Geological Survey of Canada

Geology of Canada, no. 4

GEOLOGY OF THE CORDILLERAN OROGEN IN CANADA

edited by

H. Gabrielse and C.J. Yorath

1992

This is volume G-2 of the Geological Society of America's Geology of North America series produced as part of the Decade of North American Geology project.

© Minister of Supply and Services Canada 1992

Available in Canada through
authorized bookstore agents and other bookstores
or by mail from

Canada Communications Group – Publishing
Ottawa, Canada K1A 0S9

and from

Geological Survey of Canada offices:
601 Booth Street
Ottawa, Canada K1A 0E8

3303-33rd Street N.W.,
Calgary, Alberta T2L 2A7

100 West Pender Street
Vancouver, B.C. V6B 1R8

A deposit copy of this publication is also available for reference
in public libraries across Canada

Cat. No. M40-49/4E
ISBN 0-660-13132-3

Price subject to change without notice

Cette publication est aussi disponible en français

Technical editor
P.J. Griffin

Design and layout
P.A. Melbourne

Cartography
T. Oliveric, GSC, Vancouver
GSC Cartography Sections in Ottawa and Calgary

Cover

View west to Mounts Serra I to V (centre), Asperity, and Tiedemann (far right) in the Mount Waddington area, Coast Mountains. Granitic rock in the foreground is granodiorite and tonalite, typical of the Coast Plutonic Complex. Photo by G.J. Woodsworth. KGS-2300

Printed in Canada

PREFACE

The Geology of North America series has been prepared to mark the Centennial of The Geological Society of America. It represents the efforts of more than 1000 individuals from academia, state and federal agencies of many companies, and industry to prepare syntheses that are as current and authoritative as possible about the geology of the North American continent and adjacent oceanic regions.

This series is part of the Decade of North American Geology (DNAG) Project which also includes eight wall maps at a scale of 1:5 000 000 that summarize the geology, tectonics, magnetic and gravity anomaly patterns, regional stress fields, thermal aspects, seismicity, and neotectonics of North America and its surroundings. Together the synthesis volumes and maps are the first coordinated effort to integrate all available knowledge about the geology and geophysics of a crustal plate on a regional scale.

The products of the DNAG Project present the state of knowledge of the geology and geophysics of North America in the 1980s, and they point the way toward work to be done in the decades ahead.

From time to time since its foundation in 1842 the Geological Survey of Canada has prepared and published overviews of the geology of Canada. This volume represents a part of the seventh such synthesis and besides forming part of the DNAG Project series is one of the nine volumes that make up the latest *Geology of Canada*.

J.O. Wheeler
General Editor for the volumes
published by the
Geological Survey of Canada

A.R. Palmer
General Editor for the volumes
published by the
Geological Society of America

ACKNOWLEDGMENTS

Although the *Geology of Canada* is produced and published by the Geological Survey of Canada, additional support from the following contributors through the Canadian Geological Foundation assisted in defraying special costs related to the volume on the Appalachian Orogen in Canada and Greenland.

Alberta Energy Co. Ltd.
Bow Valley Industries Ltd.
B.P. Canada Ltd.
Canterra Energy Ltd.
Norcen Energy Resources Ltd.
Petro-Canada
Shell Canada Ltd.
Westmin Resources Ltd.
J.J. Brummer
D.R. Derry (deceased)
R.E. Folinsbee

Frontispiece.

View northwesterly over the town of Cassiar, northern British Columbia. Dark weathering rocks at the upper right are oceanic sediments, volcanics and ultramafic rocks of the Sylvester Allochthon, Slide Mountain Terrane, which host the Cassiar chrysotile asbestos deposit near its contact with the underlying light grey weathering carbonates of the Middle Devonian McDame Group of the Cassiar Terrane. Rocks in the background beyond Cassiar consist mainly of mid-Cretaceous granite of the Cassiar Plutonic Suite and flanking metamorphosed strata of Late Proterozoic and Early Cambrian age. Photo by H. Gabrielse. 1991-588

ERRATA

The publication date for the volume should be 1991.

Some page numbers in the index should be changed as follows:

pages		
0 - 242	- no change required	
243 - 418	- add 2 pages to existing numbers	
419 - 434	- no change required	
435 - 490	- subtract 2 pages from existing numbers	
491 - to end	- subtract 4 pages from existing numbers	

Figure 17.62 on page 640

The asterisk located along 57° latitude should be positioned 1° north on the same longitude.

CONTENTS

Foreword 1

CHAPTER

1.	Introduction 3
2.	Tectonic framework 15
3.	Paleomagnetism: review and tectonic implications 61
4.	Precambrian basement rocks of the Canadian Cordillera 87
5.	Middle Proterozoic assemblages 97
6.	Upper Proterozoic assemblages 125
7.	Cambrian to Middle Devonian assemblages 151
8.	Upper Devonian to Middle Jurassic assemblages 219
9.	Upper Jurassic to Paleogene assemblages 329
10.	Neogene assemblages 373
11.	Physiographic evolution of the Canadian Cordillera 403
12.	Quaternary glaciation and sedimentation 419
13.	Modern plate tectonic regime of the continental margin of western Canada 435
14.	Volcanic regimes 457
15.	Plutonic regimes 491
16.	Metamorphism 533
17.	Structural styles 571
18.	Tectonic synthesis 677
19.	Regional metallogeny 707
20.	Energy and ground resources of the Canadian Cordillera 769
21.	Natural hazards 803
22.	Outstanding problems 817

Index 825

FOREWORD

Since 1909 the Geological Survey of Canada has published five editions of the Geology and Economic Minerals of Canada. The most recent of these, edited by R.J.W. Douglas and published in 1970, represented a significant advance in the synthesis of many aspects of Canadian geology, largely due to the vastly increased data base accumulated during the previous fifteen years. Though greatly expanded, it was, like the four editions that preceded it, published in only one volume. This new edition, prepared in conjunction with the Decade of North American Geology project, launched in 1980 by the Geological Society of America, consists of nine volumes (six regional, two topical and one summary volume) thus permitting a much broader discussion of the geology of the several regions of Canada than has been possible heretofore.

The present volume also has benefited from an additional fifteen years of research in the Cordilleran Orogen and, as well, from the application of new concepts relating to plate tectonic processes that were not widely considered before 1970. It is divided into twenty-two chapters, each dealing with specific aspects of stratigraphy, arranged chronologically, or with the various disciplines that constitute the spectrum of geological studies in the Cordillera. The chapters have been written by authorities on the various topics and include contributions from many researchers at universities, the British Columbia Geological Survey and the Geological Survey of Canada.

The chapters have been compiled from the efforts of more than sixty contributors, but are, to a large degree, mutually interdependent; few can stand alone. During the early stages of planning the editors chose a somewhat democratic authorship policy because the considerable complexity of Cordilleran geology demanded a wide diversity of expertise. Thus the editors dealt largely with content and format rather than with style. In instances where more than one opinion on a given issue is supported by factual information, all of these are expressed. All chapters passed through several stages of critical evaluation including those applied by the editors. Important in this process were the views expressed by an "outside" authority on Cordilleran geology and an "outside" generalist who represented the opinion of the non-specialist.

This volume has been designed to serve an audience of broad interest and background. It is hoped that it will provide information and useful reference for a wide range of earth scientists including university students, professors and exploration geologists.

H. Gabrielse
Cordilleran Division
Geological Survey of Canada
Vancouver, British Columbia

C.J. Yorath
Pacific Geoscience Centre
Geological Survey of Canada
Sidney, British Columbia

Chapter 1
INTRODUCTION

Regional character
Scope of the volume
Concepts and terminology
History of geological exploration
Acknowledgments
Reference

Chapter 1

INTRODUCTION

H. Gabrielse and C.J. Yorath

REGIONAL CHARACTER

The Canadian sector of the Cordilleran Orogen encompasses an area of over 1.6×10^6 km^2 and extends from the base of the continental slope on the west to the western limit of undeformed strata underlying the Interior Plains. Its northern boundary is the Beaufort Sea and its southern boundary, 2000 km to the south, is the International Boundary, for the most part at 49°N (Map 1701A, in pocket).

The great diversity in the geology of the region is reflected in its varied physiography. Spectacular exposures of layered sedimentary strata characterize the British, Richardson, Ogilvie, Wernecke, Mackenzie and Rocky mountains of the Foreland Belt in the eastern part of the Cordillera. To the west equivalent rocks have been intensely folded, metamorphosed and intruded by granitic rocks to form the rugged Selwyn, Kaska and Columbia mountains of the Omineca Belt (Morphogeological Belt map - inside front cover). Transecting the Omineca Belt in the Yukon Territory and separating the Foreland and Omineca belts in British Columbia is one of the world's most remarkable lineaments represented by the Tintina, Northern Rocky Mountain and Southern Rocky Mountain trenches. The straight valleys of the Tintina and Northern Rocky Mountain trenches coincide with the traces of dextral transcurrent faults, whereas the three arc-like segments of the Southern Rocky Mountain Trench, although everywhere associated with faults, are not known to be the loci of transcurrent displacements.

The Intermontane Belt of central British Columbia and south-central Yukon Territory is underlain by a wide variety of igneous and sedimentary rocks. Its generally subdued relief results from modest Late Cretaceous and Cenozoic uplift and, locally, the construction of flat, nearly horizontal lava plateaus on Cenozoic erosion surfaces. The plateaus of the Yukon Territory and northern British Columbia are separated from those to the south by the rugged Skeena Mountains formed mainly of intensely folded Mesozoic sedimentary rocks.

Bordering the Intermontane Belt to the west is the Coast Belt comprising the rugged and ice-capped Coast and Cascade mountains cored by greatly uplifted plutonic and metamorphic rocks. Contributing to great local relief in the region are numerous fiords that penetrate deeply into and across the belt from the Pacific Ocean.

Gabrielse, H. and Yorath, C. J.
1991: Introduction, Chapter 1 in Geology of the Cordilleran Orogen in Canada, H. Gabrielse and C. J. Yorath (ed.); Geological Survey of Canada, Geology of Canada, no. 4, p. 3-11 (also Geological Society of America, The Geology of North America, v. G-2)

In the Insular Belt the relatively subdued but rugged mountains of the Queen Charlotte Islands and Vancouver Island contrast with the lofty, ice-capped scenic grandeur of the Saint Elias Mountains of northwestern British Columbia and southwestern Yukon Territory. Although both are underlain by similar rocks, the spectacular relief of the latter is believed to be due to recent and continuing interaction between the North American and Pacific crustal plates.

Superimposed on the late Cenozoic physiographic elements of the Cordillera are erosional and constructional features associated with widespread and local glaciation. An exception is expressed by subdued topography of the Klondike area of the western Yukon Territory which escaped glaciation due to its position in the rainshadow of the Saint Elias Mountains.

SCOPE OF THE VOLUME

The volume begins with a discussion of the tectonic framework of the Cordillera in terms of its main morphogeological belts, tectonostratigraphic rock assemblages and component terranes. The application of paleontology to terrane biostratigraphy and timing of assembly is described as is the general nature of the crust as revealed by geophysical measurements. The third chapter is an extensive review of paleomagnetic studies undertaken in the Cordillera and their implications regarding translations and rotations of the principal component terranes.

The following seven chapters (4-10) describe the stratigraphy of the Cordillera in terms of tectonic assemblages as they occur on the miogeocline of western cratonal North America or in one or more of the component terranes. Correlation charts (in pocket) illustrate the stratigraphy at 90 locations throughout the region. Critical to these discussions is the accompanying Tectonic Assemblage Map and Legend (Map 1712A, in pocket) which illustrates the distribution of these assemblages and defines them according to their lithologies, origins and tectonic settings. Moreover, this map was used as the primary base for the construction of the Terrane Map (Map 1713A, in pocket) so that direct comparisons can be made and so that the reader can see the degrees of stratigraphic contrast among the several terranes and between these and cratonal North America.

The following seven chapters (11-17) are concerned with specific aspects of Cordilleran geology. The evolution of the modern land surface and its subsequent glacial modification is supported by a new physiographic map superimposed on a Landsat image of the Cordillera (Map 1701A, in pocket). The development and dynamics of the modern plate tectonic regime, including deformation on the continental margin, is followed by a chapter on volcanic regimes as they evolved through ancient and modern plate tectonic processes. Two chapters on plutonic and metamorphic regimes are supported by maps illustrating the distribution of the several compositional and temporal suites of granitic rocks (see Fig. 15.1, in pocket) and a new comprehensive metamorphic map that includes data from a wide variety of sources including vitrinite reflectance and conodont colouration indices (Map 1714A, in pocket). The many different structural styles of the Cordillera are described with reference to both the terranes and the morphogeological belts in which they occur; regional structural cross-sections (see Fig. 17.1, in pocket) support this chapter. The concluding chapter to these first two sections of the volume presents a tectonic synthesis of the Cordillera and comparisons with other orogens.

Chapters 19 and 20 discuss the economic geology of the Cordillera. The nature, distribution and origin of mineral deposits is described with reference to their host terranes. Petroleum, coal and groundwater resources are discussed as are the comparatively recent efforts towards geothermal energy exploitation. A chapter on natural hazards includes discussions of potential dangers associated with earthquakes, landslides, volcanism, floods, tsunamis and erosion.

The volume is concluded by a short chapter listing the many outstanding problems that remain for further research in the Canadian Cordillera.

CONCEPTS AND TERMINOLOGY

The organization of material throughout the volume was influenced by the relationship of the geology to the five morphogeological belts and to the various terranes which form the Cordilleran collage. Only during the past decade has it been realized that much of the Cordillera comprises an assembly of disparate crustal fragments, or terranes, some of which originated at distant paleolatitudes and which, during the Mesozoic and Cenozoic, became incorporated with cratonal North America. The terranes are differentiated primarily by their stratigraphy, but, in many cases they also have distinctive structural styles and volcanic and plutonic histories. As a consequence of terrane amalgamation, the Cordillera is segregated into five distinct belts each of which show marked contrasts in physiography, metamorphism, volcanism, plutonism and structural style, which, to some degree, were inherited from the original component terranes. The boundaries of the belts are somewhat arbitrary because their definition combines both geological and physiographic factors. The terrane boundaries, on the other hand, are geological contacts only, and, to some extent, are collinear with those between belts. These concepts are expanded in Chapter 2.

Throughout the volume considerable reference is made to "Ancestral North America" and the "miogeocline". The former applies to the craton, which throughout the Phanerozoic, remained relatively stable. The western edge of Ancestral North America, from mid-Proterozoic to early Mesozoic time, is thought to have approximately coincided with the western part of the Omineca Belt and southern British Mountains in northern Yukon Territory (Fig. 1.1). Since the early Mesozoic, successive episodes of terrane accretion have led to the modern outline of the western part of the continent. The "miogeocline" refers to the westward expanding then tapering wedge of supracrustal rocks that accumulated upon the westerly sloping Precambrian crystalline basement of Ancestral North America from mid-Proterozoic to Middle Jurassic time. Although many digressions from geometrical uniformity exist, the overall regional concept employed herein is that the miogeocline represents a broad, flat to gently west-sloping depositional surface upon which mainly shallow water carbonates and associated terrigenous clastics were deposited.

Two significant departures from geological terminology, formerly in common use, have been made throughout the text and in the correlation charts (in pocket). The first of these is the abandonment of Precambrian era terms "Aphebian", "Helikian" and "Hadrynian" (Stockwell, 1961) for Cordilleran rocks and their replacement by the terms "Early", "Middle" and "Late" Proterozoic with lower age limits of 2500 Ma, 1600 Ma and 800 Ma respectively. This threefold subdivision is based upon recent stratigraphic, geochronological and paleomagnetic data and is more suitable to Cordilleran Proterozoic assemblages than the former terms established for time-stratigraphic subdivision in the Canadian Shield.

A second departure is the discarding of formal names for Cordilleran orogenies. During the past few decades numerous terms have been applied to local or regional Cordilleran orogenies on the basis of different and often not widely accepted criteria. The two most commonly used terms have been "Columbian" and "Laramide" for deformational and/or metamorphic events of roughly mid-Jurassic to mid-Cretaceous and Late Cretaceous to Eocene ages respectively. In many parts of the western Foreland Belt the assignment of deformation to either of these is arbitrary and it is probable that the two events are part of an orogenic continuum. Terms used for local deformational events such as the Middle Triassic Tahltanian, Early Jurassic Inklinian and Middle Jurassic Nassian orogenies have not been used extensively in Cordilleran literature. Definition of the mid-Paleozoic Cariboo orogeny remains confused and the term is seldom used. The Precambrian Racklan, Hayhook, East Kootenay and Goat River orogenies represent local deformations of arguable significance and which lack regional counterparts. As a consequence of these and other criticisms, the many tectonic events evident in the Cordillera are herein described in terms of their age and the nature of the phenomena involved.

Amalgamation of the several terranes and their accretion to Ancestral North America are examined separately through studies of paleomagnetic signature on the one hand and the geological record of accretion on the other. Paleomagnetic data suggest that much of the western Cordillera moved more than 2000 km to the south in post-Middle Jurassic time and then moved north to its present position in mid-Cretaceous to Paleogene time. Late Jurassic to Late Cretaceous faunas of the accreted terranes, however, show no hint of residence in equatorial waters. Structural, metamorphic and stratigraphic data point to successive accretions in the Middle Jurassic and mid-Cretaceous with continued or pulsatory intraplate deformation in the Intermontane, Omineca and Foreland belts from Middle Jurassic to Paleogene time. The disposition of magmatic arcs, structural and metamorphic belts and foredeep assemblages reflect a strong component of contraction within the Cordillera during this interval, although large dextral

INTRODUCTION

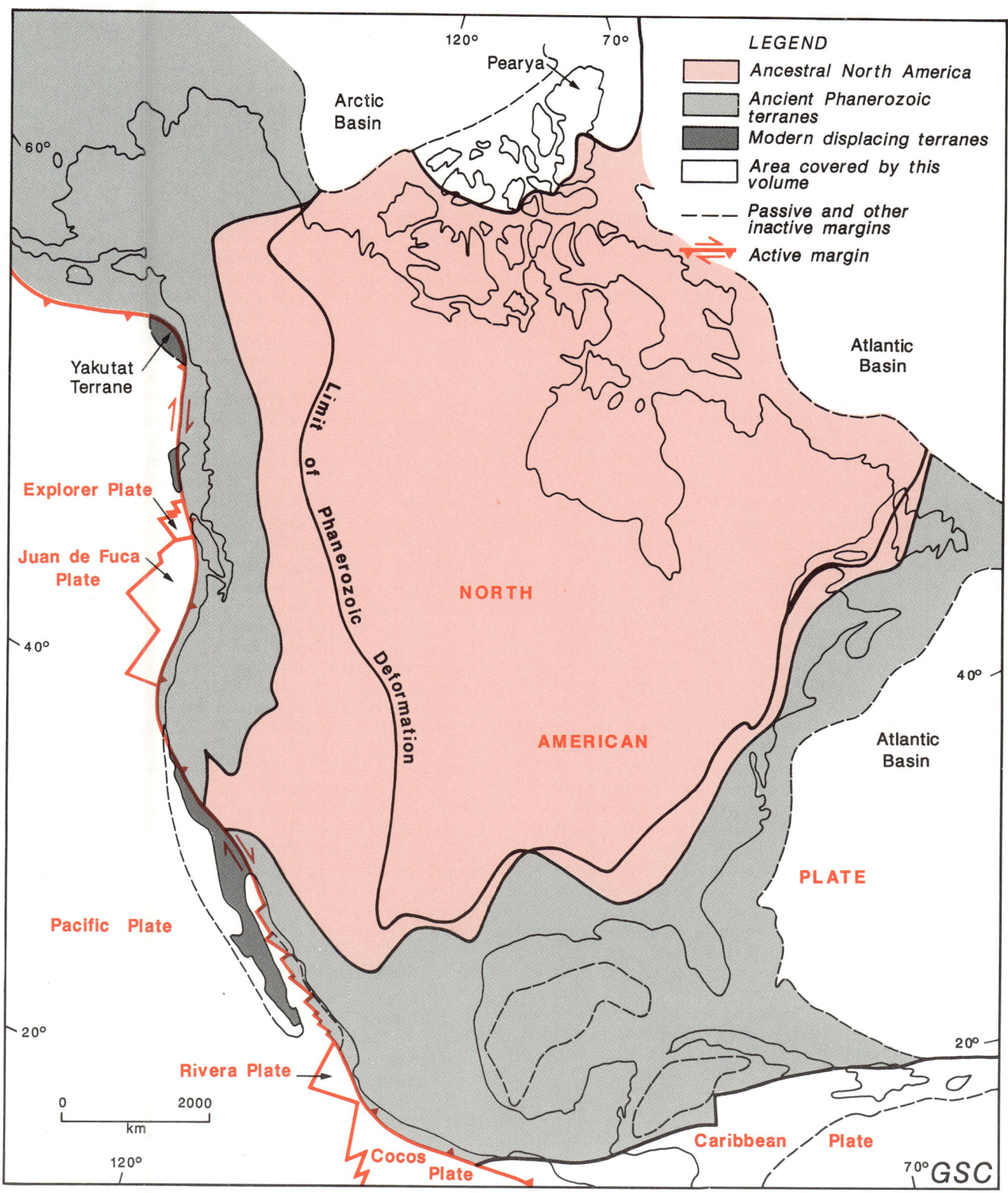

Figure 1.1. Tectonic components of North America showing the region covered by this volume (modified from R.C. Speed, pers. comm., 1987).

transcurrent displacements are also indicated. The extent to which these paleomagnetic and geological data can be reconciled remains for future studies.

HISTORY OF GEOLOGICAL EXPLORATION

The history of geological exploration in the Canadian Cordillera began with a British expedition led by Captain John Palliser from 1857 to 1860. The Palliser group conducted explorations in the Prairies and Rocky Mountains between latitudes 49°N and 54°N and as far west as the Pacific coast during which James Hector made observations on the stratigraphy of the southern Rocky Mountains and the coal deposits on Vancouver Island. The first exclusively geological surveys, however, took place in the 1870s with reconnaissance and local detailed studies in southern British Columbia by A.R.C. Selwyn, G.M. Dawson, R.G. McConnell and James Richardson of the Geological Survey of Canada. Particularly noteworthy was the work of Dawson whose perceptions of the geology of the southern Cordillera were remarkably insightful and remain currently valuable.

Comprehensive studies of the Cordillera began in the 1880s by Dawson and McConnell. The latter's geological cross-section of the Rockies became a standard reference, and the McConnell Fault, one of the great thrust faults of the southeastern Cordillera, was subsequently named for him. A far ranging reconnaissance journey by these two men in 1887-88 provided the initial geological information on the northern Cordillera including much of the Yukon Territory and northern British Columbia. McConnell's observations on parts of the northern Foreland Belt served as the only source of geological information until the 1950s. Work in the Cariboo region in the 1880s by Amos Bowman set the foundation for subsequent mapping in a geologically complex area, still the subject of much debate.

The gold rush to the Klondike in the late 1890s led to several projects by R.G. McConnell who produced important reports on the placer deposits of that and other regions of the Yukon. Near the turn of the century the efforts of McConnell, Joseph Keele and D.D. Cairnes in the Yukon Territory and those of J.C. Gwillim in the Atlin area of northwestern British Columbia led to a general understanding of the geology that, in several areas, was not superseded until three decades ago. Important surveys were conducted in support of mineral exploration including those in the Kootenay district by McConnell, R.W. Brock and W.W. Leach and in the coastal area north of Vancouver by O.E. Le Roy and J.A. Bancroft. Interest in coal and oil resources in the southeastern Cordillera provided stimulus for work in that region. D.B. Dowling's study of the coal-bearing formations of the Rockies resulted in publication of a monograph that remained the principal reference for many years.

Two investigations in southern British Columbia during the early 1900s are notable in relationship to modern tectonic concepts. One was R.A. Daly's International Boundary Survey between 1900 and 1907 during which he recognized a fundamental division of the Cordillera into an eastern part dominated by sedimentary rocks and a western part containing abundant volcanic, plutonic and metamorphic rocks. The other was C.W. Drysdale's study of the Kootenay region from which he proposed that the various types of mineral deposits were related to "terranes" encompassing different host rocks.

Charles Camsell's reconnaissance expedition in the District of Mackenzie and Yukon Territory led to the recognition of the coal-bearing Bonnet Plume Basin. Joseph Keele's traverse in 1908 along the river that bears his name provided the only transect of the Mackenzie Mountains available until extensive mapping projects were undertaken in the middle of the century.

Between 1910 and 1915 mapping in southern British Columbia was conducted by many geologists who, in later years, were to attain international recognition as researchers and teachers, men such as N.L. Bowen, R.A. Daly and A.M. Bateman. During this period C.D. Walcott drew attention to the remarkably well preserved, soft-bodied fauna in the Lower Cambrian Burgess Shale near Field in the southern Rocky Mountains, a locality which has since been designated by the United Nations as a World Heritage Site. F.H. McLearn contributed greatly to the understanding of Mesozoic biostratigraphy in the western Cordillera and L.D. Burling made significant advances in establishing the stratigraphic framework of the Cambrian rocks in the southeastern Cordillera. During these same years J.D. Mackenzie and C.H. Clapp respectively described the geological framework of the northern Queen Charlotte Islands and southern Vancouver Island.

Early investigations of important mineral districts and hydrocarbon resources provided the focus for Cordilleran exploration until the mid-1940s. C.O. Swanson's research on the origin of the magnetite deposits on Texada Island and C.S. Evans' and J.F. Walker's work in the Kootenays provided much important information on local stratigraphy and structure. In the 1920s V. Dolmage, C.E. Cairnes and C.H. Crickmay published lasting references on the southern Coast Belt. At the same time F.A. Kerr was involved in regional studies in the Stikine region of northwestern British Columbia, a mapping project that produced information, useful to the present, on the various characteristics of the plutonic rocks and complex Mesozoic stratigraphy. Several of Kerr's assistants, including C.S. Lord, G.W.H. Norman and S.S. Holland later became prominent figures in the geology of the Cordillera and other regions of Canada. Farther north W.E. Cockfield carried out diverse studies in the Yukon Territory mainly within mineral districts such as Keno Hill and Galena Hill, and W.A. Johnson examined surficial deposits critical to the search for placer gold in the Cariboo and Cassiar areas.

After 1930 H.S. Bostock became the best-known geologist in the Yukon Territory as a consequence of his 1:250 000 mapping program, his reports on the mineral industry and through his contacts with mining and exploration personnel. He greatly advanced the knowledge of the physiographic development of the northern Cordillera and his maps and reports served as guides to placer as well as lode exploration.

G.S. Hume's and E.M. Kindle's substantial contributions to petroleum exploration in the northern Foreland Belt in the early 1920s were based upon their stratigraphic, structural and paleontological studies along and near the Mackenzie River valley. Much later, in the early years of the Second World War, in recognition of the strategic importance of the Norman Wells oil field which had been discovered in 1920, a co-operative exploration and development project to increase the known limited reserves was undertaken by the Geological Survey of Canada, Imperial Oil Ltd., and the United States Government (CANOL) under the direction of T.A. Link. Forty reports and several maps resulted from this work which, following acquisition of additional subsurface information from subsequent drilling were summarized in an important report by Hume in 1954. Similarly, in the southern Foreland Belt the work of B.R. MacKay, C.O. Hage, H.H. Beach, F.H. McLearn, W.A. Bell, R.J.W. Douglas and D.K. Norris substantially aided the exploration efforts of the petroleum industry following the major discoveries of oil in Devonian reefs beneath the Alberta plains in the late 1940s and early 1950s. Moreover, it was these studies in the northern and southern Foreland Belt that provided the groundwork for the large, helicopter-supported reconnaissance mapping projects conducted by petroleum companies and the Geological Survey of Canada in the late 1950s and 1960s.

Construction of the Alaska Highway and the CANOL road to Fairbanks in the early 1940s provided access to the Mackenzie, Selwyn, and Pelly mountains for E.D. Kindle and the northern Rocky and Cassiar mountains for C.O. Hage, M.Y. Williams and C.S. Lord. Although areally restricted these investigations supplemented the meager data available to that time. A reconnaissance study of part of the northern Rocky and Cassiar mountains by M.S. Hedley and S.S. Holland of the British Columbia Geological Survey (a branch of the British Columbia Ministry of Energy, Mines and Petroleum Resources) emphasized the profound lithological and structural discordance between the two areas and illustrated the areal extent of Mesozoic granitic rocks. In 1943, K. de P. Watson and W.H. Mathews, under the auspices of the same agency, conducted a reconnaissance study of the Tuya-Teslin area of northwestern British Columbia. Their report contains descriptions of the unusual variety of granitic rocks in the region and its glacial history. The study of Pliocene and Pleistocene volcanic rocks led to Mathew's concept of "tuyas" which are flat-topped volcanic edifices erupted into lakes thawed into glacial ice by the heat of the volcano. Detailed structural studies by M.S. Hedley between 1946 and 1950 in the Slocan area of southern Omineca Belt identified large, recumbent folds involving the entire stratigraphic assemblage, the first of such structures to be recognized west of the Foreland Belt.

The late 1940s and early 1950s saw a great increase in the number of field parties engaged in 1:250 000 scale mapping. W.E. Cockfield, H.M.A. Rice and C.E. Cairnes outlined the general framework of the southern Intermontane Belt. Farther north S.S. Holland established the stratigraphic succession in part of the Cariboo Mountains and J.E. Armstrong, C.S. Lord and E.F. Roots conducted the first systematic surveys of the Omineca Mountains. In addition to Armstrong's and J.S. Grey's discovery of the Pinchi Lake mercury deposits, Armstrong was the first to attempt correlation of stratigraphic assemblages throughout the full length of the Intermontane Belt.

The 1950s saw the beginning of a new era of mapping when pack-horses and canoes were replaced by helicopters (Fig. 1.2). Several of the major oil companies carried out regional, helicopter-supported reconnaissance projects throughout the Foreland Belt. At the same time the Geological Survey of Canada began several projects in western Canada, one of the largest being "Operation Stikine" in northwestern British Columbia, which, in 1956, covered six 1:250 000 scale map areas and employed eight staff members, 45 support staff, three pack-horse groups, two helicopters and several fixed-wing aircraft. The introduction of helicopters not only greatly increased the rate at which geological mapping could be carried out but much enhanced the potential for multidisciplinary studies. The consequent liaison among stratigraphers, structural geologists, paleontologists, petrologists, geochemists and other specialists has contributed to a much more comprehensive view of the geological framework of several regions in the Cordillera than would not have been otherwise possible.

The British Columbia Geological Survey continued its detailed studies of mineral districts but also conducted investigations of regional scope. The work of G.G.L. Henderson in the Stanford Range of the Foreland Belt in the early 1950s resulted in the first clear picture of stratigraphy and structure in the western part of the southern Rocky Mountains. In a study of the Shulaps ultramafic body G.B. Leech recognized the significance of, and named the Yalakom Fault, later realized to be one of the most important terrane bounding faults in the Cordillera.

The influence on Cordilleran geological research and mineral exploration by the University of British Columbia during the first half of the 20th century was considerable. Following the appointment of R.W. Brock (formerly Director of the Geological Survey of Canada) in 1914 as the Dean of the Faculty of Applied Science, many outstanding teachers and researchers joined the Department of Geology. These included M.Y. Williams as professor of paleontology, C.O. Swanson, an excellent mathematician and geologist, H.C. Gunning, one of the greatest professors of economic geology of his time, R.M. Thompson, an outstanding teacher and mineralogist, H.V. Warren, a pioneer in the study of trace elements in plants and V.J. Okulitch, an internationally respected expert on the phyllum *Archaeocyatha*.

The modern era of geological research in the Canadian Cordillera has produced important syntheses on tectonic framework, evolution of structural styles and development of sedimentary basins. W.H. White's 1959 benchmark paper on "Cordilleran tectonics in British Columbia" was the first attempt to outline its complete evolution. This paper provided much stimulus for further research and led the way for subsequent regional treatises. R.J.W. Douglas was particularly influential in syntheses concerning the Foreland Belt. His career of stratigraphic and structural studies in the southern Rockies and Mackenzie Mountains culminated in his appointment as editor for the fifth edition of "Geology and Economic Minerals of Canada", published in 1970.

Geophysical studies in the Canadian Cordillera have been slow in developing. Although the first seismograph station was established at Victoria in 1898 other types of geophysical measurements did not begin until the latter half of this century. One of the most significant geophysical surveys ever conducted was by A.D. Raff and R.G. Mason of the Scripps Institute of Oceanography, who, in the late 1950s, conducted a marine magnetic survey off the west coast of North America between latitudes 40° and 52°N. In 1963 Vine and D.H. Mathews, as well as L.W.

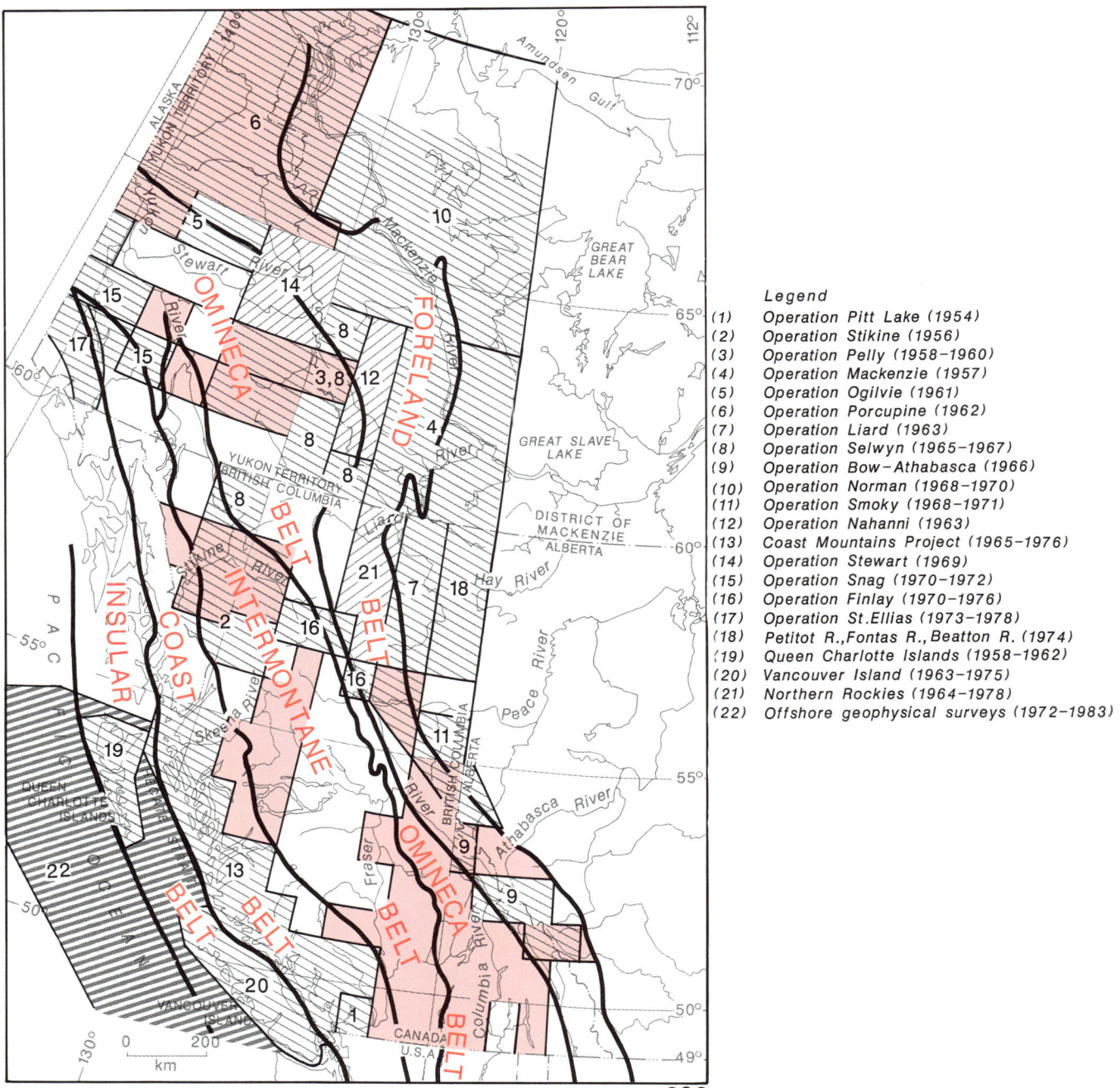

Figure 1.2. Regional geological reconnaissance operations in the Canadian Cordillera by the Geological Survey of Canada and British Columbia Geological Survey since 1954. Areas in which detailed, second phase 1:250 000 scale geological mapping have been completed are shown in pink.

Morley and A. Larochelle, proposed a model to explain the origin of the linear symmetrical anomalies observed, and, in 1965, Vine and J. Tuzo Wilson applied the model to the Juan de Fuca Ridge off Vancouver Island. It was this survey and these interpretations that led to the development of the modern hypotheses of seafloor-spreading and plate tectonics.

Since the early 1960s an increasingly broad range of geophysical measurements has been obtained in the Cordillera. Measurements of seismicity, heat flow, potential fields, conductivity and active deformation as well as paleomagnetic and reflection and refraction seismological studies by the Geological Survey of Canada, the University of British Columbia, the University of Alberta and the University of Calgary have contributed much to the knowledge of crustal structure and dynamics. In 1978 the Pacific Geoscience Centre on southern Vancouver Island was established and since then geophysical studies both on land and offshore have become a substantial component of geoscience research in the Cordillera.

The basic geological framework of the Canadian Cordillera was established by the end of the 1970s. This was achieved largely from 1:250 000 scale mapping of the Geological Survey of Canada carried out mainly by pre-helicopter reconnaissance surveys and by large helicopter operations in the 1950s and 1960s in which two or three map areas were mapped as one project coupled with simultaneous surveys of other individual map areas which received partial, shared-helicopter support. Significant contributions also came from some regional but mainly detailed mineral-district investigations by the Mineralogical Branch, predecessor of the British Columbia Geological Survey. In the last decade mapping by the Canadian and British Columbia geological surveys has involved more detailed work supported partly by helicopter. This has led to the participation of many disciplinary specialists and to contributions from innumerable detailed university thesis projects, especially in the less remote southern Cordillera.

The recent multidisciplinary approach combined with more detailed mapping has facilitated the formulation and testing of the concepts, expressed in this volume, on the plate-tectonic evolution of the Cordillera. Because of the enhanced data base, new concepts and the demands for increasingly sophisticated information by the mineral industry and land-use planners, the more recent 1:250 000 survey projects have produced much greater detail on all aspects of geology than the earlier reconnaissance work. Current studies, therefore, are concerned mainly with areas in which the geological data base has become obsolete by present standards, including many first covered by the larger helicopter operations. This obsolescence has resulted partly from new information and partly from changing concepts since the areas were mapped.

The combination of much enhanced access, the use of aircraft, the rapidly expanding and improving topographic map and air photo coverage and the application of geophysical techniques have enabled geoscientists to examine large tracts of the Cordillera in three dimensions. Reconnaissance and detailed investigations by the British Columbia Geological Survey, the Geology Section of the federal Department of Indian and Northern Affairs and the Geological Survey of Canada, together with a host of university-directed projects, as well as information provided by industrial exploration all have contributed greatly to this current synthesis of Cordilleran geology. The application of modern concepts such as plate tectonics and terrane accretion have dramatically altered former stabilist views but, as always, differences in approach and experience lead to differing interpretations. Doubtless new concepts yet unformulated will lead to a different interpretation of Cordilleran geology.

ACKNOWLEDGMENTS

The editors particularly express their thanks to A.J. Brookfield who, in addition to compiling many diagrams, provided invaluable editorial support during all stages of the project. Likewise, P. McFeely's considerable efforts in compiling the Tectonic Assemblage Map are gratefully acknowledged. B.D. Bornhold also generously gave editorial assistance.

Special thanks are extended to A. Oliveric for her patience and energy in preparing final drafts for most of the figures. The contributions of the drafting units of the Geological Survey at the Pacific Geoscience Centre in Sidney, British Columbia, the Institute of Sedimentary and Petroleum Geology in Calgary, and Geoscience Information Division in Ottawa also deserve much credit. Graphic assistance was provided by P.E. Tercier, S.J. Friday, M.D. McPherson, R.B. McFarlane and E.G. Yorath. Typing was done by B.E. Vanlier, L.G. Fox, and W. Chiu. To all of these the editors express their gratitude.

The volume was critically read by P.J. Coney, S.E. Jenness and J.O. Wheeler. Their comments on scientific content, organization and presentation are greatly appreciated and influenced the format adopted.

Finally, the editors thank the many contributors without whose efforts this volume could not have been assembled.

REFERENCE

Stockwell, C.H
1961: Structural provinces, orogenies, and time-classification of rocks of the Canadian Precambrian Shield; Geological Survey of Canada, Paper 61-17.

Authors' Addresses

H. Gabrielse
Cordilleran Division
Geological Survey of Canada
100 West Pender Street
Vancouver, British Columbia
V6B 1R8

C.J. Yorath
Pacific Geoscience Centre
9860 West Saanich road
P.O. Box 6000
Sidney, British Columbia

Printed in Canada

Chapter 2

TECTONIC FRAMEWORK

Introduction
Part A. Morphogeological belts, tectonic assemblages, and terranes
H. Gabrielse, J.W.H. Monger, J.O. Wheeler, and C.J. Yorath
Morphogeological belts
    Foreland Belt
    Omineca Belt
    Intermontane Belt
    Coast Belt
    Insular Belt
Tectonic assemblages and plutonic suites
    Tectonic settings of assemblages
    Plutonic suites and metamorphic rocks
Terranes
    Characteristics of Cordilleran terranes
    Terrane amalgamation, accretion, and dispersion
Part B. Paleontological signatures of terranes
E.S. Carter, M.J. Orchard, C.A. Ross, J.R.P. Ross, P.L. Smith, and H.W. Tipper
Fusulinids
Conodonts
Radiolaria
Ammonites
Part C. Crustal geophysics
J.F. Sweeney, R.A. Stephenson, R.G. Currie, and J.M. DeLaurier
Geophysical studies
    Deep refraction and reflection seismology
    Heat flow
    Seismicity
    Gravity and isostasy
    Geomagnetism
    Electromagnetic depth sounding
Cordilleran structure from geophysics
References

Chapter 2

TECTONIC FRAMEWORK

INTRODUCTION

The first part of this chapter deals with the concepts of morphogeological belts, tectonic assemblages and terranes. These concepts are fundamental to the understanding of Canadian Cordilleran evolution and play an important role in the organization of the volume.

Part B is mainly devoted to recent paleontological studies and their significance relative to the characterization of Cordilleran terranes. No attempt is made to synthesize the full scope of paleontology in the context of Cordilleran geology, a subject beyond the purview of this volume, but the importance of this discipline will be apparent in the chapters on stratigraphy.

Part C presents a synthesis of crustal geophysical surveys including refraction and reflection seismology, seismicity, heat flow, geomagnetism, gravity, isostasy and magnetotellurics. Although many of these studies are in their infant stages they have already provided a wealth of data on the nature of the deep crust and processes contributing to its evolution.

PART A. MORPHOGEOLOGICAL BELTS, TECTONIC ASSEMBLAGES AND TERRANES

H. Gabrielse, J.W.H. Monger, J.O. Wheeler, and C.J. Yorath

In 1959 W.H. White published the first modern tectonic synthesis of the Canadian Cordillera in which he described the region in terms of classical geosynclinal theory. At that time substantial areas remained unmapped and modern concepts of seafloor-spreading and plate tectonics had not yet been formulated. During the past two and a half decades the understanding of Cordilleran geology has grown and changed substantially as a consequence of (1) continued reconnaissance systematic mapping at 1:250 000 scale and, locally, at larger scales; (2) the completion of many detailed topical studies, many in the form of graduate research theses; (3) significant advances in the understanding of contemporary plate tectonic processes on a global scale and the application of modern models to the ancient rocks of the Cordillera; and (4) the use of new and more effective ways of acquiring and analyzing data. These factors, coupled with significant advances in micropaleontology, geochronology, geophysics, geochemistry and paleomagnetism provide a more comprehensive view of Cordilleran architecture and tectonic evolution than was possible only a few decades ago.

Until the general acceptance of the plate tectonic hypothesis in the early 1970s, the Canadian Cordillera was characterized as consisting of two geosynclinal belts (Daly, 1912; Kay, 1951; White, 1959; King, 1969): an eastern belt, comprising sedimentary and minor volcanic and intrusive rocks (miogeosyncline) and a western belt containing abundant volcanic, intrusive and metamorphic rocks (eugeosyncline). As the amount of information increased the Cordillera was divided longitudinally into five morphogeological belts (Fig. 2.1), each characterized by unique lithology, structural style and morphology (Sutherland Brown et al., 1970; Wheeler and Gabrielse, 1972). Sufficient knowledge had also been acquired on plate tectonic processes to allow Monger et al. (1972) to construct a second synthesis which identified five belts made up of combinations of tectonic settings including ancient volcanic arcs, oceanic crust, foredeep, shelf and slope deposits, successor basins and plutonic/metamorphic complexes (Fig. 2.2). Thus the old geosynclinal model was replaced by one which embodied concepts derived from observations on processes affecting modern continental margins. For example, many analogies are apparent between the present interaction of Australia with the Indonesian archipelago and New Guinea and the Mesozoic interaction of North America with the various terranes that lay to the west (Fig. 2.3).

In the 1970s the stratigraphy of the Cordillera became known in great enough detail to permit Tipper et al. (1981) to group lithostratigraphic units into tectonic assemblages. Each assemblage, commonly bounded by regional unconformities or by faults, represents a specific depositional or volcanic setting and/or response to one or more tectonic events. The tectonic assemblages illustrated in the Tectonic Assemblage Map (Map 1712A, in pocket) reflect the current interpretation of the many tectonic and depositional settings represented by the rock units of the Cordillera.

Recognition and establishment of tectonic assemblages developed concurrently with the idea that many parts of the western Cordillera are "suspect" as to their paleogeographic relationship with one another and with

---

Gabrielse, H., Monger, J.W.H., Wheeler, J.O., and Yorath, C.J.
1991: Part A. Morphogeological belts, tectonic assemblages, and terranes; in Chapter 2 of Geology of the Cordilleran Orogen in Canada, H. Gabrielse and C. J. Yorath (ed.); Geological Survey of Canada, Geology of Canada, no. 4, p. 15-28 (also Geological Society of America, The Geology of North America, v. G-2)

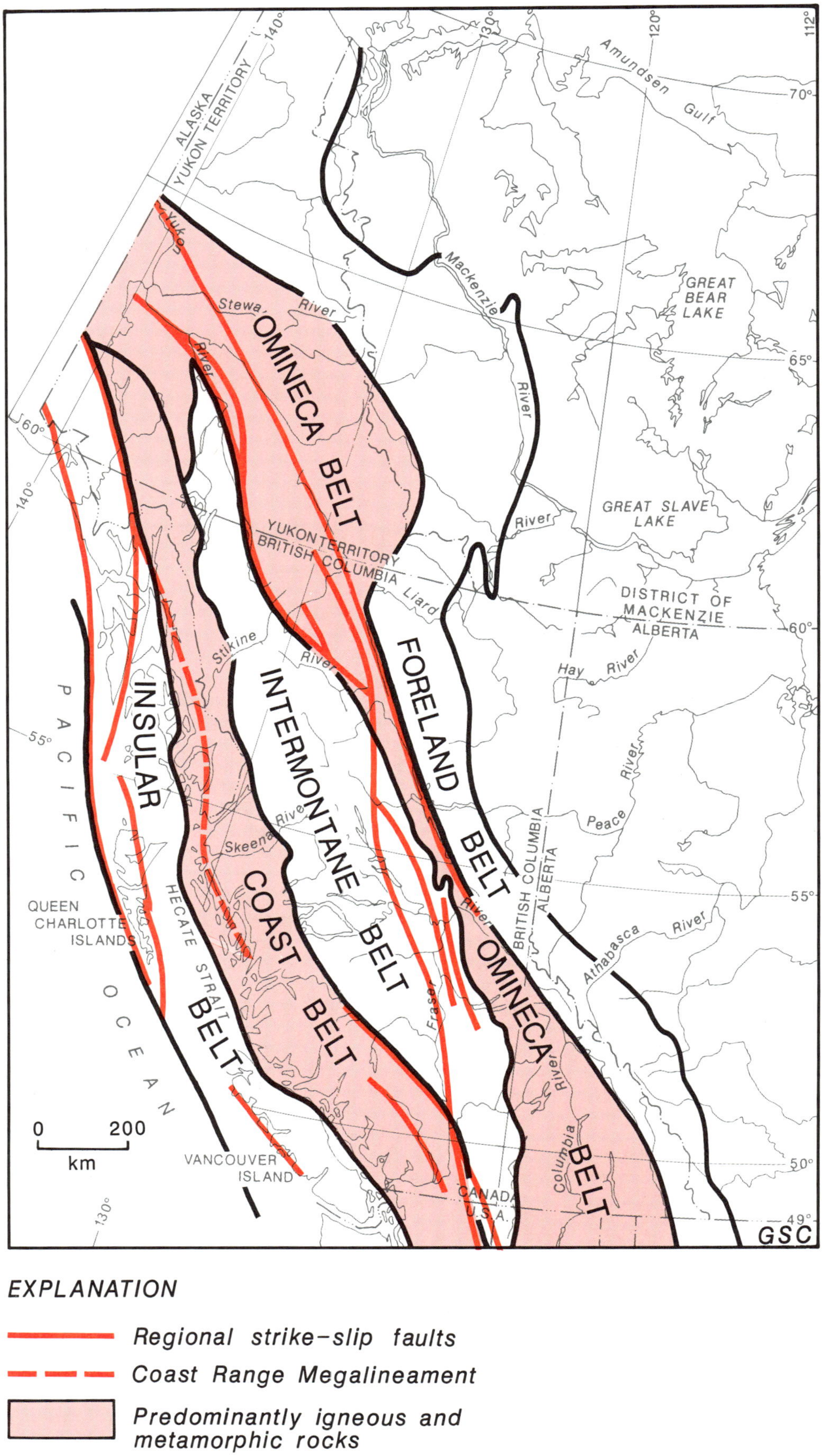

CHAPTER 2

EXPLANATION

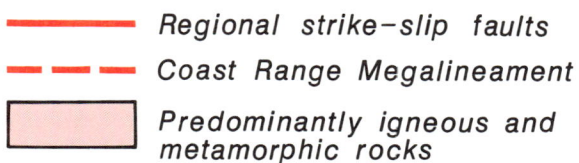

— Regional strike-slip faults
-- Coast Range Megalineament
▨ Predominantly igneous and metamorphic rocks

Figure 2.1. Morphogeological belts and regional strike-slip faults of the Canadian Cordillera.

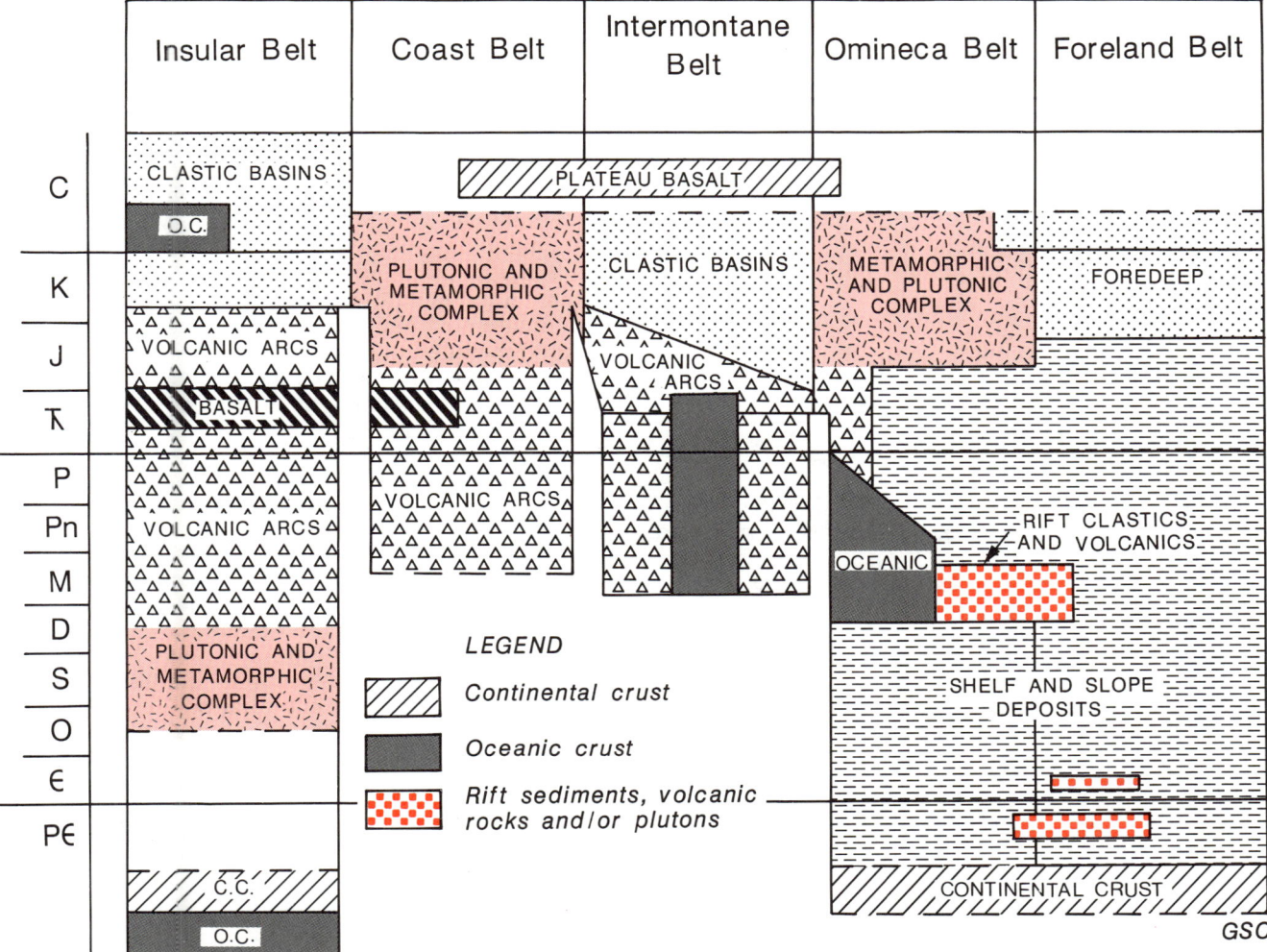

Figure 2.2. Temporal and spatial distribution of lithological assemblage types in the five morphogeological belts (modified from Monger et al., 1972).

Ancestral North America (Coney et al., 1980). Each of these so-called "terranes" possess unique tectonic assemblages that differ from those of adjacent terranes. The boundaries between them are faults. It is known from paleomagnetic and paleontological studies that some terranes originated externally to Ancestral North America and that they became accreted to the continent during the Mesozoic and Cenozoic. In some cases terranes were joined together, i.e. amalgamated, outboard of Ancestral North America to form superterranes which were later attached to the continent. In other cases their paleogeographic relationships to Ancestral North America or to one another are unknown. In some cases, subsequent to accretion, the terranes became disrupted, or "smeared out" northwestward along the continental margin by movements along dextral transcurrent faults. Because of the accretionary and disruptive style of tectonic development it is not surprising that the geometry of the morphogeological belts reflects that of the major terranes (Fig. 2.4).

In addition to the terranes and morphogeological belts, several other tectonic elements are important to the discussion of Cordilleran geology. These include basins and miogeoclinal platforms wherein much of the sedimentary record of the region is preserved (Fig. 2.5).

In summary, the architecture of the Canadian Cordillera is described in terms of tectonic assemblages, terranes, and morphogeological belts. Each tectonic assemblage reflects a specific tectonic and/or depositional environment regardless of its place of origin, and, singly or in groups, characterize the several terranes and Ancestral North America. The amalgamation, accretion and subsequent disruption of the terranes, largely through plate tectonic processes, led to the establishment of the five morphogeological belts and the modern character of the orogen.

Morphogeological belts

The five morphogeological belts of the Canadian Cordillera (Fig. 2.1) reflect the sum of geological processes which, from Middle Proterozoic to Cenozoic time, interacted to produce the geological framework of the orogen. Their description and characteristics which follow provide a synopsis of Canadian Cordilleran geology. From east to west

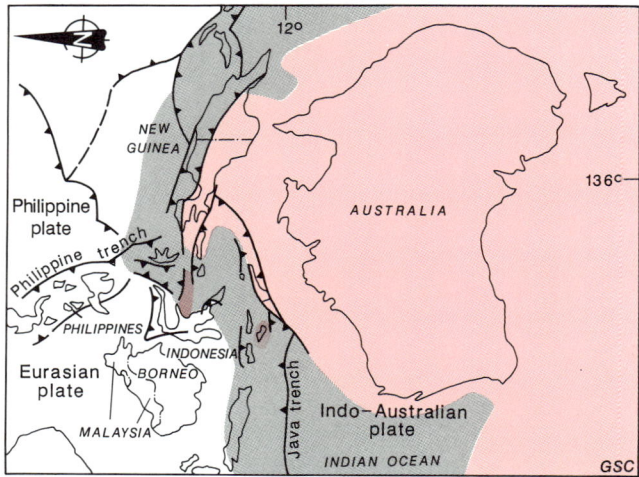

Figure 2.3. Present plate interactions involving the collision of Australia on the Indo-Australian Plate, with the Indonesian archipelago and New Guinea. Orientation is rotated 90° counter clockwise to facilitate comparison with North America during the Mesozoic.

they comprise the Foreland, Omineca, Intermontane, Coast and Insular belts, each of which is defined by a combination of lithological, structural, tectonic and physiographic attributes (see Map 1701A, and Fig. 17.1, in pocket). Together the five belts form the Cordilleran continental crust which is about 45 to 50 km thick beneath the Foreland Belt, 30 km thick under the Omineca, Intermontane and Coast belts and, in the Insular Belt, thins from nearly 30 km in the east to less than 3 km at the toe of the continental slope.

The nomenclature of the belts evolved through time (Wheeler, 1970; Sutherland Brown et al., 1970; Douglas et al., 1970; Wheeler and Gabrielse, 1972). It developed from the recognition that the Cordillera comprises five major physiographic divisions reflecting five distinct geological domains. These relationships were first apparent in the mid-1950s and led to the notion that the five belts were components of two orogens: an eastern "Columbian Orogen" (miogeosyncline) and a western "Pacific Orogen" (eugeosyncline). Although these concepts were influenced by, and applied to geosynclinal theory, the five belts are equally explicable in plate and terrane tectonic terms, specifically to the interaction between the westward moving North American Plate and variously moving Pacific plates carrying displaced crustal fragments or terranes that ultimately accreted to Ancestral North America.

For brevity the revised nomenclature uses a single word for the name of each belt. The previous well established names, or parts of them, that link geology and physiography are retained, whereas geological modifiers are excluded. The names "Insular" and "Intermontane" are unaltered whereas the names "Coast" and "Omineca", representing belts containing abundant granitic and metamorphic rocks, do not include the heretofore applied modifiers "plutonic" or "crystalline", respectively. The name "Coast Belt" is preferred for brevity over the hyphenated name "Coast-Cascade Belt", even though it includes the northern Cascade Mountains. Moreover, the belt essentially coincides with the Coast Mountains, largely underlain by the Coast Plutonic Complex. The name "Intermontane Belt" reflects its physiographic character, dominated by subdued relief (excepting the Skeena Mountains) between the two rugged and high relief Coast and Omineca belts. The term "Omineca Belt" is derived from the Omineca Mountains, one of several mountain ranges that form segments of the belt; no single name embraces all ranges of the belt. Similarly no single name embraces all of the mountain ranges of the Foreland Belt. Instead, the term "Foreland" is intended to convey its thrust-faulted and folded structural style, the dominantly miogeoclinal character of its stratigraphic succession and its position above and adjacent to the stable western margin of the North American craton.

Foreland Belt

The Foreland Belt forms the eastern mountain ranges and foothills of the Canadian Cordillera and is composed of imbricated and folded miogeoclinal and clastic wedge assemblages deposited on and adjacent to the stable craton of Ancestral North America. It extends southward from the British Mountains near the Arctic coast, swings sharply eastward through the Wernecke, Ogilvie, Mackenzie and Franklin mountains and continues southeastward in a broad re-entrant and salient along the Rocky Mountains (Map 1701A, in pocket). Its eastern boundary is the eastern limit of deformation between deformed strata of the foothills and mountain systems on the west and flat-lying, undeformed strata underlying the Interior Plains to the east. Its western boundary in the north is the eastern slope of the Selwyn Mountains and, in the south, coincides with the Rocky Mountain trenches.

The Foreland Belt contains an easterly tapering prism of mainly sedimentary rocks composed of several tectonic assemblages. These include: Precambrian crystalline basement gneiss adjacent to the Rocky Mountain trenches; clastic rift sequences of Middle and Late Proterozoic and earliest Paleozoic ages; carbonate-clastic passive continental margin and thin platform assemblages of Middle and Late Proterozoic, Paleozoic and early Mesozoic ages; and clastic wedges of Devono-Mississippian, latest Jurassic-Early Cretaceous, and latest Cretaceous-Early Tertiary ages. The clastic wedges of Mesozoic age were deposited in a foredeep on the continent in front of the easterly advancing deformation which uplifted much of the Cordillera. The amount of shortening and eastward displacement over the underlying Canadian Shield is estimated to be between 50 km in the northern Foreland Belt to over 200 km in the south.

Omineca Belt

The Omineca Belt is an uplifted region, extensively underlain by metamorphic and granitic rocks, which straddles the boundary between the accreted terranes and Ancestral North America (Fig. 2.4). It extends from the Yukon Plateau and Pelly and Selwyn mountains in Yukon Territory and Northwest Territories southeastward through the Cassiar, Omineca and Columbia mountains in British Columbia. Its eastern boundary in the Yukon and Northwest Territories is placed immediately east of the easternmost

EXPLANATION

Accreted superterranes

Pericratonic and displaced terranes

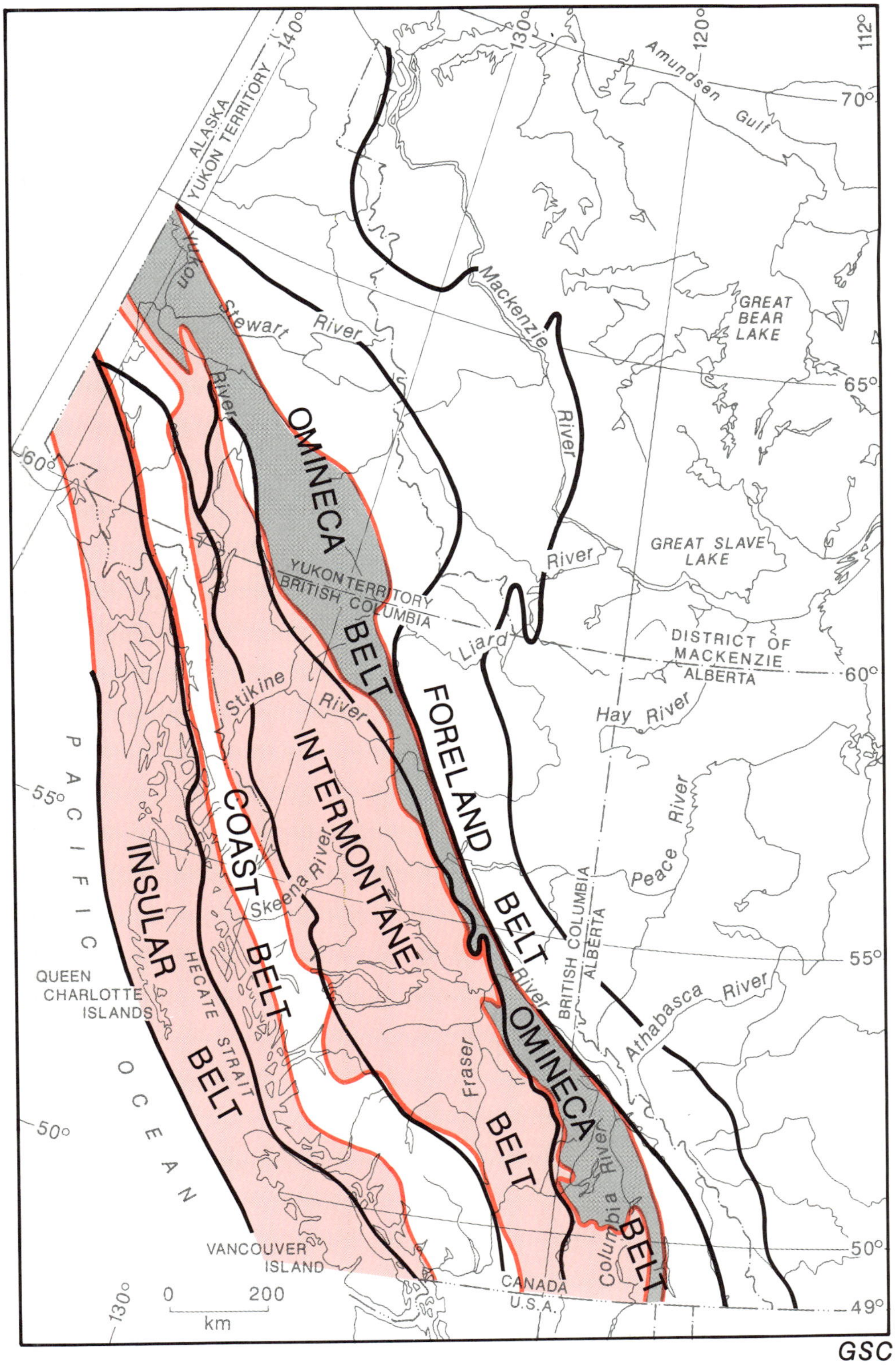

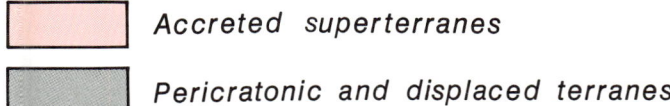

Figure 2.4. Comparison of the distributions of the five morphogeological belts with the pericratonic and displaced terranes and the accreted superterranes.

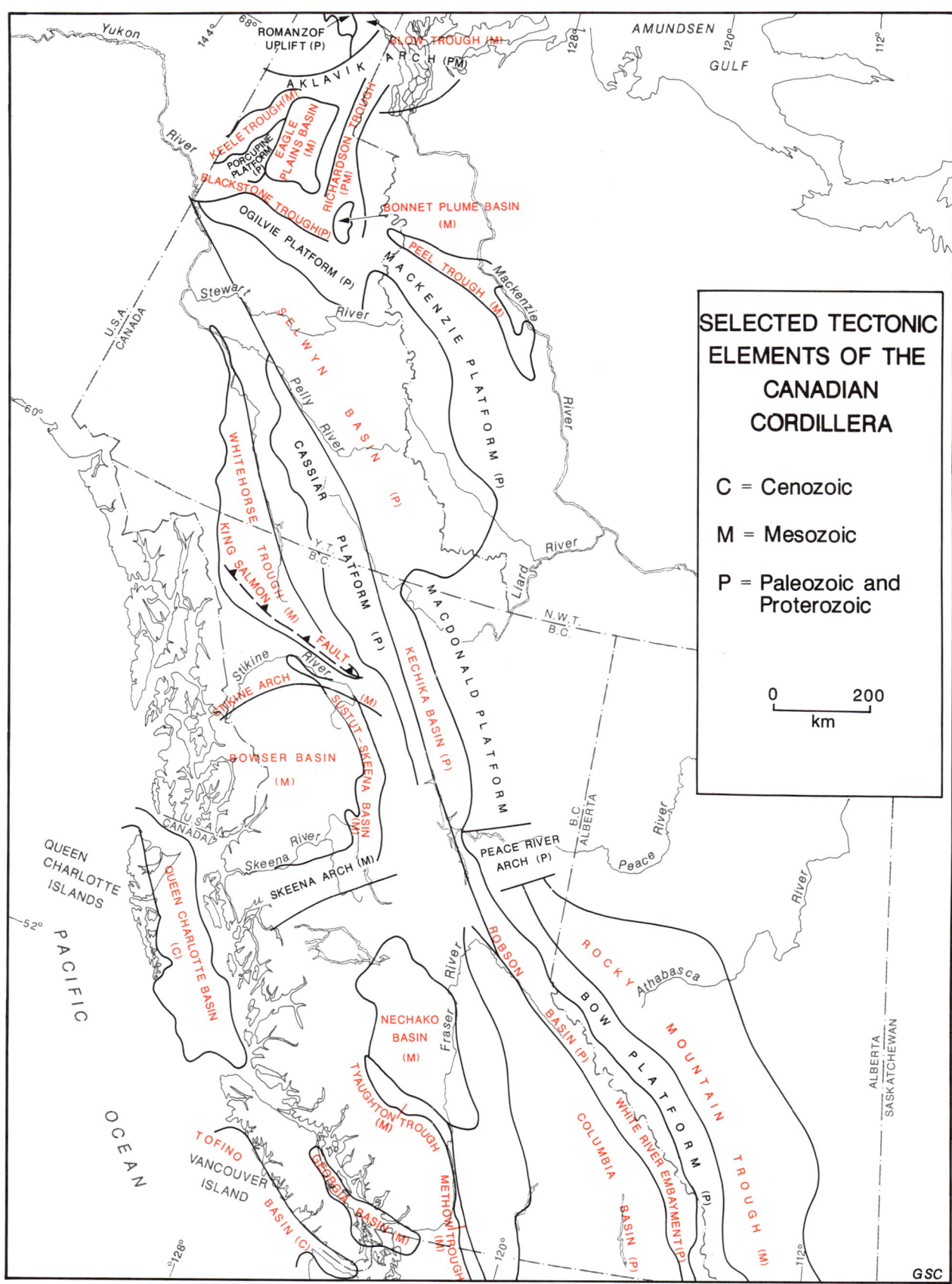

Figure 2.5. Distribution of principal basins, arches and platforms in the Canadian Cordillera.

granitic plutons and in British Columbia it coincides with the Rocky Mountain trenches. In the south the western boundary of the belt is the western limit of metamorphic and granitic rocks of the Shuswap Metamorphic Complex and throughout much of the remainder of British Columbia and southern Yukon it lies within or follows the western slope of the Kaska Mountains coinciding with the eastern boundary of the Quesnellia terrane. In southwestern Yukon the belt surrounds the northernmost Intermontane Belt where it includes the Klondike, Lewes, Teslin and Taku plateaus, parts of which are the physiographic expressions of the Whitehorse Trough. The western boundary of the Omineca Belt approximately corresponds with the line representing the initial $^{87}Sr/^{86}Sr$ ratio of 0.704. To the east the ratio is greater, expressing the recycled strontium of continental crust commonly associated with S-type plutons presumably derived from or intruded through continental crust; to the west in the volcanic terranes and associated I-type plutons, lower ratios originate from depleted mantle sources characteristic of oceanic regions.

Throughout much of its length the Omineca Belt comprises a series of northwesterly aligned structural culminations and depressions, which, in contrast to the linear uniformity of the Coast Belt, give a discontinuous character to the trend of its surface expression. The depressions are largely underlain by low-grade sedimentary rocks which extend westwards from strata in the Foreland Belt from which they differ little except where overlain by oceanic and island arc assemblages of accreted terranes. The structural culminations include core zones of moderate- to high-grade metasedimentary rocks which locally enclose slices and wedges of Lower Proterozoic basement gneiss. The core zones are overlain by, and commonly grade into less deformed cover rocks which in turn are overlain by easterly directed, folded thrust sheets containing volcanic, sedimentary and plutonic rocks. Younger gneisses, which probably represent deformed Late Proterozoic, Devonian, Mesozoic and Tertiary intrusions, are distributed throughout the belt. The plutons are mainly S-type granites of mid-Cretaceous age, although those of Middle Jurassic age are abundant in the south.

Omineca Belt rocks are more penetratively deformed than those of adjacent belts. The structurally lowest and more intensely metamorphosed rocks sustained several phases of ductile deformation and are characterized by isoclinal and recumbent folds; structurally higher levels display simple, upright, open concentric folds. Zones of intense shear and cataclasis, locally gently dipping, are associated with high-level nappes in Yukon Territory, the Northern Rocky Mountain Trench and in the outer parts of the core zones.

Regionally the belt comprises an imbricated succession of folded thrust sheets of Mesozoic age which express a dominant easterly direction of tectonic transport toward the craton. The internal folds and faults show evidence of symmetrical transport away from an irregular axis of structural divergence which generally, but not everywhere, coincides with the core zone. In Late Cretaceous and Early Tertiary time the belt was further disrupted by several northwest-trending dextral faults which, in the case of the Tintina Fault, repeated the belt along hundreds of kilometres of overlap length and doubled its width in Yukon Territory (Fig. 2.1). Finally, the southernmost part of the belt in the Okanagan region sustained Early Tertiary stretching, uplift and erosion, apparently facilitated by lateral transport of cover rocks along listric normal faults located above the metamorphic core zone.

Accreted terranes occur as large folded thrust sheets emplaced upon Upper Triassic and older miogeoclinal rocks along the continental margin of Ancestral North America. The structurally lowest is the Kootenay Terrane composed of Proterozoic to Upper Triassic assemblages of siliceous clastics, felsic and mafic volcanics and carbonate. In Yukon Territory, a segment of the Kootenay Terrane, the Nisutlin Allochthon, clearly has been thrust over the ancient continental margin (Tempelman-Kluit, 1979), however, to the south the base of the Barkerville Subterrane, another component, is not exposed. Above the Kootenay Terrane successive thrust sheets represent the Slide Mountain Terrane and Quesnellia, components of the larger Intermontane Superterrane which accreted to the continental margin in Early to Middle Jurassic time (Fig. 2.6). The process of accretion, resulting in crustal thickening, contraction, intense metamorphism, intrusion and uplift formed the Omineca Belt.

Intermontane Belt

Generally the Intermontane Belt is topographically low and physiographically subdued in comparison to the Omineca and Coast belts. A prominent exception is the Skeena Mountains comprising a series of rugged, high-relief ranges in northwestern British Columbia. In southern British Columbia its eastern boundary with the Omineca Belt is largely expressed by changes in physiography and lithology whereas in northern British Columbia and southern Yukon, regional strike-slip faults such as the Kutcho and Teslin faults separate the two belts. Its western boundary with the Coast Belt in the south coincides with the traces of the Pasayten, Fraser and Hungry Valley faults. Northward, its boundary with the Coast Belt is defined by a dramatic change in relief and lithology from mainly subdued topography underlain by volcanic and sedimentary rocks to rugged, high relief edifices largely supported by granitic rocks. In southwestern Yukon the Intermontane Belt is enclosed by the western extension of the Omineca Belt where the latter includes abundant plutons to the north and west of Whitehorse Trough.

In contrast to adjacent belts the stratified rocks of the Intermontane Belt rarely are metamorphosed as high as greenschist facies. Plutonic rocks are predominantly of I-type and associated with Late Triassic, Early Jurassic, Late Cretaceous and Early Tertiary volcanism. Volcanic and sedimentary rocks are widespread, the latter occurring in thick successions in the Skeena Mountains and Spatsizi and Nechako plateaus where they respectively define the Bowser, Sustut-Skeena and Nechako basins (Fig. 2.5).

The style and geometry of structures in the Intermontane Belt vary with competency of bedded rocks and their position within the belt. Broad folds prevail in thick volcanic sequences, whereas tighter folds, commonly associated with thrust faults, predominate in sedimentary sequences in the northern and central parts of the belt. Westward directed structures of Middle and Late Jurassic ages are conspicuous in the easternmost Intermontane

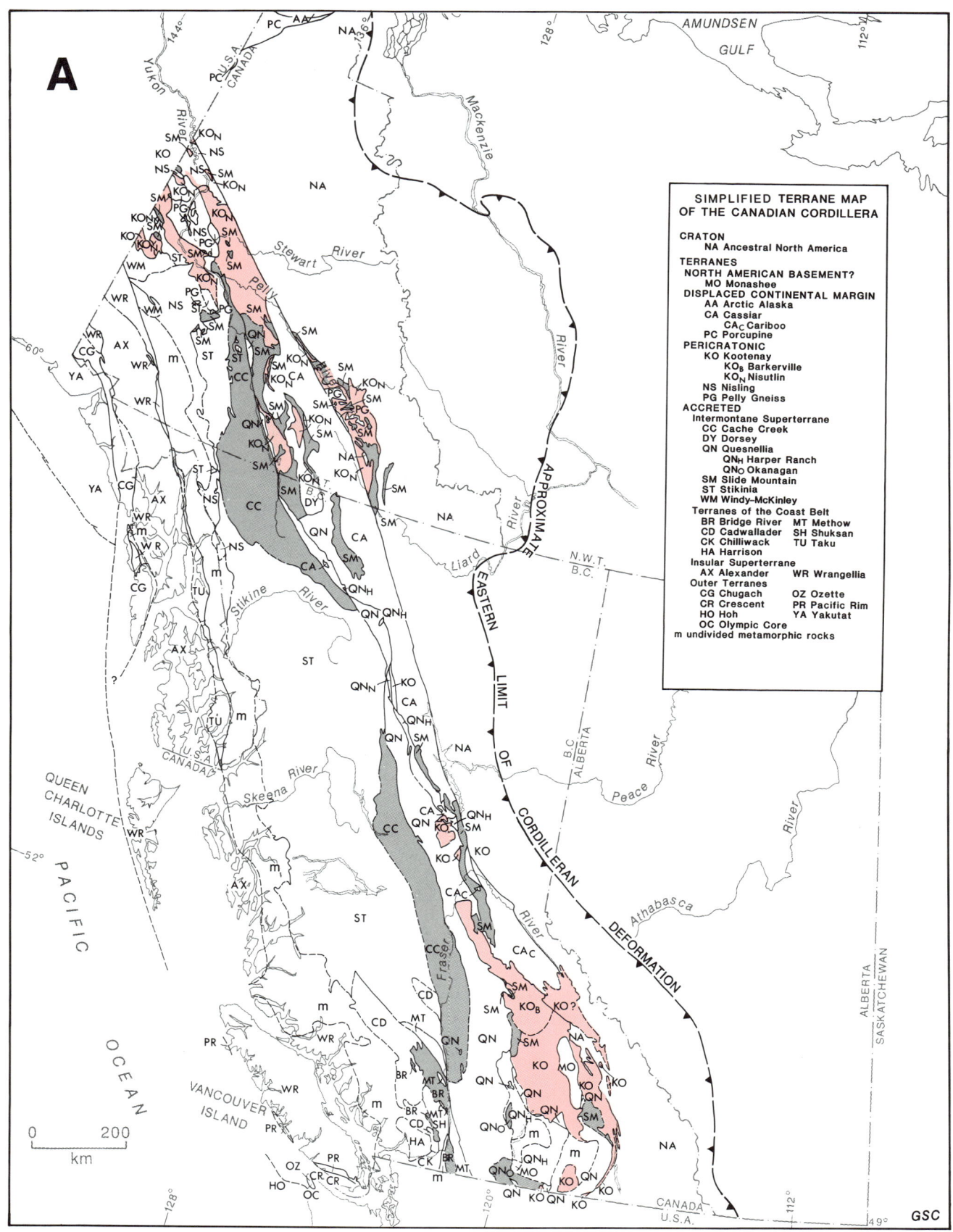

Figure 2.6. (A) Simplified terrane map of the Canadian Cordillera. Dark grey shading represents oceanic terrane and pink shading represents pericratonic terrane. (B) Block diagram illustrating structural relationship of terranes in the Quesnel Highlands and Cariboo Mountains.

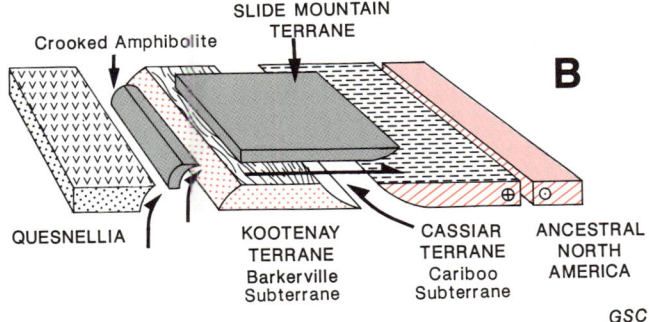

Belt whereas eastward-verging structures of Cretaceous and Tertiary ages occur in the western part of the belt.

Like the Omineca Belt the Intermontane Belt is traversed by northwesterly trending Late Cretaceous and Tertiary dextral strike-slip faults. The aggregate displacement on these faults, when combined with linked right-lateral faults in the Omineca Belt, amounts to more than 1000 km.

The Intermontane Belt is largely coextensive with the Intermontane Superterrane (Fig. 2.4), an amalgam of terranes including Stikinia, Quesnellia and the Slide Mountain and Cache Creek terranes. These terranes are overlapped by sediments derived from bordering uplifts resulting from contraction during terrane amalgamation and accretion; examples include the Bowser and Sustut basins. Flat-lying Tertiary volcanic rocks overlie the terranes and basins producing the widespread uniformly low relief in the southern half of the belt. Only the youngest eruptions gave rise to the mountainous shield volcanoes of Mount Edziza, Level Mountain and those of the westerly trending Anahim Belt. Late Cretaceous uplift and inversion of the Bowser Basin formed the Skeena Mountains, thereby interrupting the low relief of the belt at that latitude.

Coast Belt

The Coast Belt is a rugged, high relief region forming the Coast and Cascade mountains and composed largely of granitic and metamorphic rocks of the Coast Plutonic Complex. In the south its boundary with the Insular Belt lies just off the coast of mainland British Columbia and roughly follows the western margin of the Coast Plutonic Complex. Farther north, in the Yukon its western boundary follows the Denali Fault and in southeastern Alaska coincides with the eastern margin of the Gambier Assemblage. It is separated from the southern Intermontane Belt by the Pasayten, Fraser and Hungry Valley faults. Northwards the eastern boundary is the margin of the Coast Plutonic Complex and, in the Yukon, where it occurs adjacent to the western Omineca Belt, its eastern boundary is drawn at the eastern margin of dominantly Cenozoic plutonic rocks.

Metamorphism in the Coast Belt ranges from greenschist to upper amphibolite facies and reflects deeper burial and substantially greater uplift during Cretaceous and Tertiary time than that experienced by adjacent belts. The Coast Plutonic Complex consists of numerous I-type granitic plutons of average quartz diorite composition, many of which are separated from one another by high grade gneiss. Plutons in the western part of the belt are mid-Cretaceous and older, whereas those to the east are mainly Late Cretaceous and Tertiary. The southernmost part of the Coast Belt includes upper Paleozoic volcanic-arc and upper Paleozoic and Mesozoic oceanic and clastic rocks of the Cascade Fold Belt together with its plutonic and metamorphic core.

Like the Omineca Belt the Coast Belt is a symmetrical orogen with mid- to Late Cretaceous folds and thrust faults which verge outwards from a central core zone. Westward verging structures of mid-Cretaceous age may reflect eastward underthrusting of the Insular Superterrane beneath the Intermontane Superterrane, and the eastward directed structures of Late Cretaceous age possibly are due to subsequent uplift of the two superterranes.

The Coast Belt is a metamorphic and plutonic welt that appears to have been created by either or both long-term subduction and the accretion of the Insular Superterrane to the Intermontane Superterrane. The former comprises the Alexander Terrane and Wrangellia which are seen in roof pendants and inter-pluton septa, respectively in the northern and southern parts of the belt. Upper Jurassic to Lower Cretaceous volcanic and sedimentary rocks of the Gambier Assemblage overlap these two terranes. Several other smaller terranes can be identified along the southeastern margin of the Coast Belt and, in the southwest, the belt is overlapped by Upper Cretaceous and Tertiary clastic wedge of the Nanaimo Assemblage, part of which was derived from the adjacent Insular Belt and the remainder from the Coast Belt.

Insular Belt

The Insular Belt includes the modern Pacific continental margin comprising Vancouver Island, the Queen Charlotte Islands, the Alexander Archipelago of southeastern Alaska and the Saint Elias Mountains of southwestern Yukon and northwestern British Columbia. Its western boundary lies at the base of the continental slope, which off Vancouver Island is the locus of convergence between the Juan de Fuca and North American plates and, adjacent to the Queen Charlotte Islands, is represented by the Queen Charlotte Transform Fault. Its eastern boundary is hidden beneath the waters covering shelf basins between the mainland and Vancouver Island and the Queen Charlotte Islands. Northwards its boundary with the Coast Belt is the Denali Fault.

On Vancouver Island and in the Queen Charlotte Islands the belt comprises Paleozoic, Mesozoic and Cenozoic volcanic arc, oceanic and clastic wedge assemblages which are disrupted by northwesterly trending dextral strike-slip faults, westerly verging thrust faults, plutons and anticlinoria. The adjacent interior and outer continental shelves are underlain by thick clastic successions which, off western Vancouver Island, enclose detached fragments of oceanic crust within a subduction complex above the modern descending Juan de Fuca Plate. In contrast with its northern continuation in the Saint Elias Mountains and with the Coast Mountains to the east, the mountains of Vancouver Island and the Queen Charlotte Islands, though locally rugged, are topographically lower.

The Saint Elias Mountains are the highest and most scenically spectacular in Canada. Arc, back-arc, oceanic, mélange, platform and offshelf assemblages of Paleozoic, Mesozoic and Cenozoic ages are intruded by northwesterly aligned plutons and disrupted by right-lateral strike-slip faults, northeast- and southwest-dipping thrust faults and northwesterly trending folds. Vast areas of the mountains are covered by permanent ice which extends outwards as glaciers through narrow valleys to the sea. Local relief is great, with Mount Logan, the highest in Canada, exceeding 6000 m.

The Insular Belt comprises several terranes, the largest of which are Alexander Terrane and Wrangellia. The former is contained largely within Saint Elias Mountains and is believed to extend beneath Tertiary sediments and volcanics of the northeastern Queen Charlotte Islands and northern Hecate Strait. Wrangellia forms most of the remainder of the belt. Beneath western and southern Vancouver Island, the Pacific Rim and Crescent terranes have been emplaced beneath Wrangellia along prominent westerly and northwesterly trending thrust faults. Beneath these young terranes, multichannel seismic reflection profiles reveal the presence of additional material accreted to the base of Wrangellia and support the hypothesis that at least part of the Cordillera is composed of vertically stacked terranes of disparate origin.

Tectonic assemblages and plutonic suites

During the 130 years of geological exploration and research in the Canadian Cordillera vast amounts of data have been collected on the lithology, age and distribution of its component rock units (see Correlation charts in pocket). Extensive detailed descriptions of lithostratigraphic, intrusive and highly metamorphosed rocks have enabled a synthesis by grouping related units into assemblages which reflect specific tectonic or depositional environments. Thus established, assemblages allow comparisons of the tectonic behaviour of regions in the Cordillera during specific intervals of time. An assemblage may comprise one or more formations from a single region or from several separate regions. Adjacent assemblages are separated either by unconformities or by faults; in the latter case original associations between them may be unknown. Suites of plutonic rocks are defined mainly by age and subdivided on the basis of composition, and, are not considered to be integral parts of stratified assemblages which host them. Metamorphic assemblages of known stratified protolith are included among the tectonic assemblages; those of unknown protolith are unnamed.

The Tectonic Assemblage Map (Map 1712A, in pocket) is an attempt to group the rock components of the Cordillera into an optimum number of assemblages which best reflect its tectonic history. They are, in effect, the fundamental components of Cordilleran geology. Most assemblages are named for an important constituent formation or group, although a few are named after the region in which the assemblage is best developed. Each assemblage is identified, for the most part, according to the age range of its components and characterized in terms of its tectonic or depositional setting, the latter illustrated by descriptions of its principal lithologies, facies variations, source areas and other criteria. The assemblages and their constituent formations can be related readily to the terranes and morphogeological belts to which they belong.

The degrees of confidence in identifications of associated tectonic or depositional settings varies considerably and, in some cases, are controversial. Most assemblages are categorized in terms of settings currently observable on modern continental margins, island arcs and in ocean basins. Others are defined with reference to their orogenic position (foredeep clastic wedge) or their cratonic position (passive continental margin sediments). Regardless of their morphogeological belt or terrane associations, the assemblages provide an effective means of describing the tectonic and depositional history of the many regions of the Canadian Cordillera.

Tectonic settings of assemblages

The Middle Proterozoic to Middle Jurassic succession of Ancestral North America comprises several assemblages which formed on or near the ancient continental margin. Within the Foreland and eastern Omineca belts, where these sequences are exposed, the assemblages mainly represent passive continental margin and rift settings. The former includes widespread carbonate platform sequences, e.g. the Cambrian to Devonian Rocky Mountains Assemblage, characteristic of the Paleozoic miogeocline.

The terrane assemblages predominantly represent volcanic arc, e.g. the Mesozoic Stuhini, Nicola and Hazelton assemblages, and oceanic settings, e.g. Cache Creek and Slide Mountain assemblages. Regardless of its place of origin each terrane includes assemblages which, by analogy with modern examples, developed above subduction zones of differing polarity, in oceanic rift zones or marginal basins and in transform environments.

The effects of collision between amalgamating terranes and between the superterranes and Ancestral North America are recorded by accretionary and overlap assemblages derived, respectively from subduction processes and by erosion of uplifted rocks within the zones of accretion (Fig. 2.7). The mélanges of the Cache Creek Assemblage and the modern subduction complex along the Pacific continental margin are accretionary prism assemblages that resulted from subduction of oceanic lithosphere. In Bowser Basin an overlap assemblage was deposited upon a single terrane (Stikinia) in response to the amalgamation of Stikinia and Cache Creek Terrane. Other overlap assemblages were deposited across two or more terranes providing an upper age limit on the time of accretion (Gambier Assemblage on Wrangellia and the Alexander Terrane). Post-accretionary volcanic and sedimentary assemblages such as the Kamloops and Skeena assemblages, underlie extensive areas and, for the most part, developed from a variety of arc, rift and transform-related settings.

Plutonic suites and metamorphic rocks

Plutonic and metamorphic rocks underlie extensive areas in all but the Foreland Belt. Many of the plutonic rocks are assigned to suites characterized by distinctive compositions and ages. Some are of uncertain affinity to the various suites and thus are unassigned. Some plutonic rocks are closely associated spatially and temporally with island arc volcanic assemblages and, in the Intermontane, Coast and Insular belts, are of I-type and include distinctive

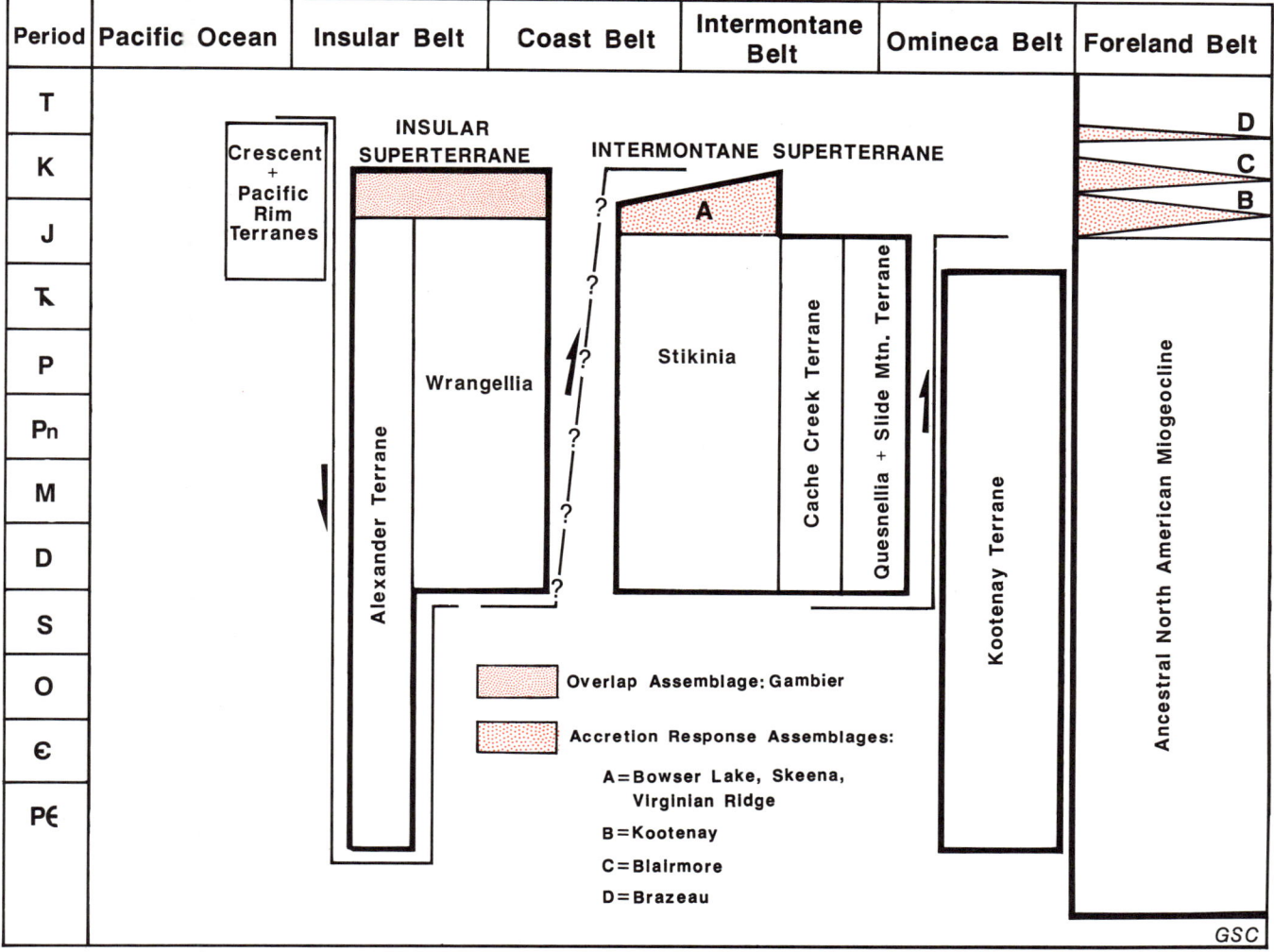

Figure 2.7. Temporal and spatial history of amalgamation and accretion of the major terranes with ancestral North America.

zoned ultramafic plutons. S-type granitic rocks occur in the Omineca Belt, the metamorphic core zone and cover rocks of which are believed to lie above the ancient margin of cratonal North America. In the Coast Belt, which includes the suture between the Insular and Intermontane superterranes, the Coast Plutonic Complex contains pre-, syn-, and post-accretionary granitic rocks which developed in response to either or both long-lived subduction and accretion. As in the Omineca Belt the metamorphic rocks of the Coast Belt are thought to have formed in response to deep burial and tectonic thickening during and following accretion between the two superterranes.

Terranes

The diverse terranes are the main building blocks of the Canadian Cordillera and account for four-fifths of its width (Fig. 2.6A; Terrane Map - in pocket). The application of tectonic assemblages to studies of Cordilleran geology has aided the recognition of terranes and has provided information on when they came together, how they were accreted to the western margin of Ancestral North America to form the five morphogeological belts, and their subsequent post-accretionary compressive deformation and dispersion by transcurrent faulting.

Terranes are parts of the earth's crust which preserve a geological record different from those of neighbouring terranes (Jones et al., 1983; Monger and Berg, 1984). The term has no genetic significance nor does it imply an origin far removed from neighbouring terranes or its present position. Because most terranes have unknown paleogeographic relationships with respect to North America, Coney et al. (1980) referred to them as "suspect" terranes. In this volume, however, only the term "terrane" is used. The boundaries between terranes are faults, though, in places, these may be concealed by younger cover rocks, intrusions, alluvium or water. The attitudes and motions on the bounding faults indicate that terranes may have been thrust onto (obducted), wedged into, subducted beneath or may have slid horizontally past adjacent terranes

(Fig. 2.6B). Terrane thickness is variable; some appear to be comparatively thin nappe-like sheets whereas others appear to be at least 18 to 20 km thick. Paleomagnetic and paleontological evidence suggests that some currently juxtaposed and overlapping terranes were originally separated from one another and/or from Ancestral North America by distances of up to thousands of kilometres (northern part of Slide Mountain and Cassiar terranes). In other instances great displacements have not been demonstrated (Quesnellia and Kootenay Terrane). Although considerable differences in stratigraphic and structural characteristics are evident, relative displacements between some terranes are thought to be small (Kootenay Terrane and Ancestral North America).

Terranes are defined only on the basis of their internal assemblage composition. In cases of those that originated as fragments of crust at distant paleolatitudes and were accreted to Ancestral North America at much later stages in their tectonic history, only the pre-accretionary plutonic suites and assemblages, and accretionary mélanges define the terrane. Post-accretionary plutonic rocks and overlap assemblages are excluded. However, in those instances where two or more terranes amalgamated prior to their accretion to Ancestral North America to produce a "superterrane" the post-accretionary and overlap assemblages overlying the component terranes are included in the definition of the superterrane (Fig. 2.8A). For example, the Lower Jurassic Inklin Assemblage is considered to be part of the Cache Creek Terrane although it has been derived, at least in part, from Quesnellia.

Characteristics of Cordilleran terranes

Terranes in the North American Cordillera range in size from Stikinia in British Columbia and Yukon Territory, which has an area of approximately 375 000 km^2, to small bodies a few kilometres in extent such as those reported from the Franciscan Complex of California (Silberling et al., 1984). Wrangellia presently extends over 16 degrees of latitude from Vancouver Island to southern Alaska, however, paleomagnetic data suggest that its original spread was less than five degrees of latitude (Yole and Irving, 1980).

Terranes display a variety of characteristics. Although dislocated by imbricate thrust faults in the Foreland Belt, Ancestral North America is comparatively coherent and displays a "layer cake" stratigraphy with laterally uninterrupted facies changes. On the other hand disrupted terranes such as the Cache Creek Terrane are, at least locally, intensely deformed and contain a wide variety of oceanic and arc-volcanic assemblages commonly in the form of mélange and broken formation. Metamorphic terranes such as the Nisling Terrane have a metamorphic overprint that largely conceals the original stratigraphy but retain sufficient protolith characteristics to distinguish them from adjacent terranes. Some terranes such as the Cache Creek and Nisling terranes contain single assemblages: respectively, the Cache Creek and Nisling assemblages. "Subterranes" are divisions of terranes and are identified so as to indicate affinity, but not necessarily stratigraphic continuity with a given terrane. "Superterranes" are composed of two or more component terranes that were amalgamated prior to their accretion to North America.

From east to west the six major terranes recognized to date are Slide Mountain Terrane, Quesnellia, Cache Creek Terrane, Stikinia, Alexander Terrane and Wrangellia (Fig. 2.5 and Map 1713A, in pocket). These "accreted" terranes became incorporated with North America at a late stage in their tectonic histories. Each has a distinctive stratigraphy and characteristic paleontological and/or paleomagnetic signatures which indicate that they originated at varying distances from and along the continental margin of Ancestral North America. All contain assemblages indicative of ensimatic volcanic arc and/or oceanic to marginal basin settings.

Between the accreted terranes and rocks that formed on and adjacent to the ancient continental margin are the "pericratonic" terranes, all of which form components of the Omineca Belt. Parts have stratigraphic affinities with the margin of Ancestral North America and therefore may not have been greatly displaced. Other parts, however, have stratigraphic and structural characteristics that are not seen in the rocks of the margin of Ancestral North America. For example, the pericratonic Kootenay Terrane is composed of rocks that traditionally have been regarded as metamorphosed sedimentary and/or volcanic equivalents of successions that accumulated on or near the continental margin.

As a consequence of Late Cretaceous and Tertiary dextral motions along transcurrent faults, parts of the ancient continental margin were displaced as were superimposed terranes which had previously accreted to the edge of the continent. An example is the Cassiar Terrane containing miogeoclinal assemblages, including platform carbonate, which was displaced northwards hundreds of kilometres from its original position (Fig. 2.6B). The Arctic Alaska Terrane of northern Yukon Territory also is regarded as a displaced fragment of the continent but its sense and amount of displacement relative to Ancestral North America is uncertain. The Monashee Terrane comprises metasedimentary rocks containing slices of paragneiss and orthogneiss of Early Proterozoic age which may be part of the North American continental crust.

Several small terranes, some the size of large nappes, occur in the southeastern part of the Coast Belt where their definition and recognition is in an active state of revision, as is the understanding of their relationship to other terranes. Small terranes beneath Wrangellia on the western and southern sides of Vancouver Island and in northwestern British Columbia probably have not travelled far and were emplaced by motion along Early Tertiary dextral faults and through underplating processes related to plate convergence.

Terrane amalgamation, accretion and dispersion

The timing and nature of terrane amalgamation and accretion are identified by various stratigraphic associations and/or structural, metamorphic and intrusive effects. The importance of "overlap" assemblages relative to the timing of amalgamation is described in the previous section on tectonic assemblages. A record of terrane accretion is contained in the foredeep clastic assemblages. For example, the Kootenay Assemblage of the Foreland Belt is a clastic wedge that is thought to have been derived from uplift of the Omineca Belt as a consequence of the accretion of the

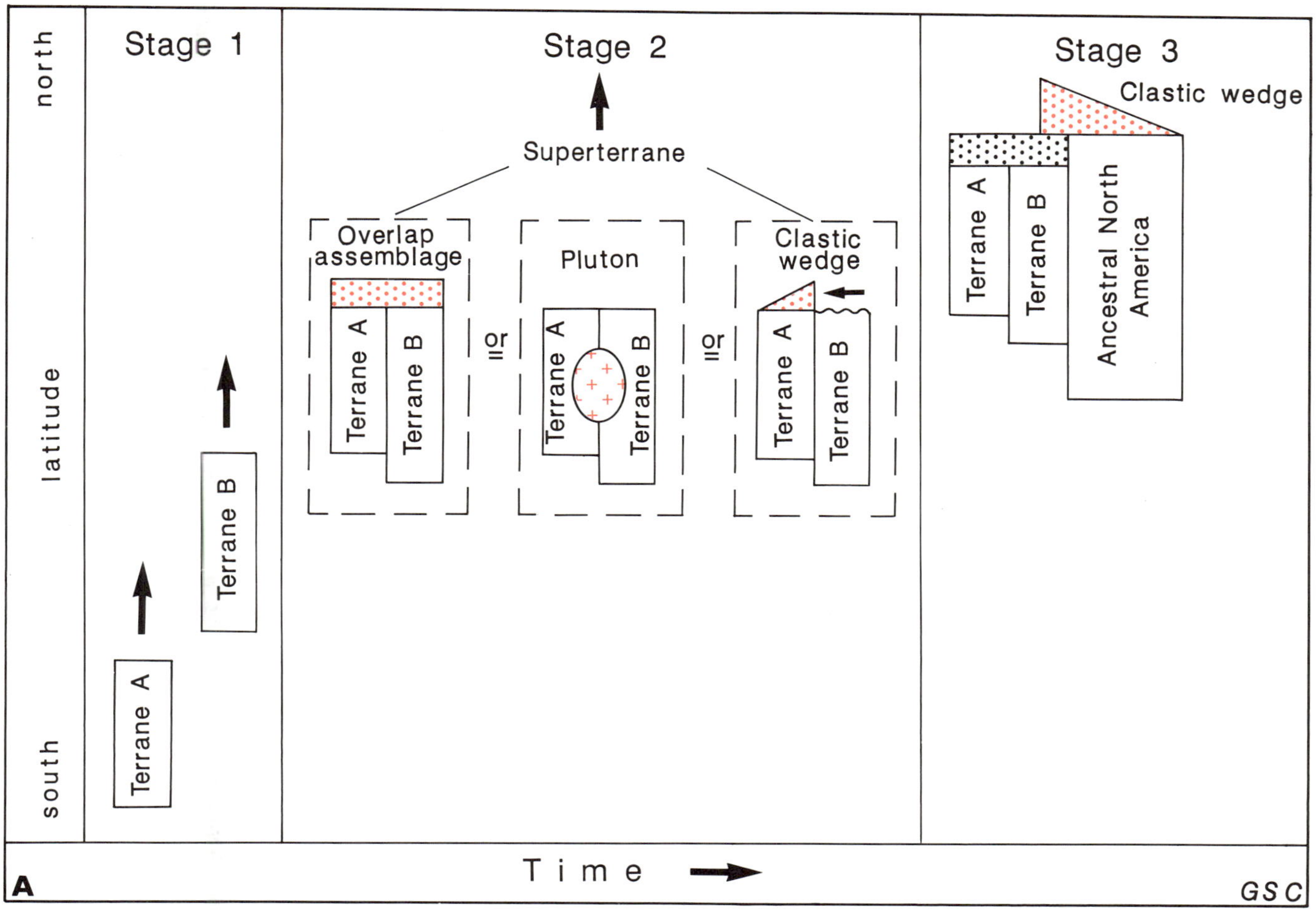

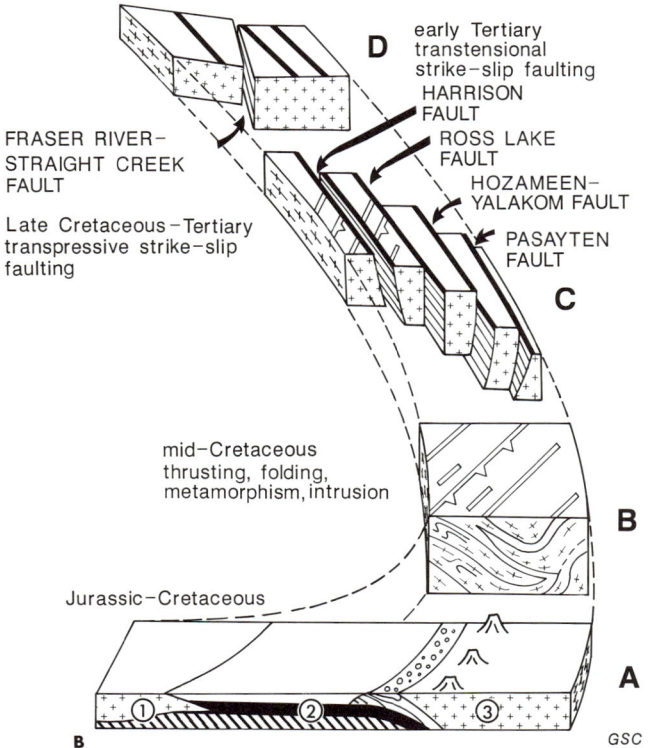

Figure 2.8. A. Schematic diagram showing stages of terrane amalgamation and accretion with ancestral North America. B. Speculative depiction of evolution of structures in the Cascade Fold Belt and southeastern Coast Plutonic Complex. An oceanic/marginal basin (A) was closed in Early Cretaceous time, with the late Early Cretaceous Spences Bridge arc, Jackass Mountain fore-arc (?) on its eastern margin. Final closure/collision (B) produced mid-Cretaceous metamorphic rocks and northeasterly trending structures. These were cut by north- northwest-trending strike-slip faults (C), with mainly dextral offsets, that were cut in turn by the north-trending, dextral early Tertiary Fraser River-Straight Creek Fault system (D). 1. Insular Superterrane (Wrangellia + Alexander + Chilliwack) overlapped by lower Lower Cretaceous Peninsula/Brokenback Hill=Gambier. 2. "Oceanic/marginal basin" Bridge River, Hozameen, Darrington, Shuksan + Cogburn, Settler; Jackass Mountain forearc?. 3. Cretaceous North America + upper Lower Cretaceous Spences Bridge continental arc.

Intermontane Superterrane with the continental margin of Ancestral North America. Finally, terranes may be "welded" by plutons whereby the age of the pluton again provides a minimum age for amalgamation or accretion; the Coast Plutonic Complex, at least in part, is thought to be due to the mid-Cretaceous accretion of the Insular Superterrane to the Intermontane Superterrane.

The converse of these processes is the disruption and dispersion of formerly continuous terranes (Fig. 2.8B). Most apparent are Late Cretaceous and Tertiary displacements that occurred along dextral transcurrent faults such as the Tintina, Northern Rocky Mountain Trench, Fraser River, Denali and Chatham Strait faults and the currently active Queen Charlotte-Fairweather Fault system. More difficult to document are episodes of extension, rifting and dispersion that probably took place in the early stages of Cordilleran evolution when, during the Middle and Late Proterozoic, the western continental margin of Ancestral North America was established.

Paleomagnetic and paleontological data indicate when accreted terranes reached their present latitude. Paleolatitude positions at the time of rock formation and at the time of a widespread mid-Cretaceous paleomagnetic overprint have been determined for Wrangellia, Stikinia, Quesnellia, the Cache Creek Terrane and from rocks in the Foreland and Coast belts. Paleomagnetic and geological data broadly agree with respect to terrane amalgamation into the Insular and Intermontane superterranes and the mid-Cretaceous time of superterrane amalgamation. Paleontological and other geological data, however, have not been fully reconciled with paleomagnetic results with respect to the amount and nature of post-mid-Cretaceous northward movement of the accreted terranes. Nor have they been reconciled regarding timing of accretion of the Intermontane Superterrane and Ancestral North America.

PART B. PALEONTOLOGICAL SIGNATURES OF TERRANES

E.S. Carter, M.J. Orchard, C.A. Ross, J.R.P. Ross, P.L. Smith, and H.W. Tipper

During the past decade, advances in Cordilleran biostratigraphy and biogeography have been made through studies of conodonts, foraminifera, radiolaria and ammonites, particularly with regard to the comparison of biostratigraphic components of Ancestral North America with the fine grained clastic and chert sequences in the accreted and pericratonic terranes. In addition to their biochronological value, fusulinacean and ammonite faunas have proven useful in indicating paleogeographic affinities of upper Paleozoic and Mesozoic sequences and thus have added support and constraints on paleolatitude origins suggested by paleomagnetic results.

Following the suggestion by Tozer (1970), Monger and Ross (1971) provided the first definitive statement of the exotic origin of part of the North American Cordillera based upon the endemism (geographical restriction) of certain fusulinacean faunas. Strong support for this conclusion was provided by Irving and Yole (1972) who determined that the ancient paleolatitude of Upper Triassic rocks of Vancouver Island, relative to cratonal North America, lay far to the south. Studies of ammonite-bearing sequences, particularly in the Intermontane and Insular superterranes, established previously unrecognized nearly complete biostratigraphic records of Lower and Middle Jurassic strata in northwestern British Columbia (Tipper and Richards, 1976) and on the Queen Charlotte Islands (Cameron and Tipper, 1985). Comparisons of the faunas suggest a resolution as to which hemisphere a terrane originated from as determined from paleomagnetic studies (Tipper, 1984). Studies of Cretaceous ammonites have been primarily of biostratigraphic value, nevertheless, there are substantial differences in the Cretaceous faunal realms of the Foreland Belt and the accreted terranes.

Fusulinids
Charles A. Ross and June R.P. Ross

Fusulinacean foraminifers, because of their abundance and wide distribution, have proven to be most useful in Cordilleran paleogeographic reconstructions. The Tethyan Permian faunal province contains abundant and diverse members of the fusulinacean family Verbeekinidae, whereas the other faunal provinces lack verbeekinids except for one rare species (Ross, 1967, 1973; Fig. 2.9). A precursor of this distinctive Permian fauna was present in both the Tethyan and Ural areas during the mid-Carboniferous. Some ubiquitous genera include distinctive species lineages or subgenera that were restricted to the Tethyan province (Fig. 2.9; Table 2.1).

Early Carboniferous (particularly Tournaisian and Viséan) faunas in the western Cordilleran terranes included a larger percentage of cosmopolitan genera than did the faunas of the mid-Carboniferous (Mamet, 1976; Ross and Ross, 1985). This trend towards increased provinciality was worldwide and reflected continental reorganizations that resulted in the closure of an equatorial seaway near the end of the Early Carboniferous (Fig. 2.10A). A further increase in the degree of endemism occurred in the Early Permian when the marine connection between the Tethyan and Ural seaways closed (Fig. 2.10C).

Common fusulinacean genera in western Canada are mid-Carboniferous through Late Permian in age and show the considerable degree of generic diversity among terranes

---

Carter, E.S., Orchard, M.J., Ross, C.A., Ross, J.R.P., Smith, P.L., and Tipper, H.W.
1991: Part B. Paleontological signatures of terranes; in Chapter 2 of Geology of the Cordilleran Orogen in Canada, H. Gabrielse and C.J. Yorath (ed.); Geological Survey of Canada, Geology of Canada, no. 4, p. 28-38 (also Geological Society of America, The Geology of North America, v. G-2).

TECTONIC FRAMEWORK

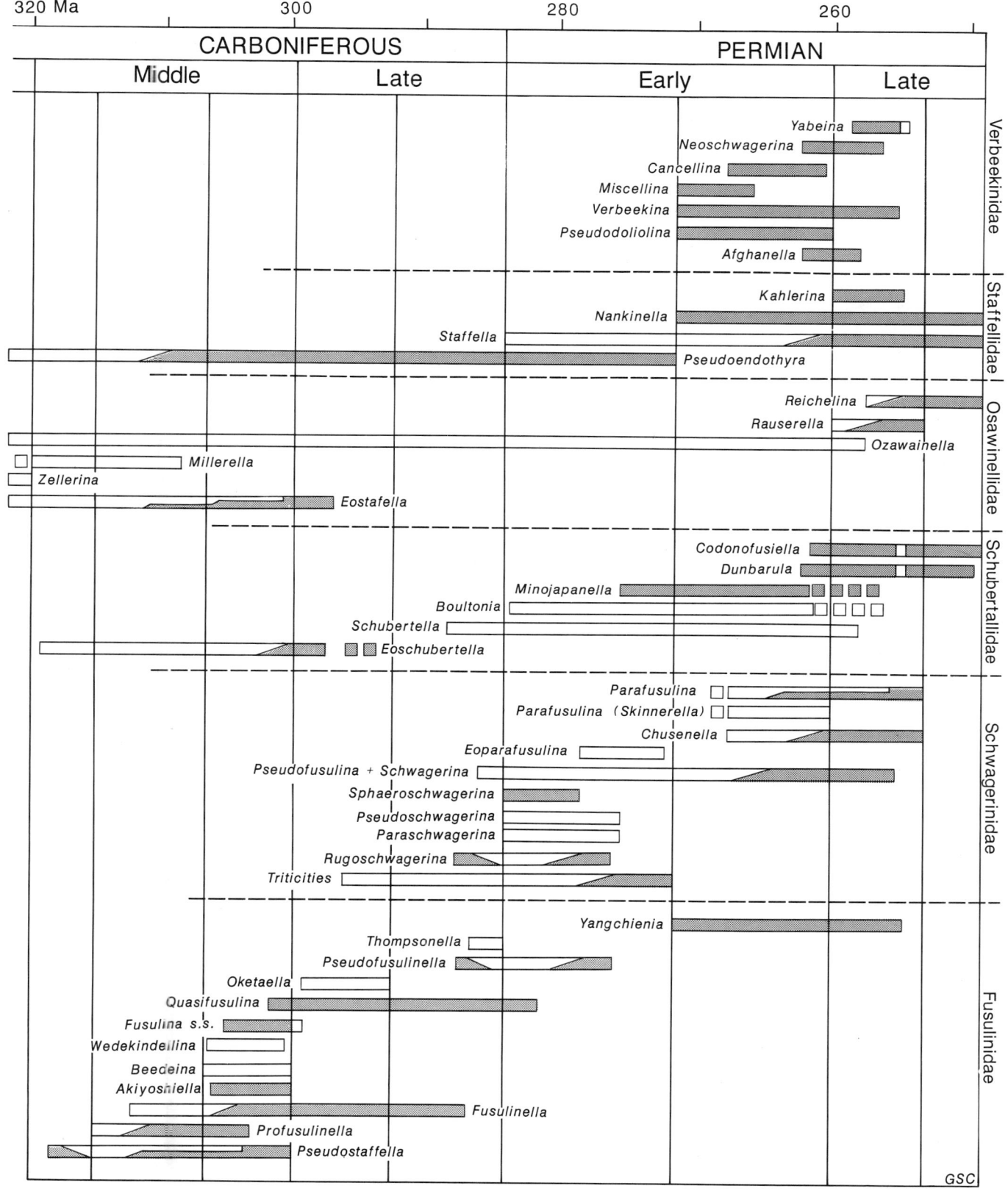

Figure 2.9. Stratigraphic ranges of fusulinacean genera listed in Table 2.1. Shaded ranges indicate those parts of a generic range that are associated only with a Verbeekinidae fauna in mid- and Upper Permian strata or are part of a Tethyan-Uralian fauna in Middle Carboniferous strata. In some, distinctive subgenera or species complexes are associated with Verbeekinidae, but not the entire genus. (Data from Ross, 1967; Rozovskaya, 1975; Kanmera et al., 1976).

CHAPTER 2

Table 2.1 Fusulinaceans and other foraminiferans reported from representative localities in the western cratonic shelf and miogeocline

	Cratonic Shelf		Slide Mountain				Cache Creek				Stikinia		Alexander		Wrangellia			
	1	2	3	4	5	6	7	8	9	10	11	12	13	14	15	16	17	18
Verbeekinindae																		
late Early to early Late Permian																		
Yabeina							X	X	X	X								
Afghanella							X	X										
Neoschwagerina							X	X	X									
Cancellina							X	X	X									
Pseudodoliolina							X	X		X								
Verbeekina							X		X									
Miscellina							X		X									
Staffellidae																		
mid-Early Carboniferous to Late Permian																		
Kahlerina							X	X										
Nankinella						X	?		?	?					X			
Staffella	?						X	X	X	?	?				X			
Pseudoendothyra	X	X			X													
Ozawainellidae																		
mid-Early Carboniferous to Late Permian																		
Reichelina							X	X		X								
Rauserella										X								
Ozawainella						?									X			
Millerella				X	X	?	?		X						X			
Zellerina							?											
Eostaffella				X	X	X	X	X	X									X
Schubertellidae																		
mid-Carboniferous to Late Permian																		
Codonofusiella							X			X								
Dunbarula										?								
Minojapanella										X								
Boultonia							X	X		X	X			?				
Schubertella	X		X				X	X		X	X					X	X	X
Eoschubertella							X	X										
Schwagerinidae																		
Late Carboniferous to early Late Permian																		
Parafusulina		X			X		X	X	X	X								
Parafusulina (Skinnerella)											X			X				
Chusenella							X	X	X									
Eoparafusulina							X	X			X		X		X	X		
Schwagerina	X	X	X			X	X	X		X	X	X	X	X		X		X
Pseudofusulina	X						X	X	X	X								
Sphaeroschwagerina	X							X										
Pseudoschwagerina							X		X	X			X					
Paraschwagerina							?											
Rugosofusulina							X											
Triticites							X	X		X			X					
Fusulinidae																		
mid-Carboniferous to early Late Permian																		
Yangchienia							X											
Thompsonella			X					X										
Pseudofusulinella	X		X			X	X			X	X	X	X	X		X		
Oketaella										X								
Quasifusulina						X	X		X									X
Susulina							X	?									X	
Wedekindellina					X													
Beedeina							X		?						X			
Akiyoshiella									X									
Fusulinella					X	X	X		X						X			
Profusulinella	X	X			X								X		X			
Pseudostaffella					X			X							X			

(Column 1 - Belcourt and unnamed formation, northeastern British Columbia; Column 2 - southwestern Alberta), Slide Mountain Terrane (Column 3 - Tay River map area, Yukon; Column 4 - Sylvester Group, Cassiar District, British Columbia; Quesnellia, Column 5 - Lay Range, Omineca Mountains, British Columbia; Column 6 - "Harper Ranch Group," Kamloops, British Columbia), Cache Creek Terrane (Column 7 - Nakina area, northwestern British Columbia; Column 8 - Dease Lake area, northern British Columbia; Column 9 - Fort St. James, central British Columbia; Column 10 - Marble Canyon, south-central British Columbia), Stikinia Terrane (Column 11 - Scud River, British Columbia; Column 12 - Telegraph Creek, British Columbia; Column 13 - Asitka Group, Omineca Mountains), Alexander Terrane (Column 14 - Saint Elias Mountains, Yukon; Column 15 - Kuiu and Prince of Wales Islands, southeast Alaska), and Wrangellia Terrane (Column 16 - east-central Alaska Range; Column 17 - east coast Vancouver Island; Column 18 - Horn and Cowichan lakes, central Vancouver Island). (See Ross and Ross, 1983, for individual references to these fusulinacean localities). Within each family, genera are listed in ascending order of their first appearances; however, there is considerable overlap in the ranges of the families (see Fig. 2.9).

and between these and the miogeocline of Ancestral North America to the east (Table 2.1). Relatively few species are known from the miogeocline, and most of these have been recovered from the Arctic archipelago. Quesnellian faunas are diverse and show some similarities to those from the miogeocline; the associated faunal assemblages also are diverse, suggesting origins in somewhat warmer waters than those on the craton.

The Slide Mountain Terrane contains a fusulinacean fauna characterized by a giant species of *Parafusulina* (Ross, 1969). The microspheric stage of this species commonly reaches 10 cm in length. This species also occurs in the Kettle Falls area of Washington, in the eastern Klamath Mountains of northern California and at El Antimonio, Sonora and in the upper part of the Word Formation in west Texas. Except for the Texas locality which occurs in the miogeocline, all are in accreted terranes, which suggests that the Slide Mountain Terrane and other localities have been displaced from southern paleolatitudes.

The most diverse fusulinacean faunas in the Canadian Cordillera are in thick limestone successions in the Cache Creek Terrane. The Cache Creek includes verbeekinid faunas which are associated with many species lineages in several other families. It is this assemblage, characterized by the distinctive Verbeekinidae, that is of Permian Tethyan origin. Its faunal characteristics suggest that the terrane originated in the tropical part of the Paleozoic Panthalassa ocean basin (Fig. 2.10C,D).

Although the faunas from Stikinia are incompletely known, they lack verbeekinids and have a considerably lower species diversity than Cache Creek faunas. The fusulinids are generally similar to those of the McCloud Limestone of Sonomia in northern California and northwestern Nevada.

Wrangellia and the Alexander Terrane show lower species diversity than the more eastern terranes. The Alexander Terrane fusulinacean faunas are physically larger and, therefore, suggest warmer paleoenvironments than do those of Wrangellia. Fusulinaceans from Wrangellia comprise several genera, some of which are common to the Franklinian miogeocline which, during the Carboniferous and Permian, is considered to have occupied a temperate marine setting.

Conodonts
M.J. Orchard

During the last decade conodonts have become increasingly valuable for determining the extent of the stratigraphic record in the accreted, pericratonic and displaced terranes (Fig. 2.11-2.14). The usefulness of conodonts lies in their cosmopolitan distribution, short ranges and resistance to metamorphism. Precise biostratigraphic correlation within and between terranes is possible, as is a degree of paleoenvironmental reconstruction such as water depth. Their widespread distribution does not favour their use for determining paleolatitudes although most post-Ordovician conodonts appear to have inhabited a region bounded by 40° North and South (Sweet, 1985), and some known geographic endemism may reflect different paleolatitudes.

Early Paleozoic conodonts are comparatively rare in the Western Cordillera (Fig. 2.11). Ordovician conodonts are known from the Alexander Terrane in the Insular Belt and from the Cassiar Terrane and Quesnellia in the Omineca Belt. The most diagnostic fauna from the Alexander Terrane is of Middle Ordovician age and is reported from the Descon Formation of southeast Alaska (A. Harris, pers. comm., 1979). Identical "North Atlantic Province" *Pygodus - Periodon* faunas occur in the Selwyn Basin and the Kechika Basin where they represent a relatively cool, deep-water biofacies. In the Cassiar Terrane, as on the miogeocline to the east, relatively shallow, warm-water "American Province" faunas occur in coeval strata. A similar pattern is evident in the conodont faunas of Silurian age; those of the Alexander Terrane (Savage, 1985) and the Selwyn Basin are closely similar and indicative of basinal environments whereas those of the Cassiar Terrane are less distinctive, platformal taxa.

Devonian conodonts occur in several areas of the Cordillera, particularly in the miogeoclinal rocks of the Foreland Belt, in the Selwyn Basin and in the Alexander Terrane of the Insular and Coast belts. In each of these regions most of the Devonian stages are readily identified (Fig. 2.11) with reference to a standard conodont zonation that is applicable throughout the Cordillera in both autochthonous and allochthonous strata. Observed variations may be due to differing biofacies. As with the lower Paleozoic, platformal sediments are characterized by specific taxa, notably *Icriodus*, whereas sediments deposited in relatively deeper waters contain diverse *Polygnathus - Palmatolepis* faunas. In contrast to older faunas, Middle and Late Devonian conodonts from the Alexander Terrane more closely resemble those from the platformal sediments rather than those of Selwyn Basin.

The discovery of Early and Late Devonian species in Stikinia and Quesnellia respectively was the first evidence that strata of those ages occurred therein; they represent the oldest parts of those terranes currently recognized. The Stikinia fauna has counterparts in both the Alexander Terrane and Selwyn Basin. Quesnellia contains two contrasting Famennian faunas: a fauna from the Harper Ranch Group represents a shallow biofacies and has no known counterpart elsewhere in the Cordillera, whereas a faunule from chert of the Shoemaker Formation correlates directly with those from the Selwyn Basin and the Kechika Basin.

Carboniferous and Permian conodont faunas are much more abundant and widespread than those of earlier periods (Fig. 2.12). Unlike the Devonian zonal scheme, however, those of global applicability are not yet established for the late Paleozoic and correlation of different biofacies is equivocal. Conodont faunules have been assigned to 17 faunas (Fig. 2.13). The faunal succession and the inferred contemporaneity of biofacies is based upon associations described largely from outside the Canadian Cordillera, but the sequence provides a useful framework for paleobiogeographic comparisons. As with older faunas, conodont taxa from upper Paleozoic rocks are generally cosmopolitan or are widespread in western North America. In the Canadian Cordillera, Faunas 2 and 3 (Fig. 2.13) are associated with barite deposits and occur only in the east in areas marginal to the craton, whereas Fauna 1 occurs farther afield although in relatively deep-water settings. The occurrence of Fauna 4, typified by *Mestognathus*, in Stikinia was thought to be anomalous, but the genus appears to be more widespread than was originally recognized (Von Bitter et al., in press). Pennsylvanian *Gondolella* (Fauna 9), which usually has been regarded as typical of

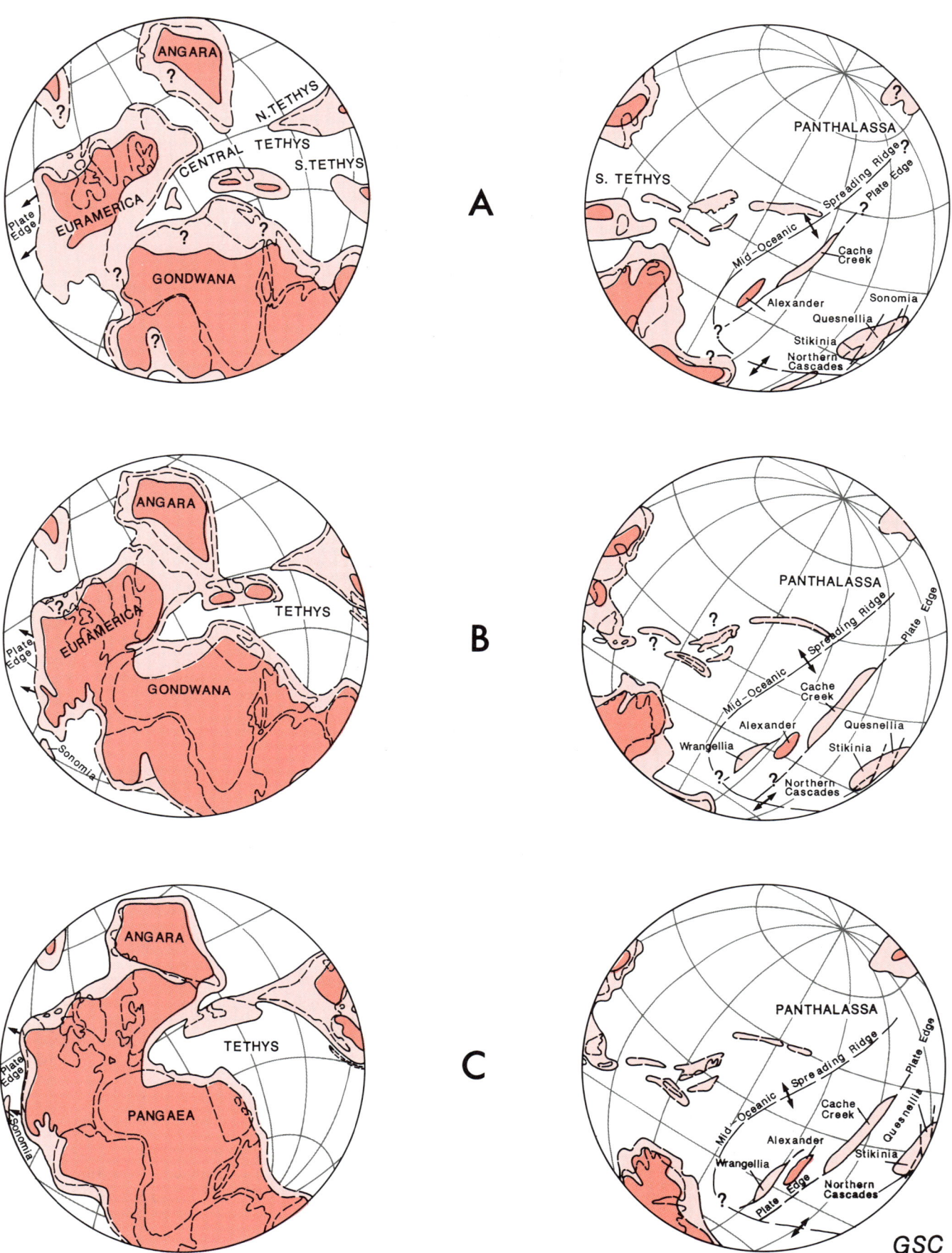

Figure 2.10. Schematic outline of possible origins for terranes bearing late Paleozoic Fusulinacean faunas. A, Early Carboniferous (Mississippian); B, Middle and Late Carboniferous (Pennsylvanian); C, Early Permian; D, Late Permian (modified from Ross and Ross, 1981, 1983, 1985).

TECTONIC FRAMEWORK

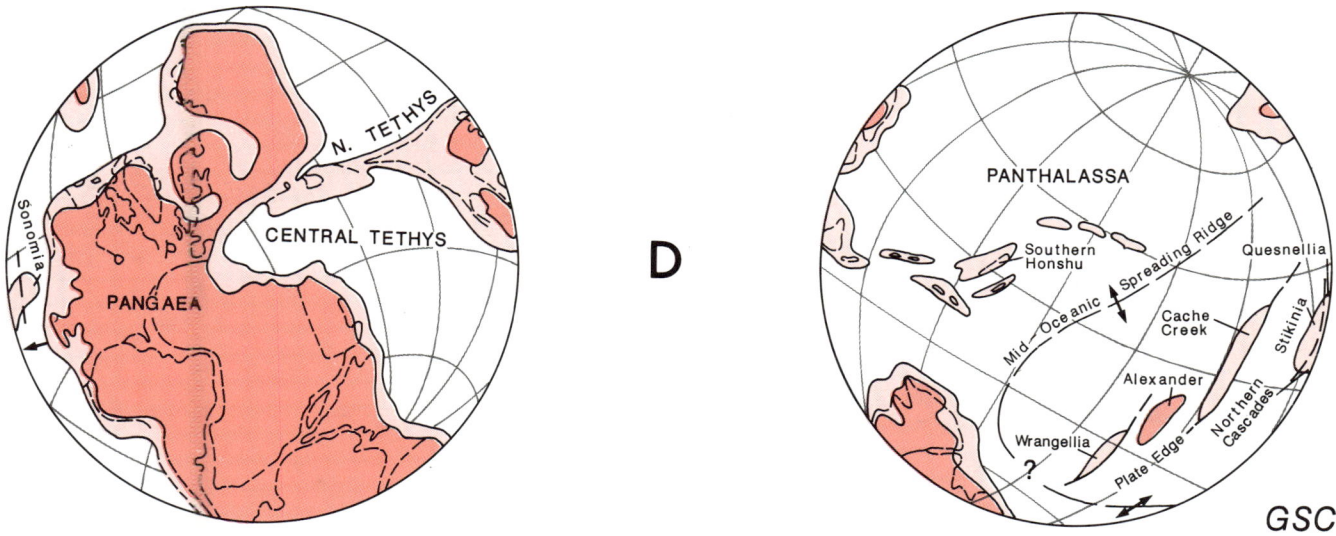

TERRANE Period/Age			NORTH AMERICAN AUTOCHTHON	CASSIAR	QUESNELLIA	STIKINIA	ALEXANDER
DEVONIAN	Late	FAMENNIAN	•	•	•		
		FRASNIAN	•	•	•		•
	Middle	GIVETIAN	•	•			•
		EIFELIAN	•	•		•	•
	Early	EMSIAN	•				•
		PRAGIAN	•			•	•
		LOCHOVIAN	•				•
SILURIAN			•	•			•
ORDOVICIAN			•	•	•		•

Figure 2.11. Ordovician through Devonian conodont record in Western Canada.

the North American midcontinent, occurs as far west as the eastern belt of the Cache Creek Terrane (in carbonate olistoliths, see Chapter 8), but has yet to be found within the Marble Canyon Formation of the central Cache Creek belt, the host of the Tethyan fusulines. This may be crucial to the understanding of relationships within the Cache Creek Terrane. Endemism is suggested for Fauna 17 from the Marble Canyon Formation, the best comparison for which is in the Late Permian of Eurasia. The uniqueness of Fauna 17 may also suggest that it is younger, and/or from a different sedimentary environment, than strata preserved elsewhere in North America.

The Triassic conodont record (Fig. 2.14) is one of relative uniformity. The Cache Creek Terrane and Stikinia are known to have similar Early Triassic faunas, although strata of this age are rare or absent elsewhere. The Marble Canyon Formation of the Cache Creek Terrane has a unique occurrence of Spathian *Platyvillosus*, a genus that otherwise is known only from low latitudes. Quesnellia and the Cassiar Terrane have several important Late Ladinian to Early Carnian elements in common and can be precisely correlated with cratonic sequences. The classic Norian succession in the Peace River area (Orchard, 1983) similarly provides a zonation that can be applied throughout the western Cordillera, although the sequence has not yet been completely studied. The youngest Late Norian conodonts from autochthonous strata in British Columbia probably predate a conodont fauna that characterizes Cadwallader, Sinwa and Lewes River rocks to the west. The 'Rhaetian' conodont *Misikella* has been found in the uppermost Triassic strata of the Cadwallader Terrane, Cache Creek Terrane and Wrangellia.

Radiolaria
E.S. Carter

Radiolarian research is just beginning in Western Canada. Radiolarian faunal zonation, calibrated with ammonoids, conodonts and foraminifers, holds great promise for correlating within and among terranes of the Canadian Cordillera. Excellent faunas from relatively uninterrupted sequences of Mesozoic strata in the Queen Charlotte Islands provide a wealth of material for future studies. These assemblages are associated with rich ammonoid and foraminiferal faunas which provide excellent stratigraphic control. Studies of latest Triassic and Early to early Middle Jurassic faunas concentrate primarily on descriptions of new taxa (Pessagno and Blome, 1980; Pessagno and Whalen, 1982; Carter et al., 1988) and on development of zonal schemes for the Jurassic (Carter, 1988; Pessagno et al., 1984) and Late Triassic (Blome, 1984).

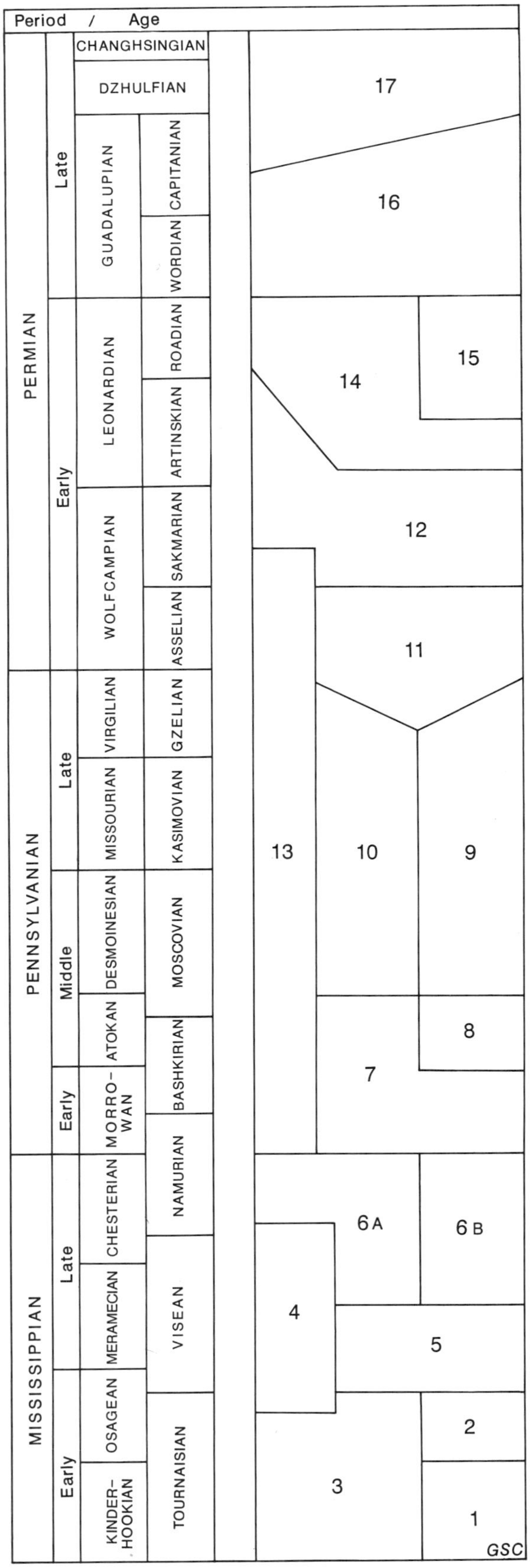

CHAPTER 2

Figure 2.13. Carboniferous-Permian conodont record in Western Canada. Faunas 1 to 17 identified in Figure 2.12. ? indicates uncertain occurrence (poor preservation).

Figure 2.12. Subdivision of Carboniferous and Permian conodont faunas 1-17 arranged with shallow-water biofacies to the left and those of deeper water affinity to the right. Characteristic or definitive taxa are: 1. *Siphonodella*. 2. '*Hindeodella*' *segaformis*. 3. *Pseudopolygnathus*. 4. *Mestognathus*. 5. '*Spathognathodus*'. 6. A. *Cavusgnathus*, B. *Gnathodus bilineatus*. 7. *Declinognathodus - Idiognathoides*. 8. *Neogondolella clarki - Gondolella laevis*. 9. *Gondolella* ex. gr. *magnus*. 10. *Gondolella* ex. gr. *gymna*. 11. *Streptognathodus elongatus*. 12. *Sweetognathus*. 13. *Adetognathus*. 14. *Neostreptognathodus*. 15. *Neogondolella* ex. gr. *idahoensis*. 16. *Neogondolella serrata* complex. 17. *Diplognathodus*.

For the Late Triassic, genera of the Capnuchosphaeridae and Pantanelliidae plus a few others of uncertain taxonomic affinity appear to be the most reliable age indicators (Fig. 2.15). These include *Capnuchosphaera*, *Sarla*, *Capnodoce*, *Betraccium*, *Kahlerosphaera*, *Ferresium* and *Laxtorum*. In addition, a few species of longer ranging genera such as *Pantanellium*, *Canoptum* and *Triassocampe* are useful. The majority of these forms are sturdy enough to be preserved in a variety of lithologies.

Jurassic radiolarians are distinct from Triassic forms, and, particularly in Lower Jurassic strata, there are many short-ranging taxa (Fig. 2.16). All occur abundantly in well preserved assemblages from the Queen Charlotte Islands and several have been observed elsewhere in the Cordillera. In the Queen Charlotte Islands (Wrangellia), Jurassic radiolarians occur in association with a diverse Tethyan ammonoid fauna. The radiolarian fauna also is diverse and has affinity with Tethyan assemblages from

Period / Age			Terrane: NORTH AMERICA AUTOCHTHON	CASSIAR	SLIDE MOUNTAIN	QUESNELLIA	CACHE CREEK	STIKINIA	BRIDGE RIVER	CADWALLADER	ALEXANDER	WRANGELLIA	PACIFIC RIM
TRIASSIC	Late	NORIAN L	●	●		●	?	●		●	●	●	
		NORIAN M	●	●	●	●		●		●	●	●	
		NORIAN E	●	●		●	●	●	●	●	●	●	●
		CARNIAN	●	●	●	●	●	●	●	●	●	●	●
	Middle	LADINIAN	●	●	●	●	?	●					
		ANISIAN	●	●		●	●	●					
	Early	SPATHIAN	●			?	●	●					
		SMITHIAN	●				●	●					
		DIENERIAN					●						
		GRIESBACHIAN					●						

GSC

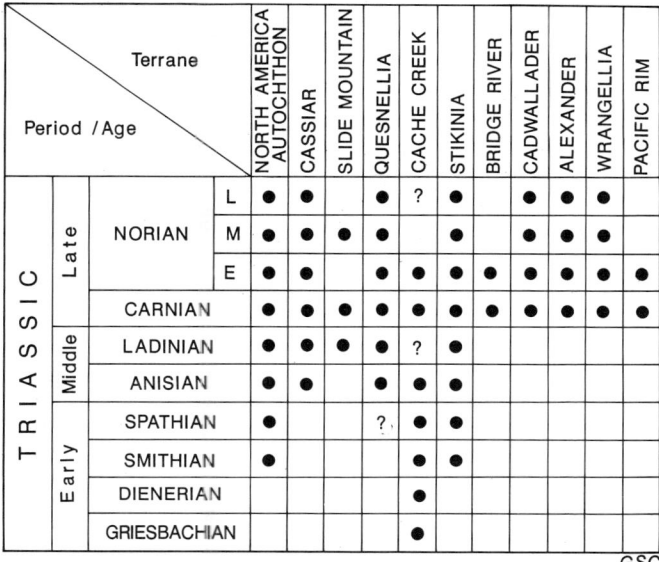

Figure 2.14. Triassic conodont record in Western Canada. ? indicates uncertain occurrence (poor preservation).

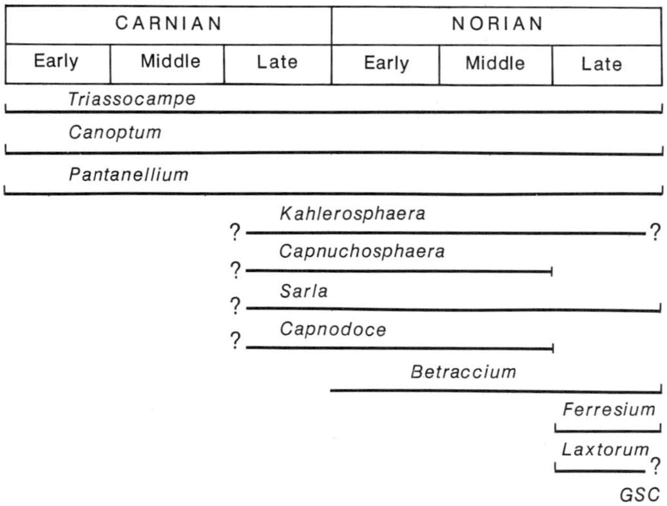

Figure 2.15. Range zones for selected Late Triassic radiolarian genera (Blome, 1984). Range of *Kahlerosphaera* from Kozur and Mostler (1979).

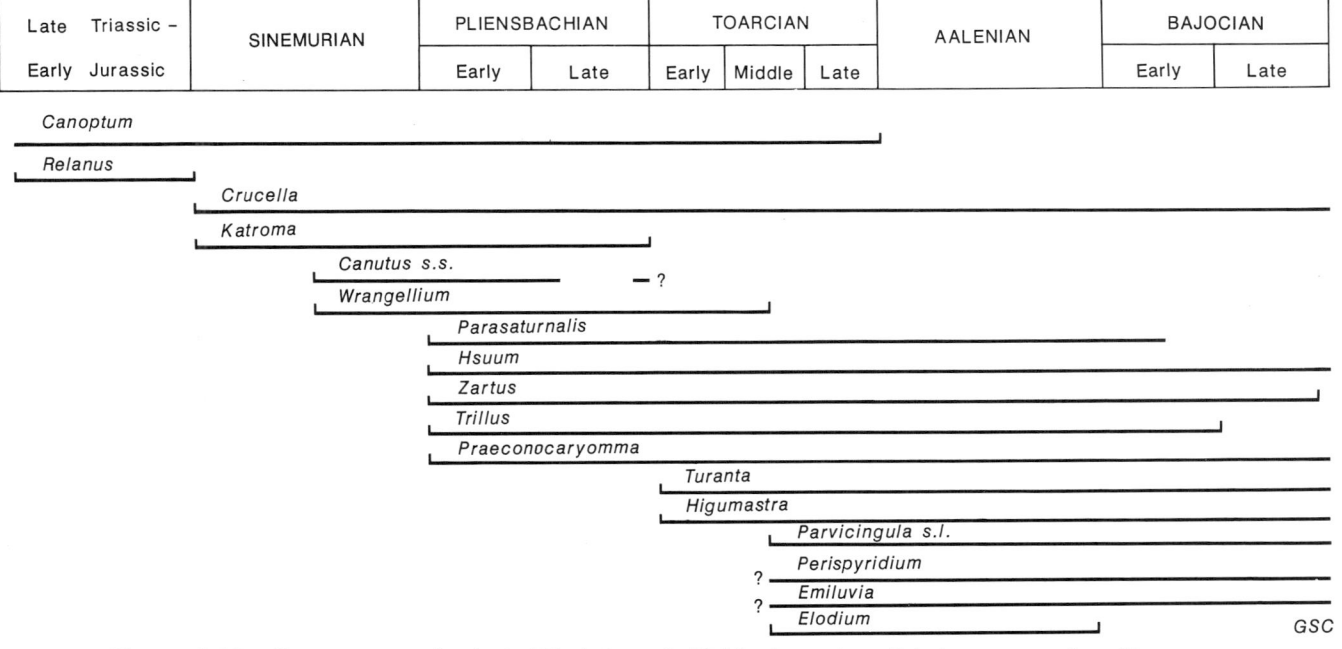

Figure 2.16. Range zones of selected Early to early Middle Jurassic radiolarian genera, from Pessagno and Blome (1980), Pessagno and Whalen (1982), Pessagno et al. (in press) and Carter (1988).

California (Franciscan Complex), east-central Oregon (Pessagno and Blome, 1980; Pessagno and Whalen, 1982), Turkey (Pessagno and Poisson, 1981; De Wever, 1982) and Japan (Yao, 1983). Pessagno et al. (1984) developed simple criteria for differentiating Jurassic radiolarian assemblages of the Tethyan and Boreal faunal realms and for subdividing each realm into provinces. Based on these criteria they assigned Hettangian to Middle Toarcian faunas in the Queen Charlotte Islands to the Central Tethyan Province of the Tethyan Faunal Realm; Middle Toarcian to Upper Bajocian faunas were compared with the Northern Tethyan Province. Central Tethyan faunas are characterized by high pantanelliid abundance/diversity, by the presence of *Ristola* and the absence of *Parvicingula* (sensu Pessagno and Whalen, 1982). Northern Tethyan faunas have the same high pantenellid abundance/diversity but possess common *Parvicingula*.

Radiolarians are known from the Bridge River, Cache Creek, Slide Mountain and Pacific Rim terranes, as well as from Quesnellia, Stikinia and Wrangellia (Fig. 2.17). With the exception of the Wrangellian Queen Charlotte Islands fauna, all collections are from isolated localities. Chert in the Sylvester Allochthon in the Slide Mountain Terrane contains radiolaria ranging from Middle(?) Devonian through Permian age (Harms, 1986). In this region a few thin, unfaulted units appear to contain fairly complete successions of radiolarian chert of Mississippian through Permian age. D.L. Jones (pers. comm., 1981, cited from Cordey, 1986) collected Permian and Triassic radiolarians from the Cache Creek Terrane and recognized Early Jurassic forms in the Bridge River Terrane (writ. comm. to E.A. Pessagno, 1977). Recent work by Cordey (1986) confirmed the presence of Early Permian to Late Triassic radiolarians in the former, and Middle Triassic to Early Jurassic forms in the latter. Igo et al. (1985) reported Permian and Late Triassic (probably Norian) radiolarians in chert of the Cache Creek mélange unit and probable Anisian taxa in the Marble Canyon Formation of the middle Cache Creek belt. They also noted possibly Middle Triassic forms strata near Kamloops. Mesozoic radiolarians have been obtained from chert in the Pacific Rim Terrane (Rusmore and Cowan, 1985; Brandon, 1985). In Wrangellia, a chert unit in the Paleozoic Sicker Group yielded Mississippian radiolarians (Brandon et al., 1986) and excellent faunas of Late Triassic to Late Cretaceous age are present in the Queen Charlotte Islands.

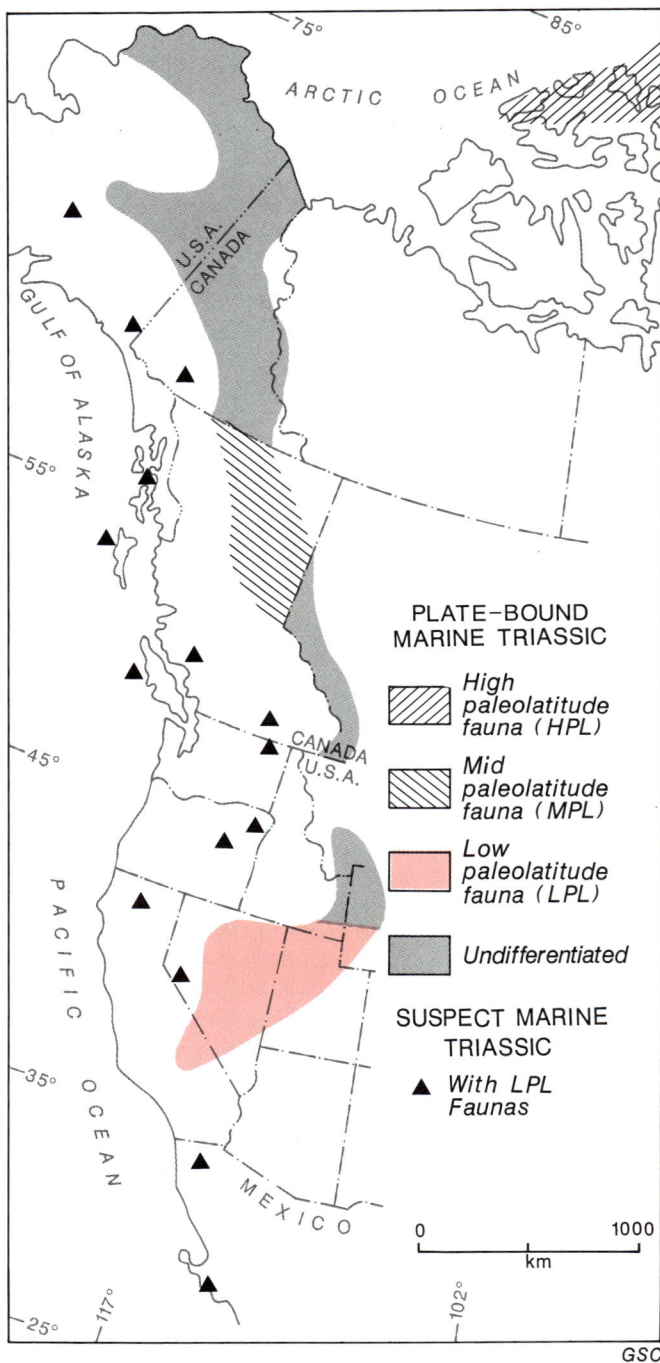

Figure 2.18. Triassic paleobiogeography of western North America (from Tozer, 1982).

Period / Terrane	CRETACEOUS L M E	JURASSIC L M E	TRIASSIC L M E	PERMIAN L M E	PENNSYLVANIAN L M E	MISSISSIPPIAN L M E
Slide Mountain				●●●	●●	●●●
Quesnellia			●			
Cache Creek			●●●	●●●		
Bridge River		●●	●			
Stikinia			●●			
Wrangellia	●●	●●●				●
Pacific Rim Complex	●	●	●●			

Figure 2.17. Radiolarian occurrences in terranes of the western Cordillera.

TECTONIC FRAMEWORK

Although still in its infancy, radiolarian biostratigraphy affords one of the most promising tools for discriminating among complexly juxtaposed terranes and for determining source areas for overlap and successor assemblages of the Cordillera. The latter is demonstrated by the occurrence of radiolarian-bearing chert clasts in the Middle to Upper Jurassic Bowser Lake Group which were derived from the Cache Creek Terrane to the north (Currie, 1984). Moreover, radiolarian-bearing chert clasts in the Upper Jurassic to Lower Cretaceous Minnes Group of the Foreland Belt suggest sources within the Slide Mountain Terrane in the eastern Intermontane and Omineca belts (M.E. McMechan, pers. comm., 1985).

Ammonites
Paul L. Smith and H.W. Tipper

Marine Triassic rocks on the craton presently span 46 degrees of latitude from Nevada to Ellesmere Island. On the basis of coral, cephalopod and pelecypod faunas (particularly *Monotis*, Silberling and Jones, 1983), Tozer (1982) recognized three cratonic faunal associations characteristic of low, mid- and high paleolatitudes (Fig. 2.18). Low-paleolatitude faunas occur west of the cratonic margin as far north as central Alaska. Another example of ammonites showing dislocated patterns of diversity and endemism has been demonstrated by Nichols and Silberling

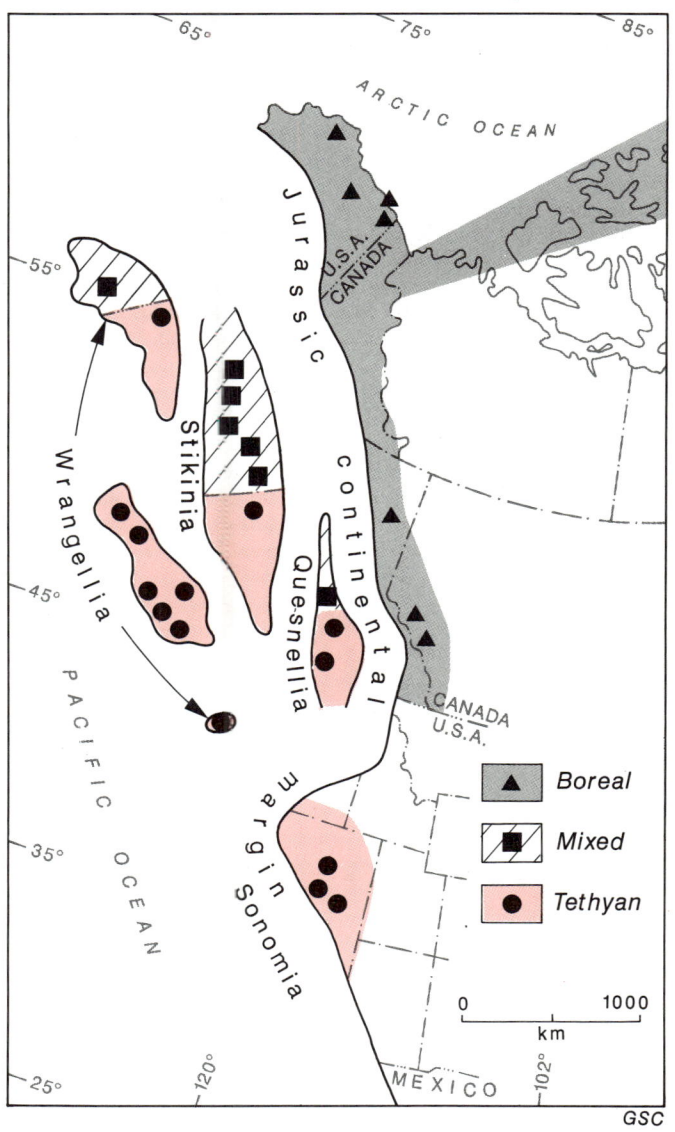
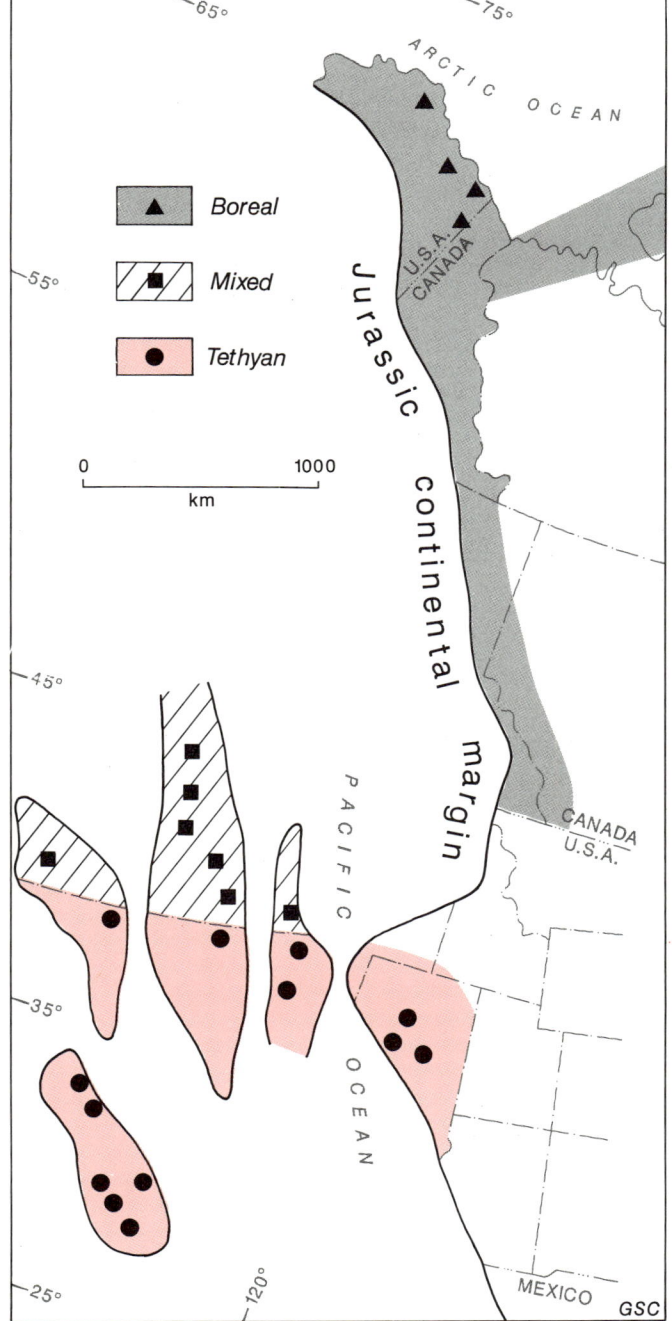

Figure 2.19. Jurassic paleobiogeography of western North America. (A) Late Pliensbachian ammonite faunas of the Canadian terranes and Sonomia. (B) Suggested restoration of latitudinal displacements.

(1979) in a study of the Smithian Stage. Twenty-four genera have been found in cratonic Smithian rocks within 20 degrees (north) of the Triassic paleoequator whereas the rocks farther north yielded only 12 genera. A single limestone bed in the Chulitna Terrane of south-central Alaska yielded 13 genera, all of which are present in low-paleolatitude faunas but only seven of which are present in high-paleolatitude faunas. At the species level the affinity with low-paleolatitude faunas is much more marked. All 13 of the Chulitna species are found at low-paleolatitudes whereas only two species occur at high-paleolatitudes.

Taylor et al. (1984) summarized the biogeography of Jurassic ammonoid and some bivalve genera. Ammonoid endemism as a function of latitude first became marked during the Pliensbachian in both North America and Europe. A low-diversity cratonic boreal fauna is found as far south as southern Alberta, and a Tethyan fauna occurs in Sonomia (Fig. 2.19A) which was probably accreted to the craton during the Triassic (Speed, 1979). The terranes of the Canadian Cordillera yield high-diversity Tethyan faunas in southern regions and mixed Boreal/Tethyan faunas in the north. Post-Pliensbachian northward displacements of approximately 2500 km, 1800 km and 500 km are indicated for Wrangellia, Stikinia and Quesnellia respectively (Fig. 2.19B). These are minimum values based on the assumption that no significant post-Triassic northward displacement of Sonomia occurred. Furthermore, these data help to resolve the ambiguity of the hemispheric origin of terranes inherent in paleomagnetic results; the presence of Boreal elements in each of the terranes indicates that they were in the northern hemisphere during the Pliensbachian. Some organisms such as the ammonite *Fanninoceras* and the bivalve *Weyla* are present in Sonomia and are characteristic of the eastern Pacific including the South American craton (Hillebrandt, 1981; Damborenea and Mancenido, 1979). Their presence in all of the Canadian Cordilleran terranes shows that the terranes were in the eastern Pacific by the Early Jurassic.

Faunal data for the early Middle Jurassic indicate that terrane relationships were essentially the same as in the Early Jurassic. By Bajocian time the Boreal Realm still extended as far south as southern Alberta on the craton and Stikinia had moved far enough north to have a mixed fauna in its southern region and a boreal fauna in the north. For the remainder of the Jurassic however, there is no definitive evidence of significant latitudinal displacement (Callomon, 1985).

Two ammonite faunal realms existed during both the Jurassic and Cretaceous periods: a North Pacific Realm dominated by marine faunas closely related to those of the Mexican, Tethyan and Andean geosynclines and the Boreal Realm, characterized by faunas almost identical to those of northern Eurasia (Jeletzky, 1965, 1970). The North Pacific Realm was restricted to the Western Cordillera throughout the Cretaceous and the Boreal Realm to the northern Yukon during the Early Cretaceous and, thereafter, to the entire Foreland Belt region. The principal difference between the two realms is in their diversity. The Boreal Realm faunas of the Foreland Belt are characterized by abundant specimens of only a few species whereas the faunas of the Pacific Realm in the Nanaimo Group of the Insular Belt are both diverse and populous (Jeletzky, 1970). Although paleolatitude endemism in Cretaceous faunas is not as well developed as in those faunas of earlier periods, there are close similarities in the Campanian faunas of the Nanaimo Group with ammonite assemblages in parts of the Great Valley sequence of California (P.D. Ward, pers. comm., 1986). After Aptian time at least, a physical barrier (the nascent Cordillera) existed between the Foreland and Insular belts. This barrier had a profound effect on both endemism and diversity by influencing the physical environment (climate, run-off etc.) and the course of evolution through the isolation of populations.

PART C. CRUSTAL GEOPHYSICS

J.F. Sweeney, R.A. Stephenson, R.G. Currie, and J.M. DeLaurier

With ever greater sensitivity and resolving power, geophysical measurements are revealing the subsurface structure of the Canadian Cordillera. The latest data (e.g. Clowes et al., 1987a) make it possible to extrapolate features exposed at the Earth's surface deep into the crust and upper mantle and also to detect structures at depth apparently unrelated to those observed at the surface. Current concepts of plate interaction and mountain building in western Canada can only be properly discussed in three-dimensional terms. In this section the record of studies of the regional geophysical character of the Cordillera in Canada is described and assessments are made on the geophysical constraints on the crustal and upper mantle architecture of the region.

Although seismicity has been monitored since the beginning of the 20th century and widespread gravity measurements were made between 1915 and 1926, systematic geophysical studies of the Canadian Cordillera began in the 1950s (Garland and Tanner, 1957; Berry et al., 1971). Investigations have focused on the southern Cordillera (49°N to 60°N), the continental margin and adjacent oceanic plates (Keen and Hyndman, 1979; Gough, 1986). Comparatively fewer geophysical data have been obtained from the northern Cordillera (north of 60°N) but new work, including regional gravity, aeromagnetic, refraction and reflection seismology, is underway.

Geophysical studies
Deep refraction and reflection seismology

Crustal seismic refraction studies in the Canadian Cordillera have been underway since 1951 (Table 2.2), mainly in southern British Columbia and the adjacent offshore area (Fig. 2.20). The types of studies undertaken include: (1) reversed and unreversed, large offset refraction and wide-angle reflection profiles interpreted using first arrivals only and/or, in more recent work, arrivals and amplitudes of secondary and other phases, as well as model ray-tracing techniques; (2) refractions along profiles, or in two-dimensional arrays, interpreted using time-term methods, basically a least-squares crustal model reduction of first arrival travel-time residuals at recording sites and shot points; (3) crust/upper mantle models based on earthquake-generated surface-wave spectra attenuation characteristics between standard seismograph stations; and (4) near vertical deep reflections, calibrated with crustal velocities from refraction analyses. The latter include deep multichannel reflection studies in southwestern British Columbia including Vancouver Island and the adjacent continental shelf and slope, carried out as part of the LITHOPROBE and Frontier Geoscience Program projects. The geological implications of these data are discussed in Chapters 13 and 17. They provide a remarkable documentation of the geometry of a modern subduction zone. Figure 2.20 does not include numerous marine, single-channel reflection profiles mostly from the waters surrounding Vancouver Island and the Queen Charlotte Islands.

Depths to the Mohorovicic discontinuity (Moho) from reversed refraction profiles in southern British Columbia (Fig. 2.21) are probably accurate to within a few kilometres. Those from unreversed data can be viewed as averages along the observed profiles. Although there is considerable variation within individual morphogeological belts, particularly in the Coast Belt (Fig. 2.21), there are systematic changes in crustal thickness across the entire southern Cordillera (e.g., Berry and Forsyth, 1975). In the Foreland Belt the average depth to the Moho is about 45 km; this is somewhat thicker than average for continental crust (33 km). Moho depths in the Omineca and Intermontane belts are more typical, averaging 33 to 34 km. The transition occurs below or a short distance west of the Southern Rocky Mountain Trench (Fig. 2.21; Berry et al., 1971).

The northern Coast Belt also exhibits Moho depths between 30 and 35 km with the thinnest crust measured in its western part. East of Vancouver Island the thickness of the Coast Belt is poorly constrained but appears to be greater than 40 km. Directly south, however, the crust thins dramatically to 20 to 25 km below the Fraser Delta. The crust of the Insular Belt thins across southern Vancouver Island from approximately 40 km in the east to about 26 km below the west coast. Across the continental margin the crust further thins to where the thickness of accreted sediments above subducting oceanic crust at the base of the continental slope is about 3 km (see Fig. 17.1, Profile N, in pocket). Within the Insular Belt near the Queen Charlotte Islands and adjacent parts of southeast Alaska Moho depths appear much more uniform and average about 26 km.

From the Foreland Belt to the eastern part of the Coast Belt a locally prominent regional P-wave velocity discontinuity exists within the crust (Monger et al., 1985). It is about 35 km deep east of the Southern Rocky Mountain Trench and occurs at depths between 20 and 25 km to the west. Above the discontinuity velocities range from 6.0 to 6.8 km/s with lower velocities measured within sediments above the basal detachment zone east of the Purcell Mountains. Below the discontinuity crustal velocities vary between 6.8 and 7.4 km/s.

The average crustal velocity apparently increases from 6.2-6.3 km/s beneath Vancouver Island to 6.5 km/s west of the Rocky Mountains (Berry and Forsyth, 1975). Uppermost mantle velocities are generally 8.0-8.1 km/s throughout the southern Cordillera except in the anomalous southwestern region where they may be as low as 7.8 km/s.

Sweeney, J.F., Stephenson, R.A., Currie, R.G., and DeLaurier, J.M.
1991: Part C. Crustal geophysics; in Chapter 2 of Geology of the Cordilleran Orogen in Canada, H. Gabrielse and C.J. Yorath (ed.); Geological Survey of Canada, Geology of Canada, no. 4, p. 39-59 (also Geological Society of America, The Geology of North America, v. G-2).

Table 2.2. Deep seismic studies in the Canadian Cordillera

Location (in Fig. 2.20)	Date	Reference
C 1	1951	Tatel and Tuve (1955)
C 2	1955	Tatel and Tuve (1955)
C 3	1956-57	Shor (1962)
C 4	1957-63	Milne and White (1960) White and Savage (1965)
C 5	1958	Milne and White (1960) Cumming et al. (1962) White and Savage (1965)
C 6	1962-65 1969	Cumming and Kanasewich (1966) Chandra and G. Cumming (1972)
C 7	1964-65	White et al. (1968)
C 8	1966-67	White et al. (1968)
	1969	Jacoby (1970) Berry and Forsyth (1975)
C 9	1967	Berry and Forsyth (1975)
C10	1969	Forsyth et al. (1974)
C11	1969	Mereu et al. (1977)
C12	1969	Hales and Nation (1973)
C13	1969	Hill (1972)
C14	1969	Johnson and Couch (1970)
C15	1969-70	Forsyth et al. (1974)
C16	1969-70	Berry and Forsyth (1975)
C17	1970	Johnson et al. (1972)
C18	1972	Mair and Lyons (1976)
C19	1972-73	Bennett et al. (1975)
C20	1972-74	Wickens (1977)
C21	1973-75	Cumming et al. (1979)
C22	1979	Horn et al. (1984)
C23	1980	Ellis et al. (1983) McMechan and Spence (1983) Spence et al. (1985) Clowes et al. (1986)
C24	1980	Clowes et al. (1983) Ellis et al. (1983)
C25	1983	Ellis and Clowes (1984) Mackie et al. (1985)
C26	1984	Clowes et al. (1984, 1986, 1987a) Yorath et al. (1985a,b) Green et al. (1986a,b; 1987)
C27	1984	Cook (1985)
C28	1985	Zelt et al. (1987) Brown et al. (1987)
C29	1985	Clowes et al. (1987b)
C30	1985	Cook et al. (1987)
C31	1987	Stephenson et al. (pers. comm., 1987)

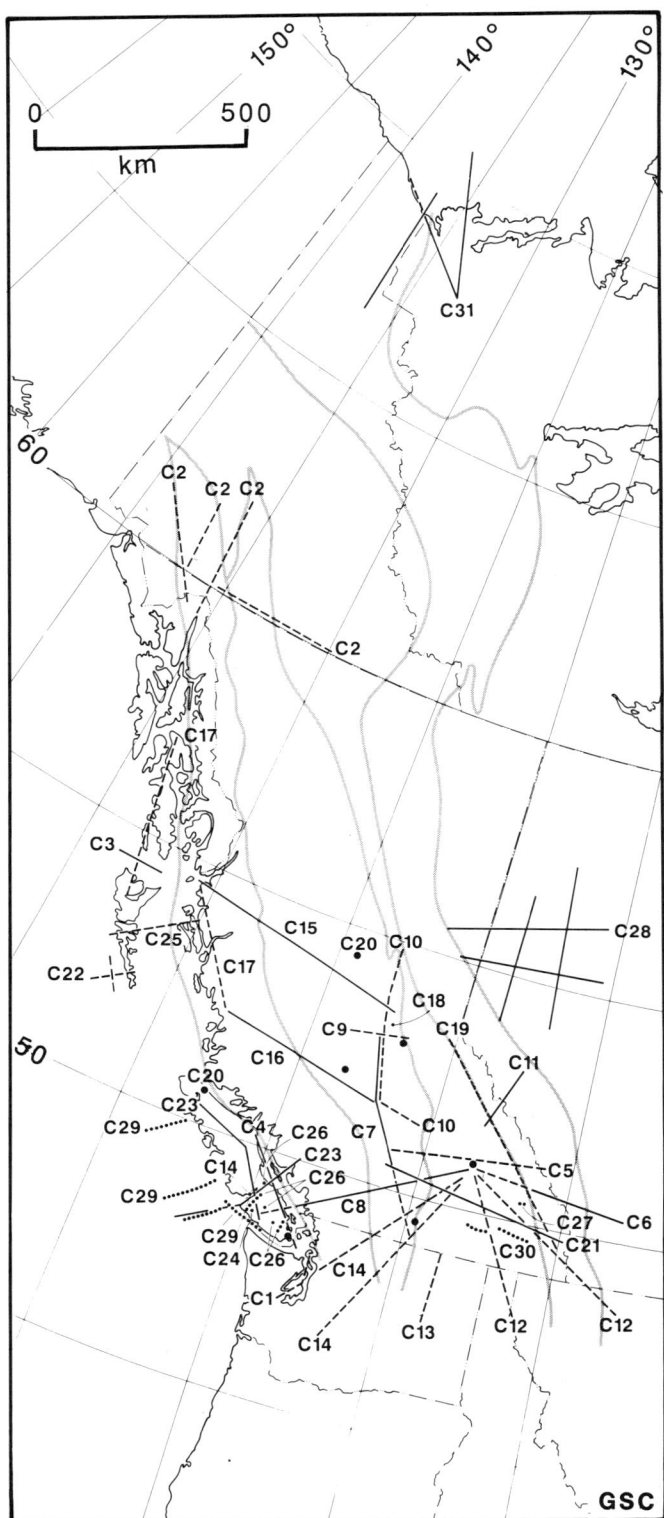

Figure 2.20. Deep seismic studies in the Canadian Cordillera: solid and dashed lines are reversed and unreversed refraction profiles, respectively; small dots and dotted lines represent deep reflection surveys; large dots represent a study of attenuation of earthquake-generated surface waves. Table 2.2 provides references. Morphogeological belt boundaries indicated by screened lines.

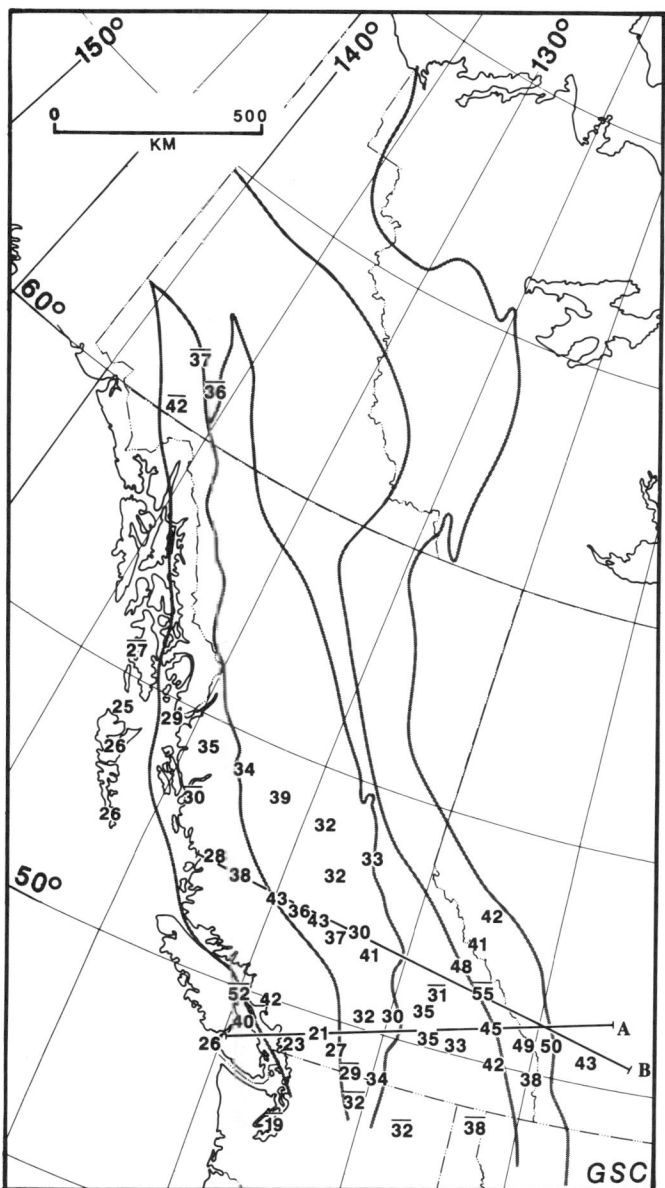

Figure 2.21. Moho depths in kilometres from selected deep seismic studies shown in Figure 2.20. Depths from unreversed refraction profiles are shown with a bar over the value. Cross-sections along Profiles A and B are shown in Figures 2.28 and 2.29. Morphogeological belt boundaries are indicated.

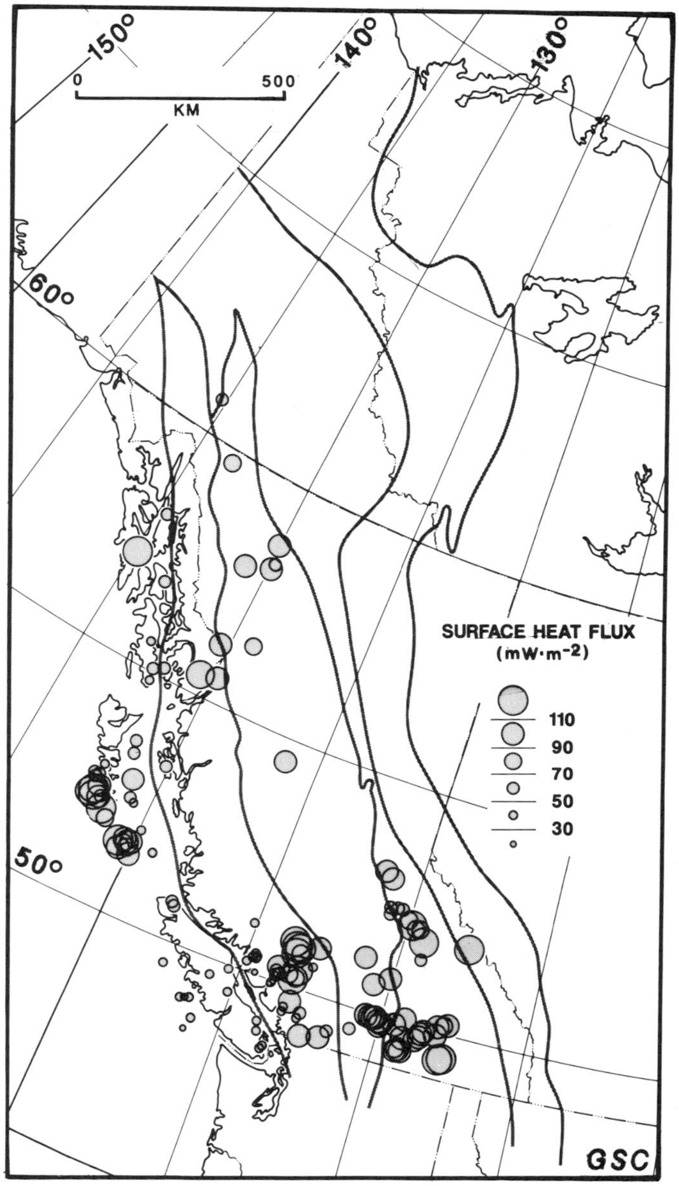

Heat flow

In the Canadian Cordillera the majority of heat flow measurements have been made in southern British Columbia and in the bordering marine areas (Fig. 2.22). Published values for the Foreland Belt are from the eastern foothills of the Rockies where heat flow is low and appears largely related to known patterns of fluid flow within sedimentary rocks of the adjacent Interior Platform (Majorowicz and Jessop, 1981; Majorowicz et al., 1984). In that region heat generation within the Precambrian basement averages 2.5 $\mu W \cdot m^{-3}$, which is within the normal range for shield/

Figure 2.22. Distribution of heat flow observations in the Canadian Cordillera.

basement rocks elsewhere on the continent. This value probably applies throughout the Foreland Belt as its basement is part of cratonic North America (Monger et al., 1985). Elsewhere in the Foreland Belt high elevations are associated with thick crust (Fig. 2.21); there is little evidence of thermal expansion. Moderate heat flow values, uncorrected for glacial disturbance, average 39 ± 4 $mW \cdot m^{-2}$ in the Foreland Belt of northern Montana (Roy et al., 1968; Blackwell, 1969).

To the west, corrected heat flow in the Intermontane and Omineca belts is about twice as high as Foreland Belt values, averaging 82 ± 7 $mW \cdot m^{-2}$ near 50°N (Davis and Lewis, 1984; Jessop and Judge, 1971) (Fig. 2.22). Similar values have been obtained within the Intermontane Belt in both the northern United States (Blackwell, 1974) and in Canada between 56°N and 61°N where heat flows average 79 ± 14 mW·m^{-2} (Jessop et al., 1984).

In the southern Cordillera the Intermontane and Omineca belts belong to the same heat flow province, as indicated by the linear relationship between heat flux and crustal heat generation (Fig. 2.23). The region is characterized by a relatively high reduced heat flux (63 mW·m^{-2}; Lewis et al., 1985) and may be part of the "Cordilleran Thermal Anomaly Zone" described by Blackwell (1969) in the northwestern United States. Data from the Intermontane Belt to the north show a considerably lower component of heat flow from the mantle (Fig. 2.23).

In the eastern part of the Coast Belt heat flow is high with large local variations (Lewis et al., 1985). Values average about 84 mW·m^{-2} in southern British Columbia (Hyndman, 1976) compared with 100 ± 5 mW·m^{-2} in northern Oregon (Blackwell et al., 1982). In both areas heat flow declines sharply to between 30 and 40 mW·m^{-2} at the western edge of the youthful Garibaldi (Cascade) Volcanic Belt in both the Coast Mountains of southern British Columbia and the Cascade Mountains of Oregon (Lewis et al., 1985). In each area the transition occurs over about 20 km and, on southern Vancouver Island, heat flow increases gradually westward across the Insular Belt to about 50 mW·m^{-2} at the edge of the continental shelf (T.J. Lewis, pers. comm., 1986).

In the intervening area of Washington and southwesternmost British Columbia, heat flow is likewise low (30 to 50 mW·m^{-2}) but eastward the transition to higher values in the Coast Belt is less pronounced. Farther north in the Insular Belt heat flow is generally much higher with values averaging 66 mW·m^{-2} on northern Vancouver Island and 86 mW·m^{-2} at the entrance of Queen Charlotte Sound (Hyndman et al., 1982). Heat flow in Hecate Strait declines southward from between 60 and 78 mW·m^{-2} near Graham Island to between 45 and 50 mW·m^{-2} in Queen Charlotte Sound (Yorath and Hyndman, 1983) whereas it increases sharply from 47 mW·m^{-2} on the west coast of the Queen Charlotte Islands to over 250 mW·m^{-2} on the young Pacific Plate; the maximum gradient lying seaward of the Queen Charlotte Fault (Hyndman et al., 1982).

In the southern Coast Belt, high heat flows are associated with average crustal heat generations that are lower (0.8 µW·m^{-3}) than those observed in the Intermontane and Omineca belts (Lewis et al., 1985). The high heat flows are thought to be produced by advective cooling of magmatic intrusions emplaced during Late Cretaceous to Recent time (T.J. Lewis, pers. comm., 1986). Crustal heat generation in the Insular Belt is likewise low (1.1 µW·m^{-3}); associated low heat flows are probably due to absorption of heat by the subducting Juan de Fuca Plate (Lewis et al., 1985). The abrupt change in heat flow between the Insular and Coast belts requires a source within 10 km of the surface, and presumably reflects upward convective transfer of heat to shallow levels beneath the Garibaldi Volcanic Belt by the ascent of magmas generated by partial melting of subducted oceanic crust at a depth of about 100 km (Blackwell et al., 1982).

Within the Insular Belt, north of the Juan de Fuca Plate system, generally higher heat flows reflect the absence of a heat-absorbing downgoing slab (T.J. Lewis, pers. comm., 1986). Very young, hot oceanic crust next to the continent in this region may produce the high heat flows measured near the shelf edge in Queen Charlotte Sound and off the west coast of the Queen Charlotte Islands (Hyndman et al., 1982). In the latter area Hyndman et al. (1982) and Yorath and Hyndman (1983) proposed that oblique underthrusting of the Pacific Plate, beginning about 6 Ma, has caused a heat sink which could create the sharp transition to low heat flow values across the Queen Charlotte Terrace. Lower heat flows in southern Queen Charlotte Basin may also be caused by the heat absorbing effects of the underthrust oceanic plate (Yorath and Hyndman, 1983).

Seismicity

Seismic activity occurs throughout much of the Canadian Cordillera with concentrations across southern British Columbia and Alberta (Milne et al., 1978) and in the Yukon and District of Mackenzie area (Basham et al., 1977) (Fig. 2-24). In the north seismicity is concentrated northeast of the Tintina Fault, particularly in the Richardson Mountains, and along the Denali Fault. By far the greatest frequency and magnitude of events are associated with the plate margin system off the west coast, especially along the Queen Charlotte-Fairweather Transform Fault system (Milne et al., 1978). Cordilleran seismological studies have focused on this tectonically active region and particularly in the populated areas of southwestern British Columbia (Rogers, 1983).

As the number of seismograph stations in the Cordillera increased, beginning in 1898 with the facility in Victoria, the threshold of complete detection declined from magnitude 7 at the close of the 19th century to

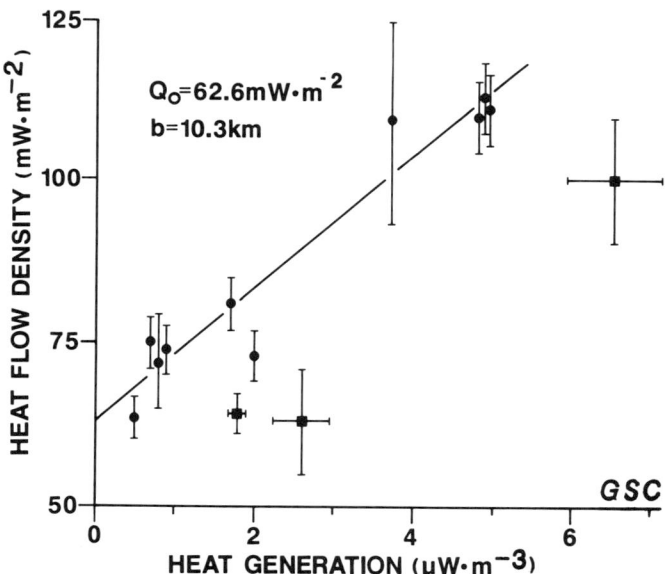

Figure 2.23. Heat flux, uncorrected for glacial effects, and crustal heat generation for sites in the Omineca and Intermontane belts at about 50°N (circles; Lewis et al., 1985) and in the Intermontane Belt between 56°N and 61°N (squares; Jessop et al., 1984). Best-fitting line for 50°N data shown. Modified from Lewis et al. (1985).

TECTONIC FRAMEWORK

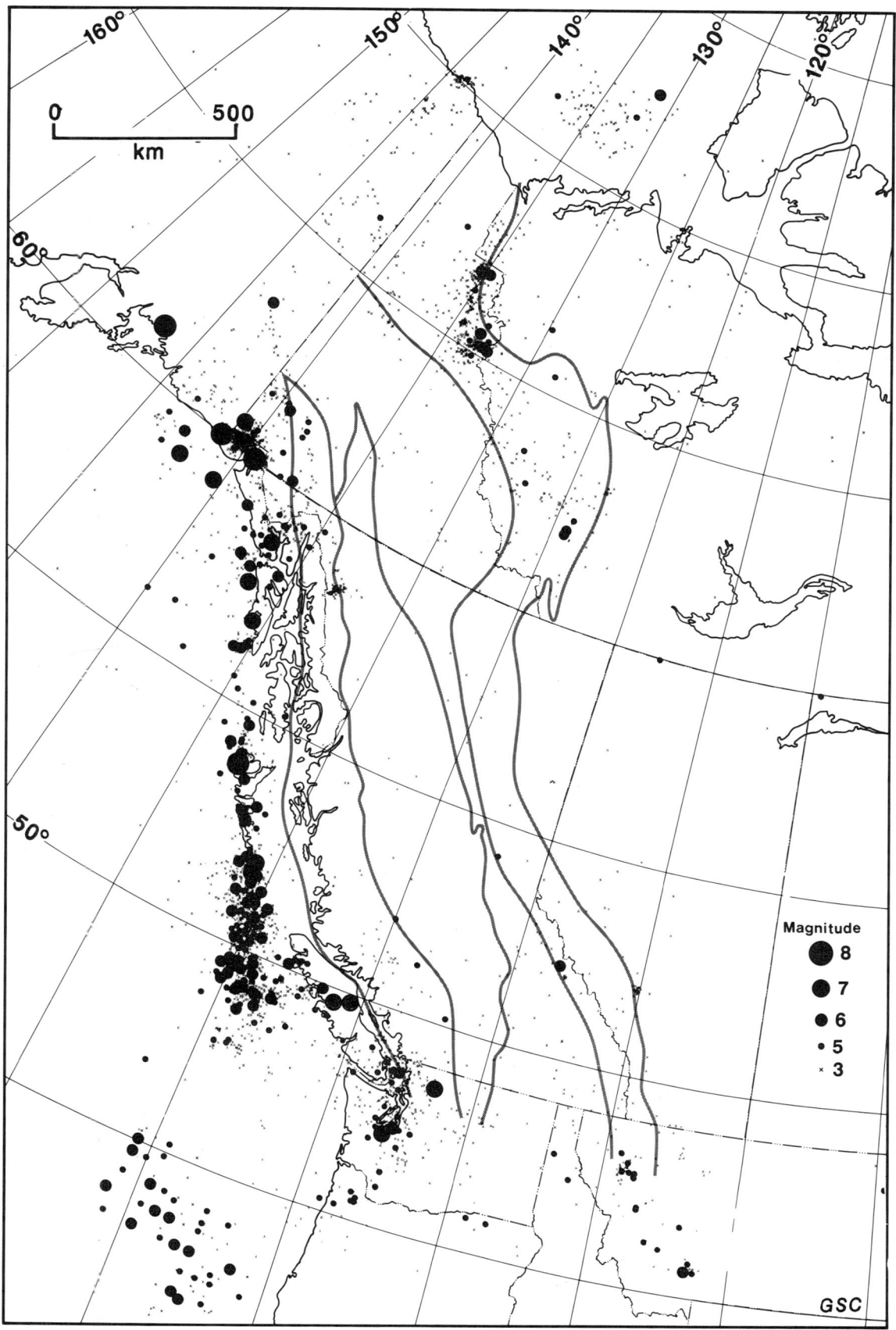

Figure 2.24. Historical seismicity in western Canada up to and including 1986. Morphogeological belt boundaries are indicated.

magnitude 4 (including the offshore) by the early 1970s and, in the coastal region of southwestern British Columbia, to magnitude 2 by the early 1980s (Basham et al., 1977; Milne et al., 1978; Horner, 1983; Rogers, 1983) (Fig. 2.25). Correspondingly, the accuracy of epicentre locations increased in the southern Cordillera from about 100 km before the early 1950s to about 20 km since that time, with a further increase in southwestern British Columbia to about 5 km in the late 1970s (Milne et al., 1978; Rogers, 1983). In the offshore, as well as in the northern Cordillera, location accuracies better than 50 to 100 km have been possible only since 1962 (Milne et al., 1978; Basham et al., 1977).

Along the Queen Charlotte-Fairweather Fault system well constrained focal mechanism solutions indicate nearly pure horizontal right-lateral strike-slip motion on a vertical plane (Milne et al., 1978; Rogers, 1983). Thrusting is noted west of where the fault joins the Aleutian Trench off southeast Alaska (Milne et al., 1978). Directly south a small component of convergence is found along the transform boundary between the Yakutat Block and the North America Plate (Perez and Jacob, 1980; Horner, 1983). In the southern Queen Charlotte Islands area, where focal depths are uncertain but are thought to be generally less than 30 km, Rogers (1983) determined thrust components for some earthquakes indicating that the Pacific Plate may be moving obliquely to the northeast beneath the North America Plate as suggested by Yorath and Hyndman (1983). Based upon the contemporary pattern of uplift near the plate boundary and subsidence farther toward the continent, Riddihough (1982) inferred that oblique convergence is taking place along the entire Queen Charlotte-Fairweather transform system.

In the deep ocean farther south, fault-plane solutions indicate mainly northwesterly strike-slip motion, presumably along transform offsets of the Juan de Fuca Ridge system off Vancouver Island; epicentre uncertainties for most events preclude association with specific transform segments (Milne et al., 1978). East of the accreting margin there is little offshore seismicity except along the Nootka Fault zone (Fig. 2.24). Onshore the maximum depth of earthquakes increases from about 30 km beneath the west coast of Vancouver Island and Washington to about 90 km just east of Georgia Strait and Puget Sound (Rogers, 1983; Taber and Smith, 1985). Shallow events (<30 km depth) show mainly strike-slip motion with generally north-south pressure axes; the preferred mechanism is right lateral displacement along northwest-striking faults (Rogers, 1983); deeper events show tension axes dipping to the east. About 200 km farther inland there is a sharp reduction in low-level seismicity, more than can be explained by reduced detectability (Milne et al., 1978). The reduced activity may be associated with the steep heat flow gradient along the Cascade and Garibaldi volcanic belts (Lewis et al., 1985; Blackwell et al., 1982).

The Quaternary Anahim and Stikine volcanic belts are likewise nearly aseismic (Fig. 2.24). The former may be the trace of an eastward propagating hot spot (Bevier et al., 1979). Seismicity on line with, and about 100 km east of the Anahim Belt, is similar to activity at intraplate hotspots elsewhere in that it appears unrelated to a nearby young volcano but instead is suggestive of reactivation of local shallow-dipping faults (Rogers, 1983). The Stikine belt is a large continental rift, apparently an artifact of shearing between the Pacific and North America plates (Rogers, 1983). Seismic activity is currently concentrated

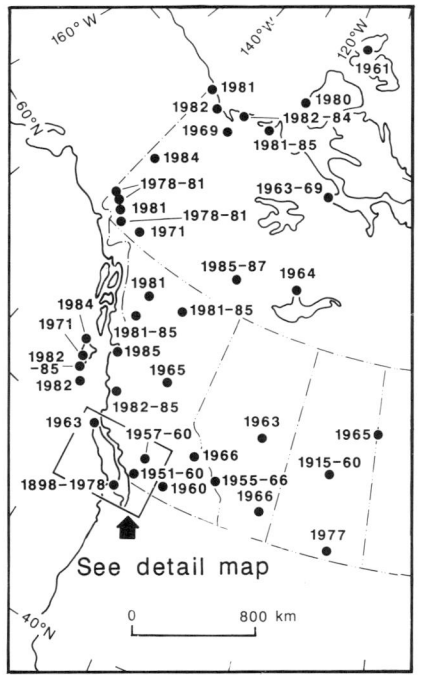
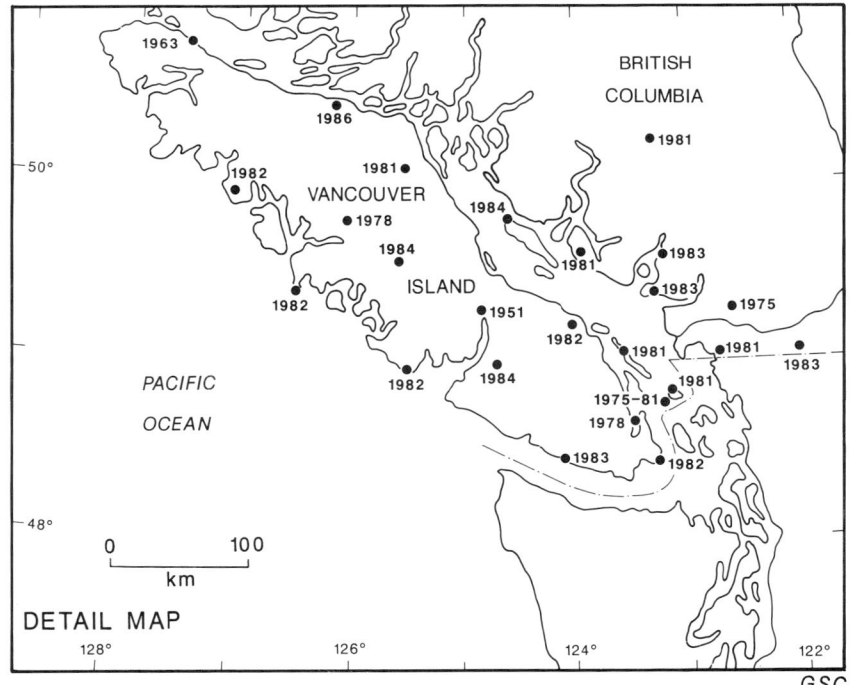

Figure 2.25. Distribution of seismograph stations in western Canada and their commencement (- closure) dates.

to the northwest along the Denali Fault system. These events indicate strike-slip motion at focal depths of less than 15 km in either an east-west or a north-south direction, oblique to the trend of the Denali system (Horner, 1983).

Seismicity in the northern Yukon occurs in areas that have sustained intense faulting (Basham et al., 1977). The most pronounced earthquake activity is in the eastern half of the Richardson Mountains (Fig. 2.24) where Leblanc and Wetmiller (1974) suggested right-lateral movement on a nearly vertical plane subparallel with the northwest trending surface faults. Another Richardson event yielded a focal depth estimate between 20 and 30 km (Leblanc and Wetmiller, 1974).

The pattern of strain release in western Canada is dominated by the distribution of large earthquakes and is thereby greatest along the Queen Charlotte-Fairweather Fault system (Milne, 1967; Milne et al., 1978). This zone of high strain release extends southward along the Pacific-Juan de Fuca Plate boundary system to about 47°N where it shifts northeastward to the mainland and continues along Georgia Strait and Puget Sound (Milne et al., 1978). Elsewhere in the southern Cordillera little significant strain release occurs. Quantitative estimates of strain release in the northern Cordillera are unavailable but significant rates are likely in areas of moderately intense seismicity such as in the Richardson Mountains and, to a lesser extent, in the Mackenzie Mountains region. The Nahanni earthquakes of October 5 and December 23, 1985, of Ms (surface wave magnitudes) 6.6 and 6.9 respectively, which occurred in the southern Mackenzie Mountains, yielded data providing a compressional axis trend oriented orthogonally to the structural grain (R.B. Horner, pers. comm., 1986) and probably closely parallel with the Late Cretaceous to Early Tertiary principal stress direction. The epicentre could not be identified with any of the mapped faults in the area, nor was an associated surface break found.

Studies of strain release with time in the plate boundary areas indicate moderate, relatively nonvarying rates over the last 80 years near the spreading axes (Milne et al., 1978). Considerably greater and significantly more step-like rates occur along both the Queen Charlotte Fault and the convergence zone in southwestern British Columbia east of the Juan de Fuca Plate and northwestern Washington demonstrating that greater strain accumulates in these areas and that it is released in periodic large events (Rogers, 1983). Milne et al. (1978) estimated the recurrence period for a potentially damaging event (M>6) to be about 10 years along the convergence zone and 5 to 10 years along the Queen Charlotte Fault.

Gravity and isostasy

The gravity field has been measured at over 20 000 localities in the Canadian Cordillera and at more than 47 000 points (about 91 000 line km) off the west coast of British Columbia (Fig. 2.26; Garland and Tanner, 1957; Buck, 1967; Walcott, 1967; Stacey and Stephens, 1969; Stacey and Steele, 1970; Stacey et al., 1973; MacLeod et al., 1977; Tiffin and Riddihough, 1977; Currie et al., 1983). The regional anomaly field is best known south of 54°N (Fig. 2.27). In general, Bouguer anomaly values within the southern Cordillera are inversely related to elevation with the most negative values (less than -200 mGal) observed in the Foreland Belt. Lower elevations in the Omineca and Intermontane belts have more positive Bouguer anomalies (generally between -100 and -150 mGal). A relatively steep regional gravity gradient separates the Foreland and Omineca belts (Stacey, 1973; Cady, 1980).

Bouguer values decline westward over the higher elevations of the Coast Belt then rise sharply (at more than 2 mGal/km) to positive anomalies west of the Cascade Garibaldi volcanic arc (Fig. 2.27). The high (up to 50 mGal) extends over the Insular Belt in a zone parallel with a linear offshore free-air anomaly low (locally exceeding -100 mGal) along the zone of convergence between the Juan de Fuca and North America plates (Riddihough, 1979). North of the convergence zone the parallel high-low free air anomaly pair is more pronounced and localized with a steep intervening gradient over the Queen Charlotte Fault (Riddihough and Seemann, 1982).

Few regional analyses of the Cordilleran gravity field have been undertaken. For Canada in general Goodacre

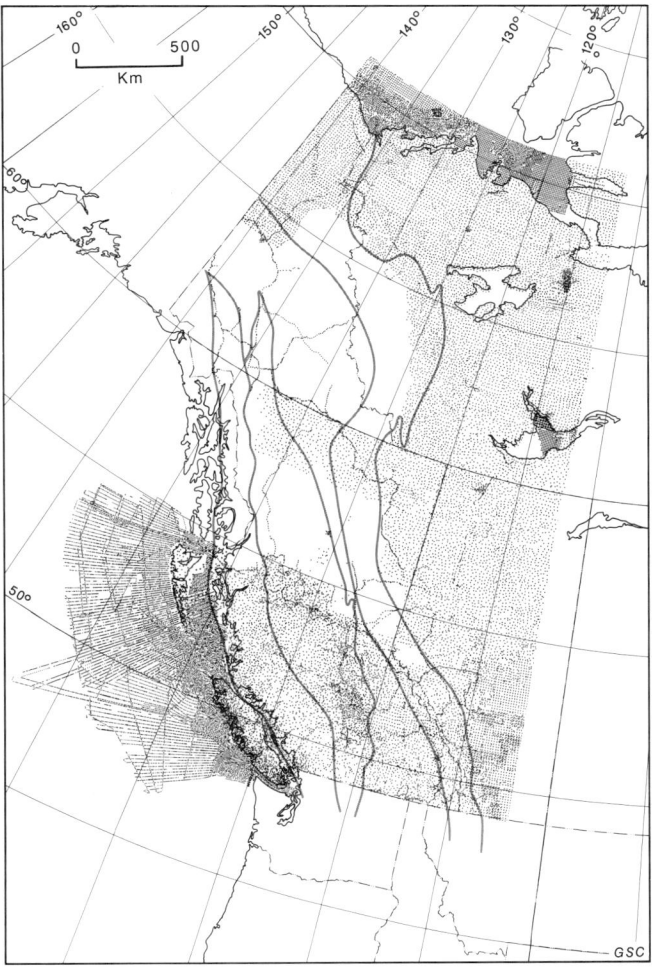

Figure 2.26. Distribution of gravity observations in western Canada and adjacent offshore regions. Shown is most central station in each 5 km x 5 km cell. Morphogeological belt boundaries are indicated.

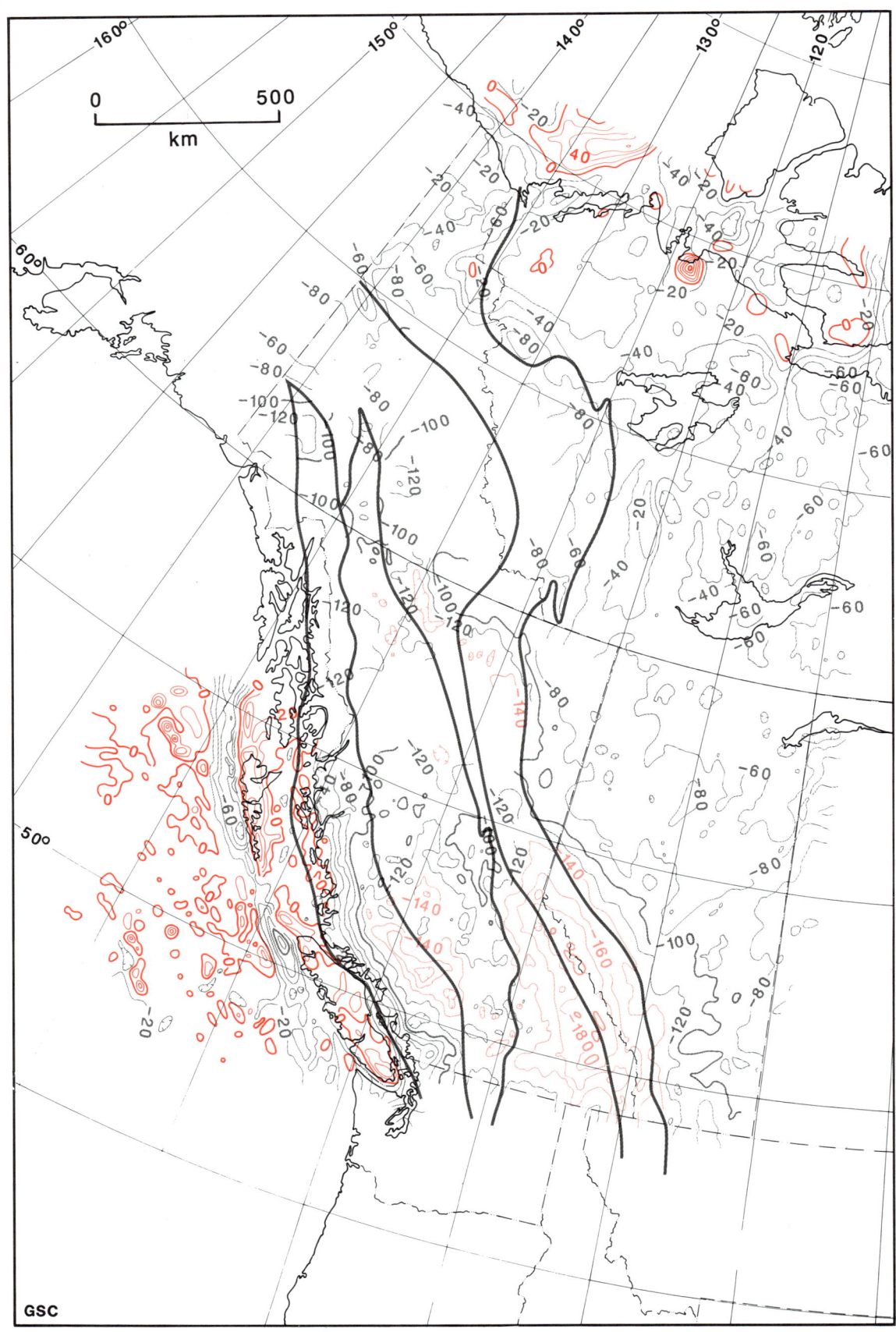

Figure 2.27. Gravity anomaly map of western Canada, terrain-corrected Bouguer onshore, free-air offshore. Contour interval 20 mGal. Positive anomalies in bold red. Morphogeological belt boundaries indicated.

(1972) showed that regional changes in Bouguer anomaly values are caused mainly by changes in crustal thickness. Moreover he noted that the source of extremely low Bouguer values in the Foreland Belt must be partly, if not entirely, within the upper mantle. Stacey (1973) determined that the Bouguer anomalies correlate well with regional changes in elevation in the southern Cordillera and that therefore the surface elevation is largely in local (Airy) isostatic balance.

In light of the variation in Moho depth revealed by deep seismic studies (Fig. 2.21), simple regional density-structure models of the southern Cordillera require relatively reduced crustal thickness in the interior region between the Foreland and Insular belts (Stacey, 1973; Berry and Forsyth, 1975). Stacey suggested that the thicker crust of the Foreland Belt reflected the presence of underlying Precambrian crystalline basement and, in the Insular Belt, the existence of underlying dense oceanic lithosphere. Pursuing the latter concept, Riddihough (1979) incorporated compositional and crustal thickness data in density-structure models to show that convergence is accompanied by subduction of the oceanic plate beneath the continent.

The regional isostatic character of the southern Canadian Cordillera can be illustrated by considering the relationships between crustal thickness measurements, Bouguer anomalies and topographic heights along two cross-sections across the orogen (Fig. 2.28, 2.29). The close correspondence between the long wavelength Bouguer gravity and the long wavelength topography along both profiles, indicates that the topography is probably in a state of local, Airy isostasy as suggested by Stacey (1973). Also, with the exception of the western end of Profile A, the variation in Moho depths, based on the refraction data, is consistent with the Airy compensation model in that regional increases in crustal thickness are apparently associated with relatively more negative Bouguer anomalies. Simple calculations suggest that the source of the long wavelength Bouguer anomalies, and hence the depth of isostatic compensation of the regional topography, is at the crust-mantle interface. At the western end of Profile A, near the convergent plate boundary, the relationship between Bouguer gravity and topography remains consistent with the Airy model, but the compensation surface must lie considerably above the surface identified by seismic refraction as the Moho. This may be related to the complex crust and upper mantle structure inferred for this region (Riddihough, 1979; Spence et al., 1985; Clowes et al., 1986).

Whereas the refraction data appear to define a Moho which, for the most part, is consistent with the long wavelength gravity and topography observations, they cannot resolve Moho depth variations of shorter wavelengths. Where shorter wavelength "bumps" occur along the Moho depth profiles in Figures 2.28 and 2.29 they generally correlate with the higher frequency (residual) gravity profiles as expected. Otherwise, the relative contributions to the residual gravity profiles of unresolved Moho depth variations and of lateral density variations within the crust cannot be judged. There is no consistent, obviously isostatic relationship, between the residual Bouguer gravity and the residual topographic profiles. A probable contributing factor is that the isostatic response of the Earth generally varies quite rapidly within wavelengths of between 80 km to 240 km (Cochran, 1980). At these wavelengths in the southern Cordillera there is possibly a relationship opposite to that of conventional isostasy, i.e. Bouguer gravity highs associated with topographic highs

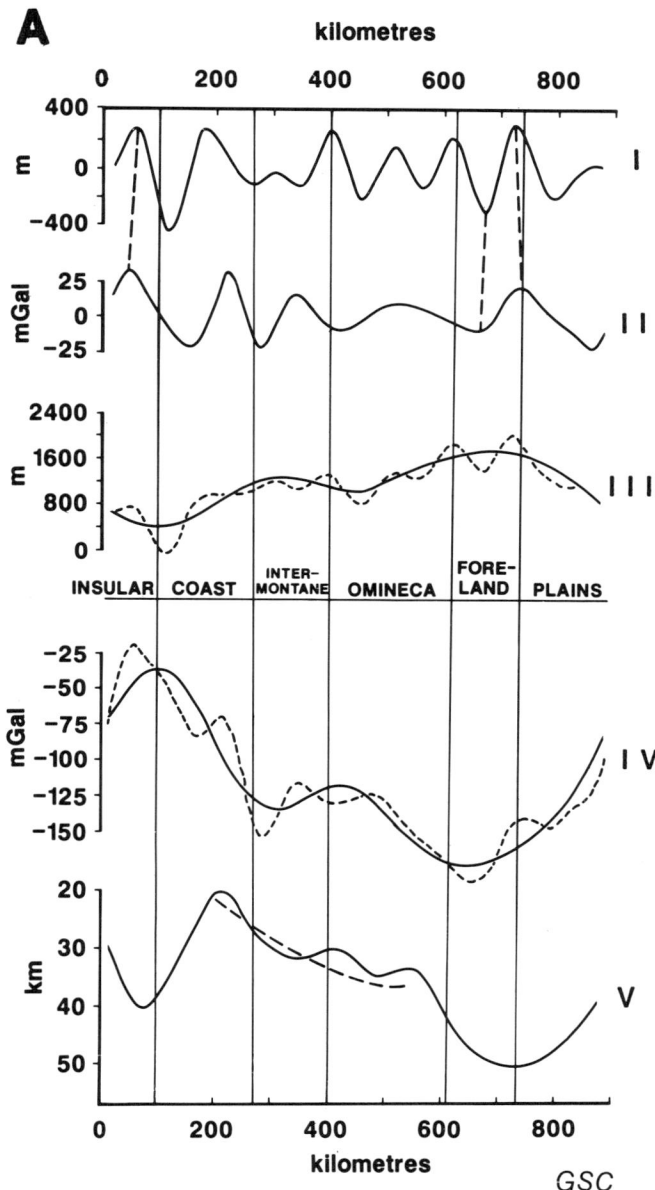

Figure 2.28. Profile A, from top to bottom: (i) topographic heights (m); (ii) Bouguer gravity anomalies (mGal) in the spectral band 240 km to 80 km wavelength; (iii) topography (m) in the spectral band ≥240 km wavelength (solid line) and ≥80 km wavelength (dashed line); (iv) Bouguer gravity anomalies (mGal) as in (ii) plus calculated values for the local isostasy model discussed in the text (dashed line); and (v) Moho depths (km) from Figure 2.21 (solid line), from Berry and Forsyth (1975; dashed line), and calculated values for the local isostasy model discussed in the text (dashed line). Location of Profile A in Figure 2.21.

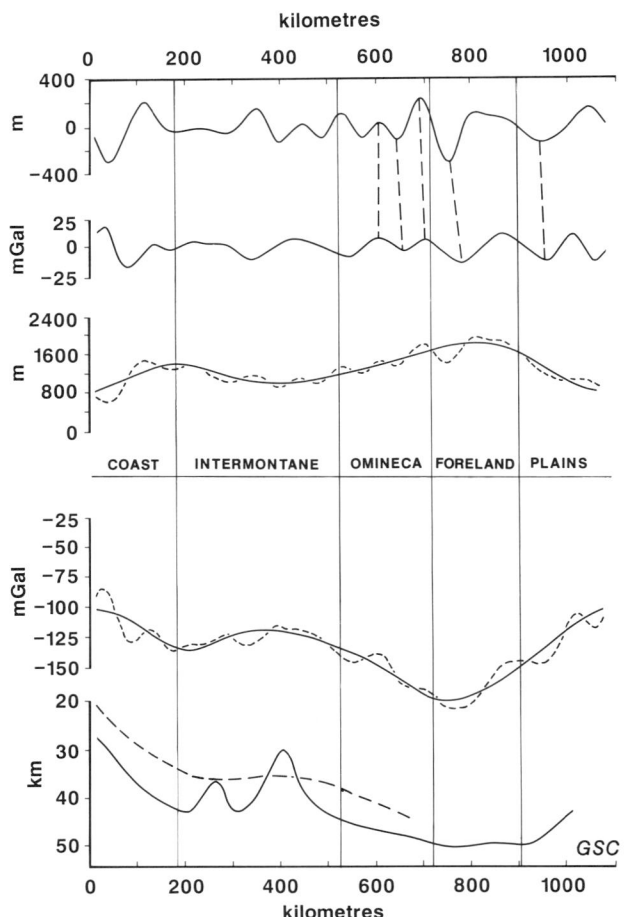

Figure 2.29. Profile B, from top to bottom as for Figure 2.28. Location in Figure 2.21.

B Geomagnetism

Only modest progress has been made in completing the aeromagnetic coverage of the Canadian Cordillera since the review of Berry et al. (1971). The Federal-Provincial program to conduct low level (typically 600 m mean terrain clearance, 1.5 to 2.0 km line spacing) magnetic surveys began in the late 1950s and to date approximately 60 per cent of the region has been flown (Fig. 2.30). The most recent surveys have been over the Insular Belt. The Queen Charlotte Islands were flown in 1985 and coverage of Vancouver Island was completed in 1986. The only comprehensive data set that exists other than satellite data is from high level (5.5 km average altitude), regional (37 km line spacing) surveys flown south of 60°N in 1969 (Haines and Hannaford, 1972) and from 60°N to the Beaufort Sea in 1972 (Haines and Hannaford, 1976). In contrast, the pioneering marine magnetic survey by Raff and Mason (1961) off the west coast of North America has been substantially augmented by systematic marine geophysical surveys (10 km average line spacing) so that the Canadian Exclusive Economic Zone (i.e. to 200 nautical miles/366 km offshore) has now been completely surveyed at a regional scale (Currie et al., 1983; Fig. 2.30).

Figure 2.30. Distribution of low-level aeromagnetic observations onshore and shipborne magnetic surveys offshore in western Canada with schematic magnetic anomalies at 500 mT contour interval. Morphogeological belt boundaries are indicated.

and vice versa, particularly toward the eastern ends of the profiles. Possible correlations are indicated with dashed lines in Figures 2.28 and 2.29. Similarly, in northern Yukon the topographically high Richardson Mountains coincide with a Bouguer gravity high (Fig. 2.27). Stephenson and Lambeck (1985) recently showed that horizontal compressional stresses within the crust in the order of 10^2 MPa, as might be expected in lithosphere close to a convergent plate boundary (e.g., west end of Profile A, Fig. 2.28), can produce non-conventional isostatic relationships between gravity and topography, especially at wavelengths between about 125 and 250 km. Although the same interpretation cannot be applied to the eastern side of the Cordillera solely on the basis of isostatic calculations, Bell and Gough (1979) determined from borehole measurements in the Western Canada Basin that large horizontal compressive stresses are present and oriented normal to fold axes in the adjacent Foreland Belt. They suggested that the stress field responsible for the deformation there could still be present, a notion that has been confirmed by the mechanism solutions of the recent Nahanni earthquake (R.B. Horner, pers. comm., 1986).

On a broad regional basis, not only is the western Cordillera easy to distinguish magnetically from the Canadian Shield and Foreland Belt to the east (Coles et al., 1976) but first order geological features within the western Cordillera have a clear magnetic expression. Cordilleran anomalies are highly irregular and complex with a predominantly northwest-trending grain that parallels tectonic strike. The Canadian Shield exhibits high amplitude, northeasterly trending magnetic anomalies that, south of 60°N, extend southwesterly across the Foreland Belt to terminate close to, but east of, the Southern Rocky Mountain Trench (Caner, 1969; Fig. 2.30). North of 60°N the orientation of cratonal anomalies is more northerly and their termination follows the eastern edge of deformation, tracing a line between the Shield and the Mackenzie Mountains (Coles et al., 1976). In this area the Foreland and Omineca belts cannot be distinguished magnetically.

Within the Omineca and Intermontane belts of the southern Cordillera the field becomes more complex for shorter wavelengths (<100 km) and smoother for long wavelength (>150 km) features. The Omineca Belt has a relatively smoothed, subdued magnetic field. In areas where more intense anomalies exist they generally have less consistent northwesterly trends than the other belts. The Intermontane Belt exhibits, on average, the most complex, short wavelength, magnetic signature of all the five belts. The anomalies tend to be intense except where the sources are deeper such as near the Bowser Basin (about 56°N, 128°W; Fig. 2.30). The Intermontane Belt is characterized by a magnetically heterogeneous upper crust and a lack of magnetic sources in the lower crust possibly due to a more sialic lower crust or to relatively elevated Curie isotherms with respect to the geological provinces to the east.

This zone of low amplitude, long wavelength anomalies, characteristic of the southern Intermontane Belt, continues south into the United States Cordillera where it appears to be coincident with a region of thinner crust, lower mantle and crustal seismic velocities, and higher electrical conductivity (Pakizer and Zeitz, 1965).

The Coast Belt has north-trending positive magnetic anomalies including a particularly intense high and associated steep gradient along its west side (Haines et al., 1971; Hannaford and Haines, 1974). Coles and Currie (1977) concluded that the west coast high cannot be explained by the measured magnetizations of the rocks. They proposed a thick (up to 40 km) magnetic crust for the western Coast Belt and suggested that it may be produced by the thermal reworking of crustal rocks above a dehydrating downgoing oceanic plate. This is in contrast to the less magnetic crust beneath the Insular Belt to the west and even less magnetic crust beneath the Intermontane Belt to the east.

The magnetic anomalies of the Insular Belt have northwesterly trends and tend to have a longer wavelength and lower amplitude than those of the other morphogeological belts. The data are of particular value in this region as they provide a basis for extending the onshore geology into the offshore (MacLeod et al., 1977).

Farther west the oceanic crust exhibits characteristic lineated magnetic anomalies, up to 1000 nT in amplitude, mostly subparallel with and approximately symmetric about the active spreading ridges (Fig. 2.30). The magnetic stripes appear to be offset across former transform segments (fracture zones). On the Juan de Fuca Plate the magnetic pattern is attenuated along the zone of convergence with the North America Plate. To the north, the Pacific Plate and its magnetic anomalies are truncated at the Queen Charlotte-Fairweather Fault system.

In the southern Cordillera the magnetic signatures of component terranes are clearly expressed. Quesnellia displays a complex pattern of irregular and closely juxtaposed positive and negative anomalies, commonly separated by steep gradients. Although much of this field is undoubtedly due to widespread Middle Jurassic to mid-Cretaceous plutons, several of the anomalies are clearly related to other sources, perhaps in the underlying miogeocline of ancestral North America. The nearby Kootenay and Monashee terranes have a much more uniform field as do the Methow and Bridge River terranes to the west.

A similar qualitative discussion of the regional magnetic patterns over a part of the Cordillera in the western United States has been presented by Mabey et al. (1978). They also observed a predominantly north-trending grain over the Cordillera with prominent northeast trends to the east over the Precambrian basement.

Quantitative analysis of magnetic data in the Canadian Cordillera is scant. Marine magnetic anomalies off western Canada have received by far the greatest analysis. It is difficult to exaggerate the importance of marine magnetic surveys and the Juan de Fuca Ridge system in the evolution of the hypothesis of plate tectonics. Riddihough et al. (1983) emphasized the unique place that the Juan de Fuca Plate occupies in these studies. Wilson (1965) and Vine and Wilson (1965) demonstrated the existence of transform faults and the seafloor spreading concept by interpreting magnetic data from the northeast Pacific. The model was refined by Vine (1968) who provided detailed seafloor ages for the region. Atwater (1970) described the plate motions and established that the Juan de Fuca Ridge is a spreading centre with a separate plate to the east, the Juan de Fuca Plate, that is probably subducting beneath the western margin of North America. Riddihough (1977) showed there has been a general decrease in spreading rate along the accreting margin over the last 10 Ma and that spreading often has been asymmetric with faster spreading on the eastern side. He showed that there has been a progressive clockwise rotation of the axes of all segments of the Juan de Fuca Ridge and that, east of the spreading axes, the plate(s) cannot be moving as a unit. The presence of multiple plates east of the ridge system was first suggested by Barr and Chase (1974) on the basis of earthquake epicentre locations. This was later substantiated by Hyndman et al. (1979).

Electromagnetic depth sounding

Variations of the electrical conductivity in the Earth's crust and upper mantle are produced by both lateral and vertical changes in rock temperature, porosity and fluid content, and/or in the concentration and continuity of conducting minerals. The electrical conductivity distribution can be obtained by measuring the fluctuations of the Earth's natural electromagnetic field on its surface. Geomagnetic Depth Sounding (GDS) is the survey practice of measuring only the three orthogonal components of the geomagnetic field. Magnetotelluric Depth Sounding (MTDS) records only the two horizontal magnetic components plus the two electric field components. Current practice is to combine these two techniques at each survey site. Magnetotelluric depth sounding provides the vertical resistivity distribution beneath each recording station, whereas, geomagnetic depth sounding gives information about the lateral variations in electrical conductivity.

The first geomagnetic depth sounding observations in the Canadian Cordillera were made by Hyndman (1963) along a profile near 49.5°N, where he concluded that the ratios of the vertical to horizontal magnetic variations were 2 to 3 times higher east of Kootenay Lake than those to the west. This was followed by several multi-site, three component experiments (Camfield et al., 1970; Caner et al., 1967, 1971; Cochran and Hyndman, 1970; Dragert, 1973; Dragert and Clark, 1977; Lambert and Caner, 1965; Miller, 1973), culminating in the recent deployment of 33 magnetometers (Gough et al., 1982) in southern British Columbia and Alberta between 55°N to 49°N. All of these studies were exploratory in nature, with site separations greater than 80 km, and capable of resolving only large scale horizontal conductivity variations beneath the Cordillera (Fig. 2.31).

Two regions have been defined (Fig. 2.32): a western conductive region, characterized by low vertical to horizontal magnetic amplitude ratios, and an eastern zone with high magnetic ratios. The transition between them follows the western side of the Rocky Mountain Trench between latitudes 55°N to 52°N and then trends southward along the Columbia River fault zone and between Arrow and Kootenay lakes (the Valhalla shear zone?) to 49.3°N. The discontinuity then changes to an easterly direction, cutting across the trend of the Kootenay Arc into southern Alberta (Lajoie and Caner, 1970). The western region has been modelled by a 15 km thick, highly conductive layer beginning at a depth of 15 ± 5 km in the crust; the eastern region lacks such a conductor (Caner et al., 1969; Lajoie and Caner, 1970). An upwelling of isotherms westward across this transition boundary could account for the higher heat flow over the Omineca and Intermontane belts, and the high conductivities in the middle to lower crust, provided these parts of the crust are hydrated (Caner, 1971; Caner et al., 1971; Hyndman and Hyndman, 1968). Filtered aeromagnetic profiles are more attenuated and featureless over the western than over the eastern region, presumably reflecting a shallower Curie point isotherm (Caner, 1970). It appears that this transition zone marks a fundamental change in crustal properties, perhaps those associated with Ancestral North America and the accreted terranes. The change to an east-west direction near the International Boundary is the result of an additional conductor (discussed below) the effect of which masks the change from low to high geomagnetic ratios between the Foreland and Omineca belts. Law and Riddihough (1971) noted similar transitions in many other parts of the world.

The western limit of the conductive region appears to underlie the western part of the Coast Belt (Kurtz et al., 1986; Dragert et al., 1980). The name 'Canadian Cordilleran Regional conductor (CCR)' was proposed by Gough (1986) who argued that the Omineca and Intermontane belts overlie a region of upper mantle upwelling. Consequently, temperatures are high in the upper mantle and lower crust, so that partial melting and heated saline fluids may

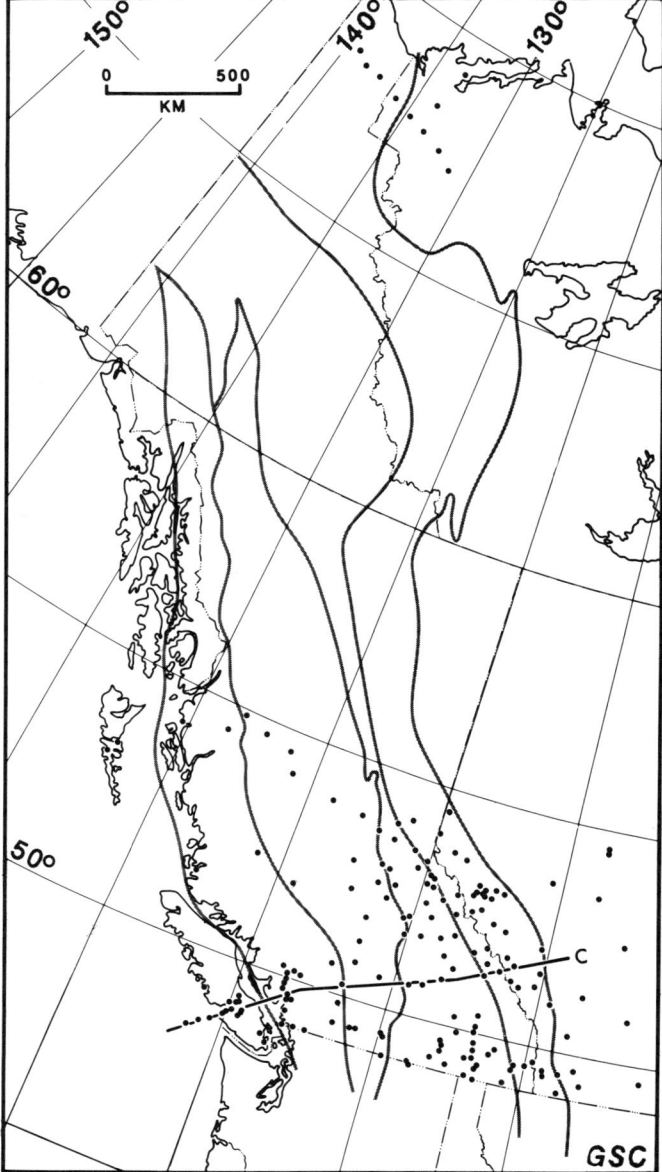

Figure 2.31. Geomagnetic depth sounding and magnetotelluric depth sounding sites in the Canadian Cordillera. Cross-section along Profile C shown in Figure 2.32.

account for the high conductivities of the Canadian Cordilleran Regional conductor (Gough, 1986).

The Foreland Belt, between 52°N-54°N, has high geomagnetic amplitudes in the form of circular anomalies parallel with and abutting the Rocky Mountain Trench (Bingham et al., 1985). A conductive ridge rising to a depth of 10 km beneath the main ranges was suggested as the source. A more recent magnetotelluric depth sounding study confirmed the high conductivities near that depth (Hutton et al., 1987). It is unlikely that high conductivities can be explained in terms of partial melting at mid- to upper crustal depths beneath the main ranges; rather, hot saline fluids filling the pore space of materials near this depth may be responsible (Ingham et al., 1987).

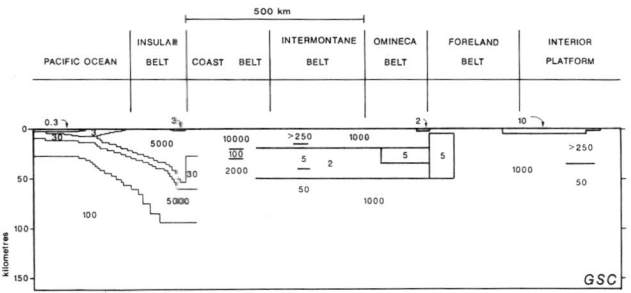

Figure 2.32. Conductivity structure model across southern Cordillera (Profile C, Fig. 2.31) showing extreme range in resistivities (in ohm.metres) proposed by various workers (see text).

In the northern Foreland Belt local lateral variations in conductivity have been inferred in the upper crust. A study in northern Yukon and the Mackenzie Delta region by DeLaurier et al. (1981) attributed conductive, fluid saturated porous materials, extending to a depth of 20 km, to be the result of fracturing associated with the Rapid Fault Array (see Chapter 17).

A linear electrical current system within an elongated, conductive zone coincident with a proposed rift zone in Precambrian basement rocks (Kanasewich et al., 1969) has been identified across the Foreland Belt from southern Alberta into southeastern British Columbia as far as Kootenay Lake (Camfield and Gough, 1975; Dragert and Clarke, 1977; Gough et al., 1982). The low magnetic ratios associated with this conductor dominate the regional pattern in the Kootenay Lake area, causing the transition between the low and high magnetic ratios to strike eastward.

The limited data available from the Coast Belt makes it difficult to distinguish conductive structure from that beneath the Intermontane Belt. Dragert et al. (1980) concluded that the upper 20 km of crust in the Garibaldi volcanic belt beneath Pemberton is very resistive, reflecting the presence of extensive crystalline rocks, underlain by a conductor, which is thinner and more resistive than the Canadian Cordilleran Regional conductor. Magnetotelluric depth sounding and DC resistivity studies indicate that highly conductive dipping structures exist in the shallow (<1 km) crust beneath Meager Mountain where drilling has detected convective flows of hot water in fractures (Flores et al., 1985; Pham van Ngoc, 1978, 1980; Fairbank et al., 1981). Similarly, magnetotelluric depth sounding data from Mount Cayley also could be the electrical response to convective hot fluids in the shallow crust, or, if topographic effects are ignored, of a larger high temperature body at middle to upper crustal depths.

In the southern Insular Belt, geomagnetic and magnetotelluric depth sounding surveys have begun to delineate details of the electrical structure beneath Vancouver Island. DeLaurier et al. (1983) concluded that dipping conductive structures beneath the island can accurately reproduce the observed geomagnetic ratios obtained at three land and three ocean bottom sites. The dipping conductive slab model is consistent with the subduction of the Juan de Fuca Plate beneath Vancouver Island. From a set of 18 magnetotelluric depth sounding sites across Vancouver Island, Kurtz et al. (1986) calculated accurate depths to a dipping conductor coincident with the lower group of reflectors shown in Figure 2.33, a package of seismic reflectors. At these depths (23-28 km), the high conductivity is believed to be the result of a saline fluid filling pore space in the materials above the descending slab. Such fluids may originate as a consequence of dehydration reactions at or near the blueschist field of metamorphism. These results explain the 65 ± 5 km depth to a conductor beneath Victoria (Caner and Auld, 1968), the current estimate to the depth to the top of subducting plate beneath that location. Farther north in the Insular Belt, sparse geomagnetic depth sounding data were modelled as an extension of the Canadian Cordilleran Regional conductor westward beneath the Queen Charlotte Islands (Miller, 1973).

Cordilleran structure from geophysics

Two zones in the southern Cordillera, identified from the available data, indicate where major regional changes in geophysical properties occur. Both zones are subparallel with tectonic strike. The eastern zone is nearly coincident with the Southern Rocky Mountain Trench and, from east to west, is characterized by a 10 to 15 km shallowing of the Moho as defined by deep refraction profiles, an apparent 40 mW·m^{-2} increase in measured surface heat flow, a 50 to 100 mGal increase in Bouguer anomaly values, a smoothing of long-wavelength magnetic anomalies and a pronounced increase in electrical conductivity.

The western zone, which lies between the Garibaldi (Cascade) volcanic arc and the Pacific Coast, is marked by a more pronounced transition in geophysical properties. From east to west, deep refraction profiles indicate that the crust thickens by about 20 km, that surface heat flow values decline by about 40 mW·m^{-2} and low-level seismicity rises sharply within a horizontal distance of 20 km and that the Bouguer anomaly rises by up to 150 mGal over a distance of about 80 km. A linear magnetic anomaly high of about 1000 nT is associated with the western zone.

The regions interior to the geophysical gradient zones, the Omineca Belt, the Intermontane Belt and the eastern part of the Coast Belt, have distinctly different geophysical properties than the Foreland, Insular and western Coast belts. The southern interior region is in local isostatic balance and, from studies of regional structure, sustained substantial tensional strain during the Eocene (Souther, 1970; Ewing, 1980; see Chapter 17). It has a relatively thin, less dense crust, high heat flow, smoother long-wavelength magnetic anomalies and high conductivity.

Caner (1970) postulated that these geophysical signatures could be produced by a hydrated, partially melted lower crust/uppermost mantle. One mechanism is convective transport of heat from deep within the Earth. Gough (1984, 1986) suggested that this may be part of a global system whereby upward diverging flow within a mantle convection cell, presently below the East Pacific Rise, continues under western North America, including the southern Canadian Cordillera. Alternatively, a warm thin crust below southwestern Canada may be a consequence of regional processes, back-arc asthenospheric flow related to convergence and subduction of the Juan de Fuca Plate system beneath the western edge of the continent (Davis and Lewis, 1984; Lewis et al., 1985). Presumably such

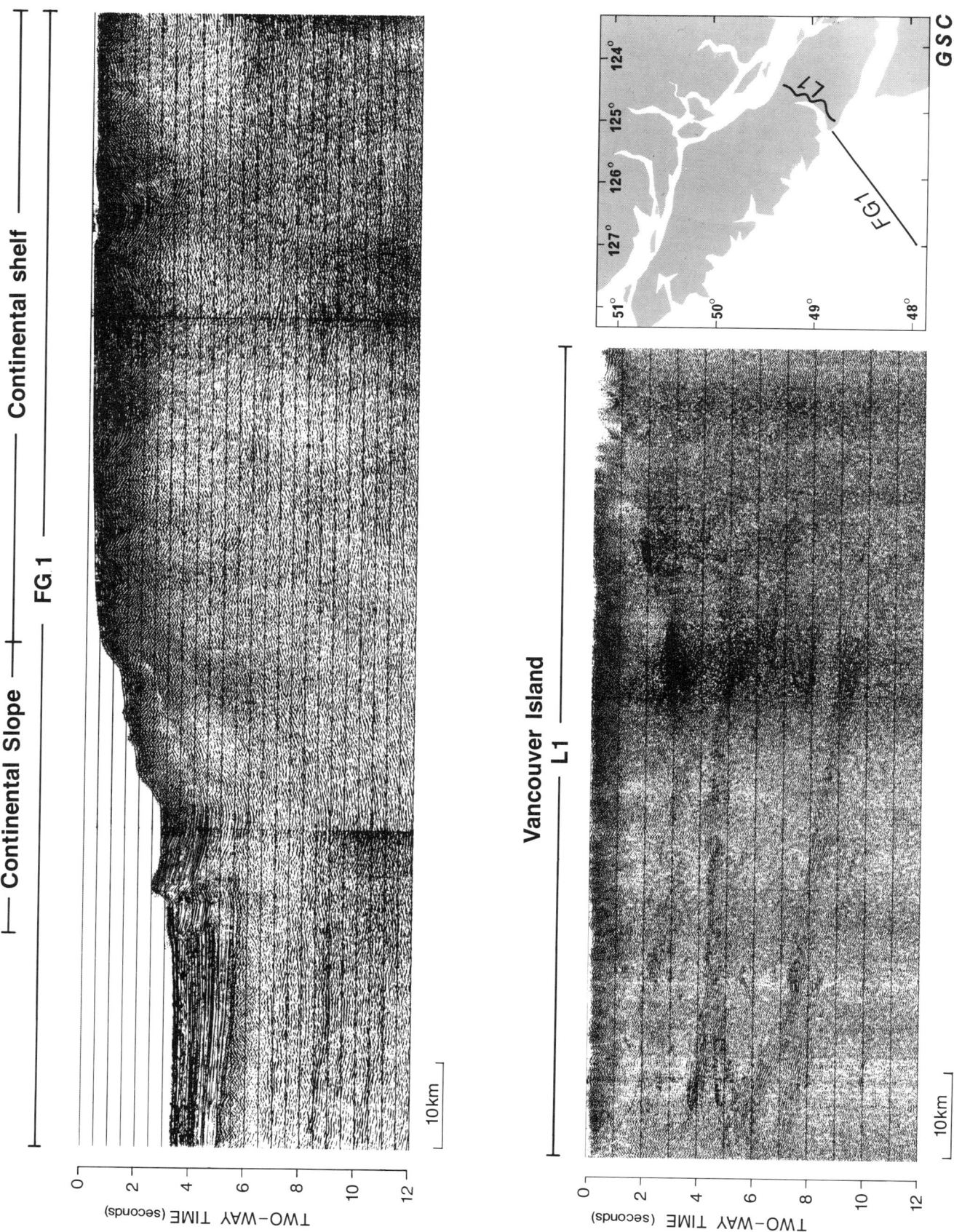

Figure 2.33. Multichannel seismic reflection profiles across Vancouver Island (L1) and the adjacent continental shelf and slope (FG1).

processes have resulted in extension and related tectonic denudation structures which are well displayed in the southern Omineca Belt (see Chapter 17).

The Foreland Belt is considered to be in a northeast-southwest compressive regime and, as a consequence, may be out of isostatic equilibrium at topographic wavelengths between about 125 and 250 km. Its crust is electrically resistive relative to the interior region and its magnetic character, crustal thickness and surface heat flow are shieldlike. It has been suggested by several researchers (e.g., Stacey, 1973) that cratonic basement rocks underlie the thrust belt, a notion that is confirmed by deep reflection profiling in southern British Columbia (Cook, 1985; Cook et al., 1987) and in nearby northern Montana (Potter et al., 1987). Crystalline shield rocks are imaged below the detachment zone along a west-dipping interface below the Foreland and eastern Omineca belts.

The western region likewise could be out of isostatic balance for the topographic wavelength range 125 to 250 km. It appears to be under northeast-southwest compression resulting from a combination of northward translation and the nearby normal (Juan de Fuca Plate) to highly oblique (Pacific Plate) convergence of oceanic crust relative to North America. In the south, geophysical measurements reflect the presence of the subducting Juan de Fuca Plate. The top part of the plate, as defined by seismicity (Rogers, 1983; Taber and Smith, 1985) and by a strong electrical conductivity anomaly (Kurtz et al., 1986), descends eastward across southern Vancouver Island from a depth of about 30 km below the west coast to between 80 and 90 km just east of Georgia Strait-Puget Sound. The plate becomes aseismic at that depth but it is thought, from petrological considerations, that the subducting plate extends to at least 100 km below the Garibaldi (Cascades) magmatic arc where it becomes partially melted, and from where the magma ascends and is emplaced into the upper crust (Blackwell et al., 1982; T.J. Lewis, pers. comm., 1986).

The pattern of surface heat flux and deep seismicity suggests that a relatively cool (about 375° to 450°C), nearly isothermal, hydrated (?) zone is present above the slab beginning at about 30 km depth; it thickens eastward to about 30 km where it is truncated by hot ascending fluids associated with the magmatic arc. This deep cool wedge is considered to be more dense and more magnetic than the surrounding rock, as required by gravity (Riddihough, 1979) and magnetic (Coles and Currie, 1977) models of the subduction. The estimate of three to four per cent saline fluids in the descending plate may account for the zone of multiple reflectors thought to represent the zone of detachment between the North America and Juan de Fuca plates (Fig. 2.33). The temperature and pressure (>850 Pa) in this zone suggest metamorphic reactions near the blueschist field are occurring and that dehydration provides a source for the fluids. Consequently this suggests a depth of origin for the widespread occurrence of high pressure/low temperature metamorphic rocks such as glaucophane schists (Kurtz et al., 1986).

Across the southern Canadian Cordillera the geophysical gradient zones appear to mark, in the east, the transition from relatively stable cratonic basement to an interior region undergoing steady-state infusion of heat from deep within the earth. In the west, the transition coincides approximately with the boundary between the interior region and the area of active plate convergence.

North of the Juan de Fuca Plate convergence, scanty geophysical data indicate a different tectonic regime across the Cordillera. The Insular and Coast belts have shallower, more uniform Moho depths (Fig. 2.21), the Intermontane Belt is underlain by a cooler mantle (Jessop et al., 1984; Fig. 2.23) and, magnetically at least, the Foreland Belt ceases to appear shieldlike (Coles et al., 1976). These changes could reflect the absence of subduction and associated back-arc asthenospheric flow in the north and could be taken as support for the Davis and Lewis (1984) concept of subduction-related regional heat transfer below the southern Cordillera. This suggestion must be regarded as tentative as geophysical results from the north are very limited.

Unique geophysical characterization of Cordilleran terranes has not been widely attempted. In the southern Cordillera Rogers and Auld (1985) drew attention to the lack of correlation between gravity and magnetic signatures across terrane boundaries. This implies that such boundaries are not vertical, a suggestion confirmed by seismic reflection profiles across the contacts between Wrangellia and the Pacific Rim and Crescent terranes on southern Vancouver Island and off the west coast (Fig. 2.33; Clowes et al., 1987a; Yorath et al., 1987).

REFERENCES

Atwater, T.
1970: Implications of plate tectonics for the Cenozoic tectonic evolution of western North America; Geological Society of America Bulletin, v. 81, p. 3513-3536.

Barr, S.M. and Chase, R.L.
1974: Geology of the north end of Juan de Fuca Ridge and sea-floor spreading; Canadian Journal of Earth Sciences, v. 11, p. 1384-1406.

Basham, P.W., Forsyth, D.A., and Wetmiller, R.J.
1977: The seismicity of northern Canada; Canadian Journal of Earth Sciences, v. 14, p. 1646-1667.

Bell, J.S. and Gough, D.I.
1979: Northeast-southwest compressive stress in Alberta: Evidence from oil wells; Earth and Planetary Science Letter, v. 45, p. 475-482.

Bennett, G.T., Clowes, R.M., and Ellis, R.M.
1975: A seismic refraction survey along the southern Rocky Mountain Trench, Canada; Bulletin of the Seismological Society of America, v. 65, p. 37-54.

Berry, M.J. and Forsyth, D.A.
1975: Structure of the Canadian Cordillera from seismic refraction and other data; Canadian Journal of Earth Sciences, v. 12, p. 182-208.

Berry, M.J., Jacoby, W.R., Niblett, E.R., and Stacey, R.A.
1971: A review of geophysical studies in the Canadian Cordillera; Canadian Journal of Earth Sciences, v. 8, p. 788-801.

Bevier, M.L., Armstrong, R.L., and Souther, J.G.
1979: Miocene peralkaline volcanism in west-central British Columbia — its temporal and plate-tectonics setting; Geology, v. 7, p. 389-392.

Bingham, D.K., Gough, D.I., and Ingham, M.R.
1985: Conductive structures under the Canadian Rocky Mountains; Canadian Journal of Earth Sciences, v. 22, p. 384-498.

Blackwell, D.D.
1969: Heat flow determinations in northwestern United States; Journal of Geophysical Research, v. 74, p. 992-1007.
1974: Terrestrial heat flow and its implications on the location of geothermal reservoirs in Washington; Washington Division of Geological and Earth Resources Information Circular, v. 50, p. 24-33.

Blackwell, D.D., Bowen, R.G., Hull, D.A., Riccio, J., and Steele, J.L.
1982: Heat flow, arc volcanism, and subduction in northern Oregon; Journal of Geophysical Research, v. 87, p. 8735-8754.

Blome, C.D.
1984: Upper Triassic Radiolaria and radiolarian zonation from Western North America; Bulletin of American Paleontology, v. 85 (318), 88 p., 17 pls.

Brandon, M.T.
1985: Mesozoic mélange of the Pacific Rim Complex, western Vancouver Island; in Field Guides to Geology and Mineral Deposits in the Southern Canadian Cordillera, D.J., Tempelman-Kluit (ed.), Geological Society of America Cordilleran Section Meeting, Vancouver, B.C.

Brandon, M.T., Orchard, M.J., Parrish, R.R., Sutherland Brown, A., and Yorath, C.J.
1986: Fossil ages and isotopic dates from the Paleozoic Sicker Group and associated intrusive rocks, Vancouver Island, British Columbia; in Current Research, Part A, Geological Survey of Canada, Paper 86-1A, p. 683-696.

Brown, J., Hajnal, Z., Zelt, C., Ellis, R.M., and Stephenson, R.
1987: Crust under the Peace River Arch from a refraction investigation; Geological Association of Canada/Mineralogical Association of Canada Joint Annual Meeting, Program with Abstracts, v. 12, p. 27.

Buck, R.J.
1967: The gravity anomaly field in Western Canada with maps, Part 1. Gravity Map Series 39 to 43; Observatories Branch, Department of Energy, Mines and Resources, Ottawa.

Cady, J.W.
1980: Gravity highs and crustal structure, Omineca Crystalline Belt, northeastern Washington and southeastern British Columbia; Geology, v. 8, p. 328-332.

Callomon, J.H.
1985: A review of the biostratigraphy of the post-lower Bajocian Jurassic ammonites of western and northern North America; Geological Association of Canada, Special Paper 27.

Cameron, B.E.B. and Tipper, H.W.
1985: Jurassic stratigraphy of the Queen Charlotte Islands, British Columbia; Geological Survey of Canada, Bulletin 365, 49 p.

Camfield, P.A. and Gough, D.I.
1975: Anomalies in daily variation magnetic fields and structure under northwestern United States and southwestern Canada; Geophysical Journal of the Royal Astronomical Society, v. 41, p. 193-218.

Camfield, P.A., Gough, D.I., and Porath, H.
1970: Magnetometer array studies in the northwestern United States and southwestern Canada; Geophysical Journal of the Royal Astronomical Society, v. 22, p. 201-221.

Caner, B.
1969: Long aeromagnetic profiles and crustal structure in western Canada; Earth and Planetary Science Letters, v. 7, p. 3-11.
1970: Electrical conductivity structure in western Canada and its petrological interpretation; Journal of Geomagnetism and Geoelectricity, v. 22, p. 113-129.
1971: Quantitative interpretation of geomagnetic depth-sounding data in western Canada; Journal of Geophysical Research, v. 76, p. 7202-7216.

Caner, B. and Auld, D.R.
1968: Magneto-telluric determination of upper mantle conductivity structure at Victoria, British Columbia; Canadian Journal of Earth Sciences, v. 5, p. 1209-1220.

Caner, B., Auld, D.R., Dragert, H., and Camfield, P.A.
1971: Geomagnetic depth sounding and crustal structure in western Canada; Journal of Geophysical Research, v. 76, p. 7181-7201.

Caner, B., Camfield, P.A., Anderson, F., and Niblett, E.R.
1969: A large scale magnetotelluric survey in western Canada; Canadian Journal of Earth Sciences, v. 6, p. 1245-1261.

Caner, B., Cannon, W.H., and Livingston, C.E.
1967: Geomagnetic depth sounding and upper mantle structure in the Cordillera region in Western North America; Journal of Geophysical Research, v. 72, p. 6335-6351.

Carter, E.S.
1988: Part 2: Systematic Paleontology; in Lower and Middle Jurassic Radiolarian Biostratigraphy and Systematic Paleontology, Queen Charlotte Islands, British Columbia, E.S. Carter, B.E.B. Cameron and P.L. Smith (ed.), Geological Survey of Canada, Bulletin 386, p. 26-109.

Carter, E.S., Cameron, B.E.B., and Smith, P.L.
1988: Part 1: Lithostratigraphy and biostratigraphy; in Lower and Middle Jurassic Radiolarian Biostratigraphy and Systematic Paleontology, Queen Charlotte Islands, British Columbia; Geological Survey of Canada, Bulletin 386, p. 5-25.

Chandra, N.N. and Cumming, G.L.
1972: Seismic refraction studies in western Canada; Canadian Journal of Earth Sciences, v. 9, p. 1099-1109.

Clowes, R.M., Brandon, M.T., Green, A.G., Yorath, C.J., Sutherland Brown, A., Kanasewich, E.R., and Spencer, C.
1987a: LITHOPROBE - southern Vancouver Island: Cenozoic subduction complex imaged by deep seismic reflections; Canadian Journal of Earth Sciences, v. 24, p. 31-51.

Clowes, R.M., Ellis, R.M., Hajnal, Z., and Jones, I.F.
1983: Seismic reflections from the subducting lithosphere?; Nature, v. 303, p. 668-670.

Clowes, R.M., Green, A.G., Yorath, C.J., Kanasewich, E.R., West, G.F., and Garland, G.D.
1984: LITHOPROBE - a national program for studying the third dimension of geology; Journal of the Canadian Society of Exploration Geophysicists, v. 20, p. 23-39.

Clowes, R.M., Spence, G.D., Ellis, R.M., and Waldron, D.A.
1986: Structure of the lithosphere in a young subduction zone: Results from reflection and refraction studies; in Reflection Seismology: The Continental Crust; M. Barazangi and L. Brown (ed.), American Geophysical Union Geodynamics Series, v. 14, p. 313-321.

Clowes, R.M., Yorath, C.J., and Hyndman, R.D.
1987b: Reflection mapping across the convergent margin of Western Canada; Geophysical Journal of the Royal Astronomical Society, v. 89, p. 79-84.

Cochran, J.R.
1980: Some remarks on isostacy and the long-term behavior of the continental lithosphere; Earth and Planetary Science Letters, v. 46, p. 266-274.

Cochran, N.A. and Hyndman, R.D.
1970: A new analysis of goemagnetic depth-sounding data from western Canada; Canadian Journal of Earth Sciences, v. 7, p. 1208-1218.

Coles, R.L. and Currie, R.G.
1977: Magnetic anomalies and rock magnetizations in the southern Coast Mountains, British Columbia: possible relation to subduction; Canadian Journal of Earth Sciences, v. 14, p. 1753-1770.

Coles, R.L., Haines, G.V., and Hannaford, W.
1976: Large scale magnetic anomalies over western Canada and the Arctic: a discussion; Canadian Journal of Earth Sciences, v. 13, p. 790-802.

Coney P.J., Jones, D.L., and Monger, J.W.H.
1980: Cordilleran suspect terranes; Nature, v. 288, p. 329-333.

Cook, F.A.
1985: Deep basement seismic reflection profiling of the Purcell Anticlinorium using a land air gun source; Journal of Geophysical Research, v. 90, p. 651-662.

Cook, F.A., Green, A.G., Simony, P.S., Price, R.A., Parrish, R., Milkereit, B., Gordy, P.L., Brown, R.L., Coflin, K.C., and Patenaude, C.
1987: LITHOPROBE - Southern Canadian Cordilleran Transect: Rocky Mountain Thrust Belt to Valhalla Gneiss Complex; Geophysical Journal of the Royal Astronomical Society, v. 89, p. 91-98.

Cordey, F.
1986: Radiolarian ages from the Cache Creek and Bridge River complexes and from chert pebbles in Cretaceous conglomerates, southwestern British Columbia; in Current Research, Part A, Geological Survey of Canada, Paper 86-1A, p. 595-602.

Cumming, G.L. and Kanasewich, E.R.
1966: Crustal structure in western Canada: Final report, Contract AF19 (628)-2835, Project 8652; Department of Physics, University of Alberta, Edmonton, Project Vela Uniform, 185 p.

Cumming, G.L., Garland, G.D., and Vozoff, K.
1962: Seismological measurements in southern Alberta: Final report, Contract AF19 (604)-8470; Department of Physics, University of Alberta, Project Vela-Uniform.

Cumming, W.B., Clowes, R.M., and Ellis, R.M.
1979: Crustal structure from a seismic refraction profile across southern British Columbia; Canadian Journal of Earth Sciences, v. 16, p. 1024-1040.

Currie, L.
1984: The provenance of chert clasts in the Ashman conglomerates of the northeastern Bowser Basin; Bachelors Thesis, Queen's University, Kingston, Ontario, 59 p.

Currie, R.G., Cooper, R., Riddihough, R.P., and Seemann, D.A.
1983: Multiparameter geophysical surveys off the west coast of Canada: 1973-1982; in Current Research, Part A, Geological Survey of Canada, Paper 83-1A, p. 253-261.

Daly, R.A.
1912: Geology of the North American Cordillera at the forty-ninth parallel; Geological Survey of Canada, Memoir 38, 857 p.

Damborenea, S.E. and Mancenido, M.O.
1979: On the Palaeogeographical distribution of the pectinid genus Weyla (Bivalvia, Lower Jurassic); Palaeogeography, Palaeoclimatology, Palaeoecology, v. 27, p. 85-102.

Davis, E.E. and Lewis, T.J.
1984: Heat flow in a back-arc environment: Intermontane and Omineca Crystalline belts, southern Canadian Cordillera; Canadian Journal of Earth Sciences, v. 21, p. 715-726.

DeLaurier, J.M., Auld, D.R., and Law, L.K.
1983: The geomagnetic response across the continental margin off Vancouver Island: comparison of results from numerical modelling and field data; Journal of Geomagnetism and Geoelectricity, v. 35, p. 571-528.

DeLaurier, J.M., Plet, F.C., and Drury, M.J.
1981: A geomagnetic depth sounding profile across the northern Yukon and the Mackenzie delta region; Canadian Journal of Earth Sciences, v. 18, p. 1092-1100.

De Wever, P.
1982: Radiolaires du Trias et du Lias de la Tethys; Societé Géologique du Nord, No. 7, 599 p.

Douglas, R.J.W., Gabrielse, H., Wheeler, J.O., Stott, D.F., and Belyea, H.R.
1970: Geology of Western Canada; Chapter 7; in Geology and Economic Minerals of Canada; Geological Survey of Canada, Economic Geology Report 1, p. 367-488.

Dragert, H.
1973: A transfer function analysis of a geomagnetic depth sounding profile across central British Columbia; Canadian Journal of Earth Sciences, v. 10, p. 1089-1098.

Dragert, H. and Clarke, G.K.C.
1977: A detailed investigation of the Canadian Cordillera geomagnetic transition anomaly; Journal of Geophysics, v. 42, p. 373-390.

Dragert, H., Law, L.K., and Sule, P.O.
1980: Magnetotelluric soundings across the Pemberton volcanic belt, British Columbia; Canadian Journal of Earth Sciences, v. 17, p. 161-167.

Ellis, R.M. and Clowes, R.M.
1984: Earthquake risk is investigated; Offshore Resources, v. 2, p. 12-13.

Ellis, R.M., Spence, G.D., Clowes, R.M., Waldron, D.A., Jones, I.F., Green, A.G., Forsyth, D.A., Mair, J.A., Berry, M.J., Mereu, R.F., Kanasewich, E.R., Cumming, G.L., Hajnal, Z., Hyndman, R.D., McMechan, G.A., and Loncarevic, B.D.
1983: The Vancouver Island seismic Project: a CO-CRUST onshore-offshore study of a convergent margin; Canadian Journal of Earth Sciences, v. 20, p. 719-741.

Ewing, T.E.
1980: Paleogene tectonic evolution of the Pacific Northwest; Journal of Geology, v. 88, p. 619-638.

Fairbank, B.D., Reader, J.F., Openshaw, R.E., and Sadlier-Brown, T.L.
1981: 1980 Drilling and Exploration Program: Meager Creek Geothermal area, upper Lillooet River, British Columbia; Report to B.C. Hydro and Power Authority.

Flores, C., Kurtz, R.D., DeLaurier, J.M.
1985: Magnetotelluric exploration in the Meager Mountain geothermal area, Canada; Acta. Geodaet. Geophys. et Montanist. Hung., v. 20, p. 165-171.

Forsyth, D.A., Berry, M.J., and Ellis, R.M.
1974: A refraction survey across the Canadian Cordillera at 54o; Canadian Journal of Earth Sciences, v. 11, p. 533-548.

Garland, G.D. and Tanner, J.G.
1957: Investigations of gravity and isostasy in the southern Canadian Cordillera; Publications of the Dominion Observatory (Department of Energy, Mines and Resources, Ottawa), v. 19, p. 169-222.

Goodacre, A.K.
1972: Generalized structure and composition of the deep crust and upper mantle in Canada; Journal of Geophysical Research, v. 77, p. 3146-3161.

Gough, D.I.
1984: Mantle upflow under North America and plate dynamics; Nature, v. 311, p. 428-432.
1986: Mantle upflow tectonics in the Canadian Cordillera; Journal of Geophysical Research, v. 91, p. 1909-1919.

Gough, D.I., Bingham, D.K., Ingham, M.R., and Alabi, A.O.
1982: Conductive structures in southwestern Canada: A regional magnetometer study; Canadian Journal of Earth Sciences, v. 19, 1680-1690.

Green, A.G., Clowes, R.M., Yorath, C.J., Spencer, C., Kanasewich, E.R., Brandon, M.T., and Sutherland Brown, A.
1986a: Seismic reflection imaging of the subducting Juan de Fuca Plate; Nature, v. 319, p. 210-213.

Green, A.G., Berry, M.J. Spencer, C.P., Kanasewich, E.R., Chiu, S., Clowes, R.M., Yorath, C.J., Stewart, D.B., Unger, J.D., and Poole, W.H.
1986b: Recent seismic reflection studies in Canada: in Reflection Seismology: A Global Perspective; M. Barazangi and L. Brown (ed.), American Geophysical Union Geodynamics Series, v. 13, p. 85-97.

Green, A.G., Milkereit, B., Mayrand, L., Spencer, C., Kurtz, R., and Clowes, R.
1987: LITHOPROBE seismic reflection profiling across Vancouver Island: results from reprocessing; Geophysical Journal of the Royal Astronomical Society, v. 89, p. 85-90.

Haines, G.V., Hannaford, W., and Riddihough, R.P.
1971: Magnetic anomalies over British Columbia and the adjacent Pacific Ocean; Canadian Journal of Earth Sciences, v. 8, p. 387-391.

Haines, G.V. and Hannaford, W.
1972: Magnetic anomaly maps of British Columbia and the adjacent Pacific Ocean; Publications of the Earth Physics Branch, Department of Energy, Mines and Resources, Ottawa, v. 42(7), p. 211-228.
1976: A three-component aeromagnetic survey of Saskatchewan, Alberta, Yukon and the District of Mackenzie; Earth Physics Branch Geomagnetism Series No. 8, Department of Energy, Mines and Resources, Ottawa, 34 p.

Hales, A. and Nation, J.B.
1973: A seismic refraction survey in the Northern Rocky Mountains: more evidence for an intermediate crustal layer; Geophysical Journal of the Royal Astronomical Society, v. 35, p. 381-399.

Hannaford, W. and Haines, G.V.
1974: A three-component aeromagnetic survey of British Columbia and the adjacent Pacific Ocean; Publications of the Earth Physics Branch, Department of Energy, Mines and Resources, Ottawa, v. 44(14), p. 323-379.

Harms, T.A.
1986: Structural and tectonic analysis of the Sylvester Allochthon, northern British Columbia: implications for paleogeography and accretion; Ph.D. thesis, University of Arizona.

Hill, D.P.
1972: Crustal and upper-mantle structure of the Columbia Plateau from long range seismic-refraction measurements; Geological Society of America Bulletin, v. 83, p. 1639-1648.

Hillebrandt, A. Von
1981: Faunas de ammonites del Liásico Inferior y Médio (Hettangiano hasta Pliensbachiano) de America del sur (excluyendo Argentina); in Cuencas Sedimentarias del Jurásico y Cretácico de América del Sur, W. Volkheimer and E.A. Musacchio (ed.), II Congreso Latinamericano de Paleontologia, v. 2, p. 499-538.

Horn, J.R., Clowes, R.M., Ellis, R.M., and Bird, D.N.
1984: The seismic structure across an active oceanic/continental transform fault zone; Journal of Geophysical Research, v. 89, p. 3107-3120.

Horner, R.B.
1983: Seismicity in the St. Elias region of northwestern Canada and southeastern Alaska; Bulletin of the Seismological Society of America, v. 73, p. 1117-1137.

Hutton, V.R.S., Gough, D.I., Dawes, G.J.K., and Travassos, J.
1987: Magnetotelluric soundings in the Canadian Rocky Mountains; Geophysical Journal of the Royal Astronomical Society, v. 90, p. 245-263.

Hyndman, R.D.
1963: Electrical conductivity inhomogeneities in the Earth's upper mantle; M.Sc. thesis, University of British Columbia, Vancouver.
1976: Heat flow measurements in the inlets of southwestern British Columbia; Journal of Geophysical Research, v. 81, p. 337-349.

Hyndman, R.D. and Hyndman, D.W.
1968: Water saturation and high electrical conductivity in the lower continental crust; Earth and Planetary Science Letters, v. 4, p. 427-432.

Hyndman, R.D., Lewis, T.J., Wright, J.A., Burgess, M., Chapman, D.S., and Yamano, M.
1982: Queen Charlotte fault zone: heat flow measurements; Canadian Journal of Earth Sciences, v. 19, p. 1657-1669.

Hyndman, R.D., Riddihough, R.P., and Herzer, R.
1979: The Nootka fault zone - a new plate boundary off western Canada; Geophysical Journal of the Royal Astronomical Society, v. 58, p. 667-683.

Igo, H. et al.
1985: Biostratigraphical studies of conodonts and radiolarians in chert formations of the Cordilleran geosyncline; in Report of Research carried out by the Oversea Scientific Research Fund of the Ministry of Education, Science and Culture, H. Igo (ed.), Japanese Government Nos. 58041013 and 59043013. March 25, 1985.

Ingham, M.R., Gough, D.I., and Parkinson, W.D.
1987: Models of conductive structure under the Canadian Cordillera; Geophysical Journal of the Royal Astronomical Society, v. 88, p. 477-485.

Irving, E. and Yole, R.W.
1972: Paleomagnetism and kinematic history of mafic and ultramafic rocks in fold mountain belts; Publications of the Earth Physics Branch, Energy, Mines and Resources, Ottawa, v. 42, p. 87-95.

Jacoby, W.R.
1970: A refraction profile across the southern Cordillera; Abstract, EOS, Transactions of the American Geophysical Union, v. 51, p. 356.

Jeletzky, J.A.
1965: Late Upper Jurassic and Early Lower Cretaceous fossil zones of the Canadian western Cordillera, British Columbia; Geological Survey of Canada, Bulletin 103, 70 p.
1970: Cretaceous macrofaunas; in Geology and Economic Minerals of Canada, R.J.W. Douglas (ed.), Geological Survey of Canada, Economic Geology Report No. 1, p. 649-662.

Jessop, A.M. and Judge, A.S.
1971: Five measurements of heat flow in southern Canada; Canadian Journal of Earth Sciences, v. 8, p. 711-716.

Jessop, A.M., Souther, J.G., Lewis, T.J., and Judge, A.S.
1984: Geothermal measurements in northern British Columbia and southern Yukon Territory; Canadian Journal of Earth Sciences, v. 21, p. 599-608.

Johnson, S.H. and Couch, R.W.
1970: Crustal structure in the North Cascade Mountains of Washington and British Columbia from seismic refraction measurements; Bulletin of the Seismological Society of America, v. 60, p. 1259-1269.

Johnson, S.H., Couch, R.W., Gemperle, M., and Banks, E.R.
1972: Seismic refraction measurements in southeast Alaska and western British Columbia; Canadian Journal of Earth Sciences, v. 9, p. 1756-1765.

Jones, D.L., Howell, D.G., Coney, P.J., and Monger, J.W.H.
1983: Recognition, character and analysis of tectonostratigraphic terranes in western North America; in Accretion Tectonics in the Circum-Pacific Regions; H. Hashimoto and S. Uyeda (ed.), Terra Scientific Publishing Co., Tokyo, p. 21-35.

Kanasewich, E.R., Clowes, R.M., and McCloughan, C.H.
1969: A buried Precambrian rift in western Canada; Tectonophysics, v. 8, p. 513-527.

Kanmera, K., Ishii, K., and Toriyama, R.
1976: The evolution and extinction patterns of Permian fusulinaceans; in Geology and Palaeontology of southeast Asia, Teiichi Kobayashi, and Wataru Hashimoto (ed.), v. 17, p. 129-154.

Kay, M.
1951: North American geosynclines; Geological Society of America, Memoir 48, 143 p.

Keen, C.E. and Hyndman, R.D.
1979: Geophysical review of the eastern and western continental margins of Canada; Canadian Journal of Earth Sciences, v. 16, p. 712-747.

King, P.B.
1969: Tectonic Map of North America, 1:5,000,000; United States Geological Survey.

Kozur, H. and Mostler, H.
1979: Beiträge zur Erforschung der mesozoischen Radiolarien. Teil 3: Die Oberfamilien Actinommacea Haeckel 1862, emend., Artiscacea Haeckel, 1882, Multiarcusellacea nov. der Spumellaria und triassische Nassellaria; Geologisch - Paläontologische Mitteilungen Innsbruck, v. 9 (1/2), p. 1-132.

Kurtz, R.D., DeLaurier, J.M., and Gupta, J.C.
1986: A magnetotelluric sounding across Vancouver Island detects the subducting Juan de Fuca Plate; Nature, v. 321, p. 596-599.

Lajoie, J.L. and Caner, B.
1970: Geomagnetic induction anomaly near Kootenay Lake - a strike slip feature in the lower crust?; Canadian Journal of Earth Sciences, v. 7, p. 1568-1579.

Lambert, A. and Caner, B.
1965: Geomagnetic "depth sounding" and the coast effect in western Canada; Canadian Journal of Earth Sciences, v. 2, p. 485-509.

Law, L.K. and Riddihough, R.
1971: A geographical relation between geomagnetic variation anomalies and tectonics; Canadian Journal of Earth Sciences, v. 8, p. 1094-1106.

Leblanc, G. and Wetmiller, R.J.
1974: An evaluation of seismological data available for the Yukon Territory and the Mackenzie valley; Canadian Journal of Earth Sciences, v. 11, p. 1435-1454.

Lewis, T.J., Jessop, A.M., and Judge, A.S.
1985: Heat flux measurements in southwestern British Columbia: the thermal consequences of plate tectonics; Canadian Journal of Earth Sciences, v. 22, p. 1262-1273.

Mabey, D.R., Zeitz, I., Eaton, G.P., and Kleinkopf, M.D.
1978: Regional magnetic patterns in part of the Cordillera in the western United States; in Cenozoic Terranes and Regional Geophysics of the Western Cordillera, R.B. Smith and G.P. Eaton (ed.), Geological Society of America, Memoir 152, p. 93-106.

Mackie, D.J., Dehler, S.A., Clowes, R.M., and Ellis, R.M.
1985: Lithospheric structure of the Queen Charlotte Islands - Hecate Strait region from onshore-offshore seismic studies; Abstract, Canadian Society of Exploration Geophysicists/Canadian Geophysical Union Joint Annual Meeting, p. 69.

MacLeod, N.S., Tiffin, D.L., Snavely, P.D., and Currie, R.G.
1977: Geological interpretation of magnetic and gravity anomalies in the Strait of Juan de Fuca, U.S.-Canada; Canadian Journal of Earth Sciences, v. 14, p. 223-238.

Mair, J.A. and Lyons, J.A.
1976: Seismic reflection techniques for crustal structure studies; Geophysics, v. 41, p. 1272-1290.

Majorowicz, J.A. and Jessop, A.M.
1981: Regional heat flow patterns in the western Canadian sedimentary basin; Tectonophysics, v. 74, p. 209-238.

Majorowicz, J.A., Jones, F.W., Lam, H.L., and Jessop, A.M.
1984: The variability of heat flow both regional and with depth in southern Alberta, Canada - effect of groundwater flow; Tectonophysics, v. 106, p. 1-29.

Mamet, B.L.
1976: An atlas of microfacies in Carboniferous carbonates of the Canadian Cordillera; Geological Survey of Canada, Bulletin 255, 131 p.

McMechan, G.A. and Spence, G.D.
1983: P-wave velocity structure of the Earth's crust beneath Vancouver Island; Canadian Journal of Earth Sciences, v. 20, p. 742-752.

Mereu, R.F., Majumdar, S.C., and White, R.E.
1977: The structure of the crust and upper mantle under the highest ranges of the Canadian Rockies from a seismic refraction survey; Canadian Journal of Earth Sciences, v. 14, p. 196-208.

Miller, H.G.
1973: An analysis of geomagnetic variations in western British Columbia; Ph.D. thesis, University of British Columbia, Vancouver.

Milne, W.G.
1967: Earthquake epicentres and strain release in Canada; Canadian Journal of Earth Sciences, v. 4, p. 1-18.

Milne, W.G. and White, W.R.H.
1960: A seismic survey in the vicinity of Vancouver Island, British Columbia; Dominion Observatory Publication, No. 24, p. 145-154.

Milne, W.G., Rogers, G.C., Riddihough, R.P., McMechan, G.A., and Hyndman, R.D.
1978: Seismicity of western Canada; Canadian Journal of Earth Sciences, v. 15, p. 1170-1193.

Monger, J.W.H. and Berg, H.C.
1984: Lithotectonic terrane map of western Canada and southeastern Alaska; in Lithotectonic Terrane Maps of the North American Cordillera; N.J. Silberling and D.L. Jones (ed.), United States Geological Survey, Open File Report 84-523.

Monger, J.W.H. and Ross, C.A.
1971: Distribution of fusulinaceans in the western Canadian Cordillera; Canadian Journal of Earth Sciences, v. 8, p. 259-278.

Monger, J.W.H., Clowes, R.M., Price, R.A., Simony, S.P., Riddihough, R.P., and Woodsworth, G.J.
1985: Corridor B-2: Juan de Fuca plate to Alberta plains; Geological Society of America, Centennial Continent/Ocean Transect #7.

Monger, J.W.H., Souther, J.G., and Gabrielse, H.
1972: Evolution of the Canadian Cordillera: a plate tectonic model; American Journal of Science, v. 272, p. 577-602.

Nichols, R.M. and Silberling, N.J.
1979: Early Triassic (Smithian) ammonites of paleo-equatorial affinity from the Chulitna terrane, south-central Alaska; United States Geological Survey, Professional Paper 1121-B.

Orchard, M.J.
1983: *Epigondolella* populations and their phylogeny and zonation in the Norian (Upper Triassic); Fossils and Strata, no. 15, p. 177-192.

Pakizer, L.C. and Zeitz, I.
1965: Transcontinental and upper mantle structure; Reviews of Geophysics, v. 3, p. 505-520.

Perez, O.J. and Jacob, K.H.
1980: Tectonic model and seismic potential of the eastern Gulf of Alaska and Yakataga seismic gap; Journal of Geophysical Research, v. 85, p. 7132-7150.

Pessagno, E.A. Jr. and Blome, C.D.
1980: Upper Triassic and Jurassic Pantanellinae from California, Oregon and British Columbia; Micropaleontology, v. 26(3), p. 255-273, pl. 1-11.

Pessagno, E.A. Jr. and Poisson, A.
1981: Lower Jurassic radiolaria from the Gumuselu Allochthon of southwest Turkey (Taurides Occidentales); Bulletin of the Mineral Research and Exploration Institute of Turkey, no. 92, 1979, p. 47-69, pls. 1-15.

Pessagno, E.A. Jr. and Whalen, P.A.
1982: Lower and Middle Jurassic radiolaria (Multicyrtid Nassellariina) from California, east-central Oregon and the Queen Charlotte Islands, B.C.; Micropaleontology, v. 28(2), p. 111-169.

Pessagno, E.A. Jr., Blome, C.D., and Longoria, J.F.
1984: A revised radiolarian zonation for the Upper Jurassic of western North America; Bulletin of American Paleontology, v. 87(320), 51 p.

Pessagno, E.A. Jr., Blome, C.D., Carter, E.S., MacLeod, N., Whalen, P.A., and Yeh, K.
in press: Part II, Preliminary Radiolarian Zonation for the Jurassic of North America; in Studies of North American Jurassic Radiolaria, Cushman Foundation for Foraminiferal Research, Special Publication No. 23, p. 1-18.

Pham van Ngoc
1978: Magnetotelluric prospecting in the Mount Meager geothermal region (British Columbia); Earth Physics Branch, Open File 78-6E.
1980: Magnetotelluric survey of the Mount Meager region and of the Squamish Valley (British Columbia); Earth Physics Branch, Open File 80-80E.

Potter, C.J., Allemdinger, R.W., Hauser, E.C., and Oliver, J.E.
1987: COCORP deep seismic reflection traverses of the U.S. Cordillera; Geophysical Journal of the Royal Astronomical Society, v. 89, p. 99-104.

Raff, A.D. and Mason, R.G.
1961: Magnetic survey of the west coast of North America, 40°N to 52°N; Geological Society of America Bulletin, v. 72, p. 1267-1270.

Riddihough, R.P.
1977: A model for recent plate interactions off Canada's west coast; Canadian Journal of Earth Sciences, v. 14, p. 384-396.
1979: Gravity and structure of an active margin - British Columbia and Washington; Canadian Journal of Earth Sciences, v. 16, p. 350-363.
1982: Contemporary movements and tectonics on Canada's west coast: a discussion; Tectonophysics, v. 86, p. 319-341.

Riddihough, R.P. and Seemann, D.A.
1982: Juan de Fuca Plate Map: JFP-8 Gravity anomaly 1:2,000,000; Earth Physics Branch, Department of Energy, Mines and Resources (Pacific Geoscience Centre), Sidney, British Columbia.

Riddihough, R.P., Beck, M.E., Chase, R.L., Davis, E.E., Hyndman, R.D., Johnson, S.H., and Rogers, G.C.
1983: Geodynamics of the Juan de Fuca Plate; in Geodynamics of the Eastern Pacific Region, Caribbean and Scotia Arcs, R.S.J. Cabre (ed.), Volume 9, Geodynamics Series, American Geophysical Union, Washington, D.C., 5-21.

Rogers, G.C.
1983: Seismotectonics of British Columbia; Ph.D. thesis, University of British Columbia, Vancouver, 247 p.

Rogers, G.C. and Auld, D.R.
1985: The Juan de Fuca Terrane Map; EOS, Transactions, American Geophysical Union, v. 66, p. 1361.

Ross, C.A.
1967: Development of fusulinid (Foraminiferida) faunal realms; Journal of Paleontology, v. 41, p. 1341-1354.
1969: Upper Paleozoic fusulinacea: *Eowaegingella* and *Wedekindellina* from Yukon Territory and giant *Parafusulina* from British Columbia; Contributions to Canadian Paleontology, Geological Survey of Canada, Bulletin 182, p. 129-134.
1973: Carboniferous Foraminiferida; in Atlas of Palaeobiogeography, A. Hallam (ed.), Amsterdam, Elsevier Science Publishing Company, p. 127-132.

Ross, C.A. and Ross, J.R.P.
1981: Late Paleozoic faunas around the Paleopacific margin; in Evolution Today, G.G.R. Scudder and J.L. Reveal (ed.), Proceedings of the Second International Congress of Systematic and Evolutionary Biology, p. 425-440.
1983: Late Paleozoic accreted terranes of western North America; in Pre-Jurassic Rocks in Western North American Suspect Terranes, C.H. Stevens (ed.), Pacific Section of Society of Economic Paleontologists and Mineralogists, Los Angeles, p. 7-22.
1985: Carboniferous and Early Permian biogeography; Geology, v. 13, p. 27-30.

Roy, R.F., Blackwell, D.D., and Birch, F.
1968: Heat generation of plutonic rocks and continental heat flow provinces; Earth and Planetary Science Letters, v. 5, p. 1-12.

Rozovskaya, S.E.
1975: Sostav, sistema i filogeniya otryada Fusulinida; Akademiya Nauk SSSR, Trudy Paleontologicheskogo Instituta, v. 149, 267 p.

Rusmore, M.E. and Cowan, D.S.
1985: Jurassic-Cretaceous rock units along the southern edge of the Wrangellia terrane on Vancouver Island; Canadian Journal of Earth Sciences, v. 22, p. 1223-1232.

Savage, N.M.
1985: Silurian (Llandovery - Wenlock) conodonts from the base of the Heceta Limestone, southeastern Alaska; Canadian Journal of Earth Sciences, v. 22, p. 711-727.

Shor, G.G., Jr.
1962: Seismic refraction studies off the coast of Alaska: 1956-1957; Bulletin of the Seismological Society of America, v. 52, p. 37-57.

Silberling, N.J. and Jones, D.L.
1983: Paleontologic evidence for the northward displacement of Mesozoic rocks in accreted terranes of the western Cordillera; Geological Association of Canada, Program with Abstracts, v. 8, p. 62.

Silberling, N.J., Jones, D.L., Coney, P.J., and Plafker, G.
1984: Lithotectonic terrane maps of the North American Cordillera; N.J. Silberling and D.L. Jones (ed.), United States Geological Survey, Open File Report 84-523.

Souther, J.G.
1970: Volcanism and its relationship to recent crustal movements in the Canadian Cordillera; Canadian Journal of Earth Sciences, v. 7, p. 553-568.

Speed, R.C.
1979: Collided Paleozoic microplate in the western United States; Journal of Geology, v. 87, p. 279-292.

Spence, G.D., Clowes, R.M., and Ellis, R.M.
1985: Seismic structure across the active subduction zone of western Canada; Journal of Geophysical Research, v. 90, p. 6754-6772.

Stacey, R.A.
1973: Gravity anomalies, crustal structure, and plate tectonics in the Canadian Cordillera; Canadian Journal of Earth Sciences, v. 10, p. 615-628.

Stacey, R.A. and Steele, J.
1970: Gravity Map Series 120 and 121 - Strait of Georgia and Juan de Fuca Strait; Earth Physics Branch, Department of Energy, Mines and Resources, Ottawa.

Stacey, R.A. and Stephens, L.E.
1969: An interpretation of gravity measurements on the west coast of Canada; Canadian Journal of Earth Sciences, v. 6, p. 463-474.

Stacey, R.A., Boyd, J.B., Stephens, L.E., and Burke, W.E.F.
1973: Gravity measurements in British Columbia; Gravity Map Series 152-155; Publications of the Earth Physics Branch, Department of Energy, Mines and Resources, Ottawa.

Stephenson, R. and Lambeck, K.
1985: Isostatic response of the lithosphere with in-plane stress: application to central Australia; Journal of Geophysical Research, v. 90, p. 8581-8588.

Sutherland Brown, A., Cathro, R.J., Panteleyev, A. and Ney, C.S.
1970: Metallogeny of the Canadian Cordillera; Canadian Institute of Mining Transactions, LXXIV, p. 121-145.

Sweet, W.C.
1985: Conodonts: those fascinating little whatzits; Journal of Paleontology, v. 59, p. 484-494.

Taber, J.J. and Smith, S.W.
1985: Seismicity and focal mechanisms associated with the subduction of the Juan de Fuca Plate beneath the Olympic Peninsula, Washington; Bulletin of the Seismological Society of America, v. 75, p. 237-249.

Tatel, H.E. and Tuve, M.A.
1955: Seismic exploration of a continental crust; Geological Society of America, Special Paper 62, p. 35-50.

Taylor, D.G., Callomon, J.H., Hall, R., Smith, P.L., Tipper, H.W., and Westermann, G.E.G.
1984: Jurassic ammonite biogeography of western North America, the tectonic implications; in Jurassic-Lower Cretaceous Biochronology and Biogeography of North America, G.E.G. Westermann (ed.), Geological Association of Canada, Special Paper 27, p. 121-124.

Tempelman-Kluit, D.J.
1979: Transported cataclasite, ophiolite and granodiorite in Yukon: evidence of arc-continent collision; Geological Survey of Canada, Paper 79-14, 27 p.

Tiffin, D.L. and Riddihough, R.P.
1977: Gravity and magnetic survey off Vancouver Island, 1975; in Report of Activities, Part A, Geological Survey of Canada Paper 77-1A, p. 311-314.

Tipper, H.W.
1984: The allochthonous Jurassic-Lower Cretaceous terranes of the Canadian Cordillera and their relation to correlative strata of the North American craton; in Jurassic-Cretaceous Biochronology and Biogeography of North America, G.E.G. Westermann (ed.), Geological Association of Canada, Special Paper 27, p. 113-141.

Tipper, H.W. and Richards, T.A.
1976: Jurassic stratigraphy and history of north-central British Columbia; Geological Survey of Canada, Bulletin 270, 71 p.

Tipper, H.W., Woodworth, G.J., and Gabrielse, H.
1981: Tectonic Assemblage Map of the Canadian Cordillera; Geological Survey of Canada, Map 1505A.

Tozer, E.T.
1970: Marine Triassic faunas; Geological Survey of Canada, Economic Geology Report No. 1, p. 663-640.
1982: Marine Triassic faunas of North America: their significance for assessing plate and terrane movements; Geologische Rundschau, v. 71, p. 1077-1104.

Vine, F.J.
1968: Magnetic anomalies associated with mid-ocean ridges; in History of the Earth's Crust, R.A. Phinney (ed.), Princeton University Press, Princeton, p. 73-89.

Vine, F.J. and Wilson, J.T.
1965: Magnetic anomalies over a young oceanic ridge off Vancouver Island; Science, v. 150, p. 485-489.

Von Bitter, P.H., Sandberg, C.A., and Orchard, M.J.
in press: Phylogeny, speciation and palaeoecology of the Early Carboniferous (Mississippian) conodont genus Mestognathus; Royal Ontario Museum, Life Sciences Contributions.

Walcott, R.I.
1967: The Bouguer anomaly map of southwestern British Columbia; University of British Columbia, Institute of Earth Sciences, Scientific Report 15.

Wheeler, J.O.
1970: Summary and discussion; in Structure of the southern Canadian Cordillera, J.O. Wheeler (ed.), Geological Association of Canada, Special Paper 6, p. 155-166.

Wheeler, J.O. and Gabrielse, H.
1972: The Cordilleran structural province; in Variations in Tectonic Styles in Canada, R.A. Price and R.J.W. Douglas (ed.); Geological Association of Canada, Special Paper 11, p. 1-81.

White, W.H.
1959: Cordilleran tectonics in British Columbia; Bulletin of the American Association of Petroleum Geologists, v. 43, p. 60-100.

White, W.R.H. and Savage, J.C.
1965: A seismic refraction and gravity study of the Earth's crust in British Columbia; Bulletin of the Seismological Society of America, v. 55, p. 463-486.

White, W.R.H., Bone, M.N., and Milne, W.G.
1968: Seismic refraction surveys in British Columbia, 1964-1966: a preliminary interpretation; American Geophysical Union, Geophysical Monograph 12, p. 81-93.

Wickens, A.J.
1977: The upper mantle of southern British Columbia; Canadian Journal of Earth Sciences, v. 14, p. 1100-1115.

Wilson, J.T.
1965: Transform faults, oceanic ridges and mantle anomalies southwest of Vancouver Island; Science, v. 150, p. 482-485.

Yao, A.
1983: Late Paleozoic and Mesozoic Radiolarians from Southwest Japan; in Siliceous Deposits in the Pacific Region, A. Iijima, J.R. Heim and R. Siever (ed.), Elsevier Publishing Company; Amsterdam, Oxford, New York, p. 361-375.

Yole, R.W. and Irving, E.
1980: Displacement of Vancouver Island: paleomagnetic evidence from the Karmutsen Formation; Canadian Journal of Earth Sciences, v. 17, no. 9, p. 1210-1228.

Yorath, C.J. and Hyndman, R.D.
1983: Subsidence and thermal history of Queen Charlotte Basin; Canadian Journal of Earth Sciences, v. 20, p. 135-159.

Yorath, C.J., Clowes, R.M., Sutherland Brown, A., Brandon, M.T., Massey, N.W.D., Green, A.G., Spencer, C., Kanasewich, E.R., and Hyndman, R.D.
1985a: LITHOPROBE - Phase I: Southern Vancouver Island: Preliminary analyses of reflection seismic profiles and surface geological studies; in Current Research, Part A, Geological Survey of Canada Paper 85-1A, p. 543-554.

Yorath, C.J., Green, A.G., Clowes, R.M., Sutherland Brown, A., Brandon, M.T., Kanasewich, E.R., and Spencer, C.
1985b: LITHOPROBE, southern Vancouver Island: Seismic reflection sees through the Juan de Fuca Plate; Geology, v. 13, p. 759-762.

Yorath, C.J., Hyndman, R.D., and Clowes, R.M.
1987: Structure and accretion style of the modern convergent margin of western Canada; Joint Annual Meeting, Geological Association of Canada, Mineralogical Association of Canada, Saskatoon, Program with Abstracts, 12, p. 102.

Zelt, B.C., Ellis, R.M., Brown, J., Hajnal, Z., and Stephenson, R.
1987: Crustal structure of the Peace River Arch from a refraction survey; Abstract, Canadian Society of Exploration Geophysicists National Convention, p. 47.

Authors' addresses

H. Gabrielse
J.W.H. Monger
J.O. Wheeler
M.J. Orchard
H.W. Tipper
Cordilleran Division
Geological Survey of Canada
100 West Pender Street
Vancouver, British Columbia
V6B 1R8

J.F. Sweeney
Geological Survey of Canada
601 Booth St
Ottawa, Ontario
K1A 0E8

C.J. Yorath
R.G. Currie
J.M. DeLaurier
Pacific Geoscience Centre
9860 West Saanich Road
P.O. Box 6000
Sidney, British Columbia
V8L 4B2

P.L. Smith
Department of Geological Sciences
University of British Columbia
Vancouver, British Columbia
V6T 2B4

R.A. Stephenson
Institute of Sedimentary and Petroleum Geology
Geological Survey of Canada
3303-33rd Street N.W.
Calgary, Alberta
T2L 2A7

C.A. Ross
J.R.P. Ross
Chevron USA, Incorporated
P.O. Box 1635
Houston, Texas
U.S.A. 77251

E.S. Carter
58335 Timber Road
Vernonia, Oregon
U.S.A. 97064

ADDENDUM

Re: Radiolaria

Based on radiolaria, the age range of the type Cache Creek rocks in southern British Columbia is now known to extend from Carboniferous to Early or Middle Jurassic. Similarly, radiolaria indicate on age range for the Bridge River Terrane of Mississipian to Middle Jurassic, one of the longest spans for the deposition of radiolarian-bearing strata known (F. Cordey, pers. comm. 1991)

Printed in Canada

Chapter 3

PALEOMAGNETISM: REVIEW AND TECTONIC IMPLICATIONS

Summary
Introduction
Terranes
  Cenozoic
  Cretaceous
  Early Jurassic and Late Triassic
  Late Paleozoic
  Summary and discussion of Phanerozoic data
    Cenozoic
    Baja British Columbia, a possible miniplate
    Post mid-Cretaceous tilt or translation
    Rotations
    Displacements and the hemispheric ambiguity
    Tectonic attenuation
    Overprinting
Foreland Belt
  Precambrian
  Cretaceous
References

Chapter 3

PALEOMAGNETISM: REVIEW AND TECTONIC IMPLICATIONS

E. Irving and P.J. Wynne

SUMMARY

Paleomagnetic studies in the Canadian Cordillera began in the late 1950s. During the past two decades efforts to determine displacements and rotations have concentrated largely on the terranes of the Intermontane and Insular belts, and, to a lesser extent, on Proterozoic and Cretaceous strata of the Foreland Belt.

Results obtained from Neogene rocks confirm the average dipolar nature of the ancient geomagnetic field and indicate that since the mid-Tertiary, except for the Queen Charlotte Islands, no paleomagnetically detectable tilting, rotations or horizontal motions relative to the craton have occurred throughout most of the Cordillera. Paleogene rocks commonly show significant rotations but no detectable latitudinal displacements.

Cretaceous paleopoles from the western Cordillera mostly are clustered in the North Atlantic whereas those from the craton occur northwest of Alaska. Many paleopoles have been determined from well-dated older rocks in the accreted terranes, all from overprints of probable mid-Cretaceous age. These data, when applied to Pacific plate motion reconstructions, suggest that the Intermontane and Insular superterranes may have been one coherent crustal fragment ("Baja British Columbia") by mid-Cretaceous time and located along the eastern edge of the Kula Plate some 2000 km south of its present position. Northward motion between mid-Cretaceous and Eocene time, together with clockwise rotation, terminated with the accretion of Baja British Columbia to ancestral North America when the Kula Plate ceased to exist as a separate plate and when the approximate position of the present Pacific-America plate boundary was established. Results from Cretaceous rocks of the Foreland Belt are indicative of insufficient sampling of the paleosecular variation.

Lower Jurassic and Triassic rocks have yielded results from Wrangellia, Stikinia and Quesnellia; large and variable rotations are ubiquitous. Due to reversals of the paleofield the hemisphere of origin is ambiguous. Assuming northern paleolatitudes, determinations from Lower Jurassic and Upper Triassic rocks of Wrangellia are indistinguishable and implied displacements are considerably less than those indicated from Cretaceous magnetizations.

Moreover, the data suggest that during the early Mesozoic, Wrangellia was a much more compact terrane than at present, and furthermore, that prior to accretion most terranes were approximately as close together as they are now.

Lower Permian rocks of Stikinia have yielded paleolatitudes not significantly different from their present position although large rotations are implied. These data, in conjunction with Mesozoic results, indicate that the Intermontane and Insular superterranes originated in the northern hemisphere, at/or somewhat south of their present position relative to North America. Paleomagnetic data can neither support nor refute their collision with or accretion to North America at that time. Between the Early Jurassic and mid-Cretaceous the terranes moved as a group ("Baja British Columbia") to the south, prior to their motion northwards and final accretion to North America in the Eocene.

Proterozoic rocks of the northern Foreland Belt yield paleopoles which may form part of a polar wander loop and which suggest that no wholesale rotations of thrust sheets occured during emplacement.

INTRODUCTION

The study of the remanent magnetization of rocks (paleomagnetism) can be used to illustrate specific gross tectonic aspects of the Canadian Cordillera. In particular, it can be used to define motions among the various terranes, and between them and the craton. If large motions occurred between two places then there will be differences in the paleomagnetic poles (paleopoles) observed from rocks of comparable age (Runcorn, 1956; Irving, 1956). The fundamental assumption is that when averaged over 10^4 years or more the geomagnetic field averages to that of a geocentric axial dipole, as demonstrated originally for the recent past by Hospers (1955). For any single study, the method is capable of determining relative rotations of more than $10°$, and differences in latitude of 800 km or more; experimental errors and uncertainties generally preclude determinations of smaller movements. When data are available from many rock units, smaller motions may be detectable.

The first paleomagnetic observations from the Canadian Cordillera by Du Bois (1959), who studied late Cenozoic volcanics from the Yukon and northern British Columbia, indicated a paleopole concordant with the present geographic pole. He confirmed that in this region the geomagnetic field averaged over the recent past had been that of an axial geocentric dipole. In the Cordillera of

---

Irving, E. and Wynne, P. J.
1991: Paleomagnetism: review and tectonic implications, Chapter 3 in Geology of the Cordilleran Orogen in Canada, H. Gabrielse and C. J. Yorath (ed.); Geological Survey of Canada, Geology of Canada, no. 4, p. 61-86 (also Geological Society of America, The Geology of North America, v. G-2)

CHAPTER 3

North America the first discordant paleomagnetic result was obtained by Cox (1957) from the Eocene Siletz volcanics of Oregon. Cox did not recognize the effect as having a tectonic cause, the realization of which came some years later (Irving, 1964). The present surge of work dates from the early 1970s when discordant results were obtained from several laboratories (Beck and Noson, 1972; Packer and Stone, 1974; Irving and Yole, 1972). This review presents results that have been obtained since that time from the Canadian Cordilleran Orogen, and includes data from Wrangellia, Stikinia, Quesnellia and the Cache Creek Terrane and the Coast and Foreland belts (Fig. 3.1A,B). It reassesses them using the most recent time scale (Kent and Gradstein, 1985), and presents new interpretations.

Successful paleomagnetic studies depend upon several requirements. Firstly, the remanent magnetizations must be dated; magnetizations may be primary and acquired when the rock was formed, or secondary (overprints) and acquired later. Primary magnetizations are usually well-dated, secondary magnetizations are not. Just as many radiometric ages do not reflect the age of rock formation, about one half the magnetizations described from the Canadian Cordillera are probably secondary, acquired during a later thermal, metasomatic or metamorphic event. Secondly, the directions of remanent magnetizations must be related to the paleohorizontal at the time of acquisition. In the predominantly igneous terranes of the western part of the Canadian Cordillera original attitude is frequently difficult to determine, and reliance often must be placed on the consistency among several determinations made over a wide area. Secondary magnetizations, however, are commonly post-deformational and tilt corrections may be unnecessary. Thirdly, paleomagnetic surveys should extend through a time se-

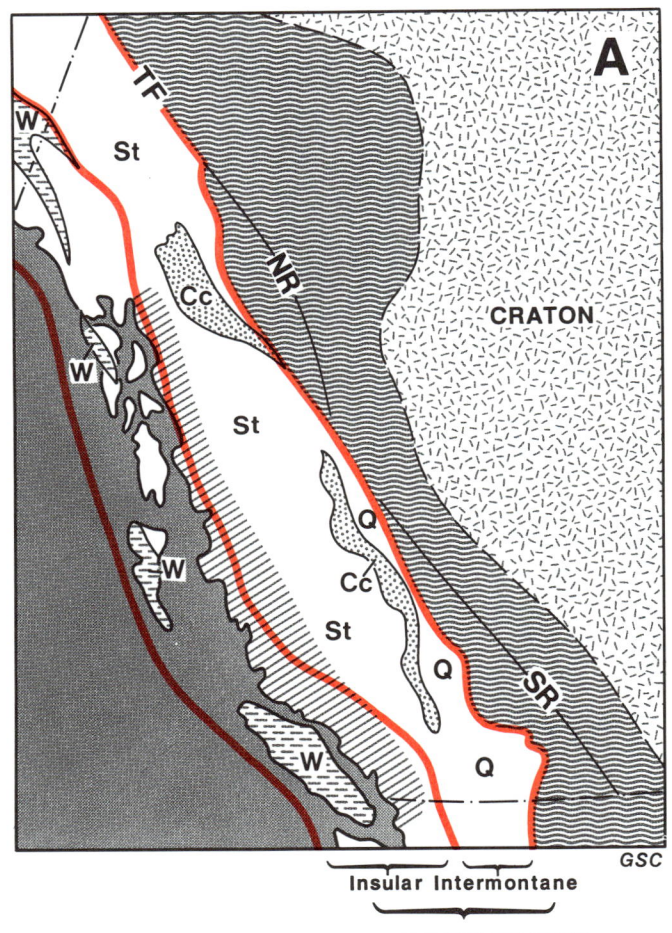

LEGEND

BAJA BRITISH COLUMBIA

- W WRANGELLIA
- St STIKINIA
- Cc CACHE CREEK
- Q QUESNELLIA
- TF TINTINA FAULT
- NR NORTHERN } ROCKY MOUNTAIN
- SR SOUTHERN } TRENCH
- COAST PLUTONIC BELT
- FORELAND AND OMINECA BELTS

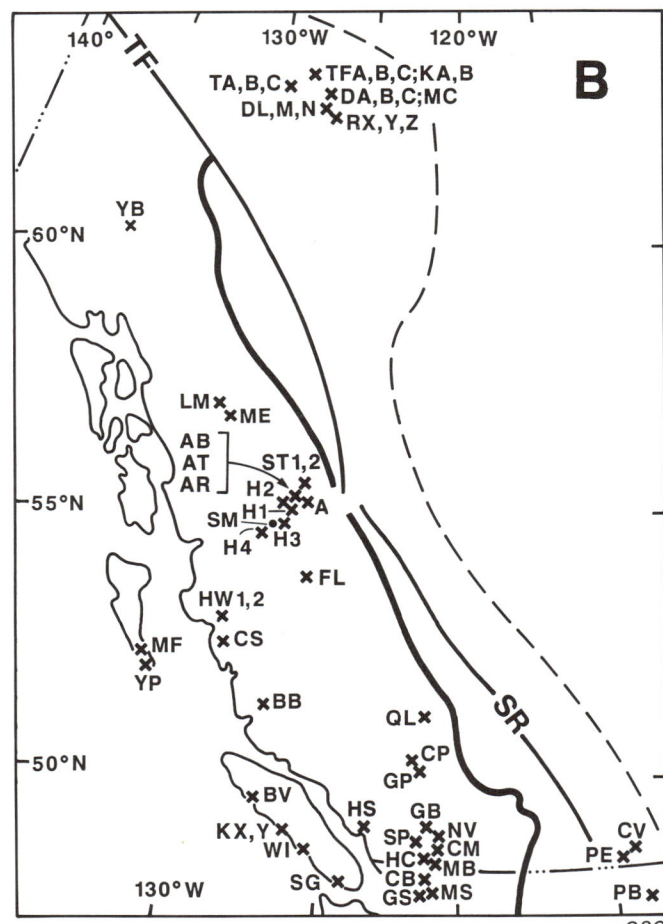

Figure 3.1. A) Major tectonic units of the Canadian Cordillera. To the west is the Pacific Plate, to the east the craton and its deformed margin. In between is the composite superterrane of Baja British Columbia, which comprises two superterranes which themselves are made up of smaller terranes. B) Paleomagnetic localities listed in Table 3.1. Geological boundaries are identified in Figure 3.1A.

quence of sufficient length to average out the paleosecular variation - preferably 10^4 years or more. Fourthly, accurate, well-dated results from the adjacent craton are needed for comparison.

Two procedures are used to obtain reference paleopoles. By the first method cratonic paleopoles representative for a given interval (usually 10-30 Ma) are averaged and their mean calculated. Single intervals may be considered, or a sequence of intervals can be assembled and running means calculated to generate a calibrated cratonic apparent polar wander (apw) reference path (Irving and Irving, 1982). This method, when applied to data from intervals of rapid apparent polar wander, may introduce undesirable smoothing, but only if the interval averaged is too great. By the second method paleopoles spread over much longer intervals (about 100 Ma) are fitted to small circles and corresponding Euler (a pole of rotation on the surface of a sphere) poles calculated, the assumption being that the plate on which the continent rode had trajectories that were constant for 100 Ma, so that paleopoles would fall sequentially on small circles (Gordon et al., 1984). Apparent polar wander paths obtained by the second method are "model" paths, being obliged by a tectonic model to lie on small circles. The procedure could be questioned because significant changes in plate trajectories occur more frequently than every 100 Ma (every 25 Ma according to Engebretson, 1982). It could also be questioned on philosophical grounds: firstly, it is, perhaps, not reasonable to expect that information about a specific interval can be obtained from summaries that incorporate data from rocks laid down much earlier or much later; secondly, the assumption prejudges the nature of the processes that are being investigated. In this review, the basic idea of the first method is retained (the need to average results for a given length of time) and smoothing is minimized by shortening the interval as far as is possible within the accuracy of geochronological and biostratigraphic dating. The approach is phenomenological, and tectonic models are eschewed for the reasons just given. Cordilleran data will simply be compared directly with determinations from cratonic rocks deposited during the same time interval.

Figure 3.2 is an early attempt to depict schematically the postulated displacements that have taken place in the Cordillera. Geophysical studies of the ocean floor to the west have provided global schemes of plate motions within which these displacements can be understood (Fig. 3.3). Both schemes are speculative, but nonetheless they provide a background against which the more recent paleomagnetic evidence can be discussed.

Four stages in the evolution of terrane relationships were suggested as shown in Figure 3.2. Initially an unknown distance separated Wrangellia in the Insular Superterrane and Stikinia in the Intermontane Superterrane. By mid-Cretaceous time these two superterranes were welded together (Fig. 3.2B). Following the concept and terminology of Stone (1977), the combined superterranes are referred to herein as "Baja British Columbia" (Irving, 1985), expressing the idea that during the Late Cretaceous and Paleocene Baja British Columbia might have had much the same relationship to North America as Baja California does today. The paleomagnetic evidence suggests that the two superterranes were close together in the Triassic and Jurassic. In the mid-Cretaceous they were welded together into a recognizable lithospheric block or minicontinent, an event that was marked by intense metamorphism and the emplacement of granitoid batholiths along their common margin (the Coast Plutonic Complex). Baja British Columbia is believed to have existed as a separate tectonic entity from its inception in the mid-Cretaceous until the Paleocene, when it became welded to the craton of North America (Fig. 3-2C). This hypothesis was based on paleomagnetic evidence and also on the geological evidence that intense deformation continued in the Foreland Belt until the Paleocene. It did not, however, address the problem of the pre-mid-Cretaceous origin of the Omineca Belt.

Irving et al. (1980) noted the possible existence of this larger entity and described it as a "combined terrane". More recently Chamberlain and Lambert (1985) used the approximately synonymous term "Cordilleria" which includes, in addition to the Insular and Intermontane superterranes, some components of the Omineca Belt and "Baja Alaska". "Baja Alaska" is the displaced superterrane of southern Alaska comprising the Peninsular Terrane, Wrangellia and the Chugach Terrane (Stone, 1977). They considered all of these elements to have been emplaced against North America together and at much the same time. The notion adopted herein is that Baja Alaska and Baja British Columbia, although they may have been initially one, separated in the earliest Tertiary. Baja Alaska subsequently moved northwards and was emplaced alongside North America later than Baja British Columbia (Fig. 3.2D).

These four stages of development can be related, in a tentative and qualitative way, to a corresponding sequence of interactions between the North American and oceanic plates to the west (Engebretson, 1982; Engebretson et al., 1985). The formation of the constituent terranes (stage A) and their later assembly into Baja British Columbia (stage B) are related to the dominantly eastward motion of the Farallon Plate and its subduction beneath the craton during the Jurassic and Early Cretaceous (Fig. 3.3A). Terranes could have been carried eastward by the Farallon Plate, or created by its interaction with the craton. The northward movement and attachment of Baja British Columbia to the craton may have been due to the rapid and oblique convergence of the Kula Plate relative to North America during latest Cretaceous and earliest Tertiary time (Fig. 3.3B). Subsequent northward displacement of Baja Alaska and the associated truncation and smearing-out of westerly elements of Baja British Columbia resulted from the northwesterly motion of the Pacific Plate relative to North America during the Tertiary (Fig. 3.3C). The following review of paleomagnetic data from the Cordillera is presented to assess the model presented in Figure 3.2.

Since submission of this chapter in July, 1986, many more recently acquired data and analyses have been presented; for reviews of those the reader is referred to Irving and Wynne (1990) and Irving et al. (1990).

TERRANES
Cenozoic

The early result of Du Bois (YB, Fig. 3.4, Table 3.1) and more recent results spanning the past 6 Ma from the Level Mountain and Mount Edziza volcanoes (LM and ME of Fig. 3.4 and Table 3.1) all show reversals of magnetic

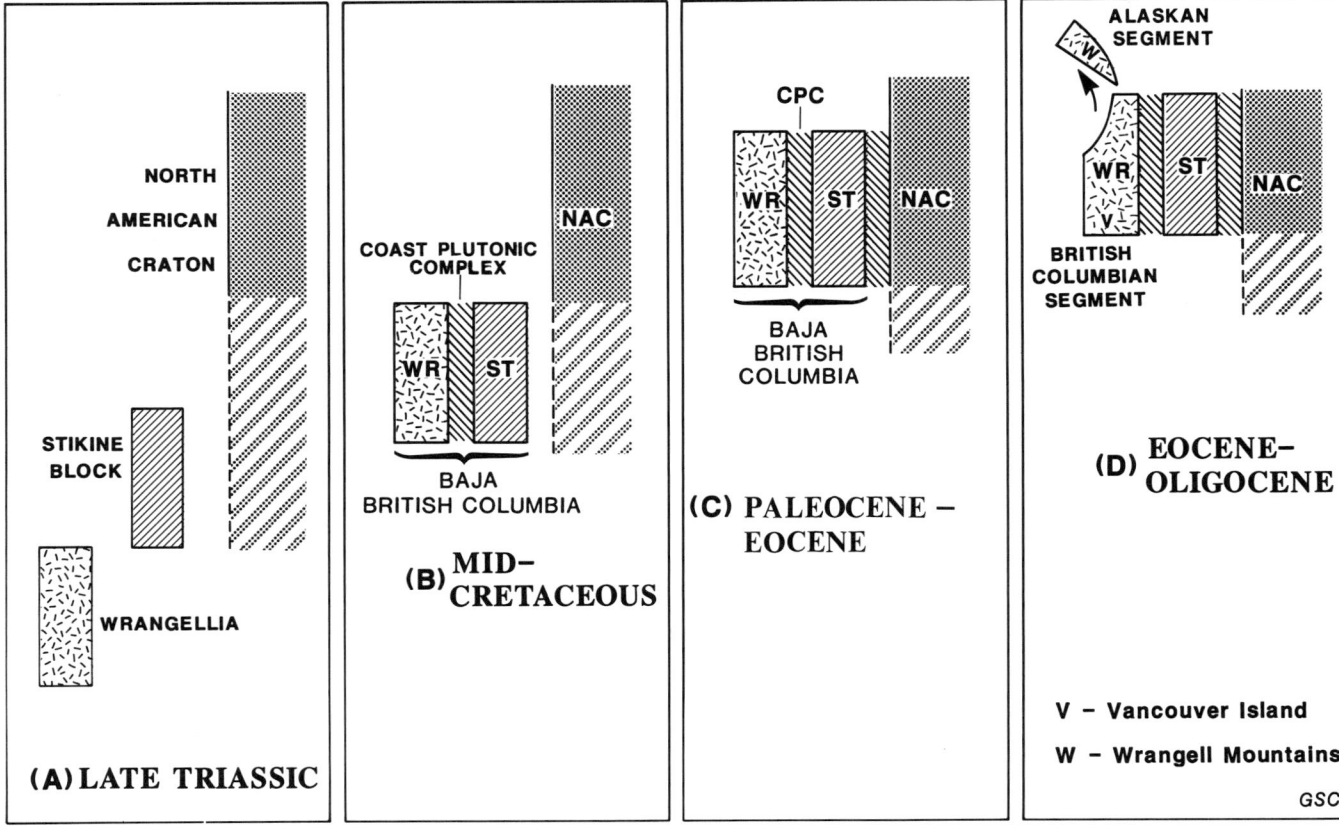

Figure 3.2. Speculative scheme showing four stages (A-D), based on paleomagnetic evidence, for assembly of major tectonic elements of the Canadian Cordillera as proposed in 1980 (Irving et al., 1980). Compare with Figure 3.22.

polarity. Paleopoles are not significantly different from the geographical pole, confirming the average dipolar nature of the paleofield in the recent past.

Except for one result of low accuracy from the Mt. Barr Complex (MB of Table 3.1), paleopoles obtained from lavas and intrusive rocks of mid-Tertiary age from mainland British Columbia and adjacent Washington State cluster around the expected position obtained from cratonic rocks (CB, HC, GS, GP, CP, BB in Fig. 3.5). Mid-Tertiary overprints in Cretaceous rocks yield a concordant paleopole (SPO). This means that throughout most of the Canadian Cordillera there has been no paleomagnetically detectable tilting, rotation or horizontal motion relative to cratonic North America since the mid-Tertiary. The divergences observed from the Queen Charlotte Islands (Masset Formation MF, and post-tectonic plutons YP) are considerable and appear to be significant because reversals are present. No corrections for tilt have been made and the divergences could reflect undetected tilting and/or rotations.

Paleopoles derived from magnetizations (a probable early Tertiary overprint) from the north-northwest limb of the 'Hawkesbury warp' of the Coast Plutonic Complex and from the Francois Lake Granite Suite (formerly designated the Topley Intrusions) in north-central British Columbia are consistent with those observed from the craton (compare paleopoles HW2 and FLO with those from craton in Fig. 3.6). The paleopole (HW1) from the more westerly trending limb of the Hawkesbury warp, although inaccurate, indicates an anticlockwise rotation of Early Tertiary age consistent with the change in structural trend (Symons, 1977a). The paleopole from the East Sooke gabbro of southern Vancouver Island (SG, Fig. 3-6) is to the left of the main group, reflecting either anticlockwise rotation, or post-emplacement tilting (Symons, 1973c). The paleohorizontal cannot be determined at the sampling localities, but dips of 20 to 30° are common in the area, so tilting cannot be excluded.

Although data from Lower Tertiary rocks of the Canadian Cordillera may indicate rotations there are, at present no accurate determinations of paleolatitude. However, limits to possible latitudinal motions have been set by data obtained in the United States. Extensive data from Eocene rocks of the Coast Ranges of Washington and Oregon (Siletzia) indicate little or no paleolatitudinal shift (but large rotations) relative to the craton (Simpson and Cox, 1977; Globerman et al., 1982; Wells and Coe, 1985). Siletzia was not firmly coupled to the Insular Belt to the north until the Oligocene, so these data only approximately constrain Early Tertiary latitudinal motions in the Canadian Cordillera. More definitive are data from Eocene volcanics (55 to 48 Ma) immediately south of the International Boundary in eastern Washington. These volcanics accumulated on Quesnellia of the Intermontane Belt (Fox and Beck, 1985). Data from five grabens agree with cratonic

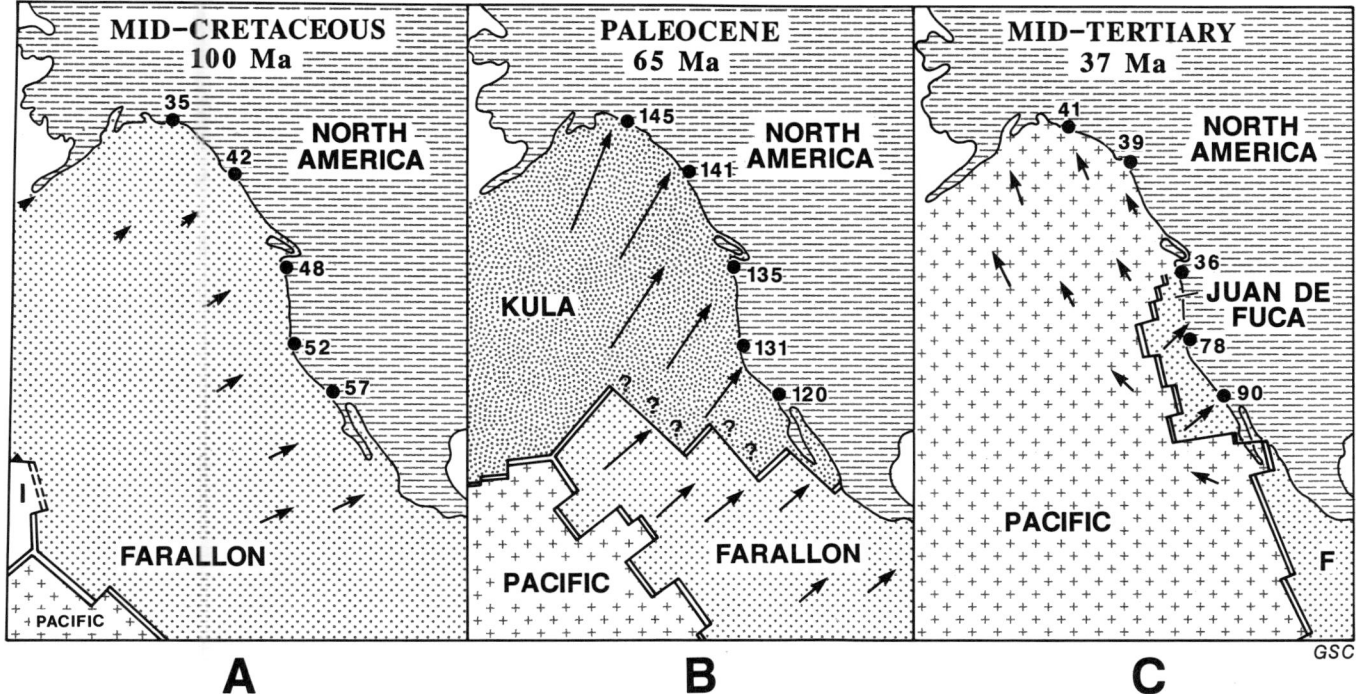

Figure 3.3. Speculative plate configuration in the mid-Cretaceous, Paleocene, and mid-Tertiary compiled from Engebretson et al. (1985). North America is fixed, and arrows give relative directions of motion of oceanic plates. Arrows lengths are proportional to velocity. Numbers show velocities (in mm/a) of oceanic plates relative to the western (fixed) margin of North America at 5 localities. I is the Izanagi Plate. The position of the boundary between the Farallon and Kula Plate is uncertain.

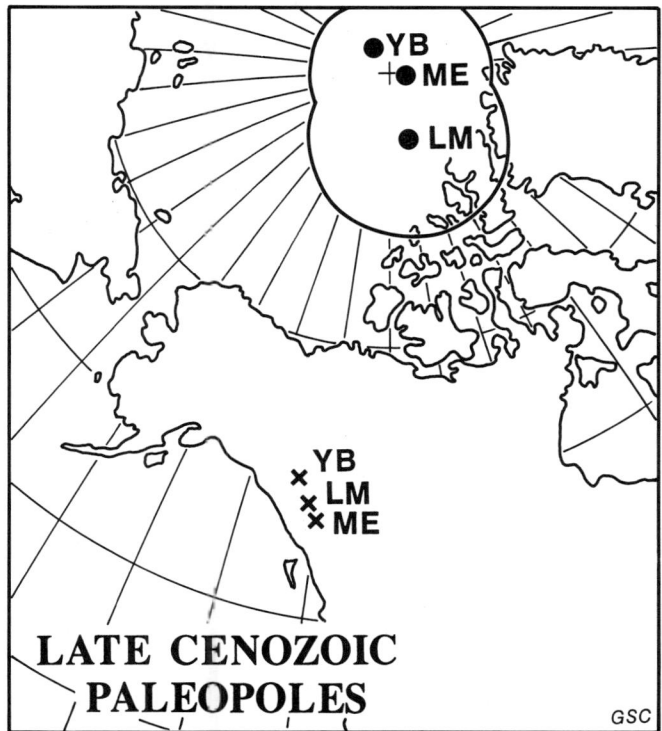

Figure 3.4. Late Cenozoic paleopoles. YB = late Tertiary basalts of Yukon and northern British Columbia; LM = Level Mountain; ME = Mount Edziza. Details in Table 3.1.

paleolatitudes, and their mean is in excellent agreement with that expected (Fig. 3.7). Some determinations, those from the south Republic graben in particular, show rotations; there is no paleolatitudinal change but variable clockwise rotation between grabens, as might be expected in the extensional tectonics of that time. These data show that the Intermontane Belt has not moved latitudinally relative to the craton (within limits determinable paleomagnetically) since the Eocene.

Cretaceous

Magnetizations of Cretaceous age differ from those of the Tertiary. Paleopoles of mid-Cretaceous age from the craton cluster just northwest of Alaska (Fig. 3.8). With the exception of some reversed magnetizations from the Monteregian Intrusion (Q), all have normal magnetization characteristic of the Cretaceous Normal Superchron (118 to 83 Ma). With the exception of a result (HS) from plutons along Howe Sound north of Vancouver (Symons, 1973e) paleopoles from the western Canadian Cordillera are strongly aberrant, indicating that they were affected by later overprinting (see SPO Table 3.1; Symons et al., 1986); these paleopoles cluster in the North Atlantic region. The number of results and their geographical range preclude the suggestion that their disagreement with cratonic paleopoles is caused by insufficient averaging of the paleosecular variation. Some of the results are well dated, however, others are not. Data from rocks that have been radiometrically dated (see Table 3.1) include those from the Mount Stuart batholith of Washington (MS), the Captains Cove and

Table 3.1. List of paleopoles. Radiometric ages are given if available. (h) and (b) indicate that the paleopole has been calculated before or after correction for geological tilt

Rock unit		lat., long. (degrees)	$A_{95}=$ (deg.)	ref.
RESULTS OF CENOZOIC AGE FROM THE CORDILLERA *				
YB	Late Tertiary basalts of Yukon and B.C. (h)	85N, 150E	6	55
LM	Level Mountain shield volcano, 6.5 to 1.0 Ma (h)	85N, 104W	7	1
ME	Mount Edziza shield volcano, 61 to 0.9 Ma (h)	89N, 032W	7	2
CB	Chilliwack batholith, 36 to 30 Ma (h)	88N, 093E	2	3
HC	Hope Complex, 41-35 Ma (h)	88N, 152W	6	4
GS	Grotto and Snoqualmie batholiths 27-15 Ma (h)	86N, 152W	5	5
SPO	Spuzzum-Porteau plutons overprint, (h)	84N, 051W	7	22
MB	Mount Barr Complex, 21-16 Ma (h)	72N, 086W	14	4
CP	Cariboo Plateau lavas, 13 to 10 Ma (h)	84N, 140W	5	6
GB	Gabbroic plug (intrudes CP) (h)	85N, 147W	5	6
BB	Brown basaltic dykes, post-Upper Miocene (h)	82N, 112W	16	7
MF	Masset Formation, Miocene (h)	72N, 097W	11	8
YP	Younger Plutons QC Islands, Miocene (h)	70N, 154W	8	8
SG	East Sooke stock (h)	70N, 151E	6	9
HW1	Eocene plutons, Hawkesbury warp, WNW trend (h)	44N, 139E	27	10
HW2	Eocene plutons, Hawkesbury warp, NNW trend (h)	81N, 134W	9	10
FLO	Francois Lake Suite, overprint (h)	75N, 176W	10	11,12
NVO	Nicola volcanics, overprint (h)	82N, 057E	8	51
RESULTS OF CRETACEOUS AGE FROM THE CRATON				
D	Lamprophyre dykes, Notre Dame Bay, mean 129 Ma (h)	71N, 154W	4	41,13
Q	Monteregian Hills, 120-100 Ma (h)	72N, 169W	3	14,54
I	Isachsen diabase, mean 101 Ma (h)	65N, 180	14	15
A	Alkalic intrusions, Arkansas, 105-95 Ma (h)	65N, 173W	11	16
B	Niobrara Fm., Montana, Coniacian-Maastrichtian, (b)	66N, 168W	6	17,18
RESULTS OF CRETACEOUS AGE FROM THE CORDILLERA				
CV	Crowsnest Formation, Albian (b)	78N, 108E	6	19
MS	Mount Stuart batholith, 95 Ma (h)	68N, 035E	5	20
HS	Howe Sound plutons, 100 Ma (h)	82N, 129W	12	30
CS	Captains Cove & Stephens plutons, 109-102 Ma (h)	67N, 0	8	21
SP	Spuzzum & Porteau plutons, 103-81 Ma (h)	65N, 015W	6	22
AX2	Axelgold intrusion, 125-115 Ma (h)	64N, 012W	8	12,39
AX1	Axelgold intrusions, 125-115 Ma (b)	76N, 033W	8	22,39
KY	Karmutsen Formation, overprint (h)	70N, 015W	11	23
WI	see Triassic-Jurassic section below			
CMA	Copper Mountain intrusion (AF), 192 Ma, overprint (h)	68N, 009W	8	24
GBA	Guichon Batholith (AF), 198 Ma, overprint (h)	66N, 013E	12	25
QL	Quesnel Lake volcaniclastics, overprint (h)	62N, 015W	12	26
FL	Francois Lake Suite (TH), 143-133 Ma (h)	58N, 122W	10	40
RESULTS OF LATE TRIASSIC AND EARLY JURASSIC AGE FROM THE CRATON				
C1	Chinle Formation, New Mexico, Late Triassic (b)	58N, 079E	3	27
M	Manicouagan Impact Crater, 212 Ma, Quebec (h)	60N, 089E	5	28,29
RT1	C1 and M combined	59N, 084E	12	
L	Los Rostras and Ischigualasto formations, Argentina, 224 Ma (b)	79N, 059E	15	46
P	Paramillos Formation, Argentina, M-L Triassic, (b)	74N, 086E	14	47
RT2	C1, M, L and P combined	64N, 099E	13	
RESULTS OF LATE TRIASSIC AND EARLY JURASSIC AGE FROM COLORADO PLATEAU				
K	Kayenta Formation, Utah, Early Jurassic (b)	62N, 073E	3	33
W	Wingate Formation, Utah, Early Jurassic (b)	59N, 064E	3	32
C2	Chinle Formation, Utah, Late Triassic (b)	61N, 064E	3	32

Table 3.1. Continued

Rock unit		lat., long. (degrees)	$A_{95}=$ (deg.)	ref.
RESULTS OF PERMIAN-TRIASSIC-JURASSIC AGE FROM THE WESTERN CORDILLERA				
H1	Hazelton Group (1), Early Jurassic (b)	40N, 144E	29	12
H2	Hazelton Group (2), Early Jurassic (b)	17N, 175W	21	12
H3	Hazelton Group (3), Early Jurassic (b)	70N, 057E	19	12
BV	Bonanza Group, Early Jurassic (b)	22N, 154E	8	53
WI	West Coast Complex and Island Intrusions, Early Jurassic (h)	74N, 013E	15	52
GBT	Guichon batholith, 200 Ma, (h)	52N, 012E	7	43
CMT	Copper Mountain intrusion, 192 Ma, (h)	57N, 012E	4	42
ST1	Takla Group (1), Late Triassic (b)	38N, 133E	13	12
ST2	Takla Group (2), Late Triassic (b)	24N, 146E	12	12
NV	Nicola volcanics (A), Late Triassic (b)	53N, 011E	6	51
KX1	Karmutsen Formation (1), Late Triassic (b)	26N, 072E	11	23
KX2	Karmutsen Formation (2), Late Triassic (b)	20N, 041E	5	23
KX	Karmutsen Fm., grand average, Late Triassic (b)	23N, 052E	6	23
AB	Asitka Group basalt, Early Permian (b)	40N, 123E	7	31
AR	Asitka Group rhyolite, Early Permian (b)	15N, 159E	3	31
AT	Asitka Group tuff, Early Permian (b)	56N, 063E	8	31
RESULTS OF PRECAMBRIAN AGE FROM THE MACKENZIE MOUNTAINS				
TFA	Tsezotene Formation A	12N, 146E	5	44
TFB	Tsezotene Formation B	23N, 162E	5	44
TFC	Tsezotene Formation C	63N, 141W	6	44
TA	Tsezotene diabase sills A, 770 Ma (b)	01N, 139E	4	34,39
TB	Tsezotene diabase sills B (b)	01S, 141E	4	34,35
DA	Little Dal Group limestone A (b)	16S, 141E	3	35
DB	Little Dal Group limestone B (b)	03N, 138E	8	35
DC	Little Dal Group limestone C (b)	60N, 170E	9	35
MC	Mudcrack Formation	09S, 143E	9	36
KA	Katherine Group A	09N, 150W	6	45
KB	Katherine Group B	17N, 164W	5	45
KC	Katherine Group C	77N, 122W	6	45
CL	Coates Lake Group lavas L, (b)	24N, 115E	6	37
CM	Coates Lake Group lavas M, (b)	24N, 159E	13	37
CN	Coates Lake Group lavas N, (b)	47N, 175E	6	37
RX	Rapitan Group X, (?)	01S, 106W	4	38
RY	Rapitan Group Y, (?)	66N, 163W	11	38
RZ	Rapitan Group Z, (?)	05N, 149E	3	38
RESULTS OF PRECAMBRIAN AGE FROM THE ROCKY MOUNTAINS				
PE	Purcell Supergroup, Alberta (b)	24S, 145W	6	48
PB	Purcell Supergroup, Montana (b)	17S, 157W	0	49
PV	Purcell Supergroup, Spokane Formation, Montana (b)	15S, 135W	5	50
PVO	Purcell, Spokane Fm., overprint, Montana (b)	05N, 163E	8	50

* Reconnaissance study of scattered Tertiary igneous bodies of Vancouver Island (Symons, 1971b) are not included. References:

(1) Hamilton and Evans (1983). (2) Souther and Symons (1973). (3) Beck et al. (1982). (4) Symons (1973d). (5) Beske et al. (1973). (6) Symons (1969). (7) Symons (1968). (8) Hicken and Irving (1977). (9) Symons (1973c). (10) Symons (1977a). (11) Symons (1973a). (12) Monger and Irving (1980). (13) Lapointe (1979). (14) Foster and Symons (1979). (15) Larochelle and Black (1963). (16) Scharon and Hsu (1969). (17) Shive and Frerichs (1974). (18) Gordon et al. (1984). (19) Irving et al. (1986). (20) Beck et al. (1981a). (21) Symons (1977b) see revisions in Symons et al. (1986). (22) Irving et al. (1985). (23) KX1 is from the NNW directions of upper Karmutsen, KX2 is from the N directions from Schwartz et al. (1980) and Yole and Irving (1980) see Figure 2.34, KX given here is the grand average. (24) Symons (1973b). (25) Symons (1971a). (26) Rees et al. (1985). (27) Reeve and Helsley (1972). (28) Robertson (1967). (29) Larochelle and Currie (1967). (30) Symons (1973e). (31) Irving and Monger (1987). (32) Reeve (1975) in Gordon et al. (1984). (33) Steiner and Helsley (1974). (34) Park (1981a). (35) Park (1981b). (36) Park (1984). (37) Morris and Aitken (1982). (38) Morris (1977). (39) Armstrong et al. (1982). (40) Symons (1983a). (41) Prasad (1981). (42) Symons and Litalien (1984). (43) Symons (1983b). (44) Park and Aitken (1986b). (45) Park and Aitken (1986a). (46) Valencio et al. (1975). (47) Valencio (1972). (48) Evans et al. (1975). (49) Elston and Bressler (1980) mean pole giving unit weight to each of 10 formations studied. (50) Vittorelli and van der Voo (1977). (51) Symons (1985a). (52) Symons (1985b) mean of 4 magnetizations observed from these two units. (53) Irving and Yole (1987) data from 1800 m section BV, two other directions with same inclination but very different declinations (BV2 and BV3 of Fig. 3.20) also observed at two other localities. (54) Larochelle (1962). (55) Du Bois (1959).

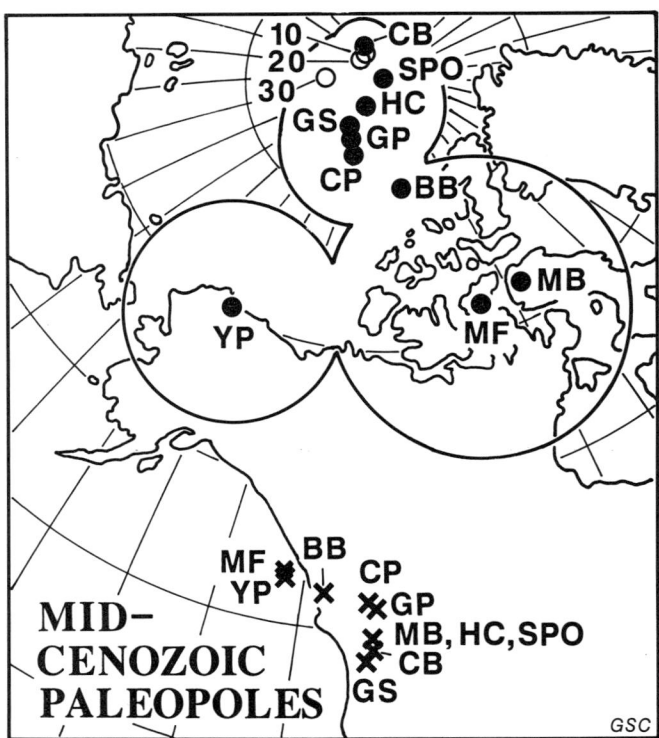

Figure 3.5. Mid-Cenozoic paleopoles. CB = Chilliwack batholith; HC = Hope Complex; GS = Grotto and Snoqualmie batholiths; MB = Mount Barr Complex; CP = Cariboo Plateau basalts; GP = gabbroic plug; BB = brown basalt dykes; SPO = overprint in Spuzzum and Porteau plutons and aureole; YP and MF = younger plutons and Masset volcanics of Queen Charlotte Islands. Errors for MB and BB are large and are not plotted. The open circles are the mean cratonic paleopoles for 10, 20 and 30 Ma taken from Irving and Irving (1982). Details in Table 3.1.

Stephens plutons in the western Coast Belt (CS), the Spuzzum and Porteau plutons of the southern Coast Plutonic Complex (SP), and the Axelgold intrusion in the Cache Creek Terrane (AX1). These bodies range in age from about 125 to 100 Ma, much the same time interval as for cratonic samples. If no attitudinal correction is applied to the Axelgold data the paleopole (AX1) falls within the main group. The small divergence of MS from the main group can be explained by local tilting, which may be several tens of degrees to the southeast (Tabor in Beck et al., 1981b). The remainder of the paleopoles are based on what are considered to be secondary magnetizations from Wrangellia (overprint KY from the Karmutsen Formation, and WI from the West Coast Complex and Island Intrusions of Vancouver Island) and from Quesnellia (Copper Mountain intrusion CMA, Guichon Batholith GBA, Quesnel Lake volcaniclastics QL). The overprints are probably mid-Cretaceous because of their close agreement with other determinations, and, because reversals which characterize older and younger rocks are almost entirely absent (Rees et al., 1985). However, the age assignment is neither certain nor generally accepted. For example, Symons (1985b) believed WI to be primary and of Early Jurassic age.

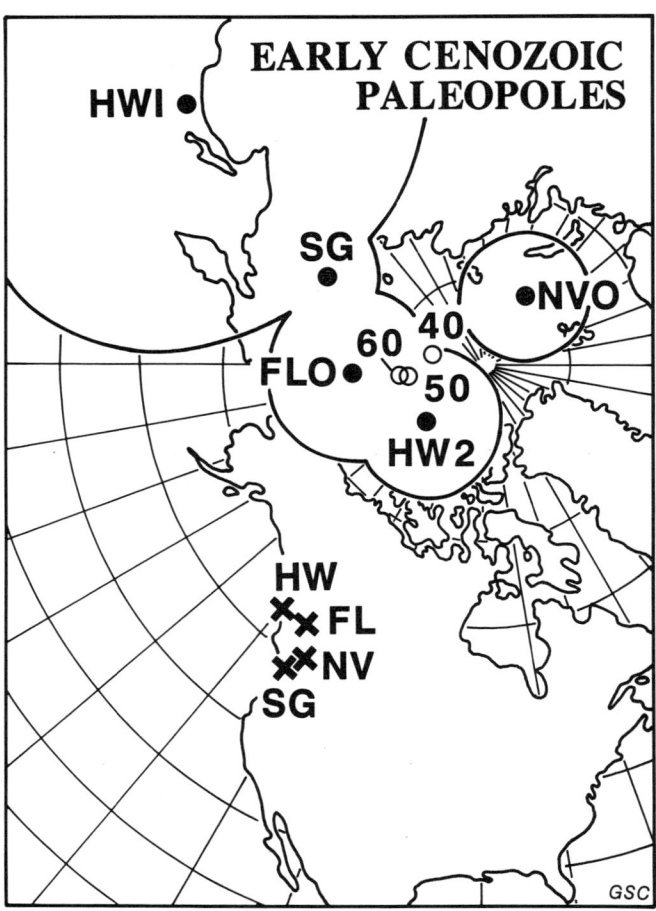

Figure 3.6. Early Cenozoic paleopoles. SG = East Sooke gabbro; HW = Eocene plutons of the Hawkesbury warp: HW1 from the west-northwest trend and HW2 from the north-northwest trend; FLO = Francois Lake Suite, overprint; NVO = Nicola volcanics, overprint. Cratonic paleopoles for 40 to 60 Ma from Irving and Irving (1982) are shown by open circles. Details in Table 3.1.

The Cretaceous determinations have been obtained from localities throughout the western Cordillera from Vancouver Island to Quesnel Lake and from northern Washington to north-central British Columbia. The general coherence of paleopoles provides paleomagnetic evidence for the existence of a Baja British Columbia superterrane in mid-Cretaceous time. Geological evidence indicates that Quesnellia was welded to the Omineca Belt in the Jurassic, and if the paleopole QL is correctly dated, then, perhaps at least some part of the Omineca Belt was attached to Baja British Columbia (Rees et al., 1985).

No tilt corrections have been made for the mid-Cretaceous Cordilleran paleopoles of Figure 3.8 (except for AX2), so the difference between Cordilleran and cratonic paleopoles could be caused either by a consistent regional tilting towards the south-southwest of about 30°, or by displacement (roughly 2000 km) northward and rotation (about 60° clockwise) of Baja British Columbia (Beck et al., 1981b; Irving et al., 1985). Systematic regional tilting

Figure 3.9 is consistent with the data. Apparent rotations are shown in the inset. The paleolatitude of Mount Stuart (MS) falls south of other paleopoles and local tilting could be responsible for the divergence. All apparent displacements are from the south, and, because the paleofield had normal polarity, there is no ambiguity as to whether paleolatitudes were north or south. The average displacement is about 2000 km, and all rotations are about 60° clockwise.

Cordilleran paleopoles are in good agreement, indicating that during the Cretaceous Baja British Columbia behaved as a single plate and not as a number of platelets acting as "ball-bearings". Internal motions may have occurred but they were second-order. The question then arises: was Baja British Columbia a separate small plate, or was it attached to the margin of another larger plate? Paleopoles from Baja British Columbia are similar to paleopoles determined from mid-Cretaceous seamounts of the Pacific Plate (Fig. 3.10). This similarity does not mean that Baja British Columbia was part of the Pacific Plate, however, because the Pacific Plate was not in contact with

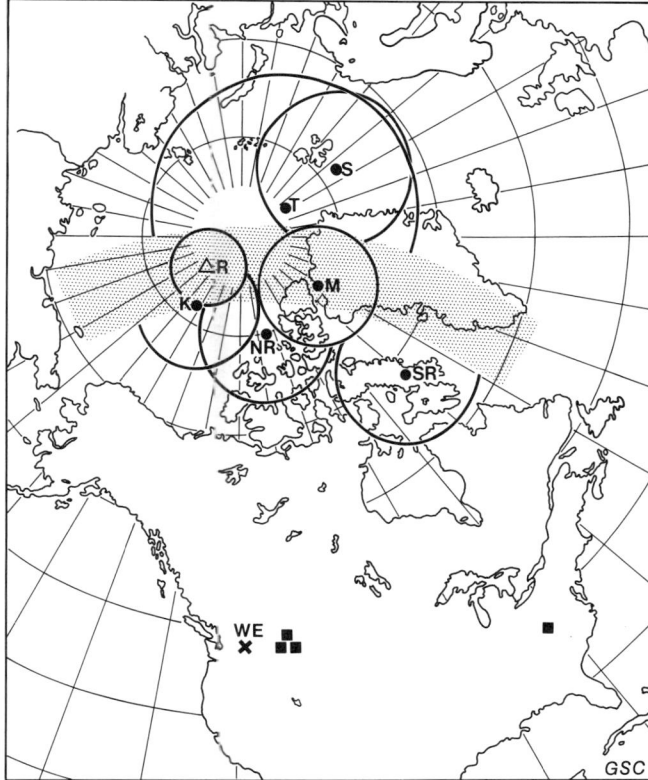

Figure 3.7. Paleopoles from Eocene (55-48 Ma) volcanics of eastern Washington (Quesnellia) compared with cratonic reference paleopole (R, derived from localities indicated by solid squares) for same time interval. The sampling region is labelled WE. Data from five grabens spread over 120 km centred at 48.5°N,119°W (Fox and Beck, 1985): K = Keller 82°N,150°W, NR = north Republic graben 80°N,106°W, S = Spokane-Enterprise lineament 78°N,6°E, SR = south Republic graben 68°N,60°W, T = Toroda 84°N,2°E, M = grand average 80°N,52°W. R = (84°N,171°W, k=203, α95=7°) calculated from locations 29 to 32 (shown as squares) spanning 50 to 47 Ma in Table I of Irving and Irving (1982). Paleopoles should lie in the shaded belt if no latitudinal displacement has occurred. Standard errors are given, so the absence of overlap indicates significant difference at P = 0.05.

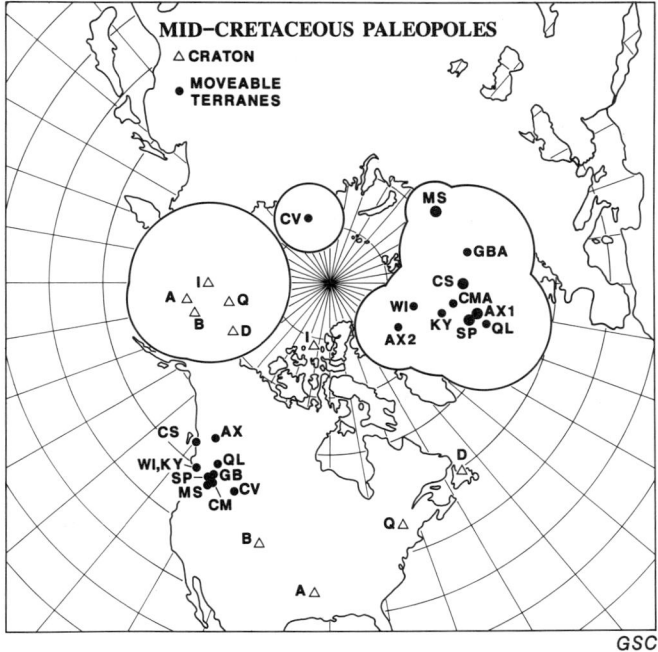

Figure 3.8. Mid-Cretaceous paleopoles. Results from the craton: D = lamprophyre dykes, Newfoundland; Q = intrusions of Monteregian Hills; I = Isachsen diabase; A = alkalic intrusions, Arkansas; B = Niobrara Formation. Results from the Cordillera: CV = Crowsnest Formation; MS = Mount Stuart batholith; CS = Captains Cove and Stephens plutons; SP = Spuzzum and Porteau plutons; AX = Axelgold intrusion: AX1 with respect to present horizontal, AX2 with respect to layering; KY = Karmutsen Formation, overprint; WI = West Coast Complex and Island Intrusions, overprint; CMA = Copper Mountain Intrusion, overprint; GBA = Guichon batholith, overprint; QL = Quesnel Lake volcaniclastics, overprint. Note that of the Cordilleran data MS, CS, SP and AX1 have larger symbols to denote that they are radiometrically dated. Details in Table 3.1.

seems unlikely but at present cannot be discounted. Displacement and rotation could have been separate processes (the "ball-bearing" hypothesis of Beck, 1980), or they could have occurred in a unified way through motion of a single quasi-coherent plate. Separate displacements and rotations are shown in Figure 3.9 in which the results of Figure 3.8 are converted into paleolatitude and plotted along the modern coast of North America. Although the mid-Cretaceous edge of the continent would have been different from that at present, this is a convenient way of expressing paleolatitudinal displacements because the present coastline was essentially paleomeridional during the Mesozoic. A displacement is simply the difference between present latitude and paleolatitude measured along the coast. Paleolongitude is not determined, and any position parallel with the lines of paleolatitudes shown in

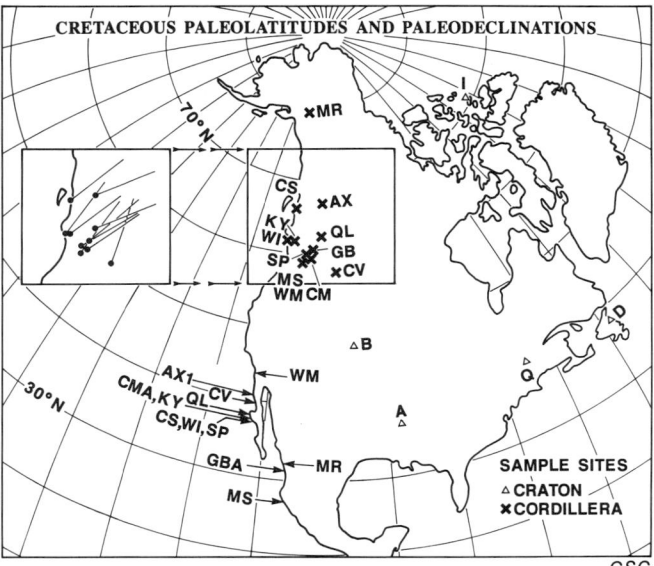

Figure 3.9. Paleolatitudinal displacements, marked by arrows along the coast, and rotation, shown in the inset, relative to cratonic North America calculated from mid-Cretaceous or presumed (overprints) mid-Cretaceous magnetizations (listed in Irving et al., 1985). Results from the Upper Cretaceous Winthrop and Midnight Peak formations of the Methow Trough in Washington (WM, Granirer et al., 1985) and the latest Cretaceous MacColl Ridge Formation of the Wrangell Mountains are included for comparison (MR, Panuska, 1985; reference paleopole used 75°N,170°W which is the mean for 70 Ma in Irving and Irving, 1982). Labels as in Figure 3.8 and Table 3.1.

North America in the Cretaceous, the similarity implies, rather, that the northward motion of Baja British Columbia took place in a fashion roughly comparable to that of the Pacific Plate and to that of all other plates of the Pacific Ocean; that is, it had a strong northerly component, in contrast to North America, which has moved generally southward since the mid-Cretaceous.

Figure 3.11 shows the plate rotation required to reconcile Cordilleran paleopoles with cratonic paleopoles and produce the latitudinal displacements and rotations of Figure 3.9. Because longitude cannot be fixed, there is no unique solution. If the motion of Baja British Columbia was coastwise, then the Euler pole was situated near Chicago (EPA). As the starting point is moved farther west, the Euler pole moves progressively southward (EPB), but its longitude is little changed. There is no evidence of oceanic sedimentation in the Cordillera in mid-Cretaceous time, and no clear evidence of an intervening ocean having closed in the interim. It is reasonable to assume that the motion has been essentially coastwise (Fig. 3.12). The oldest age limit for the displacement is set by the youngest ages of magnetizations, about 90 Ma (Table 3.1). The younger limit is certainly older than 30 or 40 Ma (Fig. 3.5, 3.6, and Table 3.1 paleopoles CB, HC). Studies of Eocene rocks in the adjacent United States, indicate that most if not all the latitudinal motion was completed by then (Fig. 3.6). Hence, the northward motion of Baja British Columbia and its emplacement in its present position relative to the craton (stage B, Fig. 3.2B) probably took place between 85 Ma and 50 Ma, that is, between mid-Cretaceous and Eocene time. The motion was broadly consistent with that of the Kula Plate (Fig. 3.12), and particularly so if the starting point is placed west of the present coast. Thus Baja British Columbia was attached

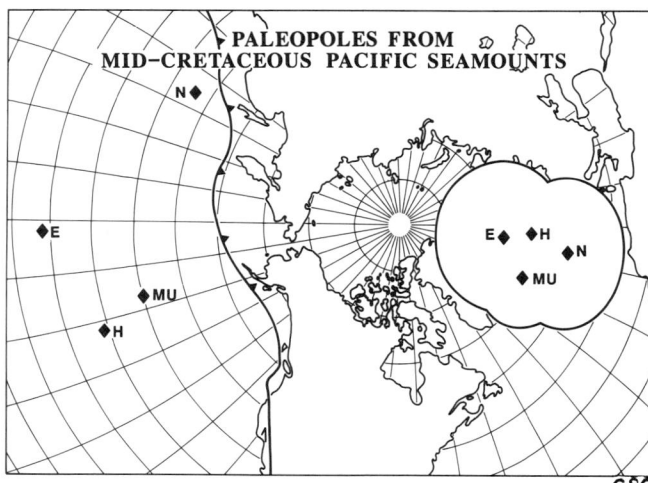

Figure 3.10. Paleopoles from mid-Cretaceous seamounts of the Pacific Ocean compiled from Harrison et al. (1975). E = Equatorial seamounts, location 8°N,178°W, paleopole 67°N,5°W; H = Hawaiian seamounts, 19°N,158°W, 60°N,3°W; MU = Musicians seamounts, 30°N,163°W, 60°N,21°W; N = North Japanese, 36°N,146°E, 51°N,8°W.

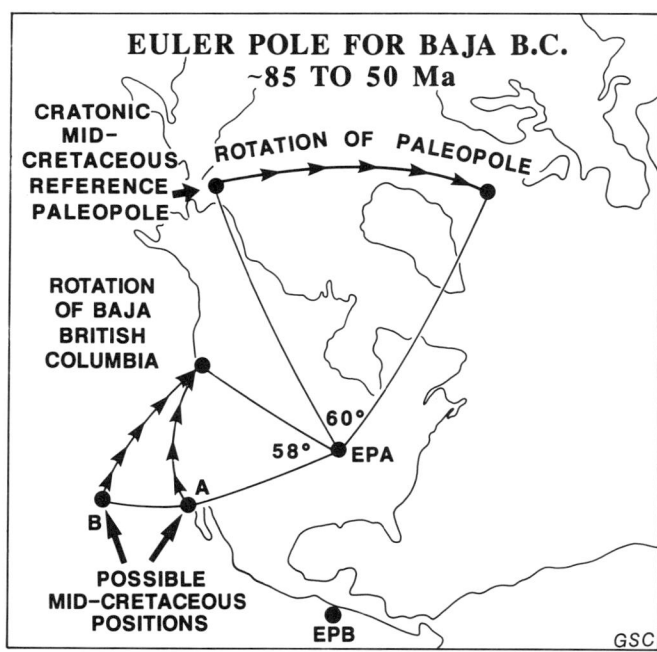

Figure 3.11. Possible Euler poles EPA and EPB for Baja British Columbia from about 85 to 50 Ma.

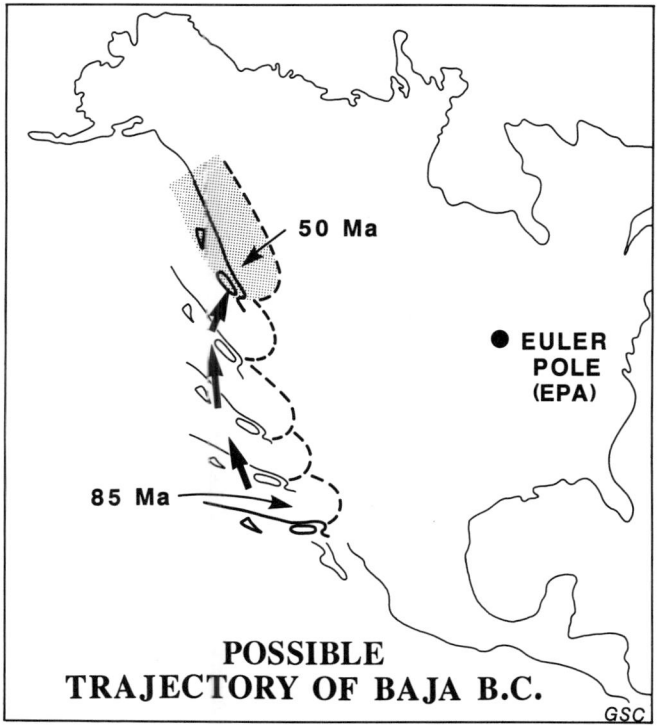

Figure 3.12. Latest Cretaceous and earliest Tertiary trajectory of Baja British Columbia deduced from the Euler pole EPA of Figure 3.11. The choice of the coastwise migration is arbitrary. Other trajectories, commencing farther west, are equally consistent with the data.

to the eastern margin of the Kula Plate during the latest Cretaceous and earliest Tertiary, much as Baja California is attached to the Pacific Plate today. The Kula Plate broke from its parent Farallon Plate at about 85 Ma (Woods and Davies, 1982) and became amalgamated with the Pacific Plate at about 50 Ma (Byrne, 1979). Hence, the northward motion of Baja British Columbia occurred during the lifetime of the Kula Plate. It is within these age limits that much of the deformation of the Foreland Belt occurred (Fig. 3.1A) and it is reasonable, as Chamberlain and Lambert (1985) suggested, to relate the deformation to the oblique collision of Baja British Columbia with the craton.

Work by Symons (1973a, 1983a) on the Early Cretaceous Francois Lake Suite in Stikinia (FLO) indicated some displacement from the south but no rotation (FL, Table 3.1). Results also have been obtained from Cretaceous rocks of the Rocky Mountains (CV) and these will be discussed later in this chapter, with other data from the Foreland Belt.

The "tilt or displacement" ambiguities inherent in Cretaceous results from the Canadian Cordillera will only be settled satisfactorily when observations from bedded sequences become available. Two recent results from sedimentary rocks in the adjacent United States shed light on the problem. The first is from Upper Cretaceous sedimentary rocks of the Methow Trough in Washington (Granirer et al., 1985). Their magnetization is post-folding, at least in part, and the result shown in Figure 3.9 (WM) was obtained by restoring beds by one half of the apparent tilt.

There is some uncertainty as to the age of the magnetization, and the possibility of regional tilt, although perhaps diminished, is not excluded. The second result (MR) is from latest Cretaceous sedimentary rocks of the Wrangell Mountains (Panuska, 1985). Both WM and MR are in agreement with data from the Canadian Cordillera and provide support for the displacement hypothesis. Both determinations indicate large clockwise rotations. It should be noted that WM and MR are from Upper Cretaceous rocks, whereas the other results of Figure 3.9 are derived from probable mid-Cretaceous magnetizations.

Early Jurassic and Late Triassic

Paleopoles from Lower Jurassic rocks are compared with the reference pole (RJ) in Figure 3.13. The reference field is based on results from radiometrically dated peri-Atlantic tholeiites. The rocks fall in the interval between 185 and 200 Ma, and are comparable in age to the Lower Jurassic rocks studied in the Cordillera. Paleopoles determined from Lower Jurassic rocks of the Colorado Plateau (K and W of Table 3.1), are displaced to the west with respect to those from the craton, and therefore Steiner (1984) suggested that the Colorado Plateau may have been rotated clockwise relative to the rest of North America. They are not used to estimate the cratonic reference field (Irving and Yole, 1987).

Determinations have been obtained from the Lower Jurassic Bonanza Group (BV) of Vancouver Island (Wrangellia), the Hazelton Group (H1, H2, H3) in Stikinia, and from two intrusive complexes (Guichon batholith, GBT and Copper Mountain intrusions, CMT of Quesnellia). Their paleopoles are widely scattered and very large and variable rotations are implied (Fig. 3.13). Reversal of the geomagnetic field occurred in the Early Jurassic, so their paleolatitudes could be north or south of the paleoequator. The paleolatitudes, assumed northern, are scattered over about 20° (Fig. 3.14). Results BV, CMT and GBT (the latter two are intrusions for which there is no attitude control) imply displacements from the south of between 10 and 15°. Paleolatitudes of the Hazelton Group (H1, H2, H3) imply no significant displacement but their accuracies are low (Table 3.1). The displacement obtained from some units may be as much as 1000 km, but never exceeds one half those observed for the mid-Cretaceous (compare Fig. 3.9 and 3.14 and Table 3.2). If the southern hemisphere is chosen the post-Early Jurassic displacements would be correspondingly larger (Fig. 3.14).

The reference field for the Early Jurassic is not very well established, and the quality and number of determinations from the Canadian Cordillera are insufficient for precise comparison. For example, one result, from the Bonanza volcanics of Vancouver Island illustrates the kinds of ambiguities inherent in the discussion of an individual datum. In Figure 3.14 northern and southern paleolatitudinal belts for the Bonanza volcanics are plotted relative to Pangea, and indicate displacement from the south of 9° ± 7 or 57° ± 7. Mid-Cretaceous data indicate a position near present Baja California (Fig. 3.9). Hence many possible paths of motion for Vancouver Island may be envisaged for the interval Early Jurassic to mid-Cretaceous (approximately 145 to 85 Ma). It could have moved along the coast from the southern belt (CWS) or from the northern belt (CWN) to the mid-Cretaceous position

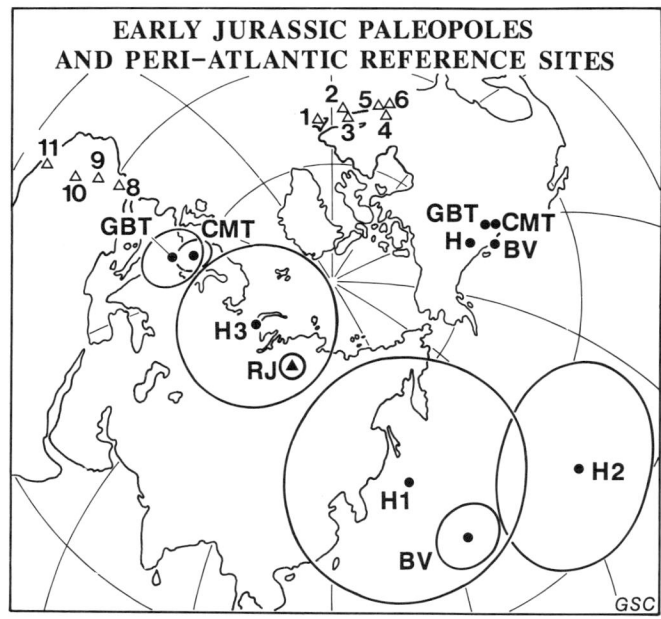

Figure 3.13. Paleopoles from Lower Jurassic rocks of the Canadian Cordillera compared with reference paleopole RJ for cratonic North America. H1, H2, H3 = Hazelton Group; BV = Bonanza volcanics; GBT = Guichon batholith; CMT = Copper Mountain intrusions; details in Table 3.2. The cratonic reference paleopole RJ (68°N, 095°E $A_{95}=3°$) is from peri-Atlantic tholeiitic intrusions and lavas in the age range 200 to 185 Ma. Localities from which cratonic data were obtained are shown by open triangles and are as follows: (1) trans-Avalon dyke, Newfoundland (Hodych and Hayatsu, 1980); (2) Shelburne dyke, Nova Scotia (Larochelle and Wanless, 1966); (3) North Mountain basalt, Nova Scotia (Larochelle, 1967; Carmichael and Palmer, 1968); (4) Connecticut valley rocks (de Boer and Snider, 1979); (5) pre-folding intrusions New England (Smith and Noltimier, 1979); (6) Newark intrusives, older group (Beck, 1972); (8) intrusives of stable Morocco; (9) northern Mauritanian dykes; (10) southern Mauritanian dykes (8,9,10, Sichler et al., 1980) (11) Liberian dykes (Dalrymple et al., 1975). Data from Africa have been merged with North American data by closing the north Atlantic. Details in Table 3.2 and Irving and Yole (1987).

(Fig. 3.15). It may have been carried in from the Pacific by the Farallon Plate, as the latter moved toward North America (Fig. 3.3) and from where trajectories may be calculated using the rotations given by Engebretson et al. (1985). The trajectory cannot be extended back to the Early Jurassic because older magnetic anomalies are not present, so a direct comparison with Bonanza paleolatitudes is not possible. The "tilt or translation" ambiguity in the Cretaceous data means that two trajectories are possible, one assuming a Late Cretaceous position near Baja California, and the other near the present location of northern Vancouver Island. With increasing age, the first trajectory moves toward the southern paleolatitudinal belt. The second is within the northern belt. Hence the data from this individual rock unit can be explained in at least four ways. Laber in this review it will be argued that the route CWN of Figure 3.15 is the most probable.

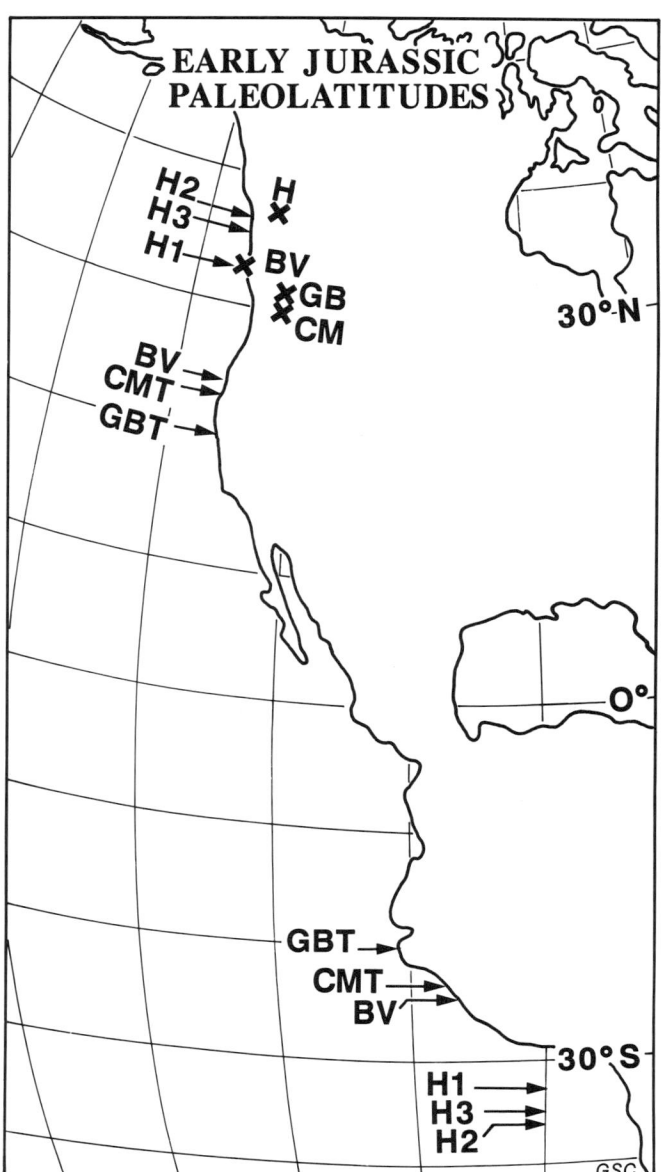

Figure 3.14. Paleolatitude map for the Early Jurassic (using reference paleopole RJ, Irving and Yole, 1987 and Figure 3.13) showing displacements observed from Cordilleran terranes. The symbols are labelled as in Figure 3.13 and Table 3.2. Possible northern and southern hemisphere positions are shown along the western coast of Pangea.

The Late Triassic reference paleofield is not well-determined, although until recently it was considered reasonably well fixed. The principal reason for this current uncertainty is the recent downward revision of the timescale, which assigns to the Early Jurassic those paleopoles (notably those from the peri-Atlantic tholeiites) that were formerly regarded as Late Triassic (Irving and Yole, 1987). Consequently, only two Late Triassic results are available from cratonic North America: one from the Chinle Formation of New Mexico (C1) and the other from the Manicouagan Impact Crater (M) of Quebec (Table 3.1).

Table 3.2. Displacements and rotations observed from Jurassic, Triassic and Permian rocks

Rock unit		D°, I°	$\alpha°_{95}$	$\lambda°_p$	North RPD°	North RR°	South RPD°	South RR°
JURASSIC (RJ 68°N, 95°E, $A_{95}=3°$)								
H1	Hazelton Group	114,-52	25	33	5 ± 18	-132 ± 22	70 ± 18	48 ± 22
H2	Hazelton Group	242, 56	18	37	1 ± 14	100 ± 18	74 ± 14	-80 ± 18
H3	Hazeltmx Group	359, 55	16	6	2 ± 12	-17 ± 15	73 ± 12	163 ± 15
BV1	Bonanza Group	276, 42	6	24	9 ± 5	66 ± 5	57 ± 5	-114 ± 5
GBT	Guichon Batholith	028, 36	7	20	12 ± 5	-43 ± 5	52 ± 5	137 ± 5
CMT	Copper Mtn. intrusion	030, 41	4	24	7 ± 3	-45 ± 4	54 ± 3	135 ± 4
TRIASSIC (RT1, n=2, 59°N, 84°E, $A_{95}=12°$)								
ST1	Stuhini Group	300, 44	6	26	2 ± 12	43 ± 14	54 ± 12	-137 ± 14
ST2	Stuhini Group	281, 38	7	21	6 ± 11	62 ± 13	49 ± 11	-118 ± 13
KX	Karmutsen Formation	003,-33	6	18	3 ± 10	161 ± 11	39 ± 10	-19 ± 11
NV	Nicola volcanics	041, 47	6	29	- 8 ± 10	-54 ± 11	49 ± 10	126 ± 11
TRIASSIC (RT2, n=4, 64°N, 99°E, $A_{95}=13°$)								
T1	Stuhini Group	300, 44	6	26	10 ± 12	38 ± 14	61 ± 12	-143 ± 14
T2	Stuhini Group	281, 38	7	21	14 ± 11	57 ± 14	57 ± 11	-133 ± 14
KX	Karmutsen Formation	003,-33	6	18	11 ± 11	± 13	47 ± 11	157 ± 13
NV	Nicola volcanics	041, 47	6	29	-1 ± 11	59 ± 12	57 ± 11	121 ± 12
PERMIAN (RP, n=12, 43°N, 126°E, $A_{95}=4°$)								
AB	Asitka Group basalt	129,-40	8	23	4 ± 5	1 ± 6	49 ± 5	179 ± 6

D,I mean declination, $\alpha°_{95}$ its error (P=0.05), $\lambda°p$ the paleolatitude. RPD and RR are the relative paleolatitudinal displacement (positive from the south) and relative rotation (positive anticlockwise). The errors have been corrected by the method of Demarest (1983). Cratonic reference paleopoles are given. "North" and "South" are values for the northern and southern hemisphere options.

Data elsewhere in Pangea are from the Los Rostras, Ischigualasto (L) and Paramillos (P) formations of Argentina (Table 3.1, Fig. 3.16). A paleopole from Upper Triassic rocks of the Colorado Plateau (C2, Table 3.1) is displaced to the west of those from the craton, and, because of possible rotation of the plateau, is not used. The South American paleopoles (P and L), after rotation into the North American frame of reference, are displaced relative to North American paleopoles. The difference appears real, but the meaning is not clear. It could be caused by errors in dating, paleomagnetic analyses, and reconstructions, or by the presence of long-term non-dipole components in the geomagnetic field. The imperfect fit of paleomagnetic observations from the northern and southern hemispheres is a long-standing problem, for which presently there is no satisfactory solution. Thus, two reference paleopoles have been calculated: RT1 uses the North American data (C1 and M), and RT2 is based upon the North and South American data combined (Table 3.1). Displacements and rotations are given in Table 3.2 and Figure 3.17.

Triassic data are available from the Karmutsen Formation of Vancouver Island (KX), and from two sections of the Stuhini Group (ST1, ST2) (formerly Takla Group) of Stikinia in north-central British Columbia (Fig. 3.1B). The paleopoles diverge widely from the reference paleopole indicating very large rotations (Table 3.2). Using the reference paleopole RT2 and the northern option the corresponding paleolatitudes are lower than expected (Fig. 3.17A). The displacements are statistically indistinguishable from those for the Early Jurassic for the northern option (compare Fig. 3.14 and 3.17). All are less than for the mid-Cretaceous (compare Fig. 3.9). If the southern option is adopted, the displacements are less than for the Jurassic (Fig. 3.14, Table 3.2). Using the reference pole RT1, the mean displacements, with the exception of NV, are all from the south, but are not individually significant. Agreement with Early Jurassic results is better for the northern than for the southern option.

Late Paleozoic

Data from the Lower Permian Asitka Group (Fig. 3.18) are based on 25 hand samples for a limited stratigraphic coverage, but are sufficient to show that very large and variable rotations occurred (Table 3.2). The paleolatitude (22°) is not significantly different from that expected (27°) but the mean is displaced to the south (5° ± 6°). Both polarities occur, but one, with negative inclination, predominates. Assuming this to be reversed, as the Permian field usually was, a northern paleolatitude is indicated (Irving and Monger, 1987). The Asitka displacement can be compared with that observed from Upper Permian rocks of the eastern Klamath mountains (Dekkas, Fig. 3.19, Mankinen et al., 1984). It indicates that Stikinia and the eastern Klamath mountains were closer together than at present, but the errors are such that considerable leeway is allowable. In any case large displacement is not required.

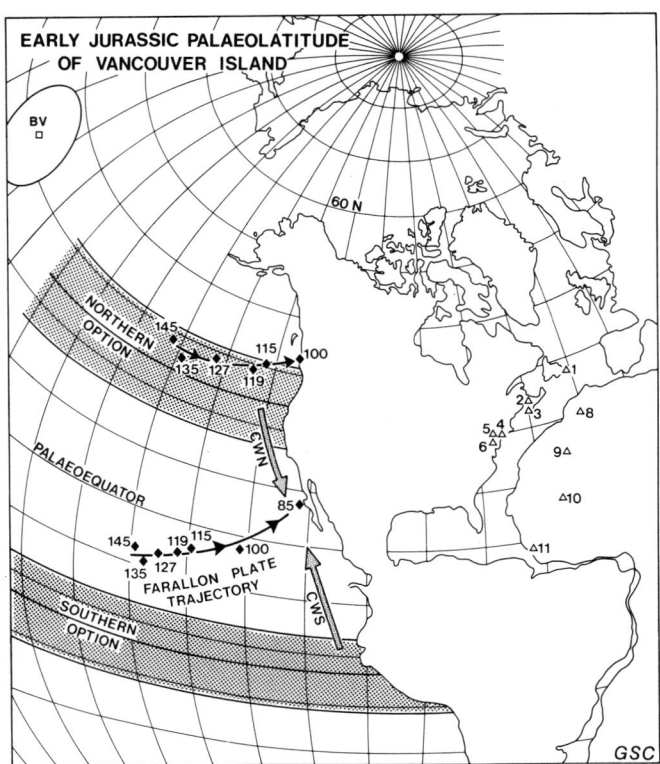

Figure 3.15. Possible motions of Vancouver Island between Early Jurassic and mid-Cretaceous time (Irving and Yole, 1987). Early Jurassic grid, drawn relative to RJ of Table 3.2 and based on data listed in legend to Figure 3.13. Late Jurassic-Early Cretaceous trajectories of Vancouver Island have been calculated assuming that the island was carried across the Pacific towards North America on the Farallon plate. Alternative coastwise trajectories also are shown, extending from the northern latitude belt (CWN) and from the southern latitude belt (CWS) to the mid-Cretaceous position determined by the data in Figure 3.8.

Summary and discussion of Phanerozoic data

Data presented above suggest the revisions to Figure 3.2 that are shown in Figure 3.22. For the early Mesozoic the Intermontane and Insular superterranes are now placed close together at a higher paleolatitude, and in the Cretaceous at a lower paleolatitude. The Cenozoic reconstruction is not changed. The former steady movement from the south relative to North America (Fig. 3.2) is replaced by a southerly motion in the Middle Jurassic to Early Cretaceous, and a larger northerly motion in the Late Cretaceous to earliest Tertiary (Fig. 3.21). Figure 3.22 is not unique and other schemes are possible.

Cenozoic

Cenozoic results show that the major tectonic elements of British Columbia were in place relative to North America by the mid-Tertiary. There is no evidence for significant latitudinal motion since Eocene time. Divergent results are confined to the western margin and could be related

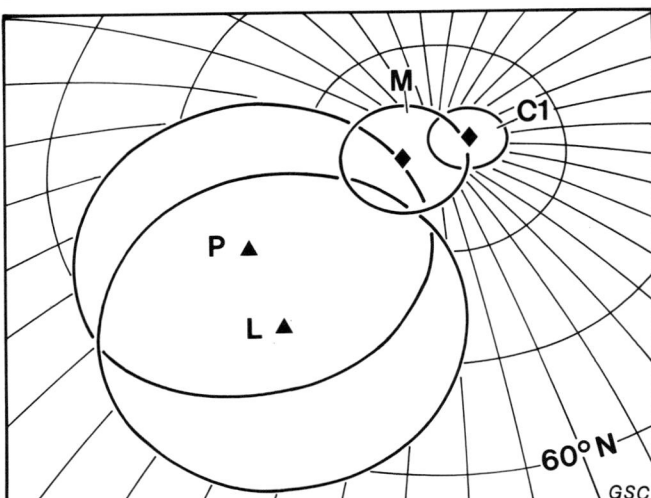

Figure 3.16. Reference paleopoles for the Triassic. The grid is drawn relative to the mean of the M (Manicouagan) and C1 (Chinle paleopoles). P (Paramillos) and L (Los Rostras and Ischigualasto) from South America have been rotated into the African frame of reference by rotation of 57.5° about 45.5°N,32.2°W (Rabinowitz and LaBrecque, 1979) and then by -78.8° about 66.2°N,12.4°W in the North American frame (Le Pichon et al., 1977).

either to undetected tilts or to rotations, perhaps as a consequence of the truncation of Baja British Columbia during stage D (Fig. 3.22D).

Baja British Columbia, a possible miniplate

The good agreement among results obtained from mid-Cretaceous rocks of the Western Cordillera of Canada and adjacent United States, and from overprints that reasonably can be assumed to be of this age, is consistent with the idea that the major terranes were assembled by then into a coherent plate (Baja British Columbia).

Post mid-Cretaceous tilt or translation

Data from Cretaceous intrusive rocks from terranes west of and including the Cache Creek Terrane (Fig. 3.1) indicate displacement from the south by distances of about 2000 km and clockwise rotation of about 60°. Data from lower Mesozoic rocks of Quesnellia can be interpreted as overprints whose age appears to be mid-Cretaceous. If this age assignment is correct, then at least part of Quesnellia has undergone a similar displacement. Geometrically these results can be explained by assuming that Baja British Columbia rotated as a single lithospheric plate about a Euler pole to the east (Fig. 3.11). There is no ambiguity as to the hemisphere of origin because the magnetizations were acquired, for the most part, during the Cretaceous normal superchron. Such motion is consistent with the idea that part of the deformation along the margin of the craton (Fig. 3.1) was formed by the oblique collision of the mini-plate with the craton, as Chamberlain and Lambert (1985) have suggested. This motion could have occurred when Baja British Columbia was attached to the Kula

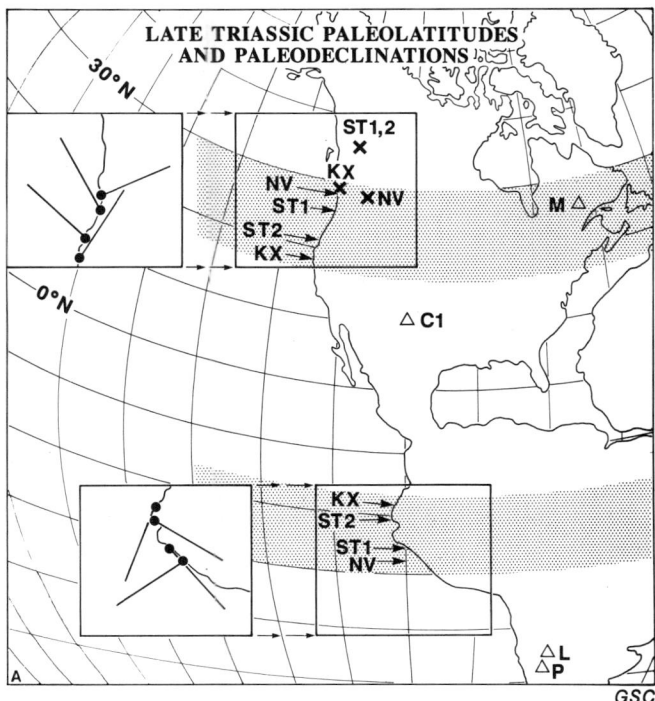

Figure 3.17. Alternative paleolatitude maps for the Late Triassic. In A the four reference paleopoles used are M, C1, L and P of Table 3.1 and Figure 3.16. Their mean is 69°N,108°E, k = 61, A_{95}=12° and has been used to calculate the grid. Paleolatitudes for Cordilleran localities are plotted along the west coast of Pangea for both northern and southern options. They are as follows: ST1, ST2 = Stuhini Group; KX = Karmutsen Formation; NV = Nicola Volcanics (see Table 3.1). Rotations are shown in the insets. In B the paleopole used is the mean of M and C1 only.

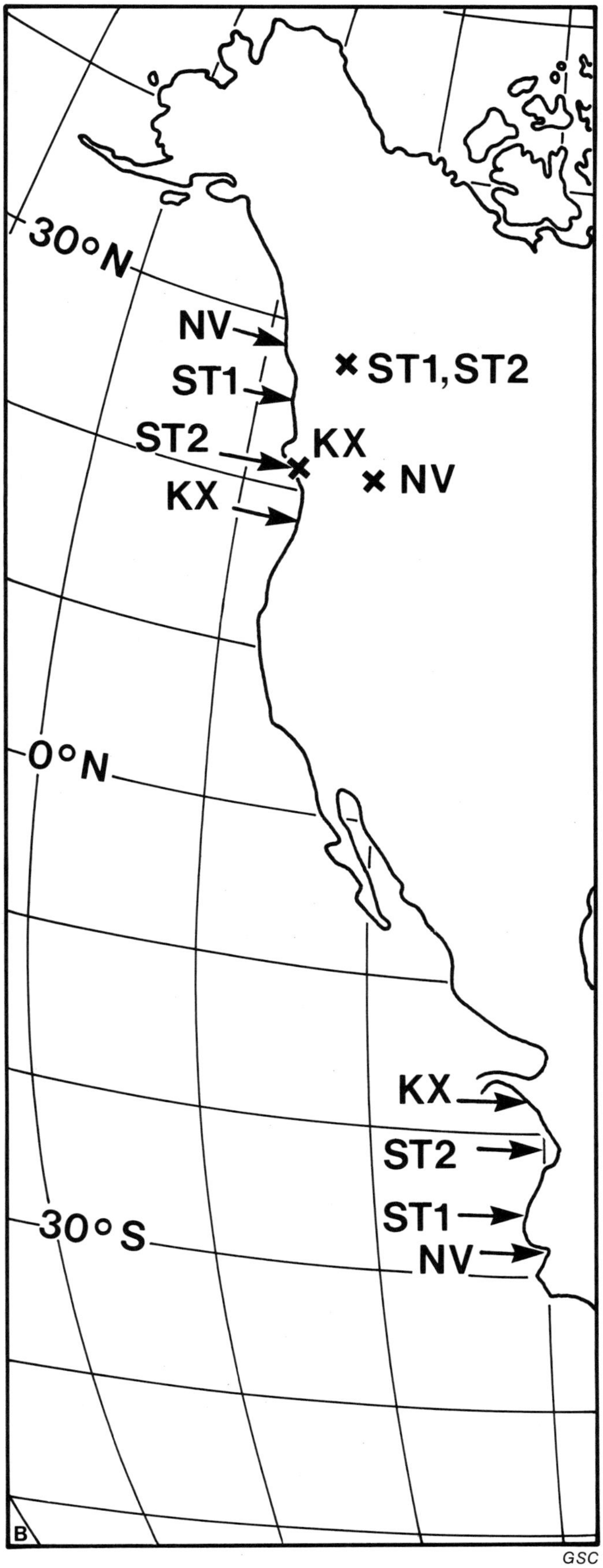

Plate (Fig. 3.3). The motion occurred between about 85 and 50 Ma ago. The Canadian data contain no direct estimates of paleohorizontal which is the main weakness of the argument; it is the agreement over a wide area that supports the hypothesis. However, recent data from Upper Cretaceous bedded sedimentary rocks in Alaska and Washington are consistent with the displacement hypothesis. Alternatively it may be supposed that Baja British Columbia tilted to the southwest by about 30°. No current geological evidence unequivocally supports the tilt hypothesis. In cases where paleohorizontal can be estimated, the paleopole is brought into agreement not with the cratonic paleopole, but with those from Baja British Columbia (MS of Fig. 3.8). Many of the bodies studied are large, presumably deeply rooted batholiths, but the possibility that tilting could have occurred in these tectonically disrupted terranes will remain until there are definitive determinations to the contrary from stratified sequences.

Rotations

Rotations, relative to the craton, observed in Lower Jurassic, Upper Triassic and Permian rocks can be compared with those observed in Cretaceous rocks (Fig. 3.20). The rotations for the former are large and variable, even within the same rock unit sampled at localities separated by no more than a few tens of kilometres. The rotations are so large and so variable that the question of whether they are real or not is essentially independent of uncertainties in the reference paleofield. The rotations obtained from presumed mid-Cretaceous magnetizations are systematic, averaging 60° clockwise, showing that the large and variable rotations observed in older rocks occurred prior to mid-Cretaceous time. Relative to the Cretaceous "zero" (0°K, Fig. 3.20B), the majority appear to be anticlockwise, consistent with sinistral strike-slip, and therefore with southerly movement such as tract CWN in Figure 3.15. Much more data are required to substantiate that the rotations are systematically anticlockwise.

The systematic clockwise rotation observed in Cretaceous rocks (Fig. 3.20A) is the typical paleomagnetic signature for motion of a single plate. The more variable rotations observed in older rocks (Fig. 3.20B) indicate that the region was not then a single coherent plate. One theory is that the rotations could have been produced soon after the rocks were formed. Some units, such as volcanics of the Bonanza Group, are the products of island arcs which may have been contorted as result of horizontal bending (Carey, 1958). Large rotations have been observed paleomagnetically in a modern island arc in the Phillipines (Fuller, 1985). Similar rotations have been observed in large nappes in Sicily (Channel et al., 1980), and in the Prospect thrust sheet of the Rocky Mountains of Wyoming (Grubbs and van der Voo, 1976). The second theory is that

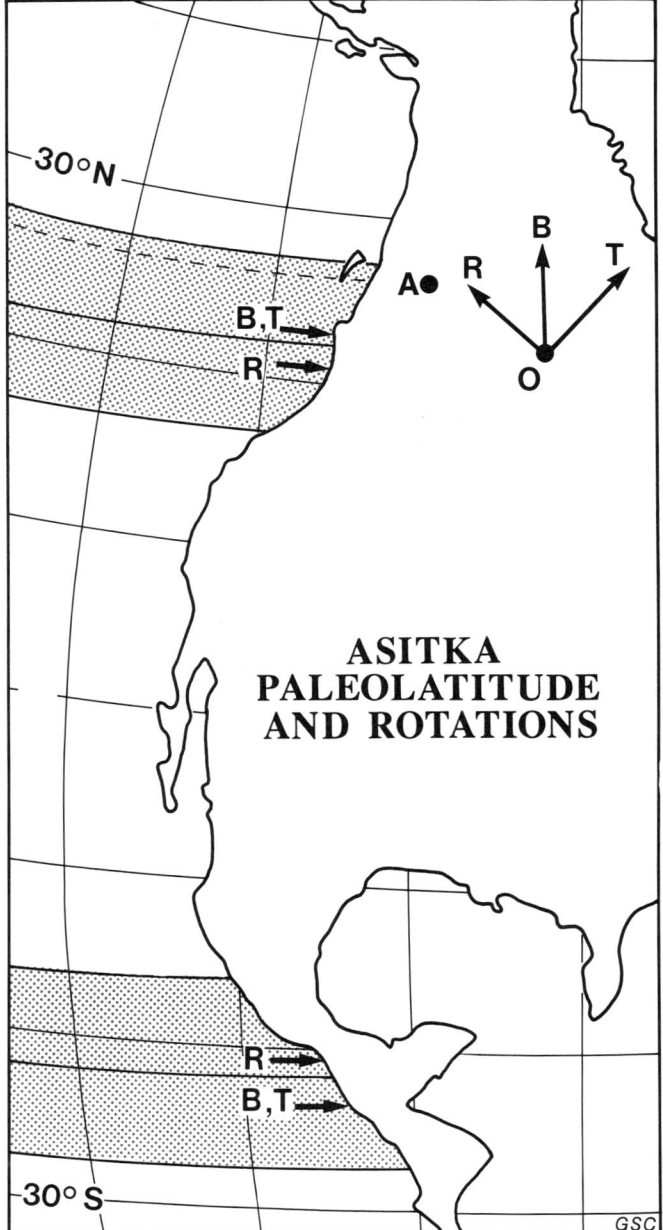

Figure 3.18. Paleolatitudes and rotations observed from Lower Permian rocks of Stikinia (Irving and Monger, 1987). A = present location of Asitka Group. O = average paleolatitude. R,B,T = horizontal components of magnetization (= paleonorth) observed in rhyolites (R), basalts (B) and tuffs (T) at three, fault-separated localities in the Asitka Group. The short arrows indicate paleolatitudes determined for the three units separately.

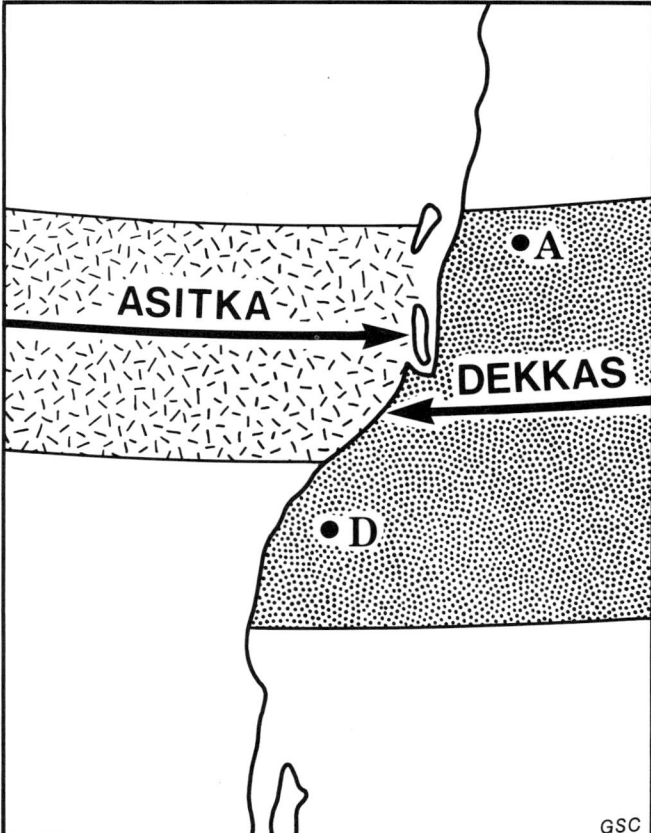

Figure 3.19. Displacements observed from Lower Permian rocks of Stikinia (Asitka Group of Fig. 3.18) compared with observations from Upper Permian rocks of the eastern Klamath Mountains. The latter data are mainly from the Dekkas andesite (Mankinen et al., 1984). Correction has been made for the age difference (Irving and Monger, 1987).

the rotations occurred during the assembly of Baja British Columbia in the Late Jurassic and Early Cretaceous, long after the rocks were deposited. In this latter view, Baja British Columbia could represent an assemblage of thin, overlapping crustal sheets. This could have occurred by overthrusting of terranes over older lithosphere, or by underplating (underthrusting and subsequent imbrication of terranes through subduction of oceanic crust). Such deformation may have occurred at an accretionary margin during underthrusting of the Farallon Plate beneath the craton in the mid-Mesozoic (Fig. 3.3). Whatever the cause of the rotations, the process by which Baja British Columbia was assembled seems to have been very different from that by which it was emplaced against the margin of the craton in its present position (Fig. 3.19).

Displacements and the hemispheric ambiguity

Uncertainties in the reference paleofield make the estimation of displacements of earlier Mesozoic rocks difficult, and the presence of paleomagnetic reversals results in ambiguity of as to hemisphere origin. If the southern hemisphere option is chosen, the displacements are in the order of 60°. If the northern is chosen the displacements are 10° or less (Fig. 3.15). Small or negligible displacements were observed in Permian rocks. There appears to be no significant difference between the displacements inferred for Stikinia and Wrangellia. Displacements calculated for the northern option are always much less than those inferred from Cretaceous rocks. Indeed, it has been argued by Gordon et al. (1984) and Butler and May (1985) that no latitudinal displacements are required for Stikinia. Where detailed studies are available and comparison between observations from essentially coeval rocks on and off the craton is possible (Fig. 3.15), displacement from the south, although somewhat less than originally believed (Fig. 3.2), appears to be real and consistent with paleontological evidence of disjunct faunas (Tipper, 1981; Taylor et al., 1984; see Chapter 2). When all data are considered, a systematic shift to the south becomes apparent (Fig. 3.21). The mean displacement for results obtained from rocks of Late Triassic and Early Jurassic age, using the RT1 reference paleopole for the Triassic, is 5° from the south. The mean using RT2 is 9° from the south. These averages do not incorporate results GBT, CMT and NV for which there is no paleohorizontal control. If they are included, the values become 4° and 7° respectively.

If the northern option is adopted, the similarity between displacements observed from Permian (Fig. 3.18), Upper Triassic and Lower Jurassic rocks (Fig. 3.21) implies that Wrangellia, Stikinia and Quesnellia, and by implication, the Insular and Intermontane superterranes, need not have been very far apart during this time interval. This is in disagreement with the dispositions shown in Figure 3.2. It is possible to separate the terranes, either longitudinally or by placing them in different hemispheres, but these solutions are indeterminable paleomagnetically. If the southern option is adopted, however, the overall agreement is considerably less and Early Jurassic displacements were generally greater than those of Upper Triassic rocks (Fig. 3.21). This difference could be an artifact of uncertainty in the reference paleofield. A renewed attempt to improve the reliability of the reference paleofield, both magnetically and chronologically, is required. Nonetheless, the analysis indicates that the terranes are better grouped paleolatitudinally for the northern rather than for the southern option. Therefore the former is regarded as the more probable, inasmuch as it requires least movement.

In summary, three pieces of evidence favour the northern option. Firstly, Permian magnetizations acquired during a time of predominantly reversed polarity indicate a northern hemisphere position. Secondly, Permian and early Mesozoic results from both the Insular and Intermontane superterranes are in better agreement for

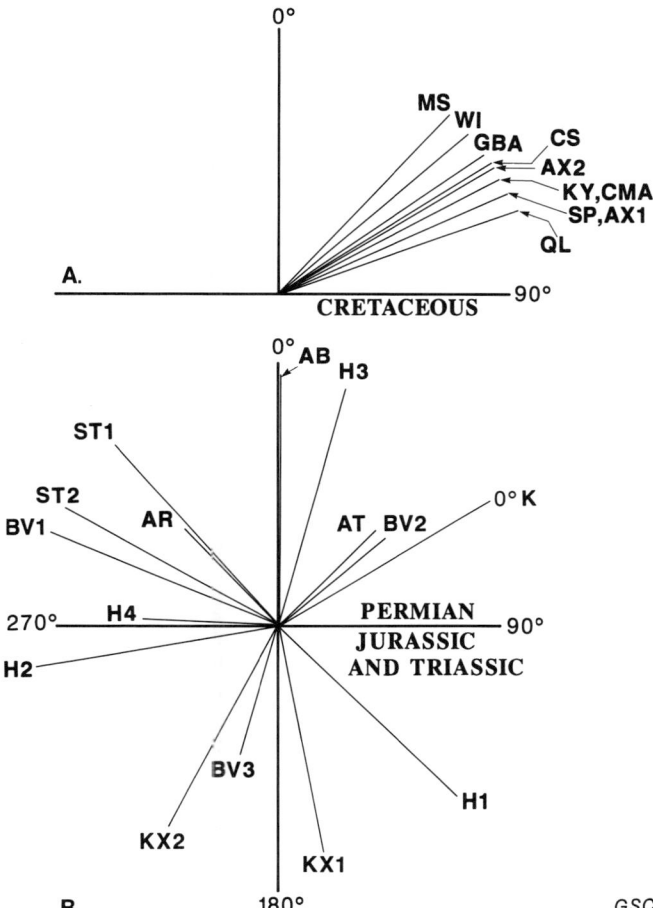

Figure 3.20. Rotation inferred from presumed mid-Cretaceous magnetizations (A) compared with those from Permian, Late Triassic and Early Jurassic magnetizations (B). The rotations of Table 3.2 are reduced to one locality (Smithers 54.8°N, 127.2°W) located in Central British Columbia. If no rotations had occurred the lines would coincide with zero azimuth (0°). In (B) the magnitude of the pre-Cretaceous rotations can be determined by viewing them relative to the "Cretaceous zero" (0°K). Errors are listed in Table 3.2 and in Irving et al. (1985). Long lines are based on observation from five or more sampling sites, short lines on fewer sites. The lines are labelled as in Table 3.1. Information from intrusions is not included in (B). H4 (RR = 88°) is from unpublished data obtained from the Lower Jurassic Telkwa Formation near Smithers, British Columbia.

the northern than for the southern option (Fig. 3.21). Thirdly, pre-Cretaceous rotations (Fig. 3.20B) seem to be predominantly anticlockwise, consistent with a southerly movement of the superterranes relative to the craton during the interval Middle Jurassic through Early Cretaceous. None of the evidence is definitive, but a northerly option is more consistent. Hence the most probable scheme of motion is north to south to north (Fig. 3.22).

A comparable scheme for Baja Alaska is implicit in the work of Stone et al. (1982). They plotted paleolatitude (not displacement relative to North America) and suggested an initial northerly motion from low (~25°) paleolatitude in the mid- to Late Jurassic, followed by a decrease to low paleolatitude (~15°) in the mid-Cretaceous, and in turn followed by an increase to present values. If relative displacements had been calculated, the Triassic and Jurassic values would be similar because North America is moving northward. This would have resulted in displacement to the south as indicated for the Late Jurassic and Early Cretaceous, followed by a displacement to the north in post mid-Cretaceous time (Fig. 3.22).

Tectonic attenuation

Some terranes show a spread in paleolatitude less than that at present. This is particularly notable among the fragments of Wrangellia (Fig. 3.21, Yole and Irving, 1980; Hillhouse and Grommé, 1984). It also seems to occur among mid-Cretaceous plutons (Irving et al., 1985). The agreement between Permian results from Stikinia and the eastern Klamaths could represent a third example (Fig. 3.19). Apparently some terranes were formally more compact and have been attenuated or smeared out lengthwise along the Cordillera.

Overprinting

Several periods of widespread magnetic overprinting are evident. Some overprints were probably acquired thermally, for example those in the Spuzzum and Porteau batholiths and related rocks are carried by pyrrhotite with low (~400°C) blocking temperatures (Table 3.1, SPO). Others, such as those in the Karmutsen Formation (Table 3.1, KY) have high blocking temperatures (up to 500°C) and were probably acquired by alteration of existing magnetic minerals or deposition of secondary minerals (chemical remanent magnetization CRM). Particularly prominent are secondary magnetizations of mid-Cretaceous age, which occur throughout the Canadian Cordillera and which correspond to the later stage of assembly of Baja British Columbia. If assembly occurred by imbrication of comparatively thin rotating crustal sheets and widespread underplating by oceanic crust, then water could have been flushed from the underplating oceanic crust into the fractured upper crust, causing the widespread occurrence of secondary CRM.

FORELAND BELT
Precambrian

Following the work by Collinson and Runcorn (1960) and Norris and Black (1962), paleomagnetic data are available from several detailed studies of the Precambrian Purcell (Belt) Supergroup of the Rocky Mountains of Alberta and Montana. The paleopole PE (Table 3.1; Fig. 3.23) was calculated as an average of results from the Kintla and Sheppard formations and the Purcell Lava. Paleopole PB represents an average of studies based on 10 or more samples from each of 10 formations. Paleopole PV is based on detailed studies of the Spokane Formation of Montana, which identified an overprint with a paleopole (PVO) close to that from the Franklin intrusions of northern Canada (FP of Fig. 3.23; 5°N, 165.3E $A_{95}=3$, Irving and Hastie, 1975

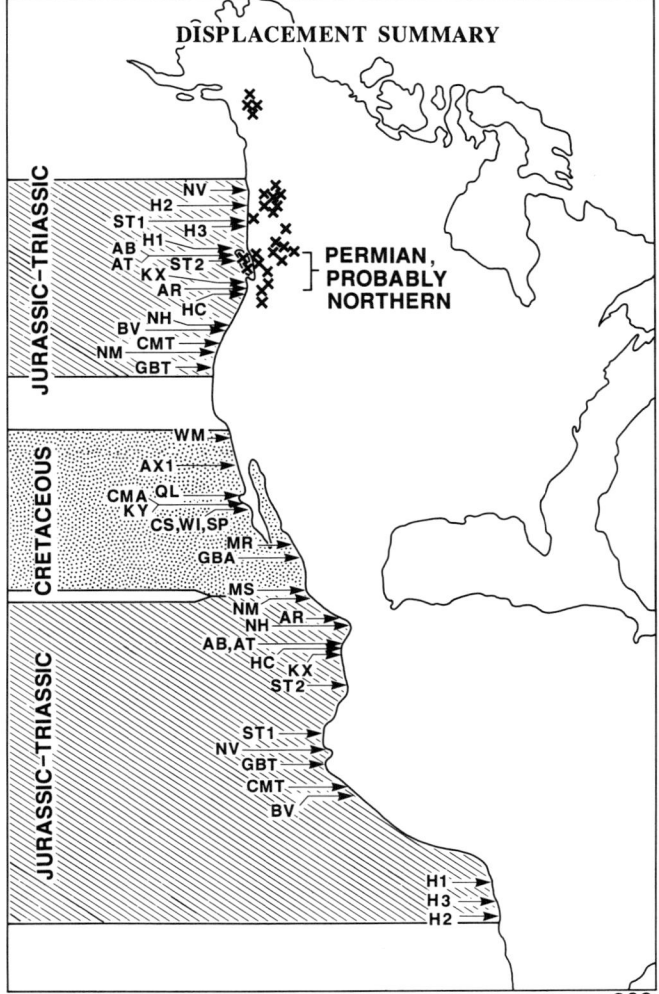

Figure 3.21. Arrows indicate the location of the estimated parallels of paleolatitude along the western coast of Pangea (and are identified in Table 3.1). Because the latter has been essentially N-S for the late Paleozoic and Mesozoic the paleolatitudinal displacement is the distance between the sampling locality (crosses, see Fig. 3.1B for details) and the arrow. Pangea configuration is relevant to the Late Triassic and Early Jurassic results for which both northern and southern hemisphere options are shown, but not to Cretaceous results because Pangea, by then, had begun to disintegrate. Results from the Nikolai greenstone (Karmutsen equivalent in Alaska, NM, NH; Hillhouse, 1977; Hillhouse and Grommé, 1984) and Hell's Creek Canyon in Oregon (HC, Hillhouse et al., 1982, compared with RJ) are added for comparison.

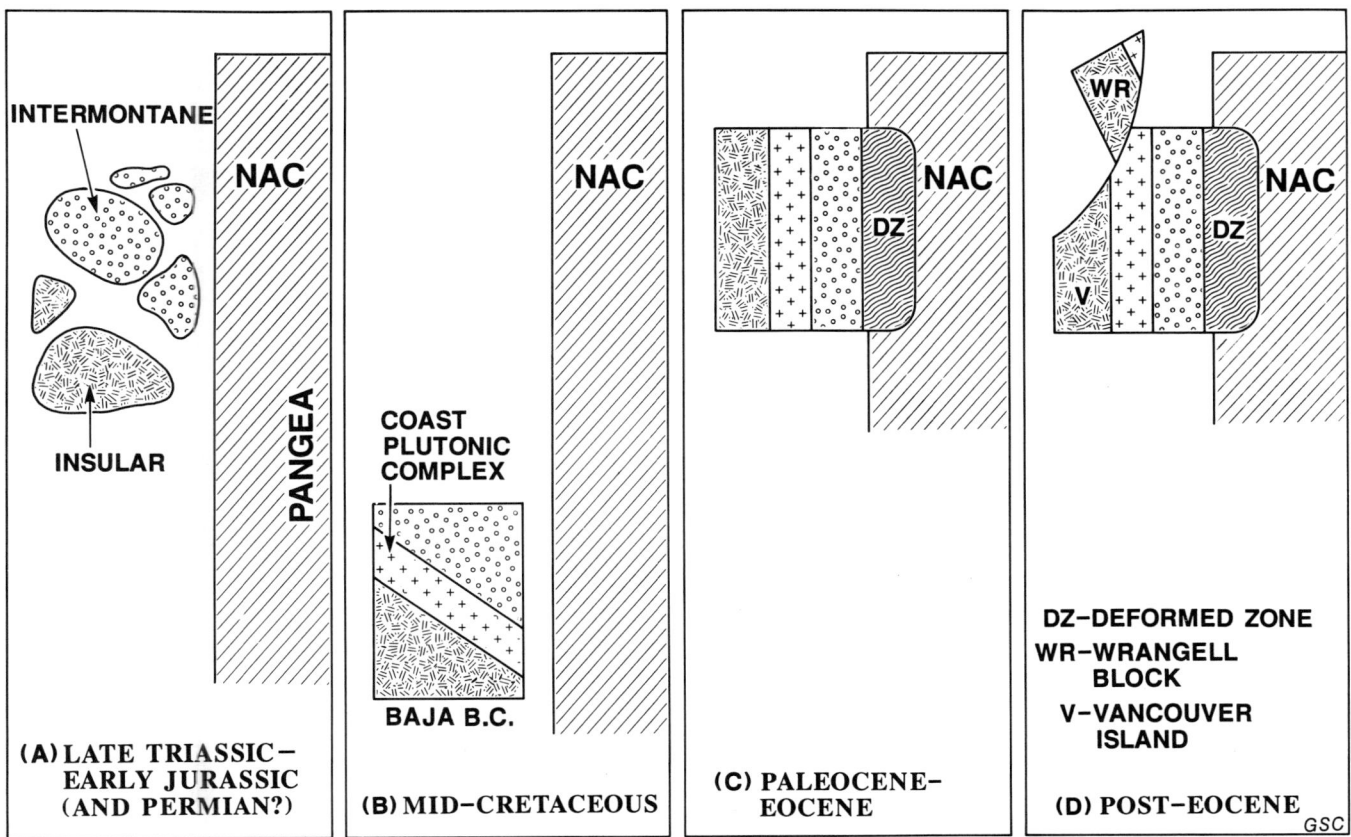

Figure 3.22. Speculative scheme showing four stages (A-D) for assembly of the western Cordillera in Canada, updated from Figure 3.2.

entry 450) dated at about 675 Ma. Alternatively, this overprint may, as Vittorelli and van der Voo (1977) suggested, mark a metamorphic event which has been dated at 750 Ma by K-Ar studies of illite. The Purcell (Belt) paleopoles PV, PE, PB and PVO lie on a crude arc, and the magnetization from which PB, and, to a lesser extent PE were derived, may have been affected by the overprint PVO. The effect on PE, however, would be small.

Purcell (Belt) rocks are of Middle Proterozoic age and in Figure 3.23 their paleopoles are compared with those from cratonic rocks formed in the interval 1350 to 1475 Ma. Obradovich et al. (1984) argued that Belt (Purcell) rocks fall within this age range. Most of the cratonic results have, by Precambrian standards, good age control. Purcell (Belt) paleopoles are more southerly than those from the craton. The difference is systematic and probably real. It could have been caused by motion of the entire Laurentian Shield (that is, apparent polar wander relative to the craton) because it is possible that Belt rocks fall entirely outside the time range of rocks from the craton. Alternatively, the difference could be attributed to relative motion between the Foreland Belt and the craton, in which case movement of the Rockies from the north is implied. The motion could have happened at any time since the Middle Proterozoic.

Younger radiometric ages have been obtained from the higher parts of the sequence from which most of the paleomagnetic data have been obtained, and these could reflect more accurately the age of the remanent magnetism. If this were so then the comparisons made here would not apply. For example, if the Belt paleopoles are 1150 Ma old they would be coeval with Keweenawan rocks and displacement from the south would be implied.

Paleopoles from the Mackenzie Mountains are shown in Figure 3-24 and represent three major stratigraphic units, all younger than the Purcell (Belt) Supergroup (Aitken et al., 1982). These are, from oldest to youngest, the Mackenzie Mountains Supergroup (results from the Tsezotene Formation, Katherine and Little Dal groups), lavas from the Coates Lake Group, and the Rapitan Group. Determinations in the Foreland Belt also are available from sills in the Tsezotene Formation, which are dated at 770 Ma and from dykes of similar composition (Armstrong et al., 1982). There is no geological evidence either for translations greater than about 20 km or of large rotations relative to the craton.

Detailed studies of the first two units (Park and Aitken, 1986a) show that four groups of magnetizations are present. Group 1 magnetizations (TFA and KA) are from the Tsezotene Formation and Katherine Group and were probably acquired as primary magnetizations at the time the rocks were deposited. Group 2 magnetizations represent primary or near primary magnetization of the Little Dal Group (DA), the Tsezotene sill (TA, TB) and overprints of

PALEOMAGNETISM FROM LATE PRECAMBRIAN ROCKS

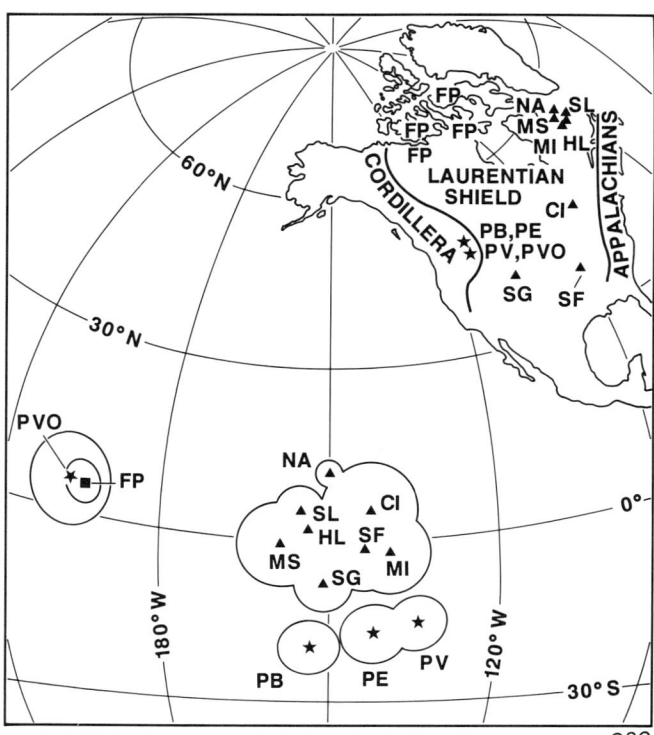

Figure 3.23. Mid-Proterozoic paleopoles from the Rocky Mountains of Alberta and Montana compared with Elsonian paleopoles from the Laurentian Shield. Paleopoles from the Rockies are as follows: PE = Purcell (Belt) Supergroup, Alberta; PB = Purcell (Belt) Supergroup, Montana; PV = Purcell (Belt) Supergroup, Spokane Formation, Montana, and its overprint PVO. Pole FP from Franklin diabase is added for comparison (see text). Paleopoles from the Laurentian Shield are, CI = Croker Island Complex (Palmer, 1969) 1475 Ma Rb-Sr isochron; HL = Harp Lake anorthosite (Irving et al., 1977) 1450 Ma U-Pb zircon; MI = Michikamau anorthosite (Emslie et al., 1976) 1460 Ma U-Pb zircon; MS = Mistastin pluton (Fahrig and Jones, 1976) 1346 Ma Rb-Sr whole-rock isochron; NA = Nain anorthosite (Murthy, 1978) 1418 Ma Rb-Sr mineral age; SF = St. Francois rocks (Hayes and Scharon, 1966) 1400 Ma Rb-Sr; SG = Sherman granite (Eggler and Larson, 1968) 1410 Ma Rb-Sr isochron; SL = Seal Lake Group redbeds (Roy and Fahrig, 1973) 1350 Ma Rb-Sr isochron from associated volcanics.

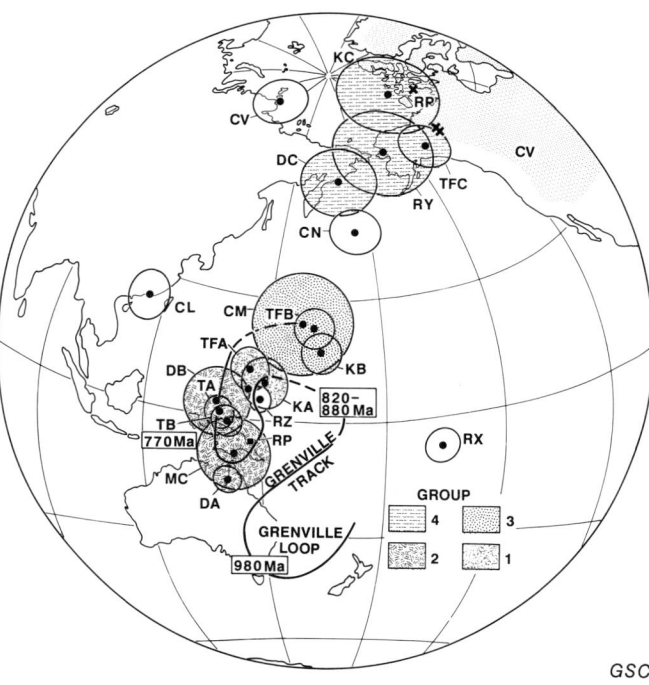

Figure 3.24. Paleopoles from Precambrian rocks of the Mackenzie Mountains. TA and TB = Tsezotene diabase sills; DA, DB and DC = Little Dal Group; MC = Mudcrack Formation; CL, CM and CN = Coates Lake Group lavas; RX, RY, RZ = Rapitan Group; RP = Reynolds Point Formation. The second letters refer to different magnetizations isolated from the same rock-unit, as detailed in Table 3.2. The Crowsnest paleopole (CV) of Cretaceous age is added for comparison. The four groups of paleopoles are detailed in the text.

comparable age (DB). Group 3 magnetizations (CM, TFB, KB) are overprints in several rock units that may have been acquired during rifting and volcanism associated with deposition of the Coates Lake Group. Park and Aitken (1986a) suggested that these three groups of paleopoles form part of a polar wander loop. Group 4 paleopoles (DC, CN, KC, TFC) are derived from much younger post-folding magnetizations. They agree generally with Cretaceous and Early Tertiary cratonic paleopoles.

Some paleopoles from the Mackenzie arc do not conform to this general scheme. Morris and Aitken (1982) obtained a paleopole from the Coates Lake lavas (CL) that falls to the east of the Mackenzie Loop. Morris (1977) obtained a paleopole (RZ) from the younger Rapitan Group that is close to those that Park and Aitken (1986a) regarded as the oldest of the Mackenzie Mountain Supergroup. The Rapitan also yielded a paleopole (RY) in group 4, and a third, (RX), far to the east (Fig. 3.24).

The origin of these apparently aberrant paleopoles is uncertain. Perhaps they reflect local rotations of thrust sheets, although no supporting evidence was found (Park and Aitken, 1986a,b). Indeed most of the paleopoles from the Mackenzie Mountains fall in an area elongated in the wrong sense to have been caused by relative rotations. If rotations had occurred, the paleopoles would have been spread in a west-northwest - east-southeast direction, not in the north-northeast - south-southwest direction observed (Fig. 3.24). Park and Aitken (1986a) maintained that not only have internal rotations been negligible but also there has been no rotation of the arc as a whole relative to the craton, based on the fact that the polar loop from the Mackenzie arc can be linked sequentially with the (almost certainly) older Grenville Loop. The paleopole from the Reynolds Point Formation of Victoria Island (Palmer et al., 1983) falls in the loop as expected from its probable age of 800-750 Ma based on correlation with the lower Coates Lake Group (RP, Fig. 3.24). Both results have been obtained from rocks on the craton. The Grenville Track (the younger, western limb of the Grenville Loop) has been dated in the range 980-820 Ma (Berger et al., 1979) and 980-880 Ma (Dallmeyer and Sutter, 1980). The Mackenzie Loop may simply be a segment of the Grenville Track which has been rotated about 20° clockwise, perhaps caused by wholesale rotation of the thrust sheets during emplacement. The Mackenzie paleopoles seem to be distinctly younger, however, so there is little foundation for such speculation.

In Figure 3.25 paleopoles derived from post-folding group 4 magnetizations observed from the Mackenzie Mountains can be compared with the paleopole (BR) derived from post-folding magnetizations observed from the Brooks Range of Alaska (Hillhouse and Grommé, 1983), and with the paleopole (EC) from syn-folding magnetizations in the fold and thrust belt of Idaho and Wyoming (Schwarz and van der Voo, 1984). With the exception of paleopole KC from the Mackenzie Mountains, all paleopoles cluster around the expected Cretaceous paleopole (compare Fig. 3.8). They have normal polarity, and hence probably acquired their magnetization during the Cretaceous Normal Superchron (118-83 Ma). These secondary magnetizations may have been acquired later in the latest Cretaceous or Tertiary, but if this were so at least a few reversed magnetizations would have been expected because the paleofield was then undergoing frequent reversals. If the magnetizations summarized in Figure 3.24 are truly mid-Cretaceous in age, then much of the deformation in the area sampled must have occurred within or prior to the Cretaceous Normal Superchron.

Cretaceous

The paleopole from the Crowsnest Formation (Fig. 3.24, CV Table 3.1), a volcanic unit of Albian age from the Rocky Mountains of Alberta, can be compared with the Cretaceous cratonic paleopole (Fig. 3.8). It indicates clockwise rotation and an original position almost 2000 km to the south of that at present. The Crowsnest Formation was probably deposited in a short interval of time, so the apparent aberrance could have been caused by incomplete averaging of the paleosecular variation. The rotation could have been real.

REFERENCES

Aitken, J.D., Cook, D.G., and Yorath, C.J.
1982: Upper Ramparts River (106G) and San Sault Rapids (106 H) map-areas, District of Mackenzie; Geological Survey of Canada, Memoir 388.

Armstrong, R.L., Eisbacher, G.H., and Evans, P.D.
1982: Age and stratigraphic-tectonic significance of Proterozoic diabase sheets, Mackenzie Mountains, northwestern Canada; Canadian Journal of Earth Sciences, v. 19, p. 316-323.

Armstrong, R.L., Monger, J.W.H., and Irving, E.
1985: Age of the magnetization of the Axelgold gabbro, north central British Columbia; Canadian Journal of Earth Sciences, v. 22, p. 1217-1222.

Beck, M.E.
1972: Paleomagnetism of Upper Triassic Diabase from southeastern Pennsylvania: Further results; Journal of Geophysical Research, v. 77, p. 5673-5687.
1980: Discordant paleomagnetic pole positions as evidence of regional shear in the western Cordillera of North America; American Journal of Science, v. 276, p. 694-712.

Beck, M.E. and Noson, L.
1972: Anomalous palaeolatitude in Cretaceous granitic rocks; Nature, v. 235, p. 11-13.

Beck, M.E., Burmester, R.F., Engebretson, D.C., and Schoonover, R.
1981a: Northward translation of Mesozoic batholiths western North America: paleomagnetic evidence and tectonic significance; Geofisica International, v. 20, p. 144-162.

Beck, M.E. Burmester, R.F., and Schoonover, R.
1981b: Paleomagnetism and tectonics of the Cretaceous Mt. Stuart Batholith of Washington: translation or tilt?; Earth and Planetary Science Letters, v. 56, p. 336-342.
1982: Tertiary paleomagnetism of the North Cascade Range, Wa.; Geophysical Research Letters, v. 9, p. 515-518.

Berger, G.W., York, D., and Dunlop, D.J.
1979: Calibration of Grenvillian paleopoles by $^{40}Ar/^{39}Ar$ dating; Nature, v. 277, p. 46-48.

Beske, S.J., Beck, M.E., and Noson, L.
1973: Paleomagnetism of the Miocene grotto and Snoqualmie batholiths, central Cascades, Washington; Journal of Geophysical Research, v. 78, p. 2601-2608.

Butler, R.F. and May, S.R.
1985: Implications of a revised Triassic-Jurassic APW path for western Cordilleran terrane displacements; EOS, v. 66, p. 862.

Byrne, T.
1979: Late Paleocene demise of the Kula-Pacific spreading center; Geology, v. 7, p. 341-344.

Carey, S.W.
1958: A tectonic approach to continental drift; in Continental Drift, a symposium; University of Tasmania, p. 177-355.

Carmichael, C.M. and Palmer, H.C.
1968: Paleomagnetism of the late Triassic, North Mountain basalt of Nova Scotia; Journal of Geophysical Research, v. 73, p. 2811-2822.

Chamberlain, V.E. and Lambert, R.St.J.
1985: Cordilleria, a newly defined Canadian microcontinent; Nature, v. 314, p. 707-713.

Channel, J.E.T., Catalano, R., and D'Argenio, B.
1980: Paleomagnetism and deformation of the Mesozoic continental margin in Sicily; Tectonophysics, v. 61, p. 391-407.

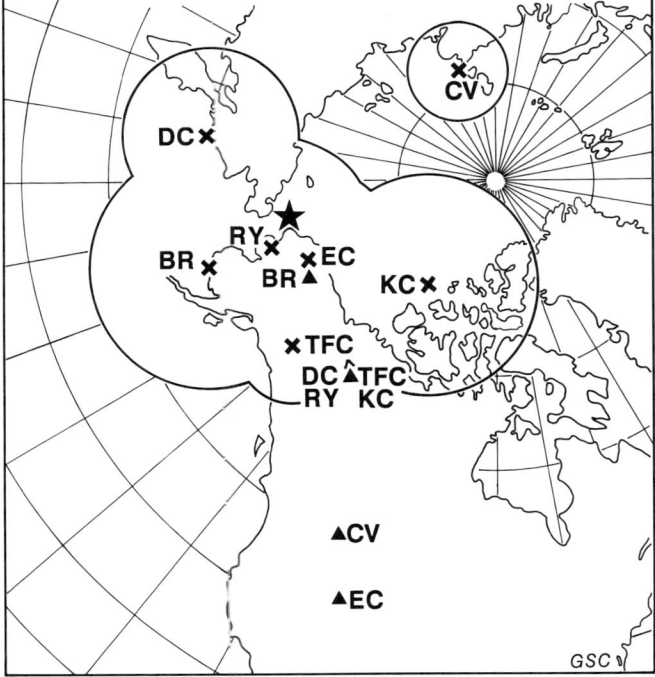

Figure 3.25. Paleopoles from primary and secondary magnetizations of Cretaceous or earliest Tertiary age observed in the Foreland Belt. They are as follows: CV = Crowsnest Volcanics, DC = Little Dal Group overprint; BR = Kanayut Formation, Brooks Range overprint; RY = Rapitan Group overprint; EC = syn-folding magnetization in Jura-Cretaceous sedimentary rocks; KC = Katherine Group overprint; TFC = Tsezotene Formation overprint; details in Table 3.2. The average paleopole for the mid-Cretaceous is shown by a star.

Collinson, D.W. and Runcorn, S.K.
1960: Polar wandering and continental drift: evidence from paleomagnetic observations in the United States; Geological Society of America Bulletin, v. 71, p. 915-958.

Cox, A.
1957: Remanent magnetism of Lower to Middle Eocene basalt flows from Oregon; Nature, v. 179, p. 685-686.

Dallmeyer, R.P. and Sutter, J.F.
1980: Acquisition chronology of remanent magnetization along the "Grenville polar path": evidence from 40Ar/39Ar ages of hornblende and biotite from the Whitestone diorite, Ontario; Journal of Geophysical Research, v. 85, p. 3177-3186.

Dalrymple, G.B., Grommé, C.S., and White, R.W.
1975: Potassium-argon age and paleomagnetism of diabase dikes in Liberia: Initiation of central Atlantic rifting; Geological Society of America Bulletin, v. 86, p. 399-411.

de Boer, J. and Snider, F.G.
1979: Magnetic and chemical variations of Mesozoic diabase dikes from eastern North America: Evidence for a hotspot in the Carolinas; Geological Society of America Bulletin, v. 90, p. 185-198.

Demarest, H.H.
1983: Error analysis for the determination of tectonic rotation from paleomagnetic data; Journal of Geophysical Research, v. 88, p. 4321-4328.

Du Bois, P.M.
1959: Late Tertiary geomagnetic field in northwestern Canada; Nature, v. 183, p. 1617-1618.

Eggler, D.H. and Larson, E.D.
1968: Paleomagnetic study of late Precambrian rocks of the Front Range, Colorado-Wyoming; Geophysical Journal of the Royal Astronomical Society, v. 14, p. 497-504.

Elston, D.P. and Bressler, S.L.
1980: Paleomagnetic poles and polarity zonations from the Middle Proterozoic Belt Supergroup, Montana and Idaho; Journal of Geophysical Research, v. 85, p. 339-355.

Emslie, R.F., Irving, E., and Park, J.K.
1976: Further paleomagnetic results from the Michikamau Intrusion, Labrador; Canadian Journal of Earth Sciences, v. 13, p. 1052-1057.

Engebretson, D.C.
1982: Relative motions between oceanic and continental plates in the Pacific Basin; Ph.D. thesis, Stanford University, Stanford, California, 211 p.

Engebretson, D.C., Gordon, R.G., and Cox, A.
1985: Relative motions between oceanic and continental plates in the Pacific Basin; Geological Society of America, Special Paper 206, p. 1-58.

Evans, M.E., Bingham, D.K., and McMurray, E.W.
1975: New paleomagnetic results from the Upper Belt-Purcell Supergroup of Alberta; Canadian Journal of Earth Sciences, v. 12, p. 52-61.

Fahrig, W.F. and Jones, D.L.
1976: The paleomagnetism of the Helikian Mistastin pluton, Labrador, Canada; Canadian Journal of Earth Sciences, v. 13, p. 832-837.

Foster, J. and Symons, D.T.A.
1979: Defining a paleomagnetic polarity pattern in the Monteregian intrusives; Canadian Journal of Earth Sciences, v. 16, p. 1716-1725.

Fox, K.F. and Beck, M.E.
1985: Paleomagnetic results from Eocene volcanic rocks from northeastern Washington and the Tertiary tectonics of the Pacific Northwest; Tectonics, v. 4, p. 323-341.

Fuller, M.
1985: Paleomagnetism in an accretionary margin, northern Phillipines; Journal of Geodynamics, v. 2, p. 141-158.

Globerman, B.R., Beck, M.E., and Duncan, R.A.
1982: Paleomagnetism and tectonic significance of Eocene basalts from the Black Hills, Washington Coast Range; Geological Society of America Bulletin, v. 93, p. 1151-1159.

Gordon, R.G., Cox, A., and O'Hare, S.
1984: Paleomagnetic Euler poles and the apparent polar wander and absolute motion of North America since the Carboniferous; Tectonics, v. 3, no. 5, p. 499-537.

Granirer, J.L., Beck, M.E., and Burmester, R.F.
1985: Paleomagnetic evidence for northward transport of the Methow-Pasayten Belt, north-central Washington; EOS Transactions, American Geophysical Union, v. 66, p. 863.

Grubbs, K.L. and van der Voo, R.
1976: Structural deformation of the Idaho-Wyoming overthrust belt (USA), as determined by Triassic paleomagnetism; Tectonophysics, v. 33, p. 321-336.

Hamilton, T.S. and Evans, M.
1983: A magnetostratigraphic and secular variation study of Level Mountain, northern British Columbia; Geophysical Journal of the Royal Astronomical Society, v. 73, p. 39-49.

Harrison, C.G.A., Jarrard, R.D., Vacquier, V., and Larson, R.L.
1975: Paleomagnetism of Cretaceous Pacific seamounts; Geophysical Journal of the Royal Astronomical Society, v. 42, p. 859-882.

Hayes, W.W. and Scharon, L.
1966: A paleomagnetic investigation of some of the Precambrian igneous rocks of southeast Missouri; Journal of Geophysical Research, v. 71, p. 553-560.

Hicken, A. and Irving, E.
1977: Tectonic rotation in western Canada; Nature, v. 268, p. 219-220.

Hillhouse, J.W.
1977: Paleomagnetism of the Triassic Nikolai greenstone, McCarthy triangle; Canadian Journal of Earth Sciences, v. 14, p. 2578-2592.

Hillhouse, J.W. and Grommé, C.S.
1983: Paleomagnetic studies and the hypothetical rotation of Arctic Alaska; Journal of the Alaska Geological Society, v. 2, p. 27-39.
1984: Northward displacement and accretion of Wrangellia: new paleomagnetic evidence from Alaska; Journal of Geophysical Research, v. 89, p. 4461-4477.

Hillhouse, J.W., Grommé, C.S., and Vallier, T.L.
1982: Paleomagnetism and Mesozoic tectonics of the Seven Devils volcanic arc in northeastern Oregon; Journal of Geophysical Research, v. 87, p. 3777-3749.

Hodych, J.P. and Hayatsu, A.
1980: K-Ar isochron age and paleomagnetism of diabase along the trans-Avalon aeromagnetic lineament - evidence of Late Triassic rifting in Newfoundland; Canadian Journal of Earth Sciences, v. 17, p. 491-499.

Hospers, J.
1955: Rock magnetism and polar wandering; Journal of Geology, v. 63, p. 59-74.

Irving, E.
1956: Paleomagnetic and palaeoclimatological aspects of polar wandering; Geofisica pura et applicata, v. 33, p. 23-41.
1964: Paleomagnetism and its application to geological and geophysical problems; John Wiley and Sons, New York, 399 p.
1985: Whence British Columbia?; Nature, v. 314, p. 673-674.

Irving, E. and Hastie, J.
1975: Catalogue of paleomagnetic directions and poles; Earth Physics Branch, Geomagnetic Series no. 3, p. 1-42.

Irving, E. and Irving, G.A.
1982: Apparent polar wander paths Carboniferous through Cenozoic and the assembly of Gondwana; Geophysical Surveys, v. 5, p. 141-188.

Irving, E. and Monger, J.W.H.
1987: Preliminary paleomagnetic results from the Permian Asitka Group, British Columbia; Canadian Journal of Earth Sciences, v. 24, p. 1490-1497.

Irving, E. and Yole, R.W.
1972: Paleomagnetism and kinematic history of mafic and ultramafic rocks in fold mountain belts; Publications of the Earth Physics Branch, Department of Energy, Mines and Resources, Ottawa, vol. 42, p. 87-95.
1987: Tectonic rotations and translations in western Canada; new evidence from Jurassic rocks of Vancouver Island; Geophysical Journal of the Royal Astronomical Society, v. 91, p. 1025-1048.

Irving, E., Emslie, R.F., and Park, J.K.
1977: Paleomagnetism of the Harp Lake complex and associated rocks; Canadian Journal of Earth Sciences, v. 14, p. 1187-1201.

Irving, E., Monger, J.W.H., and Yole, R.W.
1980: New paleomagnetic evidence for displaced terranes in B.C.; Geological Association of Canada, Special Paper 20, p. 441-456.

Irving, E., Woodsworth, G.J., Wynne, P.J., and Morrison, A.
1985: Paleomagnetic evidence for displacement from the south of the Coast Plutonic Complex, British Columbia; Canadian Journal of Earth Sciences, v. 22, p. 584-598.

Irving, E., Wheadon, P., and Horel G.
1990: Catalogue of paleomagnetic directions and paleopoles, sixth Issue, Geological Burrey of Canada, Open File 2247.

Irving, E., and Wynne, P. J.
1990: Poleomagnetic evidence bearing on the evolution of the Canadian Cordillera; Philosophical Transactions of the Royal Society, 331A, p. 487-509.

Irving, E., Wynne, P.J., Evans, M.E., and Gough, W.
1986: Anomalous paleomagnetism of the Crowsnest Formation of the Rocky Mountains; Canadian Journal of Earth Sciences, v. 23, p. 591-598.

Kent, D.V. and Gradstein, F.M.
1985: A Cretaceous and Jurassic geochronology; Geological Society of America Bulletin, v. 96, p. 1419-1427.

Lapointe, P.L.
1979: Paleomagnetism of the Notre Dame Bay lamprophyre dikes, Newfoundland, and the opening of the North Atlantic Ocean; Canadian Journal of Earth Sciences, v. 16, p. 1823-1831.

Larochelle, A.
1962: Paleomagnetism of the Monteregian Hills, southeastern Quebec; Geological Survey of Canada, Bulletin 79, 44 p.
1967: Preliminary data on the paleomagnetism of the North Mountain Basalt; Geological Survey of Canada, Paper 67-39, p. 7-12.

Larochelle, A. and Black, R.F.
1963: An application of palaeomagnetism in estimating the age of rocks; Nature, v. 198, p. 1260-1262.

Larochelle, A. and Currie, K.L.
1967: Paleomagnetic study of igneous rocks from the Manicouagan structure, Quebec; Journal of Geophysical Research, v. 72, p. 4163-4169.

Larochelle, A. and Wanless, R.K.
1966: The paleomagnetism of a Triassic diabase dike in Nova Scotia; Journal of Geophysical Research, v. 71, p. 4949-4953.

LePichon, X., Sibuet, J.C., and Francheteau, J.
1977: The fit of the continents around the North Atlantic Ocean; Tectonophysics v. 38, p. 169-209.

Mankinen, E.A., Irwin, W.P., and Grommé, C.S.
1984: Implications of paleomagnetism for the tectonic history of the eastern Klamath and related terranes in California and Oregon; Society of Economic and Petroleum Mineralogists, Pacific Section Field-Trip Guide to the Hornibrook Formation, 1984, T.H. Nilsen (ed.), v. 42, p. 221-229

Monger, J.W.H. and Irving, E.
1980: Northward displacement of north-central British Columbia; Nature, v. 285, no. 5763, p. 289-294.

Morris, W.A.
1977: Paleolatitude of glaciogenic upper Precambrian Rapitan Group and the use of tillites as chronostratigraphic marker horizons; Geology, v. 5, p. 85-88.

Morris, W.A. and Aitken, J.D.
1982: Paleomagnetism of the Little Dal lavas, Mackenzie Mountains, Northwest Territories, Canada; Canadian Journal of Earth Sciences, v. 19, no. 10, p. 2020-2027.

Murthy, G.S.
1978: Paleomagnetic results from the Nain anorthosite and their tectonic implications; Canadian Journal of Earth Sciences, v. 15, p. 516-525.

Norris, D.K. and Black, R.F.
1962: Paleomagnetism and differential rotation in the Lewis Thrust Plate; Journal Alberta Society of Petroleum Geologists, v. 10, p. 13-21.

Obradovich, J.D., Zattman, R.E., and Peterman, Z.E.
1984: Update of the geochronology of the Belt Supergroup; United States Geological Survey, Special Publication 90, p. 82-84.

Packer, D.R. and Stone, D.B.
1974: Paleomagnetism of Jurassic rocks from southern Alaska, and their tectonic implications; Canadian Journal of Earth Sciences, v. 11, p. 976-997.

Palmer, H.C.
1969: The paleomagnetism of the Croker Island Complex, Ontario, Canada; Canadian Journal of Earth Sciences, v. 6, p. 213-218.

Palmer, H.C., Baragar, W.R.A., Fortier, M., and Forster, J.H.
1983: Paleomagnetism of Late Proterozoic rocks, Victoria Island, Northwest Territories, Canada; Canadian Journal of Earth Sciences, v. 20, p. 1456-1469.

Panuska, B.C.
1985: Paleomagnetic evidence for post-Cretaceous accretion of Wrangellia; Geology, v. 13, p. 880-883.

Park, J.K.
1981a: Paleomagnetism of the Late Proterozoic sills in the Tsezotene Formations, Mackenzie Mountains, Northwest Territories, Canada; Canadian Journal of Earth Sciences, v. 18, no. 10, p. 1572-1580.
1981b: Analysis of the multicomponent magnetism of the Little Dal Group, Mackenzie Mountains, Northwest Territories, Canada; Journal of Geophysical Research, v. 86, no. B6, p. 5134-5146.
1984: Paleomagnetism of the Mudcracked Formation of the Precambrian Little Dal Group, Mackenzie Mountains, Northwest Territories, Canada; Canadian Journal of Earth Sciences, v. 21, p. 371-375.

Park, J.K. and Aitken, J.D.
1986a: Paleomagnetism of the Katherine Group in the MacKenzie Mountains: implications for post-Grenville (Hadrynian) apparent polar wander; Canadian Journal of Earth Sciences, v. 23, no. 3, p. 308-323.
1986b: Paleomagnetism of the late Proterozoic Tsezotene Formation of the Mackenzie Mountains, northwestern Canada; Journal of Geophysical Research, v. 91, p. 4955-4970.

Prasad, J.N.
1981: Paleomagnetism of the Mesozoic lamprophyre dikes in north-central Newfoundland; M.Sc. thesis, Memorial University of Newfoundland.

Rabinowitz, P.D. and LaBrecque, J.
1979: The Mesozoic South Atlantic Ocean and evolution of its continental margins; Journal of Geophysical Research, v. 84, p. 5973-6002.

Rees, C.J., Irving, E., and Brown, R.L.
1985: Secondary magnetization of Triassic-Jurassic volcaniclastic rocks of Quesnel terrane; Geophysical Research Letters, v. 12, p. 498-501.

Reeve, S.C.
1975: Paleomagnetic studies of sedimentary rocks of Cambrian and Triassic age; Ph.D. thesis, University of Texas, 426 p.

Reeve, S.C. and Helsley, C.E.
1972: Magnetic reversal sequence in the upper portion of the Chinle Formation, Montoya, New Mexico; Geological Society of America Bulletin, v. 83, p. 3795-3812.

Robertson, W.A.
1967: Manicougan, P.Q., paleomagnetic results; Canadian Journal of Earth Sciences, v. 4, p. 641-649.

Roy, J.L. and Fahrig, W.F.
1973: The paleomagnetism of Seal and Croteau rocks from the Grenville Front, Labrador: Polar wandering and tectonic implications; Canadian Journal of Earth Sciences, v. 10, p. 1279-1301.

Runcorn, S.K.
1956: Paleomagnetic comparisons between Europe and North America; Proceedings of the Geological Association of Canada, v. 8, p. 77-85.

Scharon, L. and Hsu, I.
1969: Paleomagnetic investigation of some Arkansas alkalic igneous rocks; Journal of Geophysical Research, v. 74, p. 2774-2779.

Schwartz, E.J., Muller, J.E., and Clark, K.R.
1980: Paleomagnetism of the Karmutsen basalts from southeast Vancouver Island; Canadian Journal of Earth Sciences, v. 17, p. 389-399.

Schwarz, S.Y. and van der Voo, R.
1984: Paleomagnetic study of thrust sheet rotation during foreland impingement in the Wyoming-Idaho thrust belt; Journal of Geophysical Research, v. 89, p. 10077-10086.

Shive, P.N. and Frerichs, W.E.
1974: Paleomagnetism of Niobrara Formation in Wyoming, Colorado, and Kansas; Journal of Geophysical Research, v. 79, p. 3001-3007.

Sichler, B., Olivet, J.L., Auzende, J.M., and Jonquet, H.
1980: Mobility of Morocco; Canadian Journal of Earth Sciences, v. 17, p. 1546-1558.

Simpson, R.W. and Cox, A.
1977: Paleomagnetic evidence for tectonic rotation of the Oregon Coast Range; Geology, v. 5, p. 585-589.

Smith, T.E. and Noltimier, H.C.
1979: Paleomagnetism of the Newark trend igneous rocks of the north central Appalachians and the opening of the Atlantic Ocean; American Journal of Science, v. 279, p. 778-807.

Souther, J.G. and Symons, D.T.A.
1973: Stratigraphy and paleomagnetism of Mount Edziza volcanic complex, northwestern British Columbia; Geological Survey of Canada, Paper 73-32.

Steiner, M.B.
1984: Is the Colorado Plateau rotated?; EOS (abstract), v. 65, p. 864.

Steiner, M.B. and Helsley, C.E.
1974: Magnetic polarity sequence of the Upper Triassic Kayenta Formation; Geology, v. 2, p. 191-194.

Stone, D.B.
1977: Proceedings of Sixth Annual Symposium; Geological Society of Alaska, p. 1.

Stone, D.B., Panuska, B.C., and Packer, D.R.
1982: Paleolatitude versus time for southern Alaska; Journal of Geophysical Research, v. 87, p. 3697-3707.

Symons, D.T.A.
1968: Geological implication of paleomagnetic studies in the Bella Coola and Laredo Sound map areas, British Columbia; Geological Survey of Canada, Paper 68-72, 15 p.
1969: Paleomagnetism of the Late Miocene plateau basalts in the Cariboo Region of British Columbia; Geological Survey of Canada, Paper 69-43, 16 p.
1971a: Paleomagnetism of the Triassic Guichon Batholith and rotation in the Interior Plateau, British Columbia; Canadian Journal of Earth Sciences, v. 8, no. 11, p. 1388-1396.
1971b: Paleomagnetic results on some minor Tertiary igneous bodies, Vancouver Island, British Columbia; Geological Survey of Canada, Paper 71-24, p. 1-8.
1973a: Paleomagnetic results from the Jurassic Topley intrusions near Endako, British Columbia; Canadian Journal of Earth Sciences, v. 10, p. 1099-1108.
1973b: Unit correlations and tectonic rotation from paleomagnetism of the Triassic Copper Mountain Intrusions, British Columbia; Geological Survey of Canada, Paper 73-19, p. 11-28.
1973c: Paleomagnetic zones in the Oligocene East Sooke Gabbro, Vancouver Island, British Columbia; Journal of Geophysical Research, v. 78, p. 5100-5109.
1973d: Paleomagnetic results from the Tertiary Mount Barr and Hope Plutonic Complexes, British Columbia; Geological Survey of Canada, Paper 73-19, p. 1-10.
1973e: Concordant Cretaceous paleolatitudes from felsic plutons in the Canadian Cordillera; Nature, v. 241, p. 59-61.
1977a: Geotectonics of Cretaceous and Eocene plutons in British Columbia, a paleomagnetic fold test; Canadian Journal of Earth Sciences, v. 14, p. 1246-1262.
1977b: Paleomagnetism of Mesozoic plutons in the westernmost Coast Complex of British Columbia; Canadian Journal of Earth Sciences, v. 14, p. 2127-2139.
1983a: Further paleomagnetic results from the Jurassic Topley intrusives in the Stikinia subterrane of British Columbia; Geophysical Research Letters, v. 10, p. 1065-1068.
1983b: New paleomagnetic data for the Triassic Guichon batholith of south-central British Columbia and their bearing on Terrane I tectonics; Canadian Journal of Earth Sciences, v. 10, p. 1340-1344.
1985a: Paleomagnetism of the Triassic Nicola volcanics and geotectonics of the Quesnellia subterrane of Terrane I, British Columbia; Journal of Geodynamics, v. 2, p. 229-244.
1985b: Paleomagnetism of the Westcoast Complex and the geotectonics of the Vancouver Island segment of the Wrangellian subterrane; Journal of Geodynamics, v. 2, p. 211-228.

Symons, D.T.A. and Litalien, C.R.
1984: Paleomagnetism of the Lower Jurassic Copper Mountain intrusions and geotectonics of Terrane I, British Columbia; Geophysical Research Letters, v. 11, p. 685-688.

Symons, D.T.A., Chodla, G.R., and Donner, K.
1986: Paleomagnetism of Mid-Cretaceous granitic plutons in the Coast Plutonic Complex, B.C.; M.A., C.G.U. Joint Annual Meeting, Abstract, from Program with Abstracts, p. 133.

Taylor, D.G., Callomon, J.H., Hall, R., Smith, P.L., Tipper, H.W., and Westermann, G.E.G.
1984: Jurassic ammonite biogeography of western North America, the tectonic implications-in Jurassic-Lower Cretaceous Biochronology and Biogeography of North America, G.E.G. Westermann, (ed.), Geological Association of Canada, Special Paper 27, p. 121-124.

Tipper, H.W.
1981: Offset of an upper Pliensbachian geographic zonation in the North American Cordillera by transcurrent faulting; Canadian Journal of Earth Sciences, v. 18, p. 1788-1792.

Valencio, D.A.
1972: Relative upper Palaeozoic - Mesozoic positions of South America and Africa from paleomagnetic data; Conference on Solid Earth Problems, Buenos Aires, p. 303-317.

Valencio, D.A., Mendía, J.E., and Vilas, J.F.
1975: Paleomagnetism and K-Ar ages of Triassic igneous rocks from the Ischigualastos-Ischichuca Basin and Puesto Viejo Formation, Argentina; Earth and Planetary Science Letters, v. 26, p. 319-330.

Vittorelli, I. and van der Voo, R.
1977: Late Hadrynian and Helikian pole positions from the Spokane Formation, Montana; Canadian Journal of Earth Sciences, v. 14, p. 67-73.

Wells, R. and Coe, R.S.
1985: Paleomagnetism and geology of Eocene volcanic rocks of southwest Washington, implications for mechanism of tectonic rotations; Journal of Geophysical Research, v. 90, p. 1925-1947.

Woods, M.T. and Davies, G.F.
1982: Late Cretaceous genesis of the Kula plate; Earth and Planetary Science Letters, v. 58, p. 161-166.

Yole, R.W. and Irving, E.
1980: Displacement of Vancouver Island: paleomagnetic evidence from the Karmutsen Formation; Canadian Journal of Earth Sciences, v. 17, no. 9, p. 1210-1228.

Authors' address

E. Irving
P.J. Wynne
Pacific Geoscience Centre
9860 West Saanich Road
P.O. Box 6000
Sidney, British Columbia
V8L 4B2

Chapter 4

PRECAMBRIAN BASEMENT ROCKS OF THE CANADIAN CORDILLERA

Summary
Introduction
Basement rocks of the Omineca Belt
    Yukon and Northwest Territories
    Deserters Range
    Sifton and Cormier ranges
    Malton Gneiss Complex
    Monashee Complex
    Other areas of granitoid gneissic rock
Relationship of Cordilleran basement rocks to the North American craton
Age of Proterozoic rifting events
References

Chapter 4

PRECAMBRIAN BASEMENT ROCKS OF THE CANADIAN CORDILLERA

Randall R. Parrish

SUMMARY

The metamorphic and plutonic rocks upon which the miogeocline was developed are considered to be the basement rocks of the Canadian Cordillera. They are recognized by an unconformable relationship with Proterozoic or younger bedded rocks and/or on the basis of reliable geochronological data. According to these criteria, basement rocks in the Cordillera are exposed only in fold and thrust nappes of the Omineca Belt, and, at one locality, unconformably beneath the Windermere Supergroup in the Foreland Belt. Basement rocks sampled in diatremes and drill holes provide additional geochronological data.

The principal basement exposures comprise mainly granitic paragneiss and orthogneiss, with minor amphibolite and metasedimentary rocks. U-Pb zircon dating provides the most reliable age assignment. These age determinations suggest that most Precambrian crystalline rocks of the Cordillera fall into three groups, 1.85-2.1 Ga, 1.1-1.2 Ga, and 0.7-0.8 Ga; the first and third of these categories are probably represented throughout the full length of the Canadian Cordillera. Relationships with the structurally or stratigraphically associated Windermere Supergroup suggest but do not prove that all of these rocks are part of the North American craton. The earlier Proterozoic ages (1.85-2.1 Ga) probably represent those of one or more Proterozoic tectonic provinces of the western Canadian Shield. Granitic rocks from a diatreme in the Mackenzie Mountains (1.1-1.2 Ga) may be associated with a Middle Proterozoic orogenic event in the northern Cordillera. The latest Proterozoic rocks may be related to magmatism caused by pre-Windermere rifting, eventually resulting in the development of the passive continental margin.

INTRODUCTION

Basement rocks of the Cordillera are considered to be the crystalline metamorphic and plutonic rocks forming the foundation upon which the Late Proterozoic-Paleozoic miogeocline developed. Late Middle and Upper Proterozoic miogeoclinal rocks include the Purcell, Wernecke, Mackenzie Mountains and Windermere supergroups, and perhaps the mantling metasedimentary strata of the Monashee Complex. All of these rocks occur in the Omineca and western Foreland belts (Fig. 4.1).

To the east, well and seismic data indicate basement rocks of the Canadian Shield underlie platform Phanerozoic rocks of the Interior Platform and Foreland Belt and continue westward to at least the Rocky Mountain Trench (Price and Mountjoy, 1970). The magnetic signature of the Shield rocks beneath the Alberta Basin corroborates this observation.

Precambrian crystalline rocks have not been identified in the Intermontane, Coast or Insular belts (Fig. 4.1). Only in the Omineca Belt and westernmost Foreland Belt has Precambrian crystalline basement been documented geochronologically (Fig. 4.1). A line marking the western limit of initial $^{87}Sr/^{86}Sr$ ratios of greater than 0.706 coincides roughly with the western margin of the Omineca Belt and is assumed to reflect the westernmost limit of substantial cratonal North American basement.

Much controversy has surrounded the gneiss complexes. Many, particularly in southern British Columbia, have been interpreted variously as strongly metamorphosed and mobilized Upper Precambrian sediments (Reesor, 1970), or as crystalline rock of earlier Precambrian age (Jones, 1959; Ross, 1970; Campbell, 1968; Wheeler, 1970). The modest amount of published U-Pb and Rb-Sr geochronology demonstrates that these granitic gneisses include rocks of late Mesozoic-early Cenozoic (Parrish and Ryan, 1983; Parrish, 1984; Parkinson, 1985), Paleozoic (Tempelman-Kluit and Wanless, 1980; Mortensen, 1991; Okulitch, 1985), Late Precambrian (Evenchick et al., 1984; Chamberlain and Lambert, 1985a), and Proterozoic-Archean(?) ages (Wanless and Reesor, 1975; Duncan 1978; Chamberlain et al., 1979; Evenchick et al., 1984; Chamberlain and Lambert, 1985a add: McDonough and Parrish 1991; Armstrong et al. 1991; Murphy et al. 1991). Field studies indicate that the granitic gneisses occur in a variety of structural settings, including large folds, nappes, and wedge-like thrusts, generally bounded by fault, shear, and/or mylonite zones. Penetrative deformation is widespread, but variable in its intensity; the most deformed basement rocks are generally near zones of detachment from adjacent metasediments.

Most basement complexes of the Cordillera are granitic in composition. Amphibolites, other mafic rocks, and metasedimentary or metavolcanic rocks are comparatively minor. Ultramafic rocks have not been recognized.

Latest Cretaceous to Early Tertiary K-Ar and Rb-Sr dates on micas are characteristic in basement complexes of the Cordillera. These young dates are the result of cooling and closure of isotopic systems consequent upon

---

Parrish, R. R.
1991: Precambrian basement rocks of the Canadian Cordillera, Chapter 4 in Geology of the Cordilleran Orogen in Canada, H. Gabrielse and C. J. Yorath (ed.); Geological Survey of Canada, Geology of Canada, no. 4, p. 87-96 (also Geological Society of America, The Geology of North America, v. G-2).

CHAPTER 4

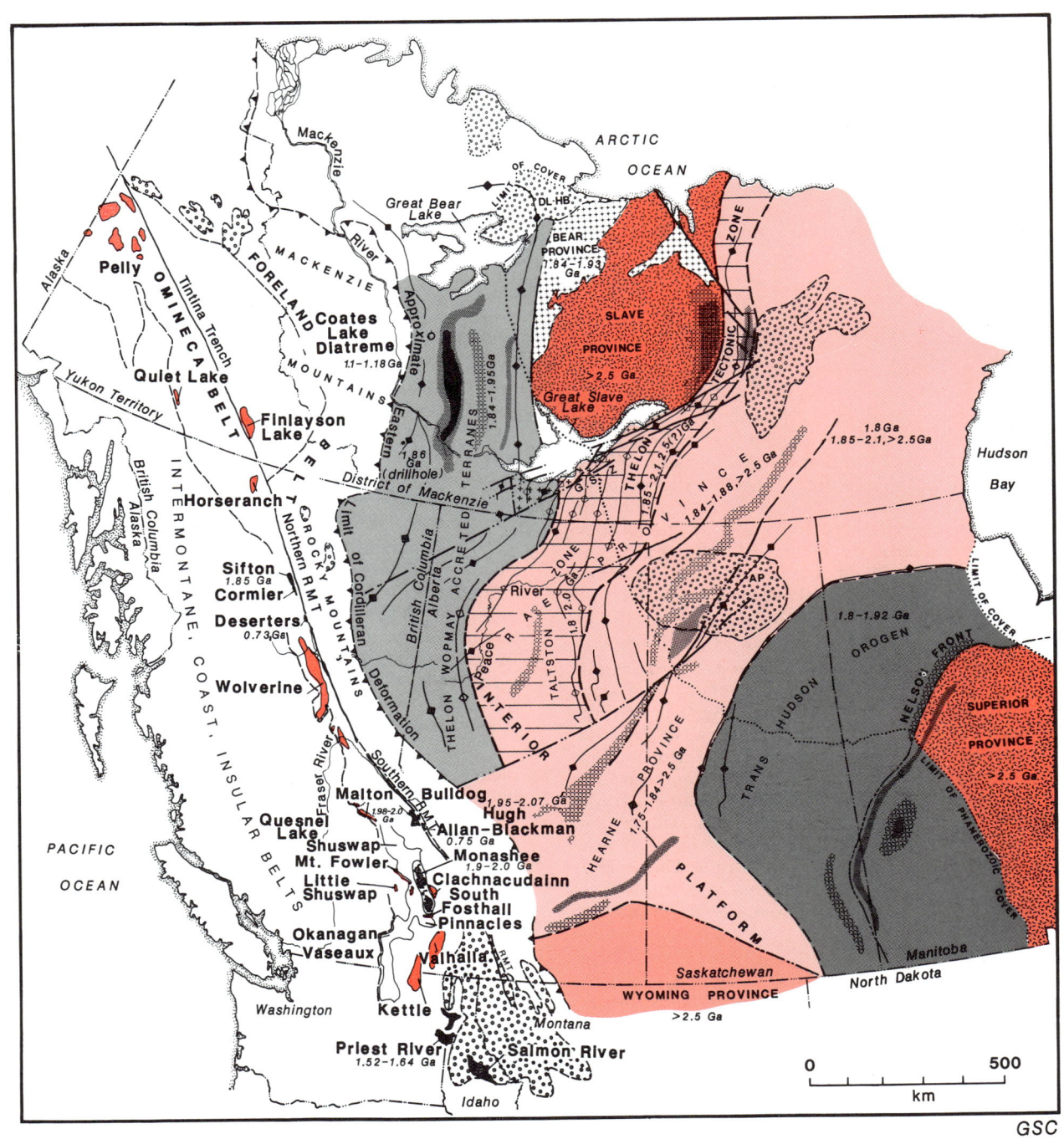

LEGEND

- Gneiss of Phanerozoic or uncertain age
- Cordilleran Precambrian crystalline basement showing name and U/Pb zircon age
- Middle Proterozoic, post-1.75 Ga sedimentary rocks
- Lower Proterozoic sedimentary cover on and adjacent to Slave Province
- Stable Archean craton
- Major faults; exposed (solid), inferred
- Tectonic belt or subprovince boundary
- G.S.Z. Great Slave Lake Shear Zone
- Linear aeromagnetic high
- Linear aeromagnetic low
- Positive Bouguer gravity anomaly
- Negative Bouguer gravity anomaly

uplift and erosion, or, in some cases, widespread tectonic denudation and quenching resulting from crustal extension (Parrish, 1979; Ewing, 1980; Price, 1979; Parrish et al. 1988). Dates from known or presumed Precambrian rocks which have been variably deformed and metamorphosed by younger events in the Cordillera are subject to a variety of interpretations. Some dates record original igneous ages, others the ages of a source terrane for paragneiss or metasediments, and some indicate variable resetting by subsequent events. The approach taken in this chapter is to rely on data from U-Pb analysis of zircons mainly in orthogneiss because of the closed system behaviour of the mineral, its high resistance to resetting, and its high level of precision. These factors make it more reliable for precise age determination than other isotope systems which rely on a multi-sample approach (Rb-Sr and Sm-Nd) and which are much more subject to disturbances caused by alteration and/or metamorphic events.

BASEMENT ROCKS OF THE OMINECA BELT

Yukon and Northwest Territories

There are no documented exposures of Precambrian crystalline rock in the northern Cordillera of the Yukon or District of Mackenzie.

Possible Precambrian basement granitic rocks are known from three widely separated localities. Basement gneiss from the subsurface east of the deformed belt near Fort Simpson yielded a U-Pb zircon age of 1.85 Ga (Villeneuve et al. 1991), suggesting the existence of basement rocks beneath the eastern MacKenzie Mountains with an age similar to that of the Great Bear magmatic zone in Wopmay Orogen to the east (Hoffman and Bowring, 1984). Granitic clasts occur in the Ordovician Coates Lake diatreme (Fig. 4.1), which intrudes the Upper Proterozoic Mackenzie Mountains Supergroup of the eastern Mackenzie Mountains. Zircon contained in these clasts has yielded

---

Figure 4.1. Major tectonic elements of both the Cordillera and the Canadian Shield which are relevant to the distribution of Precambrian basement. Within the Cordillera, the solid black pattern signifies exposures of gneissic rock of Precambrian age. Other, mainly younger gneisses, are labelled and shown in unpatterned areas within the Omineca Belt. Tectonic elements in the Canadian Shield are mainly from Stockwell (1982), Lewry and Sibbald (1980), and Hoffman and Bowring (1984); curvilinear aeromagnetic anomalies are from Dods et al. (1984) and prominent, curvilinear gravity features are modified from Earth Physics Branch (1980). Many of the prominent gravity and magnetic features in the Canadian Shield parallel prominent geological trends and tectonic boundaries and the subsurface extensions of such geological trends are inferred to correlate with these geophysical features. The ages of orogenic and pre-orogenic rocks within provinces of the Canadian Shield are shown and are from Stockwell (1982), Baldwin et al. (1985), Hoffman and Bowring (1984), Hildebrand et al. (1983), Lewry and Sibbald (1980), T.M. Gordon (pers. comm., 1984). The age assignments for the Precambrian gneissic rocks in northeast Washington and west-central Idaho are from Armstrong et al. (1987).

---

a U-Pb age of ca. 1150 Ma, and the zircons in the clasts contain inherited zircons at least 1750 Ma old (Jefferson and Parrish 1989). The granite that supplied these clasts, presumed to exist at depth, may be related to an orogenic event that affected rocks of the Wernecke Supergroup in the Wernecke Mountains (Eisbacher, 1978). Paleozoic granitic rocks of the Nisutlin Subterrane of the Kootenay Terrane contain inherited zircons that have yielded an Early Proterozoic age (Mortensen 1991), but the analytical results may represent a mixture of several ages. None of the sedimentary sequences north of 60 degrees are known to be exposed in unconformable or tectonic contact with Precambrian crystalline rocks.

Deserters Range

At latitude 56.8°N, just east of the Northern Rocky Mountain Trench (Fig. 4.1), the Deserters Range anticlinorium exposes leucocratic gneissic granite (Deserters Gneiss) with a U-Pb zircon age of 728 +8/-7 Ma nonconformably overlain by basal Windermere Supergroup strata of the Misinchinka Group (Evenchick, 1983b). Together, these rocks can be confidently linked to the North American craton, in contrast to other basement complexes of the Cordillera where original relationships have been destroyed by deformation. The granite, containing about equal amounts of quartz and potassium feldspar with somewhat less plagioclase, is strongly foliated. The foliation, steeply dipping and parallel with the Northern Rocky Mountain Trench, is cut by veinlets of ultracataclasite.

Sifton and Cormier ranges

Directly west of the Northern Rocky Mountain Trench, between latitudes 57°N and 58°N, deformed Precambrian orthogneiss (Tochieka Gneiss) is exposed in the narrow Sifton and Cormier ranges on either side of Finlay River (Fig. 4.1). In the former, granitic K-feldspar augen orthogneiss (Fig. 4.2) with a U-Pb zircon age of 1.85 ± 0.13 Ga, is structurally overlain and interleaved with amphibolite, quartzite, and chlorite-muscovite schist of uncertain age (Evenchick et al., 1984; Evenchick, 1983a). No unconformity has been recognized. In the more southerly Cormier Range, several attenuated repetitions of the gneiss unit are present as deformed mylonites. The orthogneiss which is strongly potassic, containing almost twice as much K_2O as Na_2O, is interpreted as the basal part of a thrust sheet transported southeastward over strata of the Windermere Supergroup during a time of dextral transcurrent displacement along the Northern Rocky Mountain Trench (Evenchick 1985).

Malton Gneiss Complex

A large area of well-exposed granitoid and layered quartzofeldspathic gneiss forms the Malton Gneiss Complex and straddles the southern Rocky Mountain Trench south of Valemount, British Columbia (Fig. 4.1; Campbell, 1968). The gneiss is commonly interpreted as part of the upper plate of the Purcell Thrust Fault. Reconnaissance geochronological studies indicate multiple ages of basement rocks, some dated as possibly Archean (Chamberlain et al., 1979), and others as young as 740 Ma (Parrish and Armstrong, 1983).

Figure 4.2. Augen gneiss (1.85 Ga) in Sifton Ranges, southeastern Cassiar Mountains. Photo by H. Gabrielse. GSC205235-E.

Present data indicate that mylonite zones and faults separate these basement gneisses from latest Proterozoic metamorphic rocks of the Windermere Supergroup (Campbell, 1968; Giovanella, 1967; Morrison, 1982). The gneisses are present in three areas (Fig. 4.1): the main Malton Gneiss west of the Southern Rocky Mountain Trench, the Hugh Allan Creek-Mt. Blackman gneiss on the east side of the Trench southeast of the Malton gneiss, and the Bulldog-Yellowjacket gneiss just east of the trench opposite the Malton Gneiss in the Bulldog Creek area. Significantly, in all of these areas a relatively pure but mylonitic quartzite discontinuously marks the contact between basement gneiss and structurally overlying strata of the Windermere Supergroup. The similarity of the granitic rocks in these three areas, and their close proximity to each other suggest that they represent a single, although disrupted basement block of complex internal history and variable age. Alternatively, the three gneissic bodies may be unrelated, as suggested by Chamberlain and Lambert (1983, 1985b), who proposed that the three gneissic blocks east of the Southern Rocky Mountain Trench are different in age, lithology, geochemistry and isotope composition from the block to the west (Chamberlain and Lambert, 1985b). The contact between the Windermere Supergroup and the structurally underlying Malton Gneiss is an important décollement of potentially large Jurassic-Paleogene displacement (Morrison, 1982). Several other detachments within the gneiss complex represent sheared, attenuated infolds of metamorphic rocks and may predate the main décollement. The Malton Gneiss Complex appears to be one of the most structurally and lithologically complex of all of the basement gneisses of the Canadian Cordillera. Intense late Mesozoic infolding and interleaving have obscured the original relationships between the various Precambrian basement rock units.

Morrison (1979, 1982) subdivided the main Malton Gneiss into three units, all of predominantly granitic composition. A layered biotite quartzofeldspathic paragneiss with minor quartzite is probably the oldest, and contains para-amphibolite as well as younger deformed amphibolitic dykes and intrusions. Leucocratic orthogneiss forms a major unit which, in places, apparently intruded the paragneiss; it is well lineated and relatively homogeneous. A second orthogneiss unit is a ca. 2.0 Ga (McDonough and Parrish, 1991) K-feldspar biotite augen gneiss with a very well-developed lineation and annealed mylonitic fabric. Similar lithologies have been described by Oke and Simony (1981) and McDonough and Simony (1984) in Mt. Blackman and Bulldog Creek areas. In Bulldog Creek area, these gneisses are 1.87 Ga old (McDonough and Parrish, 1991). Farther north Chamberlain et al. (1979) described alkaline granitic tonalitic and mafic gneiss of variable age and geochemistry, but the relationship of these gneisses to those farther south is uncertain.

Granitic rocks of the Hugh Allan Creek area are approximately 740 Ma old (McDonough and Parrish, 1991), similar to Rb-Sr ages from the northern Malton Range (Blenkinsop, 1972; Chamberlain et al., 1979). Archean rocks may exist in the Malton Gneiss Complex because imprecise Rb-Sr errorchrons as old as 3.2 Ga have been determined (Chamberlain et al., 1979). Rb-Sr work also suggests a complex history of multiple intrusion and isotopic homogenization, and possibly open-system Rb-Sr behaviour.

Monashee Complex

The Monashee Complex is an extensive area of high-grade Precambrian granitic gneiss and overlying stratified rocks of uncertain age in the Monashee Mountains, where it is bounded by the Monashee Décollement on the west (Fig. 4.1; Read and Brown, 1981) and the Columbia River Fault on the east. The area includes the Frenchman Cap (Høy and Brown, 1980), and Thor-Odin-Cranberry (Reesor and Moore, 1971; Duncan, 1984; see also Plate 1) "domal" complexes which expose Proterozoic crystalline basement. Neither the Pinnacles area, also part of the Monashee Complex as defined by Read and Brown (1981), nor the Valhalla Complex (Reesor, 1965; Parrish et al., 1985) exposes basement gneiss. The protolith of the core gneiss in the Monashee Complex was earlier thought to be either Hudsonian or Archean basement (Jones, 1959; Ross, 1970) or alternatively, highly metamorphosed and mobilized Upper Precambrian metasediments (Reesor, 1970). Not until recently, when U-Pb zircon dates of approximately 2.0 Ga were obtained, did their basement origin become clarified (Wanless and Reesor, 1975).

Reesor and Moore (1971), Fyles (1970), McMillan (1973), and Journeay (1986) described the basement gneisses in detail. Rock units resembling the Malton Gneiss Complex comprise layered quartzofeldspathic paragneiss (Fig. 4.3), dated as Early Proterozoic and, in the Thor-Odin area, augen orthogneiss of possible Archean to Early Proterozoic ages. Various leucogranitic gneisses, foliated sills, amphibolite, and other minor metamorphic rocks are components of the complex. The basement complex is, in places, unconformably overlain by pure quartzite and rare conglomerate which forms the base of a neritic succession of stratified rocks, including marble and diopsidic marble of uncertain age. In many cases, particularly in the Thor-Odin area, the contact is intensely sheared and repeated (Plate 1); there, the original relationship between basement and sedimentary cover is obscure.

Figure 4.3. Paragneiss (~2.0 Ga) of the Monashee Complex, core gneisses near Bourne Glacier, northwest of Revelstoke, British Columbia. The block is 1.7 m long. Photo by R.R. Parrish. GSC 204048-W.

Various interpretations of the structural origin of the domes of the Monashee Complex have been proposed. Reesor (1965, 1970) and Reesor and Moore (1971) suggested that the core rocks, because of their high metamorphic grade and granitoid composition, are diapirs. Duncan (1984) argued that the domes formed by cross-folding following earlier thrust and fold deformation. Brown and Read (1983) and Price (1985) interpreted the complex as a Mesozoic compressional feature with superimposed Eocene extensional strain. Brown et al. (1986) and Monger et al. (1985) interpreted part of the Monashee Complex as a large Jurassic-Cretaceous crustal-scale duplex, exposing the upper slabs of basement rock. Brown et al. (1991) have modified and corroborated this view.

Reconaissance U-Pb and Rb-Sr geochronology on the core gneisses of the Monashee Complex indicates ages of ca. 1.95-2.1 Ga and 1.85 Ga (Armstrong et al. 1991; Wanless and Reesor, 1975). Also, a nepheline syenite complex near Mount Copeland is about 740 Ma old (Parrish and Scammell, 1988).

A Proterozoic minimum age for foliated granitic gneiss of the Vaseaux Formation east of Okanagan Valley and south of Okanagan Lake is suggested by Sr and Nd isotope analyses (Armstrong et al. 1991). These rocks represent the westernmost exposure of basement or basement derived rocks in the southern Omineca Belt.

Other areas of granitoid gneissic rock

Precambrian crystalline basement has yet to be unambiguonsly identified in the gneissic rocks known to occur in several other structural and metamorphic culminations in the Omineca Belt (Fig. 4.1; Evenchick et al., 1984). The gneissic granitoid rocks of the Pelly Gneiss (included in Yukon Cataclastic Complex of Tempelman-Kluit (1979)), so far have yielded only Paleozoic ages. Granitic gneiss, locally flanking mid-Cretaceous granites in the Pelly Mountains, and in high-grade metamorphic rocks of the Horseranch Range (Gabrielse, 1963), and Wolverine Complex (Armstrong, 1949; Roots, 1954; Parrish, 1979) may locally include Precambrian basement. The age of granitoid rocks of the Kettle-Grand Forks Complex (Cheney, 1980; Preto, 1970; Orr, 1985) appears to be late Mesozois. Gneissic rocks of the Yukon are mostly Paleozoic (Mortensen, 1991; Tempelman-Kluit and Wanless, 1980) and those of the Valhalla Gneiss Complex of southeastern British Columbia are late Mesozoic and Paleogene (Parrish and Ryan, 1983; Parrish et al., 1985; Carr et al., 1987), considerably younger than surrounding metasediments.

RELATIONSHIP OF CORDILLERAN BASEMENT ROCKS TO THE NORTH AMERICAN CRATON

The age of the cratonic crystalline basement beneath Phanerozoic rocks of Alberta and the Northwest Territories is moderately well known. K-Ar ages on drill cuttings obtained from the subsurface basement generally have yielded Hudsonian dates (1.6-1.9 Ga; Burwash et al., 1962); however these are minimum ages. Recent U-Pb dating by Ross et al. (1991) on more than 50 cores from Alberta reveals age domains which range from 1.78 to 2.8 Ga. Detrital zircons from the Windermere Supergroup and other Late Precambrian metasedimentary rocks range in age from 1.75 to 2.9 Ga (Ross and Parrish 1991), and mimic this age pattern.

As shown in Figure 4.1, the western Canadian Shield is represented by the Churchill (Keewatin), Slave, and Bear structural provinces. The structural trends of these provinces, beneath cover rocks, are revealed by patterns of linear aeromagnetic and gravity anomaly trends which can be traced nearly to the edge of the deformed belt (Dods et al., 1984; Earth Physics Branch, 1980; Ross et al. 1991).

The age of basement rocks of the Cordillera, excepting latest Proterozoic (0.75 Ga) igneous events, can be grouped as 1.85 Ga to 2.1 Ga; this age span is similar to that of the combined Bear Province (Wopmay orogen, Hoffman and Bowring, 1984) and the Thelon Tectonic Zone of the northwest Churchill Province (Van Breemen et al., 1987) which also are 1.84 to about 2.1 Ga with a few older Archean ages (Fig. 4.1). No Archean rocks are known to occur beneath the Canadian Cordillera except perhaps in the southernmost part. Although precise correlation is not possible at present, the similarity of ages between the exposed and subsurface rocks of the northwestern Canadian Shield and basement rocks of the Cordillera argues against an exotic origin for the latter as suggested by Chamberlain and Lambert (1985b), but does not preclude considerable dextral translation as implied by paleomagnetic data of Irving (1985) and Irving et al. (1985).

AGE OF PROTEROZOIC RIFTING EVENTS

Rocks of latest Precambrian age (730-780 Ma), intrusive into older basement or metasediments of uncertain age or unconformably overlain by basal Windermere strata, are found the length of the Cordillera, including the Deserters gneissic granite, part of the Malton Complex in at least two localities, and the Mount Copeland nepheline syenite of the Monashee Complex (Fig. 4.1).

The Mount Harper volcanic complex, developed in lower Windermere strata in the western Yukon

(Roots, 1983), yielded a zircon age of approximately 780 Ma (Roots and Parrish, 1988). A diorite interpreted to be unconformably overlain by the Coates Lake Group in the Mackenzie Mountains has given a U-Pb age on zircon of 778 Ma (Jefferson and Parrish, 1989). This age is similar to that of basaltic dykes and sills which intrude the upper Mackenzie Mountains Supergroup (Armstrong et al., 1982). The magmatic event may have been related to initial rifting that formed the Late Proterozoic continental margin (see Chapter 6).

REFERENCES

Armstrong, J.E.
1949: Fort St. James map-area, Cassiar and Coast district, British Columbia; Geological Survey of Canada, Memoir 252, 210 p.

Armstrong, R.L., Eisbacher, G.H., and Evans, P.D.
1982: Age and stratigraphic-tectonic significance of Proterozoic diabase sheets, Mackenzie Mountains, northwest Canada; Canadian Journal of Earth Sciences, v. 19, p. 316-323.

Baldwin, D.A., Syme, E.C., Zwanzig, H.V., Gordon, T.M., Hunt, P.A., and Stevens, R.D.
1985: U/Pb zircon ages from the Lynn Lake and Rusty Lake Metavolcanic belts, Manitoba: Two ages of Proterozoic magmatism; Geological Association of Canada, Program with Abstracts, v. 10, p. A3.

Blenkinsop, J.
1972: Computer assisted mass spectrometry and its application to Rb-Sr geochronology; Ph.D. thesis, University of British Columbia, Vancouver, 109 p.

Boyer, S.E. and Elliot, D.
1982: Thrust systems; American Association of Petroleum Geologists Bulletin, v. 66, p. 1196-1230.

Brown, R.L. and Read, P.B.
1983: Shuswap terrane of British Columbia: A Mesozoic "core complex"; Geology, v. 11, p. 164-169.

Brown, R.L., Journeay, J.M., Lane, L.S., Murphy, D.C., and Rees, C.J.
1986: Obduction, backfolding, and piggyback thrusting in the metamorphic hinterland of the southeastern Canadian Cordillera; Journal of Structural Geology, v. 8, no. 3/4, p. 255-268.

Burwash, R.A., Baadsgaard, H., and Peterman, Z.E.
1962: Precambrian K-Ar dates from the western Canada sedimentary basin; Journal of Geophysical Research, v. 67, p. 1617-1625.

Campbell, R.B.
1968: Canoe River, British Columbia; Geological Survey of Canada, Map 15-1967.

Carr, S.D., Parrish, R.R., and Brown, R.L.
1987: Eocene structural development of the Valhalla complex, southeastern British Columbia; Tectonics, v. 6, p. 175-196.

Chamberlain, V.E. and Lambert, R.St.J.
1983: Tectonic implications of the geochronology and geochemistry of the Malton Gneiss Complex, Valemount, British Columbia; in Program with Abstracts, Geological Association of Canada Joint Annual Meeting, v. 8, p. A11.
1985a: Geochemistry and geochronology of the gneisses east of the Southern Rocky Mountains Trench, near Valemount, British Columbia; Canadian Journal of Earth Sciences, v. 22, p. 980-991.
1985b: Cordillera, a newly defined Canadian microcontinent; Nature, v. 314, p. 707-713.

Chamberlain, V., Lambert, R.St.J., Baadsgaard, H., and Gale, N.H.
1979: Geochronology of the Malton Gneiss Complex of British Columbia; in Current Research, Part B, Geological Survey of Canada, Paper 79-1B, p. 45-50.

Cheney, E.S.
1980: Kettle dome and related structures of northeastern Washington; Geological Society of America, Memoir 153, p. 463-482.

Dods, S.D., Hood, P.J., Teskey, D.J., and McGrath, P.H.
1984: Magnetic Anomaly Map of Canada; Geological Survey of Canada, Map 1255A.

Duncan, I.J.
1984: Structural evolution of the Thor-Odin gneiss dome; Tectonophysics, v. 101, p. 87-130.

Earth Physics Branch
1980: Gravity Map of Canada; Department of Energy, Mines and Resources, Gravity Map 80-1.

Eisbacher, G.H.
1978: Two major Proterozoic unconformities, northern Cordillera; in Current Research, Part A, Geological Survey of Canada, Paper 78-1A, p. 53-58.

Evenchick, C.A.
1983a: Stratigraphy, structure and metamorphism in the Sifton Ranges, Cassiar Mountains, northern British Columbia; in Current Research, Part A, Geological Survey of Canada, Paper 83-1A, p. 221-224.
1983b: Nonconformity at the base of upper Proterozoic Misinchinka Group, Deserters Range, northern Rocky Mountains; in Current Research, Part A, Geological Survey of Canada, Paper 83-1A, p. 475-476.
1985: A south-southeasterly directed contraction fault related to dextral transcurrent faulting in the northern Canadian Cordillera; Geological Society of America, Abstracts with Programs, v. 17, p. 354.

Evenchick, C.A., Parrish, R.R., and Gabrielse, H.
1984: Precambrian gneiss and late Proterozoic sedimentation in north-central British Columbia; Geology, v. 12, p. 233-237.

Ewing, T.
1980: Paleogene tectonic evolution of the Pacific Northwest; Journal of Geology, v. 88, p. 619-638.

Fyles, J.T.
1970: Jordan River Area; British Columbia Department of Mines and Petroleum Resources, Bulletin 57.

Gabrielse, H.
1963: McDame map-area, Cassiar District, British Columbia; Geological Survey of Canada, Memoir 319, 138 p.

Giovanella, C.A.
1967: Structural studies of the metamorphic rocks along the Rocky Mountain Trench at Canoe River, British Columbia; in Report of Activities, Part A, Geological Survey of Canada, Paper 68-1A, p. 27-31.

Hildebrand, R.S., Bowring, S.A., Steer, M.E., and Van Schmus, W.R.
1983: Geology and U-Pb geochronology of parts of the Leith Peninsula and Riviere Grandin map areas, District of Mackenzie; in Current Research, Part A, Geological Survey of Canada, Paper 83-1A, p. 329-342.

Hoffman, P.F. and Bowring, S.A.
1984: Short-lived 1.9 Ga continental margin and its destruction, Wopmay orogen, northwest Canada; Geology, v. 12, p. 68-72.

Høy, T. and Brown, R.L.
1980: Geology of the eastern margin of the Shuswap Complex - Frenchman Cap area, British Columbia; British Columbia Ministry of Energy, Mines and Petroleum Resources, Preliminary Map 43.

Irving, E.
1985: Paleomagnetic evidence for northward motion of the western Cordillera in latest Cretaceous to early Tertiary time; in Abstracts with Programs, Geological Society of America, v. 17, no. 6, p. 363.

Irving, E., Woodsworth, G.J., Wynne, P.J., and Morrison, A.
1985: Paleomagnetic evidence for displacement from the south of the Coast Plutonic Complex, British Columbia; Canadian Journal of Earth Sciences, v. 22, p. 584-598.

Jones, A.G.
1959: Vernon map area, Geological Survey of Canada, Memoir 296.

Journeay, J.M.
1986: Stratigraphy, internal strain and thermo-tectonic evolution of the northern Frenchman Cap dome: An exhumed duplex structure, Omineca hinterland, S.E. Canadian Cordillera; Ph.D. thesis, Queen's University, Kingston, Ontario, 350 p.

Lewry, J.F. and Sibbald, T.I.I.
1980: Thermotectonic evolution of the Churchill Province in northern Saskatchewan; Tectonophysics, v. 68, p. 45-82.

McDonongh, M.R. and Simony, P.S.
1984: Basement gneisses and Hadrynian metasediments near Bulldog Creek, Selwyn Range, British Columbia; in Current Research, Part A, Geological Survey of Canada, Paper 84-1A, p. 99-102.

McMillan, W.J.
1973: Petrology and structure of the west flank, Frenchman's Cap Dome, near Revelstoke, British Columbia; Geological Survey of Canada, Paper 71-29, 87 p.

Monger, J.W.H., Clowes, R.M., Price, R.A., Simony, P.S., Riddihough, R.P., and Woodsworth, G.J. with contributions from Currie, R.R., Høy, T., Snavely, P.D., and Yorath, C.J.
1985: B2, Juan de Fuca Plate to Alberta Plains; Geological Society of America, Centennial Continent/Ocean Transect #7.

Morrison, M.L.
1979: Structure and petrology of the southern portion of the Malton Gneiss, British Columbia; in Current Research, Part B, Geological Survey of Canada, Paper 79-1B, p. 407-410.
1982: Structure and petrology of the Malton gneiss complex; Ph.D. thesis, University of Calgary, Alberta, 314 p.

Oke, C. and Simony, P.S.
1981: Basement gneisses of the western Rocky Mountains, Hugh Allan Creek area, British Columbia; in Current Research, Part A, Geological Survey of Canada, Paper 81-1A, p. 181-184.

Okulitch, A.V.
1985: Paleozoic plutonism in southeastern British Columbia; Canadian Journal of Earth Sciences, v. 22, no. 10, p. 1409-1424.

Orr, K.
1985: Structural features along the margin of Okanagan dome, Tenas Mary Creek area, NE Washington; Geological Society of America, Abstracts with Programs, v. 17, p. 398.

Parkinson, D.
1985: Geochronology of the western side of the Okanagan metamorphic core complex, southern B.C.; in Abstracts with Programs, Geological Society of America, Cordilleran Section, v. 17, p. 399.

Parrish, R.R.
1979: Geochronology and tectonics of the northern Wolverine Complex, British Columbia; Canadian Journal of Earth Sciences, v. 16, p. 1429-1438.
1984: Slocan Lake Fault: a low angle fault zone bounding the Valhalla Gneiss Complex, Nelson map-area, southern British Columbia; in Current Research, Part A, Geological Survey of Canada, Paper 84-1A, p. 323-330.

Parrish, R.R. and Armstrong, R.L.
1983: U-Pb zircon age and tectonic significance of gneisses in structural culminations of the Omineca Crystalline Belt, British Columbia; Geological Society of America, Abstracts with Programs, v. 15, no. 5, p. 324.

Parrish, R.R. and Ryan, B.
1983: Pb-U zircon dates reflecting late Cretaceous-early Tertiary plutonism, deformation, and isotopic resetting, Valhalla Complex, southeast B.C.; Geological Association of Canada, Program with Abstracts, v. 8, p. A53.

Parrish, R.R., Carr, S.D., and Brown, R.L.
1985: Valhalla gneiss complex, southeast British Columbia, 1984 fieldwork; in Current Research, Part A, Geological Survey of Canada, Paper 85-1A, p. 81-87.

Preto, V.A.
1970: Structure and petrology of the Grand Forks Group, British Columbia; Geological Survey of Canada, Paper 69-22, 80 p.

Price, R.A.
1979: Intracontinental ductile crustal spreading linking the Fraser River and northern Rocky Mountain trench transform fault zones, south-central British Columbia and northeast Washington; Geological Society of America, Abstracts with Programs, v. 11, p. 499.
1985: Metamorphic core complexes of the first and second kind in the Cordillera of southern Canada and northern USA; Geological Society of America, Abstracts with Programs, v. 17, p. 401.

Price, R.A. and Mountjoy, E.W.
1970: Geologic structure of the Canadian Rocky Mountains between Bow and Athabasca Rivers - A progress report; Geological Association of Canada, Special Paper 6, p. 7-25.

Read, P.B. and Brown, R.L.
1981: Columbia River fault zone: southeastern margin of the Shuswap and Monashee complexes, southern British Columbia; Canadian Journal of Earth Sciences, v. 18, p. 1127-1145.

Reesor, J.E.
1965: Structural evolution and plutonism in Valhalla gneiss complex, British Columbia; Geological Survey of Canada, Bulletin 129, 128 p.
1970: Some aspects of structural evolution and regional setting in part of the Shuswap Metamorphic Complex; in Structure of the Canadian Cordillera, Geological Association of the Canada, Special Paper 6, p. 73-86.

Reesor, J.E. and Moore, J.M. Jr.
1971: Petrology and structure of Thor-Odin gneiss dome, Shuswap Metamorphic Complex, British Columbia; Geological Survey of Canada, Bulletin 195, 149 p.

Roots, E.F.
1954: Geology and mineral deposits of Aiken Lake map-area, British Columbia; Geological Survey of Canada, Memoir 274, 246 p.

Ross, J.V.
1970: Structural evolution of the Kootenay Arc, southeastern British Columbia; in Structure of the Canadian Cordillera, Geological Association of Canada, Special Paper 6, p. 53-65.

Stockwell, C.F.
1982: Proposals for time classification and correlation of Precambrian rocks and events in Canada and adjacent areas of the Canadian Shield; Part 1: A time classification of Precambrian rocks and events; Geological Survey of Canada, Paper 80-19, 135 p.

Tempelman-Kluit, D.J.
1979: Transported cataclasite, ophiolite, and granodiorite in Yukon: evidence of arc-continent collision; Geological Survey of Canada, Paper 79-14, 27 p.

Tempelman-Kluit, D.J. and Wanless, R.R.
1980: Zircon ages for the Pelly gneiss and Klotassin granodiorite in western Yukon; Canadian Journal of Earth Sciences, v. 17, p. 297-306.

Van Breemen, O., Henderson, J.B., Loveridge, W.D., and Thompson, P.H.
1987: U-Pb zircon and monazite geochronology and zircon morphology of granulites and granite from the Thelon Tectonic Zone, Healey Lake and Artillery Lake map areas, N.W.T.; in Current Research, Part A, Geological Survey of Canada, Paper 87-1A, p. 783-801.

Wanless, R.K. and Reesor, J.E.
1975: Precambrian zircon age of orthogneiss in the Shuswap Metamorphic complex, British Columbia; Canadian Journal of Earth Sciences, v. 12, p. 326-333.

Wheeler, J.O.
1970: Summary and discussion; in Structure of the Southern Canadian Cordillera, Geological Association of Canada, Special Paper 6, p. 155-166.

Author's address

R.R. Parrish
Geological Survey of Canada
601 Booth Street
Ottawa, Ontario
K1A 0E8

/ CHAPTER 4

ADDENDUM

THE PRECAMBRIAN BASEMENT BENEATH THE ALBERTA BASIN

Recent analysis of aeromagnetic and gravity potential field data, in conjunction with a program of U-Pb dating of the drilled basement beneath the Alberta Basin, has greatly clarified the age and regional tectonic elements of the Precambrian basement of western Canada. These new data and ideas are summarized by Ross et al. (1991) and Villeneuve et al. (1991). The ages of crystalline rocks range from 1.7-3.2 Ga, and demonstrate that unlike the exposed Canadian Shield to the east and north, a large expanse of 2.1-2.4 Ga crust is present in northern Alberta. With few exceptions, Archean rocks are restricted to the southeastern one third of the Province and represent continuations of the Churchill (Rae and Hearne) Province of the Shield on the west side of the Trans-Hudson orogen as well as Archean rocks traceable northward from the Wyoming craton.

The orogenic belts delineated by potential field data and U-Pb geochronology are mostly 1.9-1.75 Ga in age, are numerous, and trend northerly to northeasterly for the most part. A coherent area of 2.1-2.4 Ga rocks, the largest block of which has been termed the Buffalo Head Terrane, is contained in this orogenic collage and is one of the largest areas of this age of crust documented in Canada. The Archean Slave Province terminates north of latitude 60°N and does not extend into Alberta.

One of the interesting aspects of this part of Canada is that many of the geophysical trends cannot be clearly extended into the Cordillera. Because 1.85 Ga plutonic rocks occur in all major exposures of basement in the Cordillera, it is possible that most of the older (> 1.9 Ga) belts of the buried shield were truncated or modified by a major 1.85 Ga northwest-trending tectonic belt near the edge of the Cordillera. This proposed belt may have contributed to the geometry of the development of the Cordilleran miogeocline in later Proterozoic time.

REFERENCES

Armstrong, R.L., Parrish, R.R., van der Heyden, P., Reynolds, S.J., and Rehrig, W.A.
1987: Rb-Sr and U-Pb chronology of the Priest River metamorphic complex: Precambrian X basement and its Mesozoic-Cenozoic plutonic-metamorphic overprint near Spokane, Washington; in Geology of Washington; Washington Division of Geology and Earth Resources, Bulletin 77, p. 15-40.

Armstrong, R.L., Parrish, van der Heyden, P., Scott, K., Runkle, D., and Brown, R.L.
1991: Early Proterozoic basement exposures in the southern Canadian Cordillera: Core gneiss of Frenchman Cap, Unit I of the Grand Forks gneiss, and the Vaseaux formation; Canadian Journal of Earth Sciences, v. 28.

Brown, R.L., Carr, S.D., Johnson, B.J., Coleman, V.J., Cook, F.A., and Varsek, J.L.
1991: The Monashee decollement of the southern Canadian Cordillera: A crustal scale shear zone linking the Rocky Mountain foreland belt to lower crust beneath acreeted terranes; in Thrust Tectonics, K. McClay (ed.), Geological Society of London Special Publication.

Jefferson, C.W. and Parrish, R.R.
1989: Late Proterozoic stratigraphy, U-Pb zircon ages and tectonic implications, Mackenzie Mountains, Northwestern Canada; Canadian Journal of Sciences, v. 26, p. 1784-1801.

Mortensen, J.K.
1991: Pre-mid-Mesozoic tectonic evolution of the Yukon-Tanana terrane, Yukon and Alaska; Tectonics, v. 10.

Murphy, D.C., Walker, R.T., and Parrish, R.R.
1991: Age and geological setting of Gold Creek gneiss, crystalline basement of the Windermere Supergroup, Cariboo Mountains, British Columbia; Canadian Journal of Earth Sciences, v. 28.

McDonough, M.R. and Parrish, R.R.
1991: Proterozoic gneisses of the Malton Complex, near Valemount, British Columbia: U-Pb ages and Nd isotopic signatures; Canadian Journal of Earth Sciences, v. 28.

Parrish, R.R., Carr, S.C., and Parkinson, D.
1988: Eocene extensional tectonics and geochronology of the southern Omineca belt, British Columbia and Washington; Tectonics, v. 7, p. 181-212.

Parrish, R.R. and Scammell, R.J.
1988: The age of the Mount Copeland syenite gneiss and its metamorphic zircons, Monashee complex, southeastern British Columbia; in Radiogenic Age and Isotopic Studies: Report 2; Geological Survey of Canada, Paper 88-2, p. 21-28.

Roots, C.F. and Parrish, R.R.
1988: Age of the Mount Harper volcanic complex, southern Ogilvie Mountains, Yukon; in Radiogenic Age and Isotopic Studies, Report 2; Geological Survey of Canada Paper 88-2, p. 29-36.

Ross, G.M., Parrish, R.R., Villeneuve, M.E., and Bowring, S.A.
1991: Geophysics and geochronology of the crystalline basement of the Alberta Basin, western Canada; Canadian Journal of Earth Sciences, v. 28.

Ross, G.M. and Parrish, R.R.
1991: Detrital zircon geochronology of metasedimentary rocks in the southern Omineca Belt, Canadian Cordillera; Canadian Journal of Earth Sciences, v. 28.

Villeneuve, M.E., Theriault, R.J., and Ross, G.M.
1991: U-Pb ages and Sm-Nd signature of two subsurface granites from the Fort Simpson magnetic high, northwest Canada; Canadian Journal of Earth Sciences, v. 28.

Printed in Canada

Chapter 5

MIDDLE PROTEROZOIC ASSEMBLAGES

Summary
Introduction
Principal divisions of the Cordilleran Proterozoic
Sequence A, ≈1.7 to ≈1.2 Ga
    Purcell (Belt) Supergroup (Purcell-Wernecke Assemblage)
        Basal division
        Lower division
        Middle carbonate division
        Upper division
    Monashee Complex
    Wernecke Supergroup (Purcell-Wernecke Assemblage)
    Muskwa Ranges Succession (Muskwa Assemblage)
    Cap Mountain Succession (Cap Mountain-Hornby Bay Assemblage)
        Tectonic environment
    Age and tectonic setting of Sequence A
Sequence B, ≈1.2 to ≈0.78 Ga
    Mackenzie Mountains Supergroup (Mackenzie Mountains Assemblage)
        Age and tectonic environment
References

Chapter 5

MIDDLE PROTEROZOIC ASSEMBLAGES

J.D. Aitken and M.E. McMechan

SUMMARY

The ages of formations and larger groupings of Precambrian stratified rocks in the Cordillera are, in general, poorly constrained. Only two useful macrofossil assemblages are known: an Ediacaran, non-skeletal fauna found only in the youngest formations, and an older, *Chuaria-Tawuia* association whose duration of several hundred million years provides age control little more constrained than that provided by the few available radiometric dates. Stromatolite biostratigraphy to date has failed to provide useful chronocorrelation with the Riphean of the U.S.S.R. Only the youngest formations have been characterized palynologically.

Few reliable radiometric dates have been published, but bracketing dates and structural relationships permit the assignment of most major successions of Cordilleran Proterozoic strata to one or other of the sequences recognized by G.M. Young and co-workers:

Sequence A: ≈1.7 to ≈1.2 Ga

Sequence B: ≈1.2 to ≈0.78 Ga

Sequence C: ≈0.78 to ≈0.57 Ga

These sequences, of which A and B are the subject of this chapter, provide a useful, though imprecise, framework for discussion. Sequence C unconformably overlies Sequence B and is discussed in Chapter 6.

Sequence A (Purcell-Wernecke, Cap Mountain-Hornby Bay, Muskwa assemblages) conspicuously lacks the longitudinal Cordillera continuity of Upper Proterozoic strata (Sequence C) and is exemplified by the classical Purcell (Belt) Supergroup of southeastern British Columbia, southwestern Alberta, Montana and Idaho. Its patterns of sedimentary facies and thickness (up to 20 km) have generally been interpreted as those of a continental-margin succession, but for the Belt-Purcell Basin at least, work in the northwestern United States is yielding increasing evidence suggestive of a two-sided, and hence intracratonic basin. In the southern Canadian Cordillera this margin truncates structural trends of the Churchill Province of the Canadian Shield, and itself trends more westerly than the continental margin upon and across which Sequence C (Pinguicula, Rapitan, Windermere, Neruokpuk assemblages) was deposited. The Purcell margin appears to be truncated by the latter.

The older (pre-Coppermine lavas) formations of the Coppermine Homocline, namely, the Hornby Bay and Dismal Lakes groups, rest on crystalline basement (>1.8 Ga) and belong also to Sequence A. They outcrop in the Cordillera at Cap Mountain, Northwest Territories, and may have been deposited in an aulacogen-like rift-basin opening to the southwest. Structural and regional relations, and a few radiometric dates, suggest that the Wernecke Supergroup also belongs to Sequence A. It is known only from the Wernecke and Ogilvie mountains of the Yukon, and like the Purcell (Belt) Supergroup, was folded during the Precambrian. Its aspect is plausibly distal to that of the proximal, Hornby Bay-Dismal Lakes succession, and its thickness, over 14 km, is suggestive of a continental-margin accumulation.

The older Proterozoic succession of the Muskwa Ranges of northeastern British Columbia, over 6 km thick, is undated, but older than Sequence C, and is similar in aspect to both the Hornby Bay-Dismal Lakes and the classical Purcell. It occurs in a structural salient, and by analogy with the Waterton salient at the 49th parallel, probably occupies a more or less local, Sequence A 'thickening' at the mountain front. The nature of the 'thickening', whether an embayment in the continental margin or the southeast flank of the Coppermine Homocline rift-basin, is unclear.

Sequence B (Mackenzie Mountains Assemblage) is preserved only north of 60° and includes the Mackenzie Mountains Supergroup and the post-lavas, Rae Group of the Coppermine Homocline. These strata, in excess of 4 km thick in the Mackenzie Mountains, are platformal in facies character and thickness distribution. They may be epicratonic basin deposits, yet bear a spatial relationship to the Sequence C continental margin that is unexplained.

The source of detritus for Sequence B formations was the Canadian Shield. No evidence for a source to the west is known.

The Purcell (Belt) Supergroup is host to one giant and numerous smaller Pb-Zn (Ag) deposits. Clastic formations of Sequence A generally, and the Grinnell Formation of the classical Purcell in particular, have anomalously high Cu content, although no deposits have yet been exploited in Canada. The Wernecke Supergroup hosts numerous showings of Cu, Pb, Zn, Ag, Co and U, all of which are economically inaccessible and have not been exploited. The potential of the Hornby Bay-Dismal Lakes succession appears to lie mainly in uranium deposits, but anomalously high values of stratabound copper occur at least locally. The Muskwa Assemblage of Sequence A is host to numerous showings of copper in vein deposits associated with mafic dykes; one of these has been mined.

Aitken, J. D. and McMechan, M. E.
1991: Middle Proterozoic assemblages, Chapter 5 *in* Geology of the Cordilleran Orogen in Canada, H. Gabrielse and C. J. Yorath (ed.); Geological Survey of Canada, Geology of Canada, no. 4, p. 97-124 (*also* Geological Society of America, The Geology of North America, v. G-2)

The Little Dal Group (Mackenzie Mountains Supergroup; Sequence B) is host to a large, unexploited Zn-Pb deposit in the Mackenzie Mountains. Its remote setting, and the existence of large, more accessible deposits in Paleozoic strata of the Selwyn Basin, prohibit its development at present.

INTRODUCTION

The Middle Proterozoic rocks of the Canadian Cordillera include thick successions of essentially unaltered strata. Apart from the metamorphic culminations of the orogen, the preservation of primary sedimentary structures, even at a microscopic scale, permits thorough sedimentological analysis, limited only by the absence of fossils other than simple microfossils and algal stromatolites.

Study of the Cordilleran Proterozoic near the 49th parallel commenced near the turn of the 20th century, and there the physical stratigraphy was well established by the 1940s. On the other hand, in northeastern British Columbia study of the Proterozoic in any detail did not commence until the 1960s, and not until a decade later in the Mackenzie Mountains of the Northwest Territories and the Wernecke Mountains of the Yukon. First accounts, at anything beyond a reconnaissance level, of thick and important successions in the Wernecke Mountains were not published until the early 1980s. Thus, especially in the north, important problems of Proterozoic geology of the Cordillera await further study.

PRINCIPAL DIVISIONS OF THE CORDILLERAN PROTEROZOIC

Pre-Windermere formations of the Cordillera (Fig. 5.1, 5.2) have been studied to the degree that local lithostratigraphy and sedimentology are generally well known. A paucity of reliable radiometric dates has bedevilled chronocorrelation, however, and rarely can the succession exposed in one structural culmination (Purcell Anticlinorium, Tuchodi Lakes Anticlinorium) be related with confidence to that in another. Many published 'correlations' cannot be objectively substantiated.

The treatment adopted herein follows, with adjustments dictated by new data, the natural divisions of the Middle and Upper Proterozoic of western North America proposed by Young (1978) and Young et al. (1979):

Sequence A: ≈1.7 to ≈1.2 Ga

Sequence B: ≈1.2 to ≈0.78 Ga

Sequence C: ≈0.78 to ≈0.57 Ga

This chapter is mainly concerned with Sequences A and B, broadly the Middle Proterozoic, although some of the rocks of Sequence B may extend into the Upper Proterozoic.

SEQUENCE A, ≈1.7 TO ≈1.2 GA

Sequence A (Purcell-Wernecke, Cap Mountain-Hornby Bay, Muskwa assemblages) is underlain by basement rocks of the Churchill Province in the south, and by sedimentary, volcanic, metamorphic and plutonic rocks of the western Bear Province in the north. It is bounded above by widespread, intrusive and extrusive, mafic igneous rocks yielding dates near 1.2 Ga, of which the Coppermine lavas

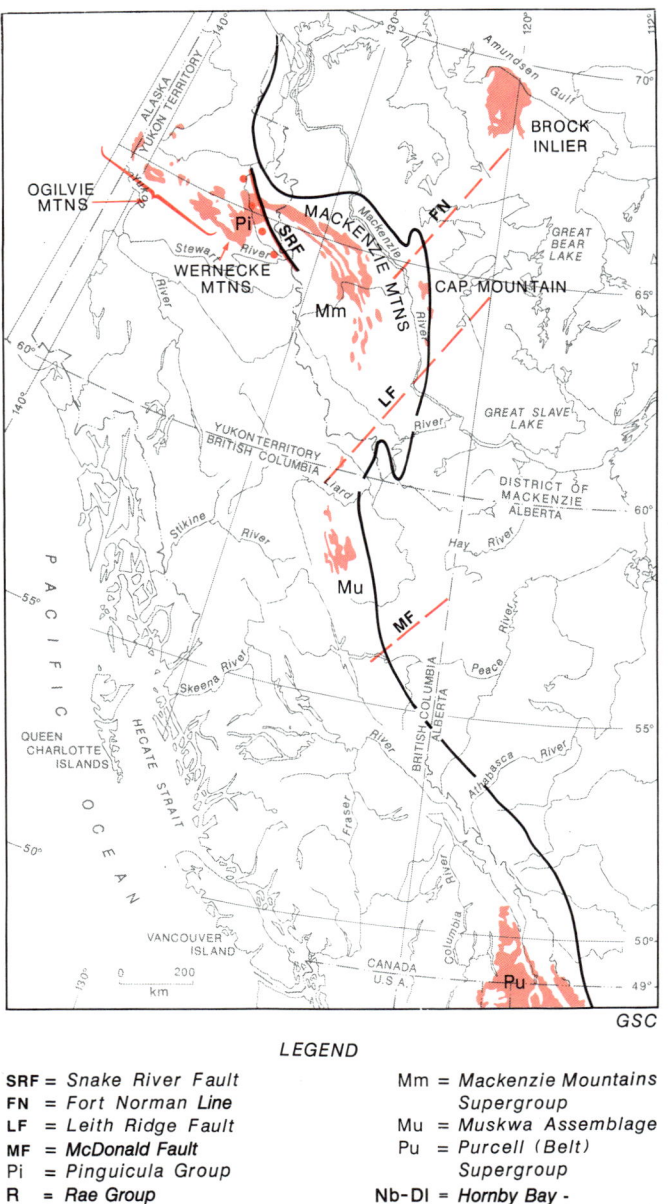

Figure 5.1. Distribution of Middle Proterozoic rocks in the Canadian Cordillera and Coppermine Homocline. Black line shows eastern limit of Cordilleran deformation.

(Copper Creek Formation of Baragar and Donaldson, 1973) are representative. On the basis of new U-Pb zircon dates, Bowring and Ross (1985) questioned the validity of the near 1.2 Ga dates, suggesting that the widespread, mafic igneous activity might be older by 'several hundred million years'. In this treatment, the widely accepted 1.2 Ga date is followed, pending further research (see addendum).

Sedimentary rocks assigned (some tentatively) to Sequence A occur in four areas of the Canadian Cordillera: the classical Purcell (Belt) Supergroup in southeastern British Columbia and southwestern Alberta; the Wernecke Supergroup in the Wernecke and Ogilvie mountains of the

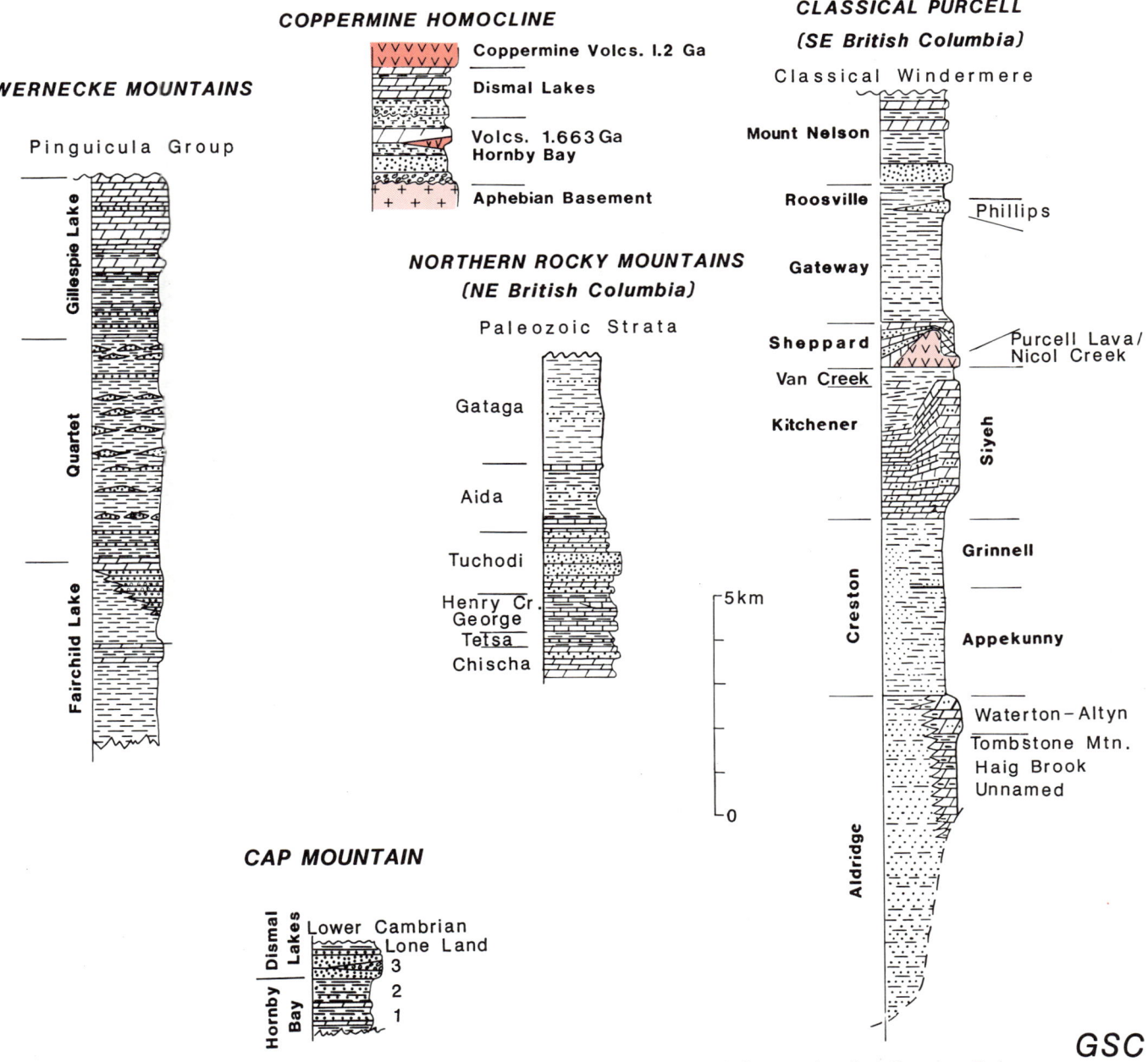

Figure 5.2. Comparison of Middle Proterozoic assemblages. No correlations are implied. Sources: Delaney (1981), Kerans et al. (1981), Bell (1968), McMechan (1981).

Yukon; the pre-Windermere strata of the northern Rocky Mountains (Muskwa Assemblage); and the sub-Cambrian rocks exposed at Cap Mountain, southern Franklin Mountains, District of Mackenzie (Figs. 5.1, 5.2).

Purcell (Belt) Supergroup (Purcell-Wernecke Assemblage)

The Purcell (Belt) Supergroup of southeastern British Columbia, southwestern Alberta, western Montana, Idaho and eastern Washington reaches reported maximum thicknesses of more than 11 km in Canada (Reesor, 1958), and 20 km in the United States (Harrison, 1972). In the Canadian part of the Belt-Purcell Basin, the Purcell (Belt) Supergroup (Fig. 5.3, 5.4) comprises a sequence of shallow-marine and nonmarine rocks and an underlying sequence expressed by fine grained, basinal clastic rocks in the west and platformal, clastic and carbonate rocks in the east. The base is not exposed. Facies changes are generally distinct near the northeastern limit of exposure, but become more subtle to the southwest. These changes of facies, the relatively large distances between areas of best exposure of various parts of the succession, and the lack of geological investigations straddling the Canada-United States boundary have resulted in complicated stratigraphic nomenclature for the northern part (north of 48°) of the Belt-Purcell Basin (Fig. 5.3).

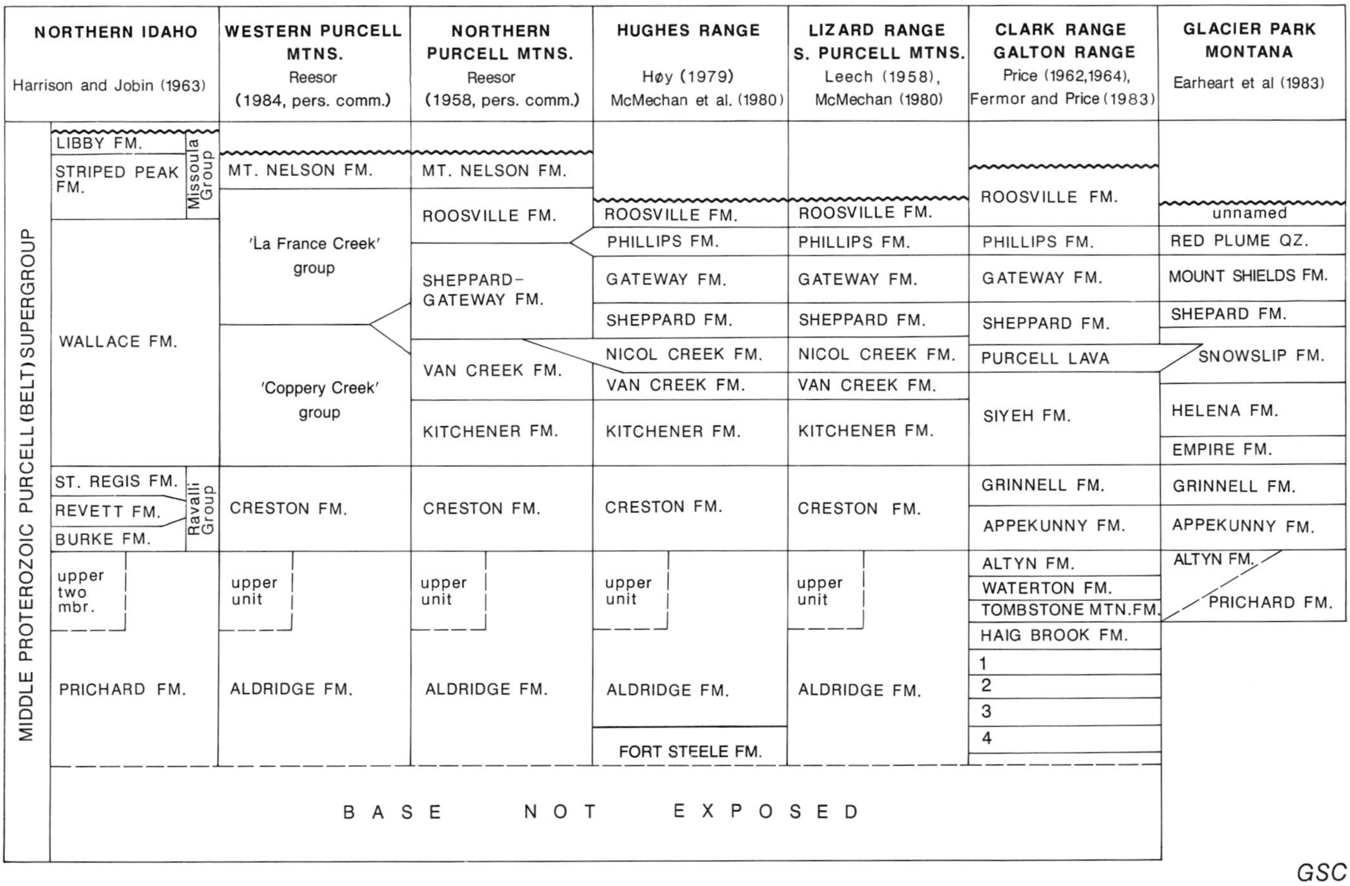

NORTHERN IDAHO Harrison and Jobin (1963)		WESTERN PURCELL MTNS. Reesor (1984, pers. comm.)	NORTHERN PURCELL MTNS. Reesor (1958, pers. comm.)	HUGHES RANGE Høy (1979) McMechan et al. (1980)	LIZARD RANGE S. PURCELL MTNS. Leech (1958), McMechan (1980)	CLARK RANGE GALTON RANGE Price (1962, 1964), Fermor and Price (1983)	GLACIER PARK MONTANA Earheart et al (1983)
LIBBY FM.	Missoula Group	MT. NELSON FM.	MT. NELSON FM.				
STRIPED PEAK FM.						ROOSVILLE FM.	unnamed
			ROOSVILLE FM.	ROOSVILLE FM.	ROOSVILLE FM.		
		'La France Creek' group		PHILLIPS FM.	PHILLIPS FM.	PHILLIPS FM.	RED PLUME QZ.
			SHEPPARD-GATEWAY FM.	GATEWAY FM.	GATEWAY FM.	GATEWAY FM.	MOUNT SHIELDS FM.
WALLACE FM.				SHEPPARD FM.	SHEPPARD FM.	SHEPPARD FM.	SHEPARD FM.
			VAN CREEK FM.	NICOL CREEK FM.	NICOL CREEK FM.	PURCELL LAVA	SNOWSLIP FM.
		'Coppery Creek' group		VAN CREEK FM.	VAN CREEK FM.	SIYEH FM.	HELENA FM.
			KITCHENER FM.	KITCHENER FM.	KITCHENER FM.		EMPIRE FM.
ST. REGIS FM.	Ravalli Group	CRESTON FM.	CRESTON FM.	CRESTON FM.	CRESTON FM.	GRINNELL FM.	GRINNELL FM.
REVETT FM.							
BURKE FM.						APPEKUNNY FM.	APPEKUNNY FM.
upper two mbr.		upper unit	upper unit	upper unit	upper unit	ALTYN FM.	ALTYN FM.
						WATERTON FM.	PRICHARD FM.
						TOMBSTONE MTN. FM.	
						HAIG BROOK FM.	
PRICHARD FM.		ALDRIDGE FM.	ALDRIDGE FM.	ALDRIDGE FM.	ALDRIDGE FM.	1	
						2	
						3	
				FORT STEELE FM.		4	
BASE NOT EXPOSED							

Figure 5.3. Nomenclature for classical Purcell (Belt) Supergroup (after McMechan, 1981).

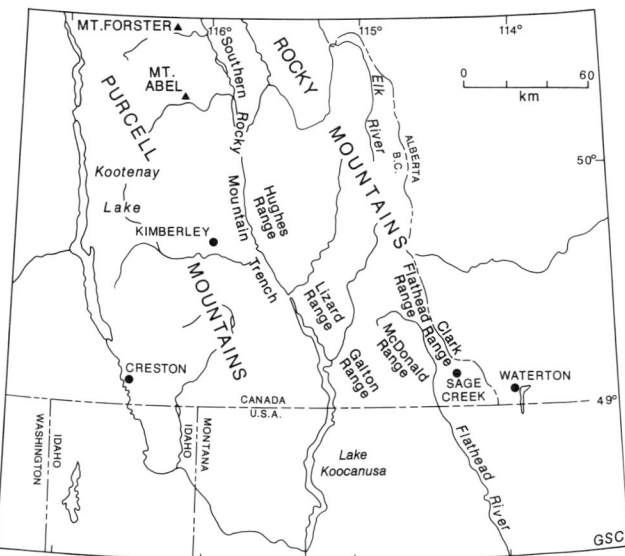

Figure 5.4. Location map for areas of Purcell (Belt) Supergroup.

Two different interpretations of the setting and tectonic character of the Belt-Purcell Basin have been proposed from time to time. Canadian geologists, impressed by the great thickness of the Purcell (Belt) Supergroup and southwestward proximal-to-distal facies changes, have generally opted for a continental-margin setting and a miogeoclinal interpretation (see McMechan, 1981). Some geologists in the United States, commencing with C.D. Walcott, have perceived the Belt Basin as intracratonic, while more recently, Harrison (1972) and Harrison et al. (1974) have perceived it to be a miogeocline or possibly an aulacogen. Recently, increasing evidence for a two-sided, intracratonic basin (Winston et al., 1984) has been published (e.g. Winston et al., 1984; Roberts, 1986). Despite extensive study of the basin, the case remains undecided.

With the exception of highly deformed and metamorphosed rocks exposed at the western margin of the Purcell Mountains, the Purcell (Belt) Supergroup can be coarsely divided into four divisions, on the basis of distinctive rocks and readily correlatable sequences: basal, lower, middle carbonate, and upper. The boundaries of these divisions do not everywhere correspond to those of established formations. Variations in thickness and facies are most pronounced in the basal division, represented by a relatively thin, platformal succession along the northeastern limit of exposure in the Clark Range of the Rocky Mountains (Fig. 5.4), and by a thick, basinal assemblage elsewhere.

Variations in the lower division reflect the embayed configuration of the continental margin and the interplay of sources of detritus to the east and south. The middle carbonate division is in a thin, platformal facies along the northeastern limit of exposure and a thicker, more basinal facies elsewhere. Important lithological changes in the upper division occur near the western and northern limits, rather than near the northeastern limits of exposure.

Like other assemblages of strata assigned to Sequence A, the Purcell (Belt) Supergroup lacks the linear, almost continuous outcrop distribution along the length of the Cordillera shown by the Windermere Supergroup (Sequence C, Fig. 6.1). Thickness and facies variations in the lower division, the middle carbonate and the lower part of the upper division show that the westward offsets in the northeastern limit of Purcell exposure reflect the original shape of the northeastern margin of the basin.

Mafic sills and, less commonly, dykes intrude the Purcell throughout its area of exposure. Sills occur in all units of the Purcell; lavas are present in the upper division. The thickest (50-400 m) and most abundant sills occur in the lower part of the basinal facies of the basal division, where they form about one quarter of the section (Hamilton, 1984). In western and southeastern exposures, sills and dykes also occur above the lavas (Rice, 1941; Hunt, 1962). The intrusions probably belong to more than one episode of emplacement. Most of the extensive sills in the lower part of the basal division are probably related to synsedimentary stretching and rifting of the sub-Purcell basement (T. Høy, pers. comm., 1985), but sills and dykes also were emplaced during upper division volcanism (Daly, 1912; Schofield, 1915; Rice, 1937; Hunt, 1962; McMechan, 1981), and in post-Purcell times (Rice, 1941; Reesor, 1958). Many of the later sills and dykes may be related to rifting and volcanism associated with Sequence C.

The age of the Purcell (Belt) Supergroup is bracketed by those of the basement beneath, about 1.7 Ga (Burwash et al., 1962; Giletti, 1966; Obradovich and Peterman, 1973), and the unconformably overlying Windermere Supergroup. An often-quoted K-Ar date for basal Windermere volcanics of about 850 Ma is suspect, in view of new U-Pb dates of about 730 Ma on zircons from granitic rocks nonconformably beneath Windermere successions north of the Belt-Purcell Basin (Evenchick et al., 1984; see also Devlin et al., 1985). Geochronologic studies (Hunt, 1962; Obradovich and Peterman, 1968; Zartman et al., 1982) suggest that sedimentation occurred in the Belt-Purcell Basin between about 1.5 Ga and 850 Ma, whereas paleomagnetic studies (Evans et al., 1975; Vitorello and Van der Voo, 1977; Elston and Bressler, 1980) indicate that sedimentation took place between about 1.5 and 1.2 Ga. The latter dates are reasonably consistent with a recalculated Rb-Sr whole-rock date of about 1300 Ma for the Hellroaring Creek stock (Ryan and Blenkinsop, 1971; McMechan and Price, 1982), which cuts folded and metamorphosed Purcell strata. Regional considerations of metamorphic fabrics, intrusive relationships and radiometric age determinations that are open to question (as dates of rock formation) suggest that the younger (<1.2 Ga) isotopic ages were updated during a later thermal event whose nature and timing are unclear (e.g., McMechan and Price, 1982). Not all students of the Purcell (Belt) Supergroup agree that it suffered pre-Windermere folding, and the absence of a clastic wedge resulting from that folding is difficult to explain.

Basal division

Major changes of thickness and facies occur in the basal division of the Purcell (Belt) Supergroup. Throughout the Purcell Mountains and in the Lizard Range of the Rockies, this division consists of a thick sequence of basinal turbidites and argillites assigned to the Aldridge Formation. Only in the Hughes Range are fluvial or littoral quartzite, mudcracked argillite and siltite of the Fort Steele Formation exposed beneath the Aldridge (Fig. 5.3).

In the Purcell Mountains the Aldridge Formation, up to 4.2 km thick, is divided into three units (Reesor, 1958; Edmunds, 1973, 1977; Høy and Diakow, 1982). The lower unit consists of rusty-weathering, laminated, thin-bedded, very fine grained quartzite, argillaceous quartzite and siltite (all interpreted as distal turbidites), with minor black argillite partings.

The middle unit is characterized by thin to thick beds of fine grained quartzite and argillaceous quartzite that are interbedded with laminated and ripple crosslaminated siltite and laminated dark argillite (Fig. 5.5). The sandstones are interpreted as A-E turbidites of the Bouma (1962) model, and are more proximal than those of the lower unit. The middle Aldridge contains remarkable, thin, marker units that provide the only means of chronocorrelation within the Aldridge/Prichard Formation (Fig. 5.3). These units, consisting of parallel-laminated, dark grey carbonaceous and pale grey non-carbonaceous siltite, contain sequences of laminae that can be matched, lamina for lamina, over distances approaching 300 km (Huebschman, 1973; Edmunds, 1977; Finch and Baldwin, 1984). The middle unit forms the base of exposure in the Lizard Range. In the Purcell Mountains the lower two divisions of the Aldridge Formation include abundant metadiorite and meta-quartz diorite sills of the Moyie Intrusions.

The upper unit consists of thin-bedded, rusty-weathering, dark and medium grey argillite, with thin, parallel,

Figure 5.5. Sandstone and siltstone of the middle Aldridge Formation (interpreted as turbidites), interbedded with argillite and cut by meta-diorite and meta-quartz diorite sills of the Moyie Intrusions. View to east side of Mount Evans about 5 km west of the Middle Proterozoic Hellroaring Creek stock. Photo by J.E. Reesor. GSC 205235-D

pale grey siltite and dark grey carbonaceous laminae, and lesser laminated siltite and argillite. Minor thin- to thick-bedded quartzite occurs locally. Towards the top, siltite becomes more abundant, laminae become discontinuous, and syneresis (shrinkage) cracks and scour-and-fill structures appear, the first signs of a gradual transition to shallow-water deposits of the overlying Creston Formation. Dolomite and dolomitic siltite occur locally.

In the Hughes Range, the Aldridge Formation also is divisible into three parts. The lower part consists of a thick (2600-1500 m), northward-thinning succession of intertidal and subtidal argillite and siltite, and a prominent cryptalgal carbonate unit (Høy, 1979). Correlation of marker sequences (Finch and Baldwin, 1984) suggests that these strata are equivalent to the proximal and distal turbidites of the lower and middle Aldridge of the Purcell Mountains and Lizard Range. Quartz sand turbidites, that become finer grained as the unit thins northward from 400 to 240 m, characterize the middle unit (Høy, 1979). Laminated argillite typical of the upper Aldridge forms the upper unit. The subaerial to shallow-marine deposits of the Fort Steele Formation and the lower part of the Aldridge Formation in the Hughes Range are platformal deposits. The upward transition to basinal sediments of the middle and upper Aldridge indicates subsidence and marine transgression. The great thickness (3-4 km) of fine grained quartz sand and silt turbidites that accumulated elsewhere in the basin suggest that this subsidence was basin-wide.

The northeasterly limit of Purcell exposures is along the leading edge of the Lewis thrust-sheet in the Clark Range. There, the basal division is mainly a platformal sequence, but to the south and west, equivalent strata acquire a more basinal aspect.

The Waterton Formation, 170 to 248 m thick, was long considered to be the base of the exposed Purcell (Belt) Supergroup, because it forms the base of the Lewis thrust-sheet in the eastern Clark Range. There, it consists of very fine crystalline, banded and laminated limestone and dolomite, green argillite and argillaceous, very fine crystalline dolomite, varicoloured brownish red and light green laminated, very fine crystalline dolomite, dense white limestone, and grey stromatolitic dolomite and cherty dolomite (Price, 1964). Fermor (1980) and Fermor and Price (1983) identified similar strata in structural windows in the western Clark Range. The strata become more argillaceous southward, and, at Sage Creek, equivalent strata are greenish grey, laminated, dolomitic and calcareous argillite with some limestone and argillite interbeds. A similar westward change is indicated between Waterton and Sage Creek (Fermor and Price, 1983). Along the western Clark Range two new mappable units, the Tombstone Mountain and Haig Brook formations, occur. Four additional units, numbered 1 to 4, penetrated by the Pacific-Atlantic Flathead No. 1 well at Sage Creek, are interpreted as a normal stratigraphic succession below the Haig Brook Formation (Fig. 5.3 and 5.6).

Following Fermor and Price (1983), the new, pre-Waterton units, described in <u>downward</u> order, are as follows:
– The Tombstone Mountain Formation (175 m), consists of dark grey argillite and argillaceous dolomite and limestone. Carbonate content decreases to the south.

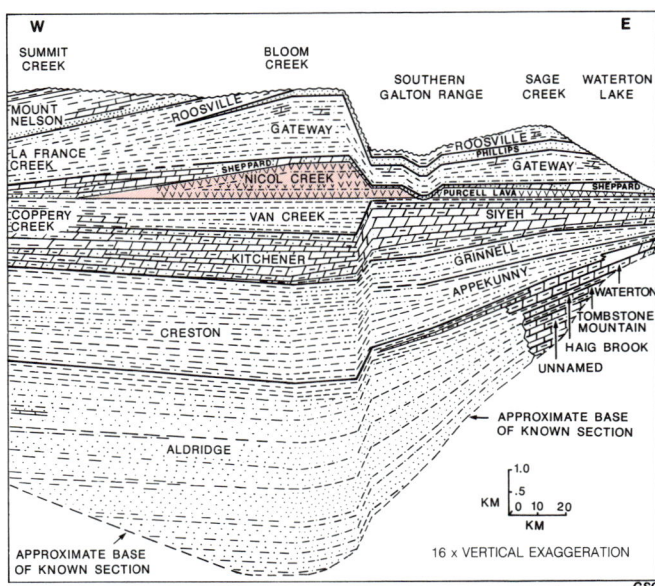

Figure 5.6. Stratigraphic cross-section of the classical Purcell (Belt) Supergroup (modified after McMechan, 1981).

– The Haig Brook Formation (150 m+), is grey dolomite and limestone and grey, red and green argillaceous limestone and dolomite.
– Unit 1 (107 m), consists of dark, calcareous and dolomitic argillite and black argillite.
– Unit 2 (314 m), is pale brownish grey to medium grey argillaceous dolomite and black argillite, with red and green dolomite toward the base.
– Unit 3 (219 m), is pale grey, green, red and white, very fine crystalline and cryptocrystalline dolomite and limestone.
– Unit 4 has a drilled thickness of 170 m, with base unknown. It consists of medium to dark grey, dense, argillaceous dolomite.

The Altyn Formation (145-375 m) consists of dolomite and subordinate limestone with varying contents of argillaceous and arenaceous impurities, and black to grey argillite, which conformably overlie the Waterton Formation (Plate 2). It is gradational upward to greenish argillite of the Appekunny Formation. Division into lower, middle and upper parts is possible in eastern but not western exposures, and the lower part becomes noticeably less argillaceous southward (Fermor and Price, 1983; Hill, 1983). The formation thickens markedly to the west across Waterton Park (Douglas, 1952), largely as a consequence of thickening (from 60-250 m) in the lower part, accompanied by increased clay content.

In the western part of Clark Range, dolomitic argillite overlies strata correlated with the Waterton Formation and underlies green argillite typical of the Appekunny. The dolomitic argillite thickens southward from 135 to 210 m as it loses its sand content and changes from dominantly green to grey. Fermor and Price (1983) proposed that this argillite should be included in the Appekunny, rather than the Altyn Formation. Price (1964) interpreted the Altyn and its equivalents as a nearshore, carbonate platform that passed westward into deeper water in which terrigenous muds accumulated.

Strata of the basal division of the Purcell (Belt) Supergroup exposed in the western Clark Range can be correlated with the upper unit of the Aldridge Formation of the Purcell and western Rocky Mountains on the basis of lithological aspect and stratigraphic position (Price, 1964; McMechan, 1981). These correlations are supported by a southward increase in argillite content, parallel lamination and grey colour in the Altyn, Waterton and Tombstone Mountains formations. In Glacier National Park, Montana, equivalent strata are rusty-weathering, laminated, carbonaceous argillite typical of the upper Aldridge (Prichard) Formation (Earhart et al., 1983).

Changes of thickness and facies in the basal division of the Purcell exposed in the Clark Range are consistent with basin margins to the north and east, as outlined by isopachs of the overlying divisions of the Purcell (see Fig. 5.7).

Lower division

The lower division of the Purcell (Belt) Supergroup consists mainly of clastic sediments laid down in shallow water. These strata gradationally overlie basinal or platformal strata of the basal unit and underlie relatively carbonate-rich strata of the middle unit. In the Purcell Mountains and the Lizard and Hughes ranges of the Rockies, they belong to the Creston Formation and the lower part of the Kitchener Formation; in Clark Range, they belong to the Appekunny and Grinnell formations and the lower member of the Siyeh Formation.

The Creston Formation (920-2350 m) consists of green, grey and purple siltite and argillite and lesser quartz arenite. Sun cracks, ripple marked surfaces and rip-up debris layers are abundant locally. Three main units are recognized: lower and upper units comprising graded siltite-argillite couplets and a middle unit characterized by laterally continuous thin beds of coarse grained siltite and very fine grained sandstone. Sun cracks occur in siltite-argillite couplets and at the top of quartz arenite lenses; they are more common in eastern than in western exposures and are generally absent at the base of the formation. Lenses of fine- to coarse-grained quartz arenite occur in the upper part of the lower unit and in the middle and upper units. Quartz arenite beds are common east of the Rocky Mountain Trench, where they display bi- or trimodal patterns of crossbedding orientations. They become less common to the west, and appear to be derived from the craton to the north and east. In contrast, the coarse grained siltites or very fine grained quartzites, characteristic of the middle unit, thin and become finer grained northward and eastward. This variation suggests the same southwestern source as that inferred for the Revett quartzites in the United States (McMechan, 1981). In the Purcell Mountains and the Hughes and Lizard ranges, coarse grained siltite of the middle unit occurs in extensive, parallel-bedded sheets. Except for grading at the top, they are internally structureless. Parallel laminations occur locally, and the bases are sharp. These siltites probably were deposited in a lowest intertidal to shallow subtidal environment, whereas sequences of interbedded sun-cracked and ripple-marked siltite-argillite and quartz arenite typical of the upper part of the lower unit and the upper unit are intertidal (McMechan, 1981). The latter, intertidal, lithofacies becomes less common westward, and is rare in the western Purcell Mountains (Reesor, 1983). Thus, the Creston Formation was deposited in environments that ranged from intertidal to shallow subtidal, with subtidal sedimentation increasing away from the northeastern limit of exposure.

Argillite and siltite of the Creston Formation grade into dolomitic argillite and siltite of the lower member of the Kitchener Formation. In most areas, the lower member (17-500 m) comprises green, or locally, grey dolomitic siltite and argillite, lesser nondolomitic siltite and argillite, and minor quartzite; pods of micritic limestone occur locally. The unit is characterized by dolomitic, graded siltite-argillite couplets. Climbing ripples, lenticular bedding and sun cracks are common in eastern exposures, and occur also in central exposures. Sedimentation appears to have taken place in the lower intertidal zone (McMechan, 1981). Rocks of this character are not recognized in highly deformed and metamorphosed equivalent strata in the western part of the Purcell Mountains (Reesor, 1983).

In the Clark Range, the Appekunny Formation overlies strata of the basal division of the Purcell. The following description excludes the previously described dolomitic strata of the western Clark Range that Fermor and Price (1983) proposed be treated as a member of the Appekunny rather than as a facies of the Altyn Formation. The Appekunny in its type area consists of green, grey and minor red argillite interbedded with siltstone and green and pale grey quartzitic sandstone. From eastern to western Clark Range it thickens from 300 to 600 m, and displays marked changes. Whereas the Appekunny argillites in the east are green with common red interbeds, redbeds are rare or absent in the western part of the range. In southwestern Clark Range, the argillites are dominantly grey rather than green, and toward the base acquire fine lamination not seen in the east. In the east, sun cracks and ripple marks occur throughout, but in the west they occur only near the top. Price (1964) concluded that the eastern facies of the Appekunny probably accumulated as flood plain or tidal flat deposits. The lenticular quartz sandstones are channel-fill deposits, whereas the persistent, thick, quartzite marker units of the lower Appekunny may be transgressive beach deposits. The laminated grey argillite facies to the southwest may represent deeper-water deposits; it is lithologically identical to the basal part of the Creston Formation in the Rocky Mountains. Green argillites of the Appekunny grade into red argillites of the Grinnell Formation. The contact is probably markedly diachronous (Price, 1964).

The Grinnell Formation consists of red argillite and argillaceous siltstone, with lenticular interbeds of white and greenish grey quartzitic sandstone that increase in amount upward. The argillite commonly shows sun cracks; ripple marks are abundant in the siltstone and sandstone. Conglomerates of red argillite rip-up clasts are characteristic. The contact with the overlying Siyeh Formation is gradational. The Grinnell increases in thickness from about 100 m in northeastern to 230 m in southeastern and 520 m in southwestern Clark Range. Price (1964) suggested that the Grinnell accumulated in a nonmarine flood plain or brackish tidal-flat environment. The overlying lower member of the Siyeh Formation contains the transition

CHAPTER 5

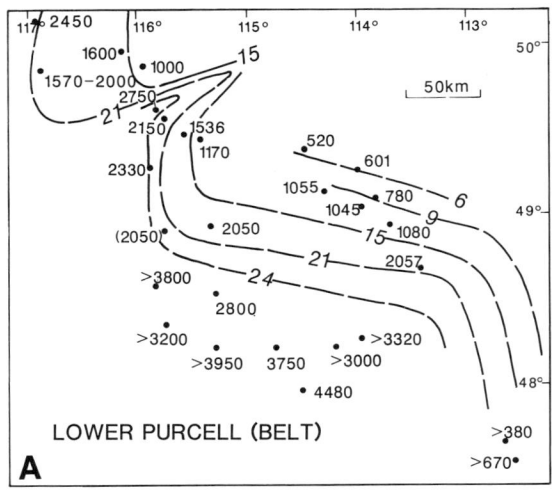
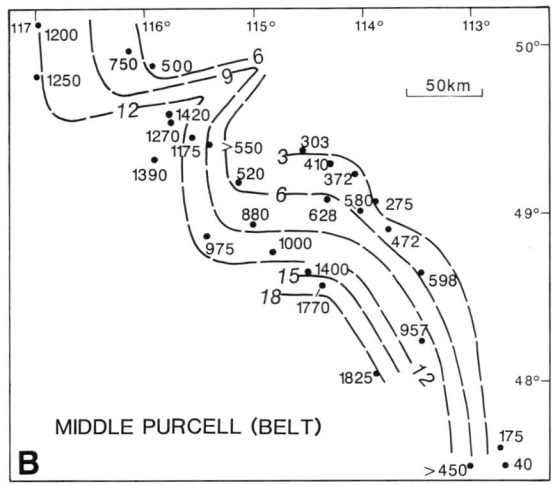
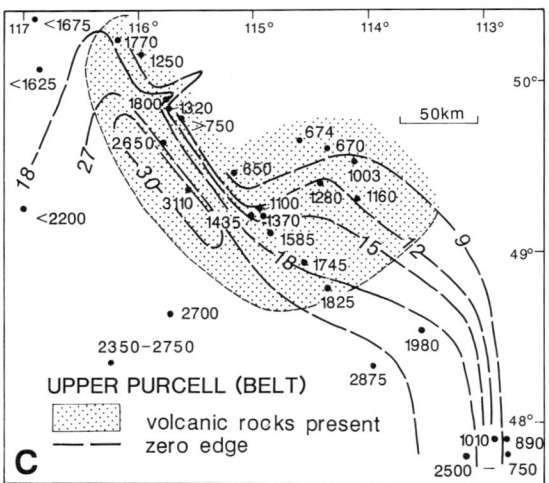
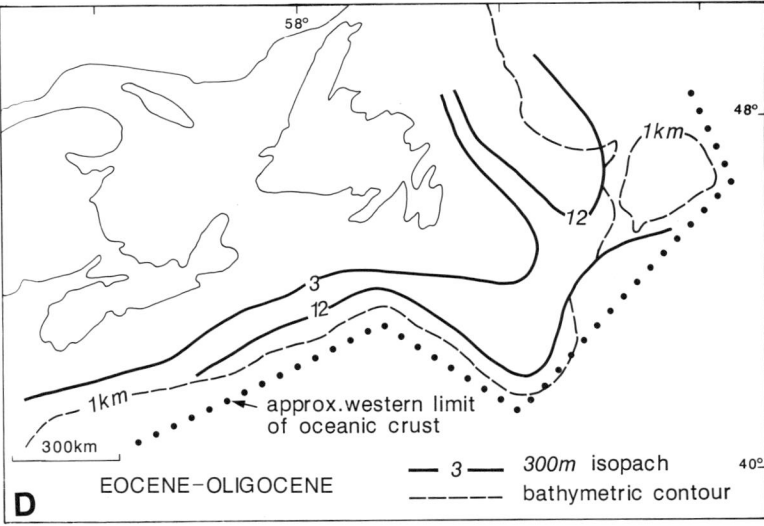
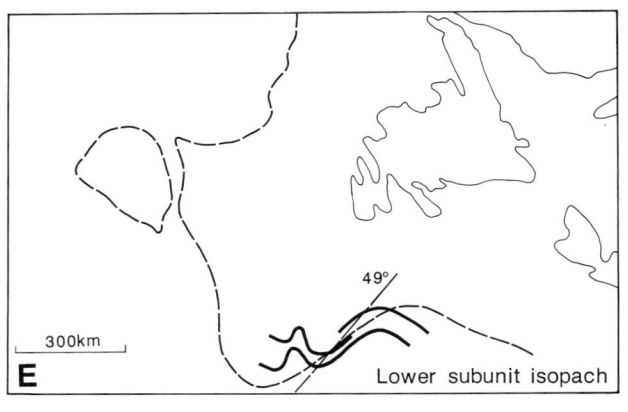

Figure 5.7. Configurations of isopachs of the three divisions of the Purcell (Belt) Supergroup (A,B,C) compared with the configuration of the Eocene-Oligocene, Atlantic passive margin (D). E, isopachs from Figure A superimposed on reversed geometry of the Atlantic margin. Locality thicknesses in metres and isopachs in hundreds of metres, (modified after McMechan, 1981).

from clastic Grinnell strata beneath to the carbonate-dominated middle member of the Siyeh above. It consists of green, grey and black argillite, green and grey argillaceous and arenaceous dolomite, and dolomitic quartz sandstone. To the northeast it thickens from 9 to 150 m at the expense of the overlying carbonate unit.

Thickness variations in the lower division of the Purcell outline a rectilinear basin margin (Fig. 5.7) whose shape is reflected by the locus of the northeastern limit of Purcell exposure. Gradual facies changes within the division, described above, also reflect this shape.

Middle carbonate division

The middle carbonate division of the Purcell (Belt) Supergroup forms a distinctive unit throughout the Belt-Purcell Basin except in the northwestern Purcell Mountains, where metamorphism and intense deformation hamper its separation from strata of the overlying, upper division of the Purcell (Reesor, 1973). The middle carbonate division is of platformal character along its northeastern limit of exposure (middle Siyeh, Plate 2), and of more basinal character elsewhere. In the Clark and Galton ranges of the Rockies, both facies are represented (Fig. 5.6). In the central and northern Purcell Mountains, and in the Hughes and Lizard ranges of the Rockies, both facies are included in the upper member of the Kitchener Formation. In the western Purcell Mountains, these strata are included in the 'Coppery Creek group', which probably also incorporates strata belonging to the upper division of the Purcell.

The platformal facies includes thin-bedded, internally laminated argillaceous dolomite and dolomite, sandy quartzose dolomite, dolomitic quartz sandstone and graded, grey to black, siltite-argillite couplets. Intraformational conglomerate with tabular or disc-shaped pebbles of dolomite is common; algal stromatolites, oolites, sun cracks and ripple marked surfaces occur locally (Price, 1964; Høy, 1979; McMechan, 1981). The thicker, deeper-water facies includes thin-bedded, silty and argillaceous dolomite and limestone, and graded, grey to black, carbonate and non-carbonate-bearing siltite-argillite couplets in laminae to very thin beds. Black argillite partings and large 'molar tooth' structures are common in the carbonates. Small 'molar tooth' structures, load casts, flame structures and irregular syneresis cracks are abundant in the siltite-argillite. Thin beds of quartzose, sandy dolomite, limestone or dolomite intraformational conglomerate, oolitic limestone, algal stromatolites and ripple-marked surfaces occur locally. These sediments were deposited mainly in shallow subtidal environments (McMechan, 1981).

In the western Purcell Mountains, strata in the more basinal facies are complexly deformed and metamorphosed. They have been included in the 'Coppery Creek group'. Reesor (1983, 1984) recognized three units within the Coppery Creek. A thick (1400 m) lower unit consists of dolomite interbedded with green, grey or black phyllite which grades upward to silvery and green phyllite, siltite and some carbonate. A middle unit (200 m) is thinly laminated black phyllite and grey siltite. An upper unit (300 m) comprises silvery phyllite, calcareous dark grey phyllite and dolomite, with a sequence of interbedded dolomite and quartzite at the top. The middle carbonate division of the Purcell is probably represented by the thick, lower unit of the 'Coppery Creek'.

A distinctive marker horizon, 15 to 25 m thick, consisting mainly of stromatolites, occurs near the top of the middle carbonate division in the Clark, Galton and southern Lizard ranges and in the central Purcell Mountains (Price, 1964; McMechan, 1981). The variable, mostly columnar stromatolites include *Conophyton* in some areas. Strata at the top of the middle carbonate division become more argillaceous, and the transition to argillite of the upper division of the Purcell generally occurs over a few metres. Locally in the Galton and eastern Lizard ranges, strata belonging to the middle part of the upper division (Sheppard Formation) disconformably overlie the middle carbonate division (Price, 1964).

The transition from platformal to more basinal facies is accompanied by westward thickening: from the southern Lizard Range (700 m) to the central Purcell Mountains (1390 m), and from the Hughes Range (500 m) to the northern Purcells (1250 m). A thick succession of more basinal facies occurs in the northern Lizard Range. In the Clark Range, the middle carbonate division thickens to the south and west, from the thin platformal succession at its northern (300 m) and southeastern (275 m) limits of exposure to the mixed facies in the southwestern Clark Range (630 m). Eastward thinning in the Clark Range may be due partly to the equivalence of beds in the middle carbonate division in the west with beds at the top of the lower Purcell in the east (Price, 1964). Regional variations in the thickness of the middle carbonate division outline a rectilinear basin margin similar to that of the lower division of the Purcell (Fig. 5.7). The facies belts within the middle division follow the isopach trends.

Upper division

Detrital, carbonate, and locally, volcanic rocks make up the upper division of the Purcell (Belt) Supergroup. The central and eastern exposures share a common succession of shallow-water to subaerial sediments with one volcanic unit, but important changes appear near the western and northern limits of Purcell exposure. In the following descriptions, central and eastern exposures are treated jointly, as are western and northern exposures.

Central and eastern exposures: The lowest part of the upper division of the Purcell in central and eastern exposures consists of green, locally red, purple or black, laminated to very thin-bedded, graded siltite-argillite couplets. Lenses of fine grained quartzite and dolomitic siltite and argillite occur locally. Sun cracks, ripple marks, ripple crosslamination, climbing ripples, rip-up debris beds, pull-apart and scour-and-fill structures are abundant in some places. The depositional environment probably varied from tidal flat to nonmarine flood plains (Price, 1964; McMechan, 1980). These strata are included in the Siyeh Formation of the Clark and Galton ranges and in the Van Creek Formation of the Lizard and Hughes ranges and the southern to northern Purcell Mountains. In all these areas, the strata are abruptly overlain by volcanic rocks.

Thickness of the upper Siyeh increases southward and eastward in the Clark Range, from 30 m in the northwest to 105 m in the southwest and 180 m in the southeast. The eastward thickening probably occurs in part at the expense of the underlying middle Purcell carbonate. The thickness of the Van Creek increases westward from the central Lizard Range (200 m) to the southern Purcell

Mountains (750 m), and from the Hughes Range (200 m) to the northern Purcell Mountains (600 m). As with the underlying formations of the Purcell, the Van Creek is thicker in the northern Lizard Range (420 m) than in the immediately adjacent Hughes and central Lizard ranges.

Volcanic rocks form an important chronostratigraphic marker in central and eastern exposures of the Purcell (Belt) Supergroup. In the Clark, southern Flathead, and Galton ranges, a succession of green, amygdaloidal, chloritized flows of basaltic to andesitic composition (Hunt, 1962) is called the Purcell Lava (60 m to 150 m). In the Clark and southern Flathead ranges, a basal unit, 10 to 20 m thick, is pillowed lava. In the southern Galton Range the lava was locally removed by erosion prior to deposition of the overlying Sheppard Formation (Price, 1962, 1964).

In the Lizard and Hughes ranges and the southern to northern Purcell Mountains, the volcanic unit is called the Nicol Creek Formation. Near its depositional, western edge, the Nicol Creek consists entirely of volcanic tuff. Elsewhere, it consists of interlayered green and maroon, massive, amygdaloidal volcanic flows, derived green volcanic sandstone and siltstone, tuff, and green and purple, locally dolomitic, nonvolcanic siltite and argillite. Based on chemical analyses the volcanic rocks are basaltic to andesitic (Daly, 1912; Hunt, 1962; McMechan et al., 1980). The local abundance of sun cracks, ripple lamination, ripples and rip-up debris beds in the sedimentary rocks suggests that they were deposited on a tidal flat or nonmarine flood plain (McMechan, 1980). The Nicol Creek is abruptly overlain by quartzose sediments. The contact is disconformable in parts of the Purcell Mountains and in the eastern Lizard Range, where the Nicol Creek and Van Creek appear to have been removed by pre-Sheppard erosion (cf. southern Galton Range). The Nicol Creek is about 750 m thick in the southern Purcell Mountains and central Hughes Range, and thins rapidly to the east and west, except for an anomalous area in the northern Lizard Range, where it is 608 m thick.

The Sheppard Formation consists of stromatolitic dolomite, dolomite, grey and dark red quartz sandstone and siltstone, pale green dolomitic sandstone and siltite, and green and red laminated argillite. Ripple marks, sun cracks, crossbedded oolites and intraformational conglomerates occur locally. A chloritized lava flow, 10 to 15 m thick, and similar to Purcell Lava flows, occurs in the lower Sheppard in the southern Clark Range. In the Clark Range, redbeds are restricted to the upper part of the Sheppard and become more abundant to the east and north (Price, 1964). A shallow-water depositional environment with an intermittent supply of terrigenous sediment is inferred (Price, 1964; McMechan, 1981). Rocks of the Sheppard Formation grade upward to siltite and argillite of the Gateway Formation. The Sheppard decreases in thickness from 275 m in the southern to 165 m in the northern Clark Range, 125 m in the southwestern Clark Range and 50 m in the southern Flathead Range. Strata included in the upper part of the Sheppard in the Clark Range may be equivalent to those assigned to the lower part of the Gateway Formation elsewhere (Price, 1964). The Sheppard is about 200 m thick in the northern Hughes Range, 115 m in the northern Lizard Range and 115 m in the southern Purcell Mountains.

The Gateway Formation consists of two members, a lower siltite and an upper siltite and dolomite member. Thin-bedded siltite with minor argillite partings, and lesser, discontinuously laminated, graded siltite-argillite couplets make up the lower member. These change from dominantly dark red or purplish red along the northeastern limit of exposure in the Clark Range to dominantly grey-green and grey elsewhere. Sun cracks, oscillation ripple marks and intraformational conglomerates are abundant locally. The common presence of casts of salt crystals are characteristic.

The upper member consists of green and greenish grey, micaceous argillite, dolomitic argillite, and coarse-grained, dolomitic sandstone with interbeds of pale grey dolomite and sandy dolomite. Oscillation ripple marks are common, and sun cracks and algal stromatolites occur locally. The upper contact is transitional to the Phillips Formation through an increase in the proportion of sandstone (Price, 1964; McMechan, 1981). The Gateway was deposited in very shallow water, possibly on a hypersaline tidal flat (Price, 1964) or in a lagoon (McMechan, 1981). In the Clark Range the Gateway thickens from 350-380 m in the northwest to 460-490 m in the northeast and south, and thence to 700-850 m in the southwest (Price, 1964). It is approximately 750 m thick in the Galton and northern Hughes ranges and thickens to 1350 m in the southern Purcell Mountains. The Gateway is truncated by sub-Cambrian and/or sub-Devonian erosion in the Lizard and southern Hughes ranges.

The Phillips Formation consists of thin-bedded, maroon and red, fine- to coarse-grained quartz sandstone and siltite. Argillite and micaceous argillite occur as partings and thin interbeds. The argillites commonly pass laterally to intraformational conglomerates composed of argillite chips in a sandstone matrix. Ripple marks and sun cracks are locally prominent (Price, 1964; McMechan, 1981). The small-scale primary structures and the sheet-like geometry of the sandstone beds suggest that the formation was deposited in very shallow water, possibly as sheet floods in a subaerial environment (McMechan, 1981). The sandstones grade upward to argillite of the Roosville Formation. Depositional thickness of the Phillips ranges from about 130 m in the northwestern Clark Range to 200 m in the southwestern Clark, southern Macdonald and southern Galton ranges. From the central Purcell Mountains (150 m) it thins northward and westward to disappearance (Leech, 1958, 1960). In the northern Hughes Range the Phillips is about 60 m thick. It is missing as a result of sub-Cambrian and/or sub-Devonian erosional truncation in the southern Hughes and Lizard ranges.

The Roosville Formation is the uppermost unit of the Purcell (Belt) Supergroup preserved in central and eastern exposures. Its thickness and distribution are controlled by sub-Windermere, sub-Cambrian and sub-Devonian erosional truncation. Maximum preservation is in the Galton Range (up to 1300 m) and in the southwestern Clark Range (600 m). The Roosville consists of green and greenish grey argillite, dolomitic argillite, siltite and sandstone, with lesser light-coloured quartzite, argillaceous and stromatolitic dolomite, and red argillite. Micaceous partings are common and sun cracks and ripple marks occur throughout. The Roosville was probably deposited in a tidal-flat setting (Price, 1964).

The overall pattern of thickness variation for the upper division of the Purcell (Belt) Supergroup (excluding the Roosville Formation) in central and eastern exposures is outlined in Figure 5.7. The isopachs trend across those of the underlying units and probably reflect the development of northwest- and west-southwest-trending epeirogenic structures. The most prominent of these, now located in the eastern Purcell Mountains, is a northwest-trending sub-basin with a steep northeast boundary, resembling a half-graben with its northeast side down-dropped.

Western and northern exposures: Towards the western and northern limits of Purcell exposure, strata of the Van Creek, Sheppard, Gateway and Roosville formations pass laterally into a succession of grey and green siltite, argillite and phyllite, quartzite, argillaceous dolomite and dolomite. With the thinning and disappearance of volcanic (Nicol Creek) and red quartzite marker (Phillips) units, subdivision of the upper division becomes impractical; assigned to it are the upper two units of the 'Coppery Creek' and 'La France Creek groups' in western exposures, and the Van Creek, Sheppard, Gateway and Roosville formations in northern sections (Reesor, 1984). In both areas, the overlying Mount Nelson Formation of dolomite, quartzite, argillite and siltite is the uppermost of the Purcell (Belt) Supergroup (Fig. 5.3).

The Sheppard-Gateway Formation of northern exposures consists of dolomite with minor argillite and crossbedded quartzite. Stromatolitic horizons commonly are associated with oolitic beds. The unit gradationally overlies argillite of the Van Creek or, locally, abruptly overlies tuff of the Nicol Creek. It is 700 m thick near Mount Abel in Lardeau (east-half) map area (Reesor, 1958, 1984).

The Roosville Formation consists of dark argillite, green siltite, and, in the lower third, dolomitic and calcareous siltite and argillite (Reesor, 1958, 1984). The formation gradationally overlies carbonate of the Sheppard-Gateway. It is 625 m thick near Mount Abel in Lardeau (east-half) map area (Reesor, 1984).

The 'La France Creek group' of the western Purcell is dominated by siltite, quartzite and phyllite. The strata commonly are intensely deformed and metamorphosed, and depositional thicknesses difficult to estimate. Reesor (1983, 1984) recognized a lower unit of thinly interbedded black phyllite and grey siltite and an upper unit of grey siltite and quartzite with black phyllite and carbonate-bearing siltite and phyllite near the top. Dolomite occurs within the unit a few hundred metres thick in northern exposures (Reesor, 1984). The 'La France Creek group' gradationally overlies silvery phyllite, calcareous dark grey phyllite, and dolomite, with a sequence of interbedded dolomite and quartzite at the top, included within the upper unit of the 'Coppery Creek group' (Reesor, 1983, 1984). Correlation of these latter with the dolomite, quartzite and argillite of the Sheppard Formation of central and eastern exposures is plausible, but uncertain. The underlying, laminated black phyllite and grey siltite of the middle 'Copper Creek group' are probably correlative with the Van Creek Formation, and thus form the base of the upper division of the Purcell. The 'La France Creek group' is about 1000 m thick. The middle and upper units of the 'Coppery Creek group' are about 300 and 200 m thick respectively (Reesor, 1983, 1984).

In most areas, strata of the 'La France Creek group' grade into thicker-bedded quartzite at the base of the Mount Nelson Formation. Locally in the northern Purcell Mountains there is an abrupt change from argillite to quartzite, and the contact may be disconformable (J.E. Reesor, pers. comm., 1985). The Mount Nelson Formation consists of a cliff-forming, basal unit of white, grey or green orthoquartzite with rare argillaceous laminae and partings, overlain by brownish red- to grey-weathering impure carbonate interbedded with black, purple or red argillite and grey siltite. Stromatolites and lenses or nodules of chert occur locally within the carbonate unit. The basal orthoquartzite, up to 70 m thick, thins gradually to the south. In northern exposures, an upper orthoquartzite up to 105 m thick commonly occurs near the top of the carbonate succession. Interbeds of green, black or red argillite are common within the upper quartzite unit, and green and black argillite and siltite form the top of the preserved formation. The carbonate unit is thicker in western exposures, where it is overlain by interbedded black phyllite and grey siltite. Cream-weathering, dark-coloured dolomite and brown-weathering, white dolomite, locally interbedded with black phyllite, occur at the top of the formation as preserved. Mud cracks in argillite, ripple marks in quartzite and solution-breccias in dolomite are locally common in both areas. The Mount Nelson was deposited in a shallow-water environment (Reesor, 1973).

The Mount Nelson Formation, whose maximum preserved thickness is about 1000 m (Reesor, 1973, 1984), is unconformably overlain by conglomerate of the Toby Formation of the Upper Proterozoic Windermere Supergroup. Evidence for small-scale, pre-Toby block faulting is found locally (Atkinson, 1975; K. Root, pers. comm., 1985), and the St. Mary's fault, in the southern Purcell Mountains, underwent up to 9 km of movement during Windermere deposition, with attendant deep erosion of Purcell strata from the uplifted block south of the fault (Lis and Price, 1976). Regionally, the unconformity cuts out progressively older Purcell strata southward, along the western Purcell Mountains. Locally, in northern Purcell exposures (Mount Forster area), a conglomerate of limestone and dolomite blocks and boulders in a dolomite matrix occurs 60 to 90 m below the Toby Conglomerate. Reesor (1973) treated the lower conglomerate as part of the Mount Nelson Formation, but it and an overlying dolomite unit are now considered part of the Windermere Supergroup, rather than the Purcell (K. Root, pers. comm., 1984; J.E. Reesor, pers. comm., 1985).

Monashee Complex

The Monashee Complex, part of the Shuswap Metamorphic Complex (Okulitch, 1984) is exposed in the Monashee Mountains of southeastern British Columbia (see Chapter 4). It consists in part of core paragneiss dated at 2.8 and 2.2 Ga, intruded by granitoid plutons dated at 2.1 and 1.96 Ga. The core is unconformably overlain by metasedimentary rocks: a basal quartzite conglomerate followed by pelitic, psammitic and calc-silicate gneisses overlain by thick psammitic gneiss. These units are intruded by syenite, zircons from which have yielded a poorly constrained U-Pb date of about 750 Ma. The various published suggestions as to the age of these metasediments, reviewed by Okulitch (1984), range from lower Paleozoic to older than 1.7 Ga. Correlation with the Purcell Supergroup is plausible but undemonstrated.

Wernecke Supergroup (Purcell-Wernecke Assemblage)

The Wernecke Supergroup is known only from the Wernecke and Ogilvie mountains of the northern Yukon (Fig. 5.1, 5.2). Although exposure of the supergroup is discontinuous, the lithic character and threefold aspect persist from the type area near Gillespie and Fairchild lakes, where the supergroup is more than 13 km thick, westward to the Coal Creek 'Dome', in the Ogilvie Mountains near the Alaska boundary.

The base of the Wernecke is not exposed. At its top, large-scale folds in cleaved, Wernecke strata are bevelled by uncleaved, basal beds of the Pinguicula Group, a stratigraphic succession of uncertain correlation which is in turn unconformable beneath the Rapitan Group of the Windermere Supergroup. Correlation of the Wernecke with the Purcell (Belt) Supergroup is defensible on the basis of evidence available, but requires further documentation.

The most thorough description of the Wernecke Supergroup in its type area is that of Delaney (1981), who divided it into three groups, which, in upward order are the Fairchild Lake, Quartet and Gillespie Lake groups, each of which he subdivided into numbered, informal units of formational rank. The rocks are regionally metamorphosed to greenschist facies, and have a weakly developed cleavage.

The Fairchild Lake Group consists of at least 4 km of generally thin-bedded, commonly laminated siltstone, mudstone, claystone and fine sandstone with minor intercalated carbonate rocks (Plate 3). Plane-parallel lamination, crosslamination, lenticular and wavy bedding, ripple marks and load structures occur in most parts of the group, whereas flute casts are more restricted in their distribution. Delaney divided the group into four units on the basis of subtle differences in bedding thickness and style, weathering colour, and amount and character of carbonate beds, which are minor in all four units. He concluded that most of the group was deposited in 'deeper' water, by predominantly southerly flowing currents, but that unit F-4 was deposited in a shallow, marginal marine environment.

The Quartet Group, gradationally succeeding the Fairchild Lake, is about 5 km thick. It is divided into two formational units. Unit Q-1, at the base, consists of about 200 m of thin-bedded, carbonaceous claystone and silty, carbonaceous mudstone. It is in part pyritic, and generally metamorphosed to slate. Unit Q-2 is up to 4800 m of dark grey- to grey-weathering siltstone, fine sandstone and claystone. The basal several hundred metres is rhythmite of pale grey-weathering siltstone and dark grey-weathering mudstone. Above this, the formation is characterized by medium to thick beds of wavy, lenticular and flaser-bedded siltstone; thin, medium and locally, thick beds of crosslaminated coarse siltstone to very fine sandstone; and medium beds of 'chaotic' carbonaceous siltstone and laminae to thin beds of mudstone and claystone. Interference ripple marks are ubiquitous; other common sedimentary structures are wavy, flaser and lenticular bedding, symmetrical and asymmetrical ripple marks, load-casted ripples, flame structures and ball-and-pillow structures. Shrinkage cracks are abundant, and a small proportion of these are interpreted as sun cracks. Slump structures are common. The medium to thick, commonly slump-folded beds of siltstone and very fine sandstone are submature to mature arkoses. Paleocurrent determinations based on crosslamination yield bimodal or polymodal patterns. Delaney (1981) interpreted unit Q-1 as a starved-basin deposit and Q-2 as deposited in a shallow, marine environment.

The Gillespie Lake Group, in gradational contact with the underlying Quartet, is at least 4 km thick, and is characterized by its predominant carbonate rocks (Plate 4). At the type locality, it is divided into seven units of potentially formational rank. At the base, the TR (transitional) unit (25-700 m) and the overlying G-2 (400-600 m) commence with strata similar to unit Q-2 of the Quartet, with an upward-increasing content of orange-weathering beds of dolomitic siltstone and silty dolomite. Minute lenses of black chert are present in the upper part. Paleocurrents, determined from crossbeds and ripple marks, where not bimodal or polymodal, indicate southerly or southeasterly flow. Units G-3 (thickness undetermined) and G-4 (about 450 m) consist of silty dolomite, dolomite, dolomitic mudstone, dolomitic siltstone and, in G-4, more or less grey to dark grey chert. Unit G-3 contains some lenses of 'edgewise' intraformational conglomerate, and upward-thickening cycles are characteristic of the middle subdivision of unit G-4. Units G-5 (thickness unreported) and G-6 (500-800 m) consist of parallel- and cross-laminated silty dolomite, with subordinate beds of dolomitic claystone and dolomitic siltstone. Both contain conglomerates of several kinds, and G-5 contains a few algal stromatolites in place, and mounds of breccia composed of stromatolitic clasts. Unit G-6 (500-800 m) contains distinctive, greenish-grey-weathering, parallel-laminated beds of limestone. The paleocurrent pattern is bimodal or polymodal, with a dominant, southward-directed mode. Unit G-7 (400-700 m) consists of orange, buff, and grey, locally pink-weathering carbonate rocks, including thin to thick beds of crinkly laminated (stromatolitic?) dolomite, oolitic and oncoidal dolomite, stromatolitic dolomite, clayey carbonaceous dolomite and parallel- to wavy-laminated and lenticular-bedded dolomite. Molar-tooth structure is ubiquitous. Some relict masses of pale grey limestone remain. Minor lenses of grey and dark grey chert are present. Paleocurrents are polymodal. The Gillespie Lake Group was deposited in shallow-marine environments; dominant, fine grained siliciclastic sediments at the base are progressively replaced upward by fine grained carbonate detritus, although a subordinate siliciclastic component persists almost to the top. Delaney (1981) inferred the appearance of a major source of carbonate detritus to the (present) north. Upward, signs of the establishment of an autochthonous carbonate platform appear in unit G-5, and G-7 represents such a platform throughout.

The areas of exposed Wernecke strata are characterized by metre- to kilometre-scale masses of monolithic and heterolithic breccia that are hosts to numerous showings of uranium, copper and cobalt. Some of these breccias have large vertical dimensions and are in part branching pipes (Delaney, 1981), whereas others are sheet-like, cutting stratification at low angles (R.I. Thompson, pers. comm.,

1985). Some bodies show little displacement of clasts relative to wall-rock, but displacements of clasts upward or downward a few hundred metres are common. The fabrics are diamict, the clasts locally rounded. Igneous material is virtually absent. A spatial association with faults and lineaments and their intersections is probable. The various kinds of alteration associated with the breccias include sodium feldspathization, silicification, hematization and carbonization. There is no agreement as to the origin of these enigmatic breccia masses; processes such as hydraulic stoping, gas streaming (true diatremes), crush-brecciation, and evaporite solution-brecciation have been suggested (Delaney, 1981; Bell, 1982; Young et al., 1982). To the extent that gases or fluids of deep origin may have been involved, as suggested by several authors, the 'breccia province' may be related to the character of sub-Wernecke basement, rather than to any process inherent to the Wernecke depositional cycle.

The fragmentary information on the relative and absolute ages of the Wernecke Supergroup leads to a broad correlation with part or all of the classical Purcell (Belt) Supergroup (Young et al., 1979; Delaney, 1981; Young et al., 1982). The correlation rests in part on the few published radiometric dates, reviewed by Delaney (1981): a uranium-lead age of 1.153 Ga and a lead-lead age of 1.249 Ga on a pitchblende sample from an intrusive breccia complex, giving an apparent minimum age; model lead ages of 1.288 and 1.440 Ga on lead minerals hosted by Gillespie Lake Group strata, with similar implications. Additionally, the great thickness of the supergroup, suggesting a continental-margin setting, and the pre-Windermere folding of the Wernecke are consistent with such an interpretation. Although not compelling, the radiometric dates suggest equivalence between the Wernecke and the pre-Coppermine Lavas succession (Hornby Bay and Dismal Lakes groups) of the Coppermine Homocline. Furthermore, both successions display (as does the classical Purcell) an overall progression from mainly clastic rocks in the lower part to mainly carbonate rocks at the top, and the relationship between the two is plausibly proximal (Hornby Bay-Dismal Lakes) to distal (Wernecke Supergroup), particularly if the Wernecke has undergone large-scale displacement northwestward from its depositional site, as speculated below (Fig. 5.8).

One of many problems regarding the Wernecke Supergroup is: where presently exposed, is it autochthonous, or allochthonous, and if the latter, to what degree? The question is raised by the striking misfit of the Proterozoic successions juxtaposed across the Snake River Fault (Fig. 5.1), at the eastern end of the Wernecke Mountains (Eisbacher, 1981; Norris, 1982, 1983). At the top of the Proterozoic succession there the Rapitan Group (Windermere) need not be displaced as the differences seen across the fault are within the marked variability of the Rapitan in the area. Beneath the Rapitan, east of the fault is the Mackenzie Mountains Supergroup, erosionally bevelled to the level of the Gypsum Formation of the Little Dal Group. West of the fault, the Rapitan rests unconformably on the Pinguicula Group. No detailed description of the Pinguicula has yet been published (for a summary, see Eisbacher, 1981), but it bears no strong resemblance to any part of the Mackenzie Mountains Supergroup and its correlation has been debated at length without resolution (Eisbacher, 1978, 1981; Yeo et al., 1978;

Young et al., 1979). Given the remarkable persistence of formations and members of the Mackenzie Mountains Supergroup, it is clear that if the Pinguicula, in its sole area of exposure, is indeed a Mackenzie Mountains correlative, as claimed by Young et al. (1979), Eisbacher (1981) and Bell (1982), it and the underlying Wernecke Supergroup are allochthonous relative to the substantially different Mackenzie Mountains Supergroup rocks east of the Snake River Fault (Fig. 5.1). Alternatively, the Pinguicula Group is not correlative with any part of the Mackenzie Mountains Supergroup, in which case the marked dissimilarity between the pre-Rapitan, Proterozoic formations juxtaposed across the Snake River Fault may or may not be an argument for an allochthonous Wernecke Supergroup. If the Wernecke is allochthonous, its position when deposited, and when folded in pre-Pinguicula (certainly pre-Windermere) time, is unknown.

If the Wernecke Supergroup is autochthonous, its detrital materials were derived from the northwestern Canadian Shield. If the hypothesis that the Wernecke is allochthonous is accepted, it is tempting to restore it to a location far to the south, where it would be closer to the speculatively contemporaneous Hornby Bay-Dismal Lakes succession of the Coppermine Homocline, and would form the distal part of a miogeoclinal wedge of which the Hornby Bay-Dismal Lakes would make up the proximal part, as suggested by Bell (1982) (Fig. 5.8). This speculation is supported by the generally feldspathic composition of Wernecke arenites, and the upward progression, in both successions, from mainly clastic to mainly carbonate strata. The hypothetical movement would be of dextral sense, consonant with the prevailing dextral displacements on documented, Phanerozoic faults of the northern Cordillera (Norris and Yorath, 1981). If the fault followed the tectonic arc of the Mackenzie Mountains, arcuate Proterozoic isopachs, and arcuate termination of Shield-type magnetic anomalies, restoration of the movement would involve clockwise rotation of the Wernecke terrane so that paleocurrents now seen as southward-directed would be westward in the restored position and consistent with the postulated proximal-distal relationship.

Muskwa Ranges Succession (Muskwa Assemblage)

The older (pre-Windermere) Proterozoic succession of the Tuchodi Lakes area in the Muskwa Ranges of northeastern British Columbia (Fig. 5.1, 5.2) has not been dated isotopically. It is here arbitrarily included in the discussion of Sequence A because its correlation with either the Purcell (Belt) Supergroup or the Mackenzie Mountains Supergroup is unknown. The sequence is assuredly Middle Proterozoic in age because at least at one locality on Mount Lloyd George it is in contact with exposed Windermere strata (Sequence C) (H. Gabrielse, pers. comm., 1985). Also, as noted by Preto (1971) dykes of probable Windermere age cut folded strata of the older succession but are not themselves folded.

The succession, over 6 km thick where exposed, has been divided into seven formations, which in upward order are the Chischa, Tetsa, George, Henry Creek, Tuchodi, Aida and Gataga (Bell, 1968; Taylor and Stott, 1973). In the subsurface it thins to the northeast and disappears to

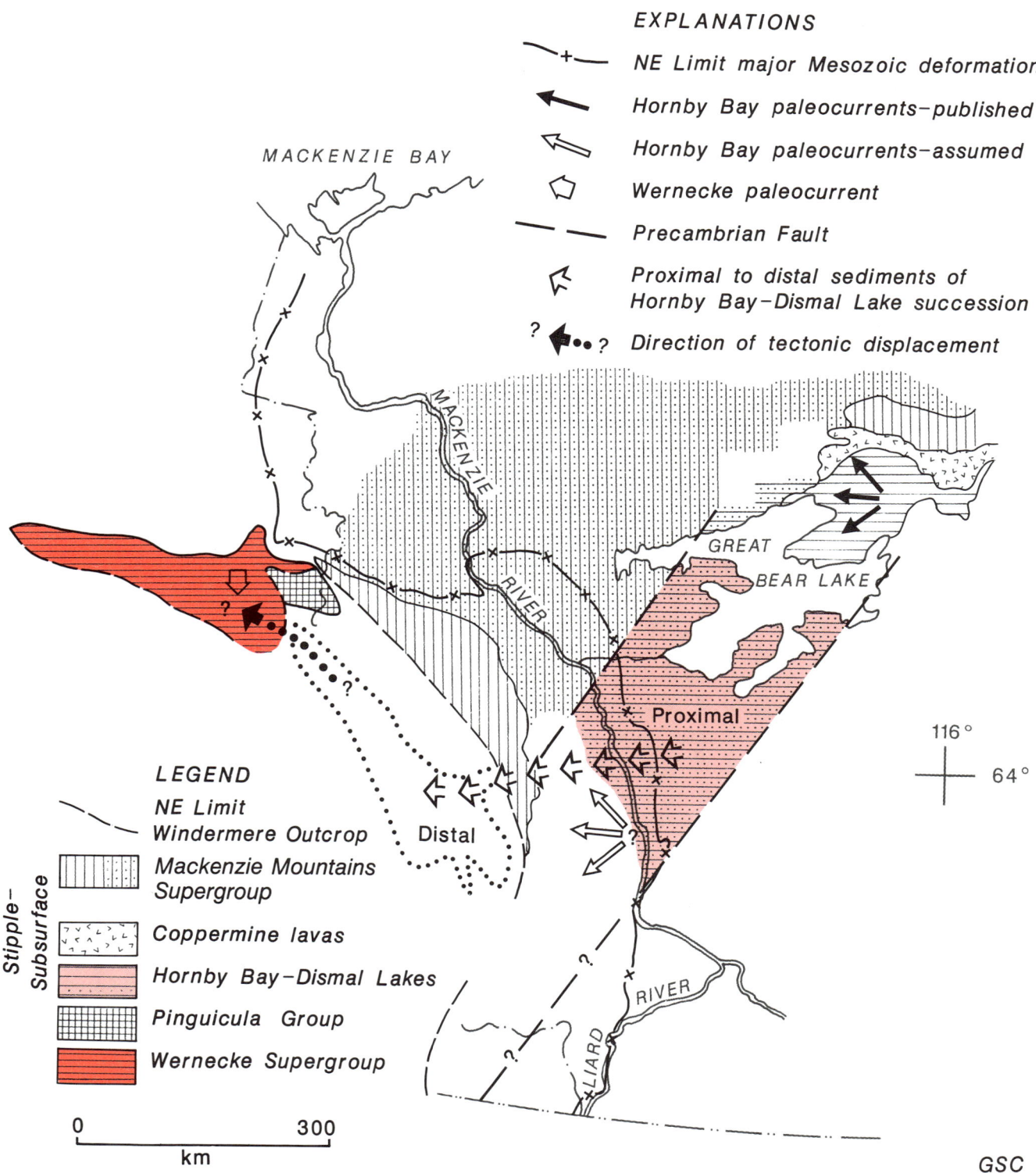

Figure 5.8. Speculations on a relationship between the Hornby Bay-Dismal Lakes succession of Coppermine Homocline, and the Wernecke Supergroup, and of the possibly displaced character of the Wernecke. These ideas were in part anticipated by Bell (1982). Dotted pattern denotes Pharnerozoic cover.

the south, north of the Peace River Arch (R.I. Thompson, pers. comm., 1985).

The Chischa Formation, with base concealed, is about 940 m thick. It is characterized by pale grey and pastel-coloured, microcrystalline dolomite, with stromatolites, abundant molar-tooth structure, flat-pebble conglomerates and ripple marks. Beds of fine grained, locally ripple marked orthoquartzite are common in the upper third of the formation. The Chischa was deposited in a shallow, marine environment.

The Tetsa Formation rests upon the Chischa with clearly erosional contact. It is about 320 m thick, and consists of thin-bedded, dark grey to black, siliceous, feldspathic, micaceous and carbonaceous mudstone, siltstone and shale. Fine, plane-parallel lamination occurs throughout, with local current ripples, load casts and flame structures. Several thin beds of orthoquartzite, some of which Bell (1968) identified as turbidites, are present in the basal 60 m. The Tetsa accumulated in a 'deep-water' environment, relative to that of the formations below and above. Its contact with the overlying George Formation is gradational.

The George Formation, 360 to 530 m thick, is a carbonate unit dominated by thick-bedded lime mudstone, generally with well-developed, molar-tooth structure, and thin-bedded, parallel-laminated and ripple-cross-laminated calcisiltite. Flat-pebble conglomerates are prominent, and generally occur in short, stout lenses; some of these display normal grading and a few are matrix-supported. Minor calcareous siltstone and very fine grained sandstone occur in the basal zone transitional from the Tetsa. Algal stromatolites, commonly of columnar-branching type, first appear about 90 m above the base, and recur at many intervals from that horizon upwards. Toward the top, 'floating' intraclasts are increasingly common in beds of lime mudstone, and some beds may be described as diamictite. Bell (1968) reported mud cracks (but these are not necessarily subaerial 'sun cracks'); based upon the stromatolites and conglomerates he concluded that the George Formation records shallow-water, partly subaerial conditions. On the other hand, a deeper-water origin is suggested by the apparent debris flow origin of at least some of the conglomerates, the diamictites and the pebbly lime mudstone beds, and the resemblance of the long-wavelength (± 30 cm) current-ripples with draping laminae to Bouma 'C' units of turbidites, albeit within the photic zone for the stromatolite-bearing parts of the formation.

The Henry Creek Formation, about 460 m thick, consists largely of grey to dark grey mudstone, and succeeds the George Formation gradationally. Bell (1968) recognized a lower, slaty-cleaved member and an upper, non-cleaved member with sandstone beds. The lower member consists of mudstone and poorly sorted, sandy and silty mudstone, with a carbonate content of 1 to 40% and feldspar in sand- and silt-sized grains up to 5%. 'Floating' chips and pebbles are widespread, and a few beds may be called diamictite. The upper member is identified by its content of sandstone beds like the dark sandstone beds of the overlying Tuchodi Formation. Bell's observations indicate that the southwestward thinning of the George Formation reflects a facies change of the upper George to Henry Creek rock-types, which he interpreted as a transition to a deeper water setting, however, if the deeper-water interpretation of the George is preferred, the facies change may simply reflect a southwestward increase in the supply of siliciclastic mud.

The Tuchodi Formation, about 1500 m thick, gradationally overlies the Henry Creek Formation. It consists typically of crossbedded quartzite, with varying amounts of argillaceous dolomite, siltstone and shale. The arenites and siltstones contain from one percent to 29% feldspar, mainly potassium feldspar; grain size and feldspar content are inversely related. The dolomites are microcrystalline, pale coloured, and normally argillaceous, silty and sandy; some contain reworked dolomitic and cherty ooids. Molar-tooth structure is widespread. Where the dolomite is laminated, the laminae are usually undulatory and lenticular; this is in part cryptalgal lamination. Algal stromatolites and flat-pebble conglomerates are rare and local. The siltstones are pale grey to black, greenish and olive grey. The shales are black to dark grey and greenish grey, slightly micaceous and carbonaceous; in the most easterly sections they are partly red. Upward-thickening, upward-coarsening, decametre-scale cycles are well developed in parts of the formation.

Bell (1966) illustrated facies of the Tuchodi Formation that form a mosaic in both the geographic and stratigraphic senses. In a generally east-to-west transect, these facies are fluvial/deltaic, mud flat/lagoonal, barrier beach/offshore bar, and offshore, shallow marine.

The Aida Formation (Fig. 5.9), estimated to be 1200 to 1800 m thick, is an assemblage of silty, sandy and calcareous, pale to medium grey mudstone and siltstone, with minor sandstone and limestone. Like the overlying Gataga Formation, it is preserved only in the western limb of the anticlinorium, that is, in relatively internal parts of the Rocky Mountains. Bell (1966, 1968) described the Aida in terms of four members: Lowest, Chamosite, Carbonaceous and Main members.

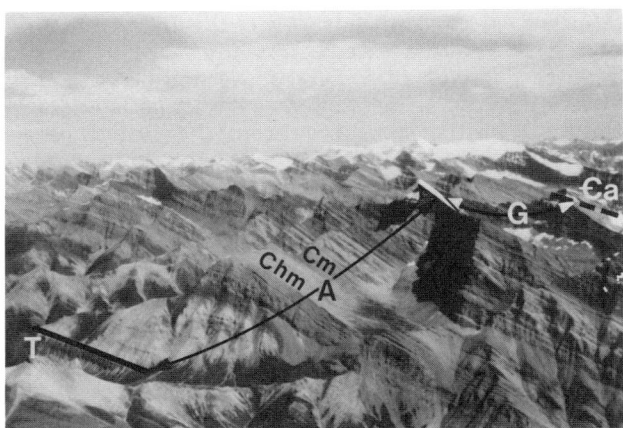

Figure 5.9. Tuchodi, Aida and Gataga formations, Tuchodi Lakes map area, British Columbia. T-Tuchodi; A-Aida (Cm-Carbonaceous Member, Chm-Chamositic Member); G-Gataga. Basal Cambrian strata at right. Photo-G.C. Taylor. GSC-1991-070

The Lowest Member consists of argillaceous limestone and dolomite, calcareous siltstone, mudstone and very fine grained sandstone, and flat-pebble conglomerate. The siltstone and mudstone have a well developed cleavage. The Chamosite Member consists of green chamositic mudstone grading upward to black, siliceous, carbonaceous mudstone and shale. The Carbonaceous Member comprises siliceous and carbonaceous shale and siltstone; Bell (1968), recorded carbon contents of 1 to 9%. The strata, cherty in places, are evenly laminated.

The Main Member, 900 to 1500 m thick, consists of calcareous and dolomitic mudstone, siltstone and shale, with minor very fine grained sandstone and argillaceous, silty limestone. Penetrative cleavage is ubiquitous, except in the basal 100 m. The member is almost entirely thin-bedded and/or laminated, and may be largely described as rhythmite. Bell (1968) listed a suite of sedimentary structures in these rocks of wacke-like texture; these include graded beds, contorted lamination, load-casted ripples, flame structures, pseudonodules and a few flute and groove casts. Scour surfaces and channels also are present. The festoon crossbedding reported by Bell is alternatively interpreted as long-wavelength, current ripples with draping laminae, strongly suggestive of Bouma 'C' units. Evidence of slumping and sliding is widespread.

The Aida Formation records the onset of prolonged basinal conditions in the Tuchodi Lakes region. The Lowest Member, transitional with the Tuchodi Formation beneath, records gradual deepening from the earlier shallow-water conditions. The Chamosite and Carbonaceous members appear to record semi-starved basin conditions, and the Main Member deep-water, turbiditic deposition of rather distal aspect. The carbonate contents of the Lowest and Main members imply the existence of either a contemporaneous carbonate platform or a nearby terrain composed partly of carbonate rocks and undergoing rapid mechanical erosion. Paleocurrents flowed northwestward (Bell, 1966).

The Gataga Formation is more than 1200 m of dark grey to black, noncalcareous slate, with minor sandy and silty mudstone, sandstone and siltstone. It records a continuation of the deep-water, turbiditic style of deposition established earlier. The contrast with the underlying Aida is marked, in that the Gataga strata are very dark grey to olive grey, relatively free of carbonate, and black to dark grey weathering. Sedimentary structures characteristic of turbiditic deposition occur throughout; they include parallel and convolute lamination, flame structures, graded beds, and minor sole markings and current-ripple crosslamination. Like those of the Aida, Gataga paleocurrents flowed northwestward.

The Gataga Formation is unconformably overlain by Lower Cambrian strata, commonly with 5 to 15° of angular discordance (Bell, 1968; Taylor and Stott, 1973). The entire Windermere Supergroup, represented by thick deposits farther west, is missing at the contact except near Mount Lloyd George where a diamictite unit is preserved between Lower Cambrian and Middle Proterozoic rocks.

Cap Mountain Succession (Cap Mountain-Hornby Bay Assemblage)

At Cap Mountain, in the southern Franklin Mountains, a faulted structure of extraordinarily high relief brings to view over 1800 m of unmetamorphosed sedimentary rocks which underlie Lower Cambrian sandstone with angular unconformity (Douglas and Norris, 1963; Aitken et al., 1973). These Precambrian rocks bear no resemblance to any of the Proterozoic formations exposed in the Mackenzie Mountains. On the other hand, as recognized by Meijer Drees (1975), they can be correlated on the basis of rock character and stratigraphic sequence with part of the Hornby Bay-Dismal Lakes succession of Coppermine Homocline (Aitken and Pugh, 1984), and accordingly, are assigned to Sequence A.

Published descriptions divide the succession at Cap Mountain into four units: Map units 1, 2, and 3, and at the top, the Lone Land Formation. Map unit 1, (508 m with base concealed) bears strong lithological resemblance to parts of Unit 9 of the Hornby Bay Group, as described by Kerans et al. (1981). It consists largely of purple mudstone, with subordinate interbeds and decametre-scale units of dolomite and silty dolomite that include flat- and crinkly-laminated, stromatolitic, nodular and breccia varieties. Protoquartzitic sandstone of "salt and pepper" aspect is a minor constituent, occurring in depositional units from laminae to stacks of very thick beds. Sun cracks and other evidence of shallow-water deposition and subaerial exposure occur throughout. Several of the dolomite sub-units bear suggestive evidence of the former presence of evaporites, such as breccias, contorted beds and apparent casts of gypsum rosettes.

Map unit 2 (525 m) resembles Unit 10 of the Hornby Bay. It is dominated by brick-red to purple, commonly sandy and silty mudstone, with sun cracks, ripple marks and ripple-drift crosslamination. Red, purple and green, muddy siltstone and sandstone, the latter with abundant rip-up clasts, are subordinate components, as are a few dolomite sub-units like those of Map unit 1. The arenites of Map unit 2 are distinctly less feldspathic than those of the apparently correlative unit of the Hornby Bay, and may have had a different source.

Map unit 3 and the Lone Land Formation, jointly, display numerous characteristics in common with units 11 and 12 of the Dismal Lakes Group. Map unit 3 (553 m) consists of prominently crossbedded, grey, orthoquartzitic sandstone, commonly in lenticular sub-units, and subequal amounts of purple-red and green mudstone and shale, with minor beds of siltstone and pebble conglomerate.

The Lone Land Formation (about 225 m), in concordant contact with the underlying Map unit 3 (Aitken et al., 1973), and in angular unconformity with overlying, Lower Cambrian sandstone, is a fining-upward, thinning-upward sequence in which white, orthoquartzitic sandstone at the base gives way gradationally upward to dominant olive-grey and brownish grey to black shale. The sandstones display trough crossbedding, ripple-drift crosslamination, abundant rip-up clasts and scour-and-fill structures. Ovoid, millimetre-scale, carbonaceous discs are common on the tops of the sandstone beds.

The lithostratigraphic correlation of the Cap Mountain section with part of the Hornby Bay-Dismal Lakes succession is strongly supported by its position, on trend with the latter and north of the 'Leith Line' of Kerans et al. (1981), where the succession would be expected to be relatively thick and complete. Furthermore, wells penetrating the Precambrian throughout a broad area north of Fort

Simpson and east of the Mackenzie River encounter a sedimentary section dominated by red mudrocks and sandstones (Meijer Drees, 1975) readily related to the Hornby Bay-Dismal Lakes to the northeast. No part of the younger, 'Mackenzie Mountains Supergroup' of Sequence B, nor of the Coppermine Lavas, is preserved in the region (Aitken and Pugh, 1984).

The significance of the Cap Mountain section is that, in the absence of evidence that the Coppermine River Group reaches the Mackenzie Valley, the Hornby Bay-Dismal Lakes succession and correlatives are the pseudo-basement of the 'Mackenzie Mountains Supergroup' (Sequence B).

Tectonic environment

The exposures at Cap Mountain and information from exploratory wells are too limited to support an interpretation of Hornby Bay-Dismal Lakes tectonics; but some insight can be gained from studies of the succession exposed in the lower part of the Coppermine Homocline (Baragar and Donaldson, 1973; Kerans et al., 1981), designated the 'Hornby basin' for present discussion. Like other Middle Proterozoic basins of the Canadian Shield (Fraser et al., 1970), the 'Hornby basin' is not delineated on reconnaissance gravity maps or on aeromagnetic maps. The predominance of northwest-side-down faults and northwestward thickening in the exposed parts might suggest the possibility of a relationship to a northeast-trending, passive continental margin. There is, however, no suggestion of such a margin, nor a later suture, in available potential field maps. The continuity of subsequent, Rae Group-Mackenzie Mountains Supergroup formations, in platformal facies, across the locus of any feasible margin or suture, similarly stands against such an interpretation.

Fraser et al. (1970) suggested that the Middle Proterozoic basins of the Canadian Shield were basins of preservation, although their repeated references to fault-bounded basins appear to modify that view. With specific reference to the 'Hornby basin', Kerans et al. (1981) clearly recognized an early, fault-trough stage. They limited their interpretation, however, to the exposed part of the succession, leaving the basin 'open' and uninterpreted to the west, beneath younger cover. The basin is possibly a bilaterally symmetrical, rift-basin or aulacogen, of which only the southeastern half is exposed. This interpretation, proposed by Burke and Dewey (1973), is supported by:

a. Synsedimentary faults aligned with the exposed basin flank, and the assumed trend of the basin axis (Kerans et al., 1981).
b. Volcanism (Narakay volcanics) during Hornby Bay deposition.
c. Alignment of the hypothetical basin axis with the cratonward bulge of the Mackenzie Mountains (Burke and Dewey, 1973).
d. The 'steer's head' relationship, characteristic of rift basins (Watts et al., 1982) mapped by Baragar and Donaldson (1973) (Fig. 5.10); the older units of the Hornby Bay Group are confined to the thickest part of the basin fill ('steer's face'), whereas the upper Hornby Bay and Dismal Lakes Group spread much farther eastward, though thinning in that direction ('steer's horn'). Note also that the Copper Creek and Husky Creek formations also are involved in the 'steer's head'. Only the southeastern half of the 'steer's head' is exposed, if the basin fill is symmetrical.

If the 'Hornby basin' is an aulacogen, Burke and Dewey (1973) were correct in postulating that it opened to a continental margin to the southwest; westward thickening and the predominance of westward-directed paleocurrents in fluvial formations (Kerans et al., 1981) require such a conclusion. The exposures at Cap Mountain provide important evidence of an elongate, southwest-trending configuration.

Although an aulacogen model appears attractive, newly available information poses difficulties. Bowring and Ross (1985) obtained a U-Pb age of 1.663 Ga on zircon from the Narakay volcanics. A life-span of nearly a half-billion years, from that date to the 1.2 Ga of the Coppermine lavas, is uniquely and implausibly long for an aulacogen. It is conceivable, however, that a significant, though unrecognized, gap in the stratigraphic record occurs beneath the Coppermine lavas.

Age and tectonic setting of Sequence A

Sequence A strata are younger than the basement of the Churchill Province of the Canadian Shield (<1.7 Ga) and apparently older than the Coppermine lavas (1.2 Ga) of the Coppermine Homocline. No radiometric age determinations are available for the Cap Mountain succession of

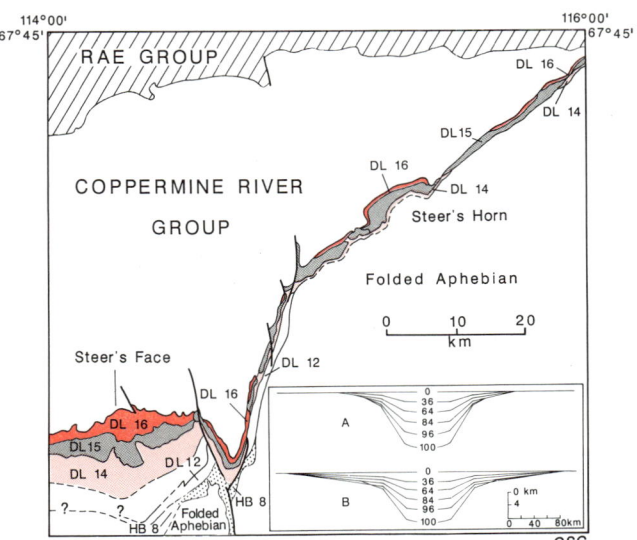

Figure 5.10. The "steer's head" configuration of the Coppermine homocline. Units are numbered continuously from HB (Hornby Bay Group) through DL (Dismal Lakes Group). This map may be seen as a structural section by viewing it down the regional dip, 3 degrees to 8 degrees northwest. Re-drawn from Geological Survey of Canada Map 1337A (Baragar and Donaldson, 1973).

Inset: calculated stratigraphy of a thermally subsiding basin. Selected horizons are flagged with their "age" in Ma. Two cases are shown: In A, the thickness of the elastic plate is the depth to the 300°C oceanic isotherm; in B to the 600°C oceanic isotherm. Re-drawn from Watts et al. (1982).

the Franklin Mountains, but it is firmly identified as part of the pre-lavas, Hornby Bay-Dismal Lakes succession of the Coppermine Homocline. The Narakay volcanics near the top of the Hornby Bay have recently been dated at 1.663 Ga (Bowring and Ross, 1985). The Muskwa Assemblage of northeastern British Columbia is undated; at present it cannot be correlated with either Sequence A or Sequence B. The available isotopic age determinations from the Wernecke Supergroup of the Yukon are from minerals formed post-depositionally, and most of these give pre-1.2 Ga ages. Geochronologic studies of the Purcell (Belt) Supergroup in southern British Columbia, Alberta and the adjacent United States suggest that sedimentation occurred between about 1.5 and 0.85 Ga, thus encompassing both Sequences A and B (onset of deposition as early as the newly dated Hornby Bay Group is not excluded). On the other hand, paleomagnetic studies of Purcell (Belt) rocks paleomagnetism suggest that sedimentation occurred between about 1.5 and 1.25 Ga, and belonged entirely to Sequence A (Evans et al., 1975; Vitorello and Van der Voo, 1977; Elston and Bressler, 1980). Regional considerations of metamorphic fabrics, intrusive relationships and Rb-Sr radiometric ages suggest that the younger (1.2 Ga) isotopic ages were updated by younger events involving heating and/or uplift, as suggested by McMechan and Price (1982), and that the Purcell belongs entirely to Sequence A.

The Purcell (Belt) and Wernecke supergroups have the configuration and thickness appropriate to passive-margin deposits; however, serious doubts about such an interpretation in the case of the Belt basin have been raised (Winston et al., 1984; Roberts, 1986). No evidence for a southwestern flank of the "Wernecke basin" is evident, and the miogeoclinal interpretation remains as yet unchallenged. The structural style of the Mackenzie Mountains requires the presence at depth, beneath Sequence B strata, of thick, layered rocks (Aitken et al., 1982) that are probably Wernecke equivalents. If this be the case, the Wernecke Supergroup also can be said to occupy an embayment in the continental margin. The Cap Mountain succession and its correlative in the Coppermine Homocline appear to have been deposited in an elongate, southwest-trending rift-depression that may have opened southwestward to the continental margin. This depression is not identified as an aulacogen because of its apparently long life, nearly a half-billion years.

The undated Muskwa succession is largely of platformal character and resembles parts of both the Hornby Bay-Dismal Lakes succession of Sequence A and the eastern (platformal) Purcell, but also resembles parts of the Mackenzie Mountains Supergroup (Sequence B). Arbitrarily placed in Sequence A, the Muskwa succession occupies a structural salient, and by analogy with the Crowsnest deflection of southern Alberta, probably forms a more or less local thickening at the mountain front. It is unclear whether deposition was along the southeastern flank of the Hornby rift-basin, or in a separate continental-margin re-entrant farther south.

The strata of Sequence A thus record a continental margin (in part, possibly an intracratonic basin) that formed 700 to 900 Ma prior to the rifting event at 700 to 800 Ma that led to deposition of Sequence C (Windermere Supergroup), and the establishment of the Cordilleran continental margin. In the south, the eastern margin of the Purcell basin had a more westerly trend (Fig. 5.7) than the continental margin upon which Sequence C was deposited. There, the younger margin appears to truncate the older northwestward. Similarly, the re-entrant(s) in which the Muskwa and Hornby Bay-Dismal Lakes successions were deposited appear to be truncated by Sequence C and the Cordilleran trend. Truncation of the Sequence A continental margin by later (Sequence C) rifting explains the discontinuous distribution of Sequence A strata as compared to the Cordillera-long continuity of Sequence C and younger, Paleozoic deposits; it also explains the absence of Sequence A strata above basement rocks exposed in the Rocky Mountains between the Muskwa re-entrant and the Purcell (Belt) basin.

SEQUENCE B, ≈1.2 TO ≈0.78 GA

Sedimentary rocks of Sequence B (Mackenzie Mountains Assemblage) occur within the Cordillera only north of the 60th parallel (Fig. 5.1). They are represented by the Mackenzie Mountains Supergroup and, possibly, by the Pinguicula Group. The absence of stratified rocks of this sequence south of the 60th parallel is one of the major, recently appreciated facts of Cordilleran Proterozoic stratigraphy, and one yet to be fully interpreted.

Mackenzie Mountains Supergroup (Mackenzie Mountains Assemblage)

The term Mackenzie Mountains Supergroup has become established in the recent literature on Mackenzie Mountains geology, although it remains formally undefined. The supergroup is widely exposed in the Mackenzie Mountains and is present in the subsurface of the northern Franklin Mountains and the Great Bear and Anderson plains. In the Mackenzies, it is at least 4 km thick, and includes all exposed strata beneath the Coates Lake Group. The Coates Lake Group (Jefferson and Ruelle, 1987) was formerly included within the Mackenzie Mountains Supergroup, but in view of the well-documented unconformity at its base (Aitken, 1981; Morris and Aitken, 1982), its harmony of tectonic setting with the overlying Rapitan Group, and its age (apparently younger than 0.78 Ga intrusions), it is interpreted herein as the basal part of the Windermere Supergroup (Sequence C) equivalents of the Mackenzie Mountains region.

The Mackenzie Mountains Supergroup, whose age is broadly bracketed by 1.2 Ga and 0.78 Ga, is divided into four main units recording dominantly platformal sedimentation: unit H_1, Tsezotene Formation, Katherine Group, and Little Dal Group. Basaltic lava, locally preserved at the top of the supergroup, now is included in the Coates Lake Group.

Isopachs for units of formation and member rank in the Katherine and Little Dal groups are north- to west-trending, arcuate and illustrate southwestward thickening (Fig. 5.11). Control for these isopachs is sparse, but their shape and position are supported by lithofacies data. With one exception, correlation panels following the trend of isopachs as drawn display marked persistence in thickness of members, and even beds (Aitken, 1981).

The oldest formation of the Mackenzie Mountains Supergroup, and the oldest exposed in Mackenzie Mountains, is informally designated map unit H_1. It is exposed only in the crestal areas of three major structures. The

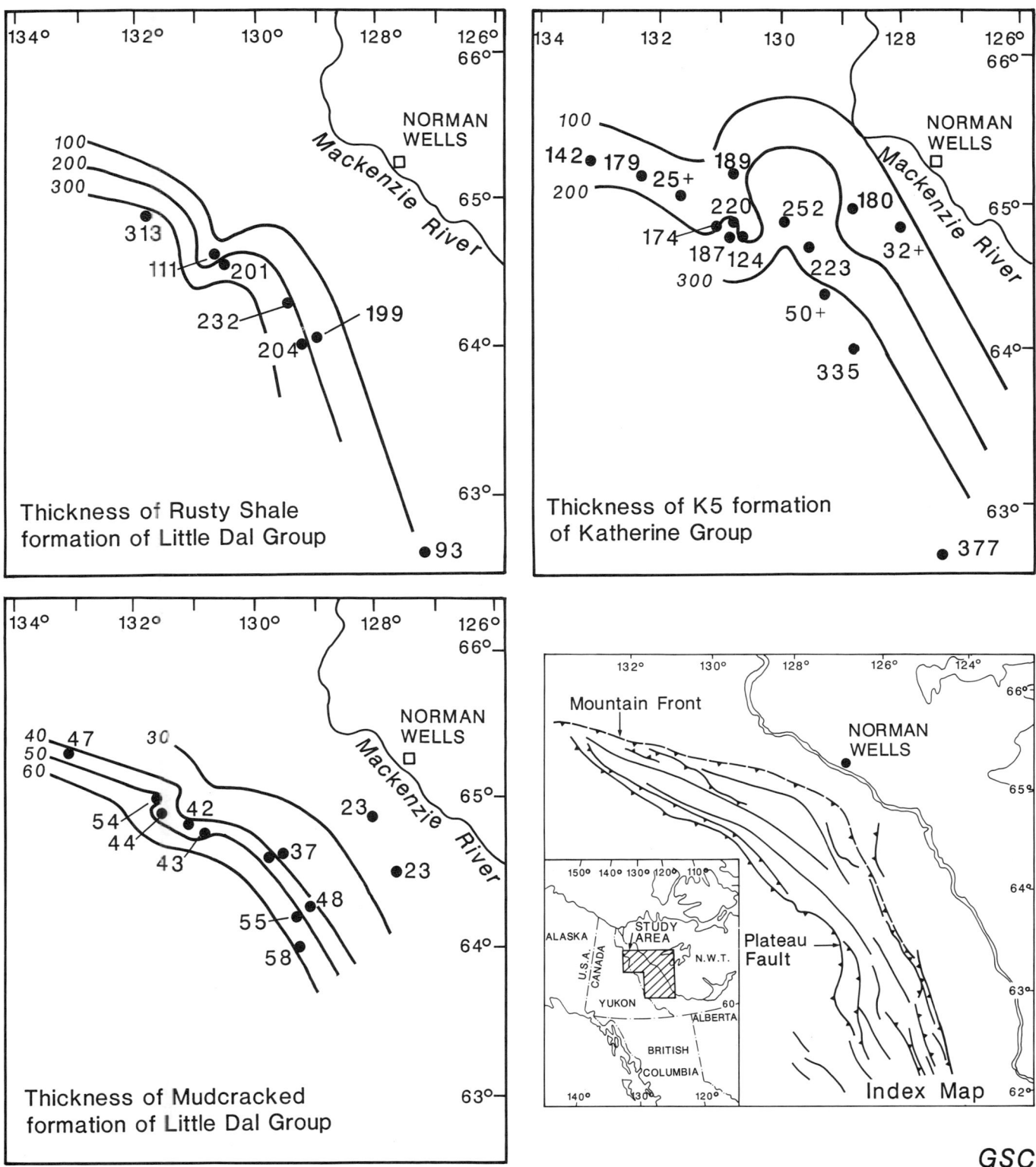

Figure 5.11. Isopach maps for selected units of the Mackenzie Mountains Supergroup. Thicknesses in metres. After Aitken and Long (1978).

profile of Tawu anticline, at the level of map unit H_1, is such that bedded rocks to the thickness of some kilometres must underlie H_1 and be involved in the folding (Aitken et al., 1982). In the type area, where over 400 m are exposed, the formation consists of three members. The lower member is largely covered, suggesting that clastic strata may be dominant; the exposed beds are silty, partly argillaceous and sandy, ripple-marked grey dolomite, and subordinate brownish grey, dolomitic shale. The middle member (183 m) consists of dark grey dolomite that weathers pale grey. It consists of thick, massive biostromes and bioherm complexes of various stromatolites, among which *Conophyton* and *Jacutophyton* are prominent. The form and height of the bioherms imply growth in water at least several metres deep. At the top of the middle member is 30 m of bedded, largely sandy (quartz) dolomite, including varieties derived from grainstone and laminated lime mudstone. Crossbedding, including 'herringbone' crossbedding, is well developed, and erosional channels are numerous. The upper member, 180 m thick, appears dark grey, largely because of shadows cast by its characteristic, resistant, silicified masses. It is dominated by cryptalgal laminite, and contains abundant, largely black, chert. Subordinate beds of sandy dolomite and dolomitic quartz sandstone are present, as are beds of flat-pebble conglomerate. Domal and rare columnar-branching stromatolites also occur. Tepee structures and channels are widespread, current-ripple marks less so. At the top, a few interbeds of dark grey shale are precursors to the abrupt appearance of the basal black shales of the Tsezotene Formation, at a non-erosional contact. At Carcajou River, the upper member, 250 m thick, is silicified and includes grey, white, and black chert. The middle member is present, but the internal structure of the bioherms obscures bedding and prohibits measurement. In the Tsezotene Range, only the uppermost 50 m of the formation are exposed and are typical of the upper member.

The Tsezotene Formation, ranging in thickness from 750 to 1500 m, represents an incursion of clastic rocks above the platform carbonates of map unit H_1, and forms the lower part of a coarsening-upward cycle that is completed by the thick sandstones of the Katherine Group. The Tsezotene, as described by D.G.F. Long (pers. comm., 1982) is divisible into two members: a lower, dominated by grey to black argillaceous rocks, and an upper member characterized by red, green and grey shales. From Keele River southward, a middle carbonate member is present. The lower, 'grey' member displays the usual pattern of thickness distribution of units of the Mackenzie Mountains Supergroup, in that isopachs are arcuate northward to northeastward, following the tectonic trend, and indicate thickening towards the southwest, from 600 to 1100 m. Dark grey to greyish-black mudrocks make up 85% of the member. Associated with these are minor, discontinuous, laminated beds of dolomite, and thin, siliciclastic beds of silt to very fine sand grade. Alternation of fissile and nonfissile mudrocks is markedly rhythmic. Sandstone units up to 200 m thick are markedly impersistent laterally; these contribute to the overall, coarsening-upward trend.

The carbonate member of the Tsezotene is best developed at Keele River, where it is 120 m thick. There, it consists mainly of flat- to wavy-laminated limestone with subordinate black shale, stromatolitic limestone and dolomite, and muddy and silty limestone and carbonate-rich shale. The stromatolitic carbonates occur in part as domal and tabular bioherms and biostromes.

The upper, 'red' member, 130 to 300 m thick, contains somewhat less than half redbeds (fissile and nonfissile mudrocks, homogeneous and laminated siltstone, muddy siltstone, muddy sandstone, sandy mudstone), and in addition, grey and green shales, thin, orange-weathering carbonates and green and white sandstone bodies. An arrangement into distinct, metre-scale, coarsening-upward cycles is characteristic. The upper member passes conformably, by interbedding, into the sandstone-dominated Katherine Group.

The Katherine Group, 700 to 1300 m thick, was divided by Long (Aitken et al., 1978a) into seven units (K1-K7) of formation and member rank, not all of which are coextensive with the group. The group commences at the base with one of the four, thick, resistant-weathering sandstone units that record progradation of a fluvial-deltaic complex and its subsequent reworking (destructional phase) by marine waves and currents. Each of the three non-resistant units, consisting of mudrocks with variable contents of carbonate strata, is at least partly marine. Certain of the units of formational rank, such as K4 (purple and maroon shale, sandstone, carbonate), and of member rank or lower, such as the two stromatolite biostromes of K6, show remarkable persistence. Casts of salt crystals have been found locally in the base of the Katherine, and widely in its uppermost beds.

D.G.F. Long (pers. comm., 1982) interpreted the Tsezotene-Katherine units in terms of basin filling (the upward-coarsening Tsezotene) followed by progradation of fluvial-deltaic sand sheets, in continuation of the upward-coarsening trend. In view of the distribution and platformal character of the H_1 dolomites and the Katherine orthoquartzites, the Tsezotene basin was a shelf basin. Paleocurrent measurements on crossbedding in the Katherine (D.G.F. Long, pers. comm., 1982) are largely bimodal or polymodal, recording tidal influence. Where a single mode, or a third mode in addition to a bimodal pair is present, it is usually directed northwestward, along the depositional strike, and is taken to record northwesterly, longshore transport of sand.

The Katherine Group is succeeded, at a concordant, erosional contact, by the Little Dal Group, 2 km thick (Fig. 5.12), which makes up the remainder of the Mackenzie Mountains Supergroup. It is divided into at most seven, informally named units of formational rank (Aitken, 1981). Lower and upper divisions are recognized, the upper commencing with the Gypsum formation. The lower Little Dal displays pronounced (though upward-diminishing) facies variation; such variation is minimal in the upper division. For some formations and members, sufficient control is available to permit the drawing of isopach maps that demonstrate a southwestward-thickening, arcuate trend congruent with the structural arc of the Mackenzie Mountains (Fig. 5.11; Aitken and Long, 1978).

The basal, Mudcracked formation is 20 to 60 m thick. It consists mainly of grey, brown, black, green and red mudstone, with subordinate fine grained sandstone, mainly in thin beds. Shallow-water structures are prominent, and include large-scale sun cracks and casts of salt-crystals. A

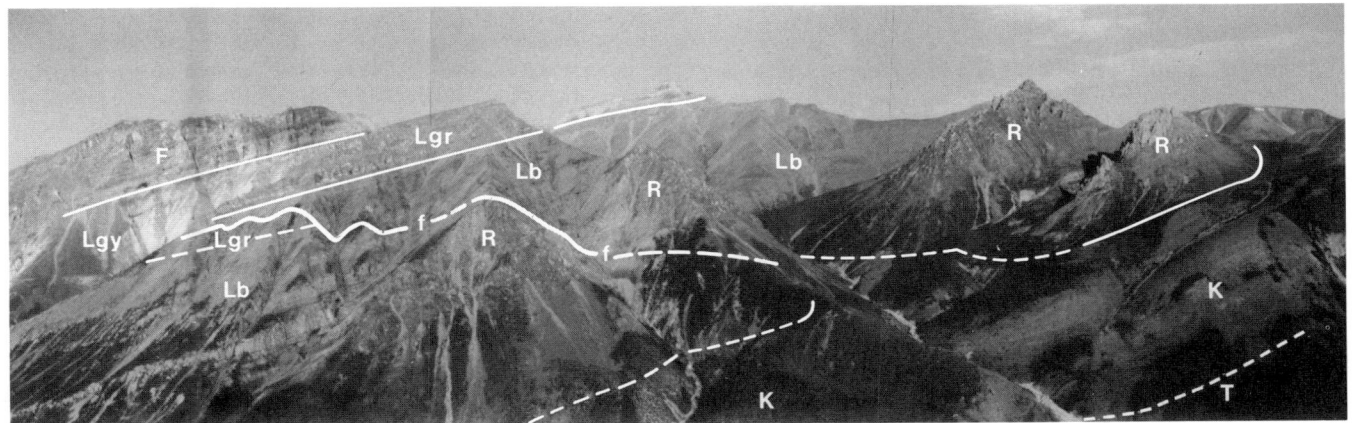

Figure 5.12. 'Mackenzie Mountains Supergroup' overlain by Franklin Mountain Formation (Cambro-Ordovician); note the low-angle discordance at the contact. Tributary of Mountain River, Mackenzie Mountains, N.W.T. T-Tsezotene Formation; K-Katherine Group; L-Little Dal Group (Lb basinal assemblage, with stromatolitic reefs (R); Lgr-Grainstone Formation; Lgy-Gypsum Formation); F-Franklin Mountain Formation; f-fault, occupied by diabase dyke GSC-1991-071.

thin, upper member of dolomitized, ooid and intraclast grainstone forms a regional, marker unit.

The strata succeeding the Mudcracked formation are divided into two, geographically exclusive, mappable units. In the southeastern extent of Little Dal exposures, which may have been slightly elevated at the onset of deposition, the platformal assemblage (450-800 m) bears the marks of deposition on a shallow, high-energy carbonate platform. The rocks are in large part intraclast and ooid grainstone, with lesser lime mudstone in which molar-tooth structure is prominent. Stromatolite bioherms and biostromes occur throughout.

In the northwest, the basinal assemblage (430-630 m), equivalent to the platformal assemblage, is of a distinct facies essentially devoid of shallow-water rocks and structures and typified by carbonate-clastic rhythmites and nodular limestone. At the base, a facies of pink limestone nodules in a matrix of red argillaceous limestone is characteristic of northern exposures; it intertongues southward with grey, basinal rhythmites. The rhythmites consist of thin beds of limestone (mainly calcisiltite) with thinner interbeds of dark grey, silty shale or siltstone. The limestone beds display current ripples of long wavelength and low amplitude with draping laminae, suggestive of Bouma C units. The detrital interbeds contain graded laminae, commonly with micro-load casts. Units of dark grey to black shale and shaley siltstone, in part with limestone nodules, also occur. This assemblage is clearly of deep-water origin. It intergrades at base and top with shallow-water strata, via intermediate facies characterized by algal stromatolites and abundant molar-tooth structures.

Within the basinal assemblage, and more than 100 km basinward of the edge of the platform, is a belt of unique stromatolitic reefs, some exceeding 300 m in height. Basinal rhythmites onlap the steep flanks of the reefs, and intertongue with coarse reef-derived talus (Fig. 5.12). The length of the talus tongues demonstrates that during their growth the reefs stood at least tens of metres above the basin floor.

Fragments of carbonaceous mats or films (*Beltina*, *Morania*) are common and locally (adjacent to reefs?) abundant in the basinal rhythmites. The millimetre-scale, carbonaceous disc *Chuaria* sp. and the centimetre-scale, carbonaceous compressions of the metaphyte *Tawuia* sp. (Fig. 5.13) have been found in both basinal and platformal facies (Hofmann and Aitken, 1979; Hofmann, 1985). Evidence is growing that these latter may serve to identify a worldwide, pre-Ediacaran biostratigraphic unit (Hofmann,

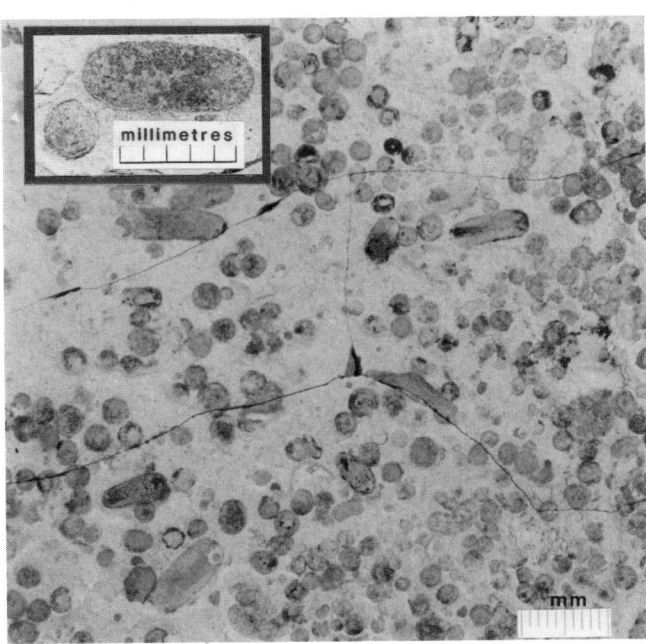

Figure 5.13. The fossils *Chuaria circularis* (compressed spheres) and *Tawuia dalensis* (compressed lozenges), from the Little Dal basinal assemblage, Mackenzie Mountains. Photos by H.J. Hofmann.

1985). A well preserved microflora also is present in the basinal facies.

The platformal and basinal assemblages are separated by a dolomitized, multi-story stromatolite biostrome complex up to 200 m thick, that appears to serve as a supporting buttress for the edge of the platform. The trend of the known part of the complex and the margin of the platform crosscut isopachs of the underlying and overlying formations at a high angle (Aitken, 1981).

The platformal and basinal assemblages are succeeded, at an intertongued contact, by the Grainstone formation, locally exceeding 425 m. The Grainstone formation consists of two principal lithofacies that intertongue in units tens of metres thick: one is dolomitized ooid grainstone, in thick, massive beds, with associated cryptalgal laminite, flakestone, stromatolites and dolomitized crosslaminated calcisiltite; the other is thin, platy and flaggy, commonly sandy beds of microcrystalline dolomite, with minor cryptalgal laminite, flakestone, simple stromatolites and pale-pellet grainstone. The 'platy dolomite' lithofacies carries abundant sun cracks, commonly sand-filled, and locally, salt-crystal casts. Masses of gypsum have been encountered in drilling the 'platy dolomites', and local breccias probably record solution of such evaporites (R. Hewton, pers. comm., 1982).

The Grainstone formation is succeeded conformably by the poorly exposed Gypsum formation, up to 530 m thick (Fig. 5.12), that has exerted an important influence on the structure of the region. It consists of at least 90% grey gypsum and anhydrite that are largely laminated, but also in part nodular, with enterolithic folds resulting from volumetric changes accompanying chemical changes. Non-sulphate material, including quartz sand, silt, and clay, occurs mainly near the base and top, and is marked by reddish and greenish tints. The amount of clastic material increases eastward and southward around the tectonic arc, as is the case with other units of the Little Dal Group, indicating a source of detritus to the southeast. A thin carbonate member in the upper half of the formation is regional in extent. Except for beds near the top and base, the environment of deposition appears to have been subaqueous. The presumed barrier that isolated the evaporitic lagoon from the open sea is buried, and its nature unknown.

The Gypsum formation is succeeded conformably by the Rusty Shale formation (180-270 m), which records an influx of muddy and sandy detritus. Shallow-water structures are prominent throughout. The formation consists of regionally persistent members of red, green and dark grey shale, very fine grained quartzite, and peritidal carbonate rocks. Small-scale cycles, from shale at the base to carbonates at the top, similar to those characteristic of the succeeding Upper Carbonate formation, are common. Sandstone members are thickest and coarsest in the southeast, recording a source of detritus in that direction. The formation contains a recurrence of the *Chuaria-Tawuia* biota first seen in the basinal and platformal assemblages.

The Upper Carbonate formation (<300-750 m) comprises the thick, cliff-forming, stromatolite-rich, largely cyclical strata that generally cap the Little Dal Group. Four members are coextensive with the formation, except where removed by erosion at subsequent unconformities. These members consist mainly of various proportions of several kinds of peritidal carbonate rocks, almost entirely dolomitized. Mudrocks occur almost exclusively as the basal beds of shallowing-upward, 'clearing-upward', metre-scale cycles. Rock colour and distinctive stromatolites provide secondary criteria for the recognition of members.

At the top of the Little Dal Group, and only locally preserved, is a series of partly pillowed, basaltic lava flows conformable with the underlying carbonate strata. A study of the paleomagnetic record across the top of these lavas demonstrates that the hiatus beneath the overlying Coates Lake Group, revealed by erosional removal of Little Dal members and formations, is of significant length (Morris and Aitken, 1982).

Age and Tectonic Environment

Much recent progress has been made toward establishing narrow limits to the age of the Mackenzie Mountains Supergroup. However, some paradoxical aspects of its distribution, thickness and lithofacies leave important unanswered questions about the tectonic environment in which it was deposited.

Radiometric determinations date the supergroup only within broad limits, but are sufficient to demonstrate a post-Purcell, pre-Windermere (Sequence C) age. The Rae Group of Coppermine Homocline lies stratigraphically above the Copper Creek basalts, which have been dated at $1.2 \pm .045$ Ga (Rb-Sr whole-rock isochron; Baragar, 1972). Because correlation of the Rae Group with the lower formations of the Mackenzie Mountains Supergroup (Fig. 5.14) is well established (Young, 1977, 1978; Aitken et al., 1978b), a maximum age for the supergroup is about 1.2 Ga. In the Mackenzie Mountains, diabase dykes that intrude formations as young as the lower Little Dal Group have yielded clasts to the Shezal (Middle Rapitan, Sequence C) diamictites. The dykes are considered comagmatic with sills dated at about 770 Ma (Rb-Sr mineral isochron; Armstrong et al., 1982) that intrude the Tsezotene Formation and probably with diorite dated at 780 Ma (U-Pb on Zircon; R.R. Parrish and C.W. Jefferson, pers. comm., 1986) that intrude the Little Dal Group. These data establish a minimum age for the Mackenzie Mountains Supergroup.

Recent paleomagnetic studies appear to limit the age of the supergroup more closely. An apparent polar wander path for the supergroup (Park and Aitken, 1986a,b) is separate from, and necessarily younger than the Grenville Loop (Irving and McGlynn, 1981). New work dates the younger end of the Grenville Loop at about 880 Ma (Dallmeyer and Sutter, 1980). This implies a post-880 Ma, pre-770 Ma age for the supergroup (see Chapter 3).

Macrofossils (*Chuaria, Tawuia*) from the Little Dal Group hold promise for future, worldwide correlation, but at present add no further refinement to the age-range given above. Intercontinental biostratigraphic correlation based on preliminary studies of algal stromatolite assemblages (Aitken et al., 1978a) apparently fails to establish correctly the age of the Mackenzie Mountains Supergroup. The suggested early Late Riphean age (950 ± 50 Ma) for the Upper Carbonate formation of the Little Dal Group can be valid only if the paleomagnetic dating is incorrect. In any event, the suggested early Middle Riphean ($1.3 \pm .050$ Ga) age for the lowest formation of the supergroup exposed in the mountains is at least 150 Ma too old, as

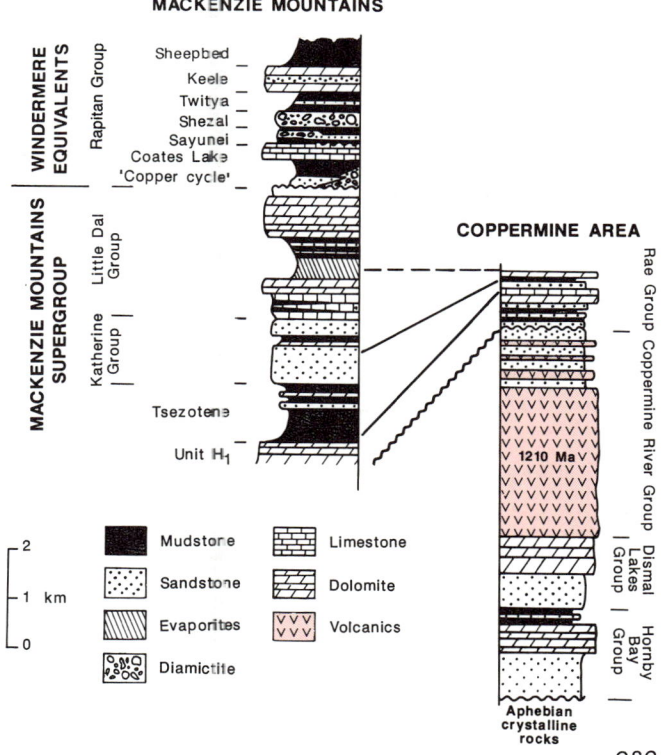

Figure 5.14. Correlation of exposed Proterozoic rocks, Mackenzie Mountains to Coppermine Homocline. Adapted in part from Young et al. (1979).

compared with the maximum age of the supergroup established by radiometric dating and lithostratigraphic correlation.

The tectonic environment in which the Mackenzie Mountains Supergroup was deposited must take account of its salient characteristics:

a. The lithological character of most of the supergroup: mature, shallow-water clastics, carbonate rocks, and evaporites.

b. Gradual basinward thickening, even within the deformed belt, as compared with abrupt rates of thickening of Windermere and lower Paleozoic units.

c. Great geographic persistence of lithostratigraphic units, both along and across the depositional strike (Fig. 5.10).

d. Isopachs congruent with tectonic arc of Mackenzie Mountains, and with the well-established, passive margin of Windermere and early Paleozoic age (Fig. 5.11).

e. Insignificant extrusive volcanic deposits, recording a single, apparently terminal event (the diabase intrusions are interpreted as dating later rifting, at the inception of Windermere sedimentation).

f. Steep faults, trending north-northwest to north-northeast, and cutting the supergroup, that influenced Coates Lake Group sedimentation (Eisbacher, 1977; Jefferson, 1978) and predate the Rapitan Group.

g. The absence of Precambrian folding.

In most of these respects, the Mackenzie Mountains Supergroup is platformal and resembles Devonian formations of the undeformed part of the miogeocline. Its breadth of exposure, from concealed western to erosional eastern limits, is about 450 km, as compared to 500 to 800 km for the Devonian example, and over this distance, it displays less variation in facies. Only in the parallelism of thickness trends to the well-established continental margin of Windermere Late Proterozoic and early Paleozoic time is there a suggestion of a relationship to a continental margin. The evidence available is insufficient to lead to a firm interpretation. The Mackenzie Mountains Supergroup was deposited either in a broad epicratonic basin, or as the eastern miogeoclinal equivalents of a passive-margin succession (post-Purcell, pre-Windermere).

The Pinguicula Group (Eisbacher, 1981) of the eastern Wernecke Mountains overlies bevelled folds in the underlying Wernecke Supergroup. It underlies the Rapitan Group unconformably, but was not folded in the Precambrian. A tentative correlation with the Mackenzie Mountains Supergroup is possible (Young et al., 1979; Eisbacher, 1981), but proof is lacking. Correlation with the Coates Lake Group also has been suggested (Eisbacher, 1978). The folding of the Wernecke Supergroup has not been dated, but is widely assumed to predate deposition of the Mackenzie Mountains Supergroup, which is not known to have undergone Precambrian folding. This is not necessarily the case. The exposed Mackenzie Mountains Supergroup may have remained unfolded in the foreland of the Wernecke deformation, just as Purcell strata of the Waterton area remained unfolded during deformation ascribed to the East Kootenay orogeny. If this argument could be proven, it would also demonstrate that the Pinguicula is younger than the Mackenzie Mountains Supergroup.

REFERENCES

Aitken, J.D.
1981:   Generalizations about Grand Cycles; in Short Papers for the Second International Symposium on the Cambrian System; M.E. Taylor (ed.), United States Geological Survey, Open File Report 81-743, p. 8-14.

Aitken, J.D. and Long, D.G.F.
1978:   Mackenzie tectonic arc - reflection of early basin configuration?; Geology, v. 6, p. 626-629.

Aitken, J.D. and Pugh, D.C.
1984:   The Fort Norman and Leith Ridge structures, major, buried Precambrian features underlying Franklin Mountains and Great Bear and Mackenzie Plains; Bulletin of Canadian Petroleum Geology, v. 32, p. 139-146.

Aitken, J.D., Cook, D.G., and Yorath, C.J.
1982:   Upper Ramparts River (106G) and Sans Sault Rapids (106H) map-areas, District of Mackenzie; Geological Survey of Canada, Memoir 388 (with maps 1452A and 1453A).

Aitken, J.D., Long, D.G.F., and Semikhatov, M.A.
1978a: Progress in Helikian stratigraphy, Mackenzie Mountains; in Current Research, Part A, Geological Survey of Canada, Paper 78-1A, p. 481-484.

1978b: Correlation of Helikian strata, Mackenzie Mountains - Brock Inlier-Victoria Island; in Current Research, Part A, Geological Survey of Canada, Paper 78-1A, p. 485-486.

Aitken, J.D., Macqueen, R.W., and Foscolos, A.E.
1973:   A Proterozoic sedimentary succession with traces of copper mineralization, Cap Mountain, southern Franklin Mountains, District of Mackenzie (95O); in Report of Activities, Part A, Geological Survey of Canada, Paper 73-1A, p. 243-246.

Armstrong, R.L., Eisbacher, G.H., and Evans, P.D.
1982:   Age and stratigraphic - tectonic significance of Proterozoic diabase sheets, Mackenzie Mountains, northwestern Canada; Canadian Journal of Earth Sciences, v. 19, p. 316-323.

Atkinson, S.J.
1975:   Surface geology of the Paradise Basin (82K/8W); British Columbia Ministry of Mines and Petroleum Resources, Geology in British Columbia 1975, p. G7-G12.

Baragar, W.R.A.
1972: Coppermine River Basalts - District of Mackenzie; in Rubidium-Strontium Isochron Age Studies, Report 1, R.K. Wanless and W.D. Loveridge (ed.), Geological Survey of Canada, Paper 72-23, p. 21-24.

Baragar, W.R.A. and Donaldson, J.A.
1973: Coppermine and Dismal Lakes map-areas; Geological Survey of Canada, Paper 71-39 (with maps 1337A and 1338A).

Bell, R.T.
1966: Precambrian rocks of the Tuchodi Lakes map-area, northeastern British Columbia; Ph.D. thesis, Princeton University, 138 p.
1968: Proterozoic stratigraphy of northeastern British Columbia; Geological Survey of Canada, Paper 67-68.
1982: Comments on the geology and uraniferous mineral occurrences of the Wernecke Mountains, Yukon and District of Mackenzie; in Current Research, Part B, Geological Survey of Canada, Paper 82-1B, p. 279-284.

Bouma, A.H.
1962: Sedimentology of some flysch deposits; Elsevier Publishing Company, Amsterdam, 168 p.

Bowring, S.A. and Ross, G.M.
1985: Geochronology of the Narakay Volcanic Complex: implications for the age of the Coppermine Homocline and Mackenzie igneous events; Canadian Journal of Earth Sciences, v. 22, p. 774-781.

Burke, K. and Dewey, J.F.
1973: Plume generated triple junctions: key indicators in applying plate tectonics to old rocks; Eos (American Geophysics Union Transactions), v. 54, no. 4, p. 239.

Burwash, R.A., Baadsgaard, H., and Peterman, Z.E.
1962: Precambrian K-Ar dates from the western Canada sedimentary basin; Journal of Geophysical Research, v. 67, p. 1617-1625.

Le Cheminant, A.N. and Heaman, L.M.
1989: MacKenzie igneous events, Canada: Middle Proterozoic hotspot magmatism associated with ocan opening; Earth and Planetary Science Letters, v. 96, p. 38-48.

Dallmeyer, R.D. and Sutter, J.F.
1980: Acquisitional chronology of remnant magnetization along the 'Grenville polar path', evidence from Ar ages of hornblende and biotite from the Whitestone Diorite, Ontario; Journal of Geophysical Research, v. 85, p. 3177-3186.

Daly, R.A.
1912: Geology of the North American Cordillera at the forty-ninth parallel; Geological Survey of Canada, Memoir 38.

Delaney, G.D.
1981: The mid-Proterozoic Wernecke Supergroup, Wernecke Mountains, Yukon Territory; in Proterozoic Basins of Canada, F.H.A. Campbell (ed.), Geological Survey of Canada, Paper 81-10, p. 1-23.

Devlin, W.J., Bond, G.C., and Brueckner, H.K.
1985: An assessment of the age and tectonic setting of volcanics near the base of the Windermere Supergroup in northeastern Washington: implications for latest Proterozoic - earliest Cambrian continental separation; Canadian Journal of Earth Sciences, v. 22, p. 829-837.

Douglas, R.J.W.
1952: Waterton, Alberta; Geological Survey of Canada, Paper 52-10.

Douglas R.J.W. and Norris, D.K.
1963: Dahadinni and Wrigley map areas, District of Mackenzie, Northwest Territories; Geological Survey of Canada, Paper 62-33.

Earhart, R.L., Raup, O.B., Corner, J.J., Carrain, P.E., Mcgimsey, D.H., Constenius, K.N., and Van Loener, R.E.
1983: Preliminary geologic map and cross sections of the northwest part of Glacier Park, Montana; United States Geological Survey, Miscellaneous Field Studies Map Mf-1604-A.

Edmunds, F.R.
1973: Stratigraphy and lithology of the lower Belt series in the southern Purcell Mountains, British Columbia; in Belt Symposium 1973, v. 1: Department of Geology, University of Idaho and Idaho Bureau of Mines and Geology, p. 230-234.
1977: Kimberley to Creston, stratigraphy and lithology of the lower Belt series in the Purcell Mountains, British Columbia; in Lead-Zinc Deposits of Southeastern British Columbia, T. Høy (ed.), Geological Association of Canada, Field Trip No. 1, Guidebook, p. 22-32.

Eisbacher, G.H.
1977: Tectono-stratigraphic framework of the Redstone Copper Belt, District of Mackenzie; in Report of Activities, Part A, Geological Survey of Canada, Paper 77-1A, p. 229.
1978: Two major Proterozoic unconformities, Northern Cordillera; in Current Research, Part A, Geological Survey of Canada, Paper 78-1A, p. 53-58.
1981: Sedimentary tectonics and glacial record in the Windermere Supergroup, Mackenzie Mountains, northwestern Canada; Geological Survey of Canada, Paper 80-27.

Elston, D.P. and Bressler, S.L.
1980: Paleomagnetic poles and polarity zonation from the middle Proterozoic Belt Supergroup, Montana and Idaho; Journal of Geophysical Research, v. 85, p. 339-356.

Evans, M.E., Bingham, D.K., and McMurray, E.W.
1975: New paleomagnetic results from the upper Belt-Purcell Supergroup of Alberta; Canadian Journal of Earth Sciences, v. 12, p. 52-61.

Evenchick, C.A., Parrish, R.R., and Gabrielse, H.
1984: Precambrian gneiss and late Proterozoic sedimentation in north-central British Columbia; Geology, v. 12, p. 233-237.

Fermor, P.R.
1980: Structural and stratigraphic analysis of rocks adjacent to the Lewis Thrust Fault around the Cate Creek and Haig Brook Windows, British Columbia and Alberta; M.Sc. thesis, Queen's University, Kingston, Ontario.

Fermor, P.R. and Price, R.A.
1983: Stratigraphy of the lower part of the Belt-Purcell Supergroup (middle Proterozoic) in the Lewis thrust sheet of southern Alberta and British Columbia; Bulletin of Canadian Petroleum Geology, v. 31, p. 169-194.

Finch, J.C. and Baldwin, D.D.
1984: Stratigraphy of the Prichard Formation, Belt Supergroup; in Belt Symposium II, S.W. Hobbs (ed.), Montana Bureau of Mines and Geology, Special Publication 90, p. 5-7.

Fraser, J.A., Donaldson, J.A., Fahrig, W.F., and Tremblay, L.P.
1970: Helikian basins and geosynclines of the northwestern Canadian Shield; Geological Survey of Canada, Paper 70-40, p. 213-239.

Giletti, B.J.
1966: Isotopic ages from southwestern Montana; Journal of Geophysical Research, v. 71, p. 4029-4036.

Hamilton, J.M.
1984: The Sullivan deposit, Kimberley, British Columbia - a magmatic component to genesis?; in Belt Symposium II, S.W. Hobbs (ed.), Montana Bureau of Mines and Geology, Special Publication 90, p. 58-59.

Harrison, J.E.
1972: Precambrian Belt basin of the northwestern United States: its geometry, sedimentation and copper occurrences; Geological Society of America Bulletin, v. 83, p. 1215, 1240.

Harrison, J.E. and Jobin, D.A.
1963: Geology of the Clark Fork quadrangle, Idaho-Montana; United States Geological Survey, Bulletin 1141-K, 38 p.

Harrison, J.E., Griggs, A.B., and Wells, J.D.
1974: Tectonic features of the Precambrian Belt Basin and their influence on post-Belt structures; United States Geological Survey, Professional Paper 866.

Hill, R.
1983: The stratigraphy and sedimentology of the Waterton and Altyn formations, Waterton Lakes National Park Alberta: a preliminary report (unpublished) in the Institute of Sedimentary and Petroleum Geology Library, Calgary, Alberta.

Hofmann, H.J.
1985: The mid-Proterozoic Little Dal macrobiota, Mackenzie Mountains north-west Canada; Palaeontology, v. 28, pt. 2, p. 331-354.

Hofmann, H.J. and Aitken, J.D.
1979: Precambrian biota of the Little Dal Group, Mackenzie Mountains, northwest Canada; Canadian Journal of Earth Sciences, v. 16, p. 150-166.

Høy, T.
1979: Geology of the Estella-Kootenay King area, Hughes Range, southeastern British Columbia; British Columbia Ministry of Energy, Mines and Petroleum Resources, Preliminary Map 36.

Høy, T. and Diakow, L.
1982: Geology of the Moyie Lake area, British Columbia; Ministry of Energy, Mines and Petroleum Resources, Preliminary Map 46.

Huebschman, R.P.
1973: Correlation of fine carbonaceous bands across a Precambrian stagnant basin; Journal of Sedimentary Petrology, v. 43, p. 688-699.

Hunt, G.
1962: Time of the Purcell eruption in southeastern British Columbia and southwestern Alberta; Journal of the Alberta Society of Petroleum Geologists, v. 10, p. 438-442.

Irving, E. and McGlynn, J.C.
1981: On the coherence, rotation and paleolatitude of Laurentia; in The Proterozoic; Precambrian Plate Tectonics, A. Kroner (ed.), Elsevier, Amsterdam, p. 561-598.

Jefferson, C.W.
1978: Correlation of middle and upper Proterozoic strata between northwestern Canada and south and central Australia (Abstract); Geological Society of America, Abstracts with Programs, 1978, p. 429.

Jefferson, C.W. and Ruelle, J.C.L.
1987: The Late Proterozoic redstone Copper Belt, Mackenzie Mountains, Northwest Territories; in Mineral Deposits of the Northern Cordillera, J.A. Morin (ed.), Canadian Institute of Mining and Metallurgy, Special Volume 37, p. 154-168.

Kerans, C., Ross, G.M., Donaldson, J.A., and Geldsetzer, H.J.
1981: Tectonism and depositional history of the Helikian Hornby Bay and Dismal Lakes Groups, District of Mackenzie; in Proterozoic Basins of Canada, F.H.A. Campbell (ed.), Geological Survey of Canada, Paper 81-10, p. 152-182.

Leech, G.B.
1958: Fernie map-area, west half, British Columbia; Geological Survey of Canada, Paper 58-10.
1960: Geology of Fernie (west half), Kootenay District, British Columbia; Geological Survey of Canada, Map 11-1960.

Lis, M.G. and Price, R.A.
1976: Large-scale block faulting during deposition of the Windermere Supergroup (Hadrynian) in southeastern British Columbia; in Report of Activities, Part A, Geological Survey of Canada, Paper 76-1A, p. 135-136.

McMechan, M.E.
1980: Stratigraphy, structure and tectonic implications of the middle Proterozoic Purcell Supergroup in the Mount Fisher area, southeastern British Columbia; Ph.D. thesis, Queen's University, Kingston, Ontario.
1981: The middle Proterozoic Purcell Supergroup in the southwestern Rocky and southeastern Purcell Mountains, British Columbia and the initiation of the Cordilleran miogeocline, southern Canada and adjacent United States; Bulletin of Canadian Petroleum Geology, v. 29, p. 583-621.

McMechan, M.E. and Price, R.A.
1982: Superimposed low-grade metamorphism in the Mount Fisher area, southeastern British Columbia - implications for the East Kootenay orogeny; Canadian Journal of Earth Sciences, v. 19, p. 476-489.

McMechan, M.E., Hoy, T., and Price, R.A.
1980: Van Creek and Nicol Creek formations (New): A revision of the stratigraphic nomenclature of the Middle Proterozoic Purcell Supergroup, southeastern British Columbia; Bulletin of Canadian Petroleum Geology, v. 28, p. 542-558.

Meijer Drees, N.C.
1975: Geology of the lower Paleozoic formations in the sub-surface of the Fort Simpson area, District of Mackenzie, Northwest Territories; Geological Survey of Canada, Paper 74-40.

Morris, W.A. and Aitken, J.D.
1982: Paleomagnetism of the Little Dal Lavas, Mackenzie Mountains, Northwest Territories, Canada; Canadian Journal of Earth Sciences, v. 19, p. 20-27.

Norris, D.K.
1982: Geology, Snake River, Yukon-Northwest Territories; Geological Survey of Canada, Map 1529A.
1983: Geotectonic correlation chart - Operation Porcupine project area; Geological Survey of Canada, Map 1532A.

Norris, D.K. and Yorath, C.J.
1981: The North American plate from the Arctic Archipelago to the Romanzof Mountains; in The Ocean Basins and Margins, v. 5, The Arctic Ocean, A.E.M. Nairn, M. Churkin, Jr. and F.G. Stehli (eds.), Plenum, New York, p. 37-104.

Obradovich, J.D. and Peterman, Z.E.
1968: Geochronology of the Belt Series, Montana; Canadian Journal of Earth Sciences, v. 5, p. 737-747.
1973: A review of the geochronology of Belt and Purcell rocks in Belt Symposium, v. 1; Department of Geology, University of Idaho and Idaho Bureau of Mines and Geology, Moscow, p. 8-9.

Okulitch, A.V.
1984: The role of the Shuswap metamorphic complex in Cordilleran tectonism: a review; Canadian Journal of Earth Sciences, v. 21, no. 10, p. 1171-1193.

Park, J.K. and Aitken, J.D.
1986a: Paleomagnetism of the Katherine Group in the Mackenzie Mountains: implications for post-Grenville (Hadrynian) apparent polar wander; Canadian Journal of Earth Sciences, v. 23, no. 3, p. 308-323.
1986b: Paleomagnetism of the late Proterozoic Tsezotene Formation of the Mackenzie Mountains, Northwestern Canada; Journal of Geophysical Research, v. 91, p. 4955-4970.

Preto, V.A.
1971: Lode copper deposits of the Racing River - Gataga River area; in Geology, Exploration and Mining in British Columbia; British Columbia Department of Mines and Petroleum Resources, p. 75-107.

Price, R.A.
1962: Fernie map-area, east half, Alberta and British Columbia; Geological Survey of Canada, Paper 61-24.
1964: The Precambrian Purcell system in the Rocky Mountains of southern Alberta and British Columbia; Bulletin of Canadian Petroleum Geology, v. 12, p. 399-426.

Reesor, J.E.
1958: Dewar Creek map-area with special emphasis on the White Creek Batholith, British Columbia; Geological Survey of Canada, Memoir 292.
1973: Geology of the Lardeau map-area, east half, British Columbia; Geological Survey of Canada, Memoir 369.
1983: Nelson map-area, east half, British Columbia; Geological Survey of Canada, Open File 929.
1984: The Purcell Supergroup in the Purcell Mountains, British Columbia; in Belt Symposium II, S.W. Hobbs (ed.), Montana Bureau of Mines and Geology, Special Publication 90, p. 33-35.

Rice, H.M.A.
1937: Cranbrook map-area, British Columbia; Geological Survey of Canada, Memoir 207.
1941: Nelson map-area, east half, British Columbia; Geological Survey of Canada, Memoir 228.

Roberts, S.M.
1986: Belt Supergroup: a guide to Proterozoic rocks of western Montana and adjacent areas; Montana Bureau of Mines and Geology, Special Publication 94.

Ryan, B.D. and Blenkinsop, J.
1971: Geology and geochronology of the Hellroaring Creek Stock, British Columbia; Canadian Journal of Earth Sciences, v. 8, no. 1, p. 85-95.

Schofield, S.J.
1915: Geology of Cranbrook map-area, British Columbia; Geological Survey of Canada, Memoir 76.

Taylor, G.C. and Stott, D.F.
1973: Tuchodi Lakes map-area, British Columbia; Geological Survey of Canada, Memoir 373.

Vitorello, I. and Van der Voo, R.
1977: Late Hadrynian and Helikian pole positions from the Spokane Formation, Montana; Canadian Journal of Earth Sciences, v. 14, p. 67-73.

Watts, A.B., Karner, G.D., and Steckler, M.S.
1982: Lithospheric flexure and the evolution of sedimentary basins; Royal Society of London, Philosophical Transactions, v. 305, p. 249-281.

Winston, D., Woods, M., and Byer, G.B.
1984: The case for an intracratonic Belt-Purcell basin: tectonic, stratigraphic and stable isotopic considerations; in Montana Geological Society, 1984 Field Conference and Symposium, J.D. McBane and P.B. Garison (ed.), p. 103-118.

Yeo, G.M., Delaney, G.D., and Jefferson, C.W.
1978: Two major Proterozoic unconformities, northern Cordillera - discussion; in Current Research, Part B, Geological Survey of Canada, Paper 78-1B, p. 225-228.

Young, G.M.
1977: Stratigraphic correlation of upper Proterozoic rocks of northwestern Canada; Canadian Journal of Earth Sciences, v. 14, p. 1771-1787.
1978: Proterozoic (<1.7 b.y.) stratigraphy, paleocurrents and orogeny in North America; Egyptian Journal of Geology, v. 22, p. 45-64.

Young, G.M., Jefferson, C.W., Delaney, G.D., and Yeo, G.M.
1979: Middle and Lake Proterozoic evolution of the northern Canadian Cordillera and Shield; Geology, v. 7, p. 125-128.

Young, G.M., Jefferson, C.W., Delaney, G.D., Yeo, G.M., and Long, D.G.F.
1982: Upper Proterozoic stratigraphy of northwestern Canada and Precambrian history of the North America Cordillera; in Society of Economic Geologists Coeur d'Alene Field Conference, Idaho-1977, R.R. Reid and G.A. Williams (ed.); Idaho Bureau of Mines and Geology, Bulletin 24, p. 73-96.

Zartman, R.E., Peterman, Z.E., Obradovich, J.D., Gallego, M.D., and Bishop, D.T.
1982: Age of the Crossport C sill near Eastport, Idaho, in Society of Economic Geologists Coeur d'Alene Field Conference, Idaho 1977, R.R. Reid and G.A. Williams (ed.),: Idaho Bureau of Mines and Geology Bulletin 24, p. 61-69.

CHAPTER 5

Authors' addresses

J.D. Aitken
2676 Jemima Rd.
Denman Island,
British Columbia, V0R 1T0

M.E. McMechan
Institute of Sedimentary and
Petroleum Geology
Geological Survey of Canada
3303-33rd Street N.W.
Calgary, Alberta
T2L 2A7

ADDENDUM

The Coppermine River basalts are now dated at 1267 ± 2 Ma (A.N. Le Cheminant and L.M. Heaman, 1989), the age of their feeder dykes (MacKenzie dykes).

Printed in Canada

Chapter 6

UPPER PROTEROZOIC ASSEMBLAGES

Summary
Introduction
Arctic Alaska Terrane
   British Mountains
Porcupine Terrane
Ancestral North America
   Yukon-Tatonduk area
   Ogilvie Mountains
   Wernecke Mountains
   Mackenzie Mountains
   Selwyn Mountains, northern Rocky Mountains, east of Tintina Trench
   Cassiar and Omineca Mountains
   North-central and central Rocky Mountains
   Southern Rocky Mountains
   Northern Columbia Mountains
   Southern Columbia Mountains
Alexander Terrane
Nisling Terrane
Nisutlin Subterrane
References

Chapter 6

UPPER PROTEROZOIC ASSEMBLAGES

H. Gabrielse and R.B. Campbell

SUMMARY

Sequences of Upper Proterozoic, dominantly clastic, sedimentary rocks, generally assigned to the Windermere Supergroup, are commonly more than 2000 m thick and are exposed almost continuously throughout the length of the eastern Cordillera (Fig. 6.1, 6.2). In the Purcell and Mackenzie mountains the supergroup unconformably overlies strata of the Purcell and Mackenzie Mountains supergroups, respectively. In and near the Omineca Belt the rocks unconformably overlie basement of granitic gneiss ranging in age from 728 Ma to more than 2 Ga (see Chapter 4). Elsewhere, the lower contact of the supergroup is not exposed.

In many places the supergroup is overlain unconformably by clastic rocks, commonly sandstone, of Early Cambrian age, but in more western areas, where Lower Cambrian rocks may be fine grained, an unconformable relationship is difficult to demonstrate.

Characteristically, in the central and southern Cordillera, the lower part of the Windermere Supergroup comprises thick, monotonous sequences of gritty, feldspathic sandstone, siltstone and shale, variably metamorphosed from low greenschist to upper amphibolite facies. The upper part of the Windermere is more calcareous and much more variable in lithology. One or more carbonate formations are of regional significance although correlations between regions are dubious. Conspicuous maroon, purple and green shale occurs in the uppermost formation of the group in the Selwyn, northern Rocky, Cassiar, Omineca and Cariboo mountains. In the Selwyn Mountains the varicoloured shales overlie thick successions of gritty sandstone that are younger than most grits assigned to the Windermere Supergroup elsewhere. Diamictite, at least locally reflecting glacial activity, occurs in scattered localities and at several stratigraphic levels.

Thick sequences of volcanic rocks are present near the Canada-U.S.A. International Boundary in the southern Cordillera, in eastern exposures near the Yukon-British Columbia boundary and in the northwestern Cordillera. Volcanic rocks in the Ogilvie Mountains and sills in the Mackenzie Mountains dated at about 770 Ma, and northerly trending dyke swarms in the northern Rocky Mountains may all be related to the same episode of volcanic activity.

Gabrielse, H. and Campbell, R. B.
1991: Upper Proterozoic assemblages, Chapter 6 in Geology of the Cordilleran Orogen in Canada, H. Gabrielse and C. J. Yorath (ed.); Geological Survey of Canada, Geology of Canada, no. 4, p. 125-150 (also Geological Society of America, The Geology of North America, v. G-2)

The easternmost exposures of Windermere rocks in the Mackenzie Mountains are unique in that they include cherty iron formation and glaciogenic deposits. The latter occur at two distinct stratigraphic levels. A similar stratigraphic succession, but lacking iron-formation, occurs in the eastern exposures of the central Rocky Mountains.

The overall character of lithology, thickness and sedimentary structures of the Windermere Supergroup suggests deposition along a rifted margin of western, cratonal North America. The early stages of rifting were accompanied by volcanism and glaciation in an environment of considerable topographic relief. Upper units reflect shallower water deposition as the rate of sedimentation exceeded that of subsidence. In general, the rocks preserved along the easternmost belt of Upper Proterozoic strata strongly indicate a rift environment, particularly in the lower formations. They are characterized by marked and abrupt facies and thickness changes from place to place. In contrast, the upper formations in the east and most of the strata farther west are more uniform in distribution and lithology, perhaps representing a prograding wedge of sediment along the western, rifted margin of the craton.

Radiometric dates ranging from 762 to 777 Ma in volcanic rocks in the basal part of the Windermere Supergroup, 780 Ma on diorite probably related to volcanic rocks, and from 728 to 750 Ma in nonconformably underlying granitic basement rocks suggest that the base of the supergroup is slightly less than 800 Ma old in the Canadian sector of the Cordillera. It is therefore practical to consider this age, which may be better constrained with future work, as the boundary between the Upper and Middle Proterozoic assemblages rather than 900 Ma as shown on the Decade of North American Geology geologic time scale. Clearly, the boundary marks a critical change in Precambrian tectonic evolution.

INTRODUCTION

The thick and widespread, dominantly clastic Windermere Supergroup, deposited between roughly 780 Ma and 570 Ma, is considered generally to represent the basal succession of the Upper Proterozoic and Paleozoic Cordilleran miogeocline which formed along and on the rifted, western margin of Ancestral North America. The source for most of the clastics was presumably the faulted and uplifted basement of Ancestral North America although external sources are possible for some of the northern Cordilleran sediments. The recognition and dating of glaciogenic diamictites, well documented in several localities, are of global significance for correlation of Upper Proterozoic successions.

Two areas with thick Proterozoic strata are poorly dated and therefore of doubtful correlation. These are the

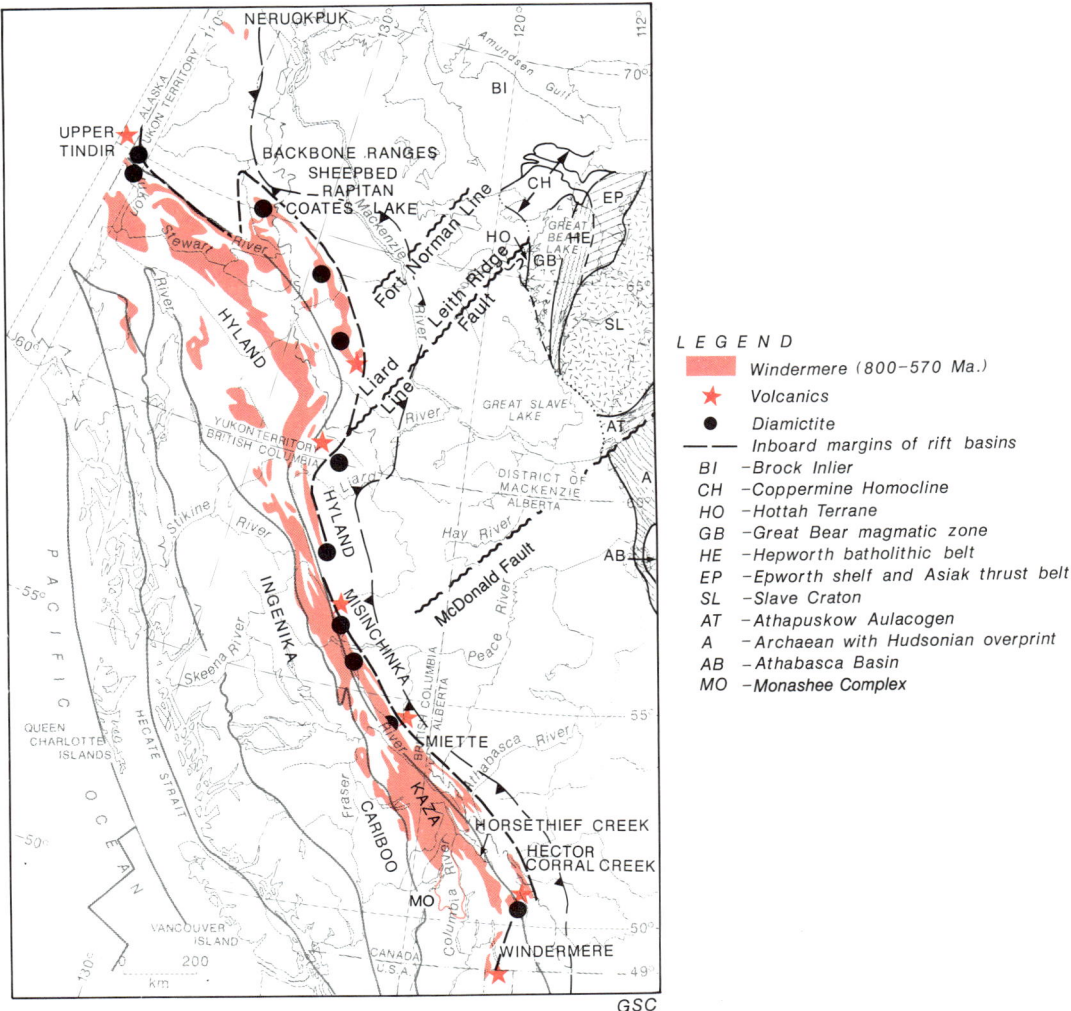

Figure 6.1. Distribution of Upper Proterozoic Windermere Supergroup in the Canadian Cordillera and the terminology used for the main stratigraphic units.

Neruokpuk Formation in the Alaska Arctic Terrane and a little studied succession in the Porcupine Terrane. The Neruokpuk Formation may have been deposited along the northern margin of Ancestral North America and therefore more closely related to the Innuitian orogen rather than to the Cordilleran orogen. The age and provenance of Proterozoic rocks in the Porcupine Terrane are critical to the understanding of possible linkages between the two orogens.

ARCTIC ALASKA TERRANE
British Mountains

A thick succession of dominantly fine grained clastic rocks in northwesternmost Yukon Territory, believed to be in large part of Late Proterozoic age, is assigned to the Neruokpuk Formation. Although as many as six different sequences and twelve members have been identified in the Neruokpuk of Alaska, which includes lower Paleozoic rocks (Dutro et al., 1972), some cannot be recognized with certainty in the Yukon where the formation has been subdivided into seven informal lithostratigraphic units (Norris, 1981a,b) (see Table 6.1). Locally the units are clearly distinct (Fig. 6.3, 6.4) but it is possible that some are lateral equivalents of others. Despite large and small scale deformation the stratigraphic succession dips generally southwest so that on a regional scale the stratigraphically uppermost of the seven units lies towards the southwest, whereas the lowest occurs towards the northeast. The total exposed thickness of the Neruokpuk Formation is estimated to be greater than 13 400 m.

Correlation of the Neruokpuk Formation with strata included in the formation in Alaska is equivocal. The formation is clearly of pre-Carboniferous (Viséan) age on the basis of its regional unconformable relationship with the overlying Endicott Group. Locally, in Alaska, an overlying, thick volcaniclastic and carbonate assemblage contains Early and Late Cambrian trilobites. These rocks, with associated shale and siltstone, rest with angular unconformity on the second youngest unit of the Neruokpuk (PN_5). The possibility exists, however, that the carbonate bodies are olistoliths, thus not constraining the age of the underlying strata. The presence of lower Paleozoic units in

UPPER PROTEROZOIC ASSEMBLAGES

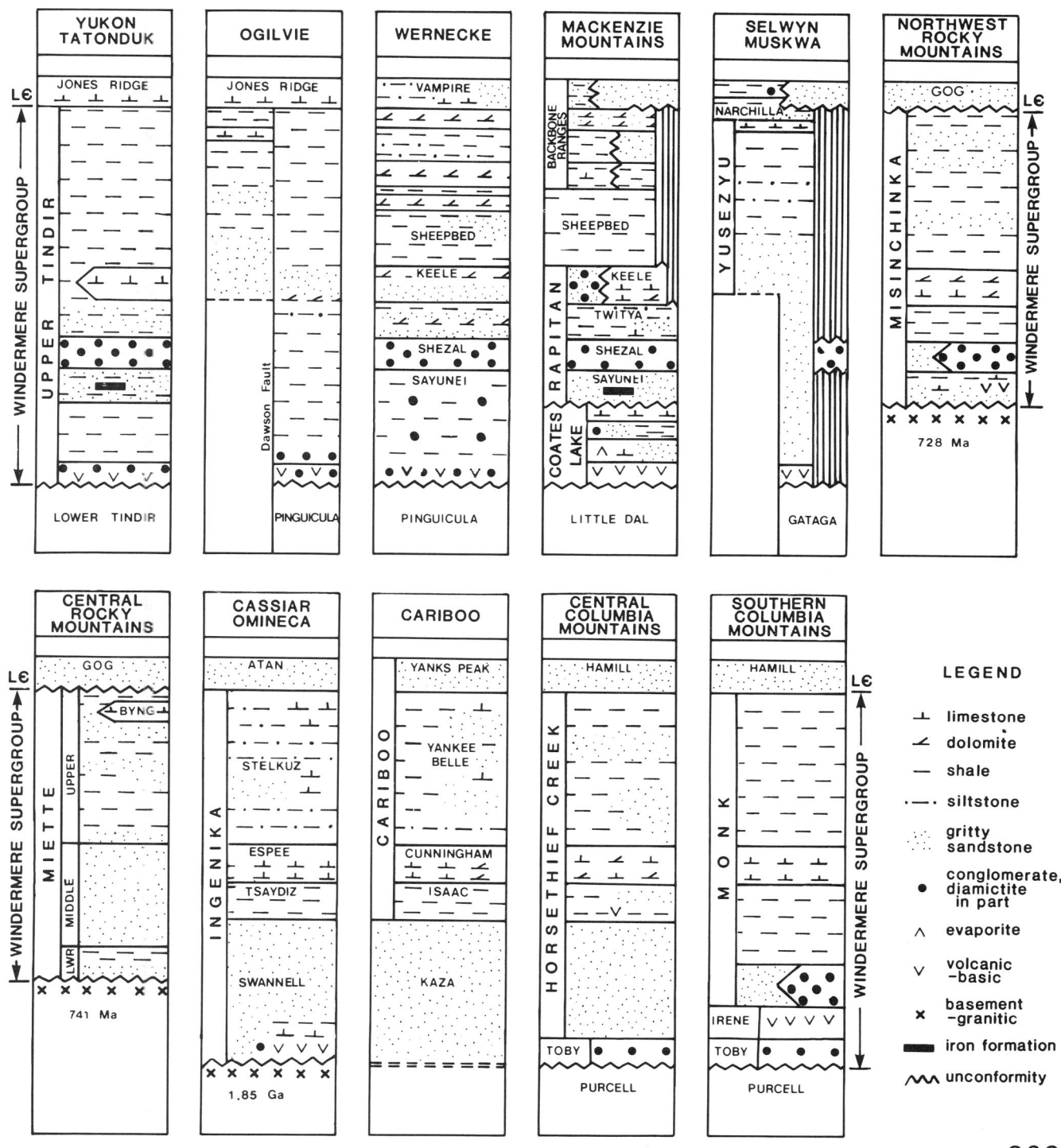

Figure 6.2. Generalized stratigraphic columns of the Windermere Supergroup showing tentative regional correlations.

Table 6.1. Tentative correlation of rock units.

Norris, 1981a,b			Dutro et al., 1972
Unit (Canada)	**Lithology**	**Thickness**	**Unit (Alaska)**
PN_6	Interbedded dark grey-weathering sandstone and olive grey and pale red, slaty argillite? Not examined on the ground.	1000 m (3000 ft)	Not recognized in Alaska
PN_5	Interbedded limestone, black, fine crystalline, yellowish weathering; and argillite, slaty, olive grey, locally red.	1300 m (4300 ft)	Calcareous Siltstone and Sandstone Member
PN_4	Interbedded sandstone, olive grey, fine to medium grained, quartz; argillite, slaty, olive grey, locally red; and chert, olive grey; top not seen.	5100 m (16 200 ft)	Quartzite and Semischist Member (Neruokpuk Schist of Leffingwell, 1919) and Ferruginous Sandstone Member
PN_3	Interbedded argillite, slaty, olive grey; limestone, dark grey, fine crystalline; and siltstone, variably calcareous, olive grey.	1000 m (3000 ft)	Limestone Member; Argillite and Limestone Member; Calcareous Siltstone and Sandstone Member (H.N. Reiser, pers. comm., 1973).
PN_2	Interbedded argillite, slaty, olive grey; sandstone, quartz, fine to coarse grained; and limestone, argillaceous, fine crystalline.	3000 m (10 000 ft)	Grey Phyllite and Chert Member
PN_1	Interbedded slaty argillite, argillaceous limestone and sandstone? Not examined on the ground.	1000 m (3000 ft)	Red and Green Phyllite Member
PN_0	Argillite, dark grey, rusty weathering and sandstone, fine grained, brown weathering; base not seen.	1000 m (3000 ft)	Slate and Quartzite Member

the Neruokpuk in Yukon Territory has been confirmed by Lane and Cecile (1989).

A thick, monotonous succession of argillite, poorly sorted quartz arenite, argillaceous limestone and chert, identified as PN_4 (Table 6.1), appears to represent a deep-water slope sequence, deposited in large part by turbidity currents. Redbeds are few, conglomerate is rare and algal biostromes and ripple marks are unknown. The succession indicates a protracted period of tectonic quiescence represented by mud, muddy limestone and silica accumulation, punctuated by pulses of quartz-rich clastics shed from a sedimentary and perhaps metamorphic provenance (Reed, 1968).

D.K. Norris (pers. comm., 1985) suggested that the paleogeographic setting for the slope deposits comprising the Neruokpuk was the extreme northwest corner of ancestral North America and that the sediments may have been deposited at least in part on oceanic crust. He further hypothesized that the Neruokpuk may be Late Precambrian and therefore, that it is the thick, distal, slope equivalent of the platformal deposits of the upper Tindir Group, now 450 km to the south, across the suture between the Arctic Alaska and North American plates. Another possibility is that the Neruokpuk rocks were deposited in the same basin as the Upper Proterozoic strata in the Innuitian orogen and were displaced in Cretaceous time by counterclockwise rotation out of the Canada Basin. They may have had no direct connection with Upper Proterozoic strata in the Cordilleran miogeocline.

There is no direct evidence for a presumed Late Precambrian age for most if not all of the Neruokpuk; there is substantial evidence for a Late Precambrian age for the upper part of the Tindir Group (Allison and Moorman, 1973). That the two successions are more or less coeval is purely speculative. Indeed, recent studies by Lane and Cecile (1989) have suggested that much of the succession described by Norris (Table 6.1) is of Early Cambrian to possibly Devonian age. In addition, they noted the similarity between the Neruokpuk rocks and correlative strata in Selwyn Basin.

Figure 6.3. Isoclinal folding in rusty weathering, slaty argillites of Unit PN_O of the Neruokpuk Formation on the northeast flank of Romanzof Uplift. View is to northwest across lower Firth River. Photo by D.K. Norris. GSC205235-JJ

Figure 6.4. Folded limestone and slaty argillite of unit PN_3 of the Neruokpuk Formation in the core of Romanzof Uplift. View is to west across Firth River, a few kilometres above Glacier Creek. Photo by D.K. Norris. GSC205235-II

PORCUPINE TERRANE

Extensive exposures of Precambrian rocks occur in the hanging wall of the Yukon Fault along and near Porcupine River where it crosses the Yukon Territory-Alaska border. A stratigraphic sequence described by Cairnes (1914) along the Porcupine River, and assigned to the Tindir Group, is more than 1200 m thick and consists of about 500 m of thinly bedded dolomite overlain by more than 100 m of shale, 350 m of thinly bedded, in part feldspathic and dolomitic, quartzite and 325 m of variably calcareous shale. South of the Porcupine River the strata include persistent units of thinly banded red and green slate, including possible trace fossils, a lithology common in the upper parts of Windermere successions, particularly the Hyland Group, elsewhere in the Cordillera. Diabase dykes cut the stratified rocks in a few places.

The base of the Proterozoic succession is unexposed. It is overlain unconformably by Upper Cambrian to Lower Devonian carbonate strata. Correlations have been suggested with the Middle Proterozoic Quartet Group in the Wernecke Mountains (Norris, 1981c) and the Windermere Supergroup in the Ogilvie Mountains (R.I. Thompson, pers. comm., 1986). On the Tectonic Assemblage Map of the Canadian Cordillera (Map 1712A, in pocket) the rocks are included tentatively in the Hyland Assemblage but correlations are essentially unknown.

The understanding of the correlation and provenance of Proterozoic rocks in the Arctic Alaska and Porcupine terranes is paramount in analyzing the paleogeography of the northwestern part of the North American craton. Until the understanding is achieved speculations on connections between the Innuitian and Cordilleran orogens will remain tenuous.

ANCESTRAL NORTH AMERICA

The rocks described below, unlike those of the Arctic Alaska North Slope and Porcupine terranes are easily correlated with Upper Proterozoic strata in the Ogilvie, Wernecke and Mackenzie mountains related to Ancestral North America.

Yukon-Tatonduk area

Proterozoic rocks of the Tindir Group outcrop in east-central Alaska (Fig. 6.1) and underlie an area of about 250 km² in the adjacent Yukon Territory where they occur in the hanging wall of a major thrust fault. The upper part of the Tindir Group is correlated with the Windermere Supergroup to the southeast in the Mackenzie, Wernecke and Ogilvie mountains (Young, 1982). The rocks have been studied in most detail along and near the Tatonduk River in east-central Alaska (Brabb and Churkin, 1969; Allison et al., 1981; Young, 1982). There, the upper part of the Tindir is divided into five informal formations, several of which show marked lateral variability in thickness and character (Young, 1982). The lowest is a discontinuous unit of green and grey amygdaloidal, tholeiitic, pillow basalt with local tuff and volcanic breccia as much as 200 m thick. Conglomerate in the upper part contains pebbles and cobbles believed to have been derived from dolostone in the lower part of the group. Overlying the volcanic rocks are generally purple-weathering rocks comprising mudstone and minor siltstone, diamictite and graded sandstone. Locally there are also lenses of pebble to boulder conglomerate and (at the top of the unit) laminated hematite-jasper iron-formation. A few clasts with deep, crosscutting striations may be of glacial origin. Elsewhere orange-weathering diamictite marks the top of this second unit. The sequence ranges from more than 700 m thick in the northwest to less than 200 m in the south. The unit coarsens upward. Slumped beds, flame structures, flute and groove casts and crossbeds indicate westward sedimentary transport. The third unit is fault bounded but may be in part related genetically and temporally to the previously described sequence. It is a crudely stratified purple and red diamictite, up to 250 m thick, with thin interbeds of purple mudstone and beds and lenses of red and orange weathering chert. The diamictite consists of clasts, as much as 60 cm across, of mainly buff dolostone and limestone and minor jasper and volcanics in a matrix of fine grained lithic fragments of dolostone and limestone with lesser amounts of microcrystalline quartz siltstone, volcanic and chert fragments and fine grained hematite. Striated clasts are present in several localities and at different stratigraphic levels. The entire diamictite assemblage may be allochthonous (Young, 1982). A fourth unit conformably overlies terrigenous mudstone of unit 2 and comprises as much as 600 m of dominantly grey shale with lesser turbiditic, quartz-dolostone-rich sandstone, siltstone and granule conglomerate. Local but minor members of diamictite and volcanic breccia are present. Paleocurrent data indicate westward transport. The uppermost unit of the upper part of the Tindir Group changes markedly in thickness and lithology from east to west. In the eastern exposures along the Alaska-Yukon boundary it is about 70 m thick and consists of a basal, pebble and boulder conglomerate with clasts of mainly dolostone in a dolostone matrix. Overlying buff dolostone contains possible tepee structures like those in the Keele Formation in the Mackenzie Mountains. Some dolostone, dark shale and fetid limestone form the upper part of the sequence. In the most western outcrops the formation is more than 700 m thick and locally has a basal, thick bedded, granule and pebble conglomerate about 12 m thick with rounded to angular clasts of dolostone and some dolomitic clast breccia. Overlying rocks comprise thinly bedded dark grey limestone, dark grey shale and limestone with black chert nodules. If suggested correlations are correct (Young, 1982), the thick diamictite and turbidite formations of the more western exposures are not present along the Yukon-Alaska boundary. The Tindir Group is concordantly overlain by Lower Cambrian carbonate rocks of the Jones Ridge and Funnel Creek formations.

The interpretation of lithologies, facies changes and thicknesses of the upper part of the Tindir Group is much like that suggested for correlative rocks of the Windermere Supergroup in the Mackenzie Mountains. Initial rifting was accompanied by volcanism and the creation of high relief, at least on the east side of the basin. Early periods of slow deposition were increasingly punctuated by the influx of turbidity currents mainly from the east and southeast. Conglomerate of mass-flow origin is common but some striated dropstones indicate a glacial influence. A prograding sequence is suggested by upward gradation from laminated mudstones, through turbidite to channel-fill conglomerate. Some of the conglomerate may have formed in the upper part of a submarine fan complex.

Local fault scarps probably were important in localizing erosion that led to the accumulation of thick, coarse sediments. The uppermost, dominantly limestone formation changes from deep water to shallow-water facies from the west to the east.

Although correlation of the upper part of the Tindir Group along the Alaska-Yukon border with Upper Proterozoic strata in the Ogilvie, Wernecke, and Mackenzie mountains is unequivocal the correlation of the lower part of the Tindir Group is problematical. In the Yukon-Tatonduk area the lower Tindir consists of six units with an aggregate thickness of 2400 m (Young, 1982). The lowest unit, with base unexposed, is more than 100 m thick and includes grey and purple mudstone and siltstone with some conglomerate. It is overlain by laminated, stromatolitic dolostone with silicified, flat-chip conglomerate about 300 m thick. The next unit, 500 m thick, consists of a grey and black shale grading upward into dolostone and partly oolitic and stromatolitic limestone. The carbonate is overlain by massive to platy quartz arenite with lesser brown and grey shale and orange-weathering, locally cryptalgal dolostone. This unit, 450 m thick, contains some mafic sills. The uppermost two units comprise a lower black, pyritic shale sequence with conglomeratic beds, locally slump folded, and possibly 600 m thick and an upper well-bedded dolostone and limestone member up to 500 m thick. Diabase sills and dykes are common in the shale.

Young (1982) interpreted the lower Tindir succession as representing two shoaling-upward sequences. Graded channel-fill conglomerate with clasts of carbonate and chert coupled with slump folds indicates deposition on unstable slopes in deeper water than the carbonate environment.

Correlation of the lower part of the Tindir Group in the Tatonduk area with the Pinguicula Group and Mackenzie Mountains Supergroup in the Wernecke and Mackenzie mountains has been suggested (Young, 1982). R.I. Thompson (pers. comm., 1986) agreed with the correlation of the lower part of the Tindir Group with the Pinguicula Group but J.D. Aitken (pers. comm., 1986) did not support the correlation of Pinguicula and 'Mackenzie Mountains Supergroup' strata.

Ogilvie Mountains

Strata of the Windermere Supergroup in the Ogilvie Mountains comprise two chrono- and litho-stratigraphic assemblages (R.I. Thompson, pers. comm., 1985). One is a succession of clastic, volcanic and carbonate rocks, as old as 775 Ma, called the Harper rift assemblage and the other is a succession of gritty sandstone, carbonate, and maroon and green shale of latest Proterozoic and Cambrian age known informally as the "maroon and green assemblage"; the two assemblages are not in stratigraphic contact. The Harper rift assemblage is overlain unconformably by Lower Cambrian dolostone and bounded by growth faults that were active during deposition; the maroon and green assemblage forms part of several large thrust slices brought northward during Late Jurassic and/or Early Cretaceous time.

Thicknesses and facies within the Harper assemblage change abruptly along and across strike. The lower part is dominated by conglomerate and volcanics; the upper part by shale, siltstone (turbidite) and carbonate. Maximum aggregate thickness is about 1800 m. It conformably overlies massive carbonate, correlated (tentatively) with the Pinguicula Group, and is overlain unconformably by uppermost Lower Cambrian shallow-water carbonate (Jones Ridge Formation). The unconformity spans about 220 Ma (i.e. 770-550 Ma).

Conglomerate occurs below, above and adjacent to the volcanics. It is massive, rarely bedded, and usually clast-supported. Clasts are angular to subrounded, and range in size from pebbles to boulders; outsized clasts are up to 10 m in diameter and olistoliths occur up to 100 m in length (Plate 6). The majority of clasts were derived from underlying dolostone of the Pinguicula Group. The matrix is dolomitic sandstone containing silt- to granule-size grains. The conglomerate is interpreted as the proximal part of a coastal fan complex: massive units represent gravity flows whereas finer grained, crossbedded and suncracked units may have had a fluvial origin. Thus far a glaciogenic component has not been recognized.

Volcanic strata, called the Mount Harper Volcanic Complex, (Roots, 1983) consist of mafic and minor felsic lava flows with related hydroclastic and pyroclastic breccias; total thickness is approximately 1200 m. The rocks were derived from two distinct magma types: the lower three members represent a calc-alkalic or transitional homogeneous basaltic suite; the upper two members represent a tholeiitic, silica-bimodal suite. The lower suite built upward from deep into shallow water and formed some emergent basaltic islands. The upper suite erupted into a shallow marine environment and consists mainly of sheet flows, tuff and hydroclastic breccia. Zircons from the upper felsic unit give an age of 777 (+47/-31) Ma.

The clastic upper part of the Harper rift assemblage is in gradational contact with the underlying Harper volcanic rocks. Its three main lithologies comprise: a basal mudstone and shale (100 m+); a middle orange weathering, silty and sandy dolomite, limestone, limestone debris flow, dolomitic quartz sandstone and quartzite (10-135 m); and an upper turbiditic succession of interbedded fine sandstone, siltstone and shale (145 m). These strata are mainly of shallow-water marine origin.

The Harper rift assemblage was deposited into a northwest-trending half graben. These rocks are the initial products of Windermere sedimentation. Structural reconstructions across the region suggest this site of crustal extension was probably more than 150 km northeast of the continental edge.

The Harper rift assemblage is lithologically similar to the upper part of the Tindir Group (Young, 1982), the Coates Lake Group (Jefferson, 1983), and possibly part of the Rapitan Group (Eisbacher, 1981).

The "maroon and green assemblage" overlying the Harper rift assemblage consists of three stratigraphic divisions: lower gritty quartz sandstone, argillite and chert; middle limestone; and upper maroon, green and grey argillite, fine sandstone, and mafic volcanics. The lower sandstone division consists of medium to thick beds of quartz sandstone and grit. Sedimentary structures are rare, normal and inverse grading are poorly developed, and contacts are sharp. The proportion of argillite interbeds increases upwards. A distinctive thin bedded, pale grey chert unit, about 50 m thick, caps the division. The middle limestone division is recrystallized and cleaved. Near the Alaska boundary it is several hundred metres thick but thins to a few metres about 60 km to the east at the Dempster Highway. The upper maroon, green and grey argillite division contains siltstone and fine grained sandstone interbeds. In places mafic, amygdaloidal flows and volcanic breccias form part of the succession. The trace fossil *Oldhamia*, together with larger forms, indicate that the upper division is probably Early to Middle Cambrian in age. The succession in general correlates lithologically and chronologically with Cambrian and Precambrian strata in the Selwyn Mountains and is typical of the Hyland Assemblage.

Flute casts along the bottom of some sandstone beds indicate south to north transport. This current direction is supported by southward coarsening of the lower sandstone division.

Wernecke Mountains

Whereas in the Mackenzie Mountains, basal strata of the Windermere Supergroup overlie older rocks with conformable, disconformable and angular unconformable contacts, in the Wernecke Mountains they overlie older rocks with marked angular unconformity (Eisbacher, 1981; Fig. 6.5). Indeed, two angular unconformities bound three distinct stratigraphic assemblages in the Wernecke Mountains. The lower one separates the Pinguicula Group from the underlying Wernecke Supergroup (Plate 7). The Pinguicula Group equivalent to the Pinguicula Assemblage has been correlated with the Mackenzie Mountains Supergroup to the east (Eisbacher, 1981) but there are some marked differences in stratigraphic succession and lithology. It may be correlative in part with the Coates Lake Group in the Mackenzie Mountains, a speculation based on a comparison of their tectonic settings. The lowest unit of the Pinguicula grades regionally from several hundred metres of mafic to intermediate volcanic flows and aquagene tuffs in the south into siliclastic, red laminite as much as 200 m thick to the north. It is overlain by about 300 m of thinly bedded, laminated and flasered limestone which grades upward into dolosiltite. A white, massive dolostone, as much as 200 m thick, overlies the dolosiltite and grades upward into about 600 m of shale, dolostone, conglomerate and limestone laminites with stromatolite bioherms and biostromes. The uppermost units consist of thin, laminated quartzite in the north and up to 2000 m of red dolosiltite, dolostone and quartzite farther south. Thinly bedded, particulate limestone comprises the uppermost unit.

The upper unconformity separates the Pinguicula Group from typical Windermere Group strata (Fig. 6.5). The basal unit, correlated with the Sayunei Formation of the Rapitan Group in the Mackenzie Mountains, is a conglomerate as much as 400 m thick. It is in a fluviatile redbed facies to the south and a buff subaqueous channelled facies to the north. The Sayunei Formation is overlain by glaciomarine diamictite of the Shezal Formation which decreases in thickness from a few hundred metres in the north to zero in the south. The diamictite comprises greenish grey greywacke and bedded siltstone with widely dispersed quartzite, greenstone and dolostone clasts. Strata correlated with the Twitya Formation in the Mackenzie Mountains are, in the Wernecke Mountains, from 400 to 800 m thick and consist of light grey, thick bedded dolostone. These strata grade northward into mass-flow breccias of

Figure 6.5. Angular unconformity between Pinguicula Group and conglomerate of Windermere Supergroup in Wernecke Mountains. Photo by G.H. Eisbacher. GSC203631-A

dolostone and turbiditic, limestone interbedded with shale (Fig. 6.6). The shale grades laterally into turbiditic, siliciclastic siltstone and sandstone at least 1000 m thick. The Keele Formation, about 40 m thick, overlies sediments correlated with the Twitya Formation and consists of distinctive light grey laminated limestone, pink dolostone, crossbedded pebbly quartzite and white dolostone at the top. The Rapitan Group in the Snake River area shows significant differences in lithology and depositional setting from equivalent units in the Mackenzie Mountains. Paleocurrents in the basal Sayunei Formation flowed southeastward in the southern redbed facies and northeastward in the subaqueous buff facies. A northeast-facing paleoslope is demonstrated by onlap of glaciomarine dolostone by off-shelf siliclastic turbidite. Slump folds, solemarks and paleocurrent flow confirm this interpretation. Sedimentary facies and thickness changes in the Snake River area suggest synsedimentary tectonics and a basin axis parallel with the dominant northwest fault trend of the region.

The uppermost strata of the Windermere Supergroup in the Wernecke Mountains are at least in part correlative with the Sheepbed and Backbone Ranges formations in the Mackenzie Mountains. They are more than 500 m thick and consist mainly of units of shale, siltstone and fine grained quartzarenite alternating with thin to thick bedded dolostone and thinly bedded limestone (Narbonne et al., 1985).

In the thickest succession so far studied a resistant grey weathering dolostone, about 60 m thick is characterized by thin, rhythmically spaced seams of coarse white dolomite spar in fine grained grey dolostone. It is correlated with the Gametrail Formation in the Mackenzie Mountains and overlies siltstone possibly correlative with the Sheepbed Formation. An overlying recessive siltstone, shale, fine grained quartzarenite and minor carbonate unit, about 350 m, thick contains Ediacaran megafossils, including *Beltanelliformis brunsae*, near the top and ventotaenid algae in the lower part. The strata are probably correlative with those of the Blueflower Formation in the Mackenzie Mountains. Three overlying units, comprising a lower dolostone and quartzite, 69 m thick, a middle shale, quartzarenite and siltstone unit, 148 m thick, containing Ediacaran megafossils and simple trace fossils and an upper dolostone, 126 m thick, are tentatively correlated with the Risky Formation in the Mackenzie Mountains. As in the Mackenzie Mountains the rocks reflect a regional, post-Keele transgression followed by fluctuating shallow and deep water environments in the upper Windermere succession. The Lower Cambrian-Precambrian boundary occurs within the strata above the Sheepbed Formation.

Mackenzie Mountains

The oldest strata included in the Windermere Supergroup in the Mackenzie Mountains are assigned to the Coates Lake Group, an assemblage bounded by unconformities, and transitional in lithology between the underlying Little Dal Group of the Mackenzie Mountains Supergroup and the overlying Rapitan Group (Fig. 6.2, 6.7; Plate 8; Jefferson and Ruelle, 1986). The Coates Lake Group occurs in an arcuate belt about 300 km long on the eastern margin of preserved Upper Proterozoic rocks (Fig. 6.8). It consists of three formations separated by gradational contacts and local unconformities.

The basal Thundercloud Formation, ranging from zero to 300 m thick, unconformably overlies stromatolitic grainstone of the Little Dal Group. Basalt at the top of the Little Dal Group (Gabrielse et al., 1973; Jefferson and Ruelle, 1986) is considered herein to be part of the Thundercloud Formation of the Coates Lake Group because it reflects local extension and the onset of a major change in the tectonic behaviour of the region. The unit gradationally becomes finer grained upward from basal volcanic conglomerate and basalt into maroon mudstone which is rhythmically interbedded with tan dolostone and a quartzarenite member. Evaporites of the Redstone River Formation overlie dolostone of the Thundercloud Formation gradationally or, in places, abruptly with a basal conglomerate. Evaporites and conglomerate grade upward into maroon mudstone rhythmite. The clastic rocks are dominantly carbonate, with fining-upward cycles. Trough

Figure 6.6. Tongues of dolomitic breccia interbedded with shale representing a facies of the Twitya Formation, eastern Wernecke Mountains. Photo by G.H. Eisbacher. GSC203631-C

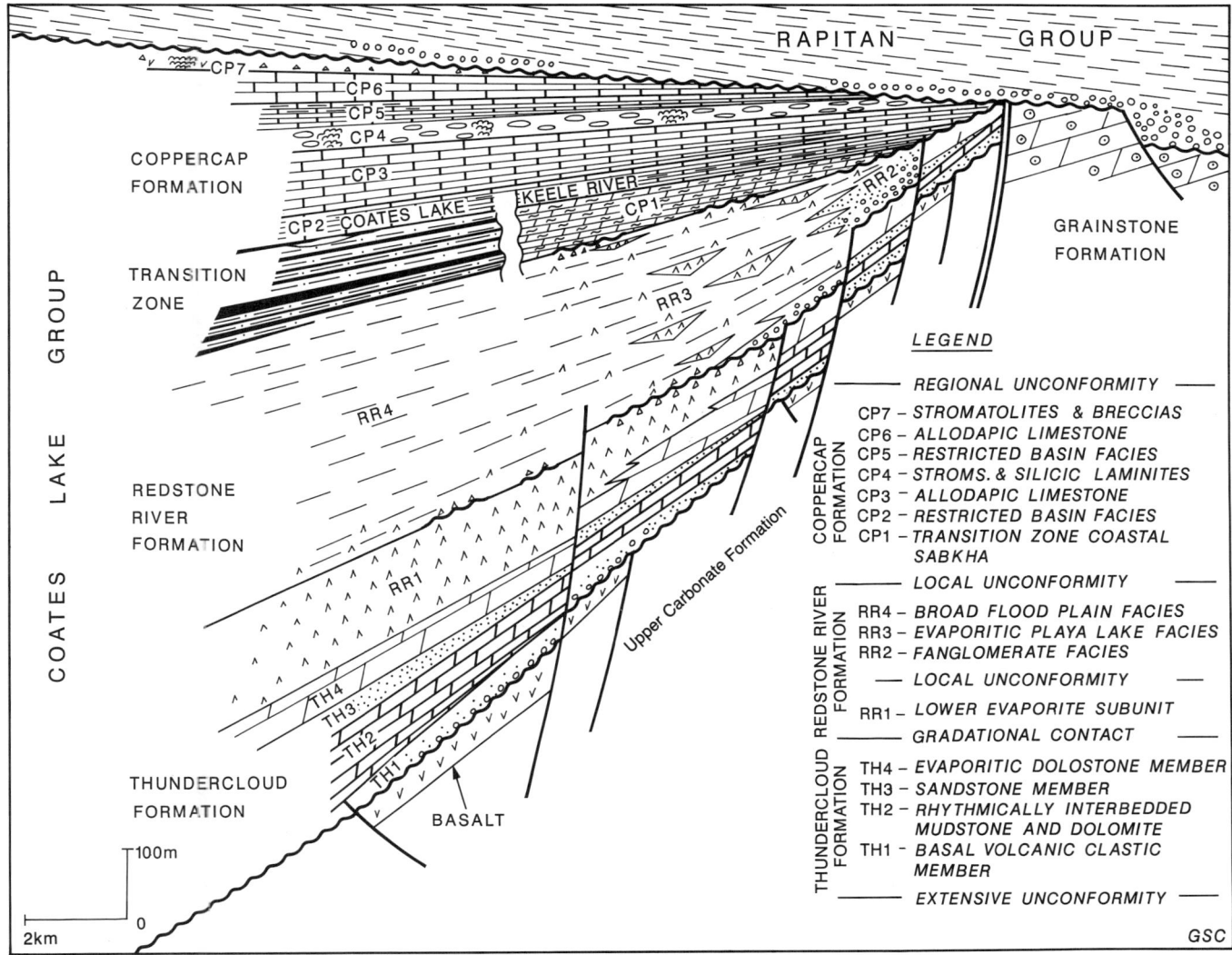

Figure 6.7. Schematic facies relationships, Coates Lake Group. After Jefferson (1983). The Upper Carbonate and Grainstone formations are respectively the main upper and lower carbonate units of the Little Dal Group (Aitken, 1981).

crossbeds and desiccation cracks are conspicuous. The formation is up to 1220 m thick. The Coppercap Formation is the uppermost formation of the Coates Lake Group. It comprises up to 300 m of coarsening-upward sequences of shaly calcisiltite to calcirudite turbidite with layers of chert and stromatolite bioherms. A transitional unit between the Redstone River and Coppercap formations consists of interbedded mudstone, evaporites and carbonate. This unit contains significant amounts of stratabound copper sulphide minerals.

The Coates Lake Group was deposited in restricted areas in an extensional tectonic setting about 760 Ma ago (Fig. 6.5; Eisbacher, 1981). Diabase sills and dykes, believed to be related to volcanic rocks of the lowermost Coates Lake Group have been dated by the Rb-Sr method at about 770 Ma (Armstrong et al., 1982). Diorite, probably related to the volcanics, has been dated by U-Pb (zircon) analyses at 780 Ma (R.R. Parrish and C.W. Jefferson, pers. comm., 1986). The Thundercloud Formation is believed to express the transition from alluvial fan to sabkha and shallow platformal marine carbonate environments following a period of extension and volcanism (Jefferson and Ruelle, 1986). Overlying evaporites of the Redstone River Formation indicate restricted marine, lagoonal or lacustrine evaporation. The associated clastics are interpreted to be components of alluvial fan complexes presumably derived from the upper part of the Little Dal Group. The transitional unit between the Redstone River and Coppercap formations reflects a coastal sabkha environment which was followed by clastic deposition during regional transgression. Facies changes and paleocurrent data throughout the Coates Lake Group delineate six main areas of deposition and preservation influenced by growth faults.

The Rapitan Group onlaps the Coates Lake and Little Dal groups with local conformity to angular unconformity (Fig. 6.9) and has been described variously as including

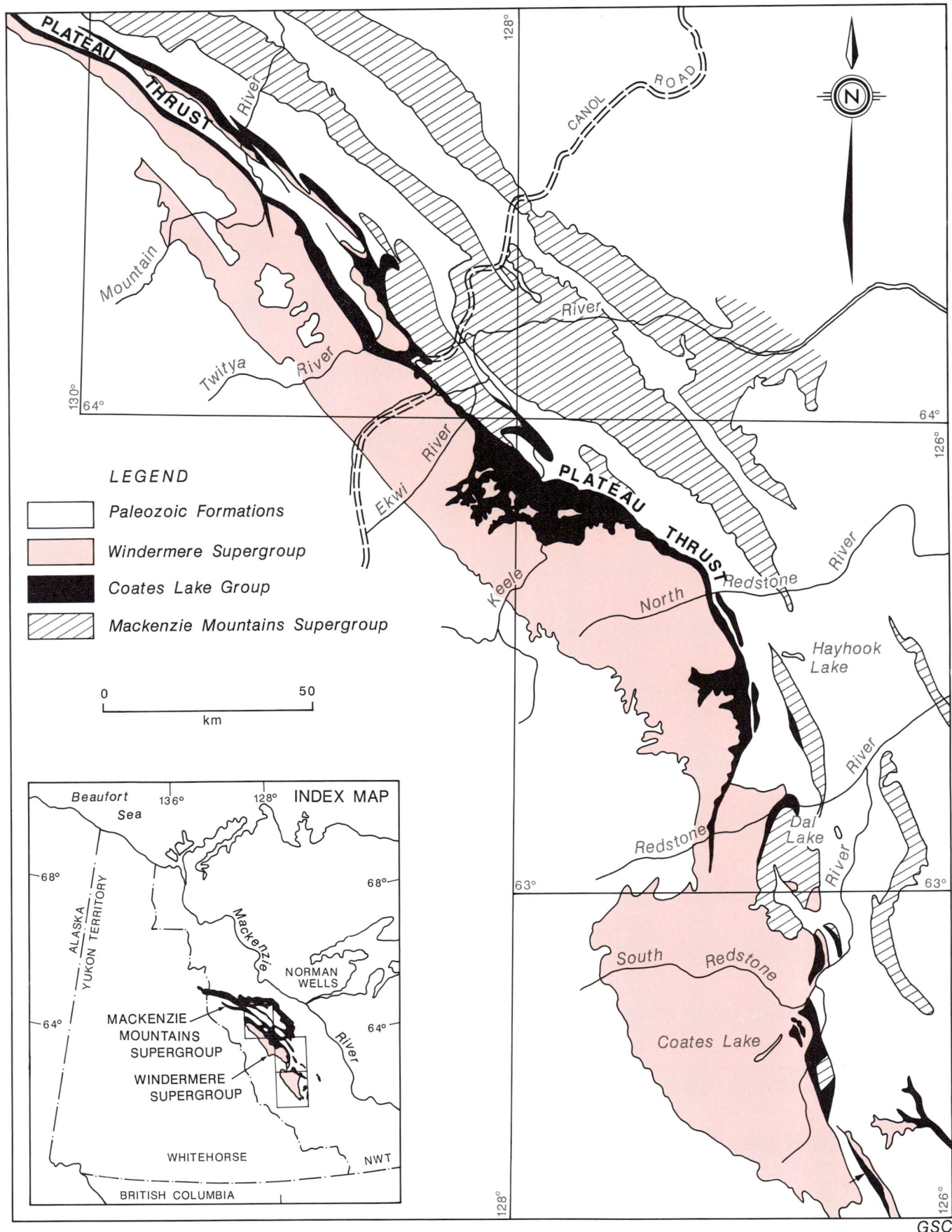

Figure 6.8. Distribution of Coates Lake Group (black). After Aitken and Cook (1974), Aitken et al. (1973), Blusson (1971), Eisbacher (1981) and Gabrielse et al. (1973). After Jefferson and Ruelle (1986).

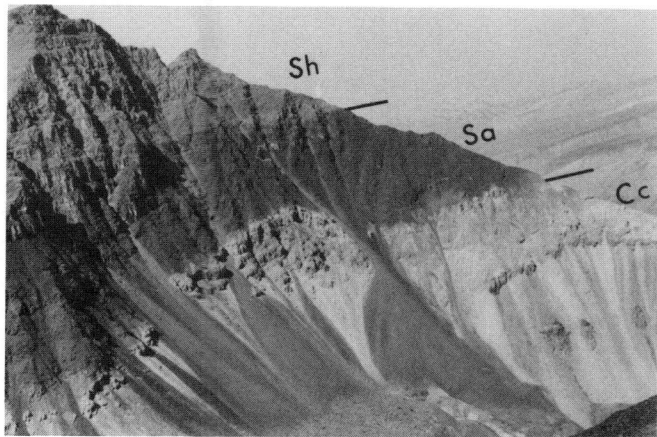

Figure 6.9. Onlap of basal Rapitan Group (Sayunei Formation, Sa) onto gently folded Coates Lake Group (Coppercap Formation, Cc), in southeastern Mackenzie Mountains. sh- Shezal Formation. Photo by G.H. Eisbacher. GSC203631-B

two (Norris, 1982), three (Gabrielse et al., 1973), or four (Eisbacher, 1978) formations all easily identified throughout the Mackenzie Mountains (Eisbacher, 1981). The nomenclature of Eisbacher is used herein because it includes the two distinctive glaciogenic units recognized in the region. The Sayunei Formation is the basal unit, with a thickness ranging from zero to 700 m (Plate 9). It is dominantly thinly bedded, graded siliciclastic laminite and turbidite alternating with dark red to maroon or green argillite. Included are local thick beds of coarse mass-flow deposits and fine grained clastic rocks with glaciogenic dropstones. The top of the formation is characterized by regionally extensive hematite-jaspilite iron formation containing abundant dropstones. In the Snake River area this iron-formation is as much as 150 m thick (Yeo, 1981). The overlying Shezal Formation ranges from a few metres to 500 m thick and in its thinner marginal facies is a massive tillite with numerous striated and glacially polished clasts (Fig. 6.10). Thicker basinal diamictite beds are interbedded with sandstone and shale. Clasts in the diamictite, commonly well rounded, are up to several metres in diameter and include coarsely crystalline dioritic boulders as well as dolomite, limestone, quartzite and dark green volcanic rocks. The Twitya Formation, as much as 800 m thick, rests with sharp but generally conformable contact on the Shezal Formation. It consists of shale, siltstone, parallel-laminated sandstone, in places arkosic, and some conglomeratic channel deposits. The base of the unit is marked locally by thinly bedded to laminated, dark grey, limestone as much as 100 m thick. The uppermost unit of the Rapitan Group is the Keele Formation, from 100 to 500 m thick, which lies in gradational contact on the Twitya Formation. It consists of shoaling-upwards carbonate quartzite cycles. The lower carbonate member ends southwestward at a breakaway scarp that shed a subregional olistostrome (chaotic breccia) into a basinal carbonate succession. The upper carbonate member pinches out and is replaced basinward by diamictite (tillite). Local occurrences of striated dropstones indicate a renewal of glacial activity. The top of the formation is defined by a laterally persistent and resistant light grey dolostone unit, about 10 m thick, characterized by 'tepee' structures.

Paleoslopes during deposition of the Rapitan Group were prograded centripetally towards Selwyn Basin (Fig. 6.11). Yeo (1981; Fig. 6.11) and Eisbacher (1981) suggested that sedimentation during deposition of the Sayunei and Shezal formations was controlled by a system of high-angle, extensional faults that fan from northeasterly trending in the southeast to northwesterly trending in the northwest. The faulting was accompanied by possible submarine volcanism and hydrothermal systems that led to the deposition of iron and silica of the iron-formations. Concurrent uplift accompanied by glaciation along the east margin of the basin may have been the result of the same extensional, rift tectonics. The Twitya Formation was probably deposited on the outer shelf and upper slope of the Rapitan clastic basin. Regional shoaling is indicated by the progradation of cyclic shallow-water marine and nonmarine clastics and carbonates of the Keele Formation. Keele deposition was influenced also by a second episode of Late Proterozoic glacial activity.

Overlying the Keele Formation of the Rapitan Group with sharp but conformable contact is the regionally extensive Sheepbed Formation comprising as much as 900 m of recessive dark weathering shale and siltstone. In places the upper part is dominantly siltstone and includes some fine grained sandstone. In its western outcrops the Sheepbed Formation is coarser grained and of more proximal aspect than farther east. The formation represents a marked and abrupt transgression following Keele deposition. Mainly in its western exposures the Sheepbed Formation is succeeded by the Gametrail Formation, consisting of 322 m of ribbon-bedded lime mudstone and grey dolostone (Aitken, 1989). The overlying Blueflower Formation comprising 450 m of turbiditic dark grey shale, mudstone and sandstone is overlain by a unit of dolostone and sandstone, 114 to 167 m thick, named the Risky Formation. The post-Sheepbed formations may be correlative with the lower two units of the Backbone Ranges Formation although these relationships are in doubt. The Blueflower Formation contains abundant well preserved

Figure 6.10. Boulders of limestone, dolomite and greenstone in diamictite of the Shezal Formation in Mackenzie Mountains. Photo by G.H. Eisbacher. GSC203632-G

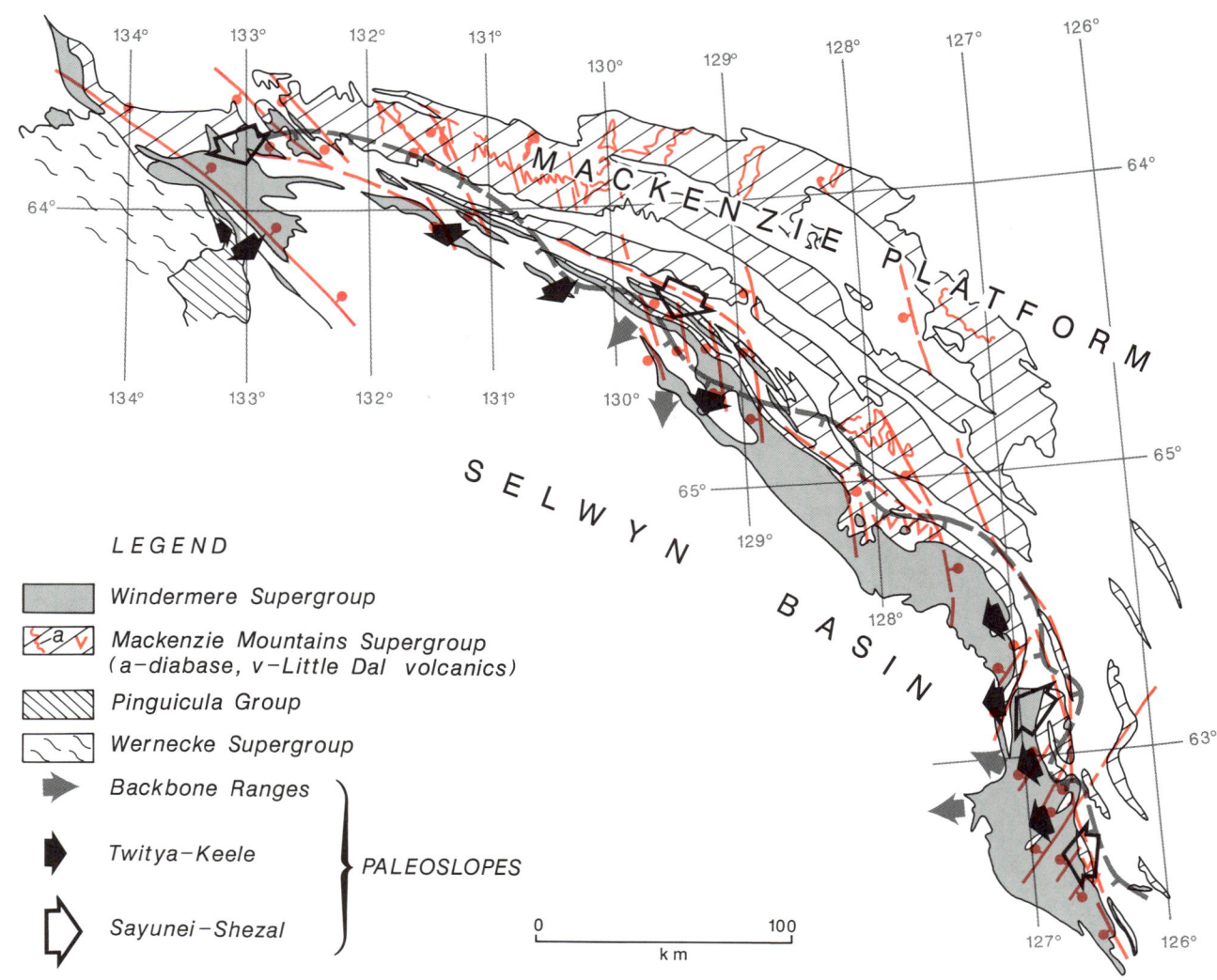

Figure 6.11. Mapped and inferred high-angle faults (dots on downthrown side) which influenced the depositional pattern of the Coates Lake Group and lower part of the Rapitan Group. Also shown are synoptic paleocurrent and slump-fold data for the Windermere Supergroup. After Eisbacher (1981).

Ediacaran megafossils and trace fossils including *Torrowangea*? sp. and *Planolites* sp. The Sheepbed Formation in its eastern exposures in Mackenzie Mountains is overlain unconformably by quartzite of the lower member of the Backbone Ranges Formation which is probably of Late Proterozoic age (Fritz et al., 1983, 1984; Aitken, 1984) but represents a pronounced change in sedimentation characterized by the overlying Lower Cambrian strata. The Backbone Ranges Formation is described with the Lower Cambrian rocks in Chapter 7.

Selwyn Mountains, northern Rocky Mountains, east of Tintina Trench

Extensive areas in the Selwyn Mountains and northern Rocky Mountains and between the Tintina Trench and the northern Selwyn Mountains are underlain by thick sequences of Upper Proterozoic to Lower Cambrian rocks characterized by gritty quartzose clastics, commonly with maroon shale in the upper part. Since they were first mapped the strata have been referred to informally as the "Grit Unit" (Gabrielse et al., 1973), but recently have been named Hyland Group divisible into two formations: a lower Yusezyu Formation and an upper, Narchilla Formation (S.P. Gordey, pers. comm., 1985). The base of the group is not exposed and the top is conformable to disconformable with Lower Cambrian fine grained clastics.

In the type region within the southern Selwyn Mountains, the Yusezyu Formation is as much as 3000 m thick and is dominated by coarse grained clastic rocks with interbedded shale and minor limestone. Monotonous sequences of well bedded, massive sandstone in sharp planar contact with shale are typical and in places amalgamated sandstone units are up to 100 m thick. Quartz-pebble conglomerate, locally feldspathic, is found only in thick sandstone members. Coarse quartz grains commonly show blue opalescence. Feldspar, including plagioclase and microcline, ranges from less than 8% to as much as 15%. Sedimentary structures, in normally graded sequences, are rare but include groove and load casts, ripple lamination, scour marks and aligned shale clasts. The upper part of the Yusezyu Formation is variably calcareous and in many places is capped by a fine grained, light to dark grey limestone member.

The Narchilla Formation conformably overlies the Yusezyu Formation and has been divided into three members. The lowest, in gradational contact with the Yusezyu, is more than 300 m thick and consists of blue-grey to green weathering slate, commonly laminated. The middle member is thin to thick bedded, fine grained quartz sandstone and siltstone about 70 m thick. The upper unit, more than 400 m thick, is mainly blue-grey slate which, in its upper part, becomes apple-green weathering. The apple-green slate grades laterally into distinctive maroon strata. Locally, sandstone beds show normal size-grading parallel lamination or ripple crosslamination. Flute and groove casts are present but not abundant. The formation is overlain by limestone conglomerate containing clasts of Lower Cambrian archaeocyathid limestone.

Strata of the Hyland Group, similar to those described above, can be traced southward into the northern Rocky Mountains as far as the area near Gataga River. Diagnostic strata include feldspathic quartz-pebble conglomerate and maroon weathering slate. In many places the top is marked by a limestone cobble conglomerate which contains archaeocyathids, quartz-pebble conglomerate and maroon weathering slate.

The easternmost rocks included in the Windermere Supergroup in the northern Rocky Mountains consist of more than 500 m of diamictite composed of locally striated clasts in a mudstone matrix. Light coloured dolostone clasts and lenses of dolostone breccia are conspicuous. The unit is believed to reflect glacial activity (Fritz, 1972; Eisbacher, 1981). East of the main belt of Hyland Group strata in southeastern Yukon Territory mafic, amygdaloidal volcanic flows, tuff and breccia range from 300 to 650 m thick. Buff, flaggy dolomite and sandy dolomite form minor interbeds. The sequence is overlain by a thick unit of feldspathic sandstone with maroon and green slate which may be in part of Late Precambrian age. The upper sandstone contains Early Cambrian trilobites.

Correlation of the Hyland Group with rocks of the Windermere Supergroup in the Mackenzie Mountains is uncertain. The Narchilla Formation is thought to be correlative with the lower part of the Backbone Ranges Formation. The upper part of the Yusezyu and lower part of the Narchilla formations are of Precambrian age based on primitive trace fossils beneath the upper limestone member of the Yusezyu formation (Fritz et al., 1983). This suggests that the Hyland Group may be correlative with the upper part of the Windermere Supergroup in the Mackenzie Mountains and mainly or entirely younger than the Rapitan Group. Although Hyland Group rocks are known to underlie large areas between the Tintina Trench and the northern Selwyn Mountains, they have been mapped only on a reconnaissance scale as yet, and further studies are necessary before the true relationship of these two major units can be resolved.

The proportion of coarse clastic sediments, amalgamation of sandstone units to form thick, shale-free members, and the presence of scour structures beneath some beds suggest that much of the Yusezyu Formation was deposited in submarine upper or mid-fan channels. Paleocurrent data, although sparse, show a dominance of southeasterly sediment transport. The formation reflects rapid deposition of sediment from a granitic and possibly high grade metamorphic terrane of considerable relief. The Narchilla Formation with its finer grain size and turbiditic structures was probably deposited in relatively deep water.

Cassiar and Omineca Mountains

Upper Proterozoic rocks are exposed extensively in the Cassiar and Omineca mountains of the Omineca Belt. Only in the southeastern Cassiar and northern Omineca mountains, have underlying rocks been observed. There, quartzite and amphibolite assigned to the Windermere Supergroup overlie basement granitic gneiss (see Chapter 4). In most areas the Upper Proterozoic strata are overlain paraconformably by clastic rocks, commonly quartzite, of Early Cambrian age. The Proterozoic rocks comprise the Ingenika Group and in the northern Omineca Mountains, where they are most fully developed, are subdivided into four formations, which, in ascending order include the Swannell, Tsaydiz, Espee and Stelkuz formations (Mansy and Gabrielse, 1978).

The Swannell Formation, up to 2000 m thick, is dominantly a clastic sequence of fine grained argillaceous to medium grained quartzitic rocks. Thin, discontinuous limestone members form a minor component of the upper part of the formation. It is characterized, however, by thick-bedded to massive members of grit and feldspar-quartz pebble conglomerate, some of which show distinct normal grading. Quartz commonly occurs as opalescent blue grains and pebbles in a phyllite matrix. Feldspar may be plagioclase and/or perthite, the proportion ranging widely from one area to another. The lowest part of the Swannell Formation is mainly micaceous quartzite and locally feldspathic. At one locality in the Omineca Mountains the stratigraphically lowest unit is a conglomerate which includes rounded boulders up to 20 cm in diameter of coarse grained, foliated granitic rock of felsic to intermediate composition (Roots, 1954). In southeastern Cassiar Mountains quartzite and amphibolite overlie a basement consisting of potassium feldspar augen gneiss dated at 1.85 Ga on U-Pb (zircon) (Evenchick et al., 1984). The structurally complex overlying assemblage includes feldspathic quartzite, amphibolite, carbonate and schist, together more than 1000 m thick in the hanging wall of a thrust fault. These rocks, bounded on the east by the Northern Rocky Mountain Trench, are not obviously correlative with the Ingenika Group elsewhere in the region because of the abundance of amphibolite and the high degree of strain they have sustained. In general, strata of the Swannell Formation were regionally metamorphosed from greenschist to amphibolite facies.

The Tsaydiz Formation overlies the Swannell Formation with a gradational contact and consists of up to 185 m of thin-bedded, grey, sericitic phyllite, which is calcareous in the upper part of the formation. Minor units of fine grained gritty quartzite are present locally. The formation is incompetent and almost invariably strongly folded and cleaved. It grades upward through thin bedded limestone into the Espee Formation.

The Espee Formation is the oldest distinctive carbonate unit exposed west of the Northern Rocky Mountain Trench. It is a thin- to thick-bedded, resistant, grey-weathering limestone unit ranging from less than 100 m to 400 m thick (Fig. 6.12). Throughout the Cassiar and Omineca mountains the limestone typically contains beds with conspicuous pisoliths and ferrodolomite particles. Algal structures, generally on a scale of 10 cm or less, and limestone intraclasts, are present locally.

The Stelkuz Formation, abruptly overlying the Espee Formation is the youngest sequence of Precambrian age in the Cassiar and Omineca mountains. It is characterized by a wide variety of varicoloured lithologies including distinctive green and maroon units which alternate, more or less regularly in vertical sequence (Fig. 6.12). In much of the region, three units can be recognized. The lowest, more than 600 m thick, consists mainly of limestone, pelite, siltstone and sandstone in cyclical units; the middle, about 150 m thick, is composed essentially of green and maroon pelite and lenses of fine grained sandstone, in places capped by a thick limestone member; the upper, up to 600 m thick comprises pelite and sandstone, the latter more abundant upward in the section. Total thickness ranges from 300 m to more than 1500 m. The uppermost contact is generally marked by the appearance of clean white sandstone of the Lower Cambrian Boya Formation. Locally the Stelkuz Formation contains probable Late Precambrian trace fossils including: *Gordia*, *Helminthopsis* and *Didymaulichnus* (Fritz and Crimes, 1985). The uppermost part of the Stelkuz Formation has trace fossils suggestive of an Early Cambrian age. Crossbeds in the Omineca Mountains indicate southwestward paleocurrent flow.

Strongly metamorphosed and migmatized Upper Proterozoic strata, probably mainly correlative with the Swannell and Tsaydiz formations, are found in the Wolverine Complex of the Omineca Mountains and in the Horseranch Range in the Cassiar Mountains. Farther north, in the southern Pelly Mountains southwest of the Tintina Trench, siltstone and shale below fossiliferous Lower Cambrian rocks may include strata correlative with the Stelkuz Formation. Precambrian rocks may be included also in the Big Salmon Complex comprising metamorphic and migmatitic rocks in the northern Pelly Mountains. South of the Omineca Mountains discontinuous exposures of Upper Proterozoic strata and regionally metamorphosed correlative rocks can be followed into the northern Cariboo Mountains.

The monotonous thick assemblage of clastic rocks in the Swannell Formation was probably deposited, in part, by turbidity currents. Poor sorting of most units suggests deposition in relatively deep water, an environment that may have accompanied rifting along the western cratonal margin. Another possible manifestation of the rifting event is the presence, particularly in the lower part of the succession, of amphibolite bodies representing volcanic flows, dykes or sills. The upper part of the Ingenika Group

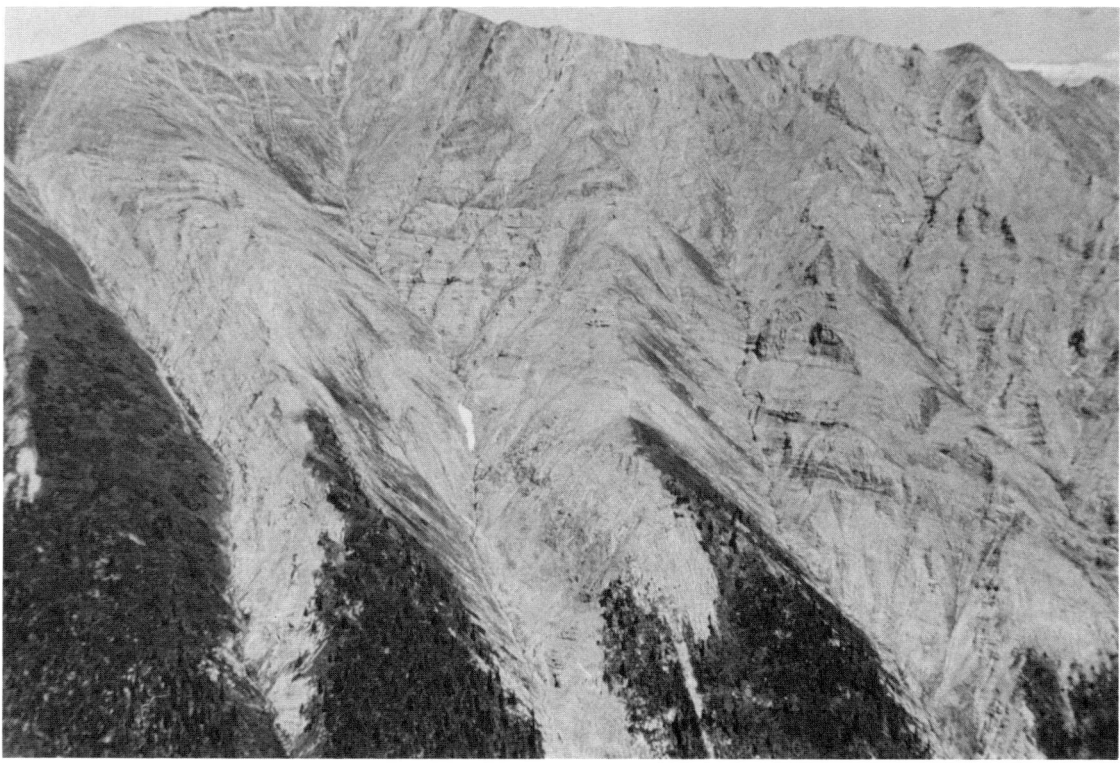

Figure 6.12. Thick- and thin-bedded limestone of Espee Formation in Espee Range overlain by dark weathering Stelkuz Formation on left. Northern Omineca Mountains. Photo by H. Gabrielse. GSC205235-F

reflects shallowing conditions of sedimentation perhaps along a prograding continental margin.

North-central and Central Rocky Mountains

Upper Proterozoic strata of the Misinchinka Group are exposed along the east side of the Northern Rocky Mountain Trench in a southeastward widening belt from its northern limit of truncation near Sifton Pass. A fairly complete sequence is present east of Williston Lake where more than 3000 m of strata nonconformably overlie a potassic granite gneiss basement dated at 728 Ma (Evenchick, 1982; Evenchick et al., 1984). The lowest unit is a thick-bedded, pure, vitreous, even grained quartzite, 200 m thick, with a thin basal member, less than 1 m thick, of locally feldspathic quartzite-pebble conglomerate. The quartzite is overlain by 400 m of fine- to coarse-grained amphibolite, intercalated with thin-bedded quartzite and pelitic schist. A heterogeneous assemblage of pelitic schist, quartzite, marble and amphibolite, also about 400 m thick, overlies the amphibolite unit. Quartzite-cobble diamictite containing well rounded clasts from a few centimetres to more than 15 cm long occurs locally. At least 300 m of amphibolite, schist, marble, grit, dolostone, phyllite and diamictite overlie the unit described above and in turn are overlain by 150 m of marble, limestone and dolostone, which form one of the most conspicuous units in the Upper Proterozoic succession. Distinctive members below the limestone include thin beds of opalescent blue quartz grit and quartzite and carbonate pebble-to-cobble diamictite. A 2- to 3-m thick chlorite schist member rich in magnetite occurs below the diamictite. The uppermost strata of the Misinchinka Group include two prominent but discontinuous limestone members 10 to 30 m thick with an intervening diamictite unit. These are overlain by about 400 m of thin-bedded dolomitic sandstone and a thick sequence of grey and green phyllite interbedded with fine grained sandstone. The Misinchinka rocks have been regionally metamorphosed, the grade increasing from lower greenschist in the east to amphibolite facies in the west, culminating adjacent to the granite gneiss basement.

Farther south to about latitude 55° a broad threefold division of the Misinchinka Group has been recognized (Stott et al., 1983; McMechan, 1987). The lower clastic unit comprises a thick succession of phyllite, siltite, diamictite, feldspathic quartzite and minor carbonate, the base of which is not exposed. Major facies changes occur within the unit across the Weston Fault (Fig. 6.13). To the east the upper 1200 m of the unit consists of green (or less commonly purple) sandy phyllite and schist with phyllite, quartzite, dolostone and greenstone clasts, which are termed diamictites, with lesser laminated green argillite and siltite and minor quartz sandstone and grey argillite. Limestone and dolostone interbeds up to tens of metres thick occur locally in the upper 200 m of the unit. Green or less commonly, grey-purple, locally laminated, silty phyllite and argillaceous siltite underlie the diamictite. Packets of quartzite and pebble conglomerate, 2 to 10 m thick, occur as interbeds in the lower part. These coarse clastics are poorly sorted, graded and locally feldspathic. They are interpreted as the deposits of turbidity flows. Many of these quartzite units have abrupt lateral terminations and appear to be tabular channels. West of the Weston Fault, grey-green pyritic phyllite forms the upper 50 m of the lower clastic unit. Units of feldspathic fine- to coarse-grained quartzite and pebble conglomerate, interbedded with rusty weathering grey argillite characterize the remaining 450 m of the exposed section. The conglomerates are probably of turbiditic origin. In general the quartzites are thicker and comprise more of the section than similar lithologies found at lower stratigraphic levels east of the Weston Fault. Along the Northern Rocky Mountain Trench, interbedded argillaceous sandstone and argillite was metamorphosed and deformed into chlorite-muscovite-(garnet)- schist. Even in eastern exposures, where deformation has not destroyed the details of sedimentation, there is no evidence for shallow water deposition. The maximum exposed thickness of the lower clastic unit is estimated at 2 km.

The middle carbonate unit comprises blue-grey weathering, fine- to medium-crystalline, platy and sandy limestone, lesser quartzite and minor argillite. The limestone has commonly been altered to tan weathering, light grey dolostone. Locally, east of the Weston Fault, a thick interbed of diamictite with a sandy limestone matrix and quartzite and dolostone clasts occurs near the base of the unit. Quartzite occurs sporadically at all stratigraphic levels within the unit. Channelled bases and crossbeds are common. The abundance of quartz sandstone and the association with channelled quartzites suggest that most of the carbonate strata composing the unit were deposited in a peritidal environment. The middle carbonate unit ranges

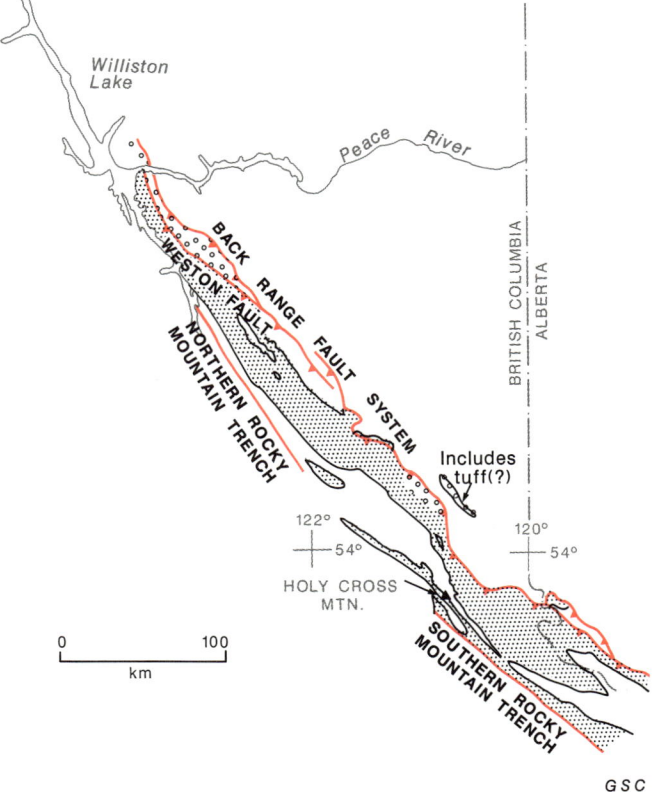

Figure 6.13. Distribution of Windermere Supergroup (stippled), areas with abundant diamictite (circles), and area of possible crystal tuff.

in thickness from approximately 600 to 900 m, owing to rapid lateral facies changes.

The upper clastic unit consists of argillite, quartzite and siltite. The argillite is dominantly grey, silty and cleaved. Rusty weathering is common. Laminae of argillaceous siltite or very fine grained sandstone occur in thin-bedded sheets that appear massive or graded. Most are probably turbidity deposits. These quartzite beds are distinct from typical 'quartzites' of the lower clastic unit because of their fine grain-size, better sorting and low feldspar content. The thickness of the upper clastic unit is at least 1100 m thick. The absence of mudcracks and ripple marks suggests that the upper clastic unit was deposited in a basinal environment.

Misinchinka strata are lithologically similar to and largely correlative with the middle and upper part of the Miette Group exposed in the McBride area 300 km to the southeast. The lower clastic unit is approximately correlative with the middle part of the Miette Group of Campbell et al. (1973) and the upper clastic unit with the upper part of the Miette. The Byng Formation (Slind and Perkins, 1966), a prominent carbonate unit in a succession containing an Ediacaran fauna in the Miette Group of the Mount Robson area, is absent in the McBride area (Campbell et al., 1973) but is almost certainly younger than the middle carbonate unit of the Mount Selwyn area.

South of latitude 56°N in the central Rocky Mountains, the Windermere Supergroup is represented by the Miette Group, a thick succession of fine to coarse clastic strata with locally important carbonate units. East of the Back Range Fault (Fig. 6.14) the easternmost exposures of the Miette assemblage are similar in general aspect to the eastern and oldest successions of Windermere rocks in the Ogilvie, Wernecke and Mackenzie mountains (Fig. 6.14; M.E. McMechan, pers. comm., 1985). The oldest unit, the base of which is unexposed, consists of more than 400 m of graded, laminated green siltite containing rare dolostone nodules and laminae and green to grey quartz-feldspar sandstone believed to be reworked crystal tuff. Locally the rocks are interlayered with an overlying unit ranging from 300 to 900 m thick of granule to boulder, commonly massive diamictite. Clasts form 10 to 30% of the rock and include fine- to medium-grained mafic extrusive rock, coarse crystalline felsic intrusive rock, intermediate volcanics, quartzite, dolostone, pebble conglomerate, green layered tuff, limestone and green and grey argillite. The matrix is brown-weathering, sandy, chloritic argillite and, locally, limestone. In the diamictite are discontinuous members, up to 40 m thick, of light and dark green laminated argillite with thin-graded siltite interbeds, fine- to coarse-grained, graded feldspathic sandstone and granule to pebble conglomerate. Overlying the diamictite unit east of the Back Range Fault is a unit, about 300 m thick, consisting mainly of quartz-feldspar sandstone, possibly redeposited crystal tuff, with laterally continuous thin beds of silty argillite in the lower part and rusty argillite in the upper part. Olistostromal blocks of tan to orange weathering dolostone from 5 to 400 m long and rare lenses of limestone and dolostone are present locally. The uppermost strata of the Miette Group east of the Back Range Fault are truncated to the north by the sub-Cambrian unconformity. They include a lower member of dolostone, locally stromatolitic and up to 300 m thick, overlain by rusty argillite up to 70 m thick.

West of the Back Range Fault (Fig. 6.13) and south to about latitude 54°15'N diamictite is an important component of middle Miette stratigraphy (Stelck et al,. 1978). At least 1000 and probably 2000 m of diamictite are present locally. Clasts are dominantly quartzite but also include argillite, bedded carbonate, meta-volcanic rock and coarsely crystalline igneous rock. Along trend to the southeast the diamictite appears to grade into more than 300 m of feldspathic sandstone and pebble conglomerate, commonly graded, in units up to 30 m thick, interbedded with argillite and siltite. An overlying argillite about 300 m thick, contains blocks or olistoliths of grey, tan and orange weathering dolostone from 5 to 400 m long. Similar blocks of carbonate overlie the thick diamictite noted above. About 250 to 500 m of feldspathic sandstone to pebble conglomerate interbedded with argillite and siltite constitute the upper part of the middle Miette assemblage. The upper Miette comprises lower and upper members of argillite, each ranging from 0 to 300 m thick, which are thickest in the northwest. An intervening carbonate unit consisting mainly of dolostone and dolostone breccia with minor limestone, is up to 400 m thick.

The succession and lithology of units described above suggests correlation of the carbonate and diamictite members with similar rocks in the Misinchinka Group to the north. The assemblage includes the southernmost exposures in the eastern Cordillera of poorly sorted, probably rift-related facies associated, at least locally, with volcanism and possibly glaciogenic environments.

A broad structural culmination in the Rocky Mountains between latitudes 52° and 54°N exposes classical sequences of the Miette Group (Charlesworth et al., 1967; Mountjoy, 1970; Campbell et al., 1973). A broad, threefold division has been used to describe the characteristically thick and monotonous, predominantly clastic strata. The lower Miette strata are highly deformed, probably more than 380 m thick, and occur in the Cushing Creek area mainly in the eastern half of the McBride map area (Carey and Simony, 1985). The lower limestone and calcareous grit member consists of coarse grained gritty sandstone to granule conglomerate, finely crystalline carbonaceous black limestone, silty and sandy limestone and rusty weathering silty black slate. The upper slate member consists of rusty weathering pyritic, locally silty black slate with minor sandstone and siltstone. Graded bedding is present in the grit and rare ripple crosslamination occurs in the sandy limestone. The environment of deposition is interpreted as an euxinic basin, perhaps distal deltaic, into which there were influxes of turbiditic clastics.

The middle part of the Miette Group, as much as 3000 m thick, is characterized by interlayered composite grit and sandstone beds in units up to 385 m thick but averaging about 140 m thick in the lower part and 24 m thick in the upper part (Carey and Simony, 1985). Interbedded with the sandstone is rusty weathering silty dark green to grey pyritic slate in sequences up to 210 m thick but averaging 70 m thick in the lower part and 21 m thick in the upper part. The sandstone is immature and contains from 0 up to 25% plagioclase; little orthoclase is present. Coarse grained beds commonly are graded and have scoured bases. Rip-up clasts of shale are abundant. Finer grained beds are locally ripple crosslaminated. Near Cushing Creek a pebbly mudstone member 30 m thick occurs about 2100 m above the base and contains grey

UPPER PROTEROZOIC ASSEMBLAGES

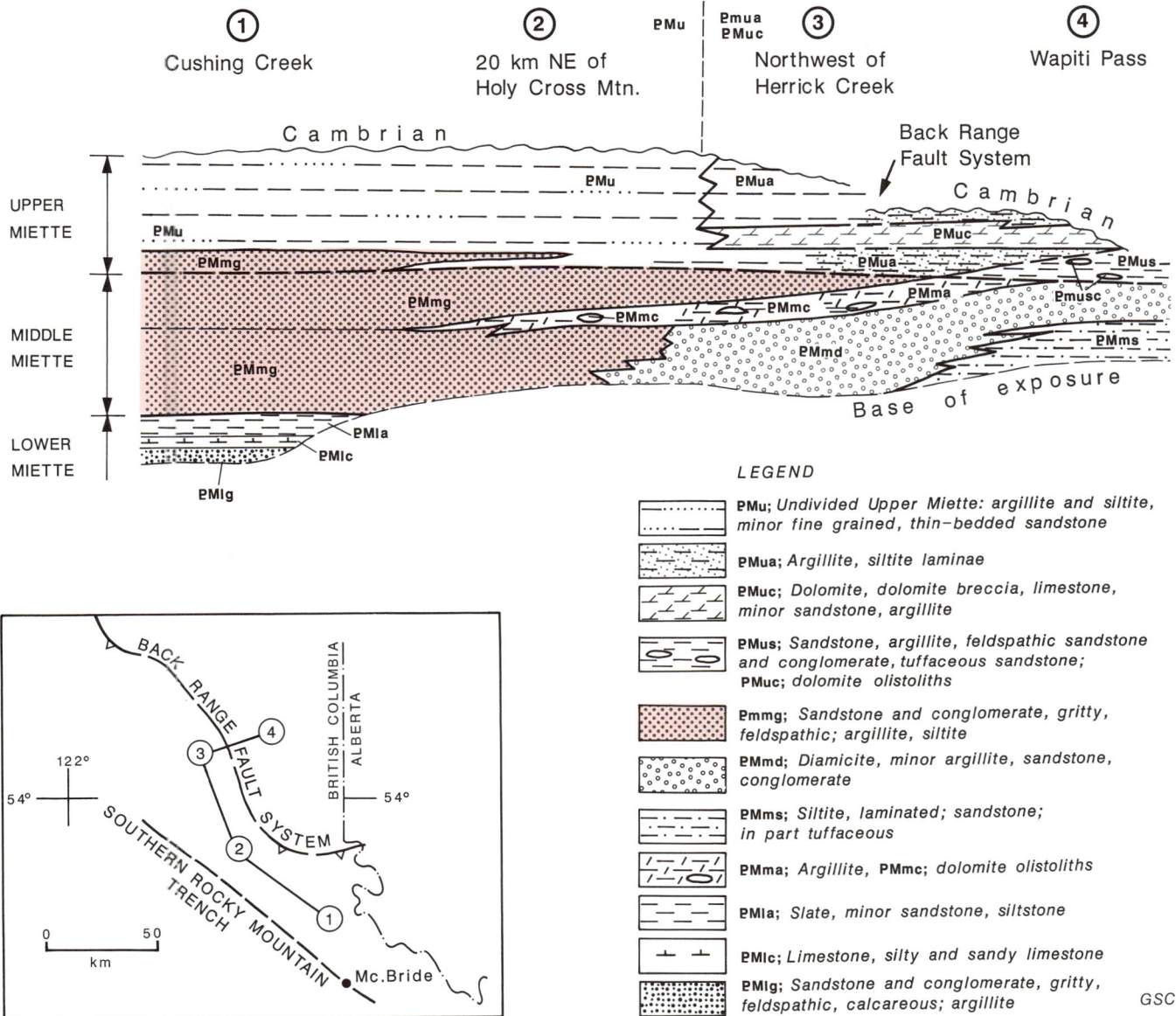

Figure 6.14. Facies relationships in the Miette Group, Hart Ranges. After M.E. McMechan, pers. comm., 1985.

micritic limestone clasts up to 40 cm diameter. This facies is perhaps a distal equivalent of much thicker diamictites to the north. Carey and Simony (1985) interpreted the middle Miette rocks as representing a submarine fan complex that accumulated during an episode of active tectonism.

The upper part of the Miette Group is approximately 1800 m thick and consists predominantly of recessive grey to dark grey mudstone and silty argillite with significant amounts of interbedded fine grained sandstone and siltstone in the lower and upper parts to the west and coarser grained clastics to the east (Campbell et al., 1973). Small scale ripples and fine silt laminae are characteristic as are coarsening-upward cycles of argillite to fine grained sandstone about 50 m thick. Lenticular interbeds of coarse grained immature sandstone and conglomeratic sandstone are sparse but widespread. Near Holy Cross Mountain on the east side of the Southern Rocky Mountain Trench (Fig. 6.13) the uppermost part of the Miette Group consists of limestone, sandy limestone, sandy dolostone, mudstone and an upper sandstone, possibly in part correlative with the carbonate and related members in the upper part of the Miette and Misinchinka groups to the north. Locally, the upper Miette is capped by carbonate members with a maximum thickness of 400 m consisting of thin to thick beds of buff grey weathering stromatolitic and pisolitic dolostone, sandstone and pelite. The rocks are believed to have formed on platforms in shallow water (Teitz and Mountjoy, 1985). These rocks, generally included in the Byng Formation, are enclosed in strata that contain an

Ediacaran fauna (Hofmann et al., 1985) and thus are almost certainly younger than the more persistent carbonates that occur in the Misinchinka and upper Miette sequences north of latitude 55°. The upper Miette rocks reflect a relatively stable sedimentary environment, possibly on a marine shelf during a period of tectonic quiescence.

Near the Southern Rocky Mountain Trench east of McNaughton Lake about 2400 m of Miette Group strata are present between a nonconformity on basement granitic gneiss and basal Lower Cambrian Gog quartzite (see Chapter 4). Locally the base is marked by foliated micaceous quartzite, 50 to 200 m thick, which in places includes a unit of quartzite-pebble conglomerate 40 m thick at the top. A continuous member of black biotite schist separates the quartzite from an overlying granule conglomerate (grit) interbedded with pelite and psammite more than 1000 m thick. The grit unit is overlain by about 1000 m of aluminosilicate-rich pelite. The uppermost rocks are meta-sandstone pelite and some marble forming a sequence approximately 200 m thick. The entire succession has been regionally metamorphosed, the grade increasing from garnet in the east to sillimanite in the west.

Southern Rocky Mountains

In the Lake Louise-Mount Assiniboine area of the southern Rocky Mountains rocks assigned to the Windermere Supergroup are correlated with those of the Miette Group in the Jasper area and are lithologically similar (Mountjoy, 1962). The group consists of a lower Corral Creek Formation and an upper Hector Formation. These two formations are similar in gross lithology, comprising grey and greenish grey slate and some thin siltstone beds interbedded with coarse grained sandstone and grit, pebbly grit and quartz-pebble conglomerate (Aitken, 1969). They are separable, however, on the basis of a unit of purple and green slate with associated limestone-bearing conglomerate, interpreted as a debris flow, which is about 100 m thick, and makes up the lower part of the Hector Formation. Locally a well defined channel more than 150 m deep in slate and turbidite is filled with coarse clastic sediment. It probably represents a submarine canyon (Arnott and Hein, 1984). The maximum exposed thickness for the Miette Group in the area is about 850 m preserved beneath a significant sub-Gog (Lower Cambrian) unconformity (Aitken, 1969).

Northern Columbia Mountains

Upper Proterozoic strata within the Omineca Belt in the northern Columbia Mountains (Cariboo and northern Monashee, Selkirk and Purcell mountains and adjacent highlands) comprise two distinct stratigraphic sequences, the Kaza Group and part of the Cariboo Group in the northwest and the Horsethief Creek Group in the southeast. In the southern Cariboo Mountains the Horsethief Creek Group is interpreted to be conformably overlain by the Kaza Group (Fig. 6.15; Pell and Simony, 1984, 1987; Pell, 1984). In the southwestern part of Cariboo Mountains and the adjacent Quesnel Highlands the Snowshoe Formation is thought to include strata equivalent to and

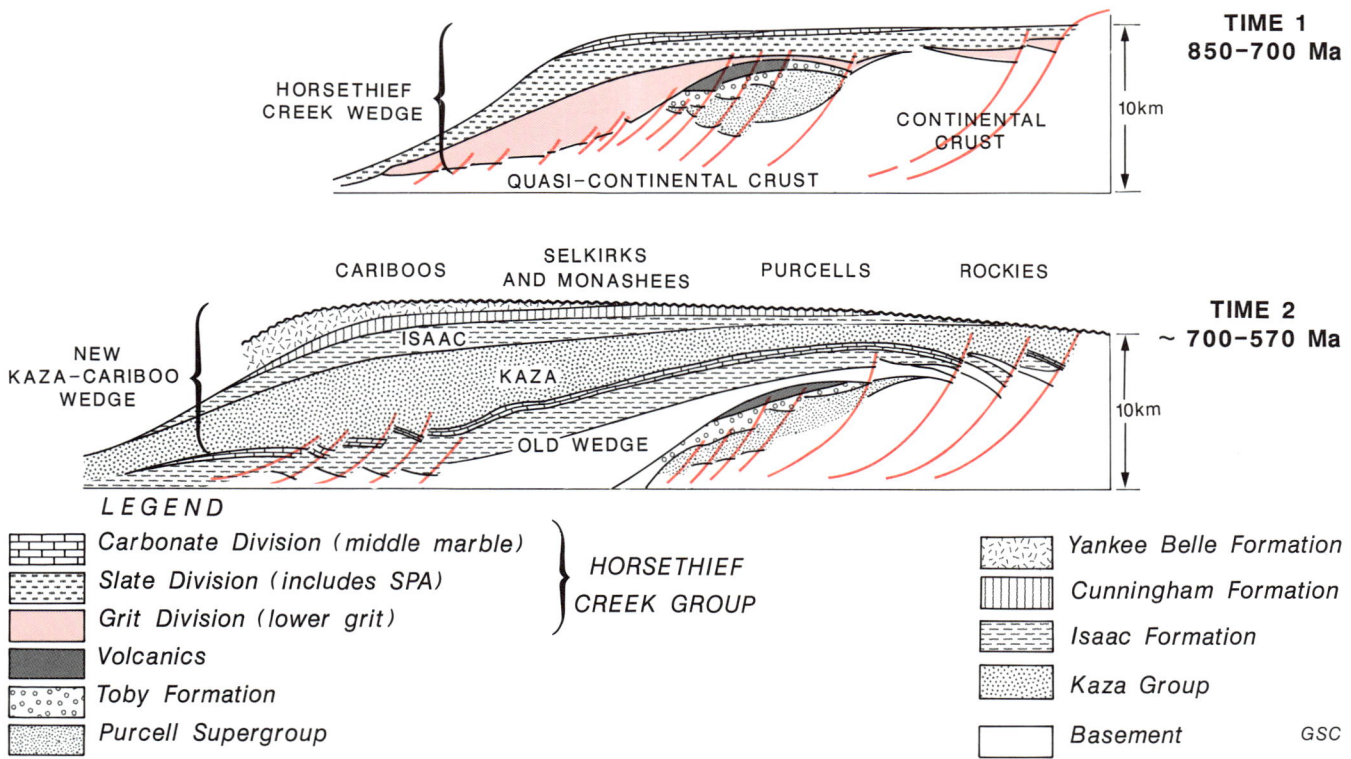

Figure 6.15. Suggested relationship between Horsethief Creek and Kaza-Cariboo strata in the Columbia Mountains. After Pell and Simony, 1987.

coextensive with the Horsethief Creek Group (Struik, 1986). The Snowshoe Formation may include rocks that range in age from Late Proterozoic to late Paleozoic, correlative with the Horsethief Creek Group and overlying Paleozoic rocks in the Selkirk Mountains (e.g. near Rogers Pass). Struik (1986) believed that the Kaza and Cariboo groups are thrust southwestward over the Snowshoe Formation and the Horsethief Creek Group, including Paleozoic rocks that may be presently included with both. Such a relationship would negate the interpretation that the Kaza Group stratigraphically overlies the Horsethief Creek Group.

The Kaza Group and the Proterozoic part of the Cariboo Group are well exposed in the Cariboo Mountains where they comprise four units totalling at least 6200 m in thickness (Campbell et al., 1973). The succession in the Cariboo Mountains is remarkably similar to the equivalent sequence in the Cassiar and Omineca mountains (Mansy and Gabrielse, 1978). Both sequences are overlain by quartzite followed by Lower Cambrian carbonate.

The Kaza Group, about 3500 m thick, is dominantly a well-bedded sequence of argillite, slate and phyllite interlayered with gritty sandstone and pebble conglomerate (Fig. 6.16); the gritty sandstone is characteristic. Where these rocks are weakly metamorphosed a greenish grey colour is imparted by the chlorite-rich matrix that encloses glassy quartz and chalky white plagioclase granules and pebbles; opalescent blue quartz is present but not abundant. Where more highly metamorphosed the matrix of chlorite and micas, which are at least partly detrital, are recrystallized to brown biotite and muscovite and the rocks assume a brownish grey colour. Garnet and aluminosilicate minerals are abundantly developed throughout the Kaza Group in rocks of appropriate composition and metamorphic grade. Even in rocks of upper amphibolite facies prominent bedding, graded beds and other sedimentary features are retained. Where the metamorphic grade is higher the strata are more strongly folded, and, because of the prominent bedding, the Kaza Group displays spectacular folds in much of Cariboo Mountains.

The Kaza Group is correlated with the Swannell Formation (Mansy and Gabrielse, 1978) in the Cassiar and Omineca mountains and with the middle part of the Miette Group to the east in the Rocky Mountains (Campbell et al., 1973).The Kaza Group passes gradationally upward into the dark coloured, fine grained phyllite, slate and argillite of the recessive-weathering Isaac Formation, the basal unit of the Cariboo Group. The base of the Isaac Formation is defined at the top of the highest prominent gritty sandstone, although the formation contains some grit beds. The fine grained rocks commonly are calcareous but distinct limestone beds are rare. The rocks are characteristically pyritic, particularly in the lower part of the formation. The strata of the Isaac Formation, which are more than 1300 m thick, are incompetent, and, in contrast to the prominent large folds in the underlying and more competent Kaza Group, are deformed into multitudes of inconspicuous small folds in highly cleaved rocks. Near the top of the formation thin interbeds of limestone become more prevalent, produce a strong striped pattern in the outcrops and are succeeded by the basal thick carbonate of the Cunningham Formation.

The Isaac Formation is correlated with the Tsaydiz Formation in the Omineca and Cassiar mountains (Mansy

Figure 6.16. Chevron folds in turbiditic clastics of the Kaza Group, Cariboo Mountains. Amplitudes of folds are about 500 m. Photo by R.B. Campbell. GSC205235-W

and Gabrielse, 1978) and with part of the upper Miette Group to the east in the Rocky Mountains (Campbell et al., 1973).

The Cunningham Formation reaches a maximum thickness of about 550 m in the northern Cariboo Mountains but the thickness varies substantially from place to place. Dominantly grey weathering, thick bedded limestone and locally dolostone, the formation is somewhat variable in lithology and locally contains beds of dark argillaceous limestone and sandstone. Compared to the underlying Isaac Formation the limestone is competent where weakly metamorphosed and is deformed into broad concentric folds. Numerous faults cutting the formation do not seem to penetrate deeply into the underlying strata. The formation is equivalent to the Espee Formation to the northwest (Mansy and Gabrielse, 1978) and like the Espee commonly displays conspicuous pisoliths, ferrodolomite grains, intraclasts and local algal structures. Skeletal fossils have not been found.

The Yankee Belle Formation, equivalent to the Stelkuz Formation (Mansy and Gabrielse, 1978) to the northwest, rests with abrupt but conformable contact on the Cunningham Formation. The thickness and lithologies of the formation vary substantially from place to place. The maximum known thickness of 900 m is attained in the central part of the northern Cariboo Mountains from where it becomes thinner both to the northeast (400 m) and southwest (500 m). At the top of the section shale and sandstone predominate. The Yankee Belle Formation is prominently and thinly bedded; characteristic but not necessarily dominant lithologies are green and less abundant maroon shale that weather a distinctive rusty brown colour. The proportion of quartz sandstone increases at the top and intertongues with the overlying Yanks Peak Formation. Rhythmic cycles of limestone, siltstone and shale are characteristic of much of the formation.

Evidence that organisms existed during deposition of the Yankee Belle Formation is rare; the stratigraphically younger Mural Formation is Early Cambrian hence the Cambrian-Precambrian boundary is likely within the intervening Yanks Peak or Midas formations. Common practice in the Cordillera has been to place the boundary at the base of thick quartzite successions, like the Yanks Peak Formation, that underlie known Lower Cambrian strata. In the Rocky Mountains it is generally placed at the base of the quartzite of the McNaughton Formation. The Yankee Belle Formation may be equivalent to the basal McNaughton Formation or to the uppermost part of the upper Miette Group or it may have no counterpart in the Rocky Mountains to the east; its age is thus uncertain but is tentatively assigned to the Proterozoic and the formation is included within the Windermere Supergroup.

Rocks of the Kaza Group, in common with other thick Windermere "grit" sequences, reflect deep water deposition, perhaps in submarine fans. The detritus was derived from crystalline rocks with high relief, possibly on the rifted continental margin. The supply of coarse clastics ended abruptly with initiation of deposition of the Isaac Formation in euxinic, still water followed by the development of a widespread, shallow water carbonate platform (Cunningham Formation). If the tentative correlations shown in Figure 6.2 are valid, a widespread shoaling of depositional environments in the miogeocline took place in Cunningham-Espee time, perhaps in response to a marked eustatic lowering of sea level. The distribution of lithologies at the base of the Yankee Belle Formation indicates a change from an alluvial environment (sandstone, maroon mudstone with mudcracks) in the northeastern Cariboo Mountains to a deeper marine environment in the south and southwest (rhythmic cycles of limestone, siltstone and shale). The Yankee Belle grades upward into the littoral and alluvial sandstone of the Yanks Peak Formation.

In the southeastern Cariboo Mountains the Horsethief Creek Group is divided into three units (Pell, 1984). Deformation has been intense and the rocks are highly strained hence thickness estimates are little more than approximations. The lowest unit is meta-semipelite and amphibolite, "SPA", which ranges from 500 m to 1000 m in thickness; the base is not exposed. The thickest and lowest part of the unit consists of micaceous pelitic and psammitic schist in which aluminosilicates are rare, and dark garnetiferous amphibolite. Amphibolites with sharp contacts probably represents basaltic flows or high level sills whereas others, with gradational contacts, may be metamorphosed tuffs. This is overlain by a rusty, variable carbonate-calcsilicate zone a few tens of metres thick followed by 100 m or more of pelitic schist with prominent aluminosilicate minerals at higher metamorphic grades.

The "SPA" is succeeded by the middle marble unit which consists of upper and lower grey marbles and associated calcareous rocks with an intervening clastic layer of micaceous schist. The middle marble unit is variable in lithology and in thickness, ranging from 21 m to more than 500 m, and the clastic unit from 5 m to 100 m.

The upper clastic unit is thought to pass gradationally upward into typical Kaza Group beds (Pell and Simony, 1984; Pell, 1984) whereas Struik (1986) argued that the contact must be structural with Kaza Group rocks thrust southwestward over the Horsethief Creek Group.

The basal unit of the upper clastic division, 150 m to 1000 m in thickness, consists of meta-sandstone and alternating layers of pelitic schist (Pell and Simony, 1984; Pell, 1984), with minor orthoquartzite and some gritty meta-sandstone beds. Graded bedding is rare. This unit is capped by a thin carbonate member followed by strata of typical Kaza Group lithologies as described above.

In the northern Purcell Mountains (Dogtooth Range) and in the eastern part of the northern Selkirk Mountains (Poulton and Simony, 1980; Brown et al., 1978) the Horsethief Creek Group, below the Hamill Group orthoquartzite, or its equivalent, is more than 6000 m thick, of which 1800 m lies below the "SPA" unit and hence below the lowest strata exposed in southeastern Cariboo Mountains. Units of the group can be traced from the Cariboo Mountains through the northern Monashee Mountains (Raeside and Simony, 1983) and the northern Selkirk Mountains, (Poulton and Simony, 1980; Brown et al., 1978) to the Dogtooth Range (Poulton and Simony, 1980). Along this distance the "SPA" unit changes character and merges with the upper slate division (1200 m) of Poulton and Simony in Dogtooth Range, where it is predominantly grey slate with minor calcareous clastics; amphibolite is absent. The upper slate is separated from the lower bluish grey slate (700 m-800 m) by an impure carbonate, phyllite and rusty schist layer. The two slate units are difficult to distinguish from one another in the southern part of the region where, together with the impure carbonate, they total about 2000 m in thickness.

The basal unit of the Horsethief Creek Group in the northern Selkirk and Purcell mountains is the lower grit division, consisting of about 1000 m of meta-sandstone and granule conglomerate, pelitic rocks and minor limestone. Opalescent blue quartz granules are prominent.

The upper clastic division, above the middle or main carbonate, is up to 3300 m thick in the northern Selkirk Mountains. It becomes more calcareous from east to west (Brown et al., 1978) and consists of feldspathic sandstone and granule and cobble conglomerate (rare blue quartz) with pelitic rocks below an upper unit of well bedded siltstone and pelitic rocks as much as 1200 m thick. These rocks are overlain, apparently conformably, by the Hamill Group and succeeding Mohican and Badshot formations. The latter three units appear to thin to the northwest and cannot be traced beyond the northern end of the Selkirk Mountains. Thus, if the Paleozoic Lardeau Group continues to the northwest, it may directly overlie strata of the Horsethief Creek Group or unrecognized equivalents of the Hamill Group and Badshot Formation, and be coextensive with the deformed and metamorphosed Snowshoe Formation in and southwest of the Cariboo Mountains as suggested by Struik (1986).

The Horsethief Creek Group displays, in a general way, a succession not unlike the Kaza-Cariboo groups of the Cariboo Mountains. Deposition was initiated with coarse feldspathic sand presumably in a basinal submarine fan followed by still and probably shallower water accumulation of pelitic, calcareous and basaltic material capped by a prominent limestone (middle marble) in turn succeeded by a clastic unit, with local coarse sediments, which grades westward to a more calcareous facies. Pell and Simony (1984) and Pell (1984) believed that the two successions are superposed clastic wedges whereby the sequence of events that produced the Horsethief Creek Group are repeated in some detail by those that produced the Kaza-Cariboo groups. Struik (1986) postulated that the two successions result from similar and roughly synchronous events in separate localities and are now tectonically juxtaposed. If such is the case then the correlations of Brown et al. (1978), equating the upper Horsethief Creek and Cariboo groups may be correct.

The Snowshoe Formation described above and the Eagle Bay Formation and perhaps other units near Adams Lake are believed to include Upper Proterozoic rocks (Schiarizza and Preto, 1984). In addition rocks of this age must be included within the highly metamorphosed rocks of the Shuswap Metamorphic Complex.

Southern Columbia Mountains

In its type area the Windermere Supergroup unconformably overlies strata of the Purcell Supergroup along the west limb and northward plunging nose of the Purcell Anticlinorium (Little, 1960; Reesor, 1973). The basal Toby Formation, best developed around the northern end of the anticlinorium ranges in thickness from a few metres to 500 m and is characterized by extreme variability in composition, grain size and sorting. Typically the Toby is a polymictic conglomerate containing pebbles, cobbles and boulders of mainly dolostone, quartzite and slate derived from underlying Purcell rocks (Fig. 6.17). Cobbles of granitic gneiss occur locally in easternmost exposures. Clasts, from a fraction of a centimetre to more than a metre in length, are generally angular to subangular but well rounded varieties occur. Classic examples of diamictite comprise poorly sorted dispersed pebbles, cobbles and boulders in commonly comminuted sandy mudstone or muddy, fine sandstone (Aalto, 1971). Sandstone and argillite containing dropstones are most abundant in the upper part of the Toby. Interbedded with the conglomerate are units of argillite, slate and graded and massive sandstone as much as 50 m thick. Vesicular andesitic volcanics locally form the matrix of the conglomerate and elsewhere occur as cobbles in the conglomerate. At least part of the Toby Formation has been interpreted as a glaciogenic deposit.

The Toby Formation is generally less than several tens of metres thick from east of the north end of Kootenay Lake to west of the south end of Kootenay Lake. There the thickness increases markedly to about 700 m and at the International Boundary it may be as much as 2000 m thick.

The Irene Volcanic Formation consisting of up to 700 m of fine grained, dark bluish green mafic tuff and massive schistose greenstone with minor intercalations of light grey phyllite overlies the Toby Formation west of the south end of Kootenay Lake and appears to thicken southward towards the Canada-U.S.A. Boundary. A Nd-Sm age obtained on the correlative Leola volcanics in Washington State is 762 ± 40 Ma (W.J. Devlin, pers. comm., 1985), in excellent agreement with ages on volcanic rocks near the base of the Windermere Supergroup in the northern Canadian Cordillera.

The Horsethief Creek Group and its correlative west of Kootenay Lake, the Monk Formation, conformably overlie the Toby Formation and consist of great thicknesses of slate, argillite and phyllite with lesser amounts of quartzite, greywacke, pebble conglomerate and limestone. In a general way east of Kootenay Lake the lower part of the Horsethief Creek Group consists dominantly of argillite and slate with some limestone, the middle part is characterized by quartzite, grit and pebble conglomerate and the upper part in its eastern and southeastern limits contains much purple and red slate and siltstone with very minor limestone (Reesor, 1973). The upper part farther west is dominated by dark slate and phyllite.

Distinctive pebble conglomerate contains pebbles of white vein quartz, feldspar, chert, quartzite, opalescent blue quartz, dolostone and slate. Irregular lenses of pebble conglomerate interfinger with quartzite or grit beds and locally occur as channel fill in finer grained clastic rocks (Fig. 6.18). Graded beds are common in sequences containing pebble conglomerate and crossbedding is locally conspicuous in quartzite. North and east of Kootenay Lake total thicknesses of the Horsethief Creek Group increase from about 1000 m near the Rocky Mountain Trench to more than 3000 m farther west.

The Monk Formation west of Kootenay Lake is about 1000 m thick and comprises two phyllite members separated by laminated grey limestone. A basal polymictic

Figure 6.17. Conglomerate of Toby Formation, southern Columbia Mountains. Photo by J.E. Reesor. GSC158448

Figure 6.18. Pebble conglomerate in Horsethief Creek Group, southern Columbia Mountains. Photo by J.E. Reesor. GSC158446

conglomerate about 60 m thick (Little, 1960) probably is correlative with strata of the Toby Formation elsewhere.

The regional unconformity at the base of the Windermere Supergroup in the southern Columbia Mountains, the great variability in thickness, sorting and composition of the basal Toby Formation, the turbiditic character of the Horsethief Creek Group, and the presence of the Irene Volcanic Formation document an episode of rifting, uplift, erosion and volcanism in Late Proterozoic time. As elsewhere in the Cordillera these events probably were accompanied by glaciation.

The rifted-margin tectonic setting attending deposition of the Windermere Supergroup in Canada extended far to the south in the Cordillera of the western United States as shown by thick successions of similar lithologies in Washington, Idaho and Utah (Stewart, 1972).

ALEXANDER TERRANE

A few, relatively small areas in the Saint Elias Mountains may include Upper Proterozoic rocks. A thick sequence of volcanic and volcaniclastic strata east of the Alsek River across from the Lowell Glacier contain Late Cambrian fossils in the upper part. Paragneiss, granitoid gneiss, amphibolite, quartzite, quartzose schist and minor marble apparently underlie strata as old as Ordovician in an area along the Alaska-British Columbia border southwest of the Tweedsmuir Glacier. Rocks in these two areas could be correlative with part of the Wales Group in southeast Alaska.

NISLING TERRANE

A locally thick sequence of regionally metamorphosed strata consisting of schist, impure, in part graphitic, quartzite and marble occurs intermittently from the Omineca Belt in southwestern Yukon Territory into the Coast Belt of northern British Columbia. On the basis of regional correlations some of the rocks are believed to be of early Paleozoic age but direct evidence is lacking and Upper Proterozoic rocks may be included.

NISUTLIN SUBTERRANE

Upper Proterozoic rocks may form part of the Nisutlin Subterrane in western Yukon Territory. The rocks consist of muscovite-biotite schist, micaceous quartzite, marble and minor amphibolite all characteristically, strongly cataclastized. These rocks, and less strongly deformed rocks of the Nisling Terrane, are assigned to the Yukon-Tanana Terrane in Alaska.

REFERENCES

Aalto, K.R.
1971: Glacial marine sedimentation and stratigraphy of the Toby Conglomerate (Upper Proterozoic) southeastern British Columbia, northwestern Idaho and northeastern Washington; Canadian Journal of Earth Sciences, v. 8, p. 755-787.

Aitken, J.D.
1969: Documentation of the sub-Cambrian unconformity, Rocky Mountains Main Ranges, Alberta; Canadian Journal of Earth Sciences, v. 6, p. 193-200.
1981: Stratigraphy and sedimentology of the Upper Proterozoic Little Dal Group, Mackenzie Mountains, Northwest Territories; in Proterozoic basins of Canada, F.H.A. Campbell (ed.), Geological Survey of Canada, Paper 81-10, p. 47-71.
1984: Strata and trace fossils near the Precambrian-Cambrian boundary, Mackenzie, Selwyn and Wernecke Mountains, Yukon and Northwest Territories: Discussion; in Current Research, Part B, Geological Survey of Canada, Paper 84-1B, p. 401-407.
1989: Uppermost Proterozoic formations in Central Mackenzie Mountains, N.W.T.; Geological Survey of Canada, Bulletin 368.

Aitken, J.D. and Cook, D.G.
1974: Geology of parts of Mount Eduni (106A) and Bonnet Plum Lake map-areas, District of MacKenzie; Geological Survey of Canada, Open File 221.

Aitken, J.D., Macqueen, R.W., and Usher, J.L.
1973: Reconnaissance studies of Proterozoic and Cambrian stratigraphy, lower Mackenzie River area (Operation Norman), District of Mackenzie; Geological Survey of Canada, Paper 73-9, 173 p.

Allison, C.W. and Moorman, M.A.
1973: Microbiota from the Late Proterozoic Tindir Group, Alaska; Geology, v. 1, no. 2, p. 65-68.

Allison, C.W., Young, G.M., Yeo, G.M., and Delaney, G.D.
1981: Glaciogenic rocks of the upper Tindir Group, east-central Alaska; in Earth's Pre-Pleistocene Glacial Record, M.J. Hambrey and W.B. Harland (ed.), Cambridge University Press, p. 720-723.

Armstrong, R.L., Eisbacher, G.H., and Evans, P.D.
1982: Age and stratigraphic-tectonic significance of Proterozoic diabase sheets, Mackenzie Mountains, northwestern Canada; Canadian Journal of Earth Sciences, v. 19, p. 316-323.

Arnott, R.W. and Hein, F.J.
1984: Proximal submarine channel deposits, Hector Formation (Hadrynian) Lake Louise, Alberta; in Program with Abstracts, v. 9, Geological Association of Canada, Annual Meeting 1984, p. 43.

Blusson, S.L.
1971: Sekwi Mountain map area, Yukon Territory and District of Mackenzie; Geological Survey of Canada, Paper 71-22, 17 p.

Brabb, E.E. and Churkin, M., Jr.
1969: Geologic map of the Charley River Quadrangle, east-central Alaska; United States Geological Survey, Map I-573.

Brown, R.L., Tippet, C.R., and Lane, L.S.
1978: Stratigraphy, facies changes and correlations in the northern Selkirk Mountains, southern Canadian Cordillera; Canadian Journal of Earth Sciences, v. 15, p. 1129-1140.

Cairnes, D.D.
1914: The Yukon-Alaska International Boundary between Porcupine and Yukon rivers; Geological Survey of Canada, Memoir 67, 161 p.

Campbell, R.B., Mountjoy, E.W., and Young, F.G.
1973: Geology of McBride map-area, British Columbia; Geological Survey of Canada, Paper 72-35.

Carey, A.J. and Simony, P.S.
1985: Stratigraphy, sedimentology and structure of Late Proterozoic Miette Group, Cushing Creek Area, B.C.; Bulletin of Canadian Petroleum Geology, v. 33, no. 2, p. 184-203.

Carey, S.W.
1959: Continental Drift, A symposium; Geology Department, University of Tasmania, Hobart, 1958.

Charlesworth, H.A.K., Weiner, J.L., Akehurst, A.J., Bielenstein, H.U., Evans, C.R., Griffiths, R.E., Remington, D.B., Stauffer, M.R., and Steiner, J.
1967: Precambrian geology of the Jasper region, Alberta; Research Council of Alberta, Bulletin 23, 74 p.

Dutro, J.T., Jr., Brosge, W.P., and Reiser, H.N.
1972: Significance of recently discovered Cambrian fossils and reinterpretation of the Neruokpuk Formation, northeastern Alaska; American Association of Petroleum Geologists Bulletin, v. 56, no. 4, p. 808-815.

Eisbacher, G.H.
1978: Re-definition and subdivision of the Rapitan Group, Mackenzie Mountains; Geological Survey of Canada, Paper 77-35, 21 p.
1981: Sedimentary tectonics and glacial record in the Windermere Supergroup, Mackenzie Mountains, northwestern Canada; Geological Survey of Canada, Paper 80-27, 40 p.

Evenchick, C.A.
1982: Stratigraphy, structure and metamorphism in Deserters Range, northern Rocky Mountains, British Columbia; in Current Research, Part A, Geological Survey of Canada, Paper 82-1A, p. 325-328.

Evenchick, C.A., Parrish, R.R., and Gabrielse, H.
1984: Precambrian gneiss and late Proterozoic sedimentation in north-central British Columbia; Geology, v. 12, p. 233-237.

Fritz, W.H.
1972: Cambrian biostratigraphy western Rocky Mountains, British Columbia (83E, 94C, 94F); in Report of Activities, Geological Survey of Canada, Paper 72-1, Part A, p. 209-211.

Fritz, W.H. and Crimes, T.P.
1985: Lithology, trace fossils, and correlation of Precambrian-Cambrian boundary beds, Cassiar Mountains, north-central British Columbia; Geological Survey of Canada, Paper 83-13.

Fritz, W.H., Narbonne, G.M., and Gordey, S.P.
1983: Strata and trace fossils near the Precambrian- Cambrian boundary, Mackenzie, Selwyn, and Wernecke Mountains, Yukon and Northwest Territories; in Current Research, Part B, Geological Survey of Canada, Paper 83-1B, p. 365-375.
1984: Strata and trace fossils near the Precambrian- Cambrian boundary, Mackenzie, Selwyn, and Wernecke Mountains, Yukon and Northwest Territories: Reply; in Current Research, Part B, Geological Survey of Canada, Paper 84-1B, p. 409-412.

Gabrielse, H.R., Blusson, S.L., and Roddick, J.A.
1973: Geology of Flat River, Glacier Lake and Wrigley Lake map-areas, District of Mackenzie and Yukon Territory; Geological Survey of Canada, Memoir 366, Part 1, 153 p, Part 2, 269 p.

Gordey, S.P. and Anderson, R.G.
in press: Evolution of the northern Cordilleran miogeocline, Nahanni map-area (105I), Yukon Territory and District of Mackenzie; Geological Survey of Canada, Memoir 428.

Hofmann, H.J., Mountjoy, E.W., and Teitz, M.W.
1985: Ediacaran fossils in the Miette Group, Rocky Mountains, British Columbia; in Program with Abstracts, v. 10, Geological Association of Canada, Annual Meeting, 1985, p. A28.

Jefferson, C.W.
1983: The Upper Proterozoic Redstone Copper Belt, Mackenzie Mountains, Northwest Territories; Ph.D. thesis, University of Western Ontario, 445 p.

Jefferson, C.W. and Ruelle, J.C.L.
1986: The Late Proterozoic Redstone Copper Belt, Mackenzie Mountains, Northwest Territories; in Mineral Deposits of the Northern Cordillera, J.A. Morin (ed.), Canadian Institute of Mining and Metallurgy, Special Volume 37, p. 154-168.

Lane, L.S. and Cecile, M.P.
1989: Stratigraphy and structure of the Neruakpuk Formation, northern Yukon; in Current Research, Part G, Geological Survey of Canada, Paper 89-1G, p. 57-62.

Leffingwell, E. de K.
1919: The Canning River region; northern Alaska; United States Geological Survey, Professional Paper 109, 251 p.

Little, H.W.
1960: Nelson map-area, west half, British Columbia (82F W 1/2); Geological Survey of Canada, Memoir 308.

Mansy, J.L. and Gabrielse, H.
1978: Stratigraphy, terminology and correlation of Upper Proterozoic rocks in Omineca and Cassiar Mountains, north-central British Columbia; Geological Survey of Canada, Paper 77-19.

McMechan, M.E.
1987: Stratigraphy and structure of the Mount Selwyn area, Rocky Mountains, northeastern British Columbia; Geological Survey of Canada, Paper 85-28, 34 p.

Mountjoy, E.W.
1962: Mount Robson (southeast) map-area; Geological Survey of Canada, Paper 61-31, 114 p.
1970: Geology of the Main Ranges between Tete Jaune Cache and Jasper; in Field Conference Guide Book, Edmonton Geological Society.

Narbonne, G.M., Hofmann, H.J., and Aitken, J.D.
1985: Precambrian-Cambrian boundary sequence, Wernecke Mountains, Yukon Territory; in Current Research, Part A, Geological Survey of Canada, Paper 85-1A, p. 603-608.

Norris, D.K.
1981a: Geology, Herschel Island and Demarcation Point, Yukon Territory; Geological Survey of Canada, Map 1514A.
1981b: Geology, Blow River and Davidson Mountains, Yukon Territory-District of Mackenzie; Geological Survey of Canada, Map 1516A.
1981c: Porcupine River, Yukon Territory; Geological Survey of Canada, Map 1522A.
1982: Snake River, Yukon-Northwest Territories; Geological Survey of Canada, Map 1529A.
1984: Post-Valanginian restructuring of the northern Cordillera and contiguous Canada Basin; in Abstracts with Programs 1984, 80th Annual Meeting Cordilleran Section Geological Society of America, Anchorage, v. 16, no. 5, p. 326.

Pell, J.
1984: Stratigraphy, structure and metamorphism of Hadrynian strata in the southeastern Cariboo Mountains, B.C.; Ph.D. thesis, University of Calgary, Alberta.

Pell, J. and Simony, P.S.
1984: Stratigraphy of the Hadrynian Kaza Group between the Azure and North Thompson rivers, Cariboo Mountains, British Columbia; in Current Research, Part A, Geological Survey of Canada, Paper 84-1A, p. 95-98.
1987: New correlations of Hadrynian strata, south-central British Columbia; Canadian Journal of Earth Sciences, v. 24, p. 302-313.

Poulton, T.P. and Simony, P.S.
1980: Stratigraphy, sedimentology and regional correlation of the Horsethief Creek Group (Hadrynian, late Precambrian) in the northern Purcell and Selkirk Mountains, British Columbia; Canadian Journal of Earth Sciences, v. 17, p. 1708-1724.

Raeside, R.P. and Simony, P.S.
1983: Stratigraphy and deformational history of the Scrip Nappe, Monashee Mountains, British Columbia; Canadian Journal of Earth Sciences, v. 20, p. 639-650.

Reed, B.L.
1968: Geology of the Lake Peters Area northeastern Brooks Range, Alaska; United States Geological Survey, Bulletin 1236.

Reesor, J.E.
1973: Geology of the Lardeau map-area, east half, British Columbia; Geological Survey of Canada, Memoir 369.

Roots, C.F.
1983: Mount Harper Complex, Yukon; early Paleozoic volcanism at the margin of the Mackenzie Platform; in Current Research, Part A, Geological Survey of Canada, Paper 83-1A, p. 423-427.

Roots, E.F.
1954: Geology and mineral deposits of Aiken Lake map-area, British Columbia; Geological Survey of Canada, Memoir 274.

Schiarizza, P. and Preto, V.A.
1984: Geology of the Adams Plateau-Clearwater Area; British Columbia Ministry of Energy, Mines and Petroleum Resources, Preliminary Map 56.

Slind, O.L. and Perkins, G.D.
1966: Lower Paleozoic and Proterozoic sediments of the Rocky Mountains between Jasper, Alberta and Pine River, British Columbia; Bulletin of Canadian Petroleum Geology, v. 14, no. 4, p. 442-468.

Stelck, C.R., Burwash, R.A., and Stelck, D.R.
1978: The Vreeland High: a Cordilleran expression of the Peace River Arch; Bulletin of Canadian Petroleum Geology, v. 26, p. 87-104.

Stewart, J.N.
1972: Initial deposits of the Cordilleran geosyncline: Evidence for a late Precambrian (850 m.y.) continental separation; Geological Society of America Bulletin, v. 83, p. 1345-1360.

Stott, D.F., McMechan, M.E., Taylor, G.C., and Muller, J.E.
1983: Geology, Pine Pass (Mackenzie), British Columbia (93O); Geological Survey of Canada, Open File 925.

Struik, L.C.
1986: Imbricated terranes of the Cariboo gold belt with correlations and implications for tectonics in southeastern British Columbia; Canadian Journal of Earth Sciences, v. 23, p. 1047-1061.

Teitz, M. and Mountjoy, E.W.
1985: The Yellowhead and Astoria carbonate platforms in the Late Proterozoic Miette Group, Jasper, Alberta; in Current Research, Part A, Geological Survey of Canada, Paper 85-1A, p. 341-348.

Yeo, G.M.
1981: The Late Proterozoic Rapitan glaciation in the northern Cordillera; in Proterozoic Basins of Canada, F.H.A. Campbell (ed.), Geological Survey of Canada, Paper 81-10, p. 25-46.

Young, G.M.
1982: The late Proterozoic Tindir Group, east-central Alaska: Evolution of a continental margin; Geological Society of America Bulletin, v. 93, p. 759-783.

Authors' addresses

H. Gabrielse
Cordilleran Division
Geological Survey of Canada
100 West Pender Street
Vancouver, British Columbia
V6B 1R8

R.B. Campbell
1760 Forest Park Drive
Sidney, British Columbia
V8L 4A6

ADDENDUM

Several important papers dealing with the terminal Proterozoic succession of the northern Cordillera have gone to press since completion of the manuscript for this chapter. These document the composition and stratigraphic range of Ediacaran fossils in the region, and the deposits of a post-Rapitan glaciation.

REFERENCES:

Aitken, J.D.
in press (a): Two Late Proterozoic glaciations, Mackenzie Mountains, Northwestern Canada; Geology, v. 19 (1991).
in press (b): the Ice Brook Formation and post-Rapitan, Late Proterozoic glaciation, Mackenzie Mountains, N.W.T.; Geological Survey of Canada, Bulletin 404.

Hofmann, H.J., Narbonne, G.M., and Aitken, J.D.
1990: Ediacaran remains from intertillite beds in northwestern Canada; Geology, v. 18, p. 1199-1202.

Narbonne, G.M., and Aitken, J.D.
1990: Ediacaran fossils from the Sekwi Brook area, Mackenzie Mountains, northwestern Canada; Paleontology, v. 33, Part 4, p. 945-980.

Chapter 7

CAMBRIAN TO MIDDLE DEVONIAN ASSEMBLAGES

Summary
Introduction
Cambrian assemblages
  W.H. Fritz
  Biostratigraphy
  Lower Cambrian Placentian Series
  Lower Cambrian Waucoban Series
  Middle Cambrian Series
  Upper Cambrian Series
Ordovician and Silurian assemblages
M.P. Cecile and *B.S. Norford* (contributions on Cordilleran terranes by *J.O. Wheeler, H. Gabrielse*, and *G.E. Gehrels*)
  Ancestral North American Miogeocline
  Cordilleran terranes
    Cassiar Terrane
    Kootenay Terrane
    Accreted terranes
  Volcanic and intrusive rocks
  Biostratigraphic zonation
Lower and Middle Devonian assemblages
D.W. Morrow and *H.H.J. Geldsetzer* (with contributions on Cordilleran terranes by *H. Gabrielse*)
  Ancestral North American miogeocline
    Principal stratigraphic subdivisions
    Delorme sequence (Gedinnian-Siegenian)
    Bear Rock-Stone sequence (Emsian-Early Eifelian)
    Hume-Dunedin sequence (Late Eifelian-Early Givetian)
  Cordilleran terranes
    Displaced North America: Cassiar Terrane
    Kootenay Terrane
    Accreted terranes
  References

Chapter 7

CAMBRIAN TO MIDDLE DEVONIAN ASSEMBLAGES

W.H. Fritz, M.P. Cecile, B.S. Norford, D. Morrow, and H.H.J. Geldsetzer

SUMMARY

H. Gabrielse

Lower Cambrian to Middle Devonian miogeoclinal strata were deposited along a passive margin of western Ancestral North America which formed as a result of rifting in Late Proterozoic time. Local anomalous thicknesses and facies of sedimentary rocks and the presence, in all systems, of minor volcanic rocks suggest repeated episodes of extension. The miogeoclinal sedimentary prism thickens markedly west of hinge lines which vary slightly in position for different rock units. Farther west thick carbonate sequences grade abruptly into much thinner argillaceous strata which, perhaps, reflect deposition on significantly attenuated continental crust. Thus, the main characteristics of the Lower Cambrian to Middle Devonian rocks in the miogeocline may be attributed to a complex interplay of continental rifting, attenuation, drifting, thermal subsidence and flexuring along the western margin of the continent. Rocks of similar ages, in part volcanogenic, mainly in the Alexander Terrane and forming minor components of several other terranes seem to have had both volcanic-island arc and possibly miogeoclinal affinities.

The Cambrian System of the Cordillera can be subdivided into four series, the Lower Cambrian Placentian and Waucoban Series, Middle Cambrian, and Upper Cambrian. The Placentian Series is strikingly different from the other three in its predominance of clastic sediments. Various water depths are indicated, but no major carbonate body was deposited, which suggests deposition in relatively cool waters. Placentian basinal shale and siltstone were probably deposited in British-Barn Basin and are present in much of Selwyn Basin. Between the two basins was the ancestral Yukon Platform, which lacks equivalent strata and was probably exposed in early Early Cambrian time. Shallow-water quartzite fringes the northeastern margin of Selwyn Basin, but is poorly represented or absent in southeastern Selwyn Basin and along the eastern margin of Kechika Basin. Abundant shallow-water quartzite filled Robson Basin, whereas fine sand and siltstone were deposited on Cassiar Platform. The Cassiar strata resemble those in the slope deposits of the same age in Selwyn Basin. Quartzite and some associated greenstone accumulated in Columbia Basin.

The boundary between Lower Cambrian Placentian and Waucoban Series is marked by an abrupt increase in carbonate and shelly fossils, suggesting an onset of the warmer conditions that dominated most of the Cambrian Period. The boundary coincides with that between the *Fallotaspis* and *Nevadella* zones. It is also near the top of the lowest half-cycle of three grand cycles (clastic-carbonate depositional pairs) that comprise the two lowest Cambrian series. The two series represent one major transgression which reached a maximum just before the end of Early Cambrian time.

Abundant Waucoban carbonate occurs in northeastern Selwyn Basin, on Yukon and Cassiar platforms, in part of Robson Basin, and in northwestern Columbia Basin. Its presence is essential for the recognition of the upper two and one half grand cycles. Clastics from Mackenzie Platform were trapped in Richardson Trough which allowed pure white carbonate to be deposited farther west on Yukon Platform. Similar carbonate on Cassiar Platform may have been deposited under the same conditions. Waucoban basinal shale with some volcanic rocks occupied British-Barn Basin, much of Selwyn Basin, there locally associated with basalt, and parts of northern Kechika Basin. Columbia Basin is dominated by a mixture of quartzite, siltstone, and carbonate, except at the northwestern end where carbonate occurs.

A regional disconformity separates Lower and Middle Cambrian strata. The disconformity represents a regression followed by a transgression that moved slowly at first but increased in rate. The pattern was locally interrupted by syndepositional block faulting in eastern Kechika Basin, northeastern Selwyn Basin and on southeastern Yukon Platform. Middle Cambrian slope and basinal dark shale and platy limestone predominate in British-Barn Basin(?), Richardson Trough, most of Selwyn Basin, most of Kechika Basin, Cassiar Platform(?), and Columbia Basin. The original lateral extent of a narrow fringe of shallow-water carbonate, locally preserved along the northeastern margin of Selwyn Basin and northeastern Kechika Basin and partly removed by post-Cambrian erosion, is unknown. A shallow-water carbonate succession occupied southeastern Kechika Basin and Robson Basin. Four grand cycles were recognized in the latter area, where carbonate half-cycles abruptly terminate basinward against predominately clastic deposits. Yukon Platform was probably exposed in late Middle Cambrian time.

Fritz, W. H., Cecile, M. P., Norford, B. S., Morrow, D., and Geldsetzer, H. H. J.
1991: Cambrian to Middle Devonian assemblages; in Geology of the Cordilleran Orogen in Canada, H. Gabrielse and C. J. Yorath (ed.); Geological Survey of Canada, Geology of Canada, no. 4, p. 151-218 (also Geological Society of America, The Geology of North America, v. G-2).

Upper Cambrian strata are widespread, but they register facies shifts that were more rapid than those that occurred during earlier epochs. Dark grey to black basinal shale and platy limestone occupy British-Barn Basin(?), Richardson Trough, north-northeastern Selwyn Basin, middle Kechika Basin, and Cassiar Platform(?). Medium grey shale and interbedded limestone with a significant clastic content unconformably overlie the south-southeastern and southwestern parts of Selwyn Basin and extend into the Kechika Basin and a small part of the Robson Basin near the Rocky Mountain Trench. Basinal shale and limestone were deposited in Columbia Basin, but strata there are dated in only one region. Throughout the Canadian Cordillera dark grey shale and platy limestone exhibit slump breccias in slope environments, whereas the medium grey equivalent strata containing more terrigenous clasts do not. Clean platform carbonate of medial Late Cambrian to Devonian age overlies unconformity-bounded Cambrian strata of variable composition of Yukon Platform. Moderately clean cratonal carbonate on Mackenzie Platform extends to the northeast edge of Richardson Trough and along part of northeastern margin of adjoining Selwyn Basin. Along the southeastern margin of Selwyn Basin and the Kechika Basin, these strata were removed by Early Ordovician erosion. In Robson Basin Upper Cambrian carbonate half-cycles, unlike similar half cycles in the Middle Cambrian, extend into the basinal clastic facies because of shallower water depths.

Ordovician and Silurian strata of the miogeocline accumulated under similar depositional settings as those established during late Early Cambrian time. Broad carbonate platforms were bordered on the west by basins, troughs and embayments in which deeper water facies were deposited. Regressions of epeiric seas caused periodic erosion of large areas of the craton, but the general lack of terrigenous clastics in platform carbonates attests to low cratonal relief. Exceptions occur in the central and southern Cordillera where platform carbonates contain significant terrigenous detritus, probably derived from the Peace River Arch.

On the northern miogeocline arches, basins and embayments may have developed as a consequence of early Paleozoic crustal extension between the Cordilleran and Innuitian regions. The Liard Depression, a depocentre that was absent during the Cambrian, contains in excess of 5500 m of Ordovician to Middle Devonian strata that could have accumulated in a rift setting.

On Mackenzie Platform, Ordovician and Silurian carbonate strata are divisible into three transgressive-regressive cycles referred to the upper Sauk, Tippecanoe and lower Kaskaskia sequences, respectively. Similar strata are preserved on nearby platforms although expansion of intervening trough settings modified the stratigraphic records such that basinal facies locally overlie platform carbonates.

Intraplatform and platform-margin basins and embayments received a variety of sediments characterized by argillaceous limestone, calcareous shale, siltstone and chert. Richardson Trough, which may have developed as an aulacogen during late Early Cambrian time, persisted as a basinal depocentre until Middle Devonian time. Misty Creek, Meilleur River and Prairie Creek embayments developed as platform-encroaching extensions from Selwyn Basin in late Early Cambrian, Early Ordovician and Late Silurian time, respectively. The Misty Creek Embayment was a distinct rift basin, the south rim of which was the site of Middle Ordovician and Late Silurian to Middle Devonian volcanism and from which volcaniclastic and alkalic basalt flow rocks extended as tongues into the basinal sediments of the embayment. To the south several centres of Late Precambrian to Devonian mafic and alkalic volcanic rocks occur throughout Selwyn Basin, particularly close to the edge of adjacent carbonate platforms. From Middle Ordovician to Late Silurian time the eastern margin of Kechika Basin was a zone of abrupt transition from basinal to platform deposition; associated sediments include foreslope breccia, conglomerate and debris flows, some of which are composed entirely of rocks derived from the adjacent MacDonald Platform.

In Robson Basin, Ordovician and Silurian strata are discontinuously preserved beneath the sub-Devonian unconformity. Only Lower Ordovician strata are preserved in the north whereas in the southern part of the basin Upper Cambrian to Lower Silurian basinal facies accumulated prior to and following late Middle Ordovician to Early Silurian encroachment of platformal carbonate. On adjacent MacDonald, Kakwa and Bow platforms the same cycles are developed as on Mackenzie Platform but with less clarity and completeness.

Lower and Middle Devonian strata are widespread in the northern Foreland Belt whereas in the south they are preserved only locally beneath the sub-Devonian unconformity. As with older successions they thicken westward and then thin abruptly where they change facies from dominantly platform and shelf carbonates to fine grained basinal shale, argillaceous limestone and chert.

Devonian miogeoclinal strata are referred to four sequences: Delorme (Gedinnian to Sieginian), Bear Rock-Stone (Emsian), Hume-Dunedin (Eifelian) and Fairholme (Givetian-Frasnian). The bulk of the last sequence is described in Chapter 8. These sequences are separated by discontinuities across which inferred tectonostratigraphic relationships change significantly. The Delorme sequence records widespread transgression following Late Silurian regression and erosion. Arenaceous carbonate first accumulated on parts of Mackenzie Shelf and later on MacDonald Shelf and Ogilvie Platform while basinal shale and argillaceous limestone continued to be deposited in Selwyn Basin and Blackstone and Richardson troughs. In Gedinnian to earliest Siegenian time Delorme sequence evaporites accumulated in widely separated cratonic platform basins such as Root Basin and Godlin Salient.

The Bear Rock-Stone sequence records a two-phase marine inundation of those areas of the northern miogeocline that were subaerially exposed at the close of Delorme sequence deposition. Throughout the Emsian and Early Eifelian shelf areas were less mobile and thus local Bear Rock-Stone thickness variations are less than those of the Delorme sequence except where transgression occurred across previously exposed paleogeographic highs. During the Emsian phase pure carbonate was deposited over Mackenzie and MacDonald shelves and Root Basin which, for the most part, ceased to be a depocentre for basinal sediments; evaporite deposition expanded on the eastern parts of the shelves and adjacent platforms. Hemipelagic sedimentation continued in Prairie Creek

Embayment and in Selwyn Basin. In northern Selwyn Basin dark grey shale and limestone is locally interbedded with volcaniclastics which originated from volcanic centres along the south rim of the Misty Creek Embayment. Richardson Trough re-established a connection between Selwyn Basin and northern seaways thus separating Porcupine Platform from Peel Platform. The second phase began following a brief interval of Early Eifelian regression when western deeper water settings expanded onto parts of the adjacent platforms. Pelletal lime wackestone and packstone accumulated on the western parts of Mackenzie and Peel platforms whereas on the edge of MacDonald Platform and in southern Selwyn Basin crinoidal and reefoid limestone accumulated. Deposition of the Bear Rock-Stone sequence was terminated by widespread regression from the platforms.

The Hume-Dunedin sequence developed during a widespread regional transgression which extended over a much broader region than did previous Early Devonian transgressions. Both the Elk Point Basin and Golden Embayment were established and many of the paleotopographic elements that had controlled deposition in the northern miogeocline ceased to have significant influence. On Peel, Mackenzie and MacDonald platforms fossiliferous and argillaceous open marine as well as peritidal carbonates pass westward into basinal, interbedded limestone and shale in Selwyn Basin. During the Late Eifelian shallow-water, restricted evaporite deposits accumulated at the southern end of the Golden Embayment. A western source for Eifelian and Early Givetian strata in the central and northern parts of Golden Embayment is suggested by the presence of clastics which become coarser toward the Purcell Landmass. In Early Givetian time the Peel, Mackenzie and MacDonald platforms became settings for deep-water shale deposition, bordered on the east by clinoform carbonate slope deposits and barrier reefs. Deposition of the Hume-Dunedin sequence terminated with widespread regression accompanied by the dramatic change from platform carbonate deposition to clastic deposition associated with rifting along the western margin of the craton. In the southern miogeocline, carbonate deposition returned following the widespread Taghanic Onlap (see Chapter 8).

Lower and middle Paleozoic strata in the western Cordillera occur mainly in the Cassiar, Kootenay and Alexander terranes. In Cassiar Terrane rocks of Cambrian to Devonian age are largely of miogeoclinal character whereas those in Kootenay Terrane both resemble and differ from temporally equivalent strata of ancestral North America. Arc volcanic, clastic and carbonate strata occur throughout Alexander Terrane where, in southeast Alaska they are associated with polymictic conglomerate, a derivative of the Klakas Orogeny which is unrecorded in the Canadian sector of the terrane. Clastic, calcareous and volcanic sequences of Early Devonian age occur locally in Quesnellia and possibly in Stikinia. The lower part of a thick volcaniclastic and epiclastic succession of Wrangellia include strata of Devonian and older age.

INTRODUCTION

This chapter is primarily concerned with Cambrian to Middle Devonian strata on the miogeocline of ancestral North America and neighbouring pericratonic and displaced terranes. Brief treatment is given to strata of this interval in the accreted terranes. As in the other assemblage chapters stratigraphic units are identified with the appropriate tectonic assemblage on the Tectonic Assemblage Map (Map 1712A, in pocket).

The Cambrian System of the miogeocline is discussed in terms of its biostratigraphic zonation and, where possible, with reference to its widespread cyclical character. Its facies are related to the well known inner, middle and outer detrital belts which are widely recognized around the rim of cratonic North America. Cambrian strata of the Cassiar Terrane, a displaced fragment of the miogeocline, are referred to the "Cassiar Platform" and those of the Porcupine and Arctic Alaska terranes of the northern Yukon are treated under the heading "British-Barn Basin" and included with the miogeocline pending identification of their proper paleogeographic setting. Because much of the data for Cambrian stratigraphy have not been synthesized previously and published reference material is not easily available, the Cambrian System is dealt with at greater length than succeeding Paleozoic systems.

In contrast to the comparatively detailed descriptive account of the Cambrian System, the Ordovician to Middle Devonian successions are given a more general and broadly interpretive treatment. The latter are described in relation to their paleogeographic settings (i.e. platforms, basins, embayments, etc.) and, moreover, with reference to several sequences broadly identifiable with all or parts of the Sauk, Tippecanoe and Kaskaskia sequences. Many paleogeographic elements with complex histories, particularly in the northern miogeocline, significantly controlled the distribution of facies and variations in thickness of Ordovician to Middle Devonian strata, whereas during the Cambrian, such features were fewer in number, particularly in platform areas.

To some extent the employment of paleogeographic terms reflects the bias of the authors. In some instances the same paleogeographic or paleotectonic elements are given different names in the three sections of this chapter. In others the editors have imposed uniformity of terminology for which they take responsibility. Some paleotectonic elements of arguable value are retained from long-standing usage. The editors have made no attempt to define such terms as "trough", "basin", "platform", "miogeocline" etc., believing these to be largely self-explanatory.

CAMBRIAN ASSEMBLAGES
W.H. Fritz

The best known work on the Cambrian System in the Canadian Cordillera derives from studies in the southern Rocky Mountains. There Walcott (1912a, 1928, etc.) described numerous formations and fossils, including the remarkable soft-bodied forms in the Burgess Shale. Walcott's stratigraphy was improved by Burling (1923, 1955), Deiss (1939, 1940), Rasetti (1951), North and Henderson (1954), and Aitken and Greggs (1967). From these studies it became apparent that the Middle and Upper Cambrian succession consisted of thick intervals of shale alternating with thick intervals of mainly unfossiliferous carbonate, each of which were recognized as a formation. In 1966 Aitken focused attention on the cyclical nature of this succession by pairing the shale and carbonate units to erect eight "grand cycles" in 3600 m of Middle Cambrian to Middle Ordovician strata. Aitken's grand cycles lie within an extensive middle carbonate belt that was first described by Robison (1960) and Palmer (1960) in the Great Basin region of the southwestern United States. They noted that it once extended around the North American craton. The middle carbonate belt is flanked by an inner detrital belt comprising light coloured clastics derived from the adjacent craton, and by an outer detrital belt comprising finer, dark coloured clastics.

In southeastern Rocky Mountains, Middle Cambrian strata of the middle carbonate belt were shown by Rasetti (1951), Ney (1954), Cook (1970, 1975), Fritz (1971) and McIlreath (1977b) to terminate abruptly against sediments of the outer detrital belt. Aitken (1971) recognized a local thickening or "rim" at the outer margin of the Middle Cambrian part of the middle carbonate belt. Balkwill (1969) demonstrated that Upper Cambrian carbonate of the middle carbonate belt extends into the outer detrital belt.

The inner detrital, middle carbonate, and outer detrital belts were shown to occur discontinuously throughout the Canadian Cordillera by Ziegler (1969), Douglas et al. (1970), and North (1964, 1971). The Middle Cambrian to Middle Ordovician grand cycles have not been recognized in the Canadian Cordillera north of southeastern Rocky Mountains. Fritz (1975) applied the grand cycle concept to Lower Cambrian strata, and recognized all or parts of three grand cycles in the Cordillera from Canada to northern Mexico. The lowest Lower Cambrian clastic half-cycle differs from younger clastic half-cycles in being thicker and having less carbonate. The greater thickness, general lack of carbonate, and a lack of shelly fossils, combine to warrant special treatment of these older Cambrian strata.

Outcrop and facies distributions for each of four major Cambrian divisions are shown in Figures 7.2, 7.6, 7.11, and 7.15.

On the Tectonic Assemblage Map (Map 1712A, in pocket) Cambrian miogeoclinal strata in the Foreland Belt, together with their displaced equivalents in the Cassiar Terrane of the Omineca Belt, are assigned to the Gog, Hyland and Rocky Mountains passive continental margin assemblages. Exceptions include Middle Cambrian clastic rocks of north-central Yukon and northeastern British Columbia (Kechika Basin); these have been identified as an unnamed rift assemblage. In northernmost Yukon Cambrian strata in the Porcupine and Arctic Alaska terranes are included with miogeoclinal assemblages of ancestral North America, however, their paleogeographic relationships are unknown.

Biostratigraphy

Until recently the Cambrian zonal succession (Fig. 7.1) in the Cordillera was considered to contain a large gap in its lower portion because of a paucity of shelly fossils. In Canada the base of the Cambrian was traditionally placed at the first formational (or group) contact below the lowest occurrence of *Olenellus* (Little, 1960). A later demonstration that the lowest occurrence of *Olenellus* equated with the base of the *Bonnia-Olenellus* Zone, and that two trilobite zones occur below that level (Fritz, 1972), had no effect on the previously selected and much lower lithostratigraphic boundary.

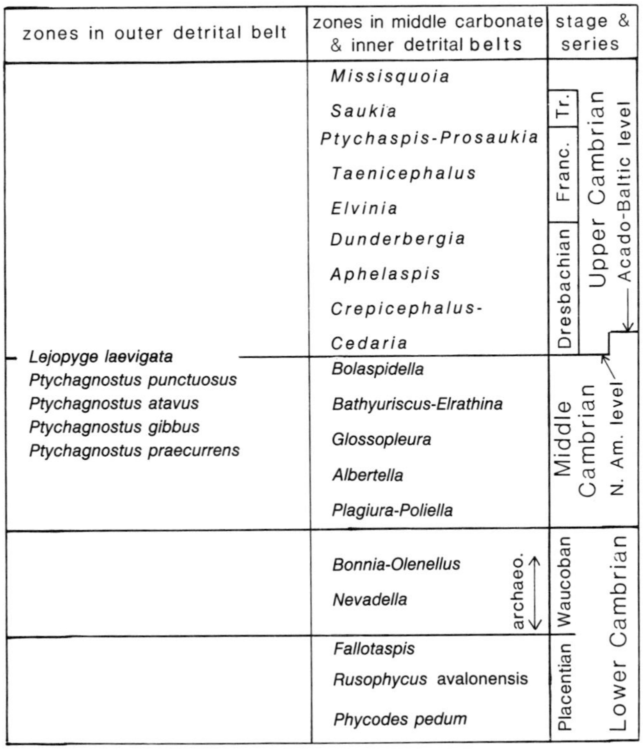

zones in outer detrital belt	zones in middle carbonate & inner detrital belts	stage & series	
	Missisquoia	Tr.	Upper Cambrian
	Saukia		
	Ptychaspis-Prosaukia	Franc.	
	Taenicephalus		
	Elvinia		
	Dunderbergia	Dresbachian	
	Aphelaspis		
	Crepicephalus-		Acado-Baltic level
	Cedaria		
Lejopyge laevigata Ptychagnostus punctuosus Ptychagnostus atavus Ptychagnostus gibbus Ptychagnostus praecurrens	Bolaspidella		Middle Cambrian
	Bathyuriscus-Elrathina		
	Glossopleura		N. Am. level
	Albertella		
	Plagiura-Poliella		
	Bonnia-Olenellus	Waucoban	Lower Cambrian
	Nevadella	archaeo.	
	Fallotaspis	Placentian	
	Rusophycus avalonensis		
	Phycodes pedum		

Figure 7.1. Trilobite zonation in the right-hand column is used in Cambrian biostratigraphy of inner detrital and middle carbonate belt strata in the Canadian Cordillera. Abbreviated stages are Franconian (Franc.) and Trempealeauan (Tr.). Zonation in the left-hand column was suggested by Robison (1984) for use in the outer detrital strata of the North American Faunal Province. This zonation was a modification of Westergard's (1946) agnostoid zonation in the Acado-Baltic Province.

The boundary between the two sub-*Olenellus* zones (*Nevadella* and *Fallotaspis* zones) is here considered as the base of the Lower Cambrian Waucoban Series, which is in turn underlain by the Lower Cambrian Placentian Series. This definition for the base of the Waucoban is close to the original but vague definition by Walcott (1912b), who defined the Waucoban in two sections, both with *Nevadia* near the lowest described beds. In this chapter the base of the Waucoban is placed at the first occurrence of *Nevadia*.

The lowest Cambrian series used in this chapter, the Placentian Series, has recently been described by Landing et al. (1989). At the type Placentian in Newfoundland, the base of the series was placed at the first occurrence of the trace fossil *Phycodes pedum* and the top at the base of the *Callavia* Zone. Because the Waucoban was erected in the North American Faunal Province and the Placentian in the Acadian Faunal Province, problems remain in correlating their mutual boundary. However the tentative correlation of the *Nevadella* Zone with the *Callavia* Zone (Fritz, 1972, p. 6) suggests this correlation may be reasonably close. In the type Placentian, all of the strata are below the first occurrence of trilobites, and the series is zoned by small shelly fossils and trace fossils. In the Canadian

Cordillera, only a few metres of strata at the top of the Placentian contain trilobites, small shelly fossils are very rare (Nowlan et al., 1985), and the series is defined almost exclusively by trace fossils.

The top of the Lower Cambrian in the Canadian Cordillera is at the uppermost occurrence of olenellid trilobites (Rasetti, 1951). The immediately overlying earliest Middle Cambrian faunas are poorly known, and their correlation has not been rigorously tested.

The Middle and Upper Cambrian zonation used herein (Fig. 7.1) is modified from that described by Lochman-Balk and Wilson (1958). The *Dicanthopyge* and *Prehousia* zones, which Palmer (1965) recognized in the Great Basin and placed between adjusted *Aphelaspis* and *Dunderbergia* zones, have not been added, as they have not yet proven useful in Canada. The *Taenicephalus* Zone is substituted for the *Conaspis* Zone to reflect the more common occurrence of *Taenicephalus* within this interval (see Longacre, 1970). The *Ptychaspis-Prosaukia* Zone is retained, although several pairs of zones (e.g. *Idahoia* and *Ellipsocephaloides* zones) elsewhere provide refinements in this interval; these have only been recognized at a few localities in Canada. The *Missisquoia* Zone is added above the *Saukia* Zone to raise the top of the Cambrian to a level presently favoured by the Cambrian-Ordovician Boundary Working Group (IUGS, IGCP). The base of the Ordovician is placed at the first occurrence of *Symphysurina brevispicata*, which, in the richly fossiliferous beds in the Cow Head Group in Newfoundland, begins at the basal level for the conodont *Cordylodus linstromi*, and slightly below the first occurrence of the graptolite *Rhabdinopora flabelliforme* (C.R. Barnes, pers. comm., 1985). Sections in which the first occurrence of *S. brevispicata* is recorded in a described faunal succession are at localities 18, 40 and 41 (see Fig. 7.15). Recently proposed, major changes in the North American biostratigraphic framework by Ludvigsen and Westrop (1985) have not been applied in this chapter. One of their new stages (see also Westrop, 1986) is based on fossils from a locality in the southern Canadian Rocky Mountains (see Fig. 7.15, loc. 40).

The zones shown in Figure 7.1 need future modification and refinement because the faunal change from one depositional belt to the next is too great for satisfactory correlation (Stitt, 1971). Lochman-Balk and Wilson (1958) demonstrated this problem by showing different ranges of genera in different areas of their shallow shelf and deeper depositional environments. These environments coincide with the inner detrital and middle carbonate belts, respectively. Their zonation for a third environment, the euxinic basin, however, cannot be used as intended. It was adapted from the the Acado-Baltic Province before the plate tectonics concept was fully appreciated and when that province was believed to represent the deep water faunas of the North American Faunal Province. Robison (1984) has shown, however, that some Acado-Baltic agnostoid zones can be used within the North American outer detrital belt. These zones are in the upper Middle Cambrian (Fig. 7.1).

Using agnostoids, Daily and Jago (1975) and Robison (1984) showed that the Middle-Upper Cambrian boundary in the North American Faunal Province is below that used in the classical Acado-Baltic Province, where it occurs between the *Lejopyge laevigata* and *Agnostus pisiformis* zones. In the North American Faunal Province the boundary has been defined as the base of the *Cedaria* Zone, and that definition is used in this chapter.

Lower Cambrian Placentian Series
British-Barn Basin

In Barn Mountains (Fig. 7.2, loc. 1) and Buckland Hills 80 km to the northwest, broadly dated Cambrian clastics underlie Ordovician chert and Silurian argillite. The Cambrian succession comprises grey, maroon and green argillite, interbedded quartzite and some limestone. In both areas the basinal Early Cambrian trace fossil *Oldhamia* is present (Cecile, 1988; Lane and Cecile, 1989), but no effort has yet been made to separate these strata into a Placentian and/or Waucoban series.

On the Canadian side of the Yukon-Alaska border, two map units in the British Mountains (Fig. 7.2, loc. 2) have been placed in the Cambrian (Norris, 1976, p. 457). The upper unit, comprising agglomerate, limestone and mafic flows, can be traced across the border into Alaska where it contains Late Cambrian fossils and the Early Cambrian trilobite cf. *Olenellus praenuntius* Cowie (Palmer in Dutro et al., 1972). No fossils were found in the lower unit which consists of dark grey argillite (150 m). The lower unit is reported to rest unconformably on the Neruokpuk Formation. If present, Placentian strata would be below the cf. *O. praenuntius* fossil horizon and probably below the dark grey argillite unit, as the latter unit is underlain by echinoderm-bearing calcareous siltstone and sandstone in nearby outcrops in Alaska (Reiser et al., 1980). Despite the unconformity at the top of the Neruokpuk, it cannot be ruled out that it does not contain Placentian strata.

Yukon Platform and Richardson Trough

No Lower Cambrian Placentian strata occur on Yukon Platform or in Richardson Trough. At the northeastern margin of Selwyn Basin (Fig. 7.2, loc. 3-5), upper Proterozoic and Placentian strata exhibit depositional thinning and disconformities as they approach Yukon Platform. Because these strata comprise fine clastics, it is likely that the adjacent Yukon Platform was barely emergent during early Early Cambrian time. Regional relationships suggest that the area of Yukon Platform and Richardson Trough were originally part of Mackenzie Platform, and did not come into existence as separate elements until late Early Cambrian time (during *Bonnia-Olenellus* Zone time).

Selwyn Basin

Near the eastern margin of Selwyn Basin, Upper Proterozoic carbonate of the Risky Formation (Gog Assemblage, see Map 1712A) is useful in locating the base of the Cambrian System (Fig. 7.3). The formation occurs a short distance below the base of the Cambrian, and has been correlated with the middle member of the Backbone Ranges Formation, and tentatively correlated with carbonate at the top of the Yusezyu Formation (Fritz et al., 1983). The Risky Formation overlies Precambrian strata containing Ediacaran fossils in northern Selwyn Basin (Fig. 7.2, loc. 3-6 and 10) and is in turn closely overlain by Placentian rocks containing small shelly fossils of the *Anabarites-Circotheca-Protohertzina* Zone (Fig. 7.2, loc. 5).

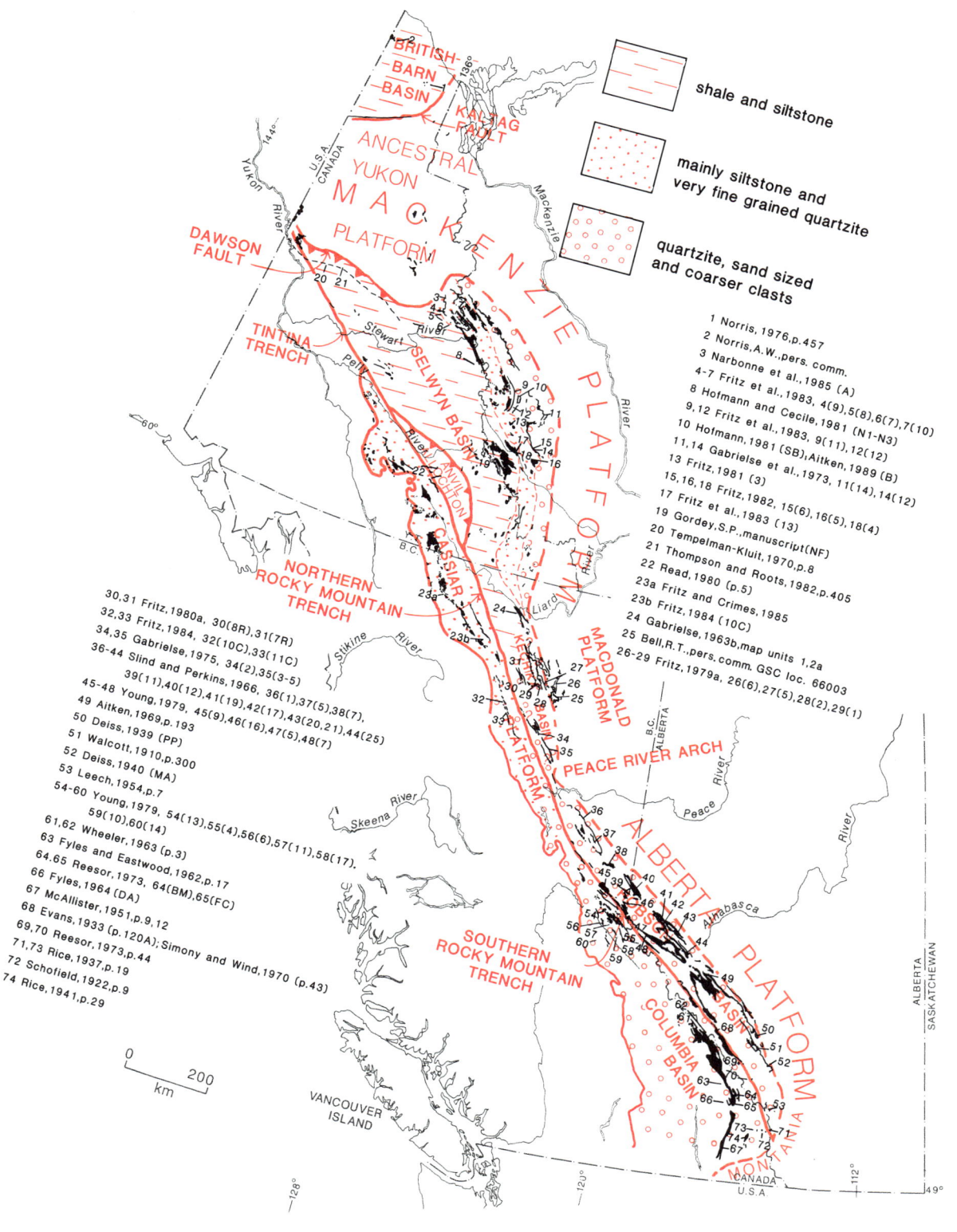

Figure 7.2. Distribution of lower Lower Cambrian out crops and Placentian localities and facies. Important out crops are numbered and each number is Keyed to a list of references. Relevant stratigraphic sections by the author are placed in parenthesies after the author's name and are indicated either by the initials of the name or number given to the section. If neither a name or number was assigned, the page number of the author's description is entered.

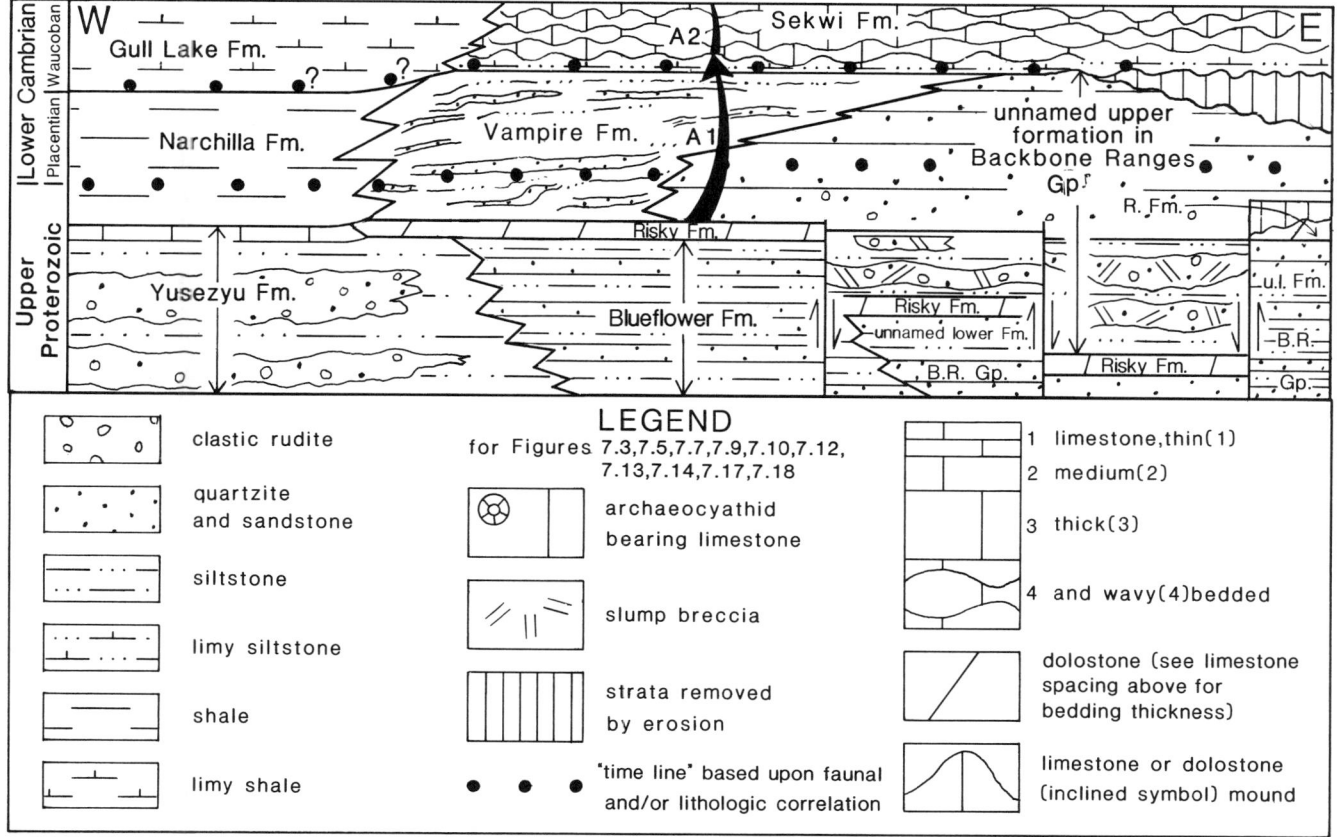

Figure 7.3. Schematic cross-section showing lateral relationships of Upper Proterozoic and Lower Cambrian strata at the northeast side of Selwyn Basin. Latest Proterozoic (post-Risky Formation) block faulting is indicated at the right side of the figure. Lower and upper rows of solid circles approximately locate the base and top of the Lower Cambrian Placentian Series, respectively. The Risky Formation is a unit formerly referred to as the middle member of Backbone Ranges Formation (Gabrielse et al., 1973). Grand cycle (half-cycles A1, A2, arrows) is from Fritz (1975).

The widespread occurrence of the Risky carbonate suggests a measure of stability during late Precambrian time. Karst(?) breccia plus suspected minor disconformities at the top of the Risky Formation in several areas indicate mild uplift just before Cambrian time. Risky karst(?) breccia in the northern Selwyn Basin near locality 6 (Fig. 7.2) contains the Goz Creek zinc deposit (1.75 million metric tonnes proven, see Reeve, 1977).

Above the Risky Formation is a thick succession of uppermost Proterozoic and Placentian strata that can be separated into three regional map units. To the east, on the western margin of Mackenzie Platform and in Selwyn Basin, is fine to coarse, light-coloured quartzite which is crossbedded, locally channelled and which contains sparse trace fossils (Plate 10). These strata belong to the unnamed, upper (post-Risky) part of the Backbone Ranges Formation (Fig. 7.4). At the Backbone Ranges type section (Fig. 7.2, loc. 14) and at nearby locality 16, the unnamed strata are 850 m and 1740 m thick, respectively, the latter figure being close to the maximum. At locality 16, a 600 m interval immediately above the Risky Formation contains massive quartzite interbedded with siltstone. The quartzite contains unsorted clasts of up to pebble size, and disoriented slabs of siltstone up to 1.5 m long. The thick interbeds may represent mass flow deposits near faults that were active just before and perhaps during earliest Cambrian time.

The next basinward map unit is the Vampire Formation, which at its type locality (Fig. 7.2, loc. 17) is 930 m thick. The Vampire comprises dark grey siltstone and shale interbedded with light brown, very fine grained quartzite. Trace fossils are common, and abundant slump folds suggest a slope environment.

Farther basinward is maroon and locally apple-green shale and siltstone comprising the Narchilla Formation. Thin, planar laminae suggest that the Narchilla was deposited in a low-energy environment. Trace fossils are sparse, but *Oldhamia* has been reported near the top of the formation at locality 8 (Fig. 7.2) and from unknown horizons at locality 21; *Gordia*, *Palaeophycus* and *Planolites* are known from the base at locality 19. The Narchilla Formation has an estimated thickness of 300 m near the northwestern margin of Selwyn Basin (loc. 20), and an estimated thickness of 830 m near the middle of the basin (loc. 19). A local member within the Backbone Ranges Formation (between localities 9 and 12), the Ingta Member, may be a tongue of the Narchilla.

Figure 7.4. Typical uppermost Proterozoic and Placentian quartzite in unnamed upper formation of Backbone Ranges Group. View is of outcrop at locality 16 in Figure 7.2. Approximately 1200 m of strata shown on ridge skyline. Figure from Fritz, 1982. GSC 203775-A,I,J

In Selwyn Basin the distribution of the upper Backbone Ranges Formation, Vampire Formation and Narchilla Formation coincides with the three lower Lower Cambrian facies shown in Figure 7.2. To the east, strata of the upper part of the Backbone Ranges Formation once overlapped Mackenzie Platform for an unknown distance but have mostly been removed by early Paleozoic erosion. The Vampire Formation lies mainly basinward of the Backbone Ranges Formation and the northeastern margin of Selwyn Basin, and, along with underlying Upper Proterozoic strata, thins against the margin of Yukon Platform. In northern Selwyn Basin, the Narchilla Formation is juxtaposed against the southwestern margin of Yukon Platform along the Dawson Fault and terminates to the southwest against the Tintina Fault. All three units probably continue southeastward into the northern part of Kechika Basin, but their mutual contacts are difficult to trace because of limited outcrop.

Kechika Basin

Placentian strata are either missing or thin across west-central MacDonald Platform and central Kechika Basin (Fig. 7.2, loc. 25-31). There, Waucoban quartzite unconformably overlies Middle and/or Upper Proterozoic strata (loc. 25 = 60 m; 26 = 240 m; 27 = 170 m; 28 = 560 m; 29 = 750 m). Two sections farther west (loc. 30 and 31, both = 540 m+) have unexposed bases, but surrounding outcrops suggest a Lower Cambrian depositional contact on Upper Proterozoic strata.

No Placentian strata have been reported in southern Kechika Basin. Sections containing Waucoban strata (Fig. 7.2, loc. 34, 35) have their bases either covered or faulted, but their positions, close to the Peace River Arch, suggest that Placentian strata are either thin or missing.

The Narchilla Formation probably extends from Selwyn Basin into northern Kechika Basin where a Waucoban archaeocyathid bioherm overlies maroon and green Narchilla(?) shale (loc. 24) in a relationship similar to that observed in Selwyn Basin (Gordey, 1979). The Narchilla(?) is exposed west of the Gundahoo Fault; no transitional units are known to the east where the Placentian is presumed to be missing. The abrupt appearance of basinal Placentian strata west of the fault is not yet understood; dextral displacement has been suggested by Taylor and Stott (1973).

Cassiar Platform

Placentian strata on Cassiar Platform (Cassiar Terrane) are best known near its central part (Fig. 7.2, loc. 23a), where they are included in the uppermost Stelkuz Formation (190 m), Boya Formation (400 m), and lowermost Rosella Formation (1.5 m). At the base of the succession is a minor disconformity tentatively considered as the Precambrian-Cambrian boundary (Fritz and Crimes, 1985). Below the disconformity are the trace fossils *Didymaulichnus miettensis, Neonereites* sp., *Skolithos* sp., and *Taphrhelminthopsis circularis*. Above are *Diplocraterion* sp., *Skolithos* sp., and *Teichichnus* sp.

A minor disconformity occurs near the top of the Placentian at locality 23a (Fig. 7.2), which elsewhere also is minor or lacking except at locality 23b, where the whole of the Boya Formation is missing.

The Placentian succession comprises dark grey siltstone and interbedded, very fine grained quartzite in the upper Stelkuz and overlying Boya formations. Strata in the latter formation resemble those of the Vampire Formation in Selwyn Basin. To the south, at locality 32, the Boya is 955 m thick, and there the Precambrian-Cambrian boundary is tentatively placed at the contact between the Boya and Stelkuz formations. No disconformity was noted at or near the contact. Still farther south, beginning at locality 33, and extending to the south end of Cassiar Platform, the Boya-Stelkuz contact is abrupt, but no erosion surface is reported (Mansy and Gabrielse, 1978). In that area there is a southward increase in quartzite within the Boya Formation.

On northern Cassiar Platform, at locality 22, the Placentian comprises at least 300 m of rust weathering, green argillite and interbedded quartzite.

Robson Basin

In the northern part of Robson Basin, between Peace River and locality 49 (Fig. 7.2), Placentian strata are contained within the McNaughton Formation, the basal unit of the Gog Group. Along strike between localities 36 and 44, the McNaughton comprises light-coloured, thick-bedded quartzite with an average thickness of 800 m. Along a parallel but more westward trend near the Southern Rocky Mountain Trench (loc. 45 to 48), thickness changes are rapid (550 m to 2200 m). There an upper McNaughton unit (Holmes Creek Member, maximum thickness 700 m) is present, comprising interbedded dark grey siltstone and white quartzite. The rapid lateral changes in thickness and lithology suggest syndepositional faulting (Young, 1979).

A disconformity below the McNaughton Formation and equivalent strata to the southeast separate these beds from the underlying Miette Group (Slind and Perkins, 1966; Aitken, 1969). Ediacaran fossils and the trace fossil *Didymaulichnus miettensis* occur within the upper part of the Miette (Young, 1979; Hofmann et al., 1985). Cambrian small shelly fossils *Campitius* sp. and *Volborthella* spp. were reported from the upper Miette (Young, 1979), but these identifications could not be substantiated following a careful study by J.W. Durham (pers. comm., 1984).

No diagnostic fossils have been reported from the base of the McNaughton Formation, thus it is suspected that the Precambrian-Cambrian boundary lies within the thicker, basinward sections of the formation, and may pass into the disconformity beneath the McNaughton at thinner sections toward the craton (Fig. 7.5).

Overlying the McNaughton is the Mural Formation, comprising limestone and siltstone (Fig. 7.5). Trilobites from the base of the Mural belong to the lower part of the *Nevadella* Zone (Fritz and Mountjoy, 1975). Because the Mural is critical to differentiating quartzite of the McNaughton Formation from overlying quartzite of the Gog Group, and, as the Mural is absent southeast of locality 49, the Gog to the southeast is undivided except for a thin Peyto Member at the top.

The Gog quartzite to the southeast, like that to the northwest, is thick bedded, contains shallow-water structures, and has a low shelly fossil content. Between localities 49 and 51 (Fig. 7.2) the entire Gog Group near the Alberta-British Columbia provincial boundary averages 670 m in thickness (Palonen, 1976). At locality 50 the group is 520 m thick and at locality 52 is 357 m thick. In the Hughes Range (loc. 53), where it is locally called Cranbrook Formation, it is 75 to 245 m thick. Trilobites within the Gog have been reported 610 m below the top at locality 51 (*Nevadella* Zone), at least 70 m below the top at locality 52 (*Bonnia-Olenellus* Zone), and 30 m below the top at locality 53 (*Nevadella* Zone).

Regional data suggest that the top of the Placentian lies a short distance below the contact between the McNaughton and Mural formations in the northwestern part of Robson Basin (loc. 49 and to the northwest), and lies well down in the Gog Group to the southeast.

Columbia Basin

In Columbia Basin, intense deformation and metamorphism have rendered the study of Cambrian strata difficult. Nonetheless, distinctive Lower Cambrian units, such as the Badshot Formation and the Hamill Group (in part Precambrian?) provide broad Lower Cambrian control.

In the Cariboo Mountains (Fig. 7.2, loc. 54 to 60) Placentian strata are included in the Yanks Peak and Midas formations. The Yanks Peak averages 400 m in thickness and comprises mainly light-coloured quartzite and, in the northern part of the mountains, some redbeds (Young, 1979). The overlying Midas Formation comprises siltstone and shale, and averages 300 m in thickness. The base of the lower Lower Cambrian is tentatively placed at

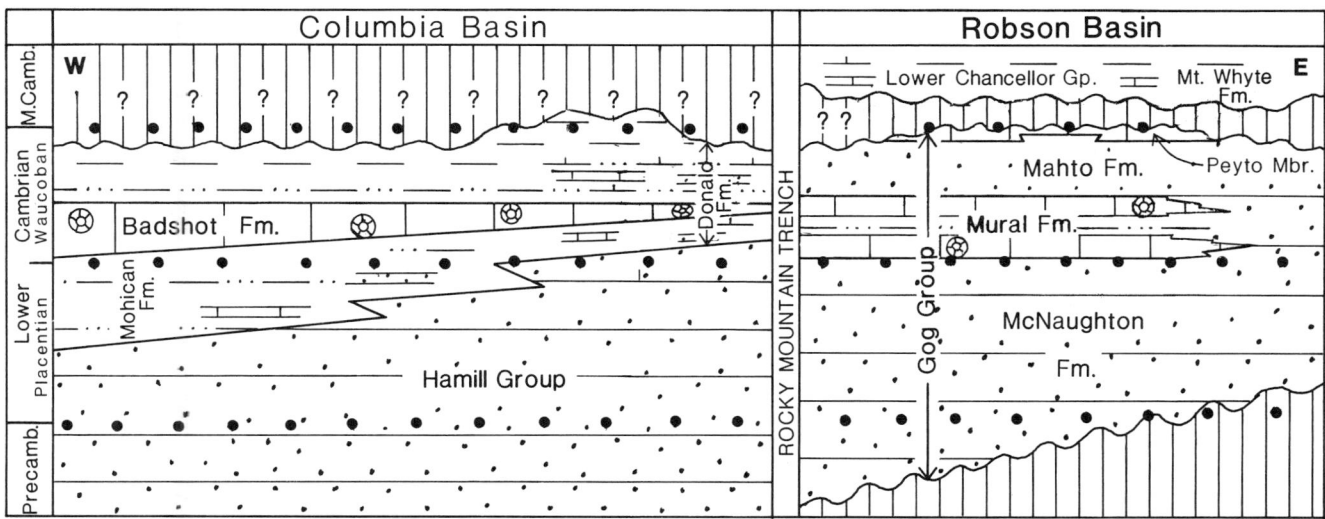

Figure 7.5. Schematic cross-section showing Lower Cambrian units in Robson and Columbia basins. Solid circles mark approximate level of the base of the Lower Cambrian, the boundary between the Placentian and Waucoban series and the top of the Lower Cambrian. Legend is in Figure 7.3.

the base of the Yanks Peak, a position founded upon correlation with similar strata on Cassiar Platform. The top of the Placentian coincides with the base of the overlying Mural Formation.

In the northernmost Monashee and Selkirk Mountains Lower Cambrian strata are either metamorphosed beyond recognition or missing. Still farther south, in the Kootenay Arc (loc. 61 to 67), the Placentian is contained within the Hamill Group and partially equivalent Mohican Formation, each of which is dated by its position below the Waucoban, archaeocyathid- bearing Badshot Formation (Fig. 7.5).

The contact between the Hamill Group and underlying Proterozoic Horsethief Creek Group of the Windermere Supergroup may be an unconformity that represents a considerable amount of missing strata (Pell and Simony, 1981, 1984). Basal Hamill beds comprise clean, light-coloured quartzite with sand- to cobble-sized clasts; younger Hamill strata are similar, but finer grained. The Hamill is 1675 m thick at locality 62, and 1300 and 2000 m thick, respectively, at localities 64 and 65. At locality 61 the group contains a significant amount of greenstone. The presence of greenstone and coarse clastic sedimentary rocks has been attributed to rifting (Devlin, 1984). Close to the Southern Rocky Mountain Trench (loc. 68), the Hamill (1650 m) is clearly unconformable on the Horsethief Creek Group and contains *Nevadella* Zone fossils in its uppermost strata.

In the western part of Kootenay Arc, the upper Hamill Group laterally changes to phyllite and interbedded limestone of the Mohican Formation. This formation varies from 610 to less than 300 m in thickness at locality 63, and from a few metres to 300 m at locality 66. Near the International Boundary (loc. 67) the Hamill Group and Mohican equivalents have thicknesses of 1460 and 225 m, respectively. These variations in thickness are possibly tectonic in origin (J.O. Wheeler, pers. comm., 1987).

On either side of the Southern Rocky Mountain Trench (loc. 53, 69-74) small Lower Cambrian outcrops are assigned to the Cranbrook, Donald and Eager formations. The Cranbrook contains uppermost Proterozoic(?), Placentian, and some Waucoban strata, whereas the Donald and Eager contain Waucoban and younger(?) strata. Trilobites of the *Nevadella* Zone have been found at the top of the Cranbrook at locality 53, and those of the *Bonnia-Olenellus* Zone closely overlie the Cranbrook at locality 72. An unconformity beneath the Cranbrook is indicated by a coarse basal conglomerate, and by the formation's position above various Middle Proterozoic formations of the Purcell (Belt) Supergroup.

A variable thickness of the basal Cranbrook conglomerate led Rice (1937) to postulate an Early Cambrian transgression over a surface of high relief. At locality 74 this conglomerate continues throughout an unusually thick outcrop of the formation. Elsewhere, above the conglomerate, the Cranbrook comprises light coloured, shallow-water quartzite. At locality 69 the Cranbrook is 760 m thick, at locality 70 it is 300 m thick, and at locality 72 it is 180 m thick. In addition to relief on the sub-Cranbrook surface, erosional bevelling at the top locally contributed to abrupt changes in thickness. This erosional surface is below the Upper Cambrian Jubilee Formation.

A magnesite member, 45 m thick, occurs near the top of the Cranbrook Formation at locality 73 (Rice, 1937).

Summary of Placentian deposition

In British-Barn Basin, sparse data suggest at least some deposition of lower Lower Cambrian sediments; the general basinal aspect of the region indicates these deposits were mainly of a shale and siltstone facies.

Richardson Trough and Yukon Platform were probably emergent during the first half of Early Cambrian time. A thin layer of clastics at the base of the overlying Waucoban succession suggests a low relief surface across which a late Early Cambrian sea transgressed. Fine grained Placentian clastics near the southeastern edge of Yukon Platform also indicate a surface of low relief.

The best record of Placentian nearshore to basinal deposits is exposed in Selwyn Basin. From the northeast edge of the basin, facies changes to the southwest, are represented by belts of shallow-water quartzite, slope deposits, and basinal shales. Strata of the latter facies extend southeastwards into the northern part of Kechika Basin. In the middle and southern part of Kechika Basin, only thin or no Placentian strata are present. This thinning to the south is attributed to the influence of the Peace River Arch.

Placentian strata on Cassiar Platform resemble the interbedded siltstone and very fine grained quartzite (Vampire Formation) in Selwyn Basin. A lateral introduction of shallow-water, coarse grained quartzite at the southeastern end of the platform is consistent with an abundance of similar equivalent strata in Cariboo Mountains (northern Columbia Basin) farther to the southeast.

A general southeastward increase in the amount of sand-sized material within the Placentian succession is shown in Figure 7.2. The wedge of quartzite in Robson Basin and Columbia Basin greatly exceeds that to the northwest and is believed to reflect significant uplift of the adjacent craton. At least locally the character of the Hamill Group suggests tectonism, perhaps related to rifting (Devlin, 1984).

The almost total lack of carbonate in the Placentian, despite different depositional environments, implies cooler water temperatures than those during later Paleozoic time.

Lower Cambrian Waucoban Series
British-Barn Basin

No fossils restricted to the Waucoban have been found in the Canadian part of British-Barn Basin, but at locality 1 (Fig. 7.6) Norris (1976) correlated a volcanic and carbonate unit with a similar unit in easternmost Alaska, which contains Waucoban and Late Cambrian fossils (Dutro et al., 1972). Below the unit at locality 1 is 150 m of argillite which, together with part of the volcanic and carbonate unit, is tentatively placed in the upper Lower Cambrian assemblage. *Oldhamia*, a trace fossil commonly reported from the Lower Cambrian, is present in the British and Barn mountains (Cecile, 1988; Lane and Cecile, 1989).

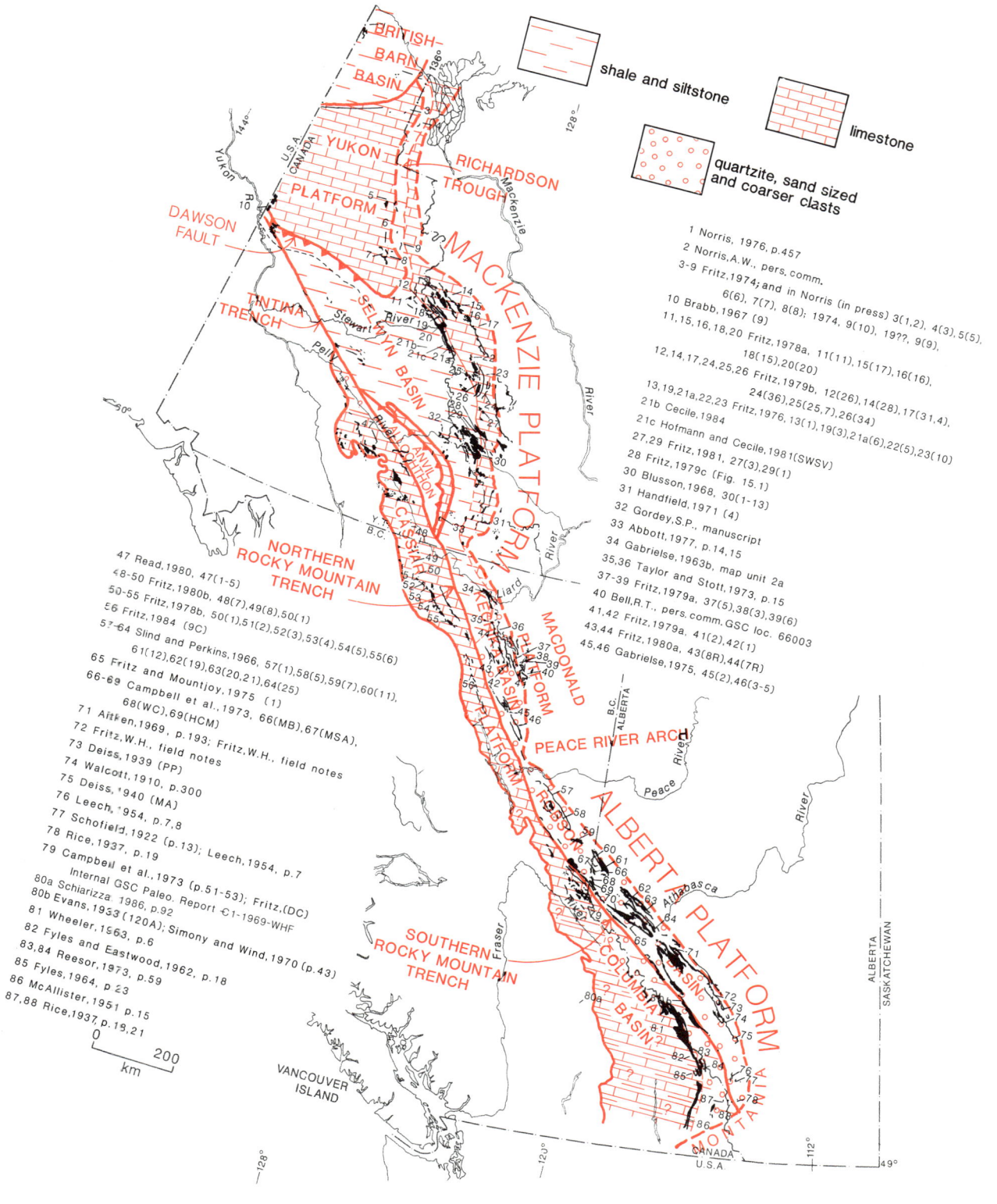

Figure 7.6. Distribution of upper Lower Cambrian outcrops and Waucoban localities and facies. See Figure 7.2 for note on references.

Yukon Platform and Richardson Trough

Waucoban strata of the Illtyd Formation unconformably overlie Middle and Upper Proterozoic rocks on Yukon Platform (Fig. 7.7). At locality 6 the formation comprises a lower unit (60 m thick, base not exposed) of limy siltstone and an upper unit (550 m) of white, finely crystalline and dense limestone in massive, cliff-forming beds alternating with light grey lime packstone in thin and medium beds. At the southeastern margin of the Yukon Platform (loc. 8), adjacent to Richardson Trough, the Illtyd is nearly twice as thick (950 m), and includes large (175 m thick), white weathering carbonate buildups. Within Richardson Trough (loc. 9), the lower half of the Illtyd is like that on eastern Yukon Platform (loc. 6), and the upper half contains considerable platy limestone and some argillaceous, nodular limestone similar to that of the Sekwi Formation in Selwyn Basin.

The remarkably pure limestone within the Illtyd Formation resembles Waucoban strata on Cassiar Platform, and is rare among equivalent carbonates in the middle carbonate belt that borders the eastern edge of the Canadian Cordillera. The purity of the Illtyd carbonate is attributed to the intervening role of Richardson Trough, which prevented clastics from the Mackenzie Platform to the east from reaching Yukon Platform.

The top of the Illtyd Formation is near the Lower-Middle Cambrian boundary, but is slightly diachronous. Near the southeastern margin of Yukon Platform (loc. 7) the boundary is at least 47 m above the Illtyd, and near the northern end of Richardson Trough (loc. 4) it is at least 160 m below the top of the formation.

On the southwestern margin of Yukon Platform near the International Boundary (loc. 10), strata equivalent to the Illtyd Formation are represented within a white, thick-bedded carbonate succession assigned to the Jones Ridge Formation of Early Cambrian through Early Ordovician age (Brabb, 1967). In the Yukon, the Jones Ridge is underlain by the same unconformity that underlies the Illtyd Formation to the east (Norris, 1982b; R.I. Thompson, pers. comm., 1984). On the Alaska side, the unconformity probably extends westward below the Jones Ridge and below its westward directed tongue, the Funnel Creek Formation (see also Young, 1982). Early Cambrian chancelloriid fossils have been reported (Payne and Allison, 1981; Allison et al., 1981) from below the suspected unconformity at a locality that is isolated from the Lower Cambrian carbonates; the true stratigraphic position of these fossils relative to the base of the Cambrian carbonates therefore remains unknown. An absence of trace fossils below the sub-Jones Ridge unconformity suggests Placentian and uppermost Proterozoic strata are missing.

Selwyn Basin

Waucoban strata in the middle carbonate belt are best known along the northeastern margin of Selwyn Basin where they have been assigned to the Sekwi Formation. The formation, which averages 700 m in thickness, comprises lime mudstone and wackestone in thin, wavy to nodular beds, thick-bedded dolostone, and limy siltstone and shale. Inner detrital variegated siltstone and sandstone that are lateral equivalents of Sekwi carbonate also have been assigned to the Sekwi. Most of these strata were removed by early and middle Paleozoic erosion. The outer slope and basinal siltstone and shale equivalent of the Sekwi have been assigned to the Gull Lake Formation.

In Figures 7.3 and 7.8 Sekwi and underlying strata, have been grouped into grand cycles. Half-cycle A1 contains the Vampire Formation, the laterally equivalent upper part of the Backbone Ranges Formation (post-Risky part) and the lower part (averaging 175 m thick) of the Sekwi Formation. Sekwi strata in this half-cycle comprise thin-bedded siltstone, shale and limestone with a basal several metres of rose to light orange-weathering limy siltstone containing *Parafallotaspis grata*, a guide fossil for the uppermost *Fallotaspis* Zone. Overlying A1 strata extend well into the *Nevadella* Zone. Half-cycle A1, therefore, ranges from uppermost Precambrian into the Waucoban *Nevadella* Zone.

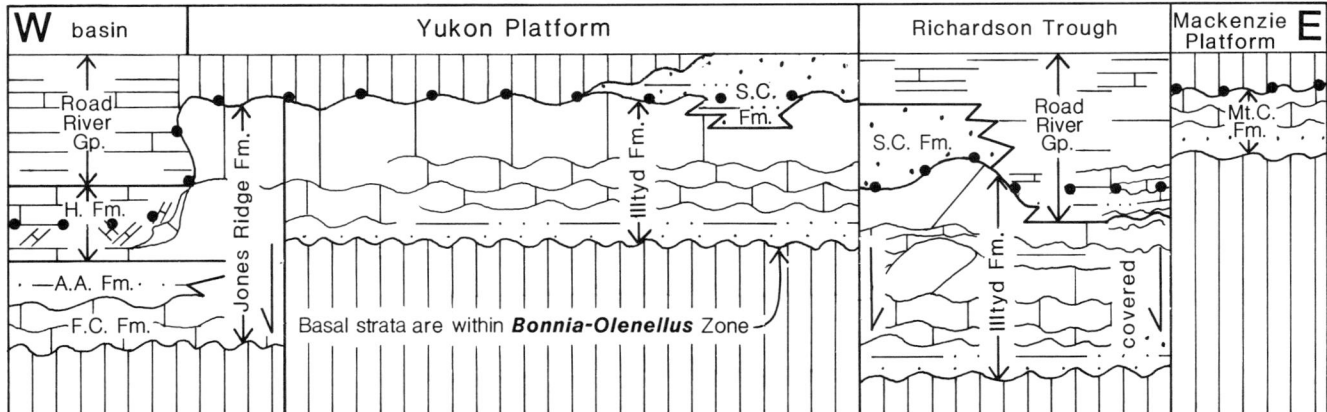

Figure 7.7. Schematic cross-section showing distribution of Waucoban Series (between lower hiatus and solid circles) in Richardson Trough and on Yukon Platform and adjacent areas. Formation name abbreviations: Hillard (H.), Adams Argillite (A.A.), Funnel Creek (F.C.), Slats Creek (S.C.), and Mount Cap (Mt. C.). Figure modified from Fritz (1974). Legend is in Figure 7.3.

CAMBRIAN TO MIDDLE DEVONIAN ASSEMBLAGES

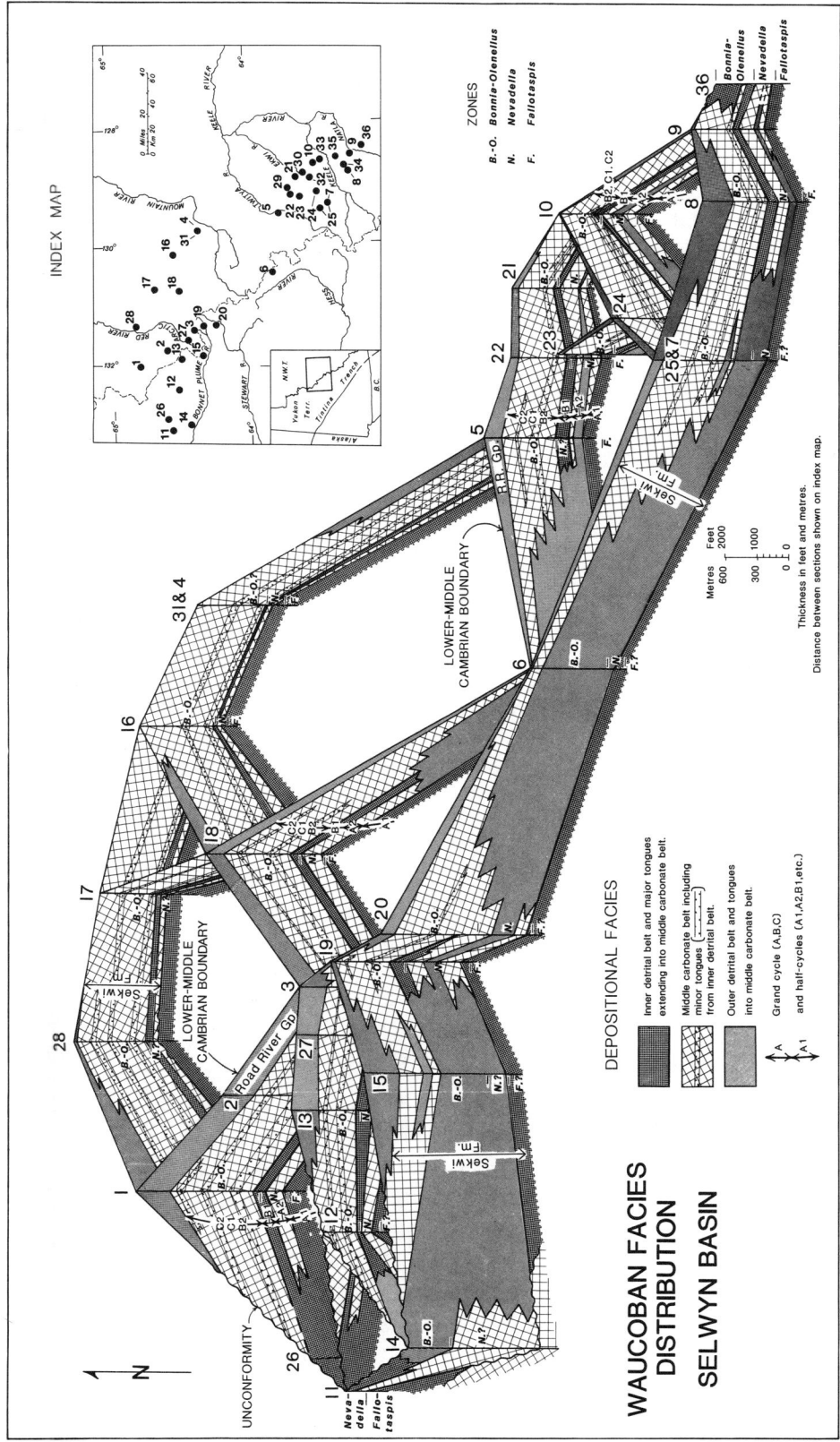

Figure 7.8. Fence diagram showing depositional facies and grand cycles within Sekwi Formation, northeastern part of Selwyn Basin. Figure modified after Figure 4 in Fritz (1979b). Locations of sections in this figure are identified in the inset and do not coincide with localities in Figure 7.6.

Half-cycle A2 is entirely within the *Nevadella* Zone and averages 100 m in thickness. It consists mainly of thick-bedded, cream-coloured, fine- to coarsely crystalline platform dolostone. Minor "floating" fine- to grit-sized quartz grains are arranged in laminae that parallel parting surfaces and outline crossbeds. Quartzite-filled caves occur near the top of the half-cycle at locality 24. Basinwards from the carbonate platform are upper slope slump deposits comprising blocks of light-coloured carbonate intermixed with broken plates of dark, thin-bedded limestone. On the craton-facing side of the platform is thin, wavy bedded to nodular limestone interbedded with siltstone.

The overlying half-cycle B1 averages 100 m in thickness and comprises light orange, maroon, and light greenish grey weathering siltstone and shale, together with some interbedded carbonate and quartzite. Near the middle of the half-cycle are several metres of quartzite with *Skolithos* burrows. This "pipe rock" constitutes one of the best regional markers within the Sekwi Formation. The boundary between the *Nevadella* and *Bonnia-Olenellus* zones is near or within the "pipe rock".

The remaining weakly expressed half-cycles (B2, C1, C2) are within the Sekwi Formation (Fig. 7.9) and have an aggregate average thickness of 325 m. They comprise mainly thin, wavy bedded limestone in thick units that are separated by thick siltstone and shale. A few locally developed platforms composed of thick to thin bedded dolostone are present (i.e., Blusson, 1971, map unit 14b). Sekwi carbonate overlying half-cycle B1 extends farther basinward than does that in A2. At the transition from typical Sekwi platform carbonate to slope deposits, both above and below half-cycle B1, there is a marked increase of otherwise very sparse carbonate buildups with archaeocyathids.

Diachronously overlying the Sekwi Formation are black shale and dark, platy limestone of the Road River Group. At the basin-facing margin of the middle carbonate belt the contact between the two formations is within the middle part of the *Bonnia-Olenellus* Zone, and at the craton-facing margin it is at the top of that zone. Post-Sekwi Lower Cambrian strata belong to the Lower and Middle Cambrian Hess River Formation (see Middle Cambrian).

The lower part of the Sekwi Formation remains relatively unchanged southward (Fig. 7.6, loc. 24, 27-31) within Selwyn Basin. At locality 30 half-cycle A2 is well developed, and there it hosts the main ore body of the Canada Tungsten Mine (Blusson, 1968). Farther south, at locality 31, half-cycle A2 becomes more variable in thickness and forms carbonate buildups up to 90 m thick (Handfield, 1971). From locality 24 southward, the upper part of the Sekwi Formation (above half-cycle B1) lacks the typical limestone to the north, and is composed of siltstone, bright orange-weathering dolostone and quartzite.

Numerous, small lead-zinc deposits occur in the Sekwi Formation, but thus far none has been judged large enough for exploitation (Brock, 1975).

The basinward equivalent of the Sekwi Formation is the Gull Lake Formation, which is 1050 m thick at its type section in southern Selwyn Basin (loc. 32). There the lower 665 m contain orange-brown to rust-brown weathering, slightly limy slate, siltstone, and very fine-grained sandstone. The upper 385 m contain grey-weathering, thick-bedded mudstone, part of which is limy. At the base of the Gull Lake is a discontinuous layer of limestone conglomerate that attains a maximum thickness of several tens of metres; archaeocyathids are present in the clasts.

A regional unconformity, the "sub-Franconian" unconformity of Gabrielse et al. (1973), at the base of the Upper Cambrian Rabbitkettle Formation reflects erosion that, in many areas, completely removed the Gull Lake Formation. The contact between the Gull Lake and the underlying Narchilla Formation is abrupt, but no significant hiatus is suspected.

Figure 7.9. View of Sekwi Formation and overlying lower Road River Group near locality 22 (Fig. 7.6). Interval within Sekwi Formation from base of half-cycle B1 (a) to top of formation (b) is 550 m thick. Strata above point "b" belong to the Hess River Formation. Typical Sekwi strata shown here can be compared with strata of the same age on Cassiar Platform (Fig. 7.10) for contrast between type 1 and type 2 limestone. GSC 203343-R,V

Near locality 21c approximately 500 m of green argillite, black shale, and minor platy limestone represent the Gull Lake Formation (Hofmann and Cecile, 1981).

In the area around locality 21a undifferentiated argillites are in part equivalent to the siltstones and shale of the Gull Lake Formation. The argillites weather buff, green, and pale green and are interstratified with black shale, siliceous argillite and chert. Basaltic breccias associated with Waucoban archaeocyathids interfinger with those argillites that correlate with the Gull Lake Formation.

Strata which correlate with the Gull Lake Formation are exposed near locality 33. There, lenses of archaeocyathid bearing limestone up to 70 m thick and 1.6 km long are enclosed by silvery greenish grey and brownish grey phyllite.

Kechika Basin

Waucoban quartzite of the Gog Group occur in central Kechika Basin (Fig. 7.6, loc. 36 to 44). To the east (loc. 37 to 40), fine- and medium-grained quartzite averages 150 m thick. At locality 40 the Gog is 60 m thick and *Nevadella* Zone trilobites are 16 m above the base. To the west (loc. 41, 42) the Gog is 620 m thick; *Nevadella* Zone fossils occur near the middle of the group and *Bonnia-Olenellus* Zone fossils are found near the top. There and farther west (loc. 43, 44) the Gog contains some interbedded carbonate and shale. At locality 44 uppermost *Bonnia-Olenellus* Zone fossils are present a short distance above the top of the group in a thick, unnamed carbonate unit.

Strata in the above areas, except for the uppermost carbonate unit, belong to the inner detrital belt, and demonstrate a uniform basinward increase in both thickness and shale and carbonate content. Small archaeocyathid-bearing carbonate mounds (average 10 cm thick) occur at the westernmost localities.

In southern Kechika Basin, sections at localities 45 (650 m) and 46 (615 m) contain Waucoban Gog strata resembling strata at localities 37 to 44. In the southern sections archaeocyathids are rare, and no carbonate buildups have been reported.

Farther north at localities 35 and 36, large archaeocyathid buildups are present. These localities are close to the northern part of Kechika Basin, which is mainly unstudied. In northern Kechika Basin (loc. 34), a large Waucoban bioherm is surrounded by dark clastic strata that overlies basinal strata of the Narchilla(?) Formation, and therefore suggests deeper water conditions than do equivalent strata farther south.

Cassiar Platform

Most of the Waucoban strata on Cassiar Platform are contained within the Rosella Formation (average thickness 760 m). The Rosella, like the Sekwi Formation, includes a basal, thin, orange-weathering or rose-weathering silty limestone or limy quartz sandstone that lies just above (Fig. 7.6, loc. 49) or contains (loc. 50) the boundary between the *Fallotaspis* and *Nevadella* zones. The same grand cycles recognized within the Sekwi are found in the Rosella.

Typically, the lowermost Rosella comprises, in ascending order, limy siltstone and interbedded limestone (upper part of half-cycle A1, 100 m), light grey to white, thick-bedded limestone with archaeocyathids (half-cycle A2, 100 m), olive green to light brown shale and siltstone with some interbedded limestone (half-cycle B1, 70 m), thick-bedded, light grey limestone with archaeocyathids (half-cycle B2, 150 m), light grey to light brown shale and siltstone (half-cycle C1, 90 m), and thin- to thick-bedded, light grey limestone (half-cycle C2?, 250 m). The uppermost unit within the Rosella Formation (Fig. 7.10) has not been dated by fossils and therefore its inclusion with the Lower Cambrian grand cycles is not fully substantiated. However, the presence of *Proliostracus* within underlying half-cycle C1 indicates that the strata immediately below C2 are older than latest Early Cambrian.

On southern Cassiar Platform (loc. 54) the depositional contact between the Rosella Formation and the overlying Kechika Group is sharp, with light-coloured, thick-bedded Rosella carbonate beneath light brown Kechika shale. The regional nature of this contact is not known, because it is nearly everywhere faulted. A Late Cambrian fossil collection from stratigraphically low in the Kechika Group near locality 54 provides a minimum age for the uppermost Rosella Formation (Gabrielse, 1963a, 1969).

Near the northern end of Cassiar Platform (loc. 47), only the lower half of the Rosella Formation is preserved. There the upper part of half-cycle A1 is 150 to 165 m thick, half-cycle A2 120 to 180 m thick, and half-cycle B1 40 to 90 m thick (Read, 1980). Disconformably overlying half-cycle B1 is calcareous argillite and thin-bedded argillaceous limestone assigned to the Upper Cambrian Series.

Within half-cycle A2 are large carbonate buildups consisting of archaeocyathid-*Renalcis*? boundstone with a carbonate mudstone matrix. Strata in half-cycles A2 and B2 to the south have the same clotted appearance, but are more stratified (reworked?) and might be better classified as floatstone or rudstone.

Figure 7.10. View of white Waucoban limestone (half-cycles B2, C1, C2?) of Rosella Formation at locality 55 (Fig. 7.6) on Cassiar Platform. Half-cycle C1 is at "a". Figure from Fritz, 1978b. GSC 203167-F

Many small, epigenetic base-metal deposits, locally associated with tungsten, occur in the Rosella carbonate throughout the Cassiar Platform.

Robson Basin

In the northern part of Robson Basin, the base of the Waucoban coincides with the base of the Mural Formation in the Gog Group. Fossils belonging to the lower part of the *Nevadella* Zone lie near the base of the Mural, but no diagnostic *Fallotaspis* Zone fossils have been found in the underlying McNaughton Formation.

Nearly everywhere the top of the Gog Group coincides with a regional disconformity separating Lower from Middle Cambrian strata. However, at locality 72 (Fig. 7.6) Early Cambrian fossils extend at least 6.5 m above the Gog into overlying siltstone of the Mount Whyte Formation.

Waucoban strata in the northern part of Robson Basin (loc. 57 to 59) average 380 m in thickness and comprise mainly thick-bedded quartzite. Half-cycles A2, B1, and B2 are not developed, and are expressed as an undivided, thin succession of dolomitic sandstone. Carbonate half-cycle C2 has either been removed by erosion or is laterally replaced by inner detrital quartzite.

Farther to the southeast (loc. 60 to 67), the Waucoban averages 800 m in thickness. A typical section, 8 km west of Mount Robson (loc. 65), includes half-cycle A2 (145 m), comprising light grey, thin- to thick-bedded limestone with archaeocyathids near the top (lower member of Mural Formation); half-cycle B1 (62 m), containing dark grey shale (62.5 m) and the boundary between the *Nevadella* and *Bonnia-Olenellus* zones (middle member of Mural Formation); half-cycle B2 (62.5 m), comprising buff to light pink-weathering, thick-bedded dolostone (upper member of Mural Formation); and half cycle C1 (410 m), containing fine- and medium-grained, thin- to medium-bedded light grey-weathering quartzite (Mahto Formation). At the top of the Lower Cambrian, half-cycle C2 (113 m), is an uncommon facies comprising dark grey lime mudstone in thin, wavy beds (Hota Formation). Elsewhere in the region this interval contains a considerable amount of bioclastic material and is commonly dolomitized.

Still farther southeast, at locality 71, the Waucoban is about 320 m thick. There the Mural Formation (half-cycle B2 only?) is 22 m thick, the Mahto is 214 m thick (Aitken, 1969), and the Peyto (=Hota) is 82 m thick.

In the southern Rocky Mountains at localities 73 and 75, the Gog Group (entire Lower Cambrian) is 520 m and 375 m thick, respectively. At both localities the succession lacks carbonate except for 8.3 m of limestone (Peyto Formation, half-cycle C2) at locality 73.

Near the southern end of the Robson Basin, near locality 76, the Lower Cambrian comprises 245 m of quartzite (Cranbrook Formation) and an overlying 235 to 460 m of interbedded quartzite, shale and minor limestone (Eager Formation). The variable thickness of the interbedded succession is attributed to erosion predating deposition of the Upper Cambrian Jubilee Formation. At locality 77 (Schofield, 1922) 370 m of white quartzite is assigned to the Cranbrook Formation (=Gog Group), and 145 m of interbedded sandstone, shale, and some limestone to the Eager Formation. Medial *Bonnia-Olenellus* Zone fossils, such as *Wanneria* and *Salterella*, have been reported from the Eager. Leech (1954) found *Nevadella* Zone fossils close to Schofield's section and only 30 m below the top of the Cranbrook. Thus all of the Eager and at least the upper part of the Cranbrook are late Early Cambrian in age. No attempt is made to assign these strata to grand cycles.

At locality 78 the Cranbrook contains a dolostone unit that has been correlated with a magnesite unit west of the Southern Rocky Mountain Trench (loc. 87) (Rice, 1937).

Columbia Basin

In Cariboo Mountains, at the northwest end of Columbia Basin, strata assigned to the Waucoban belong to the Mural Formation (150-730 m) and the lower part of the overlying Dome Creek Formation. At locality 79 (Fig. 7.6), the Mural (203 m), as in the Mount Robson area, contains half-cycles A2 (59 m), B1 (91 m) and B2 (53 m); the boundary between the *Nevadella* and *Bonnia-Olenellus* zones is within half-cycle B1. In Cariboo Mountains both half-cycles A2 and B2 contain more archaeocyathids and bioclastic material than in the Mount Robson area.

Above the Mural Formation at locality 79 is 73 m of dark grey, "papery" shale belonging to the lowest member within the Dome Creek Formation. This lithology, plus the abundance of *Ogygopsis* and other undescribed Early Cambrian genera common to slope deposits, suggests that the Mural-Dome Creek contact represents a change from a shallow to a moderately deep water environment.

The relationship between the lowest and next overlying member of the Dome Creek Formation is speculative, because the second member is undated. It comprises shallow-water, khaki and light greenish grey, highly burrowed siltstone 98 m thick, and is overlain by strata containing Late Cambrian fossils.

South of Cariboo Mountains, the Waucoban is either missing or indistinguishable from other metasediments in the Shuswap Metamorphic Complex. Still farther south (loc. 81-86), the Waucoban is represented in part by the Badshot Formation, a distinctive carbonate unit within a mainly clastic succession. The Badshot typically comprises light to dark grey limestone in thin to massive beds and is locally altered to white marble or buff dolostone. Archaeocyathid localities to the north (loc. 81) and south (loc. 86) date the Badshot as Waucoban, and suggest a close lateral relationship with the archaeocyathid-bearing Mural Formation. A short distance south of the International Boundary, the Badshot (Old Dominion Limestone) also contains the trilobite *Wanneria* sp. (Resser, 1934; Deiss, 1940) that belongs to the medial part of the *Bonnia-Olenellus* Zone. A *Nevadella* Zone fauna is present in the underlying quartzite (Okulitch, 1951). The Badshot is everywhere deformed within the Kootenay Arc, where uncorrected thicknesses are as follows: locality 81, 320 m; locality 82, 305 m; localities 83 and 84, 30 m; locality 85, 85 m; and locality 86, 50 m. Northwest of the Kootenay Arc (80b), in the Kootenay Terrane, archaeocyathids are present in a limestone up to several hundred metres thick that is associated with greenstone and dark green calcareous chlorite schist.

Lithological correlations suggest that the base of the Waucoban lies at some undesignated horizon within the underlying Mohican Formation or laterally equivalent quartzite of the Hamill Group (Fig. 7.5). The position of the Lower-Middle Cambrian boundary is tentatively placed at the top of or not far above the Badshot Formation. The Badshot Formation hosts numerous small lead-zinc-silver deposits.

East of Kootenay Arc, the Waucoban succession is assigned either to the Hamill Group and overlying Donald Formation or to the Cranbrook and overlying Eager Formation. At locality 80b the Hamill Group (280 m) mainly comprises light-coloured, thick-bedded quartzite. The Donald Formation comprises a basal 150 m of green and grey slate, 150 m of limestone and calcareous sandstone, and at least 300 m (top not exposed) of grey calcareous slate and sandstone. Fossils from locality 80 suggest that the Hamill-Donald contact is at, or near, the boundary between the *Fallotaspis* and *Nevadella* zones. If this is so, then the base of the Waucoban lies within the Hamill Group. Within the second Donald member is a cream coloured limestone 3 to 12 m thick, containing archaeocyathids and other fossils including *Salterella* sp. This limestone might represent a thin development of the Badshot Formation and/or half-cycle B2 in the Mural Formation. *Olenellus* is reported from the uppermost Donald, and therefore all of the exposed Donald is no younger than the Lower Cambrian *Bonnia-Olenellus* Zone.

At localities 87 and 88 the light-coloured, thick-bedded quartzite of the Cranbrook Formation (180 m) is overlain by dark grey to dark brownish grey argillite of the Eager Formation (1830 m, top not exposed). The Cranbrook there is similar to that in Robson Basin, but the Eager Formation is finer grained. Fossils from the lower part of the Eager at locality 88 include rare fragments attributed to *Bonnia*, one species of *Wanneria*, and an abundance of several species of *Olenellus* (Best, 1952). This assemblage indicates the medial part of the *Bonnia-Olenellus* Zone, and therefore suggests that at least a part of the underlying Cranbrook belongs to the Waucoban Series.

Interpretation of Waucoban strata

Waucoban strata are the most widespread of the Cambrian System in the Canadian Cordillera, and they contain the best regional developed grand cycles.

Carbonates assigned to the Waucoban can be divided into two major types. Type 1 fits well into the classical concept of the middle carbonate belt, as the proportion of carbonate gradually increases away from the craton. The carbonate is a somewhat argillaceous lime mudstone, commonly in thin, nodular to wavy beds (Fig. 7.9). Carbonate buildups are sparse, but where present, are best developed at the outer (basinward) margin of the middle carbonate belt. Type 1 carbonate is best represented in the Sekwi Formation. It is also interbedded within the mainly quartzitic Gog Group, in the mixed lithologies of the Donald Formation, and in the Eager Formation of Robson Basin.

Type 2 carbonate forms pure, white, irregular beds that are locally thick to massive (Fig. 7.10). The interval embraced by the *Nevadella* to middle *Bonnia-Olenellus* zones contains archaeocyathids. Algae may have played an important role during deposition, and small to large carbonate buildups are common. Large bodies of this type of carbonate do not fit well into the classic middle carbonate belt model. Their deposition may have taken place basinward from the middle carbonate belt, separated from it by an intervening trough. Examples of type 2 carbonate are the Illtyd and Jones Ridge formations on the Yukon Platform, the large carbonate buildups surrounded by basinal shale at locality 33 in southern Selwyn Basin, and similar buildups west of the Gundahoo Fault (loc. 34) in Kechika Basin. Examples west of the Rocky Mountain and Tintina trenches are the Rosella Formation on Cassiar Platform and the Mural Formation in northern Columbia Basin. Part of the Badshot Formation farther south may be similar to type 2 carbonate, but alteration nearly everywhere has masked primary features.

Middle Cambrian Series
British-Barn Basin

No Middle Cambrian fossils have been found within British-Barn Basin, but Middle Cambrian strata may be present in a basinal facies.

Yukon Platform and Richardson Trough

On Yukon Platform the Middle Cambrian is represented by the Slats Creek Formation which overlies Middle Proterozoic to Lower Cambrian strata. Block faulting on the southeastern part of the platform played an important role during Slats Creek deposition, resulting in both marine and alluvial deposits. At its type section (Fig. 7.11, loc. 4), the Slats Creek contains a lower siltstone member (800 m) with fossils ranging in age from latest Early Cambrian into the Middle Cambrian *Glossopleura* Zone. The upper member (770 m) comprises siltstone, massive conglomerate beds, and brick-red sandstone. *Glossopleura* occurs a short distance above conglomerates at the base of the member, but no fossils have been found at a higher stratigraphic level. At nearby locality 5 the Slats Creek (780 m) comprises siltstone, thick-bedded dolostone and rare brick-red sandstone. There the formation is entirely of early Middle Cambrian age, with *Plagiura-Poliella*(?) Zone fossils at the base and *Glossopleura* Zone fossils at the top. Farther southwest at locality 6, the Slats Creek comprises brick-red sandstone and abundant, massive beds of unsorted conglomerate 605 m thick. Greenstone sills and interbedded volcanics in the Slats Creek Formation indicate Middle Cambrian igneous activity (Green, 1972).

Above the Slats Creek Formation at locality 4 is a sub-*Cedaria-Crepicephalus* Zone disconformity, and above the formation at locality 3 is a sub-*Elvinia* Zone disconformity. It is not known whether these disconformities represent erosion of younger and/or more laterally extensive Middle Cambrian strata or if Middle Cambrian deposition on Yukon Platform terminated at the end of early Middle Cambrian time.

At the southwestern margin of Yukon Platform (loc. 7), no Middle Cambrian fossils are known from the fossiliferous Lower Cambrian to Middle Ordovician Jones Ridge Formation (Brabb, 1967; Palmer, 1968). Nearby slope deposits in easternmost Alaska (Hillard Formation) have yielded fossils representing only the uppermost part of the Middle Cambrian.

Within Richardson Trough (loc. 1) Middle Cambrian, medium-bedded, light brown Slats Creek sandstone is over 715 m thick, and is probably the product of block faulting on Yukon Platform. This sandstone locally replaces dark

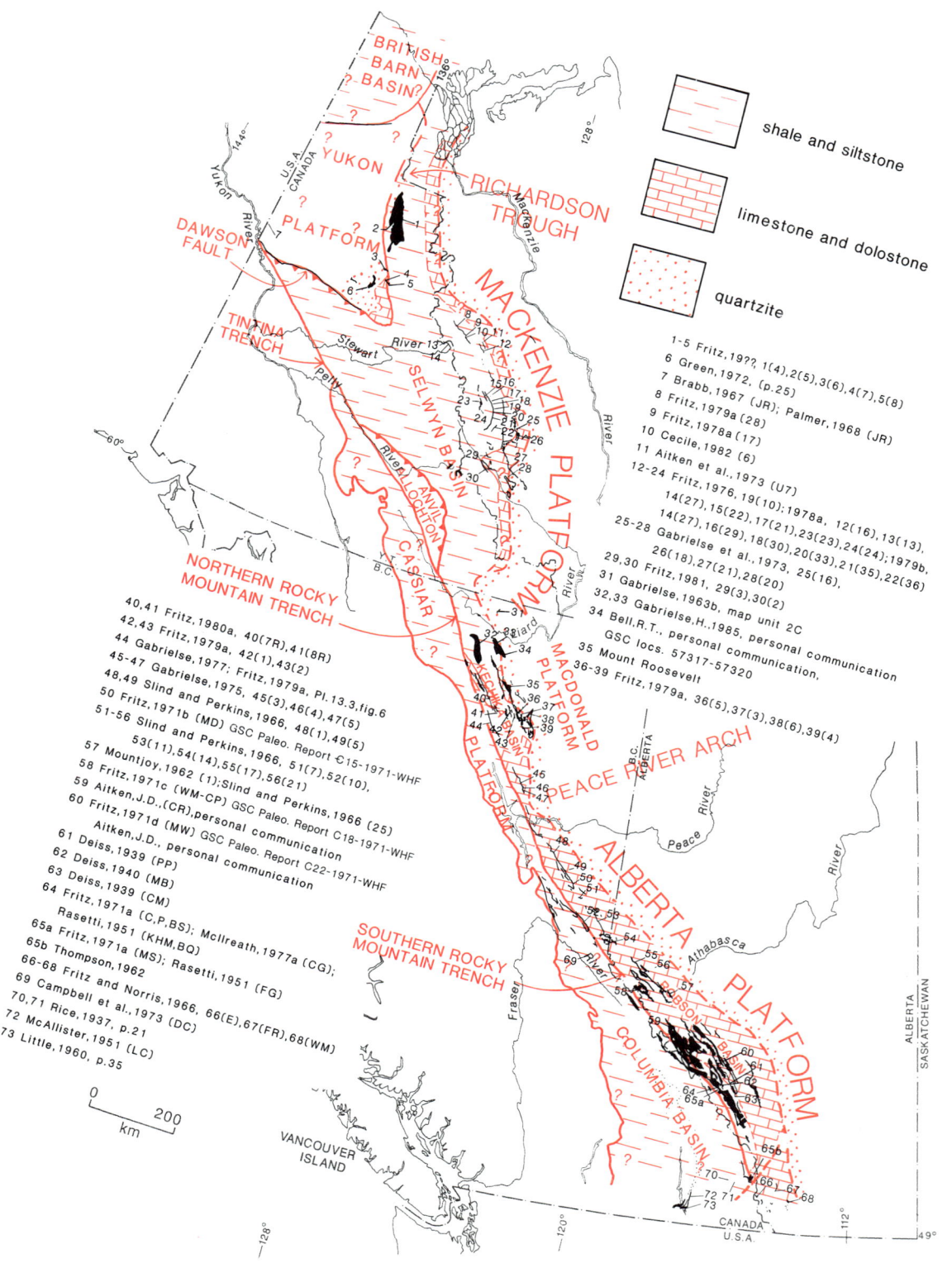

Figure 7.11. Distribution of Middle Cambrian outcrops, localities and facies. See Figure 7.2 for note on references.

shale and platy limestone in the lower part of the Road River Formation (Fig. 7.7). Approximately 150 m of Middle Cambrian(?), black Road River shale with *Protospongia* sp. spicules overlie the Slats Creek at that locality. At locality 2 lower Slats Creek sandstone interfingers with dark Road River shale that contains *Protospongia*. No diagnostic fossils are known from the Slats Creek sandstone within Richardson Trough, but at locality 2 latest Early Cambrian fossils are present in the underlying Illtyd Formation, and early Late Cambrian fossils are present 150 m above the sandstone at locality 1 (Fritz, 1985).

Two localities in Selwyn Basin (Fritz, 1976; Cecile, 1982) contain local, thick successions of Middle Cambrian? sandstone similar to that in the Richardson Trough; this sandstone may likewise have been derived from block faults on Yukon Platform.

In summary, only the lower part of the Middle Cambrian succession has been recognized on Yukon Platform. Clastics were shed into Richardson Trough to the east and possibly into Selwyn Basin to the southeast, but have not been recorded to the west. Late Middle Cambrian outer detrital strata west of Yukon Platform and suspected upper Middle Cambrian strata to the east lack these clastics, and it seems likely therefore that differential movement on the platform had ceased by middle Middle Cambrian (*Bathyuriscus-Elrathina* Zone) time.

Selwyn Basin

In northeasternmost Selwyn Basin, the Middle Cambrian, together with uppermost Lower Cambrian and lowermost Upper Cambrian strata, are represented by the Hess River Formation (Fig. 7.12). There the Hess River occurs within the upper Lower Cambrian to Devonian Road River Group, and comprises dark grey to black shale and dark grey, platy limestone. The Hess River has a diachronous contact with the underlying Sekwi Formation, and in areas where deposition was rapid, the Lower and Middle Cambrian part of the formation locally exceeds 2500 m in thickness (Cecile, 1982). Along the northeastern margin of Selwyn Basin, the thickness of Lower and Middle Cambrian within the Hess River is highly variable. In both marginal and more basinal areas, it is difficult to subdivide late Lower, Middle and Upper Cambrian strata without the aid of fossils. At basinward localities 13 and 14 (Fig. 7.11), only a few metres of Middle Cambrian beds are present. In southern and western Selwyn Basin, the Middle Cambrian is mainly absent below the Upper Cambrian to Lower Ordovician Rabbitkettle Formation and Kechika Group (Fig. 7.12; see also Fritz, 1985). The Rabbitkettle, like the Kechika, contains abundant medium grey siltstone and lacks black shale and basinal faunas.

In east-central Selwyn Basin, a Middle Cambrian tongue of the Road River Group, the Rockslide Formation, locally underlies the Rabbitkettle Formation (Fig. 7.12), and the Rabbitkettle's shallow-water equivalent, the Broken Skull Formation. The Rockslide, consisting of platy dark limestone and shale, changes laterally into shallow-water dolostone of the Avalanche Formation. As a consequence of pre-Rabbitkettle erosion, both the Rockslide and Avalanche are preserved only locally.

The Middle-Upper Cambrian boundary in northeastern Selwyn Basin was reported by Cecile (1982) to be near the contact between the Hess River Formation and an overlying succession of dark grey, platy limestone and subordinate dark grey to black grey shale. Cecile placed the overlying succession in the Rabbitkettle Formation, but in the Cambrian section of this chapter it will be referred to as the "Rabbitkettle Formation" to reflect its different lithology from that of the type Rabbitkettle to the south.

In the northeasternmost Selwyn Basin, the Hess River Formation passes eastward from a sub-"Rabbitkettle" position to a sub-Franklin Mountain Formation (craton carbonate) position, just as the Rockslide Formation to the south passes from a sub-Rabbitkettle position to a sub-Broken Skull Formation position. In the following paragraphs Middle Cambrian lithologies are traced along the northeastern and east-central margin of Selwyn Basin from the type section of the Hess River Formation (Fig. 7.11, loc. 10) southward to the type section of the Rockslide Formation (loc. 25).

At its type section, the Hess River Formation comprises 60 m of Lower(?) Cambrian black shale overlain by 360 m of dark grey, argillaceous, platy limestone and interbedded dark grey siltstone. A major part of the unit is within the Upper Cambrian, as *Cedaria* sp. and

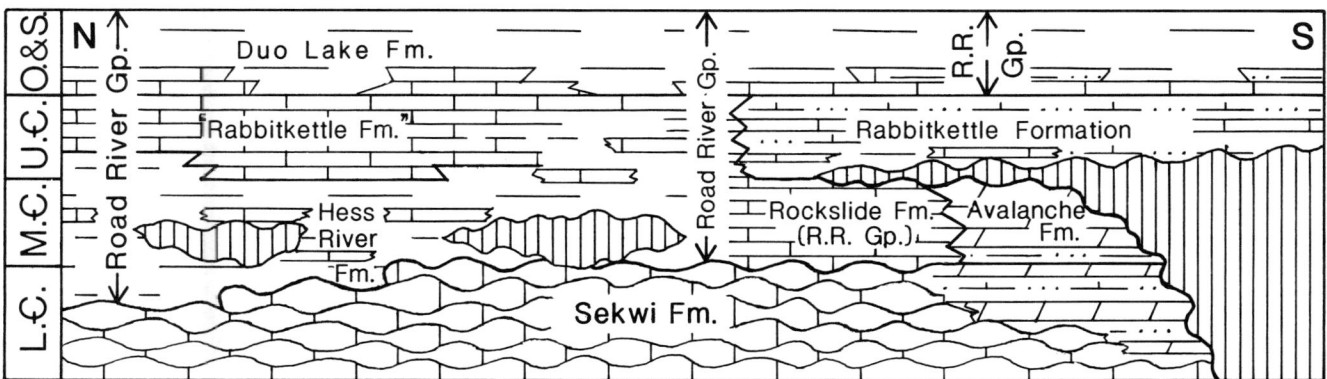

Figure 7.12. Schematic cross-section showing lateral relationships between Waucoban through Upper Cambrian units on the east side of Selwyn Basin. Platform carbonates, except for Avalanche Formation, are not shown. Legend accompanies Figure 7.3.

Glyptagnostus stolidotus? are found 140 to 150 m above the unit's base at adjacent locality 9.

Northwest of the type section of the Hess River Formation (loc. 8), the formation contains 48 m of black shale, 240 m of dark grey, platy limestone, and 110 m of light brown shale. No fossils were found in the two lower units, but the upper brown shale contains *Bathyuriscidella* sp., *Brassicicephalus* sp., *Cedaria* sp., *Orria* sp., and *Lejopyge calva* which indicate a position near the Middle-Upper Cambrian boundary.

At locality 11, east of the type section of the Hess River, the formation underlies the Franklin Mountain Formation and comprises 365 m of dark grey to black shale and interbedded, platy limestone. Fossils from the uppermost 103 m of the Hess River belong to the *Bathyuriscus-Elrathina* Zone. At locality 12 the Hess River (120 m) is also below the Franklin Mountain and has the same lithology as at locality 11. It contains earliest Middle Cambrian fossils (*Plagiura-Poliella* Zone) 10 m above its base and *Bathyuriscus-Elrathina* Zone fossils in the interval 20 to 85 m above the base. Twenty-two kilometres southeast of locality 12, the Hess River and "Rabbitkettle" are missing beneath a sub-Lower Ordovician unconformity.

Near localities 15 and 24 (Fig. 7.11) strata equivalent to the Hess River and "Rabbitkettle" formations cannot be differentiated, which led Blusson (1971) to place them in a single unit (unit 16) now recognized as the Road River Formation. At locality 15, the lowest 200 m of the Road River comprises black shale with the Early Cambrian trilobite *Goldfieldia* sp. at its base. Within the overlying dark shale and platy limestone unit, approximately 420 m belongs to the Middle and Upper Cambrian, with *Ptychagnostus richmondensis* of the *Bolaspidella* Zone at the base and *Cedaria* and *Blountia* 140 m higher. The latter collection suggests that the Middle-Upper Cambrian boundary is below the 140 m level. At locality 24, dark shale and siltstone predominate with Lower Cambrian fossils 15 m above the base and highest Middle(?) Cambrian fossils 650 m above the base.

At locality 16, platy limestone rather than shale predominates throughout the lower part of the Road River Formation. There the lower 30 m is questionably Lower Cambrian, the overlying 120 m contains fossils representing most of the Middle Cambrian zones; no lithological change takes place at the Middle-Upper Cambrian boundary.

From locality 17 southwards along an elongate outcrop, (loc. 17-21) Blusson (1971) mapped a Middle Cambrian unit (map unit 15) and an Upper Cambrian unit (map unit 17) that somewhat resemble the Hess River and overlying "Rabbitkettle" formations at locality 10. Sections along this belt indicate that the post-Sekwi Lower Cambrian is either very thin or absent, and that the Middle Cambrian, if present, is thin and represented only by the *Bolaspidella* Zone. The thickest Middle Cambrian succession is at locality 21 (140 m) and 110 m occurs at nearby locality 22.

East of localities 17 to 21, the type section of the Rockslide Formation (Fig. 7.11, loc. 25) contains only Middle Cambrian strata, and its thickness exceeds 500 m. There the Rockslide is disconformably underlain by the Backbone Ranges Formation, and conformably overlain by the Broken Skull Formation. The lower half of the type Rockslide comprises varicoloured calcareous siltstone and silty limestone. At that locality the *Plagiura-Poliella*, *Albertella* and *Glossopleura* zones are well represented by fossils. The upper half of the type section comprises dark grey, thin-bedded limestone containing *Bathyuriscus-Elrathina* Zone fossils in the lower part. The upper part remains undated, but is probably no younger than the *Bolaspidella* Zone.

At basinward locality 29, the Rockslide Formation (430 m) represents the entire Middle Cambrian Series, and conformably overlies the Sekwi Formation. Dark grey, platy limestone predominates throughout the Rockslide, but a subordinate amount of interbedded black shale with locally abundant sponge spicules also is present. Fossils in the basal 40 m correlate with those of the *Plagiura-Poliella*, *Albertella*, and *Glossopleura* zones. The *Bathyuriscus-Elrathina* Zone is poorly documented, but may occupy 100 m of section, whereas the *Bolaspidella* Zone is at least 220 m thick, and may be as thick as 290 m. *Lejopyge calva* first appears 90 m below the top of the Rockslide, and sparse, thin quartzite interbeds occur within the uppermost 60 m. These thin beds herald a basal 40 m-thick quartzite unit in the overlying Rabbitkettle Formation. As *Baltagnostus* sp., *Elrathina* sp., *Hypagnostus* and *Ptychagnostus aduleatus*? of latest Middle Cambrian age are 7 m below the 40 m-thick quartzite unit, and *Agnostus* sp., *Blountia* sp., *Cedaria* sp., *Kormagnostus* sp. and *Welleraspis* of early Late Cambrian age are 4 m below the top of the quartzite unit, the sub-Rabbitkettle unconformity is less important at locality 29 than to the south.

South of localities 25 and 29 (Fig. 7.11), the Rockslide Formation changes laterally to dolostone (loc. 27, 28, 30) of the Avalanche Formation (Fig. 7.12). The type section of the Avalanche (loc. 28, 400 m) is located near the boundary between the middle carbonate and inner detrital belts, and comprises buff weathering thin- to thick-bedded dolostone interbedded with an equal amount of greenish grey siltstone. At locality 30 the Avalanche is at the outer edge of the middle carbonate belt and comprises 325 m of thick-bedded to massive, dull grey-weathering dolostone. There it overlies 100 m of Rockslide slope deposits in a relationship that is consistent with the distribution of Avalanche platform carbonate prograding over Rockslide slope strata. The Avalanche is commonly unfossiliferous except for *Plagiura-Poliella* and *Albertella* Zone fossils in the lower, more argillaceous beds.

Middle Cambrian strata are known to extend only 45 km south of locality 28, and farther south in Selwyn Basin they are absent below the regional sub-Rabbitkettle unconformity.

To summarize, it is difficult to separate Middle Cambrian strata from Lower and Upper Cambrian strata within the Road River Formation. Local erosion and redeposition of slope and basinal strata may account for the rapid lateral changes within the Middle Cambrian, but this remains to be tested by detailed studies. Basinwards from the localities mentioned above, the Middle Cambrian probably thins. The absence of Middle Cambrian in southern and western Selwyn Basin is attributed to pre-Rabbitkettle and pre-Kechika erosion.

Kechika Basin

Within Kechika Basin the Middle Cambrian exhibits extreme facies changes attributed to syndepositional block faulting. Near Mount Roosevelt (Fig. 7.11, loc. 35) the "Roosevelt facies" comprises more than 1525 m of fanglomerate and redbeds with silt- to boulder-sized clasts (Taylor and Stott, 1973; Gabrielse and Taylor, 1982). Strata in this part of the graben are barren of fossils, but at the south end, at localities 38 (850 m+) and 36 (760 m+), the early Middle Cambrian *Plagiura-Poliella* Zone genus *Fieldaspis* is present in the lower part of the succession, the lowest occurrence being 150 m above the base (Fritz, 1979a).

South of "Roosevelt facies", in the same or a related graben, is turbidite comprising dark grey siltstone interbedded with medium-grey quartzite (see Fig. 7.17, unit A). At locality 39 Middle Cambrian strata (900 m+) are unfossiliferous, but at locality 37 (400 m) *Elrathina* sp., *Ptychagnostus punctuosus* and *Zacanthoides* sp. 120 m above the base equate with the *Bathyuriscus-Elrathina* Zone. The overlying 280 m belongs to the *Bolaspidella* Zone, with *Lejopyge laevigata*, *Bathyuriscidella* sp. and *Metisella*? near the top. The quartzite-bearing turbidite continues above the Middle-Upper Cambrian boundary into the *Dunderbergia* Zone, and therefore block faulting is believed to have continued into Late Cambrian time.

West of "Roosevelt facies", and west of the adjacent turbidite facies to the south, the Middle Cambrian Series at localities 43 (100 m) and 42 (40 m) is in a condensed, basinal facies comprising black shale and platy limestone (see Fig. 7.17, unit B). At both localities the equivalent of the *Albertella* Zone is suggested by the presence of *Yohoaspis*? near the base, and fossils equivalent to the *Bolaspidella* Zone are present near the top. Farther basinwards (loc. 40, 41), 200 m of undated, light grey carbonate may be Middle Cambrian in age (see Fig. 7.17, part of unit C). In similar carbonate below and above the 200 m interval are late Early and early Late Cambrian fossils, respectively. Still farther basinwards (loc. 44) is a northwesterly aligned row, 60 km long, of white carbonate buildups several of which have yielded Middle Cambrian fossils belonging to the *Bathyuriscus-Elrathina* Zone (Fig. 7.13). The buildups are many tens of metres thick and form peaks separated by undated dark grey shale and siltstone.

In southeastern Kechika Basin (Fig. 7.11, loc. 45-47) the Middle Cambrian is within the northernmost part of the well developed middle carbonate belt which in Robson Basin forms spectacular mountains in Jasper, Banff, and Yoho National parks. As in Robson Basin, Kechika Basin carbonate comprises mainly medium- to thick-bedded light grey, buff and medium brownish grey weathering dolostone. The total succession is thinner (loc. 45, 350 m; loc. 46, 520 m; loc. 47, 250 m) than to the southeast, because of both primary thinning and pre-Late Cambrian erosion. Below the carbonate is approximately 60 m of interbedded siltstone, quartzite, and carbonate belonging to the *Plagiura-Poliella* Zone and part of the *Albertella* Zone.

The least known Middle Cambrian strata in Kechika Basin are in the northern part. Extensive outcrops of dark brownish grey weathering siltstone, shale and conglomerate at localities 32 and 33, and a smaller exposure of

Figure 7.13. One of several aligned, discrete carbonate buildups in Middle Cambrian strata of Kechika Basin (Fig. 7.11, loc. 44). White limestone belongs to *Bathyuriscus-Elrathina* Zone. Average dip is nearly vertical. Photo by H. Gabrielse. GSC 205235-H

conglomerate at locality 31 may belong to the Middle Cambrian Series (H. Gabrielse, pers. comm., 1986).

Cassiar Platform

No Middle Cambrian strata have been identified on Cassiar Platform. Strata at the top of the Rosella Formation have been nearly everywhere faulted against the overlying Kechika Group. Near the northwestern end of Cassiar Platform (Fig. 7.2, loc. 22), a disconformity separates medial(?) Upper Cambrian Kechika Group strata from underlying Rosella Formation strata that is low in the Lower Cambrian *Bonnia-Olenellus* Zone.

Present hypotheses on the Middle Cambrian are (1) that it was deposited as a condensed, basinal succession and was removed by pre-medial Late Cambrian or pre-Ordovician erosion, or (2) that a thin Middle Cambrian succession is at least locally present within the Kechika Group and has yet to be recognized.

Robson Basin

In northwestern Robson Basin, the Middle Cambrian (Fig. 7.11, loc. 48-58, 300 m to 1100 m) disconformably overlies either quartzite of the Mahto Formation or thin carbonate of the Hota Formation, both units being near the top of the Gog Group (Fig. 7.14). The lower half of the Middle Cambrian Series (loc. 48-51, average 230 m; loc. 53-57, average 610 m) is represented by the Snake Indian Formation, which at locality 53 and to the northwest, comprises yellow-weathering, sandy dolostone with some red-weathering shale. At locality 54 and to the southeast, the Snake Indian comprises greenish grey shale and siltstone in thick intervals interbedded with equal amounts of thin- to thick-bedded carbonate. At the base of the Snake Indian Formation, the *Plagiura-Poliella* Zone is mainly thin or missing, but at locality 58 it is 275 m thick. The *Albertella* and *Glossopleura* zones are present within the Snake Indian, as is the *Bathyuriscus-Elrathina* Zone, the latter being shared with the overlying Titkana Formation.

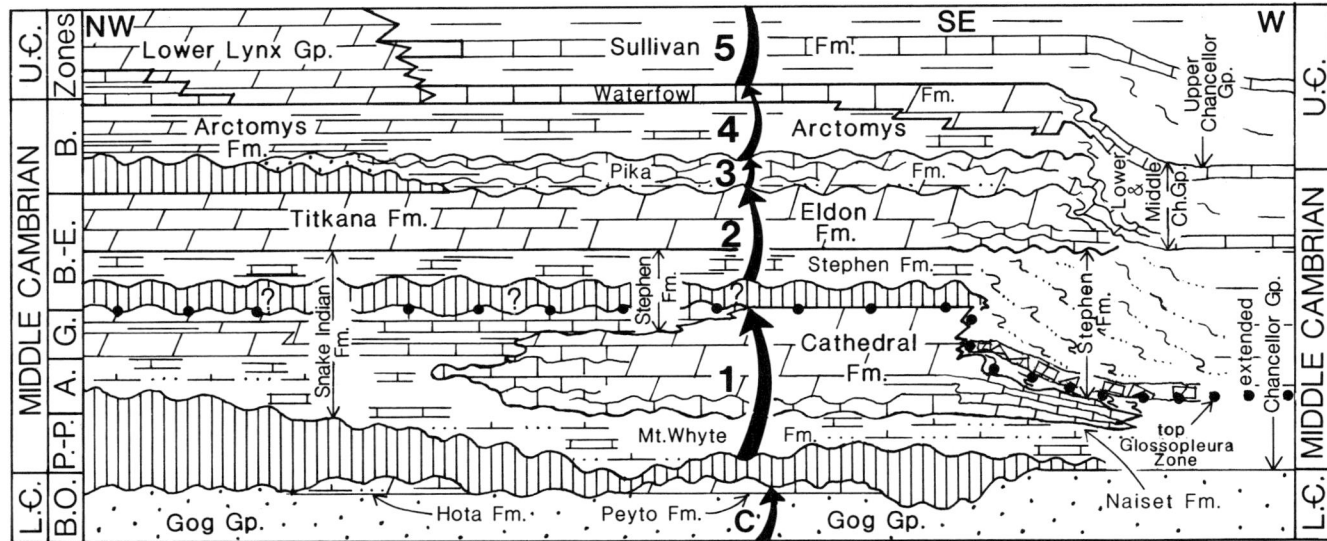

Figure 7.14. Schematic cross-section showing Middle Cambrian lateral relationships within middle carbonate belt and adjacent margin of outer detrital belt. Arrows indicate grand cycles of Fritz (1975, cycle C) and Aitken (1966b, cycles 1-5). Solid circles mark top of *Glossopleura* Zone. Legend accompanies Figure 7.3. Zones are defined in Figure 7.1.

The upper half of the Middle Cambrian is represented by the Titkana and Arctomys formations underlying the Upper Cambrian Lynx Group (Plate 11). The Titkana comprises cliff-forming dolostone and limestone in thick to thin beds, which extend southeastward to become the massive carbonate of the Eldon Formation. At locality 57 a northwestward extension of the Pika Formation (thin-bedded limestone and shale) replaces the lower Arctomys, and locally overlies the Titkana Formation (Slind and Perkins, 1966; Fig. 7.14). The Titkana ranges from 60 m thick in the northwest (loc. 48) to 300 m in the southeast (loc. 57).

The overlying Arctomys Formation is thinner than the Titkana, and comprises red and yellow shales with subordinate platy limestone and sandstone. Ripple marks, mud cracks and salt casts are common. Physical correlations from the southeast suggest that the boundary between the *Bathyuriscus-Elrathina* and *Bolaspidella* zones is near the Titkana-Arctomys contact. Similar correlations suggest the contact between the *Bolaspidella* and *Cedaria* zones may be near the Arctomys-Lynx contact. The report of *Cedaria?* sp. 30 m below the top of the Arctomys (Slind and Perkins, 1966) near section 54 suggests a younger upper Arctomys there than in the central part of Robson Basin.

The Middle Cambrian Series in the National Parks of the Canadian Rockies is described in greater detail by J.D. Aitken in Geology of Canada, no. 5 (Sedimentary Cover of the Craton in Canada), and therefore only highlights are given here. Most of the Middle Cambrian strata within the National Parks are in the middle carbonate belt, and belong to seven formations representing four grand cycles (Aitken, 1966; Fig. 7.14). The same basal disconformity noted to the northwest extends into the parks area, and the Middle-Upper Cambrian boundary lies within the Waterfowl Formation.

At the basinward edge of the middle carbonate belt, the carbonate half-cycles rapidly thicken across a narrow region known as the Kicking Horse Rim (Aitken, 1971). At the rim's western edge, the half-cycles abruptly change in facies to outer detrital belt strata comprising slope and basinal shales and platy limestone of the "thin" Cathedral Formation (see below), Stephen Formation, and Chancellor Group. Near Mount Field and Mount Stephen (Fig. 7.11, loc. 64, 65a), this change and its relationship to grand cycles can be best explained in terms of a reciprocal sedimentation model suggested by Halley (1974) to describe a similar but less spectacular change in Waucoban and lower Middle Cambrian strata in the southern Great Basin of southwestern U.S.A. During the first phase of Halley's model, bypassing or slow clastic deposition takes place in the inner detrital belt and over the area subsequently occupied by the middle carbonate belt while rapid clastic deposition takes place in the basin. During the second phase, rapid clastic deposition takes place in the inner detrital belt, rapid carbonate deposition occurs in the middle carbonate belt, and starvation takes place in the basin.

In applying the first phase of the model to strata in the Mount Field-Mount Stephen area, the Mount Whyte Formation represents the thin clastics in the area subsequently to become the middle carbonate belt, and the Naiset Formation represents the thick clastics in the basin (Fig. 7.14).

The second phase in the middle carbonate belt is represented by thick (600 m) carbonate of the Cathedral Formation, rapidly deposited during *Albertella* and *Glossopleura* Zone time. During the latter part of this phase, platy limestone and some shale belonging to the lower member of the Stephen Formation were deposited east of the rim in the middle carbonate belt. In the basin, slow deposition resulted in the accumulation of platy Cathedral limestones on the steep west-bounding slope of the rim, and, near the end of the second phase, an escarpment, 100-300 m in relief, developed at the outer edge of the rim. The end of the second phase left a unique bottom topography that strongly influenced the succeeding stratigraphic record in the Mount Field-Mount Stephen area, and that contributed to the preservation of some of the world's most famous fossils in the Burgess Shale.

Applying the reciprocal model to the next succeeding shale-carbonate pair, first-phase shale and minor bioclastic limestone belonging to the upper member of the Stephen Formation were slowly deposited in the area to again become the middle carbonate belt. These strata contain the *Ehmaniella burgessensis* fauna of the *Bathyuriscus-Elrathina* Zone. Deiss (1940) and Rasetti (1951) noted that older *Bathyuriscus-Elrathina* Zone faunas are missing below this zone, which suggests a disconformity at the contact between the upper (*E. burgessensis* fauna) and underlying (*Glossopleura* Zone) members of the Stephen Formation (see also North and Henderson, 1954). A disconformity is recognized at this level on the Kicking Horse Rim between the upper member of the Stephen Formation and the underlying massive carbonate of the Cathedral Formation (Fritz, 1971; McIlreath, 1977a).

The related first-phase sedimentation in the basin immediately west of the rim is represented by a much thicker (320 m) expression of the Stephen Formation (Fritz, 1990). Deposition began with a shale unit (45 m) belonging to the upper part of the *Glossopleura* Zone. The overlying units and their *Bathyuriscus-Elrathina* Zone faunas are as follows: (1) an apron of shallow-water carbonates comprising mainly fragments that originated from the adjacent rim (20 m; see McIlreath, 1977b), uppermost *Glossopleura* Zone and *Kootenia* sp. 1 fauna; (2) brownish grey shale of the well known "Fossil Beds" on Mount Stephen (30 m), *Ogygopsis klotzi* fauna; (3) dark grey shale (25 m), *Pagetia bootes* fauna (includes "Phyllopod Beds", 2.3 m, containing the celebrated Burgess Shale fossils (see Walcott, 1912; Whittington, 1980; etc.); (4) grey shale (200 m), the upper part of which can be traced into the upper member of the Stephen Formation overlying the rim, *Ehmaniella burgessensis* fauna.

The overlying carbonate of the Eldon Formation represents the second phase in this application of the model. The Eldon, like the Cathedral, thins rapidly basinward from the rim; this relation has not been studied in detail.

West of localities 64 and 65a (Fig. 7.11) and extending to the Southern Rocky Mountain Trench, Middle Cambrian strata reflect a basinal facies. Cook (1970, 1975) assigned all of these strata to a lower (1000 m) and middle (600 m) divisions of the Chancellor Group. The two units comprise dark grey, greenish grey and brownish grey shale and argillaceous limestone, all of which are cleaved and generally barren of fossils. *Lejopyge*? sp. at the top of the middle division supports the concept that this horizon is near the Middle-Upper Cambrian boundary, and is equivalent to the top of the Waterfowl Formation of the middle carbonate belt (Fig. 7.14). McIlreath (1977a) restricted the lower and middle Chancellor to the lateral equivalent of the Eldon through Waterfowl formations and thereby greatly reduced the age and thickness of the Chancellor. In the present interpretation (Fig. 7.14) McIlreath's two Chancellor units are accepted for the area immediately west of the rim at localities 64 and 65, but an "extended Chancellor" is recognized farther southwest, so that the base of the Chancellor there coincides with the top of the Gog Group.

In southeastern Robson Basin (loc. 65b) Middle Cambrian and lower Upper Cambrian strata have been informally assigned to the "Tanglefoot Formation" by Thompson (1962). He believed these strata (average 1200 m) represent sedimentation in a graben at the north end of Montania. Benvenuto and Price (1979) interpreted the "Tanglefoot Formation's" south-bounding fault to be a post-depositional tear fault. Because the "Tanglefoot" lithology resembles the Chancellor, it is likely that the "Tanglefoot" was deposited as part of that unit, and was later thrust to the east. The "Tanglefoot Formation" is in gradational contact with both the underlying Lower Cambrian Eager Formation and the overlying Upper Cambrian to Lower Ordovician McKay Group. Fossils collected by G.B. Leech from an undesignated horizon in the "Tanglefoot" include *Ehmaniella*, *Parkaspis*, *Protospongia* and *Ptychagnostus*. These fossils belong to the *Bathyuriscus-Elrathina* Zone and correlate with the basinal part of the Stephen Formation. The presence of the late Dresbachian-early Franconian trilobite *Pterocephalia* cf. *P. sanctisabae* (A.R. Palmer in Thompson, 1962) in the upper part of the "Tanglefoot Formation" strengthens the supposition that the resistant Upper Cambrian Jubilee Formation terminates at the north end of locality 65b by means of a lateral facies change into the McKay facies.

In southeasternmost Robson Basin (loc. 66-68) the sub-Middle Cambrian erosional surface noted to the northwest cuts deep into Middle Proterozoic strata of Montania (Norris and Price, 1966). There the overlying Flathead Formation (probably diachronous and including upper *Plagiura-Poliella* and lower *Albertella* zones) overlies a surface eroded 1500 m below the top of the Purcell (Belt) Supergroup. The Flathead comprises quartzite that is up to 45 m thick. At locality 66 (Fig. 7.11) uppermost *Plagiura-Poliella* Zone fossils are present a short distance above a gradational contact with the overlying Gordon Formation. The Gordon (45 to 90 m) comprises soft, greenish grey shale bearing an *Albertella* Zone fauna. A tongue of the Gordon that contains this fauna extends to the northwest into the Cathedral Formation where it is known as the Ross Lake Member. The uppermost Gordon and the overlying cliff-forming dolostone of the Elko Formation (0 to 155 m) belong to the *Glossopleura* Zone. A narrow interval containing siltstone separates the Elko from the overlying, generally thinner bedded, less resistant dolostone of the Windsor Formation (70 m). A regional unconformity separates Elko and Windsor strata from Middle Devonian rocks.

Columbia Basin

Little is known about Middle Cambrian strata in Columbia Basin. A possible sub-Ordovician erosional surface has been reported near the International Boundary (Park and Cannon, 1943), and sub-Devonian erosional surfaces occur in the Purcell Mountains (Root, 1983) and Cariboo Mountains (Campbell et al., 1973). In the latter area *Glossopleura* cf. *stenorhacis* in deformed beds indicates some Middle Cambrian deposition took place (Sutherland Brown, 1963). In the same area burrowed Upper? Cambrian siltstone (100 m) disconformably overlies Lower Cambrian strata indicating still another interval of erosion.

In southeastern Columbia Basin (Fig. 7.11, loc. 70, 71), a thickness of 1800 m reported for the Eager Formation suggests either tectonic thickening or that the formation may contain Middle as well as Lower Cambrian strata.

The fossiliferous Middle Cambrian Metaline Formation (900 m) immediately south of the International Boundary (Park and Cannon, 1943) comprises thick bedded to massive carbonate known in Canada as the Nelway Formation (loc. 73). Physical correlation of the Metaline with the unfossiliferous Nelway gives assurance that at least some Middle Cambrian strata extend into that part of Columbia Basin. The Metaline fossils *Elrathina*, *Kootenia*, *Ogygopsis*, *Olenoides* and *Parkaspis* correlate with the *Ogygopsis klotzi* fauna known from the Stephen Formation immediately basinward of Kicking Horse Rim (McLaughlin and Embysk, 1950). This fauna, plus the abrupt termination of the Nelway to the north against greenish grey basinal slates of the Laib Formation (mapped everywhere as a fault contact by Little, 1960), suggests that the Nelway represents the outer edge of a carbonate belt, just as do the Cathedral and Eldon formations at localities 64 and 65a in Robson Basin. The Laib may contain strata that are the Nelway's lateral equivalent and could, like the expanded interpretation of the Chancellor Group, contain the entire Middle Cambrian Series. At the only measured section of the Laib (loc. 72) the formation comprises mostly grey-green argillite and phyllite. The total thickness there, including 130 m of probable Lower Cambrian strata, is 970 m. Little (1960) stated that 1500 m is a more appropriate thickness for the Laib Formation.

An abrupt contact south of the International Boundary between the Metaline carbonate and the overlying black slate of the Ordovician Leadbetter Formation may represent an erosional surface (Park and Cannon, 1943). To the north the Active Formation (=Leadbetter Formation) is reported to be everywhere faulted against adjacent formations (Little, 1960).

Summary of Middle Cambrian deposition

The Middle Cambrian Series of the Canadian Cordillera is thickest along a zone that parallels the eastern margin of the miogeocline. Most of the strata belong to the *Bathyuriscus-Elrathina* and *Bolaspidella* zones or their equivalents. The northern half of the zone comprises slope and basinal dark shale and platy limestone of the outer detrital belt. An expected middle carbonate belt to the east was either not developed or was removed by pre-Late Cambrian erosion. An exception is in a small, irregular area underlain by the Avalanche Formation on the east-central margin of Selwyn Basin. In north and central Kechika Basin, basinal clastics grade eastward into inner detrital clastics with only a narrow intervening middle carbonate belt.

The southern half of the zone of relatively thick Middle Cambrian strata begins north of the Peace River Arch, and, from there southward mainly belongs to the middle carbonate belt. It is much thicker than the northern half. Four grand cycles are recognized plus a narrow rim of massive carbonate at the western margin. The rim juxtaposes an abrupt facies change to slope and basinal strata that disappear within a short distance into the subsurface.

In Columbia Basin Middle Cambrian strata are documented only locally. A thin unit of Middle Cambrian shale may have been deposited over the basin but, if so, was largely removed by subsequent erosion. The same history is suggested for Cassiar Platform, southern and western Selwyn Basin, and British-Barn Basin.

Middle Cambrian block faulting took place at the east-central margin of Kechika Basin and on the southeastern part of Yukon Platform. Sands from the latter area were locally trapped in Richardson Trough and Selwyn Basin(?). A lack of Middle Cambrian strata elsewhere on the platform and the limited distribution of platform sands spread elsewhere, suggests the sum of the block faulting was nearly neutral, and the platform as a whole remained close to sea-level.

Upper Cambrian Series
British-Barn Basin

Late Cambrian fossils have yet to be found in the Canadian portion of the British-Barn Basin, but strata at locality 1 (Fig. 7.15) have been correlated on the basis of lithology and stratigraphic position to a nearby unit in eastern Alaska that contains a Late Cambrian fauna (Dutro et al., 1972). On both sides of the border this unit comprises mafic volcanics and carbonates (Reiser et al., 1980), and on the Alaska side it contains both late Early and middle(?) Late Cambrian fossils. The aspect of lower Paleozoic strata in British-Barn Basin suggests that Late Cambrian deposition also was basinal.

Yukon Platform and Richardson Trough

Upper Cambrian strata on the southeastern part of Yukon Platform are contained within two unconformity-bounded sequences. The lower (locally preserved) sequence is the Taiga Formation which overlies Precambrian to Middle Cambrian strata but is only locally preserved and, in turn, is disconformably overlain by a thick, widespread carbonate succession of middle Late Cambrian through Early Devonian age. The latter succession is commonly referred to as the "unnamed carbonate formation".

At its type locality (Fig. 7.15, loc. 6) on the southeast corner of Yukon Platform, the Taiga Formation (275 m) comprises light grey weathering, thick- to thin-bedded limestone and several cyclic units of thick-bedded limestone, bright orange weathering dolostone, and light yellow-grey weathering, platy, argillaceous limestone. Strata in the cyclic units give the formation a striped appearance. *Bolaspidella?*, *Talbotina*, and *Tricrepicephalus?* near the base and *Densonella*, *Coosella* and *Meteoraspis* near the top indicate the presence of the *Cedaria* and possibly also the *Crepicephalus* zones. Farther west (loc. 8) the Taiga overlies brick-red sandstone of the Slats Creek Formation and contains a thin-bedded silty limestone member (220 m) overlain by medium- to thick-bedded, dark grey- to buff-weathering carbonate (210 m).

Strata equivalent to the Taiga Formation may be absent on the southwestern edge of Yukon Platform (loc. 9), as no fossils belonging to the *Cedaria* or *Crepicephalus* zones have been found in the Lower Cambrian through Middle Ordovician Jones Ridge Formation (Brabb, 1967; Palmer, 1968). In adjacent Alaska, a *Cedaria-Crepicephalus*

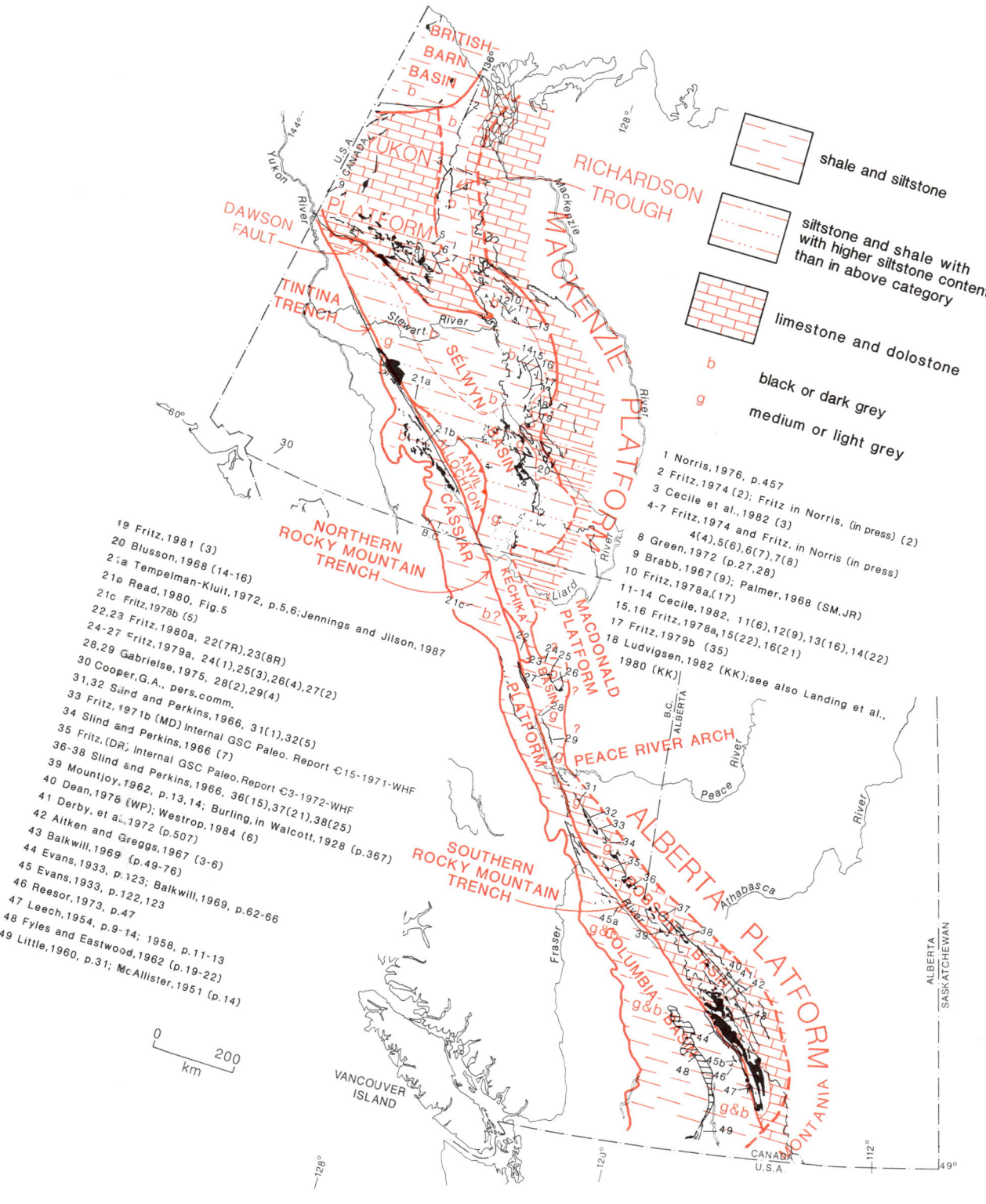

Figure 7.15. Distribution of Upper Cambrian outcrops, localities and facies. See Figure 7.2 for note on references.

Zone fauna is present within slope and basinal deposits of the upper Lower Cambrian to Lower Ordovician Hillard Formation (Palmer, 1968). The Hillard Formation is unconformably overlain by Middle Ordovician strata belonging to the Road River Formation.

At locality 5 the Taiga Formation is missing, and the unnamed carbonate formation disconformably overlies Middle Cambrian maroon conglomerate of the Slats Creek Formation. The trilobites *Buttsia* and *Iddingsia*? that belong to the *Elvinia* Zone are present 3 and 8 m above the disconformity. No fossils were found between these two horizons and Early Ordovician fossils 560 m higher in the section, and therefore the Cambrian-Ordovician boundary is not accurately located. Near localities 5 and 6 the unnamed carbonate formation is over 1860 m thick and comprises thick-bedded, white pelloidal limestone that in numerous areas has been altered to light grey dolostone (Macqueen, 1975).

In the northern Richardson Trough (loc. 2) 110 m of karst breccia is tentatively assigned to the Upper Cambrian (Fritz, 1974). The breccia directly underlies the Lower(?) Ordovician through Upper(?) Silurian Vunta Formation (Norford, 1964).

In Richardson Trough the Upper Cambrian Series is represented by a unit of platy limestone with some interbedded dark grey to black shale. The unit is part of the upper Lower Cambrian through Lower Devonian basinal succession assigned to the Road River Formation (Fritz, 1985). At locality 3 it is 2000 m thick, which is near its maximum. There the Cambrian-Ordovician boundary is close to the top of the unit, as channel samples 0 to 155 m below the top contain earliest Ordovician conodonts of the *Paltodus bassleri* group, and channel samples 155 to 255 m below the top contain the latest Cambrian conodonts *Cordylodus proavus* and *Eoconodontus notchpeakensis*.

At locality 4 a thin, faulted section, believed to represent strata from the lower part of the Road River Formation, contains the trilobites *Cedaria* cf. *prolifica*, *Crepicephalus*, *Kingstonia*, *Lonchocephalus*?, *Shickshockia*? and *Tricrepicephalus* along with the graptolites *Callograptus*, *Dendrograptus* and *Dictyonema* (Berry and Norford, 1976). The trilobites indicate the *Crepicephalus* Zone and possibly part of the *Cedaria* Zone. All of these fossils suggest that most of the unit belongs to the Upper Cambrian Series.

At the eastern edge of Richardson Trough the Road River Formation grades laterally into the Franklin Mountain Formation and at the western edge of the trough it passes into the unnamed carbonate formation. The relationship between the Taiga Formation and Richardson Trough strata has not been observed, nor has the disconformity between the Taiga and overlying unnamed carbonate been traced into Richardson Trough.

Selwyn Basin

In the northeastern Selwyn Basin, Upper Cambrian strata occur mainly in the "Rabbitkettle Formation", which interfingers to the north with the unnamed carbonate formation on Yukon Platform and to the east with dolostone of the Franklin Mountain Formation on Mackenzie Platform. The slope and basinal, dark grey platy limestone of the "Rabbitkettle" is overlain by black graptolite bearing shale and chert of the Ordovician Duo Lake Formation (Fig. 7.12).

Locality 11 (Fig. 7.15) is considered by Cecile (1982) to be a local reference section for the "Rabbitkettle Formation" (435 m). There as much as 175 m of the underlying Hess River Formation may belong to the Upper Cambrian Series. *Elvinia* Zone fossils occupy the middle of the formation and Tremadoc fossils *Caryocaris* and *Clonograptus* occur 26 m above the top. Nearby (loc. 10), the "Rabbitkettle Formation" contains *Glyptagnostus stolidotus* of the *Crepicephalus* Zone 80 m above the base, and *Crenuolimbus*?, *Dunderbergia*, *Elbergia*, and *Kindbladia* of the *Elvinia* Zone 300 m above the base. No fossils have been reported from the "Rabbitkettle" at locality 12 (785 m), but the Arenig graptolite *Glyptograptus* is present in the overlying Duo Formation 120 m above its base. At locality 13 the "Rabbitkettle" and the Franklin Mountain formations interfinger, with lower and upper Franklin Mountain tongues separated by a medial "Rabbitkettle" tongue. The lower Franklin Mountain (190 m) contains the *Cedaria* Zone. At the base of the "Rabbitkettle" (465 m thick) is the *Taenicephalus* Zone. *Eurekia* and *Hungaia*? occur in the middle and at the top, respectively. The base of the upper Franklin Mountain tongue (230 m thick) is near the Tremadoc-Arenig boundary. At the top of the upper tongue is a Middle Ordovician conodont fauna that closely underlies a disconformity between the Franklin Mountain Formation and overlying Lower Silurian strata of the Mount Kindle Formation. The interfingering relationship between the "Rabbitkettle" and Franklin Mountain formations takes place along a narrow belt, to the east of which the Franklin Mountain typically comprises light grey, brown, and orange dolostone. *Blountia*, *Coosina*?, *Kingstonia* and *Lonchocephalus* near the base of the Franklin Mountain indicate a position near the *Cedaria* Zone-*Crepicephalus* Zone boundary and *Billingsella* near the top of the lower third of the formation indicates a medial Late Cambrian age for that horizon (Norford and Macqueen, 1975). The Cambrian-Ordovician boundary has not been closely located within the Franklin Mountain Formation.

Farther southeast (Fig. 7.15, loc. 15, 16), the "Rabbitkettle Formation" is difficult to recognize as the siltstone and shale predominate over platy limestone. At locality 15 the Upper Cambrian succession is approximately 360 m thick with *Bienvillia*, cf. *Euloma*, *Nanorthis* and *Plethopeltis* below the Cambrian-Ordovician boundary and *Symphysurina* above. At locality 16 the Upper Cambrian Series is 370 m thick but there it may overlie an unconformity and rest directly on the Lower Cambrian succession.

Still farther southeast, the Upper Cambrian Series is again represented by platy limestone (Blusson, 1971, unit 17). At locality 17 the platy limestone (Fig. 7.16) is 480 m thick and the Middle-Upper Cambrian boundary is approximately 100 m above the base of the limestone. Below the boundary are *Lejopyge*, *Centropleura* and *Elrathina* which represent the uppermost *Bolaspidella* Zone. Above the boundary are *Agnostus*, *Baltagnostus* and *Cedaria* which are coeval with the *Cedaria-Crepicephalus* Zone. One hundred metres above the Middle-Upper Cambrian boundary the equivalent of the *Aphelaspis* Zone is represented by *Acmarhachis*, *Aphelaspis* and *Glyptagnostus reticulatus reticulatus*. At 160 m above the

Figure 7.16. Outcrop of typical "Rabbitkettle Formation"-like strata comprising dark grey, platy limestone and dark grey to black shale and siltstone. Geologist is standing short distance above Middle-Upper Cambrian boundary. Location is 25 km south of locality 17 (Fig. 7.15) in Selwyn Basin. GSC 203167-O

boundary are *Proceratopyge*, *Pseudagnostus* and *Taenicephalus* of the *Taenicephalus* Zone, and at 260 m are *Hedinaspis* and *Charchaqia*. The Cambrian-Ordovician boundary is near the top of the platy limestone unit and may be within a 12 m soft shale unit that separates *olenid*?-bearing platy limestone at the top from similar, but asaphid-bearing limestone of the next overlying unit.

Ludvigsen (1982) described the trilobites and lithology of the upper 180 m of the "Rabbitkettle Formation" at locality 18. The trilobites represent the lowest Ordovician *Symphysurina* Zone and the temporal equivalents of the Cambrian *Missisquoia* and *Saukia* zones. The tribolite assemblages in the Cambrian zones differ significantly from typical Cambrian trilobites in the inner detrital and middle carbonate belts, and therefore were placed by Ludvigsen in his new *Yukonaspis* and *Parabolinella* zones. The Cambrian-Ordovician boundary is 25 m below the contact between the "Rabbitkettle Formation" and an overlying unnamed unit within the Road River Formation.

The "Rabbitkettle" lithology at locality 18 reflects its intermediate position between three facies of Upper Cambrian rocks. The formation contains platy, dark grey limestone, typical of the "Rabbitkettle" to the north, medium grey, limy siltstone and nodular to platy, silty limestone resembling the type Rabbitkettle Formation to the south, and medium to light grey limestone reflecting close proximity to the Broken Skull Formation that comprises platform carbonate to the east. The Broken Skull Formation is the southern extension of the Franklin Mountain Formation but contains a higher percentage of clastics. North of locality 18 slump breccias are common in the "Rabbitkettle Formation", and are believed to reflect relatively steep slopes adjacent to shallow-water Franklin Mountain platform carbonates. To the south slope breccias are sparse or absent, and an abundance of clastics is believed to have inhibited rapid shelf-edge carbonate deposition, and created a more gradual shelf to slope transition.

At locality 19 the Rabbitkettle is over 490 m thick, and, in ascending order, comprises a quartzite and sandy siltstone unit (120 m) belonging to the *Cedaria-Crepicephalus* Zone, a platy limestone unit (140 m) with *Aphelaspis* Zone fossils at the base, a medium to light grey weathering shale and siltstone unit (165 m) of unknown age, and a thin, wavy bedded limestone unit (65 m+) containing *Saukia* Zone fossils at the top. The basal unit overlies the distal edge of an unconformity which to the south both cuts downsection and is overlain by younger Rabbitkettle strata.

In its type area (Fig. 7.15, loc. 20) the Rabbitkettle Formation is more than 1200 m thick and consists of intercalated platy limestone, silty-limestone and siltstone. The formation weathers light grey to light brownish grey, and is characterized by an irregular, undulatory or wavy layering that has prompted many to call the formation the "wavy bedded, silty limestone unit".

A mid-Late Cambrian *Elvinia* Zone through Early Ordovician age is tentatively assigned to the Rabbitkettle Formation in the type area based upon correlation with the laterally equivalent(?) and more fossiliferous Broken Skull Formation. Both formations are separated from underlying strata by a sub-Franconian unconformity. Gabrielse et al. (1973) noted that the Rabbitkettle strata are generally correlative with those of the Kechika Group.

In the Anvil Mining District of western Selwyn Basin (loc. 21a), a Middle? and/or Upper? Cambrian graphitic and quartz rich phyllite unit hosts one of Canada's most important lead-zinc-silver deposits. The succession (300 m) is underlain by strata assigned to the Lower(?) Cambrian, and overlain by a tuff and andesitic greenstone bearing phyllite (900 m) that resembles other basinward outcrops of the Kechika Group, and therefore may be latest Cambrian to Early Ordovician in age (Tempelman-Kluit, 1972). The unit is in turn overlain by black shale (120 m+) of the Road River Formation which contains Middle Ordovician to Early Silurian graptolites. Pre-mining reserves in the Anvil District are 120 million tonnes of ore containing 3.7 percent lead, 5.6 percent zinc, and 45 to 50 grams per tonne of silver (Jennings and Jilson, 1986).

In summary, the Upper Cambrian in Selwyn Basin to the northwest of locality 18 is mostly within the "Rabbitkettle Formation" that comprises mainly but variable amounts of slope and basinal platy limestone. Southeast of locality 18 the Upper Cambrian is within the more argillaceous Rabbitkettle Formation that overlies a "sub-Franconian" erosional surface. The Rabbitkettle grades westward and southward into strata of the Kechika Group and eastward into platform carbonate of the Franklin Mountain and Broken Skull formations.

Kechika Basin

In Kechika Basin Upper Cambrian strata are mainly thin-bedded limestone and shale (Fig. 7.17). At the south end of Roosevelt Graben (Fig. 7.15, loc. 26) a Middle Cambrian turbidite unit of shale and interbedded quartzite extends into the Upper Cambrian, indicated by a *Dunderbergia* Zone fauna 20 m below the faulted top. Above the fault is an Upper Cambrian unit (280 m+) of dark grey platy limestone which locally exhibits penecontemporaneous slump structures and which encloses a *Taenicephalus* Zone

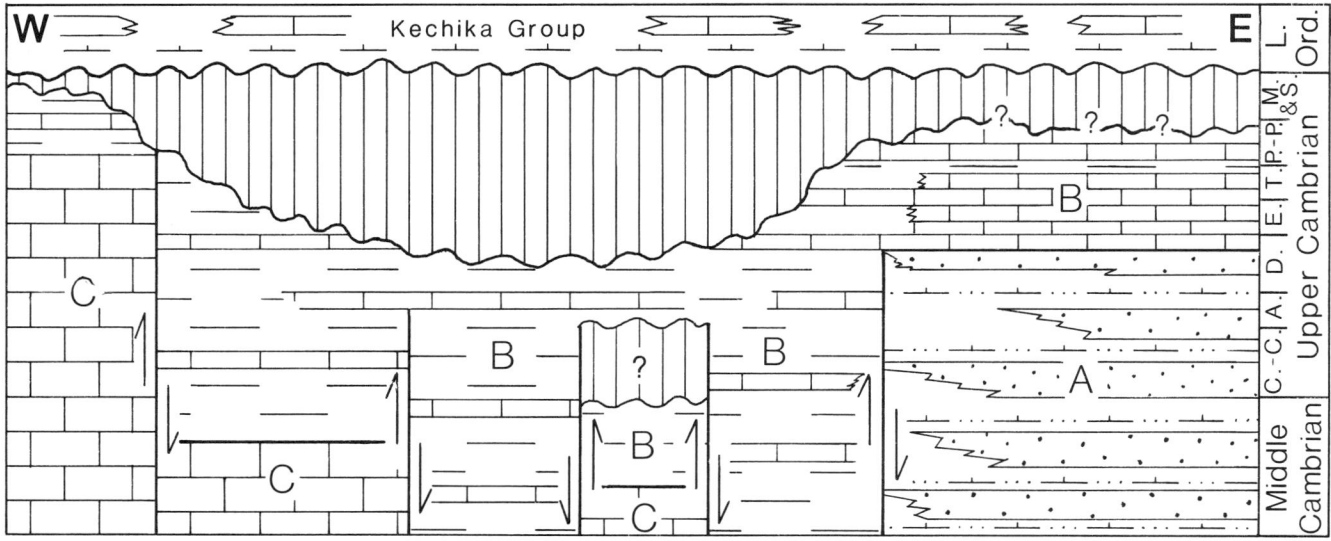

Figure 7.17. Schematic cross-section showing lateral relationship of Upper Cambrian strata across central Kechika Basin. Zone boundaries at right of figure match horizons only in adjacent strata and are offset by indicated faults farther to the west. See text for explanation of lettered units. Legend accompanies Figure 7.3, abbreviated zones written in full in Figure 7.1.

fauna. The succession is terminated at the top by a second fault that juxtaposes the platy limestone unit against light brown weathering siltstone of the Kechika Group. An earliest Early Ordovician trilobite, *Symphysurina*, is present in the lowermost Kechika beds. Basinward at locality 25 the Middle Cambrian turbidite succession extends a shorter distance into the Upper Cambrian, as *Cedaria-Crepicephalus* Zone fossils span the boundary between the turbidite unit and the overlying platy limestone unit (125 m+). *Meteoraspis* occurs 10 m below the contact the *Cedaria?* and *Kingstonia?* occur 45 m above.

At basinal locality 27, strata equivalent to both the upper part of the Middle Cambrian turbidite unit and the Upper Cambrian platy limestone unit exist in a single, condensed unit (147 m) of black shale (two-thirds) and black, platy limestone (one-third). The Middle-Upper Cambrian boundary is within the interval 85 to 110 m above the base, with *Ptychagnostus punctuosus* below and *Agnostus inexpectans* and *Glyptagnostus reticularis reticularis* above. Light brown- to buff-weathering, limy shale of the Kechika Group abruptly overlies the much darker Middle and Upper Cambrian strata. The black unit at nearby locality 24 is only 40 m thick. There the Middle-Upper Cambrian boundary is within a 4 m interval 34.5 to 38.5 m above the base, with *Hypagnostus* and *Modocia* below and *Bromella*, *Dunderbergia*, *Minupeltis*, *Olenaspella?* and *Pseudagnostus* above. At locality 24 the *Cedaria-Crepicephalus* Zone is either very thin or missing. At the top of the formation is chert-rich limestone overlain disconformably(?) by light brown to silvery weathering Kechika Group shale.

To the west (loc. 23), the Upper Cambrian is within a lower unit (65 m) of light grey limestone containing *Acmarhachis*, *Comachia?* and *Olenaspella* near the top, and an upper unit (27 m) containing dark grey, platy limestone that yields *Glyptagnostus?*, *Stenambon?*, *Crenuolimbus?*, *Pseudagnostus?* and *Tholifrons?*. Fossils from both units are of Late Dresbachian age. Within the upper part of the upper unit are *Brabbia?* and *Hungaia* which indicate a Franconian age. These rocks are overlain disconformably(?) by light brownish grey, limy shale of the Kechika Group. Nearby (loc. 22), the Upper Cambrian Series comprises at least 180 m of medium grey, clean limestone in medium and thick beds, many of which contain packstone and conglomerate. *Plethometopus* near the base and top, plus *Euptychaspis* near the base, provide a *Saukia* Zone age for the unit.

To the southwest (loc. 28), the Upper Cambrian is included in a 510 m-thick succession (Gabrielse, 1975), comprising a lower unit (265 m) of dark shale and platy limestone similar to that farther north (Fig. 7.17, unit B; Gabrielse, 1975). *Blountia*, *Cedaria* cf. *C. prolifica* and *Deiracephalus* are present 110 m above the base. An upper unit (245 m) contains thin, dark grey, platy to nodular bedded limestone that is interbedded with silvery brown-weathering, limy shale. Trilobites within the upper unit include *Bienvillia*, *Drumaspis*, *Geragnostus*, *Loganellus* and *Pseudagnostus* near the base and *Geragnostus*, *Pseudagnostus* and *Elkanaspis* near the top, therefore the unit contains the equivalents of the *Ptychaspis-Prosaukia* and *Saukia* zones. Farther south (loc. 29) Upper Cambrian strata (105 m) are represented by a unit of medium to dark, blue-grey-weathering, nodular and platy limestone containing minor silvery brown weathering shale. Trilobites within the unit belong to the *Taenicephalus*, *Ptychaspis-Prosaukia*, and *Saukia* zones. A disconformity is suspected between this unit and an underlying Middle(?) Cambrian medium and thick-bedded dolostone. Overlying the unit is silvery light brown shale. Upper Cambrian strata have not been documented in the northern part of Kechika Basin.

In summary, the Upper Cambrian Series in the area of localities 22 to 27 forms several laterally distinct units (Fig. 7.17A,B,C). At the east-central margin of the basin strata of *Dunderbergia* Zone age and older are within a turbidite unit (A) with quartzite interbeds reflecting movement along the margin of Roosevelt Graben. Basinwards is a partly equivalent black shale unit (B) lacking quartzite interbeds but instead containing some limestone. This unit has a lower Middle Cambrian diachronous base, which, when traced eastward, moves upsection to overlie an increasing thickness of the turbidite unit. The basinal unit B also increases in platy limestone content to the east.

In a limited area, it also can be demonstrated that the basinal unit climbs upsection to the west and overlies a light coloured unit (C) representing either large carbonate mounds or a carbonate platform of unknown extent. The light coloured unit contains a clean carbonate suggesting a depositional site free from contaminating terrestrial clastics. This site may have been created by block faulting within the basin. The block faulting concept may also explain the variable degree of erosion at the top of the Upper Cambrian units prior to the spreading of lower Lower Ordovician terrestrial clastics of the Kechika Group.

Basinal unit B can be traced as far south as locality 27, where it is Late Cambrian in age and overlain by a younger unit of Upper Cambrian terrestrial clastics that resemble those in the Kechika Group. Still farther south are Upper Cambrian strata similar to those in the younger unit.

Cassiar Platform

Near the northern part of Cassiar Platform (Fig. 7.15, loc. 21b), an unknown thickness of phyllite, calcareous argillite and thin-bedded limestone containing Late(?) Cambrian trilobites disconformably overlies Lower Cambrian strata belonging to the lower part of the *Bonnia-Olenellus* Zone. Near locality 21c the Late Cambrian trilobite *Hedinaspas*? was found in Kechika Group strata a short distance above the Lower Cambrian Rosella Formation (Gabrielse, 1963a, 1969). At locality 21c a rare stratigraphic contact between commonly fault-juxtaposed Lower Cambrian Rosella Formation and the Kechika Group is visible. Light grey, thick-bedded carbonate of the Rosella Formation is abruptly overlain by medium light brown siltstone. These strata are overlain by the more typical Kechika silvery grey shales and interbedded platy limestone.

Given the above meagre data, it seems likely that most of the Middle and Upper Cambrian strata on Cassiar Platform were either completely eroded, or are thin because of slow deposition and have been included with other fault slivers which have been mapped as Kechika Group. The general lack of Middle and Late Cambrian fossils suggests the former concept is the most likely.

Robson Basin

In northwestern Robson Basin (Fig. 7.15, loc. 31-39) Upper Cambrian strata occur within the Lynx Group, which lies between the Arctomys and Survey Peak formations (Fig. 7.18). Near its type section (loc. 39), the Lynx Group (1065 m) comprises moderately resistant, mainly thin-bedded, commonly crossbedded dolostone and limestone containing flat-pebble conglomerate (Mountjoy, 1962). The Early Ordovician trilobite *Symphysurina* occurs a short distance above the top, suggesting the Lynx-Survey Peak contact is near the Cambrian-Ordovician boundary. The colour and recessive weathering of the basal "putty shale" member within the Survey Peak serves as an excellent aid in locating the top of the Lynx Group.

Northwestward from the Lynx Group type section, two unmapped subunits of the group can be differentiated as far as locality 33, and farther northwest only the lower subunit can be recognized; the upper subunit blends with strata of the overlying Survey Peak Formation (Slind and Perkins, 1966; Fig. 7.18). At locality 35 the Lynx Group is 705 m thick, and the lower subunit (425 m) forms a continuous carbonate succession of resistant limestone and dolostone. Fossils of the *Cedaria-Crepicephalus* Zone are present in the interval 210 to 265 m above the base, and others of the *Aphelaspis* Zone occur 280 to 295 m above the base. The overlying beds belong to the *Dunderbergia* and/or *Elvinia* Zone. The upper subunit (280 m) comprises light brown-weathering shale with dispersed limestone nodules. Resistant, medium to thick, light grey-weathering limestone interbeds (20%) serve to lithologically differentiate this subunit from the overlying Survey Peak Formation. *Elvinia* Zone fossils are present at the base of the subunit, and *Saukia* Zone fossils are present 30 m below the top.

At locality 33 the Lynx Group is 730 m thick. There the lower carbonate subunit (220 m) resembles that at locality 35. Fossil occurrences and their distance above the base are as follows: *Aphelaspis* Zone, 75 m; *Dunderbergia* Zone, 140 to 150 m; *Taenicephalus* Zone, 155 m; and *Ptychaspis-Prosaukia* Zone fossils occur in numerous intervals between 175 m and top of the subunit. The upper subunit (510 m) can be divided into three parts. The lowest (125 m) comprises medium to light brown-weathering shale with abundant limestone pods. The boundary between the *Ptychaspis-Prosaukia* and *Saukia* Zone is within this part of the subunit. The next part (210 m) contains thin- to thick-bedded limestone and the boundary between the *Saukia* and *Missisquoia* Zone is at the top of the lower one-third. The upper part (175 m) comprises light brown-weathering shale and an equal amount of interbedded limestone. The limestone is in medium and thick beds containing flat-pebble conglomerate, and in resistant, light grey weathering beds displaying carbonate buildups up to 2 m in thickness. Similar beds with carbonate buildups are present in the uppermost 50 m of the middle part of the upper subunit. An Early Ordovician trilobite assemblage of *Bellefontia* and *Clelandia* occurs 6 m above the top of the Lynx Group. The Cambrian-Ordovician boundary therefore is either within the upper part of the upper subunit or the upper half of the medial part.

The Upper Cambrian succession in the southeastern half of the middle carbonate belt in Robson Basin is represented by four formations with type sections at or near locality 42 (Fig. 7.15). In ascending order they are as follows: Sullivan Formation (425 m), greenish and brownish grey shale and subordinate interbedded limestone, *Cedaria* Zone; Lyell Formation (345 m), resistant peritidal dolostone and limestone, *Crepicephalus, Aphelaspis, Dunderbergia* and part *Elvinia* zones; Bison Creek Formation (190 m), grey and green shale plus interbedded limestone in thin beds and nodules, part *Elvinia* Zone, *Taenicephalus, Ptychaspis-Prosaukia* zones, and part *Saukia* Zone; Mistaya Formation (100 m), resistant limestone in thick stromatolitic and bioclastic beds, part *Saukia* Zone. The base of the Upper Cambrian succession is a

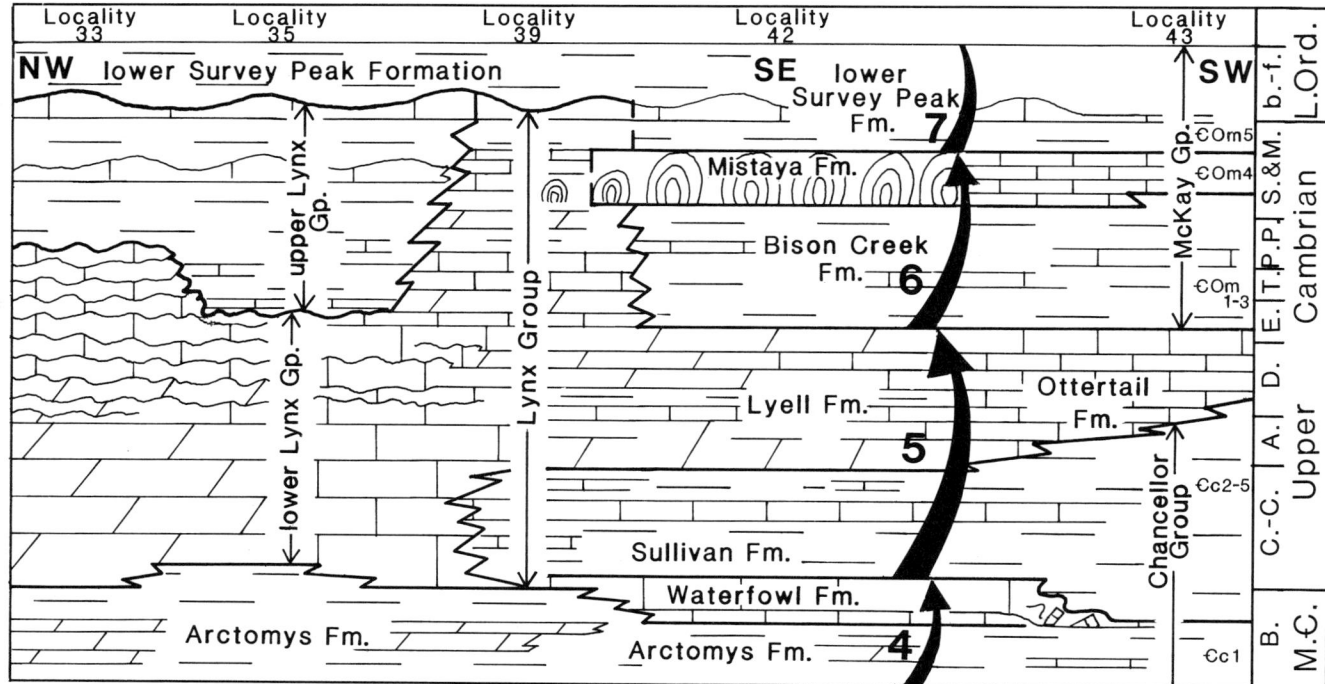

Figure 7.18. Schematic cross-section in Robson Basin showing Upper Cambrian lateral relationships within middle carbonate belt and adjacent margin of outer detrital belt. Arrows indicate grand cycles of Aitken, (1966b). Older grand cycles are shown in Figure 7.14. Layered, dome-like symbols indicate abundant occurrence of large, vertically orientated stromatolites. Locality numbers at top of figure can be found in Figure 7.15. Legend accompanies Figure 7.3, zones are identified in Figure 7.1.

short distance below the Sullivan in the underlying Waterfowl Formation, and the top of the Upper Cambrian Series is a short distance above the Mistaya in the Survey Peak Formation. At locality 40 it is 20 m above the Mistaya-Survey Peak contact and at locality 41, it is 33 m above the contact.

Aitken (1966b) placed the Sullivan and Lyell formations in one grand cycle and the overlying Bison Creek and Mistaya formations in another (Fig. 7.18). The Sullivan half-cycle extends farthest northwest and can be recognized at locality 39 where the overlying half-cycles merge with the remainder of the Lynx Group (Fig. 7.18). Farther northwest (loc. 31-37) strata equivalent to the Sullivan and Lyell formations are within the lower, carbonate subunit of the Lynx Group. Strata within the overlying shale and interbedded limestone subunit do not resemble the Bison Creek and Mistaya half-cycles. They lack abundant bioclastic limestone beds rich in echinodermal debris and closely packed, vertically oriented stromatolites. Instead they comprise shale and sparse interbeds of limestone exhibiting broad carbonate buildups. Interbeds of that type are in the lower Survey Peak Formation to the southwest, but not in the underlying Bison Creek or Mistaya half-cycle.

The basinward (southwestward) expression of the Upper Cambrian grand cycles differs from that of the Middle Cambrian, because the carbonate half-cycles extend into the outer detrital belt as mappable units, whereas the Middle Cambrian carbonate half-cycles do not. The existence of Upper Cambrian carbonate in the outer detrital belt is attributed to rapid Upper Cambrian clastic half-cycle deposition that maintained relatively shallow depths, as opposed to slow Middle-Cambrian clastic half-cycle sedimentation that resulted in deeper water conditions unfavourable to later carbonate growth. Estimated thickness for the entire Upper Cambrian succession in the outer detrital belt near locality 43 is 3000 m, whereas it is 1100 m thick in the middle carbonate belt near locality 42.

Near locality 43 the lateral equivalent of the Sullivan and Lyell half-cycle is a greenish grey slate (1600 m) assigned to the upper Chancellor Group, and a resistant limestone that locally displays shallow-water features (300 to 500 m) assigned to the Ottertail Formation (Fig. 7.18). Fossils recovered from just below the Chancellor-Ottertail contact belong to the *Crepicephalus* and *Aphelaspis* zones (Balkwill, 1969; Cook, 1975). Locally the contact is still younger with *Dunderbergia* Zone fossils just above the boundary (Gardner, 1977).

The Bison Creek and Mistaya half-cycles near locality 43 are within the lower part of a thick (1800 to 2100 m) recessive succession of light greenish grey shale assigned to the McKay Group (Fig. 7.18). Small limestone nodules are abundant within the McKay and minor limestone interbeds with flat-pebble conglomerates are conspicuous. The Bison Creek half-cycle near locality 43 comprises light greenish grey to buff-weathering shale (COm1-3 Fig. 7.18, 930 m), and the Mistaya half-cycle is represented by micritic to bioclastic limestone (COm4 Fig. 7-18, 60 m). Fossils collected a short distance above the latter half-cycle suggest that the Cambrian-Ordovician boundary is not far above its upper contact.

Near the Southern Rocky Mountain Trench (loc. 45), the Ottertail Formation (there called Jubilee Formation) unconformably overlies Lower Cambrian strata. Above the unconformity is 1180 m of Jubilee dolostone overlain by 800 m of McKay Group. The lower part of the McKay, which is equivalent to the Bison Creek half-cycle, comprises light greenish grey shale with minor limestone in nodules and thin interbeds (360 m). The next overlying part of the McKay Group equates with the Mistaya half-cycle and comprises thin-bedded limestone (130 m) and is reported to contain the Cambrian-Ordovician boundary (Leech, 1954).

Along the southeastern margin of Robson Basin (i.e., loc. 44) various outcrops of medium and light green phyllite have been assigned previously to the Middle(?) Cambrian Canyon Creek Formation (Evans, 1933). New fossil collections representing the *Ptychaspis-Prosaukia, Saukia,* and *Symphysurina* zones indicate that these strata should be assigned to the McKay Group (Balkwill, 1969).

Columbia Basin

The only areas within Columbia Basin where Upper Cambrian strata have been documented by fossils are near the Southern Rocky Mountain Trench at localities 45a and 45b (Fig. 7.15). At locality 45a in the Cariboo Mountains the fossils occur within the Dome Creek Formation (1750 m±), which comprises 74 m of basal, black Lower Cambrian shale, 100 m of Upper Cambrian(?), highly burrowed siltstone, and 1680 m+ of dark grey, tectonically thickened shale that is terminated at the top by a fault. Upper Cambrian *Taenicephalus* Zone fossils occur in folded strata 290 to 1030 m above the base of the dark grey shale unit.

At locality 45b the Upper Cambrian unconformably overlies Lower Cambrian and Upper Proterozoic strata on the Purcell Arch (see Fig. 7.19). Above the unconformity is barren dolostone of the Upper Cambrian Jubilee Formation (400-610 m), which in turn is overlain by the Upper Cambrian to Lower Ordovician McKay Group. Late Cambrian trilobites reported by Evans (1933) from undesignated stratigraphic horizons at locality 45b are *Agnostus pisiformis, Glyptagnostus reticulatus, Housia* sp., *Olenus* sp., and *Pseudagnostus* sp.

In the south-central Columbia Basin (loc. 48, 49) elongate outcrops of undated argillite and slate may in part be of Late Cambrian age. The southernmost outcrops are of greenish grey argillite that has been assigned to the Laib Formation (1500 m). This formation is in part the lateral equivalent of the Middle Cambrian Nelway Formation (=Metaline Formation) that comprises dolostone, and is overlain by Ordovician black slate of the Active Formation (=Ledbetter Formation).

In Kootenay Arc, north of outcrops mapped as Laib Formation, is a structurally complex belt of phyllite, grit, mafic volcanics and carbonate belonging to the Lardeau Group (loc. 48). This group overlies the Lower Cambrian Badshot Formation and, in turn, is overlain unconformably by the Carboniferous Milford Group. Because its upper part, at least, underwent deformation, metamorphism and granitic intrusion prior to the Late Mississippian and clearly has a history different from coeval ancestral North American strata to the east, the Lardeau Group was placed in the Kootenay Terrane (Monger and Berg, 1984). Fyles and Eastwood (1962) have recognized six formations within the Lardeau Group at locality 48, the lowest being the Index Formation. Based upon regional lithological correlations, the Index Formation (450-750 m) is the most likely formation within the group to contain Upper Cambrian strata. This is supported by recent studies (G.E. Gehrels and M.T. Smith, pers. comm., 1987) which have shown that the Index Formation lies depositionally on the Badshot Formation. The Index consists of dark grey, locally black, and greenish grey, rhythmically bedded phyllite with interlayers of limestone, minor calcite-cemented quartzite and near the top, local green mafic volcanics. The uppermost part of the Index Formation, however, contains a zone of talc schist lenses intermittently exposed over 100 km along strike (Zwanzig, 1973; G.E. Gehrels and M.T. Smith, pers. comm., 1987). The talc schist was probably derived from ultramafic rocks that may represent vestiges of early Paleozoic oceanic crust. Thus the talc schist zone may mark a major fault or suture that forms the true base of the Kootenay Terrane.

Cordilleran Terranes

Cambrian strata in the Cassiar and Cariboo terranes have been described above because of their similarity with those of ancestral North America. Elsewhere they are poorly known, reflecting the lack of biostratigraphic control and complications resulting from deformation and metamorphism.

Kootenay Terrane

Black siliceous argillite, chert, and quartzite above the talc schist zone in the Lardeau Group of Kootenay Arc may contain Cambrian strata but on lithological grounds they are similar to Ordovician strata elsewhere in the Cordillera.

In the Adams Lake area, farther northwest, the Eagle Bay Formation, occurring in southwest- and west-verging thrust sheets, includes a limestone unit (Tshinakin) with an Early Cambrian archaeocyathid fauna (Schiarizza, 1986). The limestone is intercalated with green calcareous chlorite schist and mafic and intermediate volcanics with volcaniclastic fragmental schist derived from the volcanics and volcaniclastics. An overlying unit comprises grey and black graphitic, siliceous and calcareous phyllite similar to the Index Formation of the Kootenay Arc and thus may be of Late Cambrian age.

Alexander Terrane

In the southern part of the Alexander Terrane in Yukon Territory (Fig. 7.15, loc. 30) the informally named "Field Creek volcanics", at least 1600 m thick, comprise a succession of andesite and basalt flows, breccia and pillow lava interbedded with carbonates, pebble conglomerate and minor tuffaceous sandstone. Brachiopods cf. *Billingsella* sp. and cf. *Ocnerorthis* sp. of probable Late Cambrian age have been recovered from these strata (G.A. Cooper, pers. comm., 1978).

Summary of the Late Cambrian deposition

Upper Cambrian strata along part of the northeast margin of Selwyn Basin and at the edges of Yukon Platform comprise relatively clean carbonate that pass basinwards over a short distance into dark, platy limestone with slump breccia. The platy limestone represents slope deposits that interfinger basinward with various amounts of dark shale and siltstone. Southeast of locality 18 on the northeast margin of Selwyn Basin, the platform carbonate contains more terrigenous material, the basinward change to platy limestone is less abrupt, and slope breccias are absent. These strata traced farther into the basin are medium grey rather than dark grey as to the northwest.

East-central Kechika Basin received Middle Cambrian and lower Upper Cambrian terrestrial clastics that were trapped near the basin margin. Dark, platy limestone, similar to that in northern Selwyn Basin, was deposited during the latter half of Late Cambrian time at the margin with associated shale in the basin. Uplifted blocks within the basin were sites of local, pure, light coloured limestone deposition. A brief, broad uplift at the end of Cambrian time was followed by widespread deposition of the Lower Ordovician Kechika Group.

Southern Kechika Basin and Robson Basin received a moderate and relatively evenly dispersed amount of clastics during the Late Cambrian. As in southern Selwyn Basin, slope breccias are absent. Clastic input twice abated to permit the deposition of one prominent limestone body (lower Lynx Group, Lyell, Ottertail, Jubilee formations) and another lesser limestone body (Mistaya and COm_4 Formation, Fig. 7.18).

In Columbia Basin, it is likely that basinal deposition locally accompanied by volcanism took place during Late Cambrian time. A similar environment is envisioned for the Cassiar Platform and British-Barn Basin.

ORDOVICIAN AND SILURIAN ASSEMBLAGES

M.P. Cecile and B.S. Norford

(with contributions on Cordilleran terranes by J.O. Wheeler, H. Gabrielse and G.E. Gehrels)

Ordovician and Silurian basinal and platform sedimentary and minor volcanic strata are extensively preserved in the Foreland and the northeastern Omineca belts where they form part of the miogeoclinal wedge of Ancestral North America. In the Omineca Belt, miogeoclinal basinal and platformal facies strata are also preserved in the Cassiar Terrane, a displaced segment of the ancient continental margin. In the southern Omineca Belt the pericratonic Kootenay Terrane hosts a thick sequence of clastics, carbonates and mafic volcanics, part of which is Early Ordovician. In the Insular Belt the Ordovician and Silurian systems are represented by a wide range of strata included in the Alexander Terrane of the Saint Elias Mountains.

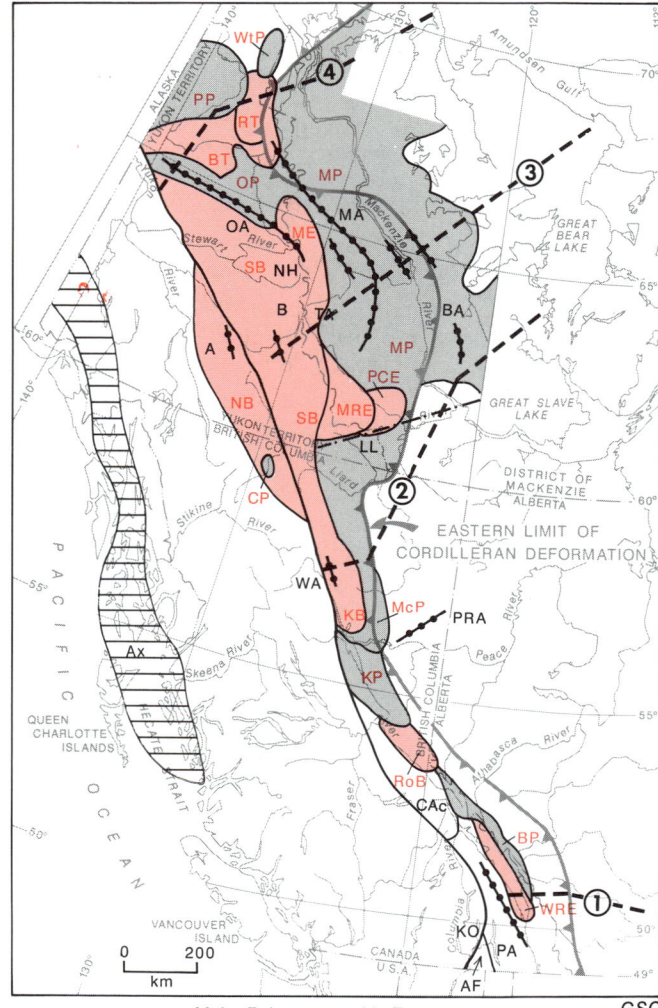

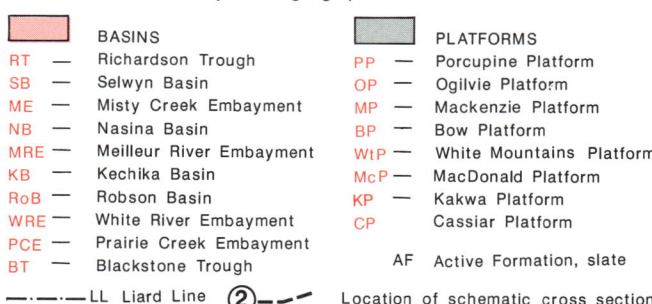

Figure 7.19. Paleogeographic features present from Early Ordovician (Tremadocian) to Late Silurian (Early Pridolian) time.

Ancestral North American Miogeocline

Within the Foreland Belt and in the Selwyn Basin part of the Omineca Belt, Lower Ordovician strata continued depositional patterns established during the Late Cambrian. Throughout most of the Northern Rocky Mountains and Mackenzie Mountains, Lower Devonian strata rest conformably upon Upper Silurian rocks. In the southern Rockies, the Upper Silurian is missing beneath Middle and Upper Devonian strata.

The miogeocline shoaled eastward with broad, carbonate platforms developing in supratidal to outer neritic shelf settings. Westward the carbonate platforms were bordered by a series of embayments, basins and troughs within which deeper water facies accumulated (Fig. 7.19, 7.20). Throughout the Ordovician and Silurian periods the depositional patterns periodically were deflected by arcuate, northwesterly trending arches. Platformal facies typically are thickly bedded carbonates. These commonly show abrupt transitions into basinal facies with dark graptolitic shale, chert, thinly bedded limestone and calcareous siltstone (Fig. 7.21). Most platform-edge and transitional facies no longer preserve clear evidence of framework-building organisms, because many of the rocks show extensive diagenesis and dolomitization. However, silicified rocks locally display abundant framework-building organisms and perhaps biostromes and bioherms were prevalent close to the facies transition.

Throughout Ordovician and Silurian time transgressions and regressions of epeiric seas caused large areas of the craton periodically to be exposed to erosion. Strata preserved adjacent to, and as outliers, on Precambrian rocks of the Canadian Shield contain little or no terrigenous material. The Shield apparently had low relief and was unimportant as a sediment source. Exceptions occur in the central and southern Cordillera where Middle Ordovician to Lower Silurian platform carbonates contain significant components of terrigenous detritus. The most likely source for this material is a broad region centred on the Peace River Arch (Fig. 7.19) from which clastics probably were dispersed through long-shore drift currents (Ketner, 1968).

An anomalous aspect of the basinal facies is the distribution of orange-brown Silurian siltstone and mudstone. These sediments are preserved only in the central miogeocline (northeastern British Columbia and southeastern Yukon Territory) with the exception of some mudstone that occurs as far north as the Richardson Trough. The distribution roughly complements an area of sub-Llandovery erosion on the adjacent platform. Unlike other regional unconformities the sub-Silurian (sub-Llandoverian) hiatus is not recognized on the Mackenzie and Bow platforms, moreover its erosional effects are most pronounced on the MacDonald Platform where much of the subjacent Cambrian and Ordovician succession is missing and Llandoverian strata locally rest directly on Precambrian rocks. Most of the orange-brown siltstone and mudstone is younger than the sub-Llandoverian erosional event, thus it is unlikely that the material was a product of this erosion but rather was derived from other source areas including the Peace River Arch.

Ordovician and Silurian platform carbonates hosted a wide variety of shallow-water flora (algae) and fauna of which trilobites, conodonts, brachiopods, corals, cephalopods, and gastropods are most significant biostratigraphically. However, there are significant variations in the relative abundance of different fossil taxa. Lower Ordovician platform carbonates are dominated by stromatolites which occur as thin biostromes or as scattered domes and biscuits. Middle Ordovician strata contain brachiopods, sponges and stromatolites and are particularly characterized by large gastropods such as *Maclurites* and *Palliseria* as well as algal pisoliths. Some Middle Ordovician carbonates represent environments hostile to most animal life and thus are barren. Upper Ordovician to Lower Silurian rocks contain the most diverse and abundant assemblages which locally form biostromes of colonial and solitary corals, large gastropods and large cephalopods. Upper Silurian strata are sparsely fossiliferous with fish and brachiopods (Thompson, 1989). Fish debris also is found in Upper Ordovician strata of the MacDonald Platform (Cecile and Norford, 1979).

Basinal and transitional strata contain abundant fossils. Transitional rocks host the most varied and abundant biota including trilobites, brachiopods, cephalopods, gastropods, sponges, solitary and colonial corals, stromatolites, graptolites and conodonts. Basinal facies contain graptolites, trilobites, the arthropod *Caryocaris*, sponge spicules, radiolarians and inarticulate brachiopods. Some Lower Silurian strata contain solitary corals and patches of colonial corals.

Some interesting faunal contrasts are found between transitional and platformal facies. Transitional Lower Ordovician strata host conodont faunas of both Mid-Continent and North Atlantic type (Tipnis et al., 1978; Kennedy and Barnes, 1981). Upper Ordovician transitional facies contain large irregular dolostone mounds and reefs which include scattered solitary corals, gastropods and brachiopods that are typical of the *Bimuria-Dicoelosia* Fauna known from terranes in northern California and Alaska (Potter and Cecile, 1985). In contrast, macrofossils and conodonts, typical of the Late Ordovician western North American *Bighornia-Thaerodonta* Fauna are found in adjacent platform carbonates.

Northern miogeocline (north of 60°N)

Paleogeographic elements of the northern miogeocline comprised the White Mountains, Mackenzie, Porcupine and Ogilvie platforms which, in the north, bordered the Richardson and Blackstone troughs (Fig. 7.19). Recent work by Lane and Cecile (1989) indicated that an area in the British and Barn mountains of northern Yukon Territory was a site of basinal deposition during Ordovician and Silurian time. Farther south the Mackenzie Platform was flanked on the west by the Selwyn Basin which to the north and south extended into the platform as the Misty Creek, Meilleur and Prairie Creek embayments. Several arches, including the Mackenzie Arch, Ogilvie Arch, Twitya Arch, Bulmer Lake Arch and Niddery High as well as unnamed features (including Positive area B, Fig. 7.19) played varying roles in the distribution of sedimentary facies.

A prominent aspect of Ordovician and Silurian, as well as Devonian stratigraphy, is the substantial and localized thickness of strata in the Liard Depression, a depocentre that includes both the Meilleur River Embayment and adjacent platform carbonate in the southern Mackenzie Mountains (compare Fig. 7.22 and 7.38 and

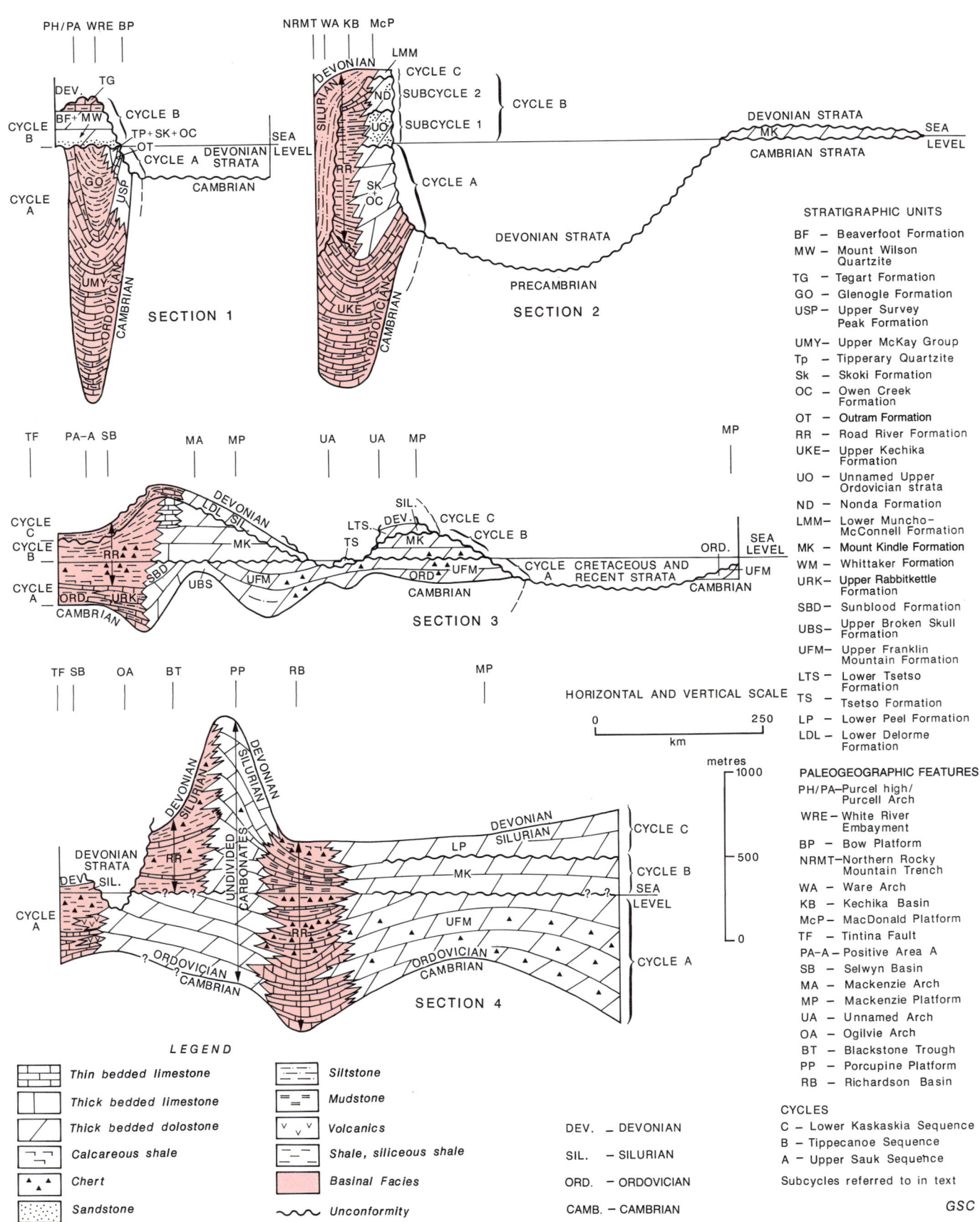

Figure 7.20. Stratigraphic cross-sections for Ordovician and Silurian strata. See Figure 7.19 for locations.

Figure 7.21. Transitional breccias in the Road River Formation along the east side of the Kechika Basin. Light coloured beds in the lower part of the photograph consist of thinly bedded limestone or stratiform breccias. The upper cliff has a lower unit with disrupted slabs of limestone to 1 m long, a middle unit of thinly bedded limestone, and an upper unit, cutting through the middle limestone of platformal dolomite debris. The upper cliff is about 40 m high. Photo by M.P. Cecile. GSC 205235-AA

7.39). On its southeast side, the depression is bounded by the Liard Line (Fig. 7.19; Aitken and Pugh, 1984) which might reflect a transform fault which offset the Proterozoic passive margin and which was reactivated during the Ordovician to form the depression. Cambrian strata show no anomalous increase in thickness in this region. The Liard Depression, together with the arches, depressions, basins and embayments of the northern miogeocline possibly developed as a consequence of crustal extension in the early Paleozoic between the Cordilleran and Innuitian regions (Cecile, 1986). Extension is suggested by the local occurrences of mafic volcanic rocks. Several authors (e.g. Miall, 1976) have suggested on the basis of facies boundary trends in lower Paleozoic rocks that the Richardson Trough was linked to the Hazen Trough of the Arctic Archipelago during early Paleozoic time.

On the Mackenzie Platform, Ordovician and Silurian strata are divisible into three transgressive-regressive cycles assigned to the uppermost third of the Sauk, the Tippecanoe and the lower Kaskaskia sequences respectively (Sloss, 1963). The lowermost, Cycle A, (Fig. 7.23, 7.24) began with widespread transgression in the Late Cambrian and is represented by the Upper Cambrian to Lower Ordovician Franklin Mountain Formation. In the northeastern Mackenzie Mountains and northern Interior Plains this formation can be divided into three members; the fourth, uppermost member, is preserved locally. The basal member is between 45 and 120 m thick and consists of interbedded (1) olive grey, argillaceous and shaly dolostone, (2) pale yellowish orange, finely crystalline dolostone, (3) conglomeratic dolostone and (4) stromatolitic dolostone. In places, the basal unit instead consists of medium- to fine-grained, thinly to medium-bedded sandstone and some conglomerate which weather purplish red. Casts of halite crystals are known from red shale interbeds.

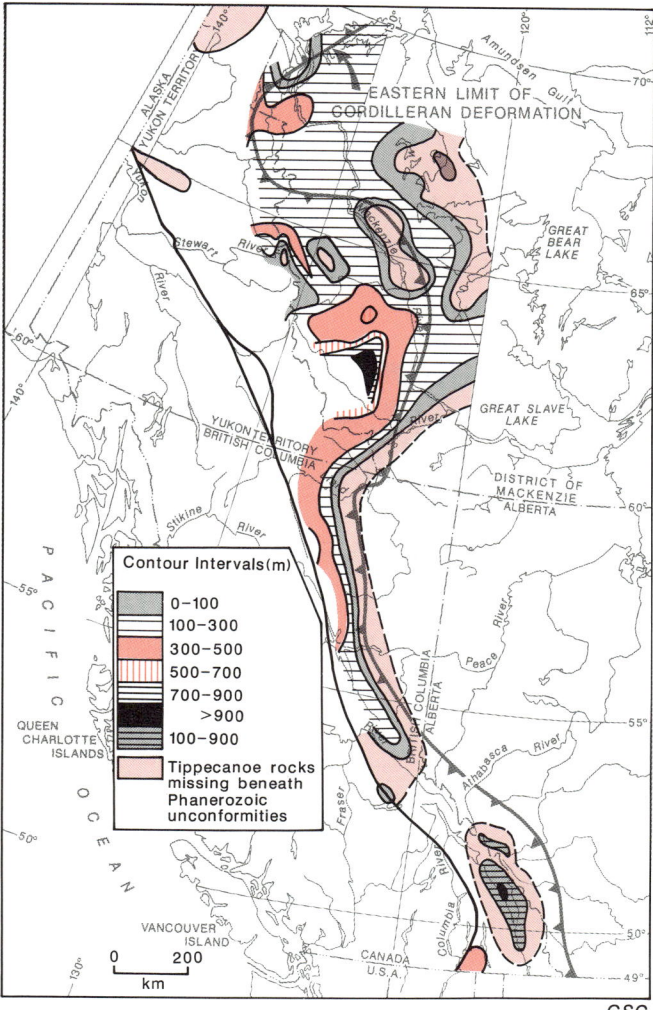

Figure 7.22. Unrestored Tippecanoe isopachs and local thicknesses.

The second unit consists of 290 to 330 m of resistant, well bedded silty dolostone; outcrops characteristically weather in brownish grey and greyish orange stripes. The third member, 40 to 170 m thick, is characterized by grey white, coarsely crystalline thickly bedded dolostone, locally with abundant silica as drusy quartz and white, tripolitic chert. Algal stromatolites and oolites are preserved in places. A fourth unit of porous dolomites is known only from a small region (subsurface) of the Mackenzie Platform, elsewhere the sub-Mount Kindle unconformity cuts lower in the stratigraphic sequence. On the southern Mackenzie Platform the equivalent of the Franklin Mountain Formation is the Broken Skull Formation with a lower member (275 m thick) of buff, orange, yellow, brown and grey weathering, variably sandy dolostone and limestone and an upper member (550 m) of grey weathering dolostone and limestone including basal sandy dolomitic limestone. Following an episode of regression and erosion during the late Middle Ordovician, widespread transgression initiated Cycle B and deposition of the Upper Ordovician to Lower Silurian Mount Kindle Formation (Fig. 7.25-7.27). The

Mount Kindle consists of dark grey and brownish grey, thinly to thickly bedded, fine to medium crystalline dolostone. Locally corals are abundant. In the platform-to-basin transitional zone of the west-central Mackenzie Mountains the Mount Kindle can be divided into a lower member (330 m thick) of well bedded, dark grey cherty dolostone and an upper member (275 m) of light grey, massive, medium to coarse grained reefoid dolostone. On the southern Mackenzie Platform similar rocks have been termed the Whittaker Formation (up to 1000 m thick) of distinctly laminated, dark grey, thinly bedded, finely crystalline dolostone with abundant discontinuous lenses of dark grey chert. Cycle B ended with mid-Silurian regression and erosion. Cycle C (Fig. 7.28) began in the latest Silurian and resulted in the accumulation of arenaceous

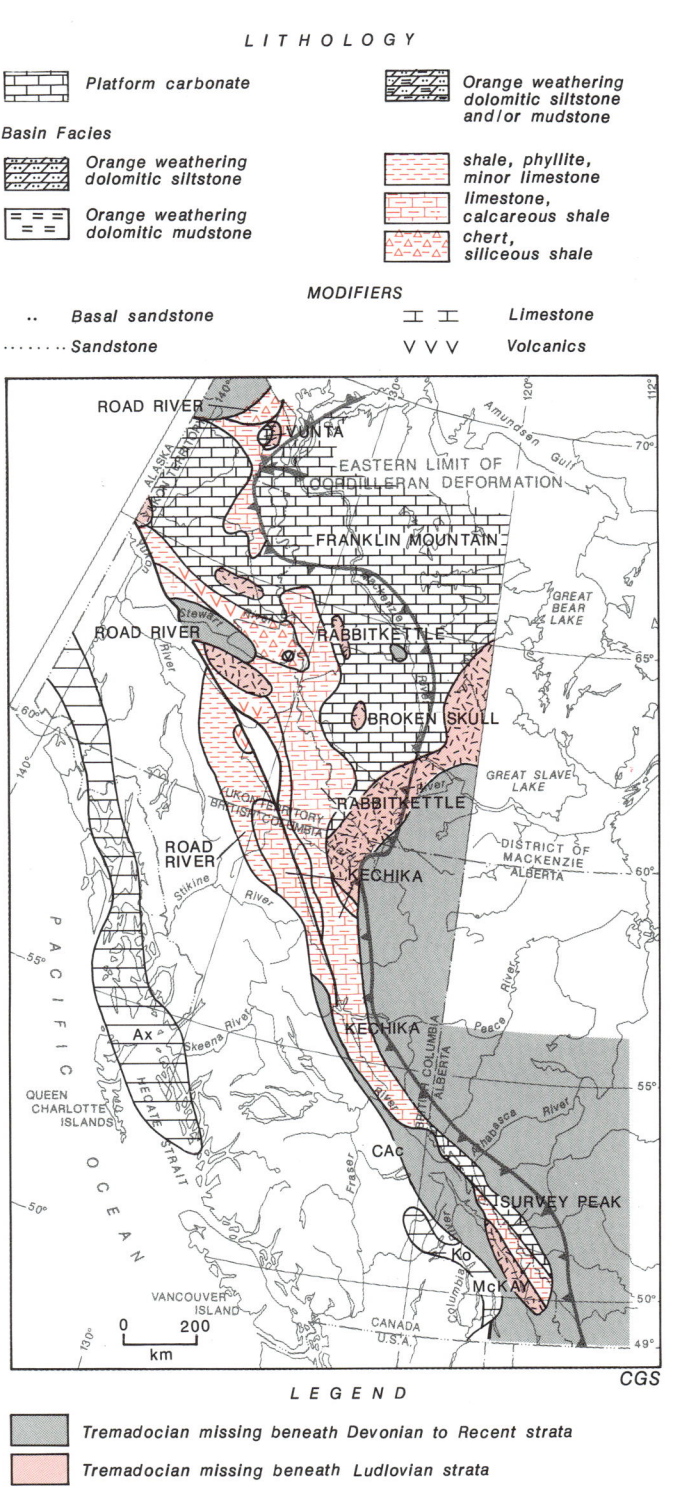

Figure 7.23. Lower Ordovician (Tremadocian) facies (Cycle A) unrestored. Lithological symbol, for Fig. 7.24, 7.25, 7.27, 7.28.

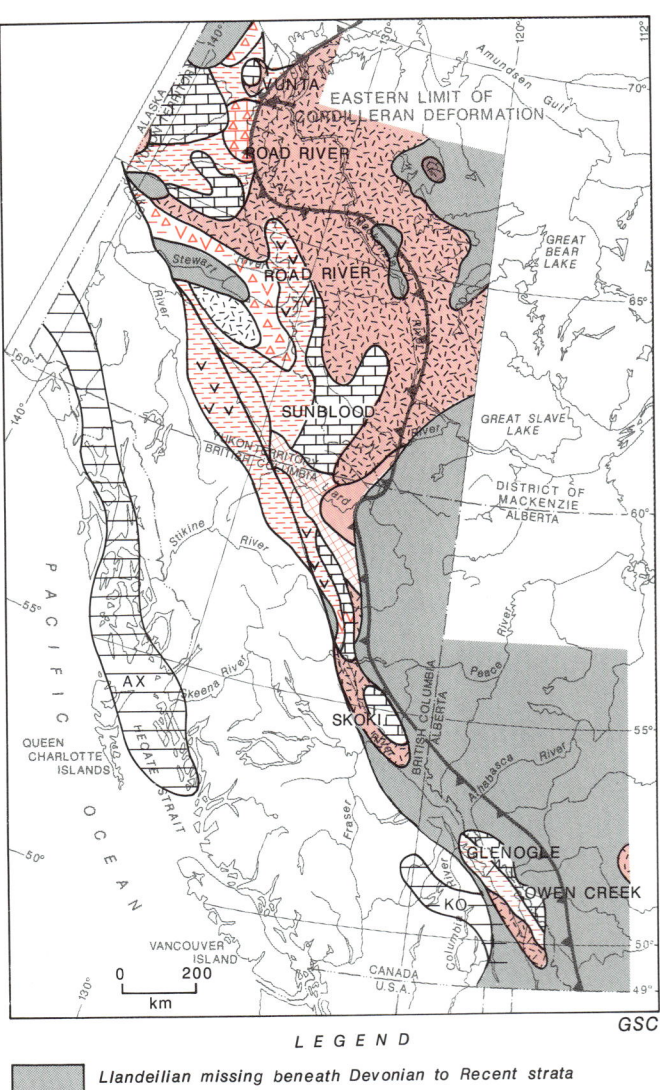

Figure 7.24. Lower Middle Ordovician (Llandeilian) facies (Cycle A) unrestored. Lithological symbols as for Fig. 23.

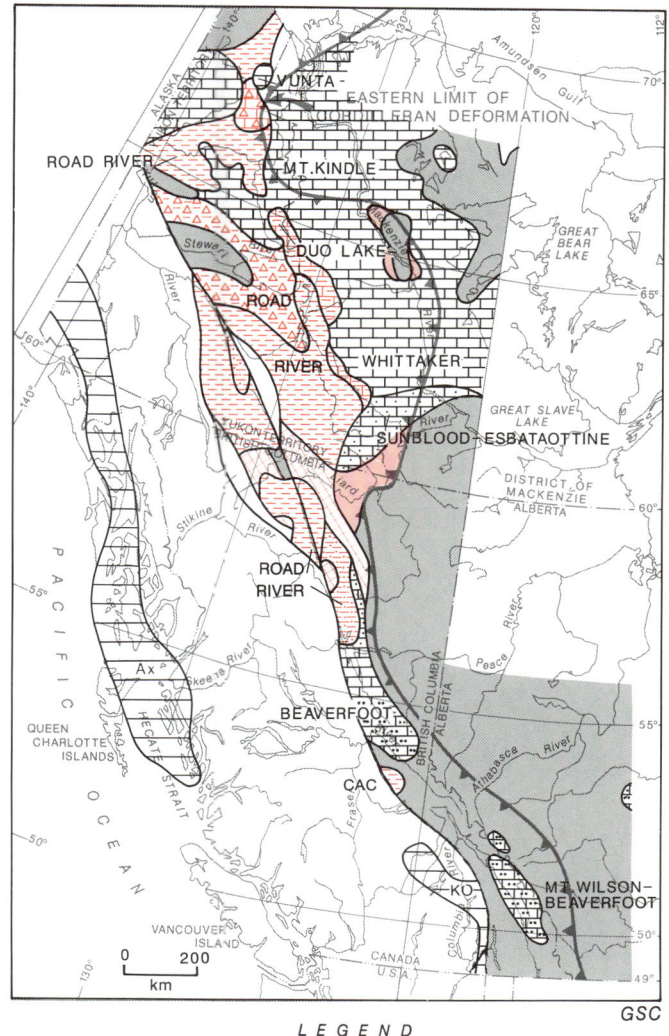

Figure 7.26. Upper Ordovician to Lower Silurian platform carbonate of the Mount Kindle Formation (MK) unconformably overlying Lower Ordovician transitional dolostone of the Franklin Mountain Formation (FM), northwestern Mackenzie Arch. Photo by M.P. Cecile. GSC 199537

LEGEND

- Upper Caradocian to Ashgillian missing beneath Devonian to Recent strata
- Upper Caradocian to Ashgillian missing beneath Ludlovian strata
- Upper Caradocian to Ashgillian missing beneath Llandoverian strata

Figure 7.25. Upper Middle to Upper Ordovician (Upper Caradocian to Ashgillian) facies (Cycle B) unrestored. Lithological symbols as for Fig. 23.

carbonates, the lowermost components of which include yellow-weathering, arenaceous dolostone of the uppermost Silurian and Lower Devonian Tsetso Formation which, at the Mackenzie Mountains front, is up to 1200 m thick.

To the northwest, beyond the Richardson Trough, strata broadly coeval with the Franklin Mountain and Mount Kindle formations were deposited on the Porcupine Platform (Fig. 7.19). The deposition of pale grey microsucrosic dolostone and minor limestone (Fig. 7.23, 7.25) continued throughout the Ordovician but most of the platform subsided during the Silurian with deposition of basinal facies. To the south, the Ogilvie Platform persisted as a region of platform carbonate deposition throughout the Ordovician and Silurian.

In the north-central part of the Richardson Trough the small White Mountains Platform apparently developed from a Cambrian and Lower Ordovician promontory of the Mackenzie Platform (Fig. 7.19). More than 850 m of light grey to greyish cream, thickly bedded micritic and biogenic limestone of the Vunta Formation accumulated throughout the Ordovician and Early Silurian, followed by thinly bedded Upper Silurian limestones and Devonian carbonate strata. A trough of basinal facies strata lay between the White Mountains and Mackenzie platforms in Late Ordovician and Silurian times (Fig. 7.23 to 7.27).

Within the platform regions of the northern miogeocline the Mackenzie (Aitken et al., 1973; Pugh, 1983), Ogilvie (Gabrielse, 1967) and Bulmer Lake (Meijer-Drees, 1974) arches are each cored by Proterozoic strata which in turn are overlain unconformably by condensed Franklin Mountain Formation (on the Mackenzie Platform) or by unnamed dolostone (Wernecke Mountains of the Ogilvie Platform). These platform carbonates are truncated by the sub-Upper Ordovician unconformity.

The intraplatform basins and platformal margin basins of the Ordovician and Silurian miogeocline received a variety of sediments characterized by argillaceous limestone, calcareous shale, siltstone and chert. Slope deposits were prominent adjacent to the platforms. The most widespread unit is the Upper Cambrian to Lower Devonian Road River Formation.

The Richardson Trough was an elongate basin that formed during the late Early Cambrian, perhaps as an aulacogen (Churkin, 1975; Norris and Yorath, 1981) and apparently linked the Cordilleran and Innuitian miogeoclines. The trough was surrounded on three sides by platform carbonates during the Late Cambrian and Early Ordovician when it was the site of a thick succession of yellow-weathering basinal limestone (lower Road River Formation, about 2000 m thick) (Fig. 7.20-Sec. 4, Fig. 7.23). More varied and deeper basinal shale, chert and thin-bedded limestone developed in late Early Ordovician to

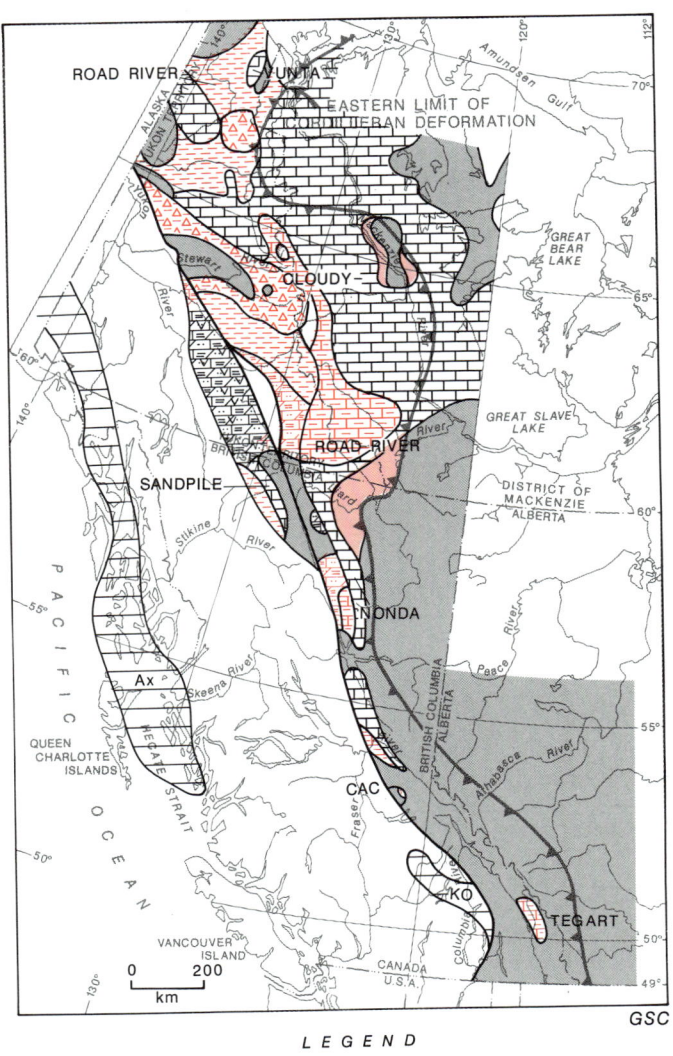

Figure 7.27. Lower Silurian (Upper Llandoverian) facies (Cycle B) unrestored. Lithological symbols as for Fig. 23.

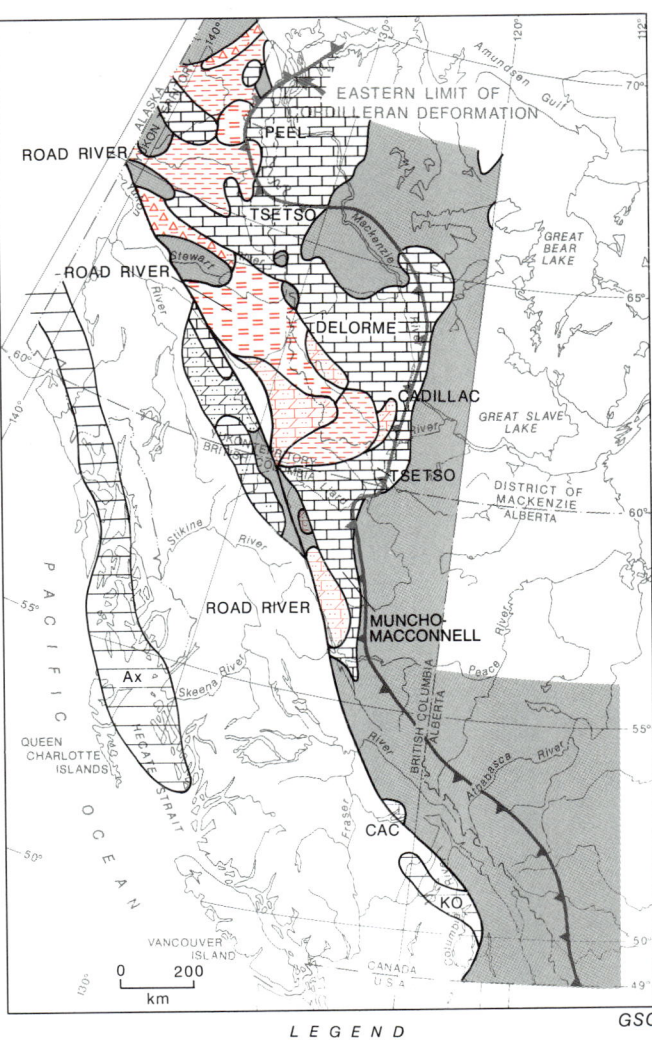

Figure 7.28. Upper Silurian (Ludlovian to Lower Pridolian) facies (Cycle C) unrestored. Lithological symbols as for Fig. 23.

Late Silurian time (upper Road River Formation, about 800 m thick). During the Silurian, both the Richardson Trough and the Blackstone Trough expanded so that shaly facies spread over much of the Porcupine Platform and the Ogilvie Arch (Fig. 7.27). The Blackstone Trough developed as a relatively shallow, westerly trending extension of the southern Richardson Trough in the late Early Ordovician (Fig. 7.24) and received a thin succession of calcareous shale and limestone (upper Road River Formation).

Basinal areas of the northwestern miogeocline were the Selwyn Basin and its extensions into the Mackenzie Platform, namely the Misty Creek, Meilleur River and Prairie Creek embayments (Fig. 7.19). To the south, the Selwyn Basin was connected to the Kechika Basin of the southern miogeocline.

The southern and northern parts of the Selwyn Basin have distinctly different lithofacies. In the north, shale, minor limestone and shaly limestone accumulated during the Early Ordovician (Fig. 7-23). From Middle Ordovician through Late Silurian time the area was the site of deposition of a condensed and monotonous succession of siliceous shale and chert (Fig. 7-25, 7-27, 7-28). Laterally discontinuous lenses of basic volcanic rocks (basalt and tuff up to 500 m thick) accumulated along the northern margin of the basin in Early and Middle Ordovician time (Fig. 7.24). The lack of detritus derived from adjacent platforms and land areas of the Ogilvie Arch might be explained by the volcanics acting as a barrier to basinward transport of sediment. In the southern Selwyn Basin, a variety of basinal sediments were deposited (Rabbitkettle Formation, Kechika Group and Road River Formation (Fig. 7.23)) in Early Ordovician to Early Silurian time. The Rabbitkettle (Upper Cambrian and Lower Ordovician, locally more than 1000 m thick) consists of grey-weathering, silty limestone and distinctly graded calcareous siltstone. Westernmost exposures are more argillaceous and darker weathering and include considerable components of argillaceous limestone and calcareous shale. The correlative Kechika Group (500 to 1500 m thick) includes yellowish grey and pale weathering, thinly bedded silty and argillaceous limestone, nodular calcareous siltstone, shale and calcareous phyllite; both units are overlain diachronously by the Road River Formation. A widespread Upper Silurian, orange-brown, carbonate-rich mudstone and siltstone (up to 140 m thick) is one of the few distinctive members of the Road River Formation in the southern Selwyn Basin. Lower Silurian shales of the Road River Formation in the east-central Selwyn Basin host a large and relatively rich zinc-lead deposit at Howards Pass (60 million tonnes drilled and 350 million tonnes inferred at 7% Pb-Zn) (E.M.R., 1985; Morganti, 1979; Norford and Orchard, 1985).

The Misty Creek Embayment (Cecile, 1982) was a rectangular depression, bounded on three sides by carbonate platforms, and extended across the submarine Niddery High from the northern Selwyn Basin into the western Mackenzie Platform (Fig. 7.19). The Niddery High is characterized by a lower Paleozoic basinal succession that is substantially thinner than elsewhere within the Misty Creek Embayment. The embayment developed in late Early Cambrian time and persisted until the mid-Early Silurian (Fig. 7.29). During the Late Cambrian and Early Ordovician it received yellowish grey weathering, thinly bedded silty limestone (Rabbitkettle Formation 200 to 900 m thick, Cecile, 1982; Fig. 7.23) with spectacular slope-breccias and slump features in transitional platform to basin successions. Subsequently the embayment was the site of a moderately thick sequence of shale, siliceous shale, limestone and chert, (Duo Lake Formation, 100-400 m thick; Fig. 7.24). During the Middle Ordovician to Middle Devonian, volcanic and volcaniclastic rocks (Marmot Formation, 0 to 500 m thick; Cecile, 1982) erupted from a centre or centres in the south-central part of the embayment. The eruptive rocks include lapilli tuff, fine grained breccia, massive amygdaloidal alkalic basalt flow rocks, sills and coarse breccia. These accumulated along with sandstone, siltstone and argillite. The Marmot Formation occurs as tongues within the upper part of the Duo Lake Formation and the succeeding Cloudy Formation (Upper Ordovician to mid-Lower Silurian, 0 to 400 m thick; Fig. 7.27) that consist of thinly bedded shaly limestone and minor shale and chert. From Late Silurian into Devonian time the site of the Misty Creek Embayment became a shallow-water platform (Fig. 7.28) which was, at times, emergent.

The Meilleur River Embayment (Bassett and Stout, 1966) was a large semicircular embayment that extended into the southern Mackenzie Platform from the southern Selwyn Basin (Fig. 7.19). During the Early Ordovician it was a minor recess into the Mackenzie Platform. In the Middle Ordovician, it was replaced by a platform on which carbonates of the Sunblood and related formations were deposited (Fig. 7.25). The Sunblood (up to 1430 m thick) consists of conspicuously banded grey and greyish yellow weathering dolostone, greyish pink and grey mottled limestone, greyish orange weathering sandstone and siliceous dolostone (Plate 12). Locally, mafic volcanic flows form a conspicuous member up to 50 m thick. The overlying and partly equivalent Esbataottine Formation (Ludvigsen, 1975) consists of between 80 and 200 m of grey, thinly bedded argillaceous and silty limestone. In the Late Ordovician to Early Silurian, basinal settings returned when Road River thin-bedded shaly limestone and shale accumulated. In the Late Silurian, the embayment expanded eastward where a satellitic feature, the Prairie Creek Embayment (Morrow, 1984) and parts of the Meilleur River Embayment received up to 700 m of carbonate-rich, orange-weathering siltstone assigned to the Cadillac Formation (Fig. 7.28). Road River shale continued to be

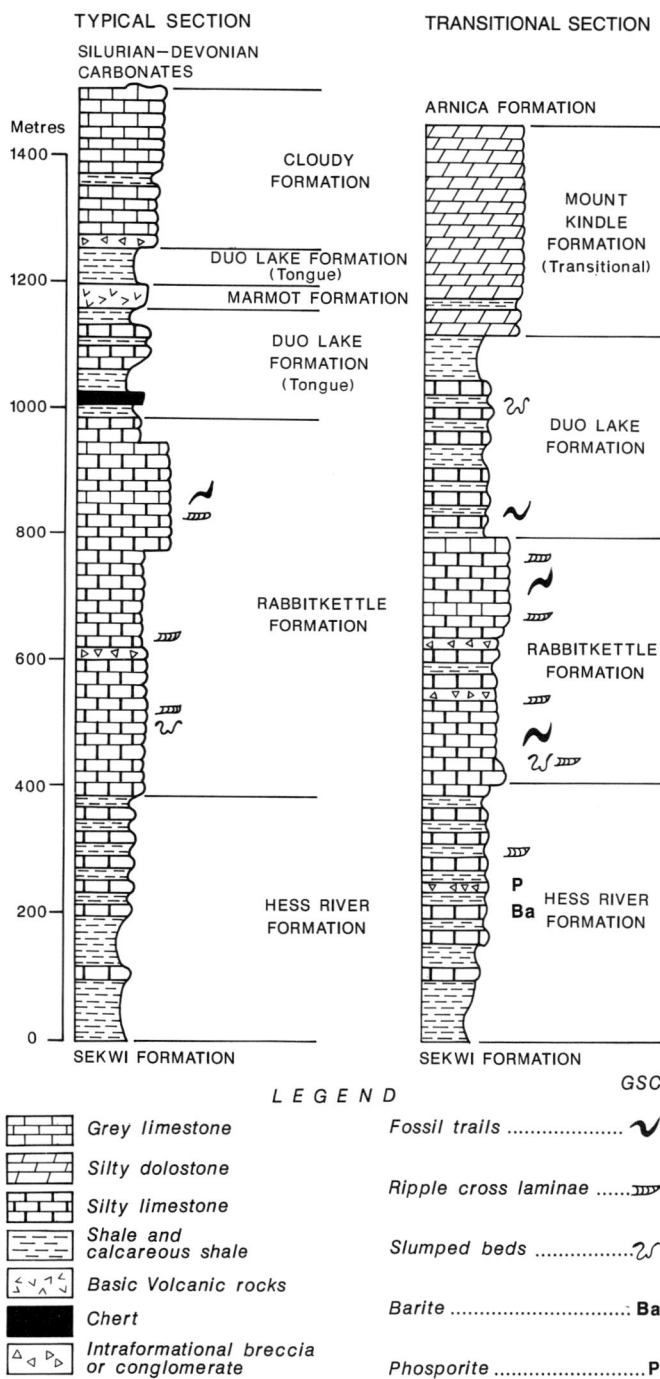

Figure 7.29. Typical basinal facies and transitional facies sections, central and northern Misty Creek Embayment (from Cecile, 1982).

deposited elsewhere in the Meilleur River Embayment. Both the Meilleur River and Prairie Creek embayments were located in the central part of the Liard Depression, a long-lived depocentre that developed in the Ordovician and continued into Devonian time.

Southern miogeocline (south of 60°N)

Ordovician and Silurian paleogeographic elements of the southern miogeocline comprised the MacDonald (Douglas et al., 1970), Kakwa and Bow platforms (introduced herein) to the east and White River Embayment which were flanked to the west by the Kechika and Robson basins (Fig. 7.19). The platforms and troughs currently are represented by discontinuous outcrop areas because of subsequent erosion. Most probably they were connected depositionally with each other and with basinal and platform areas of the northern miogeocline.

On the MacDonald, Kakwa and Bow platforms the same three cycles that developed on Mackenzie Platform are represented but less completely. On the Bow Platform the upper part of Cycle A (Fig. 7.23, 7.24) is expressed by the Upper Cambrian to lower Middle Ordovician Survey Peak, Outram, Skoki and Owen Creek formations comprising shallow-water limestones, dolostones and limy shales. At their standard Ordovician section at Mount Wilson (Fig. 7.30, 7.31) their combined thickness is about 1000 m (Norford, 1969; Aitken and Norford, 1967). On the Kakwa and MacDonald platforms, Cycle A carbonate strata are absent, but, because equivalent strata in the Kechika Basin thin dramatically and show evidence of increasing intrastratal erosion and truncation eastward, Lower Ordovician carbonate probably was present on the platforms but later was removed by erosion. On the southern Bow Platform and on the Kakwa Platform, the Lower Ordovician Tipperary (175 m) and Monkman (500 m) quartzites respectively reflect erosion of a broad land area that included Peace River Arch.

On the Bow Platform, Cycle B followed Middle Ordovician regression and erosion and is represented by the Upper Ordovician to Lower Silurian Mount Wilson, Beaverfoot and Tegart formations (Fig. 7.25, 7.27). The Mount Wilson (up to 500 m thick) is massive and thickly bedded, white sandstone and quartzite. A break is present below recessive dolomitic quartz sandstone, shale, arenaceous and argillaceous dolostone (Whiskey Trail Member, up to 35 m thick) and resistant, medium- to thickly bedded carbonates of the Beaverfoot Formation (up to 550 m). The Beaverfoot Formation hosts several barite deposits in southeastern British Columbia only one of which (Brisco) is currently in production (Manson, 1984). Ordovician and Silurian rocks are host to many diatremes in the same region. Two episodes of diatreme emplacement are indicated, late Middle Ordovician and Middle Devonian (Mott, 1989). This platformal sequence was followed by basinal facies argillaceous limestone, limestone and shale of the Tegart Formation (75 m). On the MacDonald and Kakwa platforms, Cycle B strata are represented by two sub-cycles separated by an unconformity representing latest Ordovician to earliest Silurian regression and erosion. The first documents Late Ordovician transgression and deposition of dolostone and quartzite (up to 660 m thick); and the second, Early Silurian deposition of dolostone, quartzite and dolomitic sandstone (Nonda Formation 300-700 m; Fig. 7.27). The prevalence of quartz sand in Cycle B strata reflects continued contributions from cratonic source areas. As noted earlier, it is unlikely that these included the presently exposed Canadian Shield but rather a broad area situated between the Williston Basin and Mackenzie Platform and including the Peace River Arch where Devonian strata rest unconformably upon Precambrian rocks.

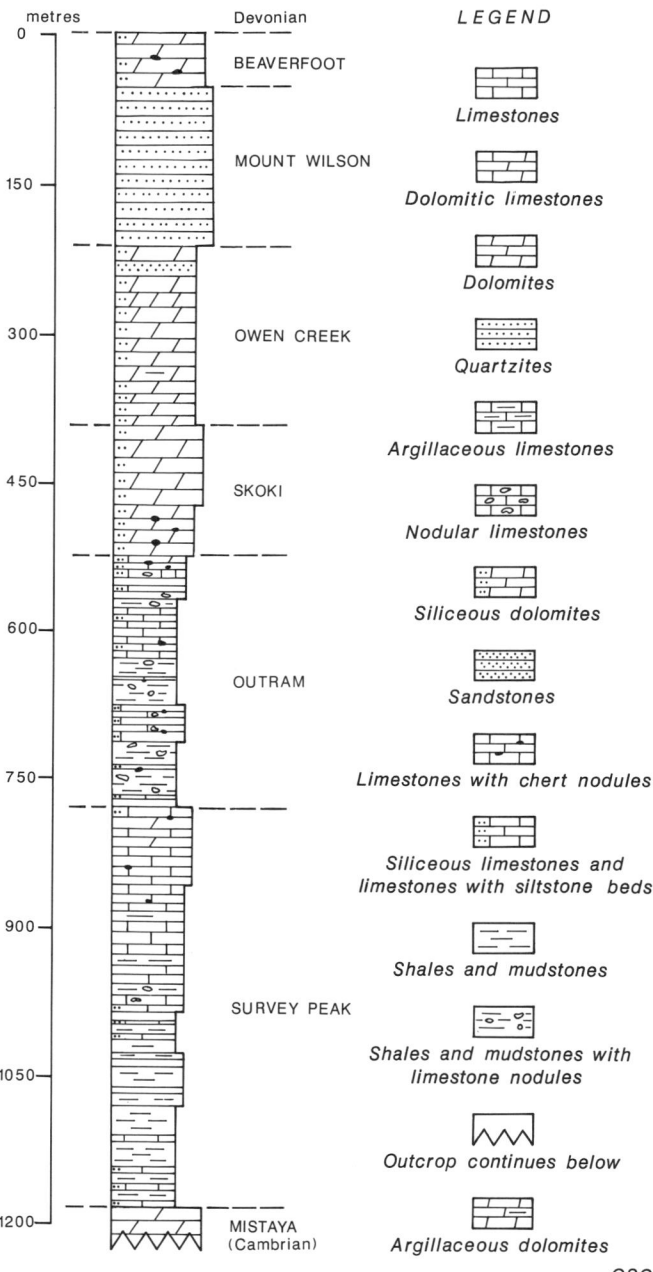

Figure 7.30. Ordovician stratigraphic section at Mount Wilson, Banff National Park (from Norford, 1969). See also Figure 7.31.

Figure 7.31. Ordovician strata exposed on Mount Wilson along the boundary of the White River Embayment and the Bow Platform, Banff National Park B, Beaverfoot Formation; MW, Mount Wilson Formation; S, Skoki Formation; O, Outram Formation; SP, Survey Peak Formation. Photo by B.S. Norford. GSC 118948

Figure 7.32. View northerly on west side of MacDonald Platform to shale-out of Nonda Formation carbonate. Photo by R.I. Thompson. GSC 205235-G

On the Bow Platform, no Ordovician and Silurian strata younger than Cycle B are known. On the MacDonald and Kakwa platforms, Cycle C essentially is identical to that preserved farther north and is represented by the Upper Silurian to Lower Devonian yellowish grey, thick-bedded dolostone of the Muncho-McConnell Formation (120 m thick; Fig. 7.28) and later Early Devonian regression and erosion.

The platform-flanking basinal regions of the southern miogeocline are represented by the Kechika and Robson basins and the White River Embayment, now separated by areas of older outcrops.

The Kechika Basin (Douglas et al., 1970) adjoins the southern end of the Selwyn Basin at the arbitrary boundary of 60° N latitude. To the east it is bordered by the MacDonald Platform and to the west by the present location of the Cassiar Platform. The southern limit is the flank of the Peace River Arch. During the Early Ordovician, most of the Kechika Basin was filled by a westward thickening succession (500 to 1500 m) of thinly bedded to nodular, yellowish grey and grey weathering, silty and argillaceous limestone (Kechika Group). Easternmost exposures display abundant thinly bedded intraformational breccia, and conglomerate and evidence for intrastratal erosion. Western exposures are more argillaceous and clearly basinal. From Middle Ordovician to Late Silurian time, the eastern margin of the basin was a zone of abrupt transition from platformal to basinal facies (Fig. 7.32) with abundant foreslope breccia conglomerate and debris flows. Some debris flows are composed entirely of platformal rocks. The foreslope breccias show a wide variety of textures including those associated with *in situ* brecciation and coarse grained carbonate turbidite. An extensive, massive microdioritic flow (up to 50 m thick) overlies Middle Ordovician shale and itself is overlain by interbedded shale and orange to brown weathering, carbonate rich, vitric, crystal and lapilli tuff (MacIntyre, 1981). A conspicuous member of the Road River strata in the Kechika Basin is a platy, greyish orange weathering dolomitic siltstone succession (up to 800 m thick) that can be traced northward into the southern Selwyn Basin. Wenlockian (late Early Silurian) fossils are known from part of this unit.

Phosphorite pavements occur in Lower Ordovician basinal limestone facies of the Kechika Group (Cecile and Norford, 1979), barite is found in Ordovician shale and tuff, and zinc is locally present in Silurian thin-bedded calcareous and argillaceous strata.

Farther south, Ordovician and Lower Silurian basinal facies are preserved as remnants beneath the sub-Devonian unconformity (Fig. 7.33). In the Robson Basin, only Lower Ordovician strata remain. In White River Embayment, the McKay Group (Upper Cambrian to Lower Ordovician, more than 2000 m thick) consists of light grey, pale greenish grey and yellowish grey to buff, calcareous slate, shale and phyllite, silty limestone, calcareous siltstone and argillaceous limestone (Fig. 7.23). These strata are overlain by recessive shale, limestone, argillaceous limestone and dolostone of the Glenogle Formation (Lower to Middle Ordovician up to 700 m; Fig. 7.24). The carbonate content of the Glenogle increases dramatically eastward towards the laterally equivalent Outram, Tipperary, Skoki and Owen Creek formations on the Bow Platform. Shallowing and an influx of quartz silt took place throughout the White River Embayment in late Middle Ordovician time. Next the area became land and was eroded prior to deposition of the upper Middle to Upper Ordovician Mount Wilson Quartzite. The White River Embayment was suppressed during Mount Wilson time and also during Beaverfoot time when platformal carbonates reached as far west as the Purcell Arch (Norford, 1981). Toward the close of the Early Silurian, a deeper environment returned with deposition of thin, wavy bedded limestone and shale of the Tegart Formation which is preserved only as isolated remnants (Fig. 7.27). No Upper Silurian rocks are preserved beneath the sub-Devonian unconformity.

In southernmost central British Columbia near the International Boundary, a small area of Active Formation (300 to 1500 m thick; Fig. 7.19) preserves outer basinal

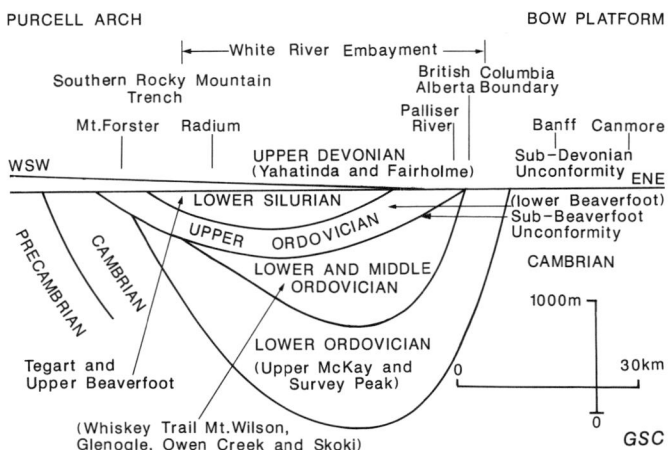

Figure 7.33. Sub-Devonian unconformity and cross-section, Bow Platform-southern Robson Basin.

Figure 7.34. Upper Ordovician carbonate of the Beaverfoot Formation overlying Middle Ordovician quartzite of the Mount Wilson Formation (MW) which in turn overlies Lower to Middle Ordovician graptolitic shale and argillaceous limestone of the Glenogle Formation. Strata occur along the boundary between the Bow Platform and White River Embayment. Photo by B.S. Norford. GSC 119004

facies rocks that are considered to be part of autochthonous North America. These rocks are sooty black argillite and slate with units of limestone, arenaceous slate, argillaceous quartzite and quartzite (Little, 1960). The Active correlates with the Ledbetter Formation in adjacent Washington State where early Ordovician (late Arenig) and Middle Ordovician (Llanvirn and Caradoc) graptolite faunas are present (Dings and Whitebread, 1965; Barnes et al., 1981).

Cordilleran terranes
Cassiar Terrane

The most complete and continuous Ordovician and Silurian sequence in the Cordilleran terranes is in the displaced Cassiar Terrane (Fig. 7.19, 7.27). The stratigraphic succession has many features in common with those in the Rocky Mountains and thus are assigned to the Rocky Mountains Assemblage (see the Tectonic Assemblage Map, Map 1712A, in pocket). The upper part of the Kechika Group in the Cassiar Mountains consists of thinly bedded, wavy banded, argillaceous limestone, calcareous phyllite and shale, increasingly thinner and more argillaceous to the west. Unlike younger strata, facies changes are relatively gradual. This assemblage is probably in part Early Ordovician in age. Two distinct facies in younger strata reflect platformal and basinal environments of deposition. Upper Ordovician platform carbonate of limited lateral extent is overlain by widespread Lower Silurian platformal carbonate of the Sandpile Group (up to 500 m thick). In places these Sandpile rocks overlie Middle Ordovician shale disconformably.

In the basinal environment, dark grey, pyritic shale (up to 150 m thick, Lower to Upper Ordovician) is followed by platy, Silurian siltstone and shale (up to 50 m thick, Lower Silurian, Upper Llandovery and Lower Wenlockian). In places the Silurian siltstone is disconformable on the Kechika Group. The facies changes between the platform carbonates and the adjacent basinal shale and siltstone are not exposed because of regional faults but the western margin of the Cassiar Terrane is entirely in basinal facies, with a thin Ordovician and Silurian sequence overlain by fine grained clastic rocks of probable Early Devonian age. Elsewhere in the basinal facies and on the platform, distinctive crossbedded 'tapioca' sandstone and sandy dolomite of possible Early Devonian age disconformably overlie the Silurian siltstone and shale unit and the Sandpile Group.

In the Cariboo Subterrane of the Cassiar Terrane the Black Stuart Formation includes greyish black pyritic mudstone and graptolitic pelite of Ordovician age with a laterally discontinuous unit of Upper Silurian to Lower Devonian platform carbonate (Struik, 1981).

Kootenay Terrane

Lower Paleozoic rocks of the Kootenay Terrane in Kootenay Arc include much of the Lardeau Group along a belt 20 to 50 km wide by 250 km long within southernmost Omineca Belt (Fig. 7.35). As discussed in the Cambrian sub-chapter the base of the Kootenay Terrane probably lies along a talc schist zone near the top of the Index Formation (Zwanzig, 1973; G.E. Gehrels and M.T. Smith, pers. comm., 1987). The Index (up to 750 m) is the lowest formation of the Lardeau Group (Fyles and Eastwood, 1962) and, since it lies almost entirely beneath the talc schist zone, it is composed largely of distal sediments related to the ancestral margin of North America. The succession above the talc schist zone is in ascending order: (1) the Triune Formation (up to 250 m) is greyish black siliceous argillite; (2) the Ajax Formation (about 250 m) is thickly bedded quartzite with limy concretions; (3) the Sharon Creek Formation (about 250 m) is greyish black siliceous argillite and recrystallized ribbon chert (Fyles and Eastwood, 1962). These three units can be interpreted as outer basinal sediment related to the margin of ancestral North America. The fourth unit, the Jowett Formation (about 800 m), consists of massive and pillowed tholeiitic lava, aquagene tuff and breccia overlain by, and intertongued with a fifth unit of foliated quartzite, quartz-feldspar grit, and grey, maroon, and green phyllite and limestone assigned to the

CAMBRIAN TO MIDDLE DEVONIAN ASSEMBLAGES

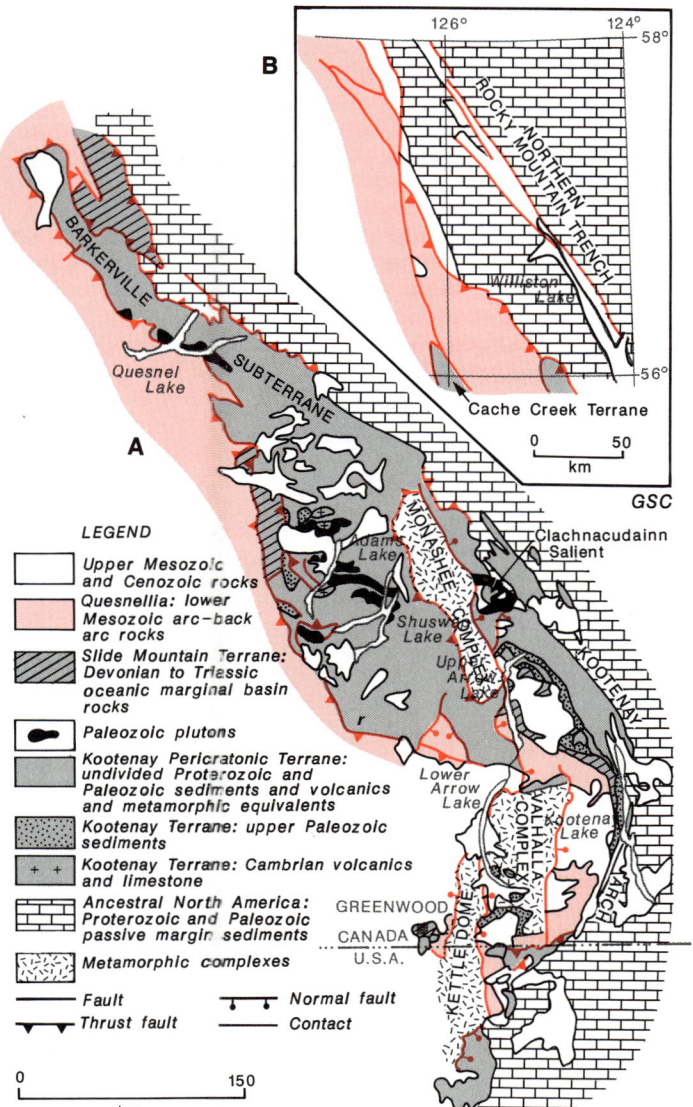

Figure 7.35. Distribution of Kootenay Terrane and its relationship to adjacent terranes in (A) southern Omineca Belt and (B) in Swannell Ranges of central Omineca Belt.

Broadview Formation (more than 1000 m). The grits contain blue quartz and are not unlike those of the upper Proterozoic Windermere Supergroup. Quartzite and phyllite units commonly are graded and resemble similar rocks in the younger Milford Group; near the Kuskanax batholith, Broadview rocks can be distinguished by their content of quartz lenses and veins. West of the Battle Range batholith, the Broadview is intimately folded with Index phyllites indicating that the Broadview also is a western facies of the Triune to Sharon Creek succession (Read and Wheeler, 1977). Although it is possible that the sands of the Broadview Formation could have been derived from the craton, bypassing the continental margin sediments (presumably the Beaverfoot to Fairholme interval), a western source of crystalline rocks outboard of ancestral North America is more likely.

No fossils have been recovered from the Lardeau Group. However, boulders of the Broadview Formation identified within the basal conglomerate of the overlying Carboniferous Milford Group have yielded an Early to Middle Ordovician Rb-Sr whole rock isochron date of 479 ± 17 Ma (Read and Wheeler, 1977). Orthogneiss which intrudes correlatives of the Lardeau Group in the Clachnacudainn salient east of Revelstoke has been dated at 367 to 422 Ma (i.e. Late Silurian to Late Devonian; Okulitch, 1985). Accordingly, the Lardeau Group is probably largely of Early Ordovician age.

Farther west, the Kootenay Terrane of the Adams Lake region (Fig. 7.35) southwest-verging thrust sheets contain strata which may be in part correlative with the Lardeau rocks. A succession of grey and greyish black, graphitic, siliceous and calcareous phyllite (Schiarizza and Preto, 1984) is similar to the Index Formation of the Kootenay Arc. In addition, a unit of grey and green phyllitic sandstone, grit, phyllite and quartzite with minor limestone and greenstone containing a distinctive metabasalt at the top is lithologically similar to the upper Lardeau Group. A.V. Okulitch (pers. comm., 1986) restudied outcrops of unfossiliferous grey limy phyllite and limestone, formerly included in the Sicamous Formation and considered to be early Mesozoic (Okulitch, 1979). He now suggests that they may be equivalent to the lower part of the Lardeau Group. Similarly Okulitch speculates that greenstones (formerly included either as part of the Nicola Assemblage or the Tsalkom Formation; Okulitch, 1979) may be equivalent to the Jowett Formation of the Kootenay Arc. Around the Chase Antiform, a sharp metamorphic transition takes place at the base of the Tsalkom Greenstone. The underlying rocks are of higher metamorphic grade and are intruded by numerous granitic sills related to the Ordovician Little Shuswap orthogneiss dated by a U-Pb zircon date at 452 +17/-19 Ma (Okulitch, 1985).

Farther north the upper part of the Snowshoe Group, underlying the Barkerville Subterrane of the Kootenay Terrane, is more than 1000 m thick. It contains the following succession in ascending stratigraphic order: dark grey micaceous quartzite and interbedded dark grey phyllite and dark grey limestone (an assemblage similar to the Index Formation of the Kootenay Arc and to rocks within the Eagle Bay Formation of the Adams Lake Region and above its Lower Cambrian limestone unit); coarse feldspathic grit, phyllite, marble and quartzitic conglomerate that are similar to parts of the Broadview Formation; grey quartzite, phyllite, limestone, and mafic pyroclastic which could be equivalent to either the Jowett Formation of Kootenay Arc or to the mafic volcanics within the Eagle Bay Formation. The highest unit of the Adams Lake succession (?Snowshoe Group) is a marble which can be traced for over 100 km beneath the Pleasant Valley Thrust Fault. Poorly preserved algae, echinoderm, bryozoan and possible ostracode fragments in the marble indicate a Paleozoic age.

Accreted terranes

Ordovician and Silurian rocks occur in several other terranes of the Canadian Cordillera. In the Alexander Terrane biostratigraphic control is sparse and insufficient for detailed paleogeographic reconstructions. Nevertheless, it appears that Paleozoic strata can be divided into two major groups. A lower succession (Descon Assemblage) includes Lower Silurian, Ordovician and older mudstone and greywacke-turbidite with interstratified units of chert and limestone. These strata commonly are juxtaposed laterally with basaltic and andesitic volcanics and volcaniclastics. An upper succession in southeastern Alaska consisting of Silurian to Lower Devonian(?) carbonate rocks, both shoal and basinal facies (Kaskawulsh Assemblage), is interstratified and laterally juxtaposed with polymictic conglomerate (including volcanic clasts), mudstone-greywacke turbidite, and mafic to intermediate volcanic rocks (Karheen Assemblage; Gehrels and Berg, 1984; Campbell and Dodds, 1979; Jones et al., 1972). The polymictic conglomerate was derived from uplifted Ordovician-Lower Silurian volcanic and plutonic rocks and reflects a significant tectonic event in the southern part of the Alexander Terrane that is not recorded elsewhere in the Canadian Cordillera. This event, the Klakas Orogeny (Gehrels et al., 1983) produced imbrication along west-verging thrust faults and local metamorphism to greenschist to amphibolite facies.

Elsewhere in the Cordilleran terranes Ordovician and Silurian strata have not been documented. They may be represented within fine grained siliceous clastic rocks of the Nisling Terrane in Yukon Territory (Nasina Quartzite, in part), in the lower part of the volcanic Nitinat Formation of Wrangellia on Vancouver Island and in a volcanic unit west of Work Channel in the central Coast Mountains.

Volcanic and intrusive rocks

Mafic volcanic rocks occur within the Kootenay and Alexander terranes. In the Lardeau Group of the Kootenay Terrane, two major volcanic units are composed of greenstone, green phyllite and calcareous green phyllite, within which rare volcanic breccia and pillow structures have been recognized (Read and Wheeler, 1977). Volcanic strata are abundant within Ordovician and older strata of the Alexander Terrane, as well as some volcanic lenses within Silurian carbonate successions. Most of these rocks have been identified in the field as andesitic and basaltic volcaniclastics, breccias and tuffs, although flows and some pillowed flows have been recognized (R.B. Campbell, pers. comm., 1985; Campbell and Dodds, 1979).

Numerous occurrences of volcanic rocks are known within miogeoclinal Ordovician and Silurian strata of western Canada. Most are in the north, firstly along the western fringe of the Mackenzie, MacDonald and Ogilvie platforms and on the Ogilvie Arch, and secondly in the Selwyn Basin, Kechika Basin and the Nasina Basin, near the Tintina-Northern Rocky Mountain Trench. The only well documented occurrence south of latitude 56° is a volcanic and intrusive complex interstratified with Lower Silurian strata on the west edge of the Kakwa Platform (Taylor et al., 1972). Mafic dykes, sills, volcanics and diatremes are present within Robson Basin and the White River Embayment but only one episode of diatreme emplacement is dated as an Ordovician event.

The autochthonous volcanic rocks typically occur as small widely spaced lenses units a few tens to hundreds of metres thick and a few kilometres long. They are highly altered and commonly contain abundant secondary minerals such as carbonate, sericite, chlorite, and leucoxene.

Many of the successions consist of, in order of abundance, lapilli tuffs, tuffs, volcaniclastics, flows, rare pillowed flows and rare breccias. In areas where accumulations are relatively thick, volcanic breccia, volcaniclastic conglomerate, and a capping succession of shallow-water carbonate usually are present. In three areas where geochemical analyses have been made, the volcanics are subalkalic to alkalic basalt and, at one location, andesite. This alkalic chemistry is illustrated in some localities by the presence of euhedral to anhedral biotite phenocrysts (Goodfellow et al., 1980a,b; Cecile, 1982). Highly altered, lenticular, alkalic volcanics of this type are found interstratified with sediments as old as Proterozoic and as young as Middle Devonian; the majority of occurrences are late Early to Middle Ordovician in age.

Four different types of Ordovician and Silurian intrusive rocks are known in western Canada. Dykes and sills are found mainly in association with volcanic strata. Very small mafic to ultramafic, highly altered diatremes are present in the east-central Misty Creek Embayment and in the adjacent platform (Oldershaw, 1981; McArthur et al., 1980). One of these, named the Mountain diatreme, is a kimberlite containing microdiamonds. It has given K-Ar and Rb-Sr dates of 445 ± 17 Ma and 427 Ma, respectively (Godwin and Price, 1986). Small sodalitic syenite intrusives are found with Silurian volcanics in the western Kakwa Platform (Taylor et al., 1972). Large granitic plutons in the Kootenay Terrane yield Ordovician isotopic ages (Okulitch, 1985).

Biostratigraphic zonation

Graptolites are the key fossils for intercontinental correlation of Ordovician and Silurian strata (Fig. 7.36). The succession of faunas found in the Richardson Trough and Selwyn Basin is one of the most continuous sequences in the world, extending from almost the base of the Ordovician to near the top of the Lower Devonian (Jackson, 1966; Jackson and Lenz, 1962; Norford, 1964; Jackson et al., 1965). Shorter sequences are present in the Kechika Basin and in the White River Embayment. Documentation of conodont and shelly fauna has begun and conodont zonation has been established in many localities (Tipnis et al., 1978; Kennedy and Barnes, 1981). Of the macrofauna, trilobites appear to be the most biostratigraphically sensitive (Ludvigsen, 1979; Chatterton and Perry, 1983). The interstratification of basinal and platformal facies in transitional belts provides good calibration between macrofossil, conodont, and graptolite zones.

LOWER AND MIDDLE DEVONIAN ASSEMBLAGES

D.W. Morrow and H.H.J. Geldsetzer
(with contributions on Cordilleran terranes by H. Gabrielse)

Ancestral North American miogeocline

Distinctive Lower and Middle Devonian strata of the miogeocline, consisting dominantly of carbonate, form an important component of the Rocky Mountains Assemblage north of latitude 56°N (Fig. 7.37). Farther south they are preserved only locally beneath a sub-Upper Devonian unconformity. As with older Paleozoic successions they

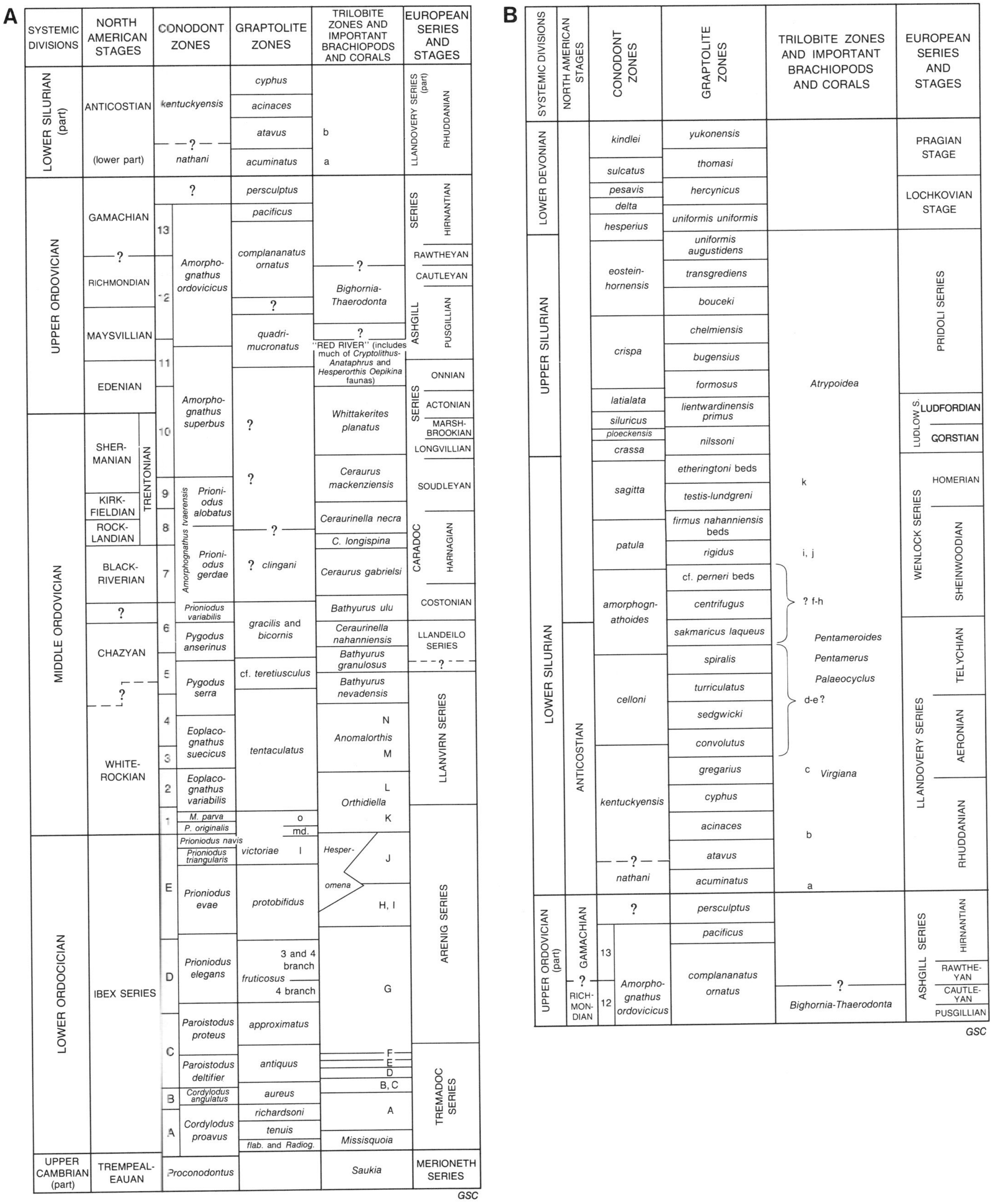

CAMBRIAN TO MIDDLE DEVONIAN ASSEMBLAGES

Figure 7.36. (A) Ordovician and (B) Silurian biostratigraphic zones. The Ordovician zonal terminology is that of Barnes, Norford and Skevington (1981) with modifications from Lenz and McCracken (1982). That for the Lower Silurian follows Lenz (1980, 1982) but uses the name *cyphus* Zone for his *gregarious* Zone, the name *gregarious* Zone for his interval *triangulatus* Zone to *argenteus* Zone. That for the Upper Silurian and Lower Devonian follows Jackson, Lenz and Pedder (1987) modified by the placement of the *formosus* Zone in the basal Pridoli. For the macrofossil zones a-k are the informal divisions of Chatterton and Perry (1984). A-N is the Nevada terminology after Ross (1949,1951) and Hintze (1953). A-E, 1-13 are the customary North American assemblages for the conodont zones.

Devonian rocks of the Foreland Belt are the source of a wide variety of economically important resources, many of which remain undeveloped. They contain oil and gas reservoirs in the Foothills of British Columbia, Alberta and in the Yukon. Sulphur is an important commodity derived from sour gas produced from deep carbonate reservoirs in the Foothills belt. Many lead-zinc deposits occur both in platform carbonates and in laterally equivalent basinal shale throughout the eastern part of the Cordillera. Limestones, such as those of the Palliser Formation, are an important source of raw material for the manufacture of cement.

Principal stratigraphic subdivisions

The most striking stratigraphic subdivision within Devonian strata in the northern Cordillera is that between dominantly carbonate Lower and lower Middle Devonian sediments and mainly siliciclastic upper Middle and Upper Devonian strata (Fig. 7.41). The boundary between them is approximately coincident with the unconformity marking the base of the mid-Givetian Taghanic Onlap

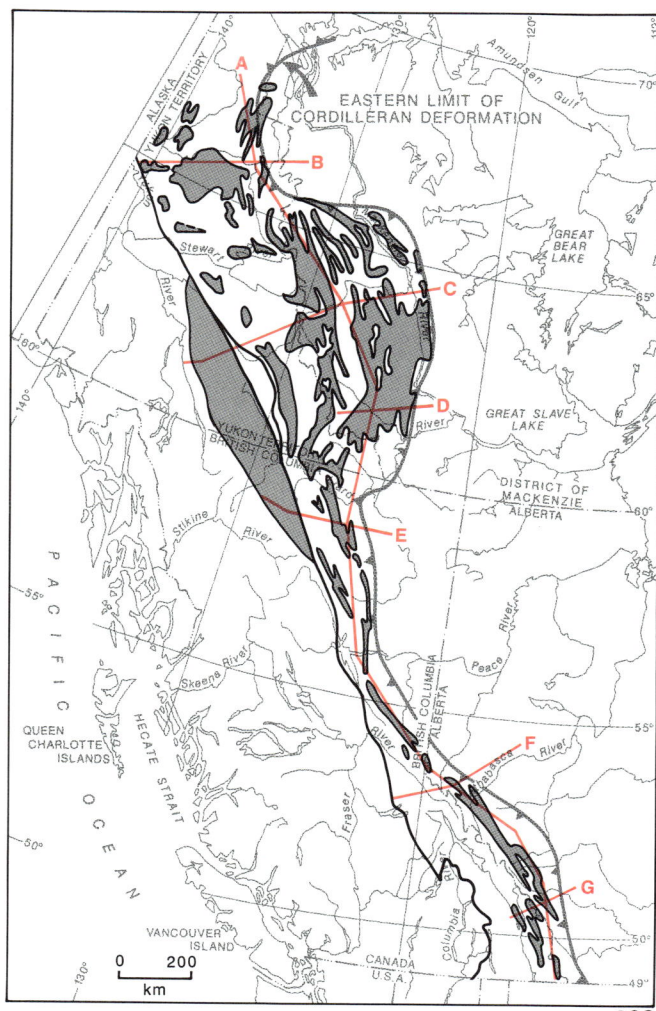

Figure 7.37. Distribution of miogeoclinal Devonian rocks in outcrop in the Canadian Cordillera showing locations of stratigraphic cross-sections.

thicken markedly westward and change facies abruptly into fine grained basinal rocks along a boundary between the inner and outer shelves of the miogeocline (Fig. 7.38, 7.39, 7.40). Facies and thickness changes are related to a variety of tectonic elements including basins, troughs, embayments, arches, highs, ridges and uplifts.

In the context of worldwide Devonian paleogeography, the Cordilleran Orogen is widely recognized to fall within tropical to subtropical latitudes based on predominant content of carbonate and evaporite. More controversial, however, is the placement of the Devonian paleoequator. Most recent studies (Chalmer and Lawson, 1985; Ettensohn and Barron, 1981; Ziegler et al., 1979) place the paleoequator approximately at the latitude of the Great Lakes with an inclination of about 30° to the present equator so that the Canadian part of the Cordillera lay between 20° and 30° north paleolatitude. Some studies, however, place the orogen in the southern hemisphere (Heckel and Witzke, 1979). The former interpretation is more consistent with paleogeographical and sedimentological data.

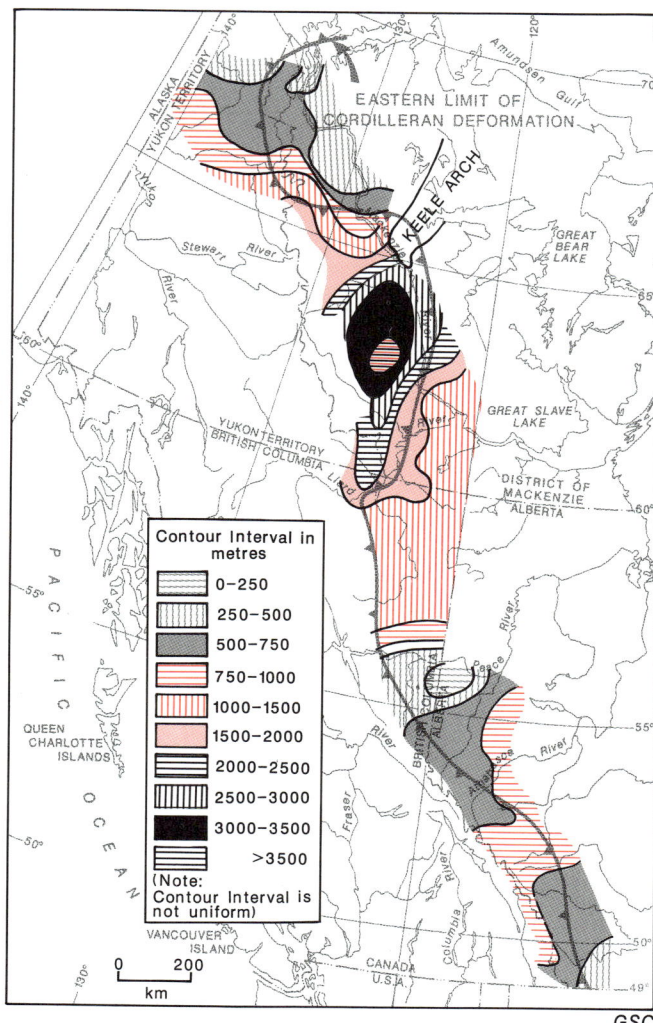

Figure 7.38. Isopach map of all Devonian strata in the eastern Cordillera.

CAMBRIAN TO MIDDLE DEVONIAN ASSEMBLAGES

bounding discontinuities (North American Commission on Stratigraphic Nomenclature, 1984). However, some of these sequences may not be the product of single overall transgressive-regressive events as are the sequences formally defined by Vail et al. (1977). Instead, they are objectively defined stratigraphic packages of laterally equivalent formations that are separated by thin intervals or discontinuities across which tectonostratigraphic relationships are markedly different.

Delorme sequence (Gedinnian-Siegenian)

The Delorme sequence (Table 7.1) records the earliest stage of widespread Devonian transgression on the continent. Sedimentation began in latest Pridolian to Gedinnian time immediately following the period of widespread regression and subaerial erosion in the Late Silurian (Fig. 7.43). This sequence is composed of carbonate strata characterized by a considerable content of mud and sand

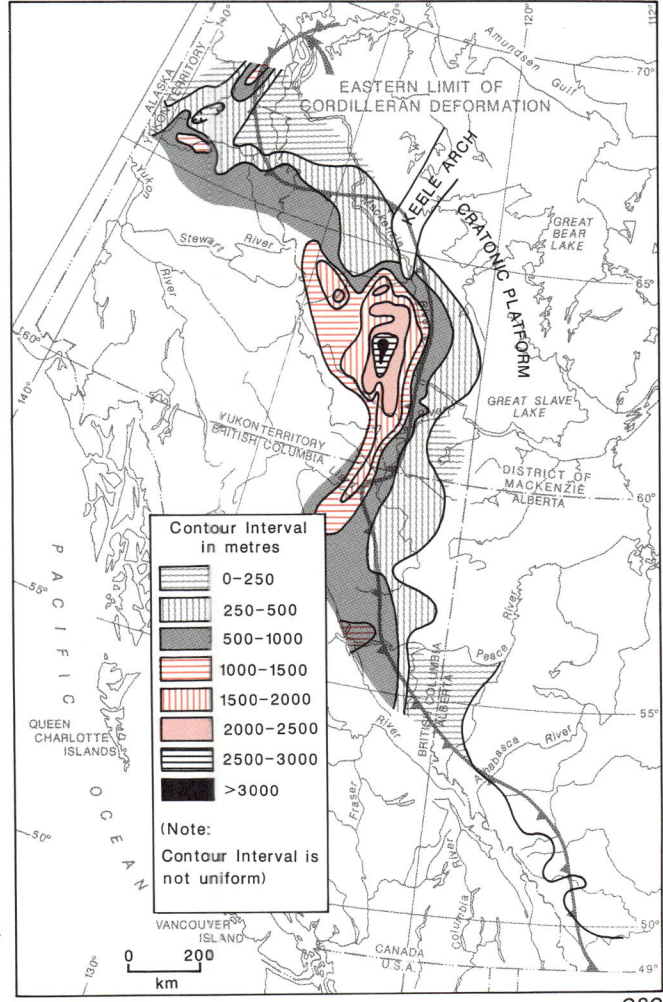

Figure 7.39. Isopach map of the 'sub-Taghanic' or Pridoli to mid-Givetian interval of the Devonian System in the eastern Cordillera.

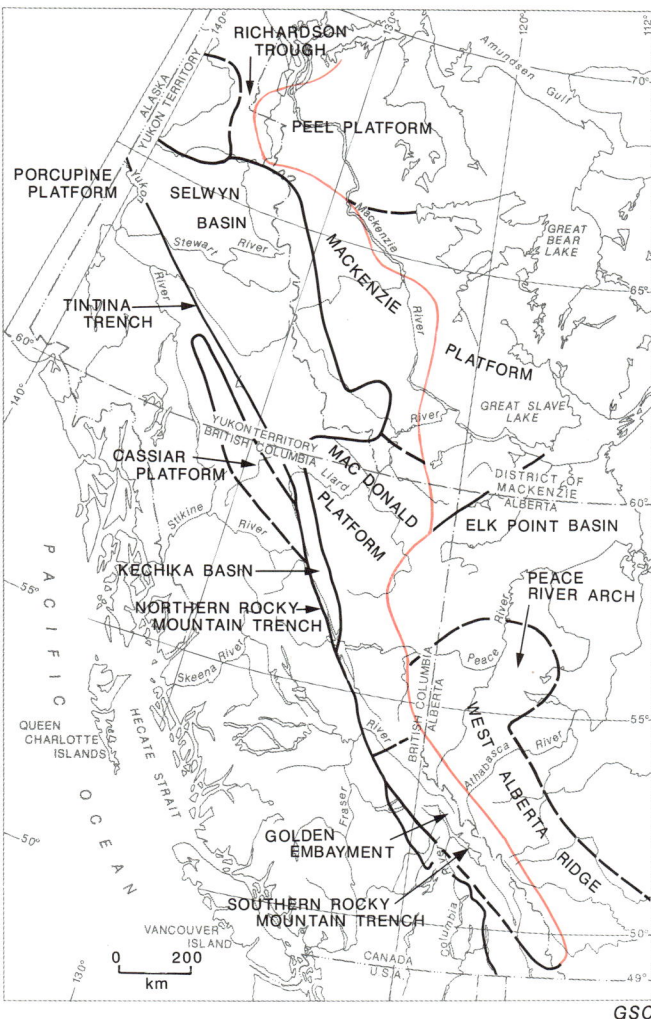

Figure 7.40. Early and Middle Devonian tectono-sedimentological elements of the eastern Cordillera. The red line represents the approximate position of flexure that separated condensed, platformal sequences to the east from thicker subsiding shelf sequences to the west.

which marked the beginning of a widespread continent-wide transgression (Johnson, 1970). In the south the preserved pre-Taghanic record lies east of the Foreland Belt and entirely within the cratonic platform where it is represented by evaporite-dominated sediments in what were restricted intracratonic basins and embayments. The Taghanic transgression in the southern Cordillera and adjacent craton is featured by widespread carbonate and evaporite deposits.

Lower and Middle Devonian strata of the miogeocline comprise three "sequences" and part of a fourth. Names for these are taken from the dominant formation or groups within each sequence (Fig. 7.42; Table 7.1); in ascending stratigraphic order these are: Delorme sequence (Gedinnian-Siegenian); Bear Rock-Stone sequence (Emsian); Hume-Dunedin sequence (Eifelian); and Fairholme sequence (Givetian-Frasnian, Chapter 8).

These sequences are similar to those described by Moore (in Geology of Canada, no. 5, Sedimentary Cover of the Craton in Canada) and, like them, could be termed "allogroups" i.e. map-units identified on the basis of their

199

CHAPTER 7

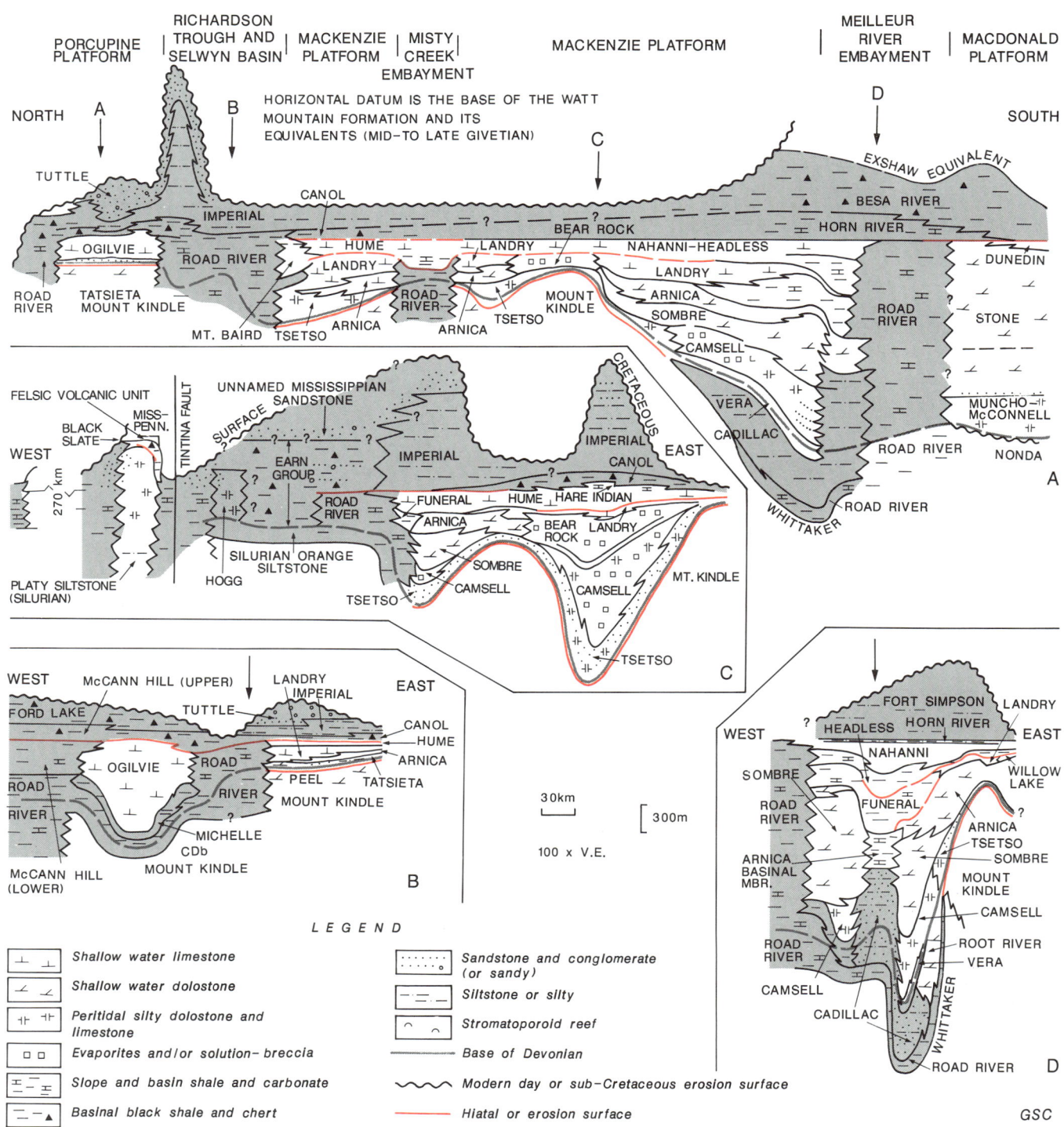

CAMBRIAN TO MIDDLE DEVONIAN ASSEMBLAGES

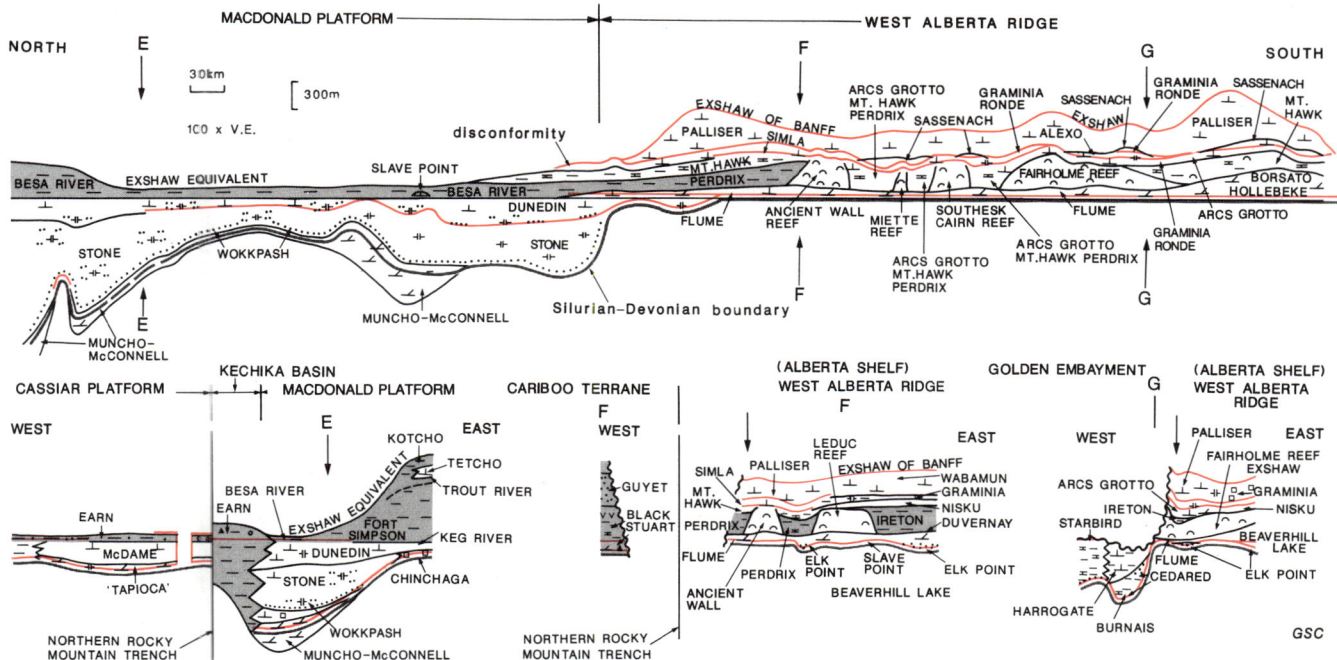

Figure 7.41. Stratigraphic cross-sections of the Devonian System in the Cordillera. See Figure 7.37 for locations. Grey shading indicates predominantly clastic rocks.

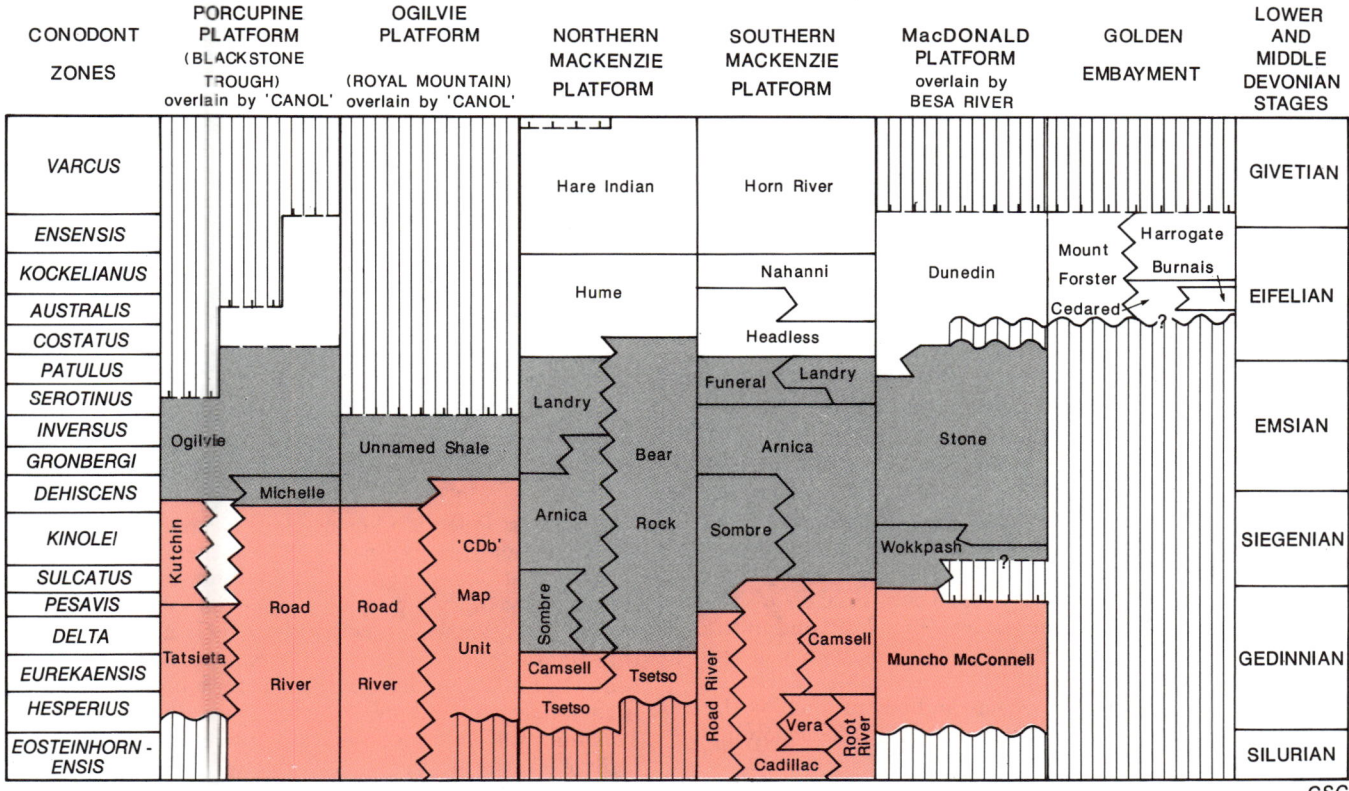

Figure 7.42. Stratigraphic chart of the Lower and Middle Devonian System. Uncoloured-sequence Hume-Dunedin, grey-sequence Bear Rock-Stone, pink-sequence Delorme.

Table 7.1. Tectonic subdivisions of Lower, Middle and Upper Devonian strata in the Foreland Belt

SEQUENCE	TECTONIC PROVINCE					
	WESTERN MIOGEOCLINE		EASTERN MIOGEOCLINE		CRATONIC PLATFORM	
	North	South	North	South	North	South
						Subsurface Nomenclature / Surface Nomenclature
Palliser	Nation River	Tuttle Ford Lake Imperial	Earn¹	Imperial	Fort Simpson Besa River	Kotcho Tetcho Trout River	Wabamun	Palliser Stettler Crossfield Sassenach
Ronde Kakisa	Nation River	Ford Lake Imperial	Earn¹	Imperial	Fort Simpson Besa River	Kakisa	Graminia Blueridge Calmar	Simla Ronde
Fairholme	Ford Lake Nation McCann Hill	Imperial 'Canol' Unnamed shale	Earn¹	Imperial Canol Ramparts Hare Indian	Fort Simpson Muskwa Besa River Horn River	Redknife Jean Marie Hay River Muskwa Waterways Slave Point Watt Mountain Gilwood	Nisku Leduc Ireton Duvernay Beaverhill Borsato	Arcs Grotto Mount Hawk Perdrix Maligne Peechee Cairn Flume Starbird Hollebeke Yahatinda
Hume- Dunedin	McCann Hill	Unnamed shale Ogilvie CDb (map unit)	Earn¹ Road River	Hume Nahanni Headless	Horn River Horn Plateau Besa River Lonely Bay Dunedin	Lonely Bay Willow Bay Ebbutt	Sulphur Point Keg River Muskeg Chinchaga Elk Point¹ Contact Rapids	Yahatinda Mount Forster Harrogate Burnais Cedared
Bear Rock- Stone	CDb (map unit)	Ogilvie Michelle	Earn¹ Road River	Mount Baird Funeral Bear Rock	Landry Arnica Snatla Grizzly Bear Sombre	Dunedin Stone Wokkpash	Landry Bear Rock	Chinchaga Elk Point¹	
Delorme	Kutchin Tatsieta	CDb (map unit)	Road River Cadillac Delorme	Road River Kutchin Tatsieta	Camsell Tsetso Vera Root River	Muncho- McConnell	Tatsieta	Tsetso

¹Group

and a yellowish-orange colour which reflects the onlap of the pre-latest Pridolian-Gedinnian subaerial erosion surface (Plate 13). Deposition was confined to the northern part of the miogeocline. Total thicknesses range from a feather edge in regions near the Norman Wells High (Fig. 7.44) to a maximum of more than 1000 m at the western edge of Mackenzie Platform (Fig. 7.39) and in the Root Basin (Fig. 7.43; Morrow, 1984). Sediments of the Delorme sequence accumulated on unstable platforms which were interrupted by arches and basins such as the Redstone Arch and the Root Basin (Fig. 7.43). The mobility of platform areas at this time is indicated by sedimentation contemporaneous with the generation of fault scarps within the Peel Platform (Pugh, 1983) and by dramatic variations in thickness of the Delorme sequence across the Mackenzie Platform (Morrow, in press).

In latest Pridolian to earliest Gedinnian time carbonate deposition was confined largely to the MacDonald and Ogilvie platforms (Fig. 7.43). Medium-bedded, faintly laminated yellowish-grey silty dolostone of the Muncho-McConnell Formation (about 400 m thick) accumulated on the MacDonald Platform and cream-coloured, medium-bedded, faintly laminated, fenestral fabric-bearing dolomitic lime mudstone and finely crystalline dolostone of the Tatsieta and Kutchin formations were deposited on the Ogilvie Platform (Fig. 7.43). The Mackenzie, Peel and Porcupine platforms were largely exposed (Pugh, 1983; Norris, 1985) and in some places even older buried unconformities were exhumed. Dark recessive brownish-grey basinal shale and argillaceous lime mudstone of the Road River Formation, and bright orange-weathering basin-slope dolomitic platy siltstone and sandstone of the Cadillac and Delorme formations (up to 800 m thick) were deposited in the Selwyn Basin (Plate 14) and in the Blackstone Trough during this time of shelf exposure (Gordey, 1980; Pugh, 1983; Morrow, 1984). Root Basin in the southern Mackenzie Shelf was filled with 150 m of brightly coloured, argillaceous, skeletal lime wackestone of Vera Formation (Fig. 7.43), which was deposited below wave base but in the photic zone. This distinctive unit contains a diverse shelly fauna (brachiopods, trilobites) that is dated as earliest Gedinnian (Morrow and Cook, 1987). Black graptolitic shale, thin-bedded limestone and chert, possibly 300 m thick, constitutes the upper part of the Road River Formation in the Richardson Trough. The uppermost strata contain *Monograptus yukonensis* the youngest species of *Monograptus* known in Canada (Jackson et al., 1978). In the Selwyn Basin east of the Tintina and Northern Rocky Mountain trenches dark grey shale of possible Gedinnian age forms a condensed succession probably less than 75 m thick.

In later Gedinnian and Siegenian times the previously exposed northern platform region was largely inundated but local areas such as the Norman Wells High (Williams, 1975; Aitken et al., 1982), Bonnet Plume High (Lenz, 1972), Dave Lord High (Jackson et al., 1978) and part of the Godlin Salient (Morrow, in press) remained exposed (Fig. 7.44). Argillaceous and orange, silty and

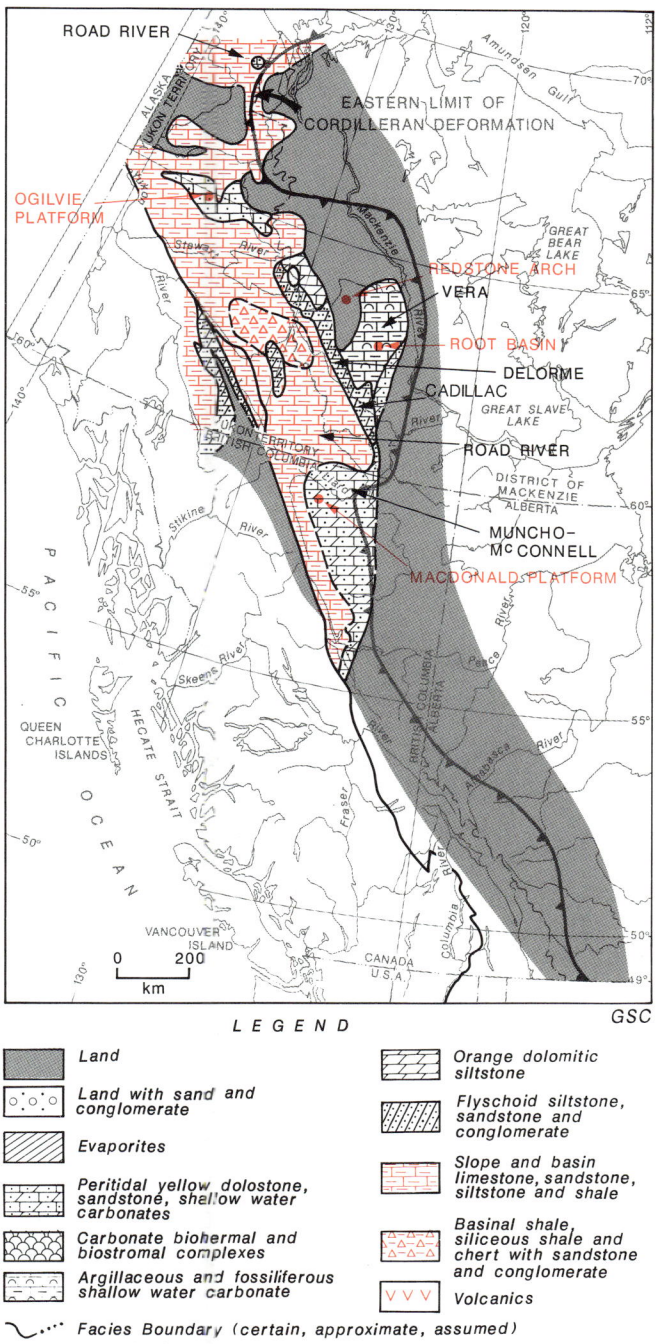

Figure 7.43. Distribution of facies in latest Silurian time (Pridoli).

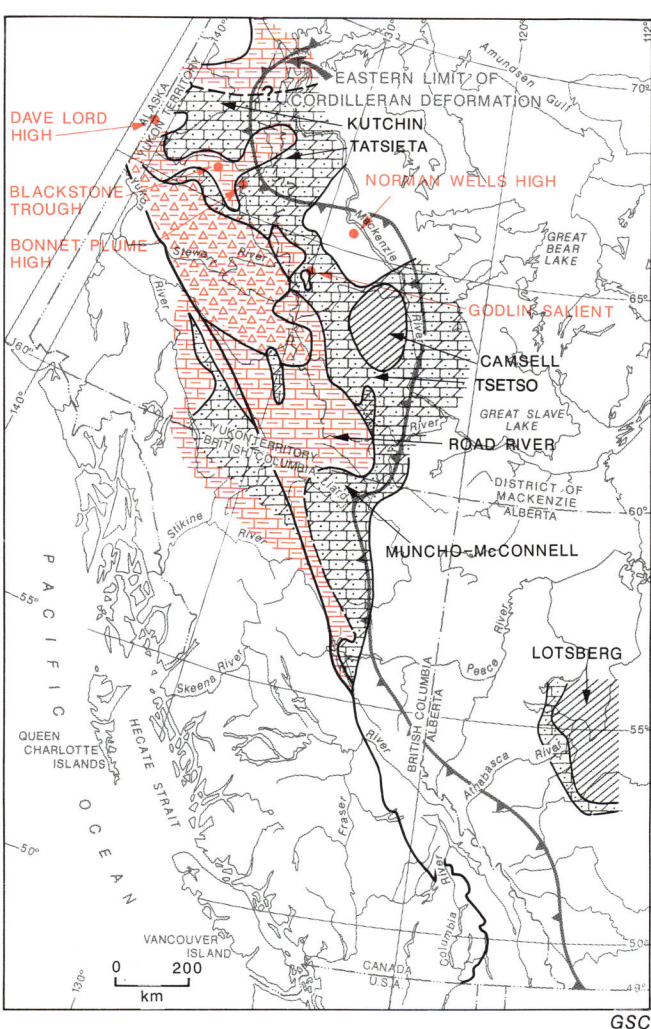

Figure 7.44. Distribution of facies during the Gedinnian and Siegenian. Legend accompanies Figure 7.43.

sandy, mudcracked intertidal and supratidal carbonate of the Tsetso and Tatsieta formations, about 100 m thick (Meijer Drees, in press; Pugh, 1983) covered large parts of the Mackenzie, Porcupine and Peel platforms while deposition of the medium grey silty laminated peritidal Muncho-McConnell dolostone continued to accumulate on the MacDonald Platform.

Deposition of Delorme sequence evaporites occurred in widely separated cratonic platforms and basins (Fig. 7.44). In the Root Basin up to 1500 m of interbedded shallow-water anhydrite and dolostone of the Camsell Formation accumulated during Gedinnian to earliest Siegenian time. Farther west, near the Godlin Salient, a thinner succession of Camsell evaporites filled a small basin that was separated from the Root Basin by the Redstone Arch (Fig. 7.44). In both regions, spectacular limestone solution-collapse breccias occur where Camsell evaporites have been exposed to surface or near-surface weathering (Morrow, in press). Minor amounts of anhydrite occur within the Tsetso and Tatsieta formations.

The yellowish-orange colouration and terrigenous siliciclastic content in the Delorme sequence on the MacDonald, Mackenzie and Peel platforms is much less evident on the Ogilvie and Porcupine platforms. This reflects the relatively great distance separating the latter two areas from the cratonic interior, where silt and sand-bearing, hematitic older strata were exposed (Morrow, in press). Consequently, the Delorme sequence is much thinner across the Porcupine Platform than on the Mackenzie and Peel platforms and is almost absent on the Ogilvie Platform. Indeed, across the Ogilvie Platform the base of this sequence can be placed only tentatively at the base of a monotonous succession of shallow-water clean carbonates immediately above beds containing abundant silicified halysitid corals and orthoconic cephalopods, characteristic of the 'Arctic Ordovician Fauna' at the top of the Upper Ordovician to Lower Silurian Mount Kindle Formation.

Remarkably abrupt facies and thickness changes in Siegenian to Emsian strata occur along the eastern margin of the Selwyn Basin. East of the Tintina Trench and west of the thick carbonate units as much as 150 m of black shale and chert form the lower part of an undivided succession included in the Lower Devonian to Lower Mississippian Earn Group. East of the Northern Rocky Mountain Trench the easternmost basinal facies consists of interbedded black, graptolitic shale, crinoidal limestone, argillaceous limestone, dolomitic sandstone and massive, fine-grained, pure quartzite as much as 100 m thick. Except for the shale all lithologies are regarded as turbiditic or mass flows derived from an eastern platform (MacIntyre, 1981). The quartzite members may be correlative with quartzite of the Wokkpash Formation east of the facies boundary.

Bear Rock-Stone sequence (Emsian-early Eifelian)

Deposition of the Bear Rock-Stone sequence (Table 7.1) accompanied marine inundation of most areas that remained subaerially exposed in the northern Cordillera at the close of deposition of the Delorme sequence (Fig. 7.45). The transgression over these areas led to the deposition of clean carbonate and evaporite successions uncontaminated by terrigenous material. The yellowish-orange colour of

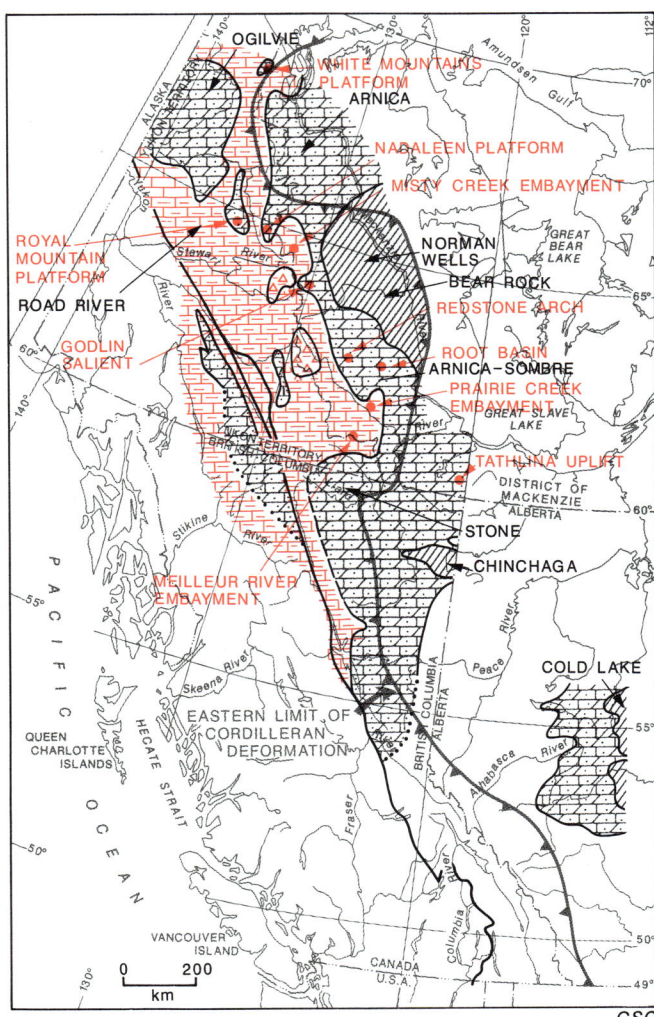

Figure 7.45. Distribution of depositional facies during Emsian time. Legend accompanies Figure 7.43.

the siliciclastic sediments of the Delorme sequence contrasts sharply with the uniformly grey and bluish grey weathering 'clean' carbonate of the Bear Rock-Stone sequence (Plate 14).

The Bear Rock-Stone sequence is thicker than the Delorme and also is confined to the northern part of the Cordillera (Fig. 7.45). It ranges in thickness from a feather edge where it onlaps the West Alberta Ridge and Peace River Arch (Fig. 7.41), to a maximum of 1700 m at the edge of the Prairie Creek Embayment (Fig. 7.45) on the Mackenzie Platform (Morrow, 1984). Moreover, the sequence characterizes deposition on carbonate platforms that underwent more uniform subsidence than those on which the Delorme sequence accumulated. This is reflected by the more regionally continuous lithofacies of the Bear Rock-Stone sequence and by its more subdued thickness variations (Morrow, in press).

The contact between the Delorme and Bear Rock-Stone sequences is generally conformable although a local unconformity may occur beneath the Wokkpash Formation, the basal unit of the Bear Rock-Stone sequence in northeast British Columbia (Thompson, 1989; others consider the contact there to be conformable also (Taylor and Mackenzie, 1970)). Along the western edge of the Mackenzie Platform the basal unit of the Bear Rock-Stone sequence is the Sombre Formation which intertongues with the Camsell Formation, the uppermost unit of the Delorme sequence (Morrow, 1984; Morrow and Cook, 1987; see also Gabrielse et al., 1973). This indicates that there was no significant regression at the close of deposition of the Delorme sequence in the Cordilleran region. There is evidence, however, that the beginning of Bear Rock-Stone deposition was initiated by a rapid transgression that abruptly altered Devonian paleogeography. Transgression across exposed areas such as the Norman Wells High allowed for the accumulation of 50 m of dark, open marine crinoidal dolostone, the 'Arnica Platform Dolomite' (Williams, 1975), which occurs at the base of the Bear Rock-Stone sequence in the Root Basin. This basal member of the Arnica Formation separates the anhydrite-dolostone sequence of the Bear Rock Formation from the underlying Camsell Formation evaporites. The transgression also is expressed by the abrupt contact between the Delorme and Bear Rock-Stone sequences across the Mackenzie and Peel platforms (Fig. 7.41; e.g. the Camsell-Sombre, Camsell-Arnica, Tsetso-Arnica and Tatsieta-Arnica contacts; Morrow and Cook, 1987; Morrow, in press; Pugh, 1983).

Deposition of the Bear Rock-Stone sequence occurred in two discrete phases, an Emsian phase (Fig. 7.45) during which the bulk of the sequence was deposited and a shorter early Eifelian phase (Fig. 7.46), the latter marked primarily by marine regression across much of the cratonic platform. The regression is marked by a disconformity or unconformity that coincides with the base of the Headless and Ebbutt formations (Meijer Drees, in press; Law, 1971) and with the middle detrital zone of the Chinchaga Formation (Belyea, 1970; Law, 1971).

On the MacDonald Platform, the deposition of crossbedded, dolomitic, nearshore quartzarenite, of the thin Wokkpash Formation (commonly less than 100 m thick; Taylor and Mackenzie, 1970) marked the beginning of Bear Rock-Stone deposition. The siliciclastics of the Wokkpash pass abruptly upwards into the thick 'clean' dolostone succession of the Stone Formation which thins southeastward towards the Peace River Arch from a maximum of about 2000 m near the northern edge of the MacDonald Platform. Rhythmically interbedded intervals of laminated, light grey intertidal dolostone and darker intervals of locally fossil-bearing, medium- to thick-bedded subtidal dolostone form the Stone Formation. The large 'Mississippi Valley type' lead-zinc deposit at Robb Lake occurs in dolomite-spar cemented, brecciated strata of the Stone Formation (Macqueen and Thompson, 1978). Northwards, on the southern Mackenzie Platform, Emsian deposits are represented by light and medium grey dolostone of the Sombre and Arnica formations, which, like the Stone, are largely shallow-water peritidal dolostone. The dominantly light grey intertidal laminites and local detrital, partly dolomitized lime mudclast beach deposits of the Sombre are confined largely to the western border of the Mackenzie Platform, whereas the darker, brownish grey, more fossiliferous (crinoids, amphiporids), medium- to thick-bedded slightly vuggy, mainly subtidal dolostones of the Arnica predominates on the central and eastern

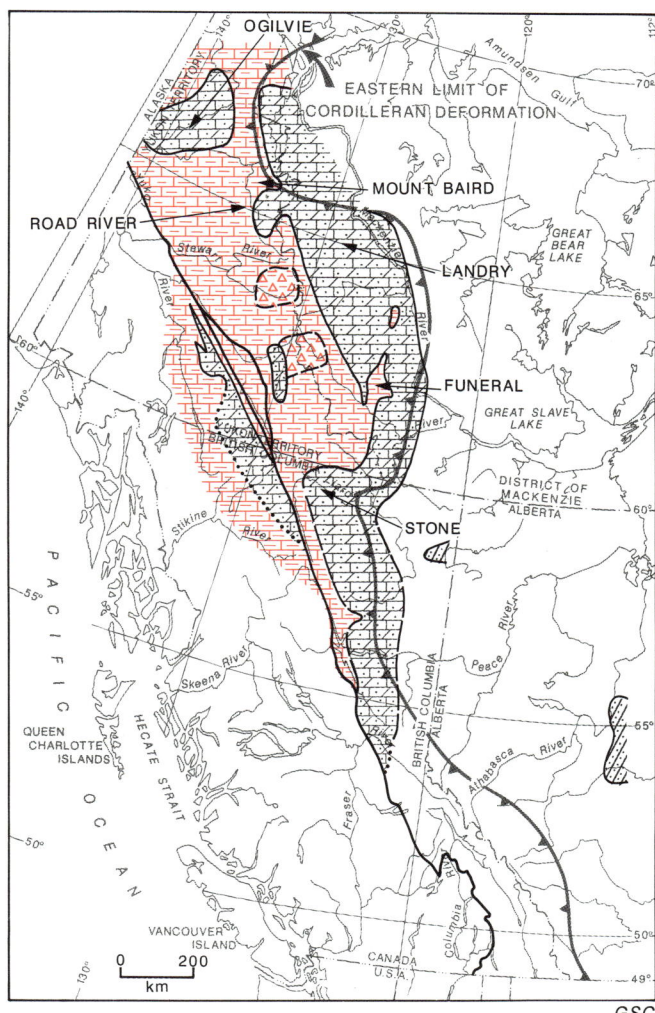

Figure 7.46. Distribution of facies during early Eifelian time. Legend accompanies Figure 7.43.

parts of the platform (Fig. 7.45). The Sombre Formation was probably deposited on a broad carbonate shoal that bordered the shelf and inshore of this shoal belt the dominantly subtidal deposits of the Arnica Formation accumulated in lagoons. At this time, the Root Basin was covered by shallow-water Arnica and Sombre sediments and no longer received basinal deposits except in the Prairie Creek Embayment where it formed an eastern expansion of the larger Meilleur River Embayment (Morrow, 1984). Deposition of deep-water, hemipelagic sediments continued within the Prairie Creek Embayment during Emsian time and included a variety of fragmental limestone gravity flows which were shed from the shallow rim of the embayment. A lack of organic nutrients caused the deposition of deep-water, pink and red calcareous shale and siltstone which are interbedded with the limestone turbidite and debris flows within the embayment. West of the Redstone Arch, towards the Selwyn Basin, the upper part of the Arnica dolostone passes laterally into deeper-water argillaceous, crinoidal limestone of the Grizzly Bear Formation, more than 200 m thick, and a thinly and rhythmically bedded turbiditic shale and limestone sequence, the Natla Formation, as much as 450 m thick (Gabrielse et al., 1973). Farther west correlative strata in the northern Selwyn Basin are relatively thin dark grey shale interbedded with limestone and locally with volcanogenic siltstone, volcanic conglomerate and lapilli tuff (Cecile, 1982).

On the northern Mackenzie Platform and on the Peel Platform the evaporite deposits of the Bear Rock Formation (about 300 m thick) extend much farther westward than on the southern Mackenzie Platform and also occur on the Norman Wells High (Fig. 7.45). These evaporites, preserved in the subsurface as repetitive peritidal 'cycles' of thin anhydritic and mud-cracked dololaminite, commonly less than 2 m thick (Meijer Drees, 1980), are the eastern facies equivalent of the Arnica-Sombre succession. Almost all surface exposures of these evaporites have undergone solution-collapse and comprise locally spectacular, calcareous, massive carbonate-rubble breccias (Morrow, in press; Morrow and Meijer Drees, 1981). Meijer Drees (in press) proposed the term "Fort Norman" for the unbrecciated Bear Rock evaporites preserved in the subsurface.

During early Emsian time the Porcupine Platform sustained major paleogeographic changes. The Blackstone Trough and Dave Lord High (Fig. 7.44) disappeared and the Richardson Trough was re-established as a deep water passageway connecting the Selwyn Basin with northern seaways (Fig. 7.45). The bluish grey-weathering, resistant, medium- and thick-bedded, shallow-water limestone of the Ogilvie Formation, up to 600 m thick, and the CDb map unit accumulated across the Porcupine Platform and on the Royal Mountain and White Mountain platforms (Norris, 1985; Pugh, 1983; Perry et al., 1974; Fig. 7.45). Across the Blackstone Trough Siegenian Road River shales are overlain by the Michelle Formation, a thin (<150 m) earliest Emsian fossiliferous and calcareous mudstone and shale slope facies, which passes abruptly upwards into the Ogilvie Formation.

The geographic and stratigraphic facies differentiation of limestone, dolostone and evaporites evident on the Peel and Mackenzie platforms is not expressed on the Porcupine Platform. Instead the Ogilvie Formation is a much more homogenous unit of open marine, skeletal lime wackestone (crinoids, solitary and ramose colonial corals, amphiporids and small lamellar and hemispheroidal stromatoporoids predominate) with minor intercalations of thin-bedded to laminated intertidal dolomitic limestone which display fenestral fabric. Like the Emsian carbonates farther south, the Ogilvie is not characterized by well developed organic shelf-edge facies, however, in contrast to the stable southern shelf edges the Ogilvie shelf edge was fairly mobile during deposition (Dubord et al., 1986). The long term stability of the Emsian Mackenzie Platform edge implies a degree of tectonic control on its position much different than that which may have influenced the position of the edge of the Porcupine Platform.

Deep-water, calcareous, dark grey shale and limestone of the Road River Formation and dark siliceous shale of the Earn Group were deposited in the Selwyn Basin and in the Richardson Trough in both Emsian and early Eifelian time (Norris, 1985; Gordey et al., 1982). Emsian basinal strata in the Selwyn Basin and in its adjoining embayments into the Mackenzie Platform are only a few hundred metres thick, much less than that of equivalent strata on the shelves (Gordey et al., 1982; Morrow, 1984).

The regional regression that marked early Eifelian deposition was coincident, somewhat paradoxically, with expansion of the area of deep-water deposition onto parts of the adjacent carbonate platforms (Fig. 7.46). Dolomitic, slightly pyritic and nonfossiliferous platy, buff shale of the Funeral Formation, as much as 500 m thick, accumulated in Root Basin and on the Mackenzie Platform west of the Redstone Arch. Farther north similar shale of the Mount Baird Formation, more than 300 m thick, was deposited in a small unnamed platform-edge embayment at the junction of the Mackenzie and Peel platforms (Fig. 7.46). At this time the Royal Mountain Platform, a large offshore carbonate bank, was covered with basinal deposits.

Across the western parts of the Mackenzie and Peel platforms up to 500 m of sparsely fossiliferous, ostracode-bearing, pelletal lime wackestone and packstone of the Landry Formation (Braun, 1978; Chatterton, 1978) accumulated. Farther south, on the MacDonald Platform a crinoidal limestone belt near the platform edge, tentatively assigned to the Dunedin Formation, may have developed during late Emsian to early Eifelian time (see Morrow, 1978; Chatterton, 1978). Farther east, the Dunedin Formation overlies the Stone Formation unconformably above the westward continuation of the sub-Ebbutt or mid-Chinchaga unconformity (Taylor and Mackenzie, 1970). The regression implied by this widespread unconformity marked the end of Bear Rock-Stone sequence deposition.

In southern Selwyn Basin (Kechika Basin) east of the Northern Rocky Mountain Trench northwest-trending belts of reefoid Emsian to Eifelian carbonate with a maximum thicknesses of 200 m are separated by thin successions of chert, argillite, carbonaceous black shale and minor limestone. The disposition of quartzose and calcareous turbidite and limestone debris flows indicates deposition of limestone on the edges of west-tilted fault blocks (MacIntyre, 1981).

Hume-Dunedin sequence (late Eifelian-early Givetian)

The Hume-Dunedin sequence (Table 7.1) was deposited during a widespread regional transgression which extended over a much broader region than did previous Early Devonian transgressions (Fig. 7.47; Moore, in press). Both the Elk Point Basin and Golden Embayment (Fig. 7.40) were established at this time. Farther north, deposition of this sequence marked the end of carbonate deposition on the Peel, Mackenzie and Porcupine platforms and the beginning of deeper water siliciclastic sedimentation (Fig. 7.48). This was followed by the even broader transgression that accompanied the Taghanic Onlap and resulted in the accumulation of the Fairholme Sequence (Rundle Assemblage) (see Chapter 8).

The Hume-Dunedin sequence is the thinnest of the three pre-Taghanic assemblages in the northern Cordillera. Typically it is about 200 m thick north of 60° latitude. In northeast British Columbia, in the Presqu'ile Complex of the Elk Point Group, it is up to 400 m thick (Griffin, 1967; Morrow, 1973) but thins southward to a feather edge where it onlaps the West Alberta Ridge (Fig. 7.40).

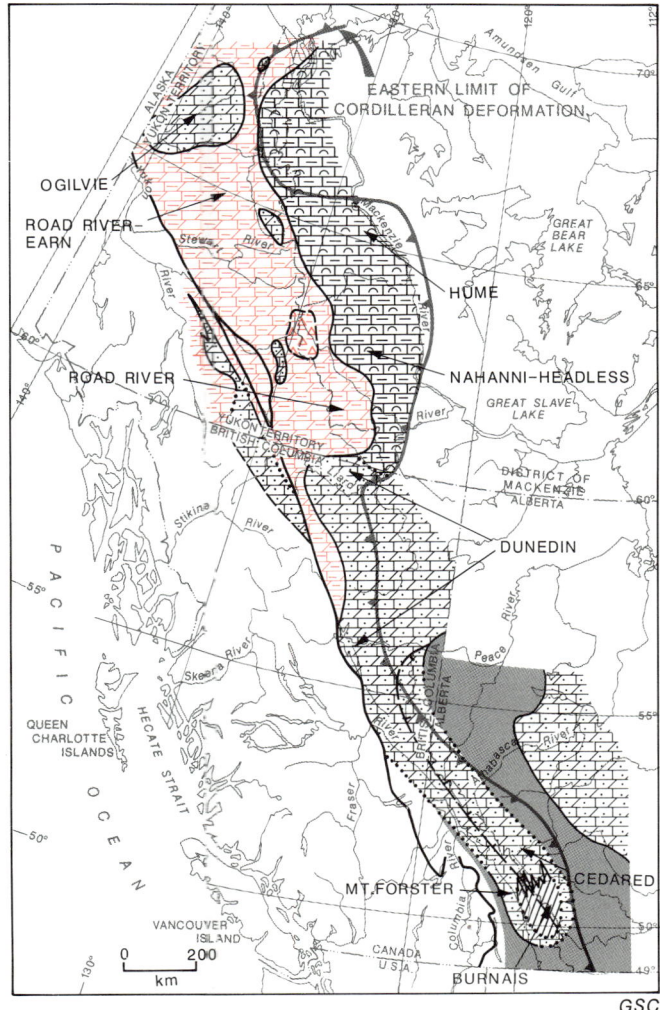

Figure 7.47. Distribution of facies during late Eifelian time. Legend accompanies Figure 7.43.

sequences of the Horn River and Hare Indian formations. In the eastern part of the northern Cordillera the base of the *disparilis* conodont zone approximately coincides with the base of the reefal buildup represented by the stromatoporoidal and coralline limestone of the Ramparts Formation (Braun, 1977; Muir and Dixon, 1984; Lenz and Pedder, 1972) which forms the reservoir for the Norman Wells oil field. Thus, the boundary between the Hume-Dunedin sequence and the overlying Upper Devonian Fairholme sequence probably coincides with the contact between the Hare Indian and Ramparts formations (see also Moore, in press).

In mid- to late Eifelian time epicontinental seas spread into the Golden Embayment (Fig. 7.40, 7.47). About 200 m of yellowish to reddish weathering, sparsely fossiliferous thin-bedded peritidal dolostone, dolomitic sandstone and laminated gypsum of the Cedared and Burnais formations rest unconformably on the Silurian Beaverfoot Formation at the southern extremity of the embayment (Norford, 1981). Based on charophyte (green algae) floras the succession is imprecisely dated as Eifelian, although it is possible that part or all of this succession may belong to the underlying Bear Rock-Stone sequence (see Moore, in press). Eifelian deposition in the interior of the Golden Embayment ended with deposition of the more open marine, argillaceous, thick-bedded limestone and dolostone of the Harrogate Formation (130 m thick) which contains abundant corals, brachiopods, crinoids and trilobites (Fig. 7.48). These strata pass westward into relatively unfossiliferous limy and dolomitic sandy mudstone of the Mount Forster Formation, 430 m thick, which borders the west side of the embayment. The yellowish- and reddish-brown colouration and westward coarsening of some of the Mount Forster sediments suggest a terrestrial origin consistent with its paleogeographical position flanking the emergent Purcell Landmass farther to the west (Norford, 1981).

On the MacDonald Platform the Dunedin Formation is a transgressive unit of medium-bedded, silty, shoreface dololaminite and carbonate sand and gravel beach-ridge deposits that accumulated on a subaerially exposed surface underlain by the Stone Formation (Morrow, 1978). These strata are overlain by more resistant thick-bedded, bluish grey-weathering, open marine, fossiliferous limestone that contains brachiopod faunas indicating that the top of the formation is as old as middle Eifelian (*Eoschuchertella adoceta* zone) near the northwest edge of the platform, but is as young as late Givetian near the Keg River Barrier to the southeast (Fig. 7.48; Taylor and Mackenzie, 1970; Geldsetzer, 1982). The influence of the Purcell Landmass in late Eifelian to earliest Givetian time may be reflected by the easterly directed paleocurrent indicators in quartz siltstone and sandstone that are abundant in the lower 50 to 100 m of the Dunedin Formation in the southern part of the MacDonald Platform and in the Golden Embayment (Geldsetzer, 1982).

On the Mackenzie and Peel platforms the Headless, Nahanni and Hume formations (Plate 15) were deposited in Eifelian time following the mid-Chinchaga or Ebbutt hiatus (Noble and Ferguson, 1971; Chatterton, 1978). Unlike the Dunedin Formation in which the vertical sequence represents an upward progression of successively deeper-water depositional environments, the Hume Formation and its southern equivalent, the Headless-Nahanni

Unlike the underlying Lower Devonian sequences that are more fully developed on the Mackenzie and Peel platforms, the Hume-Dunedin sequence is more complete in the Elk Point Basin where it occupies the interval between the mid-Chinchaga unconformity (i.e. the Ebbutt break - see Moore, in press) and the unconformity beneath the Watt Mountain Formation (Torrie, 1973; Williams, 1984). Deposition of fluvial and deltaic, waxy, green shale and thin sandstone of the Watt Mountain Formation across the Elk Point Basin marked a period of widespread emergence after which patterns of sedimentation were profoundly altered by the transgression associated with the Taghanic onlap in mid-Givetian time, i.e. post-mid-*varcus* time (Fig. 7.42; Johnson, 1970; Johnson et al., 1985).

The lower part of the sequence on the Mackenzie and Peel platforms is marked by the recessive character of the base of the Hume and Headless formations and the base of the Dunedin Formation across most of the MacDonald Platform. Strata in northern shelf regions that are correlative with the Watt Mountain hiatus (i.e. *varcus* age), however, are contained within the largely undated shale

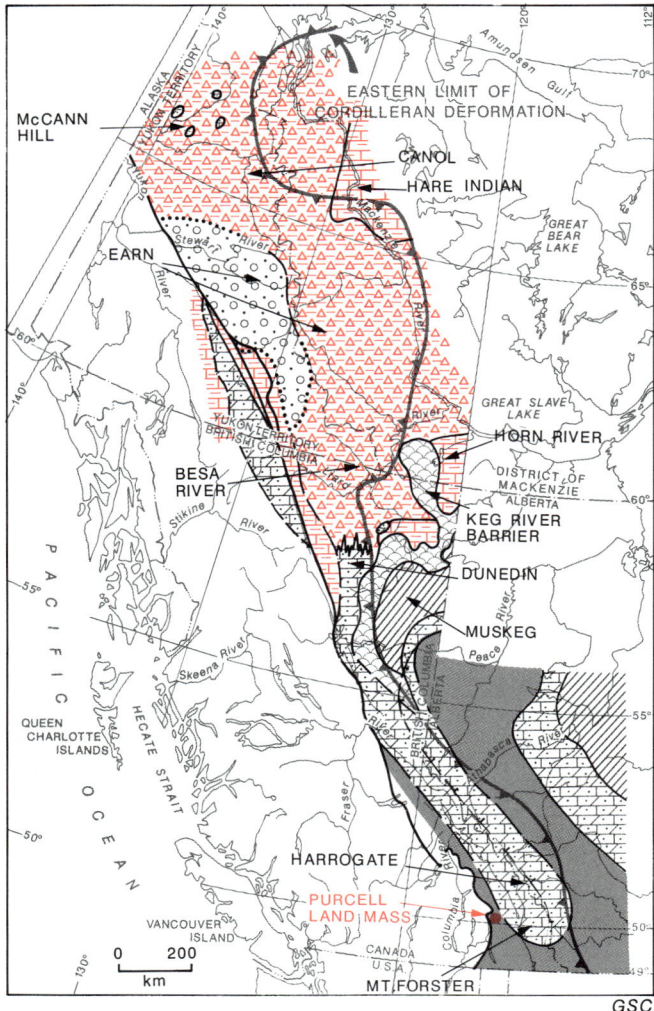

Figure 7.48. Distribution of facies during early to mid-Givetian time. Legend accompanies Figure 7.43.

progressively decreased as the platform edge retreated eastward (Dubord et al., 1986).

Early Givetian time marked the final stage of development of the Elk Point Basin behind the Presqu'ile or Keg River Barrier (Fig. 7.48; Moore, in press; Bebout and Maiklem, 1973). The western limit of the Keg River Barrier is well exposed in northeast British Columbia at Mount Bertha near the Sikanni Chief River. There the Keg River Barrier facies ('Dunedin' of Taylor and Mackenzie, 1970) is nearly 300 m thick; immediately west it is only 100 m thick. At Mount Bertha the Keg River Barrier is divided into two subequal parts: a lower sequence of dark grey, thin- to medium-bedded, crinoid and amphiporid- bearing dolostone and an upper more resistant 'reefal' part composed of thick-bedded, light grey, stromatoporoidal and coralline limestone and dolostone.

The presence of Givetian strata in the Golden Embayment is uncertain (Norford, 1981) although brightly coloured, laminated platy, yellow intertidal, dolostone and detrital, plant-bearing, channel-filling dolostone of the Yahatinda Formation (Aitken, 1966a) may be a shoreline facies that developed on the West Alberta Ridge in Givetian time.

On the MacDonald Platform crinoidal limestone of the Dunedin Formation grades into dark, silicified, rusty, basal shale of the Besa River Formation (Taylor and Mackenzie, 1970; Fig. 7.47). Pelzer (1966) suggested that a thin portion (30 m) of the basal Besa River Formation overlying most of the Dunedin Formation is equivalent to about 210 m of Givetian shale (Otter Park Member of the Horn River Formation; Griffin, 1967) and argillaceous limestone in front of the Keg River Barrier. Grey calcareous shale of the Horn River Formation (200 m thick) and its equivalents formed a seaward-sloping ramp of clinoform strata that bordered the entire Keg River Barrier during Givetian time (Williams, 1983; Fig. 7.48). Near the Keg River Barrier the shale encloses large oil- and gas-bearing patch reefs (e.g. Evie Reef). The Horn River thins rapidly to the northwest and across most of the Mackenzie Platform it is indistinguishable from the basal radioactive shales of the Upper Devonian Fort Simpson Formation.

Farther north on the Peel Platform an analogous situation exists where recessive greenish grey, slope-deposited or clinoform shale and fossiliferous, argillaceous limestone of the Hare Indian Formation in the Norman Wells area thins westward from 400 m to less than 20 m near the Snake River where equivalent strata are dark brown, radioactive, basal, spore-bearing Bluefish Member which contains a pelagic fauna of tentaculitids and styliolinids. Farther west the Hare Indian merges with part of the black, siliceous, rusty-weathering shale of the Canol Formation (Pugh, 1983; Williams, 1983). Norris (1985) considered that the Hare Indian Formation was truncated to the west by an unconformity at the base of the Canol Formation. On the Porcupine Platform, Givetian shallow-water carbonate deposition (Ogilvie Formation) was restricted to a few small remnant islands such as at Mount Burgess (Norris, 1985). It is not clear whether these sites of shallow-water deposition were surrounded by deposits of siliceous shale of the McCann Hill chert or of the Canol Formation (or the unnamed shale unit of Norris, 1985) (Fig. 7.48) or by undated deeper-water slope-deposited

succession, record an upward shoaling, progradational and regressive carbonate shelf and slope sequence that built westward over the underlying recessive subtidal argillaceous and fossiliferous limestones and calcareous shales of the Landry and Funeral formations (Williams, 1984). Consequently, the greenish and yellowish brown-weathering, brachiopod-rich (e.g. *Eoschuchertella adoceta*), argillaceous, subtidal lime wackestone of the Headless and lower Hume thickens westward at the expense of the coralline (e.g. *Hexagonaria, Favosites*), relatively clean, resistant, bluish grey-weathering, thick to massive bedded, lime wackestone and packstone of the Nahanni and upper Hume (Noble and Ferguson, 1971); the upper carbonate disappears altogether at the western edge of the Mackenzie Platform (Fig. 7.47). At the southern end of Mackenzie Platform Middle Devonian strata assigned to the Nahanni Formation form the reservoirs at Beaver River, Pointed Mountain and Kotaneelee gas fields. On the Porcupine Platform Eifelian deposition was essentially a continuation of that established during the Emsian (Ogilvie Formation) except that the area of the Porcupine Platform

limestone facies that occur at the top of the Ogilvie Formation (Dubord et al., 1986; Fig. 7.49).

In early Givetian time the Mackenzie, Peel and MacDonald platforms became sites of deeper-water, condensed-shale deposition fringed by clinoform or slope deposits along their southeastern (Keg River Barrier) and northeastern (Norman Wells area) sides. At that time the west-central Selwyn Basin may have been a region occupied by irregular, fault-bounded uplifts (Gordey, 1978; Gordey et al., 1982). In the absence of accurate biostratigraphic data for the coarse siliciclastic and chert-pebble deposits of the lower part of the Earn Group of the Selwyn Basin the timing of the initial uplift that exposed underlying chert-bearing and sandy lower Paleozoic strata to erosion was probably post-Eifelian and pre-early Frasnian (Gordey et al., 1982). Early Givetian time marked the transition from carbonate to siliciclastic deposition throughout the northern Cordillera.

Cordilleran terranes

Lower and Middle Devonian strata are best exposed in the Cassiar and Alexander terranes but are recognized only locally in Stikinia, Quesnellia and in the Kootenay Terrane. They may be present also in the metamorphic strata of the Nisling Terrane.

Displaced North America: Cassiar Terrane

In the Cassiar Mountains a sequence of probable Early Devonian age comprises from 100 to 200 m of well-bedded, light grey, laminated dolostone with a lower unit of crossbedded and ripple-marked sandstone, sandy dolostone and dolomitic sandstone (Rocky Mountains Assemblage; Gabrielse, 1963a). The rocks disconformably overlie mid-Silurian graptolitic siltstone. They are overlain unconformably by more than 100 m of fetid dark grey limestone and dolostone of the McDame Group which contains *Stringocephalus* sp. Locally a chert-pebble to cobble conglomerate less than one metre thick forms the basal unit. The fetid carbonate is overlain by light grey, well bedded to massive limestone as much as 100 m thick and in part or possibly entirely of Frasnian age (Rundle Assemblage). Locally a karst surface marks the top of the carbonate succession. Although somewhat thinner the Devonian carbonates in the Cassiar Mountains are analogous in lithology and succession to those in the Northern Rocky Mountains.

In the Pelly Mountains the Lower and Middle Devonian carbonate succession, the Askin Group, is similar to that in the Cassiar Mountains but at least part of the group on the northeast side of the Cassiar Platform appears to grade into dolostone with thin-bedded shaly limestone interbeds. On the west side of the Cassiar Platform the carbonates grade abruptly into dark grey shale, siltstone, fine grained sandstone, and calcarenite.

Dolostone breccia in a sandy and calcareous matrix of Early Devonian age forms part of the Black Stuart Formation (Rocky Mountains Assemblage) in the Cariboo Subterrane. The unit is about 30 to 40 m thick and is associated with black shale and calcareous, argillaceous sandstone. In places the top of the unit is a karst surface. The lithologies suggest a depositional environment similar

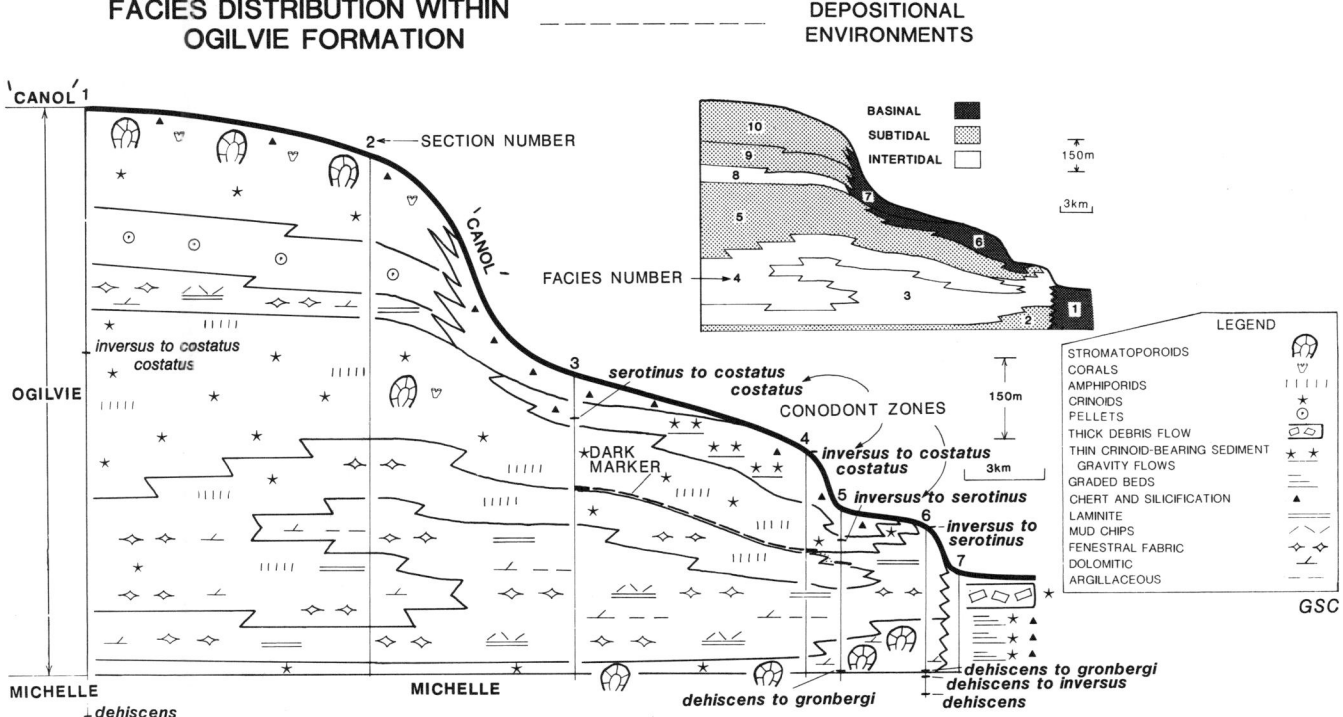

Figure 7.49. Facies distribution within the Ogilvie Formation.

to that along the eastern margin of the Selwyn Basin and the Kechika Basin and the west side of the Cassiar Platform.

Kootenay Terrane

Radiometric age determinations indicate an Early Devonian age for felsic tuff and granitic plutons in the Kootenay Terrane (Okulitch, 1985; R.L. Armstrong, pers. comm., 1985). The tuffs are included in the Eagle Bay Formation which contains Cambrian to Mississippian strata (Schiarizza and Preto, 1984).

Accreted terranes

Dark grey argillaceous limestone, siliceous argillite and phyllite and felsic volcanics of Early Devonian (Pragian) age occur in Stikinia near the Iskut River (Anarchist Assemblage). They are less than 200 m thick and appear to be faulted against Pennsylvanian and Permian strata on one side and are cut by a granitic pluton on the other (P.R. Read, pers. comm., 1983).

Poorly dated fine grained clastic and locally calcareous sequences such as the Nasina Formation in the metamorphic rocks of the Yukon and the Harper Ranch Group in Quesnellia northeast of Kamloops, may include Lower and Middle Devonian strata.

Intensely deformed Devonian rocks (Kaskawulsh and Cedar Cove assemblages) are widespread in the Alexander Terrane of the Saint Elias Mountains. Most conspicuous are discontinuous well-bedded to massive, possibly in part reefoid limestone, dolostone and marble units up to 300 m thick. They are interbedded with phyllitic, calcareous argillite, siltstone and thin-bedded argillaceous limestone. Late Early Devonian and Middle Devonian faunas have been recognized but relationships with older and younger rocks are unknown. Local occurrences of undated greywacke, sandstone and conglomerate with clasts of limestone, quartz, chert and slate are associated with the carbonates.

No direct evidence of Late Silurian tectonism and the related Lower Devonian clastic wedge, documented in the southern part of the Alexander Terrane of Alaska has been recognized in the Saint Elias Mountains.

Metamorphic rocks in the northern Coast Belt represent fine-grained siliceous, argillaceous and calcareous protoliths which in part could be of Devonian age.

The lower part of the Sicker Group (Sicker Assemblage) on Vancouver Island comprises an assemblage of marine, calc-alkaline arc volcanics overlain by Carboniferous and Lower Permian bioclastic limestone, chert and fine clastics. The volcanic sequence comprises two thick units, the lower consisting of more than 3000 m of augite-bearing agglomerate and tuff of the Nitinat Formation (Muller, 1980) and an upper, dominantly epiclastic succession, more than 1000 m thick, of volcanic sandstone, conglomerate, local pillow lavas and felsic pyroclastic rocks assigned to the McLaughlin Ridge Formation (correlative(?) with the Myra Formation of Muller, 1980; Yorath and Sutherland Brown, pers. comm., 1985). The pyroclastic pile is heterogeneous in texture, moderately uniform in composition, variably deformed and, where discernable, bedding is lenticular. It is dominated by originally scoriaceous, augite and plagioclase phyric explosive agglomerate and equivalent tuff of basaltic and andesitic composition. At many localities these rocks occur as chlorite-epidote schist. The epiclastic succession that overlies and interfingers with the agglomerate is characterized by massive- to well bedded green volcanic sandstone that coarsens upward to polymictic volcanic conglomerate. Intercalated argillite, cherty tuff, pillow lava and pyroclastic pale green dacite and rhyolite occur locally. In places the augite phyric agglomerate is intruded by Late Devonian quartz porphyry and granodiorite named the "Saltspring Intrusions" (Muller, 1980, 1983; Brandon et al., 1986). Ages based on U-Pb analyses of zircons from the quartz porphyry range between 362 Ma and 364 Ma (R.R. Parrish, pers. comm., 1987).

REFERENCES

Abbott, J.G.
1977: Structure and stratigraphy of the Mt. Hundere area, southeastern Yukon; Masters thesis, Queen's University, 105 p.
1981a: A new geological map of Mount Hundere and the area north; in Yukon Geology and Exploration 1979-1980; Department of Indian and Northern Affairs, Whitehorse, Yukon.
1981b: Geology of Seagull Tin district; in Yukon Geology and Exploration 1979-1980; Department of Indian and Northern Affairs, Whitehorse, Yukon.

Aitken, J.D.
1966a: Sub-Fairholme Devonian rocks of the eastern Front Ranges, southern Rocky Mountains, Alberta; Geological Survey of Canada, Paper 64-33.
1966b: Middle Cambrian to Middle Ordovician cyclic sedimentation, southern Rocky Mountains of Alberta; Bulletin of Canadian Petroleum Geology, v. 14, no. 4, p. 405-441.
1969: Documentation of the sub-Cambrian unconformity, Rocky Mountains Main Ranges, Alberta; Canadian Journal of Earth Sciences, v. 6, no. 2, p. 193-200.
1971: Control of lower Paleozoic sedimentary facies by the Kicking Horse Rim, southern Rocky Mountains, Canada; Bulletin of Canadian Petroleum Geology, v. 19, no. 3, p. 557-569.
1989: Uppermost Proterozoic formations in central Mackenzie Mountains, Northwest Territories; Geological Survey of Canada, Bulletin 368, 26 p.

Aitken, J.D. and Greggs, R.G.
1967: Upper Cambrian formations, southern Rocky Mountains of Alberta, an interim report; Geological Survey of Canada, Paper 66-49, 91 p.

Aitken, J.D. and Norford, B.S.
1967: Lower Ordovician Survey Peak and Outram Formations, Southern Rocky Mountains of Alberta; Bulletin of Canadian Petroleum Geology, v. 15, p. 150-207.

Aitken, J.D. and Pugh, D.C.
1984: The Fort Norman and Leith Ridge structures: major, buried Precambrian features underlying Franklin Mountains and Great Bear and Mackenzie Plains; Bulletin of Canadian Petroleum Geology, v. 32, p. 139-146.

Aitken, J.D., Cook, D.G., and Yorath, C.J.
1982: Upper Ramparts River and Sans Sault Rapids (106H) map areas, District of Mackenzie; Geological Survey of Canada, Memoir 388.

Aitken, J.D., Macqueen, R.W., and Usher, J.L.
1973: Reconnaissance studies of Proterozoic and Cambrian stratigraphy, lower Mackenzie River area (Operation Norman), District of Mackenzie; Geological Survey of Canada, Paper 73-9, 178 p.

Allison, C.W., Young, G.M., Yeo, G.M., and Delaney, G.D.
1981: Glacigenic rocks of the Upper Tindir Group, east-central Alaska; in Earth's Pre-Pleistocene Glacial Record, Hambrey, M.J. and Harland, W.B. (ed.), Cambridge University Press, p. 720-723

Armstrong, R.L.
1979: Sr isotopes in igneous rocks of the Canadian Cordillera and the extent of Precambrian rocks; in Evolution of the cratonic margin and related mineral deposits; Geological Association of Canada, Cordilleran Section, Programme and Abstracts, p. 7.

Balkwill, H.R.
1969: Structural analysis of the Western Ranges, Rocky Mountains, near Golden, British Columbia; Ph.D. thesis, University of Texas, 166 p.

Bally, A.W., Cook, T.D., Buffler, R.T., Clark, D.K., Framer, R.E., and Milner, S.
1975: Stratigraphic Atlas of North and Central America; Shell Oil Company, Exploration Department, Houston, Texas.

Barnes, C.R., Norford, B.S., and Skevington, D.
1981: The Ordovician system in Canada; International Union of Geological Sciences, Publication no. 8, 27 p.

Bassett, H.C. and Stout, J.G.
1966: Devonian of western Canada; in International Symposium on the Devonian System, v. 1, D.H. Oswald (ed.); Alberta Society of Petroleum Geologists, p. 717-752.

Bebout, D.G. and Maiklem, W.R.
1973: Ancient anhydrite facies and environments, Middle Devonian Elk Point Basin, Alberta; Bulletin of Canadian Petroleum Geology, v. 21, p. 287-343.

Belyea, H.R.
1970: Significance of an unconformity within the Chinchaga Formation, northern Alberta and northeastern British Columbia; in Report of Activities, Part A, Geological Survey of Canada, Paper 70-1A, p. 76-79.

Benvenuto, G.L. and Price, R.A.
1979: Structural evolution of the Hosmer thrust sheet, southeastern British Columbia; Bulletin of Canadian Petroleum Geology, v. 27, no. 3, p. 360-394.

Berry, W.B.N. and Norford, B.S.
1976: Early Late Cambrian dendroid graptolites from the northern Yukon; Geological Survey of Canada, Bulletin 256, p. 1-12.

Best, R.V.
1952: Two new species of *Olenellus* from British Columbia; Transactions of the Royal Society of Canada; v. 46, ser. 3, p. 13-22.

Blusson, S.L.
1968: Geology and tungsten deposits near the headwaters of Flat River, Yukon Territory and southwestern District of Mackenzie, Canada; Geological Survey of Canada, Paper 67-22, 77 p.
1971: Sekwi Mountain map-area, Yukon Territory and District of Mackenzie; Geological Survey of Canada, Paper 71-22, 17 p.
1976: Selwyn Basin, Yukon and District of Mackenzie; in Report of Activities, Part A, Geological Survey of Canada, Paper 76-1A, p. 131-132.

Bond, G.C. and Kominz, M.A.
1984: Construction of tectonic subsidence curves for the early Paleozoic miogeosyncline, southern Canadian Rocky Mountains: Implications for subsidence mechanisms, age of breakup and crustal thinning; Geological Society of America Bulletin, v. 95, p. 155-173.

Brabb, E.E.
1967: Stratigraphy of the Cambrian and Ordovician rocks of east-central Alaska; United States Geological Survey, Professional Paper 559-A, 30 p.

Brandon, M.T., Orchard, M.J., Parrish, R.R., Sutherland Brown, A., and Yorath, C.J.
1986: Fossil ages and isotopic dates from the Paleozoic Sicker Group and associated intrusive rocks, Vancouver Island, British Columbia; in Current Research, Part A, Geological Survey of Canada, Paper 86-1A, p. 683-696.

Braun, W.K.
1977: Usefulness of ostracodes in correlating Middle and Upper Devonian rock sequences in Western Canada; in Western North American, Devonian, M.A. Murphy et al. (ed.), University of California, Riverside Campus Museum Contribution, v. 4, p. 65-79.
1978: Devonian ostracodes and biostratigraphy of western Canada; in Western and Arctic Canadian biostratigraphy, C.R. Stelck and B.D.E. Chatterton (ed.), Geological Association of Canada, Special Paper 18, p. 259-288.

Brock, J.S.
1975: Mining: Yukon's first industry, review of current exploration activity; Western Miner, v. 48, no. 2, p. 63-66.
1976: Recent developments in the Selwyn-Mackenzie zinc-lead province, Yukon and Northwest Territories; Western Miner, March, p. 9-16.

Burling, L.D.
1923: Cambro-Ordovician section near Mount Robson, British Columbia; Geological Society of America Bulletin, v. 34, p. 721-748.

1928: in Pre-Devonian Paleozoic Formations of the Cordilleran Provinces of Canada, C.D. Walcott (ed.), Smithsonian Miscellaneous Collections, v. 75, no. 5, p. 367.
1955: Annotated index to the Cambro-Ordovician of the Jasper Park and Mount Robson region; Alberta Society of Petroleum Geologists, 5th Annual Field Conference Guidebook, no. 15, p. 15-51.

Campbell, R.B. and Dodds, C.J.
1979: Operation Saint Elias, British Columbia; in Current Research, Part A, Geological Survey of Canada, Paper 79-1A, p. 17-20.

Campbell, R.B., Mountjoy, E.W., and Young, F.G.
1973: Geology of McBride map-area, British Columbia; Geological Survey of Canada, Paper 72-35, 104 p.

Cecile, M.P.
1982: The Lower Paleozoic Misty Creek Embayment, Selwyn Basin, Yukon and Northwest Territories; Geological Survey of Canada, Bulletin 335, 78 p.
1984: Geology of the northwest Niddery Lake map-area, Yukon; Geological Survey of Canada, Open File 1006.
1986: Lower Paleozoic embayments, troughs and arches, Northern Canadian Cordillera; Geological Survey of Canada, "Forum on Oil and Gas in Canada", Abstract, p. 5, 6.
1988: Corridor traverse through Barn Mountains, northernmost Yukon; in Current Research, Part D, Geological Survey of Canada, Paper 88-1D, p. 99-103.

Cecile, M.P. and Cook, D.G.
1981: Structural cross-section northern Selwyn and Mackenzie Mountains; Geological Survey of Canada, Open File 807.

Cecile, M.P., Cook, D.G., and Snowdon, L.R.
1982: Plateau overthrust and its hydrocarbon potential, Mackenzie Mountains, Northwest Territories; in Current Research, Part A, Geological Survey of Canada, Paper 82-1A, p. 89-94.

Cecile, M.P. and Norford, B.S.
1979: Basin to platform transition, Paleozoic strata of Ware and Trutch map-areas, northeastern British Columbia; in Current Research, Part A, Geological Survey of Canada, Paper 79-1A, p. 219-226.

Cecile, M.P., Hutcheon, I.E., and Gardner, D.
1982: Geology of the northern Richardson Anticlinorium (map areas 106L/12, 13, 116I/9, 16); Geological Survey of Canada, Open File 875.

Cecile, M.P., Shakur, M.A., and Krouse, H.R.
1983: The isotopic composition of western Canadian barites and the possibility of a barite isotopic composition age curve; Canadian Journal of Earth Sciences, v. 20, p. 1528-1535.

Chalmer, W.G. and Lawson, J.D. (editors)
1985: Evolution and environment in the Late Silurian and Early Devonian; Philosophical Transactions of the Royal Society of London, B. Biological Sciences, v. 309, p. 1-342.

Chatterton, B.D.E.
1978: Conodont biostratigraphy of western and northwestern Canada; in Western and Arctic Canadian Biostratigraphy, C.R. Stelck and B.D.E. Chatterton (ed.), Geological Association of Canada, Special Paper 18, p. 161-232.

Chatterton, B.D.E. and Perry, D.
1983: Silicified Silurian odontopleurid trilobites from the Mackenzie Mountains; Palaeontographica Canadiana, no. 1; Canadian Society of Petroleum Geologists-Geological Association of Canada, 127 p.

Churkin, M. Jr.
1975: Basement rocks of Barrow Arch, Alaska and circum-Arctic Paleozoic mobile belt; American Association of Petroleum Geologists Bulletin, v. 59, p. 451-456.

Cook, D.G.
1970: A Cambrian facies change and its effect on structure, Mount Stephen-Mount Dennis area, Alberta-British Columbia; Geological Association of Canada, Special Paper 6, p. 27-39.
1975: Structural style influenced by lithofacies, Rocky Mountain Main Ranges, Alberta-British Columbia; Geological Survey of Canada, Bulletin 233, 73 p.

Cook, D.G. and Aitken, J.D.
1978: Twitya uplift: A pre-Delorme phase of the Mackenzie Arch; in Current Research, Part A, Geological Survey of Canada, Paper 78-1A, p. 383-388.

Daily, B. and Jago, J.B.
1975: The triobite *Lejopyge* Hawle and Corda and the middle-upper Cambrian boundary; Palaeontological Association, v. 18, pt. 3, p. 527-550.

Dawson, K.M.
1978: Map of mineral deposits and principle occurrences in the Canadian Cordillera; Geological Survey of Canada, Open File 573.

Dean, W.T.
1978: Preliminary account of the trilobite biostratigraphy of the Survey Peak and Outram formations (Late Cambrian, Early Ordovician) at Wilcox Pass, southern Canadian Rocky Mountains, Alberta; Geological Survey of Canada, Paper 75-34, 10 p.

Deiss, C.
1939: Cambrian formations of southwestern Alberta and southeastern British Columbia; Bulletin of the Geological Society of America, v. 50, p. 951-1026.
1940: Lower and Middle Cambrian stratigraphy of southwestern Alberta and southeastern British Columbia; Bulletin of the Geological Society of America, v. 51, p. 731-794.
1941: Cambrian geography and sedimentation in the central Cordilleran region; Bulletin of the Geological Society of America, v. 52, p. 1085-1115.

Derby, J.R., Lane, H.R., and Norford, B.S.
1972: Uppermost Cambrian-basal Ordovician faunal succession in Alberta and correlation with similar sequences in the western United States; 24th International Geological Congress, Section 7, Paleontology, p. 503-512.

Devlin, W.J.
1984: Syn-depositional tectonism related to continental break-up in the Hamill Group, Northern Selkirk Mountains, British Columbia; Geological Association of Canada, Program with Abstracts, v. 9, p. 57.

Dings, M.G. and Whitebread, D.H.
1965: Geology and ore deposits of the Metaline zinc-lead District, Pend Oreille County, Washington; United States Geological Survey, Professional Paper 489, 109 p.

Douglas, R.J.W., Gabrielse, H., Wheeler, J.O., Stott, D.F., and Belyea, H.R.
1970: Geology of Western Canada; in Geology and Economic Minerals of Canada, R.J.W. Douglas (ed.), Geological Survey of Canada, Economic Geology Report 1, 5th edition, p. 367-546.

Dubord, M.P., Morrow, D.W., and Macqueen, R.W.
1986: A shelf-to-basin transition of the Devonian Ogilvie Formation, Yukon Territory; in Current Research, Part A, Geological Survey of Canada, Paper 86-1A, p. 603-608.

Dutro, J.T., Jr., Brosge, W.P., and Reiser, H.N.
1972: Significance of recently discovered Cambrian fossils and reinterpretation of Neruokpuk Formation, northeastern Alaska; The American Association of Petroleum Geologists Bulletin, v. 56, p. 808-815.

Energy, Mines and Resources Canada
1985: Canadian mineral deposits not being mined in 1983; Mineral Bulletin MR198, 308 p.

Ettensohn, F.R. and Barron, L.S.
1981: Depositional model for the Devonian-Mississippian black shale sequence of North America; a tectono-climatic approach; United States Department of Energy DOE/METC/12040-2.

Evans, C.S.
1933: Brisco-Dogtooth map-area, British Columbia; Geological Survey of Canada, Summary Report, 1932, Part A/11, p. 106-176.

Fritz, W.H.
1971: Geological setting of the Burgess Shale; Proceedings of the North American Paleontological Convention, September, 1969, Part 1, p. 1155-1170.
1972: Lower Cambrian trilobites from the Sekwi Formation type section, Mackenzie Mountains, northwestern Canada; Geological Survey of Canada, Bulletin 212, 90 p.
1974: Cambrian biostratigraphy, northern Yukon Territory and adjacent areas; in Report of Activities, Part A, Geological Survey of Canada, Paper 74-1, Part A, p. 309-313.
1975: Broad correlations of some Lower and Middle Cambrian strata in the North American Cordillera; in Report of Activities, Part A, Geological Survey of Canada, Paper 75-1, Part A, p. 533-540.
1976: Ten stratigraphic sections from the Lower Cambrian Sekwi Formation, Mackenzie Mountains, Northwestern Canada; Geological Survey of Canada, Paper 76-22, 42 p.
1978a: Fifteen stratigraphic sections from the Lower Cambrian of the Mackenzie Mountains, northwestern Canada; Geological Survey of Canada, Paper 77-33, 18 p.
1978b: Upper (carbonate) part of Alan Group, Lower Cambrian, north-central British Columbia; in Current Research, Part A, Geological Survey of Canada, Paper 78-1A, p. 7-16.
1979a: Cambrian stratigraphy in the northern Rocky Mountains, British Columbia; in Current Research, Part B, Geological Survey of Canada, Paper 79-1B, p. 99-109.
1979b: Eleven stratigraphic sections from the Lower Cambrian of the Mackenzie Mountains, northwestern Canada; Geological Survey of Canada, Paper 78-23, 19 p.
1979c: Cambrian stratigraphic section between South Nahanni and Broken Skull rivers, southern Mackenzie Mountains; in Current Research, Part B, Geological Survey of Canada, Paper 79-1B, p. 121-125.
1980a: Two Cambrian stratigraphic sections near Gataga River, Northern Rocky Mountains, British Columbia; in Current Research, Part C, Geological Survey of Canada, Paper 80-1C, p. 113-119.
1980b: Two new formations in the Lower Cambrian Atan Group, Cassiar Mountains, north-central British Columbia; in Current Research, Part B, Geological Survey of Canada, Paper 80-1B, p. 217-225.
1980c: International Precambrian-Cambrian Boundary Working Group's 1979 field study to Mackenzie Mountains, Northwest Territories, Canada; in Current Research, Part A, Geological Survey of Canada, Paper 80-1A, p. 41-45.
1981: Two Cambrian stratigraphic sections, eastern Nahanni map area, Mackenzie Mountains, District of Mackenzie; in Current Research, Part A, Geological Survey of Canada, Paper 81-1A, p. 145-156.
1982: Vampire Formation, a new Upper Precambrian(?)/Lower Cambrian formation, Mackenzie Mountains, Yukon and Northwest Territories; in Current Research, Part B, Geological Survey of Canada, Paper 82-1B, p. 83-92.
1984: Uppermost Precambrian and Lower Cambrian strata, northern Omineca Mountains, north-central British Columbia; in Current Research, Part B, Geological Survey of Canada, Paper 84-1B, p. 245-254.
1985: The basal contact of the Road River Group - a proposal for its location in the type area and in other selected areas in northern Canadian Cordillera; in Current Research, Part B, Geological Survey of Canada, Paper 85-1B, p. 205-215.
1990: In defense of the escarpment near the Burgess Shale fossil locality; Geoscience Canada, v. 17, no. 2, p. 106-110.

Fritz, W.H. and Crimes, T.P.
1985: Lithology, trace fossils, and correlation of Precambrian-Cambrian boundary beds, Cassiar Mountains, north-central British Columbia; Geological Survey of Canada, Paper 83-13, 24 p.

Fritz, W.H. and Mountjoy, E.W.
1975: Lower and early Middle Cambrian formations near Mount Robson, British Columbia and Alberta; Canadian Journal of Earth Sciences, v. 12, no. 2, p. 119-131.

Fritz, W.H. and Norris, D.K.
1966: Lower Middle Cambrian correlations in the east-central Cordillera; Geological Survey of Canada, Paper 66-1, p. 105-110.

Fritz, W.H., Narbonne, G.M., and Gordey, S.P.
1983: Strata and trace fossils near the Precambrian-Cambrian boundary, Mackenzie, Selwyn, and Wernecke mountains, Yukon and Northwest Territories; in Current Research, Part B, Geological Survey of Canada, Paper 83-1B, p. 365-375.

Fyles, J.T.
1964: Geology of the Duncan Lake area, Lardeau District, British Columbia; British Columbia Department of Mines and Petroleum Resources, Bulletin 49, 87 p.

Fyles, J.T. and Eastwood, G.E.P.
1962: Geology of the Ferguson Area, Lardeau District, British Columbia; British Columbia Department of Mines and Petroleum Resources, Bulletin 45, 92 p.

Gabrielse, H.
1963a: McDame map-area, Cassiar District, British Columbia; Geological Survey of Canada, Memoir 319, 138 p.
1963b: Rabbit River, British Columbia; Geological Survey of Canada, Map 46-1962.
1967: Tectonic evolution of the northern Canadian Cordillera; Canadian Journal of Earth Sciences, v. 4, p. 271-298.
1969: Geology of Jennings River map-area, British Columbia (104-O); Geological Survey of Canada, Paper 68-55, 37 p.
1975: Geology of Fort Grahame E-1/2 map-area, British Columbia; Geological Survey of Canada, Paper 75-33, 28 p.
1977: Geology of Toodoggone (NTS 94E) and Ware (NTS 94F, W-1/2) map-areas, British Columbia; Geological Survey of Canada, Open File 483.
1981: Stratigraphy and structure of Road River and associated strata in Ware (west half) map-area, northern Rocky Mountains, British Columbia; in Current Research, Part A, Geological Survey of Canada, Paper 81-1A, p. 201-207.

1985: Major dextral transcurrent displacements along the Northern Rocky Mountain Trench and related lineaments in north-central British Columbia; Geological Society of America Bulletin, v. 96, p. 1-14.

Gabrielse, H., Blusson, S.L., and Roddick, J.A.
1973: Geology of Flat River, Glacier Lake and Wrigley Lake map-areas, District of Mackenzie and Yukon Territory; Geological Survey of Canada, Memoir 366, 153 p.

Gabrielse, H. and Taylor, G.C.
1982: Geological maps and cross-sections of the northern Canadian Cordillera from southwest of Fort Nelson, British Columbia to Gravina Island, southeastern Alaska; Geological Survey of Canada, Open File 864.

Gardner, D.A.C.
1977: Structural geology and metamorphism of calcareous Lower Paleozoic slates, Blaeberry River-Redburn Creek area, near Golden, British Columbia; Ph.D. thesis, Queen's University, 224 p.

Gehrels, G.E. and Berg, H.C.
1984: Geologic map of southeastern Alaska; United States Geological Survey, Open File Report 84-886.

Gehrels, G.E., Saleeby, J.B., and Berg, H.C.
1983: Preliminary description of the Klakas orogeny in the southern Alexander terrane, southeastern Alaska; in Pre-Jurassic Rocks in Western North American Suspect Terranes, C.H. Stephens (ed.), Pacific Section, Society of Economic Paleontologists and Mineralogists, p. 131-141.

Geldsetzer, H.H.J.
1982: Depositional history of the Devonian succession in the Rocky Mountains southwest of the Peace River Arch; in Current Research, Part C, Geological Survey of Canada, Paper 82-1C, p. 55-64.

Gilbert, D.L.F.
1973: Anderson, Horton, northern Great Bear and Mackenzie Plains, northwest Territories; in Future Petroleum Provinces of Canada, R.G. McCrossan (ed.), Canadian Society of Petroleum Geologists, Memoir 1, p. 213-244.

Godwin, C.I. and Price, B.J.
1986: Geology of the Mountain diatreme kimberlite, north-central Mackenzie Mountains, District of Mackenzie, Northwest Territories; in Mineral Deposits of Northern Cordillera, J.A. Morin (ed.), Canadian Institute of Mining and Metallurgy, Special Volume 37, p. 298-310.

Goodfellow, W.D., Jonasson, I.R., and Cecile, M.P.
1980a: Nahanni integrated multidisciplinary pilot project geochemical studies Part 1: Geochemistry and mineralogy of shales, cherts, carbonates and volcanic rocks from the Road River Formation, Misty Creek Embayment, Northwest Territories; in Current Research, Part B, Geological Survey of Canada, Paper 80-1B, p. 149-161.
1980b: Nahanni integrated multidisciplinary pilot project geochemical studies Part 2: Some thoughts on the source, transportation and concentration of elements in shales of the Misty Creek Embayment, Northwest Territories; in Current Research, Geological Survey of Canada, Paper 80-1B, p. 163-171.

Gordey, S.P.
1978: Stratigraphy and structure of the Summit Lake area, Yukon and Northwest Territories; in Current Research, Part A, Geological Survey of Canada, Paper 78-1A, p. 43-48.
1979: Stratigraphy of southeastern Selwyn Basin in the Summit Lake Area, Yukon Territory and Northwest Territories; in Current Research, Part A, Geological Survey of Canada, Paper 79-1A, p. 13-16.
1980: Stratigraphic cross-section Selwyn Basin to Mackenzie Platform, Nahanni map area, Yukon Territory and District of Mackenzie; in Current Research, Part A, Geological Survey of Canada, Paper 80-1A, p. 353-355.
1981a: Stratigraphy, structure and tectonic evolution of southern Pelly Mountains in the Indigo Lake area, Yukon Territory; Geological Survey of Canada, Bulletin 318, 44 p.
1981b: Structure section areas south-central Mackenzie Mountains, Northwest Territories; Geological Survey of Canada, Open File 809.
1983: Thrust faults in the Anvil Range and a new look at the Anvil Range Group, south-central Yukon Territory; in Current Research, Part A, Geological Survey of Canada, Paper 83-1A, p. 225-227.

Gordey, S.P. and Anderson, R.G.
in press: Evolution of the northern Cordilleran Miogeosyncline, Nahanni map area, Yukon Territory and District of Mackenzie; Geological Survey of Canada, Memoir 428.

Gordey, S.P., Abbott, J.C., and Orchard, M.J.
1982: Devono-Mississippian (Earn Group) and younger strata in east-central Yukon; in Current Research, Part B, Geological Survey of Canada, Paper 82-1B, p. 93-100.

Gordey, S.P., Wood, D., and Anderson, R.G.
1981: Stratigraphic framework of southeastern Selwyn Basin, Nahanni map area, Yukon and District of Mackenzie; in Current Research, Part A, Geological Survey of Canada, Paper 81-1A, p. 395-398.

Green, L.H.
1972: Geology of Nash Creek, Larsen Creek and Dawson map areas, Yukon Territory (106D, 116A,B,C); Geological Survey of Canada, Memoir 364, 157 p.

Griffin, D.L.
1967: Devonian of northeastern British Columbia; in International Symposium on the Devonian System, Calgary, D.H. Oswald (ed.), Alberta Society of Petroleum Geologists, v. 1, p. 803-326.

Halley, R.G.
1974: Early Middle Cambrian shelf and basin stratigraphic relations in southern Nevada and southeastern California; Geological Society of America, Abstracts with programs, v. 6, no. 3, p. 186, 187.

Handfield, R.C.
1971: Archaeocyatha from the Mackenzie and Cassiar Mountains, Northwest Territories, Yukon Territory and British Columbia; Geological Survey of Canada, Bulletin 201, 119 p.

Heckel, P.H. and Witzke, B.J.
1979: Devonian world paleogeography determined from distribution of carbonate and related lithic paleoclimatic indicators; Special Papers in Paleontology 23, p. 99-123.

Hintze, L.F.
1953: Lower Ordovician trilobites from western Utah and eastern Nevada; Utah Geological and Mineralogical Survey, Bulletin 48.

Hofmann, H.J.
1981: First record of a Late Proterozoic faunal assemblage in the North America Cordillera; Lethaia, v. 14, p. 303-310.

Hofmann, H.J. and Cecile, M.P.
1981: Occurrence of Oldhamia and other trace fossils in Lower Cambrian(?) argillites, Naddery Lake map area, Selwyn Mountains, Yukon Territory; in Current Research, Part A, Geological Survey of Canada, Paper 81-1A, p. 281-290.

Hofmann, H.J., Mountjoy, E.W., and Teitz, M.W.
1985: Ediacaran fossils from the Miette Group, Rocky Mountains, British Columbia, Canada; Geology, v. 13, p. 819-821.

Jackson, D.E.
1966: Graptolitic facies of the Canadian Cordillera and Arctic Archipelago, A Review; Bulletin of Canadian Petroleum Geology, v. 14, p. 469-485.

Jackson, D.E. and Lenz, A.C.
1962: Zonation of Ordovician and Silurian graptolites of the northern Yukon, Canada; American Association of Petroleum Geologists Bulletin, v. 46, p. 30-45.

Jackson, D.E., Lenz, A.C., and Pedder, A.E.H.
1978: Late Silurian and Early Devonian graptolite, brachiopod and coral faunas from northwestern and Arctic Canada; Geological Association of Canada, Special Publication 17, 159 p.

Jackson, D.E., Steen, G., and Sykes, D.
1965: Stratigraphy and graptolite zonation of the Kechika and Sandpile groups in northeastern British Columbia; Bulletin of Canadian Petroleum Geology, v. 13, p. 139-154.

Jennings, D.S. and Jilson, G.A.
1987: Geology and sulphide deposits of Anvil Range, Yukon; in Mineral Deposits of the Northern Cordillera, J.A. Morin (ed.), Canadian Institute of Mining and Metallurgy, Special Volume 37.

Johnson, J.G.
1970: Taghanic onlap and the end of North American Devonian provinciality; Geological Society of America Bulletin, v. 81, p. 2077-2106.

Johnson, J.G., Klapper, G., and Sandberg, C.A.
1985: Devonian eustatic fluctuations in Euramerica; Geological Society of America Bulletin, v. 96, p. 567-587.

Jones, D.L., Irwin, W.P., and Owenshire, A.T.
1972: Southeastern Alaska - a displaced continental fragment?; in Geological Survey Research, United States Geological Survey, Professional Paper 800-B, p. B211-B217.

Kennedy, D.J. and Barnes, S.R.
1981: Conodont biostratigraphy of the Survey Peak Formation, southern Rocky Mountains, Alberta; Geological Association of Canada, Abstracts, v. 6, p. A31.

Ketner, K.B.
1968: Origin of Ordovician quartzite in the Cordilleran miogeosyncline; in Geological Survey Research, United States Geological Survey, Professional Paper 600B, p. 169-177.

Landing, E., Ludvigsen, R., and von Bitter, P.H.
1980: Upper Cambrian to Lower Ordovician conodont biostratigraphy and biofacies, Rabbitkettle Formation, District of Mackenzie; Royal Ontario Museum, Life Sciences Contributions no. 126, 42 p.

Landing, E., Myrow, P., Benus, A.P., and Narbonne, G.M.
1989: The Placentian series: appearance of the oldest skeletalized faunas in southeastern Newfoundland; Journal of Paleontology, v. 63, no. 6, p. 739-769.

Lane, L.S. and Cecile, M.P.
1989: Stratigraphy and structure of the Neruokpuk Formation, northern Yukon; in Current Research, Part G, Geological Survey of Canada, Paper 89-1G, p. 57-62.

Law, J.
1971: Regional Devonian geology and oil and gas possibilities, upper Mackenzie River area; Bulletin of Canadian Petroleum Geology, v. 19, p. 437-486.

Leech, G.B.
1954: Preliminary account, Canal Flats, British Columbia; Geological Survey of Canada, Paper 54-7, 32 p.
1958: Fernie map-area, west half, British Columbia, 82G W 1/2; Geological Survey of Canada, Paper 58-10, 40 p.

Lenz, A.C.
1972: Ordovician to Devonian history of the northern Yukon and adjacent District of Mackenzie; Bulletin of Canadian Petroleum Geology, v. 20, p. 321-361.
1980: Wenlockian graptolite reference section, Clearwater Creek, Nahanni National Park, Northwest Territories, Canada; Canadian Journal of Earth Sciences, v. 17, p. 1075-1086.
1982: Llandoverian graptolites of the northern Canadian Cordillera: *Petalograptus, Cephalograptus, Rhaphidograptus, Dimorphograptus*, Retiolitidae and Monograptidae; Royal Ontario Museum, Life Sciences Contribution 130, 154 p.

Lenz, A.C. and McCracken, A.D.
1982: The Ordovician-Silurian boundary, northern Canadian Cordillera: graptolite and conodont correlation; Canadian Journal of Earth Sciences, v. 19, p. 1308-1322.

Lenz, A.C. and Pedder, A.E.H.
1972: Lower and Middle Paleozoic sediments and paleontology of Royal Creek and Peel River, Yukon and Powell Creek, N.W.T.; International Geological Congress 24th Session, Guidebook Field Excursion A14.

Lenz, A.C. and Perry, D.G.
1972: The Neruokpuk Formation of the Barn Mountains and Driftwood Hills, Northern Yukon: its age and graptolite fauna; Canadian Journal of Earth Sciences, v. 9, p. 1129-1138.

Little, H.W.
1960: Nelson map-area, west half, British Columbia (82F W1/2); Geological Survey of Canada, Memoir 308, 205 p.

Lochman-Balk, C. and Wilson, J.L.
1958: Cambrian biostratigraphy in North America; Journal of Paleontology, v. 32, no. 2, p. 312-350.
1967: Stratigraphy of Upper Cambrian-Lower Ordovician subsurface sequence in Williston Basin; American Association of Petroleum Geologists Bulletin, v. 51, p. 883-917.

Longacre, S.A.
1970: Trilobites of the Upper Cambrian *Ptychaspid* Biomere, Wilberns Formation, central Texas; Paleontological Society, Memoir 4, 70 p.

Ludvigsen, R.
1975: Ordovician formations and faunas, southern Mackenzie Mountains; Canadian Journal of Earth Sciences, v. 12, no. 4, p. 663-697.
1979: Upper Cambrian and Lower Ordovician trilobite biostratigraphy of the Rabbitkettle Formation, western District of Mackenzie; Royal Ontario Museum, Life Sciences Contribution 134, 188 p.
1982: Upper Cambrian and Lower Ordovician trilobite biostratigraphy of the Rabbitkettle Formation, western District of Mackenzie; Royal Ontario Museum, Life Sciences Contribution 134; 188 p.

Ludvigsen, R. and Westrop, S.R.
1985: Three new Upper Cambrian stages for North America; Geology, v. 13, p. 139-143.

MacIntyre, D.G.
1981: Akie River Project; in Geological Fieldwork 1980, British Columbia Ministry of Energy, Mines and Petroleum Resources, Paper 1981-1, p. 33-45.

Macqueen, R.W.
1975: Lower and middle Paleozoic sediments, northern Yukon Territory; in Report of Activities, Part C, Geological Survey of Canada, Paper 75-1C, p. 291-301.

Macqueen R.W. and Thompson, R.I.
1978: Carbonate-hosted lead-zinc occurrences in northeastern British Columbia with emphasis on the Robb Lake deposit; Canadian Journal of Earth Sciences, v. 15, p. 1737-1762.

Manson, G.R.
1984: Brisco Barite Mine; in The Geology of Industrial Minerals in Canada, C.R. Guilet and W. Martin (ed.), Canadian Institute of Mining and Metallurgy, Special v. 29, p. 264-265.

Mansy, J.L. and Gabrielse, H.
1978: Stratigraphy, terminology and correlation of Upper Proterozoic rocks in Omineca and Cassiar Mountains, north-central British Columbia; Geological Survey of Canada, Paper 74-19, 17 p.

McAllister, A.L.
1951: Ymir map-area, British Columbia; Geological Survey of Canada, Paper 51-4, 58 p.

McArthur, M.L., Tipnis, R.S., and Godwin, C.I.
1980: Early and Middle Ordovician conodont fauna from the Mountain Diatreme, northern Mackenzie Mountains; in Current Research, Part A, Geological Survey of Canada, Paper 80-1A, p. 363-368.

McIlreath, I.A.
1977a: Stratigraphic and sedimentary relationships at the western edge of the Middle Cambrian carbonate facies belt, Field, British Columbia; Ph.D. thesis, University of Calgary, Alberta, 259 p.
1977b: Accumulation of a Middle Cambrian, deep-water limestone debris apron adjacent to a vertical, submarine carbonate escarpment, southern Rocky Mountains, Canada; in Deep-water Carbonate Environments, H.E. Cook and P. Enos (ed.), Society of Economic Paleontologists and Mineralogists, Special Publication 25, p. 113-124.

McLaughlin, K.P. and Embysk, B.B.
1950: Middle Cambrian trilobites from Pend Oreille Country, Washington; Journal of Paleontology, v. 24, no. 4, p. 466-471.

Meijer Drees, N.C.
1974: Geology of the "Bulmer Lake High", a gravity feature in the southern Great Bear Plain, District of Mackenzie; in Report of Activities, Part B, Geological Survey of Canada, Paper 74-1B, p. 274-277.
1975a: Geology of the lower Paleozoic formations in the subsurface of the Fort Simpson area, District of Mackenzie; Geological Survey of Canada, Paper 74-40, 65 p.
1975b: The Little Doctor sandstone (new sub-unit) and its relationship to the Franklin Mountain and Mount Kindle Formations in the Nahanni Range and nearby subsurface, District of Mackenzie; in Report of Activities, Part C, Geological Survey of Canada, Paper 75-1C, p. 51-57.
1980: Description of the Hume, Funeral, and Bear Rock formations in the Candex et al. Dahadinni M43A well, District of Mackenzie; Geological Survey of Canada, Paper 78-17.
in press: The Devonian succession in the subsurface of the Great Slave and Great Bear Plains, Northwest Territories; Geological Survey of Canada, Memoir.

Miall, A.D.
1976: Proterozoic and Paleozoic geology of Banks Island, Arctic Canada; Geological Survey of Canada, Bulletin 258, 77 p.

Monger, J.W.H. and Berg, H.C.
1984: Lithotectonic terrane map of western Canada and southeastern Alaska; in Lithotectonic Terrane Maps of the North American Cordillera, N.J. Silberling and D.L. Jones (ed.), United States Geological Survey, Open File Report 84-523.

Moore, P.F.
in press: Devonian strata of the Interior Plains region of western Canada; in Sedimentary Cover of the Craton in Canada, D.F. Stott and J.D. Aitken (ed.), Geological Survey of Canada, Geology of Canada no. 5 (also Geological Society of America, The Geology of North America, Vol. D-1).

Morganti, J.M.
1979: The geology and ore deposits of the Howards Pass area, Yukon and Northwest Territories: The origin of basinal sedimentary stratiform sulphide deposits; Ph.D. thesis, University of British Columbia, Vancouver, 327 p.

Morin, J.A., Grapes, K.J., and Debicki, R.L.
1983: Yukon mineral industry, 1982, an overview; in Yukon Exploration and Geology, 1982; Department of Indian and Northern Affairs, Canada, Whitehorse, p. 4-17.

Morrow, D.W.
1973: The Presqu'ile Complex - implications for Middle Devonian paleogeography of northern Alberta and British Columbia; Bulletin of Canadian Petroleum Geology, v. 21, p. 421-434.
1978: The Dunedin Formation - a transgressive shelf carbonate sequence; Geological Survey of Canada, Paper 76-12.
1984: Sedimentation of Root Basin and Prairie Creek Embayment - Siluro-Devonian, Northwest Territories; Bulletin of Canadian Petroleum Geology, v. 32, p. 162-189.
in press: The Silurian-Devonian sequence of the northern part of the Mackenzie Shelf, Northwest Territories; Geological Survey of Canada, Bulletin.

Morrow, D.W. and Cook, D.G.
1987: The Prairie Creek Embayment and Lower Paleozoic stratigraphy of the southern Mackenzie Mountains; Geological Survey of Canada, Memoir 412.

Morrow, D.W. and Meijer Drees, N.C.
1981: The Early to Middle Devonian Bear Rock Formation in the type section and other surface sections, District of Mackenzie; in Current Research, Part A, Geological Survey of Canada, Paper 81-1A, p. 107-114.

Mott, J.A., Dixon, J.M., and Helmstaedt, H.
1986: Ordovician stratigraphy and the structural style at the Main Ranges - Front Ranges boundary near Smith Peak, British Columbia; in Current Research, Part B, Geological Survey of Canada, Paper 86-1B, p. 457-465.

Mountjoy, E.W.
1962: Mount Robson (southeast) map-area, Rocky Mountains of Alberta and British Columbia, 83E/SE; Geological Survey of Canada, Paper 61-31, 114 p.

Muir, I. and Dixon, D.
1984: Facies analysis of a Middle Devonian sequence in the Mountain River - Gayna River area, N.W.T.; Progress Report submitted to geology office of Department of Indian Affairs and Northern Development.

Muller, J.E.
1980: The Paleozoic Sicker Group of Vancouver Island, British Columbia; Geological Survey of Canada, Paper 79-30.
1983: Geology, Victoria, British Columbia; Geological Survey of Canada, Map 1553A, scale 1:100 000.

Narbonne, G.M., Hofmann, H.J., and Aitken, J.D.
1985: Precambrian-Cambrian boundary sequence, Wernecke Mountains, Yukon Territory; in Current Research, Part A, Geological Survey of Canada, Paper 85-1A, p. 603-608.

Narbonne, G.M., Myrow, P.M., Landing, E., and Anderson, M.M.
1987: A candidate stratotype for the Precambrian-Cambrian boundary, Fortune Head, Burin Peninsula, southeastern Newfoundland; Canadian Journal of Earth Sciences, v. 24, p. 1277-1293.

Nelson, S.J.
1963: Ordovician paleontology of the northern Hudson Bay Lowland; Geological Society of America, Memoir 90, 152 p.

Ney, C.S.
1954: Monarch and Kicking Horse mines, Field, British Columbia; Alberta Society of Petroleum Geologists, 4th Annual Field Conference Guidebook, Banff-Golden-Radium, p. 119-136.

Noble, J.P.A. and Ferguson, R.D.
1971: Facies and faunal relations at edge of early mid-Devonian Carbonate Shelf, south Nahanni River area, Northwest Territories; Bulletin of Canadian Petroleum Geology, v. 19, p. 570-588.

Norford, B.S.
1964: Reconnaissance of the Ordovician and Silurian rocks of the northern Yukon Territory; Geological Survey of Canada, Paper 63-39, 139 p.
1969: Ordovician and Silurian stratigraphy of the southern Rocky Mountains; Geological Survey of Canada, Bulletin 176, 90 p.
1981: Devonian stratigraphy at the margins of the Rocky Mountain Trench, Columbia River, southeastern British Columbia; Bulletin of Canadian Petroleum Geology, v. 29, p. 540-560.

Norford, B.S., Gabrielse, H., and Taylor, G.C.
1966: Stratigraphy of Silurian carbonate rocks of the Rocky Mountains, northern British Columbia; Bulletin of Canadian Petroleum Geology, v. 14, p. 504-519.

Norford, B.S. and Macqueen, R.W.
1975: Lower Paleozoic Franklin Mountain and Mount Kindle formations, District of Mackenzie: Their type sections and regional development; Geological Survey of Canada, Paper 74-34, 37 p.

Norford, B.S. and Orchard, M.J.
1985: Early Silurian age of rocks hosting lead-zinc mineralization at Howards Pass, Yukon Territory and District of Mackenzie: local biostratigraphy of Road River Formation and Earn Group; Geological Survey of Canada, Paper 83-18.

Norris, A.W.
1985: Stratigraphy of Devonian outcrops belts in northern Yukon Territory and northwestern District of Mackenzie (Operation Porcupine Area); Geological Survey of Canada, Memoir 410.

Norris, D.K.
1974a: Structural geometry and geological history of the northern Canadian Cordillera; in Proceedings of the First National Convention, A.E. Wren and R.B. Cruz (ed.), Canadian Society of Exploration Geophysicists, p. 18-45.
1976: Structural and stratigraphic studies in the northern Canadian Cordillera; in Report of Activities, Part A; Geological Survey of Canada, Paper 76-1A, p. 457-466.
1982a: Ogilvie River, Yukon Territory; Geological Survey of Canada, Map 1526A.
1982b: Wind River, Yukon Territory; Geological Survey of Canada, Map 1528A.
in press: The geology, mineral and hydrocarbon potential of Northern Yukon Territory and north-western District of Mackenzie (Operation Porcupine); Geological Survey of Canada.

Norris, D.K. and Price, R.A.
1966: Middle Cambrian lithostratigraphy of southeastern Canadian Cordillera; Bulletin of Canadian Petroleum Geology, v. 14, no. 4, p. 385-404.

Norris, D.K. and Yorath, C.J.
1981: The North American Plate from the Arctic Archipelago to the Romanzof Mountains; in the Arctic Ocean, Ocean Basin Series, v. 5, p. 37-103.

North American Commission on Stratigraphic Nomenclature
1984: North American Stratigraphic Code; American Association of Petroleum Geologists Bulletin, v. 65, p. 841-875.

North, F.K.
1964: Cambrian, Part II-Cordillera; in Geological History of Western Canada; Alberta Society of Petroleum Geologists, p. 28-33.
1971: The Cambrian of Canada and Alaska; in Cambrian of the New World, C.H. Holland (ed.), Wiley-Interscience, London, p. 219-324.

North, F.K. and Henderson, G.G.L.
1954: Summary of the geology of the southern Rocky Mountains of Canada; Alberta Society of Petroleum Geologists, 4th Annual Field Conference Guidebook, Banff-Golden-Radium, p. 15-81.

Nowlan, G., Narbonne, G.M., and Fritz, W.H.
1985: Small shelly fossils and trace fossils near the Precambrian-Cambrian boundary in the Yukon Territory, Canada; Lethaia, v. 18, p. 233-256.

Okulitch, A.V.
1979: Lithology, stratigraphy, structure and mineral occurrences of the Thompson-Shuswap-Okanagan area, British Columbia; Geological Survey of Canada, Open File 637.
1985: Paleozoic plutonism in southeastern British Columbia; Canadian Journal of Earth Sciences, v. 22, p. 1409-1424.

Okulitch, V.J.
1951: A Lower Cambrian fossil locality near Addy, Washington; Journal of Paleontology, v. 25, no. 3, p. 405-407.

Oldershaw, A.E.
1981: A preliminary analysis of the Mountain and Keele Diatremes; in Department of Indian and Northern Affairs, Mineral Industry Report, 1977, Northwest Territories, EGS 1981-11, p. 148-154.

Palmer, A.R.
1960: Some aspects of the early Upper Cambrian stratigraphy of White Pine County, Nevada and vicinity; in Guidebook to the Geology of East-Central Nevada, W.W. Sloan, Jr. and J.W. Boettcher (ed.), Intermountain Association of Petroleum Geologists and Eastern Nevada Geological Society's 1960 Guidebook, p. 53-58.
1965: Trilobites of the Late Cambrian Pterocephaliid Biomere in the Great Basin, United States; United States Geological Survey, Professional Paper 493, 105 p.
1968: Cambrian trilobites of east-central Alaska; United States Geological Survey, Professional Paper 559-B, 115 p.

Palonen, P.A.
1976: Sedimentary and stratigraphy of Gog Group sandstones in southern Canadian Rockies; Ph.D. thesis, University of Calgary, Alberta, 197 p.

Park, C.F., Jr. and Cannon, R.S., Jr.
1943: Geology and ore deposits of the Metaline Quadrangle, Washington; United States Geological Survey, Professional Paper 202, 81 p.

Payne, M.W. and Allison, C.W.
1981: Paleozoic continental-margin sedimentation in east-central Alaska; Geology, v. 9, p. 274-279.

Pell, J. and Simony, P.
1981: Stratigraphy, structure, and metamorphism in the southern Cariboo Mountains, British Columbia; in Current Research, Part A, Geological Survey of Canada, Paper 81-1A, p. 227-230.
1984: Stratigraphy of the Hadrynian Kaza Group between the Azure and North Thompson rivers, Cariboo Mountains, British Columbia; in Current Research, Part A, Geological Survey of Canada, Paper 84-1A, p. 95-98.

Pelzer, E.E.
1966: Mineralogy, geochemistry and stratigraphy of Besa River Shale, British Columbia; Bulletin of Canadian Petroleum Geology, v. 14, p. 273-321.

Perry, D.G., Klapper, G., and Lenz, A.C.
1974: Age of the Ogilvie Formation (Devonian) northern Yukon, based primarily on the occurrence of brachiopods and conodonts; Canadian Journal of Earth Sciences, v. 11, p. 1055-1097.

Porter, J.W., Price, R.A., and McCrossan, R.G.
1982: The Western Canada Sedimentary Basin; Philosophical Transactions of the Royal Society, London, A, v. 305, p. 169-192.

Potter, A.W. and Cecile, M.P.
1985: Paleobiogeographic significance of two Late Ordovician brachiopod faunules from the Misty Creek Embayment, Selwyn Basin, Northwest Territories, Canada; Geological Society of America, Abstracts with Program, v. 17, p. 401.

Price, R.A. and Fermor, P.R.
1985: Structure section of the Cordilleran Foreland Thrust and Fold Belt west of Calgary, Alberta; Geological Survey of Canada, Paper 84-14.

Pugh, D.C.
1983: Pre-Mesozoic geology in the subsurface of Peel River map area, Yukon Territory and District of Mackenzie; Geological Survey of Canada, Memoir 401, 61 p.

Rasetti, F.
1951: Middle Cambrian stratigraphy and faunas of the Canadian Rocky Mountains; Smithsonian Miscellaneous Collections, v. 116, no. 5, 270 p.

Read, B.C.
1980: Lower Cambrian archaeocyathid buildups, Pelly Mountains, Yukon; Geological Survey of Canada, Paper 78-18, 54 p.

Read, P.B. and Wheeler, J.O.
1977: Geology Lardeau, west-half map area and marginal notes, revised edition; Geological Survey of Canada, Open File 432, 1:125 000 scale map.

Reesor, J.E.
1973: Geology of the Lardeau map-area, east-half, British Columbia; Geological Survey of Canada, Memoir 369, 129 p.

Reeve, A.F.
1977: The Goz Creek zinc prospect, Yukon Territory, Canada; Barrier Reef Resources, Limited, Internal Report, 19 p.

Reiser, H.N., Brosgé, W.P., Dutro, J.T., Jr., and Detterman, R.L.
1980: Geologic map of the Demarcation Point Quadrangle, Alaska; United States Geological Survey, Miscellaneous Investigations Series, Map I-1133.

Resser, C.E.
1934: Recent discoveries of Cambrian beds in the northeastern United States; Smithsonian Miscellaneous Collections, v. 92, no. 10, 10 p.

Rice, H.M.A.
1937: Cranbrook map-area, British Columbia; Geological Survey of Canada, Memoir 207, 67 p.
1941: Nelson map-area, east half, British Columbia; Geological Survey of Canada, Memoir 228, 86 p.

Robison, R.A.
1960: Lower and Middle Cambrian stratigraphy of the eastern Great Basin; in Guidebook to the Geology of East-Central Nevada, W.W. Sloan, Jr. and J.W. Boettcher (ed.); International Association of Petroleum Geologists and Eastern Geological Society's 1960 Guidebook, p. 43-52.

1984: Cambrian Agnostida of North America and Greenland, part 1, Ptychagnostidae; University of Kansas Paleontological Contributions, Paper 109, 59 p.

Roddick, J.A.
1967: Tintina Trench; Journal of Geology, v. 73, p. 23-33.

Root, K.G.
1983: Upper Proterozoic and Paleozoic stratigraphy, Delphine Creek area, southeastern British Columbia: implications for the Purcell Arch; in Current Research, Part B, Geological Survey of Canada, Paper 83-1B, p. 377-380.
1985: Reinterpretation of the age of a succession of Paleozoic strata Delphine Creek, southeastern British Columbia; in Current Research Part A, Geological Survey of Canada, Paper 85-1A, p. 727-730.

Ross, R.J. Jr.
1949: Stratigraphy and trilobite faunal zones of the Garden City Formation, northeastern Utah; American Journal of Science, v. 247, p. 472-491.
1951: Stratigraphy of the Garden City Formation in northeastern Utah, and its trilobite faunas; Yale University, Peabody Museum, Bulletin.

Schiarizza, P.
1986: Geology of the Eagle Bay Formation between the Raft and Baldy batholiths (82M/5,11,12); in Geology Fieldwork, British Columbia Ministry of Energy, Mines and Petroleum Resources, Paper 1986-1, p. 89-94.

Schiarizza, P. and Preto, V.A.
1984: Geology of the Adams Plateau-Clearwater area; British Columbia Ministry of Energy, Mines and Petroleum Resources, Preliminary Map No. 56.

Schofield, S.J.
1922: Relationship of the Precambrian (Beltian) terrain to the Lower Cambrian strata of southeastern British Columbia; Geological Survey of Canada, Museum Bulletin 35, 13 p.

Simony, P.S. and Wind, G.
1970: Structure of the Dogtooth Range and adjacent portions of the Rocky Mountain Trench; Geological Association of Canada, Special Paper 6, p. 41-51.

Slind, O.L. and Perkins, G.D.
1966: Lower Paleozoic and Proterozoic sediments of the Rocky Mountains between Jasper, Alberta and Pine River, British Columbia; Bulletin of Canadian Petroleum Geology, v. 14, p. 442-468.

Sloss, L.L.
1963: Sequences in the cratonic interior of North America; Geological Society of America Bulletin, v. 74, p. 93-114.
1972: Synchrony of Phanerozoic sedimentary-tectonic events of North American craton and the Russian Platform; 24th International Geological Congress, Montreal, Section 6, p. 24-32.

Stitt, J.H.
1971: Late Cambrian and earliest Ordovician trilobites, Timber Hills and Lower Arbuckel Groups, western Arbuckle Mountains, Murry County, Oklahoma; Oklahoma Geological Survey, Bulletin 110, 83 p.

Struik, L.C.
1980: Geology of the Barkerville-Cariboo River area, central British Columbia; Ph.D. thesis, University of Calgary, Alberta, 330 p.
1981: A re-examination of the type area of the Devono-Mississippian Cariboo Orogeny, central British Columbia; Canadian Journal of Earth Sciences, v. 18, p. 1767-1775.

Sutherland Brown, A.
1963: Geology of the Cariboo River area, British Columbia; British Columbia Department of Mines and Petroleum Resources, Bulletin 47, 60 p.

Taylor, G.C., Campbell, R.B., and Norford, B.S.
1972: Silurian igneous rocks in the western Rocky Mountains, northwestern British Columbia; in Report of Activities, Part A, Geological Survey of Canada, Paper 72-1A, p. 228-229.

Taylor, G.C. and Mackenzie, W.S.
1970: Devonian stratigraphy of northeastern British Columbia; Geological Survey of Canada, Bulletin 186.

Taylor, G.C. and Stott, D.F.
1973: Tuchodi Lakes map area, British Columbia; Geological Survey of Canada, Memoir 373, 37 p.

Tempelman-Kluit, D.J.
1970: Stratigraphy and structure of the "Keno Hill Quartzite" in Tombstone River-Upper Klondike River map-areas, Yukon Territory; Geological Survey of Canada, Bulletin 180, 102 p.
1972: Geology and origin of the Faro, Vangorda, and Swim concordant zinc-lead deposits, central Yukon Territory; Geological Survey of Canada, Bulletin 208, 73 p.

1977a: Stratigraphic and structural relations between the Selwyn Basin, Pelly-Cassiar platform and Yukon crystalline terrane in the Pelly Mountains, Yukon; in Report of Activities, Part A, Geological Survey of Canada, Paper 77-1A, p. 223-227.
1977b: Geology of Quiet Lake (105F) and Finlayson Lake (105G) map areas, Yukon Territory (2 maps scale 1:250 000); Geological Survey of Canada, Open File 486.
1979: Transported cataclastite, ophiolite and granodiorite in Yukon: Evidence of arc-continent collision; Geological Survey of Canada, Paper 79-14, 27 p.

Thompson, R.I.
1981: The nature and significance of large 'blind' thrusts within the Northern Rocky Mountains of Canada; in Thrust and Nappe Tectonics, K.R. McClay and N.J. Price (ed.); Geological Society of London, Special Publication 9, p. 449-462.
1989: Stratigraphy, tectonic evolution and structural analysis of the Halfway River map area (94B), Northern Rocky Mountains, British Columbia; Geological Survey of Canada, Memoir 425.

Thompson, R.I. and Roots, C.F.
1982: Ogilvie Mountains Project, Yukon, Part A: A new regional mapping programs; in Current Research, Part A, Geological Survey of Canada, Paper 81-1A, p. 403-411.

Thompson, T.L.
1962: Stratigraphy, tectonics, structure, and gravity in the Rocky Mountain Trench area, southeastern British Columbia, Canada; Ph.D. thesis, Stanford University, 203 p.

Tipnis, R.S.
1981: Early Ordovician conodont biostratigraphy and zonation of the Kechika Formation, northeastern British Columbia; Geological Association of Canada, Abstracts, v. 6, p. A-56.

Tipnis, R.S., Chatterton, B.D.E., and Ludvigsen, R.
1978: Ordovician conodont biostratigraphy of the southern District of Mackenzie, Canada; in Western and Arctic Canadian Biostratigraphy, C.R. Stelck and B.D.E. Chatterton (ed.), Geological Association of Canada, Special Paper 18, p. 39-91.

Tipper, H.W., Woodsworth, G.J., and Gabrielse, H. (co-ordinators)
1981: Tectonic assemblage map of the Canadian Cordillera and adjacent parts of the United States; Geological Survey of Canada, Map 1505A (1:2 000 000 scale).

Torrie, J.E.
1973: Northeastern British Columbia; in Future Petroleum Provinces of Canada, R.G. McCrossan (ed.), Canadian Society of Petroleum Geologists, Memoir 1, p. 151-186.

Vail, P.R., Mitchum, R.M. Jr., and Thompson, S. III
1977: Seismic stratigraphy and global changes in sea level; in Seismic Stratigraphy - Applications to Hydrocarbon Exploration, C.E. Payton (ed.), American Association of Petroleum Geologists, Memoir 26.

Walcott, C.D.
1910: Olenellus and other genera of the Mesonacidae; Smithsonian Miscellaneous Collections, v. 53, no. 6, p. 231-420.
1912a: Middle Cambrian Brachiopoda, Malacostraca, Trilobita, and Merostomata; Smithsonian Miscellaneous Collections, v. 57, no. 6, p. 145-244.
1912b: Group terms for the Lower and Upper Cambrian series of formations; Smithsonian Miscellaneous Collections, v. 57, no. 10, p. 305-307.
1928: Pre-Devonian Paleozoic formations of the Cordilleran provinces of Canada; Smithsonian Miscellaneous Collections, v. 75, no. 5, p. 175-368.

Westergard, A.H.
1946: Agnostidea of the Middle Cambrian of Sweden; Sveriges Geologiska Undersokning, Series C, no. 477, 140 p.

Westrop, S.R.
1984: Late Cambrian and earliest Ordovician trilobites, southern Canadian Rocky Mountains, Alberta; Ph.D. thesis, University of Toronto, Ontario, 990 p.
1986: Trilobites of the Upper Cambrian Sunwaptan Stage, southern Canadian Rocky Mountains, Alberta; Palaeontographica Canadiana, no. 3, 178 p.

Wheeler, J.O.
1963: Rogers Pass map-area, British Columbia and Alberta (82N W1/2); Geological Survey of Canada, Paper 62-32, 32 p.

Wheeler, J.O. and Gabrielse, H. (co-ordinators)
1972: The Cordilleran Structural Province; in Variation in Tectonic Styles in Canada; R.A. Price and R.J.W. Douglas (ed.), Geological Association of Canada, Special Paper 11, p. 1-82.

Whittington, H.B.
1980: The significance of the fauna of the Burgess Shale, Middle Cambrian, British Columbia; Proceedings of the Geologist's Association, v. 91, pt. 3, p. 127-148.

Williams, G.K.
1975: "Arnica platform dolomite", District of Mackenzie; in Report of Activities, Part C, Geological Survey of Canada, Paper 75-1C, p. 31-35.
1981: Middle Devonian carbonate barrier complex of Western Canada; Geological Survey of Canada, Open File 761.
1983: What does the term "Horn River Formation" mean?; Bulletin of Canadian Petroleum Geology, v. 31, p. 117-122.
1984: Some musings on the Devonian Elk Point Basin, Alberta; Bulletin of Canadian Petroleum Geology, v. 32, p. 216-232.

Wilson, J.L.
1975: Carbonate Facies in Geologic History; Springs-Verlag, New York, 417 p.

Young, F.G.
1979: The lowermost Paleozoic McNaughton Formation and equivalent Cariboo Group of eastern British Columbia: piedmont and tidal complex; Geological Survey of Canada, Bulletin 288, 60 p.

Young, G.M.
1982: The late Proterozoic Tindir Group, east-central Alaska: evolution of a continental margin; Geological Society of America Bulletin, v. 93 p. 759-783.

Ziegler, A.M., Scotese, C.R., McKerrow, W.S., Johnson, M.E., and Bambach, R.K.
1979: Paleozoic paleogeography; Annual Review of Earth and Planetary Sciences, v. 7, p. 473-502.

Ziegler, P.A.
1969: The development of sedimentary basins in Western and Arctic Canada; Alberta Society of Petroleum Geologists, 89 p.

Zwanzig, H.V.
1973: Structural transitions between the foreland zone and the core zones of the Columbian Orogen, Selkirk Mountains, British Columbia; Ph.D. thesis, Queen's University, Kingston, Ontario.

Authors' addresses

W.H. Fritz
Geological Survey of Canada
601 Booth Street
Ottawa, Ontario
K1A 0E8

M.P. Cecile
B.S. Norford
D.W. Morrow
H.H.J. Geldsetzer
Institute of Sedimentary and
 Petroleum Geology
Geological Survey of Canada
3303-33rd Street N.W.
Calgary, Alberta
T2L 2A7

J.O. Wheeler
H. Gabrielse
Cordilleran Division
Geological Survey of Canada
100 West Pender Street
Vancouver, British Columbia
V6B 1R8

G.E. Gehrels
Department of Geosciences
University of Arizona
Tucson, Arizona
U.S.A. 85721

CHAPTER 7

ADDENDUM

The Nasina Basin (Tempelman-Kluit, 1979) is a general basinal area west of Selwyn Basin and the Tintina Fault (Fig. 7.19). It received a thick succession of thinly bedded limestone and shale during the Late Cambrian and Early Ordorician (mainly shale in the north and limestone in the south). In Middle and Late Ordovician and Early Silurian times, shale deposition persisted throughout the basin and was succeeded during the remainder of the Silurian by mainly orange-brown weathering green siltstone, mudstone and some quartz sandstone (D.J. Tempelman-Kluit, pers. comm., 1985).

REFERENCE: Re emplacement of diatremes

Mott, J.A.
1989: Structural and stratigraphic relations in the White River region, eastern Main Ranges, southern Canadian Rocky Mountains, British Columbia; Ph.D. Thesis, Queen's University, Kingston, Ontario.

Printed in Canada

Chapter 8

UPPER DEVONIAN TO MIDDLE JURASSIC ASSEMBLAGES

PART A. ANCESTRAL NORTH AMERICA
Summary
Introduction
Upper Devonian carbonate strata of the Foreland Belt (Rundle Assemblage)
H.H.J. Geldsetzer and D.W. Morrow
    Fairholme sequence (Upper Givetian to mid-Frasnian)
    Ronde-Kakisa sequence (uppermost Frasnian)
    Palliser sequence (Lower and Middle Famennian)
Devonian-Mississippian clastics of the Foreland and Omineca belts
S.P. Gordey
    Imperial Assemblage
    Earn Assemblage
    Besa River Assemblage and the carbonate shelf edge
    Tectonic summary
Carboniferous and Permian stratigraphy of the Foreland Belt
E.W. Bamber, C.M. Henderson, B.C. Richards and A. McGugan
    Carboniferous
    Permian
Triassic strata of the Foreland Belt
D.W. Gibson
    Southern Rocky Mountain Foothills and Front Ranges
    Northern Rocky Mountain Foothills and Front Ranges
    Northern Foreland Belt, Yukon Territory and District of Mackenzie
    Paleoenvironments and history of deposition
Lower and Middle Jurassic strata of the Foreland Belt
T.P. Poulton
    Northern Foreland Belt
    Southern Foreland Belt

PART B. CORDILLERAN TERRANES
J.W.H. Monger, J.O. Wheeler, H.W. Tipper, H. Gabrielse, T. Harms, L.C. Struik, R.B. Campbell, C.J. Dodds, G.E. Gehrels, and J. O'Brien

Summary
Introduction
Terrane stratigraphy
  Cassiar Terrane
  Arctic Alaska Terrane
  Kootenay Terrane
  Windy-McKinley Terrane
  Slide Mountain Terrane
  Tectonic history of the Slide Mountain Terrane
  Quesnellia
  Dorsey Terrane
  Cache Creek Terrane
  Stikinia
  Relationships between Stikinia, Cache Creek Terrane and Quesnellia
  Metamorphic rocks and Nisling Terrane in the northern Coast Belt
  Terranes of the southern Coast Belt
  Alexander Terrane
  Wrangellia
  Relationship of Wrangellia with other terranes
References

Chapter 8

UPPER DEVONIAN TO MIDDLE JURASSIC ASSEMBLAGES

PART A. ANCESTRAL NORTH AMERICA

S.P. Gordey, H.H.J. Geldsetzer, D.W. Morrow, E.W. Bamber, C.M. Henderson,
B.C. Richards, A. McGugan, D.W. Gibson, and T.P. Poulton

Summary

The pre-Late Devonian Cordilleran miogeocline consisted of extensive shallow-water platforms upon which carbonate-clastic deposits accumulated. They were flanked to the west by deep-water environments where shale and carbonate accumulated (Rocky Mountains Assemblage). Clastic sediments were largely craton-derived. During the Late Devonian sedimentation patterns changed dramatically as turbiditic, chert-rich clastics, derived from the west and north, flooded the northern Cordillera (Earn and Imperial assemblages). Shale (Besa River Assemblage) was deposited far out onto the miogeocline and Interior Platform; the carbonate front of the Rundle Assemblage retreated far to the east and south of its Middle Devonian position. By mid-Mississippian time the clastic influx waned and normal marine shelf carbonate and clastic sedimentation resumed, once again with clastics derived from the craton. Devono-Mississippian plutonism occurred only in northernmost Yukon Territory, and volcanism was restricted to central Yukon and south-central British Columbia. Pre-Late Mississippian folding occurred in northern Yukon but elsewhere deformation is expressed only by local high-angle faults and disconformities.

Devono-Mississippian tectonism in the northern Yukon involved uplift and granitic intrusion in Frasnian to Early Mississippian time, resulting in an upward shoaling and southward-prograding clastic wedge. The sequence consists of shale at the base, flyschoid sediments near the middle, and partly fluvial-deltaic strata at the top. Deformation migrated southward from the area of uplift until the clastics themselves were folded prior to the mid-Carboniferous.

The source of Devono-Mississippian sediments in the central Cordillera was uppermost Precambrian quartzose clastics and lower Paleozoic chert from the western miogeocline. Western coarse clastics are typified by chert-quartz wacke and arenite and chert-pebble conglomerate; feldspar and volcanic detritus locally amount to a few per cent. All of the clastics were deposited in submarine fan environments. Numerous facies changes occur over short intervals. The lack of compressional structures, the presence of local volcanics of possible rift type, and the occurrence of syn-sedimentary faults that controlled thickness and facies, favour local block uplift of the outer miogeocline as a consequence of regional extension or strike-slip faulting. It is also possible, however, that the block uplifts and clastic wedges were related to tectonism farther west expressed by the Devono-Mississippian granitic plutons and volcanics in the Kootenay Terrane. The eastern fine clastic facies consists of shale and siltstone. Its southeastern limit, the carbonate-shale boundary, fluctuated widely, possibly reflecting tectonic activity to the west.

In the southern Cordillera (south of 60°N) the Devono-Mississippian clastic record is minimal; the westernmost preserved miogeoclinal strata are shallow-water carbonate. After an initial period of late Middle Devonian erosion, shallow water settings within a southward-spreading epicontinental sea led to the establishment of a broad carbonate platform (Rundle Assemblage). During the mid-Frasnian this became the foundation for extensive, northeasterly trending linear reefs that appear to have developed as a result of differential rates of deposition after partial suffocation of organic growth by anoxic terrigenous clastics. A westward prograding apron of shale gradually buried the reefs, but as this clastic supply waned and the seas regressed in the late Frasnian impure carbonates developed over a wide area. The brief late Frasnian regression led to local emergence and the influx of silt into briefly-established peritidal carbonate environments. At the beginning of Famennian transgression, westerly derived sand filled in topographic irregularities and was succeeded by the development of a broad carbonate platform which persisted until a latest Devonian influx of black euxinic shale.

Carboniferous rocks occur in two belts. An eastern facies of marine platform and ramp carbonates and shallow marine to continental detrital clastics is succeeded to the west by a basinal shale facies. The eastern facies comprises several assemblages (Rundle, Lisburne and Mattson assemblages), each separated by regional unconformities marking Late Carboniferous and Permian episodes of

---

Gordey, S.P., Geldsetzer, H.H.J., Morrow, D.W., Bamber, E.W., Henderson, C.M., Richards, B.C., McGugan, A., Gibson, D.W., and Poulton, T.P.
1991: Part A. Ancestral North America; in Upper Devonian to Middle Jurassic assemblages, Chapter 8 of Geology of the Cordilleran Orogen in Canada, H. Gabrielse and C.J. Yorath (ed.); Geological Survey of Canada, Geology of Canada, no. 4, p. 219-327 (also Geological Society of America, The Geology of North America, v. G-2).

erosion. A sub-Permian unconformity reflects marked tectonism along the ancestral Aklavik Arch.

Carboniferous deposits of the eastern facies in the Rocky and southern Mackenzie mountains consist of Middle Tournaisian to Upper Viséan miogeoclinal carbonate platform and ramp deposits (Rundle Assemblage), overlain by widespread, Upper Viséan and Lower Namurian, supratidal, deltaic and shallow-marine siliciclastics (Mattson Assemblage). This succession is disconformably overlain by Bashkirian and Lower Moscovian carbonate and siliciclastics (lower Ishbel Assemblage) which are incompletely preserved beneath a regional sub-Permian disconformity. In northern Yukon the eastern facies consists of uninterrupted Middle Viséan to Moscovian carbonate (Lisburne Assemblage) and minor basal siliciclastics (Mattson Assemblage). Contemporaneous deltaic and shallow-marine deposits (Mattson Assemblage) are restricted to the eastern part of the belt, along the flanks of the Richardson and Barn mountains. The eastern carbonate-clastic facies overlies and passes westward into a succession of basinal deposits (CPo Assemblage) across a broad, transitional zone which temporally varied in position. Within the basinal succession, marked thinning occurs toward the west, because of starved basin conditions.

In contrast to the Carboniferous, Permian strata (upper Ishbel Assemblage) consist mainly of shallow-marine to basinal siliciclastics and silty to sandy carbonate. Extensive carbonate buildups are lacking, and neither continental deposits, nor well-defined basinal shale facies have been recognized. A regional unconformity marks the base of the Permian succession and separates it from strata ranging in age from Early Silurian to latest Carboniferous. In the Rocky and Mackenzie mountains a regional, intra-Permian disconformity separates a thin, Upper Artinskian to Wordian assemblage of chert and siliciclastics from underlying Asselian to Lower Artinskian siliciclastics and carbonate. In northern Yukon and northwestern District of Mackenzie, a thick possibly uninterrupted Asselian to Wordian clastic succession (Jungle Creek Assemblage) is present on the flanks of the ancestral Aklavik Arch. No Permian deposits younger than Wordian are preserved beneath widespread sub-Triassic and younger disconformities.

The Triassic rocks of the miogeocline comprise a westward thickening sequence of marine siliciclastics, of cratonal origin, and lesser carbonate (Spray River Assemblage). Many paralic and marine environments developed during several transgressive-regressive episodes and their deposits are preserved extensively in a belt extending from the southeast Yukon to the International Boundary (49°N). Similar strata in the central and northern Yukon are only locally preserved. The Triassic ended with regression and probable erosion prior to Jurassic marine transgression.

The Jurassic record, dominated by shallow marine shelf sandstone and shale (also Spray River Assemblage), is one of progressive but intermittent flooding, a phenomenon noted on a global scale. Extensive marine transgressions preceded Sinemurian, Toarcian, late Early Bajocian, Bathonian, Callovian (approximately), and Late Oxfordian or Kimmeridgian deposition in both the northern Yukon and southwestern Alberta. In both regions sediments were shed from the craton. The first indication of westerly derived clastics related to the onset of Mesozoic orogeny are of Late Jurassic age, and most younger Jurassic rocks (Kootenay Assemblage) of western Alberta and eastern British Columbia were shed eastward into the rapidly subsiding foredeep flanking the developing Rocky Mountain orogen. In the Yukon and District of Mackenzie there is no indication of western clastic input until the Early Cretaceous.

Introduction

In the following discussion the Upper Devonian to Middle Jurassic assemblages of the Canadian Cordillera are treated with respect to their association with ancestral North America (Part A) and with each of the Cordilleran terranes (Part B). Like the other assemblage chapters the rock stratigraphic sequences are identified according to their assignment to the tectonic assemblages (see the Tectonic Assemblage Map, Map 1712A, in pocket).

In Part A Upper Devonian carbonate and Upper Devonian to Mississippian clastic assemblages are discussed separately insofar as each represents fundamentally different tectonic settings. The former reflects continued, dominantly platform and reef carbonate settings which were established on the miogeocline during the Middle Cambrian. With significant interruption in the northern Cordillera during the Late Devonian and Mississippian, these environments continued to the end of the Paleozoic era, following which Triassic to Middle Jurassic miogeoclinal sedimentation became increasingly clastic. In the northern Cordillera important western sources supplied thick sequences of terrigenous clastics to the miogeocline in Late Devonian and Early Mississippian time.

The terranes of the western Cordillera contain assemblages which are dominantly oceanic to island-arc in origin. It was during this interval when most of the assemblages for many of the terranes accumulated and thus the definition and characterization of the terranes is an important part of this chapter. Stratigraphic evidence for the time and type of terrane linkages is described as are alternative views when such evidence is equivocal.

Upper Devonian carbonate strata of the Foreland Belt (Rundle Assemblage)
H.H.J. Geldsetzer and D.W. Morrow

Upper Devonian strata of the southern Foreland Belt (Fig. 8.1) and in the subsurface beneath the adjacent western Interior Plains comprise carbonate and mixed carbonate and siliciclastic sediments. Along the eastern margin of the belt the thickness of mid-Givetian to Famennian rocks varies between 250 and 1000 m (Fig. 8.2). Several important petroleum reservoirs occur within Upper Devonian reefs of the Interior Platform; moreover, Upper Devonian carbonates are among the dominant contributors to the magnificent scenery of the Canadian Rockies.

Like the discussion of Lower and Middle Devonian strata (see Table 7.1) the Upper Devonian succession of the southern Foreland Belt is divided into distinct sequences separated by thin intervals or discontinuities across which stratigraphic character changes markedly (Fig. 8.3). In ascending order these are the Fairholme sequence of late Givetian to mid-Frasnian age, the Ronde-Kakisa sequence

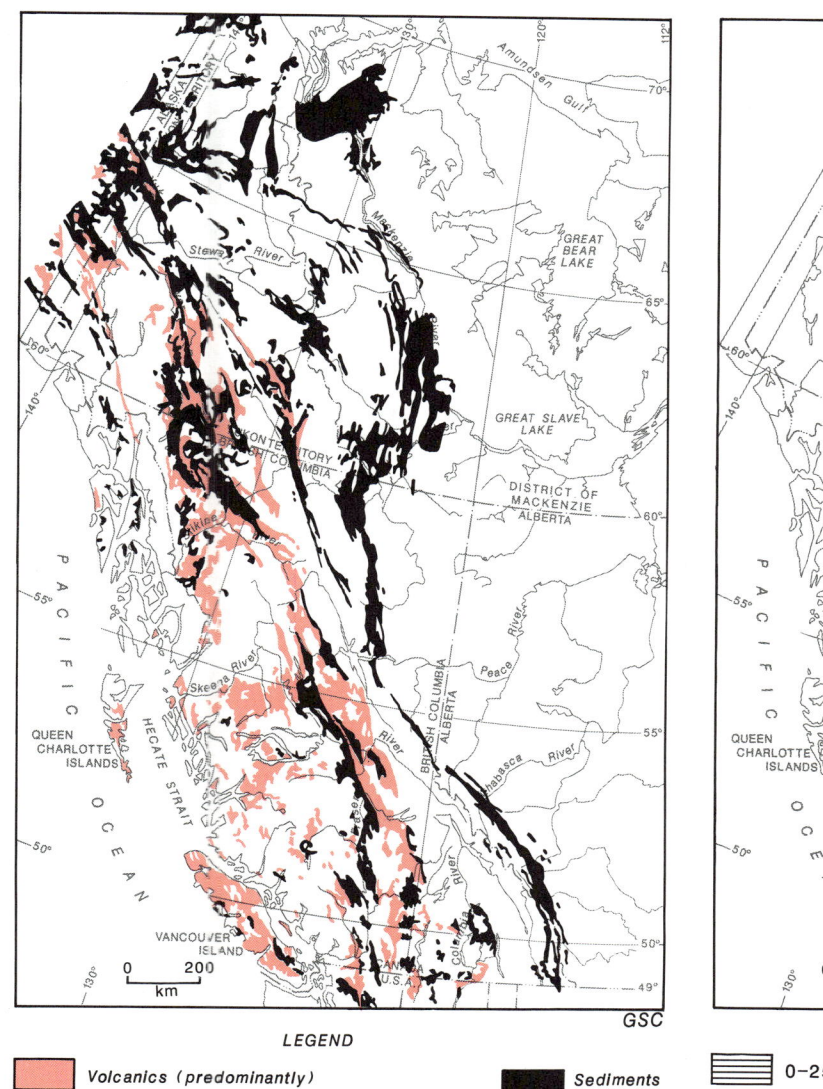

Figure 8.1. Distribution of Upper Devonian to Middle Jurassic sedimentary and volcanic rocks of the Canadian Cordillera.

of late Frasnian age and the Famennian Palliser sequence. The Fairholme sequence is separated from the Eifelian-lower Givetian Hume-Dunedin sequence by the Watt Mountain hiatus and represents the beginning of the classical "Kaskaskia Sequence" of Sloss (1963). The overlying Ronde-Kakisa sequence, bounded in the south by unconformities, represents a brief transgressive pulse within a regressive interval and was followed by deposition of the Palliser sequence, the final transgressive carbonate phase of the Devonian period. The Late Devonian cratonic platform and margin were segmented into several tectonic elements (Fig. 8.4). During most of Late Devonian time the Peace River Arch was emergent and the record of previous Paleozoic deposition there was largely removed. The Watt Mountain Formation, comprising between 20 and 75 m of nonmarine red and green shale, sandstone, limestone breccia, limestone and dolomite is the product of this erosion that formed a clastic apron, 240 km wide, about the

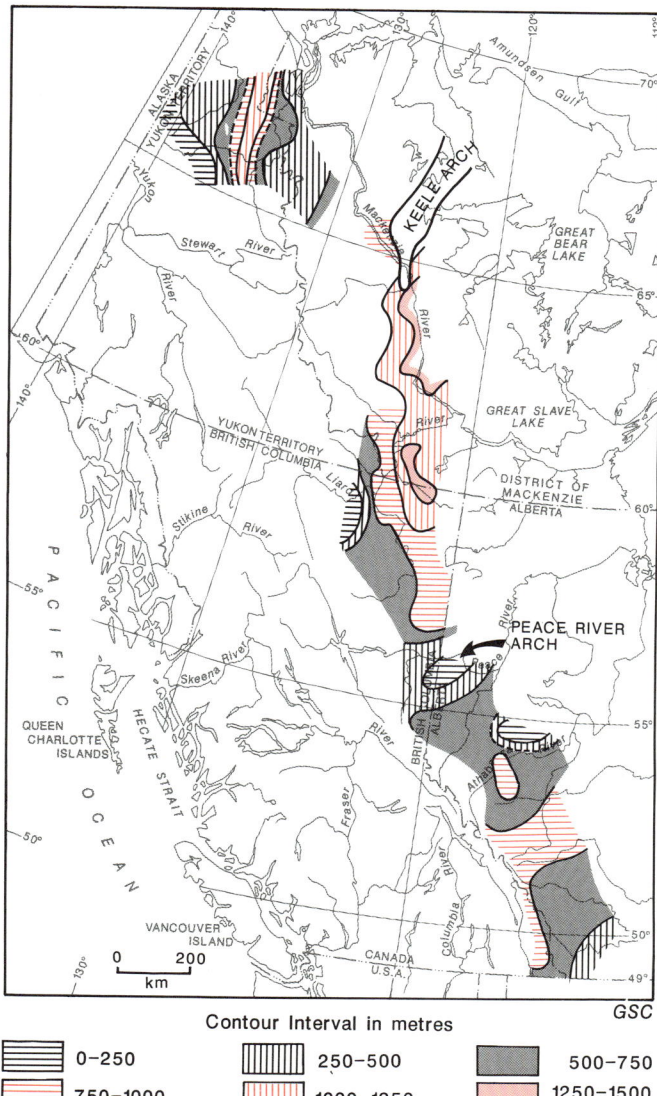

Figure 8.2. Isopach map of mid-Givetian to Famennian of the Devonian System. Thickness variations are more irregular than for Lower and Middle Devonian strata (see Chapter 7).

arch during late Givetian time. Extending southward from the Peace River Arch the West Alberta Ridge remained positive until a widespread marine transgression began in the late Givetian (Taghanic Onlap). On the crest of the ridge up to 36 m of middle or upper Givetian thinly bedded and laminated dolomite with collapse breccias of the Yahatinda Formation (Fig. 8.3) fill topographic depressions on the erosion surface underlain by Cambrian and Ordovician carbonate (Aitken, 1966). During the Late Devonian the Golden Embayment on the west side of the West Alberta Ridge ceased to exist and, south of the Peace River Arch, the cratonic platform extended westward to the limits of preservation (Fig. 8.4). During the Late Devonian an influx of external clastics from northern and western sources (see next section) limited the extent of carbonate sedimentation to areas largely south of 60°N. The only important carbonate deposition in the northern

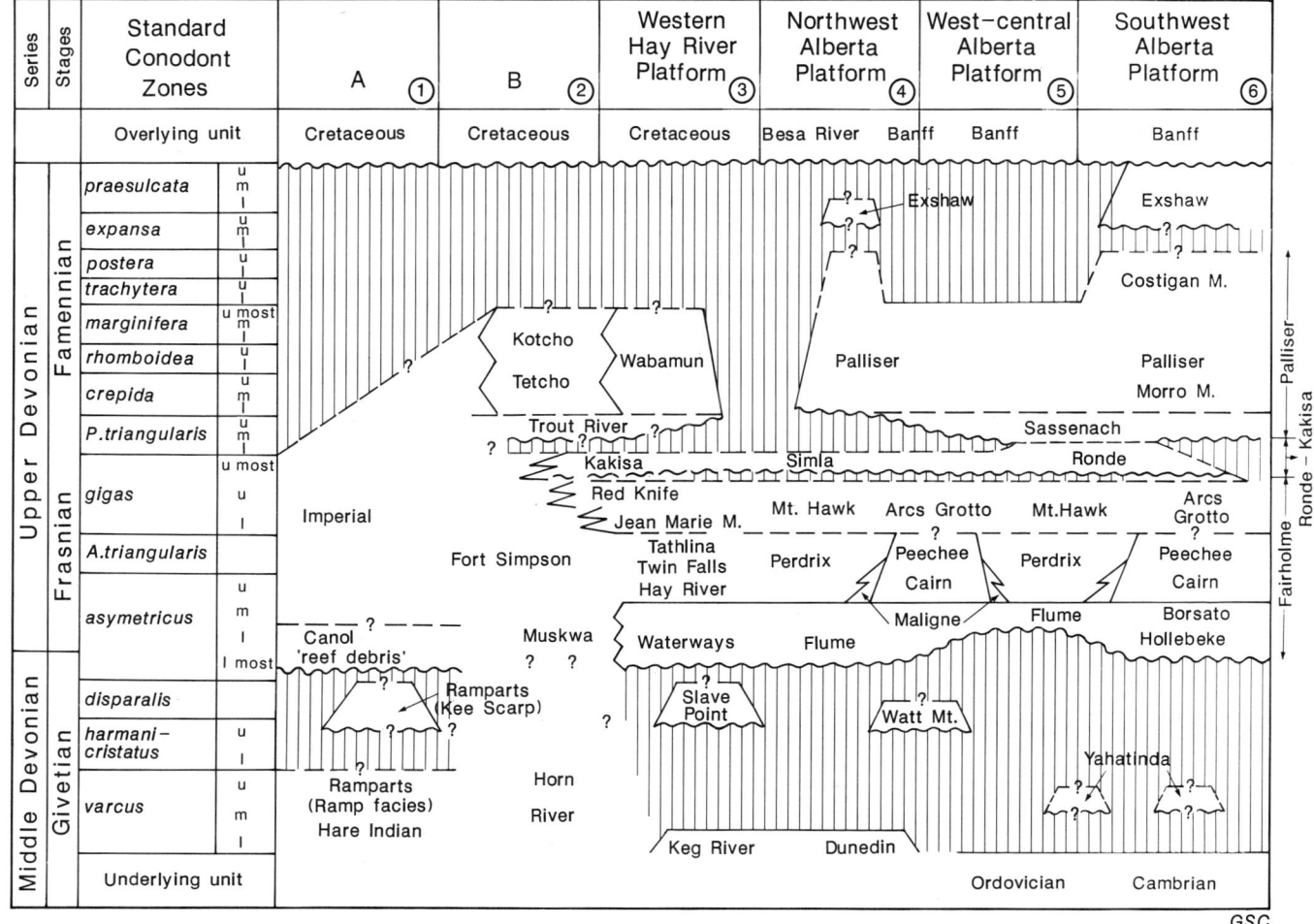

Figure 8.3. Correlation chart of upper Middle and Upper Devonian formations with standard conodont zones. These are grouped into three sequences: the Fairholme, Ronde-Kakisa, and Palliser. Locations of regions A and B are indicated in Figure 8.4.

cratonic platform after the Watt Mountain hiatus was a brief late Givetian transgression which produced reef buildups such as the Ramparts Formation (more than 200 m thick) on the Peel Platform. Time-equivalent carbonate units to the south are the Slave Point Formation (150 m thick) along the western edge of the Hay River Platform and the early stage of the Swan Hills reefs (44 m thick) south of the Peace River Arch (Fig. 8.5).

Fairholme sequence (Upper Givetian to mid-Frasnian)

The Fairholme sequence of the Rocky Mountains south of Peace River Arch (Fig. 8.3, columns 4-6) is represented by the Fairholme Group of late Givetian and Frasnian age. The group includes the western exposed part of an extensive reef domain which developed on the southern Alberta Platform (Fig. 8.5, 8.6) and embraces carbonate and clastic facies which are laterally intergradational (Fig. 8.3, 8.7; Plate 16).

The carbonate components can be grouped into a basal platform succession which locally extends upsection into biostromal buildups or platform reefs, and into carbonate ramps which developed above the buildups and project laterally above and into clastic facies. Minor components are clusters of bioherms or reef mounds along the edge of the carbonate ramps and on the foreslope of reef margins.

The lower platform succession, represented by the Flume Formation, comprises dark grey dolostone, commonly characterized by stromatoporoids, corals and chert nodules. It is about 30 m thick on the West Alberta Ridge and thickens to about 130 m on the western slope of the ridge and to about 250 m in the southernmost Rockies where equivalent rocks are assigned to the Hollebeke Formation (Price, 1965). The basal part of the thicker sections has a yellowish or brownish grey-weathering clastic component of quartzose sandstone and silty or argillaceous carbonate as much as 36 m thick in the central Rockies and 57 m thick in the southern Rockies.

Along the Southern Rocky Mountain Trench up to 60 m of grey fossiliferous dolomite and quartzose sandstone of the Starbird Formation conformably overlie Middle Devonian mixed carbonate and clastics of the Harrogate and Mount Forster formations. The Starbird is equivalent to the Flume-Hollebeke platform succession (Norford, 1981);

its terrigenous components may have been derived from western sources (Reesor, 1973).

The biostromal buildups or platform reefs rest directly on the carbonate platform succession. Thicknesses increase generally from 200 m in the Front Ranges westward to 365 m. The lower part, assigned to the Cairn Formation or Cairn and Borsato formations, consists largely of dark grey dolomitic stromatoporoid biostromes (Fig. 8.7) and may include the underlying Flume Formation where the Flume and Cairn cannot be differentiated (Plate 16). The Cairn grades upward into light grey carbonate of the Southesk Formation which, in turn, includes the upper part of the reef (Peechee Member), post-reef carbonate ramps (Grotto and Arcs members) and post-Fairholme sequence carbonate (Ronde Member, uppermost member of the Fairholme Group). The upper light grey part of the reefoid Peechee Member of the Southesk Formation is

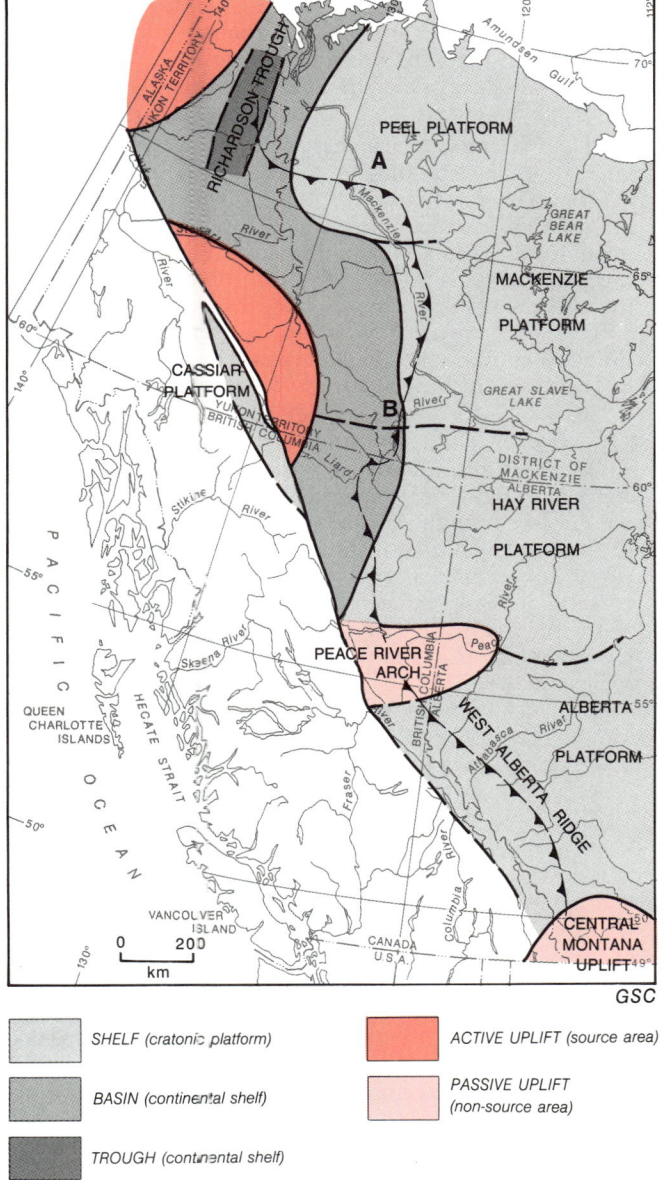

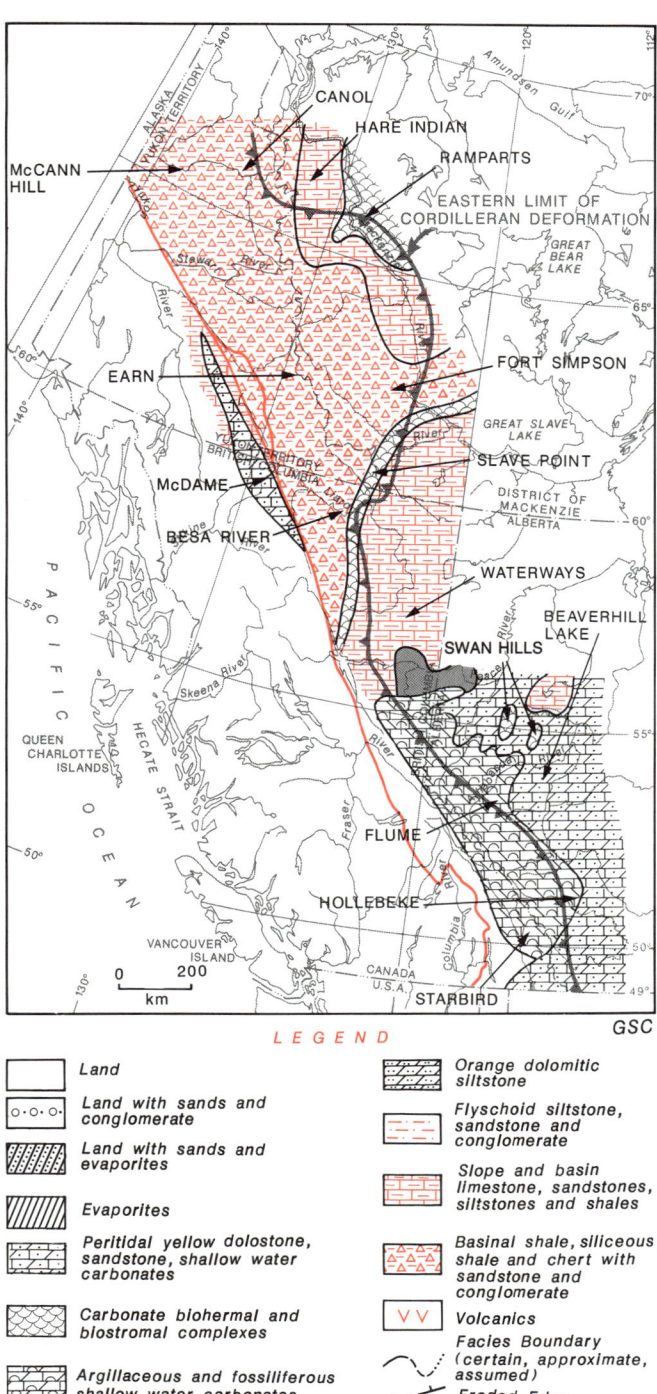

Figure 8.4. Late Devonian tectono-sedimentological elements of the eastern Cordillera. The western margin of the Hay River and Alberta platforms are outlined by the western edge of carbonate units and westward thinning of clastic units. Locations A and B designate areas for columns in Figure 8.3.

Figure 8.5. Distribution of facies during Late Givetian and Early Frasnian time (Correlations among named units not necessarily implied). The West Alberta Ridge was inundated and reefs such as the Swan Hills Reefs began to develop in south-central Alberta.

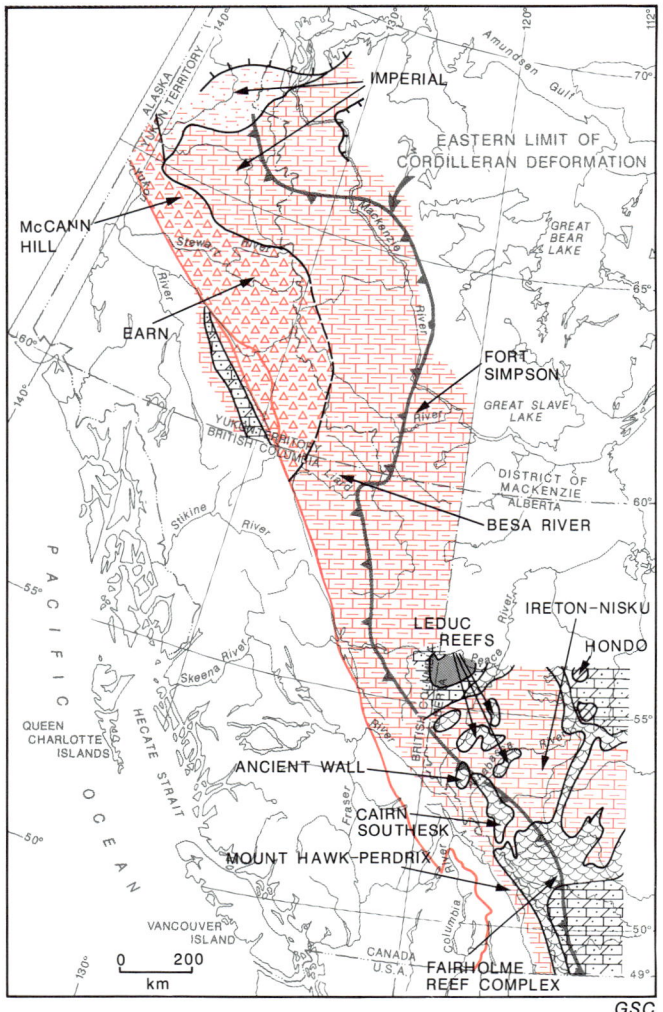

Figure 8.6. Distribution of facies during Middle Frasnian time (Correlations among named units not necessarily implied). This was a time of maximum reef development in the southern Rocky Mountains and Interior Plains. Clastics shed from northerly (Imperial) and westerly (Earn) sources were dominant in the north. Legend accompanies Figure 8.5.

dominated by laminated limestone grading laterally into stromatoporoid biostromes at the reef margin. Commonly, along the foreslope of reef margins, carbonate tongues of the lower reef (Cairn Formation) project outward across the underlying platform succession (Fig. 8.7) and change facies to dark grey laminated lime-mudstone of the Maligne Formation which, in turn, becomes argillaceous and nodular basinward; the Maligne Formation reaches up to 40 m in thickness near reef complexes (Fig. 8.7).

Carbonate ramps developed above most reefs and are characterized by light grey, commonly thick-bedded limestone consisting of grainstone and minor mudstone, to packstone assigned to the Arcs Member. These ramps extend outward across clastic-filled basins and grade laterally into nodular fossiliferous limestone of the upper Mount Hawk Formation which is equivalent to the Nisku Formation in the subsurface to the east. The lower part of this ramp facies occurs only locally and consists of dark grey, coral-bearing, slightly argillaceous carbonate of the Grotto Member.

Clusters of bioherms or reef mounds are found along the foreslope of reef margins and appear to be dominated by various algae (Mountjoy and Riding, 1981). Other bioherms occur along the edge of carbonate ramps (Fig. 8.8). Their dominant lithology is lime-mudstone with prominent stromatolitic structures. These buildups are probably equivalent to similar Nisku bioherms in the subsurface (Krause, 1984).

The clastic facies occurs as basal, relatively coarse clastics at the erosion surface below the Fairholme sequence, as a fine argillaceous component in the initial anoxic sediments (Perdrix Formation) above the carbonate platform succession and between the reefs, and as a thick silty and argillaceous basin fill (lower Mount Hawk Formation) which aggraded the reef topography (Fig. 8.3, 8.7).

The basal clastics of the lower part of the carbonate platform (Flume, Hollebeke) were discussed above. The anoxic argillaceous lime-mudstone or calcareous shale of the Perdrix Formation are distinguished by their black colour, high organic content and distinct radioactive signature. Toward the reef complexes the Perdrix grades into the Maligne Formation and in the subsurface to the east the unit correlates with the Duvernay Formation (Fig. 8.3, 8.7). Thickness varies from zero at reef margins to more than 100 m in distant basin areas. The Perdrix is temporally equivalent to the Cairn and Peechee reefs.

The basin fill of the lower Mount Hawk Formation consists of very thin-bedded to laminated, calcareous silty shale varying in colour from dark grey to greenish grey, suggesting mostly oxygenated conditions. The unit overlies the Perdrix, Maligne or extensions of the Cairn formations. In distant basinal settings the Perdrix and lower Mount Hawk have similar lithologies and are difficult to separate whereas in near-reef complexes the lower Mount Hawk is characterized by abundant limestone bands. Thickness of the basin fill varies from zero at reef margins to over 300 m in basinal areas. Along the margins of reef complexes reef-derived debris beds serve to separate the Perdrix and Mount Hawk formations (Fig. 8.7) and demonstrate that probably all of the lower Mount Hawk sediments represent a stage of post-reef infilling (Mountjoy and Mackenzie, 1973).

The upper Givetian-lower Frasnian carbonate facies of the Fairholme sequence resulted from clear, warm and shallow-water conditions within an epicontinental sea which advanced southward across the cratonic platform (Taghanic Onlap). The cyclic nature of reef buildups attests to repeated fluctuations in sea level. During the early stages of the transgression terrigenous material, presumably derived mostly from local sources, was reworked by the advancing sea and incorporated in the lower part of the carbonate platform. In the southeastern Foreland Belt the relative proportion of this material decreases to where the lower Hollebeke Formation is only moderately argillaceous. However, westward from there, grey, stromatoporoidal platform carbonates are interbedded with quartzite (Starbird). During the early Frasnian platform carbonates broadly coalesced and spread eastward (Flume, upper Hollebeke). The platform lithologies grade upsection

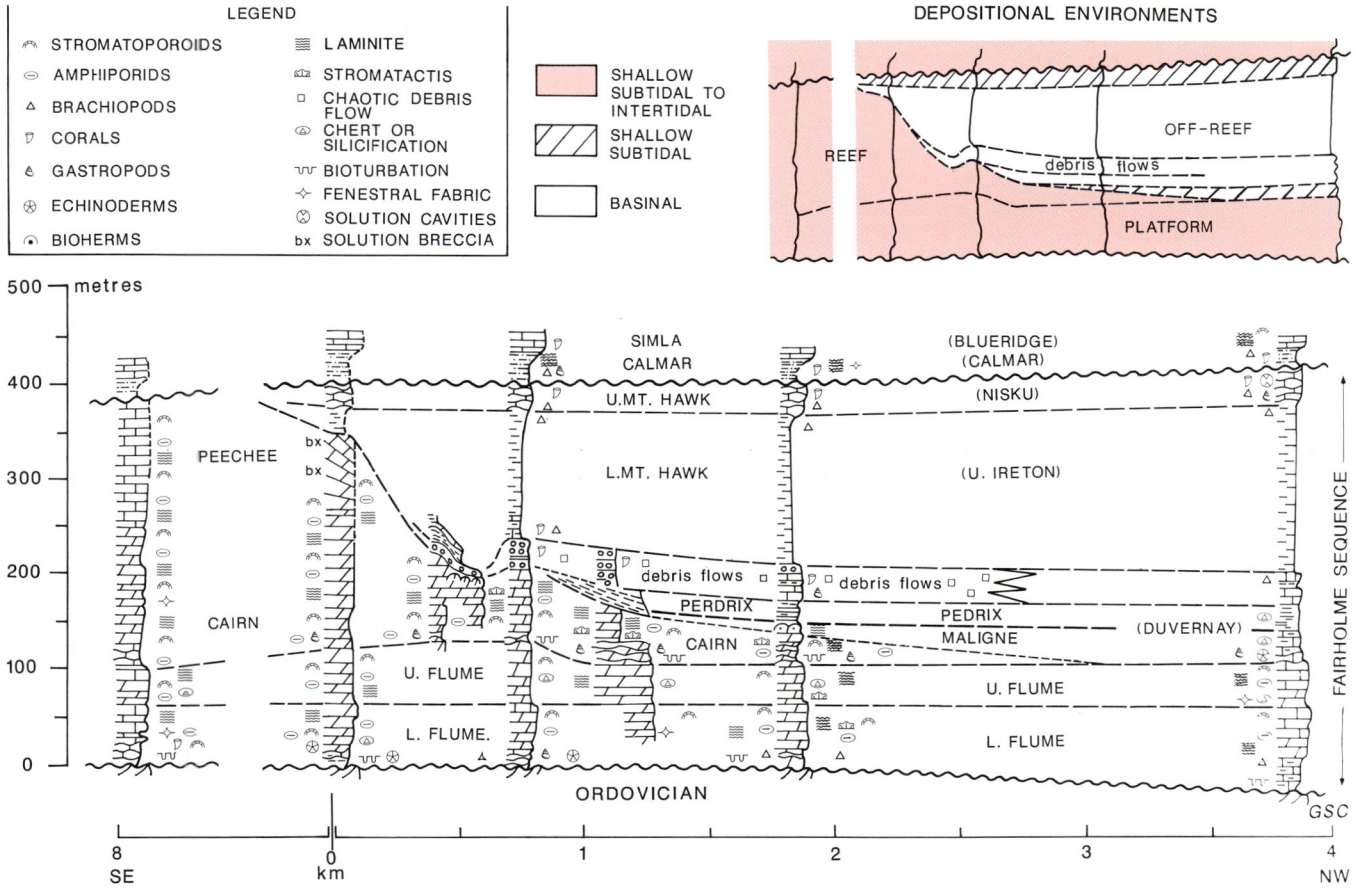

Figure 8.7. A stratigraphic cross-section of the reef to off-reef relationship of the southwest margin of the Upper Devonian Ancient Wall Reef Complex. Inferred depositional environments are shown in the inset diagram. Location of Ancient Wall Reef Complex is shown in Figure 8.6.

from brachiopod-bearing mudstone and wackestones at the base to stromatoporoid-coral biostromes above.

During mid-Frasnian time the broad carbonate platform became the foundation for extensive, northeasterly trending linear reef tracts (Fig. 8.6) which are commonly attributed to a rapid rise in sea level (Cook, 1972; Mountjoy, 1980; Moore, in press). Because the growth rate of bioconstructed, stromatoporoid-coral biostromes probably far exceeded the rate of epeirogenesis or long-term eustatic sea level change (Schlager, 1981), it is suggested that the sudden influx of a dysaerobic water mass with a minor very fine terrigenous component (Maligne and Perdrix formations) terminated organic growth, particularly stromatoporoids, in areas of comparatively low relief. Elevated areas were bypassed, thus allowing for continued upward reef growth (Geldsetzer, 1984). The much lower rate of accumulation of black, anoxic argillaceous mudstones as compared to bioconstructed carbonates led to significant topographic relief, which, along the Ancient Wall reef complex northwest of Jasper (Fig. 8.6, 8.7), exceeded 150 m.

Andrichuck (1961) suggested that the location and orientation of the northeasterly trending linear reef tracts in south-central Alberta was due to a northeasterly trending hinge zone related to faults in the Precambrian basement. These trends are closely parallel with northeasterly trending magnetic anomalies beneath the Interior Platform (see Fig. 2.30).

Reef growth continued along the western reef domain (Rocky Mountains) contemporaneously with burial of those of the eastern domain (Leduc reefs beneath the Interior Plains) by a westward prograding apron of green-grey shale (Ireton Formation). An interruption of sea level rise resulted in the exposure of the northwestern reef domain (Ancient Wall), local brecciation of the upper reef and probably numerous reef-derived debris flows. These foreslope breccias interfingered with the lowest beds of the westward-prograding basin fill which had advanced into the western reef domain (lower Mount Hawk-upper Ireton) and finally infilled the western reef topography (Fig. 8.7). In response to a gradual reduction of terrigenous supply the shale of the lower Mount Hawk changed upward into fossiliferous nodular wackestone of the upper Mount Hawk. Near and above most reef complexes these rocks grade laterally and upward into deeper water wackestone of the Grotto Member and finally into shallow-water grainstone of the Arcs Member of the Southesk Formation.

Ronde-Kakisa sequence (uppermost Frasnian)

In the central and southern Rockies the Ronde-Kakisa sequence is represented by the Ronde Member of the Southesk Formation and the Simla Formation (Fig. 8.3, 8.9). The Ronde is a variable unit, up to 55 m thick, and characterized by peritidal silty dolostone and interbedded calcareous siltstone; desiccation cracks and teepee structures are present. The Simla, up to 70 m thick, is a fossiliferous grainstone that grades upward into massive, resistant, light grey stromatoporoid limestone. Both units commonly are underlain by red, green and grey siltstone of the Calmar Formation (up to 20 m thick). In northern regions (Fig. 8.9), the Kakisa Formation, equivalent to the Simla, consists of silty dolomitic limestone with prominent coral and stromatoporoid biostromes.

The sequence developed during a short regressive interval in the late Frasnian. Following the infilling of interreef basinal areas (Mount Hawk), a lowering of sea level led to the emergence of the reef-capping grainstones (Arcs, upper Fairholme sequence) and their erosion above some reefs. Silt, probably derived from northern (Peace River Arch), eastern and southern sources, spread across the southern Alberta platform and established widespread coastal floodplain conditions (Calmar) which evolved into a peritidal setting with local salinas (Ronde). In areas of more open circulation fossiliferous limestone was deposited (Simla Formation, and its eastern subsurface equivalent, the Blue Ridge Member of the Graminia Formation; Fig. 8.9).

Figure 8.8. Upper Frasnian carbonate coral-bearing mud mounds (or bioherms) in the argillaceous limestones of the Mount Hawk Formation at Winnifred Pass near Kvass Creek in the southern Rocky Mountains of Alberta (about 40 km northwest of the Ancient Wall Reef Complex). Photo by A. Geldsetzer. GSC 205235(2)

Palliser sequence
(Lower and Middle Famennian)

In the Rocky Mountains the Palliser sequence is represented by the Sassenach and overlying Palliser formations (Fig. 8.3, 8.10; Plate 17). The Sassenach Formation (McLaren and Mountjoy, 1962) overlies an erosion surface above most of the former reef domain, infilled remnant depressions along the western margin of the reef domains, and is thus highly variable in thickness. At its type section it is more than 180 m thick and consists of a lower member (148 m) of dark grey to greenish grey, variably calcareous silty mudstone containing allochthonous reef debris in its basal beds. The upper member (32+ m) consists of sandy, medium- to coarse-grained limestone and calcareous sandstone. The Palliser overlies the Sassenach with regional conformity and is divisible into two members. The lower, Morro Member, locally as much as 530 m thick in the central Rockies (Geldsetzer, 1982) comprises a massive, cliff-forming, grey limestone, commonly bioturbated and characterized by pellet intraclast lime-mudstone. The Costigan Member consists of up to 75 m of fossiliferous, platy, and siliceous limestone and dolostone. The Palliser Formation forms the lowermost of a characteristic tripartite sequence of limestone (Palliser), recessive calcareous shale and argillaceous limestone (Exshaw-Banff) and

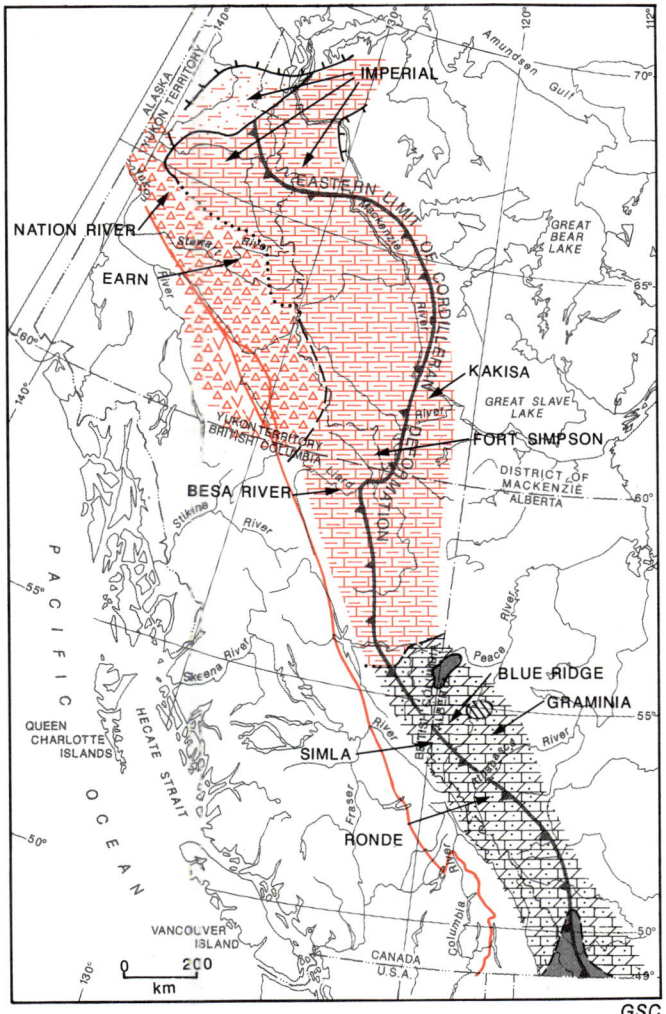

Figure 8.9. Distribution of facies during Late Frasnian time (Correlations among named units not necessarily implied). This time marked the disappearance of the reef complexes of Alberta. Legend accompanies Figure 8.5.

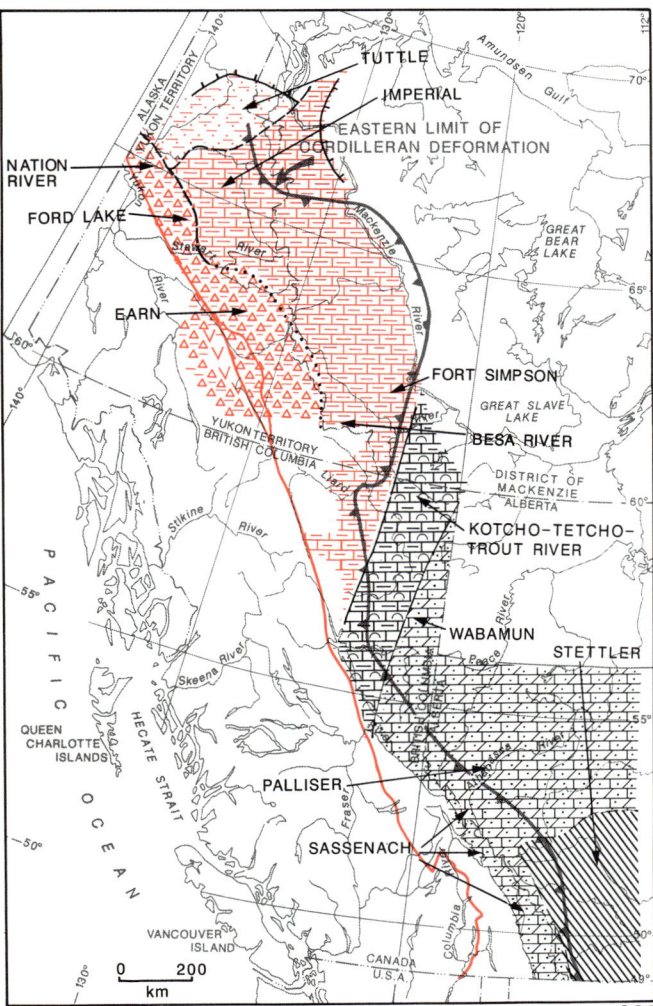

Figure 8.10. Distribution of facies during Famennian time (Correlations among named units not necessarily implied). Carbonate deposition again dominates in Alberta but reefs are not present. Legend accompanies Figure 8.5.

cliff-forming massive grey limestone (Rundle) that distinctively characterize several thrust sheets in the Front Ranges of the Rocky Mountains.

The Frasnian-Famennian boundary, which in the southern Foreland Belt occurs at or just below the Sassenach-Ronde contact (Fig. 8.3), and in the north at the Kakisa-Trout River contact, coincides with a sudden change in flora and fauna which is recognized worldwide. The cause was likely a major ocean turnover event (Geldsetzer et al., 1987) possibly as a result of a bolide impact (McLaren, 1982). The effect on carbonate sedimentation was profound, as it marked the end of most species of rugose coral (Pedder, 1982) and stromatoporoid faunas, the prime reef builders.

The Palliser sequence marks the last, widespread Devonian transgression on the cratonic platform of the miogeocline. During the early Famennian a western source supplied terrigenous clastics (lower Sassenach) to wide depressions along the western margin of the former reef domain which had not been totally infilled by upper Frasnian Mount Hawk shale (Fig. 8.10). Following aggradation of these remnant depressions silty carbonate and mudstone spread eastward onto the cratonic platform (upper Sassenach). This was followed by a period of widespread clean carbonate deposition (Palliser), and, to the southeast in the Interior Plains, the accumulation of evaporite-dominated sediments (Stettler). Towards the north carbonate passes laterally into argillaceous carbonate and shale (Kotcho, Tetcho and Trout River formations; Fig. 8.10) and thence into siliciclastics (Besa River, Fort Simpson, Earn, Ford Lake, Nation River, Imperial and Tuttle formations). The Palliser sequence was succeeded by black, bituminous shale of the lower Exshaw Formation of late Famennian age, which accumulated in an anaerobic environment during the final stage of the Palliser sequence whereas the upper Exshaw represents the regressive episode preceding the early Carboniferous transgression.

Devonian-Mississippian clastics of the Foreland and Omineca Belts

S.P. Gordey

In pre-Late Devonian time the Cordilleran miogeocline consisted of an extensive shallow-water carbonate-clastic platform flanked to the west by deep-water shale and carbonate. Until pre-Late Devonian time, clastic sediments were derived from the craton to the east. In the Late Devonian this regime abruptly changed with the influx of turbiditic, chert-rich clastic sediments of westerly and northerly derivation. By the mid-Mississippian sediment incursion from the west and north waned and marine clastic-carbonate deposition, with clastic input from the craton was re-established (Fig. 8.11).

The tectonic setting of the clastics is imperfectly understood. Plutonism and metamorphism are recorded only in northernmost Yukon (Norris and Yorath, 1981; Norris, 1981a,b) and volcanism only in central Yukon (Gordey, 1981b; Mortensen, 1982). Except in northern Yukon where pre-Late Mississippian folds and an angular unconformity are known (Bell, 1974; Norris and Yorath, 1981) structures include only local high-angle faults and associated disconformities. Parts of the clastic succession have been variously interpreted as reflecting rifting and continental separation (Tempelman-Kluit, 1979a,b), strike-slip tectonics (Eisbacher, 1983) and uplift of unknown western source terranes through thrusting or orogenesis (e.g. Gabrielse, 1976).

For description the clastics are discussed in terms of three assemblages (Fig. 8.11A): (1) the Imperial Assemblage (Givetian to mid-Tournaisian) of northern Yukon provenance, (2) the Earn Assemblage (Frasnian to Viséan), representing coarse clastic strata of westerly derivation, and (3) the Besa River Assemblage representing a mixture of the eastern and southern fine clastic fringes of the Imperial and Earn assemblages as well as fine clastics of possibly eastern cratonal derivation.

Imperial Assemblage

In northern Yukon Territory the segmented carbonate platform regime, present for much of the Early and Middle Devonian, was succeeded in Givetian time by shale and chert which hosted local carbonate reef buildups. In the Frasnian to Early Mississippian there was a profound change in depositional regime as northerly derived clastics inundated the area. The source seems to have lain within an uplifted region in northern Yukon that was cored by at least partly coeval plutons (Fig. 8.11A). The following is largely summarized from regional syntheses by Pugh (1983) and Norris (1985).

Five formations constitute the late Middle Devonian to Carboniferous Imperial Assemblage; in ascending order these are the Hare Indian (recognized in eastern areas), Canol, Imperial, Tuttle and Ford Lake formations. Equivalent clastics in the Brooks Range of Alaska include

Figure 8.11A. Facies distribution of Devono-Mississippian clastic strata in the Canadian Cordillera. Red line shows informal subdivision into three assemblages, a northern assemblage (Imperial) whose provenance was northern Yukon, a southern assemblage (Earn) of coarse clastics of generally western provenance and an eastern assemblage (Besa River) of fine clastics of mixed northerly, westerly and possibly cratonal provenance. The eastern limit of sandstone and conglomerate is shown by small circles. Large arrows indicate regional paleoflow, from non-cratonal source areas (wavy lines). Small arrow shows local paleoflow, possibly from east or northeast cratonal source area. Also shown are Devonian plutons in northern Yukon (crosses; 1-Ammerman, 2-Sedgewick, 3-Hoidahl, 4-Fitton, 5-Old Crow, 6-Schaeffer, and 7-Dave Lord plutons). Letters B to E locate sections in Figure 8.12. Barbed line shows limits of accreted terranes. Cretaceous-Tertiary offsets along Tintina Fault (450+ km) (Roddick, 1967; Tempelman-Kluit, 1979a; Gabrielse, 1985) and Kaltag Fault (100 km) (Norris and Yorath, 1981) are not restored. Five main areas of Earn sequence discussed in the text are also shown (diagonal lines). **Fig. 8.11B.** Fluctuation in position of carbonate-shale boundary during the Late Devonian and Early Mississippian (heavy wavy lines-land area; dashed line-shale; dashed pattern-carbonate). From Bamber and Waterhouse (1971), Basset and Stout (1967), Belyea (1964), Geldsetzer (1982), Grayston et al. (1964), Lenz (1972), Morrow (1984), Macauley et al. (1964), Pugh (1983), H. Gabrielse (pers. comm., 1985) and published and open file maps of the Geological Survey of Canada.

UPPER DEVONIAN TO MIDDLE JURASSIC ASSEMBLAGES

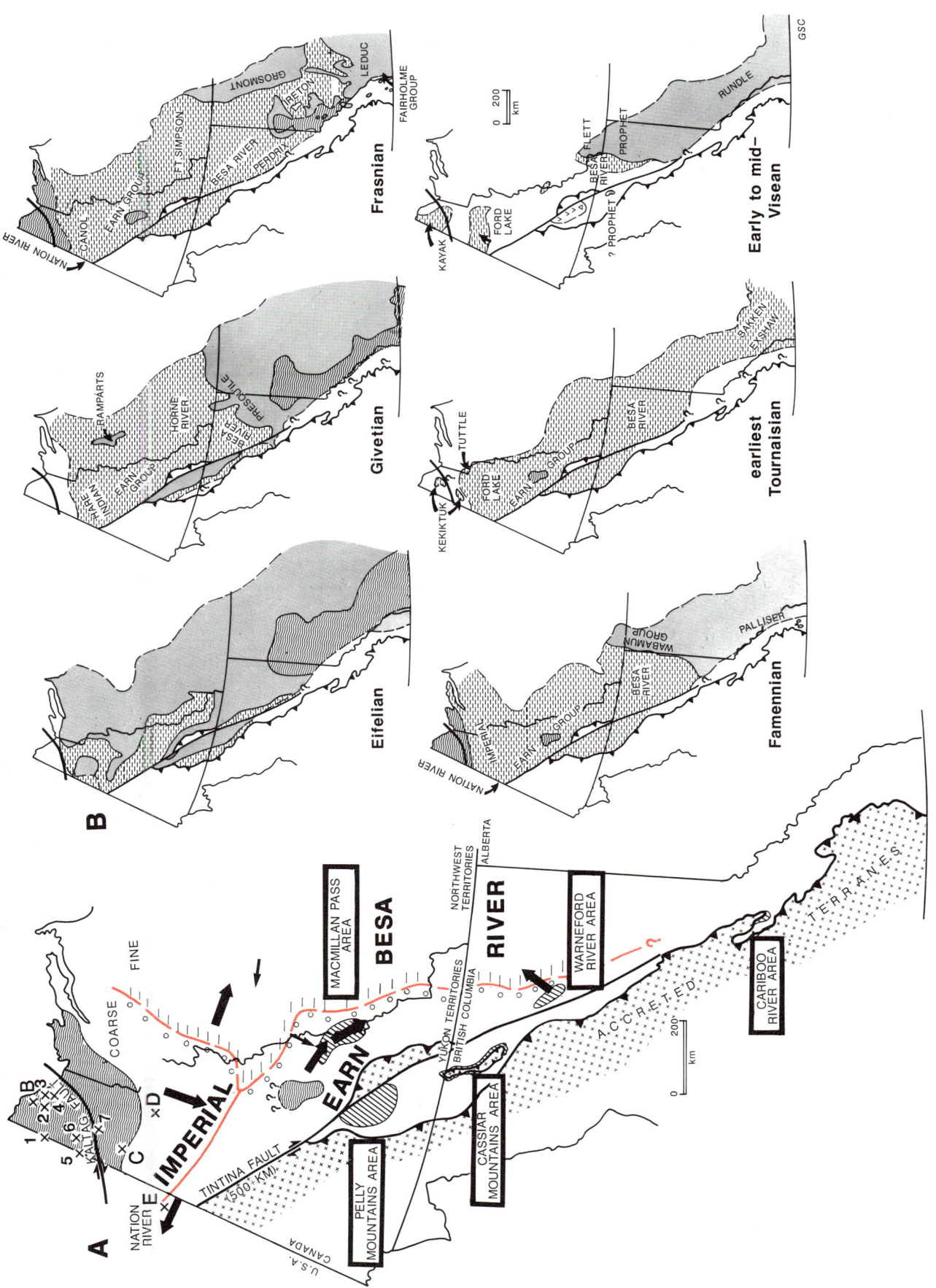

the Kanayut Conglomerate (Fig. 8.12), the overlying Kayak Formation and the underlying Noatak and Hunt Fork formations.

The basal Hare Indian Formation (150-180 m) comprises very dark brownish grey or black, bituminous, in part noncalcareous and calcareous shale, locally called the Bluefish member (10-30 m). Dark grey, calcareous shale with less than 10% interbedded calcareous siltstone and argillaceous or silty limestone form the upper part of the formation. The lower beds represent deposition in shallow euxinic water whereas upper portions may reflect open marine conditions. The lower contact of the Hare Indian is conformable with Devonian carbonate (Hume Formation).

A great variety of fossils, including conodonts, ostracodes, ammonoids and brachiopods indicate a Givetian age. Thick reefoid carbonates of the Ramparts Formation occur locally within the Hare Indian (Fig. 8.5, 8.11B).

The Canol Formation (110-225 m) consists of dark grey, noncalcareous shale which weathers grey, rust-brown, green, or jet black. This conspicuous dark shale marks the end of Devonian carbonate sedimentation throughout the northern Cordillera and reflects an abrupt change from shallow water to much deeper water marine sedimentation (Fig. 8.5). Immediately preceding deposition of the Canol, the eastern, southeastern and southern parts of the area were uplifted and eroded leading to a sub-Canol hiatus of

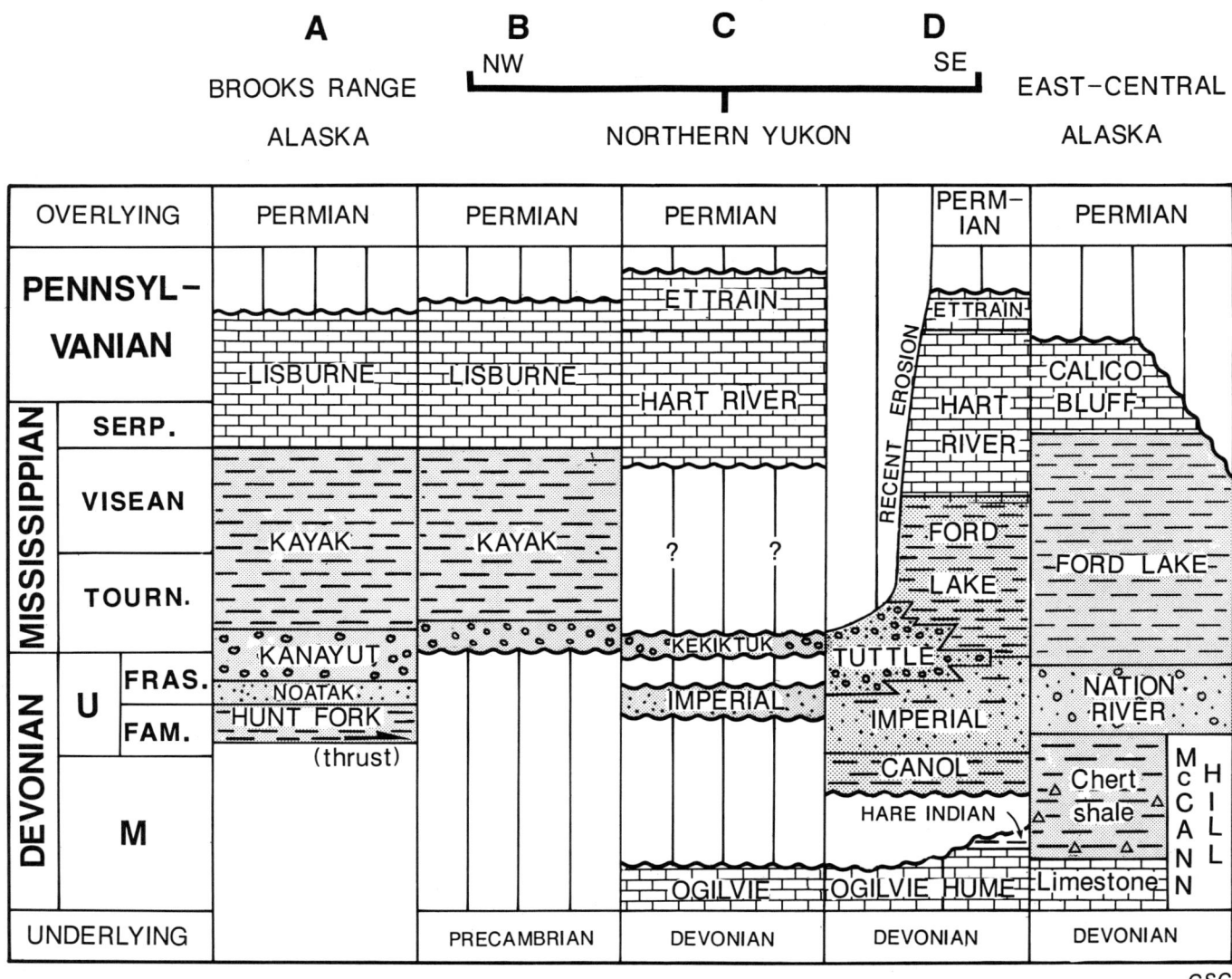

Figure 8.12. Time-stratigraphic diagram illustrating upper Paleozoic formational nomenclature in northern Yukon and parts of Alaska. Note that Nation River Formation belongs to the Earn Assemblage, Imperial and Tuttle formations to the Imperial Assemblage. Shaded areas indicate those formations mentioned in text. Approximate locations are plotted in Figure 8.10 (B - 68°55'N, 138°55'W; C - 67°05'N, 139°30'W; D - 66°28'N, 136°32'W). M=Middle, U=Upper, Fr=Frasnian, Fa=Famennian, T=Tournaisean, V=Viséan, S=Serpukhovian. Columns derived from A-Moore et al. (in press), Nilsen (1984); B-Norris (1981a, 1983), Sable (1977); C-Norris (1981b, 1983), Pugh (1983); D-Norris (1983), Norris (1985); E-Nilsen et al. (1976), Churkin (1973), Norris (1985, Fig. 3).

variable magnitude. Palynomorphs and conodonts show the Canol is Late Givetian in age.

The Imperial Formation (540-1690 m) comprises mostly brown to grey-green weathering, fine grained shale, siltstone, sandstone and minor limestone. In the area of its type section in the eastern Mackenzie Mountains Imperial strata represent alternating nearshore and offshore marine sediments derived from an eastern source area (Hills and Braman, 1978; Chi and Hills, 1974). To the north and west, however, the Imperial is predominantly turbidite (see e.g. Fig. 8.13, 8.14). There, load and flute casts and tool marks indicate southerly paleoflow. The thickest sections are near the lower Peel River area (Fig. 8.13). The lower contact is sharp, and conformable with the underlying Canol. The age of the Imperial as indicated by brachiopods, palynomorphs and conodonts ranges from Early Frasnian to Early Famennian.

An alternating succession of coarse- and fine-grained clastic rocks, the Tuttle Formation (870-1420 m) overlies the Imperial, and represents the coarsest part of this clastic assemblage. Like the Imperial the Tuttle Formation is thickest in the lower Peel River area and consists of alternating chert-pebble conglomerate, very poorly sorted quartz and chert sandstone, siltstone, dark greyish brown shale and rare coal (Fig. 8.15). The conglomeratic material consists of white, buff, grey, yellow, orange and pale green chert. In some places conglomerate clasts also include black chert, white feldspar(?) and fragments of Imperial sandstone. White, black and green chert may have been derived from northern Yukon where chert of these colours, particularly a conspicuously banded green variety, is common. The basal contact of the Tuttle with the Imperial is locally paraconformable (Braman, 1981), in other places the contact appears diachronous, becoming younger to the south. Two depositional models for the Tuttle have been proposed. Lutchman (1977) interpreted the formation as a southward advancing clastic sequence of fluvio-deltaic origin. Hills and Braman (1978) interpreted some Tuttle rocks as a turbidite sequence, within which load casts, flute casts and tool marks indicate a southerly paleoflow. Palynomorphs from the Tuttle indicate a latest Late Devonian to mid-Tournaisian age.

The Tuttle is overlain conformably by the Ford Lake shale (700 m) at one locality (Pugh, 1983), and elsewhere unconformably by Cretaceous, and locally Permian strata. Where the Tuttle is absent the Ford Lake rests directly on the Imperial Formation (Pugh, 1983). The Ford Lake consists of black pyritic shale, orthoquartzite, thinly bedded siliceous siltstone, local chert-rich sandstone, and black silty shale and black, bedded chert (Fig. 8.10, 8.11B). It ranges in age from Late Devonian to Viséan and is at least partly equivalent to the Tuttle Formation. The lower Ford Lake may represent a cycle of marine transgression and reduction in clastic supply following progradation of the Imperial-Tuttle clastic wedge. Its upper part may represent marine regression prior to deposition of overlying carbonate of the Hart River and Calico Bluff formations (Fig. 8.12).

Paleozoic igneous intrusions in the northern Yukon range from granite through quartz monzonite to monzodiorite, and occur as isolated stocks or cupolas to batholiths covering hundreds of square kilometres (Norris and Yorath 1981; Norris, 1983). Although all have been grouped as Devonian (Norris, 1981a,b), the wide range in radiometric ages (406 to 220 Ma collectively for Fitton, Sedgwick, Ammerman, Dave Lord, Schaffer, and Old Crow intrusions), intense weathering and hydrothermal alteration suggest that some of these ages have little significance. Maximum radiometric values for each pluton, which may indicate the minimum age of plutonism, range from 354-406 Ma (earliest Devonian to Early Mississippian).

The nature of late Paleozoic tectonism in the northern Yukon is obscured by poor exposure, and the lack of precise age of plutonic events. The contact between the clastic wedge sediments (Imperial, Tuttle) and younger rocks is commonly unconformable. Bell (1974) indicated that the age of the hiatus becomes younger from northwest to southeast and the time represented by it decreases in the same direction. Locally, an angular unconformity separates Imperial beds from Permian strata (Fig. 8.16; Norris, 1968), and pre-Permian broad, gentle, possibly northeast plunging folds have been documented (Bell, 1974). In northernmost Yukon (British and Barn mountains), the thin Lower Mississippian Kekiktuk Formation (20 m), a nonmarine quartzite pebble and cobble conglomerate, forms the base of the transgressive Lisburne Assemblage (discussed below) that rests with angular unconformity on deformed argillite and quartzite of the Precambrian Neruokpuk Formation, and shale as young as Early Devonian (Fig. 8.11B, 8.12; Norris and Yorath, 1981; Norris, 1981a; Lenz and Perry, 1972). Similar relationships have been described in adjacent Alaska (Romanzof Mountains), where pre-Kekiktuk strata are metamorphosed to greenschist facies (Sable, 1977). In northern Ogilvie Mountains deposition was continuous from Devonian to Carboniferous time (Fig. 8.12, column D).

In the northern Yukon Paleozoic tectonism involved uplift and granitic intrusion in Frasnian to Early Mississippian time resulting in the accumulation of an upward shoaling and southward prograding clastic wedge. The clastics, derived from pre-Kekiktuk erosion of strata in northern Yukon, consist of shale at the base, flyschoid sediments near the middle, and partly fluvial-deltaic strata at the top. Deformation migrated southward from northernmost Yukon until the clastic wedge was itself gently folded prior to the Early to mid-Carboniferous. Although compressional tectonism seems likely, other models, such as deformation related to strike-slip faulting (Bell, 1974) have been suggested.

In the Brooks Range of northern Alaska, Upper Devonian and Lower Mississippian(?) fluvial sediments of the Kanayut conglomerate form one of the coarsest, thickest (to 2600 m), and most laterally extensive nonmarine sequences in North America (Nilsen, 1984; Moore et al., in press). These proximal clastic wedge sediments prograded to the southwest above prodeltaic muds of the Hunt Fork shale and nearshore marine sands of the Noatak Formation (Fig. 8.12) from a source area in northern Alaska and northernmost Yukon; the same source as that inferred for the correlative, but dominantly flyschoid Imperial and Tuttle formations. Perhaps the more distal Imperial clastics traversed a relatively narrow highland-marginal shelf where the development of a broad delta plain was correspondingly limited.

Middle and Upper Devonian marine and nonmarine strata form a widespread clastic wedge in the Canadian

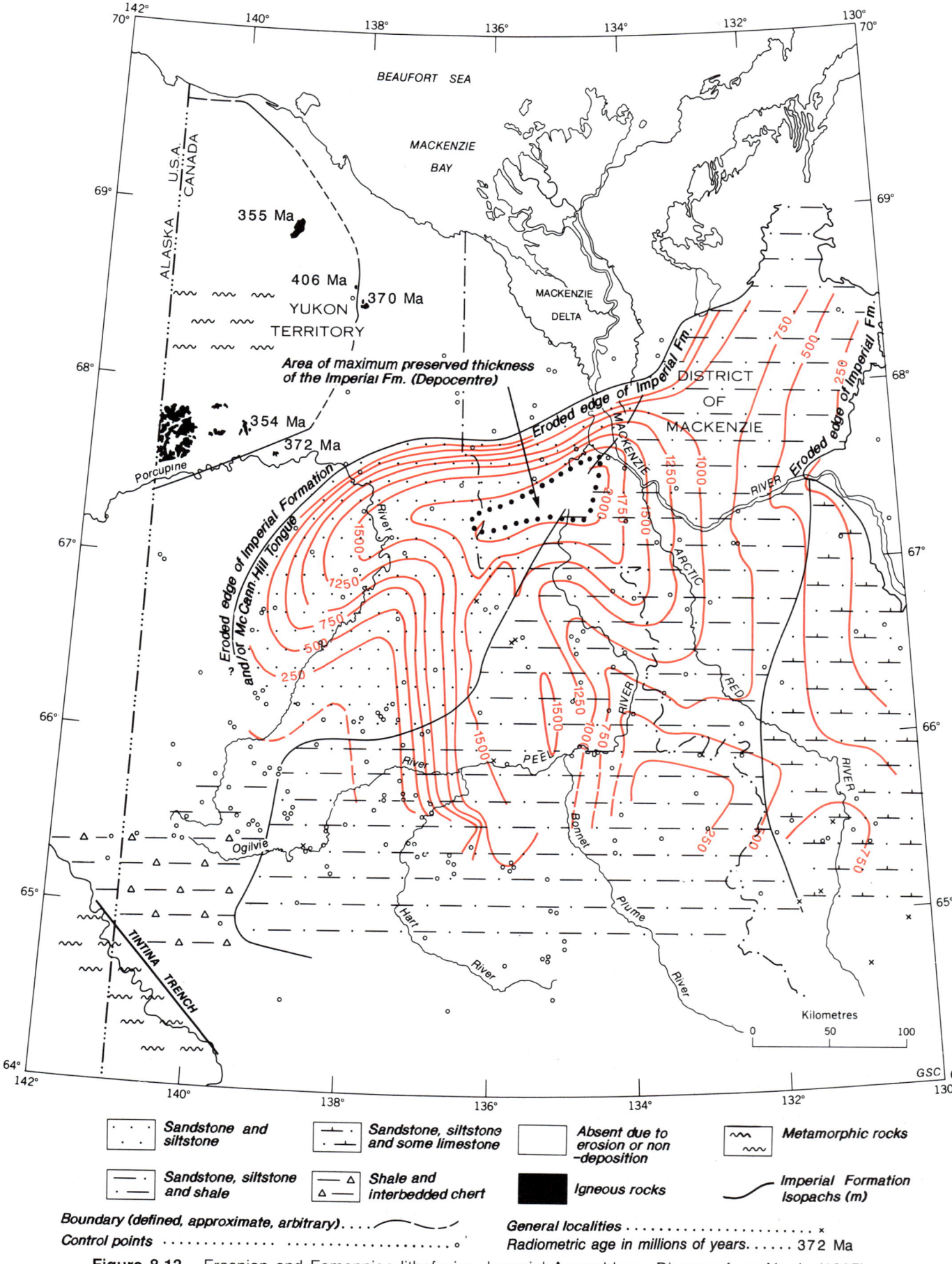

Figure 8.13. Frasnian and Famennian lithofacies, Imperial Assemblage. Diagram from Norris (1985). Isopachs (red) of the Imperial Formation from Pugh (1983).

Figure 8.14. Chevron-folded beds of sandstone, siltstone and shale of the Upper Devonian Imperial Formation in the northern Richardson Mountains. Photo by A.W. Norris. GSC 117380

Arctic Islands. These clastics, whose sedimentary transport was to the south and west, were deformed during the Early Mississippian Ellesmerian orogeny (Embry and Klovan, 1976; Kerr, 1981). Deltaic, easterly derived Imperial Formation strata in eastern Mackenzie Mountains may represent a southwestward extension of this wedge. Although the timing of development and style of Devono-Mississippian structure in the Arctic Islands is like that in the northern Yukon, a direct link is concealed beneath the Beaufort Shelf (a similar inference is drawn with respect to Ordovician and Silurian strata in Chapter 7).

Earn Assemblage

Although clastic strata of the Earn Assemblage are widespread, relatively little is known of them except in local areas. Their description is presented through synopses of five localities (Fig. 8.17). Equivalent clastic rocks in the Nation River area in east-central Alaska, also are briefly described, although the source and tectonic relationships of these to either the Earn or Imperial assemblages are unknown.

Macmillan Pass area

The Earn Group, containing the coarse clastics includes the Portrait Lake and Prevost formations (Fig. 8.17; Abbott, 1982, 1983; Carne, 1979; Gordey, 1981a; Gordey et al., 1982a; McClay, 1984). The Portrait Lake Formation (40-880 m), of Early Devonian to mid-Famennian age, comprises blue-black weathering, thin-bedded siliceous shale and chert. The base of the formation is markedly diachronous and the lower part changes facies eastward to shallow-water platform carbonate where the formation overlies a sub-Frasnian(?) unconformity of uncertain lateral extent. The upper part of the formation, probably of Frasnian age, contains the oldest westerly derived clastic detritus. It occurs in a 10-300 m thick member of massive chert-pebble conglomerate and laterally equivalent but thinner muddy chert-quartz sandstone and pebbly mudstone. The conglomerate is remarkably uniform in composition and contains rounded to angular pebbles and cobbles of grey, black, white or green chert and minor quartz sandstone in a clean matrix of quartz and chert sandstone. Contacts with enclosing rocks are sharp. Important deposits of stratiform lead-zinc (Tom, Jason properties; McClay, 1984) are associated with a widespread unit of baritic shale and barite of Frasnian age (Dawson and Orchard, 1982).

The Portrait Lake locally is overlain abruptly and unconformably by thin-bedded ripple cross-laminated siltstone, sandstone and shale of the Itzi Formation whereas at most other localities it is unconformably overlain by the Prevost Formation. Meagre conodont data suggest that the sub-Itzi unconformity is local and moreover that the Itzi may be a tongue within the upper Portrait Lake. Southerly paleoflow (J.G. Abbott, pers. comm., 1984) may indicate a cratonal or eastern source for the Itzi Formation.

The Prevost Formation (up to 900+ m), comprises brown-weathering shale, quartz-chert sandstone and chert-pebble to cobble conglomerate. Most exposures consist of up to several hundred metres of mainly shale and siltstone which commonly contain thick intervals (+200 m) of coarse clastics (Fig. 8.18). Conglomeratic mudstone is a distinctive local lithology, at one place at least 70 m thick, and containing cobbles of quartz sandstone and chert and rare large boulders of shale. Prevost conglomerate is composed of from 25% to 100% varicoloured chert and 5% to 20% fine- to coarse-grained quartz sandstone in a generally clean sandy matrix. Clasts are mostly well rounded pebbles to small cobbles, but large cobbles and boulders occur locally. Prevost sandstone consists mostly of subequal proportions of chert and quartz, a maximum of one per cent feldspar, rare siltstone and shale fragments, and trace tourmaline, zircon and detrital muscovite (Fig. 8.19A, 8.20). Wackes are slightly less abundant than arenites.

Sedimentary structures in the conglomeratic and sandy facies include thickly parted massive beds, local graded bedding, rare Bouma sequences, isolated large shale clasts, zones of chaotic (slumped) bedding, and large groove casts. Parting lineation, groove and prod casts, and rare flute casts occur locally in interbedded thin-bedded, fine grained sandstone. Sedimentary features in predominantly shale successions include thin, graded siltstone beds, and thin to thick, rare interbeds of sandstone showing graded bedding, rare Bouma sequences, and large shale clasts.

The Portrait Lake Formation was deposited in a low-energy, sub-wavebase setting with, at most times, a low rate of coarse clastic influx and accumulation of siliceous shale and chert. The chert-pebble conglomerate, chert-quartz arenite, chert-quartz wacke and pebbly mudstone were deposited from sediment gravity flows along an east-trending, possibly fault controlled submarine fan. The overlying Itzi Formation may have been deposited as a shallow marine shelf tongue of northerly and possibly craton-derived sediment. Sedimentary features of the Prevost suggest deposition of thick conglomeratic clastics as sediment gravity flows within submarine channels. The fine clastics, including relatively uncommon classical turbidite are both lateral and distal equivalents of the coarse facies.

Uplift and erosion of the outer miogeocline, including Upper Proterozoic and Cambrian gritty quartzose

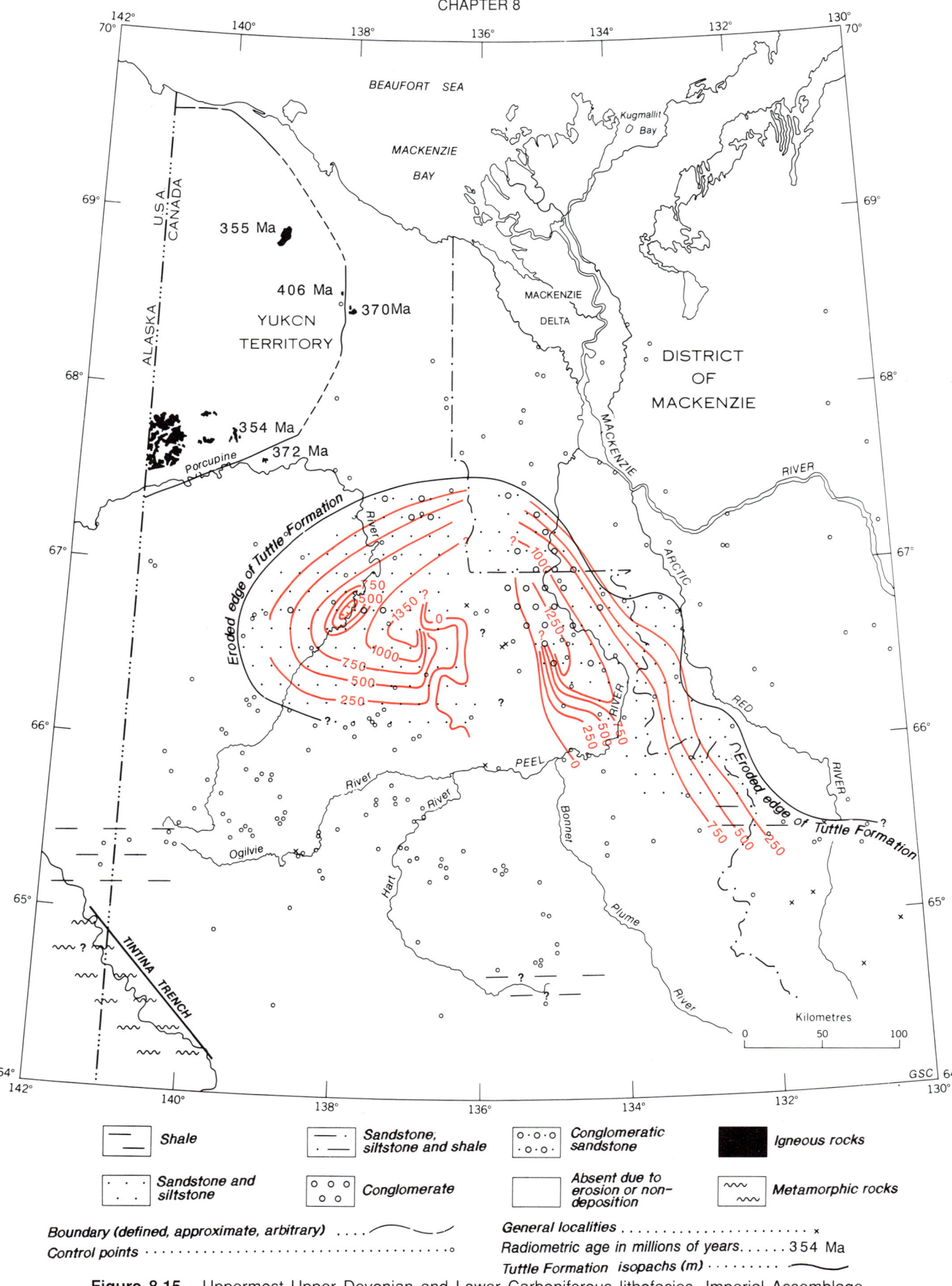

Figure 8.15. Uppermost Upper Devonian and Lower Carboniferous lithofacies, Imperial Assemblage. Diagram from Norris (1985). Isopachs (red) of the Tuttle Formation from Pugh (1983).

Figure 8.16. Angular unconformity between sandstone, siltstone and shale of the Upper Devonian Imperial Formation and overlying sandstone of Permian age, northern Richardson Mountains. Photo by A.W. Norris. GSC 205235-KK

sandstone and Ordovician and Silurian chert, provide the necessary constituents for the Prevost and Portrait Lake clastics. A recycled sedimentary source for the quartz (non-chert) detritus is indicated by the high degree of roundness and sphericity of quartz, low amounts of feldspar, and quartz sandstone cobbles within Prevost conglomerate. Given the east-southeast paleoflow for the Prevost clastics (Fig. 8.11A; Gordey, 1979) the nearest possible source where appropriate lithologies have been eroded is 200 km distant.

Syn-depositional faults with normal and possibly minor strike-slip displacement (shown diagrammatically in Fig. 8.17) occur locally (Abbott, 1982a). Such faults probably formed a graben controlling clastic dispersion of both the Portrait Lake and Prevost formations, as well as localizing important stratiform barite ± lead ± zinc mineralization. Pre-Imperial(?) folding at one locality (McClay, 1984), could be related to local faults.

Warneford River area

In northeastern British Columbia Devonian-Mississippian strata have not been formally subdivided (Jefferson et al., 1983; MacIntyre, 1983; Gabrielse et al., 1977).

In the northwest part of the area (Fig. 8.11A) blocky, well bedded pebble and cobble conglomerate with minor interbedded white, vitreous, crossbedded quartz sandstone rests unconformably on Ordovician black graptolitic shale of the Road River Formation (Fig. 8.17). Conglomerate clasts range from 5-10 cm across, are well rounded, and consist of dark to light grey chert and vitreous white quartzite. Overlying the conglomerate is well bedded calcareous sandstone capped by 10 m of clean white medium grained quartz sandstone. Overlying this succession is 25 m of conglomerate and siltstone with clasts up to 5 cm in diameter and composed of vitreous grey, medium grained quartz sandstone and lesser dark grey chert. The uppermost exposed strata is well bedded argillaceous limestone.

In the southeast part of the area strata occurring conformably beneath the lowest conglomerate consist of siliceous shale, reefal limestone, and chert which, in turn, conformably to unconformably overlie black shale and siltstone of the Road River Formation (Fig. 8.17). Farther to the southeast the conglomerate is replaced by shale and siltstone with about 25% of the sequence comprising interbeds of sandstone to conglomerate. The coarse clastic beds, commonly from 10-75 cm thick, are regularly interbedded with shale, normally graded and their bases exhibit well preserved flute and tool casts (Fig. 8.21). Similar to northwest sequences, quartz sandstone clasts are more common at higher stratigraphic levels. Easternmost strata consist entirely of shale. Deposits of bedded barite ± lead ± zinc are particularly abundant in strata of Frasnian age (MacIntyre, 1983; Jefferson et al., 1983). Barite is traceable regionally as nodular and bedded varieties within siliceous mudstone.

The southeastern clastic facies, with its graded and sole-marked turbiditic sandstone and conglomerate, represent medial to distal submarine fan deposits. Sole markings indicate east-northeast paleoflow (Fig. 8.11A; Gabrielse et al., 1977). The thick conglomerate member to the northwest may reflect a submarine channel that supplied detritus to the fan. The sub-conglomerate unconformity above non-turbiditic calcareous sandstone, together with the younger capping carbonate, suggest the channel traversed a shallow shelf.

The composition of the clastics indicates derivation from a sedimentary terrane underlain by chert and quartz sandstone. The upward increase in quartz sandstone may reflect progressive deepening of the erosion level in the source area.

Local unconformities combined with variable stratigraphic position of underlying strata suggest that high-angle faults were active during deposition (shown diagrammatically in Fig. 8.17). These also may have influenced the development and distribution of barite ± lead ± zinc deposits.

Pelly Mountains area

The Earn Assemblage in the Pelly Mountains consists of the informal "black slate", "felsic volcanic" and "cherty tuff" units (Fig. 8.17; Tempelman-Kluit, 1977, pers. comm.; Gordey, 1981b; Morin, 1977; Mortensen, 1982).

The "black slate" unit is dominated by black, fissile, laminated slate that contains rare dark greywacke (<10%) and chert-pebble conglomerate (<5%). The greywacke is well sorted, medium grained, commonly graded and rhythmically interbedded with slate in parallel-laminated beds up to 1 m thick. It is composed of subrounded chert, quartz, minor feldspar and mica and rare altered volcanic fragments. Conglomerate occurs in massive units from 10 to 60 m thick. Clasts are generally subrounded to rounded and up to 20 cm in diameter, but average about 1 cm. Most are chert, but some are very fine-grained to medium-grained

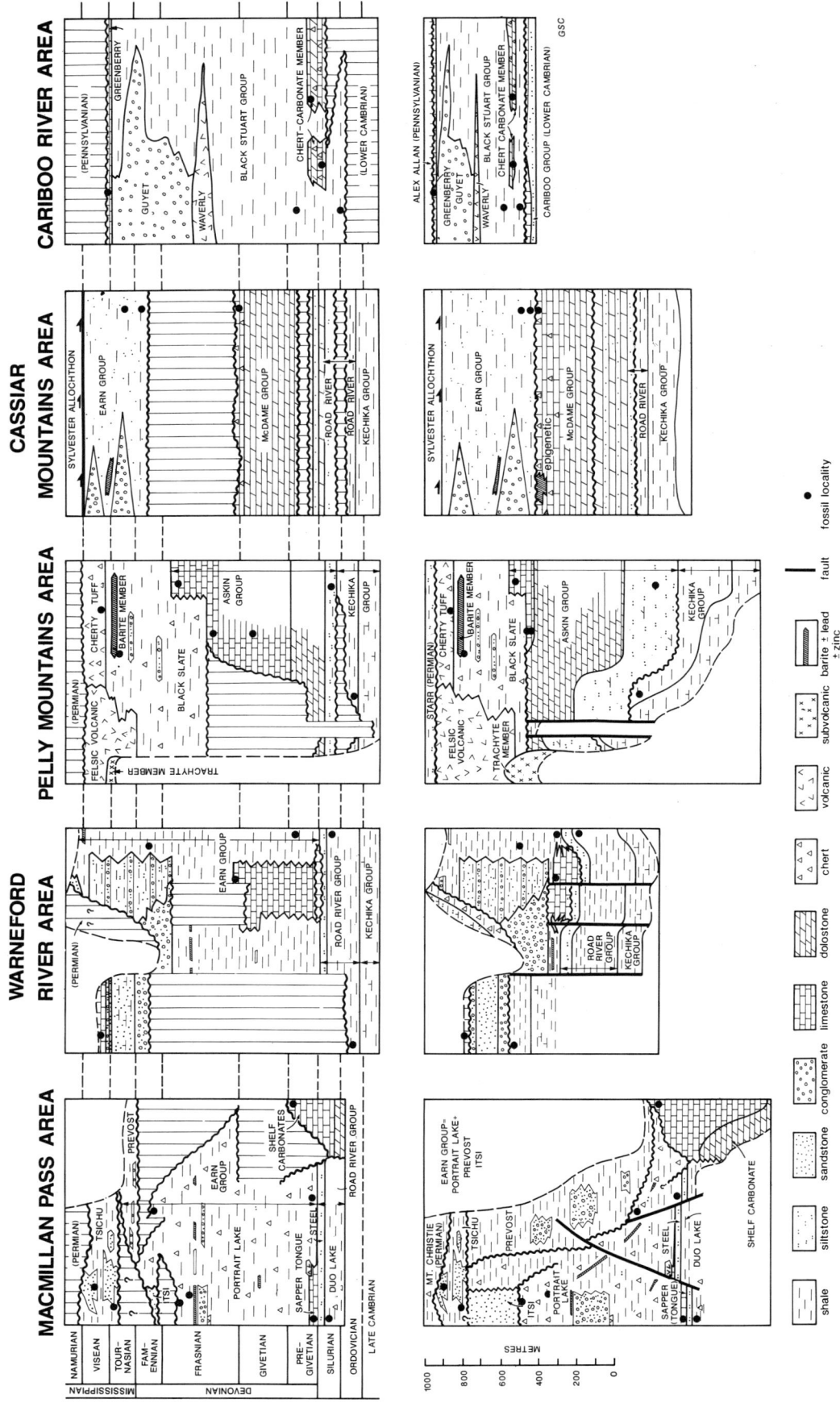

Figure 8.17. Synoptic time- and rock-stratigraphic diagrams for the five areas depicted in Figure 8.11, and described in text as typical of the coarse Earn Assemblage.

Figure 8.18. Debris flow conglomerate west of Macmillan area. Boulders are composed of gritty quartz sandstone, and chert and rest in mud matrix. Photo by S.P. Gordey. GSC 205235-T

quartz arenite. Sorting is poor to moderate and the clasts are enclosed in a clean sand-size matrix of chert and quartz. Resistant lenses of intermediate volcanic rocks, up to 50 m thick, are common within the "black slate" unit.

The "felsic volcanic" unit includes light coloured, resistant interlayered tuff, breccia and flow rocks as well as dykes, sills and plugs. Individual members are several metres to tens of metres thick and generally cannot be traced more than a kilometre. Both tuffs and flows are typically massive and composed of subhedral perthitic feldspar, albite, minor quartz, hornblende, biotite and opaque minerals. Analyses of the felsic volcanics (Fig. 8.19D) show that they are mostly alkaline trachyte or rhyolite, anomalously rich in potash and high in barium, thorium and fluorine and poor in uranium (Gordey, 1981b; Mortensen, 1982; Morin, 1977). The "cherty tuff" unit comprises thin-bedded grey to green tuffaceous chert and siliceous tuff, which may be a lateral equivalent of the felsic volcanic unit.

Siliceous shale of the "black slate" unit represents low-energy deposition below wavebase within which sandstone and conglomerate were likely deposited as sediment gravity flows. The composition of sandstone and conglomerate in the "black slate" unit indicates a source terrane underlain by chert and quartz sandstone. The trachyte plugs, felsic volcanics and cherty tuff are products of one alkaline volcanic event. The barite member of the "black slate" may be a widely dispersed chemical deposit related to the felsic volcanism.

The "black slate" unit overlies the Askin Group (Fig. 8.17), locally with marked unconformity. Steep faults cutting the Askin Group do not penetrate the slate. The unconformity reflects local uplift and erosion associated with block faulting of a regionally subsiding platform. Thus, some local fault basins received sediment continuously while companion horsts were eroded. Plugs and dykes of the "felsic volcanics" were injected along faults and fractures cutting Devonian and older strata. Mortensen (1982) concluded that the major and minor element chemistry of the "felsic volcanics" is consistent with a rift origin

and compares closely with that of the East Africa Rift Zone.

Cassiar Mountains area

A sequence of Upper Devonian (Famennian) to mid-Mississippian fine- to coarse-grained clastic rocks, more than 700 m thick, overlies Middle to Upper Devonian carbonate strata in the Cassiar Mountains. The clastics underlie oceanic rocks of the Sylvester Allochthon on the limbs of a synclinorium trending southeastward for about 175 km from near the Yukon-British Columbia boundary. The basal strata, locally up to 150 m thick, but commonly thinner, consist of grey to black, locally pyritic or calcareous shale. Ribboned porcellanite comprising interbedded siliceous mudstone and slate is present at some localities. Overlying rocks include siltstone, sandstone and conglomerate of which the latter varies markedly in thickness.

The best studied section (Fig. 8.17), near the north end of the synclinorium, rests on a karst surface at the top of Middle (Givetian) to Upper Devonian (Frasnian) carbonate. A basal unit of black argillite and fine grained siltstone, 5 to 40 m thick, of mid-Famennian age, is overlain by sandstone, conglomerate and siltstone, 35 to 280 m thick. Coarse sandstone and conglomeratic beds are commonly graded and have abundant sole markings. Clasts in the conglomerate, as much as 10 cm in diameter, are subrounded to angular and comprise silicified dolomitic sandstone, dolomite and abundant slate. Quartz is dominant in the matrix of coarser grained rocks and is commonly a characteristic opalescent mauve. The conglomeratic unit is overlain by a succession of siltstone, sandstone and calcarenite from 200 to 500 m thick, including a basal member of carbonaceous mudstone and siltstone containing silica, pyrite and barite with local sulphide. The uppermost unit, 150 to 200 m thick, is made up of sandstone, conglomerate and siltstone. Plant remains occur in several beds. Farther southeast conglomerate is rare or absent, either because of facies changes, or truncation below the basal fault of the Sylvester Allochthon.

The character of the Upper Devonian to mid-Mississippian sediments suggests abrupt deepening of the depositional basin accompanied by local uplift of nearby, older miogeoclinal carbonate and siliceous sediments. Most, if not all, siliceous clasts in greywacke and conglomerate represent silicified carbonate and sandy carbonate representative of Lower and(?) Middle Devonian miogeoclinal lithologies. Basal deposits of carbonaceous, pyritic mudstone suggest euxinic environments succeeded by influxes of turbiditic sediments. The marked relief and facies changes are indicative of a rift setting along or near the western margin of the North American miogeocline.

Cariboo River area

The southernmost outcrops of the Earn Assemblage, 360 km south-southeast of those in the Cassiar Mountains, comprise the Black Stuart Group and the Waverly, Guyet and Greenberry formations (Struik, 1980, 1981). Exposures are poor, structurally complex and metamorphism is of chlorite grade.

The Black Stuart Group (500± m) (Fig. 8.17) consists of black slate, fine grained greywacke and siltstone, minor black micritic limestone and units of chert-carbonate and

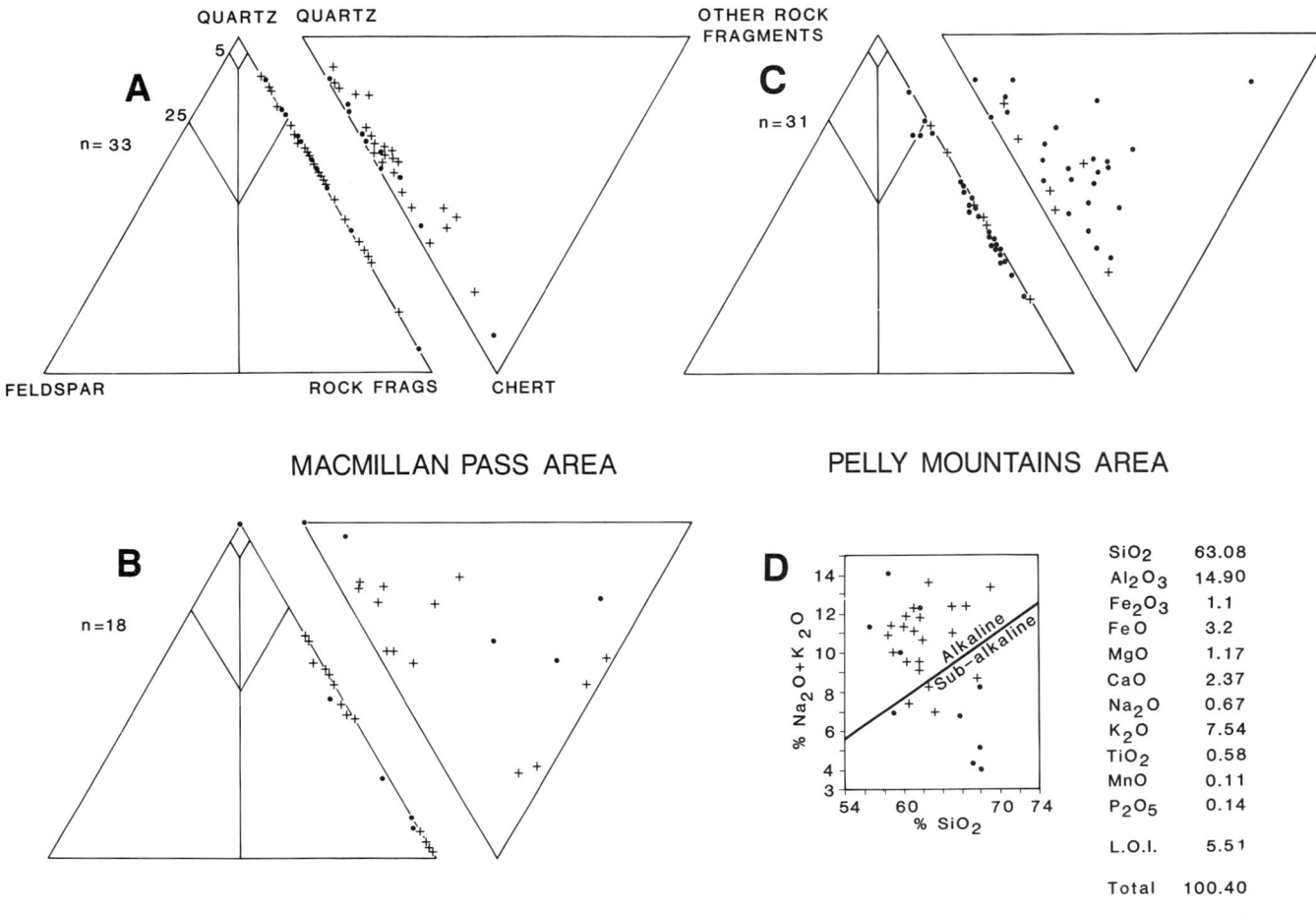

Figure 8.19. Composition of Devonian-Mississippian sandstones (A,B) and pebble conglomerate (C) for areas shown. (D) Alkali-silica diagram and a typical chemical analysis for Mississippian volcanic rocks in Pelly Mountains area (data from Gordey, 1981b (circles); Mortensen, 1983; analysis from Gordey, 1981a). Dots indicate wackes (matrix >15%); crosses indicate arenites (matrix <15%). S in lower right corner is sulphur.

sandstone. Discontinuous chert-carbonate (0-60 m thick), consists of sandy dolostone, dolostone breccia and replacement chert. In the upper Black Stuart, load and flame structures and graded beds are typical of the sandstone member which comprises quartzite, siltite and interbedded pelite.

The Waverly Formation (0-50 m) is composed of schistose, calcareous, volcanic agglomerate and flows, pyroclastics, pillow basalt, sedimentary breccias and volcaniclastics interdigitated with the Black Stuart Group.

The Guyet Formation (0-200 m) consists of sandstone-matrix and mudstone-matrix conglomerate and greywacke (Fig. 8.19B). Sandstone-matrix conglomerate consists of granule to cobble size clasts of varicoloured chert, cherty argillite, pelite, micaceous siltite and quartzite, and minor limestone and basalt. Clasts are subrounded to rounded, and are enclosed in a matrix of quartz, chert, siltite and minor feldspar. Within the massive, mudstone-matrix conglomerate clasts are of the same composition but have a greater diversity in size (up to 1 m across) and proportion (locally volcanic clasts dominate). Conglomeratic beds range from one to several tens of metres thick and are characterized by local basal scour and changes in matrix type and grain size. Coarse massive or planar laminated sandstone is locally interbedded with the conglomerate. At one locality dark grey, normally graded greywacke and shale, in beds 10-75 cm thick, form a thinning and fining upward sequence (top not exposed) above mudstone-matrix conglomerate with which it is in gradational contact.

Light grey massive bioclastic limestone comprises the Greenberry Formation (10 m), the base of which locally contains rare clasts of chert similar to that of the Guyet, and the top of which is in unconformable contact with Pennsylvanian limestone up to 5 m thick.

Figure 8.20. Photomicrograph of typical quartz chert sandstone, of the Prevost Formation, Macmillan area. Photo by S.P. Gordey. GSC 205235-U

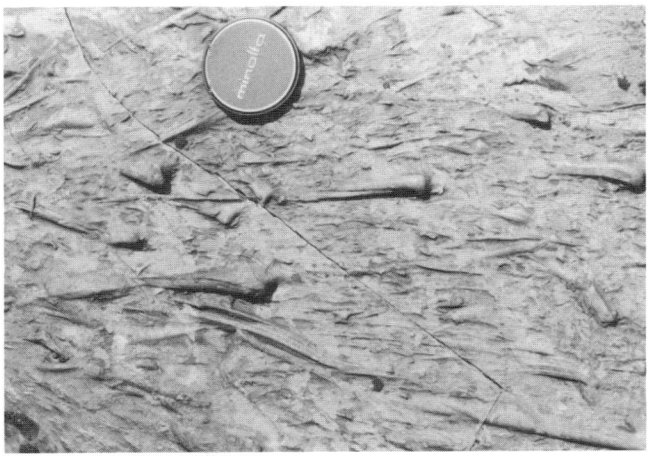

Figure 8.21. Tool marks at base of thin-bedded turbidite, Warneford area. Photo by H. Gabrielse. GSC 205235-Y

Fine clastics of the Black Stuart were deposited in low energy conditions below wave-base, an environment interrupted by shallow-water deposition of the chert-carbonate member and Waverly submarine volcanism. Clastics of the Guyet Formation and local graded sandstone of the Black Stuart were likely deposited in fairly deep water as sediment gravity flows. The Greenberry Formation is possibly a shallow-water deposit, but could represent a debris flow that carried its contained shallow-water conodont fauna into deep water.

The source terrane for the Guyet was underlain by siltstone, sandstone, and radiolarian chert. Volcanic clasts, compositionally like the Waverly volcanics, are locally abundant, however, pre- or syn-Guyet erosion of the Waverly has not been demonstrated. Complex Mesozoic deformation and poor exposure preclude identification of possible syn-sedimentary tectonic features.

Nation River Formation

In east-central Alaska straddling the Alaska-Yukon border are sediments of the Nation River Formation (Fig. 8.10, 8.11, 8.12; 610-1220 m) described as a deep-sea fan complex composed of interbedded chert-pebble conglomerate, chert-quartz arenite, siltstone, mudstone, shale and pebbly mudstone (Nilsen et al., 1976). According to these authors paleocurrents transported sediment toward the west from a source area to the east or northeast. Recent work by Howell et al. (1987), however, indicates that paleoflow was eastward. The Nation River has been dated as Late Devonian on the basis of spores. It rests conformably on the partly Upper Devonian McCann Hill chert, and is overlain conformably by the Ford Lake Shale of Late Devonian to early Late Mississippian age (Fig. 8.12).

Besa River Assemblage and the carbonate shelf edge

The fine clastics of the Besa River Assemblage, mostly Besa River (305-2200 m; Pelzer, 1966; Taylor and Mackenzie, 1970) and Perdrix (100-470? m; Geldsetzer, 1982) formations in northern British Columbia and Horn River (<320 m; Hills et al., 1981; Williams, 1977) and Fort Simpson (480-1000 m; Belyea and McLaren, 1962; Hills et al., 1981) formations in southern Northwest Territories comprise shale, mudstone, and siltstone. They represent the easterly and southerly fine clastic fringe of the Imperial and Earn assemblages, and possibly include clastic input from eastern cratonal sources. During the Late Devonian and Mississippian, the eastern limit of these clastics was a carbonate-shale boundary that fluctuated widely, at times moving far to the east and south of its pre-Late Devonian position (Fig. 8.5, 8.6, 8.9, 8.10, 8.11). Although transgression began in the Givetian, the major transgressions, punctuated by minor regression, occurred in the Frasnian, earliest Mississippian, and Early Mississippian (Fig. 8.11). In the Frasnian, shale spread as far east and south as central Alberta (Ireton Formation). In the earliest Mississippian it accumulated across northeastern British Columbia, the southern Canadian Rockies (Exshaw Formation) and far onto the continental interior (Baakan, Antrim and Chatanooga formations) (see next section). Sub-Frasnian, sub-earliest Mississippian and sub-Early Mississippian erosion occurred over many areas of the Cordillera and marked the intervening regressions. Immediately following the Early Mississippian transgression there was no distinct carbonate shelf edge as shale interdigitated with carbonate over a wide area (Banff Formation), but by the mid-Mississippian (Viséan), at the same time as clastic input from the west and north diminished, carbonates again became established along much of the eastern miogeocline (Rundle Group, Prophet Formation) (Fig. 8.11).

Tectonic summary

In the northern Yukon Devonian-Mississippian tectonism involved uplift of miogeoclinal strata and coincident granitic intrusion, resulting in an upward shoaling and southward prograding clastic wedge (Imperial Assemblage). The clastic wedge was folded prior to the mid-Carboniferous. Early Mississippian compressional tectonism, similar to that in the Franklinian geosyncline of the Arctic

Innuitian region (Ellesmerian orogeny; Kerr, 1981), seems likely, however, any evidence of structural link between the two regions is concealed beneath the Beaufort Shelf.

Coarse Devonian-Mississippian clastics in the remainder of the Cordillera (Earn Assemblage) reflect erosion of Upper Proterozoic quartzose clastics, lower Paleozoic chert, and locally platform carbonate from sources within the outer miogeocline. Lack of compressional structures, local alkaline volcanics of possible rift type, and syn-sedimentary faults controlling local unconformities, thickness, facies and exhalite mineralization favour a tensional environment. In a broader context such an environment could reflect (1) rifting and continental separation, (2) broad extension (transtension) within the miogeocline related to strike-slip faulting along the old continental margin, or (3) extensional basins in the foreland of a contractional orogen. Eisbacher (1983) related clastic deposition and uplift to pull-apart basins and compressional fault bends along a single through-going strike-slip fault. However, the location of such a structure remains enigmatic (Cecile, 1984).

The presence of Late Devonian-Early Mississippian granitic rocks in suspect terranes (Kootenay) along the western margin of the miogeocline (Gabrielse et al., 1982; Mortensen and Jilsen, 1985) raises the problem of their possible relationship to the tectonic setting of the coeval clastic rocks described herein. The displaced terranes have a lithological affinity with strata derived from the North American craton and probably represent disrupted parts of the outer miogeocline. If so, the tensional regime indicated by the Earn clastics may have been only one aspect of an orogenic event that occurred farther west.

In composition, age and facies, Devonian-Mississippian clastics in the Canadian Cordillera are remarkably like those of the westerly derived "Antler Flysch" (Upper Devonian to Upper Mississippian) in the United States Cordillera. Classical interpretations of the "Antler Flysch" (e.g. Poole, 1974) relate it to deformation and obduction of Devonian and older siliceous rocks (Roberts Mountain Allochthon) onto the outer carbonate shelf of the miogeocline, a setting not unlike the tectonism in the northern Yukon and Alaska which resulted in the southerly and easterly progradation of the Imperial clastic wedge but different from the possible rift origin of the Earn Assemblage.

Carboniferous and Permian stratigraphy of the Foreland Belt

E.W. Bamber, C.M. Henderson, B.C. Richards, and A. McGugan

Carboniferous and Permian strata in the Foreland Belt (Fig. 8.22) occur in two facies belts: an eastern succession of platform and ramp carbonates and shallow marine to continental clastics, and a correlative, western basinal shale and siltstone succession. These two facies belts, which occur from the southern Rocky Mountains to northern Yukon, were deposited on the western margin of ancestral North America. The eastern slope and shelf succession comprises several lithostratigraphic assemblages, separated by regional disconformities marking Late Carboniferous and Permian erosion. Truncation beneath these and younger, sub-Mesozoic disconformities indicates that much of the upper Paleozoic record was removed, particularly in the Rocky Mountain Foothills and Mackenzie Mountains.

Facies relationships and thickness variations within the slope, shelf, and basinal sediments are illustrated on schematic cross-sections (Fig. 8.27-8.31). Correlations (Fig. 8.23-8.25) are supported by regional zonations of Foraminifera (Mamet and Skipp, 1970), conodonts (Richards et al., in press; Henderson et al., in press), brachiopods (Bamber and Waterhouse, 1971), and corals (Sando and Bamber, 1985). Facies and faunas of Carboniferous and Permian rocks in the displaced Cassiar and Arctic Alaska terranes are similar to those in ancestral North America and are described in this section.

Carboniferous

Carboniferous deposits of the carbonate-clastic belt are preserved throughout most of the Foreland Belt where, in part, they disconformably overlie older strata. In the Rocky and Mackenzie mountains this succession, which was deposited in Prophet Trough (Fig. 8.22; Richards et al., in press) and on the westernmost cratonic platform, consists of middle Tournaisian to upper Viséan, carbonate platform and ramp deposits (Fig. 8.26A,B), overlain by widespread, upper Viséan and Serpukhovian, deltaic to nondeltaic, shallow-marine and supratidal siliciclastics (Bamber et al., 1984; Richards, 1983, 1989). This succession is disconformably overlain by Bashkirian and lower Moscovian carbonates and siliciclastics which are incompletely preserved beneath a regional sub-Permian disconformity (Fig. 8.32, 8.33; McGugan et al., 1968). In northern Yukon, where the upper part of the Carboniferous depositional record is more complete, the carbonate-clastic belt is a continuous succession of Tournaisian(?) to Moscovian carbonates and subordinate siliciclastics, with local preservation of younger Carboniferous carbonates. Contemporaneous deltaic deposits and shallow-marine siliciclastics are mainly restricted to the east, along the flanks of Richardson and Barn mountains (Bamber and Waterhouse, 1971; Pugh, 1983). The northern succession was deposited in northern Prophet Trough and onlapped the Yukon Fold Belt toward the northeast, and the cratonic platform toward the east (Fig. 8.22).

The carbonate-clastic belt overlies and passes westward into a succession of basinal deposits across a broad, transitional zone that appears to have migrated southwestward (Fig. 8.27-8.29). Within the western part of the basinal succession, marked westward thinning occurs because of reduced sedimentation rates in a starved basin (Bamber et al., 1984).

Figure 8.22. Distribution of Carboniferous deposits, locations of cross-sections and principal Carboniferous tectonic elements. 1 - Lizard Range; 2 - Hosmer Thrust Plate; 3 - Telford Thrust Plate; 4 - Bow Valley; 5 - Pine Pass; 6 - Labiche Range; 7 - Kotaneelee Range; 8 - Liard Range; 9 - Nahoni Range; 10 - Ogilvie Mountains; 11 - Barn Mountains. The lower, middle, and upper depositional units include: Lower Depositional Unit (Exshaw, Bakken, Lodgepole, Banff and Yohin formations); Middle Depositional Unit (Mission Canyon and Charles formations; Rundle Group, excluding Etherington Formation); and Upper Depositional Unit (Stoddart and Spray Lakes groups; Kibbey, Etherington, Mattson, and Golata formations).

UPPER DEVONIAN TO MIDDLE JURASSIC ASSEMBLAGES

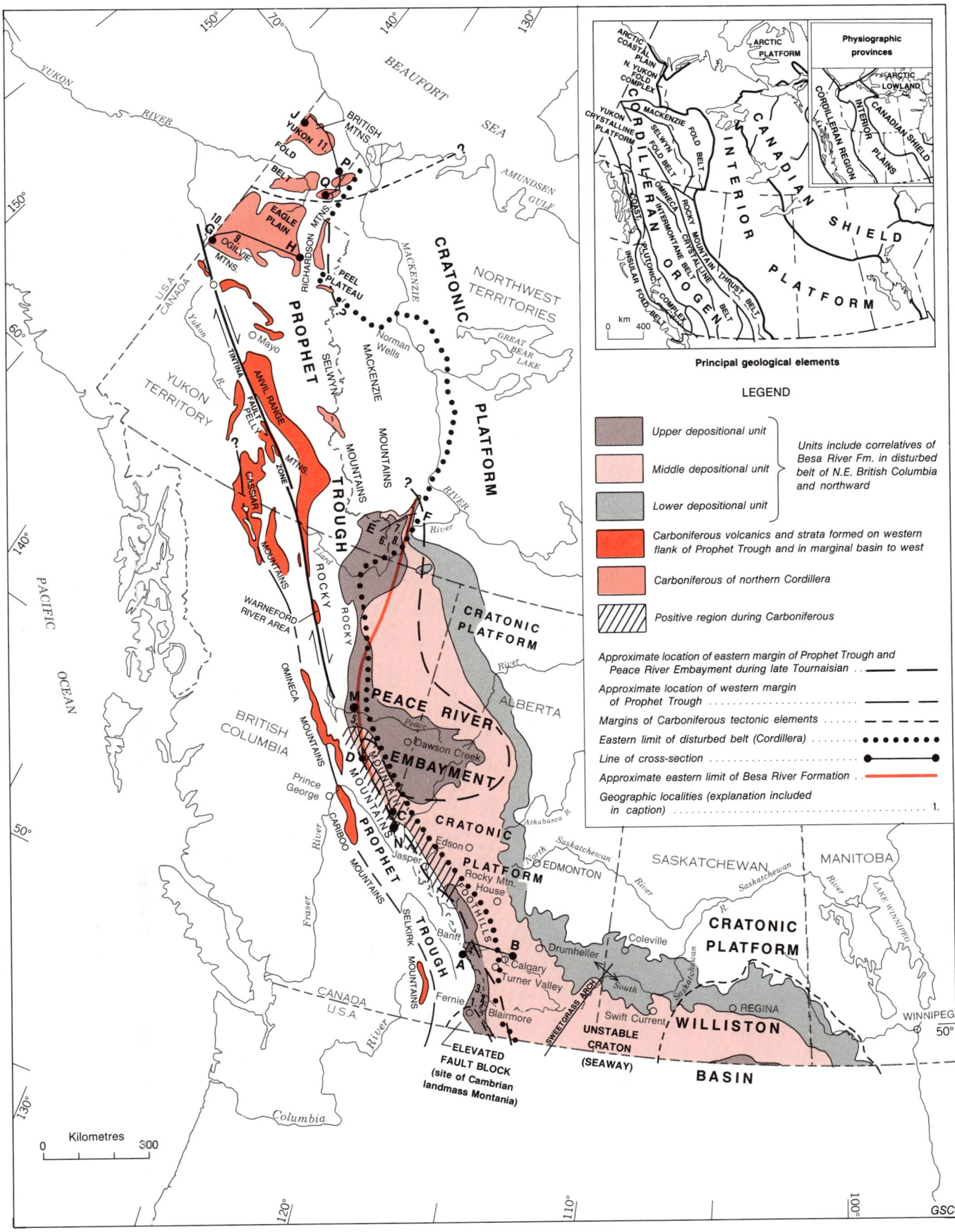

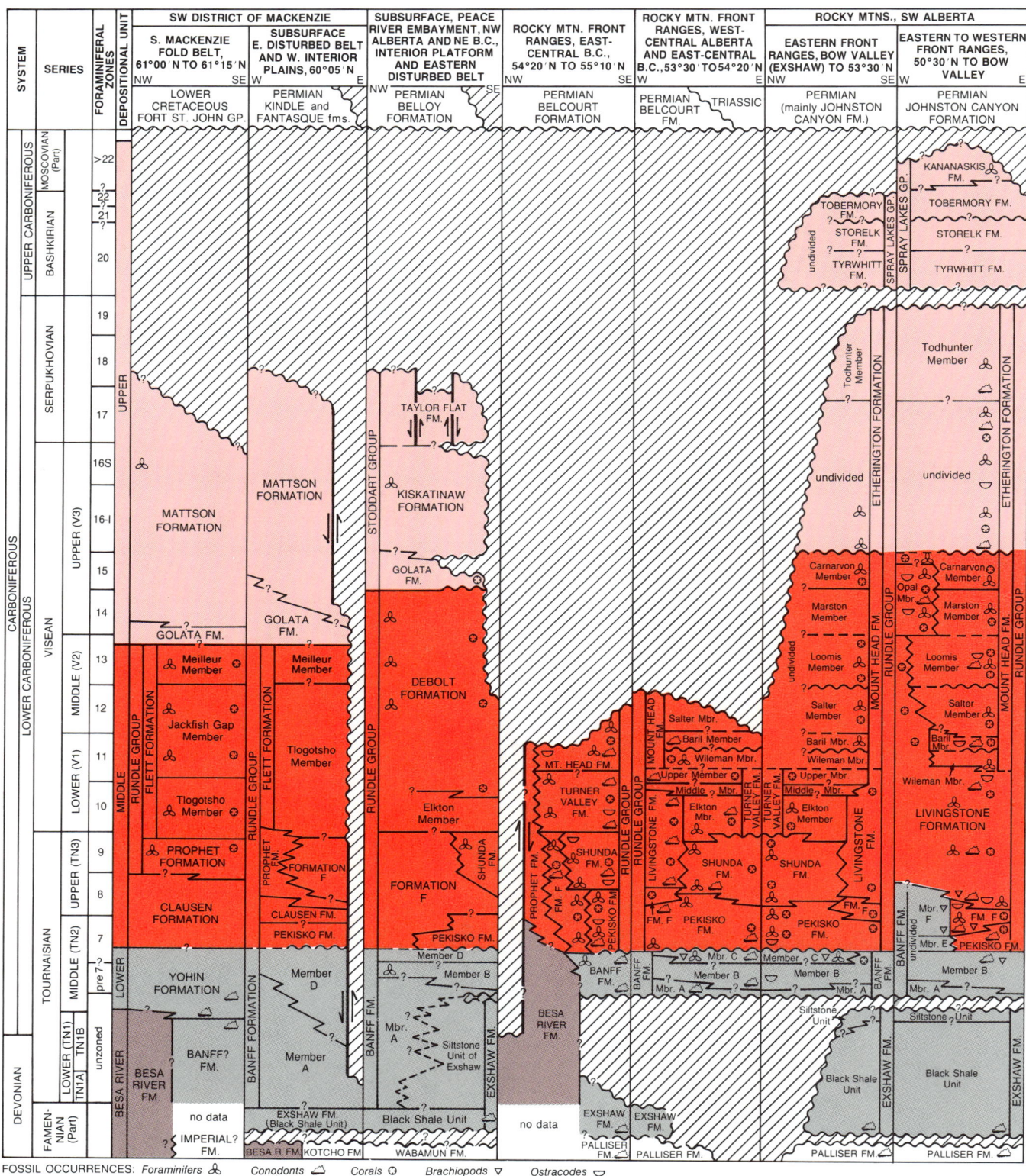

Figure 8.23. Correlation of Carboniferous lithostratigraphic units, southwestern Alberta to southwestern District of Mackenzie. The lower (grey), middle (red) and upper (reddish grey) depositional units correspond to those in Figure 8.22.

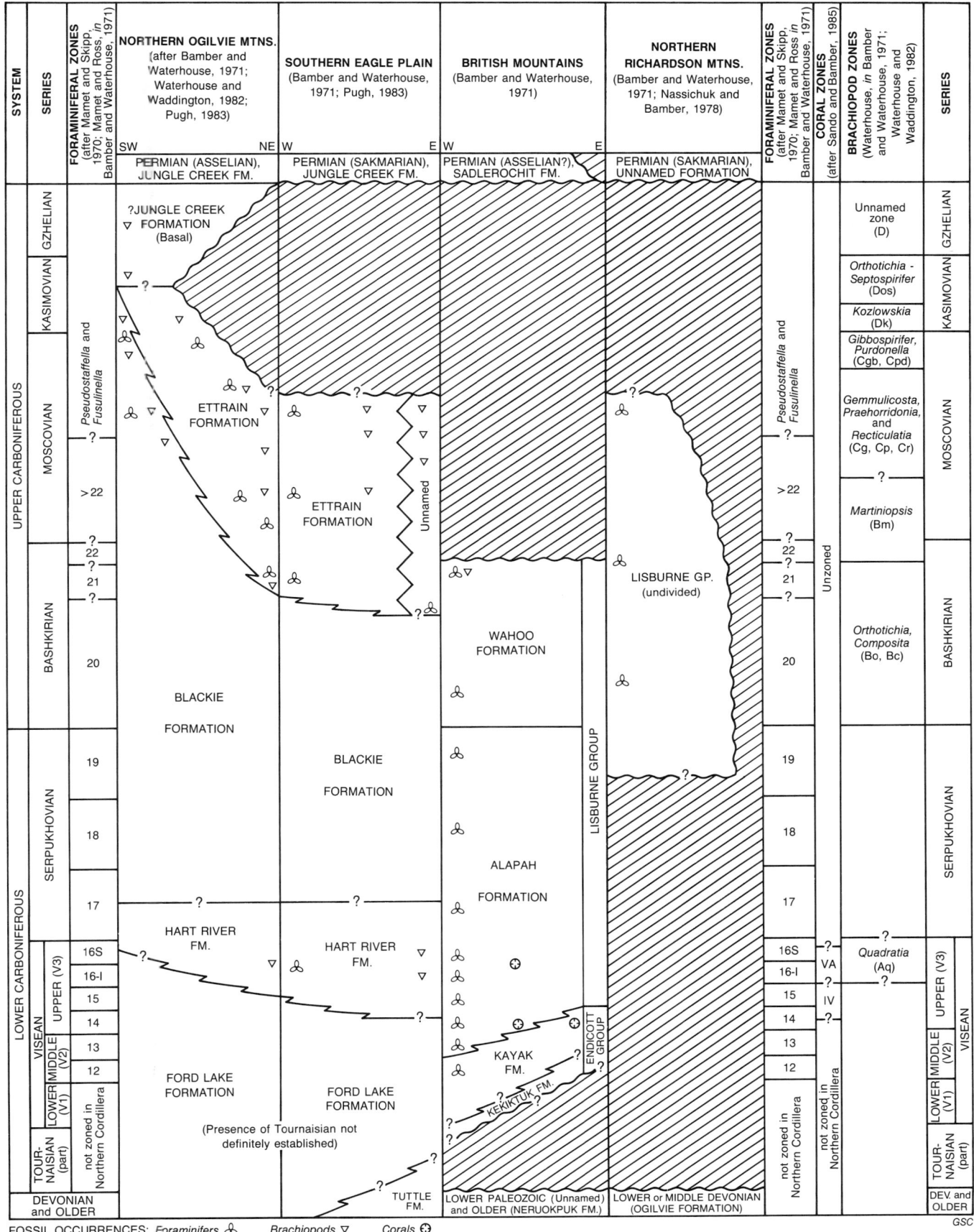

Figure 8.24. Correlation of Carboniferous lithostratigraphic units, northern Yukon Territory and northwestern District of Mackenzie.

CHAPTER 8

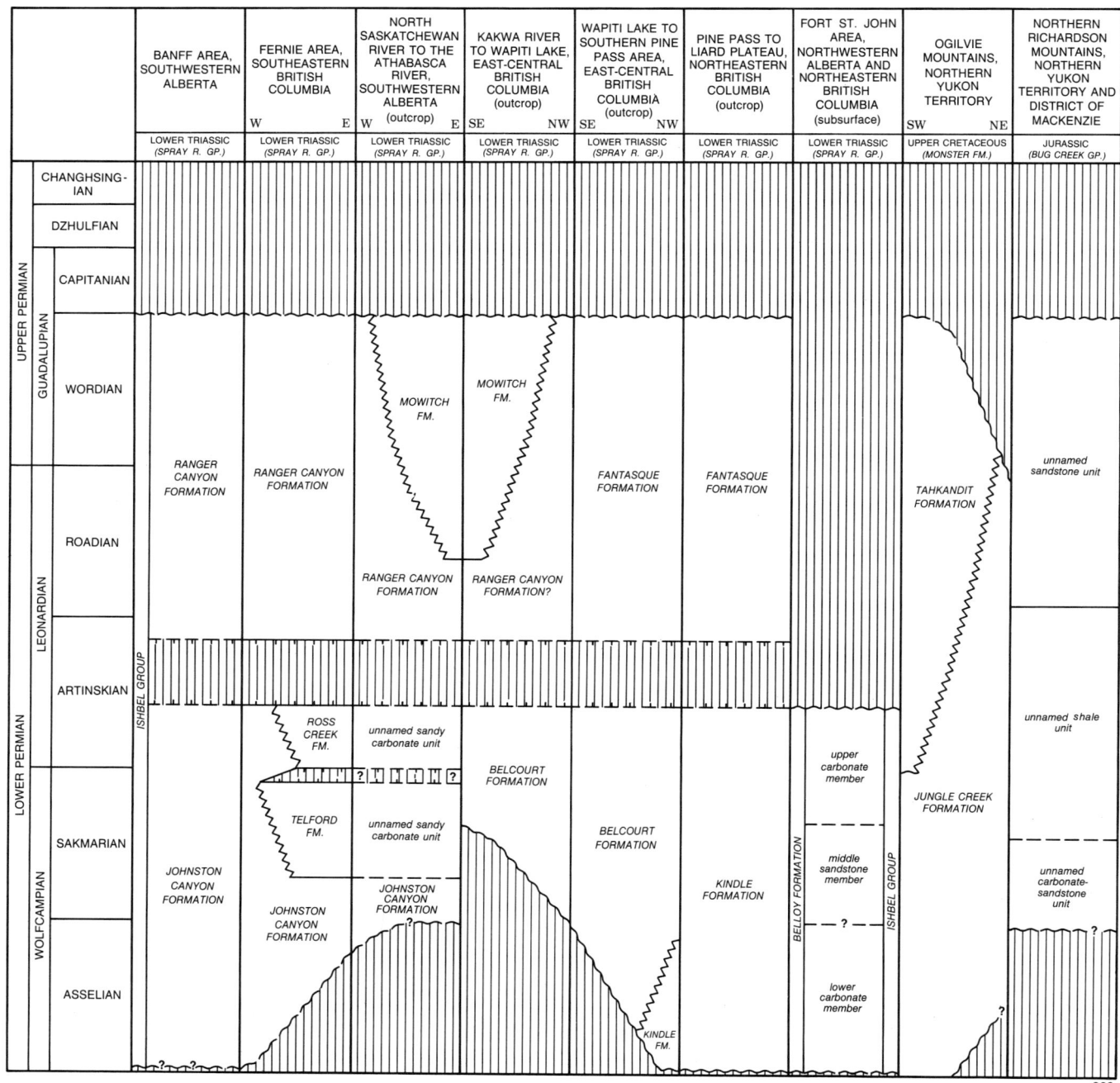

Figure 8.25. Correlation of Permian lithostratigraphic units, southeastern British Columbia to northwest District of Mackenzie.

UPPER DEVONIAN TO MIDDLE JURASSIC ASSEMBLAGES

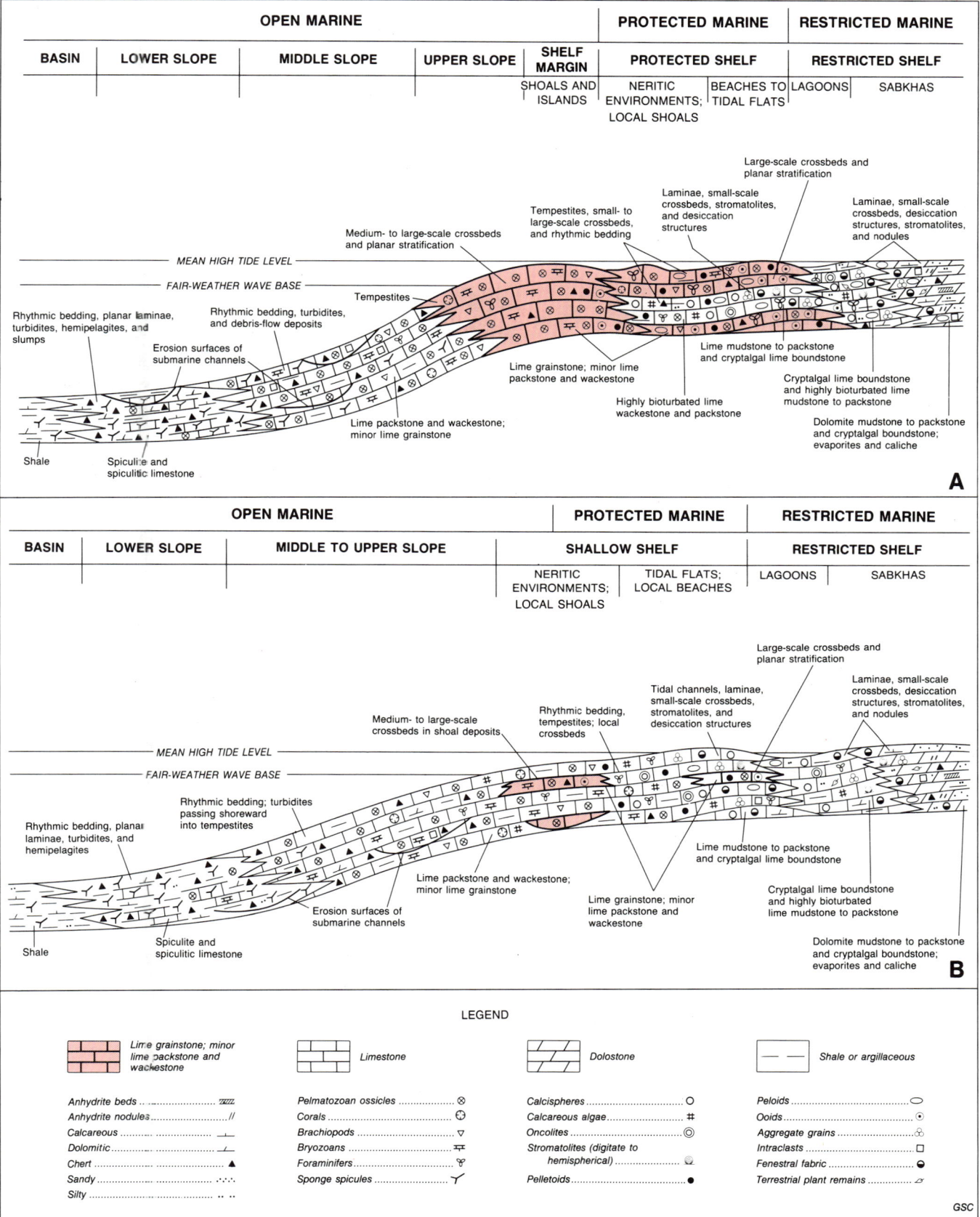

Figure 8.26 A,B. Schematic, generalized depositional model of carbonate platform (A) and ramp (B) showing environments and corresponding rock types illustrated in Figures 8.27 to 8.31.

CHAPTER 8

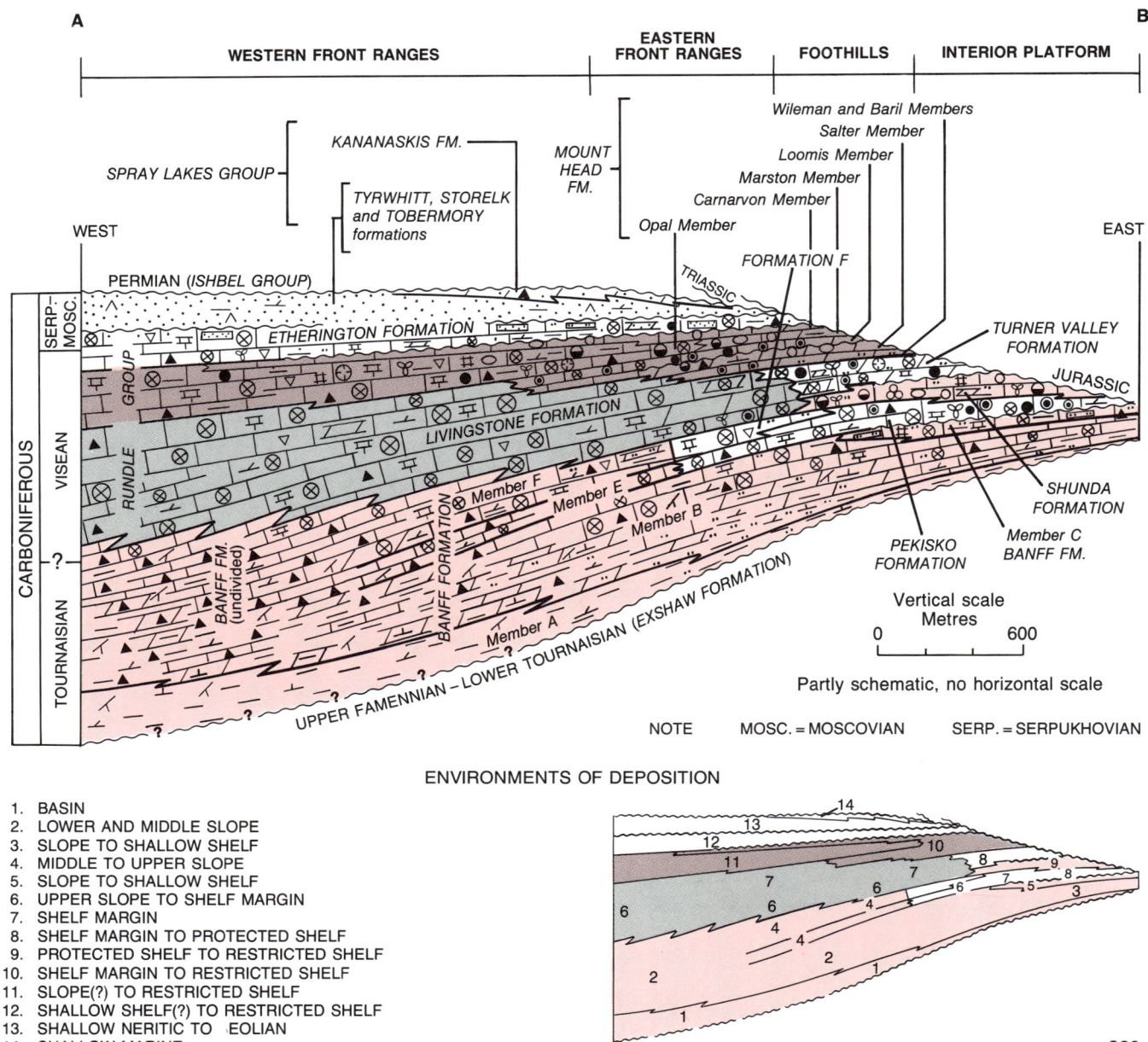

1. BASIN
2. LOWER AND MIDDLE SLOPE
3. SLOPE TO SHALLOW SHELF
4. MIDDLE TO UPPER SLOPE
5. SLOPE TO SHALLOW SHELF
6. UPPER SLOPE TO SHELF MARGIN
7. SHELF MARGIN
8. SHELF MARGIN TO PROTECTED SHELF
9. PROTECTED SHELF TO RESTRICTED SHELF
10. SHELF MARGIN TO RESTRICTED SHELF
11. SLOPE(?) TO RESTRICTED SHELF
12. SHALLOW SHELF(?) TO RESTRICTED SHELF
13. SHALLOW NERITIC TO EOLIAN
14. SHALLOW MARINE

Figure 8.27. Schematic cross-section (AB) illustrating stratigraphic relationships and lithology, Tournaisian to Moscovian, southern Rocky Mountains to western Interior Platform (southwestern Alberta - see Figure 8.22 for location).

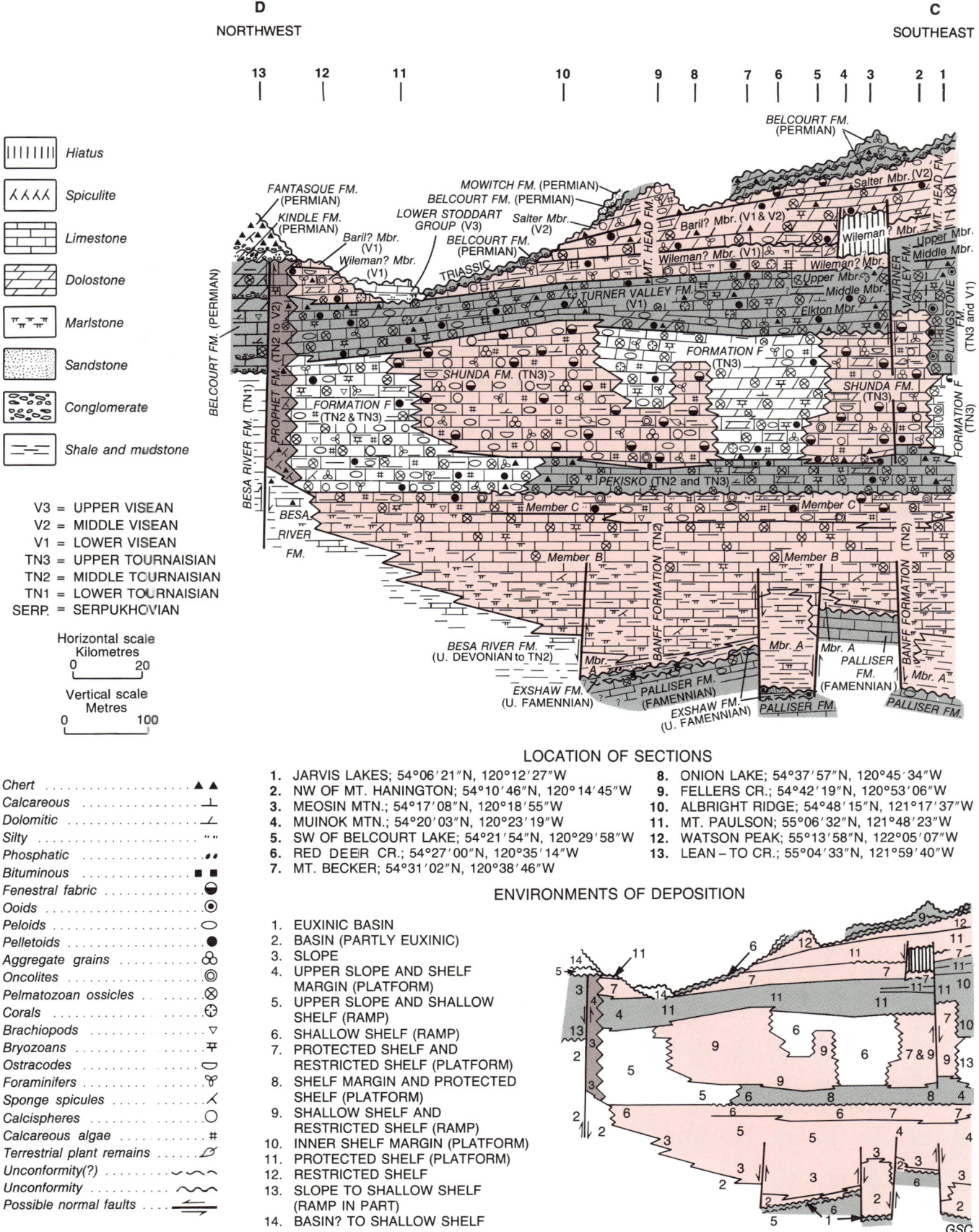

Figure 8.28. Partly schematic cross-section (CD) illustrating stratigraphic relationships and lithology, Tournaisian and Viséan, central Rocky Mountains (east-central British Columbia - see Fig. 8.22 for location).

CHAPTER 8

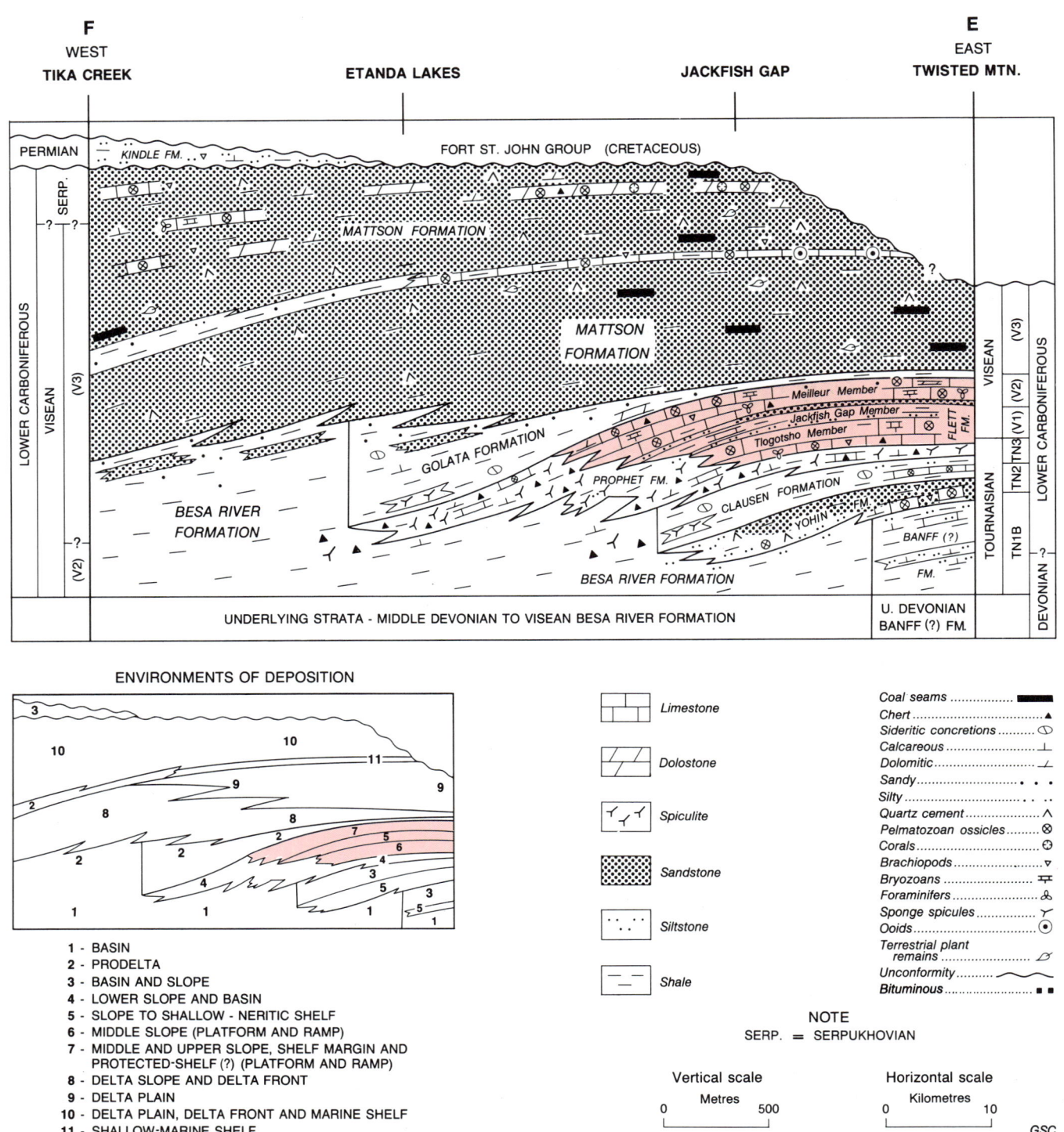

Figure 8.29. Schematic cross-section (EF) illustrating stratigraphic relationships and lithology, Tournaisian, Viséan, Serpukhovian, and Permian, Mackenzie Mountains (southwestern District of Mackenzie, southeastern Yukon - see Fig. 8.22 for location).

Southern Rocky Mountains and Foothills

In this region, only the carbonate-clastic belt occurs and is represented by components of the Rundle, Mattson, and Ishbel assemblages (Fig. 8.27, 8.32). No strata younger than early Moscovian are preserved beneath the sub-Permian disconformity. Basinal shale facies are restricted to thin units of late Famennian to middle Tournaisian age at the base of the succession (lower Exshaw Formation and lower Banff Formation). These are overlain by middle Tournaisian to lower Moscovian platform and ramp carbonates and terrigenous clastics of the middle and upper Banff Formation, Rundle Group, and Spray Lakes Group. Facies relationships (Fig. 8.27; Richards et al., in press; Bamber et al., 1984) indicate a general shallowing-upward trend which was interrupted by several episodes of regional transgression.

The basal part of the succession consists of black, bituminous shale, up to 10 m thick, of the lower Exshaw Formation (Macqueen and Sandberg, 1970). These deposits, which locally straddle the Devonian-Carboniferous boundary, accumulated in a moderately deep-water (below storm wave base), anaerobic environment during an initial, latest Devonian to earliest Carboniferous transgression. Overlying, lower to middle(?) Tournaisian siltstone, sandstone, and limestone of the upper Exshaw (up to 39 m thick) were deposited under shallowing-upward conditions in slope to shallow-marine (above wavebase) environments.

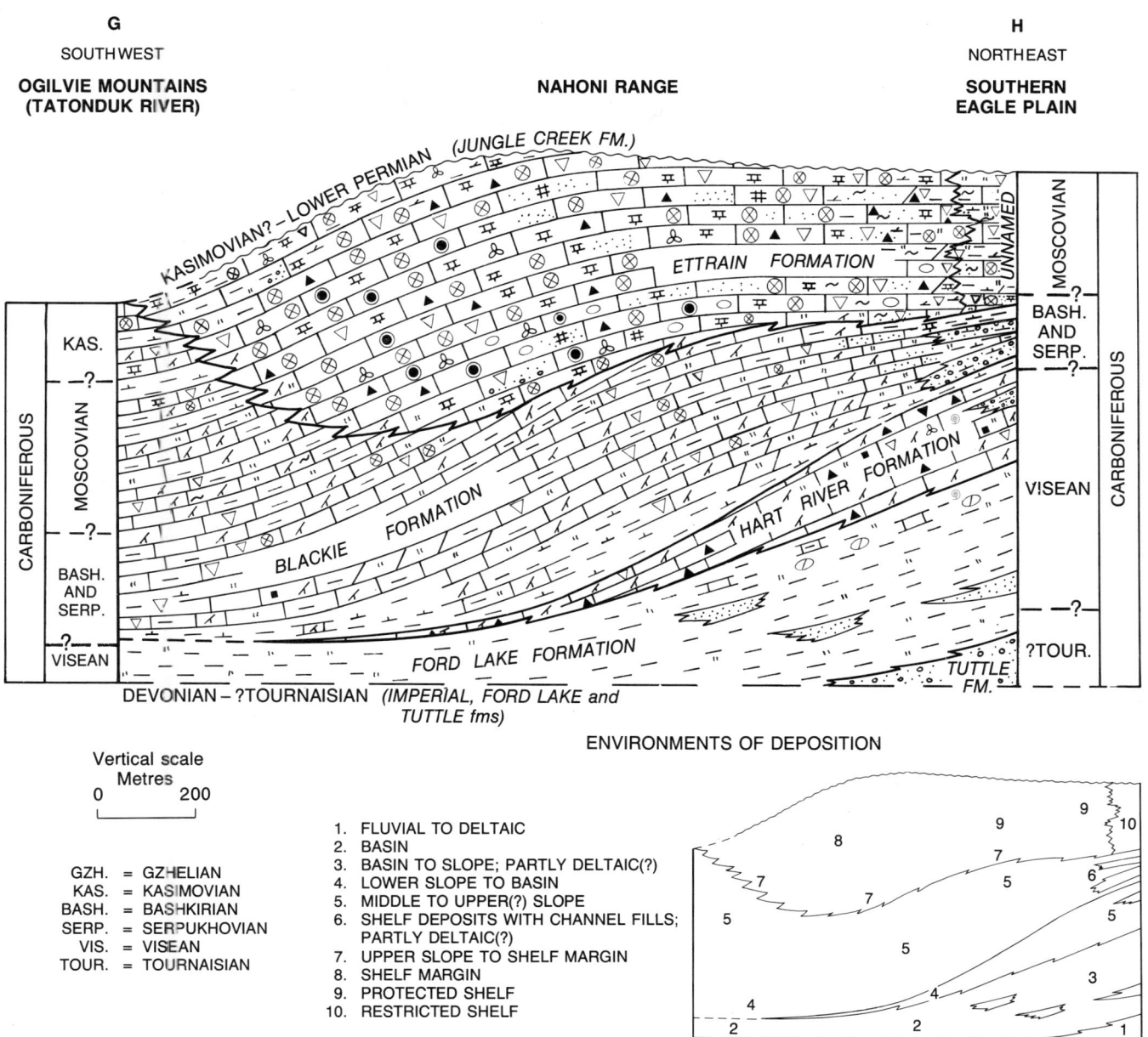

Figure 8.30. Schematic cross-section (GH) illustrating stratigraphic relationships and lithology, Carboniferous, northern Cordillera (Ogilvie Mountains, Eagle Plain - see Fig. 8.22 for location).

CHAPTER 8

The middle Tournaisian Banff Formation, which comprises several informal members (Fig. 8.32; Richards et al., in press), disconformably overlies the Exshaw Formation throughout most of the area, except in west-central Alberta and east-central British Columbia, where the Exshaw is generally absent. There, the Banff disconformably overlies Famennian carbonates of the Palliser Formation (Plate 18). The westward-thickening Banff Formation ranges in thickness from 170 to 550 m and comprises black to dark grey, basal shale grading upward into slope, shallow shelf, shelf-margin and restricted shelf carbonates. In the east, it records one regional transgression and subsequent regression, but in the west the Banff resulted from two regional transgressive-regressive events (Fig. 8.27).

The overlying Rundle Group, (Fig. 8.32, 8.34) typical of the Rundle Assemblage, is a progradational, shallowing-upward succession deposited during several transgressive-regressive hemicycles (Richards et al., in press). At the base of the group, middle to upper Tournaisian, protected-shelf to shelf-margin grainstone (Fig. 8.26A, 8.27) of the Pekisko Formation (40-152 m thick) passes southwestward into correlative slope packstone and wackestone of the Banff Formation and formation "F" (Fig. 8.27). Above the Pekisko, restricted-shelf to shelf-margin carbonates (Fig. 8.26A) of late Tournaisian to late Viséan age (Shunda, Turner Valley, and Mount Head formations) pass southwestward into coeval shelf-margin to upper-slope, skeletal grainstone of the Livingstone Formation, which ranges in thickness from 200 to 425 m (Fig. 8.27, 8.32). Farther

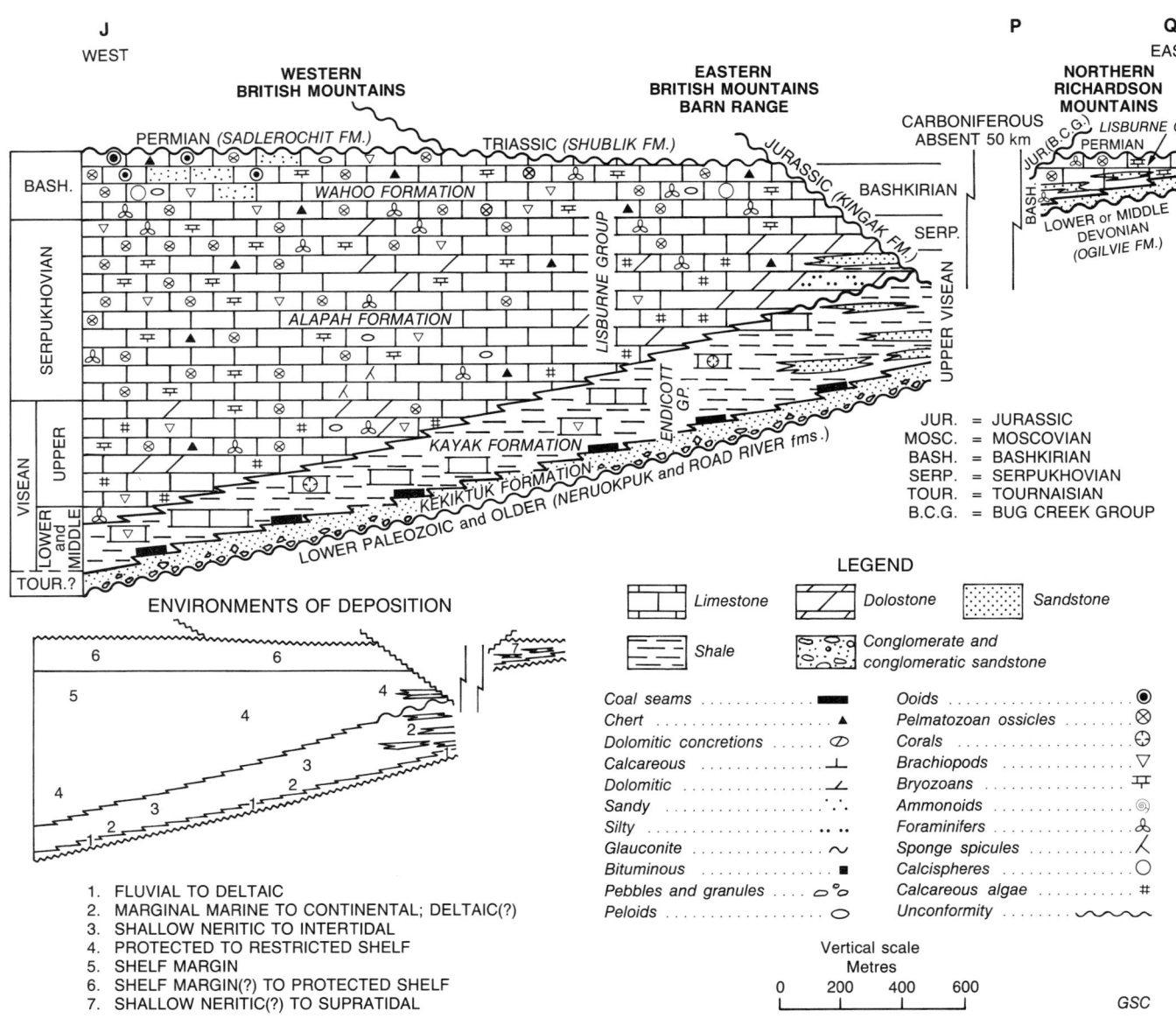

Figure 8.31. Schematic cross-section (J-P-Q) illustrating stratigraphic relationships and lithology, Viséan to Moscovian, northern Cordillera (northern Yukon and northwestern District of Mackenzie - see Fig. 8.22 for location).

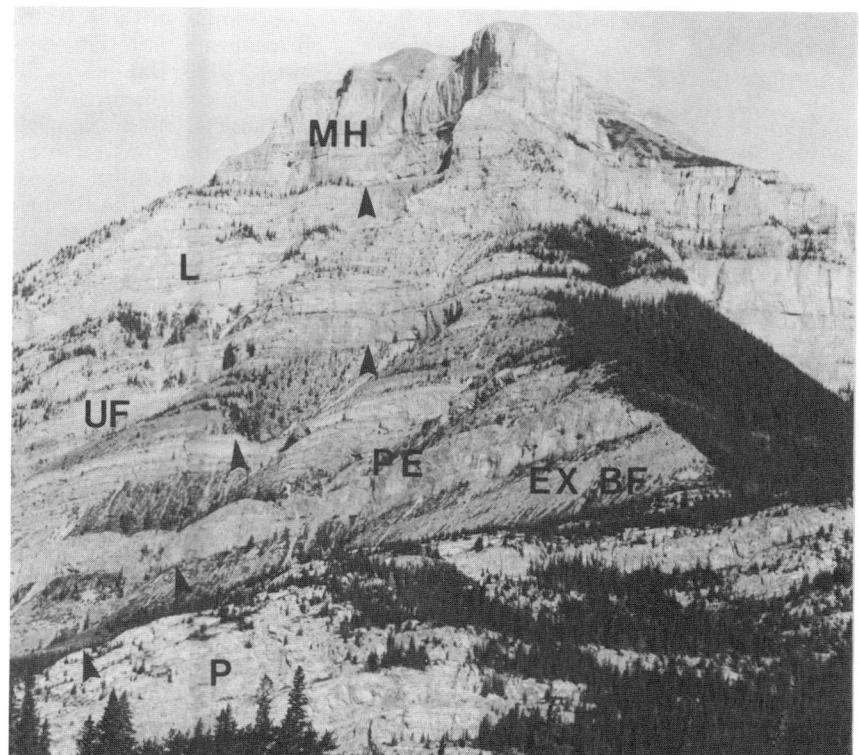

Figure 8.32. Upper Devonian and Lower Carboniferous Exshaw and Banff formations (EX,BF) and Rundle Group (Pekisko Fm. – PE, unnamed formation F – UF, Livingstone Fm. – L, Mount Head Fm. – MH), Grotto Mountain, near Banff, Alberta (51°03'N, 115°14'W); formation boundaries marked by arrows; Upper Devonian Palliser Formation (P) in foreground; view to northwest from Bow Valley. Photo by B.C. Richards. GSC 205235-C.

Figure 8.33. Permian Belcourt (B) and Mowitch (M) formations, near Kvass Creek, west central Alberta (53°37'18"N, 119°07'43"W); Belcourt Formation separated by regional disconformity from underlying, Lower Carboniferous strata (Mount Head (MH) and Turner Valley (TV) formations); Lower Triassic (T) shale and siltstone on skyline; view to southwest. GSC 205235-MM

Figure 8.34. Lower Carboniferous Banff Formation (B) and Rundle Group (Pekisko Fm.-P, Shunda Fm.-S, Turner Valley Fm.-TV, and Mount Head Fm.-MH), core of Moose Dome, cliffs north of Canyon Creek, 50 km southwest of Calgary, Alberta (lat. 50°54'30"N, long. 114°48'30"W); photo by N.C. Ollerenshaw. GSC 205235-GG

west, in the western Front Ranges of the Rockies, the lower Livingstone passes laterally into coeval slope packstone and wackestone of the upper Banff Formation. Lower slope and basinal facies, equivalent to the middle and upper Livingstone, are not preserved in the southern Rocky Mountains and Foothills.

The upper Viséan to Serpukhovian Etherington Formation (Douglas, 1958), at the top of the Rundle Group, comprises facies (58-250 m thick) that are transitional between the platform carbonates of the underlying Mount Head Formation and terrigenous clastic strata of the overlying Spray Lakes Group (Fig. 8.27; Scott, 1964). The lower Etherington, which conformably to disconformably overlies the Mount Head, is a succession of hemicycles, each comprising protected-shelf carbonates (skeletal and ooid grainstone) passing upward into restricted-shelf carbonates and siliciclastics (planar-laminated dolostone, variegated shale, algal limestone, calcrete, terra rosa breccia, and solution breccia; Fig. 8.26A; Richards et al., in press). Sandstone constitutes most of the upper Etherington (Todhunter Member), becoming more abundant towards the top of this unit. It represents a tongue of the Mattson Assemblage.

The carbonates of the Rundle Group were deposited during several transgressive-regressive episodes. Lime grainstone in the Pekisko and Turner Valley formations and the Baril and Loomis members of the Mount Head Formation (Fig. 8.27) accumulated during regional transgressions and the early phases of subsequent regressions. After deposition of each of these units, continued regression culminated with sedimentation of restricted-shelf facies preserved mainly in the Shunda Formation and in the Wileman, Salter, Marston, Carnarvon and Opal members of the Mount Head Formation. The overlying Etherington Formation, which records numerous minor transgressive-regressive events, developed during two main regressions and an intervening important transgression (Richards et al., in press).

The Bashkirian and lower Moscovian Spray Lakes Group (Ishbel Assemblage), which is approximately 250 m thick and comprises the Tyrwhitt, Storelk, Tobermory and Kananaskis formations (Scott, 1964; McGugan, 1964; =Misty Formation of Norris, 1965), disconformably overlies the Rundle Group (Fig. 8.27). Sandstone in the lower part of this interval exhibits small- to very large-scale crossbedding and is associated with subordinate sandy dolostone. The shallow-marine Tyrwhitt and overlying, eolian Storelk record a transgression and subsequent regression. Sandstone of the Tobermory disconformably overlies the Storelk and formed in shallow-neritic (above fair weather wavebase) environments during a transgression that began in the late Bashkirian and culminated with deposition of the lower Moscovian Kananaskis Formation. Sandy dolostone of the Kananaskis Formation, which gradationally overlies the Tobermory, represents re-establishment of widespread, shallow-marine carbonate deposition.

Central Rocky Mountains and Foothills (54°00'N-55°30'N)

The lower part of the carbonate-clastic belt (Rundle Assemblage) is well developed in the southeastern two-thirds of this region, where the Banff Formation and Rundle Group (Fig. 8.35) comprise facies similar to those in the Rocky Mountain Foothills and eastern Front Ranges of west-central Alberta (Mamet et al., 1986; Bamber et al., 1984; Fig. 8.28). The upper part of the carbonate-clastic belt, equivalent to the upper Mount Head Formation, Etherington Formation and the lower Spray Lakes Group, is absent beneath the sub-Permian disconformity. At the base of the succession basinal black shale of the Exshaw Formation abruptly to gradationally overlies Famennian carbonates of the Palliser Formation. The Famennian Exshaw, which is less than 5 m thick, is commonly present in the northern half of the area but generally absent in the south. The middle Tournaisian Banff Formation, which is a transgressive-regressive succession similar to that in the eastern Front Ranges of the southern Rocky Mountains, disconformably overlies the Exshaw and Palliser formations. Stratigraphic and facies relationships within the Rundle Group of this area are similar to those in southwestern Alberta, but thick intervals of shelf-margin grainstone like those in the Livingstone Formation (Fig. 8.27) are generally absent.

In the northwestern part of the region the carbonate-clastic belt consists mainly of slope, shallow-shelf and shelf-margin facies (Fig. 8.26B, 8.28) which grade westward into basinal shale of the Besa River Formation (Besa River Assemblage). The distribution of facies in this area was profoundly affected by development of Peace River Embayment (Bamber et al., 1984; Fig. 8.22), an eastward extension of deeper-water marine environments into the cratonic platform. The embayment formed during latest Devonian and Early Carboniferous time, with its axis near that of the Devonian Peace River Arch. The embayment is

Figure 8.35. Lower Carboniferous upper Banff Formation (B) and Rundle Group (R), southwest flank of Mount Becker, northeastern British Columbia (54°31'30"N, 120°39'W); disconformably overlain by Permian (P) carbonates (Belcourt Fm.) and siliciclastics (Mowitch Fm.); recessive Lower Triassic siltstone and shale (T) in upper right; view to southwest. Photo by B.C. Richards. GSC 205235-LL

defined by a marked thickening of Lower Carboniferous strata and by an increase in the proportion of deeper-water deposits in a broad area between latitudes 54°30'N and 57°30'N. Its southern margin had an easterly trend and the opposing side a northwesterly trend.

Shale and carbonate of the Exshaw and Banff formations in the central Rocky Mountains pass northwestward into Besa River shale in the Peace River Embayment. East-west facies relationships have not been determined for the Banff and Exshaw in the area. Within the overlying middle Tournaisian to middle Viséan Rundle Group, a complex carbonate assemblage of shelf to middle slope facies (Fig. 8.26A), up to 650 m thick, grades westward and northward into slope to basin facies of the Prophet Formation. The Prophet Formation, an assemblage, more than 200 m thick, of spiculite, spicule-rich carbonates, and subordinate lime grainstone, grades into basinal shale of the Besa River to the west and also to the north into Peace River Embayment (Richards et al., in press).

Northern Rocky Mountains and Foothills and southern Mackenzie Mountains

Basinal shale facies (Besa River Assemblage) and adjacent slope deposits (Fig. 8.26A,B) of the western part of the carbonate-clastic succession (Rundle Assemblage) are present along the entire outcrop belt in this area (Fig. 8.22, 8.29). Shallow-marine to supratidal carbonates and siliciclastics occur in the subsurface of the Interior Platform in northeastern British Columbia and are locally exposed north of 60°N latitude in southern Mackenzie Mountains.

Within the carbonate-clastic belt, Tournaisian to middle Viséan deposits show facies relationships similar to those in east-central British Columbia, but contain a greater proportion of terrigenous clastics. In addition, a thick sandstone-dominated succession (Mattson Assemblage) (Fig. 8.36) of late Viséan and Serpukhovian age is preserved beneath sub-Permian and sub-Mesozoic disconformities over most of the region. The oldest deposits of the carbonate-clastic succession occur in the lower and middle Tournaisian Yohin Formation of southern Mackenzie Mountains. This formation, which is coeval with the upper Banff Formation to the southeast, comprises a shallowing-upward succession, 80-160 m thick, of sandstone, siltstone, and mixed-skeletal limestone deposited in slope to shallow-marine environments. Basinal shale of the middle to upper Tournaisian Clausen Formation (30-200 m thick) conformably overlies the Yohin. Carbonate slope and shelf equivalents of the Clausen are confined almost entirely to the Prophet Formation in the Cordillera and to the Pekisko Formation and unnamed formation "F" on the Interior Platform. Above the Clausen is a thick succession of upper Tournaisian to middle Viséan, lower slope to shelf-margin facies in the Prophet and overlying Flett Formation (Richards, 1983, 1989; Richards et al., in press). The Prophet Formation (Fig. 8.37) consists of up to 770 m of spiculite and spicule-rich limestone with subordinate shale, and is continuous with similar lower slope facies in Members A and B of the Prophet Formation, which outcrops in northeastern British Columbia (Bamber et al., 1968; Bamber and Mamet, 1978). In southern Mackenzie Mountains, the Prophet thickens toward the west to occupy the entire interval equivalent to the upper Tournaisian to middle Viséan Flett Formation (Fig. 8.29). The latter, which overlies and grades westward into the Prophet Formation, comprises up to 360 m of shelf-margin to middle-slope (Fig. 8.26A), mixed-skeletal lime packstone and grainstone deposited during two main transgressive-regressive events. In its northeastern exposures the middle Flett Formation (Jackfish Gap Member; Richards, 1989) contains two thin intervals of cross-stratified sandstone

Figure 8.36. Lower Carboniferous Flett (F), Golata (G), and Mattson (M) formations, Jackfish Gap, southwestern District of Mackenzie (type section of Flett and Mattson formations - 61°05'54"N, 123°59'26"W); Golata Formation 8 m thick; base of Flett marked by arrow at left; view to north. Photo by B.C. Richards. GSC 205235-B

Figure 8.37. Lower Carboniferous Prophet (P), Golata (G), and Mattson (M) formations, Ram Creek, southwestern District of Mackenzie (61°09'50"N, 124°24'47"W); Several large-scale truncation surfaces (arrows) in slope deposits of Prophet Formation; view to southwest. Photo by B.C. Richards. GSC 205235-L

and siltstone. Grainstone occurs mainly in the upper part of the Flett (Meilleur Member, Richards, 1989) and is most abundant in eastern sections (Fig. 8.29). The upper unit appears to be continuous with similar carbonates assigned to Member C of the Prophet Formation in northeastern British Columbia (Bamber and Mamet, 1978). Large scale, submarine paleochannels are common in the slope deposits of the Yohin, Flett and Prophet formations (Richards, 1978, 1983, 1989). In the adjacent subsurface of the Interior Platform, correlatives of the Tournaisian to middle Viséan units of the outcrop belt are in the upper Banff, Pekisko, unnamed formation "F", Shunda, Debolt, Flett and Prophet formations.

The carbonate succession, both in outcrop and in the subsurface, is overlain by a thick, regressive assemblage of upper Viséan and Serpukhovian siliciclastics and subordinate carbonates (Mattson Assemblage) (Fig. 8.29). This phase of sedimentation was initiated by deposition of basinal to prodelta shale and subordinate sandstone and carbonate of the Golata Formation (8 to more than 200 m thick). It was followed by deposition of 1000-1400 m of northerly derived slope to delta-plain and shallow marine shelf sandstone of the Mattson Formation (Fig. 8.36) north of latitude 58°30'N. South of latitude 58°30'N, deltaic and marine-shelf deposits of the Kiskatinaw Formation (over 290 m thick) overlie the Golata and are overlain by slope to shallow marine carbonates and subordinate sandstone of the Taylor Flat Formation (over 280 m thick) (Richards, 1983, 1989; Bamber and Mamet, 1978). In this southern area the Golata and two overlying formations constitute the Stoddart Group.

From Peace River to southern Mackenzie Mountains, the carbonate-clastic succession passes westward into the basinal shale belt (Besa River Assemblage). The boundary between the two belts becomes younger toward the west, with sandstone of the Mattson and Stoddart formations extending farthest into Prophet Trough. Thus, the Carboniferous of the area is dominantly a progradational succession resulting from a major transgression and subsequent period of deposition, mainly under shallowing-upward conditions.

Evidence for transgressions in southern Mackenzie Mountains occurs at four levels in the succession. Basin deposits in the lower Clausen Formation (late Middle Tournaisian), shale in the uppermost Tournaisian part of the Prophet Formation, and slope deposits in the basal Meilleur Member of the Flett Formation (middle Viséan) record the first three transgressions. These three events correspond to transgresssive episodes represented in the southern Rocky Mountains by the lower Pekisko and Turner Valley formations and the Loomis Member of the Mount Head Formation (Fig. 8.27; Richards, 1989). Upper Viséan and Serpukhovian deltaic sandstone of the lower to middle Mattson Formation records the culmination of a subsequent regional regression. In Peace River Embayment, the regression is recorded by deltaic facies of the Kiskatinaw Formation. Farther south in the southern Rockies, it is indicated by the upper Mount Head and basal Etherington formations. The fourth transgression is indicated by uppermost Viséan to Serpukhovian carbonates and marine sandstone of the upper Mattson in southern Mackenzie Mountains and by the correlative Taylor Flat Formation to the south.

Farther north, in west-central Mackenzie Mountains and Selwyn Mountains (Fig. 8.22), the Carboniferous succession begins with the Tournaisian upper Earn Group (Earn Assemblage) (Gordey et al., 1982a; see previous section), consisting mainly of shale, siltstone, sandstone and conglomerate, locally more than 600 m thick. The mode of occurrence of the coarse grained siliciclastics is similar to that of coeval conglomerate and sandstone in the Tuttle Formation (Mattson Assemblage) on the flanks of the Richardson Mountains to the north. The shale and siltstone is similar in aspect and mode of occurrence to that in the basinal strata of the Besa River Formation to the south and the Ford Lake Formation in north-central Yukon (Fig. 8.30; Pugh, 1983). The upper Earn Group is abruptly, and possibly unconformably, overlain by up to 200 m of upper Viséan sandstone with shale interbeds and laminae (quartz arenite unit of Gordey et al., 1982a). This unit (Tsichu Formation, Mattson Assemblage) appears to pass westward into a dark shale facies and also is partly correlative with the Mattson Formation (Fig. 8.29). Above the sandstone unit, the remainder of the Carboniferous comprises more than 300 m of unnamed shale and interbedded limestone, grading upward into sandy, skeletal limestone with subordinate sandstone (CPo). This unit, which is of late Viséan to early Bashkirian age, is overlain unconformably by Permian strata.

Cassiar Mountains - Northern Rocky Mountain Trench

In the Cassiar Terrane near the northern end of the Northern Rocky Mountain Trench (Fig. 8.22) a Tournaisian to Viséan sequence of mainly carbonate strata, unconformably overlying Silurian and older strata, has been assigned to the Rundle Assemblage. The sequence consists of a lower member, up to 100 m thick, composed mainly of fine- to medium-grained clean sandstone interbedded with laminated dark grey shale, overlain by a thin- to thick-bedded carbonate member more than 300 m thick. The carbonate member is mainly spiculitic, argillaceous, and thin-bedded lime wackestone and packstone near the base and thick-bedded, echinoderm-rich, cherty lime grainstone in the upper part. Faunas and the vertical succession of facies are similar to those in the Prophet Formation in the Northern Rocky Mountains. Similar faunas have also been collected in the Pelly Mountains where Lower Carboniferous carbonate strata seem to be less well developed.

East of the Northern Rocky Mountain Trench (Fig. 8.22) a local occurrence of Viséan carbonate about 50 m thick is underlain by more than 50 m of clean sandstone and thin-bedded calcareous siltstone which in turn overlie, possibly conformably, pebble conglomerate and sandstone correlative with the Earn Assemblage. There, also, the faunas and succession of facies are like those in the Prophet Formation.

Northern Cordillera - Ogilvie Mountains to Beaufort Sea

In northern Yukon Territory, north of latitude 64°30'N, the Carboniferous carbonate-clastic assemblage (Mattson and Lisburne assemblages) comprises two separate, but closely related successions, disconformably overlain by Permian and Mesozoic strata (Fig. 8.30, 8.31; Bamber and Waterhouse, 1971). The most completely preserved sequence occurs mainly south of latitude 67°00'N. There, middle Viséan to Gzhelian, basinal to restricted shelf deposits, up to 1800 m thick, occur in outcrop and in the subsurface of Ogilvie Mountains and southern Eagle Plain. The second succession, comprising Viséan to Moscovian supratidal to open shelf deposits up to 1300 m thick, occurs north of latitude 68°00'N in a southeast-trending, discontinuous belt extending from the Alaska-Yukon border into northwestern District of Mackenzie (Fig. 8.22). The northern and southern successions, which originally formed part of a continuous depositional complex, are now separated by the northeast-trending ancestral Aklavik Arch (Bamber and Waterhouse, 1971). Along the arch, Carboniferous strata are absent and Permian rocks rest unconformably on Devonian and older strata.

In the southern area (Eagle Plain, Ogilvie Mountains, west flank of Richardson Mountains) the carbonate-clastic assemblage overlies and passes laterally into basinal shale of the Tournaisian? to Bashkirian Ford Lake Formation (Imperial Assemblage) and lower Blackie Formation (Fig. 8.30; Pugh, 1983). In the northeastern part of the area the lower part of the succession consists of Upper Devonian and Tournaisian conglomerate and sandstone (Tuttle Formation; Mattson Assemblage) overlain by interbedded shale and sandstone of the eastern Ford Lake Formation. The conglomerate and sandstone occur as wedges and lenses that thin southwestward into the basin (Graham, 1973) and may have been deposited in delta-slope environments. Similar sandstone and conglomerate, of Viséan and Serpukhovian age, occur in the eastern part of the overlying carbonate assemblage. The oldest carbonate accumulations occur within the upper Viséan and lower Serpukhovian slope deposits of the Hart River Formation, which consist mainly of thinly laminated spiculite and spicule-rich lime packstone. The overlying, lower Serpukhovian to Bashkirian, lower Blackie Formation comprises a second basinal shale development succeeded by a shallowing-upward succession of slope carbonates. Thick (330-550 m) upper-slope to open-shelf limestone (Fig. 8.26A,B) of the overlying Bashkirian to (?)Kasimovian Ettrain Formation (Fig. 8.30) form a shelf complex that appears to prograde westward and southward over slope carbonates and shale of the upper Blackie Formation. In eastern outcrops of Eagle Plain, near the west flank of Richardson Mountains, unnamed correlatives of the upper Ettrain contain abundant silty dolostone and siltstone. This facies suggests deposition in restricted shelf environments during regression that began in Moscovian time and culminated with regional erosion in the latest Carboniferous or earliest Permian (Bamber and Waterhouse, 1971).

The northern succession (Fig. 8.31), in British and Barn mountains of the Arctic Alaska Terrane, and in northern Richardson mountains, is a diachronous, onlapping complex of Viséan to Moscovian, continental and shallow-marine facies. No well-developed basinal shale or slope carbonates are preserved. At the base of the succession in the British and Barn mountains, conglomerate and sandstone of the Kekiktuk Formation (up to 61 m thick; Mattson Assemblage) overlies Proterozoic and lower Paleozoic rocks with angular unconformity (Norris, 1983; Bamber and Waterhouse, 1971). Sandstone and local coal seams occur at the base of the overlying Kayak Formation (Mattson Assemblage), which is up to 360 m thick and consists mainly of shallow-marine shale and subordinate limestone. Shelf carbonates up to 1200 m thick occur mainly in the overlying middle Viséan to Bashkirian Lisburne Group (Fig. 8.38). In eastern Barn Mountains the Lisburne consists of restricted to protected-shelf limestone and dolomite. To the west lime grainstone becomes abundant in shelf-margin facies of the upper Lisburne (Fig. 8.26A). Protected- to restricted-shelf facies, similar to those in the British and Barn mountains, are present in the Lisburne to the east in northwestern District of Mackenzie (northern Richardson Mountains), where up to 214 m of Bashkirian and Moscovian carbonates and sandstone rest disconformably on Middle Devonian rocks (Nassichuk and Bamber, 1978). Facies relationships within the northern Carboniferous succession suggest deposition during a regional transgression that began during early to middle Viséan time and continued toward the north and east until at least Moscovian time. No post-Moscovian Carboniferous strata are preserved beneath the sub-Permian disconformity.

Figure 8.38. Carboniferous succession, British Mountains, northern Yukon (69°23'N, 140°09'W); Endicott Group (E) (Kekiktuk and Kayak formations) conformably overlain by light grey carbonates of Lisburne Group (L); Kekiktuk Formation rests with angular unconformity on vertical to steeply dipping slate and quartzite of Precambrian Neruokpuk Formation (N); view to northwest across lower Malcolm River from top of Kekiktuk Formation (left foreground). Photo by D.K. Norris. GSC 205235-HH

Permian

In contrast to the Carboniferous, the Permian of the Foreland Belt (Ishbel, Jungle Creek, and Sadlerochit assemblages) consists mainly of shallow-marine to basinal siliciclastics and silty to sandy carbonates, lacks extensive carbonate buildups, and has neither continental deposits nor an extensive basinal shale facies (Fig. 8.39, 8.40). A regional unconformity everywhere marks the base of the Permian (Fig. 8.33), which rests on strata ranging from Lower Silurian to uppermost Carboniferous (McGugan et al., 1968; Bamber and Macqueen, 1979). In the Rocky and southern Mackenzie mountains, a regional, intra-Permian disconformity separates thin, upper Artinskian to Wordian chert and siliciclastics from underlying, Asselian to middle Artinskian siliciclastics and carbonates (Fig. 8.33; Henderson and McGugan, 1986). This disconformity also may be present in the Permian of northern Yukon and northwestern District of Mackenzie, within a thick Asselian to Wordian succession on the flanks of the ancestral Aklavik Arch (Fig. 8.41; Bamber and Waterhouse, 1971; Henderson et al., in press). No Permian deposits younger than Wordian are preserved beneath sub-Triassic and younger disconformities in the Foreland Belt and Interior Platform. Permian strata were deposited mainly in Peace River Embayment and in Ishbel Trough (Fig. 8.39; Henderson et al., in press), a down-warped and down-faulted region on the western side of the miogeocline.

Southern Rocky Mountains and Foothills (Ishbel Assemblage)

The Permian succession in the southern Rocky Mountains is generally thin, dominated by slope to basinal facies and confined to the Front Ranges from southeastern British Columbia to the Athabasca River area (Fig. 8.22, 8.42).

Biostratigraphic and facies relationships are best understood in southeastern British Columbia where the succession is thickest (Henderson and McGugan, 1983, 1986). There, the succession below the intra-Permian disconformity begins with the Johnston Canyon Formation (Fig. 8.43), which comprises a thin basal conglomerate of chert and phosphatic pebbles, overlain by approximately 60 m of phosphatic, argillaceous siltstone and nodular to bedded spicular chert of slope to basinal origin. Toward the east, the formation onlaps Bashkirian to lower Moscovian strata. The Sakmarian Telford Formation (Fig. 8.42, 8.43) progrades westward over the Johnston Canyon Formation and coarsens upward from slope to possible shallow-shelf deposits (Fig. 8.26A). The Telford consists of approximately 240 m of silty to sandy carbonates with subordinate interbedded spicular chert and pelmatozoan-brachiopod wackestone in the lower part. The upper part contains brachiopod wackestone and medium- to coarse-grained, calcareous sandstone with *Zoophycos*. The Telford is overlain by lower Artinskian (lower Leonardian) basinal to slope facies of the lower Ross Creek Formation, which consists of approximately 50 m of phosphatic and calcareous siltstone and minor, black spicular chert. The sediments of the upper Ross Creek, which prograded over the latter unit, include a succession, approximately 100 m thick, of lower to middle Artinskian silty and sandy carbonates, phosphatic siltstone and sandstone with abundant thin-shelled brachiopods, deposited on the upper slope and possibly the outer shallow shelf of a ramp. A small biostrome of solitary corals, productid brachiopods, echinoderms, bryozoans and ooids near the base of the upper Ross Creek suggests deposition in an upper slope to shallow-shelf setting. Stromatolites at the top indicate deposition of the youngest beds in the intertidal to supratidal zones on a ramp (Fig. 8.26B).

Sections containing a more westerly slope to basinal facies (Fig. 8.42) are thinner (approximately 45 m) and are characterized by black, spicular chert, platy, phosphatic siltstone and minor pelmatozoan wackestone. This facies, which ranges upward into the middle Artinskian, is referred in part to the Johnston Canyon Formation (MacRae and McGugan, 1977).

The most easterly facies represented in the area begins with the Sakmarian Johnston Canyon Formation (30 m thick), which consists of phosphatic siltstone and minor fusulinid-bearing wackestone lenses, suggesting deposition on the slope to outer shallow shelf of a ramp (Fig. 8.26B). Overlying the Johnston Canyon Formation in the area south of Jasper (Fig. 8.42 - map inset) is a thin (15-30 m) succession of sandy carbonates (McGugan and Rapson, 1963) that is similar in facies to the Telford and Ross Creek formations. The relationships between these thin

UPPER DEVONIAN TO MIDDLE JURASSIC ASSEMBLAGES

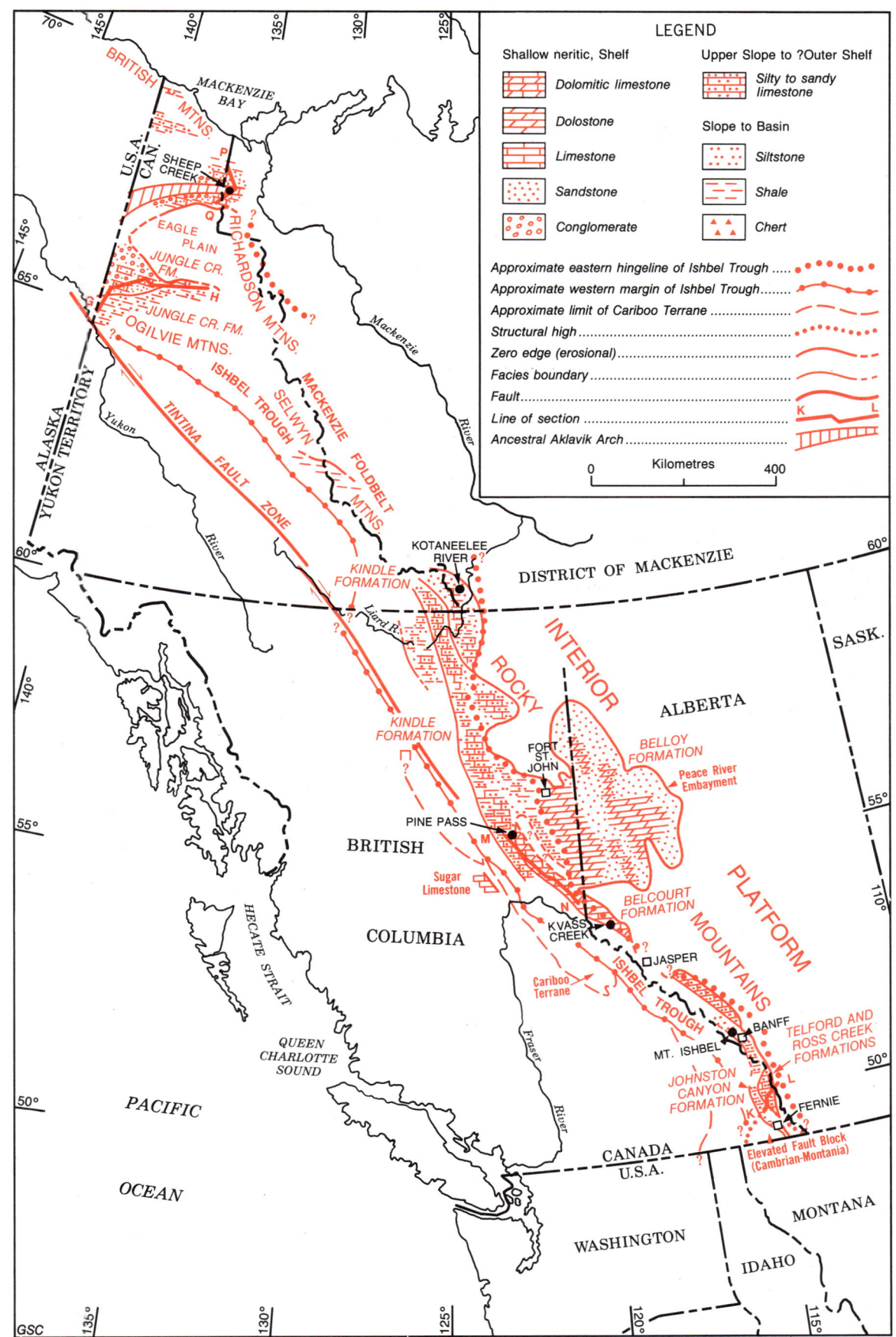

Figure 8.39. Distribution of Asselian to Lower Artinskian facies.

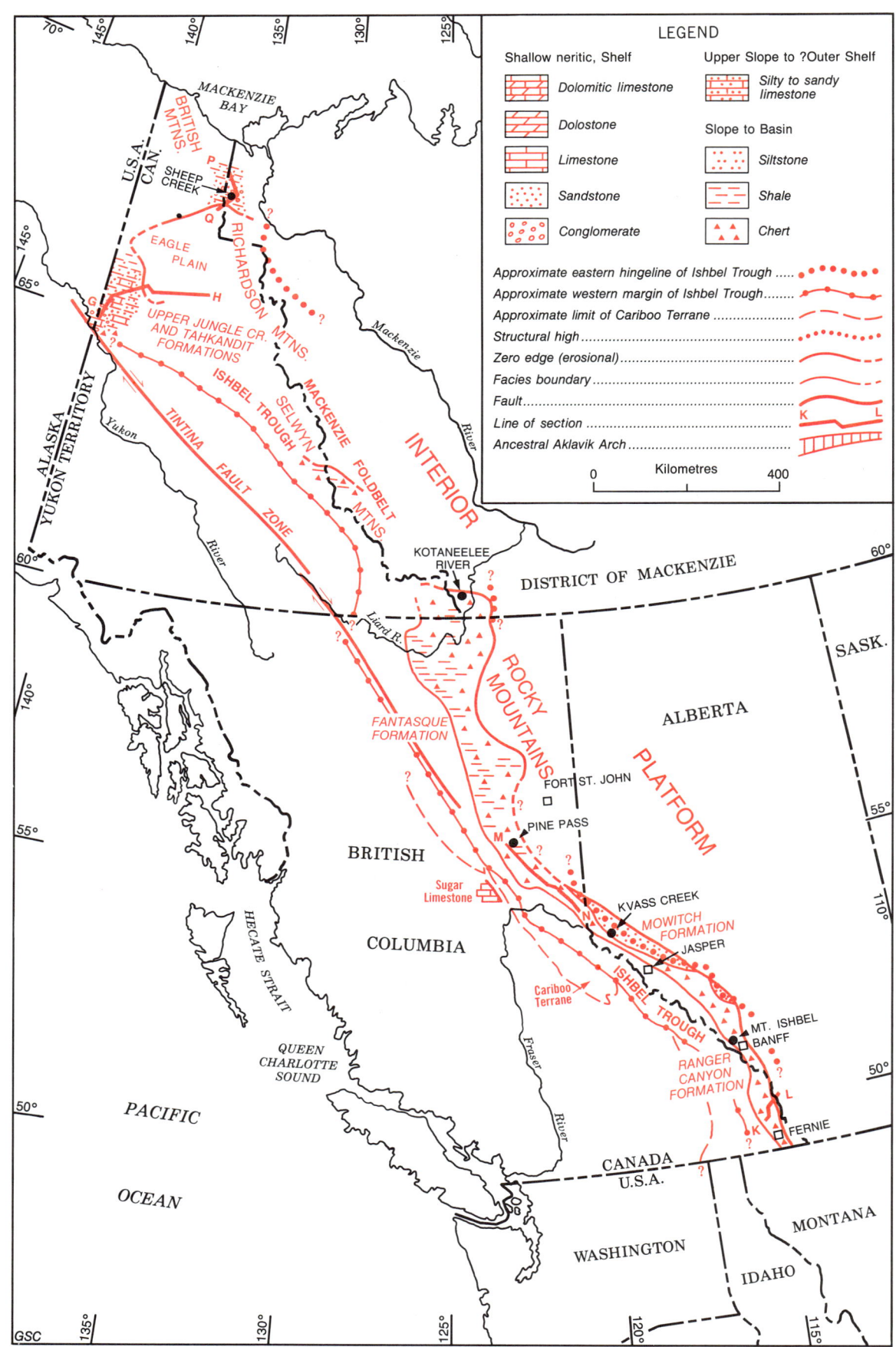

Figure 8.40. Distribution of Upper Artinskian to Wordian facies.

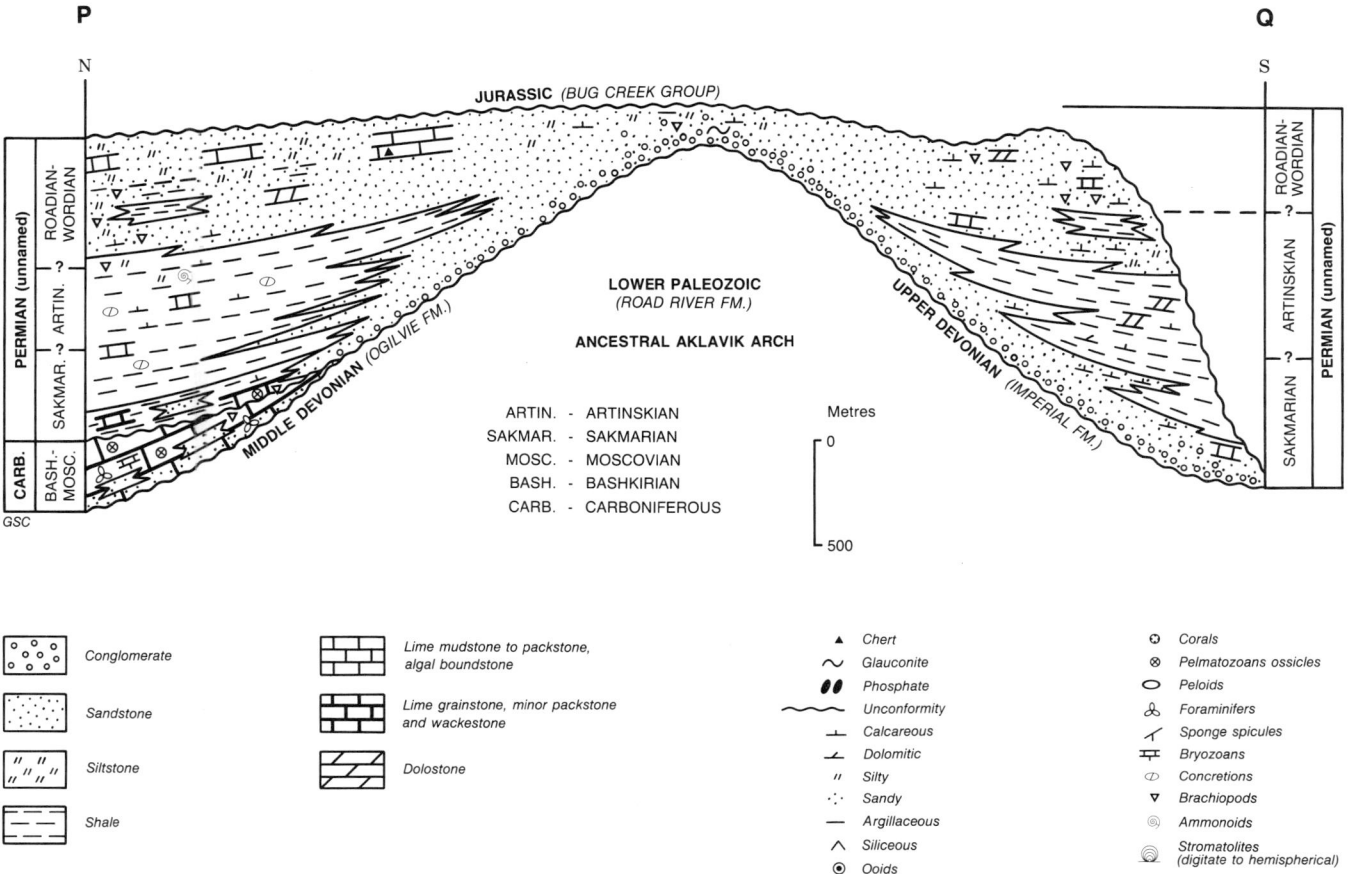

Figure 8.41. Schematic cross-section (PQ) illustrating Bashkirian, Moscovian, and Permian, stratigraphic relationships and lithology, northern Cordillera (northern Richardson Mountains - see Fig. 8.22 for location).

eastern facies and the thicker sections of southeastern British Columbia are poorly understood (Fig. 8.42) but conodonts indicate correlation with at least the Ross Creek Formation. Eastward thinning of the Permian succession is related to a combination of onlap and truncation beneath the intra-Permian disconformity.

Strata above the intra-Permian disconformity are remarkably consistent in lithology and thickness throughout the area from southeastern British Columbia to the Jasper area (Fig. 8.22; McGugan et al., 1964). A relatively thin (1 to 45 m, average) but widespread sequence of upper Artinskian to Wordian, blue-grey chert, silicified sandstone and phosphatic siltstone is referred to the Ranger Canyon Formation and interpreted as slope to basinal facies (Henderson et al., in press). A basal unit of very thin but laterally persistent, phosphatic, chert-pebble conglomerate containing inarticulate brachiopods, sponge spicules, phosphatized bone fragments and *Helicoprion* sp. onlaps the intra-Permian disconformity.

Three transgressions are represented by the deeper water sediments of the Johnston Canyon Formation (Asselian-early Sakmarian), the lower Ross Creek Formation (early Artinskian) and the Ranger Canyon Formation (?late Artinskian-Wordian).

Central and Northern Rocky Mountains and southern Mackenzie Mountains (Ishbel Assemblage)

Permian rocks of this area (Fig. 8.44) rest disconformably on Upper Devonian to Serpukhovian strata, and contain abundant thin, slope to basinal facies similar to those in the southern Rocky Mountains (Fig. 8.42). In addition, extensive shallow-neritic to supratidal facies are developed in the eastern outcrop belt and adjacent subsurface (Fig. 8.39).

In east-central British Columbia, south of Pine Pass (Fig. 8.22), the succession below the intra-Permian disconformity consists of diachronous carbonates (45-144 m thick) assigned to the Belcourt Formation (Fig. 8.35; McGugan and Rapson-McGugan, 1976; Bamber and Macqueen, 1979). Temporal and facies relationships within the formation indicate regional, eastward transgression during Early Permian time (Fig. 8.44). In the northwestern part of the area, the basal Belcourt conglomerate is overlain by Asselian to ?lower Artinskian slope deposits of silty, mixed-skeletal wackestone and packstone. The lower Belcourt becomes progressively younger along the outcrop belt toward the southeast where the basal conglomerate is

CHAPTER 8

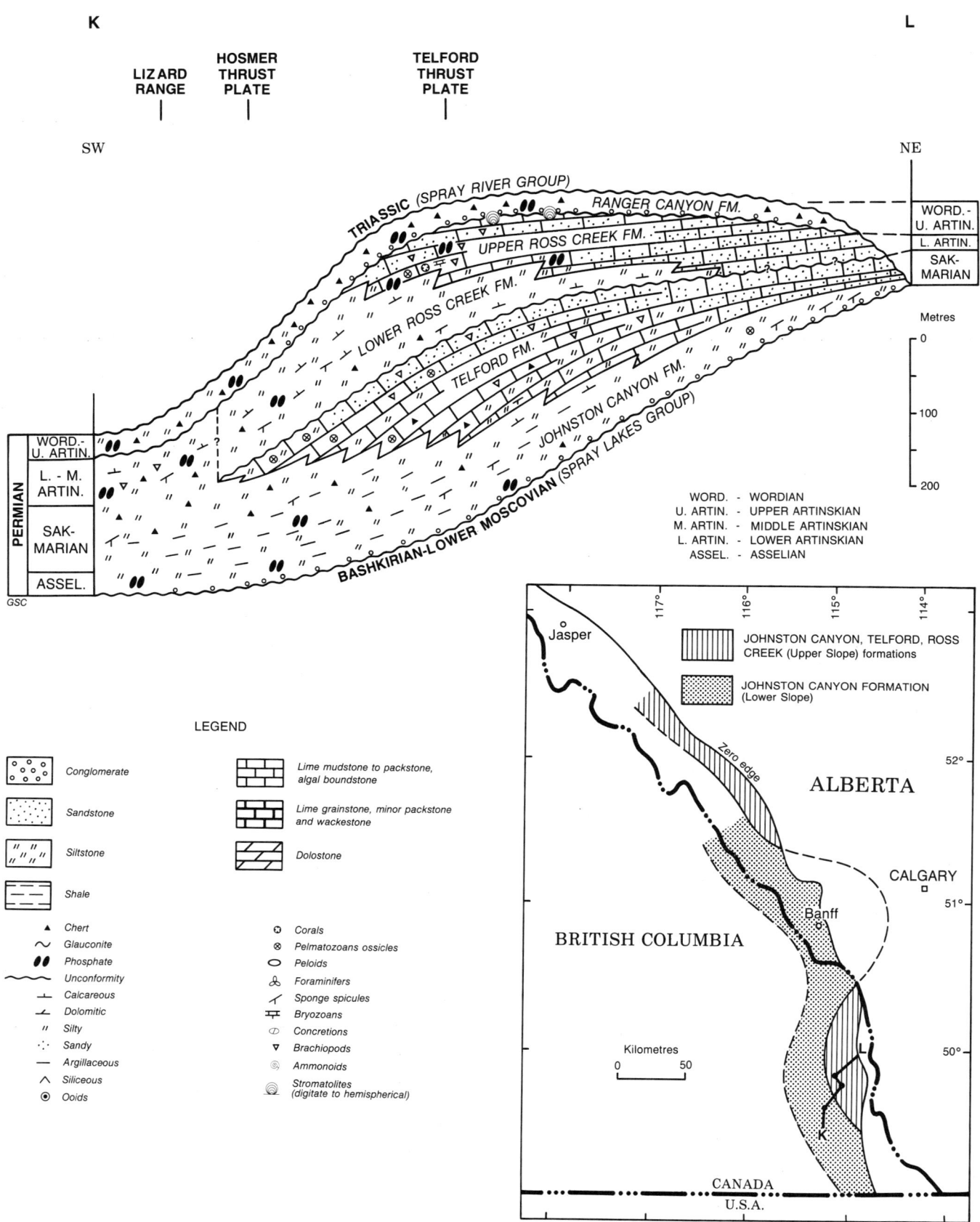

Figure 8.42. Schematic cross-section (KL) illustrating Permian stratigraphic relationships and lithology, southern Rocky Mountains. Map inset shows facies pattern for interval below intra-Permian disconformity.

Figure 8.43. Basal Permian transgressive sequence, Telford thrust plate, 40 km north of Fernie, British Columbia (49°45'N, 115°02'W); recessive, phosphatic siltstone of Johnson Canyon Formation (JC) conformably overlain by resistant, sandy carbonate of Telford Formation (T); Johnson Canyon disconformably overlies Carboniferous Spray Lakes Group (SL); view to southwest. GSC 205235-NN

directly overlain by Upper Sakmarian (Sterlitamakian), fusulinid- and coral-bearing, shallow-neritic to supratidal lime packstone and dolostone. This shallow-marine facies extends farther southeastward into west-central Alberta (Fig. 8.39; McGugan et al., 1964), where the Belcourt (Fig. 8.33) comprises planar-bedded, microcrystalline dolostone, which passes westward into cherty, dolomitized lime packstone and grainstone. Lithofacies trends within the Belcourt record deposition on a carbonate ramp (Fig. 8.26B).

In the Northern Rocky Mountains, north of Pine Pass, and in the southern Mackenzie Mountains, strata correlative with the Belcourt Formation are assigned to the Kindle Formation (Bamber et al., 1968). Slope to basinal, silty carbonates, siltstone, and shale up to 205 m thick characterize the western part of the outcrop belt. In the east, the Kindle consists of up to 133 m of rhythmically bedded siltstone, shale, and limestone, which grades farther eastward into shallow-neritic (above wave base) sandstone, shale, and carbonates in the subsurface of the Interior Platform.

Lower Permian shale and sandstone, correlative with the Kindle Formation, have been reported from the west-central Mackenzie and Selwyn mountains (Fig. 8.22; Gordey et al., 1982a,b). The contact between this northern succession and underlying Carboniferous strata may be disconformable.

The interval above the intra-Permian disconformity consists mainly of slope to basinal deposits comprising upper Artinskian to Wordian, spicular, phosphatic, locally glauconitic chert, shale, and siltstone of the Fantasque Formation (40-150 m thick). The formation extends from the Pine Pass area (Fig. 8.22, 8.40) into the southern Mackenzie Mountains and closely resembles the Ranger Canyon Formation of the southern Rocky Mountains. A thin lag(?) deposit of phosphate and chert nodules and pebbles, which accumulated during a regional eastward transgression, marks the disconformity at the base of the Fantasque. Above this deposit, the dark grey, bedded chert of eastern sections (Fig. 8.44) appears to grade westward through an intermediate facies of interbedded chert, siliceous mudstone, shale, and siltstone, into basinal shale and siliceous mudstone (Bamber et al., 1968). A unit of chert and shale resembling the Fantasque Formation caps the Permian succession to the north, in the west-central Mackenzie Mountains and Selwyn Mountains (Gordey et al., 1982a,b). In east-central British Columbia, eastern equivalents of the Fantasque, assigned to the Ranger Canyon Formation (McGugan and Rapson-McGugan, 1976; Bamber and Macqueen, 1979), comprise light grey, silicified, skeletal carbonates interbedded with fine-grained quartz sandstone. This facies grades southeastward into cherty, phosphatic, chert arenite and conglomerate of the shallow-marine Mowitch Formation (Fig. 8.33, 8.44), which occupies the entire interval (up to 25 m thick) above the Belcourt Formation in easternmost outcrops of the area and in west-central Alberta (McGugan et al., 1964).

Northern Cordillera - Ogilvie Mountains to Beaufort Sea (Jungle Creek and Sadlerochit assemblages)

Permian rocks of the northern Cordillera are preserved mainly in Ogilvie Mountains, southern Eagle Plain, northern Richardson Mountains and in British Mountains (Fig. 8.41, 8.45). Asselian to Wordian marine strata of the Jungle Creek Assemblage occur along the flanks of the ancestral Aklavik Arch (Fig. 8.22, 8.39) in facies belts trending northeast, approximately at right angles to those of the underlying Carboniferous (Bamber and Waterhouse, 1971). The arch was uplifted during latest Carboniferous and earliest Permian time, when the entire eastern Cordilleran area was affected by regional regression, local block faulting, and subaerial erosion. Permian marine transgression toward the axis of the arch continued until Wordian time, when the arch was entirely covered by siliciclastics and carbonates. This depositional pattern is best illustrated by facies relationships in the northern Richardson Mountains where similar, unnamed successions, ranging in thickness from 1050 to 1830 m occur on the north and south flanks of the arch (Fig. 8.41). These consist of basal conglomerate and sandstone, overlain by Sakmarian to Wordian sandstone, shale, siltstone, and subordinate carbonates. The basal unit which unconformably overlies Devonian strata (Fig. 8.16), becomes progressively younger toward the central part of the arch where it is overlain by ?upper Artinskian to Wordian siliciclastics and rests with angular unconformity on lower Paleozoic rocks as old as Early Silurian. At all levels coarse-grained, shallow-marine siliciclastics on both flanks of the arch pass outward into finer grained, laterally equivalent, deeper-water deposits.

To the southwest, in northern Eagle Plain, Permian strata are absent beneath a regional sub-Cretaceous unconformity (Fig. 8.39; Pugh, 1983). In the Ogilvie Mountains near the Yukon-Alaska boundary (Fig. 8.45), a

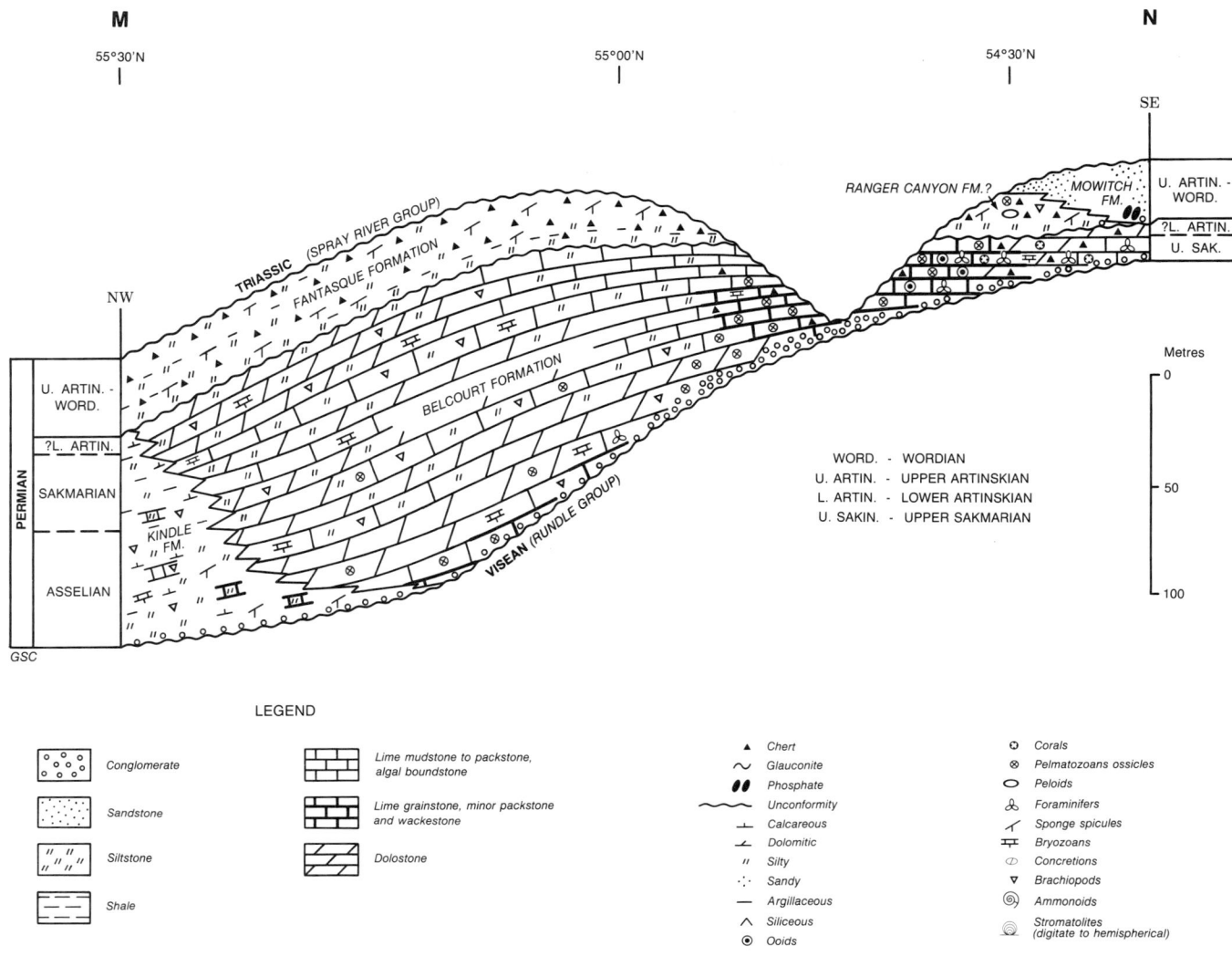

Figure 8.44. Schematic cross-section (MN) illustrating stratigraphic relationships and lithology, Permian, central Rocky Mountains and Foothills (east-central British Columbia - see Fig. 8.22 for location).

varied assemblage, more than 1100 m thick, of Asselian to Wordian siliciclastics, carbonates, and chert, assigned to the Jungle Creek and Tahkandit formations is present on the south flank of ancestral Aklavik Arch. The nature and magnitude of the sub-Permian disconformity in this area has not been clearly demonstrated (Waterhouse and Waddington, 1982). Deposition may have been locally continuous across the Carboniferous-Permian boundary. The basal part of the Tahkandit Formation contains abundant upper Artinskian chert pebbles and beds of chert pebble conglomerate. The contact with the underlying Jungle Creek Formation appears to be a disconformity, but its relationship to well-developed disconformities in the southern Cordillera is unknown (Fig. 8.42, 8.44, 8.45). Near the axis of the arch, located to the northwest in adjacent, east-central Alaska, the Permian consists of Roadian to Wordian conglomerate, sandstone, and lime grainstone of the Tahkandit and Step formations. These rest unconformably on Precambrian to Lower Carboniferous strata (Brabb and Churkin, 1969). Thus, the base of the Permian in this area becomes younger toward the axis of the arch, as it does in northern Richardson Mountains.

Toward the south and southeast, away from the arch, the shallow-neritic assemblage of northern Ogilvie Mountains grades into spicular slope carbonates, chert, and shale (?Blackie Formation; Pugh, 1983). To the east, in southern Eagle Plain, only the Jungle Creek Formation is preserved beneath the sub-Cretaceous unconformity. It consists of fine grained sandstone, siltstone, and shale, and appears to pass southward, through eastern Ogilvie Mountains, into an undifferentiated succession of slope to basinal shale, siltstone, and silty carbonates (Graham, 1973). The Permian succession is poorly known near the axis and on the north flank of the arch in the central and western Yukon. In the British Mountains (Arctic Alaska Terrane), well to the north of the axial area, more than 200 m of Lower Permian shale, sandstone, and minor carbonates of the Sadlerochit Formation (Echooka Member; Sadlerochit Assemblage) are preserved locally as erosional remnants above Bashkirian carbonate of the Lisburne

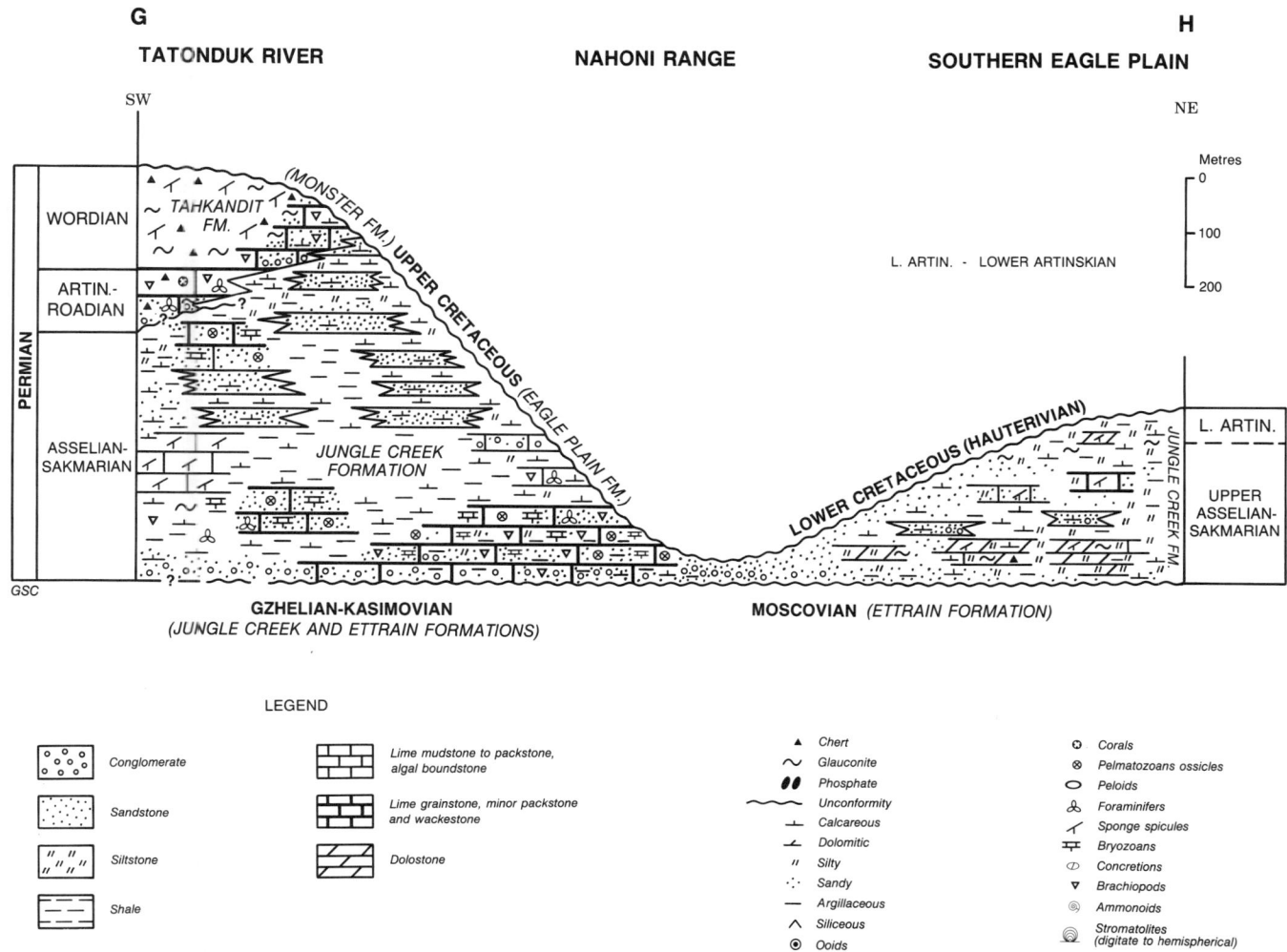

Figure 8.45. Schematic cross-section (GH) illustrating stratigraphic relationships and lithology, Permian, northern Cordillera (Ogilvie Mountains, Eagle Plain - see Fig. 8.22 for location).

Group (Bamber and Waterhouse, 1971, p. 82) and beneath Triassic siliciclastics and carbonates. No information is available on Permian facies relationships in this northern area.

Pelly Mountains

In the Cassiar Terrane southwest of the Tintina Trench local remnants of Permian (Leonardian), thin-bedded to laminated bioturbated shale and siltstone up to 100 m thick lie disconformably above Lower Carboniferous rocks and disconformably below Triassic strata (D.J. Tempelman-Kluit, pers. comm., 1986).

Triassic strata of the Foreland Belt
D.W. Gibson

The Triassic system of the Cordilleran miogeocline comprises a westward thickening marine sequence of easterly derived siliciclastics and carbonates extending from the International Boundary to near the Liard River in northeastern British Columbia and Yukon (Spray River Assemblage) (Fig. 8.46). Widely separated occurrences of Triassic strata also are reported from northern (Sadlerochit Assemblage), south-central and east-central Yukon (Mountjoy, 1967; Tempelman-Kluit, pers. comm.; Gordey, 1981a,b; Gordey et al., 1981; Fig. 8.46).

The succession was deposited from Early Triassic (Griesbachian) to Late Triassic (Norian) time and, in the Foreland Belt consists of a relatively thick (1200+ m) westward thickening sequence of siliciclastics, carbonate and lesser amounts of evaporite. Lower Triassic isopachs outline the Fort St. John Basin, flanked to the north by the Nig Creek Platform and to the south by the Wapiti Platform (Barss et al., 1964); for the remainder of the system, isopachous lines trend uniformly northwest (Fig. 8.47). Except for the presence of evaporites, the Triassic of the Yukon and western District of Mackenzie displays lithofacies like those of northeastern British Columbia.

CHAPTER 8

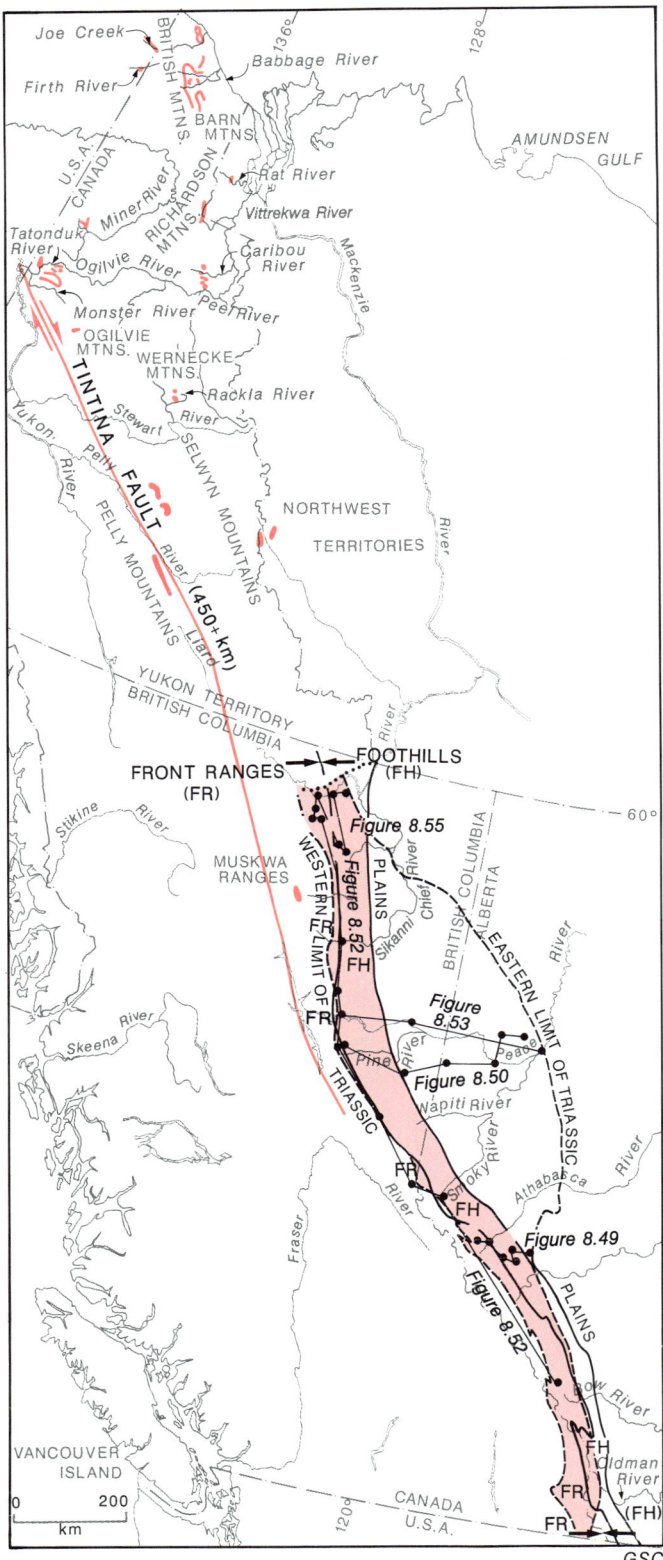

Figure 8.46. Distribution of Triassic rocks, Cordilleran Miogeocline (British Columbia and Alberta modified from Barss et al., 1964; Yukon Territory and District of Mackenzie modified from Mountjoy, 1967). Outcrop areas in Muskwa Ranges and farther north shown in red.

In the main outcrop belt of British Columbia and Alberta two nomenclatures are used. In the Rocky Mountains and Foothills between the Pine Pass-Sukunka River area of northeastern British Columbia and the International Boundary (49°N), Triassic rocks are referred to the Spray River Group (Sulphur Mountain and Whitehorse formations; Fig. 8.48, 8.49). Those in the remaining Foothills belt to the north are included in the Grayling and Toad formations and the Schooler Creek Group (Liard, Charlie Lake-Ludington, Baldonnel, Pardonet and Bocock formations in ascending order, Fig. 8.48, 8.50). The economically important oil- and gas-bearing Triassic rocks in the subsurface of the Foothills of Alberta and British Columbia also have a different nomenclature (Fig. 8.48-8.50). The Triassic system of British Columbia and Alberta has yielded outstanding ammonite faunas (Fig. 8.51-8.53) particularly near the Peace and Liard rivers.

Southern Rocky Mountain Foothills and Front Ranges

Triassic rocks of this region comprise sandstone, siltstone, shale, dolostone, limestone, intraformational breccia and rare evaporites assigned to the Spray River Group (Fig. 8.48). Two formations, a lower Sulphur Mountain (Fig. 8.54) and an upper Whitehorse are recognized, each of which is subdivided into members (Fig. 8.48, 8.49). The Spray River Group commonly disconformably overlies chert, cherty dolostone and sandstone of the Permian Ishbel Group but in parts of the eastern Foothills it rests disconformably above cherty dolostone of the Mississippian Rundle Group. The Spray River Group is disconformably overlain by the Jurassic Fernie Formation.

In most areas of the Foothills and Front Ranges of the Rockies the Sulphur Mountain Formation is subdivided into four units which, in ascending order, are the Phroso Siltstone, Vega Siltstone, Whistler and Llama members. Between Smoky River and Pine Pass, the Phroso and Vega cannot be distinguished and are grouped as a single member (Fig. 8.48).

The Phroso Siltstone Member comprises up to 240 m of recessive, shaly to flaggy weathering, grey-brown to dark grey, quartz siltstone and silty shale and less commonly very fine grained sandstone (Gibson, 1974). The contact with the overlying Vega is abrupt and conformable. The occurrence of the index fossil *Claraia stachei* indicates an Early Triassic (Griesbachian) age.

The overlying Vega Siltstone Member consists of greyish to rusty-brown weathering, cyclically bedded, dolomitic to calcareous quartz siltstone, silty limestone and shale. Locally the Vega contains very fine-grained dolomitic quartz sandstone and sandy to silty dolostone. In the Front Ranges and Foothills between Cadomin and Banff, Alberta it is characterized by light grey to buff, porous dolostone called the Mackenzie Dolomite Lentil (Fig. 8.49, Gibson, 1974). In outcrop the Vega reaches a maximum thickness of 360 m (Gibson, 1974). Throughout much of the region, it is abruptly overlain by the Whistler Member or where the Whistler is absent, possibly because of facies change, by the Llama Member. Bivalves and ammonites indicate an Early Triassic (Smithian) age.

UPPER DEVONIAN TO MIDDLE JURASSIC ASSEMBLAGES

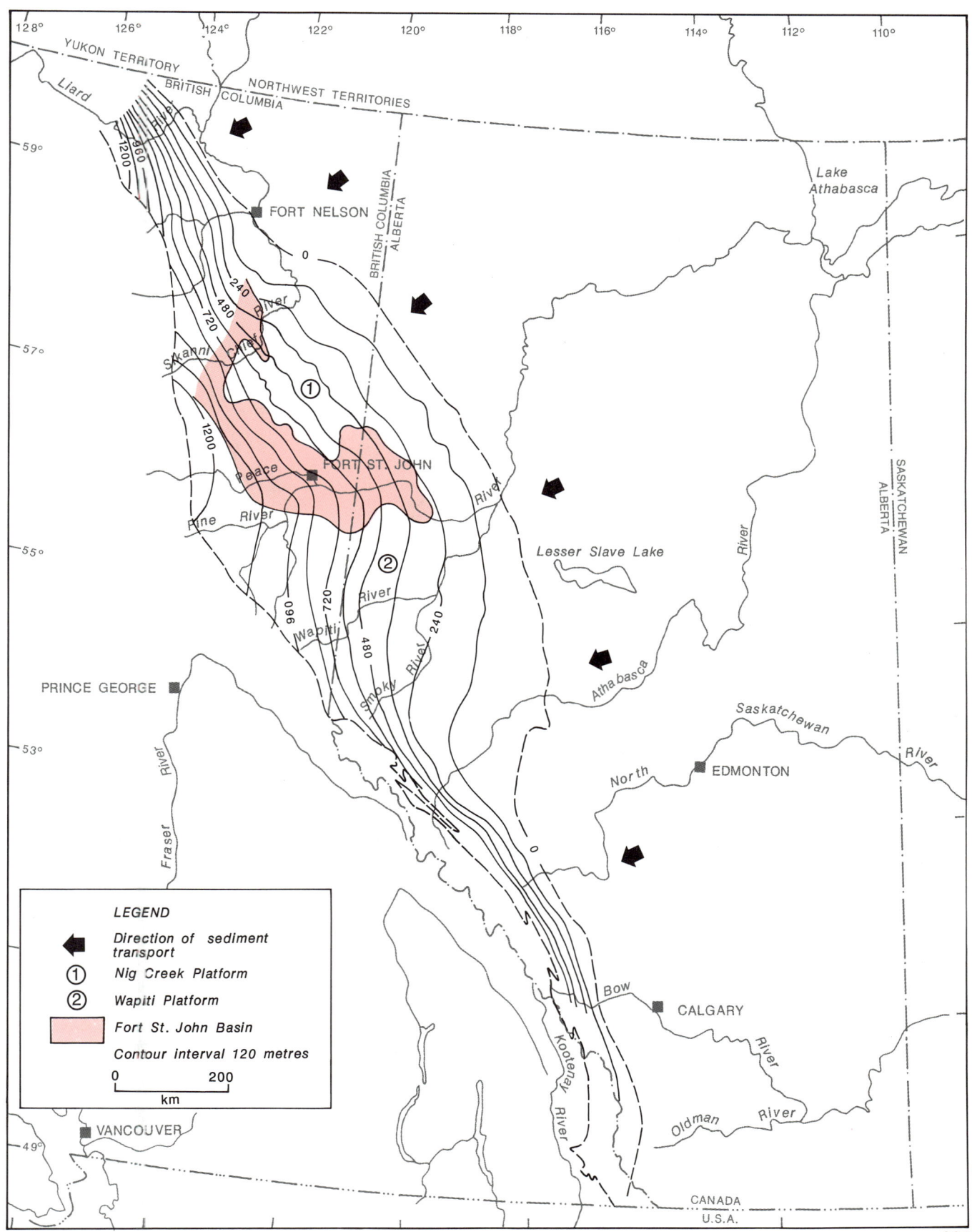

Figure 8.47. Isopach map of Triassic strata, British Columbia and Alberta (modified from Barss et al., 1964).

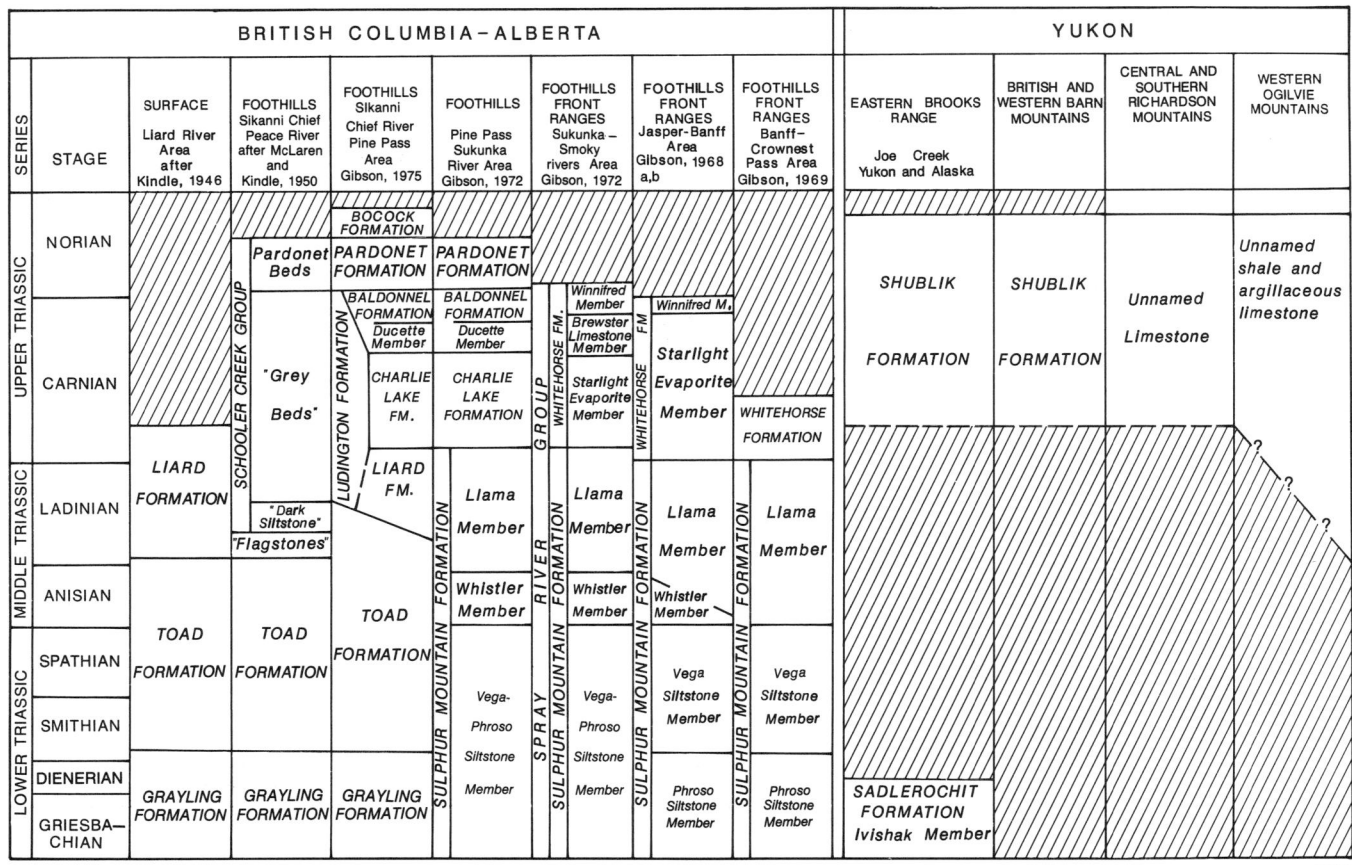

Figure 8.48. Correlation chart of Triassic strata.

The Vega-Phroso Member outcrops in the Smoky River-Pine Pass area of west-central Alberta and northeastern British Columbia and consists of dark brownish grey to rusty-brown, shaly- to flaggy-weathering, dolomitic to calcareous siltstone, finely crystalline to bioclastic limestone, silty shale, and minor very fine- to fine-grained quartz sandstone. Thicknesses range from 80 m to 270 m. Ammonites and pelecypods indicate an Early Triassic (Griesbachian to Spathian) age.

The most distinctive and easily recognized unit of the Sulphur Mountain Formation, the Whistler Member, consists of 13 to 23 m of recessive, dark grey to black weathering, carbonaceous-ferruginous silty dolostone, dolomitic quartz siltstone, silty and fossiliferous limestone, silty shale, and locally, phosphatic and quartz sandstone and phosphatic pebble conglomerate. It has a limited distribution south of Bow River and in the Foothills area east of Jasper, Alberta. Its southeastward thinning and disappearance is in part depositional and in part occurs through a facies change to the Llama Member (Fig. 8.48, 8.49). The Whistler Member is conformably overlain by thicker bedded, siltstones and silty dolostones of the Llama Member. The Whistler contains the *Beyrichites-Gymnotoceras* fauna of Middle Triassic (Anisian) age.

The Llama Member comprises cliff-forming, thin- to thick-bedded, dolomitic quartz siltstone, silty to sandy dolostone, silty and bioclastic limestone and local dolomitic quartz sandstone and silty shale. It ranges in thickness from 3 m near Athabasca River to a maximum of 356 m near Sukunka River in northeastern British Columbia. In the Pine Pass-Sukunka River region it is overlain gradationally by the Charlie Lake Formation. Elsewhere, it is overlain abruptly by the Starlight Evaporite Member of the Whitehorse Formation. The Llama ranges in age from mid-Middle to late Middle Triassic (Late Anisian to Late Ladinian).

The Whitehorse Formation is typically divisible into three members, which, in ascending order are the Starlight Evaporite Member, the Brewster Limestone Member and the Winnifred Member. In the Cadomin-Banff area the Starlight features distinctive quartzitic sandstone called the Olympus Sandstone Lentil. The Brewster is restricted to the region north of Athabasca River.

The Starlight Evaporite Member consists of buff yellow, light to medium grey to reddish brown weathering dolostone and limestone, sandstone, siltstone, intraformational and/or solution breccia, with locally intercalated beds and lenses of gypsum. Numerous abrupt facies changes are characteristic and inhibit correlation even between closely spaced sections. The member ranges in thickness from an erosional zero edge in some areas of the eastern Foothills, to over 400 m in the Front Ranges of the Rockies in Banff National Park (Gibson, 1974). It is conformably and abruptly overlain in areas north of

UPPER DEVONIAN TO MIDDLE JURASSIC ASSEMBLAGES

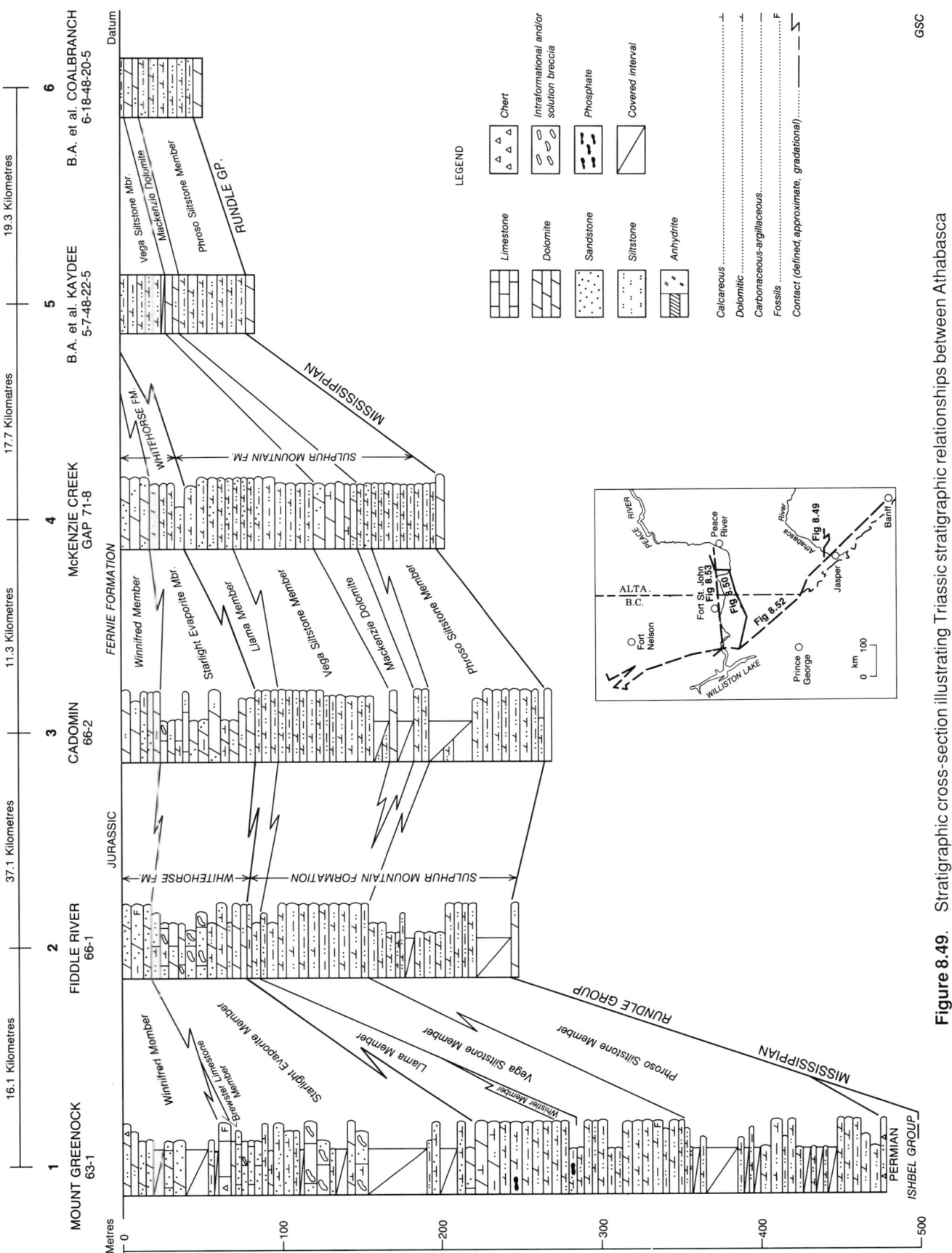

Figure 8.49. Stratigraphic cross-section illustrating Triassic stratigraphic relationships between Athabasca and Brazeau rivers, Alberta (after Gibson, 1974). Line of section shown in Figure 8.46.

Athabasca River by the Brewster Limestone Member, but in the Foothills between Athabasca and Bow rivers, is overlain by the Winnifred Member. The Starlight Member is of assumed Late Triassic (Carnian) age.

The Brewster Limestone Member comprises pale to dark grey weathering, resistant, cliff-forming, medium- to thick-bedded limestone with lesser silty to sandy dolostone and intraformational carbonate breccia that reaches up to 62 m thick. The contact with the overlying Winnifred Member is abrupt and conformable. The Brewster Member is Late Triassic (Carnian) in age.

The Winnifred Member of the Whitehorse Formation consists of yellowish to medium grey weathering, thin- to thick-bedded, sandy to silty dolostone and limestone with rare interbeds of dolomitic quartz siltstone, very fine grained quartz sandstone and intraformational conglomerate. It has a maximum thickness of 225 m. The Winnifred

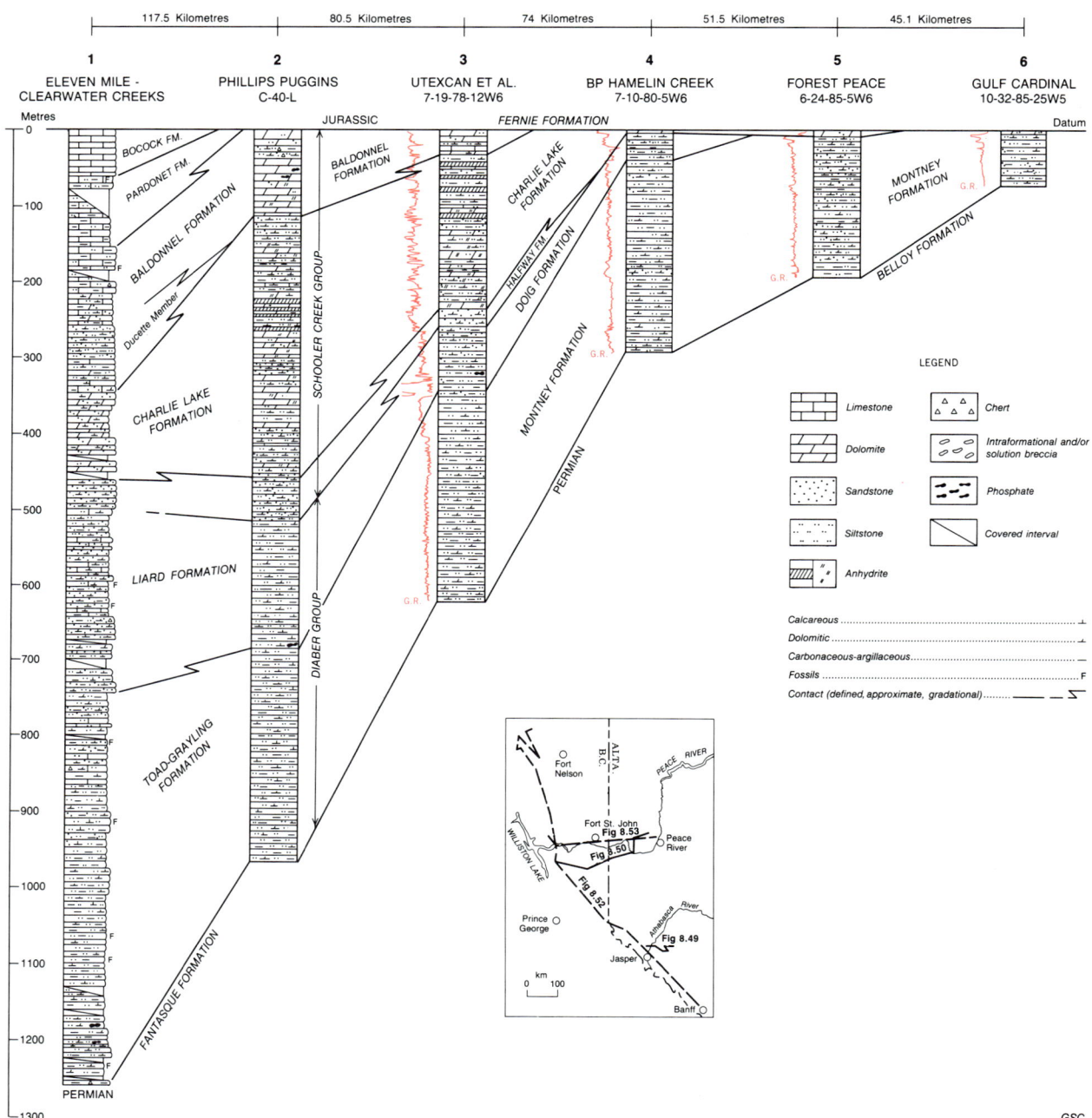

Figure 8.50. Stratigraphic cross-section illustrating lithology and facies relationships between Triassic formations of Rocky Mountain Foothills and subsurface Interior Plains, northeastern British Columbia. Line of section shown in Figure 8.46. G.R., gamma ray log.

TRIASSIC TIME SCALE

SERIES	STAGE	SUBSTAGE	ZONES
UPPER TRIASSIC	NORIAN	Upper Norian	*Choristoceras crickmayi* *Cochloceras amoenum* 27 *Gnomohalorites cordilleranus* 26
		Middle Norian (Alaunian)	*Himavatites columbianus* 25 *Drepanites rutherfordi* 24
		Lower Norian	*Juvavites magnus* 23 *Malayites dawsoni* 22 *Stikinoceras kerri* 21
	CARNIAN	Upper Carnian (Tuvalian)	*Klamathites macrolobatus* 20 *Tropites welleri* 19 *Tropites dilleri* 18
		Lower Carnian (Julian)	*Austrotrachyceras obesum* 17 *Trachyceras desatoyense* 16
MIDDLE TRIASSIC	LADINIAN		*Frankites sutherlandi* 15 *Maclearnoceras maclearni* 14 *Meginoceras meginae* 13 *Progonoceratites poseidon* 12 *Eoprotrachyceras subasperum* 11
	ANISIAN	Upper Anisian (Illyrian)	*Frechites chischa* 10 *Frechites deleeni* 9
		Middle Anisian (Pelsonian)	*Anagymnotoceras varium* 8
		Lower Anisian (Aegean)	*Lenotropites caurus* 7
LOWER TRIASSIC	SPATHIAN		*Keyserlingites subrobustus* 6 *"Olenilites" pilaticus*
	NAMMALIAN	Smithian	*Wasatchites tardus* 5 *Euflemingites romunderi* 4
		Dienerian	*Vavilovites sverdrupi* 3 *Proprychites candidus* 2
	GRIESBACHIAN	Upper Griesbachian (Ellesmerian)	*Proprychites strigatus* *Ophiceras commune* 1
		Lower Griesbachian (Gangetian)	*Otoceras boreale* *Otoceras concavum*

Figure 8.51. Triassic time scale showing ammonite zones (modified from Tozer, 1984).

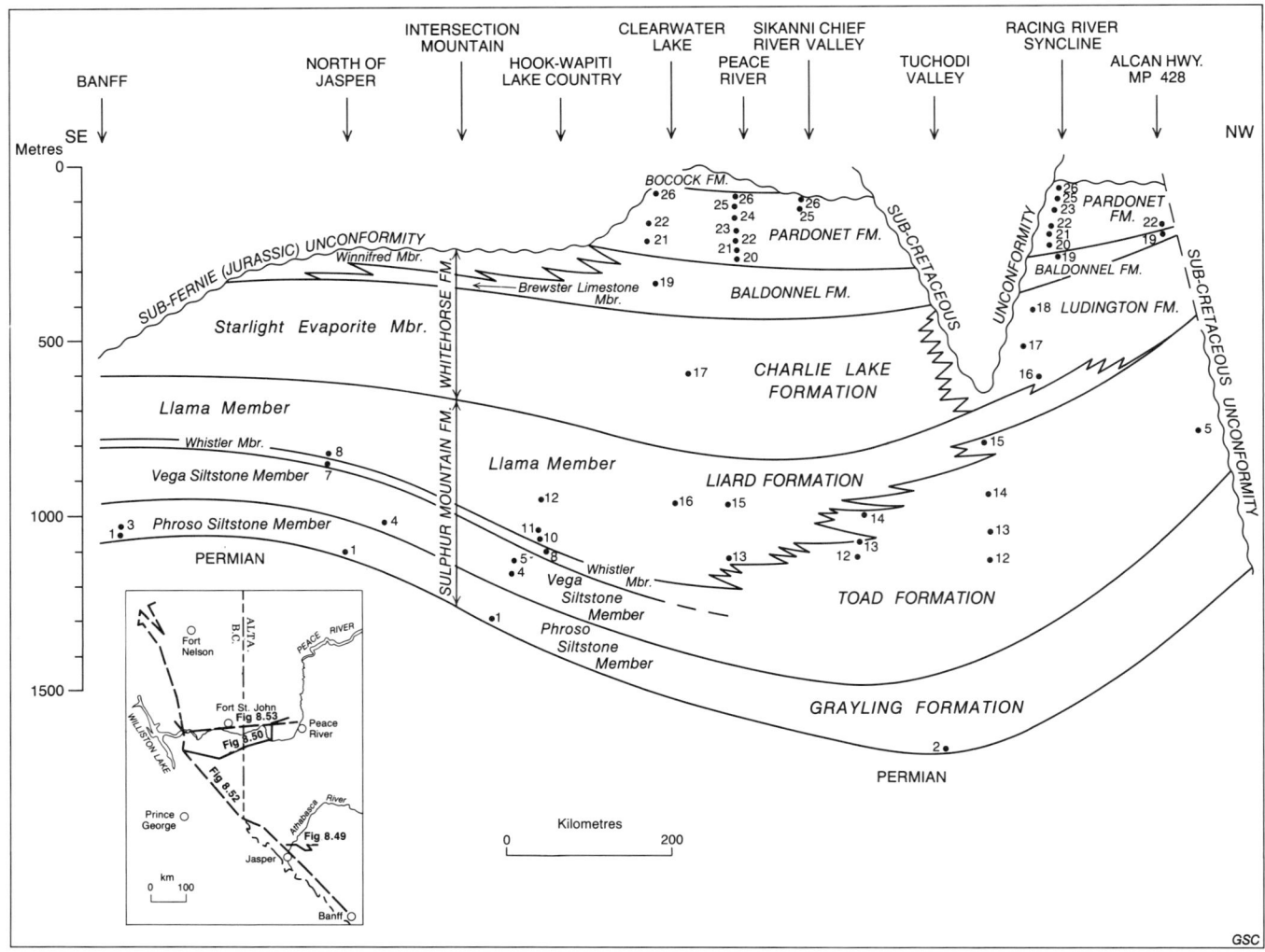

Figure 8.52. Schematic longitudinal section illustrating Triassic stratigraphic relationships and location of ammonoid faunal zones between Banff, Alberta and Milepost 428, Alaska Highway northeastern British Columbia (nos. refer to faunal zones in Fig. 8.51). Diagram by E.T. Tozer. Line of section also shown in Figure 8.46.

is interpreted to be Late Triassic (Carnian) age. It is disconformably overlain by the Jurassic Fernie Formation.

Northern Rocky Mountain Foothills and Front Ranges

In the central Foreland Belt the Triassic succession, one of the best preserved and most complete in the world, differs from that to the south and is divided into eight formations which, in ascending order, are the Grayling, Toad, Liard, Charlie Lake and laterally equivalent Ludington, Baldonnel, Pardonet and Bocock formations (Fig. 8.48, 8.49). A small area of unnamed Triassic strata, equivalent to the Ludington and Pardonet formations, is preserved in the Muskwa Ranges to the west (Fig. 8.46).

The Grayling Formation comprises recessive, dark grey to brownish grey, shaly to flaggy weathering dolomitic siltstone, silty shale, and lesser calcareous siltstone, silty limestone, dolostone and very fine grained sandstone. Its thickness ranges from 35 m near Halfway River (Gibson, 1975) to a maximum of 395 m near Liard River (Pelletier, 1963). The unit disconformably overlies chert and siliceous mudstone of the Permian Fantasque Formation and is gradationally overlain by strata of the Toad Formation. The Grayling is Early Triassic (Griesbachian to Smithian) in age and correlates with the Phroso Siltstone Member of the Sulphur Mountain Formation in the south.

The Toad Formation consists of dark grey, thin- to medium-bedded, shaly to flaggy weathering, very calcareous siltstone, silty limestone, silty shale and minor amounts of silty dolostone and calcareous sandstone. Its maximum thickness of 825 m occurs near Halfway River (Gibson, 1975). In most areas the Toad is gradationally overlain by the Liard Formation. In the extreme western Foothills north of Peace River, however, and possibly in the Liard River-southern Yukon area, it is overlain by the Ludington Formation. In the extreme eastern Foothills of the Liard River-Alaska Highway area, the Toad is disconformably overlain by the Lower Cretaceous Fort St. John Group

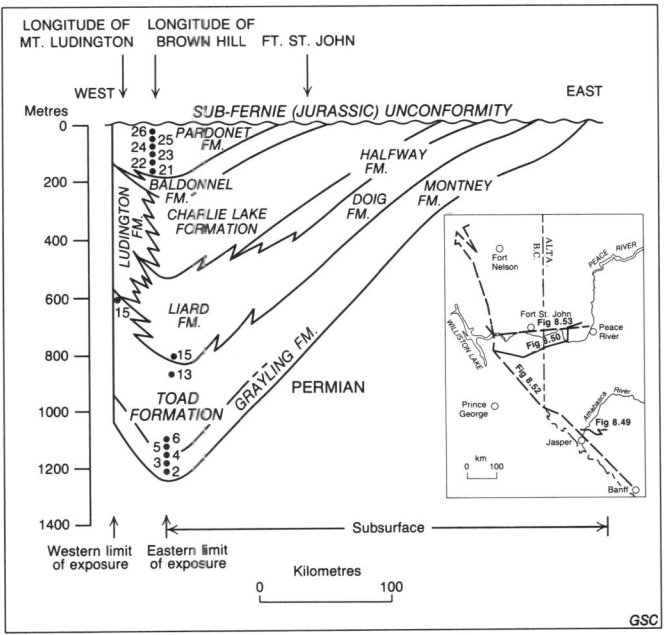

Figure 8.53. Schematic cross-section illustrating Triassic stratigraphic relationships between surface and subsurface and location of ammonoid faunal zones, between Peace River-Williston Lake and area east of Fort St. John, northeastern B.C. (nos. refer to faunal zones in Fig. 8.51). Diagram by E.T. Tozer. Line of section also shown in Figure 8.46.

Figure 8.54. Typical rhythmic bedding of Sulphur Mountain Formation at Meosin Mountain, Rocky Mountain Foothills, northeast British Columbia. GSC 205235-A

(Fig. 8.55). The fauna of the Toad Formation ranges in age from Early to Middle Triassic (Smithian to Ladinian).

The Liard Formation consists of resistant, commonly cliff-forming, very dolomitic to calcareous, fine- to coarse-grained sandstone, calcareous and dolomitic siltstone, and lesser amounts of silty to sandy dolostone. Sandy to silty bioclastic, buff-weathering limestone is sporadically interbedded within the succession at many localities. The Liard ranges in thickness from an erosional and zero depositional edge to a maximum of 417 m near Pine Pass (Gibson, 1975). Liard strata grade westward and interfinger with the Toad Formation. Throughout most of the outcrop belt, the Liard is gradationally overlain by the Charlie Lake Formation, but in the eastern Foothills it is disconformably overlain by the Lower Cretaceous Fort St. John Group (Stott, 1982). The Liard Formation is characterized by a fauna of ammonites, brachiopods and pelecypods of late Middle to early Late Triassic (Late Ladinian to Early Carnian) age.

The Charlie Lake Formation comprises variegated buff to yellow, grey to orange-brown weathering, dolomitic to calcareous sandstone, siltstone, sandy limestone, dolostone and minor amounts of intraformational and solution breccia. In the subsurface of the Interior Platform it contains anhydrite and gypsum. A maximum thickness of 405 m occurs near Williston Lake (Gibson, 1975). Toward the west the Charlie Lake grades laterally into the Ludington Formation and is gradationally overlain by limestone and dolostone of the Baldonnel Formation. In the easternmost Foothills, the Charlie Lake is unconformably overlain by the Jurassic Fernie Formation. The Charlie Lake is interpreted to be of probable early Late Triassic (Carnian) age (Gibson, 1971a,b, 1975).

The Ludington Formation in the extreme western Foothills north of Peace River, comprises medium to light grey weathering dolomitic to calcareous siltstone, very fine- to fine-grained sandstone and silty to sandy bioclastic limestone. It ranges in thickness from a minimum of 500 m to approximately 960 m. Contact relationships with overlying formations are uncertain throughout most areas as the Ludington forms the uppermost Triassic exposure. At one locality however, it is abruptly overlain by siltstone and limestone of the Pardonet Formation. The Ludington is of late Middle to early Late Triassic (Ladinian to Late Carnian) age.

The Baldonnel Formation consists of pale grey to brownish grey weathering, resistant, cliff-forming limestone and dolostone, with less common siltstone and very fine-grained sandstone. Between Williston Lake and Sukunka River however, the Baldonnel is characterized by basal dark weathering siltstone, sandstone, and dolostone called the Ducette Member. The Baldonnel Formation in outcrop attains a maximum thickness of 146 m south of Williston Lake. It is abruptly overlain by the Pardonet Formation. The Baldonnel is of Late Triassic (Carnian) age.

The Pardonet Formation comprises dark grey to dark brownish grey weathering, carbonaceous-argillaceous limestone, silty limestone, calcareous and dolomitic siltstone and minor shale, with a maximum thickness of 137 m near Peace River. In most places the Pardonet is unconformably overlain by the Jurassic Fernie Formation, but between Williston Lake and Pine Pass, it is abruptly and possibly disconformably overlain by the Bocock Formation. The Pardonet is of Late Triassic age (early to late Norian).

In the westernmost Foothills between Williston Lake and the Pine River, the Pardonet Formation is overlain by a light grey to yellowish brown weathering, medium- to thick-bedded sequence of limestone of the Bocock Formation. The Bocock is up to 63 m thick (Gibson, 1975). It is abruptly and probably unconformably overlain by the

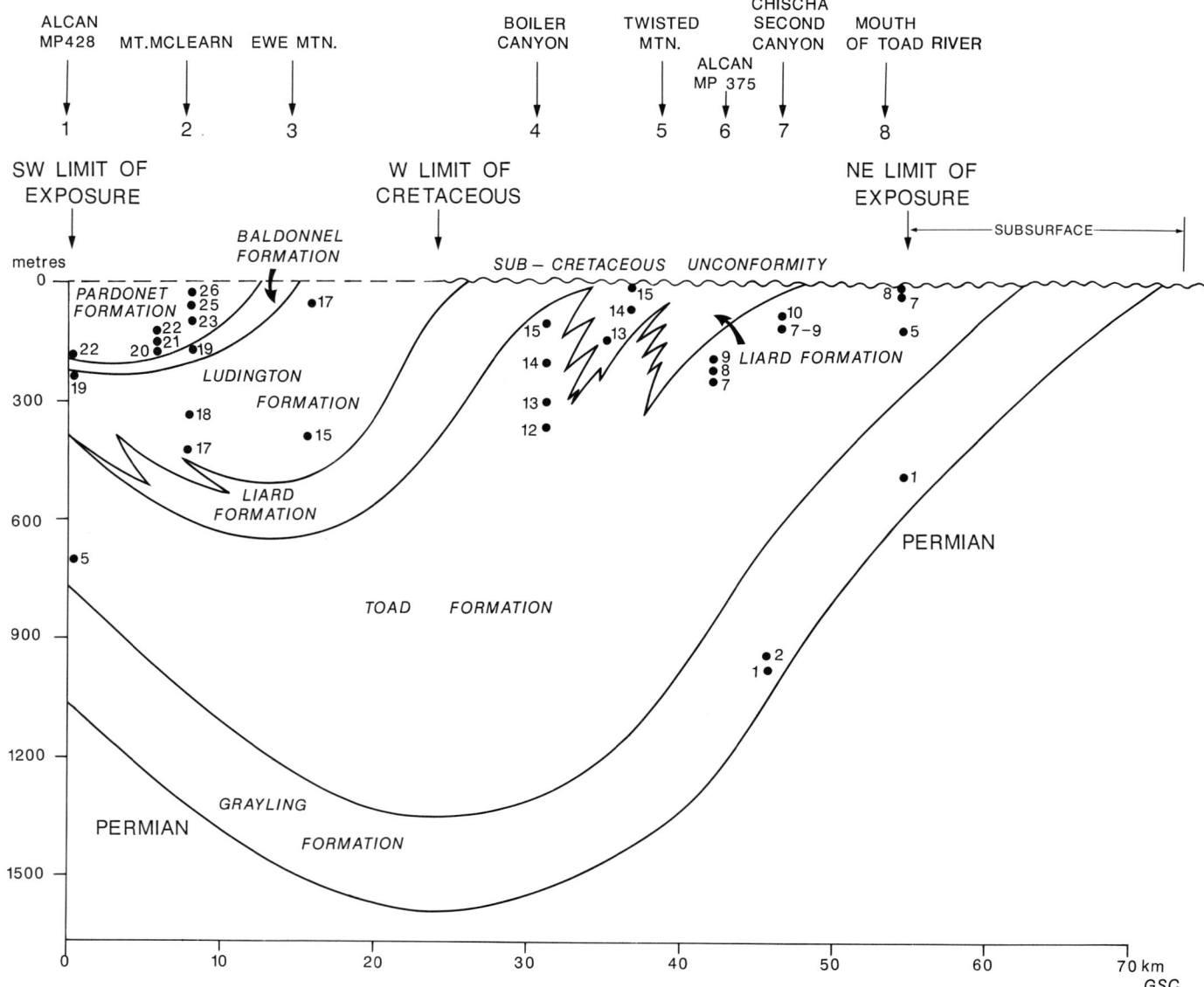

Figure 8.55. Schematic cross-section illustrating Triassic stratigraphic relationships and location of ammonoid faunal zones between Milepost 428, Alaska Highway, northeastern British Columbia and Liard River-Nelson Forks area, northeastern B.C. (nos. refer to faunal zones in Fig. 8.51). Diagram by E.T. Tozer. Line of section shown in Figure 8.46.

Jurassic Fernie Formation. Conodonts from the Bocock indicate a late Late Triassic (Late Norian) age (Orchard in Tozer, 1982a).

Unnamed Triassic strata underlie a narrow elongate area of about 60 km² in the western Muskwa Ranges. The rocks are recessive to moderately resistant, very well bedded, dark, locally markedly platy, grey weathering calcareous siltstone and shale interbedded with resistant beds of buff-orange to buff weathering silty limestone. The sequence is about 300 m thick and overlies resistant black porcellanite and cream coloured chert of Permian or Devonian-Mississippian age. Fossils indicate Early Ladinian, Carnian and Middle and Late Norian ages.

Northern Foreland Belt, Yukon Territory and District of Mackenzie

Triassic strata in the northern Foreland Belt are preserved in three widely separated areas: 1) northern Yukon-western District of Mackenzie (British, Barn, Richardson, northern Ogilvie, and Wernecke mountains), 2) west-central Yukon (southern Ogilvie Mountains), and 3) east-central Yukon (Pelly and Selwyn mountains) (Fig. 8.46).

In northern Alaska, Yukon Territory and western District of Mackenzie the Triassic comprises the Permian to Triassic Sadlerochit and the Lower to Upper Triassic Shublik formations (Sadlerochit Assemblage; Fig. 8.48).

The Sadlerochit in the British Mountains, west of the Alaska-Yukon border (Arctic Alaska Terrane) consists of two members. The upper or Ivishak Member is of Early Triassic age and consists of brown sandstone and grey shale up to 130 m thick. The abruptly and unconformably? underlying Permian Echooka Member consists of interbedded silty shale, sandstone and minor carbonate. In Yukon Territory no Triassic strata have been recognized in the Sadlerochit. The Shublik Formation occurs in the British-Barn Mountains, Richardson Mountains and northwestern Ogilvie Mountains where it comprises siltstone, sandstone, shale, silty and bioclastic limestone. Minor quartz conglomerate occurs near the base of the formation in the British Mountains. The Shublik is variable in thickness ranging up to more than 120 m thick (Mountjoy, 1967). The Shublik unconformably overlies the Lower Triassic Ivishak Member near the Alaska-Yukon border, and upper Paleozoic formations elsewhere. It is unconformably overlain by Jurassic (Kingak Formation) or Upper Cretaceous and/or Tertiary strata. The Shublik in the British and Barn mountains has yielded *Monotis ochotica* (Keyserling) of Late Norian age (Mountjoy, 1967).

In west-central Yukon (southern Ogilvie Mountains) isolated occurrences of Triassic strata (Tempelman-Kluit, 1970) consist of dark grey siltstone and fetid shaly fossiliferous limestone of probable Smithian to upper Norian age. There, the Triassic probably unconformably overlies the Permian Tahkandit Formation and is unconformably overlain by Middle Jurassic strata.

In central Yukon (Pelly Mountains) southwest of the Tintina Fault within the Cassiar Terrane, up to 500 m of Middle and Upper Triassic, thin-bedded calcareous siltstone and minor phosphatic limestone (Tempelman-Kluit, pers. comm.; Tempelman-Kluit, 1977; Gordey, 1981a) are assigned to the Spray River Assemblage. In the middle and upper parts of the succession, limestone forms a discontinuous member up to 100 m thick. The relatively uniform thickness and facies show that the unit was deposited over a larger area than where preserved. The sequence unconformably overlies Permian strata; the top is not exposed. It is the same age as and resembles the Ludington and Pardonet formations of the northern Rocky Mountains, from which it may be offset by Cretaceous-Tertiary dextral strike slip along the Tintina-Northern Rocky Mountain trench faults.

In east-central Yukon (Selwyn Mountains) northeast of the Tintina Fault Triassic strata comprise scattered exposures of thin-bedded ripple cross-laminated calcareous siltstone, sandstone, and shale of the Jones Lake Formation (Gordey, in press). Contact relations with underlying Paleozoic strata appear disconformable. Near Macmillan Pass up to 750 m of such strata are preserved (Gordey, 1981b). Meagre conodont collections indicate a range in age from Smithian to Norian.

Paleoenvironments and history of deposition

In British Columbia and Alberta Lower Triassic strata (Phroso Siltstone and Vega-Phroso Siltstone members, Grayling Formation) were deposited in an easterly transgressing sea along the western margin of a relatively deep and easterly shallowing tectonically stable, open shelf. However, in some areas of the Foothills and adjacent Plains of northeastern British Columbia, thickness variations in Lower Triassic strata suggest segmentation of the coastal margin into basins, troughs and shallow platforms (Barss et al., 1964). The Vega Siltstone Member displays a distinctive alternation of resistant and recessive weathering strata and sedimentary structures commonly found in sediments deposited from turbidity- or storm-generated offshore currents, perhaps in relatively shallow water (see e.g. Nelson, 1982). In contrast to the underlying Phroso Siltstone Member and equivalent Grayling Formation, the Vega Siltstone Member and Lower Toad Formation were deposited in a regressive or shallowing sea.

Lower Middle Triassic (Anisian) strata of the Whistler Member and middle Toad Formation were deposited during or following marine transgression. The thinness and thin bedding of the Whistler Member together with a high carbonate concentration, suggest low detrital sediment input in an oxygen-deficient, restricted depositional environment. Strata of the Middle Toad Formation in the north display features characteristic of a more open marine shelf environment.

A regional lowering of sea level along the western miogeocline occurred during deposition of the Llama Member of the Sulphur Mountain Formation, Liard and Halfway formations. The sandstones, siltstones and associated lithofacies deposited along the shoreline of the shallowing sea represent shallow marine shelf, shoreface, barrier beach, and tidal channel environments. In the westernmost Foothills and Front ranges, these strata display facies and sedimentary structures suggestive of relatively deeper water deposition.

Following Middle Triassic time, the seas continued to shoal and regress. Starlight Evaporite and Charlie Lake strata probably formed in shallow-water intertidal and tidal flat evaporitic environments. Locally, sandstones and some bioclastic carbonates may have formed barriers and offshore bars resulting in restricted water flow and consequent precipitation of gypsum-anhydrite and the development of "redbeds". Strata of the laterally equivalent Ludington Formation were deposited farther west under much less restricted and relatively deeper water conditions.

The Brewster and Winnifred members of the Whitehorse Formation and Baldonnel Formation contain lithofacies and sedimentary structures typical of relatively deeper water, and less restricted, open marine conditions. The Brewster Limestone Member and Baldonnel Formation reflect deposition in a shallow water, relatively high-energy environment, under the influence of moderate current and wave conditions. Westward, however, the Baldonnel contains increasing concentrations of organic matter and less fragmented bioclastic material suggesting a relatively deeper water, open shelf environment. Locally, however, intraformational and/or solution breccias and "redbeds" within the Winnifred Member suggest shallow-water evaporitic lagoonal settings.

Following Baldonnel deposition the seas in the north again deepened resulting in the euxinic depositional environment reflected in the Pardonet Formation. In latter Pardonet time, the sea regressed subjecting much of the region to pre-Jurassic erosion. However, in the Williston Lake-Pine Pass area in northeastern British Columbia, a shallow marine embayment remained and limestones of the Bocock Formation were deposited in a very shallow-water lagoonal environment.

The Triassic period ended with further uplift, regression and probable erosion prior to marine transgression and subsequent deposition of sediments of the Jurassic Fernie Formation.

Information on Triassic environments and history of deposition in the Yukon Territory and District of Mackenzie is sparse. The Upper Triassic Shublik Formation, characterized by rapid vertical and lateral facies variations, appears to have been deposited along a northwest-trending stable shelf in a nearshore, relatively shallow-water environment. Lithological descriptions (Mountjoy, 1967) suggest the coarser grained and conglomeratic strata represent beach and shoreface environments, and that the finer grained siltstones and shales were deposited in deeper-water restricted settings. Clastic sediments of the Shublik appear locally derived. The limestone, some of which consists of bivalve coquinas, accumulated away from the influx of terrigenous sediment, on a shallow-water shelf.

The lithology and faunal associations of Triassic strata in the Southern Ogilvie Mountains suggest low-energy marine sedimentation at moderate depth (Tempelman-Kluit, 1970).

In south-central Yukon (Pelly Mountains) the siltstone and limestone sequence in the Pelly Mountains region represents normal marine deposition on a shallow shelf at considerable water depth, but above the level of carbonate compensation. None of the rocks formed in shallow water of the intertidal zone or in agitated beach environments. Although the carbonates are detrital biogenic accumulations, none are thought to have formed reefs.

In east-central Yukon (Selwyn Mountains) Triassic strata probably represent subtidal deposition on a shallow marine shelf.

Lower and Middle Jurassic strata of the Foreland Belt
T.P. Poulton

Lower and Middle Jurassic strata of the Spray River Assemblage in the Foreland Belt are mainly shales and sandstones preserved in northern Yukon and adjacent District of Mackenzie, in central Yukon, and in the central and southern Rocky Mountains and Foothills of Alberta and British Columbia (Fig. 8.56). Preservation is incomplete due to many episodes of erosion during and following the Jurassic period.

The Lower and Middle Jurassic strata in the northern Foreland Belt are platform deposits, entirely derived from the craton. Shelf sandstone in the east and south grade to basinal shales and siltstones to the west and north. In the southern Foreland Belt, Lower and Middle Jurassic and some Upper Jurassic rocks, also are entirely craton-derived. In contrast, most of the Upper Jurassic and younger rocks in southern Cordillera reflect the early stages of Cordilleran orogenesis to the west, and were transported easterly and deposited in the foredeep adjacent to the emerging orogen.

Major transgressions preceded Sinemurian, Toarcian, late Early Bajocian, Bathonian, Callovian (approximately), and Late Oxfordian or Kimmeridgian deposition in both

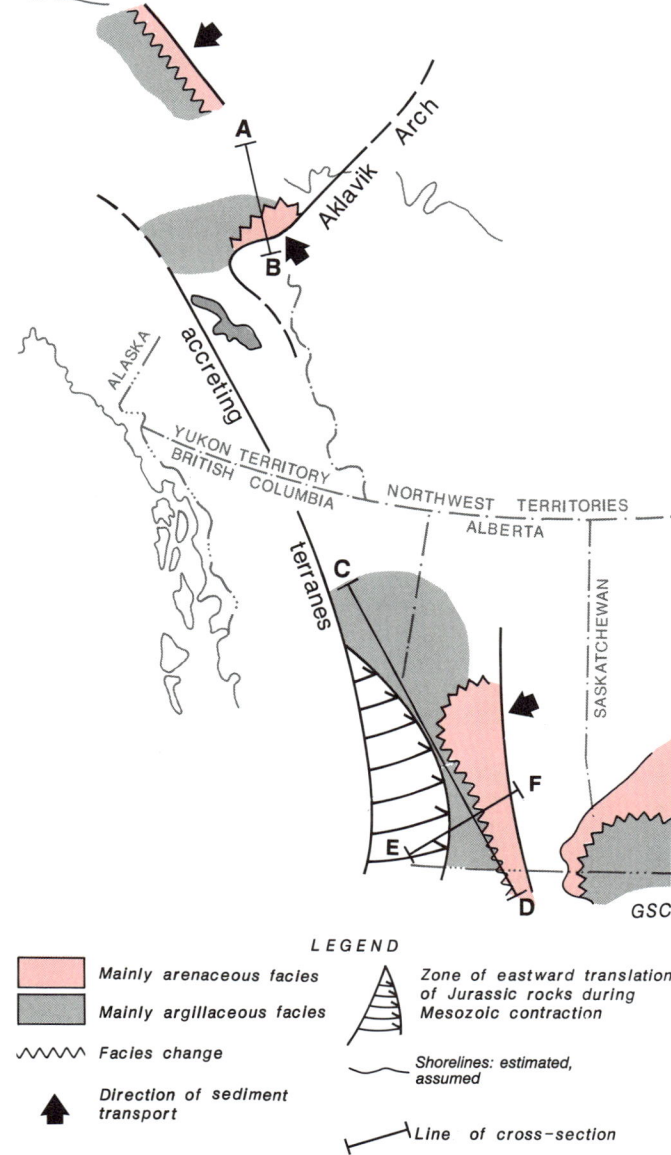

Figure 8.56. Tectonostratigraphic setting of western Canada in Early and Middle Jurassic and locations of cross-sections in Figures 8.57, 8.59, and 8.60. Generalized lithological patterns and sediment-transport directions (arrows) show the polarities of sedimentation.

northern Yukon and southwestern Alberta in response to tectonic or eustatic events along the western margin of northern North America. Until the Oxfordian, sediments were shed from the craton, after which western sources were dominant.

Thick Pliensbachian and Lower Oxfordian sandstone units in northern Yukon reflect important cratonic source regions both close by and perhaps far removed to the south and southeast. In the latter case sediments may have been delivered to the boreal region by an ancestral Mackenzie River type of drainage system east of Rocky Mountain

Trough to which it also contributed sediment (Poulton, 1984).

A progressively northerly advance of orogenic (western) sediment source areas in the Upper Jurassic may be indicated by the generally younging northward of the micaceous clastic wedges derived from them, Oxfordian in the south, Cretaceous in the northern Yukon (Poulton, 1984). This north-south variation in timing of sedimentation in the foredeep may be related to large scale mid-Cretaceous and later dextral offsets on strike-slip faults along the western side of the Omineca Belt (Eisbacher, 1981; Gabrielse, 1985).

Northern Foreland Belt

In the northern Foreland Belt Jurassic strata unconformably overlie a surface of low relief eroded into Triassic, Paleozoic, or Precambrian rocks (Fig. 8.57). Easterly or southeasterly sediment sources and shorelines are indicated by facies and thickness distributions, and by easterly increases in the magnitude of internal hiati. Depocentres and interpreted sediment sources for the Lower and Middle Jurassic were systematically displaced toward the south as a consequence of intermittent transgression of the craton. The biostratigraphic record suggests that a more or less complete sedimentary record of the Jurassic System may be preserved.

The surface on which Jurassic strata were deposited was part of a broad, unstable westward sloping cratonic platform characterized by numerous small uplifts and depressions. They were perhaps related to initial stages of Arctic Ocean rifting when crustal extension occurred beneath Beaufort Shelf (Grantz and May, 1982). There is no evidence of western sediment sources until Early Cretaceous time, nor are there indications of volcanic activity.

In the east, Jurassic strata comprise several cycles of shelf sandstone and shale that are best developed in the northern and central Richardson Mountains and western Mackenzie Delta areas (Fig. 8.57, 8.58). The Sinemurian through Middle Oxfordian Bug Creek Group includes at least five regressive cycles two of which culminate in thick sandstone units, the Almstrom Creek and Aklavik formations, of Pliensbachian and Oxfordian ages respectively (Poulton et al., 1982).

The lowest formation of the Bug Creek Group is the Murray Ridge Formation (30-80 m) comprising shale and siltstone of Sinemurian age. A basal thin sandstone and conglomerate (Scho Creek Member (3-24 m)) occurs in a fan-shaped configuration against the southern basin-marginal Aklavik Arch. The Murray Ridge grades upward into the Almstrom Creek Formation sandstone (300 m), which in turn is overlain by Toarcian and Aalenian black shales of the Manuel Creek Formation (100 m). An offshore barrier bar complex of Aalenian age assigned to the Anne Creek Member (33-72 m) occurs at the top of the Manuel Creek near the shale-out edge of the Bug Creek Group. The Murray Ridge and Manuel Creek shales thicken gradually westward and southwestward, where they merge with shale of the Kingak Formation. The Almstrom Creek sandstone thickens rapidly to over 300 m to the northwest, before it also changes facies to Kingak shale.

The present southeastern and northeastern limits of Lower and Middle Jurassic rocks are probably not far from their depositional limits, which, during the Early Jurassic were partly fault-controlled. Middle Jurassic rocks overstep those of the Lower Jurassic to the southeast (Poulton et al., 1982). A regressive episode, represented by the Anne Creek Member, was accompanied by basin-margin erosion of Lower Jurassic strata.

The Middle Jurassic succession is dominated by the Richardson Mountains Formation (50-300 m), a clastic sequence which varies from mainly sandstone and siltstone in the southeast to shale, with interbedded sandstone units, in the northwest. The Little Bell Member (30 m) of Bajocian age forms its basal sandstone unit. The Waters River Member (140 m), of earliest Callovian age, comprises a series of offshore bar deposits close to the shale-out edge of the formation. The Richardson Mountains Formation grades upward to sandstone of the Aklavik Formation (50-100 m) of Early and Middle Oxfordian age. The Aklavik thickens irregularly westward from its zero edge on the flank of Aklavik Arch to more than 700 m beyond which the sandstones merge into the Kingak shale (Poulton et al., 1982).

The outer shelf succession in the British, Barn and northern Ogilvie Mountains is the Kingak Formation (600-800 m) which includes all stages of the Jurassic except the Sinemurian. It is mostly shale and siltstone with local thin basal sandstone, conglomerate, or ironstone (Poulton, 1982, 1984). Middle and Upper Jurassic shale, siltstone and sandstone, in a narrow belt across central Yukon ("Lower Schist Division", Green, 1972; Tempelman-Kluit, 1970) are probably a southerly extension of the same succession; shoreline trends in northern Yukon indicate that they were deposited in a re-entrant separated from the Kingak

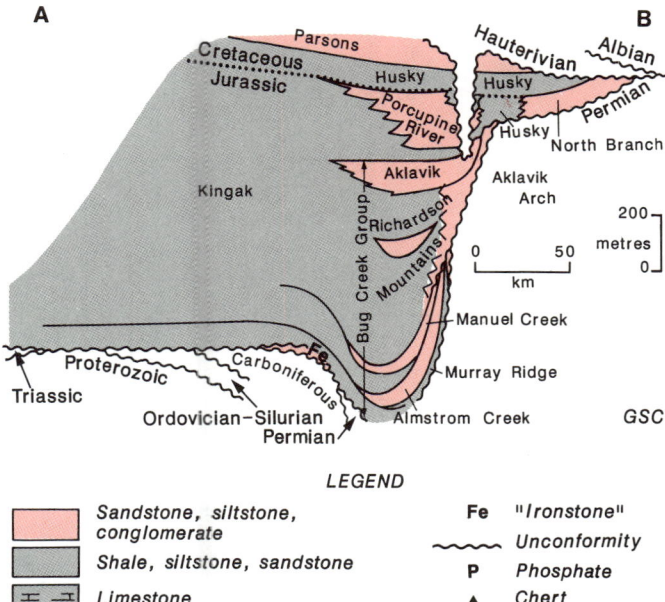

Figure 8.57. Northwest-southeast Lower and Middle Jurassic stratigraphic cross-section across northern Yukon. Location of section is shown in Figure 8.56. Legend for Figures 8.59 and 8.60, also.

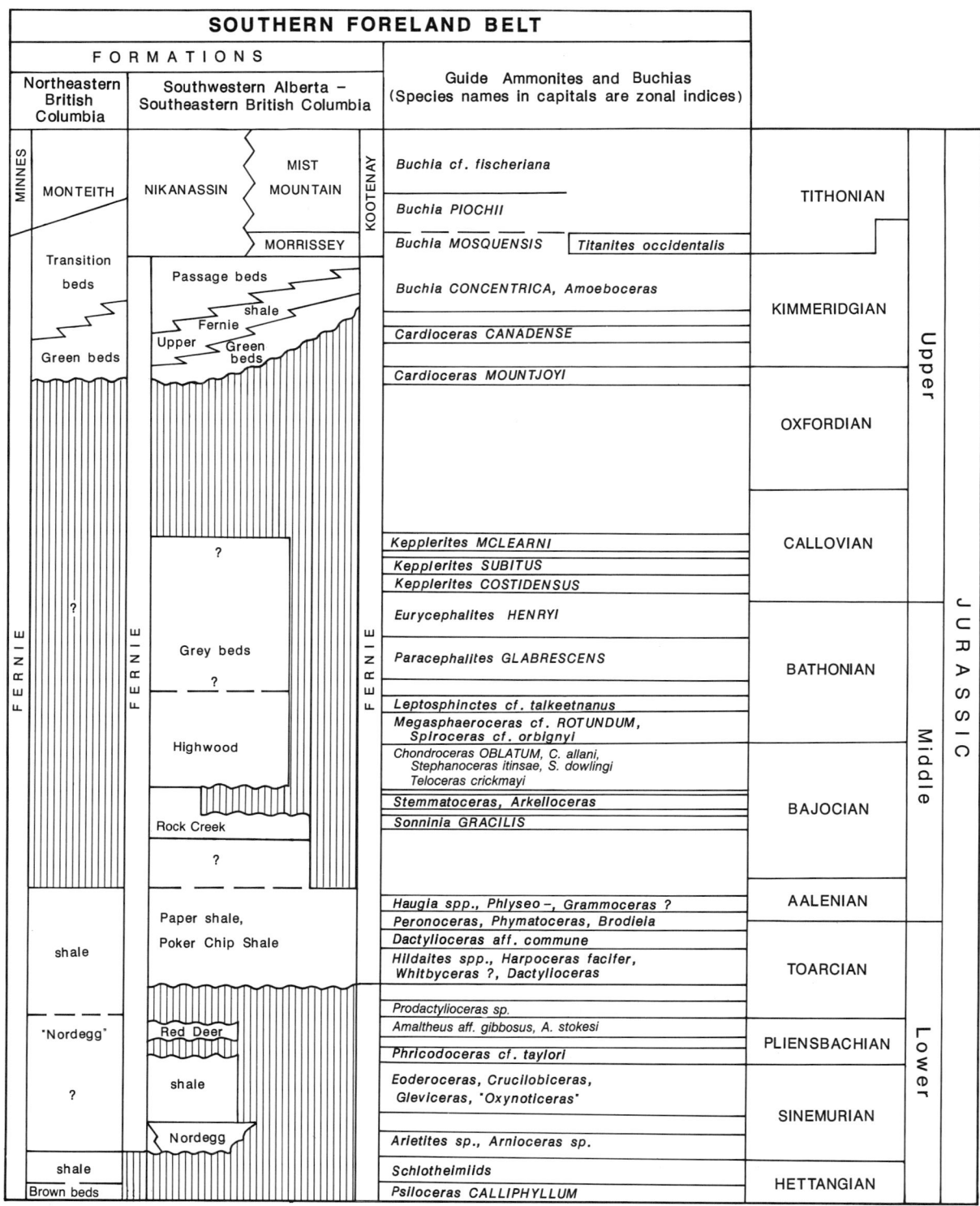

Figure 8.58. Correlation chart and key macrofossils for the Foreland Belt. Vertical lines indicate hiatuses.

UPPER DEVONIAN TO MIDDLE JURASSIC ASSEMBLAGES

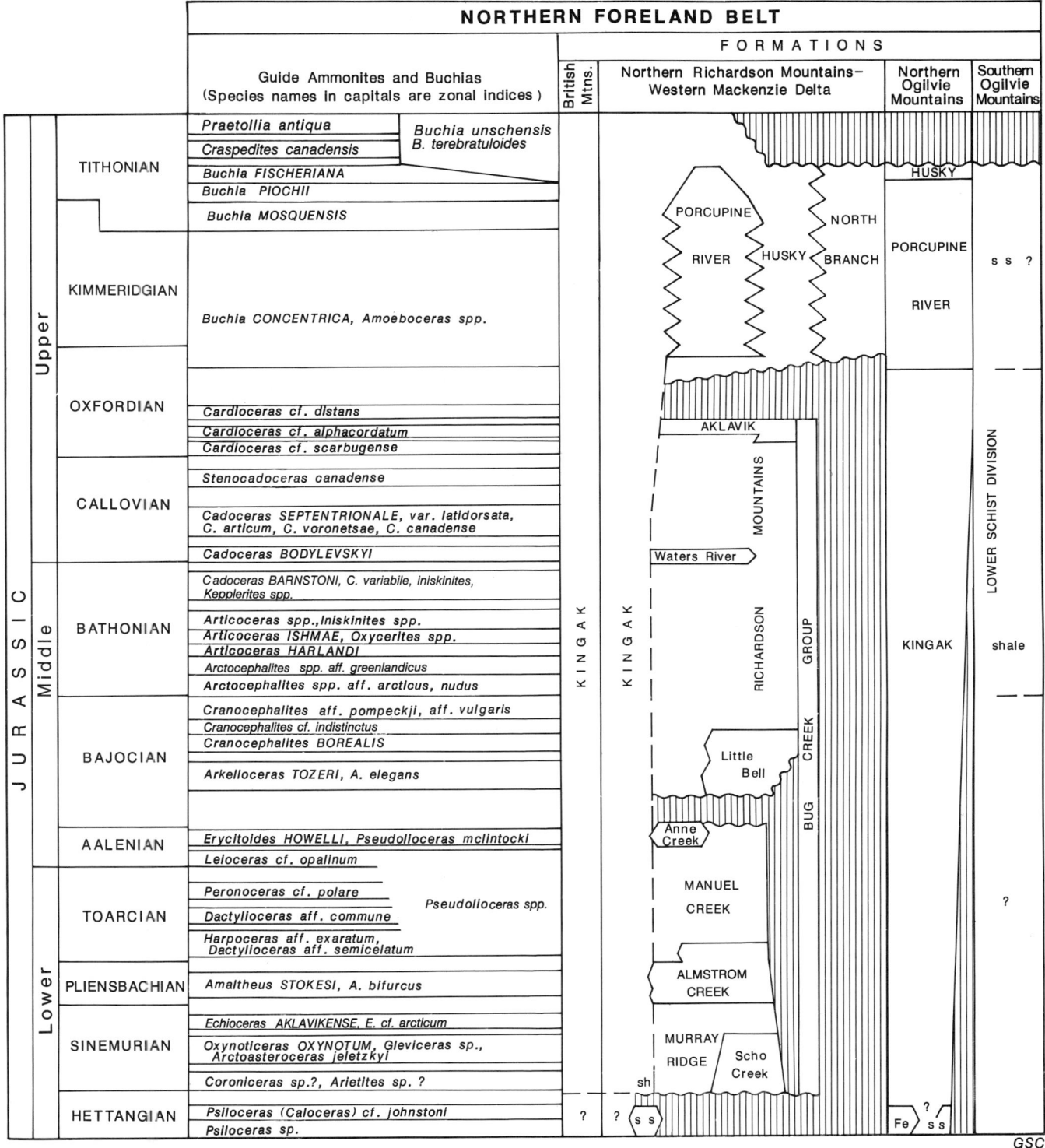

succession by a broad peninsula extending westward from the platform (Fig. 8.56; Poulton and Tempelman-Kluit, 1982).

Southern Foreland Belt

Jurassic strata of the southern Foreland Belt overlie a regional unconformity which successively truncates Triassic to Carboniferous beds eastward towards the craton (Springer et al., 1964). To the east, northeast and north, the Jurassic thins both depositionally and by erosional truncation below intra-Jurassic unconformities and beneath the Lower Cretaceous Blairmore, Bullhead, and Mannville groups (Fig. 8.59, 8.60; Stott, 1967; Poulton, 1984). Only in the westernmost part of the belt are Jurassic strata in gradational contact with those of Cretaceous age (Fig. 8.60).

Lower and Middle Jurassic rocks comprise shale, sandstone and minor bioclastic limestone of the Fernie Formation (Frebold, 1957; Poulton, 1984; Hall, 1984). They are mainly, if not entirely derived from the craton toward which they become progressively more arenaceous; shore-

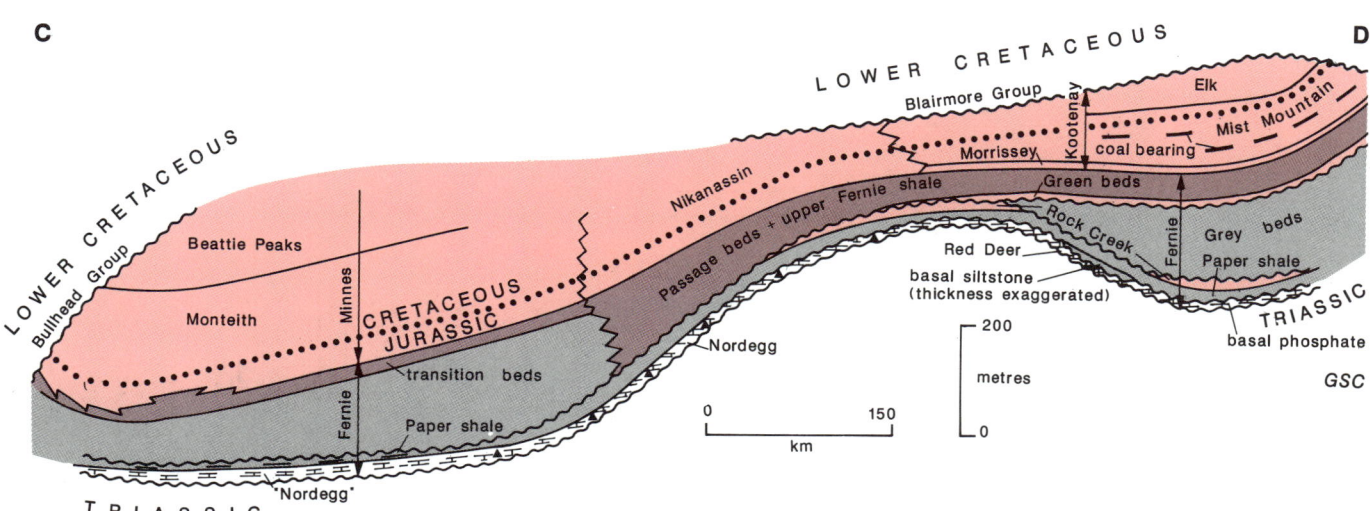

Figure 8.59. Northwest-southeast Jurassic stratigraphic cross-section along the western edge of the miogeocline. Location of section shown in Figure 8.56. Legend accompanies Figure 8.57.

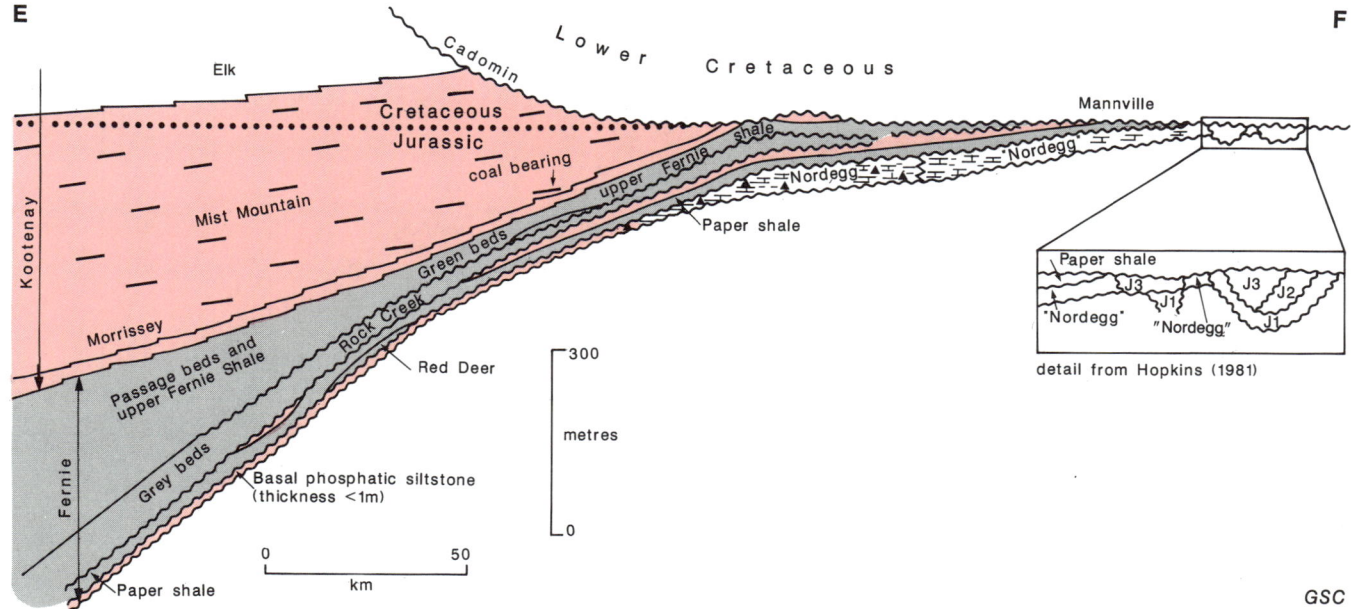

Figure 8.60. Southwest-northeast Jurassic stratigraphic cross-section across southern Foreland Belt. Location of section shown in Figure 8.56. Palinspastic restoration follows the model of Norris (1971). Legend accompanies Figure 8.57.

line facies are not preserved. The only thick sandstone unit in the dominantly shale succession is the Rock Creek Member (2-30 m) of southwestern Alberta. A thick sequence of siltstones of probable Bathonian age within the Grey beds in southeastern British Columbia may have been derived from a southern or southwestern source. The base of the Fernie varies in age from Hettangian to Toarcian. There are many internal, in part laterally impersistent gaps, some suggesting considerable erosion, thus the record is locally incomplete. Regionally however, all stages of the Jurassic are represented. The complex, partly informal, nomenclature of Lower and Middle Jurassic members of the Fernie is summarized by Poulton (1984) and Hall (1984). Thicknesses of this interval are commonly thin but reach a maximum of more than 200 m.

The thin, easterly-derived, partly chemical and highly incomplete character of the epicratonic Lower and Middle Jurassic strata contrasts with that of the thick, westerly derived Upper Jurassic clastic wedge. The first sign of the orogenic activity to the west is the enhanced rate of subsidence of the foredeep. This is indicated by the thickness of Upper Jurassic shales with minor thin turbidites(?) that, together with a locally developed, thin basal sandstone, ("Green beds") are basal to the thick sandstone sequence that culminates in the Kootenay and Minnes groups. Erratic differences in thickness of the Upper Jurassic shales, from 10 to more than 400 m, indicate variation in initial subsidence along the foredeep. The Upper Jurassic units overlie a regional disconformity which rests on various older Jurassic units (Fig. 8.57, 8.59, 8.60), perhaps a record of a passing pre-orogenic welt.

PART B. CORDILLERAN TERRANES

J.W.H. Monger, J.O. Wheeler, H.W. Tipper, H. Gabrielse, T. Harms, L.C. Struik, R.B. Campbell, C.J. Dodds, G.E. Gehrels and J. O'Brien

Summary

In comparison with the relatively well ordered and continuous stratigraphy of the miogeocline, the Upper Devonian to Middle Jurassic rocks of the Cordilleran terranes present many problems in dating, stratigraphic succession and correlation. Nonetheless, a general coherency in lithology, stratigraphic succession and fauna characterize most of the terranes.

In the terranes a wide variety of lithologies suggest oceanic, volcanic island arc and, possibly, distal miogeoclinal settings. Relationships between island arcs, subduction zones, accretionary prisms and fore-arc and back-arc basins are, for the most part, poorly understood. Therefore, at least for much of late Paleozoic time, reconstructions of paleogeography are difficult. Moreover, the distributions of the various terranes relative to one another and to ancestral North America are largely unknown.

The pericratonic Kootenay Terrane, possibly underlain by thinned continental crust, includes intensely deformed upper Paleozoic strata. In the Kootenay Arc Mississippian conglomerate, limestone, argillite and tholeiitic volcanics of the Milford Group unconformably overlie lower Paleozoic clastic and carbonate strata of the Broadview Formation. Farther northwest, in the Adams Lake and Cariboo areas, thin Mississippian and Permian limestone members are present. Throughout its length small bodies of orthogneiss and granite with Late Devonian to Early Mississippian radiometric ages are characteristic of the Kootenay Terrane. In southeastern Yukon Territory those rocks are associated with Early Mississippian mafic and felsic volcanic rocks overlain by Pennsylvanian to Permian carbonate and clastic strata. In Yukon Territory siliceous cataclastic rocks of the Nisutlin Subterrane may include upper Paleozoic strata.

Radiolarian chert in the Windy-McKinley Terrane west of the Alaska-Yukon border is of Mississippian, Triassic and possibly Jurassic age. The chert occurs with volcanic and ultramafic rocks suggesting oceanic deposition. Affinities with the Cache Creek or Slide Mountain assemblages are unknown.

Dominantly oceanic rocks of the Slide Mountain Terrane range in age from Late Devonian to Triassic and occur in thin allochthonous sheets along the western margin of the miogeocline. In the Kootenay Arc Permian tholeiitic volcanics with minor clastic rocks are cut by diorite of probable Permian age. Permian diorite also cuts previously faulted Slide Mountain rocks in the Sylvester Group in north-central British Columbia. The intrusions, pre-intrusive thrust faulting and the presence of blueschist minerals at several localities indicate significant Permian tectonism.

In the southern part of Quesnellia oceanic and volcanic-island arc terranes are unconformably overlain by Middle and Upper Triassic volcanic arc rocks. Paleozoic sedimentary rocks of the Harper Ranch Subterrane range in age from Late Devonian to Late Permian. A structurally complex assemblage of rocks in the Okanagan Subterrane, including sedimentary, volcanic and ultramafic lithologies, locally contains faunas of Ordovician to Late Triassic age. In the Omineca Mountains a sequence of sedimentary and volcanic rocks with Middle Pennsylvanian fossils has been assigned tentatively to the Harper Ranch Subterrane.

The most distinctive rocks in Quesnellia are island-arc volcanics and associated plutonic rocks of the Nicola and Takla groups of Late Triassic to Early Jurassic ages. In southern Quesnellia they grade eastwards into mixed sedimentary and volcanic rocks and finally into sedimentary strata of the Slocan Group. The plutonic rocks are host to many large copper deposits.

Monger, J.W.H., Wheeler, J.O., Tipper, H.W., Gabrielse, H., Harms, T., Struik, L.C., Campbell, R.B., Dodds, C.J. Gehrels, G.E., and O'Brien, J.
1991: Part B. Cordilleran terranes; in Upper Devonian to Middle Jurassic assemblages, Chapter 8 of Geology of the Cordilleran Orogen in Canada, H. Gabrielse and C.J. Yorath (ed.); Geological Survey of Canada, Geology of Canada, no. 4, p. 281-327 (also Geological Society of America, The Geology of North America, v. G-2).

Sedimentary rocks of the Lower to Middle Jurassic Hall Assemblage, unconformably overlie Mesozoic volcanic rocks in Quesnellia. They are mainly volcanic-arc derived clastic strata.

Upper Paleozoic, mainly oceanic sedimentary rocks of the Dorsey Terrane near the Yukon-British Columbia border include limestone units of Pennsylvanian age. Nearby cratonal source areas are indicated by abundant quartz in some sandstones and conglomerate.

The oceanic Cache Creek Assemblage ranges in age from Mississippian to Late Triassic in the northern Cordillera and from Pennsylvanian to Early Jurassic in the south. It is characterized by masses of shallow-water carbonate with fusulinids of Tethyan affinity. The carbonate in some cases, encompasses a remarkably long time span (Mississippian to Permian). The limestone units are surrounded by deep-water radiolarian chert and argillaceous sedimentary rocks. Discontinuous bodies of basalt, gabbro and ultramafic rocks are abundant. The rocks have been intensely deformed with local development of blueschist minerals during Late Triassic and Early to Middle Jurassic time. Similar lithologies occur in the Bridge River Terrane which has yielded Permian to Middle Jurassic radiolaria.

North of the Stikine Arch the oceanic rocks are overlain by Upper Triassic limestone and calc-alkaline volcanics which in turn, are overlain by Lower Jurassic greywacke, the latter perhaps derived from Quesnellia to the northeast.

Mississippian and Permian limestone formations are separated by mafic volcanic rocks and argillite in the northwestern part of Stikinia. The Permian limestone, as much as 1000 m thick, can be traced for about 500 km suggesting stable shallow water deposition, possibly on a subsiding shelf. In northeastern Stikinia bimodal Mississippian volcanics with intercalated carbonate rocks are geographically separated from sequences of Permian sediments and rhyolite.

Throughout Stikinia Upper Paleozoic rocks are overlain unconformably by Lower or Middle Triassic sediments and in places there is strong contrast in deformational style between the Paleozoic and Mesozoic successions.

Upper Triassic arc-related volcanic rocks of the Stuhini Assemblage and associated plutons extend continuously around the eastern and northern parts of Stikinia. They are commonly several kilometres thick and typically include thick flows of augite porphyry and coarse-bladed feldspar porphyry.

Almost completely surrounding the Bowser Basin are thick sequences of Early to Middle Jurassic, varicoloured, calc-alkaline volcanics and correlative sediments of the Hazelton Group. Early Jurassic granitic plutons are numerous in the volcanic successions. North of the Stikine Arch Lower and Middle Jurassic Takwahoni clastic sediments were derived from Stikinia to the west and southwest. In the northern part of the Whitehorse Trough interfingering of Lower Jurassic sediments of the Inklin and Takwahoni formations suggests at least mutual close proximity of Quesnellia, Stikinia and the Cache Creek Terrane during Early Pliensbachian time.

In the southern part of the Coast Belt the Methow, Bridge River and Cadwallader terranes are separated by a belt of high-grade metamorphic rocks from the Shuksan, Chilliwack and Harrison Lake terranes to the west. The oldest (Triassic) rocks exposed in the Methow Terrane are interpreted as mid-oceanic ridge basalts which are overlain by Lower and Middle Jurassic sediments.

Permian to Middle Jurassic oceanic rocks of the Bridge River Terrane may once have been continuous with those of the Cache Creek Terrane. Thick Upper Triassic pillow basalt in the Cadwallader Terrane is overlain by turbiditic limestone, conglomerate and sandstone. These sediments are overlain unconformably by a sequence of basal conglomerate, siltstone and shale of Late Triassic to Early Bajocian age. The Cadwallader Assemblage may represent a tholeiitic island arc environment succeeded by intra-arc and back-arc settings.

In the Shuksan Terrane one thrust sheet contains schist and amphibolite of Permian to Early Triassic age. The rocks have been subjected to high-pressure, low-temperature metamorphism in Late Jurassic to Early Cretaceous time.

A fairly complete sequence of clastic, carbonate and basaltic and dacitic volcanic rocks of Pennsylvanian to Permian age occurs in the Chilliwack Terrane. South of latitude 49°N, Devonian strata also are present. The rocks are characteristic of a volcanic, island-arc setting.

Middle Triassic argillite and siltstone of the Harrison Lake Terrane are overlain unconformably by Toarcian sediments and volcanics, Aalenian shale and a thick succession of intermediate to felsic volcanics, possibly as young as Early Bajocian in the upper part.

In the Taku Terrane, separated from Stikinia by a high-grade metamorphic belt, Permian carbonate, phyllite and metatuff are overlain by Middle and Upper Triassic carbonate, phyllite and basaltic pillow breccia.

Complexly deformed, poorly fossiliferous Upper Paleozoic and lower Mesozoic carbonate, clastic and volcanic rocks are exposed in the Alexander Terrane in the northeastern part of the Saint Elias Mountains. In the more central parts of the terrane in the Icefield and Alsek ranges Pennsylvanian to Permian limestones occur within a thick, fine grained clastic sequence. In southeast Alaska Upper Devonian to Permian carbonates, fine grained clastics and rift-related volcanics represent platformal environments of deposition. Farther south Pennsylvanian carbonate and fine grained clastic sediments are present on islands west of Prince Rupert.

Alkaline to calc-alkaline plutons of Pennsylvanian to Early Permian age in the western parts of the Alexander Terrane appear to also cut rocks assigned to Wrangellia thus demonstrating that the two terranes had been amalgamated by Pennsylvanian time or were never separated.

Upper Triassic rocks in the Alexander Terrane include thick sequences of basalt, andesite, limestone, calcareous argillite, and mudstone. Gabbro and diabase sills and plugs are common. Rhyolite is present in southeast Alaska where basal conglomerate marks a regional unconformity at the base of Upper Triassic sediments and on islands west of Prince Rupert where an unconformity is present between Pennsylvanian and Upper Triassic sediments.

Mississippian to Permian chert, tuff, sandstone and limestone occur in Wrangellia on Vancouver Island and the Queen Charlotte Islands, and in the Saint Elias Mountains. In the latter region a lower, probably Pennsylvanian volcanogenic succession is overlain by an upper, Permian sedimentary succession with conglomerate near the base.

Thick units of oceanic-rift related basalt of Late Triassic (Carnian) age overlain by Carnian and Norian limestone are the characteristic formations of Wrangellia. The limestone grades upward into fine grained sediments of Triassic and Early Jurassic ages. In the Saint Elias Mountains the Upper Triassic limestone is associated with gypsum suggesting a sabkha setting.

Lower Jurassic rocks on Vancouver Island are dominated by thick sequences of calc-alkaline volcanics which are associated with granitic plutons. Lower Jurassic sediments correlative with the volcanics occur on islands along the east side of Vancouver Island, as roof pendants in the Coast Belt and on the Queen Charlotte Islands. In the latter area they are locally rich in hydrocarbons.

Introduction

Most rocks described in this section are Upper Devonian through Jurassic marine volcanic and sedimentary strata formerly called eugeosynclinal rocks. Most can be grouped into assemblages of restricted time ranges, based on lithological and stratigraphic characteristics reflecting the tectonic settings during their deposition (Map 1712A, in pocket). At the times of their formation, however, these rocks occupied unknown or uncertain paleogeographic positions with respect to each other and with the ancient western margin of cratonic North America.

The following discussion of stratigraphy is in relation to the various terranes of the western Cordillera. The reader is referred to the Terrane Map (Map 1713A, in pocket) on which terrane boundaries have been drawn according to the distribution limits of their characteristic assemblages. A simplified version of the Terrane Map is reproduced on the inside front cover of this volume.

Terrane stratigraphy

Cassiar Terrane

Rocks of Late Devonian to Late Triassic age in the Cassiar Terrane are similar to those in ancestral North America although displaced from them by an uncertain amount along dextral transcurrent faults in the Tintina and Northern Rocky Mountain trenches. They are described in Part A of this chapter.

Arctic Alaska Terrane

Like the rocks in the Cassiar Terrane the assemblages in the Arctic Alaska Terrane have a much stronger affinity with cratonal strata than with those of the pericratonic and accreted terranes. They are described in Part A of this chapter.

Kootenay Terrane

This terrane, extending discontinuously the full length of the Canadian Cordillera, lies between ancestral North America and its displaced part, Cassiar Terrane on the east and the volcanic Slide Mountain Terrane, Quesnellia and Stikinia to the west.

The upper assemblage of the Kootenay Terrane in the Kootenay Arc (Fig. 7.37) comprises the Mississippian and Pennsylvanian Milford Group (Milford Assemblage). It lies with angular unconformity on the Broadview Formation which, prior to the Late Mississippian, had been deformed into recumbent folds, metamorphosed to greenschist, and foliated. The Milford Group includes several facies: an eastern, near-shore facies, up to 600 m thick, is composed, in ascending order, of basal conglomerate, limestone with a Viséan coralline fauna similar to that of the Rundle Group in the Rocky Mountains, sandstone and shale, cherty tuff and greenstone. A more westerly volcanic facies, about 600 m in thickness, also overlying the Broadview Formation, comprises basal conglomerate and interbedded limestone and tholeiitic volcanics. A westernmost facies, up to 600 m thick, consists, in ascending order, of limestone and calcareous sandstone, tuffaceous sandstone, boulder conglomerate, limestone, and black siliceous argillite with minor tholeiitic flows. Each of the facies contains Late Mississippian conodonts and the intermediate one also yields Early Pennsylvanian conodonts (Klepacki and Wheeler, 1985). The boulder conglomerate contains clasts of quartzite, Broadview grit, and granitic rocks yielding an Early Ordovician U-Pb zircon age of 482 +91/-49 Ma (Okulitch, 1985). Most importantly, the siliceous argillite of the westernmost facies is overlain stratigraphically by Permian tholeiite of the Kaslo Formation of the Slide Mountain Terrane, thus the Kaslo also was deposited on the distal part of the Milford Group of the Kootenay Terrane.

An isolated outcrop of calcareous phyllite in the upper plate of the Columbia River Fault zone west of Kuskanax Batholith (Fig. 7.37) has yielded Late Devonian-Early Mississippian conodonts. If this outcrop belongs to the Milford Group it implies that its base becomes younger southeastward and that it transgressed in that direction onto a mid-Paleozoic Purcell "Arch".

The Milford Assemblage of the Kootenay Terrane extends southwestwards along Kootenay Arc into Washington State (Fig. 7.37) where it includes the upper part of the Silurian to Carboniferous Pend d'Oreille sequence (Yates, 1964; Little, 1985) and the Paleozoic rocks of the Covada Group farther south along Columbia River (Snook et al., 1982). Farther west in British Columbia, between Kootenay Arc and Greenwood, clastic sediments and limestone of the Pennsylvanian Mount Roberts Formation and the Carboniferous Mount Attwood Formation near Greenwood are probably correlative with the Milford Group.

In the Adams Lake region of the Shuswap Highlands (Fig. 7.37), where the upper Proterozoic and lower Paleozoic Eagle Bay Formation (Eagle Bay Assemblage) is dislocated by five west-verging thrust sheets, Mississippian rocks occur in the middle thrust sheet (Schiarizza and Preto, 1984). They are made up of grey phyllite, siltstone, sandstone, and grit with minor limestone, conglomerate, and tuff from which Early and Late Mississippian conodonts have been obtained. Its similar lithology and temporal equivalence, therefore, makes these strata correlative with the near-shore facies of the Milford Group of Kootenay Arc. The lower Paleozoic and Proterozoic rocks in the higher and more easterly thrust sheets are intruded by Late Devonian orthogneiss dated at 372 ± 6 Ma by U-Pb zircon methods (Okulitch et al., 1975).

The Barkerville Subterrane (Fig. 7.37) to the northwest contains a few metres of Permian Sugar Limestone overlying the probably Paleozoic upper Snowshoe Group. The subterrane has been intruded by the Quesnel Lake and related orthogneiss which yields U-Pb zircon dates between 335 Ma and 375 Ma and whose emplacement was between Late Devonian and mid-Mississippian time (Mortensen et al., 1987).

In the Omineca Mountains (Fig. 7.37), Proterozoic to Lower Cambrian gritty sandstone, siltstone, shale and limestone that may belong to Kootenay Terrane are intruded by granite sills which yielded zircon dates of 429 +10/-7 Ma and 353 ± 10 Ma indicating their emplacement between Early Silurian and Early Mississippian time (Gabrielse et al., 1982).

Nisutlin Allochthon

The Nisutlin Allochthon in the Yukon Territory consists largely of siliceous cataclasites in which stratigraphic relations are unknown, and subordinately of weakly metamorphosed rocks containing a crudely coherent stratigraphy (Nisutlin Assemblage). Lithologically the allochthon ranges from quartzofeldspathic phyllitic grit, quartzite, limestone and felsic to mafic volcanic rocks, through quartz-muscovite schist, chlorite schist, marble and quartzite, to mylonite, ultramylonite and blastomylonite. The rocks are commonly strongly and pervasively sheared and their flaser fabric and schistosity masks or obliterates bedding (Tempelman-Kluit, 1979a).

The Nisutlin Allochthon forms one or more extensive thrust sheets in southern and western Yukon (Fig. 8.61). In Jura-Cretaceous time it was thrust over and imbricated with less deformed Upper Triassic and older rocks of the passive margin of ancestral North America including, according to Tempelman-Kluit (pers. comm.), Devonian-Mississippian augen orthogneiss. The latter is similar to the Pelly Gneiss of the Fifty-Mile Batholith southwest of Dawson which was dated by U-Pb methods on zircon at 375 ± 6 Ma (Tempelman-Kluit and Wanless, 1980) and 362.4 +10.2/-5.5 Ma (Mortensen, 1986). The Nisutlin Allochthon is commonly the lowest Yukon allochthon and is succeeded by the mafic volcanic-ultramafic-chert-limestone assemblage of the Anvil Allochthon, which in Yukon represents the Slide Mountain Terrane. This in turn is structurally overlain by the granitic Simpson Allochthon (Fig. 8.62). The Nisutlin Allochthon was considered by Tempelman-Kluit (1979a) to represent a Late Triassic-Early Jurassic trench mélange which formed part of an accretionary prism east of Stikinia. It contains lenses of eclogite (Erdmer and Helmstaedt, 1983) and blueschist (Erdmer, 1987) near its structural contact with the Anvil Allochthon. The eclogite is considered to have formed at depths of 40 km (Erdmer and Helmstaedt, 1983). The accretionary prism and overlying allochthons were later thrust onto ancestral North America and subsequently displaced in Late Cretaceous to Paleogene time by right-lateral movement along the Tintina Fault.

Later isotopic studies by Mortensen and Jilson (1985), revealed, however, that the Nisutlin Allochthon contains mid- and upper Paleozoic rocks that are lithologically similar to those in the Kootenay Terrane in the southern Canadian Cordillera. Accordingly, the Nisutlin Allochthon has been included in the Kootenay Terrane rather than as part of the Yukon-Tanana Terrane (Monger and Berg, 1984). As used in Alaska the latter is equivalent to the allochthonous Nisutlin and Slide Mountain terranes and the underlying craton-related Nisling Terrane consisting of micaceous quartzite, quartz-mica schist, and marble associated with Devono-Mississippian orthogneisses.

According to Mortensen and Jilson (1985) the Nisutlin Allochthon, east of Tintina Fault, contains a mappable sequence more than 3 km thick within a layered metamorphic succession. The lowest unit is a pre-Upper Devonian sequence of locally garnetiferous, quartz-mica schist, micaceous feldspathic quartzite and, near the top, marble and calcareous schist. It is overlain by a middle unit of dark grey to black siliceous and carbonaceous phyllite and quartzite, locally calcareous near its base. The carbonaceous rocks are interlayered with abundant mafic metavolcanics and lesser felsic metavolcanics. The latter yielded an Early Mississippian U-Pb zircon date on tuff-breccia of 354 +7/-5 Ma. More equivocal U-Pb zircon and sphene dating of other felsic volcanics intruded by earliest Mississippian plutons suggests they are probably of Devonian age (Mortensen, 1983). The upper part of the succession includes abundant chloritic quartz grits which locally contain blue quartz, and Lower Pennsylvanian to Lower Permian conodont-bearing carbonate and quartzite at the top.

Mortensen and Jilson (1985) referred to the granitoid rocks of the Simpson Allochthon as the Simpson Range suite. The quartz diorite-granodiorite phase has yielded an earliest Mississippian U-Pb zircon date of 359 +4/-3 Ma and the quartz monzonite phase another of 349 +3/-2 (Mortensen, 1983). Although much of the Simpson Range suite is allochthonous, Mortensen and Jilson maintained that rocks of the Nisutlin Allochthon are intruded by a dyke lithologically similar to the early phase of the Simpson Range suite and elsewhere wall rocks adjacent to the Simpson Range granitoids have been pyritized.

They also considered the augen orthogneiss, correlated with the Pelly Gneiss, to intrude the rocks of the allochthon because of a rapid increase in metamorphic grade of the country rock adjacent to the orthogneiss. U-Pb zircon dating of the orthogneiss and of related monzonitic orthogneiss is imprecise but suggests a Mississippian age. Mortensen (1983) believed that a Rb-Sr whole rock age of 342 ± 7 Ma is probably a minimum but close to the real age of Early Mississippian for the orthogneiss.

Sheared conglomerate occurs at two places along the eastern edge of the Nisutlin Allochthon. It contains detrital mica and clasts of deformed rocks of the Nisutlin and Anvil allochthons, gneissic granitic rocks, and carbonate containing Late Triassic and early Paleozoic conodonts and late Paleozoic fusulinids. Limestone and limy

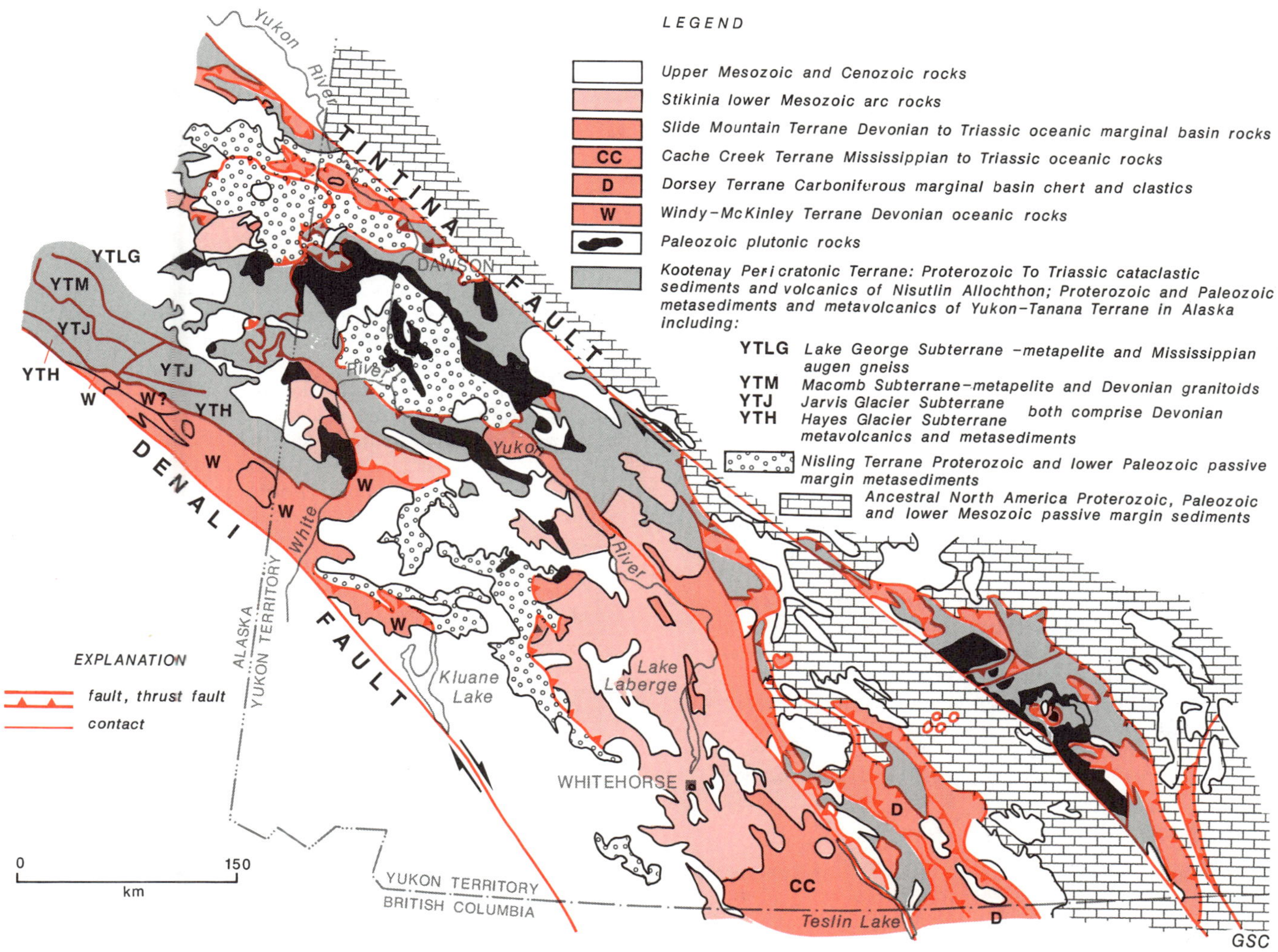

Figure 8.61. Distribution of the Kootenay Terrane and its relationships to adjoining terranes in southwestern Yukon Territory and eastern Alaska.

Figure 8.62. Early Mississippian granitic rocks of the Simpson Allochthon thrust onto oceanic rocks of the Anvil Allochthon. View northward in Simpson Ranges northeast of the Tintina Trench. Photo by P. Erdmer. GSC 205235-I

sandstone, tectonically interleaved with the conglomerate, contain Late Triassic conodonts and chips and grains of phyllonite and mylonite (Tempelman-Kluit, 1979b). It appears, therefore, that the conglomerate may have formed in front of an advancing allochthon in Late Triassic time as suggested by Tempelman-Kluit (1979a). It also supports the view of Mortensen and Jilson (1985) that the Late Devonian-Mississippian, possibly volcanic arc-plutonic succession and older rocks were deformed and metamorphosed to middle amphibolite facies by Late Triassic time. In view of its sheared nature, the conglomerate was later involved in overthrusting as the allochthon continued to be transported northeastward in Jurassic and Early Cretaceous time.

Some of the lithologies of the allochthon, notably the felsic metavolcanics, carbonaceous phyllite, and chloritic grit, have counterparts in southern Kootenay Terrane. Moreover, Devonian felsic metavolcanics occur in the northern and southern parts of the terrane. On the other hand, Mississippian felsic metavolcanics and broadly coeval Simpson Range or Simpson Allochthon granites in the north are younger than such rocks in the south. The carbonaceous phyllite and chlorite grit appear to be younger in Nisutlin Allochthon than those in similar rocks in southern Kootenay Arc. It is possible, however, in view of the high degree of shuffling and dismemberment in the Nisutlin Allochthon that the distinctive carbonaceous phyllite and chloritic grit, so characteristic of the lower part of the Kootenay Terrane in the south, may have been thrust higher in the succession in the north.

West of Tintina Fault the siliceous cataclasites of the undulating Nisutlin Allochthon extend northwestward from Pelly Mountains into the Klondike region (D.J. Tempelman-Kluit, pers. comm., 1986) (Fig. 8.61). The rocks of the allochthon are not as well understood in this region and, in areas of poor outcrop northwest of Pelly Crossing, cannot be readily differentiated from the Anvil Allochthon, Pelly Gneiss or the underlying autochthon of the Nisling Terrane. Southeast of Dawson the Nisutlin Allochthon contains the Late Devonian to Late Mississippian Mount Burnham orthogneiss yielding a U-Pb zircon age between 369 Ma and 327 Ma (J.K. Mortensen, pers. comm., 1986). Southwest of Dawson the allochthon also includes the Early Permian Sulphur Creek orthogneiss yielding a concordant U-Pb zircon age of 263.8 ± 3.2 Ma, a marginal quartz-eye schist giving a concordant U-Pb zircon age of 262 ± 3.6 Ma, and a nearby felsic metatuff producing a U-Pb zircon age between 267 and 239 Ma. The Nisutlin Allochthon Assemblage presumably continues northwestward into Alaska where it is represented by the Klondike schist (Foster, 1970, 1976). It is thrust over the carbonaceous and siliceous metasediments of the Nasina Formation of the Nisling Terrane which extends northwestwards from Dawson.

Windy-McKinley Terrane

The White River Assemblage underlies an area assigned to the Windy-McKinley Terrane (Jones et al., 1984) that straddles the Alaska-Yukon boundary northeast of Denali Fault. West of the boundary it consists of phyllite, metaconglomerate, quartzite, quartz-mica schist, greenstone, diorite, rare ultramafics and gabbro, and numerous discontinuous limestone bodies some of which contain Middle Devonian fossils (Richter, 1976). In Yukon the assemblage is unfossiliferous and comprises cherty argillite, chert, mafic volcanics, ultramafic and gabbroic rocks (Tempelman-Kluit, 1974) and metamorphic equivalents, including marble, formerly assigned to the Yukon Group (Muller, 1967).

The age of the White River Assemblage is uncertain but may contain mixed Paleozoic and Mesozoic assemblages. It contains lithologies, however, that are similar to those in the fossiliferous Windy and McKinley terranes of the Alaska Range. There, the Windy Terrane includes Devonian limestone blocks within a matrix of argillite, sandstone and conglomerate; chert clasts from the conglomerate have yielded Mississippian, Triassic and Jurassic or Early Cretaceous radiolaria; and structurally related argillite contains Silurian trilobites. The McKinley Terrane also contain Devonian limestone blocks; greywacke-argillite flysch and radiolarian chert containing late Paleozoic and late Mesozoic, mainly Cretaceous fossils; and pillow basalt with gabbro dykes and sills, both associated with Upper Triassic sediments.

If such correlations are reasonable then the White River Assemblage may consist of Paleozoic and early Mesozoic mafic volcanic-ultramafic-chert-argillite oceanic terranes, perhaps belonging to the Cache Creek Terrane, interleaved with the Jurassic and Cretaceous greywacke-argillite and mélange of the nearby Gambier Assemblage.

Slide Mountain Terrane

The Slide Mountain Terrane is the easternmost accreted terrane and lies almost entirely in fault contact on the Kootenay and Cassiar terranes. It forms a discontinuous belt that extends along the Omineca Belt for approximately 2000 km, from latitude 49°30'N near Kootenay Lake to latitude 65°N in the western Yukon. The terrane is characterized by the Slide Mountain Assemblage consisting of mainly mafic volcanics with related ultramafic, mafic and rarely granitic bodies, and predominantly fine grained clastic rock and chert, which is of mid- to late Paleozoic and, rarely, Triassic age. Where detailed paleontological and structural studies have been carried out, as in the Cassiar and Cariboo Mountains, it is known to consist of stacked and imbricated thrust sheets, and continuous stratigraphic sections can be determined only within individual sheets. Although originally called "Eastern Assemblage" by Monger (1977b) from its position as the easternmost upper Paleozoic eugeosynclinal assemblage, the term was discarded by Monger and Berg (1984) in favour of Slide Mountain Terrane, a name taken from Slide Mountain Group of the Cariboo Mountains where the internal stratigraphy and its stratigraphic and structural relationships with adjacent strata, are relatively well known. The rock successions carry many formal names, each applied to a more or less continuous outcrop area which could be designated a "subterrane", and are described below, from north to south.

Nisutlin Plateau

Upper Paleozoic rocks of the Nisutlin Plateau in south-central Yukon have been subdivided into two formations known informally as the Boswell and Semenof (Tempelman-Kluit, pers. comm.). The Boswell formation is about 1100 m thick and consists of laterally interfingering units of slate, phyllite, chert, greenstone and limestone. Locally, slate, possibly 100 m thick, is the lowest unit and is overlain by about 500 m of greywacke and chert-pebble conglomerate or grit with minor red mudstone and slate. The greywacke is in part calcareous and near the slate-greywacke transition contains bioclastic limestone pods as much as 30 m thick. Massive sills or flows of greenstone occur within the greywacke-conglomerate unit and a locally pillowed greenstone member 200 m thick overlies the sediments. Massive to thick bedded micritic limestone up to 300 m thick occurs stratigraphically above the greenstone but grades laterally into greywacke, conglomerate and greenstone. Conodonts and fusulinids indicate an Early to mid-Pennsylvanian age.

The Semenof formation is a massive volcanic formation, possibly 800 m thick. The rocks are mainly homogeneous and structureless but in places show faint pillow outlines, amygdules and layers of basaltic tuff. Augite phenocrysts are present locally. A carbonate lens in the formation contains fusulinids of Late Moscovian age.

In lithology and age the Boswell and Semenof formations are similar to those in the Pennsylvanian part of the Sylvester Group of the Slide Mountain Assemblage.

Anvil Range latitude 62°N

In the Anvil Range of southern Yukon, north of the Tintina Fault and south of the Anvil Batholith (Gordey, 1983), the Anvil Range Group, as redefined by Tempelman-Kluit (1972), consists of three units. The lowest comprises approximately 700 m of massive to thin-bedded, pale green, grey and locally brick-red, argillaceous and tuffaceous chert, with minor bioclastic limestone containing Early Permian fusulinids in its upper part. It is overlain by 500 m of massive dark green amygdaloidal flows and pillowed basalt, and basaltic pyroclastics. A younger unit, a few hundred metres thick of massive, buff to light grey, locally tuffaceous limestone contains Late Permian fusulinids, and is commonly separated by faults from the remainder of the Anvil Range Group. A narrow band of fault-bounded serpentinite bounds the Permian volcanics on the south side.

Further regional studies in the central Yukon by Tempelman-Kluit (1970), led to the use of the term "Anvil Allochthon" for allochthonous sheared ophiolite, which included the Anvil Range Group and other unnamed but lithologically similar associations. As used here, Slide Mountain Terrane in the Yukon is synonymous with Anvil Allochthon.

Cassiar Mountains latitude 59°N

In the Cassiar Mountains of north-central British Columbia Devonian to Upper Triassic rocks, included in the Sylvester Group, form a stack of fault-bounded slices which is emplaced on the Cassiar Terrane (Gabrielse, 1963; Gordey et al., 1982b; Harms, 1984, 1985a,b; 1986) (Fig. 8.63, Frontispiece). Characteristic lithologies are radiolarian ribbon chert, tholeiitic volcanics, variably serpentinized ultramafic rocks, and foliated hornblende diorite (Fig. 8.64). Although the structural style precludes recognition of an unbroken stratigraphic sequence, certain rock units or association of rock units have a specific age range, thus allowing reconstruction of pre-fault stratigraphy. Little can be said about facies changes across the depositional basin, which must have been extensive in view of the numerous fault slices which contain coeval rocks of different lithologies.

Clastic sediments in the Sylvester Group include ubiquitous shale and argillite commonly associated with chert, and at least one fault-bounded sequence of siliceous siltstone, sandstone and greywacke with local conspicuous gritty quartz sandstone and pebble conglomerate containing blue opalescent quartz grains.

Ribbon chert, commonly containing well-preserved radiolaria and, locally, conodonts, is abundant and ranges in age from pre-Devonian(?) through Permian. Black banded chert has yielded Late Devonian to Late Mississippian radiolaria, apple green chert is common in Pennsylvanian rocks and red chert is common in Permian strata. No Mesozoic chert has been identified.

Distinctive Upper Mississippian limestone of the Nizi Formation, at least 500 m thick, has a prolific foraminiferal fauna dated as late Viséan to early Namurian (Mamet and Gabrielse, 1969). The rocks are well-bedded bioclastic limestone and subordinate dolostone. A conspicuous maroon weathering shale and siltstone member occurs near the top. Some beds are characterized by chert nodules and others particularly near the base by large crinoid columnals. Pennsylvanian limestone generally less than 75 m thick is in many places crinoidal and contains fusulinids and conodonts of Early and Middle Pennsylvanian ages. An Upper Permian tuffaceous limestone unit about 50 m thick

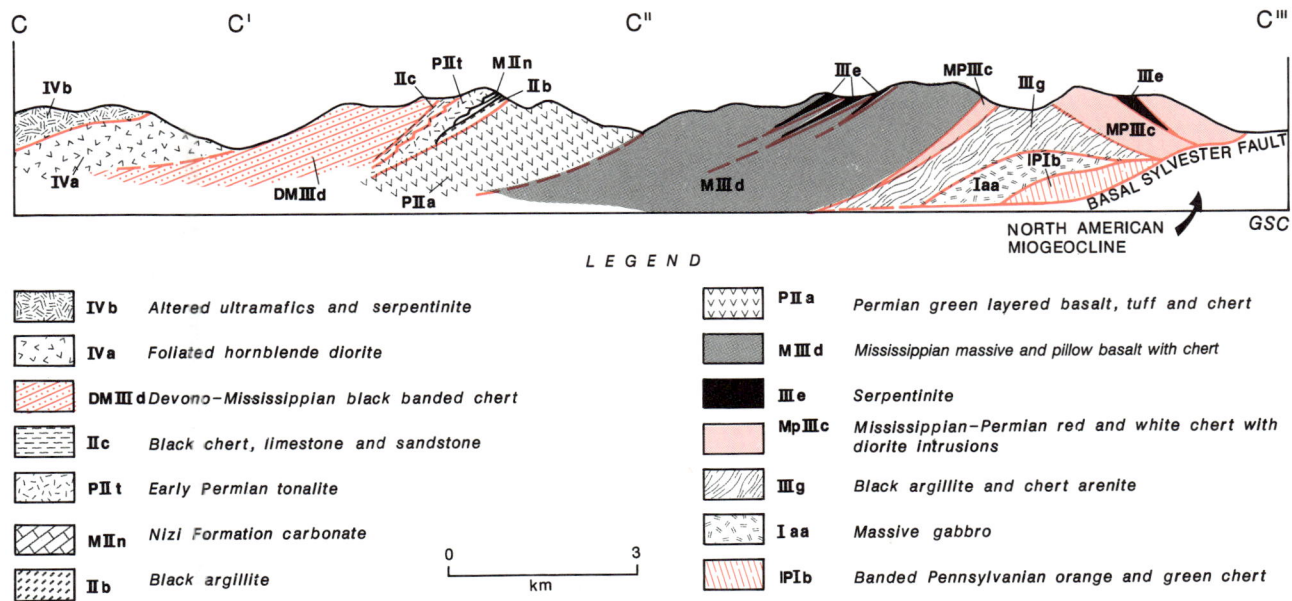

Figure 8.63. Structural cross-section through the southeastern part of the Sylvester Allochthon showing fault-bounded lithological successions. After Harms (1986).

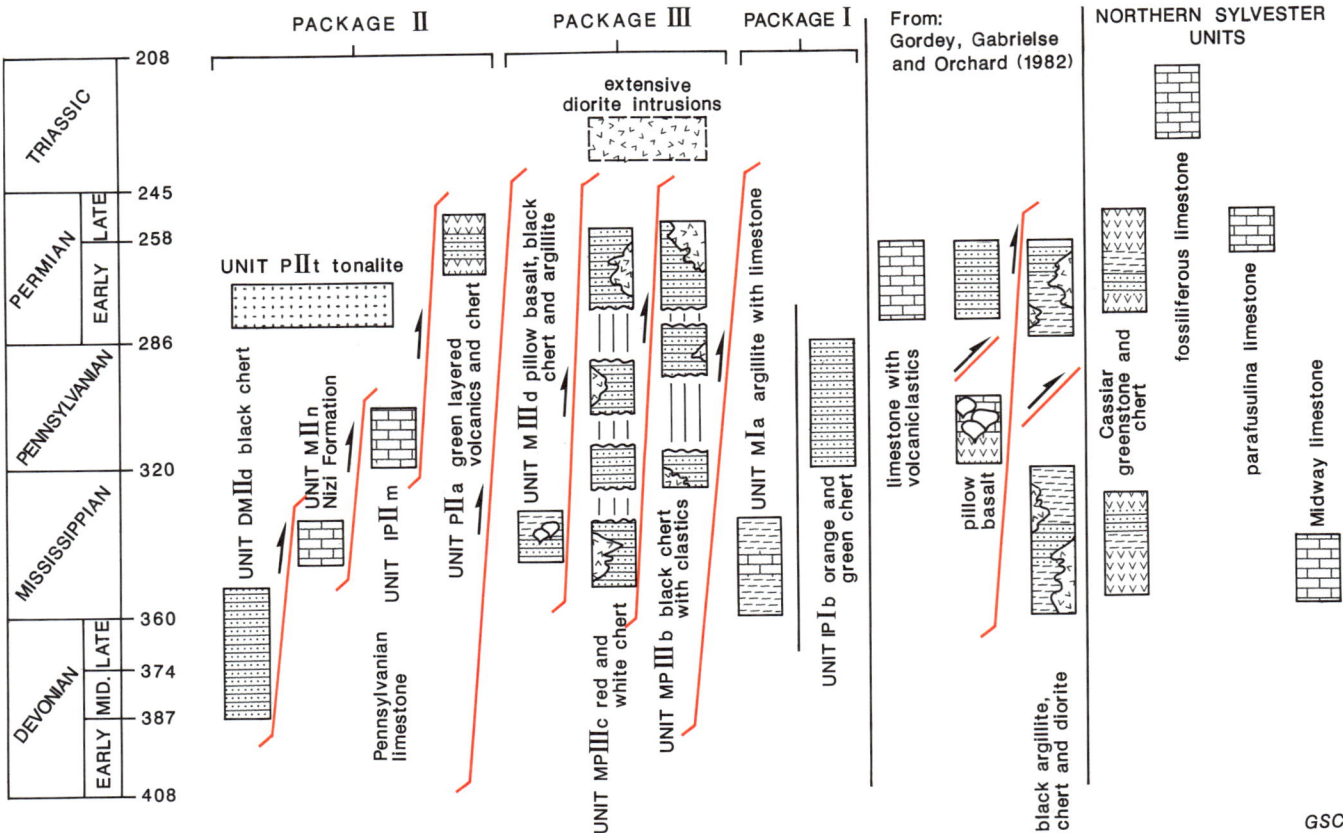

Figure 8.64. Chronostratigraphic summary chart for the Sylvester Allochthon rocks. After Harms (1986).

contains brachiopods and a distinctive giant species of *Parafusulina* similar to forms in the southwestern United States and northwestern Mexico (Ross, 1969). Fetid, platy limestone, less than 10 m thick is abundantly fossiliferous and contains Middle and Late Triassic pelecypods and ichthyosaur bones.

Thick sequences of tholeiitic basalt, locally forming spectacular pillowed units, are dated by associated sediments as Mississippian, Pennsylvanian and Permian. Pillowed basalt seems to be particularly abundant in Lower to Middle Pennsylvanian sequences. Augite basalt and associated sediments are host rocks for gold-quartz veins near Cassiar and provided the source for rich, shallow-placer gold deposits in the McDame Creek region. Tuffaceous members are present but are not conspicuous.

Ultramafic rocks occur as bodies of partly serpentinized dunite, harzburgite, peridotite and pyroxenite, as large as 25 km² or as small elongate lenses of highly sheared serpentinite along fault zones. Serpentinite is host to the asbestos deposit at Cassiar and altered nephrite bodies are hosts for important jade occurrences.

Hornblende diorite and hornblende quartz diorite plutons as much as 50 km² in extent occur within fault slices and locally may cut faults between slices. They are mainly foliated but include at least one body of granite that appears little deformed. K-Ar ages range from 261.9 to 266.5 Ma and U-Pb zircon ages range from 276 to 285 Ma. One foliated body of hornblende tonalite has given a U-Pb zircon age of 362 ± 5 Ma with a marked Early Proterozoic inheritance similar to those in the Kootenay Terrane (P. Van der Heyden, pers. comm., 1986).

Much of the Sylvester Group is of oceanic affinity. Shallow-water limestone perhaps was deposited on or around volcanic atolls or plateaus. The quartzose clastics suggest a continental or island arc source. These sediments may be part of the Devonian-Mississippian autochthonous succession interleaved with the oceanic rocks during the emplacement of the allochthon onto the miogeocline. The diorite and local granitic bodies indicate derivation from an island arc that existed during latest Pennsylvanian to Early Permian time. The latest Devonian-earliest Mississippian pluton could be a thrust slice of Kootenay Terrane juxtaposed with oceanic Sylvester rocks. Harms (1984) has argued that the Sylvester represents a variety of oceanic environments, and originally covered a vast region of the paleo-Pacific.

Crosscutting of a few faults by the Paleozoic plutons proves that some stacking of thrust sheets took place in the late Paleozoic. Other thrust faults, directed eastward, are associated with east-verging folds and reflect emplacement of the allochthonous rocks onto the miogeocline in post-Triassic and pre-mid-Cretaceous time.

Omineca Mountains latitude 56°N

In the western Omineca Mountains, upper Paleozoic strata, previously assigned to the Cache Creek Group by Armstrong (1949), are herein assigned to the Slide Mountain and Harper Ranch assemblages. The lower, Middle Pennsylvanian to earliest Permian sedimentary succession with its distinctive conglomerate facies has some similarities with the Harper Ranch Assemblage of the Harper Ranch Subterrane of Quesnellia. The Upper Pennsylvanian to Lower Permian, mafic volcanic succession is typical of the Slide Mountain Assemblage (Monger and Paterson, 1974; Monger, 1977b).

Locally, as near Nina Creek, the rocks form the upper part of an apparently homoclinal succession dipping gently westwards away from the metamorphic culmination of the Wolverine Metamorphic Complex, and are faulted on the west against Upper Triassic strata of Quesnellia. The Slide Mountain rocks comprise a succession of cherty argillite, with local thin Lower Permian carbonate beds and thick gabbro and diabase sills as much as 100 m in total thickness. The upper, mainly volcanic part comprises massive, variolitic, locally pillowed basalt, the extrusive equivalent of the sills, interbedded fine grained, calcareous lithic tuff in places with Early Permian conodonts, and small, fault-bounded bodies of sheared serpentinized peridotite.

Cariboo Mountains latitude 53°N

The Slide Mountain Terrane near Barkerville in the Cariboo Mountains consists solely of the Antler Formation of the Slide Mountain Group. It is made up of pillow basalt, diorite, chert, argillite, phyllite and minor greywacke, gabbro and ultramafic rock (Sutherland Brown, 1963; Struik, 1980). A characteristic section of Antler Formation on Sliding Mountain (Fig. 8.65A,B) consists of three thrust sheets of interbedded chert and argillite, basalt and diorite dykes and sills. The layers face stratigraphically up to the northeast as indicated by loading structures in the pillow basalt and by the order of decreasing age of conodonts extracted from the chert. Conodonts in chert at the type section (Sliding Mountain) are Early Mississippian (possibly Late Devonian) to Early Permian in age (Struik and Orchard, 1985).

Several common stratigraphic elements in thrust sheets at Sliding Mountain consist of 1) Lower Mississippian pillow basalt and minor ribbon chert (indicating possible widespread Mississippian volcanism), 2) Pennsylvanian chert and argillite, and 3) Permian pillow basalt. Upper Pennsylvanian strata, if present, are relatively thin. Diorite dykes and sills intrude the Pennsylvanian chert sequences and may have fed the Permian pillow basalt.

Elsewhere other lithologies include dykes and small pods of gabbro and ultramafic rocks, dark grey greywacke and slate, basalt breccia, tuff and agglomerate and minor volcanic-clastic conglomerate.

The structural thickness of the Antler Formation was estimated by Sutherland Brown (1963) to be about 1100 m, but the stratigraphic thickness is probably less, with 300 m as a minimum. The thickness variations in the terrane are due to the local accumulation of basalt and diorite and possibly to variations in the degree of thrust imbrication.

Typically the Slide Mountain Terrane consists of thin imbricated thrust sheets, as seen in three or possibly four sheets at Sliding Mountain. Some sheets carry much, if not all, of the known age span of the Antler Formation. Thus, approximately 100 Ma (Palmer, 1983) is recorded by less than 300 m of chert and basalt. The Antler Formation, before thrust imbrication, must have been a very thin sequence covering a large area, as has been argued by Harms (1984, 1986) for the Sylvester Group. The formation has been thrust more than 50 km eastward (Struik, 1980) and yet maintains its internal stratigraphic coherence.

Shuswap Highlands latitude 51°N

In the Adams Lake-Shuswap Lake area the mid-Carboniferous to Permo-Triassic Fennell Formation consists of a lower unit of basalt and cherty sedimentary rock and an upper unit of massive and pillowed greenstone (Campbell and Tipper, 1971; Schiarizza, 1982). The lower unit contains a wide variety of rock types that are generally discontinuous and interdigitated on a fine scale. The dominant rock is aphanitic to coarse grained, locally pillowed basalt, together with intrusive phases including concordant diorite and gabbroic bodies. In places there are small bodies of quartz feldspar porphyry. Sedimentary rocks include dominant chert and cherty argillite which contain discontinuous lenses of conglomerate with chert, basalt and argillite clasts, together with local bodies of sandstone and minor limestone.

The upper unit consists of aphanitic to fine grained, commonly pillowed basalt, with small discontinuous bodies of bedded chert.

Between latitudes 52°N and 50°N remnants of the Slide Mountain Assemblage occur sparingly across the Omineca Belt, notably northwest of Shuswap Lake, between Kootenay and Arrow lakes, and around Greenwood near the International border. Because of intense deformation and metamorphism the rocks are poorly dated but are correlated on the basis of their common content of tholeiitic basalt, metamorphosed to greenschist and amphibolite, chert, and ultramafic rocks. Near Shuswap Lake, greenstone, chlorite schist, amphibolite, serpentinized ultramafic rocks, minor shale and carbonate of the Tsalkom Formation, formerly included in the "Slide Mountain Assemblage", are now considered by A.V. Okulitch (pers. comm., 1986) to belong to the lower Paleozoic Jowett Formation. Farther southeast, Parrish (1981) suggested that sillimanite-grade pelitic and calcareous schist, amphibolite and metamorphosed ultramafic rocks of the upper part of the Nemo Lakes belt, are probably correlative in part with upper Paleozoic strata included in the Slide Mountain Terrane.

Selkirk Mountains latitude 50°N

The Kaslo Group west of Kootenay Lake consists of two parts separated by a thrust fault (Klepacki, 1983; Klepacki and Wheeler, 1985; Fig. 8.66). The upper sheet, predominantly of tholeiitic pillow lava with minor greywacke and volcanic conglomerate, is floored by ultramafic rock. The lower sheet, beneath the Whitewater Thrust Fault, includes pyroxene-porphyry pillow lava, flows and tuffaceous greenstone interbedded with green and white cherty tuff

CHAPTER 8

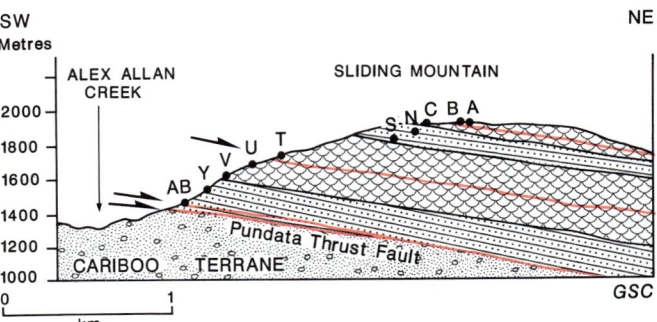

Figure 8.65A. Thrust imbricated stratigraphic section of Antler Formation along southwest ridge of Sliding Mountain. Fig. 8.65B. Structure section along ridge of Sliding Mountain. See Figure 8.65A for explanation of patterns. After Struik and Orchard, 1985.

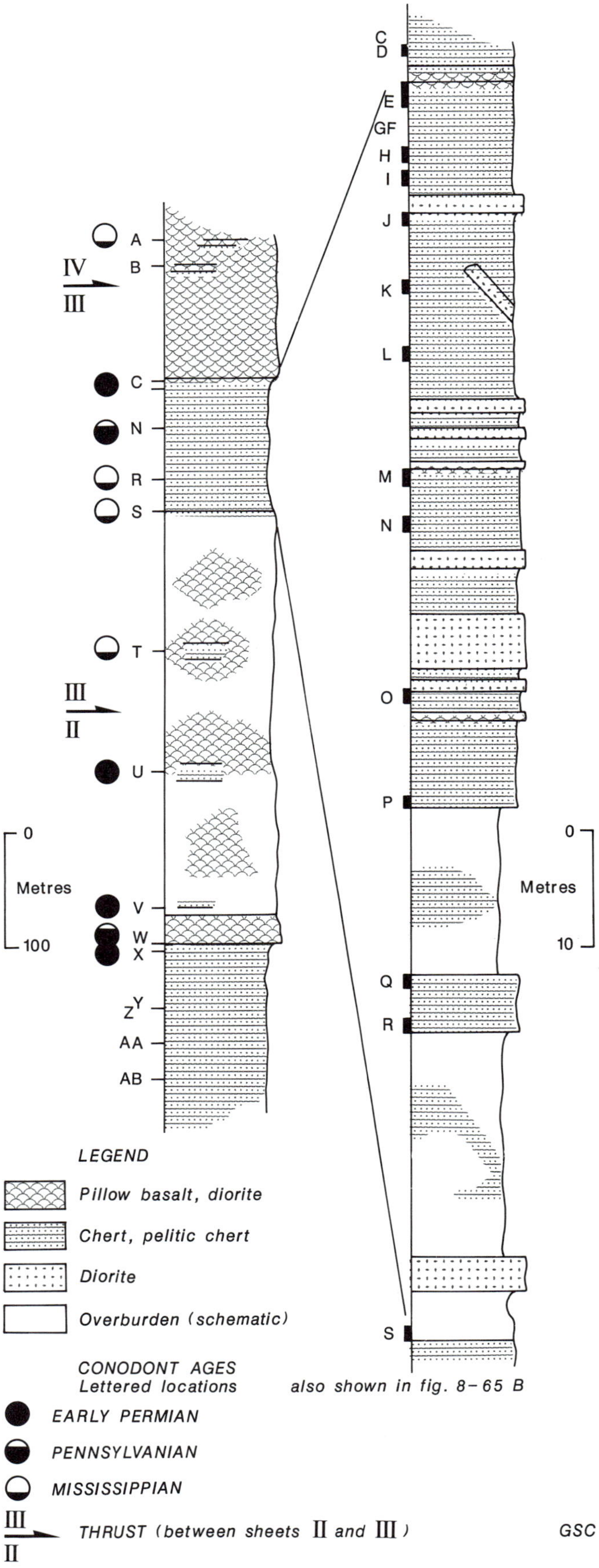

which has yielded Late Early to early Late Permian (Late Leonardian to Early Guadalupian) conodonts. The unit is stratigraphically important because it overlies fossiliferous Upper Mississippian, mainly sedimentary rocks that are correlated across the Early to early Middle Jurassic Stubbs Fault with strata of the Milford Group, which in turn lie unconformably on the lower Paleozoic Lardeau Group, a major component of the Kootenay Terrane.

Both sheets of the Kaslo Group and the Whitewater Thrust Fault are intruded by diorite of probable Permian age, a situation analogous with that in the Sylvester Allochthon in northern British Columbia described above. The upper sheet and the diorite are overlain unconformably by the Marten Conglomerate, which contains limestone lenses with conodonts of broadly Permian to Middle Triassic age.

Tectonic history of the Slide Mountain Terrane

Late Permian fusulinids in the Sylvester Group (Ross, 1969) are similar to other forms in western Cordilleran accreted terranes, such as Quesnellia, and to forms in cratonic rocks in the southwestern United States. No fusulinids are known in Permian strata of this age at comparable latitudes on the Canadian craton. This suggests that the Slide Mountain Terrane in northern British Columbia is displaced with respect to the ancient continental margin.

By Late Triassic time, rocks forming the eastern margin of Quesnellia overlap the Slide Mountain Terrane. In most places the Slide Mountain Terrane appears to be thrust eastwards on to pericratonic terranes composed of rocks of mainly Paleozoic age. One exception is in the Yukon where Upper Triassic strata occur locally in the footwalls of the thrust faults (Tempelman-Kluit, 1979a). The faults, therefore, are post-Triassic and are cut by mid-Cretaceous granitic plutons. In southern British Columbia, major eastward thrusting is clearly pre-Middle, probably pre-late Early Jurassic from crosscutting plutonic relationships. Eastward directed thrust faults separating the Fennell Formation from the Eagle Bay Formation of Kootenay Terrane predate the main, mid-Jurassic metamorphism in the region and are refolded by west-directed, syn- and post-metamorphic structures (Brown and Read,

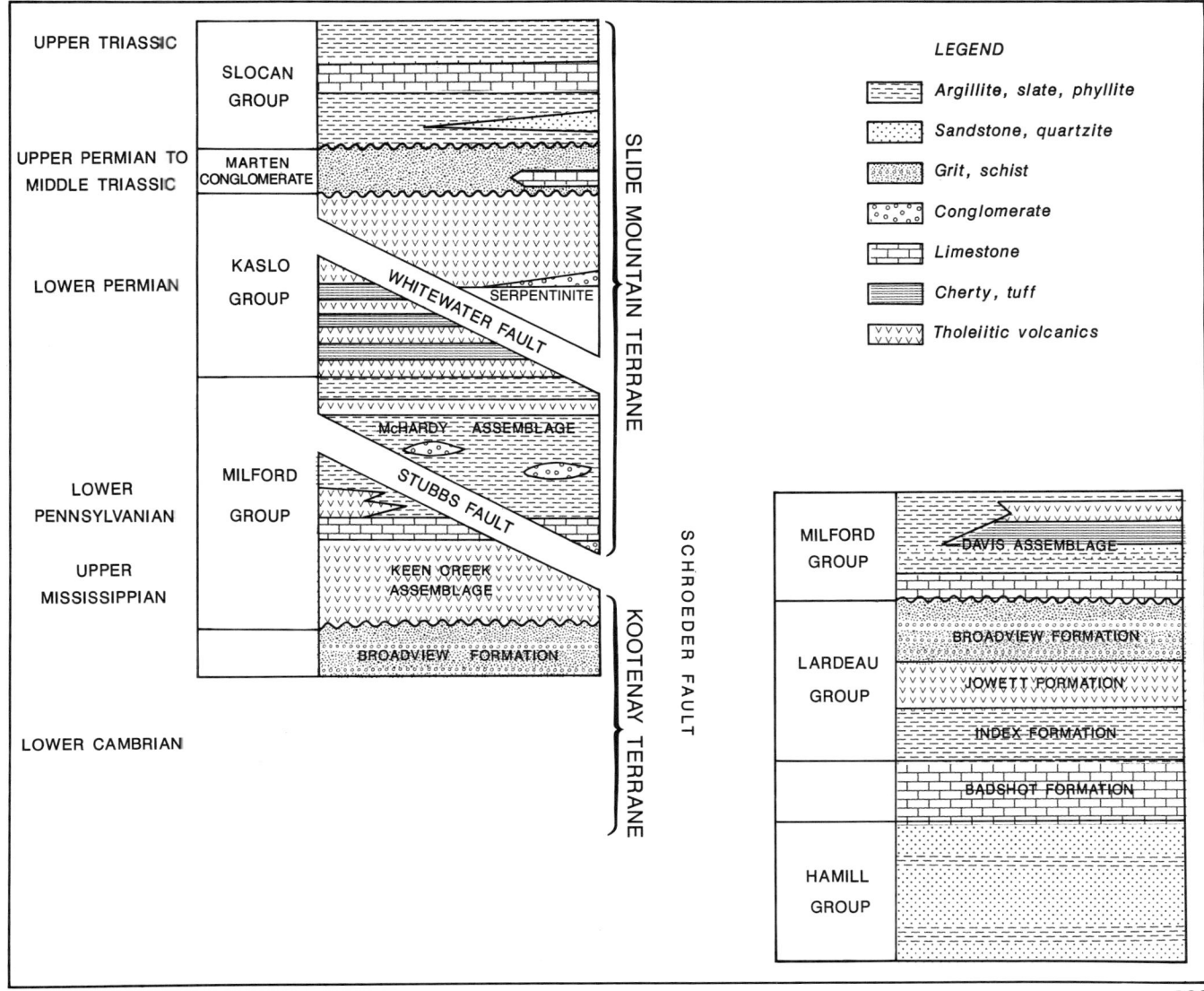

Figure 8.66. Relationships between Slide Mountain Terrane (Kaslo Group) and Kootenay Terrane (Milford Group) in southeastern British Columbia (from Klepacki and Wheeler, 1985). Rocks stratigraphically below the Kaslo Group (McHardy Assemblage) are shown as a facies of the Milford Group, which lies unconformably on strata of the Kootenay Terrane. Note: McHardy "Assemblage" is not a tectonic assemblage as used in this volume. Modified from Klepacki and Wheeler (1985).

1983; Schiarizza and Preto, 1984). The Kaslo Group and underlying rocks of the Milford Group are thrust eastwards on the pre-Middle Jurassic Stubbs Thrust Fault.

There is evidence for earlier thrust faulting within the Slide Mountain Terrane from two localities. Harms (1985) has shown that Late Permian tonalite crosscuts one thrust fault that internally imbricates the Sylvester Group in northern British Columbia, prior to its emplacement onto continental margin rocks. In southern British Columbia, Klepacki and Wheeler (1985; Fig. 8.66) documented stratigraphic overlap of the Kaslo Group, the southernmost part of Slide Mountain Terrane, on pericratonic rocks included in the Kootenay Terrane. The Kaslo Group, however, was imbricated prior to the intrusion of Permian? diorite and before the deposition of the Permo-Triassic Marten Conglomerate.

The depositional setting of much of the Slide Mountain Assemblage appears to have been in an ocean basin. However, parts of the terrane, such as some clastic rocks in the Sylvester Group, conglomerate in the Nina Creek area, and the pericratonic terranes (Klepacki and Wheeler, 1985), appear to have formed near the continent. It seems likely that the Slide Mountain Terrane includes both near-continent and distal parts of a late Paleozoic ocean basin assemblage, perhaps juxtaposed in Permian time by thrust faulting, with the whole imbricated stack thrust in Early to Middle Jurassic time over the ancient continental margin.

Quesnellia

Quesnellia comprises upper Paleozoic and lower Mesozoic volcanic and sedimentary strata and associated plutonic rocks best developed in the south-central Canadian Cordillera. The relatively wide area underlain by Quesnellia in south-central British Columbia is probably largely due to Eocene extension (see Chapter 17; Fig. 8.67). The terrane tapers to a narrow, discontinuous, region north of Prince George (lat. 54°N), where it is bounded on to the west by the Pinchi, Finlay, Kutcho, Thibert and Teslin faults (Gabrielse, 1985). On the east it stratigraphically and structurally overlies the Slide Mountain, Cassiar and Kootenay terranes. It extends into the Yukon as a narrow region lying between the Slide Mountain Terrane to the east and the Cache Creek Terrane to the west. Quesnellia is characterized by predominantly volcanogenic Upper Triassic and lowest Jurassic strata of the Nicola and Quesnel River groups, by flanking Triassic "black phyllite" on the east, by probably comagmatic plutonic rocks, and by stratigraphically overlying Lower and Middle Jurassic clastic rocks such as the Ashcroft Formation, and other unnamed units. It includes, correlative strata of the Triassic Slocan Group and Jurassic Archibald, Elise, and Hall formations of the Rossland Group in southeastern British Columbia (Little, 1982), the Triassic Takla Group and overlying Jurassic clastics east of Pinchi Fault in central British Columbia (Armstrong, 1949; Monger, 1977a; Tipper, 1976), the Triassic Shonektaw and Sinemurian Nazcha formations of north-central British Columbia (Watson and Mathews, 1944; Gabrielse, 1968), and older, upper Paleozoic strata that stratigraphically underlie them. In the Yukon, it is represented by unnamed augite porphyry volcanics and sediments east of Teslin Lake and by intrusive rocks such as the early Mesozoic Tatchun pluton.

Upper Paleozoic assemblages

Upper Paleozoic strata, stratigraphically underlying Mesozoic strata of Quesnellia in southern British Columbia (Read and Okulitch, 1977), are relatively poorly known and have a plethora of local names. They appear to comprise two distinct lithological assemblages (Fig. 8.67; Monger, 1977b; Peatfield, 1978) which characterize the Harper Ranch and Okanagan subterranes of Quesnellia. The Harper Ranch Subterrane contains possibly arc-related clastic and volcaniclastic rocks, and podiform carbonate bodies included in the Harper Ranch Assemblage. The latter comprises poorly dated chert, argillite, basalt and associated ultramafic rocks assigned to the Anarchist Assemblage and arc-related rocks of the Nicola Assemblage. The primary relationship between the Harper Ranch and Anarchist assemblages is not known and they may be facies of one another. Alternatively, they could have been independent until Triassic time when both were stratigraphically overlapped by strata of the Nicola Assemblage.

Harper Ranch Subterrane.

The Harper Ranch Group (Monger, 1975, 1977b) north and east of Kamloops (Fig. 8.67) contains rocks ranging in age from latest Devonian to Late Permian (Monger and McMillan, 1984) and is the most complete part of the Harper Ranch Assemblage. No stratigraphic base is known and the rocks are so highly folded and faulted that no continuous stratigraphic section of the complete unit has been recognized. The total thickness is not known, but is estimated to be greater than 2000 m.

The most abundant rock is thinly interbedded homogeneous argillite, laminated argillite and siltstone, which in places is silicified (cherty argillite). Graded, fine grained turbiditic sandstone is common, as is coarse grained, more massive sandstone containing local argillite rip-up clasts. The sandstones appear to be epiclastic, derived largely from volcanic sources, and contain abundant lithic volcanic rock fragments, including pumice, plagioclase minor quartz and chert (Smith, 1974). Fossils from interbedded carbonates indicate that the sections containing abundant volcanic material are Late Mississippian to Early Pennsylvanian in age, although Late Devonian and Early Mississippian fossils occur elsewhere in thin carbonate beds within the matrix and in calcareous sandstone. Chert-pebble conglomerate is a minor but distinctive part of the succession.

Carbonate is common in the argillaceous matrix, and ranges from thin beds within the predominant argillite and siltstone, to large podiform masses, some of which may be exotic blocks, one or two kilometres in length. Most carbonates contain abundant crinoidal detritus and conodonts, fusulinids and other foraminifers are widespread. Brachiopods and corals occur locally and trilobites and ammonoids have been reported. Diagnostic fossils from the carbonate are of the following ages: latest Devonian, Early Mississippian, Late Mississippian/Early Pennsylvanian, Early and Late Permian (Cockfield, 1948; Skinner and Wilde, 1966; Sada and Danner, 1974, 1976; Smith, 1974; Nelson, 1982; Nelson and Nelson, 1985; Orchard, 1984 and pers. comm., 1985).

Volcaniclastic and flow rocks occur locally and include pale green cherty tuff, and altered basalt grading into diabase. Northeast of Kamloops volcanic breccia with notable altered pyroxene grains is intermixed with argillite which appears to grade into Harper Ranch Group strata. The rocks are highly deformed, however, and may be infolded and infaulted volcaniclastics belonging to the overlying Nicola Group.

Smith (1974) concluded that the Harper Ranch Group northeast of Kamloops was probably deposited in a basin associated with an active volcanic island arc, but which lay some distance from the vent region because of the absence of much coarse volcanic material or flows.

In the Lay Range of the Omineca Mountains a structurally complicated sequence, possibly more than 1000 m thick, of argillite, siltstone, sandstone, volcaniclastics with quartz crystals, lava and local conglomerate contains limestone pods with Middle Pennsylvanian fossils. Sandstone, commonly gritty, consists mainly of quartz and is characterized by the presence of detrital mica which has yielded a preliminary K-Ar Permian age. Conglomerate is well bedded and contains rounded pebbles and boulders of granodiorite, quartz-mica schist, andesite and tuff in a gritty matrix. The source of the clasts is unknown. The assignment of this assemblage to the Harper Ranch Subterrane of Quesnellia is uncertain. The rocks occur between Quesnellia and Upper Proterozoic rocks of the Kootenay Terrane to the east and are in steep fault contact with Takla Group volcanics of Quesnellia on the west.

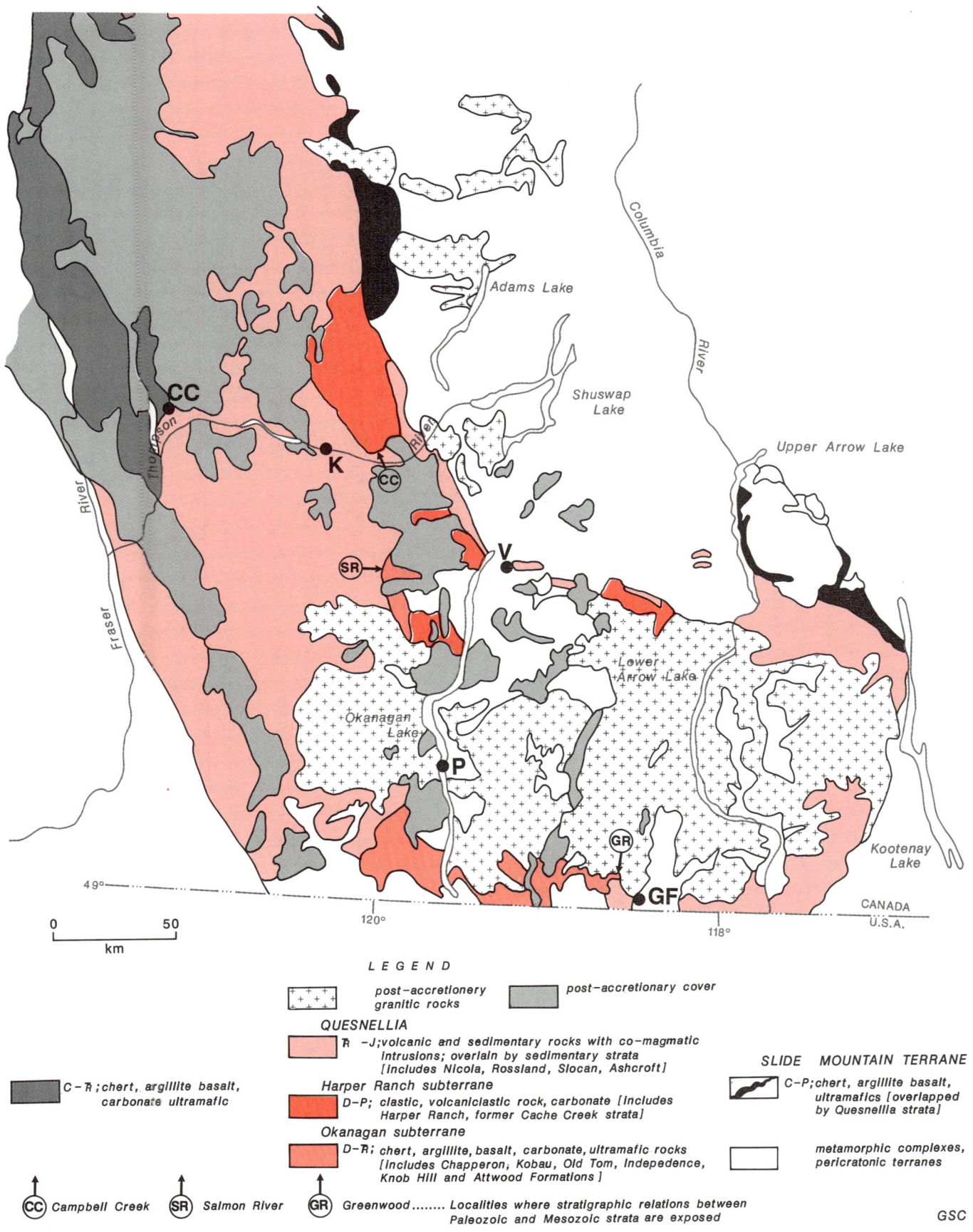

Figure 8.67. Distribution of Quesnellia and Okanagan and Harper Ranch subterranes, and location of stratigraphic contacts of subterranes with overlying, Lower Mesozoic Quesnellia strata, southern British Columbia. CC, Campbell Creek; SR, Salmon River; GR, Greenwood; GF, Grand Forks; K, Kamloops; P, Penticton; V, Vernon.

Okanagan Subterrane.

The Okanagan Subterrane (Fig. 8.67) includes limestone and argillite of the Carboniferous or Permian Attwood Formation, and chert and greenstone of the Knob Hill Group, occurring together with ultramafic bodies near Greenwood (Little, 1983). The Old Tom, Shoemaker, Independence and associated formations of the enigmatic Shoemaker Assemblage near Keremeos defined by Bostock (1940), include, according to Milford (1984), and D.J. Tempelman-Kluit and M.J. Orchard (pers. comm., 1985), a variety of structurally intermixed strata comprising greenstone, silicified tuff, minor limestone and chert breccia, locally of Ordovician to Late Triassic age and presumably correlative in part with the Upper Triassic Nicola Group to the west, but elsewhere including latest Devonian and Carboniferous radiolarian chert and pillow basalt. The pre-Upper Triassic, but otherwise undated Chapperon Group (Jones, 1959), exposed about 40 km south-southeast of the Harper Ranch Group near Kamloops, consists of metachert, phyllite, local carbonate, and greenschist and was reported by Jones (1959) to be intruded by serpentinized ultramafic dykes. The Kobau Group consisting of massive quartzite (metachert?), foliated quartzite, chloritic phyllite and schist, greenstone and marble is lithologically correlated by Okulitch (1973) with the Chapperon and Anarchist groups. The Anarchist Group in Washington, however, includes rocks similar to the Harper Ranch Formation. The last two units are assumed to be late Paleozoic on the basis of their lithological similarity to dated rocks in the region.

From its lithological association, the Okanagan Subterrane appears to have formed in an oceanic or marginal basin setting (Monger, 1977b; Peatfield, 1978; Milford, 1984). Milford (1984) suggested, on the basis of structures in rocks west of Keremeos, that parts of it may represent an ancient subduction complex. In general lithology and age it could well be Slide Mountain Assemblage strata that were structurally "mixed" with the Harper Ranch Assemblage prior to the deposition of early Mesozoic rocks across both Quesnellia and the Slide Mountain Terrane.

An angular unconformity generally separates Paleozoic strata from overlying Middle and Upper Triassic rocks of Quesnellia, (Fig. 8.67; Read and Okulitch, 1977). North of the confluence of Campbell Creek and Thompson River, 20 km east of Kamloops, Pennsylvanian limestone is unconformably overlain by breccia that in turn is overlain by augite porphyry, typical of the eastern facies of the Nicola Group. About 40 km to the south-southeast, on Salmon River, Upper Triassic sedimentary rocks, an eastern facies of the Nicola Group, lie with angular unconformity on low-grade metamorphic rocks of the Chapperon Group. In the Greenwood area, Middle Triassic strata lie unconformably on upper Paleozoic chert, greenstone, and serpentinite (Little, 1983). Recent mapping in this area suggests that the associated ultramafic rocks also lie beneath this unconformity because ultramafic detritus is found in the Triassic strata (J.T. Fyles, pers. comm., 1985). However, according to D.J. Tempelman-Kluit (pers. comm., 1985) there is little evidence for such an unconformity in rocks near Keremeos. The unconformity appears to be due to a poorly understood deformational event that juxtaposed upper Paleozoic subterranes prior to their overlapping by Triassic strata. Possibly related effects in southeastern British Columbia, are the Permian thrust faults within the Slide Mountain Terrane (Klepacki and Wheeler, 1985).

Lower Mesozoic volcanogenic assemblages of Quesnellia

The most characteristic and extensive assemblage in Quesnellia, the Nicola Group, comprises Upper Triassic and probable lowest Jurassic volcanic and sedimentary strata which are intruded by comagmatic plutons (Nicola Assemblage). The Nicola Group represents a well defined volcanic island arc. Major and trace-element chemistry (N. Mortimer, pers. comm., 1986) showed that in southern British Columbia calc-alkaline volcanic rocks occur in the west and calc-alkaline to shoshonitic rocks in the east, indicating the Nicola was a west-facing volcanic arc. Near latitude 51°N these rocks are more-or-less continuously exposed across the regional trend for over 100 km and can be subdivided into three main volcanic and one sedimentary facies belts. Westernmost, and locally sedimentologically and structurally mixed with Upper Triassic strata of the Cache Creek Terrane, are Upper Carnian and Lower Norian felsic, intermediate and, in places, mafic volcaniclastics, including local ignimbrite, minor carbonate and pelite. Farther east are intermediate volcaniclastic rocks, mainly feldspar and feldspar augite porphyries, local carbonate and volcanic sandstone of Early Norian age. The latter are locally overlain on their east side by pillow basalt and the characteristic eastern volcanic facies of augite porphyry, volcaniclastics and alkaline flows of latest Triassic (Late Norian) age. The volcanic assemblage grades eastwards into a Ladinian to Middle Norian sequence of argillite and siltstone, with local volcaniclastic lenses and limestone and breccia beds. This sequence is the lithological and temporal correlative of the Slocan Group west of Kootenay Lake at latitude 50°N.

North of latitude 51°N, the three facies belts have not been identified. The Upper Triassic (Late Norian) augite porphyry belt predominates and, in places, occurs on the western margin of Quesnellia as well as the east.

Lowermost Jurassic volcanics, lithologically similar or identical to parts of the Nicola Group, are known in several places. Sandstone, argillite, and tuff of the Sinemurian Archibald Formation grades into augite and augite feldspar porphyry flows and pyroclastics of the Elise Formation of the Rossland Group (Little, 1960). To the north, in Bonaparte and Quesnel Lake map areas, volcanic sandstone with notable pyroxene grains, and volcanic breccia containing augite porphyry fragments, contain Sinemurian fossils (Campbell and Tipper, 1971). Near latitude 56°N Sinemurian fossils are associated with augite porphyry volcanics and may date the volcanic sequences. In Quesnellia the augite porphyry volcanics apparently are Late Norian in age and younger. This contrasts with similar rocks in Stikinia (see below) which are of Late Carnian-Early Norian age.

Plutonic rocks associated with the Nicola Group are discussed in Chapter 15. They include the latest Triassic and earliest Jurassic, probably subvolcanic, alkaline, Iron Mask Batholith and Copper Mountain Stock, associated with the eastern augite porphyry facies, and the earliest Jurassic calc-alkaline Guichon Creek Batholith, which was emplaced within the western volcanic facies. If the Guichon Creek Batholith is comagmatic with an eroded, younger part of the western facies, then the plutonic rocks show the same chemical progression from calc-alkaline in the west to alkaline in the east. Other lithologies include foliated, locally mylonitic, amphibolite, metadiorite, meta-augite porphyry and felsic schist of the northern part of the Mount Lytton Complex, which lies west of the Nicola Group occurrences, and similar rocks in the Tertiary horst that contains the Nicola batholith. The rocks yield both Permo-Triassic (250 Ma) and Late Triassic (225 Ma) U-Pb dates and are cut by granodiorite plutons that give Early Jurassic K-Ar dates (P. Van der Heyden and R.R. Parrish, pers. comm., 1986). Speculatively, these rocks represent the lower crust of the early Mesozoic arc.

Near latitude 55°N, the Triassic eastern part of the Takla Group includes argillite, volcanic sandstone and tuff, together with abundant alkaline and calc-alkaline volcanics (Armstrong, 1949; Meade, 1977). These stratified rocks are associated with Late Triassic to Early Jurassic alkalic intrusive rocks of the Copper Mountain Suite ranging from gabbro, through diorite to syenite, and Early Jurassic calc-alkaline granodiorite and quartz monzodiorite (Meade, 1977) of the Guichon Suite.

Between latitudes 56°N and 57°N are numerous Alaskan-type zoned ultramafic bodies which are coeval with surrounding volcanic rocks (Irvine, 1976).

South of latitude 60°N volcanic and sedimentary rocks of the Upper Triassic Shonektaw and Lower Jurassic (Sinemurian?) Nazcha formations lie between the Cache Creek and Cassiar terranes. The Nazcha Formation comprises argillite, volcanic sandstone, and minor conglomerate and probably overlies the volcanic Shonektaw Formation with its abundant feldspar and pyroxene feldspar porphyry flows and pyroclastics (Watson and Mathews, 1944; Gabrielse, 1968; Monger, 1969, 1980). Foliated diorite, dated at 228 Ma by U-Pb (zircon) intrudes the volcanic rocks.

Jurassic sedimentary assemblages of Quesnellia

Conglomerate, sandstone, shale, and siltstone form the remnants of a Pliensbachian to Lower Bajocian overlap assemblage (Hall Assemblage) that extends from latitude 49°N to at least 56°N. In the Ashcroft area both the Nicola volcanics and early Mesozoic plutonic rocks are overlain both disconformably and with angular unconformity by the nonvolcanic Ashcroft Formation (McMillan, 1974; Grette, 1978; Travers, 1978, 1982). The Ashcroft is the youngest lithological unit in Quesnellia and ranges in age from Late Sinemurian to Callovian. It is more than 1500 m thick and consists largely of clastic rocks with minor carbonate members (Crickmay, 1930; Travers, 1982). A western facies, exposed near Ashcroft, consists of dark, carbonaceous shale with interbeds of siltstone and graded sandstone and local conglomerate with rare fossils interpreted by Travers (1982) as a submarine fan facies. Eastern facies are typically coarse grained sandstone, conglomerate and shale, and minor bioclastic limestone thought to represent a proximal, near-shore fan setting (Travers, 1982). Fossils, widespread but seldom common, include ammonites, bivalves, gastropods, and carbonaceous material. Clasts in the conglomerate include abundant granodiorite and lesser diorite and syenite cobbles, presumably derived from early Mesozoic plutonic rocks, volcanic rocks similar to various Nicola volcanics and local carbonate and chert. In undated exposures north of Kamloops Lake, quartzite pebbles suggest a source in lower Paleozoic miogeoclinal strata to the east.

In southeastern British Columbia the Lower Pliensbachian to Lower Toarcian Hall Formation (Tipper, 1984) rests with erosional unconformity (Mulligan, 1952) on the volcanic Sinemurian Elise Formation. The Hall Formation, more than 300 m thick, is dominantly black shale and buff to brown argillaceous sandstone with minor siltstone and conglomerate. The source of sediment was apparently to the north or northeast.

Between latitudes 51°N and 56°N, Lower Pliensbachian to Lower Bajocian shale, greywacke, conglomerate, and siltstone mapped within the Quesnel River and Takla groups, are distributed in isolated fault blocks throughout Quesnellia. The source of sediment and direction of transport are largely speculative but there is some suggestion locally of a western source. The thickness is unknown.

Near latitude 56°N an elongate fault block includes conglomerate and sandstone previously mapped as Lower Cretaceous Uslika Formation (Roots, 1954; Eisbacher, 1974). Alternatively the Uslika is of Early Jurassic age and underlies Lower Toarcian and (?) Upper Pliensbachian sandstone in the same fault block. The bulk of the conglomerate includes material from older Quesnellia sediments, volcanics and plutonic rocks, and from volcanics and chert of the Slide Mountain Terrane. Subordinate micaceous quartzite clasts may have had a cratonal origin consistent with a few south-southwest paleoflow indicators.

The sedimentary sequences rest unconformably on different substrates, including Upper Triassic volcanics, Sinemurian volcanics, and probably on the Cache Creek Terrane. A hiatus is suggested because earliest Pliensbachian fossils have not been found. The oldest fossils are of late Early Pliensbachian age and invariably are found along the eastern part of Quesnellia. Rocks of this age have not been seen on the western margin where Upper Pliensbachian rocks rest unconformably on Triassic or older rocks. Apparently the Late Pliensbachian sea transgressed westwards onto older rocks.

The marine sediments record the last marine incursion onto Quesnellia which for the most part ended in Early Bajocian time. Only in the Ashcroft area are younger sediments known (Callovian and Late Bajocian) but they may lie unconformably on the widespread older Jurassic marine sediments and therefore reflect a slightly later sedimentary episode.

The eastern margin of Quesnellia at one locality overlies the Slide Mountain Terrane. In southeastern British Columbia, the Slocan Group lies disconformably on the Kaslo Group (Cairnes, 1934; Klepacki and Wheeler, 1985) of the Slide Mountain Terrane, and is apparently overlain in turn by Lower Jurassic strata including the Rossland Group volcanics, typical of Quesnellia (Little, 1960). Farther north, in the Cariboo Mountains, the Slide Mountain Terrane is overlain by black phyllite, from which Middle and Late Triassic conodonts have been obtained and which grades laterally and upwards into volcanic rocks that clearly are part of Quesnellia (Struik, 1985). In the Lay Range of the Omineca Mountains (lat. 56°N), volcanic rocks of the Takla Group are so structurally mixed with volcanic rocks belonging to the Harper Ranch Subterrane that it is difficult to separate them (Monger, 1977b).

As noted above rocks herein included in the Okanagan Subterrane may be part of the Slide Mountain Terrane as they have a similar range of lithologies, and, although poorly dated, appear to be coeval. These rocks are unconformably overlain by Middle and Upper Triassic strata of Quesnellia (Read and Okulitch, 1977; Little, 1983; J.T. Fyles, pers. comm., 1985). Perhaps rocks of the Slide Mountain Terrane, its possibly correlative Okanagan Subterrane, and the lithologically different upper Paleozoic rocks of the Harper Ranch Subterrane were amalgamated by Middle Triassic time, and stratigraphically overlapped by the lower Mesozoic rocks of Quesnellia.

Dorsey Terrane

Upper Paleozoic rocks of the Dorsey Terrane overlie units of the Slide Mountain Terrane in south-central Yukon and north-central British Columbia west of the Cassiar batholith. Relationships between the terranes are unknown. In general, the Dorsey Assemblage bears some resemblance to rocks of both the Slide Mountain Terrane and the Harper Ranch Subterrane of Quesnellia but it differs in containing no volcanic rocks. Three members have been recognized by Poole (1956) but have not been mapped throughout the Dorsey Terrane. The lowest consists of in part laminated, fine grained quartzite, grey to black argillite and slate, ribbon chert and discontinuous bands of limestone, cherty limestone and dolostone. The carbonate units range from a few metres to 70 m thick and the aggregate thickness of the succession is several hundreds of metres. The middle member is more than 300 m thick, locally, and is a mainly thick bedded, cryptograined to fine grained cherty limestone of Pennsylvanian (Late Morrowan to Early Atokan) age. The uppermost member comprises a lower unit of conglomerate, grit, quartzite, sandstone and argillite as much as 300 m thick and an upper unit of ribbon chert, argillite, slate and minor limestone more than 500 m thick. Pebbles and cobbles in the conglomerate include red, pink, grey, white and green chert and, in places, orthoquartzite.

In common with the Slide Mountain Assemblage the Dorsey Terrane includes thick sequences of ribbon chert indicating deposition in deep water. Unlike the Slide Mountain, however, the Dorsey contains little if any volcanic rock and no alpine-type ultramafics. The abundance of quartz in sandstone and, in places, in conglomerate boulders suggests proximity to a cratonal source area. The age and lithology of the carbonate units are similar to those in the Sylvester Group of the Slide Mountain Terrane but the Dorsey carbonates do not appear to be related to volcanic islands or plateaus.

Cache Creek Terrane

Most rock types in the Cache Creek Terrane are of typical oceanic affinity and comprise Lower Carboniferous to Middle Jurassic radiolarian chert and argillite, distinctive, masses of shallow-water carbonate, basalt, gabbro, and alpine-type ultramafic rocks of the Cache Creek Assemblage and calc-alkaline volcanic and associated sedimentary strata of the Upper Triassic Kutcho Assemblage. Greywacke and volcaniclastic strata are local and minor components. The structural style of the terrane is commonly characterized by extreme stratigraphic and structural disruption and most internal contacts are probably faults.

The Cache Creek Assemblage occurs in three main outcrop areas. The most extensive and best exposed is in north-central British Columbia and the southernmost Yukon. There stratigraphic continuity can be established between several of the main lithological units which range in age from Early Carboniferous to Late Triassic. In central British Columbia, near Fort St. James (lat. 55°N) diagnostic fossils are Pennsylvanian and Permian. The southern and type outcrop area is well known and has excellent fossil control with diagnostic fossils ranging in age from Late Pennsylvanian to Middle Jurassic, but is structurally highly disrupted.

In north-central British Columbia the Cache Creek Assemblage comprises three distinctive "facies belts" (Fig. 8.68; Monger, 1975, 1977a), which occur in the French Range, Sentinel, and Nakina subterranes (Monger and Berg, 1984). The Nakina Subterrane provides the most complete stratigraphic section known in the Cache Creek Terrane.

The Cache Creek Assemblage in the Nakina Subterrane ranges in age from Late Mississippian to Middle Triassic. It is in fault-contact on the southwest with the Nahlin ultramafic body, the largest alpine-type ultramafic in the Canadian Cordillera. The Nahlin body contains most components of classical ophiolite sequences, including predominant foliated peridotite, minor ultramafic and mafic cumulates, local trondhjemite, and diabase dyke swarms which contain composite intrusions but no extensive tracts of sheeted dykes are known (Terry, 1977). The stratigraphically lowest unit in the Nakina Subterrane is the Nakina Formation which comprises amygdaloidal basalt, volcanic breccia, tuff and small limestone pods containing Late Mississippian (latest Viséan-earliest Namurian) foraminifera (Mamet, in Monger, 1975). It is overlain, locally gradationally, by massive, rarely tuffaceous crinoidal calcarenite of the same age. Above this is massive, crinoidal calcarenitic and/or foraminiferal calcarenitic limestone of the Horsefeed Formation ranging in age from Late Mississippian to Late Permian (Monger, 1975; Plate 19). The Horsefeed has a stratigraphic thickness of about 2000 m and surface dimensions of about 50 by 30 km. In places, it is oolitic, and many grains have an algal coating. Locally, and commonly in mid-Permian rocks, massive limestone passes upwards and laterally into algalaminate dolomitic limestone and dolostone. In places, small lenses of pillowed mafic volcanic rocks and thin tuffaceous members occur within the carbonate. The carbonate is overlain disconformably, or with slight angular unconformity, by a breccia containing carbonate clasts of mainly mid- to Late Permian age. The breccia grades upwards into radiolarian chert, which, at one locality, has yielded Middle Triassic radiolaria (D.L. Jones, pers. comm., 1978).

The carbonate is entirely of shallow water origin, and appears to have developed an atoll reef on a basaltic substrate. The reefoid character of the carbonate is shown

UPPER DEVONIAN TO MIDDLE JURASSIC ASSEMBLAGES

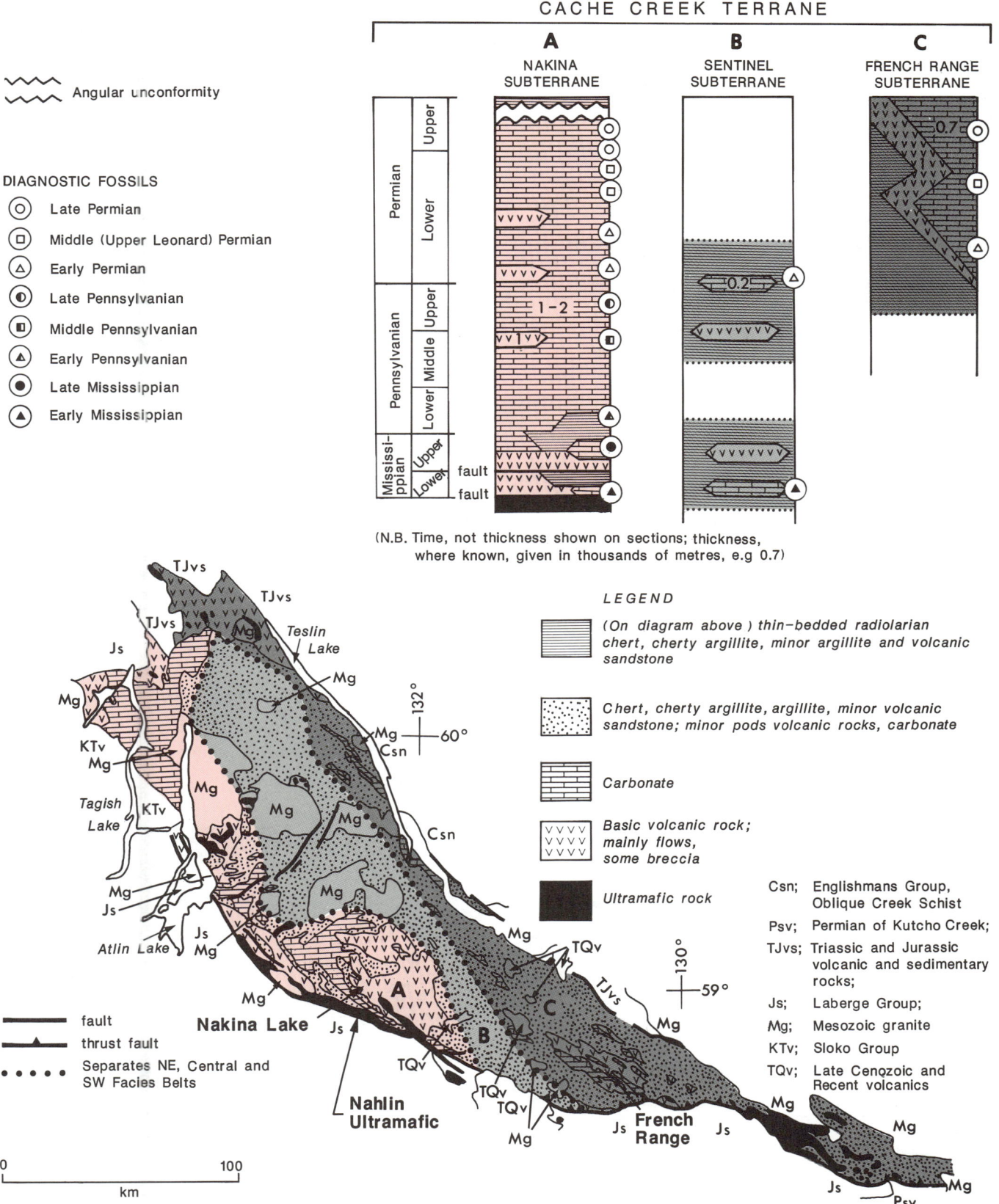

Figure 8.68. Subterranes and their facies in the northern outcrop area of the Cache Creek Terrane, northwestern British Columbia-south-central Yukon Territory.

by its lithology and lateral gradation from massive carbonate, which presumably formed the main mass of the reef, into algalaminate dolostone, the back reef facies. The reef represents an interval of about 100 Ma of remarkable tectonic stability, with an average subsidence rate of 20 m per million years, slightly less than the 25-30 m per million years reported from the Bahama Banks by Lynts (1970). Speculatively, a lowering of Permo-Triassic sea-level terminated reef growth, and subsequently only pelagic material was deposited on it.

The Cache Creek Assemblage of the Sentinel Subterrane is characterized by lithological associations that could be deeper water facies of many of the rocks in the Nakina Subterrane, although no stratigraphic continuity has been established between the two subterranes. The predominant lithologies, radiolarian chert and argillite are pelagic facies. Massive breccia with carbonate and volcanic clasts, interbeds of greywacke, distinctive aligned pods of carbonate within the bedding of enclosing chert and argillite, and local pillow basalt and diabase sills occur locally (Monger, 1977b). Fossils from the carbonate pods are mainly Pennsylvanian and Early Permian in age, although one is Early Mississippian. Chert from Sentinel Mountain has yielded Pennsylvanian and Permian radiolaria (D.L. Jones, pers. comm., 1978). The breccia and limestone blocks are submarine landslide deposits and exotic blocks derived from submarine volcanic edifices and their capping atolls, which from modern examples, could have stood 5000 m above the ocean basin floor (Fig. 8.69).

The French Range Subterrane has abundant chert and argillite, in places interbedded with greywacke together with mafic flow rocks, pyroclastic rocks including some felsic volcaniclastics, and prominent carbonate (Monger, 1969, 1975). The volcanic rocks, up to 1000 m thick, belong to the French Range Formation, and the carbonate, between 300 and 600 m thick, to the Teslin Formation. The latter contains algalaminate dolomitic limestone, similar to that of the lagoonal facies of the Nakina Subterrane. The volcanic rocks stratigraphically overlie part of the chert, and are themselves gradationally overlain by upper Lower Permian and Upper Permian carbonate of the Teslin Formation. As in the Nakina Subterrane, the limestone was deposited on a volcanic substrate. However, some radiolarian chert, locally interbedded with greywacke (Fig. 8.70) has yielded conodonts of Norian age (M.J. Orchard, pers. comm., 1980), the youngest fossils known in this part of the Cache Creek Terrane. Triassic rocks may be more extensive in northern Cache Creek Assemblage strata than formerly realized, as an abundant Norian radiolarian fauna is found in chert clasts in Middle Jurassic conglomerate of the Bowser Basin. The clasts were clearly derived from uplifted Cache Creek Terrane (Currie, 1984).

The greywacke, with its characteristic high content of hornblende, is compositionally similar to lower Mesozoic greywacke in the Lewes River Group of the Whitehorse Trough described by Wheeler (1961). Several large, variably serpentinized ultramafic bodies of peridotite, dunite, harzburgite, pyroxenite and layered gabbro occur in the French Range Subterrane, and are associated with mafic volcanics and radiolarian chert. Locally they contain lensoid bodies of nephritic jade and one body contains a significant chrysotile asbestos deposit. Many creeks within the Cache Creek Terrane of northern British Columbia have been important sites for placer gold accumulation.

In the eastern part of the French Range Subterrane Cache Creek Assemblage rocks are apparently overlain by the Upper Triassic Kutcho Formation (Kutcho Assemblage) consisting of bimodal basaltic andesite and rhyodacite,

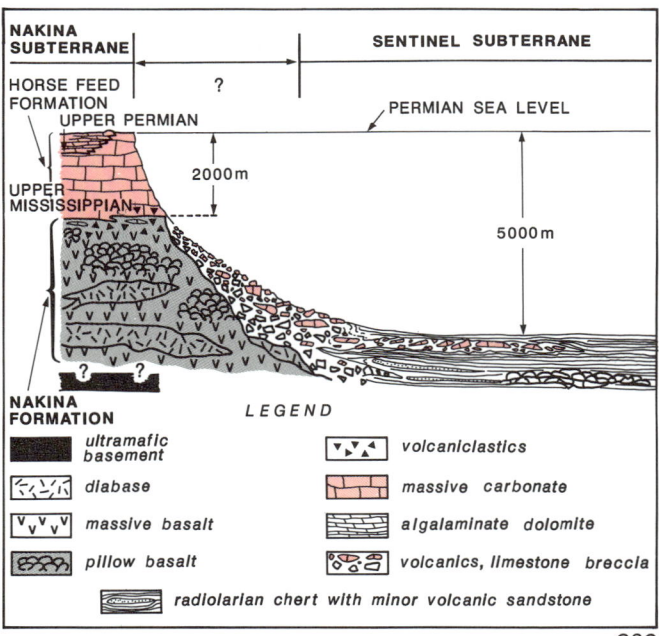

Figure 8.69. Schematic diagram showing possible relationship between the types of deposition represented by the Nakina and Sentinel subterranes of the Cache Creek Terrane; Nakina Subterrane comprises massive reefoidal carbonate, founded on a volcanic substrate (seamounts?, ridge?), Sentinel Subterrane is a deep-water equivalent and contains slump facies rocks in addition to pelagic sediments.

Figure 8.70. Interbedded Upper Triassic radiolarian chert and greywacke within the French Range Subterrane, southern Teslin map area, southern Yukon; possible link between clastic rocks of the Whitehorse Trough, founded partly on Stikinia, and Cache Creek Terrane. Photo by J.W.H. Monger. GSC 205235-P

calc-alkaline volcanics including abundant tuff, breccia and quartz-eye schist (Thorstad and Gabrielse, 1986). Overlying tuff, conglomerate and argillite may be in part or entirely of Early Jurassic age because of their intimate relationship with the Inklin Formation. Locally the Kutcho Formation is overlain by the Sinwa Formation which consists of fetid, well bedded to massive Norian limestone up to 250 m thick. Rocks similar to the Kutcho and Sinwa formations are included in the Lewes River Group in southern Yukon Territory. The Kutcho volcanics are host to a large, massive, volcanogenic, Kuroko-type copper-zinc deposit.

The central part of the Cache Creek Terrane between latitudes 56°N and 53°N, contains abundant radiolarian chert, argillite and massive Middle Pennsylvanian to Upper Permian carbonate. Basalt and alpine-type ultramafic rocks are locally important (Armstrong, 1949; Thompson, 1965; Paterson, 1973, 1977). An Upper Triassic carbonate which stratigraphically overlies mafic volcanics associated with an ophiolite sequence was included within the Triassic Takla Group by Paterson (1977) but lithologically is inseparable from the Cache Creek Group. Uppermost Triassic conglomerate of the Takla Group, which is part of Quesnellia east of the Pinchi Fault, contains abundant gabbro, basalt, and chert clasts, presumably derived from the Cache Creek Assemblage. Within the Cache Creek Terrane is a fault-bounded slice of Cache Creek Assemblage strata containing blueschist minerals. It yielded K-Ar dates of 216-223 Ma (Paterson and Harakal, 1974; Paterson, 1977). Other glaucophane and/or crossite-bearing localities are known within the Cache Creek Terrane near Dease Lake in northern British Columbia, near Takla Lake and in the type area, but in those places glaucophane or crossite is mainly associated with epidote and albite or rarely lawsonite, and belongs to the subgreenschist or perhaps lowest greenschist metamorphic facies. Jadeite occurs along the northeast margin of Cache Creek Terrane rocks along the Thibert Fault west of Dease Lake.

The type-area of the Cache Creek Terrane in southern British Columbia is divided into an eastern Bonaparte Subterrane, a central Marble Range Subterrane and a western Pavilion Subterrane (Fig. 8.71; Monger and Berg, 1984; Monger and McMillan, 1984). The Bonaparte Subterrane is represented by a mélange of highly disrupted radiolarian chert and carbonaceous argillite containing blocks of chert, carbonate, basalt, volcaniclastics and serpentinite. Fossiliferous clasts in the mélange range in age from Late Pennsylvanian to Late Permian, and the matrix ranges from Late and possibly Early Permian to Middle and possibly Late Triassic. Of note are blocks within the mélange of foliated lapilli tuff and ignimbrite (Shannon, 1982), which closely resemble some lithologies in the western felsic facies of the Nicola Group of Quesnellia. In addition, elongate, fault-bounded slices of western Nicola lithologies occur within the Bonaparte Subterrane and on the west side of the western belt (Fig. 8.71).

The central Marble Range Subterrane contains massive crinoidal, fusulinid and algal calcarenite of the Marble Canyon Formation of the Cache Creek Group, which is probably about 300 m thick, together with minor mafic flows, volcaniclastics and argillite. Fossils from the main mass of carbonate are of mid- to early Late Permian age, but fossils from other localities in unknown relationship

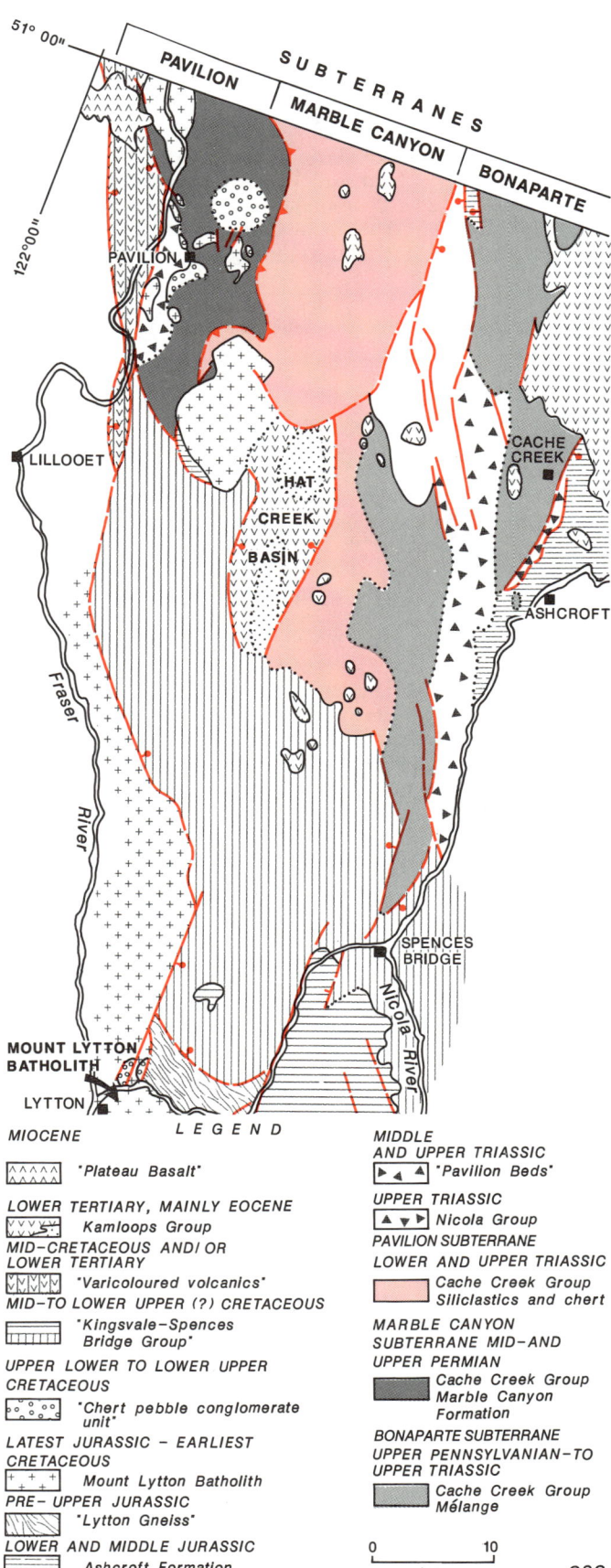

Figure 8.71. Subterranes and facies belts in the southern type area of the Cache Creek Terrane.

with the main Permian carbonate, are of Early, Middle and Late Triassic ages (M.J. Orchard, pers. comm., 1985).

The western Pavilion Subterrane is underlain by argillite, siltstone, radiolarian chert and local but notable fault slices (?) of volcaniclastic rocks, more massive volcanics, and small carbonate pods (Trettin, 1961; Monger and McMillan, 1984). Mélange and broken formation, characteristic of the assemblage in the Bonaparte Subterrane, is only locally present. The unit ranges in age from Early Triassic to Middle Jurassic (F. Cordey, pers. comm., 1987) with volcanogenic rocks in the younger part.

Stikinia

Stikinia comprises well stratified Lower Devonian to Middle Jurassic volcanic and sedimentary strata (Asitka, Stuhini, Lewes River, Hazelton and Takwahoni assemblages) and plutonic rocks which are probably comagmatic with the volcanics. Major units include unnamed upper Paleozoic rocks of the Stikine region, the Permian Asitka Group, Upper Triassic western Takla, Stuhini, and Lewes River groups, and Lower and Middle Jurassic Hazelton and Laberge groups. The rocks are best exposed around the periphery of Bowser Basin. Their continuation south of latitude 53°N is not clear. On its east side Stikinia is in fault contact with the Cache Creek Terrane. Relationships on its western margin generally have been obscured by Cretaceous and Tertiary plutonism and metamorphism in the Coast Belt, although near Prince Rupert (lat. 55°N) Stikinia appears to be structurally emplaced over the Alexander Terrane. Near Atlin (lat. 59°N), it is in fault-contact with possible late Precambrian metamorphic rocks included in the Nisling Terrane.

Upper Paleozoic assemblages

The stratigraphic relationship of the oldest rocks, which include Lower Devonian carbonate, to other Paleozoic strata in the Iskut River area is not known (P.B. Read, pers. comm., 1984). The oldest rocks, clearly in stratigraphic continuity with younger strata, are exposed between the Stikine and Iskut rivers (Souther, 1972). There, argillite and cherty argillite grade upward into mafic to intermediate tuff and breccia which is interbedded with and overlain by approximately 1000 m of massive- to medium-bedded crinoidal calcarenite and calcarenitic limestone containing foraminifera, corals, bryozoa and brachiopods of Late Mississippian age (Mamet, 1976) (Fig. 8.72). The relationship between Mississippian and overlying Permian strata is uncertain, but the disappearance of at least 1500 m of Mississippian strata over a distance of 5 km along trend, and the paucity of recognized Pennsylvanian strata, suggest a major post-Mississippian-pre-Permian unconformity. The overlying succession consists of mafic volcanic rocks grading up into argillite which, in turn, is succeeded by nearly 600 m of carbonate ranging from Early Permian (Wolfcampian) to Late Permian (Early Guadalupian) in age. The carbonate is the most distinctive Paleozoic unit in Stikinia, and, where completely preserved, consists of a lower, dark, thinly bedded, pyritic, argillaceous, micritic to calcarentic limestone containing fusulinids, small corals and bryozoa. It grades upwards into pale grey thin-bedded to medium-bedded cherty calcarenitic limestone. The uppermost member is white to cream, locally dolomitic,

Figure 8.72. View of the Mississippian succession of Stikinia, near the divide between Mess and Sphaler creeks, southern Telegraph Creek area, northwestern British Columbia. Well-bedded carbonate overlies fine grained argillite, cherty argillite and local volcaniclastic rocks and is overlain by Permian volcanic rocks. Photo by J.W.H. Monger. GSC-205235-R

calcarenite and dolomitic calcarenite containing bryozoan, brachiopods, crinoids and fusulinids. Some or all of these characteristics are retained in scattered exposures along the west side of the Intermontane Belt for a distance of over 500 km, from north of the Stikine River to south of Terrace. In the Chutine area (lat. 58°N), Permian strata are overlain with apparent conformity by calcareous and silty argillite containing Middle Triassic fossils (Souther, 1959).

The Paleozoic stratigraphic succession on the east side of the Bowser Basin is different from that on the west. The Asitka Group (Lord, 1948; Monger, 1977a) consists of a lower subdivision of argillite, chert, tuffaceous and argillaceous carbonate containing abundant Early Permian fusulinids, overlain by a probably partly subaerial basalt, brightly coloured rhyolite and breccia. An upper member of basalt, chert, tuffaceous limestone and calcareous tuff contains Early Permian sponges, corals, bryozoa and brachiopods (Rigby, 1973; Monger and Ross, 1971).

Southwest of the Kutcho Fault near latitude 58°N a polydeformed sequence of feldspathic chlorite schist, sericitic and calcareous phyllite, massive rhyolite, chert and carbonate is in part, at least, of Mississippian (Viséan) age. The succession is probably more than 500 m thick with the thickest carbonate unit about 25 m thick (Thorstad, 1980).

Along the Stikine River north of the Bowser Basin Lower Permian massive grey limestone up to 100 m thick overlies grey phyllite and is overlain by mafic metavolcanic rocks and green phyllite. Ribbon chert is present locally. The sequence is overlain unconformably by Lower and Middle Triassic rocks (Read, 1983).

The limited chemical data on Paleozoic volcanic rocks from Stikinia provide little insight into their tectonic setting. Speculatively, they represent a suite of arc-related rocks. Much carbonate, particularly that of Mississippian age in northwestern Stikinia and Asitka Group carbonate, is clearly intercalated with the volcanics and may represent fringing banks around volcanic highs. The apparently sheet-like Permian carbonate of western Stikinia, however, is remarkably uniform over a great distance, which suggests deposition on a relatively uniform and stable substrate.

Relationships between Paleozoic and younger strata differ in the various parts of Stikinia. In northwestern British Columbia an episode of deformation and metamorphism separated deposition of Paleozoic to Middle Triassic rocks from Upper Triassic and younger strata (Souther, 1971). In northwestern Spatsizi, Telegraph Creek and Tulsequah areas, upper Paleozoic strata in places are more deformed and metamorphosed than contiguous Lower and Middle Triassic rocks, and in the Tulsequah area the older rocks as well as foliated quartz diorite have northerly structural trends which differ from the later northwesterly structures produced in Middle Jurassic time (Souther, 1971). Farther south in the Oweegee Mountains Upper Triassic strata lie disconformably on Permian rocks (Monger, 1977b) and near Terrace, Lower Jurassic strata lie apparently disconformably on both Permian and Upper Triassic rocks. On the east side of the Bowser Basin, in McConnell Creek area, Lower Permian rocks of the Asitka Group are disconformably overlain by Upper Triassic argillite and siltstone comprising the basal part of the western Takla Group (Fig. 8.73).

Lower Mesozoic assemblages: Triassic rocks

Lower Mesozoic assemblages of Stikinia include Triassic volcanogenic strata of the western Takla, Stuhini and Lewes River groups, and Lower to Middle Jurassic volcanics and sediments of the Hazelton Group and the Takwahoni facies of the Laberge Group, together with comagmatic plutonic rocks.

Upper Triassic rocks extend continuously along the eastern and northern parts of Stikinia. They typically comprise mafic to felsic volcanics, predominantly volcaniclastic rocks, but include flows and associated sediments. On the east side of Stikinia, in the McConnell Creek area, the Upper Triassic (Upper Carnian to Lower Norian) western Takla Group consists of basal siltstone and argillite, overlain by about 3000 m of pillow and massive augite porphyry basalt and trachybasalt, grading upwards into subaerial flows of similar composition. The flows grade laterally into, and interfinger with, coarse- to fine-grained volcaniclastics which, distally, are intercalated with argillites. Overlying rocks are submarine to subaerial andesitic volcaniclastics and mudstone (Fig. 8.74; Monger, 1977a). The volcanics are transitional between alkaline and subalkaline basalt and andesite, and resemble volcanics in the eastern part of the Nicola Group in Quesnellia. As in Quesnellia Upper Triassic rocks contain numerous copper deposits and showings. The sequence is disconformably overlain by Lower Jurassic volcanogenic strata of the Hazelton Group.

Near the Stikine River north of the Bowser Basin basal Mesozoic sediments of Early and Middle Triassic age unconformably overlie Carboniferous and Permian rocks. The sedimentary succession, as much as 250 m thick, comprises siltstone, greywacke, pebble conglomerate and shale. Graded bedding is well developed and the rocks probably represent turbidites. Intercalated with the sediments are augite porphyry volcanics which are overlain by a thin sedimentary member. The upper sequence, of

Figure 8.73. Disconformity between Lower Permian volcanics and chert of the Asitka Group and overlying Upper Triassic argillite and siltstone of the Takla Group, on east side of Stikinia, McConnell Creek area, north-central British Columbia. Photo by J.W.H. Monger. GSC 205235-Q

Late Triassic age, is at least 1000 m thick and contains a lower augite porphyry unit, a middle coarse-bladed plagioclase porphyry and an upper maroon weathering feldspar porphyry, breccia and agglomerate unit. A pyroxenite body in the volcanic succession has been dated by the K-Ar method at about 230 Ma. Conglomerate at the base of the volcanics locally contains boulders of the underlying Cake Hill pluton, a monzodiorite body which has been dated by the K-Ar method at 220-230 Ma (Anderson, 1983).

In northwestern Stikinia, near the Taku River similar Upper Carnian and Lower Norian siltstone and argillite are interbedded with pillow basalt, forming a sequence about 1500 m thick. It is overlain, possibly unconformably, by up to 1000 m of andesite and dacite breccia and an uppermost massive limestone of Late Norian age (Souther, 1971; Monger, 1980). Possibly the younger sequence can be correlated with the foliated, volcanic-sedimentary Kutcho Formation in the Cache Creek Terrane (Fig. 8.75, Thorstad and Gabrielse, 1986). These rocks are disconformably overlain by Upper Pliensbachian conglomerate of the Takwahoni Formation of the Laberge Group.

To the north in southwestern Yukon the Lewes River Group is up to 3000 m thick. Along the eastern flank of the Coast Belt it is composed of coarse andesitic fragmentals, conglomerate, and local augite porphyry flows but grades eastward into volcanic sandstone, greywacke, argillite and, in places, prominent reefoid masses containing Carnian and Norian fossils. The western part of the Lewes River Group is overlain disconformably by Lower Jurassic conglomerate of the Laberge Group. Eastward the contact between the two groups is apparently gradational (Wheeler, 1961; Reid and Tempelman-Kluit, 1987).

Lower Mesozoic assemblages: Jurassic Strata

Lower Jurassic and lower Middle Jurassic rocks of Stikinia (Early Sinemurian to Early Bajocian) characteristically include the products of contemporaneous volcanism

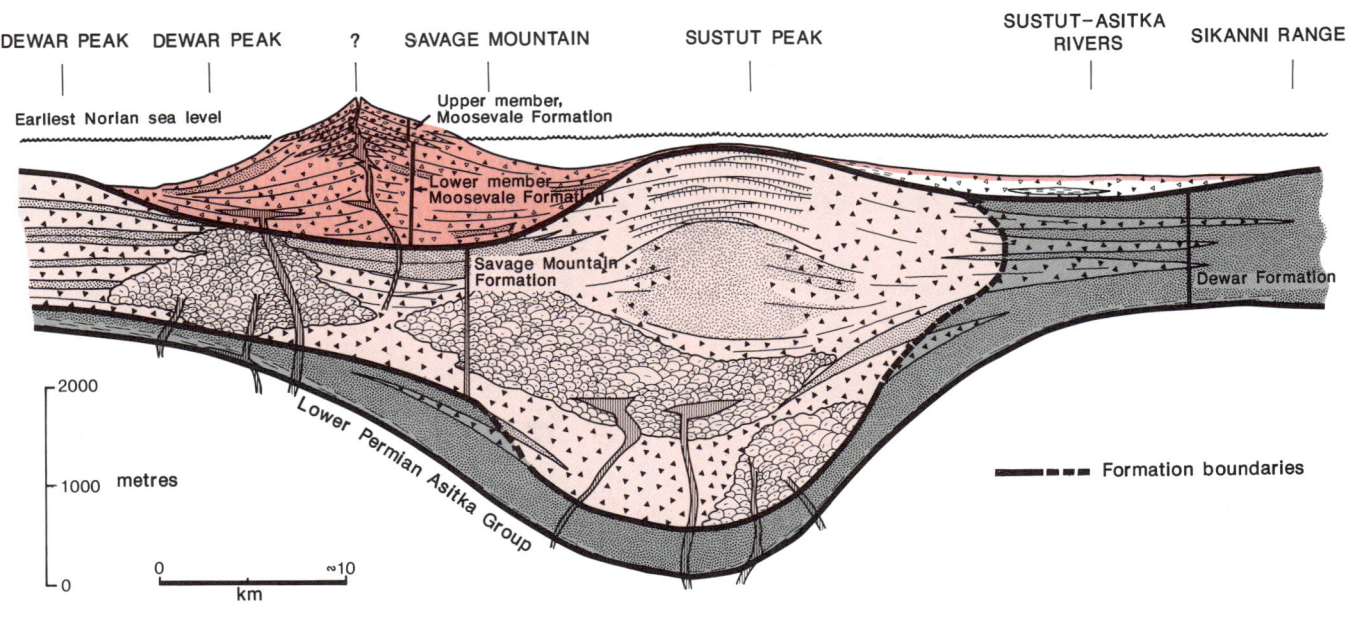

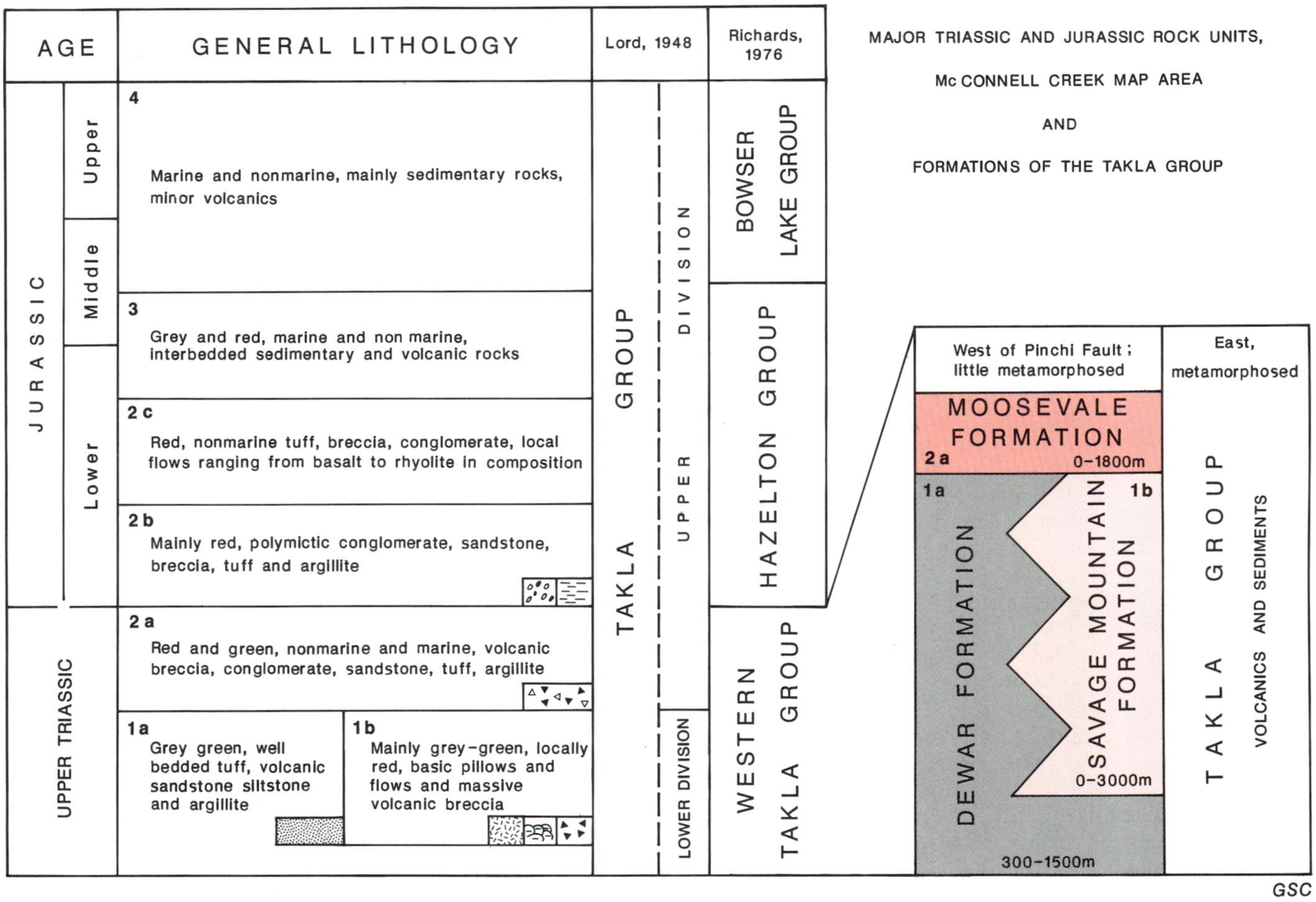

Figure 8.74. Stratigraphic relationships of the Upper Triassic western Takla Group, McConnell Creek map area (from Monger, 1977a).

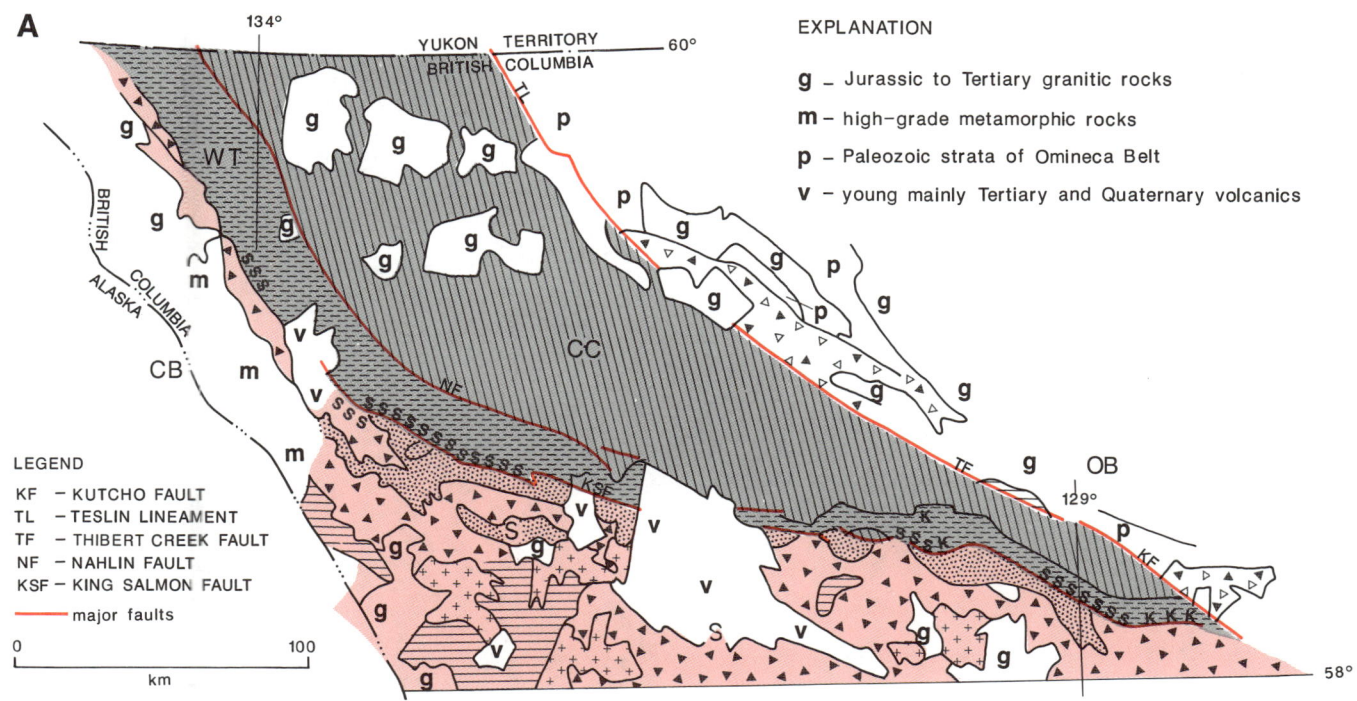

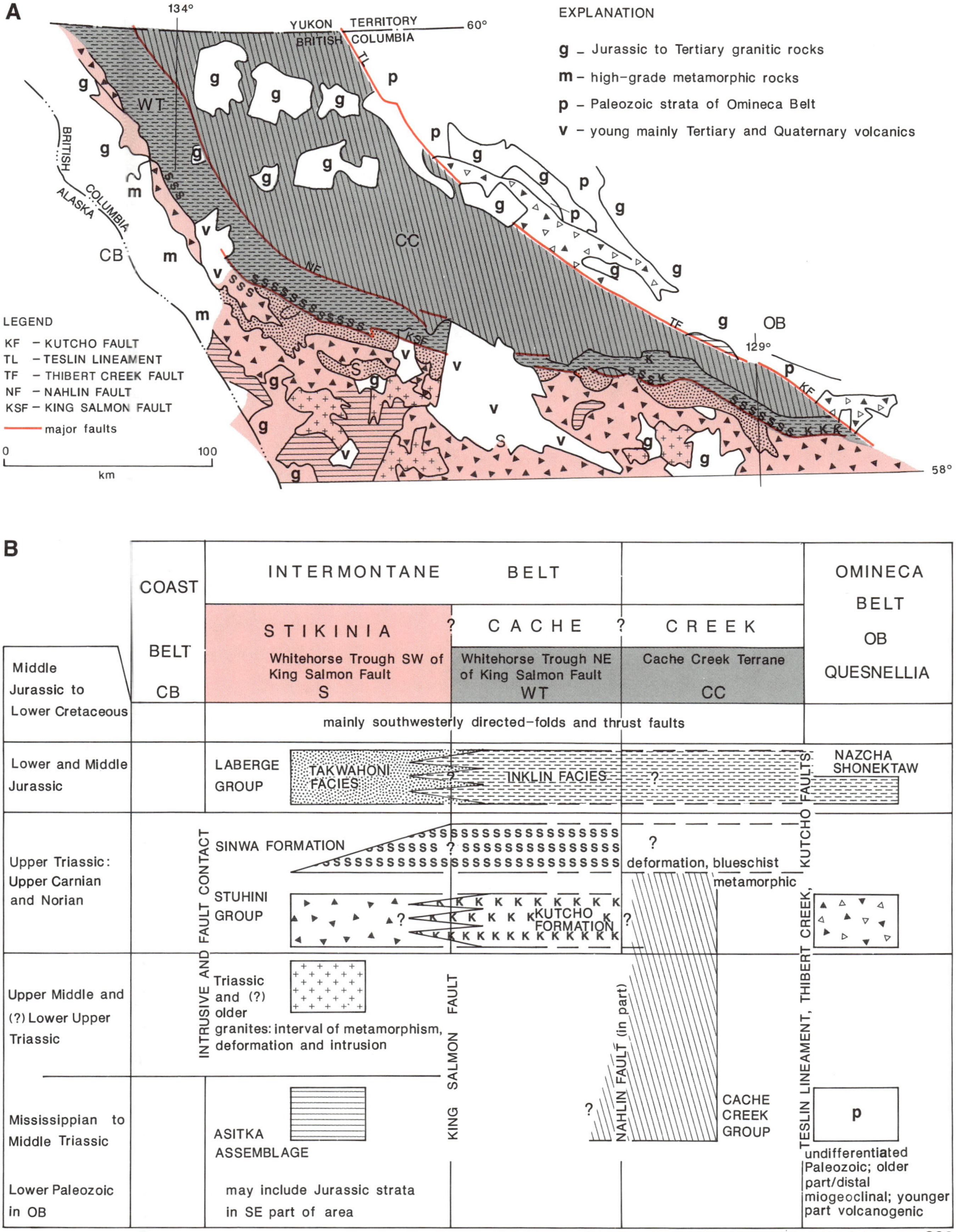

Figure 8.75. Stratigraphic relationships of lower Paleozoic to lower Mesozoic strata, northwestern British Columbia and south-central Yukon. Proximal Jurassic strata were clearly derived from Stikinia; distal Jurassic strata may overlie the Cache Creek Terrane.

(Hazelton and Takwahoni assemblages; Fig. 8.75-8.80). The volcanics are typically calc-alkaline pyroclastics deposited in both marine and nonmarine environments commonly as coarsely layered volcanic breccia, coarse poorly sorted tuffs, well bedded air-fall and reworked tuff and tuffaceous sediments. Flow rocks are generally subordinate but may predominate locally. They include rhyolite flows and rhyolite domes, minor vesicular basalt, and andesite flows. The volcanics, unless altered by later metamorphic effects or by intrusions, are commonly bright shades of red, green, mauve cream, or grey. Zeolitic altera-

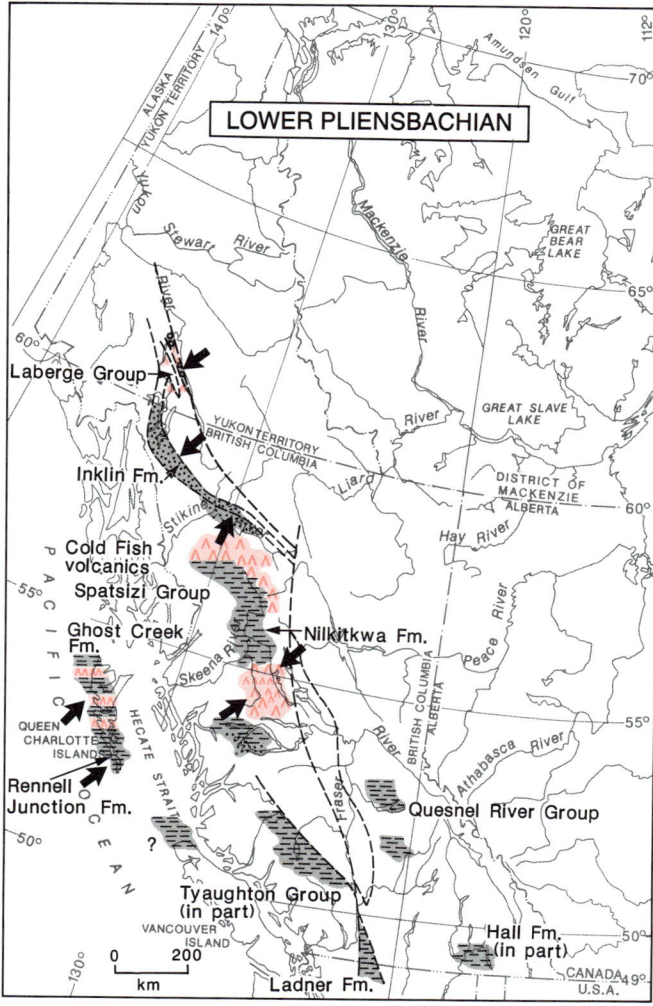

Figure 8.77. Distribution, lithology and nomenclature of Lower Pliensbachian rocks. Legend accompanies Figure 8.76.

tion is widespread but appears to be concentrated around volcanic centres.

Interbedded with the volcanic rocks are sedimentary and volcanogenic sedimentary rocks that are probably subordinate in volume but locally may represent most of the time between Sinemurian and Bajocian. Conglomerate, shale, siltstone, and tuffaceous siltstone are the most common sediments, with great variations depending on provenance and basin conditions. Greywacke and minor sandstone, although important, are only dominant in a few areas. Limestone is rare, occurring only as thin lenses and beds in shale sequences or as bioclastic lenses in volcanic piles. Massive limestone units are unknown.

Jurassic rocks define particular tectonic elements and environments of deposition (Fig. 2.5). These elements from north to south are: a) the Whitehorse Trough, mainly developed on the Cache Creek Terrane but also on bordering parts of Stikinia and Quesnellia and extending through the southern Yukon into northern British Columbia to about latitude 58°N, b) various volcanic belts lying between latitudes 58°N and 56°N along the northwest, north, and northeast margins of the younger Bowser Basin, and

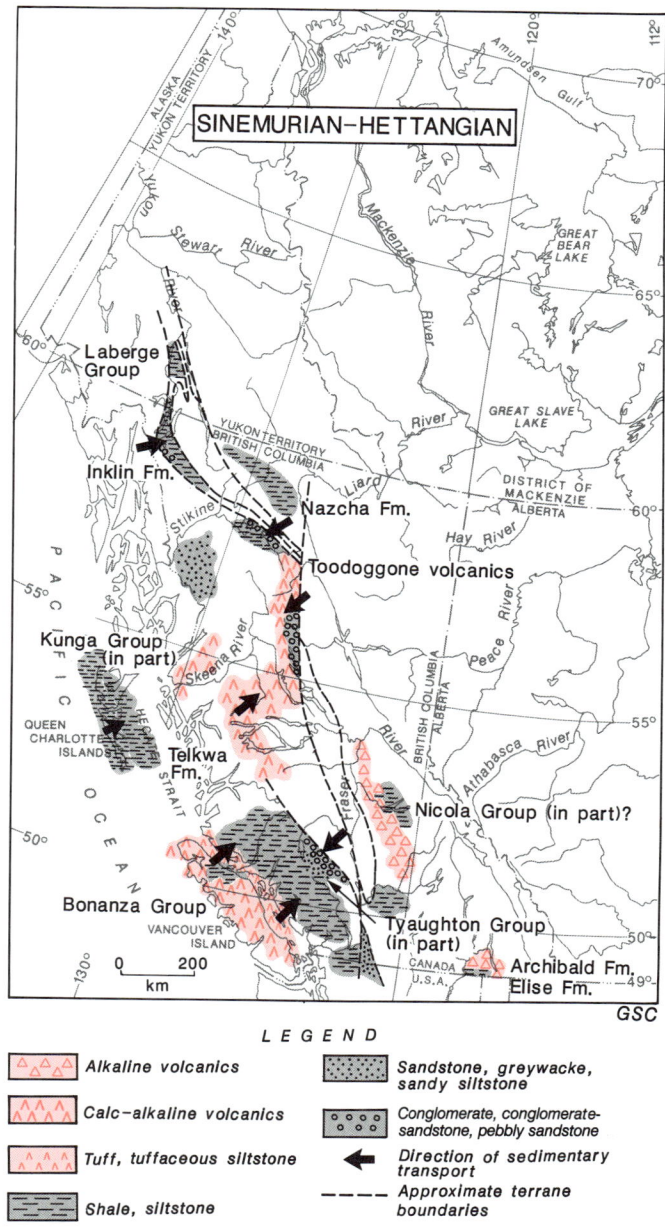

Figure 8.76. Distribution, lithology and nomenclature of Sinemurian and Hettangian rocks.

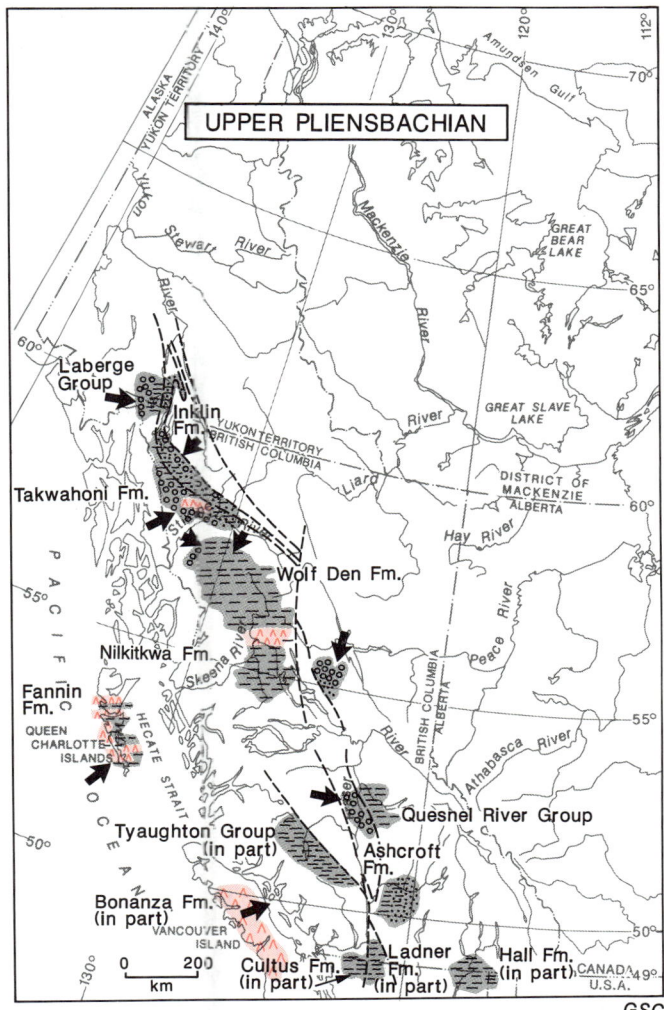

Figure 8.78. Distribution, lithology and nomenclature of Upper Pliensbachian rocks. Legend accompanies Figure 8.76.

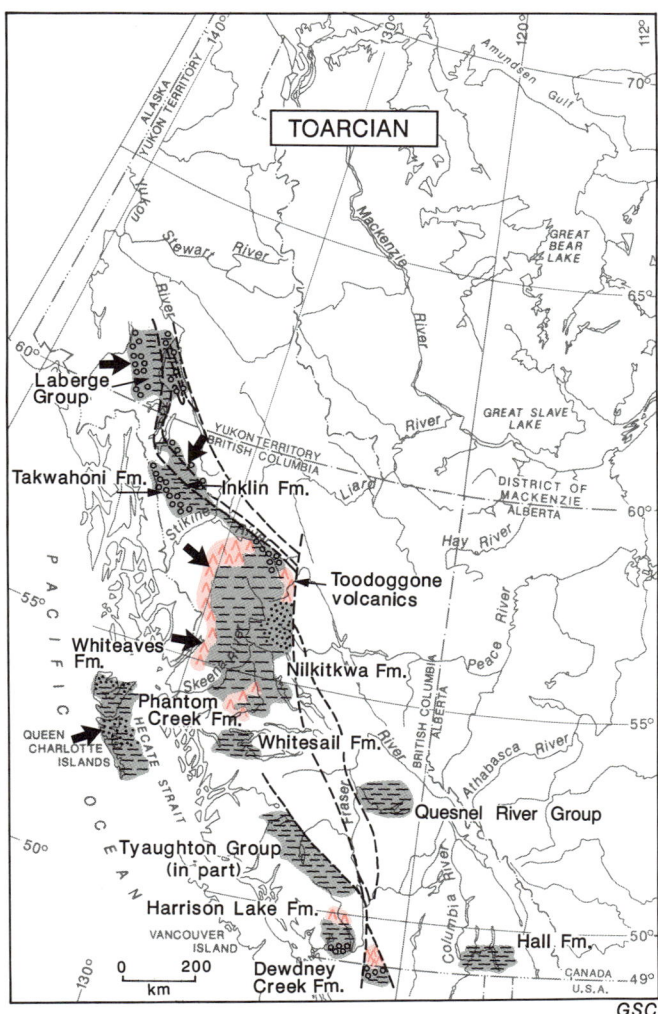

Figure 8.79. Distribution, lithology and nomenclature of Toarcian rocks. Legend accompanies Figure 8.76.

c) the type Hazelton volcanic area lying between 56°N and 53°N. The extreme southern part of Stikinia is very poorly known because of extensive Tertiary cover but apparently includes Jurassic volcanics and sediments. The Stikine and Skeena arches with related intrusive complexes and smaller plutons influenced all Jurassic tectonic and depositional history. The Whitehorse Trough lies northwest of the Stikine Arch. Through and between the two arches was a shale basin, ancestral to the younger Bowser Basin. All of the basins, arches, and volcanic belts are probably interrelated as components of one complex island arc terrane. Between the elements, however, stratigraphic continuity appears to be lacking and thus these elements may have been tectonically dislocated during or after their formation.

The Takwahoni Formation is mainly a southwesterly derived proximal facies deposited on and derived from Stikinia whereas the Inklin Formation is both a northeast- to easterly derived proximal facies (from Quesnellia, in part) and a southwesterly originating distal facies deposited on both the Cache Creek Terrane and Stikinia. Both formations are components of the Laberge Group.

The Takwahoni Formation (Fig. 8.78-8.80) consists of several kilometres of interbedded coarse cobble to pebble conglomerate, greywacke, siltstone and shale with rare flows and airfall tuff. The formation ranges in age from Late Pliensbachian to Early Bajocian and mainly overlies Upper Triassic volcanic rocks. The conglomerates, coarsest and thickest to the west and southwest, contain abundant volcanic and plutonic clasts (Fig. 8.81), some of which are more than a metre in diameter. The source area was an uplifted part of Stikinia coinciding with the Stikine Arch and the region along the boundary between the Coast and Intermontane belts (Fig. 8.78; 2.4). These areas had been the sites of Upper Triassic to Lower Jurassic island arcs southwest and west of the Whitehorse Trough. Lithologies of the granitic clasts can be related to the Stikine and Klotassin granitic suites which presumably formed the roots of the arcs.

The Inklin Formation is up to 3 km thick and consists predominantly of interbedded, turbiditic massive greywacke shale and siltstone (Fig. 8.76-8.79). In contrast to the more massive character of the Takwahoni Formation the Inklin is commonly strongly foliated, particularly in its

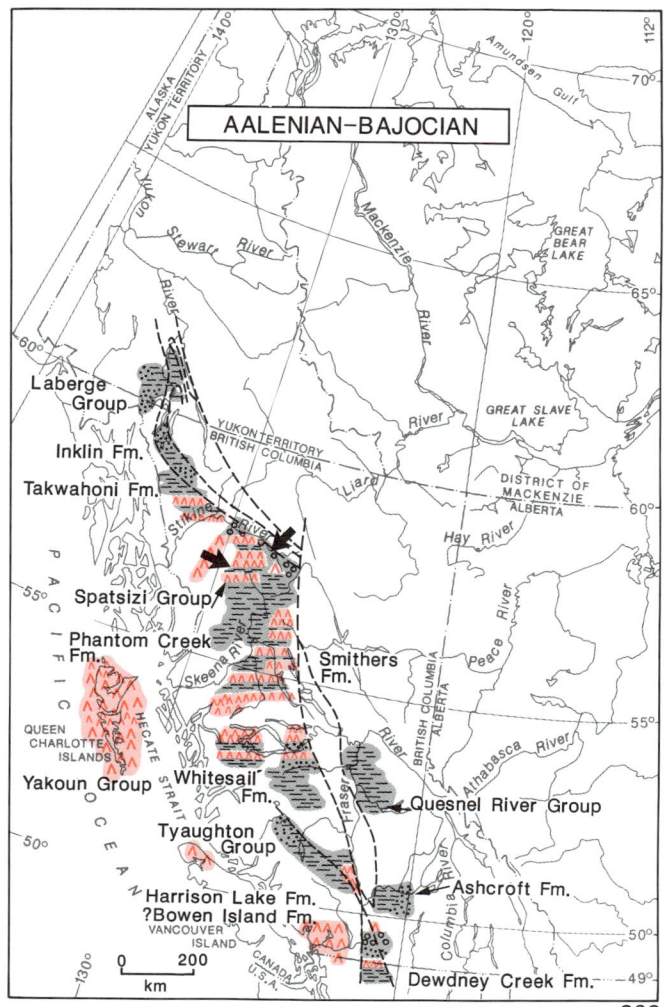

Figure 8.80. Distribution, lithology and nomenclature of Aalenian and Bajocian rocks. Legend accompanies Figure 8.76.

Figure 8.81. Conglomerate in the Takwahoni Formation on the Whitehorse Trough southwest of Whitehorse, Yukon Territory. Photo by J.O. Wheeler. GSC 205235-X

The Inklin and Takwahoni formations are everywhere separated by faults, especially along the King Salmon Fault, and it is probable that considerable shortening has taken place across the Whitehorse Trough.

Sequences of Lower to Middle Jurassic, dominantly volcanic rocks of the Hazelton Group are typically exposed along and near the Skeena Arch (Fig. 2.5). Correlative rocks, also assigned to the Hazelton Group, occur around the western, northern and eastern margins of the Bowser Basin. Except along the Skeena Arch where meagre chemical data suggest eastward subduction under an island arc (Tipper and Richards, 1976) there is little information as to the tectonic settings and relationships of the various volcanic centres. It is clear from biostratigraphic and isotopic age data and the distribution of volcanic rocks around the Bowser Basin that the group cannot represent a single, simple island arc in its present form.

The Hazelton Group on the Skeena Arch and along its flank has been subdivided into four formations with many facies and members. The most widespread and prominent volcanic unit is the Telkwa Formation of Sinemurian age (Fig. 8.76; Plate 20) and deposited unconformably on Triassic rocks; the base is marked by a coarse polymictic conglomerate with clasts of volcanic rocks, Permian limestone and, locally, granitoids. Typically the Telkwa Formation consists of reddish, maroon, purple, grey and green pyroclastic and flow rocks of marine and nonmarine origin. Zeolitic alteration has been extensive around coeval plutons and may reflect paleo-hotspring systems contemporaneous with volcanism. Thickness varies greatly but is in excess of 400 m, particularly southwest of Smithers. Lithologically similar rocks in the Terrace area, however, are of Triassic age.

The Nilkitkwa Formation comprises as much as 1000 m of interbedded shale, greywacke, andesitic to rhyolitic tuff and breccia, and minor limestone; it is mainly marine and ranges in age from Early Pliensbachian to Middle Toarcian (Fig. 8.77-8.79). Its local but distinctive basaltic flows and aquagene tuffs have been traced northward to where coeval rocks are entirely shale which disappear beneath younger strata of the Bowser Lake Group. The

southeastern exposures. Local conglomerates contain abundant clasts of limestone derived from erosion of the Upper Triassic Sinwa limestone and quartz porphyry from the Upper Triassic Kutcho Formation both in the Cache Creek Terrane (Thorstad and Gabrielse, 1986). The strata range in age from Hettangian? to Pliensbachian in its southwestern exposures and from Hettangian to Early Bajocian in the northeast. Although the formation has been interpreted as a northeast-prograding submarine fan (Bultman, 1979) and also as a distal equivalent of the Takwahoni Formation (Souther, 1971) it has now been shown that in the Atlin and Whitehorse areas most of the Inklin is older than the Takwahoni. Moreover, the Inklin Formation apparently becomes coarser towards the northeastern part of the Whitehorse Trough and thus the sediments may have been derived from Quesnellia. Some volcanics, such as the Nordenskiold dacite in the Yukon and correlative volcanics in the Atlin area of northern British Columbia are intercalated with Lower Pliensbachian and younger sediments.

formation occupies the eastern part of Stikinia between 56°N and 53°N latitude but may be thinner or covered in other areas to the south.

The Smithers Formation (Fig. 8.80), 500 m to 800 m thick, is a light grey to brown, greenish grey to drab grey succession of greywacke, siltstone, tuffaceous shale, volcanic breccia, and rare volcanic pebble conglomerate. Typically the sediments are immature and poorly sorted. The formation is widespread north and south of the Skeena Arch. Along the east side of the Bowser Basin it contains shaly, silty and sandy members. The age of the formation is variable but in most areas it is Early Bajocian; in a few areas, the more shaly members apparently are as old as Middle Toarcian and as young as Late Bajocian.

The Whitesail Formation south of the Skeena Arch in the Whitesail Lake area (Fig. 8.80), is of latest Toarcian to Aalenian age. In the type area, the formation consists of 600 m of rhyolitic tuff and breccia, shale, tuffaceous shale and siltstone that have yielded an abundant marine fauna. In the Smithers and Terrace areas rhyolite domes and rhyolitic dykes are correlated on the basis of radiometric dates. Also south of the Skeena Arch and extending to the southern extremity of Stikinia is an area of moderate to low relief in which Cretaceous and younger volcanic and sedimentary rocks cover the Jurassic strata. The meagre stratigraphic record includes Middle Bajocian strata like the Smithers Formation, Pliensbachian strata like the Nilkitkwa Formation, and coarse volcanic breccia similar to the Telkwa Formation. Sedimentary rocks of undetermined age may be Jurassic or they may be Cretaceous.

The Toodoggone volcanics along the northeast side of the Sustut Basin (Fig. 8.76, 8.79) are a succession of subaerial, intermediate, calc-alkaline to alkaline, predominantly pyroclastics, as much as 700 m thick, of Early Jurassic age. The volcanics unconformably overlie Upper Triassic volcanic rocks and are unconformably overlain by sediments of the Sustut Group. Typical lithologies include ash-flow and air-fall crystal-lithic tuffs with quartz, biotite, hornblende and plagioclase phenocrysts (Diakow, 1984). Trachyandesite flows with potassium feldspar phenocrysts form a distinctive and widespread member. Radiometric ages of volcanic rocks and related plutons range from 179 Ma to 207 Ma (Panteleyev, 1984; Diakow, 1985; Gabrielse et al., 1980) with a distinct cluster around 200 Ma. The Toodoggone volcanics may, therefore, in large part be correlative with the Telkwa Formation in the type Hazelton region. The younger radiometric dates and one collection of ammonites indicates that rocks as young as Toarcian may be present.

The Toodoggone volcanics host epithermal, amethystine quartz-gold-silver veins. Spectacular alunite-bearing alteration zones have been related to Early Jurassic, shallow hydrothermal systems.

The Cold Fish volcanics along the southwest side of the Sustut Basin (Fig. 8.77) are mainly of Early Pliensbachian age and comprise more than 700 m of subaerial to submarine felsic to mafic lava and tuff intercalated with minor shale and limestone. The volcanics are arc-related, bimodal rhyolite (mostly tholeiitic) and basalt to trachyte (transitionally tholeiitic to alkaline; D.J. Thorkelson, pers. comm., 1987).

Along the Stikine Arch Toarcian and Bajocian volcanic rocks occur but their relationships to the Cold Fish volcanics are unknown. Toarcian sediments including conglomerate, greywacke, sandstone, siltstone and minor limestone, up to 400 m thick, are overlain by volcanic breccia and flows. They rest unconformably on Pliensbachian, Sinemurian(?) and Upper Triassic rocks including granitic rocks of the northern flank of the Hotailuh batholith on Stikine Arch.

Locally, the volcanics are overlain by Aalenian(?) and Bajocian flows, breccia and tuff, commonly in distinct centres. North of Stikine River and east of the Hotailuh batholith Bajocian volcanics are intercalated with chert-pebble conglomerate, sandstone and siltstone which may represent a proximal facies of the Bowser Lake Group. South of the Stikine Arch the volcanics pass into tuffaceous sediments and eventually into shale and siltstone of the Spatsizi Group (Fig. 8.77, 8.80, 8.82; Thomson et al., 1986). Several porphyry-related copper deposits occur in probable Bajocian volcanics along the northwest margin of the Bowser Basin.

Volcanic and sedimentary rocks of Early to Middle Jurassic age are widespread along the west side of the Bowser Basin. In the Iskut River area, Lower Jurassic conglomerate with boulders and cobbles of volcanic and granitic rocks lies unconformably on Upper Triassic volcanics of the Stuhini Group (Souther, 1972). The Lower Jurassic succession comprises more than 1000 m of tuffaceous siltstone and sandstone interbedded with interbedded fragmental volcanic rocks of basalt-andesitic composition. Fossils demonstrate the presence of Hettangian, Upper Pliensbachian and Upper Toarcian stages.

About 1000 m of black shale with minor sandstone and grit of Late Toarcian to Middle Bajocian age separates the underlying Lower Jurassic rocks from an overlying sequence, as much as 2000 m thick, of Bajocian(?) submarine, basalt-andesite pillowed lavas and associated flows, dykes and sills. This unit grades laterally into sequences of tuff-breccia, tuff, volcanic sandstone and conglomerate which is overlain, possibly conformably by chert-pebble conglomerate of the Bowser Lake Group.

Farther south, in the Stewart area, a heterogeneous unit of Lower Jurassic (Pliensbachian?) intermediate to felsic flows, tuff and volcaniclastic rocks, important as a host for precious metal deposits, overlies Upper Triassic (Norian) volcanics (R.G. Anderson, pers. comm., 1986). The volcanics may be as much as 1000 m thick locally but are much thinner away from volcanic centres (Alldrick, 1985). Overlying the volcanics is a distinctive unit of buff, fossiliferous sandy limestone 1 to 3 m thick which contains a prolific Toarcian fauna. The uppermost formation of the Jurassic to Middle Jurassic succession is a widespread marker unit of alternating siliceous shale, flinty welded tuff and radiolarian chert of Bajocian age. This formation is probably correlative with similar strata at the top of the Lower and Middle Jurassic succession exposed as inliers in the northern part of the Bowser Basin and along its eastern margin. Overlying strata are Bathonian shales of the Bowser Lake Group.

The stratigraphy in several small inliers within the northern Bowser Basin and in localities along its eastern margin suggests that much of the basin may be underlain by fine grained clastic rocks of Early and Middle Jurassic age. As noted above Lower Pliensbachian volcanics of the Nilkitkwa Formation grade into shale northwest from the

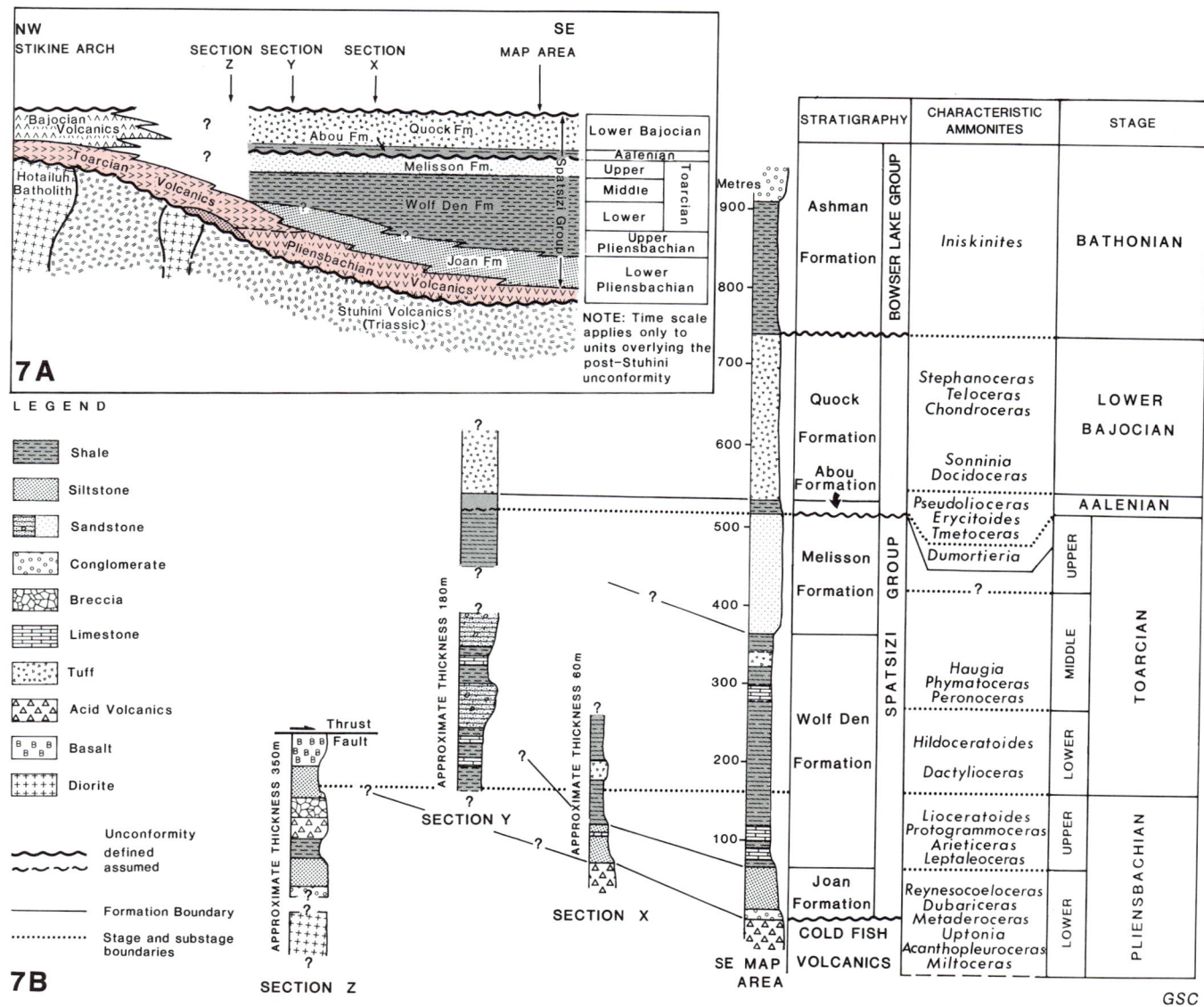

Figure 8.82. Stratigraphic relationships of Jurassic strata along Stikine Arch and in the northern Bowser Basin.

Skeena Arch. Toarcian volcanics along the southern flank of the Stikine Arch change facies to fine grained clastic sediments to the south, and, in the western part of the basin between latitudes 56°N and 57°N, no volcanic rocks are present in a Lower Jurassic succession of conglomerate volcanic sandstone and pebbly sandstone 700 m thick which unconformably overlies Upper Triassic rocks (Monger, 1977c).

At many localities around the Bowser Basin the base of the Hazelton Group and related rocks is marked by a polymictic conglomerate containing clasts of Upper Triassic volcanic rocks, Permian limestone and granitoids of mainly undetermined but presumably of Late Triassic age. These conglomerates, as much as 400 m thick record an important episode of tectonism in earliest Jurassic time.

Relationships between Stikinia, Cache Creek Terrane and Quesnellia: amalgamation forming Intermontane Superterrane

The Cache Creek Terrane was thrust southwestward over the eastern and northern margins of Stikinia by late Middle Jurassic (Bajocian) time. The upper Middle and Upper Jurassic clastic sediments of Bowser Basin contain abundant detritus derived from the Cache Creek Terrane, including rare limestone cobbles with the characteristic verbeekinid fusulinids, and chert containing Late Triassic (Norian) radiolaria. In southern British Columbia, the Cache Creek Terrane was thrust eastwards over Quesnellia in Late Jurassic time.

Earlier linkages are more tenuous, but possibly all three terranes were associated by Early Jurassic time. In northern British Columbia, the northeastern French Range Subterrane of the Cache Creek Terrane locally contains Norian chert interbedded with greywacke bearing abundant hornblende detritus (Monger, 1969). The source direction for the clastic rocks is not known, but they are compositionally similar to coeval sandstones of the Lewes River Group in the Whitehorse Trough (Wheeler, 1961). The Lower Jurassic Inklin Formation overlying the Cache Creek Assemblage was derived in part from a northeastern source, presumably in Quesnellia. In the Whitehorse area the Inklin Formation and the partly correlative Takwahoni Formation derived from Stikinia appear to interfinger. If so, Stikinia, the Cache Creek Terrane and Quesnellia were at least loosely amalgamated by Early Pliensbachian time.

In southern British Columbia the Cache Creek Terrane and Quesnellia are overlapped by the upper Lower Cretaceous Spences Bridge Group (Monger and McMillan, 1984). Travers (1978, 1982) proposed that the Cache Creek Terrane was thrust eastwards onto Quesnellia probably in Middle to Late Jurassic time although Early Cretaceous thrusting also was possible. Latest Triassic-earliest Jurassic granitic plutons occur in both the Cache Creek Terrane and Quesnellia. Near the settlement of Cache Creek, the Cache Creek Terrane and Quesnellia are separated locally by the Bonaparte Fault, a steep fault of possible Cretaceous or Tertiary age. East of the fault are overturned, well-bedded, Middle Norian volcaniclastics of the Nicola Group. They locally contain breccia composed of clasts of chert with Middle Triassic conodonts and radiolaria, as well as clasts of basalt, carbonate and serpentinite. The breccia may be derived from the Cache Creek Terrane. West of the fault, contorted and disrupted chert and argillite, at least partly of Late Triassic age and assigned to the eastern, mélange, belt of Cache Creek, contain a variety of blocks, mostly of Cache Creek Group lithologies. Some blocks of ignimbrite and lapilli tuff are typical of the western belt of the Nicola Group (Shannon, 1982). This intermixing of Quesnellian and Cache Creek lithologies together with known chemistry of the Nicola Group suggests a west-facing arc (N. Mortimer, pers. comm., 1986) and, moreover, that the Cache Creek was being subducted eastwards beneath Quesnellia in Late Triassic time. Thus the Cache Creek may represent off-scraped ocean floor material with the Nicola being the related arc. Detritus from an uplifted outer arc ridge of Cache Creek may have supplied detritus to the Nicola fore-arc basin; Nicola material possibly bypassed the ridge, to be deposited in the trench and become subsequently incorporated as blocks within the accretionary prism.

Metamorphic rocks and Nisling Terrane in the northern Coast Belt

Metamorphic rocks included in the eastern margin of the Coast Belt in northwesternmost British Columbia and southwestern Yukon originally were named the Tracy Arm Terrane (e.g. Coney et al., 1980).

In the Coast Belt west of Atlin, in decreasing order of abundance, are quartz-feldspar-muscovite to quartz-biotite schist, marble, quartz-hornblende gneiss and hornblende amphibolite (Werner, 1978). The metamorphism appears to be pre-Late Triassic in age since a granodioritic intrusion cutting metamorphic rocks gives a Rb-Sr date of about 235 Ma. Preliminary strontium isotope studies indicate a protolith age of about 900 Ma (L.J. Werner and R.L. Armstrong, pers. comm., 1979). Augite porphyry breccia, similar lithologically to Upper Triassic volcanic rocks in Stikinia, appears to stratigraphically overlie the metamorphic rocks and augite porphyry dykes cut them. Possibly the metamorphic rocks are in part basement to at least part of Stikinia. Some also may be correlative with those in the Nisling Terrane.

Terranes of the southern Coast Belt

In southwestern British Columbia, between the Intermontane and Insular superterranes in the southern Coast Belt, are six small terranes with partly coeval but contrasting stratigraphic successions (Fig. 8.83). From east to west these are: Methow, Bridge River, Cadwallader, Shuksan, Chilliwack and Harrison Lake terranes. The first three are structurally juxtaposed, and stratigraphic differences among them can be interpreted as differences in facies. The Harrison Lake and Chilliwack terranes may, in part, contain equivalent facies. Both are structurally involved with the Shuksan Terrane, and all three underwent Mesozoic high-pressure metamorphism. Largely separating the Methow, Bridge River and Cadwallader terranes from the remainder of the group is a belt of high-grade metamorphic rocks, which includes Custer and Skagit Gneiss and Settler Schist, the protoliths of which speculatively could include rocks of the Methow, Bridge River and Cadwallader terranes as well as older, stratified and plutonic rocks of unknown affinity. The six terranes appear to be stratigraphically or structurally linked with

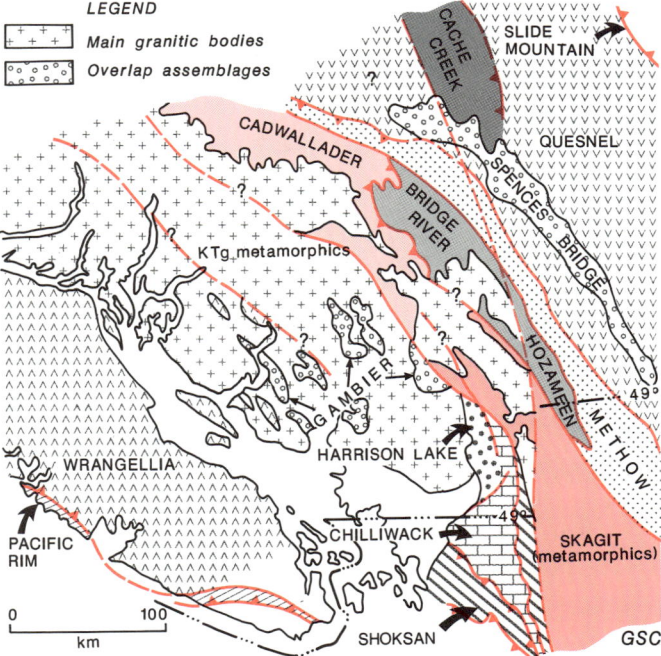

Figure 8.83. Distribution of terranes, overlap assemblages, plutonic, sedimentary and structural linkages in the southern Coast Belt-northern Cascades region prior to dextral transcurrent faulting along the Fraser River Fault zone.

one another by about mid-Cretaceous time when they were metamorphosed and cut by granitic intrusions. The mid-Cretaceous plutons yielded paleopoles which suggest that in Cretaceous time these rocks lay 2400 km to the south, with respect to North America (Chapter 3). Methow Terrane is linked to the Intermontane Superterrane by Aptian-Albian time, and Harrison Lake Terrane to the Insular Superterrane probably by the end of the Jurassic.

An alternative view held by H.W. Tipper is that the Jurassic rocks of Wrangellia, and the Cadwallader, Methow, Harrison Lake, and possibly Chilliwack terranes were part of a single, Lower Jurassic to lower Middle Jurassic basin which, judging by its fauna, lay along the cratonal margin opposite or near Sonomia in Nevada. The terrane had arc volcanics on its west margin during Sinemurian and Pliensbachian time with a central shale basin and coarser sediments on the eastern margin. During the Toarcian the west margin was uplifted and supplied sediment eastward. In Bajocian time a new volcanic arc was formed (Yakoun Group and Harrison Lake Formation?) lying to the east of the Lower Jurassic Bonanza-Kunga arc; this is clearly the case in the Queen Charlotte Islands. The Lower Jurassic-Middle Jurassic basin was destroyed by transcurrent faulting along its eastern margin after Early Bajocian time and by plutonism in the Coast Mountains. Three basins were thus created: a) Tyaughton- Methow Trough on the east margin, b) a Bathonian-Callovian basin in the Queen Charlotte Islands, and c) a Callovian-Oxfordian basin on the west side of Vancouver Island. Thus, dextral strike-slip movement was responsible for dividing the region into several fragments or subterranes.

Methow Terrane

The Methow Terrane comprises Triassic basalt, overlain stratigraphically by Lower Jurassic to Albian clastic sedimentary and subordinate volcanic rocks (Ladner, Relay Mountain and Skeena assemblages). The terrane is separated from the Intermontane Superterrane on the east by the Pasayten and Hungry Valley faults, and from the Bridge River Terrane to the west by the Jack Mountain Thrust and Hozameen and Yalakom faults.

The oldest rocks recognized belong to the Spider Peak Group consisting of Triassic(?) pillow basalt with the composition of mid-oceanic ridge basalt (Anderson, 1976; Ray, 1986). Triassic conodonts have been obtained from an intrapillow breccia at the top of the formation. The Spider Peak is overlain by the Lower and Middle Jurassic Ladner Group, comprising predominantly thin-bedded argillite and siltstone with minor beds of wacke and conglomerate near the base. Clasts in the conglomerate consist mainly of basalt, dacite, limestone and granitic rock. Thicknesses range from 300 to 700 m. Deposition probably occurred in deep water, and slump structures indicate a west-dipping paleoslope (Coates, 1974). Lower and Middle Jurassic strata in the Methow Terrane probably were deposited on oceanic crust, with local volcanic centres supplying much of the detritus.

Bridge River Terrane

The Bridge River Terrane contains radiolarian chert, argillite, basalt, local clastic rocks and small exotic blocks of carbonate assigned to the Bridge River Assemblage. The assemblage includes serpentinite and the large Shulaps ultramafic body (Leech, 1953; McTaggart and Thompson, 1967; Roddick and Hutchison, 1973; Tennyson et al., 1982; Potter, 1983; R.A. Haugerud, pers. comm., 1985). The most common fossils are of Late Triassic age, but radiolaria of Permian, and Early and Middle Jurassic ages are known also. No stratigraphic section is known because the rocks at the lowest, prehnite-pumpellyite, metamorphic grade are typically highly disrupted and those at high greenschist-lower amphibolite grades have completely transposed bedding. Basaltic rocks in the Bridge River Terrane appear to be of both island arc and tholeiitic basalt chemical affinities (Ray, 1986; Haugerud, 1985).

The Bridge River Terrane is possibly overlain by the Aptian-Albian Taylor Creek Group, which contains notable chert-pebble conglomerate, and by continental Tertiary sedimentary rocks. Metamorphosed chert, marble, basalt and ultramafic rocks, which lie on the west side of the high-grade metamorphic core of the Cascades east of Harrison Lake, are lithologically similar, but their age is uncertain.

The Bridge River Terrane appears to represent either a collapsed oceanic or marginal basin which contained arc-related volcanic rocks. The abundant pelagic sediments suggest deposition far removed from a terrigenous source. Arguably, it could be a distal facies equivalent of older parts of the Methow Terrane, or the Triassic part of the Cadwallader Terrane. Correlation of the Bridge River and Cache Creek assemblages is not ruled out.

Cadwallader Terrane

Rocks in the Cadwallader Terrane range from Late Triassic to Early Cretaceous (Albian) in age. Upper Triassic (Lower Norian) strata of the Cadwallader Group (Cadwallader Assemblage) include approximately 1300 m of pillowed basalt, basaltic breccia and minor crystal tuff which underlie and are interbedded with distinctive turbiditic limestone, volcanic and granitic conglomerate and sandstone (Jeletzky and Tipper, 1967; Rusmore, 1985). Chemically, the basalts appear to be closer to arc tholeiites than to mid-ocean ridge basalts (M. Rusmore, pers. comm., 1985). These rocks are overlain, possibly unconformably, by the Tyaughton Group (Ladner Assemblage) where the lowest part comprises a coarse, red, basal conglomerate which may be 100 m or more thick. Above the basal conglomerate is approximately 400 m of conglomerate, siltstone, and limestone that span the Triassic-Jurassic boundary, and an unknown thickness of siltstone and shale, apparently continuous to the top of the Lower Bajocian (see Fig. 17.14). Between the highest fossils of the Cadwallader Group (Early Norian) and the lowest fossils of the Tyaughton Group (Late Norian) there appears to be a gap, because the Middle Norian is not recorded. Unfossiliferous rocks, however, occur at the base of the Tyaughton Group. The red conglomerate could be the result of erosion of a deeply weathered regolith.

The Jurassic fauna of the terrane has some Tethyan affinity through most of Hettangian to Early Bajocian time; boreal components are minimal to absent.

Rusmore (1985) interpreted the lowest part of the succession as having probably formed in an intra-arc or back-arc setting. Younger rocks are basinal and contain little volcaniclastic material.

Shuksan Terrane

Much of the Shuksan Terrane is composed of actinolitic greenschist and crossite- and glaucophane-epidote schist (Shuksan Greenschist), phyllitic quartz grit and black sericitic phyllite (Darrington phyllite). Whole-rock Sr analyses indicate that the primary age of the latter unit is probably Jurassic, and the succession as a whole underwent high-pressure, blueschist-grade metamorphism in Late Jurassic to Early Cretaceous time (Armstrong, 1980). The Vedder Complex, schist and amphibolite, forms tectonic slices near the base of the Shuksan thrust sheet and has been dated as Permian to earliest Triassic by K-Ar and Rb-Sr methods suggesting the primary age of the rock and time of metamorphism are not very different (Armstrong et al., 1983). Variably retrograded clinopyroxene-plagioclase-quartz gneiss, hornblendite, and metagabbro of the Yellow Aster Complex, yield Precambrian and early Paleozoic U-Pb dates (Mattinson, 1972).

According to Vance et al. (1980) the chemistry of the Jurassic Shuksan metabasalt suggests an origin in an oceanic island arc or small ocean basin. From regional considerations they prefer an origin in a small ocean basin which was subducted in Early Cretaceous time. The terrigenous Jurassic nonvolcanic component and absence of chert supports the latter hypothesis.

Chilliwack Terrane

The oldest rocks of the Chilliwack Assemblage in the Chilliwack Terrane in Canada are of Early Pennsylvanian age, although Devonian strata are exposed south of latitude 49°N (Fig. 8.84; Monger, 1970, 1977b; Danner, 1977). In Canada, they consist of 400 m of fine grained turbiditic siltstone and sandstone, conformably overlain by 30 m of Lower Pennsylvanian carbonate. A generally coarse grained overlying sequence, about 1500 m thick, contains volcanic and local chert-grain sandstone, conglomerate, minor tuff, and pelite with plant remains. It is overlain conformably by 60 to 600 m of Lower Permian carbonate, which is gradationally overlain by up to 600 m of basalt and dacitic crystal-lithic tuff. Although the sequence is imbricated by a series of thrust faults it retains stratigraphic integrity among different thrust sheets.

The Cultus Formation (Cultus Assemblage) lies disconformably on the Chilliwack Group, and consists of 600 m of thinly bedded pelite, siltstone, volcanic sandstone and possible tuff, containing Late Triassic and Early Jurassic fossils. At one locality it is overlain by Upper Jurassic argillite.

Permian fusulinid faunas in the Chilliwack Group appear to correlate most closely with those of the McCloud Limestone of California (Sada and Danner, 1973). Permian conodont faunas are similar to those in the Harper Ranch Group of Quesnellia (M.J. Orchard, pers. comm., 1984). In general lithological association, the Chilliwack is similar to the Harper Ranch Group and to units in central Oregon, the eastern Klamath Mountains and northwestern Nevada. The Jurassic faunas of the Cultus Formation are mainly Early Jurassic in age (Sinemurian and Pliensbachian), and consist essentially of Tethyan hildoceratids and Boreal amaltheids, clearly of Late Pliensbachian age - a mixed fauna that is much farther north than would be expected from comparison with cratonal faunas (Tipper, 1981). The Cultus correlates with the Ladner Formation of the Methow Terrane and is older than the Harrison Lake Formation of the Harrison Terrane.

The general character of the unit and available chemistry from north and south of latitude 49°N (Blackwell, 1983) suggest that the Chilliwack Group was deposited in a volcanic arc. Detritus in the Cultus Formation is largely of volcanic origin, and could represent the distal, basinal part of a volcanic arc sequence.

Harrison Lake Terrane

The Harrison Lake Terrane forms the western side of Harrison Lake and occurs as isolated septa within granitic rocks of the Coast Belt to the west. At the base are silicified argillite and siltstone of the Camp Cove Formation (Cadwallader Assemblage, possibly Chilliwack Assemblage; Crickmay. 1962). Radiolaria recovered from these rocks are of Middle Triassic age (F. Cordey, pers. comm., 1985). An unconformably overlying basal Toarcian conglomerate of the Harrison Lake Assemblage, with carbonate clasts containing Permian fossils similar to those in the Chilliwack Group, suggests a pre-Toarcian linkage with the Chilliwack Terrane. Pearson (1973) considered the youngest beds to be of Toarcian (Early Jurassic) age. The Toarcian sequence includes Early Toarcian ammonites in shale that ranges upwards through Middle Toarcian(?), and Upper Toarcian shale and volcanics overlain by Aalenian shale. This section is closely allied to and overlain by the main volcanic succession of the Harrison Lake Formation, which could be as young as Early Bajocian. The sedimentary section may correlate in whole or in part with the Dewdney Creek Formation (sensu Cairnes, 1924) in the Methow Terrane (Tyaughton-Methow Trough - see Chapter 9). The overlying upper Lower and Middle Jurassic volcanic section, approximately 2500 m thick consists of predominantly intermediate to felsic pyroclastics and flows which are correlative with the Wells Creek volcanics south of latitude 49°N (Monger, 1970; Pearson, 1973). Overlying the Harrison Lake are Middle to Upper Jurassic sedimentary and volcanic rocks of, respectively, the Mysterious Creek and Billhook Creek formations, up to 1600 m thick. These strata are unconformably overlain by the Gambier Assemblage comprising locally coarse (boulders up to 1 m in diameter) conglomerate with notable granitic clasts, succeeded, in turn, by about 3000 m of sandstone of the Peninsula Formation and intermediate volcaniclastic rocks of the Brokenback Hill Formation. The last two units appear to be correlative with the Fire Lake and Gambier groups and form an overlap assemblage that extends across much of the southern Coast Belt to lie on Wrangellia to the west.

Taku Terrane

The Taku Terrane contains a polygenetic assemblage (Taku Assemblage) of highly deformed and metamorphosed strata forming the westernmost element of the Coast Belt between 54°N and 59°N (Fig. 8.85; Monger and Berg, 1984). It contains sparse fossils of Permian and Middle and Late Triassic ages (Monger and Berg, 1984). Lithologies include Permian crinoidal marble intercalated with phyllite and metatuff, Middle and Upper Triassic carbonate, phyllite and basaltic pillow breccia, and undated metagreywacke and metatuff lithologically similar to Jura-Cretaceous Gravina-Nutzotin strata. Near the southern end of the

CHAPTER 8

terrane locally conspicuous lenses of metaconglomerate of Permian (?) and Jura-Cretaceous (?) ages contain trondhjemite clasts lithologically similar to early Paleozoic trondhjemite in the Alexander Terrane.

Because of deformation and metamorphism the stratigraphic affinities of this terrane remain ambiguous. Davis and Plafker (1985) have suggested that Triassic basalt in the Taku Terrane is lithologically and geochemically similar to Triassic basalt of Wrangellia. Saleeby et al. (1985) have traced strata from the Alexander Terrane and the overlying Jura-Cretaceous Gravina-Nutzotin Belt into metamorphic rocks assigned to the Taku Assemblage. It may be that much, if not all, of the rocks of the Taku Terrane are merely the metamorphosed and deformed equivalent of those of the Insular Superterrane and the overlying Jura-Cretaceous Gambier Assemblage (Gravina-Nutzotin).

In structural setting, the Taku Terrane appears to be analogous to the small terranes in the southern Coast Belt described above, in that it lies between the Intermontane

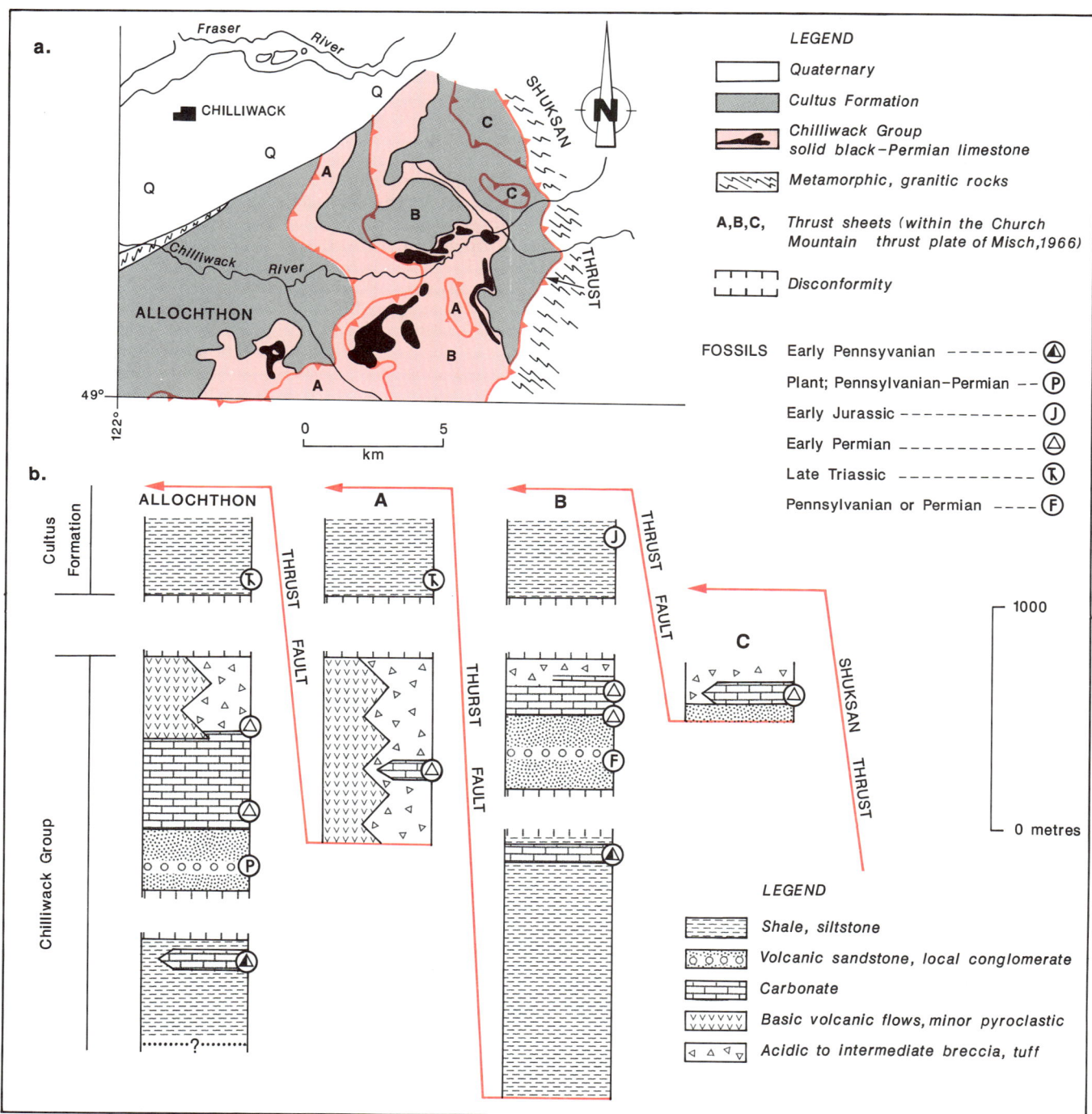

Figure 8.84. Stratigraphy of the Chilliwack Group, Chilliwack valley, southwestern British Columbia.

UPPER DEVONIAN TO MIDDLE JURASSIC ASSEMBLAGES

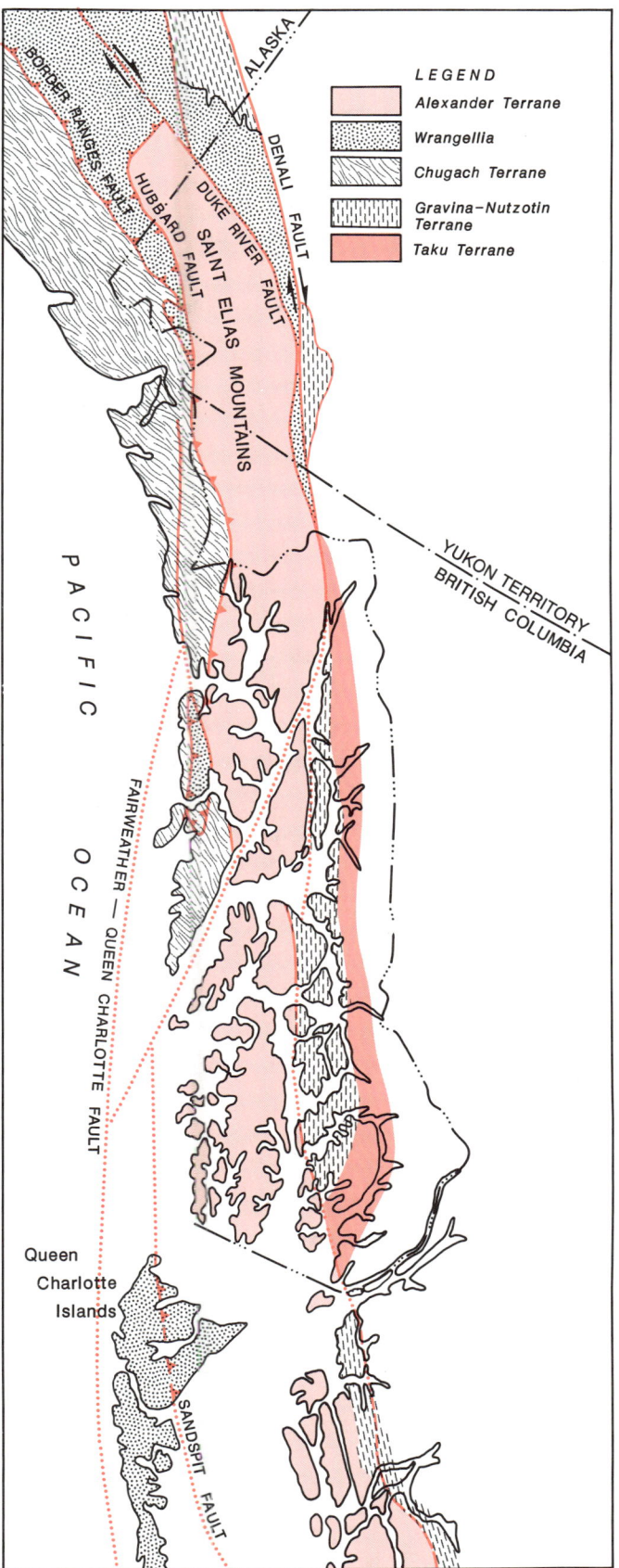

Figure 8.85. Relationship of the Alexander Terrane to neighbouring terranes.

Superterrane, accreted to North America in the Jurassic, and the Insular Superterrane, possibly not accreted until the Cretaceous. Stratigraphic similarities include the presence of Triassic basalt and Jura-Cretaceous clastic rocks that apparently overlap with the Wrangellian and Alexander terranes.

Alexander Terrane

The Alexander Terrane in Canada occurs at the northern and southern ends of a belt that includes a large part of southern Alaska (Fig. 8.85). It is underlain by heterogeneous assemblages of sedimentary, volcanic and plutonic rocks ranging in age from late Proterozoic to Late Triassic. For discussion of pre-Upper Devonian rocks see Chapter 7.

Upper Devonian to Lower Permian assemblages

Rocks within this age range occur in two belts, one along the northeastern periphery and the other in the more central parts of the Alexander Terrane in the Saint Elias Mountains (Campbell and Dodds, 1982a,b,c, 1983a,b). They are poorly documented, sparsely fossiliferous, and mostly complexly deformed and regionally metamorphosed to low greenschist facies.

Adjacent to the Duke River Fault which forms the northeastern boundary of the Alexander Terrane, the assemblage is present within a tectonic mélange. However, due largely to fault dislocations and lack of fossils, its stratigraphy, thicknesses and age relationships are poorly known. There the assemblage generally appears to be underlain by fossiliferous Middle Devonian shallow-water, "reefoid" carbonate and associated fine calcareous clastics (A.E.H. Pedder, pers. comm., 1980; M.J. Orchard, pers. comm., 1980). Overlying these carbonates with apparent conformity are sequences of shallow-water, interbedded dark calcareous fine clastics and minor thin limestones, and locally (Slims River area) marine to subaerial basaltic to andesitic flows and lesser volcaniclastics. In the latter area fault-bounded slivers also contain a complex array of small plutons, sills, and dykes of gabbro, diabase, gabbro-pegmatite with screens of mafic flows, pyroclastics, volcaniclastics, and rarer siliceous argillite and chert (collectively the informally named "Mt. Cairnes Gabbro-Greenstone Complex"), which mostly are juxtaposed with Middle Devonian carbonate. Also within this structurally complex belt are thick sequences of dark siliceous argillite, more restricted pyroclastics, and rarer gypsum-anhydrite and black serpentinite, which are strikingly similar to rocks of the immediately adjacent segment of Wrangellia (W1), and may in part represent inclusions of that terrane within the tectonic mélange.

A second poorly defined belt, believed to contain correlative rocks is exposed within the Icefield and Alsek ranges in the more central parts of the Alexander Terrane. Its stratigraphy is poorly known. The succession, of unknown thickness is dominated by thinly interbedded dark graphitic, commonly siliceous argillite and quartz-rich siltstone, which locally include dark calcareous argillite and discontinuous carbonate. In two widely separated localities a thin bedded limestone has yielded Pennsylvanian to early Early Permian conodonts (M.J. Orchard, pers. comm., 1980) and shallow-water calcareous clastics contain abundant macrofossils and scarce conodonts of probable Permian age (E.W. Bamber, pers. comm., 1980; M.J. Orchard, pers. comm., 1980). Fine clastic sediments which are widespread in the central part of the Alexander Terrane and informally named the "Icefield Ranges pelitic assemblage", may include strata of this sequence. The substrate appears to be an extensive and thick sequence of calcareous turbidite and carbonate of early Paleozoic age which locally is known to be capped by Middle Devonian limestone.

Within the Alexander Terrane in adjacent southeastern Alaska correlatives of the Saint Elias Mountains succession may include the Upper Devonian Freshwater Bay Formation (Cedar Cove Assemblage), parts of the Port Refugio Formation (Karheen Assemblage), parts of the Devonian Cedar Cove and the Saginaw Bay (Iyoukeen Assemblage) formations, the Upper Devonian and Mississippian Cannery Formation (Cannery Assemblage), the Mississippian Peratrovich and Iyoukeen formations (Iyoukeen Assemblage), the Pennsylvanian Klawak Formation (Iyoukeen Assemblage), and various Permian formations (Halleck and Pybus assemblages; Gehrels and Berg, 1984). Similar strata, including the Ducie Island limestone and Dunira Formation of the Iyoukeen Assemblage, are present on islands west of Prince Rupert, British Columbia (Woodsworth and Orchard, 1985). There, a well documented unconformity separates Upper Triassic and Pennsylvanian rocks. Much of the Upper Paleozoic stratigraphy in the southeastern part of the Alexander Terrane reflects platformal and rift-related environments of deposition.

Upper Triassic Hyd Assemblage

In the Alexander Terrane of the Saint Elias Mountains rocks of this age are present along its northeastern margin along the Duke River Fault and within its central parts in the Icefield and Alsek ranges and Tsirku Glacier area. The stratigraphy is only known in widely scattered areas. Most rocks are regionally metamorphosed to low greenschist facies.

The best documented section of the assemblage occurs near the Windy Craggy massive sulphide deposit (Cu, Co, Au) in Tatshenshini River area (Campbell and Dodds, 1983a). MacIntyre (1984) described a section which includes a basal sequence of distal to proximal turbidite and/or limestone of unknown thickness which is overlain by, and partly interfingers with, amygdaloidal andesitic flows and minor limy argillite (500-1000 m); felsic or altered porphyritic pillow lava, flows, and minor chert (100-200 m); mafic pillow lava, amygdaloidal flows, and minor chert and limy argillite (500-1000 m); limy siltstone, argillite, andesitic tuff and flows (100-1000 m); agglomerate (up to 100 m); and basaltic pillow lava (+1500 m, top unobserved). The predominantly volcanic component of this sequence has been informally named the "Tats Volcanic Complex" (J.B. Gammon and T.E. Chandler, pers. comm., 1985). Abundant conodont collections from the underlying dark thin-bedded calcareous sediments and locally within the volcanics are Early to Middle Norian in age (M.J. Orchard, pers. comm., 1982). Based on field criteria and chemistry, MacIntyre (1986) suggested that the calc-alkaline to alkaline volcanics are typical of island arcs or back arc basins developed in continental crust, and that they may have been deposited in narrow rift valleys associated with spreading centres along a transform fault system analogous to the modern day Gulf of California system.

Elsewhere within the Alexander Terrane of the Saint Elias Mountains knowledge of this assemblage is more fragmentary. Locally, near the Alsek River complexly folded, dark, thinly interbedded calcareous mudstone and shale, and discontinuous carbonate containing conodonts of Carnian and Early to Middle Norian age (M.J. Orchard, pers. comm., 1982), are juxtaposed with subaerial basaltic to andesitic flows (Campbell and Dodds, 1982c). Locally within east-central Tatshenshini area, structurally separated sequences of dark calcareous argillite and mudstone contain conodonts of Early to Middle Norian age (M.J. Orchard, pers. comm., 1982), and limestone contains megafossils of latest Norian age (E.T. Tozer, pers. comm., 1979). Locally the sediments are intruded by gabbro-diabase sills and plugs. In places massive mafic volcanics occur with gypsum-anhydrite. Apparently thick sequences, mostly of mafic volcanics, but including rarer felsic volcanics, fine clastics and carbonates, occurring immediately to the northeast of Tsirku Glacier in Tatshenshini River area (Campbell and Dodds, 1983a; MacIntyre and Schroeter, 1985) and in adjacent southeastern Alaska (Still, 1984) are believed to be in part of Late Triassic age. Sedimentary rocks of this age probably occur within parts of the "Icefield Ranges pelitic assemblage" outcropping throughout the central parts of the Alexander Terrane in the Saint Elias Mountains.

Correlatives of the Upper Triassic Hyd Assemblage (commonly including basal conglomerate) within the Alexander Terrane in southeastern Alaska overlie Paleozoic rocks with regional unconformity. However, although strongly suspected, this stratigraphic relationship has not been verified in the Saint Elias Mountains. In southeastern Alaska equivalents may include among others, felsic volcanics (Keku Volcanics and Puppets Formation), limestone and interbedded clastics (Cornwallis and Hamilton Island limestones), basaltic to andesitic pillow lavas and pyroclastics (Hound Island Volcanics and Chapin Peak Formation), and parts of the Nehenta Formation and the Hyd Group (Gehrels and Berg, 1984). Similar strata (including the Randall Formation and Moffat rhyolite) occur on islands west of Prince Rupert, British Columbia (Woodsworth and Orchard, 1985).

In southeastern Alaska the Gravina-Nutzotin Belt, an overlap assemblage (Gambier Assemblage) of Late Jurassic to mid-Cretaceous marine argillite and greywacke, and interbedded andesitic to basaltic volcanic and volcaniclastic rocks, depositionally overlies eastern parts of the Alexander Terrane (Berg et al., 1972, 1978). This relationship has not been demonstrated within the Saint Elias Mountains where Gravina-Nutzotin rocks lie on Wrangellia.

Plutonic rocks

Alkaline to calc-alkaline granitoid bodies of latest Pennsylvanian to Early Permian age (K-Ar ages range from 270 Ma to 290 Ma), informally named the "Icefield Ranges plutonic suite", are present mainly within the western parts of the Alexander Terrane in the Saint Elias Mountains (Dodds and Campbell, 1988). They comprise mostly small, elongate to rounded, fairly high-level batholithic complexes or plutons. Limited study reveals that they are predominantly agmatite and multiphase complexes, ranging from syenite to diorite, but locally include more uniform granodiorite.

In the Saint Elias Mountains no volcanics are known to be coeval with this plutonic episode. However, in adjacent southern Alaska correlative plutons (Richter et al., 1975; MacKevett, 1978; MacKevett et al., 1986) intrude extensive, coeval volcanics in Wrangellia (W1). Very few plutons within this age range are reported from southeastern Alaska (Wilson et al., 1979). The tectonic significance of the distinctive alkaline to calc-alkaline plutonism remains unknown.

Wrangellia

Most of Wrangellia is underlain by generally well stratified middle and Upper Paleozoic and lower Mesozoic volcanic and sedimentary rocks and related plutonic rocks. Exposures extend the full length of the Insular Belt from the Saint Elias Mountains to southern Vancouver Island (Fig. 8.85, 8.86).

Upper Paleozoic Sicker and Skolai assemblages

The upper sedimentary part of the Sicker Group on Vancouver Island includes Mississippian chert, tuff, argillite and sandstone overlain by 300 m of well bedded limestone of the Buttle Lake Formation which contains an abundant Middle Pennsylvanian to Early Permian brachiopod, bryozoan and conodont fauna (Yole, 1969; Sada and Danner, 1974; Muller, 1980; Brandon et al., 1986; M.J. Orchard, pers. comm., 1986). Within the Buttle Lake Formation are members of argillite, siltstone, volcanic sandstone and local beds of exotic carbonate blocks.

Rocks correlated with the Paleozoic Skolai Group of southern Alaska (Smith and MacKevett, 1970), are exposed

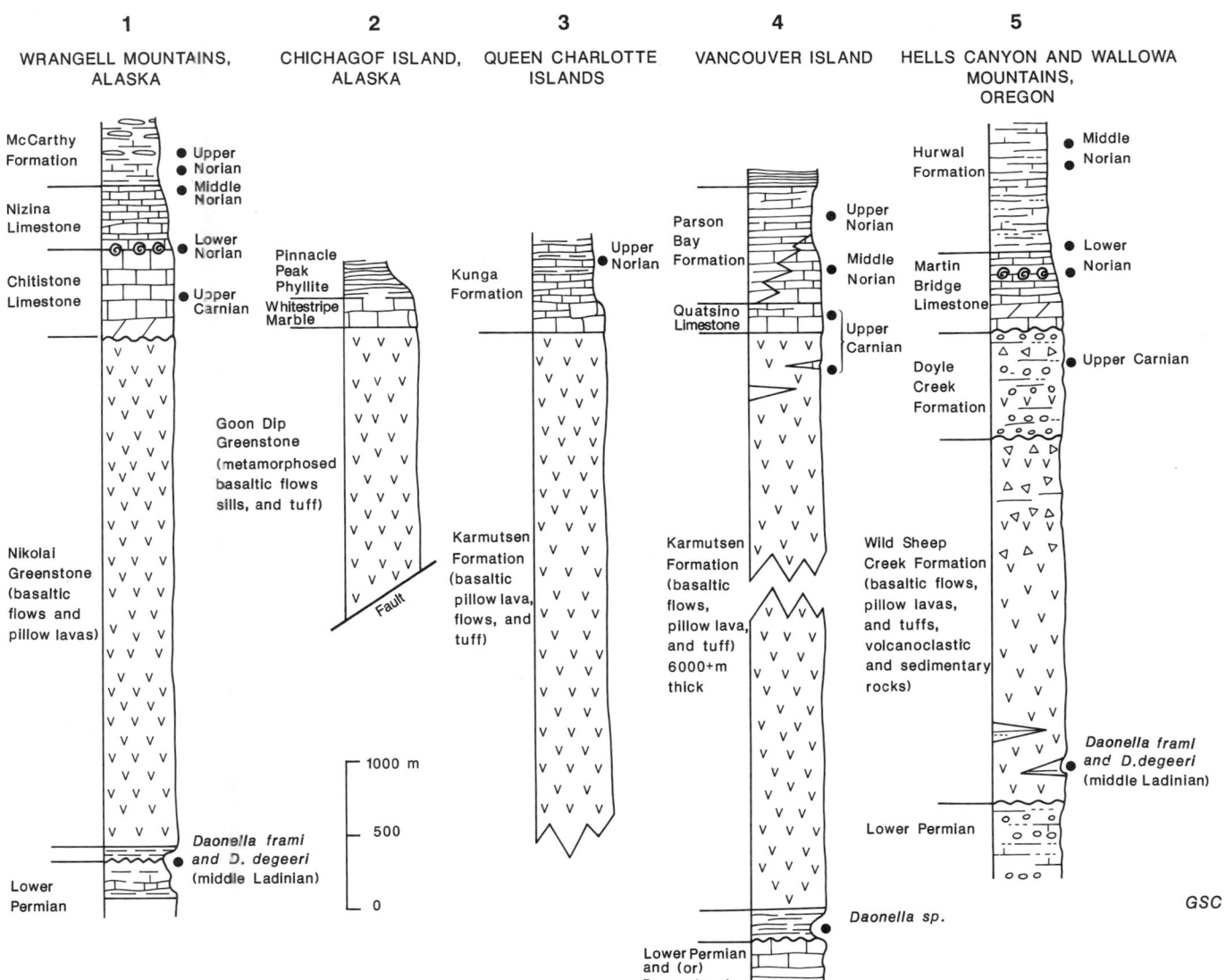

Figure 8.86. Diagrammatic stratigraphic columns in Wrangellia (modified from Jones et al., 1977).

on the east side of the Saint Elias Mountains in the southwestern Yukon (Read and Monger, 1976; Campbell and Dodds, 1982a). At one locality they nonconformably overlie a pre-Permian complex of altered gabbro and diabase of unknown age. The Skolai Group (in Canada) consists of a lower, volcanogenic Station Creek Formation and an upper, sedimentary Hasen Creek Formation. The age of the former, based on lithological correlation with rocks in Alaska, is probably Pennsylvanian, and that of the latter, based on paleontological evidence, is Early Permian. The Station Creek Formation comprises up to 1000 m of volcanic breccia containing clasts of basalt and pyroxene or plagioclase porphyry, which grades into vitric-lithic tuff and siliceous tuff. The Hasen Creek Formation consists of up to 800 m of typically siliceous argillite and siltstone with local sandstone, fossiliferous arenaceous limestone and pebble to cobble conglomerate near the base.

Lower Mesozoic Karmutsen Assemblage

Locally, and possibly gradationally in the Saint Elias Mountains and disconformably elsewhere, up to a few hundred metres of limy, dark argillite of Middle Triassic age separates the Skolai and Sicker groups from overlying basalt of the Nikolai Greenstone (in the north) and Karmutsen Formation (in the south) (Jones et al., 1977; Muller, 1977). The basalt and overlying Upper Triassic sedimentary rocks are the most characteristic units of Wrangellia stratigraphy. On Vancouver Island, the Karmutsen Formation is up to 6000 m thick and comprises pillow basalt and aquagene breccias in the lower part and massive flows in the upper part (Muller, 1977). In the Queen Charlotte Islands, it is nearly 4000 m thick, and has a similar lithology (Sutherland Brown, 1968). In the Saint Elias Mountains, the correlative Nikolai Greenstone is up to 1000 m thick, although it appears to be locally absent, and contains pillow lava and breccia in the lowest 100 m overlain by amygdaloidal flows, which are, at least in part, subaerial in setting (Read and Monger, 1976). In southeastern Alaska, the Nikolai Greenstone attains a thickness of 3500 m and in large part were erupted subaerially (MacKevett and Richter, 1974). The basalt is typically tholeiitic; Muller (1977), Jones et al. (1977) and, most recently Barker et al. (1985), concluded that they were produced by rifting of the previously established, Paleozoic arc sequences.

The upper part of the Karmutsen Formation on Vancouver Island contains locally intercalated carbonate with Upper Carnian fossils. The Karmutsen is overlain by the Upper Carnian to Middle Norian Quatsino Formation, consisting of 300 m, and locally more, of fossiliferous, shallow-water carbonate (Carlisle and Susuki, 1974). It grades upwards and laterally into dark, siliceous and carbonaceous calcareous shale, calcareous arenite and fine grained sandstone, up to 300 m thick, of the Parson Bay Formation, which is as young as Late Norian, and which, in turn is overlain gradationally by the Jurassic Harbledown Formation. In the Queen Charlotte Islands, the Kunga Formation, comprises 200 m of massive carbonate overlain by thin-bedded limestone containing latest Triassic fossils. The limestone is overlain by argillite, which contains Early Jurassic ammonites in its upper part (Sutherland Brown, 1968; Fig. 8.87). In the Saint Elias Mountains, the massive Chitistone Limestone is in places

Figure 8.87. Flaggy black argillite member (uppermost member) of the Upper Triassic to Lower Jurassic Kunga Formation exposed on northwestern Graham Island. Photo by C.J. Yorath. GSC 205235-M

associated with gypsum. It is up to 400 m thick, and is of shallow water, locally sabkha, origin (Armstrong et al., 1969). The massive limestone is succeeded by medium-bedded carbonate of the Nizina Formation and then by the McCarthy Formation, up to 400 m thick, whose lower part is distinctive light to dark grey argillaceous limestone interbedded with dark grey carbonate with uppermost Norian fossils. The McCarthy passes upwards into siliceous argillite of possible Jurassic age, in turn overlain by the Upper Jurassic to Lower Cretaceous Chisana Formation, part of the Gambier (Gravina-Nutzotin) Assemblage. As noted by Jones et al. (1977), in all of these places, the Triassic record is one of construction of a basaltic edifice, in many places to above sea-level, and of subsequent slow subsidence to deeper water basinal facies.

On Vancouver Island the Lower Sinemurian to Upper Pliensbachian Bonanza Group is a calc-alkaline dominantly pyroclastic volcanic suite with interbedded shale and siltstone as much as 2000 m thick. The Sinemurian volcanic rocks grade eastward into thinly banded shale, siltstone and tuff about 300 m thick on Harbledown Island and other islands on the east side of Vancouver Island. Farther east in the Coast Belt, shale and siltstone of Sinemurian age occur in roof pendants and are remnants of a shale sequence that may have extended eastward to the Cadwallader Terrane. Toarcian to Lower Bajocian strata are unknown on Vancouver Island.

In the Queen Charlotte Islands, the Lower and Middle Jurassic stratigraphic sequence is more complete and is included in three groups (Cameron and Tipper, 1985): the upper part of the Kunga, Maude, and Yakoun. The Jurassic part of the Kunga is a succession of mainly thinly laminated and banded shale, siltstone, and tuff more than 100 m thick, and similar to the Harbledown Formation on the east side of Vancouver Island. The Maude Group is mainly shale, shaly limestone, coquinoid sandstone, minor tuff and tuffaceous siltstone, 300 m thick. Shale in the uppermost part of the Kunga Formation and lowermost part of the Maude Formation is locally rich in hydrocarbons. The Yakoun Group comprises 480 m of volcanic breccia, lapilli tuff and agglomerate with minor lenticular sandstone and siltstone. These groups are conformable or paraconformable and span the interval from earliest Sinemurian to Early Bajocian time.

The Jurassic faunas of Wrangellia, particularly those of the Late Pliensbachian, have strong Tethyan affinities. In the Queen Charlotte Islands a Boreal influence is suggested by the presence of the genus *Amaltheus*. In Late Pliensbachian and Early Bajocian time some ammonites clearly are of east Pacific affinity.

The Early and Middle Jurassic Island Intrusions of Vancouver Island are probably comagmatic with volcanics of the Bonanza Group rocks (Muller, 1977; Isachsen et al., 1985). These intrusions grade into, and locally crosscut hornblende-plagioclase gneiss, agmatite and amphibolite, of the Early Jurassic West Coast Complex, exposed near the western margin of Wrangellia. This complex may represent the Jurassic lower crust of the Bonanza arc (Muller, 1977; Isachsen et al., 1985), and thus is an integral part of Wrangellia.

Relationship of Wrangellia with other terranes

In the north, Wrangellia and the Alexander Terrane are overlapped by Upper Jurassic to Lower Cretaceous strata of the Gravina-Nutzotin Belt (Gambier Assemblage), which lies on Wrangellia in the Nutzotin Mountains on the Yukon-Alaska boundary, and on the Alexander Terrane on Admiralty Island (Berg et al., 1972). The oldest rocks in the belt are of Oxfordian age. Recently it has been demonstrated that granitic rocks of the Pennsylvanian to Permian Icefield Ranges Suite are intrusive into rocks of both the Alexander Terrane and Wrangellia showing that these terranes were either juxtaposed by Pennsylvanian time or were never separated (MacKevett et al., 1986; R.B. Campbell and C.J. Dodds, pers. comm., 1987). As noted earlier, correlatives of the Gravina-Nutzotin Belt may be present within the Taku Terrane. On the east side of the Saint Elias Mountains, Eisbacher (1976) suggested that the Gravina-Nutzotin sequence, there called Dezadeash Group, was a deep-sea fan originating from a westerly source and shed into a marginal oceanic basin of unknown width, to the east of which lay previously accreted terranes (see Chapter 9). The basin closed after mid-Cretaceous time, possibly about 60-70 Ma ago. In the south, Wrangellia was intruded during latest Jurassic to Early Cretaceous time by granitic rocks of the Coast Belt, where if represented by a pillow basalt and overlying dacite sequence (Karmutsen and Bonanza?) in Bute Inlet map area, the terrane can be traced for at least 40 km into the Coast Belt (Roddick and Woodsworth, 1977). The Gambier Group, probably overlying Wrangellia on the west side of the southern Coast Mountains, is correlated with Middle and Upper Jurassic volcanic and clastic rocks of the Harrison Lake Assemblage, thus linking those two terranes by the end of the Jurassic. Final closure to the east took place in Cretaceous time. Wrangellia is in thrust contact with Jura-Cretaceous mélanges of the Pacific Rim Terrane near the west and south coast of Vancouver Island (Brandon, 1985; Clowes et al., 1987).

REFERENCES

Abbott, J.G.
1982: Structure and stratigraphy of the Macmillan Fold Belt: evidence for Devonian faulting; Exploration and Geological Services Division, Department of Indian and Northern Affairs, Whitehorse, Yukon, Open File, 16 p.
1983: Geology of the Macmillan Fold Belt; Exploration and Geological Services Division, Department of Indian and Northern Affairs, Whitehorse, Yukon, Open File (3 maps and legend; 1:50 000).

Aitken, J.D.
1966: Sub-Fairholme Devonian rocks of the eastern Front Ranges, southern Rocky Mountains, Alberta; Geological Survey of Canada, Paper 64-33, 88 p.

Alldrick, D.J.
1985: Stratigraphy and petrology of the Stewart mining camp; in Geological Fieldwork 1985, British Columbia Department of Energy, Mines and Petroleum Resources, Paper 1985-1, p. 316-341.

Anderson, P.
1976: Oceanic crust and arc-trench gap tectonics in southwestern British Columbia; Geology, v. 4, p. 443-446.

Anderson, R.G.
1983: Geology of the Hotailuh batholith and surrounding volcanic and sedimentary rocks north-central British Columbia; Ph.D. thesis, Carleton University, 669 p.

Andrichuk, J.M.
1961: Stratigraphic evidence for tectonic and current control of Upper Devonian reef sedimentation, Dubonel area, Alberta, Canada; Bulletin of the American Association of Petroleum Geology, v. 45, p. 612-632.

Armstrong, A.K., MacKevett, E.M., and Silberling, N.J.
1969: The Chitistone and Nizina limestones of part of the southern Wrangell Mountains, Alaska - a preliminary report stressing carbonate petrography and depositional environments; United States Geological Survey, Professional Paper 650-D, p. D49-D62.

Armstrong, J.E.
1949: Fort St. James map area, Cassiar and Coast Districts, British Columbia; Geological Survey of Canada, Memoir 252, 210 p.

Armstrong, R.L.
1980: Geochronometry of the Shuksan Metamorphic Suite, North Cascades, Washington; Geological Society of America, Abstracts with Programs, v. 12, no. 3, p. 94.

Armstrong, R.L., Harakal, J.E., Brown, E.H., Bernardi, M.L., and Rady, P.M.
1983: Late Paleozoic high-pressure metamorphic rocks in northwestern Washington and southwestern British Columbia: The Vedder Complex; Geological Society of America Bulletin, v. 94, p. 451-458.

Bamber, E.W. and Macqueen, R.W.
1979: Upper Carboniferous and Permian stratigraphy of the Monkman Pass and southern Pine Pass areas, northeastern British Columbia; Geological Survey of Canada, Bulletin 301, 27 p.

Bamber, E.W. and Mamet, B.L.
1978: Carboniferous biostratigraphy and correlations, northeastern British Columbia and southwestern District of Mackenzie; Geological Survey of Canada, Bulletin 266, 65 p.

Bamber, E.W. and Waterhouse, J.B.
1971: Carboniferous and Permian stratigraphy and paleontology, northern Yukon Territory, Canada; Bulletin of Canadian Petroleum Geology, v. 19, no. 1, p. 29-250.

Bamber, E.W., Macqueen, R.W., and Richards, B.C.
1984: Facies relationships at the Mississippian carbonate platform margin, Western Canada; 9th International Congress of Carboniferous Stratigraphy and Geology, Compte Rendu, v. 3, p. 461-478.

Bamber, E.W., Taylor, G.C., and Procter, R.M.
1968: Carboniferous and Permian stratigraphy of northeastern British Columbia; Geological Survey of Canada, Paper 68-15, 25 p.

Barker, F., Plafker, G., and Sutherland Brown, A.
1985: Karmutsen Formation, Queen Charlotte Island and Vancouver Island: an arc-rift ferrotholeiite of Wrangellia; Geological Society of America, Abstracts with Programs, v. 17, no. 6, p. 340.

Barss, D.L., Best, E.W., and Meyers, N.
1964: Chapter 9: Triassic; in Geological History of Western Canada, R.G. McCrossan and R.P. Glaister (ed.), Alberta Society of Petroleum Geologists, Calgary, Alberta, p. 113-136.

Bassett, H.G. and Stout, J.G.
1967: Devonian of western Canada; in International Symposium on the Devonian System, D.H. Oswald (ed.), Alberta Society of Petroleum Geologists, v. 1, p. 717-752.

Bell, J.S.
1973: Late-Paleozoic orogeny in the northern Yukon; in Canadian Arctic Geology, J.D. Aitken and D.J. Glass (ed.), Proceedings of Symposium on the Geology of Arctic Canada, Geological Association of Canada - Canadian Society of Petroleum Geologists, p. 23-28.
1974: Late Paleozoic orogeny in northern Yukon; in Canadian Arctic Geology, J.D. Aitken and D.J. Glass (ed.), Geological Association of Canada-Canadian Society of Petroleum Geologists Symposium, Saskatoon, Saskatchewan, p. 23-38.

Belyea, H.R.
1964: Upper Devonian (part II); in Geological History of Western Canada, R.G. McCrossan and R.P. Glaister (ed.), Alberta Society of Petroleum Geologists, p. 66-88.

Belyea, H.R. and McLaren, D.J.
1962: Upper Devonian formations, southern part of Northwest Territories, northeastern British Columbia, and northwestern Alberta; Geological Survey of Canada, Paper 61-29, 74 p.

Berg, H.C., Jones, D.L., and Coney, P.J.
1978: Map showing pre-Cenozoic tectonostratigraphic terranes of southeastern Alaska and adjacent areas; United States Geological Survey, Open-File Report 78-1085.

Berg, H.C., Jones, D.L., and Richter, D.H.
1972: Gravina-Nutzotin Belt - Tectonic significance of an Upper Mesozoic sedimentary and volcanic sequence in southern and southeastern Alaska; United States Geological Survey, Professional Paper 800D, p. D1-D24.

Blackwell, D.L.
1983: Geology of the Park Butte-Loomis Mountain area, Washington (eastern margin of the Twin Sisters Dunite); M.Sc. thesis, Western Washington University, 253 p.

Bostock, H.S.
1940: Keremeos, Similkameen District, British Columbia; Geological Survey of Canada, Map 341A.

Bostwick, D.A. and Nestell, M.K.
1967: Permian Tethyan fusulinid faunas of the northwestern United States; in Systematics Association 7, Aspects of Tethyan Biogeography, p. 92-102.

Brabb, E.E. and Churkin, M., Jr.
1969: Geologic map of the Charley River quadrangle, east-central Alaska; United States Geological Survey, Miscellaneous Geologic Investigations, Map 1-573.

Braman, D.R.
1981: Upper Devonian - Lower Carboniferous miospore biostratigraphy of the Imperial Formation, District of Mackenzie; Ph.D thesis, University of Calgary, Alberta, 377 p.

Brandon, M.T.
1985: Mesozoic mélange of the Pacific Rim Complex, western Vancouver Island; in Field Guides to Geology and Mineral Deposits in the Southern Canadian Cordillera, Geological Society of America Cordilleran Section Meeting, p. 7-1 - 7-28.

Brandon, M.T., Orchard, M.J., Parrish, R.R., Sutherland Brown, A., and Yorath, C.J.
1986: Fossil ages and isotopic dates from the Paleozoic Sicker Group and associated intrusive rocks, Vancouver Island, British Columbia; in Current Research, Part A, Geological Survey of Canada, Paper 86-1A, p. 683-694.

Brew, D.A., Ovenshine, A.T., Karl, S.M., and Hunt, S.J.
1984: Preliminary reconnaissance geologic map of the Petersburg and parts of the Port Alexander and Sumdum quadrangles, southeastern Alaska; United States Geological Survey, Open-File Report 84-405, 43 p.

Brown, R.L. and Read, P.B.
1983: Shuswap terrane of British Columbia: a Mesozoic "core complex"; Geology, v. 11, p. 164-168.

Bultman, T.R.
1979: Geology and tectonic history of the Whitehorse Trough west of Atlin, British Columbia; Ph.D. Thesis, Yale University, 284 p.

Cairnes, C.E.
1924: Coquihalla area, British Columbia; Geological Survey of Canada, Memoir 139, 187 p.
1934: Slocan Mining Camp, British Columbia; Geological Survey of Canada, Memoir 173, 137 p.

Cameron, B.E.B. and Tipper, H.W.
1985: Jurassic stratigraphy of the Queen Charlotte Islands, British Columbia; Geological Survey of Canada, Bulletin 365, 49 p.

Campbell, R.B. and Dodds, C.J.
1978: Operation Saint Elias, Yukon Territory; in Current Research, Part A, Geological Survey of Canada, Paper 78-1A, p. 35-41.
1982a: Geology of southwest Kluane Lake map-area (115G and F(E 1/2)), Yukon Territory; Geological Survey of Canada, Open File 829.
1982b: Geology of Mount St. Elias map-area (115B and C(E 1/2)), Yukon Territory; Geological Survey of Canada, Open File 830.
1982c: Geology of southwest Dezadeash map-area (115A), Yukon Territory; Geological Survey of Canada, Open File 831.
1983a: Geology of Tatshenshini River map-area (114P), British Columbia; Geological Survey of Canada, Open File 926.
1983b: Geology of Yakutat map-area (114O), British Columbia part; Geological Survey of Canada, Open File 927.
1983c: Terranes of the Saint Elias Mountains; Geological Society of America, Abstracts with Programs, 16, p. 274.

Campbell, R.B. and Tipper, H.W.
1970: Geology and mineral exploration potential of the Quesnel Trough, British Columbia; Canadian Institute of Mines and Metallurgy Transactions: Vol. LXXIII, p. 174-179.
1971: Geology of the Bonaparte Lake map-area, British Columbia; Geological Survey of Canada, Memoir 363, 100 p.

Carlisle, D. and Susuki, T.
1974: Emergent basalt and submergent carbonate-clastic sequences including the Upper Triassic Dilleri and Welleri Zones on Vancouver Island; Canadian Journal of Earth Sciences, v. 11, p. 254-279.

Carne, R.C.
1979: Geological setting and stratiform mineralization, Tom claims, Yukon Territory; Department of Indian and Northern Affairs, Canada, Report 1979-4, 30 p.

Cecile, M.P.
1984: Evidence against large-scale strike-slip displacement of Paleozoic strata along the Richardson-Hess fault system; Geology, v. 12, no. 7, p. 403-407.

Chi, B.I. and Hills, L.V.
1974: Stratigraphic and paleoenvironmental significance of Upper Devonian megaspores, type section of the Imperial Formation, Northwest Territories, Canada; in Canadian Arctic Geology, J.D. Aitken and D.J. Glass (ed.), Geological Association of Canada-Canadian Society of Petroleum Geologists Symposium, Saskatoon, Saskatchewan, p. 241-257.

Churkin, M. Jr.
1973: Paleozoic and Precambrian rocks in Alaska and their role in its structural evolution; United States Geological Survey, Professional Paper, No. 740, 64 p.

Churkin, M., Jr., Eberlein, G.D., Heuber, F.M., and Mamay, S.H.
1969: Lower Devonian land plants from graptolitic shale in southeastern Alaska; Palaeontology, v. 12, p. 559-573.

Churkin, M., Jr., Jaeger, H., and Eberlein, G.D.
1970: Lower Devonian graptolites from southeastern Alaska; Lethaia, v. 3, p. 183-202.

Clowes, R.M., Brandon, M.T., Green, A.G., Yorath, C.J., Sutherland Brown, A., Kanasewich, E.R., and Spencer, C.
1987: LITHOPROBE - southern Vancouver Island: Cenozoic subduction complex imaged by deep seismic reflections; Canadian Journal of Earth Sciences, v. 24, p. 31-51.

Coates, J.A.
1974: Geology of the Manning Park area, British Columbia; Geological Survey of Canada, Bulletin 238, 177 p.

Cockfield, W.E.
1948: Geology and mineral deposits of Nicola map-area, British Columbia; Geological Survey of Canada, Memoir 249, 164 p.

Coney, P., Jones, D.L., and Monger, J.W.H.
1980: Cordilleran suspect terranes; Nature, v. 288, p. 329-333.

Cook, H.E.
1972: Miette platform evolution and relation of overlying bank (reef) localization, Upper Devonian, Alberta; Bulletin of Canadian Petroleum Geology, v. 20, p. 375-410.

Cowan, D.S.
1982: Geologic evidence for post-40 m.y.b.p. large-scale northwestward displacement of part of southeastern Alaska; Geology, v. 10, p. 309-313.

Crickmay, C.H.
1930: The Jurassic rocks of Ashcroft, British Columbia; University of California Publications in Geological Sciences, v. 19(2), p. 23-74.
1962: Gross stratigraphy of Harrison Lake area, British Columbia; Privately published by author, available Evelyn de Mille Books, Calgary, Alberta, Article 8, 12 p.

Crockford, J.
1957: Permian Bryozoa from the Fitzroy Basin, western Australia; Australian Bureau of Mineral Resources, Geology, and Geophysics, Bulletin 34, 134 p.

Currie, L.
1984: The provenance of clasts in the Ashman conglomerate of the northeastern Bowser Basin; B.Sc. thesis, Queen's University, Kingston, Ontario, 59 p.

Danner, W.R.
1965: Limestones of the western Cordilleran eugeosyncline of southwestern British Columbia, western Washington and northern Oregon; in D.N. Wadia Commemorative Volume, Mining and Metallurgical Institute of India, p. 113-125.
1977: Paleozoic rocks of northwest Washington and adjacent parts of British Columbia; in Paleozoic Paleogeography of the Western United States, J.H. Stewart, C.H. Stevens and A.E. Fritche (eds.), Society of Economic Paleontologists and Mineralogists, Pacific Coast Section Paleogeography Symposium 1, p. 481-502.

Davis, A. and Plafker, G.
1985: Comparative geochemistry and petrology of Triassic basaltic rocks from the Taku terrane on the Chilkat Peninsula and Wrangellia; Canadian Journal of Earth Sciences, v. 22, no. 2, p. 183-194.

Davis, G.A., Monger, J.W.H., and Burchfiel, B.C.
1978: Mesozoic construction of the Cordilleran "Collage" central British Columbia to central California; in Mesozoic Paleogeography of the Western United States, D.G. Howell and K.A. McDougall (ed.), Society of Economic Paleontologists and Mineralogists, Pacific Coast Section Paleogeography Symposium 2, p. 1-32.

Dawson, K.M. and Orchard, M.J.
1982: Regional metallogeny of the northern Cordillera: biostratigraphy, correlation and metallogenic significance of bedded barite occurrences in eastern Yukon and western District of Mackenzie; in Current Research, Part C, Geological Survey of Canada, Paper 82-1C, p. 31-38.

Decker, J.E., Jr.
1980: Geology of a Cretaceous subduction complex, western Chichagof Island, southeastern Alaska; Ph.D. thesis, Stanford University, 135 p.

Diakow, L.J.
1984: Geology between Toodoggone and Chukachida rivers; in Geological Fieldwork 1983; British Columbia Ministry of Energy, Mines and Resources, Paper 1984-1, p. 139-145.
1985: Potassium-argon age determinations from biotite and hornblende in Toodoggone volcanic rocks; in Geological Fieldwork 1984; British Columbia Ministry of Energy, Mines and Resources, Paper 1985-1, p. 298-300.

Dodds, C.J. and Campbell, R.B.
1988: Potassium-argon ages of mainly intrusive rocks in the Saint Elias Mountains, Yukon and British Columbia; Geological Survey of Canada, Paper 87-16, 43 p.

Douglas, R.J.W.
1958: Mount Head map-area, Alberta; Geological Survey of Canada, Memoir 291, 241 p.

Douglas, R.J.W., Gabrielse, H., Wheeler, J.O., Stott, D.F., and Belyea, H.R.
1970: Geology of western Canada, Chapter VIII; in Geology and Economic Minerals of Canada, R.J.W. Douglas (ed.), Geological Survey of Canada, Economic Geology Report No. 1. p. 365-488.

Eberlein, G.D., Churkin, M., Jr., Carter, C., Berg, H.C., and Ovenshine, A.T.
1983: Geology of the Craig quadrangle, Alaska; United States Geological Survey, Open-File Report 83-91.

Eisbacher, G.H.
1974: Sedimentary history and tectonic evolution of the Sustut and Sifton basins, north-central British Columbia; Geological Survey of Canada, Paper 73-31, 57 p.
1976: Sedimentology of the Dezadeash flysch and its implications for strike-slip faulting along the Denali Fault, Yukon Territory and Alaska; Canadian Journal of Earth Sciences, v. 13, no. 11, p. 1495-1513.
1981: Late Mesozoic-Paleogene Bowser molasse and Cordilleran tectonics, western Canada; in Sedimentation and Tectonics in Alluvial Basins, A.D. Miall (ed.), Geological Association of Canada, Special Paper 23, p. 125-151.
1983: Devonian-Mississippian sinistral transcurrent faulting along the cratonic margin of western North America: A hypothesis; Geology, v. 11, no. 1, p. 7-10.

Embry, A.F. and Klovan, J.E.
1976: The Middle-Upper Devonian clastic wedge of the Franklinian geosyncline; Bulletin of Canadian Petroleum Geology, v. 24, no. 4, p. 485-639.

Erdmer, P.
1987: Blueschist and eclogite in mylonitic allochthons, Ross River and Watson Lake areas, southeastern Yukon; Canadian Journal of Earth Sciences, v. 24, p. 1439-1449.

Erdmer, P. and Helmstaedt, H.
1983: Eclogite from central Yukon: a record of subduction at the western margin of ancient North America; Canadian Journal of Earth Sciences, v. 20, p. 1389-1408.

Foster, H.L.
1970: Reconnaissance geologic map of the Tanacross Quadrangle, Alaska; U.S. Geological Survey Miscellaneous Geological Investigations Map I-593.
1976: Geologic map of the Eagle Quadrangle, Alaska; United States Geological Survey, Miscellaneous Geological Investigations Map I-922.

Frebold, H.
1957: The Jurassic Fernie Group in the Canadian Rocky Mountains and Foothills; Geological Survey of Canada, Memoir 287, 197 p.

Fritz, M.A.
1932: Permian Bryozoa from Vancouver Island; Royal Society of Canada Transactions, Third Series, v. 26, p. 93-109.

Gabrielse, H.
1963: McDame map area, Cassiar District, British Columbia; Geological Survey of Canada, Memoir 319, 138 p.
1968: Geology of Jennings River map-area, British Columbia (104O); Geological Survey of Canada, Paper 68-55, 37 p.
1976: Environments of Canadian Cordillera depositional basins; in Circum-Pacific Energy and Mineral Resources; American Association of Petroleum Geologists, Memoir 25, p. 492-502.
1985: Major dextral transcurrent displacements along the Northern Rocky Mountain Trench and related lineaments in north-central British Columbia; Geological Society of America Bulletin, v. 96, no. 1, p. 1-14.

Gabrielse, H., Dodds, C.J., and Mansy, J.L.
1977: Operation Finlay, British Columbia; in Report of Activities, Part A, Geological Survey of Canada, Paper 77-1A, p. 243-246.

Gabrielse, H., Wanless, R.K., Armstrong, R.L., and Erdman, L.R.
1980: Isotopic dating of Early Jurassic volcanism and plutonism in north-central British Columbia; in Current Research, Part A, Geological Survey of Canada, Paper 80-1A, p. 27-32.

Gabrielse, H., Loveridge, W.D., Sullivan, R.W., and Stevens, R.D.
1982: U-Pb measurements of zircon indicate middle Paleozoic plutonism in the Omineca Crystalline Belt, north-central British Columbia; in Current Research, Part C, Geological Survey of Canada, Paper 82-1C, p. 139-146.

Gehrels, G.E. and Berg, H.C.
1984: Geologic map of southeastern Alaska; United States Geological Survey, Open-File Report 84-886, 28 p.

Gehrels, G.E. and Saleeby, J.B.
1984: Paleozoic geologic history of the Alexander terrane in SE Alaska, and comparisons with other orogenic belts; Geological Society of America, Abstracts with Programs, v. 16, p. 516.
1985: Constraints and speculation on the displacement and accretionary history of the Alexander-Wrangellia Peninsular superterrane; Geological Society of America, Abstracts with Programs, v. 17, p. 356.

1987: Geology of southern Prince of Wales Island, southeastern Alaska; Geological Society of America Bulletin, v. 98, p. 123-137.
in press: Late Jurassic-Early Cretaceous flysch basins in western North America and the displacement and accretionary history of the Alexander-Wrangellia- Peninsular superterrane; Tectonics.

Gehrels, G.E., Saleeby, J.B., and Berg, H.C.
1983: Preliminary description of the Klakas orogeny in the southern Alexander terrane, southeastern Alaska; in Pre-Jurassic Rocks in Western North American Suspect Terranes, C.H. Stevens (ed.), Pacific Section, Society of Economic Paleontologists and Mineralogists, Los Angeles, California, p. 131-141.
1984: Geologic framework of Paleozoic rocks on Annette and Hotspur islands, southern Alexander terrane; United States Geological Survey, Circular 868, p. 113-115.

Geldsetzer, H.H.J.
1982: Depositional history of the Devonian succession in the Rocky Mountains southwest of the Peace River Arch; in Current Research, Part C, Geological Survey of Canada, Paper 82-C, p. 55-64.
1984: The role of fine-grained siliclastics in the development of organic buildups; Society of Economic Paleontologists and Mineralogists, Annual Midyear Meeting, Abstracts, p. 34.

Geldsetzer, H.H., Goodfellow, W.D., McLaren, D.J., and Orchard, M.
1987: Sulfur-isotope anomaly associated with the Frasnian-Famennian extinction, Medicine Lake, Alberta, Canada; Geology, v. 15, p. 393-396.

Gibson, D.W.
1965: Triassic stratigraphy near the northern boundary of Jasper National Park, Alberta; Geological Survey of Canada, Paper 64-9, 144 p.
1968a: Triassic stratigraphy between the Athabasca and Smoky rivers of Alberta; Geological Survey of Canada, Paper 67-65, 114 p.
1968b: Triassic stratigraphy between the Athabasca and Brazeau rivers of Alberta; Geological Survey of Canada, Paper 68-11, 84 p.
1969: Triassic stratigraphy of the Bow River-Crowsnest Pass region, Rocky Mountains of Alberta and British Columbia; Geological Survey of Canada, Paper 68-29, 48 p.
1970: Triassic stratigraphy, Pine Pass area, northeastern British Columbia; Edmonton Geological Society, Field Conference Guidebook, p. 23-38.
1971a: Triassic stratigraphy of the Sikanni Chief River-Pine Pass region, Rocky Mountain Foothills, northeastern British Columbia; Geological Survey of Canada, Paper 70-31, 105 p.
1971b: Triassic petrology of Athabasca - Smoky River region, Alberta; Geological Survey of Canada, Bulletin 194, 59 p.
1972: Triassic stratigraphy of the Pine Pass-Smoky River area, Rocky Mountain Foothills and Front Ranges of British Columbia and Alberta; Geological Survey of Canada, Paper 71-30, 108 p.
1974: Triassic rocks of the southern Canadian Rocky Mountains; Geological Survey of Canada, Bulletin 230, 65 p.
1975: Triassic rocks of the Rocky Mountain Foothills and Front Ranges of northeastern British Columbia and west-central Alberta; Geological Survey of Canada, Bulletin 247, 61 p.

Gordey, S.P.
1979: Stratigraphy of southeastern Selwyn Basin in the Summit Lake area, Yukon Territory and Northwest Territories; in Current Research, Part A, Geological Survey of Canada, Paper 79-1A, p. 13-16.
1981a: Stratigraphy, structure and tectonic evolution of southern Pelly Mountains in the Indigo Lake area, Yukon Territory; Geological Survey of Canada, Bulletin 318, 44 p.
1981b: Geology of Nahanni map area, Yukon Territory and District of Mackenzie; Geological Survey of Canada, Open File 780.
1983: Thrust faults in the Anvil Range and a new look at the Anvil Range Group, south-central Yukon Territory; in Current Research, Part A, Geological Survey of Canada, Paper 83-1A, p. 225-227.
in press: Evolution of the northern Cordilleran miogeocline, Nahanni map area (105I), Yukon Territory and District of Mackenzie; Geological Survey of Canada, Memoir.

Gordey, S.P., Abbott, J.G., and Orchard, M.J.
1982a: Devono-Mississippian (Earn group) and younger strata in east-central Yukon; in Current Research, Part B, Geological Survey of Canada, Paper 82-1B, p. 93-100.

Gordey, S.P., Gabrielse, H., and Orchard, M.J.
1982b: Stratigraphy and structure of the Sylvester Allochthon, southwest McDame map area, northern British Columbia; in Current Research, Part B, Geological Survey of Canada, Paper 82-1B, p. 101-106.

Gordey, S.P., Wood, D., and Anderson, R.G.
1981: Stratigraphic framework of southeastern Selwyn Basin, Nahanni map area, Yukon Territory and District of Mackenzie; in Current Research, Part A, Geological Survey of Canada, Paper 81-1A, p. 395-398.

Graham, A.D.
1973: Carboniferous and Permian stratigraphy, southern Eagle Plain, Yukon Territory, Canada; in Proceedings of the Symposium on the Geology of the Canadian Arctic, J.D. Aitken and D.J. Glass (ed.), Geological Association of Canada/Canadian Society of Petroleum Geologists, p. 159-180.

Grantz, A. and May, S.D.
1982: Rifting history and structural development of the continental margin north of Alaska; in Hollis Hedberg Conference, Studies on Continental Margin Geology, J.S. Watkins, and C.L. Drake (ed.), American Association of Petroleum Geologists, Memoir 34, p. 77-100.

Grayston, L.D., Sherwin, D.F., and Allan, J.F.
1964: Middle Devonian; in Geological History of Western Canada, R.G. McCrossan and R.P. Glaister (ed.), Alberta Society of Petroleum Geologists, p. 49-59.

Green, L.H.
1972: Geology of Nash Creek, Larson Creek and Dawson map-areas, Yukon Territory; Geological Survey of Canada, Memoir 364, 157 p.

Grette, J.F.
1978: Cache Creek and Nicola Group near Ashcroft, British Columbia; M.Sc. thesis, University of British Columbia, Vancouver, 88 p.

Hall, R.L.
1984: Lithostratigraphy and biostratigraphy of the Fernie Formation (Jurassic) in the southern Canadian Rocky Mountains; in The Mesozoic of Middle North America, D.F. Stott and D.J. Glass (ed.); Canadian Society of Petroleum Geologists, Memoir 9, p. 233-247.

Hall, R.L. and Stronach, N.J.
1982: A guidebook to the Fernie Formation of southern Alberta and British Columbia, 1st Field Conference, Calgary, Circum-Pacific Jurassic Research Group; International Geological Correlation Program No. 171.

Harms, T.A.
1984: Structural style of the Sylvester Allochthon, northeastern Cry Lake map area, British Columbia; in Current Research, Part A, Geological Survey of Canada, Paper 84-1A, p. 109-112.
1985a: Pre-emplacement thrust faulting in the Sylvester Allochthon, northeast Cry Lake map area, British Columbia; in Current Research, Part A, Geological Survey of Canada, Paper 85-1A, p. 301-304.
1985b: Cross-sections through Sylvester Allochthon and underlying Cassiar Platform, northern British Columbia; in Current Research, Part B, Geological Survey of Canada, Paper 85-1B, p. 109-112.
1986: Structural and tectonic analysis of the Sylvester Allochthon, northern British Columbia: implications for paleogeography and accretion; Ph.D. thesis, University of Arizona, Tuscon.

Haugerud, R.
1985: The geology of the Hozameen Group and Ross Lake shear zone, Maselpanik area, northern Cascades, southwestern British Columbia; Ph.D. thesis, University of Washington, Seattle, 263 p.

Henderson, C.M. and McGugan, A.
1983: Permian conodonts from the Canadian Rocky Mountains; Geological Society of America, Abstracts with Programs, v. 15, no. 5, p. 410.
1986: Permian conodont biostratigraphy of the Ishbel Group, southwestern Alberta and southeastern British Columbia; Contributions to Geology, University of Wyoming, v. 24, p. 219-235.

Henderson, C.M., Bamber, E.W., Higgins, A.C., Richards, B.C., and McGugan, A.
in press: Permian (Geological Description, Chapter 6); in Sedimentary Cover of the Craton in Canada; D.F. Stott and J.D. Aitken (ed.), Geological Survey of Canada, Geology of Canada, No. 5 (also Geological Society of America; The Geology of North America, v. D-1).

Hillhouse, J.W.
1977: Paleomagnetism of the Triassic Nikolai Greenstone, McCarthy Quadrangle, Alaska; Canadian Journal of Earth Sciences, v. 14, p. 2578-2592.

Hillhouse, J.W. and Grommé, C.S.
1980: Paleomagnetism of the Triassic Hound Island Volcanics, Alexander terrane, southeastern Alaska; Journal of Geophysical Research, v. 85, p. 2594-2602.

Hills, L.V. and Braman, D.R.
1978: Sedimentary structures of the Imperial Formation; in Display Summaries, Core and Field Sample Conference, A.F. Embry (compiler), Canadian Society of Petroleum Geologists, p. 35-37.

Hills, L.V., Sangster, E.V., and Suneby, L.B. (eds.)
1981: Lexicon of Canadian Stratigraphy, Volume 2, Yukon Territory and District of Mackenzie; Canadian Society of Petroleum Geologists, Calgary, Alberta, Canada.

Hopkins, J.C.
1981: Sedimentology of quartzose sandstones of Lower Mannville and associated units, Medicine River area, central Alberta; Bulletin of Canadian Petroleum Geology, v. 29, no 1, p. 12-41.

Howell, D.G., Murray, R.W., Wiley, T.J., Boundy-Sanders, S., and Kauffman-Linam, L.
1987: Sedimentology and tectonics of Devonian Nation River Formation, Alaska, part of yet another allochthonous terrane; Bulletin of the American Association of Petroleum Geologists, v. 71, p. 569.

Hudson, T., Plafker, G., and Dixon, K.
1982: Horizontal offset history of the Chatham Strait fault; United States Geological Survey, Circular 844, p. 128-132.

Irvine, T.N.
1976: Alaskan-type ultramafic-gabbroic bodies in the Aitken Lake, McConnell Creek, and Toodoggone map areas; in Report of Activities, Part A, Geological Survey of Canada, Paper 76-1A, p. 76-81.

Irving, E. and Yole, R.W.
1972: Paleomagnetism and the kinematic history of mafic and ultramafic in fold mountain belts; Canada Department of Energy, Mines and Resources, Publication of the Earth Physics Branch, v. 42, p. 87-95.

Irving, E., Woodsworth, G.J., and Wynne, P.J.
1985: Paleomagnetic evidence for displacement from the south of the Coast Plutonic Complex, British Columbia; Canadian Journal of Earth Sciences, v. 22, no. 4, p. 584-598.

Isachsen, C., Armstrong, R.L., and Parrish, R.R.
1985: U-Pb, Rb-Sr, and K-Ar geochronometry of Vancouver Island igneous rocks; in A Symposium on the Deep Structure of Southern Vancouver Island: results of LITHOPROBE Phase I, Geological Association of Canada, Lithoprobe Project Publication No. 10, Programme and Abstracts, Paper 13, p. 21-22.

Jefferson, C.W., Kilby, D.B., Pigage, L.C., and Roberts, W.J.
1983: The Cirque barite-zinc-lead deposits, northeastern British Columbia; in Sediment-Hosted Stratiform Lead-Zinc Deposits, D.F. Sangster (ed.), Mineralogical Association of Canada, Short Course Handbook, v. 9, p. 121-140.

Jeletzky, J.A.
1977: Porcupine River Formation: a new Upper Jurassic sandstone unit, northern Yukon Territory; Geological Survey of Canada, Paper 76-27, 43 p.

Jeletzky, J.A. and Tipper, H.W.
1967: Upper Jurassic and Cretaceous rocks of Taseko Lakes map-area and their bearing on the geological history of southwestern British Columbia; Geological Survey of Canada, Paper 67-54, 218 p.

Jones, A.G.
1959: Vernon map-area, British Columbia; Geological Survey of Canada, Memoir 296, 186 p.

Jones, D.L., Berg, H.C., Coney, P.J., and Harris, A.
1981: Structural and stratigraphic significance of Upper Devonian and Mississippian fossils from the Cannery Formation, Kupreanof Island, southeastern Alaska; United States Geological Survey, Circular 823-B, p. 109-112.

Jones, D.L., Irwin, W.P., and Ovenshine, A.T.
1972: Southeastern Alaska — a displaced continental fragment?; United States Geological Survey, Professional Paper 800-B, p. B211-B217.

Jones, D.L., Silberling, N.J., and Hillhouse, J.
1977: Wrangellia - A displaced terrane in northwestern North America; Canadian Journal of Earth Sciences, v. 14, p. 2565-2577.

Jones, D.L., Silberling, N.J., Gilbert, W., and Coney P.J.
1982: Character, distribution, and tectonic significance of accretionary terranes in the central Alaska Range; Journal of Geophysical Research, v. 87, p. 3709-3717.

Jones, D.L., Silberling, N.J., Coney, P.J., and Plafker, G.
1984: Part A - Lithotectonic terrane map of Alaska (west of the 141st Meridian); in Lithotectonic Terrane Maps of the North American Cordillera; D.L. Jones and N.J. Silberling (ed.), United States Geological Survey, Open-File Report 84-523, p. A1-A12.

Karl, S.M., Decker, J.E., and Johnson, B.R.
1982: Discrimination of Wrangellia and the Chugach terrane in the Kelp Bay Group on Chichagof and Baranof islands, southeastern Alaska; United States Geological Survey, Circular 844, p. 124-128.

Kerr, J.W.
1981: Evolution of the Canadian Arctic Islands: A transition between the Atlantic and Arctic Oceans; in The Ocean Basins and Margins, volume 5, The Arctic Ocean, A.E.M. Nairn, M. Churkin, Jr., and F.G. Stehli (ed.), p. 105-199.

Kindle, E.D.
1946: The Middle Triassic of Liard River, British Columbia; Geological Survey of Canada, Paper 46-1, p. 21-23.

Klepacki, D.W.
1983: Stratigraphic and structural relations of the Milford, Kaslo and Slocan groups, Rosebury Quadrangle, Lardeau map-area, British Columbia; in Current Research, Part A, Geological Survey of Canada, Paper 83-1A, p. 229-233.

Klepacki, D.W. and Wheeler, J.O.
1985: Stratigraphic and structural relations of the Milford, Kaslo and Slocan groups, Goat Range, Lardeau and Nelson map areas, British Columbia; in Current Research, Part A, Geological Survey of Canada, Paper 85-1A, p. 277-286.

Krause, F.F.
1984: Nisku reefs and Devonian lithoherms; in Carbonates in Subsurface and Outcrop, 1984 C.S.P.G. Core Conference; Canadian Society of Petroleum Geologists, p. 171-189.

Lathram, E.H., Pomeroy, J.S., Berg, H.C., and Loney, R.A.
1965: Reconnaissance geology of Admiralty Island, Alaska; United States Geological Survey, Bulletin 1181-R, 48 p.

Leech, G.B.
1953: Geology and mineral deposits of the Shulaps Range, southwestern British Columbia; British Columbia Department of Mines, Bulletin 32, 54 p.

Lenz, A.C.
1972: Ordovician to Devonian history of northern Yukon and adjacent District of Mackenzie; Bulletin of Canadian Petroleum Geology, v. 20, no. 2, p. 321-361.

Lenz, A.C. and Perry, D.G.
1972: The Neruokpuk Formation of the Barn Mountains and Driftwood Hills, northern Yukon; its age and graptolite fauna; Canadian Journal of Earth Sciences, v. 9, p. 1129-1138.

Little, H.W.
1960: Nelson map-area, west half (82F W1/2); Geological Survey of Canada, Memoir 308, 205 p.
1982: Geology of the Rossland-Trail map area, British Columbia; Geological Survey of Canada, Paper 79-26, 39 p.
1983: Geology of the Greenwood map-area, British Columbia; Geological Survey of Canada, Paper 79-29, 37 p.
1985: Geological notes and map, Nelson west half (82F W1/2) map area; Geological Survey of Canada, Open File 1195.

Loney, R.A., Brew, D.A., Muffler, L.J.P., and Pomeroy, J.S.
1975: Reconnaissance geology of Chichagof, Baranof, and Kruzof islands, southeastern Alaska; United States Geological Survey, Professional Paper 792, 105 p.

Lord, C.S.
1948: McConnell Creek map-area, Cassiar District, British Columbia; Geological Survey of Canada, Memoir 251, 72 p.

Lowes, B.E.
1972: Metamorphic petrology and structural geology of the area east of Harrison Lake, British Columbia; Ph.D thesis, University of Washington, 162 p.

Luken, M., Stevens, C.H., and Maginetti, R.
1985: Permian faunal relationships across California; Geological Society of America, Abstracts with Programs, v. 17, no. 6, p. 367.

Lutchmann, M.
1977: Mississippian clastic wedge; in Lower Mackenzie Energy Corridor Study, Geological Component; Geochem Laboratories (Canada) Ltd. and AGAT Consultants Ltd., p. M1-M10.

Lynts, G.W.
1970: Conceptual model of the Bahamian platform for the last 135 million years; Nature, v. 225, p. 1226-1228.

Macauley, G., Penner, D.G., Procter, R.M., and Tisdall, W.H.
1964: Carboniferous; in Geological History of Western Canada, R.G. McCrossan and R.P. Glaister (eds.), Alberta Society of Petroleum Geologists, p. 89-102.

MacIntyre, D.G.
1983: Geology and stratiform barite-sulphide deposits of the Gataga District, northeast British Columbia; in Sediment-Hosted Stratiform Lead-Zinc Deposits, D.F. Sangster (ed.), Mineralogical Association of Canada, Short Course Handbook, v. 9, p. 85-119.
1984: Geology of the Alsek-Tatshenshini Rivers area, British Columbia; British Columbia Ministry of Energy, Mines and Petroleum Resources, Geological Fieldwork, 1983, Paper 1984-1, p. 173-184.
1986: The geochemistry of basalts hosting massive sulphide deposits, Alexander Terrane northwest British Columbia; British Columbia Ministry of Energy, Mines and Petroleum Resources, Geological Fieldwork, 1985, Paper 1986-1, p. 197-210.

MacIntyre, D.G. and Schroeter, T.G.
1985: Mineral occurrences in the Mount Henry Clay area (114P/7, 8); British Columbia Ministry of Energy, Mines and Petroleum Resources, Geological Fieldwork, 1984, Paper 1985-1, p. 365-379.

MacKevett, E.M., Jr.
1978: Geologic map of the McCarthy quadrangle, Alaska; United States Geological Survey, Miscellaneous Inventory Series Map I-1032.

MacKevett, E.M., Jr. and Richter, D.H.
1974: The Nikolai greenstone in the Wrangell Mountains, Alaska, and nearby areas; Geological Association of Canada, Cordilleran Section, Program with Abstracts, p. 13-14.

MacKevett, E.M., Jr., Gardner, M.C., Bergman, S.C., Cushing, G., and McClelland, W.D.
1986: Geological evidence for Late Pennsylvanian juxtaposition of Wrangellia and the Alexander Terrane, Alaska; Geological Society of America, Abstracts with Programs, v. 18, no. 2, p. 128.

Macqueen, R.W. and Bamber, E.W.
1967: Stratigraphy of Banff Formation and lower Rundle Group (Mississippian), southwestern Alberta; Geological Survey of Canada, Paper 67-47, 37 p.
1968: Stratigraphy and facies relationships of the Upper Mississippian Mount Head Formation, Rocky Mountains and Foothills, southwestern Alberta; Bulletin of Canadian Society of Petroleum Geologists, v. 16, no. 3, p. 225-287.

Macqueen, R.W. and Sandberg, C.A.
1970: Stratigraphy, age, and interregional correlation of the Exshaw Formation, Alberta Rocky Mountains; Bulletin of Canadian Petroleum Geology, v. 18, no. 1, p. 32-66.

Macqueen, R.W., Bamber, E.W., and Mamet, B.L.
1972: Lower Carboniferous stratigraphy and sedimentology of the southern Rocky Mountains; 24th International Geological Congress, Montreal, Quebec, D.J. Glass (ed.), Guidebook, Field Excursion 17, p. 62.

MacRae, J. and McGugan, A.
1977: Permian stratigraphy and sedimentology - southwestern Alberta and southeastern British Columbia; Bulletin of Canadian Petroleum Geology, v. 25, no. 4, p. 752-766.

Mamet, B.E.
1976: An atlas of microfacies in carboniferous carbonates of the Canadian Cordillera; Geological Survey of Canada, Bulletin 255, 131 p.

Mamet, B.L. and Gabrielse, H.
1969: Foraminiferal zonation and stratigraphy of the type section of the Nizi Formation (Carboniferous System, Chesteron Stage), British Columbia; Geological Survey of Canada, Paper 69-16, 19 p.

Mamet, B.L. and Skipp, B.A.
1970: Preliminary foraminiferal correlations of Early Carboniferous strata in the North American Cordillera; in Colloque sur la Stratigraphie du Carbonifere; Les Congres et Colloques de l'universite de Liege, v. 55, p. 327-348.

Mamet, B.L., Bamber, E.W., and Macqueen, R.W.
1986: Microfacies of the Lower Carboniferous Banff Formation and Rundle Group, Monkman Pass map-area, northeastern British Columbia; Geological Survey of Canada, Bulletin 353, 93 p.

Mattinson, J.M.
1972: Ages of zircons from the northern Cascade Mountains, Washington; Geological Society of America Bulletin, v. 83, p. 3769-3784.

McClay, K.R.
1984: Tom barite-lead-zinc deposit, Yukon, Canada; in Structural Geology of Stratiform Lead-Zinc Deposits: case histories, K.R. McClay (ed.), Cordilleran Section, Geological Association of Canada, Short Course No. 2 (Part II), p. 60-96.

McGugan, A.
1964: New Permo-Carboniferous nomenclature, western Alberta and adjacent regions; Bulletin of the American Association of Petroleum Geologists, v. 48, no. 3, p. 357-359.

McGugan, A. and Rapson, J.E.
1963: Permo-Carboniferous stratigraphy between Banff and Jasper, Alberta; Bulletin of Canadian Petroleum Geology, v. 11, no. 2, p. 150-160.

McGugan, A. and Rapson-McGugan, J.E.
1976: Permian and Carboniferous stratigraphy, Wapiti Lake area, northeastern British Columbia; Bulletin of Canadian Petroleum Geology, v. 24, no. 2, p. 193-210.

McGugan, A., Rapson-McGugan, J.F., Mamet, B.L., and Ross, C.A.
1968: Permian and Pennsylvanian biostratigraphy, and Permian depositional environments, petrography and diagenesis, southern Canadian Rocky Mountains; in Canadian Rockies, Bow River to North Saskatchewan River, Alberta, H. Hornford (ed.), Alberta Society of Petroleum Geologists, 16th Annual Field Conference Guidebook, September, 1968, p. 48-66.

McGugan, A., Roessingh, H.K., and Danner, W.R.
1964: Permian; in Geological History of Western Canada, R.G. McCrossan and R.P. Glaister (ed.); Alberta Society of Petroleum Geologists, p. 103-112.

McLaren, D.J.
1982: Frasnian-Famennian extinctions in Geological Society of America, Special Paper 190, p. 477-484.

McLaren, D.J. and Mountjoy, E.W.
1962: Alexo equivalents in the Jasper region, Alberta; Geological Survey of Canada, Paper 62-23, 36 p.

McLearn, F.H. and Kindle, E.D.
1950: Geology of northwestern British Columbia; Geological survey of Canada, Memoir 259.

McMillan, W.J.
1974: Stratigraphic section from the Jurassic Ashcroft Formation and Triassic Nicola Group contiguous to the Guichon Creek Batholith; in Geological Fieldwork, British Columbia Department of Mines and Petroleum Resources, p. 27-34.

McTaggart, K.C. and Thompson, R.M.
1967: Geology of part of the northern Cascades in southern British Columbia; Canadian Journal of Earth Sciences, v. 4, p. 1199-1228.

Meade, H.D.
1977: Petrology and metal occurrences of the Takla Group and Hogem and Germansen batholiths, north-central British Columbia; Ph.D. thesis, University of Western Ontario, 355 p.

Milford, J.C.
1984: Geology of the Apex Mountain Group, north and east of the Similkameen River, south-central British Columbia; M.Sc. thesis, University of British Columbia, Vancouver, 108 p.

Mills, J.W. and Davis, J.R.
1962: Permian fossils of the Kettle Falls area, Stevens County, Washington; Contributions from the Cushman Foundation for Foraminiferal Research, Vol. XIII, Part 2, p. 41-51.

Misch, P.
1966: Tectonic evolution of the northern Cascades of Washington State; in Tectonic History and Mineral Deposits of the Western Cordillera, H.C. Gunning (senior ed.), Canadian Institute of Mining and Metallurgy, Special Volume 8, p. 101-148.

Monger, J.W.H.
1969: Stratigraphy and structure of Upper Paleozoic rocks, northeast Dease Lake map-area, British Columbia (104J); Geological Survey of Canada, Paper 68-48, 41 p.
1970: Hope map-area, west-half, British Columbia; Geological Survey of Canada, Paper 69-47, 75 p.
1975: Upper Paleozoic rocks of the Atlin terrane, northwestern British Columbia and south-central Yukon; Geological Survey of Canada, Paper 74-47, 63 p.
1977a: The Triassic Takla Group in McConnell Creek map-area, north-central British Columbia; Geological Survey of Canada, Paper 76-29, 45 p.
1977b: Upper Paleozoic rocks of the western Canadian Cordillera and their bearing on Cordilleran evolution; Canadian Journal of Earth Sciences, v. 14, p. 1832-1859.
1977c: Upper Paleozoic rocks of northwestern British Columbia; in Report of Activities, Part A, Geological Survey of Canada, Paper 77-1A, p. 255-262.
1980: Upper Triassic stratigraphy, Dease Lake and Tulsequah map areas, northwestern British Columbia; in Current Research, Part B, Geological Survey of Canada, Paper 80-1B, p. 1-9.

Monger, J.W.H. and Berg, H.C.
1984: Lithotectonic terrane map of western Canada and southeastern Alaska; in Lithotectonic Terrane Maps of the North American Cordillera, Part B, N.J. Silberling and D.L. Jones (ed.), United States Geological Survey, Open-File Report 84-523.

Monger, J.W.H. and Irving, E.
1980: Northward displacement of north-central British Columbia; Nature, v. 285, p. 289-293.

Monger, J.W.H. and McMillan, W.J.
1984: Bedrock geology of Ashcroft (92I) map-area, British Columbia; Geological Survey of Canada, Open File 980, Scale 1:125 000.

Monger, J.W.H. and Paterson, I.A.
1974: Upper Paleozoic and lower Mesozoic rocks of the Omineca Mountains; in Report of Activities, Part A, Geological Survey of Canada, Paper 74-1, Part A, p. 19-20.

Monger, J.W.H. and Ross, C.A.
1971: Distribution of Fusulinaceans in the western Canadian Cordillera; Canadian Journal of Earth Sciences, v. 8, p. 259-278.

Monger, J.W.H., Clowes, R.M., Price, R.A., Simony, P.S., Riddihough, R.P., and Woodsworth, G.J.
1985: Continent-Ocean Transect #7, B2; Juan de Fuca Plate to Alberta Plains; Geological Society of America, Centennial Continent/Ocean Transect No. 7.

Moore, P.F.
in press: Devonian strata of the Interior Plains region of western Canada; in Sedimentary Cover of the Craton in Canada, D.F. Stott and J.D. Aitken (ed.), Geological Survey of Canada, Geology of Canada, no. 5 (Geological Society of America, The Geology of North America, v. D-1.

Moore, T.E., Nilsen, T.H., and Brosge, W.P.
in press: Sedimentology of the Kayanut Conglomerate; in Dalton Highway, Yukon River to Prudhoe Bay, Alaska, Guidebook to Bedrock Geology of the eastern Koyukuk Basin, central Brooks Range and east-central Arctic slope; C.G. Mull (ed.), Alaska Geological Society, Guidebook 7 (published by Division of Geological and Geophysical Surveys, Department of Natural Resources, State of Alaska).

Morin, J.A.
1977: Ag-Pb-Zn mineralization in the MM deposit and associated Mississippian felsic volcanic rocks in the St. Cyr Range, Pelly Mountains; in Mineral Industry Report, 1976, Yukon Territory, Canada Department of Indian Affairs and Northern Development, Report EGS 1977-1, p. 83-97.

Morrow, D.W.
1984: Sedimentation in Root Basin and Prairie Creek Embayment - Siluro-Devonian, Northwest Territories; Bulletin of Canadian Petroleum Geology, v. 32, no. 10, p. 162-189.

Mortensen, J.K.
1982: Geological setting and tectonic significance of Mississippian felsic metavolcanic rocks in the Pelly Mountains, southeastern Yukon Territory; Canadian Journal of Earth Sciences, v. 19, no. 1, p. 8-22.
1983: Age and evolution of the Yukon-Tanana Terrane, southeastern Yukon Territory; Ph.D. thesis, University of California, Santa Barbara, 155 p.
1986: U-Pb ages for granitic orthogneiss from western Yukon Territory: Selwyn Gneiss and Fiftymile Batholith revisited; in Current Research, Part B, Geological Survey of Canada, Paper 86-1B, p. 141-146.

Mortensen, J.K. and Jilson, G.A.
1985: Evolution of the Yukon-Tanana terrane: Evidence from southeastern Yukon Territory; Geology, v. 13, no. 11, p. 806-810.

Mortensen, J.K., Montgomery, J.R., and Fillipone, J.
1987: U-Pb zircon, monazite and sphene ages for granitic orthogneiss of the Barkerville Terrane, east-central British Columbia; Canadian Journal of Earth Sciences, v. 24, p. 1261-1266.

Mountjoy, E.W.
1967: Triassic stratigraphy of northern Yukon Territory; Geological Survey of Canada, Paper 66-19, 44 p.
1980: Some questions about the development of Upper Devonian carbonate buildups (reefs), Western Canada; Bulletin of Canadian Petroleum Geology, v. 28, p. 315-344.

Mountjoy, E.W. and Mackenzie, W.S.
1973: Stratigraphy of the southern part of the Devonian Ancient Wall Complex, Jasper National Park, Alberta; Geological Survey of Canada, Paper 72-20, 121 p.

Mountjoy, E.W. and Riding, R.
1981: Foreslope stromatoporoid-renalcid bioherm with evidence of early cementation, Devonian Ancient Wall reef complex, Rocky Mountains; Sedimentology, v. 28, p. 299-319.

Muller, J.E.
1967: Kluane Lake map-area, Yukon Territory; Geological Survey of Canada, Memoir 340, 137 p.
1977: Evolution of the Pacific Margin, Vancouver Island, and adjacent regions; Canadian Journal of Earth Sciences, v. 14, p. 2062-2085.
1980: The Paleozoic Sicker Group of Vancouver Island, British Columbia; Geological Survey of Canada, Paper 79-30, 23 p.

Mulligan, R.
1952: Bonnington map-area, British Columbia; Geological Survey of Canada, Paper 52-13, 37 p.

Nassichuk, W.W. and Bamber, E.W.
1978: Site 8: Pennsylvanian and Permian stratigraphy at Little Fish Creek; in Geological and Geographical Guide to the Mackenzie Delta area, F.G. Young (ed.), Canadian Society of Petroleum Geologists, International Conference - Facts and Principles of World Oil Occurrence, June, 1978, Calgary, p. 85-89.

Nelson, C.H.
1982: Modern shallow-water graded sand layers from storm surges, Bering Shelf: a mimic of Bouma sequences and turbidite systems; Journal of Sedimentary Petrology, v. 52, no. 2, p. 537-545.

Nelson, S.J.
1982: New Pennsylvanian(?) syringoporid corals from Kamloops area, British Columbia; Canadian Journal of Earth Sciences, v. 19, p. 376-380.

Nelson, S.J. and Nelson, E.R.
1985: Allochthonous Permian micro- and macrofauna, Kamloops area, British Columbia; Canadian Journal of Earth Sciences, v. 22, no. 3, p. 442-451.

Newton, C.R.
1983: Paleozoogeographic affinities of Norian bivalves from the Wrangellian Peninsular and Alexander terranes, western North America; in Pre-Jurassic Rocks in Western North American Suspect Terranes, C.H. Stevens (ed.), Society of Economic Paleontologists and Mineralogists Pacific Section, Los Angeles, California, p. 37-48.

Nilsen, T.H.
1984: Stratigraphic nomenclature for the Upper Devonian and Lower Mississippian(?) Kanayut Conglomerate, Brooks Range, Alaska; United States Geological Survey, Bulletin 1529-A, 64 p.

Nilsen, T.H., Brabb, E.E., and Simoni, T.R.
1976: Stratigraphy and sedimentology of the Nation River Formation, a Devonian deep-sea fan deposit in east-central Alaska; in Recent and Ancient Sedimentary Environments in Alaska, T.P. Miller (ed.), Proceedings of the Alaska Geological Society Symposium, Alaska Geological Society, Alaska, p. E1-E20.

Nokleberg, W.J. and Aleinikoff, J.N.
1985: Summary of stratigraphy, structure, and metamorphism of Devonian igneous-arc terranes, northeastern Mount Hayes Quadrangle, eastern Alaska Range; in The United States Geological Survey in Alaska; accomplishments during 1984, S. Bartsch-Winkler (ed.), United States Geological Survey, Circular 967, p. 66-71.

Norford, B.S.
1981: Devonian stratigraphy at the margins of the Rocky Mountain Trench, Columbia River, southeastern British Columbia; Bulletin of Canadian Petroleum Geology, v. 29, p. 540-560.

Norris, A.W.
1968: Reconnaissance Devonian stratigraphy of northern Yukon Territory and northwestern District of Mackenzie; Geological Survey of Canada, Paper 67-53, 287 p.
1985: Stratigraphy of Devonian outcrop belts in northern Yukon Territory and northwestern District of Mackenzie (Operation Porcupine Area); Geological Survey of Canada, Memoir 410, 81 p.

Norris, D.K.
1965: Stratigraphy of the Rocky Mountain Group in the southeastern Cordillera of Canada; Geological Survey of Canada, Bulletin 125, 82 p.
1971: The geology and coal potential of the Cascade Coal Basin; in A Guide to the Geology of the Eastern Cordillera along the Trans-Canada Highway between Calgary, Alberta and Revelstoke, British Columbia; I.A.R. Halliday and D.K. Mathewson (ed.), Alberta Society of Petroleum Geologists, Calgary, p. 25-39.
1981a: Geology, Blow River and Davidson Mountains, Yukon Territory-District of Mackenzie; Geological Survey of Canada, Map 1516A.
1981b: Geology, Old Crow, Yukon Territory; Geological Survey of Canada, Map 1518A.
1983: Geotectonic correlation chart 1532A - Operation Porcupine project area; Geological Survey of Canada, Map 1532A.

Norris, D.K. and Yorath, C.J.
1981: The North American plate from the Arctic archipelago to the Romanzof Mountains; in The Ocean Basins and Margins, Volume 5, The Arctic Ocean, A.E.M. Nairn, M. Churkin, Jr. and F.G. Stehli (ed.), p. 37-103.

Okulitch, A.V.
1973: Age and correlation of the Kobau Group, Mount Kobau, British Columbia; Canadian Journal of Earth Sciences, v. 10, p. 1508-1518.
1979: Thompson-Shuswap-Okanagan map area; Geological Survey of Canada, Open File 637.
1985: Paleozoic plutonism in southeastern British Columbia; Canadian Journal of Earth Sciences, v. 22, p. 1409-1424.

Okulitch, A.V., Wanless, R.K., and Loveridge, W.J.
1975: Devonian plutonism in south-central British Columbia; Canadian Journal of Earth Sciences, v. 12, p. 1760-1769.

Orchard, M.J.
1984: Early Permian conodonts from the Harper Ranch beds, Kamloops area, southern British Columbia; in Current Research, Part B, Geological Survey of Canada, Paper 84-1B, p. 207-215.

Ovenshine, A.T.
1975: Tidal origin of parts of the Karheen Formation (Lower Devonian), southeastern Alaska; in Tidal Deposits, A Casebook of Recent Examples and Fossil Counterparts, R.N. Ginsburg (ed.), Springer-Verlag, p. 127-133.

Ovenshine, A.T., Eberlein, G.D., and Churkin, M., Jr.
1969: Paleotectonic significance of a Silurian-Devonian clastic wedge, southeastern Alaska; Geological Society of America, Abstracts with Programs, 1, p. 50.

Palmer, A.R., compiler
1983: Decade of North American Geology 1983 geological time scale; Geological Society of America Map and Chart Series MC-50.

Panteleyev, A.
1984: Stratigraphic position of the 'Toodoggone volcanics'; in Geological Fieldwork 1983; British Columbia Ministry of Energy, Mines and Petroleum Resources, Paper 1984-1, p. 136-138.

Parrish, R.R.
1981: Geology of the Nemo Lakes Belt, northern Valhalla Range, southeast British Columbia; Canadian Journal of Earth Sciences, v. 18, p. 944-958.

Paterson, I.A.
1973: The geology of the Pinchi Lake area, central British Columbia; Ph.D. thesis, University of British Columbia, Vancouver, 260 p.
1977: The geology and evolution of the Pinchi Fault zone at Pinchi Lake, central British Columbia; Canadian Journal of Earth Sciences, v. 14, p. 1324-1342.

Paterson, I.A. and Harakal, J.E.
1974: Potassium-argon dating of blueschists from Pinchi Lake, central British Columbia; Canadian Journal of Earth Sciences, v. 11, p. 1007-1011.

Pavlis, T.L.
1982: Origin and age of the Border Ranges fault of southern Alaska and its bearing on the late Mesozoic tectonic evolution of Alaska; Tectonics, v. 1, p. 343-368.

Pearson, D.A.
1973: Harrison, Lucky Jim Properties; in British Columbia Department of Mines and Petroleum Resources, Geology, Exploration and Mining in British Columbia, p. 125-131.

Peatfield, G.R.
1978: Geologic history and metallogeny of the "Boundary District", southern British Columbia and Northern Washington; Ph.D. thesis, Queen's University, Kingston, Ontario, 247 p.

Pedder, A.E.H.
1982: The rugose coral record across the Frasnian/Famennian boundary; Geological Society of America, Paper 190, p. 485-489.

Pelletier, B.R.
1963: Triassic stratigraphy of the Rocky Mountains and Foothills, Peace River District, British Columbia; Geological Survey of Canada, Paper 62-26.

Pelzer, E.E.
1966: Mineralogy, geochemistry, and stratigraphy of the Besa River Shale, British Columbia; Bulletin of Canadian Petroleum Geology, v. 14, no. 2, p. 273-321.

Petocz, R.G.
1970: Biostratigraphy and Lower Permian fusulinidae of the Upper Delta River area, east-central Alaska range; Geological Society of America, Special Paper 130, 94 p.

Poole, F.G.
1974: Flysch deposits of the Antler foreland basin, western United States; in Tectonics and Sedimentation, W.R. Dickinson (ed.), Society of Economic Paleontologists and Mineralogists, Special Publication 22, p. 58-82.

Poole, W.H.
1956: Geology of the Cassiar Mountains in the vicinity of the Yukon-British Columbia boundary; Ph.D. thesis, Princeton University, 247 p.

Potter, C.J.
1983: Geology of the Bridge River Complex, southern Shulaps range, British Columbia: A record of Mesozoic convergent tectonics; Ph.D thesis, University of Washington, 169 p.

Poulton, T.P.
1982: Paleogeographic and tectonic implications of Lower and Middle Jurassic facies patterns in northern Yukon Territory and adjacent Northwest Territories; in Arctic Geology and Geophysics, A.F. Embry and H.R. Balkwill (ed.), Canadian Society of Petroleum Geologists, Memoir 8, p. 13-27.
1984: The Jurassic of the Canadian Western Interior, from 49_N Latitude to Beaufort Sea; in The Mesozoic of Middle North America, D.F. Stott and D.J. Glass (ed.), Canadian Society of Petroleum Geologists Memoir 9, p. 15-41.

Poulton, T.P. and Tempelman-Kluit, D.J.
1982: Recent discoveries of Jurassic fossils in the Lower Schist division of central Yukon; in Current Research, Part C, Geological Survey of Canada, Paper 82-1C, p. 91-94.

Poulton, T.P., Leskiw, K., and Audretsch, A.P.
1982: Stratigraphy and microfossils of the Jurassic Bug Creek Group of northern Richardson Mountains, northern Yukon and adjacent Northwest Territories; Geological Survey of Canada, Bulletin 325.

Price, R.A.
1965: Flathead map area, British Columbia and Alberta; Geological Survey of Canada, Memoir 336, 221 p.

Price, R.A., Monger, J.W.H., and Muller, J.E.
1981: Cordilleran cross-section - Calgary to Victoria; in Field Guides to Geology and Mineral Deposits; Geological Association of Canada, Mineralogical Association of Canada, and Canadian Geophysical Union, p. 261-334.

Pugh, D.C.
1983: Pre-Mesozoic geology in the subsurface of Peel River map-area, Yukon Territory and District of Mackenzie; Geological Survey of Canada, Memoir 401.

Ray, G.E.
1986: The Hozameen fault system and related Coquihalla serpentine belt of southwestern British Columbia; Canadian Journal of Earth Sciences, v. 23, p. 1022-1041.

Read, P.B.
1983: Geology, Classy Creek (104J/2E) and Stikine Canyon (104J/1W); Geological Survey of Canada, Open File 940, 1 sheet.

Read, P.B. and Brown, R.L.
1981: Columbia River fault zone: southeast margin of the Shuswap and Monashee complexes, southern British Columbia; Canadian Journal of Earth Sciences, v. 18, p. 1127-1145.

Read, P.B. and Monger, J.W.H.
1976: Pre-Cenozoic volcanic assemblages of the Kluane and Alsek ranges, southwestern Yukon Territory; Geological Survey of Canada, Open File 381, 96 p.

Read, P.B. and Okulitch, A.V.
1977: The Triassic unconformity of south-central British Columbia; Canadian Journal of Earth Sciences, v. 14, p. 606-638.

Rees, C.J., Irving, E., and Brown, R.L.
1985: Paleomagnetism of Triassic-Jurassic volcaniclastic rocks of Quesnellia from Quesnel Lake, British Columbia; Geological Society of America, Abstracts with Programs, v. 17, no. 6, p. 403.

Reesor, J.E.
1973: Geology of the Lardeau map-area, east-half, British Columbia; Geological Survey of Canada, Memoir 369.

Reid, R.P. and Tempelman-Kluit, D.J.
1987: Upper Triassic Tethyan-type reefs in the Yukon; Canadian Society of Petroleum Geologists, v. 35, no. 3, p. 316-332.

Richards, B.C.
1978: Submarine channels in the Flett Formation: a deeper water carbonate succession, District of Mackenzie, Canada; Geological Association of Canada, Abstracts with Programs, Annual Meeting, v. 3, p. 478.

1983: Uppermost Devonian and Lower Carboniferous stratigraphy, sedimentation, and diagenesis, southwestern District of Mackenzie and southeastern Yukon Territory (NTS 95B, C, F, and G); Ph.D. thesis, University of Kansas.

1989: Uppermost Devonian and Lower Carboniferous stratigraphy, sedimentation, and diagenesis, southwestern District of Mackenzie and southeastern Yukon Territory (NTS 95B, C, and G); Geological Survey of Canada, Bulletin 390, 135 p.

Richards, B.C., Bamber, E.W., Higgins, A.C., and Utting, J.
in press: Carboniferous, (Geological Description, Chapter 6); in Sedimentary Cover of the Craton: Canada, D.F. Stott and J.D. Aitken (ed.); Geological Survey of Canada, Geology of Canada, no. 5 (also Geological Society of America, The Geology of North America, v. D-1).

Richards, T.A.
1976: Takla project (Reports 10-16): McConnell Creek map-area (94D, East Half) British Columbia; in Report of Activities, Part A, Geological Survey of Canada, Paper 76-1A, p. 43-50.

Richter, D.H.
1976: Geologic map of the Nabesna Quadrangle, Alaska; United States Geological Survey, Miscellaneous Geological Investigations Map I-932.

Richter, D.H., Lanphere, M.A., and Matson, N.A., Jr.
1975: Granitic plutonism and metamorphism, eastern Alaska Range, Alaska; Geological Society of America Bulletin, v. 86, p. 819-829.

Rigby, K.
1973: Permian sponges from northwestern British Columbia; Canadian Journal of Earth Sciences, v. 10, p. 1600-1606.

Rinehart, C.D. and Fox, K.F.
1972: Geology and mineral deposits of the Loomis Quadrangle, Okanogan County, Washington; Washington Division of Mines and Geology, Bulletin 64, 124 p.

Roddick, J.A.
1967: Tintina Trench; Journal of Geology, v. 75, p. 23-33.

Roddick, J.A. and Hutchison, W.W.
1973: Pemberton (east half) map-area, British Columbia; Geological Survey of Canada, Paper 73-17, 21 p.

Roddick, J.A. and Woodsworth, G.J.
1977: Coast Mountain Project; in Report of Activities, Part A, Geological Survey of Canada, Paper 77-1A, p. 271-272.

Roots, E.F.
1954: Geology and mineral deposits of Aiken Lake map-area, British Columbia; Geological Survey of Canada, Memoir 274, 246 p.

Ross, C.A.
1969: Upper Paleozoic fusulinacea: *Eowaegingella* and *Wedekindellina* from Yukon Territory and giant *Parafusulina* from British Columbia; in Contributions to Canadian Paleontology, Geological Survey of Canada, Bulletin 182, p. 129-134.

Ross, C.A. and Nassichuk, W.W.
1970: Yabeina and Waagenoceras from the Atlin Horst area, northwestern British Columbia; Journal of Paleontology, v. 44, no. 4, p. 779-781.

Ross, C.A. and Monger, J.W.H.
1977: Carboniferous and Permian fusulinaceans from the Omineca Mountains; Geological Survey of Canada, Bulletin 267, 75 p.

Ross, C.A. and Ross, J.R.P.
1983: Late Paleozoic accreted terranes of western North America; in Pre-Jurassic rocks in western North American suspect terranes, C.H. Stevens (ed.), Society of Economic Paleontologists and Mineralogists Pacific Section, Los Angeles, California, p. 7-22.

Rossman, D.L.
1959: Geology and ore deposits of northwestern Chichagof Island, Alaska; United States Geological Survey, Bulletin 1058-E, p. E139-E216.

Rusmore, M.E.
1987: Geology of the Cadwallader Group and the Intermontane-Insular superterrane boundary, southwestern British Columbia; Canadian Journal of Earth Sciences, v. 24, p. 2279-2291.

Sable, E.G.
1977: Geology of the western Romanzof Mountains, Brooks Range, northeastern Alaska; United States Geological Survey, Professional Paper 897, 84 p.

Sada, K. and Danner, W.R.
1973: Early Permian *Para Fusulina* and *Pseudo Fusulinella* from the Chilliwack Group, southwestern British Columbia, Canada; Transects and Proceedings, Palaeontological Society of Japan, New Series, no. 90, p. 72-80.

1974: Early and Middle Pennsylvanian Fusulinids from southern British Columbia, Canada and northwestern Washington, U.S.A.; Transects and Proceedings, Paleontological Society of Japan, New Series, no. 93, p. 249-265.

1976: *Pseudoschwagerina* from Harper Ranch area near Kamloops, British Columbia; Commemoration Volume of the Founding of the Faculty of Integrated Arts and Sciences, Hiroshima University, p. 213-228.

Saleeby, J.B., Gehrels, G.E., and Berg, H.C.
1985: Character of the Alexander-Taku Terrane Boundary - Cape Fox to Cleveland Peninsula Region - SE Alaska; Geological Society of America, Abstracts with Programs, v. 17, no. 6, p. 406.

Sando, W.J. and Bamber, E.W.
1985: Coral Zonation of the Mississippian System in the Western Interior Province of North America; United States Geological Survey, Professional Paper 1334, 61 p.

Schiarizza, P.
1982: Clearwater area; in Geological Fieldwork, 1981, British Columbia Ministry of Energy, Mines and Petroleum Resources, p. 59-67.

Schiarizza, P. and Preto, V.A.
1984: Geology of the Adams Plateau-Clearwater area; British Columbia Ministry of Energy, Mines and Petroleum Resources, Preliminary Map 56.

Schlager, W.
1981: The paradox of drowned reefs and carbonate platforms; Geological Society of America Bulletin, v. 92, Part 1, p. 197-211.

Schuchert, C.
1923: Sites and nature of the North American geosynclines; Geological Society of America Bulletin, v. 34, p. 151-239.

Scott, D.L.
1964: Pennsylvanian stratigraphy; Alberta Society of Petroleum Geologists, Guidebook, 14th Annual Field Conference; Bulletin Canadian Petroleum Geologists, v. 12, p. 460-493.

Selwyn, A.R.C.
1872: Journal and report of preliminary explorations in British Columbia; Geological Survey of Canada, Report of Progress for 1871-72, p. 16-72.

Shannon, K.R.
1982: Cache Creek Group and contiguous rocks, near Cache Creek, British Columbia; M.Sc. thesis, University of British Columbia, 72 p.

Skinner, J.W. and Wilde, G.L.
1966: A new Permian fusulinid from southern British Columbia; Academy of Sciences of U.S.S.R., Problems in Micropaleontology, v. 10, p. 105-108.

Sloss, L.L.
1963: Sequences in the cratonic interior of North America; Geological Society of America Bulletin, v. 74, p. 93-113.

Smith, J.G. and MacKevett, E.M., Jr.
1970: The Skolai Group in the McCarthy B-4, C-4, C-5 Quadrangles, Wrangell Mountains, Alaska; United States Geological Survey, Bulletin 1274-Q, p. Q1-Q26.

Smith, R.B.
1974: Geology of the Harper Ranch Group (Carboniferous- Permian) and Nicola Group (Upper Triassic), northwest of Kamloops, British Columbia; M.Sc. thesis, University of British Columbia, Vancouver, 211 p.

Snook, J.R., Ellis, M.A., Mills, J.W., and Watkinson, A.J.
1982: The Kootenay Arc in N.E. Washington; Geological Society of America, Abstracts with Programs, 1982, v. 14, p. 235.

Souther, J.G.
1959: Chutine, Cassiar District; Geological Survey of Canada, Map 7-1959.

1971: Geology and mineral deposits of Tulsequah map-area, British Columbia; Geological Survey of Canada, Memoir 362, 84 p.

1972: Telegraph Creek map-area, British Columbia; Geological Survey of Canada, Paper 71-44, 38 p.

Springer, G.D., MacDonald, W.D., and Crockford, M.B.B.
1964: Jurassic; in Geological History of Western Canada, Chapter 10, R.G. McCrossan and R.P. Glaister (ed.), Alberta Society of Petroleum Geologists, Calgary, p. 137-155.

Still, J.C.
1984: Stratiform massive sulphide deposits of the Mt. Hendry Clay area, southeast Alaska; United States Bureau of Mines, Open File Report 118-184.

Stone, D.B., Panuska, B.C., and Packer, D.R.
1982: Paleolatitudes versus time for southern Alaska; Journal of Geophysical Research, v. 87, p. 3697-3708.

Stott, D.F.
1967: Jurassic and Cretaceous stratigraphy between Peace and Tetsa rivers, northeastern British Columbia; Geological Survey of Canada, Paper 66-7, 73 p.
1982: Lower Cretaceous Fort St. John Group and Upper Cretaceous Dunvegan Formation of the Foothills and Plains of Alberta, British Columbia, District of Mackenzie and Yukon Territory; Geological Survey of Canada, Bulletin 328.

Struik, L.C.
1980: Geology of the Barkerville-Cariboo River area, east-central British Columbia; Ph.D. thesis, University of Calgary, Alberta, 335 p.
1981: A re-examination of the type area of the Devono-Mississippian Cariboo Orogeny, central British Columbia; Canadian Journal of Earth Sciences, v. 18, p. 1767-1775.
1985: Pre-Cretaceous terranes and their thrust and strike-slip contacts, Prince George (east half) and McBride (west half) map areas, British Columbia; in Current Research, Part A, Geological Survey of Canada, Paper 85-1A, p. 267-272.

Struik, L.C. and Orchard, M.J.
1985: Upper Paleozoic conodonts from ribbon chert of the Antler Formation of Slide Mountain Terrane denote imbricate thrust sheets, central British Columbia; Geology, v. 13, no. 11, p. 794-798.

Sutherland Brown, A.
1963: Geology of the Cariboo River area, British Columbia: British Columbia Department of Mines and Petroleum Resources, Bulletin 47, 113 p.
1968: Geology of the Queen Charlotte Islands, British Columbia; British Columbia Department of Mines and Petroleum Resources, Bulletin 54, 226 p.

Symons, D.T.A.
1971: Paleomagnetic notes on the Karmutsen basalts, Vancouver Island, British Columbia; Geological Survey of Canada, Paper 71-24, p. 10-24.
1976: Paleomagnetism of the Triassic Guichon Batholith and rotation in the Interior Plateau, British Columbia; Canadian Journal of Earth Sciences, v. 8, p. 1388-1396.

Taylor, G.C. and Mackenzie, W.S.
1970: Devonian stratigraphy of northeastern British Columbia; Geological Survey of Canada, Bulletin 186, 62 p.

Tempelman-Kluit, D.J.
1970: Stratigraphy and structure of the "Keno Hill Quartzite" in Tombstone River-Upper Klondike River map-areas, Yukon Territory (116B/7, B/8); Geological Survey of Canada, Bulletin 180, 102 p.
1972: Geology and origin of the Faro, Vangorda and Swim concordant lead-zinc deposits, central Yukon Territory; Geological Survey of Canada, Bulletin 208, 73 p.
1974: Reconnaissance geology of Aishihik Lake, Snag and part of Stewart River map-areas, west-central Yukon; Geological Survey of Canada, Paper 73-41, 97 p.
1977: Geology of Quiet Lake and Finlayson Lake map areas, Yukon Territory (105F and G); Geological Survey of Canada, Open File 486.
1979a: Transported cataclasite, ophiolite and granodiorite in Yukon: evidence for arc-continent collision; Geological Survey of Canada, Paper 79-14, 27 p.
1979b: Five occurrences of transported synorogenic clastic rocks in Yukon Territory; in Current Research, Part A, Geological Survey of Canada, Paper 79-1A, p. 1-12.

Tempelman-Kluit, D. and Wanless, R.K.
1980: Zircon ages for the Pelly Gneiss and Klotassin granodiorite in western Yukon; Canadian Journal of Earth Sciences, v. 17, p. 297-306.

Tennyson, M.E. and Coles, M.R.
1978: Tectonic significance of Upper Mesozoic Methow-Pasayten sequence, northeastern Cascade Range, Washington and British Columbia; in Mesozoic Paleogeography of the Western United States, D.G. Howell and K.A. McDougall (ed.), Society of Economic Paleontologists and Mineralogists, Pacific Coast Section Paleography Symposium 2, p. 499-508.

Tennyson, M.E., Jones, D.L., and Murchie, B.L.
1982: Age and nature of chert and mafic rocks in the Hozameen Group, north Cascade Range, Washington; Geological Society of America, Abstracts with Programs, v. 14, p. 239.

Terry, J.
1977: Geology of the Nahlin ultramafic body, Atlin and Tulsequah map-areas, northwestern British Columbia; in Report of Activities, Part A, Geological Survey of Canada, Paper 77-1A, p. 263-266.

Thompson, M.L.
1965: Pennsylvanian and Early Permian fusulinids from Fort St. James area, British Columbia, Canada; Journal of Paleontology, v. 39, p. 224-234.

Thomson, R.C., Smith, P.L., and Tipper, H.W.
1986: Lower to Middle Jurassic (Pliensbachian to Bajocian) stratigraphy of the northern Spatsizi area, north-central British Columbia; Canadian Journal of Earth Sciences, v. 23, no. 12, p. 1963-1973.

Thorstad, L.E.
1980: Upper Paleozoic volcanic and volcaniclastic rocks in northwest Toodoggone map-area, British Columbia; in Current Research, Part B; Geological Survey of Canada, Paper 80-1B, p. 207-211.

Thorstad, L.E. and Gabrielse, H.
1986: The Upper Triassic Kutcho Formation Cassiar Mountains, north-central British Columbia; Geological Survey of Canada, Paper 86-16.

Tipper, H.W.
1976: Biostratigraphic study of Mesozoic rocks in Intermontane and Insular belts of the Canadian Cordillera, British Columbia; in Report of Activities, Part A, Geological Survey of Canada, Paper 76-1A, p. 57-59.
1981: Offset of an upper Pliensbachian geographic zonation in the North American Cordillera by transcurrent movement; Canadian Journal of Earth Sciences, v. 18, p. 1788-1792.
1984: The age of the Jurassic Rossland Group of southeastern British Columbia; in Current Research, Part A, Geological Survey of Canada, Paper 84-1A, p. 631-632.

Tipper, H.W. and Richards, T.A.
1976: Jurassic stratigraphy and history of north-central British Columbia; Geological Survey of Canada, Bulletin 270, 73 p.

Tozer, E.T.
1967 :A standard for Triassic time; Geological Survey of Canada, Bulletin 156, 103 p.
1979: Latest Triassic ammonoid faunas and biochronology, western Canada; in Current Research, Part B, Geological Survey of Canada, Paper 79-1B, p. 127-135.
1982a: Late Triassic (upper Norian) and earliest Jurassic (Hettangian) rocks and ammonoid faunas, Halfway River and Pine Pass map areas, British Columbia; in Current Research, Part A, Geological Survey of Canada, Paper 82-1A, p. 385-391.
1982b: Marine Triassic faunas of North America: their significance for assessing plate and terrane movements; Geologische Rundschau, v. 71, no. 3, p. 1077-1104.
1984: The Triassic and its ammonoids; evolution of a time scale; Geological Survey of Canada, Miscellaneous Report 35, 171 p.

Travers, W.B.
1978: Overturned Nicola and Ashcroft strata and their relation to the Cache Creek Group, southwestern Intermontane Belt, British Columbia; Canadian Journal of Earth Sciences, v. 15, p. 99-116.
1982: Possible large-scale overthrusting near Ashcroft, British Columbia: Implications for petroleum prospecting; Bulletin of Canadian Petroleum Geology, v. 30, no. 1, p. 1-8.

Trettin, H.
1961: Geology of the Fraser River Valley between Lillooet and Big Bar Creek; British Columbia Department of Mines and Petroleum Resources, Bulletin 44, 109 p.

Turner, D.L., Herreid, G., and Bundtzen, T.K.
1977: Geochronology of southern Prince of Wales Island, Alaska; Alaska Division of Geological and Geophysical Surveys Geologic Report 55, p. 11-16.

United States Geological Survey
1982: Interpretation of the heterogeneous rocks in the Duncan Canal area as a Cretaceous mélange; United States Geological Survey, Professional Paper 1275, 90 p.

Vance, J.A., Dungan, M.A., Blanchard, D.P., and Rhodes, J.M.
1980: Tectonic setting and trace element geochemistry of Mesozoic ophiolitic rocks in western Washington; American Journal of Science, v. 280-A, p. 359-388.

Van der Voo, R., Jones, M., Grommé, C.S., Eberlein, G.D., and Churkin, M., Jr.
1980: Paleozoic paleomagnetism and northward drift of the Alexander terrane, southeastern Alaska; Journal of Geophysical Research, v. 85, no. B10, p. 5281-5296.

Waterhouse, J.B. and Waddington, J.
1982: Systematic descriptions, paleoecology and correlations of the Late Paleozoic subfamily Spiriferellinae (Brachiopoda) from the Yukon Territory and the Canadian Arctic Archipelago; Geological Survey of Canada, Bulletin 289, 72 p.

Watson, K.DeP. and Mathews, W.H.
1944: The Tuya-Teslin area, northern British Columbia; British Columbia Department of Mines, Bulletin 19, 52 p.

Werner, L.J.
1978: Metamorphic terrane, northern Coast Mountains west of Atlin Lake, British Columbia; in Current Research, Part A, Geological Survey of Canada, Paper 78-1A, p. 69-70.

Wheeler, J.O.
1961: Whitehorse map-area, Yukon Territory, 105D; Geological Survey of Canada, Memoir 312, 156 p.

Williams, G.K.
1977: The Hay River Formation and its relationship to adjacent formations, Slave River map-area, N.W.T.; Geological Survey of Canada, Paper 75-12, 17 p.

Wilson, F.H., Dadisman, S.V., and Herson, P.L.
1979: Map showing radiometric ages of rocks in southeastern Alaska; United States Geological Survey, Open File Report 79-594.

Wilson, J.T.
1968: Static or mobile earth: the current scientific revolution; Proceedings of the American Philosophical Society, v. 112, no. 5, p. 309-320.

Woodsworth, G.J. and Orchard, M.J.
1985: Upper Paleozoic to lower Mesozoic strata and their conodonts, western Coast Plutonic Complex, British Columbia; Canadian Journal of Earth Sciences, v. 22, p. 1329-1344.

Woodsworth, G.J., Crawford, M.L., and Hollister, L.S.
1983: Metamorphism and structure of the Coast Plutonic Complex and adjacent belts, Prince Rupert and Terrace areas, British Columbia; Geological Association of Canada Field Trip Guide 14, 62 p.

Yates, R.G.
1964: Geological map and sections of the Deep Creek areas, Stevens and Pend Oreille counties, Washington; United States Geological Survey, Miscellaneous Geological Investigations Map I-412.

Yole, R.W.
1963: An early Permian fauna from Vancouver Island, British Columbia; Canadian Petroleum Geology Bulletin, v. 11, no. 2, p. 138-149.
1969: Upper Paleozoic stratigraphy of Vancouver Island, British Columbia; Proceedings, Geological Association of Canada, v. 20, p. 30-40.

Yole, R.W. and Irving, E.
1980: Displacement of Vancouver Island: Paleomagnetic evidence from the Karmutsen Formation; Canadian Journal of Earth Sciences, v. 17, p. 1210-1288.

Yorath, C.J. and Chase, R.L.
1981: Tectonic history of the Queen Charlotte Islands and adjacent areas - a model; Canadian Journal of Earth Sciences, v. 18, p. 1717-1739.

ADDENDUM

Detailed accounts of the biostratigraphy and tectonic settings of Upper Triassic and Lower Jurassic rocks in the Queen Charlotte Islands are given in several papers contained in the following volume:

1991: Evolution and hydrocarbon potential of the Queen Charlotte Basin, British Columbia; G.J. Woodsworth (ed.); Geological Survey of Canada, Paper 90-10, 569 p.

Re: Lower and Middle Jurassic strata of the Foreland Belt

Significant changes in the Lower and Middle Jurassic from south to north within the southern Foreland Belt have been elucidated in recent reports, indicating instability in depositional, and perhaps tectonic regimes, of the craton prior to Cordilleran orogeny. The basal Jurassic (Sinemurian) strata of southwesternmost Alberta and southeastern British Columbia include a phosphorite unit as thick as 10 m. The requirement of "West coast type" phosphorite depositional models for access to upwelling ocean currents places limits on how close allochthonous terranes could have lain off the west coast of the continent in Sinemurian times (Poulton and Aitken, 1989).

The basal Jurassic Nordegg black chert and limestone member of west central Alberta is absent in northwestern Alberta and northeastern British Columbia, where a platy argillaceous limestone facies occupies the same position, although most of it may be somewhat younger (Pliensbachian rather than Sinemurian). The limits and general character of the carbonate platform and of the argillaceous facies to its north have been outlined by Poulton et al. (1991).

Similarly, the Middle Jurassic Rock Creek sandstone, and associated sandstones and shales of west central Alberta are apparently absent in northwestern Alberta and northeastern British Columbia, below an unconformity at the base of the superficially similar Upper Jurassic sequence (Poulton et al., 1991).

Most recent summaries of the Jurassic, including access to references, are by Poulton (1989a) for the northern Foreland Belt and Poulton (1989b) for the southern Foreland Belt.

REFERENCES

Poulton, T.P.
1989a: Current status of Jurassic biostratigraphy and stratigraphy, northern Yukon and adjacent Mackenzie Delta; in Current Research, Part G, Geological Survey of Canada, Paper 89-1G, p. 25-30.
1989b: Upper Absaroka to Lower Zuni: the transition to the foreland basin; in Western Canada Sedimentary Basin. A case history, B.D. Ricketts (ed.), Canadian Society of Petroleum Geologists, p. 233-247.

Poulton, T.P. and Aitken, J.D.
1989: The Lower Jurassic phosporites of southeastern British Columbia and terrane accretion to western North America; Canadian Journal of Earth Sciences, v. 26, p. 1612-1616.

Poulton, T.P., Tittemore, J. and Golby, G.
1991: Jurassic strata, northwestern (and west central) Alberta and northeastern British Columbia; Bulletin of Canadian Petroleum Geology, v. 38A, p. 159-175.

Authors' addresses

S.P. Gordey, J.W.H. Monger, J.O. Wheeler,
H.W. Tipper, H. Gabrielse, L.C. Struik, C.J. Dodds
Cordilleran Division
Geological Survey of Canada
100 West Pender Street
Vancouver, British Columbia
V6B 1R8

H.H.J. Geldsetzer, D.W. Morrow, E.W. Bamber,
C.M. Henderson, B.C. Richards, D.W. Gibson T.P. Poulton
Institute of Sedimentary and Petroleum Geology
Geological Survey of Canada
3303-33rd Street N.W.
Calgary, Alberta
T2L 2A7

A. McGugan
University of Calgary
Department of Geology and Geophysics
Calgary, Alberta
T2N 1N4

G.E. Gehrels, J. O'Brien
Department of Geosciences
University of Arizona
Tucson, Arizona
U.S.A. 85721

R.B. Campbell
1760 Forest Park Drive
Sidney, British Columbia
V8L 4A6

T. Harms
Department of Geology
Amherst College
Amherst, Massachusetts
U.S.A. 01002

Plate 1. View northerly toward the core of the Thor-Odin Gneiss Complex near Cariboo Alp, Monashee Mountains. Grey massive gneiss in the distance has a U-Pb age of about 2.0 Ga. The foreground is a southwesterly dipping imbricated zone, structurally overlying the basement massif, and consisting of grey basement slivers and rusty weathering metasediments of uncertain age. Photo by R.R. Parrish. KGS 2288

Plate 2. Typical exposures of the northeastern platformal facies of the Purcell (Belt) Supergroup at Waterton Lake, Alberta. Al=Altyn Formation, Ap=Appekunny Formation, G=Grinnell Formation, S=Siyeh Formation. Photo by D.K. Norris. KGS 2269

Plate 3. Exposures of fine grained sedimentary rocks of the Fairchild Lake Group of the Purcell-Wernecke Assemblage, Wernecke Mountains. Photo by J.D. Aitken. KGS 2260

Plate 4. Gillespie Lake Group (Wernecke Supergroup) near its type section. Nearly all of the rock in view is dolomite; the dark-weathering unit forming the saddle contains much silty and argillaceous dolomite and chert, and consists in large part of metre-scale, shallowing upward cycles. Algal stromatolites are prominent in the thick-bedded, resistant, ridge-forming dolomites at the right. Photo by J.D. Aitken. KGS 2270

Plate 5. View northwest towards Coates Lake in the central Mackenzie Mountains, showing the contact between dark brown quartzite of the Katherine Group to the right and carbonate of the Little Dal Group on the left. The thin cream-weathering unit at the base of the Little Dal Group consists of mud-cracked siltstone and dolomitic siltstone. Pink-weathering strata in the middle ground are in the Redstone River Formation. Photo by J.D. Aitken. KGS 2263

Plate 6. View westward in southern Ogilvie Mountains to two north-dipping faults that formed during extension at the beginning of Late Proterozoic time. The southern fault (1) is a reactivated Mesozoic thrust fault separating lower Paleozoic dolostone in the footwall from siltstone and sandstone of the Middle Proterozoic Wernecke Supergroup in the hanging wall. The northern fault (2) is a Late Proterozoic normal fault with dolomitic clastic rocks and slide blocks (white masses) in the hanging wall. The poorly sorted clastics represent erosion of shelf carbonates which overlie the Wernecke Supergroup and which were eroded before carbonate deposition in early Paleozoic time. Photo by R.I. Thompson. KGS 2289

Plate 7. Purple mudrocks and greenstone (Kohse Creek volcanics) of the Pinguicula Group overlying, with angular unconformity, steeply dipping dolomite beds of the Gillespie Lake Group near Pinguicula Lake, Mackenzie Mountains. Photo by J.D. Aitken. KGS 2259

Plate 8. Coates Lake Group near Mountain River, Mackenzie Mountains, showing basal lava (bottom right) successively overlain by grey resistant carbonate of the Thundercloud Formation, recessive siltstone and white gypsum of the Redstone River Formation, resistant light grey carbonate of the Coppercap Formation and dark weathering siltstone of the Sayunei Formation. Photo by C.W. Jefferson. KGS 2279

Plate 9. View northwesterly in north-central Mackenzie Mountains showing buff-weathering diamictite of the Shezal Formation separating the generally fine grained clastics of the overlying Twitya Formation from the dark red to maroon siliciclastic laminites and turbidites of the Sayunei Formation. Photo by J.D. Aitken. KGS 2282

Plate 10. View northwesterly in west-central Mackenzie Mountains. Resistant quartzite of the upper formation of the Backbone Ranges Group is overlain at the extreme left by buff-weathering dolomite of the Middle Cambrian Avalanche Formation. Photo by H. Gabrielse. KGS 2271

Plate 11. View northwesterly to Mount Robson, west of Jasper National Park. About 2900 m of Middle and Upper Cambrian, mainly carbonate strata are exposed. The upper, resistant, dark-weathering rocks represent a nearly complete section of the Upper Cambrian Lynx Formation. The uppermost Middle Cambrian Arctomys Formation forms the recessive buff-brown-weathering band about halfway down the mountain side. The lower part of the mountain is underlain by Middle Cambrian strata, mainly Titkana Formation. Photo by R.B. Campbell. KGS 2274

Plate 12. View northerly near southeastern margin of Selwyn Basin showing typically striped exposures of the Middle Ordovician Sunblood Formation (left) with orange-brown-weathering sandstone forming the uppermost unit; relatively thin (<100 m) Middle Ordovician limestone of the Esbataottine Formation; Middle Ordovician dolomite and limestone of the lower Whittaker Formation and recessive dark grey calcareous shale of the Road River Formation on the extreme right. Photo by H. Gabrielse. KGS 2257

Plate 13. A thin, north-dipping Lower and Middle Devonian sequence, 500 m thick, exposed along the northern flank of the Mackenzie Mountains. The Vera, Camsell and Sombre formations of the southern Mackenzie Mountains are absent and are represented by unnamed argillaceous, peritidal dolostone equivalent to the Tsetso Formation, unconformably overlying the Ordovician-Silurian Mount Kindle Formation. Photo by D.W. Morrow. KGS 2287

Plate 14. View northwesterly near the eastern margin of Selwyn Basin showing orange-weathering dolomitic siltstone and silty dolomite of the Delorme Formation overlying reefoid carbonate of the Mount Kindle Formation and underlying dull grey-weathering dolostone of the Sombre Formation (Bear Rock-Stone sequence). Photo by H. Gabrielse. KGS 2258

Plate 15. Anticline on Mackenzie Platform in the central Mackenzie Mountains exposing resistant cap of Nahanni Formation limestone, underlain by recessive, argillaceous limestone of the Headless Formation. Valleys are underlain by black shale, possibly correlative with the Horn River Formation. Photo by J.D. Aitken. KGS 2266

Plate 16. Lower part of the Upper Devonian Fairholm sequence, unconformably overlying Upper Cambrian carbonate of the Lyell Formation. L=Lyell Formation; F=Flume Formation; C=Cairn Formation (reef); I=Ireton (Mount Hawk) Formation (offreef). Burnt Timber Creek area, Alberta. Photo by H. Geldsetzer. KGS 2286

Plate 17. View looking south to Roche Miette, east of Jasper in Jasper National Park. The peak is composed of Palliser Formation carbonate which successively is underlain by a thin, dark band of Sassenach Formation, sandstone, moderately recessive shale, argillaceous carbonate and calcareous sandstone of the Mount Hawk Formation and recessive calcareous shale of the Perdrix Formation. Photo by R.B. Campbell. KGS 2273

Plate 18. View north to Cascade Mountain near Banff in Banff National Park. Recessive weathering rocks along the highway are fine grained clastics of the Fernie Formation and Kootenay Group. Carbonates of the cliff-forming Upper Devonian Palliser Formation form the lower part of the mountain. These are overlain by recessive limestone and shale of the Mississippian Banff Formation which, in turn are overlain by resistant carbonate of the Mississippian Livingston Formation of the Rundle Group. Photo by T.J. Lewis. KGS 2256

Plate 19. Massive limestone, as much as 2000 m thick, of the Upper Mississippian to Upper Permian Horsefeed Formation, Nakina Subterrane of the Cache Creek Terrane. Junction of Nakina and Silver Salmon rivers, Atlin area, northwestern British Columbia. Photo by J.W.H. Monger. KGS 2278

Plate 20. View northerly to upper part of the Telkwa Formation in the Hazelton Group. The Seven Sisters peaks in the background are underlain by a fault-bounded block of Middle and Upper Jurassic Bowser Lake Group. About 60 km northeast of Terrace, B.C. Photo by G.J. Woodsworth. KGS 2261

Plate 21. Bowser Lake Group, Bowser Basin, northern British Columbia. The gradation from dark-weathering marine siltstone and shale with sparse buff-orange-weathering fine-grained sandstone beds upwards into buff-brown-weathering siltstone, sandstone and conglomerate containing much plant material is common in the northern part of the basin. View looking west to Coast Mountains in the background. Photo by C.A. Evenchick. KGS 2277

Plate 22. Large scale, southward-verging folds in the Dezadeash Formation of the Gambier Assemblage (Gravina-Nutzotin). The folds are possibly the result of transpressional stress along the Denali Fault system. Kluane Ranges, Saint Elias Mountains, Yukon Territory. Photo by W. Dekur. KGS 2264

Plate 23. Quartz monzonite of the Salal Creek stock (buff weathering), dated at about 8 Ma, overlain by unconsolidated sediments and ash, in turn overlain by basalt flows. View looking northerly between Bridge and Lillooet rivers, southern Coast Mountains. Photo by G.J. Woodsworth. KGS 2265

Plate 24. Typical tuya, consisting of a cap of gently dipping basalt flows overlying aquagene breccias in the southern Coast Mountains about 90 km north of Vancouver. The edifice was emplaced into a lake thawed in a glacier by the volcano's heat. Photo by J.G. Souther. KGS 2303

Plate 25. View northerly to Mount Edziza in northwestern British Columbia. Basalt flows in the foreground are of Late Miocene age whereas the upper part of the edifice is as young as Pleistocene. Photo by J.G. Souther. KGS 2280

Plate 26. Feeder plug to Miocene-Pliocene (about 5.0 Ma) basaltic lavas in the northwestern part of the Bowser Basin. Well bedded rocks in the background are Upper Jurassic clastics of the Bowser Lake Group. Photo by J.G. Souther. KGS 2285

Plate 27. View westerly up Lowell Glacier in the Saint Elias Mountains, Yukon Territory. Mounts Hubbard, Kennedy and Alverstone all exceed 4300 m in elevation. Photo by W. Dekur. KGS 2290

Plate 28. View south along the Fraser River south of Big Bar showing postglacial dissection of alluvial fill. Photo by G.J. Woodsworth. KGS 2275

Plate 29. View south to a rusty-weathering Jurassic metasedimentary pendant in grey, slabby, jointed alkali feldspar syenite (Mount Monolith) of the mid-Cretaceous Tombstone Plutonic Suite. Locality is 58 km northeast of Dawson City, Yukon Territory. Photo by R.G. Anderson. KGS 2276

Plate 30. Granitoid gneiss which forms a zone up to 39 km wide in the axis of the Coast Plutonic Complex, may be the parent for many plutons in the Coast Mountains. This northwestward areal view shows a ridge of steeply dipping, banded granodiorite megagneiss, about 8 km northwest of Mount Waddington, southern Coast Belt. Photo by J.A. Roddick. KGS 2268

Plate 31. With a peak elevation in excess of 5951 m, Mount Logan is the highest mountain in Canada and the second highest in North America. Named after Sir William Logan, the first Director of the Geological Survey of Canada, Mount Logan is one of the largest mountain massifs in the world. The main mass is composed of grey Jurassic and Cretaceous granitic rock which intrudes Wrangellia, whereas the lower southwestern slopes are underlain by sedimentary strata of the Cretaceous Valdez Group of the Chugach Terrane. Separating these two rock types is the Border Ranges Fault, clearly shown as the sharp change from grey to dark brown colours. At the base of the mountain is the broad Seward Ice Field that lies between Mount Logan and Mount Saint Elias on the Yukon-Alaska border. Photo by C.J. Yorath. KGS 2262

Plate 32. Exploratory well, Shell Waterton #1, southern Foreland Belt, Alberta. The well, which produced gas from Cretaceous rocks in the footwall of the Lewis Thrust Fault, was spudded into Proterozoic strata of the Purcell Supergroup which are exposed in the mountains in the background. Photo by Shell Canada Resources, Limited.

Chapter 9

UPPER JURASSIC TO PALEOGENE ASSEMBLAGES

Summary
Introduction
The Foreland Belt
D.F. Stott, C.J. Yorath, and J. Dixon
    Upper Jurassic to Lower Cretaceous assemblages (Kootenay and Parsons)
    Mid-Cretaceous assemblages (Blairmore and Trevor)
    Upper Cretaceous to Paleogene assemblages (Brazeau, Smoky, Fish River, Reindeer)

The Omineca Belt
H. Gabrielse and R.G. Anderson
    Volcanic and plutonic rocks
    Sedimentary rocks

The Intermontane Belt
H. Gabrielse, J.G. Souther, G.J. Woodsworth, H.W. Tipper, and J.W.H. Monger
    Volcanic and plutonic rocks
    Sedimentary rocks
        Bowser Basin
        Nechako Basin
        Sustut Basin
        Tuya and Nahlin basins

The Coast Belt
G.J. Woodsworth and J.W.H. Monger
    Tyaughton-Methow Trough
    Plutonic rocks

The Insular Belt
C.J. Yorath, A. Sutherland Brown, R.B. Campbell, and C.J. Dodds
    Plutonic rocks
    Sedimentary rocks
    Saint Elias Mountains
        Upper Jurassic to Lower Cretaceous assemblages
        Upper Cretaceous to Paleogene assemblages
    Queen Charlotte Islands
        Upper Jurassic to Lower Cretaceous assemblages
        Mid-Cretaceous assemblage
    Vancouver Island

      Upper Jurassic to Lower Cretaceous assemblages
      Mid-Cretaceous assemblage
      Upper Cretaceous to Paleogene assemblage

Fault controlled basins
H. Gabrielse
    Tintina and Northern Rocky Mountain trenches
    Ross River area
    Liard Plain region
    Sifton Basin
    Bowron Basin
    Fraser River and North Thompson River areas
    Hat Creek
    Nicola-Princeton basins
    Tantalus basins

Conclusions
References

Chapter 9

UPPER JURASSIC TO PALEOGENE ASSEMBLAGES

C.J. Yorath

SUMMARY

The Upper Jurassic (Oxfordian) to Paleogene (Oligocene) assemblages record the effects of Mesozoic and early Cenozoic terrane amalgamations and collisions which, with consequent orogenesis and sedimentation, created the modern framework of the Canadian Cordillera. During this interval the five geological belts were established, more or less, but not entirely concordantly with major terrane boundaries. The sedimentary basins which developed as a result of and subsequent to terrane accretion reflect the full range of continental and oceanic plate interactions including orthogonal, oblique and transform motions. The assemblages are important hosts to several episodic suites of copper, copper-molybdenum and molybdenum porphyry deposits as well as all of the economic coal deposits of the Cordillera. Additionally, Cretaceous strata in the Foothills of the Foreland Belt serve as reservoirs for many hydrocarbon accumulations.

The tectonic setting within which the assemblages developed was dominated by the accretion of large, composite crustal fragments to the continental margin. The Intermontane Superterrane, comprising the amalgamated components of Stikinia, Quesnellia, the Slide Mountain and Cache Creek terranes probably collided with and subsequently was thrust over the pericratonic terranes at the western edge of North America in Early to Middle Jurassic time. Likewise, the Insular Superterrane, composed of Wrangellia and the Alexander Terrane, together with several smaller terranes in the Coast Belt, accreted to the Intermontane Superterrane no later than mid-Cretaceous time and possibly as early as latest Jurassic to earliest Cretaceous time. The great width of the Canadian Cordillera is a reflection of the dimensions of the successively accreted superterranes, the dynamics of accretion and subsequent intraplate contraction and dextral transcurrent displacement as well as extension in the southernmost Cordillera. Each of these processes resulted in fundamental changes to the organization and geological architecture of the western part of the continent.

The Upper Jurassic to Paleogene assemblages of the Foreland Belt (Kootenay, Blairmore and Brazeau assemblages) represent the principal responses to accretion and deformation in the western Cordillera. During this interval the eastern part of the miogeocline, for the first time received clastic detritus from a western source. The assemblages comprise synorogenic foreland basin-fill sediments, which, for the most part, were derived from shallow-water miogeoclinal strata in the Rocky Mountains and adjacent parts of the Omineca Belt. The older strata were deformed when the miogeocline was compressed, detached from its basement, thickened and displaced 100 to 200 km northeastward onto the craton. In front of the eastward advancing thrust and fold succession a linear moat developed in which a thick succession of clastic sediments was deposited in a wide variety of environments ranging from paralic settings, particularly along the western margin of the trough, to outer shelf and upper slope regimes in the trough axis. On the north side of Mackenzie arc, clastic sediments of Bonnet Plume, Eagle Plains and Keele-Kandik basins and Blow Trough were derived from cratonic sources until latest Aptian to Albian time (late Early Cretaceous); thereafter they were eroded from the developing northern Cordillera.

Temporal and spatial differences in phases of orogenic activity within the Cordillera as well as epeirogenic movements on the adjacent craton and global eustatic changes in sea level led to a wide range of facies variations within the foreland basin-fill during Late Jurassic and Cretaceous times. The extent and depth of epicontinental and pericratonic seas were variable; bordering paralic environments and local intrabasin source areas and paleotopographic culminations were transgressed and exposed in a complex manner. During part of Cretaceous time extensive Boreal and Gulfian marine embayments joined and formed an elongate seaway through the entire mid-continent region of North America. A wide variety and large number of stratigraphic units can be grouped into three main clastic wedges or tectonostratigraphic assemblages, each of which can be described in terms of several transgressive-regressive cycles. The Upper Jurassic to Lower Cretaceous (Kootenay and Parsons), mid-Cretaceous (Blairmore and Trevor) and Upper Cretaceous to Paleogene (Brazeau, Smoky, Moose Channel and Reindeer) assemblages each reflect phases of orogenic activity in the Cordillera and each are bounded above and below by regional unconformities.

The Omineca and Coast belts are regional tectonic welts within which were concentrated intense deformation, regional metamorphism, granitic magmatism, uplift and erosion as a consequence of tectonic overlap and/or compressional thickening of crustal rocks during and following collision of the two superterranes with the pre-Middle Jurassic and pre-mid-Cretaceous continental margins. The effect of accretion of the Insular Superterrane perhaps is further illustrated by the appearance of a western source for coarse clastic deposits in the Tyaughton-Methow

---

Yorath, C.J.
1991: Upper Jurassic to Paleogene assemblages, Chapter 9 in Geology of the Cordilleran Orogen in Canada, H. Gabrielse and C.J. Yorath (ed.); Geological Survey of Canada, Geology of Canada, no. 4, p. 329-371 (also Geological Society of America, The Geology of North America, v. G-2)

Trough, derived from the rising Coast and Cascade mountains between Aptian and Cenomanian time (early Late Cretaceous). Rocks of possibly latest Jurassic and Early Cretaceous age, however, form an overlap assemblage across several smaller terranes in the southern Coast Belt which may form parts of either superterrane. A pre-Early Cretaceous time of superterrane accretion could thus be inferred.

Internal effects of terrane accretion are illustrated in Stikinia. In Middle and Upper Jurassic strata of Bowser Basin the upward gradation of fine grained, marine to coarse nonmarine sediments reflects rapid filling by centripetal, high-gradient delta systems, deeply incised prodelta channel deposits and basinal turbidites in the lower part succeeded by deposition of a thick wedge of alluvial and paralic deposits which prograded towards the basin centre (Bowser Lake Assemblage). Paleocurrent trends and the composition of clasts in the northern part of the basin demonstrate a source in nearby Cache Creek rocks where uplift of chert, basalt and limestone occurred in response to contraction as a consequence of accretion of Stikinia with the Cache Creek Terrane. In contrast to the northern and eastern margins, the proximal facies in the southern part of Bowser Basin contain abundant volcanic and minor granitic material derived from the emergent Skeena Arch.

Although the relationship of the present geography of the Bowser Basin to the Skeena Arch and northern Cache Creek source areas is probably similar to paleogeography during deposition of the Middle Jurassic to ?Lower Cretaceous Bowser Lake Group, its relationship to the paleogeography of the eastern Intermontane Belt is more complex. There, dextral transcurrent displacements, perhaps in the order of more than 300 km disrupted the original distributions of the Cache Creek source area which may have been more or less continuous along the east side of the basin. There was no contribution from the Omineca or Coast belts during Bowser Lake deposition.

During Albian time, widespread marine transgression occurred throughout much of the Intermontane Belt south of Skeena Arch (Skeena and Gambier assemblages). This transgression linked deposition of the Taylor Creek and Jackass Mountain groups in the Tyaughton-Methow Trough and possibly the Gambier Group in the Coast Belt. The surface across which the transgression occurred is a widespread pediment throughout the Intermontane Belt which has a correlative counterpart in the Foreland Belt. These erosion surfaces reflect a lull in deformation which coincides with a dramatic decrease in magmatism throughout the Cordillera during much of the Neocomian (early Early Cretaceous). Although disrupted by transcurrent faults, the disposition of the Dezadeash, Gravina-Nutzotin, Gambier and Tyaughton-Methow successions suggests the possibility of a continuous basin connecting through the Bowser Basin, that extended the full length of the Cordillera from Late Jurassic to Albian (late Early Cretaceous) time.

In contrast to the effects of accretion seen in the Bowser Basin (Stikinia to the Cache Creek Terrane) the Cretaceous to Tertiary sediments of the Sustut Basin provide a record of uplift of the Omineca Belt during Albian to Campanian time (mainly Skeena Assemblage) and uplift in Campanian to Maastrichtian time of Bowser Lake Group strata to the west. Characteristic of both lower Sustut and Skeena Group sediments is ubiquitous muscovite derived from metamorphic source terranes in the Omineca Belt.

In the Insular Belt the Pacific Rim Terrane, comprising Lower Cretaceous mélanges enclosing exotic blocks of Upper Triassic volcanics, together with the Leech River metavolcanic and metasedimentary rocks (Pacific Rim Assemblage), was emplaced against and beneath the western and southern margins of Wrangellia in latest Cretaceous or Early Tertiary time. In the Queen Charlotte Islands the mid-Cretaceous Queen Charlotte Group (Honna Assemblage) developed as an epicontinental and clastic wedge sequence on the Alexander Terrane and Wrangellia. The Georgia Basin, enclosing Upper Cretaceous and Paleogene marine and nonmarine sediments (Nanaimo Assemblage), has been variously described as a forearc basin or as a pull-apart basin within a broad zone of normal and transcurrent faulting in response to oblique convergence between the Kula and North American plates. However, some characteristics, such as easterly to westerly changes in dominant source areas and syndepositional westward verging folds and thrust faults suggest a foreland basin origin for the Georgia Basin, perhaps related to the developing Cascade and Coast mountains. Likewise, the Dezadeash Formation (Gambier Assemblage) of Late Jurassic to Early Cretaceous age represents a foredeep flysch accumulation adjacent to the Saint Elias Mountains. On southernmost Vancouver Island the Eocene Crescent Terrane, comprising an ophiolitic sequence of ridge-centred oceanic island origin (Metchosin Assemblage) was emplaced beneath the Leech River Formation during Late Eocene time. In Tofino Basin off the west coast of Vancouver Island, chemically similar rocks occur beneath the Paleogene Carmanah Group and overlie a subduction complex which has developed above the modern descending Juan de Fuca Plate.

Many elongate, nonmarine, clastic sequences of Late Cretaceous to Paleogene age occur in and along major fault-controlled lineaments in the Intermontane and Omineca belts (Sifton Assemblage). Several of these faults are the loci of large dextral transcurrent displacements that markedly disrupt regional paleogeographic elements. Characteristically the rocks are highly deformed although significant sections of homoclinal, uniformly dipping beds are present locally. Some of these sequences are thick and clearly reflect syndepositional faulting. Most have minor to significant coal accumulations. Typical nonmarine environments deduced from the character of the sediments include lake, pond, delta, alluvial fan, colluvium and various stream settings evincing high- to low-energy regimes. In addition, the tensional stress regimes generated by dextral transcurrent faulting controlled the emplacement of numerous alkalic volcanic complexes and related intrusions throughout the Cordillera.

Widespread plutonic and volcanic activity occurred in distinct pulses during the Late Jurassic to Paleogene interval. The processes are variously ascribed to terrane collision in the Coast Belt, intraplate thickening in the Omineca Belt, subduction and/or tensional environments related to dextral strain in the Intermontane and Omineca belts and to changes in relative Pacific-America plate motions giving rise to Eocene plutons and related volcanics in the southern Insular Belt. Mid-Cretaceous granitic plutons are rarely associated with volcanic rocks except in the

Yukon Territory. However, classic examples of spatially and temporally related plutons and volcanics are common in Upper Cretaceous to Paleogene rocks. Metamorphic and plutonic assemblages suggest enormous uplifts in the core zones of the Coast and Omineca belts following burial to depths of 20 km or more. In the Coast Belt, uplift occurred mainly after mid-Cretaceous time whereas in the Omineca Belt, possibly half of the uplift occurred between Middle Jurassic and mid-Cretaceous time.

The aforementioned cause-and-effect hypothesis pertaining to the pre-Late Jurassic accretion of the Intermontane Superterrane to the continental margin is further complicated by reassessment of, and new paleomagnetic data obtained from volcanic rocks from Wrangellia, plutonic rocks in the Coast Belt and the Cache Creek Terrane and inferred from Cretaceous magnetic overprints in plutonic rocks in Quesnellia (see Chapter 3). Mid-Cretaceous paleopoles determined from these regions suggest that the amalgamated Insular and Intermontane superterranes of central and southern British Columbia ("Baja British Columbia") were 2400 km south of their present position at about 100 Ma, and that this "megaterrane" moved northward with the Kula plate at about 8 cm/a and arrived at its present position between 80 and 50 Ma. On the other hand, evidence from faunal distributions suggest considerably smaller displacements in post-Bajocian time. The implications of these results are substantial to the understanding of Cordilleran evolution, particularly with regard to the origin of the Foreland Belt, the degree of displacement on major known and presently unknown transcurrent faults, and the fate or location of that portion of the Cordillera that must have been displaced northward prior to Eocene time. Clearly, lateral stratigraphic correlation from one belt to another must be reviewed with caution.

INTRODUCTION

The Upper Jurassic to Paleogene assemblages of the Canadian Cordillera record the effects of successive accretion of the Intermontane and Insular superterranes with the western margin of ancestral North America. Moreover, it was during this interval when the five morphogeological belts and the modern plate tectonic regime were established. In this chapter the assemblages are treated with reference to the five belts rather than with the several terranes.

The accumulation of Upper Jurassic to lower Tertiary sedimentary and volcanic rocks (Fig. 9.1) can be described in terms of three groups of assemblages (Fig. 9.2-9.4), each of which developed in response to specific tectonic events summarized in Figure 9.5. The several sedimentary basins which developed as a consequence of, or successor to these events are illustrated in Figure 9.6.

The accumulation of thick clastic wedges in the Insular and Foreland belts began as a consequence of Middle to Late Jurassic tectonism in adjacent belts. In the Intermontane and northern Foreland belts, however, the sedimentary accumulations in the Bowser Basin and Rapid Depression began in Middle and Early Jurassic times respectively and are included in this chapter to provide a complete perspective of their depositional history. Similarly,

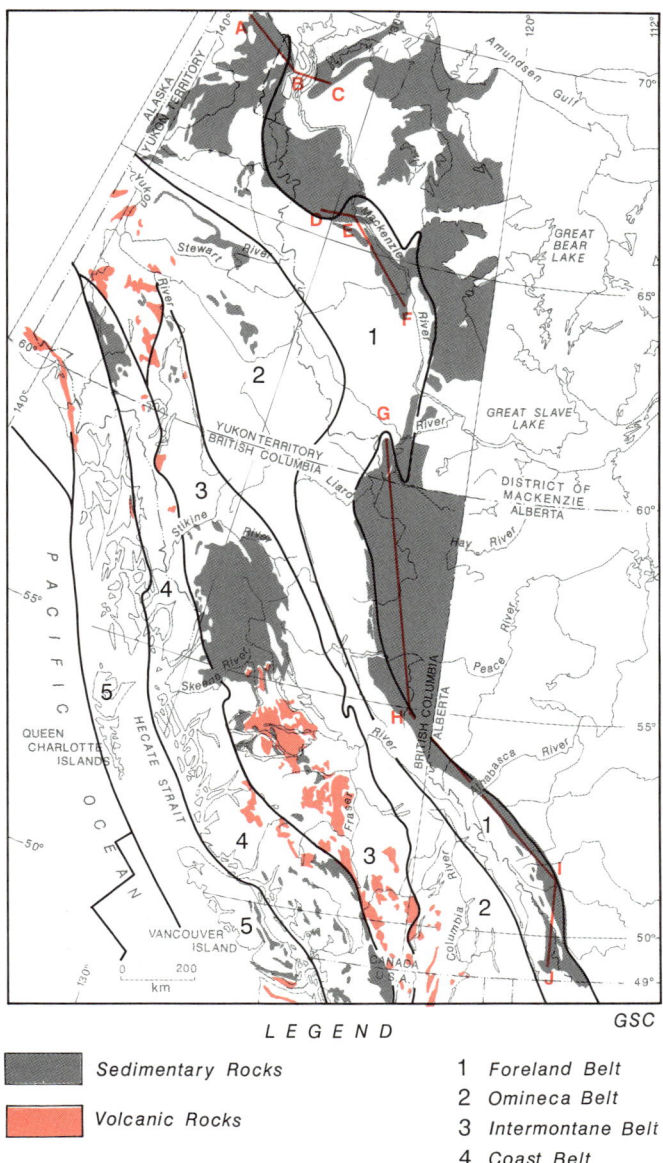

Figure 9.1. Distribution of Middle Jurassic to Paleogene sedimentary and volcanic rocks. Cross-sections A-C, D-F, and G-J are illustrated in Figures 9.7, 9.8, and 9.11.

in the Tyaughton-Methow Trough of the southern Coast Belt laterally intertonguing volcanic, volcaniclastic and terrigenous clastic sediments attest to a complex tectonic history that began in late Early Jurassic time and continued into Late Cretaceous time; again, the total fill of the trough is described in this chapter.

During the Late Jurassic to Paleogene interval the Canadian Cordillera achieved its present form. Most researchers agree that the Intermontane and Insular superterranes were amalgamated by mid-Cretaceous time; however, from there opinions differ widely on how and when the Cordillera was constructed. Paleomagnetic data from a few localities suggest that the Insular and Intermontane superterranes formed a composite "megaterrane" (Baja B.C. - see Chapter 3) which, during mid-Cretaceous time was located far to the south and which did not dock with North America until the Paleogene. On the other hand, a wide variety of geological and paleontological data support the hypothesis that the Intermontane Superterrane collided with ancestral North America during the Middle Jurassic, resulting in deformation, plutonism, and metamorphism in the Omineca Belt and the consequent accumulation of a foredeep clastic wedge in the Foreland Belt. The subsequent collision of

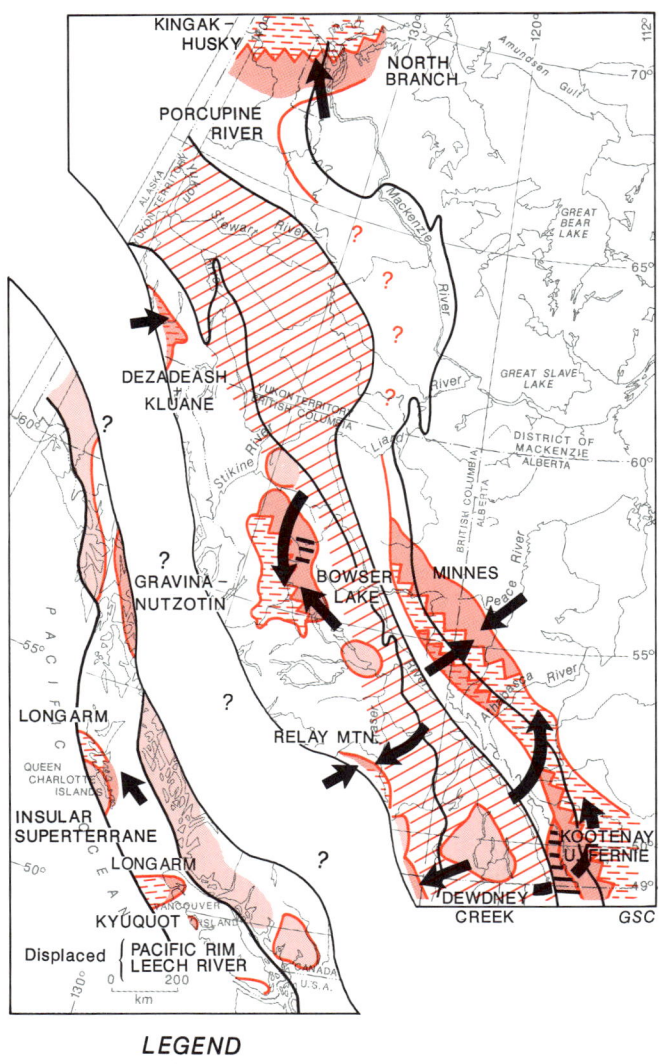

Figure 9.2. Upper Jurassic (Oxfordian) to Lower Cretaceous (approx. Valanginian) lithofacies. The position of the Insular Superterrane with respect to the remainder of the Cordillera is unknown.

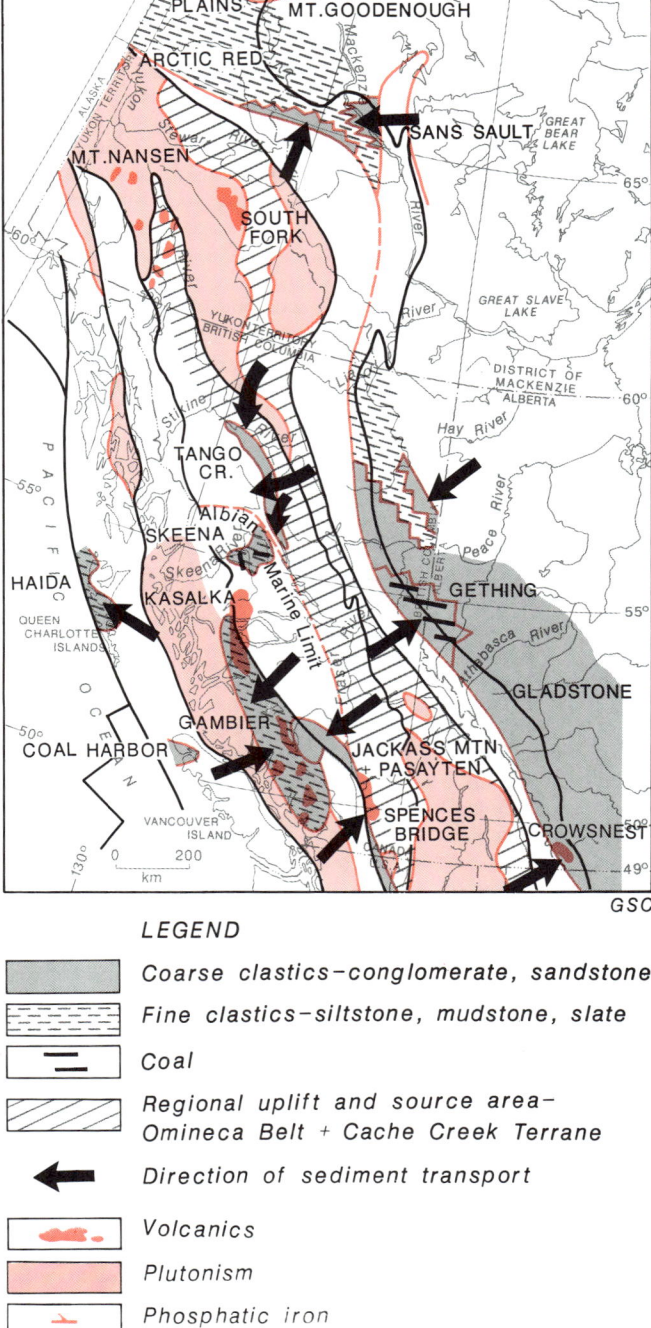

Figure 9.3. Mid-Cretaceous (approx. Hauterivian to Turonian) lithofacies.

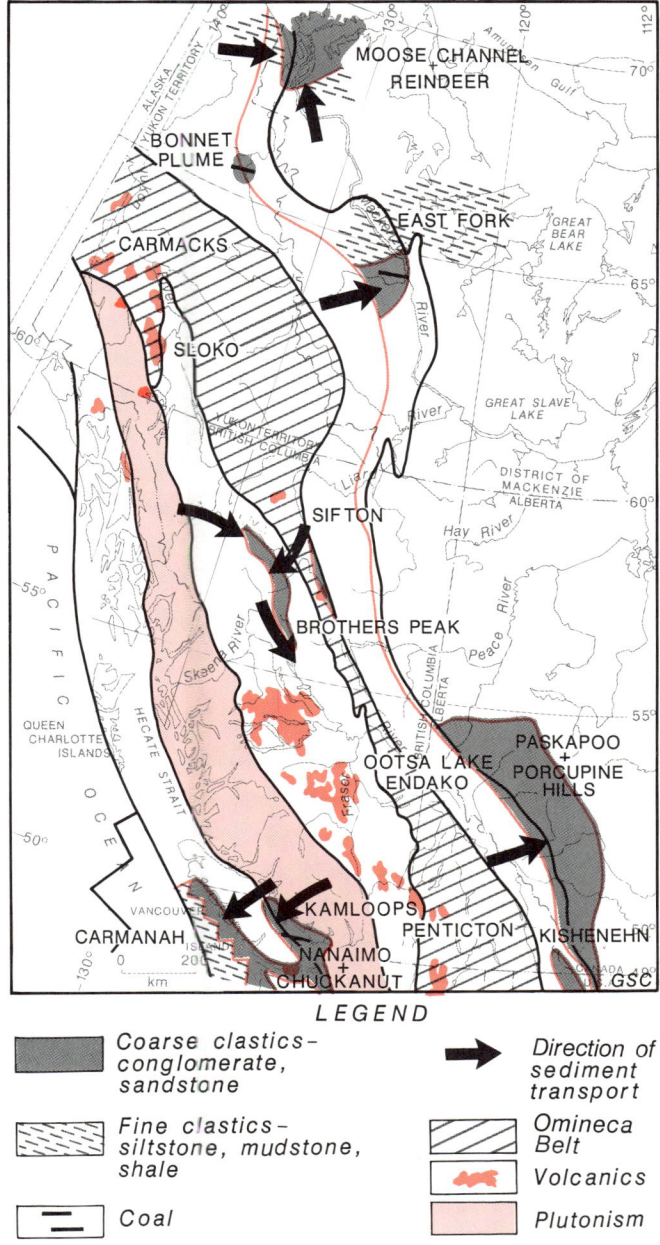

Figure 9.4. Upper Cretaceous (approx. Campanian) to Paleogene (Oligocene) lithofacies.

the Insular Superterrane has been suggested to have resulted in widespread plutonism and uplift in the Coast Belt and the development of a second foredeep clastic wedge in mid-Cretaceous time (Fig. 9.5). Overlap relationships in the southern Coast Belt, however, indicate a pre-Early Cretaceous time of superterrane amalgamation. Clearly these lines of evidence are incompatible; much work yet needs to be done to resolve the manner and time of assembly of the western Cordillera.

THE FORELAND BELT

D.F. Stott, C.J. Yorath and J. Dixon

The Upper Jurassic to Paleogene foredeep deposits in the Foreland Belt contain an excellent record of mountain building in the eastern Canadian Cordillera. Because of their importance for hydrocarbon and coal resources the assemblages in the southern part of the belt have been studied extensively both in surface exposures and in the subsurface. Although some pulses of clastic wedge deposition in the Foreland Belt can be roughly correlated with those in the Intermontane Belt, suggesting drainage both to the east and west from the uplifted Omineca Belt, their paleogeographic relationships are not yet fully understood.

Sedimentation in the southern part of the Foreland Belt indicates uplift in the Cordillera beginning in Late Jurassic time. In the northern part of the Foreland Belt, however, sedimentation was not influenced by Cordilleran uplift until Late Aptian to Early Albian time. Prior to the Late Aptian, extensional tectonics related to rifting in the area that was to become the Arctic Ocean, was dominant. Sediments were derived principally from the craton and rift margins, not from a compressional orogen.

The following account is brief; for a more extensive treatment the reader is referred to Stott and Aitken (in press).

Upper Jurassic to Lower Cretaceous assemblages (Kootenay and Parsons)

Northern Foreland Belt

Upper Oxfordian to Berriasian rocks of the Parsons Assemblage are well exposed in the northern part of the Richardson Mountains, in the Ogilvie Mountains, and in northernmost Yukon Territory where they unconformably overlie sandstone and siltstone of the Sinemurian to Lower Oxfordian Bug Creek Group. Progradational, coarsening upward, shallow marine sandstone of the North Branch, Husky, Martin Creek, and Porcupine River formations (Jeletzky, 1967, 1977) pass and thicken basinward to the north, west and southwest into open marine concretionary shale of the Husky Formation and upper part of the Kingak Formation (Fig. 9.2, 9.7). On the eastern slopes of Richardson Mountains the Martin Creek consists of hummocky cross-stratified sandstone units interpreted as representing a shallow neritic environment; to the west these sediments grade into bioturbated fine grained sandstone of a mid- to outer-shelf setting.

Valanginian to Middle(?) Hauterivian strata include the McGuire Formation (100 m) and Kamik Formation (800 m) which have wide distribution throughout Porcupine Plateau and Mackenzie Delta. The McGuire Formation consists dominantly of bioturbated shale with subordinate interbedded siltstone and fine grained sandstone, reflecting outer to mid-shelf deposits which grade upward to inner neritic sand and mud. The lower third of the Kamik Formation in the Mackenzie Delta area consists of thick, nonmarine sandstone beds identified as alluvial deposits (Myhr and Young, 1975; Dixon, 1982a,b). On the western slope of the northern Richardson Mountains equivalent strata are marine and contain sedimentary structures suggesting strandline to nearshore environments. To the

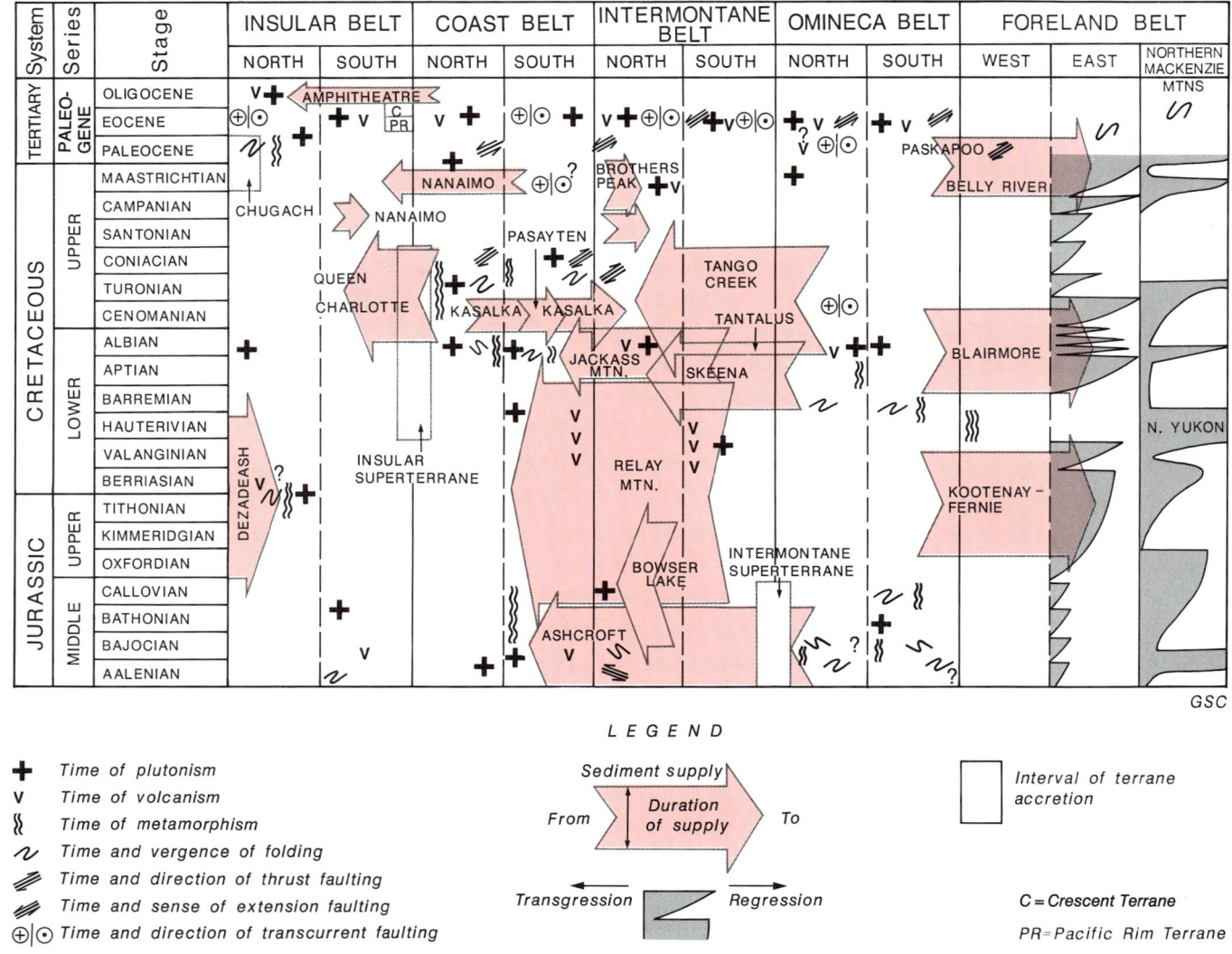

Figure 9.5. Summary diagram of principal Middle Jurassic to Paleogene tectonic events in the Canadian Cordillera.

west very fine- to fine-grained, laminated, hummocky cross-stratified and bioturbated sandstone is interpreted as representing inner to mid-shelf deposits. The overall setting during deposition of the lower Kamik was that of a broad shelf over much of northern Yukon Territory with a delta-complex situated in the Mackenzie Delta area. The upper two-thirds of the Kamik Formation is thought to be predominantly marine and consists of several coarsening upward cycles. An east-to-west change from littoral to shelf sediments has been noted in the upper part of the Kamik Formation (Dixon, 1986a).

Regional facies and isopach trends indicate that the primary source region for McGuire and Kamik strata lay to the south and southeast on the craton and that the depositional edge was probably not far removed from the preserved edge. Major thickening occurs immediately basinward of Eskimo Lakes Arch and Cache Creek Uplift (Fig. 9.2, 9.6).

Within the Foreland Belt between northern Yukon and northeastern British Columbia, no Upper Jurassic to Barremian rocks are known, although Jurassic palynomorphs have been recovered from Aptian and younger formations in the outer Mackenzie arc (Fig. 9.2, 9.8; Yorath and Cook, 1981).

Uplift in latest Middle to earliest Late Hauterivian time led to a regionally extensive unconformity that occurs at the base of the Mount Goodenough Formation (Fig. 9.7). This formation comprises about 530 m of dark grey to black concretionary shale, siltstone and sandstone of Late Hauterivian to Barremian and possibly Aptian age (Jeletzky, 1958, 1960, 1961). The overlying Rat River Formation of probable Late Barremian and Aptian age consists of about 200 m of neritic, bioturbated and crossbedded, fine- to medium-grained sandstone which becomes finer grained and increasingly argillaceous towards the northwest (Fig. 9.3).

Southern Foreland Belt

In the Rocky Mountain Foothills, extending from Peace River to southwestern Alberta, and in the Fernie Basin of southeastern British Columbia, Upper Jurassic to Valanginian sediments of the Kootenay Assemblage reflect the earliest phase of uplift and deformation in the Cordillera (Fig. 9.2). The upper part of the Fernie Formation is separated from its lower part by a poorly dated unconformity, which may reflect the change of the dominant source area from the eastern craton to the emerging Cordillera. Stott (1984) has suggested that the significant quantities of Oxfordian siltstone and sandstone in westernmost exposures of the upper part of the Fernie Formation may be derived from western sources. The upper Fernie Formation, approximately 300 m thick, comprises green, glauconitic sandstone (Green Beds) overlain by dark grey, concretionary shale which in turn passes upward into concretionary shale with interbedded mudstone and siltstone (Passage Beds; Fig. 9.9) displaying abundant ripple marks, bioturbation structures, and plant debris including large logs. In northeastern British Columbia the Fernie is transitionally overlain by the uppermost Jurassic to lowermost Cretaceous Minnes Group (Fig. 9.10). This succession grades southeastward into the Nikanassin Formation of the central Alberta Foothills and thence into the Kootenay Group of the southern Foothills and Rocky Mountains (Fig. 9.11).

The Minnes Group of Tithonian to Late Valanginian age (Ziegler and Pocock, 1960; Stott, 1967a, 1975) is most complete in the Foothills of northern Alberta and northeastern British Columbia where it is more than 2 km thick. Its four formations are predominantly marine sandstone and interbedded mudstone north of Pine River but southward, the upper three formations grade laterally into nonmarine, coarse channel sandstone, conglomerate and coal-bearing beds of the Gorman Creek Formation (1030 m). In general Minnes Group lithologies reflect fluvial-deltaic conditions on the trough margins to deep water environments represented by turbidites in the basin centre (Fig. 9.2). The presence of green, radiolarian chert clasts in the Monteith Formation (M.E. McMechan, pers. comm., 1985) points to Slide Mountain or Cache Creek source terranes in the Omineca Belt. This is significant as evidence for the emplacement of oceanic terranes in the Omineca Belt by latest Jurassic or earliest Cretaceous time. Throughout the region the sequence is bevelled by the pre-Bullhead (Hauterivian-Barremian) unconformity and is overlain by the Cadomin conglomerate of the Lower Cretaceous Bullhead and Blairmore groups (Fig. 9.11).

In the Athabasca River region, the Nikanassin Formation is poorly dated, fossils of Kimmeridgian to Tithonian/Volgian age (latest Jurassic) having been reported only from the basal beds (Mountjoy, 1962). The formation is 300 to 600 m thick in western exposures and is truncated eastward by the aforementioned unconformity. The Nikanassin consists of alternating fine grained sandstone and dark mudstone which, in its lower part, probably represents a shaly offshore facies; thin carbonaceous beds occur in the upper part, perhaps suggesting an estuarine setting.

The Kootenay Group, also of Late Jurassic to Early Cretaceous age, is thickest in the Fernie Basin (1.1 km) and comprises inversely graded basal sandstone overlain by interbedded sandstone, mudstone, chert-pebble conglomerate and coal. The group is bevelled by the sub-Cadomin unconformity. Paleocurrent data and facies analyses of uppermost Fernie and lower Kootenay strata in the southern Rocky Mountain Trough support a northward prograding, stream-fed beach complex and

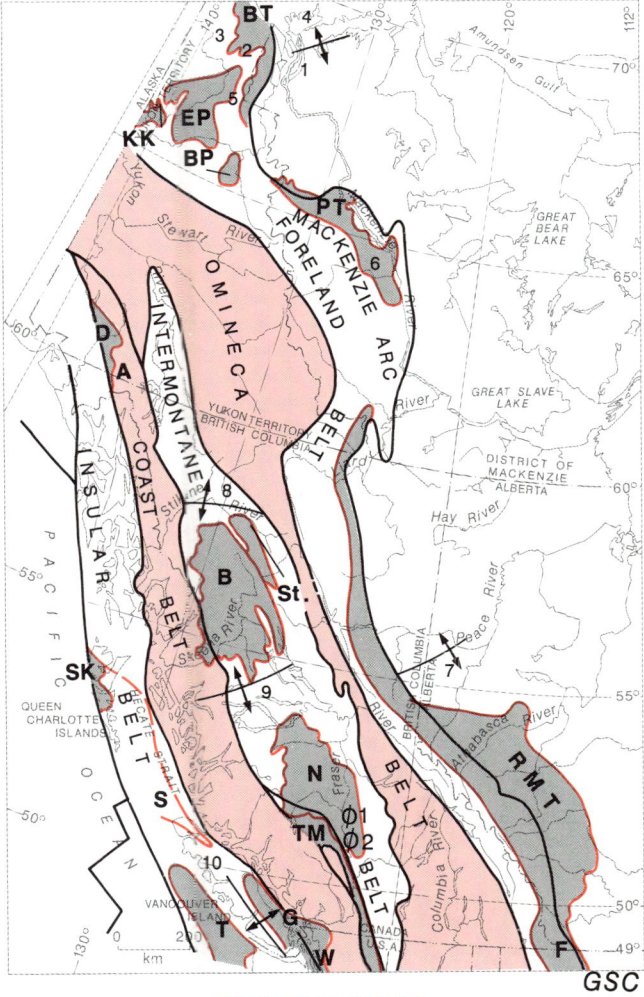

Figure 9.6. Morphogeological belts and principal Middle Jurassic to Paleogene sedimentary basins and arches.

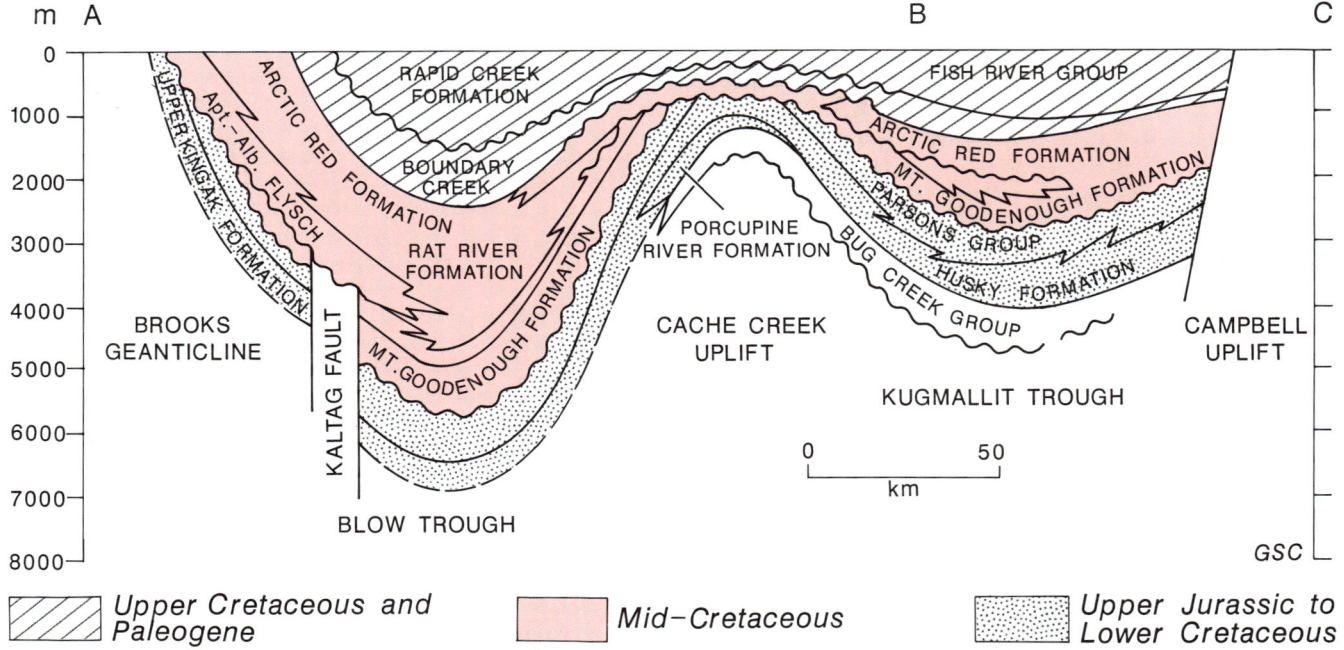

Figure 9.7. Schematic diagram illustrating the three clastic assemblages in Upper Jurassic to Paleogene strata of Blow and Kugmallit troughs (after Balkwill et al., 1983). For section location see Figure 9.1.

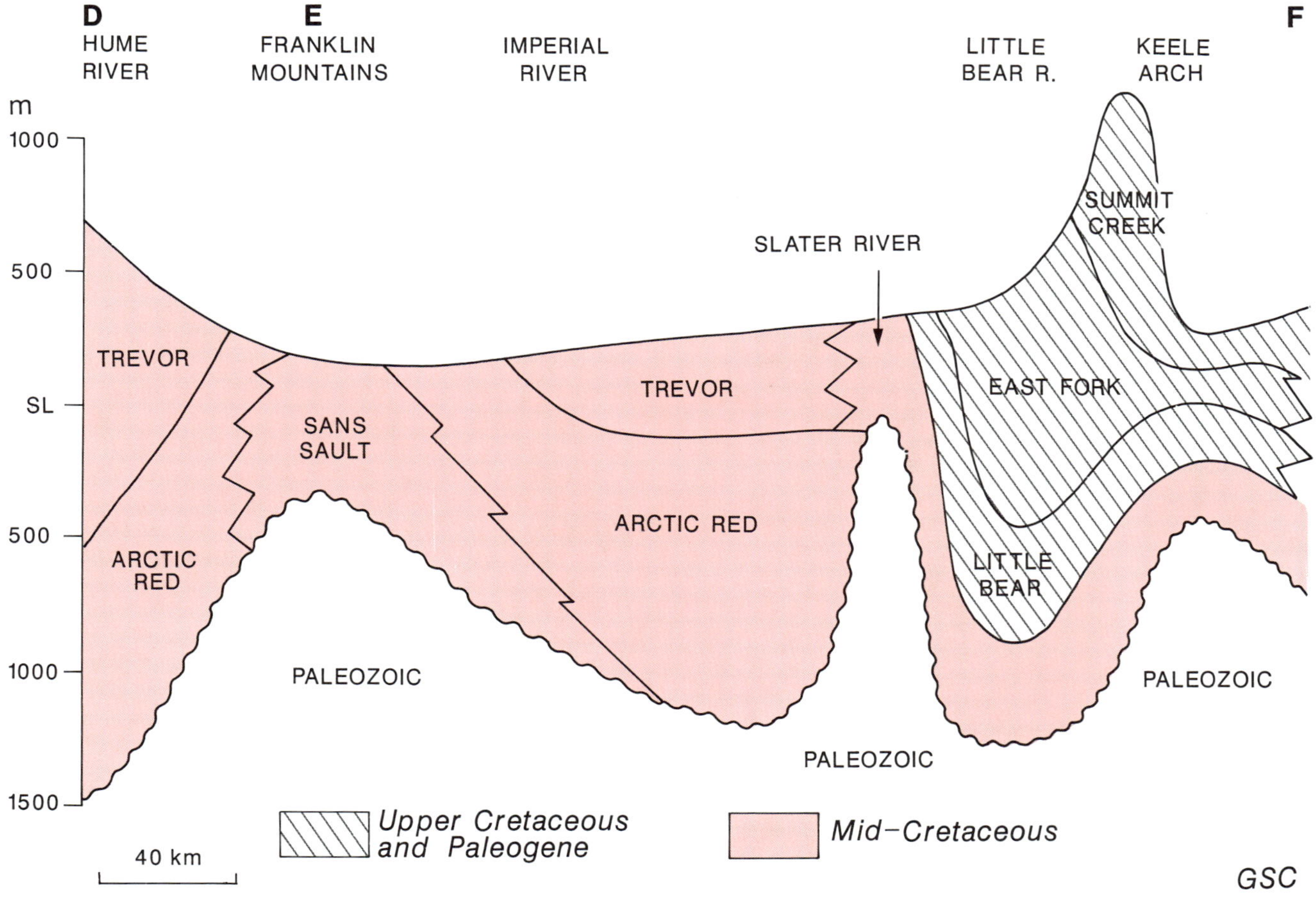

Figure 9.8. Schematic diagram illustrating the two clastic wedges in Cretaceous and Tertiary strata of the axial part of Peel Trough (modified from Yorath and Cook, 1981). For section location see Figure 9.1.

Figure 9.9. Passage Beds of the Fernie Formation, Banff, Alberta. These sediments represent the earliest phase of westerly derived orogenic clastics in the Foreland Belt. Photo by W.R. Price. GSC 205235-O

Figure 9.10. Kootenay Assemblage in the foothills of northeastern British Columbia. Light coloured exposures at lower left are Triassic. The overlying dark recessive shales of the Fernie Formation grade transitionally into sandstones of the Minnes Group forming the central peak. Photo by D.F. Stott. GSC 205235-DD

storm-dominated shallow-marine setting within which sediment transport was axial and towards the northwest (Fig. 9.2; Hamblin and Walker, 1979). A similar dispersal pattern of the strata in the Kootenay Group is shown by Eisbacher et al. (1974).

Mid-Cretaceous assemblages (Blairmore and Trevor)

Northern Foreland Belt

In response to Aptian to Albian tectonic activity, southerly and westerly source areas in the northern Cordillera began to shed sediment into several foreland troughs and basins that developed in front of the rising orogen. In the northern part of Yukon Territory and extending southwestward into Alaska, Blow Trough and Keele-Kandik Basin developed on the east and south margin of the Brooks Range Geanticline (Fig. 9.3, 9.6; Young et al., 1976). In Peel Trough, the foredeep that developed in front of the Mackenzie Mountains, latest Aptian to earliest Albian time is represented by the Glauconite Member of the Martin House Formation, the basal glauconitic sandstone of the Arctic Red Formation, and the lower sandstone unit of the Sans Sault Formation (Fig. 9.3, 9.8).

Steeply dipping and folded unnamed nonmarine Cretaceous strata are preserved as an isolated fault-bounded panel in western Mackenzie Mountains (Blusson, 1971). Their thickness, about 1315 m, is a minimum as the lowest beds observed are in fault contact with Devonian rocks; moreover, these fossiliferous strata are the youngest exposed. They comprise a succession of sandstone, conglomerate and shale arranged in upward-fining cycles from 2 to 15 m thick. Most of the deposits represent meandering stream settings. Coaly shale and/or coal commonly caps each cycle. Scant paleocurrent data suggest southeasterly paleoflow, parallel with the long axis of the present outcrop trend. At the base of the sequence chert is the predominant clast type in sandstone and conglomerate, but stratigraphically upward becomes secondary in abundance to quartz sandstone. Source beds were probably the Mississippian Mattson and Permian Fantasque formations. The age of these deposits is not older than Barremian and not younger than Albian (A.R. Sweet, pers. comm., 1986) which, together with the Late Aptian age of the earliest foredeep deposits suggests that the source area was uplifted about Aptian time.

Thicknesses of Albian strata in the Northwest Territories and Yukon Territory are variable. In Blow Trough as much as 4 km of Lower Albian strata have been recorded (Young, 1973), yet directly to the east, on the flanks of Cache Creek Uplift (Fig. 9.6) they are about 400 m thick (Young, 1972). Southward in Peel Trough, Albian strata are up to 1600 m thick. In Eagle Plain, Albian shale is up to 1500 m thick and in Kandik Basin, the Kathul Greywacke is at least 450 m thick (Brabb, 1969) in Alaska and at least 900 m thick in Yukon Territory.

In Blow Trough the Albian flyschoid sequence includes a lower shale unit, a conglomerate and sandstone unit, an upper shale unit and a turbiditic sandstone and shale unit. On the western side of Blow Trough and Keele-Kandik Basin, conglomeratic and sandy gravity-flow deposits, 2100 m thick, (Young, 1972, 1973; Young et al., 1976) appear to have been deposited as submarine fans. Lateral equivalents to the north are finer grained and have Bouma sequences, typical of outer fan or basin-plain environments. The Lower Albian coarse grained basin deposits become finer grained upwards, indicating either curtailment of sediment supply or a change in the locus of fan accumulation. The location of the Lower Albian Sharp Mountain Formation sandstones between the deep water sediments of Blow Trough to the north and the Keele-Kandik Basin to the southwest suggest that they may be a continuation of those deep water sediments as suggested by Dixon (1986b). If they are shallow marine as suggested by Jeletzky (1975) then they probably represent shelf deposits adjacent to a western land area.

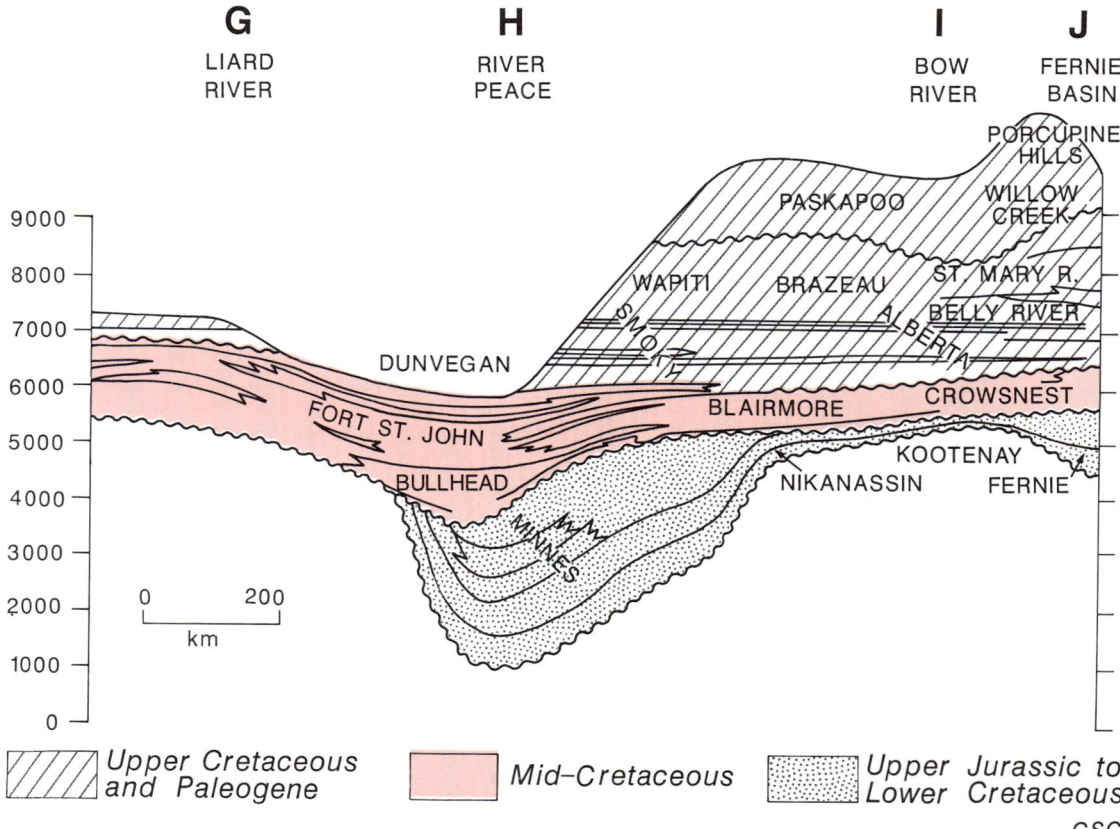

Figure 9.11. Schematic diagram illustrating the three clastic wedges in Jurassic, Cretaceous and Tertiary strata in the Rocky Mountains and Foothills between Fernie Basin and Liard River (from Stott, 1984). For section location see Figure 9.1.

To the east, the younger and upper part of the flyschoid sequence thins and grades eastward towards Cache Creek Uplift into the Rapid Creek Formation, a sedimentary phosphatic iron deposit, 60 m thick (Fig. 9.3; Young and Robertson, 1984). Phosphate grains within the deposit are composed of rare minerals such as satterlyite, arrojadite and ormanite which reflect an original calcium-deficient composition. The coexistence of iron and magnesium phosphates as well as apatite is unusual. This most northerly known phosphorite probably resulted from cold, northeast-flowing currents upwelling on the western flank of Cache Creek Uplift.

During Early Albian time much of the northern platform area was the site of mud and silt accumulation, represented by the Arctic Red and Whitestone River formations (Fig. 9.3). The rapidly deepening Peel Trough was filled with a basal glauconitic sandstone and overlying thick concretionary, fossiliferous and silty mudstone of the Arctic Red Formation (Fig. 9.12), which disconformably overlie strata of the Upper Devonian Imperial Formation between the Mackenzie Mountains front and Mackenzie River (Fig. 9.8). The Sans Sault Formation, composed of sandstone and mudstone, developed on the western flank of Keele Arch (Fig. 9.6) and extends westerly into Peel Trough. In southern Peel Trough, bentonitic shales of the lowermost Slater River Formation form the basal unit of the Cretaceous succession (Yorath and Cook, 1981, 1984).

By late Middle Albian time Peel Trough was receiving coarse clastic marine sediments from the south and southwest, represented by the Trevor Formation (Yorath and Cook, 1981). The coarse clastics of the Trevor Formation and unnamed conglomerates on Keele Arch grade laterally southward along the Mackenzie Valley into shelf mudstone and siltstone of the upper Slater River Formation (Fig. 9.8).

Orogenic activity in the Cordillera brought to a close the extensive flooding of earlier Albian time. Throughout much of the Northwest Territories, a regional unconformity separates Albian from Upper Cretaceous strata, except in Peel Trough where this hiatus has not been recognized. In Bonnet Plume Basin, Middle or Upper Albian sandstone, conglomerate and shale of the lower part of the Bonnet Plume Formation form part of a thick sequence of sediments overlying Paleozoic miogeoclinal rocks of the Richardson Anticlinorium (Norris and Hopkins, 1977). These are unconformably overlain by upper Bonnet Plume Formation strata of Maastrichtian to Paleocene age.

Southern Foreland Belt

As uplift continued in the Cordillera, coarse clastic material accumulated in the rapidly subsiding Rocky Mountain Trough; subsidence allowed the gradual incursion of marine water from boreal regions. The clastic wedge, developed

Figure 9.12. Trevor Formation (luKt) overlying Arctic Red Formation (lKa) in Peel Trough, Imperial River, Northwest Territories. Photo by C.J. Yorath. GSC 205235-N

Figure 9.13. Massive conglomerate of the Cadomin Formation (basal) Blairmore Assemblage, Monkman Pass area, northeast British Columbia. Photo by D.F. Stott. GSC 205235-EE

between Barremian to Early Cenomanian time, comprises the Blairmore, Luscar, Bullhead and Fort St. John groups and Dunvegan Formation (Fig. 9.3, 9.11). It is 450 m to 2.1 km thick in westernmost exposures but thins eastward to less than 300 m beneath the plains. In northeastern British Columbia, the wedge is bounded at its base by the pre-Bullhead unconformity and at its top by a post-Dunvegan unconformity. In southwestern Alberta the wedge is bounded by pre- and post-Blairmore unconformities. The paleogeographic characteristics of the Blairmore-Fort St. John succession indicate extensive flooding of the Interior Plains region and lateral changes in shoreline positions.

The Aptian-earliest Albian embayment recorded in the Gething Formation of the Bullhead Group, did not extend much farther south than Peace River (Fig. 9.3). A large delta prograded eastward into the seaway from its southwestern shore. Other deltas were present on the eastern side, flanked by ridges of Paleozoic rocks (Rudkin, 1964). The seaway was bordered to the south by an extensive, poorly drained alluvial plain.

In northeastern British Columbia, the basal succession, presumably of Barremian to earliest Albian age, is included in the Bullhead Group (Stott, 1968, 1973, 1982). The sequence is about 1 km thick in the western Foothills at Peace River and its base is marked by the distinctive Cadomin Formation (Fig. 9.13), predominantly a chert-pebble conglomerate, which, in northeastern British Columbia, was deposited in a piedmont-alluvial plain setting. Like the underlying sequence, the Cadomin was deposited from a northeasterly thence northwesterly directed paleodrainage regime (McLean, 1977). The overlying coal-bearing facies in the Gething Formation, which forms extensive deltas between Athabasca and Peace rivers, grades northward into fine grained sandstone, in turn passing into marine mudstone and siltstone (Fig. 9.14). These beds represent an Aptian to possibly earliest Albian transgression expressing the initial incursion of a boreal sea following a period of erosion. Significantly, the Gething sandstone contains the first suite of metamorphic minerals, including garnet, staurolite and sillimanite, in the Foreland Belt; these were derived from uplift in the Omineca Belt to the west (Fig. 9.3; Stott, 1968) which, at

Figure 9.14. Interbedded sandstone, siltstone and mudstone at the type section of the Gething Formation (Trevor Assemblage), Peace River Canyon, British Columbia. Photo by D.F. Stott. GSC 205235-CC

the same time, was the source also for abundant mica in the Skeena Formation in the Intermontane Belt.

In the central Foothills, the Gething coal-bearing facies gives way to alluvial sand with point-bar and flood-plain deposits included in the lower part of the Gladstone Formation (Fig. 9.3; Mellon, 1967; McLean, 1982). The upper part of the Gladstone Formation in the southern Foothills includes calcareous shale and limestone (Loranger, 1951; Glaister, 1959), which contain an abundant fresh-water fauna replaced to the north by a brackish-water fauna (Mellon and Wall, 1963; Mellon, 1967; McLean and Wall, 1981).

During the Early Albian a new phase of Cordilleran orogenesis was initiated and the foredeep, including extensive coal deposits, migrated eastward (Fig. 9.15). The boreal embayment advanced southward and eventually joined, in latest Middle Albian time, with the embayment extending northward from the Gulf of Mexico (Stelck et al.,

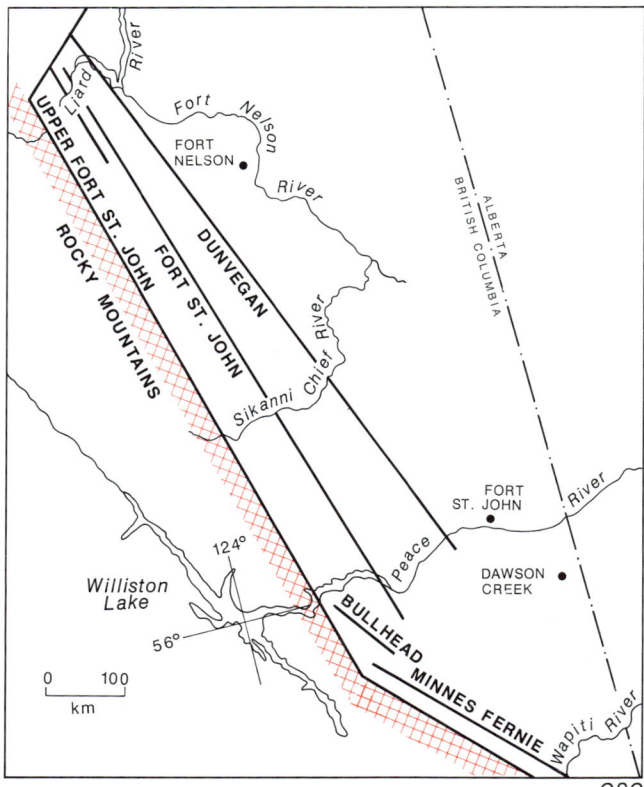

Figure 9.15. Positions of foredeep axes during deposition of Upper Jurassic to Lower Cretaceous and mid-Cretaceous assemblages (from Stott, 1984).

1956; Stelck, 1958; Jeletzky, 1971a,b). Variations in the rate of subsidence, deposition and uplift of the source region, and global eustatic changes in sea level resulted in major transgressions and regressions.

The Albian succession in northeastern British Columbia, represented by the Fort St. John Group, comprises four transgressive-regressive cycles, each of which is expressed by marine shale and overlying paralic coarse clastic couplets including extensive coal deposits (Stott, 1968, 1982). The pattern of deposition was influenced by renewed activity along the ancient Peace River Arch where deltaic and shoreline sandstone and associated facies of the first two cycles show easterly trends. In the central and southern Foothills, equivalent strata are represented by the Luscar Group (Langenberg and McMechan, 1984) and the Blairmore Group. The former comprises fining upward sequences of sandstone shale and coal which reflect neritic, estuarine, fluvial and backswamp settings. Blairmore strata consist of varicoloured shales and green feldspathic sandstones that contain abundant volcanic detritus and chloritic cement. Several members of conglomerate containing igneous pebbles occur throughout the Blairmore. Those from the McDougall-Segur conglomerate (Douglas, 1950) have yielded K-Ar ages between about 149 and 178 Ma (Norris et al., 1965). The source for these clasts was almost certainly in the Omineca Belt.

During the third cycle, in Late Albian time, marine transgression from the north and south linked the Boreal and Gulfian seas. Along the southwestern margin of the seaway volcanic activity is recorded by trachytic tuff and agglomerate of the Crowsnest Formation (see Chapter 14) which underlies the Fish-Scale marker generally thought to approximately mark the Albian-Cenomanian (Early-Late Cretaceous) boundary. The regressive phase of the last cycle is represented by the Cenomanian Dunvegan Formation which consists mainly of piedmont and alluvial plain quartz, quartzite and chert pebble conglomeratic sandstone in the Liard region (Fig. 9.16) alluvial-deltaic sandstone between there and Peace River, and delta-front and prodeltaic fine grained sandstone and mudstone to the south (Stott, 1982). In Cenomanian time the connection between the Boreal and Gulfian seaways was restricted by the broad Dunvegan fluvial-deltaic system which developed between the marine troughs in the Mackenzie River and central Foothills regions.

In the Liard region of British Columbia, the unconformity at the top of the Dunvegan Formation (Fig. 9.11) represents part of the Cenomanian and all of the Turonian and Coniacian stages (Stott, 1963). Although

Figure 9.16. Massive conglomerate of the Dunvegan Formation (Blairmore Assemblage), Dunedin River, British Columbia. Photo by D.F. Stott. GSC 205235-FF

much of the Cenomanian and Turonian are represented by marine shale in the northern Foothills of Alberta, the unconformity at the top of the Blairmore Group in the central and southern Foothills appears to represent most of Cenomanian and possibly some part of Albian time.

Upper Cretaceous to Paleogene assemblages (Brazeau, Smoky, Fish River, Reindeer)

Northern Foreland Belt

In the western Ogilvie Mountains the Cenomanian Monster Formation records the northward to northeastward progradation of a coastal fan-delta complex. A thin marine mudstone at the base of the formation is succeeded by littoral and nonmarine strata. On the southern edge of Eagle Plain, the lower part of the Eagle Plain Group, over 700 m thick, consists of interbedded sandstones and shales. The presence of both marine and nonmarine fossils and the facies suggests deposition in a transitional lacustrine to fluviatile and deltaic environment near an oscillating shoreline (Mountjoy, 1967; Dixon, in press). Farther north, marine strata of the Eagle Plain Group are interpreted as inner neritic deposits. In Rapid Depression, organic-rich shelf muds of the Boundary Creek Formation are dated as Cenomanian to Turonian and lie unconformably on Albian shales. In Peel Trough the possibility remains that the Trevor Formation, as described by Yorath and Cook (1981), includes an unconformity representing some part of Late Albian to Cenomanian time. This regional unconformity, which occurs throughout most of the Mackenzie River region forms an important tectonostratigraphic boundary.

The paleogeographic evidence of Coniacian to Campanian time indicates that an extensive shallow-water shelf extended over most of the Northwest Territories and northern Yukon Territory. The main source of coarse detritus was from the rising Cordillera to the south, with deposition of nonmarine sandstone and siltstone of the upper part of the Eagle Plain Group along its northern flank.

Upper Coniacian to Lower Campanian strata of the upper part of the Eagle Plain Group and Little Bear Formation are intermittently exposed and are of limited areal extent. Thicknesses are not great; the Little Bear succession over most of its distribution is less than 300 m thick at the surface and about 500 m thick in subsurface basin centre sections (Yorath and Cook, 1981). The Little Bear Formation of southern Peel Trough, which has yielded Santonian to Campanian palynomorphs (Brideaux, 1971; Aitken and Cook, 1974) is a unit of interbedded sandstone, siltstone, and mudstone. The formation occurs mainly along the southwest margin of Peel Trough and grades northward into marine shale of the East Fork-Slater River succession in the central part of the trough (Fig. 9.8).

Maastrichtian to Paleocene strata in northern Rapid Depression include about 1800 m of terrigenous molassoid clastics of the Fish River Group overlain by approximately 575 m of delta plain sediments of the Reindeer Formation (Reindeer Assemblage; Fig. 9.4; Young, 1975). In Bonnet Plume Basin the upper Bonnet Plume Formation comprises sandstone, shale and lignite units of Maastrichtian to Paleocene age. In central Peel Trough accumulation of marine shale of the East Fork Formation continued; during the Maastrichtian the shale became interbedded with fine grained, argillaceous sandstone. A succession of Upper Maastrichtian to Paleocene conglomerate, sandstone, tuff and low grade coal included in the Summit Creek Formation overlies the East Fork Formation in southern Peel Trough. These beds are part of a large alluvial fan derived from uplift of the Mackenzie Mountains (Fig. 9.4) and exhibit fluvial features such as point bars and channels. The ash beds are thought to represent periodic deposition by fallout from high-altitude ash columns originating from volcanic sources in the central part of Yukon Territory (Rickets, 1985).

Southern Foreland Belt

The Late Cenomanian to Paleogene history of the Rocky Mountains Trough and adjacent platform is marked by widespread marine flooding of the continental interior and expansion of Boreal and Gulfian seas. Several transgressive-regressive couplets occur within a dominantly shale sequence with the associated development of epineritic sands during Turonian and Santonian time. Phases of Cordilleran orogenesis, reactivated during Campanian to Maastrichtian time, resulted in widespread molassoid, eastward-prograding alluvial-deltaic deposits with associated coal beds and dinosaur remains (Brazeau Assemblage).

Upper Cretaceous marine strata of the Smoky Assemblage attained a maximum thickness of 4 km in the western Foothills. The succession lies unconformably on the Blairmore Group in the southern and central Foothills of Alberta and on the Dunvegan Formation in northeastern British Columbia (Fig. 9.11). The top of the Dunvegan is marked by a regional unconformity that represents all of Turonian and probably Coniacian time. In the northern Foothills of Alberta and extending into the Peace River region, the succession is more continuous and the lower contact is transitional and conformable. Throughout the Foothills the Upper Cretaceous beds are overlain by Paleocene sediments assigned to the upper Willow Creek, Porcupine Hills and Paskapoo formations (Fig. 9.4).

At the beginning of Late Cretaceous time the continental interior was a low-lying region bounded on the west by the rising Cordillera, on the east by the Precambrian Canadian Shield and on the south and north by the embayments of the Gulf and Boreal seas. Deposition of marine sediments was initiated as a widespread epeirogenic subsidence in the mid-continent region that resulted in the invasion of bordering seaways. The boreal marine incursion was a continuation of the advance that began in the northern Foothills near the end of Late Albian time. This invasion spread southward into the Western Interior Basin of the United States in Late Albian time but did not extend westward into the Foothills region of southern Alberta until Late Cenomanian to Turonian time.

The marine succession, of the Alberta and Smoky groups, of Late Cenomanian to Campanian age, includes sideritic and calcareous shale, siltstone and fine grained sandstone (Stott, 1963, 1967b). The thickness of the succession which occurs along the entire length of the Foothills to a northern erosional limit near Peace River ranges from 600 to 1400 m. The upper part of the marine succession and its overlying alluvial beds reappear in the broad Liard Syncline north of the Alaska Highway. Two megacycles are recognized, the first of Cenomanian to Turonian age and the second of Turonian to Campanian age. The regressive phase of the first megacycle is found in the relatively thin marine sandstone and mudstone of the Cardium Formation which, at Pembina, Alberta, forms one of Canada's most important oil fields. Walker (1983) suggested that Cardium sediment transport in west-central Alberta was southeasterly, more or less along the axis of the Rocky Mountain Trough. The second megacycle marks the greatest Cretaceous epeiric flooding of western North America (Jeletzky, 1971b).

The extensive and prolonged marine deposition during the Late Cretaceous Epoch was brought to an end as alluvial, coarse grained sandstone and interbedded mudstone, variously assigned to the Belly River, St. Mary River, Willow Creek, Brazeau and Wapiti formations, were deposited. These nonmarine sediments in the west grade eastward into marine shale and are conformably to unconformably overlain by nonmarine shale and sandstone of the Tertiary Porcupine Hills and Paskapoo formations (Fig. 9.4). This total succession reaches 1060 m thick in the western Foothills.

The Belly River Formation and its equivalents (Tozer, 1956; Carrigy, 1971) form a wedge, over 600 m thick, of clastic sediments consisting of green sandstone, shale and some coal. It is overlain by the Campanian to Maastrichtian Bearpaw marine shale (Wall and Rosene, 1977; Caldwell et al., 1978) which records a significant readvance of the seaway during Late Campanian time. Although this seaway extended to the west in the Crowsnest region, it did not extend into the central and northern Foothills of Alberta and did not reach northwestward beyond Edmonton.

In the southern Alberta Foothills, the Bearpaw Formation is overlain by 250 m of alluvial sandstone and shale of the St. Mary River Formation. The younger Willow Creek Formation, is about 1.2 km thick and consists of sandstones with varicoloured shale which grade upward into buff-weathering sandstone and brown shale. No hiatus is recognized within the Willow Creek although the upper part contains a typical Paleocene fauna. The Paleocene Porcupine Hills Formation of the southern Foothills includes about 1 km of coarse grained, massive, crossbedded sandstone and calcareous, bentonitic shale.

In the central Foothills, the lower beds of the Brazeau Formation include pebble conglomerate, and the upper 1.6 km consists of greenish sandstone, shale, some tuff and thin coal seams. The Brazeau is overlain by the Entrance conglomerate, coals of the Coalspur Formation and thick deposits of the Paleocene Paskapoo Formation (Jerzykiewicz and McLean, 1980). The Wapiti farther north is similar to the Brazeau Formation (Kramers and Mellon, 1972). These latest Cretaceous clastic sediments, together with Paleocene strata, are very thick and record the final phase of Cordilleran orogenesis. Mountain building along the orogen, together with uplift of the northern craton, resulted in the retreat of the seaway from the continental interior.

In the upper Flathead Valley of the southern Rocky Mountains, the Kishenehn Formation, of latest Eocene to earliest Oligocene age (Russell, 1954; McMechan, 1981), comprises about 3600 m of alluvial deposits. They accumulated in an asymmetric graben which formed during westward listric normal movement on the Flathead Fault (Price, 1965). The Kishenehn Formation lies with angular unconformity on strata deformed by the last phase of Cordilleran orogenesis and establishes an upper limit for dating the underlying structures of the southern Rocky Mountains (Fig. 9.4).

THE OMINECA BELT

H. Gabrielse and R.G. Anderson

During the interval between Late Jurassic and Paleogene time the Omineca Belt was a locus of deformation, metamorphism, plutonism, volcanism and uplift. As a consequence the stratigraphic record is sparse, consisting of widely scattered, fault-controlled basins which contain Upper Cretaceous and Paleogene nonmarine sediments locally associated with volcanic rocks; these are discussed under the heading "Fault-controlled basins" at the end of this chapter. Nonetheless the belt was of fundamental importance as a source region for clastic detritus shed into the adjacent Foreland and Intermontane belts. Mid- to Upper Cretaceous volcanic rocks occur mainly in the northern part of the belt, and a number of Cretaceous and early Cenozoic plutonic suites are exposed throughout its length.

Volcanic and plutonic rocks

Three Cretaceous to Paleogene assemblages of volcanic rocks occur in the northern Cordillera. These are the mid-Cretaceous Mount Nansen Group and South Fork volcanics (South Fork Assemblage) of western part of Yukon Territory and Selwyn Basin respectively (Fig. 9.3), the Upper Cretaceous Carmacks and Hutshi groups of western Yukon Territory (Carmacks Assemblage) and the Eocene Sloko and Skukum groups (Kamloops Assemblage) of northwestern British Columbia and southwestern Yukon Territory (Fig. 9.4). Although they occur mainly in the northern Omineca Belt, some are present as overlap assemblages between the northernmost Intermontane and Coast belts. The assemblages are spatially related to temporally equivalent plutonic rocks (see Chapters 14 and 15).

Outcrops of the mid-Cretaceous South Fork volcanics (total area 2610 km^2) are distributed along a 210-km northwest-trending belt northeast of Tintina Trench and north, west and east of Ross River, in southeastern Yukon Territory (Fig. 9.3). They are predominantly densely welded crystal tuffs and are preserved within five discrete calderas, the largest about 30 km in diameter. Extra-caldera ashfall or ashflow deposits that may once have covered the region have been removed by Cretaceous to Recent erosion. In the northern Omineca Belt the South Fork succession is the only suite of volcanic rocks known to be coeval with and related to the belt's extensive and characteristic mid-Cretaceous plutons.

The South Fork volcanics are typically composed of broken plagioclase, hornblende, quartz and biotite crystals and rare lithic fragments in a very fine felsic crystalline matrix. They are predominantly subalkaline, high-K, calc-alkaline dacite but include mafic and basaltic compositions (Roddick and Green, 1961a,b; Wood and Armstrong, 1982; Pigage and Anderson, 1985; S.P. Gordey and R.G. Anderson, pers. comm., 1987). Typically the volcanics are structureless, but in places where bedding is observed, they dip gently and are undeformed. Minimum thickness estimates range up to 1500 m. Hornblende and biotite K-Ar ages range from 86 to 109 Ma (Wood and Armstrong, 1982; S.P. Gordey, pers. comm., 1987) and are similar to those from the geochemically similar comagmatic plutonic rocks in the Anvil plutonic suite (Pigage and Anderson, 1985).

The Mount Nansen Group consists of relatively uniform, commonly structureless, medium to dark greenish grey, aphanitic volcanics, locally with textures of tuff and tuff breccia (Tempelman-Kluit, 1974). In places the rocks overlie or are cut by feldspar porphyry dykes which are believed to be about the same age. The group rests on a gently dipping unconformable surface and appears to have been extruded from several isolated centres prior to the establishment of the present drainage.

The Carmacks Group is a sheet of volcanic rocks of relatively uniform composition, 600 to 900 m thick, whose thickness and distribution was influenced by the present topography (Tempelman-Kluit, 1974). The sequence consists of well defined flows of brown weathering andesite and basalt. Some tuff breccias and intrusive plugs and dykes of diabase are present. Locally the volcanics are underlain by a conformable sequence of carbonaceous, conglomeratic sediments.

The Hutshi Group comprises flat-lying to gently dipping flows and flow breccias of basalt and rhyolite with variable amounts of agglomerate, tuff and minor clastic sedimentary rocks (Wheeler, 1961). It is locally more than 1000 m thick, lies with angular unconformity on the Lower and Middle Jurassic Laberge Group, and is cut by a variety of granitic plutons.

The Sloko and Skukum volcanic rocks are characteristically brightly coloured andesitic, dacitic, rhyolitic and basaltic breccias, tuffs and lavas in which pyroclastic rocks are dominant (Wheeler, 1961). A detailed study of one of the volcanic centres by Lambert (1974) revealed an elliptical complex of 570 km^2 consisting of two nested calderas, a central dome, concentric and radial fracture systems and a subelliptical ring dyke. Shattered granitic breccia, derived from underlying rocks is common. In northwestern British Columbia Sloko Group volcanics occur at several localities close to the boundary between the Coast and Intermontane belts. Locally they contain units of sedimentary breccia, grit and conglomerate (Aitken, 1959; Souther, 1971, 1972). In exceptional examples pyroclastic deposits can be traced downward into homogeneous aphanitic felsite which in turn grades downward into finely crystalline and ultimately medium crystalline granite. Everywhere the contact of the Sloko and Skukum volcanics with the substrate is unconformable.

Exposures of Eocene rhyolite and basalt occur along the Tintina Fault (Jackson et al., 1986) and minor outcrops of Eocene dacitic volcanics occur in the Northern Rocky Mountain Trench between the Kechika and Spinel faults. An Eocene volcanic centre consisting of rhyolite flows, tuff and breccia lies along the Kechika Fault about 120 km northwest of the trench (Fig. 9.4).

Evidence of Late Cretaceous (Campanian-Maastrichtian) volcanism is present in the form of widespread tuff in the upper part of the Sustut Group in the Intermontane Belt. Although much of this material is believed to have been derived from volcanism in the Intermontane Belt, some may have come from the Omineca Belt. Also, in the Summit Creek Formation (Maastrichtian-Paleocene) adjacent to the Mackenzie Mountains front, numerous tuffs are thought to represent periodic deposition by fallout from high-altitude ash columns originating from volcanic centres in the central part of Yukon Territory (Yorath and Cook, 1981; Rickets, 1985).

Plutonic rocks in the Omineca Belt, of both S- and I-type, fall into four distinct suites (see Chapter 15). These are: (1) Middle Jurassic granitic plutons, widespread across the southern Cordillera but not identified north of Prince George; (2) mid-Cretaceous granites, the most widespread and voluminous plutonic rocks in the Omineca Belt; (3) Late Cretaceous granite, commonly occurring as small, high-level intrusive bodies; and (4) Eocene, high-level plutons ranging from calc-alkaline to alkaline in composition. Middle Jurassic granitic rocks are typically syntectonic to post-tectonic, whereas the mid-Cretaceous and younger plutons are characteristically post-tectonic relative to contractional structures in the Omineca Belt. Strongly alkaline plutonic rocks of the Coryell Suite and related volcanics of the Marron Formation in the southernmost Omineca Belt were emplaced during an episode of Eocene extension. The Marron volcanics are described with the Penticton Group in the section on the Intermontane Belt.

Sedimentary rocks

From Late Jurassic to Paleogene time the Omineca Belt was mainly a source area for sediments deposited in basins in the Foreland and Intermontane belts. Most of the sedimentary rocks, mainly of Paleocene and Eocene ages, occur in fault-controlled basins, locally associated with volcanic rocks, and are discussed in a later section.

The Sophie Mountain Formation in the Rossland area of the Monashee Mountains, possibly more than 100 m thick, consists of quartzitic conglomerate and minor argillite containing plant fossils of Cenomanian to Campanian age. The formation appears to rest conformably on other units but contains no clasts of the underlying nearby granitic rocks.

THE INTERMONTANE BELT

H. Gabrielse, J.G. Souther, G.J. Woodsworth, H.W. Tipper, and J.W.H. Monger

Collision of Stikinia with the Cache Creek Terrane and uplift of the Skeena Arch in Middle Jurassic time led to the formation of the Bowser Basin which persisted until Early Cretaceous time and, at least in its northern part, had many characteristics of a foredeep (Fig. 9.2). Only the most proximal parts of the foredeep clastic wedge, however, were deformed in response to the southward and westward tectonic transport of the dominantly Cache Creek source rocks. Mid-Cretaceous uplift of the Omineca Belt, locally associated with westward verging structures, provided a source area for the Hauterivian to Santonian sediments deposited in the Sustut-Skeena Basin. In the Tyaughton-Methow Trough Upper Jurassic and Lower Cretaceous sediments also were derived mainly from uplifts to the east. Thus the polarity of sedimentation and tectonic transport along the west side of the Omineca Belt was opposite to that in the eastern part of the Omineca Belt and in the Foreland Belt.

Mid- and late Late Cretaceous uplift and eastward tectonic transport in the eastern Coast Belt resulted in deposition of westerly derived sediments in the Tyaughton-Methow Trough and in the Sustut Basin, respectively. This tectonism coincided with mid-Cretaceous magmatism along the eastern flank of the Coast Belt and late Late Cretaceous magmatism in southern Yukon Territory and in Skeena Arch. Early Paleogene magmatism was widespread throughout the Intermontane Belt.

Many elongate areas underlain by Paleogene clastic rocks were either fault controlled or were closely associated with regional strike-slip faults that may be genetically related to those in the Omineca and Coast belts. These are discussed under the heading "Fault-controlled basins" at the end of this chapter. In the following discussion Middle Jurassic strata in the Bowser Basin are included so as to avoid fragmentary treatment of this important tectonic element.

Volcanic and plutonic rocks

Nonmarine volcanic rocks of Late Cretaceous to Paleogene age are widespread in west-central British Columbia where they overlie Jura-Cretaceous strata and older rocks. The age of volcanism varies areally suggesting that it was restricted to several centres of igneous activity, all of which were not necessarily active at any one time.

The oldest and best studied unit is the Kasalka Group (South Fork Assemblage; Fig. 9.3; MacIntyre, 1976; G.J. Woodsworth and P. van der Heyden, pers. comm., 1986). These rocks unconformably overlie marine, Middle Albian sediments of the Skeena Group in the Whitesail Lake area (Fig. 9.17; Woodsworth, 1979a, 1980). The base of the unit is commonly marked by a coarse conglomerate, up to 50 m thick, and containing granitic clasts derived from the Coast Plutonic Complex. This is overlain by about 1000 m of calc-alkaline, felsic pyroclastic rocks, porphyritic andesite, rhyolite flows and domes, and volcanic mudflows. Small subvolcanic intrusions (Kasalka intrusions) are spatially associated with the volcanic rocks. K-Ar dates from the Kasalka Group and related intrusions range from 87 to 100 Ma (MacIntyre, 1976; R.L. Armstrong, pers. comm., 1987), suggesting that the group is mainly latest Early Cretaceous (Middle Albian or younger) to early Late Cretaceous in age.

Significantly younger than the Kasalka Group are the Brian Boru Formation and Tip Top Hill volcanics (Carmacks Assemblage). The Brian Boru Formation (Sutherland Brown, 1960) is best exposed near Hazelton where it is up to 1800 m thick and unconformably overlies the Skeena Group (Fig. 9.3). Compositions range from basalt to rhyolite with a pink to cream coloured hornblende porphyry, dacite and rhyolite being the most common. K-Ar ages of 70 and 72 Ma from these volcanic rocks are substantially younger than those from the Kasalka Group and indicate a latest Cretaceous age. The Tip Top Hill volcanics are similar in age and composition to the Brian Boru Formation (Church, 1973).

Plutonic rocks roughly coeval with the Brian Boru and Tip Top Hill successions are abundant in west-central British Columbia (see Chapter 15). These plutons, the Bulkley intrusions, give K-Ar ages ranging from about 70 to 84 Ma. MacIntyre (1976) suggested that the plutons are deep-seated equivalents of the volcanic piles. He regarded

Figure 9.17. Well bedded Middle Albian marine sandstone of the Skeena Group overlain by massive nonmarine volcanics of the Kasalka Group (upper third of mountain) near Tahtsa Lake. Paleocurrent data from the Skeena Group indicate a northeasterly source whereas paleocurrent data from the base of the Kasalka Group at this locality indicate a westerly source. Photo by G.J. Woodsworth. GSC 205235-S.

the mid-Cretaceous Kasalka Group as a fault-bounded cauldron subsidence complex. An alternative hypothesis, not inconsistent with the former, is that the volcanic suites and related plutons are high-level equivalents of deep-seated, Late Cretaceous plutons of the Coast Belt. In this hypothesis, Late Cretaceous plutonism and volcanism in the Coast Belt and western Intermontane Belt represent a wrench-related orogen, with caldera complexes developing in local pull-apart basins in the east and later uplift exposing deeper crustal levels on the west.

The mid-Cretaceous Spences Bridge Group (South Fork Assemblage) in south-central British Columbia comprises approximately 2400 m of equal proportions of calc-alkaline lava and clastic rocks (Thorkelson, 1985). The group is conformably overlain by about 600 m of altered andesite and minor sediments which were formerly referred to the Kingsvale Group. The components of the Spences Bridge probably formed as a stratovolcano above plutonic rocks and metamorphosed volcanics of the Triassic Nicola Group (Fig. 9.3).

Lower Tertiary volcanic rocks of the Kamloops Assemblage are locally abundant in west-central British Columbia (Fig. 9.4; Church, 1971, 1973). Generally they occur farther east than the Upper Cretaceous volcanic suites. Most commonly they consist of a lower, relatively felsic unit overlain by a basalt-andesite pile. The lower unit, generally called the Ootsa Lake Group, consists mainly of calc-alkaline, dacite to rhyolite flows, tuff and breccia with minor intercalated sediments. The subvolcanic Nanika intrusions are spatially associated with the Ootsa Lake volcanic rocks. The Goosly Lake volcanic and intrusive rocks have alkaline compositions. The Ootsa Lake Group grades upwards into the Endako Group, a unit consisting mainly of andesite and olivine basalt and related hypabyssal intrusions. A few K-Ar dates suggest that the Ootsa Lake and Endako groups are mainly Eocene in age but may be as old as Late Cretaceous and as young as Oligocene (Fig. 9.4).

MacIntyre (1976) viewed the early Tertiary magmatic activity in the Whitesail Lake area as a reactivation of an older cauldron subsidence complex. Alternatively, the volcanic rocks are high-level manifestations of tectonism, plutonism and limited extension within the Coast Belt to the west. Godwin (1975) proposed the development of imbricate subduction along and near the Skeena Arch, to account for its Late Cretaceous to early Tertiary plutonic and metallogenic history.

The Kamloops Group (Kamloops Assemblage) is a succession of Lower to Middle Eocene alkali-rich calcalkaline volcanic and sedimentary rocks which occurs widely throughout south-central British Columbia (Fig. 9.4; Ewing, 1981a,b). In the type area west of Kamloops, the basal Tranquille Formation consists of 500 m of lacustrine and deltaic sediments, pillowed flows and hyaloclastites. Elsewhere, basal coal-bearing nonvolcanic fluvial and lacustrine units occur, such as the Coldwater Formation at Merritt, the Chu Chua Formation at Barriere and the Shorts Creek Formation west of Vernon. Overlying these formations are dominantly volcanic units. At the type area the Dewdrop Flats Formation includes over 1000 m of interstratified basaltic andesite flows, andesitic flow-breccia sheets and cones, basaltic tuff rings and an andesitic composite cone. In places, flat-lying basaltic andesite flows about 600 m thick with local flow breccias are common. Chemical data support a subduction-related continental arc origin of the volcanics (Ewing, 1981b). Ewing (1981a) interpreted the basal sediments as having accumulated in separate fault-bounded basins initiated immediately before the onset of volcanism. The volcanic rocks filled the basins and formed a widespread volcanic blanket which was disrupted by a regional fault network.

In the Okanagan region, and occurring in both the Intermontane and Omineca belts, the Penticton Group (Kamloops Assemblage) comprises volcanic and sedimentary rocks of Eocene age (Fig. 9.4; Church, 1982; Church et al., 1983). The group consists of five formations with an aggregate thickness of between 2500 m and 3500 m in the type area near Penticton (Fig. 9.18). At the base is polymictic conglomerate and breccia of the Springbrook Formation and coeval beds of the Kettle River Formation consisting of granite-boulder conglomerate, rhyolite breccia and tuffaceous sedimentary rocks. The overlying Marron Formation comprising thick andesite, trachyte and phonolitic lava flows is succeeded by dacitic domes of the Marama Formation. The overlying White Lake Formation consists of volcanic breccia, fluvial and lacustrine sedimentary rocks which are overlain by fanglomerates of the Skaha Formation. K-Ar ages obtained from samples of the Penticton Group are 48.4 Ma (whole-rock analysis) and 53.1 Ma (biotite analysis). To the north and west of Penticton the constituent formations interfinger with and are replaced by units of the Kamloops Group. In the Vernon area such terms as Naswhito Creek Formation, Bouleau rhyolite, Attenborough Creek Formation and Shorts Creek Formation are applied to Eocene volcanic and sedimentary units that only partly correlate with the Penticton Group (Church et al., 1983). The potassic porphyries of the Penticton Group are chemically similar to, and are believed to be comagmatic with the nearby Coryell syenites, a suite

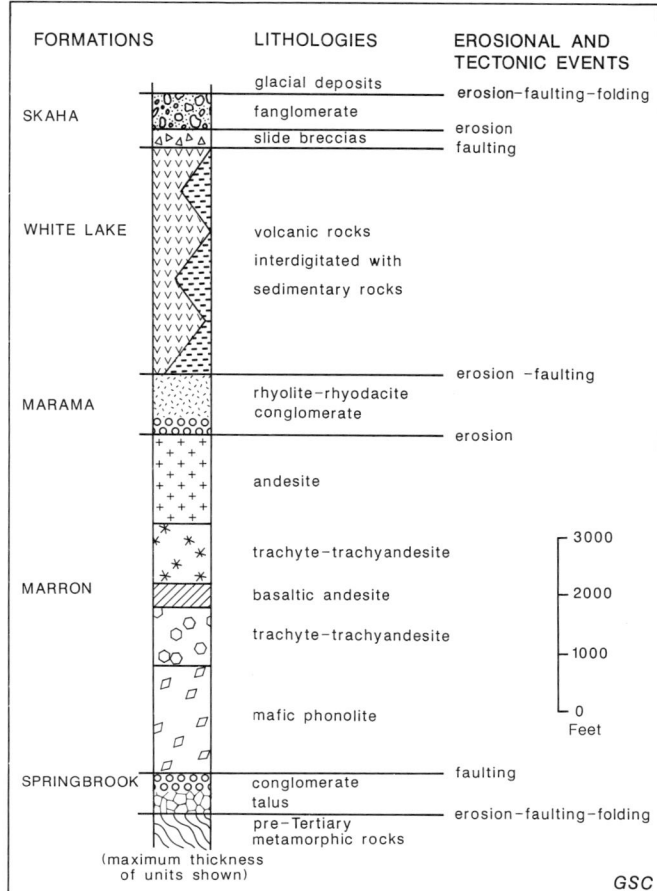

Figure 9.18. Generalized columnar section of the Penticton Group (from Church, 1982).

of epizonal plutons in the southwestern Omineca Belt (see Chapter 15).

The widespread Paleogene calc-alkaline volcanism in the Okanagan region has been related to ductile crustal spreading which linked dextral movement on the Tintina-Northern Rocky Mountain Trench Fault zone and the Fraser River Fault zone (Price et al., 1981).

Sedimentary rocks
Bowser Basin

Clastic sedimentary rocks of the Middle Jurassic to Lower(?) Cretaceous Bowser Lake Group (Bowser Lake Assemblage), locally as much as 3500 m thick, underlie the Bowser Basin in north-central British Columbia (Fig. 9.6). The basin encloses an area of about 4900 km^2 and is bounded by the Omineca Belt on the east, the Coast Belt on the west, the Stikine Arch to the north, and the Skeena Arch in the south. Except for local, proximal, nonmarine deposits, sediments in the lower part are of marine origin and range in age from Late Bathonian to Early Oxfordian in the north, Late Bajocian to Early Oxfordian in the south, and Early Bathonian to Early Oxfordian in the east. The upper and thickest part of the group comprises Upper

Jurassic and possibly Lower Cretaceous nonmarine sediments and locally includes anthracite coal in the Groundhog coal measures (Eisbacher, 1974, 1981; Jeletzky, 1976; Richards and Jeletzky, 1975; Tipper and Richards, 1976; Bustin and Moffat, 1983).

The sedimentary sequence has many attributes of a foredeep assemblage reflected by the progradation of coarse grained, nonmarine molasse over fine grained basinal flysch (Fig. 9.19). The stratigraphic units express a record of uplift in perimeter areas to the north, east, and southeast beginning in Middle Jurassic time and culminating in the Late Jurassic Epoch but possibly continuing into the Early Cretaceous (Fig. 9.2, 9.20).

Where observed the base of the Bowser Lake Group is marked by an unconformity, which, in places, is angular. The group is overlain by mid-Albian nonmarine strata of the Sustut Group on the northern and eastern margins of the basin and by Hauterivian marine and nonmarine strata intercalated with minor volcanic rocks assigned to the Skeena Group along the southern margin of the basin.

The basal unit of the Bowser Lake Group is the Ashman Formation. Unlike younger strata, the Ashman can be traced southwards into the Nechako Basin south of Skeena Arch (Fig. 9.6). The Ashman consists of grey to black shale with varying amounts of feldspathic to quartzose sandstone, greywacke, chert-pebble conglomerate and greywacke conglomerate (Tipper and Richards, 1976). In the Groundhog Coalfield, in the northwestern part of the basin, partly correlative rocks have been named informally the "Jackson Unit" (Fig. 9.21; Bustin and Moffat, 1983). In many parts of the basin, particularly in the north, black shale and siltstone contains sparse but conspicuous beds of buff to buff-orange weathering, rusty, locally calcareous, fine grained sandstone beds ranging from less than 1 cm to more than 1 m thick. Graded bedding is common in sandy units. Near-shore facies of the Ashman Formation contain much sandstone and variable amounts of conglomerate. North of the Stikine River a conglomerate overlying Lower Bajocian siltstone and sandstone grades eastward into a sequence of nonmarine conglomerate, sandstone, siltstone and shale intercalated with maroon and purple andesitic volcanic rocks. Abundant woody material and coarse-shelled pelecypods are present in crossbedded sandstone near the southwestern margin of the basin. The sandstone and associated pebble conglomerate grade basinward into black to dark grey shale and grey to greenish grey, fine grained sandstone.

South of Skeena Arch the Ashman Formation consists of mainly thin bedded pyritic siltstone, greywacke, shale, sandstone, minor conglomerate and tuff. Clasts are predominantly feldspar and volcanic fragments. Sedimentary structures are few, possibly due to bioturbation; calcareous concretions are common. Fossils include belemnites, pelecypods and ammonites. The Ashman Formation is between 300 and 760 m thick in the southern Bowser Basin and up to 1800 m thick in the Groundhog Coalfield. Its age ranges from Late Bajocian to Early Oxfordian. The Ashman Formation and at least part of the Jackson Unit are interpreted as representing submarine fan deposits in the lower and distal sequences and a variety of deltaic and paralic environments in the upper and proximal successions (Fig. 9.19).

Three distinct pebble- to cobble-conglomeratic units of the Bowser Lake Group occur along the margin of the northern part of Bowser Basin (Currie, 1984). The lowest, in the Ashman Formation, forms a sheet-like deposit comprising interfingering lenses of conglomerate, less than 1 m to 4 m thick, with associated shale and sandstone

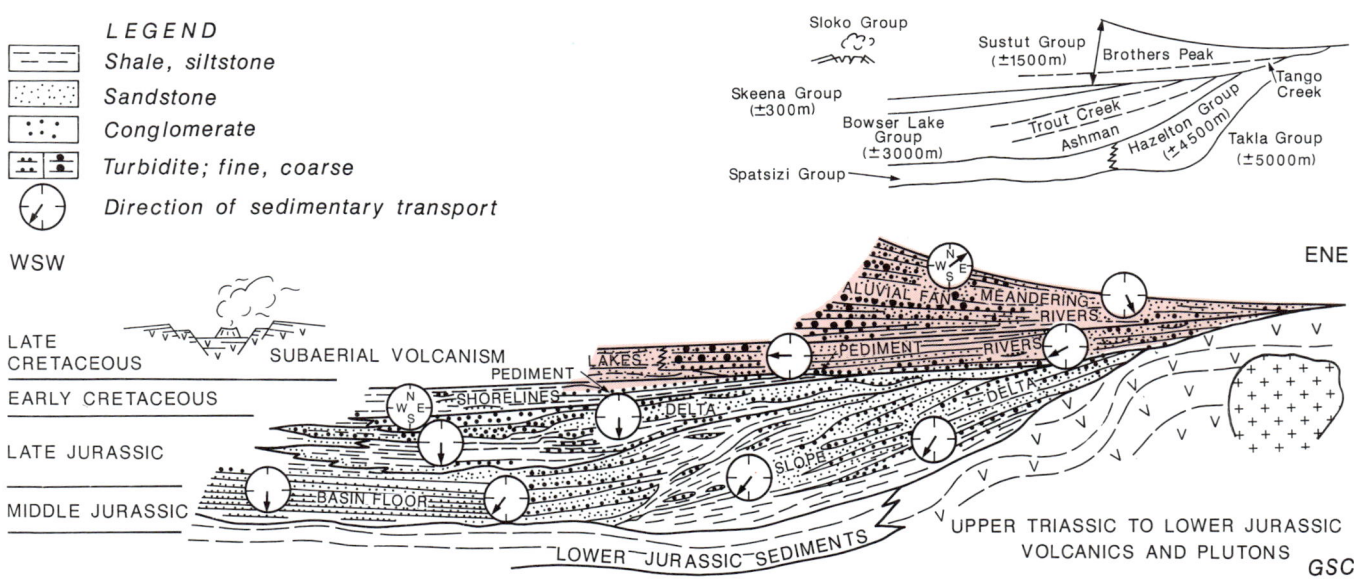

Figure 9.19. Model of the Bowser Basin (horizontal distance approx. 250 km). Main rock units are identified in the inset. Paleocurrent directions are shown (after Eisbacher, 1981). Grey, marine; pink, nonmarine.

comprising a continuous coarsening upward sequence tens of metres thick. The middle conglomerate, immediately above the Ashman, is made up of discrete conglomerate lenses tens of metres thick and hundreds of metres wide, separated by black shale sequences (Plate 22). In places sandstone is present at the base of the lenses and a few thin beds of sandstone or shale occur within the conglomerate. The upper conglomerate, forming the uppermost unit in the group, is massive, rusty weathering, and consists of fining-upward sequences. It is up to 200 m thick and is much more widespread than lower conglomerates. Clasts in the upper conglomerate range from 1 to 20 cm in diameter. In this region all of the conglomerates include abundant green, grey and maroon radiolarian chert derived from Cache Creek rocks to the north (Fig. 9.22) as well as felsic volcanic rocks and polycrystalline quartz.

Conformably overlying the Ashman Formation in the southern part of the basin is an assemblage of sandstone, conglomerate, siltstone and minor coal with a maximum thickness in proximal facies of 300 m. This sequence is informally named the "Trout Creek assemblage" and is Late Oxfordian in age (Tipper and Richards, 1976). A lower marine part grades upward into nonmarine units containing pebble and cobble conglomerate members as much as 50 m thick. Clasts of volcanic, plutonic and sedimentary rocks indicate derivation from the Skeena Arch to the south.

The Trout Creek assemblage is overlain locally by, and in part intercalated with, basaltic and andesitic subaerial volcanic rocks informally termed the "Netalzul volcanics". In places these rocks occur as volcanic mudflows in units 50 to 100 m thick. Some mud flows contain carbonized tree trunks, a few in growth position.

The youngest strata of the Bowser Lake Group in the southern part of the basin represent alluvial shoreline and

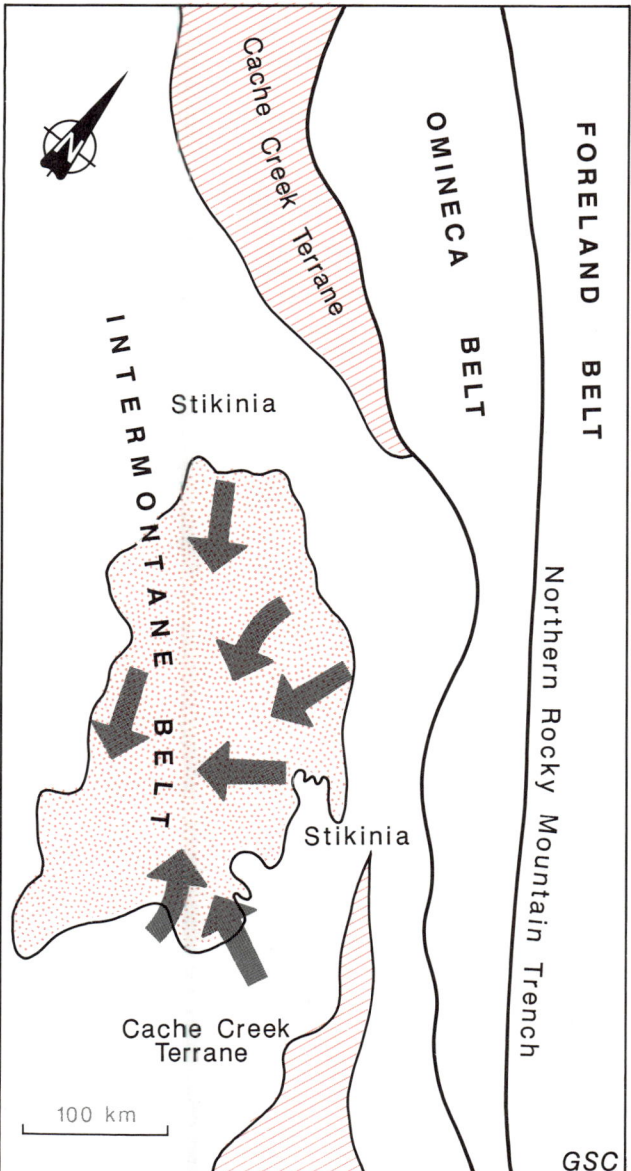

Figure 9.20. Paleocurrent trends and structural setting of the Bowser Lake Group (modified from Eisbacher, 1981).

		Eisbacher, 1974 NORTHERN BOWSER BASIN	Tipper & Richards, 1976 SOUTHERN BOWSER BASIN	Bustin & Moffat, 1983 Groundhog Coalfield
CRETACEOUS	Upper	Sustut-Sifton Assemblage	Sustut Group	
	Lower	Bowser Assemblage: Jenkins Creek Facies; Gunanoot-Groundhog Facies	Skeena Group	Devils Claw Unit; McEvoy Unit
JURASSIC	Upper	Bowser Assemblage: Duti River Slamgeesh Facies	Bowser Lake Group	Currier / Prudential
	Middle			Jackson Unit
	Lower	Takla-Hazelton Assemblage	Hazelton Group	
TRIASSIC	Upper		Takla group	
	Middle			

GSC

Figure 9.21. Stratigraphy and regional correlations in Bowser Basin (after Bustin and Moffat, 1983).

Figure 9.22. Interbedded shale and conglomerate of the Ashman Formation, northern Bowser Basin, British Columbia. The resistant and laterally discontinuous lenses of conglomerate contain chert pebbles derived from the Cache Creek Terrane. Photo by C.A. Evenchick. GSC 205235-V

shallow marine facies. Monotonous sequences of siltstone, mudstone, sandstone and shale are interrupted by minor conglomerate and marl. Conglomerate clasts are predominantly feldspar and volcanic fragments with lesser quartz, shale and siltstone. Poorly preserved plant fossils are common but ammonites and bivalves are rare; shoreline coquinas are locally conspicuous. Thicknesses are difficult to estimate because of faulting and the lack of distinctive markers, but are probably less than 100 m. The rocks range in age from Late Oxfordian to Early Kimmeridgian and probably younger.

In the Groundhog region the Ashman Formation is overlain by a sequence consisting of four informal units (Bustin and Moffat, 1983; Fig. 9.21). The lowest is the "Currier unit", composed of thin- to thick-bedded sandstone with minor shale, siltstone and coal between 400 and 600 m thick. Its lower part comprises a series of upward-coarsening sequences from 5 to 25 m thick. The sandstone contains clasts of volcanic rocks, chert and feldspar. The middle part is predominantly fine- to medium-grained sandstone, mudstone, siltstone and coal which locally constitute upward-fining units. At least eight coal seams between 0.5 and 4.0 m thick are present. The upper part of the Currier unit consists of siltstone and shale, thin- to thick-bedded sandstone and coal in seams up to 2.5 m thick. Most of the unit is fluvial in origin but some of it is regarded as paralic and deltaic.

The "Prudential unit", consisting of 400 to 500 m of alluvial and paralic interbedded conglomerate, sandstone, siltstone and shale with minor coal seams up to 1.5 m thick, occurs in the eastern part of the Groundhog area and is believed to be, at least in part, a correlative, coarser grained facies of the Currier unit. Volcanic and chert pebbles and cobbles are dominant in the conglomerates.

The "McEvoy unit", ranging from 400 to 800 m thick, overlies the "Currier unit" and comprises thick interbedded sequences of siltstone, shale, minor limestone, sandstone, conglomerate and coal seams up to 0.5 m thick. The conglomerates include clasts of chert, volcanic rocks, and widely disseminated granodiorite. The McEvoy contains marine strata with belemnites in its lower part and nonmarine, partly lacustrine sediments, including limestone, in the upper part.

The uppermost sequence in the Groundhog area is the "Devil's Claw unit", which is between 300 and 500 m thick. It consists dominantly of pebble to cobble conglomerate and rare sandstone. The homogeneous conglomerates display large-scale crossbeds and contain well rounded and well sorted clasts of chert, volcanic rocks, quartz, and sparse granodiorite. The lower part includes thin sequences of siltstone, shale, minor sandstone and coal. The age of the "Devil's Claw" is, at least in part, Albian (Moffat, 1985) and therefore is probably correlative with the lower part of the Sustut Group and the upper part of the Skeena Group. Several characteristics of the conglomerates suggest that they were deposited in alluvial-fan to alluvial-plain environments.

Note: Recent stratigraphic and biostratigraphic studies in north-central Bowser Basin have resulted in the establishment of formal stratigraphic nomenclature and improved ages and correlations with Bowser Lake and Skeena groups strata (Cookenboo and Bastin, 1989).

Following the Early Kimmeridgian to (?)Early Cretaceous period of nonmarine deposition in the southern part of Bowser Basin, a marine transgression advanced eastward and spread southward across Skeena Arch in Albian time, perhaps connecting with the Tyaughton-Methow Trough. The basal unit of the Skeena Group (Skeena Assemblage) is represented by the "Kitsuns Creek" sediments, which are up to 2000 m thick and comprise Hauterivian to Aptian paralic and neritic conglomerate, pebbly sandstone, shale and minor coal. The unit rests unconformably upon rocks of the Hazelton Group and locally is paraconformable on the Bowser Lake Group. The overlying 'Rocky Ridge' volcanic rocks comprise up to several hundred metres of basaltic and andesitic augite and feldspar porphyry, subaerially erupted during Barremian to Early Albian time. The unit forms a series of volcanic piles, possibly originating from separate vents. Away from the vents the volcanics are represented by hematitic tuff and breccia which locally is interbedded with conglomerate. The next youngest unit is a sequence of marine shale and chert-pebble conglomerate of variable thickness and Middle Albian age. The unit occurs widely throughout British Columbia and is recognized by the common presence of the ammonite *Cleoniceras*. The uppermost unit of the Skeena Group is the Red Rose Formation, about 300 m thick, and consisting of conglomerate, siltstone mudstone and coal of Albian and (?)younger age.

Nechako Basin

South of Skeena Arch, in southern Nechako Basin, strata temporally equivalent to the Skeena Group were penetrated by two exploratory wells (Fig. 9.6). In the Honolulu Nazko well the uppermost 1612 m comprises marine and nonmarine sandstone, shale and conglomerate (Koch, 1973). These strata are underlain by about 910 m of variegated red, green, grey and brown shale, tuff, chert and minor nonmarine sandstone, which in turn rests upon a Permian chert sequence. The nearby Hudson's Bay Oil and Gas Redstone well penetrated a similar section 637 m thick (Koch, 1973). Elsewhere scattered outcrops throughout the region comprise chert-pebble conglomerate, black shale and siltstone, which resemble components of the Skeena Group.

Sustut Basin

Nonmarine Cretaceous and (?)Tertiary clastic rocks of the Sustut Basin (Fig. 9.6), ranging in thickness from 300 to 2600 m, occupy a narrow belt between the Skeena and Omineca mountains and extend from the Stikine River in the northwest to Takla Lake in the southeast (Eisbacher, 1974, 1981). The sedimentary assemblage, named the Sustut Group, has been divided into two formations recognizable throughout the basin and a number of informal members of local importance.

The Tango Creek Formation (Fig. 9.3), increasing in thickness from 500 m on the east to about 1400 m in the west, generally rests unconformably on strata of the Bowser Lake Group in western exposures and upon older volcanic, sedimentary and granitic terranes along the northern and eastern margins of the basin. Locally a polymictic basal conglomerate, up to 100 m thick, overlies a clearly defined and widespread pediment surface. Along the northeastern margin of the basin basal rocks consist locally of poorly sorted, coarse conglomerate with angular to subangular clasts of hematite-stained volcanic rocks. Elsewhere in eastern outliers, they include clasts of marble and granitic rocks. However, most of the Tango Creek Formation consists of interbedded sandstone, mudstone and pebble conglomerate, locally enclosing thin seams of lignite. In an isolated area of Sustut rocks along the Grand Canyon of Stikine River, strata correlative with the Tango Creek are less than 250 m thick and consist of interbedded sandstone, shale and carbonaceous shale with lenses of chert- and quartz-pebble conglomerate (Read, 1983). An abundance of detrital mica and quartz in the lower Tango Creek Formation (Skeena Assemblage) markedly contrasts with their paucity in the Albian Devil's Claw unit of northern Bowser Basin (Moffat, 1985) and, further, suggests differing provenances for the two successions. Based upon abundant spores and pollen, the age of the Tango Creek Formation spans an interval from mid-Albian to Campanian (late Early to middle Late Cretaceous).

Sediments of the lower Tango Creek Formation are believed to have been deposited by two principal river systems in the northern part of the basin (Eisbacher, 1974). These streams entered the basin from the Omineca Belt to the north and east and thence merged to flow southwesterly (Fig. 9.23). Similarly, in the southern part of the basin the main rivers entered from the northeast and probably flowed towards the southwest. A change from this early centripetal drainage pattern to a longitudinal system occurred during deposition of the upper part of the Tango Creek Formation (Virginian Ridge Assemblage) in response to tectonic uplift along the southwestern part of the basin.

The basal part of the overlying Brothers Peak Formation (Virginian Ridge Assemblage; Fig. 9.4) consists of a widespread, coarse, thick conglomerate which rests conformably, and unconformably on, and locally oversteps the Tango Creek Formation. Numerous ash-fall tuffs are interbedded with the conglomerates. Along the Grand Canyon of the Stikine River chert-pebble conglomerate,

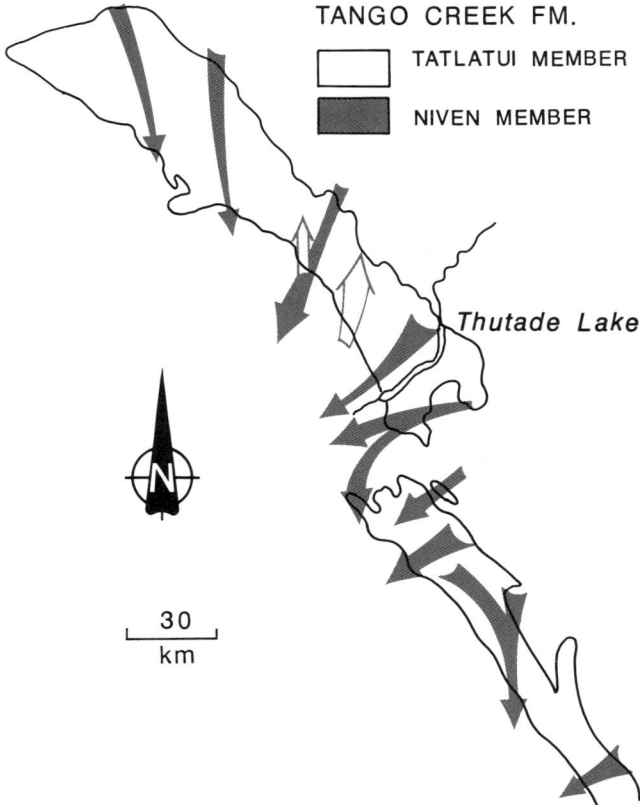

Figure 9.23. Paleocurrent directions in the Tango Creek Formation of the Sustut Group in Sustut Basin (from Eisbacher, 1974). Figure 9.6 shows location of Sustut Basin.

more than 300 m thick onlaps the northern margin of the Sustut Basin and overlies Triassic volcanic rocks into which were cut southerly trending valleys. The upper part of the Brothers Peak Formation is made up of pebbly sandstone interlayered with tuff and mudstone. The total thickness of the formation ranges from 300 to 1500 m and its age is mainly Campanian and Maastrichtian (middle Late to late Late Cretaceous). The influx of thick, coarse conglomerate at the base of the Brothers Peak Formation documents a significant pulse of uplift along the west side of the Sustut Basin in Campanian time which was probably linked with deformation in the eastern Coast Belt and western part of Bowser Basin.

The dominant clastic input into the northern part of the basin during deposition of the Brothers Peak Formation was from the west, with transport mainly towards the southeast (Fig. 9.24). This drainage was joined farther south by drainage from the east, and the combined systems thence flowed southerly. The Brothers Peak reflects uplift of Bowser Basin sediments which became sources for a massive conglomerate complex deposited as alluvial fans which prograded over swampy floodplains (Eisbacher, 1981). The final phase of deposition consisted of high gradient streams entering the basin from the northwest and thence swinging to the south and southwest, parallel with the main axis of the basin.

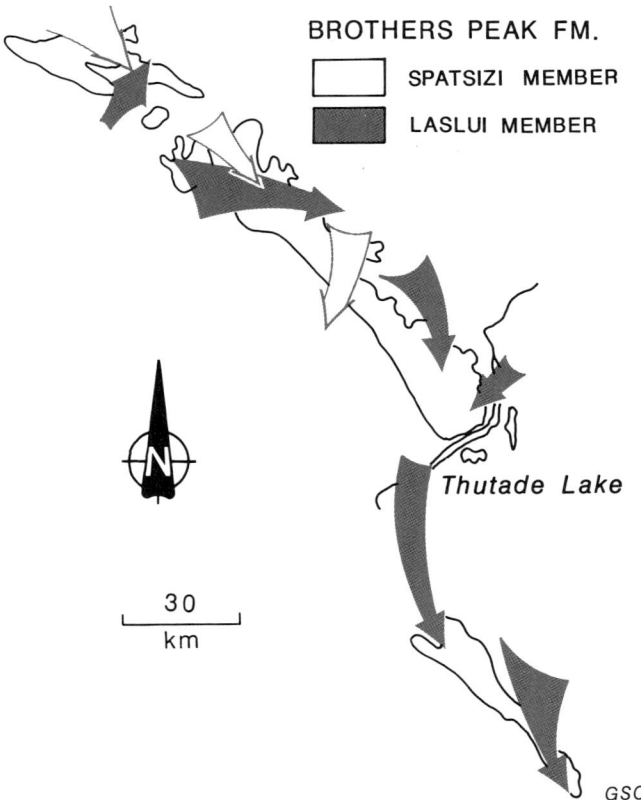

Figure 9.24. Paleocurrent directions in the Brothers Peak Formation of the Sustut Group in Sustut Basin (from Eisbacher, 1974). Figure 9.6 shows location of Sustut Basin.

Tuya and Nahlin basins

Coarse, nonmarine, poorly sorted Eocene clastic rocks, tentatively assigned to the Sifton Assemblage, are exposed along the lower reaches of the Tuya and Tanzilla rivers and along the Nahlin River, roughly on trend with exposures of the Sustut Group in the Grand Canyon of the Stikine River. Strata along the Tuya River are separated from the Cretaceous Sustut rocks by a northeast-trending fault; no direct paleogeographic connection between the two areas is evident. The sequence consists of locally derived pebble to cobble conglomerate sandstone, shale and carbonaceous shale. Lignite is common.

THE COAST BELT

G.J. Woodsworth and J.W.H. Monger

About 85% of the Coast Belt is composed of plutonic rocks, and the original nature of much of the remainder has been obscured by intense metamorphism and deformation. Nevertheless, some tentative conclusions can be drawn about the nature of the surviving strata within the Coast Belt. As noted in Chapter 8, the pre-Upper Jurassic stratigraphic framework of the Coast Belt in the Prince Rupert-Terrace region consists mainly of rocks assigned to the Alexander Terrane on the west and Stikinia to the east; to the south they are included mainly in Wrangellia and the Chilliwack Terrane. These two regions have different post-Upper Jurassic stratigraphic and plutonic assemblages.

In the Alexander Terrane near Prince Rupert, Woodsworth and Orchard (1985) have suggested that flyschoid Upper Jurassic to Lower Cretaceous rocks correlate with those in the Gravina-Nutzotin Assemblage in southeastern Alaska (Berg et al., 1972). Gravina-Nutzotin strata and associated ultramafic rocks may also be a component of the Central Gneiss Complex and related metamorphic rocks (mainly metamorphosed Stikinia) at this latitude (Douglas, 1983). This relationship indicates that the Gravina-Nutzotin succession is an overlap assemblage deposited on both the Alexander Terrane and Stikinia and that the two terranes may have been juxtaposed by Late Jurassic time. A similar relationship attends the overlap of the Gravina-Nutzotin sequence with the Alexander Terrane and Wrangellia. Strata of the Bowser Lake Group (Middle Jurassic to ?Lower Cretaceous) also may be a component protolith of the Central Gneiss Complex in the Prince Rupert-Terrace region (Hill et al., 1985).

In the southern Coast Belt most of the post-Middle Jurassic pendants appear to belong to the Upper Jurassic(?) to Lower Cretaceous Gambier Group and correlative rocks (Fig. 9.3) (Gambier Assemblage). The Gambier Group extends in a belt from near Vancouver north into the Intermontane Belt in the Mount Waddington, Bella Coola and Whitesail Lake areas. The group consists of a lower volcanic unit, mainly Hauterivian-Barremian in age, with perhaps some older strata, and an upper sedimentary unit of Middle Albian age. The lower unit consists mainly of marine and nonmarine basaltic to rhyolitic fragmental volcanic rocks and associated flyschoid sediments. Most of the volcanic members are probably calc-alkaline and dacitic to andesitic in composition. An uppermost Jurassic (Kimmeridgian-Tithonian) conglomerate containing 1.5 m blocks of hornblende granodiorite occurs in a fault block in the Tyaughton-Methow Trough between the Tchaikazan and Yalakom faults (Tipper, 1969). This conglomerate may be an easterly extension of the Gambier Group. Elsewhere, near Harrison Lake, granite-clast conglomerate at or near the base of the Fire Lake Group and Peninsula Formation underlie strata bearing Neocomian (Berriasian-Valanginian) fossils.

The Fire Lake Group (Roddick, 1965) is similar in age to the lower part of the Gambier Group but contains a greater proportion of sedimentary rocks. The basal conglomerate of the Peninsula Formation (Arthur, 1986) unconformably overlies Upper Jurassic (Oxfordian) volcaniclastics of the Billhook Creek Formation, the uppermost unit of the Harrison Lake Assemblage of the Harrison Lake Terrane. The conglomerate is succeeded by 3000 m of feldspathic sandstone, in turn overlain by volcaniclastics, pyroclastics and intermediate flows of the Valanginian to Hauterivian Brokenback Hill Formation. Thus, the age of the basal Gambier Group and related units is Early Cretaceous and may be latest Jurassic. These beds are host to important base-metal deposits of which the Britannia massive sulphide deposit is the largest.

The upper unit of the Gambier Group consists mainly of greywacke and siltstone with minor rhyolitic tuff which paraconformably overlies the lower unit. Broadly correlative strata include a thick sequence of greywacke, argillite, conglomerate and volcanic rocks assigned to the Helm, Empetrum and Cheakamus formations in the Mt. Garibaldi area (Mathews, 1958). The Middle Albian fauna from the upper unit of the Gambier Group is identical to that from the Skeena and Taylor Creek groups of north-central British Columbia and the Tyaughton-Methow Trough, respectively; these temporal as well as lithological similarities suggest depositional continuity.

The lower unit of the Gambier Group may represent arc-derived material deposited along the southwest side of the Tyaughton-Methow Trough (Fig. 9.6) and possibly also on the Hazelton Assemblage of Stikinia in the Bella Coola and Whitesail areas. Thus the Gambier Group and related strata in the southern Coast Belt may represent an overlap assemblage, correlative with the Gravina-Nutzotin Assemblage, therefore indicating the amalgamation of the Insular and Intermontane superterranes, certainly in pre-Hauterivian time and probably during the latest Jurassic.

Tyaughton-Methow Trough

The Tyaughton-Methow Trough of southwestern British Columbia, lies along the southeastern margin of the Coast Belt (Fig. 9.6) where it occurs between Quesnellia, Stikinia, and the Cache Creek Terrane to the northeast and Wrangellia, the Bridge River and Cadwallader terranes to the southwest (Fig. 9.25). The basin is dislocated and/or bounded by the Fraser, Straight Creek, Yalakom, Pasayten, Hozameen and Ross Lake faults (Fig. 9.25). Within the basin, the succession includes the Dewdney Creek (Ladner Assemblage), Relay Mountain (Relay Mountain Assemblage), Taylor Creek, Jackass Mountain, and Pasayten groups (Skeena Assemblage), which, like the assemblage in the Foreland Belt, expresses the time and effects of terrane amalgamation in the southern Cordillera (Kleinspehn, 1982, 1985).

The oldest clastic rocks in the Tyaughton-Methow Trough are assigned to the Ladner Group (Ladner Assemblage) of Early to Middle Jurassic age. The group occurs in the southern part of the basin where it ranges between 1800 and 3600 m thick. Lithologies include andesitic and dacitic volcanic conglomerate and sandstone, breccia-conglomerate, breccia, shale and argillite, most of which appear to have been deposited under a turbidity flow regime (Coates, 1974) upon Upper Triassic oceanic basalt and ultramafic rocks of the Spider Peak Group. The Ladner grades eastwards and upwards into between 500 and 1000 m of Lower to Middle Jurassic (upper Toarcian to middle Bajocian) crystal-lithic tuff, tuffaceous siltstone and wacke, thin conglomerate beds bearing rare granitic clasts, volcanic breccia and andesite flows of the Dewdney Creek Group. This usage of "Dewdney Creek" corresponds more closely to the original definition by Cairnes (1924) than to its common application to about 300 m of Oxfordian to Tithonian sandstone, sandy argillite and minor conglomerate that, in the southern or Methow portion of the trough, disconformably overlies the Ladner and Dewdney Creek groups. In the northern, or Tyaughton part of the trough, the Callovian to Barremian Relay Mountain Group consists of shale and siltstone in the central part of the trough which grade laterally into greywacke and conglomerate towards each side of the basin (Jeletzky and Tipper, 1968). The Dewdney Creek and Relay Mountain groups possibly represent a westward-facing forearc sedimentary wedge which was supplied by detritus from volcanic sources mainly to the east. Alternatively, the Relay Mountain Group may have accumulated on the eastern

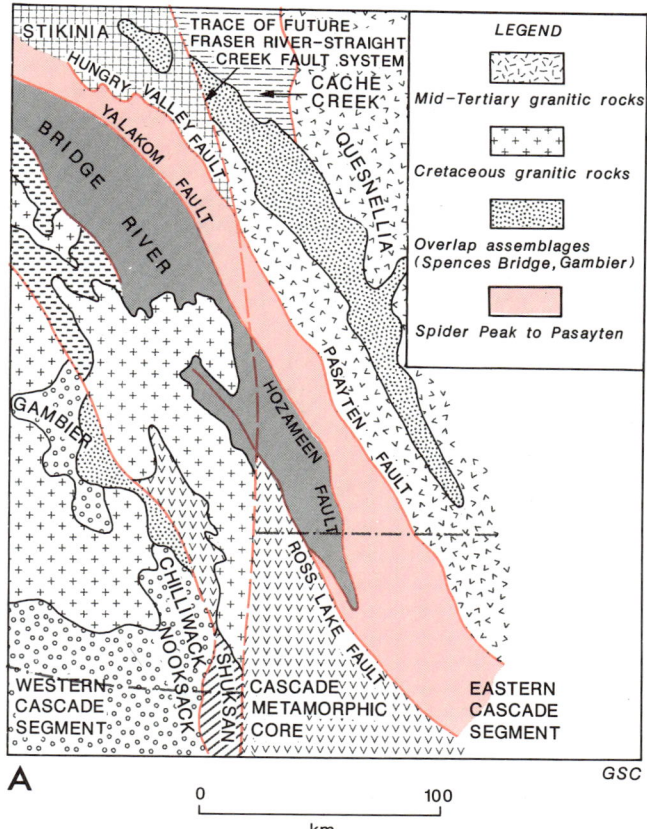

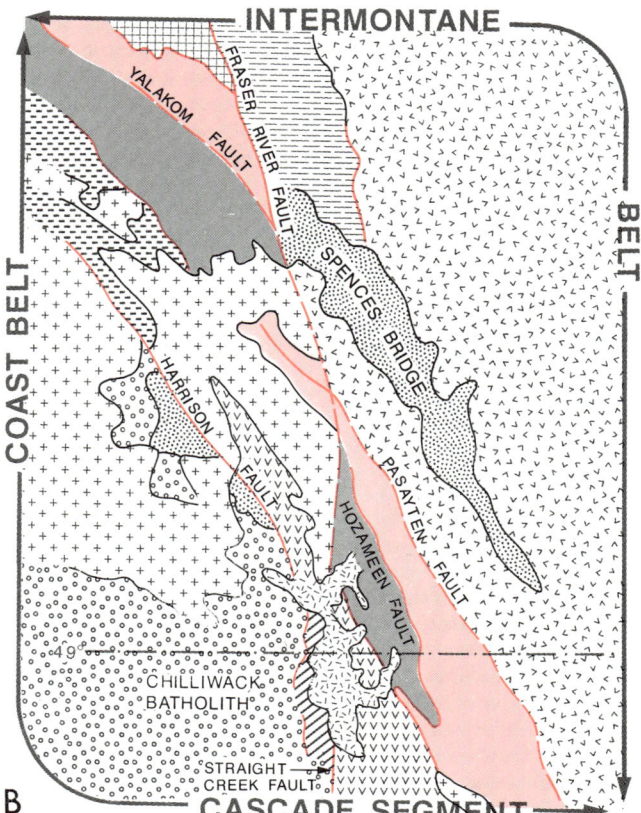

Figure. 9.25. Structural setting and reconstruction of the Tyaughton-Methow Trough, southeastern Coast Belt (after Monger, 1985).

side of a trough, the western side of which is represented by the lower, volcanic succession of the Gambier Group (H.W. Tipper, pers. comm., 1986).

The Jackass Mountain Group, variably of Barremian to Albian age, conformably and unconformably overlies the Dewdney Creek and Relay Mountain groups and interfingers with the Taylor Creek Group. The Jackass Mountain is thought to be in part coeval with Middle to Upper Albian porphyritic andesite, dacite, basalt and rhyolite of the Spences Bridge Group which have yielded K-Ar ages of 90 Ma to 112 Ma. The Jackass Mountain Group consists of up to 5000 m of polymictic, mainly normal to inversely graded, clast-supported boulder conglomerate, volcanic-rich lithic wackes and shale which are displayed in upward-coarsening, somewhat structureless sequences interpreted as submarine debris flow and turbidite components of fan complexes (Kleinspehn, 1985). Plutonic clasts constitute 43 to 67% of the conglomerates with volcanic components making up 30 to 40%. Sediment dispersal data indicate source areas lay to the east and northeast (Fig. 9.3). The group may reflect a late phase of forearc sedimentation within the Tyaughton-Methow Trough in which easterly derived volcanic debris in part derived from the Spences Bridge Group was mixed with increased quantities of plutonic clasts, the latter probably derived from the adjacent Mount Lytton Complex (Coates, 1974).

The Aptian and Albian Taylor Creek Group in the Tyaughton part of the Tyaughton-Methow Trough is as much as 700 m thick and consists of chert-pebble conglomerate, sandstone, shale and siltstone. East of the Tchaikazan Fault the sediments were apparently derived from the east whereas west of the fault they may have had a westerly source although this contention is still in doubt.

Approximately 1000 m of westerly derived Albian to Cenomanian, mostly normally graded sequences of chert-rich pebble conglomerate, volcanic-rich lithic arenites and volcaniclastic rocks conformably to unconformably overlie the Jackass Mountain Group in the Tyaughton part of the trough (Virginian Ridge Assemblage). These rocks probably are correlative with easterly derived clastics of the Virginian Ridge and Winthrop formations in the Methow Basin of northern Washington (Trexler, 1985; Trexler and Bourgeois, 1985). The succession is interpreted to represent alluvial fans or a low-sinuosity fluvial system developed near a volcanic terrain (Spences Bridge) which supplied detritus to the basin and which ultimately extruded lava sheets adjacent to and above the sedimentary section (Kleinspehn, 1985). Paleocurrent data support a western source provided during Cenomanian time by uplift of the southern part of the Coast Belt. This uplift terminated open marine conditions to the west.

In the southern part of Tyaughton-Methow Trough the Pasayten Group of probable Late Albian to possibly early Late Cretaceous age overlies the Jackass Mountain Group with gradational, unconformable and fault contact (Coates, 1974). The various components of the group comprise nonmarine sandstone of variable immaturity, lesser siltstone, minor conglomerate and a wide range of these lithologies in a coarse redbed facies. Both eastern and western sources are indicated.

The similarity of granitic and gneissic cobbles in the Jackass Mountain Group with rocks of the Mount Lytton Complex suggest that by at least Albian time, the Methow Terrane was linked to terranes to the east. Moreover, the occurrence of Jackass Mountain conglomerate interfingering with westerly derived Taylor Creek rocks of the Cadwallader Terrane west of the Yalakom Fault suggests that terranes to the west and east of the Tyaughton-Methow Trough were together prior to the Albian. Cenomanian uplift of Bridge River and Hozameen rocks due to emplacement of Coast Belt plutons provided a western source for the coarse Albian to Cenomanian clastics (Kleinspehn, 1985; Coates, 1974).

The intergradational relationship between the Jackass Mountain and Taylor Creek groups as opposed to the paleogeographically separate older sequences in the northern and southern parts of the trough suggests that the trough may not have existed as a single feature until Albian time; earlier, as many as three separate basins may have been widely separated along the western margin of the Intermontane Superterrane. Such an interpretation adds complexity to, but is not inconsistent with the suggestion of a longitudinally continuous trough extending the full length of the Cordillera during Late Jurassic to Early Cretaceous time.

Plutonic rocks

The Late Jurassic to Eocene plutonic suites in the Coast Belt are discussed in detail in Chapter 15. Late Jurassic to Early Cretaceous commonly foliated granitic rocks along the western side of the Coast Belt may be the deep-seated equivalents of the Gambier and Gravina-Nutzotin volcanic rocks (Fig. 9.2). Mid-Cretaceous granitic rocks generally occur in a matrix of foliated and metamorphosed plutonic rocks and are in part post-tectonic in the southern part of the Coast Belt and syntectonic in the Skeena River region; they may be the plutonic equivalents of the Spences Bridge and Kasalka groups (Fig. 9.3). Late Cretaceous and early Tertiary plutons in the eastern part of the Coast Belt may have been related to calc-alkaline volcanism in the Intermontane Belt (Fig. 9.4).

THE INSULAR BELT

C.J. Yorath, A. Sutherland Brown, R.B. Campbell, and C.J. Dodds

Upper Jurassic to Paleogene assemblages of the Insular Belt developed largely upon foundations comprising rocks of Wrangellia and the Alexander Terrane. Plutonic, metamorphic, volcanic and sedimentary rocks are included in the assemblages which are well exposed on Vancouver Island, on the Queen Charlotte Islands and in the Saint Elias Mountains. Early Tertiary and older strata have been penetrated in offshore exploratory wells in Tofino and Queen Charlotte basins.

Plutonic rocks

Throughout the Insular Belt the foundation rocks (mostly Wrangellia) are intruded by northwesterly and northerly aligned intermediate to felsic plutons (Sutherland Brown, 1968b; Muller, 1977; Yorath and Chase, 1981). On Vancouver Island, where they are subvolcanic to the Bonanza Group (see Chapter 8), these rocks range in age from 180 to 200 Ma (Early to early Middle Jurassic) (Isachsen et al.,

1985). An alternative interpretation is that, although Early Jurassic plutons do exist, the majority of plutons on Vancouver Island are Middle to Late Jurassic in age (see Chapter 15). On the Queen Charlotte Islands the plutons give K-Ar (hornblende analysis) ages ranging from 119 to 156 Ma; the mean of the several ages available is 144 Ma (latest Jurassic; Young, 1981). The plutons may have been subvolcanic to the calc-alkaline volcanics of the Yakoun Formation.

On the Queen Charlotte Islands the Late Jurassic plutons ("Syntectonic Plutons" of Sutherland Brown, 1968b) comprise tabular to lensoid, coarsely crystalline, foliated, mesozonal bodies that range in composition from hornblende diorite to quartz diorite. For the most part they dip steeply eastward as does their foliation, and, on the basis of structural, petrological and chemical criteria, were believed by Sutherland Brown (1968b) to have been derived through anatexis of basaltic lavas of the Karmutsen Formation. The enclosing host rocks (mostly Karmutsen Formation) have been dynamothermally metamorphosed to fine black amphibolites in fairly compact aureoles up to 1.5 km wide.

The Eocene Catface Intrusions of Vancouver Island comprise several comparatively small hypabyssal bodies of a broadly quartz diorite composition. They occur in two belts (Andrew, 1987). Those of the western belt near the coast yield K-Ar ages of 50 to 55 Ma and may be associated with the nearby "Flores volcanics" of approximately the same age (see below). The eastern belt includes small bodies ranging in age between 38 and 42 Ma intrusive into the Upper Cretaceous Nanaimo Group, the older limit of which closely corresponds to the time of a significant change in Pacific-America plate motion (Engebretson, 1982). A concordant zircon age of 42 Ma is probably the best estimate of the time of intrusion (Isachsen et al., 1985).

The Saint Elias Plutonic Suite in the Icefield Ranges of the Saint Elias Mountains comprises a varied assemblage of mainly discordant bodies of tonalite to quartz monzonite composition. They range in age between 160 and 130 Ma (Oxfordian to Hauterivian), intrude both Wrangellia and the Alexander Terrane and are approximately coeval with the Gravina-Nutzotin overlap assemblage. The Kluane Ranges Plutonic Suite comprise mesozonal quartz diorite, diorite and granodiorite bodies that range in age between 117 and 106 Ma (Aptian to Albian). Along the southwest periphery of the Saint Elias Mountains a varied assemblage of calc-alkaline epizonal plutons which yield radiometric ages between 52 and 41 Ma (Eocene) have intruded the Chugach Terrane. In Canada there is no evidence for Paleocene crustally derived granites which are abundant north of the Gulf of Alaska (Hudson and Plafker, 1982).

Sedimentary rocks
Saint Elias Mountains

Upper Jurassic to Lower Cretaceous assemblages

The Dezadeash Formation (Gambier Assemblage), in the Saint Elias Mountains (Plate 23), contains Late Tithonian to Valanginian fossils and underlies an elongate, elliptical area in the Kluane Ranges where it is bounded on the southwest by the Denali Fault (Eisbacher, 1976). To the north, regionally metamorphosed clastic strata of the Kluane Schist in the northern Coast Belt occur on the northeast side of the Denali Fault and may be correlative (see Fig. 17.83). The Dezadeash, about 3000 m thick, is a typical foredeep flysch sequence consisting of turbidite, mass flow deposits and argillite. Abundant paleocurrent indicators and orientations of slump folds document eastward to northeastward inclined paleoslopes possibly related to a deep-sea fan system with a granitoid and volcanogenic source terrain to the west (Fig. 9.2). Reconstruction of the basin involving juxtaposition of a proximal Nutzotin Mountains Sequence with a mid-fan Dezadeash Formation suggests 300 km of dextral displacement along the Denali Fault during mid-Tertiary time (Fig. 9.26).

The Valdez Group (Valdez Assemblage) of possible Early to Late Cretaceous age underlies two rugged areas on the southwestern side of the Saint Elias Mountains. In Alaska the group or its equivalents are collectively included in the Chugach Terrane which extends from Baranof Island in the southeast to Kodiak Island in the west. On its northern and eastern side the Chugach Terrane is bounded by the Border Ranges Fault along which it is juxtaposed against a variety of older terranes (Berg et al., 1978; Campbell and Dodds, 1983; Monger and Berg, 1984).

South from the lower Alsek River the Valdez Group consists of well bedded, weakly metamorphosed greywacke and argillite near the Border Ranges Fault. The metamorphic grade increases progressively toward the southwest and reaches mid- to upper amphibolite facies where the rocks have been converted to schist, gneiss and minor amphibolite.

Between Mount Logan and Mount Saint Elias the Valdez Group is contained in two thrust sheets separated by the north-dipping Columbus Fault (Campbell and Dodds, 1982). The northern sheet consists of black, locally limy argillite, greywacke, and biotite and andalusite schist. At two localities the argillites have yielded *Buchias* of Berriasisan (Sharp and Rigsby, 1956) and Early Valanginian and Late Berriasian ages (J.A. Jeletzky, pers. comm., 1975, 1980). Metamorphic grade increases away from the Border Ranges Fault and, in the mid- to upper amphibolite facies the rocks are converted to brown-weathering quartzo-feldspathic gneiss with interlayered schist. Along the Columbus Fault the metamorphic rocks are thrust southwestward over the lower plate which comprises low-grade tuff, breccia and pillowed basaltic flows. Within the lower plate the metamorphic grade also increases downward, or to the south, and the basaltic rocks are converted to crystalline amphibolite, and related sedimentary layers to schist. The age of the volcanic rocks is unknown.

In Alaska the Valdez Group contains Maastrichtian fossils (Jones and Clark, 1973) which, together with Canadian collections indicate the group may range in age from Early to Late Cretaceous. The principal episode of deformation, metamorphism and associated plutonism, presumably related to large displacement on the Border Ranges Fault, is early Tertiary (50 Ma; Hudson and Plafker, 1982).

Upper Cretaceous to Paleogene assemblages

Oligocene and (?)older conglomerate, pebbly sandstone, sandstone and shale and some coal of the Amphitheatre

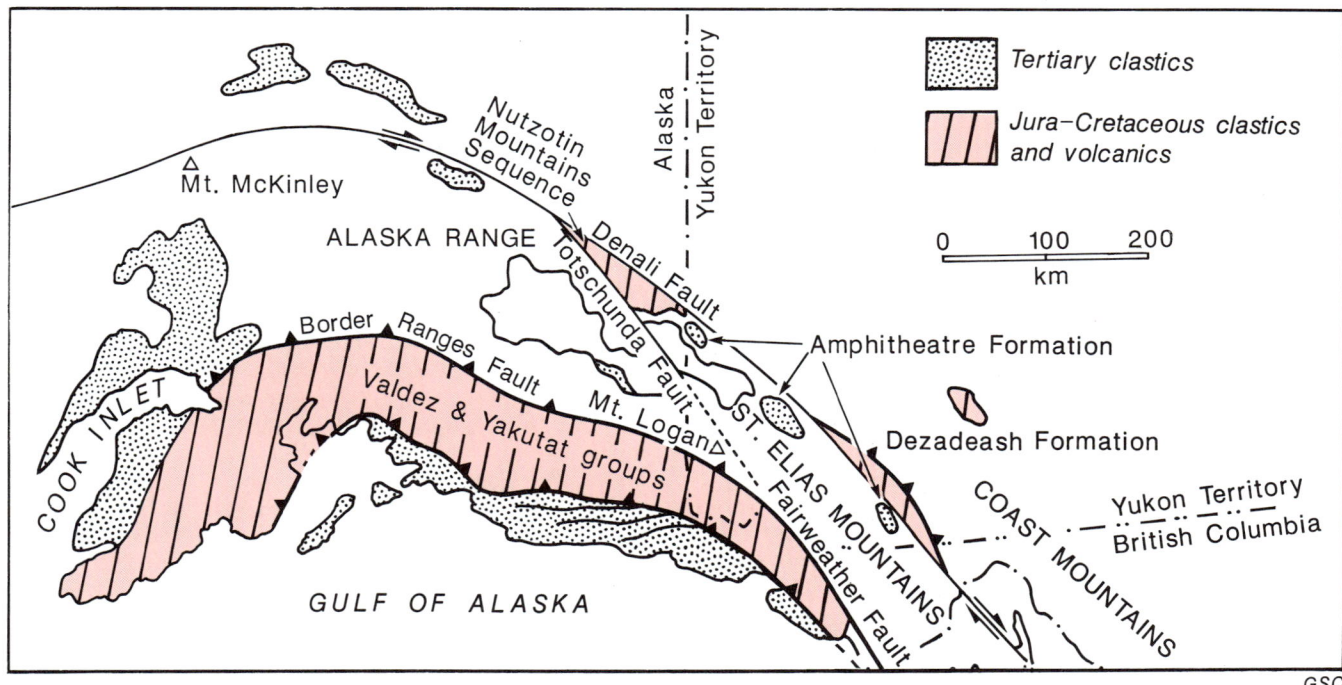

Figure 9.26. Structural setting of Jurassic-Cretaceous and Tertiary assemblages in the Saint Elias Mountains, northern Insular Belt (after Eisbacher and Hopkins, 1977).

Formation (Amphitheatre Assemblage) underlie elongate areas that form a belt more than 250 km long along the Denali Fault system in the southwestern part of Yukon Territory and adjacent British Columbia (Fig. 9.26; Eisbacher and Hopkins, 1977). The strata, locally more than 500 m thick, rest with angular unconformity on a wide variety of older rocks. Granitic and metamorphic clasts, yielding Eocene metamorphic ages, and paleocurrent data indicate an eastern source, perhaps from the northern Coast Belt which also has yielded Eocene metamorphic ages. It is inferred that a broad pediment extended eastward from the base of the Amphitheatre Formation onto Yukon Plateau. Deformation of the formation is clearly related to movements on regional boundary faults that have been active to Recent time. Coarse conglomerate, including debris flow and landslide material, was deposited towards the close of Amphitheatre deposition, indicating the control on sedimentation by active faults.

In Canada the Yakutat Group (Yakutat Assemblage) of presumed Late Cretaceous and older age is exposed in a small, remote and little known area on the southwestern side of the Saint Elias Mountains (Fig. 9.26). In Alaska it consists of more than 3000 m of flysch and associated mélange. The flysch facies, of Campanian to Maastrichtian(?) age, consists of grey-brown, lithofeldspathic greywacke and conglomerate in thick channel deposits or interbedded with grey to black siltstone, argillite or slate. The mélange facies contains blocks up to several kilometres in size composed of greenstone, oolitic limestone, marble, granitic plutonic rocks, chert and greywacke in a matrix of black, cherty, tuffaceous pelite. The clasts contain fossils of Tithonian to Valanginian age (G.P. Plafker, pers. comm., 1986).

Queen Charlotte Islands
Upper Jurassic to Lower Cretaceous assemblages

On the Queen Charlotte Islands the oldest sedimentary rocks of this assemblage are a comparatively thin and intermittently exposed sequence of unnamed *Buchia*-bearing sandstone and fine pebble conglomerate of probable Late Tithonian age (Cameron and Tipper, 1986). These rocks, included in the Bonanza Assemblage, were not observed in contact with older strata of the Moresby Group (see Chapter 8), nor were they observed in contact with the Lower Cretaceous Longarm Formation. In the adjacent offshore beneath Hecate Strait, Tertiary sedimentary rocks overlie basalt porphyry and altered basalt providing poorly constrained K-Ar dates between 118 and 165 Ma (Shouldice, 1971, 1973; Young, 1981; Yorath and Chase, 1981).

The Longarm Formation (Longarm Assemblage) of Neocomian age (Early Cretaceous) has been divided into a shoreline or proximal facies and a trough or distal facies (Sutherland Brown, 1968b; Fig. 9.27). The proximal facies, about 180 m thick, comprises very large boulder, pebble and granule conglomerate and coarse sandstone exposed on several small islands near the southeastern part of Moresby Island where this facies locally overlies the Upper Triassic to Lower Jurassic Kunga Formation with pronounced angular unconformity. The conglomerate clasts are composed of andesitic volcanic rocks, similar to Yakoun dykes which intrude the Kunga Formation. The pebble conglomerates contain both normal and inversely graded successions and all sedimentary structures, including crossbedding, imbrication, and sole marks, suggest northwesterly transport (Yorath and Chase, 1981; Fig. 9.27).

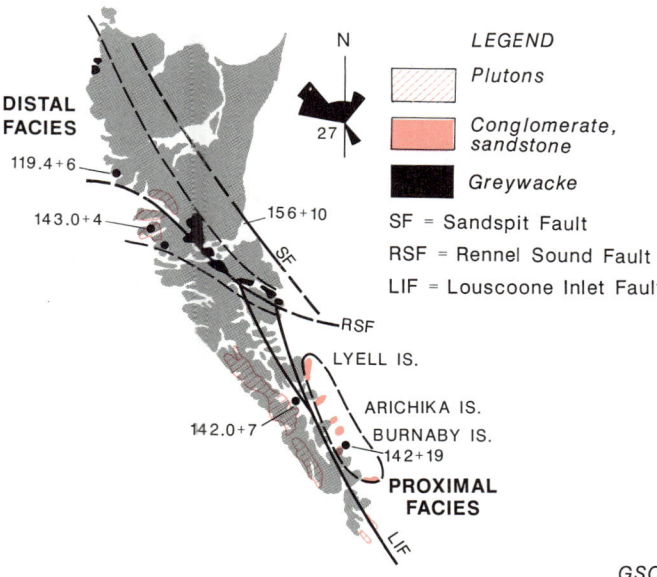

Figure 9.27. Structural setting, paleocurrent directions and facies distribution of the Longarm Formation on the Queen Charlotte Islands. Distribution of "Syntectonic Plutons" (Sutherland Brown, 1968b) is also shown (after Yorath and Chase, 1981).

The distal facies, possibly as much as 1200 m thick, lies largely within the Rennell Sound Fault Zone and consists of calcareous siltstone, fine- to medium-grained greywacke and argillite. The latter two components form couplets of well displayed uniform turbidites where original sedimentary structures are observable. Minor andesitic agglomerate and porphyritic flows occur locally in the distal facies. A characteristic feature of both facies of the Longarm Formation is the presence of very large *Inoceramus* remains, both as shell material but more commonly as moulds. The distal facies contains abundant broken fragments and calcite prisms of this fauna. The Longarm Formation and Late Jurassic plutons were included in the "Suture Assemblage" by Yorath and Chase (1981) insofar as the coarse conglomerates were interpreted as having been derived through erosion of high-relief edifices caused by the collision of Wrangellia with the Alexander Terrane in latest Jurassic to earliest Cretaceous time. The distal facies turbidites were thought to represent deep-water marine sediments delivered to an ocean floor via a trough that traversed the Queen Charlotte Islands. As noted elsewhere in this volume, in the Icefield Ranges of the Saint Elias Mountains Late Paleozoic syenites have been observed to be intrusive into both the Alexander Terrane and Wrangellia implying a) that the two terranes were together by Late Pennsylvanian time, or b) that they were never separated and that they are separate tectonostratigraphic components of a single terrane (Insular Superterrane).

Mid-Cretaceous assemblage

Overlying the Longarm Formation, probably with significant unconformity, is the Mid- to Upper Cretaceous Queen Charlotte Group (Skeena and Honna assemblages), which is divided into the Haida, Skidegate and Honna formations (Sutherland Brown, 1968b; Sutherland Brown et al., 1983; Yorath and Chase, 1981). This group is exposed in three basinal areas: in Skidegate Inlet where the component formations have their type sections and where they are thickest; on northwestern Graham Island; and about Sewell Inlet on central Moresby Island. In the latter two basins only the lower two formations are represented. The Haida Formation (Skeena Assemblage) is divisible into a lower sandstone member (820 m) and an upper shale member (330 m) which, in the Skidegate Inlet area, thicken markedly toward the west. The sandstone member consists of a basal unit of black and white, coarse, pebbly sandstone, overlain by an interbedded sequence of grey-green and green, fine- to medium-grained, immature glauconitic arkose and more mature grey-brown, argillaceous and silty quartzose sandstone (Fig. 9.28). Ironstone, large, oblate calcareous "onion ring" concretions, coal pebbles, carbonized woody debris, pelecypod coquinas and ammonites are common. Low-angle crossbedding occurs in the lower half of the member and grazing trails and convolute laminations are present in the upper part. Small logs occur in the lower 200 m of the member. The interpreted depositional environment of the Haida sandstone is that of both locally and easterly derived clastics deposited in a low wave energy, high tidal energy, estuarine embayment surrounded by subdued topography underlain by the Yakoun Formation. The sandstone passes gradationally upwards into dark grey, locally calcareous, argillaceous siltstone and mudstone of the shale division which contain grazing trails and burrows. Fossiliferous concretions are common. The age of the Haida Formation is well documented by many ammonite collections as Early Albian to Early Turonian in age (Sutherland Brown, 1968b; Haggart, 1986).

The Skidegate Formation (600 m), of Cenomanian to Early Coniacian age, appears to be the lateral facies equivalent of the upper part of the shale division of the Haida Formation (Haggart, 1986). The Skidegate (Skeena Assemblage) occurs only in the western Skidegate Inlet area where it is composed of thinly interbedded flyschoid sandstone, siltstone, mudstone and shale. Foraminifera indicate temporal equivalence but a deeper water setting than those within the upper Haida (B.E.B. Cameron, pers.

Figure 9.28. Typical shore-line exposure of Haida Formation sandstone, Skidegate Inlet, Queen Charlotte Islands. Photo by C.J. Yorath. GSC 205235-Z

comm., 1985). The Skidegate is overlain by Neogene volcanics of the Masset Formation. Beneath Hecate Strait interbedded sediments and volcanics of poorly constrained mid- to Late Cretaceous age underlie Tertiary sediments (Shouldice, 1971, 1973; Yorath and Chase, 1981)

The Honna Formation (Honna Assemblage; 400 to 1200 m) rests upon the Haida with disconformable and sharp contact, but fossils indicate that the hiatus is probably not significant. The unit is composed of polymictic conglomerate and coarse sandstone (Fig. 9.29) which form westward thickening sequences towards the Rennell Sound Fault Zone. Clast compositions reflect the total Wrangellian succession both local and probably from the Coast Belt. Primary sedimentary structures, including imbrication, crossbedding, clast-size distribution and sorting indicate a dominant eastern source for the formation in all three of its basin localities (Sutherland Brown, 1968b; Sutherland Brown et al., 1983; Yorath and Chase, 1981; C.J. Yorath, unpub. data). The conglomerate commonly occurs in poorly to well graded packages; eastern outcrops show normal grading whereas western exposures display both normal and inverse grading and resedimented components characteristic of submarine debris flows. Fossils collected from the base and middle of the unit indicate a general Coniacian age (Haggart, 1986; B.E.B. Cameron, pers. comm., 1987).

The Queen Charlotte Group was included in the "Post-suture Assemblage" by Yorath and Chase (1981). Its coarse clastic components reflect a dominant westward transport direction, and the conglomerate of the Honna Formation may represent continued response to accretion of the Insular and Intermontane superterranes (Fig. 9.3).

Vancouver Island
Upper Jurassic to Lower Cretaceous assemblages

On Vancouver Island sedimentary rocks of this assemblage unconformably overlie Wrangellia on the west and east coasts and in isolated areas on the northern part of the island. On the southern and southwestern parts of Vancouver Island they occur within a thrust panel beneath Wrangellia.

In a small area on the central west coast of Vancouver Island the Kyuquot Group (Bonanza and Longarm assemblages) of Callovian to Valanginian age comprises a composite shallow marine clastic wedge which unconformably overlies the Lower Jurassic Bonanza Group (Muller et al., 1981). The succession consists of a lower Kapoose Formation (thickness unknown) composed dominantly of a basal conglomerate and succeeding fine grained lithic sandstone, siltstone and minor sandy limestone and a disconformably overlying calcareous sandstone assigned to the One Tree Formation (about 200 m). The disconformity represents the Upper Tithonian stage. The uppermost component of the group is the Longarm Formation (Longarm Assemblage) of Valanginian to Barremian age. It is exposed in the Quatsino Sound area of northern Vancouver Island and consists of approximately 75 to 275 m of conglomerate and fine grained greywacke (Muller et al., 1974) and thus is similar in lithology and depositional environment to the proximal facies of the formation on the Queen Charlotte Islands. Similar rocks bearing a sparse Early Cretaceous (Barremian-Albian) microfauna were recovered from the

Figure 9.29. Sea stack (Height = 29 m) of Honna Formation conglomerate and coarse sandstone, Pillar Bay, Graham Island, Queen Charlotte Islands. Photo by C.J. Yorath. GSC 205235-J

continental slope off northwestern Vancouver Island (Yorath et al., 1977). The Longarm and One Tree formations have not been observed in contact.

Along the southwest coast of Vancouver Island a narrow, fault-bounded outcrop belt exposes the Pacific Rim Complex (Pacific Rim Assemblage), a sequence of Lower Cretaceous mélanges which overlie uppermost Triassic fragmental calc-alkaline arc volcanic rocks and subordinate diorite intrusions and interbedded limestone, ("Ucluth Volcanics") all of which are included in the Pacific Rim Terrane. The mélanges consist of chaotic assemblages of mudstone, sandstone and chert containing slide blocks of the basement volcanic rocks (Fig. 9.30). The heterogeneous structural style of the mélanges suggests an origin through down-slope mass movement processes such as submarine

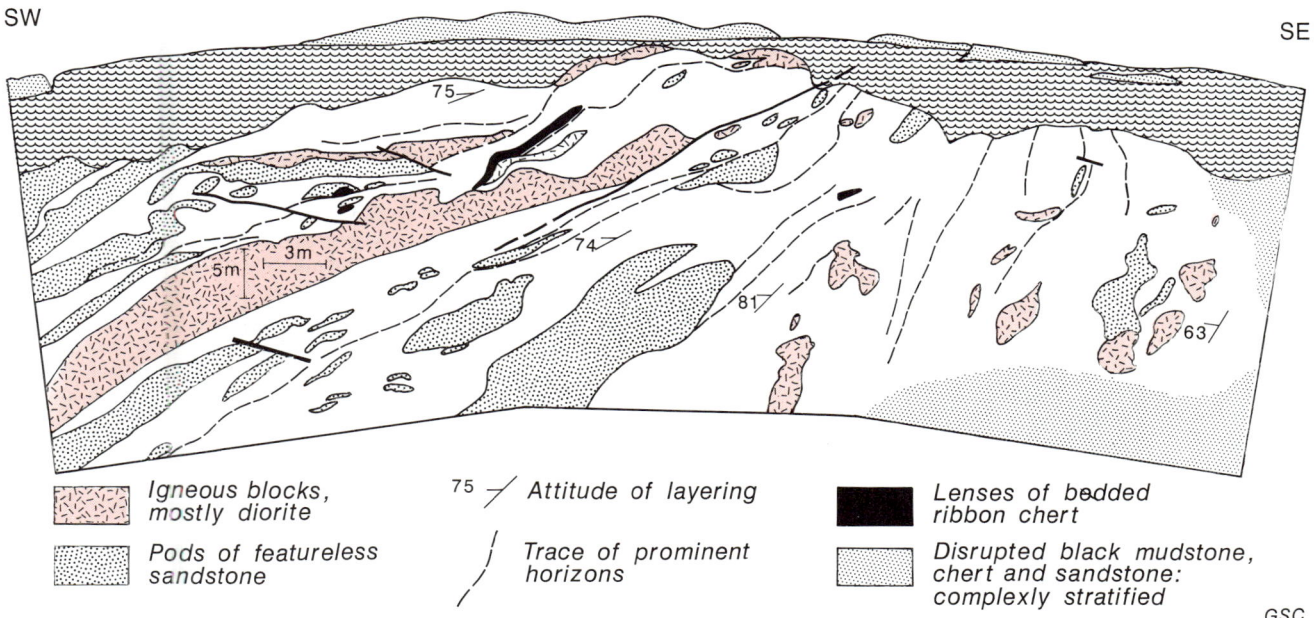

Figure 9.30. Detailed map of mélange fabric, Pacific Rim Complex, Vancouver Island (from Brandon, 1985). Wavy pattern is water.

slides, rock falls, debris flows and in situ liquefaction (Brandon, 1985). Based upon similarities in stratigraphy and metamorphism with the Constitution Formation, Lopez Complex and Decatur Terrane rocks of the San Juan Islands, Brandon and Cowan (1983) suggested that the Pacific Rim Terrane is a displaced fragment within a large transform fault system which truncated the west side of Vancouver Island during latest Cretaceous or early Tertiary time, which would mean that the Upper Triassic volcanic rocks are not part of Wrangellia.

Farther to the southeast are the Jura-Cretaceous Leech River Formation and so-called "Pandora Peak Unit", each of which are components of the Pacific Rim Terrane and are separated from Wrangellia by the San Juan Fault. The Leech River Formation, bounded by the San Juan, Survey Mountain and Leech River faults, is composed of poorly dated Jura-Cretaceous metasedimentary and metavolcanic rocks which are thought to be higher grade equivalents of the Pacific Rim Complex (Muller, in Brandon et al., 1983). The metasediments range from schistose greywacke and argillite through slate and phyllite to quartz-staurolite-andalusite-biotite-garnet schist (Muller, 1977); the protoliths are considered to have been turbidites and minor volcanic rocks (Rusmore and Cowan, 1985). The rocks are intensely deformed and locally intruded by small quartz diorite and granite bodies. Rb-Sr data suggest only that the depositional age of the complex is Jurassic-Cretaceous; abundant K-Ar dates indicate that deformation and metamorphism culminated between 41 and 39 Ma (Rusmore and Cowan, 1985; Fairchild and Cowan, 1982). The Pandora Peak Unit (Rusmore and Cowan, 1985) comprises mudstone, argillite, greywacke, radiolarian chert, tuff and basaltic metavolcanic rocks of general Late Jurassic to Early Cretaceous age. Unlike the high temperature-moderate pressure amphibolite grade metamorphism of the Leech River Formation the Pandora Peak Unit contains a distinctive low temperature-high pressure lawsonite + prehnite + quartz assemblage similar to that observed in the Pacific Rim Complex and Constitution Formations (Brandon, 1985). Although several interpretations of tectonic setting are possible, Rusmore and Cowan (1985) appeared to favour a marginal slope or forearc setting adjacent to a subaerial volcanic arc associated with transform-related rifting.

Mid-Cretaceous assemblage

The Coal Harbour Group (Skeena Assemblage) of Aptian to Cenomanian age occurs on northern Vancouver Island where it was assigned to the Queen Charlotte Group by Muller et al. (1974). Measured and studied in detail by Jeletzky (1976), the group comprises an Aptian "Coarse arenite unit" (240 to 1650 m) composed of greywacke, shale, minor limestone, conglomerate and coal, a middle nonmarine Blumberg Formation (900+ m) composed dominantly of conglomerate but with minor siltstone, shale and coal and an upper informal "Upper shale unit" (200+ m) consisting of siltstone, shale and minor conglomerate and limestone. Like the Queen Charlotte Group the Coal Harbour Group appears to represent molassoid accumulation consequent upon uplift and erosion of older Wrangellian rocks.

Upper Cretaceous to Paleogene assemblage

The Nanaimo Group (Nanaimo Assemblage) of Late Santonian to Maastrichtian, and possibly early Tertiary age, underlies the coastal plain of southeastern Vancouver Island, the adjacent Gulf Islands and the Cowichan and Alberni valleys of south-central Vancouver Island (Muller and Jeletzky, 1970). The group is enclosed mainly in two subdivisions of the Georgia Basin, the Comox Basin in the north and the Nanaimo Basin in the south where locally the base of the group may be as old as Turonian

(P.D. Ward, pers. comm., 1985). Its nine component formations are, in upwards stratigraphic order: Comox-350 m; Haslam-200 m; Extension-Protection-300 m; Cedar District-300 m; De Courcy-350 m; Northumberland-250 m; Geoffrey-150 m; Spray-200 m; and Gabriola-350 m. These formations comprise variously alternating sequences of coarse- and fine-grained terrigenous clastics (Fig. 9.31). In the Comox Basin these are dominantly of neritic to outer shelf affinity, whereas, in the Nanaimo Basin, turbidites with high-energy sedimentary structures suggest the presence of deeper water slope and fan environments adjacent to a narrow shelf zone (Sliter, 1973; Ward and Stanley, 1982; Pacht, 1980, 1984). During the Santonian and earliest Campanian the dominant source regions for Nanaimo Basin clastic sediments probably were local areas to the west and to the southeast, whereas during the remainder of the Late Cretaceous Epoch, Coast Belt and Cascade sources are important (Fig. 9.32; Pacht, 1984). In the Comox Basin the sediments are more or less undeformed. However, thrust faults apparently dislocate the lower part of the Nanaimo Group in the southern part of the basin and in Alberni Valley (Sutherland Brown and Yorath, 1985). In the Nanaimo Basin all components of the group are deformed into linear, northwesterly trending cylindrical folds and possibly cut by thrust faults. Along the northeast coast of Vancouver Island the small Suquash Basin contains Campanian sandstone and shale that are assigned to the Nanaimo Group.

Towards the centre of the Georgia Basin the Richfield Pure Point Roberts well penetrated approximately 2750 m of Middle and Upper Eocene nonmarine sandstone, mudstone and minor coal overlying the Nanaimo Group and assigned to the Nanaimo Assemblage (Fig. 9.31). Hopkins (1966) assigned the Eocene strata to the Burrard and Kitsilano formations of the Vancouver area which in turn are equivalent to the Chuckanut Formation of northwestern Washington (Johnson, 1984).

In the Vancouver area, the nonmarine, Middle Eocene Burrard Formation is represented by about 600 m of conglomerate, feldspathic sandstone, shale and minor lignite; the formation rests upon granitic rocks. The overlying nonmarine Kitsilano Formation is perhaps more than 760 m thick and consists of a basal conglomerate overlain by coarse, crossbedded sandstone and shale. The unit is thought to be Late Eocene in age (Hopkins, 1966). The Burrard and Kitsilano, together with younger sediments in Washington, are enclosed within the Whatcom Basin which appears to be a Tertiary embayment of Georgia Basin (Fig. 9.6).

On the southwest coast of Vancouver Island the Middle Eocene to Oligocene Carmanah Group (Carmanah Assemblage) is exposed in the Nootka Sound area and on the north shore of the Strait of Juan de Fuca. In the former region the group is in both depositional and fault contact with the Lower Jurassic Bonanza Group and West Coast Crystalline Complex (Muller et al., 1981) and in the latter area it is in stratigraphic contact with the Jura-Cretaceous Leech River Formation and the Lower Eocene Metchosin volcanics (Muller, 1980). The Carmanah Group comprises the Escalante, Hesquiat and Sooke formations, in upwards stratigraphic order. The Escalante Formation (140 m) consists of outer shelf to bathyal calcareous sandstone and minor polymictic conglomerate which gradationally passes upward into irregularly interbedded deep water shale, sandstone and conglomerate of the Hesquiat Formation (1100 m; Fig. 9.33; Cameron, 1980; Muller et al., 1981). The Sooke Formation (<200 m) consists of coarse, littoral sandstone and conglomerate. Beneath the continental shelf, Tofino Basin contains Eocene and Oligocene immature sandstone and mudstone which are believed to be a distal facies of the Carmanah Group.

On Vancouver Island two volcanic sequences of Eocene age but different lithology occur along the west and south coasts. On the west coast the informally named "Flores Volcanics" (Kamloops Assemblage) consists of intermediate to silicic calc-alkaline laharic flows and breccias which have yielded zircon ages of between 54 and 56 Ma (Brandon, 1985).

On the southern part of Vancouver Island the Metchosin Volcanics and associated Sooke Gabbro of Early Eocene age (Muller, 1977, 1980) occur in the footwall of the Leech River Fault. More properly termed the "Metchosin Igneous Complex" (Metchosin Assemblage) the rocks have an ophiolitic stratigraphy. Massive and layered gabbros pass upward into and are intruded by a well developed sheeted dike complex. This in turn is succeeded by subaqueous pillow and sheet flow basalts with minor pyroclastic rocks overlain by subaerial(?) amygdaloidal flows. The general stratigraphy suggests that the succession developed in a ridge-centred oceanic island setting (Massey, 1986). Muller (1980) suggested a seamount origin based upon chemical studies.

The basalts of the Metchosin Igneous Complex are chemically identical to basalts that were penetrated by exploratory wells in Tofino Basin (Shouldice, 1973), and which have been mapped magnetically by MacLeod et al. (1977). Moreover, they are also similar in chemistry and age to the Eocene Crescent Formation of the Olympic Mountains of northwestern Washington. The Metchosin Igneous Complex and Crescent Formation together comprise the Crescent Terrane which was emplaced beneath the Pacific Rim Terrane by the end of the Eocene (Clowes et al., 1987), possibly as a consequence of a change in the Pacific Plate tectonic regime at about 42 Ma (Engebretson, 1982).

FAULT-CONTROLLED BASINS
H. Gabrielse

Numerous fault-controlled basins of Late Cretaceous to early Tertiary age occur throughout the Canadian Cordillera, principally in the Omineca and Intermontane belts (Fig. 9.34). Whereas those in the northern Omineca Belt are clearly associated with the major dextral strike-slip faults, the basins in the southern Omineca Belt and southeastern Intermontane Belt more probably resulted from regional tensional stress associated with strain transfer between the Northern Rocky Mountain Trench and Fraser-Straight Creek Fault systems; volcanic rocks of the Kamloops and Penticton groups also may be linked to this extension.

Additional information on these basins is given in Chapter 20 (Part B).

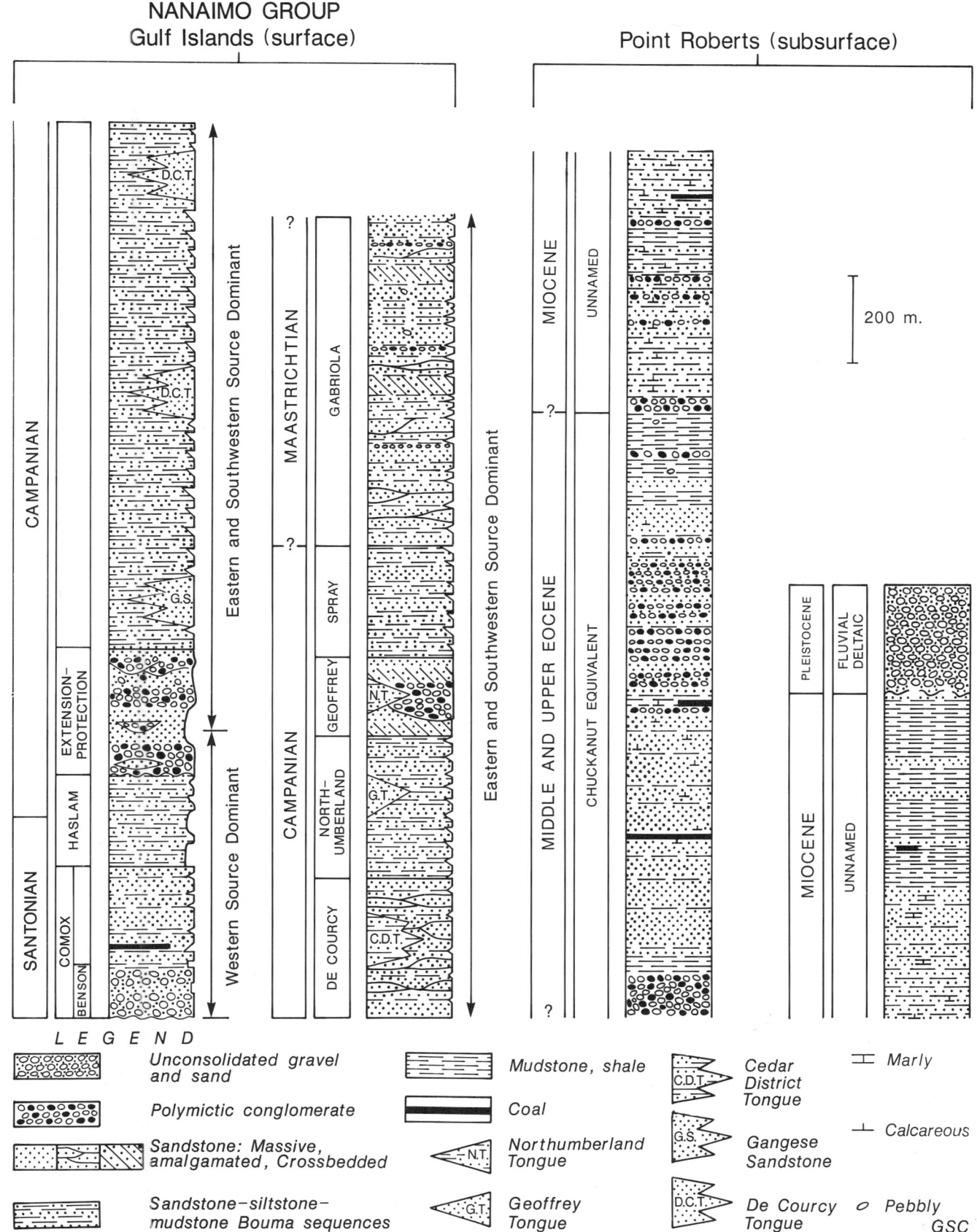

Figure 9.31. Generalized stratigraphy of the Nanaimo Group and Tertiary sediments of southern Georgia Basin (based on data from Muller and Jeletzky, 1970; Hanson, 1976; Carter, 1977; Hopkins, 1966; Pacht, 1984; Johnson, 1984; Yorath, unpubl. data).

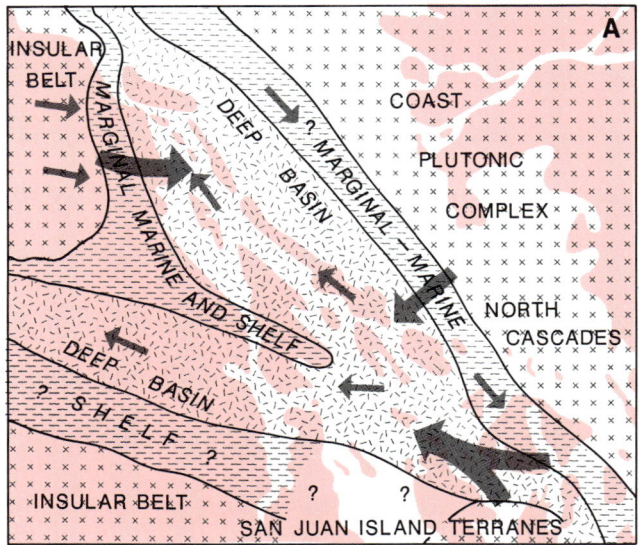

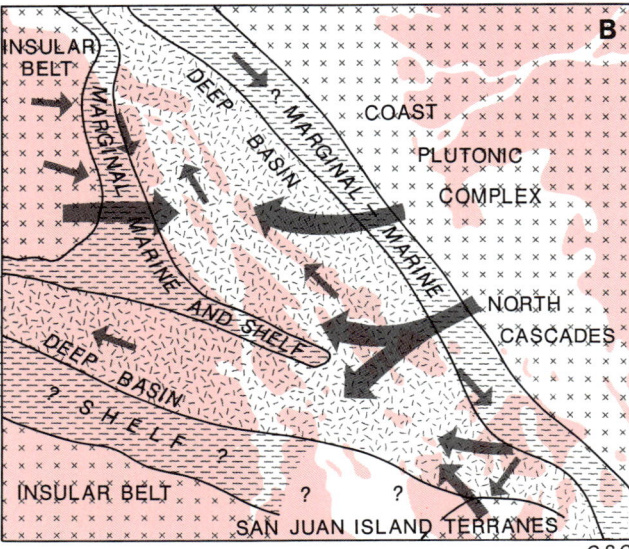

Figure 9.32. Paleogeography and sediment transport directions in southern Georgia Basin during (A) Santonian to earliest Campanian time (Comox, Haslam and lower Extension-Protection formations) and (B) mid-Campanian and Maastrichtian time (upper Extension-Protection, Cedar District, DeCourcy, Northumberland, Geoffrey, Spray and Gabriola formations) (from Pacht, 1984).

Tintina and Northern Rocky Mountain trenches

Scattered exposures of coal-bearing Paleogene rocks occur along the Tintina Trench southeastward from the Alaska-Yukon boundary for more than 200 km. Typical sequences consist of interstratified, poorly sorted, massive and indistinctly bedded pebble conglomerate, organically rich mudstone and minor thin beds of medium- to coarse-grained sandstone and pebbly mudstone. Lignitic coal with claystone partings is present at several localities. Bimodal volcanic rocks also occur locally. Thicknesses of these

Figure 9.33. Fluxoturbidite conglomerate overlying pebbly mudstone, lower Hesquiat Formation, Le Clair Point, Hesquiat Peninsula, Vancouver Island, British Columbia. Photo by B.E.B. Cameron. GSC 205235-K

sediments are difficult to determine owing to poor exposure and intense deformation. Locally as much as 700 m has been estimated. Reported ages, based on macroflora and pollen, range from Late Cretaceous to, more commonly, Eocene.

Ross River area

Two fault blocks, one trending northwest, parallel with the Tintina Trench, and another trending east-west contain successions of conglomerate, sandstone, shale, claystone and coal up to 400 m thick. In this region, as in the Dawson area to the northwest, the sediments generally coarsen upward with coal being more conspicuous in the lower members. Vitrinite reflectance values suggest an abnormally high geothermal gradient during the burial period of the coals (Hughes and Long, 1980). The succession is probably of Eocene age.

Liard Plain region

Several occurrences of Eocene strata along and near Liard River are similar in lithology to those in the Ross River and Dawson areas. The extent of the basin is obscured by widespread thick surficial deposits which also conceals their structural style as it might be related to the Tintina and Northern Rocky Mountain trenches. Along the southwestern margin of Liard Plain immediately south of the Yukon-British Columbia boundary, a sequence of micaceous greywacke, pebbly greywacke and pebble conglomerate, possibly 200 to 300 m thick, underlies a northwesterly trending area about 2 km wide and 20 km long. Pebbles of chert are most abundant but others include much creamy feldspar in clasts to 1 cm long and abundant slate chips. The lithologies indicate a western source including the Sylvester Group and the Cassiar Suite. A Late Cretaceous or Paleogene age is inferred.

Two localities of coal-bearing Paleogene clastic rocks are known east of the Liard Plain in the Hyland Highland. A small area is underlain by coal and claystone on Coal

UPPER JURASSIC TO PALEOGENE ASSEMBLAGES

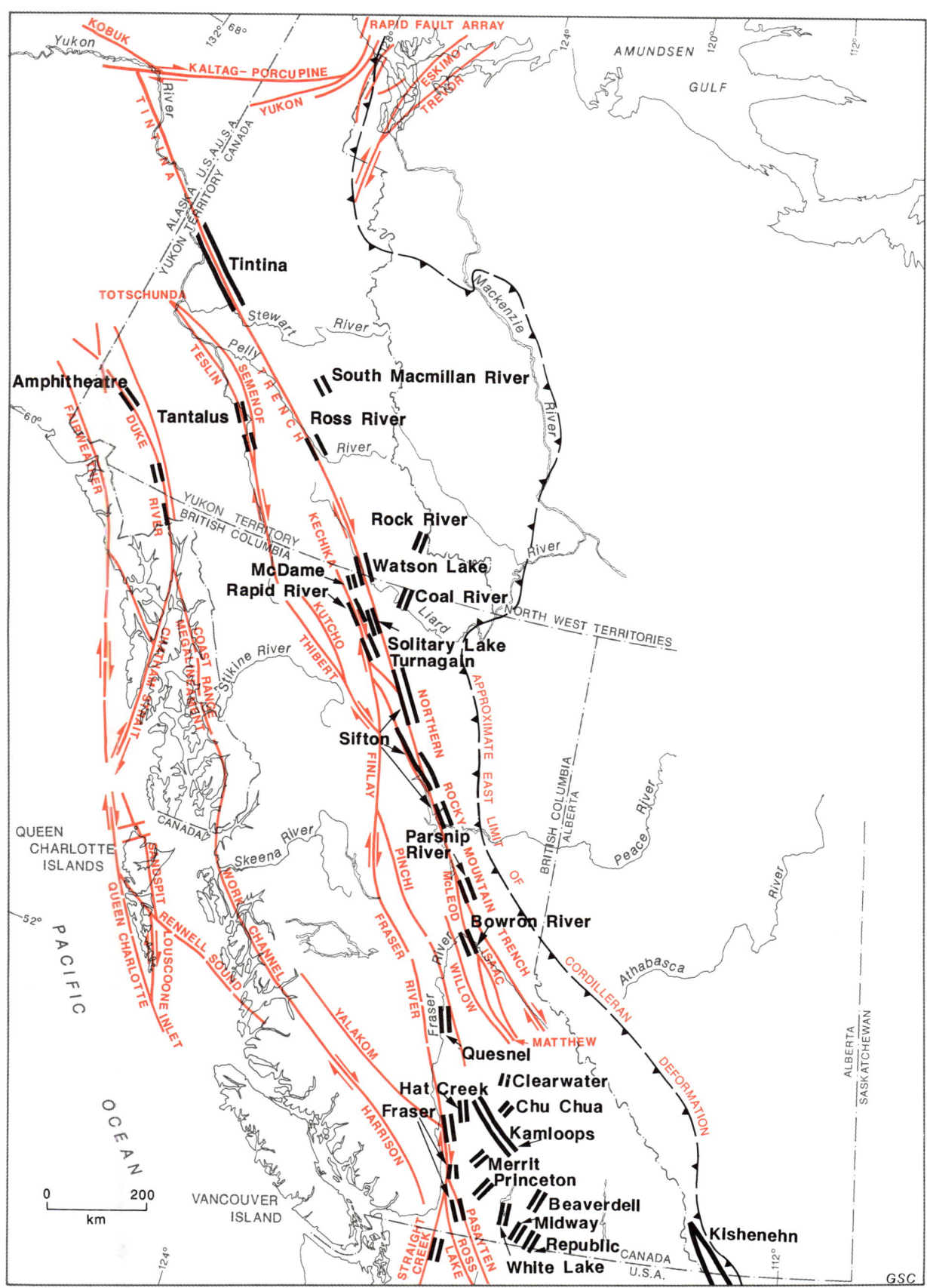

Figure 9.34. Fault-controlled basins of the Canadian Cordillera.

River near its confluence with Liard River and more extensive outcrops occur along Rock River, about 100 km to the north. There, a basin more than 40 km long has been outlined by geophysical and surface exploration. Substantial lignitic coal resources are contained in a sequence of clastic rocks, possibly more than 1000 m thick (Wright and Miller, 1983). The suggested thickness of sediment indicates depositional control by north-trending faults.

Sifton Basin

Upper Cretaceous and Paleogene nonmarine clastic sediments of the Sifton Formation are exposed in and along the Northern Rocky Mountain Trench (Fig. 9.34) and along several other prominent fault-related valleys to the west (Eisbacher, 1974; Gabrielse, 1985). The rocks consist of a varied assemblage of boulder conglomerate, breccia, conglomeratic sandstone, sandstone, siltstone and mudstone and local lignite. The strata are poorly exposed and, in places, much deformed but locally display unbroken sections greater than 150 m thick. Dacitic volcanic rocks of Eocene age occur within the sediments in a few areas in and near the Northern Rocky Mountain Trench. Lamprophyre dykes of about the same age cut Sifton sediments in the Sifton Ranges. Several areas along the Kechika Fault near the Rapid River (Fig. 9.34) are underlain by coal-bearing sediments of probable Paleogene age. One locality has yielded a Santonian-Campanian flora. Flows and tuff of rhyolitic composition have been dated as Eocene.

Paleocurrent data (Eisbacher, 1974) indicate that along the Northern Rocky Mountain Trench major drainage was directed to the south (Fig. 9.35) although vigorous local alluvial fans from the east may have caused a shift of the principal valley towards the southwest. From the southerly directed paleocurrent data it appears that the Sifton Basin constituted a much broader alluvial plain than the present floor of the trench. Drainage in the basin was parallel with the regional structure and directed towards the south.

Bowron Basin

Coal-bearing strata, which outcrop along the Bowron River, southeast of Prince George, are more than 600 m thick. They consist predominantly of sandstone and fine conglomerate with a coarse basal conglomerate (Campbell et al., 1973). Coal occurs only in the basal 60 m where mudstone is interbedded with the coarser clastic sediments. The structure of the basin suggests deposition in a graben, bounded on the northeast by a westward-dipping normal fault that was probably active during deposition.

Fraser River and North Thompson River areas

Scattered exposures of Paleogene to possibly early Neogene clastic rocks, locally containing coal, occur intermittently along the Fraser River between Prince George and Hope. In places the sequences are more than 500 m thick, and,

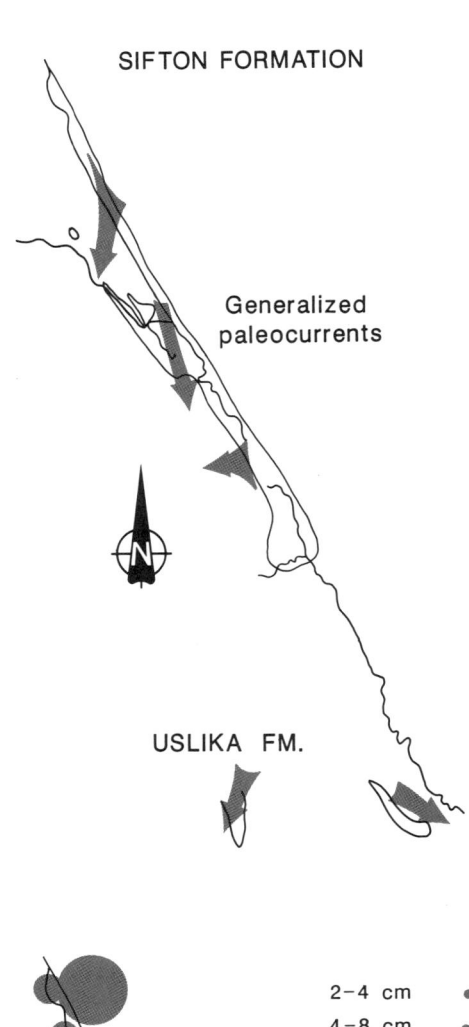

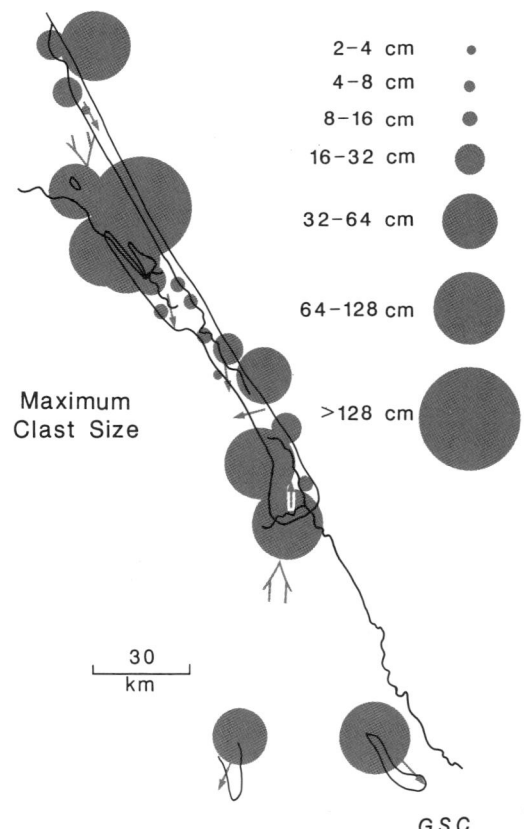

Figure 9.35. Paleocurrents and maximum clast sizes of the Sifton Formation (for location see Fig. 9.23). After Eisbacher, 1974.

north of Lytton, a steeply folded Eocene sequence comprising a basal boulder conglomerate overlain by pebble and cobble conglomerate, arkose and shale is more than 900 m thick. Similar strata nearby include basalt, volcanic breccia and volcanic conglomerate (Duffell and McTaggart, 1952). On the North Thompson River, fluvial and lacustrine sediments, as much as 700 m thick, consist of basal conglomerate overlain by coarse pebble sandstone and fine conglomerate and a considerable thickness of arkose and sandy shale (Campbell and Tipper, 1971).

Hat Creek

An assemblage of Eocene clastic strata in the Hat Creek area, part of the Kamloops Group, is remarkable because of its great thickness, in excess of 2300 m, and for the considerable thicknesses of its several coal seams (see Chapter 20; Church, 1975). The lowest unit, referred to as the Coldwater beds, is about 1300 m thick and consists of sandstone and conglomerate. A middle unit, more than 400 m thick, comprises four coal seams with an aggregate thickness of 370 m; these are separated by siltstone, sandstone and conglomeratic sandstone. The upper unit is more than 600 m thick and is made up of a monotonous, massive siltstone and claystone. The Hat Creek basin occurs in a graben bounded by north-trending faults which are locally offset by northwest- and northeast-striking conjugate shear faults.

Nicola-Princeton basins

A north-trending belt of mainly Eocene(?) sandstone, shale, conglomerate and coal is represented by several discontinuous exposures underlain by the Nicola Group or younger rocks and, in many places, overlain by Miocene volcanic rocks. Locally the strata are tightly folded, possibly adjacent to faults that controlled basin development. Thicknesses range up to a few hundred metres.

The most prominent faults active in Early Paleogene time trend north-northeast and are believed to be extension faults related to a regional, north-northwest trending shear regime (Ewing, 1981a; Monger, 1985).

Tantalus basins

Several northwest-trending, at least in part fault-bounded, exposures of nonmarine, coal-bearing fluvial clastic rocks assigned to the Tantalus Formation occur within the northern Intermontane Belt in the southwestern part of Yukon Territory (Cairnes, 1910; Bostock, 1936; Wheeler, 1961; Tempelman-Kluit, 1974). Possibly correlative rocks occur about 40 km south-southeast of Dawson (Lowey, 1984). The lower part of the formation is characteristically made up of recessive, massive, medium- to coarse-grained sandstone and shale, the latter containing abundant detrital muscovite. The shale is locally carbonaceous and, in places, includes discontinuous coal seams several metres thick. The base of the sequence rests unconformably on a variety of older rocks. Locally, as much as 100 m of red-weathering volcanic breccia marks the base of the formation. The upper part of the Tantalus Formation typically consists of chert-pebble conglomerate with well rounded clasts, mainly of black, grey, green and white chert and white or grey quartz. Unlike underlying strata of the Laberge Group the Tantalus rocks contain little or no K-feldspar or volcanic fragments. The correlative rocks southeast of Dawson, however, include significant amounts of orthoclase as well as plutonic, volcanic and metamorphic detritus. Thicknesses range from 200 m to more than 1000 m.

Deposition of the Tantalus Formation was probably influenced by contemporaneous faulting, which created small pull-apart basins in part related to the Teslin fault system (D.J. Tempelman-Kluit, pers. comm., 1986). Southerly and southwesterly flowing fluvial systems drained source areas underlain probably by deeply weathered rocks.

The age of the Tantalus Formation has been considered generally as Late Jurassic and Early Cretaceous on the basis of enclosed macroflora and palynological material. Most recent determinations on spores and pollen suggest an Early Cretaceous age, in part Albian. The lithology of the formation, with its abundant mica and quartz is like that of the Skeena Formation and Sustut Group in north-central British Columbia and unlike that of the older Bowser Lake Group with which it has been correlated.

CONCLUSIONS

The interval between the Late Jurassic and the end of the Paleogene witnessed a complicated succession of events that led to the modern framework of the Canadian Cordillera. The geological record contains evidence for major plate collisions giving rise to structural, metamorphic, magmatic and sedimentary phenomena whose relationships are still not fully understood. The following are the main elements considered:

1. Between Middle Jurassic and late Paleogene time the five morphogeological belts were established.

2. In Early to Middle Jurassic time the Intermontane Superterrane collided with the western margin of ancestral North America resulting in plutonism, metamorphism and uplift in the Omineca Belt and, eventually, the development of the first of three foredeep clastic wedges in the Foreland Belt.

3. The time of accretion of the Insular Superterrane is uncertain. An overlap assemblage earliest Cretaceous or possibly older age bridges the Insular and Intermontane superterranes across the Coast Belt, indicating accretion or close proximity of the superterrane to North America at least by latest Jurassic time. Mid-Cretaceous deformation, metamorphism and plutonism in the Coast Belt and the change from eastern to western source areas for sediments in the Tyaughton-Methow Trough suggest additional convergence and final accretion of the two superterranes. These events were synchronous with plutonism, metamorphism and uplift in the Omineca Belt and the related development of a second clastic wedge in the Foreland Belt. Together, these mid-Cretaceous events reflect a time of rapid convergence between ancestral North America and newly accreted superterranes.

4. Bowser Basin developed as a foredeep upon Stikinia as a consequence of contraction and uplift of the Cache Creek Terrane probably during accretion of the Intermontane Superterrane with the ancient continental margin. The subsequent accumulation of the first clastic wedge of the Foreland Belt was in response to continued

deformation and uplift which resulted in the development of the Omineca Belt.

5. During and following a widespread lull in magmatism in Early Cretaceous time a broad marine transgression occurred throughout the central and northern Intermontane Belt. Although subsequently disrupted by faults a single depositional continuum may have linked the Dezadeash, Gravina-Nutzotin, Bowser, Skeena-Sustut and Gambier 'basins' with the Tyaughton-Methow Trough in Albian time.

6. Upper Cretaceous rocks in the Sustut Basin record a reversal in the directions of sediment transport. Omineca Belt sources were important during Albian to Campanian time whereas from Campanian to Maastrichtian time the sediment source was dominantly from the inverted Bowser Basin (Skeena Mountains) to the west. Deformation and uplift of the Bowser Basin was linked to uplift and the development of eastward verging structures in the eastern part of the Coast Belt.

7. The Tyaughton-Methow Trough, on the perimeter of the southeastern Coast Belt, contains a record of opposed clastic sources. Until the close of the Albian, sediment supply to the trough was mainly from the east. At the beginning of Cenomanian time, however, the newly uplifting Coast Belt to the west became an important source area. The uplift may have been a response to continued convergence of the Insular Superterrane with the Intermontane Superterrane.

8. In the Insular Belt the Pacific Rim and Crescent terranes were emplaced beneath Wrangellia by the end of the Eocene, possibly as a consequence of a change in the Pacific Plate tectonic regime at about 42 Ma. In the Saint Elias Mountains the Chugach Terrane was accreted to the continental margin in mid-Paleocene time.

9. Numerous fault-controlled sedimentary basins of Late Cretaceous to early Tertiary age occur throughout the Intermontane and Omineca belts. Those in the central and northern parts of the region are related to major dextral motions. In the south, several sedimentary and volcanic sequences owe their origin to an early Tertiary tensional stress regime that resulted in extension of the southern Omineca Belt and southeastern Intermontane Belt.

10. Whereas Wrangellia has been considered to be a separate terrane that amalgamated with the Alexander Terrane to form the Insular Superterrane in latest Jurassic to earliest Cretaceous time, recent observations of intrusive relationships in the Icefield Ranges of the Saint Elias Mountains indicate that these two terranes were amalgamated prior to the Late Pennsylvanian or were never separated.

11. Paleomagnetic data indicate that the combined Insular and Intermontane superterranes (Baja British Columbia) were at southerly paleolatitudes during the mid-Cretaceous and that they did not accrete to the continental margin in their present position until the Eocene. The character and distribution of Jurassic faunas support the contention of northward movement of Baja British Columbia, but perhaps from not as far south as suggested by paleomagnetic data. The eastern Boundary of Baja British Columbia has not been recognized nor have structural events related to its emplacement in Late Cretaceous to Eocene time. Considerable additional effort is required to reconcile geological and paleomagnetic data.

REFERENCES

Aitken, J.D.
1959: Atlin map-area, British Columbia; Geological Survey of Canada, Memoir 307.

Aitken, J.D. and Cook, D.G.
1974: Carcajou Canyon map-area, District of Mackenzie, Northwest Territories; Geological Survey of Canada, Paper 74-13.

Anderson, R.G.
1984: Late Triassic and Jurassic magmatism along the Stikine Arch and the geology of the Stikine batholith, north-central British Columbia; in Current Research, Part A, Geological Survey of Canada, Paper 84-1A, p. 67-73.

Andrew, A.
1987: Lead and strontium isotope study of five volcanic and intrusive rock suites and related mineral deposits, Vancouver Island, British Columbia; Ph.D. thesis, University of British Columbia, Vancouver.

Arthur, A.J.
1986: Stratigraphy along the west site of Harrison Lake, southwestern British Columbia; in Current Research, Part B, Geological Survey of Canada, Paper 86-1B, p. 715-720.

Balkwill, H.R., Cook, D.G., Detterman, R.L., Embry, A.F., Hakansson, E., Miall, A.D., Poulton, T.P., and Young, F.G.
1983: Arctic North America and northern Greenland; in The Phanerozoic Geology of the World, II The Mesozoic B; Elsevier Science Publishers B.V., Netherlands.

Berg, H.C., Jones, D.L., and Richter, D.H.
1972: Gravina-Nutzotin belt - tectonic significance of an upper Mesozoic sedimentary and volcanic sequence in southern and southeastern Alaska; United States Geological Survey, Professional Paper 800-D, p. 1-24.

Berg, H.C., Jones, D.L., and Coney, P.J.
1978: Map showing pre-Cenozoic tectonostratigraphic terranes of southeastern Alaska, and adjacent areas; United States Geological Survey, Open File Report 78-1085, 2 sheets.

Blusson, S.L.
1971: Sekwi Mountain map-area, Yukon Territory and District of Mackenzie; Geological Survey of Canada, Paper 71-22.

Bostock, H.S.
1936: Carmacks District, Yukon; Geological Survey of Canada, Memoir 307.

Brabb, E.E.
1969: Six new Paleozoic and Mesozoic formations in east-central Alaska; United States Geological Survey, Bulletin 1274-1, 25 p.

Brandon, M.T.
1985: Mesozoic melange of the Pacific Rim Complex, western Vancouver Island; Geological Society of America, Cordilleran Section Meeting, Vancouver B.C., Guidebook, 28 p.

Brandon, M.T. and Cowan, D.S.
1983: Mesozoic terrane convergence and dispersion within the Fraser block, Pacific Northwest; Geological Society of America, Annual Meeting, Program with Abstracts, v. 15, p. 295.

Brandon, M.T., Cowan, D.S., Muller, J.E. and Vance, J.A.
1983: Pre-Tertiary geology of San Juan Islands, Washington and southeast Vancouver Island, British Columbia; Geological Association of Canada/Mineralogical Association of Canada/ Canadian Geophysical Union, Annual Meeting, Victoria. Field Trip Guidebooks, Field Trip No. 5, 65 p.

Brandon, M.T., Cowan, D.S., and Vance, J.A.
in press: Stratigraphic and structural framework of the mid-Cretaceous San Juan thrust system, San Juan Islands, Washington; in Geology of Washington State: Washington Division of Geology and Earth Resources Bulletin.

Brideaux, W.W.
1971: Palynology of the lower Colorado Group, central Alberta, Canada: I, Introductory remarks, geology and microplankton studies; Palaeontographica, Section B, v. 135, p. 53-114.

Bustin, R.M. and Moffat, I.
1983: Groundhog Coalfield, central British Columbia: Reconnaissance stratigraphy and structure; Bulletin of the Canadian Society of Petroleum Geologists, v. 31, p. 231-245.

Cairnes, C.E.
1924: Coquihalla area, British Columbia; Geological Survey of Canada, Memoir 139.

Cairnes, D.D.
1910: The Lewis and Nordenskiold rivers coal district; Geological Survey of Canada, Memoir 5.

Caldwell, W.G.E., North, B.R., Stelck, C.R., and Wall, J.H.
1978: A foraminiferal zonal scheme for the Cretaceous system in the Interior Plains of Canada; in Western and Arctic Canadian biostratigraphy; C.R. Stelck and B.D.E. Chatterton (ed.), Geological Association of Canada, Special Paper 18, p. 495-575.

Cameron, B.E.B.
1980: Biostratigraphy and depositional environment of the Escalante and Hesquiat formations (Early Tertiary) of the Nootka Sound area, Vancouver Island, British Columbia; Geological Survey of Canada, Paper 78-9.

Cameron, B.E.B. and Tipper, H.W.
1986: Jurassic stratigraphy of the Queen Charlotte Islands; Geological Survey of Canada, Bulletin 365.

Campbell, R.B. and Dodds, C.J.
1982: Geology of Mount Saint Elias map area (115B and C), Yukon Territory; Geological Survey of Canada, Open File 830.
1983: Terranes and major faults of the Saint Elias Mountains, Yukon Territory, British Columbia and Alaska; Geological Association of Canada/Mineralogical Association of Canada/Canadian Geophysical Union, Annual Meeting, Victoria, Program with Abstracts, v. 8, p. A10.

Campbell, R.B. and Tipper, H.W.
1971: Geology of the Bonaparte Lake map-area, British Columbia; Geological Survey of Canada, Memoir 363.

Campbell, R.B., Mountjoy, E.W., and Young, F.G.
1973: Geology of the McBride map-area, British Columbia; Geological Survey of Canada, Paper 72-35.

Carrigy, M.A.
1971: Lithostratigraphy of the uppermost Cretaceous (Lance) and Paleocene strata of the Alberta plains; Research Council of Alberta, Bulletin 27.

Carter, J.M.
1977: The stratigraphy, structure, and sedimentology of the Cretaceous Nanaimo Group, Galiano Island, British Columbia; M.Sc. thesis, Oregon State University.

Christopher, P.A. and Carter, N.C.
1976: Metallogeny and metallogenic epochs for porphyry mineral deposits; in Porphyry Deposits of the Canadian Cordillera, A. Sutherland Brown (ed.), Canadian Institute of Mining and Metallurgy, Special Volume 15, p. 64-71.

Church, B.N.
1971: Geology of the Owen Lake, Parrott Lakes, and Goosley Lake area; in Geology, Exploration and Mining in British Columbia 1970; British Columbia Department of Mines and Petroleum Resources, p. 119-125.
1973: Geology of the Buck Creek area; in Geology, Exploration and Mining in British Columbia 1972, British Columbia Department of Mines and Petroleum Resources, p. 353-363.
1975: Geology of the Hat Creek coal basin (92I/13E); British Columbia Ministry of Energy, Mines and Petroleum Resources, Geology in British Columbia, 1975, p. G99-G118.
1982: Notes on the Penticton Group: A progress report on a new stratigraphic subdivision of the Tertiary, south-central British Columbia; British Columbia Ministry of Energy, Mines and Petroleum Resources, Geological Field Work, 1981, Paper 1982-1, p. 12-16.

Church, B.N., Ewing, T.E., and Hora, Z.D.
1983: Volcanology, structure, coal and mineral resources of Early Tertiary outliers in south-central British Columbia; Geological Association of Canada/Mineralogical Association of Canada/Canadian Geophysical Union Annual Meeting, Victoria, British Columbia; Field Guide Book, Trip #1.

Clowes, R.M., Brandon, M.T., Green, A.G., Yorath, C.J., Sutherland Brown, A., Kanasewich, E.R., and Spencer, C.
1987: LITHOPROBE - southern Vancouver Island: Cenozoic subduction complex imaged by deep seismic reflections; Canadian Journal of Earth Sciences, v. 24, p. 31-51.

Coates, J.A.
1974: Geology of the Manning Park area, British Columbia; Geological Survey of Canada, Bulletin 238.

Cookenboo, H.O. and Bastin, R.M.
1989: Jura-Cretaceous (Oxfordian to Cenomanian) stratygraphy of the north-central Bowser Basin, northern British Columbia; Canadian Journal of Earth Sciences, 26, p. 1001-1012.

Currie, L.
1984: The provenance of chert clasts in the Ashman conglomerates of the northeastern Bowser Basin; B.Sc. thesis, Queen's University, Kingston, Ontario.

Dixon, J.
1982a: Jurassic and Lower Cretaceous subsurface stratigraphy of the Mackenzie Delta - Tuktoyaktuk Peninsula, N.W.T.; Geological Survey of Canada, Bulletin 349.
1982b: Sedimentology of the Neocomian Parsons Group in the subsurface of the Mackenzie Delta area, Arctic Canada; Bulletin of the Canadian Society of Petroleum Geologists, v. 30, p. 9-28.
1986a: Cretaceous to Pleistocene stratigraphy and paleogeography, northern Yukon and northwestern District of Mackenzie; Bulletin of Canadian Petroleum Geology, v. 34, p. 49-70.
1986b: Comments on the stratigraphy, sedimentology and distribution of the Albian Sharp Mountain Formation, northern Yukon; in Current Research, Part B, Geological Survey of Canada, Paper 86-1B, p. 375-381.
in press: Mesozoic stratigraphy, Eagle Plain area, northern Yukon; Geological Survey of Canada, Bulletin.

Douglas, B.J.
1983: Structural and stratigraphic analysis of a metasedimentary inlier within the Coast Plutonic Complex, British Columbia, Canada; Ph.D. thesis, Princeton University, New Jersey.

Douglas, R.J.W.
1950: Callum Creek, Langford Creek and Gap map-areas, Alberta; Geological Survey of Canada, Memoir 255.

Duffell, S. and McTaggart, K.C.
1952: Ashcroft map-area, British Columbia; Geological Survey of Canada, Memoir 262.

Eisbacher, G.H.
1974: Sedimentary history and tectonic evolution of the Sustut and Sifton basins, north-central British Columbia; Geological Survey of Canada, Paper 73-31.
1976: Sedimentology of the Dezadeash flysch and its implications for strike-slip faulting along the Denali Fault, Yukon Territory and Alaska; Canadian Journal of Earth Sciences, v. 13, p. 1495-1513.
1981: Late Mesozoic - Paleogene Bowser Basin molasse and Cordilleran tectonics, western Canada; in Sedimentation and Tectonics in Alluvial Basins, A.D. Miall (ed.), Geological Association of Canada, Special Paper 23, p. 125-151.

Eisbacher, G.H. and Hopkins, S.L.
1977: Mid-Cenozoic paleogeography and tectonic setting of the St. Elias Mountains, Yukon Territory; in Report of Activities, Part B, Geological Survey of Canada, Paper 77-1B, p. 319-335.

Eisbacher, G.H., Carrigy, M.A., and Campbell, R.B.
1974: Paleodrainage pattern and late-orogenic basins of the Canadian Cordillera; in Tectonics and Sedimentation, W.R. Dickinson (ed.), Society of Economic Paleontologists and Mineralogists, Special Publication No. 22, p. 143-166.

Engebretson, D.C.
1982: Relative motions between oceanic and continental plates in the Pacific Basin; Ph.D. thesis, Stanford University, California, 211 p.

Ewing, T.E.
1981a: Regional stratigraphy and structural setting of the Kamloops Group, south-central British Columbia; Canadian Journal of Earth Sciences, v. 18, p. 1464-1477.
1981b: Petrology and geochemistry of the Kamloops Group volcanics, British Columbia; Canadian Journal of Earth Sciences, v. 18, p. 1478-1505.

Fairchild, L.H. and Cowan, D.S.
1982: Structure, petrology and tectonic history of the Leech River Complex, northwest of Victoria, Vancouver Island; Canadian Journal of Earth Sciences, v. 19, p. 1816-1835.

Gabrielse, H.
1985: Major dextral transcurrent displacements along the Northern Rocky Mountain Trench and related lineaments in north-central British Columbia; Geological Society of America Bulletin, v. 96, p. 1-14.

Glaister, P.
1959: Lower Cretaceous of southern Alberta and adjoining areas; American Association of Petroleum Geologists, Bulletin, v. 43, p. 590-640.

Godwin, C.I.
1975: Imbricate subduction zones and their relationship with Upper Cretaceous to Tertiary porphyry deposits in the Canadian Cordillera; Canadian Journal of Earth Sciences, v. 12, p. 1362-1378.

Haggart, J.W.
1986: Stratigraphic investigations of the Cretaceous Queen Charlotte Group, Queen Charlotte Islands, British Columbia; Geological Survey of Canada, Paper 86-20.

Hamblin, A.P. and Walker, R.G.
1979: Storm-dominated shallow marine deposits: the Fernie - Kootenay (Jurassic) transition, southern Rocky Mountains; Canadian Journal of Earth Sciences, v. 16, p. 1673-1690.

Hanson, W.B.
1976: Stratigraphy and sedimentology of the Cretaceous Nanaimo Group, Saltspring Island, British Columbia; Ph.D. thesis, Oregon State University.

Hill, M.L., Woodsworth, G.J., and van der Heyden, P.
1985: The Coast Plutonic Complex near Terrace, British Columbia: a metamorphosed western extension of Stikinia; Geological Association of America, Program With Abstracts, v. 17, p. 362.

Hopkins, W.S. Jr.
1966: Palynology of Tertiary rocks of the Whatcom Basin, southwestern British Columbia and northwestern Washington; Ph.D. thesis, University of British Columbia, Vancouver.

Hudson, T. and Plafker, G.
1982: Paleogene metamorphism of an accretionary flysch terrane, eastern Gulf of Alaska; Geological Society of America Bulletin, v. 93, p. 1280-1290.

Hughes, J.D. and Long, D.G.F.
1980: Geology and coal resource potential of Early Tertiary strata along Tintina Trench, Yukon Territory; Geological Survey of Canada, Paper 79-32, 22 p.

Irving, E. and Wynne, P.J.
1985: Possible Late Cretaceous - Early Tertiary trajectory of "Baja B.C."; in Programme and Abstracts, A symposium on deep structure of southern Vancouver Island: Results of Lithoprobe Phase 1, Geological Association of Canada, Victoria Section.

Isachsen, C., Armstrong, R.L., and Parrish, R.R.
1985: U-Pb, Rb-Sr, and K-Ar geochronometry of Vancouver Island igneous rocks; in Programme and Abstracts, A symposium on deep structure of southern Vancouver Island: Results of Lithoprobe Phase 1, Geological Association of Canada, Victoria Section.

Jackson, L.F., Gordey, S.P., Armstrong, R.L., and Harakal, J.E.
1986: Bimodal Paleogene volcanics near Tintina Fault, east-central Yukon, and their possible relationship to placer gold; in Yukon Geology, v. 1, Department of Indian and Northern Affairs, p. 139-147.

Jeletzky, J.A.
1958: Uppermost Jurassic and Cretaceous rocks of Aklavik Range, northeastern Richardson Mountains, Northwest Territories; Geological Survey of Canada, Paper 58-2.
1960: Uppermost Jurassic and Cretaceous rocks, east flank of Richardson Mountains between Stony Creek and lower Donna River, Northwest Territories; Geological Survey of Canada, Paper 59-14.
1961: Uppermost Jurassic and Lower Cretaceous rocks, west flank of Richardson Mountains between the headwaters of Blow River and Bell River, Yukon Territory; Geological Survey of Canada, Paper 61-9.
1967: Jurassic and (?)Triassic rocks of the eastern slopes of the Richardson Mountains, northwestern District of Mackenzie; Geological Survey of Canada, Paper 66-50.
1971a: Marine Cretaceous biotic provinces and paleogeography of western and Arctic Canada, illustrated by a detailed study of ammonites; Geological Survey of Canada, Paper 70-22.
1971b: Marine Cretaceous biotic provinces of western and Arctic Canada; Proceedings of the North American Paleontological Convention, September, 1969, pt. 1, p. 1638-1659, Allen Press, Kansas.
1975: Sharp Mountain Formation (new): A shoreline facies of the Upper Aptian-Lower Albian Flysch Division, eastern Keele Range, Yukon Territory (117-O); in Report of Activities, Part B, Geological Survey of Canada, Paper 75-1B, p. 237-244.
1976: Mesozoic and ?Tertiary rocks of Quatsino Sound, Vancouver Island, British Columbia; Geological Survey of Canada, Bulletin 242.
1977: Causes of Cretaceous oscillations of sea level in western and Arctic Canada and some general geotectonic implications; in Mid-Cretaceous Events, K. Kanmera (ed.), Paleontological Society of Japan, Special Paper 21, p. 233-246.

Jeletzky, J.A. and Tipper, H.W.
1968: Upper Jurassic and Cretaceous rocks of Taseko Lakes map-area and their bearing on the geological history of southwestern British Columbia; Geological Survey of Canada, Paper 67-54.

Jeletzky, O.L.
1976: Preliminary report on stratigraphy and depositional history of Middle to Upper Jurassic strata in McConnell Creek map-area (94D West Half), British Columbia; in Report of Activities, Part A, Geological Survey of Canada, Paper 76-1A, p. 63-67.

Jerzykiewicz, T. and McLean, J.R.
1980: Lithostratigraphical and sedimentological framework of coal-bearing Upper Cretaceous and Lower Tertiary strata, Coal Valley area, central Alberta Foothills; Geological Survey of Canada, Paper 79-12.

Johnson, S.Y.
1984: Stratigraphy, age and petrography of the Eocene Chuckanut Formation, northwest Washington; Canadian Journal of Earth Sciences, v. 21, p. 92-106.

Jones, D.L. and Clark, S.H.B.
1973: Upper Cretaceous (Maastrichtian) fossils from the Kenai-Chugach Mountains, Kodiak and Shumagin Islands, southern Alaska; United States Geological Survey, Journal of Research, v. 1, no. 2, p. 125-136.

Kleinspehn, K.L.
1982: Cretaceous sedimentation and tectonics, Tyaughton - Methow Basin, southwestern British Columbia; Ph.D. thesis, Princeton University.
1985: Cretaceous sedimentation and tectonics, Tyaughton - Methow Basin, southwestern British Columbia; Canadian Journal of Earth Sciences, v. 22, p. 154-174.

Koch, N.G.
1973: The central Cordilleran region; in The Future Petroleum Provinces of Canada, R.G. McCrossan (ed.), Canadian Society of Petroleum Geologists, Memoir 1, p. 37-71.

Koo, J.
1984: Coal geology of the Mount Klappan area in northwestern British Columbia (104H/2,3,6,7); in Geological Field Work 1984; British Columbia Ministry of Energy, Mines and Petroleum Resources, Paper 1985-1, p. 343-351.

Kramers, J.W. and Mellon, G.B.
1972: Upper Cretaceous - Paleocene coal-bearing strata, northwest-central Alberta plains; Proceedings, First Geological Conference on Western Canadian Coal; Research Council of Alberta, Information Series, No. 60, p. 109-124.

Lambert, M.B.
1974: The Bennett Lake cauldron subsidence complex, British Columbia and Yukon Territory; Geological Survey of Canada, Bulletin 227.

Langenberg, C.W. and McMechan, M.E.
1984: Lower Cretaceous Luscar Group of the northern and north-central Foothills of Alberta; Bulletin of Canadian Petroleum Geology, v. 33, p. 1-11.

Long, D.G.F.
1981: Dextral strike-slip faults in the Canadian Cordillera and depositional environments of related fresh-water intermontane coal basins; Geological Association of Canada, Special Paper 23, p. 153-186.

Loranger, D.M.
1951: Useful Blairmore microfossil zone in central and southern Alberta, Canada; American Association of Petroleum Geologists Bulletin, v. 35, p. 2348-2367.

Lowey, G.W.
1984: The stratigraphy and sedimentology of siliciclastic rocks, west-central Yukon, and their tectonic implications; Ph.D. thesis, University of Calgary.

MacIntyre, D.G.
1976: Evolution of Upper Cretaceous volcanic and plutonic centres and associated porphyry copper occurrences, Tahtsa Lake area, British Columbia; Ph.D. thesis, University of Western Ontario, London.

MacLeod, N.S., Tiffin, D.L., Snavely, P.D. Jr., and Currie, R.G.
1977: Geologic interpretation of magnetic and gravity anomalies in the Strait of Juan de Fuca, U.S. - Canada; Canadian Journal of Earth Sciences, v. 14, p. 223-238.

McLean, J.R.
1977: The Cadomin Formation: stratigraphy, sedimentology and tectonic implications; Bulletin of Canadian Petroleum Geology, v. 25, p. 792-827.
1982: Lithostratigraphy of the Lower Cretaceous coal-bearing sequence, Foothills of Alberta; Geological Survey of Canada, Paper 80-29.

McLean, J.R. and Wall, J.H.
1981: Early Cretaceous Moosebar sea in Alberta; Bulletin of Canadian Petroleum Geology, v. 29, p. 334-377.

McMechan, R.D.
1981: Stratigraphy, sedimentology, structure and tectonic implications of the Oligocene Kishenehn Formation, Flathead valley graben, southeastern British Columbia; Ph.D. thesis, Queen's University, Kingston, Ontario.

Massey, N.W.D.
1986: The Metchosin igneous complex, southern Vancouver Island: an Eocene ocean-island ophiolite; Geology, v. 14, p. 602-605.

Mathews, W.H.
1958: Geology of the Mount Garibaldi map-area, southwestern B.C., Canada; Geological Society of America Bulletin, v. 69, p. 161-178.

Mellon, G.B.
1967: Stratigraphy and petrology of the Lower Cretaceous Blairmore and Mannville groups, Alberta Foothills and plains; Research Council of Alberta, Bulletin 21.

Mellon, G.B. and Wall, J.H.
1963: Correlation of the Blairmore Group and equivalent strata; Bulletin of Canadian Petroleum Geology, v. 11, p. 396-409.

Moffat, I.W.
1985: The nature and timing of depositional events and organic and inorganic metamorphism in the northern Groundhog coalfield: implications for the tectonic history of the Bowser Basin; Ph.D. thesis, University of British Columbia, Vancouver, 205 p.

Monger, J.W.H.
1985: Structural evolution of the southwestern Intermontane Belt, Ashcroft and Hope map-areas, British Columbia; in Current Research, Part A, Geological Survey of Canada, Paper 85-1A, p. 349-358.

Monger, J.W.H. and Berg, H.C.
1984: Lithotectonic terrane map of western Canada and southeastern Alaska; in Lithotectonic Terrane Maps of the North American Cordillera, N.J. Silberling and D.L. Jones (ed.), United States Geological Survey, Open File Report 84-523.

Monger, J.W.H., Price, R.A., and Tempelman-Kluit, D.J.
1982: Tectonic accretion and the origin of the two major metamorphic and plutonic welts in the Canadian Cordillera; Geology, v. 10, p. 70-75.

Mountjoy, E.W.
1962: Mount Robson (southeast) map-area, Rocky Mountains of Alberta and British Columbia; Geological Survey of Canada, Paper 61-31.
1967: Upper Cretaceous and Tertiary stratigraphy, northern Yukon and northwestern District of Mackenzie; Geological Survey of Canada, Paper 66-16.

Muller, J.E.
1977: Evolution of the Pacific margin, Vancouver Island and adjacent regions; Canadian Journal of Earth Sciences, v. 14, p. 2062-2085.
1980: Chemistry and origin of the Eocene Metchosin volcanics, Vancouver Island, British Columbia; Canadian Journal of Earth Sciences, v. 17, p. 199-209.

Muller, J.E. and Jeletzky, J.A.
1970: Geology of the Upper Cretaceous Nanaimo Group, Vancouver Island and Gulf Islands, British Columbia; Geological Survey of Canada, Paper 69-25.

Muller, J.E., Northcote, K.E., and Carlisle, D.
1974: Geology and mineral deposits of Alert Bay - Cape Scott map-area, Vancouver Island, British Columbia; Geological Survey of Canada, Paper 74-8.

Muller, J.E., Cameron, B.E.B., and Northcote, K.E.
1981: Geology and mineral deposits of Nootka Sound map-area, Vancouver Island, British Columbia; Geological Survey of Canada, Paper 80-16.

Myhr, D.W. and Young, F.G.
1975: Lower Cretaceous (Neocomian) sandstone sequence of Mackenzie Delta and Richardson Mountains area; in Report of Activities, Part C, Geological Survey of Canada, Paper 75-1C, p. 247-266.

Norris, D.K. and Hopkins, W.S. Jr.
1977: The geology of Bonnet Plume Basin, Yukon Territory; Geological Survey of Canada, Paper 76-8.

Norris, D.K. and Yorath, C.J.
1981: The North American plate from the Arctic Archipelago to the Romanzof Mountains; in The Ocean Basins and Margins, Volume 5, The Arctic Ocean, A.E.M. Nairn, M. Churkin, Jr., and F.G. Stehli (ed.), p. 37-103, Plenum Press, New York and London.

Norris, D.K., Stevens, R.D., and Wanless, R.K.
1965: K-Ar age of igneous pebbles in the McDougall-Segur conglomerate, southeastern Canadian Cordillera; Geological Survey of Canada, Paper 65-26.

Pacht, J.A.
1980: Sedimentology and petrology of the Late Cretaceous Nanaimo Group deposited in the Nanaimo Basin, western Washington and British Columbia: Implications for Cretaceous tectonics; Ph.D. thesis, Ohio State University.
1984: Petrologic evolution and paleogeography of the Late Cretaceous Nanaimo Basin, Washington and British Columbia: Implications for Cretaceous tectonics; Geological Society of America Bulletin, v. 95, p. 766-778.

Pigage, L.C. and Anderson, R.G.
1985: The Anvil plutonic suite, Faro, Yukon Territory; Canadian Journal of Earth Sciences, v. 22, p. 1204-1216.

Price, R.A.
1965: Flathead map-area, British Columbia and Alberta; Geological Survey of Canada, Memoir 336.
1980: The Cordilleran thrust and fold belt in the southern Canadian Rocky Mountains; Geological Society of London, Special Volume.

Price, R.A., Monger, J.W.H., and Muller, J.A.
1981: Cordilleran cross-section - Calgary to Victoria; in Field Guides to Geology and Mineral Deposits, R.I. Thompson and D.G. Cook (ed.), Annual Meeting, Geological Association of Canada, Mineralogical Association of Canada, Canadian Geophysical Union, Calgary, Alberta, p. 261-334.

Read, P.B.
1983: Geology, Classy Creek (104 J/2E) and Stikine Canyon (104 J/1W), British Columbia; Geological Survey of Canada, Open File 940.

Richards, T.A. and Jeletzky, O.L.
1975: A preliminary study of the Upper Jurassic Bowser assemblage in the Hazelton, west half, map-area, British Columbia; in Report of Activities, Part A: April to October, 1974, Geological Survey of Canada, Paper 75-1A, p. 31-36.

Rickets, B.D.
1985: Possible plinian eruptions of Paleocene ash in central Yukon: evidence from volcanic ash, Norman Wells area, N.W.T.; Canadian Journal of Earth Sciences, v. 22, p. 473-479.

Roddick, J.A.
1965: Vancouver North, Coquitlam and Pitt Lake map-areas, British Columbia; Geological Survey of Canada, Memoir 335.

Roddick, J.A. and Green, L.H.
1961a: Sheldon Lake, Yukon Territory; Geological Survey of Canada, Map 12-1961.
1961b: Tay River, Yukon Territory; Geological Survey of Canada, Map 13-1961.

Rudkin, R.A.
1964: Lower Cretaceous; in Geological History of Western Canada, Chapter 11, R.G. McCrossan and R.P. Glaister (ed.), Alberta Society of Petroleum Geologists, p. 156-168.

Rusmore, M.E. and Cowan, D.S.
1985: Jurassic-Cretaceous rock units along the southern edge of the Wrangellia terrane on Vancouver Island; Canadian Journal of Earth Sciences, v. 22, p. 1223-1232.

Russell, L.S.
1954: Mammalian fauna of the Kishenehn Formation, southeastern British Columbia; National Museum of Canada, Bulletin 132, p. 92-111.

Sharp, R.P. and Rigsby, G.P.
1956: Some rocks of the central St. Elias Mountains, Yukon Territory, Canada; American Journal of Science, v. 254, p. 110-122.

Shouldice, D.N.
1971: Geology of the western Canadian continental shelf; Canadian Society of Petroleum Geologists Bulletin, v. 19, p. 405-436.
1973: Western Canadian continental shelf; in Future Petroleum Provinces of Canada, R.G. McCrossan (ed.), Canadian Society of Petroleum Geologists, Memoir 1, p. 7-35.

Sliter, W.V.
1973: Upper Cretaceous foraminifers from the Vancouver Island area, British Columbia, Canada; Journal of Foraminiferal Research, v. 2, p. 167-183.

Souther, J.G.
1971: Geology and mineral deposits of the Tulsequah map-area, British Columbia; Geological Survey of Canada, Memoir 362.
1972: Telegraph Creek map-area; Geological Survey of Canada, Paper 71-44.

Stelck, C.R.
1958: Stratigraphic position of the Viking Sand; Alberta Society of Petroleum Geologists, Journal, v. 6, p. 2-7.

Stelck, C.R., Wall, J.H., Bahan, W.G., and Martin, L.J.
1956: Middle Albian foraminifera from Athabasca and Peace drainage areas of western Canada; Research Council of Alberta, Report 75.

Stott, D.F.
1963: The Cretaceous Alberta Group and equivalent rocks, Rocky Mountain Foothills, Alberta; Geological Survey of Canada, Memoir 317.
1967a: The Fernie and Minnes strata north of Peace River, Foothills of northeastern British Columbia; Geological Survey of Canada, Paper 67-19, Pt. A.
1967b: The Cretaceous Smoky Group, Rocky Mountain Foothills, Alberta and British Columbia; Geological Survey of Canada, Bulletin 132.
1968: Lower Cretaceous Bullhead and Fort St. John Groups between Smoky and Peace Rivers, Rocky Mountain Foothills, Alberta and British Columbia; Geological Survey of Canada, Bulletin 152.
1973: Lower Cretaceous Bullhead Group between Bullmoose Mountain and Tetsa River, Rocky Mountain Foothills, northeastern British Columbia; Geological Survey of Canada, Bulletin 219.
1975: The Cretaceous System in northeastern British Columbia; in The Cretaceous System in the Western Interior of North America, W.G.E. Caldwell (ed.), Geological Association of Canada, Special Paper 13, p. 441-467.
1982: Lower Cretaceous Fort St. John Group and Upper Cretaceous Dunvegan Formation of the Foothills and Plains of Alberta, British Columbia, District of Mackenzie and Yukon Territory; Geological Survey of Canada, Bulletin 328.
1984: Cretaceous sequences of the foothills of the Canadian Rocky Mountains; in The Mesozoic of Middle North America, D.F. Stott and D.J. Glass (ed.), Canadian Society of Petroleum Geologists, Memoir 9, p. 85-107.

Stott, D.F. and Aitken, J.D. (editors)
in press: Sedimentary cover of the Craton in Canada; Geological Survey of Canada, Geology of Canada, no. 5 (also Geological Society of America, The Geology of North America, v. D-1).

Sutherland Brown, A.
1960: Geology of the Rocher Deboule Range; British Columbia Department of Mines and Petroleum Resources, Bulletin 43.
1968a: Bowron River; British Columbia Department of Mines and Petroleum Resources, Annual Report, 1967, p. 459-460.
1968b: Geology of the Queen Charlotte Islands, British Columbia; British Columbia Department of Mines and Petroleum Resources, Bulletin 54.

Sutherland Brown, A., Yorath, C.J., and Tipper, H.W.
1983: Geology and tectonic history of the Queen Charlotte Islands; Field Trip Guidebook 8; Geological Association of Canada, Mineralogical Association of Canada, Canadian Geophysical Union, Annual Meeting, Victoria.

Sutherland Brown, A. and Yorath, C.J.
1985: LITHOPROBE profile across southern Vancouver Island: Geology and tectonics; Geological Society of America, Cordilleran Section Annual Meeting, Vancouver, British Columbia; Field Guidebook.

Tempelman-Kluit, D.J.
1970: Stratigraphy and structure of the "Keno Hill Quartzite" in Tombstone River - Upper Klondike River map-areas, Yukon Territory (116B/7, B/8); Geological Survey of Canada, Bulletin 180.
1974: Reconnaissance geology of the Aishihik Lake, Snag and part of Stewart River map-areas, west central Yukon; Geological Survey of Canada, Paper 73-41.

Tempelman-Kluit, D.J. and Wanless, R.K.
1975: Potassium-argon age determinations of metamorphic and plutonic rocks in the Yukon crystalline terrane; Canadian Journal of Earth Sciences, v. 12, p. 1895-1909.

Thorkelson, D.J.
1985: Geology of the mid-Cretaceous volcanic units near Kingsvale, southwestern British Columbia; in Current Research, Part B, Geological Survey of Canada, Paper 85-1B, p. 333-339.

Tipper, H.W.
1969: Mesozoic and Cenozoic geology of the northeast part of Mount Waddington map-area (92N), Coast District, British Columbia; Geological Survey of Canada, Paper 68-33.

Tipper, H.W. and Richards, T.A.
1976: Jurassic stratigraphy and history of north-central British Columbia; Geological Survey of Canada, Bulletin 270.

Tozer, E.T.
1956: Uppermost Cretaceous and Paleocene nonmarine molluscan faunas of western Alberta; Geological Survey of Canada, Memoir 280.

Trexler, J.H., Jr.
1985: Sedimentology and stratigraphy of the Cretaceous Virginian Ridge Formation, Methow Basin, Washington; Canadian Journal of Earth Sciences, v. 22, p. 1274-1285.

Trexler, J.H., Jr. and Bourgeois, J.
1985: Evidence for mid-Cretaceous wrench faulting in the Methow Basin, Washington: Tectonostratigraphic setting of the Virginian Ridge Formation; Tectonics, v. 4, p. 379-394.

Walker, R.G.
1983: Cardium Formation 3. Sedimentology and stratigraphy in the Garrington - Caroline area, Alberta; Bulletin of Canadian Petroleum Geology, v. 31, p. 213-230.

Wall, J.H. and Rosene, R.K.
1977: Upper Cretaceous stratigraphy and micropaleontology of the Crowsnest Pass - Waterton area, southern Alberta Foothills; Bulletin of Canadian Petroleum Geology, v. 25, p. 842-867.

Ward, P.D.
1978: Revisions to the stratigraphy and biochronology of the Upper Cretaceous Nanaimo Group, British Columbia and Washington State; Canadian Journal of Earth Sciences, v. 15, p. 405-423.

Ward, P. and Stanley, K.O.
1982: The Haslam Formation: A Late Santonian - Early Campanian forearc basin deposit in the Insular Belt of southwestern British Columbia and adjacent Washington; Journal of Sedimentary Petrology, v. 52, p. 975-990.

Wheeler, J.O.
1961: Whitehorse map-area, Yukon Territory; Geological Survey of Canada, Memoir 312.

Wood, D.H. and Armstrong, R.L.
1982: Geology, chemistry, and geochronometry of the Cretaceous South Fork Volcanics, Yukon Territory; in Current Research, Part A, Geological Survey of Canada, Paper 82-1A, p. 309-316.

Woodsworth, G.J.
1979a: Geology of Whitesail Lake map-area, British Columbia; in Current Research, Part A, Geological Survey of Canada, Paper 79-1A, p. 25-29.
1979b: Metamorphism, deformation, and plutonism in the Mount Raleigh pendant, Coast Mountains, British Columbia; Geological Survey of Canada, Bulletin 295.
1980: Geology of Whitesail Lake (93E) map-area, British Columbia; Geological Survey of Canada, Open File 708.

Woodsworth, G.J. and Orchard, M.J.
1985: Upper Paleozoic to lower Mesozoic strata and their conodonts, western Coast Plutonic Complex, British Columbia; Canadian Journal of Earth Sciences, v. 22, p. 1329-1344.

Wright, J. and Miller, D.
1983: Rock River coal basin; Mineral Deposits of the Northern Cordillera, Symposium, Whitehorse, Yukon, Abstract.

Yorath, C.J., Tiffin, D.L., and Cameron, B.E.B.
1977: Submersible operation on the Pacific continental margin; in Report of Activities, Part A, Geological Survey of Canada, Paper 77-1A, p. 301-310.

Yorath, C.J. and Chase, R.L.
1981: Tectonic history of the Queen Charlotte Islands and adjacent areas - a model; Canadian Journal of Earth Science, v. 18, p. 1717-1739.

Yorath, C.J. and Cook, D.G.
1981: Cretaceous and Tertiary stratigraphy and paleogeography, northern Interior Plains, District of Mackenzie; Geological Survey of Canada, Memoir 398.
1984: Mesozoic and Cenozoic depositional history of the northern Interior Plains of Canada; in The Mesozoic of Middle North America, D.F. Stott and D.J. Glass (ed.), Canadian Society of Petroleum Geologists, Memoir 9, p. 69-83.

Young, F.G.
1972: Cretaceous stratigraphy between Blow and Fish Rivers, Yukon Territory; in Report of Activities, Part A, Geological Survey of Canada, Paper 72-1A, p. 229-235.
1973: Jurassic and Cretaceous stratigraphy between Babbage and Blow Rivers, Yukon Territory; in Report of Activities, Part A, Geological Survey of Canada, Paper 73-1A, p. 277-281.
1975: Upper Cretaceous stratigraphy, Yukon coastal plain and northwestern Mackenzie Delta; Geological Survey of Canada, Bulletin 249.

Young, F.G. and Robertson, B.T.
1984: The Rapid Creek Formation: An Albian flysch-related phosphatic iron formation in northern Yukon Territory; in The Mesozoic of Middle North America, D.F. Stott and D.J. Glass (ed.), Canadian Society of Petroleum Geologists, Memoir, 9, p. 361-372.

Young, F.G., Myhr, D.W., and Yorath, C.J.
1976: Geology of the Beaufort-Mackenzie Basin; Geological Survey of Canada, Paper 76-11.

Young, I.F.
1981: Geological development of the western margin of the Queen Charlotte Basin; M.Sc. thesis, University of British Columbia, Vancouver.

Ziegler, W.H. and Pocock, S.A.J.
1960: The Minnes Formation; Second Annual Field Conference Guidebook, Edmonton Geological Society, p. 43-71.

Authors' addresses

D.F. Stott
J. Dixon
Institute of Sedimentary and
Petroleum Geology
Geological Survey of Canada
3303-33rd Street N.W.
Calgary, Alberta
T2L 2A7

C.J. Yorath
Pacific Geoscience Centre
9860 West Saanich Road
P.O. Box 6000
Sidney, British Columbia
V8L 4B2

H. Gabrielse
R.G. Anderson
J.G. Souther
G.J. Woodsworth
H.W. Tipper
J.W.H. Monger
C.J. Dodds
Cordilleran Division
Geological Survey of Canada
100 West Pender Street
Vancouver, British Columbia
V6B 1R8

A. Sutherland Brown
546 Newport Avenue
Victoria, British Columbia
V8S 5C7

R.B. Campbell
1760 Forest Park Drive
Sidney, British Columbia
V8L 4A6

ADDENDUM

Important biostratigraphic and basin analysis papers dealing with Middle Jurassic and Cretaceous rocks of the Queen Charlotte Islands appear in the following volume:

1991: Evolution and hydrocarbon potential of the Queen Charlotte Basin, British Columbia; G.J. Woodsworth (ed.); Geological Survey of Canada, Paper 90-10, 569 p.

Printed in Canada

Chapter 10

NEOGENE ASSEMBLAGES

Summary
Sedimentary assemblages
J.G. Souther and C.J. Yorath
    Queen Charlotte Islands and Queen Charlotte Basin
    Northwestern Vancouver Island continental shelf and slope
    Tofino Basin
    Georgia Basin
    Intermontane Belt

Igneous assemblages
J.G. Souther
    Southern British Columbia subduction complexes - related arc volcanics
        Pemberton Volcanic Belt
        Franklin Glacier Complex
        Mount Silverthrone Complex
        Masset Formation
        Alert Bay Volcanic Belt
        Garibaldi Volcanic Belt
            Southern segment
            Central segment
            Northern segment
    Southern and central British Columbia plateau lavas
        Chilcotin Group
    Central and northern British Columbia plume and rift complexes
        Anahim Volcanic Belt
            King Island-Bella Bella Complex
            Central shield volcanoes
            Postglacial monogenetic cones
        Stikine Volcanic Belt
    Northwestern British Columbia and southwestern Yukon subduction and transform-related arc volcanics
        Wrangell Volcanic Belt
    Volcanic rocks of east-central British Columbia
        Clearwater-Quesnel Volcanic Province
        McConnell Creek Volcanic Province
References

Chapter 10

NEOGENE ASSEMBLAGES

J.G. Souther and C.J. Yorath

SUMMARY

By the beginning of Neogene time, some 24 Ma ago, the Kula Plate had disappeared and the Pacific-Farallon spreading ridge had collided with the western margin of North America north of Queen Charlotte Islands. During the early Neogene the Pacific-North America-Farallon triple junction migrated southeastward to a position adjacent to the northern end of Vancouver Island where it has remained relatively fixed for the past 10 Ma. The early Neogene shift in the position of the triple junction was accompanied by a rapid decrease in the size of the easterly-subducting Farallon Plate. The present small remnants of the Farallon Plate have been renamed the Juan de Fuca and Explorer plates, and the active spreading centre (Farallon-Pacific spreading ridge) which bounds them on the west has been renamed the Juan de Fuca Ridge system. Motion vectors relative to the absolute (hotspot) framework indicate that the western Canadian part of the North American Plate moved southwestward at about 22 mm per year for most of Neogene time.

Dextral transcurrent faulting, which dominated the early Tertiary tectonics of the Intermontane Belt, was greatly reduced during the Neogene, and confined to faults at or near the continental margin. Movement on the Totschunda and Border Ranges fault systems accompanied profound Neogene uplift, folding and northeasterly directed thrusting in the Saint Elias Mountains. During this time the Intermontane Belt remained relatively stable whereas the axis of the Coast Belt was greatly uplifted and deeply dissected.

Neogene sedimentary deposits occur extensively throughout the offshore regions of the Insular Belt, on parts of Vancouver Island and the Queen Charlotte Islands, and in isolated basins within the Intermontane Belt of British Columbia (Fig. 10-1). Neogene and Quaternary igneous rocks are concentrated along five principal volcano-tectonic belts that are closely related to the modern tectonic regime (Fig. 10-1). The Pemberton and Garibaldi volcanic belts in southern British Columbia are volcanic fronts related to eastward subduction of the Juan de Fuca Plate whereas the transverse Alert Bay Volcanic Belt is the product of a descending-plate-edge effect where disruption of steady state plate consumption triggered magma genesis

Souther, J.G. and Yorath, C.J.
1991: Neogene assemblages, Chapter 10 in Geology of the Cordilleran Orogen in Canada, H. Gabrielse and C.J. Yorath (ed.); Geological Survey of Canada, Geology of Canada, no. 4, p. 373-401 (also Geological Society of America, The Geology of North America, v. G-2)

along the subducting slab. The Wrangell Volcanic Belt of southwestern Yukon is an extension of the Aleutian Arc, and is related to the underflow of Pacific oceanic crust along its converging margin with northwestern North America. The east-trending Anahim and north-trending Stikine volcanic belts of central and northern British Columbia are hotspot traces and zones of incipient rifting, respectively that lie inland from the dextral transcurrent boundary between the Pacific and North American plates.

Neogene igneous rocks in the Clearwater-Quesnel and McConnell Creek areas appear to lie outside the principal volcanic belts but may be localized along a major, continental suture zone.

Because the age of many young volcanic piles in the Canadian Cordillera spans the Neogene-Quaternary boundary all Miocene and younger volcanic and plutonic assemblages are included in this chapter.

SEDIMENTARY ASSEMBLAGES

C.J. Yorath and J.G. Souther

Queen Charlotte Islands and Queen Charlotte Basin

Queen Charlotte Basin was first described by Shouldice (1971) to include marine and nonmarine sedimentary rocks that underlie northeastern Graham Island, Hecate Strait and Queen Charlotte Sound (Fig. 10.2). The basin is approximately 450 km long and 130 km wide at its widest point in Queen Charlotte Sound where its western boundary is defined by the Queen Charlotte Fault. Within the basin, sediments of the Skonun Formation reach a maximum thickness of slightly over 4500 m, however, their thickness rarely exceeds 3000 m. The isopachs (Fig. 10.2) show a linear trough extending through Hecate Strait but interrupted in its southern part by a prominent northeasterly trending basement ridge over which the Tertiary sediments are thin. Yorath and Chase (1981) interpreted the ridge as a pre-Tertiary structural culmination in Wrangellian rocks. The foundation of the basin consists mostly of volcanic rocks of the Masset Formation which are exposed widely on the Queen Charlotte Islands (Sutherland Brown, 1968) and which were penetrated in several exploratory wells drilled during the 1960s by Atlantic Richfield Ltd. and Shell Canada Resources Ltd. The Skonun Formation conformably overlies, and locally is interbedded with, the volcanic Masset Formation. It comprises a succession of dominantly nonmarine immature sandstone, minor shale, lignite and pebble conglomerate (Fig. 10.3). Northern sections of the formation are mainly nonmarine but minor marine units occur near the top of the Tow Hill, Gold Creek, Tlell and Cape Ball wells and near the base of

CHAPTER 10

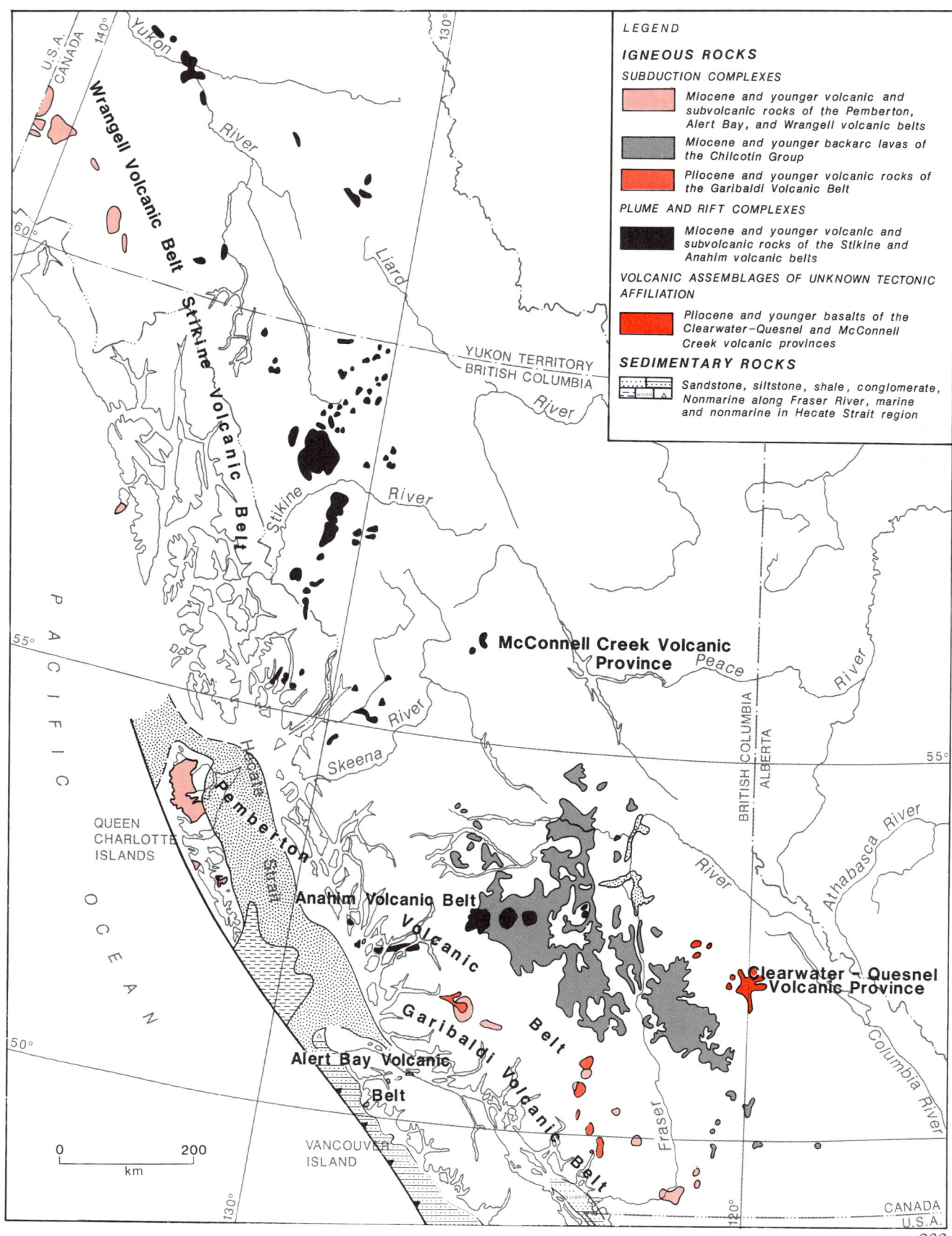

Fig. 10.1. Distribution of Neogene sedimentary assemblages and Neogene and Quaternary igneous assemblages in the Canadian Cordillera.

the formation in the last well where marine sediments are interbedded with volcanics of the Masset Formation. To the south, beneath southern Hecate Strait and Queen Charlotte Sound, the succession is partly marine in the Murrelet well and entirely marine in Harlequin and Osprey wells (Fig. 10.2). In the Osprey well the underlying Masset volcanics are interbedded with nonmarine sandstone (Shouldice, 1971). On northeastern Graham Island olivine basalt of the Tow Hill sill occurs as outcrop and in the subsurface where it was encountered in the Tow Hill well as an intrusion in the upper part of the Skonun Formation

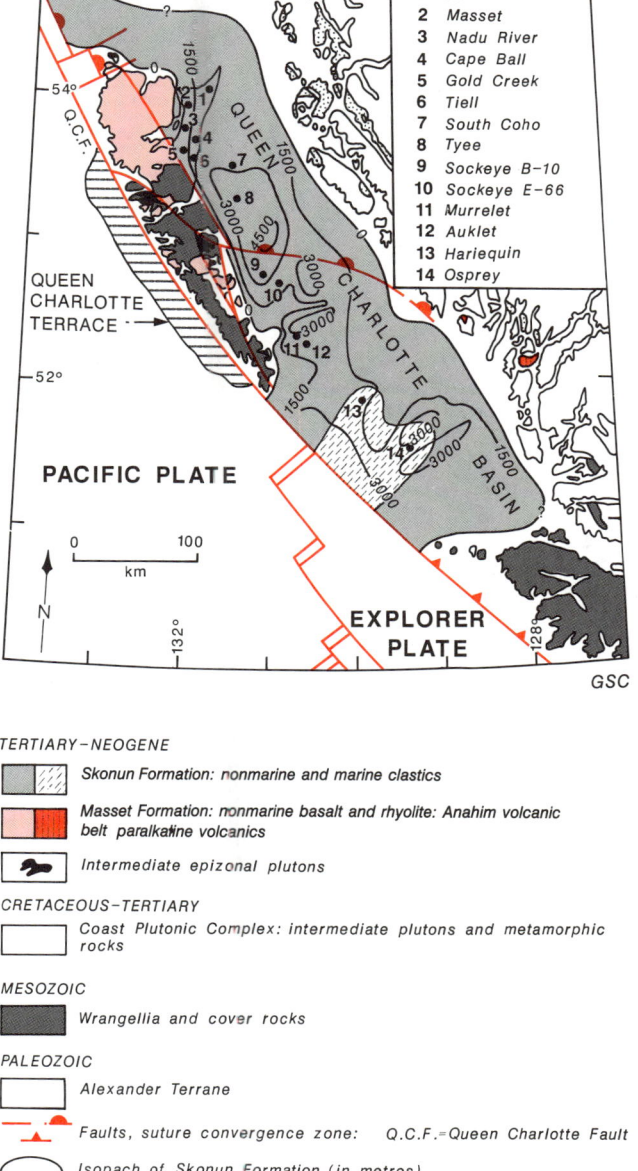

Fig. 10.2. Distribution of sedimentary facies and locations of boreholes on Queen Charlotte Islands and adjacent Hecate Strait and Queen Charlotte Sound.

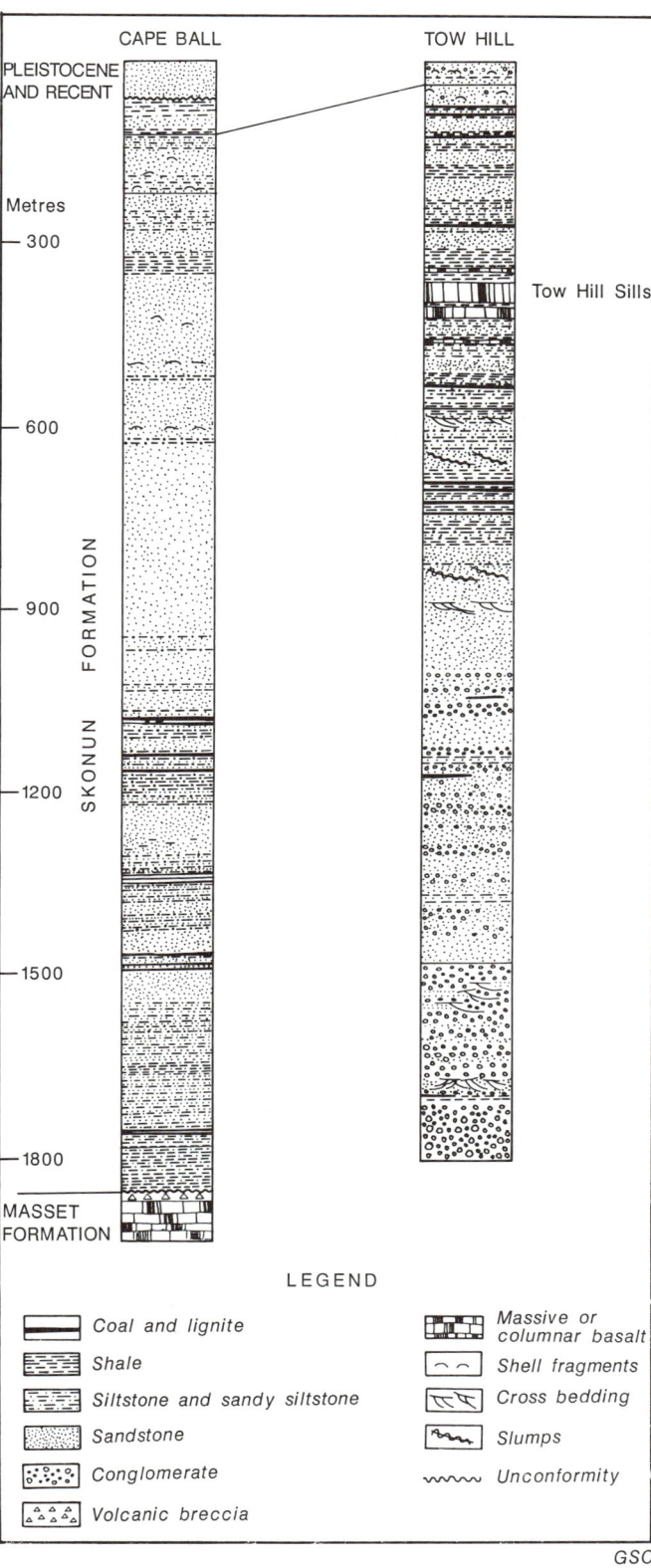

Fig. 10.3. Columnar sections of the Skonun Formation based on subsurface data from the Cape Ball and Tow Hill wells on northern Graham Island (see Fig. 10.2 for locations; after Sutherland Brown, 1968).

(Fig. 10.3). The age of the sill is poorly dated as Late Pliocene or earliest Pleistocene (Sutherland Brown, 1968).

The age of the Skonun Formation varies throughout Queen Charlotte Basin. The oldest beds are found at the base of the Harlequin well in Queen Charlotte Sound where they are Early Miocene (Burdigalian) in age (B.E.B. Cameron, pers. comm., 1982). On Graham Island basal sections range from late Early to Middle Miocene and in all remaining subsurface sections the oldest beds are Late Miocene. Throughout the basin the uppermost beds are Late Pliocene in age (Sutherland Brown, 1968; Addicott, 1978; Champigny et al., 1981; Shouldice, 1971).

Northwestern Vancouver Island Continental Shelf and Slope

The northwestern Vancouver Island continental shelf and slope is underlain by an assemblage of argillite, volcanics and fine grained clastics of presumed Neogene age (Fig. 10.4) (Bornhold and Yorath, 1984). The distribution and lithology of the several rock types have been determined by submersible observations, augmented by dredging, grab sampling and geophysical techniques such as side-scan-sonar and single- and multi-channel seismic profiling.

A volcano-sedimentary succession forms the foundation of the Neogene succession which appears to be separated from the adjacent Bonanza Group of Vancouver Island by a continuation or splay of the West Coast Fault. The sequence comprises argillite, greywacke and volcanics which are probably intimately interbedded. A sample of the volcanics yielded a whole-rock K-Ar age of 18.3 Ma (R.L. Armstrong, pers. comm., 1979). No fossils have been recovered from the sediments which are low to mid-greenschist facies. The relationship of small quartz diorite bodies to the presumed Neogene sequence is unknown. The volcano-sedimentary sequence has been deformed into southerly and southwesterly directed recumbent folds, probably related to thrust faults which, on the basis of multi-channel seismic data, outcrop on the continental slope. Miocene shale and mudstone (B.E.B. Cameron, pers. comm., 1980), Lower Cretaceous greywackes (Yorath et al., 1977), and volcanics of unknown age, have been recovered in dredge hauls from the continental slope. Unconformably overlying the volcano-sedimentary sequence is an undeformed, westward thickening wedge of Upper Miocene to Upper Pliocene siltstone, mudstone and minor spherulitic basalt(?) beds (Yorath et al., 1977). Stratigraphic studies on the upper part of the continental slope show the unit to be about 400 m thick, crossbedded and intruded by several narrow siltstone dykes.

The tectonic implications of this succession are poorly understood. At least the lower part of the sequence probably developed, and was deformed, in an accretionary and compressive setting in pre-Late Miocene time. However, the lack of knowledge of stratigraphic relationships among the components of the volcano-sedimentary sequence, and their relationship to older rocks precludes tectonic analysis.

Tofino Basin

The Neogene component of the Tofino basin is an extension of the Paleogene succession of marine siltstone, shale and minor, immature feldspathic sandstone (Fig. 10.4). Local unconformities partition the Middle Miocene and Upper Pliocene rocks (Shouldice, 1971).

Georgia Basin

No Neogene strata have been confirmed in Georgia Basin, however, the upper part of the section penetrated by the Point Roberts well could include Upper Tertiary strata (see Chapter 9).

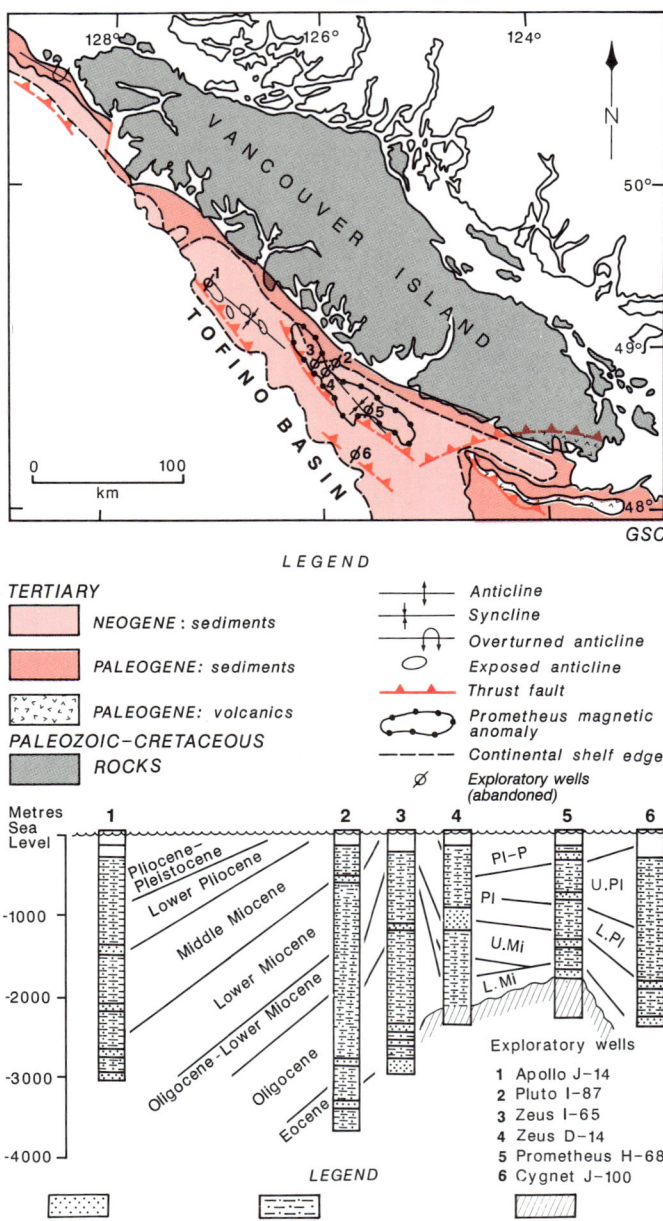

Fig. 10.4. Distribution, structure, and borehole data from Tertiary sedimentary rocks of Tofino Basin and the northern continental slope west of Vancouver Island.

Intermontane Belt

Neogene sediments outcrop locally throughout the central interior of British Columbia where they are commonly overlain, and protected from erosion by younger plateau basalts of the Chilcotin Group. The beds, which rest unconformably on older rocks, are essentially undeformed but dips of a few degrees suggest that they have been locally affected by tilting and minor faulting. The thickness of the sedimentary sequence varies from a few centimetres to more than 150 m and includes both lacustrine and fluvial deposits which locally include small amounts of carbonaceous shale and lignite.

Middle to upper Tertiary beds of coarse, well sorted lithic arenite, conglomerate, and minor calcareous shale outcrop in a northeasterly trending zone that extends for roughly 8 km along the Fraser River near Pavilion (Trettin, 1961). These sediments, which are locally up to 150 m thick, are probably flood plain deposits derived mainly from Cache Creek strata.

In the Bonaparte Lake area (Campbell and Tipper, 1971), upper Tertiary lacustrine sediments of the Deadman River Formation comprise up to 150 m of interbedded tuff, breccia, diatomite, diatomaceous siltstone, pebbly arenite and conglomerate. The beds are correlative, at least in part, with those in assemblages near Hanceville in the Taseko Lakes area (Tipper, 1963), and in the Anahim Lake (Tipper, 1957), Prince George (Tipper, 1961), Nicola (Cockfield, 1948), and Quesnel (Tipper, 1959) areas. Near Quesnel the Crownite Formation, a succession of poorly consolidated diatomite 12 m thick, is overlain by plateau basalt and underlain by 200 m of gravel, sand and silt of the Fraser Bend Formation. Fossil plants from the diatomite indicate a late Middle Miocene age, between 11 and 13 Ma (Rouse and Mathews, 1979). Sediment deposition likely took place in a wide valley under fan, floodplain, swamp and lacustrine conditions.

In the Yukon clastic sedimentary rocks of probable Miocene age are interlayered with lavas of the Wrangell Volcanic Belt. They are locally as much as 100 m thick and comprise sandstone, coaly siltstone, tuff and conglomerate in which all clasts are volcanic and are believed to have been shed from constructional volcanic edifices (Souther and Stanciu, 1975). Conglomerate and sandstone underlying parts of the volcanic succession may also be of Neogene age but they have not been separated from older (mainly Oligocene) sediments of the Amphitheatre Formation (Eisbacher and Hopkins, 1977). In the northern part of the St. Clare province, Wrangell lavas are overlain unconformably by ancient deposits of polymict fluvial and glacial gravel, till, and lacustrine deposits which contain both Wrangell and pre-Wrangell lava clasts. These deposits have been tilted and faulted along with the underlying Wrangell strata and may be as old as Late Miocene (Denton and Armstrong, 1969).

IGNEOUS ASSEMBLAGES
J.G. Souther

Miocene and younger igneous activity in the Canadian Cordillera was concentrated along five volcano-tectonic belts and in three relatively small igneous provinces that lie outside the main belts (Fig. 10.1). In southwestern British Columbia the Garibaldi and Pemberton volcanic belts, and the Chilcotin Group plateau basalts define an arc-backarc pair related to subduction of the Juan de Fuca and Explorer plates under the continental margin. The Anahim Volcanic Belt (Bevier et al., 1979; Souther, 1984), which extends from the coast, near Bella Bella, easterly across central British Columbia to the eastern boundary of the Intermontane Belt (Fig. 10.1), is interpreted to be the trace of a mantle hotspot. The Stikine Volcanic Belt (Souther, 1977), which forms a broad zone that, curving through northwestern British Columbia and the southern Yukon (Fig. 10.1), is interpreted to be a zone of extension developed in response to shear along the adjacent, transcurrent boundary between the continent and Pacific crust. The Wrangell Volcanic Belt (Souther, 1977), which extends from Alaska into southwestern Yukon, is interpreted to be a continental arc related to convergence between the Pacific and northern North American plates.

Neogene volcanic activity outside the five main volcano-tectonic belts was limited to the eruption of small basaltic centres in the Clearwater-Quesnel and McConnell Creek areas, near the suture bounding the eastern edge of Quesnellia and to the eruption of basaltic to rhyolitic volcanics and hypabyssal rocks of the Alert Bay Belt possibly coincident with the boundary between the Explorer and Juan de Fuca plates.

Southern British Columbia subduction complexes - related arc volcanics

At least four volcanic elements are associated with the converging plate boundary between the Juan de Fuca-Explorer plate system and the continental margin. Calc-alkaline volcanic fronts of the Pemberton and Garibaldi volcanic belts and alkaline lavas of the Chilcotin Group form an arc-backarc pair which persisted through Neogene and Quaternary time (Fig. 10.5). The Masset volcanics on the Queen Charlotte Islands are believed to be an early manifestation of this same subduction complex and the Alert Bay Volcanic Belt on Vancouver Island is probably a zone of basaltic and salic volcanism related to the subducted Juan de Fuca-Explorer plate edge.

Pemberton Volcanic Belt

The southern Pemberton Volcanic Belt (Souther, 1975, 1977; Berman and Armstrong, 1980) is defined by a group of epizonal plutons and a few erosional remnants of eruptive rock that lie along a linear zone extending west-northwest from the Chilliwack Batholith to the Salal Creek stock (Fig. 10.1). The Mount Silverthrone and Franklin Glacier complexes, in the central Coast Mountains, and the Masset Formation on Queen Charlotte Islands, though widely separated from one another, are of similar age and calc-alkaline affinity and are believed to be the northern extremity of Pemberton arc volcanism related to subduction of the Farallon Plate.

The Mount Barr Batholith (Richards, 1971) is the youngest intrusive body (21 to 16 Ma) in a group of mid- to late Tertiary epizonal stocks and batholiths that cut the Fraser River-Straight Creek fault system in southwestern British Columbia and northern Washington. Other bodies in the group include the Silver Creek Stock (35 Ma), and the Chilliwack Batholith (29 to 26 Ma). The range of ages

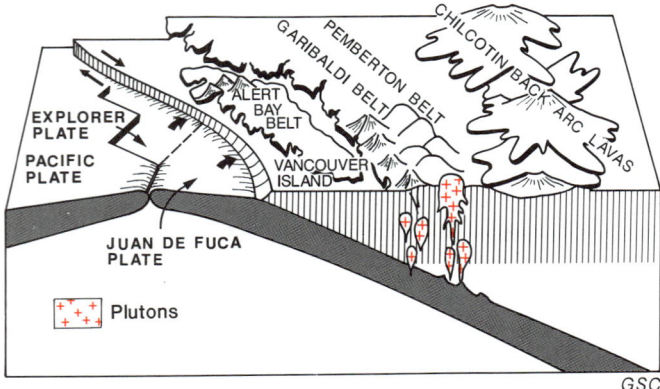

Fig.10.5. Schematic diagram showing the relationship of the Pemberton, Garibaldi, and Alert Bay volcanic belts and the Chilcotin back-arc lavas to offshore plate geometry.

suggests that episodic igneous activity occurred throughout late Tertiary time and culminated with emplacement of the Mount Barr Batholith in the Early Miocene.

Mount Barr Batholith is roughly circular, about 16 km in diameter, and composed of four intrusive phases ranging in composition from tonalite to quartz monzonite. The oldest, Conway phase, is a medium-grained tonalite concentrated in discrete areas along the margin of the body. Hornblende-plagioclase-phyric granodiorite of the Mount Barr phase forms the main mass of the batholith and two younger phases, quartz monzonite (leucocratic stocks) and fine grained tonalite (Wahleach Lake phase) intrude the central part. Contacts between phases are commonly gradational suggesting that successive intrusions took place in fairly rapid succession, while many of the phases were still hot and possibly plastic. The older Conway and Mount Barr phases contain layered sequences, probably cumulates, that suggest a sill-like structure. The younger stocks are characterized by granophyric textures and miarolitic cavities, indicative of an epizonal environment. The phases of the Mount Barr Batholith are believed to have been emplaced as successive pulses of magma undergoing crystal fractionation in a deep chamber.

The Coquihalla Complex (Berman, 1981; Berman and Armstrong, 1980) covers an area of about 30 km² near Hope, British Columbia (Fig. 10.6). It consists of calc-alkaline, felsic to intermediate extrusive and intrusive rocks including rhyolitic pyroclastic deposits, andesite and dacite domes, dykes, sills, and a central, diorite to quartz diorite stock. Three K-Ar dates average 21.4 ± 0.7 Ma, and are concordant with a Rb-Sr isochron (22.3 ± 4 Ma) based on seven whole-rock samples which span the entire compositional range of the suite. The complex rests unconformably on the mid-Cretaceous Eagle Pluton and Lower Cretaceous Pasayten Group rocks.

The lower, pyroclastic assemblage of the Coquihalla Complex includes rhyolitic tuff, ash flows, and breccia with a combined thickness of about 1600 m. The pyroclastic succession is overlain and intruded by numerous andesite to dacite domes, dykes and sills all cut by a diorite to quartz diorite stock which forms the core of Coquihalla Mountain. The stock is coarser grained than other andesitic intrusions in the complex, suggesting that it crystallized at a later time, under a thicker accumulation of pyroclastic rocks.

Volcanic rocks of the Coquihalla Complex are commonly separated from surrounding rocks by faults associated with distinctive, monolithological avalanche breccias. The breccias, which are cut locally by andesite dykes, are characterized by poorly sorted angular clasts of either Eagle granodiorite or Pasayten sediments up to 2 m in size, set in a matrix of smaller clasts. They are believed to have formed through large scale avalanching from uplifted fault-blocks bounding the complex. Locally, Coquihalla ash flows rest unconformably on Pasayten rocks outside the bounding faults. The implication of these overlapping relationships is that the Coquihalla Complex was deposited in a subsiding basin. Subsidence must have been rapid and contemporaneous with faulting, tilting of the basal unconformity, and basin filling volcanism. Toward the end of the volcanic cycle, the rate of pyroclastic production exceeded the rate of basin subsidence and late ash flows spilled over and rest unconformably on Pasayten Group rocks.

The Salal Creek Pluton (Stephens, 1972) is an epizonal granitic body with an exposed area of about 40 km². It is near the younger volcanic edifice of Meager Mountain, where the Pemberton and younger Garibaldi Volcanic belts converge. The pluton is intrusive into regionally metamorphosed rocks of the Coast Belt and its deeply dissected surface is locally overlain by erosional remnants of Garibaldi Group volcanics (Plate 23). The body is elliptical in plan with a northeasterly elongation and sharp discordant contacts with the older foliated rocks.

The Salal Creek rocks vary from granodiorite to granite but the majority of samples are quartz monzonite. The body exhibits a crude, eccentric, textural zonation from a marginal coarse grained, biotite-rich phase, through an intermediate phase of medium grained, biotite-poor rock, to a central leucocratic, fine grained phase with less than 1% biotite. In addition, quartz-feldspar porphyry occurs in pods and lenses that are gradational with the other phases and as crosscutting dykes. The internal zonation is the result of crystal fractionation and changing pressure conditions during intrusion by piecemeal stoping.

Comagmatic silicic dykes, major joint-sets and shear zones trend northeast, parallel with the long axis of the body. Their orientation was controlled by a northeasterly trending axis of uplift that accompanied the emplacement and crystallization of the pluton. Younger, post-Pliocene basalt dykes have a more northerly orientation, parallel with the regional trend of the Garibaldi Group volcanic centres.

Franklin Glacier Complex

The Franklin Glacier Complex (McKnight, 1965; Ney, 1968) underlies an elliptical, northwesterly trending area about 20 km long and 6 km wide with local relief exceeding 2000 m. Within it the Mesozoic to early Tertiary granitic and metamorphic rocks of the Coast Plutonic Complex are fractured, hydrothermally altered, and cut by dyke swarms and high-level plutons of biotite-quartz-porphyry, biotite quartz-feldspar-porphyry and quartz monzonite. At least some of the younger intrusive phases appear to be feeders for an overlying succession of eruptive rocks comprising

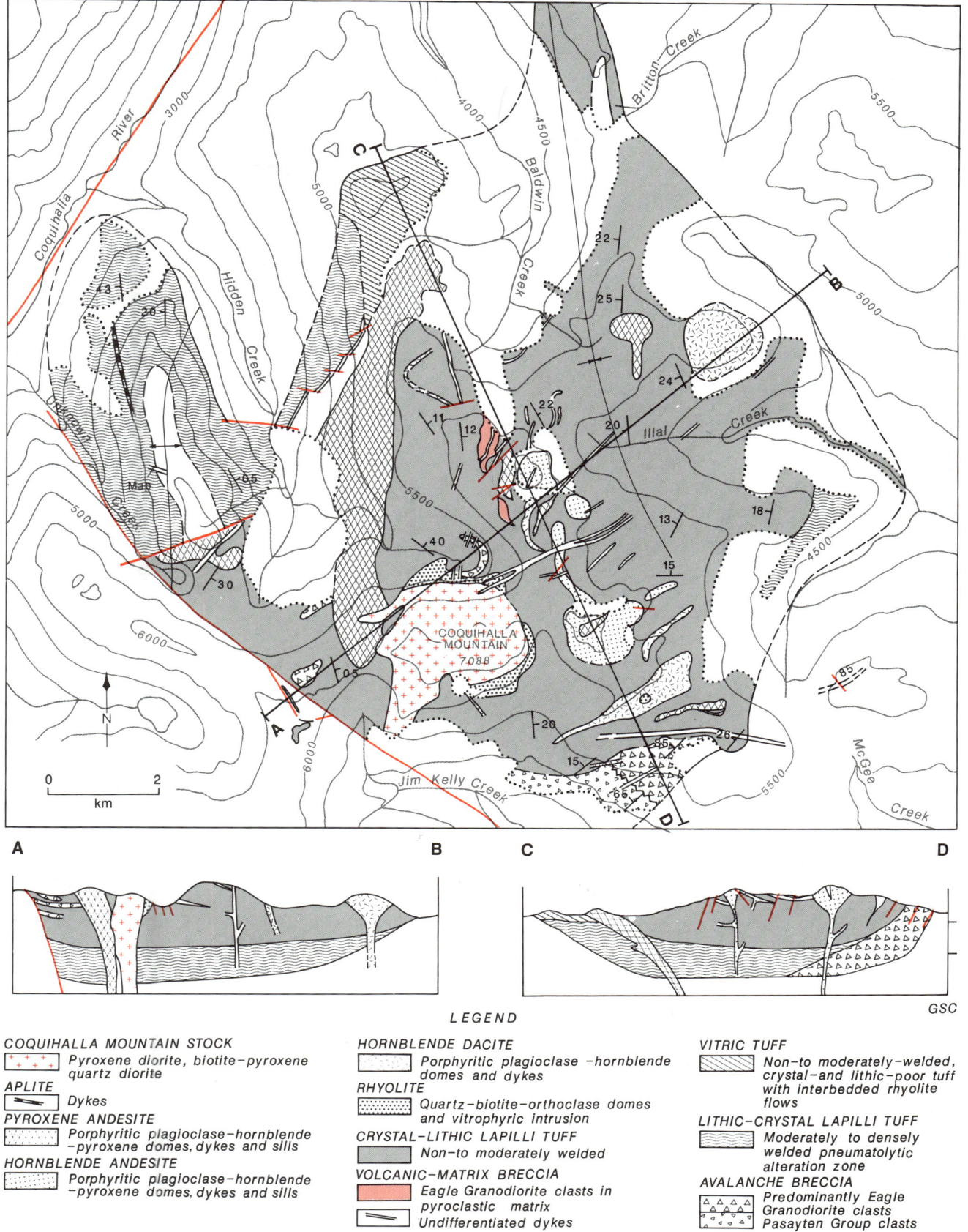

Fig. 10.6. Sketch map and cross-sections of the Coquihalla Complex. After Berman and Armstrong (1980).

mainly dacite breccia, minor dacite flows and a few remnants of hornblende andesite at the top of the succession. The breccia includes clasts of porphyritic dacite and basement rocks up to a metre across. Both types of clasts are angular to subangular and randomly suspended in a gritty matrix of comminuted rock debris. The absence of pumice, glassy blocks or bombs in these breccias suggests that they may be of epiclastic origin, possibly formed during collapse and infilling of a cauldron subsidence structure.

Medium grained quartz monzonite forms the largest of the subvolcanic plutons in the Franklin Glacier Complex. It yielded a K-Ar (biotite) age of 6.8 ± 0.2 Ma, which is only slightly younger than the Salal Creek Pluton (7.9 Ma, 8.1 Ma; Wanless et al., 1978) and the adjacent Fall Creek stock (10.1 Ma; Stevens et al., 1982). Both of the latter granitic bodies are overlain unconformably by Upper Pliocene to Holocene volcanic rocks near Meager Mountain (Read, 1977; Lawrence et al., 1984) and are considered to be high-level, possibly subvolcanic plutons belonging to the pre-Garibaldi, Pemberton Volcanic Belt (Souther, 1975). A porphyritic dacite dyke, cutting the Franklin Glacier quartz monzonite, yielded a K-Ar (biotite) age of 3.9 ± 0.1 Ma and a fine grained dacite dome or flow yielded a K-Ar (biotite) age of 2.2 ± 0.1 Ma. The ages are only slightly older than the basal Meager Mountain rocks (1.9 Ma). Thus, the Franklin Glacier Complex, like the Meager Mountain and Salal complexes, appears to be the product of two discrete thermal events - the first between 6 and 8 Ma and the second between 2 and 3 Ma.

Mount Silverthrone Complex

The Mount Silverthrone Complex (Green et al., 1988) lies about 55 km west-northwest of the Franklin Glacier Complex. Between the two, several small remnants of andesitic and rhyolitic lava cap ridge crests and occupy valleys. Although deeply dissected and equally as rugged as the Franklin Glacier Complex, the roughly circular Mount Silverthrone Complex is composed mostly of eruptive rocks (Fig. 10.7). Many of the volcanic products postdate the present topography and are clearly younger than those associated with the Franklin Glacier centre.

The lowest unit is a breccia containing angular to subangular clasts of granitic, metamorphic and volcanic rocks in a dense, well indurated matrix of comminuted rock debris. Locally, welded shards give the matrix a fluidal, eutaxitic (banded) texture. The breccia is exposed up to 1200 m above valley bottoms and contacts with the older crystalline rocks of adjacent peaks are steep, suggesting a fault-bounded subvolcanic structure. The breccia is overlain by thick, lenticular flows of rhyolite, dacite and andesite having a composite thickness of about 900 m. Rhyolite glass about 100 m stratigraphically above the top of the basal breccia and a thick andesite flow overlying rhyolite in the central part of the complex yielded whole rock K-Ar dates of 0.75 ± 0.08 Ma and 0.40 ± 0.10 Ma, respectively. Both dates are compatible with the degree of dissection.

The early flows and underlying breccia were deeply eroded and much of the present topography was established before the culminating eruption of basaltic andesite from numerous centres around the periphery of the complex. Remnants of pyroclastic cones, built during the final surge of activity, project through the glacial ice which mantles the eastern half of the complex. Distal flows ex-

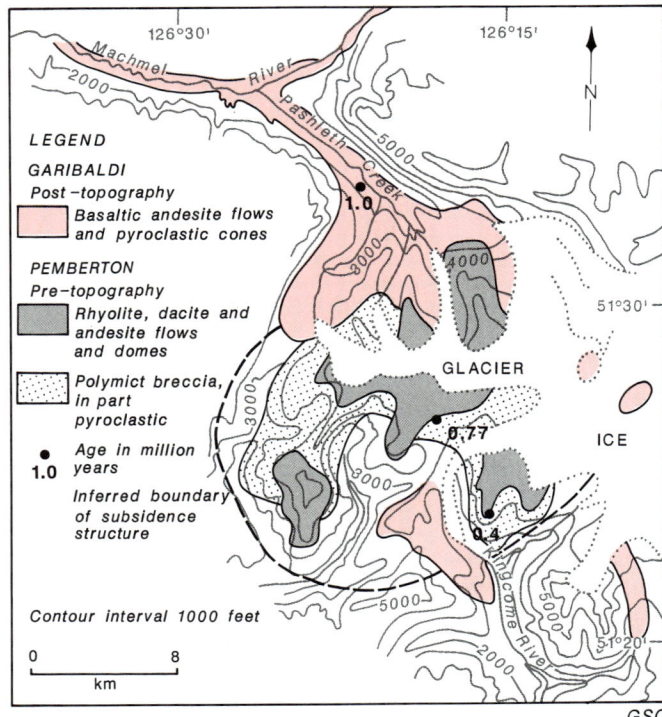

Fig. 10.7. Sketch map of the Mount Silverthrone Complex showing deeply dissected (Pemberton Belt) phase on which younger eruptive rocks of the Garibaldi Group are superimposed.

tend south into the headwaters of Kingcome River, and north along Pashleth Creek into Machmell River valley. The latter flows are continuously exposed for more than 25 km, decreasing in elevation from 2000 m above sea level to less than 100 m above sea level. They rest on several metres of unconsolidated, crossbedded tuff breccia and fluvial gravel. The basaltic andesite flow which occupies the Pashleth Creek and Machmell River valleys, yielded a K-Ar date of 0.95 ± 0.2 Ma. The lava may be much younger than indicated by the K-Ar date, because the high energy, glacier-fed stream has only begun to etch a channel along the edge of the flow. Inclusions and xenocrysts from the underlying, Mesozoic granitic basement are widely distributed in the lava and their incorporation could explain the probably erroneous age.

Masset Formation

Basalt and rhyolite flows, pyroclastic rocks and related intrusions of the Masset Formation are exposed on Graham and Moresby islands (Fig. 10.8) and basalt, believed to be correlative with the lower Masset, has been encountered in boreholes within southern Hecate Strait and Queen Charlotte Sound. The Masset rests unconformably on the Karmutsen and younger formations and is both overlain unconformably and locally interbedded with the Neogene Skonun Formation. K-Ar ages range from 63 to 11 Ma on Queen Charlotte Islands and from 57 to less than 10 Ma in the subsurface, beneath Queen Charlotte Sound.

Sutherland Brown (1968) recognized three distinct facies: the Tartu Facies, which underlies most of Graham

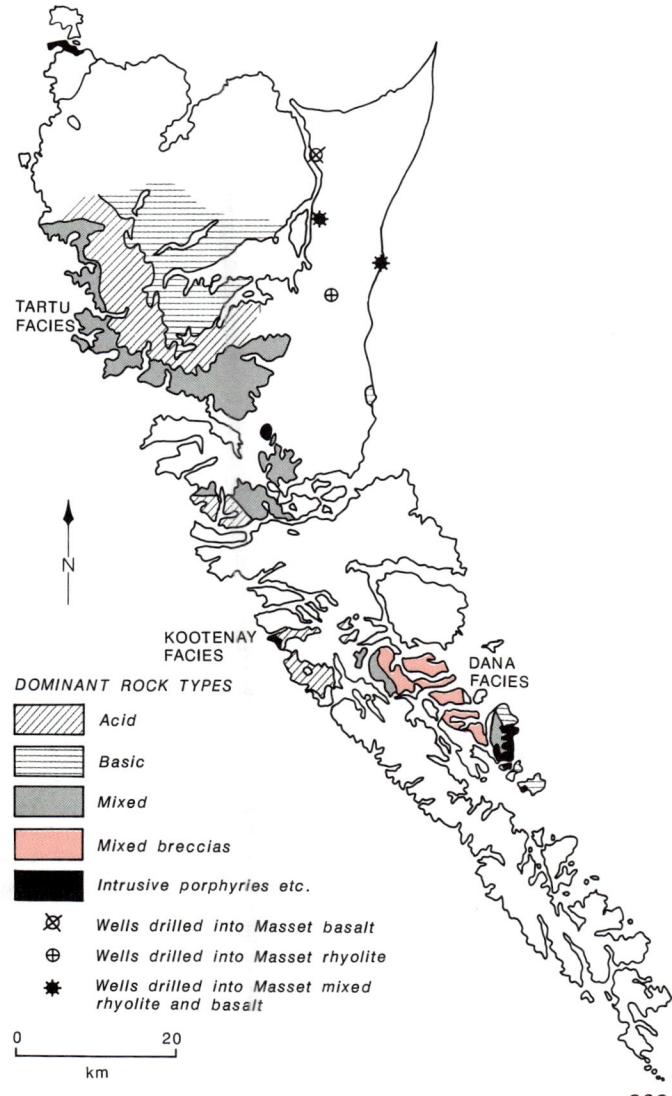

Fig. 10.8. The three facies of the Masset Formation on the Queen Charlotte Islands (after Sutherland Brown, 1968).

succession, more than 1500 m thick, of columnar basalt flows associated with minor basaltic and rhyolitic pyroclastic rocks.

The Kootenay Facies, exposed on the west coast and near Dawson Inlet, is characterized by rhyolite tuff breccias, welded ash flows and massive spherulitic rhyolite. The pyroclastic rocks form beds from 5 to 120 m thick. They are typically eutaxitic and contain cognate lithic and collapsed pumice fragments in the 5 to 8 cm size range. Accidental clasts of basalt are present in some flows. Associated aphanitic, subhorizontally foliated rhyolite is commonly spherulitic and may contain up to 50% drusy lithophysae (concentrically structured bubbles). These units are gradational with welded ash flows and may themselves be intensely welded pyroclastic flow units that have sintered and sufficiently recrystallized to mask their original vitroclastic textures. The type Kootenay section consists entirely of crudely columnar flows of sodic rhyolite, welded tuff breccia and ash flows of varying thickness and fragment size. The rocks grade laterally into spherulitic rhyolite and are locally intercalated with columnar basalt flows.

The Dana Facies, on the east coast of Moresby Island, is at least 1500 m thick. It is mainly tuff breccia containing rhyolite and basalt clasts. The units are characteristically fine grained and unbedded, although some contain perceptible graded bedding. Sorting is generally poor; angular blocks from 30 cm to 6.5 m being randomly suspended in a matrix of lapilli-sized clasts. The graded sequences show evidence of reworking and some contain fragments of charred wood, suggesting deposition in a fluvial-lacustrine or lagoonal environment.

Eruptive rocks of the Dana Facies are cut by numerous feldsparphyric rhyolite dykes and sills. Elsewhere the Masset Formation includes intrusive hypabyssal rocks of gabbro and diabase as well as salic feldspar porphyry. These are believed to represent plug-like, dyke-like and sill-like orifices or upper magma chambers within the volcanic succession that cut older rocks beneath the Masset pile.

The Masset complex is chemically bimodal, containing about equal proportions of subalkaline sodic rhyolite and high-alumina basalt with MORB affinities (Sutherland Brown, 1968; T.S. Hamilton, pers. comm., 1985). Intermediate rocks are rare and, at least in the Dana tuff breccia, appear to be mechanical mixtures of the two fundamental rock types. Sutherland Brown (1968) considered the basalt to be upper mantle material slightly modified by contamination and fractionation and suggested that the rhyolites may be early melting material remobilized from deeply buried older plutons in Wrangellia. Yorath and Chase (1981) suggested that Masset volcanism began with the initiation of rifting in Queen Charlotte Sound and migrated northwestward along the Louscoone Inlet-Sandspit Fault system and later along the Rennell Sound Fault. A subsidence and thermal model supports this hypothesis (Yorath and Hyndman 1983). Unpublished chemical data indicate that the salic rocks, like those elsewhere in the Pemberton Belt, are calc-alkaline whereas the basic suite has MORB affinities (T.S. Hamilton, pers. comm., 1985). This bimodal, arc-MORB affinity suggests that Masset activity may have been initiated where the spreading ridge between Pacific and Farallon plates intersected the continent.

Island north of Rennell Sound, comprises microporphyritic to aphanitic, commonly columnar jointed basalt flows associated with oxidized pyroclastic and flow-top breccias. A smaller proportion of the Tartu facies consists of sodic rhyolite flows and domes, commonly with varicoloured banding and irregular vesicles which are either open or partly lined with chalcedony and zeolites. Most of the rhyolite is porphyritic with phenocrysts of feldspar, pyroxene and rarely quartz in an aphanitic to glassy groundmass. Welded ash flows, with pronounced eutaxitic texture and air-fall lapilli tuff are present as minor components of the Tartu facies. The type section near Tartu Inlet includes three members: 1. a lower member comprising 1800 to 2000 m of mixed basalt breccia and flows and rhyolite ash flow tuff; 2. a middle rhyolite member composed of very thick (30 to 120 m) units of rhyolitic ash flow tuff and minor columnar basalt and; 3. an upper

Alert Bay Volcanic Belt

The Alert Bay Volcanic Belt (Armstrong et al., 1985) extends from Brooks Peninsula northeastward across Vancouver Island to Port McNeil (Fig. 10.1, 10.5). It encompasses several separate remnants of late Neogene volcanic piles and related intrusions ranging in composition from basalt to rhyolite and in age from about 8 Ma in the west to about 3.5 Ma in the central and eastern part. On the Brooks Peninsula highly deformed argillite of the Pacific Rim Complex is cut by coarse grained plagiophyric basalt dykes which yielded K-Ar (feldspar) dates of about 8 Ma. Farther east, in central Vancouver Island, dykes and sills of basalt and dacite cut Cretaceous Nanaimo Group sandstone. The largest remnant in the belt is preserved at Twin Peaks where surfaces above 770 m within an area about 4 km by 8 km are underlain by flat-lying columnar basalt flows, basaltic tuff breccia, and volcanic conglomerate containing clasts of vesicular and feldsparphyric basalt. The flat-lying deposits suggest the former presence of a volcanic edifice large enough to have generated a drainage system capable of producing coarse fluvial and laharic deposits. The eastern end of the Alert Bay Volcanic Belt is defined by a cluster of remnants near Port McNeil. These consist of coarsely feldsparphyric basalt flows, tuff breccia and volcanic conglomerate and more felsic rocks. The latter include massive, light-coloured dacite that has been quarried on Haddington Island for use as a building stone. Sills and dykes of dacite and less abundant microporphyritic rhyolite are associated with the Twin Peaks and Port McNeil remnants.

Major element analyses of Alert Bay volcanic and hypabyssal rocks suggest two different basalt-andesite-dacite-rhyolite suites with divergent fractionation trends. The first coincides with the typical calc-alkaline, Cascade trend, whereas the other is more alkaline and more Fe-enriched, following a trend analogous to that of Mull in the Hebrides which straddles the calc-alkaline-tholeiite boundary.

The western end of the Alert Bay Volcanic Belt is now about 80 km northeast of the Nootka Fault Zone which separates the Explorer and Juan de Fuca plates. However, at the time of its formation the volcanic belt may have been coincident with the subducted plate edge between Juan de Fuca and Explorer plates. Also, the timing of volcanism corresponds to shifts of plate motion and changes in the locus of volcanism along the Pemberton and Garibaldi volcanic fronts. This brief interval of plate motion adjustment at about 3.5 Ma may have triggered the generation of basaltic magma along the descending plate edge. These mantle-derived magmas appear to have undergone minor crustal contamination and shallow fractionation along divergent paths to create the variety of observed rock types.

Garibaldi Volcanic Belt

The Garibaldi Volcanic Belt (Mathews, 1958; Souther, 1977) is a northern extension of the Cascades Volcanic Belt in the western United States (Fig. 10.1). It is offset to the west of the main Cascade trend and is a composite of at least three en échelon, north-trending segments, referred to here as the southern, central, and northern segments. The Garibaldi Volcanic Belt intersects the older Pemberton Volcanic Belt at a low angle near Meager Mountain where Garibaldi Group lavas rest on uplifted and deeply eroded remnants of Pemberton Volcanic Belt subvolcanic plutons (see Salal Creek Stock, Plate 24). North of Meager Mountain the Garibaldi and Pemberton volcanic belts appear to merge into a single belt and Mount Silverthrone, farther to the northwest, was episodically active during both Pemberton and Garibaldi stages of volcanism.

Southern segment

The principal volcanoes in the southern segment of the Garibaldi Volcanic Belt, are Mount Garibaldi, Mount Price and the Black Tusk (Fig. 10.9). Black Tusk, the oldest volcano in the southern segment, is a composite pile formed during two distinct stages of magmatic activity. The initial eruption of hornblende-dacite flows and minor ash flow tuffs, between 1.1 and 1.3 Ma, built a composite cone that was deeply eroded before the onset of the second stage. During the second stage remnants of the ancestral cone were buried by hypersthene andesite flows and pyroclastic breccia between 0.17 and 0.21 Ma. The culminating event of Black Tusk volcanism was the extrusion of a mushroom-shaped plug dome near the summit of the second stage cone.

Mount Garibaldi is a moderately dissected, composite Pelean cone, built during the waning stages of the last major glaciation (Mathews, 1952, 1958; Green, 1981). It rests on erosional remnants of an older, Round Mountain assemblage of andesite breccias, polymict conglomerate, minor dacite and basalt flows and lahar deposits that occupy preglacial valleys beneath the western part of the Mount Garibaldi edifice. Similar breccias south of Mount Garibaldi outcrop on Paul Ridge and form most of Round Mountain which is believed to be one of several pre-Garibaldi eruptive centres. All of the Round Mountain lavas, like those of the overlying Garibaldi pile, are normally magnetized and probably of Brunhes age (± 0.75 Ma).

The edifice of Mount Garibaldi was built during at least three principal stages of activity and construction of the central cone was accompanied by episodic flank eruptions. The initial, Cheekye Stage (ca 20 ka), began after considerable degradation of the Round Mountain pile, when the ice surface lay at about 1340 m ASL. Eruption of dacite from vents near the eastern end of Brohm Ridge formed steep-sided, coalescing domes flanked by tuff-breccia and lava flows which formed a broad composite cone. A period of quiescence followed, during which the flanks of the cone were incised by radial drainage gulleys. The second, Atwell Stage of activity began while ice still filled the surrounding valleys but after the Wisconsin climax. A central, Pelean spine of dacite erupted near the present summit of Mount Garibaldi and repeated avalanches of hot dacite debris poured radially outward onto the flanks of the growing cone. Locally the distal edge of the pyroclastic pile rested on glacier ice and crossbedded volcanic sandstone was deposited in flanking meltwater ponds. The third, Dalton Stage of activity began after the ice sheet had begun its final retreat. As the supporting ice melted the western part of the pyroclastic cone began to collapse, producing landslides and mudflows that spilled into Cheekye valley and built thick deposits of fanglomerate. Dacite lava of the Dalton Stage flowed down one of the major landslide scars. The flow itself was partly destroyed by additional sapping of the underlying tuff-breccia, suggesting that this final

NEOGENE ASSEMBLAGES

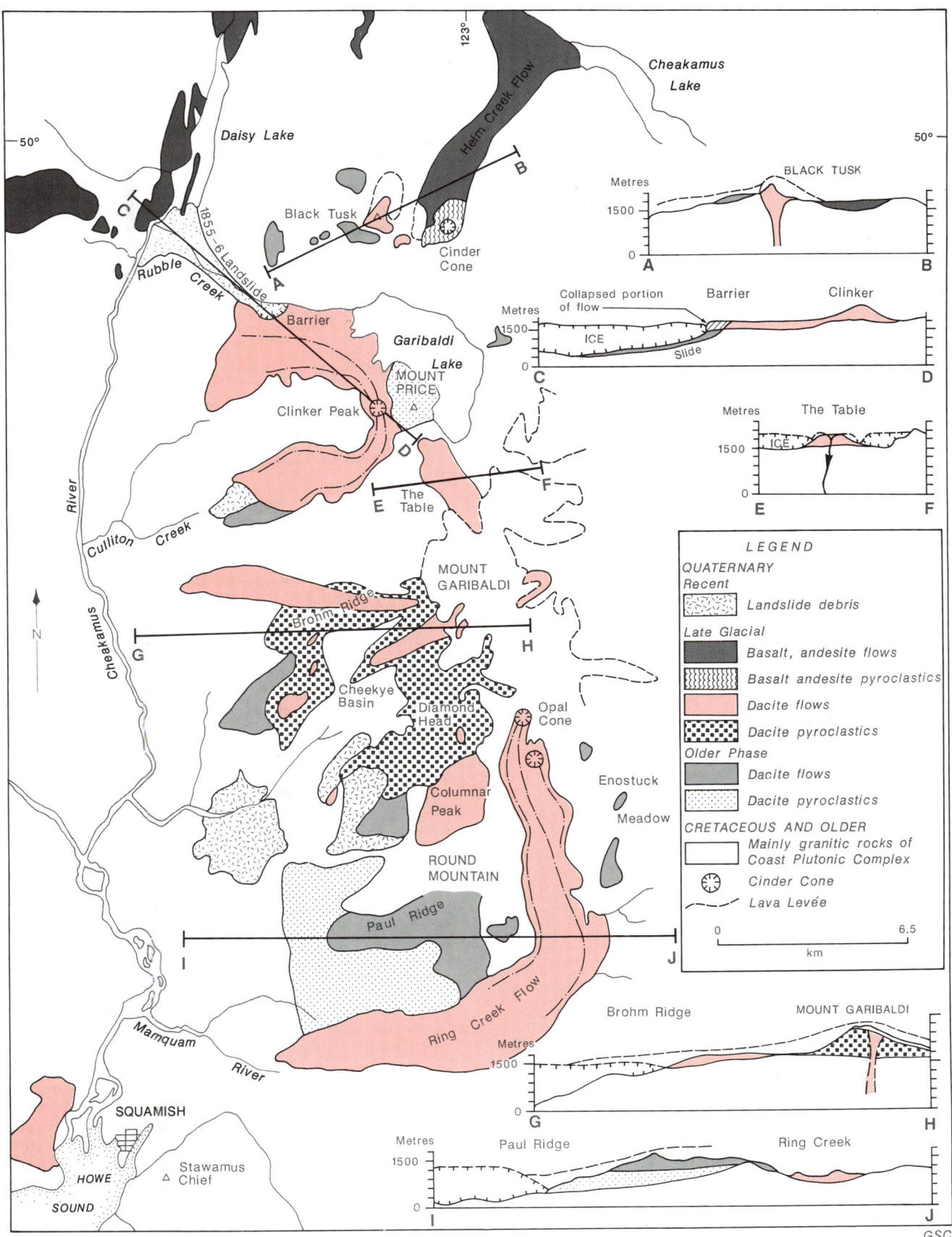

Fig. 10.9. The principal volcanoes of the central Garibaldi Belt. Modified from Green et al. (1988).

eruption from the summit area of Mount Garibaldi occurred soon after withdrawal of the ice. Mazama ash on Brohm Ridge indicates that the ice sheet was gone before 6670 years ago.

Eruption of lava from satellitic vents near the flanks of Mount Garibaldi occurred both before and after withdrawal of the ice. The Table (Mathews, 1951) on the north slope of Mount Garibaldi is a steep-sided, flat-topped tuya - a pile of hornblende andesite flows that erupted into and were contained by a vertical pipe thawed into glacier ice. At the time of its eruption the ice surface must have stood at least as high as the 2020 m summit of the Table. A similar, but smaller, subglacial pile of hornblende basaltic andesite forms a steep sided mass at Enostuck Meadow on the southeast flank of Mount Garibaldi. The postglacial eruption of Opal Cone, on the south flank of the mountain, marked the final episode of Garibaldi volcanism. A massive effusion of dacite issued from the vent low on the south flank of the mountain and formed the Ring Creek Flow which is as much as 2 km wide and extends more than 17 km from its source into Mamquam Valley. Opal cone, a composite of three nested, annular ridges, was probably constructed by explosive surges from the vent during the waning stages of activity.

Mount Price a composite volcano much smaller than Mount Garibaldi, formed during three distinct periods of activity (Green, 1981). The initial eruptions produced a small stratovolcano of interlayered hornblende andesite flows and pyroclastic material that rest on glacial drift and, based on K-Ar dates of about 1.2 Ma, are approximately coeval with the early stage of Black Tusk activity. Following a period of quiescence and erosion the focus of volcanic activity shifted westward and the present symmetrical cone of Mount Price was constructed on, and partly buried, the older edifice. The second stage culminated with the summit eruption of two andesite flows which yielded a K-Ar date of 0.3 ± 0.2 Ma (Green, et al, 1988). During the final stage of activity a breached lava ring, Clinker Peak, was built on the western shoulder of Mount Price. Hornblende andesite lava from this vent formed the Barrier and Culliton Creek flows which postdate the disappearance of the Cordilleran ice-sheet from higher altitudes but predate the disappearance of valley glaciers from Cheakamus River Valley. The oversteepened front of the Barrier flow, formed by contact of the flow with a remnant of glacier ice, has been the source of repeated landslides. Garibaldi and Lesser Garibaldi lakes are ponded behind the Barrier flow and drain through subterranean channels along its base. Future collapse of the barrier and catastrophic release of Garibaldi Lake into the head of Rubble Creek is considered to pose sufficient risk to warrant legislation excluding the area from development (Moore and Mathews, 1978).

In addition to the large, central andesite-dacitic volcanoes the southern Garibaldi Belt includes numerous remnants of basalt and basaltic andesite flows and pyroclastic rocks that are distributed from the head of Howe Sound, along Cheakamus River and Helm Creek valleys. The source of the Helm Creek flow is exposed at the head of Desolation valley where the initial activity produced a broad tuff ring and an associated basaltic andesite flow which yielded a K-Ar date of 0.11 ± 0.03 Ma (Green et al, 1988). Subsequent eruption of olivine basalt built a cinder cone on the eastern rim of the older tuff ring and produced the main body of the Helm Creek flow which yielded a minimum K-Ar date of 0.04 ± 0.04 Ma. The source of olivine basalt flow-remnants in Cheakamus River valley is unknown, but at least two distinct episodes of volcanism are represented. The lowermost basalt rests on ice-scoured basement rock and locally on fluvio-glacial sand and gravel. The flows have also been glaciated and, where several flows are present, they are commonly separated by layers of baked till. Wood contained in a silty interbed yields a date of ca 34,000 years. This date is within the Olympic Interstade, the nonglacial period preceding the last (Fraser) glaciation. The younger flows in the Cheakamus Valley are characterized by fanned entablature columns and blocky, unglaciated flow tops. They commonly stand as anastomosing ridges above the older flow surfaces and are believed to have formed during the waning stages of Fraser glaciation, when lava entered meltwater caverns thawed beneath the ice (Mathews, 1958).

Central Segment

The central segment of the Garibaldi Belt (Souther, 1980) is defined by a group of eight volcanoes lying along the height of land east of Squamish River, and by remnants of basaltic flows preserved in the adjacent Squamish valley (Fig. 10.10).

Mount Cayley, the largest and most long-lived centre (3.8 to 0.31 Ma; Green et al, 1988) is a multiple plug dome of dacite and minor rhyodacite from which most of the original, outer cone of pyroclastic material has been eroded away. Pyroclastic deposits and dacite flows, erupted during the initial, Mount Cayley stage, rest on a preglacial surface of high relief that slopes steeply westward into Squamish valley. The Mount Cayley stage culminated with the intrusion of a central dacite spine after which the locus of volcanism shifted south to the Vulcan's Thumb centre. Vulcan's Thumb activity produced a thick, asymmetrical pile of dacite flows, domes and pyroclastic deposits that partially buried the southwestern flank of the older Mount Cayley edifice. Westerly flowing streams dissected the composite central pile and cut the deeply incised valleys of Turbid and Shovelnose creeks prior to the final, Shovelnose stage. This final stage of activity began with the effusion of dacite lava that flowed west into the Squamish valley from a vent in upper Shovelnose Creek and culminated with the emplacement of two small domes, each about 300 m high and a kilometre in diameter. Glass-rimmed, subvolcanic cupolas are exposed in the deeply incised valley of Turbid Creek where they cut basement rocks and are associated with north-trending fractures. Seeps of warm water issue from some of the fractures and the surrounding rock is intensely altered, suggesting that an active hydrothermal system is associated with the volcano.

Mount Fee is a narrow elliptical spine of rhyodacite about 1 km long and 0.25 km across at its widest point. The massive, sparsely jointed rock forms a series of near-vertical towers that rise 100 to 150 m above the ridge. Most of the summit ridge is a denuded neck of intrusive, pale grey porphyritic rhyodacite with abundant phenocrysts of glassy feldspar and sparse biotite. The mantle of pyroclastics that must once have enclosed it has been stripped away except for a small remnant on the western side and along the northern end where the lip of the conduit is exposed in cross-section. Contacts between the spine and older rocks of the Coast Plutonic Complex are

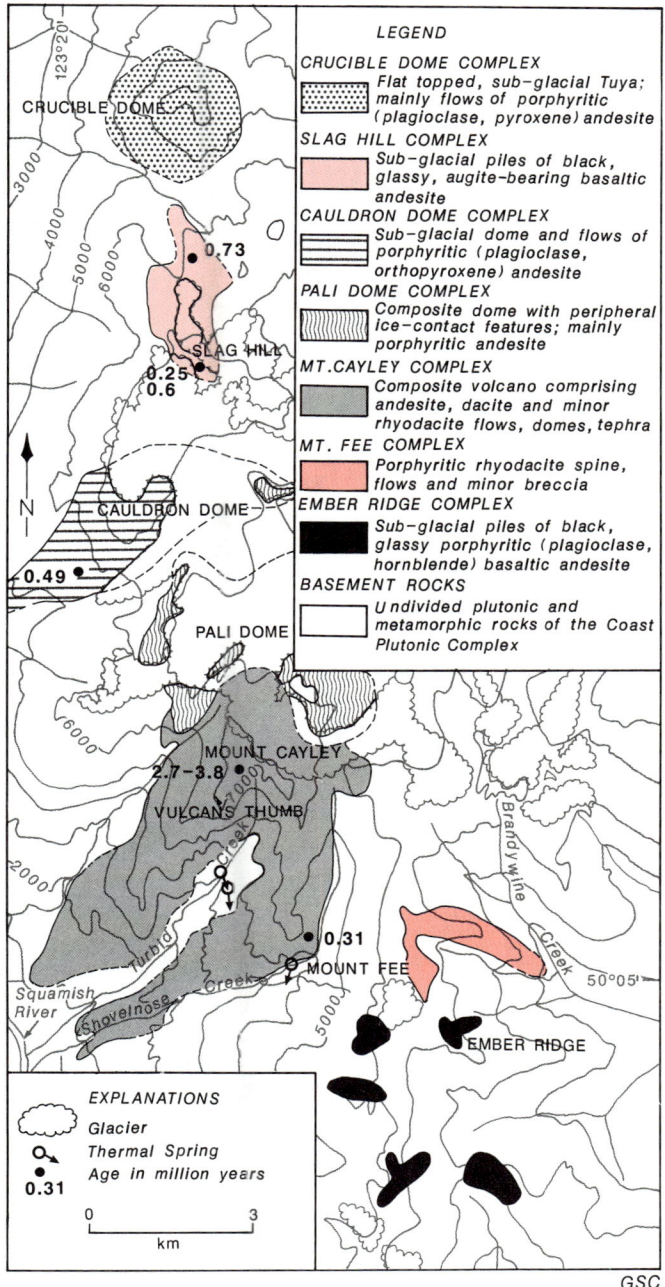

Fig. 10.10. The distribution of volcanic complexes in the central segment of the Garibaldi Belt.

2 km. The flows are underlain by steep, northerly-dipping beds of blocky tuff-breccia which are probably part of a steep-sided pyroclastic cone formed during the early stages of activity and across which the flows were ramped. The existing towers and spines of Mount Fee are erosional remnants bounded by joint planes. Complete denudation of the central spine as well as the absence of a till layer under the Mount Fee flows suggest a pre-glacial age.

The other volcanoes of the central Garibaldi Belt (Ember Ridge, Pali Dome, Cauldron Dome, Slag Hill and Crucible Dome) are intraglacial, tuya-like forms with oversteepened, ice-contact margins. The Ember Ridge complex comprises five separate remnants of aphanitic to vitreous, hornblende-bearing basalt which forms steeply inclined exogenous domes. Individual flows, up to 60 m thick, are separated by deep furrows which locally enlarge into small elliptical glass-lined caverns. Most exposures are characterized by complex, small diameter, columnar and closely spaced polygonal joints, typical of quenched lavas. The volcanism probably occurred during a period of extensive ice cover, causing the rapidly quenched lava to pile up in steep-sided domes directly over the vents.

Pali Dome, at least 4 km in diameter, is a composite, largely ice-covered lava dome overlapping the northern edge of the Mount Cayley Complex. It consists entirely of coarsely porphyritic (plagioclase, hypersthene, ± hornblende) andesite. The proximal portion of most flows appears to be subaerial. Large diameter columns form well developed vertical colonnades which are underlain by scoriaceous, oxidized flow breccia. In contrast the distal flows have small diameter radiating and subhorizontal columns characteristic of rapid cooling. Most flows terminate in near vertical ice-contact cliffs from 100 to 200 m high. Locally these are flanked by thick piles of granular glass which must have accumulated in the moat between lava and ice.

Cauldron Dome is a nearly flat-topped elliptical pile of thick porphyritic andesite flows. It is about 2 km in diameter and has the classical form of a tuya. Erosion has removed the outer, quenched portion of the edifice, exposing tiers of thick nearly vertical columnar flows separated by oxidized scoria. Two very thick (100 to 130 m) complexly jointed lava flows of similar andesite extend 2 km southwest and 300 m below the summit. Cauldron Dome formed as a subglacial pile whereas the lower flows were probably directed into a meltwater channel which breached the enclosing barrier of ice during the latter stages of activity.

The Slag Hill pile comprises steep-sided bulbous masses of black, glassy basaltic andesite with small diameter, curved and radiating columns. It is clearly a quenched, subglacial pile, but except for a small flat-topped bluff at the top, it lacks the classical tuya form. The Slag Hill lava rests on a steep northwesterly sloping basement surface along which outflow channels probably developed shortly after the eruption started. Thus only a small amount of lava was ice-ponded to form the flat-topped bluff in the source area whereas most of it was channelled, along with meltwater, into subglacial caverns where it was quenched and solidified into its present bulbous form.

Crucible Dome is a circular flat-topped pile of porphyritic (plagioclase, hypersthene) andesite (Plate 24). It is flanked by talus and its outer margins have been greatly modified by erosion. The flat upper surface is a

nearly vertical. Along them the granitic wall rock has been shattered and granulated for a few metres and the adjacent rhyodacite is quenched to a narrow (40-50 cm) selvage of granular, porphyritic glass. The exposed northern rim of the main conduit consists of blocky tuff-breccia overlain successively by angular colluvium comprising blocks of basement rock, about 2 m of lapilli ash, about 6 m of lightly welded rhyodacite tuff-breccia and finally by columnar-jointed, rhyodacite lavas. Three relatively thin flows are overlain by a massive upper flow at least 150 m thick that drops more than 1200 m in elevation within less than clinkery red scoria strewn with bomb-like chunks of vesicular oxidized lava. It is probably a modified tuya on which at least the upper flows were subaerial.

Northern Segment

The northern segment of the Garibaldi Volcanic Belt includes the Meager Mountain complex and several remnants of basaltic piles north of Meager Mountain.

Meager Mountain (Read, 1978; Lewis and Souther, 1978) is a complex of at least four overlapping composite volcanoes that become progressively younger from south to north (Fig. 10.11). The earliest eruptions of rhyodacite lavas and pyroclastic materials built the Pylon Peak cone on a basement surface with up to 400 m of relief. Basal breccias up to 300 m thick are overlain by subhorizontal quartz dacite flows which yielded a K-Ar date of about 2 Ma. A period of quiescence and deep dissection preceded the subsequent eruption of hornblende andesite flows and pyroclastic material which overlap the Pylon Peak edifice and form most of the southern and western parts of the complex. The second-stage flows are commonly flow-banded, interbedded with thin oxidized breccia and tuff lenses, and have initial dips up to 15° away from The Devastator where a concentration of hypabyssal intrusions and the presence of clasts up to several metres in length suggest a major vent. Whole rock K-Ar dates of 0.9 to 0.5 Ma (Green et al, 1988) from the lavas indicate a long period of andesitic volcanism. After a second period of quiescence the focus of volcanism again shifted north to the Job, Capricorn, and Meager centres. Eruption of biotite-rhyodacite and hornblende-biotite quartz dacite flows, domes and pyroclastic material from these centres built a steep-sided edifice on the north slope of the older andesitic cone. K-Ar dates of about 0.09 to 0.1 Ma suggest that this activity was short-lived. The final eruptive phase of the Meager Mountain Complex, and the youngest volcanic activity in the Garibaldi Volcanic Belt, produced the Bridge River tephra and rhyodacite flows from a vent on the northern margin of the complex. Near their source crudely stratified breccia and ash deposits are up to 20 m thick and distal Bridge River ash has been detected as far east as southern Alberta. Scoriaceous rhyodacite lava and pyroclastic block and ash flows up to 145 m thick extend for 6 km along the adjacent Lillooet valley and are believed to have issued from the same vent. A charred tree in living position in tephra beneath one of the valley flows gave a ^{14}C date of 2490 ± 50 BP (Read, 1978). An active hydrothermal system associated with the Meager Mountain Volcanic Complex has been extensively explored as a potential geothermal energy source (see Chapter 20).

Several occurrences of Neogene volcanic rocks lie north of Meager Mountain but only those in the Salal Creek area have been studied in detail (Lawrence et al., 1984). The Salal Glacier Volcanic Complex, with an area of about 2 km², is the largest remnant. It contains two flow units which rest on the peneplaned surface of the Late Miocene, Salal Creek stock (Plate 25). The lower flow is a columnar-jointed, coarsely feldsparphyric hawaiite (K-Ar 0.97 ± 0.05 Ma, Green et al, 1988) underlain by 20 m of breccia containing both basement and basalt clasts. The upper flow is alkali olivine basalt (K-Ar 0.59 ± 0.05 Ma) containing euhedral phenocrysts of olivine, clinopyroxene and plagioclase. The relatively thin proximal parts of the flow

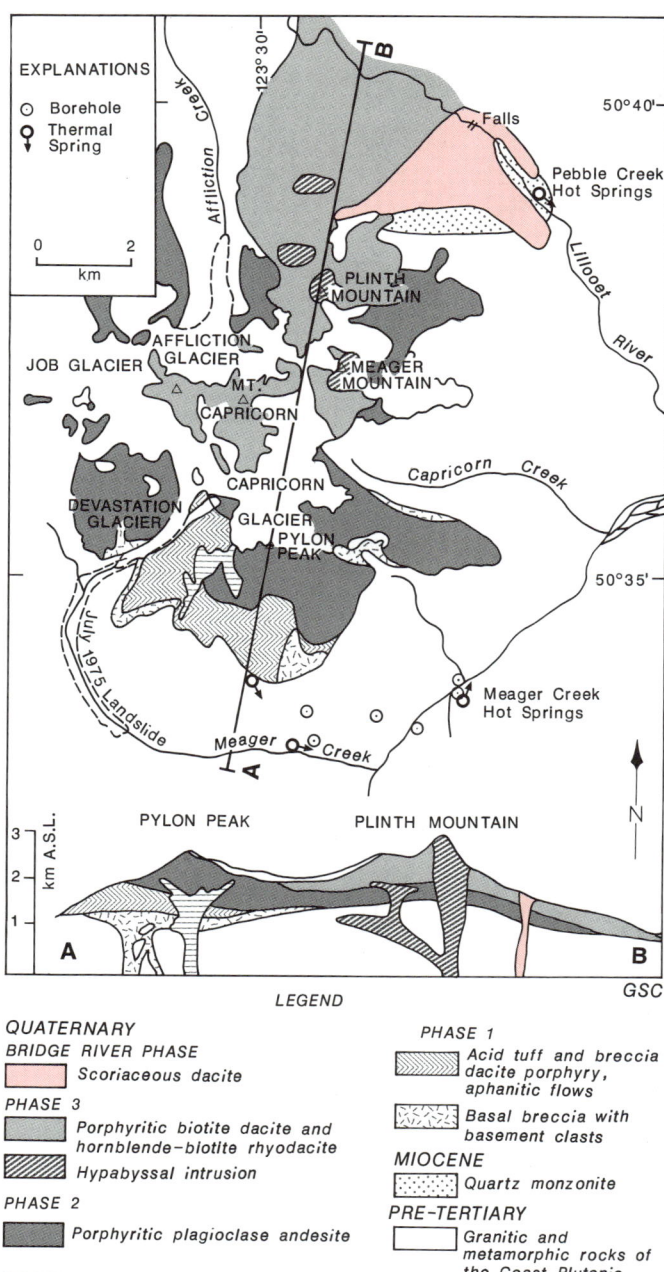

Fig. 10.11. Sketch map and schematic cross-section of Meager Mountain (after Read, 1978).

are strewn with scoriaceous, subaerial bombs and cinders whereas the southern terminus forms a vertical ice-contact face over 100 m high. Columnar joints radiating outward toward the exposed cliff face indicate close proximity of the original cooling surface.

All other flow remnants in the Salal Creek area are basaltic and most contain some evidence of ice contact. The alkaline affinity of many of these lavas contrasts with the calc-alkaline character of the larger central volcanoes of the Garibaldi Volcanic Belt. Their proximity to the volcanic front suggests a discontinuity in the subducted plate, possibly a subducted plate edge analogous to that proposed for the Alert Bay Volcanic Belt, or magma generation at the northern end of the active arc.

Southern and central British Columbia plateau lavas

Flat-lying Mio-Pliocene basalt flows cover 50 000 km^2 of British Columbia's interior plateau. Since they were first described by Dawson (1879) they have been informally referred to as the "Plateau Lavas". Tipper (1978) introduced the name Chilcotin Group to which he assigned both the volcanic rocks and locally underlying sediments. In a later study the name Chilcotin Group was restricted to the volcanic rocks (Bevier, 1983), a convention that is followed herein. Flat-lying basalt flows similar in age and composition but geographically separated from the lavas of central British Columbia are not included in the Chilcotin Group.

Chilcotin Group

The Chilcotin Group (Bevier, 1983) consists of thin, crudely columnar-jointed pahoehoe flows, some thick, tiered flows, pillow lava and pillow breccia, and rare silicic tephra layers. The flows are predominantly flat-lying or dip less than 2°, and are inferred to form a series of coalesced, low shield volcanoes erupted from central vents. Most sections include from 2 to 20 cooling units. The average thickness of the Chilcotin Group is 67 m; the thickest known exposure is 141 m. In many places the entire section consists of only one flow. The Bull Canyon and Deadman River exposures, among the thickest and most complete sections of Chilcotin Group lavas, are described by Bevier (1983) as follows:

"The section at Bull Canyon, located along the Chilcotin River, consists of at least 17 pahoehoe flow lobes with intercalated lenses of pillow lava. Individual flow lobes are thin (3-14 m), crudely columnar jointed, and have platy, vesicular zones at the base (up to 0.5 m thick) and top (up to 2 m thick). Tops and bottoms of flows are oxidized to a red-brown. Ropy pahoehoe texture is found on well preserved flow tops, and collapsed pahoehoe toes are common. Thin pipe vesicles (4-10 cm long) occur 1-3 cm above the base of the flows. Vesicle sheets and cylinders (1-10 cm wide, up to 15 m long) are abundant and are confined to the massive interiors of flow lobes. The basalt is dark grey, orange-brown weathering, aphyric to olivine and plagioclase microphyric (up to 10%), and contains 5-20% vesicles. In places the vesicles are filled with chabazite. Pillow lava and broken pillow breccia make up about 25% of the stratigraphic section and thin layers of unconsolidated silt occur between some flows, but pillow interstices are empty, suggesting that the pillows formed where sediment could not accumulate."

"The Deadman River section, near the southeastern margin of the lava plateau, contains three silicic air-fall tephra layers intercalated with at least 16 lava flow lobes. The lava flows, 2-30 m thick, are pahoehoe lobes some of which are collapsed, with red, ropy tops and bases, and crudely to moderately well developed columnar-jointed interiors. Chisel marks are common on the better developed columns. Vesicle sheets and cylinders are found in most flows; zeolites are prominent only in the vesicular parts of some flows. The basalt is medium to dark grey, orange-brown weathering, and aphyric to olivine microphyric. All three tephra layers mantle the topography of the underlying basalt flows. The oldest of the three tephra layers is 9.3 m thick and consists of five beds. The tephra consists of buff, light-brown weathering, well sorted angular pumice fragments, less than 5% subrounded basaltic lithic fragments (<2 mm), and 1-2% fragments of plagioclase, biotite, sanidine, and quartz phenocrysts. Different beds in this layer can be distinguished on the basis of size of the pumice fragments, which range from 1-3 mm to 0.5-1 cm in diameter. In the finer grained beds, up to 10% matrix of fine ash and shards is present. The middle tephra layer is a 1.6 m thick 'popcorn' textured layer consisting of light grey, buff weathering, 1-10 mm rounded accretionary lapilli and broken accretionary lapilli, with 2% rounded basaltic scoria fragments (2-5 mm) and 1-2% fragments of plagioclase, biotite, and quartz. The top 50 cm of this layer is baked red by the overlying basalt flow. The youngest tephra layer is a 1.2 m thick bed consisting of light grey, buff weathering fine ash and shards with rare accretionary lapilli, and it bears 2-4% plagioclase, hornblende, and biotite fragments. No vents for these tephras have been identified within the plateau lava field. Their most likely source is Pemberton belt volcanoes 150 km to the west (Bevier, 1981a)."

The surface of the Chilcotin lava plain was intensely glaciated and deeply incised by major streams, many of which have cut steep-walled canyons through the basalt and into the underlying basement rocks. At least 50% of the original volcanic pile is believed to have been stripped off by glacial and fluvial erosion. Pillow lava is found only in sections along major river canyons, suggesting that the present drainage pattern was well established by the Late Miocene. Post-Miocene uplift of the Coast Mountains has tilted basalt remnants along the southwestern margin of the plateau as much as 15° toward the northeast.

Chilcotin lava flows are intruded by several gabbroic and basaltic plugs (Farquharson, 1973) which are presumed to be vents for the flows. They define a northwest trend along the axis of the volcanic plateau, parallel with the Pemberton Volcanic Belt centres. Four plugs of olivine-bearing diabasic gabbro near Williams Lake, British Columbia are elliptical bodies, 82-250 m across, that exhibit a steep, upwardly diverging foliation. Pegmatoidal segregations and crystal-lined cavities occur near their margins and the adjacent basalt has been metamorphosed to hornfels. Anahim Peak, near the eastern flank of the Rainbow Range, was the source for a series of basalt flows that are the same age as lavas on the plateau and may be included in the Chilcotin Group. The Anahim Peak centre lies near the intersection of the basalt plateau with the Anahim Volcanic Belt and its vent is occupied by a plug dome of trachyte similar in age and composition to flows and domes in the adjacent shield volcanoes of the Anahim Volcanic Belt. North of the Anahim belt, in Whitesail Lake area, Woodsworth (1979) reported many small necks of columnar basalt cutting 20 Ma, flat-lying basalt flows in the extreme northern part of the Chilcotin Group lava plain.

K-Ar dates from Chilcotin Group flows suggest that the lavas issued from many different centres in several eruptive episodes, predominantly between 2-3 and 6-10 Ma. However, locally the lavas and outliers of similar lavas in the Whitesail Lake and Kamloops areas yielded K-Ar dates as old as Late Oligocene. Chilcotin Group lavas and plugs exhibit both normal and reverse magnetic polarity but in most sections all flows have the same polarity, suggesting that thick sections of flows accumulated rapidly near discrete vents at different times.

Chilcotin Group basalts are coeval with, and approximately 150 km inland from arc volcanoes of the Pemberton-Garibaldi volcanic front. Most of the them are coeval with volcanics and subvolcanic plutons of the Pemberton Volcanic Belt but some of the youngest flows are coeval with early stages of Garibaldi volcanism. They are mainly olivine-bearing, transitional basalts believed to have been generated by partial melting in the upper mantle, due to asthenospheric upwelling in a back-arc setting, above the subducting Juan de Fuca Plate. Silicic tuff, interbedded with the basalt, probably originated from calc-alkaline, arc volcanoes in the Pemberton Volcanic Belt and was preserved between successive flows of basalt in the Chilcotin, back-arc lava plain.

Central and northern British Columbia plume and rift complexes

Two well defined belts of Neogene volcanics extend into the continent from near the transcurrent boundary between the North American and Pacific plates (Fig. 10.1). The east-trending Anahim Volcanic Belt is defined by a chain of Neogene volcanoes, plutons and dyke swarms which extend across the Coast Belt and flat-lying basaltic lavas of the Chilcotin Group in central British Columbia. The Stikine Volcanic Belt of northern British Columbia and southern Yukon encompasses a series of en échelon, north-trending segments which occupy a broad zone curving northwestward and subparallel with the continental margin. A province of basaltic volcanism in the Quesnel-Shuswap Highlands includes the Clearwater Valley flows and numerous postglacial monogenetic cones.

Anahim Volcanic Belt

The Anahim Volcanic Belt (Souther, 1977, 1986; Bevier et al., 1979) extends from coastal British Columbia, near Bella Bella, across the Coast Mountains into the Interior Plateau. Its western end lies adjacent to Queen Charlotte Sound which probably formed by crustal rifting initiated in latest Oligocene or earliest Miocene time (Yorath and Chase, 1981; Yorath and Hyndman, 1983). The rift zone is filled with sediments of the Neogene, Skonun Formation which rests on basalt considered to be an early phase of Masset volcanism. East of Queen Charlotte Sound, intrusive and comagmatic volcanic rocks of the Bella Bella-King Island complex (Souther, 1986; Fig. 10-12) are exposed in fiords and islands of the western Coast Range. The central part of the Anahim Belt consists of three moderately dissected, peralkaline shield volcanoes, the Rainbow, Ilgachuz, and Itcha ranges which lie along an east-west zone between the Chilcotin and Nechako plateaus. A cluster of postglacial basaltic cones in the Nazko area west of Quesnel comprise the eastern part. In addition to the major eruptive centres several small postglacial basaltic cones are distributed along the entire length of the belt.

A systematic decrease in the age of volcanism from west to east along the Anahim Belt suggests that it may be the trace of a mantle hotspot over which the continent moved during the past ten to fifteen million years (Bevier

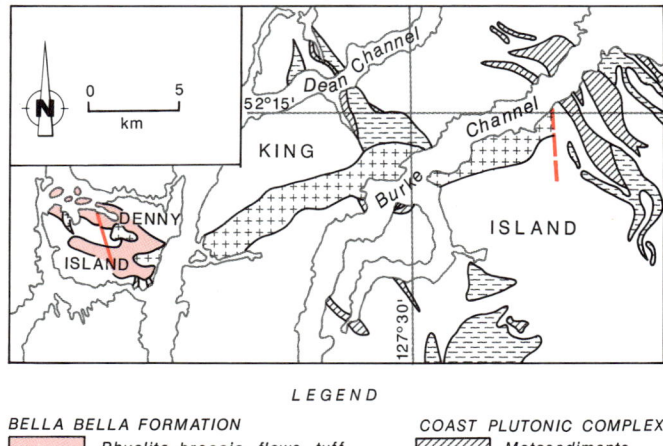

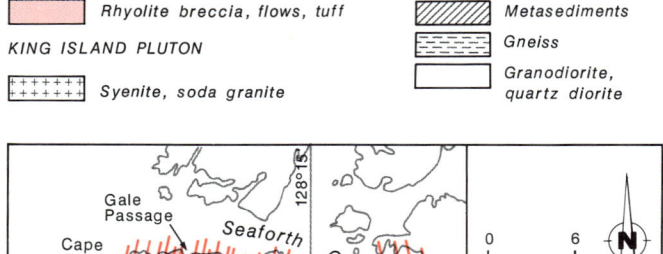

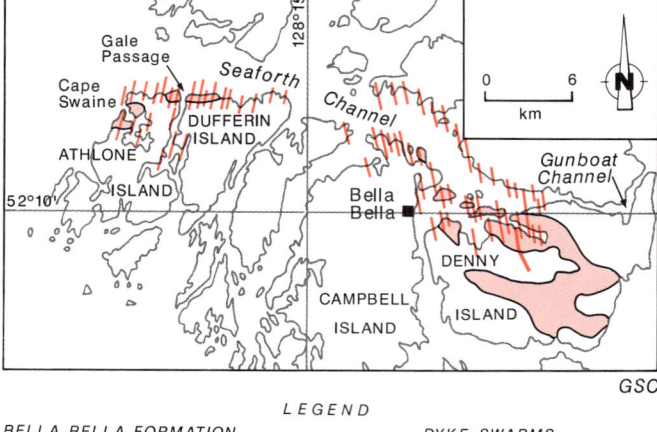

Fig. 10.12. The distribution of dyke swarms and volcanic and plutonic rocks at the western end of the Anahim Belt.

et al., 1979; Souther, 1986). The major element composition of rocks from volcanic, hypabyssal and plutonic environments is remarkably similar, each form a bimodal suite of highly alkaline rocks. Petrographic modelling (Souther, 1986) suggests that primitive mantle-derived alkali basalt fractionated in crustal reservoirs analogous to King Island Pluton to produce the oversaturated peralkaline end members.

King Island-Bella Bella Complex

King Island Pluton (Souther, 1986; Fig. 10.12) is exposed along Dean and Burke channels, west of Bella Coola. It is more than 50 km long and 5 km wide and trends roughly east across the predominantly northwest structural trend of the Coast Plutonic Complex. The core of the body is coarse grained alkaline to peralkaline syenite consisting almost entirely of coarsely perthitic alkali feldspar and a small proportion of intergranular mafic minerals, including fayalitic olivine, aenigmatite and riebeckite. With the appearance of quartz the syenite grades into soda granite, a phase that is best developed as a marginal zone and in satellitic stocks west of King Island Pluton. The granite is medium to fine grained, commonly with myrmekite and clusters of alkali amphibole and pyroxene interstitial to subhedral grains of quartz and alkali feldspar. Numerous open, quartz- and sodic amphibole-lined miarolitic cavities indicate that the granite was emplaced at low enough pressures to permit the separation of an alkaline vapour phase.

Eruptive rocks in the western Anahim Volcanic Belt are assigned to the Bella Bella Formation (Dolmage, 1922; Baer, 1973; Souther, 1986). They are exposed as flat-lying to gently dipping erosional remnants in the type area on Denny and adjacent islands, and in wave-cut benches along Seaforth Channel as far west as Cape Swaine on the outermost coast. The dominant lithology is coarse, dark purple rhyolite breccia containing numerous subangular accidental clasts of basement granodiorite, suggesting that they are the product of explosive volcanic activity. Locally the breccia is associated with flow-banded rhyolite flows and bedded tuffaceous sandstone containing leaf impressions.

Each of the remnant volcanic piles lies near the centre of, and is cut by, a major northerly trending dyke swarm representing an apparent extension of the host rocks between 20 and 40%. The western, Gale Passage swarm, and the eastern, Bella Bella swarm are each over 10 km wide and comprise mainly basalt and comendite dykes up to 20 m thick and a smaller proportion of relatively thin trachyte dykes. The basaltic dykes include both fine grained to aphanitic alkali olivine basalt and coarsely porphyritic hawaiite with phenocrysts of plagioclase up to a centimetre across. The comendite dykes, many with vitreous selvages, are fine equigranular to moderately porphyritic rocks with subhedral phenocrysts of alkali feldspar up to a centimetre long. Mafic minerals include sodic amphiboles, aegirine, ferrohedenbergite and lesser aenigmatite. Like the granitic phases of King Island Pluton they commonly have open miarolitic cavities lined with quartz, aegirine, and arfvedsonite. Relatively sparse trachyte dykes are commonly fine grained to aphanitic and consist almost entirely of alkali feldspar, opaques, and small interstitial crystals of sodic amphibole and/or pyroxene.

Central shield volcanoes

The central Anahim Volcanic Belt includes three large shield complexes and at least one smaller eruptive centre which form a linear, east-west-trending chain of loci extending across the Interior Plateau. The belt crosses the northern end of the area flooded by Chilcotin plateau lava, and distal flows at the margins of the shields merge imperceptibly with flat-lying flows of the Chilcotin Group. Unlike the Chilcotin basalts, which are not associated with any salic derivatives, the volcanoes of the central Anahim Volcanic Belt are markedly bimodal, comprising a mixed assemblage of basalt and peralkaline silicic rocks.

The Rainbow Range (Bevier, 1981b), with a diameter of about 25 km, is the largest and most westerly of the shield volcanoes. Its northern flank is a simple, concordant pile which from oldest to youngest, comprises comenditic-trachyte, mugearite, comendite and hawaiite; but the structure of the central part is unknown. About half a million years after cessation of volcanic activity on the shield, a series of hawaiite flows issued from a vent on its northeast flank, building Anahim Peak. A trachyte plug subsequently filled the vent from which these flows issued and most of the flanking pyroclastic deposits have been removed by erosion.

The Ilgachuz Range (Souther, 1984) is slightly smaller and less deeply dissected than the Rainbow Range. The shield is a composite of many overlapping domes and thick flows of comendite and comenditic trachyte, overlain by a relatively thin mantle of hawaiite, some of which issued from satellitic vents on the flanks of the shield. A circular, fault-bounded caldera, filled with epiclastic debris, lacustrine tuff and thick ponded trachyte flows, occupies the central part of the shield (Fig. 10.13).

The Itcha Range (Stout and Nicholls, 1983) is the smallest and most easterly of the three shield volcanoes in the central Anahim Volcanic Belt. Trachyte, comendite and comenditic-trachyte in the central core of the shield are overlain by a mantle of hawaiite flows forming most of the outer flanks.

Postglacial monogenetic cones

Small postglacial basaltic cinder cones are scattered along the Anahim Belt. In the western part of the belt slightly eroded pyroclastic cones, clustered a short distance north of the Bella Bella-King Island Complex, form Kitasu Hill, Lake Island, and several smaller hills on Lady Douglas, Price, and Dufferin Islands (Fig. 10.12). Similar basaltic cones, slightly modified by erosion, occur on the flanks of the central shield volcanoes. A composite pyroclastic cone and associated basalt lava flow near Nazko, 75 km east of the Itcha Range, is believed to be the youngest and most easterly manifestation of Anahim Belt volcanism. Tephra from the Nazko centre is intercalated with peat having a ^{14}C date of ca 7200 BP (Souther et al., 1987).

Stikine Volcanic Belt

Volcanics in the northerly trending Stikine Belt formed in an extensional regime related to northwesterly trending dextral transcurrent faults to the west which were active during Neogene time. The central part of the Stikine Volcanic Belt includes three large, compositionally diverse volcanic complexes; Level Mountain in the north, the Edziza-Spectrum Complex in the centre, and Hoodoo Mountain in the south (Fig. 10.14). South of Mount Edziza the belt is defined by several relatively small pyroclastic cones and associated basalt flows within a fairly narrow north-trending zone that traverses diagonally across the northwesterly trend of the Coast Mountains. North of Mount Edziza the Stikine Belt is less clearly defined. Level Mountain shield volcano and the Neogene part of the adjacent Heart Peaks dome complex lie east of the projected trend of the southern part of the belt and small pyroclastic cones are distributed throughout a broad zone that curves northwesterly from Mount Edziza across Level Mountain and into the southern Yukon. The volcanic centres within this zone appear to lie along six north-trending, en échelon segments. From south to north these are: the Aiyansh,

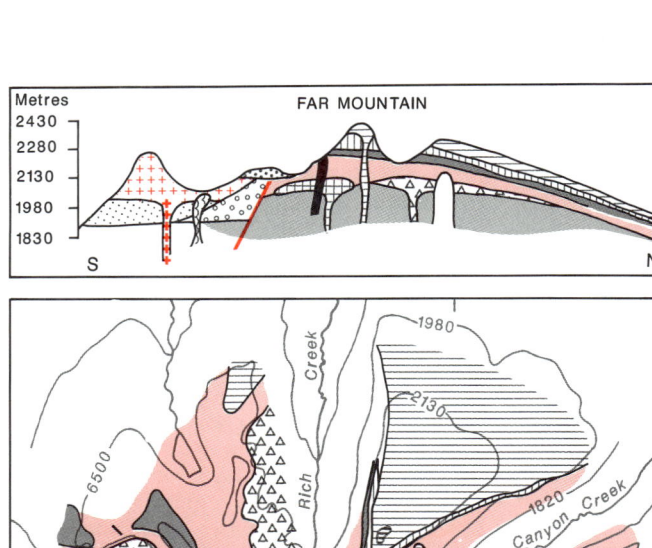

LEGEND

LATE SHIELD-FORMING ASSEMBLAGE
Blue Canyon Basalt
  Thick, columnar-jointed flows of plagioclase-olivine-phyric basalt
FAR MOUNTAIN BASALT
  Dark grey, very fine grained, aphyric basalt flows, flow breccia and related intrusions
EARLY SHIELD-FORMING ASSEMBLAGE
  Flows(a), and intrusions (b) of flaggy, fine grained, pale green, feldspar-phyric, flow-layered rhyolite and trachyte.
North Rift Basalt
  Coarsely porphyritic, plagioclase-olivine-pyroxene-phyric basalt flows, scoria, lapilli tuff and pyroclastic breccia (a), and dykes (b)
INTRACALDERA ASSEMBLAGE
  Massive, very coarse-grained trachyte flows and blocky flow-breccia (a), and related intrusions (b)
  Fine grained, thin-bedded, laminated, vitric basaltic tuff; locally containing large drop-stones of basalt
  Crudely bedded, epiclastic boulder and block deposits, debris flows, landslides and fossil talus
LATE DOME-FORMING ASSEMBLAGE
Ilgachuz Comendite
  Flaggy green-weathering, pale green, flow-layered, feldspar-phyric, rhyolite and trachyte flows and overlapping domes. Locally, thick flow-top and pyroclastic breccia
EARLY DOME-FORMING ASSEMBLAGE
Rich Creek Rhyolite
  Flows and domes of feldspar-phyric rhyolite: minor pumice and obsidian
Blue Canyon Rhyolite
  Flows and related intrusions of coarsely feldspar-phyric rhyolite, breccia and granular obsidian
PRE-CALDERA ASSEMBLAGE
  Rusty-weathering, hydrothermally altered flows, chaotic breccias, cupolas, dykes and pyroclastic deposits of greenish to purplish-brown, feldspar-phyric trachyte and rhyolite
– – – Inferred Caldera Margin

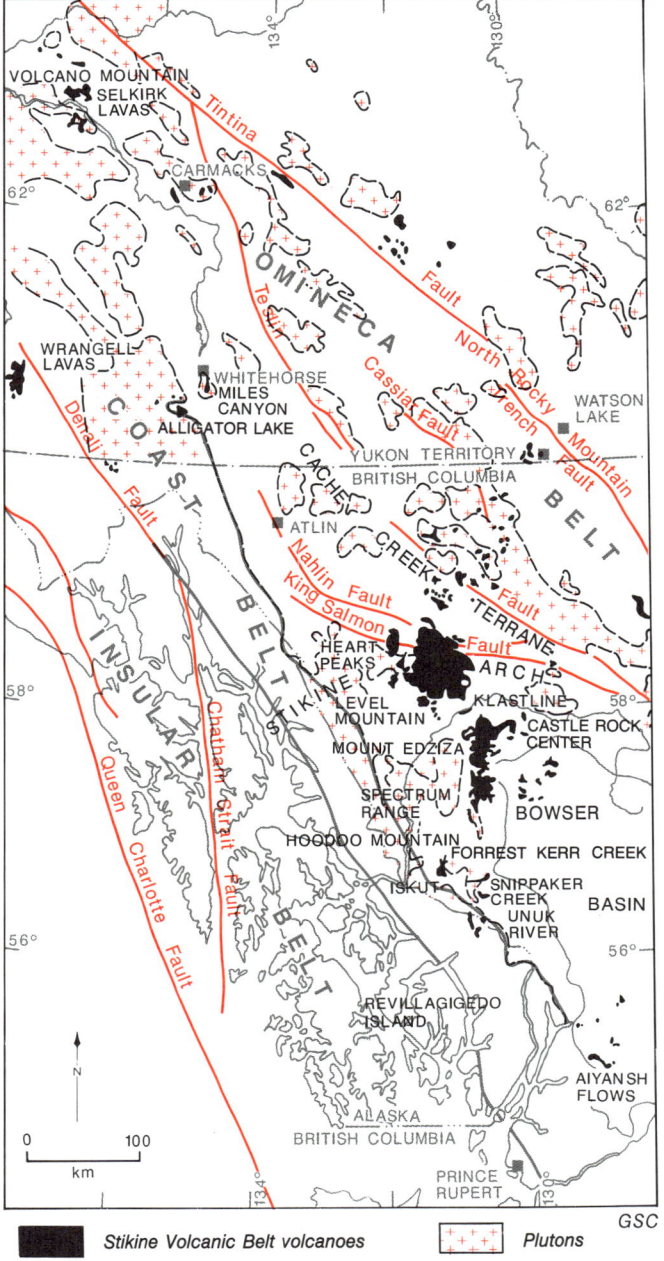

Fig. 10.14. The distribution of volcanic centres in the Stikine Volcanic Belt and their relationship to major transcurrent faults, terranes, and plutons.

Fig. 10.13. Sketch map of the central Ilgachuz Range showing the shield and central, caldera-filling facies. Cross section shows the relative positions of map units projected onto a north-south plane through Far Mountain (after Souther, 1984).

Edziza, Level Mountain, Atlin, Miles Canyon, and Carmacks segments.

The Aiyansh segment at the southern end of the belt includes the Aiyansh valley-filling flows and several smaller flow remnants farther north along Nass River. The Aiyansh flows issued from a cluster of well preserved nested pyroclastic cones and flooded the Nass River valley about 200 ^{14}C years ago (Sutherland Brown, 1969). Remnants in the northern part of this segment are underlain by unconsolidated river gravel and a single date of 1.6 ± 0.3 Ma indicates that they are also of Quaternary age.

The Edziza segment includes the Mount Edziza-Spectrum Complex, Hoodoo Mountain and scores of smaller eruptive centres that occur along a narrow zone extending from Revillagigedo Islands in southeastern Alaska to the Tanzilla Plateau. The remnants on Revillagigedo and adjacent islands (Berg et al., 1977) comprise pyroclastic cones, columnar-jointed lava flows, pumice and scoria that range in composition from alkali olivine basalt to trachyandesite. The volcanic rocks both predate and postdate glacial features and K-Ar dates (5.9 to 0.39 Ma) confirm the presence of both Late Tertiary and Quaternary volcanic events (Berg et al., 1978).

North of the Alaska Border, a cluster of basaltic pyroclastic cones and related flows occupy tributary valleys of the Iskut and Unuk rivers. The youngest of these erupted from a vent in granitic basement rocks high on the east side of Lava Fork valley. Tephra from this eruption still blankets the surrounding terrain and blocky basalt flows that spilled down the steep slope into the valley extend along the gently sloping valley floor south into Alaska almost to the Unuk River estuary at the head of Burroughs Bay where carbonized wood on the flow yielded a ^{14}C date of 360 ± 60 BP (Elliott et al., 1981). Farther north, a cluster of cones at the head of Snippaker Creek include both subglacial piles of hyaloclastite and subaerial flows that descend Snippaker Creek almost to Iskut River. Near its junction with Forrest Kerr Creek, the Iskut River valley is occupied by at least 10 flat-lying basalt flows which issued from vents within the valley and from a well preserved pyroclastic cone on its south side. Several small, monogenetic basalt cones are exposed on Arctic Lake Plateau, south of the Spectrum Range. Similar intraglacial and postglacial cones and basalt flows are scattered along the length of the Edziza-Spectrum Complex (see below) and north, across the Cache Creek Terrane, into the southwestern ranges of the Cassiar Mountains. The Klastline Plateau basalts, particularly the Castle Rock centre, contain numerous lherzolite inclusions (Littlejohn and Greenwood, 1974).

Hoodoo Mountain, a composite, central volcano, lies east of the main trend of the central Stikine Volcanic Belt. Its ice-capped, nearly circular, steep-sided edifice has undergone little dissection except for superficial gullying of the upper slopes and truncation of its lower slopes by the alpine glaciers surrounding it on three sides. It consists mostly of thick flows, domes and pyroclastic breccia of coarsely porphyritic, comenditic trachyte. A till or glacial-fluvial layer near the base of the pile is underlain by a thick trachyte flow that gave a K-Ar date of 0.09 Ma and overlain by similar trachyte dated at 0.11 Ma. A group of younger, hawaiite flows that issued from a vent on the south flank of the volcano rest on glacial-fluvial gravel filling the present valley of Iskut River.

Mount Edziza and the adjacent Spectrum Range (Fig. 10.15) form the longest-lived, and lithologically most diverse, Neogene volcanic complex in this segment of the Stikine Volcanic Belt. Together they form a volcanic terrane, the Mount Edziza Volcanic Complex (Plate 25), that covers an area of about 1000 km^2 and ranges in age from Late Miocene (7-10 Ma) to Holocene (Souther et al., 1984). The complex is strongly bimodal, consisting of overlapping basaltic shields and a group of four moderately dissected, salic, mainly peralkaline, central volcanoes (Armadillo Peak, Spectrum Range, Ice Peak, Mount Edziza). The flat-lying, relatively thin basalt flows have spread out to form an intermontane plateau on which the central volcanoes are superimposed (Fig. 10.15). Armadillo Peak (ca 6 Ma) and Spectrum Range (ca 3 Ma) are associated with small calderas, each about 3 km in diameter, whereas the younger, composite volcanoes of Ice Peak (ca 1 Ma) and Mount Edziza (ca 0.9 Ma) rise to relatively narrow summit craters. The central volcanoes are built mainly of trachyte and comendite flows, domes and minor pyroclastic deposits. These salic assemblages have spread onto the surrounding shield where they are separated by thick sequences of alkali olivine basalt and hawaiite flows. Five magmatic cycles are recognized. Each cycle began with the effusion of a relatively large volume of alkali olivine basalt and culminated with the eruption of a smaller volume of salic lava from a central vent. The basaltic sequences commonly grade up from aphyric alkali olivine basalt to highly feldsparphyric hawaiite which is overlain abruptly by the upper salic assemblage.

The cyclical repetition of basalt, trachyte, comendite sequences and the smooth compositional variation between end member groups suggests a genetic relationship between basic and salic members of the complex. Major element modelling (Souther and Hickson, 1984) indicates that the entire range of intermediate and salic rocks in the Mount Edziza Volcanic Complex can be derived by crystal fractionation of a common alkali olivine basalt parent. The stratigraphic repetition of magmatic cycles beginning with basalt and followed by a lesser volume of salic lava is consistent with the repeated injection of primitive, mantle-derived alkali olivine basalt into crustal chambers. The tectonic setting of the Mount Edziza Volcanic Complex, in a region of crustal extension, would favour the development of high-level reservoirs of sufficient size and thermal capacity to sustain prolonged fractionation.

East of the Edziza-Spectrum Complex, flat-lying erosional remnants of basalt, which cap concordant summits in the western Skeena Mountains, rest unconformably on deformed Upper Jurassic sediments of the Bowser Lake Group. The flows are associated with a group of volcanic necks (Plate 26), believed to be the central conduits through which the basalt issued to form several low, shield volcanoes. Although the lavas and necks are deeply dissected, K-Ar dates of about 5 Ma indicate that they are coeval with basalt in the lower, shield portion of the Mount Edziza Complex. The surface on which they rest is above 2000 m in elevation whereas the moderately rolling surface that underlies the Edziza flows varies in elevation from about 1700 m in the east to about 1200 m in the west where the surface is truncated by the eastern ranges of the

Coast Mountains. Depression of the prevolcanic erosion surface beneath the Edziza-Spectrum Complex relative to both the Skeena Mountains on the east and the Coast Mountains on the west suggests that eruption of the Mount Edziza Volcanic Complex was associated with crustal subsidence and extension.

The Level Mountain segment of the Stikine Volcanic Belt includes the Level Mountain composite volcano part of the adjacent Heart Peaks shield volcano and scores of smaller intra- and postglacial tuyas, pyroclastic cones and valley flows distributed within a broad north-northeast-trending zone which is offset to the north and west of the Edziza segment. The segment begins south of Tahltan River, where remnants of basaltic pyroclastic cones and flows are exposed on the upland west of Stikine River, and extends north through Level Mountain, across the Slide Mountain and Cassiar terranes, to Liard River valley, near Watson Lake where basalt flows and tephra yielded K-Ar dates of about 0.5 Ma (J.A. Westgate, pers. comm., 1984). Level Mountain is a composite volcano comprising a lower shield overlain by a broad stratocone (Hamilton, 1981). It covers an area of more than 1800 km^2 and contains at least 860 km^3 of lava and pyroclastic rocks. The lower shield rests on a regional Tertiary erosion surface of about 750 m elevation and forms a cliff-bounded constructional basaltic plateau that rises to an average elevation of 1400 m. The deeply dissected stratocone, with an area of about 1000 km^2, rises to a maximum elevation of 2200 m above the central part of the shield.

The shield consists almost wholly of transitional alkali basalt-tholeiite and silica saturated hawaiite in thin columnar flows separated by ropy flow-top breccia. In contrast the relatively complex stratocone comprises at least five bimodal packages of flows and pyroclastic ejecta that form overlapping piles erupted from several adjacent vents. Felsic components, mostly peralkaline trachyte and comendite, make up about 80% of the stratocone (Hamilton, 1981). Associated basalts are mostly alkaline and include hawaiite, mugearite, and benmoreite. Evidence of contemporaneous glaciation and volcanism is widespread. Till, lahar, glacial erratics and volcano-glacial tuff-breccia and pillow lava are interlayered with the flows, and tuyas are present both on the uppermost surface of the shield and as outliers.

K-Ar dates (14.9 to 5.8 Ma) from the Level Mountain pile span a large part of Middle and Late Miocene time and suggest that there was no hiatus between the shield-building and stratocone stages of activity. Level Mountain is the largest Neogene pile in the Stikine Volcanic Belt and probably the first centre to have erupted.

The Late Miocene, Heart Peaks basaltic shield (Souther, 1971; Casey and Scarfe, 1980) underlies an area of about 275 km^2 west of the Level Mountain composite volcano. It is separated from Level Mountain by the deeply incised valley of Dudidontu River but flows from the two

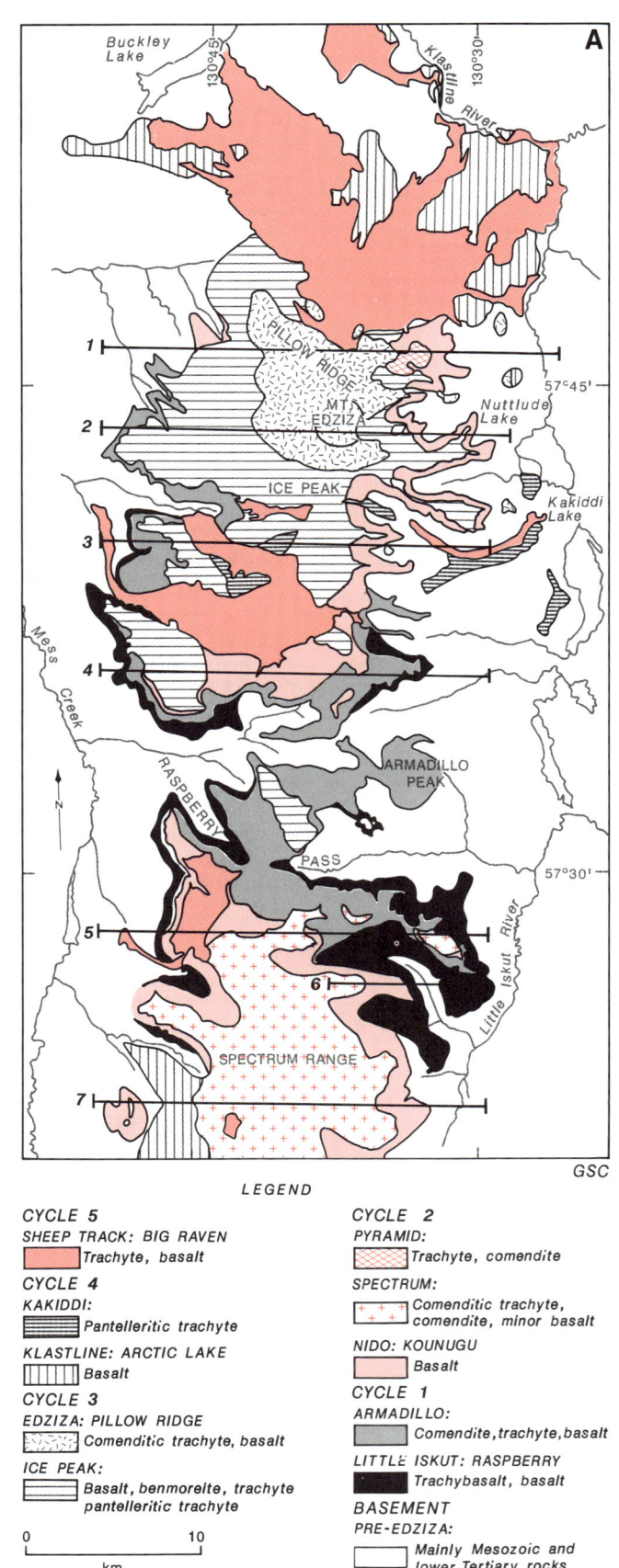

Fig. 10.15. Sketch map (A) of the Mount Edziza Volcanic Complex showing the relationship of the volcanic plateau to the central, composite volcanoes and schematic cross-sections, (B) showing the relationship between stratigraphy and K-Ar dates (Ages in Ma). The postglacial Big Raven Formation is omitted from the cross-sections.

NEOGENE ASSEMBLAGES

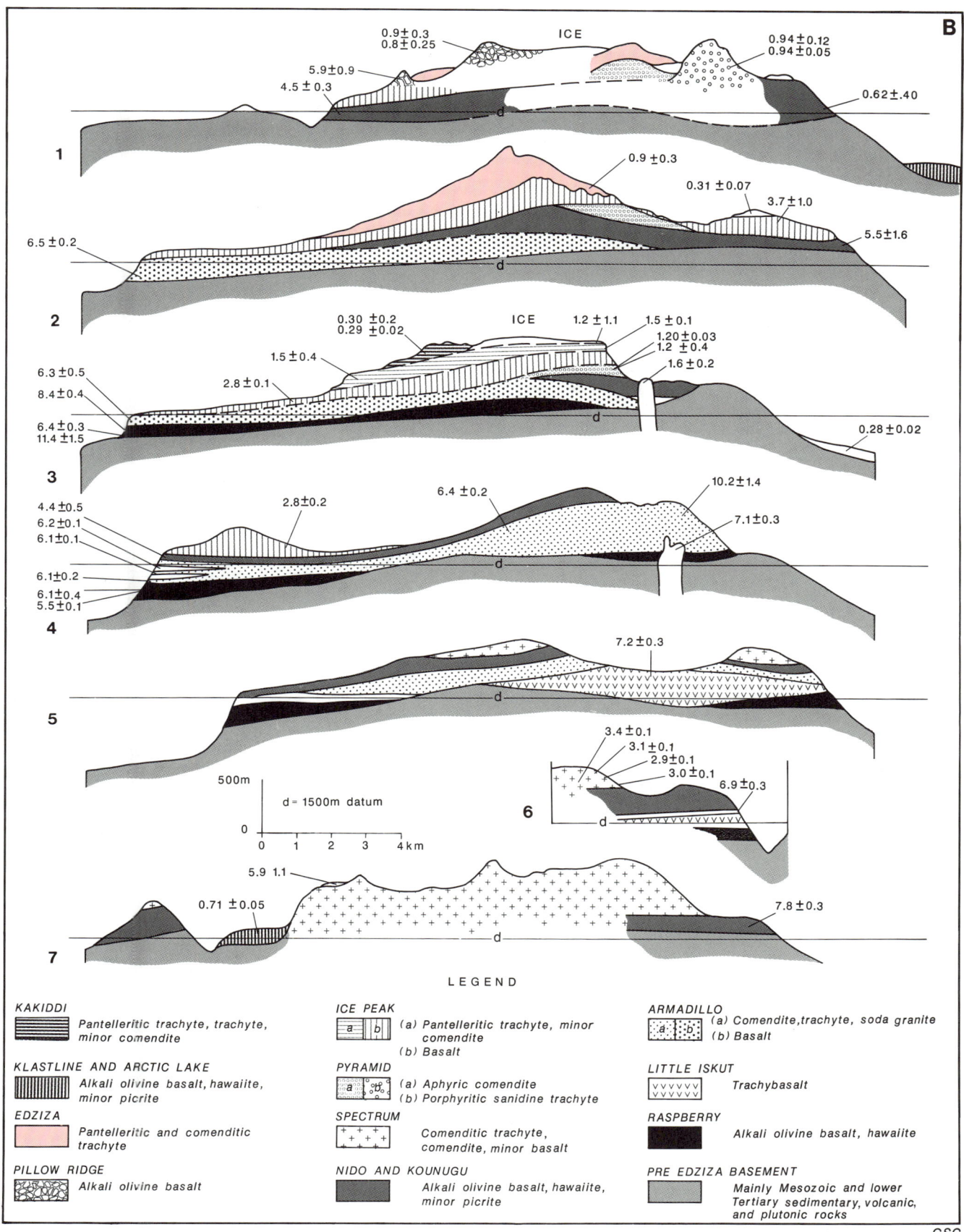

centres may once have been coextensive. An escarpment on the western side of the Heart Peaks pile exposes about 25 flows of alkali olivine basalt with a composite thickness of more than 450 m. Basalt flows at the northwestern edge of the shield appear to overlie or interdigitate with a cluster of rhyolite domes (Heart Peaks Formation) that were formerly thought to be coeval with the basalt. However, the rhyolite has yielded Eocene K-Ar dates, suggesting that it is correlative with the Sloko Formation and that it formed a pre-existing pile against which the younger basalt flows lapped.

The Atlin, Whitehorse, and Fort Selkirk segments of the Stikine Volcanic Belt are relatively short. They comprise discrete clusters of a few eruptive centres and flow remnants that lie progressively farther to the northwest but they lack the internal, north-south linearity of the larger segments. The Atlin segment includes postglacial cinder cones east of Atlin Lake (Aitken, 1959). The Whitehorse segment includes the Miles Canyon and Alligator Lake basalt flows and scattered flow remnants north and east of Ibex Mountain.

The Alligator Lake pile, with an estimated volume of 0.5 km^3, is believed to be representative of the relatively small centres in the Stikine Volcanic Belt. Basaltic lava issued from two composite cinder cones which, though deeply eroded, retain their circular form. The succession of flows has been subdivided into five chemically distinctive units (Eiche, 1986). Two of these comprise alkali olivine basalt and basanite which have high Mg contents (13.5-19.5 cation %) and host spinel lherzolite xenoliths, suggesting that they are primary magmas derived directly from the mantle. Differences in rare earth element and large-ion lithophile-element abundances indicate that the two primary magmas may have been derived by different degrees of partial melting at different depths in a garnet-bearing mantle source. The other three units are hypersthene, quartz, and nepheline normative respectively. Their chemical diversity suggests a complex history involving the differentiation of their parent magmas after leaving their mantle source. Trace element abundances in the differentiated lavas indicate that they are not derived from the same magmas as the primitive, lherzolite-bearing alkali olivine basalt and basanite units. It appears rather that each unit is the product of a small batch of compositionally distinct primary magma.

Volcano Mountain (Bostock, 1936), a composite basalt cone on the north side of Yukon River valley, is the principal centre in the Fort Selkirk segment. Flows issuing from it flooded a large area around Fort Selkirk, and remnants of similar basalt from nearby satellitic cones occupy tributary valleys of the Stewart River, north of Volcano Mountain. North of Fort Selkirk five flows with a combined thickness of more than 150 m are exposed in the banks of the Yukon River. Individual flows of crudely columnar, porphyritic alkali olivine basalt range from 6 to as much as 65 m thick. Locally the basalt is underlain by unconsolidated river gravel, glacial fluvial outwash and loess. The charred trunks of small trees enveloped by the lower flow yielded ^{14}C dates of 38 000 BP (Bostock, 1966). A nearby cinder cone, built on the flows, is mantled by 1500 year-old White River ash, suggesting that the Selkirk lavas and the cinder cone postdate the Reid (Nisling) glacial advance and may be younger than the McConnell advance. Basaltic tuff on the flanks of the cinder cone contains abundant nodules of spinel lherzolite (Sinclair et al., 1978).

Northwestern British Columbia and southwestern Yukon subduction and transform-related arc volcanics

Scattered remnants of upper Tertiary subaerial lavas and pyroclastic rocks are preserved along the entire eastern fringe of the Saint Elias Mountains. This calc-alkaline assemblage is believed to be the product of arc volcanism along a volcanic front related to the convergence of Pacific crust with the northern, North American continent. Over large areas extrusive rocks lie in flat undisturbed piles on a Tertiary surface of moderate relief, but locally, strata of the same age have been affected by a late pulse of tectonism during which they were faulted, contorted into tight symmetrical folds, or overridden by pre-Tertiary basement rocks along southwesterly dipping thrust faults. Considerable recent uplift, accompanied by rapid erosion has reduced once vast areas of upper Tertiary volcanic rocks to small isolated remnants. In places the eruptive rocks have been stripped away completely, leaving only subvolcanic plutons, necks, dyke swarms and zones of hydrothermal alteration.

Wrangell Volcanic Belt

In Canada volcanic rocks of the Wrangell Volcanic Belt are divided into three stratigraphically and structurally distinct subprovinces (Souther and Stanciu, 1975). The Canyon Mountain Province in northwestern Kluane Lake area is coextensive with a much larger area of Wrangell Lava in adjacent Alaska (Fig. 10.14). Approximately 610 m of gently tilted basalt and basaltic andesite lava and pyroclastic rocks rest conformably on Tertiary sandstone and conglomerate. Coarse pyroclastic breccia comprising loosely welded clasts of black vesicular lava form the lower third of the section. The lower unit is cut by numerous dykes and cupolas and overlain by a layer of volcanic conglomerate, pumiceous tuff, mud-flows and sedimentary breccia that separate it from an overlying sequence of thin irregular olivine basalt flows and scoria beds. The predominance of coarse pyroclastic debris, thick irregular flows, and masses of randomly oriented dykes, suggest a central, near vent environment of a large composite basaltic-andesite dome on which numerous satellitic, basalt cones developed.

The St. Clare Province extends from White River south to Slims River (Fig. 10.16). It includes a broad belt of rugged volcanic terrane that flanks the Icefield Ranges as far south as the Steele Glacier and isolated remnants farther south along the Kluane Range. Thicknesses vary from a few metres in some of the residual caps to more than 1800 m along the St. Clare syncline (Fig. 10.17). Flows of basalt and basaltic andesite, generally less than 12 m thick, are several times more abundant than related pyroclastic rocks. Intrusive bodies of white, hornblende-biotite rhyolite and dacite cut the lower Wrangell eruptive rocks. Contacts are commonly sharp but the outer part of the intrusion and the adjacent wall rock may be intensely brecciated and hydrothermally altered. Interbedded with the flows are thin layers of felsic pumice, locally redeposited by water, minor welded ash flows, pods of coaly siltstone and mudstone, and discontinuous lenses of volcanic

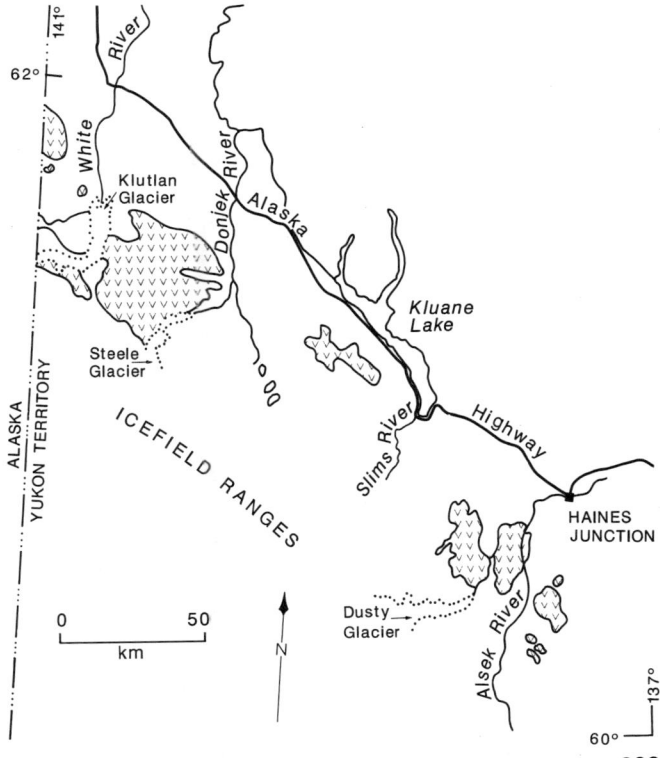

Fig. 10.16. Outline map showing the distribution of Neogene volcanic rocks in the Saint Elias Mountains.

conglomerate. The clasts are derived entirely from Wrangell lavas and are believed to have been shed from constructional volcanic edifices rather than from tectonically uplifted terrane. The absence of thick, westerly-derived, polymict clastic wedges within the Wrangell succession, plus the presence of interlayered coal deposits suggest that the Wrangell lavas were erupted onto a stable surface of low relief prior to uplift of the Saint Elias Mountains. Evidence of uplift is found in the northern St. Clare Province where the volcanic rocks are overlain unconformably by ancient deposits of polymictic fluvial and glacial gravels, tills and lacustrine deposits which contain clasts of both Wrangell and pre-Wrangell rocks.

The Alsek Province includes three principal piles of eruptive rocks and numerous small isolated flow remnants and subvolcanic intrusions. Although they are locally tilted, the Neogene rocks of the Alsek Province have not undergone the extensive post-volcanic deformation observed in the St. Clare Province. Conversely, the sequence of eruptive events is extremely complex, involving at least four distinct pulses of activity. The pre-volcanic surface is mantled by a discontinuous veneer of upper Tertiary sediments comprising sandstone, quartz-pebble conglomerate, carbonaceous shale and thin coal seams. These are overlain by the products of a rhyolitic eruption (Phase 1) that was centred east of the Alsek River where thick proximal flows grade laterally into welded ash flows and distal beds of air-fall pumice. A very large composite basaltic-andesite cone (Phase 2) overlaps the older rhyolite pile east of Dusty River. A second phase of rhyolitic volcanism (Phase 3) produced a great volume of pyroclastic rocks from vents west of Dusty River where eruption was accompanied by collapse and partial destruction of the phase 2 basaltic andesite cone. This phase of activity was accompanied by intrusion of large subvolcanic bodies and by extensive hydrothermal alteration and veining of the overlying strata. Distal phase 3 ash flows and air-fall pumice are widespread north and south of the vent area. They are commonly preserved beneath relatively thin caps of phase 4 basaltic andesite and may be correlative with distal tuff layers in the St. Clare succession. During phase 4, basaltic andesite lava issued from many small central vents on the surface of the older, phase 3, rhyolite pile.

The Wrangell Volcanic Belt of northern Canada is coextensive with active Quaternary volcanoes in adjacent Alaska. Although no centres younger than Late Miocene are known in Canada the eruption of rhyolite pumice, White River Ash, from a vent near the head of Klutlan Glacier, 24 km west of the Alaska-Yukon border, blanketed large areas of northwestern Canada with tephra (Lerbekmo and Campbell, 1969). The ash is distributed in two lobes, a northern lobe resting on peat that yielded ^{14}C dates of 1990 to 1750 BP and an easterly trending lobe resting on peat that yielded ^{14}C dates of 1460 to 1200 BP

Volcanic Rocks of East-Central British Columbia

Late Neogene and Quaternary basaltic cones, necks, and associated intravalley lava flows are distributed in the highlands east of the Interior Plateau. These centres lie outside the main volcanic belts and their tectonic affiliation is not clear. They lie near and parallel to Slide Mountain rocks along the eastern edge of Quesnellia and may be a manifestation of small magma batches generated in a zone of incipient extension related to normal faulting (Hickson, 1986).

Clearwater-Quesnel Volcanic Province

Basalt flows in the Wells Grey-Clearwater area (Hickson and Souther, 1984; Hickson, 1986) issued from a number of separate centres both during periods of regional glaciation and during ice-free interglacial periods (Fig. 10.18). A glacial assemblage including circular, flat-topped tuyas from 2 to 4 km across is composed of pillow lava and hyaloclastite as well as ponded subaerial flows. Also included are cone-shaped subglacial piles of hyaloclastite, such as Pyramid Mountain, that may not have risen above the ice, and two transitional ice-contact piles. The latter erupted onto high ground near the upper ice-surface where a broad lower edifice of pillow lava and sideromelane tuff-breccia is overlain by a superstructure of subaerial flows and pyroclastic deposits. These structures are adjacent to and surrounded by an assemblage of valley-filling flows that coalesced into the Clearwater Valley from undefined vents. The resulting intermontane lava field was deeply dissected along the Clearwater valley and its tributaries where sections up to 180 m thick are exposed. Most of the flows are subaerial, from 1 to 10 m thick, and separated by thin layers of ropy or scoriaceous flow-top breccia. Locally between the columnar flows are small lenses of pillow lava and pockets of fluvial gravel, the relicts of small streams and lakes that formed briefly on the lava surface before being displaced by yet another flow. Twig and tree molds

CHAPTER 10

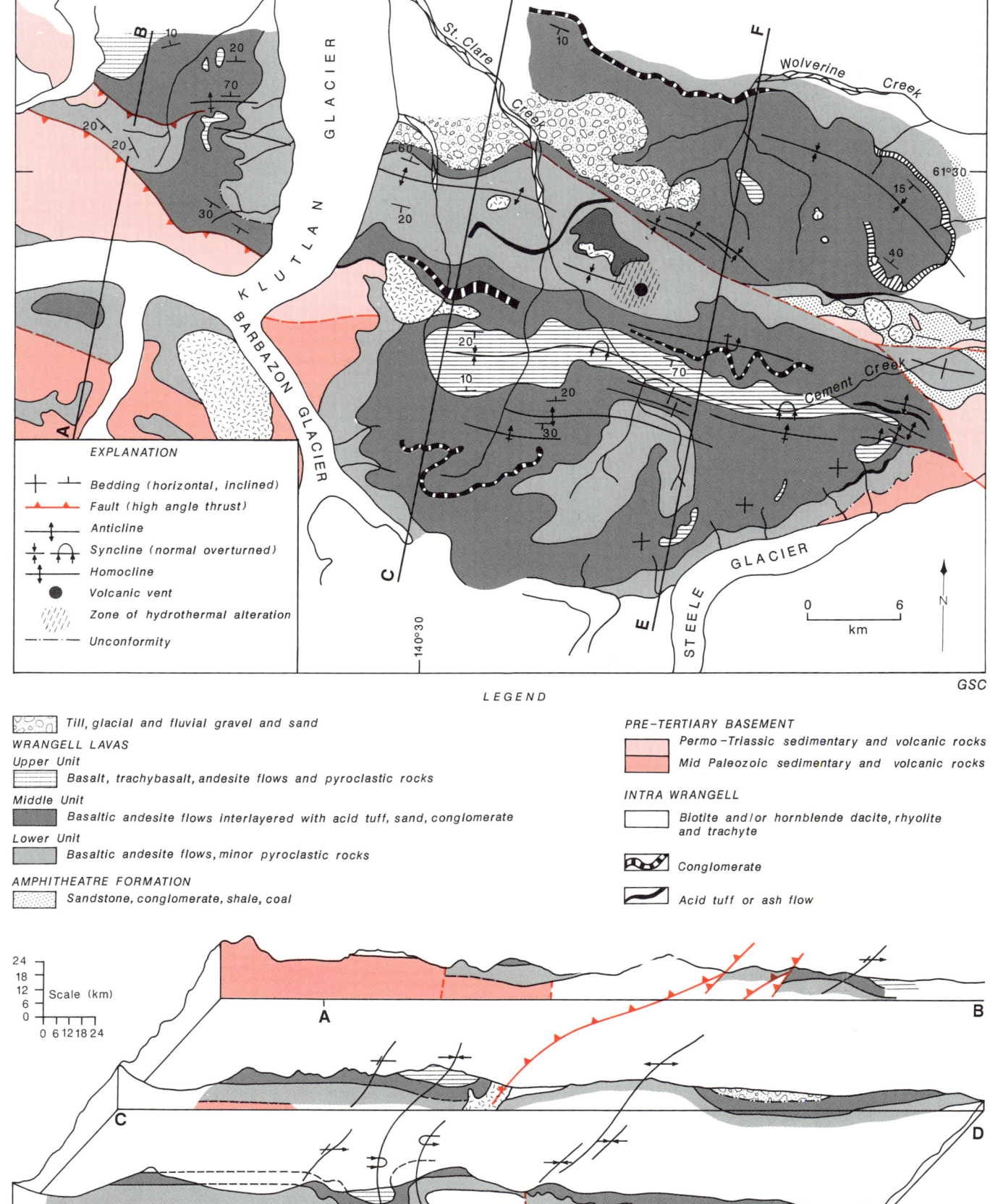

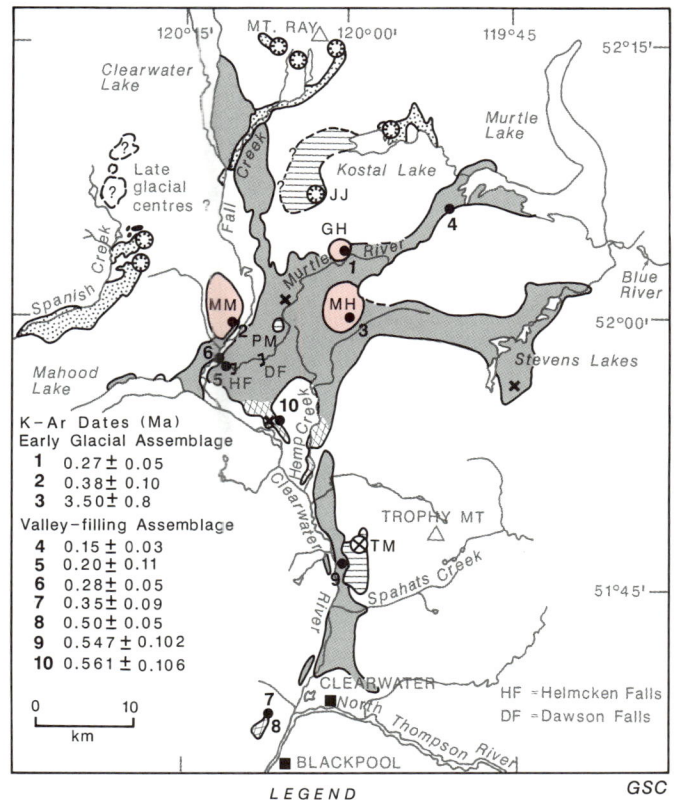

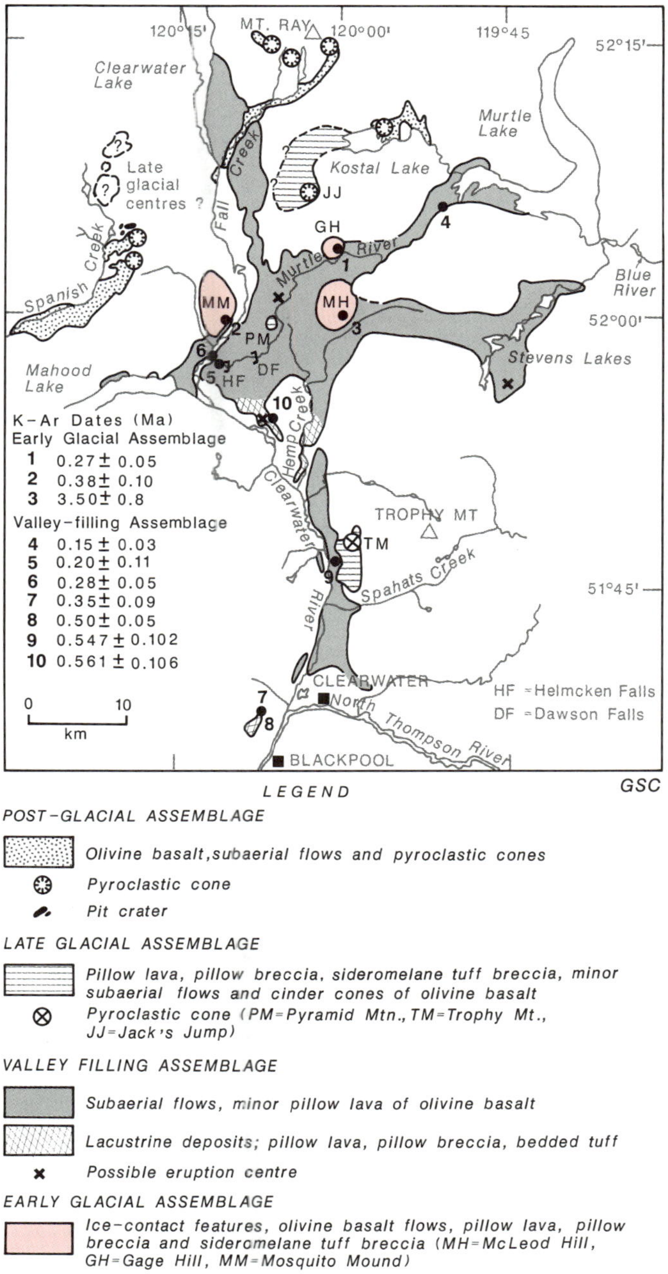

Fig. 10.18. Generalized map showing the distribution of preglacial, intraglacial and postglacial facies of the Clearwater basalt. After Hickson (1986).

Fig. 10.17. Map (A) and schematic cross-sections, (B) showing the distribution of Wrangell Lavas and variation in structural style across part of the St. Clare Province (after Souther and Stanciu, 1975).

indicate that mature forest thrived on or near the lava field during its growth and rare paleosols indicate periods of quiescence.

Small postglacial pyroclastic cones and related flows, many of them containing lherzolite inclusions, are scattered throughout the Clearwater area and as far north as Quesnel Lake. Perfectly preserved pyroclastic cones in the Spanish Creek region and Kostal Cone in Murtle River valley are among the most recently active eruptive centres in British Columbia.

McConnell Creek Volcanic Province

Basaltic necks, dykes and lavas in the eastern Omineca Mountains of McConnell Creek area are coextensive with and may be extrusive equivalents of the Kastberg intrusions (Lord, 1948). The necks, which are up to 300 m in diameter and over 100 m high, are the eroded remnants of cylindrical, lava-filled conduits that cut Upper Cretaceous Sustut Group sandstone. Associated basalt dykes, mostly less than 3 m wide but locally more than 60 m wide, cut Sustut strata and Kastberg porphyry intrusions. The lavas form horizontal or gently inclined caps on interfluves more than 300 m above adjacent valleys. They are believed to be remnants of formerly more extensive lava fields which were extruded onto a middle Tertiary erosion surface. Associated pyroclastic cones do not appear to have been glaciated, suggesting that the period of volcanism extended, with or without significant interruptions, from about middle Tertiary to Recent time.

REFERENCES

Addicott, W.D.
1978: Late Miocene molluscs from the Queen Charlotte Islands, British Columbia, Canada; Journal of Research of the United States Geological Survey, v. 6, p. 667-690.

Aitken, J.D.
1959: Atlin map-area, British Columbia (104 N); Geological Survey of Canada, Memoir 307, p. 69-70.

Armstrong, R.L., Muller, J.E., Harakal, J.E., and Muehlenbachs, K.
1985: The Neogene Alert Bay Volcanic Belt of northern Vancouver Island, Canada: descending-plate-edge volcanism in the arc-trench gap; Journal of Volcanology and Geothermal Research, v. 26, p. 75-97.

Baer, A.J.
1973: Bella Coola-Laredo Sound map-areas, British Columbia; Geological Survey of Canada, Memoir 372.

Berg, H.C., Elliott, R.L., Smith, J.G., Pittman, T.L., and Kimball, A.L.
1977: Mineral resources of the Granite Fiords Wilderness Study Area, Alaska; United States Geological Survey, Bulletin 1403.

Berg, H.C., Elliott, R.L., Smith, J.G., and Koch, R.D.
1978: Geological map of the Ketchikan and Prince Rupert quadrangles, Alaska; United States Geological Survey Open-File Report 78-73A.

Berman, R.G.
1981: Differentiation of calc-alkaline magmas: evidence from the Coquihalla volcanic complex, British Columbia; Journal of Volcanology and Geothermal Research, v. 9, p. 151-179.

Berman, R.G. and Armstrong, R.L.
1980: Geology of the Coquihalla Volcanic Complex, southwestern British Columbia; Canadian Journal of Earth Sciences, v. 17, p. 985-995.

Bevier, M.L.
1981a: Stratigraphy and petrology of the Miocene plateau lavas of British Columbia; Geological Association of Canada, Program with Abstracts, v. 6, p. A-4.

1981b: The Rainbow Range, British Columbia: a Miocene peralkaline shield volcano; Journal of Volcanology and Geothermal Research, v. 11, p. 225-251.

1983: Implications of chemical and isotopic composition for petrogenesis of Chilcotin Group basalts, British Columbia; Journal of Petrology, v. 24, p. 207-226.

Bevier, M.L., Armstrong, R.L., and Souther, J.G.
1979: Miocene peralkaline volcanism in west-central British Columbia - its temporal and plate-tectonic setting; Geology, v. 7, p. 389-392.

Bornhold, B.D. and Yorath, C.J.
1984: Surficial geology of the continental shelf, northwestern Vancouver Island; Marine Geology, v. 57, p. 98-112.

Bostock, H.S.
1936: Carmacks District, Yukon; Geological Survey of Canada, Memoir 189.
1966: Notes on glaciation in central Yukon Territory; Geological Survey of Canada, Paper 65-36.

Campbell, R.B. and Tipper, H.W.
1971: Bonaparte Lake map-area, British Columbia; Geological Survey of Canada, Memoir 363.

Casey, J.J. and Scarfe, C.M.
1980: Summary of the petrology of the Heart Peaks volcanic centre, northwestern British Columbia; in Current Research, Part A, Geological Survey of Canada, Paper 80-1A, p. 356.

Champigny, N., Henderson, C.M., and Rouse, G.E.
1981: New evidence for the age of the Skonun Formation, Queen Charlotte Islands, British Columbia; Canadian Journal of Earth Sciences, v. 18, p. 1900-1903.

Cockfield, W.E.
1948: Geology and mineral deposits of Nicola map-area, British Columbia; Geological Survey of Canada, Memoir 249.

Dawson, G.M.
1879: Preliminary report on the physical and geological features of the southern portion of the interior of British Columbia; Geological Survey of Canada, Report of Progress 1877-1878, p. B1-B173.

Denton, G.H. and Armstrong, R.L.
1969: Miocene-Pliocene glaciations in southern Alaska; American Journal of Science, v. 267, p. 1121-1142.

Dolmage, V.
1922: Coast and islands of British Columbia between Burke and Douglas Channels; Geological Survey of Canada, Summary Report for 1921, Part A, p. 22-49.

Eiche, G.
1986: Petrology of Quaternary alkaline lavas from the Alligator Lake Volcanic Complex, Yukon Territory, Canada; M.Sc. thesis, McGill University, Montreal.

Eisbacher, G.E. and Hopkins, S.L.
1977: Mid-Cenozoic paleogeomorphology and tectonic setting of the St. Elias Mountains, Yukon Territory; in Report of Activities, Part B, Geological Survey of Canada, Paper 77-1B, p. 319-335.

Elliott, R.L., Koch, R.D., and Robinson, S.W.
1981: Age of basalt flows in the Blue River valley, Bradfield Canal quadrangle; in The United States Geological Survey in Alaska: Accomplishments during 1979, Geological Survey Circular 823-B, p. B115-B116.

Farquharson, R.B.
1973: The petrology of late Tertiary dolerite plugs in the South Cariboo region, British Columbia; Canadian Journal of Earth Sciences, v. 10, p. 205-225.

Green, N.L.
1981: Geology and petrology of Quaternary volcanic rocks, Garibaldi Lake area, southwestern British Columbia; Geological Society of America Bulletin, v. 92, Part 1, p. 697-702, and Part 2, p. 1359-1470.

Green, N.L., Armstrong, R.L., Harakal, J.E., Souther, J.G., and Read, P.B.
1988: Eruptive history and K-Ar geochronology of the Garibaldi volcanic belt, southwestern British Columbia; Geological Society of America Bulletin, v. 100, p. 563-579.

Hamilton, T.S.
1981: Late Cenozoic alkaline volcanics of the Level Mountain Range, northwestern British Columbia: Geology, petrology and paleomagnetism; Ph.D. thesis, University of Alberta, Edmonton.

Hickson, C.J.
1986: Quaternary volcanics of the Wells Gray-Clearwater area, east central British Columbia; Ph.D. thesis, University of British Columbia, Vancouver.

Hickson, C.J. and Souther, J.G.
1984: Late Cenozoic volcanic rocks of the Clearwater-Wells Gray area, British Columbia; Canadian Journal of Earth Sciences, v. 21, p. 267-277.

Lawrence, R.B., Armstrong, R.L., and Berman, R.G.
1984: Garibaldi Group volcanic rocks of the Salal Creek area, southwestern British Columbia: alkaline lavas on the fringe of the predominantly calc-alkaline Garibaldi (Cascade) volcanic arc; Journal of Volcanology and Geothermal Research, v. 21, p. 255-276.

Lerbekmo, J.F. and Campbell, F.A.
1969: Distribution, composition, and source of the White River Ash, Yukon Territory; Canadian Journal of Earth Sciences, v. 6, p. 109-116.

Lewis, T.J. and Souther, J.G.
1978: Meager Mountain, B.C. - a possible geothermal energy resource; Earth Physics Branch, Geothermal Series No. 9.

Littlejohn, A.L. and Greenwood, H.J.
1974: Lherzolite nodules in basalts from British Columbia, Canada; Canadian Journal of Earth Sciences, v. 11, p. 1288-1308.

Lord, C.S.
1948: McConnell Creek Map-Area, Cassiar District, British Columbia; Geological Survey of Canada, Memoir 251, p. 43-45.

Mathews, H.W.
1951: The Table, a flat-topped volcano in southwestern British Columbia; American Journal of Science, v. 9, p. 830-841.
1952: Mount Garibaldi, a supraglacial Pleistocene volcano in southwestern British Columbia; American Journal of Science, v. 250, p. 81-103.
1958: Geology of the Mount Garibaldi map-area southwestern British Columbia, Canada; Geological Society of America Bulletin, v. 69, p. 179-198.

McKnight, B.
1965: Tertiary igneous activity in the Franklin Glacier area, British Columbia; B.A.Sc. thesis, University of British Columbia, Vancouver.

Moore, D.P. and Mathews, W.H.
1978: The Rubble Creek landslide, southwestern British Columbia; Canadian Journal of Earth Sciences, v. 15, p. 1039-1052.

Ney, C.S.
1968: Geological and geochemical report on the VAN claims, British Columbia; British Columbia Department of Mines, unpublished Assessment Report.

Read, P.B.
1977: Meager Creek volcanic complex: southwestern British Columbia; in Report of Activities, Part A, Geological Survey of Canada, Paper 77-1A, p. 277-281.
1978: Geology, Meager Creek geothermal area, British Columbia; Geological Survey of Canada, Open File 603.

Richards, T.
1971: Plutonic rocks between Hope, B.C. and the 49th parallel; Ph.D. thesis, University of British Columbia, Vancouver.

Rouse, G.E. and Mathews, W.H.
1979: Tertiary geology and palynology of Quesnel area, British Columbia; Bulletin of Canadian Petroleum Geology, v. 27, p. 418-445.

Shouldice, D.H.
1971: Geology of the western Canadian continental shelf; Bulletin of Canadian Petroleum Geology, v. 19, p. 405-436.

Sinclair, P.D., Tempelman-Kluit, D.J., and Medaris, L.G. Jr.
1978: Lherzolite nodules from a Pleistocene cinder cone in central Yukon; Canadian Journal of Earth Sciences, v. 15, p. 220-226.

Souther, J.G.
1971: Geology and mineral deposits of Tulsequah map-area, British Columbia (104K); Geological Survey of Canada, Memoir 362.
1975: Geothermal Potential of Western Canada, in Proceedings Volume, 2nd United Nations Symposium on the Development and Use of Geothermal Resources, San Francisco, May 1975, p. 259-267.
1977: Volcanism and tectonic environments in the Canadian Cordillera - a second look; in Volcanic Regimes in Canada, W.R.A. Baragar, L.C. Coleman, and J.M. Hall (ed.), Geological Association of Canada, Special Paper 16, p. 3-24.
1980: Geothermal reconnaissance in the central Garibaldi Belt, British Columbia; in Current Research, Part A, Geological Survey of Canada, Paper 80-1A, p. 1-11.
1984: The Ilgachuz Range, a peralkaline shield volcano in central British Columbia; in Current Research, Part A, Geological Survey of Canada, Paper 84-1A, p. 1-10.
1986: The western Anahim Belt, rootzone of a peralkaline magma system; Canadian Journal of Earth Sciences, v. 23, p. 895-908.

Souther, J.G. and Stanciu, C.
1975: Operation Saint Elias, Yukon Territory: Tertiary volcanic rocks; in Report of Activities, Part A, Geological Survey of Canada, Paper 75-1A, p. 63-70.

Souther, J.G., Armstrong, R.L., and Harakal, J.
1984: Chronology of the peralkaline, late Cenozoic Mount Edziza Volcanic Complex, northern British Columbia, Canada; Geological Society of America Bulletin, v. 95, p. 337-349.

Souther, J.G. and Hickson, C.J.
1984: Crystal fractionation of the basalt comendite series of the Mount Edziza Volcanic Complex, British Columbia: major and trace elements; Journal of Volcanology and Geothermal Research, v. 21, p. 79-106.

Souther, J.G., Clague, J.J., and Mathewes, R.W.
1987: Nazko Cone, a Quaternary volcano in the eastern Anahim Belt; Canadian Journal of Earth Sciences, v. 24, p. 2477-2485.

Stephens, G.C.
1972: The geology of the Salal Creek Pluton, southwestern British Columbia; Ph.D. thesis, Lehigh University.

Stevens, R.D., Delabio, R.N., Lachance, G.R.
1982: Age determinations and geological studies, K-Ar Isotopic Ages, Report 15; Geological Survey of Canada, Paper 81-2, p. 15.

Stout, M.Z. and Nicholls, J.
1983: Origin of the hawaiites from the Itcha Mountain Range, British Columbia; Canadian Mineralogist, v. 21, p. 575-581.

Sutherland Brown, A.
1968: Geology of the Queen Charlotte Islands, British Columbia; British Columbia Department of Mines and Petroleum Resources, Bulletin 54.
1969: Aiyansh lava flow, British Columbia; Canadian Journal of Earth Sciences, v. 6, p. 1460-1468.

Tipper, H.W.
1957: Anahim Lake, British Columbia; Geological Survey of Canada, Map 10-1957.
1959: Quesnel, British Columbia; Geological Survey of Canada, Map 12-1959.
1961: Prince George, British Columbia; Geological Survey of Canada, Map 49-1960.
1963: Taseko Lakes, British Columbia; Geological Survey of Canada, Map 29-1963.
1978: Taseko Lakes map-area, British Columbia; Geological Survey of Canada, Open File 534.

Trettin, H.P.
1961: Geology of the Fraser River Valley between Lillooet and Big Bar Creek; British Columbia Department of Mines and Petroleum Resources, Bulletin 44.

Wanless, R.K., Stevens, R.D., Lachance, G.R., and Delabio, R.N.
1978: Age determinations and geological studies, K-Ar Isotopic Ages, Report 13; Geological Survey of Canada, Paper 77-2.

Woodsworth, G.J.
1979: Geology of Whitesail Lake map-area, British Columbia; in Current Research, Part A, Geological Survey of Canada, Paper 79-1A, p. 25-29.

Yorath, C.J. and Chase R.L.
1981: Tectonic history of the Queen Charlotte Islands and adjacent areas - a model; Canadian Journal of Earth Sciences, v. 18, p. 1717-1739.

Yorath, C.J. and Hyndman, R.D.
1983: Subsidence and thermal history of Queen Charlotte Basin; Canadian Journal of Earth Sciences, v. 20, p. 135-159.

Yorath, C.J., Tiffin, D.L., and Cameron, B.E.B.
1977: Submersible operation on the Pacific continental margin; in Report of Activities, Part A, Geological Survey of Canada, Paper 77-1A, p. 301-310.

Authors' addresses

J.C. Souther
Cordilleran Division
Geological Survey of Canada
100 West Pender Street
Vancouver, British Columbia
V6B 1R8

C.J. Yorath
Pacific Geoscience Centre
9860 West Saanich Road
P.O. Box 6000
Sidney, British Columbia
V8L 4B2

ADDENDUM

For recent studies of Neogene Assemblages in the Queen Charlotte Islands the reader is referred to the following publication:

1991: Evolution and hydrocarbon potential of the Queen Charlotte Basin, British Columbia; G.J. Woodsworth, (ed.); Geological Survey of Canada, Paper 90-10, 569 p.

Printed in Canada

Chapter 11

PHYSIOGRAPHIC EVOLUTION OF THE CANADIAN CORDILLERA

Summary
Introduction
Paleogeomorphology
    Mid-Mesozoic
    Early Cretaceous
    Late Cretaceous to Paleocene
    Early to Middle Eocene
    Oligocene
    Miocene
    Pliocene to Holocene
Other geomorphic developments
References

Chapter 11

PHYSIOGRAPHIC EVOLUTION OF THE CANADIAN CORDILLERA

W.H. Mathews

SUMMARY

The geomorphic development of the Canadian Cordillera is here considered as starting from Middle or Late Jurassic time when ancestral North America collided with the Intermontane Superterrane along its western margin. This, and a similar event later in the Mesozoic, produced two metamorphic and plutonic complexes, each of which are loci of high-grade metamorphism, rapid uplift, and vigorous erosion. These complexes, the Omineca Belt in the east and the Coast Belt to the west, constitute two of the five morphogeological belts in the Cordillera.

Development of the Foreland Belt was an additional outcome of plate collision. With the growth of this belt in mid-Mesozoic to earliest Cenozoic time, a trellis drainage pattern developed, simulating that in the present Cordillera. Individual river courses of that time, however, bear little or no relationship to present drainage. The early erosion-products, together with material from the Omineca Belt to the west, contributed to an almost continuous apron of alluvial fans and associated deltaic deposits in the foredeep, east of the rising orogen.

Widespread Early Cretaceous pediplanation, recorded by pre-Hauterivian to Albian erosion surfaces, coincided with a lull in magmatic activity throughout the Cordillera. Contemporaneous uplift and volcanism in the western Cordillera provided sources for rapid degradation and alluvial sedimentation in and adjacent to contemporaneous narrow seaways.

Although transcurrent faulting and arc (Andean?) magmatism prevailed in part of the western Cordillera erosional and depositional activity west of the Foreland Belt subsided in Late Cretaceous and Paleocene time only to become vigorously rejuvenated in Early to Middle Eocene time. The earlier, dominantly compressional deformation gave way to rotation and extension. Mafic to felsic volcanism, both calc-alkaline and alkaline, was widespread in the interior of the Cordillera. High heat flow led to extensive resetting of isotope systems in metamorphic and plutonic rocks and was accompanied by extension and followed by rapid uplift and denudation. Local grabens, half-grabens and basins became the sites of accumulation of sedimentary and pyroclastic rocks which, by virtue of their low resistance to erosion, are preserved only in localized present-day topographic lows associated with older structural depressions. Thus, Eocene tectonism indirectly contributed to modern topography; indeed, many Eocene faults localized modern valleys. However, other valleys clearly crosscut Eocene deposits and structures.

In the interior of the Cordillera a relatively inactive interval followed the mid-Eocene event and led to the development of subdued topography characterized by broad valley floors. Within these broad valleys brief rejuvenation of streams in the Middle Miocene led to the incision of narrow channels which were soon refilled with fluvial and tuffaceous sediments. Extrusion of widespread fluid basaltic lavas in mid-Miocene to Pliocene time mantled both the broad valley floors and the infilled channels with a thin resistant capping. In the eastern Cordillera pediplanes developed; throughout the Cordillera relicts of this late Cenozoic landscape, only slightly modified, can still be recognized.

Post-Miocene uplift not only re-established the main mountain belts but also caused stream incision to depths of hundreds, and locally, thousands of metres below the late Cenozoic surface. Several periods of glaciation, each including a major ice sheet and several satellite ice caps covering most of the Cordillera, aided in the enlargement of the stream-cut valleys. Glacial ice undercut the few summits standing above the ice and rounded those which it overrode. Plateaus and lowlands became mantled with drift. Drift dams and the ice itself obstructed many of the valleys and disrupted pre-existing drainage; river systems, such as seen today, may have come into being only after this glacial derangement, even though the valleys they now occupy may be traced back to preglacial time.

INTRODUCTION

Paleogeomorphology of the Canadian Cordillera deals not only with the shape of ancient land surfaces, their position relative to the present surface and their temporal changes in shape and position, but also with the processes that brought about these changes. In classical studies 'fossil' land surfaces, or unconformities, and the character and attitude of the bounding strata are used to interpret paleogeography and environment. Sedimentary and volcanic deposits yield information on land form and rates of supply from source areas. Geobarometry, geothermometry and geochronology of rock suites and mineral deposits have been used to infer depths of cover at various times in the past (e.g. Høy, 1976, 1980; Ghent et al., 1983; Pigage

---

Mathews, W. H.
1991: Physiographic evolution of the Canadian Cordillera, Chapter 11 in Geology of the Cordilleran Orogen in Canada, H. Gabrielse and C. J. Yorath (ed.); Geological Survey of Canada, Geology of Canada, no. 4, p. 403-418 (also Geological Society of America, The Geology of North America, v. G-2).

and Greenwood, 1982; Hollister, 1979). Locally fossil thermal gradients and heat flow have been inferred from mineral assemblages or fission-track dates (Parrish, 1981a,b); cooling histories have been reconstructed based in part on calculated erosion rates (Harrison et al., 1979).

The paleogeomorphology of the Canadian Cordillera is herein considered from the time of accretion of various terranes to ancestral North America in the mid- and late Mesozoic. Before then the distribution of land and sea, of high and low ground, with respect to present geography can not be reliably determined. However, as a consequence of plate collisions which created the Omineca and Coast belts and the adjacent mountain systems, the Cordillera took on a more familiar aspect. The original assembly may have taken place considerably south of present locations; transcurrent faulting in Cretaceous and Paleogene time brought the terranes into their current positions (Gabrielse, 1985). Erosion has since obliterated the Mesozoic mountain belts, but their loci can be identified even though continued deformation has distorted the frame of geographic reference.

PALEOGEOMORPHOLOGY
Mid-Mesozoic

The earliest known record of continental accretion is found in the high-grade metamorphic rocks of the Omineca Belt. The rocks were metamorphosed at pressures of 5 to 7 kb (about 20 km depth) and at temperatures in excess of 600°C (Høy, 1976; Ghent et al., 1983). A Middle Jurassic time of metamorphism is suggested from integrated structural, metamorphic and geochronological studies. Subsequently the Omineca Belt was greatly uplifted, particularly in mid-Cretaceous and Eocene times and was deeply eroded. It became a principal source for Cretaceous and lower Cenozoic sediment which accumulated in the foredeep of the Foreland Belt and in basins to the west (Eisbacher, 1974; Bustin and Moffat, 1983).

Early Cretaceous

Paleogeographic models identifying the distribution of land and sea of mountainous source-areas and sites of sediment deposition in coastal plains or seaways during Late Jurassic to Early Cretaceous time were proposed by Eisbacher et al. (1974), McLean (1977), Hamblin and Walker (1979), Gibson and Hughes (1981), Stott (1984), Jeletzky and Tipper (1968) and Coates (1974). These models portray an elongate mountain system, centred about the present Omineca Belt, which shed detritus easterly across an alluvial apron and coastal plain to a epicontinental boreal sea (e.g., the Kootenay and Blairmore clastic wedges of the Foreland Belt), and westerly into one or more local marine basins or troughs including the Tyaughton-Methow Trough and Skeena Basin. The present width of the source area, as inferred from the distance between known bounding occurrences of Upper Jurassic or Lower Cretaceous sediments, is 370 km near latitude 49°N (Fernie to the Cascade Mountains) and 240 km near latitude 58°N (Rocky Mountain Foothills to Bowser Basin). However, these figures are only crude indications of the original width because of significant crustal contraction in Cretaceous-Paleocene time, crustal extension in Eocene time and displacement along longitudinal strike-slip faults. A north-south orientation and relatively high paleolatitude (55° to 70°N) for the miogeocline is indicated by paleomagnetic data from the craton (Couillard and Irving, 1975), although the fossil record gives no hint of an arctic or cool temperate climate. The central and western Cordillera, however, formed at significantly lower latitudes (Monger and Irving, 1980).

In its structure and morphology the Cretaceous Cordillera may have resembled the present mountain belt of New Guinea, or more properly, that belt before Quaternary volcanism (Fig. 11.1). The New Guinea belt, like the Cordillera, is bounded on one side, the south, by a craton locally covered by an epicontinental sea, into which a broad coastal plain has been constructed. On the north side is a deep ocean. The watershed is centrally disposed between ocean on the north and coastal plain on the south. The southern slope is underlain by dominantly sedimentary rocks in a fold and thrust belt directed towards the craton (D'Addario et al., 1975). The northern slope, on the other hand, exposes low-grade metamorphic, intrusive and extrusive rocks which provide a very different suite of clastic sediments to a bordering longitudinal trough and adjacent ocean basin.

Though these structural and lithological comparisons may apply to the Cretaceous Cordillera, the climates and small-scale topography of the two regions must have contrasted markedly. Tropical weathering and reef development, conspicuous in New Guinea, have no counterparts in the Cretaceous record of the Cordillera except perhaps in the Insular Belt where Upper Cretaceous marine sandstone overlies deeply weathered Lower Jurassic granodiorite (grus) which yield high δ^{18} values (Sutherland Brown and Yorath, 1985). Lying athwart the high-latitude zone of westerly winds the Lower Cretaceous Cordillera seems to have produced a significant rain shadow on its eastern side. Redbeds occur in Aptian coastal-plain sediments in southwestern Alberta and Montana, and pedimentation, generally associated with arid and semi-arid environments, has been suggested as the mechanism that developed the thin but extensive Cadomin conglomerate (McLean, 1977). Farther north in the Richardson Mountains and Mackenzie Delta the widespread occurrence of coal suggests a more humid Early Cretaceous environment (Young et al., 1976).

The denudation rate of at least the eastern Cordillera may have changed significantly with time judging from variations in rates of sedimentation in the foredeep. Stott (1984) recorded peaks of sedimentation in the Valanginian, Albian, and Maastrichtian. These changes probably, in part, reflect pulses of diastrophism in the Cordilleran belt, but also may be influenced by shifts in the loci of sedimentation.

Pedimentation surfaces overlain by strata ranging in age from Hauterivian to Albian have been recognized in several regions of the Cordillera (Eisbacher, 1981). The development of these surfaces coincided with a distinct lull in tectonic and magmatic activity (see Chapters 9 and 15).

During much of Cretaceous time the eastern front of the Cordillera was marked by an almost continuous series of alluvial fans, several of which are distinguished by bodies of either very thick or very coarse conglomerate (McLean, 1977; Stott, 1968). Some, like the Belcourt fan (Stott, 1968), characterized by abundant coarse

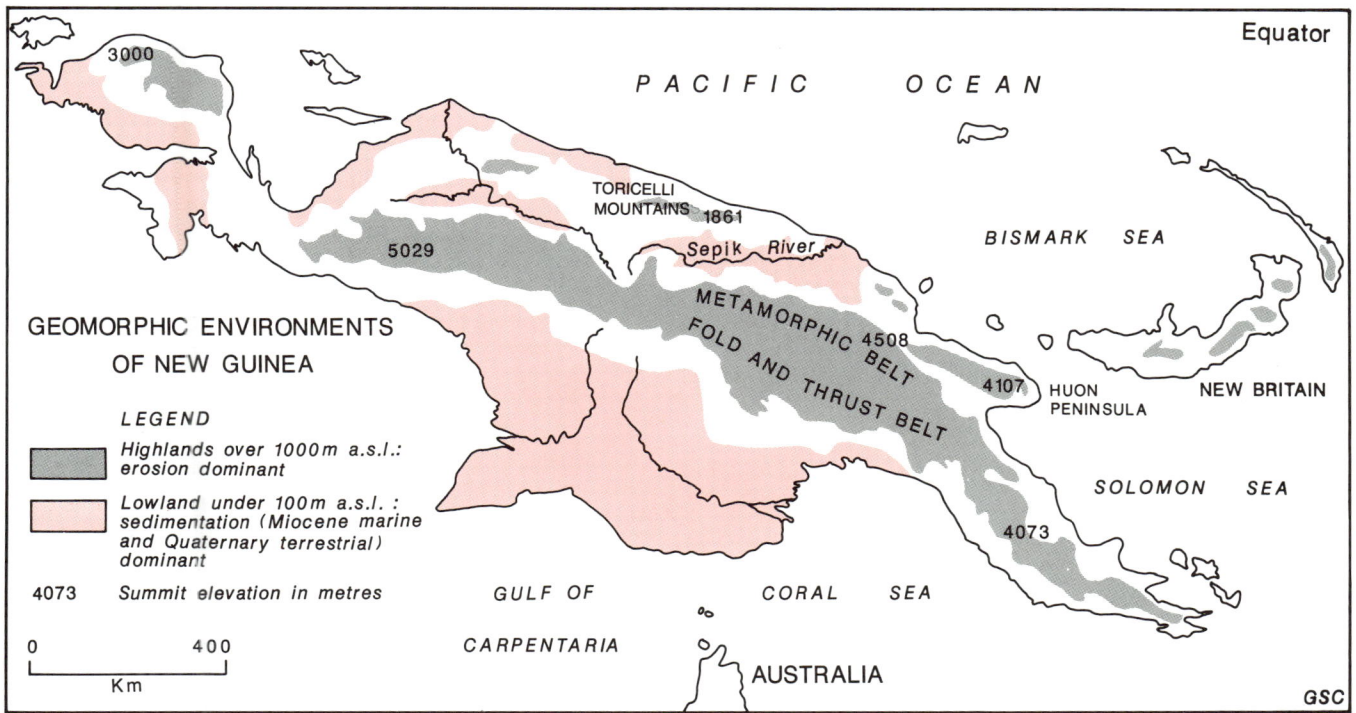

Figure 11.1. Geomorphic environments of New Guinea, possibly a modern geomorphic analogue of the Canadian Cordillera during the Cretaceous Period.

conglomerate, seem to be the product of short, high-gradient streams feeding easterly from the rising Cordillera and containing clasts of quartzite, chert, dolomite and silicified limestone derived from Paleozoic strata. Deposits from low-gradient streams flowing parallel with the trend of the newly developing structures in the hinterland before breaking through to the eastern front should be revealed by (1) finer sediment in their fans or deltas, (2) their greater diversity in clast lithology and mineralogy, including representation from metamorphic and igneous rocks, and (3) their much greater volumes. The last criterion, however, may be difficult to recognize owing to the potential complexity of geomorphic processes and because of tectonic dislocations which influenced sites and rates of accumulation. Re-entrants may be favoured as sites of emergence of major streams from the hinterland (Eisbacher et al., 1974) but many exceptions occur in other young mountain ranges.

Several sites have been found where Cretaceous rivers may have emerged from the hinterland. Granitic and volcanic clasts from the McDougall-Segur conglomerate of the Blairmore Group, collected from four sites between latitudes 49° and 50°N, have yielded K-Ar ages ranging from 148 Ma to 178 Ma (Norris et al., 1965). The clasts presumably were derived from the Omineca Belt. At Mt. Allan, near Banff, Rapson (1965) reported volcanic detritus also probably derived from the southern Omineca Belt. Another is near the Peace River where sediment facies indicate an east-flowing fluvial system in earliest Cretaceous time (Stott, 1984) and where sillimanite, staurolite and chlorite, eroded from metamorphic terranes, are reported in Lower Albian beds (Stott, 1968). Fans derived from the rising mountains were built easterly onto a northwesterly sloping surface, at times covered by a seaway, and at other times crossed by northwesterly to northerly-flowing streams (Taylor and Walker, 1984).

The western margin of the Lower Cretaceous Cordillera is less well known by virtue of more complex structure and stratigraphy. Great thicknesses of nearshore to terrestrial sands and gravels of the Jackass Mountain and Pasayten groups were deposited in the Tyaughton-Methow Trough (Jeletzky and Tipper, 1968; Coates, 1974; see also Chapter 9). Thicknesses of coarse detritus, including boulder conglomerate, reach 4.5 km and imply not only an actively rising source area but also a basin deep enough to retain such thicknesses of sediment. Again the New Guinea analogy is instructive. The Huon peninsula, on the northeast coast of the island (Fig. 11.1) is rising at rates approaching 3 mm/a (Bloom et al., 1974); streams descending the steep slope from an elevation of about 2500 m in a horizontal distance of about 40 km have deposited a gravel mantle up to 100 m thick (Chappell, 1974). The nearshore relief of the Cordilleran terrain also must have been great to support the continuing transport of coarse gravels. On the other hand the possibility of lengthy, longitudinal, low-gradient streams such as the Sepik River of northern New Guinea, draining the central and western Cordillera is not precluded. Clast lithology in the upper Lower Cretaceous sediments is diverse, and includes a significant proportion of granitic (20-90%) and volcanic (25-50%) detritus and lesser amounts of chert, schist and gneiss (average <10% each).

The history of the Tyaughton-Methow Trough was repeated, with variations, farther to the north in the Bowser Basin and, slightly later, in the Sustut Basin. The Bowser

Basin records infilling by southward and westward prograding, high-gradient delta systems during Middle Jurassic to Early Cretaceous time in response to collision of Stikinia and the Cache Creek Terrane. The Skeena Basin and lower part of the Sustut Basin contain the first evidence in the north-central Intermontane Belt of uplift in the Omineca Belt which shed metamorphic and plutonic debris westward.

West of the Tyaughton-Methow Trough in Early Cretaceous time was another, probably low landmass whose western limit is considered to lie across the northwestern end of Vancouver Island and along or immediately east of Queen Charlotte Islands (Jeletzky and Tipper, 1968 - Fig. 8 and 9). The Toricelli Mountains of New Guinea may be analogous.

In the Insular Belt mid-Cretaceous marine and nonmarine sandstone and conglomerate were deposited upon comparatively steep, westward sloping surfaces. These ultimately connected with narrow, marine troughs along the edge of the continental margin adjacent to northern Vancouver Island and the Queen Charlotte Islands (Yorath and Chase, 1981).

Vigorous tectonic and erosional activity subsided in mid- to Late Albian time and was succeeded in the mid-Cretaceous by intense volcanism, which produced mafic to felsic lavas, breccias and tuffs and associated sedimentary rocks of the Spences Bridge Group in southwestern British Columbia, the Kasalka volcanics in central British Columbia and the South Fork and Mount Nansen volcanics in Yukon Territory. A thickness of 3 km of these volcanic accumulations is preserved in a synclinal trough immediately east of the Tyaughton-Methow Trough.

The record of the Lower Cretaceous Cordillera north of latitude 58°N is fragmentary. Terrestrial sediments of the Tantalus Formation are sparingly exposed in southern Yukon in small basins possibly related to transcurrent faulting. Marine clastics of Cordilleran derivation occur north of latitude 64° in the Mackenzie Lowland, Peel Plain and Peel Plateau (Yorath and Cook, 1981). Terrestrial beds at one locality in the Mackenzie Mountains are intimately involved in late Mesozoic to Cenozoic deformation. Lower and Upper Cretaceous marine strata underlie much of the area of the Richardson Mountains where they thin over the axis of the anticlinorium suggesting the presence of syndepositional relief (Jeletzky, 1975).

Late Cretaceous to Paleocene

By the end of the Cretaceous Period the Cordillera began to acquire its present character. The coastal plain on the east became emergent but still received copious alluvium. To the west the rising Rocky Mountains of the Foreland Belt probably displayed trellis drainage like the modern corresponding area in New Guinea. This drainage may have been complicated by numerous antecedent and deeply incised rivers such as are present in New Guinea (Jenkins, 1974). Farther west, in the Omineca Belt, metamorphic rocks, sources of chlorite, staurolite and sillimanite, were already exposed. As in New Guinea the drainage may have been in large part fault-controlled. Farther west the remains of the mid-Cretaceous volcanic highlands possibly supported a radial drainage pattern. Very little is known of the paleogeomorphology of the Coast and Insular belts except that a trough along the present Strait of Georgia received fluviatile and marine sediments from local source areas to the west, east and southwest and that local deep saprolitic weathering had affected granitic rocks overlapped by these sediments.

The Sustut and Sifton basins in northern British Columbia (Eisbacher, 1974, 1981), which developed between Albian and Paleogene time, received copious, coarse alluvial sediments, mostly from northern and eastern sources within the Omineca Belt, but locally, in Maastrichtian time and possibly earlier, from a landmass to the west which also supplied pyroclastic debris.

Marine conditions, which persisted throughout the Late Cretaceous in the Mackenzie River area, gave way to deltaic and fluviatile sedimentation as a consequence of uplift in the Mackenzie and Richardson mountains (Yorath and Cook, 1981).

Much of the interior of the Cordillera, however, has little or no record of Late Cretaceous to Paleocene sedimentation, probably reflecting a time of extensive erosion. The earlier concept, that this erosion led to a 'Cretaceous peneplain' preserved in the present concordance of summit levels or on the surface of the Interior Plateau is unjustified. On the contrary, the record of the Early Eocene is one of an irregular land surface contributing coarse debris to local basins. The sub-Eocene unconformity, moreover, has local structural relief 2000-3000 m (e.g. at Kamloops Lake) and bears no relationship to the present summit level or plateau surface.

Early to Middle Eocene

In both the Omineca and Coast belts resetting of the Rb-Sr, K-Ar and fission-track systems at about 50 to 60 Ma was widespread. This recorded a thermal pulse followed by rapid uplift, erosion and probable tectonic denudation along extension faults, which in turn led to chilling of the radioactive minerals below their blocking temperatures. Harrison et al. (1979) suggested a cooling history for the Quottoon pluton in the Coast Belt east of Prince Rupert that began about 50 Ma when rocks, then at a depth of as much as 15 km (Hollister, 1979, p. 6; Woodsworth et al., 1983), cooled below 500°C. Similar data for the nearby Ecstall Pluton suggest that cooling began about 80 Ma, but was interrupted by a heat pulse about 60 Ma. The cooling curves of both plutons indicate the rocks cooled below 200°C by Late Eocene time. Hollister (1979) suggested that about 10 km of erosion occurred between about 52 and 47 Ma at a denudation rate of about 2 to 3 m/ka, and perhaps by as much as 8 m/ka immediately prior to 52 Ma. Such rates imply tectonic in addition to normal geomorphic denudation processes.

In considering the disposition of the former cover Hollister (1979) proposed that most was carried westerly as clastic sediment and deposited on the floor of the Pacific Ocean near a spreading ridge. From there it could have been subsequently carried northwesterly and southeasterly some 2000 km to form the Early Tertiary marine sediments of the present Alaskan and continental shelves. An indeterminate volume may also have been subducted under the western edge of the America Plate and into the Aleutian Trench.

In the southern Omineca Belt metamorphic rocks which had been at or above 500°C and buried to depths of 5 to 13 km, depending on assumed temperature gradients of between 100°C/km and 37°C/km, were unroofed in the brief span of 6 to 9 Ma during Early Eocene time (Mathews, 1981). The volume of cover rocks removed is estimated to have approached 100 000 km³. Where did this material go? One possible answer is suggested by Middle Eocene extensional structures in southern British Columbia (Parrish et al., 1985; Tempelman-Kluit and Parkinson, 1986), from where the cover rocks may have moved outward on detachment faults and remain, more or less intact, adjacent to exposures of the high-grade metamorphic complexes which they formerly buried. The problem of disaggregating, eroding, and disposing of this huge volume of cover rocks in some distant basin all in an extremely short interval of time, is eliminated by this suggested mechanism.

Conventional explanations involve the cover rocks being moved as clastic sediment to more remote sites. Johnson (1984), for example, suggested that the Eocene Chuckanut Formation of northwestern Washington, consisting of up to 6 km of alluvial strata, much of it of appropriate mineral content, may represent some of the redistributed cover. Deep seismic reflection profiles across Vancouver Island suggest that the southern Insular Belt was uplifted as a consequence of post-Eocene and pre-Late Miocene underplating of oceanic materials (Yorath et al., 1985). Thus, Tofino Basin, which contains Eocene sediments, and the adjacent Eocene sea floor (Kula Plate) could have been the depositional site for much of this material. Another possible depositional area could have been the eastern Foreland Belt. Estimates of the amount of post-Paleocene erosion of strata in western Alberta range between 1 and 9 km (Nurkowski, 1984; Beaumont, 1981; England and Bustin, 1986), part of which could have been sediments derived from the Eocene pulse in the Omineca Belt; no post-Paleocene clastic-wedge sediments are known in the Foreland Belt (see Chapter 9).

A study in the Kootenay Lake area, also in the southern Omineca Belt (Mathews, 1983), reveals a similar history: a record of rapid cooling, probably reflecting erosion in Paleocene to Early Eocene time when the ground surface was perhaps as much as 6 km above the present land level.

In the Cordillera the Early to Middle Eocene was a time of pronounced block faulting accompanied by local terrigenous sedimentation and extensive volcanism (Ewing, 1981). The southern Cordillera was cut by a series of right-lateral northwesterly trending transcurrent faults between which there was northwest-southeast extension (Fig. 11.2) (Price et al., 1985; Ewing, 1981). A mosaic of grabens and half-grabens developed, commonly with northerly to northeasterly long axes, which locally terminated against easterly trending reverse faults. Some grabens formed early, receiving coarse sediments of local derivation. Elsewhere, the stratigraphy suggests a widespread and nearly continuous cover of volcanic rocks which were later involved in the block faulting (Ewing, 1981; Monger, 1968). In a few places local basins along the Fraser River and in the Princeton and Penticton areas which received fine grained sediment and coal, developed late in the history of deformation (Church, 1973; McMechan, 1983; Mathews and Rouse, 1984). The thickness of Eocene sedimentary and volcanic rocks exceeds 1.5 km in several grabens and may reach 3 km in as many as three. The topographic relief during this period of tectonism may well have exceeded

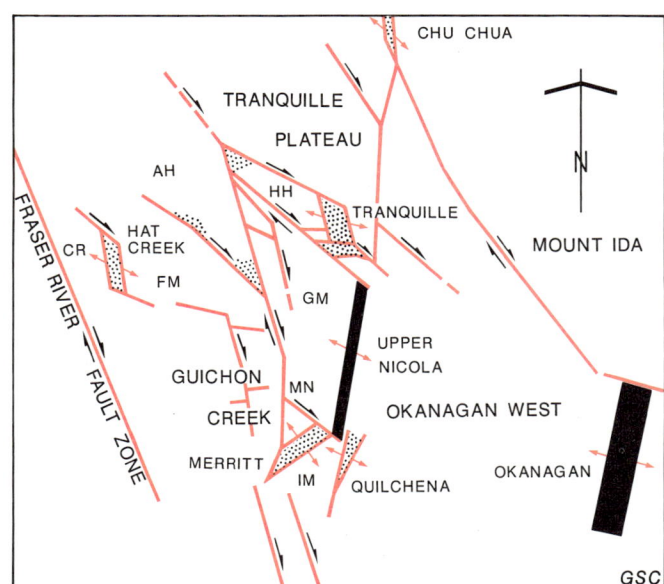

Figure 11.2. Eocene faults, grabens (stippled) and uplifts (black) in south-central British Columbia (from Ewing, 1981) (HH = Hardie Hill; GM = Greenstone Mountain; AH = Arrowstone Hills; FM = Forge Mountain; CR = Clear Range; MN = Merritt North; IM = Iron Mountain).

3 km. Notwithstanding this active tectonism and volcanism, elevations in the central and western Cordillera may not have been as great as today even though only a small proportion of summits now exceed 3 km above sea level. There is no sign in the Eocene pollen record of a rain shadow such as exists at present (Rouse, 1967).

Regardless of the total relief at any one time in the Eocene, local slopes were sufficiently high and steep to shed rockslide debris (Church, 1973; Mathews, 1981). For example, megabreccia in the White Lake area in the southern Okanagan valley (Church, 1973) includes crackled chert cut by continuous alkali syenite dykes hundreds of metres in length, which together appear to be rafted slabs. It is possible that these mega-breccias are associated with a regional décollement involving tectonic denudation of the adjacent metamorphic terrane.

The thick deposit of Upper Eocene (Rouse, 1977) low-rank coal and associated sediments at Hat Creek is noteworthy (see Chapters 9 and 20). The coal is more than 350 m thick and represents a local basin or rift zone which, for perhaps a quarter of a million years, was neither filled to its limit by peat and fine grained sediment, nor was it emptied by erosion at the outlet. Notwithstanding the continuing tectonic activity required to maintain the basin, no coarse debris was introduced to its centre to interrupt the peat accumulation.

Oligocene

Tectonism, so active in Early to Middle Eocene time, dwindled into and through Oligocene time. The rare dated beds of this age mostly show only gentle warping, but local tight folds are found near Quesnel (Rouse and Mathews, 1980). In the southernmost Rocky Mountains the Flathead graben developed concurrently with the accumulation of nonmarine Kishenehn sediments (Jones, 1969).

Miocene

By Early to Middle Miocene time the Eocene fault blocks were reduced to a landscape characterized by broad, low valleys with gentle slopes rising to narrow divides with summits 300 to 500 m above the valley floors. Though hardly a classical peneplain, this late mature erosion surface is in sharp contrast to the rugged terrain that probably existed in the Early and mid-Cretaceous, Eocene and Holocene. The Middle Miocene surface, slightly modified, persists today in many places capped by Upper Miocene lavas (Fig. 11.3).

Middle Miocene uplift and stream entrenchment is marked locally by fluviatile deposits inset below the level of the mature surface. This infilling includes, in several places, coarse, well rounded and sorted gravel of distant provenance which mark the sites of large streams. Vegetation in the interior of the Cordillera, judging from palynological analyses (Rouse and Mathews, 1979), consisted of an oak-beech forest in floodplains and backswamps and an assemblage of hemlock and cedars on the higher slopes. The climate is interpreted as having been warmer and more humid than at present and without the clearly defined rain shadow marked by extensive grasslands that now exists in the lee of the Coast Mountains (Mathews and Rouse, 1963).

Locally (e.g. in the Okanagan Highlands) the Middle Miocene stream courses show no relationship to the present drainage (Fig. 11.4) particularly where they are covered by younger volcanic rocks; in other places (e.g. Quesnel) they coincide with modern rivers (Fig. 11.5). Survival of the Miocene valleys may have been a matter of chance; the presence of easily eroded sediment in the old valleys may have helped to localize later streams.

An analysis of the Miocene drainage system of the Yukon Territory by Tempelman-Kluit (1980; Fig. 11.6) is based largely on the continuity and changing dimensions of existing valleys. It suggests a general southward drainage from the Yukon plateaus through the rising Saint Elias Mountains via the Alsek valley. An alternative explanation, however, is that the southward widening and deepening of the Dezadeash-Alsek valley may have been caused by more intense Pleistocene glaciation in and near the Saint Elias Mountains. Thus, in pre-Pleistocene time, its drainage may have been northward and eastward into the Yukon River.

The Early to Middle Miocene shoreline of the Pacific coast was close to that of modern southern Vancouver Island and northeastern Queen Charlotte Islands in both plan position and elevation. Data from the latter area (Martin and Rouse, 1966) indicate an environment of coastal swamps and lagoons vegetated with swamp cypress and groves of oak, alder and pine. A frost-free climate, some-

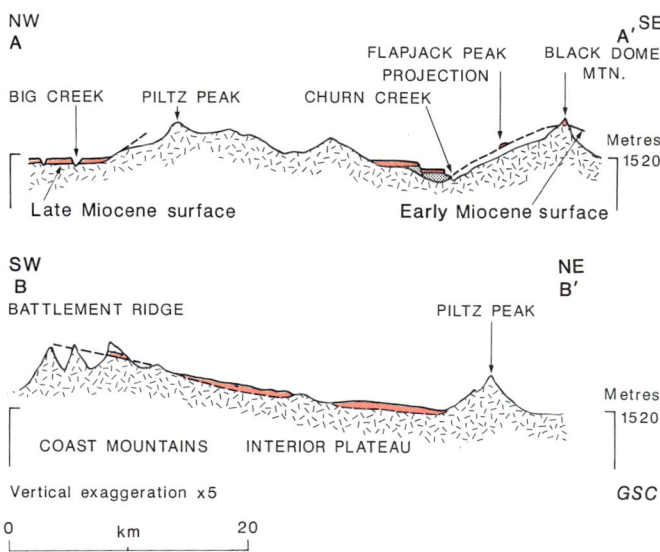

Figure 11.3. Physiographic cross-sections in the south Chilcotin area, south-central British Columbia, showing the relationship between Early and Late Miocene land surfaces, the pre-Miocene bedrock (random pattern), Lower Miocene sediments (stippled), Lower Miocene basalt (heavy, black, dashed line) and Upper Miocene basalt (red shading). For location see Figure 11.9.

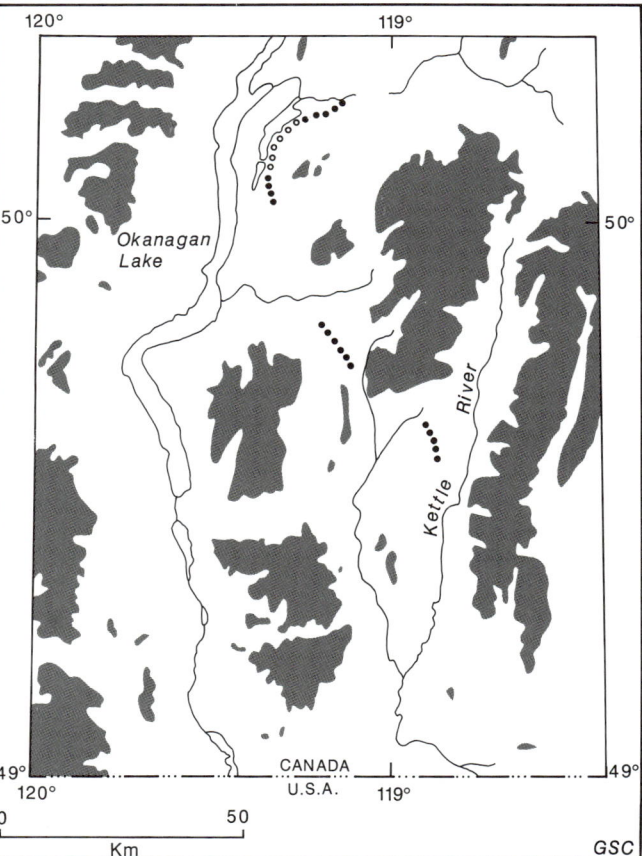

Figure 11.4. Deposits of Miocene river channels (solid dots) and inferred link (open dots) in the Okanagan Highland, south-central British Columbia. Shaded area is over 1500 m in elevation.

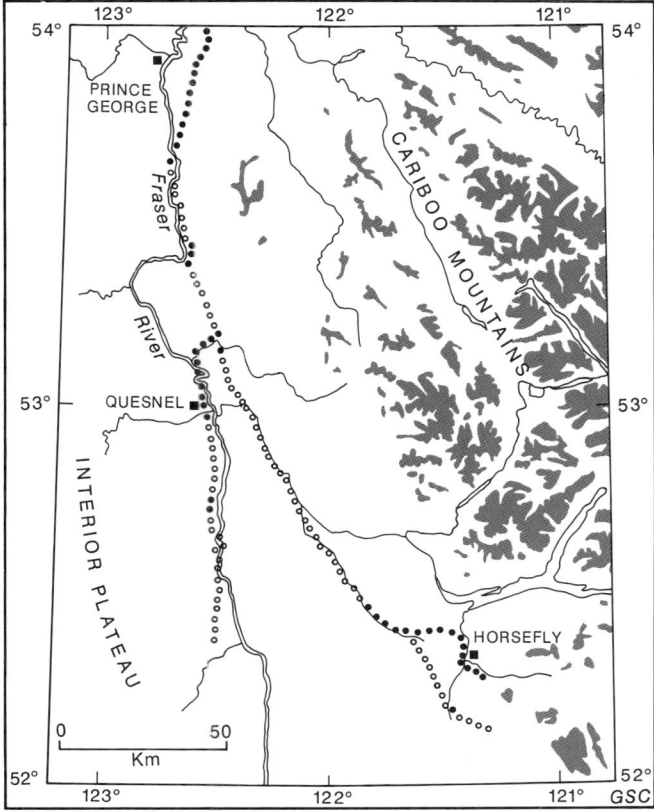

Figure 11.5. Deposits of Miocene river channels (solid dots) and inferred links or extensions (open dots) Cariboo District, central British Columbia. Shaded area is over 1500 m in elevation.

what more temperate than that of the present in Queen Charlotte Islands, is inferred.

The Coast Mountains of Early to Middle Miocene time were not as lofty as they are now (Mathews and Rouse, 1963; Rouse and Mathews, 1979; Parrish, 1981b; Fig. 11.7). In the south, outliers of Miocene lavas cap a few of the summits up to 2.5 km above present sea level, and it is clear that much of the height of the system (locally up to 4 km above sea level) was developed in post-Miocene time. In contrast, at and north of latitude 52°N, Miocene lavas, or the surface on which these lavas were deposited, extend into mountain valleys between peaks rising 1000 m above the valley floors. In this more northerly area a significant part of the elevation may have originated in pre-Miocene time. Apatite-dating (Parrish, 1981b) indicates that the Coast Mountains were regionally elevated by as much as 4 km in the last 10 Ma (Fig. 11.8). Tilting of Pliocene lavas along the Fraser River immediately east of the Coast Mountains indicates that uplift continued into Quaternary time (Mathews and Rouse, 1984).

In southeastern British Columbia Clague (1975) found evidence for 600 m of uplift of the east wall of the Rocky Mountain Trench in post-Miocene time. In an adjacent area Ford et al. (1981) concluded that cutting of valleys averaging 1.3 km deep required between 1.2 and 12 Ma.

A thin (150 m) mantle of Upper Miocene basalt (Bevier, 1983) buries mid-Miocene fluviatile sediments, channel fills and the adjacent mature erosion surface throughout much of the southern interior of the Cordillera (see Chapter 14). Pre-existing ridges and summits locally protruding above the lava partly or completely isolated one lava field from its neighbour in an adjacent valley. The regional consistency in elevation of the base of the lava among the numerous valleys suggests that the valley floors were not only broad and level, but also did not deviate greatly from a common base level. Accordingly, regional variations in elevation of the lava (Fig. 11.9) can be attributed in large part to later tectonism. The gradual but persistent southward rise in elevation of the lava fields from about 1000 m above sea level near the Chilcotin River to almost 2500 m at the northeastern front of the Coast Mountains, 80 km away (Fig. 11.3), records post-Miocene uplift of the Coast Belt. Similarly, a decline of 400 m from the Chilcotin River to near Prince George, 200 km in the opposite direction, may record the same but less pronounced uplift. Older units of the basalt succession (e.g. Bonaparte Lake area) locally show more tilt than later units, suggesting that uplift may already have begun in the Miocene.

Remnants of formerly more extensive basalt flows cap isolated summits and plateaus in an area 30 by 40 km east of the Iskut Trench in north-central British Columbia (Geological Survey of Canada, 1957). The basalts have limited thickness (about 150 m or less) and rest on a low-relief erosion surface which rises in elevation from about 1400 m in the west and east, to 2100 m in the central part of the area (Fig. 11.10; see also Fig. 11.12). This arching of the basalt sheet corresponds to the arch in summit levels and points to uplift of the mountains since deposition of the lavas at about 5.0 Ma.

Other elongated topographic highs occur in the Cordillera and are identified on the Physiographic Map (Map 1701A, in pocket). However, for all but a few, young arched caprock is lacking as is supporting evidence of neotectonic activity. In the Okanagan Highlands a young caprock exists with significant range in elevation of its base but it remains unclear how much of this relief may be due to pre-existing topography and how much to later flexing.

Broad shield volcanoes of Late Miocene to Pliocene age, still retaining much of their original form, occur in the Anahim Volcanic Belt which extends across British Columbia approximately at 52°N. These volcanoes, together with young intrusive rocks to the west, are believed to mark a mantle hot spot over which the North America Plate has moved westerly (Bevier et al., 1979). Other similarly dissected shield volcanoes in the Stikine River area (Stikine Volcanic Belt) show a general north-south alignment and may be related to Late Tertiary intraplate extension and/or plate interactions (see Chapter 14).

Pliocene to Holocene

The Pliocene and early Pleistocene stratigraphic record in the Cordillera is meagre but includes scattered areas of basaltic lava and indications of an early glaciation in the Cariboo Plateau, the Stikine region of northwestern British Columbia, southern Yukon Territory, and in the Saint Elias Mountains.

Uplift in the Saint Elias Mountains, estimated by Parrish (1981a) to have averaged 3 m/ka for the past 15 Ma, was sufficient to create not only the highest

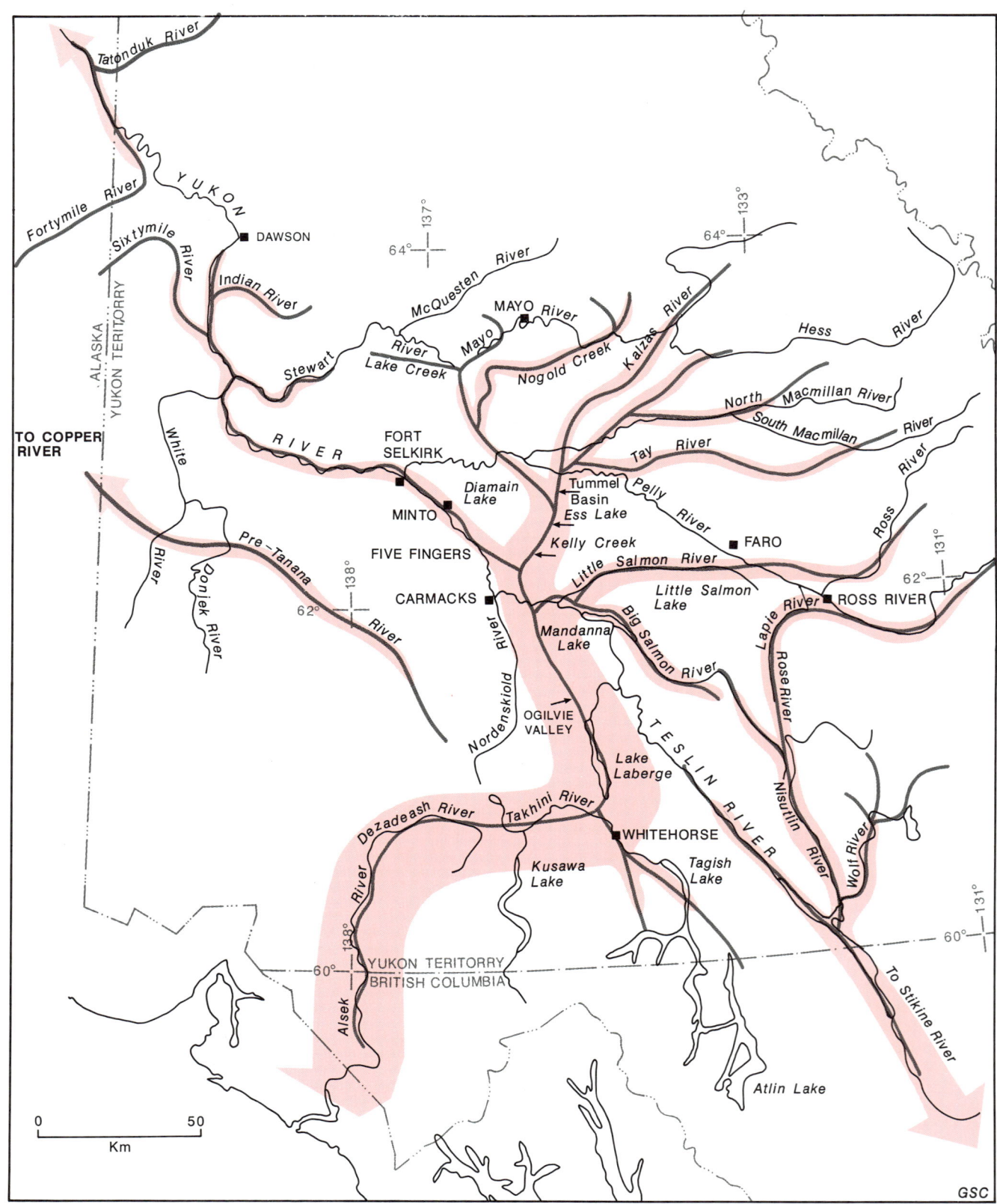

Figure 11.6. Possible Miocene drainage pattern in southwestern Yukon, after Tempelman-Kluit (1980).

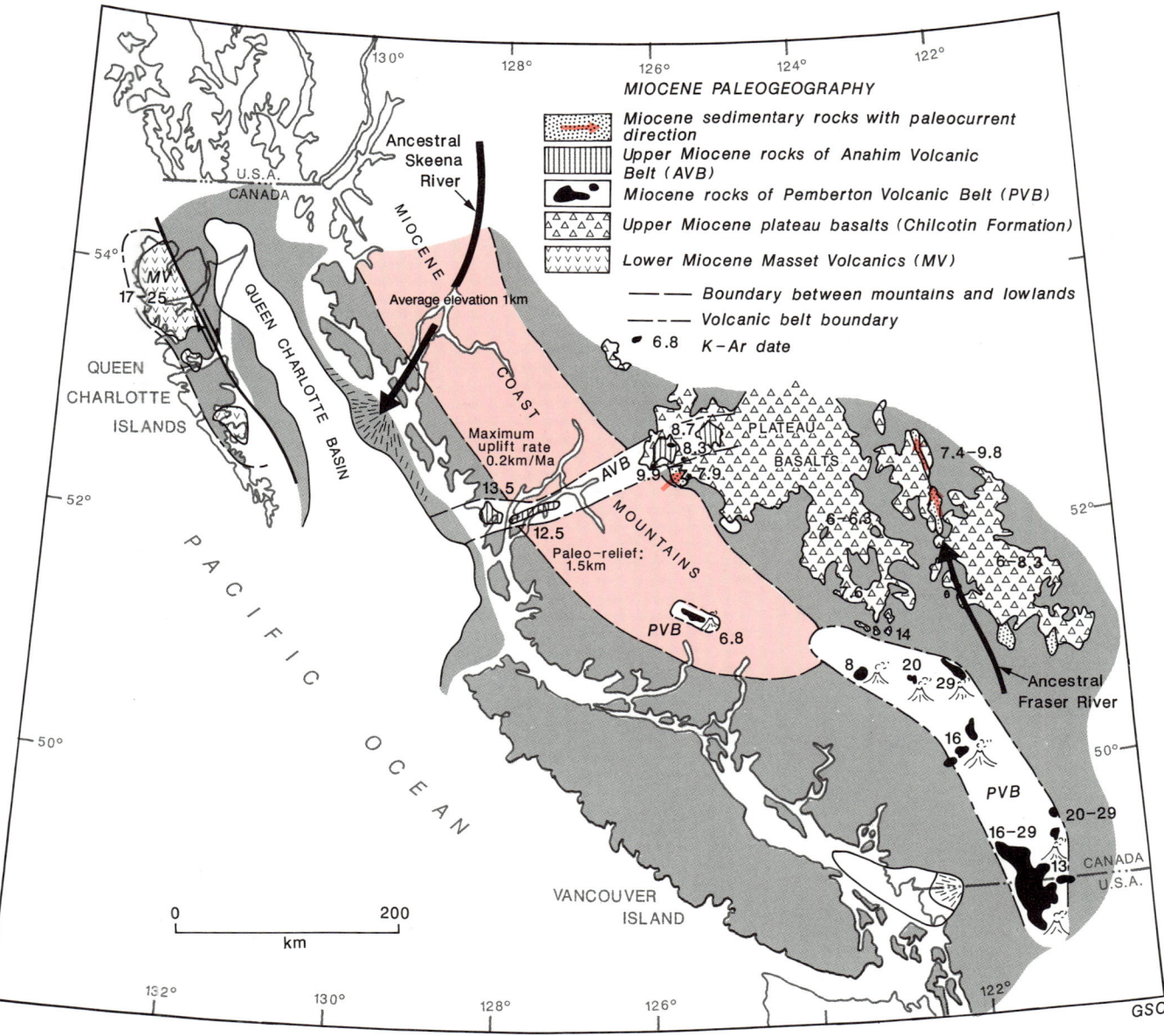

Figure 11.7. Miocene paleogeography in southwestern British Columbia since 10 Ma (modified after Parrish, 1981b).

mountains in Canada (Plate 27) but to severely deform the initially near-horizontal lava sheets which were extruded in the Miocene along the eastern margin of the mountain belt (Souther and Stanciu, 1975). In contrast, in the plateau of central Yukon, the slight divergence in slope of Plio-Pleistocene terrace gravels from those along the adjacent Yukon River has been explained in two ways. Hughes et al. (1972) attributed it to differential uplift of preglacial gravels in the northwest (near Dawson) with respect to those in the southeast; conversely Tempelman-Kluit (1980) proposed a reversal in stream flow as a result of glacial damming in the southeast.

Though glaciers, at their maxima, covered only the upper reaches of the Yukon drainage basin, glaciation had, nonetheless, a marked influence on the Yukon River and its tributaries. In the Klondike goldfields, the highly auriferous "White Channel gravels", dominated by vein-quartz detritus, are overlain by less-auriferous gravel containing, in addition to vein-quartz, much quartzite and schist (Hughes et al., 1972). The former are interpreted as being a product of intense and perhaps prolonged Late Tertiary weathering, and the latter may represent the aggradational detritus from an early Pleistocene glaciation.

The Late Tertiary landscape of the eastern Rocky, Mackenzie and Richardson mountains is marked by a series of gravel-capped pediplains. Though most lack firm dates, they record easterly drainage from the mountains to the plains at levels hundreds of metres above those of the present streams. Streams leaving the mountains and entering the plains, such as the Bow, Saskatchewan Athabasca and Peace rivers, probably became established on these pediplains and have since maintained their

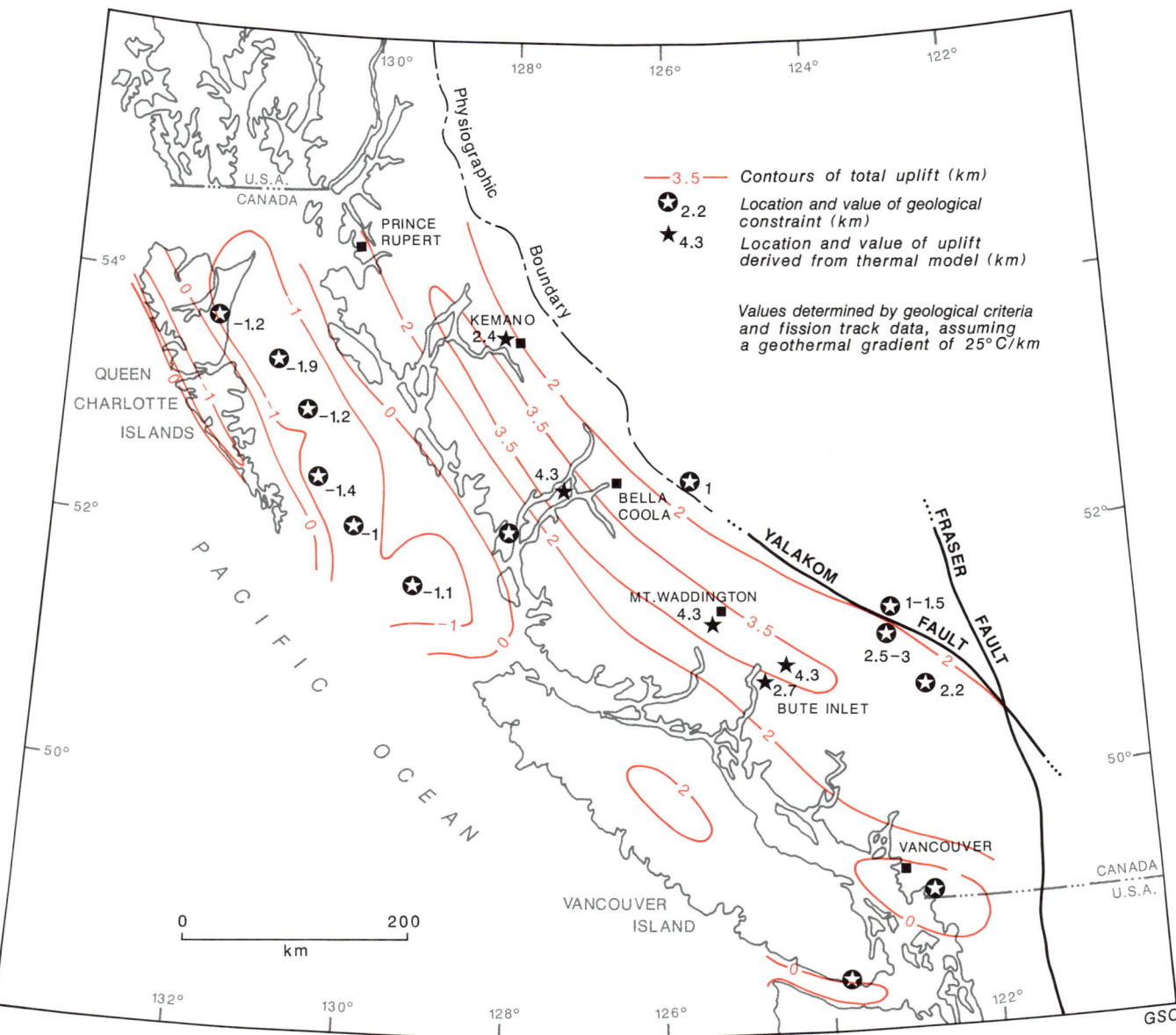

Figure 11.8. Total uplift, with respect to sea level in the last 10 Ma in the southern Coast Mountains. After Parrish (1981b).

easterly or northeasterly courses during subsequent incision. In the Intermontane Belt post-Miocene uplift encouraged stream incision to depths of hundreds of metres and perhaps much more in the adjacent mountain belts. The incised valleys served to guide ice-streams in subsequent glaciations during which they were widened, deepened and their courses straightened so as to create the fiords and fiord-lakes of the Pacific Coast as well as the linear trenches of the major mountain system.

Drainage diversions were probably commonplace during the advancing and retreating stages of each of several glaciations. Dams of glacial drift or marginal ice diverted streams into new courses across former low divides and in many instances these new routes were sufficiently incised to become the established drainage paths. Fraser River at the southeastern end of the Coast Mountains may be an example of an ice-diverted stream (Mathews and Rouse, 1984). Kootenay Lake may have formerly drained south into Idaho; its present outlet, westward to Columbia River, could be an example of a drift-diverted stream. Peace River valley at the eastern front of the Rocky Mountains was blocked by a kame-moraine during the last glaciation and a new course was developed in a spectacular canyon which bypasses the obstruction (Mathews, 1980).

Superposition is another method of stream diversion. For example, Capilano River, in North Vancouver, leaves a broad valley occupied by a storage reservoir and enters a narrow bedrock canyon at Cleveland Dam (Fig. 11.11). At the end of the last ice age the river became established on the western side of an aggraded outwash plain, sloping gently southward, and in the degradational stage that followed ice retreat it cut through the outwash into the

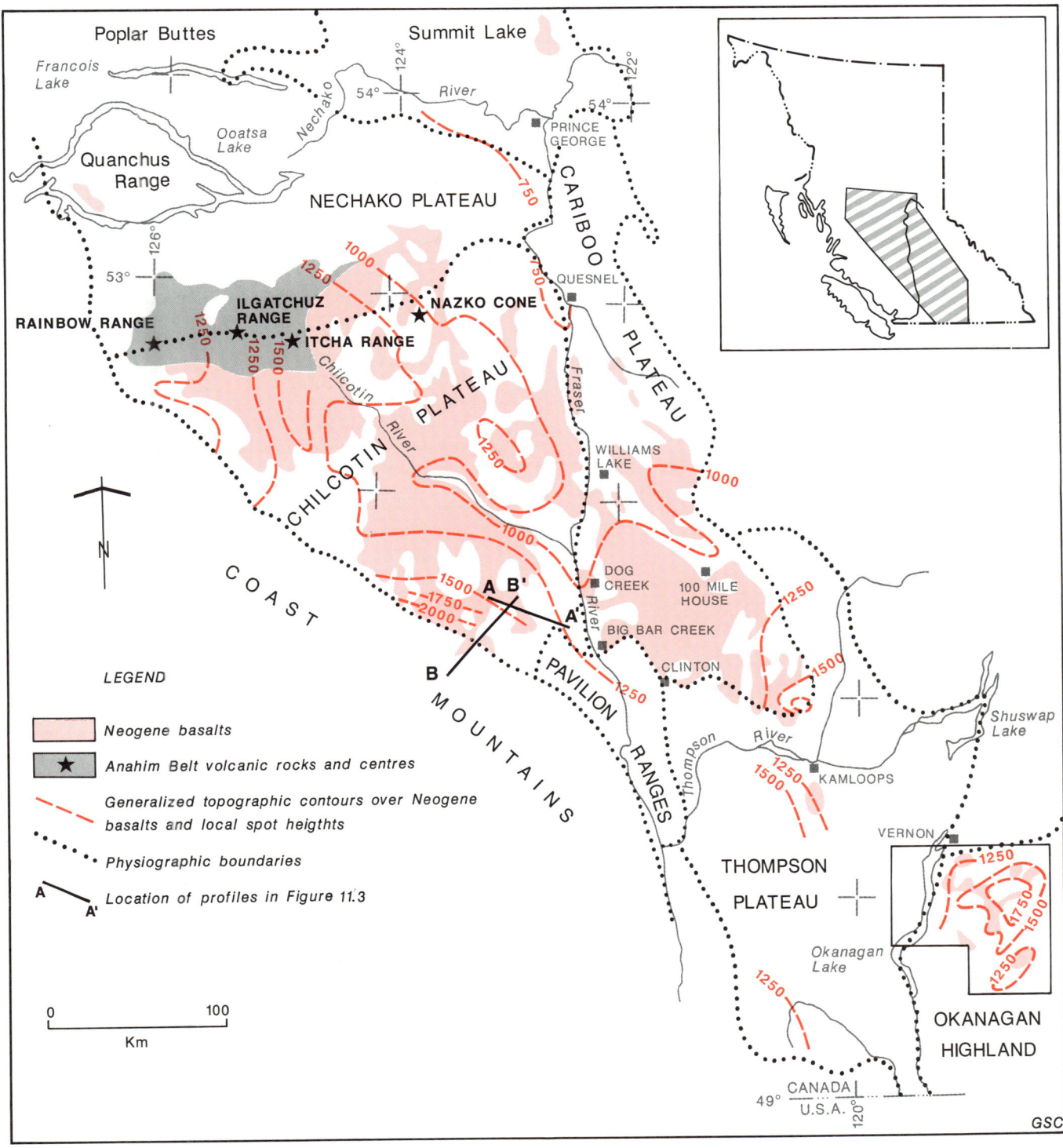

Figure 11.9. Distribution and elevation of Miocene basalts in southern British Columbia (after Mathews, 1988, 1989).

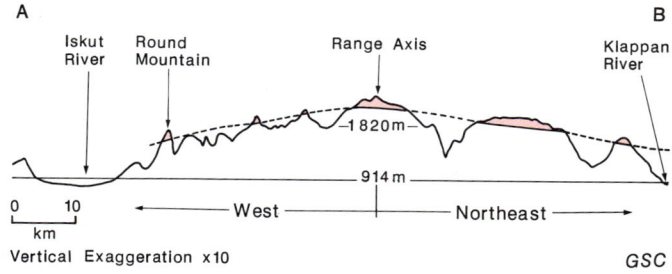

Figure 11.10. Physiographic cross-section, western Skeena Mountains, Cassiar District, north-central British Columbia. For location see Figure 11.12.

bedrock underlying the valley wall at an elevation well above that of the old valley floor to the east. Two other streams a few kilometres to the east, Lynn Creek and Seymour River, exhibit similar characteristics, alternating between narrow bedrock canyon and open valley cut in unconsolidated materials.

Other diversions are less clearly associated with glaciations. The upper Stikine River, for example, flows along a broad valley west to longitude 130° W where it enters its spectacular and obviously youthful Grand Canyon (Fig. 11.12). It is probable that the river formerly flowed southwesterly from the head of this canyon to rejoin the present lower Stikine Valley either near the mouth of Mess Creek or that of Iskut River. Whether the diversion was caused by an ice dam, a blockage by glacial debris, or by growth of the Mount Edziza volcanic complex remains undetermined.

Numerous volcanic centres were active during the Quaternary (Map 1701A, in pocket) though they did not affect as large an area as did the Miocene cones. Some centres erupted under, into, through, or onto glacier ice producing highly anomalous land forms (Mathews, 1947, 1951, 1952a,b). The Garibaldi chain of volcanoes in the southern Coast Mountains, erupting mainly andesitic to dacitic products, is considered to represent the active arc related to the subducting Juan de Fuca Plate. A small basaltic centre in the Kitimat Trench, active only about two centuries ago, is the youngest known in Canada (Sutherland Brown, 1969).

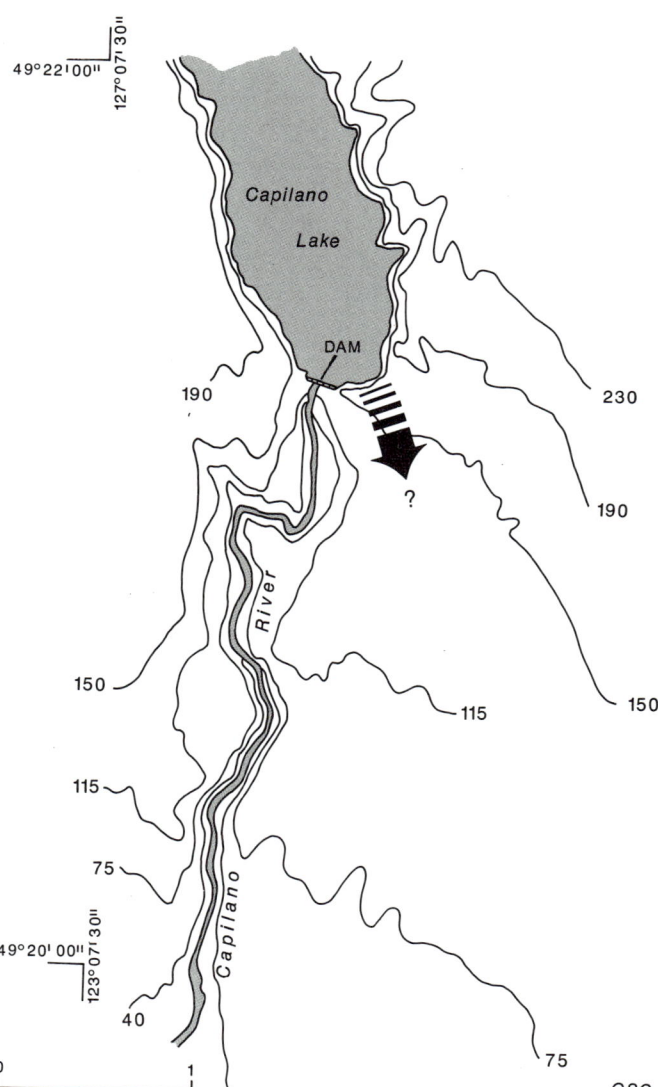

Figure 11.11. Present course of the Capilano River incised in a bedrock canyon. Axis of former valley farther east (broken lines and arrow) is unknown. Contours in metres.

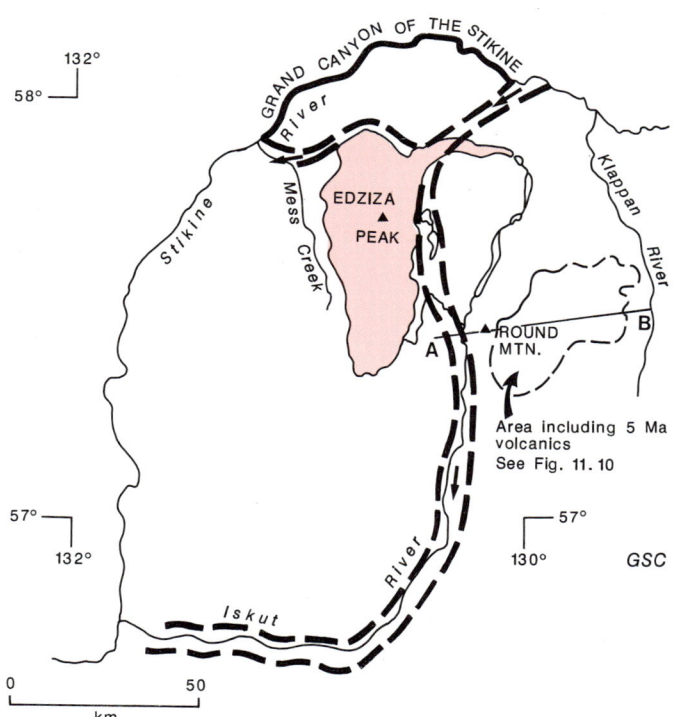

Figure 11.12. Possible former courses of the Stikine River (broken lines and arrows). Tertiary and Quaternary volcanics shown in pink.

OTHER GEOMORPHIC DEVELOPMENTS

Some significant composite geomorphic developments cannot be closely dated hence fall outside the chronological framework of the foregoing discussion, yet deserve comment. Although juxtaposition of resistant and non-resistant rocks took place in the distant past it is their differential erosion in relatively recent time that has created well defined topography. For example Paleozoic carbonate rocks of the Rocky Mountains were thrust over much less resistant shale and minor sandstone of Cretaceous age during the Late Cretaceous to Paleogene deformations, but the present contrasts in relief did not come about until after late Cenozoic pedimentation and glaciation. The Northern Rocky Mountain Trench is underlain by relatively weak Paleogene sediments, now fault-bounded. Differential erosion seems to be at least partly responsible for this long linear topographic depression. However, the topographic difference in elevation between the floor and east side of the Southern Rocky Mountain Trench was substantially increased by Late and post-Miocene faulting and this raises the question whether normal faulting is largely responsible for the relief in other parts of the Rocky Mountain Trench. A similar problem exists in the Vancouver area where resistant plutonic rocks form the mountains to the north and relatively weak Upper Cretaceous to Miocene sediments underlie the lowland at and south of the city. Differential erosion has clearly played a part in controlling the topography, indeed the mountain front is a stripped surface dipping southward at about 15° with local patches of Cretaceous sediment still surviving on it.

REFERENCES

Beaumont, C.
1981: Foreland Basins; Geophysical Journal of the Royal Astronomical Society, v. 65, no. 2, p. 291-329.

Bevier, M.L.
1983: Regional stratigraphy and age of Chilcotin Group basalts, south-central British Columbia; Canadian Journal of Earth Sciences, v. 20, p. 515-514.

Bevier, M.L., Armstrong, R.L., and Souther, J.G.
1979: Miocene peralkaline volcanism in west-central British Columbia - its temporal and plate-tectonic setting; Geology, v. 7, p. 389-392.

Bloom, A.L., Broecker, W.S., Chappell, J.M.A., Matthews, R.K., and Mesolella, K.J.
1974: Quaternary sea level fluctuations on Tectonic Coast: new ^{230}Th ^{234}U dates from the Huon Peninsula, New Guinea; Quaternary Research, v. 4, p. 185-205.

Bustin, R.M. and Moffat, I.
1983: Groundhog coalfield, central British Columbia; reconnaissance stratigraphy and structure; Bulletin of Canadian Petroleum Geology, v. 31, p. 231-245.

Chappell, J.H.
1974: Geology of coral terraces, Huon Peninsula, New Guinea: a study of Quaternary tectonic movements and sea level changes; Geological Society of America Bulletin, v. 85, p. 553-570.

Church, B.N.
1973: Geology of the White Lake Basin; British Columbia Department of Mines and Petroleum Resources, Bulletin 61, 120 p.

Clague, J.J.
1975: Late Quaternary sediments and geomorphic history of the southern Rocky Mountain Trench, British Columbia; Canadian Journal of Earth Sciences, v. 12, p. 595-605.

Coates, J.A.
1974: Geology of the Manning Park area, British Columbia; Geological Survey of Canada, Bulletin 238, 177 p.

Couillard, R. and Irving, E.
1974: Paleolatitude and reversals: evidence from the Cretaceous Period in Cretaceous system; in The Western Interior of North America, W.G.E. Caldwell (ed.), Geological Association of Canada, Special Publication no. 13, p. 21-29.

D'Addario, G.W., Dow, D.B., and Swoboda, R.
1975: Geology of Papua-New Guinea, 1:2,5000,000; Bureau of Mineral Resources, Canberra, Australia.

Eisbacher, G.H.
1974: Sedimentary history and tectonic evolution of the Sustut and Sifton basins, north-central British Columbia; Geological Survey of Canada, Paper 73-31, 57 p.
1981: Late Mesozoic-Paleogene Bowser Basin molasse and Cordilleran tectonics in alluvial basins; A.D. Miall (ed.), Geological Association of Canada, Special Paper 23, p. 125-151.

Eisbacher, G.H., Carrigy, M.A., and Campbell, R.B.
1974: Paleodrainage pattern and late-orogenic basins in the Canadian Cordillera; in Tectonics and Sedimentation, W.R. Dickinson (ed.), Society of Economic Paleontologists and Mineralogists, Special Paper 22, p. 143-166.

England, T.D.J. and Bustin, R.M.
1986: Thermal maturation of the western Canadian sedimentary basin south of the Red Deer River; Alberta Plains Bulletin of Canadian Petroleum Geology, v. 34, no. 1, p. 71-90.

Ewing, T.E.
1981: Regional stratigraphy and structural setting of the Kamloops Group, south-central British Columbia; Canadian Journal of Earth Sciences, v. 18, p. 1464-1477.

Ford, D.C., Schwarez, H.P., Drake, J.J., Gascoyne, M., Harmon, R.S., and Latham, A.G.
1981: Estimates of the age of the existing relief of the Rocky Mountains of Canada; Arctic and Alpine Research, v. 13, p. 1-10.

Gabrielse, H.
1985: Major dextral transcurrent displacements along the Northern Rocky Mountain Trench and related lineaments in north-central British Columbia; Geological Society of America Bulletin, v. 96, p. 1-14.

Geological Survey of Canada
1957: Stikine River area, Cassiar District, British Columbia; Geological Survey of Canada, Map 9-1957.

Ghent, E.D., Stout, M.Z., and Raeside, R.P.
1983: Plagioclase-clinopyroxene-quartz equilibria and the geobarometry and geothermometry of garnet amphibolites from Mica Creek, British Columbia; Canadian Journal of Earth Sciences, v. 20, p. 699-706.

Gibson, D.W. and Hughes, J.D.
1981: Structure, stratigraphy, sedimentary environments and coal deposits of the Jura-Cretaceous Kootenay Group, Crowsnest Pass area, Alberta and British Columbia; in Field Guides to Geology and Mineral Deposits, R.I. Thompson (ed.), Geological Association of Canada, 1981 Annual Meeting, Calgary, p. 1-40.

Hamblin, A.P. and Walker, R.G.
1979: Storm-dominated shallow marine deposits: the Fernie-Kootenay (Jurassic) transition, southern Rocky Mountains; Canadian Journal of Earth Sciences, v. 16, p. 1673-1690.

Harrison, T.M., Armstrong, R.L., Naeser, C.W., and Harakal, J.E.
1979: Geochronology and thermal history of the Coast Plutonic Complex near Prince Rupert, British Columbia; Canadian Journal of Earth Sciences, v. 16, p. 400-410.

Hollister, L.S.
1979: Metamorphism and crustal displacements: new insights; Episodes, 1979, no. 3, p. 3-8.

Høy, T.
1976: Calc-silicate isograds in the Riondel Area, southeastern British Columbia; Canadian Journal of Earth Sciences, v. 13, no. 8, p. 1093-1104.
1980: Geology of the Riondel area, central Kootenay Arc, southeastern British Columbia; British Columbia Ministry of Energy, Mines and Petroleum Resources, Bulletin 73, 89 p.

Hughes, O.L., Rampton, V.N., and Rutter, N.W.
1972: Quaternary geology and geomorphology, southern and central Yukon (northern Canada); XXVI International Geological Congress, Guidebook to excursion A11, 59 p.

Jeletzky, J.A.
1975: Jurassic and Lower Cretaceous paleogeography and depositional tectonics of Porcupine Plateau, adjacent areas of northern Yukon, and those of Mackenzie District; Geological Survey of Canada, Paper 74-16, 52 p.

Jeletzky, J.A. and Tipper, H.W.
1968: Upper Jurassic and Cretaceous rocks of Taseko Lakes map area and their bearing on the geological history of southwestern British Columbia; Geological Survey of Canada, Paper 67-54, 218 p.

Jenkins, D.A.L.
1974: Detachment tectonics in Western Papua-New Guinea; Geological Society of America Bulletin, v. 85, p. 533-548.

Johnson, S.Y.
1984: Stratigraphy, age and paleogeography of the Eocene Chuckanut Formation, northwestern Washington; Canadian Journal of Earth Sciences, v. 21, p. 92-106.

Jones, P.B.
1969: The Tertiary Kishenehn Formation, British Columbia; Bulletin of Canadian Petroleum Geology, v. 17, p. 234-246.

Martin, H.A. and Rouse, G.E.
1966: Palynology of Late Tertiary sediments from Queen Charlotte Islands, British Columbia; Canadian Journal of Botany, v. 44, p. 171-208.

Mathews, W.H.
1947: 'Tuyas,' flat-topped volcanoes in northern British Columbia; American Journal of Science, v. 245, p. 560-570.
1951: The Table, a flat-topped volcano in southwestern British Columbia; American Journal of Science, v. 249, p. 830-841.
1952a: Mount Garibaldi, a supraglacial Pleistocene volcano in southwestern British Columbia; American Journal of Science, v. 250, p. 81-103.
1952b: Ice-dammed lavas from Clinker Mountain, southwestern British Columbia; American Journal of Science, v. 250, p. 553-565.
1980: Retreat of the last ice sheets in northeastern British Columbia and adjacent Alberta; Geological Survey of Canada, Bulletin 331, 22 p.
1981: Early Cenozoic resetting of potassium-argon dates and geothermal history of north Okanagan area, British Columbia; Canadian Journal of Earth Sciences, v. 18, p. 1310-1319.
1983: Early Tertiary resetting of potassium-argon dates in the Kootenay Arc, southeastern British Columbia; Canadian Journal of Earth Sciences, v. 20, p. 867-872.
1988: Neogene geology of the Okanagan Highland, British Columbia; Canadian Journal of Earth Sciences, v. 25, p. 725-731.
1989: Neogene Chilcotin basalts in south-central British Columbia; geology, ages and geomorphic history; Canadian Journal of Earth Sciences, v. 26, p. 969-982.

Mathews, W.H. and Rouse, G.E.
1963: Late Tertiary volcanic rocks and plant-bearing deposits in British Columbia; Geological Society of America Bulletin, v. 74, p. 55-60.
1984: The Gang Ranch-Big Bar area, south-central British Columbia: stratigraphy, geochronology, and palynology of the Tertiary beds and their relationship to the Fraser fault; Canadian Journal of Earth Sciences, v. 21, p. 1132-1144.

McLean, J.R.
1977: The Cadomin Formation: stratigraphy, sedimentology, and tectonic implications; Bulletin of Canadian Petroleum Geology, v. 25, p. 792-827.

McMechan, R.D.
1983: Geology of the Princeton Basin; British Columbia Ministry of Energy, Mines, and Petroleum Resources, Paper 1983-3, 52 p.

Monger, J.W.H.
1968: Early Tertiary stratified rocks, Greenwood map-area, (82 E/2) British Columbia; Geological Survey of Canada, Paper 67-42, 39 p.

Monger, J.W.H. and Irving, E.
1980: Northward displacement of north-central British Columbia; Nature, v. 285, p. 289-294.

Norris, D.K., Stevens, R.D., and Wanless, R.K.
1965: K-Ar age of igneous pebbles in the McDougall-Segur conglomerate, southeastern Canadian Cordillera; Geological Survey of Canada, Paper 65-26, 11 p.

Nurkowski, J.R.
1984: Coal quality, coal rank variation and its relation to reconstructed overburden, Upper Cretaceous and Tertiary Plains coals, Alberta, Canada; American Association of Petroleum Geologists, Bulletin, v. 68, no. 5, p. 285-295.

Parrish, R.R.
1981a: Uplift rates of Mt. Logan, Yukon Territory and British Columbia's central Coast Mountains using fission track dating methods; Eos, v. 62, p. 59-60.
1981b: Cenozoic thermal and tectonic history of the Coast Mountains of British Columbia as revealed by fission track and geological data and quantitative thermal models; Ph.D. thesis, University of British Columbia, Vancouver, 116 p.

Parrish, R.R., Carr, S., and Parkinson, D.
1985: Metamorphic complexes and extensional tectonics, southern Shuswap Complex, southeastern British Columbia; in Field Guides to Geology and Mineral Deposits in the southern Canadian Cordillera, D.J. Tempelman-Kluit (ed.), p. 12-1 to 12-15.

Pigage, L.C. and Greenwood, H.J.
1982: Internally consistent estimates of pressure and temperature: the staurolite problem; American Journal of Science, v. 282, p. 943-969.

Price, R.A., Monger, J.W.H., and Roddick, J.A.
1985: Cordilleran cross-section, Calgary to Vancouver; Geological Society of America, Cordilleran Section Meeting, Vancouver, Field Trip 3 Guide Book.

Rapson, J.E.
1965: Petrography and derivation of Jurassic-Cretaceous clastic rocks, southern Rocky Mountains, Canada; American Association of Petroleum Geologists Bulletin, v. 49, p. 1426-1452.

Rouse, G.E.
1967: A late Cretaceous plant assemblage from east-central British Columbia: Part 1, fossil leaves; Canadian Journal of Earth Sciences, v. 4, p. 1185-1212.
1977: Paleogene palynomorph ranges in western and northern Canada; in Contributions of Stratigraphic Palynology, vol. 1 (Cenozoic Palynology), W.C. Elsik (ed.), American Association of Stratigraphic Palynologists, Contributions Series 5A, p. 48-65.

Rouse, G.E. and Mathews, W.H.
1979: Tertiary geology and palynology of the Quesnel area, British Columbia; Bulletin of Canadian Petroleum Geology, v. 27, p. 418-445.

Souther, J.G. and Stanciu, C.
1975: Operation Saint Elias, Yukon Territory: Tertiary volcanic rocks; in Report of Activities, Part A, Geological Survey of Canada, Paper 75-1A, p. 63-70.

Stott, D.H.
1968: Lower Cretaceous Bullhead and Fort St. John groups between Smoky and Peace rivers, Rocky Mountain Foothills, Alberta and British Columbia; Geological Survey of Canada, Bulletin 152, 279 p.
1984: Cretaceous sequences of the foothills of the Canadian Rocky Mountains; in The Mesozoic of Middle North America, D.F. Stott and D.J. Glass (ed.), Canadian Society of Petroleum Geologists, Memoir 9, p. 85-107.

Sutherland Brown, A.
1969: Aiyansh lava flow, British Columbia; Canadian Journal of Earth Sciences, v. 6, p. 1460-1467.

Sutherland Brown, A. and Yorath, C.J.
1985: Lithoprobe profile across southern Vancouver Island: geology and tectonics; Geological Society of America, Cordilleran Section Meeting, Vancouver, B.C. Field Trip 8, Guide Book.

Taylor, D.R. and Walker, R.G.
1984: Depositional environments and paleogeography in the Albian Moosebar Formation and adjacent fluvial Gladstone and Beaver Mines formations, Alberta; Canadian Journal of Earth Sciences, v. 21, p. 698-714.

Tempelman-Kluit, D.
1980: Evolution of physiography and drainage in southern Yukon; Canadian Journal of Earth Sciences, v. 17, p. 1189-1203.

Tempelman-Kluit, D.J. and Parkinson, D.
1986: Extension across the Eocene Okanagan crustal shear in southern British Columbia; Geology, v. 14, p. 318-321.

Woodsworth, G.J., Crawford, M.L., and Hollister, L.S.
1983: Metamorphism and structure of the Coast Plutonic Complex and adjacent belts, Prince Rupert and Terrace area, British Columbia; Guidebook to Field Trip 14, Geological Association of Canada, 1983 National Meeting, Victoria, B.C.

Yorath, C.J. and Chase, R.L.
1981: Tectonic history of the Queen Charlotte Islands and adjacent areas - a model; Canadian Journal of Earth Sciences, v. 18, p. 1717-1739.

Yorath, C.J. and Cook, D.G.
1981: Cretaceous and Tertiary stratigraphy and paleogeography, northern Interior Plains, District of Mackenzie; Geological Survey of Canada, Memoir 398.

Yorath, C.J., Green, A.G., Clowes, R.M., Sutherland Brown, A., Brandon, M.T., Kanasewich, E.R., Hyndman, R.D., and Spencer, C.
1985: Lithoprobe-southern Vancouver Island: Seismic reflection sees through Wrangellia to the Juan de Fuca Plate; Geology, v. 13, no. 11, p. 759-762.

Young, F.G., Myhr, D.W., and Yorath, C.J.
1976: Geology of the Beaufort-Mackenzie Basin; Geological Survey of Canada, Paper 76-11.

Author's address

W.H. Mathews
Department of Geological Survey
University of British Columbia
Vancouver, British Columbia
V6T 2B4

Chapter 12

QUATERNARY GLACIATION AND SEDIMENTATION

Summary
Introduction
Cordilleran Ice Sheet
Record of glaciation
Record of nonglacial periods
Distribution and character of Quaternary sediments
Quaternary sedimentation and erosion
    Effects of glaciation on sedimentation and erosion
    Spatial constraints on sedimentation
Crustal deformation related to glaciation
References

Chapter 12

QUATERNARY GLACIATION AND SEDIMENTATION

John J. Clague

SUMMARY

The Quaternary Period, encompassing the last two million years of geological time, is noteworthy for major climatic perturbations that resulted in episodic growth and decay of continental ice sheets in middle latitudes of the Northern Hemisphere. One such ice sheet and smaller independent satellite glaciers repeatedly enveloped most of the Canadian Cordillera with the exception of the northern Yukon and parts of the western District of Mackenzie.

Global cooling at the beginning of each glaciation led to the expansion of cirque and valley glaciers in the high mountains of western Canada. As climate deteriorated, glaciers advanced and coalesced to form piedmont complexes and mountain ice sheets. Eventually, glaciers from separate mountain ranges joined to cover most of British Columbia, southern Yukon, and parts of westernmost Alberta, Alaska, and the northwestern conterminous United States. During most Quaternary glaciations, the Cordilleran Ice Sheet was continuously nourished from source areas in high mountain ranges, and ice flow was controlled mainly by topography. However, at the climaxes of a few glaciations, ice in the interior of British Columbia became sufficently thick for one or more ice domes to develop with surface flow radially away from their centres.

Glaciations ended with rapid climatic amelioration. Deglaciation occurred by complex frontal retreat and by downwasting accompanied by widespread stagnation. In areas of moderate relief, uplands appeared through the ice sheet first, dividing it into a series of tongues that decayed in response to local conditions.

Glaciers existed in the Canadian Cordillera during late Tertiary and early Quaternary time, but little is known of the character and chronology of early glacial events. Reasonable stratigraphic and landform evidence is available only for the last three glaciations. The youngest of these (Late Wisconsinan) is by far the best documented because its deposits and landforms occur at the surface throughout the glaciated Cordillera. In contrast, deposits of older glaciations are less common and, where present, are generally covered by younger sediments. Landforms assignable to these older glaciations are found only beyond Late Wisconsinan glacial limits in parts of the Yukon, western Alberta, and northwestern United States. Available evidence indicates that glacial styles during the last three glaciations were broadly similar, although the Late Wisconsinan Cordilleran Ice Sheet apparently was less extensive than its predecessors.

Glaciations alternated with periods of restricted ice cover (i.e., glaciers were largely limited to mountain ranges). Stratigraphic, paleoecological, and sedimentological data are available for the last three nonglacial periods, the most recent of which began about 10 000 years ago and continues to the present. These data indicate that sedimentary environments and the geomorphic framework were similar during each of these periods, at least in the southern part of the Canadian Cordillera.

The Quaternary stratigraphic record of the Canadian Cordillera is mainly a product of brief sedimentation events separated by long periods of nondeposition and erosion. Thick stratified sediments lie mainly in valleys and coastal lowlands and were deposited in proglacial and ice-contact environments during periods of growth and decay of the Cordilleran Ice Sheet. At glacial maxima, till was deposited over large areas of low and moderate relief. However, at the same time, much of the landscape was eroded by glaciers.

Sedimentation was more restricted and occurred at lower rates during nonglacial periods than during glaciations. Although large amounts of sediment accumulated offshore, especially in fiords and basins, the only important nonglacial sediment-accumulation sites on land were lakes, floodplains, and fans. As a result, the terrestrial stratigraphic record of nonglacial periods is meagre. Where present, true nonglacial units generally are thin and discontinuous. Commonly, a nonglacial period is recorded only by an unconformity produced when streams incised valley fills shortly after the end of the preceding glaciation.

Growth and decay of the Cordilleran Ice Sheet triggered isostatic adjustments in the crust and mantle of the Canadian Cordillera. As glaciers expanded during the early part of each glaciation, an increasing part of the Cordilleran landmass was depressed isostatically. At the climax of each major glaciation, the entire glaciated Cordillera was depressed, with areas near the centre of the ice sheet, in general, displaced downward more than areas near the periphery. The presence on the British Columbia coast of relict shorelines up to about 200 m above present sea level indicates that the crust probably was displaced downward several hundred metres in areas of thickest ice at the climax of glaciation.

Isostatic adjustments during deglacial intervals were opposite in direction to those that occurred during periods of ice-sheet growth. Studies of elevated shorelines on the

---

Clague, J.J.
1991: Quaternary glaciation and sedimentation, Chapter 12 in Geology of the Cordilleran Orogen in Canada, H. Gabrielse and C. J. Yorath (ed.); Geological Survey of Canada, Geology of Canada, no. 4, p. 419-434 (also Geological Society of America, The Geology of North America, v. G-2)

coast of British Columbia indicate that isostatic uplift resulting from the most recent decay of the Cordilleran Ice Sheet occurred very rapidly and was largely complete within a few thousand years of deglaciation. However, because of the diachronous nature of glacier retreat, uplift occurred at different times in different places; in general, regions that were deglaciated first rebounded earlier than those deglaciated at a later date.

INTRODUCTION

The Quaternary Period, comprising the last two million years of Earth history, is perhaps most noteworthy for cyclic climatic fluctuations that led to the repeated growth and decay of continental ice sheets in North America and Europe. Isotopic and magnetic studies of deep-sea sediments show that there have been eight major climatic cycles in the last 800 ka, each about 100 ka in duration and each marked by sharp fluctuations in climate on shorter timescales; similar cycles of lesser magnitude but greater frequency occurred from 800 ka B.P. to before the beginning of the Quaternary Period (Shackleton and Opdyke, 1973, 1976). The colder parts of most of these climatic cycles probably were times of widespread glaciation in the Canadian Cordillera.

Glaciation profoundly altered the landscape and biota of the Canadian Cordillera and, to a large extent, controlled Quaternary sedimentation and erosion in the region (Clague, 1986). Pleistocene glaciers also deformed the crust (Clague, 1983) and may have triggered some volcanic eruptions (Grove, 1974).

CORDILLERAN ICE SHEET

The Cordilleran Ice Sheet was a large mass of confluent valley and piedmont glaciers and mountain ice sheets that at its maximum enveloped almost all of British Columbia and the southern Yukon, as well as parts of Alaska, Alberta, Montana, Idaho, and Washington (Fig. 12.1, 12.2; Flint, 1971; Clague et al., 1989). The ice sheet was confined mainly between the high mountains of the western and eastern Cordillera, although large areas west of the Coast Mountains and, to a lesser extent, east of the Rocky Mountains also were covered. Glaciers in several ranges bordering the core area of the Cordilleran Ice Sheet, for example the Ogilvie and Mackenzie mountains and Queen Charlotte Ranges, were more or less independent of the ice sheet, even at the climax of glaciation.

Growth of the Cordilleran Ice Sheet was initiated during periods of global cooling by the advance of glaciers in the high mountains of western Canada (Davis and Mathews, 1944). With continued cooling and perhaps increased precipitation, these glaciers advanced beyond mountain fronts and coalesced to form piedmont complexes over plateaus and lowlands (Fig. 12.3). Eventually, piedmont complexes from separate mountain ranges joined to cover most of British Columbia and adjacent areas. Throughout this period, mountains were the main sources of glacier ice, and the pattern of ice flow was controlled largely by topography. Occasionally, however, ice thickened to such an extent over the interior of British Columbia (up to about 2.5 km) that one or more domes with surface flow radially away from their centres became established (Flint, 1971).

Each glacial cycle terminated with rapid climatic amelioration. Deglaciation occurred by complex frontal retreat in peripheral glaciated areas and by downwasting accompanied by widespread stagnation throughout much of the interior (Fig. 12.3; Fulton, 1967; Clague, 1981). Along the western periphery of the ice sheet, glaciers calved back in contact with eustatically rising seas. In areas of moderate relief, the pattern of deglaciation was even more complex, with uplands becoming ice free first and dividing the ice sheet into a series of valley tongues that decayed in response to local conditions. During each glacial cycle, the time of deglaciation varied from region to region (Fulton, 1971). The first areas to become ice free were those near the periphery of the ice sheet. Active glaciers probably persisted longest in some mountain valleys, however, these glaciers may have coexisted with large masses of dead ice on interior plateaus.

Lobes of the Cordilleran Ice Sheet and independent valley glaciers in the Rocky and Mackenzie mountains at times coalesced with the Laurentide Ice Sheet along the eastern margin of the Canadian Cordillera (Rutter, 1984). During most glaciations, coalescence was restricted to a relatively small portion of the Cordilleran margin, mainly east of the central and northern Rocky Mountains. During minor glaciations, a continuous ice-free zone separated Keewatin ice on the east from the Cordilleran Ice Sheet and satellite mountain glaciers on the west.

RECORD OF GLACIATION

Recent studies of ice-contact volcanic landforms and glacial deposits interstratified with lavas and pyroclastics at Mount Edziza, British Columbia indicate that there was glaciation before 1 Ma BP and perhaps before 3 Ma BP (Souther et al., 1984). Similar evidence from Fort Selkirk in the Yukon (Naeser et al., 1982) and Dog Creek, British Columbia (Mathews and Rouse, 1986) shows that one or more of the largest continental glaciations in the Canadian Cordillera is older than 1 Ma BP. Finally, Miocene and Pliocene glaciations have been documented in the Wrangell Mountains in easternmost Alaska (Denton and Armstrong, 1969). Unfortunately, the deposits of these early glaciations are fragmentary, thus the glacial history of the Cordillera during Late Tertiary and early Quaternary time is poorly known.

The earliest periods of ice-sheet glaciation for which there is reasonable stratigraphic and landform evidence predate the Sangamonian Stage (Fig. 12.4, Table 12.1) and consequently are older than 128 ka BP (Shackleton and Opdyke, 1973). Deposits of these glaciations occur beneath Wisconsinan and Sangamonian(?) sediments at several sites in south coastal British Columbia (Armstrong, 1981) and the southern Yukon (Denton and Stuiver, 1967; Klassen, 1978, 1987) and also are exposed at the surface beyond Wisconsinan glacial limits in the central Yukon (Bostock, 1966; Hughes et al., 1969) (Fig. 12.5). Little information exists on the extent of ice cover during these glaciations, except that in the Yukon the Nansen and Klaza advances (Early and Middle Pleistocene) apparently were more extensive than younger advances.

It is generally believed that there have been two major glaciations in the Canadian Cordillera during the last 75-100 ka, one during Early Wisconsinan time and the other during the Late Wisconsinan (Fulton, 1984). The

QUATERNARY GLACIATION AND SEDIMENTATION

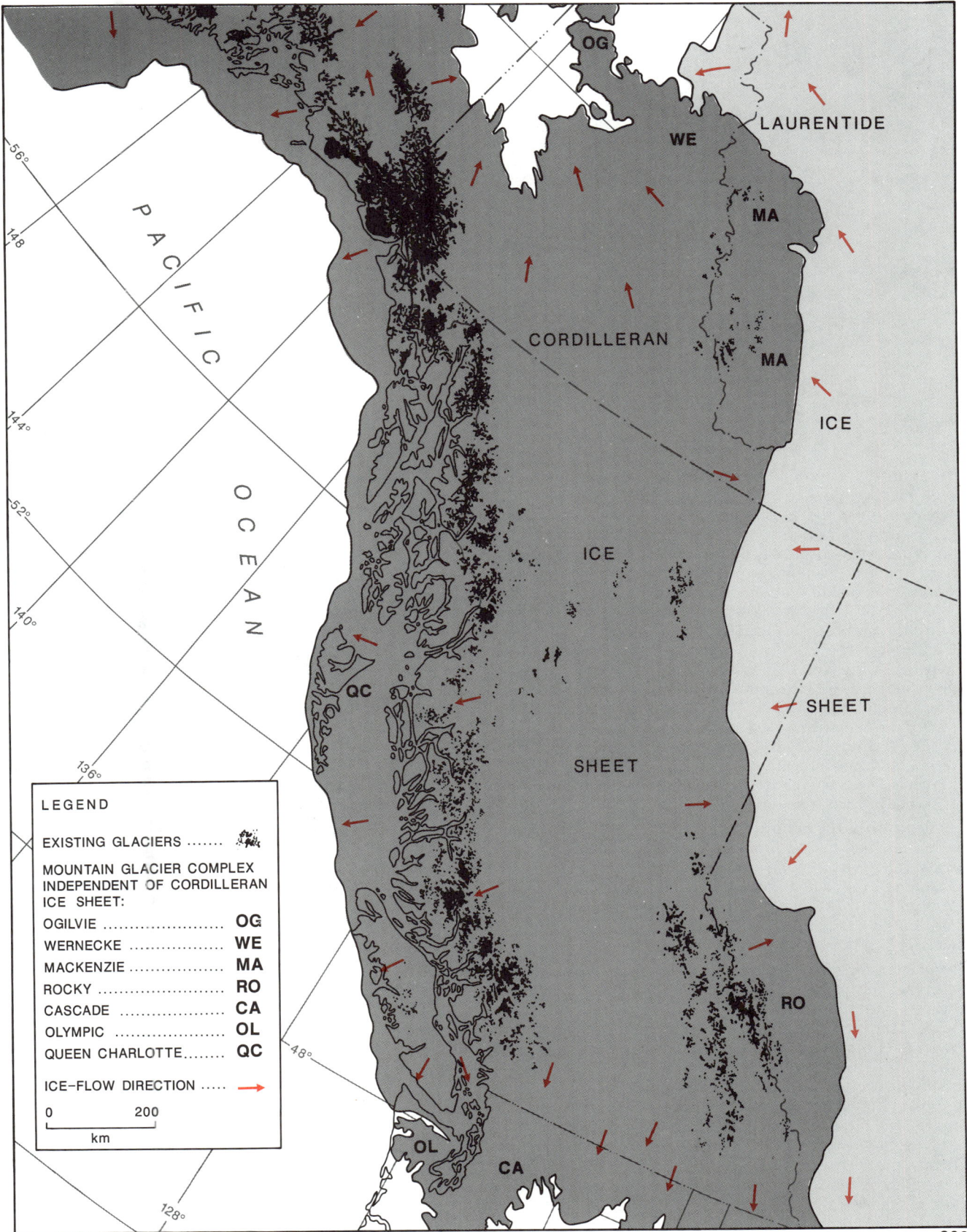

Figure 12.1. The Cordilleran glacier complex at its maximum extent. Some mountain ranges in peripheral areas supported valley glaciers that were independent or semi-independent of the Cordilleran Ice Sheet. Extent of glaciation, in part, from Crandell (1965), Lemke et al. (1965), Richmond et al. (1965), Prest et al. (1968), Hamilton and Thorson (1983), Porter et al. (1983), and Prest (1984).

CHAPTER 12

Figure 12.2. Glaciated landscape near Mount Logan: a modern analogue of the Cordilleran Ice Sheet. Photo by Austin Post, United States Geological Survey.

et al., 1982) and western Vancouver Island apparently escaped glaciation during Late Wisconsinan time.

Radiocarbon dates and paleoecological studies indicate that climatic deterioration and glacier growth marking the onset of the last glaciation began about 25-30 ka BP (Clague, 1980, 1981). Glacier growth was slow at first,

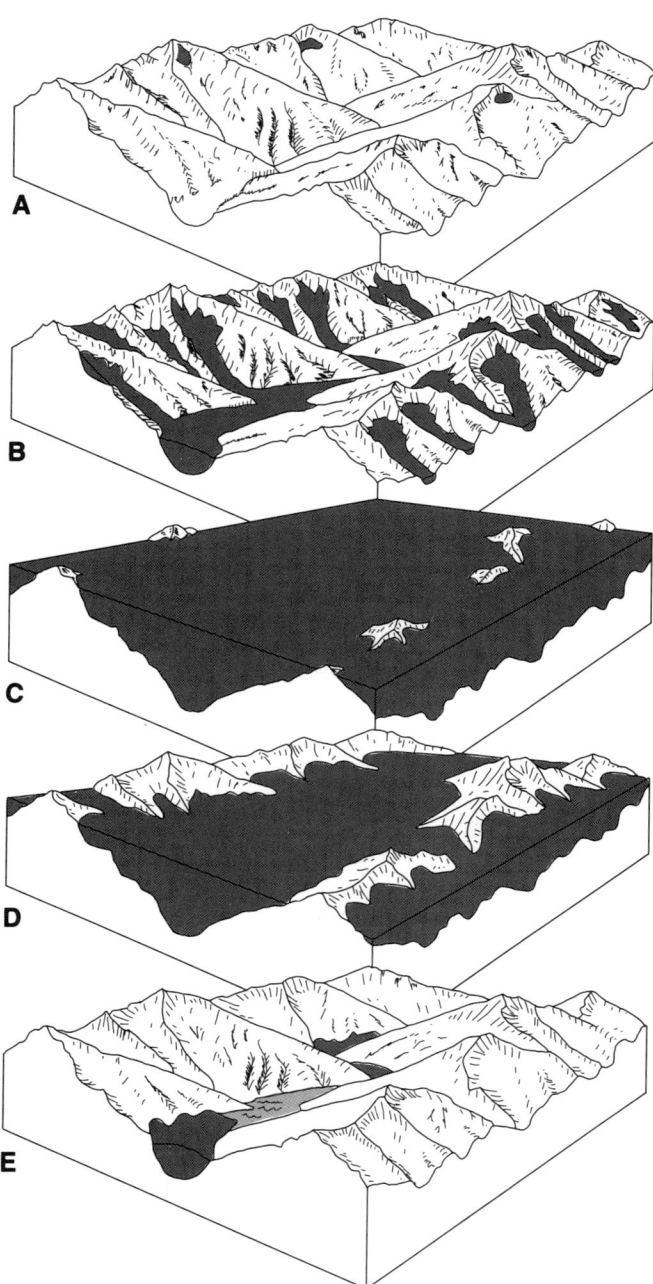

Figure 12.3. Growth and decay of the Cordilleran Ice Sheet. A. Mountain area at the beginning of a glaciation. B. Development of a network of valley glaciers. C. Coalescence of valley and piedmont lobes to form an ice sheet. D. Decay of the ice sheet by downwasting; upland areas are deglaciated before adjacent valleys. E. Residual dead ice masses confined to valleys.

penultimate glaciation in this region has been assigned an Early Wisconsinan age mainly on inferential grounds. Because deposits of this glaciation are very poorly dated, this age assignment is here considered to be provisional; an Illinoian age cannot be ruled out. Drift of the penultimate glaciation is present at the surface beyond Late Wisconsinan glacial limits in the central Yukon (Fig. 12.5; Bostock, 1966; Hughes et al., 1969) and in the Rocky Mountain Foothills (Rutter, 1972, 1977; Alley, 1973; Roed, 1975; Stalker and Harrison, 1977; Jackson, 1980). It also underlies Middle and Late Wisconsinan sediments in some valleys and coastal lowlands in British Columbia and the Yukon (Denton and Stuiver, 1967; Fulton and Smith, 1978; Hicock and Armstrong, 1983; Hicock, 1984). This drift is similar in character and complexity to drift of Late Wisconsinan age, indicating a similar pattern of glaciation. However, ice cover apparently was more extensive during the earlier of these two episodes.

Deposits of the penultimate glaciation are older than the radiocarbon dating limit, but, by how much is uncertain. Middle Wisconsinan nonglacial sediments in British Columbia have yielded radiocarbon dates as old as 58 800 + 2900/-2100 BP (QL-195; Armstrong and Clague, 1977; Clague, 1980), consequently the penultimate glaciation in the Canadian Cordillera is older than 59 ka BP.

Glacial deposits and landforms of the last glaciation are present at or near the surface in almost all parts of British Columbia and the southern Yukon (Fig. 12.6). These materials and landforms record in detail the pattern of ice flow (Prest et al., 1968), and further indicate that all of the Cordillera was covered by ice at the Late Wisconsinan glacial maximum, with the exception of the far north, some fringing areas on the east and west, and scattered nunataks (Fulton, 1971; Clague, 1981; Clague et al., 1989). Along the east side of the Cordillera, ice flowed from the central and northern Rocky Mountains onto the Interior Plains and coalesced with the Laurentide Ice Sheet (Rutter, 1984). In contrast, glaciers in the Mackenzie Mountains terminated well inside the mountain front at the climax of the last glaciation. Along the western margin of the Cordillera, parts of the Queen Charlotte Islands (Warner

424

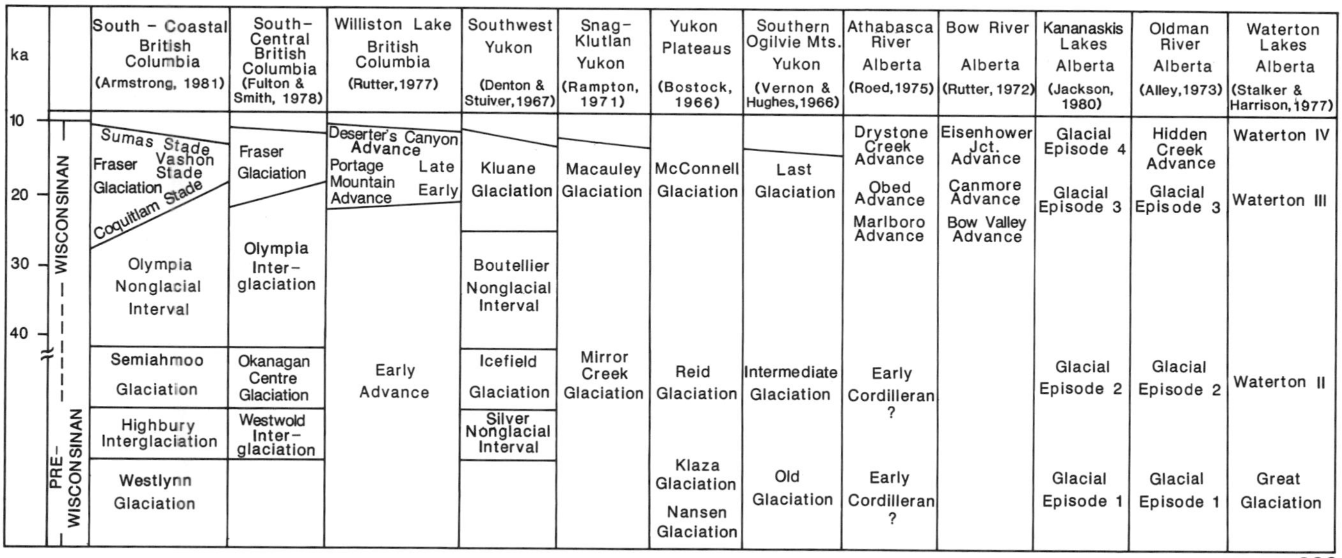

QUATERNARY GLACIATION AND SEDIMENTATION

ka	South - Coastal British Columbia (Armstrong, 1981)	South-Central British Columbia (Fulton & Smith, 1978)	Williston Lake British Columbia (Rutter, 1977)	Southwest Yukon (Denton & Stuiver, 1967)	Snag-Klutlan Yukon (Rampton, 1971)	Yukon Plateaus (Bostock, 1966)	Southern Ogilvie Mts. Yukon (Vernon & Hughes, 1966)	Athabasca River Alberta (Roed, 1975)	Bow River Alberta (Rutter, 1972)	Kananaskis Lakes Alberta (Jackson, 1980)	Oldman River Alberta (Alley, 1973)	Waterton Lakes Alberta (Stalker & Harrison, 1977)
10 —	Sumas Stade / Fraser Vashon Stade / Glaciation Coquitlam Stade	Fraser Glaciation	Deserter's Canyon Advance / Portage Mountain Advance Late / Early	Kluane Glaciation	Macauley Glaciation	McConnell Glaciation	Last Glaciation	Drystone Creek Advance / Obed Advance / Marlboro Advance	Eisenhower Jct. Advance / Canmore Advance / Bow Valley Advance	Glacial Episode 4 / Glacial Episode 3	Hidden Creek Advance / Glacial Episode 3	Waterton IV / Waterton III
20 —												
30 —	Olympia Nonglacial Interval	Olympia Interglaciation		Boutellier Nonglacial Interval								
40 —	Semiahmoo Glaciation	Okanagan Centre Glaciation	Early Advance	Icefield Glaciation	Mirror Creek Glaciation	Reid Glaciation	Intermediate Glaciation	Early Cordilleran ?		Glacial Episode 2	Glacial Episode 2	Waterton II
	Highbury Interglaciation	Westwold Interglaciation		Silver Nonglacial Interval								
	Westlynn Glaciation					Klaza Glaciation / Nansen Glaciation	Old Glaciation	Early Cordilleran ?		Glacial Episode 1	Glacial Episode 1	Great Glaciation

(Left margin: WISCONSINAN / PRE-WISCONSINAN)

Figure 12.4. Quaternary events in the Canadian Cordillera and their proposed correlations. Ages and correlations are provisional; they differ, in some cases, from those of the original authors.

with ice confined to mountain ranges until 20-25 ka BP depending on the locality. Some areas in British Columbia remained ice-free until after 17 ka BP, and the Cordilleran Ice Sheet did not attain its maximum extent in the south until 14-14.5 ka BP (Clague et al., 1980). The ice sheet began to decay shortly after 14 ka BP. Parts of the coastal lowlands of southwestern British Columbia were ice-free by 13 ka BP, and the ice sheet and most satellite glaciers had completely disappeared by 10 ka BP or shortly thereafter (Fulton, 1971; Clague, 1980, 1981). Detailed studies in a few areas, mainly near the periphery of the ice sheet, have shown that glacier growth and decay were complex; at least locally, there were intervals of retreat prior to the climax of this glaciation, and there were stillstands and readvances during deglaciation (Fig. 12.7; Armstrong et al., 1965; Clague, 1975; Rutter, 1977; Hicock and Armstrong, 1981; Saunders et al., 1987).

RECORD OF NONGLACIAL PERIODS

Glaciations alternated with periods during which ice was restricted largely or entirely to major mountain ranges. The record of early and middle Quaternary nonglacial periods is sparse, and little is known of their ages or environments. However, three periods of speleothem deposition in caves in the Rocky and Mackenzie mountains are thought to correspond to major nonglacial periods, and these have been dated by the Th^{230}-U^{234} method as >350, 275-320, and 185-235 ka BP (Harmon et al., 1977).

The oldest nonglacial period for which there is good stratigraphic evidence is thought to correspond to the Sangamonian Stage (ca. 115-128 ka BP; oxygen isotope stage 5e; Shackleton and Opdyke, 1973). However, most nonglacial sediments in the Canadian Cordillera that are assigned to the Sangamonian Stage are poorly dated; some, in fact, may be older. Paleoecological studies of sediments deposited during this nonglacial period indicate a climate

at times as warm as, or warmer than, the present (Hicock and Armstrong, 1983; Hicock, 1984; Alley and Hicock, 1986); however, cooler conditions apparently prevailed during the later part of this period (Fulton and Smith, 1978).

Late Wisconsinan glaciation was preceded by a nonglacial period that began sometime before 59 ka BP and ended about 25-30 ka BP (Fig. 12.4; Clague, 1980, 1981). Plateaus and lowlands probably were continuously ice-free throughout this period, and sedimentary environments and the geomorphic framework were similar to those of the present. Paleoecological analyses of Middle Wisconsinan nonglacial sediments indicate a climate at times similar to and at times cooler than the present (Clague, 1978; Alley, 1979; Warner et al., 1984; Alley et al., 1986). Middle Wisconsinan sediments underlie drift of the last glaciation and, in general, are not exposed at the surface in the Cordillera.

The present nonglacial period (i.e., Holocene), which spans the last 10 ka, has been a time of adjustment of geomorphic systems to changed environmental conditions resulting from deglaciation (see, for example, Church and Ryder (1972) for a discussion of adjustments of fluvial systems to deglaciation). Climatic changes during the Holocene have been pronounced, although much smaller in magnitude than those that occurred during the Pleistocene. Holocene climatic changes have affected the distribution of plants and animals in some parts of the Cordillera and have caused cirque and valley glaciers to fluctuate (for reviews, see Clague et al., 1989). In general, the coolest and wettest conditions during the last 10 ka occurred during the last few centuries, and it is at this time that most alpine glaciers in at least the southern Cordillera attained their maximum Holocene extent (Mathews, 1951; Luckman and Osborn, 1979; Clague, 1981).

CHAPTER 12

Table 12.1. Selected Quaternary stratigraphic units in the Canadian Cordillera.

Region	Unit	Reference
Middle and Early Pleistocene		
southwest British Columbia	Westlynn Drift	Armstrong, 1975
southwest Yukon	Shakwak Drift	Denton and Stuiver, 1967
central Yukon	Nansen Drift	Bostock, 1966
	Klaza Drift	Bostock, 1966
southwest Alberta	Albertan Till	Stalker and Harrison, 1977
Sangamonian or older		
southwest British Columbia	Muir Point Formation	Hicock and Armstrong, 1983
	Highbury Sediments	
south-central British Columbia	Westwold Sediments	Hicock and Armstrong, 1983
		Fulton and Smith, 1978
Early Wisconsinan or Illinoian		
southwest British Columbia	Semiahmoo Drift	Hicock and Armstrong, 1983
	Muchalat River Drift	
	Older Drift	Howes, 1981
south-central British Columbia	Okanagan Centre Drift	Howes, 1983
east-central British Columbia	Early Advance Drift	Fulton and Smith, 1978
southwest Yukon	Icefield Drift	Rutter, 1977
west-central Yukon	Mirror Creek Drift	Denton and Stuiver, 1967
central Yukon	Reid Drift	Rampton, 1971
	Waterton II Drift	Bostock, 1966
southwest Alberta	Maycroft Till	Stalker and Harrison, 1977
		Alley, 1973
Middle Wisconsinan		
southwest British Columbia	Cowichan Head Formation	Armstrong and Clague, 1977
south-central British Columbia	Bessette Sediments	
		Fulton and Smith, 1978
Late Wisconsinan		
southwest British Columbia	Quadra Sand	Clague, 1976
	Coquitlam Drift	Hicock and Armstrong, 1981
	Vashon Drift	
	Capilano Sediments	Armstrong, 1981
	Fort Langley Formation	Armstrong, 1981
	Sumas Drift	Armstrong, 1981
	Gold River Drift	Armstrong, 1981
	Port McNeill Drift	Howes, 1981
south-central British Columbia	Kamloops Lake Drift	Howes, 1983
east-central British Columbia	Early Portage Mountain Drift	Fulton and Smith, 1978
	Late Portage Mountain Drift	Rutter, 1977
	Deserter's Canyon Till[1]	Rutter, 1977
southwest Yukon	Kluane Drift	Rutter, 1977
west-central Yukon	Macauley Drift	Denton and Stuiver, 1967
eastern Yukon	Hungry Creek Till[2]	Rampton, 1971
central Yukon	McConnell Drift	Hughes et al., 1981
southwest Alberta	Waterton III Drift[1]	Bostock, 1966
	Waterton IV Drift	Stalker and Harrison, 1977
	Ernst Till[1]	Stalker and Harrison, 1977
	Hidden Creek Till	Alley, 1973
	Bow Valley Drift	Alley, 1973
	Canmore Drift	Rutter, 1972
	Eisenhower Junction Drift	Rutter, 1972
	Erratics Train Till[3]	Rutter, 1972
	Marlboro Till	Jackson, 1980
	Obed Till	Roed, 1975
	Drystone Creek Till	Roed, 1975
		Roed, 1975

[1]Stratigraphic position modified from cited work.
[2]Drift of Keewatin provenance.
[3]Drift of mixed Cordilleran and Keewatin provenance.

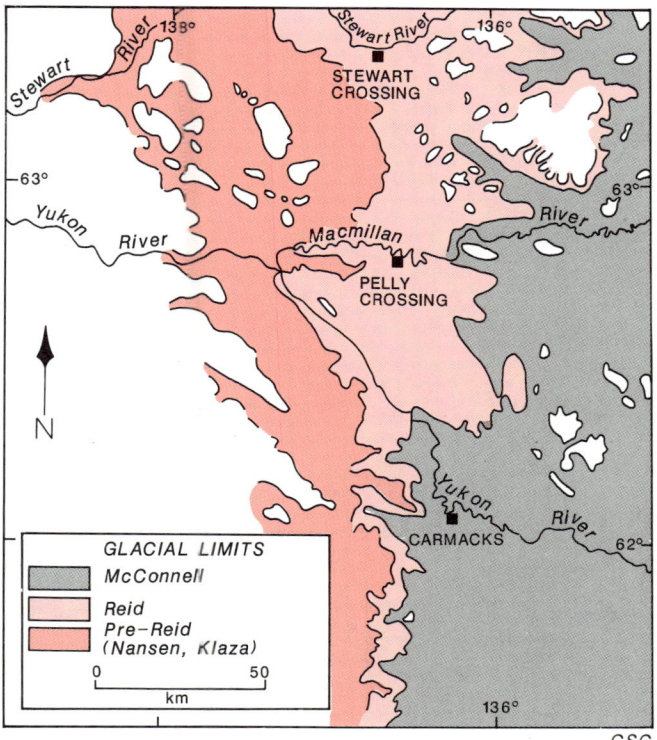

Figure 12.5. Glacial limits in central Yukon Territory. Note that older advances reached farther than younger ones. Adapted from Hughes et al. (1969).

Figure 12.6. Drumlinized plateau surface northeast of Prince George, British Columbia (view east towards Rocky Mountains). These landforms were produced during the last glaciation (Late Wisconsinan); ice flowed from lower right towards upper left. Province of British Columbia photo BC761-70.

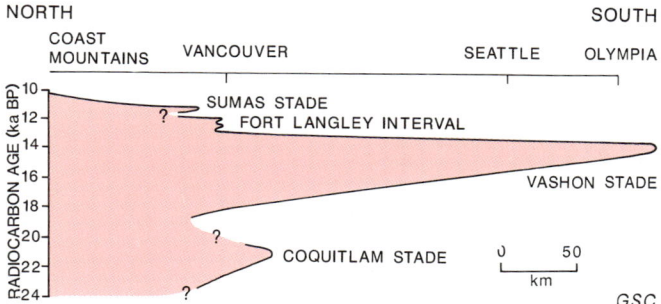

Figure 12.7. Time-distance diagram showing fluctuations of the margin of the Cordilleran Ice Sheet in southwestern British Columbia and northwestern Washington during the last glaciation.

DISTRIBUTION AND CHARACTER OF QUATERNARY SEDIMENTS

The Quaternary stratigraphic record in the Canadian Cordillera is mainly a product of episodic growth and decay of the Cordilleran Ice Sheet (Clague, 1986). Most Quaternary units are of glacial origin and consist of (1) stratified sediments deposited in proglacial and ice-contact environments (glaciofluvial, glaciolacustrine, and glaciomarine sediments), and (2) till deposited subglacially and supraglacially (Fig. 12.8, 12.9). The thickest and most varied glacial sediments occur mainly in valleys and in some coastal lowlands. In these places, the deposits of each glaciation comprise several units laid down in different sedimentary environments. Each of these units may have sharp or gradational contacts with bounding strata, and most are characterized by extremely variable lithofacies. Unconformities within glacial sequences are common and are products of glacial and, to a lesser extent, fluvial erosion.

Glacial sequences are separated from one another by nonglacial sediments or by an unconformity. Where present, nonglacial sediments generally are thin and discontinuous. For example, in south coastal British Columbia, thin Middle Wisconsinan nonglacial sediments, which range in age from >60 to about 25 ka BP, occur between much thicker drift units deposited over periods of hundreds of years to at most several thousand years (Fig. 12.8, 12.9; Armstrong and Clague, 1977). Although there are thick fluvial and lacustrine deposits in some areas, most of these occur high above present-day base levels and probably were deposited during periods of glacier growth and decay (see below). Some, however, may be truly nonglacial in origin. More commonly, unconformities separate major glacial sequences. These define former land surfaces similar in morphology and relief to the present.

Most Quaternary stratigraphic units in the Canadian Cordillera are time-transgressive. This is largely a result of diachronous glacier growth and decay during Pleistocene glaciations. Sedimentation and erosion during periods of glacier growth began first in the mountains and propagated outward onto plateaus, lowlands, and the continental shelf. In contrast, sedimentation during deglacial phases commenced earlier in peripheral glaciated areas than in the core area of the Cordilleran Ice Sheet.

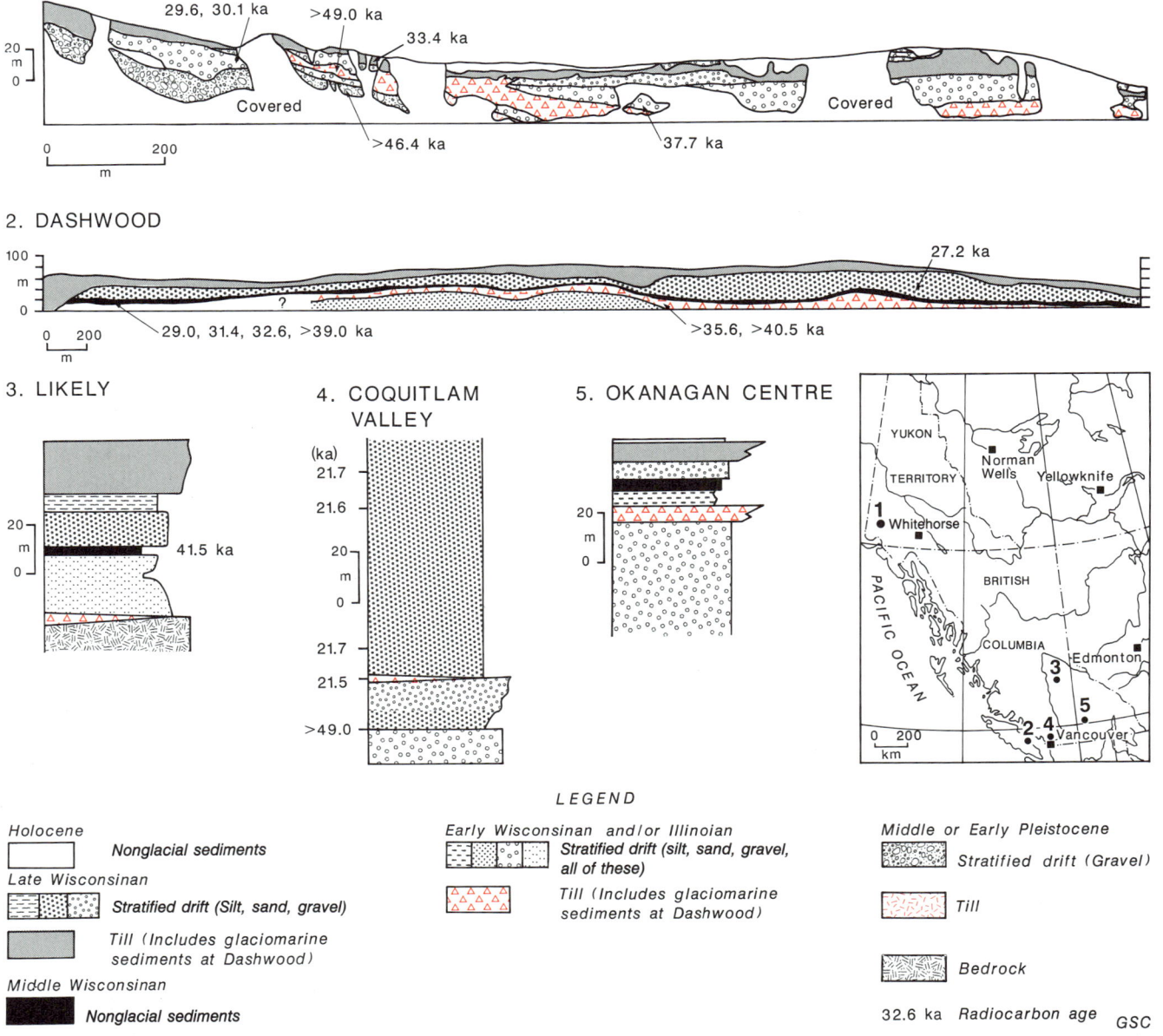

Figure 12.8. Representative Quaternary stratigraphic sections showing the relationships of glacial and nonglacial deposits of Holocene, Wisconsinan, and pre-Wisconsinan age. Sources of information: sections - Fyles (1963), Denton and Stuiver (1967), Fulton and Smith (1978), Clague and Luternauer (1982); radiocarbon dates - Denton and Stuiver (1967) and Clague (1980).

Most terrestrial Quaternary deposits in the glaciated part of the Canadian Cordillera record Wisconsinan and Holocene events. The pre-Wisconsinan stratigraphic record in this region is poor for two reasons: (1) Sangamonian and older deposits are covered by thick Wisconsinan sediments and thus are not well exposed; (2) most of these old deposits were obliterated by glacial erosion during the Wisconsinan Stage. It seems likely, however, that significant early and middle Quaternary sediments underlie parts of the continental shelf at the periphery of the former Cordilleran Ice Sheet (Fig. 12.10; Herzer and Bornhold, 1982; Luternauer and Murray, 1983). In addition, old Quaternary sediments occur near or below present base level in some formerly glaciated valleys and in lowlands bordering the Strait of Georgia. They also are widespread in parts of the Porcupine River basin in unglaciated northern Yukon (Hughes, 1972; Hughes et al., 1972, 1983).

Figure 12.9. Exposure of Quaternary sediments in Coquitlam River valley, British Columbia. Glaciofluvial gravel of Early Wisconsinan or older age (1) is overlain successively by thin Middle Wisconsinan sand and silt (2), upward-coarsening glaciofluvial or fluvial sediments (3), Late Wisconsinan till (4), and Late Wisconsinan gravelly and sandy subaqueous (?) outwash (5). See Figure 12.8 for a sketch of this section. Photo by J.J. Clague (GSC 205437).

QUATERNARY SEDIMENTATION AND EROSION

The framework of Quaternary sedimentary deposits in the Canadian Cordillera is a product of relatively brief intervals of deposition separated by long periods of nondeposition and erosion (Clague, 1986). The thickest and most extensive stratigraphic units were laid down in valleys and coastal lowlands during periods of ice-sheet growth and decay. In comparison, true nonglacial units are thin and very restricted in distribution.

Effects of glaciation on sedimentation and erosion

Climatic deterioration and glacier expansion during the early phase of each glaciation led to an increase in sediment production, especially in alpine areas. Initially, much of this sediment accumulated as outwash in mountain valleys. However, as glaciers advanced out of the mountains, large amounts of this material, as well as colluvial and fluvial sediments deposited during the preceding nonglacial period, were flushed from these staging areas into intermontane valleys and fiords. Streams, unable to transport the large amounts of sediment made available to them, aggraded their valleys, and as a result substantial amounts of outwash accumulated in short periods of time. Loci of deposition shifted as glaciers advanced, thus aggradation occurred at different times in different places. As glaciers continued to expand, they increasingly disrupted the drainage, ponding large, rapidly evolving lakes in which significant quantities of clay, silt, and sand accumulated.

Aggradation on a similar scale also occurred at the close of each major glaciation (Church and Ryder, 1972). At these times, large amounts of sediment were released from wasting ice masses and discharged onto floodplains and into glacial lakes. In addition, newly deposited, unstable drift was transported by running water and mass-wasting processes from slopes into valleys where it accumulated on floodplains and fans. Along the coast, glaciomarine sediments were laid down on isostatically depressed lowlands and in offshore areas (Armstrong, 1981; Conway and Luternauer, 1984; Clague, 1985). As a result of these processes, thick fills accumulated in most interior and mountain valleys, on lowlands bordering the Strait of Georgia, in fiords, and in some offshore basins. Remnants of fills dating to the end of the last glaciation and to the end of the penultimate glaciation are important stratigraphic units in the Canadian Cordillera. They are disproportionately large in size and volume considering the extremely short periods of time in which they accumulated.

The pattern of sedimentation and erosion at glacial maxima, when most of the Cordillera was covered by ice, is more difficult to characterize. Most areas, especially mountains, fiords, and valleys parallel to the direction of ice flow, were eroded by glaciers and their sediment cover stripped away. However, glaciers accomplished little erosion in some areas, and significant remnants of older sediments remained following deglaciation. In addition, many glacially eroded surfaces were covered by till before becoming deglaciated. Till is more widely distributed than the stratified valley and lowland fills deposited during periods of ice-sheet growth and decay, but generally is thinner. It consists of detritus both eroded directly from bedrock and recycled from older Quaternary sediments.

Significant sedimentation has been much more limited during nonglacial periods than during glaciations. The only important nonglacial sedimentation sites on land were large lakes in river valleys, some floodplains, and fans at the mouths of some streams. However, large amounts of sediment accumulated in marine deltas and on the floors of fiords and coastal marine basins.

Early during each nonglacial period, valley fills in most areas were deeply incised by streams (Fig. 12.11; Plate 28). This resulted mainly from a significant reduction in the amount of sediment supplied to the fluvial system as slopes stabilized and became vegetated. In addition, base level in coastal areas was lowered up to 200 m due to an isostatically induced fall in the level of the sea relative to the land. The effects of this base-level lowering probably were propagated slowly inland and may have contributed to valley incision by streams. However, other local base-level controls such as rock ledges and constrictions probably played an equal or more important role in controlling downcutting. Whatever the causes, streams deeply incised deglacial and older fills before achieving quasi-equilibrium at much lower levels. Barring tectonic uplift, subsidence, or renewed glaciation, streams tended to flow at or near these levels for long periods. In the case of the present nonglacial period, many streams in British Columbia were flowing near their present levels within several thousand years of the close of Late Wisconsinan glaciation; subsequent changes in floodplain levels have been relatively minor.

Spatial constraints on sedimentation

Physiography has played a major role in determining patterns of sedimentation in the Canadian Cordillera

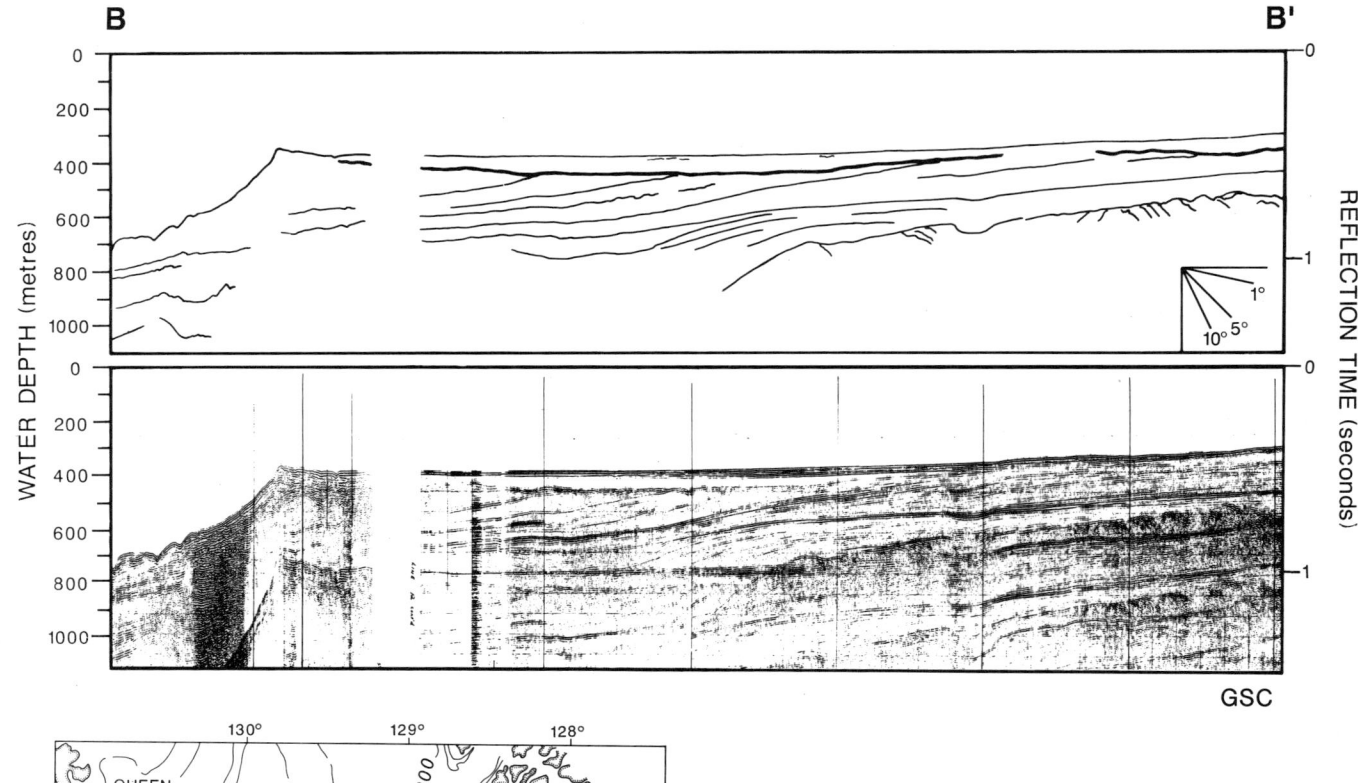

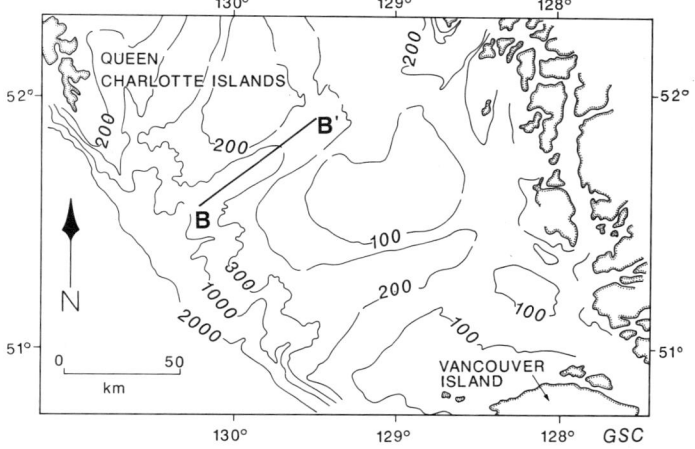

Figure 12.10. Continuous seismic reflection profile across Queen Charlotte Sound showing a thick succession of parallel-bedded sediments overlying folded(?) "bedrock" and unconformably overlain by irregularly stratified sediments. The uppermost unit and perhaps the thick parallel-bedded succession are Quaternary in age. The map shows the location of the profile. Adapted from Luternauer and Murray (1983, Fig. 22 and 24).

during the Quaternary Period. On the timescale of a glacial-interglacial cycle (10-100 ka), the major terrestrial sediment sinks have been large valleys and coastal lowlands. In addition, some plateaus have been repositories of large quantities of till and ice-contact sediments. Mountains and other highlands, although serving as temporary sediment storage sites, have been areas of erosion on the timescale considered here. Sediments accumulated locally in these areas during nonglacial periods, but were removed during glaciations, accompanied by significant erosion of rock surfaces by glaciers. Sediment fills in valleys and lowlands were partially eroded during nonglacial periods and the erosion products transferred to floodplains, lakes, and the sea. The large plateau areas of British Columbia and the southern Yukon were dominantly passive surfaces during nonglacial periods, experiencing neither significant deposition nor erosion, except in the vicinity of streams and in areas of high relief. However, during each glaciation, large amounts of till and glaciofluvial sediments were removed from plateau surfaces by ice before a new mantle of drift was laid down.

The ultimate sink for Cordilleran Quaternary sediments is the Pacific Ocean. Over the long term (1-10 Ma), uplift of the Canadian Cordillera due to lithospheric plate interactions in the northeast Pacific Ocean will ensure the destruction of much of the existing Quaternary cover. This material will be carried to the sea by streams, perhaps after a number of cycles of erosion, transportation, and deposition, and will accumulate in deltas, fiords, on parts of the continental shelf, and on abyssal plains and basins beyond the continental margin. However, if the Cordilleran Ice Sheet should ever form again, some of this sediment will be eroded by grounded ice.

Figure 12.11. Incised Quaternary fill, Fraser River valley west of Clinton, British Columbia (view south). Incision occurred during and shortly after deglaciation at the end of the Pleistocene. Province of British Columbia photo BC1087-46.

CRUSTAL DEFORMATION RELATED TO GLACIATION

Growth and decay of the Cordilleran Ice Sheet triggered isostatic adjustments in the crust and mantle of western Canada. In combination with related eustatic and diastrophic effects, these adjustments produced complex sea-level changes along the coast of British Columbia (Mathews et al., 1970; Clague et al., 1982b; Clague, 1983). Gradual growth of glaciers at the beginning of each glacial cycle led to progressive isostatic depression of the land surface. At first, this depression was localized beneath mountain ranges that served as loci of glacier growth. Lateral movement of material in the asthenosphere away from these areas probably produced outward-migrating forebulges. Initially, shorelines may have fallen as these forebulges passed through coastal areas and as water was transferred from oceans to expanding ice sheets. Eustatic sea-level lowering, in turn, may have caused hydro-isostatic uplift of the continental shelf, resulting in a further lowering of the sea relative to the land. However, as glaciers advanced onto plateaus and lowlands, isostatically depressed areas grew in size and the coastal region began to subside. Eventually, glacio-isostatic depression became dominant in most areas, and the sea rose far above its present level relative to the land.

At the climax of each major glaciation, the entire glaciated Cordillera was isostatically depressed, with areas near the centre of the ice sheet displaced downward more than areas near the periphery. Although the magnitude of isostatic depression at such times is unknown, limits are provided by levels of shorelines that formed at the end of the Pleistocene soon after the Late Wisconsinan Cordilleran Ice Sheet achieved its maximum extent. Relict shorelines are found on the British Columbia coast up to about 200 m above sea level (Fig. 12.12). Taking into account a eustatic sea-level lowering of perhaps 50-100 m at the time the highest shorelines formed, local glacio-isostatic depression apparently was more than 250 m (Clague, 1983). In fact, isostatic depression in areas of thickest ice was probably much greater than this, because the Cordilleran Ice Sheet had decreased in size before the highest shorelines formed, and consequently isostatic rebound probably had commenced earlier.

The elevation of the Late Wisconsinan marine limit on the British Columbia coast varies in relation to distance from the main centres of ice accumulation and, to a lesser extent, to the timing of retreat. In general, the marine limit is highest on the mainland and declines towards the west and southwest (Fig. 12.12; Mathews et al., 1970; Clague, 1981). Late Pleistocene shorelines on the Queen Charlotte Islands were lower than at present, indicating that glacio-isostatic depression was relatively minor there (Fladmark, 1975; Clague et al., 1982a,b).

Isostatic adjustments during periods of deglaciation were opposite in direction to those that occurred during periods of ice-sheet growth. Isostatic uplift in most coastal areas was greater than the coeval eustatic rise, thus the sea fell from its highest shorelines as deglaciation progressed. The sea fell rapidly relative to the land, and within a few thousand years isostatic uplift was largely complete (Mathews et al., 1970; Clague, 1983).

Uplift occurred at different times along the British Columbia coast during deglaciation due to diachronous retreat of the Cordilleran Ice Sheet. In general, regions that were deglaciated first rebounded earlier than those deglaciated at a later date. For example, the sea had fallen to its present level at Victoria by about 11.5 ka BP, about 1.5 ka after initial deglaciation of the area. In contrast, the sea was at its upper limit at Kitimat about 11 ka BP when that area was first deglaciated; most isostatic uplift there occurred between 10.5 and 9 ka BP. Such data indicate that there were substantial regional differences in isostatic response to deglaciation and show that the crust deformed in a complex nonuniform manner as a result of the stresses imposed upon it.

Even within a region, the sea probably did not fall uniformly relative to the land at the end of each glaciation. There are at least three reasons for this:(1) the rate and direction of eustatic change varied as deglaciation progressed; (2) the rate of isostatic uplift also varied due to local stillstands and readvances of glaciers; and (3) some vertical displacements may have occurred instantaneously and sporadically along faults during earthquakes.

The record of crustal movements on the Queen Charlotte Islands at the close of the last glaciation is different from that for other parts of the British Columbia coast (Clague et al., 1982a,b). Shorelines on the Queen Charlottes were lower than at present from at least 16 ka BP until 9.5-10 ka BP, whereas on the mainland coast directly to the east, shorelines were higher during deglaciation than they are today. The period of low sea levels on the Queen Charlotte Islands was followed by a marine transgression

CHAPTER 12

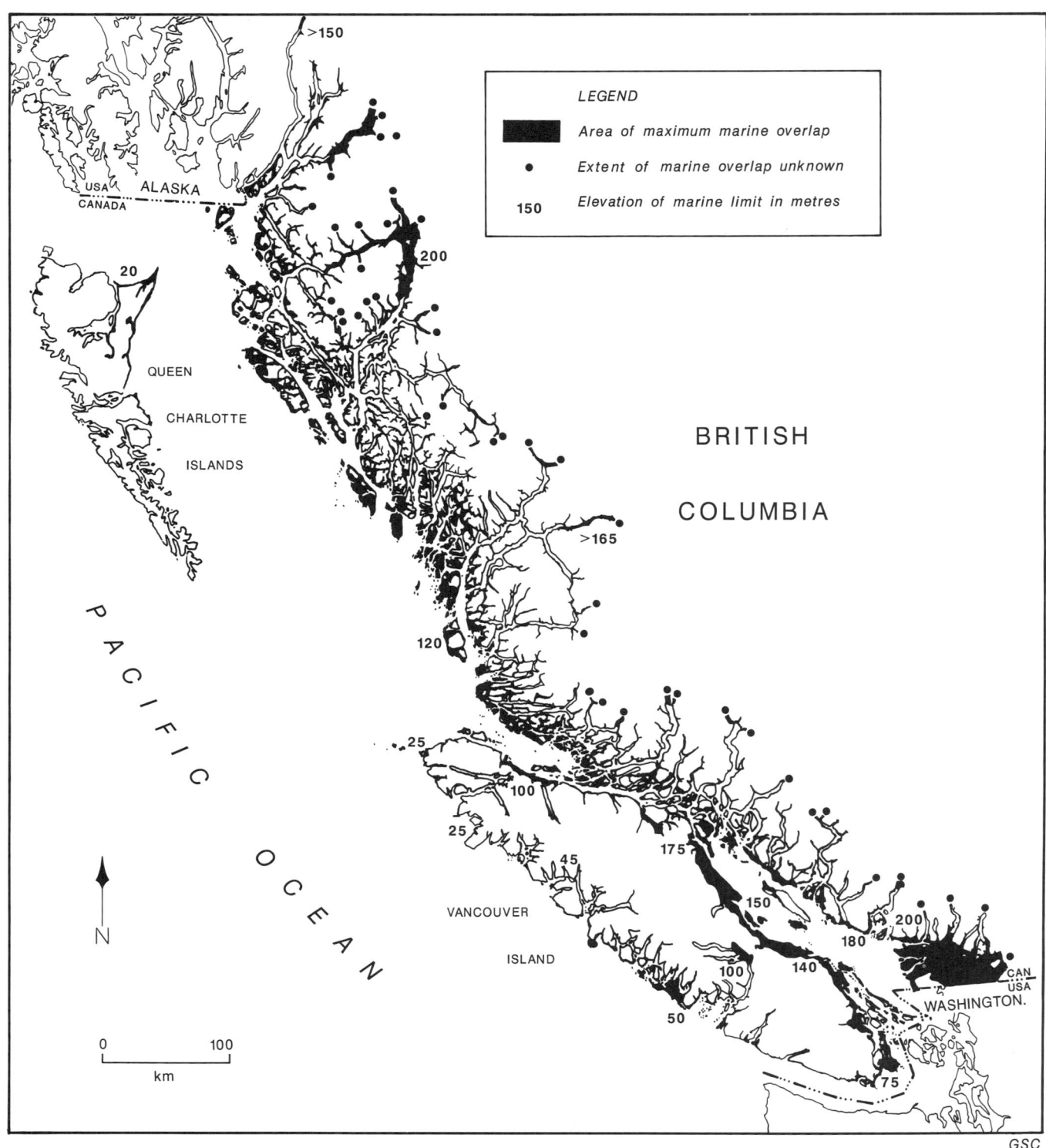

Figure 12.12. Extent of marine transgression in British Columbia during latest Pleistocene and early Holocene time. From Clague (1981, Fig. 5).

that culminated about 7.5-8.5 ka BP when shorelines in some areas were about 15 m higher than at present.

The opposing character of sea-level changes on the Queen Charlotte Islands and the adjacent mainland is best explained in terms of differing ice loads and forebulge migration during deglaciation. Late Pleistocene shorelines in the former area were low because ice loads there were insufficient to depress the crust below lowered eustatic water levels prevailing at that time. As the Cordilleran Ice Sheet thinned and retreated at the end of the Pleistocene, matter flowed in the asthenosphere from peripheral glaciated areas towards the centre of the ice sheet. As a result, mantle material was depleted beneath the Queen Charlotte Islands, the crust subsided, and lowlands were transgressed by the sea. In contrast, eastward forebulge migration beneath the mainland coast caused uplift and a marine regression.

REFERENCES

Alley, N.F.
1973: Glacial stratigraphy and the limits of Rocky Mountain and Laurentide ice sheets in southwestern Alberta, Canada; Bulletin of Canadian Petroleum Geology, v. 21, p. 153-177.
1979: Middle Wisconsin stratigraphy and climatic reconstruction, southern Vancouver Island, British Columbia; Quaternary Research, v. 11, p. 213-237.

Alley, N.F. and Hicock, S.R.
1986: The stratigraphy, palynology, and climatic significance of pre-middle Wisconsin Pleistocene sediments, southern Vancouver Island, British Columbia; Canadian Journal of Earth Sciences, v. 23, p. 369-382.

Alley, N.F., Valentine, K.W.G., and Fulton, R.J.
1986: Paleoclimatic implications of middle Wisconsinan pollen and a paleosol from the Purcell Trench, south central British Columbia; Canadian Journal of Earth Sciences, v. 23, p. 1156-1168.

Armstrong, J.E.
1975: Quaternary geology, stratigraphic studies and revaluation of terrain inventory maps, Fraser Lowland, British Columbia (92 G/1, 2, and parts of 92 G/3, 6, 7, and H/4); in Report of Activities, Part A, April to October 1974; Geological Survey of Canada, Paper 75-1A, p. 377-380.
1981: Post-Vashon Wisconsin glaciation, Fraser Lowland, British Columbia; Geological Survey of Canada, Bulletin 322, 34 p.

Armstrong, J.E. and Clague, J.J.
1977: Two major Wisconsin lithostratigraphic units in southwestern British Columbia; Canadian Journal of Earth Sciences, v. 14, p. 1471-1480.

Armstrong, J.E., Crandell, D.R., Easterbrook, D.J., and Noble, J.B.
1965: Late Pleistocene stratigraphy and chronology in southwestern British Columbia and northwestern Washington; Geological Society of America Bulletin, v. 76, p. 321-330.

Bostock, H.S.
1966: Notes on glaciation in central Yukon Territory; Geological Survey of Canada, Paper 65-36, 18 p.

Church, M.A. and Ryder, J.M.
1972: Paraglacial sedimentation:consideration of fluvial processes conditioned by glaciation; Geological Society of America Bulletin, v. 83, p. 3059-3072.

Clague, J.J.
1975: Late Quaternary sediments and geomorphic history of the Southern Rocky Mountain Trench, British Columbia; Canadian Journal of Earth Sciences, v. 12, p. 595-605.
1976: Quadra Sand and its relation to the late Wisconsin glaciation of southwest British Columbia; Canadian Journal of Earth Sciences, v. 13, p. 803-815.
1978: Mid-Wisconsin climates of the Pacific Northwest; in Current Research, Part B, Geological Survey of Canada, Paper 78-1B, p. 95-100.
1980: Late Quaternary geology and geochronology of British Columbia.Part 1: radiocarbon dates; Geological Survey of Canada, Paper 80-13, 28 p.
1981: Late Quaternary geology and geochronology of British Columbia.Part 2: summary and discussion of radiocarbon-dated Quaternary history; Geological Survey of Canada, Paper 80-35, 41 p.

Clague, J.J. Cont.
1983: Glacio-isostatic effects of the Cordilleran Ice Sheet, British Columbia, Canada; in Shorelines and Isostasy, D.E. Smith and A.G. Dawson (ed.), Academic Press, London, p. 321-343.
1985: Deglaciation of the Prince Rupert-Kitimat area, British Columbia; Canadian Journal of Earth Sciences, v. 22, p. 256-265.
1986: The Quaternary stratigraphic record of British Columbia - evidence for episodic sedimentation and erosion controlled by glaciation; Canadian Journal of Earth Sciences, v. 23, p. 885-894.

Clague, J.J. and Luternauer, J.L.
1982: Excursion 30A:late Quaternary sedimentary environments, southwestern British Columbia; International Association of Sedimentologists, 11th International Congress on Sedimentology (Hamilton), Field Excursion Guide Book, 167 p.

Clague, J.J., Armstrong, J.E., and Mathews, W.H.
1980: Advance of the late Wisconsin Cordilleran Ice Sheet in southern British Columbia since 22,000 yr B.P.; Quaternary Research, v. 13, p. 322-326.

Clague, J.J., Mathews, R.W., and Warner, B.G.
1982a: Late Quaternary geology of eastern Graham Island, Queen Charlotte Islands, British Columbia; Canadian Journal of Earth Sciences, v. 19, p. 1786-1795.

Clague, J.J., Harper, J.R., Hebda, R.J., and Howes, D.E.
1982b: Late Quaternary sea levels and crustal movements, coastal British Columbia; Canadian Journal of Earth Sciences, v. 19, p. 597-618.

Clague, J.J., Hughes, O.L., Jackson, L.E., Jr., MacDonald, G.M., Mathews, W.H., Matthews, J.V., Jr., Rutter, N.W., and Ryder, J.M.
1989: Quaternary geology of the Canadian Cordillera; Chapter 1 in Qua-ternary Geology of Canada and Greenland, R.J. Fulton (ed.), Geological Survey of Canada, Geology of Canada, no. 1 p. 15-96 (also Geo-logical Society of America, The Geology of North America, v. K-1).

Conway, K.W. and Luternauer, J.L.
1984: Longest core of Quaternary sediments from Queen Charlotte Sound: preliminary description and identification; in Current Research, Part A, Geological Survey of Canada, Paper 84-1A, p. 647-649.

Crandell, D.R.
1965: The glacial history of western Washington and Oregon; in The Quaternary of the United States, H.E. Wright Jr. and D.G. Frey (ed.), Princeton University Press, Princeton, p. 341-353.

Davis, N.F.G. and Mathews, W.H.
1944: Four phases of glaciation with illustrations from southwestern British Columbia; Journal of Geology, v. 52, p. 403-413.

Denton, G.H. and Armstrong, R.L.
1969: Miocene-Pliocene glaciations in southern Alaska; American Journal of Science, v. 267, p. 1121-1142.

Denton, G.H. and Stuiver, M.
1967: Late Pleistocene glacial stratigraphy and chronology, northeastern St. Elias Mountains, Yukon Territory, Canada; Geological Society of America Bulletin, v. 78, p. 485-510.

Fladmark, K.R.
1975: A paleoecological model for Northwest Coast prehistory; National Museums of Canada, National Museum of Man, Mercury Series, Archaeological Survey of Canada, Paper 43, 328 p.

Flint, R.F.
1971: Glacial and Quaternary geology; John Wiley and Sons, New York, 892 p.

Fulton, R.J.
1967: Deglaciation studies in Kamloops region, an area of moderate relief, British Columbia; Geological Survey of Canada, Bulletin 154, 36 p.
1971: Radiocarbon geochronology of southern British Columbia; Geological Survey of Canada, Paper 71-37, 28 p.
1984: Quaternary glaciation, Canadian Cordillera; in Quaternary Stratigraphy of Canada—a Canadian Contribution to IGCP Project 24, R.J. Fulton (ed.), Geological Survey of Canada, Paper 84-10, p. 39-48.

Fulton, R.J. and Smith, G.W.
1978: Late Pleistocene stratigraphy of south-central British Columbia; Canadian Journal of Earth Sciences, v. 15, p. 971-980.

Fyles, J.G.
1963: Surficial geology of Horne Lake and Parksville map-areas, Vancouver Island, British Columbia; Geological Survey of Canada, Memoir 318, 142 p.

Grove, E.W.
1974: Deglaciation - a possible triggering mechanism for recent volcanism; International Association of Volcanology and Chemistry of the Earth's Interior, Symposium on Andean and Antarctic Volcanology Problems (Santiago), Proceedings, p. 88-97.

Hamilton, T.D. and Thorson, R.M.
1983: The Cordilleran ice sheet in Alaska; in Late-Quaternary Environments of the United States, Volume 1, the Late Pleistocene, S.C. Porter (ed.), University of Minnesota Press, Minneapolis, p. 38-52.

Harmon, R.S., Ford, D.C., and Schwarcz, H.P.
1977: Interglacial chronology of the Rocky and Mackenzie Mountains based upon ^{230}Th-^{234}U dating of calcite speleothems; Canadian Journal of Earth Sciences, v. 14, p. 2543-2552.

Herzer, R.H. and Bornhold, B.D.
1982: Glaciation and post-glacial history of the continental shelf off southwestern Vancouver Island, British Columbia; Marine Geology, v. 48, p. 285-319.

Hicock, S.R.
1984: Southwest British Columbia:Pleistocene chronology, stratigraphy and correlation; in Correlation of Quaternary Chronologies, W.C. Mahaney (ed.), Geo Books, Norwich, p. 479-489.

Hicock, S.R. and Armstrong, J.E.
1981: Coquitlam Drift:a pre-Vashon Fraser Glaciation formation in the Fraser Lowland, British Columbia; Canadian Journal of Earth Sciences, v. 18, p. 1443-1451.
1983: Four Pleistocene formations in southwest British Columbia:their implications for patterns of sedimentation of possible Sangamonian to early Wisconsinan age; Canadian Journal of Earth Sciences, v. 20, p. 1232-1247.

Howes, D.E.
1981: Late Quaternary sediments and geomorphic history of north-central Vancouver Island; Canadian Journal of Earth Sciences, v. 18, p. 1-12.
1983: Late Quaternary sediments and geomorphic history of northern Vancouver Island, British Columbia; Canadian Journal of Earth Sciences, v. 20, p. 57-65.

Hughes, O.L.
1972: Surficial geology of northern Yukon Territory and northwestern District of Mackenzie, Northwest Territories; Geological Survey of Canada, Paper 69-36, 11 p.

Hughes, O.L., Campbell, R.B., Muller, J.E., and Wheeler, J.O.
1969: Glacial limits and flow patterns, Yukon Territory, south of 65 degrees north latitude; Geological Survey of Canada, Paper 68-34, 9 p.Includes Map 1319A.

Hughes, O.L., Rampton, V.N., and Rutter, N.W.
1972: Quaternary geology and geomorphology, southern and central Yukon (northern Canada); 24th International Geological Congress (Montreal), Guidebook, Field Excursion A11, 59 p.

Hughes, O.L., Harington, C.R., Janssens, J.A., Matthews, J.V., Jr., Morlan, R.E., Rutter, N.W., and Schweger, C.E.
1981: Upper Pleistocene stratigraphy, paleoecology, and archaeology of the northern Yukon interior, eastern Beringia.1. Bonnet Plume Basin; Arctic, v. 34, p. 329-365.

Hughes, O.L., van Everdingen, R.O., and Tarnocai, C.
1983: Regional setting—physiography and geology; in Guidebook to Permafrost and Related Features of the Northern Yukon Territory and Mackenzie Delta, Canada, H.M. French and J.A. Heginbottom (ed.), Alaska Division of Geological and Geophysical Surveys, Fairbanks, p. 5-34.

Jackson, L.E., Jr.
1980: Glacial history and stratigraphy of the Alberta portion of the Kananaskis Lakes map area; Canadian Journal of Earth Sciences, v. 17, p. 459-477.

Klassen, R.W.
1978: A unique stratigraphic record of late Tertiary-Quaternary events in southeastern Yukon; Canadian Journal of Earth Sciences, v. 15, p. 1884-1886.
1987: The Tertiary-Pleistocene stratigraphy of the Liard Plain, southeastern Yukon Territory; Geological Survey of Canada, Paper 86-17, 16 p.

Lemke, R.W., Laird, W.M., Tipton, M.J., and Lindvall, R.M.
1965: Quaternary geology of northern Great Plains; in The Quaternary of the United States, H.E. Wright Jr. and D.G. Frey (ed.), Princeton University Press, Princeton, p. 15-27.

Luckman, B.H. and Osborn, G.D.
1979: Holocene glacier fluctuations in the middle Canadian Rocky Mountains; Quaternary Research, v. 11, p. 52-77.

Luternauer, J.L. and Murray, J.W.
1983: Late Quaternary morphologic development and sedimentation, central British Columbia continental shelf; Geological Survey of Canada, Paper 83-21, 38 p.

Mathews, W.H.
1951: Historic and prehistoric fluctuations of alpine glaciers in the Mount Garibaldi map-area, southwestern British Columbia; Journal of Geology, v. 59, p. 357-380.

Mathews, W.H. and Rouse, G.E.
1986: An Early Pleistocene proglacial succession in south-central British Columbia; Canadian Journal of Earth Sciences, v. 23, p. 1796-1803.

Mathews, W.H., Fyles, J.G., and Nasmith, H.W.
1970: Postglacial crustal movements in southwestern British Columbia and adjacent Washington state; Canadian Journal of Earth Sciences, v. 7, p. 690-702.

Naeser, N.D., Westgate, J.A., Hughes, O.L., and Péwé, T.L.
1982: Fission-track ages of late Cenozoic distal tephra beds in the Yukon Territory and Alaska; Canadian Journal of Earth Sciences, v. 19, p. 2167-2178.

Porter, S.C., Pierce, K.L., and Hamilton, T.D.
1983: Late Wisconsin mountain glaciation in the western United States; in Late-Quaternary Environments of the United States, Volume 1, the Late Pleistocene, S.C. Porter (ed.), University of Minnesota Press, Minneapolis, p. 71-111.

Prest, V.K.
1984: Late Wisconsinan glacier complex; Geological Survey of Canada, Map 1584A.

Prest, V.K., Grant, D.R., and Rampton, V.N.
1968: Glacial map of Canada; Geological Survey of Canada, Map 1253A.

Rampton, V.N.
1971: Late Pleistocene glaciations of the Snag-Klutlan area, Yukon Territory; Arctic, v. 24, p. 277-300.

Richmond, G.M., Fryxell, R., Neff, G.E., and Weis, P.L.
1965: The Cordilleran Ice Sheet of the northern Rocky Mountains, and related Quaternary history of the Columbia Plateau; in The Quaternary of the United States, H.E. Wright Jr. and D.G. Frey (ed.), Princeton University Press, Princeton, p. 231-242.

Roed, M.A.
1975: Cordilleran and Laurentide multiple glaciation, west-central Alberta, Canada; Canadian Journal of Earth Sciences, v. 12, p. 1493-1515.

Rutter, N.W.
1972: Geomorphology and multiple glaciation in the area of Banff, Alberta; Geological Survey of Canada, Bulletin 206, 54 p.
1977: Multiple glaciation in the area of Williston Lake, British Columbia; Geological Survey of Canada, Bulletin 273, 31 p.
1984: Pleistocene history of the western Canadian ice-free corridor; in Quaternary Stratigraphy of Canada—a Canadian Contribution to IGCP Project 24, R.J. Fulton (ed.), Geological Survey of Canada, Paper 84-10, p. 49-56.

Saunders, I.R., Clague, J.J., and Roberts, M.C.
1987: Deglaciation of Chilliwack River valley, British Columbia; Canadian Journal of Earth Sciences, v. 24, p. 915-923.

Shackleton, N.J. and Opdyke, N.D.
1973: Oxygen-isotope and palaeomagnetic stratigraphy of equatorial Pacific core V28-238:oxygen isotope temperatures and ice volumes on a 10^5 year and 10^6 year scale; Quaternary Research, v. 3, p. 39-55.
1976: Oxygen-isotope and paleomagnetic stratigraphy of Pacific core V28-239, late Pliocene to latest Pleistocene; in Investigation of Late Quaternary Paleoceanography and Paleoclimatology, R.M. Cline and J.D. Hayes (ed.), Geological Society of America, Memoir 145, p. 449-464.

Souther, J.G., Armstrong, R.L., and Harakal, J.
1984: Chronology of the peralkaline, late Cenozoic Mount Edziza Volcanic Complex, northern British Columbia, Canada; Geological Society of America Bulletin, v. 95, p. 337-349.

Stalker, A.MacS. and Harrison, J.E.
1977: Quaternary glaciation of the Waterton-Castle River region of Alberta; Bulletin of Canadian Petroleum Geology, v. 25, p. 882-906.

Vernon, P. and Hughes, O.L.
1966: Surficial geology, Dawson, Larsen Creek, and Nash Creek map-areas, Yukon Territory (116B and 116C E1/2, 116A and 106D); Geological Survey of Canada, Bulletin 136, 25 p.

Warner, B.G., Clague, J.J., and Mathewes, R.W.
1984: Geology and paleoecology of a mid-Wisconsin peat from the Queen Charlotte Islands, British Columbia, Canada; Quaternary Research, v. 21, p. 337-350.

Warner, B.G., Mathewes, R.W., and Clague, J.J.
1982: Ice-free conditions on the Queen Charlotte Islands, British Columbia, at the height of late Wisconsin glaciation; Science, v. 218, p. 675-677.

Author's address

J.J. Clague
Terrain Sciences Division
Geological Survey of Canada
100 West Pender Street
Vancouver, British Columbia
V6B 1R8

Printed in Canada

Chapter 13

MODERN PLATE TECTONIC REGINE OF THE CONTINENTAL MARGIN OF WESTERN CANADA

Summary
Introduction
Plate motions
   Pacific-America motion
   Juan de Fuca Plate system motions - ridge and margin
Juan de Fuca Ridge system
Pacific-America margin
   Queen Charlotte Fault System
Pacific-America-Juan de Fuca triple junction
   Nootka Fault Zone
Juan de Fuca-America boundary, the subduction zone
   The filled trench
   Margin seismicity
   The subducted slab
   Vertical and horizontal motions of the margin
   Thermal and structural models
America Plate intraplate deformation
References

Chapter 13

MODERN PLATE TECTONIC REGIME OF THE CONTINENTAL MARGIN OF WESTERN CANADA

R.P. Riddihough and R.D. Hyndman

SUMMARY

The western margin of the Cordillera and adjacent offshore areas of western Canada exhibit pronounced tectonic activity through a variety of plate tectonic interactions, including ocean-ridge spreading, transform faulting, and subduction. Interactions occur among three principal plates or plate systems: the Pacific Plate, the America Plate and the intervening small Juan de Fuca Plate system.

The en échelon Juan de Fuca spreading ridge system is the accretionary boundary between the Pacific and Juan de Fuca plates. The ridge system spreads at rates of between 40 and 60 mm/a and is fragmented and complex. Detailed surveys show that the ridge morphology and tectonics are extremely variable. The Juan de Fuca Plate system east of the ridge is equally complex and has apparently responded to varying resistance at the subduction zone along the continental margin by plate breakup, ridge jumping and re-orientation.

The Pacific-America interaction, which extends along the continental margin northwards from Queen Charlotte Sound, is predominantly right-lateral transform at a rate of 50 to 60 mm/a. However, there is evidence for a small component of convergence along the Queen Charlotte Islands that causes underthrusting. To the north, off Dixon Entrance, there appears to be pure transcurrent motion. Farther north in southeast Alaska and the western Yukon, the plate boundary becomes more complex and through studies of seismicity, it seems likely that several faults in the region transfer strike-slip motion along the Fairweather system into thrust motion in the Chugach-Saint Elias system and the Aleutian Trench.

Convergence beneath the western margin of Vancouver Island from plate models is 40 to 45 mm/a. Convergence and underthrusting is evident in compression of margin sediments, uplift, seismicity, heat flow, and volcanism in the Cascade-Garibaldi volcanic chain. The descending plate has been identified by earthquake hypocentre determinations, seismic refraction, seismic reflection, and magnetotellurics, all of which are compatible with gravity models. The nature of the strain at the margin is, however, unclear. There have been no thrust earthquakes identified, but crustal shortening has been measured by repeated surveys on the land near the coast, notably in the Puget Sound area.

Intraplate activity within the British Columbia interior of the America Plate is limited to deformation rates of at least an order of magnitude less than at the plate boundaries. However, the east-west Anahim Volcanic Belt may be associated with the motion of the America Plate over a "hot spot" and the north-south Stikine volcanic belt in northern British Columbia may indicate some form of extensional faulting or rifting associated with Pacific-America relative movement.

INTRODUCTION

Modern plate tectonic interactions in the Canadian Cordillera occur primarily along the continental margin. There, a variety of ocean ridge, transform fault, and subduction zone boundaries occur among three principal plates or plate systems: the Pacific Plate, the America Plate and the small Juan de Fuca Plate system (Fig. 13.1). North of Vancouver Island, the predominantly oceanic Pacific Plate interacts with the America Plate in a largely right-lateral transform along the Queen Charlotte Fault zone. Farther northwards, the boundary splays into the Fairweather-Chatham Strait-Denali Fault systems of southeast Alaska. At least part of the continental shelf and the Alexander Archipelago may be partly coupled to the Pacific Plate. Offshore, the boundary between the Pacific and Juan de Fuca plates is the accretionary Juan de Fuca Ridge system that extends from a point off northern California to the south of the Queen Charlotte Islands. The ridge is highly segmented and is interrupted by several transform faults. The Juan de Fuca Plate system which lies between the ridge and the America Plate appears to be broken into a number of subplates. Off Vancouver Island, this plate system converges northeastwards and subducts beneath the continental margin of the America Plate. The convergence zone is characterized by seismicity, volcanism and contemporary tectonism. Away from the immediate effects of this convergence, the America Plate is generally stable. However, a few tectonically active features such as the Stikine and Anahim volcanic belts, together with minor seismic activity in areas such as the Rocky Mountain Trench area suggest that slow intraplate deformation is taking place.

This chapter begins with a review of estimates of contemporary plate motions in the region, then discusses the tectonic interactions and structure of five parts of the plate system: the Juan de Fuca Ridge system, the Pacific-America transform margin, the Pacific-America-Juan de

---

Riddihough, R. R. and Hyndman, R. D.
1991: Modern plate tectonic regime of the continental margin of western Canada, Chapter 13 in Geology of the Cordilleran Orogen in Canada, H. Gabrielse and C. J. Yorath (ed.); Geological Survey of Canada, Geology of Canada, no. 4, p. 435-455 (also Geological Society of America, The Geology of North America, v. G-2)

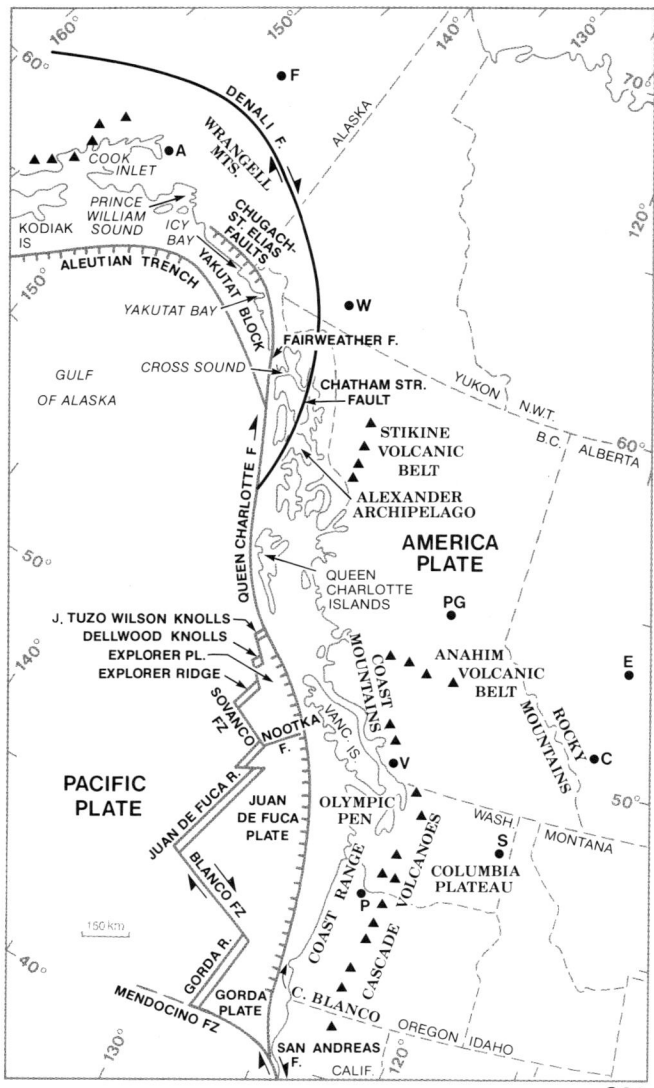

Figure 13.1. Location map showing modern plate tectonic regime of the northeast Pacific. A, Anchorage; F, Fairbanks; W, Whitehorse; PG, Prince George; V, Vancouver; P, Portland; S, Spokane; C, Calgary; E, Edmonton. Triangles represent Miocene to Recent volcanic centres.

Fuca triple junction, the Pacific-America subduction boundary, and the deformation within the America Plate. Previous reviews covering many of these subjects are given in Keen and Hyndman (1979) and Riddihough et al. (1983).

PLATE MOTIONS

Since the original hypothesis of seafloor spreading (e.g. Vine and Matthews, 1963; Vine and Wilson, 1965) and the subsequent development of plate tectonic theory, the determination of contemporary plate motion has generally been based on a combination of marine magnetic-anomaly interpretation (both geometry and reversal chronology), transform fault orientation (from magnetic anomalies, bathymetry and seismicity), earthquake fault-plane solutions, seismicity rates; and "hot spot" trace geometry and chronology. The geological time interval represented by the results of each of these methods of plate-movement analysis varies from a few decades for seismicity, to tens of millions of years for "hot spot" identification. Magnetic-anomaly analysis generally has a time resolution of about 0.5 Ma.

Pacific-America motion

The two major plates in the region, the Pacific and America plates, are among the largest on the globe. However, there are few direct measures of motion between them. The most widely accepted measure of their contemporary (e.g. the past few million years) relative motion is the global plate motion solution of Minster and Jordan (1978) which used a combination of global fault-plane solutions, transform fault orientations and the recent magnetic anomaly spreading rate data. The resultant relative motion in the Queen Charlotte Islands region (RM2) is 56 mm/a at 20° west of north (Fig. 13.2A), with 95% confidence limits of approximately 2 mm/a and 4°. The direction is significantly different from the orientation of the Queen Charlotte Fault in this region, which is close to 40° west of north. Thus a component of convergence along the margin is predicted. The convergence appears to have been initiated at about 5 to 6 Ma with a small change in Pacific-America relative plate motion (Cox and Engebretson, 1985; Yorath and Hyndman, 1983). North of the Queen Charlotte Islands off Dixon Entrance, the Queen Charlotte-Fairweather Fault system curves eastwards to be parallel with the calculated present Pacific-America motion direction. The rate of Pacific-America motion has recently been estimated on much shorter time scales (tens of years), on the San Andreas Fault system by laser ranging experiments and trilateration and triangulation resurveying (e.g. Minster and Jordan, 1984; Savage, 1983), and on both the San Andreas and Queen Charlotte systems using seismic moment release rates (e.g. Hyndman and Weichert, 1983). These short-term local methods generally have much lower resolution than global solutions, but they agree within the measurement accuracy and indicate that the plate models are generally applicable for the past few tens of years. One of the difficulties with the local methods, both along the Canadian continental margin and in the western United States, is that although motion may be primarily on a single fault strand, some motion may be distributed over a considerable area.

"Absolute" motions of the Pacific and America plates are defined as motions relative to the "hot spot" framework. This framework is based on the conclusion that, whereas hot spots (e.g. Hawaii, Iceland etc.) may slowly move with respect to each other, the rates are generally much less than the relative plate motion rates. Motions relative to hot spots can only be obtained on a longer time scale (>5 Ma) and are less accurate than relative motions. A number of absolute Pacific Plate motions have been proposed based on volcanic and island arc chains within the Pacific Plate (e.g. Minster and Jordan, 1978; McDougall and Duncan, 1980; Turner et al., 1980; Clague and Jarrard, 1973). The global plate motion solution AM1-2 of Minster

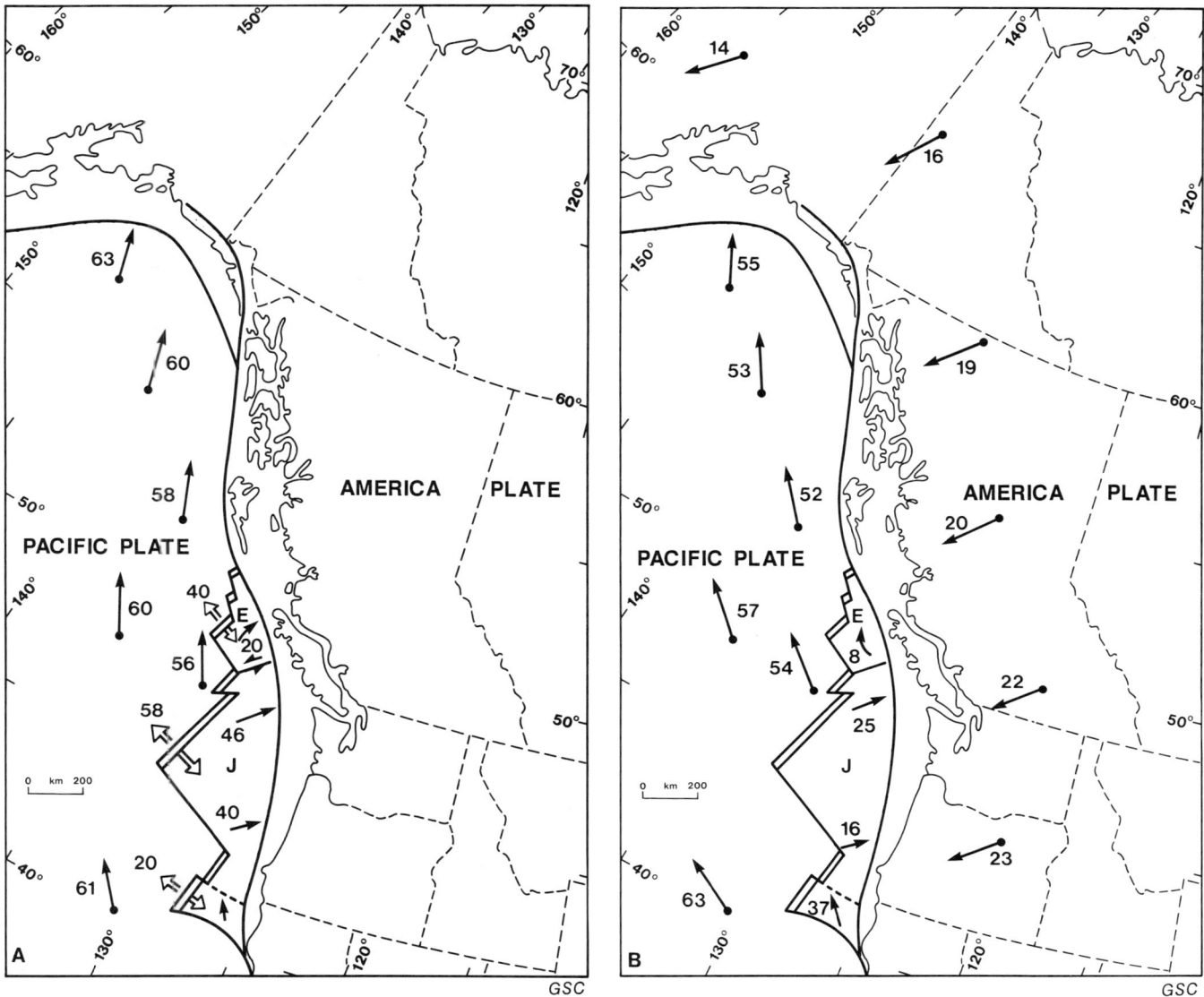

Figure 13.2. Contemporary plate motions in western Canada and southeast Alaska from Minster and Jordan (1978) and Riddihough (1984) (in mm/a). E is Explorer Plate and J is Juan de Fuca Plate - (A) motions relative to the America Plate; (B) motions relative to an assumed fixed "hot spot" framework.

and Jordan (1978), calculates a present northwesterly absolute motion for the Pacific Plate off the Queen Charlotte Islands of 51 mm/a (± 10 mm/a) at 320° (± 11°) (Fig. 13.2B). Absolute motion for the America Plate is considerably less certain as hot spot traces within the plate are few and poorly defined (Coney, 1971, 1977; Duncan, 1982; Morgan, 1981; Suppe et al., 1975). America Plate absolute motion is normally determined by a combination of Atlantic Ocean spreading and absolute motions of the African or other plates (e.g. Duncan, 1982). Plate motion solution AM1-2 of Minster and Jordan (1978) predicts a southwesterly motion of 21 mm/a (± 12 mm/a) at 230° (± 30°) for southwestern British Columbia.

Juan de Fuca Plate system motions - ridge and margin

Relative motions for the Juan de Fuca Plate system have been determined by analysis of the magnetic anomaly data of the Juan de Fuca Ridge and of transform fault data. The motions are described by poles of motion relative to the Pacific Plate. For the Juan de Fuca Plate proper, such a relative motion pole lies near Tahiti (Riddihough, 1984; Nishimura et al., 1984). Spreading rates at the Juan de Fuca Ridge proper vary from 58 to 60 mm/a and predicted convergence rates with North America are northeasterly and vary from 37 mm/a at 044° near Cape Blanco in southern Oregon, to 47 mm/a at 056° off Vancouver Island

(Fig. 13.2A). The estimated errors are ± 7 mm/a and ± 7° (Riddihough, 1984). Absolute motions for the Juan de Fuca Plate also are northeasterly, although at a slower rate of approximately 25 mm/a (± 10 mm/a) at 062° (± 20°) near Vancouver Island (Fig. 13.2B).

Riddihough (1977, 1984) showed that the Explorer Plate (first named by Barr, 1972) must move independently of the Juan de Fuca Plate proper. The calculated motions relative to North America near Brooks Peninsula on northern Vancouver Island are 21 mm/a (± 15 mm/a) at 050° (± 12°). Absolute motions at the same location are small, but are extremely uncertain as the Explorer Plate lies very close to the 95% confidence limit field of its absolute pole of rotation at 50°N, 127.4°W. The Nootka transform fault which separates the Juan de Fuca and Explorer plates (Hyndman et al., 1979) has an estimated present left-lateral motion of approximately 25 mm/a in a direction of approximately 050°.

Other lower accuracy estimates of Juan de Fuca-America Plate convergence are provided by sediment deformation at the foot of the continental slope (von Heune and Kulm, 1973; Barnard, 1978). The estimated rates are lower than those from plate models. However, continental crustal shortening along the margin of northern Washington appears to be of the order of 25 mm/a from repeated surveying data (Savage et al., 1981). This amount combined with sediment deformation rates gives a total convergence comparable to that predicted from the plate motion calculations (see Hyndman and Weichert, 1983).

JUAN DE FUCA RIDGE SYSTEM

The Juan de Fuca Ridge system off western Canada consists of an en échelon series of ridge segments with offsetting transform faults (Fig. 13.1). The central Juan de Fuca Ridge proper is connected to the north through a series of short transform fault zones: the Sovanco fracture zone, the Revere-Dellwood fracture zone, and the Dellwood-Wilson zone. The structure becomes progressively more complex towards the triple junction off northern Vancouver Island. This progression seems to be reflected in the seismicity which shows little or no activity on the sections of the ridge south of 48°N (low seismicity is characteristic of most ocean ridges), but shows increased activity along the spreading segments and transform segments near the triple junction (Rogers, 1983a; Riddihough et al., 1983; Chandra, 1974). The main scatter of epicentres shown on summary plots such as Figure 13.3 is thought to be due to mislocation from land-based observations, but relocations from ocean-bottom seismographs (Hyndman and Rogers, 1981) show that seismicity is closely associated with the plate boundaries. However, towards the triple junction, even these more accurate epicentres do not distinguish between an orthogonal en échelon ridge-transform system and a broader zone of general fracturing.

Spreading rates on the ridge system from magnetic anomaly analyses (e.g. Carlson, 1976; Riddihough, 1977, 1984) show a variation from 60 mm/a near the northern intersection of the Juan de Fuca Ridge with the Sovanco fracture zone, to about 40 mm/a on the northern Explorer Ridge, and perhaps lower at the Dellwood Knolls (Riddihough et al., 1980). On a global scale, these rates are intermediate between fast and slow spreading ridges, and the resulting morphology of the ridge system exhibits

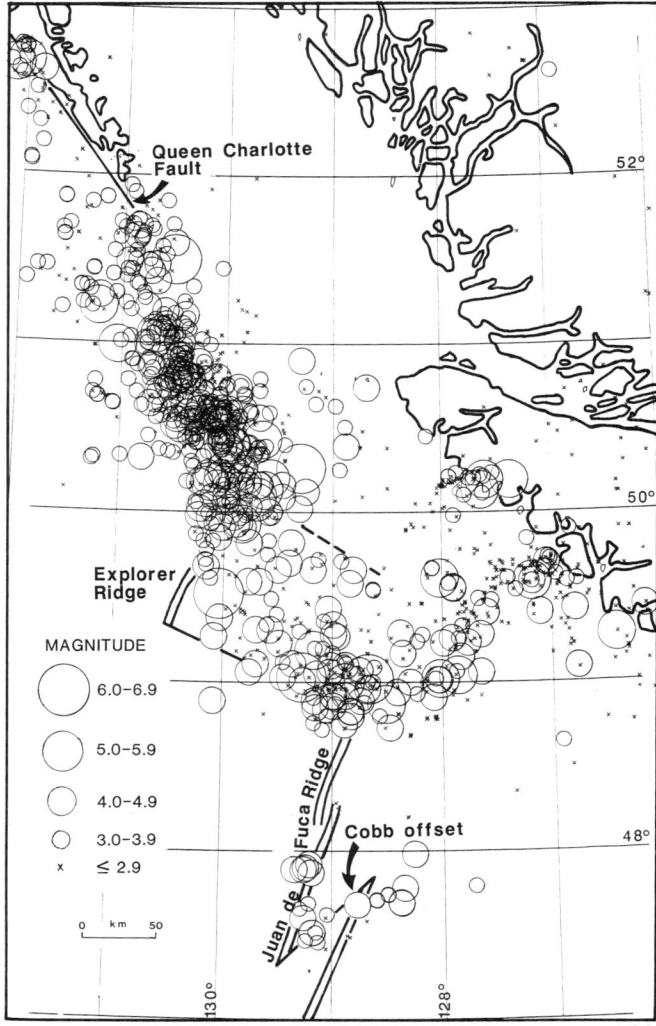

Figure 13.3. Seismicity of the northern Juan de Fuca Ridge and triple junction region. Magnitude of earthquakes (Richter Scale) shown by diameter of circles. Epicentres from 1972 to 1984 from the Geological Survey of Canada Canadian seismicity data files.

features which are characteristic of both. Recent acoustic imaging (side-scan) and swath bathymetry surveys (Malahoff et al., 1984; Davis et al., 1984) show, in detail, morphology varying from symmetrical ridge-trough topography (i.e. Endeavour segment of the northern Juan de Fuca Ridge; Karsten et al., 1984), a deep flat-floored valley with steep rifted walls (i.e. West Valley, northern Juan de Fuca Ridge; Davis and Lister, 1977; Barr and Chase, 1974), a valley with a prominent central ridge (i.e. Explorer Ridge) and scattered areas of magmatism (i.e. the Dellwood and Tuzo Wilson Knolls; Riddihough et al., 1980; Chase et al., 1975). Computer generated 3-dimensional views of the first three examples are shown in Figure 13.4. The genesis of these different spreading ridge structural styles is not yet well understood, but it is generally concluded that they resulted from a combination of adjustments to variable plate motions, particularly reorientations, and temporal and spatial variations in magma supply along the ridge. A

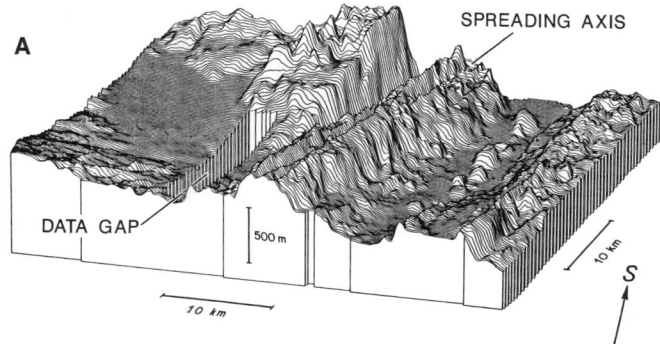

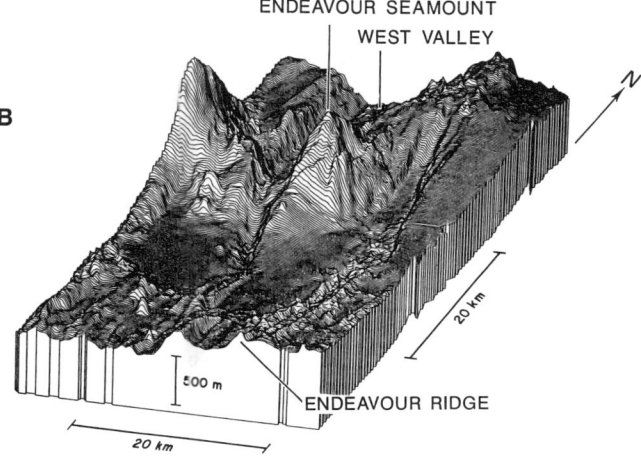

Figure 13.4. Three-dimensional perspective views of the spreading axes on Explorer Ridege (A) and northern Juan de Fuca Ridge (B). Data from Malahoff et al. (1984).

conjugate or overlapping rift system, probably a response to ridge reorientation, occurs toward the northern end of the Juan de Fuca Ridge (Fig. 13.4B) centred around Endeavour Seamount, and may occur in other areas.

The oldest magnetic anomalies on the Juan de Fuca Plate have been dated at about 8 to 9 Ma (Riddihough, 1977), and the record of ridge spreading back to this time can be reconstructed with some accuracy. Atwater and Menard (1970) first noted that during this period the ridge system had rotated clockwise 10 to 15° and that the spreading rates had decreased. They suggested that this rotation which resulted in the ridge becoming more nearly perpendicular to the Queen Charlotte and San Andreas faults, was part of an adjustment to the diminishing size of the Juan de Fuca Plate and its interaction with the continental America Plate. Carlson (1976) and Riddihough (1977) showed that the margin convergence rates have generally decreased during the last 10 Ma.

The calculation of rotation poles (Riddihough, 1984; Wilson et al., 1984) suggests that the changes in the pattern of the spreading ridge segments may have been a result of varying resistance at the subduction zone to the downgoing Juan de Fuca Plate. The youngest, warmest and thus most buoyant part of the plate shows the most resistance to subduction (Atwater, 1970) and produces "pivoting subduction" (Menard, 1978). This effect is clearly seen in the change in Juan de Fuca-America convergence at the time of the separation of the Explorer Plate. Prior to 5 Ma, the youngest crust of the Juan de Fuca Plate with the greatest resistance to subduction lay beneath Vancouver Island. As a consequence, the Juan de Fuca-America pole of rotation lay to the north, and convergence rates increased southwards along the margin. Between 5 and 3 Ma, the northern, youngest portion of the plate broke off to form the independent Explorer Plate. The youngest part of the plate was then at the south end of the Juan de Fuca Plate where the Gorda Ridge was closest to the margin. Motions adjusted so that convergence became slowest at the south end of the plate and increased northwards. The Explorer Plate then began to move about its own local pole so that its convergence at the margin was greatly reduced. There also is evidence of a small piece of the southern end of the Juan de Fuca Plate breaking off (i.e. the Gorda Block; Riddihough et al., 1983), but this appears to have been too small to significantly affect the Juan de Fuca Plate convergence. A model of this suggested plate history is shown in Figure 13.5.

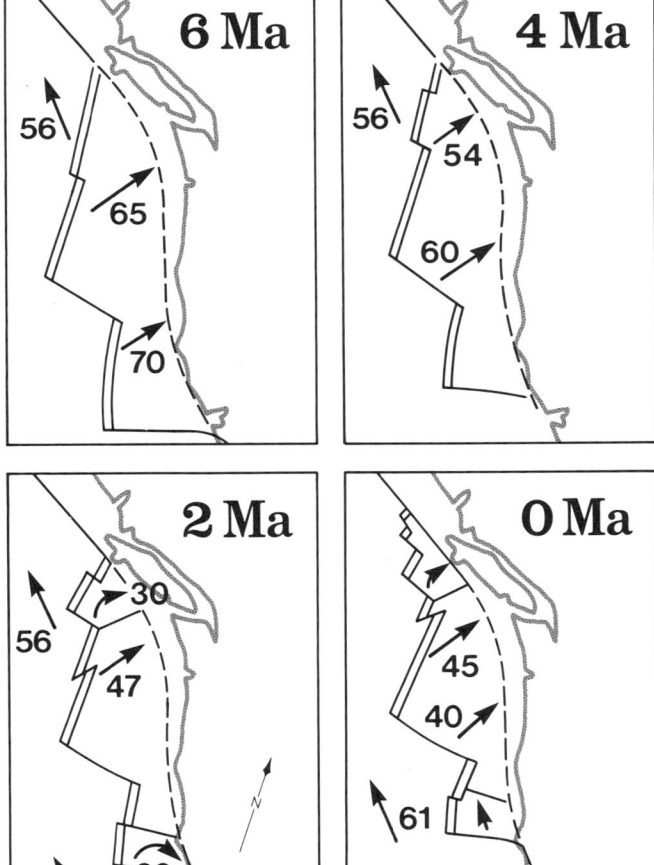

Figure 13.5. Model of the Juan de Fuca Plate motions and break-off of the Explorer Plate. The motions in mm/a (or km/Ma) are relative to the America Plate.

Throughout the complex adjustment of plate breakup, ridge reorientations, changes in spreading rates and poles of motion, the Juan de Fuca Ridge remained approximately the same distance offshore from the continental margin (e.g. Engebretson, 1982). This may have been the result of increasing resistance to subduction as the ridge approached the margin. All plate and ridge rotations also were clockwise in conformity with the general right-lateral shear imposed by the much larger Pacific and America plates (Fig. 13.2). A general picture of continual adjustment and interaction between the plates of the region is emerging, confirming the view that the ridge and spreading orientation adjusts passively in response to plate motions.

Hey (1977) and Wilson et al. (1984) suggested that the adjustment of the ridge to changes in spreading has been achieved through the mechanism of ridge propagation. Hey (1977) showed how such a process results in V-shaped "pseudo-fault" intersections at the ridge, marked as discontinuities in the magnetic anomaly pattern. Such discontinuities in the Juan de Fuca area were noted by Vine (1966) and Silver (1971), and detailed reconstructions were made by Hey and Wilson (1982) and Wilson et al. (1984). A major contemporary "propagator", the Cobb offset (Fig. 13.3) which has apparently moved northward was described by Johnson et al. (1983).

The detailed mapping, analysis and data coverage of the Juan de Fuca Ridge system make it an ideal laboratory for study of oceanic spreading systems. Ridge hydrothermal systems and the resulting sulphide minerals have received particular attention recently. From detailed heat flow, sampling and seismic investigations (e.g. Davis and Lister, 1977; Hyndman et al., 1978; Davis et al., 1980), an understanding of the role of seawater circulation through fractured oceanic crust near the ridge system has developed (e.g. Lister, 1977). Bottom photography and dredging have located a series of active high-temperature hydrothermal vent systems on the ridge with associated sulphide deposits and exotic biological communities (e.g. Karsten et al., 1984; Hammond et al., 1984; Tunnicliffe et al., 1984; Crane et al., 1985). The recognition that such metal-enriched hydrothermal fluids were the source of many sulphide deposits now found on land, and that these fluids may play an important role in seawater chemistry is giving considerable impetus to such investigations.

PACIFIC-AMERICA MARGIN
Queen Charlotte Fault System

The predominantly transform Pacific-America margin extends north from the region of the Tuzo Wilson Knolls, north of Vancouver Island, to the Aleutian Trench system (Fig. 13.1). Along the west coast of the Queen Charlotte Islands the margin is prominently lineated with steep scarps and no continental shelf or rise (Fig. 13.6A). From high cliffs onshore, the seafloor drops steeply directly offshore to a terrace, 30 km wide at a depth of approximately 1500 m (Fig. 13.6B). The outer edge of the terrace is similarly linear and descends steeply to depths of 2500 to 3000 m (Chase and Tiffin, 1972; Srivastava et al., 1971; Srivastava, 1973; Chase et al., 1975). Seismic reflection and refraction profiling, heat flow measurements, and gravity interpretation (Hyndman et al., 1982; Horn et al., 1984) suggest that the terrace is composed of folded and indurated sediments up to 5 km thick which overlie

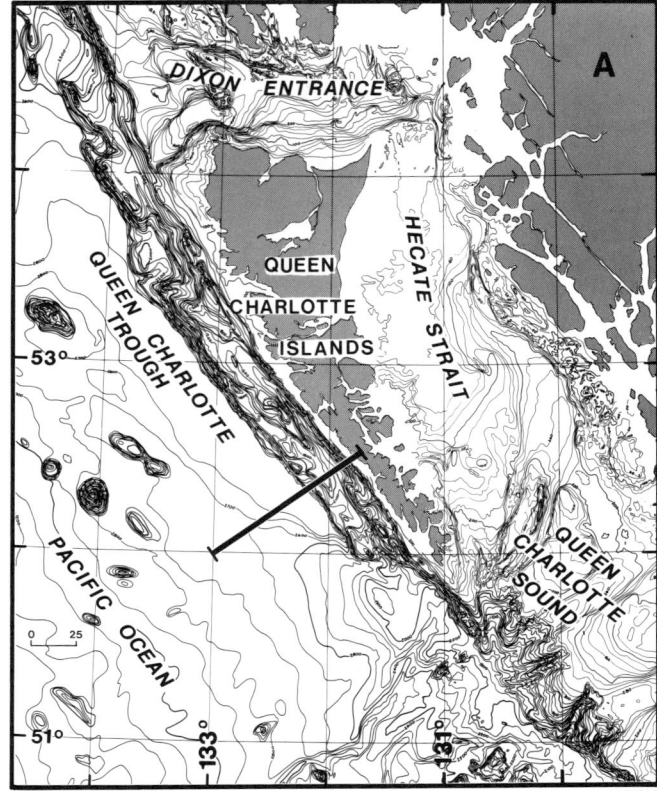

Figure 13.6A. Bathymetry of the continental margin off the Queen Charlotte Islands from Currie et al. (1983). Contour interval for Dixon Entrance, Hecate Strait, and Queen Charlotte Sound is 20 m and for Queen Charlotte Trough is 100 m. **B.** A sample seismic reflection profile across the trough and terrace (Davis and Seemann, 1981). Location of profile is shown in Figure 13.6A.

depressed oceanic crust. Magnetic anomalies of the Pacific Ocean floor (Currie et al., 1983) terminate at the outer scarp of the terrace, although microseismicity indicates that current faulting is primarily along its inner flank (Hyndman and Ellis, 1981). Composite sections across the margin (Horn et al., 1984) suggest relatively recent faulting on both sides of the terrace.

The Queen Charlotte margin exhibits many characteristics of both transform and underthrusting margins. Strike-slip faulting on a vertical fault plane is indicated by most earthquake mechanism solutions (Wickens and Hodgson, 1967; Rogers, 1983a; Bostwick, 1984) and by microearthquake hypocentres (Hyndman and Ellis, 1981; G.C. Rogers, pers. comm., 1985). Convergence and oblique underthrusting are suggested collectively by: the 20° difference between the orientation of the fault trace and the Pacific-America motion determined from global plate models (e.g. Minster and Jordan, 1978); the Queen Charlotte Terrace, which appears to be an accretionary prism of sediments exhibiting compressive structures; a shallow seafloor trough and depression near the margin and associated bathymetric high some 100 km offshore that suggest flexural bending of the oceanic crust; the characteristic parallel low and high gravity anomaly pair (Riddihough, 1979); the heat-flow pattern (Hyndman et al., 1982); and

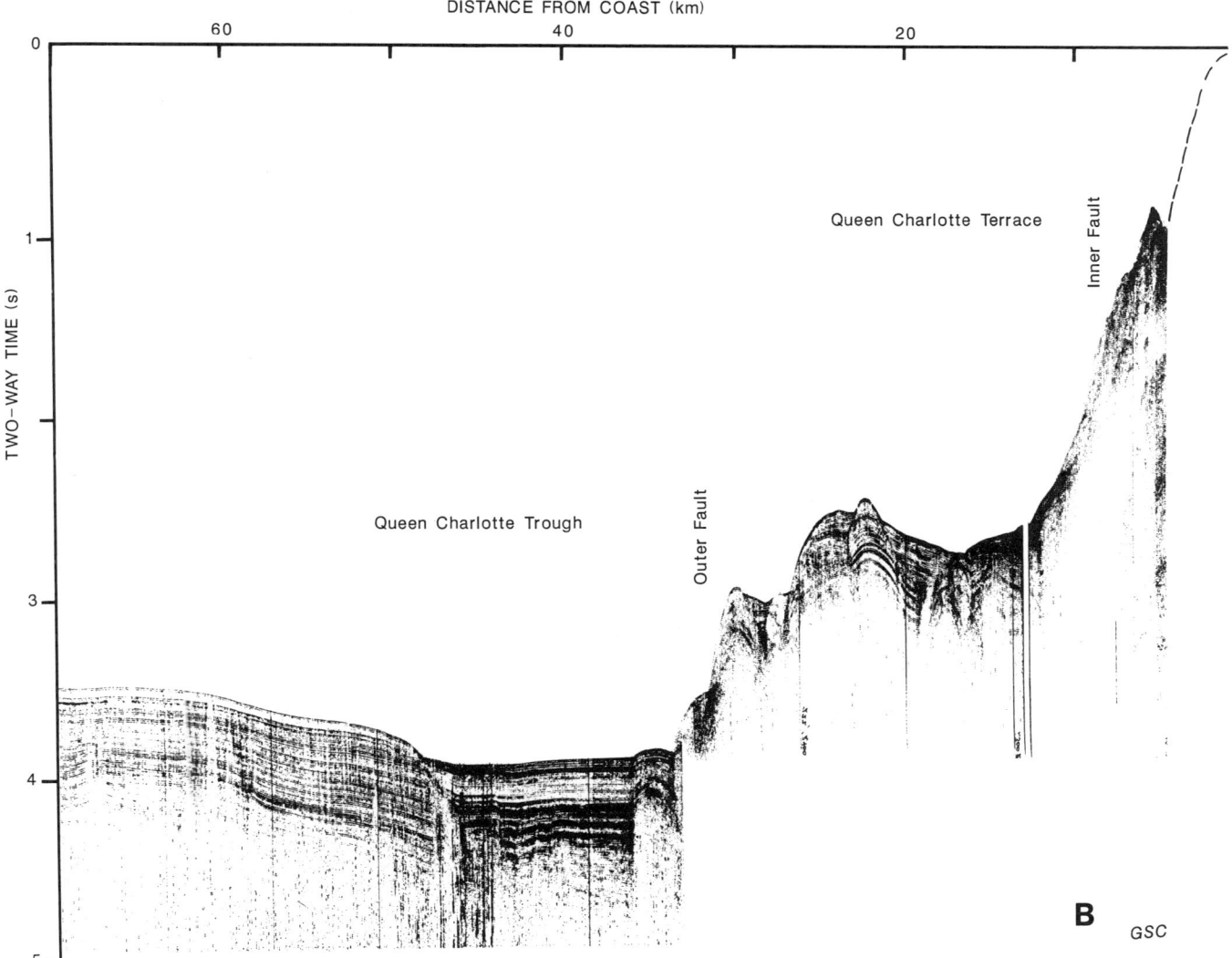

contemporary and recent uplift along the margin of the Queen Charlotte Islands (Sutherland Brown, 1968; Riddihough, 1982a). At the southernmost end of the Queen Charlotte Islands, one earthquake mechanism supports oblique thrusting on a steeply dipping plane (Rogers, 1982). The model required by earthquake data shows the Queen Charlotte Fault as vertical with right-lateral motion parallel with the margin. However, the deeper section of the fault, bounded on both sides by oceanic lithosphere, is being thrust beneath the margin (Hyndman and Ellis, 1981). Presumably, when it is some distance beneath and inboard of the margin, a new fault would be initiated at the edge of the continent. In this model, the piece of oceanic lithosphere beneath the Queen Charlotte Islands is moving inland, orthogonally to the coast. This piece also could be connected to the Winona Block to the south (Yorath and Hyndman, 1983). Alternatively, the space problem posed by convergence may be solved by the periodic translation to the northwest of slivers of the continent or ocean floor. The Queen Charlotte Terrace could be one such sliver now being transported to the north.

Yorath and Hyndman (1983) modelled the continental lithosphere flexure generated by the underthrusting with at least 5 km of uplift along the margin and a smaller amount of flexural subsidence in Hecate Strait (Fig. 13.7). The Hecate Strait subsidence history, estimated from exploratory well data, indicates that the flexure and presumed underthrusting was initiated from 5 to 6 Ma, so that a perpendicular component of motion of 10 mm/a would have resulted in underthrust Pacific Ocean crust extending only some 60 km inboard beneath the margin. This short distance may explain the observed paucity of both major thrust and Benioff-Wadati type earthquakes, and also the paucity of arc-type volcanism to the east. Plate models suggest a change in Pacific-America motion to a more northerly direction beginning about 5 to 6 Ma (Cox and Engebretson, 1985).

North of the Queen Charlotte Islands, opposite Dixon Entrance, Queen Charlotte Terrace becomes wider and less linear. In that area, the global plate solution RM2 of Minster and Jordan (1978) agrees well with the strike of the margin, suggesting that there is at most a minor amount of compression and perhaps even some extension. That the underthrusting does not continue to the north also is suggested by the apparent lack of recent margin uplift northward from Dixon Entrance. Seismic profiles

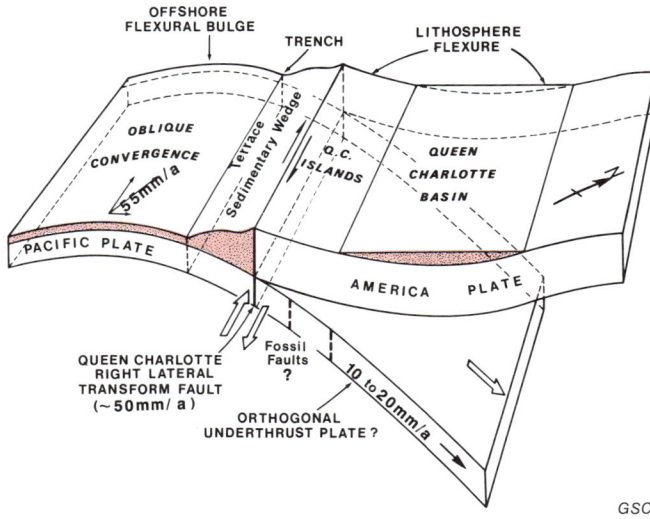

Figure 13.7. Tectonic model of oblique subduction beneath the Queen Charlotte Islands (modified from Yorath and Hyndman, 1983).

across the margin off Dixon Entrance show the basin containing up to 5 km of largely undisturbed sediments and the ocean crust dipping gently toward the continent (von Heune et al., 1978).

North of Dixon Entrance, contemporary seismic activity (Fig. 13.8; Horner, 1983) and seismic reflection profiles indicate that the Queen Charlotte Fault system is continuous with the Chatham Strait Fault (e.g. von Heune et al., 1978) and Fairweather Fault system. From Dixon Entrance, the fault system extends along the continental margin as far as Cross Sound (Fig. 13.1). From Cross Sound, the boundary divides at what is a form of triple junction between the Pacific Plate, the America Plate and the Yakutat Block (Perez and Jacob, 1980; Bruns, 1983). Bruns (1983) considered that this block is currently moving with the Pacific Plate so that the Fairweather Fault represents the Pacific-America Plate boundary. However, Perez and Jacob (1980) calculated seismic slip vectors which suggest 5 to 20 mm/a 030° convergence between the Pacific Plate and the Yakutat Block. Moreover, Plafker et al. (1980) considered the Yakutat Block to be separated from the Pacific Plate by a thrust fault as shown on the Tectonic Assemblage map (Map 1712A, in pocket). Strike-slip motion along the Fairweather Fault from geological data (Plafker et al., 1978) is estimated to be 48 to 58 mm/a for the last 1000 years, and from seismicity data to be of a similar order (R.G. Horner, pers. comm., 1984); both are close to the rate predicted from global plate motion studies.

The Chugach-Saint Elias Fault system is interpreted as the primary collision and thrust zone beneath which the Yakutat Block is being thrust under the America Plate, where it reaches a depth of about 200 km below the Wrangell volcanoes (Perez and Jacob, 1980). Vertical movements associated with this process are very great. Rapid vertical movements also have been measured at Glacier Bay (Hicks and Shofnos, 1965), and although they may be at least partly attributable to glacial unloading (Clark, 1977; Hudson et al., 1982), some of the motion is

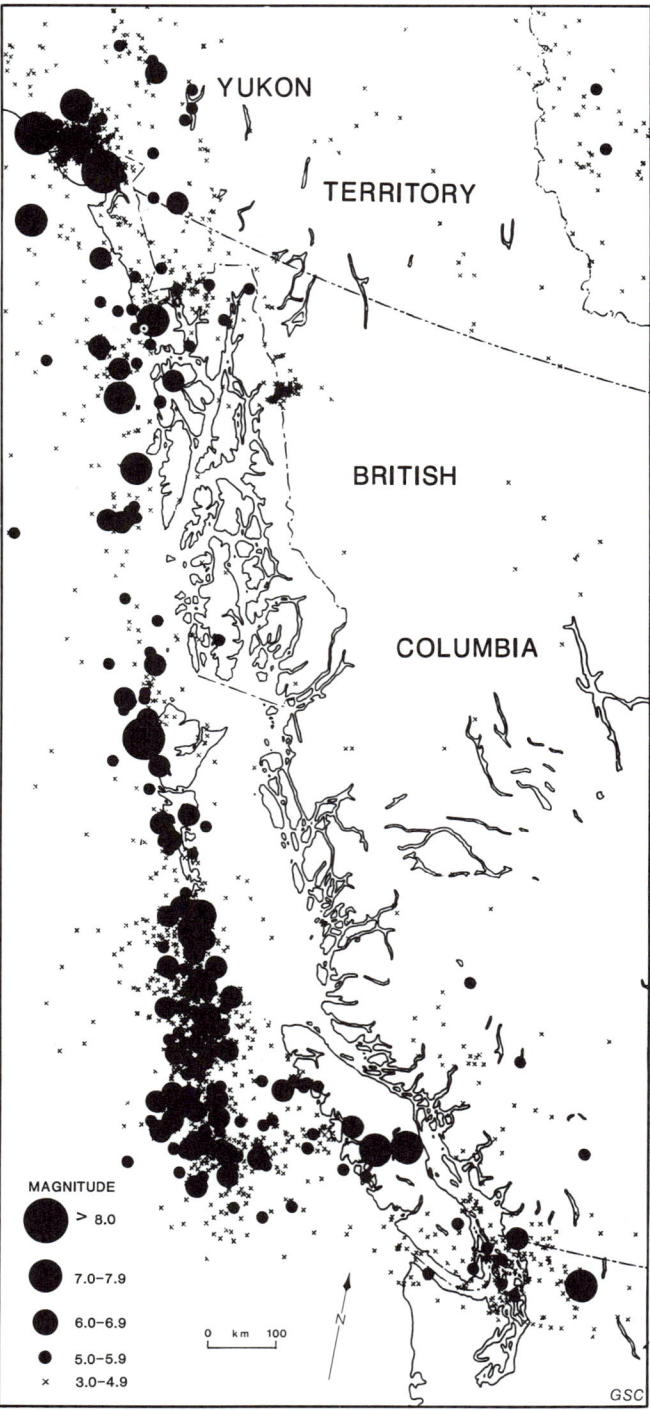

Figure 13.8. Seismicity of the Queen Charlotte-Fairweather and Denali fault systems. Relative magnitude of earthquakes (Richter scale) shown by diameter of circles.

probably tectonic in origin (Horner, 1983; Riddihough, 1982a). Inland of the main plate boundaries, the Denali Fault system in the Yukon exhibits a lower, but significant seismic activity (Fig. 13.8; Rogers, 1976; Horner, 1983). The seismicity data suggest slip rates of up to 1 mm/a although there is no geological evidence showing recent strike-slip movement (Richter and Matson, 1971; Clague, 1979). The connection of this motion to the Pacific-America boundary is unclear because the Chatham Strait Fault shows no current seismic activity. The connection may be made along a hitherto unrecognized zone near Glacier Bay (Horner, 1983).

PACIFIC-AMERICA-JUAN DE FUCA TRIPLE JUNCTION

The Pacific-America-Juan de Fuca triple junction is poorly defined and, as noted by Chase et al. (1975), consists of a complex area of small plates, short ridge segments and faults (Fig. 13.9). Prior to 1 Ma, the junction with a nearly linear single northern Juan de Fuca Ridge seems to have been simpler and to have been stable for at least 5 Ma near Brooks Peninsula (Riddihough, 1977; Davis and Riddihough, 1982). A change in the tectonic style of the continental shelf and slope to the north and south was first noted by Tiffin et al. (1972). The shelf to the northwest is characterized by vertical faults trending parallel with the slope, that to the southeast by compression folds and thrust faults (Clowes et al., 1987). One of the implications of a stable triple junction at this location is that the northwestern edge of the subducting ocean crust would lie beneath Vancouver Island along an approximately northeast trend (Riddihough, 1977). A series of very young extrusive cones, the Alert Bay volcanic belt (2.4 to 7.9 Ma), occurs along this trend and may reflect the plate edge (Bevier et al., 1979). Margin volcanic activity close to a triple junction has been observed elsewhere (e.g. Marshak and Karig, 1977).

The triple junction near Brooks Peninsula began to migrate northward at about 1 Ma, establishing new spreading centres at the Dellwood Knolls and Tuzo Wilson Knolls (Riddihough et al., 1980; Chase, 1977; Davis and Riddihough, 1982). This migration probably involved the detachment, subsidence and tilting of the Winona Block, the youngest part of the Pacific Plate. The uplifted edge of this block forms the Paul Revere Ridge. The depressed Winona Block to the northeast became the Winona Basin which is filled with over 5 km of Pleistocene sediments (Davis and Clowes, 1986). Several causes for the recent change in geometry have been proposed. It may have been due to the regional plate breakup process described earlier, or a response to the influence of a "hot spot" near the Tuzo Wilson Knolls (Chase, 1977). Davis and Riddihough (1982) suggested that the migration of the spreading centres and the detachment of the Winona Block was in response to resistance associated with oblique subduction. In a related model, Yorath and Hyndman (1983) suggested that very oblique underthrusting beneath the Queen Charlotte islands started about 6 Ma. Increasing resistance to this motion resulted in the initiation of the present Queen Charlotte transform fault at about 1 Ma. The Winona Block may be part of the piece of oceanic lithosphere, then detached from the Pacific Plate, which extends beneath the Queen Charlotte margin. The present convergent motion of this block is slow, approximately orthogonal to the North America margin.

Results from SEABEAM swath bathymetry mapping (Malahoff et al., 1984) indicate that recent changes in the Explorer Ridge system, immediately southwest of the Paul Revere Ridge, occurred at about the time as separation of the Winona Block, as suggested by Srivastava et al. (1971). A further complication is evident in the mosaic of SeaMarc II acoustic imaging surveys (Davis et al., 1984). They show a fault trace, affecting the most recent sediments, which connects the Paul Revere fracture zone with the southwest end of the Tuzo Wilson Knolls (approximately aligned with the outer edge of the Queen Charlotte Terrace) and bypassing the Dellwood Knolls. However, clear extensional faults affecting recent sediments at the eastern end of the Dellwood Knolls indicate that spreading occurs at the Dellwood Knolls.

The complexity of the present triple junction and the adjacent margin interaction is instructive for the interpretation of the older margin geological history. The margin of western North America has been characterized by several temporally and spatially distinct triple junctions which separated right-lateral transform from convergent margins (e.g. Engebretson, 1982; Riddihough, 1982b) the effects of these changing plate motion regimes should be evident in the geological record.

Nootka Fault zone

The Nootka Fault zone (Fig. 13.9), separating the Explorer Plate from the main Juan de Fuca Plate, extends from the north end of the Juan de Fuca Ridge to the margin of north-central Vancouver Island. Its presence was first suggested by Barr and Chase (1974) based on earthquake data. Riddihough (1977, 1984) showed that the marine magnetic anomaly data off western Canada requires the independence of the Explorer Plate and left-lateral relative motion between it and the Juan de Fuca Plate during at least the last 4 Ma. Hyndman et al. (1979) found that the fault has a complex structure and history. Its intersection with the margin has probably moved northwestwards along the coast during the past 5 Ma, in response to the general northerly movement of the Juan de Fuca Plate system (Fig. 13.5). The present rate of motion on the fault is estimated to be some 25 mm/a, which results in convergence rates of approximately 20 mm/a to the north and 45 mm/a to the south of the fault zone.

The triple junction formed by the Sovanco and Nootka faults and the Juan de Fuca Ridge (Fig. 13.9) may not be stable. The west end of the Nootka Fault may migrate southward along the ridge some distance, then disappear, and a new fault might be initiated from the northern end of the ridge. This process, plus the rotation of the Explorer Plate about a local pole, may explain the broad complex zone of faults and irregular magnetic anomalies. The triple junction at the margin is a trench-fault-trench type with approximately collinear convergence zones to the north and south. Its configuration may thus remain stable, even though it may have migrated northwestwards along the coast.

JUAN DE FUCA-AMERICA BOUNDARY, THE SUBDUCTION ZONE
The filled trench

The contemporary convergence across the continental margin of southern British Columbia is indicated by a wide range of geological and geophysical data (e.g.

CHAPTER 13

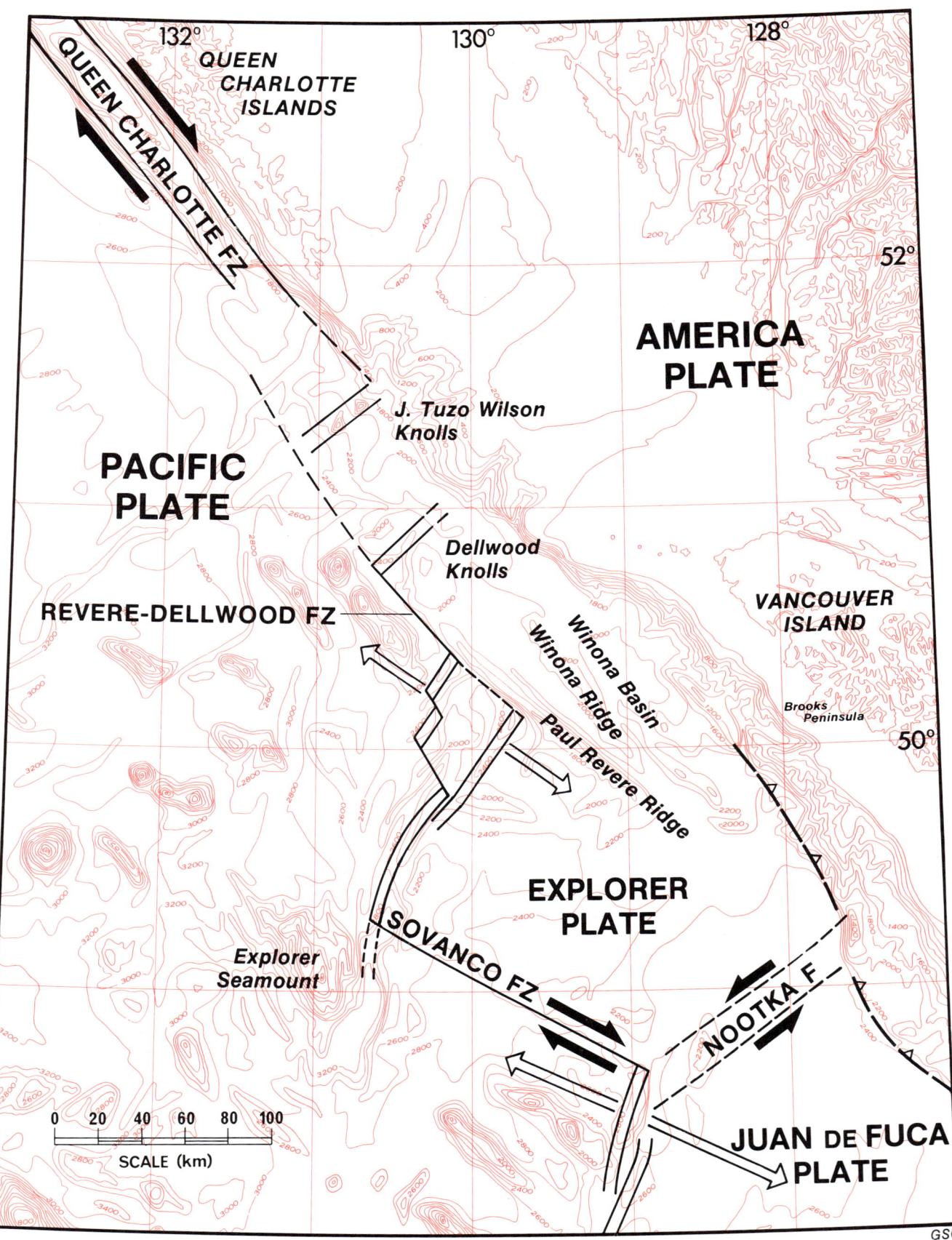

Figure 13.9. Detailed bathymetry of the northern Juan de Fuca-Explorer Ridge system and triple junction region.

Riddihough and Hyndman, 1977). Nonetheless, the absence of a bathymetric trench along the foot of the continental slope beneath which the Juan de Fuca Plate is converging is a major difference between this and most other subduction zones. However, multichannel seismic reflection profiles, seismic refraction studies, and gravity interpretations (e.g. Barr, 1974; Chase et al., 1975; Davis and Seemann, 1981; Ellis et al., 1983; Srivastava, 1973; Davis and Clowes, 1986; Clowes et al., 1987) show that the ocean crust dips towards and beneath the margin at 5° or more, so that there is a pronounced basement trench. Rapid sedimentation during the Pleistocene filled any topographic depression that may have existed earlier (e.g. von Heune and Kulm, 1973; Kulm and Fowler, 1974). Locally in the Winona Basin, sedimentation rates have exceeded 6 km in 1 Ma (Davis and Riddihough, 1982; Davis and Clowes, 1986).

Extending from the foot of the continental slope to the Juan de Fuca Ridge is a blanket of Pleistocene terrigenous sediments forming the Cascadia Basin and Nitinat and Astoria fans (Kulm and Fowler, 1974). The ridge has acted as a sediment barrier, broken only by sea channels through the Sovanco and Blanco fracture zones (Fig. 13.1). Sediment supply has apparently been high enough to maintain the Cascadia Basin filled to these overflow levels for much of its recent history.

There is clear evidence for deformation and accretion at the foot of the continental slope from a number of detailed studies along the continental margin off Vancouver Island (Tiffin et al., 1972; Yorath, 1980; Yorath et al., 1988; Clowes et al., 1987) and off Washington (Silver, 1972; Barnard, 1978; Carson et al., 1974), that have shown evidence of compression, imbricate thrusting, and uplift in Upper Tertiary and younger sediments (Fig. 13.10 and Fig. 2.33). The geometry of an anticlinal ridge at the foot of the slope (deformation front) was used by Carson et al. (1974), von Heune and Kulm (1973), and Barnard (1978) to estimate convergence rates that are about half those predicted from magnetic anomalies.

Margin seismicity

Little seismic activity has been recorded beneath the continental shelf and slope, but the region of the continent that lies between the coast and the Cascade Volcanic Belt is characterized by general seismicity with events up to magnitude 7.3 (Fig. 13.11; Milne et al., 1978; Crosson, 1972, 1983; Chandra, 1974; Rogers, 1983a). The events fall into two groups, those associated with the downgoing Juan de Fuca Plate at depths of up to 90 km (Benioff-type events), and those occurring within the overlying crust.

The absence of an easily identifiable Benioff Zone of earthquakes extending to hundreds of kilometres beneath the Pacific northwest continental margin was at one time advanced as a significant argument against the existence of subduction. However, as suggested by Atwater (1970), Riddihough and Hyndman (1977) and Keen and Hyndman (1979), the maximum depth of earthquakes is limited by the rapid reheating of the young Juan de Fuca Plate as it descends relatively slowly into the mantle. The whole plate exceeds the maximum temperature for seismic failure (i.e. brittle-type fracture) at a shallow depth. Some recent hypocentres have been located as deep as 90 km (Taber and Smith, 1985) which is in general agreement with the predicted maximum depth of 70 km based on global seismicity studies (Deffeyes, 1972).

Seismicity within the downgoing Juan de Fuca Plate has generally been identified as tensional, with some events indicating movement on eastward-dipping normal faults (McKenzie and Julian, 1971; Chandra, 1974; Rogers, 1983a). Rogers (1983a) noted that the seismicity within the slab is concentrated near the bend between the lower and more steeply dipping plate segments and may be associated with a phase change occurring in the descending oceanic crust and mantle. The observed seismic moment is comparable with the magnitude of such a phase change, the thickness of the Juan de Fuca Plate, and its rate of descent into the mantle.

The seismic activity in the overlying crust at the arc-trench gap occurs mainly in the Puget Sound area (Milne et al., 1978; Rogers, 1983a), with the northern limit near Texada Island, British Columbia in the Strait of Georgia and the southern limit near the Columbia River in Oregon (Fig. 13.11). Fault-mechanism solutions (Chandra, 1974; Crosson, 1972; Milne et al., 1978; Rogers, 1979a,b) suggest dominantly northwest-oriented, right-lateral, strike-slip motion. The earthquakes have been interpreted to be a consequence of north-south compression. Recent precisely located epicentres have identified at least one north-northwest-trending zone which is active in the Puget Sound region (Weaver and Smith, 1983). A possible cause, suggested by Rogers (1983a) and Weaver and Smith (1983) may be a kinematic association between continental faults subparallel with the margin, and the oblique northeasterly convergence of the Juan de Fuca Plate beneath the Washington margin.

The subducted slab

The geometry of the dipping oceanic plate beneath Washington and Oregon has been determined from both hypocentre locations (Taber and Smith, 1985; Crosson, 1983; Michaelson, 1983; Smith and Knapp, 1980) and from observations of travel-time delays for teleseisms (Michaelson and Weaver, 1986; Langston, 1981). The slab seems to be characterized by an upper gently dipping section extending to a depth of about 40 to 60 km, then a more steeply inclined section that extends to a depth of 100 to 200 km beneath the volcanic arc. Along the whole Juan de Fuca Plate subduction zone, the dips seem to decrease from south to north (Riddihough, 1979). In addition, there is possible segmentation and folding in the descending plate (Michaelson, 1983; Weaver and Michaelson, 1985; Hughes et al., 1980; Keen and Hyndman, 1979; Rogers, 1983a,b). One of the most significant segmentations off southern British Columbia has resulted in the independent convergence of the Explorer Plate as a consequence of the development of the Nootka Fault (Hyndman et al., 1979; Riddihough, 1977, 1984). The complex motions of the Explorer Plate may result in a markedly different character of contemporary compression, uplift, and seismicity between the southern Vancouver Island margin and that of northern Vancouver Island northwest of the Nootka Fault (Fig. 13.1, 13.9). The presence of this fault in the oceanic crust beneath Vancouver Island may be reflected by several earthquakes that have occurred in south-central Vancouver Island (e.g. Rogers, 1983a,b). Farther north, the Winona Block may be

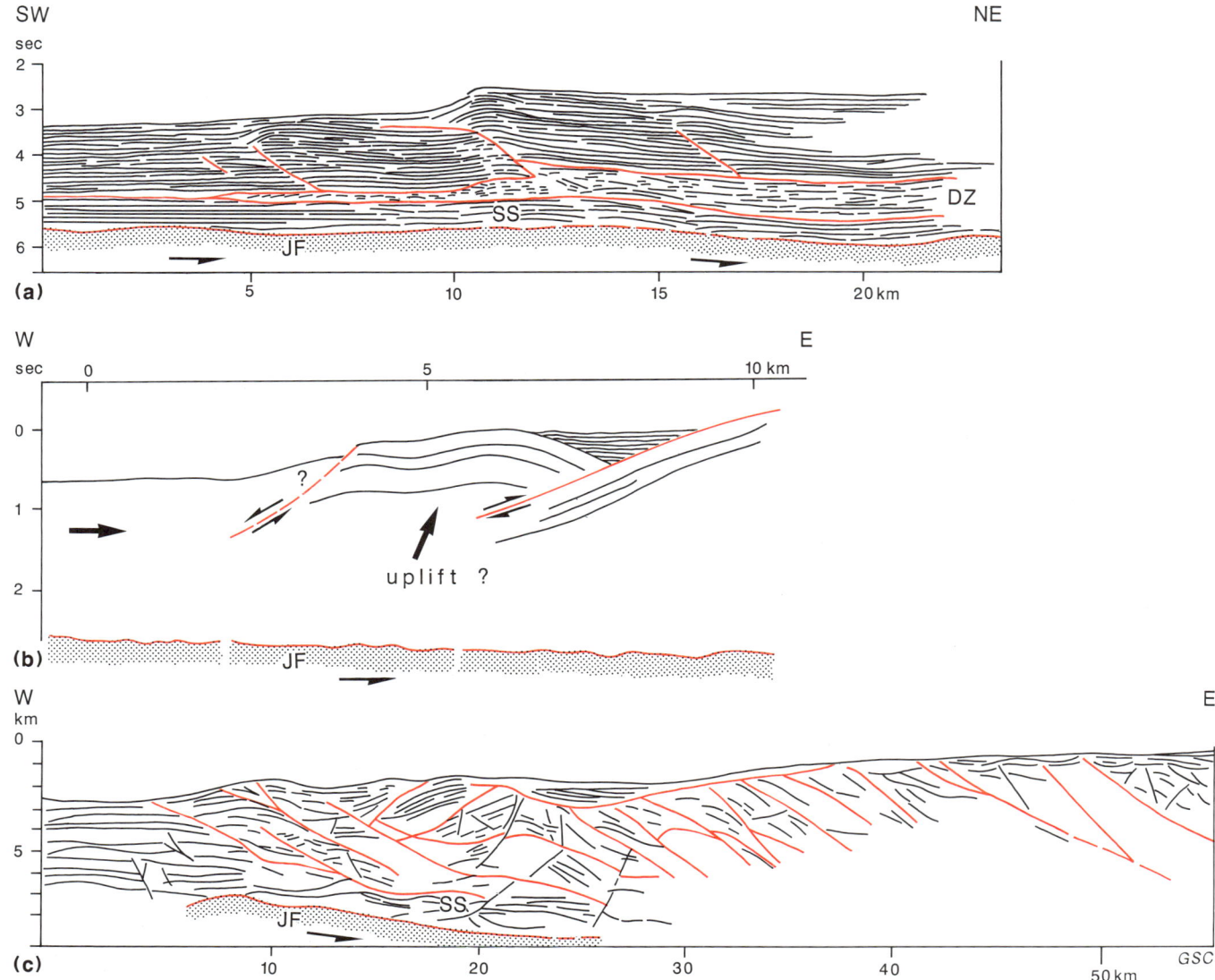

Figure 13.10. Line drawings of sedimentary structure at the foot of the continental slope in the Juan de Fuca Plate convergence zone. (a) southwest Vancouver Island: JF is Juan de Fuca Plate, SS is subducted sediments, DZ is disturbed sediments of proposed décollement zone; (b) northern Washington (Lewis, 1984) showing seaward dipping thrust and ?extensional faults; (c) central Oregon margin (Snavely et al., in press).

underthrusting the margin as another independent dipping slab of crust (Davis and Riddihough, 1982; Yorath and Hyndman, 1983). In addition to the breaks in the subducting plate, there may be bends or folds generated by the transverse shortening required by subduction into a corner of the margin near the Strait of Juan de Fuca (Rogers, 1983b).

The geometry of the subducting slab beneath Vancouver Island was first defined by seismic refraction (Fig. 13.12b) and gravity (Fig. 13.12a) models summarized in Riddihough, 1979). Subsequent seismic reflection and refraction studies on Vancouver Island and in the offshore confirmed the geometry (Ellis et al., 1983). More recently, multichannel reflection data, together with seismic and magnetotelluric studies conducted during the comprehensive Lithoprobe program (Clowes, 1984) have defined the downgoing slab beneath Vancouver Island with remarkable clarity (Fig. 13.12d) (Yorath et al., 1985a,b). The reflecting horizon that is taken to be near the top of the oceanic crust dips at 10 to 12° and lies at depths of about 25 km near the coast to about 30 km beneath central Vancouver Island (Fig. 2.33). Benioff-Wadati type seismicity defines a similar dip at a slightly greater depth, as does a dipping electrically conductive zone. Above the present underthrusting oceanic lithosphere is a second horizon that may represent the top of older underplated oceanic crust.

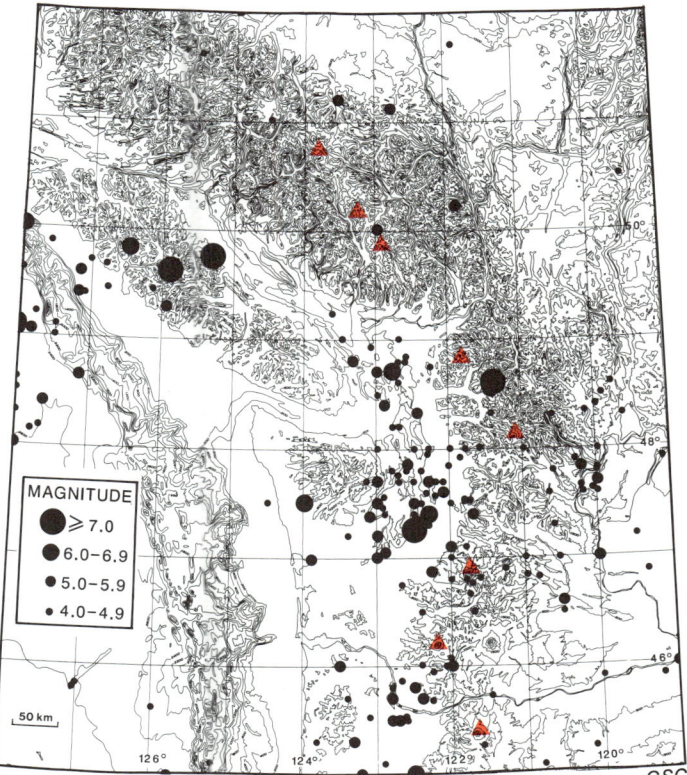

Figure 13.11. Seismicity in the volcanic arc-trench gap of the Juan de Fuca-America Plate boundary (from Rogers, 1979b). Volcanic centres of the Cascade-Garibaldi arc are shown by solid triangles.

Vertical and horizontal motions of the margin

Subduction of the Juan de Fuca Plate system is reflected by data on contemporary and recent vertical movement. Long-term mean sea level trends suggested that the outer coast of Vancouver Island is rising at a rate of between 1 and 2 mm per year and that the inner coast is subsiding at the same rate (Riddihough, 1982b). Relevelling surveys perpendicular to the coasts of Oregon and Washington confirm the mean sea level trends. Across central Vancouver Island, however, repeated levelling during the past decade suggests an opposite sense of movement at about the same rate (Adams and Reilinger, 1980; Adams, 1984; Ando and Balazs, 1979; Dragert, 1986, 1987a,b). The longterm mean vector is characteristic of co-seismic deformation in other subduction zones. However, as no major thrust earthquakes have been recorded, Ando and Balazs (1979) suggested that this indicates aseismic deformation during underthrusting. Alternatively, it may imply elastic response within a very long period earthquake cycle (H. Dragert, pers. comm., 1987; Heaton and Kanamori, 1984).

Horizontal strain measurements in the Puget Sound part of the arc-trench gap (Savage et al., 1981), show significant rates of shortening, generally in conformity with the calculated northeasterly convergence between the Juan de Fuca and America plates. However, the motion reflected in the seismicity is small (Hyndman and Weichert, 1983). These data have been interpreted as indicating that strain could be accumulating for a major earthquake (Savage, 1983). Savage argued that the conformity of strain direction with that predicted by plate convergence is significant and that the return period of great thrust earthquakes is likely to be about 250 years. Across Johnstone Strait between northern Vancouver Island and the mainland, Slawson and Savage (1979) and H. Dragert (pers. comm., 1985) have suggested small right-lateral shear parallel to the margin.

A fore-arc depression lies between the Coast Range of Oregon and Washington and the Cascade volcanic arc (Fig. 13.1). It extends north from the Willamette Valley in Oregon, through Puget Sound to Georgia Strait in southern British Columbia. Rogers (1983a) speculated that the depression is associated with phase changes and shrinkage in the downgoing oceanic plate. Alternatively, it may reflect lithosphere flexure subsidence generated from subduction uplift along the margin. Whatever its origin, it seems to be a characteristic feature of many similar active subduction zones and arc-trench regions.

Thermal and structural models

The arc-trench region is characterized by very low heat flow across the continental margin and sharply higher values near the volcanic arc (Hyndman, 1976; Jessop et al., 1984; Davis and Lewis, 1984; Blackwell et al., 1982; Hyndman, 1984; Fig. 13.13). This pattern is characteristic of many subduction zones, indicating that the low heat flow may be the result of the downgoing slab acting as a thermal sink, both because it is comparatively cold and because of endothermic phase changes in the descending oceanic material. The higher values of heat flow farther inland may be the result of convective upwelling in the vicinity of the volcanic arc.

The crustal structure of the fore-arc region, from seismic refraction investigations, indicates that crustal thicknesses beneath Vancouver Island are 30 to 40 km (White et al., 1961; Ellis et al., 1983; Spence, 1984). These thicknesses are in apparent conflict with gravity interpretations (Stacey, 1974; Dehlinger et al., 1971; Riddihough, 1979) which indicate mantle densities occurring at depths as shallow as 25 km. Riddihough (1979) suggested that low velocity-high density material may occur above the downgoing subducted slab, and Spence (1984), following Keen and Hyndman (1979), proposed that both sets of data can be satisfied if the material beneath Vancouver Island is a detached slab of oceanic lithosphere (see Fig. 13.12a,b). The structure and tectonic history of central Vancouver Island region has been defined in considerable detail by the Lithoprobe program involving four multichannel seismic reflection lines and a wide range of other geological and geophysical studies (Yorath et al., 1985a,b; Clowes et al., 1987; DeLaurier and Kurtz, 1985). The new seismic reflection data appear to confirm the presence of older underplated material above the present descending slab. From the surface to the top of the underplated slab, a series of easterly dipping faults appear to have resulted in significant shortening within the Vancouver Island crust (Yorath et al., 1985a,b).

Note: More recent discussions of the structure and tectonic history of the Cascadia subduction zone and Vancouver Island appear in Hyndman et al. (1990) and Yorath et al. (in press).

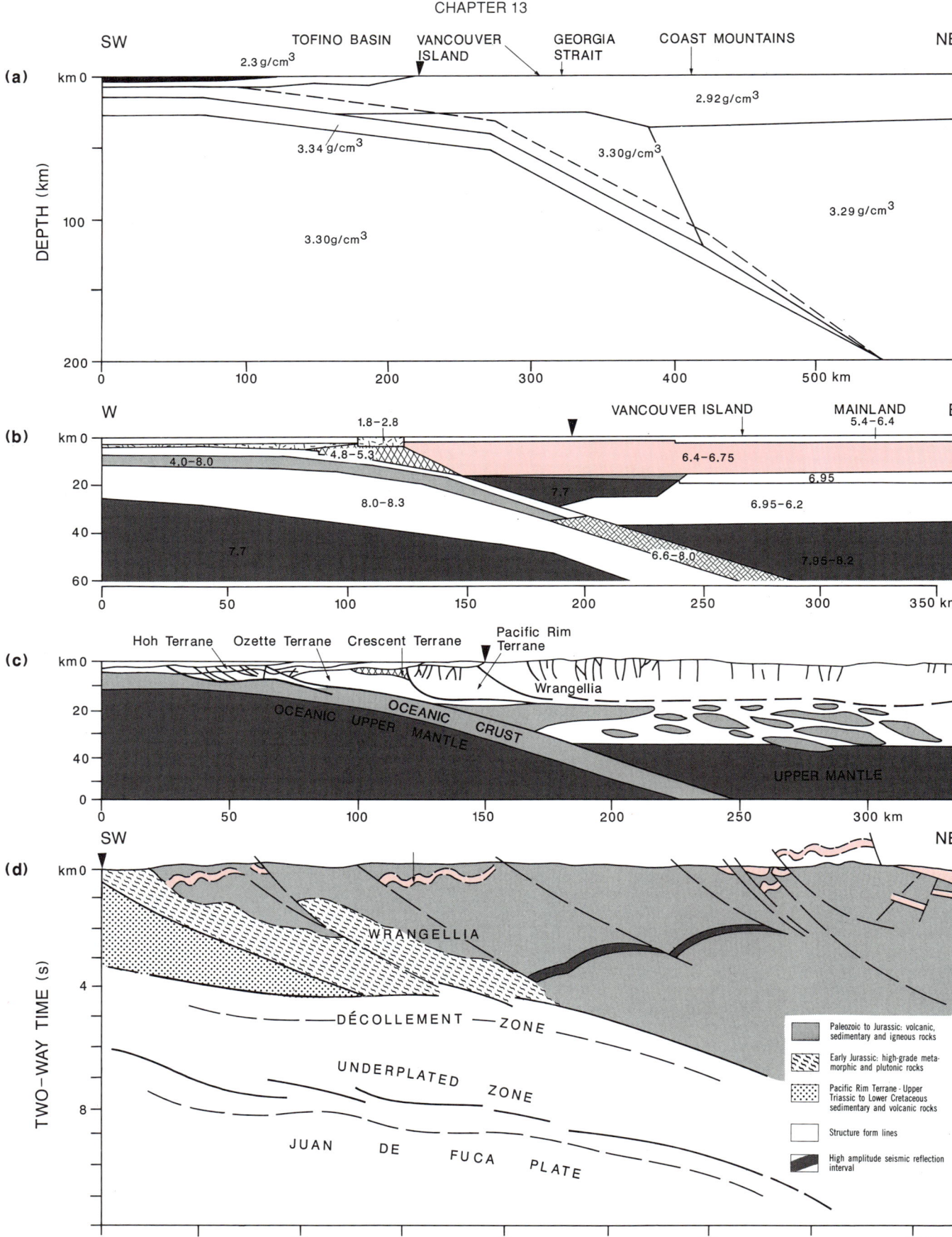

Figure 13.12. Interpreted sections of the subducting Juan de Fuca Plate across south-central Vancouver Island emphasizing: (a) gravity (Riddihough, 1979); (b) seismic refraction-units in km/sec. (Spence, 1984); (c) geology and geophysics (Monger et al., 1985); (d) multichannel reflection seismic data (Yorath et al., 1985a,b). Note scales differ, particularly section d.

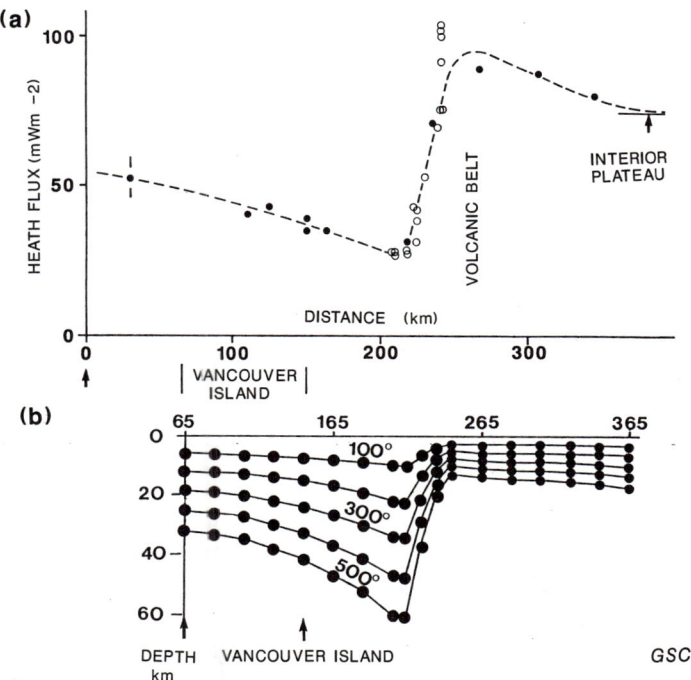

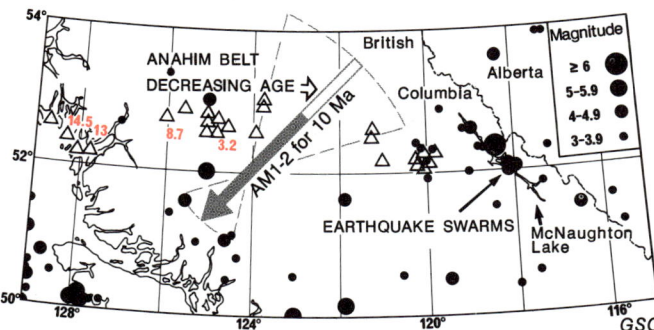

Figure 13.13. Heat flow and temperatures across the southwestern Canadian Cordillera. (a) Heat flow values from Hyndman (1976, 1984) and Lewis et al. (1985). Solid dots are on line of section, open circles are projected. (b) Isotherms computed from heat flow data by T. Lewis (pers. comm., 1986).

Figure 13.14. Volcanic centres (triangles, ages in Ma) and seismic epicentres of the Anahim Volcanic Belt from Rogers (1981). The arrow shows the direction of America Plate motion (AM1-2) with confidence limits (Minster and Jordan, 1978). The lengths of the shaded (210 km) and shaded + unshaded (340 km) arrow are the mean and maximum displacement estimates for 10 Ma movement.

AMERICA PLATE INTRAPLATE DEFORMATION

The effects of the modern plate tectonic regime are largely confined to the plate boundaries, but minor intraplate deformation and other tectonic activity occurs within other parts of the Canadian Cordillera. The level of seismic activity within the Cordillera is relatively low, suggesting deformation and fault motion rates of less than 1 mm/a. In central and eastern Washington and Oregon evidence of widespread extensional faulting, folding and recent volcanism is observed within and east of the Cascade volcanic chain (Eaton, 1979; Suppe et al., 1975; Smith, 1977). The origin of these phenomena has variously been proposed to be due to one or more of back-arc spreading of the Cascade arc, widespread distributed shear as a consequence of Pacific-America Plate motion, a developing spreading system in the Basin and Range Province of the western United States, or the influence of the suggested Yellowstone-Snake River hot spot and its associated volcanism. It seems likely that similar activity may occur in southern British Columbia, particularly in the back-arc region of the Cascade-Garibaldi volcanic belt (Fig. 13.1).

The Anahim Volcanic Belt has been suggested to be the trace of a hot spot (Bevier et al., 1979; Rogers, 1981). The age progression of volcanic centres is appropriate to the amount of America Plate absolute motion (Minster and Jordan, 1978), although the east-west orientation of the belt differs somewhat from the predicted southwesterly azimuth (AM1-2 of Fig. 13.14). Seismic activity near McNaughton Lake (Rogers, 1981) may be associated with the hot spot; a correspondence of contemporary vertical movements was noted by Riddihough (1982b).

The north-trending Stikine volcanic belt in northwestern British Columbia (Souther, 1977) is associated with collinear normal faults that appear to be the result of intraplate rifting, possibly a consequence of a dextral wrench couple provided by the relative movements of the Pacific and America plates. Stacey (1974) suggested that the belt formed as a result of oblique convergence between the Pacific and America plates. Rogers (1976) noted, however, that there was little seismic activity associated with the volcanic belt.

REFERENCES

Adams, J.
1984: Active deformation of the Pacific Northwest continental margin; Tectonics, v. 3, p. 449-472.

Adams, J. and Reilinger, R.
1980: Time behavior of vertical crustal movements measured by releveling in North America; a geological perspective; in Proceedings, Second International Symposium on Problems Related to the Redefinition of the North American Vertical Geodetic Networks (NAD), Canadian Institute for Surveying, Ottawa, Ontario, p. 327-339.

Ando, M. and Balazs, E.I.
1979: Geodetic evidence for aseismic subduction of the Juan de Fuca Plate; Journal of Geophysical Research, v. 84, p. 3023-3028.

Atwater, T.
1970: Implications of plate tectonics for the Cenozoic tectonic evolution of western North America; Geological Society of America Bulletin, v. 81, p. 3513-3536.

Atwater, T. and Menard, H.W.
1970: Magnetic lineations in the Northeast Pacific; Earth and Planetary Science Letters, v. 7, p. 445-450.

Barnard, W.D.
1978: The Washington continental slope: Quaternary tectonics and sedimentation; Marine Geology, v. 27, p. 79-114.

Barr, S.M.
1972: Geology of the north end of Juan de Fuca Ridge and adjacent continental slope; Ph.D. thesis, University of British Columbia, Vancouver.
1974: Structure and tectonics of the continental slope west of Vancouver Island; Canadian Journal of Earth Sciences, v. 11, p. 1187-1199.

Barr, S.M. and Chase, R.L.
1974: Geology of the northern end of Juan de Fuca Ridge and sea-floor spreading; Canadian Journal of Earth Sciences, v. 11, p. 1384-1406.

Bevier, M.L., Armstrong, R.L., and Souther, J.G.
1979: Miocene peralkaline volcanism in west central British Columbia - its temporal and plate tectonics setting; Geology, v. 7, p. 389-392.

Blackwell, D.D., Bowen, R.G., Hull, D.A., Riccio, J., and Steele, J.L.
1982: Heat flow, arc volcanism and subduction in northern Oregon; Journal of Geophysical Research, v. 87, p. 8735-8754.

Bostwick, T.K.
1984: Re-examination of the August 22, 1949 Queen Charlotte earthquake; M.Sc. thesis, University of British Columbia, Vancouver, 115 p.

Bruns, T.R.
1983: Model for the origin of the Yakutat block, an accreting terrane in the northern Gulf of Alaska; Geology, v. 11, p. 718-721.

Carlson, R.L.
1976: Cenozoic plate convergence in the vicinity of the Pacific Northwest: a synthesis and assessment of plate tectonics in the northeastern Pacific; Ph.D. thesis, Department of Geological Sciences, University of Washington, Seattle.

Carson, B.J., Juan, P.B., Myers, P.B., and Barnard, W.D.
1974: Initial deep-sea sediment deformation at the base of the Washington continental slope: a response to subduction; Geology, v. 3, p. 561-564.

Chandra, U.
1974: Seismicity, earthquake mechanisms and tectonics along the western coast of North America from 42°N to 61°N; Bulletin of Seismological Society of America, v. 64, p. 5129-5149.

Chase, R.L.
1977: J. Tuzo Wilson Knolls: Canadian Hot Spot; Nature, v. 266, p. 344-346.

Chase, R.L. and Tiffin, D.L.
1972: Queen Charlotte Fault Zone, British Columbia; in Marine Geology and Geophysics, Section 8, 24th International Geological Congress Proceedings, p. 17-27.

Chase, R.L., Tiffin, D.L., and Murray, J.W.
1975: The western Canadian Continental Margin; in Canada's Continental Margins and Offshore Petroleum Exploration, C.J. Yorath, E.R. Parker, and D.J. Glass (ed.), Canadian Society of Petroleum Geology, Memoir 4, p. 701-722.

Clague, D.A. and Jarrard, R.D.
1973: Tertiary Pacific plate motion deduced from the Hawaiian-Emperor Chain; Geological Society of America Bulletin, v. 84, p. 1135-1154.

Clague, J.J.
1979: The Denali fault system in southwest Yukon Territory - geological hazard?; in Current Research, Part A, Geological Survey of Canada, Paper 79-1A, p. 169-178.

Clark, J.A.
1977: An inverse problem in glacial geology: the reconstruction of glacier thinning in Glacier Bay, Alaska, between AD 1910 and 1960 for relative sea-level data; Journal of Glaciology, v. 18, p. 481-503.

Clowes, R.M.
1984: Phase 1 LITHOPROBE, A coordinated National Geoscience Project; Geoscience Canada, v. 11, p. 122-126.

Clowes, R.M., Yorath, C.J., and Hyndman, R.D.
1987: Reflection mapping across the convergent margin of western Canada; Geophysical Journal of the Royal Astronomical Society, v. 87, p. 79-84.

Coney, P.J.
1971: Cordilleran tectonic transitions and motion of the North American plate; Nature, v. 233, p. 462-465.
1977: Mesozoic-Cenozoic Cordilleran plate tectonics; in Cenozoic Tectonics and Regional Geophysics of the Western Cordillera, R.B. Smith and G.P. Eaton (ed.), Geological Society of America Memoir, v. 152, p. 33-50.

Cox, A. and Engebretson, D.
1985: Change in motion of Pacific plate at 5 Myr BP; Nature, v. 313, p. 472-474.

Crane, K., Aikman, F., Embley, R., Hammond, S., Malahoff, A., and Lupton, J.
1985: The distribution of geothermal fields on the Juan de Fuca Ridge; Journal of Geophysical Research, v. 90, p. 727-744.

Crosson, R.S.
1972: Small earthquakes, structure and tectonics of the Puget Sound Region; Bulletin, Seismological Society of America, v. 62, p. 1133-1171.
1983: Review of seismicity in the Puget Sound Region from 1970 through 1978; in Earthquake Hazards of the Puget Sound Region, Washington, Proceedings of Workshop IV, United States Geological Survey, Open File Report 83-19, p. 6-18.

Currie, R.G., Cooper, R.V., Riddihough, R.P., and Seemann, D.A.
1983: Multi-parameter geophysical surveys off the west coast of Canada: 1973-1982; in Current Research, Part A, Geological Survey of Canada, Paper 83-1A, p. 207-212.

Currie, R.G., Davis, E.E., Riddihough, R.P., and Hussong, D.M.
1984: Acoustic imagery of the Pacific-America-Explorer triple junction; Transactions, American Geophysical Union (EOS), v. 65, p. 1110.

Davis, E.E. and Clowes, R.M.
1986: High velocities and seismic anisotropy in Pleistocene turbidites off Western Canada; Geophysical Journal of the Royal Astronomical Society, v. 84, p. 381-399.

Davis, E.E. and Lewis, T.J.
1984: Heat flow in a back-arc environment: Intermontane and Omineca belts, southern Canadian Cordillera; Canadian Journal of Earth Sciences, v. 21, p. 715-726.

Davis, E.E. and Lister, C.R.B.
1977: Tectonic structures on the Juan de Fuca Ridge; Geological Society of America Bulletin, v. 88, p. 346-363.

Davis, E.E. and Riddihough, R.P.
1982: The Winona Basin: structure and tectonics; Canadian Journal of Earth Science, v. 19, p. 767-788.

Davis, E.E. and Seemann, D.A.
1981: A compilation of seismic reflection profiles across the continental margin of western Canada; Geological Survey of Canada, Open File 751.

Davis, E.E., Currie, R.G., Sawyer, B.S., and Hussong, D.M.
1984: Juan de Fuca Ridge Atlas: SeaMarc II Acoustic Imagery; Earth Physics Branch, Department of Energy, Mines and Resources, Open File 84-17.

Davis, E.E., Lister, C.R.B., Wade, U.S., and Hyndman, R.D.
1980: Detailed heat flow measurements over the Juan de Fuca ridge system and implications for the evolution of hydrothermal circulation in young ocean crust; Journal of Geophysical Research, v. 85, p. 299-310.

Deffeyes, K.S.
1972: Plume convection with an upper mantle temperature inversion; Nature, v. 240, p. 539-544.

Dehlinger, P., Couch, R.W., McManus, D.A., and Gemperle, M.
1971: Northeast Pacific Structure; in The Sea, A.E. Maxwell (ed.), v. 4, Part II, John Wiley and Sons, p. 133-189.

DeLaurier, J.M. and Kurtz, R.D.
1985: A magnetotelluric survey across Vancouver Island; in A Symposium on the Deep Structure of Southern Vancouver Island: Results of LITHOPROBE Phase I, Geological Association of Canada, Pacific Section, Program with Abstracts, p. 12.

Dragert, H.
1986: A summary of recent geodetic measurements of surface deformation on central Vancouver Island, British Columbia; Royal Society of New Zealand Bulletin, v. 24, p. 29-37.
1987a: Contemporary crustal deformation measurements on central Vancouver Island; in Recent Crustal Movements in the Pacific Northwest, Programme and Abstracts, Pacific Section, Geological Association of Canada, p. 15-16.
1987b: The fall (and rise) of central Vancouver Island: 1930-1985; Canadian Journal of Earth Sciences, v. 24.

Duncan, R.A.
1982: A captured island chain in the Coast Range of Oregon and Washington; Journal of Geophysical Research, v. 87, p. 10,827-10,837.

Eaton, G.P.
1979: A plate tectonic model for late Cenozoic crustal spreading in the western United States; in The Rio Grande Rift: Tectonics and Magmatism, R.F. Riecker (ed.), American Geophysical Union, Washington, D.C., p. 7-32.

Ellis, R.M., Spence, G.D., Clowes, R.M., Waldron, D.A., Jones, I.F., Green, A.G., Forsyth, D.A., Mair, J.A., Berry, M.J., Mereau, R.F., Kanasewich, E.R., Cumming, G.L., Hajnal, Z., Hyndman, R.D., McMechan, G.A., and Loncarevic, B.D.
1983: The Vancouver Island seismic project: a co-crust onshore-offshore study of a convergent margin; Canadian Journal of Earth Sciences, v. 20, p. 719-741.

Engebretson, D.C.
1982: Relative motions between oceanic and continental plates in the Pacific Basin; Ph.D. thesis, Stanford University, 211 p.

Hammond, S.R., Lee, J.S., Malahoff, A., Feely, R., Embley, R.W., and Franklin, J.
1984: Discovery of high temperature hydrothermal venting on the Endeavour Segment of the Juan de Fuca Ridge; Transactions, American Geophysical Union (EOS).

Heaton, T.H. and Kanamori, H.
1984: Seismic potential associated with subduction in the northwestern United States; Bulletin of the Seismological Society of America, v. 74, p. 933-941.

Hey, R.N.
1977: A new class of "pseudofaults" and their bearing on plate tectonics: a propagating rift model; Earth and Planetary Science Letters, v. 37, p. 321-325.

Hey, R.N. and Wilson, D.S.
1982: Propagating rift explanation for the tectonic evolution of the northeast Pacific - the pseudomovie; Earth and Planetary Science Letters, v. 58, p. 167-188.

Hicks, S.D. and Shofnos, W.
1965: The determination of land emergence from sea-level observations in southeast Alaska; Journal of Geophysical Research, v. 70, p. 3315-3320.

Horn, J.R., Clowes, R.M., Ellis, R.M., and Bird, D.N.
1984: The seismic structure across an active oceanic/continental transform fault zone; Journal of Geophysical Research, v. 89, p. 3107-3120.

Horner, R.G.
1983: Seismicity in the Saint Elias region of northwestern Canada and southwestern Alaska, Bulletin of the Seismological Society of America, v. 73, p. 1117-1137.

Hudson, T., Dixon, T., and Plafker, G.
1982: Regional uplift in southeastern Alaska; in The U.S. Geological Survey in Alaska: Accomplishments During 1980, W.L. Coonrad (ed.), United States Geological Survey Circular 844, p. 132-135.

Hughes, J.M., Stoiber, R.E., and Carr, M.J.
1980: Segmentation of the Cascade volcanic chain; Geology, v. 8, p. 15-17.

Hyndman, R.D.
1976: Heat flow measurements in the inlets of southwestern British Columbia; Journal of Geophysical Research, v. 81, p. 337-349.
1984: Juan de Fuca Plate Map JFP-10: Geothermal Heat Flux; Open File Map, Pacific Geoscience Centre, Sidney, B.C.

Hyndman, R.D. and Ellis, R.M.
1981: Queen Charlotte fault zone: Microearthquakes from a temporary array of land stations and ocean bottom seismographs; Canadian Journal of Earth Sciences, v. 18, p. 776-788.

Hyndman, R.D. and Rogers, G.C.
1981: Seismicity surveys with ocean bottom seismographs off western Canada; Journal of Geophysical Research, v. 86, p. 3867-3880.

Hyndman, R.D. and Weichert, D.H.
1983: Seismicity and rates of relative motion on the plate boundaries of western North America; Geophysical Journal of the Royal Astronomical Society, v. 72, p. 59-82.

Hyndman, R.D., Lewis, T.J., Wright, J.A., Burgess, M., Chapman, D.S., and Yamano, M.
1982: Queen Charlotte fault zone: heat flow measurements; Canadian Journal of Earth Sciences, v. 19, p. 1657-1669.

Hyndman, R.D., Riddihough, R.P., and Herzer, R.
1979: The Nootka Fault Zone - A new plate boundary off western Canada; Geophysical Journal of the Royal Astronomical Society, v. 58, p. 667-683.

Hyndman, R.D., Rogers, G.C., Bone, M.N., Lister, C.R.B., Wade, U.S., Barrett, D.L., Davis, E.E., Lewis, T., Lynch, S., and Seemann, D.
1978: Geophysical measurements in the region of the Explorer Ridge off western Canada; Canadian Journal of Earth Sciences, v. 15, p. 1508-1525.

Hyndman, R.D., Yorath, C.J., Clowes, R.M., and Davis, E.E.
1990: The northern Cascadia subduction zone at Vancouver Island: seismic structure and tectonic history; Canadian Journal of Earth Sciences, v. 27, p. 313-329.

Jessop, A.M., Lewis, T.J., Judge, A.S., Taylor, A.E., and Drury, M.J.
1984: Terrestrial heat flow in Canada; Tectonophysics, v. 103, p. 239-261.

Johnson, H.P., Karsten, J.L., Delaney, J.R., Davis, E.E., Currie, R.G., and Chase, R.L.
1983: A detailed study of the Cobb offset of the Juan de Fuca Ridge: evolution of a propagating rift; Journal of Geophysical Research, v. 88, p. 2297-2315.

Karsten, J., Delaney, J., Johnson, P., Goldfarb, M., McDuff, R., Kingston, M., Tivey, M., Dymond, J., Kadko, D., Taghan, G., Leinen, M., Lupton, J., Rhodes, M., and Tunnicliffe, V.
1984: Regional setting and local character of a hydrothermal field sulfide deposit on the Endeavour segment of the Juan de Fuca Ridge; Transactions of the American Geophysical Union (EOS), v. 65, p. 1111.

Keen, C.E. and Hyndman, R.D.
1979: Geophysical review of the continental margins of eastern and western Canada; Canadian Journal of Earth Sciences, v. 16, p. 712-747.

Kulm, L.D. and Fowler, G.A.
1974: Cenozoic sedimentary framework of the Gorda-Juan de Fuca Plate and adjacent continental margin - a review; in Modern and Ancient Geosynclinal Sedimentation, R.H. Dott and R.H. Shave (ed.), Society of Economic Paleontologists, Special Publication No. 19, p. 212-229.

Langston, C.A.
1981: Evidence for the subducting lithosphere under Vancouver Island and western Oregon from teleseismic P wave conversions; Journal of Geophysical Research, v. 86, p. 3857-3866.

Lewis, B.T.R.
1984: Deep-tow seismic reflection experiment, Washington margin; in Western North American Continental Margin and Adjacent Ocean Floor off Oregon and Washington, L.D. Kulm et al. (ed.), Ocean Margin Drilling Program, Regional Atlas Series, Atlas 1, p. 27.

Lewis, T.J., Bentkowski, W., Davis, E.E., Hyndman, R.D., Wright, J., and Souther, J.G.
1985: A heat flux profile across southern Vancouver Island to the Garibaldi Volcanic Belt; in A Symposium on the Deep Structure of Southern Vancouver Island: Results of LITHOPROBE Phase I, Geological Association of Canada, Pacific Section, Programme with Abstracts.

Lister, C.R.B.
1977: Qualitative models of spreading center processes including hydrothermal penetration; Tectonophysics, v. 37, p. 203-218.

Malahoff, A., Hammond, S.R., Embley, R.W., Currie, R.G., Davis, E.E., Riddihough, R.P., and Sawyer, B.S.
1984: Juan de Fuca Ridge Atlas: Preliminary SEABEAM bathymetry; Earth Physics Branch, Open File 84-6, Department of Energy, Mines and Resources, Ottawa, Ontario, Canada.

Marshak, R.S. and Karig, D.E.
1977: Triple junctions as a cause for anomalously near trench igneous activity between the trench and volcanic arc; Geology, v. 5, p. 233-236.

McDougall, I. and Duncan, R.A.
1980: Linear volcanic chains - recording plate motions?; Tectonophysics, v. 63, p. 275-295.

McKenzie, D.P. and Julian, B.
1971: The Puget Sound, Washington, earthquake and mantle structure beneath the northwestern U.S.; Geological Society of America Bulletin, v. 82, p. 3519-3524.

Menard, H.W.
1978: Fragmentation of the Farallon Plate by pivoting subduction; Journal of Geology, v. 86, p. 99-110.

Michaelson, C.A.
1983: Three-dimensional velocity structure of the crust and upper mantle in Washington and northern Oregon; M.Sc. thesis, University of Washington, Seattle.

Michaelson, C.A. and Weaver, C.S.
1986: Upper mantle structure from teleseismic P-wave arrivals in Washington and northern Oregon; Journal of Geophysical Research, v. 91, p. 2077-2094.

Milne, W.G., Rogers, G.C., Riddihough, R.P., McMechan, G.A., and Hyndman, R.D.
1978: Seismicity of western Canada; Canadian Journal of Earth Sciences, v. 15, p. 1170-1193.

Minster, J.B. and Jordan, T.H.
1978: Present day plate motions; Journal of Geophysical Research, v. 83, p. 5331-5354.
1984: Vector constraints on Quaternary deformation of the western United States east and west of the San Andreas Fault; in Tectonics and Sedimentation along the California Margin, J.K. Crouch, and S.B. Bachman (ed.), Pacific Section, Society of Economic Paleontologists and Mineralogists, v. 38, p. 1-16.

Monger, J.W.H., Clowes, R.M., Price, R.A., Riddihough, R.P., Simony, P., and Woodsworth, G.J.
1985: Continent-Ocean Transect B2: Juan de Fuca Plate to Alberta Plains; Geological Society of America.

Morgan, W.J.
1981: Hotspot tracks and the opening of the Atlantic and Indian oceans; in The Sea, C. Emiliani (ed.), Wiley, New York, v. 7, p. 443-488.

Nishimura, C., Wilson, D.S., and Hey, R.N.
1984: Pole of rotation analysis of present day Juan de Fuca Plate motion; Journal of Geophysical Research, v. 89, p. 10,283-10,290.

Perez, O.J. and Jacob, K.H.
1980: Tectonic model and seismic potential of the eastern Gulf of Alaska and Yakatoga seismic gap; Journal of Geophysical Research, v. 85, p. 7132-7150.

Plafker, G., Hudson, T., Bruns, T.R., and Rubin, M.
1978: Late Quaternary offset along the Fairweather fault and crustal plate interactions in Southern Alaska; Canadian Journal of Earth Sciences, v. 15, p. 805-816.

Plafker, G., Winkler, G.R., Coonrad, W.L., and Claypool, G.
1980: Preliminary geology of the continental slope adjacent to OCS Lease sale 55, Eastern Gulf of Alaska; petroleum resource implications; United States Geological Survey, Open File Report 80-1089.

Richter, D.H. and Matson, N.A., Jr.
1971: Quaternary faulting in the eastern Alaska Range; Geological Society of America Bulletin, v. 82, p. 1529-1540.

Riddihough, R.P.
1977: A model for recent plate interactions off Canada's west coast; Canadian Journal of Earth Sciences, v. 14, p. 384-396.
1979: Structure and gravity of an active margin - British Columbia and Washington; Canadian Journal of Earth Sciences, v. 16, p. 350-363.
1982a: Contemporary movements and tectonics on Canada's west coast: a discussion; Tectonophysics, v. 86, p. 319-341.
1982b: One hundred million years of plate tectonics in western Canada; Geoscience Canada, v. 9, p. 28-34.
1984: Recent movements of the Juan de Fuca Plate system; Journal of Geophysical Research, v. 89, p. 6980-6994.

Riddihough, R.P. and Hyndman, R.D.
1977: Canada's active western margin: the case for subduction; Geoscience Canada, v. 3, p. 269-278.

Riddihough, R.P., Currie, R.G., and Hyndman, R.D.
1980: The Dellwood Knolls and their role in triple junction tectonics off northern Vancouver Island; Canadian Journal of Earth Sciences, v. 17, p. 577-593.

Riddihough, R.P., Beck, M.E., Chase, R.L., Davis, E.E., Hyndman, R.D., Johnson, S.H., and Rogers, G.C.
1983: Geodynamics of the Juan de Fuca Plate; in Geodynamics of the Eastern Pacific Region, Caribbean and Scotia Arcs, R. Cabre (ed.), American Geophysical Union, Geodynamics Series, v. 9, p. 5-21.

Rogers, G.C.
1976: A microearthquake survey in northwest British Columbia and southeast Alaska; Bulletin Seismological Society of America, v. 66, p. 1643-1655.
1979a: Earthquake fault plane solutions near Vancouver Island; Canadian Journal of Earth Sciences, v. 16, p. 523-531.
1979b: Juan de Fuca Plate Map: fault plane solutions; JFP-6, Earth Physics Branch, Department of Energy, Mines and Resources, Open File.
1981: McNaughton Lake seismicity - more evidence for an Anahim hot-spot?; Canadian Journal of Earth Sciences, v. 18, p. 826-828.
1982: Revised seismicity and revised fault plane solutions for the Queen Charlotte Islands region; Earth Physics Branch, Open File Report No. 82-23.
1983a: Seismotectonics of British Columbia; unpublished Ph.D. thesis, University of British Columbia, Vancouver.
1983b: Some comments on the seismicity of the northern Puget Sound - southern Vancouver Island region; in Earthquake Hazards of the Puget Sound Region, Washington, Proceedings of Workshop IV, United States Geological Survey, Open File Report 83-19, p. 19-39.

Savage, J.C.
1983: Strain accumulation in western United States; Annual Reviews of Earth and Planetary Science, v. 11, p. 11-43.

Savage, J.C., Lisowski, M., and Prescott, W.H.
1981: Geodetic strain measurements in Washington; Journal of Geophysical Research, v. 86, p. 4929-4040.

Silver, E.A.
1971: Small plate tectonics in the northeastern Pacific; Geological Society of America Bulletin, v. 82, p. 3491-3495.
1972: Pleistocene tectonic accretion of the continental slope off Washington; Marine Geology, v. 13, p. 239-249.

Slawson, W.F. and Savage, J.C.
1979: Geodetic deformation associated with the 1946 Vancouver Island, Canada, earthquake; Bulletin Seismological Society of America, v. 69, p. 1487-1496.

Smith, R.B.
1977: Intraplate tectonics of the western North America plate; Tectonophysics, v. 37, p. 323-336.

Smith, S.W. and Knapp, J.S.
1980: The northern termination of the San Andreas Fault; in Studies of the San Andreas Fault Zone in Northern California, R. Sherborne (ed.), California Division of Mines and Geology, Special Report 140, p. 153-164.

Snavely, P.D., Miller, J., von Heune, R., and Mann, D.
in press: Central Oregon margin lines WO76-4 and 5; in Seismic Images of Modern Convergent Margin Tectonic Structure, R. von Heune, (ed.), American Association of Petroleum Geologists Memoir.

Souther, J.G.
1977: Volcanism and tectonic environments in the Canadian Cordillera - a second look; in Volcanic Regimes in Canada, W.R.A. Baragar, L.C. Coleman, and J.M. Hall (ed.), Geological Association of Canada Special Paper, v. 16, p. 3-24.

Spence, G.D.
1984: Seismic structure across the active subduction zone of western Canada; Ph.D. thesis, University of British Columbia, Vancouver.

Spence, G.D., Clowes, R.M., and Ellis, R.M.
1985: Seismic structure across the active subduction zone of western Canada; Journal of Geophysical Research, v. 90, no. B8, p. 6754-6772.

Srivastava, S.P.
1973: Interpretation of gravity and magnetic measurements across the continental margin of British Columbia, Canada; Canadian Journal of Earth Sciences, v. 10, p. 1664-1677.

Srivastava, S.P., Barrett, D.L., Keen, C.E., Manchester, K.S., Shih, K.G., Tiffin, D.L., Chase, R.L., Thomlinson, A.G., Davis, E.E., and Lister, C.R.B.
1971: Preliminary analysis of geophysical measurements north of Juan de Fuca Ridge; Canadian Journal of Earth Sciences, v. 8, p. 1265-1281.

Stacey, R.A.
1974: Plate tectonics, volcanism and the lithosphere in British Columbia; Nature, v. 250, p. 133-134.

Suppe, J., Powell, C., and Berry, R.
1975: Regional topography, seismicity, Quaternary volcanism and the present day tectonics of the western United States; American Journal of Science, v. 275-A, p. 397-436.

Sutherland Brown, A.
1968: Geology of the Queen Charlotte Islands, British Columbia; British Columbia Department of Mines and Petroleum Resources, Bulletin 54.

Taber, J.J. and Smith, S.W.
1985: Seismicity and focal mechanisms associated with the subduction of the Juan de Fuca plate beneath the Olympic Peninsula, Washington; Bulletin of the Seismological Society of America, v. 75, p. 237-249.

Tiffin, D.L., Cameron, B.E.B., and Murray, J.W.
1972: Tectonic and depositional history of the continental margin off Vancouver Island, B.C.; Canadian Journal of Earth Sciences, v. 9, p. 280-296.

Tunnicliffe, V., Johnson, H.P., and Botros, M.
1984: Along strike variation in hydrothermal activity on the Explorer Ridge, N.E. Pacific; American Geophysical Union Fall Meeting, Additional Abstracts, p. 75.

Turner, D.L., Jarrard, R.D., and Forbes, R.B.
1980: Geochronology and origin of the Pratt-Welker seamount chain, Gulf of Alaska: a new pole of rotation for the Pacific Plate; Journal of Geophysical Research, v. 85, p. 6547-6556.

Vine, F.J.
1966: Spreading of the ocean floor: new evidence; Science, v. 154, p. 1405-1415.

Vine, F.J. and Matthews, D.H.
1963: Magnetic anomalies over ocean ridges; Nature, v. 199, p. 947-949.

Vine, F.J. and Wilson, J.T.
1965: Magnetic anomalies over a young oceanic ridge off Vancouver Island; Science, v. 150, p. 485-489.

von Heune, R. and Kulm, L.D.
1973: Tectonic summary of Leg 18; in Initial Reports of the Deep Sea Drilling Project, v. 18, p. 961-976.

von Heune, R., Shor, G.O., and Wageman, J.
1978: Continental margins of the eastern Gulf of Alaska and boundaries of tectonic plates; in Geological and Geophysical Investigations of Continental Margins, J.S. Watkins, L. Montadert, and P.W. Dickerson (ed.), American Association of Petroleum Geologists Memoir 29, p. 273-290.

Weaver, C.S. and Smith, S.W.
1983: Regional tectonic and earthquake hazard implications of a crustal fault zone in southwestern Washington; Journal of Geophysical Research, v. 88, p. 10,371-10,383.

Weaver, C.S. and Michaelson, C.A.
1985: Seismicity and volcanism in the Pacific Northwest: evidence for the segmentation of the Juan de Fuca plate; Geophysical Research Letters, v. 12, p. 215-218.

White, W.R.H., Bone, M.N., and Milne, W.G.
1961: Seismic refraction surveys in British Columbia, 1964-1966: a preliminary interpretation; American Geophysical Union Monograph, v. 12, p. 81-93.

Wickens, A.J. and Hodgson, J.H.
1967: Computer re-evaluation of earthquake mechanism solutions 1922-1962; Publications of the Dominion Observatory, v. 33, 560 p.

Wilson, D.S., Hey, R.N., and Nishimura, C.
1984: Propagation as a mechanism of reorientation of the Juan de Fuca Ridge; Journal of Geophysical Research, v. 89, p. 9215-9225.

Yorath, C.J.
1980: The Apollo structure in Tofino Basin, Canadian Pacific continental margin; Canadian Journal of Earth Sciences, v. 17, p. 758-775.

Yorath, C.J. and Hyndman, R.D.
1983: Subsidence and thermal history of Queen Charlotte basin; Canadian Journal of Earth Sciences, v. 20, p. 135-159.

Yorath, C.J., Clowes, R.M., Green, A.G., Sutherland Brown, A., Brandon, M.T., Massey, N.W.D., Spencer, C., Kanasewich, E.R., and Hyndman, R.D.
1985a: LITHOPROBE - Phase 1: Southern Vancouver Island: Preliminary analyses of reflection seismic profiles and surface geological studies; in Current Research, Part A, Geological Survey of Canada, Paper 85-1A, p. 543-554.

Yorath, C.J., Green, A.G., Clowes, R.M., Sutherland Brown, A., Brandon, M.T., Kanasewich, E.R., Hyndman, R.D., and Spencer, C.
1985b: Lithoprobe, southern Vancouver Island: Seismic reflection sees through Wrangellia to the subducting Juan de Fuca plate; Geology, v. 13, p. 759-762.

Yorath, C.J., Clowes, R.M., Macdonald, R.D., Spencer, C., Davis, E.E., Hyndman, R.D., Rohr, K., Sweeney, J.F., Currie, R.G., Halpenny, J.F., and Seemann, D.A.
1988: Marine multichannel seismic reflection, gravity and magnetic profiles - Vancouver Island continental margin and Juan de Fuca Ridge; Geological Survey of Canada, Open File 1661.

Yorath, C.J., Hyndman, R.D., Sutherland Brown, A. and Massey, N.W.D.
in press: Lithoprobe- Southern Vancouver Island: A summary; Geological Society of America Bulletin.

Authors' addresses

R.P. Riddihough
Geological Survey of Canada
601 Booth Street
Ottawa, Ontario
K1A 0E8

R.D. Hyndman
Pacific Geoscience Centre
9860 West Saanich Road
P.O. Box 6000
Sidney, British Columbia
V8L 4B2

Printed in Canada

Chapter 14
VOLCANIC REGIMES

Summary
Volcanic rocks of the miogeocline
    Proterozoic assemblages
        Purcell(Belt) and Wernecke assemblages
        Windermere Assemblage
    Paleozoic volcanic assemblages
        Lower Paleozoic
        Upper Paleozoic
        Volcanic rocks of pericratonic Kootenay Terrane
Volcanism in the accreted terranes
    Proterozoic-lower Paleozoic assemblages
    Volcanism in Alexander Terrane
    Upper Paleozoic assemblages
        Slide Mountain Terrane
        Quesnellia
        Cache Creek Terrane
        Stikinia
        Wrangellia
        Alexander Terrane
    Triassic assemblages
        Quesnellia
        Cache Creek Terrane
        Stikinia
        Wrangellia
        Alexander Terrane
    Jurassic Assemblages
        Quesnellia
        Stikinia
        Wrangellia
    Upper Mesozoic and Tertiary assemblages
        Chugach Terrane
        Pacific Rim Terrane
        Crescent Terrane

Post-accretionary volcanism
    Cretaceous magmatism
      Early and Mid-Cretaceous
      Mid-Cretaceous
        Foreland Belt
        Northern Intermontane and Omineca belts
        Insular and south-central Intermontane belts
      Late Cretaceous
    Early Tertiary magmatism
      Southern Insular Belt
      Southern Intermontane Belt
      Central Intermontane Belt
      Northern Intermontane Belt
    Late Tertiary and Quaternary magmatism
      Farallon-Juan de Fuca subduction complexes
        Pemberton Belt
        Alert Bay Volcanic Belt
        Garibaldi Volcanic Belt
        Chilcotin Group, back-arc lavas
      Pacific Plate subduction complexes
        Wrangell Volcanic Belt
      Rift and plume complexes within the accreted terranes
        Stikine Volcanic Belt
        Anahim Volcanic Belt
        Clearwater-Quesnel and McConnell Creek volcanic provinces
  Conclusions
  References

Chapter 14

VOLCANIC REGIMES

J.G. Souther

SUMMARY

Volcanic rocks are found in most of the major rock assemblages in the western Cordillera. In-situ models of their origin (Souther, 1977) have been superseded by tectonic models based on evidence that the Cordillera is a collage of separate, once distant terranes that have been accreted to the continental margin. This concept is the basis for the three-fold subdivision of volcanic regimes in this chapter: (1) volcanic rocks of the miogeocline, (2) volcanism in the accreted terranes, (3) post-accretionary volcanism.

Following the separation of ancestral North America from a Precambrian continental mass in mid-Proterozoic time the deposition of a prograded terrace wedge along the western continental margin was accompanied by episodes of crustal extension and minor igneous activity. Tectonism about 1.2 Ga produced a widespread unconformity in the northern Cordillera between the Wernecke and Mackenzie Mountains supergroups and was accompanied by local effusions of alkali-tholeiite lavas and the emplacement of diabase dyke swarms and sills throughout the entire western half of North America. A second tectonic event about 780 Ma initiated deposition of the Windermere Supergroup in a relatively narrow rift depression that truncated the older terrace wedge deposits. Rifting was accompanied by widespread eruption of mafic volcanic rocks and, at least locally, by the explosive eruption of intermediate to salic magma. Deposition of the Windermere Supergroup was followed by a relatively stable regime that lasted through the early Paleozoic. Deposits of quartzite and carbonate reefs on the continental shelf and thick deep-water shale deposits on the outer slope show little evidence of contemporaneous tectonic or igneous activity. The episodic eruptions of small batches of alkaline, mafic lava, particularly during the Ordovician, Silurian and possibly also in the Devonian periods, were probably the result of minor extension accompanying thermal subsidence and contraction of the lower crust. A major rifting event in Devono-Mississippian time initiated deposition of thick wedges of clastic rocks, intrusion of syenite plutons, and eruption of highly evolved, oversaturated alkaline and peralkaline lavas.

The miogeocline is bounded on the west by a collage of transported terranes that include arc and oceanic volcanic complexes of distant origin as well as volcanic rocks generated during accretion. A fairly continuous belt of highly deformed rocks (pericratonic terranes) was thrust eastward over the miogeoclinal succession, and the pericratonic terranes were themselves overthrust by younger terranes to the west. Alkaline lavas within the miogeocline probably originated in minor rifts within the outer part of the continental terrace wedge. West of the pericratonic terranes the accreted, Slide Mountain Terrane comprises a series of easterly transported allochthons each consisting of numerous thin imbricated thrust slices that represent as much as 500 km of shortening. The predominance of upper Paleozoic oceanic basalt, chert, diabase and ultramafic rocks suggests an ophiolitic assemblage derived from oceanic crust that lay west of the miogeoclinal wedge at the continental margin. The chemistry of the basalts and their association with ultramafic rocks suggests that they include both mid-ocean ridge tholeiites and more alkaline rocks characteristic of off-ridge seamounts. The oceanic Slide Mountain Terrane was bounded on the west by Quesnellia where a succession of calc-alkaline to shoshonitic arcs was active during late Paleozoic and most of Mesozoic time. The polarity of the pre-Mesozoic arcs on Quesnellia is unknown but the Triassic Nicola and eastern Takla volcanics and the Jurassic Rossland volcanics have a pronounced easterly increase in alkalis, consistent with an arc-back-arc pair above an east-dipping subduction zone. West of Quesnellia the Cache Creek Terrane comprises a discontinuous belt of mainly oceanic basalts and ultramafic rocks associated with chert, argillite and minor limestone ranging in age from late Paleozoic to Triassic. Most of the Cache Creek volcanic rocks are typical of mid-ocean ridge basalts. A lesser number are more alkaline, probably related to off-ridge seamounts, and a small percentage with calc-alkaline affinities probably were derived from adjacent arc complexes and tectonically mixed into the Cache Creek mélange. The Cache Creek oceanic terrane is bounded on the west by Stikinia, where local eruption of acid volcanic rocks in the Permian and Mississippian periods was followed by widespread volcanism in the Late Triassic and throughout much of the Jurassic Period, producing a mixed assemblage of calc-alkaline and mildly alkaline submarine and subaerial volcanic and volcaniclastic deposits. The great breadth and variable chemical polarity that characterizes the Mesozoic Stuhini and Hazelton volcanic belts on Stikinia suggests that they are not the product of simple subduction-related arcs. The presence of alkaline rocks and deep, local basins of accumulation suggests a transtensional environment in which subduction was accompanied by pull-apart structures developed in response to penetrative transcurrent faulting. The suture between Stikinia and the Insular Superterrane to the west is

---

Souther, J. G.
1991: Volcanic regimes, Chapter 14 in Geology of the Cordilleran Orogen in Canada, H. Gabrielse and C. J. Yorath (ed.); Geological Survey of Canada, Geology of Canada, no. 4, p. 457-490 (also Geological Society of America, The Geology of North America, v. G-2)

obscured by uplift, intrusion, and metamorphism in the Coast Belt, but there is compelling faunal and paleomagnetic evidence to show that they were widely separated until latest Jurassic time. The mid- and upper Paleozoic Skolai and Sicker volcanic and volcaniclastic assemblages of Wrangellia are interpreted as arc assemblages developed on oceanic crust. Subsequent rifting of the arc complex in the Triassic Period was accompanied by the effusion of Karmutsen-Nikolai, marine and nonmarine tholeiitic basalt. The resulting Karmutsen volcanic platform became the site of Jurassic arc volcanism that produced the calc-alkaline lavas and caldera complexes of the Bonanza Group.

By the end of Jurassic time several components of the Cordilleran collage had amalgamated into two superterranes. The more easterly Intermontane Superterrane had already docked against the western edge of North America. Its accretion was accompanied by crustal thickening, metamorphism, and plutonism along the suture to form the Omineca Belt. At that time the Insular Superterrane lay an unknown distance to the west, separated from the Intermontane Superterrane by the Dezadeash-Tyaughton-Methow Trough. The contraction of this Trough during Late Jurassic and Cretaceous time was accompanied by the eruption of Gravina-Nutzotin and Gambier volcanic rocks from an arc along its western margin and by the deposition of distal submarine fans that overlap the western edge of the Intermontane Superterrane. Mid-Cretaceous contraction of southern Tyaughton Trough coincided with declining volcanism in the Gambier arc to the west and with the initiation of Spences Bridge volcanism in an arc along its eastern margin related to subduction of crust flooring the trough beneath the Intermontane Superterrane. Elsewhere, in the northern Omineca Belt and north-central Intermontane Belt mid- and Late Cretaceous volcanism at widely separated centres appears to have been closely related to relaxation or to a transpressive regime accompanying the onset of transcurrent faulting. The South Fork and Kasalka volcanic piles are probably cauldron subsidence complexes, and correlative assemblages appear to occupy local, fault-bounded basins. There is little record of volcanism in the Canadian Cordillera during the Paleocene. However, early in the Eocene Epoch volcanism resumed, and within the short interval between 55 and 36 Ma the northern Cordillera was the site of intense and widespread magmatic activity. The Nisling Range and Skukum-Sloko igneous complexes of Yukon and northern British Columbia, the Ootsa-Endako volcanic complexes in central British Columbia, and the Kamloops, Marron-Coryell volcanic-plutonic assemblages of southern British Columbia comprise a broad zone of Eocene calc-alkaline to alkaline volcanism that has been related to an east-dipping subduction zone beneath the Challis-Absaroka arc. Chemical polarity in the Kamloops-Marron assemblages supports such an interpretation, but the association between Eocene magmatism and pervasive transcurrent faulting suggests that transtensional regimes may have had an important, if not the dominant role in magma generation in the northern Cordillera. The Eocene magmatic culmination was followed by a period of relative tectonic quiescence that lasted through Oligocene time. By early Neogene time the current tectonic relationship between the continent and Pacific margin began to evolve and exert a controlling influence on the distribution, and chemical affiliation of Neogene volcanoes on the adjacent continent. In southern British Columbia the Pemberton-Garibaldi volcanic fronts and the Chilcotin plateau basalts comprise an arc-back-arc pair that developed above the subducting Juan de Fuca Plate and was intermittently active throughout Miocene and early Quaternary time. In southwestern Yukon Territory and adjacent Alaska Miocene and younger basaltic andesite, dacite, and rhyolite magmas were erupted from volcanoes of the Wrangell volcanic arc, which developed inland from the Aleutian Trench. Between the northern triple junction of the Juan de Fuca Plate and the southern end of the Aleutian Trench the continent is separated from Pacific crust by the Queen Charlotte transform fault. Two Late Miocene to Quaternary volcanic belts, both characterized by alkaline to peralkaline bimodal complexes, developed on the continent adjacent to this transform boundary. The Stikine Volcanic Belt, which trends northerly across the Coast and northern Intermontane belts, is related to incipient rifting. Volcanoes of the Anahim Volcanic Belt, which trends easterly across central British Columbia, become progressively younger toward the east and are believed to define the trace of a hotspot.

VOLCANIC ROCKS OF THE MIOGEOCLINE

From its inception in the Middle Proterozoic until the onset of accretion in the early Mesozoic accumulation of continental terrace wedge deposits was the dominant feature of Cordilleran tectonics. This long interval of sedimentation was interrupted by several periods of uplift, faulting, and volcanic activity: about 1200 Ma ago (Purcell Lavas); about 780 Ma ago (Irene volcanics, Coates Lake volcanics, Mount Harper volcanics); Middle Cambrian volcanics and sills on Ogilvie Platform; Cambrian and Middle Ordovician (Selwyn Basin and Kechika Trough volcanics); Middle Ordovician to Devonian (Marmot volcanics, Misty Creek Embayment); Silurian volcanics in the northern part of the Robson Basin and Late Devonian to Early Carboniferous volcanics on the Cassiar Platform.

Proterozoic Assemblages

Proterozoic rocks of the miogeocline can be grouped into three principal assemblages (Fig. 14.1), separated in most places by major disconformities and locally by angular unconformities. They are: the Purcell (Belt), Muskwa(?) and Wernecke assemblages (1700 to 1200 Ma), the Mackenzie Assemblage (1200 to 780 Ma) and the Windermere Assemblage (780 to 600 Ma). Strata older than 780 Ma have limited north-south continuity, whereas the Windermere Assemblage extends almost the entire length of the Cordillera. Volcanic rocks are known only in the Purcell (Belt) and Windermere assemblages (see Chapters 5 and 6).

Purcell (Belt) and Wernecke assemblages

Volcanic rocks comprising the "Purcell Lava" (Daly, 1912; Høy, 1979) and the Nicol Creek Formation (McMechan et al., 1980) are interbedded with dolomitic siltite and argillite near the top of the Purcell (Belt) Assemblage. The volcanic rocks include massive to amygdaloidal, locally pillowed

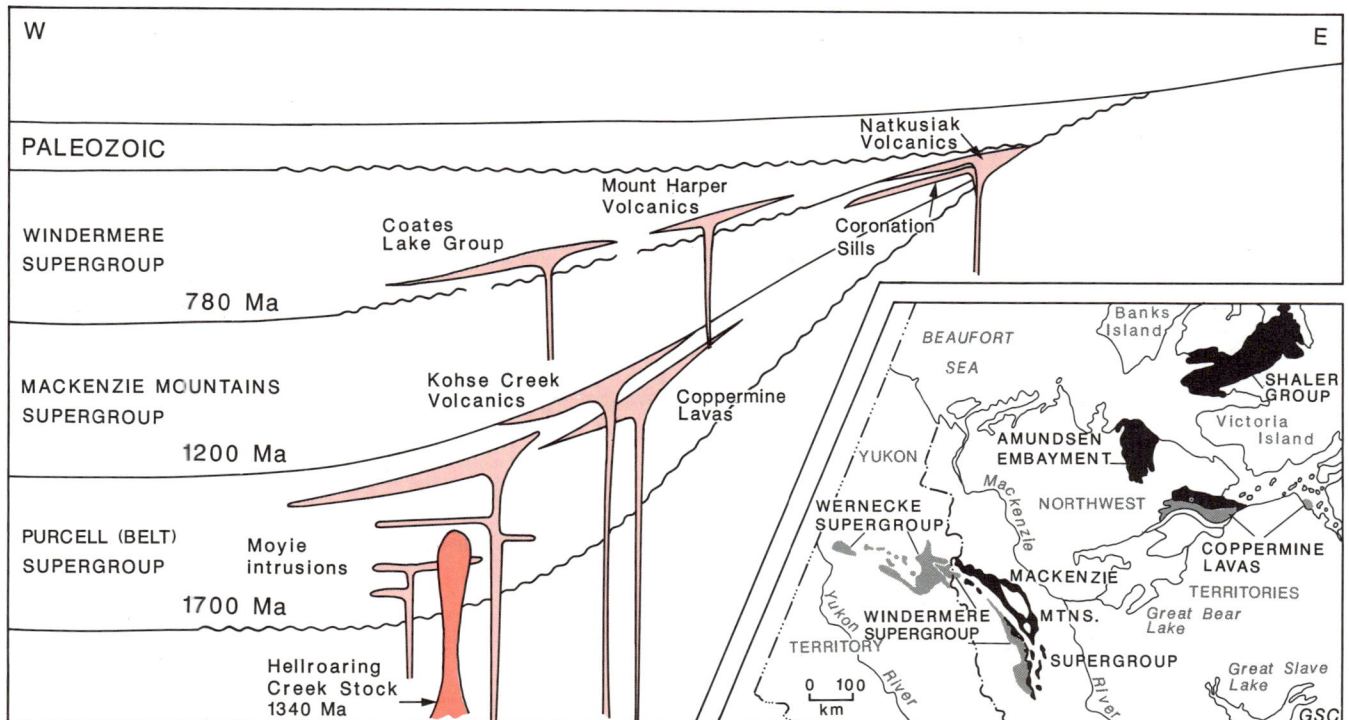

Figure 14.1. Schematic section showing the stratigraphic relationships of some Proterozoic igneous rocks of western North America. Index map shows the distribution of Proterozoic rocks in northwestern Canada. Modified from Young et al. (1982).

lava flows interlayered with volcanic sandstone and tuff. The lavas are chloritized and sericitized plagioclase microporphyritic basalts and andesites. Limited chemical data (McMechan et al., 1980) suggest an alkaline, K-rich affinity. The flows, which erupted about 1075 to 1200 Ma, occur over a wide area in the central Purcell Mountains, where they have a cumulative thickness of up to 750 m. Sedimentary strata below the lavas are cut by numerous diabase dykes and thick, extensive sills of the Moyie Intrusions (Reesor, 1958). Some of these may be subvolcanic equivalents of the Purcell lavas, but others are clearly older (1433 Ma) and may be coeval with the emplacement of granodiorite plutons in the western part of the basin (Reynolds, 1984).

In the Wernecke Mountains, the Wernecke Assemblage, a thick sedimentary succession, believed to be correlative with Purcell (Belt) rocks, is overlain unconformably by the Kohse Creek Volcanics at the base of the Pinguicula Group (Fig. 14.2; Eisbacher, 1981). The volcanic succession grades from several hundred metres of basic to intermediate flows and aquagene tuffs in the south, northward to siliciclastic laminites. Their age is not known and they could be as young as basal Windermere volcanic rocks in the eastern Mackenzie Mountains. The emplacement of intrusive mega-breccias within the Wernecke Supergroup (Bell, 1986) may have been accompanied by volcanic and/or diatreme activity.

East of the Wernecke Mountains, strata of Purcell (Belt) age overlap the cratonic interior in Amundsen Embayment (Fig. 14.1), where Middle Proterozoic clastic sediments and dolomite are conformably overlain by basalt flows of the Coppermine Group. These aphyric lava flows have a composite thickness of at least 3600 m. Individual flows are from 15 to 60 m thick and show no apparent lithological variation throughout the pile. The narrow range of compositions has been confirmed by a large number of chemical analyses that are transitional between tholeiite and alkali basalt, suggesting that they originated from a relatively undepleted mantle source.

The basic lavas of the Purcell (Belt), Wernecke, and Coppermine assemblages are broadly coeval with the Muskox Intrusion and Keweenawan lavas (1.07-1.1 Ga), with the unconformity at the base of the Mackenzie Mountains Supergroup, and with the Mackenzie Dyke Swarm that trends northwest-southeast across the full width of the Canadian Shield. This widespread array of basic intrusions, extrusions, and associated structures may have developed over a period as long as 200 Ma. It was clearly not a single event but rather a protracted period during which repeated episodes of extension were accompanied by the rise of primitive, mantle-derived tholeiitic and alkaline basic magma. The absence of salic volcanic rocks and intrusions suggests that the rate of extension and magma generation was slow and never resulted in asthenospheric upwelling of sufficient magnitude to melt lower crustal rocks or produce well defined rifts.

Windermere Assemblage

Extensional faulting and minor associated volcanism marked the onset of Windermere deposition (Fig. 14.1, 14.2). In the Mackenzie Mountains altered basalt flows, up

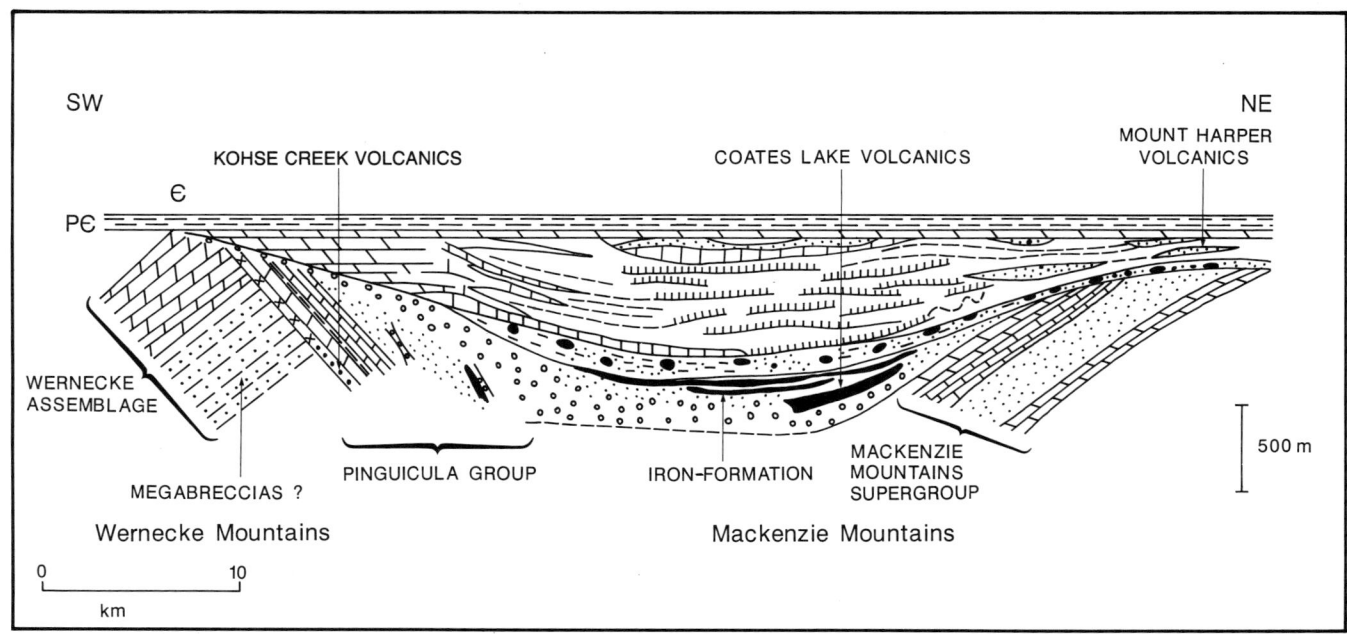

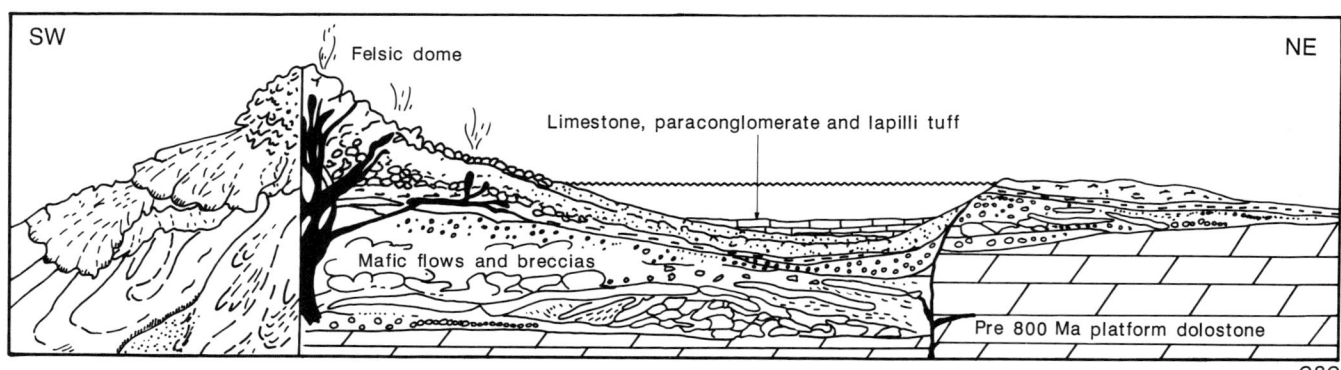

Figure 14.2. Schematic reconstruction showing the relative positions of the Kohse Creek, Coates Lake and Mount Harper volcanics on a transect of the Windermere Supergroup between the eastern Wernecke and Mackenzie mountains. Detailed section (below) is an interpretation of basin development during a late stage of Mount Harper volcanism (from Roots and Moore, 1982).

to 50 m thick, in the Coates Lake Group are thought to be coeval with north-northwesterly trending diabase dykes and sills that intrude underlying strata of the Mackenzie Mountains Supergroup and yield Rb-Sr isochron dates of about 780 Ma (Armstrong et al., 1982). The dykes occupy pre-existing faults that probably were associated with the rifting event that initiated Windermere deposition.

In east-central Alaska strata equivalent to the Windermere are included in the upper part of the Tindir Group (Young, 1982). The discontinuous, basal unit consists of up to 200 m of amygdaloidal pillow basalts and locally cupriferous tuffs and volcanic breccia. Chemical analyses suggest a tholeiitic affinity.

In the Ogilvie Mountains of west-central Yukon Territory Upper Proterozoic rocks include the Mount Harper volcanic complex (Roots, 1982, 1983; Thompson, in press), one of the best preserved records of Windermere volcanism in the Cordillera. The complex (Fig. 14.2) covers an area of 105 km^2 and consists of unaltered and undeformed volcanic rocks, up to 1200 m thick, underlain by "Pinguicula-like" dolostone and overlain by Lower Cambrian Jones Ridge dolostone. The stratigraphy indicates three periods of mafic volcanism separated by two periods of explosive activity that produced intermediate and felsic eruptive products. The mafic rocks comprise most of the pile and include both subaqueous and, less abundant, subaerial facies. Pillow lava and hyaloclastite of the subaqueous facies are associated with epiclastic deposits derived from nearby scarps, whereas massive and columnar jointed lavas of the subaerial facies are associated with scoriaceous oxidized flow-top breccia, cinder and pyroclastic beds containing cored bombs. The pile evolved from a broad subaqueous shield into an emergent cone that was broadened by the deposition of epiclastic deposits eroded from the slopes of the growing edifice and by the formation of hyaloclastite deposits at the waters edge. Intermediate and felsic volcanic rocks are exposed only in the southern part of the complex where light grey, spherulitic, quartz and/or feldsparphyric flows and domes are associated with thick successions of welded pyroclastic flow deposits. Epiclastic boulder deposits associated with the volcanic rocks suggest that volcanism was contemporaneous with northwest-trending normal faulting. A 777 Ma U-Pb zircon date on quartz-feldsparphyric Harper flows (Thompson, in press) is in close agreement with the ±770 Ma Rb-Sr age of diabase sheets in the Mackenzie Mountains (Armstrong et al., 1982).

In the Columbia-Purcell Mountains of southeastern British Columbia, basal conglomerate of the Windermere Supergroup (Toby Formation) is locally overlain by volcanic rocks of the Irene Formation. The Irene is a fine grained, sheared, andesitic greenstone with interbeds of conglomerate and limestone. It is correlative with mafic metavolcanics of the Huckleberry Formation in northeastern Washington. The volcanics yield K-Ar dates between 827 and 918 Ma, but Rb-Sr data suggest that these dates may be erroneously old (Devlin et al., 1985). A Nd-Sm age of about 762 Ma has been obtained from the Huckleberry Formation (W.J. Devlin, pers. comm., 1985).

Proterozoic strata of Windermere age are exposed in several structural and metamorphic culminations in the Omineca Belt (Evenchick et al., 1984). Metamorphosed clastic successions in the Sifton and Deserters ranges are intercalated with thick sequences of amphibolite that are thought to be derived from mafic volcanic rocks. The scarcity of associated salic volcanic rocks may be due to erosion rather than nondeposition. If significant salic volcanism occurred during Windermere time it was probably localized above the evolving igneous and metamorphic culminations where the chance of surviving erosion was negligible. The salic volcanic rocks of the Mount Harper Complex may be unique simply because they were not uplifted and cannibalized by syntectonic erosion that removed the infrastructure of larger and more vigorous igneous centres along the length of the Windermere belt.

Paleozoic volcanic assemblages

Paleozoic basic volcanic rocks occur throughout the miogeoclinal succession. However, much of the activity was concentrated in two long intervals: Late Cambrian to Early Devonian and Late Devonian to Early Carboniferous.

Lower Paleozoic

Volcanic rocks of the Marmot Formation (Fig. 14.3) are interbedded with Middle Ordovician to Lower Devonian clastic and minor carbonate rocks in the Misty Creek Embayment, a northerly trending embayment in northern Selwyn Basin (Cecile, 1982). The formation consists of basaltic lapilli tuff and breccia, massive amygdaloidal flows, sills, minor coarse breccia, and pillow breccia. Commonly the entire formation is represented by a single cooling unit from one to several tens of metres thick. However, in the southeastern part of the embayment an anomalously thick (500 m) section of proximal breccia and volcanic conglomerate is associated with sills and dykes. The lower part of the pile is mainly hyaloclastite comprising crudely stratified tuff breccia, pillow lava, and blocky amygdaloidal basalt flows. The devitrified breccias are commonly

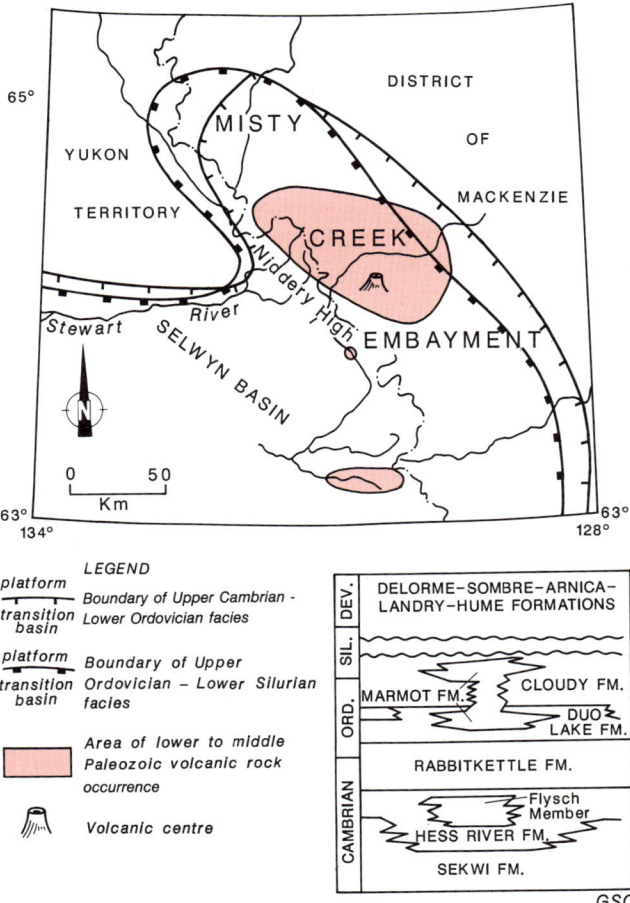

Figure 14.3. Sketch map showing the distribution of lower to middle Paleozoic volcanic rocks within the Misty Creek Embayment on the northern boundary of Selwyn Basin. Inset shows the stratigraphic relationship of Marmot volcanics to basin sediments. After Cecile (1982).

cemented by calcareous or argillaceous ooze, typical of peperites associated with seamounts where basalt erupted into marine basins. At the volcanic centre the upper half of the formation is mostly volcanic conglomerate and fossiliferous limestone, indicating shoaling and erosion of the volcanic edifice. The Marmot basalts contain primary biotite phenocrysts, and chemical analyses (Goodfellow et al., 1980) confirm that they are alkalic, suggesting a rift origin.

Farther west, in the Ogilvie Mountains, the Selwyn Basin volcanics at least locally of Early Ordovician age comprise up to 2000 m of sericitized basic lava flows, tuff breccia, stratified ash and lapilli deposits and debris flows (Roots, 1982). The fragmental rocks include both peperite, deposited with lime mudstone, and subaerial cinder deposits. Vesicular lava forms massive flows with related flow-top breccia or, more commonly, discontinuous pillow accumulations. The high density of vesicles and abundant chlorite alteration suggest deposition within a shallow marine environment. Areal distribution suggests narrow, fissure-like source areas associated with deep-seated crustal extension. Rapidly built volcanic piles locally may have risen above the surface.

In the southern Pelly Mountains, Gordey (1981) reported 450 m of basalt in the Upper Cambrian to Ordovician Kechika Group. The succession comprises flows, pyroclastic rocks, and derived sediments. The basalt includes augite-bearing and augite-free varieties, locally with calcite- or chlorite-filled amygdules. Chemical analyses show small scatter on an AFM diagram and, on an alkali-silica diagram, plot in the alkaline field. According to Gordey (1981) "the basalt and tuff were erupted from local centres and intertongue with contemporaneous marine pelite. The basalt forms thick local accumulations that may have been built to above sea level."

The above are only three of many local basaltic assemblages that occur as minor lithologies within the lower and middle Paleozoic strata of Yukon Territory and northeastern British Columbia. Thicknesses generally range from a few metres to several tens of metres and the units are laterally discontinuous. Basic volcanic assemblages have been reported in Middle Ordovician and Lower Cambrian strata of the southern Mackenzie Mountains (Gabrielse et al., 1973), within the Upper Cambrian-Lower Ordovician Rabbitkettle Formation in the Nahanni area (Gordey et al., 1982); in Lower Cambrian rocks in the Coal River area of southeastern Yukon (Gabrielse and Blusson, 1969); in Middle Ordovician platform and basin strata of northeastern British Columbia (Thompson, 1976; Cecile and Norford, 1979), associated with exhalative barite in Middle Ordovician shale in the Akie River area (MacIntyre, 1981); with a shale-dolostone-limestone sequence of Silurian age in the western Rocky Mountains east of Prince George (Campbell et al., 1973); in Cambrian limestones in the British Mountains of northern Yukon (Norris, 1974); associated with lower Paleozoic basin strata of the northwestern Selwyn Basin (Green, 1972); with Middle Cambrian Ordovician and Silurian carbonates on the Ogilvie Platform (Green, 1972); associated with lower Paleozoic basin strata throughout the Anvil Range near Ross River (Tempelman-Kluit, 1977a; Gordey, 1983); and southwest of the Tintina fault in the Pelly Mountains (Tempelman-Kluit, 1977a,b; Gordey, 1981). Volcanic rocks are scarce in the miogeoclinal lower Paleozoic strata of the southeastern Cordillera. Pillowed, amygdaloidal, basic metavolcanic rocks occur in the Lower Cambrian Hamill Group at several localities in the northern Selkirk Mountains (Lane, 1977). Chlorite-magnetite schist and meta-tuff are present locally. The association of the volcanic rocks with feldspathic sandstone and conglomerate suggest a syndepositional rift environment (Devlin and Bond, 1984; Bond et al., 1985).

Diatremes of alkalic ultrabasic composition in the "White River Embayment" of the southern Rocky Mountains have been emplaced mainly into Ordovician and Silurian carbonate and sandstone and locally grade into pyroclastic and volcaniclastic strata (Pell, 1987). The rocks are rich in mafic minerals including clinopyroxene and augite. All were emplaced near the shelf to off-shelf hinge line of the lower Paleozoic Cordilleran Miogeocline and suggest a period of extension between 435 and 440 Ma.

The Cassiar Terrane, considered to be a displaced slice of ancestral North America, includes local members of basalt and agglomerate in argillaceous limestone and calcareous shale of the Upper Cambrian to Lower Ordovician Kechika Group. In the Pelly Mountains where the volcanoes are best developed they are as much as 450 m thick. The rocks are generally much altered but analyses of relatively fresh samples suggests a K-poor alkali olivine basalt composition (Gordey, 1981). The volcanics appear to have formed in a rift setting along the outer margin of the miogeocline.

Upper Paleozoic

The final and most widespread rifting events in the miogeocline occurred in Devono-Mississippian time. A thick clastic wedge of chert-bearing grit, the Earn Group, was locally deposited unconformably on older rocks of the miogeocline. Volcanic assemblages associated with the Earn Group are reported from several places within the Pelly Mountains.

In the southern Pelly Mountains, Gordey (1981) described a succession of acid volcanic flows, tuffs, and intercalated shale at least 1700 m thick. The volcanics overlie Upper Silurian to Middle Devonian dolomite along a northwesterly trending belt within the Cassiar Terrane. Shale intercalated with the volcanics is similar to that in the Earn Group. More than 60% of the volcanic unit consists of tuff and fine breccia, with clast size commonly less than 5 cm. Flow rocks are aphanitic with small phenocrysts of alkali feldspar. The chemistry of this suite is distinctly different from that of the older Ordovician and Silurian basalts. Silica ranges from 60 to 75% and on both the AFM and alkali-silica diagrams analyses plot in fields that are typical of over-saturated, alkaline to peralkaline suites associated with continental rift zones.

Mortensen (1981) described a highly variable sequence of felsic tuff, breccia, and flows in the Earn Group of the central Pelly Mountains. Detailed isotopic studies of flows and associated intrusions (Chronic, 1979) yielded a Rb-Sr mineral isochron of 333.0 ± 10 Ma and a K-Ar date of 319 ± 10 Ma. The volcanic succession is discontinuous. Individual flows and fragmental units are from 1 to 20 m thick, and the total thickness of volcanic sections ranges from 100 to 600 m.

Volcanic units of intermediate composition form the lower part and the bulk of the Pelly Mountains succession. They comprise massive, locally amygdaloidal trachyte flows, and tuffs. K-feldspar and albite-twinned plagioclase are the predominant, and commonly the only silicate minerals, but a distinctive "dark trachyte" contains 12 to 15% riebeckite phenocrysts in a trachytic groundmass of K-feldspar, plagioclase, and minor biotite. The intermediate succession is overlain by a widespread unit of felsic, crystal-lithic tuffs erupted during an episode of explosive volcanism. These deposits contain fragments of sanidine, minor plagioclase and, locally, partly resorbed euhedral quartz grains. The related volcanic centres are syenite intrusive bodies flanked by coarse proximal breccia that becomes finer grained outward. The syenite and trachyte stocks, domes, and sills are mineralogically identical to the felsic tuffs. In the largest syenite stock (12 by 4 km) K-feldspar grains reach 1 cm in diameter. The presence of miarolitic cavities indicates emplacement at shallow depths. The hypabyssal nature of coarse grained syenite intrusions also has been noted by Morin (1977) in the nearby St. Cyr Range, where contacts between trachyte flows and syenite plutons are gradational.

The felsic volcanics are host to, or are immediately overlain by several types of exhalite deposits, formed by hydrothermal convection through the volcanic pile (Morin, 1977; Mortensen and Godwin, 1982). Pyritic chert is widespread, whereas barite horizons and pyritic iron formation are much more restricted.

Alkali ultrabasic diatreme breccias and dykes intrude Upper Cambrian to Permian miogeoclinal rocks in the Western and Main ranges of the Rocky Mountains (Pell, 1987; Ijewliw, 1986). They were emplaced prior to Jura-Cretaceous deformation and appear to be related to rifting along the continental margin during deepening of the miogeoclinal basin. At least one of the diatremes is a true kimberlite, containing xenoliths of garnet and spinel lherzolite as well as granitic xenoliths which indicate the presence of continental basement beneath the miogeoclinal succession.

Volcanic rocks of pericratonic Kootenay Terrane

The pericratonic, Kootenay Terrane forms a fairly continuous belt of highly deformed rocks that were thrust eastward over the displaced miogeoclinal succession and are themselves overthrust by younger terranes to the west. It includes volcanic sequences in the lower Paleozoic Lardeau Group (Jowett Formation), and the Lower and Upper Carboniferous Milford Group (Keen Creek assemblage). Each of these volcanic successions consists of tholeiitic pyroxene-porphyry pillow lava, flows, tuffaceous greenstone, and volcanic conglomerate associated with chert. The volcanic units are interstratified with fine grained clastics and carbonates and although the rocks are more highly deformed the association is analogous to the rift-related volcanics within the miogeoclinal succession (Marmot Formation and Earn Group). However, basic lavas near the top of the Lardeau Group may be correlative with clastic, arc-like volcanic rocks farther west in the Eagle Bay Formation which gave a U-Pb age of 387 Ma. Remnants of a late Paleozoic magmatic arc also are found farther west in the Slide Mountain Terrane (Kaslo Group) and Quesnellia. There the Permo-Carboniferous, Mount Roberts Formation includes pebble conglomerate with clasts of basic and salic volcanics as well as quartz and coarse feldspar, which suggest contributions from an island archipelago or cratonic source.

VOLCANISM IN THE ACCRETED TERRANES

Many of the Devono-Mississippian to Upper Jurassic rocks of the western Cordillera were not deposited on the continental margin but rather in arcs and basins that lay far to the south and west. The amalgamation of these distant terranes into larger blocks, the accretion of the resulting superterranes to western North America, and their subsequent fragmentation by pervasive transcurrent faulting, produced the present collage of the western Cordillera. Mantle-derived volcanic rocks on these terranes are far removed from the root zones in which their parent magmas were generated, and salic volcanic rocks of crustal origin, though possibly still coupled to their protoliths, are isolated from their original tectonic setting.

Proterozoic-lower Paleozoic assemblages

With few exceptions the oldest exposed rocks in the terranes are late Paleozoic or Mesozoic in age, but upper Precambrian rocks are exposed on the Alexander Terrane, Devonian rocks are reported from a single locality in Stikinia, the Sicker Group on Vancouver Island (Wrangellia) includes rocks possibly as old as Silurian.

Volcanism in Alexander Terrane

Volcanic rocks of Late Cambrian to Early Ordovician age are interlayered with the oldest fossiliferous rocks in the Saint Elias Mountains. In the Dezadeash area the "Field Creek" volcanics (Campbell and Dodds, 1982c) comprise at least 1600 m of thick-bedded plagio-phyric andesite and basalt flows, locally with red oxidized tops and, more rarely, varicoloured volcanic breccias and pillow lava. They are interbedded with carbonate rocks, pebble conglomerate, and minor tuffaceous sandstone containing Late Cambrian brachiopods. The assemblage suggests a mixed shallow marine and subaerial depositional environment.

In the Donjek and Alsek ranges of Saint Elias Mountains widespread early Paleozoic volcaniclastic sediments, at least 2000 m thick, are interlayered with thin carbonate beds, locally associated with lenses of basic volcanics and intruded by basic dykes, sills, and small plutons (Campbell and Dodds, 1982a,b,c, 1983a). Some of the basic volcanics may be facies equivalents of the "Field Creek" volcanics, but others appear to be younger, grading upward into, and locally interfingering with, the basal part of fossiliferous, Lower Ordovician carbonates. The Lower Paleozoic assemblage in the Saint Elias Mountains of western Yukon Territory is in part correlative with the Wales Group and Descon Formation in southeastern Alaska and is believed to be a marine volcanic arc complex.

Upper Paleozoic assemblages

Volcanic rocks comprise a major part of the upper Paleozoic successions in all of the accreted terranes. Oceanic volcanics predominate in the Slide Mountain and Cache Creek terranes, whereas rocks with arc affinities characterize Quesnellia, Stikinia, and Wrangellia.

Slide Mountain Terrane

The Slide Mountain Terrane (Fig. 14.4) comprises three principal allochthons: the Anvil Allochthon of Yukon Territory and northern British Columbia, the Sylvester Allochthon of north-central British Columbia, and the Slide Mountain Allochthon of south-central British Columbia.

The upper Paleozoic Anvil Allochthon (Tempelman-Kluit, 1979) is a complex of sheared ophiolite including locally pillowed basalt, serpentinized peridotite, and altered gabbro, probably representing a slice of structurally disrupted oceanic crust. It rests on the extensive Nisutlin Allochthon and is overlain in many places by the Simpson Allochthon, although in places the various allochthons are completely interleaved. The Nisutlin rocks include siliceous cataclasite, slate, and both intermediate and basic volcanics. The uppermost, Simpson Allochthon, includes granitic cataclasite that may be the deformed root of Devono-Mississippian volcanic arc (Mortensen, 1981).

The Anvil and Nisutlin allochthons are probably equivalent to less metamorphosed sedimentary and volcanic rocks in western Yukon Territory and adjacent Alaska, where the presence of early Mississippian augen gneiss bodies in the Yukon-Tanana Terrane (Dusel-Bacon and Aleinikoff, 1985) suggests that an early Paleozoic magmatic arc was associated with the Anvil ophiolite succession. According to Dusel-Bacon, correlation of the Alaskan augen gneiss with similar bodies in Yukon Territory provides evidence for an early Mississippian plutonic belt 400 km long.

In north-central British Columbia the Sylvester Allochthon (Gordey et al., 1982; Harms, 1984) comprises Upper Devonian to Triassic chert, greenstone, gabbro, clastic and ultramafic rocks thrust over autochthonous or parautochthonous strata of ancestral North America in post-Triassic time. Gabbro and basalt are commonly intrusive into chert and argillite members, suggesting that the Sylvester is not a simple, tectonically disturbed ophiolite but rather a complex assemblage of oceanic sediments, volcanics, and subvolcanic intrusions (layer 1) whose emplacement was contemporaneous with tectonic shortening. The volcanic rocks are typically saussuritized, but diabasic textures and the presence of relict andesine and labradorite in specimens containing secondary oligoclase or albite suggest that the Sylvester greenstones were originally pyroxene andesites or basalts (Gabrielse, 1963).

A tonalite intrusion yielding a Permian zircon date (288 Ma) cuts at least one of the Sylvester thrust sheets. A body of granite in one of the thrust sheets has been dated by the U-Pb method on zircon at 277.5 Ma (R.R. Parrish, pers. comm., 1986). These dates not only confirm that some Sylvester thrust faults significantly predate emplacement of the allochthon, but also suggests that a late Paleozoic, calc-alkaline magmatic arc lay west of, or within, the Sylvester ocean basin (Fig. 14.4). The presence of quartz-bearing greywacke as a minor facies in some of the Sylvester thrust slices suggests a source for detrital quartz from an arc or island archipelago to the west or the North American craton to the east.

In the Nina Creek area of the Omineca Mountains an ophiolitic succession of probable Permian age includes massive variolitic, locally pillowed basalt, fine grained tuff, and small, sheared serpentinized peridotite and gabbro bodies (Gabrielse, 1975).

At the north end of the Shuswap Metamorphic Complex the Slide Mountain Group consists of an imbricated succession of basic pillowed flows, chert, argillite, and diabase sills (Antler Formation), which are thought to be correlative with basalts and associated marine sediments of the Fennell Formation of south-central British Columbia and with basic volcanic rocks in the nearby Milford Group. Similar rocks are included in the Kobau, Knob Hill, Chapperon and Anarchist groups in the southern part of the Intermontane Belt.

The Slide Mountain Assemblage of pillow basalt, chert, argillite, diabase, and serpentinized peridotite suggests an ophiolitic affiliation, although no complete ophiolite successions are known. According to Read (1976) some of the ultramafic rocks may be parts of differentiated sills. Also, the chemistry of the lavas has been variously interpreted. According to Monger (1977a) major-element analyses from the Sylvester and Anvil Range groups show a predominance of basalts that plot on the AFM diagram within the onland ophiolite field of Bailey and Blake (1974). In contrast, diabase in the Nina Creek area, presumed to be comagmatic with the flows, is notably low in potash and relatively high in titanium, typical of low-K abyssal basalt (Gabrielse, 1975). The inherent problem of using alkalies as chemical discriminants in altered rocks was addressed by Hall-Beyer (1976), who demonstrated that almost all of the Slide Mountain analyses plot within the oceanic field of the TiO_2-K_2O-P_2O_5 diagram developed by Pearce et al. (1975). This, plus relatively low, bimodal, potash levels (0.13% and 0.30% compared to an average of 0.14% for oceanic and 0.40% for island-arc tholeiites) support the contention that the suite as a whole reflects the geochemistry of ocean floor rather than island arcs. Aggarwal et al. (1984) suggested that correlative greenstones of the Fennell Formation are characteristic of alkalic and transitional basalts. They contain abundant phenocrysts of Al- and Ti-enriched augite and titanium-rich amphibole, which are typical of modern alkaline and transitional basalts of oceanic islands or seamounts.

Volcanic rocks of the Kaslo Group (Klepacki and Wheeler, 1985) are divided by the Whitewater Fault. Units in the lower thrust plate comprise tholeiitic pyroxene-porphyry pillow lava, flows, and tuffaceous greenstone interbedded with cherty tuff. The upper thrust plate is floored by ultramafic rock in turn overlain by clastic sediments and a thick succession of pillow lava. The ultramafic unit is interpreted as oceanic floor basement to the upper plate of the Kaslo Group.

Quesnellia

In south-central British Columbia and northern Washington the oceanic rocks of the Slide Mountain Terrane are flanked on the west by Quesnellia where upper Paleozoic rocks may be remnants of an arc and subduction complex (Okulitch and Peatfield, 1977). The sequences include parts of the Harper Ranch and Anarchist groups, which contain abundant clastic detritus, including volcanic sandstone as well as minor andesite, basalt flows, and pyroclastics. According to Smith (1974) the sandstones have an andesitic to dacitic provenance, and the presence of pumice indicates that volcanism was probably contemporaneous with clastic deposition.

In the Lay Range of the Omineca Mountains a succession of clastic and volcanic rocks, at least in part of Pennsylvanian age, is as much as 1000 m thick and includes thick units of andesitic to basaltic lava, breccia and tuff. The rocks are tentatively assigned to the Harper Ranch Subterrane.

If these upper Paleozoic rocks in Quesnellia are the remnants of a volcanic arc that lay west of the Sylvester ocean its original extent is unknown. In the south it may have been the source of the lenses of salic volcanic rock that are interbedded with sediments in the lower part of the Slide Mountain Assemblage. A northern continuation of the arc may be represented by tonalite and other upper Paleozoic plutonic rocks of Quesnellia. Also, tonalite of Permian age (280-290 Ma) within the Sylvester Terrane of north-central British Columbia may be related to a Quesnel arc that lay west of the Sylvester ocean.

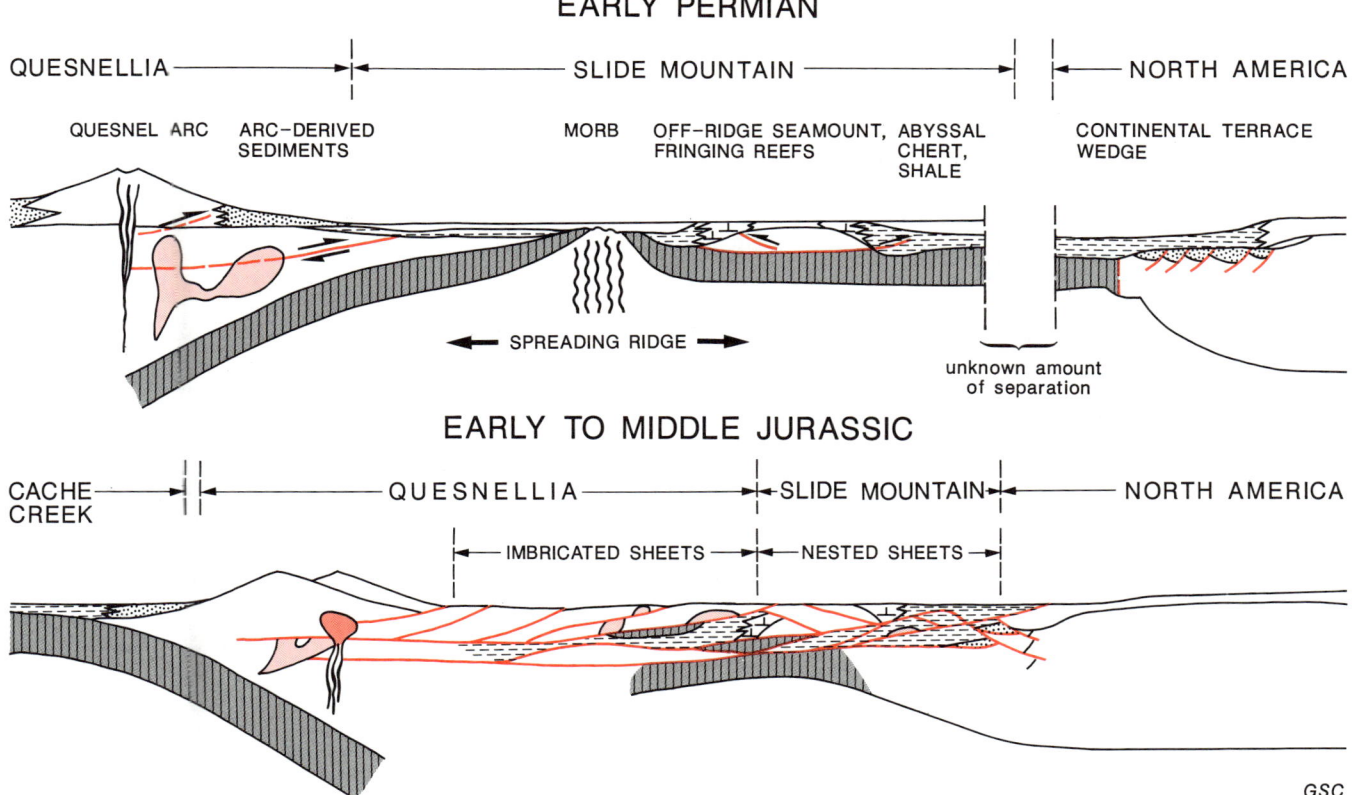

Figure 14.4. Schematic sections across the north-central Cordillera showing an interpretation of the origin of the Slide Mountain Terrane and its relationship to Quesnellia and the continental margin.

Cache Creek Terrane

The Cache Creek Terrane (Monger, 1977a,b; Shannon, 1981) ranging in age from Late Mississippian to Late Triassic in the northern Cordillera and from Middle Pennsylvanian to Middle Jurassic in the southern Cordillera is characterized by discontinuous pods of metabasalt and small to large bodies of ultramafic rock associated with chert, argillite, and local shallow-water limestone. The volcanic rocks are mainly basaltic flows that are locally pillowed, but include massive diabase and local breccia. Major-element analyses plot in a field that is transitional between alkaline basalt and tholeiite. The ultramafic rocks are variably serpentinized peridotite and dunite, commonly associated with cumulate ultramafic and gabbroic rocks, trondhjemite, and diabase, all mainly of subgreenschist grade. In places they have undergone multiple deformations, at least one of which is associated with blueschist metamorphism.

The Cache Creek assemblage is clearly of oceanic origin. Most of the volcanics are tholeiitic, similar in composition to typical mid-ocean ridge basalts. A smaller proportion of the lavas are relatively enriched in alkalies and may have originated as seamounts. A thick basaltic succession in the Atlin region (Monger, 1977a) is overlain by carbonate, which is thought to be a large bank, reef and lagoon complex. The carbonate assemblage spans a time interval of about 100 Ma, from Late Mississippian to Late Triassic, and represents a period of tectonic stability. It seems likely that the underlying volcanic platform was deposited as large, off-ridge seamounts locally capped by atolls. Limestone with primary textures indicative of shallow-water deposition is commonly associated with volcanic breccia and chert. These are believed to be products of submarine slumps from carbonate-volcanic atolls into deep-water chert basins.

Stikinia

The oldest rocks in Stikinia are Paleozoic (Early Devonian to Permian) and occur around the periphery of, and probably beneath much of the Bowser Basin. The succession consists of basalt to rhyolite flows and pyroclastic rocks, interbedded with carbonate, shale, volcanic sandstone, and minor chert. East of Bowser Basin carbonate rocks have been dragged up along the margins of Mesozoic intrusions and are consequently sheared and metamorphosed. The Asitka Group in McConnell Creek area is a sequence of fossiliferous Lower Permian tuffaceous and argillaceous carbonate overlain by basalt and rhyolite, the latter giving a Rb-Sr age of 250 ± 20 Ma. West of Bowser Basin and on Oweegee Peak, an inlier within the basin, the upper Paleozoic sediments are characterized by abundant tuff. The associated flows and breccias include basalt and andesite, but much less rhyolite than in the Asitka succession.

Analyses of upper Paleozoic lavas from Stikinia plot mostly in the calc-alkaline field of the AFM diagram. This, combined with the high proportion of tuff and coarse pyroclastic deposits, subaerial lavas, and volcanic sandstone, suggest that the sequence accumulated in an arc environment. Stikinia is entirely bounded by tectonic contacts, thus the original extent of the parent arc is unknown. However, Lower Pennsylvanian and older strata exposed in thrust sheets in the Chilliwack Terrane of southern British Columbia and northern Washington include basaltic to dacitic volcanic rocks. Palinspastic restoration (Monger, 1977a,b) suggests that these may be part of a Permian arc that was coextensive with the southern part of Stikinia.

Wrangellia

Paleozoic rocks are exposed only in the extreme northern and southern parts of Wrangellia. In Alaska, upper Paleozoic rocks of the Skolai Group consist of a lower volcanic sequence of basalt, basaltic andesite, minor andesite and dacite flows and pyroclastic rocks, overlain by interbedded sandstone, argillite, chert, and limestone. Correlative rocks in most parts of adjacent Yukon Territory and northern British Columbia comprise a lower sequence of basic volcanic breccia up to 1000 m thick, interbedded locally with siliceous, vitric-lithic tuff, and overlain by up to 800 m of fine clastic, Lower Permian sediments.

No volcanic Paleozoic rocks are known in the Queen Charlotte Islands, but near the southern end of Wrangellia, volcanics occur in mid- to upper Paleozoic rocks of the Sicker Group exposed on Vancouver Island (Muller, 1980b; Sutherland Brown and Yorath, 1985; Brandon et al., 1986). The Sicker Group rocks are generally of epidote-actinolite grade, but primary textures as well as sparse microfossils are preserved. The lowest unit comprises at least 3000 m of augite-bearing agglomerate with intercalated lapilli tuff, lesser pillow lava, epiclastic breccia, and minor chert. In central Vancouver Island this basal unit is overlain by 700 to 1200 m of volcanic sandstone, conglomerate, cherty argillite, and argillite intercalated with andesitic pillow lavas, lapilli tuff, and flow-layered rhyolite. The uppermost unit of the group comprises between 100 and 400 m of crinoidal limestone and minor chert and argillite. Correlative units farther east, in the Nanoose Uplift of eastern Vancouver Island, are dominated by epiclastic rocks, turbidites, and chert with some crinoidal limestone, a relationship that suggests a polarity from volcanic centres in the west to an epiclastic apron to the east. No volcanic rocks are interbedded with sediments in the upper 200 to 750 m of the Sicker Group.

The base of the Sicker has not been observed, but zircon dates (362 and 364 Ma) from synvolcanic quartz porphyry plutons in the lower volcanic succession suggest that it is at least as old as Late Devonian. Early Carboniferous to Early Permian conodonts and radiolaria occur in the upper sedimentary succession (Brandon et al., 1986).

Both the Skolai and Sicker assemblages are interpreted as mid- to late Paleozoic arcs developed on oceanic crust. The volcanic piles and underlying gabbroic basement along the axis of the Skolai arc appear to have been uplifted and eroded one or more times during late Paleozoic time, producing wedges of epiclastic, graded sandstones that prograded from the arc onto oceanic crust in adjacent deep-water basins. Chemical differences between the Skolai and Sicker assemblages are apparent in the greater range of silica and alkalis in the Sicker rocks, which is consistent with the greater variety of rocks reported from field observation. Extreme iron enrichment suggests that they represent a highly fractionated tholeiitic suite. In contrast, many of the Skolai analyses have alkaline affinities, and the suite appears to be transitional between alkaline and tholeiitic rocks, possibly reflecting the presence of back-arc lavas. Despite their present wide separation the Skolai and Sicker assemblages were probably once coextensive.

Alexander Terrane

Volcanic rocks and associated basic intrusions of late Paleozoic age in the Saint Elias Mountains of western Yukon Territory are confined to the Kluane Ranges between the Duke River and Denali faults where they occur within a complex tectonic mélange arising from combined dextral strike-slip movements along both the faults (Campbell and Dodds, 1982a,b,c, 1983a,b). As a consequence the stratigraphy, thickness, and lateral extent are unknown. The assemblage may include rocks of more than one age and terrane affinity. The rocks are regionally metamorphosed from sub- to mid-greenschist facies and include gabbro, diabase, rare peridotite, and both marine and subaerial varicoloured flows and pyroclastic rocks. The volcanic assemblage is associated with shallow-water marine sediments containing Middle Devonian to Early Permian macrofossils and Devonian conodonts.

Triassic assemblages

In the Intermontane Superterrane the Nicola and Rossland groups of southern Quesnellia and the eastern belt of Takla Group rocks in central and northern Quesnellia range in age from latest Triassic to Early Jurassic. The Cache Creek Terrane contains a mixed assemblage of oceanic and arc lithologies that include rocks as young as Late Triassic. Volcanic rocks of Stikinia include the western Takla, and the Stuhini and Lewes River groups, which range in age from Late Triassic to middle Jurassic. In the Insular Superterrane Upper Triassic volcanic rocks comprise the Karmutsen Formation and Nikolai Greenstone of Wrangellia and part of the Alexander Terrane.

Quesnellia

In southern Quesnellia Upper Triassic volcanic rocks and sediments of the Nicola group (Preto, 1977; Mortimer, 1986; Church, 1973) rest unconformably on upper Paleozoic strata that were deposited west of the Sylvester oceanic plate during an earlier (Permian) episode of arc volcanism.

Calc-alkaline acidic to intermediate flows, breccia, and volcaniclastic rocks as well as basic volcaniclastic rocks in the western facies of the Nicola appear to be intercalated, on the west, with oceanic rocks of the Cache Creek Terrane. The western facies is a typical arc assemblage and the presence within it of local ignimbrite and subaerial flows indicates that at least part of the arc was emergent. The arc assemblage is flanked on the east by a predominantly intermediate, feldspar and feldspar-augite porphyry volcaniclastic assemblage, probably deposited in a back-arc basin. Still farther east the volcaniclastic wedge is overlain by an easterly thickening pile of relatively alkaline, augite porphyry pillow lavas and related volcaniclastic rocks that comprise the widespread eastern facies of the Nicola Group.

The distinct sedimentological and chemical polarity of the Nicola assemblage suggests that it is a calc-alkaline arc complex developed above an east-dipping subduction zone and flanked on the east by a back-arc basin in which a large volume of contemporaneous alkaline lava was erupted.

Cache Creek Terrane

Suspended within the mélange of the Cache Creek Group in southern British Columbia are blocks that resemble lithologies in the western, calc-alkaline facies of the adjacent Nicola arc. Late Triassic dates on the enclosed blueschist facies rocks suggest that the mélange represents an accretionary wedge that was contemporaneous with Nicola volcanism above an east-dipping subduction zone. Farther north, in the Cassiar Mountains, volcanic and volcaniclastic rocks are interlayered with clastic and carbonate sediments of the Upper Triassic Kutcho Formation (Thorstad and Gabrielse, 1986). These rocks, which include flowbanded rhyolite and silicic tuff, are commonly metamorphosed to greenschist or higher grades. Lithologically similar, correlative rocks occur also in the Omineca Mountains (Sitlika assemblage of Monger et al., 1978) and a unit of andesite and basalt of the same age overlies Cache Creek strata in the Whitehorse Trough, Yukon Territory (D.J. Tempelman-Kluit, pers. comm., 1985).

Stikinia

Upper Triassic volcanic assemblages are associated with sedimentary and intrusive rocks in the Lewes River, Stuhini, and western Takla groups. The Lewes River and Stuhini groups are best developed in the northern and north-central part of Stikinia, whereas the western belt of the Takla Group is confined to its eastern margin. A suture, or intervening belt of Cache Creek rocks between the eastern and western Takla belts suggests that the rocks are parts of two distinct volcanic provinces that were separated by oceanic crust during at least part of Late Triassic time (Fig. 14.5).

In central Yukon Territory volcanic rocks in the lower part of the Lewes River Group are interbedded with clastic sediments in the western facies of Whitehorse Trough (Wheeler, 1961). From 30 to more than 300 m of black, purple, grey, and green basalt and andesite flows, breccia, and tuff are associated with coarse conglomerate and greywacke that were derived from a western source. The breccias, which contain angular blocks as much as a metre across, are believed to be proximal to a volcanic terrane that lay a short distance to the west. In Yukon Territory, the source region of the western volcanic facies has been obscured by deformation and plutonism, but farther south, in the Tulsequah area of northwestern British Columbia, clastic rocks equivalent to Upper Triassic strata in Whitehorse Trough are flanked on the west by thick sections of proximal lava flows, pyroclastic and volcaniclastic rocks of the Stuhini Group. Varicoloured andesites and basaltic-andesites with phenocrysts of augite and/or feldspar, and bladed feldspar porphyry with radiating feldspar clusters up to a centimetre across are characteristic of many Stuhini volcanic and related subvolcanic intrusions. At least 3600 m of mainly subaerial andesitic flows, pyroclastic, and volcaniclastic rocks are exposed in the western Tulsequah area, whereas sections of similar age and composition in the central Tulsequah region include as much as 1200 m of pillow lava interlayered with marine volcaniclastic rocks deposited in a southern extension of Whitehorse Trough. South of the Tulsequah area the belt of Upper Triassic volcanics trends easterly along the axis of the Stikine Arch, where thick sequences of greywacke, siltstone, and volcanic sandstone are overlain by as much

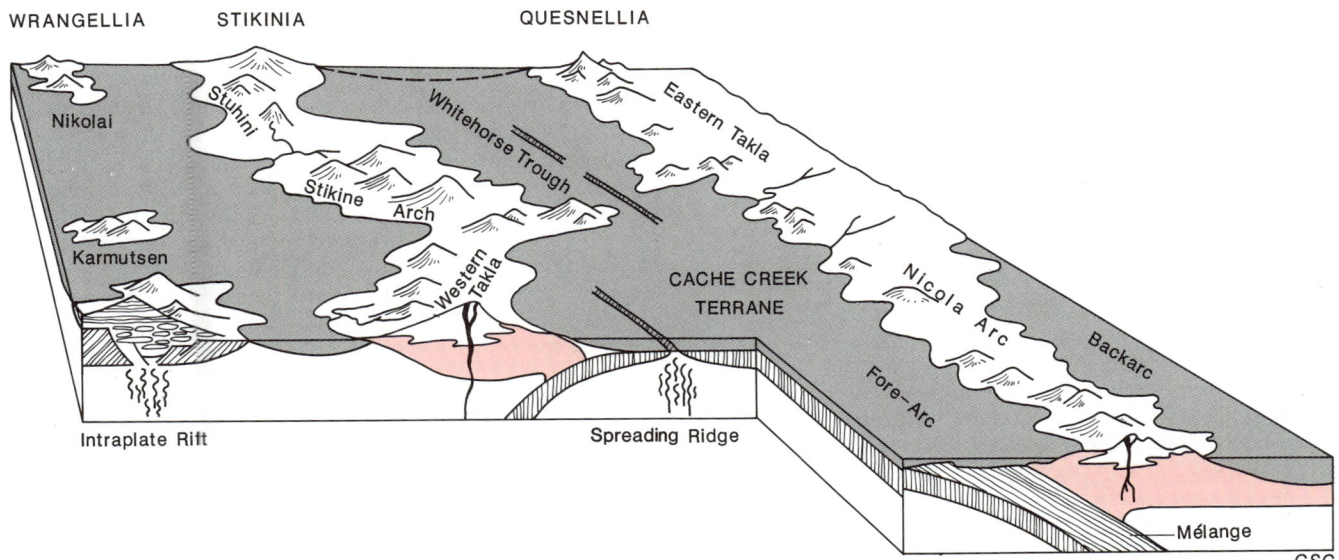

Figure 14.5. An interpretation showing the possible relationship between volcanic belts and basins of deposition in the north-central Cordillera during Late Triassic time. The separation between Wrangellia and Stikinia was probably much greater than shown in the sketch.

as 1200 m of varicoloured augite andesite and basaltic andesite flows, breccias, heterolithic conglomerate, tuff, and minor ignimbrite. Several large Late Triassic plutons on Stikine Arch are probably coeval with and, at least in part, comagmatic with the surrounding Stuhini volcanics. At the eastern end of the Stikine Arch, Upper Triassic rocks of the Stuhini Group are coextensive with western Takla Group rocks of similar age and lithology. The western Takla succession is at least 600 m thick and occupies southeasterly-trending basins that are truncated by the Vital Fault and other faults that separate Stikinia from the adjacent Cache Creek Terrane and Quesnellia. In McConnell Creek area the western Takla Group is divided into three formations (Monger, 1977c). The basal Dewar Formation comprises 700 to 1500 m of mainly submarine volcaniclastic rocks, sandstone, siltstone and, locally, graphitic shale. The overlying Savage Mountain Formation contains about 3000 m of mostly augite, and bladed feldspar porphyry flows and pyroclastic rocks. It includes thick successions of pillow lava, but the upper part is predominantly subaerial and the overlying Moosevale Formation includes 1600 m of subaerial volcaniclastic rocks with predominantly fine grained feldspar porphyry clasts.

Elsewhere in Stikinia Upper Triassic rocks vary widely in composition and thickness. Relatively thin successions of bladed feldspar porphyry breccias and flows have been reported along Babine Lake, in Whitesail and Nechako areas, and along the eastern side of the Coast Belt. In Terrace and Bowser Lake areas, south of the Stikine Arch, the Upper Triassic has no volcanic component, whereas basalt flows, tuff, and breccia are associated with clastic sediments in the Taseko Lake area of southernmost Stikinia.

Stratigraphic relationships suggest that the Stuhini-western Takla belt originated as a volcanic arc bounded on the northeast by Whitehorse Trough, into which volcaniclastic sediments were shed from volcanic islands and submarine vents. The axis of the arc has a broad sigmoidal trace that runs southeasterly in Yukon Territory, then easterly along the crest of the Stikine arch, and again southeasterly into the Takla belt of eastern Stikinia. The most intense igneous activity appears to have been along the central Stikine Arch segment, where the thickest volcanic successions are associated with Late Triassic batholiths.

The Stuhini volcanics are entirely calc-alkaline, whereas the western Takla belt includes rocks that have both calc-alkaline and alkaline affinities. This apparent chemical difference suggests that the Stuhini and Takla volcanics erupted in different tectonic environments. It is reasonable to speculate that the Stuhini-western Takla arc developed above a southwesterly dipping subduction zone related to closing of the Cache Creek ocean basin. The sigmoidal trace of the magmatic axis would result in different angles of convergence along different segments of the arc. T. Richards (pers. comm., 1985) proposed that the voluminous calc-alkaline volcanism along the central part of the arc resulted from rapid, orthogonal convergence, whereas oblique convergence along the southeasterly trending portions of the arc would result in pull-apart basins associated with thick clastic successions and more alkaline volcanism.

Wrangellia

Triassic volcanics of the Karmutsen Formation on Vancouver Island and Queen Charlotte Islands are correlative with the Nikolai greenstone of Alaska (Muller, 1977a,b; Muller et al., 1974; Sutherland Brown, 1968). On Vancouver Island the succession comprises up to 6000 m of pillowed, brecciated, and layered tholeiitic lavas, which rest on a substratum of Paleozoic volcanic and sedimentary rocks and locally on Middle Triassic argillite. Numerous basaltic sills and dykes in the substratum are thought to be subvolcanic equivalents. This thick basaltic sequence was produced in the interval between Late Ladinian and Late Carnian time, a span of possibly 10 Ma. The change from pillowed to layered lavas indicates temporary emergence of the basaltic shield. The lavas are overlain by thick- and thin-bedded, commonly bioclastic limestone of the Upper Carnian Quatsino Formation.

Similar sequences of Upper Triassic basalt, overlain by shelf sediments, are known throughout Wrangellia. In the Queen Charlotte Islands Sutherland Brown (1968) described the Karmutsen Formation as an accumulation of submarine basic lavas, related clastic rocks, dykes, sills, and minor limestone as much as 4300 m thick. In the Saint Elias Mountains the correlative Nikolai greenstone is 1000 m thick and in adjacent southeastern Alaska it reaches a thickness 3500 m. Unlike the Karmutsen, much of the Nikolai pile consists of subaerial flows but, like the Karmutsen, it is overlain by reefoid limestone. Throughout Wrangellia Karmutsen-Nikolai volcanism recorded the construction of vast submarine platforms of tholeiitic basalt, which rose locally above sea level and then subsided. As the volcanic edifices were submerged limestone was deposited on their surfaces and was later covered by fine clastic sediment.

The tholeiitic composition and great thickness of the Karmutsen-Nikolai Assemblage suggests that it may have been an oceanic plateau, possibly produced by rifting of older, Paleozoic arc sequences (Muller, 1977b; Jones et al., 1977; Barker et al., 1985).

Alexander Terrane

Upper Triassic volcanic rocks occur sporadically in the Saint Elias Mountains of western Yukon Territory and adjacent Alaska. In northwestern Tatshenshini River area they are assigned informally to the "Tats Volcanic Complex" (Gammon and Chandler, 1986). According to MacIntyre (1984) the succession is underlain by and interbedded with Upper Triassic turbidites and limestone. It comprises mafic pillow lava, a lesser volume of amygdaloidal andesite flows, minor limy argillite, chert, siltstone, and andesitic tuff, with a composite thickness of 2700 to 4700 m. The succession contains Early to Middle Norian conodonts (M.J. Orchard, pers. comm., 1985). The volcanic rocks have a calc-alkaline to alkaline affinity. This, plus the close association with turbidites and other coarse clastic sediments, suggests an arc, back-arc environment of deposition. MacIntyre (1986) speculated that the succession was deposited in narrow rift valleys associated with spreading centres along a transform fault system analogous to that in the modern day Gulf of California. The nearby Windy Craggy massive sulphide deposit (Cu, Co, Au, Ag, Zn) may

Co, Au, Ag, Zn) may be the product of a hydrothermal system related to eruption of the "Tats Volcanics".

In southeastern Tatshenshini River area a thick sequence of mostly basic volcanics is isolated from, but probably at least partly the same age as the "Tats Volcanic Complex". These volcanics have been regionally metamorphosed to chlorite grade and include basalt and andesite flows, rhyolite, tuff, and minor interbedded limy slate, siltstone, and carbonate (MacIntyre and Schroeter, 1985). The sediments have yielded both (?)Late Triassic and (?)Devonian conodonts. The volcanic succession is bimodal and includes a relatively greater proportion of acid rocks, some of which host polymetallic massive sulphide and barite deposits. Like the "Tats Volcanics" they have both calc-alkaline and alkaline affinities and are believed to have erupted in a rift environment associated with transform faulting. The acid rocks are believed to have been erupted as volatile-rich differentiates formed in subvolcanic, crustal reservoirs (MacIntyre, 1986).

Jurassic assemblages

In Quesnellia Jurassic volcanic rocks comprise the Rossland volcanics and part of the Nicola Group. In Stikinia diverse assemblages of Jurassic volcanic and volcaniclastic rocks are assigned collectively to the Hazelton Group, and in Wrangellia the principal Jurassic volcanic assemblages comprise parts of the Bonanza Group and somewhat younger Yakoun Group of Vancouver Island and Queen Charlotte Islands, respectively. In each of these terranes a hiatus at the beginning of the Jurassic Period was followed successively by an Early and Middle Jurassic episode of volcanism, a widespread episode of shale deposition and, finally, by uplift accompanied by marine and nonmarine clastic deposition in marginal basins (Tipper and Richards, 1976; Tipper, 1984) (Fig. 14.6).

Quesnellia

Following a hiatus in latest Triassic to Early Jurassic time volcanism in southern Quesnellia resumed with eruption of the Sinemurian Rossland volcanics (Beddoe-Stephens, 1982). The succession consists of variably metamorphosed agglomerate, conglomerate, lavas and volcaniclastics intercalated with marine shale and siltstone. The lavas range from augite-rich ankaramites to plagioclase-phyric andesite which are believed to have evolved in an island arc setting. The suite, which includes silica undersaturated rocks, is not strictly calc-alkaline. The parent magma is thought to have originated in the upper mantle. Melting at about 20 kbar followed by crystal fractionation could account for the highly porphyritic ankaramitic compositions. Like the underlying Nicola, these lavas are characteristically augite porphyry basalt, but the Jurassic volcanic front shifted eastward out of the Nicola belt toward the continent. By Middle Jurassic time the axis of magmatism had shifted still farther east; Middle Jurassic (173-182 Ma) plutons cut the Slide Mountain Terrane and locally encroach on North America. Although no eruptive equivalents of these plutons are preserved, it seems likely that they are the final expression of a magmatic front that migrated eastward across Quesnellia between Early Triassic and Middle Jurassic times.

Mafic, mildly alkaline, augite porphyry volcanics, similar to those of the Nicola and Rossland groups, are present in the Upper Triassic to Sinemurian, eastern Takla belt of central and northern British Columbia. This suggests that the early Mesozoic arc extended along the Quesnel Terrane at least the length of British Columbia.

Stikinia

Jurassic volcanic rocks in Stikinia are assigned to the Hazelton Group. Their distribution mimics that of the Lewes River, Stuhini, and western Takla rocks that were erupted during Late Triassic time but, except for a small area in the southwestern Coast Mountains, the two assemblages are separated by a widespread hiatus. As in the Triassic Period the Stikine Arch, which was a site of Triassic magmatism, and the Skeena Arch remained as relatively positive features, along which volcanic activity was concentrated throughout most of the Jurassic Period and from which much of the clastic and volcaniclastic debris was shed into the Whitehorse and Hazelton troughs (Fig. 14.6).

The Hazelton Group volcanics range in age from Early Sinemurian to Early Bajocian. North of latitude 57°, near the Stikine Arch, rocks of Hazelton age are assigned to the Toodoggone volcanics and to the Takwahoni Formation.

The Upper Pliensbachian to Lower Bajocian Takwahoni Formation is mainly a clastic sedimentary assemblage, locally 3350 m thick, that includes boulder conglomerates containing clasts of Triassic volcanic and granitic rocks. The detritus was derived from uplifted Triassic arc rocks along the Stikine Arch, and deposited in the western and southwestern part of Whitehorse Trough. Minor andesitic and basaltic andesite flows and pyroclastic rocks interbedded with coarse, proximal conglomerates near the base of the Takwahoni Formation indicate that volcanism resumed along the axis of the Stikine Arch shortly after uplift and erosion of the Upper Triassic Lewes River-Stuhini arc assemblages.

Southeast of Stikine Arch Jurassic volcanic rocks are assigned to the Toodoggone volcanics, which include a great variety of volcanic rocks that were erupted from a multitude of vents between Early Pliensbachian and Early Bajocian time. In the Toodoggone River area (Diakow, 1984) subaerial and shallow marine dacite and latite porphyry flows and pyroclastic rocks are associated with Early Jurassic granitic plutons and numerous small subvolcanic bodies. Tuffaceous sediments have yielded fossils of Toarcian age, and biotite and hornblende from the volcanics gave K-Ar dates ranging from 182 to 204 Ma (Diakow, 1985). The Sinemurian to Lower Bajocian assemblage comprises at least 700 m of subaerial lava flows, ash flow tuffs, and pyroclastic air-fall deposits. Interbedded conglomerate and finer grained clastic sediments indicate that volcanism was accompanied by periodic uplift and reworking of volcanic members. Also, tuffaceous carbonate rocks, fine grained clastic rocks, and siliceous tuffs flanking the subaerial deposits suggest periods of short-term tectonic stability.

Farther west the Toodoggone volcanics include at least three distinct volcanic assemblages: (1) an Early Pliensbachian assemblage that occupies a northwesterly trending basin near Cold Fish Lake in Spatsizi area, where

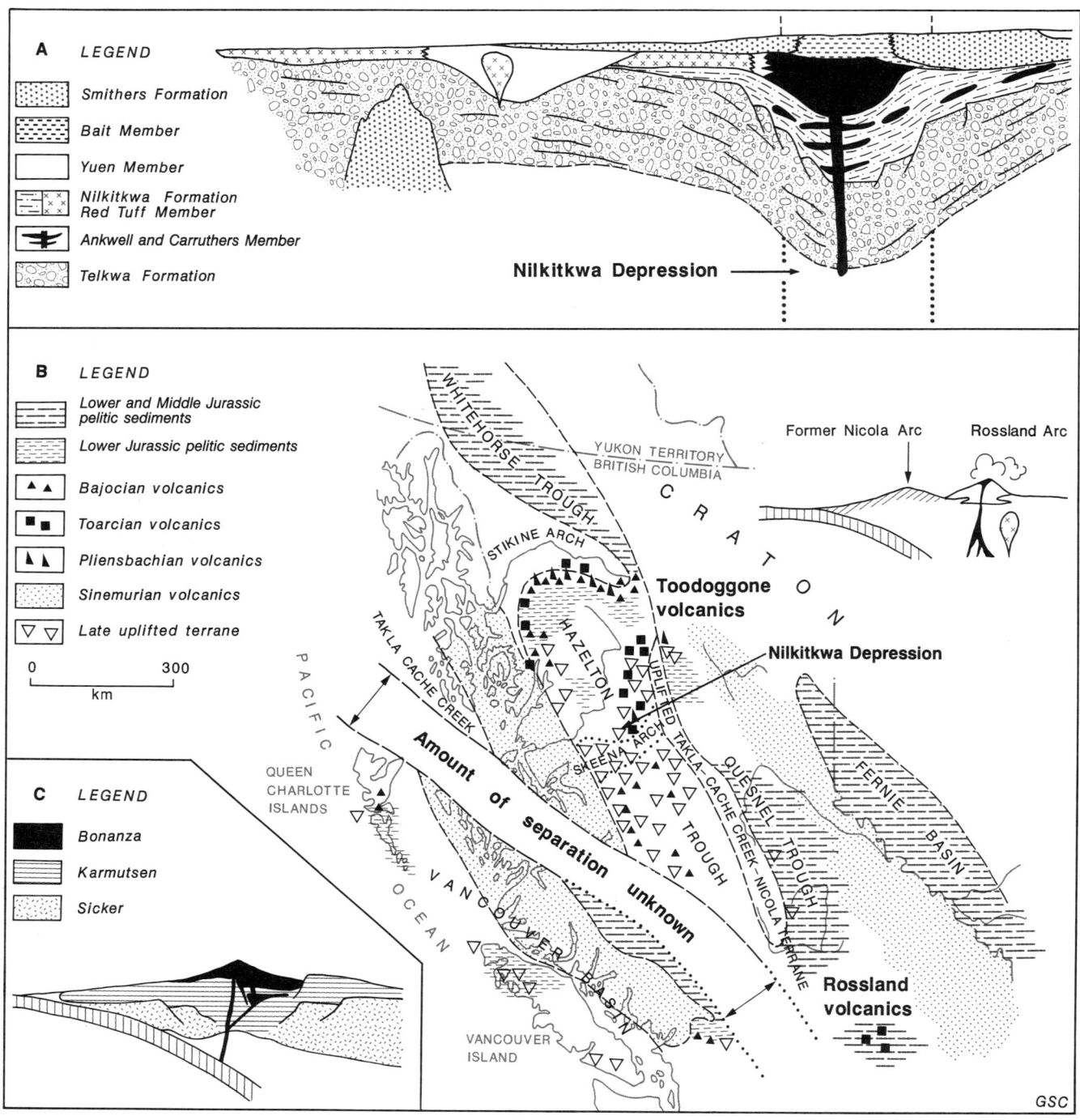

Figure 14.6. Sketch map (B) and interpretive sections (A and C) showing the distribution and tectonic setting of Jurassic volcanic rocks in British Columbia. Modified from Tipper and Richards (1976).

at least 600 m of rhyolite, dacite, andesite, and minor basalt flows and pyroclastic rocks are interbedded with fossiliferous sediments; (2) a Toarcian assemblage of andesite, basalt, and rhyolite tuff, breccia, and flows with interbedded fossiliferous sediments that lap onto the central part of Stikine Arch; and (3) a Bajocian assemblage of basalt and basaltic andesite that forms a discontinuous girdle around the northern and western part of Bowser Basin and is commonly associated with fossiliferous Lower Bajocian shale.

The volcanic succession records a decrease in the magnitude of pyroclastic eruptions with time: from an early major eruptive stage that produced widespread ash flow sheets and air-fall deposits, through an intermediate stage of sporadic volcanic activity characterized by the well layered calcareous tuffs, to a final stage marked by major lava flow eruptions.

The predominantly andesitic, Lower Pliensbachian succession in Spatsizi area includes andesitic to rhyolitic flows, breccia, and pyroclastic rocks including welded ash flows. The eruption of these rocks appears to have been coeval with the subsidence of northwesterly trending, fault-bounded basins in which both volcanic and epiclastic sediments accumulated. The stratigraphy in adjacent basins is commonly different and, in places, the bounding fault zones are occupied by tabular, probably comagmatic bodies of rhyolite.

The record of proximal Toarcian volcanics in the Spatsizi area is fragmentary, but the predominance of basalt and basaltic-andesite flows of this age on the central Stikine Arch suggests a correlation with the upper part of the Toodoggone pile.

The Lower Bajocian assemblage comprises basalt and basaltic-andesite pillow lavas, tuff breccia, tuff and tuffaceous siltstone. The finer grained tuffs and tuff breccia are exposed over wide areas and at several stratigraphic levels within the Bajocian sediments of the Hazelton Trough, whereas the pillow lavas are confined to a relatively narrow zone around its margins. West of Iskut River, in the Telegraph Creek area, Lower Bajocian shale is underlain by at least 1200 m of tuffaceous sandstone and brownish grey peperite consisting of quenched basaltic clasts in a limy mudstone matrix. Above the shale is a pile of pillow lava 2600 m thick that thins rapidly northward and grades into tuff breccia and peperite similar to the lower unit. The pillow lava and flanking aprons of tuff breccia and peperite are interpreted to be the proximal and distal facies, respectively, of seamounts that were episodically erupted into the margin of a subsiding shale basin. The distribution of the seamounts around the margins of the basin may reflect the locus of bounding faults along which subsidence of the Bowser Basin continued during latest Jurassic time.

Each of these three proximal volcanic assemblages is flanked by a distal tuff and volcaniclastic facies that was deposited in the northern part of Hazelton Trough (Fig. 14.6) and interbedded with the Spatsizi sediments.

South of latitude 57°, near the Skeena Arch, the Hazelton Group is subdivided (Tipper and Richards, 1976; Woodsworth, 1979a) into four major formations: the Telkwa, Nilkitkwa, Whitesail, and Smithers, each of which includes a variety of volcanic facies (Fig. 14.6A). The Hazelton Group represents an island arc and back-arc volcanogenic basin assemblage that accumulated on uplifted and eroded remnants of the Takla Group, and subsequently evolved into the basin in which the Bowser Lake Group accumulated.

The greater volume of the southern Hazelton Group is represented by the Telkwa Formation, which is exposed from the Coast Mountains eastward across Stikinia along the Skeena Arch. The Howson, Babine, Kotsine, Bear Lake, and Sikanni facies of the Telkwa represent calc-alkaline volcanism within different environments. Chemically the Howson facies is a low-alkali, high-iron, calc-alkaline suite, whereas the other facies are more alkalic. Volcanic rocks are most abundant in the southwest where local volcanoes of the Howson facies comprise up to 2500 m of reddish pyroclastic breccia. The locus of volcanism probably was centred over the Topley Intrusions (Fig. 14.6B; see Fig. 15.1, in pocket), resulting in the formation of an ancestral Skeena Arch. Air-fall tuff and breccia, epiclastic deposits, and interlayered flows of basalt and ignimbritic flows indicate explosive volcanism. Volcanic highlands are commonly surrounded by extensive areas of zeolitized rock formed by hydrothermal activity adjacent to vents. The volcanic arc was flanked on the east and northeast by a broad, shallow basin, with islands where marine and nonmarine volcanics and sediments of the Babine shelf facies were deposited. Farther northeast, in the Nilkitkwa Depression, the shelf facies pass abruptly into the correlative Kotsine facies, which includes more than 1500 m of submarine volcanics. The marine succession thins northward and laps onto a thick succession of red, subaerial volcanics of the Bear Lake facies. Subsidence of the Nilkitkwa Depression during Telkwa time may account for the onlapping of marine strata onto nonmarine volcanics. Along the eastern margin of the Hazelton Trough faulting along the Pinchi-Two Lake Creek fault systems was accompanied by volcanism. There a graben developed, traceable for 80 km, in which the Sikanni facies was deposited.

The abrupt transition from Telkwa Formation to Nilkitkwa Formation reflects a sudden Early Pliensbachian marine transgression. Only the southwestern part of the area, underlain by the Howson facies and the coeval Topley Intrusions, remained emergent. Elsewhere shallow-water pelitic and acidic tuffaceous sediments accumulated to thicknesses of 100 to 150 m above the shelf and thickened to more than 1200 m in the Nilkitkwa Depression. Eruption of acid tuffs, characteristic of volcanism within the Nilkitkwa Depression, was interrupted by episodic effusion of basic lavas of the Carruthers Member and terminated by eruption of more than 1000 m of alkali-olivine basalt of the Ankwell Member, an emergent pile at the top of the Nilkitkwa Formation. Distal, air-fall lapilli tuff (Red Tuff Member), related to eruptions in or near the Nilkitkwa Depression, was deposited to thicknesses of as much as 300 m across much of the Hazelton Trough (Fig. 14.6), including the emergent Howson and Babine areas.

A westward-directed marine transgression began in Late Toarcian time with overlap of the Red Tuff Member by fine grained clastic sediments of the mid-Toarcian to Lower Callovian Smithers Formation. Deposition of these sediments was accompanied by minor contemporaneous volcanism, leading to the formation of basaltic seamounts and flanking breccia and tuff deposits. A widespread, distinctively banded, siliceous tuff unit of Bajocian age is an excellent stratigraphic marker around the western and northern parts of the Bowser Basin. The close of Early Callovian time marks the end of widespread Mesozoic volcanism and the beginning of deposition of sediments derived from erosion of older terranes exposed by the rise of Skeena and Stikine arches. Evidence of volcanic activity during Middle and Late Jurassic deposition of the Bowser Lake Group is found in tuff beds interbedded with shale and siltstone of Early Callovian age in the Smithers and Whitesail Lake areas. The only other volcanic unit in the Bowser Lake Group is the Netalzul volcanics, a succession of mid-Oxfordian to earliest Late Oxfordian basaltic to andesitic, subaerial to shallow marine breccia, tuff, and thick flows exposed along the northern margin of the Skeena Arch in the Smithers and Hazelton areas (Tipper and Richards, 1976). The volcanic succession is up to 300 m thick and may be related to faulting along the margins of the Skeena Arch during its rapid uplift in the mid-Oxfordian.

Wrangellia

Jurassic volcanic successions are present in strata that conformably overlie the Upper Triassic (Karmutsen Formation) on both Vancouver Island and Queen Charlotte Islands (Fig. 14.6C).

On Vancouver Island the Sinemurian to Upper Pliensbachian Bonanza Group (Muller, 1977a,b) comprises a varied assemblage of intermediate to acid volcanic and volcaniclastic rocks including hawaiite, minor basalt of various types, tholeiitic andesite, dacite, and rhyolite as well as calc-alkaline dacite and rhyolite. Extensive pyroclastic deposits suggest that much of the volcanism was explosive. The volcanic rocks are associated with the Early to Middle Jurassic Island Intrusions (Northcote and Muller, 1972). These high-level plutons and dykes of quartz monzonite and quartz-feldspar porphyry yield Pb-Pb dates between 180 and 190 Ma. This and other age data suggest that the Island intrusions may represent magma chambers and subvolcanic feeders of the Bonanza volcanics (Isachsen et al., 1985).

South of the Cowichan Uplift the Bonanza Group contains a higher proportion of acid rocks compared with areas farther north (A. Sutherland Brown, pers. comm., 1985). The basal units, from 100 to 1000 m thick, are redbeds composed of microbreccia to angular lapilli-sized breccia. These are overlain by dacitic to rhyolitic, feldspathic vitrophyres or, in flanking areas, by rounded lapilli tuff breccias or fine andesitic agglomerates. Vitrophyres, up to 1500 m thick, are commonly overlain by well laminated welded feldspar-phyric ash flows intercalated with air-fall crystal-lithic tuffs. Where the Bonanza Group is thick, dykes of green feldspar porphyry form 30 to 50% of the volcanic pile, but they are notably absent in the Cowichan Uplift. On southern Vancouver Island the Bonanza succession has a large component of subaerially deposited pyroclastic rocks, probably related to caldera collapse. The area from Alberni Canal to Cowichan Lake may once have formed a single volcanic edifice rooted in a marine basin but largely deposited subaerially.

On Queen Charlotte Islands Jurassic volcanic rocks occur in the Sinemurian Kunga Formation, and in the Early Bajocian Yakoun Group (Sutherland Brown, 1968). The volcanic components of the Kunga comprise interbedded, varicoloured tuff and shale with local interbeds of fine breccia. No proximal Lower Jurassic rocks are known on Queen Charlotte Islands. The Yakoun Group (Cameron and Tipper, 1985), consists of volcanic members with a composite thickness of up to 850 m. Pyroclastic rocks predominate over flows and include feldspar-pyroxene-phyric andesite breccia and epiclastic conglomerate, interbedded with waterlain lapilli tuff and finely laminated volcaniclastic sediments (Sutherland Brown, 1968). Some of the lapilli tuffs have a calcareous matrix, indicative of eruption into submarine carbonate ooze.

Both the Bonanza Group and Kunga tuffs, and the somewhat younger Yakoun Group represent calc-alkaline arc sequences that were erupted from submarine and subaerial volcanic centres in an island archipelago.

Upper Mesozoic and Tertiary assemblages

By the end of Jurassic time most of the terranes of the western Cordillera had been either accreted to North America or amalgamated into large superterranes. Exceptions are the Chugach Terrane of southeastern Alaska, the Pacific Rim Complex of western Vancouver Island, and the Crescent and Olympic terranes of northern Washington and southern Vancouver Island. These four small terranes were either accreted to, or developed on the western margin of the Insular Superterrane in latest Cretaceous or Tertiary time.

Chugach Terrane

The Chugach Terrane (Helwig and Emmet, 1981) is a narrow wedge of upper Mesozoic and Tertiary rocks west of, and separated from, the Alexander Terrane by the Border Ranges Fault. It is believed to have been deposited much farther south, off the coast of present British Columbia, and transported on the northward moving Kula Plate to its present site during Paleogene time. The succession comprises Cretaceous to Tertiary mélange, flysch, and basalt. The igneous rocks include sheeted mafic intrusions that probably constitute an ophiolite sequence formed in a near-shore ridge complex, as well as small volcanic masses that are not related to a seafloor spreading centre and may have erupted above a transform fault.

Pacific Rim Terrane

The Pacific Rim Terrane (Brandon, 1985) is a sequence of Lower Cretaceous mélanges exposed in a fault-bounded slice along the west coast of Vancouver Island. It has been interpreted as an accretionary wedge, formed within a late Mesozoic subduction complex and correlated with the Chugach Terrane and Franciscan Complex. However, the Pacific Rim rocks overlie a lower Mesozoic volcanic arc succession that is not associated with oceanic crust. This and other evidence suggest that the Pacific Rim mélange is not an accretionary wedge but rather an accumulation of disrupted sediment and locally derived slump blocks. The material was transported by a variety of mass-movement processes into small slope-basins that formed within a complex, active transcurrent margin.

Crescent Terrane

Most of the Crescent Terrane is in northwestern Washington, where it includes a succession of tholeiitic pillow lava (Crescent Formation) overlain by a sequence of Middle Eocene to Oligocene clastic sediments.

On Vancouver Island basalts correlative with the Crescent Formation are assigned to the Metchosin Igneous Complex, which includes basic intrusions comagmatic with the Sooke Gabbro (Muller, 1980a; Massey, 1986). The Metchosin Igneous Complex is underthrust beneath Vancouver Island along the northerly dipping Leech River Fault. The Metchosin volcanics consist of an estimated 3000 m of tholeiitic pillow basalt, breccia, and minor siliceous tuff, succeeded by about 1000 m of layered amygdaloidal flows. Massive and layered gabbros and leucogabbros pass upward into, and are intruded by, a well developed sheeted dyke complex (Massey, 1986). Minor shallow-marine limestone near the top of the pillow-lava succession contains Early Eocene fossils. The pseudo-ophiolitic stratigraphy and tholeiitic chemical affiliation of the Metchosin lavas suggest an emergent island setting within a transform marginal basin (Massey, 1986).

POST-ACCRETIONARY VOLCANISM

By the end of Jurassic time the Intermontane Superterrane had docked against the western edge of North America. The Insular Superterrane lay an unknown distance to the west and, through much of latest Jurassic and earliest Cretaceous time, the two superterranes were separated by a sedimentary basin. A discontinuous volcanic belt of unknown polarity was active along the axis of the Insular Superterrane.

Mid-Cretaceous magmatic events are closely linked, in the east, to the initiation of transcurrent faulting, and in the west, to the final closing of the Dezadeash and Tyaughton troughs. By mid-Cretaceous time accretion was essentially complete and subsequent, Late Cretaceous and Tertiary magmatic events are related to dislocation and deformation of the accreted terranes and to their interaction with the underlying mantle and adjacent oceanic plates. The oldest unsubducted ocean rocks in the Pacific basin are Early Jurassic. Plate reconstructions based on magnetic striping (Engebretson et al., 1984) suggest that western North America was bounded by the northerly moving Kula Place through most of Cretaceous and early Tertiary time, corresponding to a time of intermittent arc volcanism in the northern Canadian Cordillera. A major change in the direction and rate of Kula Plate motion at about 56 Ma corresponds closely with the onset of pervasive Eocene magmatism throughout the Cordillera.

Cretaceous magmatism

Cretaceous magmatism was concentrated in two major episodes of volcanic and plutonic activity concentrated in the Coast and Omineca belts, along the western edge of the Intermontane Belt, and in the Foreland Belt of southern Alberta (Fig. 14.7).

Early and Mid-Cretaceous

Calc-alkaline volcanic rocks and associated sediments of the Gravina-Nutzotin Belt in the north and the Gambier Group in the south include rocks that were deposited on and adjacent to the Insular Superterrane prior to its amalgamation with the Intermontane Superterrane. The Gravina-Nutzotin assemblage can be traced more than 1000 km through southeastern Alaska, and Gambier rocks occur almost continuously from near Prince Rupert through the Coast and western Intermontane Belts to southern British Columbia. Because of their similar age and lithology, the Gravina-Nutzotin and Gambier rocks are considered to be part of the same volcano-sedimentary succession. Interruption of the belt southeast of Prince Rupert probably reflects younger and more intense uplift and plutonism in that part of the Coast Belt. The Gravina-Nutzotin succession (Berg et al., 1972) overlaps the boundary between Wrangellia and the Alexander Terrane. The assemblage includes flysch-like argillite, greywacke, and channel conglomerate and minor nonmarine sediments intertongued with lenses of andesitic volcanic rocks, including augite-phyric pillow lava. Fossils indicate that sedimentation in the Gravina-Nutzotin Belt may have begun in Middle Jurassic (Bajocian) time and continued with minor interruptions into the late Early Cretaceous (Albian). Volcanism was episodic but the bulk of volcanic activity occurred during post-Valanginian time (post-131 Ma). Most of the lavas are classed as andesite, but silica values from 26 analyzed specimens range from 44.0 to 59.8% (Berg et al., 1972).

The Gravina-Nutzotin layered rocks are closely associated with and intruded by numerous small stocks and plutons, principally granodiorite but also quartz monzonite, quartz diorite, diorite, and monzodiorite. Zoned ultramafic complexes that grade outward from a core of dunite through successive shells of peridotite and pyroxenite to peripheral gabbro also may be comagmatic with the eruptive rocks.

The Gravina-Nutzotin Belt is interpreted to be a basinal arc (Berg et al., 1972). The proximal volcanogenic rocks are correlative with distal submarine fans that were deposited in the Dezadeash-Tyaughton Trough on the east and with deep marine sediments outboard of Alexander Terrane on the west which may now be represented by clasts in the Chugach mélange. According to Berg et al. (1972) the Gravina-Nutzotin arc was genetically linked with subduction that resulted in accretion of the Chugach trench facies (Fig. 14.8).

South of Prince Rupert, rocks assigned to the Gambier Group (Heah et al., 1986) are exposed as pendants within the Coast Belt, and as moderately deformed stratified sequences that unconformably overlie pre-Cretaceous Hazelton Assemblage rocks in the adjacent Intermontane Belt. Volcanic units are most abundant in the lower part of the group, where most K-Ar dates are of Hauterivian or Barremian age. In roof pendants within the Coast Mountains the volcanic rocks are overlain unconformably by a predominantly sedimentary succession. In the Mount Raleigh pendant (Woodsworth, 1979b), dacite tuffs, flows, and coarse volcaniclastic rocks containing clasts of granitic as well as felsic and basic volcanic rocks appear to have been deposited in an unstable basin adjacent to a volcanic source. In the southern Coast Mountains (Roddick, 1965) the lower part of the Gambier Group is estimated to be 1800 m thick and consists of a complex assemblage of andesitic pyroclastic rocks, flows, pelitic sediments, and lesser coarse conglomerate containing granitic debris. The ratio of volcanics to sediments decreases in the more easterly pendants, and correlative rocks east of the Coast Mountains (Relay Mountain Group) are predominantly marine clastic sequences in which minor volcanic units are interpreted to be distal equivalents of volcanic piles preserved as pendants in the Coast Mountains.

The Gambier Group rocks are believed to be the products of Early Cretaceous, calc-alkaline volcanism in an arc associated with active, fault-bounded basins accumulating both clastic and pyroclastic rocks. The arc was bounded on the east by the Tyaughton-Methow Trough, which separated the Insular Superterrane from the newly accreted western margin of North America added in early Middle Jurassic time.

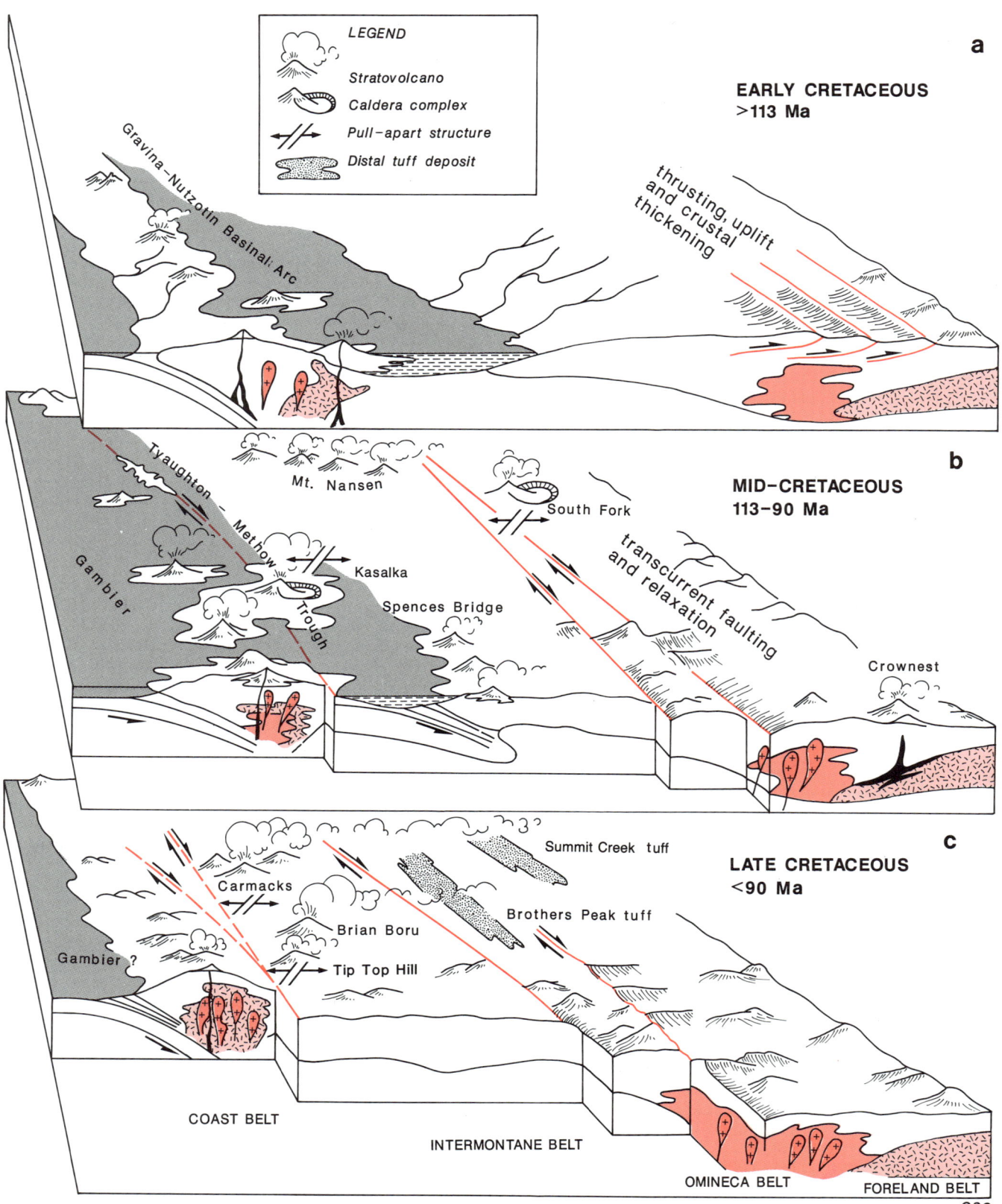

Figure 14.7. An interpretation of volcanic environments in the north-central Cordillera during Cretaceous time: (a) Early Cretaceous calc-alkaline volcanism along the Gravina-Nutzotin-Gambier basinal arc accompanied by crustal thickening along the suture between North America and the Intermontane Superterrane, (b) mid-Cretaceous pervasive transcurrent faulting accompanied by arc volcanism related to convergence between Pacific crust and the Insular Superterrane (Gambier), to the closing of Tyaughton Trough between the Insular and Intermontane superterranes (Spences Bridge), and to pull-apart structures in transtensional regimes within the thickened continental root zones, (c) late Late Cretaceous closure and uplift of Tyaughton Trough and eruption of discrete volcanic centres in response to transtensional, pull-apart regimes in the Intermontane Belt. (Age ranges are approximate, there may be overlap of Kasalka, Brian Boru, and Tip Top Hill activity).

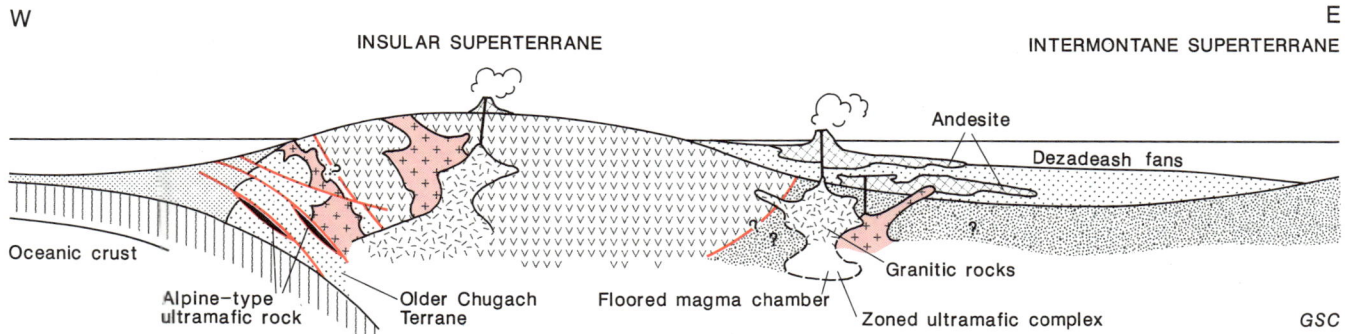

Figure 14.8. Schematic section across the Gravina-Nutzotin basinal arc as it may have appeared during the mid-Cretaceous, prior to amalgamation of the Insular and Intermontane superterranes.

Mid-Cretaceous

Early to mid-Cretaceous Gravina-Nutzotin and Gambier volcanism was followed in the Albian by widespread marine transgression (Fig. 14.7). Several widely separated volcanic belts were active either during or shortly following the deposition of these sediments. In south-central Yukon Territory and western Northwest Territories the Mount Nansen, parts of the Hutshi Formation, and the South Fork volcanics form a broad, discontinuous belt that extends across the Intermontane and Omineca belts. In southern British Columbia approximately coeval volcanic activity produced the Spences Bridge-Kingsvale volcanics near the southeastern margin of the Tyaughton Trough. In west-central Intermontane Belt eruption of the Kasalka Group volcanics in the mid-Cretaceous was followed in the Late Cretaceous by intrusion of the Bulkley and related intrusions. In the southwestern Foreland Belt of Alberta eruption of the alkaline Crowsnest volcanics was approximately coeval with the emplacement of nearby trachyte intrusions.

Foreland Belt

The Crowsnest Formation (Ricketts, 1982) comprises an assemblage of alkaline volcanic rocks that is restricted to the eastern part of Fernie Basin, in southwestern Alberta. From a maximum thickness of 425 m the volcanic pile thins rapidly, becoming interbedded with and grading into sandstone and shale of the upper Blairmore Group. The presence of a typical upper Blairmore flora, and a K-Ar date of 96 Ma confirms an Albian age. The volcanic rocks are mainly pyroclastic and epiclastic deposits, associated with rare flows and intrusive rocks. The lavas are believed to have been extruded as domes, which brecciated during cooling and collapsed to form the lahars.

The major lithology in both clasts and lavas is sanidine-rich trachyte, associated with lesser amounts of analcime phonolite and blairmorite (a phonolite with analcime phenocrysts). The analcime is believed to be a primary igneous phase and the parent magmas may have originated by crustal melting at a depth of about 25 km.

Small alkalic intrusions in the Foreland Belt of southeastern British Columbia cut Lower Cretaceous strata and yield K-Ar ages that range from 72 to 112 Ma (Gordy and Edwards, 1962). They include syenite and trachyte which are chemically similar to, and may be comagmatic with, the Crowsnest Volcanics.

Northern Intermontane and Omineca belts

The Mount Nansen volcanic rocks in Carmacks and Laberge areas (Tempelman-Kluit, 1980) consist of felsic flows and pyroclastic rocks, which occur in isolated remnants as much as 2000 m thick. The volcanics are closely associated, and considered to be coeval with, granitic plutons that yield Albian (109 to 116 Ma) K-Ar dates. K-Ar dating has shown that some of the volcanic rocks are of Late Cretaceous age but the name Mount Nansen was retained (Grond et al., 1984). The Mount Nansen is restricted here to the Lower Cretaceous volcanic and subvolcanic rocks as originally defined by Tempelman-Kluit (1980) and the Upper Cretaceous rocks of Grond et al. (1984) are assigned to the Carmacks Group. Most of the Mount Nansen flows and breccias are hornblende-plagioclase-phyric andesite. Those in the upper part of the pile commonly contain carbonate-filled vesicles and a relatively low phenocryst content compared to nonvesicular, highly porphyritic lower flows. The Mount Nansen eruptive rocks are cut by a variety of dykes, some of which were comagmatic feeders, and others that are probably related to the younger Carmacks Group and Nisling Range alaskite. Subvolcanic intrusions and breccia pipes include coeval granitic stocks of several square kilometres, implying that an extensive mid-Cretaceous volcanic cover was removed.

In the Whitehorse area, thick, moderately-dipping sequences of volcanics formerly mapped as Hutshi are now believed to be less deeply eroded equivalents of the Mount Nansen. The relatively steep dips may be the result of synvolcanic faulting related to graben formation or caldera collapse. The lithology and morphology of the Mount Nansen volcanics are similar to thick mid-Cretaceous (about 100 Ma) pyroclastic sequences in the Yukon-Tanana Terrane of eastern Alaska (Bacon et al., 1985) and to the South Fork Volcanics east of the Tintina Fault.

South Fork volcanics, which occupy a zone northeast of Tintina Fault 200 km long, are associated with mid-Cretaceous (ca 100 Ma) epizonal plutons that occur throughout the Omineca Belt. However, eruptive rocks are preserved only in the north. The South Fork succession is nearly flat-lying and from 580 to 1520 m thick. Locally the lower 50 to 600 m of the pile contains hornblende-bearing basaltic andesite flows and hornblende andesite ash flow tuffs, which are overlain by as much as 150 m of volcanic sandstone, mudstone, conglomerate, lapilli tuff, and epiclastic, chert-bearing breccia (Wood and Armstrong,

1982). Most of the tuffs are crystal-rich, with abundant unsorted fragments of alkali and plagioclase feldspar, hornblende, and smoky quartz, as well as pumice lapilli and lithic clasts. Individual flow units are as much as 250 m thick and display irregularly spaced columnar jointing typical of ash-flow tuffs. The dense welding, fine grain size and association with epiclastic breccias, as well as the great thickness of both individual cooling units and the succession as a whole suggest that it is a ponded, intracaldera assemblage.

Seven analyses of South Fork volcanics (Wood and Armstrong, 1982) plot in the subalkaline, calc-alkaline field of the Irvine and Baragar (1971) classification. They are richer in calcium, iron, and magnesium and lower in alumina and alkalis than most calc-alkaline rocks of similar silica content. Also an initial $^{87}Sr/^{86}Sr$ ratio of 0.7160 ± 0.0002 is believed to be due to contamination with radiogenic crustal strontium. Such mixing of siliceous crustal melt with mafic, mantle-derived magma may explain the distinctive chemistry of this suite.

The South Fork volcanics are closely associated with and geochemically similar to epizonal plutons in the Anvil Range, which include peraluminous, muscovite-bearing granite having I- and S-type affinities (Pigage and Anderson, 1985). Plutons of similar age and lithology extend the length of the Omineca Belt and are believed to have been generated by crustal melting in the root zone of tectonically thickened continental crust along the suture between North America and the Intermontane Superterrane. Melting, intrusion, and caldera formation may have been initiated as a consequence of the shift from Late Jurassic-Early Cretaceous compression, which accompanied accretion, to mid- and Late Cretaceous relaxation, which accompanied subsequent dextral movement on transcurrent faults of the northern Rocky Mountain Trench-Tintina system.

Insular and south-central Intermontane belts

Thick sections of the mainly pyroclastic, mid-Cretaceous Kasalka Group rest unconformably or disconformably on Albian marine sediments of the Skeena Group. In the type area near the southwestern end of the Skeena Arch (Hodder and MacIntyre, 1977; MacIntyre, 1976) the basal unit comprises pebble conglomerate and sandstone 5 to 50 m thick. It is overlain by as much as 600 m of interlayered flows of rhyodacite, ash-flow tuff, lapilli tuff, minor volcanic breccia, and discontinuous andesite-latite flows. Overlying the flows are 800-1000 m of lahar comprising heterolithic clasts in a matrix of fine lithic fragments, broken crystals, and clay. In most places the lahar is overlain by at least 300 m of prominently jointed latite-andesite flows.

At Tahtsa Lake as much as 1500 m of Kasalka strata are confined to a circular area about 40 km in diameter interpreted to be a caldera (Fig. 14.9). The intracaldera succession is cut by diorite and rhyodacite intrusions that occupy radial and concentric fractures and are considered to be feeders. The succession also is spatially related to and cut by two or more phases of the Bulkley Intrusions. These porphyritic granodiorite stocks are more siliceous and more potassic than the Kasalka intrusions and may be a late, more differentiated, resurgent phase of Kasalka magmatism. Most Kasalka dates are greater than 100 Ma but a K-Ar date of 87 ± 4 Ma from the volcanic pile compares to a date of 83.8 ± 2.8 Ma for crosscutting Bulkley granodiorite. It is not clear whether this is truly a genetic relationship or whether the young date of the volcanics is partially reset.

The Kasalka volcanics lie near a welt of greatly thickened salic crust, the Coast Belt, along the suture between the Insular and Intermontane superterranes. Like the Mount Nansen and related Anvil-type plutons, the Kasalka and Bulkley Intrusions are relatively rich in potassium. It is reasonable to speculate that the Kasalka magmas were generated by crustal melting in a root zone of greatly thickened crust during a period of crustal relaxation.

Albian bentonite layers in the Fort St. John Group (Stott, 1982) of northeastern British Columbia may be distal ash layers related to Kasalka volcanism.

The Spences Bridge Group (Thorkelson, 1985) is confined to a narrow belt east of the Fraser Fault in south-central British Columbia. It comprises at least 2400 m of calc-alkaline, basaltic to rhyolitic lavas and clastic rocks

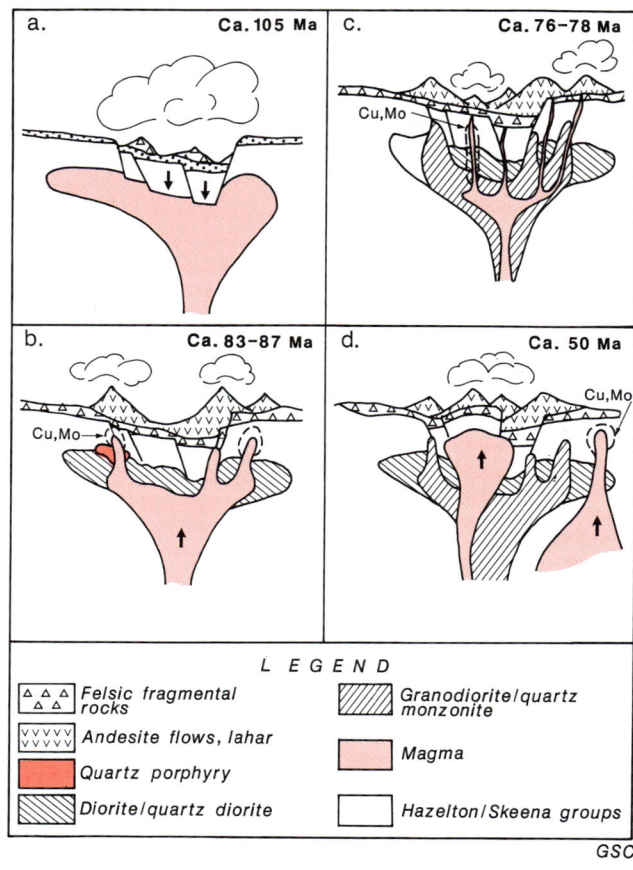

Figure 14.9. Interpretative cross-sections showing four stages in the evolution of the Tahtsa Lake caldera: (a) explosive rhyodacite volcanism and caldera collapse; (b) construction of andesitic cones and emplacement of intrusions related to the Kasalka volcanics; (c) magmatic resurgence and intrusion of porphyritic granodiorite stocks; (d) magmatic resurgence and intrusion of large diorite and quartz monzonite batholiths and late dyke swarms. After MacIntyre (1976).

that rest on a surface of moderate relief. Plagioclase-augite-hypersthene phyric andesite lavas predominate in the lower part and are commonly overlain by coarse conglomerate containing basement clasts. The upper part of the group is a complex of felsic and mafic lavas interlayered with welded and nonwelded ignimbrite, air-fall tuff, lahar, and epiclastic breccia. Felsic and mafic dykes, believed to have been feeders, are the only intrusions that cut the Spences Bridge.

The Spences Bridge Group is overlain by at least 600 m of highly altered, amygdaloidal to dense andesite flows that were formerly assigned to the Kingsvale Group. Both conformable and unconformable relationships are reported, but the unconformities are believed to be local and of synvolcanic origin rather than the result of regional uplift and erosion. K-Ar dates of 94.4 ± 3.4 Ma for the Spences Bridge and 91.7 ± 3.3 Ma for the overlying andesite suggest no significant hiatus between the units. Both the Spences Bridge and the overlying andesite flows appear to be part of the same volcano-sedimentary assemblage, probably a chain of stratovolcanoes associated with subsiding, fault-bounded basins. The Spences Bridge lavas are predominantly calc-alkaline, low to medium K, basalt and andesite. However, silica values range from 49% to more than 70%, and iron-enrichment in some of the acidic Spences Bridge lavas suggests a possible tholeiitic affinity.

Spences Bridge volcanism was contemporaneous with the filling of Tyaughton-Methow Trough and with uplift of the Mount Lytton Batholithic Complex along its eastern margin (Monger, 1982, 1985). Thick wedges of conglomerate, the Jackass Mountain Formation, were deposited in the Tyaughton-Methow Trough west of the Mount Lytton Complex and are believed to be coeval with, and comprise a facies of, the Spences Bridge volcaniclastic rocks. The Spences Bridge arc may thus be linked to eastward subduction of complex older crust flooring the Tyaughton-Methow Trough (Fig. 14.7). Age relationships suggest that the onset of Spences Bridge volcanism corresponded closely in time to the decline of Gambier volcanism along the western margin of the closing trough.

Late Cretaceous

Late Cretaceous volcanism was confined to isolated centres in the Insular Belt and the Intermontane Belt of Yukon Territory and north-central British Columbia (Fig. 14.7). The principal volcanic successions are the Carmacks Group, which underlies large areas in southwestern Yukon, and the Brian Boru and Tip Top Hill formations in north-central British Columbia. Both of these assemblages are post-Albian in age and yield K-Ar dates between 80 and 64 Ma.

The Carmacks Group (Churchill, 1980; Grond et al., 1984) is exposed in isolated piles of gently dipping volcanic and volcaniclastic strata that lie between Tintina and Denali faults in central Yukon Territory. Sections as much as 200 m thick consist of lava flows, breccias, sintered tuff, and immature volcanic sandstone that were deposited on a surface of high local relief. The base of the pile is commonly marked by a thin conglomerate, above which massive lavas grade upward into interbedded vesicular lavas and epiclastic breccia. The lower part of the succession is relatively acid, including hornblende-phyric andesite flows, pyroclastic rocks, dyke swarms, and subvolcanic laccoliths. The lavas are mostly andesite, but range in composition from rhyolite through trachyte and dacite to basalt. Epiclastic units containing clasts as much as a metre in diameter are commonly interbedded with sintered tuff and lava flows. The upper part of the succession is dominated by flat-lying columnar basalt flows. The chemistry of the Carmacks Group lavas is typical of potassic, alkaline suites.

Upper Cretaceous volcanics in central British Columbia are assigned to the Brian Boru and Tip Top Hill formations. The Brian Boru (Sutherland Brown, 1960; Richards, 1980) comprises at least 1800 m of porphyritic andesite flows, breccias, and minor interlayered tuff. At the type locality on Brian Boru Peak 30 m of tuff at the base of the formation are overlain by porphyritic andesite flows and breccia with phenocrysts of calcic plagioclase, hornblende, and augite. The eruptive rocks are similar in composition and closely associated with the granodiorite-quartz monzonite Rocher Deboule Stock, a 70 km^2 body that may be genetically related to the volcanics.

The Tip Top Hill Formation (Church, 1972) consists mostly of andesite and dacite flows and pyroclastic rocks dated at 75.8 ± 2.7 Ma, which are associated with microdiorite feeder sills and dykes that yielded a 74.0 ± 2.0 Ma date. The andesite is underlain by alkali rhyolite (76.0 ± 3.0 Ma) and an associated granitic pluton (76.0 ± 2.0 Ma). Like the Brian Boru, the Tip Top Hill volcanics appear to be remnants of a large central, mainly andesitic volcanic-plutonic complex. Although they are widely separated the rocks from these isolated centres of late Late Cretaceous volcanism exhibit a strong chemical polarity suggestive of an east-dipping subduction zone. The tectonic setting in which their parent magmas were generated was probably one of oblique subduction accompanied by dextral movement on major faults such as Tintina and Denali. It seems probable that some of this strain was translated to tensional fractures along which the Late Cretaceous, Carmacks, Brian Boru, and Tip Top Hill magmas rose into shallow crustal chambers and locally erupted, both passively to form small basalt shields, and explosively, to form composite piles with a large pyroclastic component. A distal facies of this explosive volcanism is preserved as fine airfall tuff interbedded with sediments far to the east and northeast of the eruptive centres (Fig. 14.7). The upper Upper Cretaceous, Brothers Peak Formation, in the Sustut Group of central British Columbia contains abundant fine, air-fall tuff that probably originated from the Brian Boru and/or Tip Top Hill eruptive centres. Similarly, the Maastrichtian and Lower Paleocene, Summit Creek Formation (Yorath and Cook, 1981; Ricketts, 1985), south and west of Fort Norman, Northwest Territories, contains up to 17 m of fine tuff representing between 100 and 300 eruption-deposition events. The tuff is believed to have been deposited from airborne ash erupted from Carmacks Group volcanic centres 600 km to the southeast, in central Yukon Territory.

Early Tertiary magmatism

Early in the Eocene Epoch, during the short interval between 55 and 36 Ma, the northern Cordillera was the site of intense and pervasive magmatic activity. The Challis-Kamloops Volcanic Belt extends from Yellowstone northwest through Idaho, Wyoming, and Montana and into central British Columbia, as far north as the central part of Yukon Territory. Igneous rocks emplaced during this event include the Nisling Range and Skukum-Sloko assemblages of Yukon and northern British Columbia, the Ootsa-Endako assemblage in central British Columbia, the Kamloops, Penticton-Coryell volcanic-plutonic assemblages of south-central British Columbia, and the contemporaneous but possibly unrelated Flores Volcanics of western Vancouver Island.

Southern Insular Belt

The Flores Volcanics (Brandon, 1985) and associated Catface Intrusions of western Vancouver Island are exposed close to the West Coast Fault which separates Wrangellia from the Pacific Rim Complex. The volcanics, which comprise dacite breccias, welded ash flows, and laharic breccia, yield early Eocene zircon fission-track dates (54, 55, 56 Ma). The eruptive rocks are associated with dyke swarms and plutons. The latter include a younger suite (U-Pb zircon ± 42 Ma) and an older suite (K-Ar 50, 52, 60 Ma), both of which may be comagmatic with the volcanics.

Southern Intermontane Belt

In south-central British Columbia Eocene volcanism was coeval with dextral, transcurrent movement on the Northern Rocky Mountain Trench fault systems and in part with the younger Fraser River Fault. Magma generated above an east-dipping subduction zone and injected into this complex crustal environment was of two distinct types, characterized on the one hand by the calc-alkaline Kamloops Group and, on the other, by the alkaline rhomb porphyries and syenites of the Penticton-Coryell assemblage.

The Kamloops Group (Ewing, 1981a,b) is a widespread assemblage of volcanic and sedimentary rocks preserved in north-northeasterly trending, fault-bounded basins east of the Fraser Fault. Individual basins are linked by large throughgoing faults and, although subsidence and volcanism may have been contemporaneous in different basins, the stratigraphic succession and thickness varies from one to another. The basins are not simple grabens but rather a complex of slumped blocks that are the surface manifestation of high-angle reverse faults at depth. Strike-slip motion on major faults is believed to be the controlling mechanism. The small sedimentary and volcanic depressions occupy triangular areas between splaying faults, or pull-apart basins between transcurrent faults.

Clastic sediments commonly form the basal members of the Kamloops Group. In the type area the Tranquille Formation comprises as much as 450 m of lacustrine sediments and andesitic bedded tuffs that grade upward into a complexly interlayered succession of andesitic and basaltic tuffs and flows, tuffaceous lacustrine sediments, phreatic breccias, mudflows, and pillowed andesite flanked by an apron of hyaloclastic breccia. The Tranquille beds are overlain by at least 1000 m of mainly basaltic andesite flows and pyroclastic rocks of the Dewdrop Flats Formation. The lower part of this succession includes palagonitic breccia, mudflows, and phreatic breccia indicative of an aqueous environment, whereas the upper part is dominated by basaltic to andesitic, subaerial flows and pyroclastics.

The Kamloops Group is an alkali-rich, calc-alkaline suite showing little or no Fe enrichment. Initial $^{87}Sr/^{86}Sr$ ratios increase from 0.7040 in the west to 0.7060 in the east, and the content of K_2O at 60% SiO_2 increases regularly eastward across southern British Columbia. According to Ewing (1981b) the chemical data support a subduction-related, continental arc origin. The basalts appear to form a continuous comagmatic series with the more felsic rocks, and the continuity indicates that they represent a differentiated cogenetic suite.

Southeast of the axis of Kamloops volcanism, Eocene sedimentary and volcanic rocks of the Penticton Group are exposed in Okanagan valley. Near Kelowna the assemblage occupies the White Lake Basin (Church, 1973), which is bounded by normal faults and contains at least 2400 m of sedimentary and volcanic deposits. A discontinuous basal conglomerate and breccia, derived from the underlying basement, is overlain by more than 1500 m of highly alkaline, mainly rhomb porphyry lavas and related breccias of the Marron Formation. These lavas are overlain unconformably by more than 200 m of rhyolite and rhyodacite of the Marama Formation which, in turn, is overlain by conglomerate and epiclastic volcanic breccia that Church (1973) interpreted to be slide deposits from nearby terrane underlain by pre-Tertiary rock. The White Lake succession is cut by at least three intrusive phases of rhomb-porphyry or augite porphyry sills and dykes, which were probably feeders.

The succession near Penticton is similar to the volcanic rocks of White Lake Basin and is probably correlative with the Marron and Marama formations.

The highly potassic, rhomb porphyries of the Marron and correlative volcanics are chemically similar, and peripheral to the Coryell, syenitic epizonal plutons. They are believed to be coeval and comagmatic - the intrusive and extrusive phases of a chemically distinctive, alkaline assemblage that lies east of the main axis of Kamloops activity. The more alkaline affiliation of the Marron-Coryell suite may reflect its position relative to the east-dipping subduction zone and the proportionately greater thickness of underlying continental crust. Emplacement of the syenite was accompanied by regional thermal upwelling that reset K-Ar systems in the older, peripheral rocks to about 50 Ma. Doming culminated in the development of low-angle detachment surfaces along which the upper part of the dome slid to the west and north, carrying the Marron volcanics, now exposed near Kelowna and Penticton, as much as 100 km from the subvolcanic Coryell plutons above which they were erupted (Tempelman-Kluit and Parkinson, 1986).

Central Intermontane Belt

Eocene volcanic and plutonic activity east of the Fraser Fault, in the central Intermontane Belt, is poorly understood. Salic volcanic rocks commonly have been assigned to the Ootsa Lake Group and basic flows to the Endako Group. However, some rocks formerly thought to be of Eocene (Ootsa) age have been shown to be correlative with the mid-Cretaceous Kasalka Group and some basalts formerly assigned to the Endako Group are outliers of Miocene Chilcotin Group lavas. Nevertheless, a large number of K-Ar dates ranging from 54 to 42 Ma from the Ootsa and Endako lavas confirm that Eocene rhyolitic and basaltic volcanism was widespread in the west-central Intermontane Belt. In the Buck Creek area in central

British Columbia, Church (1972) reported Eocene volcanic rocks that are predominantly andesitic, but range in composition from basalt to dacite and may include minor rhyolite flows and dykes. The Buck Creek volcanics are closely associated with the Eocene Goosly Lake and nearby granitic stocks, which may be subvolcanic equivalents.

Northern Intermontane Belt

The products of Eocene volcanism include the Nisling Range alaskite and the predominantly salic to intermediate rocks of the Skukum Group in western Yukon Territory (Fig. 14.10), and the Sloko Group in northwestern British Columbia.

The Nisling Range alaskite (Tempelman-Kluit, 1977a,b, 1979) forms discordant, high-level leucocratic plutons, which intrude Paleozoic and older rocks of the Yukon Crystalline Terrane. Associated dyke swarms are localized by north-trending fractures cutting the regional northwesterly trend of older structures. Explosive acid volcanics, associated with the alaskite, lie on a surface of low relief, which was later tilted toward the northeast. Deeper erosion and consequent removal of eruptive rocks from the southwest part of the belt exposed the subvolcanic alaskite plutons and dyke swarms. The alaskite is a medium grained miarolitic granite comprising perthitic potash feldspar, smoky quartz, interstitial albite, and accessory biotite and fluorite. Extrusive equivalents of the Nisling Range alaskite include hornblende-feldspar porphyry flows, breccia, and tuff. Both the alaskite and related volcanics yield Eocene K-Ar and Sr-Rb dates of 52 to 62 Ma. A similar range of K-Ar ages has been obtained from granodiorite of the Ruby Range Batholith, which may be a deeper, comagmatic phase of the Nisling Range igneous suite.

In southern Yukon Territory, Eocene volcanic rocks of the Skukum Group (Wheeler, 1961; Smith, 1983) are exposed in down-faulted blocks within the Intermontane and Coast belts. The Eocene succession includes more than 1200 m of andesite, dacite, rhyolite, and minor basaltic volcanic and volcaniclastic rocks that commonly have moderate dips. The group is divided into a basal division of mixed, mainly andesitic rocks, a middle division of mainly felsic rocks, and an upper division of mainly basaltic rocks. In the Mount Skukum area (Fig. 14.10) the lower division is from 120 to 450 m thick and includes coarse, granite-bearing breccia and conglomerate units interlayered with andesite and dacite flows and pyroclastic deposits up to 20 m thick. The overlying middle division comprises an estimated 450 m of varicoloured dacite and rhyolite breccia, feldspar porphyry andesite flows, and breccia, welded and nonwelded tuffs, and ash flows. The basaltic upper division is absent.

As much as 420 km^3 of mainly pyroclastic deposits and lesser lava flows and epiclastic rocks lie within a cauldron subsidence complex (Fig. 14.10) about 20 km in diameter near Bennett Lake (Lambert, 1974). The intracaldera rocks include moderately to densely welded ignimbrites comprising volcanic, granitic, and metamorphic lapilli in a matrix of devitrified pumice, shards and dust, and ash-sized lithic and crystal fragments. Individual ash-flow units, which are as much as 250 m thick, are interbedded with dacite and rhyolite flows, and wedges of epiclastic, granite-bearing breccia and conglomerate that were shed from the bounding scarps of the caldera. Along much of its circumference the caldera is bounded by steeply dipping arcuate, ring-fracture intrusions of leucocratic granite and porphyritic rhyolite. Granitic basement rocks around the complex are commonly shattered and brecciated.

In northwestern British Columbia, Eocene volcanic rocks correlative with the Skukum Group are assigned to the Sloko Group (Souther, 1971), which is preserved in down-faulted blocks and erosional remnants on many of the higher uplands along the eastern flank of the Coast Mountains. The succession is predominantly pyroclastic, varying from coarse explosion breccia and agglomerate to fine grained, banded vitric tuff and ignimbrite. Andesite and local basalt flows are present in most sections, but they are everywhere subordinate to rhyolitic and dacitic pyroclastic rocks. Sediments within the Sloko Group consist almost entirely of angular to subangular debris derived from adjacent volcanic accumulations, but wedges of epiclastic breccia containing basement clasts are interbedded with the volcanogenic rocks along some of the bounding faults. This implies that subsidence of fault-bounded basins was contemporaneous with volcanism and sedimentation. Many of the interbedded sediments consist of shards, ash, and lapilli, erupted directly into bodies of water where they settled to form graded sequences. The presence, locally, of coalified plant debris and carbonized logs in some of these beds indicates episodic volcanism.

The Sloko eruptive rocks are spatially associated with quartz monzonite intrusive bodies that also yield Eocene K-Ar dates. At many places, particularly in the eastern Coast Belt, pyroclastic deposits grade imperceptibly downward into homogeneous aphanitic felsite and finally into medium grained quartz monzonite. Intrusion of the quartz monzonite appears to have accompanied periodic explosive eruptions, block foundering, and stoping, which in some places brought the intrusive magma into contact with the lower part of the comagmatic, Sloko volcanic accumulation.

Farther east, rhyolitic volcanic rocks are exposed along Kechika Fault north of Turnagain River, andesitic volcanics are exposed in the Northern Rocky Mountain Trench near Sifton Pass (Gabrielse, 1985), and a bimodal, rhyolite-olivine basalt suite is exposed north of Tintina Fault in east-central Yukon (Jackson et al., 1986). The eruptive rocks are associated with high-level, fluorite-bearing, miarolitic granitic stocks and slightly younger, northeasterly trending lamprophyre dykes. The volcanics, dykes, and associated plutons all yield Eocene K-Ar dates and are probably comagmatic.

In the southern Cordillera the Eocene magmatic culmination has been linked to the Challis-Absaroka arc (Ewing, 1981a). However, it is difficult to rationalize the great breadth of Eocene volcanic and plutonic activity in the Canadian Cordillera with a simple subduction model. Ductile spreading and thinning of the deep crust accompanied by high heat flow in a tensional and/or transcurrent stress regime have been proposed as important elements of the Eocene tectonic setting, which produced widespread magmatism in the southern Cordilleran (Ewing 1981a,b; Price, 1979). In the northern Cordillera (Gabrielse, 1985) the coincidence of early Cenozoic plutonism, volcanism, high heat flow, and rapid uplift with structures related to transcurrent faulting suggests that the plutonic and

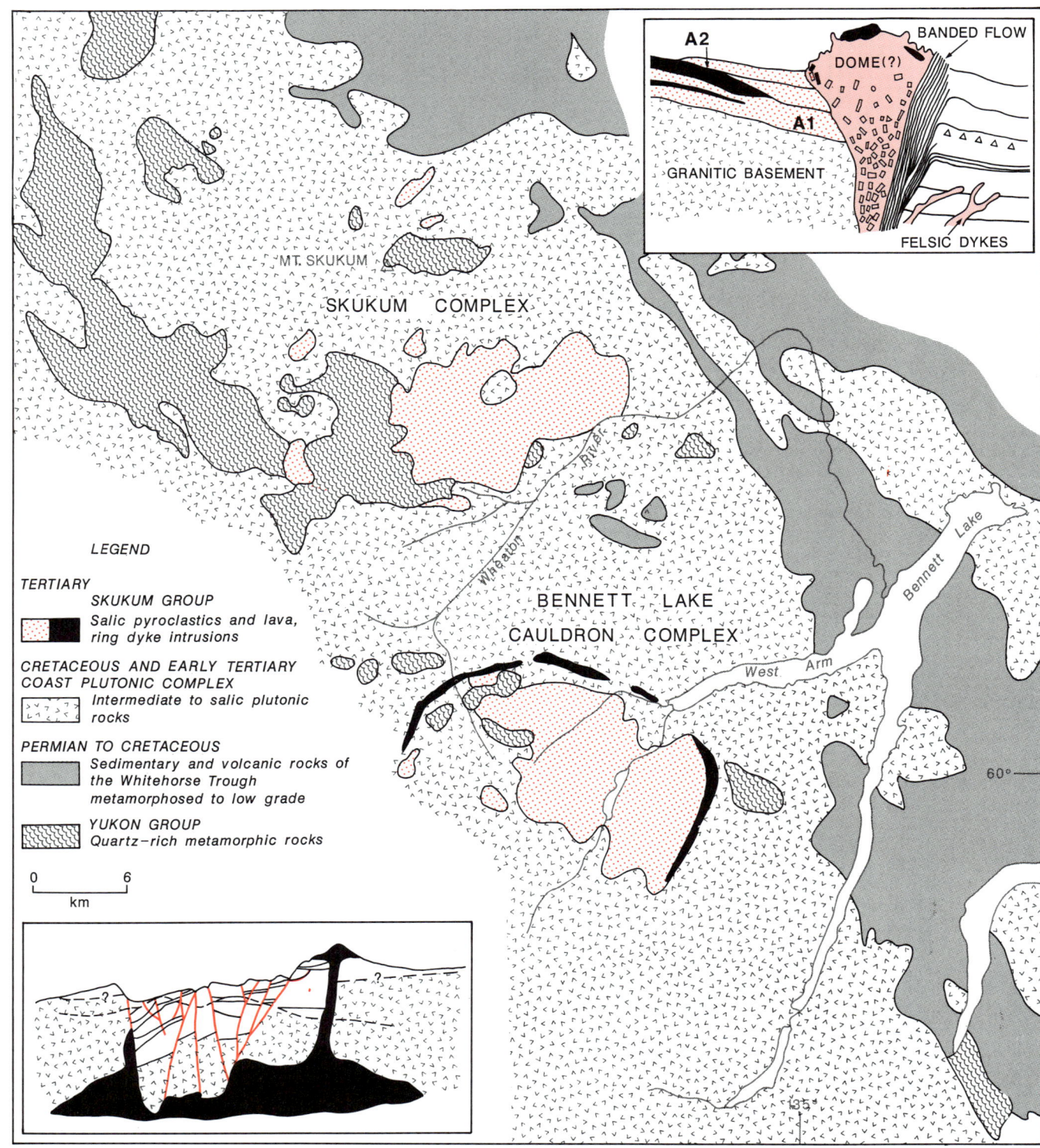

Figure 14.10. Sketch map and schematic sections of the Skukum Volcanic Complex (upper right) (after Smith, 1983) and the Bennett Lake Cauldron Complex (lower left), (after Lambert, 1974).

volcanic events may not have been directly associated with subduction but rather resulted from thinning of continental crust accompanied by a rapid rise in geothermal gradient. Whether such a process could, by itself, produce significant crustal melting is open to question. It is possible that local extension related to transcurrent faulting was contemporaneous with subduction and the transfer of heat to the crust by mantle-derived magma.

Late Tertiary and Quaternary magmatism

During early Neogene time the Farallon-Pacific ridge intersected the continental margin north of Queen Charlotte Islands. By about 10 Ma the triple junction had migrated south to its present position off the west coast of Vancouver Island and the Farallon Plate had been reduced to a small remnant (the Juan de Fuca Plate). The changing geometry of offshore plate motion exerted a controlling influence on the distribution, history, and chemical affiliation of Neogene volcanism on the adjacent continent (Fig. 14.11).

Farallon-Juan de Fuca subduction complexes

Subduction of the Farallon-Juan de Fuca Plate was associated with development of the Pemberton and Garibaldi (Cascade) volcanic arcs in the Coast and Insular belts, effusion of Chilcotin Group back-arc basalt in the Intermontane Belt, and with formation of the transverse Alert Bay Belt across northern Vancouver Island.

Pemberton Belt

In the southern Coast Mountains the Pemberton Volcanic Belt (Souther, 1975; Berman, 1981) is defined by deeply eroded epizonal stocks and a few erosional remnants of associated volcanic rocks that are believed to be vestiges of a Neogene, calc-alkaline arc. It includes younger phases of the Chilliwack Batholith, Salal Creek Stock, the Coquihalla Complex, and older phases of the Franklin Glacier and Silverthrone complexes. K-Ar dates from the Pemberton Belt range from 21 Ma for Chilliwack Batholith through 8 Ma for Salal Creek Stock to 6.8 Ma for Franklin Glacier Complex. A gap of 400 km, with no known Neogene centres, separates the Franklin Glacier Complex, in the central Coast Mountains, from the Masset volcanics of Queen Charlotte Islands. Although the Masset Formation has not been considered part of the Pemberton Belt it is approximately coeval (21 Ma) and is probably related to early Neogene interaction between North America and the subducting Farallon Plate. The Masset comprises a bimodal succession of subalkaline rocks (Sutherland Brown, 1968; T.S. Hamilton, pers. comm., 1985). The salic rocks, like those in other parts of the Pemberton Belt, are calc-alkaline whereas the basic Masset suite has mid-ocean ridge basalt affinities. The Masset volcanics may be the most northerly manifestation of volcanism related to early Neogene subduction of the Farallon Plate and thus comprise the northern limit of the Pemberton Belt. Their bimodal, calc-alkaline-MORB affinity suggests that Masset activity may have been initiated where the spreading ridge between Pacific and Farallon plates intersected the edge of the continent (T.S. Hamilton, pers. comm., 1985).

Alert Bay Volcanic Belt

The Alert Bay Group volcanic rocks are exposed in a belt extending from Brooks Peninsula northeasterly across northern Vancouver Island. Its trend is coincident with the trace of the subducted Juan de Fuca-Explorer plate edge. According to Armstrong et al. (1985), "Volcanism began in the west, at Brooks Peninsula, about 8 Ma ago, but occurred in most centres 3.5 ± 1 Ma ago. There is a suggestion of eastward migration of activity and a shift from basalt to dacite or rhyolite with time. Most of the volcanism was coincident with a time of rapid changes in the geometry of subduction, as inferred from offshore magnetic patterns, and with a hiatus in mainland, Cascade

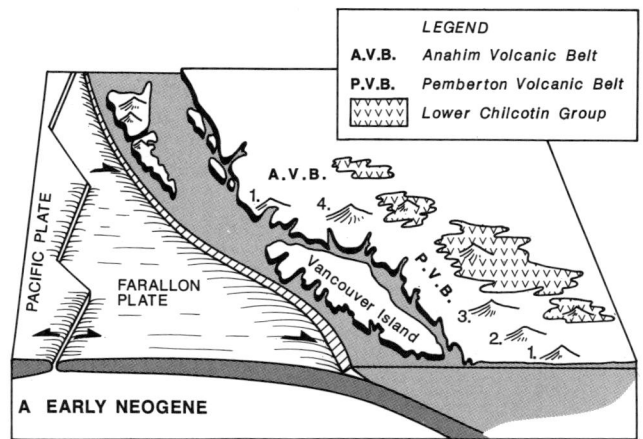

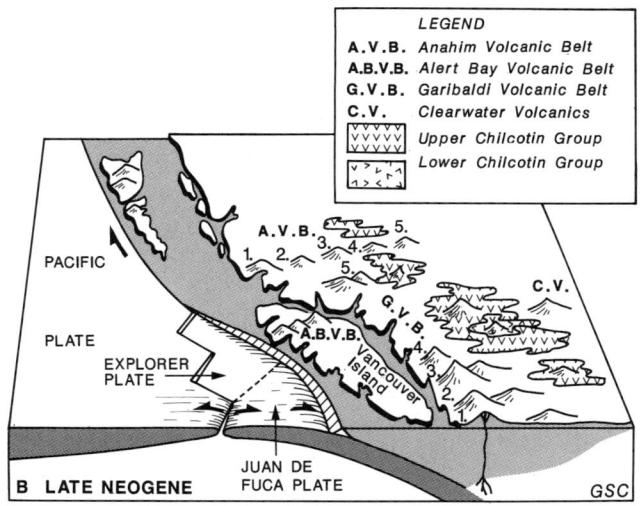

Figure 14.11. Map and schematic sections showing the distribution and possible tectonic setting of Neogene volcanic belts in the northern Cordillera. In A, the Anahim Volcanic Belt includes: 1. Bella Bella Complex; and the Pemberton Volcanic Belt includes: 1. Chilliwack Batholith, 2. Coquihalla Complex, 3. Salal Creek Stock, and 4. Silverthrone Complex. In B, the Anahim Volcanic Belt includes: 1. Bella Bella Complex, 2. Rainbow Range, 3. Ilgachuz Range, 4. Itcha Range, and 5. Nazko Cone; and the Garibaldi Volcanic Belt includes: 1. Mount Baker, 2. Mount Garibaldi, 3. Mount Cayley, 4. Meager Mountain, and 5. Mount Silverthrone.

volcanic arc activity. Geometry and chronometry suggest that this is a descending-plate-edge volcanic belt, where disruption of steady-state plate-consumption patterns triggered magma genesis."

Garibaldi Volcanic Belt

The Juan de Fuca-Pacific spreading ridge has remained relatively fixed for the last 7 to 10 Ma during which time minor adjustments in plate geometry were accompanied by shifts in the orientation of onshore volcanic fronts (Riddihough, 1984). Older centres of the Pemberton Belt, which lay inland from the Juan de Fuca Plate, were strongly uplifted and deeply eroded prior to the eruption of Garibaldi Belt volcanoes in Pleistocene and late Quaternary times. The Garibaldi Belt (Mathews, 1952) is a northern extension of the Cascade Volcanic Belt of the western United States. It is offset to the west of the main Cascade trend and is itself a composite of three north-trending segments (Souther, 1980). The most westerly segment intersects the older, northwest-trending Pemberton Belt where the 8 Ma Salal Creek Stock is overlain unconformably by Garibaldi Group lavas from Meager Mountain.

The volcanoes of the Garibaldi Belt (Green, 1981; Lawrence et al., 1984) include preglacial, intraglacial, and postglacial piles that range in age from 3.8 Ma to dacite pumice and block flows that erupted from Meager Mountain 1340 years BP. The Garibaldi Belt lavas include a complete spectrum of calc-alkaline rocks as well as mildly alkaline, high-alumina basalt from peripheral centres.

Chilcotin Group, back-arc lavas

Calc-alkaline volcanism along the Pemberton-Garibaldi fronts was contemporaneous with the effusion of basalt from a multitude of centres that extend northwesterly through the Intermontane Belt from southern British Columbia to the Bowser Basin. They are most extensive in central British Columbia where they form a volcanic plateau of 50 000 km^2. The lavas are assigned collectively to the Chilcotin Group (Bevier, 1983a,b), which yields K-Ar dates ranging from 15 to 2 Ma. Chemical variation is restricted to a small range transitional between alkaline and tholeiitic basalt. The only salic rocks interlayered with Chilcotin Group lavas are fine tuffs, which are believed to be distal, airborne ash from the calc-alkaline volcanoes of the Pemberton-Garibaldi arc. Consistent differences in magnetic polarity, and minor differences in trace-element and isotopic chemistry indicate that the Chilcotin volcanic plateau is a composite of separate, small overlapping shields, each shield-forming sequence of flows having issued in a relatively short interval of time from a separate vent. The Chilcotin, transitional basalts are believed to have erupted in a back-arc environment, above a zone of asthenospheric upwelling behind the Pemberton-Garibaldi arc.

Pacific Plate subduction complexes

Throughout most of early Neogene time, northerly-moving Pacific crust, north of the Farallon-Pacific Ridge, was being consumed in a subduction zone beneath the western hook of North America. The Wrangell Volcanic Belt developed inland from the trench along what is now the outer ranges of the Saint Elias Mountains.

Wrangell Volcanic Belt

In southwestern Yukon Territory Late Miocene effusion of calc-alkaline basaltic andesite and andesite lava built several composite volcanoes and flooded a vast area with flat-lying flows having a composite thickness of more than 1500 m (Souther and Stanciu, 1975). Locally the effusion of basaltic andesite was followed by voluminous eruptions of dacite and rhyolite accompanied by block foundering and intrusion of felsic stocks into the volcanic pile. Post-Miocene uplift accompanied by northeast-directed thrusting and dextral transcurrent faulting has severely deformed part of the Wrangell succession in southern Yukon. Following deformation and uplift the locus of volcanism shifted west into Alaska, where many centres are still active. The White River ash, 1500 years old, which covers 324 000 km^2 of southern Yukon and eastern Alaska, issued from a vent a short distance west of the Alaska-Yukon border (Lerbekmo and Campbell, 1969).

Rift and plume complexes within the accreted terranes

By 10 Ma the Farallon Plate had been reduced to its present small remnant (Juan de Fuca) and the Queen Charlotte transform boundary extended from northern Vancouver Island to Alaska. The late Neogene to Quaternary Stikine and Anahim Volcanic belts developed on the continent adjacent to the transform boundary and, farther east, basaltic cones erupted along the eastern margin of Quesnellia in the Clearwater and McConnell Creek areas.

Stikine Volcanic Belt

The Stikine Volcanic Belt (Souther, 1977) includes a multitude of late Neogene volcanoes ranging from small, usually lherzolite-bearing, monogenetic cones of alkali olivine basalt to large, bimodal shield and dome complexes comprising alkaline basalt and a peralkaline salic suite. Major centres include Level Mountain, Mount Edziza, Spectrum Range, and Hoodoo Mountain. The eruptive centres occur along several north-trending, en échelon segments, which define a broad zone extending diagonally across the northern Coast Belt into the Intermontane Belt of northern British Columbia and central Yukon. The belt evolved over a period of 10 to 15 Ma and has remained active into post-Pleistocene time. Many of the volcanoes in the Stikine Volcanic Belt are characterized by ice-contact features such as tuff breccia rings, tuyas and subglacial mounds, formed where lava was erupted beneath glacial ice. The oldest lavas in the Stikine Volcanic Belt erupted near its centre, and younger events propagated to en échelon segments to the north and south. Mount Edziza, one of the largest complexes, is clearly related to synvolcanic movement on north-trending normal faults, suggestive of incipient rifting. The segments of the Stikine Volcanic Belt lie between the northwesterly trending Denali and Tintina transcurrent fault systems. On the former there is evidence for Neogene displacement. Pull apart structures that controlled the segments of the Stikine Volcanic Belt may be wrench faults related to dextral movement on the bounding transcurrent faults (see Chapter 17).

Anahim Volcanic Belt

The Anahim Volcanic Belt (Souther, 1984, 1986) trends easterly across the Coast and Intermontane belts of central British Columbia to the western edge of the Omineca Belt. It includes both monogenetic cones of alkali olivine basalt and bimodal, basalt-hawaiite-comendite, peralkaline complexes. Three large shield volcanoes, the Rainbow, Ilgachuz, and Itcha ranges, dominate the central part. They lie along a trajectory that projects eastward to a group of postglacial basaltic cones near Nazko and westward into a zone of high-level, syenitic to soda granite plutons and bimodal basalt-comendite dyke swarms in the Coast and Insular belts. Unlike the Stikine Belt, the onset of volcanism in the Anahim Belt decreases systematically in age from about 15 Ma in the west to only a few hundred years at Nazko in the east, suggesting that it may define the trace of a mantle hotspot beneath the westerly-moving North American Plate (Souther, 1986).

Clearwater-Quesnel and McConnell Creek volcanic provinces

Although they are separated by more than 500 km the late Neogene and Quaternary volcanoes of the Clearwater-Quesnel and McConnell Creek areas occupy a similar tectonic setting along the eastern edge of Quesnellia.

The Clearwater-Quesnel volcanoes (Hickson and Souther, 1984; Hickson, 1986) form a poorly defined belt that extends northwesterly from Wells Gray Park to Quesnel Lake. They lie east of and are similar to the extensive Chilcotin Lavas, but the Clearwater flows are younger and originated from local, intravalley eruptive centres. Four morphological assemblages are recognized. An early glacial assemblage, characterized by tuya-like forms, gives K-Ar dates of 0.27 to 3.5 Ma. They are surrounded by valley-filling flows and tuff breccia that rest locally on lag gravel and till. Subaerial flows in this assemblage give K-Ar dates of 0.15 to 0.56 Ma. A late interglacial assemblage is composed of subaerial pyroclastic material, transitional deposits, and deposits that are clearly subaqueous. The youngest assemblage comprises pyroclastic cones, blocky lava flows, and pit craters that postdate the last Cordilleran glaciation. Abundant lherzolite nodules and other accidental clasts are a common feature of the young centres in both the Clearwater area and farther north near Quesnel Lake.

The Clearwater basalts are less altered but otherwise similar to the transitional (tholeiite-alkali basalt) Chilcotin lavas. Most flows are porphyritic, with phenocrysts of olivine, and less abundant plagioclase in a groundmass of plagioclase, titaniferous augite, olivine, opaque oxides, and interstitial glass.

In McConnell Creek area Lord (1948) described a group of intrusive and extrusive basaltic rocks that intrude or overlie Sustut strata in the Connelly Range. The associated dykes, volcanic necks, lava flows, and pyroclastic deposits have undergone differing amounts of dissection. The oldest flows form flat-lying erosional remnants, up to 160 m thick, that may be eruptive equivalents of the coextensive, middle Tertiary Kastberg intrusions, whereas young pyroclastic cones in adjacent valleys are postglacial. The assemblage probably records a period of volcanism that extended from about mid-Tertiary to Recent time.

The Clearwater-Quesnel and McConnell Creek volcanic provinces appear to be related to fault zones bounding the eastern edge of the Quesnel Terrane. Dextral shear on these faults may have induced incipient gash fractures along which basaltic magma rose to the surface.

CONCLUSIONS

The history of the Canadian Cordillera has been punctuated by repeated eruption and accretion of volcanic rocks which now form a large proportion of the total mass of the orogen west of the Foreland Belt. Nowhere is there a greater diversity of volcanic assemblages than in western Canada where the products of multiple episodes of oceanic and continental volcanism are juxtaposed in relatively narrow belts along the western edge of the North American craton. Despite the complex character and association of the various volcanic suites some general trends in the nature of volcanism are clearly related to the evolution of the orogen with time, to the diverse tectonic environments that prevailed in the accreted terranes, and to the interaction of continental crust with adjacent oceanic plates.

From its inception as a rifted continental terrace wedge in the Middle Proterozoic throughout most of early Paleozoic time the evolving orogenic belt was dominated by extension and thermal subsidence associated with minor volcanic activity. Volcanism was characterized by the eruption of K-rich tholeiitic basalt and lesser amounts of alkali basalt. The proportion of alkaline lavas appears to have increased toward the end of this period and locally, where rifting was most intense, the basalt is associated with cogenetic, alkaline to peralkaline rhyolite. The chemistry of this suite suggests a primary source from undepleted mantle that underlay the Proterozoic craton.

West of the rifted continental margin stretched an unknown expanse of ocean basins, arcs, and platforms which evolved independently and were later accreted to the continent. Among these accreted terranes are two belts of disrupted oceanic crust, the Upper Devonian to Triassic Slide Mountain Terrane and the Mississippian to Jurassic Cache Creek Terrane. Both of these consist of imbricated thrust slices of chert, pillowed basalt, serpentinized peridotite, and altered gabbro representing ophiolite assemblages. Most of the pillow lava is typical low-K, mid-ocean ridge basalt (MORB) but the assemblage includes a lesser amount of more alkaline basalt which probably erupted as intraplate seamounts and platforms.

Lower Mesozoic arc volcanics, which are widely distributed throughout the western Canadian Cordillera, were erupted almost entirely on distant terranes which, during the late Mesozoic were assembled into superterranes and accreted to the craton. On the western, Insular Superterrane, a thick marine platform of tholeiitic basalt is overlain by a calc-alkaline arc assemblage. Elsewhere, throughout the central Cordillera, lower Mesozoic volcanics are dominated by mainly intermediate, calc-alkaline and lesser alkaline intermediate rocks which locally display normal chemical polarity (more alkaline toward the craton) suggesting that they formed above an east-dipping subduction zone.

The latest Jurassic is an important landmark in the history of Cordilleran evolution: it saw the final amalgamation of the major terranes with the continent and it is the age of the oldest, unsubducted oceanic crust still preserved in the Pacific Basin. Thus, theoretically at least, it is possible to correlate Cretaceous and younger volcanic events on the extended continental margin with the motion of adjacent oceanic plates.

Cretaceous volcanism was associated with two magmatic culminations, the first at about 100 Ma and the second at about 80 Ma. The 100 Ma event was manifest in both the western, Coast Belt, where arc volcanism produced calc-alkaline assemblages which overlap older terrane boundaries, and in the eastern, Omineca-Selwyn Belt, where the emplacement of high level aluminous plutons was associated locally with caldera formation. The 80 Ma event produced a strongly polarized calc-alkaline to alkaline arc assemblage which is restricted to a northerly expanding belt along the eastern margin of the Coast Plutonic Complex. The western Cretaceous arcs were probably developed above subduction zones along the eastern and northern edges of the northerly moving Kula Plate. Their concentration in the north reflects the more orthogonal convergence where the continental margin of the northern Cordillera swings west toward the Alaska Peninsula. The 100 Ma felsic volcanics and coeval muscovite-bearing plutons of the eastern belt have extremely high radiogenic Sr contents, indicating assimilation of old crust. These rocks may not be related to a subduction zone but rather to melting at the base of an overthickened sialic root zone formed by duplexing of the crust along an old suture.

Widespread Eocene volcanism, which followed a brief magmatic lull in the early Tertiary, began about 55 Ma, at almost exactly the same time that magnetic anomalies record major changes in the direction and spreading rate of Kula Plate. During the intense magmatic episode that followed, the continental margin was emergent and volcanism was manifest in the construction of subaerial composite volcanoes, caldera complexes and the intrusion of numerous epizonal plutons in a broad zone through the central Cordillera. The eruptive rocks, which range in composition from highly evolved calc-alkaline to potassic alkaline suites, exhibit normal chemical polarity and probably developed above a shallow, northeast dipping subduction zone along the rapidly converging margin of Kula Plate. Subduction was probably oblique to much of the continental margin and associated with dextral transcurrent faulting. In the southern Canadian Cordillera this resulted in the formation of extensional, pull-apart basins into which bimodal, alkaline to calc-alkaline volcanic suites were erupted.

By Miocene time the Kula Plate had disappeared and subsequent magmatic events in the northern Cordillera result from interaction of the continent with the Farallon and Pacific plates. A complex calc-alkaline arc, developed above northerly subducting Pacific crust in the northern Cordillera, was initiated in the Miocene and volcanoes in its western, Alaskan, portion are still active. Farther south a late Miocene calc-alkaline arc developed above easterly subducting Farallon Plate in the south coastal region and related back-arc basalts flooded a vast area farther east, in south-central British Columbia. Unlike the rift basalts generated earlier in Cordilleran history, the chemistry of Miocene back-arc lavas indicates an origin from highly depleted mantle. As the size of the Farallon Plate decreased with time so too did the length of the adjacent volcanic front. Today a small remnant of Farallon Plate (Juan de Fuca Plate) is flanked by Pleistocene volcanoes of the Garibaldi-Cascade Volcanic Belt in southwestern British Columbia.

Throughout Miocene and later time the north-central Cordillera of British Columbia was bounded by a transcurrent plate margin and no calc-alkaline volcanoes developed on the adjacent continental margin. However, a belt of highly evolved, alkaline to peralkaline volcanic complexes erupted farther inland. They are associated with northerly trending normal faults which, like the alkaline parent magmas, are believed to be the result of incipient continental rifting. An east-trending, Miocene to Recent belt of similar, highly alkaline volcanic complexes extends through central British Columbia. An easterly decrease in the age of initial volcanism suggests that the belt may represent the trace of a mantle hotspot over which the continent has moved.

Recognition of the important role volcanism has played in Cordilleran evolution is reflected in the increased amount of volcanological research in Western Canada. On the one hand volcanic rocks provide our most direct window to the deep crustal and mantle processes that drive tectonism and, on the other, the transfer of volcanic heat to the upper crust is an essential step in the formation of many metalliferous deposits.

During the past decade a dramatic increase in the amount and quality of analytical data has defined the chemical affinity of most of the major Cordilleran volcanic suites and, in many places, established the chemical polarity of ancient volcanic arcs. Dating and stable isotope studies have established temporal and genetic links between volcanic and plutonic rocks and provided insights into the nature of the crust and mantle from which they originated. Mapping and stratigraphic work continue to refine our understanding of the geometry of volcanic terranes, their association with crustal structures, and the relationship between volcanic protoliths and derived sediments. Future researchers will face the added challenge of more rigorous interpretation of what are now mainly empirical models. The thermodynamic and chemical criteria which control the relationship between tectonic environments and the chemistry of associated volcanic suites are only beginning to be addressed. Such information is needed to provide independent constraints on plate tectonic models and on the interpretation of geophysical data from the deep crust and mantle.

Volcanogenic mineral deposits are known in almost every major volcanic assemblage in the Canadian Cordillera. Massive sulphide deposits, iron-formation, and strata-bound barite deposits are associated with rift and oceanic (MORB) basalts of Proterozoic to Mesozoic age, porphyry copper and molybdenum deposits are associated with Mesozoic subvolcanic plutons, and epithermal precious metal deposits are associated with Mesozoic and Tertiary volcanic centres. The study of modern hydrothermal systems both on the ocean floor and associated with active terrestrial volcanoes is providing new insights into the processes of metal transport and deposition but there is a need for better coordination between volcanology and metallogeny, particularly in the study of ancient hydrothermal systems and their relationship to the larger structural and stratigraphic features of host volcanic complexes.

Each new advance in the science of volcanology has the potential to sharpen our understanding of western Cordilleran geology. Conversely the western Canadian Cordillera, with its diversity of volcanic rocks and tectonic environments, is fertile ground for research into the fundamental processes of volcanism and metallogeny.

REFERENCES

Aggarwal, P.K., Fujii, T., and Nesbitt, B.E.
1984: Magmatic composition and tectonic setting of altered volcanic rocks of the Fennell Formation, British Columbia; Canadian Journal of Earth Sciences, v. 21, p. 745-752.

Armstrong, R.L., Eisbacher, G.E., and Evans, P.D.
1982: Age and stratigraphic-tectonic significance of Proterozoic diabase sheets, Mackenzie Mountains, northwestern Canada; Canadian Journal of Earth Sciences, v. 19, p. 316-323.

Armstrong, R.L., Muller, J.E., Harakal, J.E., and Muehlenbachs, K.
1985: The Neogene Alert Bay Volcanic Belt of northern Vancouver Island, Canada: descending-plate-edge volcanism in the arc-trench gap; Journal of Volcanology and Geothermal Research, v. 26, p. 75-97.

Bacon, C.R., Foster, H.L., and Smith, J.G.
1985: Cretaceous calderas and rhyolitic welded tuffs in the Yukon-Tanana Terrane, east-central Alaska; Abstracts with Programs, 81st Annual Meeting, Cordilleran Section, Geological Society of America with Pacific Coast Section of the Paleontological Society, May, 1985, University of British Columbia, Vancouver, v. 17, no. 6.

Bailey, E.H. and Blake, M.C.
1974: Major chemical characteristics of Mesozoic Coast Range ophiolite in California; Journal of Research of the United States Geological Survey, v. 2, p. 637-656.

Barker, F., Plafker, G., and Sutherland Brown, A.
1985: Karmutsen Formation, Queen Charlotte Is. and Vancouver Is.: an arc-rift ferrotholeiite of Wrangellia; Geological Society of America, Abstracts with Programs, Annual Meeting Cordilleran Section, p. 340.

Beddoe-Stephens, B.
1982: The petrology of the Rossland volcanic rocks, southern British Columbia; Geological Society of America Bulletin, v. 93, p. 585-594.

Bell, R.T.
1986: Megabreccias in northeastern Wernecke Mountains, Yukon Territory; in Current Research, Part A, Geological Survey of Canada, Paper 86-1A, p. 375-384.

Berg, H.C., Jones, D.L., and Richter, D.H.
1972: Gravina-Nutzotin Belt - Tectonic significance of an Upper Mesozoic sedimentary and volcanic sequence in southern and southeastern Alaska; United States Geological Survey, Professional Paper 800-D, p. D1-D24.

Berman, R.G.
1981: Differentiation of calc-alkaline magmas: evidence from the Coquihalla volcanic complex, British Columbia; Journal of Volcanology and Geothermal Research, v. 9, p. 151-179.

Bevier, M.L.
1983a: Regional stratigraphy and age of Chilcotin Group basalts, south-central British Columbia; Canadian Journal of Earth Sciences, v. 20, p. 515-524.
1983b: Implications of chemical and isotopic composition for petrogenesis of Chilcotin Group basalts, British Columbia; Journal of Petrology, v. 24, p. 207-226.

Bond, G.C., Christie-Blick, N., Kominz, M.A., and Devlin, W.J.
1985: An early Cambrian rift to post-rift transition in the Cordillera of western North America; Nature, v. 315, p. 742-746.

Brandon, M.T.
1985: Mesozoic melange of the Pacific Rim Complex, western Vancouver Island; Geological Society of America, Cordilleran Section Meeting, Field Guidebook, p. 7-1 - 7-28.

Brandon, M.T., Orchard, M.J., Parrish, R.R., Sutherland Brown, A., and Yorath, C.J.
1986: Fossil ages and isotopic dates from the Paleozoic Sicker Group and associated intrusive rocks, Vancouver Island, British Columbia; in Current Research, Part A, Geological Survey of Canada, Paper 86-1A, p. 683-696.

Cameron, B.E.B. and Tipper, H.W.
1985: Jurassic stratigraphy of the Queen Charlotte Islands; Geological Survey of Canada, Bulletin 365, 49 p.

Campbell, R.B. and Dodds, C.J.
1982a: Geology of S.W. Kluane Lake map-area (115G and F (E1/2)), Yukon Territory; Geological Survey of Canada, Open File 829.
1982b: Geology of Mount St. Elias map-area (115B and C (E1/2)), Yukon Territory part; Geological Survey of Canada, Open File 830.
1982c: Geology S.W. Dezadeash map-area (115A), Yukon Territory; Geological Survey of Canada, Open File 831.
1983a: Geology of Tatshenshini river map-area (114P), British Columbia; Geological Survey of Canada, Open File 926.
1983b: Terranes and major faults of the Saint Elias Mountains, Yukon Territory, British Columbia and Alaska; Geological Association of Canada, Annual Meeting, Victoria, Program with Abstracts, p. A10.

Campbell, R.B., Mountjoy, E.W., and Young, F.G.
1973: Geology of McBride map area, British Columbia; Geological Survey of Canada, Paper 72-35.

Cecile, M.P.
1982: The Lower Paleozoic Misty Creek Embayment, Selwyn Basin, Yukon and Northwest Territories; Geological Survey of Canada, Bulletin 335, 78 p.

Cecile, M.P. and Norford, B.S.
1979: Basin to platform transition, Paleozoic strata of Ware and Trutch map-areas, northeastern British Columbia; in Current Research, Part A, Geological Survey of Canada, Paper 79-1A, p. 219-226.

Chronic, F.
1979: Geology of the Guano-Guayes rare earth element-bearing skarn porphyry, Pelly Mountains, Yukon Territory; M.Sc. thesis, University of British Columbia, Vancouver, 122 p.

Church, B.N.
1972: Geology of the Buck Creek area; in British Columbia Ministry of Mines and Petroleum Resources, Geology, Exploration and Mining in British Columbia, 1972, p. 353-363.
1973: Geology of the White Lake Basin; British Columbia Department of Mines and Petroleum Resources, Bulletin 61, 120 p.

Churchill, S.
1980: Geochronometry and chemistry of the Cretaceous Carmacks Group, Yukon; B.Sc. thesis, University of British Columbia, Vancouver, April 1980.

Daly, R.A.
1912: Geology of the North American Cordillera at the forty-ninth parallel; Geological Survey of Canada, Memoir 38, p. 443-506.

Devlin, W.J. and Bond, G.C.
1984: Syn-depositional tectonism related to continental breakup in the Hamill Group, Northern Selkirk Mountains, British Columbia; Geological Association of Canada - Mineralogical Association of Canada, Program with Abstracts, v. 9, Joint Annual Meeting, London, Ontario.

Devlin, W.J., Bond, G.C., and Hannes, K.B.
1985: An assessment of the age and tectonic setting of volcanics near the base of the Windermere Supergroup in northeastern Washington: implications for latest Proterozoic-earliest Cambrian continental separation; Canadian Journal of Earth Sciences, v. 22, p. 829-837.

Diakow, L.J.
1984: Geology between Toodoggone and Chukachida Rivers; Geological Fieldwork, 1983, British Columbia Ministry of Energy, Mines and Petroleum Resources, Paper 1984-1, p. 139-145.
1985: Potassium-argon age determinations from biotite and hornblende in Toodoggone volcanic rocks; Geological Fieldwork, 1984, British Columbia Ministry of Energy, Mines and Petroleum Resources, Paper 1985-1, p. 298-300.

Dusel-Bacon, C. and Aleinikoff, J.N.
1985: Petrology and tectonic significance of augen gneiss from the belt of Mississippian granitoids in the Yukon-Tanana terrane, east-central Alaska; Geological Society of America Bulletin, v. 96, p. 411-425.

Eisbacher, G.H.
1981: Sedimentary tectonics and glacial record in the Windermere Supergroup, Mackenzie Mountains, northwestern Canada; Geological Survey of Canada, Paper 80-27, 40 p.

Engebretson, D.C., Cox, A., and Gordon, R.G.
1984: Relative motions between oceanic plates of the Pacific Basin; Journal of Geophysical Research, v. 89, no. B12, p. 10,291-10,310.

Evenchick, C.A., Parrish, R.R., and Gabrielse, H.
1984: Precambrian gneiss and late Proterozoic sedimentation in north-central British Columbia; Geology, v. 12, p. 233-237.

Ewing, T.E.
1981a: Petrology and geochemistry of the Kamloops Group volcanics, British Columbia; Canadian Journal of Earth Sciences, v. 18, p. 1478-1491.
1981b: Regional stratigraphy and structural setting of the Kamloops Group, south-central British Columbia; Canadian Journal of Earth Sciences, v. 18, p. 1464-1477.

Gabrielse, H.
1963: McDame map-area, Cassiar District, British Columbia; Geological Survey of Canada, Memoir 319, 96 p.
1975: Geology of Fort Grahame E1/2 map-area, British Columbia; Geological Survey of Canada, Paper 75-33.
1985: Major dextral transcurrent displacements along the Northern Rocky Mountain Trench and related lineaments in north-central British Columbia; Geological Society of America Bulletin, v. 96, p. 1-14.

Gabrielse, H. and Blusson, S.L.
1969: Geology of Coal River map-area, Yukon Territory and District of Mackenzie (95D); Geological Survey of Canada, Paper 68-38.

Gabrielse, H., Blusson, S.L., and Roddick, J.A.
1973: Geology of Flat River, Glacier Lake, and Wrigley Lake map-areas, District of Mackenzie and Yukon Territory; Geological Survey of Canada, Memoir 366, 153 p.

Gammon, J.B. and Chandler, T.E.
1986: Exploration of the Windy Craggy massive sulphide deposit, British Columbia, Canada; in Geology in the Real World, Kingsley Dunham Volume, Institution of Mining and Metallurgy, p. 131-141.

Goodfellow, W.D., Jonasson, I.R., and Cecile, M.P.
1980: Geochemistry and mineralogy of shales, cherts, carbonates and volcanic rocks from the Road River Formation, Misty Creek embayment, Northwest Territories, Part I; in Current Research, Part B, Geological Survey of Canada, Paper 80-1B, p. 149-161.

Gordey, S.P.
1981: Stratigraphy, structure and tectonic evolution of southern Pelly Mountains in the Indigo Lake Area, Yukon Territory; Geological Survey of Canada, Bulletin 318, 44 p.
1983: Thrust faults in the Anvil Range and a new look at the Anvil Range Group, south-central Yukon Territory; in Current Research, Part A, Geological Survey of Canada, Paper 83-1A, p. 225-227.

Gordey, S.P., Gabrielse, H., and Orchard, M.J.
1982: Stratigraphy and structure of Sylvester Allochthon, southwest McDame map area, northern British Columbia; in Current Research, Part B, Geological Survey of Canada, Paper 82-1B, p. 101-106.

Gordy, P.L. and Edwards, G.
1962: Age of the Howell Creek Intrusives; Journal of the Alberta Society of Petroleum Geologists, v. 10, no. 7, p. 369-372.

Green, L.H.
1972: Geology of Nash Creek, Larsen Creek, and Dawson map-areas, Yukon Territory (106D, 116A, 116B, and 116E (E1/2)), Operation Ogilvie; Geological Survey of Canada, Memoir 364, p. 103-106.

Green, N.L.
1981: Geology and petrology of Quaternary volcanic rocks, Garibaldi Lake area, southwestern British Columbia; Geological Society of America Bulletin, v. 92, Part 1, p. 697-702, and Part 2, p. 1359-1470.

Grond, H.C., Churchill, S.J., Armstrong, R.L., Harakal, J.E., and Nixon, G.T.
1984: Late Cretaceous age of the Hutshi, Mount Nansen, and Carmacks groups, southwestern Yukon Territory and northwestern British Columbia; Canadian Journal of Earth Sciences, v. 21, p. 554-558.

Hall-Beyer, B.
1976: Geochemistry of some ocean-floor basalts of central B.C.; M.Sc. thesis, University of Alberta, Fall 1976.

Harms, T.
1984: Structural style of the Sylvester Allochthon, northeastern Cry Lake map area, British Columbia; in Current Research, Part A, Geological Survey of Canada, Paper 84-1A, p. 109-112.

Heah, T.S.T., Armstrong, R.L., and Woodsworth, G.J.
1986: The Gambier Group in the Sky Pilot area, southwestern Coast Mountains, British Columbia; in Current Research, Part B, Geological Survey of Canada, Paper 86-1B, p. 685-692.

Helwig, J. and Emmet, P.
1981: Structure of the Early Tertiary Orca Group in Prince William Sound and some implications for the plate tectonic history of Southern Alaska; Journal of the Alaska Geological Society, v. 1, p. 12-35.

Hickson, C.J.
1986: Quaternary volcanics of the Wells Gray-Clearwater area, east-central British Columbia; Ph.D. thesis, University of British Columbia, Vancouver, 357 p.

Hickson, C.J. and Souther, J.G.
1984: Late Cenozoic volcanic rocks of the Clearwater-Wells Gray area, British Columbia; Canadian Journal of Earth Sciences, v. 21, p. 267-277.

Hodder, R.W. and MacIntyre, D.G.
1977: Place and time of porphyry-type Cu-Mo mineralization in Upper Cretaceous caldera development, Tahtsa Lake, British Columbia; Fifth IAGOD Quadrennial Symposium, p. 175-183.

Høy, T.
1979: Geology of the Estella-Kootenay King area, Hughes Range, southeastern British Columbia; British Columbia Ministry of Energy, Mines and Petroleum Resources, Preliminary Map 36.

Ijewliw, O.J.
1986: Comparative mineralogy of three ultramafic breccia diatremes in southeastern British Columbia; in Geological Fieldwork, 1986, British Columbia Ministry of Energy, Mines and Petroleum Resources, Paper 1987-1.

Irvine, T.N. and Baragar, W.R.A.
1971: A guide to the chemical classification of the common volcanic rocks; Canadian Journal of Earth Sciences, v. 8, p. 523-548.

Isachsen, C., Armstrong, R.L., and Parrish, R.R.
1985: U-Pb, Rb-Sr, and K-Ar geochronometry of Vancouver Island igneous rocks; in Geological Association of Canada, Symposium on the Deep Structure of Southern Vancouver Island: Results of LITHOPROBE Phase I, Program and Abstracts, LITHOPROBE Project Publication No. 10, p. 21-22.

Jackson, L.E., Gordey, S.P., Armstrong, R.L., and Harakal, J.E.
1986: Bimodal Paleogene volcanics near Tintina Fault, east-central Yukon, and their possible relationship to placer gold; in Yukon Geology, v. 1; Exploration and Geological Services Division, Yukon, Indian and Northern Affairs Canada, p. 139-147.

Jones, D.L., Silberling, N.J., and Hillhouse, J.W.
1977: Wrangellia - A displaced terrane in northwestern North America; Canadian Journal of Earth Sciences, v. 14, p. 2565-2577.

Klepacki, D.W. and Wheeler, J.O.
1985: Stratigraphic and structural relations of the Milford, Kaslo and Slocan groups, Goat Range, Lardeau and Nelson map areas, British Columbia; in Current Research, Part A, Geological Survey of Canada, Paper 85-1A, p. 277-286.

Lambert, M.B.
1974: The Bennett Lake Cauldron Subsidence Complex, British Columbia and Yukon Territory; Geological Survey of Canada, Bulletin 227, 213 p.

Lane, L.S.
1977: Structure and stratigraphy, Goldstream River-Downie Creek area, Selkirk Mountains, British Columbia; M.Sc. thesis, Carleton University, Ottawa, Ontario, 139 p.

Lawrence, R.B., Armstrong, R.L., and Berman, R.G.
1984: Garibaldi Group volcanic rocks of the Salal Creek area, southwestern British Columbia: alkaline lavas on the fringe of the predominantly calc-alkaline Garibaldi (Cascade) volcanic arc; Journal of Volcanology and Geothermal Research, v. 21, p. 255-276.

Lerbekmo, J.F. and Campbell, F.A.
1969: Distribution, composition, and source of the White River Ash, Yukon Territory; Canadian Journal of Earth Sciences, v. 6, p. 109-116.

Lord, C.S.
1948: McConnell Creek map-area, Cassiar District, British Columbia; Geological Survey of Canada, Memoir 251, p. 43-45.

MacIntyre, D.G.
1976: Evolution of Upper Cretaceous volcanic and plutonic centres and associated porphyry copper occurrences, Tahtsa Lake area, British Columbia; Ph.D. thesis, University of Western Ontario, London, Ontario.
1981: Akie River Project (94F); in Geological Fieldwork 1980, British Columbia Ministry of Mines and Petroleum Resources, Paper 1981-1, p. 33-45.
1984: Geology of the Alsek-Tatshenshini Rivers area (114P); Geological Fieldwork, 1983, British Columbia Ministry of Energy, Mines and Petroleum Resources, Paper 1984-1, p. 173-184.
1986: The geochemistry of basalts hosting massive sulphide deposits, Alexander Terrane northwest British Columbia; Geological Fieldwork, 1985, British Columbia Ministry of Energy, Mines and Petroleum Resources, Paper 1986-1, p. 197-210.

MacIntyre, D.G. and Schroeter, T.G.
1985: Mineral occurrences in the Mount Henry Clay area (114P/7,8); Geological Fieldwork, 1984, British Columbia Ministry of Energy, Mines and Petroleum Resources, Paper 1985-1, p. 365-379.

McMechan, M.E., Høy, T., and Price, R.A.
1980: Van Creek and Nicol Creek Formations (new): a revision of the stratigraphic nomenclature of the Middle Proterozoic Purcell Supergroup, southeastern British Columbia; Bulletin of Canadian Petroleum Geology, v. 28, p. 542-558.

Massey, N.W.D.
1986: Metchosin Igneous Complex, southern Vancouver Island: Ophiolite stratigraphy developed in an emergent island setting; Geology, v. 14, p. 602-605.

Mathews, H.W.
1952: Mount Garibaldi, a supraglacial Pleistocene volcano in southwestern British Columbia; American Journal of Science, v. 250, p. 81-103.

Monger, J.W.H.
1977a: Upper Paleozoic rocks of the western Canadian Cordillera and their bearing on Cordilleran evolution; Canadian Journal of Earth Sciences, v. 14, p. 1832-1859.
1977b: Upper Paleozoic rocks of northwestern British Columbia; in Report of Activities, Part A, Geological Survey of Canada, Paper 77-1A, p. 255-262.
1977c: The Triassic Takla Group in McConnell Creek map-area, north-central British Columbia; Geological Survey of Canada, Paper 76-29.
1982: Geology of Ashcroft map area, Southwestern British Columbia; in Current Research, Part A, Geological Survey of Canada, Paper 82-1A, p. 293-297.
1985: Structural evolution of the southwestern Intermontane Belt, Ashcroft and Hope map areas, British Columbia; in Current Research, Part A, Geological Survey of Canada, Paper 85-1A, p. 349-358.

Monger, J.W.H., Richards, T.A., and Paterson, I.A.
1978: The hinterland belt of the Canadian Cordillera: new data from northern and central British Columbia; Canadian Journal of Earth Sciences, v. 15, no. 5, p. 823-830.

Morin, J.A.
1977: Ag-Pb-Zn mineralization in the MM deposit and associated Mississippian felsic volcanic rocks in the Ct. Cyr Range, Pelly Mountains; Department of Indian and Northern Affairs, 1976 Mineral Inventory Report, p. 83-97.

Mortensen, J.K.
1981: Geological setting and tectonic significance of Mississippian felsic metavolcanic rocks in the Pelly Mountains, southeastern Yukon Territory; Canadian Journal of Earth Sciences, v. 19, p. 8-22.

Mortensen, J.K. and Godwin, C.I.
1982: Volcanogenic massive sulfide deposits associated with highly alkaline rift volcanics in the southeastern Yukon Territory; Economic Geology, v. 77, p. 1225-1230.

Mortimer, N.
1986: Late Triassic, arc-related, potassic igneous rocks in the North American Cordillera; Geology, v. 14, p. 1035-1038.

Muller, J.E.
1977a: Geology of Vancouver Island; in Geological Association of Canada; Mineralogical Association of Canada; Canadian Geophysical Union; Society of Economic Geologists; Field Trip Guidebook, v. 7.
1977b: Evolution of the Pacific Margin, Vancouver Island and adjacent regions; Canadian Journal of Earth Sciences, v. 14, p. 2062-2085.
1980a: Chemistry and origin of the Eocene Metchosin Volcanics, Vancouver Island, British Columbia; Canadian Journal of Earth Sciences, v. 17, p. 199-209.
1980b: The Paleozoic Sicker Group of Vancouver Island, British Columbia; Geological Survey of Canada, Paper 79-30.

Muller, J.E., Northcote, K.E., and Carlisle, D.
1974: Geology and mineral deposits of Alert Bay-Cape Scott map-area. Vancouver Island, British Columbia; Geological Survey of Canada, Paper 74-8.

Norris, D.K.
1974: Structural and stratigraphic studies in the northern Canadian Cordillera; in Report of Activities, Part A, Geological Survey of Canada, Paper 74-1A, p. 343-349.

Northcote, K.E. and Muller, J.E.
1972: Volcanism, plutonism and mineralization: Vancouver Island; Canadian Institute of Mining and Metallurgy Bulletin, v. 65, p. 49-57.

Okulitch, A.V. and Peatfield, G.R.
1977: Geologic history of the late Paleozoic-early Mesozoic eugeocline in southern British Columbia and eastern Washington; Abstract, Geological Association of Canada, Program with Abstracts, 1977, Annual Meeting, Vancouver, B.C., p. 40.

Pearce, J.A., Gorman, B.E., and Birkett, T.C.
1975: The TiO_2-K_2O-P_2O_5 diagram: a method of discrimination between oceanic and non-oceanic basalts; Earth and Planetary Science Letters, v. 24, 419 p.

Pell, J.
1987: Alkalic ultrabasic diatremes in British Columbia: petrology, geochronology and tectonic significance; in Geological Fieldwork 1986, Ministry of Energy, Mines and Petroleum Resources, Geological Survey Branch, Paper 1987-1, p. 259-272.

Pigage, L.C. and Anderson, R.G.
1985: The Anvil plutonic suite, Faro, Yukon Territory; Canadian Journal of Earth Sciences, v. 22, p. 1204-1216.

Preto, V.A.
1977: The Nicola Group: Mesozoic volcanism related to rifting in southern British Columbia; in Volcanic Regimes in Canada, W.R.A. Baragar, L.C. Coleman and J.M. Hall (ed.), Geological Association of Canada Special Paper 16, p. 39-57.

Price, R.A.
1979: Intracontinental ductile crustal spreading linking the Fraser River and northern Rocky Mountain Trench transform fault zones, south-central British Columbia and northeast Washington; Abstract, Geological Society of America, Abstracts with Programs, v. 11, p. 499.

Read, P.B.
1976: Operation Saint Elias, Yukon Territory: pre-Cenozoic volcanic assemblages in the Kluane Ranges; in Current Research, Part A, Geological Survey of Canada, Paper 76-1A, p. 187-193.

Reesor, J.E.
1958: Dewar Creek map-area with special emphasis on the White Creek Batholith, British Columbia; Geological Survey of Canada, Memoir 292, 78 p.

Reynolds, M.W.
1984: Tectonic setting and development of the Belt Basin, northwestern United States; in The Belt, S. Warren Hobbs (ed.), Montana Bureau of Mines and Geology, Special Publication 90, p. 44-46.

Richards, T.A.
1980: Geology, Hazelton map-area, British Columbia (93M); Geological Survey of Canada, Open File 720.

Ricketts, B.D.
1982: Laharic breccias from the Crowsnest Formation, Southern Alberta; in Current Research, Part A, Geological Survey of Canada, Paper 82-1A, p. 83-87.
1985: Possible plinian eruptions of Paleocene age in central Yukon: evidence from volcanic ash, Norman Wells area, N.W.T.; Canadian Journal of Earth Sciences, v. 22, p. 473-479.

Riddihough, R.
1984: Recent movements of the Juan de Fuca Plate system; Journal of Geophysical Research, v. 89, no. B8, p. 6980-6994.

Roddick, J.A.
1965: Vancouver North, Coquitlam, and Pitt Lake map-areas, British Columbia; Geological Survey of Canada, Memoir 335, p. 70-71.

Roots, C.F.
1982: Ogilvie Mountains Project, Yukon; Part B: volcanic rocks in north-central Dawson map area; in Current Research, Part A, Geological Survey of Canada, Paper 82-1A, p. 411-414.
1983: Mount Harper complex, Yukon: Early Paleozoic volcanism at the margin of the Mackenzie Platform; in Current Research, Part A, Geological Survey of Canada, Paper 83-1A, p. 423-427.

Roots, C.F. and Moore, J.M., Jr.
1982: Proterozoic and Early Paleozoic volcanism in the Ogilvie Mountains: an example from Mount Harper, west-central Yukon; Yukon Exploration and Geology 1982, p. 55-62.

Shannon, K.R.
1981: The Cache Creek Group and contiguous rocks near Cache Creek, British Columbia; in Current Research, Part A, Geological Survey of Canada, Paper 81-1A, p. 217-221.

Smith, M.J.
1983: The Skukum Volcanic Complex, 105D SW: Geology and comparison to the Bennett Lake Cauldron Complex; Indian and Northern Affairs Canada, Yukon Exploration and Geology 1982, p. 68-72.

Smith, R.B.
1974: Geology of the Harper Ranch Group (Carboniferous-Permian) and Nicola Group (Upper Triassic), northwest of Kamloops, British Columbia; M.Sc. thesis, University of British Columbia, Vancouver, 211 p.

Souther, J.G.
1971: Geology and mineral deposits of Tulsequah map-area, British Columbia (104K); Geological Survey of Canada, Memoir 362, 84 p.
1975: Geothermal potential of western Canada; in Proceedings, 2nd United Nations symposium on the development and use of geothermal resources, v. 1, p. 259-267.
1977: Volcanism and tectonic environments in the Canadian Cordillera - a second look; in Volcanic Regimes in Canada, W.R.A. Baragar, L.C. Coleman and J.M. Hall (ed.), Geological Association of Canada, Special Paper 16, p. 3-24.
1980: Geothermal reconnaissance in the central Garibaldi Belt, British Columbia; in Current Research, Part A, Geological Survey of Canada, Paper 80-1A, p. 1-11.
1984: The Ilgachuz Range, a peralkaline shield volcano in central British Columbia; in Current Research, Part A, Geological Survey of Canada, Paper 84-1A, p. 1-10.
1986: The western Anahim Belt: root zone of a peralkaline magma system; Canadian Journal of Earth Sciences, v. 23, no. 6, p. 895-908.

Souther, J.G. and Stanciu, C.
1975: Operation Saint Elias, Yukon Territory: Tertiary volcanic rocks; in Report of Activities, Part A, Geological Survey of Canada, Paper 75-1A, p. 63-70.

Stott, D.F.
1982: Lower Cretaceous Fort St. John Group and Upper Cretaceous Dunvegan Formation of the Foothills and Plains of Alberta, British Columbia, District of Mackenzie and Yukon Territory; Geological Survey of Canada, Bulletin 328.

Sutherland Brown, A.
1960: Geology of the Rocher Deboule Range; British Columbia Department of Mines and Petroleum Resources, Bulletin 43, 78 p.
1968: Geology of the Queen Charlotte Islands, British Columbia; British Columbia Department of Mines and Petroleum Resources, Bulletin 54, p. 127-128.

Sutherland Brown, A. and Yorath, C.J.
1985: Lithoprobe profile across southern Vancouver Island: geology and tectonics; Geological Society of America, Cordilleran Section Annual Meeting, Field Guidebook 8.

Tempelman-Kluit, D.J.
1977a: Stratigraphic and structural relations between the Selwyn Basin, Pelly Cassiar Platform and Yukon Crystalline Terrane in the Pelly Mountains, Yukon Territory; in Report of Activities, Part A, Geological Survey of Canada, Paper 77-1A, p. 223-227.
1977b: Geology of Quiet Lake (105F) and Finlayson Lake (105G) map-areas, Yukon Territory (2 maps; Scale: 1:250,000); Geological Survey of Canada, Open File 486.
1979: Transported cataclastite, ophiolite and granodiorite in Yukon: evidence of arc-continent collision; Geological Survey of Canada, Paper 79-14.
1980: Highlights of field work in Laberge and Carmacks map areas; Scientific and Technical Notes in Current Research, Part A, Geological Survey of Canada, Paper 80-1A, p. 357-362.

Tempelman-Kluit, D.J. and Parkinson, D.
1986: Extension across the Eocene Okanagan crustal shear in southern British Columbia; Geology, v. 14, p. 318-321.

Thompson, R.I.
1976: Some aspects of stratigraphy and structure in the Halfway River map-area (94B), British Columbia; in Report of Activities, Part A, Geological Survey of Canada, Paper 76-1A, p. 471-477.
in press: Rifting and its influence on Canadian Cordilleran miogeocline evolution; Geological Society of London Special Publication.

Thorkelson, D.J.
1985: Geology of the mid-Cretaceous volcanic units near Kingsvale, southwestern British Columbia; in Current Research, Part B, Geological Survey of Canada, Paper 85-1B, p. 333-339.

Thorstad, L.E. and Gabrielse, H.
1986: The Upper Triassic Kutcho Formation Cassiar Mountains, north-central British Columbia; Geological Survey of Canada, Paper 86-16.

Tipper, H.W.
1984: The allochthonous Jurassic-Lower Cretaceous terranes of the Canadian Cordillera and their relation to correlative strata of the North American Craton; in Jurassic-Cretaceous Biochronology and Paleogeography of North America, G.E.G. Westermann (ed.), Geological Association of Canada, Special Paper 27, p. 113-120.

Tipper, H.W. and Richards, T.A.
1976: Jurassic stratigraphy and history of north-central British Columbia; Geological Survey of Canada, Bulletin 270, 73 p.

Wheeler, J.O.
1961: Whitehorse map-area, Yukon Territory; Geological Survey of Canada, Memoir 312, 156 p.

Wood, D.H. and Armstrong, R.L.
1982: Geology, geochronometry of the Cretaceous South Fork volcanics, Yukon Territory; in Current Research, Part A, Geological Survey of Canada, Paper 82-1A, p. 309-316.

Woodsworth, G.J.
1979a: Geology of Whitesail Lake map area, British Columbia; in Current Research, Part A, Geological Survey of Canada, Paper 79-1A, p. 25-29.
1979b: Metamorphism, deformation, and plutonism in the Mount Raleigh Pendant, Coast Mountains, British Columbia; Geological Survey of Canada, Bulletin 295, 58 p.

Yorath, C.J. and Cook, D.G.
1981: Cretaceous and Tertiary stratigraphy and paleogeography, northern interior plains, District of Mackenzie; Geological Survey of Canada, Memoir 398.

Young, G.M.
1982: The late Proterozoic Tindir Group, east-central Alaska: Evolution of a continental margin; Geological Society of America Bulletin, v. 93, p. 759-783.

Young, G.M., Jefferson, C.W., Delaney, G.D., Yeo, G.M., and Long, D.G.F.
1982: Upper Proterozoic stratigraphy of northwestern Canada and Precambrian history of the North American Cordillera; Idaho Bureau of Mines, Bulletin 24, p. 73-96.

Author's address

J.G. Souther
Cordilleran Division
Geological Survey of Canada
100 West Pender Street
Vancouver, British Columbia
V6B 1R8

ADDENDUM

Papers on Neogene to Recent volcanics and dyke swarms in the Queen Charlotte Islands and east of Hecate Strait are included in the following publication:

1991: Evolution and hydrocarbon potential of the Queen Charlotte Basin, British Columbia; G.J. Woodsworth (ed.); Geological Survey of Canada, Pager 90-10, 569 p.

Printed in Canada

Chapter 15

PLUTONIC REGIMES

Summary
Introduction
Proterozoic and Paleozoic
    Ancestral North America
    Arctic-Alaska Terrane
    Kootenay Terrane
    Monashee Terrane
    Slide Mountain Terrane
    Coast and Insular belts
Late Triassic
    Polaris Ultramafic Suite
    Stikine Suite
Predominantly Early Jurassic
    Guichon Suite in Quesnellia
    Copper Mountain Suite
    Topley and Blake Lake suites in Stikinia
    Klotassin and Long Lake suites
    Westcoast Suite
Predominantly Middle Jurassic
    Kuskanax and Nelson suites, and Galena Bay stock
    Other Jurassic plutons in southern British Columbia
    Three Sisters Suite
    Teslin Crossing stock
    Island Suite
    Bokan Mountain Granite
Late Jurassic to Early Cretaceous
    Saint Elias Suite
    Francois Lake Suite
Mid-Cretaceous
    Howell Creek Suite
    Bayonne Suite
    Cassiar Suite
    Axelgold Gabbro
    Selwyn Suite

    Tombstone Suite
    Whitehorse Suite
    Kluane Suite
    Union Bay Ultramafic Suite
    Cascade Suite
    Southern Intermontane and Coast belts
Late Cretaceous
    Bulkley Suite
    Surprise Lake Suite
Early Tertiary (mainly Eocene)
    Insular Belt
    Southern British Columbia
    West-central British Columbia
    Northern British Columbia and Yukon Territory
    Foreland Basin
Coast Plutonic Complex
    Paleozoic and Late Triassic plutons
    Early Jurassic plutons
    Middle to Late Jurassic plutons
    Late Jurassic to early Early Cretaceous plutons
    Late Early to mid-Cretaceous plutons
    Late Cretaceous plutons
    Early Tertiary plutons
Late Tertiary
    Chilliwack Suite
    King Island Pluton
    Kano Suite
    Tkope, Wrangell, and La Perouse suites
Regional trends and tectonic classification
References

Chapter 15

PLUTONIC REGIMES

G.J. Woodsworth, R.G. Anderson, and R.L. Armstrong
with contributions by L.C. Struik (Proterozoic and Paleozoic) and P. van der Heyden (Coast Plutonic Complex)

SUMMARY

The greatest concentrations of plutonic rocks in the Canadian Cordillera are in the Coast and Omineca belts but significant amounts also occur in adjacent belts. Most Cordilleran plutons are Late Triassic to Paleogene in age, and are coeval and comagmatic with volcanic rock suites.

Proterozoic and Paleozoic plutons of ancestral North America consist of Early and Middle Proterozoic granodiorite, Late Proterozoic alkalic plutons, early Paleozoic alkalic to carbonatitic suites, and Proterozoic and Paleozoic mafic sills and diatremes. The pericratonic Kootenay Terrane contains granite to quartz diorite intrusions of mainly Ordovician to Mississippian age. The Monashee Terrane has Proterozoic and Paleozoic(?) alkaline intrusions. The Slide Mountain Terrane contains a variety of Paleozoic plutons, mostly diorite, quartz porphyry, and tonalite. The Alexander Terrane includes Ordovician to Early Silurian calc-alkaline plutons; mid- to Late Silurian sodic plutons emplaced during the Klakas orogeny; and, in the Saint Elias Mountains, late Paleozoic calc-alkaline stocks and batholiths. Wrangellia has small mafic to ultramafic plutons in the Saint Elias Mountains and Devonian quartz-feldspar porphyry in southwestern British Columbia.

Late Triassic plutons are largely restricted to small, Alaskan-type ultramafic bodies in Quesnellia and Stikinia, and to a belt of tholeiitic to calc-alkaline granitoid rocks that intrude Stikinia along the Stikine Arch. Both suites are spatially and probably genetically related and are associated with Middle to Upper Triassic volcanic rocks.

In the Early Jurassic, plutonic activity occurred in Quesnellia, Stikinia, and Wrangellia. Calc-alkaline batholiths in Quesnellia and alkaline bodies there and in Stikinia show close spatial and temporal affinities with Upper Triassic to Lower Jurassic volcanic rocks. Calc-alkaline plutons in Stikinia occur on the west, north and south sides of the Bowser Basin and may be related to Early and Middle Jurassic Hazelton volcanism. A poorly dated suite of plutons in Yukon Territory may in large part be related to volcanism in the Whitehorse Trough. In Wrangellia, Early to Late Jurassic, calc-alkaline, I-type plutons form a well-defined belt related to Early Jurassic Bonanza and Middle Jurassic Yakoun volcanism. A belt of plutons of Early to Middle Jurassic age seems to extend along the east side of the Coast Plutonic Complex for much of its length and may be coeval with Early to Middle Jurassic volcanism of the Hazelton Group.

Middle Jurassic plutons in southern British Columbia consist of an early alkaline suite that was followed by widespread calc-alkaline plutonism and then by areally restricted two-mica granites. Volcanic equivalents of these rocks are rarely preserved. Middle Jurassic plutons also are present in the Coast Belt, and in small, distinctive groups east and north of the Bowser Basin.

Late Jurassic to early Early Cretaceous plutons are common in the northern Insular and Coast belts but uncommon elsewhere in the Canadian Cordillera. Plutons of this age and related to arc volcanics of the Gambier Group occur in the Coast Plutonic Complex. Late Jurassic to Early Cretaceous plutons in the Saint Elias Mountains may be the northern extension of those in the Coast Plutonic Complex. In the Intermontane Belt, the earliest Cretaceous Francois Lake Intrusions are confined to Stikinia and the Cache Creek Terrane along the Skeena Arch.

Mid-Cretaceous granitic rocks are concentrated in two broad geographic belts that coincide, roughly, with the two great zones of metamorphic rocks in the Canadian Cordillera. Those in the eastern zone, roughly along the Omineca Belt, are predominantly S-type, felsic, and have initial $^{87}Sr/^{86}Sr$ ratios greater than 0.710; those in the western belt, mainly in the Coast Belt, are I-type, felsic to mafic, and have initial $^{87}Sr/^{86}Sr$ ratios less than 0.706. In the eastern zone, the numerous discordant batholiths and stocks are mainly granite and granodiorite in composition. These appear to have been emplaced after or, in some cases, during the main periods of metamorphism and deformation in the region. Many of these plutons have no known volcanic counterparts, nor do contemporaneous syenite and quartz monzonite plutons in the western Yukon. In the Whitehorse Trough, a suite of predominantly granodiorite plutons may be subvolcanic to the mid-Cretaceous Mount Nansen volcanics. In the Foreland Belt, small alkaline intrusions are comagmatic with the mid-Cretaceous Crowsnest volcanics.

Mid-Cretaceous plutonic rocks of the western zone are concentrated mainly in the Coast Belt, where they were emplaced before, during, and after regional deformation and metamorphism. In the northern Insular Belt,

---

Woodsworth G. J., Anderson, R. G., and Armstrong, R. L.
1991: Plutonic regimes, Chapter 15 in Geology of the Cordilleran Orogen in Canada, H. Gabrielse and C. J. Yorath (ed.); Geological Survey of Canada, Geology of Canada, no. 4, p. 491-531 (also Geological Society of America, The Geology of North America, v. G-2)

calc-alkaline plutons cut rocks of Wrangellia and the Alexander Terrane and the overlapping Gravina-Nutzotin Assemblage. Alaskan-type ultramafics of mid-Cretaceous age form a linear belt extending from the Saint Elias Mountains through southeastern Alaska and possibly into the Coast Plutonic Complex of British Columbia. These may be related to volcanic rocks of the Gravina-Nutzotin Assemblage.

Late Cretaceous plutons are abundant in the Intermontane Belt and, from north to south, the western to central to eastern parts of the Coast Belt. Those in the Intermontane Belt are generally small, high-level, calc-alkaline bodies that may be comagmatic with widespread nonmarine Upper Cretaceous volcanics. In the Coast Belt, Late Cretaceous plutons range from high-level, post-tectonic intrusions to deep-seated and syn-metamorphic bodies.

Most Paleogene plutons are Early and Middle Eocene in age. This time marks the end of extensive plutonism in the Canadian Cordillera, as well as uplift and unroofing of metamorphic complexes in the Omineca and Coast belts. The bulk of Eocene plutons are in the central and eastern Coast Plutonic Complex where they may be deep-seated counterparts of high-level Intermontane Belt plutons and volcanics; most of the remainder are in the Intermontane and Insular belts. A few are in the Foreland Belt. Most Eocene plutons are small, high level, discordant stocks; many are spatially and genetically related to Eocene volcanic rocks.

Oligocene and Miocene felsic plutons are restricted to the Coast and Insular belts. Most are small, were emplaced at shallow depths, and may be genetically related to nearby volcanic rocks of the same general age and composition. Most are probably subduction related, and those in the Saint Elias Mountains are associated with regional transcurrent faults.

INTRODUCTION

Plutonic rocks form a substantial proportion of the Canadian Cordillera, particularly in the Coast and Omineca belts. Most are calc-alkaline granitoid rocks of Mesozoic and Tertiary age that range in composition from diorite to granite. In the dozen years since the publication of the reviews of Cordilleran plutonism by Gabrielse and Reesor (1974) and Roddick and Hutchison (1974), there has been nearly an order of magnitude increase in the number of isotopic dates from plutons and volcanic rocks in the Cordillera. Armstrong (1988) reviewed some 3000 dates, about a third of them from plutons. These are mostly K-Ar but include significant numbers of Rb-Sr dates and a rapidly growing body of U-Pb data. However, detailed structural and petrological studies have not kept pace with the ever-growing isotopic data base, and the last decade has seen the publication of few thorough, modern petrological and structural studies of plutonic rocks in the Canadian Cordillera.

As with other rocks, plutonic rocks may be pre-, syn- or post-accretionary. Like volcanic rocks, the mineralogy, texture, chemical and isotopic composition, structure, and mode of emplacement of plutons reflect the tectonic environment in which the pluton was formed and emplaced. However, as most plutonic rocks are associated with regional metamorphic belts in which the accretionary history of the terranes (and even the identity of the terranes themselves) is uncertain, and because the mode of origin of many plutons is unclear, it has seemed best to treat Cordilleran plutonism primarily descriptively and chronologically. The time intervals chosen here differ slightly from those of Armstrong (1988), although the same data base is used. This review emphasizes the chemical and petrographic characteristics, subdivision into suites, and tectonic setting of the rocks rather than the isotopic data and is thus complementary to Armstrong (1988). The exception to the chronological approach is the Coast Plutonic Complex, which is treated as an entity although it comprises rocks of many different ages. In this chapter, for descriptive purposes, a number of plutonic suites, each having distinctive lithological and structural characteristics, are given informal names. The distribution of these suites and individual plutons is shown in Figure 15.1 (in pocket).

PROTEROZOIC AND PALEOZOIC

The limited knowledge of Proterozoic and Paleozoic plutons in the Canadian Cordillera permits no comprehensive, Cordillera-wide synthesis. Paleozoic terranes were probably widely scattered and the plutons are mainly deformed and metamorphosed. Granitoid Precambrian basement rocks are confined to the Foreland and Omineca belts and are discussed in Chapter 4.

Pre-Mesozoic plutonic rocks of ancestral North America consist of Precambrian granodiorite and gabbro, early Paleozoic alkalic to carbonatitic suites (except north of the Kaltag Fault where they consist mainly of granite), and Proterozoic and Paleozoic mafic sills and diatremes. Within the pericratonic Kootenay Terrane granite to quartz dioritic intrusions are mainly Ordovician to Mississippian in age; syenitic rocks are in part coeval with the granitic rocks and in part of uncertain age. The Monashee Terrane has Proterozoic and Paleozoic(?) alkaline intrusions. The Slide Mountain Terrane has a variety of small Paleozoic plutons, mostly diorite, quartz porphyry, and tonalite. Alexander Terrane includes Ordovician-Silurian plutons and late Paleozoic complexes of diorite to granite. Wrangellia contains Paleozoic mafic plutons and Devonian quartz-feldspar porphyry.

Ancestral North America

In ancestral North America, Precambrian plutonic rocks include Early, Middle, and Late Proterozoic granitoid gneisses (Chapter 4). In the Wernecke and Ogilvie mountains, the Wernecke Supergroup is intruded by numerous sills and lensoid bodies of gabbro, diorite, and minor peridotite (Green, 1972), here called the Rackla dykes. These are rare in the overlying Pinguicula Group, suggesting a post-Wernecke-pre-Pinguicula age. Numerous breccia bodies and diatremes of uncertain origin, the Coal Creek Dome and Wernecke Mountains diatremes, also are post-Wernecke and pre-Pinguicula in age and gave a U-Pb date on monazite of about 1270 ± 40 Ma (Parrish and Bell, 1987).

In the Mackenzie Mountains, diabase sills and dykes, the Thundercloud dykes, gave Rb-Sr isochron dates of about 770 Ma (Armstrong et al., 1982). Quartz diorite spatially associated with flows at the base of the Coates Lake Group gave a U-Pb date on zircons of 778 ± 2 Ma (Jefferson and Parrish, 1989). Farther west in the Ogilvie

Mountains, volcanics of the Mount Harper complex have been dated (U-Pb on zircons) at 751 Ma (Roots, 1987). These igneous events are associated with stratigraphy and structures that suggest continental extension (Thompson et al., 1987).

In northeastern British Columbia, the MacDonald Suite of northerly trending mafic dykes intrude the Muskwa Assemblage but are younger than the Windermere Assemblage (Taylor and Stott, 1973) and are assumed to be Late Proterozoic in age.

In southeastern British Columbia, a suite of sills and dykes, variously called the Moyie sills, Purcell sills, and Purcell intrusives (Schofield, 1915; Rice, 1937; Reesor, 1958) and here called the Moyie Suite, have received considerable attention. The sills, which reach 700 m in thickness, intrude all units of the Purcell Supergroup. Zartman et al. (1982) obtained near-concordant U-Pb dates of 1433 Ma from zircons of the Crossport sill in Idaho; however, the sills probably represent more than one age of emplacement (Chapter 5). Many of the sills are differentiated and range in composition from gabbro to quartz diorite granophyre (Bishop, 1974). Chemically, the sills are tholeiitic; the more differentiated members show extreme Fe and Ti enrichment (Hunt, 1964; Bishop, 1974). As with the Rackla and Thundercloud dykes, the Moyie Suite probably represents rifting and thinning of the continental crust in the Middle Proterozoic.

Also in southeastern British Columbia, the Hellroaring Creek Stock is a coarse grained, S-type, tourmaline-muscovite granodiorite that intrudes the Purcell Supergroup. A Rb-Sr age of 1320 ± 50 Ma at an initial $^{87}Sr/^{86}Sr$ ratio of 0.81 ± 0.06 (Ryan and Blenkinsop, 1971) and a U-Pb age of 1342 ± 7 Ma (J.K. Mortensen, pers. comm., 1987) clearly indicate a Middle Proterozoic age.

Alkalic, ultramafic diatremes occur in several areas in the Foreland Belt (Pell, 1987a,b). Except for the Cross kimberlite, all are hosted in Cambrian to Silurian platformal carbonate and clastic sediments. The Bull River cluster consists of 40 or more breccia pipes and related rocks. Most are hosted and interbedded with carbonates of the Ordovician-Silurian Beaverfoot Formation. The Mountain diatremes in the Mackenzie Mountains are kimberlites by most criteria and gave K-Ar and Rb-Sr dates indicating a Late Ordovician to Silurian age (Godwin and Price, 1986). These pipes are in part similar in age to mafic volcanics of the Marmot Formation found nearby (Chapter 7). Diatremes in the Golden cluster consist of macrocryst-rich breccias and dykes containing macrocrysts of titaniferous augite, phlogopite and other minerals (Ijewliw, 1987). Rb-Sr and K-Ar dates suggest a Devono-Mississippian age for this suite (Pell, 1987b). The youngest mafic diatreme is the Cross kimberlite in southeastern British Columbia. This body is a deeply eroded pipe containing olivine, phlogopite, pyroxene, garnet, and spinel megacrysts (Pell, 1987a,b; Ijewliw, 1987). Rb-Sr dates from 240 to 250 Ma on phlogopite (Smith et al., 1988) indicate a Late Permian or Early Triassic age for the kimberlite.

The Aley carbonatite complex is intrusive into Cambrian sediments in the Foreland Belt. The complex consists of an outer shell of altered syenite and a core of carbonatite; ferrocarbonatite dykes rich in Nb and rare earth elements intrude the contact aureole (Mader, 1987). A small ultramafic lamprophyre breccia pipe, the Ospika diatreme, is exposed a few hundred metres from the carbonatite complex, but the relations between the carbonatite and diatreme are unclear (Pell, 1986). Rb-Sr and K-Ar dates on micas from both the Aley complex and the Ospika diatreme range from 323 to 349 Ma, suggesting a Mississippian age of intrusion.

Paleozoic syenitic intrusions are characteristic of ancestral North America as far north as the Kaltag Fault. In the Yukon, these include the Dave Lord Pluton (which gave 372 ± 15 Ma on hornblende (Wanless et al., 1979) and a 352 ± 5 Rb-Sr whole-rock isochron (K. Bell, pers. comm., 1986)) and three plutons from which Cambrian to Permian K-Ar dates have been obtained (Baadsgaard et al., 1961). In the central Rocky Mountains, the Bearpaw Ridge sodalite syenite intrudes Silurian volcaniclastic rocks and may be synvolcanic (Pell, 1985, 1987b). The Ice River Complex is the largest of several alkaline complexes in the southern Rocky Mountains. It intrudes Upper Cambrian carbonate and consists of nepheline and pyroxene-rich rocks (jacupirangite to urtite) and a carbonatite body cut by nepheline and sodalite-bearing syenite (Currie, 1975). The complex has proven difficult to date because of disturbances to all isotopic systems. However, recent work indicates that its age is between 356 and 372 Ma, and most probably 368 ± 4 Ma (Late Devonian) (Parrish et al., 1987). In the Monashee Mountains, the Blue River cluster consists of a number of carbonatite layers intrusive into the Horsethief Creek Group (Pell, 1985, 1987b); some contain economically interesting concentrations of rare earth elements.

Arctic-Alaska Terrane

Poorly exposed and inadequately studied Paleozoic plutons in the Arctic-Alaska Terrane of the Yukon, north of the Kaltag Fault range from granite and granodiorite to monzodiorite. Most are medium- to coarse-grained, peraluminous biotite granite. Plutons are from one to 1700 km^2 in area, the largest being the Old Crow Batholith. K-Ar dates, mostly on biotite, range from 95 to 431 Ma and are mostly from 250 to 350 Ma (Norris and Yorath, 1981; Brosge and Reiser, 1969) but strong hydrothermal alteration in many of the bodies makes these dates questionable. The Mount Fitton and Mount Sedgwick plutons generally are regarded as Devonian, an age in accord with widespread Late Devonian tectonism in the area. Similar plutons from the eastern Brooks Range in Alaska give U-Pb upper intercepts with concordia ranging from 365 to 410 Ma (Dillon et al., 1987) and were interpreted by them as indicating widespread Devonian plutonism in the Arctic-Alaska Terrane.

Kootenay Terrane

In the Yukon, all granitic orthogneisses in the Kootenay Terrane were originally included in the Pelly Gneiss, but current usage favours more specific names. Mortensen (1983, 1986) recognized three distinct intrusive events based on lithology and isotopic data. The rocks occur both as crosscutting intrusions and as foliated sills up to several kilometres thick. The most widespread is the Mink Creek Suite, which includes large bodies of granite to quartz monzonite augen orthogneiss. These rocks are early Mississippian in age and are S-type in character. A second plutonic suite, the Simpson Range Suite (Mortensen, 1983;

Mortensen and Jilson, 1985) includes rocks of the same name in the southeastern Yukon and the Selwyn Gneiss (Tempelman-Kluit and Wanless, 1980) in west-central Yukon. The suite consists of hornblende-biotite quartz diorite to quartz monzonite, gave Late Devonian to early Mississippian U-Pb dates on zircons, and is typically I-type in nature. A third, somewhat younger unit, the Sulphur Creek orthogneiss, forms a batholith-sized body in the Klondike region in west-central Yukon (J.K. Mortensen, pers. comm., 1987). It is a gneissic, biotite quartz monzonite that gave a concordant, Early Permian U-Pb date on zircons.

In northern British Columbia, folded and foliated biotite-muscovite granite sills in Upper Proterozoic strata gave discordant U-Pb dates suggesting a Silurian or possibly Devonian age (Gabrielse et al., 1982). In southern British Columbia, a number of bodies that include the Mount Fowler Batholith, the Quesnel Lake Orthogneiss, and the Clachnacudainn Gneiss have been interpreted as Paleozoic on the basis of U-Pb dates on zircons (Okulitch et al., 1975; Okulitch, 1985; Getsinger, 1985; P. van der Heyden, pers. comm., 1986). However, the isotopic systematics of the zircons are invariably complex and interpretations are complicated by problems of lead loss and possible inherited zircons. Most of the plutons are well foliated and/or lineated; all are interpreted as orthogneisses. All have been metamorphosed and deformed in the Mesozoic and/or Tertiary, and all have given early Tertiary K-Ar dates. It is probable that these bodies are Paleozoic but the data do not permit a more refined age assignment. The Quesnel Lake Orthogneiss, however, is certainly Late Devonian to early Mississippian in age, based on discordant U-Pb dates on zircons (Mortensen et al., 1987).

Monashee Terrane

Several bodies of extrusive and intrusive carbonatites and gneissic syenites occur in the mantling gneisses structurally overlying the Frenchman Cap gneiss dome (Pell, 1987b). The best-studied of the extrusive bodies is the Mount Grace carbonatite (Höy and Kwong, 1986). The Mount Copeland syenite gneiss, which may lie structurally beneath the Mount Grace carbonatite, is an alkaline igneous complex emplaced as a sill or series of sills and dykes and later metamorphosed to amphibolite facies (Currie, 1976). Okulitch et al. (1981) obtained Late Proterozoic U-Pb dates from the Mount Copeland syenite. Further study of the same samples (Parrish and Scammell, 1988) gave U-Pb dates of 740 ± 36 Ma from the cores of the zircons, which may represent the age of the syenite. However, the isotopic systematics are complex and unusual, and a younger age for the syenite cannot be ruled out.

Slide Mountain Terrane

The Slide Mountain Terrane hosts a variety of small Paleozoic plutons. For example, in northern British Columbia, a thick, Early Permian (276 ± 6 Ma U-Pb dates on zircon and sphene) tonalite sill cuts a thrust fault in the Sylvester Group (Harms, 1986), and bodies of Permian hornblende diorite and granite occur within thrust sheets. A Late Devonian foliated diorite with inherited Early Proterozoic zircons forms a thrust slice in the Sylvester Group. In southern British Columbia, the syntectonic Permian Whitewater diorite intrudes Kaslo Group strata (Klepacki and Wheeler, 1985).

Coast and Insular belts

In the Bridge River Terrane, altered diorite, quartz diorite and granite within the Bralorne fault zone gave Permian K-Ar, Rb-Sr and U-Pb dates (Leitch and Godwin, 1987; Leitch, et al., 1990).

In southeastern Alaska, the Alexander Terrane includes Ordovician to Lower Silurian arc volcanics spatially and temporally associated with a compositionally similar, cogenetic calc-alkaline suite of diorite to granodiorite and subordinate gabbro and ultramafics, granite and quartz syenite. In mid- to Late Silurian time, trondhjemite and sodic leucodiorite were emplaced during the main phase of deformation and metamorphism of the essentially compressive Klakas orogeny (Gehrels and Saleeby, 1987; Gehrels et al., 1983; Armstrong, 1985). In the Saint Elias Mountains, the Icefield Ranges Suite consists of high-level, multi-phase, batholithic complexes and smaller bodies ranging in composition from syenite and granodiorite to diorite. Other, fairly homogeneous bodies of megacrystic granite and granodiorite also are known. Most K-Ar dates on biotite and hornblende range from 270 to 290 Ma (Late Pennsylvanian to Early Permian) (Dodds and Campbell, 1988). In eastern Alaska, plutons of the Icefield Ranges Suite are not confined to the Alexander Terrane, but also intrude strata assigned to Wrangellia (e.g. MacKevett, 1978). In the Wrangell Mountains of eastern Alaska, the Barnard Glacier pluton crosscuts both the Alexander Terrane and Wrangellia. U-Pb dates on zircons indicate a middle Pennsylvanian age for the pluton, suggesting that Wrangellia and the Alexander Terrane were amalgamated by middle Pennsylvanian time (Gardner et al., 1988).

In the Saint Elias Mountains, Wrangellia contains many small bodies of gabbro and diorite of possible Paleozoic age (Dodds and Campbell, 1988). The Steele Creek and Mount Constantine gabbro complexes and associated serpentinite may form the basement to Wrangellia. The former body is overlain by Lower Permian strata, and the latter gave a hornblende K-Ar date of 364 ± 37 Ma (Dodds and Campbell, 1988). In Wrangellia of southern British Columbia the Saltspring Intrusions gave discordant U-Pb dates on zircons suggesting an Early Devonian age (Muller, 1980; Brandon et al., 1986). These intrusions are metamorphosed and foliated quartz-feldspar porphyry and may be related to the Paleozoic quartz diorite of the Turtleback Complex in the San Juan Islands.

In the North Cascades, the Yellow Aster Complex is a crystalline unit consisting of gabbroic to trondhjemitic rocks that range in age from Precambrian to early Paleozoic (Mattinson, 1972; Misch, 1977). Gabbro on Vedder Mountain in the northern Cascades east of Vancouver is of unknown age but is lithologically like both the early Paleozoic and Jurassic gabbros on the San Juan Islands (Armstrong et al., 1983).

LATE TRIASSIC

Most Late Triassic plutonic rocks in the Canadian Cordillera occur in two main suites. The Polaris ultramafic suite is a belt of small, Alaskan-type ultramafic bodies that are spatially associated with Upper Triassic volcanic rocks of Quesnellia and Stikinia. The Stikine Suite consists of tholeiitic to calc-alkaline granitoid rocks that intrude Stikinia along the Stikine Arch. Both suites are spatially and probably genetically associated with Middle to Upper Triassic volcanics and, therefore, probably with each other. Indeed, in the Hotailuh, Hickman and Stikine batholiths and Kaketsa stock, the two suites occur together. Both suites are hornblende-rich and tend to have alkaline affinities.

In the southern Ogilvie Mountains, a thick, differentiated gabbro sill intrudes the Mississippian Keno Hill Quartzite (Tempelman-Kluit, 1970). Preliminary U-Pb dating suggests a Middle Triassic age for the sill (J.K. Mortensen, pers. comm., 1987).

Several Late Triassic gabbro bodies are known from Wrangellia and the Alexander Terrane. On Vancouver Island, a hornblende gabbro gives concordant 217 to 222 Ma U-Pb dates on zircons and is thought to be related to Upper Triassic Karmutsen basalts characteristic of Wrangellia (Isachsen et al., 1985). In Wrangellia of southwestern Yukon, mafic to ultramafic bodies give 224-225 Ma K-Ar dates on phlogopite (Campbell, 1981). On Duke Island in southeastern Alaska, a pyroxene gabbro has traditionally been thought to be Jurassic in age, but a U-Pb date of 226 ± 3 Ma on zircon clearly indicates a Late Triassic age (Gehrels et al., 1987).

In the North Cascades of Washington, the Marblemount Quartz Diorite and Dumbell Mountain Pluton are elongate tonalitic plutons that gave relatively concordant U-Pb dates of about 220 Ma (Tabor et al., 1987; Cater, 1982; Mattinson, 1972).

Polaris Ultramafic Suite

Numerous ultramafic to mafic intrusions form an arcuate belt extending around the eastern and northern margins of the Bowser Basin. Most are similar in lithology, petrography, chemistry and structure to the mid-Cretaceous "zoned" ultramafic complexes of southeastern Alaska (Clark, 1980; Irvine, 1976) and are only a few square kilometres in area. The largest, the Polaris Complex, is about 14 km long and 3 km wide. Most are circular or elongated in a northwesterly direction. They are characterized by dunite, wehrlite, clinopyroxenite, hornblendite, and hornblende gabbro or diorite. Orthopyroxene is typically absent and plagioclase is rare and interstitial. In several of the larger bodies, particularly the Turnagain complex (Clark, 1980), concentric zonation of lithologies is shown by a dunite or peridotite core surrounded by clinopyroxene and hornblende-rich rocks and a gabbro or diorite margin (Irvine, 1976).

Ultramafics of the Polaris Suite are most abundant near the north and east margins of the Hogem Batholith, where they intrude rocks of Quesnellia. The Polaris Complex to the east is separated from the Hogem cluster by the Lay Range fault and intrudes rocks also thought to be part of Quesnellia. North of the main cluster, the Menard Creek ultramafic body intrudes volcanics of the Stuhini Assemblage, a unit diagnostic of Stikinia. Still farther north, the Lunar Creek and Turnagain plutons lie along the Kutcho Fault. The Gnat Lakes Ultramafite is near the northwestern end of the Polaris chain and forms an early phase of the Hotailuh Batholith (Fig. 15.2) (Anderson, 1983a). This ultramafite consists of hornblende clinopyroxenite and hornblendite and is coeval with most of the Upper Triassic Stuhini Group (Anderson, 1983b).

K-Ar dates on hornblende from the Gnat Lakes, Wrede Creek, and Johanson Lake range from 215 to 232 Ma (Anderson, in Stevens et al., 1982b; Wong et al., 1985; Woodsworth, in Stevens et al., 1982a), suggesting a Middle to Late Triassic age. Middle to Late Jurassic dates from the Polaris Complex appear to reflect extensive later alteration and plutonism (Irvine, 1976). Irvine (1976), on the basis of a close spatial association between the ultramafic bodies and Upper Triassic volcanics, suggested a genetic relationship between the two. Clark (1980) suggested that the Turnagain body represents an alkaline crystal cumulate that formed in a magma chamber beneath a Late Triassic volcano. In the Hotailuh Batholith, Anderson (1983a) suggested that clinopyroxene and plagioclase porphyry basalt dykes form the hypabyssal link between the Gnat Lakes Ultramafite and the calc-alkaline to tholeiitic Stuhini Group basalts.

The similarities in lithology and age of the ultramafic bodies north and east of the Bowser Basin, regardless of terrane, suggest that the volcanic rocks of Quesnellia and Stikinia were amalgamated by Late Triassic time. The linearity of the ultramafic belt further suggests that large amounts of strike-slip faulting on the Lay Range, Ingenika, and Pinchi fault systems did not occur after this time. Alternatively, the assignment of the Menard Creek body to Stikinia may be erroneous, or the linear distribution of the ultramafic rocks coincidental. In this case, Quesnellia and Stikinia could have evolved in separate arcs until Early Jurassic time.

In southern British Columbia, the Late Triassic Tulameen gabbro and ultramafic complex (Findlay, 1969; St. Louis et al., 1986; Nixon and Rublee, 1988) intrudes Upper Triassic Nicola Group rocks and is similar in most respects to the other Alaskan-type ultramafics of the Polaris ultramafic suite. K-Ar dates on hornblende range from about 197 Ma to 215 Ma (Roddick and Farrar, 1971).

Stikine Suite

Late Triassic granitoid plutons are confined to the Stikine Arch on the north and west flanks of the Bowser Basin. Unlike the Polaris Ultramafic Suite of similar age, plutons of the Stikine Suite occur only in Stikinia.

Part of the Hotailuh Batholith (Anderson, 1983a) is typical of the Stikine Suite (Fig. 15.2). This batholith, one of the best-studied in the Canadian Cordillera, is a large (1100 km^2) composite body consisting of a Late Triassic plutonic suite that is cut by a Jurassic plutonic suite. (The Jurassic suite, mainly Middle and Late Jurassic, is described elsewhere in this chapter.) The Triassic suite consists of four units. From oldest to youngest, based on field relations, these are: Latham Pluton (biotite-hornblende diorite); Cake Hill Pluton (mainly quartz monzodiorite and monzodiorite); Gnat Lakes Ultramafite (described above); and Beggerlay Creek Pluton (altered hornblende gabbro and diorite). Parts of the suite are overlain by granitoid-cobble volcanic conglomerate forming part of the basal Upper Triassic Stuhini Group (Anderson, 1983a). The Beggerlay Creek Pluton clearly intrudes and metamorphoses the Stuhini Group. K-Ar dates range from 215 to 230 Ma and are consistent with the Late Triassic age

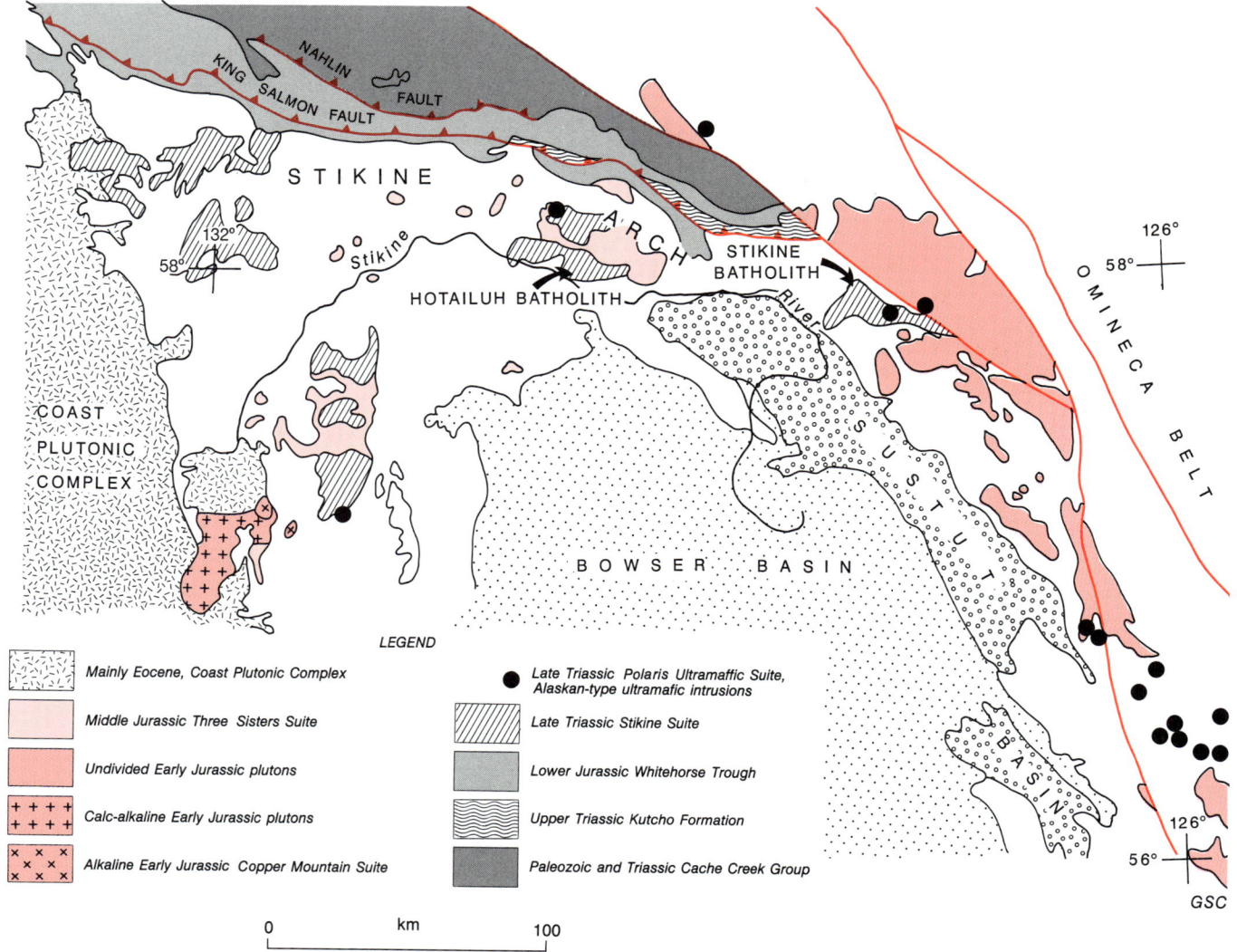

Figure 15.2. Distribution of Stikine, Polaris, unnamed Early Jurassic calc-alkaline, Copper Mountain and Three Sisters plutonic suites north and west of the Bowser Basin. Distribution of plutons in the west partly based on unpublished isotopic dates (D.A. Brown and M. Gunning, written communication, 1989 and M.L. Bevier, written communication, 1990). Unadorned red lines east of the Bowser Basin are mainly Cretaceous-Tertiary transcurrent faults.

inferred from stratigraphy, but are only roughly consistent with the inferred relative order of intrusion (Anderson, 1983a; Anderson, in Stevens et al., 1982b).

The Latham and Cake Hill plutons are moderately to strongly foliated; the other plutons are unfoliated. Hornblende is the sole mafic mineral in most plutons. Magnetite, apatite, and sphene are characteristic accessory minerals. The Beggerlay Creek Pluton contains coarse, poikiloblastic biotite and is pervasively altered. Subalkaline chemical compositions are most common, but the Beggerlay Creek Pluton falls in the alkaline field on an alkali-silica plot. The mafic phases are generally tholeiitic and the felsic plutons are mainly calc-alkaline. In general, the plutons have chemical characteristics typical of I-type granitoids. Compositionally similar pyroxenes in the plutons and nearby Upper Triassic volcanics support field and isotopic data which indicate that the Gnat Lakes Ultramafite and the Beggerlay Lake Pluton may be comagmatic and coeval with the Stuhini Group volcanics.

Several other batholiths along the Stikine Arch are comparable with the Hotailuh Batholith. The Stikine Batholith near the east end of the arch (Fig. 15.2) (Anderson, 1984) contains a suite of Late Triassic plutons similar to Late Triassic plutons of the Hotailuh and is cut, also, by a Middle(?) Jurassic plutonic suite. The Hickman Batholith on the northwest side of the Bowser Basin (Souther, 1972) contains both Late Triassic and Jurassic plutonic suites (Holbek, 1988). All three of these large bodies appear to have been emplaced at relatively shallow levels and were uplifted and unroofed within a few million years. The Kaketsa Stock (McMillan et al., 1975), dated as Late Triassic (K-Ar on hornblende) also appears to belong to this suite, as may the Mount Caplice Stock on the east edge of the Coast Belt near Atlin Lake (Werner, 1978).

PREDOMINANTLY EARLY JURASSIC

In contrast to the restricted extent of Late Triassic plutonism, the time from about 210 to 187 Ma was marked by extensive but scattered plutonism. The emplacement of highly alkaline plutons in the suspect terranes is distinctive of this time. Most plutonic activity took place in the terranes west of ancestral North America, particularly in Quesnellia (the Guichon Suite and most of the Copper Mountain Suite), and Stikinia (Copper Mountain and Topley suites). In the Yukon, the poorly dated and understood Klotassin and Long Lake suites form part of Stikinia and Quesnellia. In the Insular Belt of Vancouver Island, the Westcoast Suite may be largely Early Jurassic in age. A zone of plutons of Early to Middle Jurassic age seems to extend along the east side of the Coast Plutonic Complex for much of its length; the southern half of this belt is described in the section on the Coast Plutonic Complex. Based on regional setting, field characteristics, and geochronometry, several distinct plutonic suites can be recognized in the Coast Plutonic Complex; however, the distinction between plutons of Early and Middle Jurassic age is commonly blurred by resetting of isotopic systems.

Guichon Suite in Quesnellia

The western part of Quesnellia, particularly in the southern Canadian Cordillera, hosts numerous large plutons. These plutons are mainly Early Jurassic in age (200 to 210 Ma), although some may be as old as latest Triassic (up to 217 Ma), and form one of the characteristic and distinctive units of Quesnellia. Most of these plutons, like the alkalic suite discussed below, show close spatial and temporal affinities with Upper Triassic to Lower Jurassic volcanic rocks of the Nicola Assemblage. Many of the larger plutons, such as the Guichon Creek, Hogem, and Pitman batholiths, have an elongate shape, suggesting emplacement may have been controlled by pre- or synplutonic faults. In general, the plutons are hornblende-rich and mesocratic; granodiorite is the most abundant lithology.

The Guichon Creek Batholith (Fig. 15.3), southeast of Cache Creek, is typical of Early Jurassic calc-alkaline plutons in Quesnellia. It is one of the most-studied intrusions in the Canadian Cordillera because it hosts rich porphyry Cu and Mo deposits and is the principal Cu reserve for British Columbia (McMillan, 1985). Gravity data suggest that the north-northwesterly elongate, 1000 km² plan of the intrusion is a section through the top of a 4 km thick "lip" of an inverted champagne glass shape which leads downwards to a steeply east-northeast plunging, conical "stem" more than 8 km deep (Ager et al., 1973).

The batholith intrudes Lower Norian volcanics of the Nicola Group and is overlain by the Upper Pliensbachian to Callovian clastics of the Ashcroft Formation (Monger and McMillan, 1984). Reliable K-Ar isotopic ages range from 196 to 213 Ma for mineralization and all plutonic phases (e.g. Northcote, 1969; McMillan, 1985) and cluster around 205 Ma. A well-defined Rb-Sr whole-rock isochron date of 205 ± 10 Ma agrees with the K-Ar dates (Preto et al., 1979) and with U-Pb dates on zircon of 210 ± 3 Ma (Mortimer et al., 1990). The discrete periods of intrusion suggested by interphase intrusive relations (Northcote, 1969) are not reflected in the isotopic dates, and the batholith was apparently emplaced in a single brief period of intrusion in the latest Triassic.

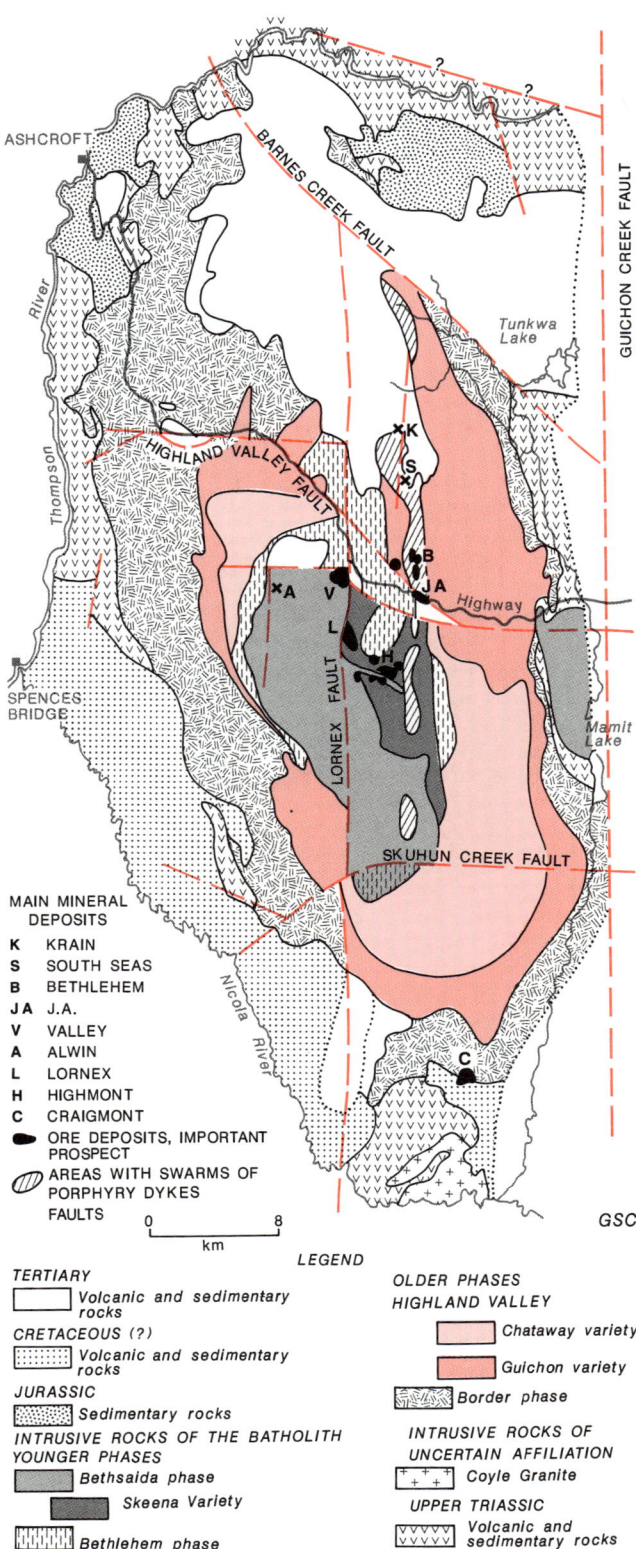

Figure 15.3. Simplified geology of the Guichon Creek Batholith, south-central British Columbia (after McMillan, 1985).

Sequential, radially inward emplacement of four heterogeneous phases is indicated by field relations (Northcote, 1969). Interphase contacts are generally gradational and rarely intrusive. Mafic-rich, well-foliated, heterogeneous tonalite and minor hornblendite, diorite, quartz diorite and quartz monzodiorite are oldest phases. The next two units comprise several varieties of granodiorite with lesser quartz monzodiorite and tonalite. The youngest, innermost, and most leucocratic phase are of unfoliated, porphyritic granodiorite and granite. Mineral showings are dispersed throughout the batholith but the important deposits are associated with late dyke swarms, faults, or the youngest phase.

All phases contain amphibole (ranging from magnesian hornblende in older phases to tremolite compositions in younger units) and biotite (predominantly Mg-rich and Al-poor in all phases; Johan et al., 1980). Older phases are more mafic, rarely hypersthene-bearing, contain augite as cores in hornblende and have higher hornblende/biotite ratios (Northcote, 1969). Alkali feldspar is interstitial and commonly perthitic. Quartz is interstitial and exhibits undulatory extinction except in the youngest phase where it is commonly clear, coarse and euhedral and apparently formed before plagioclase.

The batholith is calc-alkaline and both hypersthene- and diopside-normative. Smooth trends for cafemic elements, Mn and Ti on most variation diagrams and the distribution of Mn between amphibole and biotite support derivation of the younger, felsic phases from older, comagmatic mafic phases through fractionation of mineral phases similar to those in the mafic inclusions (e.g., Olade, 1976, McMillan, 1985; Johan et al., 1980).

Femic trace elements were incorporated into biotite and hornblende during crystal fractionation and their abundance decreases smoothly with intrusive sequence (Olade, 1976; Johan et al., 1980). Large discontinuities in large-ion trace element contents and smaller discontinuities in U, Eu and Sc with intrusive sequence suggest the importance of an evolving volatile phase in trace element partitioning during genesis of the later phases (Johan et al., 1980). High Sr values (average 530-725 ppm) and low Rb/Sr ratios (average 0.05-0.09) support indications of a primitive protolith and crustal environment, consistent with low $^{87}Sr/^{86}Sr$ initial ratios of 0.7034-0.7037 (Preto et al., 1979).

In contrast to the Guichon Creek Batholith, in which multiple phases were emplaced in a very short time, the Hogem Batholith in central British Columbia consists of several plutons that were emplaced over about 100 Ma (Fig. 15.4). An altered diorite to monzodiorite suite gave K-Ar dates ranging from 206 to 175 Ma (Garnett, 1978; Eadie, 1976). These rocks grade into and are cut by granodiorite plutons with K-Ar dates of 189-190 Ma. The Middle Jurassic Duckling Creek Syenite, described below, cuts the Lower Jurassic rocks and is in turn intruded by mid-Cretaceous granite and granodiorite of the Mesilinka Pluton and Osilinka Intrusions. The Early Jurassic plutons are calc-alkaline to slightly alkaline, and have low (less than 0.705) initial $^{87}Sr/^{86}Sr$ ratios. The batholith and its constituent plutons are elongate northwesterly, suggesting long-lived structural control of plutonism. Farther north, the northwest-trending Pitman Batholith, the largest pluton of this suite in northern British Columbia, consists of well foliated quartz diorite and granodiorite with characteristic large, prismatic hornblende grains and white plagioclase. Biotite is everywhere less abundant than hornblende. Much of the pluton is altered to greenschist-facies assemblages, particularly along and near faults.

In southern British Columbia, the Similkameen and Pennask batholiths represent Early Jurassic parts of the poorly-studied and complex Okanagan Composite Batholith (described below). The Mount Lytton Complex east of the Pasayten Fault, is an elongate, north-northwest trending, composite body that is continuous with the Eagle Plutonic Complex to the south. The Mount Lytton Complex is diorite to granodiorite in composition and may represent the roots of the western part of the Triassic part of the Nicola Group (Monger, 1985).

The Guichon and other Early Jurassic plutons in southern Quesnellia are closely associated in time, space and origin with the Upper Triassic to Lower Jurassic Nicola Assemblage (Mortimer, 1986). In this view, the plutons represent subvolcanic roots of a Late Triassic to Early Jurassic island arc (e.g. Chapter 8; Gabrielse and Reesor, 1974; Preto et al., 1979, Monger and McMillan, 1984). The alkalic, Late Triassic to Middle Jurassic Copper Mountain Suite (discussed below) also appears spatially and temporally linked with the Nicola Assemblage. Near the Guichon Creek Batholith, the Nicola volcanics show a progression from calc-alkaline on the west to alkaline on the east, suggesting that the volcanics represent a west-facing arc (Mortimer, 1986). Monger (Chapter 8) suggested that the Guichon Creek Batholith represents the eroded roots of the western, calc-alkaline facies, whereas the alkaline Iron Mask Batholith is thought to be comagmatic with the eastern, alkaline, augite porphyry facies. However, the presence of the calc-alkaline, Early Jurassic Wildhorse Batholith in the same belt as the Iron Mask Batholith is unexplained. In other parts of Quesnellia, close correlations between volcanic and plutonic chemistry have not been found, and the genetic relationships between volcanism, calc-alkaline plutons, and alkaline plutons are neither simple nor well understood.

Copper Mountain Suite

Numerous small alkaline bodies of the Copper Mountain Suite form a roughly linear, northwest-trending belt extending from the International Border to the Stikine Arch and are characteristic of Quesnellia. A shorter, northwest-trending belt extends along the east side of the Coast Plutonic Complex in northwestern British Columbia; these plutons intrude Stikinia in British Columbia and the Yukon. Most are Early Jurassic, but latest Late Triassic and Middle Jurassic representatives are also known. Most are small, roughly equant stocks, only a few kilometres in diameter. The largest (about 30 by 5 km) is probably the northwesterly elongate, Middle Jurassic Duckling Creek Syenite Complex, part of the Hogem Batholith (Fig. 15.4). The emplacement of individual plutons may have been controlled by faults and fault intersections (Barr et al., 1976).

Many of the alkaline plutons contain important porphyry-type mineralization; examples include the Copper Mountain-Ingerbelle and Afton mines in southern British Columbia, Cariboo-Bell in the Cariboo area, and the Galore

Creek deposits in the Stikine River area. Unlike the Cu-Mo porphyry deposits in Early Jurassic calc-alkaline plutons, alkaline porphyries contain Cu, significant Au and Ag, and only minor Mo (Barr et al., 1976). At least one of the bodies in Stikinia has been explored for magnetite.

Syenite, monzonite, and monzodiorite are typical of most plutons of the Copper Mountain Suite but diorite, monzogranite and clinopyroxenite are important locally. Nepheline and pseudoleucite are present in some bodies, particularly Galore Creek (Barr, 1966; Allen et al., 1976) and Cariboo-Bell (Hodgson et al., 1976). Many are extremely complex lithologically and texturally, with multiple phases of intrusion and K-feldspar metasomatism of both host rocks and early intrusive phases. Clinopyroxene and biotite are the main mafic minerals. Hornblende is common in the Duckling Creek body (Garnett, 1978), and epidote is conspicuous at Galore Creek. Garnet is a minor constituent of several plutons. All intrusions in this suite are characterized by abundant magnetite and apatite. The Ten Mile Creek body near Telegraph Creek on the Stikine River consists of a pyroxenite rim surrounding a syenite core (Morgan, 1976; Souther, 1972) and has some resemblance to ultramafics of the Late Triassic Polaris Suite. Ultramafic phases in most of the syenites appear to represent cumulates (e.g. Duckling Creek).

The plutons are alkaline but not peralkaline. They are characterized by high K_2O/Na_2O ratios, up to 24:1 (Fig. 15.4). Some (e.g. Galore Creek) are nepheline and leucite normative; others (e.g. Duckling Creek) contain both quartz-saturated and undersaturated compositions. Low initial $^{87}Sr/^{86}Sr$ ratios (e.g. 0.7035 for the Iron Mask (Preto et al., 1979) and for the Duckling Creek Syenite (Woodsworth, E.T. Eadie, and Armstrong, unpublished data) are similar to initial ratios for calc-alkaline plutons of the Guichon Suite and volcanics of the Nicola Assemblage of Quesnellia (Preto et al., 1979).

K-Ar dates for these plutons range from 209 to 170 Ma, with the oldest dates coming from Galore Creek, Iron Mask, and Copper Mountain (Preto et al., 1979; White et al., 1968). Most plutons are Early Jurassic in age, but some appear to be entirely Middle Jurassic (e.g. Duckling Creek, Kruger) and others are close to the Early-Middle Jurassic boundary (e.g. Olalla; Parkinson, 1985 and Cariboo-Bell). The alkaline plutons are spatially associated with Upper Triassic and Lower Jurassic alkaline volcanics. Plutons in the eastern belt intrude volcanic rocks of the Nicola Assemblage. In the western belt, most plutons intrude volcanics of the Upper Triassic Stuhini Assemblage. Fragments of alkaline plutonic rocks are locally abundant in volcanic breccia. The general similarities in age, chemistry, $^{87}Sr/^{86}Sr$ initial ratios, and setting of the volcanics and plutons suggest that the eastern alkalic plutons are subvolcanic equivalents of the Nicola volcanics (e.g. Preto, 1972; Barr et al., 1976), perhaps resulting from extensive differentiation of the Nicola magmas by pyroxene fractionation (Woodsworth, unpublished data). The genetic link, if any, between the alkaline suite and the Early Jurassic, calc-alkaline plutonic suite is unknown. The two appear to overlap in time and space within both Quesnellia and Stikinia.

Topley and Black Lake suites in Stikinia

In contrast to the widespread, large batholiths of the Guichon Suite in Quesnellia, Early Jurassic calc-alkaline plutons in Stikinia are relatively small and areally restricted. The largest concentration, the Topley suite, occurs along the Skeena Arch; a second group, the Black Lake Suite, outcrops along the east and north margins of the Bowser Basin and southeast of Teslin Lake. Both groups are relatively felsic in composition in contrast to the Guichon Suite.

The term "Topley Intrusions" was originally applied to batholiths and stocks along the Skeena Arch, particularly in the Francois Lake-Babine Lake area. Most of these plutons are currently grouped with the Late Jurassic to Early Cretaceous Francois Lake Intrusions. Following Carter (1982), the term Topley Intrusions is retained for Early to Middle Jurassic plutons in the western part of the Intermontane Belt. All dated plutons of this suite are within Stikinia, but undated plutons northeast of Babine Lake, in the Cache Creek Terrane, also may belong to this group. The plutons define a northeast-trending belt along the Skeena Arch and south of the Bowser Basin, in a tectonic setting analogous to the Late Triassic plutonic suite along the Stikine Arch.

The Topley Intrusions are epizonal, calc-alkaline stocks and batholiths (Tipper and Richards, 1976). Most are massive to weakly foliated, biotite- and hornblende-bearing granite to quartz diorite. K-feldspar megacrysts are common in the Tachek batholith. K-Ar dates from the Topley Intrusions range from 178 to 210 Ma. Most are between 199 and 210 Ma (Early Jurassic). Tipper and Richards (1976) noted that the Topley Intrusions are coincident with the thickest and most felsic piles of the Lower Jurassic (mainly Upper Sinemurian) Telkwa Formation of the Hazelton Group. They suggested that the intrusions are roughly the same age as the volcanics and represent root or feeder zones for Telkwa volcanism. Younger, late Early to Middle Jurassic K-Ar dates (178-185 Ma) from the Topley Intrusions may reflect resetting of the isotopic system. However, in the Whitesail Lake area, one pluton with a K-Ar date of 181 Ma is spatially associated with extensive felsic volcanism of the Middle Jurassic Whitesail Formation (Woodsworth and van der Heyden, unpublished data). It thus appears that most Early and Middle Jurassic plutons along the Skeena Arch broadly represent plutonic equivalents of the Hazelton Group.

The Black Lake Suite consists of several plutons that intrude Stikinia along the east and north margins of the Bowser Basin and Slide Mountain and Dorsey terranes southeast of Teslin Lake. The largest of those rimming the Bowser Basin, the Black Lake stock, consists of biotite-hornblende granodiorite to quartz monzodiorite. It is elongate northwesterly and cuts the Lower Jurassic Toodoggone volcanics. K-Ar dates from the pluton average about 190 Ma. Numerous smaller plutons associated with Sinemurian volcanics occur on the north rim of the Bowser Basin. The close spatial association of all these bodies with Lower Jurassic volcanics suggests that these plutons are analogous to the Topley Intrusions and represent the subvolcanic equivalent of a broadly-defined Hazelton arc.

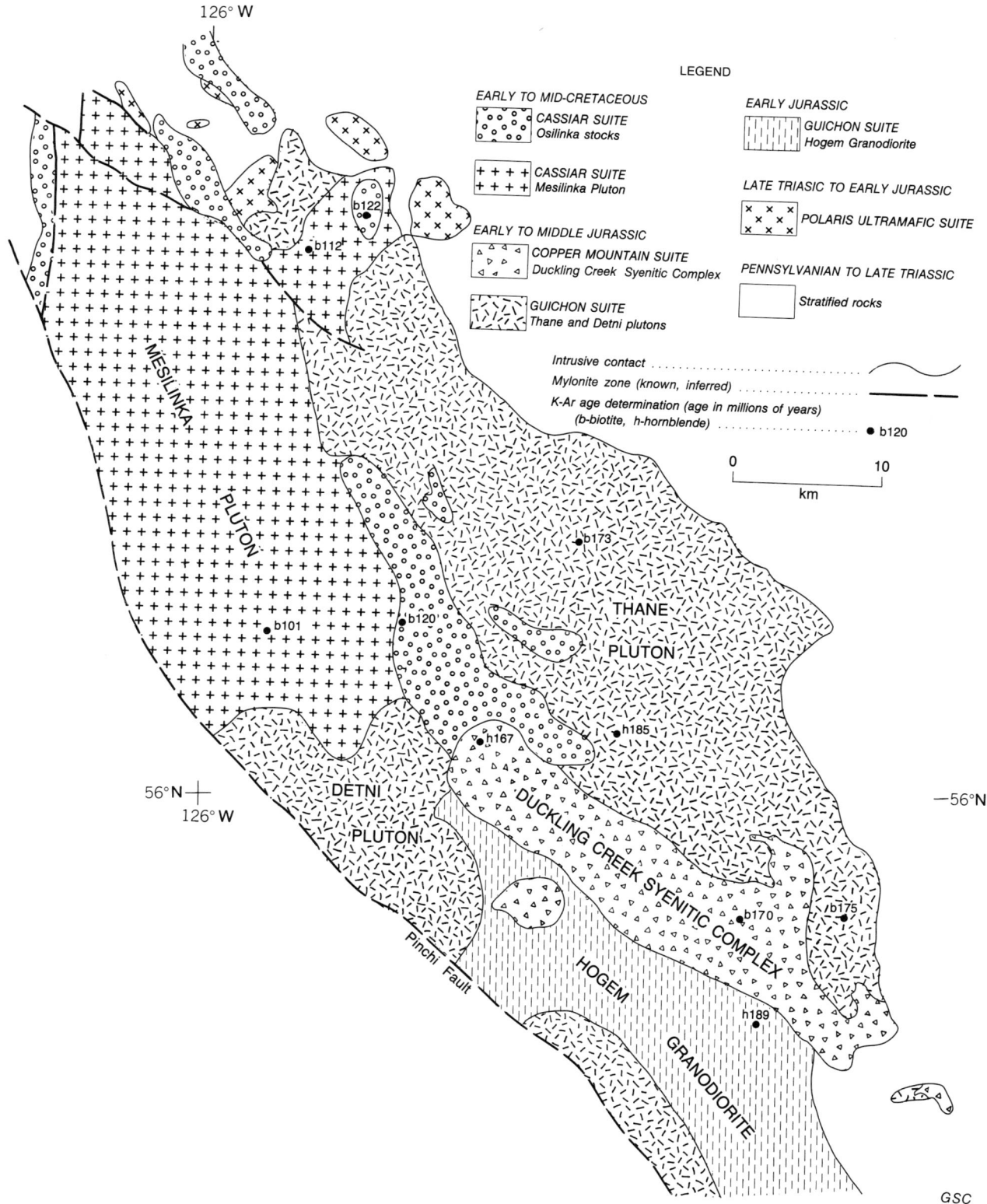

Figure 15.4A. Distribution of Polaris, Guichon, Copper Mountain and Cassiar plutonic suites in the Hogem Batholith (after Woodsworth, 1976 and unpublished data).

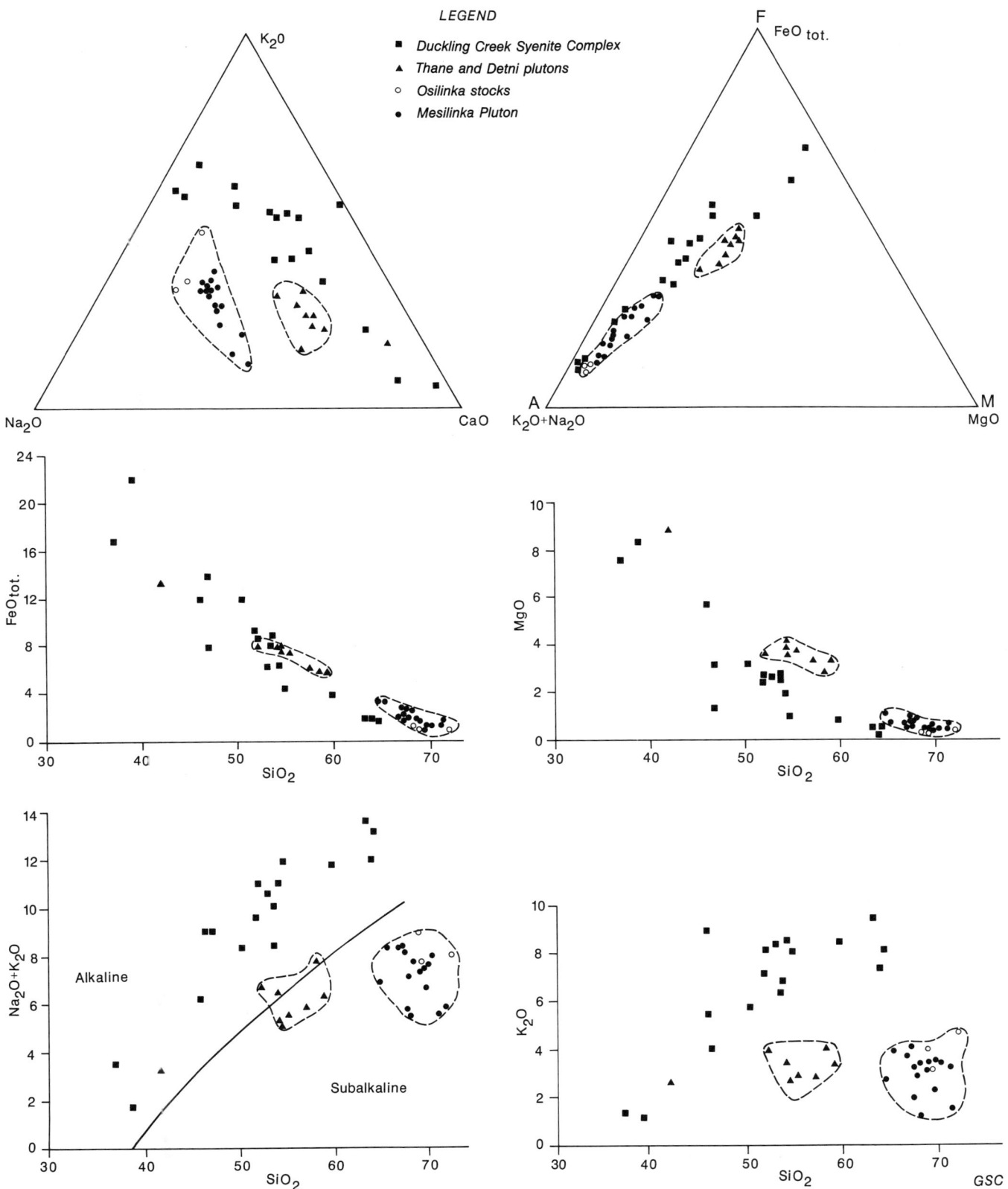

Figure 15.4B. Chemistry of Duckling Creek Syenitic Complex, Thane and Detni plutons, Osilinka stocks and Mesilinka Pluton.

Just south of the British Columbia-Yukon border, the easterly trending Early Jurassic Nome Lake and Simpson Peak batholiths lie athwart the regional, northwest grain, intrude strata of the Slide Mountain and Dorsey terranes and have no obvious regional structural control. These plutons are mainly granodiorite and granite; biotite is about equal to or more abundant than hornblende.

Klotassin and Long Lake suites

In the Yukon Territory, numerous stocks and batholiths in a broad belt extending west-northwest from Little Salmon and Bennett lakes are here informally grouped into the Klotassin Suite of older, darker rocks and the Long Lake Suite of younger, more felsic plutons. Most of the bodies are poorly dated and studied, and detailed work may allow subdivision into several suites. Except for the Lokken, Tatchun, and Tatlmain batholiths, all the large plutons lie southwest of the Teslin Fault. Most are elongate in a west-northwesterly direction; granodiorite and granite are the main lithologies. Many may be genetically related to Early Jurassic calc-alkaline volcanism in the Whitehorse Trough.

Representative bodies of the Klotassin Suite include the Klotassin Batholith, Minto Pluton, and Granite Batholith. These are medium grained, heterogeneous, equigranular granodiorite, generally containing more hornblende than biotite. Most rocks are foliated; many are intensely deformed. K-feldspar megacrysts are locally abundant. Granodiorite boulders, presumably derived from these plutons, occur in the Sinemurian to Toarcian part of the Laberge Group. K-Ar dates on the boulders range from 144 to 201 Ma (Bultman, 1979; Tempelman-Kluit and Wanless, 1975; Morrison et al., 1979). K-Ar and Rb-Sr dates on the plutons range from about 140 to 190 Ma and reflect cooling, resetting, and alteration but not necessarily the ages of the bodies (Tempelman-Kluit and Wanless, 1975; Le Couteur and Tempelman-Kluit, 1976; Godwin, 1975). Nearly concordant U-Pb dates of 192 Ma from zircons of the Minto Pluton appear to be closer to the age of emplacement (Tempelman-Kluit and Wanless, 1980). Tentatively, an Early Jurassic age is suggested for all plutons in this suite, but a Late Triassic age cannot be ruled out.

The Long Lake Suite is more felsic and, based on field relations, younger than the Klotassin Suite. The Carmacks and Long Lake batholiths are composed of pink, unfoliated, coarse grained biotite granite with conspicuous K-feldspar megacrysts. These rocks intrude the granodiorite plutons of the Klotassin Suite and also occur as boulders in the Laberge Group. K-Ar dates from the Carmacks Batholith range from 160 to 165 Ma. However, Tempelman-Kluit (pers. comm., 1985) suggested that these plutons are the intrusive equivalents of the Sinemurian to Toarcian Nordenskiold Dacite. A granitoid boulder from the Laberge Group that resembles rocks of the Carmacks Batholith gave an Early Jurassic K-Ar date (Tempelman-Kluit, pers. comm., 1985), strongly supporting an Early Jurassic age for the Carmacks Batholith.

Northeast of the Teslin Fault are the large Tatchun Batholith and the smaller Tatlmain and Lokken batholiths. The Tatchun and Lokken bodies are lithologically similar to the Minto and Granite Mountain plutons of the Klotassin Suite and are presumed to be Early Jurassic in age. The Tatlmain Batholith, dominantly unfoliated, leucocratic, biotite granite with large K-feldspar phenocrysts may be correlative with the Long Lake Suite.

Westcoast Suite

The Westcoast Complex forms a discontinuous belt along the west side of Vancouver Island. It is composed mainly of heterogeneous amphibolitic country rocks and plutonic rock of trondhjemitic to gabbroic composition (Isachsen, 1984), similar to the less potassic parts of the Island Suite. The Wark and Colquitz complexes on southern Vancouver Island are comparable in texture, composition and age to the Westcoast Complex. Excluding one anomalously old date reported by Muller (1977), U-Pb dates on zircons from the Westcoast, Wark, and Colquitz complexes indicate that the rocks are early Early Jurassic in age, or that Early to Middle Jurassic plutons have assimilated small amounts of older material (Isachsen et al., 1985). Based on chemistry and low (0.70329-0.70360) initial $^{87}Sr/^{86}Sr$ ratios, Isachsen (1984) argued that the Westcoast Suite represents the deeper crustal equivalent of the more evolved Bonanza volcanics. The rocks show a subalkaline tholeiitic to calc-alkaline compositional trends similar to those produced in magmatic arc settings.

PREDOMINANTLY MIDDLE JURASSIC

Plutons of Middle Jurassic age are common in the southern part of the Intermontane Belt. They are less abundant in the northern Intermontane Belt and the Omineca and Foreland belts. Middle Jurassic plutons, generally deformed and metamorphosed, are present in the Coast Belt, particularly in the region southeast of Terrace, but their abundance and distribution elsewhere in the belt are poorly known. East of the Coast Belt, most Middle Jurassic plutons occur in southern British Columbia, where alkaline plutonism was followed by calc-alkaline and then by two-mica granite. A small, distinctive group of Middle Jurassic plutons, the Three Sisters Suite, lies east, north and west of the Bowser Basin. In the Insular Belt, Middle to Late Jurassic plutons of the Island Suite form the great bulk of granitoid rocks on Vancouver Island and Queen Charlotte Islands.

Kuskanax and Nelson suites, and Galena Bay stock

The largest concentration of plutons of roughly Middle Jurassic age in the Canadian Cordillera occurs in a broad area between the Pasayten Fault and the Rocky Mountain Trench in southern British Columbia. Unlike the Jurassic plutons in the Insular Belt, this zone of plutons cuts across the regional grain of the Cordillera and across all terranes from Quesnellia to ancestral North America. Also unlike those in the Insular Belt, most plutons have abundant quartz; silica undersaturated phases are rare. Volcanic equivalents are common in the Insular Belt but are absent or rare in southern British Columbia. The ages, distribution, and characteristics of many of these plutons are poorly known, particularly in the west. However, mid-Jurassic plutons that intrude metamorphic and other rocks of the southwestern Omineca and southern Intermontane belts fall into four suites. Together these record a change from alkaline plutonism from about 180 to 170 Ma

(Kuskanax Suite) to calc-alkaline from about 170 to 160 Ma (Nelson Suite and Adamant pluton) to two-mica granites from about 160 to 150 Ma (Galena Bay stock).

The Kuskanax Suite, the oldest group of Middle Jurassic plutons in southeastern British Columbia, consists of the Kuskanax Batholith and related stocks. They intrude metamorphosed and deformed rocks of the Kootenay Arc and are highly distinctive mineralogically. The Kuskanax, 1200 km^2 in area, is a homogeneous batholith of leucocratic, aegerine-augite quartz monzonite (Read, 1973). The rock is composed mainly of microcline megacrysts (40-70 per cent) with lesser quartz and albite. Aegerine-augite with 10-30 per cent aegirine is the main mafic mineral. Hydrous minerals are uncommon. The few chemical analyses indicate that the rock is strongly alkaline to peralkaline. Satellite stocks are similar in mineralogy and may contain epidote and andradite (Read, 1973). Satellite stocks cut the regional metamorphic fabric in the Lardeau Group, but the Kuskanax Batholith was emplaced during the main, second phase of compressive deformation in the region (Read, 1973). The minimum age of the Kuskanax, based on U-Pb analyses of zircons, is 173 ± 5 Ma (Parrish and Wheeler, 1983), and this age may in effect be the age of second phase folding in the Kootenay Arc. However, volcanic rocks with chemistry similar to the Kuskanax Batholith are generally characteristic of an anorogenic, extensional setting and, to some extent, an early or late arc setting. The contradiction between tectonic setting suggested by chemistry and that inferred from the structure cannot be resolved with the available data.

Alkalic stocks described by Hyndman (1968) from the southwest side of the Kuskanax Batholith commonly contain primary(?) epidote, as do nearby calc-alkaline stocks that closely resemble parts of the Nelson Batholith. Both suites of stocks appear to have been emplaced at pressures of less than about 5 kbar (5×10^5 kPa), based on the presence of andalusite in contact aureoles.

The Nelson Suite, consisting of the Nelson Batholith, its satellites and compositionally similar plutons, is slightly younger than the Kuskanax Suite. Areally and economically, the Nelson is the most important Jurassic suite in southeastern British Columbia. The Ag-rich veins of the Slocan area, the Au-bearing veins of the Sheep Creek district, and some of the mineralization in the Rossland area are spatially and probably genetically related to the Nelson intrusions (Little, 1960; Fyles, 1984; Cairnes, 1934).

Nelson rocks are typically medium- to coarse-grained granite to granodiorite (Little, 1960). Megacrysts of K-feldspar are common in much of the Nelson Batholith, particularly in the granite phases. Hornblende commonly predominates over biotite, although some areas of leucogranite contain no hornblende and only a little biotite and/or muscovite (Gabrielse and Reesor, 1974). The few chemical analyses available indicate that the plutons are calc-alkaline.

Contact relations between plutons of the Nelson Suite and the country rocks are in many places controversial and not well understood. Structurally, these plutons occur in the hanging wall of the Slocan Lake and Valkyr shear zones (Parrish, 1984). Close to the shear zones, the plutonic rocks are altered and affected by ductile and brittle deformation. The Trail Pluton is intrusive into and has hornfelsed Lower Jurassic strata of the Rossland Group (Little, 1982; Simony, 1979; Fyles, 1984).

There is general agreement that plutons of the Nelson Suite were intruded after or during the waning phases of regional, Middle Jurassic metamorphism and deformation. For example, the post-tectonic Mount Carlyle stock, with a K-Ar date of 165 Ma on hornblende, has superimposed a contact metamorphic aureole on low grade Upper Triassic strata of the Slocan Group. On the other hand, the Mine Stock in the southern Kootenay Arc, which yielded dates of 166-171 Ma by U-Pb on zircon and K-Ar on biotite, is one of several late synkinematic plutons that have structurally concordant and deformed margins (Archibald et al., 1983). U-Pb and reliable K-Ar dates from the Nelson Suite cluster around 160-172 Ma (Armstrong, 1988), indicating a Middle Jurassic age of emplacement. Younger K-Ar dates (e.g. Archibald et al., 1983) may reflect overprinting by unrecognized and unrelated mid-Cretaceous plutonism and metamorphism (Archibald et al., 1983) or partial resetting of the K-Ar system by deformation related to the Slocan Lake, Valkyr, and other fault systems.

The Adamant Pluton and Mount Toby stock represent a third, distinctive type of plutonism in the southern Omineca Belt. The Adamant Pluton, by far the largest (150 km^2) and best-studied (Fox, 1969; Shaw, 1980a) of this group, intrudes Upper Proterozoic rocks of the Horsethief Creek Group and trends east-west, oblique to the predominant regional structural grain. The pluton is a composite body with a hypersthene-augite monzonite core that grades outwards to hornblende granite. Lineation and foliation parallel those in the enclosing Horsethief Creek Group and indicate that the pluton was emplaced prior to or during regional metamorphism and deformation. Fox (1969) concluded that the pluton was emplaced as a single intrusion of pyroxene monzonite. The transitional and marginal zones resulted from hydration of the pyroxene monzonite during metamorphism and deformation. The age of the Adamant Pluton is controversial. Shaw (1980b) obtained concordant U-Pb dates of 167-170 Ma from the hydrated, outer zones of the pluton. He argued that the zircons nucleated and grew during hydration accompanying metamorphism and do not reflect the age of the pluton. However, an equally plausible interpretation is that the zircons were originally concentrated in the outer parts of the pluton and that the concordant dates reflect the age of the pluton, i.e. Middle Jurassic and coeval with the Nelson Suite.

The Galena Bay stock, Oliver Pluton, and other small bodies consist of unfoliated, leucocratic, biotite-muscovite granite. The relationship of the Galena Bay stock to the Columbia River fault zone is controversial. Read and Brown (1981) thought that the stock intruded the fault and argued for pre-Middle Jurassic movement on the Monashee Décollement. Recent workers (Parrish et al., 1988), however, concluded that the Galena Bay stock is confined to the hanging wall of the Columbia River fault and that movement on the fault was entirely Eocene in age. U-Pb dates from zircon and monazite indicate that the age of the pluton is 162 Ma, slightly older than a 157 ± 2.4 Ma Rb-Sr whole-rock isochron (Parrish and Armstrong, 1987).

Other Jurassic plutons in southern British Columbia

The large region between the Nelson Suite and the Pasayten fault contains numerous, poorly-studied plutonic complexes, mapped as "granitic rocks of uncertain age" on the Tectonic Assemblage Map (Map 1712A, in pocket). East of Okanagan Lake, many plutonic rocks are lithologically similar to the Nelson suite and may be roughly the same age, although some may be early Tertiary in age (R.R. Parrish, pers. comm., 1987). Most of the few isotopic dates are K-Ar dates, and these must be assumed to minima unless corroborated by other methods, although an Early Jurassic age is probable for deformed plutons near Osoyoos Lake on the basis of U-Pb results on zircons and complex Rb-Sr data (Ryan, 1973; Parkinson, 1985).

Between Okanagan Lake and the Pasayten Fault, the largest plutonic complex of general Jurassic age has been variously called the Similkameen, Pennask, and Okanagan Batholith (Petö and Armstrong, 1976; Gabrielse and Reesor, 1974) and is here called the Okanagan Composite Batholith. The batholith, crudely zoned both spatially and temporally (Petö, 1973), consists of at least seven plutonic units that intrude the Upper Triassic Nicola Group and are overlain by Tertiary volcanics. The margin consists of older granodiorite to quartz diorite called the Pennask Batholith in the north and the Similkameen Intrusions to the south. These rocks are characteristically equigranular and contain more hornblende than biotite. The marginal Similkameen Batholith gave a preliminary Early Jurassic U-Pb date (R.R. Parrish, pers. comm., 1986) which suggests that the Similkameen and Pennask bodies are part of the Guichon Suite. The core of the batholithic complex, here called the Osprey Lake Pluton, consists of characteristically pink granodiorite to granite that intrudes the typically greenish to grey Similkameen and Pennask intrusions. Abundant K-feldspar megacrysts are characteristic of the Osprey Lake Pluton. Biotite generally predominates over hornblende. Based on Rb-Sr studies and a review of the K-Ar data, Petö and Armstrong (1976) thought that the Osprey Lake Pluton was emplaced at about 156 Ma. This conclusion is confirmed by U-Pb dates on zircons of about 162.5 Ma (R.R. Parrish, pers. comm., 1987).

The Eagle plutonic complex is a narrow, northwest-trending plutonic belt bounded on the west by the Pasayten Fault. Rocks include foliated, gneissic biotite tonalite and granodiorite (Greig, 1989). U-Pb dating indicates that earlier Late Jurassic phases are cut by a mid-Cretaceous, crosscutting muscovite granite (Greig, 1989). K-Ar and Rb-Sr dates range from 100 to 130 Ma and may reflect resetting during mid-Cretaceous deformation along the west margin of the Intermontane Belt and intrusion of younger granite.

Three Sisters Suite

Except for a few alkaline bodies of the Copper Mountain Suite, Middle Jurassic plutons in north-central British Columbia and the Yukon tend to be calc-alkaline and felsic. Most plutons belong to the Three Sisters Suite and occur along the Stikine Arch north of the Bowser Basin.

The best-studied pluton of the Three Sisters Suite is the Three Sisters Pluton (Fig. 15.2), one of several intrusions that together make up the younger intrusions comprising the Hotailuh Batholith (Anderson, 1983a). The Middle Jurassic intrusions intrude the Late Triassic Stikine Suite as well as Middle Triassic to Pliensbachian sediments and volcanics. Most plutons of the Three Sisters Suite are heterogeneous bodies that range from diorite through quartz monzodiorite or granodiorite to quartz monzonite. Some plutons are homogeneous granodiorite. Biotite and hornblende are present in all units. In the Three Sisters Pluton, all phases (including the most felsic) contain clinopyroxene cores in hornblende. U-Pb dates on zircons and K-Ar dates indicate that most of the Jurassic suite was intruded about 170 Ma ago, but there is considerable scatter in the data and some plutons may be as old as 180-186 Ma or as young as 150 Ma (Anderson, 1983a; Anderson et al., 1982).

The predominantly Late Triassic Stikine Batholith also contains Middle Jurassic(?) intrusions (Anderson, 1984). These range from an early, biotite-hornblende quartz monzodiorite and diorite to a younger, pink weathering, hornblende-biotite quartz syenite. Pink weathering, leucocratic granite is characteristic of the McNamara, Albert Dease, and other plutons that are spatially associated with Pliensbachian to Bajocian volcanics on the northeast rim of the Bowser Basin. Many of these plutons are similar in age and composition to the younger parts of the Topley Suite on the south rim of the Bowser Basin.

Teslin Crossing stock

In the Yukon, the Teslin Crossing stock and nearby small intrusions in the Whitehorse Trough are the only plutons known with reasonable certainty to be Middle Jurassic in age. They are mostly fine- to medium-grained, equigranular monzonite with lesser syenite and granite; hornblende is the main mafic mineral. The stock was emplaced in local pull-apart basins in Toarcian and Bajocian strata of the Laberge Group (Tempelman-Kluit, pers. comm., 1985) and gave K-Ar dates of 173-186 Ma. Many of the K-Ar dates in the 165-170 Ma range reported by Tempelman-Kluit and Wanless (1975) may reflect resetting of Early Jurassic dates. However, parts of the pink quartz monzonite plutonic suite near Carmacks which intrude the Early Jurassic Klotassin Suite, may indeed be Middle Jurassic in age.

Island Suite

Most plutons in the Canadian part of the Insular Belt are Jurassic. These plutons intrude Wrangellian strata on Vancouver Island where they are Early to Middle Jurassic in age. In the Queen Charlotte Islands most plutons are of Middle to Late Jurassic age.

About 10 per cent of Vancouver Island is underlain by what Carson (1973) and Muller et al. (1974) called the Island Intrusions. Along the west coast of the island is a complex of plutonic and volcanic rocks, the Westcoast Complex, which appears genetically related to the Island Intrusions. The Island Intrusions intrude volcanics of the Bonanza Group, as well as all older units, and are unconformably overlain by Upper Cretaceous sediments of the Nanaimo Group. They occur as batholiths and stocks, commonly aligned northwesterly and are most abundant in the central and western parts of the island. Contacts between plutons and Sicker and Bonanza assemblages are commonly concordant and gradational across zones of gneiss and migmatite. Contacts with Karmutsen volcanics are generally sharp, with a narrow hornfels aureole. Larger bodies are, in general, less felsic, more mafic-rich, and emplaced at deeper levels than the smaller stocks. The most abundant rock types are tonalite, quartz diorite, and granodiorite, with lesser quartz monzodiorite, granite, diorite, and gabbro. Hornblende and biotite are the main mafic minerals in all rock types. K-Ar, Rb-Sr, and U-Pb data indicate that the Island Intrusions are late Early to early Middle Jurassic (Isachsen et al., 1985) and are genetically related to the Bonanza Group.

Jurassic plutons on the Queen Charlotte Islands have been recently studied by Anderson (1988a) and Anderson and Reichenbach (1991). They recognized two groups of eastward-younging, Middle to Late Jurassic plutons based on contact relations, composition, texture, intraplutonic dykes, and geophysical signature. The San Christoval plutons (172-171 Ma) occur along the west side of the islands and are massive to foliated quartz diorite and diorite with conspicuous prismatic hornblende and aligned mafic inclusions. The Burnaby Island plutons (168 to $\geq$158 Ma) are the most heterogeneous (gabbro, quartz monzodiorite, quartz monzonite and rare leucogranite) and have suffered the most intense brittle fracturing and alteration (locally to endoskarn) of all the Queen Charlotte Islands Jurassic plutons.

Most of the economically important Fe-Cu and Pb-Zn skarn deposits on Vancouver, Texada, and Queen Charlotte islands are genetically related to Jurassic intrusions (Sangster, 1969; Muller and Carson, 1969). The largest producing mine on Vancouver Island, Island Copper, is a porphyry deposit in volcanics of the Bonanza Group and is related to the emplacement of the Island Intrusions (Cargill et al., 1976). Tasu and Jedway were important, past producing Fe-Cu mines on Moresby Island of the Queen Charlotte Islands (Sutherland Brown, 1968). They are skarn deposits developed in the Triassic and Lower Jurassic Kunga Group limestone and Upper Triassic Karmutsen Formation greenstone near the San Christoval and Burnaby Island plutons.

Bokan Mountain Granite

Of several Jurassic plutons in the Alexander Terrane of southeastern Alaska, the Bokan Mountain Granite has received the most attention because of its U-Th mineralization and peralkaline chemistry. The pluton is a circular ring-dyke complex about 5 km in diameter that intrudes Ordovician to Silurian sedimentary, volcanic, and plutonic rocks (MacKevett, 1963). The pluton is zoned, and the main lithologies from core to border are fine grained, albitic aegerine granite; albitic arfvedsonite granite; and albitic aegirine granite (Thompson et al., 1982; Saint-Andre et al., 1983).

Dating of the complex has given conflicting results. A U-Pb date on zircons of 171 ± 5 Ma (Saint-Andre et al., 1983) is questioned on the basis of analytical and inherited zircon problems but may represent a maximum age of the pluton (Armstrong, 1985). A Rb-Sr whole-rock isochron date of 151 ± 5 Ma with an initial ratio of 0.711 ± 0.001 (Armstrong, 1985), although plagued with discordant samples, gives an minimum age for the complex. A Middle to Late Jurassic age of emplacement is most probable.

The Bokan Mountain Granite is unique in the western Cordillera. The peralkaline chemistry and ring-dyke nature of the complex suggest that it formed in an extensional setting.

LATE JURASSIC TO EARLY CRETACEOUS

Plutons of Late Jurassic to early Early Cretaceous age (about 155 to about 130 Ma) are uncommon in the Canadian Cordillera. Some plutons of this age are known from the Coast Plutonic Complex and are discussed under that heading. Elsewhere, two suites are reasonably well dated. Both are areally restricted and are confined to one or two of the western suspect terranes. The Saint Elias Suite occurs mainly in the Alexander Terrane of the Saint Elias Mountains. The Francois Lake Suite intrudes Cache Creek Terrane, Quesnellia, and Stikinia along the eastern part of the Skeena Arch. Late Jurassic to Early Cretaceous volcanic rocks are unknown in both regions.

Saint Elias Suite

Late Jurassic plutons are abundant in the Saint Elias Mountains of the Insular Belt (Dodds and Campbell, 1988). They are restricted to the area northeast of the Border Ranges Fault and southwest of the Denali Fault. Most plutons intrude rocks of the Alexander Terrane, but near the Border Ranges Fault and in southern Alaska they intrude Wrangellian strata. The plutons vary from large, elongate batholithic complexes to small single intrusions and generally decrease in size, abundance, and possibly depth of emplacement from southwest to northeast. On the southwest, large relatively mafic bodies such as the Mount Logan Batholith (Plate 31) are elongated northwesterly, parallel with the regional tectonic grain and are weakly to moderately foliated. On the northeast, most plutons are small and relatively felsic. They are commonly unfoliated, circular and discordant with contact metamorphic aureoles. Granodiorite and quartz diorite with subordinate diorite and quartz monzonite are the main lithologies; all are calc-alkaline. Both biotite and hornblende are present in most plutons.

Almost all the numerous K-Ar dates on biotite and hornblende from the Saint Elias Suite fall between 160 and 130 Ma, spanning the Late Jurassic and early Early Cretaceous. Coeval plutons continue into southeastern Alaska, perhaps as far south as Baranof Island, as what Hudson (1983) called the Tosina-Chichagof Belt. The plutons may be the deep-seated expression of Late Jurassic to Early Cretaceous volcanism of the Gambier Assemblage which overlaps both Wrangellia and the Alexander Terrane. The Saint Elias Suite may represent a northward extension of the poorly-defined Late Jurassic to Early Cretaceous belt of plutons in the Coast Plutonic Complex south of latitude 55°N. Late Jurassic to Early Cretaceous plutonism and volcanism therefore may have extended more or less continuously from northern Washington into southern Alaska.

Francois Lake Suite

An economically important cluster of Late Jurassic to Early Cretaceous plutons underlies much of the area between

Francois and Stuart lakes in central British Columbia. These plutons, originally included with the Topley Intrusions, were named the Francois Lake Intrusions by Carter (1982). They host the Endako molybdenum deposit, the largest molybdenum deposit known in Canada (Kimura et al. 1976). Most of the Francois Lake Intrusions are northwesterly-elongate batholiths with a moderately to strong northwest-trending foliation. Biotite granite to quartz monzonite, commonly with conspicuous K-feldspar megacrysts, is the dominant lithology. In the Endako area, a foliated, diorite to granodiorite suite is cut by unfoliated, megacrystic biotite granite and granodiorite. Most K-Ar dates on biotite range from 135 to 145 Ma; a few are older (White et al., 1970). Rb-Sr whole-rock dates range from 138 to 146 Ma (R.L. Armstrong, unpublished data).

All but one of the Francois Lake bodies are in the Cache Creek Terrane or rocks that may be part of Stikinia. The exception is the "Jean Marie" stock, in Quesnellia on the northeast side of the Pinchi Fault. The stock is mainly grey, medium grained, hornblende-biotite granodiorite and hosts porphyry copper and molybdenum mineralization (Garnett, 1978). K-Ar dates on biotite and hornblende are concordant at about 135 Ma. The presence of the Francois Lake Suite in an areally restricted, roughly north-trending belt across Stikinia, Cache Creek Terrane, and Quesnellia suggests that dextral transcurrent faulting within and between these terranes was unimportant after early Early Cretaceous time.

MID-CRETACEOUS

Plutons intruded between about 85 and 130 Ma are concentrated in two broad geographic belts that coincide, roughly, with the Coast and northern Insular belts and with the Omineca Belt.

Mid-Cretaceous granitic rocks are areally the most important plutons in the eastern Canadian Cordillera and are grouped into six suites. In the Foreland Belt, small, alkaline intrusions of the Howell Creek Suite are comagmatic with volcanics of the mid-Cretaceous Crowsnest Formation. In the southern Omineca Belt, large, discordant granite to granodiorite plutons of the Bayonne suite are abundant, particularly in a broad, arcuate belt extending from east of Kootenay Lake to south of Quesnel Lake. The Cassiar Suite is similar lithologically and forms a linear belt in northern British Columbia and the Yukon southwest of the Northern Rocky Mountain-Tintina trench. The Selwyn Suite includes the Anvil Batholith as well as numerous plutons in the Selwyn Mountains. These are mostly granite, granodiorite and quartz syenite, and contain, or are associated with, much of the important Sn and W mineralization in the Cordillera. In the western Yukon, the Tombstone Suite is an areally restricted, bimodal suite of syenite and quartz monzonite north of the Tintina Trench. In the Whitehorse Trough the Whitehorse Suite of predominantly granodiorite may be the subvolcanic equivalent of the mid-Cretaceous Mount Nansen Group volcanics.

The numerous calc-alkaline plutons in the Coast Plutonic Complex of the Coast Belt are discussed elsewhere in this chapter. In the northern Insular Belt, mid-Cretaceous, calc-alkaline plutons are present in Wrangellia and the Gambier Assemblage in the Saint Elias Mountains (Kluane Suite) and in Alexander Terrane in southeastern Alaska. Alaskan-type ultramafic bodies of the Union Bay Ultramatic Suite form a linear belt extending from the Saint Elias Mountains through southeastern Alaska and possibly into the Coast Plutonic Complex of British Columbia. The Cascade Suite in southwest British Columbia and northern Washington consists mainly of large, foliated tonalite plutons.

The two main belts of mid-Cretaceous plutons are markedly different in lithology and chemistry. Those in the Coast Plutonic Complex are, with few exceptions, I-type, commonly contain hornblende in addition to biotite, and are poor in large-ion lithophile elements. Initial $^{87}Sr/^{86}Sr$ ratios are commonly low (less than 0.706 and generally less than 0.7045). In contrast, plutons in the Selwyn and Bayonne suites of the eastern belt are generally S-type, felsic, rich in large-ion lithophile elements, and have initial $^{87}Sr/^{86}Sr$ ratios between 0.710 and 0.740 (Anderson, 1988b; Armstrong, 1988). The Tombstone Suite is similar to A-type plutons in linear distribution of plutons, subcircular pluton shape, alkaline mineralogy and geochemistry and texture. Plutons in the Coast and Insular belts may be subduction-related; those in the eastern belts may be related to A-type subduction of the miogeocline and transtensional tectonics.

Howell Creek Suite

Small, high-level stocks and dykes of alkaline plutonic rock in the southern Rocky Mountains consist mainly of trachytic-textured syenite and porphyritic trachyte (Price, 1965). These variously have phenocrysts of sanidine, plagioclase, hornblende and pyroxene; the matrix locally contains leucite and epidote. Based on stratigraphic setting, composition, and texture, Price (1965) concluded that these plutons are comagmatic with alkaline volcanic rocks of the Albian Crowsnest Formation to the east. K-Ar dates from the intrusions range from about 72 to 128 Ma (Gordy and Edwards, 1962; Price, in Leech et al., 1963) and, although showing much scatter, support a mid-Cretaceous age.

Bayonne Suite

Mid-Cretaceous plutons in the southeastern Cordillera intrude most terranes west of ancestral North America and ancestral North America itself. Most are batholiths and large stocks that postdate regional metamorphism and are strongly discordant with the country rocks. Many are roughly equant, although those in the Shuswap Lake region are commonly elongate in an east-west direction, across the regional structural grain. The main lithologies are biotite and biotite-muscovite leucogranite and granodiorite, and biotite-hornblende granodiorite and granite. K-feldspar megacrysts are locally abundant. Coeval volcanic rocks are absent.

Typical and the best-studied of these plutons is the White Creek Batholith (Reesor, 1958). This is a roughly oval body that intrudes Proterozoic sediments of the Purcell Supergroup along the Hall Lake fault system. The batholith has a rim of biotite granodiorite which grades inwards to hornblende-biotite granodiorite and K-feldspar megacrystic biotite granite. A core of muscovite-biotite leucogranite both grades into and intrudes the outer units. Foliation in the pluton parallels the outer contact and is independent of internal compositional boundaries. The pluton disrupts structures in the surrounding strata, and a strong secondary foliation has been superimposed on the sediments. Mineral assemblages in the contact metamorphic aureole suggest the pluton was emplaced at a depth of about 15 km.

The K-Ar dates from the White Creek Batholith may be unreliable (Archibald et al., 1983). However, a Rb-Sr whole-rock isochron indicates emplacement at about 111 Ma (Wanless et al., 1968). Initial $^{87}Sr/^{86}Sr$ ratios of 0.7250 for the core of the batholith are much higher than the 0.7077 ratio for the marginal zones, suggesting different degrees of assimilation of crustal materials by the magmas. On the basis of U-Pb dates on zircons from the Kaniksu Batholith and more precise K-Ar and Rb-Sr dates from other plutons in the area, Archibald et al. (1984) concluded that most plutons were emplaced between 115 and 90 Ma. The Kaniksu Batholith was emplaced at about 94 Ma at deeper levels than most of the mid-Cretaceous plutons in the region, and was accompanied by penetrative deformation and prograde metamorphism while the remainder of the area was tectonically inactive.

Cassiar Suite

Mid-Cretaceous plutons in north-central British Columbia form a narrow, linear belt parallel with the regional tectonic grain. In contrast with plutons of the Bayonne Suite, those of the Cassiar Suite are generally elongate northwesterly, and many have been affected by post-emplacement deformation. Lithologically the plutons are similar to the Bayonne Suite: biotite granite and granodiorite predominate. Muscovite is common in some plutons and hornblende, generally subordinate to biotite is present in others. Many rocks have a pink colour, and K-feldspar megacrysts are common. Coeval volcanic rocks are unknown.

The Cassiar Batholith is the largest pluton in this suite and one of the largest plutons in the Canadian Cordillera. The batholith is markedly elongate, with a length of about 350 km and a maximum width of some 40 km. General descriptions of parts of the batholith are given in Gabrielse (1963, 1969) and Poole (1956), but there has been no overall study of this important pluton. Most of the batholith consists of pinkish-grey, medium- to coarse-grained granite and granodiorite, commonly containing megacrysts of K-feldspar. Biotite is the main mafic mineral; hornblende is locally abundant but generally subordinate to biotite. Muscovite is common in the sheared, western parts of the batholith.

A striking feature of the Cassiar Batholith, and many other plutons of the Cassiar Suite is its association with regional transcurrent faults. The western margin of the Cassiar Batholith is marked by a shear zone in which the pluton has been deformed and mylonitized over widths of up to 3 km. Overall, the degree of deformation decreases from an intensely mylonitized western border through a zone of augen and flaser gneiss to weakly foliated and unfoliated granite (Poole, 1956; Gabrielse, 1969). Subhorizontal stretching lineations in the mylonitized pluton suggest transcurrent motion on the faults.

Most K-Ar dates on biotite from the Cassiar Batholith and other plutons of the suite range from about 85 to 110 Ma. A Rb-Sr whole-rock isochron date of 109 Ma has been obtained from the Cassiar Batholith (G. Medford and R.L. Armstrong, pers. comm., 1987). Middle Cretaceous K-Ar dates on muscovite generated in the mylonitic margins of the plutons suggest that at least some of the deformation is mid-Cretaceous in age, roughly synchronous with emplacement of the plutons. However, faulting and deformation may have persisted into the Late Eocene or Oligocene (Gabrielse, 1985).

The Mesilinka and Osilinka plutons are part of the predominantly Early to Middle Jurassic Hogem Batholith. The Mesilinka Pluton (Woodsworth, 1976) is a heterogeneous, biotite granite, quartz monzodiorite, granodiorite, and quartz diorite. Like the Cassiar Batholith, the Mesilinka Pluton is strongly mylonitized on its western margin, with foliation decreasing in intensity to the east. Intrusive into the Mesilinka and the Jurassic parts of the batholith are bodies of miarolitic biotite granite and granodiorite containing conspicuous quartz eyes, the Osilinka Intrusions. These stocks are elongate northwesterly, parallel with the regional fault pattern, but are internally unfoliated and undeformed.

The Mesilinka and Osilinka plutons cannot be distinguished using the available major or trace element data. Both units are calc-alkaline, have between about 58 and 71 per cent SiO_2, and have Na_2O/K_2O ratios greater than one (Meade, 1977; Garnett, 1978; Woodsworth, E.T. Eadie, and Armstrong, unpublished data). Heat production (average about 2.5 $\mu W/m^3$) is much higher than for the Jurassic parts of the batholith (0.8-1.0 $\mu W/m^3$) (Lewis and Woodsworth, 1981).

K-Ar biotite dates of 103 ± 4 and 114 ± 4 Ma have been obtained from the Mesilinka Pluton, and 112-123 Ma from the Osilinka stocks. A Rb-Sr whole rock isochron from the Osilinka stocks of 117 ± 8 Ma and a whole rock Rb-Sr errorchron from the Mesilinka Pluton also suggest a mid-Cretaceous age for these bodies (Eadie, 1976; Woodsworth and Armstrong, unpublished data). The scatter in the Rb-Sr data and the anomalously younger (compared with the crosscutting Osilinka Intrusions) K-Ar dates from the Mesilinka Pluton may reflect partial resetting of the isotopic systems during the intense shearing of the Mesilinka Pluton. The absence of deformation in the Osilinka Intrusions indicates that significant movement on nearby fault systems ceased by mid-Cretaceous time.

The strongly elongate Cassiar and Hogem batholiths and smaller bodies such as the Thudaka and Whudzi plutons are regionally associated with strike-slip faults that is, at least in part, synplutonic. Part of the elongate nature may be simple, passive control of magma by the fault system, as with the elongate, post-faulting Osilinka intrusions. However, for greatly elongate bodies such as the Cassiar Batholith, it seems plausible that synplutonic ductile shear may have greatly stretched the plutons to several times their original length. Slightly older plutons such as the Mesilinka Pluton may also owe their shape to ductile shear and may be, in effect, giant porphyroclasts.

Axelgold Gabbro

The Axelgold gabbro body, just west of the Hogem Batholith, is only slightly older than most plutons in the Cassiar Suite and unique in the Canadian Cordillera. It is a layered gabbro that is intrusive into Cache Creek Terrane. The main rock type is olivine gabbro, with a few picritic and anorthositic layers of cumulate origin (Irvine, 1975).

The main gabbro is cut by small bodies of syenite. Based on K-Ar and Rb-Sr dates, the gabbro is almost certainly Early Cretaceous (125 ± 5 Ma) in age (Armstrong et al., 1985). Coeval volcanic rocks are unknown in the immediate area but may be represented on the south edge of the Bowser Basin as the mafic and alkaline Lower Cretaceous Rocky Ridge volcanics. The pluton has no known counterpart in the Canadian Cordillera, but in age is not unlike the Union Bay ultramafic suite in southeast Alaska and in petrology closely resembles the Crillon and La Perouse layered gabbros in the Fairweather Range of Alaska (Irvine, 1975).

Selwyn Suite

Homogeneous, siliceous stocks and batholiths are characteristic of mid-Cretaceous plutonic rocks in the Omineca Belt in the Yukon. These plutons are mainly discordant, high-level intrusions of granite, granodiorite, and quartz syenite; many have pronounced S-type characteristics. Many are roughly equant in form, but others are elongate parallel with the regional grain. Important tin and tungsten skarns and greisens (Dick, 1979, 1980; Dick and Hodgson, 1982) are associated with the suite.

The plutons occur in two tectonically distinct settings. Those in the Anvil Ranges (Pigage and Anderson, 1986) were intruded after Buchan-type regional metamorphism accompanying second phase deformation in the Omineca Belt and before movement on the Tintina Fault. The posttectonic plutons in the Selwyn Mountains (Anderson, 1982, 1983b) were emplaced into platformal to basinal facies of a Proterozoic to Paleozoic miogeoclinal succession.

In the Selwyn Mountains, contact metamorphic aureoles anneal slaty cleavage developed in the sedimentary rocks. Andalusite hornfels, common pendants, and rare miarolitic cavities indicate epizonal emplacement of the barely unroofed plutons. Minor structures in the plutons and subcircular cross-sections suggest forcible emplacement of some plutons as nearly consolidated diapirs.

Most rocks are granite and granodiorite; quartz syenite is locally abundant. In both the Anvil and Selwyn areas, plutons contain either hornblende+biotite, or muscovite+biotite as essential minerals. Biotite is phlogopite-rich in some hornblende-bearing rocks, and aluminous and annite-rich in muscovitic varieties. Irregular clinopyroxene cores in hornblende, and sphene and magnetite are present in hornblende-bearing plutons. Fine grained, round monazite and zircon inclusions with radiation-damage haloes characterize biotite in muscovite-bearing rocks. In two-mica rocks, ragged, subhedral or anhedral muscovite predominates and is locally associated with fibrolite sprays; euhedral muscovite occurs in the most siliceous phases. Aluminous accessory minerals such as garnet, andalusite, and retrograded cordierite(?) typify marginal or satellitic phases of two-mica granites associated with tungsten skarns. Apatite, zircon, allanite and tourmaline are ubiquitous accessory minerals but primary opaque oxides are rare or absent. Alkali feldspar, commonly as megacrysts, is predominantly microperthitic. Myrmekite is common in the two-mica rocks. Tourmaline-bearing aplite and pegmatite dykes are abundant, especially along pluton margins. Mafic, hornblende-biotite-plagioclase-phyric dykes are coeval with and restricted to possible subvolcanic, hornblende-bearing plutons.

Plutons in both areas resemble evolved (rich in large ion lithophile elements), compositionally restricted, silicic, calc-alkaline plutonic suites recognized elsewhere from continental regions. In general, hornblende-bearing plutons are diopside-normative, have higher TiO_2, MnO, Cr, Ni, Sr, Ba, and Y contents; lower SiO_2, K_2O, Rb, Rb/Sr and K/Ba values than the more evolved, peraluminous muscovite-bearing plutons (Anderson, 1988b; Pigage and Anderson, 1986). Hornblende-bearing plutons tend to be slightly depleted in $\delta^{18}O$ and less radiogenic (initial $^{87}Sr/^{86}Sr$ ratios of 0.710 to 0.720) compared with the two-mica rocks (initial $^{87}Sr/^{86}Sr$ ratios of 0.730 to 0.740) (Pigage and Anderson 1986; Dagenais 1984; Anderson, 1988b).

K-Ar, U-Pb, and Rb-Sr dates range from 88-114 Ma, with the preferred age of emplacement of the plutons being from 90 to 100 Ma. The isotopic systematics are complex, no doubt partly reflecting the large amount of assimilation the magmas have undergone. South Fork Volcanics, considered cogenetic with the hornblende-bearing phases (Pigage and Anderson, 1986), also are mid-Cretaceous in age (Wood and Armstrong, 1982). No other coeval volcanics are known from the region, and there are no other stratigraphic constraints on the ages of the plutons.

Neither the Selwyn nor the Cassiar suites appear to have formed as a direct result of subduction. Rather they may have resulted from thickening and melting of the continental crust in the outer part of the miogeoclinal sedimentary wedge during regional compression.

Tombstone Suite

This suite is restricted to ancestral North America northeast of the Tintina Trench in the western Yukon. The Tombstone Pluton is the largest body and underlies about 90 km^2. The Mount Brenner, Antimony Range, and other small bodies occur near the Tombstone Pluton, and other plutons that may be part of the suite, such as the Syenite Range stock, outcrop up to 100 km southeast along the linear belt.

The Tombstone Pluton is a high-level, discordant, roughly equant body (Plate 29). It was intruded along a zone of Late Proterozoic extension into the Mississippian Keno Hill Quartzite and postdates regional deformation. The intrusion is composite, consisting mainly of alkali feldspar syenite with marginal phases of quartz monzonite and diorite (Tempelman-Kluit, 1970; Anderson, 1987). The syenite is both equigranular and K-feldspar phyric with less abundant plagioclase, aegirine-augite, arfvedsonite, biotite, and minor quartz. In the Mount Brenner Pluton, alkali feldspar foliation is concentric with pluton margins and mimics the annular distribution of phases (Lambert, 1966). Melanite garnet, fluorite, and sphene are widespread accessory minerals. Phonolite containing pseudoleucite and acmite (tinguaite) is present near the margin of the Tombstone Pluton and in its core (Tempelman-Kluit, 1969). Fluorite, calcite, chalcopyrite and galena occur in miarolitic cavities in alkaline phases, and the intrusions have been extensively explored for U-Sn-Mo-F-Sb-Au-Ag veins and skarns.

Diorite is commonly the oldest phase. Tinguaite intrusions separate two phases of syenite intrusion. Quartz monzonite was the last major phase to be intruded. The Mount Brenner (Lambert, 1966) and Deadman plutons locally contain an early hornblende clinopyroxenite phase.

Geochemical (Olade and Goodfellow, 1978) and lithological compositions suggest that the Tombstone Suite has a skewed, bimodal nature, with an alkaline, mafic to peralkaline salic compositions, and minor calc-alkaline siliceous rocks but with few intermediate compositions. Reliable K-Ar and Rb-Sr isotopic ages which range between 90 and 110 Ma (M.L. Bevier, pers. comm., 1988) suggest a minimum early Late Cretaceous age for the suite.

Whitehorse Suite

Numerous medium- to high-level plutons form a roughly linear belt that extends through the Whitehorse Trough and into the adjacent Nisling and Kootenay terranes to the north and west. Most of these plutons are stocks or small batholiths that were intruded after regional deformation. In the Whitehorse area, plutons are equant to elongate north-northwesterly and cut the regional structural grain at a small angle.

In the southern part of the belt, the plutons intrude lower Mesozoic strata of the Lewes River and Laberge groups. Most rocks are biotite-hornblende granodiorite and quartz diorite, locally with cores of either pink, granophyric granite or very coarse grained leucogranite (Morrison et al., 1979). Farther northwest, near Aishihik Lake, granodiorite in the Nisling Range (Tempelman-Kluit, 1974) forms two regionally discordant, westerly-elongate batholiths of coarse grained, equigranular granodiorite to quartz diorite. The rocks contain either hornblende or both hornblende and biotite, and have a distinctive purplish colour. The Coffee Creek intrusions form stocks and batholiths of coarse grained, equigranular biotite granite containing conspicuous, smoky quartz (Tempelman-Kluit, 1974).

K-Ar dates from the Whitehorse Suite range from 90 to 116 Ma (Morrison et al., 1979; Tempelman-Kluit and Wanless, 1975; Godwin, 1975), indicating a mid-Cretaceous age. Volcanics of the Mount Nansen Group are the same age and are spatially associated with the Whitehorse Suite in the Nisling Range and Whitehorse Trough, suggesting a comagmatic relationship between plutons and volcanics. In the Dawson Range, swarms of mid-Cretaceous feldspar porphyry dykes, presumably related to the Mount Nansen volcanics, have a north-south trend, parallel with the general trend of mid-Cretaceous plutons. The Whitehorse Suite of plutons is interpreted as the subvolcanic roots of a mid-Cretaceous continental volcanic arc superposed on the Whitehorse Trough and Lewes River arc (Tempelman-Kluit, pers. comm., 1985).

Kluane Suite

Mid-Cretaceous plutons in the Saint Elias Mountains intrude Wrangellia and its overlying cover of Gambier Assemblage rocks (Dodds and Campbell, 1988). The rocks are mainly high-level batholithic complexes and stocks of biotite-hornblende granodiorite, quartz diorite, and lesser biotite quartz monzonite. Many of the plutons are elongate northwesterly, parallel with the regional tectonic grain. Major-element chemistry suggests a calc-alkaline, I-type affinity for the suite (Dodds and Campbell, 1988). Most K-Ar dates on biotite and hornblende range from 106 to 117 Ma, suggesting a mid-Cretaceous age of emplacement.

Volcanic rocks of this age are unknown in the Saint Elias Mountains of Canada. However, coeval volcanics are present in the northwestward continuation of the Kluane Suite in eastern Alaska (Richter et al., 1975).

In southeastern Alaska, mid-Cretaceous plutons, called the Muir-Chichagof Suite by Brew and Morrell (1983), intrude Alexander and Chugach terranes, as well as Wrangellia. Lithologically, the rocks are similar to the Kluane Suite. Granodiorite, tonalite, quartz diorite and diorite containing biotite and hornblende are the main lithologies. Many plutons are large bodies with contacts parallel with regional structural trends, but plutons in the northern and western parts of the belt are mostly small, discordant stocks. Plutons in this belt and Alaskan ultramafics of the Union Bay Suite may represent the general base of the Early to mid-Cretaceous volcanic arc that has its surface expression in the Gravina-Nutzotin belt of volcanics and flysch (Berg et al., 1972; Brew and Morrell, 1983).

Union Bay Ultramafic Suite

A linear belt of Alaskan-type ultramafic complexes extends from near Haines Junction in southern Yukon through southeastern Alaska and possibly into the Coast Belt near Prince Rupert. Most are intrusive into rocks of the Alexander and Taku terranes and are closely associated with strata of the Gambier overlap assemblage.

The bodies consist mainly of dunite, pyroxenite, hornblendite, and gabbro (Taylor, 1967; Murray, 1972; Irvine, 1974). Most are small, generally less than 3 km across. The largest, the Union Bay body, is about 6 km in length; this body has been extensively explored for magnetite. Complexes containing olivine are commonly zoned; a dunite core is surrounded by successive zones of olivine pyroxenite, hornblende pyroxenite, and gabbro. The gabbro shells are commonly larger than the ultramafic bodies and older than the cores of the complexes. Clinopyroxene is exclusively diopsidic augite; orthopyroxene is absent or scarce. Plagioclase is absent from the ultramafic rocks except as late-stage gabbroic pegmatite and in the gabbroic margins. Magnetite and subordinate ilmenite form between 5 and 20 per cent of hornblende pyroxenite and hornblendite. Many bodies are surrounded by high-temperature contact aureoles.

K-Ar dates from these complexes range from 100 to 124 Ma (Brew and Morrell, 1983; Sturrock et al., 1980), suggesting an Early to mid-Cretaceous age of emplacement. Murray (1972) and Irvine (1973) suggested that the Cretaceous ultramafic rocks are genetically linked with late Early Cretaceous volcanic rocks of the Gravina-Nutzotin belt (Gambier Assemblage). Based on similarities in pyroxene compositions and abundances of alkalis, Irvine (1973) suggested that the alkaline Bridget Cove Volcanics represent the parental magma of the ultramafic complexes. In this view, the ultramafic rocks represent subvolcanic magma reservoirs in which the mafic minerals were precipitated and formed cumulate deposits. The concentric, zoned structure may have resulted from upward diapiric emplacement (Himmelberg et al., 1986).

The northernmost body in the Union Bay Ultramafic Suite is the Pyroxenite Creek ultramafite, which intrudes strata of the Dezadeash Group. The western to central interior of the body is composed of fine grained magnetite pyroxenite, whereas the eastern part of the interior is coarse grained olivine pyroxenite (Sturrock et al., 1980). Gabbro and diorite occur discontinuously around the perimeter. K-Ar and Rb-Sr mineral dates range from 113 to 124 Ma. The Haines ultramafic complex, some 200 km south-southeast of Pyroxenite Creek, gave a Rb-Sr age of 120 ± 9 Ma (Lanphere, 1968). The Haines and Pyroxenite Creek bodies are either somewhat older, or have been subject to less resetting of the isotopic systems than the ultramafic complexes farther south.

Several ultramafic bodies exposed in the Coast Plutonic Complex east of Work Channel may represent the southeastward continuation of the Union Bay suite (Douglas, 1983). These bodies contain abundant orthopyroxene, unlike the Union Bay suite, and their age and mode of origin is unknown.

Cascade Suite

Most mid-Cretaceous plutons in the northern Cascades of Washington and southern British Columbia were intruded synchronously with regional metamorphism and deformation. These bodies are northwest-elongate batholiths largely composed of tonalite, diorite and quartz diorite; hornblende generally is more abundant than biotite. Many have strongly foliated margins and some, such as the Gabriel Peak Pluton, have been mapped as orthogneiss (Misch, 1966). The Ten Peak Pluton is similar to the Ecstall Pluton of the Coast Plutonic Complex (discussed later) in that it contains primary epidote and has complex structural relations with surrounding metamorphic rocks (R.A. Haugerud, pers. comm., 1986).

The best Canadian example of the Cascade Suite is the Spuzzum Pluton, in the southern Coast Belt west of the Fraser River Fault, that was intruded synchronously with regional metamorphism. The southern part of the pluton has an irregular zonation similar to that in the Middle Jurassic Adamant Pluton. A central core of hypersthene-augite diorite and hornblende diorite is rimmed by tonalite (Richards, 1971; Vining, 1977). U-Pb dates on zircons range from about 120 to 110 Ma (Gabites, 1985). K-Ar dates range from 76 to 104 Ma (data reviewed in Irving et al., 1985) and are younger towards the Fraser River Fault. The data suggest an Early Cretaceous age for the pluton and hence for regional metamorphism and deformation and setting or resetting of the K-Ar systems during the Late Cretaceous.

The Giant Mascot ultramafic body is situated within the Spuzzum Pluton and is probably mid-Cretaceous in age (McLeod et al., 1976). The ultramafic is crudely zoned, with several peridotite to dunite cores (some with pyrrhotite-pentlandite-chalcopyrite ore bodies) surrounded by pyroxenite and diorite (Aho, 1956). Unlike the Union Bay Ultramafic Suite in southeastern Alaska of the same general age, the Giant Mascot body contains abundant orthopyroxene. It may represent a cumulate phase of crystallization of the Early to mid-Cretaceous Spuzzum Batholith. Alternatively, it may have formed by the reaction of tonalitic Spuzzum magma with mantle-derived ultramafic rocks, as was suggested by Kelemen and Ghiorso (1986) for parts of the Mount Stuart Batholith, which forms part of the Cascade Suite in the North Cascades of Washington.

Southern Intermontane and Coast belts

Most Early to mid-Cretaceous plutons in the southern Intermontane Belt are poorly studied and dated. The Verde Creek Pluton is a miarolitic leucogranite that cuts the mid-Cretaceous Spences Bridge volcanics and may be their hypabyssal equivalent. Farther northwest, between the Yalakom and Fraser faults, the Piltz Peak Pluton and parts of the Tatla Lake metamorphic complex (Friedman and Armstrong, 1988) also are Early to mid-Cretaceous in age.

LATE CRETACEOUS

Late Cretaceous plutons are abundant in the Intermontane Belt and the western and central parts of the Coast Belt. K-Ar dates from most range from 64 to 85 Ma (Armstrong, 1988). Those in the Intermontane Belt (Bulkley and Surprise Lake suites) are small, high-level, calc-alkaline bodies that may be comagmatic with widespread, nonmarine Upper Cretaceous volcanics of the Carmacks Assemblage. Plutons in the Coast Plutonic Complex, including the Bendor Suite of felsic plutons in southwestern British Columbia, are described under that heading later in this chapter.

Bulkley Suite

The greatest concentration of Late Cretaceous plutons occurs along the Skeena Arch, where they are collectively known as the Bulkley Intrusions (Carter, 1982). Individual bodies range in size from dyke swarms to small stocks and small batholiths. The larger plutons form the cores of block-faulted mountain massifs in the Intermontane Belt. Plutons are unfoliated or weakly foliated and appear to have been emplaced at relatively shallow levels; many have well developed contact-metamorphic aureoles. The intrusions are calc-alkaline, biotite- and hornblende-bearing granite to diorite; most are granodiorite and quartz diorite. Many of the smaller bodies host significant Cu-Mo stockwork/porphyry mineralization (Carter, 1982; MacIntyre, 1985; and Christopher and Carter, 1976).

The Rocher Deboule stock is the best studied of the larger Bulkley Intrusions (Sutherland Brown, 1960). The stock covers about 80 km^2 and is the core of an uplifted block of Upper Jurassic to Upper Cretaceous volcanics and sediments. It is composed of weakly porphyritic, biotite-hornblende granodiorite with lesser fine grained granite. Compositional similarity, proximity, and similar K-Ar dates (70-72 Ma) for the stock and the intruded Brian Boru volcanics suggest that the two units are comagmatic.

MacIntyre (1985) concluded that the Bulkley Intrusions near Tahtsa Lake were comagmatic with the Albian(?) to lower Upper Cretaceous Kasalka Group and were emplaced on the periphery of a large cauldron subsidence complex. However, the Kasalka Group is significantly older (mid-Cretaceous) than the Bulkley Suite; the Upper Cretaceous Brian Boru and Tip Top Hill formations are more probable volcanic equivalents of the Bulkley Suite. The Bulkley Suite and related volcanic rocks are the high-level products of deep-seated plutonism and metamorphism in the Coast Belt to the west. Late Cretaceous plutonism, volcanism and metamorphism may represent a transpressive orogen with the Bulkley Intrusions and coeval volcanics being controlled by local pull-apart structures in the Intermontane Belt.

East of the southern Coast Plutonic Complex, only a few plutons gave K-Ar dates between 70 and 80 Ma. High-level, hornblende-bearing granodiorite to quartz diorite bodies occur on both sides of the Yalakom Fault. The best studied is the Fish Lake stock, which contains significant porphyry Cu and Au mineralization (Wolfhard, 1976; McMillan, 1983).

Surprise Lake Suite

This suite consists of a northwest-trending belt of stocks and batholiths in north-central British Columbia and a north-northwest belt extending along the eastern flank of the Coast Belt from northwest British Columbia into Yukon Territory. Most are subcircular in plan; all are high-level, post-tectonic and unfoliated. The rocks are mainly hornblende-biotite and biotite-hornblende granite, alkali-feldspar granite, granodiorite, and hornblende quartz-syenite. Diorite and gabbro are common marginal phases. Miarolitic cavities, commonly lined with quartz, feldspar, fluorite, and other minerals, are present in all members of this suite (Gabrielse, 1969). The Needlepoint Pluton is unusually rich in sphene. Important U-W-Mo-Sn mineralization is associated with some of the plutons (Ballantyne and Littlejohn, 1982).

K-Ar dates cluster around 70 Ma. Plutons of the north-northwest belt may be deeper-seated equivalents of volcanics of the Upper Cretaceous Carmacks Group exposed further north (Grond et al., 1984). No Upper Cretaceous volcanic equivalents are obvious for the northwest-trending belt; however, the plutons possibly represent the roots of volcanic systems that supplied Campanian to Maastrichtian tuff to the Sustut Basin. The plutons in the two belts are similar, and it is probable that they represent continental, volcanic-plutonic magmatism on the east side of the Coast Plutonic Complex, analogous to the Bulkley Suite south of the Bowser Basin.

EARLY TERTIARY (MAINLY EOCENE)

Most of the numerous isotopic dates, mainly K-Ar, from Paleogene plutons fall in the 45 to 55 Ma range (Early to Middle Eocene). The Middle Eocene marks the end of extensive plutonism in the Canadian Cordillera, as well as uplift and unroofing of metamorphic complexes in the Omineca, Intermontane and Coast belts. The bulk of Eocene plutons are in the eastern Coast Plutonic Complex, where they may be deep-seated counterparts of high-level Intermontane Belt plutons and volcanics; most of the remainder are in the Intermontane and Insular belts. The great majority of Eocene plutons are small, high-level, discordant stocks and uncounted dykes and sills; many are spatially and genetically related to volcanic rocks. Plutons in the Coast Plutonic Complex are described later in this chapter; other suites are described below, by region.

Insular Belt

Eocene plutons west of the Coast Plutonic Complex consist of the Early Eocene Sooke and related gabbro bodies on Vancouver Island; the largely Middle Eocene Catface Suite on Vancouver Island; and the Early to Middle Eocene Seward Suite in the Saint Elias Mountains.

The Sooke Gabbro and related smaller bodies occur south of the Leech River Fault on Vancouver Island. The Sooke Gabbro forms the basement to, and grades upwards into sheeted dykes and the Eocene Metchosin Volcanics (Massey, 1986). A U-Pb date of 52 ± 2 Ma on zircons from the gabbro, a 55 ± 1 Ma K-Ar date on hornblende from the gabbro, and a $^{40}Ar/^{39}Ar$ whole rock date from the Metchosin Volcanics (Marvin and Cole, 1978, Duncan, 1982; Massey, 1986) indicate that the volcanics and gabbro are Early Eocene and contemporaneous. The gabbros and volcanics form an ophiolitic stratigraphy that formed in an area of regional extension, perhaps as an emergent island in a transform marginal basin (Massey, 1986).

The Catface Suite consist of numerous small, irregular stocks, dykes and sills (Carson, 1973) in a broad belt extending from near Nanaimo west to Ucluelet and north to Zeballos on Vancouver Island. An important cluster of plutons outcrops on east-central Vancouver Island near Courtenay. Catface Intrusions are mainly tonalite and quartz diorite and minor granodiorite and granite. The rocks contain up to 10 per cent biotite and hornblende. Many intrusions are porphyritic; plagioclase- and hornblende-phyric phases are recognized in all well-studied plutons (Carson, 1973). The rocks are calc-alkaline, with K/Na ratios slightly lower than those of the Jurassic Island Intrusions.

In the Tofino area, the Catface Intrusions grade into and are comagmatic with dacitic ignimbrites, informally called the Flores volcanics by Brandon (1985), which gave Early to Middle Eocene isotopic dates (R.R. Parrish and R.L. Armstrong, unpublished data). K-Ar dates from the Catface Intrusions range from 32 to 59 Ma; most are between 36 and 45 Ma (Armstrong, 1988). They are thus Middle to possibly Late Eocene in age, roughly synchronous with most Paleogene plutons from the Coast and Intermontane Belts in southern British Columbia. In most areas, the Catface Intrusions show no apparent association with coeval volcanism, possibly because of subsequent erosion. The plutons intrude rocks as young as the Upper Cretaceous Nanaimo Group and were emplaced at high levels. Diatremes or collapse breccias occur in and near many bodies, suggesting high volatile contents during intrusion. Many of the Catface Intrusions contain economically interesting sulphide mineralization. Porphyry Cu-Mo deposits are associated with the small intrusions at Catface Mountain (McDougall, 1976) and Mount Washington, and Au-Ag-bearing quartz veins are associated with the Zeballos stock.

The Seward Suite in the Saint Elias Mountains of Canada (Dodds and Campbell, 1988), consists of tonalite, quartz diorite, granodiorite and quartz monzonite. Most plutons lie southwest of the Border Ranges Fault and were emplaced at high levels into the Chugach Terrane. Most K-Ar dates suggest a 41-55 Ma (late Early and Middle Eocene) age.

Southern British Columbia

Numerous Paleocene and Eocene plutons occur in southern British Columbia east of the Coast Plutonic Complex. These have a wide variety of compositions and structural styles and fall into several distinct suites.

The Ladybird Suite is exposed in the footwall of the Okanagan and Valkyr shear zones (Carr et al., 1987). In the Valhalla gneiss complex (Reesor, 1965), plutonic rocks of the Ladybird Suite are largely Paleogene in age (Carr et al., 1987; Parrish et al., 1988). These include an I-type, hornblende-biotite, megacrystic quartz monzonite of Paleocene age and a homogeneous, leucocratic, biotite quartz monzonite of Early Eocene age. Both units are sheet-like bodies that intrude and structurally overlie Late Cretaceous and older plutonic and metamorphic rocks. They have a pervasive, moderate to strong mylonitic foliation.

Other granitic rocks between the Okanagan valley and the Columbia River are here included in the Ladybird Suite. These include the Sheppard Intrusions south of Trail. These rocks are generally granite, but syenitic phases also occur. Biotite is the main mafic mineral. Unlike the Coryell Suite (described below), the Sheppard Intrusions are locally intensely sheared and mylonitized (Little, 1982). No K-Ar dates are available but a U-Pb date on zircons of 47 Ma (Parrish, et al., 1988) indicates a Middle Eocene age. The Sheppard Intrusions may be related to the calc-alkaline volcanics of the upper part of the White Lake basin (Church, 1973) and the felsic volcanics of the Middle Eocene Kettle River Formation (Little, 1982). The genetic link, if any, between the Coryell Suite and Sheppard Intrusions is unclear.

The Coryell Suite consists of distinctive high-level, Eocene plutons widespread between Okanagan Lake and the Columbia River. The Coryell Suite consists of batholiths and stocks of pink and buff, variably textured, commonly porphyritic syenite and lesser granite, shonkinite, diorite, and monzonite (Little, 1960, 1982). Biotite and hornblende are the main mafic minerals, but pyroxene is present locally. Chemically the rocks are strongly alkaline and have anomalously high concentrations of U and Th (Leroux, 1980). K-Ar dates (compiled in Little, 1982) and a U-Pb date on zircons indicate a late Early to Middle Eocene age for the Coryell Intrusions. The plutons are genetically related to the widespread Marron Formation of Middle Eocene age (Church, 1973; Little, 1982). Tempelman-Kluit and Parkinson (1986) suggested that the Coryell Intrusions were contemporaneous with extension across the low-angle Okanagan shear zone. If so, then the alkaline chemistry of the rocks may reflect crustal thinning during Eocene extension; however, most of the extension postdates the alkaline intrusions.

The Shingle Creek porphyry, just west of the Okanagan shear zone, is mainly a calc-alkaline, coarsely porphyritic granite. Phenocrysts of sanidine up to 10 cm long, and smaller phenocrysts of albite and quartz are set in a fine grained matrix (Bostock, 1966). The porphyry is unconformably overlain by volcanics of the Marron Formation and may represent the westernmost member of the Coryell Suite.

The southwestern Intermontane Belt and northern Cascades in British Columbia and Washington contain numerous plutons, mostly undated, that are probably Paleogene to Eocene in age. One of the largest is the crudely circular Needle Peak Pluton. This body intrudes Mesozoic strata of the Methow Trough and cuts most structures in the trough and may be offset by northeast-trending block faults. It is a high-level, unfoliated body of coarse grained biotite-hornblende monzogranite of Eocene age (Greig, 1989). Several large Eocene plutons occur in the North Cascades of Washington; all postdate regional metamorphism and deformation. The Golden Horn Batholith is unusual among these plutons in that it is distinctly alkaline, in contrast to the calc-alkaline nature of the others.

West of the Fraser River and Yalakom faults, numerous granodiorite sills cut schists of the Bridge River Terrane. The sills are mylonitized and isoclinally folded; little remains of the original textures (Potter, 1983). Concordant U-Pb dates on zircons indicate a Middle Eocene age of emplacement and deformation (Potter, 1983; P. van der Heyden, pers. comm., 1985). Mylonitization and recumbent folding, probably related to dextral wrench faulting on the Fraser River and Yalakom faults, thus is in part Late Eocene or younger in age. Undeformed dykes and sills of felsite, feldspar porphyry, and biotite-feldspar porphyry are common southwest of the Yalakom Fault. These cut Mesozoic rocks of the Tyaughton Trough and may be feeders to volcanics of the Eocene Sheba Group. Early Tertiary granodiorite in the Tatla Lake metamorphic complex has undergone ductile deformation and mylonitization (Friedman and Armstrong, 1988).

West-central British Columbia

Early Tertiary plutons are abundant in the Intermontane Belt of west-central British Columbia, where they occur as small, high-level stocks and are known by a variety of names. Most plutons fall into one of three distinct suites: the granitic Nanika and Quanchus suites, the granodioritic Babine Suite, and the gabbroic Goosly Lake Suite. With few exceptions, K-Ar dates range from 44 to 56 Ma (Carter, 1982).

Plutons of the Nanika Suite are called the Nanika Intrusions south of the Bowser Basin, the Kastberg Intrusions east of the basin, and the Alice Arm Intrusions west of the basin. The rocks are calc-alkaline and vary from miarolitic granite to granodiorite. Texturally they vary from coarse grained, equigranular rocks to aphanitic rhyolite breccia. Many intrusions were emplaced as small plutons, dyke swarms, and sills along steeply dipping faults. Some contain important stockwork/porphyry Cu and Mo mineralization such as the Berg (Panteleyev, 1981) and Kitsault deposits (Steininger, 1985). Coeval and spatially associated volcanics are present south of the Bowser Basin, and in general the plutons appear to represent the roots of deeply eroded volcanoes (MacIntyre, 1985). The Alice Arm Intrusions just east of the Coast Plutonic Complex may occupy an intermediate structural level between deeper, Eocene plutons in the eastern Coast Plutonic Complex and Sloko-like calc-alkaline volcanics.

The Quanchus Suite consists of an arcuate chain of large stocks in west-central British Columbia. The intrusions are similar in composition to the Nanika Suite but are larger, have generally more hornblende and biotite, and lack significant Mo and Cu mineralization. The few available K-Ar dates indicate Late Paleocene to Eocene ages of emplacement.

The granodiorite suite (Babine Intrusions) is restricted to a northwest-trending belt near Babine Lake. It hosts several large porphyry Cu-Mo deposits (e.g. Granisle, Bell Copper; Carson and Jambor, 1974). Most of the rocks are a distinctive, biotite-feldspar granodiorite to diorite porphyry.

The bodies range in form from dyke swarms to small stocks, commonly showing evidence of multiple intrusion. They may be the roots of volcanic centres, and extrusive equivalents are preserved in several areas (Carter, 1982).

The Goosly Lake Intrusions consist of several small plugs south of Houston. These are alkaline gabbro and monzodiorite regarded by Church (1971) as centres of nearby Eocene volcanism.

Northern British Columbia and Yukon Territory

Numerous Tertiary plutons are exposed along the east side of the Coast Mountains in northern British Columbia and western Yukon Territory. The Ruby Range granodiorite forms the eastern part of the Coast Plutonic Complex in the southwestern Yukon. It gives mainly Eocene K-Ar dates but parts of the unit are probably older (Tempelman-Kluit and Wanless, 1975). The rock is mainly heterogeneous, grey hornblende-biotite granodiorite (Muller, 1967). Some outcrops are well foliated, and recrystallization has produced a pronounced granoblastic texture. Smoky quartz is locally present.

The Bennett Suite consists of high-level, calc-alkaline plutons on the east side of the Coast Plutonic Complex. In composition and tectonic setting, these plutons resemble those of the Nanika Suite. The rocks are mostly buff-weathering, leucogranite to alaskite which intrude Stikinia. Smoky quartz eyes and miarolitic cavities are common; biotite is the main mafic mineral and fluorite is a common accessory. In the Yukon, the Nisling Range alaskite (Muller, 1967; Tempelman-Kluit and Wanless, 1975; Ridgway, 1973) is related to extensive north-trending feldspar-porphyry dyke swarms that may be feeders to nearby intermediate to felsic pyroclastic rocks. The Pattison Pluton (Lynch and Pride, 1984) is the best-studied member of the Bennett Suite. It consists of fine-, medium- and coarse-grained phases of alaskite and contains economically interesting molybdenite mineralization.

In northwestern British Columbia, Eocene plutons of the Bennett Suite are thought to be the subvolcanic equivalents of the mainly Eocene Sloko volcanics (Bultman, 1979; Lambert, 1974). The Fourth of July Batholith (Aitken, 1959) is a large, zoned granite and granodiorite pluton near Atlin. Hornblende, commonly with pyroxene cores, is the main mafic mineral. Although this body has abundant smoky quartz suggestive of the Bennett Suite, recent dating indicates a Jurassic age (R.L. Armstrong, unpublished data).

The Bennett and Nanika suites are similar in composition and tectonic setting. Both suites consist of high-level, felsic plutons that intrude Stikinia; both gave K-Ar dates of about 45 to 57 Ma. Although the Bennett Suite can be more easily linked with correlative volcanics, it is likely that both suites represent high-level expressions of the extensive plutonism occurring in the Eocene in the Coast Belt to the west.

In southeastern Yukon Territory, the alkalic "Ting Creek" intrusion is one of several small, epizonal syenite and trachyte plutons. K-Ar dates indicate an Eocene age (Harrison, 1981).

In the Deserters Range west of the Northern Rocky Mountain Trench, the post-deformation Balourdet stock cuts structures in the Sifton Fault. This pluton is a medium grained, equigranular biotite granite that gave a K-Ar date of 42 ± 2 Ma (Evenchick, 1988). Undeformed lamprophyre (kersantite) and pegmatite dykes also cut the Sifton Fault and give K-Ar dates ranging from 37 to 52 Ma (Evenchick, 1988). Eocene dykes are common elsewhere in the Omineca Belt (e.g. Parrish, 1979).

Foreland Basin

Several clusters of alkaline igneous rocks outcrop in north-central Montana. Those in the Sweetgrass Hills are Early to Middle Eocene in age and range in composition from pyroxene syenite to aplite (Marvin et al., 1980). The Bearpaw Mountains igneous complexes outcrop over an area of 2500 km^2. Intrusive rocks related to volcanic activity are Eocene in age and alkaline to peralkaline in composition. Marvin et al. (1980) concluded that the Eocene magmas are mantle-derived and owe their origin to one or more hot spots.

COAST PLUTONIC COMPLEX

The greatest concentration of plutonic rocks in the Canadian Cordillera is in the Coast Plutonic Complex of the Coast Belt, a long and narrow zone of plutonic and lesser metamorphic rocks extending from southern British Columbia into Yukon Territory. The east and west boundaries of the complex are somewhat arbitrary as there is no sharp change in the age, composition, or plutonic style with the flanking belts. Nonetheless, the term remains useful for the main part of the Coast Belt in which plutonic and metamorphic rocks predominate over unmetamorphosed and low grade strata. This review is mainly descriptive and concentrates on the area south of 55°N latitude and particularly on the Prince Rupert to Terrace region, the best-studied cross-section of the Coast Plutonic Complex. The Coast Plutonic Complex north of 55°N, in Alaska, was reviewed by Brew and Morrell (1983).

Most reconnaissance work in the Coast Plutonic Complex has concentrated on mapping the modal compositions of plutonic rocks (e.g. Roddick and Hutchison, 1974; Roddick, 1983). The markedly different descriptions and interpretations of parts of the complex given by Roddick (1965), Souther (1971), Baer (1973), Woodsworth (1979), Hutchison (1982), Barker et al. (1986), and Crawford et al. (1987) show the lithological and structural complexity of the belt (e.g. Fig. 15.5). The Coast Plutonic Complex consists of a complex of migmatite and gneiss (Plate 30) (the Central Gneiss Complex in the Prince Rupert area); foliated to massive, discrete to coalescing plutons; and minor zones of schist and variably metamorphosed volcanics and sediments. The overall structural trend is northwesterly. Most pre-Tertiary plutons also are elongate northwesterly, suggesting that their shape and form have a strong structural control. Granitoid rocks underlie about three quarters of the Coast Plutonic Complex and range in composition from granite to gabbro. Quartz diorite forms about 40 per cent of the plutonic rock, and tonalite and diorite together constitute another 30 per cent (Roddick, 1983). Hornblende and biotite are the main mafic minerals, and clinopyroxene is common in the cores of hornblende grains, particularly in the more mafic rocks. Sphene, magnetite, apatite, and zircon are the main accessory minerals. Two-mica plutons

Figure 15.5. Mafic dyke cutting foliated quartz diorite. The dyke is dismembered and intruded by the plutonic rock, suggesting that the quartz diorite was a crystal mush when the dyke was injected. Such dykes, typical of the complexities of many rocks in the Coast Plutonic Complex, were termed "synplutonic" by Roddick and Armstrong (1959). About 25 km west of the head of Bute Inlet. Photo by G.J. Woodsworth. (GSC 205378)

are rare. Of the circum-Pacific plutonic terranes, the Coast Plutonic Complex is the largest, most mafic, and most deficient in K-feldspar.

The overall chemical composition of plutonic rocks of the Coast Plutonic Complex is similar to average continental crust, but K_2O is lower and Na_2O and Al_2O_3 are markedly higher. (Roddick, 1983). The rocks are calc-alkaline. Smith et al. (1979) reported systematic increases in K_2O, Rb, and Ba, and decreases in CaO, Na_2O, Mn, and Sr in the plutonic rocks from west to east across the belt. However, Roddick (1983), using a much larger number of analyses, found no significant increases or decreases from west to east. $^{87}Sr/^{86}Sr$ initial ratios are near or below 0.704 in the southern Coast Plutonic Complex and between 0.706 and 0.704 farther north (Armstrong, 1988; Barker et al., 1986). In summary, the granitic rocks of the Coast Plutonic Complex are fairly mafic and typical of the Cordilleran species of I-type plutons as defined by Pitcher (1983).

Early work on the age of plutons in the Coast Plutonic Complex relied mainly on K-Ar dates from the Prince Rupert and Douglas Channel region (Hutchison, 1982). These data define an eastern belt giving Eocene (43 to 55 Ma) dates, and a western belt characterized by Late Jurassic dates on the west to Late Cretaceous dates on the east. The boundary between the two belts coincides with a ductile shear zone called the Work Channel Lineament in British Columbia (see Fig. 17.8; Crawford and Hollister, 1982) and the Coast Range Megalineament in southeastern Alaska (Brew and Ford, 1978). Rocks on either side of this zone also have differing metamorphic and structural histories (Chapter 17; Crawford et al., 1987).

About 100 km southeast of Terrace, plutonic rocks along the east flank of the Coast Plutonic Complex are predominantly Middle Jurassic, Late Cretaceous and early Tertiary in age (van der Heyden, 1982). This zone may be separated from the Tertiary "plateau" to the west by a brittle(?) fault zone. Thus, at least locally, the K-Ar dates give a symmetrical pattern with a Tertiary core flanked by Cretaceous dates.

K-Ar dates in the southern Coast Plutonic Complex show a general decrease from west to east but the pattern is more complex than that in the north. In both regions, the K-Ar dates may represent final cooling of large parts of the Coast Plutonic Complex, and indicate that the western part was cooled and uplifted before the eastern. No firm generalizations can be made about the extent to which the K-Ar dates represent time of pluton emplacement; each date must be examined individually.

U-Pb dates on zircons indicate that plutonic rocks range from Silurian to Eocene in age; probably more than 95 percent are Jurassic to Eocene. Generally, the older plutons are the most highly deformed and metamorphosed; degree of deformation and metamorphism can be used to determine a crude relative plutonic chronology in a given area.

Several important U-Pb studies in the Coast Plutonic Complex were completed too late to be fully incorporated into this review. In particular, van der Heyden (1989) used 48 new U-Pb dates to develop a tectonic model for the region between latitudes 53° and 54°N. His data show that the western flank of the Coast Plutonic Complex contains three main tectonic units: syn- and post-kinematic plutons (165-155 Ma), flanked on the east by Early Cretaceon plutons (131-123 Ma), which in turn are flanked on the east by a belt of mid-Cretaceous plutons (110-84 Ma). He suggested that all Jurassic to Eocene plutons of the Coast Plutonic Complex and flanking belts are related to east-dipping subduction between a single, allochthonous Alexander-Wrangellia-Stikinia superterrane, emplaced against North America in Middle Jurassic time.

Paleozoic and Late Triassic plutons

The oldest known plutons in the Coast Plutonic Complex occur in a northwest trending belt on islands west and south of Prince Rupert. On Baron and Dunira islands, quartz diorite and tonalite intrude the lowermost units of an undated volcanic and sedimentary sequence (Hutchison, 1982). Regional greenschist-facies metamorphism has altered the original mafic minerals to actinolite and chlorite. Most of the rocks show a strong, superimposed foliation; many are in effect chlorite schist. U-Pb dates on zircons from one pluton range from 411 to 419 Ma, suggesting a Late Silurian or older age for at least part of the suite. Woodsworth and Orchard (1985) correlated upper Paleozoic and lower Mesozoic strata with those in the Alexander Terrane in southeastern Alaska. The Silurian plutons on Baron and Dunira islands are probably calc-alkaline and resemble Ordovician to Early Silurian calc-alkaline plutons in southeastern Alaska. However, Late Silurian trondhjemite and sodic leucodiorite, characteristic of the Alexander Terrane in southeastern Alaska, are unknown in the Coast Plutonic Complex in British Columbia. Paleozoic plutons are also present on Porcher and Pitt islands and in the Ecstall River area, where they form narrow northwest-trending bodies that commonly are intensely deformed.

The Dala River Pluton is a large body of metamorphosed and highly deformed quartz diorite and tonalite with conspicuous blue quartz eyes. Preliminary U-Pb dates on zircons suggest a Mississippian age, hinting that the pluton may be basement to Upper Paleozoic strata in the Atna Peak area (Chapter 16) that are tentatively assigned to the Stikine Terrane.

The Captain Cove Pluton south of Prince Rupert is a northwesterly elongated body of predominantly quartz diorite (Roddick, 1970). U-Pb dating of zircons shows that the northern part of the pluton is late Middle to early Late Triassic in age whereas the southern part is Early Cretaceous. Late Triassic tonalite is also known from southeast Alaska (Gehrels et al., 1991b).

Early Jurassic plutons

Plutons of this age form a discontinuous strip along the eastern margin of the Coast Belt and adjacent Intermontane Belt. Dated plutons lie between about 53° and 56°N, but bodies of the same character outcrop intermittently along the east side of the Coast Plutonic Complex between latitudes 52°N and 59°N. All plutons are deformed, many are mylonitized, and most are regionally metamorphosed to greenschist facies. This belt of plutons refutes any simple model which interprets the plutons of the Coast Plutonic Complex as becoming progressively younger and more felsic from west to east.

The Texas Creek Granodiorite near Stewart is the best-studied of these plutons. It is a small, irregular, homogeneous body characterized by alkali feldspar megacrysts and prismatic hornblende phenocrysts (Grove, 1971). The pluton is cut by east-trending, subvertical mylonitic fabric and zones that contain lineations plunging gently south-southwest. Contacts with regionally metamorphosed Jurassic Hazelton Group andesite and siltstone are probably intrusive; coeval, cogenetic hornblende-, plagioclase- and alkali feldspar-phyric "Premier Porphyry" dykes clearly intrude the Texas Creek Granodiorite and Hazelton Group. K-Ar and U-Pb dates indicate an Early Jurassic age for the granodiorite and dykes (Smith, 1977; Alldrick et al., 1986, 1987).

Early Jurassic plutons are present along the east side of the Coast Belt between Kitimat and Bella Coola and best exposed west of Whitesail Lake. The plutons are aligned northwesterly and consist largely of well foliated quartz diorite and diorite metamorphosed to greenschist facies.

All Early Jurassic plutons are in areas inferred to be part of Stikinia. The close spatial association of plutons and Hazelton volcanics suggests a link between volcanism and plutonism. Those between Kitimat and Bella Coola are similar in age and lithology to the Early to Middle Jurassic Topley Suite. Early Jurassic plutons in the Coast Plutonic Complex may be cogenetic with the Lower Jurassic Telkwa Formation of the Hazelton Group.

Middle to Late Jurassic plutons

Between about 53° and 55°N, the west side of the Coast Plutonic Complex is dominated by Late Jurassic plutons, whereas the east side contains Middle and Late Jurassic bodies.

On the west side of the Coast Belt, discrete and coalescing plutons range in composition from hornblende-biotite granodiorite to biotite-hornblende diorite (Roddick, 1970). In the western Coast Plutonic Complex, the roughly circular Gil Island diorite complex gave K-Ar dates of 133 (hornblende) and 139 (biotite) Ma (Roddick, 1970). Hornblende is more abundant than biotite, and much of the complex is epidotized and chloritized. About 25 km west of Gil Island, a small, circular body of biotite-hornblende granodiorite gave a 144 Ma K-Ar date on hornblende and cuts highly altered quartz diorite. The large northwest-trending Campania Pluton is unusual for the Coast Plutonic Complex in that it commonly contains abundant muscovite and biotite together with accessory amounts of garnet and gahnite. K-Ar dates from these plutons range from 106 to 148 Ma, but U-Pb dates from zircons clearly indicate Late Jurassic (155 to 165 Ma) emplacement ages.

East of Work Channel Lineament, in the high grade core of the Central Gneiss Complex, at least some metamorphosed and deformed plutonic rock gives U-Pb dates ranging from 128 to 140 Ma (Barker and Arth, 1984; Hill, 1984). In the eastern Coast Plutonic Complex, mylonitic and protomylonitic quartz diorite to granodiorite with 155 to 160 Ma U-Pb dates on zircon form part of the Gamsby Metamorphic Complex (van der Heyden, 1982, unpublished data).

On the east flank of the Coast Belt near Terrace, Middle Jurassic plutons are chloritized and were strongly affected by intense brittle deformation; mylonitization is confined to narrow, discrete zones. The rocks intrude and are thrust northeastward over Permian carbonate and Lower Jurassic volcanics of the Hazelton Group. West of Whitesail Lake, Middle Jurassic granodiorite occurs as sheets of mylonite in an imbricate thrust zone at the western margin of the Intermontane Belt (van der Heyden, 1982) and as altered and deformed granodiorite bodies intruding Hazelton Group strata along the eastern margin of the Coast Plutonic Complex. The bodies are commonly cut by swarms of altered and deformed mafic dykes which are absent from nearby, younger plutons. The Middle Jurassic Tahtsa Complex is a complex of calc-alkaline, hypersthene-normative, hornblende diorite and quartz diorite and volcanic rocks (Stuart, 1960). The complex has been metamorphosed to greenschist facies and is cut by numerous mafic dykes and brittle to ductile shear zones.

Late Jurassic to early Early Cretaceous plutons

Much of the Coast Plutonic Complex north and west of Vancouver, and southwest of the Harrison fault system and its extension to the northwest, is composed of a "matrix" of generally foliated and altered plutonic rock which is intruded by fresher, crosscutting plutons. The matrix is mostly quartz diorite, tonalite, and granodiorite, but quartz monzodiorite and granite are locally abundant. Generally, hornblende is more abundant than biotite. These rocks are at least in part altered to greenschist facies assemblages; thus relatively few have been dated by K-Ar methods. Planar fabric ranges from absent or very weak to mylonitic. Mylonitic rocks are mainly confined to steeply-dipping, northwest-trending, discrete zones, in most places of unknown width and extent. In several areas, the plutons form the basement to the Upper Jurassic to Lower Cretaceous Gambier Group, the base of which is commonly marked by conglomerate containing boulders of the underlying plutonic rock. Such clasts indicate that plutonic rocks were exposed in the Coast Plutonic Complex during the Early Cretaceous (e.g. Woodsworth, 1979). McKillop (1973) obtained Late Jurassic K-Ar dates from metamorphosed quartz diorite unconformably overlain by the Gambier Group near Vancouver. U-Pb dates from similar quartz diorite north of Squamish are equivocal but suggest a Late Jurassic age of emplacement. The widespread presence of plutonic clasts in sediments of the lower part of the Gambier Group indicate that pre-Gambier plutonic rocks may be more abundant in the Coast Plutonic Complex than is presently realized. It is significant that some plutonic clasts in the basal Gambier Group are foliated, suggesting plutonism, metamorphism and deformation prior to deposition of the Gambier Group. Plutonic-clast conglomerate is also present in Upper Jurassic strata in the southwestern part of Tyaughton Trough, immediately east of the Coast Plutonic Complex (Tipper, 1969), further indicating uplift and erosion of the plutonic terrane. These observations may imply that formation of the southern Coast Plutonic Complex, perhaps in response to amalgamation of the Insular Superterrane and the Intermontane Superterrane, was underway by Late Jurassic time.

Late Early to mid-Cretaceous plutons

Although plutons with K-Ar dates between 130 and 120 Ma are uncommon in the Coast Plutonic Complex, extremely varied plutons with K-Ar dates from about 120 to 98 Ma are abundant, particularly in the western part. Based on timing with respect to superimposed deformation and metamorphism, pre-, syn- and post-tectonic plutons were emplaced during this time interval.

Deformed and metamorphosed plutons of known or presumed Early to mid-Cretaceous age are lithologically similar to plutons of the Late Jurassic suite. The northwest-trending Pemberton Diorite Complex is heterogeneous, with diorite grading into both quartz diorite and amphibolite; all phases are intruded by numerous mafic dykes. The complex has many characteristics of the Black Dome and Tahtsa complexes farther north and emphasizes the presence of a belt of dioritic plutons and complexes of various ages along the east side of the Coast Plutonic Complex. U-Pb dates on zircons from the Pemberton diorite are discordant and range from 115 to 127 Ma; hornblende from the same outcrop gave a K-Ar date of 53 Ma. The dates indicate the difficulty of interpreting isotopic dates in the Coast Plutonic Complex; the zircons suggest but do not prove an Early Cretaceous age. The westernmost plutons of the Coast Plutonic Complex on Vancouver Island and Sechelt Peninsula are Early Cretaceous, based on U-Pb dates from zircon (Isachsen et al., 1985). South of Prince Rupert, Early Cretaceous plutons, including some on McCauley, Pitt, and Gil islands, form a narrow belt east of the Jurassic plutonic suite and west of the Ecstall belt of mid-Cretaceous plutons (van der Heyden, unpublished data).

The Ecstall Pluton, east of Prince Rupert and just west of Work Channel Lineament, is an excellent example of a synmetamorphic, mid-Cretaceous pluton that illustrates the problems of dating plutons in the Coast Plutonic Complex. The pluton is a tadpole-shaped, concentrically zoned body of diorite to granodiorite (Hutchison, 1982; Roddick, 1970). The "head" of the pluton has the same fabric as the schists it intrudes, and field relations suggest the pluton was thrust over the schist late in the metamorphic history of the area (Crawford and Hollister, 1982; Crawford et al., 1987). Attempts to date the Ecstall Pluton have given mixed results. Rb-Sr mineral dates of 78 and 79 Ma (Armstrong and Runkle, 1979) agree with conventional K-Ar and a $^{40}Ar/^{39}Ar$ plateau date of 76 Ma (Harrison et al., 1979) and indicate a Late Cretaceous age for the pluton. Rb-Sr whole-rock isotopic data for pegmatites cutting the pluton suggest an Early Jurassic age, but samples from other late, leucocratic dykes had excess radiogenic Sr. U-Pb dates on zircons from the pluton are concordant at 98 ± 4 Ma and were interpreted by Woodsworth et al. (1983) as indicating the age of emplacement of the pluton. Pb-Pb dates of 104 ± 4 Ma on zircons from the southern part of the pluton support the U-Pb dates (van der Heyden, unpublished data).

The Ecstall Pluton is the largest and most spectacular of a chain of "magmatic epidote"-bearing plutons in the western Cordillera, called the Ecstall Suite here and on the Tectonic Assemblage Map (Map 1712A, in pocket). This belt includes the mid- to Late Cretaceous Moth Bay, Bushy Point, and other plutons in southeastern Alaska (Zen and Hammarstrom, 1984b). Similar plutons occur elsewhere in the Coast Plutonic Complex (e.g. the Mount Gilbert and Butedale plutons) and the North Cascades of Washington (e.g. the Ten Peak Pluton). The common characteristic of these plutons is the presence of up to about 6 per cent epidote, not as a fine grained alteration product of plagioclase, but as discrete grains up to several millimetres across in fresh, unaltered rock. The Bushy Point Pluton contains garnet with a composition roughly intermediate between almandine and grossular (Zen and Hammarstrom, 1984a). Zen and Hammarstrom (1984b) and Zen (1985) concluded that the presence of "magmatic epidote" implies that the pluton crystallized and was emplaced in the lower crust (at pressures of at least 8 kbar, 8×10^5 kPa) and was rapidly uplifted at rates of at least 1 mm/a. Independent petrological constraints (Crawford and Hollister, 1982) suggest that the Ecstall Pluton was indeed emplaced in the lower crust. However, "magmatic" epidote also is present in plutons emplaced at shallower levels, for example the Mount Gilbert Pluton (less than 6.5 kbar, 6.5×10^5 kPa, Woodsworth, 1979) and plutons northeast of Terrace (less than 5 kbar, 5×10^5 kPa). Epidote may indeed reflect crystallization at great depths, but it also appears that it can exist metastably in plutons emplaced at fairly high levels in the crust.

In the southern Coast Plutonic Complex, a good example of a post-tectonic, late Early Cretaceous pluton is the Squamish Granodiorite. This is a roughly circular body of fresh, unfoliated biotite granodiorite. In contrast with the Ecstall Pluton, K-Ar dates on biotite of about 94 Ma are concordant within the limits of error with a whole rock Rb-Sr isochron of 100 ± 10 Ma and U-Pb dates of 101 ± 2 Ma (R.L. Armstrong, pers. comm., 1986). Also in contrast with the Ecstall Pluton, the Squamish granodiorite was emplaced at high levels into sedimentary rocks as young as Middle Albian.

The nearby Porteau Pluton is the same age as the Squamish granodiorite and was emplaced at about the same level, but has a well developed cataclastic texture (Roddick, in Price et al., 1985). The fabric may reflect deformation superimposed on the pluton by regional metamorphism and deformation or, alternatively, stresses set up in the pluton by emplacement and cooling. However, the abundance and intrusive relations of late-syn- and post-tectonic plutons of roughly mid-Cretaceous age indicate that regionally extensive metamorphism and deformation in the western Coast Plutonic Complex had concluded by about that time. Plutons similar to the Porteau and Ecstall plutons are present in the northern Cascades of British Columbia and Washington, and are discussed under the Cascade Suite.

Late Cretaceous plutons

The time corresponding roughly to the Early to Late Cretaceous boundary (about 98 Ma) was important in the evolution of the Coast Plutonic Complex in the Prince Rupert, Terrace, and Whitesail Lake region. In the western Intermontane Belt, the time marked the end of marine sedimentation, the first appearance of detritus derived from the Coast Plutonic Complex, and a resurgence of volcanism. In the west, it marked the end of extensive plutonism west of Work Channel Lineament. In the southern Coast Plutonic Complex, plutons of this age are known from all but the westernmost areas.

In the Prince Rupert-Terrace-Whitesail Lake region, the Late Cretaceous was a time of uplift and cooling west of Work Channel Lineament. East of the lineament, in the core of the Coast Plutonic Complex, Late Cretaceous upper amphibolite to granulite-facies metamorphism has reset much of the U-Pb systematics in zircons, making interpretation of other dates ambiguous (Woodsworth et al., 1983). However, it is possible that many of the plutons in the core of the belt are Late Cretaceous in age. For example, the Dubose stock near Kemano gives U-Pb dates of about 75 Ma and K-Ar dates of about 55 Ma. The Horetzky Dyke, immediately east of the Dubose stock, is a high level tonalite pluton with a K-Ar date of about 73 Ma that probably indicates the age of emplacement. The dates indicate a greater depth of emplacement of the Dubose stock and greater uplift relative to the Horetzky Dyke. The Horetzky Dyke belongs to the Late Cretaceous Bulkley Suite of the Intermontane Belt, suggesting that the Bulkley Intrusions and coeval volcanics are the high-level counterparts to metamorphism and plutonism occurring in the core of the Coast Plutonic Complex.

Northeast of Terrace, the Ponder Pluton may be largely Late Cretaceous in its western part and early Tertiary in the east (van der Heyden, unpublished data), although all parts give Eocene K-Ar dates and a large part gives and Eocene whole-rock Rb-Sr date (K. Scott, pers. comm., 1987). The pluton is composed largely of granite and granodiorite and, in its western parts, structurally overlies metamorphic rocks of the Central Gneiss Complex (Hutchison, 1982). The Late Cretaceous Lluvia Peak Pluton west of Ponder Pluton postdates metamorphism of the Central Gneiss Complex, which it structurally overlies (Hill, 1984).

In the southern Coast Belt, most Late Cretaceous plutons are post-tectonic, unfoliated to weakly foliated, granite to granodiorite. Most have biotite as the main mafic mineral. These plutons are similar to the post-tectonic, Early to mid-Cretaceous plutons such as the Squamish Granodiorite, but give younger K-Ar dates. Syntectonic plutons of presumed Late Cretaceous age in the Mount Raleigh area give K-Ar dates of about 72 Ma. The Mount Gilbert Pluton is medium grained quartz diorite and tonalite that contains roughly equal proportions of hornblende and biotite. The rock is moderately to well foliated and seems to have been emplaced synchronously with high grade regional metamorphism (Woodsworth, 1979).

An areally restricted cluster of Late Cretaceous to early Tertiary(?) plutons, the Bendor Suite, is present in the Bralorne area of the southern Coast Belt. These plutons are high-level, subcircular bodies of biotite and biotite-hornblende granodiorite, quartz diorite, and granite. They are unfoliated and appear to have been emplaced after most deformation of the host Bridge River and Cadwallader groups. K-Ar dates from the plutons range from about 57 to 77 Ma and U-Pb dates on zircons from the Copp Creek stock of 84 Ma (Rusmore, 1985) suggest a Late Cretaceous to Paleocene age for this suite.

Early Tertiary plutons

In the northern half of the Coast Plutonic Complex, particularly in the east, plutons giving Eocene isotopic dates are abundant. Although many of these plutons may be older than Tertiary, as discussed above, there is no doubt that the Paleogene, and particularly the Eocene, was the last great episode of plutonism in the Coast Plutonic Complex and, for that matter, in the Canadian Cordillera.

In the Prince Rupert-Terrace-Whitesail Lake region, early Tertiary plutons are absent or rare west of Work Channel Lineament. The Quottoon Pluton, the westernmost pluton of known early Tertiary age, is a long, narrow body that extends north through southeastern Alaska, where it is called the "foliated tonalite sill" (Brew and Ford, 1978; Gehrels et al. 1991b), and perhaps as far south as Douglas Channel. The Quottoon Pluton is predominantly tonalite with roughly equal proportions of hornblende and biotite. Its age, as given by concordant U-Pb dates of about 59 Ma (Armstrong and Runkle, 1979) and 64 and 67 Ma (Gehrels et al., 1984) and 80 Ma (van der Heyden, unpublished data) is Campanian to Paleocene. This is one of the rare examples in the Canadian Cordillera of a pluton whose age is close to the Cretaceous-Tertiary boundary. The western margin of the pluton shows moderate to strong ductile deformation, suggesting that the pluton was emplaced late in the history of movement along the Work Channel Lineament.

Between the Work Channel Lineament and the eastern margin of the Coast Plutonic Complex, several large plutons yield Eocene U-Pb, Rb-Sr and K-Ar dates. Most of these, such as the eastern part of the Ponder Pluton and the Tsaytis Pluton, are biotite or hornblende-biotite granite to granodiorite. The plutons may have been emplaced at relatively high levels, as andalusite is common in the contact aureoles on the east side of the Ponder Pluton (Hutchison, 1982; Sisson, 1985). However, some of the plutons, such as the Tsaytis Pluton and parts of the Ponder Pluton, are intimately interfolded and deformed with the enclosing Central Gneiss Complex.

Swarms of lamprophyre dykes cut Tertiary and older plutons in the Prince Rupert region, west of Whitesail Lake, and in southeastern Alaska. Most trend northeast, have steep dips and, in southeastern Alaska, are alkali-olivine basalt (Smith, 1973). They are Late Eocene to Miocene in age, and may reflect minor crustal extension during uplift of the Coast Plutonic Complex (Hill, 1984).

LATE TERTIARY

Late Tertiary plutons are restricted to the Coast (Chilliwack Suite and King Island Pluton) and Insular (Kano, Tkope, Wrangell, and La Perouse suites) belts. Most are small and were emplaced at high levels. Many appear genetically related to nearby volcanic rocks of the same general age and composition. All but the King Island syenite may be subduction related, and those in the Saint Elias Mountains are associated with regional transcurrent faults.

Chilliwack Suite

A linear belt of late Tertiary plutons extending from northern Washington into the Coast Belt is roughly coincident with the Pemberton belt of late Tertiary and Quaternary volcanics (Chapter 14). The largest body in this suite is the composite Chilliwack Batholith, exposed along and near the 49th parallel. Only the northern third is exposed in Canada, where it was dated and studied in detail by Richards and McTaggart (1976). Based on K-Ar dates, mostly on biotite, the batholith appears to have been emplaced in four stages: a late Eocene phase in northern Washington, about 35 Ma, 29 to 26 Ma, and 21 to 16 Ma. Most of the plutonic events produced several calc-alkaline intrusive phases that differ texturally and chemically from one another. In general, the more mafic rocks were emplaced early, followed by progressively more felsic units. Lithologies range from hypersthene diorite to alaskite; most are granodiorite, tonalite, and granite. Biotite and hornblende are the main mafic minerals.

The batholith intrudes and thermally metamorphoses Eocene(?) conglomerate and the Oligocene(?) Skagit volcanics. Farther south in Washington, the batholith and related volcanics cluster along and cut the Fraser River-Straight Creek transcurrent fault zone, suggesting that the fault was an important control on emplacement of the plutons. Richards and McTaggart suggested that the Chilliwack Batholith is the root of a late Tertiary volcanic complex represented in part by the Skagit volcanics. Berman and Armstrong (1980) reached a similar conclusion from a study of the nearby Coquihalla volcanic and plutonic complex of Late Oligocene to Early Miocene age.

In Canada, the Chilliwack Batholith is part of the Pemberton belt, an arcuate chain of high-level plutons and associated volcanics that extends northwest for at least 400 km from the Cloudy Pass (Cater, 1969) and Cascade Pass plutons in Washington into British Columbia (Souther, 1977; Souther, Chapter 14). K-Ar ages of plutons show a progressive northward decrease from about 16 to 35 Ma in the south to about 7 Ma in the north. Plutons are calc-alkaline granite to quartz diorite. Subvolcanic plutons in the Franklin Glacier, Salal Creek and Meager Mountain volcanic complexes, are distinctly older (7-10 Ma) than spatially associated volcanic rocks (less than 4 Ma), indicating that the complexes are the product of two discrete magmatic events. Porphyry and stockwork Mo (Salal Creek; Stephens, 1972), and Cu-Mo (Franklin Glacier) and Au (Ray, 1986) mineralization is associated with some of the intrusions.

King Island Pluton

This narrow, easterly-trending pluton lies athwart the western Coast Plutonic Complex near Bella Bella. The dominant phase is coarse grained grey syenite that grades into soda granite near the western margin of the body. The syenite is a miarolitic hypersolvus, alkaline to peralkaline rock with small amounts of mafic minerals, including fayalitic olivine, ferrohedenbergite, and sodic amphibole. K-Ar dates from the pluton and related dykes and volcanic rocks range from 10 to 13 Ma. Souther (1986) concluded that the King Island and cogenetic dyke complex represent different levels of erosion of the root zone of a peralkaline alkaline magma system. On a more regional scale, the King Island Pluton is near the western and older end of the late Cenozoic Anahim Volcanic Belt (Chapter 14) that may mark the westward motion of the North American plate over a mantle hot spot.

Kano Suite

Numerous small, post-tectonic plutons outcrop on the western and eastern coasts of the Queen Charlotte Islands and intrude all older units including part of the Tertiary Masset Formation (Sutherland Brown, 1968; Anderson, 1988a). Most are northwesterly aligned stocks of fine- to medium-grained, equigranular to seriate, augite-hypersthene diorite, quartz monzodiorite and local leucocratic granite. The mafic rocks are sub-alkaline, hypersthene normative, and chemically resemble the basaltic andesites of the Masset Formation. U-Pb dates from the plutons range from 20 to 46 Ma (Anderson and Reichenbach, 1991) and diminish in age from south to north (Young, 1981). These dates overlap those from volcanics of the nearby Masset Formation. The Carpenter Bay plutonic variety along the southeast side of Moresby Island consists of quartz monzonite, augite-hypersthene monzodiorite and miarolitic granite. Abundant, north-trending coeval and co-spatial basalt, andesite, and rhyolite dykes in some of the plutons suggest that they may be the roots of part of the Tertiary Masset volcanism. In dyke-poor plutons, the plutonic-volcanic link is more tenuous.

Tkope, Wrangell, and La Perouse suites

K-Ar dates suggest that the late Tertiary plutons of the Saint Elias Mountains fall into two groups: Late Oligocene (24-31 Ma) group, the Tkope Suite, and a Middle to Late Miocene (6-16 Ma) unit, the Wrangell Suite (Dodds and Campbell, 1988). A third suite, the La Perouse gabbroic suite, is present only in southeastern Alaska and on Mount Fairweather in adjacent British Columbia.

The compositionally diverse Tkope Suite is typified by the Tkope River Batholith which intrudes Alexander Terrane rocks in northwestern British Columbia (Jacobson et al., 1980). The batholith is an epizonal, composite, gabbro to granophyre. All phases are calc-alkaline, and the body has an initial $^{87}Sr/^{86}Sr$ ratio of 0.7031. Nearby andesitic volcanic rocks may be extrusive equivalents. Many of the Oligocene plutons are associated with coeval volcanics and may be related to large-scale transcurrent faults, but the origin and tectonic significance of the plutons remains unknown.

Most plutons of the Middle to Late Miocene Wrangell Suite were emplaced within or near the coeval, calc-alkaline Wrangell lavas. The Bock's Brook stock in southwestern Yukon Territory has been studied in some detail (Downey et al., 1980). The intrusion is a medium grained, biotite diorite with a narrow gabbroic border. Major element chemistry indicates a high-alumina, tholeiitic composition; the $^{87}Sr/^{86}Sr$ ratio of 0.70414 is within the normal range of circum-Pacific volcanic arcs and is consistent with an origin related to late Tertiary subduction of the Pacific Plate.

The La Perouse Suite consists of a chain of late Tertiary gabbro complexes in the Fairweather Range in southeastern Alaska. The largest and best studied of these is the La Perouse layered gabbro (Loney and Himmelberg, 1983). The complex consists of basal cumulates of peridotite, olivine gabbro and troctolite that are overlain by layered olivine norite and gabbro. $^{40}Ar/^{39}Ar$ dating suggests an Oligocene age for the intrusion. The layering in the intrusion forms an asymmetric funnel that is elongate northwestward. Although the pluton was probably emplaced in a transpressive regime, the synformal shape of the layering may be due to subsidence from magmatic loading at high temperatures that continued after solidification (Loney and Himmelberg, 1983).

Several Late Oligocene stocks in the Coast Belt of southeast Alaska east of Ketchikan range in composition from biotite granite to gabbro. One of these, the Quartz Hill stock, hosts a world-class porphyry Mo deposit. K-Ar dates from these stocks range from 22 to 30 Ma (Hudson et al., 1979).

REGIONAL TRENDS AND TECTONIC CLASSIFICATION

Discussion of Cordillera-wide patterns of timing, composition, geochemistry, and tectonic setting of plutonism must be tempered by the fact that the Cordillera consists of a collage of disparate terranes that were disrupted by strike-slip faults before, during, and after amalgamation. Nonetheless, a few generalities emerge that are likely to withstand the continuing growth of the database.

Plutons in any given suite share a common plutonic style characterized by similar tectonic setting, contemporaneity, and shared mineralogical, textural, geochemical and isotopic characteristics. Although the geochronometric database for Cordilleran plutonism has outstripped the complementary field, petrographic and geochemical studies, enough is known of Mesozoic and Tertiary plutons in the Canadian Cordillera to allow comparison with well-known plutonic classifications of other workers (Anderson, 1988a,b). Many of the same criteria used here were used by previous workers to develop means of categorizing plutonic rocks on a regional scale. They have led to concepts of the quartz diorite line (Moore, 1959); the $^{87}Sr/^{86}Sr$ line (Armstrong et al., 1977; magma sequences (Bateman and Dodge, 1970); super units (Cobbing et al., 1977, 1981); ilmenite- and magnetite-series granites (Ishihara, 1977); I-, S-, and A-type granitoids (Chappell and White, 1974; White and Chappell, 1983; White et al., 1986); and a four-part tectonic environmental granite classification (Pitcher, 1983).

There is no clear-cut, simple temporal migration of plutonism across the Cordillera. The most obvious diachronous feature is an eastward migration of plutonism, and volcanism, from the Insular Belt across the Coast Belt and into the western Intermontane Belt from the Early Jurassic to Eocene (Roddick and Hutchison, 1974; Armstrong, 1988). This migration is not unique to the Canadian Cordillera, but is common in the Americas from Alaska to Chile. Although its cause is in doubt, the time-transgressive nature of plutonism in the western Canadian Cordillera is difficult to reconcile with the hypothesis of Monger et al. (1982) that suggests the Jurassic to Eocene Coast Plutonic Complex resulted from a single Late Cretaceous collision between the Insular and Intermontane superterranes. Few other Cordillera-wide time-transgressive patterns emerge from the distribution maps. For example, in the mid-Cretaceous, intense plutonism occurred simultaneously in both the eastern and western Cordillera, although the plutonic styles are very different.

In the Canadian Cordillera, there is a good correlation between plutonic style and known or inferred host terrane composition until the Cretaceous, when most of the Cordillera had amalgamated but the terranes had not necessarily ceased their northward motion relative to ancestral North America. In a general way, this is reflected in the strontium isotopic composition of plutonic rocks. The initial $^{87}Sr/^{86}Sr$ data, mostly unpublished, from the Canadian Cordillera were reviewed by Armstrong (1988). Initial ratios in the Omineca and Foreland belts are invariably greater than 0.705, whereas those in the western belts are commonly below 0.705. Armstrong (1988) pointed out that initial ratios for mid-Cretaceous plutons in the eastern belts commonly exceed 0.710. The transition from low to high ratios coincides roughly with the western limit of old, radiogenic continental crust and the eastern limit of the composite, mainly volcanic terranes. Plutons west of the transition may be largely arc-related; those to the east may reflect anatexis and/or assimilation of older, sialic crust.

Most Triassic and Jurassic plutons in the Canadian Cordillera occur in volcanic and volcaniclastic host rocks. These are commonly associated with nearly coeval, compositionally similar, calc-alkaline volcanism. Examples include Polaris, Stikine, Guichon, Topley, Black Lake, Long Lake, Vancouver Island and Three Sisters suites, and plutons in the Coast Plutonic Complex. Nearby, coeval subduction complexes and constituent alpine ultramafic bodies support a subduction-related origin and mantle source for the Triassic volcano-plutonic complexes. Jurassic subduction, although reasonable to explain the large volumes of intermediate-composition magma, cannot be proven on the basis of the plutonic evidence alone. Most of the Triassic and Jurassic plutons are medium- to high-level bodies. Ultramafic, mafic, intermediate and felsic phases are characteristic. Hornblende (± biotite) is the main mafic mineral, and magnetite and sphene are common accessory minerals. Calc-alkaline, metaluminous, cafemic-rich, and less evolved and radiogenic (compared to mid-Cretaceous and younger suites) compositions are typical. All early Mesozoic plutons have $^{87}Sr/^{86}Sr$ initial ratios less than 0.707 and many less than 0.705. Early to mid-Mesozoic Cordilleran plutonism is mineralogically and geochemically akin to magnetite-series, Cordilleran I-type plutons produced and emplaced in an Andinotype, tectonic environment at the continental margin. Many of the Cordillera's rich porphyry copper and molybdenum deposits are associated with these plutons.

The Copper Mountain and parts of the Kuskanax and Nelson suites, and the Galena Bay stock, are important exceptions to the above generalizations. The Copper Mountain Suite is associated with alkaline volcanic rocks. The other three suites lack volcanic equivalents from which an independent tectonic setting may be inferred. However, these plutons were emplaced in medium- to high-grade metamorphic terranes, and coeval volcanics may have been removed by erosion or tectonic processes. The alkaline Copper Mountain and Kuskanax suites are mineralogically and compositionally similar to some Cretaceous-Tertiary suites described below; two-mica granite members of the Nelson Suite and Galena Bay stock are comparable to, but older than the mid-Cretaceous plutonic suites.

Members of the mid-Cretaceous Selwyn, Bayonne, and Cassiar plutonic suites are the closest Cordilleran analogues of S-type, ilmenite-series plutons emplaced in a Hercynotype environment. Plutonism in the Omineca and Foreland belts was syn- or post-kinematic with respect to both the middle to late Mesozoic deformation and metamorphism in the Omineca Belt and folding and thrust faulting of the miogeocline. Nonetheless, the plutons do form an integral part of the eastern of the two metamorphic welts ascribed by Monger et al. (1982) to accretion of the composite, volcanic-dominated terranes to ancestral North America continent during an arc-continent collision. Coeval volcanics are rare, clearly related to hornblende-bearing phases and only sporadically preserved. Hornblende-biotite granodiorite and two-mica granite are characteristic but generally do not coexist in the same pluton; the hornblende-bearing bodies may be slightly older than the two-mica plutons. Mafic plutonic rocks are absent. $^{87}Sr/^{86}Sr$ initial ratios for both varieties are high (>0.710) but highest in the two-mica granitoids (0.730 and greater) and are distinct from hornblende-bearing plutons. The two-mica plutons, particularly those with peraluminous accessory minerals such as andalusite, muscovite, tourmaline and garnet, are closely similar to S-type plutons (White et al., 1986). Hornblende-bearing granodiorite is geochemically similar to, although not identical with, Lachlan Fold Belt I-type plutons. Exceptions in the Selwyn plutonic suite are the most northeastern plutons which are siliceous equivalents of the alkali-feldspar syenite of the Tombstone and Howell Creek suites. These, and particularly the former, although apparently mid-Cretaceous, have many of the attributes of the Cordilleran Late Cretaceous-Tertiary Andinotype plutonism.

In contrast to mid-Cretaceous plutonism in the eastern Cordillera, Mesozoic plutonism in the Coast Belt typifies I-type, magnetite-series plutons emplaced in an Andinotype environment. Many of these plutons are deep-seated and are intimately associated with the metamorphism and deformation that produced the western of the two great metamorphic welts in the Canadian Cordillera. Coeval volcanics are abundant and appear to be arc-derived. The plutons are calc-alkaline and range in composition from granite to gabbro; tonalite, granodiorite, and diorite are the most abundant lithologies. Hornblende and biotite are the main mafic minerals. Primary muscovite and peraluminous minerals are rare. $^{87}Sr/^{86}Sr$ initial ratios for plutons and volcanics are below 0.706 and mostly below 0.704, consistent with derivation by anatexis of mantle or lower crust in an arc setting.

Cretaceous and Tertiary plutons in the central and eastern Cordillera include the closest analogues to A-type (anorogenic) plutonism in a Caledonian-style, tectonic setting. The suites consist of small, circular, high-level stocks and their cogenetic dyke swarms which, together with their cogenetic volcanics, are distributed along northerly or westerly trending belts. The plutons postdate terrane accretion and regional metamorphism, and the belts cross important terrane and structural boundaries. They are known to intrude previously extended crust (e.g., Tombstone Suite), to be coincident with a change from transpressional to transtensional movement along nearby faults (e.g., Surprise Lake Suite) or to be associated with coeval extension (e.g., Coryell Suite). The suites or groups of related suites are bimodal and comprise diorite or gabbro and alkali-feldspar syenite, granite and monzogranite. Mafic minerals clinopyroxene, hornblende and biotite are locally rich in Na, Fe, and/or F. Sphene, magnetite, fluorite, and feldspathoids are important accessory minerals. Silicic, subaluminous, met- to peraluminous, geochemically evolved compositions are typical for the felsic end members. $^{87}Sr/^{86}Sr$ initial ratios are variable and likely reflect the post-accretionary crustal composition unique to each locality.

As these selected examples indicate, only part of the Cordilleran plutonic record unequivocally fits other tectonic classification schemes proposed for plutonic rocks. This conclusion, not unexpected in such a tectonically complex orogen, in large part reflects the inadequacy of both the understanding of pluton genesis and emplacement and the current tectonic-compositional classifications of plutonic terranes, as well as the inadequacy of our knowledge of Cordilleran plutons. Nonetheless, it is clear that there are Cordilleran analogues to I-, S- and A-type plutonism. Moreover, there is a close relationship between the overall Mesozoic and Paleogene Cordilleran tectonic evolution, host terrane and pluton type. Cordilleran I-types are associated with Late Triassic to Cretaceous island arc and ocean island volcanic rocks in the allochthonous terranes of the western Cordillera. S-type plutonism within the interior and along the edge of the continental miogeocline may reflect the in part transpressive movement of these outboard terranes along and onto the continental margin of North America. A-type plutonism, occurring within the composite plate some distance from the Late Cretaceous to Tertiary continental margin, marks the change from compressional to tensional stress in different crustal environments accompanying or postdating final movement along extensive dextral strike-slip faults.

REFERENCES

Ager, C.A., Ulrych, T.J., and McMillan, W.J.
1973: A gravity model for the Guichon Creek batholith, south-central British Columbia; Canadian Journal of Earth Sciences, v. 10, p. 920-935.

Aho, A.E.
1956: Nickel-copper pyrrhotite deposits at the Pacific Nickel Property, southwestern British Columbia; Economic Geology, v. 51, p. 444-481.

Aitken, J.D.
1959: Atlin map-area, British Columbia (104 N); Geological Survey of Canada, Memoir 307, 89 p.

Alldrick, D.J., Mortensen, J.K., and Armstrong, R.L.
1986: Uranium-lead age determinations in the Stewart area (104B/1); in Geological Fieldwork 1985, British Columbia Ministry of Energy, Mines and Petroleum Resources, Paper 1986-1, p. 217-218.

Alldrick, D.J., Brown, D.A., Harakal, J.E., Mortensen, J.K., and Armstrong, R.L.
1987: Geochronology of the Stewart mining camp (104B/1); in Geological Fieldwork 1986, British Columbia Ministry of Energy, Mines and Petroleum Resources, Paper 1987-1, p. 81-92.

Allen, D.G., Panteleyev, A., and Armstrong, A.T.
1976: Galore Creek; in Porphyry Deposits of the Canadian Cordillera, A. Sutherland Brown (ed.), Canadian Institute of Mining and Metallurgy, Special Volume 15, p. 402-414.

Anderson, R.G.
1982: Geology of the Mactung pluton in Niddery Lake map area and some of the plutons in Nahanni map area, Yukon Territory and District of Mackenzie; in Current Research, Part A, Geological Survey of Canada, Paper 82-1A, p. 299-304.
1983a: Geology of the Hotailuh Batholith and surrounding volcanic and sedimentary rocks, north-central British Columbia; Ph.D. thesis, Carleton University, 669 p.
1983b: Selwyn plutonic suite and its relationship to tungsten skarn mineralization, southeastern Yukon and District of Mackenzie; in Current Research, Part B, Geological Survey of Canada, Paper 83-1B, p. 151-163.
1984: Late Triassic and Jurassic magmatism along the Stikine Arch and the geology of the Stikine batholith, north-central British Columbia; in Current Research, Part A, Geological Survey of Canada, Paper 84-1A, p. 67-73.
1987: Plutonic rocks in the Dawson map area, Yukon Territory; in Current Research, Part A, Geological Survey of Canada, Paper 87-1A, p. 689-697.
1988a: Jurassic and Cretaceous-Tertiary plutonic rocks on the Queen Charlotte Islands, British Columbia; in Current Research, Part E, Geological Survey of Canada, Paper 88-1E, p. 213-216.
1988b: An overview of some Mesozoic and Tertiary plutonic suites and their associated mineralization in the northern Cordillera; in Recent Advances in the Geology of Granite-Related Mineral Deposits, R.P. Taylor and D.F. Strong (ed.), Canadian Institute of Mining and Metallurgy, Special Volume 39, p. 96-113.

Anderson, R.G., Loveridge, W.D., and Sullivan, R.W.
1982: U-Pb isotopic ages of zircon from the Jurassic plutonic suite, Hotailuh Batholith, north-central British Columbia; in Current Research, Part C, Geological Survey of Canada, Paper 82-1C, p. 133-137.

Anderson, R.G. and Reichenbach, I.
1991: U-Pb and K-Ar framework for Middle to Late Jurassic (172-≥158 Ma) and Tertiary (46-27 Ma) plutons in Queen Charlotte Islands, British Columbia; in Evolution and Petroleum Potential of the Queen Charlotte Basin, G.J. Woodsworth (ed.); Geological Survey of Canada, Paper 90-10.

Archibald, D.A., Glover, J.K., Price, R.A., Farrar, E., and Carmichael, D.M.
1983: Geochronology and tectonic implications of magmatism and metamorphism, southern Kootenay Arc and neighbouring regions, southeastern British Columbia. Part I: Jurassic to mid-Cretaceous; Canadian Journal of Earth Sciences, v. 20, p. 1891-1913.

Archibald, D.A., Krogh, T.E., Armstrong, R.L., and Farrar, E.
1984: Geochronology and tectonic implications of magmatism and metamorphism, southern Kootenay Arc and neighbouring regions, southeastern British Columbia. Part II: Mid-Cretaceous to Eocene; Canadian Journal of Earth Sciences, v. 21, p. 567-583.

Armstrong, R.L.
1985: Rb-Sr dating of the Bokan Mountain granite complex and its country rocks; Canadian Journal of Earth Sciences, v. 22, p. 1233-1236.
1988: Mesozoic and early Cenozoic magmatic evolution of the Canadian Cordillera; in Processes in Continental Lithospheric Deformation, S.P. Clark, Jr., B.C. Burchfiel, and J. Suppe (ed.), Geological Society of America, Special Paper 218, p. 55-91.

Armstrong, R.L. and Runkle, D.
1979: Rb-Sr geochronometry of the Ecstall, Kitkiata, and Quottoon plutons and their country rocks, Prince Rupert region, Coast Plutonic Complex, British Columbia; Canadian Journal of Earth Sciences, v. 16, p. 387-399.

Armstrong, R.L., Taubeneck, W.H., and Hales, P.O.
1977: Rb-Sr and K-Ar geochronometry of Mesozoic granitic rocks and their Sr isotope composition, Oregon, Washington and Idaho; Geological Society of America Bulletin, v. 88, p. 397-411.

Armstrong, R.L., Eisbacher, G.H., and Evans, P.D.
1982: Age and stratigraphic-tectonic significance of Proterozoic diabase sheets, Mackenzie Mountains, northwestern Canada; Canadian Journal of Earth Sciences, v. 19, p. 316-323.

Armstrong, R.L., Harakal, J.E., Brown, E.H., Bernardi, M.L., and Rady, P.M.
1983: Late Paleozoic high-pressure metamorphic rocks in northwestern Washington and southwestern British Columbia: the Vedder Complex; Geological Society of America Bulletin, v. 94, p. 451-458.

Armstrong, R.L., Monger, J.W.H., and Irving, E.
1985: Age of magnetization of the Axelgold Gabbro, north-central British Columbia; Canadian Journal of Earth Sciences, v. 22, p. 1217-1222.

Baadsgaard, H., Folinsbee, R.E., and Lipson, J.
1961: Caledonian or Acadian granites of the northern Yukon Territory; in Geology of the Arctic, Volume 1, G.O. Raasch (ed.), p. 458-465.

Baer, A.J.
1973: Bella Coola-Laredo Sound map areas; Geological Survey of Canada, Memoir 372, 122.

Ballantyne, S.B. and Littlejohn, A.L.
1982: Uranium mineralization and lithogeochemistry of the Surprise Lake batholith, Atlin, British Columbia; in Uranium in Granites, Y.T. Maurice (ed.), Geological Survey of Canada, Paper 81-23, p. 145-155.

Barker, F. and Arth, J.G.
1984: Preliminary results, Central Gneiss Complex of the Coast Range batholith, southeastern Alaska: the roots of a high-K calc-alkaline arc?; Physics of the Earth and Planetary Interiors, v. 35, p. 191-198.

Barker, F., Arth, J.G., and Stern, T.W.
1986: Evolution of the Coast Batholith along the Skagway traverse, Alaska and British Columbia; American Mineralogist, v. 71, p. 632-643.

Barr, D.A.
1966: The Galore Creek copper deposits; Canadian Mining and Metallurgical Bulletin, v. 59, p. 841-853.

Barr, D.A., Fox, P.E., Northcote, K.E., and Preto, V.A.
1976: The alkaline suite porphyry deposits — a summary; in Porphyry Deposits of the Canadian Cordillera, A. Sutherland Brown (ed.), Canadian Institute of Mining and Metallurgy, Special Volume 15, p. 359-367.

Bateman, P.C. and Dodge, F.C.
1970: Variations of major chemical constituents across the Sierra Nevada Batholith; Geological Society of America Bulletin, v. 81, p. 409-420.

Berg, H.C., Jones, D.L., and Richter, D.H.
1972: Gravina-Nutzotin belt — tectonic significance of an upper Mesozoic sedimentary and volcanic sequence in southern and southeastern Alaska; in Geological Survey Research 1972, United States Geological Survey, Professional Paper 800-D, p. D1-D24.

Berman, R.G. and Armstrong, R.L.
1980: Geology of the Coquihalla Volcanic Complex, southwestern British Columbia; Canadian Journal of Earth Sciences, v. 17, p. 985-995.

Bishop, D.T.
1974: Petrology and Geochemistry of the Purcell sills in Boundary County, Idaho; in Belt Symposium 1973, Department of Geology, University of Idaho, and Idaho Bureau of Mines, v. 2, p. 15-66.

Bostock, H.H.
1966: Feldspar and quartz phenocrysts in the Shingle Creek porphyry, British Columbia; Geological Survey of Canada, Bulletin 126, 71 p.

Brandon, M.T.
1985: Mesozoic mélange of the Pacific Rim Complex, western Vancouver Island; in Field Guides to Geology and Mineral Deposits in the Southern Canadian Cordillera, D.J. Tempelman-Kluit (ed.), Geological Society of America, Cordilleran Section Meeting, Vancouver, B.C., May, 1985, p. 7-1 - 7-28.

Brandon, M.T., Orchard, M.J., Parrish, R.R., Sutherland Brown, A., and Yorath, C.J.
1986: Fossil ages and isotopic dates from the Paleozoic Sicker Group and associated intrusive rocks, Vancouver Island, British Columbia; in Current Research, Part A, Geological Survey of Canada, Paper 86-1A, p. 683-696.

Brew, D.A. and Ford, A.B.
1978: Megalineament in southeastern Alaska marks southwest edge of Coast Range batholithic complex; Canadian Journal of Earth Sciences, v. 15, p. 1763-1772.

Brew, D.A. and Morrell, R.P.
1983: Intrusive rocks and plutonic belts of southeastern Alaska, U.S.A.; in Circum-Pacific Plutonic Terranes, J.A. Roddick (ed.), Geological Society of America, Memoir 159, p. 171-193.

Brosge, W.P. and Reiser, H.N.
1969: Preliminary geologic map of the Coleen Quadrangle, Alaska; United States Geological Survey, Open File Report 69-25.

Bultman, T.R.
1979: Geology and tectonic history of the Whitehorse Trough west of Atlin, British Columbia; Ph.D. thesis, Yale University, 284 p.

Cairnes, C.E.
1934: Slocan mining camp; Geological Survey of Canada, Memoir 173, 137 p.

Campbell, S.W.
1981: Geology and genesis of copper deposits and associated host rocks in and near the Quill Creek area, southwestern Yukon; Ph.D. thesis, University of British Columbia, 215 p.

Cargill, D.G., Lamb, J., Young, M.J., and Rugg, E.S.
1976: Island Copper; in Porphyry Deposits of the Canadian Cordillera, A. Sutherland Brown (ed.), Canadian Institute of Mining and Metallurgy, Special Volume 15, p. 206-218.

Carr, S.D., Parrish, R.R., and Brown, R.L.
1987: Eocene structural development of the Valhalla Complex, southeastern British Columbia; Tectonics, v. 6, p. 175-196.

Carson, D.J.T.
1973: The plutonic rocks of Vancouver Island; Geological Survey of Canada, Paper 72-44, 70 p.

Carson, D.J.T. and Jambor, J.L.
1974: Mineralogy, zonal relationships and economic significance of hydrothermal alteration at porphyry copper deposits, Babine Lake area, British Columbia; Canadian Mining and Metallurgical Bulletin, v. 67, p. 110-133.

Carter, N.C.
1982: Porphyry copper and molybdenum deposits, west-central British Columbia; British Columbia Ministry of Energy, Mines, and Petroleum Resources, Bulletin 64, 150 p.

Cater, F.W.
1969: The Cloudy Pass epizonal batholith and associated subvolcanic rocks; Geological Society of America, Special Paper 115, 54 p.
1982: Intrusive rocks of the Holden and Lucerne quadrangles, Washington — the relation of depth zones, composition, textures, and emplacement of plutons; United States Geological Survey, Professional Paper 1220, 108 p.

Chappell, B.W. and White, A.J.R.
1974: Two contrasting granite types; Pacific Geology, v. 8, p. 173-174.

Christopher, P.A. and Carter, N.C.
1976: Metallogeny and metallogenic epochs for porphyry mineral deposits in the Canadian Cordillera; in Porphyry Deposits of the Canadian Cordillera, A. Sutherland Brown (ed.), Canadian Institute of Mining and Metallurgy, Special Volume 15, p. 64-71.

Church, B.N.
1971: Geology of the Owen Lake, Parrott Lakes, and Goosly Lake area; in Geology Exploration and Mining in British Columbia 1970, British Columbia Department of Mines and Petroleum Resources, p. 119-125.
1973: Geology of the White Lake basin; British Columbia Department of Mines and Petroleum Resources, Bulletin 61, 120 p.

Clark, T.
1980: Petrology of the Turnagain ultramafic complex, northwestern British Columbia; Canadian Journal of Earth Sciences, v. 17, p. 744-757.

Cobbing, E.J., Pitcher, W.S., and Taylor, W.P.
1977: Segments and super-units in the Coastal Batholith of Peru; Journal of Geology, v. 85, p. 625-631.

Cobbing, E.J., Pitcher, W.S., Wilson, J.J., Baldock, J.W., Taylor, W.P., McCourt, W., and Snelling N.J.
1981: The geology of the Western Cordillera of northern Peru; Institute of Geological Sciences, Overseas Memoir 5, 143 p.

Crawford, M.L. and Hollister, L.S.
1982: Contrast of metamorphic and structural histories across the Work Channel lineament, Coast Plutonic Complex, British Columbia; Journal of Geophysical Research, v. 87, p. 3849-3860.

Crawford, M.L., Hollister, L.S., and Woodsworth, G.J.
1987: Crustal deformation and regional metamorphism across a terrane boundary, Coast Plutonic Complex, British Columbia; Tectonics, v. 6, p. 343-361.

Currie, K.L.
1975: The geology and petrology of the Ice River alkaline complex, British Columbia; Geological Survey of Canada, Bulletin 245, 68 p.
1976: Notes on the petrology of nepheline gneisses near Mount Copeland, British Columbia; Geological Survey of Canada, Bulletin 265, 32 p.

Dagenais, G.R.
1984: The oxygen isotope geochemistry of granitoid rocks from the southern and central Yukon; M.Sc. thesis, University of Alberta, 168 p.

Dick, L.A.
1979: Tungsten and base metal skarns in the northern Cordillera; in Current Research, Part A, Geological Survey of Canada, Paper 79-1A, p. 259-266.
1980: A comparative study of the geology, mineralogy and conditions of formation of contact metasomatic mineral deposits in the northeastern Canadian Cordillera; Ph.D. thesis, Queen's University, 471 p.

Dick, L.A. and Hodgson, C.J.
1982: The MacTung W-Cu(Zn) contact metasomatic and related deposits of the northeastern Canadian Cordillera; Economic Geology, v. 77, p. 845-867.

Dillon, J.T., Tilton, G.R., Decker, J., and Kelly, M.J.
1987: Resource implications of magmatic and metamorphic ages for Devonian igneous rocks in the Brooks Range; in Alaskan North Slope Geology, Volume 2, I. Tailleur and P. Weimer (ed.), Pacific Section, Society of Economic Paleontologists and Mineralogists, and The Alaskan Geological Society, p. 713-723.

Dodds, C.J. and Campbell, R.B.
1988: Potassium-argon ages of mainly intrusive rocks in the Saint Elias Mountains, Yukon and British Columbia; Geological Survey of Canada, Paper 87-16, 43 p.

Douglas, B.J.
1983: Structural and stratigraphic analysis of a metasedimentary inlier within the Coast Plutonic Complex, British Columbia, Canada; Ph.D. thesis, Princeton University, 298 p.

Downey, M.E., Armstrong, R.L., and Parrish, R.R.
1980: K-Ar, Rb-Sr and fission track geochronometry of the Bock's Brook Stock, Kluane Ranges, southwestern Yukon Territory; in Current Research, Part B, Geological Survey of Canada, Paper 80-1B, p. 189-193.

Duncan, R.A.
1982: A captured island chain in the Coast Range of Oregon and Washington; Journal of Geophysical Research, v. 87, p. 10827-10837.

Eadie, E.T.
1976: K-Ar and Rb-Sr geochronology of the northern Hogem Batholith, B.C.; B.Sc. thesis, University of British Columbia, 46 p.

Evenchick, C.A.
1988: Stratigraphy, metamorphism, structure, and their tectonic implications in the Sifton and Deserters ranges, Cassiar and northern Rocky mountains, northern British Columbia; Geological Survey of Canada Bulletin 376, 90 p.

Findlay, D.C.
1969: Origin of the Tulameen ultramafic-gabbro complex, southern British Columbia; Canadian Journal of Earth Sciences, v. 6, p. 399-425.

Fox, P.E.
1969: Petrology of Adamant pluton, British Columbia; Geological Survey of Canada, Paper 67-61, 101 p.

Friedman, R.M. and Armstrong, R.L.
1988: Tatla Lake Metamorphic Complex: an Eocene metamorphic core complex on the southwestern edge of the Intermontane Belt of British Columbia; Tectonics, v. 7, p. 1141-1166.

Fyles, J.T.
1984: Geological setting of the Rossland mining camp; British Columbia Ministry of Energy, Mines and Petroleum Resources, Bulletin 74, 61 p.

Gabites, J.E.
1985: Geology and geochronometry of the Cogburn Creek-Settler Creek area, northeast of Harrison Lake, B.C.; M.Sc. thesis, University of British Columbia, 153 p.

Gabrielse, H.
1963: McDame map-area, Cassiar District, British Columbia; Geological Survey of Canada, Memoir 319, 138 p.
1969: Geology of Jennings River map-area, British Columbia (104-O); Geological Survey of Canada, Paper 68-55, 37 p.
1985: Major dextral transcurrent displacements along the Northern Rocky Mountain Trench and related lineaments in north-central British Columbia; Geological Society of America Bulletin, v. 96, p. 1-14

Gabrielse, H. and Reesor, J.E.
1974: The nature and setting of granitic plutons in the central and eastern parts of the Canadian Cordillera; Pacific Geology, v. 8, p. 109-138.

Gabrielse, H., Loveridge, W.D., Sullivan, R.W. and Stevens, R.D.
1982: U-Pb measurements on zircon indicate middle Paleozoic plutonism in the Omineca Crystalline Belt, north-central British Columbia; in Current Research, Part C, Geological Survey of Canada, Paper 82-1C, p. 139-146.

Gardner, M.C., Bergman, S.C., Cushing, G.W., MacKevett, E.M., Jr., Plafker, G., Campbell, R.B., Dodds, C.J., McClelland, W.C., and Mueller, P.A.
1988: Pennsylvanian pluton stitching of Wrangellia and the Alexander terrane, Wrangell Mountains, Alaska; Geology, v. 16, p. 967-971.

Garnett, J.A.
1978: Geology and mineral occurrences of the southern Hogem batholith; British Columbia Ministry of Mines and Petroleum Resources, Bulletin 70, 75 p.

Gehrels, G.E., Brew, D.A., and Saleeby, J.B.
1984: Progress report on U/Pb (zircon) geochronologic studies in the Coast plutonic-metamorphic complex east of Juneau, southeastern Alaska; in The United States Geological Survey in Alaska: Accomplishments During 1982, United States Geological Survey, Circular 939, p. 100-102.

Gehrels, G.E. and Saleeby, J.B.
1987: Geology of southern Prince of Wales Island, southeastern Alaska; Geological Society of America Bulletin, v. 98, p. 123-137.

Gehrels, G.E., Saleeby, J.B., and Berg, H.C.
1983: Preliminary description of the Klakas orogeny in the southern Alexander terrane, southeastern Alaska; in Pre-Jurassic Rocks in Western North American Suspect Terranes, C.H. Stephens (ed.), Society of Economic Paleontologists and Mineralogists, Pacific Section, p. 131-141.
1987: Geology of Annette, Gravina, and Duke islands, southeastern Alaska; Canadian Journal of Earth Sciences, v. 24, p. 866-881.

Getsinger, J.S.
1985: Geology of the Three Ladies Mountain/Mount Stevenson area, Quesnel Highland, British Columbia; Ph.D. thesis, University of British Columbia, 239 p.

Godwin, C.I.
1975: Alternative interpretations for the Casino Complex and Klotassin Batholith in the Yukon Crystalline Terrane; Canadian Journal of Earth Sciences, v. 12, p. 1910-1975.

Godwin, C.I. and Price, B.J.
1986: Geology of the Mountain diatreme kimberlite, north-central Mackenzie Mountains, District of Mackenzie, Northwest Territories; in Mineral Deposits of Northern Cordillera, J.A. Morin (ed.), Canadian Institute of Mining and Metallurgy, Special Volume 37, p. 298-310.

Gordy, P.L. and Edwards, G.
1962: Age of the Howell Creek intrusives; Journal of the Alberta Society of Petroleum Geologists, v. 10, p. 369-372.

Green, L.H.
1972: Geology of Nash Creek, Larsen Creek, and Dawson map-areas, Yukon Territory (106D, 116A, 116B and 116C (E 1/2)); Geological Survey of Canada, Memoir 364, 157 p.

Greig, C.J.
1989: Geology and geochronometry of the Eagle plutonic complex, Coquihalla area, southwestern British Columbia; M.Sc. thesis, University of British Columbia, Vancouver.

Grond, H.C., Churchill, S.J., Armstrong, R.L., Harakal, J.E., and Nixon, G.T.
1984: Late Cretaceous age of the Hutshi, Mount Nansen, and Carmacks groups, southwestern Yukon Territory and northwestern British Columbia; Canadian Journal of Earth Sciences, v. 21 p. 554-558.

Grove, E.W.
1971: Geology and mineral deposits of the Stewart area, British Columbia; British Columbia Department of Mines and Petroleum Resources, Bulletin 58, 219 p.

Harms, T.
1986: Structural and tectonic analysis of the Sylvester Allochthon, northern British Columbia: implications for paleogeography and accretion; Ph.D. dissertation, University of Arizona, 80 p.

Harrison, J.C.
1981: Petrology of the 'Ting Creek' alkalic intrusion, southeast Alaska; M.Sc. thesis, University of Toronto, 299 p.

Harrison, T.M., Armstrong, R.L., Naeser, C.W., and Harakal, J.E.
1979: Geochronology and thermal history of the Coast Plutonic Complex near Prince Rupert, British Columbia; Canadian Journal of Earth Sciences, v. 16, p. 400-410.

Hill, M.L.
1984: Geology of the Redcap Mountain area, Coast Plutonic Complex, British Columbia; Ph.D. thesis, Princeton University, 216 p.

Himmelberg, G.R., Loney, R.A., and Craig, J.T.
1986: Petrogenesis of the ultramafic complex at the Blashke Islands, southeastern Alaska; United States Geological Survey, Bulletin 1662, 14 p.

Hodgson, C.J., Bailes, R.J., and Verzosa, R.S.
1976: Cariboo-Bell; in Porphyry Deposits of the Canadian Cordillera, A. Sutherland Brown (ed.), Canadian Institute of Mining and Metallurgy, Special Volume 15, p. 388-396.

Holbek, P.M.
1988: Geology and mineralization of the Stikine Assemblage, Mess Creek area, northwestern British Columbia; M.Sc. thesis, University of British Columbia, Vancouver, 184 p.

Höy, T. and Kwong, Y.T.
1986: The Mount Grace carbonatite—an Nb and light rare earth element enriched marble of probable pyroclastic origin in the Shuswap Complex, southeastern British Columbia; Economic Geology, v. 81, p. 1374-1386.

Hudson, T.
1983: Calc-alkaline plutonism along the Pacific Rim of southern Alaska; in Circum-Pacific Plutonic Terranes, J.A. Roddick (ed.), Geological Society of America, Memoir 159, p. 159-169.

Hudson, T.L., Smith, J.G., and Elliott, R.L.
1979: Petrology, composition, and age of intrusive rocks associated with the Quartz Hill molybdenite deposit, southeastern Alaska; Canadian Journal of Earth Sciences, v. 16, p. 1805-1822.

Hunt, G.
1964: Chemical correlation of the Purcell igneous rocks; Bulletin of Canadian Petroleum Geology, v. 12, p. 544-555.

Hutchison, W.W.
1982: Geology of the Prince Rupert-Skeena map area, British Columbia; Geological Survey of Canada, Memoir 394, 116 p.

Hyndman, D.W.
1968: Petrology and structure of Nakusp map-area, British Columbia; Geological Survey of Canada, Bulletin 161, 95 p.

Ijewliw, O.J.
1987: Comparative mineralogy of three ultramafic diatremes in southeastern British Columbia: Cross, Blackfoot and HP (82J, 82G, 82N); in Geological Fieldwork 1986, British Columbia Ministry of Energy, Mines and Petroleum Resources, Paper 1987-1, p. 273-282.

Irvine, T.N.
1973: Bridget Cove Volcanics, Juneau area, Alaska: Possible parental magma of Alaskan-type ultramafic complexes; Carnegie Institution of Washington, Year Book 72, p. 478-491.
1974: Petrology of the Duke Island ultramafic complex, southeastern Alaska; Geological Society of America, Memoir 138, 240 p.

1975: Axelgold layered gabbro intrusion, McConnell Creek map-area, British Columbia; in Report of Activities, Part B, Geological Survey of Canada, Paper 75-1B, p. 81-88.
1976: Studies of Cordilleran gabbroic and ultramafic intrusions, British Columbia. Part 2: Alaskan-type ultramafic-gabbroic bodies in the Aiken Lake, McConnell Creek, and Toodoggone map-areas; in Report of Activities, Part A, Geological Survey of Canada, Paper 76-1A, p. 76-81.

Irving, E., Woodsworth, G.J., Wynne, P.J., and Morrison, A.
1985: Paleomagnetic evidence for displacement from the south of the Coast Plutonic Complex, British Columbia; Canadian Journal of Earth Sciences, v. 22, p. 584-598.

Isachsen, C.
1984: Geology, geochemistry, and geochronology of the Westcoast Crystalline Complex and related rocks, Vancouver Island, British Columbia; M.Sc. thesis, University of British Columbia, 144 p.

Isachsen, C., Armstrong, R.L., and Parrish, R.R.
1985: U-Pb, Rb-Sr, and K-Ar geochronometry of Vancouver Island igneous rocks; in Programme and Abstracts, A symposium on deep structure of southern Vancouver Island: Results of Lithoprobe Phase 1, Geological Association of Canada, Victoria Section, p. 21-22.

Ishihara, S.
1977: The magnetite-series and the ilmenite-series granitic rocks; Mining Geology, v. 27, p. 293-305.

Jacobson, B., Parrish, R.R., and Armstrong, R.L.
1980: Geochronology and petrology of the Tkope River batholith in the Saint Elias Mountains, northwestern British Columbia; in Current Research, Part B, Geological Survey of Canada, Paper 80-1B, p. 195-206.

Jefferson, C.W. and Parrish, R.R.
1989: Late Proterozoic stratigraphy, U-Pb zircon ages, and rift tectonics, Mackenzie Mountains, northwestern Canada; Canadian Journal of Earth Sciences, v. 26, p. 1784-1801.

Johan, Z., Le Bel, L., and McMillan, W.J.
1980: Evolution géologique et pétrologique des complexes granitoides fertiles: Étude comparative des batholites de La Caldera (Perou) et de Guichon Creek (Canada), deux exemples de plutonisme et mineralisations associées de la Cordillere peri-pacifique; in Mineralisations Liees aux Granitoides, Z. Johan (ed.), Bureau de Recherches Géologiques et Minières, Memoir 99, p. 21-70.

Kelemen, P.B. and Ghiorso, M.S.
1986: Assimilation of peridotite in zoned calc-alkaline plutonic complexes: evidence from the Big Jim complex, Washington Cascades; Contributions to Mineralogy and Petrology, v. 94, p. 12-28.

Kimura, E.T., Bysouth, G.D., and Drummond, A.D.
1976: Endako; in Porphyry Deposits of the Canadian Cordillera, A. Sutherland Brown (ed.), Canadian Institute of Mining and Metallurgy, Special Volume 15, p. 444-454.

Klepacki, D.W. and Wheeler, J.O.
1985: Stratigraphic and structural relations of the Milford, Kaslo and Slocan groups, Goat Range, Lardeau and Nelson map areas, British Columbia; in Current Research, Part A, Geological Survey of Canada, Paper 85-1A, p. 277-286.

Lambert, M.B.
1966: Geology of the Mount Brenner stock near Dawson City, Yukon Territory; M.Sc. thesis, University of British Columbia, 64 p.
1974: The Bennett Lake cauldron subsidence complex, British Columbia and Yukon Territory; Geological Survey of Canada, Bulletin 227, 213 p.

Lanphere, M.A.
1968: Sr-Rb-K and Sr isotopic relationships in ultramafic rocks, southeastern Alaska; Earth and Planetary Science Letters, v. 4, p. 185-190.

Le Couteur, P.C. and Tempelman-Kluit, D.J.
1976: Rb/Sr ages and a profile of initial Sr^{87}/Sr^{86} ratios for plutonic rocks across the Yukon Crystalline Terrane; Canadian Journal of Earth Sciences, v. 13, p. 319-330.

Leech, G.B., Lowdon, J.A., Stockwell, C.H., and Wanless, R.K.
1963: Age determinations and geological studies (including isotopic ages - Report 4); Geological Survey of Canada, Paper 63-17, 140 p.

Leitch, C.H.B., Dawson, K.M., and Godwin, C.I.
1989: Early Late Cretaceous-early Tertiary gold mineralization: a galena-lead isotope study of the Bridge River mining camp, southwestern British Columbia, Canada; Economic Geology, v. 84, p. 2226-2236.

Leitch, C.H.B. and Godwin, C.I.
1987: The Bralorne gold vein deposit: an update (92J/15); in Geological Fieldwork 1986, British Columbia Ministry of Energy, Mines and Petroleum Resources, Paper 1987-1, p. 35-38.

Leroux, J.
1980: Geothermal potential of the Coryell Intrusions, Granby River area, British Columbia; in Current Research, Part B, Geological Survey of Canada, Paper 80-1B, p. 213-215.

Lewis, T.J. and Woodsworth, G.J.
1981: Heat generation in the northern Hogem Batholith and nearby plutons of McConnell Creek map-area, British Columbia; in Current Research, Part B, Geological Survey of Canada, Paper 81-1B, p. 163-164.

Little, H.W.
1960: Nelson map-area, west half, British Columbia (82F W 1/2); Geological Survey of Canada, Memoir 308, 205 p.
1982: Geology of the Rossland-Trail map-area, British Columbia; Geological Survey of Canada, Paper 79-26, 38 p.

Loney, R.A. and Himmelberg, G.R.
1983: Structure and petrology of the La Perouse gabbro intrusion, Fairweather Range, southeastern Alaska; Journal of Petrology, v. 24, p. 377-423.

Lynch, G.V. and Pride, C.
1984: Evolution of a high-level, high-silica magma chamber: the Pattison pluton, Nisling Range alaskites, Yukon; Canadian Journal of Earth Sciences, v. 21, p. 407-414.

MacIntyre, D.G.
1985: Geology and mineral deposits of the Tahtsa Lake district, west central British Columbia; British Columbia Ministry of Energy, Mines and Petroleum Resources, Bulletin 75, 82.

MacKevett, E.M., Jr.
1963: Geology and ore deposits of the Bokan Mountain uranium-thorium area, southeastern Alaska; United States Geological Survey, Bulletin 1154, 125 p.
1978: Geologic map of the McCarthy quadrangle, Alaska; United States Geological Survey, Miscellaneous Investigations Series, Map I-1032.

Mader, U.K.
1987: The Aley carbonatite complex, northern Rocky Mountains, British Columbia; in Geological Fieldwork 1986, British Columbia Ministry of Energy, Mines and Petroleum Resources, Paper 1987-1, p. 283-288.

Marvin, R.F. and Cole, J.C.
1978: Radiometric ages: compilation A; United States Geological Survey, Isochron/West, no. 22, p. 3-14.

Marvin, R.F., Hearn, B.C., Jr., Mehnert, H.H., Naeser, C.W., Zartman, R.E., and Lindsey, D.A.
1980: Late Cretaceous-Paleocene -Eocene igneous activity in north-central Montana; Isochron/West, no. 29, p. 5-25.

Massey, N.W.D.
1986: Metchosin Igneous Complex, southern Vancouver Island: Ophiolite stratigraphy developed in an emergent island setting; Geology, v. 14, p. 602-605.

Mattinson, J.M.
1972: Ages of zircons from the Northern Cascade Mountains, Washington; Geological Society of America Bulletin, v. 83, p. 3769-3783.

McDougall, J.J.
1976: Catface; in Porphyry Deposits of the Canadian Cordillera, A. Sutherland Brown (ed.), Canadian Institute of Mining and Metallurgy, Special Volume 15, p. 299-310.

McKillop, G.R.
1973: Geology of southwestern Gambier Island, Howe Sound, British Columbia; B.Sc. thesis, University of British Columbia, 24 p.

McLeod, J.A., Vining, M., and McTaggart, K.C.
1976: Note on the age of the Giant Mascot ultramafic body, near Hope, B.C.; Canadian Journal of Earth Sciences, v. 13, p. 1152-1154.

McMillan, W.J.
1983: Fish Lake deposit (92O/5E); in Geology in British Columbia, 1976, British Columbia Ministry of Energy, Mines and Petroleum Resources, p. 84-103.
1985: Geology and ore deposits of the Highland Valley camp; Geological Association of Canada, Mineral Deposits Division, Field Guide and Reference Manual Series, No. 1, 121 p.

McMillan, W.J., Panteleyev, A., and Preto, V.A.
1975: Geochemical sampling, geology, and magnetics of the Kaketsa stock; in Geological Fieldwork 1974, British Columbia Department of Mines and Petroleum resources, p. 63-68.

Meade, H.D.
1977: Petrology and metal occurrences of the Takla Group and Hogem and Germansen batholiths, north-central British Columbia; Ph.D. thesis, University of Western Ontario, 355 p.

Misch, P.
1966: Tectonic evolution of the northern Cascades of Washington State; in A Symposium on the Tectonic History and Mineral Deposits of the Western Cordillera in British Columbia and Neighbouring Parts of the United States, Canadian Institute of Mining and Metallurgy, Special Volume 15, p. 101-148.
1977: Bedrock geology of the North Cascades: Field trip no. 1; in Geological Excursions in the Pacific Northwest, E.H. Brown and R.C. Ellis (ed.), Geological Society of America 1977 Annual Meeting, p. 1-62.

Monger, J.W.H.
1985: Structural evolution of the southwestern Intermontane Belt, Ashcroft and Hope map-areas, British Columbia; in Current Research, Part A, Geological Survey of Canada, Paper 85-1A, p. 349-358.

Monger, J.W.H. and McMillan, W.J.
1984: Bedrock geology of Ashcroft (92I) map area; Geological Survey of Canada, Open File 980.

Monger, J.W.H., Price, R.A., and Tempelman-Kluit, D.J.
1982: Tectonic accretion and the origin of two major metamorphic and plutonic welts in the Canadian Cordillera; Geology, v. 10, p. 70-75.

Moore, J.G.
1959: The quartz diorite boundary line in the western United States; Journal of Geology, v. 67, p. 198-210.

Morgan, J.T.
1976: Geology of the Ten Mile Creek syenite-pyroxenite pluton, Telegraph Creek, B.C.; B.Sc. thesis, University of British Columbia, 42 p.

Morrison, G.W., Godwin, C.L., and Armstrong, R.L.
1979: Interpretation of isotopic ages and $^{87}Sr/^{86}Sr$ initial ratios for plutonic rocks in the Whitehorse map area, Yukon; Canadian Journal of Earth Sciences, v. 16, p. 1988-1997.

Mortensen, J.K.
1983: Age and evolution of the Yukon-Tanana Terrane, southeastern Yukon Territory; Ph.D. thesis, University of California (Santa Barbara).
1986: U-Pb ages for granitic orthogneiss from western Yukon Territory: Selwyn Gneiss and Fiftymile Batholith revisited; in Current Research, Part B, Geological Survey of Canada, Paper 86-1B, p. 141-146.

Mortensen, J.K. and Jilson, G.A.
1985: Evolution of the Yukon-Tanana terrane: evidence from southeastern Yukon; Geology, v. 13, p. 806-809.

Mortensen, J.K., Montgomery, J.R., and Fillipone, J.
1987: U-Pb zircon, monazite and sphene ages for granitic orthogneiss of the Barkerville terrane, east-central British Columbia; Canadian Journal of Earth Sciences, v. 24, p. 1261-1266.

Mortimer, N.
1986: Late Triassic, arc-related, potassic igneous rocks in the North American Cordillera; Geology, v. 14, p. 70-75.

Mortimer, N., van der Heyden, P., Armstrong, R.L., and Harakal, J.
1990: U-Pb and K-Ar dates for the timing of magmatism and deformation in the Cache Creek terrane and Quesnellia, southern British Columbia; Canadian Journal of Earth Sciences, v. 27, p. 117-123.

Muller, J.E.
1967: Kluane Lake map-area (115G, 115F E 1/2), Yukon Territory; Geological Survey of Canada, Memoir 340, 137 p.
1977: Evolution of the Pacific margin, Vancouver Island, and adjacent regions; Canadian Journal of Earth Sciences, v. 14, p. 2062-2085.
1980: The Paleozoic Sicker Group of Vancouver Island, British Columbia; Geological Survey of Canada, Paper 79-30, 23 p.

Muller, J.E. and Carson, D.J.T.
1969: Geology and Mineral Deposits of Alberni map-area, British Columbia (92F); Geological Survey of Canada, Paper 68-50, 52 p.

Muller, J.E., Northcote, K.E., and Carlisle, D.
1974: Geology and mineral deposits of Alert Bay-Cape Scott map-area (92L-102I) Vancouver Island, British Columbia; Geological Survey of Canada, Paper 74-8, 77 p.

Murray, C.G.
1972: Petrologic studies of zoned ultramafic complexes in Venezuela and Alaska; Ph.D. thesis, Princeton University, 188 p.

Nixon, G.T. and Rublee, V.J.
1988: Alaskan-type ultramafic rocks in British Columbia: new concepts of the structure of the Tulameen complex; in Geological Fieldwork 1987, British Columbia Ministry of Energy, Mines and Petroleum Resources, Paper 1988-1, p. 281-294.

Norris, D.K. and Yorath, C.J.
1981: The North American plate from the Arctic Archipelago to the Romanzof Mountains; in The Ocean Basins and Margins, Volume 5: The Arctic Ocean, A.E.M Nairn, M. Churkin, Jr. and G.G. Stehli (ed.), Plenum Press, p. 37-103.

Northcote, K.E.
1969: Geology and geochronology of the Guichon Creek batholith; British Columbia Department of Mines and Petroleum Resources, Bulletin 56, 73 p.

Okulitch, A.V.
1985: Paleozoic plutonism in southeastern British Columbia; Canadian Journal of Earth Sciences, v. 22, p. 1409-1424.

Okulitch, A.V., Loveridge, W.D., and Sullivan, R.W.
1981: Preliminary radiometric analyses of zircons from the Mount Copeland syenite gneiss, Shuswap Metamorphic Complex, British Columbia; in Current Research, Part A, Geological Survey of Canada, Paper 81-1A, p. 33-36.

Okulitch, A.V., Wanless, R.K., and Loveridge, W.D.
1975: Devonian plutonism in south-central British Columbia; Canadian Journal of Earth Sciences, v. 12, p. 1760-1769.

Olade, M.A.
1976: Geochemical evolution of copper-bearing granitic rocks of the Guichon Creek Batholith, British Columbia, Canada; Canadian Journal of Earth Sciences, v. 13, p. 199-209.

Olade, M.A. and Goodfellow, W.D.
1978: Lithogeochemistry and hydrogeochemistry of uranium and associated elements in the Tombstone batholith, Yukon, Canada; in Proceedings of the 7th International Geochemical Symposium, Golden, Colorado, J.R. Watterson and T.K. Theobold (ed.), Association of Exploration Geochemists, p. 407-428.

Panteleyev, A.
1981: Berg porphyry copper-molybdenum deposit; British Columbia Ministry of Energy, Mines, and Petroleum Resources, Bulletin 66, 158 p.

Parkinson, D.L.
1985: U-Pb geochronometry and regional geology of the southern Okanagan Valley, British Columbia: the western boundary of a metamorphic core complex; M.Sc. thesis, University of British Columbia, 149 p.

Parrish, R.R.
1979: Geochronology and tectonics of the northern Wolverine Complex, British Columbia; Canadian Journal of Earth Sciences, v. 16, p. 1428-1438.
1984: Slocan Lake fault: a low angle fault zone bounding the Valhalla gneiss complex, Nelson map-area, southern British Columbia; in Current Research, Part A, Geological Survey of Canada, Paper 84-1A, p. 323-330.

Parrish, R.R. and Armstrong, R.L.
1987: The ca. 162 Ma Galena Bay stock and its relationship to the Columbia River fault zone, southeast British Columbia; in Radiogenic Age and Isotopic Studies, Report 1, Geological Survey of Canada, Paper 87-2, p. 25-32.

Parrish, R.R. and Bell, R.T.
1987: Age of the NOR breccia pipe, Wernecke Supergroup, Yukon Territory; in Radiogenic Age and Isotopic Studies, Report 1, Geological Survey of Canada, Paper 87-2, p. 39-42.

Parrish, R.R. and Scammell, R.J.
1988: The age of the Mount Coipeland Syenite Gneiss and its metamorphic zircons, Monashee Complex, southeastern British Columbia; in Radiogenic Age and Isotope Studies: Report 2, Geological Survey of Canada, Paper 88-2, p. 21-28.

Parrish, R.R. and Wheeler, J.O.
1983: A U-Pb zircon age from the Kuskanax batholith, southeastern British Columbia; Canadian Journal of Earth Sciences, v. 20, p. 1751-1756.

Parrish, R.R., Carr, S.D., and Parkinson, D.L.
1988: Eocene extensional tectonics and geochronology of the southern Omineca Belt, British Columbia and Washington; Tectonics, v. 7, p. 181-212.

Parrish, R.R., Heinrich, S., and Archibald, D.
1987: Age of the Ice River complex, southeastern British Columbia; in Radiogenic Age and Isotopic Studies, Report 1, Geological Survey of Canada, Paper 87-2, p. 33-37.

Pell, J.
1985: Carbonatites and related rocks in British Columbia (82L, 83D, 93I, 93N); in Geological Fieldwork 1984, British Columbia Ministry of Energy, Mines and Petroleum Resources, Paper 1985-1, p. 84-94.
1986: Nepheline syenite gneiss complexes in British Columbia (82M, N, 83D, 93I); in Geological Fieldwork 1985, British Columbia Ministry of Energy, Mines and Petroleum Resources, Paper 1986-1, p. 255-260.
1987a: Alkalic ultrabasic diatremes in British Columbia: petrology, geochronology and tectonic significance (82G, J, 83C, 94B); in Geological Fieldwork 1986, British Columbia Ministry of Energy, Mines and Petroleum Resources, Paper 1987-1, p. 259-267.
1987b: Alkaline ultrabasic rocks in British Columbia: carbonatites, nepheline syenites, kimberlites, ultramafic lamprophyres and related rocks; British Columbia Ministry of Energy, Mines and Petroleum Resources, Open File 1987-17, 109 p.

Petö, P.
1973: Petrochemical study of the Similkameen batholith, British Columbia; Geological Society of America Bulletin, v. 84, p. 3977-3983.

Petö, P. and Armstrong, R.L.
1976: Strontium isotope study of the composite batholith between Princeton and Okanagan Lake; Canadian Journal of Earth Sciences, v. 13, p. 1577-1583.

Pigage, L.C. and Anderson, R.G.
1986: The Anvil plutonic suite, Faro, Yukon Territory; Canadian Journal of Earth Sciences, v. 22, p. 1204-1216.

Pitcher, W.S.
1983: Granite type and tectonic environment; in Mountain Building Processes, K.J. Hsu (ed.), Academic Press, London, p. 19-40.

Poole, W.H.
1956: Geology of the Cassiar Mountains in the vicinity of the Yukon-British Columbia boundary; Ph.D. thesis, Princeton University, 247 p.

Potter, C.J.
1983: Geology of the Bridge River complex, southern Shulaps Range, British Columbia: a record of Mesozoic convergent tectonics; Ph.D. thesis, University of Washington, 192 p.

Preto, V.A.
1972: Geology of Copper Mountain; British Columbia Department of Mines and Petroleum Resources, Bulletin 59, 87 p.

Preto, V.A., Osatenko, M.J., McMillan, W.J., and Armstrong, R.L.
1979: Isotopic dates and strontium isotopic ratios for plutonic and volcanic rocks in the Quesnel Trough and Nicola Belt, south-central British Columbia; Canadian Journal of Earth Sciences, v. 16, p. 1658-1672.

Price, R.A.
1965: Flathead map-area, British Columbia and Alberta; Geological Survey of Canada, Memoir 336, 221 p.

Price, R.A., Monger, J.W.H., and Roddick, J.A.
1985: Cordilleran cross-section: Calgary to Vancouver; in Field Guides to Geology and Mineral Deposits in the Southern Canadian Cordillera, D.J. Tempelman-Kluit (ed.), Geological Society of America, Cordilleran Section, Vancouver, p. 3-1 to 3-85.

Ray, G.E.
1986: Gold associated with a regionally developed mid-Tertiary plutonic event in the Harrison Lake area, southwestern British Columbia (92G/9, 92H/3, 4, 5, 6, 12); in Geological Fieldwork 1985, British Columbia Ministry of Energy, Mines and Petroleum Resources, Paper 1986-1, p. 95-98.

Read, P.B.
1973: Petrology and structure of Poplar Creek map-area, British Columbia; Geological Survey of Canada, Bulletin 193, 144 p.

Read, P.B. and Brown, R.L.
1981: Columbia River Fault zone: southeast margin of the Shuswap and Monashee Complex, southern British Columbia; Canadian Journal of Earth Sciences, v. 18, p. 1127-1145.

Reesor, J.E.
1958: Dewar Creek map-area with special emphasis on the White Creek Batholith, British Columbia; Geological Survey of Canada, Memoir 292, 78 p.
1965: Structural evolution and plutonism in the Valhalla Gneiss Complex, British Columbia; Geological Survey of Canada, Bulletin 129, 128 p.

Rice, H.M.A.
1937: Cranbrook map-area, British Columbia; Geological Survey of Canada, Memoir 207, 67 p.

Richards, T.A.
1971: Plutonic rocks between Hope, B.C. and the 49th parallel; Ph.D. thesis, University of British Columbia, 178 p.

Richards, T.A. and McTaggart, K.C.
1976: Granitic rocks of the southern Coast Plutonic Complex and northern Cascades of British Columbia; Geological Society of America Bulletin, v. 87, p. 935-953.

Richter, D.H., Lanphyre, M.A., and Matson, N.A. Jr.
1975: Granitic plutonism and metamorphism, eastern Alaska Range, Alaska; Geological Society of America Bulletin, v. 86, p. 819-829.

Ridgway, W.R.
1973: The petrology of Nisling and Ruby range volcanic and plutonic rocks, Yukon Territory; B.Sc. thesis, University of British Columbia, 53 p.

Roddick, J.A.
1965: Vancouver North, Coquitlam, and Pitt Lake map-areas, British Columbia; Geological Survey of Canada, Memoir 335, 276 p.
1970: Douglas Channel-Hecate Strait map-area, British Columbia; Geological Survey of Canada, Paper 70-41, 56 p.
1983: Geophysical review and composition of the Coast Plutonic Complex, south of latitude 55°N; in Circum-Pacific Plutonic Terranes, J.A. Roddick (ed.), Geological Society of America, Memoir 159, p. 195-211.

Roddick, J.A. and Armstrong, J.E.
1959: Relict dikes in the Coast Mountains near Vancouver, B.C.; Journal of Geology, v. 67, p. 603-613.

Roddick, J.A. and Hutchison, W.W.
1974: Setting of the Coast Plutonic Complex, British Columbia; Pacific Geology, v. 8, p. 91-108.

Roddick, J.C. and Farrar, E.
1971: High initial argon ratios in hornblendes; Earth and Planetary Science Letters, v. 12, p. 208-214.

Roots, C.F.
1987: Regional tectonic setting and evolution of the Late Proterozoic Mount Harper Volcanic Complex, Ogilvie Mountains, Yukon; Ph.D. thesis, Carleton University, Ottawa, 219 p.

Rusmore, M.E.
1985: Geology and tectonic significance of the Upper Triassic Cadwallader Group and its bounding faults, southwestern British Columbia; Ph.D. thesis, University of Washington, 174 p.

Ryan, B.D.
1973: Structural geology and Rb-Sr geochronology of the Anarchist Mountain area, south-central British Columbia; Ph.D. thesis, University of British Columbia, 256 p.

Ryan, B.D. and Blenkinsop, J.
1971: Geology and geochronology of the Hellroaring Creek Stock, British Columbia; Canadian Journal of Earth Sciences, v. 8, p. 85-95.

Saint-Andre, B. de, Lancelot, J.R., and Collot, B.
1983: U-Pb geochronology of the Bokan Mountain peralkaline granite, southeastern Alaska; Canadian Journal of Earth Sciences, v. 20, p. 236-245.

Sangster, D.F.
1969: The contact metasomatic magnetite deposits of southwestern British Columbia; Geological Survey of Canada, Bulletin 172, 85 p.

Schofield, S.J.
1915: Geology of Cranbrook map-area, British Columbia; Geological Survey of Canada, Memoir 76, 245 p.

Shaw, D.A.
1980a: Structural setting of the Adamant pluton, northern Selkirk Mountains, British Columbia; Ph.D. thesis, Carleton University.
1980b: A concordant Uranium-lead age for zircons in the Adamant Pluton, British Columbia; in Current Research, Part C, Geological Survey of Canada, Paper 80-1C, p. 243-246.

Simony P.S.
1979: Pre-Carboniferous basement near Trail, British Columbia; Canadian Journal of Earth Sciences, v. 16, p. 1-11.

Sisson, V.B.
1985: Contact metamorphism and fluid evolution associated with the intrusion of the Ponder Pluton, Coast Plutonic Complex, British Columbia, Canada; Ph.D. thesis, Princeton University, 345 p.

Smith, C.B., Colgan, E.A., Hawthorne, J.B., and Hutchinson, G.
1988: Emplacement age of the Cross kimberlite, southeastern British Columbia, by the Rb-Sr phlogopite method; Canadian Journal of Earth Sciences, v. 790-792.

Smith, J.G.
1973: A Tertiary lamprophyre dike province in southeastern Alaska; Canadian Journal of Earth Sciences, v. 10, p. 408-420.
1977: Geology of the Ketchikan D-1 and Bradfield Canal A-1 quadrangles, southeastern Alaska; United States Geological Survey, Bulletin 1245, 49 p.

Smith, T.E., Riddle, C., and Jackson, T.A.
1979: Chemical variation within the Coast Plutonic Complex of British Columbia between lat 53° and 55°N; Geological Society of America Bulletin, v. 90, p. 346-356.

Souther, J.G.
1971: Geology and mineral deposits of Tulsequah map-area, British Columbia; Geological Survey of Canada, Memoir 362, 84 p.
1972: Telegraph Creek map-area, British Columbia; Geological Survey of Canada, Paper 71-44, 38 p.
1977: Volcanism and tectonic environments in the Canadian Cordillera - a second look; in Volcanic Regimes in Canada, W.R.A. Baragar and others (ed.), Geological Association of Canada, Special Paper 16, p. 3-24.
1986: The western Anahim Belt: root zone of a peralkaline magma system; Canadian Journal of Earth Sciences, v. 23, p. 895-908.

Steininger, R.C.
1985: Geology of the Kitsault molybdenum deposit, British Columbia; Economic Geology, v. 80, p. 57-71.

Stephens, G.C.
1972: The geology of the Salal Creek Pluton, southwestern British Columbia; Ph.D. thesis, Lehigh University, 177 p.

Stevens, R.D., Delabio, R.N., and Lachance, G.R.
1982a: Age determinations and geological studies: K-Ar isotopic ages, Report 15; Geological Survey of Canada, Paper 81-2, 56 p.
1982b: Age determinations and geological studies: K-Ar isotopic ages, Report 16; Geological Survey of Canada, Paper 82-2, 56 p.

St. Louis, R.M., Nesbitt, B.E., and Morton, R.D.
1986: Geochemistry of platinum-group elements in the Tulameen Ultramafic Complex, southern British Columbia; Economic Geology, v. 81, p. 961-973.

Stuart, R.A.
1960: Geology of the Kemano-Tahtsa area; British Columbia Department of Mines and Petroleum Resources, Bulletin 42, 52 p.

Sturrock, D.L., Armstrong, R.L., and Maxwell, R.B.
1980: Age and Sr isotopic composition of the Pyroxenite Creek Ultramafic Complex, southwestern Yukon territory: an Alaskan-type ultramafic intrusion; in Current Research, Part B, Geological Survey of Canada, Paper 80-1B, p. 185-188.

Sutherland Brown, A.
1960: Geology of the Rocher Deboule Range; British Columbia Department of Mines and Petroleum Resources, Bulletin 43, 78 p.
1968: Geology of the Queen Charlotte Islands, British Columbia; British Columbia Department of Mines and Petroleum Resources, Bulletin 54, 226 p.

Tabor, R.W., Zartman, R.E., and Frizzell, V.A., Jr.
1987: Possible tectonostratigraphic terranes in the North Cascades crystalline core, Washington; in Selected Papers on the Geology of Washington, J.E. Schuster (ed.), Washington Division of Geology and Earth Resources, Bulletin 77, p. 107-127.

Taylor, G.C. and Stott, D.F.
1973: Tuchodi Lakes map-area, British Columbia (94K); Geological Survey of Canada, Memoir 373, 37 p.

Taylor, H.P.
1967: The zoned ultramafic complexes of southeastern Alaska; in Ultramafic and Related Rocks, P.J. Wylie (ed.), New York, Wiley, p. 96-121.

Tempelman-Kluit, D.J.
1969: A re-examination of pseudoleucite from Spotted Fawn Creek, west-central Yukon; Canadian Journal of Earth Sciences, v. 6, p. 55-62.
1970: Stratigraphy and structure of the "Keno Hill Quartzite" in Tombstone River - upper Klondike River map-areas, Yukon Territory (116 B/7, B/8); Geological Survey of Canada, Bulletin 180, 102 p.
1974: Reconnaissance geology of Aishihik Lake, Snag and part of Stewart River map-areas, west-central Yukon; Geological Survey of Canada, Paper 73-41, 97 p.
in press : Geology of Carmacks and Laberge map areas, Yukon Territory; Geological Survey of Canada, Memoir.

Tempelman-Kluit, D. and Parkinson, D.
1986: Extension across the Eocene Okanagan crustal shear zone in southern British Columbia; Geology, v. 14, p. 318-321.

Tempelman-Kluit, D.J. and Wanless, R.K.
1975: Potassium-argon age determinations of metamorphic and plutonic rocks in the Yukon Crystalline Terrane; Canadian Journal of Earth Sciences, v. 12, p. 895-1909.
1980: Zircon ages for the Pelly Gneiss and Klotassin granodiorite in western Yukon; Canadian Journal of Earth Sciences, v. 17, p. 297-306.

Thompson, R.I., Mercier, E., and Roots, C.
1987: Extension and its influence on Canadian Cordilleran passive-margin evolution; in Continental Extensional Tectonics, M.P. Coward, J.F. Dewey and P.L. Hancock (ed.), Geological Society, Special Publication 28, p. 409-417.

Thompson, T.B., Pierson, J.R., and Lyttle, T.
1982: Petrology and petrogenesis of the Bokan Granite Complex, southeastern Alaska; Geological Society of America Bulletin, v. 93, p. 898-908.

Tipper, H.W.
1969: Mesozoic and Cenozoic geology of the northeast part of Mount Waddington map-area (92N), Coast District, British Columbia; Geological Survey of Canada, Paper 68-33, 103 p.

Tipper, H.W. and Richards, T.A.
1976: Jurassic stratigraphy and history of north-central British Columbia; Geological Survey of Canada, Bulletin 270, 73 p.

van der Heyden, P.
1982: Tectonic and stratigraphic relations between the Coast Plutonic Complex and Intermontane Belt, west-central Whitesail Lake map area, British Columbia; M.Sc. thesis, University of British Columbia, 172 p.
1989: U-Pb and K-Ar geochronometry of the Coast Plutonic Complex, 53°N-54°N, and implications for the Insular-Intermontane superterrane boundary, British Columbia; Ph.D. thesis, University of British Columbia, 392 p.

Vining, M.R.
1977: The Spuzzum Pluton northwest of Hope, B.C.; M.Sc. thesis, University of British Columbia, 147 p.

Wanless, R.K., Loveridge, W.D., and Mursky, G.
1968: A geochronological study of the White Creek batholith, southeastern British Columbia; Canadian Journal of Earth Sciences, v. 5, p. 375-386.

Wanless, R.K., Stevens, R.D., Lachance, G.R., and Delabio, R.N.
1979: Age determinations and geological studies: K-Ar isotopic ages, Report 14; Geological Survey of Canada, Paper 79-2, 67 p.

Werner, L.J.
1978: Metamorphic terrane, northern Coast Mountains west of Atlin Lake, British Columbia; in Current Research, Part A, Geological Survey of Canada, Paper 78-1A, p. 69-70.

White, A.J.R. and Chappell, B.W.
1983: Granitoid types and their distribution in the Lachlan Fold Belt, southeastern Australia; in Circum-Pacific Plutonic Terranes, J.A. Roddick (ed.), Geological Society of America, Memoir 159, p. 21-34.

White, A.J.R., Clemens, J.D., Holloway, J.R., Silver, L.T., Chappell, B.W., and Wall, V.J.
1986: S-type granites and their probable absence in southwestern North America; Geology, v. 14, p. 115-118.

White, W.H., Harakal, J.E., and Carter, N.M.
1968: Potassium-argon ages of some ore deposits in British Columbia; Canadian Institute of Mining and Metallurgy Bulletin, v. 61, p. 1326-1334.

White, W.H., Sinclair, A.J., Harakal, J.E., and Dawson, K.M.
1970: Potassium-argon ages of Topley Intrusions near Endako, British Columbia; Canadian Journal of Earth Sciences, v. 7, p. 1172-1178.

Wolfhard, M.R.
1976: Fish Lake; in Porphyry Deposits of the Canadian Cordillera, A. Sutherland Brown (ed.), Canadian Institute of Mining and Metallurgy, Special Volume 15, p. 317-322.

Wong, R.H., Godwin, C.I., and McTaggart, K.C.
1985: Geology, K/Ar dates, and associated sulphide mineralization of Wrede Creek zoned ultramafic complex (94D/9E); in Geology in British Columbia, 1977-1981, British Columbia Ministry of Energy, Mines and Petroleum Resources, p. 148-155.

Wood, D.H. and Armstrong, R.L.
1982: Geology, chemistry, and geochronometry of the Cretaceous South Fork Volcanics, Yukon Territory; in Current Research, Part A, Geological Survey of Canada, Paper 82-1A, p. 309-316.

Woodsworth, G.J.
1976: Plutonic rocks of McConnell Creek (94D west half) and Aiken Lake (94C east half) map-areas, British Columbia; in Report of Activities, Part A, Geological Survey of Canada, Paper 76-1A, p. 69-73.
1979: Metamorphism, deformation, and plutonism in the Mount Raleigh pendant, Coast Mountains, British Columbia; Geological Survey of Canada, Bulletin 295, 58 p.

Woodsworth, G.J. and Orchard, M.J.
1985: Upper Paleozoic to lower Mesozoic strata and their conodonts, western Coast Plutonic Complex, British Columbia; Canadian Journal of Earth Sciences, v. 22, p. 1329-1344.

Woodsworth, G.J., Loveridge, W.D., Parrish, R.R., and Sullivan, R.W.
1983: Uranium-lead dates from the Central Gneiss Complex and Ecstall pluton, Prince Rupert map area, British Columbia; Canadian Journal of Earth Sciences, v. 20, p. 1475-1483.

Young, I.F.
1981: Structure of the western margin of the Queen Charlotte Basin, British Columbia; M.Sc. thesis, University of British Columbia, Vancouver, 380 p.

Zartman, R.E., Peterman, Z.E., Obradovich, J.D., Gallego, M.D., and Bishop, D.T.
1982: Age of the Crossport C sill near Eastport, Idaho; Society of Economic Geologists' Coeur d'Alene Field Conference, Idaho, 1977, R.R. Reid and G.A. Williams (ed.), Idaho Bureau of Mines and Geology, Bulletin 24, p. 61-69.

Zen, E-an
1985: Implications of magmatic epidote-bearing plutons on crustal evolution in the accreted terranes of northwestern North America; Geology, v. 13, p. 266-269.

Zen, E-an and Hammarstrom, J.M.
1984a: Mineralogy and a petrogenetic model for the tonalite pluton at Bushy Point, Revillagigedo Island, Ketchikan 1° x 2° quadrangle, southeast Alaska; in The United States Geological Survey in Alaska: Accomplishments During 1982, United States Geological Survey, Circular 939, p. 118-123.
1984b: Magmatic epidote and its petrologic significance; Geology, v. 12, p. 515-518.

Authors' addresses

G.J. Woodsworth
R.G. Anderson
L.C. Struik
P. van der Heyden
Cordilleran Division
Geological Survey of Canada
100 West Pender Street
Vancouver, British Columbia
V6B 1R8

R.L. Armstrong (deceased)
Department of Geological Sciences
University of British Columbia
Vancouver, British Columbia
V6T 1Z4

ADDENDA

Polaris Ultramatic Suite. The following publications deal with the geology of Alaskan - type, mafic - ultramafic rocks in British Columbia:

Hammack, J.L., Nixon, G.T., Wong, R.H., and Paterson, W.P.E.
1990: Geology and noble metal geochemistry of the Wrede Creek ultramafic complex, north-central British Columbia (94D/9); in Geological Fieldwork 1989, British Columbia Ministry of Energy, Mines and Petroleum Resources, Paper 1990-1, p. 405-415.

Nixon, G.T., Ash, C.H., Connelly, J.N., and Case, G.
1989: Alaskan-type mafic-ultramafic rocks in British Columbia: The Gnat Lakes, Hickman, and Menard Creek complexes; in Geological Fieldwork 1988, British Columbia Ministry of Energy, Mines and Petroleum Resources, Paper 1989-1, p. 429-442.

Nixon, G.T., Hammack, J.L., Connelly, J.N., Case, G., and Paterson, W.P.E.
1990: Geology and noble metal geochemistry of the Polaris ultramafic complex, north-central British Columbia (94C/5, 12); in Geological Fieldwork 1989, British Columbia Ministry of Energy, Mines and Petroleum Resources, Paper 1990-1, p. 387-403.

Nixon, G.T., Hammack, J.L., and Paterson, W.P.E.
1990: Geology and noble metal geochemistry of the Johanson Lake mafic-ultramafic complex, north-central British Columbia (94D/9); in Geological Fieldwork 1989, British Columbia Minisytry of Energy, Mines and Petroleum Resources, Paper 1990-1, p. 417-424.

Copper Mountain Suite. In northwestern Stikina, south of Iskut River recent unpublished data (R.G. Anderson, M.L. Bevier, J.K. Mortensen and R. Kirkham) suggest the presence of two, coeval, northwest-trending belts of Early Jurassic plutons: a calc-alkaline belt on the southwest and an alkaline belt (including the Galore Creek Pluton) on the northeast.

Nelson Suite. The age of Ag-rich veins in the Slocan area is controversial. The possibility of an Eocene age is suggested by new $^{40}Ar/^{39}Ar$ data (G. Beaudain and D. Sangster, pers. comm., 1990).

Three Sisters Suite. K-Ar and U-Pb dating (Holbek, 1988; M.L. Bevier and R.G. Anderson, unpublished data) indicate that the Three Sisters Suite west of the Bowser Basin extends in a south trending belt from the Stikine Arch.

Coast Plutonic Complex. Foliated granodiorite in Burroughs Bay, southeastern Alaska (Mississippian; Gehrels et al., 1991a) and in the Scotia River area southeast of Price Rupert (Middle Devonian; Gareau, 1991) are recently discovered Paleozoic plutons in the Coast Belt.

Gareau, S.A.
1991: The Scotia-Quaal metamorphic belt: a distinct assemblage with pre-early Late Cretaceous deformational and metamorphic history, Coast Plutonic Complex, B.C.; Canadian Journal of Earth Sciences, v. 28; no 6.

Gehrels, G.E., McClelland, W.C., Samson, S.D., Jackson, J.L., and Patchett, P.J.
1991a: U-Pb geochronology of two pre-Tertiary plutons in the Coast Mountains batholith near Ketchikan, southeastern Alaska; Canadian Journal of Earth Sciences, v. 28, no. 6.

Coast Plutonic Complex. Age dating suggests that the Texas Creek Granodiorite forms part of a belt of northwest-trending, calc-alkaline, Early Jurassic plutons west of Bowser Basin (M.L. Bevier and R.G. Anderson, unpublished data). Early Jurassic orthogneiss has been reported from the Scotia River area where it intrudes rocks tentatively correlated with the Nisling Terrane (Gareau, 1991). Data from the northern Coast Plutonic Complex show an evolution in pluton shape and fabric in eastwardly younging Late Cretaceous and Paleogene plutons (Gehrels et al., 1991; R.G. Anderson and M.L. Bevier, unpublished data). Plutonic styles change from highly deformed, Late Cretaceous tonalitic sheets to less elongate Paleocene bodies to large, massive, Eocene plutons which make up most of the complex at northern latitudes.

Gehrels, G.E., McClelland, W.C., Samson, S.D., Patchett, P.J. and Brew, D.A.
1991b: U-Pb geochronology of Late Cretaceous and early Tertiary plutons in the northern Coast Mountains batholith; Canadian Journal of Earth Sciences, v. 28, no. 6.

Regional trends and tectonic classification.
ENd is between 2.4 and 6.8 for the few analyzed Stikinian and Wrangellian plutons (Samson et al., 1989, 1990).

Samson, S.D., McClelland, W.C., Patchett, P.J., Gehrels, G.E. and Anderson, R.G.
1989: Evidence from neodymium isotopes for mantle contributions to Phanerozoic crustal genesis in the Canadian Cordillera; Nature, v. 337, no. 6209, p. 705-709.

Samson, S.D., Patchett, P.J., Gehrels, G.E. and Anderson, R.G.
1990: Nd and Sr isotopic characterization of the Wrangellia terrane and implications for crustal growth of the Canadian Cordillera; Journal of Geology, v. 98, p. 749-762.

Printed in Canada

Chapter 16
METAMORPHISM

Summary
Introduction
Parameters of metamorphism
Facies and zones
   Petrogenetic grid
   Metamorphism and organic maturity
Foreland Belt and adjacent Interior Platform
   Interior Platform
   Foreland Belt
Omineca Belt
   Precambrian
     Monashee Complex
     Okanagan Complex
   Paleozoic (pre-Late Mississippian)
     Kootenay Terrane
   Paleozoic (Permian)
   Late Triassic (230-214 Ma)
     Cariboo, Dorsey(?), Nisling and Nisutlin terranes
   Middle Jurassic (200-155 Ma)
     Transition between the Omineca and Foreland belts
     Slide Mountain Terrane
     Southern Omineca Belt
       Monashee Complex
       Kootenay Arc
       Cariboo Mountains
       Summary
     Northern Omineca Belt
       Swannell Ranges
       Sifton Ranges
       Finlay Ranges
       Cassiar Batholith region
       Horseranch Range
   Relationship of deformation to metamorphism in Cassiar and Omineca mountains
   Northern Omineca Belt, Yukon Territory

    Late Jurassic-Early Cretaceous (155-138 Ma)
    Early to mid-Cretaceous
    Late Paleocene-mid-Eocene (65-45 Ma)
Intermontane Belt
    Cache Creek Assemblage
    Asitka Assemblage
    Nicola and Stuhini assemblages
    Hazelton Assemblage
    Bowser Lake Assemblage
    Cretaceous assemblages
        Tatla Lake Metamorphic Complex
    Tertiary and younger rocks
Coast Belt
    Northern Coast Belt
    Central Coast Belt
        West of Work Channel
        East of Work Channel
        East margin of the Coast Belt
    Southern Coast Belt
        Pendants west of Lillooet Lake and Lillooet River
    Cascade core zone
    Northeast margin of the southern Coast Belt
        Bridge River region
Insular Belt
    Wrangellia
    Alexander Terrane
    Pacific Rim and Crescent terranes
Conclusions
References

Chapter 16

METAMORPHISM

H.J. Greenwood, G.J. Woodsworth, P.B. Read, E.D. Ghent and C.A. Evenchick

SUMMARY

All pre-Miocene rocks in the Canadian Cordillera have been regionally metamorphosed. The highest grade rocks, reflecting deep burial and high temperatures, form core zones in the Coast and Omineca belts whereas lower grade rocks, suggestive of burial metamorphism, characterize most of the Insular, Intermontane, and Foreland belts. Regional metamorphism reached its peak in the Omineca Belt in Middle Jurassic time and in the Coast Belt in Late Cretaceous time. Both episodes correlate with periods of intense crustal contraction and thickening and were followed by great and rapid uplift.

Except for metamorphic culminations in the Deserters Range east of the Northern Rocky Mountain Trench and a local area east of the Southern Rocky Mountain Trench most of the regional metamorphism in the Foreland Belt is of low-grade burial type. Precambrian rocks are commonly in greenschist facies, Paleozoic and some Mesozoic strata are mainly in prehnite-pumpellyite facies, and most Mesozoic strata are in zeolite facies. Although there is a general westward increase in coal rank with increasing stratigraphic burial, several east- to northeast-trending belts of anomalous organic maturation parallel present geothermal gradients. These belts may be related to faults in the Precambrian basement.

Rocks in the Omineca Belt received their main metamorphic imprint in Middle Jurassic time, presumably as a result of collision between ancestral North America and the Intermontane Superterrane. Locally, there is evidence for Precambrian metamorphism in the Monashee Complex, pre-Late Mississippian metamorphism in the Kootenay Arc, Late Permian(?) high-pressure and low-temperature metamorphism in accreted terranes in the southern part of Yukon Territory, and Late(?) Triassic to Early Jurassic metamorphism in accreted terranes of southwestern Yukon Territory. Widespread mid-Cretaceous metamorphism accompanied emplacement of voluminous granitic rocks in Columbia, Kaska, and Selwyn mountains. High heat flow, high-level plutonism, and great uplift occurred in Eocene time. In the northern Cordillera allochthons in the Omineca Belt show evidence for pre-emplacement deformation and high-pressure, low-temperature metamorphism. Geothermometric and geobarometric studies on mineral assemblages in the southern Omineca Belt indicate burial to depths in excess of 20 km in Middle Jurassic time and rapid uplift to less than 14 km of cover in Early Cretaceous time. The Eocene event was associated with extension and tectonic denudation.

Characteristic metamorphic assemblages in the Intermontane Belt are of low-temperature and low-pressure facies. Exceptions include an amphibolitic core complex near Tatla Lake and local, high-pressure, blueschist- and jadeite-bearing rocks in the Cache Creek Group of the central and northern Cordillera, which indicate subduction related events. Zeolite-facies rocks are well developed in Mesozoic volcanics and in several places grade into prehnite-pumpellyite facies at depth. In several localities a zonation of zeolite to prehnite-pumpellyite to subgreenschist facies near granitic plutons can be demonstrated. Zeolite facies are common in clastic Jurassic and Cretaceous sediments of the Intermontane Belt.

In some areas in northern British Columbia, upper Paleozoic strata of greenschist facies are overlain by Upper Triassic rocks of prehnite-pumpellyite facies with no evidence of a transition between the two. More work is needed to show whether or not this relationship is a local or regional phenomenon. High-pressure low-temperature metamorphism of the Cache Creek Terrane in the Pinchi Lake area and the development of a transitional blueschist assemblage in the Ashcroft area took place in Late Triassic time. Regional metamorphism in the thick Mesozoic volcanic successions was, in many cases, contemporaneous with volcanism and associated plutonism. Burial metamorphism in Mesozoic and volcanic assemblages probably took place mainly during the time of deposition. Metamorphism in the Tatla Lake Metamorphic Complex took place in mid- to Late Cretaceous time in the core zone and in Eocene time in the mylonitic carapace.

Metamorphic rocks in the Coast Belt occur as ortho- and paragneiss and less metamorphosed screens and pendants surrounded and intruded by plutonic rocks which make up at least 80% of the belt. In a transect across the most deeply exhumed part of the Coast Plutonic Complex at the latitude of Prince Rupert, metamorphic grade increases eastward from greenschist facies to amphibolite facies near Work Channel Lineament. East of the lineament in metamorphic rocks of the Central Gneiss Complex, amphibolite facies dominate and granulite is present locally. Peak metamorphism with temperatures of about 600°C and pressures of about 6 kbar (6x10^5 kPa) occurred

---

Greenwood, H. J., Woodsworth, G. J., Read, P. B., Ghent, E. D., and Evenchick, C. A.
1991: Metamorphism, Chapter 16 in Geology of the Cordilleran Orogen in Canada, H. Gabrielse and C. J. Yorath (ed.); Geological Survey of Canada, Geology of Canada, no. 4, p. 533-570 (also Geological Society of America, The Geology of North America, v. G-2)

in Late Cretaceous time. Rapid uplift took place in early Tertiary time. Along the eastern margin of the Coast Plutonic Complex rocks metamorphosed to amphibolite facies in Late Jurassic and Early Cretaceous time were thrust eastward over subgreenschist facies strata of the Intermontane Belt.

In the southern part of the Coast Belt pendants and screens are predominantly of prehnite-pumpellyite facies in the west and are separated from prehnite-pumpellyite and greenschist facies rocks in the east by a central belt of amphibolitic rocks extending north from the core zone of the Cascade Mountains. Blueschist-facies minerals are locally present in the area west of Harrison Lake and south of the Fraser River in the western Cascades.

Evidence for pre-Cretaceous regional metamorphism is found locally in the southern part of the Coast Belt. Early and mid-Cretaceous high-pressure and low-temperature metamorphism west of Harrison Lake and in the western Cascades has been correlated with amphibolite-facies Barrovian metamorphism in the core of the Cascades.

The main metamorphic events on Vancouver Island in the southern part of the Insular Belt involved Permian to Triassic greenschist metamorphism, an Early to Middle Jurassic prehnite-pumpellyite burial metamorphism, and a Late Cretaceous to Eocene burial metamorphism. On the Queen Charlotte Islands burial metamorphism may be as young as Pliocene. On Vancouver Island, in the Queen Charlotte Islands, and in the southern Alexander Terrane metamorphic grade and intensity of plutonism increase to the southwest. Late Cretaceous high-pressure and low-temperature metamorphism affected the Pacific Rim Terrane.

Wrangellian strata in southwestern Yukon Territory underwent Mesozoic and possibly Miocene burial metamorphism, mainly reaching prehnite-pumpellyite facies. In the adjacent Alexander Terrane low greenschist facies developed, at least locally, in pre-Late Pennsylvanian time.

INTRODUCTION

In the nearly two decades since publication of the first metamorphic map of the Canadian Cordillera (Monger and Hutchison, 1971), metamorphic petrology has advanced greatly through the increased number of reliable measurements of thermodynamic properties of minerals, and through the development of internally consistent thermodynamic data bases and realistic solution models for many metamorphic minerals. Such theoretical and experimental studies have led to the development of geologically useful geothermometers and geobarometers. Moreover, the widespread availability of high-precision electron microbeam analyses has led to the widespread application of such techniques. The last two decades also have seen the development of tools such as vitrinite reflectance, illite crystallinity, and conodont alteration indices which are particularly useful in the study of subgreenschist facies metamorphism.

Application of modern techniques to the Canadian Cordillera has shown that almost all strata older than Miocene have been metamorphosed to some extent. It has long been known that the greatest concentrations of amphibolite-facies rocks are in and near the two main belts of plutonic rocks: the Omineca and Coast belts. Monger and Hutchison (1971), in their pioneering description of metamorphism in the Canadian Cordillera, recognized that much of the strata in the Intermontane and Insular belts is in subgreenschist facies, but most of the Foreland Belt was thought to be unmetamorphosed. Perhaps the main contribution of the present review and the accompanying Metamorphic Map of the Canadian Cordillera (Map 1714A, in pocket) is the recognition of widespread subgreenschist facies metamorphism in the Foreland and Intermontane belts.

In this review, the facies and zone classification and the petrogenetic grid on which they are based are described. Metamorphism in the Cordillera is discussed sequentially from east to west, using the five morphogeological belts as a broad framework (indeed, contrasting metamorphic styles is one of the criteria for delineating the belts; see Chapter 2). The emphasis is on the southern Cordillera, particularly the Omineca Belt, reflecting the great number of detailed studies done in that region in the last twenty years and the fact that data on metamorphism in much of the rest of the Cordillera are still meagre or absent.

PARAMETERS OF METAMORPHISM
Facies and zones

A key factor in choosing the boundaries of facies and zones of regional metamorphism is the need to use data from a wide range of investigations, only some of which emphasize metamorphism. In addition, an attempt has been made to place the boundaries in positions that are compatible with the best available experimental and thermodynamic data on critical minerals. Depending upon the quality of the metamorphic information, subgreenschist, blueschist, greenschist, amphibolite, granulite, and eclogite facies have been subdivided into a total of 22 zones or combinations of zones described below and shown in Figure 16.1.

Unmetamorphosed

No evidence of regionally developed secondary minerals, and little or no sign of thermal maturation of organic materials. Specifically, vitrinite reflectance (R_{max}) is less than 0.38%, and coal is of lignite ASTM (American Society for Testing and Materials) rank.

Cryptic

No secondary minerals or textural reconstitution reported; vitrinite reflectance more than 0.38% or Conodont Alteration Indices (CAI) greater than 1.0. This category exists only because of a lack of metamorphic data.

Zeolite

One or more of analcime, laumontite, stilbite, wairakite, or heulandite-clinoptilolite. Not texturally reconstituted, the diagnostic minerals occur as cement agents, fillings of vesicles and veins, and as replacements of volcanic glass and feldspars. The rocks commonly appear to be unmetamorphosed and undeformed.

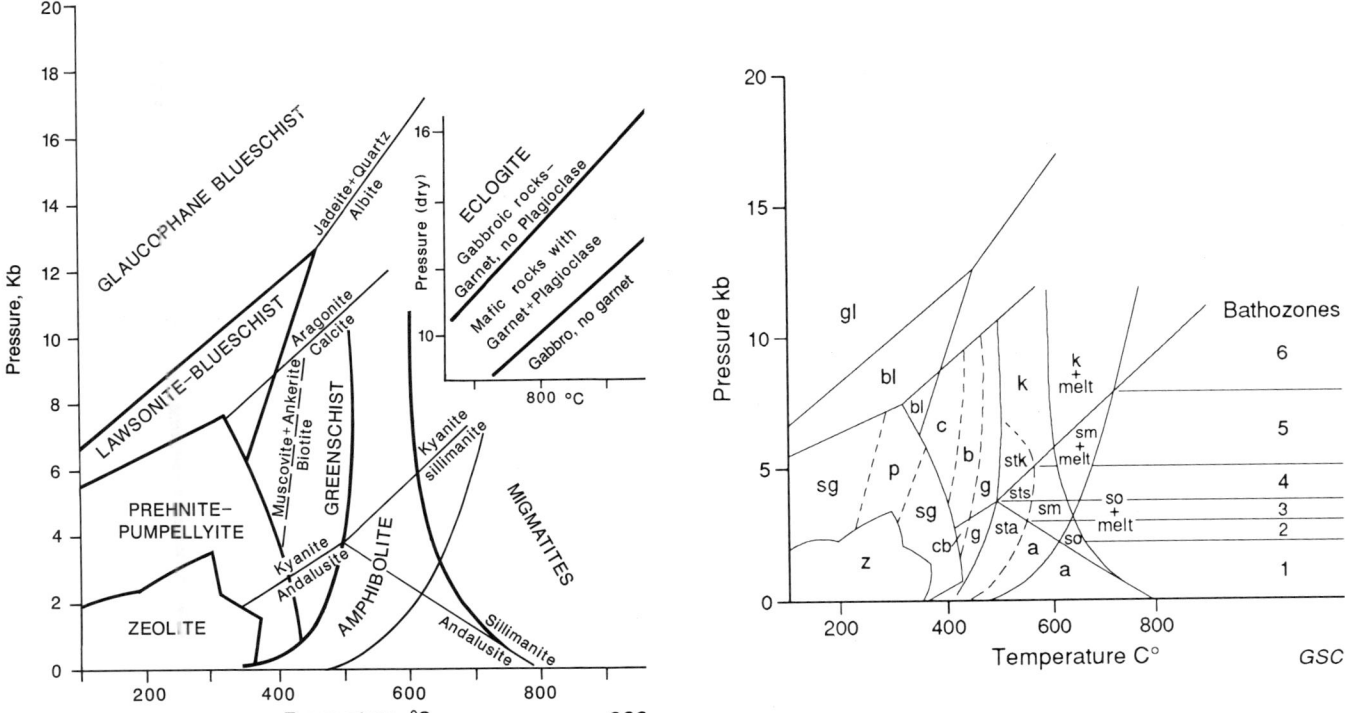

Figure 16.1. Pressure-temperature diagrams with the distribution of the major metamorphic facies, and metamorphic zones and bathozones shown on the Metamorphic Map of the Canadian Cordillera (Map 1714A, in pocket). Curves computed with the programs of Perkins et al. (1986), using the UBCDATA thermodynamic database (Berman and Brown, 1985; Berman, 1988). The stability fields of mafic granulite and eclogite are from Ringwood (1972) and Ringwood and Green (1966) for basaltic rocks in the absence of H2O vapour. A. Major metamorphic facies boundaries. B. Metamorphic zones and bathozones.

Prehnite-pumpellyite

Prehnite and/or pumpellyite; epidote and chlorite common; no lawsonite, zeolites, or actinolite. No significant textural reconstitution or foliation. Volcanogenic rocks commonly retain relict igneous compositions of pyroxene, hornblende, and plagioclase phenocrysts; matrix minerals do not.

Subgreenschist

(Undifferentiated zeolite and prehnite-pumpellyite). Combinations of zeolites, prehnite, pumpellyite, chlorite, white mica, epidote, albite, and rarely stilpnomelane. Lawsonite and actinolite absent. No textural reconstitution, relict igneous compositions of phenocrysts common.

Lawsonite blueschist

Lawsonite, with combinations of prehnite, pumpellyite, chlorite, epidote/zoisite, and either aragonite or calcite. Glaucophane is absent. Albite is common in the lower-pressure part of the facies. The assemblage jadeite + quartz is possible at higher pressures according to Figure 16.1 but is rare in glaucophane-free rocks of the Canadian Cordillera. These rocks are not texturally reconstituted and may be more abundant than reported because of difficulty in identifying lawsonite.

Glaucophane blueschist

Glaucophane, with combinations of jadeitic pyroxene, quartz, stilpnomelane, and aragonite. Albite is absent; the rocks commonly are foliated.

Greenschist facies

Chlorite zone

Combinations of chlorite, albite, epidote-zoisite, actinolite, and locally stilpnomelane, Mn-garnet, or chloritoid. Biotite and oligoclase are absent. Foliation and textural reconstitution are common; relict igneous phenocrysts are rare.

Biotite zone

Biotite plus minerals of the chlorite zone. Foliation and textural reconstitution are widespread.

Garnet zone

Almandine-rich garnet and hornblende in addition to the minerals of the chlorite and biotite zones. Plagioclase is normally albite.

Chlorite and biotite zones, undivided. Data insufficient for subdivision of rocks containing biotite, chlorite, and albite.

Greenschist, undivided

Some of garnet, biotite, chlorite, actinolite, and albite; insufficient detail for further subdivision. Epidote common, with white mica.

Amphibolite facies

Staurolite-andalusite zone. Staurolite + andalusite, possibly with cordierite; including combinations of almandine-rich garnet, biotite, muscovite, amphibole, and oligoclase or more calcic plagioclase. Generally defined in pelitic rocks. Most rocks are foliated and/or schistose.

Andalusite zone

All minerals of the staurolite-andalusite zone except staurolite.

Staurolite-kyanite zone

Staurolite + kyanite, chlorite present; rarely chloritoid or cordierite; other minerals as in rocks of the staurolite-andalusite zone.

Kyanite zone

All minerals of the staurolite-kyanite zone except staurolite.

Staurolite-sillimanite zone

Staurolite + sillimanite, with combinations of almandine-rich garnet, biotite, muscovite, amphibole, and oligoclase or more calcic plagioclase. Generally defined in pelitic rocks. Most rocks are foliated and/or schistose.

Sillimanite-kyanite zone

Kyanite + sillimanite (or fibrolite); with combinations of almandine-rich garnet, biotite, muscovite, amphibole, and oligoclase or more calcic plagioclase; staurolite absent.

Sillimanite-muscovite zone

Sillimanite (or fibrolite); with combinations of almandine-rich garnet, biotite, muscovite, amphibole, and oligoclase or more calcic plagioclase. Andalusite, kyanite, or staurolite absent or present only as metastable relicts. In high grade part partial melting and anatexis produce pegmatite and granite.

K-feldspar zone

K-feldspar + sillimanite + quartz. Muscovite + quartz do not coexist. Combinations of biotite, amphibole, and garnet are common. Common in areas of schist and gneiss, associated with abundant pegmatite and partial melting products.

Amphibolite, undifferentiated

Rocks with plagioclase at least as calcic as oligoclase, in combination with some of biotite, muscovite, chlorite, amphibole, garnet, epidote. Isolated occurrences of kyanite, sillimanite, or staurolite.

Granulite facies

Isolated occurrences of anhydrous mafic granulites. Combinations of clinopyroxene, orthopyroxene, hornblende, pyrope-almandine garnet, plagioclase, perthite, and antiperthite.

Eclogite facies

Type III eclogites, consisting of pyrope-rich garnet, omphacitic clinopyroxene; may contain orthopyroxene and/or quartz; plagioclase absent. Typically overprinted by glaucophane blueschist and chlorite zone assemblages.

Combustion metamorphism

Bocannes produced by natural combustion of organic material. Most are of Recent age.

Discussion

The division between diagenesis and metamorphism is to some extent arbitrary and the boundary depends significantly on the bulk composition of the rocks. The processes of replacement and cementation in the miogeoclinal rocks of the Foreland Belt are usually included in diagenesis by stratigraphers and sedimentologists but, under similar physical conditions, as deduced from organic maturity data, the volcanics and volcaniclastic sediments of the western Cordillera contain zeolites, prehnite, or pumpellyite, and are treated as low-grade metamorphic rocks by metamorphic petrologists. Volcanic rocks intercalated with sediments having lignite ASTM rank coals, with carbonaceous sediments having R_{max} <0.38%, and with rocks having conodont alteration indices (CAI) of 1 are unaltered. Hence these organic maturity parameters have been chosen as the dividing line between diagenesis and metamorphism (Table 16.2).

Successful subdivision of the subgreenschist facies requires data on the mineralogy of mafic volcanics and greywackes, as these are most likely to contain the critical mineral assemblages. Because these rocks are the least studied of any in the Canadian Cordillera, the position of the boundary between the prehnite-pumpellyite facies and the chlorite zone of the greenschist facies is commonly uncertain. The pumpellyite-actinolite facies has not been mapped separately even though it exists locally in the Chilliwack, Sylvester, Bridge River, Sicker, and some other groups. A few local occurrences of barroisite-garnet-epidote-biotite-albite-quartz indicate transitional greenschist-blueschist facies, but these have been included in the greenschist facies.

With the exception of the chlorite zone, pelitic and semipelitic bulk compositions contain the critical minerals most suited to placing metamorphic rocks in the appropriate zones of the greenschist or amphibolite facies. In the amphibolite facies, use of the combination of aluminosilicate polymorphs and the disappearance of staurolite relative to the andalusite, kyanite, and sillimanite isograds yields zones with geobarometric significance (Fig. 16.1B; Table 16.1). Some of the experimental work on the stability of staurolite-bearing assemblages is of uncertain applicability, and staurolite-bearing assemblages are restricted to a limited range of bulk composition. Nonetheless, staurolite-aluminosilicate zones, and staurolite-free aluminosilicate

Table 16.1. Symbols used in figures

	Metamorphic Facies or Zone		Metamorphic Minerals/Rocks
um	Unmetamorphosed	Ab	albite
y	Cryptic	Act	actinolite
z	Zeolite facies	And	andalusite
p	Prehnite pumpellyite facies	Ank	ankerite
sg	Subgreenschist facies	Ant	antigorite
bl	Lawsonite blueschist	ASK	stable aluminosilicate polymorph
gl	Glaucophane blueschist	Bio	biotite
c	Chlorite zone	Ca	calcite
b	Biotite zone	Cord	cordierite
cb	Chlorite and biotite zones	Diop	diopside
g	Garnet zone	Dol	dolomite
gs	Greenschist facies	Ep	epidote
st	Staurolite zone	Fo	forsterite
sta	Staurolite-andalusite zone	Gar	garnet
a	Andalusite zone	Hbl	hornblende
stk	Staurolite-kyanite zone	Ksp	K-feldspar
k	Kyanite zone	melt	Water-saturated silicate melt at equilibrium with solids shown
sts	Staurolite-sillimanite zone		
sk	Sillimanite-kyanite zone	Ms	muscovite
s	Sillimanite zone	Phl	phlogopite
sm	Sillimanite-muscovite zone	Plag	plagioclase
so	Sillimanite-K-feldspar zone	Q	quartz
am	Amphibolite facies	Sill	sillimanite
■	Granulite facies	Trem	tremolite
▲	Eclogite facies	Ky	kyanite
✻	Combustion metamorphism	Chl	chlorite
		St	staurolite

zones can be distinguished in well-studied areas in the Omineca and Foreland belts. Broad, 'default' categories accommodate inadequately studied rocks which cannot be placed more precisely than subgreenschist, greenschist, or amphibolite facies.

Petrogenetic grid

The facies divisions and mineral zones shown on the Metamorphic Map (Map 1714A, in pocket) have been plotted, as far as possible, on a P-T petrogenetic grid (Fig. 16.1A; see also Table 16.1). No attempt has been made to account for the effects of variations in the composition of the metamorphic pore fluid or in the solid solutions involved. In spite of this simplification, the petrogenetic grid provides a reasonably consistent framework in which to regard mineral assemblages, zones, and facies. Precise estimates of pressure and temperature are restricted to rocks that have been studied thoroughly through application of solid-solution geothermometry and geobarometry, but approximate ranges of pressure and temperature for zones and facies can be estimated from the petrogenetic grid.

The database for equilibrium univariant reactions in the subgreenschist and blueschist part of the petrogenetic grid (Fig. 16.2) comes from Berman and Brown (1985) and Berman (1988). Exceptions to the computed placement of equilibrium boundaries are the positions of the glaucophane schist facies adapted from Carman and Gilbert (1983) and the prehnite-pumpellyite field taken from Liou et al. (1985).

Figure 16.3 shows the P-T fields of the higher temperature assemblages and a few of the equilibria on which boundaries have been drawn. The boundary between greenschist and amphibolite (Fig. 16.1A) has been placed at a group of curves that all pass close to the kyanite-sillimanite-andalusite triple point. These reactions are chemically unrelated but can serve to define the boundary in rocks of very different bulk composition. The upper stability of antigorite + diopside marks the boundary in ultramafic and low-silica magnesian carbonates, the upper stability of pyrophyllite in highly aluminous but low-potassium rocks, and the disappearance of chlorite with appearance of oligoclase in the presence of albite, actinolite, and epidote in mafic rocks (Apted and Liou, 1984). Field studies of pelitic rocks show that this boundary coincides with the appearance of staurolite. Within the biotite zone of the greenschist facies the equilibrium between dolomite, talc, quartz, tremolite, and calcite is a convenient 'half-way marker', even though the composition of the CO_2-H_2O pore-fluid changes continuously along the equilibrium line. This equilibrium, which can be observed in calcareous sediments and ultramafites, lies about 50°C above the appearance of biotite in pelitic rocks.

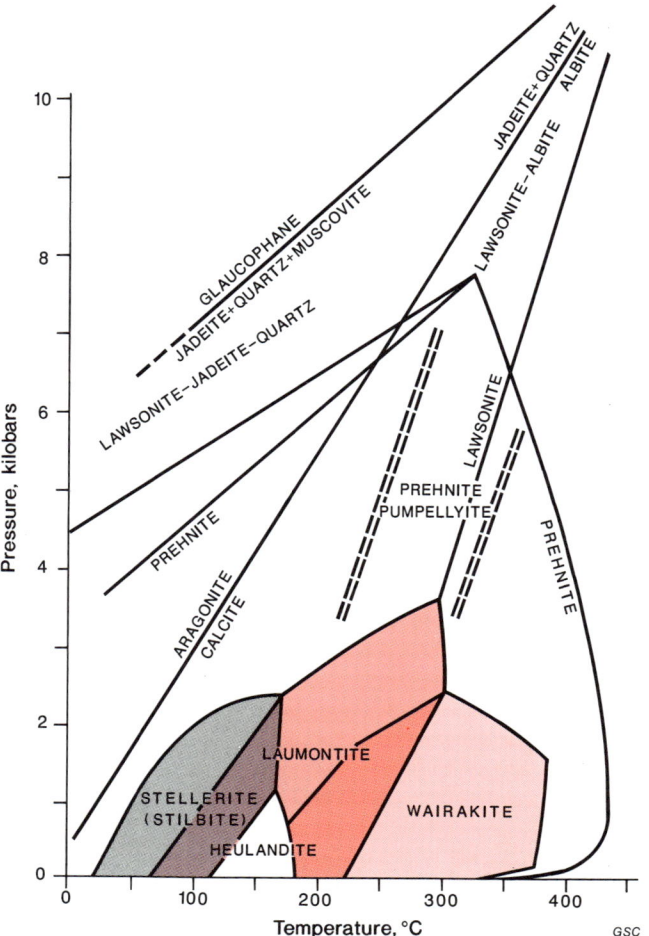

Figure 16.2. Equilibrium phase boundaries for low-temperature equilibria defining some of the metamorphic facies. Computed as noted in Figure 16.1 and in the text. The dashed boundaries for prehnite + pumpellyite are approximate, as their underlying data have not yet been incorporated into the database.

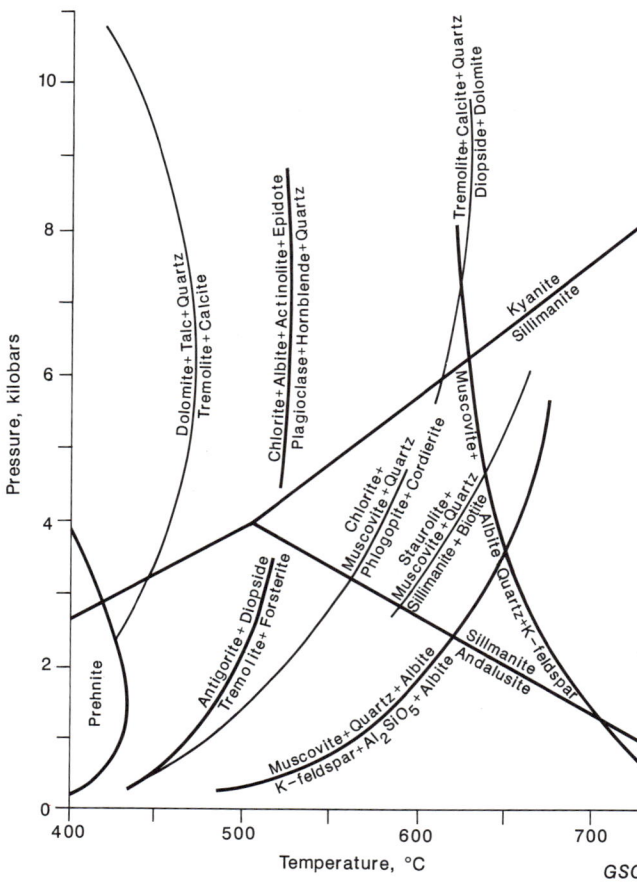

Figure 16.3. Pressure stabilities of equilibria used to define greenschist and amphibolite facies. Computed as noted in Figure 16.1 and in the text.

Within the amphibolite facies, the upper stability of clinochlore in the presence of muscovite and quartz provides another intermediate reference point. It lies in the middle of the staurolite-aluminosilicate zones (sta, sts, stk), and other 'chlorite-out' reactions in pelitic rocks are taken to be in the same general neighbourhood (Fig. 16.1B). This equilibrium coincidentally plots at almost the same position as the trace of the univariant equilibrium between tremolite, calcite, quartz, diopside, and dolomite, so that the same conditions of intermediate amphibolite facies may be recognized in both pelitic and carbonate-rich rocks. The upper stability of staurolite with quartz and muscovite is plotted in Figure 16.3 because it seems to bear a closer relationship to the deduced conditions of staurolite breakdown in pelitic rocks than the simpler staurolite + quartz breakdown. The first appearance of abundant anatectic pegmatite has been modelled in Figure 16.3 by the melting curves in the muscovite + quartz + albite + vapour system (Tracy, 1978), represented here by a single curve showing the equilibrium between muscovite, quartz, albite, K-feldspar, H_2O, and liquid.

All curves are subject to displacement solid-solution effects within the reacting minerals, and through the presence of extra components of the fluid phase. In addition, if a fluid phase was not present during metamorphism, the activity of the volatile components may have been less than unity, causing large displacements of the equilibria. Nonetheless, it is believed that the general agreement and internal consistency found upon applying the diagrams is an indication that, for the mapping of metamorphism on a Cordilleran-wide scale, such a representation is adequate.

Metamorphism and organic maturity

Estimates of the intensive variables of metamorphism such as pressure and temperature in low-grade metamorphic rocks are commonly made from mineral equilibria (Zen and Thompson, 1974, and references therein), but more recently a number of methods have been applied which depend upon the degree of conversion of material from a more metastable to a less metastable state. Although these methods are rate dependent, a number of studies, both in the experimental laboratory and in areas undergoing active metamorphism, have shown that temperature and not pressure or fluid composition is the dominant intensive variable. The methods that have been developed include: illite crystallinity (Kubler, 1967);

conodont colour or conodont alteration index (CAI) (Epstein et al., 1977); palynomorph colour or temperature alteration index (TAI) (Staplin, 1977); catagenesis of solid organic matter (Bostick, 1979), as expressed by ASTM coal rank (Hacquebard and Donaldson, 1974), vitrinite reflectance (R_{max}) (Teichmuller and Teichmuller, 1979), and alteration of solid organic matter in graptolites (F. Goodarzi and B.S. Norford, 1985).

In the southern Alberta plains, ASTM rank or vitrinite reflectance data are sufficient to contour with a 0.05% contour interval (Metamorphic Map 1714A, in pocket). The isorank and isoreflectance contours are based on data taken from samples within 100 m of the present surface in the Alberta plains. In the Foreland Belt, isoreflectance contours of 0.5, 0.7, 1.1, 1.5, and 2.0% correspond closely to major coal rank boundaries and permit contouring of coal rank and vitrinite reflectance data together. The irregular and broader contour interval permits use of data collected at depth which are projected up dip to the surface. The contours, like isograds, indicate the distribution of 'metamorphic zones' between contour intervals, and the direction of change in 'metamorphic grade' within low-grade metamorphic rocks.

Correlation of the zeolite, prehnite-pumpellyite, and greenschist facies with illite crystallinity, clay mineral polytypes, vitrinite reflectance, conodont color alteration, and palynomorph temperature alteration indices is complex (Kisch, 1987). Illite crystallinity studies, such as those of Hoffman and Hower (1979) in Montana, have been tried in a few areas in the Canadian Cordillera (e.g. McMechan and Price, 1982; Hutcheon et al., 1980). In view of the limited region covered and the difficulties in the interpretation of illite crystallinity (Hutcheon et al., 1980), however, these studies have not been used in this compilation. Large areas of the Canadian Cordillera yield organic maturity data in the form of TAI, CAI, ASTM coal rank and vitrinite reflectance data. Among the studies done, England's (1984) is the most regional in its treatment of the western Canadian sedimentary basin in the Rocky Mountain Foothills and plains of Alberta south of latitude 52°N.

Some of the several thousand organic maturity values collected during the present compilation come from rocks with suitable bulk compositions for determining the metamorphic facies. They result in correlations of maturity and metamorphism which are consistent with and expand upon those from rare studies such as Kisch (1981). Because of the limited database, the correlation of metamorphic subdivisions and organic maturity parameters given in Table 16.2 is tentative compared with the correlations among the various organic maturity parameters which have been taken largely from Teichmuller and Teichmuller (1979), and Higgins et al. (1985).

FORELAND BELT AND ADJACENT INTERIOR PLATFORM

The rocks in the Foothills part of the Foreland Belt and those of the adjacent Interior Platform do not show textural reconstitution or foliation, and mineralogical changes are commonly thought to be diagenetic. In the western part of the Foreland Belt, regional metamorphism appears to be stratigraphically controlled and related to depth of burial; pelitic rocks typically have a slaty cleavage. Precambrian rocks generally reach chlorite or biotite zones except for two small areas, one in the Deserters Range along the Northern Rocky Mountain Trench and one east of the Malton Gneiss along the Southern Rocky Mountain Trench. Rocks in these areas reach amphibolite facies and have metamorphic styles similar to those of the adjacent Omineca Belt with which they are discussed later in the chapter.

Interior Platform

Locally, Cretaceous and lower Tertiary sediments in the Alberta Syncline contain zeolites. Heulandite-clinoptilolite cements some sandstones of the Paskapoo and Bearpaw formations and Wapiti Group to as far north as 55°N (Carrigy and Mellon, 1964; Carrigy, 1971; Kramers and Mellon, 1972). Coal seams, intercalated with the zeolite-bearing sandstones, have vitrinite R_{max} between 0.45 and 0.55%, within the range of values for zeolite-facies rocks in the western Canadian Cordillera. Most Cretaceous and lower Tertiary sediments of the Alberta Syncline have carbonate, illite, montmorillonite, chlorite, and interstratified clay minerals (typical of but not diagnostic of the zeolite facies), and coal rank values ranging from sub-bituminous C to high volatile bituminous B (well within the range of zeolite facies rocks; Table 16.2).

North of the Peace River Arch and east of Mackenzie River, sparse organic maturity data from the Cretaceous and Lower Tertiary cover rocks indicate the presence of lignite, which characterizes unmetamorphosed rocks. However, coal rank and vitrinite reflectance values in the cover rise abruptly towards the Mackenzie delta and reach values typical of zeolite-facies rocks. In a sandstone in the Moose Channel Formation, analcime occurs both as replacements of quartz and feldspar and as cement (Holmes and Oliver, 1973); nearby coals are high-volatile bituminous A with R_{max} values between 0.60 and 0.76%.

The Paleozoic basement emerges from beneath the cover in a broad area around Fort Simpson. Northwest of Fort Simpson, vitrinite R_{max} values between 2.83% and 3.09% come from Devonian coal seams about 455 m below the surface (Gunther and Meijer-Drees, 1977) and imply prehnite-pumpellyite facies metamorphism of Devonian rocks now at surface. Vitrinite reflectance values decline eastward to those typical of the zeolite facies. The Paleozoic basement reemerges along and east of the Mackenzie River downstream from Fort Norman but metamorphic and organic maturity data, except for a bitumen with R_{max} of 0.96% (Norris and Cameron, 1986), are absent. This large region is thought to have undergone cryptic metamorphism.

Until recently the general view of metamorphism in the Alberta plains and adjacent Foreland Belt was that coal rank generally increased from east to west with increasing stratigraphic depth of burial (Hacquebard and Donaldson, 1974). England's study (1984) has shown that the westward increase is irregular with several east- to northeast-trending, regional perturbations in organic maturity contours. England (1984) also showed that the distribution of organic maturity values reflects Late Eocene processes. These perturbations reflect in part the present geothermal gradients in the Alberta Basin. Lam and Jones (1984) showed that present geothermal gradients are influenced by groundwater flow related to present topography, faults in the Precambrian basement, and in the amount of heat supplied from the Precambrian basement in which areas of high basement heat flow coincide with aeromagnetic highs. The partly defined, northeast-trending maturity high through Peace River lies on the south flank of a northeast-trending aeromagnetic high through Peace River. The northeasterly oriented maturity maximum ending south of Lesser Slave Lake corresponds with a number of small aeromagnetic highs between Hinton and Swan Hills south of Lesser Slave lake, and the maturity high through Edmonton coincides with an aeromagnetic high in the basement. The eastward-trending bulge in the isoreflectance contours through and north of a line through Medicine Hat and Lethbridge coincides in position and orientation with a buried Precambrian rift outlined by Kanasewich et al. (1969). The close correspondence between basement structure, aeromagnetic highs, and organic maturity implies that the properties of the Precambrian basement were more important controls of organic maturity than was present topography. A similarly irregular distribution of organic maturity values may have been present in the Foreland Belt, but the pattern has been complicated by faulting and folding.

Foreland Belt

Except for the few metamorphic culminations along the southwestern edge of the Foreland Belt, the low-grade regional metamorphism affecting most of the belt is of burial type (Kalkreuth and McMechan, 1984). Precambrian rocks commonly are in greenschist facies, and younger rocks in subgreenschist facies. In small areas, organic maturity contours in Mesozoic and Paleozoic rocks are subparallel with stratigraphy, but in broad regions they cut across stratigraphy.

The few studies of volcaniclastic Mesozoic sediments, such as those of the Blairmore Group, show that they are in zeolite facies (Ghent and Miller, 1974; Pearce, 1970).

Table 16.2. Correlations among coal rank, fixed carbon (F.C.), Btu/lb, vitrinite reflectance (RMAX), TAI, CAI, and metamorphic facies or zones

Rank	F.C.	$\overline{R}_o$MAX	TAI	CAI	Metamorphic Facies or Zones
				7.0	///////////// Biotite and garnet zones
				6.0	///////////// chlorite zone
				5.0	
Meta-anthracite				5.0	/////////////
Anthracite	98	4.00	4.0	4.0	
Semi-anthractie	92	3.00			prehnite-pumpellyite facies
Low-volatile bituminous	86	2.05	3.75		
Medium-volatile bituminous	78	1.50	3.5	3.5 3.0	/////////////
High-volatile A bituminous	69	1.10	3.0	2.0	
Btu/lb* High-volatile B bituminous	14 000	0.71	2.75		
High-volatile C bituminous	13 000	0.57	1.5		zeolite facies
Sub-bituminous A	11 500	0.47	2.5		
Sub-bituminous B	10 500				
Sub-bituminous C	9 500	0.43			
Lignite A	8 300	0.38	2.25	1.0	/////////////
Lignite B	6 300				unmetamorphosed

* Moist, mineral matter-free B.t.u./lb

Heulandite-clinoptilolite, laumontite, and analcime form cement and partly replace albitized detrital feldspar grains. Although zeolites are unknown from the remainder of the Foreland Belt, most of the R_{max} values from Mesozoic rocks lie between 0.38% and 2.05%, and on this basis most have been assigned to the zeolite facies. Exceptions to this generalization include a large area of Mesozoic rocks in the Foothills west of Fort Nelson and Dawson Creek. There, sparse vitrinite reflectance data from the Lower Cretaceous Buckinghorse Formation range from 2.20% to 2.53% (Foscolos et al., 1976); anthracite occurs in the Upper Triassic Pardonet Formation (D.F. Stott, pers. comm., 1985), and CAIs in the Triassic rocks of the Liard Canyon range from 3 to 4.5. In this region, Triassic and some Lower Cretaceous strata are probably in prehnite-pumpellyite facies. Near Banff, the lower part of the Upper Jurassic to Lower Cretaceous Mist Mountain Formation of the Cascade coal basin has R_{max} values greater than 2.0% (Gibson, 1985), which implies that these rocks also are in the prehnite-pumpellyite facies.

In the Paleozoic rocks of the Foreland Belt, CAI values range from 3.5 to 5, typical of the prehnite-pumpellyite facies elsewhere in the Canadian Cordillera. Most of the Paleozoic rocks of the Foreland Belt are probably in the prehnite-pumpellyite facies. Because of the unsuitable bulk compositions, however, prehnite- and pumpellyite-bearing assemblages are unreported except for prehnite-bearing metadiabase dykes that intrude the McKay Group and Glenogle Formation about 30 km north of Fernie (J.A. Mott, 1989). Throughout the Foreland Belt, there are scattered occurrences of mafic metavolcanics with a chlorite-epidote-albite-calcite-quartz assemblage - for example, the Ordovician Skoki Formation near the Aley carbonatite northeast of Williston Lake (Mader, 1987). This actinolite-free assemblage is typical of mafic metavolcanics in the belt and suggests the rocks probably belong to the prehnite-pumpellyite facies. However, organic maturity data outline at least three areas of low organic maturity which indicate the presence of zeolite facies. The largest of these lies west of the confluence of the South Nahanni and Liard rivers, where the Mississippian Mattson Formation and surrounding strata have R_{max} values (Read at al., 1991a), ASTM coal ranks (Hacquebard and Barss, 1957), and TAI values (J. Utting, pers. comm., 1983) in ranges associated with zeolite facies rocks. Southward in the Liard Canyon, Triassic and Paleozoic rocks have CAIs in the range 3 to 5, and northward along and north of the South Nahanni River, Paleozoic rocks have CAIs of 5, which indicate that the rocks are in prehnite-pumpellyite facies. Farther northwest, rocks of the Upper Devonian Imperial Formation with R_{max} in the range of 0.6 to 0.9% and TAI of 3 outcrop along the northeastern edge of the Mackenzie Mountains (Braman, 1981; Read et al., 1991a) and indicate that the upper part of the Paleozoic succession is in zeolite facies. Sparse data show that vitrinite reflectance rises rapidly southwards to R_{max} greater than 2, and the Paleozoic rocks probably reach prehnite-pumpellyite facies towards the core of the Mackenzie Mountains. The smallest known area of presumed zeolite-facies rocks in the Foreland Belt lies west to northwest of Hinton, where thrust slices carrying Upper Devonian and Mississippian rocks contain conodonts with CAIs in the range 1.5 to 3.5 (Higgins et al., 1985).

Precambrian rocks of the Foreland Belt typically fall in the chlorite and biotite zones in the southern part of the Canadian Cordillera and in the prehnite-pumpellyite, chlorite, and locally biotite zone in the northern part of the Cordillera. Within areas studied, the metamorphic grade increases with stratigraphic depth of burial. In Yukon Territory, the Neruokpuk Formation, which forms the Precambrian core of the British Mountains, attains the chlorite zone. In Coal Creek Dome of the Southern Ogilvie Mountains, the lowest sequence of phyllite and metasiltstone, correlated with the Wernecke Supergroup, reaches the chlorite (and locally biotite?) zone. The overlying carbonate section of the Pinguicula Group, 2500 m thick, underlies prehnite-pumpellyite-bearing volcanics of the Harper rift assemblage (Roots, 1987). Surrounding the dome, scattered conodont localities with CAIs in the range of 4 to 6 indicate that the Paleozoic succession is probably in the prehnite-pumpellyite facies. To the east in central Yukon Territory, the Wernecke Supergroup ranges from biotite zone in the lowest stratigraphic levels to chlorite zone in the highest (Delaney, 1985). Southeastwards, scattered descriptions of the Upper Proterozoic Rapitan Group indicate that it is probably in prehnite-pumpellyite facies.

OMINECA BELT

Rocks in the Omineca Belt, which include the Shuswap and Wolverine complexes, received their main metamorphic imprint in Middle Jurassic time. Locally there is evidence for Precambrian metamorphism in the Monashee Complex, pre-Late Mississippian metamorphism in the Kootenay Arc, Late Permian(?) high-pressure and low-temperature metamorphism in accreted terranes in the southern part of Yukon Territory, and Late(?) Triassic to Early Jurassic metamorphism in accreted terranes of southwestern Yukon Territory. Widespread mid-Cretaceous metamorphism accompanied the emplacement of voluminous granitic rocks in the Columbia, Omineca, Cassiar, Pelly, and Selwyn mountains. High heat flow, high-level plutonism, and great uplift occurred in Eocene time.

Detailed studies have concentrated on Barrovian events that affected the Omineca Belt in the Cariboo and Columbia mountains. Integration of metamorphic and structural studies there have played an important role in the development of concepts concerning the tectonic evolution in the eastern Cordillera.

Precambrian
Monashee Complex

The grade of metamorphism in the core gneisses of the Monashee Complex is as high as K-feldspar + sillimanite in the Thor-Odin Dome and K-feldspar + kyanite at Frenchman Cap, with widespread occurrences of kyanite, staurolite, and a later andalusite-cordierite-gedrite assemblage. The age of the sillimanite-amphibolite assemblages has not been determined with certainty because of the strong effects associated with the dominant Jurassic metamorphism and the widespread Tertiary resetting of K-Ar ages. Duncan (1982) determined that migmatization in the basement gneisses occurred about 950 Ma, and that the first phase of deformation was accompanied by sillimanite-amphibolite metamorphism. Okulitch (1984) also supported the possibility of Precambrian metamorphism. Brown et al. (1986) suggested, however, that all the structures and metamorphism can be related to Mesozoic events.

Okanagan Complex

The Okanagan Complex (Okulitch, 1984) contains Proterozoic gneissic basement rocks exposed in domal culminations (Trail Gneiss, Simony, 1979; Valhalla and Kettle domes, Reesor, 1965; Cheney, 1980). These rocks, as old as Early Proterozoic, record temperatures up to 700°C and pressures up to 6 kbar (6×10^5 kPa), accompanied in some places by migmatization. It is not clear, however, how much of the high-grade metamorphism is of Precambrian age because of the strong Jurassic overprint, nor is it proven how much of this material is of North American cratonic origin and how much represents subjacent continental crust derived from elsewhere by tectonic transport in the Mesozoic Era.

Paleozoic (pre-Late Mississippian)
Kootenay Terrane

Evidence for Paleozoic metamorphism is sparse in the Omineca Belt, but Read (1973) reported the survival of chlorite-zone mineralogy in randomly oriented foliated clasts in conglomerate at the base of the Upper Mississippian Milford Group. The matrix of the conglomerate contains chlorite zone minerals oriented parallel with foliations related to Mesozoic deformation, showing that rocks now in the northern Kootenay Arc were metamorphosed before Late Mississippian time at least to chlorite zone, and again in the Middle Jurassic Epoch to chlorite zone. According to Brown et al. (1986), Brown and Read (1983), and Read and Brown (1981), these rocks received their Paleozoic metamorphic imprint while they were located more than 80 km to the west, before the Middle Jurassic Epoch when they were transported to their present position on the Monashee Décollement.

Paleozoic (Permian)

In the Anvil Allochthon of the Slide Mountain Terrane and in spatially associated rocks on both sides of the Tintina Trench more than a dozen eclogite lenses are known at eight localities (P. Erdmer, 1987; Erdmer and Helmstaedt, 1983). K-Ar and Rb-Sr dates on blueschist minerals from these eclogite lenses indicate a Late(?) Permian age (P. Erdmer and R.L. Armstrong, pers. comm., 1987). The lenses are thought to have formed in a subduction complex at depths of 40 km or more and subsequently to have been incorporated into the presently gently dipping thrust sheets that overlie miogeoclinal rocks.

Late Triassic (230-214 Ma)
Cariboo, Dorsey(?), Nisling and Nisutlin terranes

In Triassic to earliest Jurassic time regional metamorphism occurred in the Cariboo, Nisling, and Nisutlin terranes and possibly in the Dorsey Terrane. Metamorphism in the Nisling Terrane locally reached amphibolite grade but the most widespread metamorphism produced greenschist facies in Nisutlin and possibly Dorsey terranes. Upper Triassic sediments locally contain metamorphic clasts of greenschist facies, in part derived from the Nisutlin Terrane (Tempelman-Kluit, 1979). However, the major regional metamorphism of terranes in the Yukon Territory occurred in Early Jurassic time (Tempelman-Kluit, 1976). This metamorphic episode may correspond with the Middle Jurassic emplacement of Quesnellia recorded in the southern Omineca Belt.

Middle Jurassic (200-155 Ma)

The most intense metamorphism recorded in the Omineca Belt occurred in Middle Jurassic time and was presumably linked to the arrival of the Intermontaine Superterrane at the margin of the North American continent. Rocks of the continent, and the pericratonic terranes that were caught between it and the approaching Quesnellia (Kootenay Terrane, Cassiar Terrane) were deformed, transported, metamorphosed, and uplifted.

Transition between the Omineca and Foreland belts

The boundary between the Foreland and Omineca belts locally does not lie at the metamorphic transition from low grade to high grade rocks. Several areas of high grade metamorphism, for example near the Ice River Complex, in the Park and Deserters ranges and along the Hart Highway, lie in the Foreland Belt along its western margin. A few areas in the Omineca Belt, such as in the Purcell Anticlinorium and a large area east of the Tintina Fault in the Central Yukon, have undergone only chlorite or subgreenschist facies metamorphism.

In the Foreland Belt about 40 km southeast of Golden, Currie (1975) noted that tremolite, biotite and chlorite were regionally developed in the Chancellor Formation distant from the Ice River Complex. In the southern Park Ranges Craw (1978) compared metamorphic conditions across the Purcell Fault which there forms the western boundary of the Foreland belt and demonstrated 7-10 km (2-3 Kb) post-metamorphic vertical displacement (Fig. 16.4). Based on the disappearance of staurolite within the kyanite zone west of the fault, its persistence into the sillimanite zone east of the fault and geothermometry, the throw is west side up to the southeast. Ferri (1984) traced the garnet isograd across the belt boundary with no lateral displacement along the Southern Rocky Mountain Trench.

To the northwest, the Deserters Range is one of the few structural culminations in the Rocky Mountains in which amphibolite-facies rocks are exposed (Evenchick, 1988). The grade increases with stratigraphic depth from greenschist in the highest Windermere strata (Misinchinka Group) exposed, to kyanite + garnet + biotite zone in rocks above crystalline basement. Mappable zones are chlorite, biotite, garnet + staurolite + biotite, kyanite + garnet + biotite. In the structurally lowest rocks orthoamphibole coesists with kyanite, garnet, staurolite, biotite, plagioclase, and quartz. The assemblages indicate pressure-temperature conditions in bathozone 5 or 6 (Carmichael, 1978) (Fig. 16.1B), the highest grade rocks with temperatures between 550 and 650°C, and pressure greater than 7.5 Kbar (7.5×10^5 kPa).

An Eocene thermal event reset K-Ar ages in the metamorphic rocks in ranges flanking the Northern Rocky Mountain Trench. This resetting did not affect K-Ar ages on muscovite-bearing chloritiod phyllite in the Misinchinka Group along the Hart Highway which range from 118 Ma to 121 Ma, indicating that the metamorphism was at least as old as Early Cretaceous.

In the Omineca Belt west and south of Golden the Purcell Anticlinorium exposes a small part of Ancestral North America. The metamorphic grade passes westward to chlorite and biotite zones, independant of stratigraphic depth (Root, 1987), and the area has been affected by metamorphism that may have been coeval with high grade metamorphism farther west.

In the central Yukon Territory a large segment of the Omineca Belt belonging to Ancestral North America lies east of the Tintina Fault. In this area of chlorite zone and subgreenschist-facies rocks are three metamorphic culminations: Anvil Batholith and aureole, east of Keno Hill and 100 km northeast of Watson lake. The three metamorphic culminations are spatially associated with intrusions concentrically ringed by metamorphic zones from and outer biotite through garnet to staurolite-andalusite and, for the Anvil aureole, the sillimanite zone. For the Anvil, peak metamorphic conditions of 600-620°C and 3 kbar (3×10^5 kPa) occured during intrusion and were followed immediately by 10 km of regional uplift and erosion before deposition of the mid-Cretaceous South Fork Volcanics

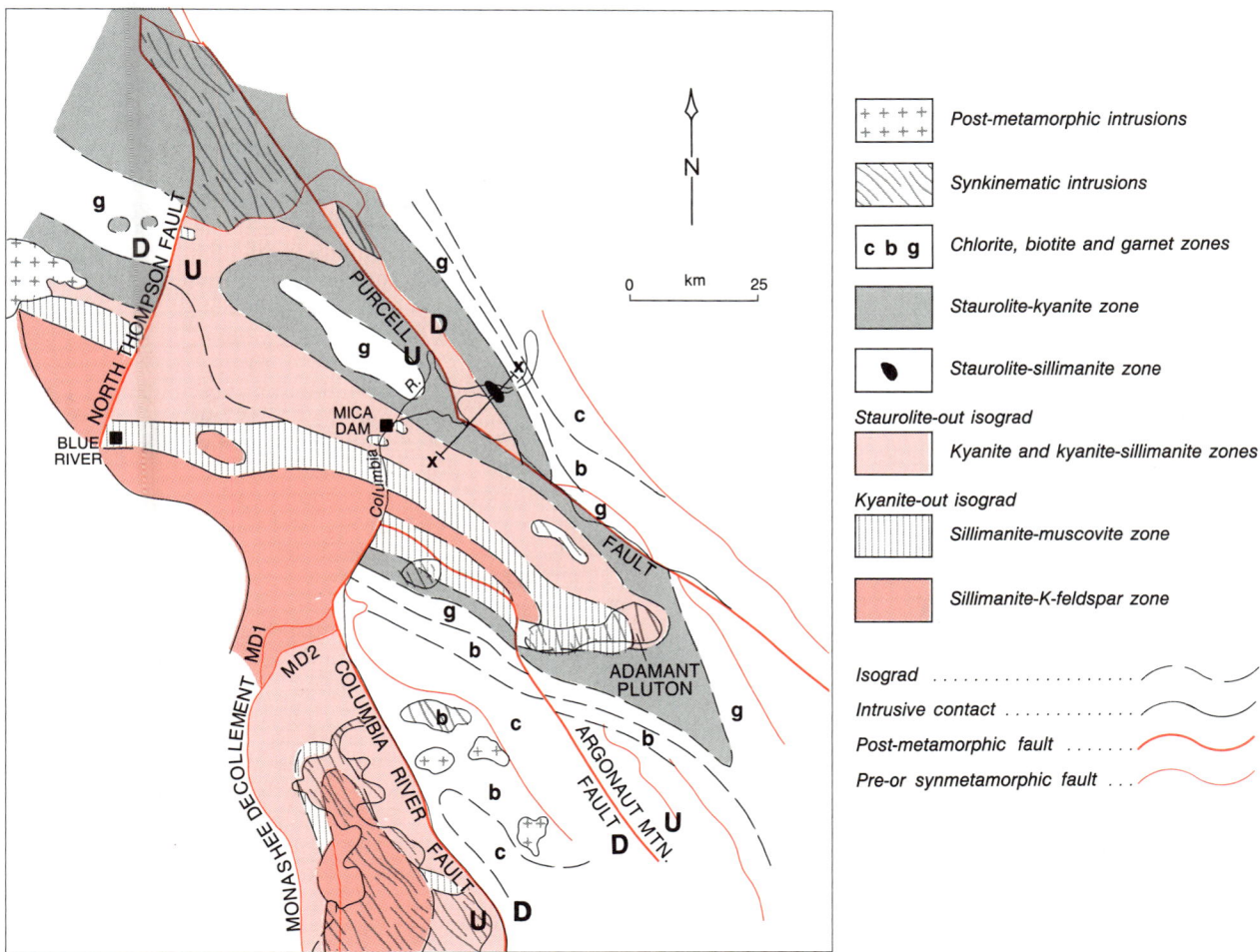

Figure 16.4. Metamorphic map of the northern Kootenay Arc and Monashee Complex. Movement of the Purcell, North Thompson and Argonaut Mountain faults creates a northwesterly elongate block elevated from 3 to 10 km vertically along post-metamorphic faults. To the southwest, Argonaut Mountain and Columbia River faults bounds a down-dropped block. Staurolite-bearing and staurolite-free but kyanite-bearing rocks are shaded to emphasize the position of the staurolite-out isograd relative to the kyanite-out isograd. Cross-section line X-X is the position of Craw's (1978) pressure difference across the Purcell Fault.

(Smith and Erdmer, 1991). The volcanics are large, down-faulted, erosional remnants of mid-Cretaceous age (Gordey, 1988), probably in zeolite facies.

Slide Mountain Terrane

In the Slide Mountain Terrane of the Cariboo Mountains, metamorphic grade ranges from subgreenschist facies to staurolite-kyanite zone (Struik, 1980, 1986a). The metamorphic isograds clearly cross the folded and metamorphosed basal detachment of the terrane indicating that metamorphism was imposed after obduction.

In the Cassiar Mountains, the Sylvester Allochthon of the Slide Mountain Terrane contains prehnite-pumpellyite, pumpellyite-actinolite, and chlorite zone assemblages which have higher conodont alteration indices than those of the underlying miogeoclinal rocks. In this region, the regional metamorphism of the Slide Mountain rocks predated their emplacement onto miogeoclinal strata.

Southern Omineca Belt

The southern Omineca Belt consists of sillimanite-bearing metamorphic rocks of the Monashee Complex exposed as culminations beneath the Monashee and other decollements. The decollement separates the complex from intensely deformed rocks ranging from chlorite to sillimanite zones which form the Kootenay Arc in the Selkirk Mountains to the east and continue along strike to the northwest into the Cariboo Mountains.

Monashee Complex

High-grade metamorphic rocks of the Monashee Complex are enveloped by Late Cretaceous and older mylonitic shear zones of the Monashee Décollement and are flanked by early Tertiary extensional faults of the Eagle River and Columbia River fault zones. Rocks exposed in the lower plate of the décollement record a two-stage history of prograde metamorphism (M1 and M2) which can be linked to major episodes of Mesozoic shortening and tectonic thickening (Journeay, 1986).

The earliest of the two metamorphic events (M1) outlasted initial eastward ovethrusting of the Selkirk Allochthon and associated folding and imbrication of the underlying Monashee Complex. It is characterized by a normal (hot-side-down) sequence of metamorphic zones ranging in grade from middle amphibolite facies (staurolite-garnet-biotite zone at 640°C and 6.4 kbar (6.4×10^5 kPa)) to lower granulite facies (kyanite-K-feldspar zone 685°C and 7.5 kbar (7.5×10^5 kPa)). M1 isotherms are antiformal and cut across the trace of folded mylonitic shear zones which define the locus of early displacement along the Monashee Décollement (MD1).

M2 metamorphism in the complex is associated with renewed overthrusting on the décollement in Late Cretaceous and early Tertiary time (MD2). Geometrical reconstructions of M2 assemblage zones in the northern end of the complex reveal an inverted (hot-side-up) metamorphic gradient which increases in grade towards MD2 from greenshcist facies (chlorite-biotite zone: 450°C and 3.5 kbar (3.5×10^5 kPa)) in the core of Frenchman Cap Dome to upper amphibolite facies (sillimanite-K-feldspar zone; 650-680°C and 2.5 kbar (2.5×10^5 kPa)) along its outward-dipping flanks. The development of this inverted metamorphic gradient is attributed to thermal relaxation associated with large-scale crustal imbrication and eastward overthrusting of high grade metamorphic and related plutonic rocks of the Shuswap Complex onto uplifted and previously metamorphosed rocks of the Monashee Complex.

Preservation of this inverted metamorphic gradient probably results from rapid uplift and cooling associated with extensional faulting and tectonic denudation of the Monashee Complex in Early Eocene time. Regional isotopic signatures in the Monashee and Shuswap complexes reflect this history of extensional unroofing and associated early Tertiary plutonism.

Kootenay Arc

In the Kootenay Arc and Purcell Anticlinorium, the transition from the Foreland Belt to the crystalline core zone is represented by a gradual westward increase in metamorphic grade across a broad slate belt in lower greenschist facies near the International Boundary to a rapid increase east of Mica Creek. The transition to the metamorphic culminations in the core of the Omineca Balt is commonly telescoped by post-metamorphic, east-dipping faults such as the Slocan Lake and Columbia River fault zones on the east side of the core, and west dipping faults such as the Eagle River and Okanagan valley fault zones on the west.

South of Kootenay Lake, mineral assemblages of the Windermere Supergroup record pressures between 5 and 6 kbars (5-6×10^5 kPa, 20-24 km depth). Farther north in the Riondel area (Crosby, 1968; 1976); Höy, 1976), Paleozoic rocks of the Lardeau Group were metamorphosed at conditions transitional between bathozone 4 and 5 (about 4.5 kbar, 4.5×10^5 kPa, 18 km depth). Kootenay Lake occupies the core of a narrow, northerly oriented metamorphic high in the sillimanite-K-feldspar zone in bathozone 5 near Riondel but only abot 5 km to the west the rocks are in the biotite or chlorite zone along the western margin of Nelson Batholith. Because neither post-metamorphic folding of faulting caused the quick succession of isograds, the closely spaced isograds may have resulted from a localized high heat flow. As a result of the high metamorphic pressure estimates in this part of the arc, it seems inescapable that both the Windermere Supergroup and the Lardeau Group were buried to depths of 18 to 24 km, far in excess of any reasonable estimate of overlying stratigraphic thickness. Tectonic burial seems to be the only available process to develop the metamorphic pressures.

The northerly trending Kootenay Lake high dissapiates north of the lake and a northwesterly trending metamorphic low, lying along Slocan syncline, passes between the Nelson and Kuskanax batholiths to its northwestern termination on the Columbia River fault zone. West of the Nelson Batholith and south of the metamorphic low is the Valhalla Complex (Reesor, 1965). This Eocene metamorphic culmination (Carr et. al, 1987) with a sillimanite-K-feldspar core permeated by leucosomes is juxtaposed by the Eocene Slocan Lake Fault (Parrish, 1981) against chlorite zone metamorphic rocks and the western side of the Nelson Batholith. The absence of kyanite in the complex is characteristic of all the metamorphic rocks east of the Columbia River fault zone for 350 km from south of the complex to south of Mica Creek. North of Valhalla Complex, the rocks pass essentially unfaulted from the staurolite-sillimanite zone at 5.0-6.8 kbar (5.0-6.8x10^5 kPa) and 630-680°C to chlorite zone in the Slocan low (Parrish, 1981). Around and north of the Kuskanax Batholith staurolite ± andalusite are sparsely developed adjacent to the batholith and the metamorphic grade decreases westward to the Columbia River Fault Zone which forms the eastern boundary of the Monashee Complex.

Southeast of Mica Creek, the Windy Range metamorphic high straddles the post-metamorphic Argonaut Mountain Fault (Leatherbarrow, 1981). Northwesterly trending zones are arranged such that metamorphic grade increases from the southwestern chlorite zone and the northeastern staurolite-kyanite zone margins towards a centrally located metamorphic culmination of sillimanite-K-feldspar. Sequences of mineral reactions and zones imply that assemblages southwest of the culmination formed at temperatures up to about 550°C at pressures of about 5 kbar (5x10^5 kPa), whereas assemblages northeast of the culmination formed at temperatures from approximately 550°C to 700° and pressures between 5 and 7 kbar (5-7x10^5 kPa). A discontinuity in metamorphic conditions of about 100°C and 2 kbar (2x10^5 kPa) across Argonaut Mountain Fault requires a 7 km throw with the northeast side up.

The rocks of the Mica Creek area (Fig. 16.4), are in the high-pressure block northeast of Argonaut Mountain Fault. They range in metamorphic grade from upper garnet zone to sillimanite-K-feldspar zone (e.g. Ghent et al., 1979, 1982; Simony et al., 1980 Leatherbarrow, 1980). Because staurolite disappears in the stability field of kyanite, pressures were in the upper range of bathozone 5. The disappearance of staurolite marks the appearance of abundant trondjhemite migmatite which is widespread in the kyanite zone up to 6 km thick. The appearance of sillimanite and kyanite occurs at the disappearance of muscovite and quartz (Ghent et al., 1982; Raeside, 1982). Clinopyroxene-garnet metabasites occur slightly upgrade from the kyanite-sillimanite isograd near Mica Creek (Ghent et all, 1983), and represent a relatively high pressure transition in mafic rocks (Newton and Perkings, 1982). In aluminous calcsilicate rocks, the high pressure assemblage epidote-kyanite-garnet-calcic plagioclase is present (Ghent et al., 1979).

At the north end of the Kootenay Arc, the distribution of metamorphic zones and mineral assemblages allows and estimation of post-metamorphic fault displacement along three major faults: Purcell Fault, North Thompson Fault and Argonaut Mountain Fault. In the Southern Rocky Mountain Trench north of Mica Creek, minor Tertiary normal movement on the Purcell Fault is west side down, but metamorphic mineral assemblages show that there was a post-metamorphic reverse displacement of about 7 km, with the west side up as shown in Fig. 16.4. The North Thompson fault, of Tertiary age, has west side down displacement (Pell, 1984). Stratigraphic throw is estimated to be about 3 to 4 km, and the pressure difference estimated with the garnet-plagioclase-Al$_2$SiO$_5$-quartz geobarometer is consistent at about 1 kbar (1x10^5 kPa). The pressure difference is expressed by the disappearance of staurolite at the sillimanite isograd or slightly within the sillimanite zone west of the fault whereas to the east of the fault it disappears in the kyanite stability field. In addition, metabasites west of the fault lack the high pressure garnet-clinopyroxene assemblage found east of the fault. Metamorphic conditions across the Argonaut Mountain Fault are estimated from garnet-biotite geothermometry and the calcite-dolomite solvus geothermometer, and pressure estimates are based on the garnet-plagioclase-Al$_2$SiO$_5$-quartz geobarometer. Southwest of the fault temperatures range from 490 to 590°C and pressures are about 5 kbar (5x10^5 kPa), and northeast of the fault temperatures range from 550°C to 650°C with pressures of about 7 kbar (7x10^5 kPa). The three faults bound a fault block 30 km wide which is more than 100 km long and retains metamorphic assemblages developed 4 to 10 km deeper than those present in the surrounding rocks. The southwest and northeast margins of the block are the Argonaut Mountain and Purcell faults respectively; North Thompson Fault forms the northwestern end.

Cariboo Mountains

Campbell et al. (1973) were the first to conclude that the northern Cariboo Mountains between the North Thompson River and Quesnel Lake and north of Shuswap Lake formed a major northwest-plunging complex of anticlinoria and synclinoria in which metamorphic grade and structural style are directly linked to structural depth. The most northwesterly locations expose rocks that are relatively high in the structural stacking and locations to the southeast expose rocks that are structurally deep.

The Matthews Fault (Struik, 1986b; Campbell, 1970, 1973) is a right-lateral transpressional feature that marks the northeastern margin of the high-grade metamorphic core for many kilometres, separating the Paleozoic Snowshoe Group on the west from the Upper Proterozoic Kaza Group on the east. Detailed work along the eastern margin of the high-grade zone has shown this fault to be syn- or post-metamorphic with a definite temperature discontinuity across it (Fig. 16.5). Fletcher and Greenwood (1979) found a discontinuity based on garnet-biotite geothermometry of 150°C, with the southwest side hotter. Engi (1984) reported a discontinuity of 75°C in the same sense. Pigage (1978), Klepacki (1981), and Getsinger (1985) all found low-angle, east-dipping, post-metamorphic, listric faults or tectonic slides at the eastern margin of the high-grade zone, all with measured or inferred temperature differences of about 100°C in the same sense, hotter on the southwest and presumably deeper by 3 or 4 km. In summary, the northeast margin of the metamorphic complex is marked by syn- to post-metamorphic faults with east side down about 4 km and a temperature discontinuity of about 100°C.

Within the central zone (Fig. 16.5), all the pressure indicators suggest pressures near the upper end of bathozone 4, with staurolite becoming unstable in the stability field of sillimanite rather than kyanite. The exception is the discovery of bathograd 4-5 by Getsinger (1985) immediately north of the north arm of Quesnel Lake (Fig. 16.5), confirming the deduction of Campbell (1970, 1973) that this arm of the lake coincides with a northeast-trending post-metamorphic Little River Fault with the northwest side up some 250°C and presumably about 8 km of throw.

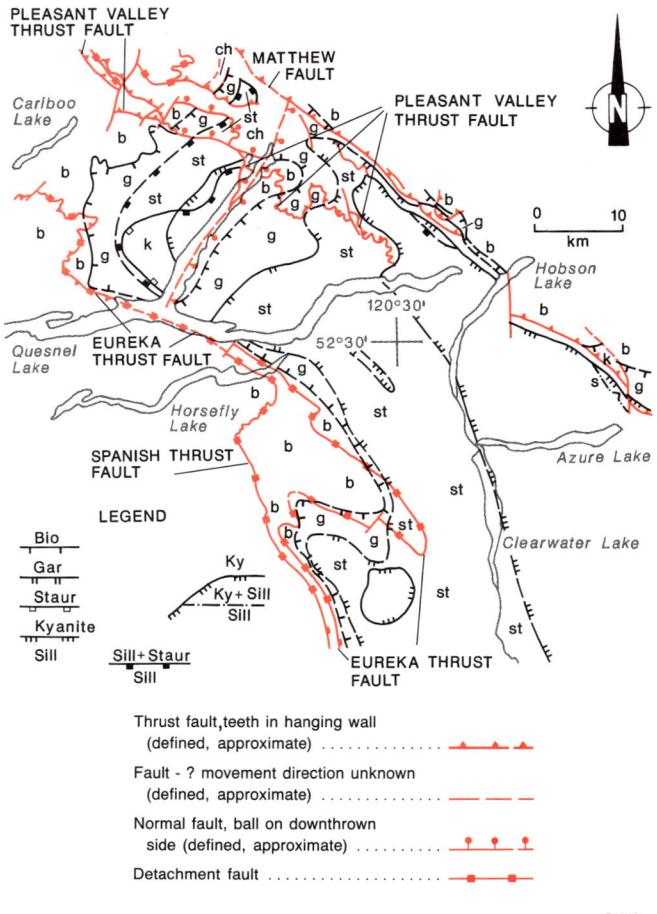

Figure 16.5. Metamorphic map of the northern end of the high-grade core zone near Quesnel Lake. Hachures are on the high-temperature side of each isograd. The location of bathograd 4-5 to the northwest of the north arm of Quesnel Lake is marked by the intersection of the staurolite-kyanite and sillimanite-kyanite isograds. The change in grade requires a post-metamorphic fault beneath the lake (Little River Fault). Note that isograds cut across terrane boundaries near the southwest limit of the map. Abbreviations are explained in Table 16.1.

Geothermometry and geobarometry by Pigage (1977), Fletcher and Greenwood (1979), Klepacki (1981), Getsinger (1985), McMullin and Greenwood (1986), Montgomery (1985), Fillipone (1985), and Engi (1984) have outlined the following general characteristics of the high-grade belt:

1. A broad central high-grade zone reached temperatures of 600°C to 650°C in pressure regimes characteristic of Carmichael's bathozone 4, near its upper range with pressures of 5 to 6 kbar (5 to 6×10^5 kPa, about 19 to 23 km in depth). Pegmatite appears at temperatures above the breakdown of staurolite.

2. Well defined isograds have been mapped extending through the normal Barrovian sequence outward from the centre to chlorite zone except along the northeast margin where the margin is cut by syn- to post-metamorphic faults.

3. Along the southwestern margin the metamorphic grade decreases rapidly but without a break, continuing into Quesnellia and the Slide Mountain Terrane, which have been metamorphosed in the same Middle Jurassic event at least up to staurolite-kyanite zone (Fig. 16.5). This has consequences for interpretation of the cause/effect relationships between the supposed underthrusting of Quesnellia and the development of the high metamorphic pressures and temperatures recorded in the Cassiar and Kootenay terranes. Because the west-verging structures which were coeval with metamorphism are supposed (Brown et al., 1986; Okulitch, 1984) to have been caused by underthrusting of Quesnellia they should, presumably, have had an uplifting effect on the superjacent terranes to the east, which were undergoing high-pressure Barrovian metamorphism. In this process, Quesnellia sustained only marginal metamorphism where in contact with the other terranes.

4. Textural evolution of the main metamorphism shows early garnet cores with kyanite inclusions, and clear, post-kinematic idiomorphic garnet rims associated with staurolite and sillimanite.

5. The latest high-temperature crystallization produced Buchan-type assemblages near Boss Mountain, including cordierite and andalusite near small pegmatite bodies (Campbell, 1971).

6. Kaza and Isaac rocks lying northeast of the Matthew fault and similar east-side-down faults contain, besides biotite, garnet, and chlorite, numerous occurrences of chloritoid and scattered occurrences of paragonite (Engi, 1984; Pigage, 1978).

In the northern Cariboo Mountains, northeast of the main high-grade core zone, a progression to tighter and more recumbent west-verging folds with depth coincides with a regular depth progression of steadily increasing metamorphic grade (Murphy, 1987; Murphy and Rees, 1983; Pell, 1984). The metamorphic zones are roughly parallel with the present topographic surface and clearly discordant with the limbs of the major F2 folds, which are, on textural criteria, the same age as the metamorphism. It seems most likely that this is a consequence of post-kinematic continuation of metamorphism due to the time lag of heat flow following cessation of strain, and may possibly be linked to the post-kinematic rims on prograde garnets.

Summary

Commonly observed relationships between regional metamorphism and structural development in the Cariboo and Colombia mountains have played a major role in models of tectonic evolution for the region (see Ch. 17). Unfortunately, only a few localities have carefully dated evidence bearing on the timing of deformation and metamorphism.

Evidence for a pre-Late Mississippian, probably mid- to Late Devonian deformation and regional metamorphism is recorded in Lardeau clasts present in conglomerates of the Upper Mississippian to Pennsylvanian Milford Group. Read (1976), and Klepacki (1985) noted angular, foliated clasts from the Lardeau Group, with the attitude of the foliation varying from clast to clast, in the basal conglomerate lenses of the Milford Group over a distance of 100 km in the central Kootenay Arc. This early deformation of the

Lardeau Group was so pervasive that it commonly obliterated bedding and the accompanying metamorphism produced chlorite zone assemblages with quartz-albite segregations along bedding and early foliation. Because assured recognition of this early phase requires the presence of the Milford Group, the extent of the mid-Paleozoic deformation and metamorphism is unknown.

In the Kootenay Arc south of Kootenay Lake, a U-Pb zircon date from the late synkinematic Mine stock of 171 Ma gives the age for the metamorphic maximum which outlasted the first and most intense phase of Mesozoic deformation (Archibald et al, 1983). This timing is supported by evidence from the plutonic rocks on either side of the Slocan syncline. In highly folded sediments of the Milford Group thin dikes of aegerine-augite leucocratic-quartz monzonite, typical of the Kuskanax, are folded and disrupted; others are undeformed and lie in the axial-plane foliation of the strongly folded rocks. A U-Pb zircon date of 173±5 Ma (Parrish and Wheeler, 1983) comes from a marginally foliated and lineated part of the Kuskanax Batholith which was involved in and partly recrystallized by the same intense phase of mid-Jurassic deformation and metamorphism recorded south of Kootenay Lake. A minimum age for the deformation and metamorphism comes from a K/Ar date on hornblende of 164±6 Ma from the northern margin of the Nelson Batholith (Nguyen et al, 1968) where it overprints a contact metamorphic aureole on rocks already regionally deformed and metamorphosed by the mid-Jurassic event. East of Mica Creek the deformed and metamorphosed Adamant Pluton yielded a U-Pb zircon age of 169±4 Ma which records the age of the peak or regional metamorphism (Shaw, 1980). To the northwest in the Cariboo Mountains, a U-Pb zircon date of 174±1 Ma on the undeformed and unmetamorphosed Hobson Lake pluton, yields a minimum age for the regional metamorphism (Gerasimoff, 1987). In summary, the radiometric data indicate that the widespread, regional metamorphism and two intense deformations of the Kootenay Arc and Cariboo Mountains are of Middle Jurassic age.

The metamorphic history of the Monashee Complex and other high-grade metamorphic culminations is much younger and spans at least Late Cretaceous to Eocene. U-Pb zircon ages from two of the orthogneiss sheets intruded into the Ladybird thrust plate of the Valhalla Complex during deformation and metamorphism date the events as lying in the range 56.5± to 62±1 Ma (late Paleocene to earliest Eocene) (Carr et al., 1987). West of the Columbia River Fault Zone, Pinnacles Dome is the southern exposure of the high grade metamorphic rocks. Four U-Pb zircon ages on metamorphic zircons from amphibolite in the Ladybird thrust plate above the Monashee Complex yield Late Paleocene to Early Eocene dates for the regional metamorphism (Carr, 1989). To the north, on the southern Flank of Thor-Odin Dome, undeformed pegmatite cutting thrusts associated with Monashee Décollement, not only indicate the pre-Eocene age for the thrusts (Coleman, 1989), but also a pre-Eocene age for the high-grade regional metamorphism.

In conclusion, a low-grade mid-Paleozoic metamorphism of unknown extent preceded two widespread periods of low to high-grade regional metamorphism affecting the southern Omineca Belt. A Middle Jurassic metamorphism affected the structurally highest rocks of the upper plate, a Paleocene to Eocene event affected rocks of the medial Ladybird plate, and a Paleocene to Late Cretaceous and (?) older metamorphism modified the lowest plate containing the Monashee Complex. The boundary between the Jurassic and Tertiary events follows major listric normal faults responsible for the tectonic denudation of the high-grade Tertiary metamorphic rocks.

Northern Omineca Belt

Metamorphism in the northern Omineca Belt was dominantly dynamo-thermal and Barrovian. The grade varies from subgreenschist to amphibolite facies, the highest grade pelites containing sillimanite, K-feldspar, and muscovite. The grade generally increases with stratigraphic depth; the highest grade rocks are exposed in the cores of post-metamorphic anticlinoria. All rocks of greenschist and amphibolite facies belong to the Upper Proterozoic Windermere Supergroup.

In the Chase Mountain area up the Wolverine Complex, Rb-Sr dates on metamorphic muscovite range from 154 to 166 Ma (Parrish, 1979). The oldest K-Ar dates on micas range from 120 to 140 Ma, and the first metamorphic detritus reached the Sustut Basin in Albian time. The basin was probably south of presently adjacent parts of the Omineca Belt. The data indicate uplift in the Early Cretaceous, probably following or overlapping with late stages of metamorphism that may have lasted from Middle or Late Jurassic to Early Cretaceous time. It was probably associated with mid-Mesozoic regional northeast contraction which was accommodated by faults and recumbent isoclinal folds. Isolated areas with Eocene K-Ar ages indicate that renewed uplift in Eocene time was local and not associated with significant new mineral growth.

Detailed studies of metamorphism are few, particularly in the lower grade rocks. The Wolverine Complex south of 56°N has not been studied other than during reconnaissance mapping by Armstrong (1949), who applied the name to high-grade metamorphic rocks with associated granites and pegmatites in which it was not possible to distinguish stratigraphic units. The Wolverine Complex appears on the Metamorphic Map (Map 1714A, in pocket) as undivided amphibolite facies. Along trend to the northwest, rocks between 56°N and 58°N are the largest exposures of high-grade rocks in the northern Omineca Belt. There, in the Swannell, Finlay, Butler, and Sifton ranges, chlorite, biotite, garnet, kyanite and sillmanite isograds were mapped by Gabrielse (1975), and Gabrielse et al. (1977); Mansy, 1986; and Evenchick, 1988; the isograds appear to be folded in regional anticlinoria.

Swannell Ranges

Much of the Swannell Ranges is underlain by rocks in greenschist facies, in which the chlorite, biotite, and garnet isograds are concordant with the anticlinal structure. Metamorphic grade increases towards the core of the anticlinorium where amphibolite facies rocks are exposed. The highest grade culminations are small areas outlined by the kyanite isograd. Although Mansy (1986) described staurolite and garnet in these rocks, no staurolite isograd has been mapped and it is assumed that the highest grade rocks contain the assemblage kyanite + staurolite + garnet + biotite, i.e. above the boundary of bathozones 4 and 5.

The assemblages indicate a temperature of at least 500°C at a pressure greater than 5 kbar (5×10^5 kPa). Geothermometry gave temperatures of about 400 to 450°C. Near the south end of the range, Roots (1954) described sillimanite in the stratigraphically lowest rocks. Twenty kilometres to the southeast, Parrish (1979) noted isolated localities of sillimanite in schist of the Wolverine Complex.

Sifton Ranges

In the Sifton Ranges, the lowest grade rocks are in greenschist facies. Across a steep fault to the east, and in the footwall of the Sifton Fault, the staurolite + biotite, kyanite + garnet + biotite, sillimanite + kyanite, sillimanite, and sillimanite + K-feldspar zones are defined (Evenchick, 1988). The kyanite + sillimanite and sillimanite zones appear to follow the form of the Sifton Antiform. Staurolite is stable (with biotite, muscovite, garnet) in the kyanite zone. Both kyanite and staurolite persist after the first appearance of sillimanite, resulting in a zone of coexisting sillimanite + kyanite + garnet + staurolite + biotite + muscovite + plagioclase + quartz. The assemblage probably resulted from sluggish dissolution of staurolite and kyanite. In the sillimanite zone, staurolite and kyanite, if present, are sheathed by wide rims of muscovite and are clearly in textural disequilibrium. A small area in the core of the antiform has the assemblage K-feldspar + muscovite + sillimanite + biotite ± garnet. The relationships suggest a pressure at about the boundary between bathozones 4 and 5. All sillimanite appears to be in textural equilibrium with other mineral phases that define each zone, and sillimanite is not present in the staurolite + biotite or kyanite + garnet + biotite zones.

The isograds in the footwall of the Sifton Fault are truncated by the gently dipping fault (Evenchick, 1988). The fault is inferred to be post-metamorphic; the hanging wall is a shear zone in which pelites are rare, the textural relations are obliterated by mylonitic textures, and the zones were probably shuffled and interleaved during faulting. The pre-fault structural and metamorphic history is inferred to be about the same as the rest of the core of the Omineca Belt. A large area with the assemblage garnet + staurolite + biotite is present. The assemblage kyanite + garnet + biotite + staurolite occurs the length of the hanging wall, almost entirely within the lowest stratigraphic unit. The only occurrence of sillimanite (with staurolite, garnet, biotite, muscovite) is in the stratigraphically lowest pelites. The grade represented by these assemblages is at about the boundary of bathozones 4 and 5, similar to the footwall of the fault. The stable mineral assemblages during faulting were chlorite + biotite + muscovite in pelites, and hornblende + plagioclase + quartz in mafic rocks. The pressure-temperature conditions inferred for the development of ductile structures in the mylonites are consistent with the grade represented by these assemblages. Geothermobarometry on kyanite + staurolite + garnet + biotite assemblages yielded: footwall temperature 600 ± 50°C, pressure 7 ± 1.5 kbar ($7 \times 10^5 \pm 1.5 \times 10^5$ kPa) and hanging wall temperature 630 ± 50°C, pressure 6 ± 1.5 kbar ($6 \times 10^5 \pm 1.5 \times 10^5$ kPa). The data show that the footwall was metamorphosed at about the same conditions as was the hanging wall.

The Sifton Ranges is one of the few areas in the northern Omineca Belt with Eocene K-Ar ages for metamorphic and plutonic rocks. The footwall has a structural and metamorphic style similar to the Swannell Ranges and is therefore inferred to have had a similar Jurassic-Cretaceous history. The Eocene ages are thought to be the result of reheating associated with displacement on the Sifton Fault in Late Cretaceous to Eocene time. In the south Sifton Ranges, rare pelites in basement gneiss 300 m from the northeast edges of the undeformed Balourdet Pluton (Eocene) has the assemblage andalusite + cordierite + muscovite + biotite + plagioclase + K-feldspar + quartz. The textures and assemblage are diagnostic of bathozone 1.

Finlay Ranges

The Finlay Ranges anticlinorium has the same stratigraphy as the Sifton and Swannell ranges but mainly at lower metamorphic grade. Lower Paleozoic rocks are in subgreenschist facies. From analysis of illite crystallinity, Mansy (1986) suggested that the anchizone was reached in the Stelkuz Formation. Chlorite appears in the Tsaydiz Formation, and only biotite and garnet are present in the Swannell Formation.

Cassiar Batholith region

Along the southeastern margin of the Cassiar Batholith the following zones indicate increasing metamorphic grade towards the batholith: chloritoid, staurolite, andalusite, sillimanite. High-grade assemblages include biotite + muscovite + cordierite, K-feldspar and sillimanite, i.e. above bathozone 1. Geothermometry on the cores of biotite and garnet in one sample gave a temperature of 580°C, whereas the rim gave a temperature of 800°C (Mansy, 1986).

Horseranch Range

The Horseranch Range is a doubly plunging anticlinorium bounded on the east and west by the Deadwood and Horseranch faults, respectively. The central schist complex consists of pelitic to psammitic schist, quartzite, marble, and minor amphibolite intruded by granitic to mafic and ultramafic rocks. On the western margin, the west dipping Horseranch Fault juxtaposes biotite zone rocks against sillimanite-bearing schist. The fault is a mylonitic zone up to 1 km thick with movement indicators reflecting a top-down-to-the-northwest sense of shear. In the pelitic schist the widespread replacement of muscovite + quartz by sillimanite-K-feldspar throughout the complex suggest that peak metamorphic temperatures exceeded 650-700°C in the range of 5-7 kbar ($5-5 \times 10^5$ kPa). Structurally overlying staurolite-kyanite schist is locally present and indicates minimum pressure temperature limits of 4 kbar (4×10^5 kPa) at 500°C (Plint and Erdmer, 1989). Muscovite and zircons from abundant pegmatite have yielded Eocene K-Ar and U-Pb ages, respectively.

Relationship of deformation to metamorphism in Cassiar and Omineca mountains

Generalizations based on observations in the Swannell and Sifton ranges are that syn-metamorphic deformations resulted in the development of a foliation parallel with axial surfaces of isoclinal folds; the foliation also is isoclinally folded. Rotational garnets resulted from syntectonic growth. Mansy (1986) and Parrish (1979) referred to the foliation parallel with the axial surfaces of isoclinal folds of bedding as S_1, and the axial surfaces of folded foliation as S_2. According to Mansy the porphyroblasts are syn- to post-S_2. Both sets of folds are the same style and orientation and are difficult to separate. An alternative view is that because of the lack of textural or structural data to suggest that S_1 and S_2 represent separate folding events, the lack of evidence to show that they occurred at significantly different grades, and the likelihood of progressive deformation occurring with progressive metamorphism, the structures are inferred to be a result of one phase of syn- to late-metamorphic deformation. In all areas the foliation is kinked and porphyroblasts are brittley deformed. These textures are inferred to be a result of post-metamorphic upright folding, which also folded isograds to roughly the present form, and resulted in the metamorphic relief within each range. The Sifton Fault truncates an upright antiform in the footwall, and displacement on the fault defines a third distinct phase of deformation in which contraction was south-southeast rather than northeast as for earlier phases.

Northern Omineca Belt, Yukon Territory

In general, the timing of regional metamorphism, the distributions of isograds in the various terranes and the relationships of metamorphism between terranes have been little studied in the northernmost part of the Omineca Belt. In Jurassic time a major metamorphic event seems to have occurred in the southwestern part of the Yukon Territory (Tempelman-Kluit, 1976). Large areas of garnet-mica-quartz schist and paragneiss, locally including kyanite, occur in the Nisling Terrane. Metamorphic foliation, commonly parallel with compositional layering is well developed and is cut by Cretaceous granitic plutons. On the other hand, foliation is generally concordant with that in adjacent Early Jurassic plutons. Metamorphic clasts are abundant in sediments of the Lower Cretaceous(?) Tantalus Formation.

Late Jurassic-Early Cretaceous (155-138 Ma)

Within and following this time period, Quesnellia, the Kootenay and Cassiar terranes and North America behaved as a single tectonic unit, at least from the metamorphic standpoint, and no major reconstitutive metamorphism is recorded. However, there may well have been elevated temperature and anomalous heat flow, suggested by the pattern of reset K-Ar dates reported by Armstrong (1988). Armstrong plotted lines of constant reset ages, or 'chrontours', which outline all of the major Middle Jurassic metamorphic zones of the Omineca Belt. A large proportion of these reset dates fall inside the 150 Ma 'chrontour' but were not fully reset by the Eocene thermal event. It is possible that these 150 Ma dates reflect the 'cooling ages' of the Jurassic metamorphism, as there is no identifiable metamorphic event with which to associate the resetting.

Early to mid-Cretaceous

The relationships of Early to mid-Cretaceous and Middle Jurassic to Late Jurassic events in the Omineca Belt are not well understood. In the Kaska and Selwyn mountains of northern British Columbia and Yukon Territory a widespread thermal event accompanied the emplacement of mid-Cretaceous granitic rocks. The regional pattern of isograds is related to broad anticlinoria that formed during this interval. In the Cassiar and Omineca mountains isotopic ages suggest that the isograds are of Middle to Late Jurassic age and that thermal overprints, locally with significant recrystallization, took place in mid-Cretaceous and early Tertiary times. No evidence supporting a Middle to Late Jurassic age for regional metamorphism is known from the Pelly and Selwyn mountains.

Late Paleocene-mid-Eocene (65-45 Ma)

This interval witnessed a general resetting of radiometric ages over wide areas in the Omineca Belt coinciding with a 50 Ma magmatic event (Armstrong, 1982, 1988; Parrish, 1979, 1981; Mathews, 1981, 1983; Archibald et al., 1984). It was a time of uplift, crustal extension, erosion (both denudational and tectonic), and presumably high heat flow. Temperatures must have reached about 500°C at least locally because of the resetting of the K-Ar ages measured on amphiboles (Medford, 1975). One of the difficult facts to explain is the existence of little metamorphosed Eocene sediments unconformably overlying the reset rocks of the Okanagan Complex (Okulitch, 1984; Medford, 1975), which had just shortly before been heated to temperatures of about 500°C. This seems to necessitate a combination of extremely steep vertical geothermal gradient and very rapid uplift and erosion. Even with an assumed geothermal gradient of 50°C/km, the amount of material removed before the end of the Eocene is 10 km, necessitating a rapid rate of erosion.

It is possible that at this time some of the marginal faulting to the Shuswap Complex was taking place. It has been noted above that the northeast margins of the high-grade zone in the Cariboo Mountains were undergoing listric normal faulting, with east side down at the Little River Fault (Klepacki, 1981), near Niagara Creek (Engi, 1984), and near Azure Lake (Pigage, 1978). Also active, but west side down, was the normal, North Thompson Fault (Fig. 16.4, Pell, 1984). This tectonic activity can be rationalized in terms of crustal extension, uplift, tectonic denudation, and related high heat flow, and thus may be another manifestation of the processes that led to resetting of radiometric ages.

INTERMONTANE BELT

Throughout most of its length the Intermontane Belt contains metamorphic assemblages characteristic of low-temperature and low-pressure facies, particularly the prehnite-pumpellyite and zeolite facies. Few detailed studies of metamorphism have been completed; conspicuous exceptions include studies of the spectacular blueschist assemblages in the Pinchi Lake area, the core complex near Tatla Lake, and the subgreenschist assemblages of the

Hazelton Group near Smithers and Terrace. The sparse data from elsewhere suggest that all rocks older than Miocene have been metamorphosed to at least subgreenschist grade.

Metamorphism in the Intermontane Belt took place from Late Triassic to early Tertiary time. In most areas, the rocks regionally never exceeded about 250°C and, except for some transitional blueschist assemblages in the Cache Creek Assemblage, were never at pressures greater than 2 kbar (2×10^5 kPa) and some were demonstrably never buried deeper than about 2.5 km.

Cache Creek Assemblage

The Cache Creek Assemblage in the Ashcroft area is in fault contact with essentially unmetamorphosed strata of the Upper Triassic to Lower Jurassic Nicola Assemblage (Grette, 1978). The difference in metamorphic grade between the two units implies that they were juxtaposed by faulting. Within the Cache Creek Assemblage, a transitional blueschist assemblage is variably developed, the intensity of which depended on the lithology. Minerals include albite, chlorite, prehnite, pumpellyite, actinolite, magnesioriebeckite, stilpnomelane, calcite, epidote, and sphene. The coexistence of prehnite, pumpellyite, actinolite, magnesioriebeckite, and epidote, together with albite and absence of analcime and lawsonite, indicate a temperature of about 225°C and a pressure of about 4 kbar (4×10^5 kPa), which must have been imposed in Late Triassic time.

In the Dease Lake region, much of the Cache Creek Assemblage is metamorphosed to prehnite-pumpellyite facies (Monger, 1975). Actinolite is locally present. Near Dease Lake, local blue amphibole and lawsonite indicates the presence of the lawsonite blueschist facies. One of the few occurrences of jadeite in the Canadian Cordillera is in the Thibert fault zone along the northern margin of the Cache Creek Terrane just west of the north end of Dease Lake.

In the Pinchi Lake area between Prince George and Manson Creek, the Cache Creek Terrane has spectacular blueschist assemblages of Late Triassic age (Paterson, 1973; Paterson and Harakal, 1974). The assemblages include combinations of lawsonite, jadeite, glaucophane, aragonite, and calcite, chlorite, quartz, and phengitic white micas. Pressures were estimated to be about 8 kbar (8×10^5 kPa) with temperatures of 300°C, corresponding to a depth of about 28 km and a geothermal gradient of about 10°C/km, typical of downgoing subducting slabs.

High-pressure assemblages occur also farther north in the Vital Range (I.A. Paterson, pers. comm., 1986) and in the Dease Lake area, reflecting the erratic but widespread distribution of high-pressure, subduction-related assemblages in the Cache Creek Terrane.

Asitka Assemblage

Little is known about the metamorphic history of the upper Paleozoic Asitka Assemblage. West of the Ingenika Fault, the rocks are in prehnite-pumpellyite facies with local occurences of actinolite implying the presence of pumpellyite-actinolite facies (Burns, 1973). In the Iskut River area, the Asitka Assemblage is in the chlorite zone and conodonts from Lower Permian carbonates and older strata generally have CAI values greater than 5 in contrast to those in the Triassic strata which are less than 5 (Anderson, 1989). These observations augment those of Read et al. (1983) who noted that a decrease in intensity of deformation, development of foliation and colour alteration index of conodonts marks the break from Lower Permian to Lower or Middle Triassic rocks in the Iskut River area and in the Grand Canyon of the Stikine River 150 km to the north. In a regional context, these data indicate that the period of deformation and low-grade regional metamorphism, occured during the Late Permian to Early Triassic.

Nicola and Stuhini assemblages

In south-central British Columbia, the age of the Nicola Assemblage overlaps that of the Cache Creek Group. Near the faulted contact between the group and the Nicola Assemblage, the Nicola is in the prehnite-pumpellyite facies (Shannon, 1982). This grade is widespread south of the Trans Canada Highway (Schau, 1968; Lefebure, 1976), and only locally drops to zeolite facies. Along the eastern edge of the Eagle Plutonic Complex (Greig, 1989) and surrounding the plutons in the Nicola horst (Moore and Pettipas, 1990), rocks of the Nicola Group rapidly pass through the garnet zone and into the lower part of the amphibolite facies. Listric normal faults bound the horst and set the amphibolite facies rocks against Nicola metavolcanics in the prehnite-pumpelyite facies.

North of the Trans Canada Highway, particularly west of Quesnel Lake, tuffaceous sediments and volcanics of the Nicola Assemblage are zeolite-bearing with heulandite, chabazite and analcime most common (Bailey, 1978). Some of the analcime occurs as igneous phenocrysts but the remainder, developed along joints and partly replacing the matrix, is secondary. Although these rocks have been placed in zeolite facies (Map 1714A), the zeolite±calcite ±sericite±clay mineral assemblages are apparently quartz-free.

In north-central British Columbia, volcanics of the Nicola Assemblage are mainly in the prehnite-pumpellyite facies (Meade, 1977; Ferri and Melville, 1988 and 1989). Near the margins of the Germansen batholith, narrow chlorite and biotite zones result from contact metamorphism. West of the Moose Valley Fault, Upper Triassic volcanics of the Stuhini Assemblage are in prehnite-pumpellyite and zeolite facies (Burns, 1973). Prehnite-pumpellyite assemblages are most common in the stratigraphically lower parts of the assemblage, and zeolite (characterized by laumontite) is most abundant in the upper parts of the unit. The Stuhini Assemblage is in part unconformably overlain by the Cretaceous Sustut Group. The Sustut Group truncates isograds in the Stuhini Assemblage, indicating that metamorphism of the Stuhini Assemblage was pre-Late Cretaceous in age.

Hazelton Assemblage

The Hazelton Group in west-central British Columbia has well developed, subgreenschist assemblages over a wide area (Tipper and Richards, 1976). Metamorphic features are particularly well displayed in the Howson subaerial facies of the Telkwa Formation. In this unit in the Smithers area, the authigenic minerals occur as amygdules and void fillings, as cements, groundmass replacement, and veins and fracture fillings (Dudley and Ghent, 1980; Dudley, 1983). Zeolites include laumontite, wairakite, heulandite, mesolite, scolecite, thomsonite, stilbite, and analcime. Other important minerals include epidote, prehnite, adularia, albite, chlorite, calcite, quartz, and less commonly, pumpellyite. The alteration, although on a regional scale, is patchy with locally extensive veining and development of zeolite-cemented breccias. The high percentage of void fillings indicates that the alteration may have occurred at relatively shallow depths, probably under locally high fluid/rock ratios. Veins of laumontite and calcite cutting zeolite-cemented clastic units suggests more than one episode of metamorphic crystallization. Mineral zoning is concentric around the Early Jurassic Topley Suite. The zone nearest the intrusions lacks zeolites and is characterized by quartz, adularia, kaolinite, chlorite, epidote, and calcite. The middle zone is dominated by zeolite-bearing assemblages with chlorite, prehnite, albite, calcite, and quartz. The third zone consists predominantly of clay-carbonate assemblages without zeolites. The apparently stable coexistence of laumontite-chlorite-quartz (Coombs, 1971) and the coexistence of epidote-chlorite-quartz (Frost, 1980) suggest that both zeolite and prehnite-pumpellyite facies assemblages are present, commonly in the same outcrop.

In the Terrace region, metamorphic grade of the mostly nonmarine Telkwa Formation generally increases from zeolite to prehnite-pumpellyite facies with increasing depth of burial (Mihalynuk, 1987). However, contemporaneous and later hydrothermal events have modified the metamorphic pattern and, on the whole, temperatures appear to have been slightly higher near Terrace than at Smithers. In both areas, the presence of low-pressure (less than 2 kbar, 2×10^5 kPa) assemblages indicates that the area was not deeply buried at any time since deposition: i.e. the region has remained within about 6 km of the surface, about the thickness of the stratigraphic pile, for the last 200 Ma. This conclusion implies that eastward-directed thrusting related to development of the Coast Plutonic Complex in the Cretaceous did not tectonically thicken or load the Mesozoic stratigraphic section to any significant extent.

The Hazelton Assemblage about 100 km northeast of Hazelton is predominantly marine. The development of secondary minerals was largely controlled by position in the stratigraphic section (T.A. Richards, pers. comm., 1985). Pumpellyite is common in the oldest unit, the Telkwa Formation; actinolite is locally present. Pliensbachian to Bajocian tuffaceous rocks of the Nilkitkwa Formation are locally intercalated by the mid-Toarcian Ankwell volcanics. Below the volcanics, the tuffs contain abundant prehnite; above, laumontite is present. The alteration in the Nilkitkwa Formation may have been controlled more by heat introduced by the Ankwell volcanics than by depth of burial (T.A. Richards, pers. comm., 1985).

Bowser Lake Assemblage

Little is known of the metamorphic condition of the rocks in the vast Bowser Basin except that most of them are subgreenschist in grade. The few detailed studies carried out to date centred on the Groundhog coal area in the northeast part of the basin. Bustin (1984) reported vitrinite reflectance values ranging from 1.70% R_{max} to 5.8% R_{max}. He concluded that these values are anomalously high for the calculated maximum depth of burial of 3500 m, and suggested that they resulted from high geothermal gradients (50-70°C/km) in Early Cretaceous time. Moffat (1985), using vitrinite reflectance data, illite crystallinity, and mineral assemblages, concluded that metamorphism preceded Late Cretaceous deformation and that paleo-geothermal gradients in the northern part of the Groundhog were about 30-40°C/km. In the northwestern part of the Bowser Basin, pumpellyite is common as cement in sandstones, and a single R_{max} value is 1.43% (T.J.D. England, pers. comm., 1983).

Near Hazelton, laumontite is common as cement in sandstones of the Bowser Lake Group, and prehnite is known from two localities (T.A. Richards, pers. comm., 1985). The basaltic Netalzul volcanics, locally interbedded with clastic sediments of the Bowser Lake Group, are intensely laumontized.

Cretaceous assemblages

Most Cretaceous strata in the Intermontane Belt were metamorphosed to subgreenschist grade. For example, the Lower Cretaceous Spences Bridge Group near Spences Bridge shows evidence of a stratigraphically controlled burial metamorphism (Mamu, 1974). Deep in the succession is a zone of chlorite, calcite, quartz, albite and epidote, which grades upsection into a laumontite zone that formed at less than about 250°C. Higher in the stratigraphic section and above the prehnite-laumontite-bearing rocks are laumontite-calcite and laumonite-albite-quartz assemblages of the laumontite zone which formed at less than 250°C. Above laumonite is a heulandite-stilbite zone with heulandite in the west (temperatures about 180°C and stilbite in the east (about 160°C); albite and montmorillonite are common. The highest zeolite zone is characterized by mordenite accompanied by celadonite in rocks which are usually vesicular rather than amygdaloidal (Mamu, 1974). Other highly hydrated zeolites such as levyne, epistilbite, natrolite, analcime (Drown, 1973) occur here and there in the mordenite zone. The wide spectrum of ages given by several whole-rock K-Ar dates ranging from 58.7 ± 2.3 Ma to 94.4 ± 3.3 Ma (Thorkelson and Rouse, 1989) results from argon loss during burial metamorphism of Late Cretaceous to Paleocene age.

In the area near Hazelton and Smithers, clastic rocks of the Lower Cretaceous Skeena Group have no obvious development of secondary minerals. In contrast, the Rocky Ridge volcanic member of the Skeena Group is locally strongly zeolitized. Laumontite, scolecite, mesolite, natrolite, and thompsonite are present in veins and, less commonly, as matrix (T.A. Richards, pers. comm., 1985).

In north-central British Columbia, the mid- to Upper Cretaceous Sustut Group is largely in zeolite facies (Read and Eisbacher, 1974). The nontuffaceous sandstones of the Tango Creek Formation have common laumontite and some secondary albite and sphene. The overlying Brothers Peak Formation contains widespread laumontite, but albite cement, analcime and heulandite are developed only within or adjacent to tuffaceous sediments which contained rhyolite shards. As a result, within the Brothers Peak Formation, laumontite-, heulandite-, albite- and analcime-bearing rocks are intercalated rather than arranged according to depth of burial. This pattern resulted from dissolution of the vitric shards which controlled the composition of the fluids permeating the rocks to the extent that it was more important than the pressure and temperature changes due to stratigraphic burial. Whole-rock K-Ar ages of crystal-free vitric tuffs from the Brothers Peak Formation are 53 ± 6 Ma and 40 ± 5 Ma (Eisbacher, 1974). Both tuffs are dominantly heulandite-clinoptilolite, which is the only potassium-bearing mineral in the rock, and the dates are the age of zeolite metamorphism. The maximum stratigraphic thickness overlying the basal tuffs of the Brother Peak could not have exceeded 1000 m yielding a lithostatic pressure of 0.4 kbar (0.4×10^5 kPa) at 65°C. The total stratigraphic thickness of 2800 m to the base of the group would only yield 0.75 kbar ($.75 \times 10^5$ kPar) at 120°C, but this estimate does not take into account any tectonic loading developed as a result of thrusting in the Skeena Fold Belt (Evenchick, 1991). In the southern part of the basin, five vitrinite reflectance values range from R_{max} 0.89% to 1.50% for samples up to 670 m above the base of the group (McKenzie, 1985).

Tatla Lake Metamorphic Complex

Most metamorphic core complexes occur in the Omineca Belt; the Tatla Lake complex lies on the southwest side of the Intermontane Belt and is bounded to the southwest by the Yalakom dextral transcurrent fault. The complex has a typical, metamorphic-core complex geometry. An anticlinorial amphibolite-facies gneissic and migmatitic core is structurally overlain by a mylonitic carapace that reaches 2 km in thickness. The mylonitic rocks are in fault contact with greenschist-facies cover rocks (Friedman and Armstrong, 1988). The latter rocks have a Late Jurassic minimum age given by a U-Pb zircon date of 157 Ma from a sill and are tentatively correlated with the Hazelton Group, as are the upper plate fossiliferous rocks of probable Bajocian age.

U-Pb zircon dating brackets the timing for the two periods of metamorphism and deformation affecting the lower plate. The gneissic core records an early period of mid- to Late Cretaceous (107-79 Ma) deformation, related to a regional compressional event (Rubin et al., 1990), and amphibolite-facies metamorphism under conditions of about 650°C and at least 8 kbar (8×10^5 kPa). Metamorphic conditions are loosely constrained by pegmatites developed from partial melts and the presence of magmatic epidote in the orthogneiss. A late period of Eocene (55-47 Ma) deformation and metamorphism developed a mylonitic fabric and metamorphic assemblages in the upper part of the lower plate and obliterated any earlier metamorphic history. Three metamorphic zones (chlorite-biotite, staurolite and staurolite-kyanite±sillimanite) developed in the upper part of the lower plate. Geothermometry and geobarometry on the latter two zones yield temperatures which increase with depth from 500 to 650 ± 50°C at 8 + 3 kbar (8×10^5 kPa). Hornblende and biotite K-Ar dates of 53.6-45.6 Ma for the lower plate record cooling through the 500-300°C interval during extensional tectonic denudation of the lower plate along a low to moderate angle normal fault (Friedman, 1988; pers. comm. 1991). The style and timing of metamorphism in the Tatla Lake metamorphic complex are similar to those of the core complexes to the east in the southern Omineca Belt.

Tertiary and younger rocks

Most rocks of Eocene age have probably undergone cryptic to zeolite facies metamorphism, but few show any obvious metamorphic minerals. A conspicuous exception is the mid-Eocene Penticton Group, particularly the White Lake Formation (Church, 1973), which locally contains amygdules and fractures spectacularly filled with thomsonite, analcime, stilbite, stevensite, mordenite, and natrolite. This mineral assemblage limits temperatures to less than 175°C and pressures to less than about 2.5 kbar (2.5×10^5 kPa). In southern British Columbia, particularly in the Princston Basin, heulandite-clinoptilolite is so extensive and pervasive in tephra that it has an industrial mineral potential; laumontite and mordenite are rare. Vitrinite reflectance values from nearby carbonaceous shales and coal seams range from 0.34 to 0.94 (Read et al., 1991a). In flows, highly hydrated zeolites are widespread as joint and amygdule fillings.

In the Tahtsa Lake area, basalts of the Oligocene Endako Group are locally heavily laumontized. In most outcrops, however, the basalt shows little evidence of alteration and contains conspicuous fresh olivine. In contrast to the block-faulted structure of the Endako Group, the overlying plateau basalts of the Chilcotin Group are undeformed. Zeolites are locally present as amygdules and fracture fillings, but to a lesser extent than in the Endako basalts. Zeolites are rare in the younger parts of the Chilcotin Group in the south-central Intermontane Belt.

COAST BELT

The Coast Belt is the western of the two great welts of high-grade metamorphic rocks in the Canadian Cordillera. Metamorphism was mainly Mesozoic and early Tertiary in age, although hints of Paleozoic metamorphism are preserved locally. Polymetamorphism was probably more common than has been recognized; intense Cretaceous and early Tertiary metamorphism, deformation, and plutonism have, in many places, obscured evidence of earlier events.

Plutonic rocks make up at least 80% of the Coast Belt; the remainder is granitoid gneiss, metasediments, and metavolcanics. The metamorphic rocks commonly occur as screens or pendants surrounded and intruded by plutonic rock. The narrow width of most screens makes it difficult to determine whether some metamorphism was regional in extent or simply contact metamorphism resulting from emplacement of plutons. However, the overall consistency of the metamorphic pattern shown on the Metamorphic Map (Map 1714A, in pocket) suggests that metamorphism reflects regional thermal gradients into which plutons were emplaced (Woodsworth, 1979), rather than classical contact metamorphism. The plutons present a related problem. Many plutons were emplaced during or before metamorphism, but many show little or no evidence of metamorphism. In low-grade terranes, this in part may be because plutons require volatiles and a fracture system to develop subgreenschist and greenschist-facies assemblages. Metamorphic minerals may be present only as fracture fillings, none of which have been examined in detail in the Coast Belt. On the other hand, the normal mineral assemblages of granitoid plutons are, in effect, amphibolite to granulite-facies assemblages, and thus plutons emplaced during amphibolite-facies metamorphism commonly show no evidence of metamorphism.

Northern Coast Belt

The Coast Belt in the Yukon Territory includes metamorphosed rocks in numerous pendants within areas dominated by plutonic rocks and a large area northeast of the Denali Fault underlain by strata assigned to the Upper Jurassic and Lower Cretaceous Gravina-Nutzotin rocks of the Gambier Assemblage. Garnetiferous quartz-mica schist is common in the pendants. The Gravina-Nutzotin rocks have been described as hornfelsed schists resulting from a thermal overprint of regionally metamorphosed layered clastic rocks (Tempelman-Kluit, 1974). They contain cordierite, andalusite, quartz, mica, plagioclase, and retrograded staurolite. The thermal overprint may be related to widespread, adjacent, and intrusive, early Cenozoic granitic rocks.

Central Coast Belt

The best understood area of the Coast Belt is in the Prince Rupert-Terrace-Tahtsa Lake area, between about 53° and 55°N, which represents the most deeply exhumed part of the Coast Plutonic Complex. Metamorphic grade increases eastward from greenschist facies on the west to kyanite-amphibolite facies and to sillimanite-amphibolite and, locally, granulite facies in the core of the belt. Farther east, andalusite-amphibolite facies rocks are juxtaposed by faulting against subgreenschist assemblages of the Intermontane Belt. This pattern of metamorphism continues north into southeastern Alaska and northwestern British Columbia.

In the Prince Rupert-Terrace-Tahtsa Lake region, the Coast Belt consists of two longitudinal segments (Fig. 16.6), each with distinctive metamorphic, structural, and plutonic styles. These belts are separated by a conspicuous structural break, the Work Channel Lineament (see Fig. 17.8, Chapter 17). In some places, the eastern belt is juxtaposed by faults against generally subgreenschist-facies rocks of the Intermontane Belt; in others the contact is obscured by large, post-tectonic plutons that have superimposed contact metamorphic aureoles on low-grade rocks of the Intermontane Belt.

West of Work Channel

Metamorphism in this region consists of a Barrovian sequence, progressing from greenschist facies in the west to amphibolite facies near Work Channel Lineament (Hutchison, 1982; Crawford and Hollister, 1982; Crawford et al., 1987). The area is subdivided by an inferred ductile thrust fault near Prince Rupert, which may represent the boundary between Alexander Terrane on the west and Taku(?) Terrane to the east.

West of, and structurally below the shear thrust zone, rocks range from chlorite and biotite zone on the islands west of Chatham Sound (Woodsworth and Orchard, 1985) through garnet and chloritoid + staurolite and kyanite zones (Crawford and Hollister, 1982). Garnet-biotite and garnet-plagioclase-kyanite geothermometry and geobarometry suggest that kyanite formed at about 560°C at about 5.6 ± 0.5 kbar ($5.6 \times 10^5 \pm 0.5 \times 10^5$ kPa) (Crawford et al., 1987). East of and structurally above the Prince Rupert shear zone, kyanite + staurolite assemblages developed during westerly directed folding and thrusting. Sillimanite is not present, except in contact aureoles. The assemblage margarite + zoisite + kyanite indicates metamorphic conditions were about 625°C at about 8 kbar (8×10^5 kPa) (Crawford et al., 1979; Crawford and Hollister, 1982). The syn-metamorphic, epidote-bearing Ecstall Pluton was emplaced at similar conditions but is now at a structurally high level.

The overall pattern of metamorphic grades west of Work Channel Lineament is inverted, with high-pressure schist and plutons on the east structurally overlying lower pressure rocks farther west. The high-pressure assemblages formed at about 98 Ma, based on U-Pb dates from the syn-metamorphic Ecstall Pluton (Woodsworth et al., 1983b). Biotite through kyanite zone metamorphism farther west is slightly younger, about 90 ± 1 Ma (Sutter and Crawford, 1985) and may have resulted in part from westward thrusting of hot rocks of Taku Terrane westward over the Alexander Terrane. Glaucophane, lawsonite, and other high-pressure minerals have not been found west of Prince Rupert. K-Ar dates from plutons west of Prince

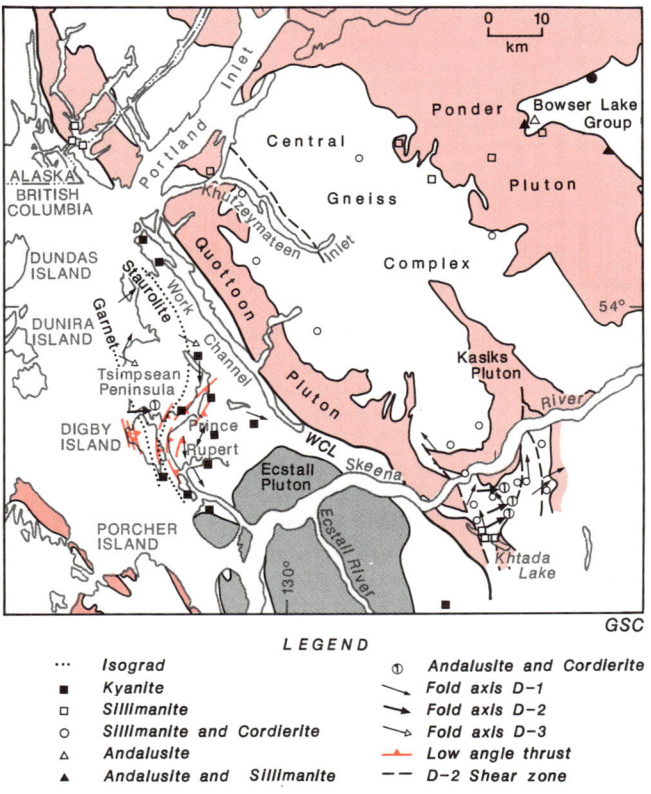

Figure 16.6. Simplified map of Prince Rupert region, after Hutchison (1982), Hollister (1982), Sisson (1985), and Woodsworth et al. (1983a). Large intrusive bodies are stippled. Gar: garnet isograd, st: staurolite isograd. Kyanite is restricted to west of Work Channel Lineament, sillimanite is restricted to the east.

Rupert indicate that the rocks had cooled to 250°C or less by about 84 Ma.

In the Douglas Channel area, regional metamorphic grade in the westernmost Coast Belt reaches upper greenschist and lower amphibolite facies. The local presence of andalusite suggests that maximum pressures were much less than in the Prince Rupert area. Metamorphism was synchronous with or slightly earlier than emplacement of Late Jurassic plutons (P. van der Heyden and G.J. Woodsworth, unpublished data), and is clearly older than the high-pressure metamorphism in the Prince Rupert area.

East of Work Channel

Amphibolite-facies metamorphism dominates the region between Work Channel and the east edge of the Coast Belt. The metamorphic rocks are known collectively as the Central Gneiss Complex (Hutchison, 1982); about half is orthogneiss and granitoid gneiss, with metasediments and metavolcanics comprising the remainder. Sillimanite is the dominant aluminosilicate. Kyanite and staurolite are absent from most areas, but local occurrences of sillimanite pseudomorphous after kyanite (Sisson, 1985) and inclusions of kyanite and staurolite in garnet indicate pre-sillimanite metamorphism (Hollister, 1977, 1982). Kyanite also is present in the belt of metamorphic rocks between the Ecstall and Quottoon plutons south of Skeena River; Krage's (1984) P-T estimates of 5-6 kbars ($5-6 \times 10^5$ kPa) at 565-600°C suggest that a drop in metamorphic temperatures may be responsible for the appearance of kyanite and staurolite-kyanite zones along part of the southwestern edge of the eastern belt. In general, peak metamorphic temperatures in the Central Gneiss Complex exceeded 650°C at about 5 to 6 kbar (5 to 6×10^5 kPa) (Hollister, 1982; Selverstone and Hollister, 1980).

Partial melting accompanied metamorphism in many parts of the Central Gneiss Complex and produced spectacular migmatites, some of which were studied in detail by Lappin and Hollister (1980) and Kenah and Hollister (1983). They concluded that partial melting of biotite + hornblende + quartz + plagioclase gneiss resulted in the removal of biotite + hornblende and the production of leucotonalite melt and the formation of hornblende as a residual mineral. Other reactions involving the partial melting of hornblende + quartz-bearing assemblages and production of granitic melt also may have occurred.

Locally, assemblages typical of low-pressure granulite facies were developed (Hollister, 1975; Selverstone and Hollister, 1980). Partial assemblages include garnet + hypersthene + quartz + plagioclase; biotite + orthoclase + orthopyroxene + garnet + quartz + plagioclase, and biotite + garnet + cordierite + sillimanite + orthoclase + plagioclase. Garnet + clinopyroxene is absent. Metamorphic conditions were constrained to about 800°C at about 5 kbar (5×10^5 kPa).

Hollister (1982) argued that kyanite-staurolite metamorphism, preserved as pseudomorphs and relicts in garnet, reached temperatures of about 600°C at about 6 kbar (6×10^5 kPa) prior to 85 Ma and speculatively correlated this metamorphism with the high-pressure metamorphism west of Work Channel Lineament. Using U-Pb dates on zircons from orthogneiss and granulites, Woodsworth et al. (1983b) suggested that peak metamorphic temperatures, preserved as sillimanite-zone rocks, were attained in Late Cretaceous time, between 83 and 65 Ma. Metamorphism was accompanied by development of large-scale recumbent folds and a gentle, north-dipping foliation (Hill, 1984).

In early Tertiary time, the Central Gneiss Complex was uplifted at the rate of about 2 mm/a (Hollister, 1982). Decompression is shown by the development of symplectites of hypersthene and plagioclase replacing garnet, by the formation of cordierite rims around garnet, and by the presence of andalusite in discordant veins. Large tonalite and diorite plutons such as the Quottoon and Kasiks were intruded during uplift; the accompanying introduction of heat may have aided in the local formation of the granulite-facies assemblages. By about 45 Ma, the region had cooled to about 250°C or less, as indicated by reset K-Ar dates on biotite from metamorphic and plutonic rocks, and was within about 5 km of the surface. Crawford et al. (1987) suggested that uplift took place along steeply-dipping ductile shears along Work Channel Lineament on the west, and ductile to brittle northeasterly directed thrusts and steep brittle faults on the east. However, the similarity in K-Ar dates between the Central Gneiss Complex and the Tatla Lake metamorphic complex immediately east of the Coast Belt, and the general absence either to the east or west of clastic sediment derived from the Coast Belt, indicate that perhaps an extensional cause for uplift and unroofing should be considered.

East margin of the Coast Belt

Along much of the Coast Belt, relationships between the Coast and Intermontane belts are obscured by large plutons. For example, the eastern part of the Ponder Pluton northwest of Terrace, was intruded at about 45 Ma into subgreenschist-facies rocks of the Bowser Lake Group (Fig. 16.6). Assemblages in the well developed contact aureole contain cordierite, sillimanite, and andalusite and indicate temperatures ranging from 400 to 700°C at 2-3 kbar ($2-3 \times 10^5$ kPa) (Sisson, 1985, 1987).

In the Tahtsa Lake area, the Coast and Intermontane belts are juxtaposed by faulting. As described by van der Heyden (1982), the boundary is a highly disrupted thrust complex. Rocks of the Central Gneiss Complex and Gamsby Metamorphic Complex were mylonitized and metamorphosed to amphibolite facies in Late Jurassic to Early Cretaceous time and later were thrust northeastward over subgreenschist facies rocks of the Intermontane Belt. The Central Gneiss Complex consists mainly of granitoid gneiss and garnet-bearing amphibolite. Structurally beneath the Central Gneiss Complex is a transition zone, the Gamsby Metamorphic Complex (van der Heyden, 1982). This unit consists largely of greenschist-facies metavolcanic rocks that may represent the metamorphosed equivalent of the Hazelton Assemblage of the Intermontane Belt. The structural and metamorphic pattern indicates that folding and easterly-directed thrusting of the Central Gneiss Complex over the Gamsby Group took place while the rocks of the former were still hot, because many of the deformational features in the Central Gneiss Complex are annealed (van der Heyden, 1982).

In the Atna Peak area, the easternmost rocks in the Coast Belt consist of metamorphosed sediments and volcanics of Stikinia. Regional metamorphic assemblages are unusual for the Central Gneiss Complex in that they are characterized by andalusite + staurolite and indicate a general Buchan-type facies sequence. Metamorphic conditions were about 550°C at about 2.5 kbar (2.5×10^5 kPa) (Ghent, Woodsworth and Evenchick, unpublished data). Metamorphism was accompanied by the development of tight, upright folds with gentle, southwest-trending axes. The intrusion of small, syn-metamorphic plutons of biotite granite plutons appears to have provided enough heat to produce sillimanite in the rocks near the plutons; pressures in the sillimanite-zone rocks are markedly higher than from the lower grade assemblages. Metamorphism occurred in early Late Cretaceous time, the same age as Buchan-type metamorphism in the Mount Raleigh area, the Chism Creek schist (both described below), and elsewhere in the eastern Coast Plutonic Complex and northern Cascade Mountains. Amphibolite-facies rocks of the Atna Peak area are in contact on the east with prehnite-pumpellyite facies rocks of the Hazelton Assemblage of the Intermontane Belt. Reconnaissance mapping suggests that this fault is a high-angle reverse fault. The metamorphism in the Atna Peak area has no known equivalent in age or style in the Prince Rupert-Terrace area. However, it is similar in many ways to metamorphism in the Mount Raleigh area about 400 km southeast, also near the east margin of the Coast Plutonic Complex.

Southern Coast Belt

In the southern Coast Belt between Vancouver and the Fraser River fault system, prehnite-pumpellyite facies assemblages are predominant in the west. East of Harrison Lake, Barrovian assemblages predominate in the northern extension of the Cascade core zone, whereas rocks of the Bridge River, Cadwallader, and Methow terranes generally have prehnite-pumpellyite to greenschist-facies assemblages.

Metamorphic rocks in the western two-thirds of the southern Coast Belt, west of Lillooet River and Harrison Lake, consist largely of greenschist and subgreenschist assemblages with local areas of Buchan-type amphibolite-facies and lawsonite blueschist assemblages. A Barrovian metamorphic sequence is well developed east of Harrison Lake, in the Canadian extension of the Cascade core zone. East of the Coast Plutonic Complex, the Coast Belt is dominated by prehnite-pumpellyite facies assemblages. The highest pressure rocks occur on both sides of the Harrison Lake-Lillooet River fault zone and its inferred extension to the northwest. Metamorphic temperatures and pressures nowhere reach those found in the Prince Rupert-Terrace region.

Pendants west of Lillooet Lake and Lillooet River

Most of the generally elongate belts of metamorphic rocks in this region lie in greenschist facies, but lower- and higher-grade rocks also occur, even in rocks of the same general age. The metamorphic grade of the Lower Cretaceous Gambier Group shows a general increase from prehnite-pumpellyite and lower greenschist facies in the west to Buchan-type amphibolite facies near the Lillooet River lineament.

For example, the Gambier Group in the Sky Pilot area (see Fig. 17.12), contains both the assemblage prehnite + pumpellyite + chlorite + quartz, typical of prehnite-pumpellyite facies, and epidote + chlorite + actinolite, characteristic of greenschist facies (Heah et al., 1986). Near the mid-Cretaceous Squamish Granodiorite, the rocks are contact metamorphosed to amphibolite facies, indicating that the prehnite-pumpellyite to greenschist metamorphism was Early to mid-Cretaceous in age.

Near the head of Jervis Inlet, pelitic rocks of probable Albian age contain biotite + andalusite and volcanic rocks contain chlorite + epidote + actinolite, suggesting temperatures of roughly 450°C and pressures of 1 to 3 kbar (1-3×10^5 kPa) (Woodsworth, unpublished data).

The highest grade exposures of probable Gambier Group occur in the Mount Raleigh area (Fig. 16.7), directly southwest of the northwestern extension of the Lillooet River fault zone. Buchan-type metamorphism is shown by staurolite, fibrolite, sillimanite, and cordierite isograds; grade increases from northeast to southwest (Woodsworth, 1977, 1979; Kerrick and Woodsworth, 1989). Andalusite is the dominant aluminosilicate throughout most of the pendant, and kyanite is absent. Temperatures during metamorphism reached about 700°C at pressures of about 2-3 kbar (2-3×10^5 kPa) (Kerrick and Woodsworth, 1989). Woodsworth (1979) concluded that metamorphism resulted from regional thermal gradients and is independent of plutonism at the present level of exposure. Pluton contacts are near-vertical, but the staurolite isograd dips about 30° southwest. Peak metamorphism occurred during uplift and was accompanied by the emplacement of syn-metamorphic plutons that deformed and modified earlier structures in the pluton. Woodsworth (1979) suggested that metamorphism was early Late Cretaceous in age; K-Ar dates indicate that the rocks had cooled to about 250°C by 70 Ma. A high-temperature, low-pressure contact aureole was superimposed on the regional metamorphism in early Tertiary time; temperatures reached about 800°C at pressures of about 2 kbar (2×10^5 kPa) (Woodsworth, 1979).

In a few areas in the Coast Belt, there is evidence for a pre-Gambier metamorphism. Foliated cobbles in conglomerate near the base of the Gambier Group (Woodsworth, 1979) indicate at least local pre-Gambier ductile deformation. Woodsworth (1979) postulated a pre-Early Cretaceous metamorphism for parts of the Mount Raleigh area, but the best evidence for this event comes from Gambier Island. There, the Gambier Group contains subgreenschist assemblages containing prehnite, chlorite, and epidote; actinolite and penetrative deformation are absent (McKillop, 1973). The underlying Bowen Island Group is foliated and contains the greenschist assemblage albite + actinolite + epidote + chlorite. The group which includes lower and Middle Jurassic rocks (Friedman et al., 1990) is cut by plutons, giving Late Jurassic K-Ar dates which imposed a contact metamorphic halo on the greenschist-facies rocks of the Bowen Island Group (McKillop, 1973).

The region west of Harrison Lake is largely underlain by low-grade Middle Triassic to mid-Cretaceous strata. All units have been metamorphosed to prehnite-pumpellyite facies. In the Lower to Middle Jurassic Harrison Lake Formation, assemblages include laumontite + prehnite +

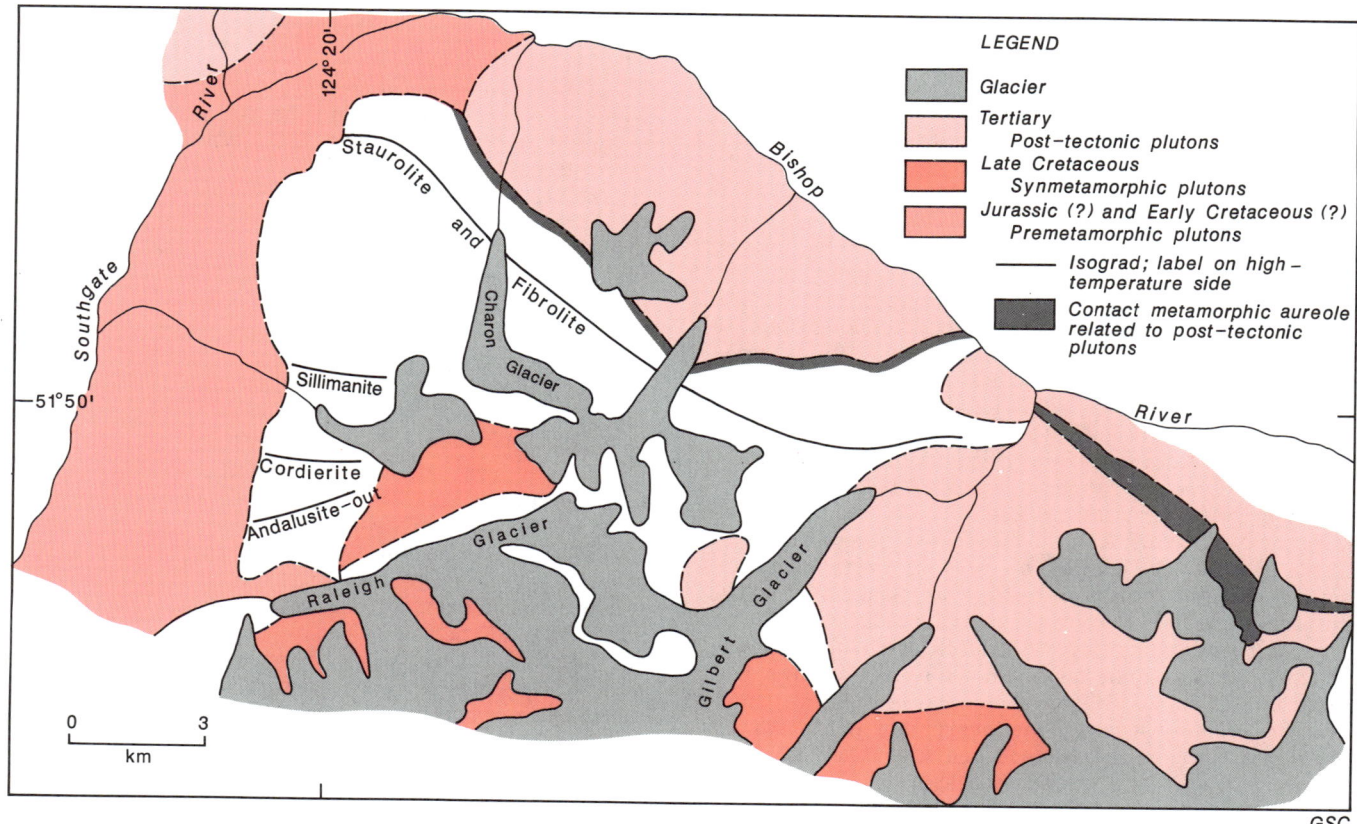

Figure 16.7. Distribution of pre-, syn-, and post-metamorphic plutons in the Mount Raleigh region, and isograds in pelitic rocks. After Woodsworth (1979) and Kerrick and Woodsworth (1989).

chlorite, prehnite + pumpellyite + epidote + chlorite, and tremolite + epidote + chlorite, all with quartz + albite + sphene (Beaty, 1974). Lawsonite has been reported by M.T. Brandon (pers. comm., 1985) from the Upper Jurassic Billhook Creek Formation. Conodonts from the Middle Triassic Camp Cove Formation have CAI values of 4 to 4.5 (M.J. Orchard, pers. comm., 1987). Assemblages characteristic of zeolite, prehnite-pumpellyite, and lawsonite blueschist all appear to be present west of Harrison Lake, but most rocks are in the prehnite-pumpellyite facies.

South of the Fraser River, in areas underlain by strata of the Chilliwack and Cultus groups, lawsonite is common in association with calcite, chlorite, albite and, rarely, pumpellyite (Beaty, 1974). Zeolites are absent. The assemblages, indicative of lawsonite blueschist facies, indicate slightly higher pressure conditions than prevailed west of Harrison Lake.

Cascade core zone

The region between Harrison Lake and Fraser River and west of the Ross Lake fault zone south of Hope represents the extension into Canada of the high-grade metamorphic rocks of the North Cascades of Washington. East of Harrison Lake, on the west side of the syn-metamorphic Spuzzum Pluton, metamorphic grade increases from west to east from garnet to garnet-staurolite, to staurolite-kyanite, to coarse sillimanite zones (Gabites, 1985; Lowes, 1972; Reamsbotton, 1971, 1974). Peak metamorphic conditions reached about 700°C at 6 to 8 kbar (6-8x10^5 kPa).

Between Fraser River and the Spuzzum Pluton, metamorphic conditions were similar (Pigage, 1976; Bartholomew, 1979; Hollister, 1969a,b). Isograds do not parallel contacts with the Spuzzum Pluton and in places are truncated by the pluton.

Rocks showing similar Barrovian metamorphic effects extend discontinuously south into the North Cascades of Washington (Haugerud, 1985). Gabites (1985) inferred an Early Cretaceous age for the metamorphism, based on U-Pb dating of the pre- to syn-metamorphic Spuzzum Pluton, an age that is in accord with Early Cretaceous blueschist metamorphism in the Shuksan Terrane in Washington (Brown, 1988). Cowan and Potter (1986) and Monger et al. (in press) suggested that metamorphic rocks in the southern Coast Mountains and North Cascades occur in a series of stacked thrust sheets whose age decreases towards the west. In this interpretation, Barrovian rocks of the Harrison Lake-Fraser River region are high in the structural sequence and structurally overlie lower-grade rocks west of Harrison Lake. Metamorphism west of Harrison Lake may reflect structural burial during mid-Cretaceous thrusting.

There is some evidence for an early Buchan-type metamorphism in the area. Reamsbottom (1974) described andalusite porphyroblasts near the Spuzzum pluton and Pigage (1976) and Hollister (1969a) described andalusite pseudomorphs that have been replaced by sillimanite. Hollister (1969a) suggested that andalusite crystallized metastably in the kyanite field and was replaced by stable assemblages. However, Pigage (1976) suggested that the andalusite formed by early, low-pressure contact metamorphism related to intrusion of the Spuzzum Pluton and was later overprinted by higher-pressure assemblages. The widespread occurrence of the andalusite pseudomorphs in the North Cascades (Evans and Berti, 1986) suggests that Pigage's interpretation is probably correct.

Northeast margin of the southern Coast Belt

Rocks east of most plutons forming the Coast Plutonic Complex but west of the Yalakom and Pasayten faults show varying styles of metamorphism. In general, the rocks are sub-greenschist facies; Triassic strata are probably mainly prehnite-pumpellyite facies.

Bridge River region

Between Lillooet and Bralorne, the Bridge River Terrane consists of two structural blocks separated by a high-angle fault (Potter, 1983). Greenstones in the southwestern block have prehnite-pumpellyite facies assemblages. Actinolite is absent; laumontite + chlorite + quartz is locally present. Metamorphism was essentially static, as few rocks show penetrative fabrics, and probably reflects burial at pressures less than about 4 kbar (4×10^5 kPa). Local amphibolite-facies assemblages may reflect early ocean-floor metamorphism close to a volcanic or intrusive centre (Potter, 1983).

The northeastern block shows an inverted metamorphic sequence ranging from lower amphibolite facies at high structural levels to prehnite-pumpellyite facies at the structural base. Metamorphic conditions were about 325 to 525°C at about 4 to 5 kbar (4-5$\times10^5$ kPa) (Potter, 1983). Potter interpreted deformation and metamorphism as resulting from Late Jurassic overthrusting by hot rocks of the Shulaps ultramafic complex.

Triassic and Jurassic strata in the Cadwallader Terrane west and north of the Bridge River Terrane are metamorphosed in prehnite-pumpellyite facies (Rusmore, 1985, 1987). Actinolite, and possibly biotite, are present in some samples that lack pumpellyite. Conodont alteration indices are mostly between 2 and 4 (Rusmore, 1985). Metamorphic grade of the overlying Cretaceous and early Tertiary strata is not known but is assumed to be zeolite facies.

A zone of Buchan-type metamorphic rocks southwest of Bralorne was called the Chism Creek schist by Rusmore (1985); the protoliths may in part be Bridge River and Cadwallader groups. The assemblage of andalusite + cordierite + biotite + garnet + quartz and others suggest peak temperatures of about 500 to 650°C at pressures of less than 3 kbar (3×10^5 kPa). U-Pb dates on zircons from both deformed and crosscutting intrusions bracket the age of metamorphism and associated deformation to between 84 and 100 Ma (Rusmore, 1985). Metamorphism was thus the same age as that of the lawsonite blueschist metamorphism on San Juan Islands, and as that in the Mount Raleigh area, but may be older than peak metamorphism of the Central Gneiss Complex in the Terrace region.

The Chism Creek schist is part of a belt of Buchan-type metamorphism of mid- to Late Cretaceous age that extends from Mount Raleigh southeast through the eastern Coast Belt into the North Cascades of Washington. In some areas (e.g. Mount Raleigh, Kwoiek Creek, and parts of the North Cascades) peak metamorphic temperatures coincided with uplift and decompression. The metamorphism was accompanied by high heat flow, deformation, and plutonism and may be related to crustal thickening related to transpression along the length of the Coast Belt. Maximum metamorphic pressures in this belt occur in the Spuzzum area, suggesting that the amount of relative uplift decreased from south, near Hope, to northwest in the Chism Creek and Mount Raleigh areas.

INSULAR BELT

Most studies of metamorphism in the Insular Belt, which in Canada is mainly underlain by Wrangellia, have been on Vancouver Island. There, the main metamorphic events were an Early Permian to Middle Triassic greenschist metamorphism, an Early to Middle Jurassic prehnite-pumpellyite burial metamorphism, and a middle to late Eocene zeolite burial metamorphism. On the Queen Charlotte Islands, possible burial metamorphism of Pliocene age may be the youngest known metamorphism of that type in the Canadian Cordillera. On Vancouver Island, metamorphic grade and intensity of plutonism increase to the southwest. A similar pattern also is present in the Queen Charlotte Islands and in the Craig subterrane of the Alexander Terrane. Eocene contact metamorphism locally has affected rocks that were metamorphosed in the zeolite and pumpellyite facies. Metamorphism in the Pacific Rim Terrane has two elements. One, Late Cretaceous in age, has blueschist affinities and the other is an Eocene Buchan-type of metamorphism. On southern Vancouver Island, Wrangellia is underthrust by the Pacific Rim Terrane, which is in turn underthrust by the Crescent Terrane of prehnite-pumpellyite facies. The high-pressure parts of the Pacific Rim Terrane are related not to subduction but to extreme tectonic thickening during plate collision in Late Cretaceous time.

Wrangellia

Away from plutons, Wrangellian rocks on Vancouver Island show metamorphic effects ranging from chlorite and biotite zones in parts of the Sicker Group through prehnite-pumpellyite and zeolite facies in the Karmutsen Formation to zeolite facies in Jurassic and Cretaceous strata (Fig. 16.8 and 16.9). Unmetamorphosed Eocene to Oligocene clastic sediments of the Carmanah Group place a minimum age on the latest period of metamorphism and latest movements on faults truncating the southern end of the terrane. A rough increase in metamorphic grade occurs from northeast to southwest, reflecting perhaps a general increase in depth of erosion and the abundance of plutonic rocks towards the west side of the island.

The age and origin of metamorphism of the Devonian to Permian Sicker Group is uncertain, but there is evidence for both pre-early Triassic and post-Late Triassic events. Metamorphic grade of the Sicker Group varies from place to place. It is probably greenschist facies in most areas but reaches amphibolite facies on the west side of Vancouver Island in areas of abundant synkinematic plutons of the Island Suite. In some regions, the Sicker Group shows little or no evidence of penetrative deformation, and igneous clinopyroxenes are well preserved (e.g.

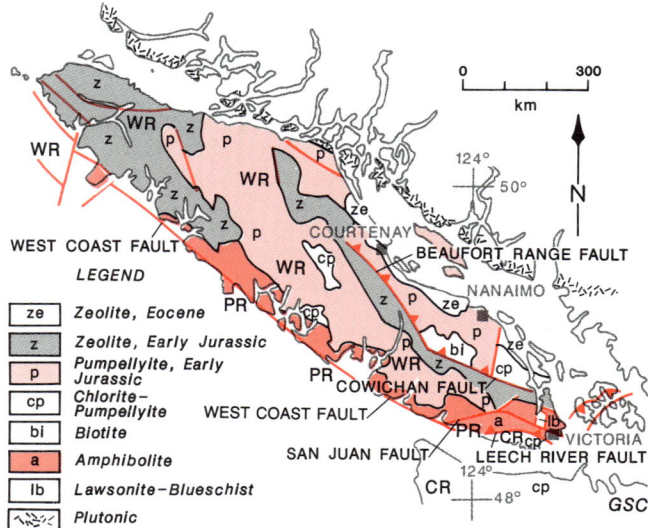

Figure 16.8. Simplified metamorphic map of Vancouver Island, showing the distribution of facies and the principal structural elements. Noteworthy features are the broad belt of prehnite-pumpellyite facies in the central portion of the island, zeolite facies at north and south, and both blueschist and Buchan-type metamorphism together in the narrow belt of Crescent Terrane near Victoria. WR=Wrangellia; CR=Crescent Terrane; PR=Pacific Rim Terrane.

Muller, 1980). In other areas the rocks have a strong schistose fabric and pyroxenes are extensively uralitized. It is possible that the schistosity was produced during Late Cretaceous faulting (N.W.D. Massey, pers. comm., 1987). Nonetheless, it seems clear that the Sicker Group was regionally deformed prior to deposition of the Upper Triassic Karmutsen Formation.

The most detailed study of stratigraphy and metamorphism in the Sicker Group is that of Juras (1987) in the Buttle Lake area. There, the Sicker Group underwent regional submarine hydrothermal metamorphism and hydrothermal alteration associated with volcanogenic sulphide mineralization, and burial metamorphism. Mineral assemblages are characteristic of the chlorite zone of the greenschist facies and pumpellyite-actinolite facies. Pumpellyite-actinolite facies assemblages are found in the uppermost parts of the sequence and chlorite zone rocks are dominant in the lower parts of the section.

Conodonts from Pennsylvanian-Permian limestone of the Buttle Lake Formation, the youngest unit in the Sicker Group, and from Early Mississippian chert lower in the succession have CAI values ranging from 5 to 7 (Brandon et al., 1986). These values are consistent with chlorite zone metamorphism. CAI values from the overlying Upper Triassic Parson Bay Formation range from 3 to 7 and on average are significantly lower than those from the Sicker Group. The CAI data strongly suggest that metamorphism of the Sicker Group took place in pre-Late Triassic time.

Basaltic volcanics of the Karmutsen Formation and the roughly 2000 m of overlying calc-alkaline volcanics of the Bonanza Group show evidence of a stratigraphically controlled burial metamorphism. Laumontite-bearing rocks of the Bonanza overlie zeolite-bearing rocks of the upper part of the Karmutsen Formation (Fig. 16.9). These pass downward into prehnite and pumpellyite-bearing Karmutsen volcanics. Although the transition is complicated in detail, as was emphasized by Surdam (1967), it is restricted to a few thousand metres vertically within Karmutsen rocks. Throughout central and northeastern Vancouver Island, plutons of the Middle to Late Jurassic Vancouver Island Suite imposed contact-metamorphic aureoles on the earlier low-grade regional assemblages (Kuniyoshi and Liou, 1976a,b; Surdam, 1967, 1973). The burial metamorphism was thus Early to Middle Jurassic in age.

On the west side of Vancouver Island, amphibolite-facies metamorphic rocks form part of the Westcoast Crystalline Complex of Early to Middle Jurassic age (Isachsen, 1984). He considered the difference between the Westcoast Complex and the Island Suite to have resulted from the deeper level of plutonic emplacement in the metamorphic complex compared with that of most of the plutons in subgreenschist facies rocks to the east.

Clastic sediments of the Upper Cretaceous Nanaimo Group unconformably overlie the Bonanza and older units. In the Nanaimo Group (Fig. 16.9), laumontite cements grains and replaces some plagioclase in volcanic-lithic arenites in all but the uppermost formation. There, heulandite occurs with laumontite as a cement and replacement mineral (Stewart and Page, 1974). Minor prehnite near the base of the group may be a detrital mineral derived from the underlying Karmutsen Formation. The distribution of coal rank and vitrinite reflectance data appears related to local heat flow patterns rather than to stratigraphic depth. In the San Juan Islands and in northwestern mainland Washington, C.W. Naeser (pers. comm., 1986) obtained Eocene fission track dates on apatite from the 6000 m thick Eocene Chuckanut Formation. These may represent the time of metamorphism not only for the laumontite-bearing Chuckanut Formation, but probably also the age of zeolite facies metamorphism of the Nanaimo Group. At Mount Washington on Vancouver Island, quartz diorite porphyry of earliest Oligocene age thermally metamorphosed sandstone and argillite near the base of the Nanaimo Group (Carson, 1960; McGuigan, 1975), restricting the zeolite-facies metamorphism to a time prior to earliest Oligocene but later than the Late Cretaceous protolith.

On northern Vancouver Island, faulted remnants of Cretaceous clastic units have been assigned to the zeolite facies category on the Metamorphic Map (Map 1714A, in pocket) on the basis of vitrinite reflectance data. The youngest remnants, part of the Nanaimo Group, contain coal of high-volatile bituminous B rank similar to that of the zeolite-bearing rocks of the Nanaimo Group elsewhere. The other remnants contain no analyzed coals but are assumed to have undergone zeolite metamorphism.

On the Queen Charlotte Islands and in the adjacent Queen Charlotte Basin, burial metamorphism has affected the oldest stratified rocks, to the youngest, the Miocene volcanics of the Masset Formation and Mio-Pliocene

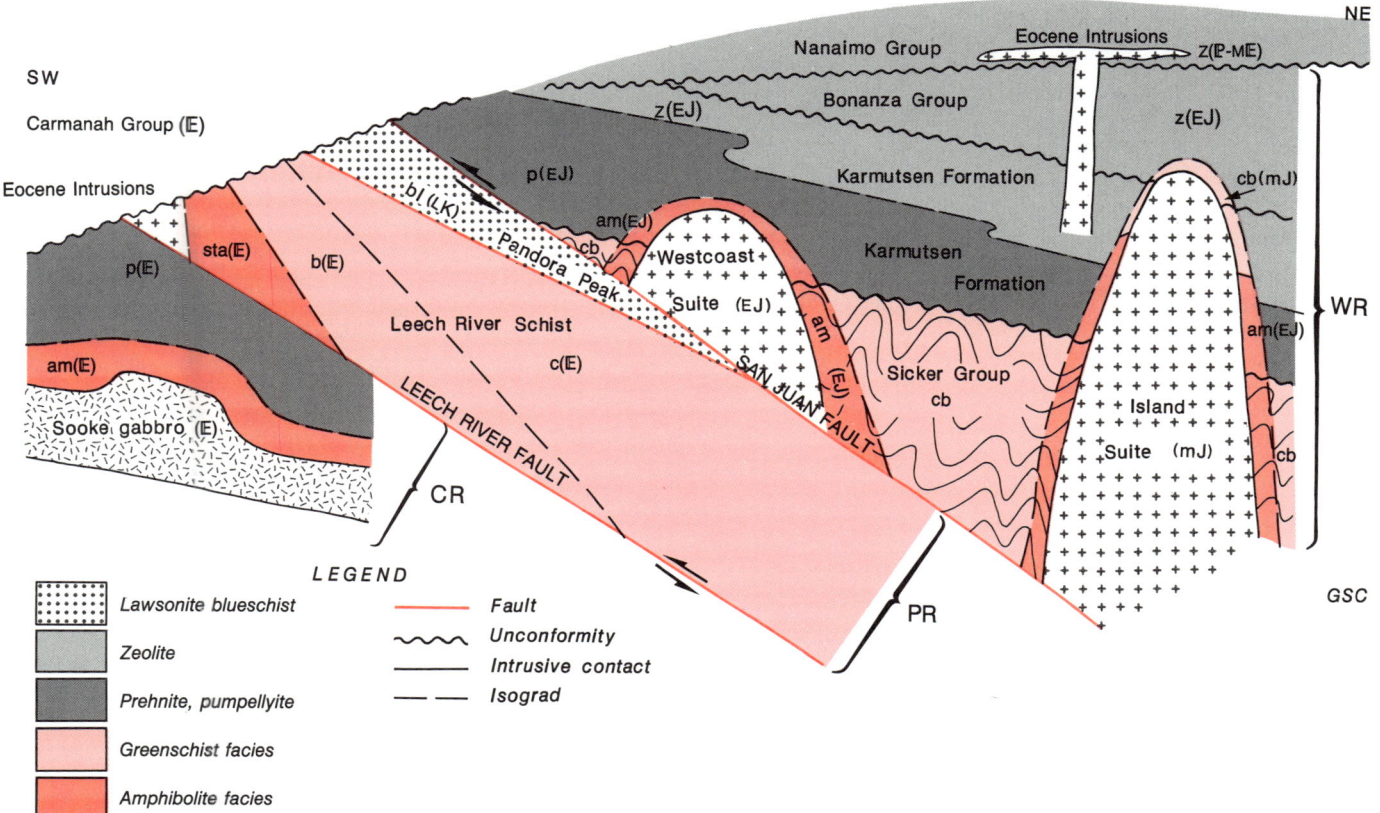

Figure 16.9. Schematic cross-section through Vancouver Island illustrating the stratigraphic, structural, and temporal relationships between lithologies and metamorphism. The diagram is not to scale, and the vertical scale is exaggerated. See Table 16.1 for metamorphic symbols. PR-Pacific Rim Terrane; CR-Crescent Terrane; WR-Wrangellia.

sediments of the Skonun Formation. The Karmutsen volcanics are in the prehnite-pumpellyite facies (Sutherland Brown, 1968). Zeolite-bearing rocks first appear near the base of the overlying Upper Triassic to Jurassic stratified rocks and continue upward into Middle Jurassic volcanics which contain stilbite. Conodonts from Upper Triassic strata have CAI values ranging from 2 to 8 (Orchard, 1988). Values from Graham Island are, on average, lower than those from Moresby Island, and the higher values are related to proximity to intrusions. Higher in the stratigraphic sequence, the mid-Cretaceous Haida and Honna formations contain laumontite and mordenite as cement in sandstones (Yagishita, 1985).

Along the southwest coast of the islands, intrusion of plutons of the Middle to Late Jurassic Island Suite(?) has developed chlorite greenschist to amphibolite facies hornfels aureoles in the Karmutsen Formation.

Volcanic rocks of the Miocene Masset Formation (about 1500 to 2000 m thick) locally are amygdaloidal, with fillings of chlorite, silica, and zeolites including chabazite, thomsonite and stilbite (Sutherland Brown, 1968; C.J. Hickson, pers. comm., 1988). However, zeolites are only locally developed and their development was controlled by permeability rather than stratigraphic depth (T. Hamilton, pers. comm., 1987). Many samples of the Masset Formation contain glass. In Queen Charlotte Basin east of the islands, volcanic litharenite of the Miocene to Pliocene Skonun Formation contains chloritized mafic fragments, laumontite cement, and laumontite replacing plagioclase from drill cores as little as 1600 m deep (Galloway, 1974). Because the Skonun Formation interfingers laterally with the Masset (Hickson, 1988) zeolite-facies burial metamorphism did not end before the Pliocene.

In southwestern Yukon Territory, Wrangellian strata underwent Mesozoic and possibly also Miocene burial metamorphism. Between the Denali and Duke River faults, upper Paleozoic Skolai Group and Upper Triassic Nikolai Greenstone are in prehnite-pumpellyite facies. The Nikolai Greenstone is structurally (and stratigraphically?) overlain by Jura-Cretaceous clastic sediments of the Gravina-Nutzotin Assemblage, which show no obvious metamorphic effects. The strata are cut by Early Cretaceous post-kinematic plutons. Some plutons have well-developed contact aureoles, thus burial metamorphism is no younger than latest Early Cretaceous (Read and Monger, 1976).

In the Alaskan part of Wrangellia between the Border Ranges and Hubbard faults, prehnite-pumpellyite facies volcanics of the Nikolai Greenstone locally contain zeolites. These rocks are overlain by about 2000 m of Jurassic and Matanuska Assemblage, Lower Cretaceous rocks. The Cretaceous strata contain laumontite-, heulandite- or analcime-bearing assemblages in the matrix of the clastic sediments (MacKevett, 1971).

In the Denali-Duke River segment of Wrangellia, 400 m of Oligocene clastic sediments with subbituminous B and C coal are overlain by about 1800 m of Miocene to Pliocene lavas. The coal rank indicates thermal maturation of probable Pliocene age, of these zeolite facies sediments. The overlying Wrangell lavas are unmetamorphosed.

Alexander Terrane

The Alexander Terrane in the Saint Elias Mountains comprises a thick, mainly sedimentary, sequence of rocks ranging in age from Late Cambrian to Late Triassic. In most places these strata have been regionally metamorphosed to chlorite and biotite zones. Conodonts from chlorite-zone rocks of Ordovician to Late Triassic age have CAI values of about 5 (M.J. Orchard, pers. comm., 1986). Locally in the western parts of the region, the metamorphic grade reaches upper greenschist to mid-amphibolite facies.

In the central and northeastern parts of the Saint Elias Mountains, epizonal plutons of the Saint Elias Suite have well developed hornfels aureoles and discordantly cut greenschist-facies strata. These plutons give K-Ar dates ranging from 130 to 160 Ma (Dodds and Campbell, 1988). Age of metamorphism was therefore post-Late Triassic and pre-Late Jurassic in age. However, plutons of the Late Pennsylvanian to Early Permian Icefield Ranges Suite locally intrude and superimpose contact aureoles on greenschist-grade strata, suggesting that some regional metamorphism was pre-Late Pennsylvanian in age (Dodds and Campbell, 1988).

Oligocene sediments underlie as much as 1900 m of Wrangell Lava. The volcanics are only locally amygdaloidal and only locally zeolite-bearing. Although the lavas are essentially unmetamorphosed, the subbituminous or higher rank of coals of the underlying sediments implies a zeolite metamorphism of Miocene to Pliocene age.

Pacific Rim and Crescent terranes

The southwest part of Vancouver Island is underlain largely by the Jura-Cretaceous Pacific Rim Terrane (separated from Wrangellia by the San Juan Fault) and the Eocene Crescent Terrane (juxtaposed against Pacific Rim Terrane along the Leech River Fault). The Pacific Rim Terrane contains two distinct lithological units, each with its own metamorphic style: Upper Cretaceous, high-pressure, low-temperature lawsonite assemblages; and fault-bounded blocks of low-pressure high-temperature assemblages metamorphosed in Late Eocene time. The Crescent Terrane is entirely in prehnite-pumpellyite facies.

The Leech River Schist, a component of the Pacific Rim Terrane, consists of pelite and wacke bearing a low-pressure, high-temperature Buchan-type overprint. Metamorphic biotite gave a K-Ar date of 39 Ma, indicating a Late Eocene age for the metamorphism (Fairchild and Cowan, 1982). Well defined isograds separate chlorite zone rocks near the San Juan Fault from biotite zone and finally andalusite + staurolite + biotite + garnet-bearing rocks near the Leech River Fault. Isograds are truncated by the Leech River Fault and are inferred to be cut by the San Juan Fault. The highest grade rocks reached temperatures of 500 to 600°C at pressures between 1.5 and 3.5 kbar (1.5-3.5x10^5 kPa) (Fairchild and Cowan, 1982).

High-pressure metamorphic rocks, the Pandora Peak unit of Rusmore and Cowan (1985), are exposed in several small fault blocks. The diagnostic assemblages are lawsonite + quartz + calcite, with or without prehnite (Rusmore and Cowan, 1985). Pumpellyite + epidote is present in some lawsonite-free rocks. Similar assemblages are found in other fault-bounded slices and in the Constitution Formation on San Juan Islands, where the assemblages lawsonite + quartz + aragonite and prehnite ± lawsonite + quartz + aragonite define a prehnite-in isograd (Brandon, 1980; Vance, 1968). On the San Juan Islands, metamorphism affected rocks as young as Late Albian, and metamorphic clasts occur in rocks of Santonian-Campanian age. Metamorphism is thus tightly constrained to an interval of about 10 Ma between 93 and 83 Ma (Brandon, 1980).

Juxtaposition of the high-pressure assemblages of the Pandora Peak and Constitution rocks against the Buchan metamorphic rocks of the Leech River Schist requires a fault of large displacement. Further, the contrast in metamorphic styles between the two units suggests that the Pacific Rim Terrane may require subdivision.

The Crescent Terrane of southern Vancouver Island consists largely of the Eocene Metchosin dismembered ophiolite suite (Massey, 1986). The rocks were subjected to seafloor metamorphism followed by prehnite-pumpellyite metamorphism. The time of metamorphism postdates the Late Eocene Buchan-type metamorphism of the Leech River Schist but is older than the unmetamorphosed, overlying clastic sediments of the Carmanah Group.

Yorath et al. (1985) concluded that the San Juan and the Leech River faults, which form the north and south boundaries of the Pacific Rim Terrane, dip gently north, indicating that both the Pacific Rim and Crescent terranes have been thrust beneath Wrangellia from the south. Rusmore and Cowan (1985) suggested development of a Late Cretaceous high-pressure metamorphism of a Jura-Cretaceous protolith of the Pacific Rim Terrane and the Constitution Formation of the San Juan Islands. These rocks were underthrust in early Tertiary time beneath Wrangellia. Buchan-type metamorphism of the Leech River Schist took place in Late Eocene time, and the rocks were subsequently underthrust beneath the lawsonite-bearing rocks of the Pandora Peak unit.

CONCLUSIONS

All rocks in the Canadian Cordillera older than Miocene, even those in the Foreland Belt that have traditionally been thought to be unmetamorphosed, show evidence of metamorphism beyond that assigned to diagenesis. Cryptic and subgreenschist metamorphism in the Foreland Belt is apparent from the patterns of organic maturity parameters (particularly CAI and reflectance data) found in these regions rather than from the more traditional mineral assemblage approach. The omnipresent nature of metamorphism in the Cordillera is a corollary of the fact that, since Late Triassic time, the region has been one of high heat flow (see Sweeney et al. in Chapter 2). Indeed, it would be surprising if any old rocks in a tectonically active orogen such as the Cordillera survived in an unmetamorphosed state.

Greenschist- and amphibolite-facies rocks reflect deep burial and high temperatures and form core zones in the Coast and Omineca belts. Lower grade rocks, suggestive largely of burial metamorphism, characterize most of the Foreland, Intermontane, and Insular belts. The core zones coincide roughly with the two main belts of plutons in the Cordillera and resulted from crustal thickening and contraction, followed by generally rapid uplift and erosion. Metamorphism in the flanking belts is at least in part a high-level reflection of deep-seated processes in the core zones.

In the broadest sense, metamorphism coincided with and probably resulted from accretion of allochthonous terranes to North America in Mesozoic and early Tertiary time. Late Triassic and Early Jurassic dates on metamorphic micas in eastern Alaska (C. Dusel-Bacon, pers. comm., 1985) and from near the Yukon-British Columbia border (Gabrielse and Reesor, 1964) offer evidence for regional metamorphism at this time, resulting perhaps from the accretion and overlap of allochthonous terranes. Farther south, however, metamorphism in the Omineca Belt peaked in late Early to Middle Jurassic time, coincident with the later accretion and overlap there of allochthonous terranes. Similarly, post-tectonic plutons in the north, such as the Tatlmain Pluton and those of the Black Lake Suite are slightly older than the Middle Jurassic syn- and post-tectonic Nelson and Kuskanax plutonic suites in the southern Omineca Belt. Accordingly, metamorphism and crustal thickening, likely associated with terrane accretion in the Omineca Belt, may be diachronous, having started significantly earlier in the north.

Metamorphism in the western high-grade welt, the Coast Belt, is younger than that in the Omineca Belt. Greenschist- to amphibolite-facies metamorphism may have begun in Middle Jurassic time, reached its peak in mid-Cretaceous time, and persisted into the early Tertiary. As in the Omineca Belt, metamorphism reflects crustal thickening and compression related to accretion. Metamorphism appears to be diachronous, but the details are poorly known.

The highest-grade parts of the Omineca Belt are characterized by Eocene K-Ar dates on metamorphic minerals. These dates, long a problem to workers in the Canadian Cordillera, are now known to record Eocene extension and crustal thinning related to collapse of the hot, over-thickened crust. Recent work in the Omineca Belt has shown that high-pressure, Jurassic and mid-Cretaceous assemblages are overprinted by low-pressure, high-temperature Eocene assemblages. In the Coast Belt, the presence of Eocene K-Ar cooling dates from the highest grade parts of the belt suggests that rapid uplift by extension and crustal thinning likely occurred there as well.

REFERENCES

Apted, M.J. and Liou, J.G.
1984: Phase relations among greenschist, epidote amphibolite and amphibolite in a basaltic system; American Journal of Science, v. 283-A, p. 328-354.

Archibald, D.A., Glover, J.K., Price, R.A., Farrar, E., and Carmichael, D.M.
1983: Geochronology and tectonic implications of magmatism and metamorphism, southern Kootenay Arc and neighbouring regions, southeastern British Columbia. Part I: Jurassic to mid-Cretaceous; Canadian Journal of Earth Sciences, v. 20, p. 1891-1913.

Archibald, D.A., Krogh, T.E., Armstrong, R.L., and Farrar, E.
1984: Geochronology and tectonic implications of magmatism and metamorphism, southern Kootenay Arc and neighbouring regions, southeastern British Columbia. Part II: Mid-Cretaceous to Eocene; Canadian Journal of Earth Sciences, v. 21, p. 567-583.

Armstrong, J.E.
1949: Fort St. James map-area, Cassiar and Coast districts, British Columbia; Geological Survey of Canada, Memoir 252, 210 p.

Armstrong, R.L.
1982: Cordilleran metamorphic core complexes - from Arizona to southern Canada; Annual Review of Earth and Planetary Sciences, v. 10. p. 129-154.

1988: Mesozoic and early Cenozoic magmatic evolution of the Canadian Cordillera; in Processes in Continental Lithospheric Deformation, S.D. Clark, Jr. et al. (ed.), Geological Society of America, Special Paper 218, p. 55-91.

Bartholomew, P.R.
1979: Geology and metamorphism of the Yale Creek area, British Columbia; M.Sc. thesis, University of British Columbia, Vancouver, 105 p.

Beaty, R.J.
1974: Low grade metamorphism of Permian to Early Cretaceous volcanic and volcanoclastic rocks near Chilliwack, British Columbia; B.Sc. thesis, University of British Columbia, Vancouver.

Berman, R.G.
1988: Internally-consistent thermodynamic data for minerals in the system $Na_2O-K_2O-CaO-MgO-FeO-Fe_2O_3-Al_2O_3-SiO_2-TiO_2-H_2O-CO_2$; Journal of Petrology, v. 29, p. 445-522.

Berman, R.G. and Brown, T.H.
1985: Heat capacity of minerals in the system $Na_2O-K_2O-CaO-MgO-FeO-Fe_2O_3-Al_2O_3-SiO_2-TiO_2-H_2O-CO_2$: representation, estimation, and high temperature extrapolation; Contributions to Mineralogy and Petrology, v. 89, p. 168-183.

Bostick, N.H.
1979: Microscopic measurement of the level of catagenesis of solid organic matter in sedimentary rocks to aid exploration for petroleum and to determine former burial temperatures - a review; Society Economic Paleontologists and Mineralogists, Special Publication 26, p. 17-43.

Braman, D.R.
1981: Upper Devonian-Lower Carboniferous microspore biostratigraphy of the Imperial Formation, District of Mackenzie and Yukon; Ph.D. thesis, University of Calgary, Alberta, 378 p.

Brandon, M.T.
1980: Structural geology of Middle Cretaceous thrust faulting on San Juan Island, Washington; M.Sc. thesis, University of Washington, Seattle, 130 p.

Brandon, M.T., Orchard, M.J., Parrish, R.R., Sutherland Brown, A., and Yorath, C.J.
1986: Fossil ages and isotopic dates from the Paleozoic Sicker Group and associated intrusive rocks, Vancouver Island, British Columbia; in Current Research, Part A, Geological Survey of Canada, Paper 86-1A, p. 683-696.

Brown, E.H.
1988: Structural geology and accretionary history of the North Cascades system, Washington and British Columbia; Geological Society of America Bulletin, v. 99, p. 201-214.

Brown, R.L. and Read, P.B.
1983: Shuswap terrane of British Columbia: a Mesozoic "core complex"; Geology, v. 11, p. 164-168.

Brown, R.L., Journeay, J.M., Lane, L.S., Murphy, D.C., and Rees, C.J.
1986: Obduction, backfolding and piggyback thrusting in the metamorphic hinterland of the southeastern Canadian Cordillera; Journal of Structural Geology, v. 8, p. 255-268.

Burns, P.J.
1973: Stratigraphy and low grade metamorphism of the "Takla-Hazelton" Group, McConnell Creek map area, north-central British Columbia; B.Sc. thesis, University of British Columbia, Vancouver, 56 p.

Bustin, R.M.
1984: Coalification levels and their significance in the Groundhog coalfield, north-central British Columbia; International Journal of Coal Geology, v. 4, p. 21-44.

Campbell, K.V.
1971: Metamorphic petrology and structural geology of the Crooked Lake area, Cariboo Mountains, British Columbia; Ph.D. thesis, University of Washington, Seattle, 192 p.

Campbell, R.B.
1970: Structural and metamorphic transitions from infrastructure to superstructure, Cariboo Mountains, British Columbia; in Structure of the Southern Canadian Cordillera, J.O. Wheeler (ed.), Geological Association of Canada, Special Paper No. 6, p. 67-72.
1973: Structural cross-section and tectonic model of the southeastern Canadian Cordillera; Canadian Journal of Earth Sciences, v. 10, p. 1607-1620.

Campbell, R.B., Mountjoy, E.W., and Young, F.G.
1973: Geology of McBride map-area, British Columbia; Geological Survey of Canada, Paper 72-35, 104 p.

Carman, J.H. and Gilbert, M.C.
1983: Experimental studies on glaucophane stability; American Journal of Science, v. 283-A, p. 414-437.

Carmichael, D.M.
1978: Metamorphic bathozones and bathograds: a measure of depth of post-metamorphic uplift and erosion on the regional scale; American Journal of Science, v. 278, p. 769-797.

Carrigy, M.A.
1971: Lithostratigraphy of the uppermost Cretaceous (Lance) and Paleocene strata of the Alberta Plains; Research Council of Alberta, Bulletin 27, 161 p.

Carrigy, M.A. and Mellon, G.B.
1964: Authigenic clay mineral cements in Cretaceous and Tertiary sandstones of Alberta; Journal of Sedimentary Petrology, v. 34, p. 461-472.

Carson, D.J.T.
1960: Geology of Mount Washington, Vancouver Island, British Columbia; MA.Sc. thesis, University of British Columbia, Vancouver, 116 p.

Cheney, E.S.
1980: Kettle dome and related structures of northeastern Washington; in Cordilleran Metamorphic Core Complexes, M.D. Crittenden, Jr., P.J. Coney and G.H. Davis (ed.), Geological Society of America, Memoir 153, p. 463-484.

Church, B.N.
1973: Geology of the White Lake Basin; British Columbia Department of Mines and Petroleum Resources, Bulletin 61, 120 p.

Coates, J.A.
1960: Analcite bearing volcanic rocks of the Quesnel River Group, Cariboo District, British Columbia; B.Sc. thesis, University of British Columbia, Vancouver, 57 p.

Coombs, D.S.
1971: Present status of the zeolite facies: molecular sieve zeolites, v. 1; in Advances in Chemistry, Series 101, American Chemical Society, p. 317-327.

Cowan, D.S. and Potter, J.
1986: B3. Juan de Fuca Spreading Ridge to Montana Thrust Belt; Geological Society of America, Decade of North American Geology, Centennial Continent/Ocean Transect #9, 3 sheets.

Craw, D.
1978: Metamorphism, structure and stratigraphy in the Southern Park Ranges, British Columbia; Canadian Journal of Earth Sciences, v. 15, p. 86-98.

Crawford, M.L. and Hollister, L.S.
1982: Contrast of metamorphic and structural histories across the Work Channel lineament, Coast Plutonic Complex, British Columbia; Journal of Geophysical Research, v. 87, p. 3849-3860.

Crawford, M.L., Kraus, D.W., and Hollister, L.S.
1979: Petrologic and fluid inclusion study of calc-silicate rocks, Prince Rupert, British Columbia; American Journal of Science, v. 279, p. 1135-1159.

Crawford, M.L., Hollister, L.S., and Woodsworth, G.J.
1987: Crustal deformation and regional metamorphism across a terrane boundary, Coast Plutonic Complex, British Columbia; Tectonics, v. 6, p. 343-361.

Crosby, P.
1968: Tectonic, plutonic, and metamorphic history of the central Kootenay arc, British Columbia, Canada; Geological Society of America, Special Paper 99, 94 p.

Currie, K.L.
1975: The geology and petrology of the Ice River alkaline complex, British Columbia; Geological Survey of Canada, Bulletin 245, 68 p.

Dechesne, R.G., Simony, P.S., and Ghent, E.D.
1984: Structural evolution and metamorphism of the southern Cariboo Mountains near Blue River, British Columbia; in Current Research, Part A, Geological Survey of Canada, Paper 84-1A, p. 91-94.

Delaney, G.D.
1985: The Middle Proterozoic Wernecke Supergroup, Wernecke Mountains, Yukon Territory; Ph.D. thesis, University of Western Ontario, 373 p.

Dodds, C.J. and Campbell, R.B.
1988: Potassium-argon ages of mainly intrusive rocks in the Saint Elias Mountains, Yukon and British Columbia; Geological Survey of Canada, Paper 87-16.

Drown, T.J.
1973: Zeolite facies metamorphism of the Kingsvale and Spences Bridge Groups; B.Sc. thesis, University of British Columbia, Vancouver, 37 p.

Dudley, J.S.
1983: Zeolitization of the Howson facies, Telkwa Formation, British Columbia; Ph.D. thesis, University of Calgary, Alberta, 311 p.

Dudley, J.S. and Ghent, E.D.
1980: Zeolite alteration of the Howson facies volcanics (Jurassic), British Columbia, Canada; in Proceedings of the Fifth International Conference on Zeolites, L.V. Rees (ed.), Heyden and Son Ltd., London, p. 129-137.

Duncan, I.J.
1982: The evolution of the Thor-Odin Gneiss Dome and related geochronological studies; Ph.D. thesis, University of British Columbia, Vancouver, 353 p.
1984: Structural evolution of the Thor-Odin Gneiss Dome; Tectonophysics, v. 101, p. 87-130.

Engi, J.E.
1984: Structure and metamorphism north of Quesnel Lake and east of Niagara Creek, Cariboo Mountains, British Columbia; M.Sc. thesis, University of British Columbia, Vancouver, 137 p.

England, T.D.J.
1984: Thermal maturation of the Western Canadian Sedimentary Basin in the Rocky Mountains Foothills and Plains of Alberta south of Red Deer River; M.Sc. thesis, University of British Columbia, Vancouver, B.C., 171 p.

Epstein, A.G., Epstein, J.B., and Harris, L.D.
1977: Conodont color alteration - an index to organic metamorphism; United States Geological Survey, Professional Paper 995, 27 p.

Erdmer, P. and Helmstaedt, H.
1983: Eclogite from central Yukon: a record of subduction at the western margin of ancient North America; Canadian Journal of Earth Sciences, v. 20, p. 1389-1408.

Evans, B.W. and Berti, J.W.
1986: Revised metamorphic history for the Chiwaukum Schist, North Cascades, Washington; Geology, v. 14, p. 695-698.

Evenchick, C.A.
1988: Stratigraphy, metamorphism, structure, and their tectonic implications in the Sifton and Deserters ranges, Cassiar and northern Rocky Mountains, northern British Columbia; Geological Survey of Canada, Bulletin 376, 90 p.

Fairchild, L.H. and Cowan, D.S.
1982: Structure, petrology, and tectonic history of the Leech River complex northwest of Victoria, Vancouver Island; Canadian Journal of Earth Sciences, v. 19, p. 1817-1835.

Ferri, F.
1984: Structure of the Blackwater Range, British Columbia; M.Sc. thesis, University of Calgary, Alberta, 143 p.

Fillipone, J.A.
1985: Structure and metamorphism at the western margin of the Omineca Belt near Boss Mountain, east-central British Columbia; M.Sc. thesis, University of British Columbia, Vancouver, 150 p.

Fletcher, C.J.N. and Greenwood, H.J.
1979: Metamorphism and structure of the Penfold Creek area, near Quesnel Lake, British Columbia; Journal of Petrology, v. 20, p. 743-794.

Foscolos, A.E., Powell, T.G., and Gunther, P.R.
1976: The use of clay minerals and inorganic and organic geochemical indicators for evaluating the degree of diagenesis and oil generating potential of shales; Geochimica et Cosmochimica Acta, v. 40, p. 953-966.

Friedman, R.M. and Armstrong, R.L.
1988: The Tatla Lake Metamorphic Complex: an Eocene metamorphic cone complex on the southwestern edge of British Columbia; Tectonics, v. 7, p. 1141-1166.

Froese, E.
1970: Chemical petrology of some pelitic gneisses and migmatites from the Thor-Odin area, British Columbia; Canadian Journal of Earth Sciences, v. 7, p. 164-175.

1973: The assemblage quartz - K-feldspar - biotite - garnet - sillimanite as an indicator of P_{H_2O}-T conditions; Canadian Journal of Earth Sciences, v. 10, p. 1575-1579.

Frost, B.R.
1980: Observations on the boundary between zeolite facies and prehnite-pumpellyite facies; Contributions to Mineralogy and Petrology, v. 73, p. 365-373.

Gabites, J.E.
1985: Geology and geochronometry of the Cogburn Creek-Settler Creek area, northeast of Harrison Lake, B.C.; M.Sc. thesis, University of British Columbia, Vancouver, 153 p.

Gabrielse, H.
1975: Geology of the Fort Grahame E 1/2 map-area, B.C.; Geological Survey of Canada, Paper 75-33, 28 p.

Gabrielse, H. and Reesor, J.E.
1964: Geochronology of plutonic rocks in two areas of the Canadian Cordillera; Royal Society of Canada, Special Publication No. 8, p. 96-138.

Gabrielse, H., Dodds, C.J., Eisbacher, G.H., and Mansy, J.L.
1977: Geology of Toodoggone (94E) and Ware W 1/2 (94F W 1/2) map areas; Geological Survey of Canada, Open File 483.

Galloway, W.E.
1974: Deposition and diagenetic alteration of sandstone in northeast Pacific arc-related basins: implications for greywacke genesis; Geological Society of America Bulletin, v. 85, p. 379-390.

Getsinger, J.S.
1985: Geology of the Three Ladies Mountain/Mount Stevenson area, Quesnel Highland, British Columbia; Ph.D. thesis, University of British Columbia, Vancouver, 239 p.

Ghent, E.D. and Miller, B.E.
1974: Zeolite and clay-carbonate assemblages in the Blairmore Group (Cretaceous), southern Alberta Foothills, Canada; Contributions to Mineralogy and Petrology, v. 44, p. 313-329.

Ghent, E.D., Robbins, D.B., and Stout, M.Z.
1979: Geothermometry, geobarometry and fluid compositions of metamorphosed calc-silicates and pelites, Mica Creek, British Columbia; American Mineralogist, v. 64, p. 874-885.

Ghent, E.D., Simony, P.S., and Knitter, C.C.
1980: Geometry and pressure-temperature significance of the kyanite-sillimanite isograd in the Mica Creek area, British Columbia; Contributions to Mineralogy and Petrology, v. 74, p. 67-73.

Ghent, E.D., Simony, P.S., and Raeside, R.P.
1981: Metamorphism and its relation to structure within the core zone west of the southern Rocky Mountains; in Field Guide to Geology and Mineral Deposits, R.I. Thompson and D.G. Cook (ed.), Calgary '81 Annual Meeting, Geological Association of Canada, p. 373-391.

Ghent, E.D., Knitter, C.C., Raeside, R.P., and Stout, M.Z.
1982: Geothermometry and geobarometry of pelitic rocks, upper kyanite and sillimanite zones, Mica Creek area, British Columbia; Canadian Mineralogist, v. 20, p. 295-305.

Ghent, E.D., Stout, M.Z., and Raeside, R.P.
1983: Plagioclase-clinopyroxene-garnet-quartz equilibria and the geobarometry and geothermometry of garnet amphibolites from Mica Creek, British Columbia; Canadian Journal of Earth Sciences, v. 20, p. 699-706.

Gibson, D.W.
1985: Stratigraphy, sedimentology and depositional environments of the coal-bearing Jurassic-Cretaceous Kootenay Group, Alberta and British Columbia; Geological Survey of Canada, Bulletin 357, 108 p.

Goldsmith, J.R. and Newton, R.C.
1969: P-T-X relations in the system $CaCO_3$-$MgCO_3$ at high temperatures and pressures; American Journal of Science, v. 267A, p. 160-190.

Grette, J.F.
1978: Cache Creek and Nicola Groups near Ashcroft, British Columbia; M.Sc. thesis, University of British Columbia, Vancouver, 88 p.

Gunther, P.R. and Meijer-Drees, N.C.
1977: Devonian coal in the subsurface of Great Slave Plain: a guide to exploration for oil and gas; in Report of Activities, Part A, Geological Survey of Canada, Paper 77-1A, p. 147-150.

Hacquebard, P.A. and Barss, M.S.
1957: A Carboniferous spore assemblage in coal from the South Nahanni River area, Northwest Territories; Geological Survey of Canada, Bulletin 40, 63 p.

Hacquebard, P.A. and Donaldson, J.R.
1974: Rank studies of coals in the Rocky Mountains and Inner Foothills Belt, Canada; Geological Society of America, Special Paper 153, p. 75-94.

Haugerud, R.A.
1985: Geology of the Hozameen Group and the Ross Lake shear zone, Maselpanik area, North Cascades, southwest British Columbia; Ph.D. thesis, University of Washington, Seattle, 268 p.

Heah, T.S.T., Armstrong, R.L., and Woodsworth, G.J.
1986: The Gambier Group in the Sky Pilot area, southwestern Coast Mountains, British Columbia; in Current Research, Part B, Geological Survey of Canada, Paper 86-1B, p. 685-692.

Hickson, C.J.
1988: Structure and stratigraphy of the Masset Formation, Queen Charlotte Islands, British Columbia; in Current Research, Part E, Geological Survey of Canada, Paper 88-1E, p. 269-274.

Higgins, A.C., Kalkreuth, W.D., and Dougherty, B.J.
1985: Correlation of conodont colour alteration indices (CAI) and vitrinite reflectance in the Upper Devonian and Lower Carboniferous rocks of N.E. British Columbia and N.W. Alberta - indicators of thermal maturity; Abstract, Canadian Society of Petroleum Geologists, Annual Meeting, Calgary, Alberta.

Hill, M.L.
1984: Geology of the Redcap Mountain area, Coast Plutonic Complex, British Columbia; Ph.D. thesis, Princeton University, 216 p.

Hoffman, J. and Hower, J.
1979: Clay mineral assemblages as low grade metamorphic geothermometers: application to the thrust faulted disturbed belt of Montana, U.S.A.; in Aspects of Diagenesis, P.A. Scholle and P.R. Schluger (ed.), Society of Economic Paleontologists and Mineralogists, Special Publication No. 26, p. 55-79.

Hollister, L.S.
1969a: Metastable paragenetic sequence of andalusite, kyanite, and sillimanite, Kwoiek area, British Columbia; American Journal of Science, v. 267, p. 352-370.
1969b: Contact metamorphism in the Kwoiek area of British Columbia: an end member of the metamorphic process; Geological Society of America Bulletin, v. 80, p. 2465-2494.
1975: Granulite facies metamorphism in the Coast Range Crystalline Belt; Canadian Journal of Earth Sciences, v. 12, p. 1953-1955.
1977: The reaction forming cordierite from garnet, the Khtada Lake Metamorphic Complex, British Columbia; Canadian Mineralogist, v. 15, p. 217-229.
1982: Metamorphic evidence for rapid (2mm/yr) uplift of a portion of the Central Gneiss Complex, Coast Mountains, British Columbia; Canadian Mineralogist, v. 20, p. 319-332.

Holmes, D.W. and Oliver, T.A.
1973: Source and depositional environments of the Moose Channel Formation, Northwest Territories; Bulletin of Canadian Petroleum Geology, v. 21, p. 435-478.

Høy, T.
1976: Calc-silicate isograds in the Riondel area, southeastern British Columbia; Canadian Journal of Earth Sciences, v. 13, p. 1093-1104.

Hutcheon, I., Oldershaw, A., and Ghent, E.D.
1980: Diagenesis of Cretaceous sandstones of the Kootenay Formation at Elk Valley (southeastern British Columbia) and Mt. Allan (southwestern Alberta); Geochimica et Cosmochimica Acta, v. 44, p. 1425-1435.

Hutchison, W.W.
1982: Geology of the Prince Rupert-Skeena map area, British Columbia; Geological Survey of Canada, Memoir 394, 116 p.

Isachsen, C.
1984: Geology, geochemistry, and geochronology of the Westcoast Crystalline Complex and related rocks, Vancouver Island, British Columbia; M.Sc. thesis, University of British Columbia, Vancouver, 144 p.

Journeay, J.M.
1986: Stratigraphy, internal strain and thermo-tectonic evolution of northern Frenchman Cap Dome: An exhumed duplex structure, Omineca Hinterland, S.E. Canadian Cordillera; Ph.D. thesis, Queen's University, Kingston, Ontario, 401 p.

Juras, S.J.
1987: Geology of the polymetallic volcanogenic Buttle Lake Camp, with emphasis on the Price Hillside, central Vancouver Island, British Columbia, Canada; Ph.D. thesis, University of British Columbia, Vancouver, 279 p.

Kalkreuth, W. and McMechan, M.E.
1984: Regional pattern of thermal maturation as determined from coal rank studies, Rocky Mountain Foothills and Front Ranges north of Grande Cache, Alberta - implications for petroleum exploration; Bulletin of Canadian Petroleum Geology, v. 32, p. 249-271.

Kanasewich, E.R., Clowes, R.M., and McCloughan, C.H.
1969: A buried Precambrian rift in western Canada; Tectonophysics, v. 8, p. 513-527.

Kenah, C. and Hollister, L.S.
1983: Anatexis in the Central Gneiss Complex, British Columbia; in Migmatites, Melting and Metamorphism, M.P. Atherton and C.D. Gribble (ed.), Shiva Publishing, Nantwich, England, p. 142-162.

Kerrick, D.M. and Woodsworth, G.J.
1989: Aluminum silicates in the Mount Raleigh pendant, British Columbia; Journal of Metamorphic Geology, v. 7, p. 547-563.

Kisch, H.J.
1981: Coal rank and illite crystallinity associated with the zeolite facies of Southland and the pumpellyite-bearing facies of Otago, southern New Zealand; New Zealand Journal of Geology and Geophysics, v. 24, p. 349-360.

Klepacki, D.W.
1981: The Little River Fault - a low angle boundary fault of the northern Shuswap Complex, Quesnel Lake, British Columbia; in Program with Abstracts, Geological Association of Canada, v. 6, p. 32.

Kramers, J.W. and Mellon, G.B.
1972: Upper Cretaceous-Paleocene coal-bearing strata, northwest-central Alberta Plains; in Proceedings First Geological Conference on Western Canadian Coal, Research Council of Alberta, Information Series No. 60, p. 109-124.

Kubler, B.
1967: Anchimetamorphisme et schistosite; Centre de Recherche Pau-SNPA Bulletin, v. 1, p. 259-278.

Kuniyoshi, S. and Liou, J.G.
1976a: Burial metamorphism of the Karmutsen volcanics, northeastern Vancouver Island, British Columbia; American Journal of Science, v. 276, p. 1096-1119.
1976b: Contact metamorphism of the Karmutsen volcanics, Vancouver Island, British Columbia; Journal of Petrology, v. 17, p. 73-99.

Lam, H.L. and Jones, F.W.
1984: Geothermal gradients of Alberta in Western Canada; Geothermics, v. 13, p. 181-192.

Lappin, A.R. and Hollister, L.S.
1980: Partial melting in the Central Gneiss Complex near Prince Rupert, British Columbia; American Journal of Science, v. 280, p. 518-545.

Leatherbarrow, R.W.
1981: Metamorphism of pelitic rocks from the northern Selkirk Mountains, southeastern British Columbia; Ph.D. thesis, Carleton University, Ottawa, 218 p.

Lefebure, D.V.
1976: Geology of the Nicola Group in the Fairweather Hills, British Columbia; M.Sc. thesis, Queen's University, Kingston, Ontario, 179 p.

Liou, J.G., Maruyama, S., and Cho, M.
1985: Phase equilibria and mineral parageneses of metabasites in low-grade metamorphism; Mineralogical Magazine, v. 49, p. 321-333.

Lowes, B.E.
1972: Metamorphic petrology and structural geology of the area east of Harrison Lake, British Columbia; Ph.D. thesis, University of Washington, Seattle, 162 p.

MacKevett, E.M., Jr.
1971: Stratigraphy and general geology of the McCarthy C-5 Quadrangle, Alaska; United States Geological Survey, Bulletin 1323, 35 p.

Mader, U.K.
1987: The Aley carbonatite complex, Northern Rocky Mountains, British Columbia; in Geological Fieldwork 1986, British Columbia Ministry of Energy, Mines and Petroleum Resources, Paper 1987-1, p. 283-288.

Mamu, R.R.D.
1974: Zeolite facies metamorphism of the Spences Bridge and Kingsvale Groups, Spences Bridge, southwestern British Columbia; B.Sc. thesis, University of British Columbia, Vancouver, 49 p.

Mansy, J.L.
1986: Géologie de la chaine d'Omineca des Rocheuses aux Plateaux Interieurs (Cordillere canadienne); Evolution depuis le Precambrien; Societe Geologique du Nord, Pub. no. 13, 718 p.

Massey, N.W.D.
1986: Metchosin Igneous Complex, southern Vancouver Island: Ophiolite stratigraphy developed in an emergent island setting; Geology, v. 14, p. 602-605.

Mathews, W.H.
1981: Early Cenozoic resetting of potassium-argon dates and geothermal history of north Okanagan area, British Columbia; Canadian Journal of Earth Sciences, v. 18, p. 1310-1319.
1983: Early Tertiary resetting of potassium-argon dates in the Kootenay arc, southeastern British Columbia; Canadian Journal of Earth Sciences, v. 20, p. 867-872.

McGuigan, P.J.
1975: Certain breccias of Mount Washington property, Vancouver Island; B.Sc. thesis, University of British Columbia, Vancouver, 66 p.

McKenzie, K.J.
1985: Sedimentology and stratigraphy of the southern Sustut Basin, north-central British Columbia; M.Sc. thesis, University of British Columbia, Vancouver, 120 p.

McKillop, G.R.
1973: Geology of southwestern Gambier Island, Howe Sound, British Columbia; B.Sc. thesis, University of British Columbia, Vancouver, 24 p.

McMechan, M.E. and Price, R.A.
1982: Superimposed low-grade metamorphism in the Mount Fisher area, southeastern British Columbia - implications for the East Kootenay orogeny; Canadian Journal of Earth Sciences, v. 19, p. 476-489.

McMullin, D.W.A. and Greenwood, H.J.
1986: Metamorphic pressures and temperatures in the Barkerville and Cariboo terranes, Quesnel Lake, British Columbia: preliminary results; in Current Research, Part B, Geological Survey of Canada, Paper 86-1B, p. 727-732.

Medford, G.A.
1975: K-Ar and fission track geochronometry of an Eocene thermal event in the Kettle River (west half) map area, southern British Columbia; Canadian Journal of Earth Sciences, v. 12, p. 836-843.

Mihalynuk, M.G.
1987: Metamorphic, structural, and stratigraphic evolution of the Telkwa Formation, Zymoetz River area (NTS 103I/8 and 93L/5), near Terrace, British Columbia; M.Sc. thesis, University of Calgary, Alberta, 128 p.

Moffat, I.W.
1985: The nature and timing of deformation events and organic and inorganic metamorphism in the northern Groundhog coalfield: implications for the tectonic history of the Bowser Basin; Ph.D. thesis, University of British Columbia, 205 p.

Monger, J.W.H.
1975: Upper Paleozoic rocks of the Atlin Terrane, northwestern British Columbia and south-central Yukon; Geological Survey of Canada, Paper 74-47, 63 p.

Monger, J.W.H. and Hutchison, W.W.
1971: Metamorphic map of the Canadian Cordillera; Geological Survey of Canada, Paper 70-33, 61 p. and Supplement, 19 p.

Monger, J.W.H., Clowes, R.M., Cowan, D.S., Potter, C.J., Price, R.A., and Yorath
in press: Continent-ocean transitions in western North America between latitudes 46 and 56 degrees: Transects B1, B2, B3; in Geological Society of America, The Decade of North American Geology, Centennial Continent-Ocean Transects, no. 7, 8, 9.

Montgomery, J.R.
1985: Structural relations of the southern Quesnel Lake Gneiss, Isosceles Mountain area, southwest Cariboo Mountains, British Columbia; M.Sc. thesis, University of British Columbia, Vancouver, 96 p.

Muller, J.E.
1980: The Paleozoic Sicker Group of Vancouver Island, British Columbia; Geological Survey of Canada, Paper 79-30, 23 p.

Murphy, D.C.
1987: Suprastructure/infrastructure transition, east-central Cariboo Mountains, British Columbia: geometry, kinematics, and tectonic implications; Journal of Structural Geology, v. 9, p. 13-29.

Murphy, D.C. and Rees, C.J.
1983: Structural transition and stratigraphy in the Cariboo Mountains, British Columbia; in Current Research, Part A, Geological Survey of Canada, Paper 83-1A, p. 245-252.

Newton, R.C. and Perkins, D., III
1982: Thermodynamic calibration of geobarometers based on the assemblages garnet-plagioclase-orthopyroxene (clinopyroxene)-quartz; American Mineralogist, v. 67, p. 203-222.

Norris, D.K. and Cameron, A.R.
1986: An occurrence of bitumen in the Interior Platform near Rengleng River, District of Mackenzie; in Current Research, Part A, Geological Survey of Canada, Paper 86-1A, p. 645-648.

Okulitch, A.V.
1984: The role of the Shuswap Metamorphic Complex in Cordilleran Tectonism: a review; Canadian Journal of Earth Sciences, v. 21, p. 1171-1193.

Orchard, M.J.
1988: Studies on the Triassic Kunga Group, Queen Charlotte Islands, British Columbia; in Current Research, Part E, Geological Survey of Canada, Paper 88-1E, p. 229.

Parrish, R.R.
1979: Geochronology and tectonics of the northern Wolverine Complex, British Columbia; Canadian Journal of Earth Sciences, v. 16, p. 1428-1438.
1981: Geology of the Nemo Lakes belt, northern Valhalla Range, southeast British Columbia; Canadian Journal of Earth Sciences, v. 18, p. 944-958.

Paterson, I.A.
1973: The geology of the Pinchi Lake area, central British Columbia; Ph.D. thesis, University of British Columbia, Vancouver, 260 p.

Paterson, I.A. and Harakal, J.E.
1974: Potassium-argon dating of blueschists from Pinchi Lake, central British Columbia; Canadian Journal of Earth Sciences, v. 11, p. 1007-1011.

Pearce, T.H.
1970: The analcite-bearing rocks of the Crowsnest Formation, Alberta; Canadian Journal of Earth Sciences, v. 7, p. 46-66.

Pell, J.
1984: Stratigraphy, structure, and metamorphism of Hadrynian strata in the southeastern Cariboo Mountains, British Columbia; Ph.D. thesis, University of Calgary, Alberta, 185 p.

Perkins, E.H., Brown, T.H., and Berman, R.G.
1986: PT-system, TX-system, PX-system: three programs which calculate pressure-temperature-composition phase diagrams; Computers & Geosciences, v. 12, p. 749-755.

Pigage, L.C.
1976: Metamorphism of the Settler Schist, southwest of Yale, British Columbia; Canadian Journal of Earth Sciences, v. 13, 405-421.
1977: Rb-Sr dates for granodiorite intrusions on the northeast margin of the Shuswap Metamorphic Complex, Cariboo Mountains, British Columbia; Canadian Journal of Earth Sciences, v. 14, p. 1690-1695.
1978: Metamorphism and deformation on the northeast margin of the Shuswap Metamorphic Complex Azure Lake, British Columbia; Ph.D. thesis, University of British Columbia, Vancouver, 289 p.

Potter, C.J.
1983: Geology of the Bridge River complex, southern Shulaps Range, British Columbia: a record of Mesozoic convergent tectonics; Ph.D. thesis, University of Washington, Seattle, 192 p.

Raeside, R.P.
1982: Structure, metamorphism and migmatization of the Scrip Range, Mica Creek, British Columbia; Ph.D. thesis, University of Calgary, Alberta, 204 p.

Read, P.B.
1973: Petrology and structure of Poplar Creek map-area, British Columbia; Geological Survey of Canada, Bulletin 193, 144 p.

Read, P.B. and Brown, R.L.
1981: Columbia River fault zone: southeastern margin of the Shuswap and Monashee complexes, southern British Columbia; Canadian Journal of Earth Sciences, v. 18, p. 1127-1145.

Read, P.B. and Eisbacher, G.H.
1974: Regional zeolite alteration of the Sustut Group, north-central British Columbia; Canadian Mineralogist, v. 12, p. 527-541.

Read, P.B. and Monger, J.W.H.
1976: Pre-Cenozoic volcanic assemblages of the Kluane and Alsek ranges, southwestern Yukon Territory; Geological Survey of Canada, Open File 381, 96 p.

Read, P.B., Psutka, J.F., Brown, R.L., and Orchard, M.J.
1983: "Tahltanian" Orogeny and younger deformations, Grand Canyon of the Stikine British Columbia; Geological Association of Canada, Program with Abstracts, v. 8, p. A57.

Reamsbottom, S.B.
1971: Geology of the Mount Breakenridge area, Harrison Lake, British Columbia; M.Sc. thesis, University of British Columbia, Vancouver, 144 p.
1974: Geology and metamorphism of the Mount Breakenridge area, Harrison Lake, British Columbia; Ph.D. thesis, University of British Columbia, Vancouver, 155 p.

Reesor, J.E.
1965: Structural evolution and plutonism in Valhalla Gneiss Complex, British Columbia; Geological Survey of Canada, Bulletin 129, 128 p.
1973: Geology of the Lardeau map-area, east-half, British Columbia; Geological Survey of Canada, Memoir 369, 129 p.

Reesor, J.E. and Moore, J.M., Jr.
1971: Petrology and structure of Thor-Odin Gneiss Dome, Shuswap Metamorphic Complex; Geological Survey of Canada, Bulletin 195, 147 p.

Ringwood, A.E.
1972: Phase transformation and mantle dynamics; Earth and Planetary Science Letters, v. 14, p. 233-241.

Ringwood, A.E. and Green, D.H.
1966: An experimental investigation of the gabbro-eclogite transformation and some geophysical implications; Tectonophysics, v. 3, p. 383-427.

Roots, C.F.
1987: Regional tectonic setting and evolution of the Late Proterozoic Mount Harper Volcanic Complex, Ogilvie Mountains, Yukon; Ph.D. thesis, Carleton University, Ottawa, 219 p.

Roots, E.F.
1954: Geology and mineral deposits of Aiken Lake map-area, British Columbia; Geological Survey of Canada, Memoir 274, 246 p.

Rusmore, M.E.
1985: Geology and tectonic significance of the Upper Triassic Cadwallader Group and its bounding faults, southwestern British Columbia; Ph.D. thesis, University of Washington, Seattle, 174 p.
1987: Geology of the Cadwallader Group and the Intermontane-Insular Superterrane boundary, southwestern British Columbia; Canadian Journal of Earth Sciences, v. 24, p. 2279-2291.

Rusmore, M.E. and Cowan, D.S.
1985: Jurassic-Cretaceous rock units along the southern edge of the Wrangellia terrane on Vancouver Island; Canadian Journal of Earth Sciences, v. 22, p. 1223-1232.

Schau, M.P.
1968: Geology of the Upper Triassic Nicola Group in south central British Columbia; Ph.D. thesis, University of British Columbia, Vancouver, 211 p.
1970: Stratigraphy and structure of the type area of the Upper Triassic Nicola Group in south-central British Columbia; in Structure of the Southern Canadian Cordillera, J.O. Wheeler (ed.), Geological Association of Canada, Special Paper No. 6, p. 123-135.

Selverstone, J. and Hollister, L.S.
1980: Cordierite-bearing granulites from the Coast Ranges, British Columbia: P-T conditions of metamorphism; Canadian Mineralogist, v. 18, p. 119-129.

Simony, P.S.
1979: Pre-Carboniferous basement near Trail, British Columbia; Canadian Journal of Earth Sciences, v. 16, p. 1-11.

Simony, P.S., Ghent, E.D., Craw, D., Mitchell, W., and Robbins, D.B.
1980: Structural and metamorphic evolution of northeast flank of Shuswap complex southern Canoe River area, British Columbia; Geological Society of America, Memoir 153, p. 445-461.

Sisson, V.B.
1985: Contact metamorphism and fluid evolution associated with the intrusion of the Ponder Pluton, Coast Plutonic Complex, British Columbia, Canada; Ph.D. thesis, Princeton University, 345 p.
1987: Halogen chemistry as an indicator of metamorphic fluid interaction with the Ponder pluton, Coast Plutonic Complex, British Columbia, Canada; Contributions to Mineralogy and Petrology, v. 95, p. 123-131.

Staplin, F.L.
1977: Interpretation of thermal history from color of particulate organic matter - A review; Palynology I, Proceedings of Eighth Annual Meeting, American Association of Stratigraphic Palynologists, Houston, 1975, p. 9-18.

Stewart, R.J. and Page, R.J.
1974: Zeolite facies metamorphism of the Late Cretaceous Nanaimo Group, Vancouver Island and Gulf Islands, British Columbia; Canadian Journal of Earth Sciences, v. 11, p. 280-284.

Struik, L.C.
1980: Geology of the Barkerville-Cariboo River area, central British Columbia; Ph.D. thesis, University of Calgary, Alberta, 350 p.
1986a: Imbricated terranes of the Cariboo gold belt with correlations and implications for tectonics in southeastern British Columbia; Canadian Journal of Earth Sciences, v. 23, p. 1047-1061.

1986b: A regional east-dipping thrust places Hadrynian onto probable Paleozoic rocks in Cariboo Mountains, British Columbia; in Current Research, Part A, Geological Survey of Canada, Paper 86-1A, p. 589-594.

Surdam, R.C.
1967: Low-grade metamorphism of the Karmutsen Group, Buttle Lake area, Vancouver Island; Ph.D. thesis, University of California, Los Angeles, California, 313 p.
1973: Low-grade metamorphism of tuffaceous rocks in the Karmutsen Group, Vancouver Island, British Columbia; Geological Society of America Bulletin, v. 84, p. 1911-1922.

Sutherland Brown, A.
1968: Geology of the Queen Charlotte Islands, British Columbia; British Columbia Department of Mines and Petroleum Resources, Bulletin 54, 226 p.

Sutter, J.F. and Crawford, M.L.
1985: Timing of metamorphism and uplift in the vicinity of Prince Rupert, British Columbia and Ketchikan, Alaska; Geological Society of America, Abstracts With Programs, v. 17, p. 411.

Teichmuller, M. and Teichmuller, R.
1979: Diagenesis of coal (coalification); in Diagenesis in Sediments and Sedimentary Rocks, G. Larsen and G.V. Chilingar (ed.), Developments in Sedimentology 25A, Elsevier Publishing Company, New York, p. 207-246.

Tempelman-Kluit, D.J.
1974: Reconnaissance geology of the Aishihik Lake, Snag and part of Stewart River map-areas, west-central Yukon; Geological Survey of Canada, Paper 73-41, 97 p.
1976: The Yukon Crystalline Terrane: enigma in the Canadian Cordillera; Geological Society of America Bulletin, v. 87, p. 1343-1357.
1979: Five occurrences of transported synorogenic clastic rocks in Yukon Territory; in Current Research, Part A, Geological Survey of Canada, Paper 79-1A, p. 1-12.

Tipper, H.W. and Richards, T.A.
1976: Jurassic stratigraphy and history of central and north-central British Columbia; Geological Survey of Canada, Bulletin 270, 73 p.

Tracy, R.J.
1978: High grade metamorphic reactions and partial melting in pelitic schist, west-central Massachusetts; American Journal of Sciences, v. 278, p. 150-178.

Vance, J.A.
1968: Metamorphic aragonite in the prehnite-pumpellyite facies, Northwest Washington; American Journal of Science, v. 266, p. 299-315.

van der Heyden, P.
1982: Tectonic and stratigraphic relations between the Coast Plutonic Complex and Intermontane Belt, west-central Whitesail Lake map area, British Columbia; M.Sc. thesis, University of British Columbia, 172 p.

Wheeler, J.O.
1970: Summary and discussion; in Structure of the Southern Canadian Cordillera, J.O. Wheeler (ed.), Geological Association of Canada Special Paper No. 6, p. 155-166.

Woodsworth, G.J.
1977: Homogenization of zoned garnets from pelitic schists; Canadian Mineralogist, v. 15, p. 230-242.
1979: Metamorphism, deformation, and plutonism in the Mount Raleigh pendant, Coast Mountains, British Columbia; Geological Survey of Canada, Bulletin 295, 58 p.

Woodsworth, G.J., Crawford, M.L., and Hollister, L.S.
1983a: Metamorphism and structure of the Coast Plutonic Complex and adjacent belts, Prince Rupert and Terrace areas, British Columbia; Geological Association of Canada, Annual Meeting, 1983, Field Trip Guidebook 14, 66 p.

Woodsworth, G.J., Loveridge, W.D., Parrish, R.R., and Sullivan, R.W.
1983b: Uranium-lead dates from the Central Gneiss Complex and Ecstall pluton, Prince Rupert map area, British Columbia; Canadian Journal of Earth Sciences, v. 20, p. 1475-1483.

Woodsworth, G.J. and Orchard, M.J.
1985: Upper Paleozoic to lower Mesozoic strata and their conodonts, western Coast Plutonic Complex, British Columbia; Canadian Journal of Earth Sciences, v. 22, p. 1329-1344.

Yagishita, K.
1985: Mid- to Late Cretaceous sedimentation in the Queen Charlotte Islands, British Columbia; lithofacies, paleocurrent and petrographic analyses of sediments; Ph.D. thesis, University of Toronto.

Yorath, C.J., Green, A.G., Clowes, R.M., Sutherland Brown, A., Brandon, M.T., Kanasewich, E.R., Hyndman, R.D., and Spencer, C.
1985: LITHOPROBE, southern Vancouver Island: seismic reflection sees through Wrangellia to the Juan de Fuca Plate; Geology v. 13, p. 759-762.

Zen, E-An and Thompson, A.B.
1974: Low-grade regional metamorphism: mineral equilibrium relations; Annual Reviews, Earth and Planetary Science Letters, v. 2, p. 179-212.

Authors' addresses

H.J. Greenwood
Department of Geological Sciences
University of British Columbia
Vancouver, British Columbia
V6T 2B4

G.J. Woodsworth
C.A. Evenchick
Cordilleran Division
Geological Survey of Canada
100 West Pender Street
Vancouver, British Columbia
V6B 1R8

P.B. Read
Geotex Consultants, Limited
100 West Pender Street
Vancouver, British Columbia
V6B 1R8

E.D. Ghent
University of Calgary
Department of Geology and Geophysics
Calgary, Alberta
T2N 1N4

ADDENDUM:

Much new information on metamorphism in the Canadian Cordillera has been published since the original compilation of this chapter. Most of the recent material has been integrated within the text and is supported by the following references:

REFERENCES

Anderson, R.G.
1989: A stratigraphic, plutonic, and structural framework for the Iskut River map area, northwestern British Columbia; in Current Research Part E, Geological Survey of Canada, Paper 89-1E, p. 145-153.

Carr, S.D. (1989):
1989: Implications of Early Eocene Ladybird granite in the Thor-Odin - Pinnacle area, southern British Columbia; in Current Research Part E, Geological Survey of Canada, Paper 89-1E, p. 69-77.

Carr, S.D., Parrish, R.R. and Brown, R.L.
1987: Eocene structural development of the Valhalla complex, southeastern British Columbia; Tectonics, v. 6, p. 175-196.

Coleman, V.J.
1989: The Cariboo duplex at the southern boundary of the Monashee Complex, southern British Columbia; in Current Research Part E, Geological Survey of Canada, Paper 89-1E, p. 89-93.

Eisbacher, G.H.
1974: Sedimentary history and tectonic evolution of the Sustut and Sifton basins, north-central British Columbia; Geological Survey of Canada, Paper 73-31, 57 p.

England, T.D.J.
1989: Late Cretaceous to Paleogene evolution of the Georgia Basin, southwestern British Columbia; unpublished Ph.D. thesis, Memorial University of Newfoundland, St. John's, Newfoundland, 481 p.

England, T.D.J. and Calon, T.J.
1991: The Cowichan fold and thrust system, Vancouver Island, southwestern British Columbia; Geological Society of America Bulletin, v. 103, p. 336-362.

Erdmer, P.
1987: Blueschist and eclogite in mylonitic allochthons, Ross River and Watson Lake areas, southeastern Yukon; Canadian Journal of Earth Sciences, v. 24, p. 1439-1449.

Evenchick, C.A.
1991: Geometry, evolution, and tectonic framework of the Skeena Fold Belt, north central British Columbia; Tectonics, v. 10, June 1991.

Ferri, F. and Melville, D.M.
1988: Manson Creek mapping project; British Columbia Ministry of Energy, Mines and Petroleum Ressources, Geological Fieldwork, 1987, Paper 1988-1, p. 169-180.

Ferri, F. and Melville, D.M.
1989: Geology of the Germansen Landing area, British Columbia; British Columbia Ministry of Energy, Mines and Petroleum Resources, Geological Fieldwork, 1988, Paper 1989-1, p. 209-220.

Fogarassy, J.A.S. and Barnes, W.C.
1991: Stratigraphy and diagenesis of the middle to Upper Cretaceous Queen Charlotte Group, Queen Charlotte Islands, British Columbia; in Evolution and Hydrocarbon Potential of the Queen Charlotte Basin, British Columbia, Geological survey of Canada, Paper 90-10, p. 279-294.

Friedman, R.M.
1988: Geology and geochronometry of the Eocene Tatla Lake metamorphic core complex, western edge of the Intermontane Belt, British Columbia,; Ph.D. thesis, University of British Columbia, Vancouver, B.C., 348 p.

Friedman, R.M., Monger, J.W.H., and Tipper, H.W.
1990: Age of the Bowen Island Group, southwestern Coast Mountains, British Columbia; Canadian Journal of Earth Sciences, v. 27, p. 1456-1461.

Gerasimoff, M.D.
1988: The Hobson Lake pluton, Cariboo Mountains, British Columbia and its significance to Mesozoic and early Cenozoic Cordilleran tectonics; unpublished M.Sc. thesis, Queen's University, Kingston Ontario, 196 p.

Goodarzi, F. and Norford, B.S.
1985: Graptolites as indicators of the temperature histories of rocks; Journal of the Geological Society of London, v. 142, p. 1089-1099.

Gordey, S.P.
1988: The South Fork Volcanics: mid-Cretaceous caldera fill tuffs in east-central Yukon; in Current Research, Part E, Geological Survey of Canada, Paper 88-1E, p. 13-18.

Greig, C.J.
1989: Geology and geochronometry of the Eagle plutonic complex, Coquihalla area, southwestern British Columbia; unpublished M.Sc. thesis, University of British Columbia, Vancouver, B.C. 423 p.

Johnson, S.Y., Zimmermann, R.A. and Naeser, C.W.
1986: Fission-track dating of the tectonic development of the San Juan Islands, Washington; Canadian Journal of Earth Sciences, v. 23, p. 1318-1330.

Journeay, J.M.
1990: A progress report on the structural and tectonic framework of the southern Coast Belt, British Columbia; in Current Research, Part E, Geological Survey of Canada, Paper 90-1E, p. 183-195.

Kisch, H.J.
1987: Correlation between indicators of very low-grade metamorphism; in Low Temperature Metamorphism (ed. Frey, M.), Blackie & Son, Glasgow, p. 301-304.

Klepacki, D.W.
1985: Stratigraphy and structural geology of the Goat Range anrea, southeastern British Columbia; unpublished Ph.D. thesis, Massachusetts Institute of technology, 268 p.

Krage, S.M.
1984: Metamorphic and fluid inclusion study of amphibolite-grade rocks, West Scotia, British Columbia; unpublished M.Sc. thesis, Bryn Mawr College, Bryn Mawr, Pennsylvania, 98 p.

Massey, N.W.D. and Friday, S.J.
1987: Geology of the Cowichan Lake area, Vancouver Island; British Columbia Ministry of Energy, Mines and Petroleum Resources, Geological Fieldwork, 1986, Paper 1987-1, p. 223-229.

Massey, N.W.D. and Friday, S.J.
1988: Geology of the Chemainus River-Duncan area, Vancouver Island; British Columbia Ministry of Energy, Mines and Petroleum Resources, Geological Fieldwork, 1987, Paper 1988-1, p. 81-91.

Massey, N.W.D. and Friday, S.J.
1989: Geology of the Alberni-Nanaimo Lakes area, Vancouver Island; British Columbia Ministry of Energy, Mines and Petroleum Resources, Geological Fieldwork, 1988, Paper 1989-1, p. 61-74.

Meade, H.D.
1977: Petrology and metal occurrences of the Takla Group and Hogem and Germansen batholiths, north central British Columbia; unpublished Ph.D. thesis, University of British Columbia, Vancouver, B.C., 355 p.

Moore, J.M. and Pettipas, A.
1990: Nicola Lake region geology and mineral deposits; British Columbia Ministry of Energy, Mines and Petroleum Resources, Open File 1990-29 Part A, p. 1-13.

Mott, A.J.
1989: Structural and stratigraphic relations in the White River region, eastern Main Ranges, southern Canadian Rocky Mountains, British Columbia; unpublished Ph.D. thesis, Queen's University, Kingston, Ontario, 405 pp.

Nguyen, K.K., Sinclair, A.J. and Libby, W.G.
1968: Age of the northern part of the Nelson batholith; Canadian Journal of Earth Sciences, v. 5, p. 955-957.

Orchard, M.J. and Forster, P.J.L.
1991: Conodont colour and thermal maturity of the Late Triassic Kunga Group, Queen Charlotte Islands, British Columbia; in Evolution and Hydrocarbon Potential of the Queen Charlotte Basin, British Columbia, Geological Survey of Canada, Paper 90-10, p. 453-464.

Parrish, R.R. and Wheeler, J.O.
1983: A U-Pb zircon age from the Kuskanax batholith, southeastern British Columbia; Canadian Journal of Earth Sciences, v. 20, p. 1751-1756.

Paterson, I.A.
1974: Geology of Cache Creek Group and Mesozoic rocks at the northern end of the Stuart Lake Belt, central British Columbia; in Report of Activities, Part B, Geological Survey of Canada, Paper 74-1, Part B, p. 31-42.

Pearson, W.N.
1974: Zeolite facies metamorphism of the Spences Bridge Group, Spences Bridge, British Columbia; unpublished B.Sc. thesis, University of British Columbia, Vancouver, B.C., 44 p.

Plint, H.E. and Erdmer, P.
1989: Structure and metamorphism in the Horseranch Range, north-central British Columbia; British Columbia Ministry of Energy, Mines and Petroleum Resources, Geological Fieldwork, 1988, Paper 1989-1, p. 347-351.

Read, P.B.
1976: Lardeau map-area (82K west half), British Columbia; in Report of Activities Part A, Geological Survey of Canada, Paper 76-1A, p. 95-96.

Read, P.B., Psutka, J.F. and Fillipone, J.
1991a: Organic maturity data for the Canadian Cordillera; Geological Survey of Canada, Open File 2341.

Read, P.B., Psutka, J.F. and Fillipone, J.
1991b: Abbreviated metamorphic data for the Canadian Cordillera: arranged by NTS sheet; Geological Survey of Canada, Open File 2373, 217 p.

Root, K.G.
1987: Geology of the Delphine Creek area, southeastern British Columbia: Implications for the Proterozoic and Paleozoic development of the Cordilleran divergent margin; unpublished Ph.D. thesis, University of Calgary, Calgary, Alberta, 446 p.

Rubin, C.M., Saleeby, J.B., Cowan, D.S., Brandon, M.T. and McGrader, M.F.
1990: Regionally extensive mid-Cretaceous west-vergent thrust system in the northwestern Cordillera; implications for continent-margin terctonism; Geology, v. 18, p. 276-280.

Shannon, K.R.
1982: Cache Creek Group and contiguous rocks, near Cache Creek, B.C.; unpublished M.Sc. thesis, University of British Columbia, Vancouver, B.C., 72 p.

Shaw, D.A.
1980: A concordant uranium-lead age for zircons in the Adamant Pluton, British Columbia; Geological Survey of Canada; in Current Research Part C, Geological Survey of Canada, Paper 80-1C, p. 243-246.

Smith, J.M. and Erdmer, P.
1991: The Anvil aureole, an atypical mid-Cretaceous culmination in the northern Canadian Cordillera; Canadian Journal of Earth Science, v. 27, p. 344-356.

Thorkelson, D.J. and Rouse, G.E.
1989: Revised stratigraphic nomenclature and age determinations for mid-Cretaceous volcanic rocks in southwestern British Columbia; Canadian Journal of Earth Sciences, v. 26, p. 2016-2031.

Chapter 17
STRUCTURAL STYLES

Summary
Part A. INSULAR BELT
R.B. Campbell, C.J. Dodds, C.J. Yorath, and A. Sutherland Brown
   Summary
   Saint Elias Mountains
   Southern Insular Belt
Part B. COAST BELT
G.J. Woodsworth, J.W.H. Monger, and H. Gabrielse
   Summary
   Northern Coast Belt
   Central Coast Belt
   Southwestern Coast Belt
   Cascade Segment of the Coast Belt
Part C. INTERMONTANE BELT
H. Gabrielse, J.W.H. Monger, D.J. Tempelman-Kluit, and G.J. Woodsworth
   Summary
   Cache Creek Terrane and Overlap assemblages
   Quesnellia
   Stikinia
   Structures related to transcurrent faults
Part D. OMINECA BELT
D.J. Tempelman-Kluit, H. Gabrielse, C.A. Evenchick, J.L. Mansy, R.L. Brown, J.M. Journeay, L.S. Lane, L.C. Struik, D.C. Murphy, C.J. Rees, P.S. Simony, J.T. Fyles, T. Høy, S.P. Gordey, R.I. Thompson, M.E. McMechan, and T.A. Harms
   Summary
   Introduction
   Core zones
      Nisling Terrane
      Cassiar Terrane
      Monashee Terrane
      Okanagan-Kootenay region
   Cover rocks
     Cassiar Terrane
      Pelly Mountains

      Cassiar and Omineca mountains
      Cariboo Mountains and Quesnel Highlands
    Ancestral North America and Kootenay Terrane
      Northeastern Columbia Mountains
      Selkirk Allochthon
      Kootenay Arc
    Ancestral North America
      Selwyn Basin
      Purcell Anticlinorium
    Accreted Terranes
      Yukon Territory
      Northern British Columbia
      Cariboo Mountains and Quesnel Highlands
      Kootenay Arc region
      Implications of Structural Styles in the Omineca Belt

Part E. FORELAND BELT
M.E. McMechan, R.I. Thompson, D.G. Cook, H. Gabrielse, and C.J. Yorath
  Summary
  The Rocky Mountains
  Mackenzie Mountains, Franklin Mountains and Colville Hills
  Wernecke and Southern Ogilvie Mountains
  Northern Yukon

Part F. TRANSCURRENT FAULTS
H. Gabrielse, J.W.H. Monger, C.J. Yorath, and C.J. Dodds
  Summary
  Tintina and Northern Rocky Mountain Trench System
  Semenof, Teslin, Thibert, Kutcho and Pinchi Fault System
  Finlay, Ingenika, Takla, Fraser River and Straight Creek System
  Yalakom, Pasayten, Harrison, and Ross Lake System
  Sandspit-Louscoone Inlet and Rennell Sound Fault Systems
  Queen Charlotte Fault
  Westcoast Fault
  Denali Fault System
  Duke River Fault
  Chatham Strait Fault
  Fairweather Fault
  Kaltag-Porcupine Fault and Related Fault Systems

Part G. EOCENE EXTENSION FAULTS
R.R. Parrish, R.M. Friedman, and R.L. Armstrong
  Summary
  Southern Cordillera
    Omineca Belt
    Intermontane Belt
  Central and northern Cordillera
    McLeod Lake area
    Horseranch Range
    Klondike Plateau
  References

Chapter 17

STRUCTURAL STYLES

SUMMARY

H. Gabrielse

The dominant elements of structural style in the Canadian Cordillera are related to the Insular, Coast, Intermontane, Omineca, and Foreland morphogeological belts, of which the Coast and Omineca belts represent greatly uplifted granitic and metamorphic orogenic core zones. Structures commonly verge outward from the core zones so that, in cross-section, the Cordilleran orogen contains two symmetrical suborogens (Fig. 17.1, in pocket). The first to develop was the Omineca Belt wherein Mesozoic deformation is attributed to the collision of the Intermontane Superterrane with ancestral North America. Orogenesis in the Coast Belt is attributed to the long-lived development of a volcanic-plutonic arc perhaps coupled with collision of the Insular and Intermontane superterranes beginning in Jurassic time. Subsequent dextral strike-slip faulting greatly modified the distribution of components of the amalgamated terranes.

Mesozoic and Cenozoic structures in the Insular Belt comprise two main elements: 1) contractional, subduction or accretion related faults and folds in the Saint Elias Mountains and Vancouver Island and 2) dextral strike-slip faults and transpressive folds in the Queen Charlotte Islands. In the Saint Elias Mountains contractional structures are cut by Late Jurassic and Early Cretaceous plutons, and, in the southern Insular Belt, both extension and contraction structures are associated with hypabyssal, felsic dykes, sills and small plutons. On Vancouver Island northwest-trending anticlinoria and northerly trending Early and Middle Jurassic plutons dominate the structural grain; on the Queen Charlotte Islands, similar plutons are of Late Jurassic age.

The structurally symmetrical Coast Belt consists of a western part with westward verging folds and thrust faults involving intensely foliated rocks, a central zone with gently dipping foliation and recumbent folds and an eastern part with eastward verging ductile and brittle structures similar to those in the western Intermontane Belt. Transpressive structures associated with dextral transcurrent faults occupy the central part of the Coast Belt and Cascade Mountains.

The Intermontane Belt south of the Stikine Arch is characterized by eastward verging, thin thrust sheets and folds. A narrow belt of southwest- and west-verging structures lies along the west side of the Omineca Belt where, in places, it is disrupted by dextral strike-slip faults. Block faults of several ages are locally conspicuous. The dominant structures appear to be due to deformation and uplift of the bordering Coast and Omineca core zones, the former in mid- to Late Cretaceous time and the latter during the Middle to Late Jurassic. Rocks in the Cache Creek Terrane are generally intensely foliated and, in places, form tectonic mélange, deformed during Mesozoic subduction and accretion. North of the Stikine Arch they, and associated Mesozoic rocks of Stikinia, occur in a wide southwest-verging thrust belt.

Two domains of different structural style characterize the Omineca Belt: ductile, recumbent folds and gently dipping foliations mark the metamorphic core zones whereas brittle faults and more upright, flexural slip folds are typical of the cover rocks. The structures are in part of Middle to Late Jurassic age. Substantial uplift of the core zones, associated with mid-Cretaceous and Eocene granitic intrusion, resulted in northwesterly trending anticlinoria. Dextral transcurrent faulting during the same interval was associated with significant crustal extension in the southern Omineca Belt and transpressive faulting and folding in the northern Cordillera.

Cratonward verging, supracrustal folds and thrust faults are the dominant elements of Foreland Belt structural style. Several levels of décollement constrain the degree of transport which took place in pulsatory phases from Early Cretaceous to Paleocene time. In the northern Yukon folds and thrust faults change trend markedly around a prominent salient in the craton. Ancient tectonic elements such as the Aklavik Arch Complex and Richardson Trough exerted a strong influence on the development of Mesozoic structures which include both dextral strike-slip and steep, dip-slip faults. This complex region occurs where the Cordilleran and Innuitian orogens meet the Brooks Range of Alaska.

Gabrielse, H. (Comp.)
1991: Structural styles, Chapter 17 in Geology of the Cordilleran Orogen in Canada, H. Gabrielse and C. J. Yorath (ed.); Geological Survey of Canada, Geology of Canada, no. 4, p. 571-675 (also Geological Society of America, The Geology of North America, v. G-2)

CHAPTER 17

PART A. INSULAR BELT

R.B. Campbell, C.J. Dodds, C.J. Yorath, and A. Sutherland Brown

Summary

Mesozoic and Cenozoic terrane collisions in the Insular Belt resulted in a complex structural style related to contractional and strike-slip displacements. In the Saint Elias Mountains of the northern Insular Belt a northwesterly structural grain is expressed by elongation of plutons, fold axes and foliation. Widespread deformation and metamorphism affected the Alexander Terrane in post-Late Triassic and pre-earliest Cretaceous time obscuring any Paleozoic deformation. Late Jurassic to Early Cretaceous plutons, at least in part, postdate regional metamorphism and folding. Northeast-verging anticlinoria and synclinoria characterize the northeastern part of the Alexander Terrane, whereas tight, in part recumbent, southeast-verging folds underlie an extensive area in its extreme northern part.

In Wrangellia (W_1), northward directed thrust faults and related folds occur between the Hubbard and Border Ranges faults. Between Duke River and Denali faults, folded and metamorphosed Wrangellian (W_2) strata are cut by elongate Early Cretaceous plutons. East of the Denali Fault in Wrangellian (W_3) strata northwest-trending folds of Cretaceous(?) age have been refolded during Cenozoic strike-slip movements. Cenozoic dextral strike-slip displacement has greatly modified both the earlier contractional structures in Wrangellia and the Alexander Terrane, and the later Tertiary overlap assemblages. Locally, a fault mélange lies between the Duke River and Denali faults.

Southwest of the Border Ranges Fault in the Chugach Terrane, thrust sheets are directed southward, and, within each sheet the intensity of deformation increases structurally downward. Subduction of the Yakutat Terrane has caused the continuing uplift of the Saint Elias Mountains and the greater deformation of Tertiary strata in the outer terranes than in Wrangellia or the Alexander Terrane.

The structural style of the Queen Charlotte Islands, Vancouver Island and the continental margin is dominated by Tertiary to Recent strike-slip faults resulting from the northwestward movement of the Pacific Plate relative to the North America Plate and by Late Cretaceous to Recent northwesterly and westerly trending contractional structures caused by subduction of oceanic rocks beneath North America. Paleozoic rocks on Vancouver Island are locally intensely deformed, but it is not known if this deformation predates gentle folds and broad warps in overlying, more competent Upper Triassic volcanic rocks. Late Cretaceous and Tertiary, northwesterly trending folds and faults are widespread.

Saint Elias Mountains
R.B. Campbell and C.J. Dodds

The Saint Elias Mountains lie along the southwestern margin of the Coast Belt within the northern extension of the Insular Belt in extreme southwestern Yukon, northwestern British Columbia and adjacent parts of Alaska. For the most part they are southwest of the Denali Fault.

The mountains are divisible into six fault bounded terranes, all of which extend into adjacent Alaska (Fig. 17.2). These comprise the Alexander Terrane, three segments of Wrangellia (W_1, W_2, W_3), the Chugach Terrane and a narrow sliver of the Yakutat Terrane; each is characterized by distinctive combinations of plutons, metamorphic rocks and structural styles (Berg et al., 1972, 1978; Coney et al., 1980; Campbell and Dodds, 1983c; Monger and Berg, 1984).

Alexander Terrane

In spite of the excellent exposures that protrude from the vast cover of ice and snow, the structural style of the Alexander Terrane is imperfectly known because of a lack of regionally distinctive stratigraphy and because of the difficulty in recognizing and tracing significant faults and macroscopic folds. The rocks generally exhibit a pervasive cleavage or foliation. The relationship of the Early Permian plutons (270-290 Ma) to metamorphism and deformation is unknown. The younger of the 130-150 Ma plutons (Late Jurassic-Early Cretaceous) seem clearly to postdate the metamorphism and deformation in the northeast whereas, in the southeast, the older plutons of this group both truncate contacts and are elongate parallel with the regional structural and metamorphic grain (see Metamorphic Map 1714A, in pocket). Although Upper Triassic strata probably rest unconformably on Paleozoic rocks they are nonetheless equally deformed and metamorphosed. Thus the widespread deformation and metamorphism seems to be post-Late Triassic and pre-earliest Cretaceous in age. Mid-Cretaceous rocks north of Mount Logan rest unconformably on the Alexander Terrane, are faulted, but are neither folded nor metamorphosed.

Complex deformation in the Alexander Terrane resulting from contraction during collision of the Insular and Intermontane superterranes produced the widely variable dips of strata and the ubiquitous and refolded cleavage. Except at a few localities the style and magnitude of individual folds have not been determined. Observations in the northeast have outlined large, but poorly defined, northwest-trending and northeast-verging synclinoria and anticlinoria. In the extreme northern part of the terrane large folds, some of which are recumbent and southeast verging, trend northeasterly. Similar structural trends occur in the Alsek River region near the British Columbia-Yukon border (Fig. 17.2).

The style of folds varies from place to place and from unit to unit depending on lithology. Folds are tight and multiphase where the metamorphic grade is high. Although supporting data are inadequate, large thrust faults may predate the youngest regional folding event. Several unconformities within the succession in southeastern Alaska suggest episodes of deformation (Gehrels et al., 1983) but they have not been documented in the Yukon.

The Alexander Terrane is bounded by the Hubbard and Duke River faults; the former, a post-Triassic to pre-Late Jurassic terrane boundary or suture (Alexander and W_1) of which the type and magnitude of displacement is unknown; the latter, a post-Triassic fault, perhaps mainly active during the Cretaceous or early Tertiary, with large dextral transcurrent displacements offsetting terrane boundaries. Later minor displacements (Late Miocene-Pliocene) along the Duke River Fault were probably mainly vertical.

Wrangellia - W_1, W_2, W_3

The segments of Wrangellia, with their Mesozoic cover sequences (Berg et al., 1972; Monger and Berg, 1984) bound the Alexander Terrane on the southwest and west (W_1) and north and northeast (W_2 and W_3). They consist predominantly of weakly metamorphosed volcanic and sedimentary rocks of late Paleozoic and Triassic age.

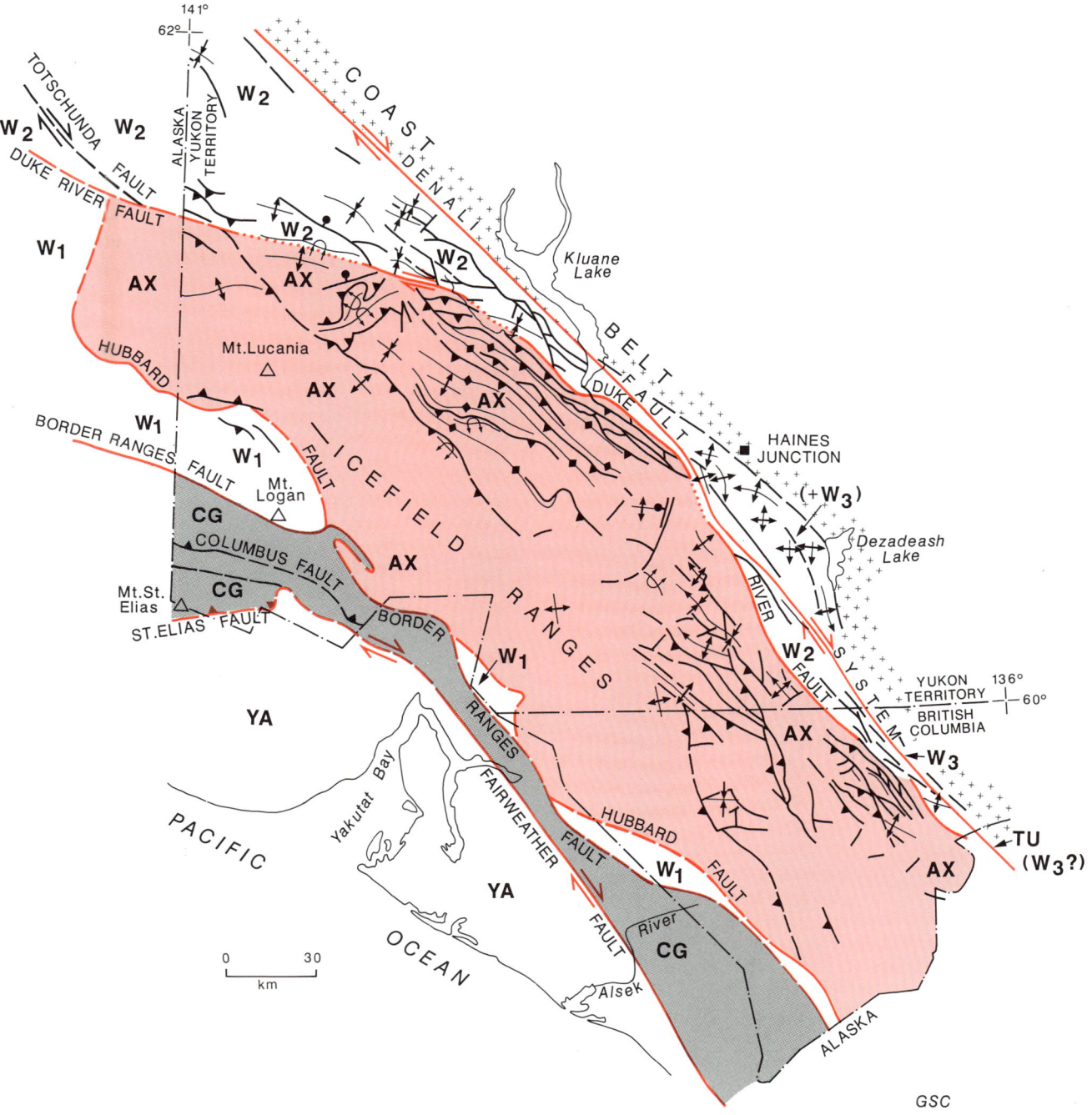

Figure 17.2. Main structures in the Saint Elias Mountains. AX-Alexander Terrane; W_1, W_2, W_3-Subterranes of Wrangellia; CG-Chugach Terrane; YA-Yakutat Terrane; TU-Taku Terrane.

W_1 is confined between the Border Ranges and Hubbard faults. Along its northern margin in Canada, W_1 rocks of greenschist to amphibolite grade are locally thrust northward over unmetamorphosed strata of W_1 which are deformed into large northward overturned folds and cut by south-dipping thrust faults. Farther north, along the Hubbard Fault, the rocks are intruded by numerous syenitic dykes in a zone apparently pervasively faulted. W_1 is intruded by large elongate batholithic complexes (mainly 130-160 Ma; Campbell and Dodds, 1983c). To the west in Alaska, W_1 strata are deformed into broad, open, northwesterly trending folds and are offset by southerly dipping thrust faults. To the southeast, rocks of proven equivalence to those of W_1 have not been recognized but may be included in the mélange of the Tarr Inlet suture zone (Brew and Morrell, 1978). The age of principal folding of W_1 is believed to be Late Jurassic or Early Cretaceous.

Rocks of W_2 and W_3, and their cover, are generally of subgreenschist grade (Muller, 1967; Read and Monger, 1976). The rocks of W_2 are deformed into upright to northeast-verging folds; except in pelitic rocks, cleavage is not strongly developed. Large elongate granitic plutons (106-117 Ma; Campbell and Dodds, 1983c) may be syn- or post-tectonic. In the narrow belt between the Duke River and Denali faults folded W_2 rocks are cut by an anastomosing network of faults. Folds are highly attenuated along the trend of the faults (Read and Monger, 1976) and commonly strata are disposed in numerous long, thin, fault-bounded lenses forming a fault mélange. The bounding faults of the lenses are subparallel with the nearby major faults and within each lens the rocks and stratigraphy are coherent.

In the cover of W_3 (Dezadeash Formation) northwest trending folds of probable Cretaceous age were refolded, presumably during Cenozoic displacements along the Denali Fault. East and southeast of the Saint Elias Mountains in the northern Coast Belt metamorphic rocks of the Taku Terrane (Berg et al., 1978) are tentatively correlated with Wrangellia and included with W_3; little is known of their internal structure.

Chugach and Yakutat terranes

The Chugach Terrane is an accretionary prism lying between the Border Ranges (Fig. 17.3) and Fairweather-Saint Elias faults. Where exposed the bounding faults appear to be narrow and sharply defined structures. On the seaward side of the Chugach Terrane large-scale late Cenozoic and presently active dextral transcurrent faults and related northerly dipping thrust faults (Fairweather Fault and related structures) are the product of continuing oblique northwestward convergence of the Pacific and North American plates and have emplaced the younger accretionary prism of the Yakutat Terrane against the Chugach Terrane (Plafker et al., 1977a, 1978).

Between Mount Logan and Mount Saint Elias the Chugach Terrane comprises two north-dipping thrust sheets which partition the Cretaceous Valdez Group. The upper sheet consists of meta-flysch (MacKevett and Plafker, 1974), and the lower of meta-volcanics and sediments (Campbell and Dodds, 1982b,c). The Columbus Fault dips gently northward but steepens to vertical along its trace toward the southeast. Within each sheet the intensity of metamorphism and deformation increases downward from subgreenschist at the top to amphibolite facies at the

Figure 17.3. View west-northwest along the Border Ranges Fault which separates dark weathering Cretaceous rocks of the Chugach Terrane from light weathering granitic rocks on the south side of Mount Logan. Photo by R.B. Campbell. (GSC 205379)

bottom. In the upper, metasedimentary thrust sheet, high temperature-low pressure metamorphism and related deformation produced a progression from low-grade fossiliferous rocks at the top, close to the Border Ranges Fault, through cleaved andalusite schist to granitoid gneiss with pegmatite dykes at the bottom. The rocks are cut by syn- or post-metamorphic plutons (ca. 50 Ma). The lower plate ranges from unsheared, low grade pillow basalt, tuff and breccia at the top to crystalline, foliated amphibolite at the base above the north dipping Saint Elias Fault. The division of the Chugach Terrane into sedimentary and volcanic thrust sheets, is recognizable far to the southeast in the Alsek River area (Plafker, pers. comm., 1979).

The Fairweather and the Saint Elias faults bound the Chugach Terrane on the southwest and separate it from the Yakutat Terrane (Plafker et al., 1977a). Beneath the Saint Elias thrust overturned folds in Cenozoic sediments verge southward and the rocks are cut by a series of northerly dipping thrust faults (Fig. 17.4).

The Chugach and Yakutat terranes were apparently successively accreted to North America, the former in the early Cenozoic by subduction along the Border Ranges Fault (MacKevett and Plafker, 1974), and the latter by a combination of transcurrent displacement and oblique subduction on a currently active family of faults including the Fairweather (Plafker et al., 1978).

Overlap assemblages

Unconformably overlying deformed Alexander Terrane and Wrangellia (W_1 and W_2) are Tertiary sedimentary strata and volcanic flows (Campbell and Dodds, 1982a,b,c) which are generally flat-lying or tilted at low angles except near fault traces where they are folded or steeply tilted (Souther and Stanciu, 1975). North of Mount Logan flat-lying mid-Cretaceous, fossiliferous marine sediments of the Valdez Assemblage lie unconformably upon deformed Alexander Terrane strata and plutonic rocks; these extend westward into southern Alaska where similar strata lie unconformably on rocks of W_1 (MacKevett, 1978). The

Figure 17.4. View westerly to east face of Hayden Peak in the Saint Elias Mountains. Paleogene(?) sedimentary rocks of the Yakutat Terrane have been deformed into southward verging folds in the footwall of the Saint Elias Fault. Photo by R.B. Campbell. (GSC 205370)

structural style is in marked contrast to that of the metamorphosed and highly deformed rocks of equivalent age in the Chugach Terrane just 20 km to the south, beyond the Border Ranges Fault. Similarly, the mildly deformed Cenozoic rocks on the Alexander Terrane and on W_1 and W_2 to the north contrast with the highly folded and faulted strata of equivalent age in the Yakutat Terrane near the coast. Within Canada clear evidence is lacking for Pleistocene or Holocene offsets on any faults but many are seismically active (R.B. Horner, pers. comm., 1986).

Oligocene nonmarine sedimentary rocks of the Amphitheatre Assemblage, locally as much as 300 m thick (Eisbacher and Hopkins, 1977), once were distributed widely over the northeastern Saint Elias Mountains but now remain only in small faulted, erosional remnants. They evidently underlay a relatively flat, bevelled surface upon which the Miocene-Pliocene Wrangell lavas and related sediments accumulated. These younger rocks were deposited well beyond the remnants of the older strata and extended far into the area underlain by the Alexander Terrane and W_2 and are now locally preserved in the high Icefield Ranges and the Kluane Ranges to as much as 3500 m above the basal lava surface in the adjacent synclines.

In addition to regional tilting and local folding the Wrangell lavas are faulted, particularly in the northern Saint Elias Mountains (Souther and Stanciu, 1975) where a family of northerly directed reverse faults may have provided a mechanism for transfer of dextral displacement from the seismically active Totschunda Fault (Plafker et al., 1977b and Richter and Matson, 1971) in Alaska to the northwest to the Denali Fault system south of Kluane Lake in the east. Such a mechanism might explain the lack of recent seismic activity on the Denali between its junction with the Totschunda Fault and Kluane Lake as indicated by the loci of microseismic events (Horner, 1983).

Rocks as young as Pliocene are cut and displaced by the Denali Fault; the displacements may be large but cannot be measured. South of Kluane Lake linear ponds and mounds along the trace of the fault suggest that some movements may be Pleistocene but evidence for displacements of latest Pleistocene or younger deposits is lacking (Clague, 1979). Northwest from Kluane Lake evidence of relatively young movements on the Denali is even more obscure; thus it seems that the most recent significant displacement on the Denali Fault was Pliocene and/or early Pleistocene in age. The principal movement, however, occurred between mid-Cretaceous and early to mid-Cenozoic time (Eisbacher, 1976).

The Duke River Fault offsets only the basal part of the Miocene-Pliocene Wrangell lavas but along its trace the upper part of the lava sequence is locally folded. In a general way, the fault trace marks the axes of two regional synclinal warps outlined by the basal lavas and associated Miocene alluvial sediments. Synclinal structure is obscure in the far northern part of the mountains but is particularly apparent along the valley of Duke River from where the base of the lava pile rises, east and west, from below the valley bottom to the crests of the ridges on either side, a change in elevation of about 1000 m. Similarly, farther to the south along an apparently en échelon synclinal warp, the base of the lava sequence and associated alluvial sediments are broadly folded. On the southwestern limb the basal surface rises more than 1000 m in a distance of 16 km; the eastern limb is truncated by the Denali Fault. These en échelon synclines together form part of the Duke Depression (Bostock, 1948) within which much of the Cenozoic sedimentary and volcanic successions in the northeastern Saint Elias Mountains are preserved.

The rise of the mountains

The Saint Elias Mountains are the highest and geologically youngest mountains in Canada. They rise, locally with spectacular relief to elevations in excess of 5500 m (e.g. Mount Saint Elias) above the narrow coastal lowland along the margin of the Gulf of Alaska. Oblique subduction of the Yakutat Terrane has caused the uplift which continues today. Beginning in the Miocene, extrusion of the Wrangell lavas also was probably in response to the subduction process; modern volcanic activity continues today in southern Alaska. The Wrangell lavas were originally deposited on a gently undulating surface of low elevation (Souther and Stanciu, 1975). The principal uplift of the mountains and the tilting, faulting and folding of the Wrangell lavas may have begun in the Miocene as a consequence of oblique subduction and transcurrent displacements along the Denali Fault; however, most of the tectonism was concentrated in the Pliocene and Pleistocene. Mount Logan rises to nearly 6000 m, about 4000 m above the adjoining ice-filled valleys (Fig. 17.3). That such spectacular relief exists within the region attests to the erosive power of the vast ice sheets and glaciers that adorn the mountains, and, perhaps to unrecognized fault displacements. The mountains provide an awe-inspiring vista of the continuing struggle between the dynamic forces of topographical construction and destruction, the most spectacular such view in Canada and one that ranks with any in the world (Plate 31).

Southern Insular Belt

C.J. Yorath and A. Sutherland Brown

The structural style of the southern Insular Belt is expressed by a dominant northwesterly to northerly aligned stratigraphic, plutonic and structural fabric. Local variations are found on southern Vancouver Island where westerly trending faults juxtapose terranes of disparate origins against and beneath Wrangellia. The contemporary crustal architecture of the region is dominated by the subducting Juan de Fuca Plate and the dextral, transform Queen Charlotte Fault which, adjacent to Queen Charlotte Sound, meet the Juan de Fuca Ridge system at a ridge-transform-trench triple junction. Beneath the continental shelf and slope, off Vancouver Island, Tofino Basin contains Tertiary sediments and volcanics which are complexly dislocated by westerly directed thrust faults and associated folds.

Queen Charlotte Islands

Three major fault systems have been recognized on the Queen Charlotte Islands (Fig. 17.1, Section M, in pocket; 17.5). The Rennell Sound Fault Zone consists of a broadly curvilinear, northwesterly trending set of vertical to steeply northeasterly dipping faults which have both dextral and east-side-down normal separation (Sutherland Brown, 1968). The zone extends from Rennell Sound to Louise Island and is physiographically most strongly expressed in the comparatively incompetent sediments of the mid- to Upper Cretaceous Queen Charlotte Group and by abrupt changes in stratigraphy across individual strands. Eastward, in the subsurface beneath Hecate Strait, some components of the zone have been recognized in single and multichannel seismic profiles where they dislocate Neogene sediments and deposits of possible Pleistocene age. The zone may coincide with the suture between the Alexander Terrane and Wrangellia (Yorath and Chase, 1981), but this contention has been disputed by Woodsworth (1988).

The Louscoone Inlet Fault Zone extends northwestward from southernmost Kunghit Island, through the island passages in and along eastern Moresby Island to Louise Island where it appears to be truncated by the Rennell Sound Fault Zone. Described by Sutherland Brown (1968) as a component of the "Rennell Sound-Louscoone Inlet Fault Zone" the major strands of the system are vertical faults with apparent right-lateral separation. Faults with this sense of displacement, however, have not been documented in subsequent detailed mapping (Thompson and Thorkelson, 1989). On the basis of geophysical and stratigraphic data, Yorath and Chase (1981) proposed that the two components of the system are temporally and dynamically distinct and that during the late Tertiary movement on the Louscoone Inlet system predated that on the Rennell Sound Zone.

The Sandspit Fault System extends northwestward from northeastern Moresby Island to central Graham Island beyond which the zone is masked by Quaternary drift and probably by the younger flows of the Neogene Masset Formation. Geophysical data support its continuation to beyond northernmost Graham Island (Yorath and Chase, 1981; Fig. 17.5). The youngest movements along the system are expressed by offset creeks, small scarplets in Pleistocene sediments, sag ponds and inactive sea cliffs (Sutherland

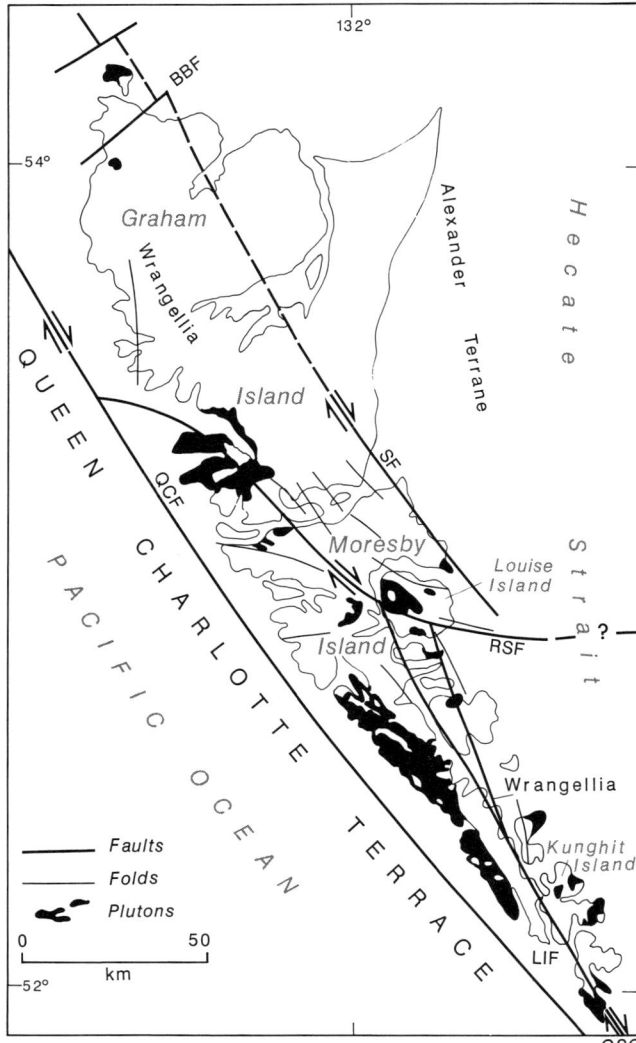

Figure 17.5 Faults, folds and plutons in the Queen Charlotte Islands. QCF= Queen Charlotte Fault; BBF= Beresford Bay Fault; SF= Sandspit Fault; RSF= Rennell Sound Fault; LIF= Louscoone Inlet Fault;

Brown, 1968). Older activity is indicated by the alignment of small, hypabyssal Tertiary plutons parallel with the system and by the dramatic increase in thickness of upper Neogene sediments of the Skonun Formation across the system into the subsurface beneath Hecate Strait. Yorath and Chase (1981) proposed that the Sandspit Fault System was once continuous and collinear with the Louscoone Inlet Fault Zone, the two components having been dislocated by late Tertiary movement along the Rennell Sound Fault Zone, and, further, that it is also collinear with the boundary between Wrangellia and the Alexander Terrane. Recent studies have cast some doubt on the significance of these structures (Thompson and Thorkelson, 1989).

A few northeasterly trending vertical faults, principal among which is the Beresford Bay Fault, cross northwestern Graham Island. These have a vertical component of separation and a significant horizontal sinistral component evinced by the dislocation of Upper Cretaceous rocks of the Queen Charlotte Group. In west central Graham Island H.W. Tipper (in Sutherland Brown et al., 1983) suggested the presence of northeasterly dipping fault-bounded panels which may reflect thrust separation.

The currently active Queen Charlotte Fault marking the edge of the North American continent is described in the subchapter on transcurrent faults.

On the Queen Charlotte Islands folds are subordinate to faults and, regionally, are of low intensity. Locally however, and particularly within the Queen Charlotte Group adjacent to the Rennell Sound Fault Zone, steep upright and overturned panels occur. Within the fault zone, the turbidite facies of the Lower Cretaceous Longarm Formation is intensely sheared and bedding is commonly completely transposed. Folds in older rocks are more westerly trending and are commonly truncated by the northwesterly trending faults. The thick pillow lavas of the Upper Triassic Karmutsen Formation have been warped into broad, low-amplitude anticlines whereas immediately overlying thinly interbedded argillite and limestone of the Kunga Formation are commonly intensely deformed into tight, chevron and disharmonic folds. At least some of this deformation took place prior to the development of an unconformity at the base of the Middle Jurassic Yakoun Formation (Thompson and Thorkelson, 1989).

The structural style within Queen Charlotte Basin is largely unknown. Given the rifting hypothesis in Queen Charlotte Sound (Yorath and Hyndman, 1983) its seems probable that significant transcurrent faults would dislocate the Tertiary volcanic and sedimentary succession and also, that associated en échelon folds would be present. The probable extension of the Rennell Sound Fault Zone across Hecate Strait suggests the presence of similar structures.

Off the west coast of the Queen Charlotte Islands, the Queen Charlotte Terrace comprises a prism of sediments that have been deformed into arcuate, high-amplitude folds. The terrace, more than 6 km thick, is bounded on its east side by the Queen Charlotte Fault and on the west by the continental slope.

Vancouver Island

The structural style of Vancouver Island is expressed by northwesterly trending structural culminations, northwesterly and northerly trending plutons and by northwesterly and southwesterly directed faults (Fig. 17.6). Minor northeasterly striking faults interrupt these trends (Muller, 1977). As in the Queen Charlotte Islands the Karmutsen Formation appears to be the most competent unit which, between important faults, has been warped into broad culminations and depressions. Less competent underlying and overlying units have been more intensely and, locally, penetratively deformed. In the Gulf Islands

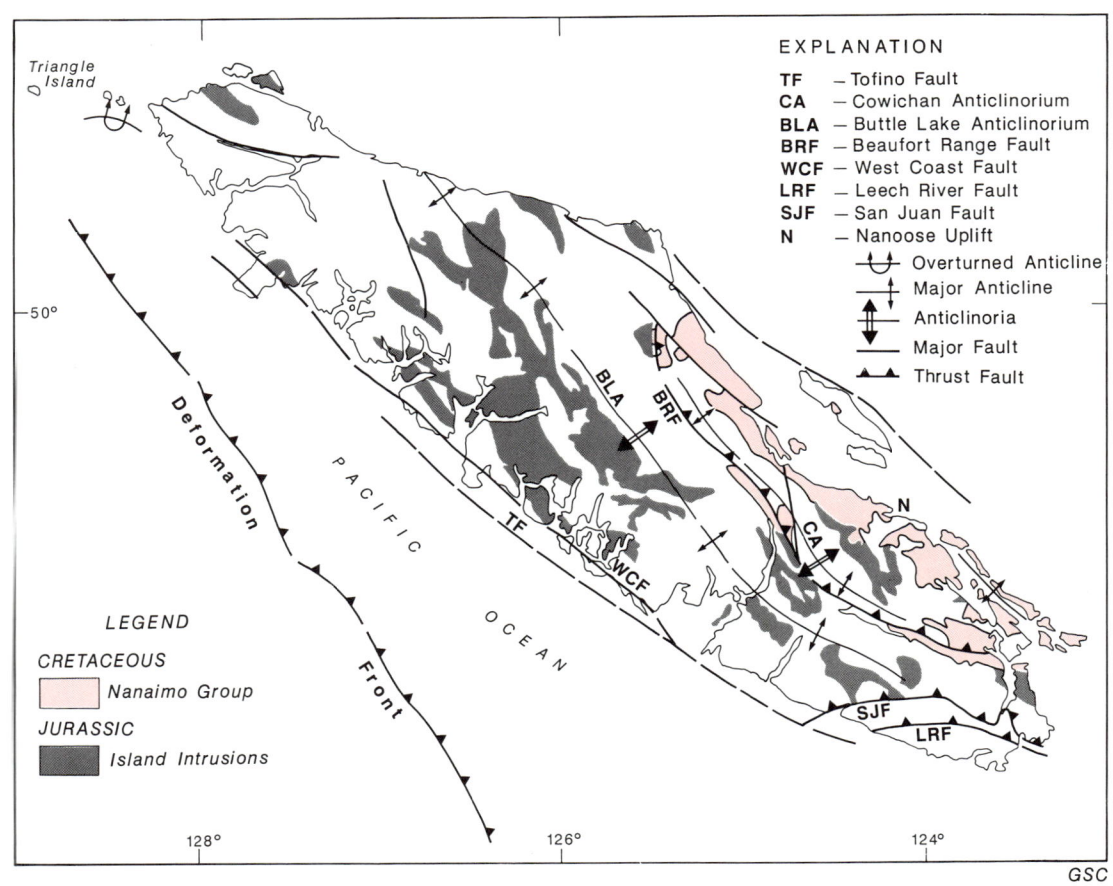

Figure 17.6. Main faults, folds and plutons on Vancouver Island.

near the island's southeast coast, the Nanaimo Group occurs in northwesterly trending cylindrical and isoclinal folds and, perhaps, is cut by westerly directed thrust faults.

The Cowichan and Buttle Lake anticlinoria and the Nanoose Uplift (Fig. 17.6) are northwesterly trending structures within which Sicker Group sedimentary and volcanic strata are commonly folded into tight to isoclinal folds. The Cowichan Anticlinorium and its collinear expression in the Beaufort Range are bounded on the southwest by two fault systems that are believed to be dynamically linked (Sutherland Brown and Yorath, 1985). These are the Cowichan Lake and Beaufort Range faults; the latter has been suggested to have been the locus of the 1946 earthquake that caused significant damage to several communities in central Vancouver Island (Rogers and Hasegawa, 1978). In the Nanoose Uplift, on the east coast of the Island, Sicker Group strata are isoclinally folded and penetrative strain occurs in argillite. Northwesterly trending faults, some of which may be thrusts, disrupt the sequence.

Subsurface structures of Vancouver Island and the adjacent continental margin (Fig. 17.1, Section N, in pocket; 17.7) are revealed by land-based and marine multichannel seismic reflection profiles (see Fig. 2.34), and by other geophysical and geological data (Shouldice, 1971; MacLeod et al., 1977; Yorath, 1980; Yorath et al., 1985a,b; Sutherland Brown and Yorath, 1985; Clowes et al., 1987; Kurtz et al., 1986).

Beneath Vancouver Island Wrangellia is partitioned by several westward verging thrust faults, mostly of probable Late Eocene age. Some high-level thrust faults of probable Late Cretaceous age occur near the east coast where they dislocate Cretaceous strata and the west limb of the Cowichan Anticlinorium (Fig. 17.7). Beneath the west coast of the island the Pacific Rim and Crescent terranes occur in the footwalls of the Westcoast and Tofino-Leech River faults, respectively. Similar relationships are observed on southern Vancouver Island (Clowes et al., 1987). These terranes were emplaced beneath Wrangellia during the Late Eocene and probably led to the development of many of the thrust faults that disrupt Wrangellia. Moreover, the emplacement and underplating of these terranes, together with the subsequent and current accretion of the modern subduction complex, has resulted in uplift of western Vancouver Island and companion subsidence of Georgia Basin. Ensuing erosion led to removal

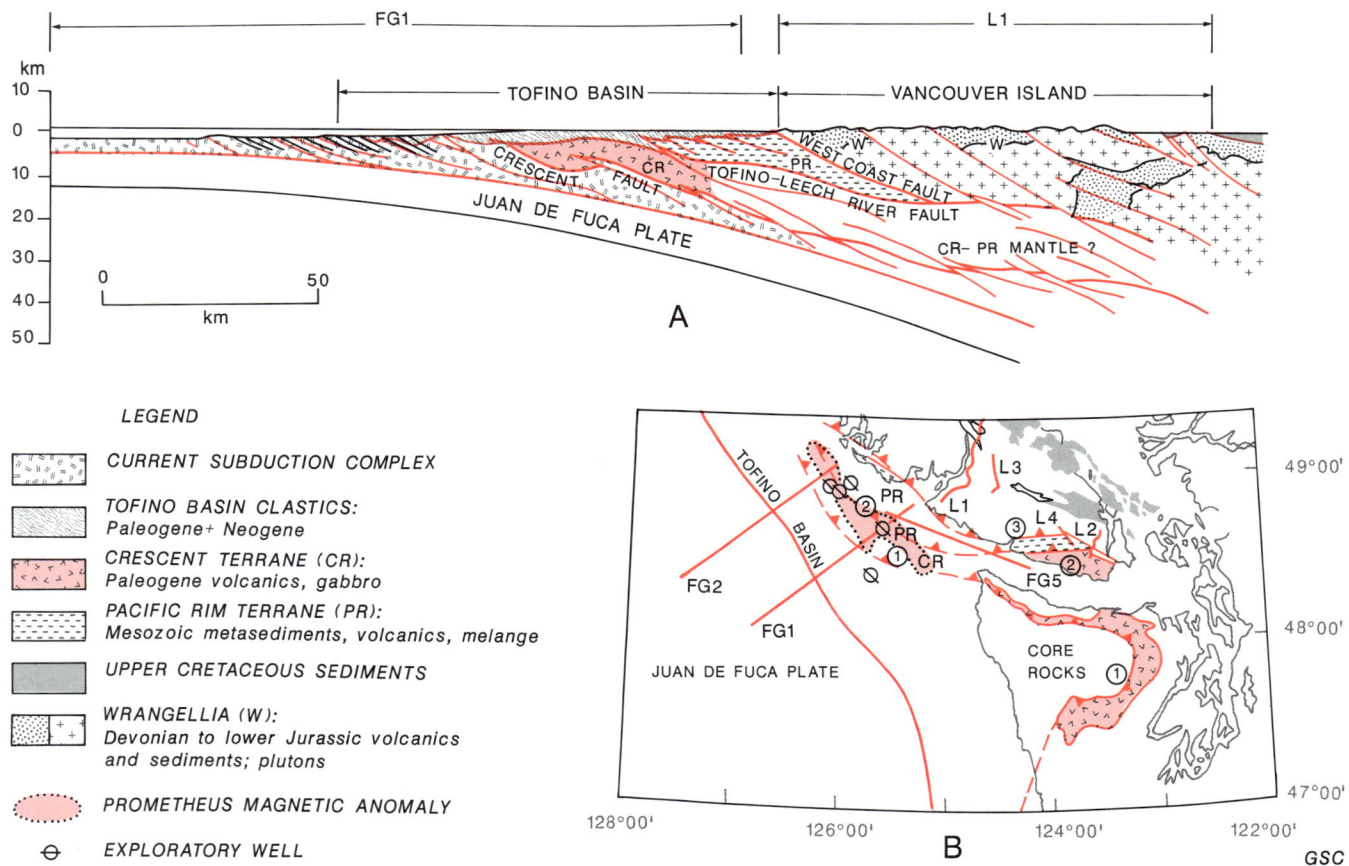

Figure 17.7A. Structural interpretation of multichannel seismic reflection profiles across the continental margin (FG1) and southern Vancouver Island (L1). **B.** Main structural elements and locations of multichannel seismic reflection profiles, southern Vancouver Island and northwestern Washington continental margin; 1-Crescent Fault; 2-Tofino-Leech River Fault; 3-San Juan-West Coast Fault.

of several kilometres of Wrangellian stratigraphy and exposure of the deeper levels of the Early Jurassic plutons which are represented by metamorphic and plutonic rocks of the West Coast Complex on western Vancouver Island.

The Crescent Terrane, penetrated by offshore exploratory wells and represented by the Prometheus magnetic anomaly (Fig. 17.5; Shouldice, 1971; MacLeod et al., 1977) comprises rocks chemically similar to basalt of the Metchosin Igneous Complex of southern Vancouver Island. The terrane occurs in the hanging wall of the Crescent Thrust, a structure extrapolated from beneath the Olympic Mountains of northern Washington to beneath the continental shelf off Vancouver Island. There the terrane structurally overlies intensely deformed Middle and Upper Eocene sedimentary rocks comprising part of the current Cenozoic subduction complex in the core of the Olympic Mountains. Although correlations of reflections between the offshore and onshore seismic profiles are uncertain (see Fig. 2.34), in Figure 17.5 the Crescent Terrane is interpreted to comprise much of the underplated material beneath Wrangellia. Other components of the underplated zone might be additional segments of the Pacific Rim Terrane and, perhaps Eocene mantle associated with the basalts (see Chapter 2, Part C). An important corollary to this interpretation is that, assuming that the original lithosphere of Wrangellia was in the order of 100 km thick, then more than 70 km of "Wrangellian" lithosphere must have been removed prior to the underplating by the Pacific Rim and Crescent terranes.

The current subduction complex extends from the limit of deformation to beneath western Vancouver Island where its sediments are perhaps equivalent to the Core Rocks of the Olympic Mountains. Landward from the deformation front, thrust faults with associated ramp folds intersect the top of the subducting oceanic crust and some extend upward to close to the seafloor. Thus since the underplating of the Crescent Terrane, the sedimentary section of the Juan de Fuca Plate has been accreting to and beneath the continental margin. The accretion process has led to folding and uplift of the Crescent Terrane and superimposed bathyal sediments (Cameron, 1980) which form the lower part of Tofino Basin. Within the Neogene sediments of the upper part of Tofino Basin a series of curvilinear folds interrupt the otherwise flat, gently sloping seafloor. One of these, the Apollo structure, may be an anticline which developed above a shallow detachment surface within the Neogene and Quaternary sediments, possibly as a consequence of Holocene earthquake activity (Yorath, 1980).

Note: More recent discussions of the structure and tectonic history of the Cascadia subduction zone and Vancouver Island appear in Hyndman et al. (1990) and Yorath et al. (in press).

A volcano-sedimentary sequence, either of Jura-Cretaceous or Cenozoic age, underlying the northwestern Vancouver Island continental shelf is complexly deformed. Submersible traverses and side-scan-sonar profiles suggest that the sequence is deformed into an arcuate series of southwesterly and southerly verging recumbent folds (Yorath and Currie, 1980; Bornhold and Yorath, 1984). On Triangle Island these are subaerially exposed within a thick sequence of argillite and greywacke, intruded by one or more felsite sills. Unconformably overlying Upper Miocene to Pliocene strata are undeformed except for a few vertical faults near the shelf edge. Beneath the continental slope, dredging by B.E.B. Cameron (pers. comm., 1980), seismic reflection profiles and submersible traverses suggest the presence of thrust faults involving Lower Cretaceous and Tertiary strata.

PART B. COAST BELT

G.J. Woodsworth, J.W.H. Monger, and H. Gabrielse

Summary

The structural styles of the Coast Belt are difficult to interpret because of extensive areas of plutonic and metamorphic rocks, a paucity of stratigraphic markers and difficulties in correlation. Structures in metamorphic and migmatitic rocks range from steeply dipping to flat-lying over considerable areas and reflect deep-seated ductile deformation overprinted by shallower level ductile to brittle deformation.

The most distinctive structures in the northern Coast Belt are commonly gently dipping foliations in the older granitic and metamorphic rocks, and swarms of early Tertiary tensional joints occupied by leucocratic dykes.

In the Prince Rupert-Terrace region of the Coast Belt two longitudinal structural belts can be recognized east and west of Work Channel Lineament, together with an eastern zone of rocks correlative with those in the westernmost part of the Intermontane Belt. The belt east of Work Channel, a core zone of high grade metamorphic and plutonic rocks, is characterized by gently dipping ductile fabrics and large sheet-like plutons. The belt of metamorphic and plutonic rocks west of Work Channel has east-dipping, west- to northwest-directed ductile to brittle fabrics and structures. The most eastern zone is characterized by unmetamorphosed to low greenschist-grade strata and by mainly east- to northeast-verging brittle to ductile shear zones related to similar structures in the western Intermontane Belt. Superimposed on these pervasive and relatively gentle structures are discrete, steep shear zones that trend northwest to north. Plutons in the central and western zones have been affected by both ductile and brittle deformation; pluton shape may have been controlled by emplacement into a dynamic regime. This style of structure and plutonism appears to indicate a component of longitudinal motion during development of the orogen.

In the southern Coast Belt, structures are dominated by steep, northwest-trending fabrics and by northwest-trending belts of pendants of metamorphosed sediments and volcanics. These belts are interpreted as graben-like structures of mid- to Late Cretaceous age. In the

Intermontane Belt and the eastern part of the southern Coast Plutonic Complex, northeast-directed thrust faults involve Early Cretaceous and older strata. Collectively, the structures indicate that the Coast Plutonic Complex and the western Intermontane Belt in both the northern and southern sections form a two-sided, perhaps in part wrench-related, orogen in which deformation spanned the time from Late Jurassic to early Tertiary.

The Cascade segment of the Coast Belt includes a granitic to high grade metamorphic core zone of polyphase deformed rocks cut by steeply dipping faults. West of the core zone are at least three major east-dipping thrust faults, one of which contains a large northwestward directed recumbent fold. East of the Cascade core zone structures are mainly eastward-overturned flexural slip folds associated with west-dipping reverse faults. The core zone of the Cascade segment is bounded on the west by the Straight Creek dextral fault zone in the south. Its extension to the north, the Fraser River dextral fault zone, occurs on the east side of the core zone. Structures in the Cascade segment are mainly of Cretaceous age and may represent closure of oceanic or marginal basins followed by Tertiary dextral wrench and associated normal and reverse faulting.

Northern Coast Belt

H. Gabrielse

Little is known about the details of the structural style in the Coast Belt north of the Prince Rupert region. The area includes large tracts of foliated granitic rocks with pendants and bordering assemblages of regionally and contact metamorphosed rocks. Weakly foliated to non-foliated granitic rocks cut the foliated bodies and associated metamorphic successions. Foliation in the metamorphic rocks, gently dipping over large areas, is commonly parallel with compositional layering which in many cases represents transposed bedding. Folds trend mainly northwest and range from open to tight with no clear, consistent sense of vergence. Two linear elements have been recognized in the metamorphic rocks of Aishihik area of southwestern Yukon Territory (Tempelman-Kluit, 1974). The most conspicuous is related to a coarse crenulation which kinks the foliation in micaceous rocks. Fold axes trend from northerly to northeasterly. The more quartzose rocks show a strong rodding or streaking of quartz. Locally, small scale folds plunge northward at gentle angles and are subisoclinal and asymmetrical in cross-section with a westward vergence. In one locality east of the Denali Fault the locus of an inferred northward dipping thrust fault is marked by altered mafic and ultramafic rock (Muller, 1967).

The only well dated structures in the northern Coast Belt are tensional joints and faults occupied by swarms of leucocratic dykes of early Tertiary age (Tempelman-Kluit, 1974). In northern British Columbia similar swarms of leucocratic dykes trend northerly and northeasterly. Eocene volcanic rocks locally are associated with cauldron subsidence structures (Lambert, 1974) or grabens.

Central Coast Belt

G.J. Woodsworth

The structural style of the central Coast Belt is best understood in the Prince Rupert-Terrace cross-section where it is intimately linked to structures in the Intermontane Belt (Fig. 17.8; 17.1, Section M (in pocket)). In the Prince Rupert-Terrace area, the Coast Belt, mainly composed of the Coast Plutonic Complex, is divided into two longitudinal segments on the basis of contrasting metamorphic, structural and plutonic styles (Crawford and Hollister, 1982; Crawford et al., 1987). The core zone of high grade metamorphic and plutonic rocks east of Work Channel is characterized by gently dipping ductile fabrics and large sheet-like plutons. The segment west of Work Channel, composed of metamorphic and plutonic rocks, has east-dipping, west- to northwest-directed ductile to brittle fabrics and structures. Superimposed on these pervasive and relatively gentle structures in both belts are discrete, steep shear zones that trend northwest to north. Plutons have been affected by both ductile and brittle deformation; pluton shape may have been controlled by emplacement into a dynamic regime. This style of structure and plutonism appears to indicate a component of transpressive motion during development of the orogen.

West of Work Channel

Rocks in this segment consist mainly of greenschist to amphibolite facies metasediments and metavolcanics intruded by plutons. The western part, in the Dunira Island area, is characterized by foliation and transposed bedding which dip moderately toward the east. Where coherent stratigraphy is preserved, bedding is upright and dips moderately east to northeast. Locally, cleavage is axial planar to small isoclinal folds, many of which have sheared limbs. Folds and foliation represent a single deformation, characterized by ductile shear, and reflect east-dipping, southwesterly-directed thrust faults.

From Prince Rupert east to Work Channel, mineral foliation and compositional layering dip east to southeast at moderate angles. Metamorphic pressures and temperatures increased from west to east, and the highest grade rocks occupy the structurally highest position. Isoclinal folds have east-plunging axes and are cut by a later fracture cleavage. The cleavage increases in intensity to the east and becomes a penetrative foliation axial planar to tight folds. Metamorphic conditions in the structurally highest rocks reached 8 ± 1 kb ($8 \pm 1 \times 10^5$ kPa) and 625°C (Crawford and Hollister, 1982); metamorphic rocks on the small islands west of Prince Rupert are in greenschist facies. Synmetamorphic plutons emplaced into the highest grade schists at about 98 Ma (Woodsworth et al., 1983) indicate that metamorphism and deformation were roughly mid-Cretaceous in age. Syn- to post-tectonic plutons on Dunira and nearby islands yield K-Ar dates of 84 and 96 Ma on biotite, suggesting that metamorphism and deformation extended into Late Cretaceous time.

Taken as a whole, the best interpretation of the metamorphic pattern and the minor structures may be that westerly-directed, ductile thrust faults (during amphibolite facies conditions), superimposing high grade rocks over lower, are the dominant structures. The locations of these faults are unknown, although Hutchison (1982) postulated one near Prince Rupert. During cooling, ductile shear was followed by brittle thrusting, some of these faults are visible near Prince Rupert.

The western belt about 50 to 100 km south of Prince Rupert is dominated by plutonic rocks. Schist and gneiss underlie less than ten per cent of the area and occur

STRUCTURAL STYLES

mostly as narrow, steeply-dipping screens elongated northwesterly. Foliation is well developed in the schist and in many plutons. In general, foliation strikes northwest and dips vertically or steeply to the east, but there are numerous exceptions. Many plutons appear to have been emplaced late in the main period of deformation, but these too are commonly elongate, parallel with the regional grain. Preliminary U-Pb dates on zircons suggest that most plutons west of Grenville Channel are of Late Jurassic age (P. van der Heyden, pers. comm., 1987). If so, then ductile deformation in the western Coast Plutonic Complex is also Late Jurassic in age.

East of Work Channel

Rocks between Work Channel and the Shames River fault zone, consist mainly of metamorphosed sediments, volcanics, and plutons, collectively called the Central Gneiss

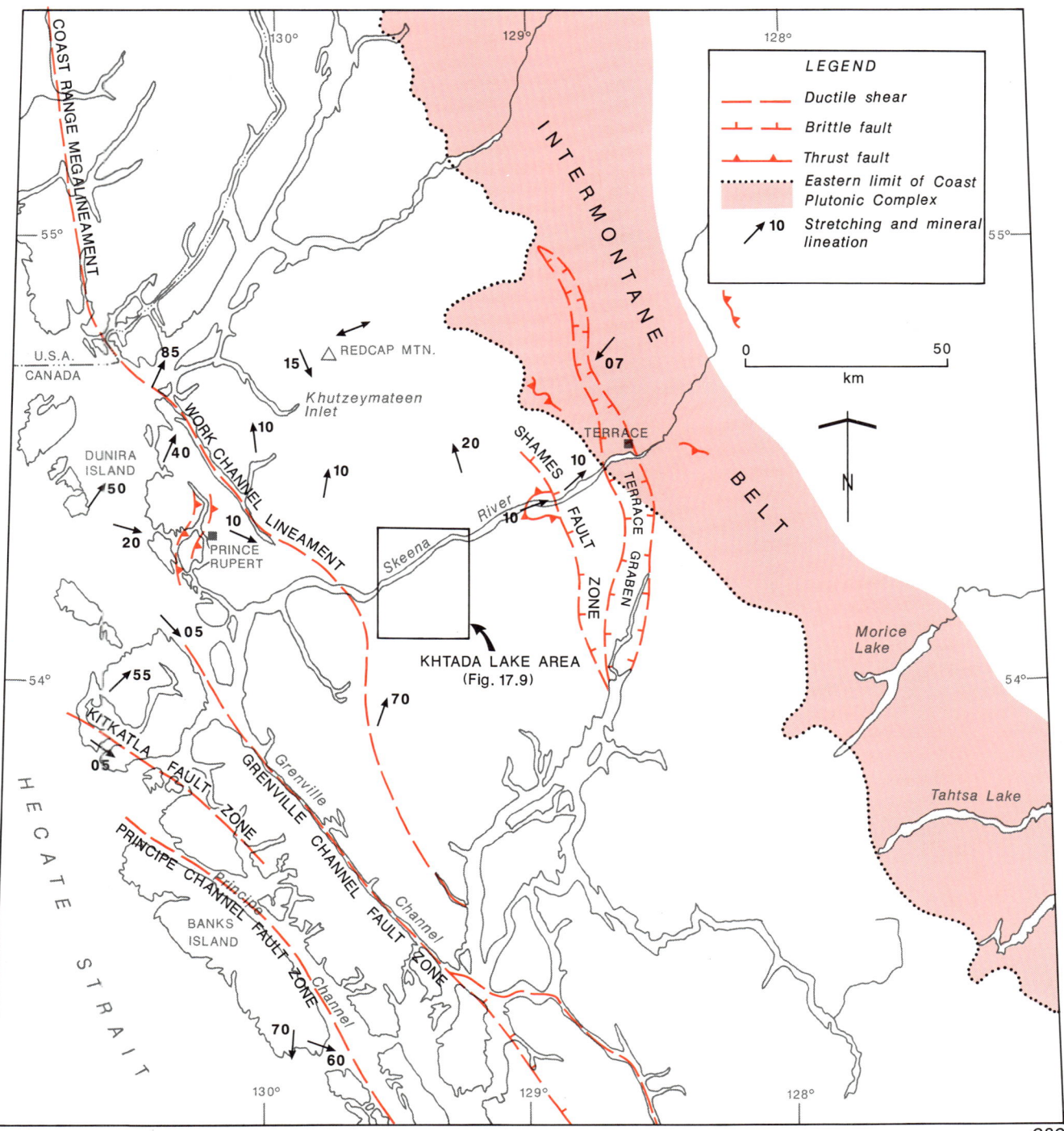

Figure 17.8. Structural elements of the central Coast Belt.

Complex. Abundant syn- to post-metamorphic plutons cut the Central Gneiss Complex, and in many places are difficult to distinguish from it. Deformation was everywhere dominated by ductile shear, but the structural histories of the rocks vary from place to place.

Throughout much of the area the overall dip of foliation is gentle to the north. Recumbent folds with north-dipping axial surfaces are common and are well illustrated in the Redcap Mountain area (Fig. 17.8; Hill, 1984). There, the Central Gneiss Complex and related metasediments have a pervasive mylonitic foliation, and may be termed blastomylonites. The rocks are exposed in large, south-verging recumbent folds; lineations and axes of tight, noncylindrical minor folds are moderately to gently plunging. The structures are interpreted as resulting from ductile shear at high temperatures. The attitude of the foliation suggests that the shear plane was roughly horizontal and that the latest motion was directed to the northwest, parallel with the overall strike of the Coast Belt. These conclusions may apply to much of the Central Gneiss Complex beyond the bounds of the Redcap Mountain area, including the Khtada Lake area farther south.

In the Khtada Lake area (Fig. 17.9), early recumbent folds with gentle east-plunging axes have been overprinted by tight, upright folds whose axes parallel those of the recumbent folds. The later folds appear to be related to the development of discrete, steep, northeast- to northwest-trending ductile shears. Both generations of folds developed while the rocks were above 650°C and probably while they were partly molten (Kenah and Hollister, 1983). In the Khutzeymateen Inlet area southwest of Redcap Mountain, the early recumbent folds have not been recognized. There folds are dominantly upright, open to tight with axes that plunge gently north-northwest (Douglas, 1983). These folds deform an earlier mylonitic foliation and may correlate with the upright folds in the Khtada Lake area.

Estimation of the timing of deformation in the Central Gneiss Complex is constrained by U-Pb dates from a variety of lithologies suggesting that high grade metamorphism of the belt was mid- to Late Cretaceous in age (Crawford et al., 1987; Woodsworth et al., 1983; Hill, 1984). High grade metamorphism was accompanied by partial melting, ductile deformation, and emplacement of syntectonic plutons. These phenomena resulted from thickening of the crust, possibly in response to accretion of the Insular Superterrane to North America. Horizontal shear involved northwestward thrusting of an upper plate (at depths of about 20 km) with a component of right-lateral transverse motion (Hill, 1984).

Ductile shear zones

Work Channel lineament is a conspicuous topographic feature that can be traced northwest from Douglas Channel through southeast Alaska to Juneau (Brew and Ford, 1978). In British Columbia it marks the boundary between the western and central parts of the Coast Belt (Crawford and Hollister, 1982). The lineament marks a complex ductile shear zone (Fig. 17.10, 17.11). Foliation is steep to vertical, and stretching lineations plunge steeply, suggesting that the latest recorded movement on the shear zone was vertical (Gareau, 1988). Foliation and mineral lineation in the Quottoon pluton, which bounds the lineament on the east, parallel those in metamorphic rocks within the shear zone.

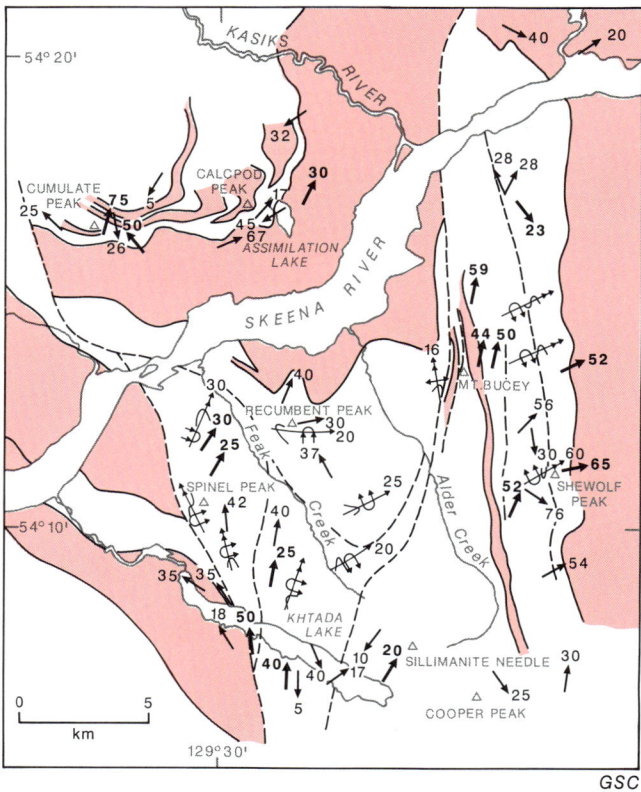

Figure 17.9. Structures in the Khtada Lake area. See Figure 17.8 for location.

Intrusion of the Quottoon pluton in Paleogene time was synchronous with the last major ductile movement on the shear zone.

Other ductile shears are common in the central part of the Coast Belt. Like the Work Channel zone, they show steep foliations and have been superimposed on the earlier, gently-dipping foliation. In the western belt of the Coast Plutonic Complex, many prominent northwest-trending waterways such as Grenville, Principe, and Kitkatla channels mark the loci of important faults (Fig. 17.8). Foliation in these zones is steep to vertical; many of the rocks are augen gneiss or mylonite. The amount, nature, and timing of movement on these faults is not known, but on the Grenville Channel Fault early ductile movement was followed by later brittle motion. At least some of these faults may represent major zones of transcurrent displacements.

Miocene extension

Northerly trending swarms of Miocene basalt and comendite dykes in the Bella Bella area on the western margin of the Coast Belt represent local extension of up to

40% (see Fig. 17.8). Their emplacement has been related to crustal expansion and pluton emplacement beneath the western part of the Anahim volcanic belt (see Chapter 14).

Southwestern Coast Belt

G.J. Woodsworth

Structures in the southern Coast Belt south of 52°N and west of Harrison Lake and Yalakom faults (Fig. 17.12) are not as well understood as in the Prince Rupert-Terrace transect. Differences exist in the stratigraphic and metamorphic framework and in the style and timing of deformation.

Southwest of the Harrison Fault

The area southwest of the Harrison Fault is dominated by plutonic rock; only about 10 to 15% consists of metamorphosed sediments and volcanics. Most of the metamorphic rocks occur as narrow, northwest-trending pendants or screens separated by large expanses of plutonic rock. Metamorphic grade of the pendants varies from subgreenschist in the west to upper amphibolite facies in the eastern part of the area. Structures in the plutons and most pendants are dominated by northwest-trending foliations that commonly dip steeply northeast. Many plutons are elongate northwesterly and have steep contacts.

The internal structures of most pendants are poorly known. The Mount Raleigh area is the best-studied of the highly metamorphosed bodies (Woodsworth, 1979). Granitoid gneiss that may be correlative with the Central Gneiss Complex farther north has a pervasive foliation cut by small, discrete ductile shears. Deformation in the granitoid gneiss is believed to be older than that in the structurally overlying Mount Raleigh pendant. Mesozoic rocks of the pendant were affected by northeast-directed thrusting that brought Upper Triassic over Lower Cretaceous rocks and were folded into an open syncline which plunges gently east-southeast. The dominant fabric in the pendant rocks and some adjacent plutons is a penetrative foliation and south- to west-southwest-plunging mineral lineation. The fabric developed during amphibolite facies metamorphism in mid- to Late Cretaceous time, and probably represents ductile shear caused by forcible emplacement of synmetamorphic, diapiric plutons.

The Sky Pilot pendant is the best studied of the lower grade pendants. The main structure of the Mesozoic rocks in the pendant is an upright homocline that dips gently south (Heah, 1982). The central part of the pendant is cut by a broad, northwest-trending, ductile to brittle shear zone. This zone dips steeply and was formed by compression along a northeast-southwest axis, with a component of dextral shear (Payne et al., 1980). The relation of the shear zone to the emplacement of nearby plutons is not known.

In both the Sky Pilot and Mount Raleigh pendants (and probably numerous others), the rocks are cut by late, northeast to northwest-trending brittle faults and dykes. Movement on each fault is probably small (a few metres) and took place in early Tertiary time.

The distribution of pendants in the southern Coast Belt is concentrated in narrow, northwest-trending belts that are semi-continuous for many kilometres along strike. These belts appear to be grabens in which the pendant strata have been downdropped relative to the neighbouring plutons (Woodsworth, 1979; Roddick and Woodsworth, 1980). The bounding faults are brittle gouge zones in some cases, but more commonly they are ductile shears in which the cooling plutonic rock recrystallized as it moved upwards. The sense of motion on the faults is not known, but their continuity and parallelism suggests that right-lateral transcurrent motion may prove to be more important than the relatively small amount of vertical movement indicated. It is suggested that movement on these faults occurred in Late Cretaceous to early Tertiary time; small displacements in the late Tertiary are also possible (Woodsworth, 1979).

Figure 17.10. Steeply dipping fabrics within and typical of the Work Channel lineament. The concordant, leucocratic material has been interpreted as an *in situ* partial melting product of amphibolite (dark layers). Along Highway 16, about 50 km east of Prince Rupert. Photo by G.J. Woodsworth. (GSC 205376)

Figure 17.11. Banded gneiss consisting of light coloured quartzo-feldspathic and dark coloured amphibolitic bands along a probable southerly continuation of the Work Channel lineament in the northeastern part of Rivers Inlet area. Photo by M.D. Dubord. (GSC 205377)

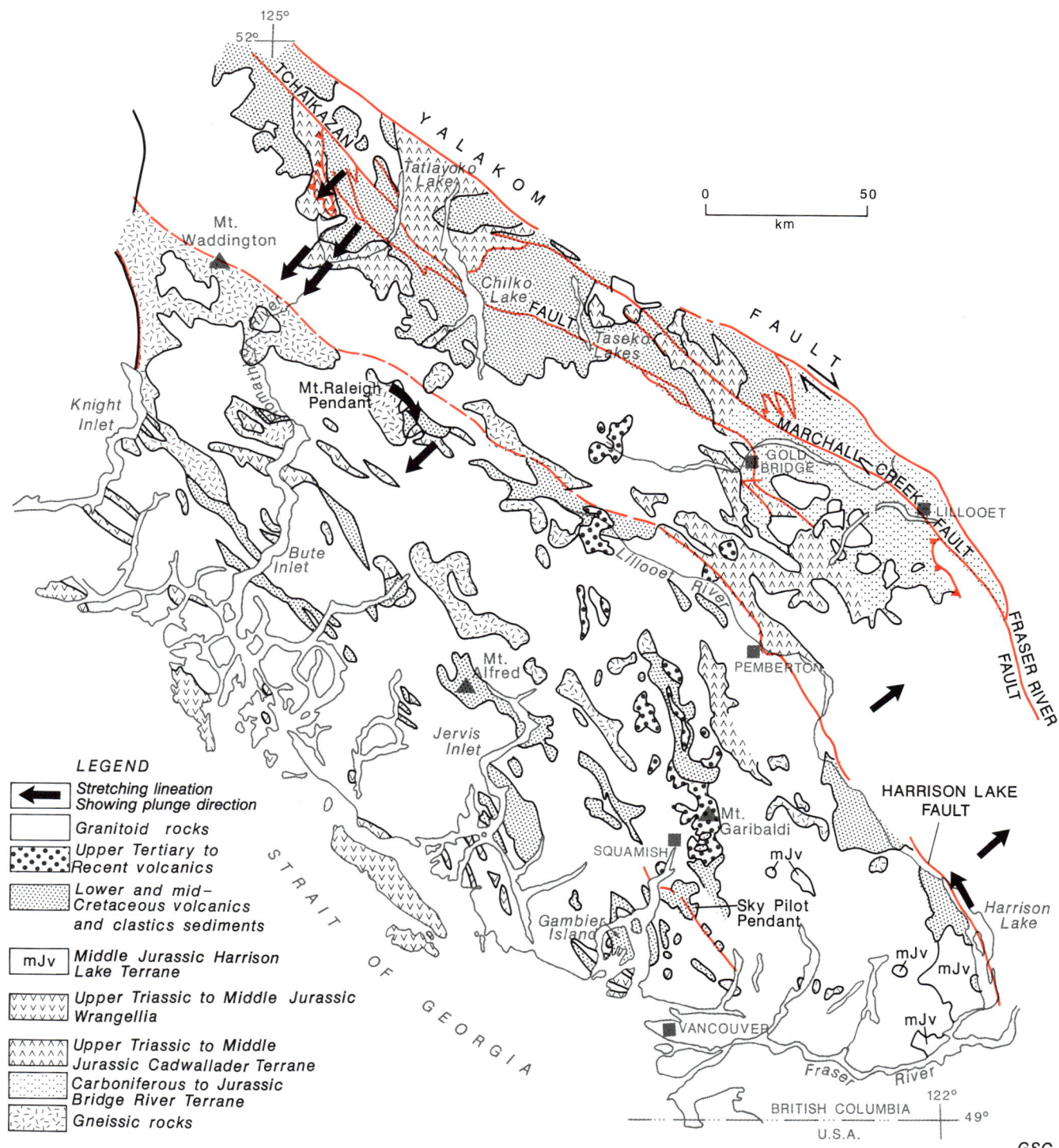

Figure 17.12. Index map for the southern Coast Belt.

Between Squamish and Pemberton, most plutons show moderate to strong ductile deformation. Stretching lineations in plutons and pendants are near vertical to steeply plunging. Foliation strikes north to northwest and dips steeply east. The ages of these rocks are poorly known, but the available data suggest deformation was Late Jurassic to Early Cretaceous in age. On Gambier Island, Late Jurassic plutons and Upper Jurassic or older strata of the Bowen Island Group are foliated and cataclastically deformed (McKillop, 1973), whereas the unconformably overlying strata of the Upper Jurassic to Lower Cretaceous Gambier Group are unfoliated.

These data suggest that Late Jurassic to Early Cretaceous penetrative deformation and uplift was succeeded by deposition of the Gambier Group and by post-Middle Albian (early Late Cretaceous(?)) deformation. The later deformation varies in style from open folds and discrete brittle to ductile shear zones (e.g. the Sky Pilot pendant) to penetrative fabrics with accompanying high grade metamorphism (e.g. Mount Raleigh). Even in the latter case, however, evidence for an earlier deformation is preserved, although its timing is uncertain (Woodsworth, 1979).

Northeast of the Harrison Fault

Dominant structures in this region are right-lateral transcurrent faults and related thrust faults (Fig. 17.12; Tipper, 1969). Ductile deformation is uncommon except near the Coast Plutonic Complex.

Between the Tchaikazan Fault and Coast Plutonic Complex west of Tatlayoko Lake, the structures are dominated by northeasterly verging thrust faults (Fig. 17.13, 17.14) and spectacular recumbent folds (Rusmore and Woodsworth, 1988). The faults generally strike northwest to north; dips are generally between 10 and 35° to the southwest or west. The faults are marked by zones of highly strained rocks in which clasts in sandstone, conglomerate, and lapilli-rich phyllite are strongly elongated. Lineations plunge gently southwest or (rarely) northeast and are remarkably consistent along the 35 km-length of the thrust belt. Rocks involved in the thrusting include Upper Triassic volcanics and sediments, Lower Cretaceous sediments, and a thick volcanic unit of uncertain but possible Early Cretaceous age.

If these volcanics are indeed Early Cretaceous in age, then the western thrusts place younger rocks over older. Rusmore and Woodsworth (1988) suggested that this relationship results from a profound angular unconformity between Upper Triassic and Lower Cretaceous strata, although out-of-sequence thrusting could also explain the structural order. It is unlikely that the faults are low angle normal faults, because they are indistinguishable from structurally lower thrusts.

Field evidence suggests that displacements were from southwest to northeast. Most fold axes lie at high angles to the elongation lineation, suggesting the transport direction is parallel to the lineation. Northeast-directed, rather than southwestward, thrusting is supported by the northeast vergence of several very large folds, and thrusts that steepen and cut upsection to the northeast.

Thrusting was Late Cretaceous in age. The youngest (and structurally lowest) rocks in the thrust system are the Upper Cretaceous Kingsvale volcanics. An upper limit for the thrusting is given by a 68 Ma U-Pb date on zircons from a pluton that cuts the thrust system (Rusmore and Woodsworth, 1988).

Figure 17.13. View northwesterly to the southwest side of Ottarasko Mountain about 20 km west of Tatlayoko Lake where Upper Triassic basalt has been thrust northeastward over Lower Cretaceous sediments. The fault, marked by a light coloured band of intensely deformed Upper Triassic limestone is typical of large northeasterly directed thrust faults along the east margin of the Coast Belt. Photo by G.J. Woodsworth. (GSC 205375)

Figure 17.14. Northeast slope of Castle Mountain in the Tyaughton Trough with an overturned succession of Norian shale, siltstone and conglomerate assigned to the Tyaughton Group (including the resistant rib and strata above) overlying Sinemurian strata exposed on the lower slopes. Photo by H.W. Tipper. (GSC 205374)

Cascade Segment of the Coast Belt
J.W.H. Monger

Structural evolution of the Cascade segment of the southeastern Coast Belt may record the final closure, during Cretaceous time, of oceanic or marginal basins, represented by the Bridge River and Shuksan assemblages, which were trapped between Chilliwack-Harrison terranes and Wrangellia to the west, and the Cretaceous margin of North America to the east. It is not clear whether this closure was by largely orthogonal or transcurrent movements.

There is fragmentary evidence of Paleozoic deformations in the Cascade segment, but the main structural characteristics appear to be related to four widespread Mesozoic to Tertiary deformations: (1) structural disruption of Middle Jurassic to Eocene age of the Bridge River and Shuksan terranes; (2) a dominant, mid- and Late Cretaceous deformation which established the metamorphic core, structures in flanking lower grade rocks, and the north-northwest-trending regional grain; (3) early Tertiary dextral wrench faults, and associated normal and reverse faults, centred on the Fraser River Fault System, and possibly on the Harrison Fault; (4) mid-Tertiary to Recent uplift accompanied in places by normal faulting.

Pre-Tertiary structures

By contrast with the southern Intermontane Belt where the predominant structures are of early Tertiary age, the dominant structures of the Cascade segment of the Coast Belt are of Cretaceous age. The segment near latitude 49°N has a central granitic and high grade metamorphic core zone (Custer and Skagit gneisses) flanked on both sides by low grade metamorphic and unmetamorphosed sedimentary and volcanic rocks that are included in the Methow and Bridge River terranes on the east, and the Shuksan, Chilliwack and Harrison terranes on the west (Fig. 17.15). The core narrows and appears to plunge to the north-northwest beneath rocks of lower metamorphic grade.

East side of Cascade core

Most structures in the Methow Terrane have a fairly simple style, with predominantly eastward-overturned flexural-slip folds a few kilometres in wavelength, and complementary west-dipping reverse faults (Fig. 17.16). In Washington, K-Ar dates of 90-60 Ma on metamorphic rocks in the core zone of the Cascade Mountains are considered to be uplift ages (McGroder and Mohrig, 1987). In the Methow Trough there, and in British Columbia, coarse clastic sediments were derived from the uplifted Cascade core zone during Middle Albian to Cenomanian time. These data indicate a mid- to Late Cretaceous time for deformation of strata in the Methow Trough. Near the Pasayten Fault in British Columbia Late Albian to Cenomanian clastic rocks locally have been thrust eastward onto Middle Eocene sediments (Greig, 1988). The relationship of the post-Middle Eocene thrusting to the Albian-Cenomanian deformation is unknown.

The Coast Belt is bounded on the east by the Pasayten Fault (Fig. 17.15). It separates Jura-Cretaceous, mainly marine, sedimentary rocks of the Methow Terrane to the west, from a mainly granitic and gneissic terrane with

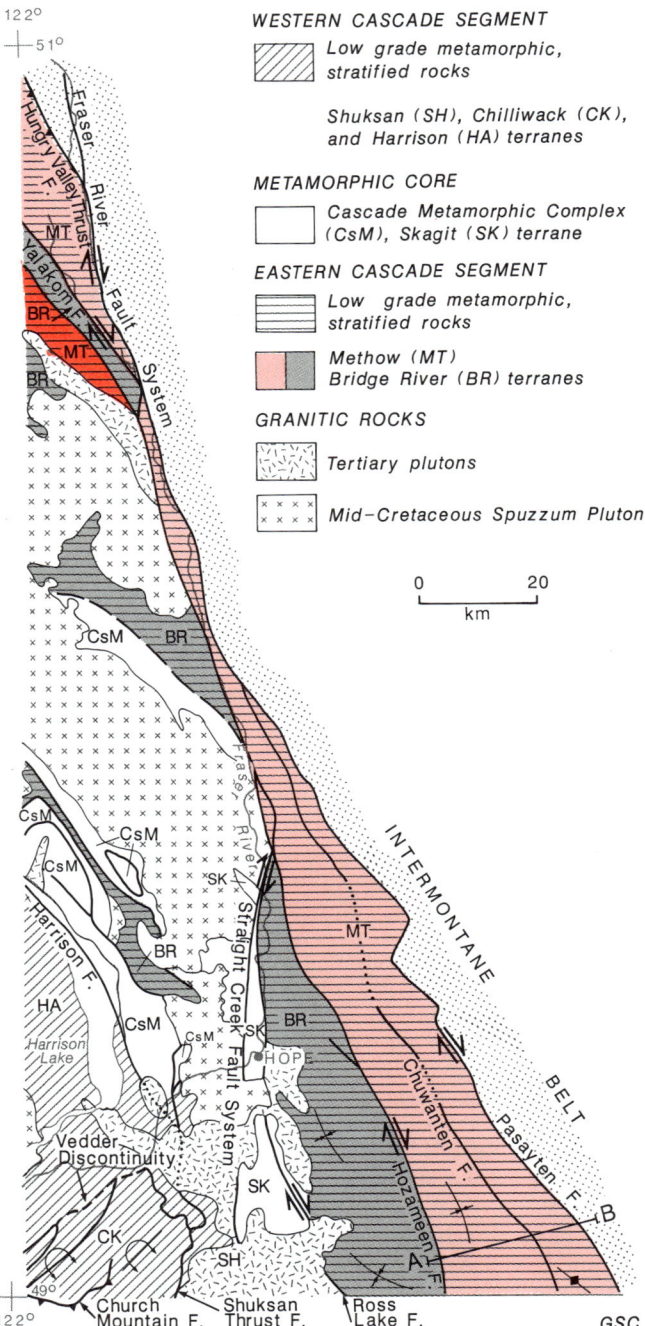

Figure 17.15. Main structures of southern Coast Belt, Cascade Segment.

Cretaceous continental volcanics to the east, possibly the Cretaceous continental margin. Superficially, the Pasayten Fault is a vertical, west-side-down normal fault, across which Albian and older strata of the Methow Terrane to the west are juxtaposed against the Mount Lytton Plutonic Complex to the east. Coates (1974) suggested possible right-lateral transcurrent movement of some tens of kilometres. The inferred extension of the Pasayten Fault to the northwest is the Hungry Valley Fault which is an Eocene thrust fault, probably related to movement in the

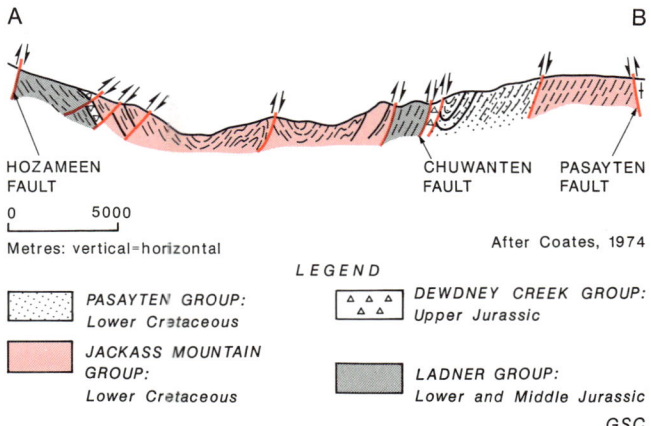

Figure 17.16. Cross-section across the Methow Terrane. Section located in Figure 17.15.

Fraser River Fault zone (Fig. 17.9; Mathews and Rouse, 1984; Monger, 1985).

The Chuwanten Fault in the Methow Terrane extends for about 120 km north of the International Boundary and appears to merge with faults of the Fraser River system in the north. It juxtaposes Jurassic rocks to the west against Cretaceous strata to the east, and is shown by Coates (1974) as a steep thrust, or reverse fault.

The Hozameen Fault, which separates the Hozameen Group in the Bridge River Terrane to the west from the Methow Terrane, is a vertical, east-side-down fault. To the south, in Washington, the Hozameen Fault links up with the east-directed Jack Mountain Thrust Fault of Misch (1966). To the north-northwest, its extension may be the Yalakom Fault, which bounds the Bridge River Terrane on the northeast. Tipper (1969) suggested dextral offset along the Yalakom Fault of 80-90 km and Kleinspehn (1985) proposed 125 km offset on the basis of displacement of the Lower Cretaceous Jackass Mountain Group.

Structures southwest of the Yalakom Fault in the Bridge River Terrane have a different sense of movement from those described by Coates (1974) in the Methow Terrane. Potter (1983) described a southwest-directed recumbent syncline, 4 km in amplitude, in the Bridge River Complex. In the same area, in Lower Cretaceous clastics imbricated with the Bridge River Complex (Monger and McMillan, 1984), Mustard (1983) recorded a smaller recumbent structure with a similar orientation cut by an Eocene pluton, dated by the U-Pb method at 47 Ma.

Extensive tracts of the Bridge River Complex (Potter, 1983), of subgreenschist grade, exhibit distinctive "broken formation" and mélange structural fabrics similar to those in the mélange belt of the Cache Creek Terrane. Ribbon chert with argillaceous interbeds is commonly broken into small fragments, some of which may be lensoidal, so that where these are abundant the rock has an overall scaly appearance. Massive basalt to pillow basalt is commonly cut by shear zones and individual pillow surfaces may have polished surfaces. Blocks of limestone, chert and basalt occur in a chert and argillite matrix and in many places are flattened. The structural fabric evidently has been superimposed on a sequence that included olistoliths.

The time of deformation is not known. The youngest rocks known in the predominantly Upper Triassic Bridge River Complex are of Early Jurassic (probably Toarcian) age, although R.A. Haugerud (pers. comm., 1986) reported Aalenian (early Middle Jurassic) fossils from the correlative Hozameen Group. Potter (1983) suggested that deformation and uplift of the Bridge River Complex preceded deposition of the Albian Taylor Creek Group, and favoured a Jurassic age for the main deformation.

The same disrupted and broken structural style is exhibited by thinly bedded, quartz-rich sandstone and argillite of unknown age, found in the upper part of the Chilliwack valley and tentatively correlated with the Darrington Phyllite, of probable Jurassic age.

Directly northwest of the Bridge River Terrane, rocks of the Cadwallader Group have undergone two phases of deformation (Rusmore, 1985, 1987). The earlier formed northeast-trending folds, with steep axial surfaces, parallel with numerous steep, brittle faults. The overall map pattern suggests that the major structure is an upright, gently plunging syncline. The younger structures consist of north-verging thrust faults and folds. Folds are open to isoclinal upright to slightly inclined to the southwest, and plunge moderately southeast. The ages of the structures are not well known, but the younger may be related to Late Cretaceous to early Tertiary transcurrent faulting.

Cascade core

The metamorphic core of the Cascade Belt is bounded on the east by the Ross Lake Fault (Fig. 17.15), and an unnamed zone about 0.5 km wide across which sillimanite and kyanite-bearing schists are juxtaposed with low greenschist facies rocks in the Kwoiek Creek area northwest of Boston Bar (Misch, 1966; Hollister, 1969; Monger and McMillan, 1984). Work by Haugerud (1985) suggested that, near latitude 49°N, the Ross Lake Fault was the locus of Eocene dextral displacement. The west side of the core, south of the bend of the Fraser River at Hope, either is cutoff and bounded by the Tertiary Fraser River-Straight Creek Fault system, or is intruded by the Oligocene Chilliwack Batholith. North of Fraser River, the metamorphic rocks are limited by the Harrison Fault (Fig. 17.15), which appears to be a ductile shear zone in rocks of low metamorphic grade crosscutting structures to the east.

In the metamorphic core north of the Fraser River deformation is polyphase, and includes rootless isoclinal early folds and open late folds, with kyanite and sillimanite forming late in the deformational history (Reamsbottom, 1974). Most faults are vertical or steeply dipping at the surface but possibly, as suggested by Lowes (1972), were originally eastward dipping thrust faults. The age of the deformation and metamorphism is given by the relationships of these rocks with the syn- to late-tectonic Spuzzum Pluton, which is probably of late Early-Cretaceous age (Irving et al., 1985). In places the Spuzzum Pluton crosscuts the bounding faults but elsewhere is concordant or locally cut by them. Zircon dating from the Washington Cascades suggests the main episode of metamorphism was mid-Cretaceous to earliest Tertiary, with hints of much older, early Proterozoic ages in some gneissic rocks (lower intercepts of 60 to 90 Ma on concordia with upper intercepts in the 1400-2000 Ma range; Mattinson, 1972).

West side of Cascade core

West-directed thrust faults on the west side of the Cascade core were recognized by Crickmay (1930) and Misch (1966). The latter identified three main structural elements separated by major east-dipping thrust faults (Fig. 17.15). From west to east these are: (1) Mesozoic strata separated by the Church Mountain Thrust Fault from (2) upper Paleozoic rocks, which are separated by the Shuksan Thrust Fault from (3) high-pressure metamorphic rocks of probable Jurassic protolith age. Limestone of the upper Paleozoic sequence outlines large northwestward directed recumbent folds, whose axes trend and plunge gently to the northeast. The folds are associated with northwest-verging thrust faults (Monger, 1970). Tight, isoclinal minor folds and cleavage parallel or subparallel with bedding are associated with this deformation. The folds and thrusts are redeformed by west-southwesterly directed reverse faults and associated chevron and local conjugate folds with northwest-trending axes, and strain-slip cleavage. Jewett and Brown (1985) suggested that the root zone of the Shuksan Thrust Fault is a plexus of high-angle dextral strike-slip faults. Speculatively, the northerly continuation of the fault system north of Fraser River is the Harrison Fault (Fig. 17.15), a low-grade but ductile shear zone, with strong subhorizontal, north-northwest-trending stretching lineations. To the west are non-foliated, openly folded Mesozoic strata.

West of Harrison Lake the Vedder discontinuity (Fig. 17.15) separates highly deformed strata to the south, from less deformed strata and abundant granitic intrusions to the north. Its significance is not known, but Davis et al. (1978) speculated that it was possibly the southern limit of Wrangellia.

Evidence of old metamorphisms and(?) deformations is found near latitude 49°N in amphibolite and mica schist of the Vedder Complex and in mafic/ultramafic tectonites correlated with the Yellow Aster Complex. The metamorphic age of amphibolite and mica schist of the Vedder Complex is late Paleozoic (Armstrong et al., 1983), and that of the Yellow Aster Complex (from south of latitude 49°N) is Silurian (Mattinson, 1972). Present evidence suggests that these old metamorphic episodes are restricted to scattered, small fault slivers.

Tertiary structures
Fraser River Fault System

The major structure active in the early Tertiary was the Fraser River Fault system, predominantly a dextral wrench system with some dip-slip displacement. Its northerly trend cuts acutely across the older, Cretaceous strata, offsetting rock units and older structures, so that estimates of displacement can be made with some confidence (Fig. 17.15). Rocks on the west side were apparently displaced 80 to 100 km to the north relative to rocks on the east. Structures probably related to the fault occur within both the Intermontane and Coast belts.

Associated with the fault system, and particularly well displayed near Lillooet, are northwest-trending thrust faults, small folds with northwest-trending axes, northeast-trending normal faults, and north-trending, possibly synthetic strike-slip faults. This family of structures fits the wrench fault strain ellipse depicted by Wilcox et al. (1973) and suggests a direction of maximum compression oriented about north 30° east.

On the regional scale, the Fraser River Fault system to the north apparently crosses the Intermontane Belt, linking up near latitude 57°N with faults associated with the Northern Rocky Mountain Trench-Tintina Trench fault system (Gabrielse, 1985). To the south, it slices into the central part of the Cascade segment of the Coast Belt, and passes into Washington, where it is known as the Straight Creek Fault.

West side of the Cascade Belt

Miller and Misch (1963) described an angular unconformity between rocks of the Paleocene and Eocene Chuckanut Formation and the later Eocene and younger Huntingdon Formation on the western front of the Cascades, a few kilometres south of Sumas on the International Boundary.

Eastern Margin, southern part of Coast Belt

On the eastern margin of the Coast Belt, near Lillooet, a narrow sliver (0-10 km wide, 80 km long) of upper greenschist-lower amphibolite facies siliceous schist, biotite and garnet-bearing pelitic schist, minor amphibolite and marble, partitioned by numerous concordant and crosscutting felsic intrusions (Fig. 17.17), lies within disrupted, subgreenschist grade rocks of the Bridge River Complex, from which it is separated by northwest-trending vertical faults (Fig. 17.15; Monger and McMillan, 1984). The composition of the schists suggests their derivation from Bridge River strata. Concordant felsic intrusions commonly have a mylonitic foliation conforming with and grading into that of the enclosing schists. The intrusions were presumably emplaced during deformation and metamorphism of the schists, which took place at about 4-5 kb ($4-5 \times 10^5$ kPa)

Figure 17.17. Concordant and crosscutting, foliated felsic sills and dykes cutting metamorphosed equivalents of the Bridge River Complex. The felsic intrusions are syn- to late deformational and give concordant U-Pb dates of about 40 Ma. View looking north, Bridge River canyon, about 20 km northwest of Lillooet. Photo by G.J. Woodsworth. (GSC 205369)

pressure (Potter, 1983). U-Pb dates of 41 Ma and 40 Ma have been obtained from both concordant and crosscutting intrusions. From granitic rocks, which are locally mylonitic, in this same structural sliver, Parrish (1983) obtained fission track ages of 27-37 Ma. These results indicate that the sliver of metamorphic rocks was uplifted extremely rapidly, at rates about 1 km per million years, in latest Eocene-earliest Oligocene time.

PART C. INTERMONTANE BELT

H. Gabrielse, J.W.H. Monger, D.J. Tempelman-Kluit, and G.J. Woodsworth

Summary

The Intermontane Belt comprises a variety of oceanic, island-arc, foredeep and clastic wedge assemblages each characterized by a distinctive structural style. The most important elements of structural style can be related to deformation in accretionary prisms and subduction zones in early Mesozoic time, exemplified by structures in the oceanic Cache Creek Terrane; block faulting during early Mesozoic time and later arc volcanism in many parts of Stikinia; folding related to uplift of the Coast and Omineca belts in late Mesozoic to early Cenozoic time, best observed in the Bowser and Sustut basins (Fig. 17.18); and, finally, dextral transcurrent strain regimes during Mesozoic to early Cenozoic time, well displayed in the southern part of the belt.

Structures generally verge eastward away from the Coast Belt and, in a narrow zone, westward away from the Omineca Belt. Except in the Cache Creek Terrane, rocks older than Middle Triassic are more highly cleaved and metamorphosed than younger strata. To what extent this reflects Permo-Triassic tectonism or different responses to later tectonism is not known. Increasing evidence in the northern Intermontane Belt suggests that detachment surfaces of regional extent may underlie Mesozoic rocks.

The main deformation of the Intermontane Belt was superimposed on pre-Triassic structures and formed during juxtaposition of Stikinia, Quesnellia and the Cache Creek Terrane during Middle and Late Jurassic time. The collisions produced the dominant southwest-verging folds and southwest-directed thrust faults along and north of the Stikine Arch (Fig. 17.18). Some dextral transcurrent movements may have accompanied the contractional deformation.

Throughout the Intermontane Belt deformation of Cache Creek Terrane rocks, resulting in the development of mélange and blueschist minerals, is interpreted to reflect Late Triassic eastward subduction of an accretionary wedge. In the Yukon and northernmost British Columbia the degree of deformation and disruption of stratigraphic units increases southeastward to east of Dease Lake where the distribution of lithological units is chaotic. Two phases of deformation are recognized there and in the Pinchi Lake area. In southern British Columbia mélange in the Cache Creek Group, faults in the Nicola Group and the Guichon Creek Batholith, and metamorphic fabric in the northern part of the Mount Lytton Plutonic Complex presumably developed prior to accretion of these rock units to the continental margin. The enigmatic northern part of the Mount Lytton Plutonic Complex may represent the roots of the early Mesozoic magmatic arc derived from deeper parts of the accretionary prism and down-dragged parts of the arc that were underplated beneath Quesnellia at that time.

Possibly coeval with deformation of the Cache Creek Terrane was northeast-directed thrusting of Quesnellia onto ancestral North America. Where competent volcanic rocks of Quesnellia are dominant, folds are generally open

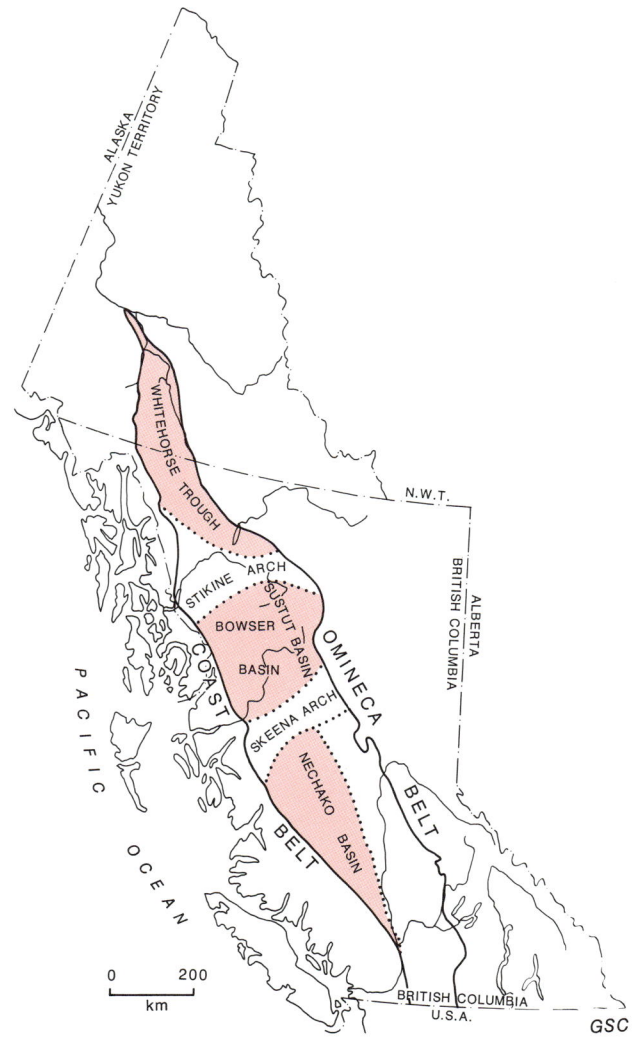

Figure 17.18. Main tectonic elements of the Intermontane Belt.

and cleavage is not well developed. Strong foliation is common only near thrust faults. On the east side of the southern Intermontane Belt, westerly directed structures in the Harper Ranch Group and the eastern sedimentary facies of the Nicola Group, probably of late Early to Middle Jurassic age, are congruent with structures in the Omineca Belt.

Late Permian to Early Triassic deformation in northern Stikinia produced tight northerly trending folds accompanied by a strong foliation. The structures were overprinted by southeasterly trending Middle to Late Jurassic folds and thrust faults.

Widespread, pre-Hauterivian erosion records a period of uplift in the Early Cretaceous. Eastward thrusts in the Bowser and Sustut basins, related to uplift of the Coast Belt, occurred during the Late Cretaceous and early Cenozoic. Westward thrusting related to uplift of the Omineca Belt was of about the same age. The great shortening of Mesozoic assemblages in the Bowser and Sustut basins from south of Skeena Arch (Fig. 17.18) to the Stikine Arch took place above décollement surfaces that probably rooted along the eastern margin of the Coast Belt. Block faults in the Skeena Arch may have been in part inherited from the Early Jurassic island arc environment, and to uplift during the Middle Jurassic. They were reactivated in Cretaceous through Cenozoic time. South of the Skeena Arch ductile shear zones in the easternmost Coast Belt are closely related to brittle, eastward directed thrust faults in the western Intermontane Belt.

A regional dextral transcurrent strain regime appears to have been important in the evolution of early Cenozoic structures in the southern part of the Intermontane Belt and in Mesozoic structures in the north (Thorstad and Gabrielse, 1986). These structures have been related to right-lateral transform motions and to regional extension.

Cache Creek Terrane and overlap assemblages

Northern Intermontane Belt

The northward dipping King Salmon Fault in northern British Columbia separates early Mesozoic Stuhini and Takwahoni assemblages of Stikinia with poorly developed cleavage to the south and southwest, from penetratively cleaved rocks of the upper Paleozoic and Mesozoic Cache Creek and Kutcho assemblages of the Cache Creek Terrane and overlying Inklin Assemblage to the north (Fig. 17.19, 17.20). The eastern part of the Cache Creek Terrane has a distinct structural asymmetry defined by south-southwest-directed thrust faults and north-northeastward dipping cleavage. The intensity of deformation increases progressively from west to east, and east of Dease Lake disruption of lithological units has produced widespread mélange. Folds in well-bedded Lower Jurassic sediments are commonly open and have wave lengths and amplitudes of several hundred metres to a kilometre. Near the King Salmon Fault, however, tight, locally isoclinal folds overturned to the south are apparent (Pearson and Panteleyev, 1975). Limestone in the hanging wall of the King Salmon Fault is intensely brecciated and clasts of feldspar porphyry in conglomerate are deformed such that flattening is in the plane of cleavage and elongation is parallel with the dip.

Although major, presumably older structures trend west-northwest, north-trending folds are present locally.

Near Atlin Lake the King Salmon Fault consists of a steeply dipping fault zone. Its northwest trend and steep dip suggests a component of dextral strike-slip displacement in accord with the north-northeast contraction documented along the more eastward trending extension of the fault farther southeast.

In the Yukon westward verging folds occur in Mesozoic overlap assemblages in the Cache Creek Terrane in the hanging wall of a westward-directed thrust fault which is a probable continuation of the King Salmon Fault. A significant change in structural style of the Intermontane Belt occurs just north of the British Columbia-Yukon boundary (Fig. 17.19, 17.21) where the Cache Creek Terrane is overlapped by Mesozoic assemblages of the Whitehorse Trough (Fig. 17.18). Typical structures are north-trending vertical faults, the most important of which may have had strike-slip displacements, and open and upright northwest-trending folds (Fig. 17.21). Near Whitehorse and west of Marsh Lake folds are slightly asymmetric with northeast dipping axial surfaces, or, are locally overturned to the southwest. Southwesterly trending folds southwest of Marsh Lake are parallel with a fault separating Cache Creek and Whitehorse Trough strata. Stratigraphic sequences containing abundant conglomerate or greywacke generally lack a well defined cleavage and are openly folded whereas sequences with considerable shale and turbidite have a slaty cleavage and are tightly folded. Steep faults, perhaps related to the strike-slip faults, cut the Mesozoic strata into numerous blocks most of which are elongate northwesterly.

Mesozoic rocks of the Inklin overlap assemblage in the Whitehorse Trough interdigitate along trend to the southeast with underlying strata of the Cache Creek Assemblage (Fig. 17.21). Contacts between the two assemblages

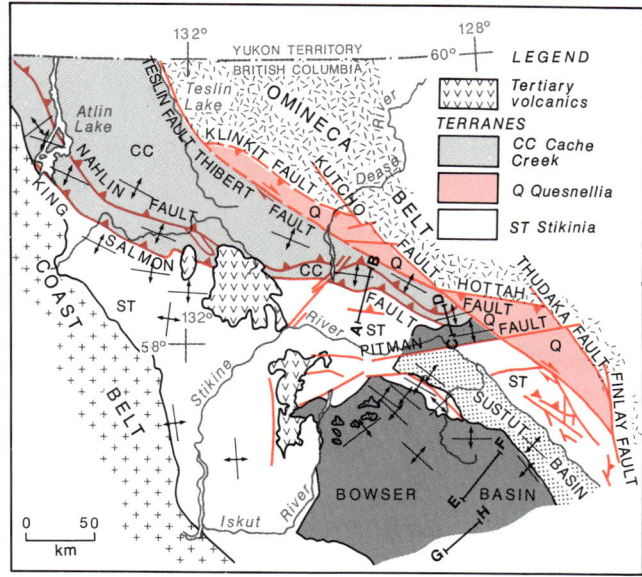

Figure 17.19. Structures in the Intermontane Belt, northwestern British Columbia. Locations of cross-sections of Figure 17.20 are indicated.

STRUCTURAL STYLES

are prominently discordant. Juxtaposition of synclines in Mesozoic strata above anticlines in the Cache Creek suggest the presence of an intervening regional detachment surface. This detachment may be continuous with mylonitic rocks at the north end of the Whitehorse Trough.

The Nahlin Fault (Fig. 17.19) juxtaposes mainly oceanic upper Paleozoic rocks in the Cache Creek Terrane to the north against dominantly Mesozoic island-arc related rocks to the south. It is generally steeply dipping where its trace trends northwest, and north-dipping where its trace trends east-west suggesting that it may have components of dextral transcurrent and contraction displacements related to a north-south principal stress. Strata north of the Nahlin Fault are commonly discontinuous and have a wide range in competency resulting in a heterogeneous structural style. Chert and argillite are tightly folded with

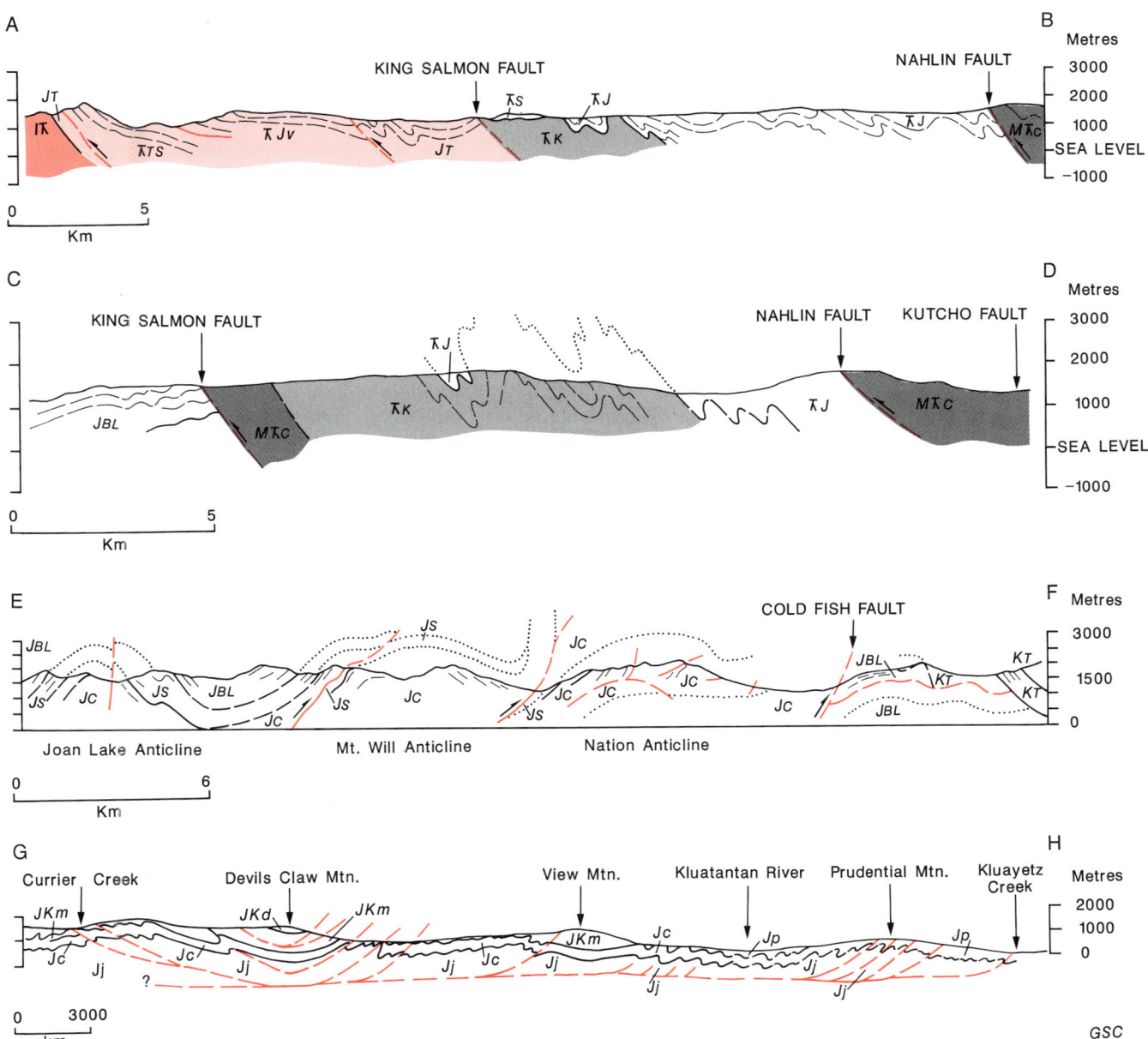

Figure 17.20. Structural cross-sections across parts of the Intermontane Belt in northern British Columbia (see Fig. 17.19 for locations). Sections A-B and C-D; Cache Creek Group-MTC, Kutcho Formation-TK, Sinwa Formation-Ts, Sinwa and Inklin formations-TJ, Tahwahoni Formation-JT, Triassic and/or Jurassic volcanics-TJv, Stuhini Group-Tst, Cake Hill Pluton-lT, Bowser Lake Group-JBL (after Thorstad and Gabrielse, 1986). Section E-F; Lower Jurassic volcanic rocks-Jc, Spatsizi Group-Js, Bowser Lake Group-JBL, Tango Creek Formation-KT, Brothers Peak Formation-KB (after Evenchick, 1986). Section G-H; Informal units of the Bowser Lake Group, Jackson-Jj, Currier-Jc, McEvoy-JKm, Devils Claw-JKd, Prudential-Jp (after Bustin and Moffat, 1983).

CHAPTER 17

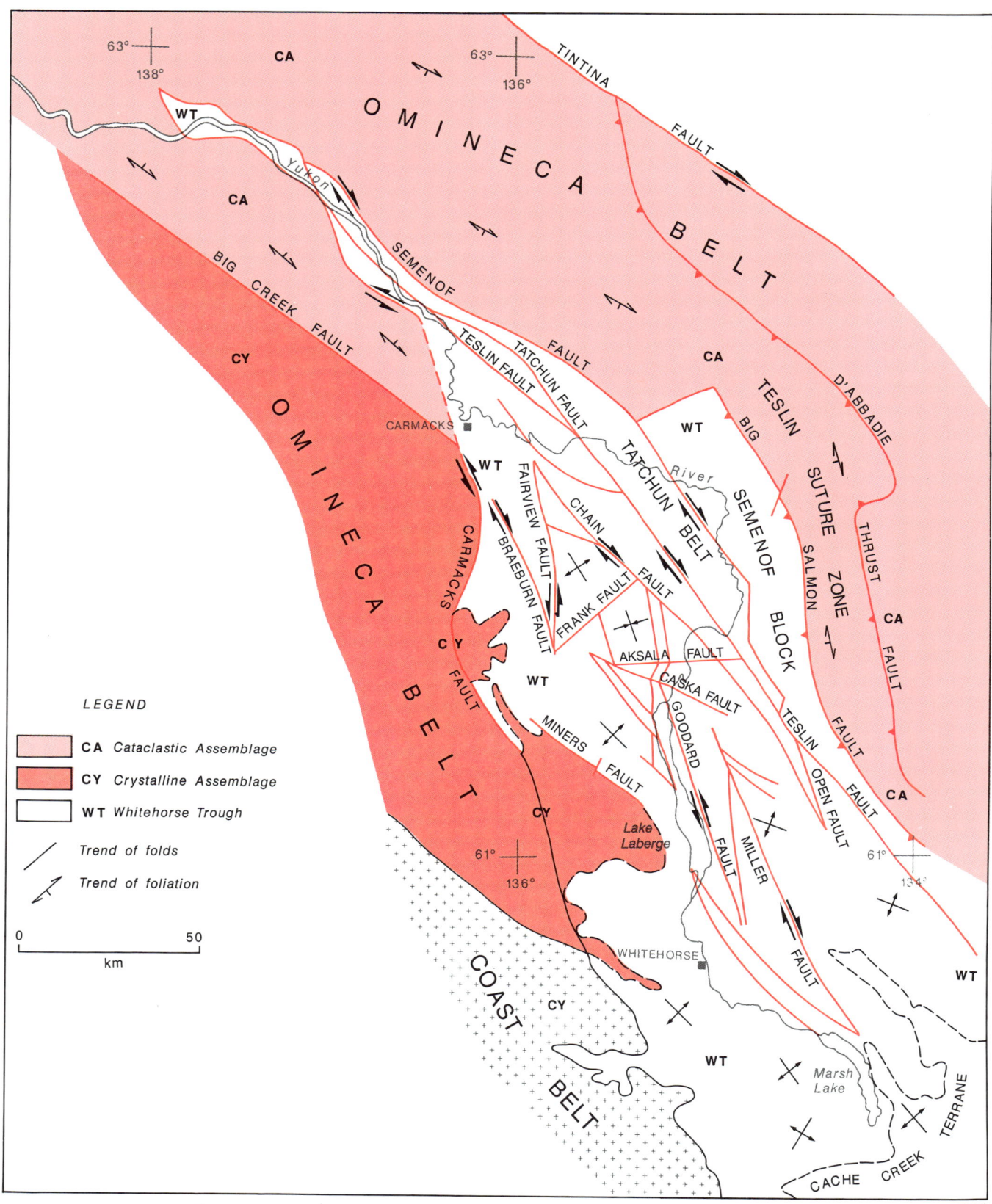

Figure 17.21. Structures in the northern Intermontane Belt, Yukon Territory.

amplitudes of a few metres to tens of metres. A few volcanic and limestone formations in the northwestern part of the region, however, outline southwestward verging folds traceable for several kilometres (Monger, 1968). In the northeastern part folds are mainly upright or with axial surfaces dipping steeply to the southwest; large recumbent folds occur locally. In the southeastern part of the area many contacts between rocks of different lithology are faults. Early, tight to isoclinal, roughly north-south trending folds are related to a penetrative foliation nearly parallel with bedding. These structures are refolded by later west-northwest- trending regional folds. The presence of crossite crystals in the plane of the early foliation indicate that the deformation took place at relatively high pressure. The later pervasive foliation and related folds are essentially parallel with the King Salmon Fault and are probably of Middle to Late Jurassic age. Uplift of the Cache Creek Terrane, presumably contemporaneous with southward directed structures, provided the principal source for Middle to Upper Jurassic sediments in the Bowser Basin.

Some deformation within the Cache Creek rocks may have occurred before southwestward thrusting along the Nahlin and King Salmon faults in Middle Jurassic time. Northwesterly trending folds and faults near Atlin Lake probably include those formed during uplift in the Coast Belt during deposition of Lower Jurassic rocks and during subsequent uplift as late as Cenozoic time.

Central Intermontane Belt

The Vital Fault in the Pinchi Lake area (Fig. 17.22) is a westerly directed thrust fault which is perhaps an offset continuation of the Nahlin Fault in north-central British Columbia (Gabrielse, 1985). The Cache Creek Group, between the Vital and Pinchi faults and farther south, is characterized by Late Triassic(?) easterly trending recumbent anticlines with axial planar foliation later deformed by flexural slip folds with westerly dipping axial surfaces (Paterson, 1977). The Pinchi Fault forms the eastern boundary of the Cache Creek Terrane and may have been the locus for early contractional faulting and folding and later dextral strike-slip faulting. The fault zone contains elongate fault-bounded blocks of contrasting lithology and metamorphic grade including ultramafic rocks and lawsonite- glaucophane- and aragonite-bearing metasedimentary and metavolcanic rocks. High pressure-low temperature metamorphism took place in Late Triassic time presumably in a subduction zone. Paterson (1977) suggested that the Pinchi Fault originated as a transform fault associated with a local northeast-trending subduction zone.

Southern Intermontane Belt

In the Cache Creek area (Fig. 17.23A) rocks in the Cache Creek Terrane and Quesnellia (Nicola Group) are structurally imbricated and thrust eastwards over the Lower and Middle Jurassic Ashcroft Formation (Fig. 17.23B). Within the Ashcroft, the shaly "offshore" facies is intensely deformed into asymmetric, eastward overturned folds that locally have well-developed slaty cleavage. Deformation probably took place before the sediment was well lithified in Late Jurassic time (Travers, 1978, 1982). The eastern part of the Cache Creek Group consists of a mélange comprising polymict clasts enclosed by a highly contorted locally phyllitic matrix which is cut by numerous shear zones. The matrix wraps around the clasts, many of which have shiny, slickensided surfaces. Some clasts are flattened and phacoidal in form. The structural fabric of the mélange appears to be superimposed on a pre-existing sedimentary, olistostromal fabric rather than being produced entirely by tectonism. The age of deformation is not known; it was possibly related to Permian and Triassic subduction. The fabric closely resembles that of many parts of the Franciscan Complex of California, and likewise may have formed within an accretionary prism.

Quesnellia
Northern Intermontane Belt

In the Yukon the Tatchun Belt, east of the Teslin Fault, is assigned to Quesnellia (Fig. 17.21; Tempelman-Kluit, in press-b). The Tatchun Belt is dominated by massive volcanic sequences; little is known of its detailed structure. It is bordered by two important dextral transcurrent faults, the Teslin on the southwest and the Semenof on the northeast. Northerly trending segments in the southeastern part of the Semenof Fault are associated with local basins containing the Lower Cretaceous(?) Tantalus Formation. The Tantalus basins in this area are interpreted as pull-apart structures related to transtension along bends in the transcurrent Semenof Fault.

In north-central British Columbia Quesnellia is in contact with the Cache Creek Terrane and Stikinia along the Thibert, Kutcho and Finlay faults (Fig. 17.19) which separate significant dextral transcurrent displacements (Gabrielse and Dodds, 1982). The Thibert Fault is a conspicuous fault zone marked by sheared and carbonatized ultramafic rocks and wide zones of gouge. It is linked to the Kutcho Fault which also contains sheared and carbonatized ultramafic rocks and zones of mylonite where it involves granitic rocks. Combined dextral strike-slip displacements on the faults appears to have been about 200 km (Gabrielse, 1985). The Kutcho Fault cuts mid-Cretaceous granite whereas the Thibert Fault, continuous with the Teslin Fault to the northwest, is cut by a Late Cretaceous granitic pluton. Along the Finlay Fault rocks are cut by numerous north-trending shear zones which locally juxtapose slices of volcanic and plutonic rocks. Pre-mid-Cretaceous(?), mid-Cretaceous and early Cenozoic movements have occurred along the southern part of the fault. The eastern margins of rock assemblages in Quesnellia are strongly sheared and appear everywhere to have been thrust onto ancestral North America.

Central Intermontane Belt

Between the Pinchi Fault and the eastern boundary of the Intermontane Belt the distribution of Upper Proterozoic(?) metamorphic rocks and nearby rocks of late Paleozoic and Triassic ages suggests the presence of northwest- and northeast-trending block faults or possibly the presence of extensional faults which facilitated tectonic denudation.

The eastern margin of the central Intermontane Belt is believed to be marked by an eastward directed thrust fault which placed Mesozoic rocks of Quesnellia over upper

CHAPTER 17

Paleozoic strata of the Slide Mountain Terrane (see Cariboo Mountains in Part D of this chapter).

Southern Intermontane Belt

In southern Quesnellia Preto (1977) noted the coincidence of intrusions and lithological variations in Upper Triassic Nicola Group volcanic rocks with the north-south-trending Summers Creek-Quilchena and Allison fault systems (Fig. 17.24), and suggested that the distribution of Nicola volcanism was controlled by crustal fractures. If so, the fault systems were reactivated in the Tertiary because they offset Cretaceous and early Tertiary strata and it is difficult to separate Tertiary deformation from older structural events. Hollister et al. (1975) and McMillan (1976) concluded that the Lornex and Highland Valley faults (Fig. 17.24) and parallel fault sets were important in localizing early Mesozoic mineralization, and were active in the late stages of cooling of the Guichon Creek batholith in Early Jurassic time.

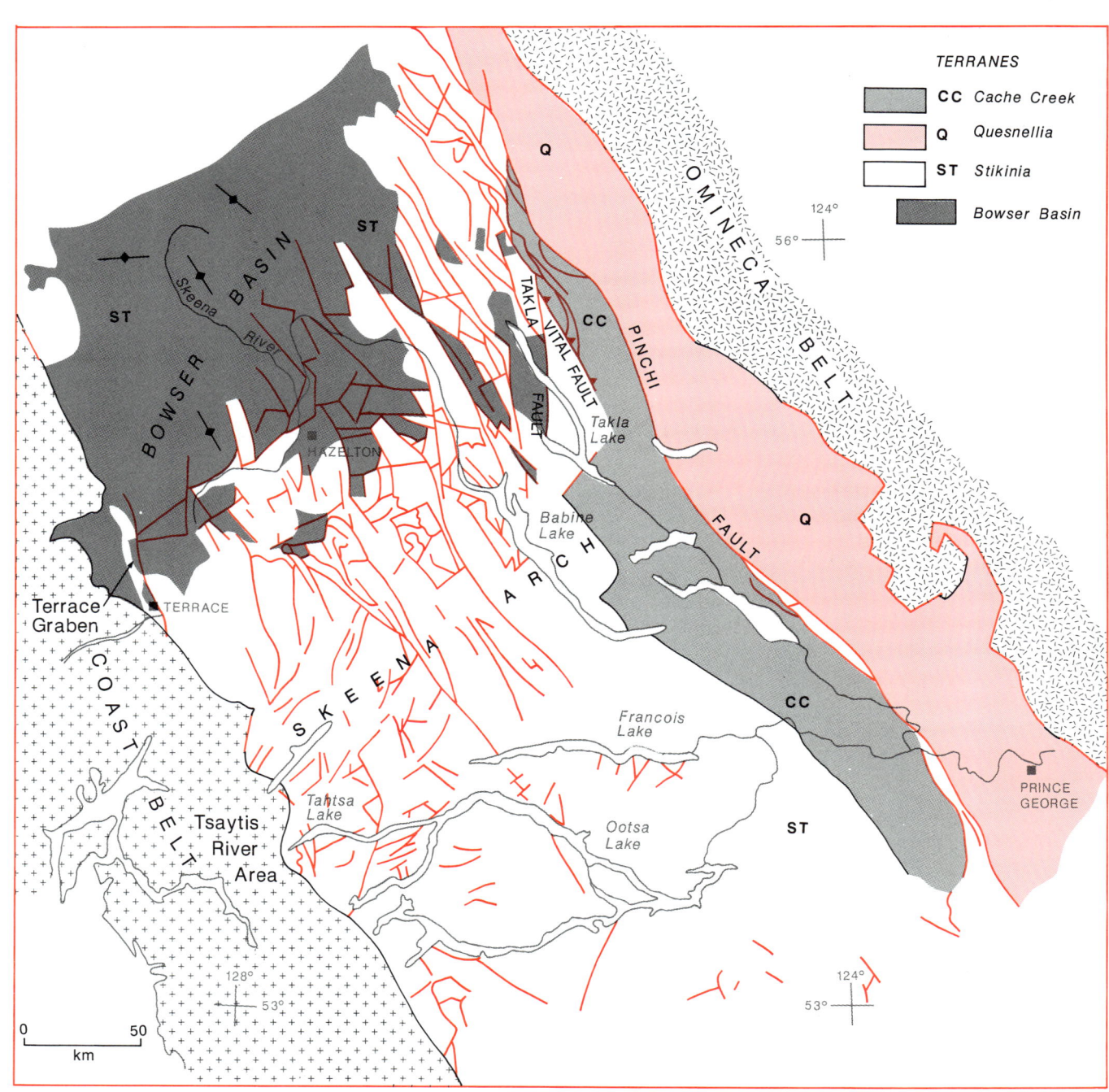

Figure 17.22. Faults along and near the Skeena Arch and structural trends in the southern Bowser Basin.

Mylonitic quartzofeldspathic schist and gneiss with locally tightly folded amphibolite layers, foliated diorite and granodiorite, and crosscutting intrusions ranging from gabbro to granodiorite constitute the northern part of the Mount Lytton Complex (Fig. 17.23A; Brown, 1981; Monger and McMillan, 1984). Orientations of foliation are variable, although northwesterly trends and dips to the northeast are most common. The age of deformation and metamorphism of the northern part of the Mount Lytton Complex appears to be early Mesozoic on the basis of U-Pb ages on zircons from metamorphic rocks of 218-230 Ma and of a K-Ar age of 186 Ma on a crosscutting pluton (P. van der Heyden and R.R. Parrish, pers. comm., 1985).

Easterly-directed structures in the Cache Creek Group, Nicola Group and Ashcroft Formation, and in the southern part of the Mount Lytton Plutonic Complex appear to be of Late Jurassic-earliest Cretaceous age (Fig. 17.23A,B). The southern end of the Mount Lytton Plutonic Complex, known as the Eagle Granodiorite (Rice, 1947; Anderson, 1974), comprises granodiorite, granodiorite and hornblende gneiss, or migmatite, and local pegmatitic quartz monzonite. On its east side, it grades into, interfingers with, and also crosscuts biotite hornblende schist, biotite schist and minor marble, whose uniformly north-northwest-striking, steeply west-dipping foliation is generally concordant with layering in the gneissic rocks. Eastwards the schist decreases in metamorphic grade, the foliation becomes less regularly orientated, and 5 to 10 km east of the contact with the Eagle Granodiorite, foliated rocks grade into poorly or non-foliated subgreenschist-grade rocks of the Nicola Group. The formation of the schist and granodioritic gneiss appear to be related. A U-Pb date of 137 Ma, obtained by P. van der Heyden (pers. comm., 1986) from migmatite of the Eagle Granodiorite, suggests that the fabric formed in earliest Cretaceous time.

The relationship between the southern end of the Mount Lytton Plutonic Complex, with its uniformly oriented fabric, and the clearly older northern end is not known because the boundary between the two is obscured by extensive granitic intrusions. Structures developed in the southern end do not appear to affect the upper Lower Cretaceous Spences Bridge and Kingsvale groups, which lie acutely across the trend of the southern part of the belt (Fig. 17.23A). However, they lie on trend with the eastward-directed structures that affect Cache Creek, Nicola and Ashcroft strata, noted above, and show a similar sense of displacement. Contiguous to the west, across the Pasayten Fault, are Jurassic and Cretaceous sedimentary rocks in the Methow Terrane which do not appear to have been deformed during this interval.

The Spences Bridge and Kingsvale groups are overlap assemblages on Quesnellia, and are composed of continental volcanics which lie in the north-northwest-trending, open Nicoamen Syncline (Fig. 17.23A). Stratigraphic relationships within the syncline suggest that it was forming during deposition of these units in latest Early Cretaceous time, coincident with uplift of the Mount Lytton Complex. The Mount Lytton Complex apparently was being elevated during extrusion of the volcanics.

On the east side of the Intermontane Belt, subgreenschist-grade rocks of the upper Paleozoic Harper Ranch Group and of the eastern, sedimentary, facies of the Nicola Group are deformed into north-northwest-trending, southwest-overturned folds and apparently congruent layer-parallel faults (Fig. 17.23A,C; Smith, 1974). To the northeast the rocks have a strong, penetrative foliation and a high greenschist metamorphic grade near the Louis Creek Fault. The same grade is observed in rocks of similar lithology across the fault; the Louis Creek Fault marks the boundary between Intermontane and Omineca belts.

The age of deformation is not known but it appears to be congruent in style, and in metamorphic continuity, with westerly-directed synmetamorphic structures in greenschist-grade metamorphic rocks on the west side of the Omineca Belt in the Shuswap region (Brown and Read, 1983). To the northeast in the Adams Plateau they are thought to be later, syn- or post-metamorphic folds and faults (Schiarizza and Preto, 1984). These structures are Middle Jurassic (171-164 Ma) in age (Brown and Read, 1983), and presumably were formed following accretion of the Intermontane Superterrane to the ancient western margin of North America (Fig. 17.25; Monger et al., 1982).

In the area near Merritt rocks of the Nicola Group are folded into open folds. The relationship of these folds to other structures is not known, but they are of pre-Tertiary age.

Stikinia
Northern Intermontane Belt

A southward widening strip of Paleozoic and Mesozoic rocks in Stikinia lies between the Cache Creek Terrane and the Coast Belt (Fig. 17.19, 17.21). In the Whitehorse-Atlin region Mesozoic strata are in contact with metamorphic rocks of the Coast Belt along steeply dipping faults marked by zones of epidotization and shear (Bultman, 1979). Rocks of the Stuhini Group, east of one of the faults which dips 65°N to the northeast, have been deformed into drag folds indicating that the southwest side moved relatively upward and to the northwest. Locally, folds east of the fault are overturned to the northeast. Some displacement may have been earlier than movement on the Middle Jurassic King Salmon Fault but relative uplift of the Coast Belt continued into the Cenozoic.

Between the Nahlin Fault and the Coast Belt in the Atlin area (Fig. 17.19) open folds with wave-lengths of 10 km or less commonly trend more westerly than the bounding faults against which they terminate in an en échelon manner. Typically the folds have a chevron profile, range from open to closed, and mainly verge to the southwest. A well developed axial-plane cleavage is characteristic in thin-bedded sediments. In a few places tight, northeast-trending synclines form a classic basin and dome interference pattern with the northwest-trending folds. In addition, several northeast-trending dextral strike-slip faults have been recognized.

Farther south pre-Upper Triassic rocks of the Asitka Assemblage near the east margin of the Coast Belt are tightly folded along northerly trends (Souther, 1971). Locally, limestone members outline complex isoclinal and fan folds with thinned limbs and thickened axial regions (Fig. 17.26). In places, extreme attenuation of the limbs has separated them from the crestal cores resulting in diapiric folds. Steeply dipping axial-plane foliation is well developed in the most westerly exposures where the grade of regional metamorphism is highest.

CHAPTER 17

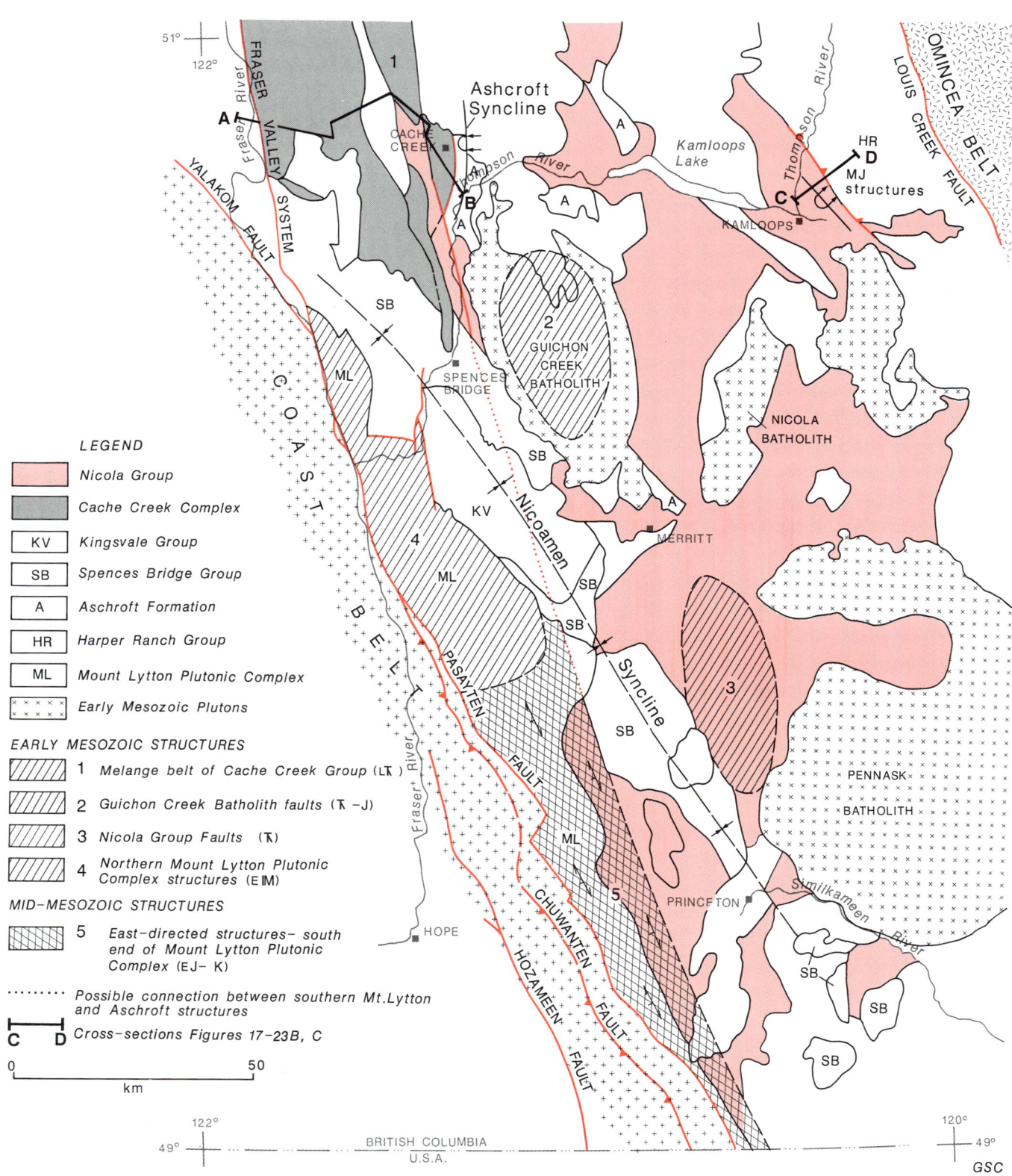

Figure 17.23A. Mesozoic structures of the southern Intermontane Belt and locations of cross-sections, **B.** Cross-section through the Cache Creek Group (see Figure 17.23A for location); after Travers (1982), **C.** Cross-section showing structural style northeast of Kamloops. Location shown in Figure 17.23A.

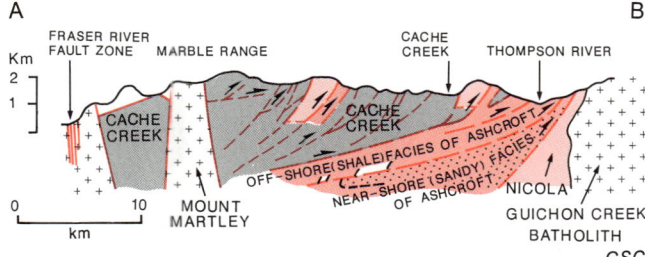

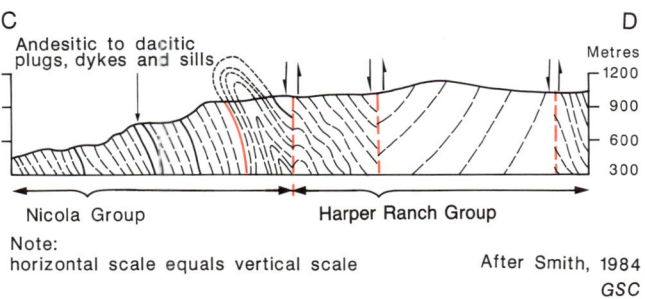

After Smith, 1984
GSC

Pre-Lower Triassic rocks in the area near the Grand Canyon of the Stikine River are tightly folded and strongly cleaved. There, and particularly in a small area in the northeastern part of Stikinia, two fold trends are evident, an earlier trend directed toward the north-northeast and a later one trending west-northwest. The first phase of deformation resulted in transposition of bedding to parallelism with penetrative foliation (Thorstad, 1980); the second phase produced isoclinal to tight upright folds. As in the Whitehorse Trough area in the Yukon a marked discordance in structural style occurs at the base of Mesozoic rocks in the Stikine River region.

In a belt up to 20 km wide, parallel with and south of the King Salmon Fault, south- and southwest-directed thrust faults involve Triassic and Lower Jurassic volcanic and sedimentary rocks (Fig. 17.19, 17.20). The zone of imbrication formed with collision of Stikinia with the Cache Creek Terrane by Middle Jurassic time, followed by further convergence and uplift in Late Jurassic time. Open, concentric folds are present in Lower Jurassic sedimentary rocks but no well defined vergence is apparent except near the King Salmon Fault. Northeasterly trending, Late Cretaceous or Cenozoic faults, are abundant. Some have had sinistral transcurrent displacements of up to 3 km. North-trending Cenozoic extensional faults extending into the Cache Creek Terrane controlled the emplacement of Miocene to Recent alkalic volcanic rocks.

A thick sequence of well bedded shale, siltstone, sandstone and conglomerate of mainly Jurassic age in northern Bowser Basin overlaps Stikinia and was subjected to major southwest-northeast contraction during pre-mid-Albian(?) and post-Maastrichtian times. Tight, laterally continuous, northeast-verging folds with amplitudes and wave lengths of 100 to 300 m are characteristic (Fig. 17.27) but, locally, southwest-verging folds are present on the southwest limbs of major synclines (Fig. 17.19, 17.20; Bustin and Moffat, 1983). Disharmonic chevron folds are widespread and fore-limb and back-limb thrust faults are common. Open to tight northwesterly overturned, east-northeast-trending folds are present in the northwest end of the basin. Northeast-trending block faults were active during pre-mid-Albian and post-Maastrichtian times (C.A. Evenchick, pers. comm., 1985). The structural style of the Bowser Basin suggests that the entire Mesozoic sedimentary sequence is underlain by a detachment surface or surfaces (Fig. 17.1, in pocket). Along the northeastern margin of the basin Lower Jurassic volcanic rocks are involved in thrusts and folds thus indicating that a detachment must lie below them. A root zone for regional detachments is difficult to define because of Tertiary plutonic rocks along the east side of the Coast Belt. Nonetheless, a relationship with uplift of the Coast Belt is inferred.

Strata in the Sustut Basin (Fig. 17.18, 17.19) occur in broad, open folds of low-amplitude where the sediments lie on a pediment surface underlain by Triassic and Jurassic volcanic rocks (Eisbacher, 1974). Farther west, however, where the lower Sustut rocks lie on predominantly pelitic rocks of the Bowser Lake Group, the structure is similar to that involving Bowser sediments, and is characterized by tight, commonly thrust-faulted, east-verging folds. An east- to east-northeast-trending set of feeder dykes to Pliocene lavas in the northwestern Bowser Basin indicate an episode of extensional faulting.

Northeastward verging folds and related gently dipping thrust faults characterize deformation in Lower Jurassic volcanic rocks between the Bowser and Sustut basins (Fig. 17.19, 17.20, Evenchick, 1986). Northeasterly trending folds and northwestward directed thrust faults occur locally. Block faulting, in part of Early Jurassic age, accompanied volcanism. The main deformation was of post-mid-Cretaceous age.

East of the Sustut Basin Lower Jurassic rocks are block faulted along mainly north and northwest trends. Westward directed thrust faults are common just west of Kutcho and Finlay faults. Some cut strata of the Sustut Group and could be of Late Cretaceous age.

The structural style in the Stikine River area is probably representative of the Intermontane Belt as far south as the Skeena Arch (Fig. 17.18). There, the dominant fold style of the Bowser Basin is subordinate to the block fault style of the arch (Fig. 17.22).

Central Intermontane Belt

The oldest principal structural feature in the central Intermontane Belt is the Skeena Arch, a paleogeographic arch trending approximately northeast from Terrace (Fig. 17.18, 17.22). The development of the arch in Late Jurassic time involved only uplift (possibly by block faulting) and was not accompanied by significant folding or plutonism (Tipper and Richards, 1976). Nonetheless, the Skeena Arch seems to have had an influence, albeit subtle and poorly understood, on later deformation in the western Intermontane Belt. The best record of widespread tectonism in the area is recorded in the deposition of mid-Oxfordian,

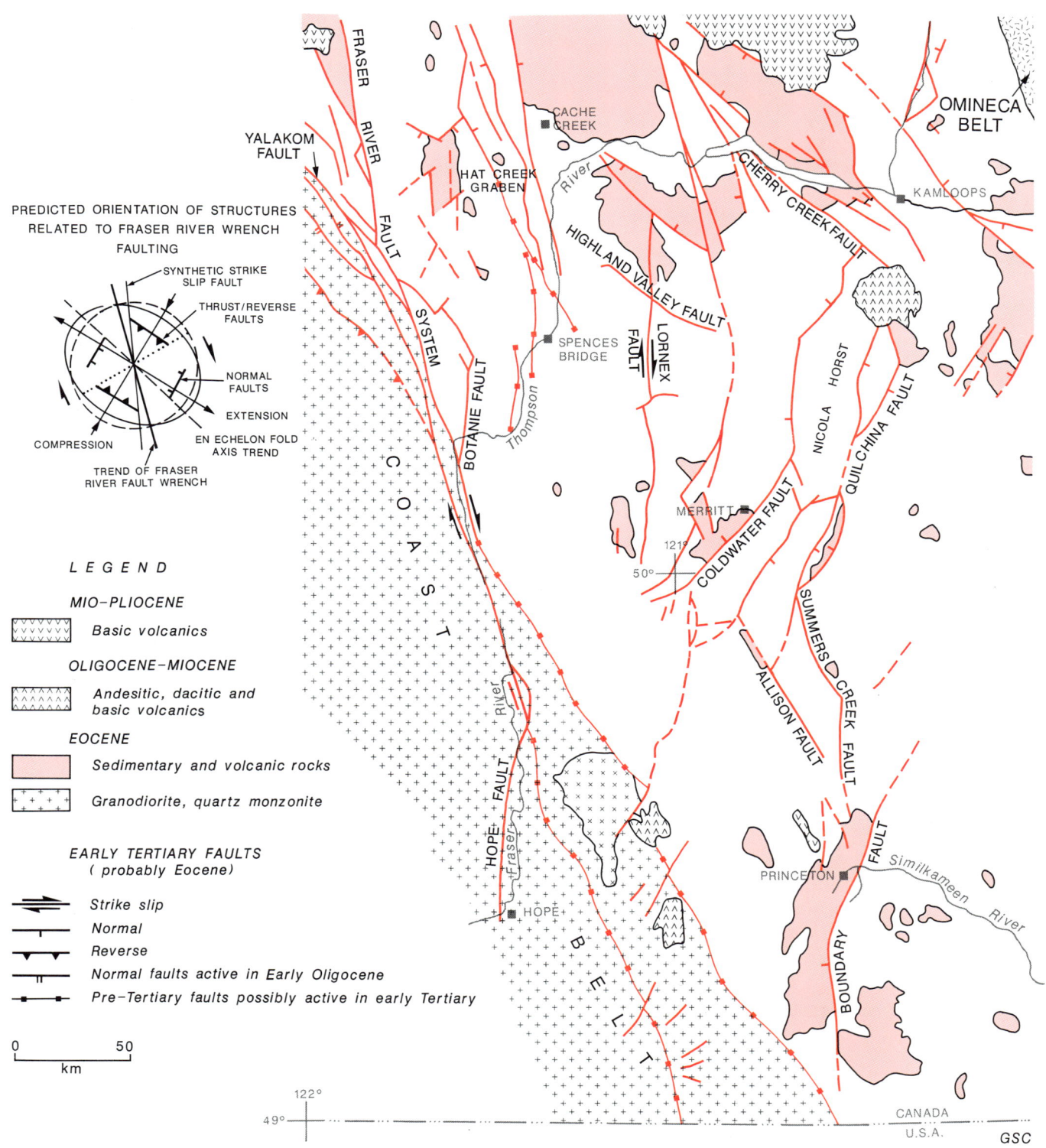

Figure 17.24. Tertiary structures of the southern Intermontane Belt.

STRUCTURAL STYLES

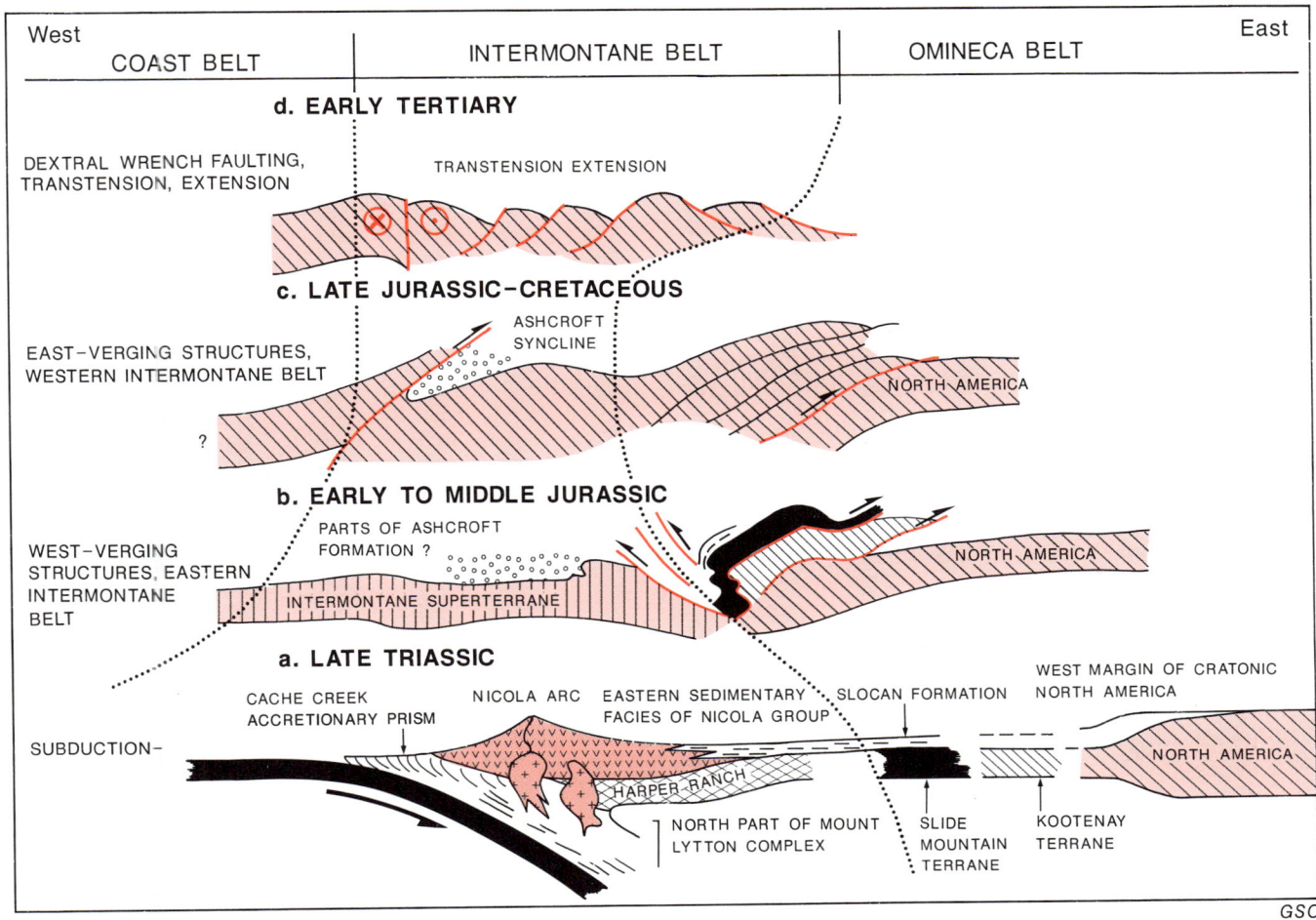

Figure 17.25. Conceptual model for the evolution of structural style in the southern Intermontane Belt.

basal conglomerate of the Bowser Lake Group. It was shed northwesterly off the arch into the Bowser Basin.

Most of the deformation in the western Intermontane Belt is Late Cretaceous and Tertiary in age. Structural style varies among the various stratigraphic units; in particular, folding is most intense in the Upper Jurassic and Lower Cretaceous sediments of the Bowser Lake and Skeena groups. Folds are variable: most are concentric to chevron in style, vary from open to closed, and most verge northeastwards. The massive, underlying volcanic rocks of the Hazelton Group form broad, open warps. In many places, the contact between the Hazelton and Bowser Lake groups is a décollement surface.

Folds in the western Intermontane Belt generally are related to thrust faults although thrusts are difficult to recognize because of the lack of good marker horizons, and because many appear to be bedding plane thrusts. Imbricate thrust faults are visible near Terrace in Lower Permian to Lower Jurassic rocks; regionally, most are directed north-northeasterly (Fig. 17.1, in pocket) and affect rocks as young as Late Cretaceous.

Ductile shear is locally important in the western Intermontane Belt, especially along the axis of Skeena Arch. Northeast of Terrace, along the Skeena River, rocks tentatively correlated with the Lower Jurassic Telkwa Formation have gently to moderately dipping foliations and stretching lineations that generally plunge gently east-northeast or west-southwest. These lineations can be traced southwestward into the high grade rocks of the Coast Belt and show that ductile deformation in the western Intermontane Belt is related to that in the Coast Belt. Deformed chert-pebble conglomerate of the Lower Cretaceous Skeena Group indicates that thrust faults and ductile deformation structures are Late Cretaceous or younger in age, a conclusion consistent with the age of deformation estimated from radiometric ages from high grade rocks of the Coast Belt.

The structural transition between the Coast and Intermontane belts can be observed in the Tsaytis River area where the boundary is a highly disrupted thrust complex (Fig. 17.8; van der Heyden, 1982). Rocks of the eastern Coast Belt were metamorphosed and mylonitized in mid- to Late Cretaceous time and later were thrust northeastward over low grade rocks of the Intermontane Belt. In both the Terrace and Tsaytis River areas, differences in structural style and grade of metamorphism between the eastern Coast Belt and western Intermontane Belt reflect different structural levels of the same tectonic

601

Figure 17.26. Isoclinal fold in Lower Permian limestone in the Iskut River area, northwestern British Columbia. The cliff face is about 365 m high. Photo by R.G. Anderson. (GSC 205373)

Figure 17.27. View northwest in northern Bowser Basin showing inclined syncline in Upper Jurassic sandstone, siltstone and conglomerate of the Bowser Lake Group. Photo by C.A. Evenchick. (GSC 205372)

events. Minor structures in both belts indicate that Late Cretaceous compression was accompanied by a component of right-lateral motion, not only between the two belts but within them (Hill, 1984).

Following the compressive events in the central Intermontane Belt, the region was disrupted by steep, brittle, high-angle faults which produced the conspicuous mosaic of polygonal mountain blocks and broad, linear valleys. Most of these faults appear to be normal fault zones whose traces approximately follow the change in slope between valley floor and hillside. Movement on many is difficult to document as they are commonly not exposed, but in most cases it is probably small (few metres to hundreds of metres). However, up to 2 km of normal displacement is evident on faults in the Skeena Arch. The uplifted areas resulting from block faulting commonly are cored by Upper Cretaceous to Paleocene intrusions whereas adjacent depressions are generally underlain by younger strata.

Block faulting may have begun as early as the Jurassic, but reached its peak in the Eocene and continued into the Oligocene. Miocene plateau basalts are unfaulted. The regular, orthogonal pattern of block faults near Hazelton (Fig. 17.22; Richards, 1980) suggests an extensional regime resulting from domal uplift whereas the rhomboid pattern near Tahtsa Lake (Woodsworth, 1980) indicates a component of transcurrent stress. In the Terrace area of the eastern Coast Belt, the Terrace graben (Fig. 17.8) trends north-south for over 100 km and has a vertical displacement of between 1300 to 1800 m. At the north end of the graben, the Aiyansh lava flow erupted about 250 years ago and together with the existence of hotsprings along the edge of the graben south of Terrace, indicate recent tectonic activity along the graben.

The imbricate zone between the Coast and Intermontane belts is disrupted by brittle, high-angle faults which trend north to north-northwest. In both the Tsaytis River and Terrace areas, these high-angle faults cut Eocene plutons. Steep, narrow, northeast- to northwest-trending brittle faults also are common in the Redcap Mountain and other areas of the Central Gneiss Complex in the core of the Coast Belt. These faults and related mafic dykes are interpreted as evidence for minor east-west extension related to cooling and uplift of the Coast Belt during the Eocene (Hill, 1984).

Due to poor exposure structural styles in the eastern part of the central Intermontane Belt are less well understood than those to the west (Fig. 17.22, Fig. 17.1, in pocket). Block faulting appears to be important as far east as the Takla and Vital faults. The Takla Fault is a dextral strike-slip fault with an offset of more than 100 km (Gabrielse, 1985) separating Stikinia from the Cache Creek Terrane.

Northeast of the Yalakom Fault the Tatla Lake Metamorphic Complex, comprising a core zone of sillimanite-grade gneiss, migmatite and amphibolite intruded by tonalite and a peripheral zone of mylonite metagranite and metavolcanic augen gneiss and metasedimentary schist, has undergone polydeformation and has many characteristics of metamorphic core complexes (R. Friedman and R.L. Armstrong, pers. comm., 1986). The structures are described in the section on Eocene extensional faulting.

Scattered exposures of Mesozoic strata in the Nechako Basin (Fig. 17.18) have been deformed into open northwest-trending folds lacking significant cleavage. Overlying Tertiary volcanic rocks are locally block-faulted.

Structures related to transcurrent faults

Contractional structures in the Intermontane Belt may be in part related to, but mainly predate, structures associated with transcurrent faults which are particularly important in the northern and southern extremities of the belt. In the Yukon, many faults associated with the Teslin dextral strike-slip fault cut Mesozoic overlap assemblages in the

Whitehorse Trough (Fig. 17.21). They may have localized deposition of the Lower Cretaceous(?) Tantalus Formation in pull-apart basins. The Teslin Fault can be traced southeastward from the north end of the Whitehorse Trough into northern British Columbia where it connects with the Thibert Fault marking the boundary of Quesnellia and the Cache Creek Terrane (Fig. 17.19).

The Teslin and Thibert faults appear to have had a long history perhaps beginning with terrane collision in the Jurassic(?) and culminating in the Early to mid-Cretaceous. The Thibert Fault is cut by a Late Cretaceous pluton but has generated mylonite as young as mid-Cretaceous.

The structural style of the southern Intermontane Belt is dominated by early Tertiary normal, reverse and dextral strike-slip faults (Fig. 17.24). On a Cordilleran-wide basis these structures have been related by Price (1979) and Ewing (1980) to right-lateral transform motions and to regional extension. The prominent system of north-, northeast- and northwest-trending faults disrupts Eocene and older strata, and is overlain by flat-lying Miocene and older basalt flows (Monger and McMillan, 1984). Ewing (1980, 1981) recognized that the fault pattern reflects regional clockwise rotation and extension which is both congruent with and intermediate between dextral wrench displacement on the Fraser River Fault System to the west, and basin-and-range type regional extension to the southeast, in and east of the Okanagan Valley.

A prominent Tertiary structure is the north-northeast-trending horst composed of the Nicola batholith and associated metamorphic rocks. Isotopic data suggest that parts of the Nicola batholith are of earliest Tertiary age, but other parts appear to be of early Mesozoic age, and isotopic systems were re-set during uplift in the early Tertiary. Faults that extend southwestwards from the horst appear to cut the Mount Lytton Complex and extend into the eastern part of the Cascade segment of the Coast Belt.

Structural depressions containing Tertiary strata are associated with normal, reverse and strike-slip faults. The Princeton Basin is an Eocene half-graben, containing all three types of faults (Fig. 17.24; McMechan, 1983). The Hat Creek Graben appears to be on trend with the strike-slip Botanie Creek Fault, which splays off the Fraser Fault System, and as proposed by Church (1977), may be due to local north-south compression. The northwest-trending structural depression containing Eocene volcanics northwest of Kamloops exhibits a complex fault pattern (Ewing, 1981) and is bounded on the southwest side by the Cherry Creek Fault which probably has both reverse and dextral strike-slip components.

In summary, Tertiary structures in the southern part of the Intermontane Belt appear to be largely transtensional, and are transitional between the dextral wrench faults along the Fraser River to the west, and basin-and-range style extension faults to the southeast.

PART D. OMINECA BELT

Summary

The Omineca Belt consists of moderate to high grade regionally metamorphosed core zones which are characterized by complex, polyphase deformation, overlain and flanked by less deformed cover rocks commonly in stratigraphic continuity with the core zones. The cover rocks are overlain in turn by accreted terranes. Throughout much of the belt south of the British Columbia-Yukon boundary the main regional structures are westward verging, although the easternmost structures in the Cassiar, Cariboo and Columbia mountains verge eastward. Easterly verging structures are dominant in the northern Cassiar, Pelly and Selwyn mountains. It is not clear that vergence of folds bears any relationship to their age except in specific regions. In the Kootenay Arc the earliest structures have been dated as pre-Late Mississippian but elsewhere they may be entirely of Mesozoic age. Invariably the structurally lowest rocks show the greatest amount of shear strain. The Cariboo Mountains provide an excellent example of a gradational change from brittle strata deformed into upright folds at high structural levels to ductile strata sheared and recumbently flow folded at low structural levels.

The core zone rocks appear to be mainly regionally metamorphosed basement and miogeoclinal cover rocks of the western margin of the North American craton which were markedly shortened and thickened by collision and convergence with the Intermontane Superterrane during Middle Jurassic and later time. All of the regionally extensive folds are cut by Early to mid-Cretaceous granitic plutons. In the Cariboo and Columbia mountains late Middle Jurassic plutons truncate regional folds and metamorphic isograds. Uplift of regional anticlinoria took place in the Omineca, Cassiar and Pelly mountains in Early to early Late Cretaceous time. Late Cretaceous and early Cenozoic transcurrent faults and Eocene extension faults, particularly evident in the southern Omineca Belt, have been superimposed on the earlier structures. Rapid uplift occurred in core zones associated with Eocene extension.

Introduction

The Omineca Belt consists of a number of northwest-trending, elongate structural culminations and depressions in which the grade of metamorphism and degree of deformation can be related, in a general way, to structural and, commonly, stratigraphic depth of burial. Some culminations expose rocks that have been uplifted more than 20 km, locally exposing crystalline basement. In contrast, the depressions are probably sites of considerably less structural thickening and are underlain by rocks whose structural style is similar to that of strata in the Foreland Belt.

Core zones

Core zones in the Omineca Belt occur in displaced and pericratonic terranes. Core zones as described herein, consist of polydeformed, high-grade metamorphic rocks that have been regionally metamorphosed and ductilely

STRUCTURAL STYLES

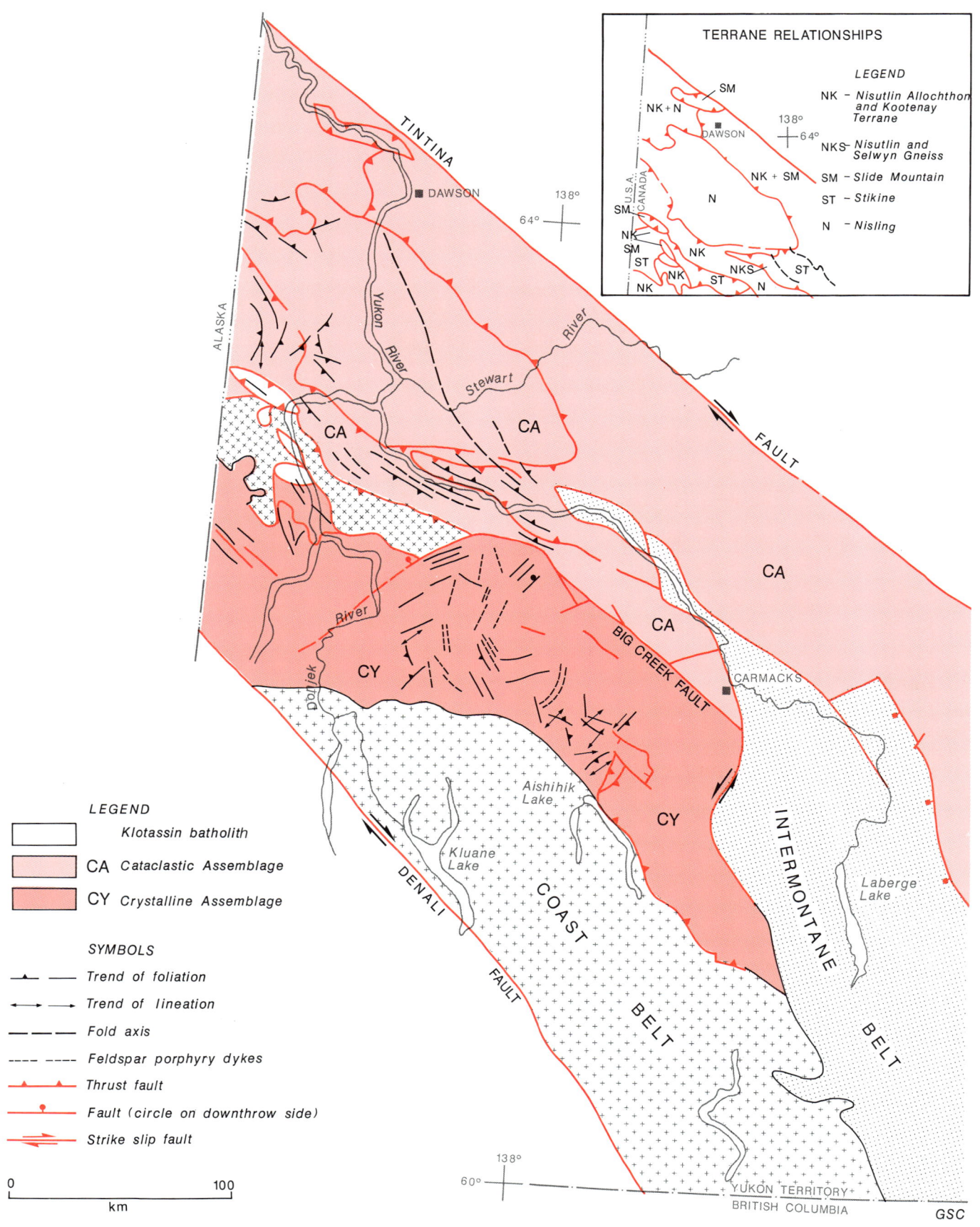

Figure 17.28. Main structural elements in the northwestern Omineca Belt, Yukon Territory.

deformed at deep structural and stratigraphic levels. They include the areas of exposed crystalline basement and represent regions of greatest uplift in the Omineca Belt. Some of the structural culminations described in the section on cover rocks have characteristics of core zones in their deepest structural levels but because their structural styles and metamorphic grades form a continuum with those of strata at much higher levels it is convenient to treat them together. Local structural culminations in the Rocky Mountains are described in Part E in this chapter.

Nisling Terrane

D.J. Tempelman-Kluit

The northernmost of the structural and metamorphic culminations in the Omineca Belt is underlain by regionally metamorphosed crystalline rocks southwest of the Big Creek Fault in the Klondike Plateau of western Yukon Territory (Fig. 17.28; Tempelman-Kluit, 1974, 1976). Large areas consist of fine grained metaquartzite (Nasina) which has a strongly developed, gently to moderately dipping foliation. Schistosity is parallel with bedding. Marble lenses in biotite schist are believed to be structurally disconnected remnants of once continuous beds. Lineations related to a late phase crinkle crenulation of the schistosity plunge north to northeast. On their southeast margin the crystalline rocks are concordant with steeply dipping foliated granodiorite.

A belt of metamorphic rocks east of the Donjek River has anomalous, northeast-trending folds and lineations almost at right angles to the more common structural grain of the region. As elsewhere the rocks, mainly Nasina Quartzite, are strongly foliated. Quartz shows a prominent crystallographic orientation parallel with a pervasive northeast-trending mullion structure.

The ages of deformation are poorly constrained. Mid-Cretaceous granitic plutons clearly cut the main regional structural fabric. Early Jurassic plutons, on the other hand, are generally foliated and are roughly concordant with foliation in adjacent rocks. Data from adjacent parts of Alaska (Dusel-Bacon and Foster, 1983) suggest that Triassic deformation and metamorphism also affected the northwest part of the Omineca Belt in Canada.

Cassiar Terrane

H. Gabrielse, C.A. Evenchick, J.L. Mansy, and D.J. Tempelman-Kluit

In the Kaska Mountains core zones define a number of regional anticlinoria in which structures and metamorphic grade are gradational into overlying cover rocks. In the Pelly Mountains northwest-trending structural culminations, each about 30 km across and more than 100 km long, are centred on the mid-Cretaceous Quiet Lake and Nisutlin batholiths. Culminations in the Cassiar and Omineca mountains coincide with areas with either mid-Cretaceous or Eocene granitic rocks or which yield mid-Cretaceous or Eocene K-Ar ages on metamorphic minerals.

The only detailed study of core zone structural style in the northern Cordillera has been done in the southern Cassiar and northern Omineca mountains. In the Sifton Ranges clastic strata with minor carbonate and amphibolite of the Windermere Assemblage have undergone an early phase of deformation which produced recumbent, probably westward verging folds with amplitudes of several kilometres (Fig. 17.29B; Evenchick, 1985, 1988). Minor, similar folds of a few to tens of metres in amplitude are characteristic. Structures developed during the first two phases of deformation are truncated by the gently southeastward-dipping Sifton Fault. The hanging wall rocks of the fault, comprising quartzite, amphibolite, carbonate and pelitic schist, are intensely and penetratively deformed. Sheath folds, shear bands, minor folds and lineations indicate relative south-southeastward transport of hanging wall rocks in a ductile environment (Fig. 17.29B; Evenchick, 1985, 1988). The strain is interpreted to reflect local transpression along a dextral transcurrent fault system caused by a bend from the Northern Rocky Mountain Trench into the Kechika Fault.

Structural culminations in the Russell, Swannell and Butler ranges are north-northwesterly trending anticlinoria in which westward verging folds and westward directed thrust faults have been refolded by more upright folds (Fig. 17.29A; Mansy and Dodds, 1976). In the Swannell and Russell ranges a pervasive almost east-west trending crinkle crenulation and associated cleavage may have been produced by mid-Cretaceous through Cenozoic stress in accord with a regional north-northwest-trending dextral transcurrent strain regime. Similar fold trends occur in Paleogene sedimentary rocks in the major transcurrent fault zones.

Great uplift of the core zone anticlinoria occurred after their early structural development. K-Ar ages on metamorphic micas and hornblende suggest culminations of uplift during mid-Cretaceous and Eocene times. Geobarometry and geothermometry studies indicate that metamorphic assemblages in the high grade rocks formed at temperatures of 550-650°C and pressures of 5-6 kb (5-6x10^5 kPa) (Evenchick, 1985, 1988; Mansy, 1986).

Southward in the Omineca Belt westward verging folds dominate the structure of the core zone rocks to south of Williston Lake. Little is known about the structural style between the south end of Williston Lake and the northern Cariboo Mountains. Structural style in the Cariboo Mountains includes elements of both core zone and cover rock geometry and is described with the latter.

Monashee Terrane

R.L. Brown, J.M. Journeay, and L.S. Lane

The Monashee Terrane, more commonly referred to as the Monashee Complex, is separated from the overlying Kootenay Terrane of the Selkirk Allochthon by the Monashee Décollement on the northwest and south, and on the east by the Standfast Creek Slide and the Columbia River Fault zone (Fig. 17.30). The Monashee Complex is interpreted to be an exhumed antiformal duplex system of basement-cored horses (Journeay, 1986; Journeay, 1983, 1986; Monger et al., 1985). Frenchman Cap and Thor-Odin domes are second order structural culminations along the axis of the antiformal duplex (Fig. 17.31) and are considered to be regional interference structures produced by the superposition of at least three generations of non-coaxial folding (Read, 1979, 1980; Journeay, 1981; Read and Klepacki, 1981; Duncan, 1984). The oldest structures are eastward verging isoclinal fold nappes and associated low-angle detachment faults. These imbricate structures were refolded by two distinct sets of reclined isoclinal folds kinematically linked to eastward and northeastward directed shear strain associated with emplacement of the overlying Selkirk Allochthon (Fig. 17.31). Penetrative, northeast-trending stretching lineations and associated flattening foliations are well developed at all structural levels but are most intense within mylonitic rock of the Monashee Décollement. Northerly trending, late- to post-metamorphic folds deformed the earlier fabrics and are consistently inclined to the east on all flanks of the terrane. They are believed to have been generated during late stage uplift and arching.

The Monashee Décollement (Read and Brown, 1981) is a fundamental stratigraphic, structural and metamorphic discontinuity within the southern Omineca Belt. It is defined by a network of mylonitic shear zones which separate the Monashee Complex from deep-level metasedimentary and related meta-plutonic rocks of the overlying Selkirk Allochthon. The fault zone is well exposed along the outward dipping flanks of both Frenchman Cap and Thor-Odin domes, and clearly predates the most recent episodes of uplift and arching within this part of the southern Omineca Belt. Along its length, the décollement is characterized by the occurrence of recrystallized and annealed mylonitic fabrics, sheath folds, and discrete ductile shear zones, which collectively define a zone of intense non-coaxial strain which extends nearly 500 m into footwall rocks of the Monashee Complex, and as much as 750-1000 m into the overlying Selkirk Allochthon.

Recent structural and petrological studies of the Monashee Décollement have shown it to be a composite shear zone across which there have been at least two major episodes of upper plate to-the-east displacement (Journeay, 1986). The older of the two shear zones (MD_1) cuts early generation (F_1) folds within Frenchman Cap Dome and predates early stage uplift and arching of the Monashee Complex. The younger shear zone (MD_2) cuts MD_1 and older structures, and is itself arched over the structural culmination of Frenchman Cap Dome. The locus of displacement for MD_2 occurs near the base of a sequence of pelitic and semi-pelitic schists and interlayered amphibolites, locally containing elongate pods of mafic and ultramafic gneiss. These rocks are contiguous with, and are presumed to be at least in part correlative with Upper Proterozoic sequences of the Horsethief Creek Group, exposed within the northern Monashee and adjacent Selkirk Mountains (Wheeler, 1965; Brown, 1981).

The upper plate of MD_2 is a metamorphic-plutonic complex composed of uniformly high-grade (bathozone 6) migmatitic gneisses, semi-concordant sheets of anatectically-derived granite, and associated syn- to late kinematic pegmatitic dykes and sills. Synkinematic granites intrude an earlier suite of granitic rocks within the Anstey Pluton, and make up at least 50% of the exposed upper plate by volume. They are intruded and overlain by a narrow belt of late- and post-kinematic granitic plutons of uncertain age and origin (Anstey Pluton and Map Unit E; Wheeler, 1965). These two belts of gneissic rocks extend southward, and are known to occur along the west and south flanks of Thor-Odin Dome (Reesor and Moore, 1971; Journeay, 1986). Together, they represent the deep level metamorphic-plutonic infrastructure of the Selkirk Allochthon.

The absence of matching hanging wall and footwall cut-offs in the segment of the décollement exposed along the flanks of Frenchman Cap Dome suggests that upper plate rocks of the Selkirk Allochthon were displaced a minimum distance of 80 km with respect to their lower plate counterparts (Read and Brown, 1981). The bulk of this differential displacement appears to have been accommodated at successively shallower levels in the crust by a combination of progressive ductile strain and dynamic recrystallization; these processes produced strongly foliated and lineated mylonitic rocks within a wide zone of strain softening along the margins of the Monashee Complex and overlying Selkirk Allochthon.

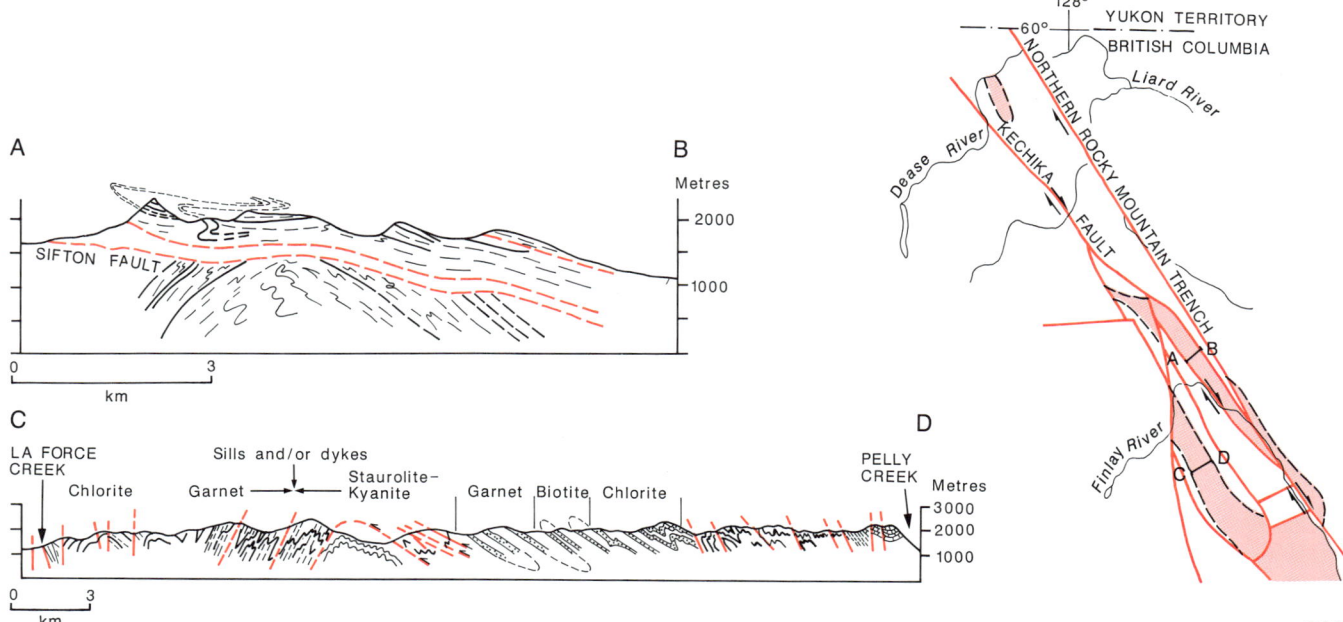

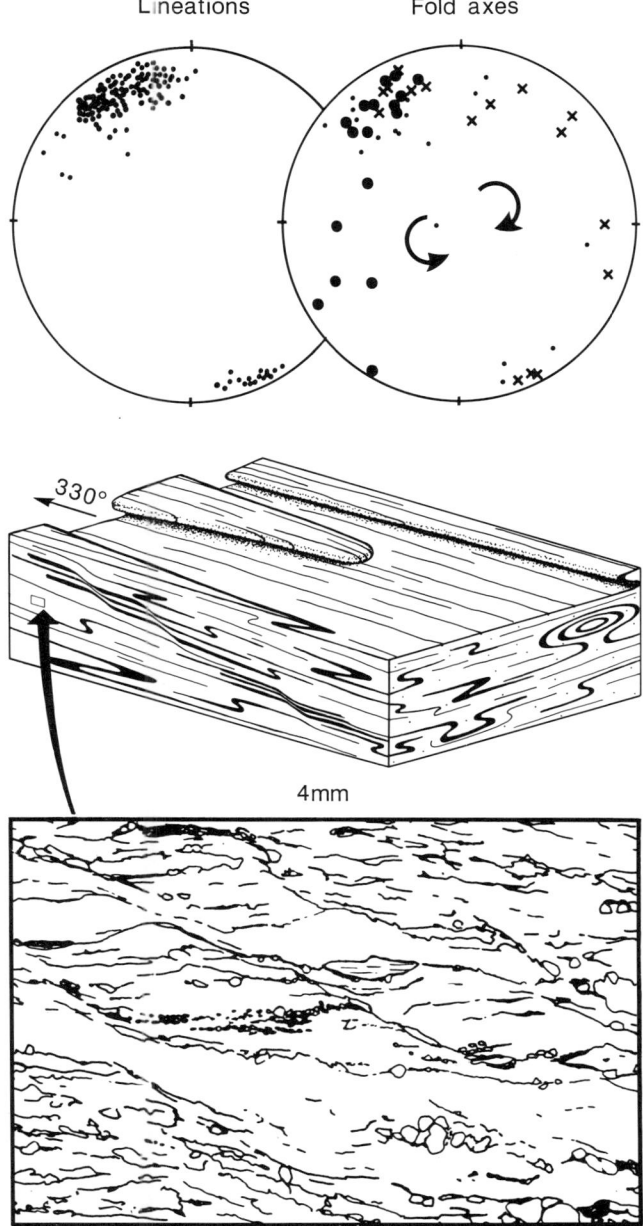

Figure 17.29A. Main structural elements of core zones in the Omineca Belt in the Cassiar and Omineca mountains. Inset map shows distribution of core zones (pink) and location of structural cross-sections. A-B) Sifton Ranges with anticlinorium of recumbently folded, core-zone rocks truncated by the transgressive Sifton Fault which marks the base of a mylonitic shear zone. Upper plate moved southeast parallel with the Northern Rocky Mountain Trench, C-D) Central Swannell Ranges showing the refolding of structures across the anticlinorium and the positions of metamorphic isograds (after Mansey and Dodds, 1976) **B.** Structures in the shear zone above the Sifton Fault (see Section A-B in Fig. 17.29A) characterized by mylonitic fabric, shear bands, intense lineation and sheath folds (after Evenchick, 1985).

Planar and linear fabric elements are gradational into synmetamorphic foliations and lineations in both upper and lower plates of the décollement, and become increasingly more flattened and rotated near the zone of detachment. Secondary bands of intense ductile strain form a network of anastomosing mylonitic shear zones that envelop lozenge-shaped domains in which mylonitic fabrics are less intensely developed.

Synmetamorphic northeast-trending lineations of associated mylonitic foliations consistently record an "upper-plate-to-the-east" sense of shear along both flanks of the Monashee Complex (Brown and Murphy, 1982; Lane, 1984a,b; Journeay, 1983, 1986).

The sequential development of retrograde sillimanite, andalusite and assemblages of muscovite-quartz and biotite-chlorite within tensional fractures in high-pressure minerals throughout the décollement indicates that overthrusting of the Selkirk Allochthon must have been synchronous with uplift and cooling of the underlying Monashee Complex. Mylonitic fabrics postdating the quenching of peak metamorphic assemblages within the décollement are virtually identical in both style and orientation to those generated during earlier stages of overthrusting.

Contractional deformation in the Monashee Complex is presumed to have begun in the Middle Jurassic and to have culminated in the Late Cretaceous or early Tertiary time (Brown and Journeay, 1987). The late stage thrusting and duplexing at deep structural levels may have been linked with late thrusting in the Foreland Belt. Estimates of crustal shortening are as much as 75% and thickening in excess of 20 km during Mesozoic to early Cenozoic time.

Okanagan-Kootenay region

H. Gabrielse

Structural styles of the southern Omineca Belt west of the Kootenay Arc are complex (Fig. 17.32). They include southwestward verging folds and thrust faults in the Kootenay Terrane north of Okanagan Lake, eastward verging structures culminating in the imbricated, duplex-style thrust sheets of the Monashee and Valhalla complexes, northwest-verging structures along the southern part of the Kootenay Arc and extension faults across the entire belt. The oldest structures in the Adams and Shuswap lakes area are pervasive foliations subparallel with bedding, rare tight to isoclinal sheared recumbent folds and intersection and mineral lineation (Okulitch, 1979). Foliation is tightly folded about axes trending north to east-northeast which are parallel with distinct mineral rodding and foliation intersection lineations. These two phases of deformation were succeeded by the west-northwest- to northwest-trending, southwest-verging folds and thrust faults which dominate the distribution of map units in four thrust sheets (Preto, 1981; Okulitch, 1979). The folds are commonly overturned and tight, and are associated with well developed axial-planar cleavage, prominent crinkle axes and intersection lineations.

The contractional structures noted above are cut by the Middle Jurassic Raft and mid-Cretaceous Baldy batholiths. Contacts are sharply discordant to the west but concordant to the east at lower structural levels where the adjacent high grade metamorphic rocks contain abundant

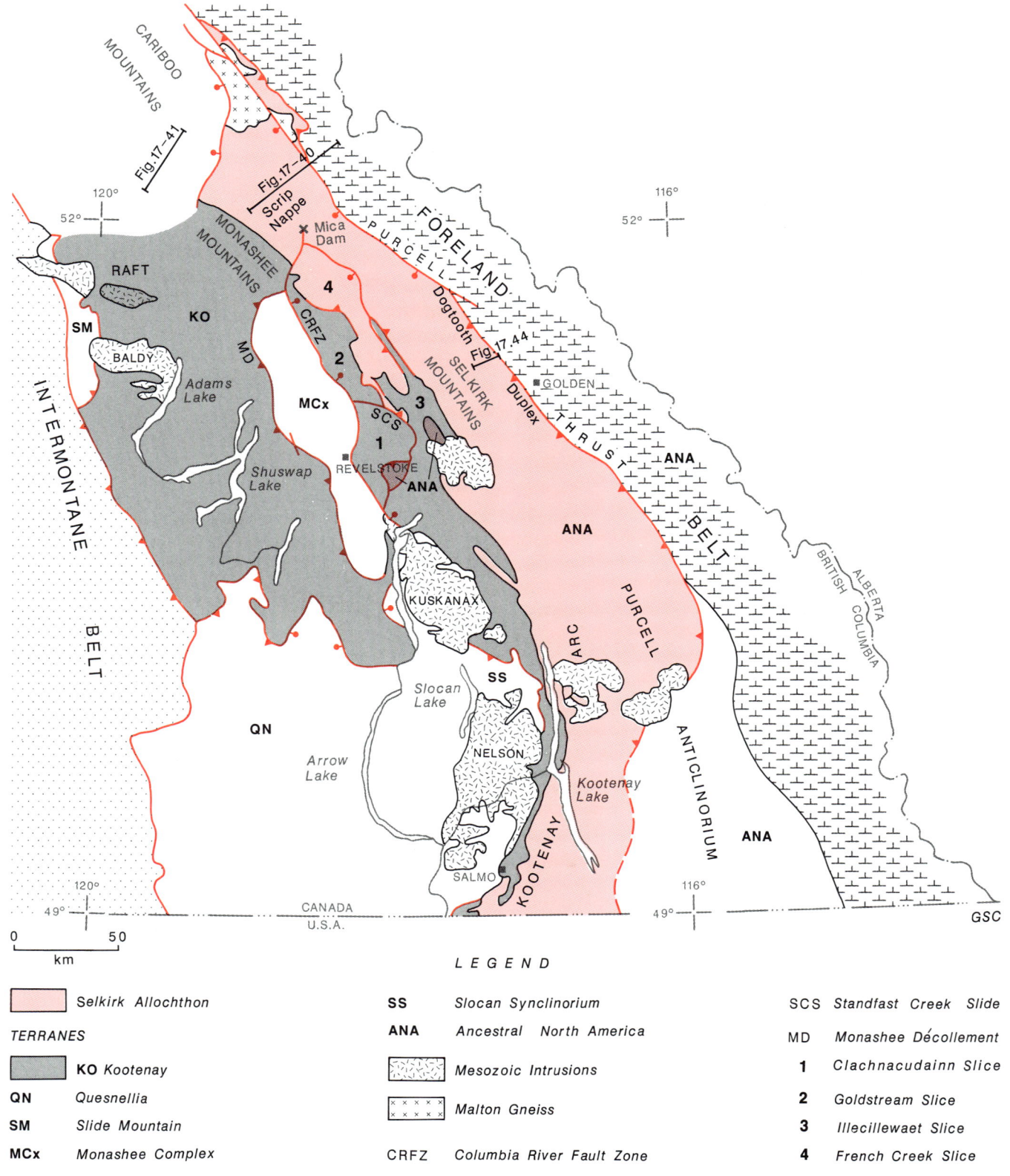

Figure 17.30. Southeastern Omineca Belt showing the distribution of terranes, some of the regional structures, and the location of structural cross-sections in Figures 17.40, 17.41 and 17.44.

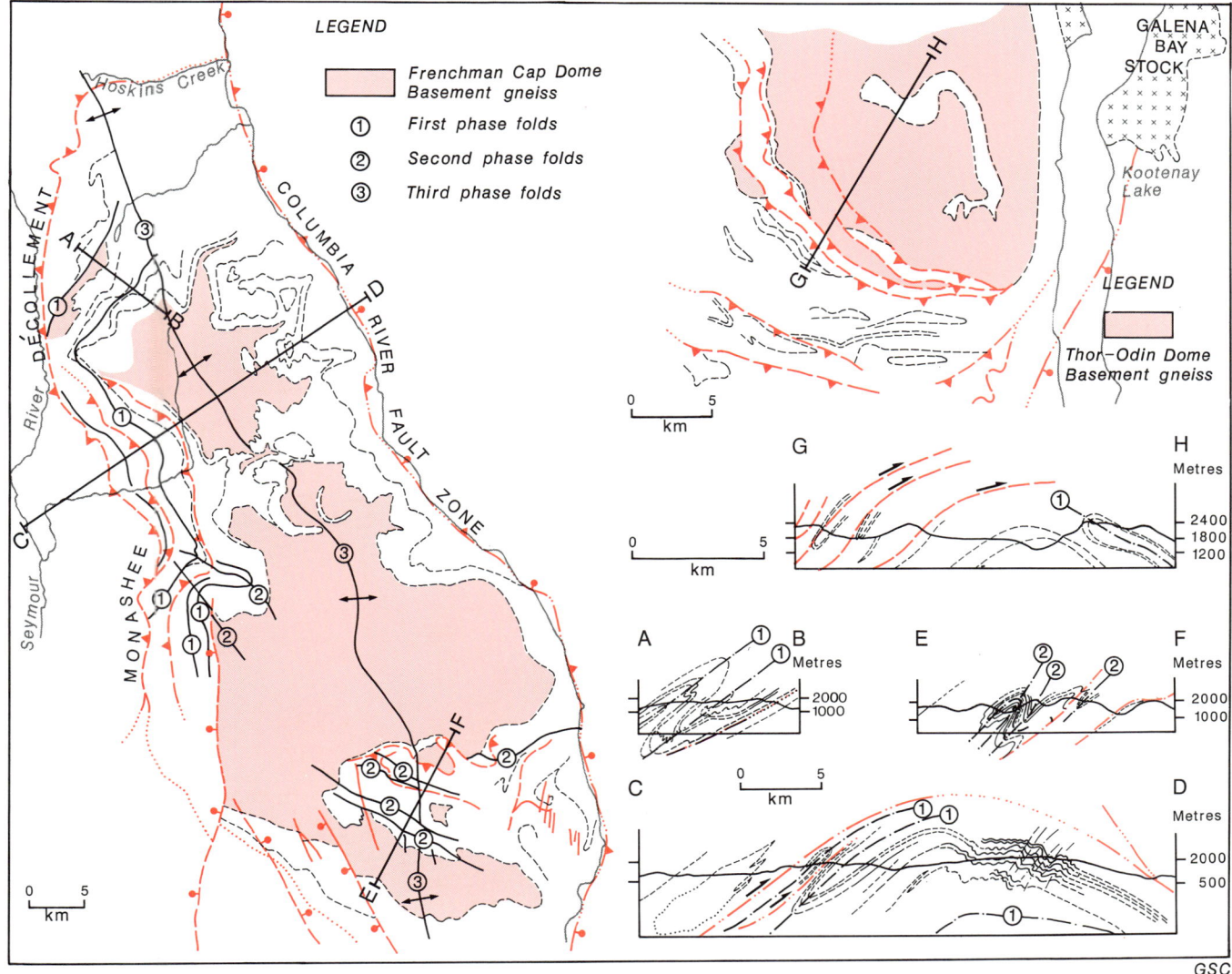

Figure 17.31. Structures of the Monashee Complex illustrating three phases of folding and duplex style of basement deformation. Structural cross-sections in Frenchman Cap Dome; (AB, CD, EF), Structural cross-section in Thor-Odin Dome, (GH).

granitoid gneiss and foliated granitic rock. The granitic bodies and host rocks are cut by north-trending faults of probable early Tertiary age (Schiarizza, 1986).

In the southern Okanagan area near Vaseaux Lake detailed studies have documented the same sequence and styles of deformational events noted above but the earlier structures are better exposed (Ross, 1981; Ross and Christie, 1979). The earliest folds are isoclinal, northerly trending and have amplitudes of 4-5 km (Fig. 17.33). Locally, narrow mylonite zones separate large recumbent folds and are the loci of numerous ultramafic bodies. The dominant, second phase structures, are also recumbent with amplitudes up to 8 km and have axes parallel with a ubiquitous east-west or northwest-southeast stretching lineation. This deformation accompanied the highest grade metamorphism. The third phase folds are similar to those in the Adams Lake-Shuswap area. They are southwestward or southward verging with axial surfaces dipping 60°-70° to the north. The age of these structures may range from as old as late Paleozoic to as young as Late Cretaceous (Parkinson, 1985).

Cover rocks

Structures in cover rocks range from ductile, polyphase and recumbent where they are gradational with those of the core zones, to brittle and dominantly upright at high structural levels. In concert with the change in structural style is a decrease in metamorphic grade and intensity of foliation although the latter is closely related to lithology.

CHAPTER 17

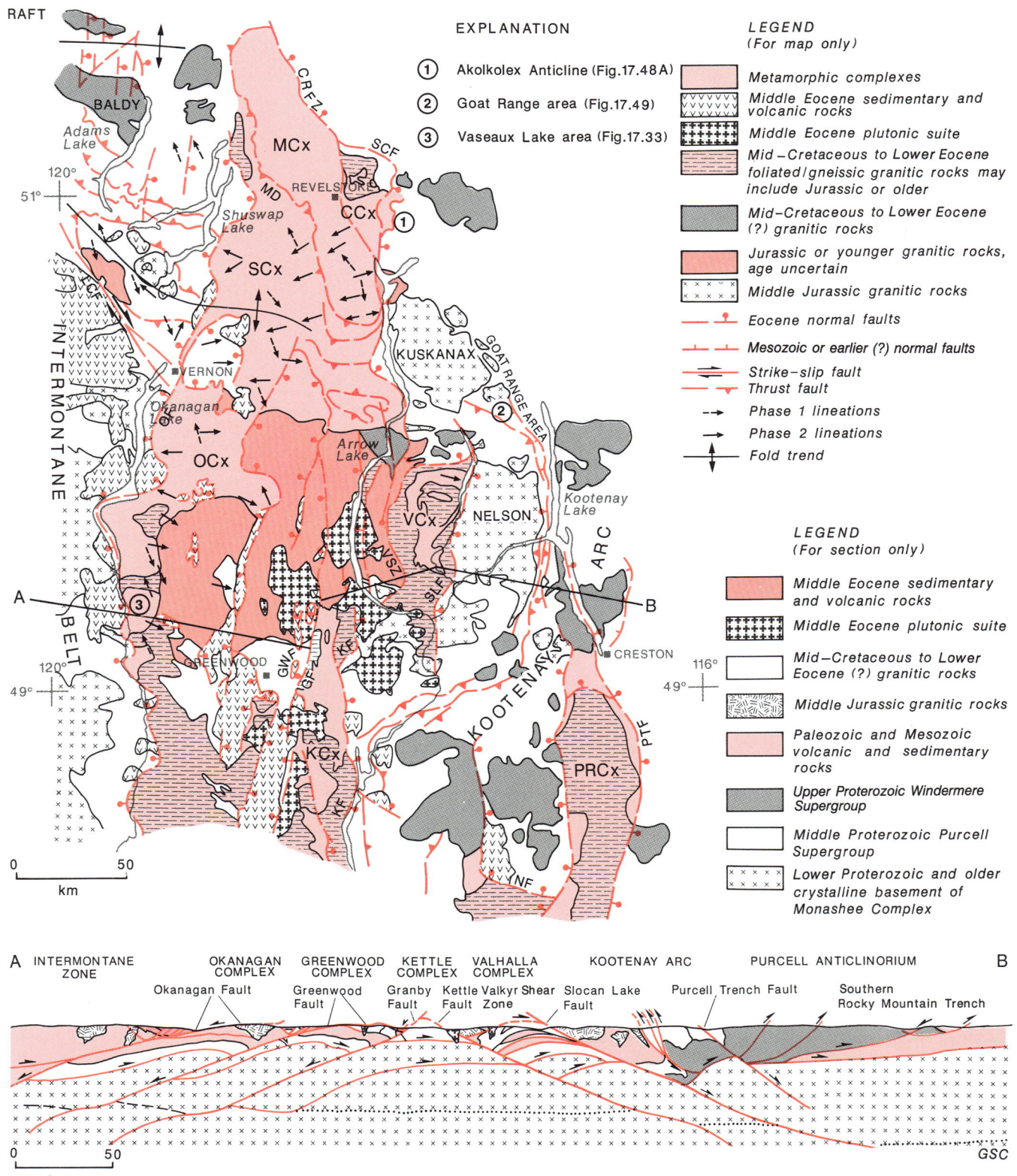

Figure 17.32. Main structural elements of the southeastern Omineca Belt. Data from R.R. Parrish (pers. comm., 1986). MCx, Monashee Complex; CCx, Clachnacudainn Complex; SCx, Shuswap Complex; OCx, Okanagan Complex; VCx, Valhalla Complex; KCx, Kettle Complex; PRCx, Priest River Complex; MD, Monashee Décollement; CRFZ, Columbia River Fault Zone; SCF, Standfast Creek Slide Fault; PTF, Purcell Trench Fault; NF, Newport Fault; SLF, Slocan Lake Fault; VSZ, Valkyr Shear Zone; KF, Kettle Fault; GF, Granby Fault; GWF, Greenwood Fault; OF, Okanagan Fault; LCF, Lewis Creek Fault.

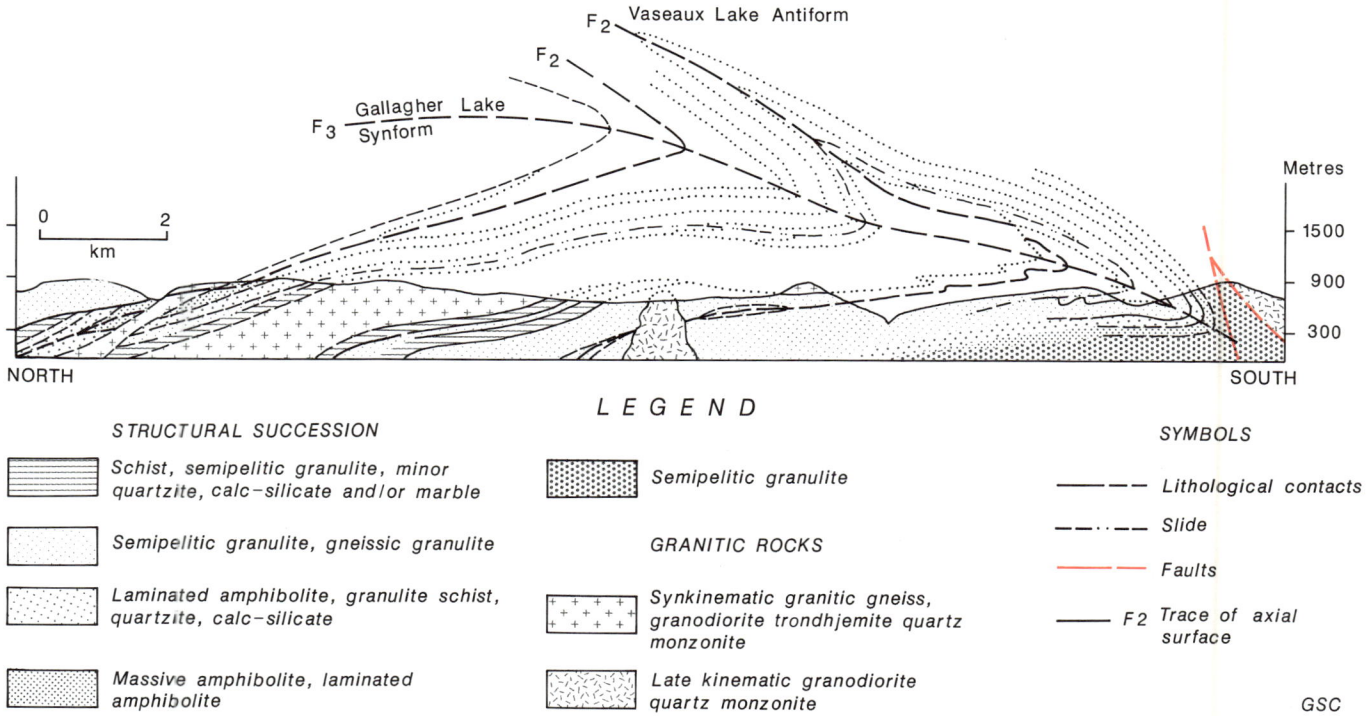

Figure 17.33. Structural cross-section in the Vaseaux Lake area, southern Okanagan Valley. After Ross and Christie, 1979. See Figure 17.32 (no. 3) for location.

Cassiar Terrane

Pelly Mountains

D.J. Tempelman-Kluit

Strata of the Cassiar Terrane in the Pelly Mountains between the core zone and the Tintina Fault are imbricated by four thrust faults each of which carries a thrust sheet 2 or 3 km thick (Fig. 17.1, in pocket and 17.34; D.J. Tempelman-Kluit, in press-a). Dip-slip on each fault is 10 km or more as demonstrated by offset facies; cumulative shortening is between 80 and 135 km. The thrust faults dip gently southwest and some are folded, so that they are exposed across strike for up to 25 km. Most have up-dip splays; some are restricted to the rocks below the next higher thrust sheet whereas others displace higher thrust sheets also. Each thrust fault probably merges down dip with the next lower. Strata in each thrust sheet are folded and tear faults above the thrust faults allowed folds to attain different shapes on either side. Some thrust faults are folded on a grand scale above a lower one that acted as the base of folding. Although most thrust faults are northeast directed, a few, considered to be splays of the main northeast-directed faults, verge southwest. Structures in the autochthonous rocks were formed before the emplacement of mid-Cretaceous granite and after deposition of Upper Triassic strata.

Cassiar and Omineca mountains

H. Gabrielse and J.L. Mansy

Upper Proterozoic to Lower Mississippian miogeoclinal strata are moderately to severely deformed throughout the northern Omineca Belt but generally are not involved in the recumbent folds typical of the early phase of deformation in the high grade metamorphic rocks of the core zones. Exceptions occur in a few localities, however, and demonstrate that the change in structural style between core and cover rocks is transitional. Southwest of the Kechika Fault folds verge southwesterly whereas to the northeast they verge northeasterly (Fig. 17.1, in pocket). Some folds and thrust faults may be related to late Mesozoic and Cenozoic strike-slip faulting.

Northwest of Dease River northeastward directed thrust faults are conspicuous; associated folds are generally upright with axial surfaces dipping steeply either to the northeast or southwest (Fig. 17.35). One major anticline has a distinct, moderately northeastward dipping axial surface but whether the asymmetry reflects original southwestward vergence or rotation above a west dipping thrust fault is not known. Northeast of the Kechika Fault, the northeastward vergence of folds is prominent. Incompetent Cambro-Ordovician strata show a well developed axial-surface cleavage, commonly moderately to steeply dipping. Cleavage fans are present locally. Isoclinal folds, overturned to the northeast, are well displayed in some areas. Locally, in autochthonous rocks along the west side and south end of the Sylvester Allochthon, a set of westerly directed faults repeat competent strata in a duplex structural style (Harms, 1985; Fig. 17.35,K-L).

Southeast of Dease River and southwest of the Kechika Fault vergence of folds and displacements on thrust faults indicate southwestward tectonic transport. In places gently dipping faults floor overturned panels of strata representing the erosional remnants of nappes (Fig. 17.35,E-F). Interstratal faults, localized by competence differences between formations, are common.

CHAPTER 17

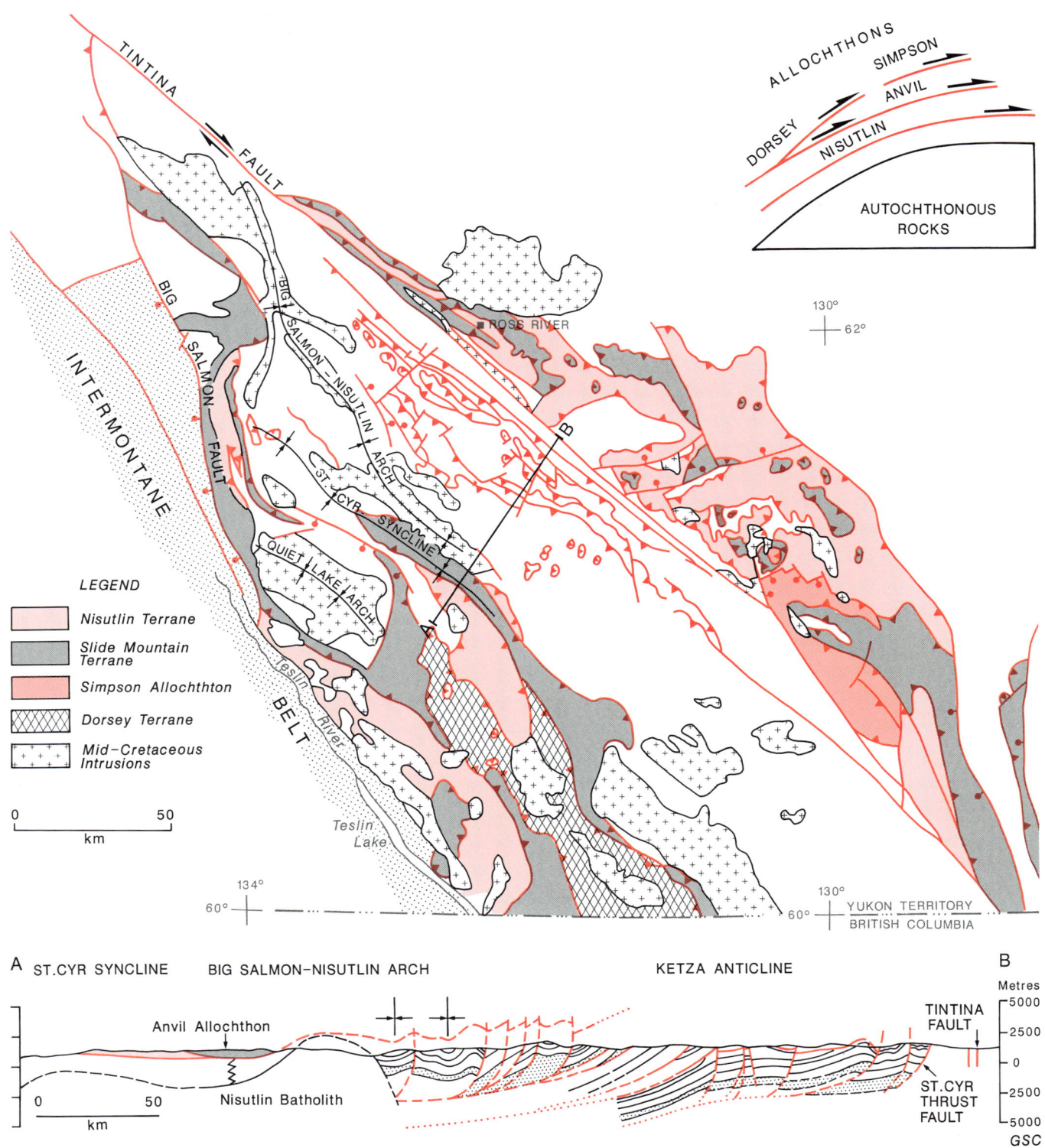

Figure 17.34. Distribution of accreted terranes and structural style in southeastern Yukon Territory.

STRUCTURAL STYLES

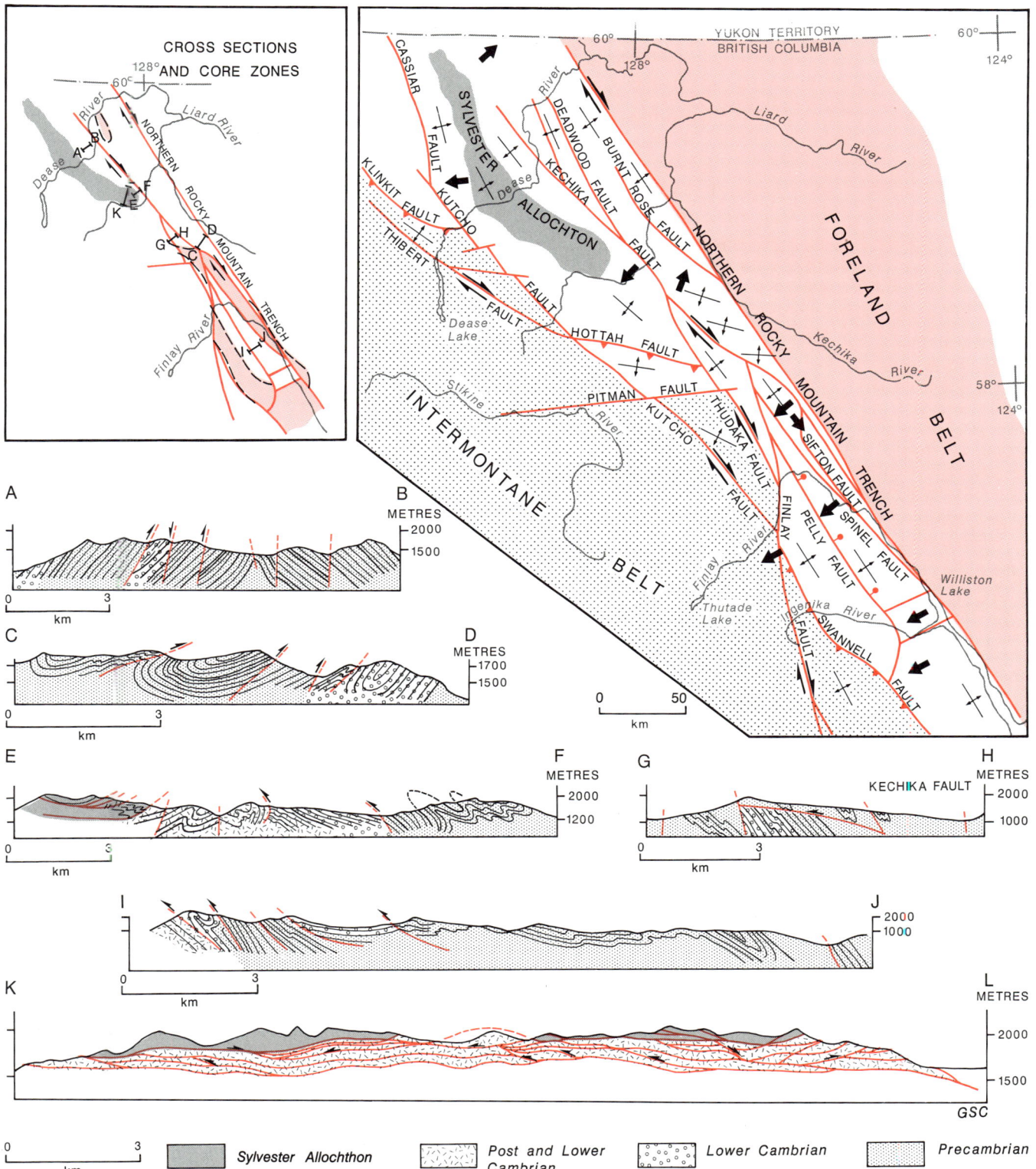

Figure 17.35. Main structural elements in the Cassiar and Omineca mountains. Location of structural cross-sections shown on inset map. Heavy black arrows indicate dominant directions of vergence.

In northern British Columbia the Omineca Belt is bounded and segmented by an array of anastomosing, dextral transcurrent faults (Gabrielse, 1985). The fault zones are loci of intense cataclasis and mylonitization and the panels between the faults commonly have east trending folds and faults, at least in part related to the dextral strain. They are described in Part F.

Cariboo Mountains and Quesnel Highlands
L.C. Struik, D.C. Murphy and C.J. Rees

The Cariboo and Barkerville subterranes were juxtaposed by thrust faulting prior to the emplacement of the overriding Slide Mountain Terrane in post-Permian time. The contact dips to the east and, assuming transport in the updip direction, records westward directed shear. The transition from westward directed ductile flow in the southern part of the Cariboo Subterrane to the average vertical, slightly compressional brittle folds and faults of the northern part implies a loss of shear strain. The evidence of eastward verging structures prior to the westward directed shear requires involvement of the Cariboo Subterrane in relative eastward directed flow before its thrust emplacement onto the Barkerville Subterrane.

The pervasive ductile shear in the Barkerville Subterrane is the result of overthrusting to the northeast by Quesnellia and the Slide Mountain Terrane and to the southwest by the Cariboo Subterrane. Thrusts directed to the east involved Quesnellia and the early shear in the Barkerville Subterrane. Subsequent westward directed shear and folds may be due to the emplacement of the Cariboo Subterrane from the northeast (Fig. 17.1, Sections J and K, in pocket).

Cassiar Terrane: Cariboo Subterrane

Rocks in the Cariboo Subterrane represent one of the best examples in the Canadian Cordillera of a gradation in structural style between ductile, recumbent flow folds characteristic of core zones to brittle, upright folds typical of relatively shallow levels of deformation (Campbell, 1970, 1973; Murphy, 1987). Thrust southwestward onto the Barkerville Subterrane, it is typified by flow folds that are transitional to the northwest and stratigraphically upwards to open folds and steep to moderately dipping faults (Fig. 17.36, 17.37, 17.38).

A ductile zone, typified by rock flow, is recorded by isoclinal folds and to a minor degree by shear zones that deform the lower part of the Kaza Group in the southern Cariboo Subterrane (Fig. 17.38, 17.39). Isoclinal folds define two oppositely verging coaxial sets; southwesterly verging superimposed on northeasterly verging. The folds are northwest-trending, recumbent to asymmetric, and have shallow to moderately dipping axial surfaces. Limb lengths are less than 3 km and generally less than 1 km. Metamorphic minerals of garnet grade and higher are mainly post-kinematic although locally they may be partly deformed during formation of the westerly verging isoclines.

Shear zones are layer parallel, less than 3 m thick, and are localized in thinly bedded, fine grained rock which is subordinate to thick-bedded, coarse grained sandstone. The shear records little displacement and may postdate the isoclinal folding.

A brittle zone is marked by open folds and steep faults that deform the Cariboo Group and the upper part of the Kaza Group in northern Cariboo Subterrane. Open folds comprise two sets; the earlier may be the same system of folds as the west-verging isoclines of the ductile zone. The two sets are northwest-trending, concentric and have steep axial surfaces the steepest of which is superimposed on the slightly asymmetric earlier folds. The first order asymmetric folds have a wavelength of 4 km and amplitude of 1.5 km. The superimposed folds are regional and include the Premier Anticlinorium and Isaac Lake Synclinorium. Metamorphic minerals, chlorite and white mica, are post-kinematic relative to the westerly verging symmetric folds.

Faults of the brittle zone crosscut and parallel the northwest trend of the folding. Cross faults generally dip steeply and offset the strike faults. Strike faults dip moderately to steeply to the northeast and southwest and are normal and reverse. Dip-slip displacement is generally less than 4 km and for the regional faults averages 1 km.

In the transition zone between the southeastern ductile structures and the brittle ones to the northwest the characteristics of flow diminish as the metamorphic grade decreases but persists in low viscosity rock such as shale. As a result folds change form from recumbent isoclines to upright and open and, with the decrease in flow, become more upright. The folds are increasingly brittle and have faulted limbs to the northwest. Layer parallel shear faults are most common within the transition zone and may reflect viscosity differences due to changes in rock type. They formed after isoclinal folding and therefore may be related to later folding recorded by the Premier Anticlinorium.

Folds including the Premier Anticlinorium and the Isaac Lake Synclinorium and the northeasterly trending crenulations affect the ductile and brittle zones in the same way and therefore are assumed to be superimposed upon them after low viscosity conditions in the ductile zone diminished. The brittle conditions throughout imply an uplifted and cold terrane.

Kootenay Terrane: Barkerville Subterrane

Rock flow, recorded by isoclinal to tight folds and shear zones, pervades rocks of the Barkerville Subterrane. Flow by layer parallel shear is characteristic and is in contrast to the Cariboo Subterrane where flow was mainly restricted to folds. Two or more sets of folds and related cleavages were formed during conditions of ductile flow; southwestward directed folds are superimposed on one or two northeastward directed structures. The folds are recumbent to asymmetrical, isoclinal to tight and have shallow to moderately dipping axial surfaces. Amplitude and wavelength are generally less than several kilometres and in the more common minor folds are from 10 to 50 m. The plunge of the southwestward directed folds is mainly to the northwest and less so to the southeast. Earlier northeastward directed folds, however, have a more variable plunge. Inconsistency of plunge is thought to be due to differential flow rates throughout the rock mass as evidenced by disparate trends in folds with the same characteristics in a single outcrop. Evidence for earlier folds being coaxial with later ones is the lack of interference patterns and the inability of the plunges of earlier folds to define a great circle surface (planar surface) of flow.

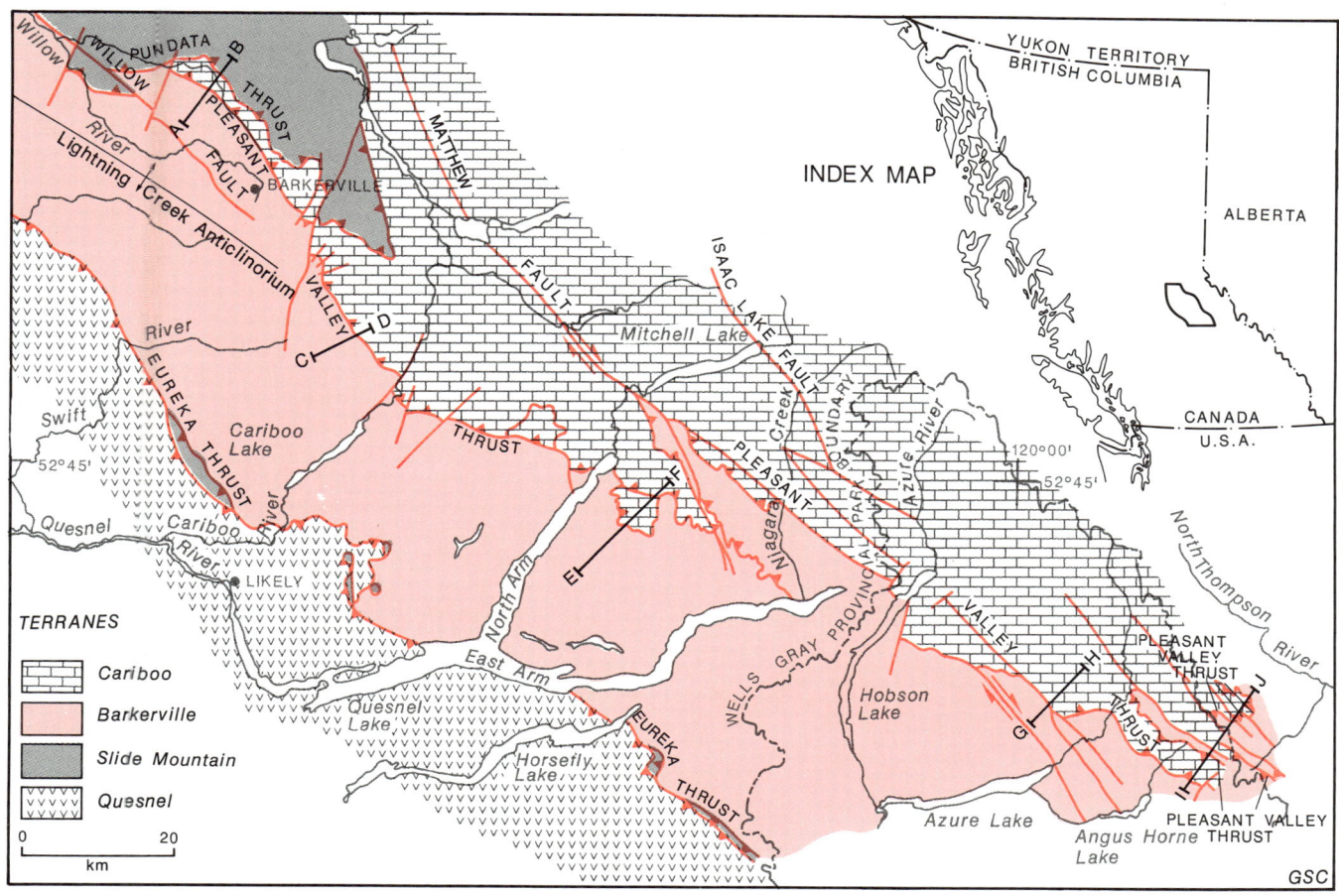

Figure 17.36. Distribution of terranes and regional faults (as well as the locations of structural cross-sections in Fig. 17.37) in the Cariboo Mountains.

Shear evidenced by cataclastic rocks is localized along the western boundary with the Crooked Amphibolite of the Slide Mountain Terrane and near the Island Mountain amphibolite of the Barkerville Subterrane. Mylonite at the boundary with these rocks is gradational down section to variably cataclastic quartzose rock for distances generally less than 500 m. Most of the terrane has zones of shear producing rootless folds, disrupted layering and sigmoidal cleavage wrapping around augen-like lenses of formerly continuous layering. In amphibolite facies shear is displayed mainly by finely and evenly crystalline quartzose rock in even, thin layering. That shear persisted throughout deformation, continuously or sporadically, is shown by disruption of young features such as quartz veins and overprinting of shear structures by folds and metamorphic minerals.

Folds and faults characteristic of brittle conditions are superimposed on ductile structures throughout the terrane. Open folds with upright cleavage, crinkles, kinks and post-metamorphic faults compose the brittle structures which are akin to those of the brittle regime of the Cariboo Subterrane. One set of open folds has moderate to upright axial-surface cleavage, regionally forming gently arching anticlinoria and synclinoria that postdate the crystallizing of metamorphic minerals. An example is the Lightning Creek Anticlinorium.

Faults of the brittle zone are shallow and steeply dipping; the earliest parallel the northwest regional fold trend and the later crosscut it. Rare strike faults are generally moderate to steeply dipping and are normal or reverse. Low-angle faults are primarily post-metamorphic slides related to extension, probably in the Tertiary. An exception is a post-metamorphic easterly directed thrust near Cariboo Lake. Cross faults are related to right-lateral strike slip or Tertiary extension. Those related to extension are common south of the North Arm of Quesnel Lake where they dip moderately to the southeast and are invariably normal. Folds, crenulations and kinks parallel the extension faults in this area and they are assumed to be related. These may be the equivalent of the northeast-trending crenulations described from the Cariboo Subterrane.

Ancestral North America and Kootenay Terrane

Northeastern Columbia Mountains

P.S. Simony

The ancestral North American strata of the northern Columbia Mountains (Fig. 17.30) exhibit a wide range of structural styles. Because of a gentle west-northwest plunge

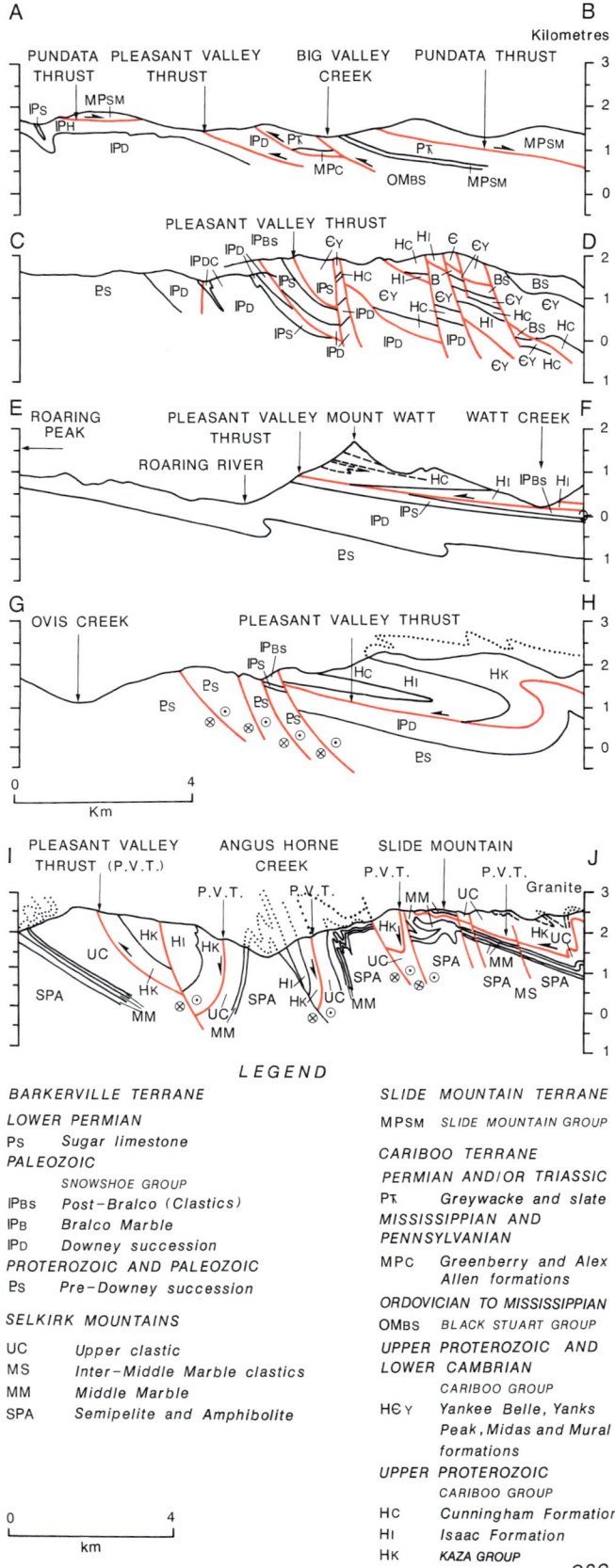

Figure 17.37. Structural cross-sections in the western part of the Cariboo Mountains showing dominant westward verging style (see Fig. 17.36 for locations).

in the Cariboo Mountains and a regional southeast plunge in the Selkirk Mountains and northern Purcell Mountains, structures are seen through a crustal depth range of more than 15 km. A broad range of lithologies, from marble and pelitic schist to quartzite, conglomerate and gneiss is involved and the structures were formed in several superimposed sets. The lowest stratigraphic level and the deepest structures occur in a culmination in the northern Monashee Mountains (the Malton Gneiss).

In the western part of the area large Phase 1, southwest-verging nappes are refolded by two sets of later folds. Basement gneisses (Malton Gneiss) are involved in the early, as well as later, folding and thrust faulting. Large fan folds and associated cleavage fans formed during Phase 2 such that belts of southwest vergence alternate with those showing northeast vergence. In general, southwest vergence is dominant on the southwest side of the region. The large southwest-verging Phase 1 fold nappes developed with the onset of metamorphism. Phase 2 folds are coeval with the metamorphic climax and may be Middle to Late Jurassic in age whereas the Phase 3 folds, cut by the pre-mid-Cretaceous Purcell Thrust Fault, could be of Early Cretaceous age. Normal and oblique slip faults, probably associated with Eocene extension, cut the area into a number of blocks and, in part, influenced the position of major valleys such as the Southern Rocky Mountain Trench.

Phase 1 nappes

The best example of the large southwest-verging nappes is the Scrip Nappe which has the greatest amplitude of any fold known in the Canadian Cordillera (Fig. 17.30, 17.40; Raeside and Simony, 1983; Simony et al., 1980). Its inverted limb is some 50 km long as outlined by a marble marker within the Windermere Supergroup. It is thus a recumbent fold of large amplitude and short (some 5 km) wavelength that must have involved much lengthening of its limbs and flattening of its fold core during intense, ductile deformation. Metamorphism during deformation apparently was at grades no higher than garnet.

Near Mica Creek in the northern Selkirk Mountains minor structures and fabrics associated with the Phase 1 Scrip Nappe are identified only locally because of the intense Phase 2 overprint. To the northwest at a higher structural level in the Cariboo Mountains, Phase 1 minor folds and early schistosity are more easily identified. Parasitic folds of a Phase 1 nappe above the Scrip Nappe on the scale of a kilometre are outlined by a Windermere Supergroup marble (Fig. 17.41). There, Phase 1 major and minor folds, as well as Phase 1 schistosity, disappear upwards within the higher units of the Upper Proterozoic sequence (Pell and Simony, 1982, 1984). Basement gneiss, exposed in the Malton Range at the north end of the Monashee Mountains, also is involved in large, southwest-verging, Phase 1 folds (Morrison, 1980, 1982). No Phase 1 folds have been identified east of a zone of thrust faults marking the northeast flank of the Selkirk and Monashee mountains and all these relationships indicate that the Phase 1 nappes were restricted to the deeper parts of the core zone.

STRUCTURAL STYLES

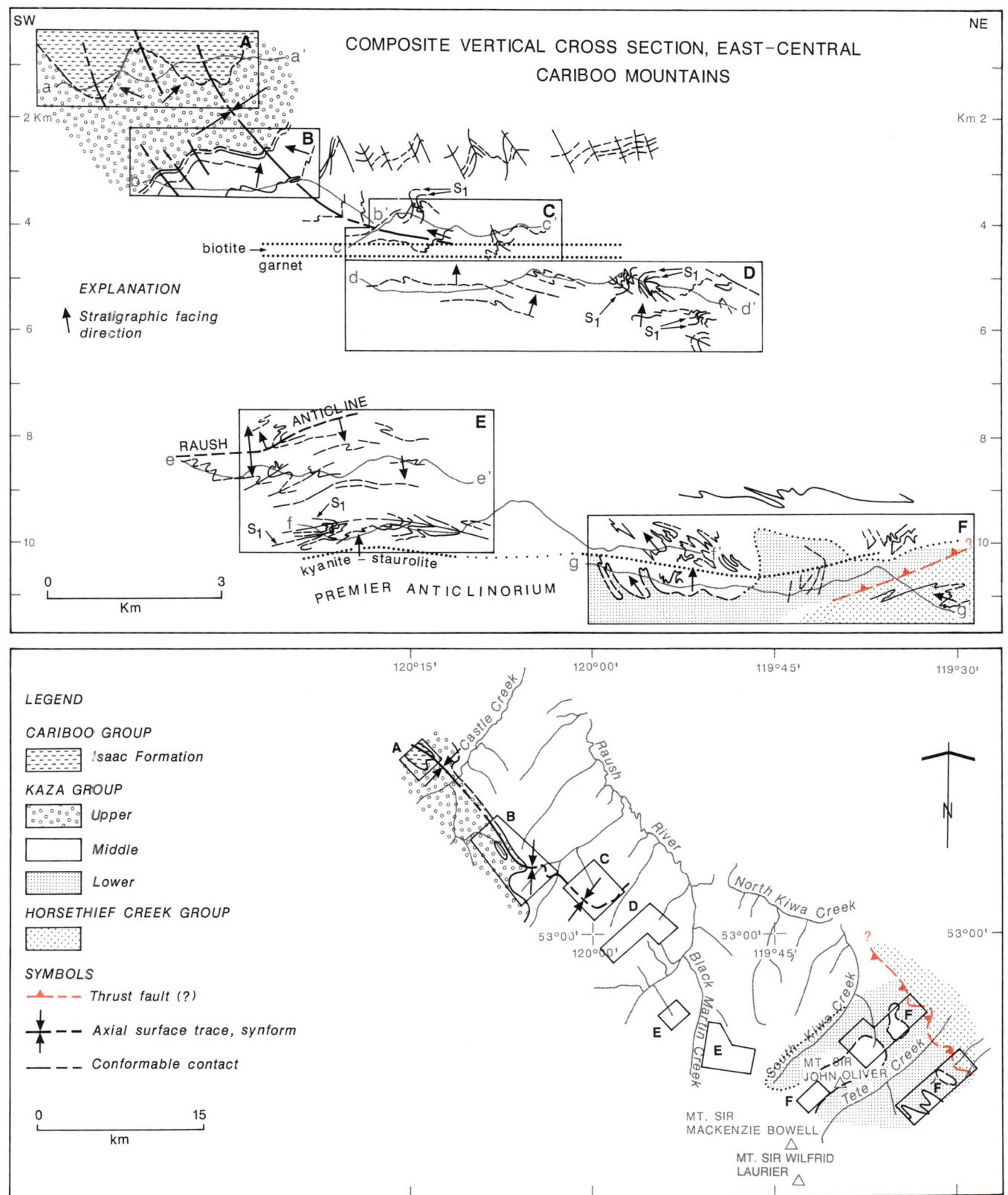

Figure 17.38. Structural cross-section in the central Cariboo Mountains showing the progressive change from recumbent, ductile structures in high grade metamorphic rocks at low structural levels to upright folds in low grade metamorphic rocks at high structural levels (after Murphy, 1987).

Figure 17.39. View westerly to isoclinally folded carbonate and quartzite layers in sillimanite gneiss in the southern part of the Cariboo Mountains. The style of folding is characteristic of early fold phases in metamorphic rocks generally included in the Shuswap Metamorphic Complex. Photo by R.B. Campbell. (GSC 205371)

Phase 2 folds and thrust faults

Phase 2 folds and related thrust faults, roughly coeval with the regional metamorphic climax, are the most widespread in the area here considered. Those in the hanging wall of the Purcell Thrust Fault verge northeast but farther southwest they comprise fans of cleavage folds in the Cariboo Mountains (Fig. 17.41; Campbell, 1968) and in the Selkirk Mountains (Fig. 17.1, Section L; Wheeler, 1963). West of the zone of fans, the Phase 2 structures verge southwest (Brown and Read, 1983).

Phase 2 folds vary in structural style with depth and metamorphic grade. Near Mica Creek, in the northern Selkirks, strata in upper amphibolite facies were buried 12 km deeper than greenschist facies rocks near the headwaters of the North Thompson River in the Cariboo Mountains. Phase 2 folds in the Selkirks are tight to nearly isoclinal with amplitudes of 4 to 6 km and wavelengths of 2 to 4 km (Fig. 17.40). A strongly developed axial-planar schistosity is associated with them (Raeside and Simony, 1983) and all earlier fabric elements are largely destroyed. As the Phase 2 folds are traced to higher structural levels they become more open, their amplitude decreases and their hinges become rounder. Their axial plane schistosity grades into strain-slip cleavage where they are superimposed on earlier schistosity. Where such earlier schistosity is lacking, Phase 2 schistosity grades into slaty cleavage. Broad upright folds are formed with associated rounded and angular mesoscopic folds, as well as kink bands, in well bedded, competent formations (Fig. 17.41; Pell and Simony, 1982).

Phase 2 folds also can be traced southward from the northern Monashees, along the eastern flank of the Selkirks and into the northern Purcell Mountains near Golden. The Proterozoic and Paleozoic facies, typical of the Selkirk Mountains, are carried northeastward on thrust faults such as the Esplanade Thrust Fault (Fig. 17.42) which may merge at depth with the Monashee Décollement. The Esplanade Thrust Fault was tightly folded during metamorphism but prior to the metamorphic climax. The tight folds are therefore interpreted as Phase 2 folds and the thrust fault as an early Phase 2 structure. Phase 1 structures are restricted to the region west of, and structurally above, the Esplanade Thrust Fault.

The tightly folded Esplanade Thrust Fault and the train of folds, thrust faults and folded thrust faults which constitute the northern Purcells, form the transition from the ductile, fold-dominated core zone to the thrust fault-dominated and less ductile Foreland Belt. Neither the Rocky Mountain Trench, nor the Purcell Thrust Fault form a southwest boundary to the Foreland Belt.

Metamorphism in the northern Purcells was in low grade greenschist facies during Phase 2 deformation, there the earliest phase, and gave rise to penetrative, slaty cleavage in argillaceous rocks. The stratigraphic thicknesses, regional structures and metamorphic mineral assemblages suggest that the rocks were buried to depths of 10-12 km (Simony and Wind, 1970). The large scale structural style is dominated by buckle folds and thrust faults particularly in the 1000 m-thick Lower Cambrian quartzite but some of the folds, including thrust faults, are very tight (Fig. 17.43). This implies considerable flattening.

The complex of thrust sheets constituting the northeastern part of the Purcell Mountains in the hanging wall of the Purcell Thrust Fault, west of Golden, forms a lens bounded on its western and upper side by a thrust fault with which the lower thrust faults merge. This is the Dogtooth Duplex (Fig. 17.44) which formed where a major thrust fault ramped from a detachment within the Windermere strata up to one above the Lower Cambrian quartzite. The Purcell Thrust Fault does not floor the duplex but later truncated the duplex and raised its western, deeper part to the present erosion level.

Synmetamorphic, small, tight folds and associated penetrative cleavage and schistosity on the western flank of the Rockies are correlated with Phase 2 folds west of the Rocky Mountain Trench. A number of thrust faults, which cut through the stratigraphy at a small angle and which are deformed by later structures, also are considered to belong to Phase 2. These structures do not extend eastward beyond the southwest flank of the Rocky Mountains.

Phase 3 folds and thrust faults

In the northern Selkirks and northeastern Monashees, large northeast-verging open folds such as the Mica Dam Antiform (Fig. 17.40; Spang et al., 1980) are clearly postmetamorphic and refold the Phase 2 structures. These Phase 3 folds generally have broad, rounded hinges with amplitude to wavelength ratios lower than those of the typical Phase 2 and Phase 1 folds. Their fold style is, however, highly variable ranging from kink folds, to parallel curved, to similar, as illustrated by parasitic folds in the core of the Mica Dam Antiform (Fig. 17.45) in the northern Selkirks. Mesoscopic folds and crenulation cleavage are widespread but large mappable Phase 3 folds are virtually confined to the northeastern Selkirks.

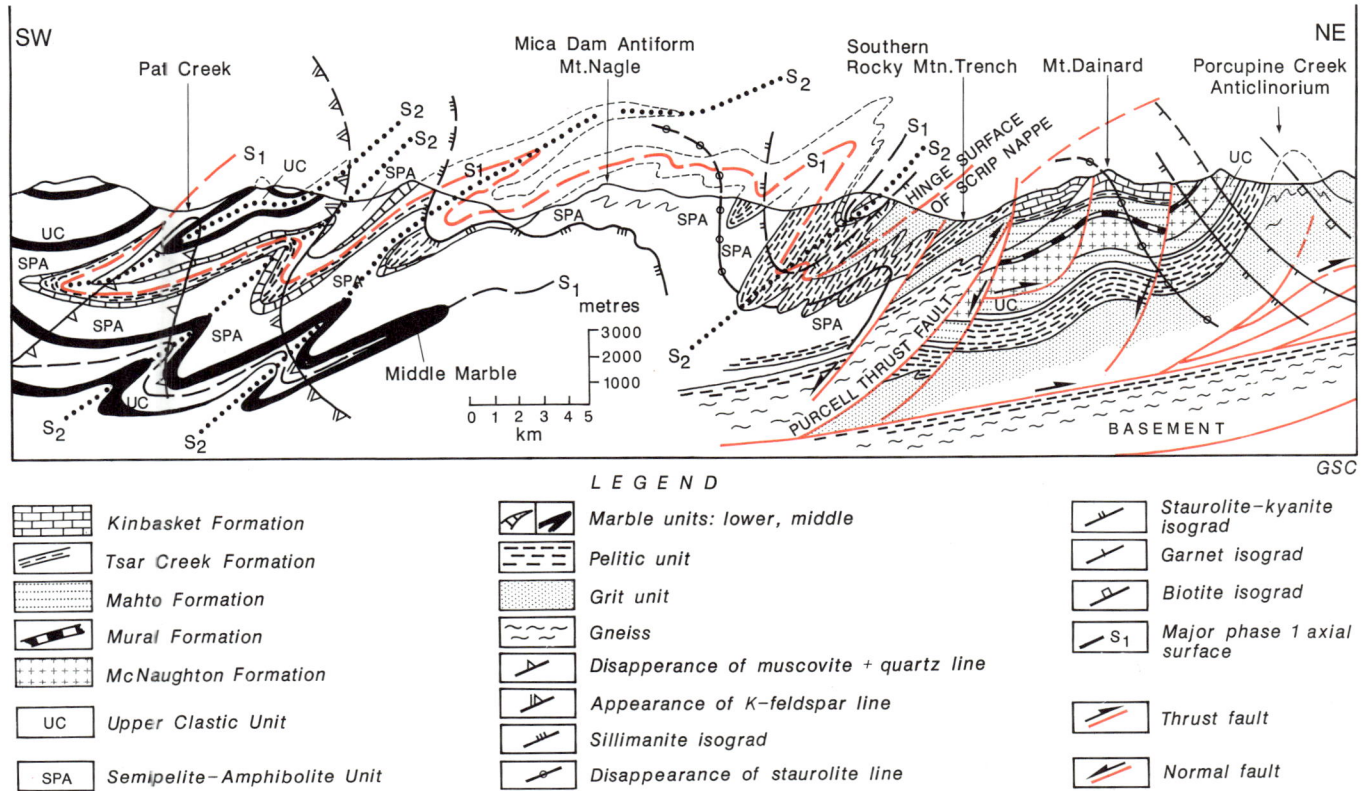

Figure 17.40. Southwest-northeast structural cross-section from the northern Selkirk to the western Rocky Mountains showing the west-verging phase 1 Scrip Nappe refolded by phase 2 and phase 3 folds and carried by the Purcell Thrust Fault over the simpler structures of the Porcupine Creek Anticlinorium. See Figure 17.30 for location.

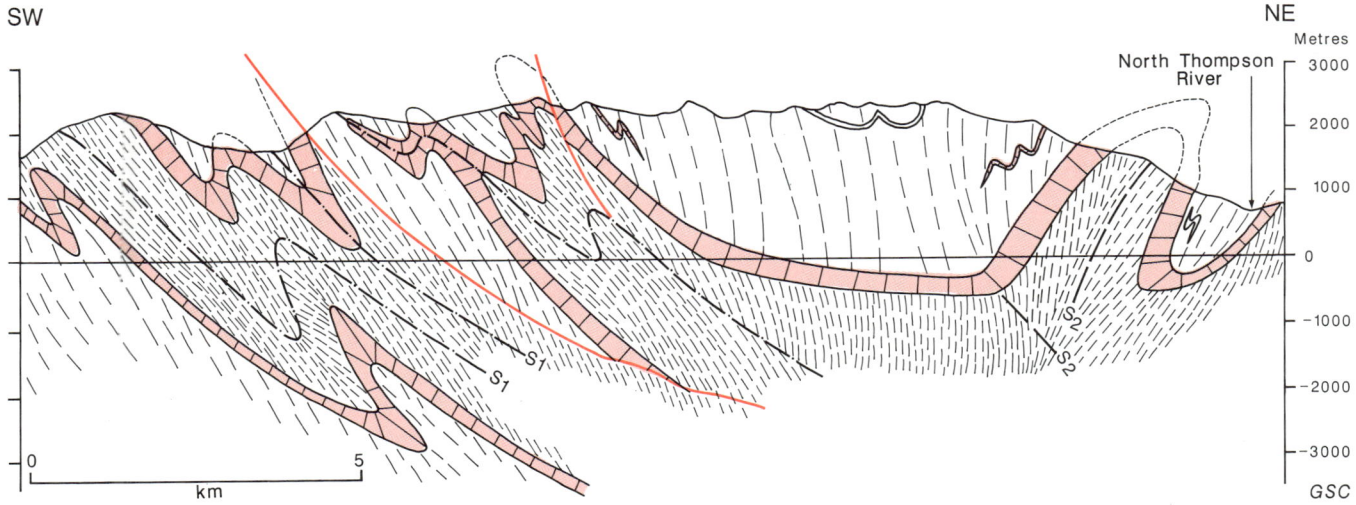

Figure 17.41. Style of phase 1 and phase 2 folds outlined by Windermere marble (pink) above the Scrip Nappe near the headwaters of the North Thompson River, Cariboo Mountains. See Figure 17.30 for location.

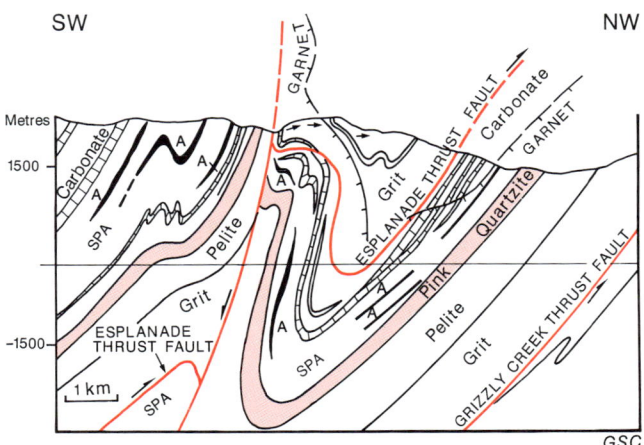

Figure 17.42. Structural cross-section through the Esplanade Range, eastern Selkirk Mountains showing relation of garnet isograd to folded Esplanade Thrust Fault. Brick symbol, marble bands of carbonate division; SPA, semipelite-amphibolite; A, thick amphibolite sheets in SPA.

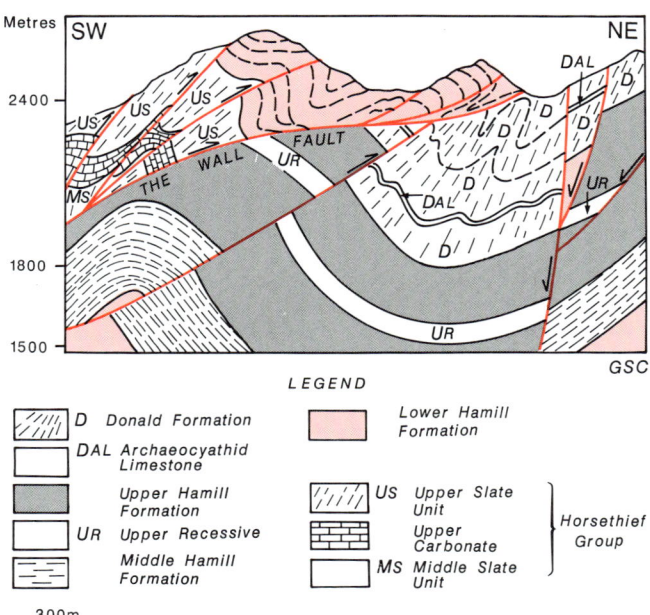

Figure 17.43. Typical east-verging structures in the Dogtooth Range of northern Purcell Mountains.

Implications

A complete gradation of styles and associations of structures can be documented from the core zone of the Omineca Belt to the Foreland Belt. No structural or metamorphic boundary can be singled out as separating them and a deformational sequence which links them can be identified (Simony et al., 1980; Brown and Read, 1983). That sequence is consistent with the west-to-east progression of thrusting in the eastern Rockies. The pattern, on the Tectonic Assemblage Map (Map 1712A, in pocket), of the eastern boundary of accreted and obducted terranes

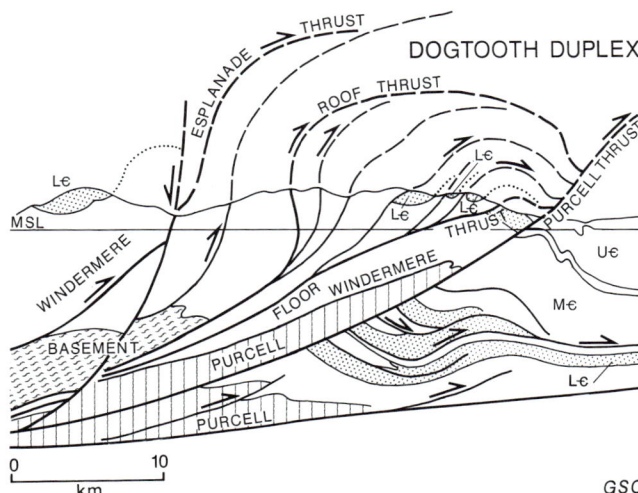

Figure 17.44. Structural cross section across northwest Dogtooth Range from the eastern flank of the Selkirk Mountains to the Southern Rocky Mountain Trench. East-verging thrust stack of the Dogtooth Duplex is cut by the Purcell Thrust Fault. The stratigraphic succession involves Proterozoic (Purcell and Windermere) and Cambrian strata (LC, MC, UC). See Figure 17.30 for location.

indicates that the obducted mass overlapped much of the core zone in the form of an eastward tapering wedge contributing perhaps 2-4 kb (2-4 x 10^5 kPa) load pressure. This, in conjunction with the great thickness of the Proterozoic and Paleozoic miogeoclinal stratigraphy and its duplication by early structures, could explain the high ductility of the rocks implied by their structural style as well as by the high pressures obtained from geobarometry (Ghent et al., 1982).

Selkirk Allochthon

R.L. Brown, J.M. Journeay, and L.S. Lane

The Selkirk Allochthon, enveloping the Monashee Complex and extending east to the Southern Rocky Mountain Trench (Fig. 17.30, 17.46A), is a composite, thick-skinned thrust nappe about 25 km thick which includes strata of the Kootenay Terrane and Ancestral North America. The earliest recognized structures are recumbent isoclinal folds in the Kootenay Arc. These structures predate an erosional unconformity of Devono-Mississippian age (Read and Wheeler, 1976) and are presently overturned to the west. They are interpreted to have initially been eastward-verging structures that were back-folded during younger generations of Mesozoic deformation (Høy, 1979). Eastward-verging isoclinal folds of equivalent magnitude, but uncertain age, are recognized in the central and northern Selkirk Mountains. At least some of the structures were generated by shear strain associated with the imbricate stacking of tectonic slivers within the Selkirk Allochthon in Early Jurassic time (Brown and Read, 1983).

Early isoclinal folds at all structural levels of the Selkirk Allochthon are refolded by westward and eastward- verging, pre- to syn-metamorphic folds. These structures become increasingly more overturned and attenuated with depth (Brown, 1978; Read and Brown, 1981;

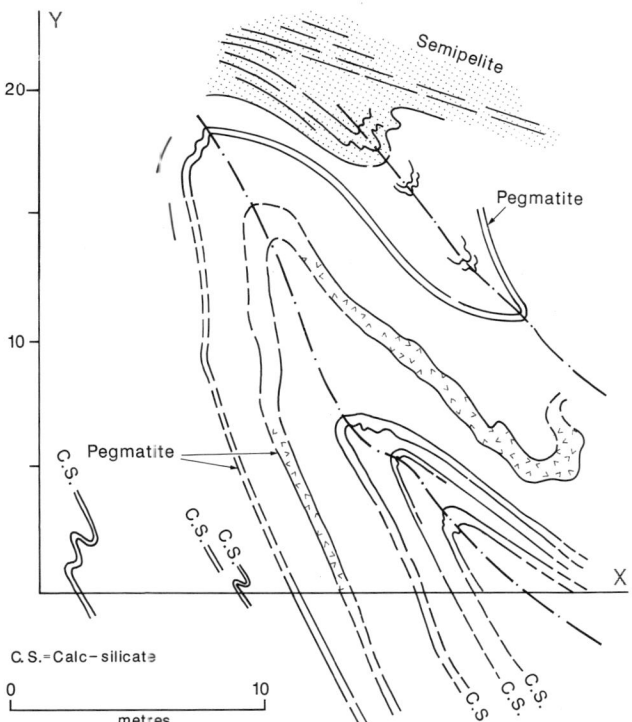

Figure 17.45. Phase 3 folds, view southeast to Mica Dam Antiform.

Murphy, 1987), and define a structural fan whose axis lies along the eastern margin of the Selkirk Allochthon (Wheeler, 1963; Franzen, 1972; Brown and Tippett, 1978; Perkins, 1983; Fig. 17.46B,C). Westward-verging folds are by far the dominant structural element within the allochthon. They are developed exclusively in the upper plate of the Monashee Décollement, and reflect an episode of backfolding and crustal thickening which predates the detachment and eastward displacement of the Selkirk Allochthon (Brown and Read, 1983; Brown et al., 1986; Murphy, 1987).

The widespread occurrence of late or post-metamorphic, eastward-verging folds in both upper and lower plates of the Monashee Décollement suggests that the Selkirk Allochthon and underlying Monashee Complex may have experienced similar histories of shear strain during uplift and eastward telescoping of the southern Omineca Belt in the Mesozoic to early Cenozoic.

Regional metamorphic isograds within the allochthon are inferred to have a pre-fold geometry characterized by a series of northward-plunging thermal culminations in the Kootenay Arc (Kootenay Lake High, Høy, 1976; Kuskanax-Clachnacudainn High, Wheeler, 1965; Leatherbarrow, 1981) and along the axis of the Monashee Mountains (Shuswap Complex, Wheeler, 1965; Reesor and Moore, 1971). These isograds are obliquely truncated along the base of the Selkirk Allochthon by the Monashee Décollement, and at higher structural levels by the Purcell Thrust Fault, along which there is an estimated 7 km of post-metamorphic dip-slip separation. The net displacement along this fault is inferred to be much greater, and depends not only upon the subsurface dip of the fault, but also on the magnitude of displacement prior to the quenching of high-pressure isograds in Middle Jurassic time. The present juxtaposition of high and low grade segments of the Selkirk Allochthon is believed to be the result of both tectonic denudation and high-angle block faulting associated with the uplift and arching of the Monashee Complex, and subsequent attenuation of the upper lithosphere in early Cenozoic time.

Kootenay Arc

J.T. Fyles and T. Høy

The Kootenay Arc is a north-trending arcuate structural zone characterized by intense polyphase deformation and local high grade regional metamorphism. It extends from near Revelstoke on the Trans-Canada Highway, southeast and south along the Lardeau Valley and Kootenay Lake, and south and southwest through Salmo into northern Washington (Fig. 17.30). Syn-kinematic and post-tectonic batholiths and stocks are prominent throughout the Kootenay Arc. The eastern edge of the arc merges with the Purcell Anticlinorium, a broad north-plunging structure in less intensely deformed Precambrian rocks. The western edge is defined by the Columbia River Fault Zone in the north, by the margins of the Nelson and Kuskanax batholiths in the centre, and by Mesozoic volcanic rocks in the south. Structures trending northwestward into the area between the Nelson and Kuskanax batholiths are discussed also because they provide critical information on timing of structural events in the Kootenay Arc.

The structures (Fig. 17.47) are developed in strata belonging partly to ancestral North America and partly to the Kootenay Terrane. The former include quartzite of the Hamill Group and Quartzite Range Formation, the distinctive Lower Cambrian carbonate (Badshot and Reeves limestone), and the lower pelitic part of the overlying Lardeau Group. The latter includes siliceous argillite, chert, mafic volcanics and grit of the upper part of the Lardeau Group and the unconformably overlying Upper Mississippian-Lower Pennsylvanian Milford Group. It comprises limestone, siltstone, sandstone, chert, mafic volcanics and conglomerate.

In general, the oldest folds in the arc are tight to isoclinal with axial planes trending parallel with the arc and dips changing from upright to recumbent depending on subsequent deformation. These folds include both Phase 1 and Phase 2 structures and culminate in the central part of the arc along Kootenay Lake. Phase 3 folds also have low plunges approximately coaxial with the earlier folds and generally are west-verging with west-dipping axial planes. A fourth phase of folding is developed as local crenulations or as broad warps in the regional strike, the orientations of which appear to be related to the curvature of the arc.

Faults are associated with many of the folds. Attenuated limbs and bedding-plane faults are typical of Phase 1; slip planes on axial-plane cleavage are typical of Phase 2. Cross faults, particularly in the southern part of the arc may be associated with Phase 4. Many normal, west-dipping strike faults along the western side of Kootenay Lake, and elsewhere, are probably Tertiary extension structures.

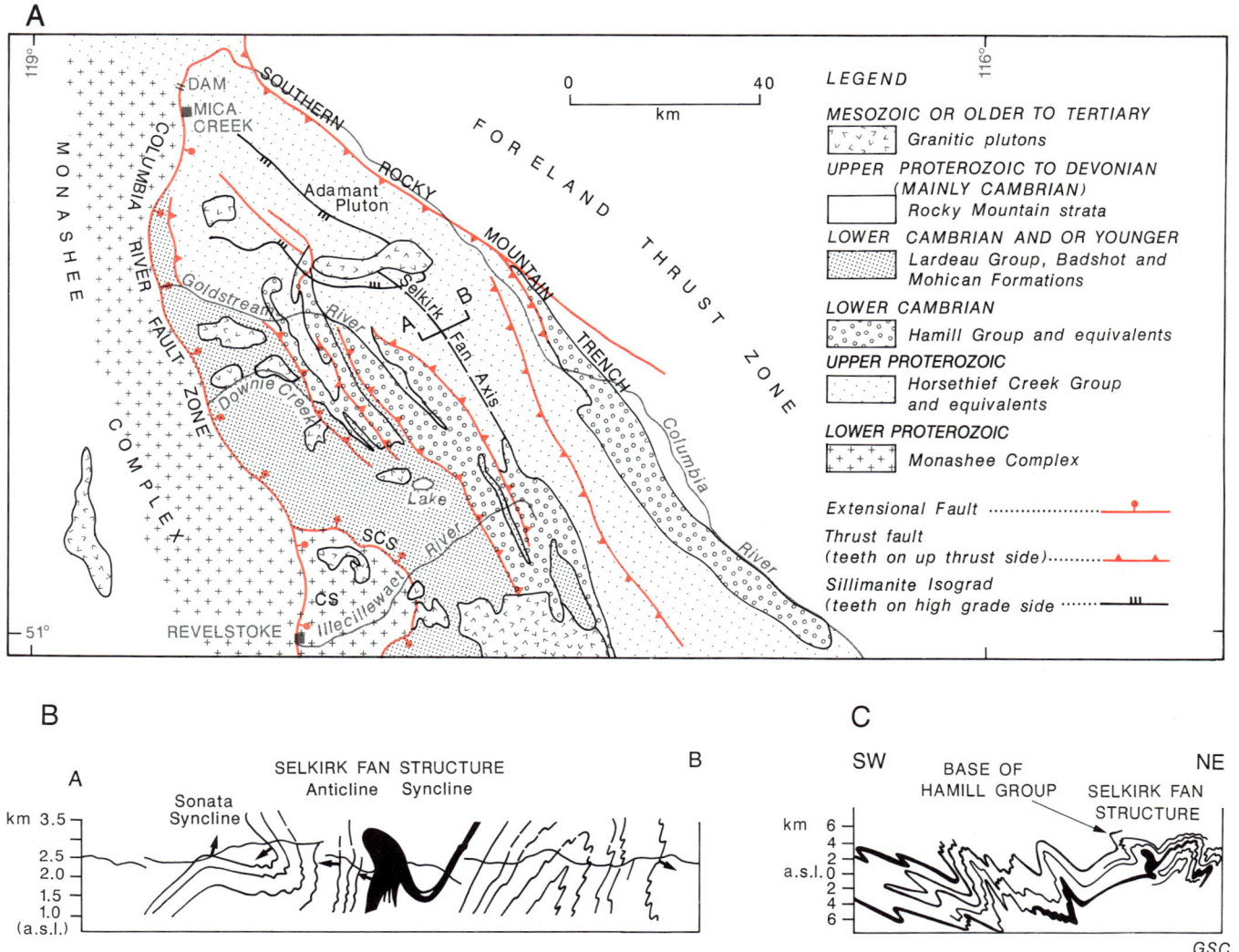

Figure 17.46. (**A**) General geology and main structures of the Northern Selkirk Mountains. CS-Clachnacudainn Salient, SCS-Standfast Creek Slide Fault, (**B**) Detailed structural cross-section of the Selkirk Fan Structure. After Brown and Tippett (1978), (**C**) Generalized structural styles across the northern Selkirk Mountains. After Brown and Tippett (1978).

Phase 1 folds

An early pre-late Paleozoic deformation is well documented in the central and northern parts of the Kootenay Arc. Small-scale structures include a penetrative mineral foliation recognized in the lower to middle Paleozoic Lardeau Group but which is absent in unconformably overlying Carboniferous Milford Group (Read, 1971; Klepacki and Wheeler, 1985). At a number of locations, conglomerate at the base of the Milford Group contains clasts of the underlying Lardeau Group in which the earliest foliation differs in orientation from clast to clast and locally from a later second phase foliation in the matrix (Wheeler, 1968; Read, 1975; Read and Wheeler, 1976). In addition, in the north-central part of the arc, rocks of the Lardeau Group, in an inverted limb of a recumbent Phase 1 syncline, are unconformably overlain by the Milford Group (Read, 1976).

Large early isoclinal folds recognized elsewhere in the arc are also probably pre-late Paleozoic structures. In the Akolkolex area near Revelstoke at the north end of the arc, a recumbent anticline-syncline pair with a fold amplitude of 8 to 9 km overlies the gentle southeast-dipping Standfast Creek Slide Fault (Fig. 17.48A; Thompson, 1978). A penetrative mineral foliation parallels the axial planes of these folds. They are the earliest structures recognized and are overprinted by upright, more open folds with an axial-planar fracture cleavage.

The oldest structures recognized to the southeast through the Lardeau area to Kootenay Lake in Hamill and Lardeau Group rocks are isoclinal folds with curving but generally upright axial planes (Fyles and Eastwood, 1962; Fig. 17.47, upper right). At the north end of Kootenay Lake they become recumbent by having being folded by structures with steep west-dipping axial planes (Fyles, 1964; Fig. 17.47, lower right). Southward along the lake folds of the two phases are almost isoclinal (Fig. 17.48B; Fyles, 1967; Høy, 1977, 1980). South of Kootenay Lake these axial planes steepen through vertical and become east-dipping near Salmo and gently southeast-dipping along the International Border (Fig. 17.48C; Fyles and Hewlett,

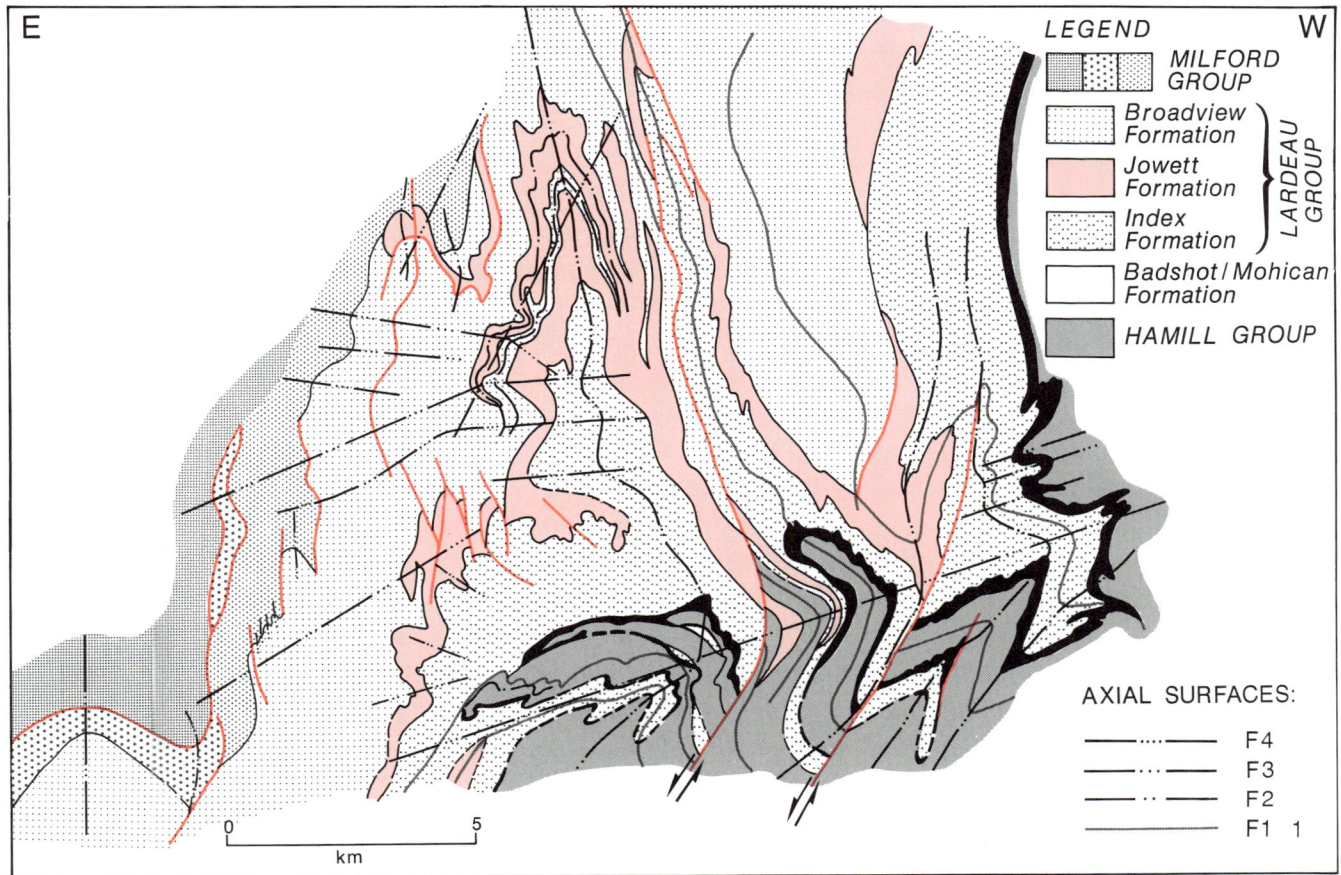

Figure 17.47. Composite tectonic cross-section of the north-central Kootenay Arc showing four phases of deformation (F1-F4). From Fyles (1964), Fyles and Eastwood (1962), Klepacki and Wheeler (1985) and Read (1973).

1959). This reversal in dip, described as a fan structure by LeClair (1983), is accentuated by the abrupt curvature in the arc south of Salmo. Pre-late Paleozoic folds, however, have not been identified in this part of the arc.

Phase 2 folds

Phase 2 folds are well defined in the central and northern parts of the arc where they generally plunge to the north and northwest but locally to the southeast (Klepacki and Wheeler, 1985). They are east-verging upright folds with associated steeply dipping strike faults. Large continuous upright folds in Milford Group rocks have been mapped by Read (1976) and Klepacki and Wheeler (1985) in the mountains east and southeast of the Kuskanax Batholith and at the head of Kootenay Lake (Fig. 17.30). These structures fold the unconformity at the base of the Milford and are cut by post-tectonic Middle Jurassic plutons.

A broad, upright, north-plunging Phase 2 antiform in the Hamill and lower Lardeau groups dominates the structure of the northern end of Kootenay Lake (Fig. 17.48C, bottom centre). South along the lake the fold tightens and its axial plane flattens, dipping at a moderate angle to the west (Fig. 17.48B). Fold limbs are sheared and commonly cut out by reverse faults. Amphibolite facies regional metamorphism accompanied deformation producing a pronounced mineral lineation and penetrative axial-plane schistosity (Høy, 1976). In southern Kootenay Arc, south-plunging isoclinal folds with vertical to east- and southeast-dipping axial planes dominate the structure (Fig. 17.48C). They are faulted and complexly refolded but at a regional scale are eastward verging. Although they are the oldest folds recognized, radiometric studies and field relationships indicate that they are Jurassic (Archibald et al., 1983) and are therefore considered to be Phase 2 structures.

Phase 3 folds

Phase 3 folds are more variable in form and attitude than the earlier folds. In general, they are west-verging, tight to open folds with moderate to gently dipping axial planes. Along Kootenay Lake they are coaxial with the older folds whereas to the northwest they plunge moderately to the south and trend southeast and occur as conjugate sets (Fig. 17.47). In the southern part of the arc, Phase 3 folds plunge gently to the south causing a warping of the axial planes of earlier structures and probably also resulting in the recumbent structures at the western side of the arc (Fig. 17.48C). Phase 3 folds superimposed on Phase 2 isoclines are very tight to isoclinal with moderate plunges to the southwest, and axial planes trending southwest.

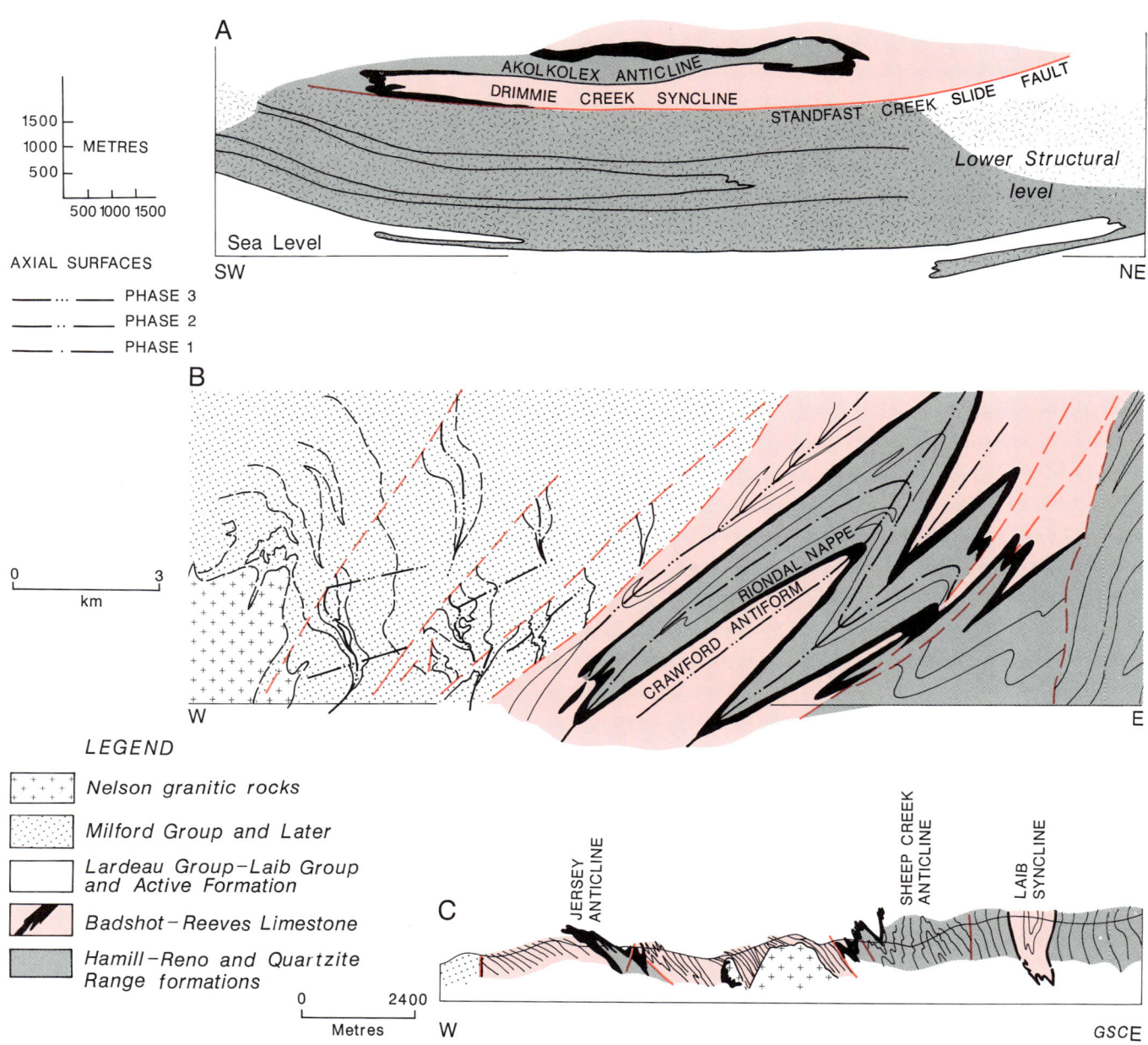

Figure 17.48. Structural cross-sections in the northern (A), central (B) and southern (C) Kootenay Arc. A, after Thompson (1978); B, after Fyles (1967) and Høy (1980); C, after Fyles (1967). See Figure 17.32 for location of A.

Phase 4 folds

Phase 4 folds are developed locally and cannot be correlated from one part of the arc to another although comparable patterns of late structures can be recognized. In the area west of northern Kootenay Lake, south- to southwest-trending Phase 4 folds plunge at steep to moderate angles to the south. They are associated with southeast-trending faults and are thought to be related to the curvature of the arc (Klepacki and Wheeler, 1985). Comparable structures along central Kootenay Lake are open to fairly tight folds that plunge westward down the dip of the earlier foliation. In the southern part of the arc, late, steeply plunging folds are associated with the emplacement of Mesozoic plutons. Southeasterly trending faults with significant right-hand offset, however, may be an expression of Phase 4 structures related to the curvature of the arc.

The structures in a northwest-trending belt of upper Paleozoic and Triassic rocks between Kuskanax and Nelson batholiths merge with those of the Kootenay Arc near Kootenay Lake (Fig. 17.30). Between the batholiths the dominant structure is a regional syncline in strata of the Upper Triassic Slocan Group, complicated by subsidiary folds and faults, which is recumbently overturned to the southwest in its eastern part and opens to a more upright but still asymmetrical structure to the northwest. This fold, recognized by the presence of overturned beds on a large scale, was the first of its type documented in the Canadian Cordillera (Hedley, 1952). Near Kootenay Lake around the northeast end of the Nelson Batholith the axial trace of the syncline swings parallel with the trend of Kootenay Arc structures where the limbs are tightly appressed. Thrust faults and foliation of pre-late Mississippian ages are the oldest recognized structures. Two foliations have been recognized locally in the older rocks (Broadview Formation) one of which is crenulated (Klepacki and Wheeler, 1985). Younger strata show only one cleavage.

The Whitewater Fault (Fig. 17.49) has placed ultramafic and volcanic rocks of the Permian and (?)Carboniferous Kaslo Group onto volcanic rocks of the same sequence and is cut by diorite uncertainly dated as Permian (Klepacki and Wheeler, 1985) but overlain unconformably by Permian-Triassic conglomerate in turn overlain by the Upper Triassic Slocan Group sediments. The principal fold in the area, the Dryden Anticline, deforms the Whitewater Fault and has a well developed axial-planar foliation as do probably related tight, east-verging smaller folds and crenulations. The Dryden Anticline is truncated by Middle Jurassic granitic plutons dated by the U-Pb zircon method at 173 +4/-5 Ma (Parrish and Wheeler, 1983) and 180 ± 7 Ma (Klepacki and Wheeler, 1985). The Stubbs Fault, also folded by the Dryden Anticline is interpreted as an easterly directed thrust fault closely related in development of the Dryden Anticline. Pre-Middle Jurassic normal faulting, down to the west, is demonstrated by the Schroeder Fault which is cut by Middle Jurassic granitic rocks. The fault cuts the Dryden Anticline which apparently formed during the peak of regional metamorphism and is therefore a post-metamorphic structure. The four structures discussed above document important pre- to Middle Jurassic deformational events in the southern Omineca Belt. The structures were later involved in two westward verging fold phases of locally undetermined age.

The structure of the Kootenay Arc records a protracted period of intermittent deformation, metamorphism, and plutonic activities from late Paleozoic to Tertiary time. Pre-late Mississippian and probably post-Ordovician folding represents the earliest significant tectonic event. At least in the upper western part of the arc, thrusting took place between Late Permian and Late Triassic time (Klepacki, 1985) which has not been related to the fold phases described in this review. The major deformation and associated regional metamorphism is Mesozoic and may have resulted from the collision in the Middle Jurassic of composite Slide Mountain and Quesnellia terranes to the west with North American Proterozoic and Paleozoic rocks to the east (Price, 1981a,b; Archibald et al., 1983). In the Late Cretaceous and Early Tertiary, rocks of the Kootenay Arc were transported eastward to their present position during the formation of the Purcell Anticlinorium and Foreland Belt (Price, 1981a,b; Archibald et al., 1984). Northerly trending dyke swarms and normal faults, probably preceded by the intrusion of high level, small alkalic plutons, record a period of extension during the Eocene.

Ancestral North America

Selwyn Basin

S.P. Gordey and R.I. Thompson

Southeast Yukon and the southwest District of Mackenzie comprise two domains of contrasting structural style, the Selwyn and Mackenzie fold belts (Fig. 17.50). The Selwyn belt coincides with Paleozoic and early Mesozoic shaly facies of the outer northern Cordilleran miogeocline and is dominated by thrust faults and open to tight similar folds. Axial-planar slaty cleavage, locally intense, small scale folds, and zones of closely spaced imbricate thrusts are characteristic. Its western boundary is marked by the leading edge of accreted terranes comprising cataclasite, ophiolite and granodiorite emplaced onto the ancient continental margin in the Mesozoic, and by the dextral transcurrent Tintina Fault. The Mackenzie belt to the northeast and east consisting of shallow water carbonate and clastics of Proterozoic to mid-Paleozoic age, lacks slaty cleavage and is characterized by large concentric folds and thrust faults. The two contrasting structural styles reflect the differing responses of incompetent (Selwyn fold belt) versus competent (Mackenzie fold belt) lithologies to largely northeast directed compression. Structural trends vary from north in southeast Yukon to east-northeast in central Yukon to east-west in the northwestern Selwyn fold belt parallel with the arcuate Paleozoic carbonate-shale facies boundary. Deformation was Early Cretaceous, older than emplacement of post-tectonic mid-Cretaceous plutons and coeval with the eruption of the South Fork Volcanics, and younger than locally deformed Lower Cretaceous (Albian?) strata (Blusson, 1971). The Mackenzie fold belt represents a transition in structural style between the Omineca Belt and the Foreland Belt, and is discussed in Part E of this chapter.

Selwyn fold belt structural styles can be grouped into four main types including large-scale thrust faults, imbricate fault zones, chevron folds, and similar folds. Structural vergence may be northward, eastward or westward. In some areas complex internal crumpling and faulting may have uniformly doubled or tripled original stratigraphic thickness, without destroying gross stratigraphic integrity. Four relatively well understood regions, Chandindu, Anvil Range, Macmillan Pass, and Nahanni areas, exemplify the main structural types (Fig. 17.50).

Chandindu area

In the Chandindu region of northwestern Selwyn Basin (Fig. 17.50) the Dawson Fault forms the boundary between the Omineca and Foreland belts. Three structurally complex, gently south-dipping thrust faults have been recognized and named, from north to south, the Dawson, Tombstone and Robert Service faults (Fig. 17.1, in pocket). The Dawson Thrust Sheet is a composite of subsidiary thrust faults and tight folds (upright or north verging; Fig. 17.51). The fault presumably represents a detachment within Windermere argillite and sandstone near a "pseudo basement" of Middle Proterozoic shelf strata. Despite a myriad of folds and thrust faults, few can be mapped because stratigraphic markers are rare.

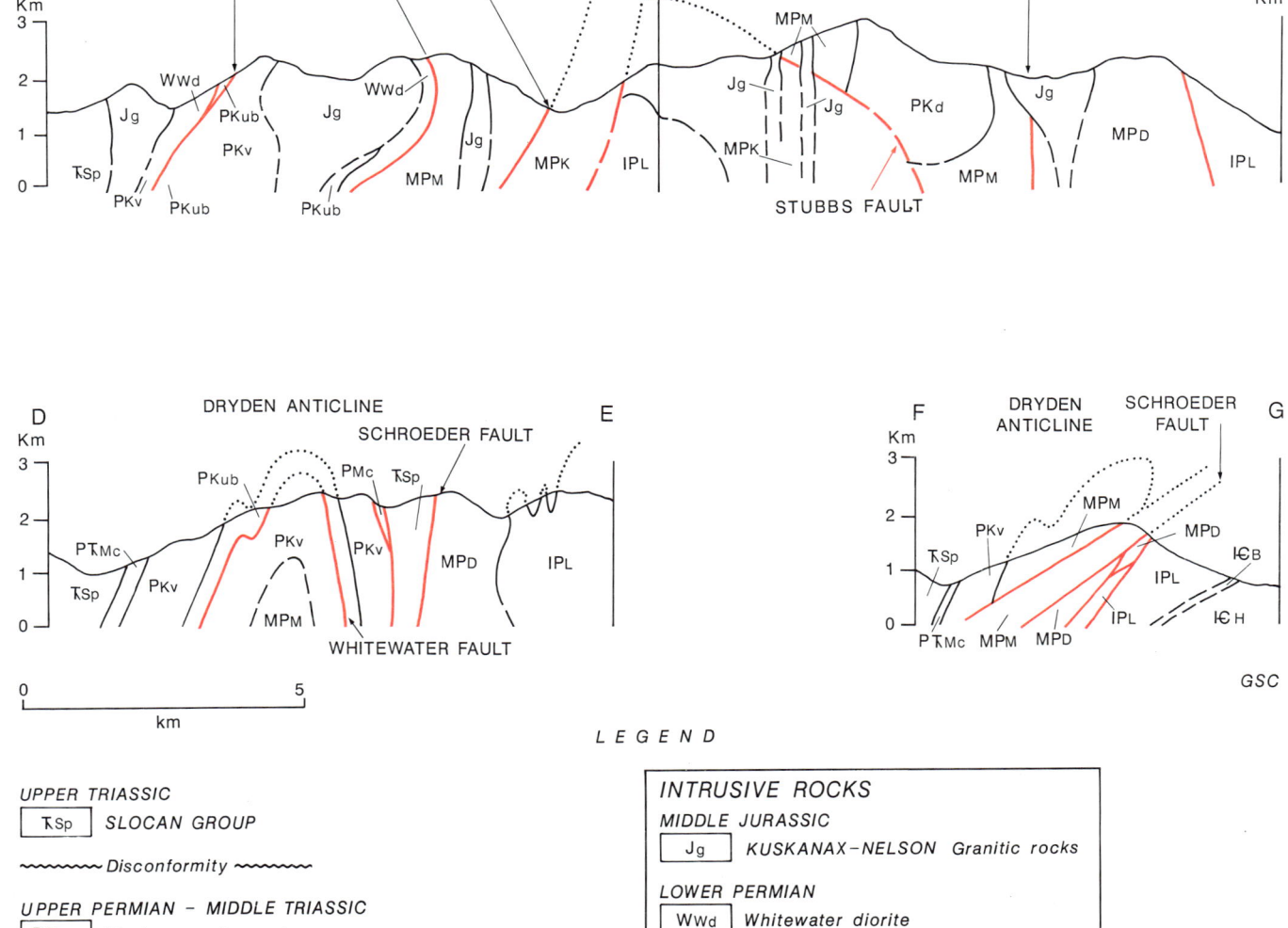

Figure 17.49. Structural cross-sections in the Goat Range area between Nelson and Kuskanax batholiths. After Klepacki and Wheeler, 1985. See Figure 17.32 for location.

STRUCTURAL STYLES

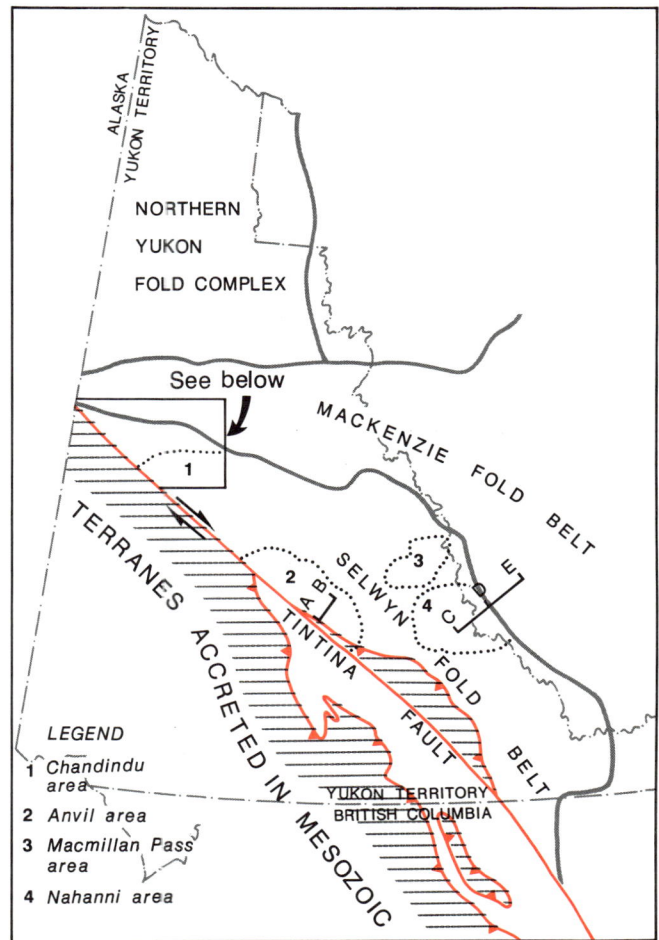

Figure 17.51. View easterly to northward verging refolded isoclinal fold in dolostone of the Upper Proterozoic to Cambrian Hyland Group. The direction of vergence is consistent with northward translation of the Dawson Thrust Sheet in the Southern Ogilvie Mountains. Photo by R.I. Thompson. (GSC 205380)

A narrow border of the Tombstone Thrust Sheet protrudes from beneath the Robert Service thrust sheet. At its northwestern limit, where exposure is best, folds are concentric, close and overturned to the northwest; bedding-parallel thrust faults are numerous (Tempelman-Kluit, 1970). Shortening within the exposed part of the thrust sheet is 50% (10+ km) or more. An important contrast between the Tombstone and Dawson thrust sheets, is the divergence of fold and fault trends between them in the Dawson City area. Trends in the Dawson Thrust Sheet are east-west, those in the Tombstone Thrust Sheet are north-south. As footwall rocks are traced eastward, around the thrust fault and into the large re-entrant near Keno Hill (Fig. 17.50), metamorphic grade increases, and folds are smaller, overturned to recumbent, and rootless; their axes remain oriented northeast-southwest (S60W, McTaggart, 1970).

The Robert Service Thrust Sheet is like the Dawson Thrust Sheet in that complex folds and thrust faults can be mapped locally, thrust faults are shallow, south-dipping and may truncate large open folds (Tempelman-Kluit, 1981a,b).

The exaggerated northward bow in the trace of the Tombstone and Robert Service faults demonstrates a shallow southward dip. Assuming south to north displacement, the two sheets represent at least 100 km of structural overlap and possibly much more. Displacement on the Dawson Thrust Fault is difficult to estimate; internal folding and thrusting probably accounts for 50% or about 20 km of shortening; total displacement of the Dawson Thrust Sheet was at least that much.

Anvil Range

In the Anvil Range and along trend to the northwest structure is dominated by moderate southwest-dipping or flat-lying strata imbricated by several large northwest-trending, northeast-directed thrust faults. The faults have strike lengths of many tens of kilometres, and one has a

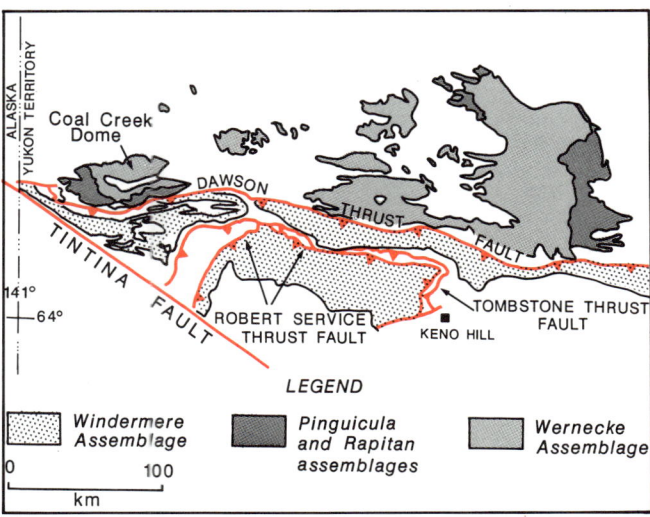

Figure 17.50. Selwyn Fold Belt and adjoining regions with locations of areas and structural cross-sections referred to in text.

627

documented overlap of at least 9-10 km (Fig. 17.52A). Stratigraphic separations probably amount to 400-500 m. Preferred horizons of bedding glide or detachment are at the base of competent Devono-Mississippian chert-pebble conglomerate or at the base of thick Ordovician volcanics.

Macmillan Pass

In the north Macmillan Pass area (Fig. 17.50) a narrow imbricate fault zone of southerly directed east-west trending thrust faults repeats Lower Cambrian to Devonian stratigraphy (Abbott, 1983). Stratigraphic separations may be less than 200 m. South of the imbricate belt open to close folds and steep faults are the dominant structures. Some of the steep faults may have been active in the Devonian, and later exerted control on development of the Mesozoic imbricate belt. Northeast from the imbricate belt structural trend enigmatically diverges or bends northerly about an arc to a northwest- southeast orientation.

In the southwest Macmillan Pass area the structure is dominated by small to intermediate scale chevron folds well-developed in thin-bedded chert of early Paleozoic age (Fig. 17.53B). Although the strata are intensely crumpled across a distance of at least 50 km perpendicular to the northwest structural grain, the stratigraphic level exposed remains nearly constant. Viewed on a large scale the chert succession has been homogeneously shortened and vertically thickened, but not tilted, or imbricated by major faults (Fig. 17.53C).

Immediately north and northwest of the Macmillan Pass area small-scale thrust faulting and isoclinal folding have internally thickened stratigraphic units (Cecile, 1984). These structurally thickened units are themselves folded in open to tight folds.

Nahanni

In western Nahanni area structural style is dominated by imbricate thrust faults (Fig. 17.1, in pocket; Fig. 17.52B, 17.53A) within Silurian siliceous shale, chert and mudstone. In one area, over a distance of 15 km beds dip 40-60° consistently to the northeast yet similarity of graptolite fauna and lithology indicate no change in gross stratigraphic level. Probably not more than 200 m of true stratigraphic thickness is exposed.

In central Nahanni area three different scales of folds seem to dominate in a probable continuum of fold size. Large, upright, close to open folds which control much of the map distribution, have lengths of up to 30-40 km, amplitudes of about 1-2 km, and half-wavelengths varying from 5-9 km (Fig. 17.53D). Hinge zones are subangular to rounded, and limb dips commonly do not exceed 60° although locally they are vertical or overturned. Smaller parasitic or second order folds have strike lengths of up to 13 km, amplitudes of up to 500 m and half-wavelengths up to about 1 km. Their geometry is similar to the larger scale folds, although they may be more appressed. Smaller outcrop scale folds may be isoclinal. All folds are congruent with a northwest-striking axial-planar slaty cleavage, that on average dips steeply to the northeast, congruent with a weak to locally strong southwest fold vergence. The slaty cleavage indicates that beds deformed internally rather than by slip along bedding and that folds are broadly of similar type.

Purcell Anticlinorium

M.E. McMechan

The Purcell Anticlinorium is a large north-plunging, asymmetric box fold cored by Middle Proterozoic (Purcell)

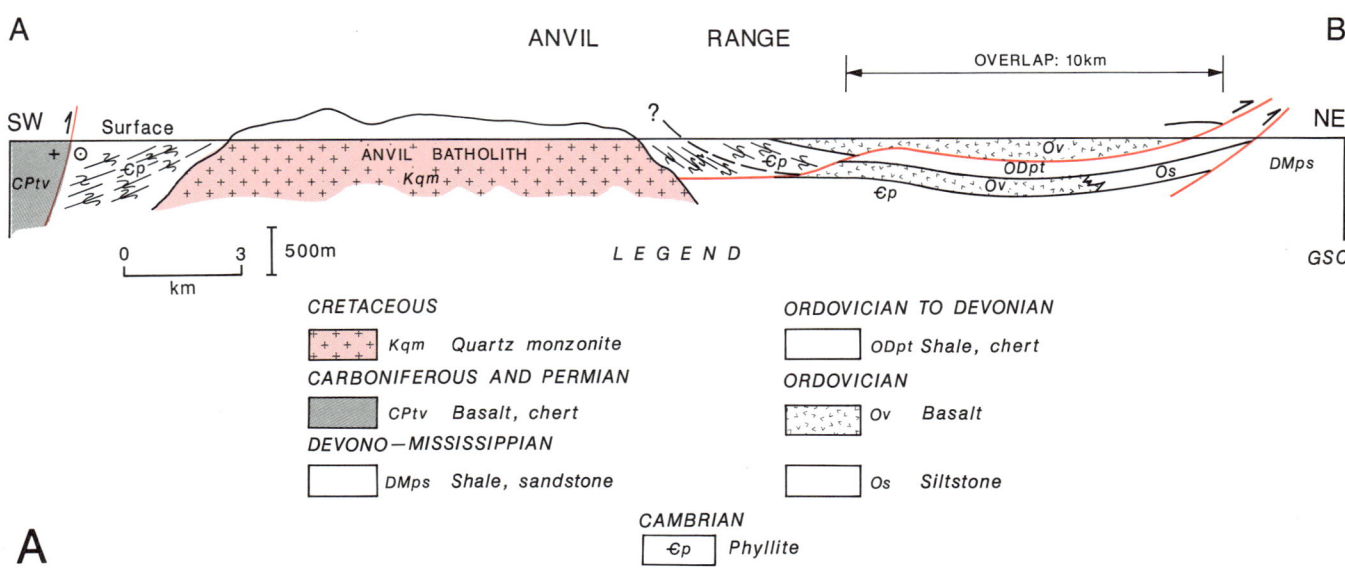

Figure 17.52. Structural sections across: A, the Anvil Range dominated by a gently dipping thrust fault east of the Anvil Batholith and B, across the boundary between Selwyn and Mackenzie fold belts. See Figure 17.50 for location.

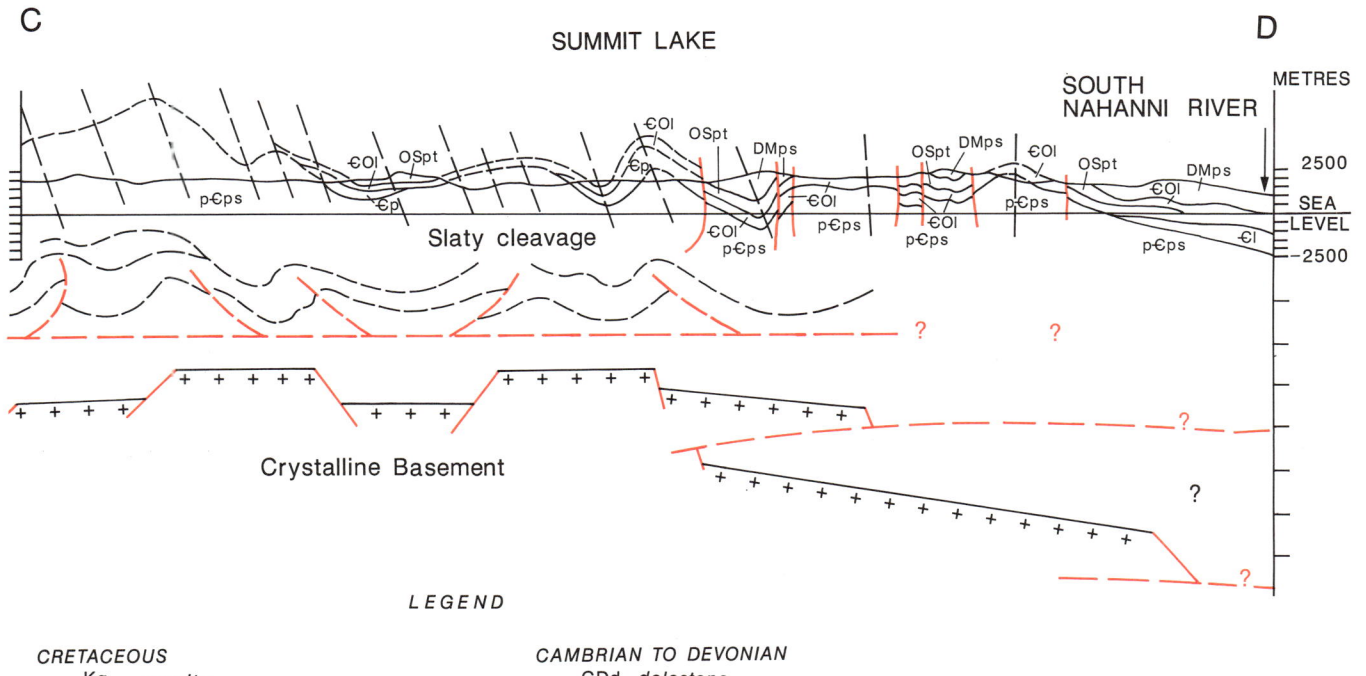

LEGEND

CRETACEOUS
Kg *granite*

DEVONIAN AND MISSISSIPPIAN
DMps *slate, sandstone, conglomerate*

SILURIAN AND DEVONIAN
SDl *limestone*

ORDOVICIAN AND SILURIAN
OSpt *shale, chert*

CAMBRIAN TO DEVONIAN
€Dd *dolostone*

CAMBRIAN AND ORDOVICIAN
€Ol *limestone*

CAMBRIAN
€p *slate, siltstone*
€l *limestone, sandstone, shale*

PRECAMBRIAN
p€ps *sandstone slate*

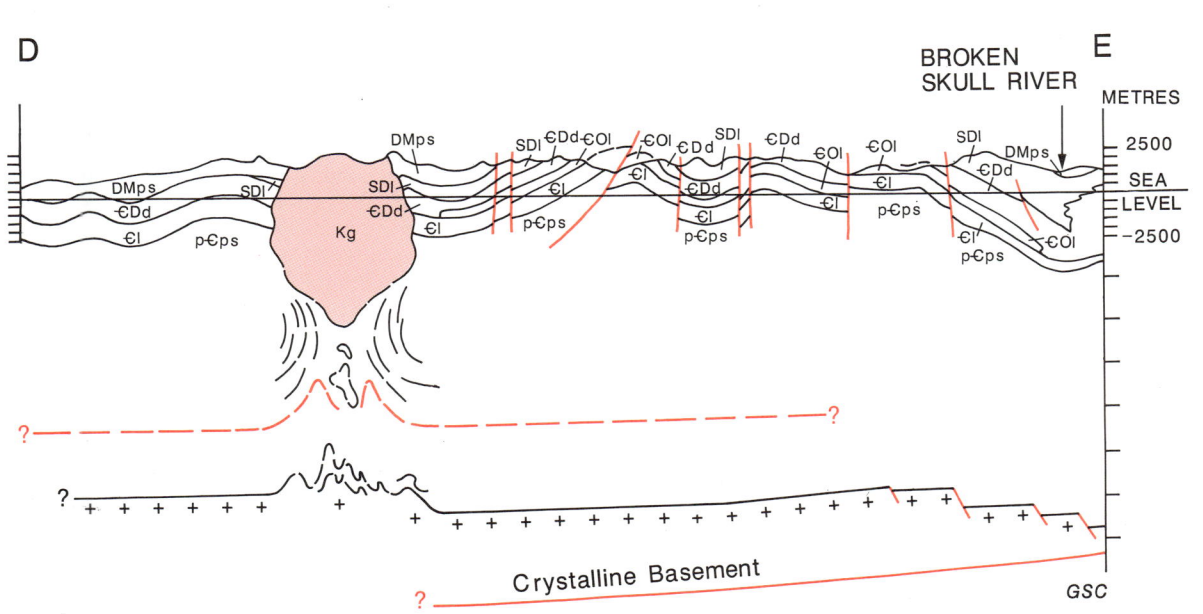

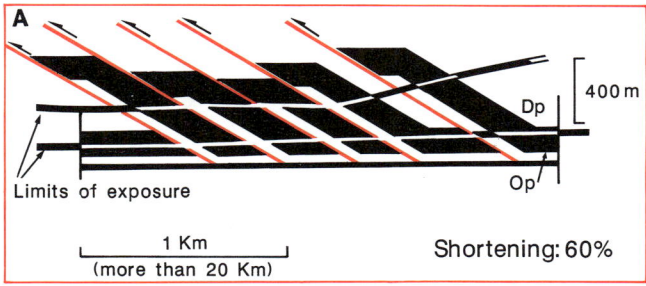
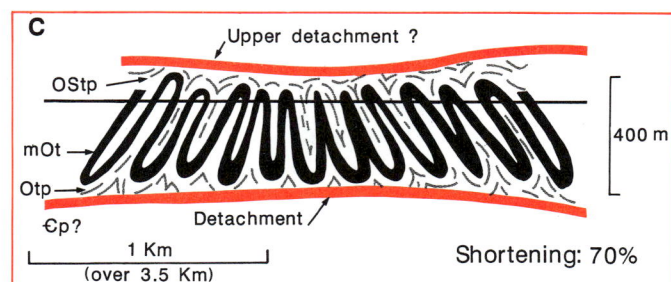
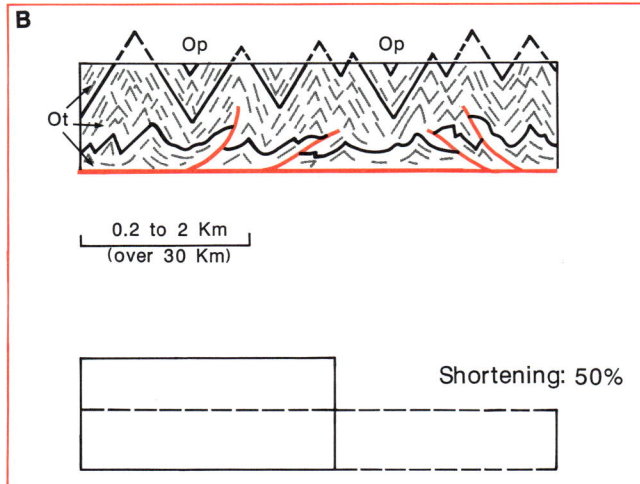
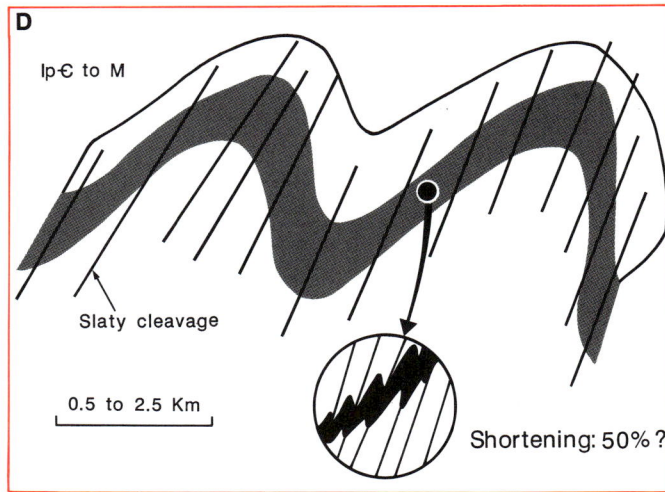

Figure 17.53. Various structural styles noted in Selwyn Basin. A, Imbricate faulting characteristic of relatively competent Silurian siltstone overlain by Devonian shale (Dp) and underlain by Ordovician shale (Op). B, Chevron folding in Ordovician chert (Ot) with infolds of Ordovician shale (Op). C, Isoclinal folding of Ordovician chert (mOt) bounded by incompetent Ordovician chert and shale (Otp) and Ordovician and Silurian chert and shale (OStp). Presumed detachment is in Cambrian shale (Cp). D, typical similar folds style in Lower Cambrian (lpC) to Mississippian strata (M).

strata which formed above a basal detachment (Fig. 17.54A). Penetrative strain increases and minor folds become more common to the west across the anticlinorium where complex structures merge with those of the Kootenay Arc. Most structures in the anticlinorium are the result of Mesozoic deformation but some of the minor folds and cleavage may have formed during Middle Proterozoic deformation and metamorphism (~1250 Ma; Leech, 1962; McMechan and Price, 1982).

The four segments of the anticlinorium are bounded by four major right-hand reverse fault systems: 1. a north-trending segment structurally concordant with the west flank of the anticlinorium, 2 and 3. transverse northeast-trending segments which cut across the anticlinorium and follow the locus of structures formed during Middle or Late Proterozoic rifting, and 4. a northwest-trending segment which extends into the Rocky Mountains on the east side of the anticlinorium (Fig. 17.54B). Overlap on the northeast segments of these fault systems cause the anticlinorium to persist northward despite its north plunge (Price, 1981). Normal faults along the Southern Rocky Mountain Trench disrupt the east limb of the anticlinorium south of 50°N.

Some of the northeasterly trending faults in the anticlinorium appear to be cut by mid-Cretaceous granitic plutons. The faults are continuous with eastward directed thrust faults in the Rocky Mountains. Therefore, these and structures of smaller scale in the anticlinorium, also cut by mid-Cretaceous plutons, are considered to be of mid-Cretaceous or earlier age. The geometry of the Purcell Anticlinorium has been interpreted as a reflection of ramping above a basal detachment in Late Cretaceous to Paleocene time (Price and Fermor, 1985).

Accreted terranes

Structural styles of the accreted terranes in the Omineca Belt are related to pre- and post-emplacement tectonic activity well documented in parts of the Slide Mountain Terrane. In other successions it is difficult to separate pre-emplacement and emplacement structures.

Yukon Territory
D.J. Tempelman-Kluit

Allochthons of the Nisutlin Subterrane (Kootenay Terrane) and the Slide Mountain Terrane (Anvil Allochthons) are important elements in the northern Omineca Belt and have distinctive structural styles. No detailed studies have been made of structures in the Windy-McKinley and Dorsey

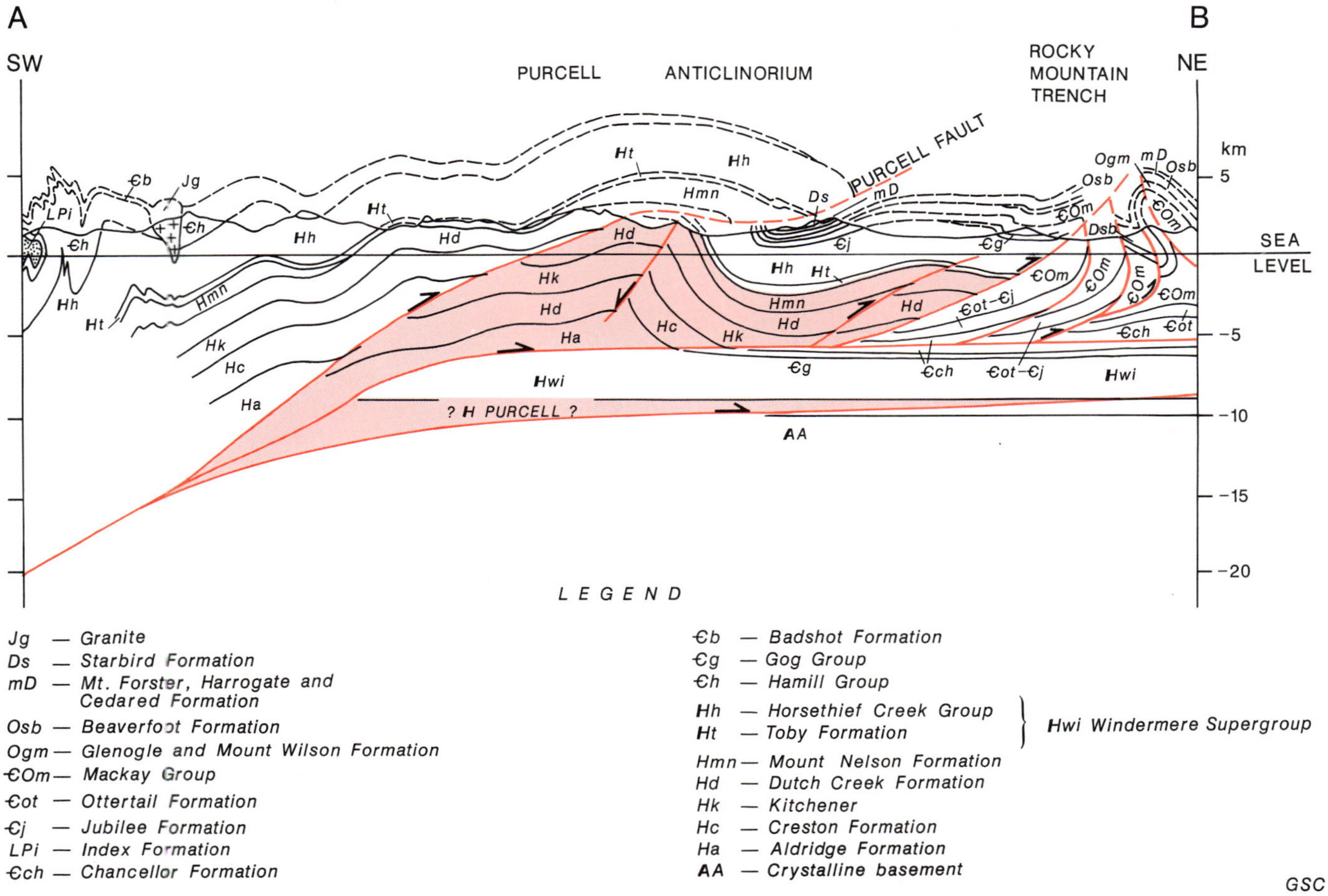

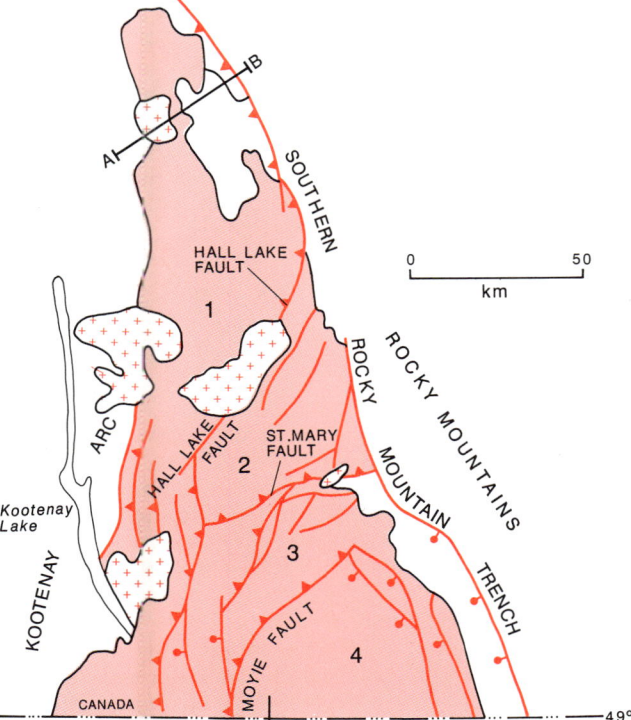

Figure 17.54. A, Structural cross-section of the southern Purcell Anticlinorium. After Price and Fermor (1985). B, Location of cross-section A-B and four segments of Purcell Anticlinorium.

terranes. Commonly but not everywhere, the allochthons occur in a consistent stacking order from the Nisutlin sedimentary and volcanic rocks at the base to oceanic rocks of the Anvil allochthons in the middle to sheared granite of the Nisutlin Subterrane at the top.

Nisutlin Subterrane

Siliciclastic, intermediate volcanic rocks and granitic rocks of the Nisutlin Subterrane have been strongly sheared and cataclastized so that primary layering commonly has been obliterated. North of the Big Creek Fault and the Klotassin batholith (Fig. 17.28) a pervasive flaser structure is evident. Banded siliceous mylonite and blastomylonite are common. Open folds and possibly related smaller folds are slightly asymmetrical and show a southwestward sense of vergence (Tempelman-Kluit, 1974). The structures are superimposed on an earlier foliation and are not accompanied by metamorphic recrystallization. Augen gneiss contains northwesterly trending quartz and quartz-feldspar boudins. The steeply dipping contact of the cataclastic rocks and the Klotassin batholith is marked by shearing and cataclasis of the granitic rocks and the presence, locally, of small ultramafic bodies.

In southern Yukon Territory penetratively deformed rocks of the Nisutlin Subterrane have a well developed flaser fabric which dips gently southwest and a strong lineation generally trending easterly. The pervasive strain contrasts with spaced thrust faults and absence of penetrative fabrics in the strata above and below. Folds and refolded folds on a scale of a metre or less commonly verge northeastward but represent a continuum of orientations resultant upon overturning, flattening and rotation towards the direction of maximum shear during deformation (Gordey, 1981).

Slide Mountain Terrane

The Anvil Allochthons of the Slide Mountain Terrane in the Yukon occur as gently dipping sheets which, although probably imbricated, retain coherent stratigraphy. On both sides of the Tintina Trench they are intimately associated with allochthons of the Nisutlin Subterrane and are separated from them by narrow zones of mylonite. Small scale folds are present in chert members and a weakly developed foliation is common. Volcanic rocks are massive.

The Semenof Block, east of the Semenof Fault (Fig. 17.21), is included in the Slide Mountain Terrane and is underlain by strata which form an open, northwest-trending syncline cut by steeply dipping northwest-trending faults. The block is bounded on the northeast by the Big Salmon Fault, a thrust which places it structurally above penetratively deformed rocks including mylonite of the Teslin Suture Zone in the Omineca Belt.

Windy-McKinley and Dorsey terranes

Little is known about structural style in the poorly exposed oceanic rocks of the Windy-McKinley Terrane. The Dorsey Terrane occurs in regional synclinoria and anticlinoria complicated by tight, in many places chevron folds with axial surfaces dipping to the southwest or northwest (Gabrielse, 1969).

Teslin Suture Zone

A root zone for the Nisutlin Subterrane, and Slide Mountain and Dorsey terranes may have been the Teslin Suture Zone (Fig. 17.21) which is interpreted as recording the evolution of a Mesozoic transpressional margin of ancestral North America from high-angle collision to dominantly dextral strike-slip displacement. There, the earliest recognized structures are ductile, variably mylonitized shear zones with west-trending stretching lineations and mineral assemblages indicating moderate temperature (575-625°C) and high pressure (8.5-11 kb; 8.5-11x10^5 kPa) metamorphic conditions (Hansen, 1986). Later structures and northwest-trending stretching lineations formed under lower temperature (425-550°C) and pressure (5-6.5 kb; 5-6.5x10^5 kPa) conditions in a dextral shear zone.

Ductile strain and metamorphism are dated isotopically, by K-Ar model ages that range between 215 and 160 Ma. The rocks may have been structurally assembled about Middle Jurassic time and thrust eastward over the autochthonous strata. Metamorphism reaching amphibolite grade and local occurrences of eclogite and blueschist minerals indicate relatively high pressure presumably related to subduction (Erdmer and Helmstaedt, 1983). K-Ar and Rb-Sr dates on metamorphic assemblages associated with blueschist minerals range from 246 to 258 (P. Erdmer and R.L. Armstrong, pers. comm., 1987). This evidence for Permian tectonism is supported by evidence for Early Permian or earlier deformation in the Sylvester Allochthon of the Slide Mountain Terrane in northern British Columbia.

Northern British Columbia
H. Gabrielse and T.A. Harms

Slide Mountain Terrane

The Sylvester Allochthon in northern British Columbia is an assemblage of Upper Devonian to Upper Triassic, dominantly oceanic lithologies forming a nested stack of gently dipping thrust slices within a regional northwest-trending synclinorium. The sole fault of the allochthon is the roof thrust of a westward verging duplex of thrust faults involving Devonian miogeoclinal strata (Harms, 1985; Fig. 17.35, 17.55). Minor folds and faults in a narrow zone near the sole fault, however, suggest northeastward emplacement of the allochthon. Within chert beds of the individual thrust slices many fold trains have northeast-dipping axial surfaces.

One thrust fault in the Sylvester Allochthon has been cut by a tonalite sill with a U-Pb date on zircon of 276 ± 6 Ma (Early Permian) showing that the rocks have undergone a pre-emplacement structural history which is not easily separated from emplacement events. Conodonts in the allochthon generally have higher alteration indices than those in underlying strata suggesting pre-emplacement metamorphism.

Included in the Sylvester Allochthon are strongly foliated diorite and granodiorite bodies and relatively massive granite. These units are interpreted as fault slices much like those that have emplaced granitic rocks of the Nisutlin Subterrane in Yukon Territory.

A combination of data from the Slide Mountain allochthons of northern British Columbia and Yukon Territory point to a Permian history of contraction, subduction, plutonism and volcanism. Obduction of the allochthons onto ancestral North America took place considerably later, possibly in Middle to Late Jurassic time.

Cariboo Mountains and Quesnel Highlands
L.C. Struik

Slide Mountain Terrane

The Slide Mountain Terrane thrust eastward onto the Barkerville and Cariboo subterranes is characterized by brittle folding and faulting with stratigraphic duplication by flat thrust faults (Fig. 17.37). Folds in the Slide Mountain Terrane are upright to slightly asymmetric, have generally east-dipping axial surfaces and trend northwest to northeast. They are open and rarely tight to near isoclinal, the tightest of which are confined to the sedimentary rocks of the predominantly volcanic terrane. Amplitudes are less than 40 m. Soft sediment folds are confined to muddy chert beds and directed to the east and northeast. Rocks of the Slide Mountain Terrane are folded with the regional Black Stuart Synclinorium.

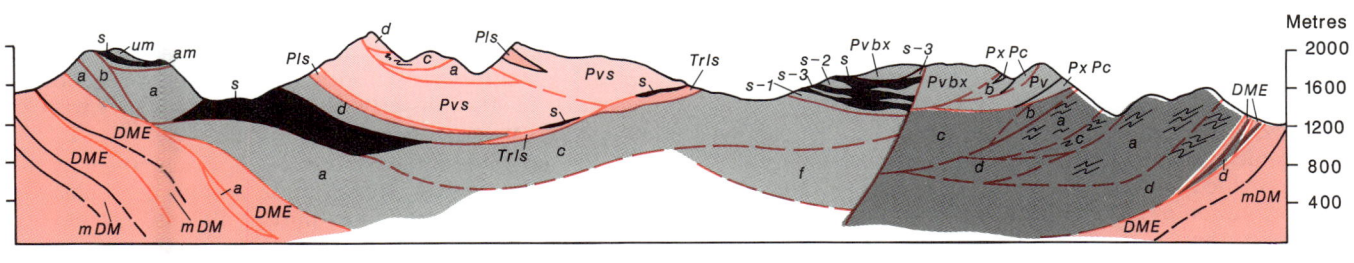

Figure 17.55. Structural cross-section across the northern part of the Sylvester Allochthon of the Slide Mountain Terrane. The allochthon consists of numerous lithotectonic slices bounded by mainly gently dipping faults. After Harms et al. (1988).

Thrust faults imbricate the entire sequence but are poorly exposed and therefore the sense of thrusting is unknown. Steep faults generally crosscut and offset previous structures.

The peak of metamorphism is confined to the upper part of the prehnite-pumpellyite zone of metamorphism and the metamorphic minerals have formed during the later stages of the folding.

The Crooked Amphibolite lies between Quesnellia and the Barkerville Subterrane on the western margin of the Quesnel Highlands and is included with the Slide Mountain Terrane. Rocks everywhere have a flow foliation interpreted to be the result of shearing. Mafic metamorphic minerals define the foliation.

Brittle structure throughout the Slide Mountain Terrane implies emplacement at shallow levels with thermal metamorphism ranging from 300° to 400°C.

Quesnellia

Quesnellia, the structurally highest terrane in the region, is characterized by open folding and shear localized near its contact with the Crooked Amphibolite. Layer-parallel shear zones have bands of concentrated flow cleavage in phyllonite 1 to 15 m thick and rootless folds outlined by distended siltite laminae. They transpose and cut bedding but parallel the regional contact of Quesnellia and the Barkerville Subterrane. The parallel shear is related to the emplacement of Quesnellia and the Slide Mountain Terrane onto the Barkerville Subterrane. Open to tight folds with wave lengths generally less than 10 m trend northwest, are upright to asymmetric to the east, and have moderately west-dipping axial surfaces. Cleavage varies from slaty to phyllitic. Part of the fold system within Quesnellia is the same age as the shear and is refolded by later somewhat more brittle structures consisting of open folds to crenulations related to regional anticlinoria and synclinoria. They are superimposed on the ductile structures of the transitional zone. Steep faults cutting most of the folds are related to Eocene extension.

Quesnellia generally remained in transitional or brittle conditions of deformation throughout its emplacement onto the Barkerville Subterrane. It locally was affected by regional amphibolite grade metamorphism. Quesnellia, upsection from the contact with the Barkerville Subterrane, displays progressively more brittle features and in concert, less metamorphism. Part of the effect is probably due to the increase in the viscosity of the terrane because of thick Triassic-Jurassic basalt.

Kootenay Arc region

The structural style of rocks assigned to the Slide Mountain Terrane in the Kootenay Arc is intimately involved and described with rocks of the Kootenay Terrane. As in the northern Cordillera the Slide Mountain sequence includes a history of Permian deformation and plutonism.

Implications of structural styles in the Omineca Belt

Detailed studies of structure, metamorphism and timing of tectonic events in the Cariboo and Omineca mountains have provided a basis for several syntheses on the structural evolution of the southern Omineca Belt. There is a general consensus that the main tectonism was initiated during Middle Jurassic time as a result of collision between the Intermontane Superterrane and ancestral North America. The origin of early formed, westerly verging structures in the western or distal part of the miogeoclinal wedge has been the subject of two leading hypotheses. One proposes eastward obduction of the Slide Mountain Terrane and Quesnellia followed by eastward decoupling and underthrusting of basement rocks (Brown et al., 1986). The other proposes obduction of the accreted terranes followed by tectonic wedging in which accreted rocks were driven eastward between parautochthonous supracrustal rocks delaminated from autochthonous basement (Price, 1986). The importance and significance of earliest formed, eastward-verging structures noted in several localities is unknown.

Eastward from the zone of westerly verging structures, on the east side of a structural fan axis, most structures verge eastwards. These structures predate the emplacement of mid-Cretaceous plutons and may have formed in part contemporaneous with, and in part subsequent to, formation of the westerly directed structures. A west to east progression of deformation is evident.

Strong uplift of the Omineca Belt accompanied by the emplacement of extensive granitic plutons took place in mid-Cretaceous time. In the northern Cordillera dextral strike-slip faulting was active at the same time. The relationship of these events to possible collision of the Insular Superterrane with the amalgamated Intermontane Superterrane and ancestral North America remains to be determined.

Finally, during Eocene time, uplift, plutonism and extensional faulting in the Omineca Belt was contemporaneous with dextral strike-slip faulting in the northern Cordillera and widespread volcanism and sedimentation in extensional basins in the southern Intermontane Belt.

The timing and sequence of structural events in the Omineca Belt are similar to those described above as far north as the British Columbia-Yukon Territory boundary. Farther north evidence for Middle Jurassic tectonism has not been found but other aspects of the structural evolution are analogous with those to the south.

PART E. FORELAND BELT

Summary

The structural style of the Foreland Belt is dominated by long, linear to curvilinear contraction structures disposed in broad salients and re-entrants from latitude 49°N to the Yukon-Alaska border. From south to north these structures occur in the northwest-trending Rocky Mountains, the north-, northwest-, and west-trending Mackenzie and Franklin Mountains, the west-trending Wernecke and Southern Ogilvie mountains, the north-trending Northern Ogilvie and Richardson mountains and the east- to southeast-trending British and Barn mountains.

In the Rocky Mountains, the structural style of the Rocky Mountains and Foothills subprovinces is expressed by thrust faults and detached folds which formed in a west-to-east progression above a shallow master décollement. In the southern Rockies basement is not involved. To the north, at two localities in the westernmost Rockies, crystalline rocks occupy the hanging walls of thrust faults. Structural style and gross stratigraphic character are closely linked; thick competent carbonate or sandstone successions of the eastern Rockies commonly form large thrust sheets whereas in interbedded argillaceous limestone and calcareous shale sequences in the western ranges detached folds and penetrative strain are characteristic. In the Foothills, closely spaced thrust faults, cylindrical folds and chevron folds lie above detachment surfaces. Northward in the Rocky Mountains and Foothills subprovinces the style changes from thrust- to fold-dominated regimes due to a change in facies in Phanerozoic rocks accompanied by a decrease in shortening from 200 km in the south to 50 km in the north.

The dominant structural style in the Mackenzie Mountains is one of easterly and westerly verging concentric folds and thrust faults. This contrasts markedly with the penetrative deformation and closely spaced imbricate thrust faults within the shaly facies of Selwyn Mountains to the west. Narrow, locally en échelon, doubly plunging folds of the southern Mackenzies contrast with the broad, regionally continuous folds to the north. Throughout the arc late stage wrench faults which cut obliquely across the folds probably reflect reactivation of structures as old as Precambrian.

The Franklin Mountains comprise thrust plates and doubly plunging anticlines disposed in easterly, northerly and southeasterly trending ranges along the outer perimeter of the Mackenzie structural arc. From one range to another and within individual ranges opposing asymmetry indicates opposing vergence.

At least four levels of detachment are inferred to have been involved in Mackenzie arc deformation. Broad concentric folds in the northeastern Mackenzies imply deep dislocation whereas thrust faults in the Franklin Mountains suggest shallow detachment. Beneath the nearby Colville Hills of the Interior Platform a deep level of décollement may extend beneath the Franklin Mountains to link with structures in the Mackenzies. Shortening across the outer Mackenzie arc is estimated to be between 45 and 55 km, whereas in the Selwyn Mountains of the inner arc, mainly in the Omineca Belt, the Paleozoic depositional width may have been reduced more than 100 km.

In the Wernecke Mountains structures arising from two episodes of Proterozoic extension as well as Proterozoic, Mesozoic and Tertiary contraction are recorded in three structural culminations extending westward from the termination of the Mackenzie Mountains to the Yukon-Alaska Boundary. Extension structures are represented by steeply dipping normal and wrench faults of comparatively small displacement that probably developed well inside of the Proterozoic margin. Contraction structures are expressed by broad, open, concentric, northwesterly trending folds and thrust faults that dip steeply to the northeast or gently to the south. Wrench faults of probable Tertiary age may represent reactivated Precambrian structures.

A marked change in structural trend takes place between the east-west en échelon folds of the Southern Ogilvie Mountains and the north-south en échelon folds of the Northern Ogilvie Mountains. The northerly trending Richardson Mountains contain the basement cored Richardson Anticlinorium and Richardson Fault Array, a set of north- and northeast-trending, near vertical, normal and dextral faults. The array traverses the Aklavik Arch Complex, a northeasterly trending composite structural culmination that was intermittently active throughout the Phanerozoic. To the northwest the British and Barn mountains display contraction structures disposed in an eastward convex arc that may reflect oroclinal bending as a consequence of the counter clockwise rotation of the Brooks Range away from the Arctic Archipelago.

The Rocky Mountains

M.E. McMechan and R.I. Thompson

The Rocky Mountains are structurally divisible into two parts, each extending from latitudes 49° to 60°N (Fig. 17.56). In the east the Foothills Subprovince comprises deformed strata of dominantly Mesozoic age whereas in the west the Rocky Mountains Subprovince consists of rocks of mainly Paleozoic and older ages.

Thrust faults and detached folds typify Rocky Mountain structures. Analysis is based upon two precepts: (1) folds and thrust faults formed in an orderly temporal and spatial progression from west to east, and (2) all detachment surfaces converge downward onto a throughgoing, shallow, west-dipping décollement surface along which Rocky Mountains strata were displaced eastward (Douglas, 1958a; Bally et al., 1966; Price and Mountjoy, 1970; Royse et al., 1975; Thompson, 1979). Basement rocks are exposed in the hanging walls of faults along the Northern Rocky Mountain Trench in the Deserters Range and along the Southern Rocky Mountain Trench adjacent to the Malton Gneiss (see Chapter 4).

In all parts of the Foreland Belt thrust faults commonly follow bedding glide zones that are linked by ramps where the faults cut across stratigraphic layering. Above the ramps strata within the hanging wall as well as those in overlying thrust sheets are folded into anticlines (Rich, 1934; Douglas, 1950).

Structural style is largely influenced by the character of stratigraphic sequences. Thick, competent carbonate or sandstone successions favoured development of large thrust plates whereas less competent interlayered shale and sandstone or carbonate sequences are disrupted by detached folds and, in western areas, penetrative strain. Facies changes within the Cambrian to Lower Cretaceous succession result in a general northward decrease in competency of the stratigraphic section and thus a consequent northward change from thrust-dominated to fold-dominated style. The transition in style occurs over a wide area between the Athabasca and Peace rivers. It is convenient, therefore, to subdivide the Rocky Mountains into southern, central (transitional) and northern segments (Fig. 17.56). Within each segment important lateral changes in structural style reflect stratigraphic changes and differences in level of exposure.

Southern Rocky Mountains

The southernmost Foreland Belt, together with the Purcell Anticlinorium of the southeastern Omineca Belt (Fig. 17.54A, 17.54B), narrows from approximately 250 km near the International Boundary to 150 km at 52°N, largely because of northward convergence between the eastern limit of deformation and the eastern boundary of the Kootenay Arc (Fig. 17.56).

Rocky Mountains Subprovince

Thrust sheets of Proterozoic, Paleozoic and minor Mesozoic strata underlie the southern Rockies (Plate 32). Some of

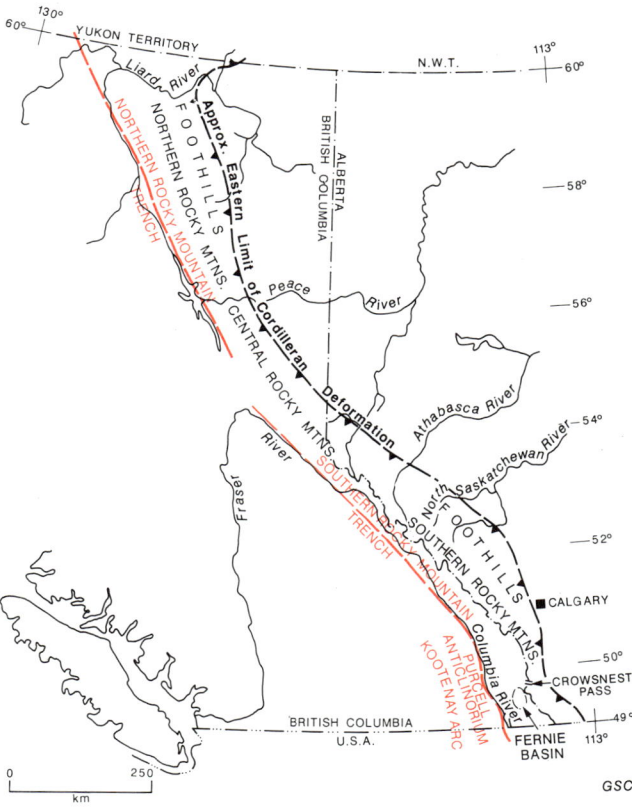

Figure 17.56. Principal tectonic elements in the Rocky Mountains.

the faults, including the well known Lewis (Plate 32), McConnell and Snake Indian thrust faults can be traced for hundreds of kilometres. Several structural styles, marked by variations in spacing, orientation and internal deformation, reflect the stratigraphic sequence and level of exposure. Where the lower Paleozoic section is thin and competent, Middle Cambrian to Lower Cretaceous strata typically form a series of stacked, west-dipping thrust sheets that locally were folded by movement on underlying thrust faults (Fig. 17.1, Section L, in pocket). This style characterizes the eastern part of the Rockies from Crowsnest Pass (49°30'N) to north of Athabasca River and probably occurs beneath open folded Jura-Cretaceous sandstone in Fernie Basin (Price, 1962; Ollerenshaw, 1981). Internal thrust-plate geometry is largely controlled by differences in facies of Upper Devonian strata (Fox, 1969; Jones, 1978; Beattie, 1984). Where the strata are in a carbonate facies the thrust sheets are generally simple tabular plates with subsidiary folds, however, where they are in argillaceous limestone or shale, large similar folds are prominent (Fig. 17.57).

North of Crowsnest Pass, the lower Paleozoic section thickens markedly to the west and comprises competent quartzitic and carbonate strata within a few thick, folded thrust sheets. Paleozoic strata therein are broadly folded and cut by predominantly west-dipping normal faults whereas underlying, Upper Proterozoic argillaceous strata with subordinate sandstone and carbonate, are cleaved and variably folded and faulted.

A marked change in structural style occurs where Middle Cambrian to Lower Ordovician carbonate strata, in thick thrust sheets, change facies westward to penetratively deformed and folded argillaceous limestone and calcareous shale (Cook, 1970, 1975). Variations in orientation of cleavage, axial planes of folds and faults define a regional, upward diverging fan structure (Porcupine Creek Anticlinorium) that extends from 49°30'N to near North Saskatchewan River at 52°N (Fig. 17.40, 17.56). Structures on the southwest side of the fan, originally northeast-verging, were rotated and overturned as the structure developed (Balkwill, 1972). Middle Ordovician to Middle Devonian strata exposed in the southwest limb of the fan outline complex synclines cut by contraction faults and an array of northwesterly trending transverse tear faults (Fig. 17.58).

Thick thrust sheets of Middle Proterozoic strata occur near the Southern Rocky Mountain Trench south of 50°N. One of these, the Lewis Thrust Sheet, extends across the width of the Rocky Mountains subprovince south of Crowsnest Pass (49°30'N; Fig. 17.59). The Middle Proterozoic strata outline broad open folds which formed as a consequence of stacking of underlying thrust sheets, or folds which formed above hanging wall ramps.

Normal faults are most common in thick competent successions. Most strike subparallel with thrust faults and folds, have west-side-down displacement and postdate thrusting. They are generally thought to flatten at depth and merge with an earlier formed thrust fault. Examples include normal faults next to the Southern Rocky Mountain Trench and Flathead Valley (Bally et al., 1966). One of these, the Flathead Fault (Fig. 17.59), forms part of a graben system which postdates contraction structures in the Lewis Thrust Sheet (Price, 1965; McMechan, 1981).

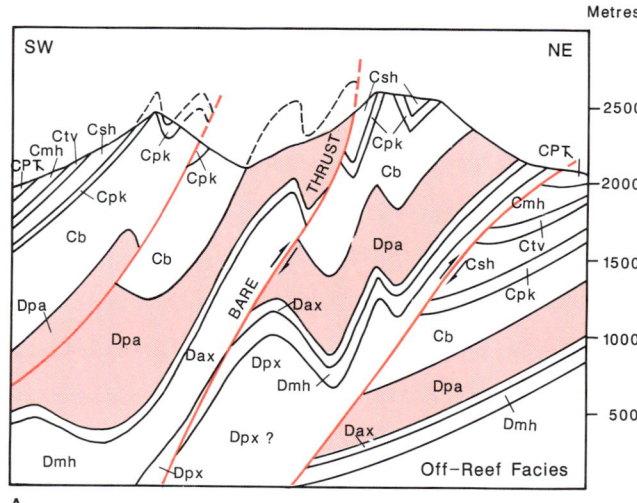

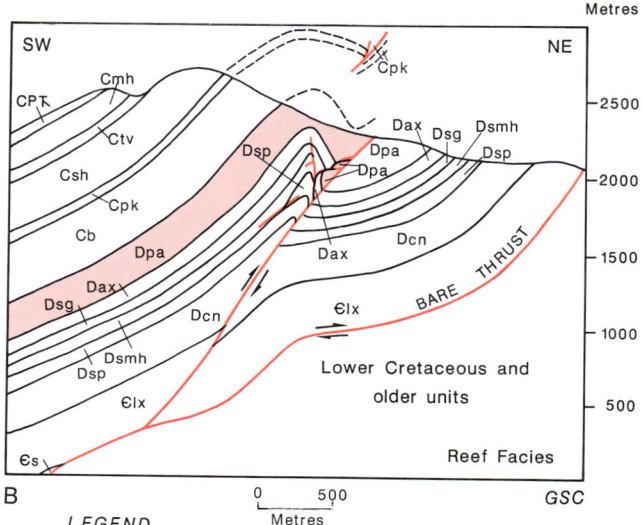

LEGEND

CPŤ	Rocky Mountain Group, Sulphur Mountain Formation
Cmh	Mount Head Formation
Ctv	Turner Valley Formation
Csh	Shunda Formation
Cpk	Pekisko Formation
Cb	Exshaw and Banff formations
Dpa	Palliser Formation
Dax	Alexo Formation
Dmh	Mount Hawk Formation } Off-Reef Facies
Dpx	Perdrix Formation
Dsg	Grotto Member Southesk Formation
Dsmh	Mount Hawk Member Southesk Formation } Reef Facies
Dsp	Pechee Member Southesk Formation
Dcn	Cairn Formation
Єlx	Lynx Formation
Єs	Sullivan Formation

Figure 17.57. Change in structural style with change in Upper Devonian Fairholme Group lithofacies, Ram Range, Alberta (near North Saskatchewan River; after Beattie, 1984). Note the predominantly tabular thrust plate in reef facies (B) in contrast with the similar fold style in the off-reef facies (A).

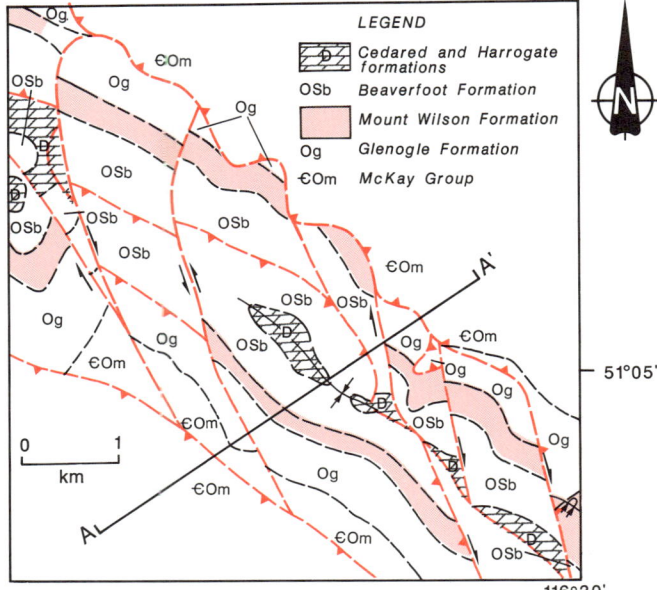

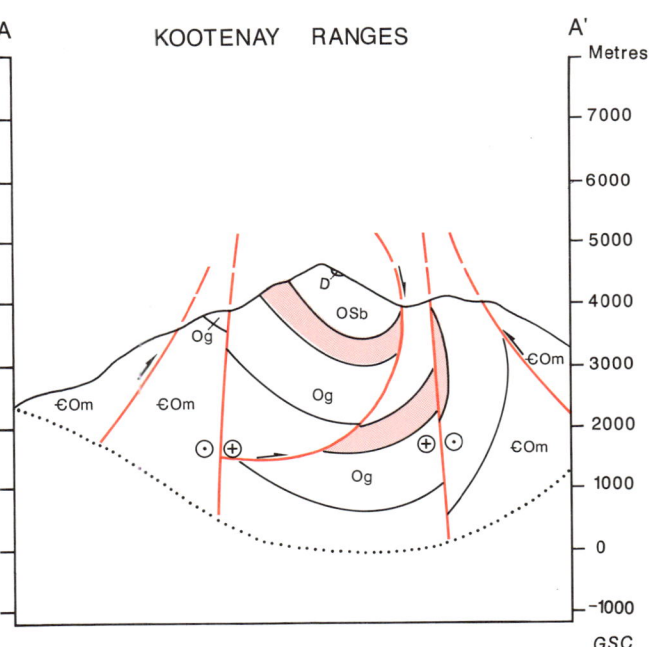

Figure 17.58. Characteristic structural style of brittley deformed Cambro-Ordovician argillaceous limestone and calcareous shales of the MacKay Group on the west limb of the Porcupine Creek fan structure (after Price et al., 1979).

Transverse structural and stratigraphic elements, such as the southern edge of the Cambrian to Ordovician shale basin extend without strike-slip offset from the Purcell Anticlinorium across the Southern Rocky Mountain Trench into the southern Rockies (Leech, 1966).

Foothills Subprovince

Northeast-verging thrust faults in Mesozoic and Paleozoic strata characterize the southern Foothills. Concentric folds locally are important near the leading edge of thrust sheets. The typical association of structures in the Foothills was used by Dahlstrom (1969) to illustrate the concept of balanced structural cross-sections. Between the Bow River (51°N) and latitude 49°N the Foothills consist of Mesozoic clastics, cut by multiple thrust fault splays from larger thrust faults which dislocate underlying Paleozoic strata (Bally et al., 1966). Stacking of thrust sheets carrying Paleozoic strata resulted in local structural culminations which are important targets for petroleum exploration (see chapter 20, Part A). Fewer faults dislocate Mesozoic clastics in the western part of the subprovince where flexural-slip folds with associated minor faults are locally more important (Douglas, 1950). Imbrication of Mesozoic strata becomes less common to the north and large synclines, cored by unfaulted Upper Cretaceous and Tertiary clastics, are found across much of the Foothills between Ram River (52°N) and Athabasca River (53°15'N) (Fig. 17.56, Tectonic Assemblage Map 1712A, in pocket). North of the North Saskatchewan River synclines in the eastern Foothills are commonly separated by zones of closely spaced faults (Fig. 17.60). The faults dip both to the east and west, many are folded (MacKay, 1943; Douglas, 1958b) and geometrically they resemble triangle zones. Western synclines are at least locally separated from imbricated faults in Lower Cretaceous clastics by a detachment in Upper Cretaceous shale (Charlesworth and Kilby, 1981). The boundary between the Southern Rocky Mountains and Foothills is commonly a major thrust fault.

A triangle zone occurs at the eastern limit of deformation along most of the southern Foothills (Fig. 17.1, Section K). Within this zone northeast-verging thrust structures were inserted beneath an east-dipping fault system and above a basal detachment to produce the east-dipping western limb of the Alberta Syncline (Fig. 17.59; Gordy et al., 1977; Jones, 1982).

Central Rocky Mountains

In the central Rocky Mountains the Foreland Belt changes from thrust-dominated to fold-dominated structural style. Faults become progressively less important northwards (Tectonic Assemblage Map 1712A, in pocket). The central Foreland Belt narrows from approximately 130 km near Athabasca River to 100 km at Peace River as a consequence of progressive northward truncation along the Southern and Northern Rocky Mountain trenches.

Rocky Mountains Subprovince

In the eastern part of the Rocky Mountains subprovince southwest-dipping thrust sheets of disharmonically folded and locally imbricated Middle Cambrian to Jurassic strata at Athabasca River (Fig. 17.1, Sections K,I, in pocket) are replaced north of 54°N by complex faulted folds in Upper Devonian to Triassic strata formed above a detachment in Upper Devonian shale (Fig. 17.61). Thick thrust sheets of folded Upper Proterozoic and lower Paleozoic strata underlie the wide western part.

In the western part of the subprovince thick, competent lower Paleozoic strata within the thrust sheets form broad folds or thrust-parallel homoclines cut by subsidiary thrust faults and normal faults (Fig. 17.1, Sections K,I, in pocket). Reversal of stratigraphic separation along strike suggests

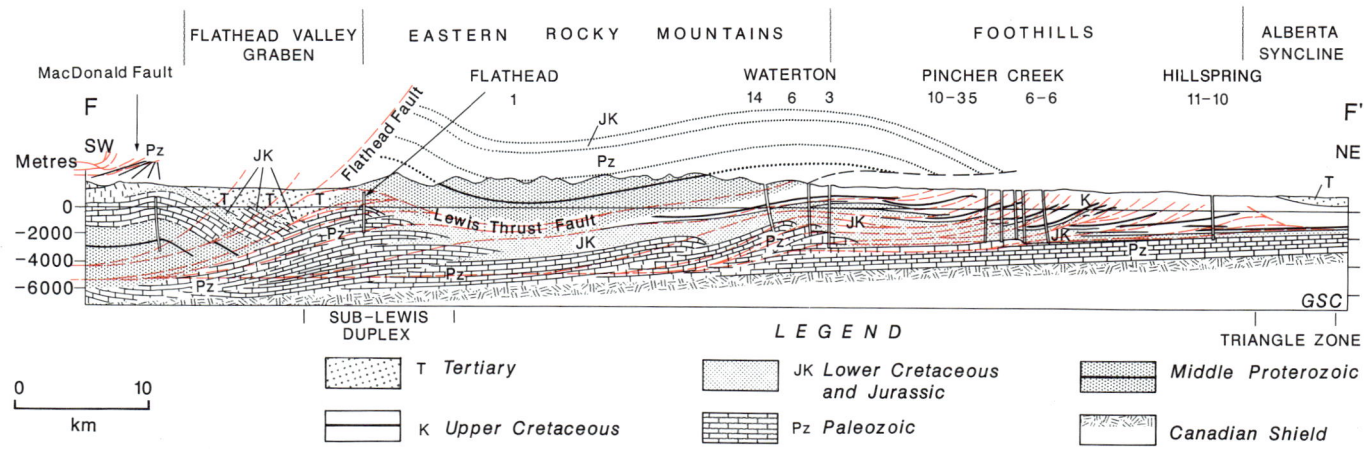

Figure 17.59. Structure section across the Foothills, eastern Rocky Mountains and Flathead Valley graben in the Waterton area (after Gordy, in Gordy et al., 1977, and McMechan, 1981).

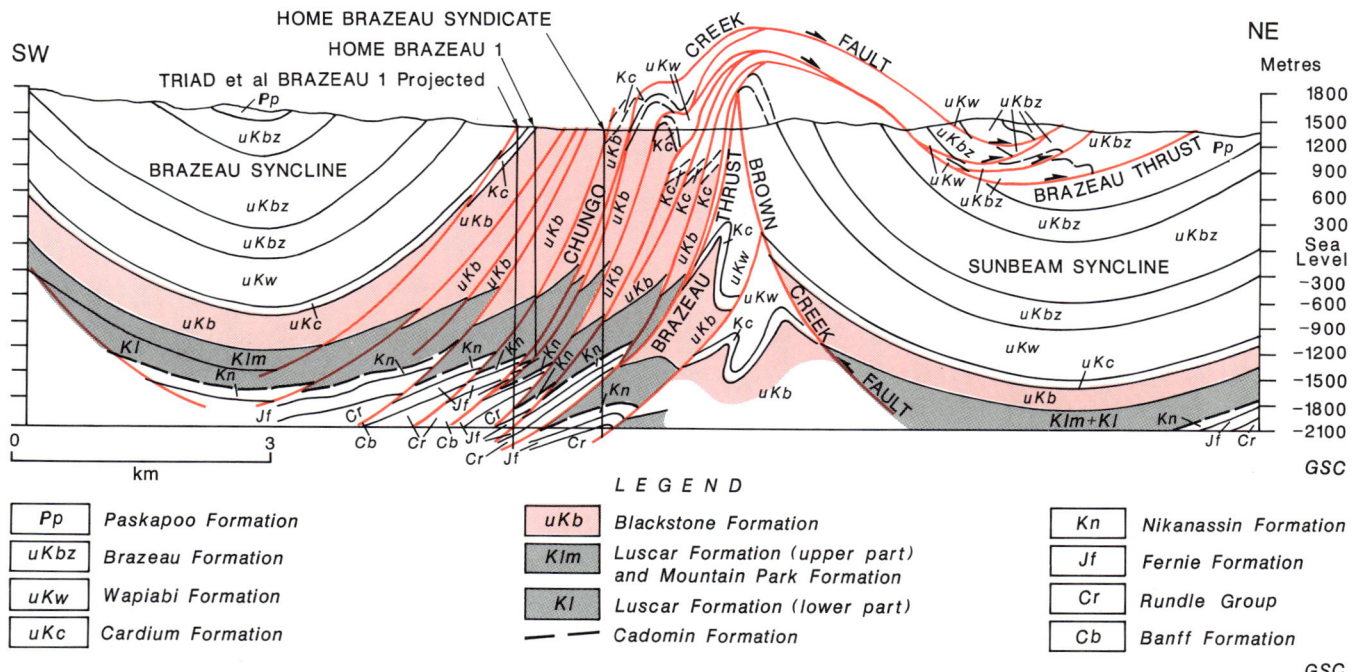

Figure 17.60. Structure section illustrating the structural style of the eastern Foothills north of North Saskatchewan River (after Douglas, 1958b). Large synclines cored by Upper Cretaceous sandstones are separated by zones of complex faulting.

that some normal faults originated as thrust faults. South of 54°30'N underlying, cleaved Upper Proterozoic strata are faulted and disharmonically folded with wavelengths and amplitudes much smaller than folds in overlying Lower Cambrian quartzite. This disharmony indicates that a detachment occurs in argillite near the top of the Upper Proterozoic succession (Charlesworth et al., 1967).

The principal bedding-glide zones for thrust faults are near the base and top of the Upper Proterozoic sequence in the west, and in the Middle Cambrian (south of 54°N) or Upper Devonian (north of 54°N) to the east. Near Peace River, Cambrian to Ordovician calcareous shale and argillaceous limestone form an intermediate detachment extending into the northern Rockies.

Blind (nonsurfacing) thrust faults carrying Middle Devonian and older strata underlie the eastern Rocky Mountains at least as far south as Pine Pass (55°30'N) where they have been documented by exploratory drilling. The small stratigraphic separation on, or lack of faults at the eastern edge of the western Rockies suggests that blind thrust faults probably occur as far south as 54°30'N.

Figure 17.61. View toward the northwest on the north side of Kakwa Lake, Monkman Pass area, central Rocky Mountains of disrupted box bold in Upper Devonian Palliser Formation which passes downward into a chevron fold of the Upper Devonian Mount Hawk Formation. The fold is in a thrust sheet of Devonian and Carboniferous carbonate and shale. Note disrupted west hinge and curvilinear trace of the axial plane of the east hinge. Photo by R.I. Thompson. (GSC 205384)

Foothills Subprovince

Northeast-verging thrust faults in Lower Cretaceous, Jurassic, Triassic and Paleozoic strata characterize the central Foothills north of Athabasca River (Fig. 17.1, Sections K,I, in pocket; Tectonic Assemblage Map 1712A, in pocket). Paleozoic inliers expose spectacular folds in Upper Devonian and Carboniferous strata. To the north such inliers are absent and thrust faults in Cretaceous strata are replaced by chevron and box folds which, at Smoky River, dominate the surface structural style. Detachments in Cretaceous shale separate folds with different wavelengths and amplitudes that deform the more competent strata. Unexposed thrust sheets of folded Paleozoic and Triassic strata are separated from overlying folds by an important detachment in Jurassic and Lower Cretaceous strata.

Structural disharmony above the detachment in Jura-Cretaceous rocks decreases to the north. At Sukunka River (55°N) broad synclines, separated by narrower anticlines and faults predominate. Between Sukunka River and 55°30'N the main zone of detachment separating Foothills structures from underlying faults and folds changes from the Jura-Cretaceous level to Lower Triassic shale (McMechan, 1985) as a consequence of a change in facies and competency. In the same area, Upper Devonian carbonate is replaced by a thick sequence of Upper Devonian to Lower Carboniferous shale which localizes a regional detachment throughout the northern Foothills separating faulted folds in Carboniferous carbonates from underlying, undeformed or thrust-faulted Middle Devonian and older strata (Fitzgerald, 1968; Thompson, 1979; McMechan, 1985).

A triangle zone occurs at the eastern margin of the Foothills south of 54°30'N where northeast-verging thrust structures were inserted beneath a detachment in Upper Cretaceous shale and above a basal detachment to produce the east-dipping west limb of the Alberta Syncline (Fig. 17.1, Sections K,I, in pocket; Jones, 1982; Teal, 1983). To the north underthrusting may have occurred across the width of the Foothills beneath an upper detachment in Jura-Cretaceous or Lower Triassic strata (McMechan, 1985; Fig. 17.1, Section I, in pocket).

Northern Rocky Mountains

The northern Rocky Mountains widen from 100 km at Peace River to 150 km at 58°N as a consequence of divergence between north-northwest-trending Foothills structures and northwest-trending Rocky Mountain structures (Fig. 17.62). At Liard River (59°N) divergence is accentuated where Foothills structures form a narrow re-entrant before they bend northward into continuity with the broad arc of the Mackenzie and Franklin mountains.

On the west, the Rocky Mountain subprovince is obliquely truncated by the Northern Rocky Mountain Trench which, unlike the southern trench, is the locus of at least 750 km of strike-slip displacement (Gabrielse, 1985).

Rocky Mountains Subprovince

Folded thrust sheets of Ordovician to Middle Devonian carbonate strata form a north-trending spine of rugged peaks along the eastern margin of the northern Rocky Mountains subprovince (Fig. 17.63). To the west, more subdued topography is underlain by folded, thrust-faulted and, in part, penetratively cleaved, mainly incompetent strata including middle Paleozoic and Upper Proterozoic fine grained clastics, Cambrian to Ordovician shale and argillaceous limestone and sandstone of Ordovician to Devonian age (Fig. 17.64). The structural style in the incompetent strata is similar to that in the Selwyn Basin. Abrupt changes in structural geometry occur where the tectonic strike crosscuts Paleozoic carbonate-to-shale facies transitions (Fig. 17.65). The northward trend of these structural regimes is interrupted between 57°30'N and 59°N by the Muskwa Anticlinorium (Fig. 17.62), about 50 km wide, comprising folded thrust sheets of Middle Proterozoic carbonates and clastics. The Gundahoo Fault truncates the western margin of the anticlinorium where Upper Proterozoic rocks are brought to the surface.

Two main levels of structural detachment occur in the northern Rockies. One, in the east between 56°N and 58°N, lies within Cambrian to Ordovician calcareous shale and argillaceous limestone and the other, to the west, occurs in Upper Proterozoic argillite. Across Muskwa Anticlinorium, detachments lie at the base of Silurian carbonate (Taylor and Stott, 1973) and within Middle Proterozoic clastics (Thompson, 1981; Gabrielse and Taylor, 1982; Fig. 17.66, Fig. 17.1, Section G, in pocket).

Many of the large thrust faults have sigmoidal traces. Each makes an abrupt step from west to east in a southward direction due to their position above underlying transverse fault ramps.

The boundary between the Rocky Mountains and Foothills subprovinces coincides with east-dipping limbs of large mountain-front anticlines cored by middle Paleozoic strata. This is in marked contrast to the southern Rockies where large thrust sheets such as the Lewis and McConnell separate the two subprovinces. This succession of large

CHAPTER 17

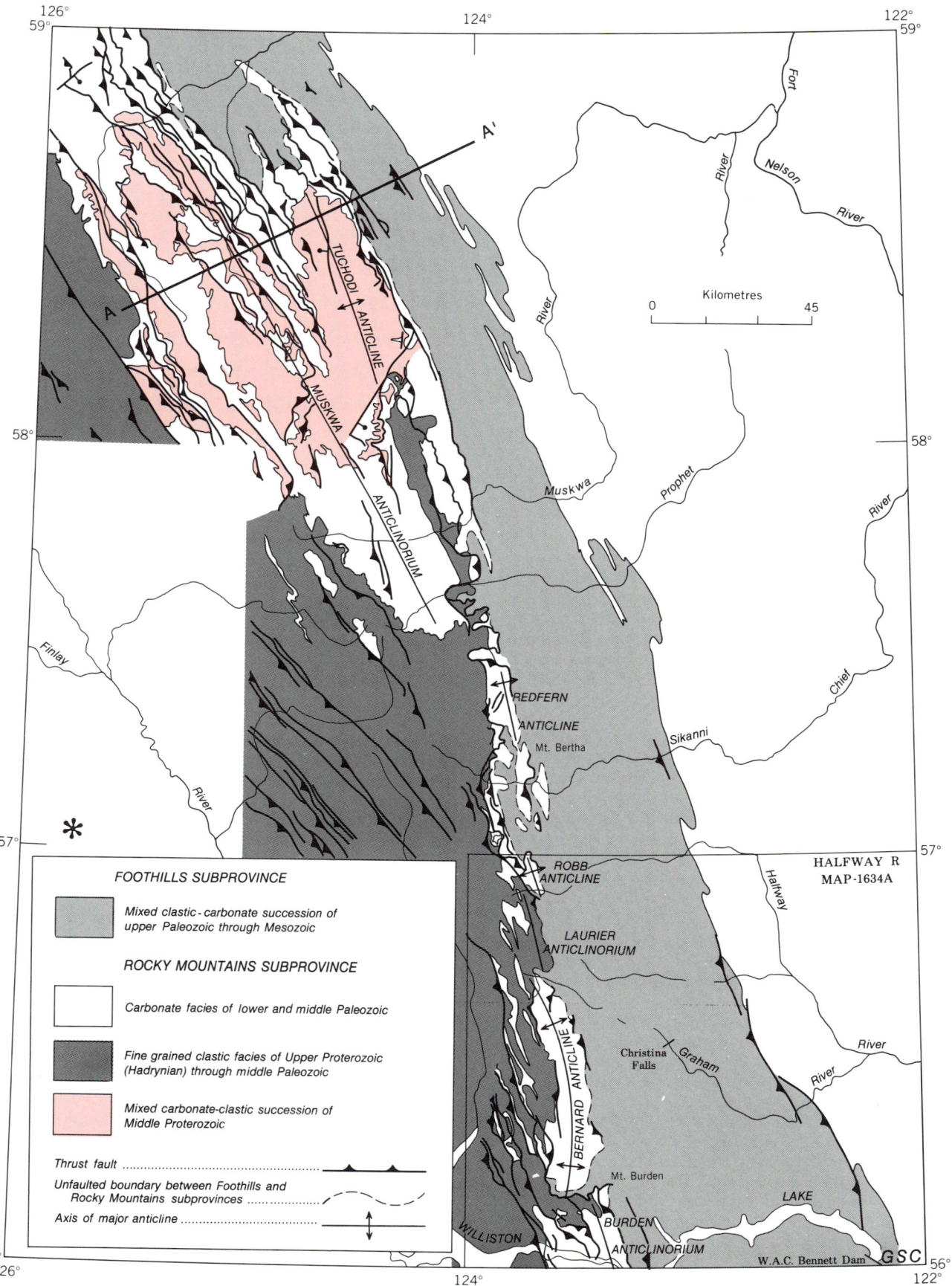

Figure 17.62. Tectonic map of the northern Rocky Mountains showing the major structural features. AA' is the location of structure cross-section illustrated in Figure 17.66. Asterisk shows location of cross-section in Figure 17.64.

anticlines expresses subsurface ramps linking deep detachments on the west with shallower ones on the east. Between 56°N and 58°N, thrust faults within Lower Ordovician calcareous shale (Kechika Group) and, between 58°N and 59°N, in the Middle Proterozoic Muskwa Assemblage, both slice upward into Devonian and Carboniferous Besa River shale; displacement over the resulting ramps produced the mountain front anticlines which, in the south are cored by Paleozoic carbonates and, in the north by carbonates of Proterozoic age. The eastern limb of each anticline dips unbroken beneath more complex Foothills folds. The thrusts that produced them continue eastward within the Besa River Formation as non-surfacing blind thrusts (Thompson, 1979, 1981). As thrust displacement climbs upward and eastward into the Besa River, overlap is compensated by an equal amount of shortening within overlying folds (Thompson, 1979, 1981; Fig. 17.1, Section H, in pocket; Fig. 17.67).

Foothills Subprovince

Large amplitude box and chevron folds in upper Paleozoic and Mesozoic strata characterize the Foothills subprovince (Fitzgerald, 1968; Thompson, 1979); local topographic relief is from 500 to 900 m, about the same as in the Rocky Mountains subprovince. The eastern margin of the Foothills is geologically and topographically abrupt and is defined by steep, east-dipping anticlinal fold limbs. Farther east smaller amplitude folds exist beneath the western part of the Alberta Plateau.

Northern Foothills folds formed above a regional detachment in thick Devonian and Carboniferous shale. Underlying Middle Devonian and older carbonates are undeformed across the eastern two-thirds of the subprovince. A second detachment within thick shale of Late Carboniferous to Triassic age accommodates structural disharmony within the fold array south of Halfway River (57°N) (Fig. 17.51; Fitzgerald, 1968).

Effect of basement features on Rocky Mountain structure

Thrust faults in the Foothills and eastern part of the Southern Rocky Mountains change in strike by as much as 80° at a prominent structural re-entrant between 49°N and 50°N (Tectonic Assemblage Map 1712A, in pocket). The re-entrant is not the result of differential rotation (Norris, 1969) and coincides with and has a similar shape to a re-entrant in the eastern limit of Middle Proterozoic (Purcell) deposition. Thus it is probable that structural trends along the re-entrant are influenced by the shape of the Mid-Proterozoic rift basin (Norris, 1969; McMechan, 1981; see Chapter 5).

In the northern Rocky Mountains a significant divergence occurs between northwest-trending thrusts and folds in the western Rocky Mountains and the north-trending folded thrust sheets and Tuchodi Anticline at the eastern edge of the Rockies (Fig. 17.50). The divergence suggests

Figure 17.63. View south to the Sidenius Thrust Fault in the Rocky Mountains Subprovince of the northern Rocky Mountains, Halfway River area. The fault separates a broad, hanging wall syncline of carbonate (Nonda and Muncho-McConnell formations, Silurian and Devonian) from shale of the Besa River Formation (Devono-Mississippian). Carbonate in overturned anticlines in centre foreground is Dunedin Formation (Middle Devonian). Displacement of the hanging wall is at least 10 km from west to east. Trace of the thrust fault passes over the small pyramidal peak on the skyline at the right side of the photo. Photo by R.I. Thompson. (GSC 205383)

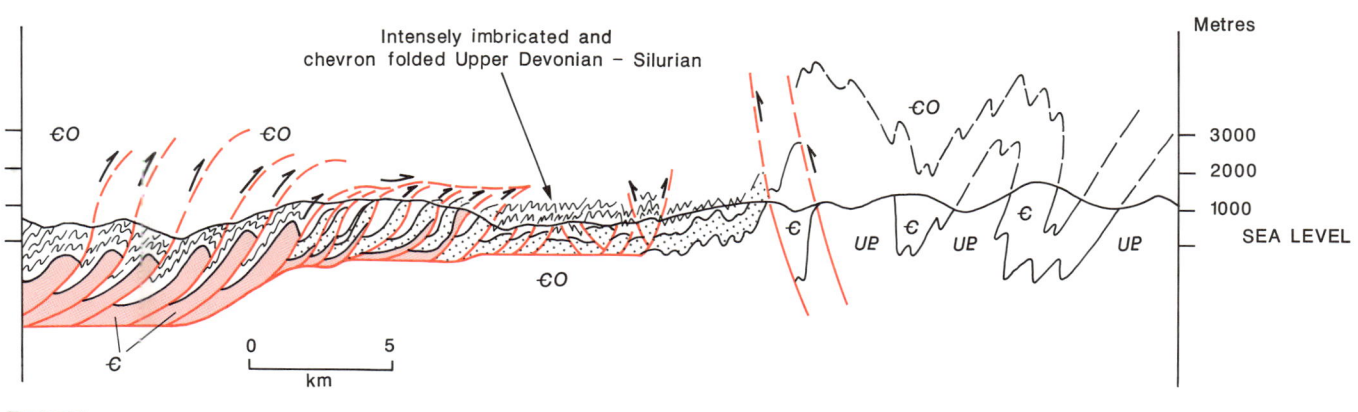

Figure 17.64. Typical structural style in incompetent lower Paleozoic strata in the Kechika Basin, northwestern Rocky Mountains. After McClay et al. (1988).

Figure 17.65. View south to detached fold complex in the Lower Carboniferous Prophet Formation, Halfway River area, northern Rocky Mountains. Resistant, light coloured units are limestone; dark recessive units are siltstone and shale. Structural disharmony is typical of those areas where the Prophet Formation changes facies from carbonate to shale. Photo by R.I. Thompson. (GSC 205385)

that early formed Proterozoic basement faults influenced Mesozoic fold and thrust development. Early continental extension is recorded in Tuchodi Anticline by a pre-Lower Cambrian north-northwest and north-trending dyke swarm, by Middle Cambrian rift facies parallel with regional structural trends, and by a large, north-northwest-trending pre-Silurian extension fault, oblique to the northwest regional trend of the Rocky Mountains. Tuchodi Anticline reflects a hanging wall ramp juxtaposed above a footwall flat. The ramp probably formed along a pre-Mesozoic basement flexure (west-side-down) or extension fault (Fig. 17.1, Section G, in pocket; Thompson, 1981; Fig. 17.66). The position and orientation of the ramp and, therefore, the orientation of the surface anticline, were controlled by basement structure. Alignment of Tuchodi Anticline with mountain front anticlines south of Peace River suggests that north-trending basement flexures and faults extend the length of the northern Rockies.

Shortening and Extension in the southern part of the Foreland Belt

Shortening across the Foreland Belt is at a maximum in the south and decreases progressively northward. Across the southern part of the southern Rockies and Purcell Anticlinorium it is estimated to be about 200 km (Price and Fermor, 1985; Monger et al., 1982). In the southern part of the central Rockies it is about 140 km (Campbell et al., 1982); in the northern part of the central Rockies at Peace River shortening is estimated to be about 70 km (McMechan, 1987); and across the northernmost Rocky Mountains at 58°30'N it decreases to a conservative estimate of about 50 km (Gabrielse and Taylor, 1982), perhaps due to major faults merging into the Northern Rocky Mountain Trench. North of the Peace River and west of the general shale-out boundary of lower Paleozoic strata the intensity of deformation increases markedly.

At least 25 km of post-orogenic extension occurred across the southern Rockies near latitude 49°N (McMechan and Price, 1984). No estimates of extension have been made for other parts of the Foreland Belt.

Timing of deformation

Compressive deformation in the southern Canadian Rocky Mountains began in the west and progressed eastward. In the Purcell Anticlinorium, reverse faults which are continuous with thrust faults in the western Rocky Mountains predate the emplacement of mid-Cretaceous plutons. To the east, Late Campanian foredeep sediments record orogenesis farther west in the Foreland Belt but are themselves cut by thrust faults in the Foothills. At the eastern edge of the Foothills thrust faults deform Paleocene strata and probably are no younger than Eocene because Oligocene extension affected the southern Rockies as recorded by graben-fill deposits in the hanging wall of the Flathead Fault.

In the central Rocky Mountains, a western fault (Herchmer-Snake Indian) postdates Late Jurassic or earlier cleavage and metamorphism (McMechan, 1987) and is older than Upper Cretaceous to Eocene conglomerate (Sifton) in the Rocky Mountain Trench. At the eastern edge of the Foothills compressive structures dislocate Upper Cretaceous and Paleocene strata.

In the northern Rocky Mountains a west-to-east progression of deformation is evident. Deformation synchronous with regional metamorphism in the western Muskwa Ranges was Early Cretaceous (ca. 120 Ma) or earlier in age (Wanless et al., 1979). Movement on the Herchmer Fault predates deposition of the Sifton conglomerate and, in the Foothills to the east, faults and folds are known only to be younger than Campanian.

Mackenzie Mountains, Franklin Mountains and Colville Hills

D.G. Cook

The Foreland Belt north of 60°N is a broad, arcuate belt comprising the Mackenzie Mountains, Franklin Mountains and Colville Hills (Fig. 17.68). The belt is bounded on the east by undeformed strata beneath the northern Interior Plains and on the west by the Selwyn Mountains of the Omineca Belt. Underlying strata consist of shallow-water carbonate and clastics of Proterozoic to mid-Paleozoic age which form the eastern or, inner part of the miogeocline. The overall structural style is one of easterly and westerly verging concentric folds and thrust faults in contrast to penetrative deformation marked by slaty cleavage, small-scale folds and closely spaced imbrication characteristic of the dominantly shaly facies of Selwyn Mountains (Selwyn Basin), in the outer part of the miogeocline. The region can be subdivided into five areas each with its own structural style: (a) central Mackenzie Mountains underlain by a single large thrust fault, the Plateau Fault, (b) the southern Mackenzie Mountains which display narrow anticlines and relatively broad synclines, (c) the northern Mackenzie Mountains characterized by broad anticlines and relatively narrow synclines, (d) the Franklin Mountains typified by isolated anticlinal and thrust bound ranges, and (e) the Colville Hills, a group of low, isolated anticlinal ranges occurring outside the Cordillera but forming an integral

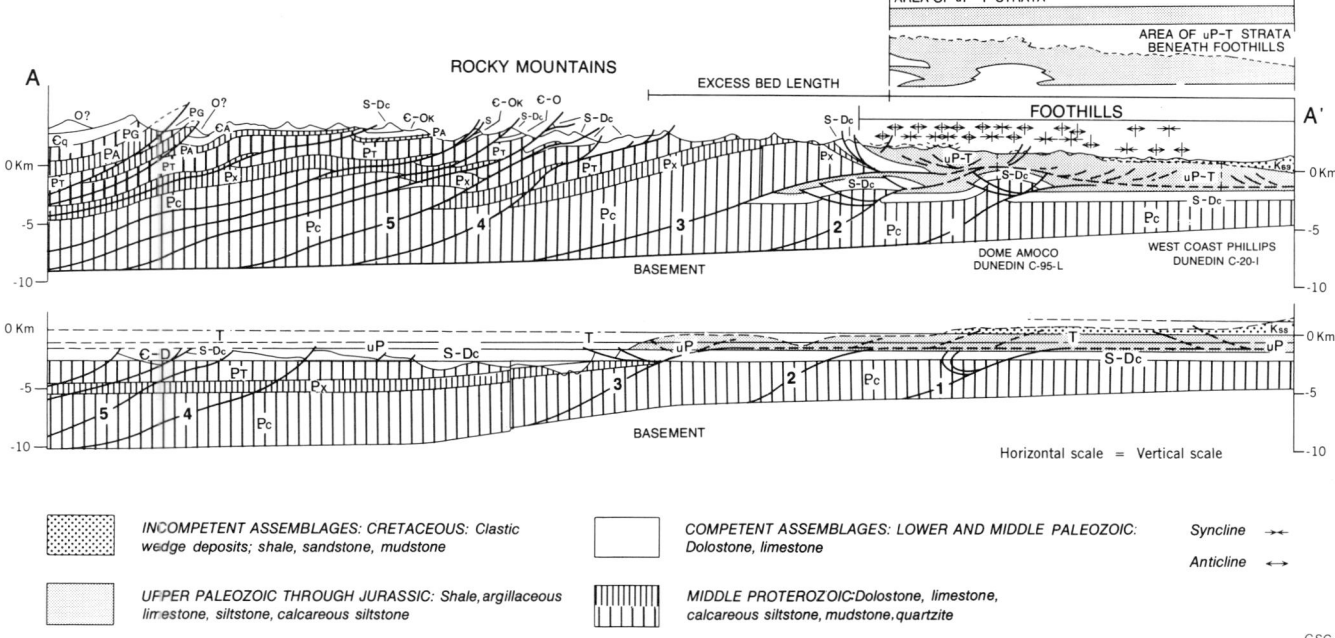

Figure 17.66. Structure cross-sections AA' (location shown in Fig. 17.62) illustrating the major elements of structural style across the Foothills and Rocky Mountains subprovinces (after Thompson, 1981). See also Figure 17.1, Section H.

part of its deformation. Variations in fold style from one region to another can be attributed to the nature of the stratigraphic succession and the depth of inferred detachment levels. Many of the dextral strike-slip faults can be documented or inferred to have been localized by earlier Precambrian faults.

West-central Mackenzie Mountains

The dominant structure of the west-central Mackenzie Mountains is the Plateau Fault (Fig. 17.1, Sections D,E, in pocket; Fig. 17.68; Gabrielse et al., 1973) which can be traced for a length of about 270 km. The thrust fault follows a zone of bedding-plane glide in Middle Proterozoic gypsum and emplaces gypsum and younger rocks over strata as young as Late Devonian. Stratigraphic throw may amount to more than 6 km. Hanging wall strata are essentially undeformed and form a plateau of structurally elevated flat-lying strata up to 30 km in width in marked contrast to the folded rocks to the west and east. The southwest margin of the plateau is marked by a syncline, the east limb of which marks a flexure within the Plateau plate where bedding changes abruptly from subhorizontal to southwest dipping (Fig. 17.1, Sections D, E, in pocket). The flexure is interpreted by Cecile and Cook (1981) to express the locus of a footwall ramp beneath the thrust fault, and by Gordey (1981) to reflect the presence of a ramp in the footwall of a deeper detachment. The Plateau Fault may have had lateral displacement of at least 35 km (Cecile and Cook, 1981).

Flat-lying to gently dipping Middle Proterozoic to lower Paleozoic strata of the Plateau Thrust Sheet underlie the Redstone Plateau which is bounded on the west by tightly folded lower Paleozoic rocks. The change in structural style coincides with an abrupt facies change in Upper Ordovician to Devonian strata from competent carbonate units in the east to incompetent argillaceous units in the west.

Southern Mackenzie Mountains

The change from the northern Rocky Mountains to the southern Mackenzies occurs near latitude 60°N and is marked by an abrupt diversion from northwest to north- and north-northeast-trending structures. Proterozoic, Paleozoic and Mesozoic strata are folded into narrow, locally en échelon, doubly plunging, faulted anticlines and broad, flat bottomed synclines. Most anticlines expose resistant Paleozoic carbonate as young as Middle Devonian with a few exposing Middle Proterozoic quartzite. Incompetent Upper Devonian shale underlies most synclines.

The limbs of many folds are cut by eastward and westward directed faults of moderate displacement. Late stage north-northeast-trending right-lateral wrench faults, such as the Spirit Fault, cut obliquely across folds (Fig. 17.69). Many of these faults are regionally aligned which suggests the presence of a basement fault reactivated during the late stages of compression (Cook, 1983).

Southern Mackenzie Mountains structures are mainly flexural folds and steep reverse faults believed to be detached from crystalline basement. Gordey (1981) calculated a detachment level at about 9 km below sea level.

Northern Mackenzie Mountains

The northern Mackenzie Mountains display broad, regionally continuous folds which markedly contrast in wave length and amplitude with the discontinuous en

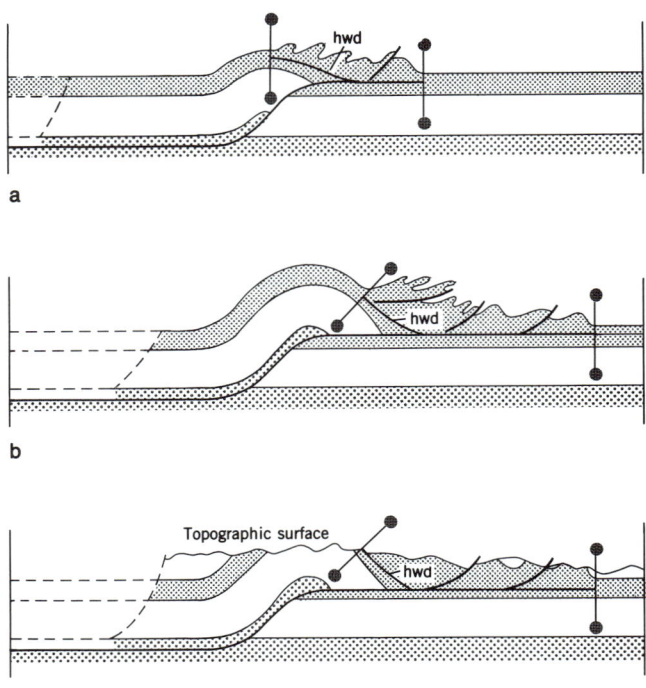

Figure 17.67. A diagrammatic representation of the blind thrust model. Stippled patterns represent mechanically incompetent strata separated by a rigid carbonate unit. a, illustrates the onset of displacement across the thrust accompanied by development of a hanging wall detachment(s) (hwd) allowing the incompetent strata within the hanging wall plate to deform disharmonically and absorb displacement on the underlying thrust. The thrust ceases to exist at the point where shortening due to folding in the hanging wall equals displacement on the thrust — hence the pin on the right side of the section. b, continued thrust fault displacement increases the width of the disharmonically deformed hanging wall succession. c, illustrates the difficulty in deciphering the detached nature of the mountain front anticline using surface exposures. The major thrust remains "blind", and much of the shortening within the disharmonically deformed incompetent unit may be difficult to assess unless good stratigraphic markers are present.

échelon folds of the southern Mackenzies (Fig. 17.1, Sections D, E, in pocket). Anticlines up to 20 km across and more than 50 km long expose Cambrian to Devonian carbonates unconformably overlying Proterozoic clastics and minor carbonate. Narrower intervening synclines are underlain by recessive Upper Devonian shale and sandstone.

Three broad anticlines dominate the structural style of the belt, about 125 km wide, from Plateau Fault to the mountain front. Steeply dipping thrust faults disrupt the folds but displacements are small such that syncline-anticline continuity is rarely destroyed. The longest of these, the Deadend Fault is 170 km long, and, near Arctic Red River, forms the mountain front. Elsewhere the mountain front is formed by the east limb of a frontal anticline.

Minor north-trending wrench faults offset Proterozoic and Paleozoic strata in the anticline cores. These are

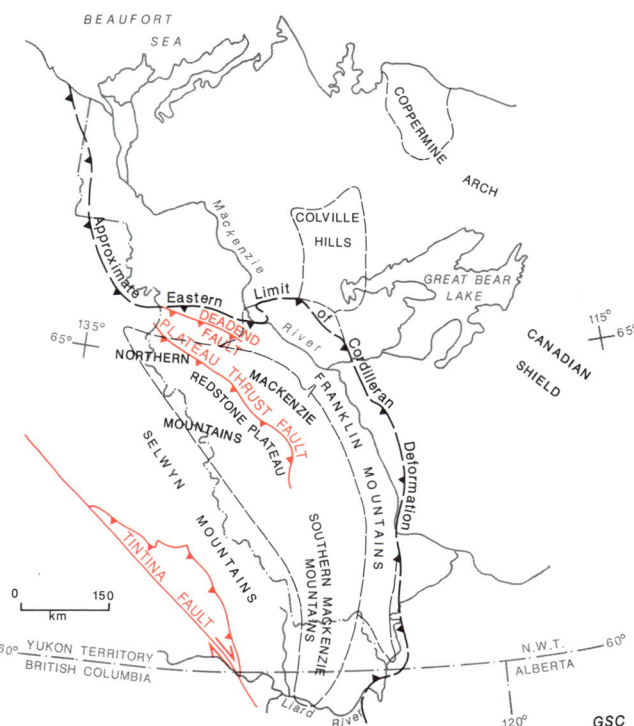

Figure 17.68. Index map for the eastern part of the Foreland Belt in the Yukon and Northwest Territories.

Precambrian faults which were favourably oriented for subsequent dextral wrench reactivation during Cretaceous and Tertiary compression (Eisbacher, 1981).

Franklin Mountains and Colville Hills

The Franklin Mountains occupy the outermost part of the southern and central Mackenzie arc (Fig. 17.68) wherein they widen northward to a maximum of 120 km in the Norman Wells region. From there, together with the Colville Hills, they form a deformed zone which extends into the Interior Plains. This outer belt terminates abruptly westward and does not continue around the northern part of the arc.

The Franklin Mountains illustrate unusual structural relationships. They comprise isolated structural ridges formed from thrust plates and doubly plunging, asymmetric anticlines and fall into three, sharply differentiated structural trends: northwest, west, and north (Fig. 17.70). In the Colville Hills the westerly trend is not represented but an additional northeast trend is evident. No one trend is superimposed on another but instead, entire ranges change from one trend to another and back again; each ridge appears to have originated with its present sinuous geometry.

From range to range the sense of asymmetry and direction of fault transport are variable (Fig. 17.70). Similarly, thrust faults exhibit opposite dips and sense of relative transport from one range to another. Moreover, faults with opposing sense of transport can occur nearly on strike with one another and within the same range.

STRUCTURAL STYLES

Detachment levels

Four levels of detachment are inferred to have been involved in Mackenzie Mountains, Franklin Mountains and Colville Hills deformation. Concentric folds imply deep detachments in Proterozoic strata of Mackenzie Mountains and their scale suggests that the detachment beneath the northern Mackenzies is much deeper than that to the south (compare Sections D and E, Fig. 17.1, in pocket). Shallower detachments are represented by the Plateau Fault and those beneath the Franklin Mountains. The Plateau Fault appears to have detached above a gypsum unit in the Middle Proterozoic Little Dal Group (Cecile et al., 1982). Eastward the northern Franklin Mountains appear to be in large part detached above Upper Cambrian salt (Saline River Formation).

The Colville Hills involve Proterozoic strata (Davis and Willot, 1978) and may require a deep level of detachment extending beneath the Franklin Mountains and linking to deep structures in the Mackenzie Mountains. Conversely, the Colville Hills may represent a shallow response to deep seated wrench faults (Cook, 1983). Similarly oriented subsurface wrench faults may have developed beneath the Upper Cambrian salt, resulting in a component of regional shear on compressive deformation and accounting for the enigmatic asymmetric structures in the northern Franklin Mountains (Fig. 17.71; Cook, 1983).

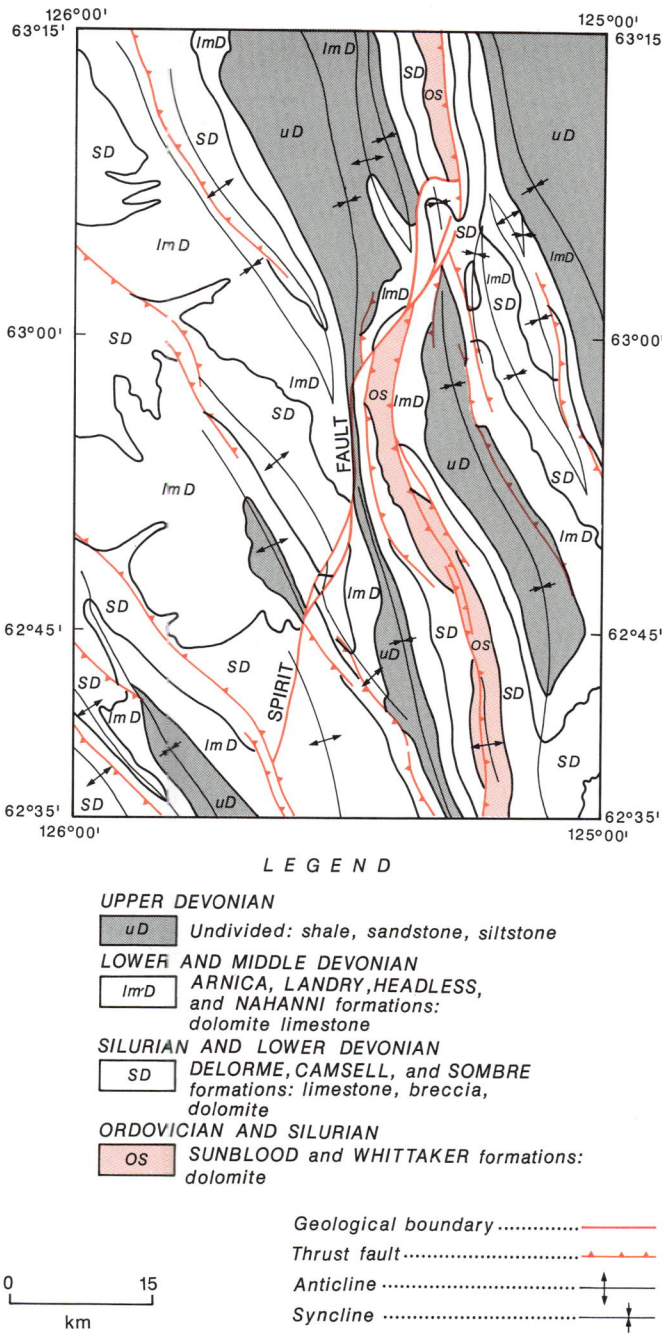

Figure 17.69. Spirit Fault, southern Mackenzie Mountains. Interrelated action of detachment folding and dextral displacement on the Spirit Fault has resulted in synchronous development of northwest-southeast and north-south structural orientations.

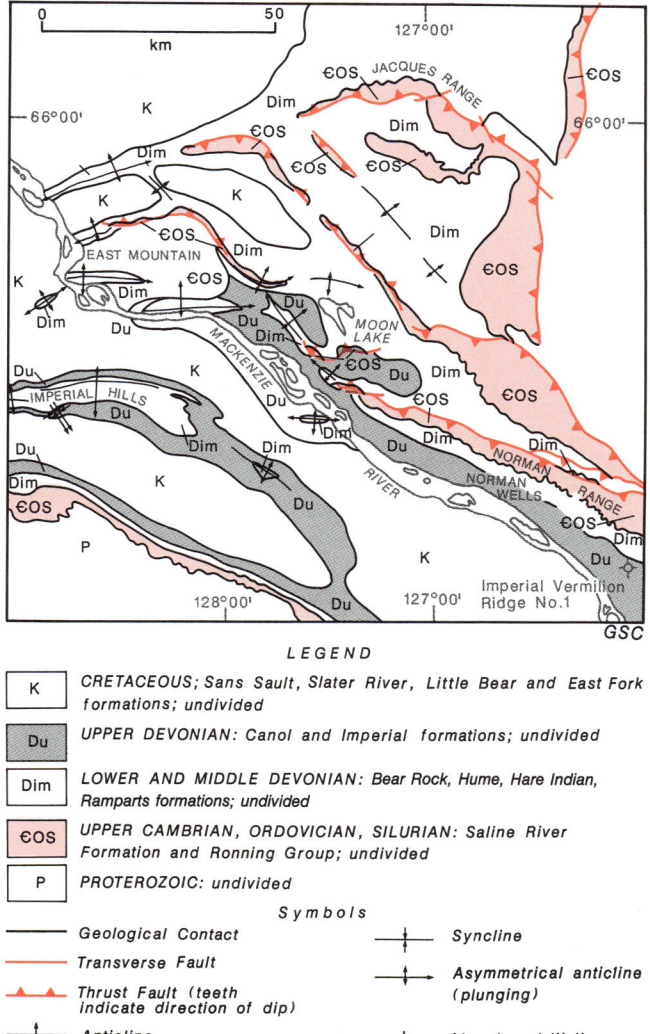

Figure 17.70. Geological sketch map of the northern Franklin Mountains.

645

CHAPTER 17

Wernecke Mountains and Southern Ogilvie Mountains

R.I. Thompson

The Wernecke Mountains are the southern part of the westward continuation of the Foreland Belt in central and western Yukon Territory (Fig. 17.72). Their boundary with the Omineca Belt is the Dawson Fault.

The Wernecke Mountains display Mesozoic and Tertiary craton-directed thrust faults and folds, typical of those in Mackenzie and Rocky mountains. Additionally at least two episodes of Proterozoic extension are recorded. Wholesale northward translation of Wernecke Mountains strata occurred in Late Jurassic and mid-Cretaceous time. An east to west oriented ramp, or succession of ramps in the detachment resulted in the Coal Creek, Hart River and Wind River structural culminations (Fig. 17.72, Fig. 17.1, Section C, in pocket).

Pre-Mesozoic structures

The Wind River culmination contains at least two generations of Proterozoic structures (Eisbacher, 1981). The first consists of tightly folded, northeastward trending Middle Proterozoic clastic and carbonate rocks (Wheeler, 1954; Gabrielse, 1967; Eisbacher, 1981) truncated by an unconformity at the base of Middle or Upper Proterozoic volcanics and siliciclastics. This so called "Racklan event" may have occurred about 1200 Ma. The second generation of structures, attributed to the "Mount Harper rift event" (Roots, 1987; = "Hayhook event" of Eisbacher, 1981), comprise steeply dipping extension and strike-slip faults which were active at the beginning of Windermere deposition. The number of older structures reactivated during the younger event is unknown.

The Coal Creek culmination contains two groups of extension faults, the younger of which is associated with the Mount Harper rift event; the older may reflect an earlier stage of Mount Harper rifting (Mercier, 1985) or may have originated during the Racklan event (Roots, 1987). The most prominent fault belongs to the older set, strikes westerly for 20 km, dips steeply to the north, and has a throw of at least 1 km; smaller faults trend northeast and northwest. Syndepositional faulting is recorded by the variety of sediment gravity flows deposited on down-faulted

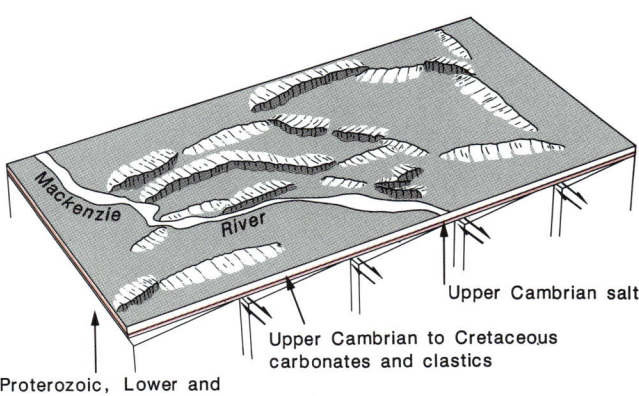

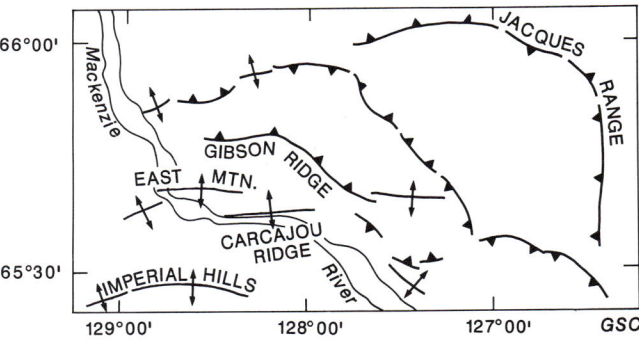

Figure 17.71. Schematic block diagram of northern Franklin Mountains showing Upper Cambrian salt (pink) as a ductile layer separating wrench-fault style below, from complex distributed shear style above (from Cook, 1983).

Shortening

Shortening across the Mackenzie and Franklin mountains has been estimated at from 45 to 55 km (Gordey, 1981; Cecile et al., 1982) in contrast to that in the Selwyn Mountains to the west where the amount of shortening has been estimated at greater than 150 km. The relatively minor shortening across the Mackenzie and Franklin mountains is consistent with paleomagnetic data which argue against a large degree of internal rotation and in favour of Mackenzie Arc reflecting original basin geometry (see Chapter 3).

Timing

As in the Rocky Mountains, deformation progressed from west to east. Deformation in the Selwyn Mountains preceded the emplacement of widespread mid-Cretaceous plutons, whereas to the east in the Franklin Mountains, rocks as young as Paleocene are faulted and folded. One area in the central Mackenzie Mountains is underlain by nonmarine Albian(?) strata which have been strongly deformed. These rocks may represent the remnants of a Lower Cretaceous foredeep which formed in response to deformation and crustal thickening in the Selwyn Mountains area to the west.

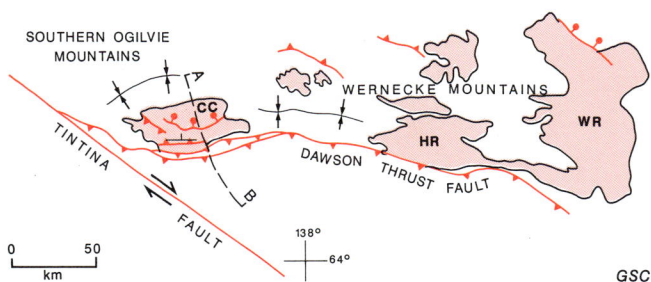

Figure 17.72. Outline map showing distribution of Coal Creek (CC), Hart River (HR) and Wind River (WR) culminations in the Wernecke and Southern Ogilvie Mountains. Precambrian rocks north of the Dawson Thrust Fault are shown in pink. AB shows line of section for Section C, Figure 17.1, in pocket. North and South Harper faults shown in red.

blocks and by rapid changes in stratigraphic thickness from one block to another. The rocks deposited during the Mount Harper rift event occur along the south side of the culmination. The North Harper and South Harper faults are the most prominent, dip northward at about 70° and have throws of between 1 and 3 km; the down-faulted blocks contain massive debris flows, turbidite, olistoliths and volcanics (Plate 6). Thicknesses and facies change rapidly along and across depositional trends. The rifting which produced the Harper faults probably affected a narrow zone of crust well inside the proto-Pacific margin.

Mesozoic and Tertiary structures

Broad folds and minor thrust faults accompanied Mesozoic northward translation of the Wernecke Mountains. In the Wind River culmination most thrust faults dip steeply to the northeast (Blusson, 1974; Eisbacher, 1981); folds are open, concentric and trend northwesterly except in the southwest where westerly trends predominate. The culmination plunges westward beneath broad, west-northwest trending folds in Paleozoic carbonate strata (Green, 1972).

Hart River culmination is a broad, low anticline, broken by steeply dipping faults of unknown displacement and dislocated by two imbricated thrust faults along its southern margin. Between Hart River and Coal Creek culminations Paleozoic clastics are tightly folded and imbricated. Coal Creek culmination is notable for the many subvertical faults that partition it in northerly and westerly trends (Mercier, 1985). Many are strike-slip faults with both dextral and sinistral displacements greater than 2 km. They cut folds and thrust faults and thus represent a later stage of Tertiary deformation; some, if not most, may be reactivated Precambrian faults. Most folds are small but locally prominent, commonly of chevron style with steep westward plunges. Thrust faults occur along the southern and western margins of the culmination where they have low southward dips and juxtapose Upper Proterozoic on lower Paleozoic strata.

Northern Yukon

H. Gabrielse and C.J. Yorath

The following account of structural styles in northern Yukon Territory is modified in large part from published and unpublished reports by D.K. Norris.

The mountain systems of northern Yukon Territory, and their interrelationships, are among the most complex in the North American Cordillera for they form part of the linkage between the Brooks Range of Alaska and the Innuitian and Cordilleran orogens. Moreover, it is in this region where a convergent margin tectonic history passes into that of a dominantly passive margin setting. Contraction and extension stress systems played equal roles in the development of the structural style of tectonic elements in the region.

Ogilvie Deflection

The Ogilvie Deflection (Norris, 1972), complementing the broad salient in Mackenzie Mountains, is expressed by an abrupt change in trend of folds and faults in Paleozoic and older rocks within the Taiga-Nahoni fold belt, the components of which are respectively embraced by the Taiga and Nahoni ranges of the Ogilvie Mountains (Fig. 17.73). The Taiga arm is truncated to the east by the Richardson Anticlinorium in the southern Richardson Mountains and the Nahoni segment against the Aklavik Arch Complex in the Dave Lord Range. Eagle Plain, containing the Eagle Fold Belt, lies in the interior angle of the deflection. The total length of the fold belt is 375 km.

The en échelon fold trains in the Taiga-Nahoni fold belt contain excellent examples of fold linkages. These are arranged characteristically in right- and left-hand patterns, depending upon their positions in the deflection. An example is seen in the Taiga segment, where, over a distance of 110 km, a bundle of folds, outlined by Lower and Middle Devonian limestone, is arranged in a zig-zag, left-hand pattern. The limestone forms the mountain front of Taiga Ranges and, because of the en échelon arrangement of folds, the mountain front steps systematically to the left (south) toward the apex of the deflection (Fig. 17.74). Within the apex of the deflection both left- and right-hand linkages occur. Northward, in the Nahoni segment, right-hand en échelon patterns occur to where the fold bundles terminate against the Aklavik Arch Complex. Axial surfaces are commonly steeply dipping and verge both towards and away from the interior angle of the deflection. Reversal of vergence along strike occurs locally. Contraction faults, both north- and south-verging in the Taiga Ranges and east- and west-verging in Nahoni Ranges, locally disrupt fold trends.

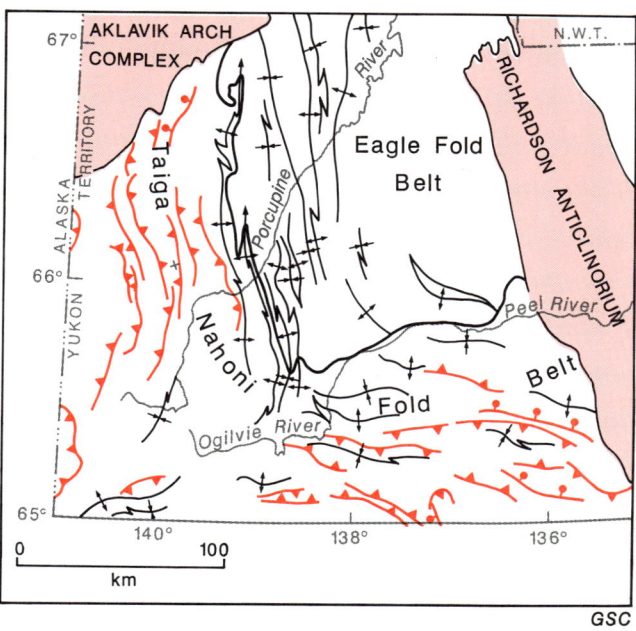

Figure 17.73. Ogilvie deflection and Wernecke Mountains structural elements at the headwaters of Ogilvie River, northern Yukon Territory, delineated by major folds and contraction faults comprising the Taiga-Nahoni fold belt. Heavy solid line is boundary among the several tectonic elements included in or adjacent to the deflection (after Norris, 1982a,b).

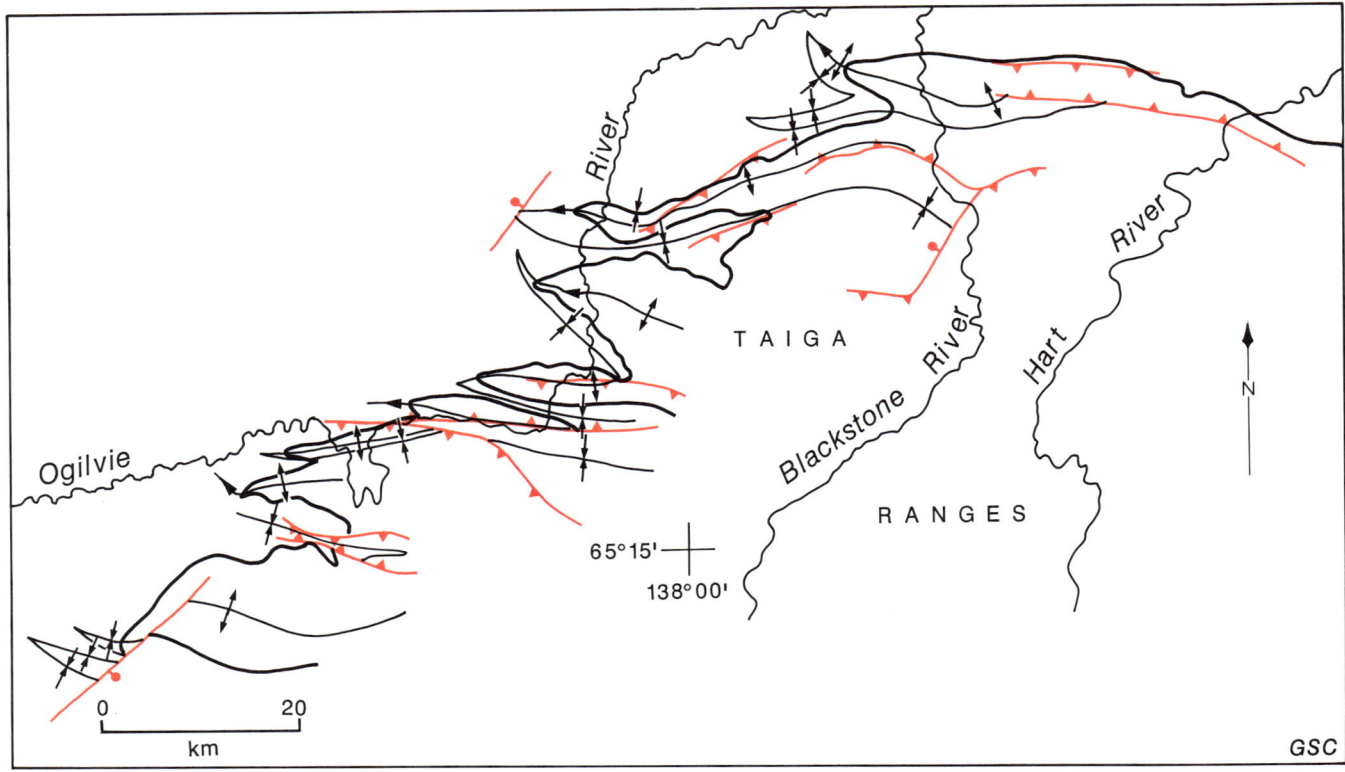

Figure 17.74. Ogilvie fold train in western Taiga Ranges, northern Yukon Territory. Heavy solid line outlines west-plunging folds in Lower and Middle Devonian Ogilvie Formation. Contraction faults with teeth in direction of hanging wall blocks offset on or both limbs of folds (after Norris, 1982a).

Eagle Fold Belt

The Eagle Fold Belt (Norris, 1974; Fig. 17.73) largely involves Upper Cretaceous clastic sediments. As in the Taiga-Nahoni fold belt, flexural-slip, cylindrical folds, commonly linked right-hand en échelon, express the structural style. Where asymmetrical their axial surfaces dip steeply eastward. Some folds can be traced for distances of up to 120 km. Locally they are cut by east- or west-dipping contraction faults of short length and throw.

Richardson Anticlinorium

The Richardson Mountains are atypical of the Foreland Belt. Their southern half is a single northwesterly trending anticlinorium which is strongly divergent from the east-west trend of the northern Mackenzie, Wernecke and Southern Ogilvie mountains (Fig. 17.73). To the east and west they are bordered by undeformed strata beneath the Northern Interior Plains and Eagle Plains respectively. The anticlinorium is 360 km long and 90 km wide. Its axis closely coincides with the depositional axis of lower Paleozoic sediments of the Richardson Trough, a possible aulacogen in the ancient northern continental margin (see Chapter 7). The anticlinorium is flanked by younger, shallow-water platform carbonate and shale. Beneath the axial lower Paleozoic strata a positive Bouguer gravity anomaly probably reflects a Precambrian crystalline basement core and may suggest the absence of a crustal root. Both the anticlinorium and the northern Richardson Mountains are cut by a series of faults assigned to the Richardson Fault Array.

Richardson Fault Array

The Richardson Fault Array (Norris and Hopkins, 1977), a family of north-trending, curviplanar, near-vertical faults, extends for some 600 km from the Mackenzie Mountains, through Richardson Anticlinorium and into the Aklavik Arch Complex to beneath Tuktoyaktuk Peninsula (Fig. 17.1, Section B, in pocket; 17.75). To the northeast it probably continues across the continental shelf to connect with the Cape Kellet Fault Zone (Norris and Yorath, 1981) west of Banks Island. From south to north, around its broad arc the faults change style from steeply dipping, high-angle reverse faults to vertical normal faults. The great length of the array, the curvature and anastomosing pattern of surface fault traces, the stratigraphic contrasts across and within the array, and the reciprocation of elevated and depressed blocks across as well as along the zone, collectively identify it as a structure of fundamental crustal magnitude and importance. Moreover, the omission of more than half of the estimated total thickness of supracrustal rocks in one block (Norris, 1980) suggests that Precambrian crystalline basement was involved, at least locally.

The Richardson Fault Array controlled the initial development of both the Richardson Trough and, later, Richardson Anticlinorium. From mid-Proterozoic until Devonian time the main components were probably dextral wrench faults (Norris and Yorath, 1981). Reactivation of the faults in the Late Cretaceous and early Tertiary was predominantly dip-slip, simultaneously resulting in inversion of the trough into the anticlinorium and subsidence of the continental margin underlying the Mackenzie delta.

Extrapolation of extension faults beneath Tuktoyaktuk Peninsula to west of Banks Island may provide a structural link between the Cordilleran and Innuitian orogens (Norris, 1983a). The link, which occurs approximately at 68°N and 137°W, marks the region where the northerly trends of the Cordillera bifurcate to a northeasterly Innuitian trend expressed by normal faults beneath Tuktoyaktuk Peninsula and a west- and northwest-trending Alaska arm expressed by thrust faults and folds in the British Mountains and Brooks Range and beneath the adjacent continental shelf. This bifurcation, termed the "Porcupine virgation" (Norris, 1983a), is perhaps a domain of extension in supracrustal rocks which increases northward beneath the continental shelf and which is complemented by a domain of synchronous contraction comprising thrust faults in the Brooks Range and British Mountains (Fig. 17.76).

Brooks Orocline

Structures in the Brooks Range and British Mountains, together comprising Romanzof Uplift (Norris, 1974) and nearby Beaufort Shelf are disposed in a broad arc from northerly trending in the east to easterly trending in the west (Fig. 17.76). The curvature occurs over an arcuate distance of 400 km between Blow River in the east and the Sagavanirktok River to the west. On the northern mainland the orocline involves sedimentary and igneous rocks, ranging in age from Proterozoic to Early Cretaceous. Beneath the continental shelf it is overlain by thick Tertiary clastics shed from the Cordillera.

In the eastern Brooks Range of Alaska, contraction faults and ancillary folds trend slightly north of east. Eastward, near the International Boundary, structures veer, first gently to the east then to the southeast and finally to the south in Barn Mountains at the apex of the orocline (Fig. 17.62).

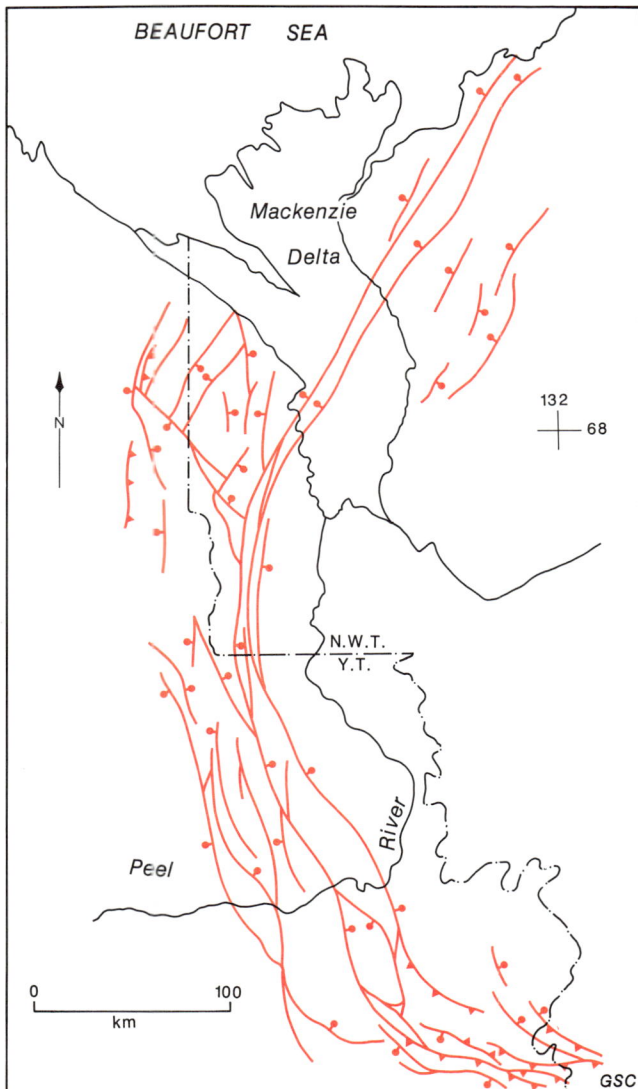

Figure 17.75. Richardson fault array, northern Yukon Territory and northwestern District of Mackenzie. Faults are shown as solid lines; solid circles are on side of downthrown segments of blocks, teeth on hanging wall side of contraction faults (after Norris, 1985).

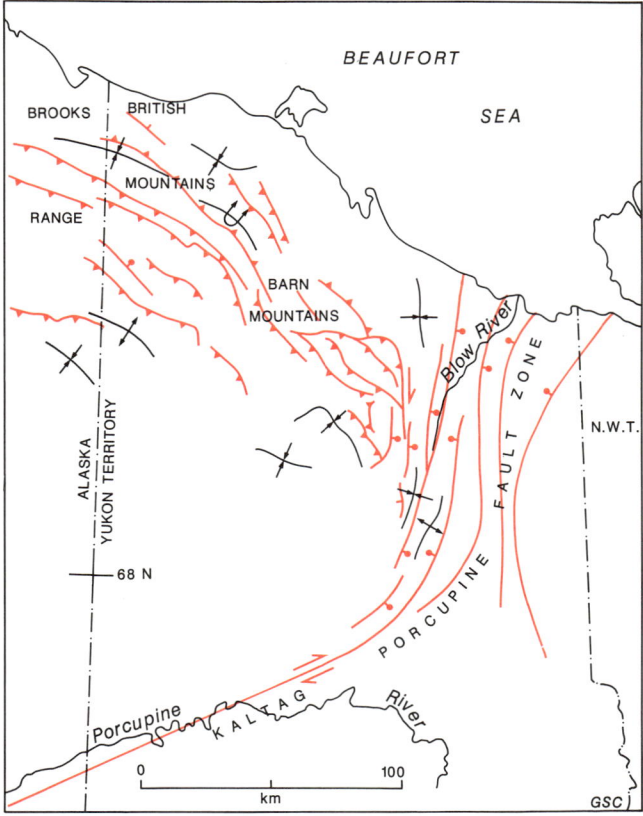

Figure 17.76. Brooks orocline, northern Yukon Territory as delineated by major faults and folds. Arrows indicate direction of displacement on strike-slip faults where known or reasonably inferred, solid circles are on side of depressed blocks, teeth on contraction faults are in direction of hanging walls (after Norris, 1981a,b).

Two structural styles are evident in the British and Barn mountains. The older, of pre-Mississippian age, is confined to the Precambrian Neruokpuk Formation and comprises open to collapsed, cylindrical flexural-slip folds with axial planes either vertical or steeply dipping to the southwest or northeast, and high-angle reverse faults. In contrast, imbricate thrust faults and open folds characterize the Cretaceous style. Eastward and southward around the orocline the faults decrease in number and stratigraphic separation and merge with high-angle strike-slip and normal faults of the Kaltag-Porcupine Fault Zone (Fig. 17.76).

An array of diapiric anticlines in the thick Tertiary succession occurs beneath the continental shelf (Fig. 17.1, Section A, in pocket). Eastward from the offshore extension of the International Boundary individual anticlines trend east-southeast, parallel with the structural grain of the British Mountains, to the approximate position of the offshore extension of the Kaltag-Porcupine Fault Zone. There, some of the anticlines veer southeasterly, subparallel with the fault zone. Within and east of the zone, there is a marked reduction in the number of large diapirs in conjunction with a change in structural style from elongate, doubly plunging anticlines with curviplanar axial surfaces to pervasively growth-faulted domes.

The structural grain of the orocline may be primarily the product of counterclockwise rotation of the Arctic Alaska plate about a pivot point located west of Mackenzie Delta at about 68°N and 137°W, near the south end of the Porcupine segment of the Kaltag-Porcupine Fault Zone (Blow Trough - see Chapter 9).

Aklavik Arch Complex

The Aklavik Arch Complex (Norris, 1974; Yorath and Norris, 1975; Norris and Yorath, 1981) is a composite, northeast-trending element extending in Canada from the Keele Range near the International Boundary to east of Mackenzie Delta (Fig. 17.77). Its various components are bounded and partitioned by northeasterly trending vertical faults and, beneath the southern Mackenzie Delta, the complex is traversed by the Richardson Fault Array. An influence on younger structures is suggested at the southwestern end of the complex where the Kaltag-Porcupine Fault comes close to but ultimately is deflected away from it. Although it truncates the northerly trending structures of the Nahoni and Eagle fold belts, some folds in the northern part of the latter are deflected subparallel with the complex. Where it intersects the north-trending projection of the Richardson Anticlinorium, the Aklavik Arch Complex is expressed by a group of small uplifts including the Rat, Scho, White and Cache Creek uplifts. To the northeast and southwest, other components of the complex respectively occur in the Campbell and Dave Lord uplifts.

Five angular unconformities and as many disconformities partition the stratigraphic succession in the Aklavik Arch Complex and attest to a prolonged tectonic history extending from Late Proterozoic to Tertiary time (Norris, 1974). The two most important deformations recorded occurred between Early Devonian and Middle Permian time and during Late Cretaceous and early Tertiary time. The former may have coincided in part with the Ellesmerian Orogeny in the Arctic Archipelago and perhaps with a probable rifting episode in Selwyn Basin and Misty Creek Embayment. Of note is the fact that the Aklavik Arch Complex lies close to the possible suture zone between the Arctic Alaska Terrane and ancestral North America.

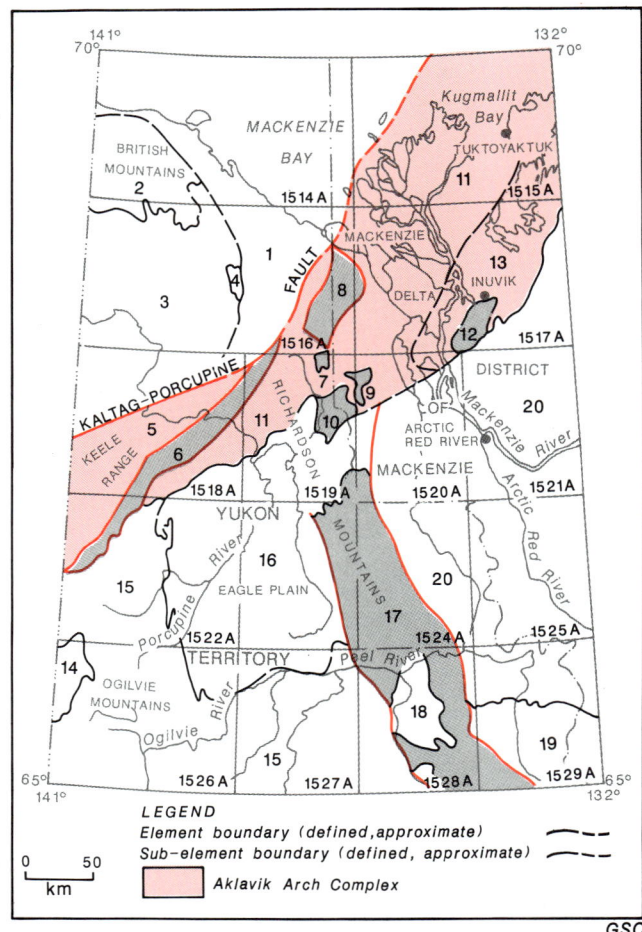

INDEX TO NUMBERED SECTORS

1. Rapid Depression
2. Romanzof Uplift
3. Old Crow-Babbage Depression
4. Barn Uplift
5. Keele Platform
6. Dave Lord Platform
7. White Uplift
8. Cache Creek Uplift
9. Scho Uplift
10. Rat Uplift
11. Canoe Depression
12. Campbell Uplift
13. Eskimo Lakes Uplift
14. Kandik Basin
15. Taiga-Nahoni Fold Belt
16. Eagle Fold Belt
17. Richardson Anticlinorium
18. Bonnet Plume Basin
19. Mackenzie Fold Belt
20. Northern Mountain Platform

Figure 17.77. Tectonic elements of northern Yukon Territory and northwestern District of Mackenzie (after Norris, 1983a,b). GSC map numbers are in bottom right corner of grids.

PART F. TRANSCURRENT FAULTS

Summary

Restorations of paleogeography in the Canadian Cordillera have been concerned mainly with the effects of contractional deformation and have been attempted primarily in the Foreland Belt where basement is known not to have been involved. It has become increasingly apparent, however, that the largest dislocations of paleogeographic elements, and the most difficult to characterize, are those which have resulted from regional, generally dextral, transcurrent faulting. Transcurrent faults form several well defined systems, each commonly including a number of anastomosing strands (Fig. 17.78). From east to west these are: Tintina and Northern Rocky Mountain Trench faults which displaced the western part of the miogeocline and overlying allochthons as much as 750 km between mid-Cretaceous and Eocene times; Semenof, Teslin, Thibert, Kutcho and Pinchi faults along or near the eastern boundary of the Intermontane Belt which offset island arc and oceanic terranes more than 200 km in post-Early Jurassic and pre-Late Cretaceous times; a family of more westerly post-mid-Cretaceous faults, namely, Finlay, Ingenika, Takla, Fraser River and Straight Creek faults with offsets across Omineca, Intermontane and Coast belts of about 100 km; Yalakom, Pasayten and Ross Lake faults with displacements in the Coast Belt of about 200 km; Denali, Duke River, Chatham Strait faults and possibly the Work Channel Lineament and Harrison Fault showing displacements across the Coast and Insular belts amounting to 300 km for the Denali Fault and between 150 and 200 km for the Chatham Strait Fault; and finally, Queen Charlotte and Fairweather faults currently active along the western continental margin with unknown displacements. Dextral displacement on the Kaltag-Porcupine Fault in the northern Yukon was possibly about 30 to 50 km, much less than the minimum estimates of 150 to 200 km suggested for the Kaltag Fault in Alaska. The fault may have been active from pre-mid-Cretaceous to early Tertiary times.

Post-mid-Cretaceous dextral movements west of the Foreland Belt, on geological evidence, suggest that the inboard terranes have been displaced about 750 km, and the most outboard terranes at least 1300 km northward, relative to ancestral North America.

Tintina and Northern Rocky Mountain Trench System

H. Gabrielse

The Tintina and Rocky Mountain trenches together form one of the world's great topographic lineaments extending for more than 2600 km from Alaska to Montana (see Map 1701A, in pocket; Fig. 17.78). The Tintina and Northern Rocky Mountain trenches are loci of major dextral transcurrent displacements but the southern part of the Southern Rocky Mountain Trench shows no indication of strike-slip faulting in supracrustal rocks.

Bedrock is sparse within the trench floors. Near Dawson, gouge and crushed rock a few hundred metres wide mask the Tintina fault zone (Aho, 1959). In many places the trenches contain Paleogene (mainly Paleocene and Eocene) nonmarine clastic rocks which are gently to tightly folded. Bimodal Eocene volcanic rocks are present along the Tintina Fault northwest and southeast of Ross River and may have been emplaced in pull-apart basins (Jackson et al., 1986; M.J. Pride, pers. comm., 1985, 1987). Along the Northern Rocky Mountain Trench and related lineaments the fold axes of these rocks trend more westerly than the related faults indicating a component of dextral transcurrent displacement (Gabrielse, 1985). In the Northern Rocky Mountain Trench Eocene dacitic volcanic rocks occur at its junction with the Spinel and Kechika faults (Fig. 17.79). Eocene rhyolite is found locally along the Kechika Fault. A swarm of north-trending Eocene lamprophyre dykes between Spinel Fault and the Rocky Mountain Trench represent a tensional fracture set resulting from dextral offsets along the faults. A set of faults, including the Kechika and Spinel, are characterized by intense cataclasis and mylonitization. Granitic rocks may be foliated over widths of several kilometres culminating in mylonite in the more highly strained rocks. Near-horizontal stretching lineations and the orientation of C/S planes demonstrate dextral strike-slip movements. East-west trending cleavage and crinkle crenulations in metamorphic rocks and west-northwest trending folds in Paleogene rocks in fault zones west of the Northern Rocky Mountain Trench are believed to be related to regional dextral strain. Some of the larger scale westerly trending folds and thrust faults may have a similar origin. A large fault south of the junction of the Kechika and Northern Rocky Mountain Trench faults is interpreted as a contraction fault related to dextral transcurrent motion. The hanging wall rocks form an intensely strained shear zone characterized by shear bands, sheath folds and stretching lineations all consistently indicating a north over south sense of displacement (Evenchick, 1985). In the Cariboo Mountains a group of dextral transcurrent faults, presumably related to the fault zone in the Northern Rocky Mountain Trench, is marked by mylonite zones which truncate isograds, thrust faults and a pluton (Struik, 1985).

Estimates of displacements on the Tintina-Northern Rocky Mountain system are based on apparent offsets of shelf to off-shelf facies boundaries of Paleozoic rocks, structural domains and plutons. They range from 425 km to 500 km for the Tintina Fault (Tempelman-Kluit, in press-a) and from 750 to more than 900 km for the Northern Rocky Mountain Trench and related faults (Fig. 17.80; Gabrielse, 1985). Mid-Cretaceous displacements are demonstrated by K-Ar ages on mica generated in shear zones in granitic rocks; Late Cretaceous to Eocene movements are suggested by the presence Santonian-Campanian, Paleocene and Eocene sediments in basins along the fault zones. Eocene volcanism and emplacement of lamprophyre dykes was restricted to the fault zones.

In the southern part of the Tintina Fault a set of more northerly trending faults intersect the Tintina at acute angles. Near the Tintina the faults are steep and are dextral strike-slip but near their southern extremities they appear to be steeply southwest-dipping thrust faults. The Tintina Fault is interpreted, therefore, as a shallowly rooted tear fault along which dextral slip took place as the supracrustal rocks were shortened above a basal

CHAPTER 17

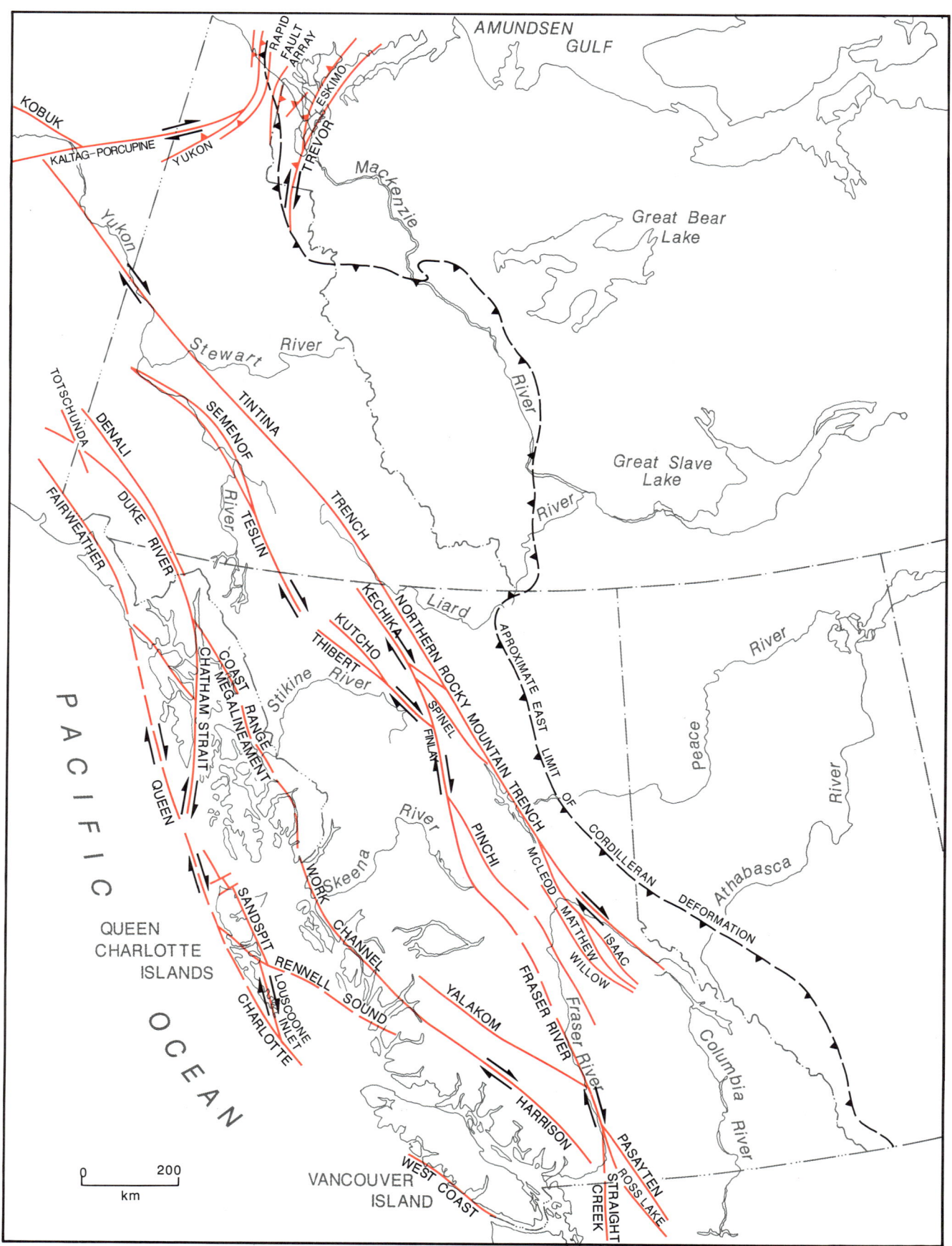

Figure 17.78. Distribution of the main transcurrent faults in the Canadian Cordillera and adjacent Alaska.

Figure 17.79. View northwest along the Northern Rocky Mountain Trench and Kechika River towards the Liard Plain in the background. This locality, about 135 km south of the boundary between British Columbia and the Yukon Territory is the farthest north where the east wall of the trench is clearly defined. Photo by H. Gabrielse. (GSC 205367)

detachment (Tempelman-Kluit, in press-b). On the other hand, the coincidence of volcanism and lamprophyre dyke emplacement with the Northern Rocky Mountain Trench and related faults suggests the fracture zones extended to considerable depth.

Semenof, Teslin, Thibert, Kutcho and Pinchi fault System

H. Gabrielse

This system of transcurrent faults occurs along or near the eastern margin of the Intermontane Belt. Except along the Teslin Trench and, locally, along the Kutcho Fault, the structures generally are not indicated by conspicuous lineaments, possibly because they have not been active since Mesozoic time.

In the Yukon the Semenof and Teslin faults bound the northeast side of the Whitehorse Trough, thus separating Mesozoic rocks having undergone brittle deformation to the southwest from openly folded Pennsylvanian rocks structurally overlying penetratively deformed mylonitic rocks of the Teslin Suture Zone to the northeast (Tempelman-Kluit, in press-b). Farther southeast the Teslin Fault is on trend with the Thibert Fault which forms the northeast boundary of the Cache Creek Terrane (Fig. 17.81). The Thibert Fault is a shear zone containing fish-scale serpentinite bodies and much quartz-carbonate alteration. It intersects the Kutcho Fault, a zone of mylonitic rocks containing serpentinite bodies and two zoned Alaskan-type ultramafic plutons. The Kutcho Fault is presumed to be an offset segment of the Pinchi Fault (Gabrielse, 1985) which lies along the northeast boundary of the Cache Creek Terrane. The Pinchi locally contains blueschist minerals. These have been dated as Late Triassic, the same age as the zoned ultramafic plutons.

Several lines of evidence suggest complex histories for the fault system. Fabrics in mylonite along the Teslin Suture Zone indicate dip-slip and strike-slip movements,

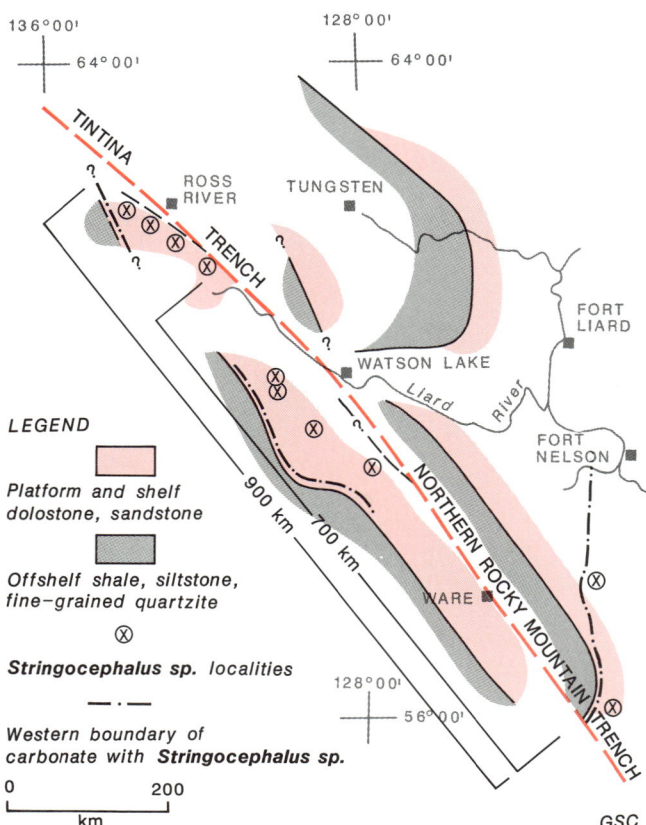

Figure 17.80. Suggested displacement of Middle Devonian carbonate (pink) to shale (grey) facies boundary along the Tintina and Northern Rocky Mountain trenches.

the latter with a dextral sense of displacement (Hansen, 1986). Displacements are constrained by the age of the youngest rocks involved in faulting (Bajocian) and the oldest overlapping strata (mid-Cretaceous). The poorly dated Tantalus Formation, possibly in part of Albian age, is interpreted to have been deposited in fault-bounded extension basins created by releasing double bends in the Teslin Fault (Tempelman-Kluit, in press-b). The Thibert Fault appears to be cut by a Late Cretaceous granitic pluton. Mylonitic mid-Cretaceous granitic rocks occur locally along the Kutcho Fault and, in places, along the Pinchi Fault. The presence of high pressure-low temperature mineral assemblages, including eclogite in the Teslin Suture Zone, jadeite in the Thibert Fault and blueschist minerals along the Pinchi Fault, imply an early history of subduction, at least in the Pinchi area, during Late Triassic time. Strike-slip faulting then became dominant perhaps culminating during the mid-Cretaceous. Dextral offset along the Thibert and Kutcho faults may be in the order of 200 km (Fig. 17.81).

Finlay, Ingenika, Takla, Fraser River and Straight Creek System

H. Gabrielse

This system of faults cuts across the Omineca and Intermontane belts and clearly offsets the Kutcho and

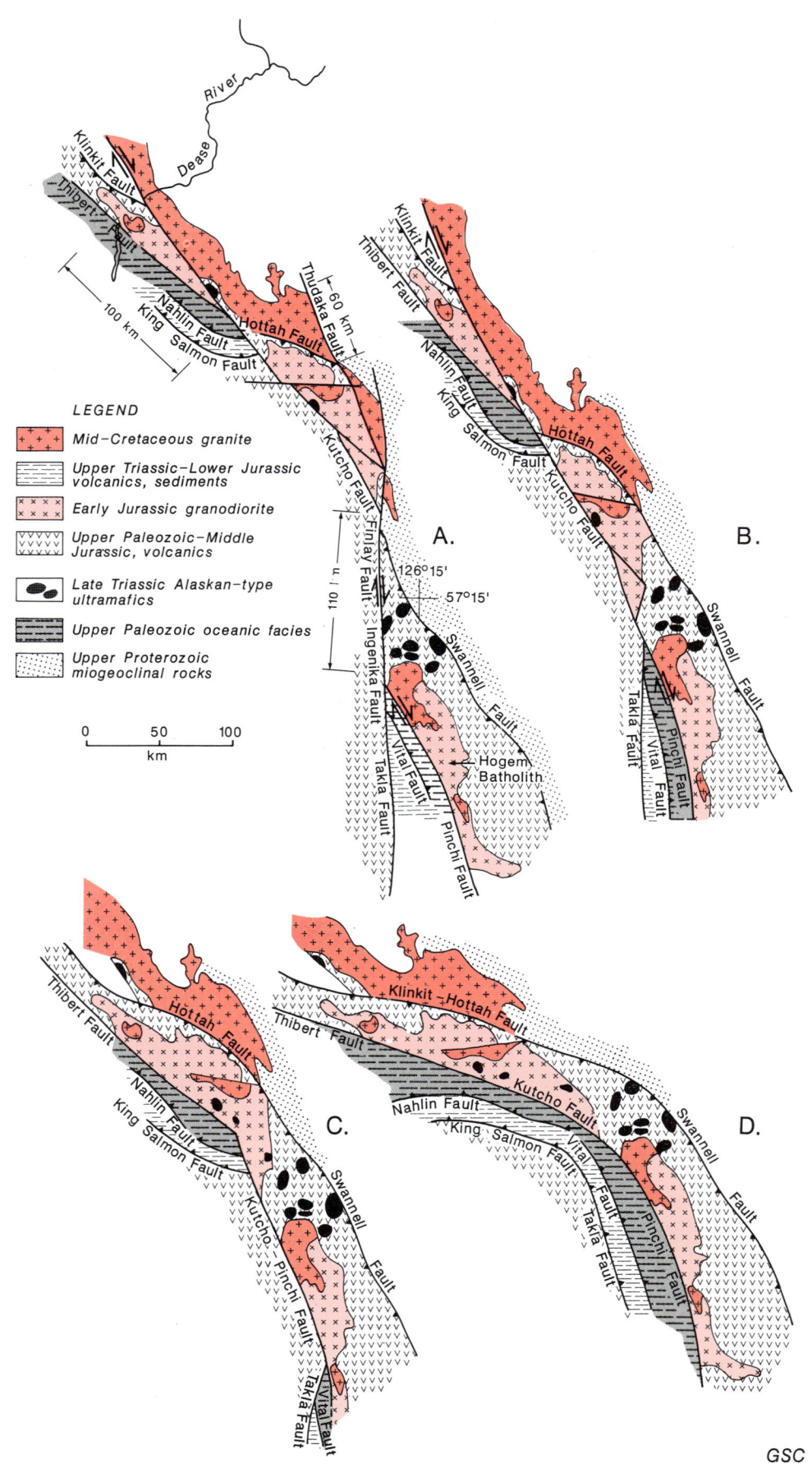

Pinchi faults and the Yalakom, Pasayten and Ross Lake faults. In the north it merges with the Kechika Fault. The connection between the northern and southern parts of the system is obscured by Tertiary volcanic rocks and may be more complex than shown in Figure 17.78. Almost everywhere the structures are marked by distinctive physiographic lineaments, a reflection of activity locally as late as Oligocene. The Finlay and Ingenika faults form shear zones, in places a few kilometres wide, in which numerous discrete, essentially vertical faults are present. The Fraser River Fault Zone consists of steeply dipping faults containing brecciated and shattered rocks (Duffell and McTaggart, 1952). For a considerable distance Cretaceous strata are restricted to the boundary faults. In its southern part the Fraser River Fault Zone and its continuation, the Straight Creek Fault, are cut by an Oligocene pluton. Eocene strata along the fault have been intensely deformed. Displacement along the Fraser River Fault Zone, determined from offsets of stratigraphic assemblages, are between 85 and 100 km. The Finlay, Ingenika and Takla faults offset stratigraphy and structures about 115 km.

Yalakom, Pasayten, Harrison, and Ross Lake System

J.W.H. Monger

Although these terrane-bounding faults are considered to have had strike-slip displacements, data concerning amounts of offset are equivocal. In southwestern British Columbia, an older set of mainly straight, north-northwest-trending faults, including the Pasayten, Yalakom and Ross Lake faults, is truncated and dextrally offset by the more northerly trending Fraser River-Straight Creek Fault System (Roddick et al., 1979). With the exception of the Ross Lake Fault, these are brittle structures, some of which have had dip-slip movement. Major problems involve correlations of older faults across the Fraser River-Straight Creek system and the senses and amounts of offsets on the older faults.

Kleinspehn (1985) suggested correlation of the Yalakom Fault with the Ross Lake Fault thus indicating offset along the Fraser system of 110 km (Fig. 17.82). Misch (1966) had previously proposed an offset of 190 km on the basis of correlation of metamorphic terranes in the Northern Cascades and southern Coast Plutonic Complex. Kleinspehn further proposed that this dislocation succeeded 150 ± 25 km of dextral displacement on the Yalakom Fault, a figure based on offset of Albian Jackass Mountain strata. This agrees well with the 160 km of dextral offset on the Ross Lake Fault which was proposed by Davis et al. (1978) to account for the apparent duplication of lithotectonic units in the northern Cascades. Restoration along these faults leaves the Tyaughton-Methow Trough as a continuous basin and the Bridge River-Hozameen complex as a continuous belt.

Alternatively Monger (1985; Fig. 17.15) proposed the following correlations and displacements: (1) Ross Lake Fault, separating high grade metamorphic rocks on the west (Custer/Skagit gneiss) from the Hozameen Group is matched, west of the Fraser system, with the boundary between high grade metamorphic rocks to the west and the Bridge River Complex; the offset is 70 km. (2) The Yalakom and Hozameen faults are matched across the Fraser River Fault System; both form the eastern/northeastern boundaries of the similar Hozameen and Bridge River units; the offset is 90 km. Finally, the Pasayten Fault is matched with the Hungry Valley Fault; both bound the Jackass Mountain Group on its northeast side; the offset is 110 km. The increased offset in the last correlation, which is the amount used by Kleinspehn (1985), may be due to Early Tertiary thrust faulting on the Hungry Valley Fault, complementary to early Tertiary wrench faulting on the Fraser River Fault System (Monger, 1985). A dextral offset of 70 to 90 km is favoured for the latter system, which agrees well with the 90 km on the Straight Creek Fault suggested by J.A. Vance (pers. comm., 1985). A second, ductile, shear zone locally exposed on the west side of Harrison Lake, is marked by a narrow (1 km) zone containing a prominent horizontal stretching lineation, orientated 340 degrees, which truncates older, pre-mid-Cretaceous structures (Monger, 1986). The northward continuation of this zone, the Harrison Fault, extends along the Harrison-Lillooet lakes valley, possibly crosses the Coast Belt and continues north to link up with either the Shakwak-Denali system or the Rennell Sound Fault.

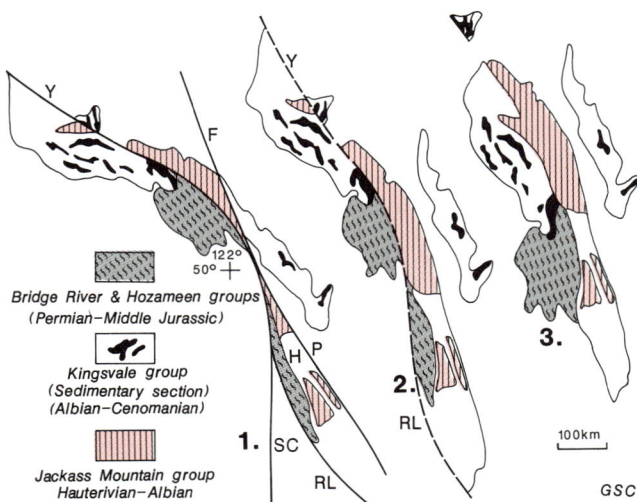

Figure 17.81. Proposed restoration of geological elements along the Thudaka, Kutcho, Finlay, and Pinchi faults. A, Generalized geology near faults. B, Restoration of 60 km on Thudaka Fault and 50 km on Finlay and Ingenika faults. C, Further restoration of 100 km on Kutcho Fault. D, Restoration juxtaposing Nahlin and Vital faults with displacements distributed on Thibert, Kutcho, Thudaka, Finlay, Ingenika, and Pinchi faults

Figure 17.82. Possible restorations of geology along the Fraser River-Straight Creek and Yalakom-Ross Lake fault systems. 1, present configuration; 2, restoration removing 110 km of dextral movement on the Fraser River-Straight Creek fault; 3, restoration removing 150 km of dextral displacement on the Yalakom-Ross Lake fault. F, Fraser Fault; Y, Yalakom Fault; H, Hozameen Fault; P, Pasayten Fault; SC, Straight Creek Fault; RL, Ross Lake Fault. Modified from Kleinspehn (1985).

Displacements on the older fault systems are difficult to determine. Tipper (1969) suggested that the distribution of distinctive Upper Triassic to Lower Jurassic stratigraphic assemblages in the Tyaughton area could be explained by 80-90 km of dextral offset on the Yalakom System. If the Ross Lake Fault is a separate fault, as suggested by Monger (1985), then the total displacement across both is at least 300 km. The Pasayten Fault is a major crustal break, and juxtaposes rocks of the Intermontane Belt to the east with rocks of the Methow Trough to the west, with no indications of a common history until late Early Cretaceous time. The fault is vertical and relatively straight. Coates (1974) proposed that it was a normal fault with downthrow to the west during the Cretaceous but entertained the possibility of dextral separation of several tens of kilometres because of a notable lack of correspondence between cobbles in Cretaceous conglomerates and potential source rocks to the east.

Sandspit-Louscoone Inlet and Rennell Sound fault systems
C.J. Yorath

The Queen Charlotte Islands are traversed by three major fault systems, the Rennell Sound Fault zone, the Louscoone Inlet Fault zone and the Sandspit Fault System.

The dynamics of the Louscoone Inlet-Sandspit Fault System were suggested to be related to rifting in Queen Charlotte Sound during the Early Miocene when the bulk of the Queen Charlotte Islands were located approximately 70 km southwest of their present position (Yorath and Hyndman, 1983). The 70 km of dextral displacement, based upon reconstructions of the distribution of mineral deposits and geophysical anomalies (Yorath and Chase, 1981), and supported by thermal and subsidence modelling (Yorath and Hyndman, 1983), agrees fairly closely with a maximum of 93 km suggested by Sutherland Brown (1968) on the basis of fold axes correlations across the Louscoone Inlet Fault. Subsequent detailed mapping, however, has shown little or no offset of the base of the Upper Cretaceous Honna Formation across this proposed fault (Thompson and Thorkelson, 1989). Following the northwesterly displacement of the Queen Charlotte Islands to near their present position, the ancient suture between the Alexander Terrane and Wrangellia may have been reactivated as the Rennell Sound Fault Zone which traversed an Early Cretaceous graben; the Louscoone Inlet-Sandspit system was offset by about 15 km. The final rotation of about 10° may have been accomplished by left lateral motion along the Beresford Bay Fault and, perhaps by other northwesterly directed sinistral faults associated with the rift zone in Queen Charlotte Sound. As noted above, recent studies have questioned these interpretations (Thompson and Thorkelson, 1989).

Queen Charlotte Fault
C.J. Yorath

The Queen Charlotte Fault defines the edge of the North American craton in the Queen Charlotte Islands area. Although the fault is the locus of primarily strike-slip motion at an average rate of about 55 mm/a (Riddihough, 1977; Keen and Hyndman, 1979), there is a discrepancy of between 10° and 20° between the strike of the fault off the southern islands and the direction of relative plate motion as predicted by global models (Minster and Jordan, 1978; Davis and Riddihough, 1982) which suggests a component of convergence along the boundary. Convergence and oblique underthrusting are supported by a wide range of data (see Chapter 13) including the presence of an apparent accretionary sedimentary prism with arcuate, high amplitude folds. The prism rests upon oceanic crust and is bounded on its landward side by the Queen Charlotte Fault.

Westcoast Fault
C.J. Yorath

On the west coast of Vancouver Island the Westcoast Fault (Muller, 1977) juxtaposes Jura-Cretaceous rocks of the Pacific Rim Complex against Jurassic plutonic rocks. Muller (1977) suggested that the fault was a subduction-related thrust fault. More recently, however, Brandon (1984) proposed that the West Coast Fault is a dextral transcurrent fault associated with the truncation of the west side of Wrangellia during the Late Cretaceous or early Tertiary.

Denali Fault System
C.J. Dodds

The Denali Fault System extends for more than 2000 km across south-central Alaska, southwestern Yukon, northwestern British Columbia and southeastern Alaska. It bounds the Saint Elias Mountains along most of its northeast side (Campbell and Dodds, 1982a,b,c, 1983a) and incorporates the Shakwak (Bostock, 1952; Muller, 1967), Dalton (Campbell and Eisbacher, 1974) and Chilkat River (MacKevett et al., 1974) faults. Along the Saint Elias Mountains the Denali Fault System is remarkably straight and steeply dipping. Northwest of the south end of Kluane Lake it occupies the prominent Shakwak Trench. Southwest of Kluane Lake it is discontinuously exposed where it transects the Kluane Ranges. At Haines, southeastern Alaska, it bends southward to either merge with or be offset by the Chatham Strait Fault.

The Denali Fault System consists of anastomosing strands of linear to curvilinear faults. Zones of major displacement comprise complex, mainly straight, vertical to steeply dipping zones of intensely foliated to highly broken, rusty stained rock with horizontal to near horizontal linear fabrics. Southwest of the main dislocation zone, transpressional stress resulted in curvilinear, generally northeast- to northward directed, open to tight, variably plunging, en échelon folds and associated contraction faults, well developed in the Gravina-Nutzotin belt. The trends of these structures commonly sweep progressively more west-northwesterly away from the zone of major breaking. Directly to the northeast of the Denali Fault System, Eisbacher (1976) reported curvilinear, variably plunging, southwest- to southward directed, open to overturned folds and associated faults within the Gravina-Nutzotin belt. These structures also are probably the result of transpressional stresses along Denali Fault System.

Northerly-trending extension faults, and belts of mafic and felisc dyke-swarm intrusions of various Cenozoic ages are probably the result of transtensional stress. Eisbacher and Hopkins (1977) noted tilting and broad local buckling of the pre-Oligocene pediment. Widespread fracturing of cobbles within the Oligocene and (?)older nonmarine clastic rocks southwest of the Denali Fault System are compatible with dextral stress. Southeast of Kluane Lake where the Denali Fault and the older Duke River Fault closely juxtapose or merge, pre-Cenozoic rocks are intensely and complexly deformed. To the southwest of the Denali Fault System folds and faults indicating dextral strain, have refolded and faulted earlier northwest-trending structures in Wrangellia W2 and the adjacent Alexander Terrane.

Locally, Miocene Pliocene Wrangell Lava and possibly Late Pliocene tillite are deformed by prominent west-northwest- trending, northerly directed cylindrical folds and related contraction faults (Souther and Stanciu, 1975). These structures merge with the southern known limit of the recently active Totschunda Fault (Richter and Matson, 1971; Plafker et al., 1977b). This compressive couple may provide a mechanism for stress transfer from the Denali Fault System south of Kluane Lake to the east to the Totschunda Fault to the west. It may also explain the lack of recent activity on the Denali Fault System between its junction with the Totschunda Fault and Kluane Lake, moreover it is consistent with the loci of current microseismicity (Horner, 1983) on these structures. High-angle linear scarps, sediment mounds and "sackung" features of latest Pleistocene to middle Holocene age are reported close to the Denali Fault System, suggesting late, mainly dip-slip related motion (Clague, 1979).

St. Amand (1957) postulated over 240 km of right-lateral offset along the Denali Fault System. Forbes et al. (1974) and Turner et al. (1974) suggested 400 km of displacement derived from the apparent dislocation of the MacLaren Metamorphic belt in Alaska from the Kluane Schist in southwest Yukon. Eisbacher (1976; Fig. 17.83) interpreted the Kluane Schist as the metamorphic equivalent of the Dezadeash flysch and, further, proposed 300 km of dextral displacement along the Denali Fault System based on the possible offset of the Jura-Cretaceous Nutzotin Mountains Sequence (Berg et al., 1972) from the Dezadeash flysch.

Most displacement along the Denali Fault System was post-Early Cretaceous, based on the youngest fossil-dated rocks within the Nutzotin Mountain Sequence and Dezadeash flysch, and the McLaren belt. Eisbacher (1976), Eisbacher and Hopkins (1977), and Lanphere (1978) considered most of the displacement to be post-latest Cretaceous and earliest Tertiary and to be largely post-approximately 55 Ma (earliest Eocene). Lanphere (1978), from evidence in Alaska, further suggested that displacements since 38 Ma (latest Eocene) on the McKinley, Shakwak and Dalton segments of the fault system cannot have been greater than 40 km. Richter and Matson (1971) indicated that the Denali Fault System southeast of its junction with the Totschunda Fault in Alaska has been essentially passive since Early and Middle Pleistocene time, and that subsequent dextral strain has occurred along the latter fault.

There is no evidence of post-latest Pleistocene to Present strike-slip displacement along the Denali Fault System in Canada. Scarps and aligned sediment mounds indicate only minor dip-slip movements during latest Wisconsin and early and possibly middle Holocene time, with no further movements since then (Clague, 1979). Microseismicity is centred along the section from north of Haines, Alaska to Kluane Lake, crudely along the trace of the Duke River Fault, and along the Totschunda Fault (Boucher and Fitch, 1969; Horner, 1983).

Duke River Fault

C.J. Dodds

The sinuous trace of the Duke River Fault (Muller, 1967) juxtaposes the Alexander Terrane to the southwest and a segment of Wrangellia (W_2) to the northeast (Fig. 17.2; Campbell and Dodds, 1982a,b,c, 1983a). Its locus mostly follows the Duke Depression (Bostock, 1948) where it closely adjoins and has been disrupted by the younger Denali Fault System. Northward it swings west-northwesterly into adjacent Alaska and is largely covered by Neogene Wrangell Lava. Farther northwest it may join with the Castle Mountain Fault. In northwest British Columbia, the Duke River Fault is truncated by the Denali Fault System. In southeastern Alaska, strands of the Duke River and Denali fault systems may be present on eastern Admiralty and Kupreanof islands, with further offset along the Chatham Strait Fault.

To the northwest, the Duke River Fault appears to be a fairly straight, steeply dipping single structure with the southwest side up. Locally it cuts the lowermost beds of the Wrangell Lava (Souther and Stanciu, 1975). Farther southeast almost to the southeastern Alaska border, it closely adjoins the Denali Fault System, and the combined displacements along these two major faults resulted in a "fault mélange" within a zone of intense deformation. There, the trace of the Duke River Fault is more difficult to define, but its locus appears to be more sinuous, possibly due to later displacements along the Denali Fault System. Dextral stress indicators are plentiful but it is difficult to distinguish between those related to motions along the Duke River Fault from those superimposed by later dextral movements along the adjacent Denali Fault System.

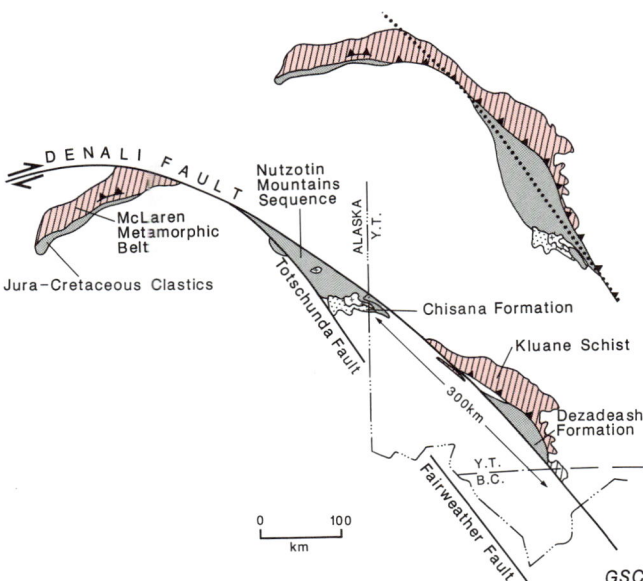

Figure 17.83. Restoration of 300 km of dextral transcurrent displacement along the Denali Fault. After Eisbacher (1976).

The Duke River Fault is interpreted as a major intracontinental dextral strike-slip fault with a considerable, but unknown displacement (Campbell and Dodds, 1978, 1979, 1983b). Time constraints for movement(s) along it are poorly known. Because only the lowermost flows of the Wrangell Lava are cut by the fault and K-Ar whole rock ages from the older parts of the lavas elsewhere yield Middle Miocene ages (Richter, 1976; Stevens et al., 1982), displacements along this part of the fault are probably small but must predate the Middle Miocene. Wrangellia W_2 is believed to have been brought inboard of the Alexander Terrane through large displacement along the Duke River Fault (Campbell and Dodds, 1983b). The youngest fossil-dated rocks within these two juxtaposed terranes are of latest Triassic age thus providing a maximum age for the large pre-Miocene motion along the fault. Moreover, segments of the Duke River Fault have undoubtedly been reactivated by movements involving the Denali Fault System.

Chatham Strait Fault
C.J. Dodds

The Chatham Strait Fault lies entirely within southeastern Alaska (Wright and Wright, 1908; Fig. 17.84). It forms a prominent, straight physiographic feature which transects the panhandle of Alaska from west of Coronation Island in the south to near Haines in the north. It is mostly concealed beneath Christian Sound, Chatham Strait, and Lynn Canal, which collectively form a linear fiord over 340 km long. To the north the fault is believed to join with the Denali Fault System, and to the south to merge with the Fairweather-Queen Charlotte faults.

The fault displaces the Alexander Terrane, the projected trace of Wrangellia W_1, the Taku Terrane, the Jura-Cretaceous Gravina-Nutzotin overlap assemblage, and the Chugach Terrane. Further, it offsets a lithologically distinctive belt of large elongate batholiths of mainly Late Jurassic to earliest Cretaceous age, mostly composed of quartz diorite to tonalite. It also appears to displace a belt of Eocene plutons which form the eastern part of the Sanak-Baranof belt (Hudson, 1983) which mostly intrude the Chugach Terrane. The latter may be further evidence of major post-Eocene movement along the Chatham Strait Fault.

Many estimates have been made of displacement along the Chatham Strait Fault based on various criteria (St. Amand, 1957-240 km; Lathram, 1964-190 km; Ovenshine and Brew, 1972-200 km; Sonnevil, 1981 - 100-180 km). Hudson et al. (1982) proposed a right-lateral separation of about 150 km, during post-mid-Cretaceous and pre-Holocene time based on the possible offset of a polydeformed metamorphic terrane, a Silurian turbidite sequence, Cretaceous flysch and mélange on Baranof Island and a Tertiary volcanic sequence. They postulated that the southern boundary of greenschist facies metasedimentary, metavolcanic and metaplutonic rocks exposed southward from near Haines for some 50 km along the west shores of Lynn Canal, could be displaced 148 km from the southern boundary of an equivalent metamorphic terrane on Admiralty Island. They further postulated that the northern boundary of a Silurian turbidite sequence (Point Augusta Formation) north of Icy Strait may have been offset some 150 km from the northern boundary of lithologically similar and age equivalent rocks (Bay of Pillars Formation) on the west side of Kuiu Island. Right-lateral shift of at least 150 km is suggested by the apparent offset of the northern boundary of a predominantly Cretaceous flysch and mélange terrane (the Chugach Terrane of Berg et al., 1972) on Baranof Island. Basaltic rocks outcropping in Icy Strait to the west of the fault may have been displaced from correlative volcanic rocks (Admiralty Island Volcanics) adjacent to the fault on southern Admiralty Island. Based on potassium-argon data from both areas, they infer that about 100 km of post-Oligocene dextral offset may have occurred. Marine geophysical data together with studies along the onshore projection of the Chatham Strait Fault reveal that it is not presently active and that the last significant displacement on it is pre-Holocene and may be pre-Quaternary.

The Chatham Strait Fault is considered by many to be an extension of the Denali Fault System. It is possible, however, that some of the Denali displacement was distributed on a family of faults east of the Chatham Strait Fault which may link with the Clarence Strait Fault and possibly also the Coast Range megalineament.

Fairweather Fault
C.J. Dodds

The Fairweather Fault (Miller, 1953) is the most important active fault on land along the Pacific margin of Canada and Alaska. The onshore part of the fault is located just inboard of the Gulf of Alaska along the southwestern periphery of the Saint Elias Mountains where it outcrops almost entirely within Alaska. The Fairweather Fault is believed to have undergone a lengthy and complex history, however the most recent displacements are dominantly dextral strike-slip (Plafker et al., 1978). Since 1899 major earthquakes (Ms 7.3 or larger) occurred along or adjoining its trace (Thatcher and Plafker, 1977; Stauder, 1960; Page, 1973, 1974; Lahr et al., 1979, 1980). It is considered to be a part of the current and late Cenozoic transform boundary between the Pacific and North American plates (Plafker et al., 1978). As suggested by Carlson et al. (1985), however, much of the latest Cenozoic transform plate motion may have occurred along the offshore Icy Point-Lituya Bay Fault, rather than along the northern, onshore part of the Fairweather Fault as proposed by Plafker et al. (1978).

The Fairweather Fault near the upper part of the Seward Glacier in southwestern Yukon trends westward where its strike-slip displacement may be taken up in whole, or in part, by a family of thrust or oblique thrust faults, the most important of which are the Chugach-Saint Elias and the Coal Glacier faults (Plafker et al., 1978). Other workers postulated, however, that the Fairweather Fault connects across the Saint Elias Mountains to the recently active, dextral strike-slip Totschunda Fault (St. Amand, 1957; Tocher, 1960; Grantz, 1966; Hamilton and Meyers, 1966; Richter and Matson, 1971; Page, 1969, 1973; Naugler and Wageman, 1973; Lahr and Plafker, 1980). Farther to the south-southeast, along the continental margin, the Fairweather Fault probably connects with the Queen Charlotte Fault (Tobin and Sykes, 1968; Morgan, 1968; Carlson et al., 1985), which is considered to be the transform boundary between the Pacific and North American plates (Atwater, 1970).

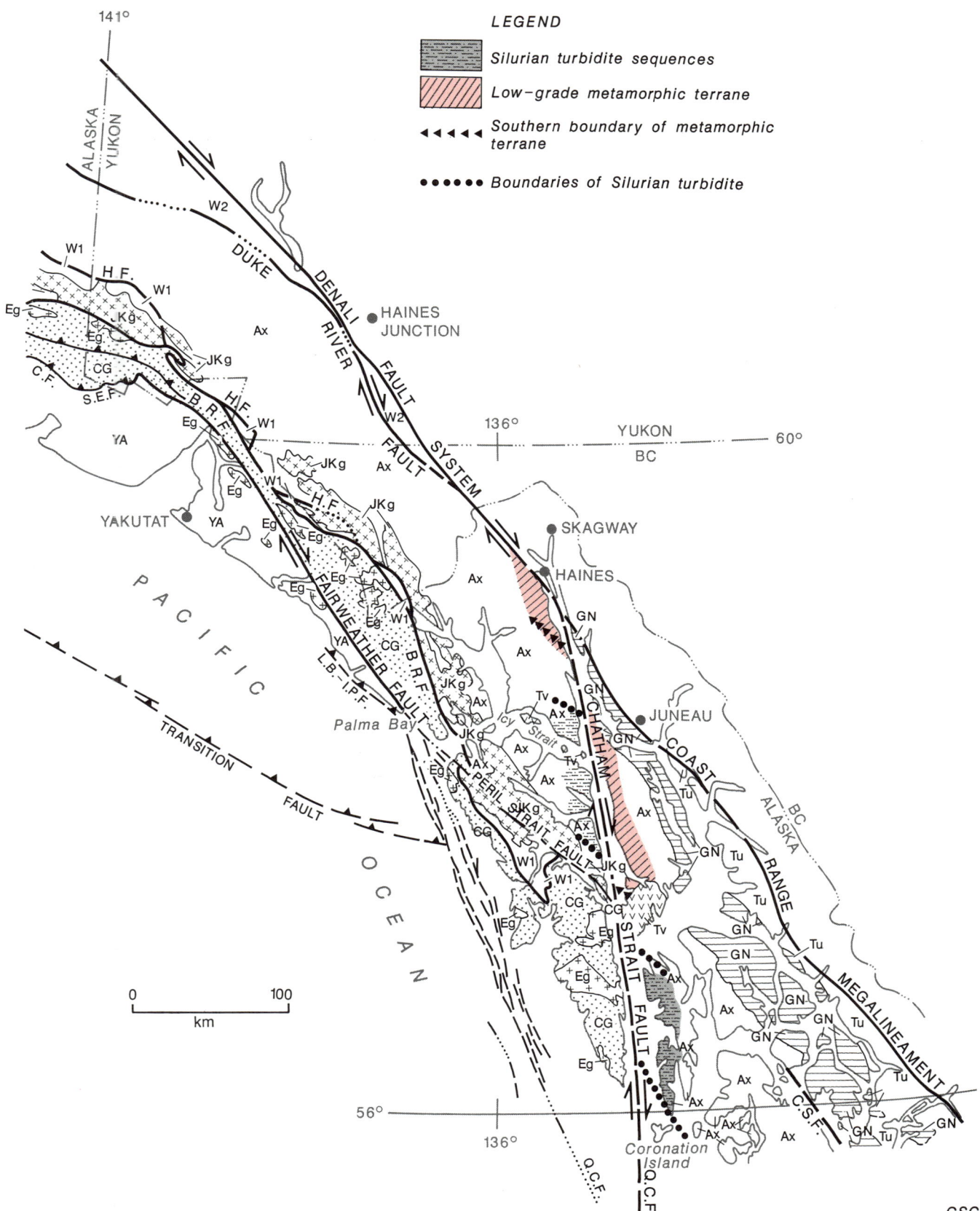

Figure 17.84. Outline map showing the location of major faults in southeastern Alaska, northwestern British Columbia and southwestern Yukon Territory. Ax, Alexander Terrane; W_1, W_2, Wrangellia; Tu, Taku; CG, Chugach Terrane; YA, Yakutat Terrane; GN, Gravina-Nutzotin Belt; JKg, Late Jurassic-earliest Cretaceous plutons; Eg, Eocene plutons; Tv, Tertiary volcanics; C.F., Chugach Fault; S.E.F., Saint Elias Fault; B.R.F., Border Ranges Fault; H.F., Hubbard Fault; L.B.-I.P.F., Lituya Bay-Icy Point Fault; C.S.F., Clarence Strait Fault.

Onshore the trace of the Fairweather Fault forms a prominent, remarkably straight, northwest-trending, mostly ice-, water- and drift-filled physiographic depression, which extends for about 280 km from Icy Point, Palma Bay in Alaska to the upper Seward Glacier area in southwesternmost Yukon. The onshore Fairweather Fault is believed to correspond to a segment of a major boundary between the Chugach Terrane to the northeast and the Yakutat Terrane to the southwest (Monger and Berg, 1984; Bruns, 1983a). The Fairweather Fault forms the northeast boundary of the Yakutat block as defined by Plafker et al. (1978).

The main fault zone of the northern Fairweather Fault is exposed at only a few localities (Plafker et al., 1978) where it comprises intensely sheared to locally crushed rocks which enclose boudin-shaped remnants of country rock. Subhorizontal mullion structures and slickensides within the shear zone indicate dextral strike-slip movement.

Most of the reliable data concerning the Fairweather Fault involve its late Quaternary displacements. In a review of movements along it during this period, Plafker et al. (1978) reported evidence for 50 to 55 m of dextral-slip since 940-1300 to 200 BP, based on stream and lateral moraine offsets, and speculated an additional 5.5 km of dextral-slip during the last 100 000 years ago (Sangamon), derived from apparent drainage offsets. Comparing displacement rates calculated from the latter movements with those deduced for the current northward relative motion of the Pacific Plate (about 5.4 cm per year, Minster et al., 1974), they postulated that the Fairweather Fault is presently, and conceivably has been since at least Sangamon time, the transform boundary along which most of, if not all, the relative motion between the Pacific and North American plates occurred. Prior to this time, they contend that the plate transform involved one or more offshore faults.

Kaltag-Porcupine Fault and related fault systems

C.J. Yorath

The Kaltag-Porcupine Fault can be traced discontinuously for approximately 1200 km from Norton Sound on the Bering Sea, across Alaska and thence into northern Yukon to the Beaufort Shelf on the southern rim of the Canada Basin. Direct connection of the Kaltag and Porcupine faults is obscured, however, by an extensive drift-covered area along the Yukon River and it is possible that they could be separate structures. From where it enters Canada on the south side of the Old Crow granite the Porcupine Fault approaches the Yukon Fault and thence turns sharply towards the north and splays into the Rapid Fault Array (Norris and Yorath, 1981), a fan-shaped family of nearly vertical faults cutting lower Tertiary and older sequences (Fig. 17.78).

In Alaska the best evidence for the timing of large-scale, dextral displacement along the Kaltag Fault probably is the lateral separation of Albian and Cenomanian rocks on the east flank of the Yukon-Koyukuk Basin and the offset of geological trends within the basin. In Canada the juxtaposition of lower Paleozoic and Upper Cretaceous (Santonian) formations along Porcupine River supports the Late Cretaceous or younger movement. The evidence is further substantiated by the presence of vertical coal measures of the lower Tertiary Reindeer Formation in the Rapid Fault Array immediately west of Mackenzie Delta.

Norris and Yorath (1981) suggested that the Kaltag-Porcupine Fault dislocated the Tintina-Kobuk Fault system by about 100 km prior to the deposition of the mid-Cretaceous clastic succession in Yukon-Koyukuk Basin. Additional slip in the early or mid-Tertiary could account for the suggested 65 to 130 km offset of the basin and for the net right-lateral separation of 200 km for the Tintina and Kobuk faults.

This net horizontal displacement is assumed to have been attenuated among the components of the Rapid Fault Array. Beneath Mackenzie Delta the more easterly strands of the array merge with vertical faults of the Eskimo Lakes Fault Zone on the seaward flank of the Aklavik Arch Complex. The Rapid Fault Array was suggested by Young et al. (1976) to comprise a series of high-angle reverse faults which, in conjunction with components of the Eskimo Lakes Fault Zone, define a structural depression beneath the western Mackenzie Delta.

An alternative view, based on the trends of Paleozoic facies boundaries in Yukon Territory is that offset on the Porcupine Fault may have been minor, perhaps a few tens of kilometres (L.S. Lane, pers. comm., 1988). If so then the Porcupine Fault and the Kaltag Fault in Alaska were not continuous throughout their history.

PART G. EOCENE EXTENSION FAULTS

Summary

The important role of Eocene extensional faulting in Canadian Cordilleran tectonics has been recognized only recently. In the southern Omineca Belt extensional shear zones occur in a southward widening wedge from north of the Monashee Complex to the International Boundary where they are present between Okanagan and Kootenay lakes. The area of Eocene extension continues southward in northeastern Washington and northern Idaho to disappear beneath volcanics of the Snake River Plain.

The Tatla Lake Metamorphic Complex in the southwestern Intermontane Belt has been interpreted as an Eocene Metamorphic Core Complex with crustal extension and denudation faulting. Another area of possibly similar structural style is the metamorphic terrane in the Omineca Belt between the Cariboo and Omineca mountains. In addition to the extension associated with tectonic denudation, widespread extension is represented by swarms of Eocene granitic and lamprophyre dykes, many probably related to dextral strike-slip regimes.

Although the distribution of metamorphic complexes in the southern Omineca Belt has been related to Eocene extension, faulting, denudation, volcanism and sedimentation, the ultimate cause of these phenomena remains obscure. The complexes may represent gravitational collapse of crust which had been overthickened during Mesozoic compression (Coney and Harms, 1984). If so, the collapse may have been initiated by a change in plate interactions perhaps expressed by the dextral-fault strain regime which resulted in the numerous sediment- and volcanic-bearing grabens in the southern Cordillera.

Southern Cordillera
Omineca Belt
R.R. Parrish

Sets of conjugate, high-angle normal faults, related shear fractures and dyke-filled extension joints of possible Eocene age record regional extension within the Monashee Complex (Fig. 17.85; Brown and Journeay, 1987). In the southern part of the complex displacements on high-angle normal faults are up to several hundred metres but in the northern part they are minor. They are regarded as scissor-type faults which, with related structures, have resulted in extension of from 4-5 km or 5%. In contrast to the faults within the complex, low-angle detachment faults along the western and eastern margins have effected displacements of about 10-15 km. On the east side, the Columbia River Fault dips from 20-30° eastward and can be traced for 150 km, well beyond the northern and southern limits of the Monashee Complex. The fault is marked by a zone of mylonite up to 1 km wide, in part intensely fractured and folded (Lane, 1984a). It has down-dropped both low- and high-grade metamorphic rocks of the Selkirk Allochthon onto high-pressure mylonitic rocks of the Monashee Décollement (Brown and Journeay, 1987). Orientation of slickensides, fibre growth and strain features in the mylonite indicate normal displacement. The fault zone truncates folds and metamorphic isograds and at least the youngest structures are of Eocene age. A low-angle detachment zone on the west side of the northern Monashee Complex juxtaposes brittley deformed metasedimentary rocks of greenschist-grade against mylonitic footwall rocks of upper amphibolite grade. The fault and/or related faults, such as the Okanagan Valley Fault, can be traced southward more than 300 km into northern Washington (Fig. 17.85).

Extensional shear zones have been studied in most detail in the region near Okanagan Lake and around the Valhalla Complex. The Okanagan Valley Fault is marked by a gently west-dipping, down-dip-thickening shear zone, 1 to 2 km thick at the top of the lower plate which consists of granodiorite, granite, syenite, amphibolite and greenstone (Tempelman-Kluit and Parkinson, 1986; Fig. 17.86). Coarsely crystalline amphibole granodiorite gneiss in the lower plate grades upward through fine grained mylonitic gneiss, augen gneiss, and finally to mylonite and microbreccia at the top. The shear zone is overlain abruptly across a clean, sharp break by Eocene undersaturated volcanic rocks, Mesozoic granite, and Mesozoic and Paleozoic metavolcanic and metasedimentary rocks. Steeply and gently dipping extension faults and gravity-slide blocks characterize the upper plate.

The mylonitic foliation, dipping west at generally less than 20°, contains a consistent, widespread west-north-west plunging lineation. Asymmetrical augen and intersecting C and S surfaces indicate that the upper plate moved down to the west. Numerous radiometric age determinations from foliated rocks and associated dykes suggest Eocene displacement. Estimates of displacements based on possible offsets of plutons and volcanic rocks range from 60-90 km.

The Valhalla Complex (Reesor, 1965) comprises Mesozoic or older, polydeformed, high grade paragneiss and variably deformed, sheets of Cretaceous to Eocene granitoid rocks disposed in an antiformal elongate culmination (Fig. 17.85; Parrish, 1984; Parrish et al., 1985b; Carr et al., 1987). Its protracted history probably includes Middle Jurassic to Paleocene compressional metamorphism and deformation as well as extensive ductile deformation of Tertiary granite (Carr, 1985). The amphibolite-grade Valkyr shear zone, 1 to 3 km thick, separates high grade, penetratively deformed Valhalla Complex from surrounding somewhat lower grade rocks on its north, west and south flanks (Carr et al., 1987). The shear zone has a consistent easterly directed displacement sense of upper plate rocks (Carr, 1985; Carr et al., 1987). The eastern side is bounded by the north-south trending ductile/brittle Slocan Lake Fault which dips 30° east and exposes the complex as a tectonic window. Lower to middle greenschist facies ductile and brittle deformation of the Slocan Lake Fault outlasted movement on the Valkyr shear zone and truncated it after it was domed and deactivated on all but the east flank. The displacement sense of the Slocan Lake Fault is easterly-directed, similar to the Valkyr shear zone (Parrish, 1984). The two faults may be part of a middle to upper crustal normal fault or shear. Seismic studies support this model (Price et al., 1986; Cook et al., 1987). Early Eocene granite is involved in the deformation and the rocks have been intruded by the 52 Ma Middle Eocene syenite (Parrish and Carr, 1986).

The Valhalla Complex has K-Ar dates less than 60 Ma, whereas dates in the upper plate of both Valkyr shear zone and Slocan Lake Fault range from 70 Ma to 165 Ma. The K-Ar data are consistent with the Eocene extraction of the complex from considerable depth and high temperature to the surface during the normal faulting event.

The relationships summarized for the Valhalla Complex are similar to internal and external structural and geochronological aspects of other metamorphic complexes of the southern Omineca Belt (Okanagan, Kettle-Grand Forks, Priest River, part of Monashee and Clachnacudainn).

The Eocene normal fault/shear zone systems of the southern Omineca Belt form a regionally extensive network of faults accommodating significant extension (Fig. 17.85). The Kettle River and Columbia River faults are also Eocene normal faults. Some segments are in places intruded by 52 Ma Coryell syenite and are thus predominantly of Early Eocene age.

The west-dipping faults of southern British Columbia (Okanagan, Granby, Greenwood fault systems) involve Middle Eocene volcanic rocks of the Kamloops Group and so must be Middle Eocene or younger in age. Thus the east- and west-dipping major fault systems are not necessarily symmetrical, contemporaneous features.

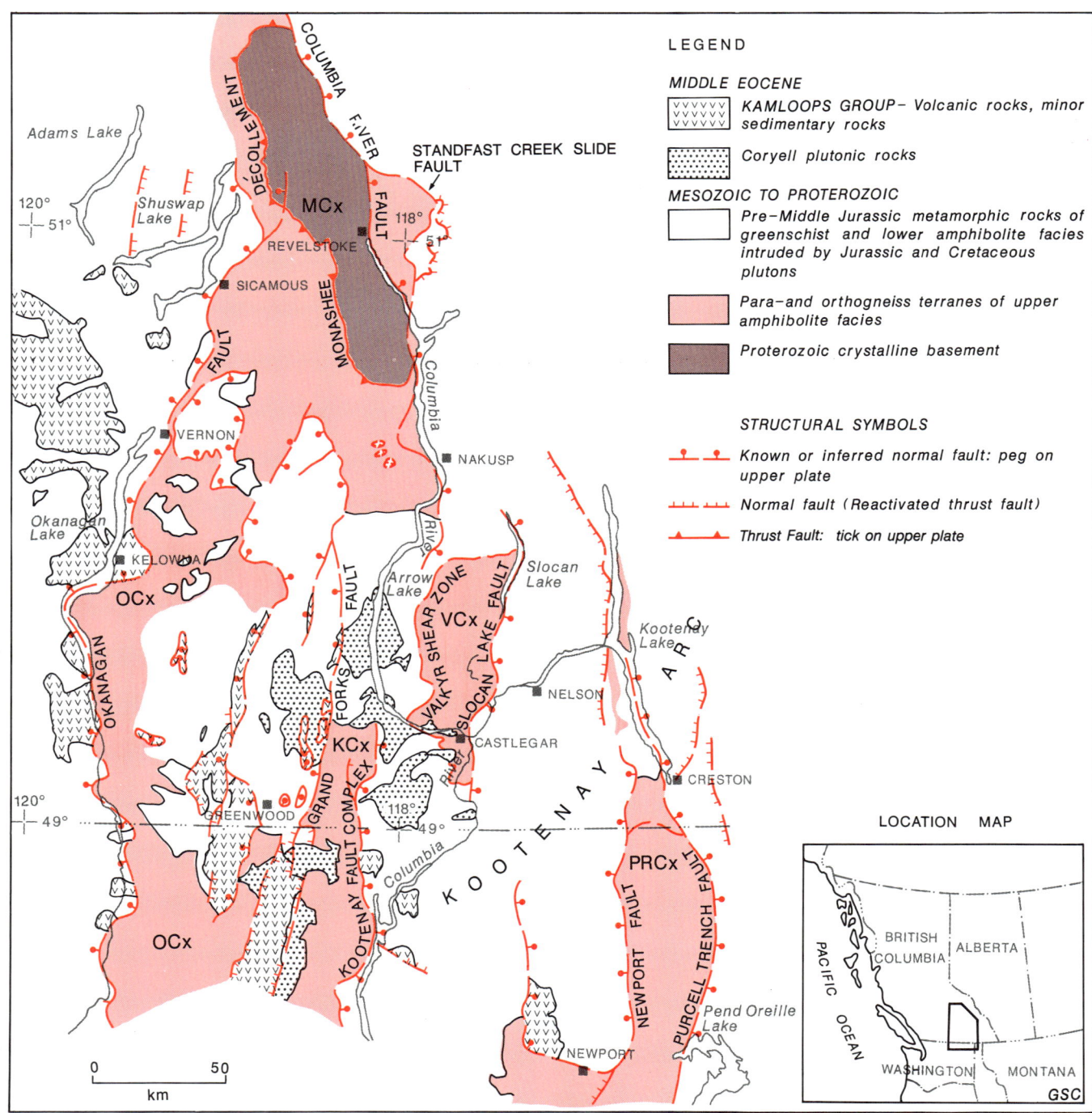

Figure 17.85. Main structural elements of the southeastern Omineca Belt showing the distribution of prominent normal faults. MCx, Monashee Complex; VCx, Valhalla Complex; OCx, Okanagan Complex; KCx, Kettle-Grand Forks Complex; PRCx, Priest River Complex; KF, Kettle Fault.

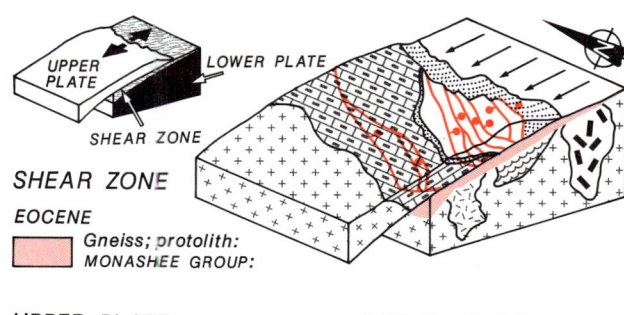
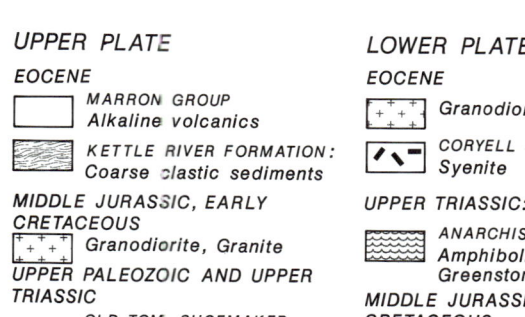

Figure 17.86. Model for Eocene extension along the Okanagan Fault Zone. The detachment is a gently west-dipping extension fault above a ductilely deformed gneiss (pink) at the top of the lower plate. After Tempelman-Kluit and Parkinson (1986).

Eocene extension has localized a number of north-northeast- trending basins of sedimentary and volcanic rocks which extend southward into northeastern Washington (Church, 1973; Ewing, 1980; Cheney, 1980; Price, 1981a,b; Harms, 1982; Parrish et al., 1985b). Eocene tectonic events appear to span an age range of 58 Ma to 47 Ma and may have resulted in an extension of about 60-80% across the width of the belt (Parrish and Carr, 1986).

Intermontane Belt

R.M. Friedman and R.L. Armstrong

The Tatla Lake Metamorphic Complex, which lies northeast of the Yalakom Fault on the southwestern side of the Intermontane Belt in British Columbia, has a typical metamorphic core complex geometry with an anticlinorial amphibolite-grade gneissic and migmatitic core underlying a ductilely sheared mylonitic zone, 1 to 2.5 km thick, in fault contact with low metamorphic grade cover rocks (Fig. 17.87). Early northwest-trending folds are observed in the granoblastic biotite and hornblende-bearing gneiss of the core. The ductilely sheared unit includes amphibolite-grade feldspathic and pelitic metasedimentary rocks structurally overlain by greenschist-grade intermediate composition metavolcanic rocks and intruded by tonalitic to granodiorite orthogneiss. A pervasive gently dipping mylonitic foliation and stretching lineation which trends east-northeast in northeastern areas to west-northwest adjacent to the Yalakom fault in the northwestern part of the area have been locally deformed by north-south trending recumbent folds. Kinematic indicators imply a tops-to-the-west sense of shear throughout the ductilely sheared unit.

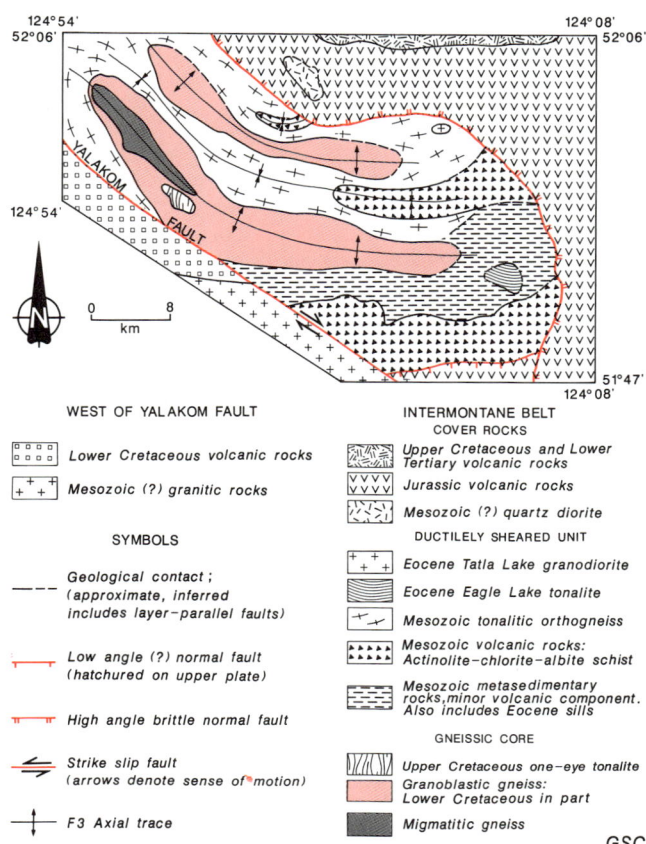

Figure 17.87. Map of the Tatla Lake Metamorphic Complex on the northeast side of the Yalakom Fault.

Both gneissic and ductilely sheared units have been warped by upright east-west to northwest- trending map-scale folds. The Yalakom strike slip fault cuts all units and forms the southwestern margin of the complex. U-Pb zircon geochronometry has bracketed Cretaceous (107 Ma to 77 Ma, in the core zone) and Eocene (55 Ma to 47 Ma, in the mylonitic zone) deformation and metamorphism. K-Ar dates for biotite and hornblende of 53.4 Ma to 45.6 Ma record final uplift and rapid cooling. The ductilely sheared unit developed during Early to Middle Eocene time when mid-crustal rocks in the structurally lower packages of the Tatla Lake Metamorphic Complex were translated to higher crustal levels, at a time of widespread magmatism and crustal extension. Similar sheared metamorphic rocks with pervasive flat foliations and interlayering of lithological units may underlie the entire southern Intermontane Belt. The complex may be a rare exposure of the middle to lower crust reflectors seen on seismic traverses in other parts of the Cordillera.

Central and northern Cordillera
McLeod Lake area

No detailed studies have been made in the region between the Cariboo and Omineca mountains where high-grade metamorphic and migmatitic rocks assigned to the Wolverine Complex, possibly belonging to the Kootenay Terrane, are juxtaposed against low-grade rocks of Slide

Mountain Terrane and Quesnellia. Exposure is poor and no contacts have been described.

Horseranch Range

In the southern Dease Plateau the Horseranch Range is underlain by paragneiss of the Windermere Assemblage. The high-grade metamorphic rocks grade abruptly upwards through mylonitic lower Paleozoic strata into essentially unmetamorphosed Siluro-Devonian carbonate strata suggesting the possibility of extension and denudation. A strong thermal event in Eocene time is documented by K-Ar and U-Pb zircon ages. No detailed structural studies have been made.

Klondike Plateau

Northerly trending swarms of Late Cretaceous and possibly Eocene feldspar porphyry dykes cut assemblages of crystalline rocks in the Omineca Belt between the Denali and Tintina faults. The east-west extension expressed by the dykes may reflect strain related to the regional, dextral strike-slip faults.

REFERENCES

Abbott, J.G.
1983: Geology of the Macmillan Fold Belt; Exploration and Geological Services Division, Department of Indian and Northern Affairs, Whitehorse, Yukon, Open File.

Aho, A.E.
1959: Similar trenchlike lineaments in Yukon; in Symposium on the Rocky Mountain Trench, Canadian Institute of Mining and Metallurgy Bulletin, v. 62, no. 565, p. 173-174.

Aitken, J.D., Macqueen, R.W., and Usher, J.L.
1973: Reconnaissance studies of Proterozoic and Cambrian stratigraphy, Lower Mackenzie River area (Operation Norman), District of Mackenzie; Geological Survey of Canada, Paper 73-9.

Anderson, P.
1974: Geology, petrology, origin and metamorphic history of the Eagle 'Granodiorite' and Nicola Group, Whipsaw Creek, Princeton area, southern British Columbia; B.Sc. thesis, University of British Columbia, Vancouver, 201 p.

Archibald, D.A., Glover, J.K., Price, R.A., Farrar, E., and Carmichael, D.M.
1983: Geochronology and tectonic implications of magmatism and metamorphism, southern Kootenay Arc and neighbouring regions, southeastern British Columbia, Part 1: Jurassic to mid-Cretaceous; Canadian Journal of Earth Sciences, v. 20, p. 1891-1913.

Archibald, D.A., Krogh, T.E., Armstrong, R.L., and Farrar, E.
1984: Geochronology and tectonic implications of magmatism and metamorphism, southern Kootenay Arc and neighbouring regions, southeastern British Columbia, Part II: Middle Cretaceous to Eocene; Canadian Journal of Earth Sciences, v. 21, p. 567-583.

Armstrong, R.L.
1982: Cordilleran metamorphic core complexes-from Arizona to southern Canada; Annual Review of Earth and Planetary Sciences, v. 10, p. 129-154.

Armstrong, R.L., Eisbacher, G.H., and Evans, P.D.
1982: Age and stratigraphic-tectonic significance of Proterozoic diabase sheets, Mackenzie Mountains, northwestern Canada; Canadian Journal of Earth Sciences, v. 19, p. 316-323.

Armstrong, R.L., Harakal, J.E., Brown, E.H., Bernardi, M.L., and Rady, P.M.
1983: Late Paleozoic high-pressure metamorphic rocks in northwestern Washington and southwestern British Columbia: The Vedder Complex; Geological Society of America Bulletin, v. 94, p. 451-458.

Atwater, T.
1970: Implications of plate tectonics for the Cenozoic tectonic evolution of Western North America; Geological Society of America Bulletin, v. 81, p. 3513-3536.

Balkwill, H.R.
1972: Structural geology, Lower Kicking Horse River region, Rocky Mountains, British Columbia; Bulletin of Canadian Petroleum Geology, v. 20, p. 608-633.

Bally, A.W., Gordy, P.L., and Stewart, G.A.
1966: Structure, seismic data and orogenic evolution of southern Canadian Rocky Mountains; Bulletin of Canadian Petroleum Geology, v. 14, p. 337-381.

Barss, D.L. and Montandon, F.A.
1981: Sukunka-Bullmoose gas fields: models for a developing trend in the southern Foothills of northeast British Columbia; Bulletin of Canadian Petroleum Geology, v. 29, p. 293-333.

Bayly, M.B.
1974: An energy calculation concerning the roundness of folds; Tectonophysics, v. 24, p. 291-316.

Beattie, E.T.
1984: Structural style affected by Devonian facies change, Ram Range, Alberta; M.Sc. thesis, University of Calgary, Calgary, Alberta.

Benedict, P.C.
1945: Structure at Island Mountain Mine; Canadian Institute of Mining and Metallurgy, Transactions, v. 48, p. 755-770.

Berg, H.C., Jones, D.L., and Richter, D.H.
1972: Gravina-Nutzotin Belt - tectonic significance of an Upper Mesozoic sedimentary and volcanic sequence in southern and southeastern Alaska; United States Geological Survey, Professional Paper 800-D, p. D1-D24.

Berg, H.C., Jones, D.L., and Coney, P.J.
1978: Map showing pre Cenozoic Tectonostratigraphic Terranes of southeastern Alaska, and adjacent areas; United States Geological Survey, Open File Report 78-1085, 2 sheets.

Blusson, S.L.
1971: Sekwi Mountain map-area, Yukon Territory and District of Mackenzie (105P); Geological Survey of Canada, Paper 71-22.
1974: Geology, Operation Stewart (northern Selwyn Basin) Yukon and District of Mackenzie, N.W.T. (106A, B, C; 105N, O); Geological Survey of Canada, Open File 205.

Bornhold, B.D. and Yorath, C.J.
1984: Surficial geology of the continental shelf, northwestern Vancouver Island; in Marine Geology, Special Volume, Sedimentation on High-Latitude Continental Shelves, A. Guilcher and B.D. Bornhold (ed.), v. 57, p. 89-112.

Bostock, H.S.
1948: Physiography of the Canadian Cordillera with special reference to the area north of the fifty-fifth parallel; Geological Survey of Canada, Memoir 247.
1952: Geology of northwest Shakwak Valley, Yukon Territory; Geological Survey of Canada, Memoir 267, 54 p.

Boucher, G. and Fitch, T.J.
1969: Microearthquake seismicity of the Denali Fault; Journal of Geophysical Research, v. 74, p. 6638-6648.

Brandon, M.T.
1984: Deformational processes affecting unlithified sediments at active margins: a field study and a structural model; Ph.D. thesis, University of Washington, 159 p.

Brew, D.A. and Ford, A.B.
1978: Megalineament in southeastern Alaska marks southwest edge of Coast Range batholithic complex; Canadian Journal of Earth Sciences, v. 15, p. 1763-1772.

Brew, D.A. and Morrell, P.M.
1978: Tarr Inlet Suture Zone, Glacier Bay National Monument, Alaska; in The United States Geological Survey in Alaska - Accomplishments During 1977; K.M. Johnson (ed.), United States Geological Survey, Circular 772-B, p. B90-B92.

Brew, D.A., Loney, R.A., and Muffler, L.J.P.
1966: Tectonic history of southeastern Alaska; in A Symposium on Tectonic History and Mineral Deposits of the Western Cordillera in British Columbia and Neighbouring Parts of the United States; Canadian Institute of Mining and Metallurgy, Special Volume 8, p. 149-170.

Brown, D.A.
1981: Geology of the Lytton area, British Columbia; B.Sc. thesis, Carleton University, 69 p.

Brown, R.L.
1978: Structural evolution of the southeast Canadian Cordillera: a new hypothesis; Tectonophysics, v. 48, p. 133-151.
1980: Frenchman Cap Dome, Shuswap Complex, British Columbia: a progress report; in Current Research, Part A, Geological Survey of Canada, Paper 80-1A, p. 47-51.

1981: Metamorphic core complex of southeast Canadian Cordillera and relationship to foreland thrusting; in Thrust and Nappe Tectonics, K. McClay and N.J. Price (ed.), Geological Society of London Special Publication 9, p. 463-474.

Brown, R.L. and Journeay, J.M.
1987: Tectonic denudation of the Shuswap metamorphic terrane of southeastern British Columbia; Geology, v. 15, p. 142-146.

Brown, R.L. and Murphy, D.C.
1982: Kinematic interpretation of mylonitic rocks in part of the Columbia River fault zone, Shuswap Terrane, British Columbia; Canadian Journal of Earth Sciences, v. 19, p. 456-465.

Brown, R.L. and Psutka, J.F.
1979: Stratigraphy of the east flank of Frenchman Cap Dome, Shuswap Complex, British Columbia; in Current Research, Part A, Geological Survey of Canada, Paper 79-1A, p. 35-36.

Brown, R.L. and Read, P.B.
1983: Shuswap terrane of British Columbia: a Mesozoic "core complex"; Geology, v. 11, p. 164-168.

Brown, R.L. and Tippett, C.R.
1978: The Selkirk fan structure of the southeast Canadian Cordillera; Geological Society of America Bulletin, v. 89, p. 548-558.

Brown, R.L., Journeay, J.M., Lane, L.S., Murphy, D.C., and Rees, C.J.
1986: Obduction, backfolding and piggyback thrusting in the metamorphic hinterland of the southeast Canadian Cordillera; Journal of Structural Geology, v. 8, p. 255-268.

Bruns, T.R.
1979: Late Cenozoic structure of the continental margin, northern Gulf of Alaska; in The Relationship of Plate Tectonics to Alaskan Geology and Resources, A. Sisson (ed.), Proceedings, Sixth Alaska Geological Society Symposium, Anchorage, Alaska, 1977, p. I1-I30.
1982: Structure and petroleum potential of the continental margin between Cross Sound and Icy Bay, northern Gulf of Alaska; United States Geological Survey, Open File Report 82-929, 64 p.
1983a: Model for the origin of the Yakutat block, an accreting terrane in the northern Gulf of Alaska; Geology, v. 11, p. 718-712.
1983b: Structure maps and petroleum potential of the Yakutat segment of the northern Gulf of Alaska continental margin; United States Geological Survey, Map MF-1480, 3 sheets, scale 1:500,000.

Bultman, T.R.
1979: Geology and tectonic history of the Whitehorse Trough west of Atlin, British Columbia; Ph.D. thesis, Yale University, 284 p.

Bustin, R.M. and Moffat, I.
1983: Groundhog Coalfield, central British Columbia: reconnaissance stratigraphy and structure; Bulletin of Canadian Petroleum Geology, v. 31, p. 231-245.

Cameron, B.E.B.
1980: Biostratigraphy and depositional environment of the Escalante and Hesquiat formations (Early Tertiary) of the Nootka Sound area, Vancouver Island, British Columbia; Geological Survey of Canada, Paper 78-9.

Campbell, R.B.
1968: Canoe River, B.C.; Geological Survey of Canada, Map 15-1967.
1970: Structural and metamorphic transitions from infrastructure to suprastructure, Cariboo Mountains, British Columbia; in Structure of the Southern Canadian Cordillera, J.O. Wheeler (ed.), Geological Association of Canada, Special Paper 6, p. 67-72.
1973: Structural Cross-section and tectonic model of the southeastern Canadian Cordillera; Canadian Journal of Earth Sciences, v. 10, p. 1607-1620.

Campbell, R.B. and Dodds, C.J.
1978: Operation Saint Elias, Yukon Territory; in Current Research, Part A, Geological Survey of Canada, Paper 78-1A, p. 35-41.
1979: Operation Saint Elias, British Columbia; in Current Research, Part A, Geological Survey of Canada, Paper 79-1A, p. 17-20.
1982a: Geology of S.W. Kluane Lake map-area (115G and F (E 1/2)), Yukon Territory; Geological Survey of Canada, Open File 829.
1982b: Geology of Mount St. Elias map-area (115B and C (E 1/2)), Yukon Territory part; Geological Survey of Canada, Open File 830.
1982c: Geology of S.W. Dezadeash map-area (115A), Yukon Territory; Geological Survey of Canada, Open File 831.
1983a: Geology of Tatshenshini River map-area (114P), British Columbia; Geological Survey of Canada, Open File 926.
1983b: Geology of Yakutat map-area (114O), British Columbia part; Geological Survey of Canada, Open File 927.
1983c: Terranes and major faults of the Saint Elias Mountains, Yukon Territory, British Columbia, and Alaska; Geological Association of Canada, Annual Meeting in Victoria, Program with Abstracts, p. A10.

Campbell, R.B. and Eisbacher, G.H.
1974: Operation Saint Elias; in Report of Activities, Part A, Geological Survey of Canada, Paper 74-1A, p. 11-12.

Campbell, R.B., Mountjoy, E.W., and Struik, L.C.
1982: Structural cross-section through south-central Rocky and Cariboo mountains to the Coast Range; Geological Survey of Canada, Open File 844.

Campbell, R.B., Mountjoy, E.W., and Young, F.G.
1973: Geology of McBride map area, British Columbia; Geological Survey of Canada, Paper 72-35.

Carey, J.A.
1984: Geology of Late Proterozoic Miette Group, southern Main Ranges, Cushing Creek area, B.C.; M.Sc. thesis, University of Calgary, Alberta, 119 p.

Carlson, P.R., Plafker, G., and Bruns, T.R.
1985: Map and selected seismic profiles of the seaward extension of the Fairweather Fault, eastern Gulf of Alaska; United States Geological Survey, Map MF-1722, 2 sheets.

Carr, S.D.
1985: Ductile shearing and brittle faulting in Valhalla gneiss complex, southeastern British Columbia; in Current Research, Part A, Geological Survey of Canada, Paper 85-1A, p. 89-96.

Carr, S.D., Parrish, R.R., and Brown, R.L.
1987: Eocene structural development of the Valhalla Complex, southeastern British Columbia; Tectonics, v. 6, p. 175-196.

Cecile, M.P.
1984: Geology of the northwest Niddery Lake map area, Yukon; Geological Survey of Canada, Open File 1066.

Cecile, M.P. and Cook, D.G.
1981: Structural cross-section, northern Selwyn and Mackenzie mountains; Geological Survey of Canada, Open File 807.

Cecile, M.P., Cook, D.G., and Snowdon, L.P.
1982: Plateau overthrust and its hydrocarbon potential, Mackenzie Mountains, Northwest Territories; in Current Research, Part A, Geological Survey of Canada, Paper 82-1A, p. 89-94.

Chapple, W.M.
1978: Mechanics of the thin-skinned fold and thrust belts; Geological Society of America Bulletin, v. 89, p. 1189-1198.

Charlesworth, H.A.K., Weiner, J.L., Akehurst, A.J., Bielenstein, H.U., Evans, C.R., Griffith, R.E., Remington, D.B., Stauffer, M.R., and Steiner, J.
1967: Precambrian geology of the Jasper region; Alberta Research Council, Bulletin 23, 74 p.

Charlesworth, H.A.K. and Kilby, W.E.
1981: Thrust nappes in the Rocky Mountain foothills near Mountain Park, Alberta; in Thrust and Nappe Tectonics, K.R. McClay and N.J. Price (ed.), Geological Society of London, Special Publication No. 9, p. 475-482.

Cheney, E.S.
1980: Kettle dome and related structures of northeastern Washington; in Cordilleran Metamorphic Core Complexes, M.D. Crittenden, Jr. et al. (ed.), Geological Society of America, Memoir 153, p. 463-483.

Church, B.N.
1973: Geology of the White Lake area; British Columbia Department of Mines and Petroleum Resources, Bulletin 61.
1975: Geology of the Hat Creek coal basin (93I/13E); in Geology in British Columbia 1975, British Columbia Ministry of Mines and Petroleum Resources, p. G99-G118.

Clague, J.J.
1974: The St. Eugene Formation and the development of the southern Rocky Mountain Trench; Canadian Journal of Earth Sciences, v. 15, p. 86-98.
1979: The Denali Fault System in southwest Yukon Territory - A geologic hazard?; in Current Research, Part A, Geological Survey of Canada, Paper 79-1A, p. 169-178.

Clapp, C.H.
1914: Geology of the Nanaimo map-area; Geological Survey of Canada, Memoir 51.

Clowes, R.M., Brandon, M.T., Green, A.G., Yorath, C.J., Sutherland Brown, A., Kanasewich, E.R., and Spencer, C.
1987: Lithoprobe - southern Vancouver Island: Cenozoic subduction complex imaged by deep seismic reflections; Canadian Journal of Earth Sciences, v. 24, p. 31-51.

Coates, J.A.
1974: Geology of the Manning Park area, British Columbia; Geological Survey of Canada, Bulletin 238, 177 p.

Coney, P.J. and Harms, T.A.
1984: Cordilleran metamorphic core complexes: Cenozoic extensional relics of Mesozoic compression; Geology, v. 12, p. 550-554.

Coney, P.J., Jones, D.L., and Monger, J.W.H.
1980: Cordilleran suspect terranes; Nature, v. 288, p. 329-333.

Cook, D.G.
1970: A Cambrian facies change and its effect on structure, Mountain Stephen-Mount Dennis area, British Columbia; in Structure of the Southern Canadian Cordillera, J.O. Wheeler (ed.), Geological Association of Canada, Special Paper No. 6, p. 27-39.
1975: Structural style influenced by lithofacies, Rocky Mountain Main Ranges, Alberta-British Columbia; Geological Survey of Canada, Bulletin 233, 73 p.
1983: The northern Franklin Mountains, Northwest Territories, Canada - a scale model of the Wyoming province; in Rocky Mountain Foreland Basins and Uplifts, J.D. Lowell (ed.), Field Conference - Rocky Mountain Association of Geologists, p. 314-338.

Cook, F.A., Green, A.G., Simony, P.S., Price, R.A., Parrish, R., Milkereit, B., Gordy, P.L., Brown, R.L., Coflin, K.C., and Patenaude, C.
in press: Lithoprobe, southern Canadian Cordilleran transect: Rocky Mountain thrust belt to Valhalla gneiss complex; Geophysical Journal of the Royal Astronomical Society.

Cook, F., Simony, P., Coflin, K., Green, A., Milkereit, B., Price, R., Parrish, R., Patenaude, C., and Gordey, P.
1987: Lithoprobe southern Canadian Cordilleran transect: Rocky Mountain thrust belt to Valhalla gneiss complex; Geophysical Journal of the royal Astronomical Society, 89, p. 91-98.

Craw, D.
1978: Metamorphism, structure and stratigraphy in the Southern Park Ranges, British Columbia; Canadian Journal of Earth Sciences, v. 15, p. 86-98.

Crawford, M.L. and Hollister, L.S.
1982: Contrast of metamorphic and structural histories across the Work Channel lineament, Coast Plutonic Complex, British Columbia; Journal of Geophysical Research, v. 87, p. 3849-3860.

Crawford, M.L., Hollister, L.S., and Woodsworth, G.J.
1987: Crustal deformation and regional metamorphism across a terrane boundary, Coast Plutonic Complex, British Columbia; Tectonics, v. 6, p. 343-361.

Crickmay, C.H.
1930: The structural connection between the Coast Range of British Columbia and the Cascades Range of Washington; Geological Magazine, v. 67, p. 27-39.

Crittenden, M.D. Jr., Coney, P.J., and Davis, G.H.
1980: Cordilleran metamorphic core complexes; Geological Society of America, Memoir 153, 490 p.

Dahlstrom, C.D.A.
1969: Balanced cross-sections; Canadian Journal of Earth Sciences, v. 6, p. 743-757.

Danner, W.R.
1977: Paleozoic rocks of northwestern Washington and adjacent part of British Columbia; in Paleozoic Paleogeography of the Western United States, J.H. Stewart, C.H. Stevens and A.E. Fritsche (ed.), Society of Economic Paleontologists and Mineralogists, Pacific Section, Pacific Coast Paleogeography Symposium 1, p. 481-502.

Davis, D.M. and Suppe, J.
1980: Critical taper in mechanics of fold and thrust belt; Geological Society of America, Programs with Abstracts, v. 12, p. 410.

Davis, D., Suppe, J., and Dahlen, F.A.
1983: Mechanics of fold and thrust belts and accretionary wedges; Journal of Geophysical Research, v. 88, p. 1153-1172.

Davis, E.E. and Riddihough, R.P.
1982: The Winona Basin: structure and tectonics; Canadian Journal of Earth Sciences, v. 19, p. 767-788.

Davis, G.A., Monger, J.W.H., and Burchfiel, B.C.
1978: Mesozoic construction of the Cordilleran "Collage" central British Columbia to central California; in Mesozoic Paleogeography of the Western United States, D.G. Howell and K.A. McDougall (ed.), Society of Economic Paleontologists and Mineralogists, Pacific Coast Section paleogeography Symposium 2, p. 1-32.

Davis, J.W., and Willet, R.
1978: Structural geology of the Colville Hills; Bulletin of Canadian Petroleum Geology, v. 26, p. 105-122.

Douglas, B.J.
1983: Structural and stratigraphic analysis of a metasedimentary inlier within the Coast Plutonic Complex, British Columbia, Canada; Ph.D. thesis, Princeton University, 298 p.

Douglas, R.J.W.
1950: Callum Creek, Langford Creek and Gap map-areas, Alberta; Geological Survey of Canada, Memoir 255, 124 p.
1985a: Mount Head map-area, Alberta; Geological Survey of Canada, Memoir 291, 241 p.
1958b: Chungo Creek map-area, Alberta (83C/9); Geological Survey of Canada, Paper 58-3, 45 p.

Duffell, S. and McTaggart, K.C.
1952: Ashcroft map-area, British Columbia; Geological Survey of Canada, Memoir 262, 122 p.

Duncan, I.J.
1984: Structural evolution of the Thor-Odin gneiss dome; Tectonophysics, v. 101, p. 87-130.

Dusel-Bacon, C. and Foster, H.L.
1983: A sillimanite gneiss dome in the Yukon Crystalline Terrane, east-central Alaska: Petrography and garnet-biotite geothermometry; United States Geological Survey, Professional Paper 1170-E.

Eisbacher, G.H.
1974: Sedimentary and tectonic evolution of the Sustut and Sifton basins, north-central British Columbia; Geological Survey of Canada, Paper 73-31.
1976: Sedimentology of the Dezadeash flysch and its implications for strike-slip faulting along the Denali Fault, Yukon Territory and Alaska; Canadian Journal of Earth Sciences, v. 13, p. 1495-1513.
1981: Sedimentary tectonics and glacial record in the Windermere Supergroup, Mackenzie Mountains, northwestern Canada; Geological Survey of Canada, Paper 80-27, 40 p.

Eisbacher, G.H. and Hopkins, S.L.
1977: Mid-Cenozoic paleogeomorphology and tectonic setting of the St. Elias Mountains, Yukon Territory; in Report of Activities, Part B, Geological Survey of Canada, Paper 77-1B, p. 319-335.

Erdmer, P. and Helmstaedt, H.
1983: Eclogite from central Yukon: a record of subduction at the western margin of ancient North America; Canadian Journal of Earth Sciences, v. 20, p. 1389-1408.

Evenchick, C.A.
1985: Stratigraphy, metamorphism, structure, and their tectonic implications in the Sifton and Deserters ranges, Cassiar and northern Rocky Mountains, northern British Columbia; Ph.D. thesis, Queen's University, Kingston, Ontario, 197 p.
1986: Structural style of the northwest margin of the Bowser Basin, Spatsizi map area, north-central British Columbia; in Current Research, Part B, Geological Survey of Canada, Paper 86-1B, p. 733-739.
1988: Stratigraphy, metamorphism, structure, and their tectonic implications in the Sifton and Deserters ranges, Cassiar and northern Rocky Mountains, northern British Columbia; Geological Survey of Canada, Bulletin 376, 90 p.

Ewing, T.E.
1980: Paleogene tectonic evolution of the Pacific Northwest; Journal of Geology, v. 88, p. 619-638.
1981: Geological and tectonic setting of the Kamloops Group, south-central British Columbia; Ph.D. thesis, University of British Columbia, Vancouver, 225 p.

Fitzgerald, E.L.
1968: Structure of British Columbia Foothills, Canada; Bulletin of America Association of Petroleum Geologists, v. 52, p. 641-664.

Forbes, R.B., Smith, T.E., and Turner, D.L.
1974: Comparative petrology and structure of the McLaren, Ruby Range and Coast Range belts: Implications for offset along the Denali Fault System; Geological Society of America, Abstracts with Programs, v. 6, p. 177.

Fox, F.G.
1969: Some principles governing interpretation of structure in the Rocky Mountain orogenic belt; in Time and Place in Orogeny, P.E. Kent, G.E. Satterthwaite and A.M. Spencer (ed.), Geological Society of London, Special Paper No. 3, p. 23-41.

Franzen, J.P.
1972: Some aspects of the petrology and structure in the Kliyul Creek area, north-central British Columbia; B.Sc. thesis, University of British Columbia, Vancouver, 46 p.

Fyles, J.T.
1960: Geological reconnaissance of the Columbia River between Bluewater Creek and Mica Creek, B.C.; Minister of Mines Annual Report 1959, p. 90-105.
1964: Geology of the Duncan Lake area, British Columbia; B.C. Ministry of Energy, Mines and Petroleum Resources, Bulletin 49, p. 87.
1967: Geology of the Ainsworth-Kaslo area, British Columbia; B.C. Ministry of Energy, Mines and Petroleum Resources, Bulletin 53, p. 125.
1970: Structure of the Shuswap Metamorphic Complex in the Jordan River area, northwest of Revelstoke, British Columbia; in Structure of the Southern Canadian Cordillera, J.O. Wheeler (ed.), Geological Association of Canada, Special Paper, Number 6, p. 87-98.

Fyles, J.T. and Eastwood, G.E.
1962: Geology of the Ferguson area, Lardeau district, British Columbia; B.C. Ministry of Energy, Mines and Petroleum Resources, Bulletin 45, p. 92.

Fyles, J.T. and Hewlett, C.G.
1959: Stratigraphy and structure of the Salmo lead-zinc area, British Columbia, B.C. Ministry of Energy, Mines and Petroleum Resources, Bulletin 41, p. 162.

Gabrielse, H.
1967: Tectonic evolution of the northern Canadian Cordillera; Canadian Journal of Earth Sciences, v. 4, p. 271-298.
1969: Geology of Jennings River map-area, British Columbia (104-O); Geological Survey of Canada, Paper 68-55.
1985: Major dextral transcurrent displacements along the Northern Rocky Mountain Trench and related lineaments in north-central British Columbia; Geological Society of America Bulletin, v. 96, p. 1-14.

Gabrielse, H. and Dodds, C.J.
1982: Faulting and plutonism in northwestern Cry Lake and adjacent map areas, British Columbia; in Current Research, Part A, Geological Survey of Canada, Paper 82-1A, p. 321-323.

Gabrielse, H. and Taylor, G.C.
1982: Geological maps and cross-sections of the Cordillera from near Fort Nelson, British Columbia to Gravina Island, southeastern Alaska: Geological Survey of Canada, Open File 864.

Gabrielse, H. and Wheeler, J.O.
1961: Tectonic framework of southern Yukon and northwestern British Columbia; Geological Survey of Canada, Paper 60-24, 37 p.

Gabrielse, H., Blusson, S.L., and Roddick, J.A.
1973: Geology of Flat River, Glacier Lake, and Wrigley Lake map-areas, District of Mackenzie and Yukon Territory; Geological Survey of Canada, Memoir 366, 153 p.

Gareau, S.A.
1988: Preliminary study of the Work Channel lineament in the Ecstall River area, Coast Plutonic Complex, British Columbia; in Current Research, Part E, Geological Survey of Canada, Paper 88-1E, p. 49-55.

Gehrels, G.E. and Berg, H.C.
1984: Geologic map of southeastern Alaska; United States Geological Survey, Open File Report 84-886.

Gehrels, G.E., Saleeby, J.B., and Berg, H.C.
1983: Preliminary description of the Klakas orogeny in the southern Alexander Terrane, southeastern Alaska; in Pre-Jurassic rocks in Western North American Suspect Terranes, C.H. Stevens (ed.), Society of Economic Paleontologists and Mineralogists, Pacific Section, Los Angeles, California, p. 131-141.

Getsinger, J.S.
1982: Metamorphism and structure of Three Ladies Mountain area, Cariboo Mountains, British Columbia; in Current Research, Part A, Geological Survey of Canada, Paper 82-1A, p. 317-320.

Ghent, E.D., Nicholl, J., Stout, M., and Rottenfusser, B.
1977: Clinopyroxene amphibolite boudins from Three Valley Gap, British Columbia; Canadian Mineralogist, v. 15, p. 269-282.

Ghent, E.D., Robbins, D.B., and Stout, M.Z.
1979: Geothermometry, geobarometry and fluid compositions of metamorphosed calc-silicates and pelites, Mica Creek, British Columbia American Mineralogist, v. 64, p. 874-885.

Ghent, E.D., Simony, P.S., and Knitter, C.C.
1980: Geometry and pressure-temperature significance of the kyanite-sillimanite isograd in the Mica Creek area, British Columbia; Contributions to Mineralogy and Petrology, v. 74, p. 67-73.

Ghent, E.D., Knitter, C.C., Raeside, R.P., and Stout, M.Z.
1982: Geothermometry and geobarometry of pelitic rocks, upper kyanite and sillimanite zones, Mica Creek area, British Columbia; Canadian Mineralogist, v. 20, p. 295-305.

Gordey, S.P.
1981: Structure section across south central Mackenzie Mountains, N.W.T.; Geological Survey of Canada, Open File 809.

Gordy, P.L., Frey, F.R., and Norris, D.K.
1977: Geological guide for the C.S.P.G. and 1977 Waterton-Glacier Park Field Conference; Canadian Society of Petroleum Geologists, Calgary, Alberta.

Grantz, A.
1966: Strike-slip faults in Alaska; United States Geological Survey, Open File Report 66-53, 82 p.

Green, A.G., Berry, M.J., Spencer, C.P., Kanasewich, E.R., Chiu, S., Clowes, R.M., Yorath, C.J., Stewart, D.B., Unger, J.D., and Poole, W.H.
1985: Recent seismic reflection studies in Canada; in Deep Structure of the Continental Crust: Results from Reflection Seismology, M. Barazargi and L. Brown (ed.), American Geophysical Union, Geodynamics Series.

Green, L.H.
1972: Geology of Nash Creek, Larsen Creek and Dawson map areas, Yukon Territory; Geological Survey of Canada, Memoir 364, 157 p.

Greig, C.J.
1988: Geology and geochronometry of the Eagle Plutonic Complex, Hope map area, southwestern British Columbia; in Current Research, Part E, Geological Survey of Canada, Paper 88-1E, p. 177-183.

Hamilton, W. and Meyers, W.B.
1966: Cenozoic tectonics of the western United States; Reviews of Geophysics, v. 4, p. 509-549.

Hansen, V.L.
1986: Metamorphic and structural evidence of a Mesozoic transpressional North American Margin; in Abstracts with Programs, 1986, Geological Society of America, v. 18, p. 627.

Harms, T.
1982: The Newport Fault: Low-angle normal faulting and Eocene extension, northeast Washington and northwest Idaho; M.Sc. thesis, Queen's University, Kingston, Ontario, 157 p.
1985: Pre-emplacement thrust faulting in the Sylvester Allochthon, northeast Cry Lake map area, British Columbia; in Current Research, Part A, Geological Survey of Canada, Paper 85-1A, p. 301-304.

Harms, T.A., Nelson, J., and Bradford, J.
1988: Geological transect across the Sylvester Allochthon north of the Blue River, northern British Columbia (104P/12); in Geological Fieldwork 1987, British Columbia Ministry of Energy, Mines and Petroleum Resources, Paper 1988-1, p. 245-248.

Haugerud, R.A.
1985: Ross Lake Fault Zone (RLZ) in the Maselpanik area, SW B.C., a major shear zone during the Eocene; Geological Society of America, Abstracts with Programs, v. 17, p. 360.

Heah, T.S.T.
1982: Stratigraphy, geochemistry and geochronology of the Lower Cretaceous Gambier Group, Sky Pilot area, southern British Columbia; B.Sc. thesis, University of British Columbia, Vancouver, 97 p.

Hedley, M.S.
1952: Geology and ore deposits of the Sandon area, Slocan Mining Camp, British Columbia; British Columbia Department of Mines, Bulletin 29, 130 p.

Hill, M.L.
1984: Geology of the Redcap Mountain area, Coast Plutonic Complex, British Columbia; Ph.D. thesis, Princeton University, 216 p.

Holland, S.S.
1954: Geology of the Yanks Peak-Roundtop Mountain area, Cariboo District, British Columbia; British Columbia Department of Mines, Bulletin 34, 102 p.

Hollister, L.S.
1969: Metastable paragenetic sequences of andalusite, kyanite and sillimanite, Kwoiek area, British Columbia; American Journal of Sciences, v. 267, p. 352-370.

Hollister, V.F., Allen, J.M., Anzalone, S.A., and Seraphim, R.H.
1975: Structural evolution of porphyry mineralization at Highland Valley, British Columbia; Canadian Journal of Earth Sciences, v. 12, p. 807-820.

Horner, R.B.
1983: Seismicity in the Saint Elias region of northwestern Canada and southeastern Alaska; Bulletin of the Seismological Society of America, v. 73, p. 1117-1137.

Høy, T.
1976: Calc-silicate isograds in the Riondel area, southeastern British Columbia; Canadian Journal of Earth Sciences, v. 13, p. 1093-1104.
1977: Stratigraphy and structure of the Kootenay Arc in the Riondel area, southeastern British Columbia; Canadian Journal of Earth Sciences, v. 14, p. 2301-2315.
1979: Cottonbelt lead-zinc deposit; in Geological Fieldwork 1978; British Columbia Ministry of Energy, Mines and Petroleum Resources, Paper 79-1, p. 18-23.
1980: Geology of the Riondel area, central Kootenay Arc, southeastern British Columbia; British Columbia Ministry of Energy, Mines and Petroleum Resources, Bulletin 73, 89 p.
1980: Geology in the Bews Creek area, southwestern margin of Frenchman Cap gneiss dome; in Geological Fieldwork 1979, British Columbia Ministry of Energy, Mines and Petroleum Resources, Paper 80-1, p. 17-22.

Høy, T. and Brown, R.L.
1981: Geology of eastern margin of Shuswap Complex, Frenchman Cap area; British Columbia Ministry of Energy, Mines and Petroleum Resources, Preliminary Map No. 43.

Høy, T. and McMillan, W.J.
1979: Geology in the vicinity of Frenchman Cap gneiss dome; in Geological Fieldwork 1979, British Columbia Ministry of Energy, Mines and Petroleum Resources, Paper 79-1, p. 24-30.

Hudson, T.
1983: Calc-alkaline plutonism along the Pacific rim of southern Alaska; Geological Society of America, Memoir 159, p. 159-169.

Hudson, T., Plafker, G., and Dixon, K.
1982: Horizontal offset history of the Chatham Strait fault; United States Geological Survey, Circular 844, p. 128-132.

Hutchison, W.W.
1982: Geology of the Prince Rupert - Skeena map area, British Columbia; Geological Survey of Canada, Memoir 394, 116 p.

Hyndman, R.D., Riddihough, R.P., and Hertzer, R.
1979: The Nootka fault zone; a new plate boundary off western Canada; Geophysical Journal, v. 58, p. 667-683.

Hydman, R.D., Yorath, C.J., Clowes, R.M., and Davis, E.E.
1990: The northern Cascadia subduction zone at Vancouver Island: seismic structure and tectonic history; Canadian Journal of Earth sciences, 27, p. 313-329.

Irving, E., Woodsworth, G.J., Wynne, P.J., and Morrison, A.
1985: Paleomagnetic evidence for displacement from the south of the Coast Plutonic Complex, British Columbia; Canadian Journal of Earth Sciences, v. 22, p. 584-598.

Jackson, L.E., Gordey, S.P., Armstrong, R.L., and Harakal, J.E.
1986: Bimodal Paleogene volcanics near Tintina Fault, east-central Yukon, and their possible relationship to placer gold; in Yukon Geology, Volume 1; Exploration and Geological Services Division, Yukon, Indian and Northern Affairs Canada, p. 139-147.

Jewett, P.D. and Brown, E.H.
1985: Fault motions in the Pleiades-Slesse Mountain area, North Cascades of Washington and British Columbia; Geological Society of America, Abstracts with Programs, v. 17, no. 16, p. 364.

Jones, P.B.
1978: Notes on the geologic structure of the Foothills and eastern Rocky Mountains between Rocky Mountain House and Saskatchewan River Crossing, Alberta; in Geologic Guide to the Central Foothills and Rocky Mountains of Alberta, P.B. Jones and R.H. Workum (ed.), Canadian Society of Petroleum Geologists, Calgary, Alberta, p. 16-25.
1982: Oil and gas beneath east-dipping underthrust faults in the Alberta Foothills; in Studies of the Cordilleran Thrust Belt, R.B. Powers (ed.), Rocky Mountain Association of Geologists, p. 61-74.

Journeay, J.M.
1981: Basement-cover relationships and structural evolution of north-central Frenchman Cap dome, Shuswap Complex, S.E. British Columbia; in Geological Association of Canada, Program with Abstracts, v. 29, p. A-29.
1982: Structural setting along the northwest flank of Frenchman Cap dome, Monashee Complex (82M/7); in Geological Fieldwork 1980, British Columbia Ministry of Energy, Mines and Petroleum Resources, Paper 81-1, p. 187-202.
1983: Progressive deformation and inverted regional metamorphism associated with Mesozoic emplacement of the Shuswap-Monashee Complex, S.E. British Columbia; in Geological Society of America, Programs with Abstracts, v. 15, p. 606.
1986: Stratigraphy, internal strain and thermo-tectonic evolution of the northern Frenchman Cap dome: An exhumed duplex structure, Omineca hinterland, S.E. Canadian Cordillera, Ph.D. thesis, Queen's University, Kingston, Ontario, 350 p.

Keen, C.E. and Hyndman, R.D.
1979: Geophysical review of the continental margins of eastern and western Canada; Canadian Journal of Earth Sciences, v. 16, p. 712-747.

Kenah, C. and Hollister, L.S.
1983: Anatexis in the Central Gneiss Complex, British Columbia; in Migmatites, Melting and Metamorphism, M.P. Atherton and C.D. Gribble (ed.), Shiva Publishing Ltd., p. 142-162.

King, P.B.
1969: The tectonics of North America - a discussion to accompany the Tectonic Map of North America Scale 1:5,000,000; United States Geological Survey Professional Paper 628, 94 p.

Kleinspehn, K.L.
1985: Cretaceous sedimentation and tectonics, Tyaughton-Methow Basin, southwestern British Columbia; Canadian Journal of Earth Sciences, v. 22, p. 154-174.

Klepacki, D.W. and Wheeler, J.O.
1985: Stratigraphic and structural relations of the Milford, Kaslo and Slocan Groups, Goat Range, Lardeau and Nelson map areas, British Columbia; in Current Research, Part A, Geological Survey of Canada, Paper 85-1A, p. 277-286.

Kurtz, R.D., DeLaurier, J.M., and Gupta, J.C.
1986: A magnetotelluric sounding acorss Vancouver Island detects the subducting Juan de Fuca Plate; Nature, v. 321, p. 596-599.

Lahr, J.C. and Plafker, G.
1980: Holocene Pacific-North America plate interaction in southern Alaska: Implications for the Yakataga seismic gap; Geology, v. 8, p. 483-486.

Lahr, J.C., Plafker, G., Stephens, C.D., Fogleman, K.A., and Blackford, M.E.
1979: Interim report on the St. Elias earthquake of 28 February 1979; United States Geological Survey, Open File Report 79-670, 35 p.

Lahr, J.C., Stephens, C.D., Hasegawa, H.S., and Boatwright, J.
1980: Alaskan seismic gap only partially filled by 28 February 1979 earthquake; Science, v. 207, p. 1351-1353.

Lambert, M.B.
1974: The Bennett Lake cauldron subsidence complex, British Columbia and Yukon Territory; Geological Survey of Canada, Bulletin 227, 213 p.

Lane, L.S.
1984a: Brittle deformation in the Columbia River fault zone near Revelstoke, southeastern British Columbia; Canadian Journal of Earth Sciences, v. 21, p. 584-598.
1984b: Deformation history of the Monashee Décollement north of Revelstoke, British Columbia; Ph.D. thesis, Carleton University, Ottawa, Ontario, 240 p.

Langenberg, C.W.
1984: Structural and sedimentological framework of Lower Cretaceous coal-bearing rocks in the Grande Cache area; Alberta; in The Mesozoic of Middle North America, D.F. Stott and D.J. Glass (ed.), Canadian Society of Petroleum Geologists, Calgary, Alberta, p. 533-540.

Lanphere, M.A.
1978: Displacement history of the Denali fault system, Alaska and Canada; Canadian Journal of Earth Sciences, v. 15, p. 817-822.

Lappin, A.R. and Hollister, L.S.
1980: Partial melting in the Central Gneiss Complex near Prince Rupert, British Columbia; American Journal of Science, v. 280, p. 518-545.

Lathram, E.H.
1964: Apparent right-lateral separation on Chatham Strait fault, southeastern Alaska; Geological Society of America Bulletin, v. 75, p. 249-252.

Leatherbarrow, R.W.
1981: Metamorphism of pelitic rocks from the northern Selkirk Mountains, southeastern British Columbia; Ph.D. Dissertation, Carleton University, Ottawa, Ontario, 218 p.

LeClair, A.D.
1983: Stratigraphy and structural implications of central Kootenay Arc rocks, southeastern British Columbia; in Current Research, Part A, Geological Survey of Canada, Paper 83-1A, p. 235-239.
1985: Stratigraphic and structural relations of the Milford, Kaslo and Slocan Groups, Goat Range, Lardeau and Nelson map areas, British Columbia; in Current Research, Part A, Geological Survey of Canada, Paper 85-1A, p. 277-286.

Leech, G.B.
1962: Metamorphism and granite intrusions of Precambrian age in southeastern British Columbia; Geological Survey of Canada, Paper 62-13, 8 p.
1966: The Rocky Mountain Trench; in The World Rift System, T.N. Irvine (ed.), Geological Survey of Canada, Paper 66-14, p. 307-329.

Loney, R.A.
1964: Stratigraphy and petrography of the Pybus-Gambier area, Admiralty Island, Alaska; United States Geological Survey, Bulletin 1178, 103 p.

Loney, R.A., Brew, D.A., and Lanphere, M.A.
1967: Post-Paleozoic radiometric ages and their relevance to fault movements, northern southeastern Alaska; Geological Society of America Bulletin, v. 78, p. 511-526.

Lowes, B.E.
1972: Metamorphic petrology and structural geology of the area east of Harrison Lake, British Columbia; Ph.D. thesis, University of Washington, 162 p.

MacKay, B.R.
1943: Foothills belt of central Alberta; Geological Survey of Canada, Paper 43-3, 1 sheet.

MacKevett, E.M., Jr.
1978: Geologic map of the McCarthy quadrangle, Alaska; United States Geological Survey, Map I-1032.

MacKevett, E.M., Jr. and Plafker, G.
1974: The Border Ranges Fault in south-central Alaska; United States Geological Survey, Journal of Research, v. 2, p. 323-329.

MacKevett, E.M., Jr., Robertson, E.C., and Winkler, G.R.
1974: Geology of the Skagway B3 and B4 quadrangles, southeastern Alaska; United States Geological Survey, Professional Paper 832, 33 p.

MacLeod, N.S., Tiffin, D.L., Snavely, P.D., Jr., and Currie, R.G.
1977: Geologic interpretation of magnetic and gravity anomalies in the Strait of Juan de Fuca, U.S.-Canada; Canadian Journal of Earth Sciences, v. 14, p. 223-238.

Mansy, J.L.
1986: Géologie de la chaîne d'Omineca des Rocheuses aux plateaux intérieurs (Cordillère canadienne), son évolution depuis le Précambrien; thèse de doctorat d'état et sciences naturelles, L'Université des Sciences et Techniques de Lille, France, 718 p.

Mansy, J.L. and Dodds, C.
1976: Stratigraphy, structure and metamorphism in northern central Swannell Ranges; in Report of Activities, Part A, Geological Survey of Canada, Paper 76-1A, p. 91-93.

Mathews, W.H.
1981: Early Cenozoic resetting of potassium-argon dates and thermal history of north Okanagan area, British Columbia; Canadian Journal of Earth Sciences, v. 18, p. 1310-1319.
1983: Early Tertiary resetting of potassium-argon dates in the Kootenay Arc, southeastern British Columbia; Canadian Journal of Earth Sciences, v. 20, p. 867-872.

Mathews, W.H. and Rouse, G.E.
1984: The Gang Ranch-Big Bar area, south-central British Columbia: stratigraphy, geochronology and palynology of the Tertiary beds and their relationship with the Fraser Fault; Canadian Journal of Earth Sciences, v. 21, p. 1132-1144.

Mattinson, J.M.
1972: Ages of zircons from the northern Cascade Mountains, Washington; Geological Society of America Bulletin, v. 83, p. 3769-3784.

McClay, K.R., Insley, M.W., Way, N.A., and Anderton, R.
1988: Tectonics and mineralization at the Kechika Trough, Gataga area, northeastern British Columbia; in Current Research, Part E, Cordillera and Pacific Margin; Geological Survey of Canada, Paper 88-1E, p. 1-18.

McGroder, M.F. and Mohrig, D.
1987: Piggyback to foreland transition in the Methow Basin, Washington, 2: balanced cross-section of the Cascade Orogen; in Geological Society of America, Abstracts with Programs, Annual Meetings, v. 19, p. 766.

McKillop, G.R.
1973: Geology of southwestern Gambier Island, Howe Sound, British Columbia; B.Sc. thesis, University of British Columbia, Vancouver, 24 p.

McMechan, M.E.
1985: Low-taper triangle zone geometry: an interpretation for the Rocky Mountain Foothills, Pine Pass-Peace River area, British Columbia; Bulletin of Canadian Petroleum Geology, v. 33, p. 31-38.
1987: Stratigraphy and structure, Mount Selwyn map area, Rocky Mountains, northeast British Columbia; Geological Survey of Canada, Paper 85-28, 34 p.

McMechan, M.E. and Price, R.A.
1982: Superimposed low-grade metamorphism in the Mount Fisher area, southeastern British Columbia - implications for the East Kootenay orogeny; Canadian Journal of Earth Sciences, v. 19, p. 476-489.

McMechan, R.D.
1981: Stratigraphy, sedimentology, structure and tectonic implications of the Oligocene Kishenehn Formation, Flathead Valley graben, southeastern British Columbia; Ph.D. thesis, Queen's University, Kingston, Ontario, v. 1 and 2, 327 p.
1983: Geology of the Princeton Basin, British Columbia; Ministry of Energy, Mines and Petroleum Resources, Paper 1983-3, 52 p.

McMechan, R.D. and Price, R.A.
1984: Crustal extension and thinning in a foreland thrust and fold belt, southern Canadian Rockies; Geological Society of America, Abstracts with Programs, v. 16, p. 591.

McMillan, W.J.
1970: West Flank, Frenchman Cap gneiss dome, Shuswap terrane, British Columbia; in Structure of the Southern Canadian Cordillera, J.O. Wheeler (ed.), Geological Association of Canada, Special Paper Number 6, p. 99-122.
1976: Geology and genesis of the Highland Valley ore deposits and the Guichon Batholith; in Porphyry Deposits of the Canadian Cordillera, A. Sutherland Brown (ed.), Canadian Institute of Mining and Metallurgy Special Volume 15, p. 85-104.

McTaggart, K.C.
1960: The geology of Keno and Galena Hills, Yukon Territory (105M); Geological Survey of Canada, Bulletin 58, 37 p.

Mercier, E.
1985: Précambrien de "Coal Creek Dome" (montagnes Ogilvie, Yukon, Canada); thèse présentée à l'Université des Sciences et Techniques de Lille, France, 246 p.

Mertie, J.B., Jr.
1931: Notes on the geography and geology of Lituya Bay, Alaska; United States Geological Survey Bulletin 836-B, p. 117-135.

Miller, D.J.
1953: Preliminary geologic map of Tertiary rocks in the southeastern part of the Lituya district, Alaska; United States Geological Survey, Open File Report 53-193.
1960: Giant waves in Lituya Bay, Alaska; United States Geological Survey, Professional Paper 354-C, p. 51-86.

Miller, G.M. and Misch, P.
1963: Early Eocene angular unconformity at western front of Northern Cascades, Whatcom County, Washington; American Association of Petroleum Geologists Bulletin, v. 47, p. 163-174.

Minster, J.B. and Jordan, T.H.
1978: Present day plate motions; Journal of Geophysical Research, v. 83, p. 5331-5354.

Minster, J.B., Jordan, T.H., Molnar, P., and Haines, E.
1974: Numerical modeling of instantaneous plate tectonics; Geophysical Journal of the Royal Astronomical Society, v. 36, p. 541-576.

Misch, P.
1966: Tectonic evolution of the Northern Cascades of Washington State; Canadian Institute of Mining and Metallurgy, Special Volume 8, p. 101-148.

Monger, J.W.H.
1968: Early Tertiary stratified rocks, Greenwood map-area; Geological Survey of Canada, Paper 67-42.
1970: Hope map-area, west-half, British Columbia; Geological Survey of Canada, Paper 69-47, 75 p.
1985: Structural evolution of the southwestern Intermontane Belt, Ashcroft and Hope map-areas, British Columbia; in Current Research, Part A, Geological Survey of Canada, Paper 85-1A, p. 349-358.
1986: Geology between Harrison Lake and Fraser River, Hope map area, southwestern British Columbia; in Current Research, Part B, Geological Survey of Canada, Paper 86-1B, p. 699-706.

Monger, J.W.H. and Berg, H.C.
1984: Lithotectonic terrane map of western Canada and southeastern Alaska; in Lithotectonic Terrane Maps of the North American Cordillera, N.J. Silberling and D.L. Jones (ed.), United States Geological Survey, Open-File Report 84-523.

Monger, J.W.H. and McMillan, W.J.
1984: Bedrock geology of Ashcroft (92) map-area, British Columbia; Geological Survey of Canada, Open File 980, Scale 1:125 000.

Monger, J.W.H., Price, R.A., and Tempelman-Kluit, D.J.
1982: Tectonic accretion and the origin of two major metamorphic and plutonic welts in the Canadian Cordillera; Geology, v. 10, p. 70-75.

Monger, J.W.H., Clowes, R.M., Price, R.A., Simony, P.S., Riddihough, R.P., and Woodsworth, G.J.
1985: B-2 Juan de Fuca Plate to Alberta Plains; Geological Society of America, The Decade of North American Geology, Centennial Continent/Ocean Transect #7, 21 p., 2 sheets.

Morgan, W.
1968: Rises, trenches, great faults, and crustal blocks; Journal of Geophysical Research, v. 73, p. 1959-1982.

Morrison, M.L.
1980: Basement involvement on the southwest flank of the southern Canadian Rockies; 26th International Geological Congress, Abstract, v. 1, p. 366.
1982: Structure and petrology of the southern portion of the Malton Gneiss, British Columbia; Ph.D. thesis, University of Calgary, Calgary, Alberta, 240 p.

Muller, J.E.
1967: Kluane Lake map-area, Yukon Territory; Geological Survey of Canada, Memoir 340, 137 p.
1977: Evolution of the Pacific Margin, Vancouver Island and adjacent regions; Canadian Journal of Earth Sciences, 14, p. 2062-2085.

Murphy, D.C.
1987: Suprastructure-infrastructure transition, east-central Cariboo Mountains, British Columbia: geometry, kinematics and tectonic implications; Journal of Structural Geology, v. 9, p. 13-29.

Murphy, D.C. and Journeay, J.M.
1982: Structural style in the Premier Range, Cariboo Mountains, southeastern British Columbia: Preliminary results; in Current Research, Part A, Geological Survey of Canada, Paper 82-1A, p. 289-292.

Murphy, D.C. and Rees, C.J.
1983: Structural transition and stratigraphy in the Cariboo Mountains, British Columbia; in Current Research, Part A, Geological Survey of Canada, Paper 83-1A, p. 245-252.

Mustard, J.F.
1983: The geology of the Mount Brew area, Lillooet, British Columbia; B.Sc. thesis, University of British Columbia, Vancouver, 33 p.

Naugler, F.P. and Wageman, J.M.
1973: Gulf of Alaska: Magnetic anomalies, fracture zones, and plate interaction; Geological Society of America Bulletin, v. 84, p. 1575-1584.

Norris, D.K.
1968: The Crowsnest Deflection of the eastern Cordillera of Canada; in Program with Abstracts, Geological Society of America Annual Meeting, Mexico City, p. 221.
1969: The Crowsnest deflection of the eastern Cordillera of Canada; Geological Society of America, Special Paper 121, p. 221.
1972: En échelon folding in the northern Cordillera of Canada; Bulletin of Canadian Petroleum Geology, v. 20, p. 634-642.
1974: Structural geometry and geological history of the northern Canadian Cordillera; Proceedings of First National Convention, Canadian Society of Exploration Geophysicists, Calgary, 1973; p. 18-45.
1980: Bedrock geology of the Dempster Lateral; in Facts and Principles of World Petroleum Occurrence, A.D. Miall (ed.), Canadian Society of Petroleum Geologists, Memoir 6, p. 535-550.
1981a: Geology, Herschel Island - Demarcation Point, Yukon Territory; Geological Survey of Canada, Map 1514A, Scale 1:250,000.
1981b: Geology, Blow River - Davidson Mountains, Yukon Territory - Northwest Territories; Geological Survey of Canada, Map 1516A, Scale 1:250 000.
1981c: Transform, contraction and extension faults in the northern cordillera of Canada - their spatial and temporal relationships since mid-Cretaceous time; in Geological Association of Canada Program and Abstracts, Cordilleran Section Meeting, Vancouver, B.C., 1981.
1982a: Geology, Ogilvie River, Yukon Territory; Geological Survey of Canada, Map 1526A, Scale 1:250 000.
1982b: Geology, Hart River, Yukon Territory; Geological Survey of Canada, Map 1527A, Scale 1:250 000.
1983a: Porcupine Virgation - the structural link among the Columbian, Innuitian and Alaskan orogens; in Program with Abstracts, v. 8, Joint Annual Meeting, Geological Association of Canada, Mineralogical Association of Canada and Canadian Geophysical Union, Victoria, B.C., May 1983, p. A51.
1983b: Geotectonic Correlation Chart 1532A - Operation Porcupine Project Area; Geological Survey of Canada, Chart 1532A.
1985: Geology of the northern Yukon and Northwestern District of Mackenzie; Geological Survey of Canada, Map 1581A, Scale 1:500 000.

Norris, D.K. and Hopkins, W.S. Jr.
1977: The geology of the Bonnet Plume Basin, Yukon Territory; Geological Survey of Canada, Paper 76-8.

Norris, D.K. and Yorath, C.J.
1981: The North American plate from the Arctic Archipelago to the Romanzof Mountains; in The Ocean Basins and Margins - The Arctic Ocean, v. 5 (3), A.E.M. Nairn, M. Churkin, Jr. and F.G. Stehli (ed.), Plenum Publishing Corporation, New York, London, Washington, Boston, p. 37-103.

Okulitch, A.V.
1979: Thompson-Shuswap-Okanagan; Geological Survey of Canada, Open File 637, 1:250 000.

Ollerenshaw, N.C.
1981: Parcel 82, Dominion Coal Block, southeastern British Columbia; in Current Research, Part B, Geological Survey of Canada, Paper 81-1B, p. 145-152.

Ovenshine, A.T. and Brew, D.A.
1972: Separation and history of the Chatham Strait Fault, southeast Alaska, North America; 24th Proceeding of the International Geological Congress, Section 3, p. 245-254.

Page, R.A.
1969: Late Cenozoic movement on the Fairweather Fault in southeastern Alaska; Geological Society of America Bulletin, v. 80, p. 1873-1878.
1973: The Sitka, Alaska, earthquake of 1972; Earthquake Information Bulletin, v. 5, p. 4-9.
1974: Evaluation of seismicity and shaking of offshore sites; Proceedings of the 7th Conference on Offshore Technology, Houston, Texas, v. 3, p. 179-190.

Parkinson, D.
1985: U-Pb geochronology and regional geology of the southern Okanagan Valley, British Columbia: the western boundary of a metamorphic core complex; M.Sc. thesis, University of British Columbia, Vancouver, 149 p.

Parrish, R.R.
1983: Cenozoic thermal evolution and tectonics of the Coast Mountains, British Columbia 1. Fission track dating, apparent uplift rates and pattern of uplift; Tectonics, v. 2, p. 601-631.
1984: Slocan Lake Fault: a low angle fault zone bounding the Valhalla gneiss complex, Nelson map area, southern British Columbia; in Current Research, Part A, Geological Survey of Canada, Paper 84-1A, p. 323-330.

Parrish, R. and Carr, S.D.
1986: Extensional tectonics of southeastern British Columbia: new data and interpretations; Geological Association of Canada, Annual Meeting, Program with Abstracts, p. 112.

Parrish, R.R. and Wheeler, J.O.
1983: A U-Pb zircon age of the Kuskanax batholith, southeastern British Columbia; Canadian Journal of Earth Sciences, v. 20, p. 1751-1756.

Parrish, R., Carr, S.D., and Parkinson, D.
1985a: Metamorphic Complexes and Extensional Tectonics, Southern Shuswap Complex, southeastern British Columbia; in Field Guides to Geology and Mineral Deposits in the Southern Canadian Cordillera, Geological Society of America, Cordilleran Section Meeting, p. 12.1-12.15.

Parrish, R., Carr, S.D., and Brown, R.L.
1985b: Valhalla gneiss complex, southeast British Columbia: 1984 Fieldwork; in Current Research, Part A, Geological Survey of Canada, Paper 85-1A, p. 81-87.

Paterson, I.A.
1977: The geology and evolution of the Pinchi fault zone at Pinchi Lake, central British Columbia; Canadian Journal of Earth Sciences, v. 14, p. 1324-1342.

Payne, J.G., Bratt, J.A., and Stone, B.G.
1980: Deformed Mesozoic volcanogenic Cu-Zn sulfide deposits in the Britannia District, British Columbia; Economic Geology, v. 75, p. 700-721.

Pearson, D.E. and Panteleyev, A.
1975: Cupriferous iron sulphide deposits, Kutcho Creek map-area (104I/1W); in Geological Field Work 1975, British Columbia epartment of Mines and Petroleum Resources, p. 86-92.

Pell, J.
1984: Stratigraphy, structure and metamorphism of Hadrynian strata in southeastern Cariboo Mountains, British Columbia; Ph.D. thesis, University of Calgary, Calgary, Alberta, 189 p.

Pell, J. and Simony, P.S.
1981: Stratigraphy, structure, and metamorphism in southern Cariboo Mountains, British Columbia; in Current Research, Part A, Geological Survey of Canada, Paper 81-1A, p. 227-230.

1982: Hadrynian Horsethief Creek Group/Kaza Group correlations in the southern Cariboo Mountains, British Columbia; in Current Research, Part A, Geological Survey of Canada, Paper 82-1A, p. 305-308.

1984: Stratigraphy of the Hadrynian Kaza Group between the Azure and North Thompson Rivers, Cariboo Mountains British Columbia; in Current Research, Part A, Geological Survey of Canada, Paper 84-1A, p. 95-98.

Perkins, M.J.
1983: Structural geology and stratigraphy, Big Bend of the Columbia River, Selkirk Mountains, B.C.; Ph.D. thesis, Carleton University, Ottawa, Ontario, 237 p.

Pigage, L.C.
1978: Geochronology and structure, Wells Gray Provincial Park, Cariboo Mountains, British Columbia; Ph.D. thesis, University of British Columbia, Vancouver, 289 p.

Plafker, G.
1967: Geologic map of the Gulf of Alaska Tertiary Province, Alaska; United States Geological Survey, Map I-484, scale 1:500,000.

Plafker, G. and Campbell, R.B.
1979: The Border Ranges fault in the St. Elias Mountains; in United States Geological Survey in Alaska: Accomplishments during 1978, K.M. Johnson and J.R. Williams (ed.), United States Geological Circular 804-B, p. B102-B104.

Plafker, G., Jones, D.H., and Pessagno, E.A., Jr.
1977a: A Cretaceous accretionary flysch and melange terrane along the Gulf of Alaska margin; United States Geological Survey, Circular 751-B, p. B41-B43.

Plafker, G., Hudson, T., and Richter, D.H.
1977b: Preliminary observations on late Cenozoic displacements along Totschunda and Denali fault systems; United States Geological Survey, Circular 751B, p. B67-B69.

Plafker, G., Hudson, T., Bruns, T., and Rubin, M.
1978: Late Quaternary offsets along the Fairweather fault and crustal plate interactions in southern Alaska; Canadian Journal of Earth Sciences, v. 15, p. 805-816.

Potter, C.J.
1983: Geology of the Bridge River Complex, southern Shulaps Range, British Columbia: A record of Mesozoic convergent tectonics; Ph.D. thesis, University of Washington, 192 p.

Preto, V.A.G.
1977: The Nicola Group: Mesozoic volcanism related to rifting in southern British Columbia; in Volcanic Regimes in Canada, W.R.A. Baragar, L.G. Coleman and J.M. Hall (ed.), Geological Association of Canada, Special Paper 16, p. 39-57.

1981: Barriere Lakes-Adams Plateau area; in Geological Fieldwork 1980, British Columbia Ministry of Energy, Mines and Petroleum Resources, Paper 1981-1, p. 15-23.

Price, R.A.
1962: Fernie map-area east half, Alberta and British Columbia (82G E1/2); Geological Survey of Canada, Paper 61-24, 65 p.

1965: Flathead map-area, British Columbia and Alberta; Geological Survey of Canada, Memoir 336, 221 p.

1979: Intracontinental ductile crustal spreading linking the Fraser River and northern Rocky Mountain transform fault zones, south-central British Columbia and northeast Washington; Geological Society of America, Abstracts with Programs, v. 11, p. 499.

1981a: Eocene stretching and necking of the crust and tectonic unroofing of the Cordilleran metamorphic infrastructure, southeastern British Columbia and adjacent Washington and Idaho; Geological Association of Canada, Program with Abstracts, p. A47.

1981b: The Cordilleran thrust and fold belt in the southern Canadian Rocky Mountains; in Thrust and Nappe Tectonics, K.R. McClay and N.J. Price (ed.), Geological Society of London, Special Publication 9, p. 427-448.

1983: The Rocky Mountain belt of Canada, thrust faulting, tectonic wedging and delamination of the lithosphere; Geological Association of Canada, Annual Meeting, Program with Abstracts, p. A55.

1986: The southeastern Canadian Cordillera: thrust faulting, tectonic wedging, and delamination of the lithosphere; Journal of Structural Geology, v. 8, p. 239-254.

Price, R.A. and Fermor, P.R.
1985: Structure section of the Cordilleran foreland thrust and fold belt west of Calgary, Alberta; Geological Survey of Canada, Paper 84-14, 1 sheet.

Price, R.A. and Mountjoy, E.W.
1970: Geologic structure of the Canadian Rocky Mountains between Bow and Athabasca rivers, a progress report; in Structure of the Southern Canadian Cordillera, J.O. Wheeler (ed.), Geological Association of Canada, Special Publication No. 6, p. 7-25.

Price, R.A., Balkwill, H.R., and Mountjoy, E.W.
1979: McMurdo (east half) British Columbia; Geological Survey of Canada, Map 1501A.

Price, R.A., Monger, J.W.H., and Muller, J.E.
1981: Cordilleran cross-section - Calgary to Vancouver; in Field Guides to Geology and Mineral Deposits, Geological Association of Canada Guide Book, Calgary '81.

Price, R.A., Parrish, R., Green, A.G., Cook, F.A., Simony, P.S., Gordy, P.L., Milkereit, B., Coflin, K.C., and Patenaude, C.
1986: Lithoprobe seismic reflection images of deep crustal structure, southeastern Canadian Cordillera; Geological Society of America, Abstracts with Programs, v. 18, p. 723.

Psutka, J.F.
1978: Structural setting of the Downie Slide, northeast flank of Frenchman Cap gneiss dome, Shuswap Metamorphic Complex, southern British Columbia; M.Sc. thesis, Carleton University, Ottawa, Ontario, 70 p.

Raeside, R.P. and Simony, P.S.
1983: Stratigraphy and deformational history of the Scrip nappe, Monashee Mountains, British Columbia; Canadian Journal of Earth Sciences, v. 20, p. 639-650.

Ramsay, J.G.
1974: Development of chevron folds; Geological Society of America Bulletin, v. 85, p. 1741-1754.

Read, P.B.
1971: Metamorphic environment and timing of deformation, plutonism, and metamorphism in part of central Kootenay Arc, British Columbia; in Metamorphism in the Canadian Cordillera, Geological Association of Canada, Cordilleran Section, Program and Abstracts, p. 27.

1973: Petrology and structure of Poplar Creek map area, British Columbia; Geological Survey of Canada, Bulletin 193, 144 p.

1975: Lardeau Group, Lardeau map-area, west half (92K, W1/2), British Columbia; in Report of Activities, Part A, Geological Survey of Canada, Canada, Paper 75-1, p. 29-30.

1976: Lardeau map area, west half (82K, W1/2), British Columbia; in Report of Activities, Part A, Geological Survey of Canada, Paper 76-1A, p. 95-96.

1979: Relationship between the Shuswap Metamorphic Complex and Kootenay Arc, Vernon East-half, southern British Columbia; in Current Research, Part A, Geological Survey of Canada, Paper 79-1A, 37-40.

1980: Stratigraphy and structure: Thor-Odin to Frenchman Cap "domes", Vernon east-half map area, southern British Columbia; in Current Research, Part A, Geological Survey of Canada, Paper 80-1A, p. 19-25.

Read, P.B. and Brown, R.L.
1981: Columbia River fault zone: southeast margin of the Shuswap and Monashee Complex, southern British Columbia; Canadian Journal of Earth Sciences, v. 18, p. 1127-1145.

Read, P.B. and Klepacki, D.W.
1981: Stratigraphy and structure: northern half of Thor-Odin nappe, Vernon east-half map area, southern British Columbia; in Current Research, Part A, Geological Survey of Canada, Paper 81-1A, p. 169-173.

Read, P.B. and Monger, J.W.H.
1976: Pre-Cenozoic volcanic assemblages of the Kluane and Alsek Ranges, southwestern Yukon Territory; Geological Survey of Canada, Open File 381, 96 p.

Read, P.B. and Thompson, R.I.
1980: Geology of the Akolkolex River area: an addendum; in Geological Fieldwork 1979, British Columbia Ministry of Energy, Mines and Resources, Paper 80-1, p. 183-187.

Read, P.B. and Wheeler, J.O.
1976: Geology of the Lardeau west-half map area, British Columbia; Geological Survey of Canada, Open File 432.

Reamsbottom, S.B.
1974: Geology and metamorphism of Mount Breaker ridge area, Harrison Lake, British Columbia; Ph.D. thesis, University of British Columbia, Vancouver, 155 p.

Rees, C.J.
1981: Western margin of the Omineca Belt at Quesnel Lake, British Columbia; in Current Research, Part A, Geological Survey of Canada, Paper 81-1A, p. 223-226.

Rees, C.J. and Ferri, F.
1983: A kinematic study of mylonitic rocks in the Omineca Intermontane Belt tectonic boundary in east-central British Columbia; in Current Research, Part B, Geological Survey of Canada, Paper 83-1B, p. 121-125.

Reesor, J.E.
1965: Structural evolution and plutonism in Valhalla gneiss complex, British Columbia; Geological Survey of Canada, Bulletin 129, 128 p.

Reesor, J.E. and Moore, J.M.
1971: Petrology and structure of Thor-Odin gneiss dome, Shuswap Metamorphic Complex, British Columbia; Geological Survey of Canada, Memoir 228, 86 p.

Rehrig, W.A. and Reynolds, S.J.
1981: Eocene metamorphic core complex tectonics near the Lewis and Clark zone, western Montana and northern Idaho; Geological Society of America, Abstracts with Programs, v. 13, p. 102.

Rice, H.M.A.
1947: Geology and mineral deposits of Princeton map-area, British Columbia; Geological Survey of Canada, Memoir 243, 136 p.

Rich, J.L.
1934: Mechanics of low-angle overthrust faulting illustrated by Cumberland Thrust block, Virginia, Kentucky and Tennessee; American Association of Petroleum Geologists, Bulletin, v. 18, p. 1584-1596.

Richards, T.A.
1980: Geology, Hazelton map area, British Columbia (93M); Geological Survey of Canada, Open File 720.

Richter, D.H.
1976: Geologic map of the Nabesna quadrangle, Alaska; United States Geological Survey, Map I-932.

Richter, D.H. and Matson, N.A., Jr.
1971: Quaternary faulting in the eastern Alaska Range; Geological Society of America Bulletin, v. 82, p. 1529-1539.

Riddihough, R.P.
1977: A model for recent plate interactions off Canada's west coast; Canadian Journal of Earth Sciences, v. 14, p. 384-396.

Robertson Research International Ltd.
1978: The People's Republic of China: Its petroleum geology and resources, v. 1 and 2.

Roddick, J.A. and Woodsworth, G.J.
1980: Geology of Bute Inlet map area, British Columbia (NTS 92K); Geological Survey of Canada, Open File 480.

Roddick, J.A., Muller, J.E., and Okulitch, A.V.
1979: Fraser River, British Columbia-Washington; Geological Survey of Canada, Map 1386A, 1:1,000,000 scale.

Rogers, G.C. and Hasegawa, H.S.
1978: A second look at the British Columbia earthquake of June 23, 1946; Bulletin of the Seismological Society of America; v. 68, p. 653-675.

Roots, C.F.
1987: Regional tectonic setting and evolution of the Late Proterozoic Mount Harper volcanic complex, Ogilvie Mountains, Yukon; Ph.D. thesis, Carleton University, Ottawa, Ontario, 219 p.

Ross, J.A. and Christie, J.S.
1979: Early recumbent folding in some westernmost exposures of the Shuswap Complex, southern Okanagan, British Columbia; Canadian Journal of Earth Sciences, v. 16, p. 877-894.

Ross, J.V.
1981: A geodynamic model for some structures within and adjacent to the Okanagan Valley, southern British Columbia; Canadian Journal of Earth Sciences, v. 18, p. 1581-1598.

Royse, F.J., Warner, M.A., and Reese, D.L.
1975: Thrust belt structural geometry and related stratigraphic problems Wyoming-Idaho-Northern Utah; Rocky Mountain Association of Geologists, Symposium on Deep Drilling Frontiers in the Central Rocky Mountains, p. 41-54.

Rusmore, M.E.
1985: Geology and tectonic significance of the Upper Triassic Cadwallader Group and its bounding faults, southwestern British Columbia; Ph.D. thesis, University of Washington, 174 p.
1987: Geology of the Cadwallader Group and the Intermontane - Insular superterrane boundary, southern British Columbia; Canadian Journal of Earth Sciences, v. 24, p. 2279-2291.

Rusmore, M.E. and Woodsworth, G.J.
1988: Eastern margin of the Coast Plutonic Complex, Mount Waddington map area (92N), B.C.; in Current Research, Part E, Geological Survey of Canada, Paper 88-1E, p. 185-190.

Sainsbury, C.L. and Twenhofel, S.W.
1954: Fault patterns in southeastern Alaska; Abstract, Geological Society of America Bulletin, v. 65, p. 1300.

St. Amand, P.
1954: Tectonics of Alaska as deduced from seismic data; Abstract, Geological Society of America Bulletin, v. 65, p. 1350.
1957: Geological and geophysical synthesis of the tectonics of proportions of British Columbia, the Yukon Territory, and Alaska; Geological Society of America Bulletin, v. 68, p. 1343-1370.

Schiarizza, P.
1986: Geology of the Eagle Bay Formation between the Raft and Baldy batholiths; in Geological Fieldwork 1985, British Columbia Ministry of Energy, Mines and Petroleum Resources, Paper 1986-1, p. 89-94.

Schiarizza, P. and Preto, V.A.G.
1984: Geology of the Adams Plateau-Clearwater area; British Columbia Ministry of Energy, Mines and Petroleum Resources, Preliminary Map 56.

Sears, J.W.
1979: Tectonic contrasts between the infrastructure and suprastructure of the Columbian orogen, Albert Canyon area, western Selkirk Mountains, British Columbia; Ph.D. thesis, Queen's University, Kingston, Ontario.

Shouldice, D.H.
1971: Geology of the western Canadian continental shelf; Bulletin of Canadian Petroleum Geology, Canadian Society of Petroleum Geologists, v. 19, p. 405-436.

Simony, P.S. and Wind, G.
1970: Structure of the Dogtooth Range and adjacent portions of the Rocky Mountain Trench; in Structure of the Canadian Cordillera, Geological Association of Canada, Special Paper 6, p. 41-51.

Simony, P.S., Ghent, E.D., Craw, D., Mitchell, W., and Robbins, D.B.
1980: Structural and metamorphic evolution of the northeast flank of the Shuswap Complex, southern Canoe River area, British Columbia; Geological Society of America, Memoir 153, p. 445-461.

Smith, R.B.
1974: Geology of the Harper Ranch Group (Carboniferous-Permian) and Nicola Group (Upper Triassic), northwest of Kamloops, British Columbia; M.Sc. thesis, University of British Columbia, Vancouver, 211 p.

Snavely, P.D., Jr. and Wagner, H.C.
1981: Geologic cross-section across the continental margin off Cape Flattery, Washington, and Vancouver Island, British Columbia; United States Geological Survey, Open File Report 81-978.

Sonnevil, R.A.
1981: The Chilkat-Prince of Wales plutonic province, southeastern Alaska; United States Geological Survey, Circular 823-B, p. B113-B115.

Souther, J.G.
1971: Geology and mineral deposits of Tulsequah map-area, British Columbia; Geological Survey of Canada, Memoir 362, 84 p.

Souther, J.G. and Stanciu, C.
1975: Operation Saint Elias, Yukon Territory: Tertiary volcanic rocks; in Report of Activities, Part A, Geological Survey of Canada, Paper 75-1A, p. 63-70.

Spang, J.H., Simony, P.S., and Mitchell, W.J.
1980: Strain and folding mechanisms in a similar style fold from the Northern Selkirks of the Canadian Cordillera; in Analytical Studies in Structural Geology, Tectonophysics, v. 66, p. 253-267.

Stauder, W.
1960: The Alaska earthquake of July 10, 1958: Seismic studies; Bulletin of the Seismological Society of America, v. 50, p. 293-322.

Stevens, R.D., Delabio, R.N., and Lachance, G.R.
1982: Age determinations and geological studies, K-Ar isotopic ages, Report 16; Geological Survey of Canada, Paper 82-2, 56 p.

Stockmal, G.S. and Chapple, W.M.
1981: Modelling accretionary wedge deformation using a rigid-perfectly plastic rheology; American Geophysical Union, Transactions, v. 62, p. 46-55.

Stout, J.H. and Chase, C.G.
1980: Plate kinematics of the Denali fault system; Canadian Journal of Earth Sciences, v. 17, p. 1527-1537.

Struik, L.C.
1980: Geology of the Barkerville-Cariboo River area, east central British Columbia; Ph.D. thesis, University of Calgary, Alberta, 335 p.
1981: A re-examination of the type area of the Devono-Mississippian Cariboo Orogeny, central British Columbia; Canadian Journal of Earth Sciences, v. 18, p. 1767-1775.

1982: Snowshoe Formation (1982), central British Columbia; in Current Research, Part B, Geological Survey of Canada, Paper 82-1B, p. 117-124.
1985: Imbricated terranes of Cariboo Gold Belt with correlations and implications for tectonics in southeastern British Columbia; Canadian Journal of Earth Sciences, v. 23, p. 1057-1061.

Sutherland Brown, A.
1957: Geology of the Antler Creek area, Cariboo District, British Columbia; British Columbia Department of Mines, Bulletin 38.
1968: Geology of the Queen Charlotte Islands, British Columbia; British Columbia Department of Energy, Mines and Petroleum Resources, Bulletin 54.

Sutherland Brown, A. and Yorath, C.J.
1985: Lithoprobe profile across southern Vancouver Island: Geology and Tectonics; in Field Guides to Geology and Mineral Deposits in the southern Canadian Cordillera, Geological Society of America, Cordilleran Section Meeting, Vancouver, B.C., p. 8-1 to 8-23.

Sutherland Brown, A., Yorath, C.J., and Tipper, H.W.
1983: Geology and tectonic history of the Queen Charlotte Islands; Geological Association of Canada/ Mineralogical Association of Canada/Canadian Geophysical Union, Annual Meeting, Field Trip #8, Guide Book.

Tarr, R.S. and Martin, L.
1912: The earthquakes at Yakutat Bay, Alaska, in September, 1899; United States Geological Survey, Professional Paper 69, 35 p.

Taylor, G.C. and Stott, D.F.
1973: Tuchodi Lakes map-area, northeastern British Columbia; Geological Survey of Canada, Memoir 373, 37 p.

Teal, P.R.
1983: The triangle zone at Cabin Creek, Alberta; in Seismic Expression of Structural Styles, A.W. Bally (ed.), American Association of Petroleum Geologists, Studies in Geology Series, No. 15, v. 3, p. 4.1-48 to 4.1-53.

Tempelman-Kluit, D.J.
1970: Stratigraphy and structure of the "Keno Hill Quartzite" in Tombstone River - Upper Klondike River map-areas, Yukon Territory; Geological Survey of Canada, Bulletin 180, 102 p.
1974: Reconnaissance geology of Aishihik Lake, Snag and part of Stewart River map-areas, west central Yukon; Geological Survey of Canada, Paper 73-41, 97 p.
1976: The Yukon Crystalline Terrane: Enigma in the Canadian Cordillera; Geological Society of America Bulletin, v. 87, p. 1343-1357.
1981a: Geology of the Craig Claims; in Yukon Geology and Exploration 1979-80, Geology Section, Department of Indian and Northern Affairs Whitehorse, p. 225-230.
1981b: Geology of the Thor Claims; in Yukon Geology and Exploration 1979-80, Geology Section, Department of Indian and Northern Affairs Whitehorse, p. 289-291.
in press-a: Geology of Quiet Lake and Finlayson Lake (105F, 105G) map-areas; Geological Survey of Canada, Memoir.
in press-b: Geology of Carmacks and Laberge map-areas, Yukon Territory; Geological Survey of Canada, Memoir.

Tempelman-Kluit, D.J. and Parkinson, D.
1986: Extension across the Eocene Okanagan crustal shear in southern British Columbia; Geology, v. 14, p. 318-321.

Thatcher, W. and Plafker, G.
1977: The Yakutat Bay, Alaska, earthquakes: seismograms and crustal deformation; Geological Society of America, Abstracts with Programs, v. 9, p. 515.

Thompson, R.I.
1978: Geology of the Akolkolex River area, B.C. Ministry of Energy, Mines and Petroleum Resources, Bulletin 60, p. 77.
1979: A structural interpretation across part of the northern Rocky Mountains, British Columbia, Canada; Canadian Journal of Earth Sciences, v. 16, p. 1228-1241.
1981: The nature and significance of large "blind" thrusts within the northern Rocky Mountains of Canada; in Thrust and Nappe Tectonics, K.R. McClay and N.J. Price (ed.), Geological Society of London Special Publication 9, p. 449-462.
1984: Late Proterozoic extension and its influence on Mesozoic deformation western Ogilvie Mountains, Yukon; Abstract, Exploration Roundup, Canadian Society of Petroleum Geologists.
in press: Stratigraphy, structural analysis, and tectonic evolution of the Halfway River map-area (94B), northern Rocky Mountains, British Columbia; Geological Survey of Canada, Memoir.

Thompson, R.I. and Eisbacher, G.H.
1984: Late Proterozoic rift assemblages, northern Canadian Cordillera; Geological Society of America, Abstracts with Programs, v. 16, p. 336.

Thompson, R.I. and Thorkelson, D.
1989: Regional mapping update, central Queen Charlotte Islands, British Columbia; in Current Research, Part H, Geological Survey of Canada, Paper 89-1H, p. 7-11.

Thorstad, L.
1980: Upper Paleozoic volcanic and volcaniclastic rocks in northwest Toodoggone map area, British Columbia; in Current Research, Part B, Geological Survey of Canada, Paper 80-1B, p. 207-211.

Thorstad, L.E. and Gabrielse, H.
1986: The Upper Triassic Kutcho Formation, Cassiar Mountains, north-central British Columbia; Geological Survey of Canada, Paper 86-16, 53 p.

Thorsteinsson, R. and Tozer, E.T.
1960: Summary account of structural history of the Canadian Arctic Archipelago since Precambrian time; Geological Survey of Canada, Paper 60-7.

Tiffin, D.L., Cameron, B.E.B., and Murray, J.W.
1972: Tectonics and depositional history of the continental margin off Vancouver Island, British Columbia; Canadian Journal of Earth Sciences, v. 9, p. 280-296.

Tipper, H.W.
1969: Mesozoic and Cenozoic geology of the northeast part of Mount Waddington map area (92N), Coast District, British Columbia; Geological Survey of Canada, Paper 68-33, 103 p.

Tipper, H.W. and Richards, T.A.
1976: Jurassic stratigraphy and history of north-central British Columbia; Geological Survey of Canada, Bulletin 270, 73 p.

Tobin, D. and Sykes, L.
1968: Seismicity and tectonics of the northeast Pacific Ocean; Journal of Geophysics Research, v. 73, p. 3821-3845.

Tocher, D.
1960: The Alaska earthquake of July 10, 1958: Movement on the Fairweather Fault and field investigations of southern epicentral region; Seismological Society of America Bulletin, v. 50, p. 267-292.

Travers, W.B.
1978: Overturned Nicola and Ashcroft strata and their relation to the Cache Creek Group, southwestern Intermontane Belt, British Columbia; Canadian Journal of Earth Sciences, v. 15, p. 99-116.
1982: Possible large-scale overthrusting near Ashcroft, British Columbia: implications for petroleum prospecting; Bulletin of Canadian Petroleum Geology, v. 30, p. 1-8.

Turner, D.L., Smith, T.E., and Forbes, R.B.
1974: Geochronology of offset along the Denali fault system in Alaska; Geological Society of America, Abstracts with Programs, v. 6, p. 268-269.

van der Heyden, P.
1982: Tectonic and stratigraphic relations between the Coast Plutonic Complex and Intermontane Belt, west-central Whitesail Lake map area, British Columbia; M.Sc. thesis, University of British Columbia, Vancouver, 172 p.

Wanless, R.K., Stevens, R.D., Lachance, G.R., and Delabio, R.N.
1979: Age determinations and geological studies, K-Ar isotopic-ages, Report 14; Geological Survey of Canada, Paper 79-2.

Wheeler, J.O.
1954: A geological reconnaissance of the northern Selwyn Mountains region, Yukon and Northwest Territories; Geological Survey of Canada, Paper 53-7, 42 p.
1963: Rogers Pass map area, British Columbia and Alberta (82N W1/2); Geological Survey of Canada, Paper 62-32, 32 p.
1965: Big-Bend map area, British Columbia; Geological Survey of Canada, Paper 64-32, 37 p.
1968: Lardeau (west-half) map area, British Columbia; in Report of Activities, 1967, Part A, Geological Survey of Canada, Paper 68-1A, p. 56-58.

Wilcox, R.E., Harding, T.P., and Seely, D.R.
1973: Basic wrench tectonics; American Association of Petroleum Geology Bulletin, v. 57, p. 74-96.

Woodsworth, G.J.
1979: Metamorphism, deformation, and plutonism in the Mount Raleigh pendant, Coast Mountains, British Columbia; Geological Survey of Canada, Bulletin 295, 58 p.
1980: Geology of Whitesail Lake (93E) map-area, B.C.; Geological Survey of Canada, Open File 708.

1988: Karmutsen Formation and the east boundary of Wrangellia, Queen Charlotte Basin, British Columbia; in Current Research, Part E, Geological Survey of Canada, Paper 88-1E, p. 209-212.

Woodsworth, G.J., Crawford, M.L., and Hollister, L.S.
1983: Metamorphism and structure of the Coast Plutonic Complex and adjacent belts, Prince Rupert and Terrace areas, British Columbia; Geological Association of Canada, Annual Meeting 1983, Field Trip Guidebook 14, 62 p.

Wright, F.E. and Wright, C.W.
1908: The Ketchikan and Wrangell mining districts, Alaska; United States Geological Survey, Bulletin 347, 210 p.

Yorath, C.J.
1980: The Apollo structure in Tofino Basin, Canadian Pacific continental shelf; Canadian Journal of Earth Sciences, v. 17, p. 758-775.

Yorath, C.J. and Chase, R.L.
1981: Tectonic history of the Queen Charlotte Islands and adjacent areas - a model; Canadian Journal of Earth Sciences, v. 18, p. 1717-1739.

Yorath, C.J. and Currie, R.G.
1980: Some aspects of the geology and structural style of the Vancouver Island continental margin; Geological Association of Canada/Mineralogical Association of Canada, Annual Meeting, Program with Abstracts, v. 8, p. 88.

Yorath, C.J. and Hyndman, R.D.
1983: Subsidence and thermal history of Queen Charlotte Basin; Canadian Journal of Earth Sciences, v. 20, p. 135-159.

Yorath, C.J. and Norris, D.K.
1975: The tectonic development of the southern Beaufort Sea and its relationship to the origin of the Arctic Ocean basin; Canadian Society of Petroleum Geologists, Memoir 4, p. 589-611.

Yorath, C.J., Clowes, R.M., Green, A.G., Sutherland Brown, A., Brandon, M.T., Massey, N.W.D., Spencer, C., Kanasewich, E.R., and Hyndman, R.D.
1985a: Lithoprobe - Phase 1: Southern Vancouver Island: Preliminary analyses of reflection seismic profiles and surface geological studies; in Current Research, Part A, Geological Survey of Canada, Paper 85-1A, p. 543-554.

Yorath, C.J., Green, A.G., Clowes, R.M., Sutherland Brown, A., Brandon, M.T., Kanasewich, E.R., Hyndman, R.D., and Spencer, C.
1985b: Lithoprobe, Southern Vancouver Island: Seismic reflection sees through Wrangellia to the Juan de Fuca Plate; Geology, v. 13, p. 759-762.

Yorath, C.J., Hyndman, R.D., Sutherland Brown, A. and Massey, N.W.D.
in press: Lithoprobe Southern Vancouver Island: A summary; Geological Society of America Bulletin.

Yorath, C.J., Woodsworth, G.J., Riddihough, R.P., Currie, R.G., Hyndman, R.D., Rogers, G.C., and Seemann, D.A.
1985c: B-1 Intermontane Belt (Skeena Mountains) to Insular Belt (Queen Charlotte Islands); Geological Society of America, The Decade of North American Geology, Centennial Continent/Ocean Transect #8, 8 p., 2 sheets.

Young, G.M., Jefferson, C.W., Delaney, G.D., and Yeo, G.M.
1979: Middle and late Proterozoic evolution of the northern Canadian Cordillera and Shield; Geology, v. 7, p. 125-128.

Younger, F.G., Myhr, D.W., and Yorath, C.J.
1976: Geology of the Beaufort-Mackenzie Basin; Geological Survey of Canada, Paper 76-11.

Ziegler, P.A.
1969: The development of sedimentary basins in western and Arctic Canada; Alberta Society of Petroleum Geologists, Calgary, Alberta, 89 p.

Authors' addresses

H. Gabrielse
C.J. Dodds
G.J. Woodsworth
J.W.H. Monger
D.J. Tempelman-Kluit
C.A. Evenchick
J.M. Journeay
L.C. Struik
D.C. Murphy
S.P. Gordey
R.I. Thompson
Cordilleran Division
Geological Survey of Canada
100 West Pender Street
Vancouver, British Columbia
V6B 1R8

C.J. Yorath
Pacific Geoscience Centre
9860 West Saanich Road
P.O. Box 6000
Sidney, British Columbia
V8L 4B2

R.M. Friedman
R.L. Armstrong
Department of Geological Services
University of British Columbia
Vancouver, British Columbia
V6T 2B4

A. Sutherland Brown
546 Newport Avenue
Victoria, British Columbia
V8S 5C7

R.B. Campbell
1760 Forest Park Drive
Sidney, British Columbia
V8L 4A6

R.L. Brown
C.J. Rees
Department of Earth Sciences
Carleton University
Ottawa, Ontario
K1S 5B6

L.S. Lane
M.E. McMechan
D.G. Cook
Institute of Sedimentary and Petroleum Geology
Geological Survey of Canada
3303-33rd Street, N.W.
Calgary, Alberta
T2L 2A7

J.L. Mansy
Laboratoire de Dynamique sédimentaire
et structurale
Université des Sciences et Techniques de
 Lille Flandres-Artois
59655, Villaneuve d'Ascq.
France

P.S. Simony
University of Calgary
Department of Geology and Geophysics
Calgary, Alberta
T2N 1N4

T.A. Harms
Department of Geology
Amherst College
Amherst, Massachusetts
U.S.A. 01002

R.R. Parrish
Geological Survey of Canada
601 Booth Street
Ottawa, Ontario
K1A 0E8

T. Høy
British Columbia Ministry of Energy,
Mines and Petroleum Resources
Geological Survey Branch
756 Fort Street
Victoria, British Columbia
V8V 1X4

J.T. Fyles
1720 Kingsberry Crescent
Victoria, British Columbia
V8P 2A7

ADDENDUM

Recent studies along and east of the eastern margin of the Coast Belt have documented the importance of eastward-verging structures in the western Intermontane Belt. These structures, of possibly, Oxfordian to latest Cretaceous or early Tertiary age, have resulted in as much as 160 km of shortening in the Skeena Fold Belt and may be linked with deformation during the same interval involving, the Coast, Omineca and Foreland belts.

REFERENCE

Evenchick, C.A.
1991: Geometry, evolution, and tectonic framework of the Skeena Fold Belt, north-central British Columbia; Tectonics, V.10

The significance of east-vergent structures along the eastern margin of the Coast Plutonic Complex during mid-Creteceous time is discussed in:

Rusmore, M.E. and Woodsworth, G.J.,
1991: Coast Plutonic Complex: a mid-Cretaceous contractional orogen; Geology vol. 19, p. 941-944.

Important papers on structural style and ages of deformation in the Queen Charlotte Islands appear in the following reference:

Woodsworth, G.J. (ed),
1991: Evolution and hydrocarbon potential of the Queen Charlotte Basin, British Columbia; Geological Survey of Canada, Paper 90-10, 569 p.

A detailed study of the Horseranch Range in north-central British Columbia documents structures indicating tectonic denudation, probably during Eocene time.

Plint, H.E. and Erdmer, P.,
1989: Structure and metamorphism in the Horseranch Range, north-central British Columbia; British Columbia Ministry of Energy, Mines and Petroleum Resources, Geological Fieldwork, 1988, Paper 1989-1, p. 347-351.

Printed in Canada

Chapter 18

TECTONIC SYNTHESIS

Summary
Introduction
Proterozoic
    Precambrian basement
    Middle Proterozoic - the early rifted and passive margin
    Late Proterozoic - the classic, rifted continental margin
Paleozoic and Triassic
    Cambrian to Middle Devonian
    Ancestral North America - the classic miogeocline
    Accreted terranes - volcanic archipelagos, related
    clastics and carbonate platforms
Late Devonian to Triassic
    Accreted terranes - oceanic basins, volcanic archipelagos
    and related clastics and carbonate platforms
    Displaced terranes and ancestral North America - miogeocline
    and Devono-Mississippian tectonismCorrelations
Early Jurassic to Paleogene
    Early Jurassic - volcanic island arcs, amalgamation
    of the Intermontane Superterrane and terminal
    phase of the miogeocline
    Middle Jurassic to Early Cretaceous (Hauterivian) - accretion and orogeny
    Pre-Hauterivian to Barremian pedimentation
    Early Cretaceous (Barremian) to Paleocene
    Eocene - magmatism, uplift, extension, dextral strike-slip faulting
Neogene
    Oligocene to Recent
Conclusions and comparisons with other orogens
References

Chapter 18

TECTONIC SYNTHESIS

H. Gabrielse and C.J. Yorath

SUMMARY

The geological architecture of the Cordilleran Orogen in Canada is the product of a long-lived evolution through a variety of tectonic processes acting upon and adjacent to the ancient continental margin. From its rift inception during the Middle Proterozoic until the present, the continental margin discontinuously moved oceanward through sediment progradation and as a result of convergence and transform-related processes involving the accretion of island arc and oceanic assemblages of distant and disparate origin. Throughout much of this time (1.44 Ga), the miogeocline of ancestral western North America evolved in a passive margin setting. One or more Precambrian folding events, accompanied by low-grade regional metamorphism, occurred in the northern and southern parts of the Cordillera but their timing and extent are poorly documented. Following Late Proterozoic to Early Cambrian rifting, broad carbonate platforms developed on a westerly sloping shelf, the edge of which was irregularly indented by numerous embayments and basins which received terrigenous clastics from the craton and intra-platform ridges and shoals. Local, episodic volcanism and graben development reflect rifting in the outer part of the miogeocline in early Paleozoic time. The northern part of the Cordillera was flooded by an Upper Devonian and Mississippian westerly derived clastic wedge succeeded by Mississippian and Pennsylvanian clastics whose source lay to the north, presumably in the Innuitian Orogen. Since the Middle Jurassic Epoch, as a consequence of collision and incorporation of the large exotic superterranes, intense metamorphism, plutonism and uplift took place in the Omineca Belt, and the miogeocline was detached from its basement, thickened through folding and telescoping along imbricate thrust faults and transported eastward over the edge of the craton. The rising metamorphic welt and prograding thrust sheets thereafter provided clastic fill for the foredeep moat that progressively migrated eastward in front of the developing orogen.

The oceanic and island arc assemblages of the central and western Cordillera comprise a variety of crustal fragments which form a collage of accreted terranes. Two groupings of terranes, the Intermontane and Insular superterranes, each consist of several smaller terranes which amalgamated "offshore" and subsequently became incorporated into the continent during Mesozoic time. Between the Intermontane and Insular superterranes the Coast Belt evolved as a long-lived island arc into the core of a two-sided orogen in mid- to Late Cretaceous time. Two small terranes were subsequently added to the Insular Superterrane during the early Tertiary. Between the accreted superterranes and ancestral North America the Omineca Belt hosts several terranes, some of which have close continental affinities whereas others have been transported long distances and have been thrust over the continental margin.

Superimposed upon the terranes and across the boundaries between them are sedimentary overlap assemblages, some of which developed in response to terrane amalgamation and intra-terrane contraction whereas others formed as a consequence of crustal attenuation, thermal subsidence and lithospheric flexure. In the southern Intermontane and Coast belts, extensive areas of young volcanic rocks reflect the effects of a changing Cenozoic plate tectonic regime. Beneath the modern continental shelf and slope the Tertiary to modern accretionary prism sediments are disrupted by folds and thrust faults to form a subduction complex in response to the Benioff (B) subduction of oceanic lithosphere. The structural style of the B-subduction complex is similar to that of the deformed clastic wedge sediments of the Foothills in the Foreland Belt produced by Ampferer (A) subduction of continental lithosphere.

The dominant northwesterly directed structural fabric of the Cordillera is expressed by imbricate thrust sheets and cylindrical folds disposed in broad salients and re-entrants in the Foreland Belt and by plutons, thrust faults, folds and regional dextral strike-slip faults throughout the remaining four belts. In the northern part of Yukon Territory, northerly and northeasterly trending oblique-slip faults are congruent with and dislocate northerly and northeasterly trending structural culminations. In northwestern British Columbia and southwestern Yukon, the great elevation of the Saint Elias Mountains is the product of oblique subduction of the Yakutat Terrane beneath the continental margin along the Fairweather Fault, which in turn is part of the modern Pacific-North America transform plate boundary. Major right-lateral strike-slip faults occur as components of numerous fault systems and together define northwestward movements of more than 1500 km. In the southern Omineca Belt, and locally in the Intermontane Belt, structures arising from crustal extension are probably related to movements on these faults.

The Canadian Cordillera can be compared to modern oceanic tectonic settings and to other orogenic belts. The Japanese Archipelago is an attractive model for the tectonic

---

Gabrielse, H. and Yorath, C. J.
1991: Tectonic synthesis, Chapter 18 in Geology of the Cordilleran Orogen in Canada, H. Gabrielse and C. J. Yorath (ed.); Geological Survey of Canada, Geology of Canada, no. 4, p. 677-705 (also Geological Society of America, The Geology of North America, v. G-2)

settings of Quesnellia and Stikinia whereas Wrangellia, which originated as a rifted arc, bears a striking resemblance to the Lau Basin behind the Tonga Trench. An Andean model of oceanic subduction beneath a continental margin was long thought to apply to the origin of the Coast Belt; the latter displays considerably greater complexity, however, which may be due to the added effects of terrane collision. In contrast to both the Alps and Himalayas, which resulted from cratonic collisions, the Cordillera is characterized by a much greater volume of plutonic rock and the more widespread occurrence of obducted oceanic crust.

INTRODUCTION

The tectonic evolution of the Canadian Cordillera encompassed a wide variety of processes including: 1) development of a miogeoclinal succession along the rifted western, passive margin of ancestral North America beginning in Middle Proterozoic time; 2) an orogenic event which resulted in rifting, volcanism and plutonism in the outer part of the miogeocline in Late Devonian and Mississippian time; 3) the amalgamation and accretion of volcanic, island-arc and oceanic terranes with deformation, metamorphism, volcanism and plutonism during Mesozoic and Cenozoic time; and 4) major displacements along dextral transcurrent faults during Cretaceous and Cenozoic time.

Many aspects of tectonic evolution in the Canadian Cordillera are common with those of Alaska and the western conterminous United States. Other significant global aspects are continental breakup and glaciation in Late Proterozoic time, the breakup of Pangaea in the Triassic Period and the interactions of crustal plates during the Mesozoic and Cenozoic eras. In the following account an attempt is made to summarize the temporal and spatial evolution of the Cordillera in terms of modern tectonic processes and to compare it with other orogens.

PROTEROZOIC
Precambrian Basement

Precambrian gneissic basement rocks occur rarely and intermittently throughout the length of the Omineca Belt and, locally, in the southern part of the Foreland Belt where they fall into three age groups: 1.85-2.1 Ga, 1.1-1.2 Ga, and 0.7-0.8 Ga (Fig. 18.1). Although the oldest group has an age span similar to that of the Wopmay Orogen and Talston and Thelon tectonic zones of the Canadian Shield, there is no unequivocal evidence that these rocks belong to ancestral North America. The middle group perhaps is related to an extension event that produced the Coppermine Lavas. The youngest are clearly related to rifting along or near the western margin of ancestral North America.

Middle Proterozoic - the early rifted and passive margin

Stratigraphic sequences assigned to the Middle Proterozoic in the Canadian Cordillera include at least two thick assemblages of miogeoclinal character (Fig. 18.1). The older comprises the Purcell (Belt) Supergroup in the southeastern Cordillera where it overlies crystalline basement greater in age than 1.7 Ga, the Wernecke Supergroup in

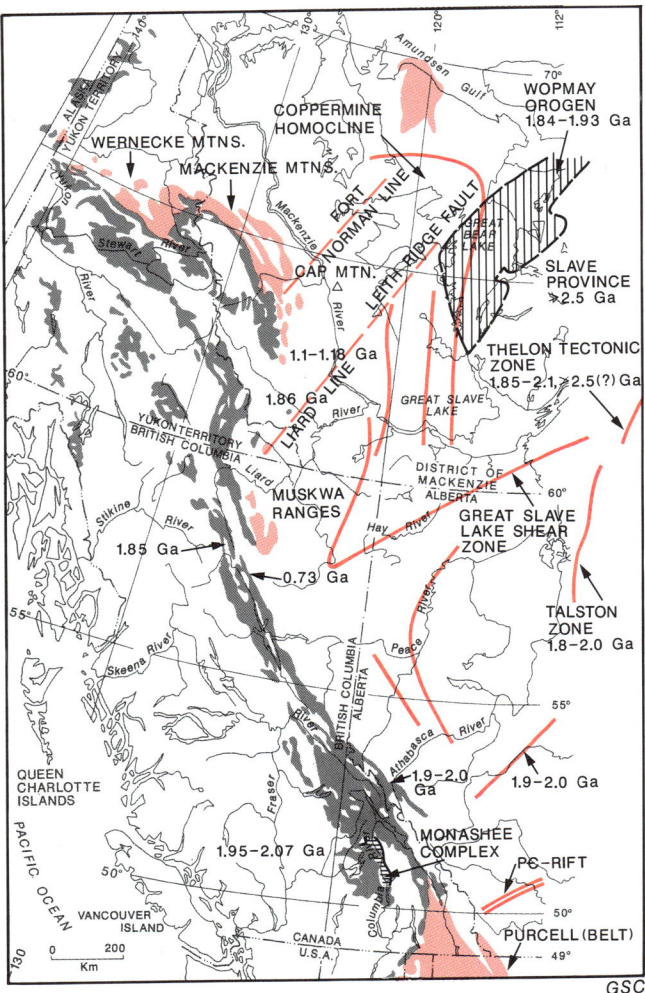

Figure 18.1. Generalized distribution of Middle (pink) and Upper Proterozoic (grey) rocks in the Canadian Cordillera with U-Pb, zircon ages of basement gneisses. Ages of various regimes in the Canadian Shield are shown for comparison. Heavy lines depict trends of aeromagnetic and Bouguer gravity anomalies and projected trends of geologically determined structures.

the Ogilvie and Wernecke mountains, possibly the Muskwa Ranges succession in the northern Rocky Mountains and a sequence exposed at Cap Mountain in the southern Franklin Mountains. The younger assemblage, the 'Mackenzie Mountains Supergroup', occurs only north of latitude 60°N and may include the Pinguicula Group in the Wernecke and Ogilvie mountains. Isotopic age determinations, paleomagnetic results, structural and stratigraphic relationships with crystalline basement and regional correlations suggest that the older succession of Middle Proterozoic strata represents an interval between 1.7 and 1.2 Ga. The 'Mackenzie Mountains Supergroup' and possibly the Pinguicula Group, on the basis of correlation with the Rae Group of the Coppermine Homocline, is younger than about 1.2 Ga and is older than 770 Ma, the age of diabase dykes which intrude the sequence in the Mackenzie Mountains.

Thicknesses of more than 10 km, as well as the character of stratigraphic units, point to a passive margin setting for Middle Proterozoic rocks. In the southern part of the Canadian Cordillera, and perhaps in the northern Rocky Mountains, facies and thickness trends are approximately orthogonal to the regional structural grain of basement rocks as deduced from magnetic and gravity anomalies, suggesting deposition along a rifted continental margin. Although outcrops of the Wernecke, Muskwa and Purcell rocks lie within the confines of the Cordilleran Orogen the configuration and orientation of the cratonal margin during their deposition is unknown. Aitken and McMechan (see Chapter 5) suggest that the older sequence of Middle Proterozoic rocks, including the Hornby Bay-Dismal Lakes succession in the Coppermine Homocline and the Mount Cap succession in the Franklin Mountains, may have been deposited in a presently southwest-trending aulacogen (Fig. 18.1). Farther south is the southwest-trending Great Slave Lake Shear Zone with early movement older than 1.7 Ma and a Precambrian rift zone of uncertain age but of similar trend in southern Alberta (Kanasewich et al., 1968). The southern margin of the Belt Basin in Montana is bounded by the east-trending Willow Creek Fault (McMannis, 1963; Fig. 18.2) and the northern boundary of the Uinta Trough in Utah is marked by an east-west trending strand line (Crittenden and Wallace, 1973). Another possible trough of the same general age but trending northwest in northwestern Arizona is represented by the depositional site of the Unkar and Chuar groups of the Grand Canyon Supergroup. Thus there is evidence that the western margin of the North American craton during deposition of the older Middle Proterozoic rocks (1.2-1.7 Ga) included a number of southwest-, west-, and northwest-trending aulacogens which opened to the southwest, west and northwest into an ocean basin.

If, as seems likely, the western margin of North America during Middle Proterozoic time resulted from rifting, the location of the cratonic rocks that were removed is unknown; Sears and Price (1978, Fig. 18.3) suggested that they may be in Siberia. In any event the Purcell (Belt) passive margin was modified or obscured by a subsequent Middle Proterozoic episode of folding and weak regional metamorphism in the Wernecke Mountains and Late Proterozoic rifting the full length of the North American Cordillera documented by the Windermere Supergroup and its equivalents.

The upper sequence of Middle Proterozoic rocks, the 'Mackenzie Mountains Supergroup', appears to have been deposited on a deeply subsiding epicratonic platform (Fig. 18.1). On the southwest side thickness trends parallel those of the younger Windermere strata suggesting a continental margin setting. The age of the Pinguicula Group is important in this analysis. If the Pinguicula is correlative with the 'Mackenzie Mountains Supergroup' rather than the basal part of the Upper Proterozoic Windermere Supergroup (see below), its basal volcanic unit would be the equivalent of the Coppermine River Group lavas

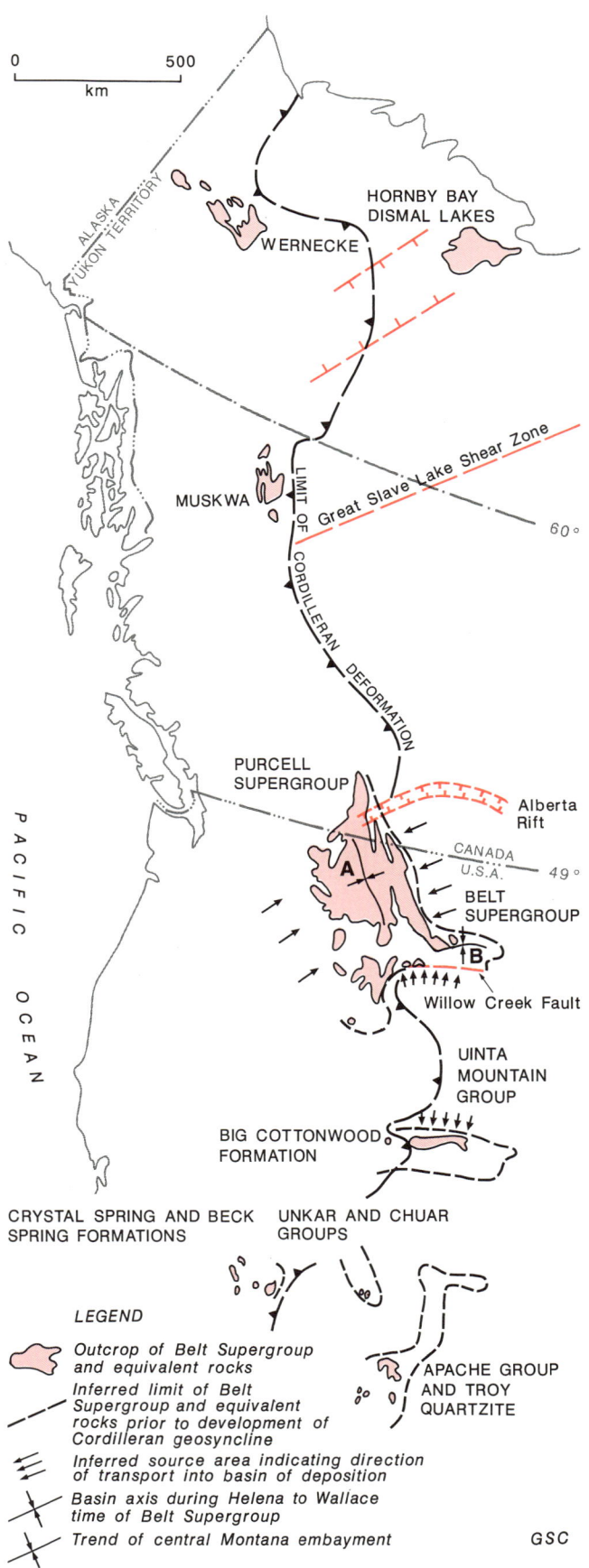

Figure 18.2. Distribution of Precambrian faults that may have affected deposition of Middle Proterozoic rocks in the North American Cordillera. Data after Stewart (1972), Kanasewich et al. (1968) and Parrish (Chapter 4).

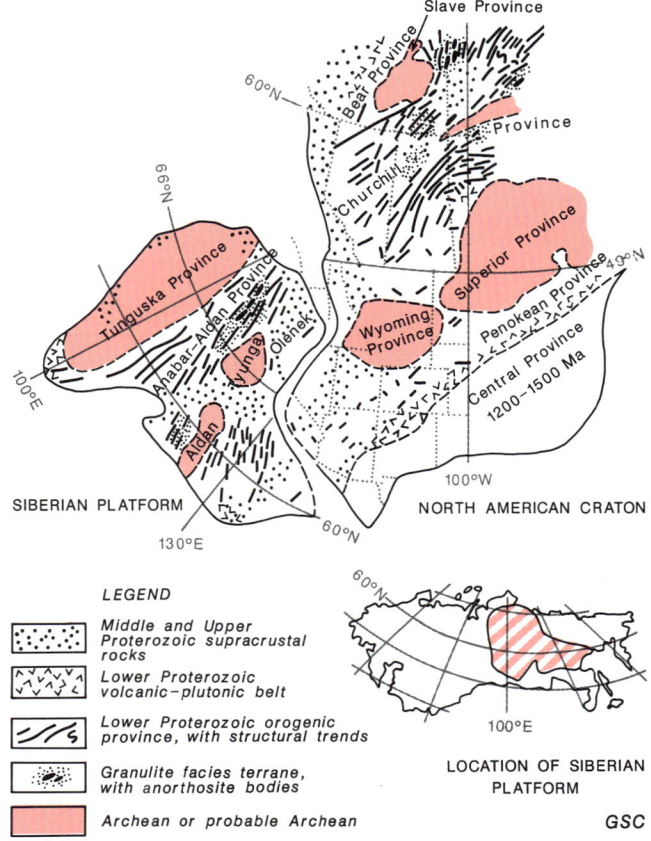

Figure 18.3. Hypothetical reconstruction of Siberian Platform and North American craton prior to Middle Proterozoic rifting. After Sears and Price (1978).

Late Proterozoic - the classic, rifted continental margin

Thick, dominantly clastic strata of the Windermere Supergroup outcrop almost continuously throughout the full length of the North America Cordillera and bear a close spatial relationship with lower Paleozoic miogeoclinal rocks (Gabrielse, 1972; Stewart, 1972; Fig. 18.4). Windermere strata lie variously on crystalline basement ranging in age from about 0.73 Ga to more than 2.0 Ga, deformed and weakly metamorphosed rocks of the older Middle Proterozoic succession (1.2 to 1.7 Ga) and, north of latitude 60°N, on little metamorphosed units of the younger Middle Proterozoic sequence (0.8 to 1.2 Ga).

The great variety of sediment types, abrupt facies changes, the widespread occurrence of mafic dykes and flows, the local presence of evaporites and abrupt thinning to disappearance towards the craton point to a rift origin for Windermere rocks. Significant local relief is indicated by abundant feldspathic sandstone and pebble conglomerate derived from crystalline basement. The upper part of the succession has characteristics typical of passive margin deposition. Cherty iron-formation in the Mackenzie Mountains and diamictite at numerous localities along the Cordillera are distinctive lithologies. Iron-formation in the Mackenzie Mountains suggests a volcanic-exhalative origin in fault-bounded basins. For some, but not all diamictites, a glacial origin has been confirmed. At two localities alkalic granite plutons are spatially related to lower Windermere rocks.

U-Pb ages on zircons from Windermere metasediments in the Yukon-Tanana Terrane of Alaska (Aleinikoff et al., 1984) and from the Pelly, Cassiar, Omineca and Cariboo mountains (P. Erdmer and H. Baadsgaard, pers. comm., 1986; Gabrielse et al., 1982; and L.C. Struik, pers. comm., 1986) cluster around 2.2 Ga with some slightly older. The ages are difficult to reconcile with those found in crystalline basement rocks of the Canadian Cordillera or the bordering North American craton. Therefore, the craton may not have been the sole source for the Windermere clastic sediments, or, alternatively, the western part of the craton includes covered terranes not yet identified.

The eastern limit of Windermere rifting, readily traceable into the southwestern United States, can be determined within narrow limits based on present distribution of facies (Fig. 18.3). Many characteristics of the sedimentation were remarkably similar along the full length of the Cordillera. Diamictites of probable glacial origin occur from the central part of Yukon Territory to southeastern California (Crittenden et al., 1983; Stewart, 1972) showing that similar depositional environments were extensive. Based mainly on data from the Canadian Cordillera rifting and glaciation began less than 0.8 Ga ago. The eastern limit of rifting in Alaska occurs near the Alaska-Yukon border north of Yukon River. In the British Mountains the initial configuration, position and depositional setting of the Neruokpuk basin within the Arctic Alaska Terrane are unknown.

Approximately 700 to 900 Ma separated the Middle and Late Proterozoic rifting events; time enough to create at least five ocean basins the size of the modern Pacific, even at moderate spreading rates. This surely implies that

perhaps indicating a common rift-aulacogen related origin for the three successions at about 1.2 Ga. South of latitude 60°N the only rocks conceivably correlative with the 'Mackenzie Mountains Supergroup' are those of the Muskwa Ranges succession in the northern Rocky Mountains. If, however, these rocks belong to the older Purcell (Belt) Assemblage, as favoured herein, much of the Cordillera in British Columbia and the United States contains little if any record of the interval between 1.2-1.3 Ga and 0.8 Ga. In the classical Purcell (Belt) areas of outcrop and in the Muskwa Ranges these thick sequences probably behaved as stable pseudocratons upon which lower Paleozoic rocks are commonly much condensed with many stratigraphic gaps. This might suggest why the 'Mackenzie Mountains Supergroup' rocks are absent in the south insofar as they would probably have been easily removed by erosion along the uplifted edge of the Windermere rifted margin.

Fragmentary evidence for deformation, metamorphism and granitic intrusion in the Purcell rocks suggests one or two episodes of tectonism between the close of Purcell deposition and the onset of Windermere rifting (McMechan and Price, 1982). The significance of these events and their regional extent are essentially unknown. No orogenic sediments associated with pre-Windermere uplifts have been identified.

the two events are unrelated, and moreover, that within that period, one or more continental configurations completely different from Pangaea could have developed. If a complete Wilson cycle occurred between the two rift events, the lack of evidence for an intervening, widespread Proterozoic collision orogen is puzzling. Deformation and metamorphism of the Wernecke Supergroup, however, is at least local evidence for a Proterozoic orogenic event.

The relationship of Windermere rifting to that during the Early Cambrian Epoch, 175 Ma later (Bond et al., 1985) is enigmatic. Clearly, the most convincing evidence for Windermere rifting is in its lower strata. It is possible that the Late Proterozoic witnessed a nearly complete cycle of rift-drift followed by a new cycle early in the Paleozoic Era.

The characteristic suite of lithologies in the Windermere Supergroup, particularly diamictite, basement-derived clastic and tholeiitic volcanic rocks, has approximately coeval equivalents in several orogens (Fig. 18.5): the Mount Rogers Formation in the south-central Appalachians (Schwab, 1981); the Conception Group in southeastern Newfoundland (Anderson and King, 1981); the Tillite Group in central east Greenland (lacking volcanic rocks; Henriksen and Higgins, 1976); the Middle Dalradian of Ireland and Great Britain where mafic volcanic rocks occur in a succession of sandstone, shale and limestone above a widespread tillite (Harris et al., 1978); the Hedmark Group and related rocks in Scandinavia (sparagmites; Bjorlykke, 1978); and the Comfortlessbreen Group and correlative strata in Svalbard. Upper Proterozoic tillites occur in many parts of Africa except in the northeast. Probable Upper Proterozoic tillites and diamictites are widespread in Brazil but their age is poorly known. In South China Upper Proterozoic (Sinian) feldspathic sandstone and shale, several hundred metres thick, separate a lower Changan tillite from an upper Nantuo tillite both clearly of glacial origin (Fig. 18.6). Elsewhere in China a third, sub-Cambrian glaciation has been suggested (Yongji, 1981; Shih-Fan, 1981). The Changan is dated at between 800 and 760 Ma whereas the Nantuo is between 740 and 700 Ma. Two main glacial episodes are recorded in Upper Proterozoic tillites of the Adelaide Basin of South Australia. The lower, Sturtian tillites are overlain by a sequence of shale, siltstone and carbonate grading upward into arkosic sandstone and an upper, Marinoan tillite. The Marinoan is overlain by dolostone, shale and siltstone grading upward into sub-Cambrian quartzite and dolostone containing an Ediacaran fauna. Von der Borch (1980) interpreted the glaciogenic rocks and underlying evaporitic dolostone, shale, quartzarenite and alkaline volcanic rocks as having been deposited in graben and half-graben basins. Still to be resolved is the apparent paradox of regional glaciations having occurred at low paleolatitudes suggested by paleomagnetic data.

The widespread occurrences of distinctive Upper Proterozoic rocks imply similarities in depositional environments. Although rift histories have not been demonstrated or suggested for all successions, many, including those in the North American Cordillera, the Caledonides and Australia, clearly have a rift affinity. A further implication is that the rifting may have been related to the breakup of a supercontinent in Late Proterozoic time perhaps similar to the breakup of Pangaea in the Triassic Period. Eisbacher (1985) proposed that the Adelaide, Sinian

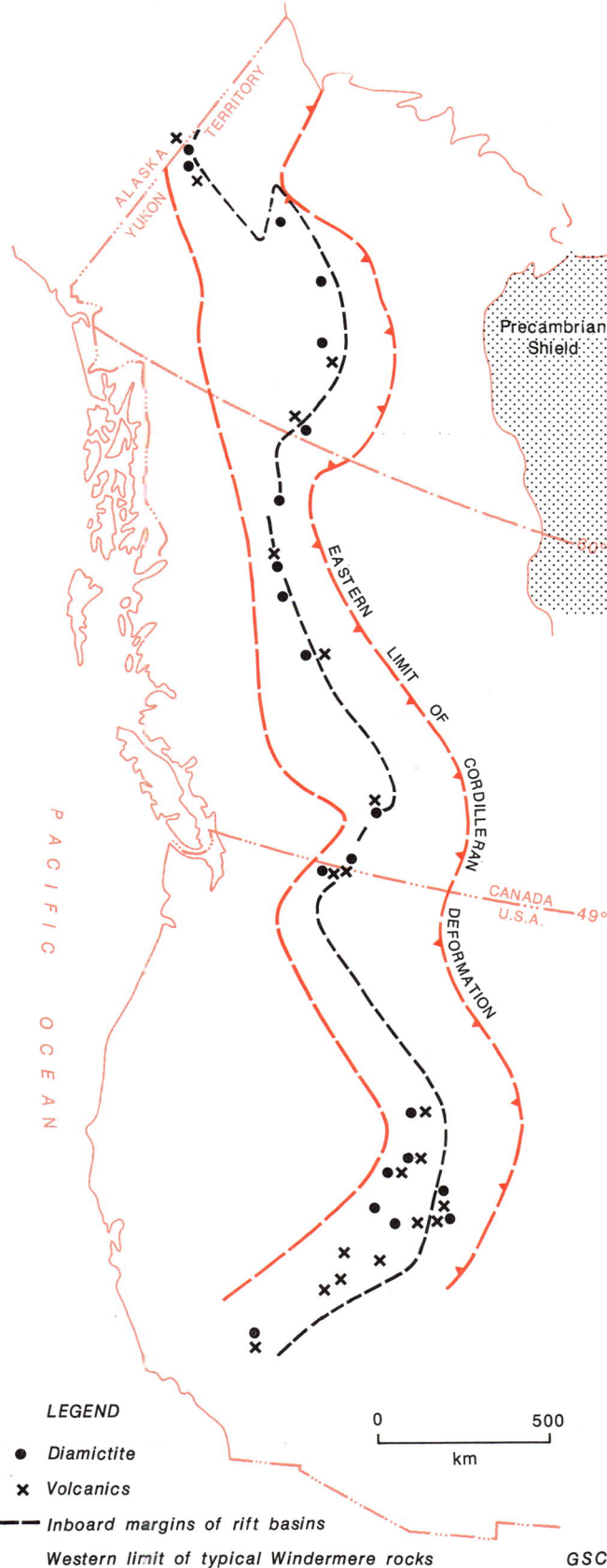

Figure 18.4. Main belt of Windermere Assemblage showing eastern margin of Windermere rifting with localities of mafic volcanics and diamictite. After Stewart (1972) and Gabrielse and Campbell (Chapter 6).

and Windermere basins were once juxtaposed and separated by a maturing rift system preceding development of a paleo-Pacific basin (Fig. 18.6). A similar argument could be made for Upper Proterozoic rift basins in the Caledonides of eastern North America, Greenland and Europe.

Many localities preserve a record of two Late Proterozoic glaciations. Better dating of these episodes coupled with the expected effects of eustatic rise and fall of sea level accompanying deglaciation and glaciation ultimately may allow a means of global correlation of Late Proterozoic events (Eisbacher, 1985).

PALEOZOIC AND TRIASSIC
Cambrian to Middle Devonian

The early Paleozoic evolution of the Canadian Cordillera is illustrated most completely by the stratigraphy of the miogeocline of ancestral North America and the Alexander Terrane (Fig. 18.7). Only a fragmentary record is present in Wrangellia and Stikinia. A fairly complete stratigraphic succession occurs in parts of the Kootenay Terrane but a lack of age control inhibits reliable correlations and analysis.

Ancestral North America - the classic miogeocline

The miogeocline, including the displaced Cassiar Terrane and possibly parts of the Kootenay Terrane, consisted of a subsiding shelf that passed abruptly into a deeper, off-shelf region to the west. The shelf contains platform carbonate and clean clastic strata whereas off-shelf areas are characterized by thin successions of fine grained argillaceous and cherty clastic rocks, olistostromal deposits near the shelf margin, local volcanic rocks and narrow carbonate reef tracts. In the northern part of Yukon Territory the early Paleozoic miogeocline was the site of platform carbonate accumulation and intraplatform trough and basin clastic deposition, the components of which were probably derived from cratonic sources to the east and northeast and, possibly from source terranes in the area of the modern Canada Basin.

The lower Paleozoic rocks of the miogeocline are generally attributed to a passive margin setting superimposed upon the rifted margin of ancestral North America. Bond et al. (1985) constructed tectonic subsidence curves for Cambrian to Early Ordovician rocks and concluded that

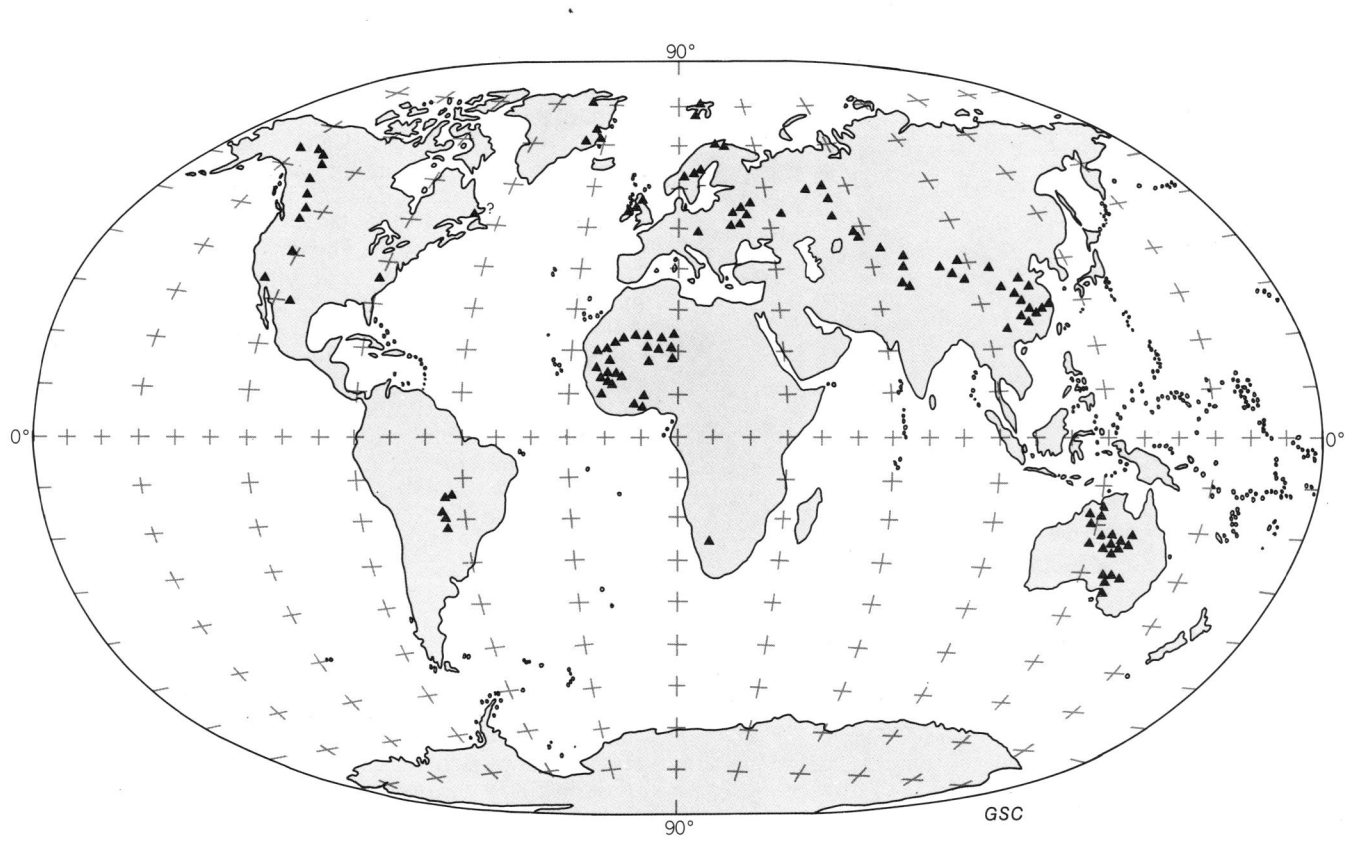

Figure 18.5. Global localities with assemblages of about the same age and lithology as those of the Windermere Supergroup. Modified from Hambrey and Harland (1981).

TECTONIC SYNTHESIS

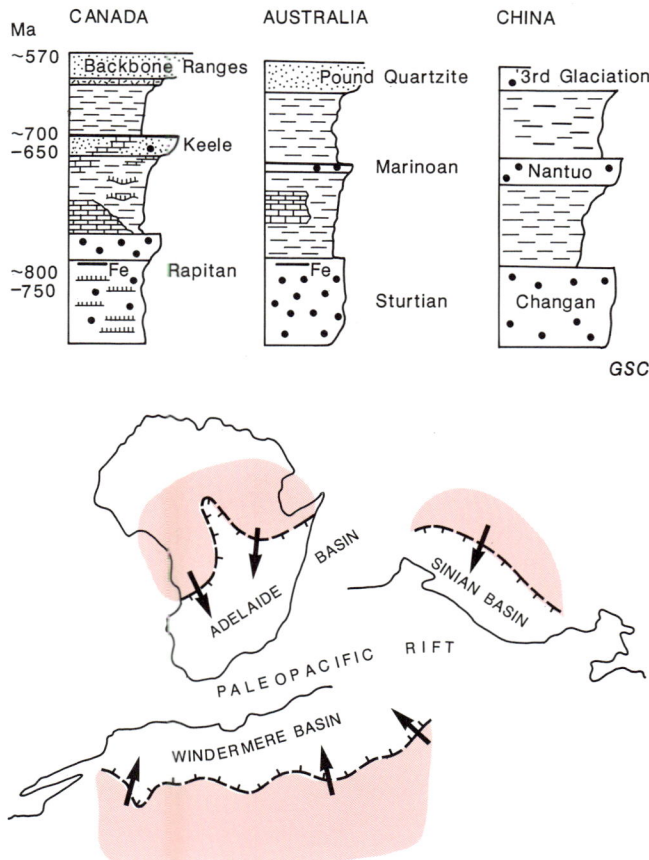

Figure 18.6. Suggested correlation at Upper Proterozoic rocks and hypothetical restoration of Adelaide, Sinian and Windermere basins prior to Late Proterozoic drift. After Eisbacher (1985). Pink areas indicate pre-Windermere cratonal rocks; solid circles indicate presence of tillite; arrows indicate relative directions of sedimentary transport.

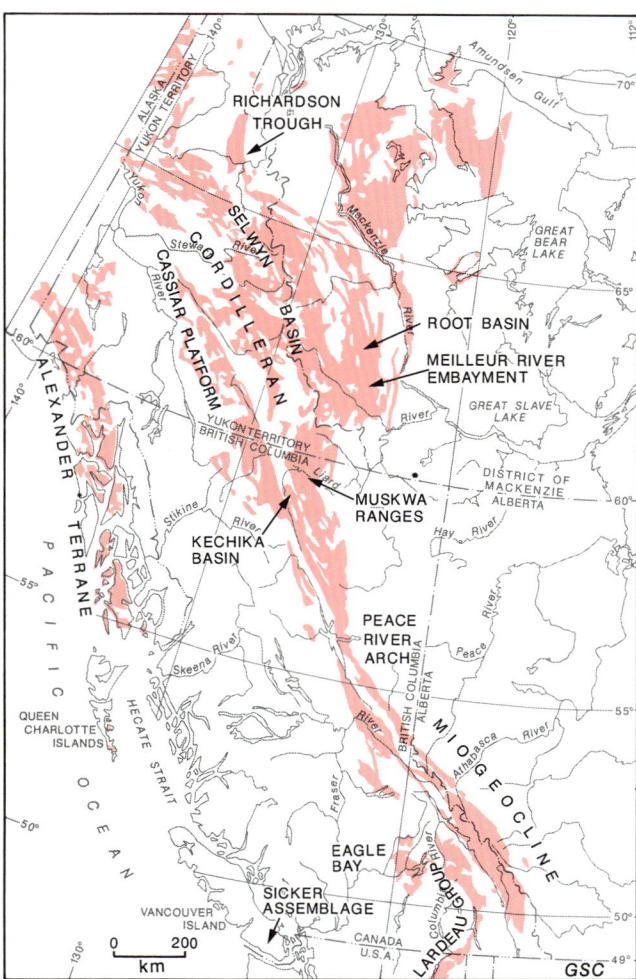

Figure 18.7. Distribution of lower Paleozoic assemblages and selected tectonic elements.

the transition from rift to post-rift cooling and subsidence occurred in latest Proterozoic or Early Cambrian time (Fig. 18.8). It is clear, however, that the concept of single stage rifting and drifting is at best a simplification of the evolution demonstrated by the stratigraphic record. The character of Lower Cambrian sediments almost everywhere implies nearly synchronous environments of deposition beginning with thick, relatively clean sandstone units near the eastern limit of deposition which define a pronounced flexure in the cratonic margin. These pass westward into finer grained clastic rocks of probable continental slope setting. Mafic volcanics are known locally in Lower Cambrian strata in the Yukon and in the uppermost Proterozoic members of the Hamill Group in the southeastern Cordillera.

Although the early Paleozoic witnessed typical passive margin development, several criteria indicate episodic rift environments, but mostly on a local scale. The Richardson Trough in the northern Yukon may be an aulacogen that developed during the Middle Cambrian and, moreover, may have connected with the Hazen Trough of the Arctic Archipelago; the Muskwa Ranges of the northern Rocky mountains (and eastern Ogilvie and Wernecke mountains) contain further evidence of Middle Cambrian rifting. In both of these areas coarse, olistostromal debris flows are restricted to grabens or half grabens on either side of which sedimentary rocks are of markedly different facies. In the Muskwa Ranges rocks west of the graben facies comprise greatly condensed basinal shale and argillaceous limestone (Fig. 18.9). Farther west are shallow water carbonates of possible Middle Cambrian age which pass westerly into a belt of spectacular, northwesterly aligned, reefoid Middle Cambrian carbonates in the Kechika Basin. Similarly, in the Kechika Basin, northwesterly aligned reefoid bodies of Middle Devonian carbonate are bounded by deeper water sediments on either side. The various facies and their distributions suggest deposition in and along the margins of rift basins. Between Early Ordovician and Middle Devonian time more than 4000 m of dominantly carbonate strata were deposited in the northerly trending Meilleur River Embayment of Selwyn Basin and Root Basin (Liard Depression). Extension, possibly associated with broad crustal attenuation between the Cordilleran and Franklinian miogeoclines, may have caused local rift basins to form.

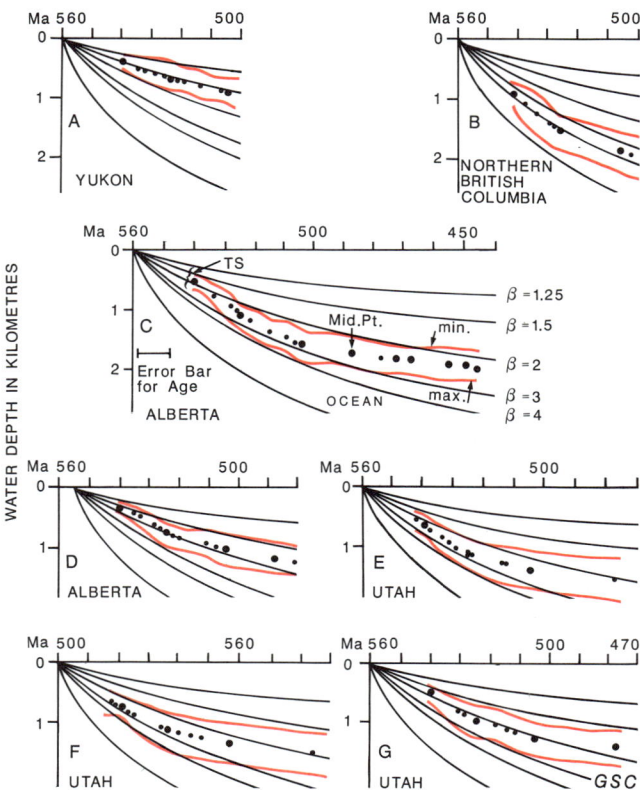

Figure 18.8. Tectonic subsidence (TS) curves for Cambrian to Ordovician rocks in the North American miogeocline. Black curves are post-rift thermal subsidence curves from McKenzie (1978) stretching model for different amounts of stretching (β). Red curves are tectonic subsidence curves for maximum delithification factors (max) and minimum delithification factors (min). Dots are mid-points between the two curves; large dots are stratigraphic boundaries with ages; small dots are boundaries without ages located by assuming constant sedimentation between dated points. Error bar for age is uncertainty in faunal correlations. After Bond et al. (1984).

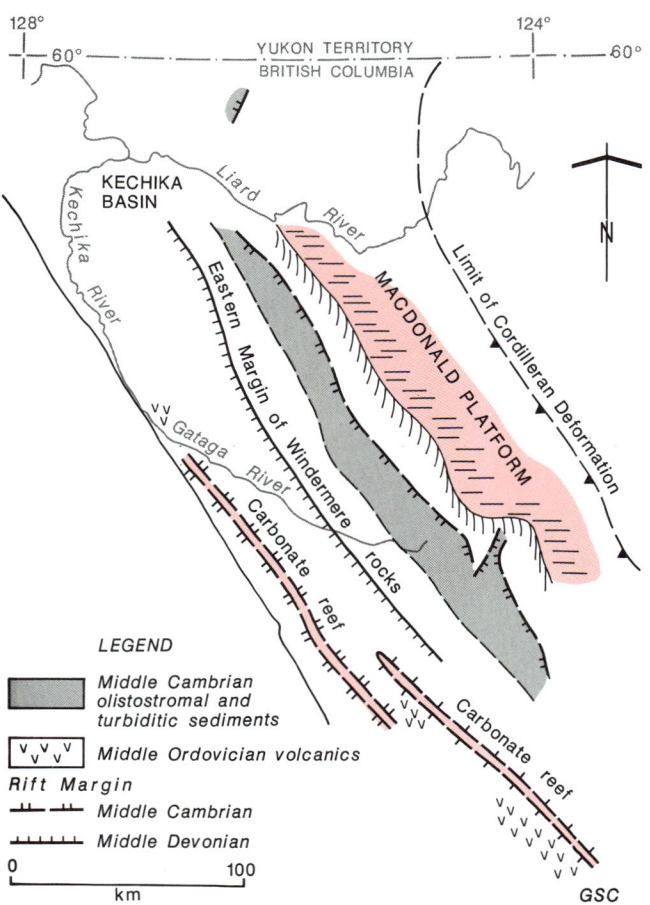

Figure 18.9. Rifts in miogeoclinal rocks of the northern Rocky Mountains shown by the distribution of volcanic rocks and sedimentary facies.

Middle Ordovician volcanic rocks, perhaps associated with rifting, occur at many localities in miogeoclinal strata north of the Peace River Arch. Not widespread but locally important was Late Cambrian-Early Ordovician volcanism on the Cassiar Platform and in the Selwyn Basin. Volcanic rocks occur in Ordovician, Silurian and Devonian strata in the Misty Creek Embayment of Selwyn Basin. Silurian alkalic volcanism occurred near the Southern Rocky Mountain Trench at latitude 54°N. Several important Ordovician and Silurian stratiform Pb-Zn-barite deposits are found in areas with rift histories.

In summary, the early Paleozoic miogeocline (passive margin) succession has several features suggestive of intermittent rifting (Fig. 18.10). Much of the tectonic activity apparently was concentrated near the shelf to off-shelf transition zone where basement may have been substantially attenuated during Late Proterozoic rifting.

Many elements of the early Paleozoic passive margin are recognized in coeval rocks of the Cordillera in the United States. Stratigraphic successions in the Great Basin of Utah, Nevada and southeastern California display the same sequences with weathering, textural and faunal characteristics remarkably similar to those in the Canadian Cordillera. These similarities attest to the continuity of like depositional environments for a length of almost 4000 km. Unlike the Caledonian-Appalachian Orogen and the Adelaide Basin of Australia, each of which was initiated by rifting at about the same time, the Cordilleran passive margin was not affected by early Paleozoic deformation.

The Mesozoic and Cenozoic Atlantic passive margin of North America provides a useful model for comparison with the Cordilleran example (Fig. 18.11). The former evolved over a span of about 190 Ma, about the same as the interval between the Early Cambrian and Middle Devonian epochs. A comparison between like margin facies suggests that the break between thick shelf carbonate strata and condensed, off-shelf facies to the west, well exposed along the eastern margin of Selwyn Basin, must have been close to a change in the character of the underlying crust. Decoupling of the supracrustal rocks from their original basement during Mesozoic deformation, however, precludes an assessment of where the crustal change occurred. The presence of continental crust beneath Selwyn Basin is indicated by mid-Cretaceous, S-type granitic rocks. Two

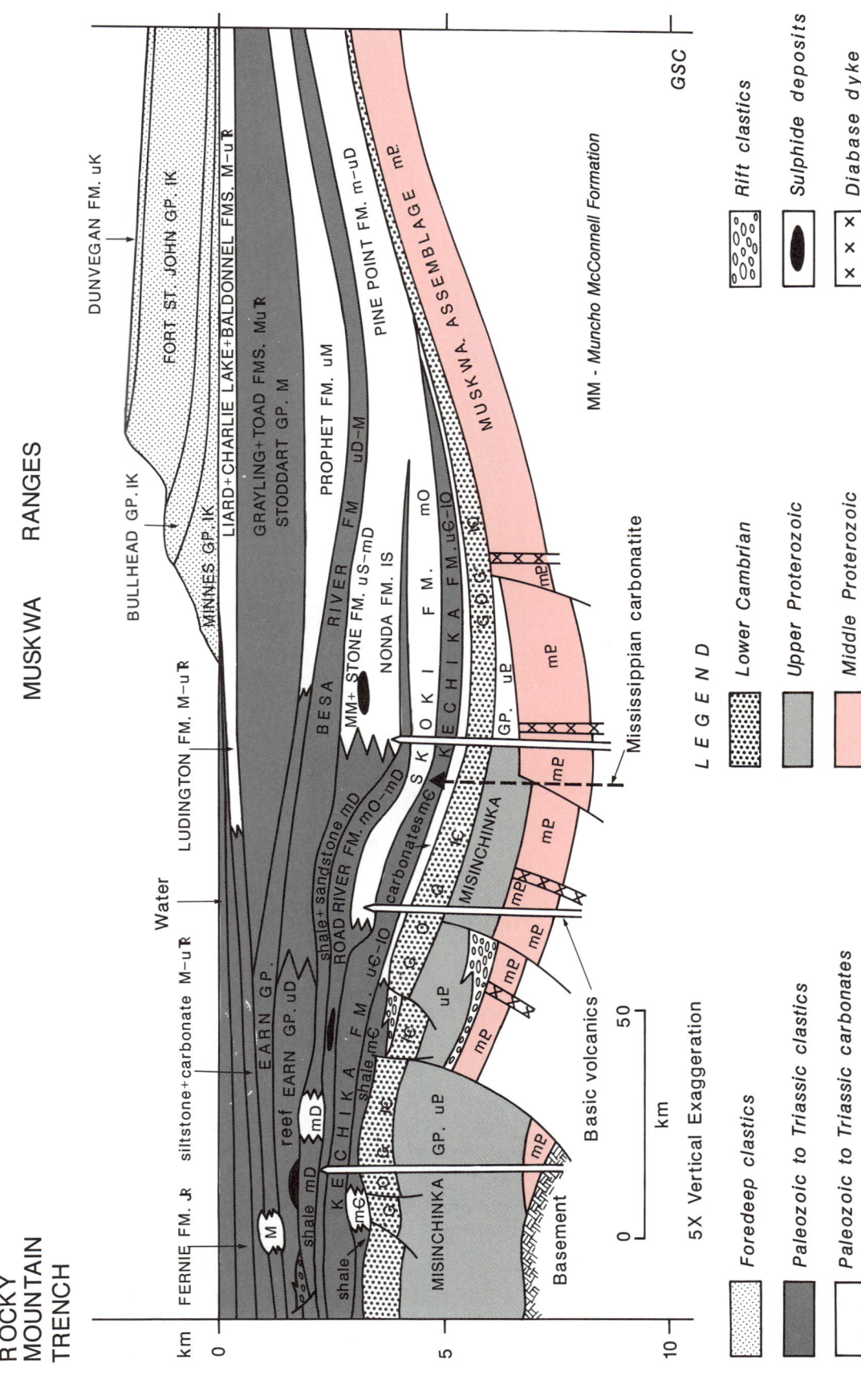

Figure 18.10. Restored section of miogeoclinal succession in the Muskwa Ranges. Datum is the base of the Jurassic Fernie Formation. Data for Muskwa Ranges provided by R.I. Thompson.

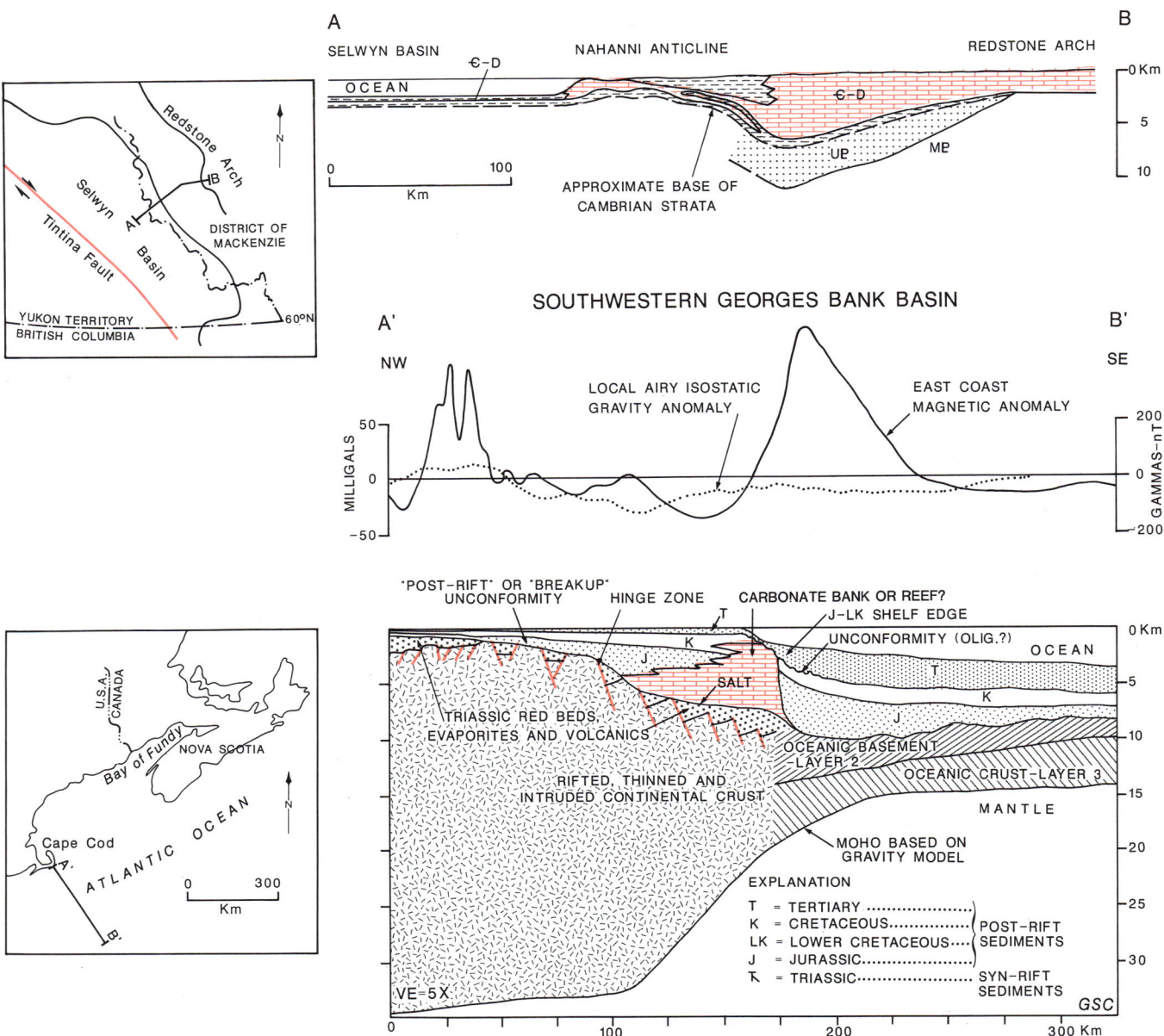

Figure 18.11. Interpretive crustal cross-section through the North American Atlantic passive margin (Grow, 1981) compared with a restored stratigraphic cross-section for Upper Proterozoic and lower to mid-Paleozoic rocks from Mackenzie Platform into Selwyn Basin (S.P. Gordey, pers. comm., 1987).

different models can be applied. The first assumes easterly dipping, mid-Cretaceous Benioff subduction of oceanic lithosphere generating granitic magma in, or strongly contaminated by, thinned sialic crust that formed the basement of the Selwyn Basin. The second assumes westerly dipping Ampferer subduction of cratonic basement during the mid-Cretaceous, with the consequent insertion of sialic material beneath the Selwyn Basin which until the Mesozoic was underlain by transitional crust. Whatever the case, palinspastic restorations of supracrustal rocks in the Selwyn Basin and the Kechika Basin imply a wide, subsiding shelf, presumably overlying thinned continental crust.

Accreted Terranes - volcanic archipelagos, related clastics and carbonate platforms

The oldest rocks recognized in the Alexander Terrane of the Saint Elias Mountains are Cambrian and possibly Upper Proterozoic strata grouped with an Upper Cambrian to Lower Ordovician sequence of volcaniclastics. An overlying succession comprises Lower Ordovician to Devonian limestone, clastic and minor volcanic and volcaniclastic rocks. In southeastern Alaska, Cambrian and possibly older volcanic, plutonic and metamorphic complexes probably form the basement of the Alexander Terrane. The pre-late Early Ordovician deformation and metamorphism that affected rocks possibly as old as Late Proterozoic in southeastern Alaska has not been noted in the Saint Elias Mountains. Ordovician to Lower Silurian rocks in southeastern Alaska include limestone, terrigenous clastics, a variety of basaltic, andesitic and rhyolitic volcanics and coeval plutons. They are overlain by Silurian volcaniclastic and quartzo-feldspathic turbidite, gradational with thick units of thick-bedded to massive limestone. In the southern part of the Alexander Terrane, polymictic conglomerate, interbedded with limestone and turbidite, was derived from uplift related to the Klakas Orogeny (Gehrels et al., 1983). This event, which extended into Early Devonian time and produced a thick redbed clastic wedge, is not recorded in the Saint Elias Mountains of the northern part of the Alexander Terrane where tectonic conditions were evidently more stable.

Correlations of inferred depositional environments and tectonic-plutonic events in the Alexander Terrane with those in other terranes and with the miogeocline are not apparent. Hints of Ordovician and Devono-Mississippian plutonism are given by isotopic ages from the Kootenay Terrane, and Late Silurian or Early Devonian plutonism has been identified in the Northern Sierra Terrane of California (Schweickert et al., 1984). Both the Kootenay and Northern Sierra terranes have island arc affinities. Early Ordovician and Late Silurian-Early Devonian tectonism, noted in the Alexander Terrane of southeastern Alaska, also occurred in parts of the Caledonides and in the Lachlan Orogen of Australia (Gehrels and Saleeby, 1984; Brown et al., 1968).

The oldest rocks identified in Wrangellia, the Sicker Assemblage, are possibly of Late Devonian age and include calc-alkaline volcanic rocks and related quartz porphyry on Vancouver Island. These are of island arc affinity and are overlain by Carboniferous to Lower Permian argillite, chert and bioclastic limestone. One locality in Stikinia is known to have marine strata of Early Devonian age.

The Kootenay Terrane, lying between ancestral North America and the accreted terranes to the west, comprises assemblages of volcanic, clastic and carbonate rocks of Late Proterozoic to Triassic ages enclosing Ordovician, Devonian and Mississippian plutons. Sedimentary units may represent off-shelf margin facies of the miogeocline of ancestral North America. Tholeiitic basaltic rocks in the early Paleozoic Lardeau Group of the southeastern Cordillera suggest an extensional setting (back-arc?) whereas felsic tuffaceous rocks in the Eagle Bay sequence to the northwest suggest an island arc environment. An unconformity at the base of the Mississippian Milford Group truncates previously folded and metamorphosed strata of the Lardeau Group, but the age of deformation is otherwise unconstrained. No evidence of early Paleozoic deformation is recognized in rocks of the miogeocline farther to the east although, in the southern Rocky Mountains, a widespread unconformity separating Upper Devonian from Ordovician and Silurian strata attests to broad epeirogenic movements. The spatial separation of early Paleozoic exposures in the Kootenay Terrane from those in ancestral North America, however, is considerable.

Aspects of sedimentation, volcanism, plutonism and deformation of early Paleozoic age in the Kootenay Terrane are like those in the Northern Sierra Terrane of California which has a similar setting relative to ancestral North America (Fig. 18.12). In the north, the Nisutlin Subterrane of the Kootenay Terrane is included in the Yukon-Tanana Terrane of Alaska. The Kootenay Terrane also has similarities with the Hammond Subterrane (Silberling and Jones, 1984) of the Arctic Alaska Terrane in the Brooks Range.

Late Devonian to Triassic

During late Paleozoic and early Mesozoic times the contrast in stratigraphy and inferred tectonism between those terranes linked to ancestral North America and the accreted terranes is most evident. In particular, Wrangellia, the Alexander Terrane, Stikinia, the Cache Creek Terrane and the Slide Mountain Terrane have fairly complete stratigraphic records allowing comparison among them and with ancestral North America. Several of the small terranes also have well documented stratigraphic histories for this interval.

Accreted terranes - oceanic basins, volcanic archipelagos and related clastics and carbonate platforms

Definition of Wrangellia is based on the widespread occurrence of Upper Triassic, tholeiitic, ocean-rift basalt and overlying carbonate. These rocks unconformably overlie a variety of mid- to upper Paleozoic volcanic, volcaniclastic and carbonate rocks of arc affinity. Permian granitic plutonism was especially important in the Saint Elias Mountains. The Upper Triassic basalt provided the first paleomagnetic data suggesting that Wrangellia moved northward a considerable distance relative to the craton since the Late Triassic Epoch.

In the Alexander Terrane a variety of volcanic, clastic and carbonate rocks, ranging in age from Middle Devonian to Late Triassic is inferred to have accumulated in island arc and, ultimately, rift environments. As in Wrangellia, a regional unconformity separates upper Paleozoic from Upper Triassic rocks. Stable conditions seem to have been widespread during Late Devonian and Mississippian times when carbonate deposition was widespread.

Island arc settings also are interpreted for most of the upper Paleozoic and Triassic rocks of Stikinia. A belt of Mississippian and Permian limestones, about 400 km long along the northwest side of the terrane, indicates that there were probably intervals when Stikinia behaved as a stable block. In the northern part of the terrane, structural and metamorphic differences distinguish Paleozoic from Mesozoic rocks but the nature of the related tectonic event is not understood. During the Triassic Period a volcanic arc extended from the southwestern Yukon along the Stikine Arch and thence south along the east boundary of the terrane. Middle to Late Triassic plutonism accompanied the volcanism. Subduction related to volcanism ultimately led to the Middle Jurassic closing of the Cache Creek ocean to the northeast and east.

The Cache Creek and Slide Mountain terranes are characterized by typical oceanic rocks ranging from mid-Paleozoic to Triassic in age. The identification of a Tethyan fusulinid fauna in the Cache Creek Group provided the first evidence for exotic terranes in the Canadian Cordillera. The Slide Mountain Terrane, on the other hand, has a

689

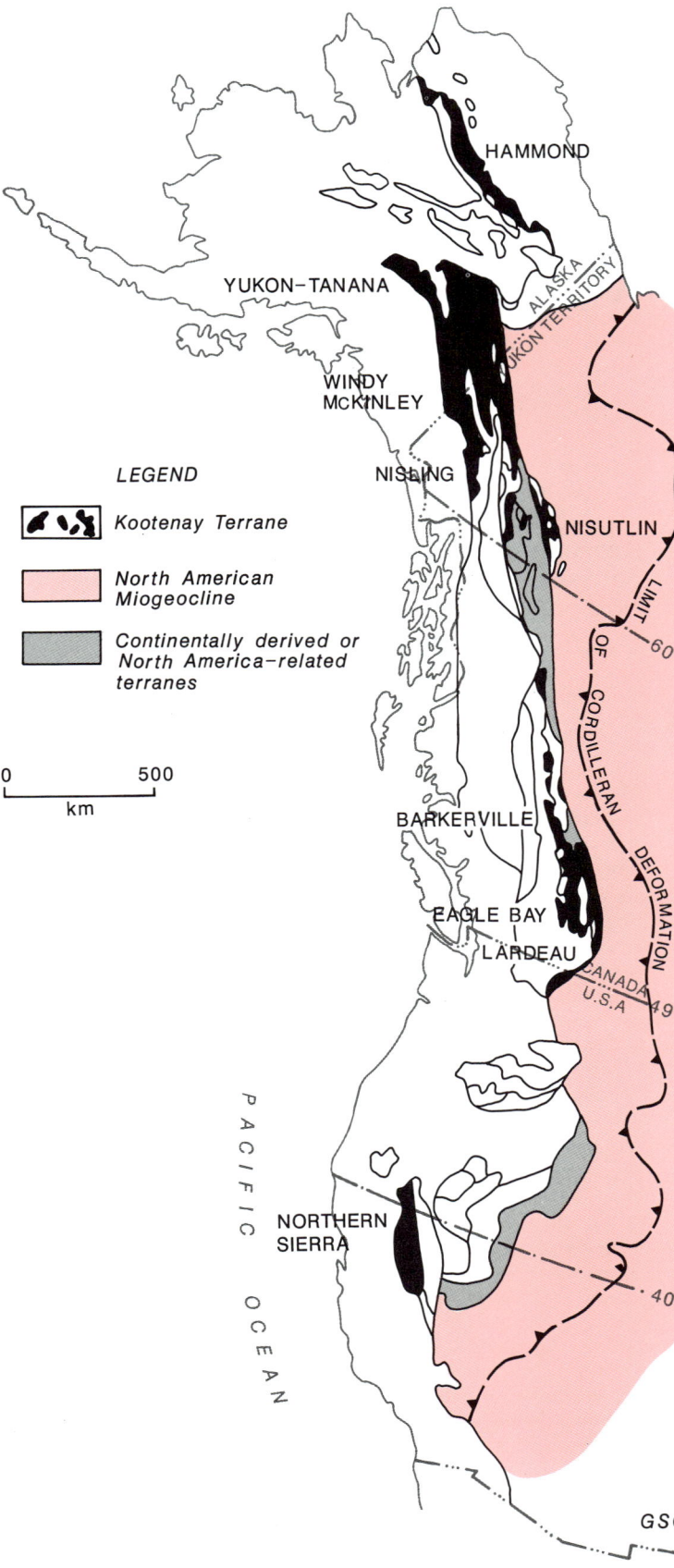

Figure 18.12. Distribution of Kootenay and similar pericratonic terranes in the North American Cordillera.

North American fauna, although possibly greatly displaced northwards relative to the craton, and includes minor sedimentary units of probable cratonal source.

The record in Quesnellia is mainly one of Late Triassic to Middle Jurassic arc volcanism, plutonism and sedimentation; local rift settings are suggested by the presence of strongly alkalic volcanic rocks. Zoned, Alaskan-type ultramafic rocks are characteristic.

The most distinguishing feature of the Kootenay Terrane is the presence of Upper Devonian to Lower Mississippian granitic rocks locally associated with calc-alkalic volcanism. In the south, evidence exists for a significant pre-Lower Mississippian unconformity as a consequence of deformation.

In the smaller terranes the most complete stratigraphic record is preserved in the island-arc Chilliwack Terrane. The Carboniferous and Permian rocks there have an affinity with those in Quesnellia.

Displaced terranes and ancestral North America - miogeocline and Devono-Mississippian tectonism

On the displaced terranes (Cassiar and North Slope) and ancestral North America, marine shelf conditions persisted in the miogeocline throughout the Late Devonian to Triassic interval. Of note was the widespread deposition of coarse- to fine-grained clastic rocks coeval with rifting and local alkaline volcanism in the northern Cordillera. Alkalic intrusive rocks and carbonatites are known in many places along a belt near the shelf to off-shelf facies boundary (Fig. 18.10). A regional unconformity occurs below Upper Devonian strata in the southern Rocky Mountains.

Uplift and erosion took place in the Permian Period along a northeasterly trending zone from Alaska to the northern Richardson Mountains (Aklavik Arch Complex) and a pronounced unconformity occurs beneath Permian strata elsewhere in the eastern Cordillera. Another regional unconformity lies at the base of Triassic strata. Conceivably, the sub-Permian unconformity which occurs throughout much of the miogeocline could have formed during a time of eustatic lowering of sea level resulting from extensive Carboniferous to Permian glaciation in the southern hemisphere.

Correlations

Within the Canadian Cordillera there is little evidence to suggest linkages during late Paleozoic to Late Triassic time among the accreted terranes and between them and the pericratonic terranes to the east. The Permo-Triassic unconformity is common to all terranes of island-arc affinity and thus appears to have resulted from a global phenomenon. The unconformity is not recognized in terranes of oceanic affinity. In the northern part of Cache Creek Terrane radiolarian chert accumulated up to Late Triassic (Norian) time whereas in the southern part chert locally contains radiolaria as young as Middle Jurassic. During uppermost Triassic time hints of linkages between Stikinia, the Cache Creek Terrane and Quesnellia appear. In the south the Cache Creek perhaps became an accretionary prism beneath the Quesnellia arc with Nicola Group volcanics forming clasts within the wedge. In the north a similar relationship occurs where Upper Triassic to Lower Jurassic greywacke, of unknown origin (possibly Quesnellia?), flooded the northern part of the Cache Creek Terrane upon which a volcanic arc was constructed in Late Triassic time. By the Early Jurassic the northern part of Stikinia in the Yukon shed clastic sediments onto the Cache Creek Terrane.

A link between the Kootenay Terrane and the miogeocline of ancestral North America is suggested by the timing of plutonism, volcanism, rifting and clastic sedimentation. During Late Devonian to Early Mississippian time, coarse clastic sediments, probably derived from uplifted rift margins in the northern Cordillera spread eastward onto the craton. In the southern Kootenay Terrane folding preceded deposition of Mississippian sediments and throughout the terrane granitic rocks were emplaced, coincident with the local eruption of calc-alkaline volcanics.

Miogeoclinal rocks of Late Devonian to Late Triassic age in the Canadian Cordillera readily correlate with those in the western United States where an Upper Devonian-mid-Mississippian flysch is attributed to the Antler Orogeny. In all respects, including westerly derivation and ubiquitous barite of Frasnian-Famennian age, these rocks are identical to those of assumed rift origin in the northern Cordillera. The Roberts Mountain Thrust is regarded as an important contractional structure resulting from the Antler Orogeny (Poole, 1974; but see Ketner and Smith, 1982 for an opposing view). No contractional structures of this age have been identified in Canada but it is conceivable that the rifting, volcanism, uplift and sedimentation in the northern Cordillera was linked to tectonism to the west in the Kootenay Terrane.

In the Arctic Archipelago a pronounced unconformity, an associated thick clastic wedge and local plutonism reflect widespread deformation between Famennian and Viséan time. This event, called the Ellesmerian Orogeny, provides a possible link between the Innuitian and Cordilleran orogens although their relationship during Devono-Mississippian time is far from clear. A widespread, northerly derived clastic wedge of Carboniferous age in northern Yukon Territory and in Selwyn and Mackenzie mountains reflects uplift of the Ellesmerian orogen. The Brooks Range of Alaska has much in common with both orogens although its metallogeny, particularly for stratiform base metal deposits, is more like that of the Cordillera.

The Kootenay Terrane and its subterranes form part of the Yukon-Tanana Terrane in Alaska and have similarities with the Hammond Terrane in the Brooks Range (Fig. 18.12; Silberling and Jones, 1984). Correlation to the south is hampered by extensive young volcanic cover in Washington, Oregon and Idaho but similar suites of rock constitute the Northern Sierra Terrane in California (Schweickert et al., 1984). A link has been suggested between tectonism in the Sierra region and the Antler Orogeny in Nevada (Schweickert et al., 1984). It appears therefore that terranes characterized by gritty and mixed clastic rocks, volcanics and Late Devonian-Early Mississippian granitic plutons lie directly west of typical miogeoclinal rocks from Alaska to the southwestern United States.

In the North Slope Terrane an unconformity separates Carboniferous carbonate and basal clastic strata from underlying rocks. Because of stratigraphic continuity of the Carboniferous carbonate and underlying transgressive clastics with those in the Ogilvie Mountains of ancestral North America the subdivision into terranes is debatable. An unresolved problem is whether these terranes are separated by a fundamental suture or simply were offset by the transcurrent Kaltag-Porcupine fault zone.

The Slide Mountain and correlative terranes occur almost everywhere as flat-soled klippen emplaced onto miogeoclinal strata. In Alaska, equivalent terranes include the Innoko, Tozitna and Angayucham (Fig. 18.13; Silberling and Jones, 1984). In Nevada a group of similar terranes comprising the Schoonover, Havallah and Pumpernickel occur in the Golconda Allochthon; these in part represent more transitional tectonic settings than those in Canada, as indicated by a greater abundance of siliciclastics, volcanic-lithic greywacke and carbonate turbidite (Miller et al., 1981, 1984; Silberling and Roberts, 1962; Stewart et al., 1977; Snyder and Brueckner, 1983). Part of the composite Caborca Terrane in northwestern Mexico includes oceanic chert. Associated carbonate contains the unusual *Parafusulina* sp., also found in the Sylvester Allochthon of the Slide Mountain Terrane.

In Nevada, emplacement of the Golconda Allochthon has been identified with the Sonoman Orogeny which is documented by the unconformity between the Koipato Formation (late Early Triassic) and the underlying upper Paleozoic Havallah sequence (Silberling and Roberts, 1962). The location of these rocks when Sonoman deformation occurred is uncertain, and the Triassic age of the Golconda thrust has been questioned (Gabrielse et al., 1983). Speed (1984) proposed that the Lower Triassic Candelaria Formation in Nevada was deposited in a foredeep related to Sonoman deformation. The Slide Mountain allochthons in the northern Canadian Cordillera overlie Upper Triassic rocks along their eastern margins. In southeastern Yukon Territory, however, local, coarse conglomerate of Late Triassic age contains clasts of mylonite similar to lithologies in the allochthons. Therefore, emplacement of allochthons onto cratonic North America may have begun earlier than Late Triassic time and could have led to deep subsidence of the Liard Basin throughout all of the Triassic.

In all well studied examples, the Slide Mountain and correlative terranes comprise a succession of thrust-bounded sheets (Harms, 1986; Struik and Orchard, 1985; Miller et al., 1982, 1984; Brueckner and Snyder, 1985; Snyder and Brueckner, 1983). In some cases, the degree of imbrication and variation in lithology between thrust sheets suggest contraction of a wide oceanic basin (Harms, 1986). A setting proximal to the Kootenay Terrane has been demonstrated in the Kootenay Arc region (Klepacki and Wheeler, 1985). In the northern Cordillera some Early Permian tonalite is contained within the Sylvester Allochthon and cuts some but not all of the thrust faults (Harms, 1986). In the Kootenay Arc some imbrication of Upper Permian strata took place in pre-Late Triassic time (Klepacki and Wheeler, 1985). Muscovite associated with blueschist minerals and eclogites in the allochthons of southern Yukon Territory have yielded mid-Permian K-Ar ages (P. Erdmer, pers. comm., 1986). One explanation is that contraction of an ocean basin, facilitated by subduction, began in late Paleozoic time and culminated with the emplacement of its contents onto the miogeocline in the Mesozoic. Perhaps related to this process was the evolution

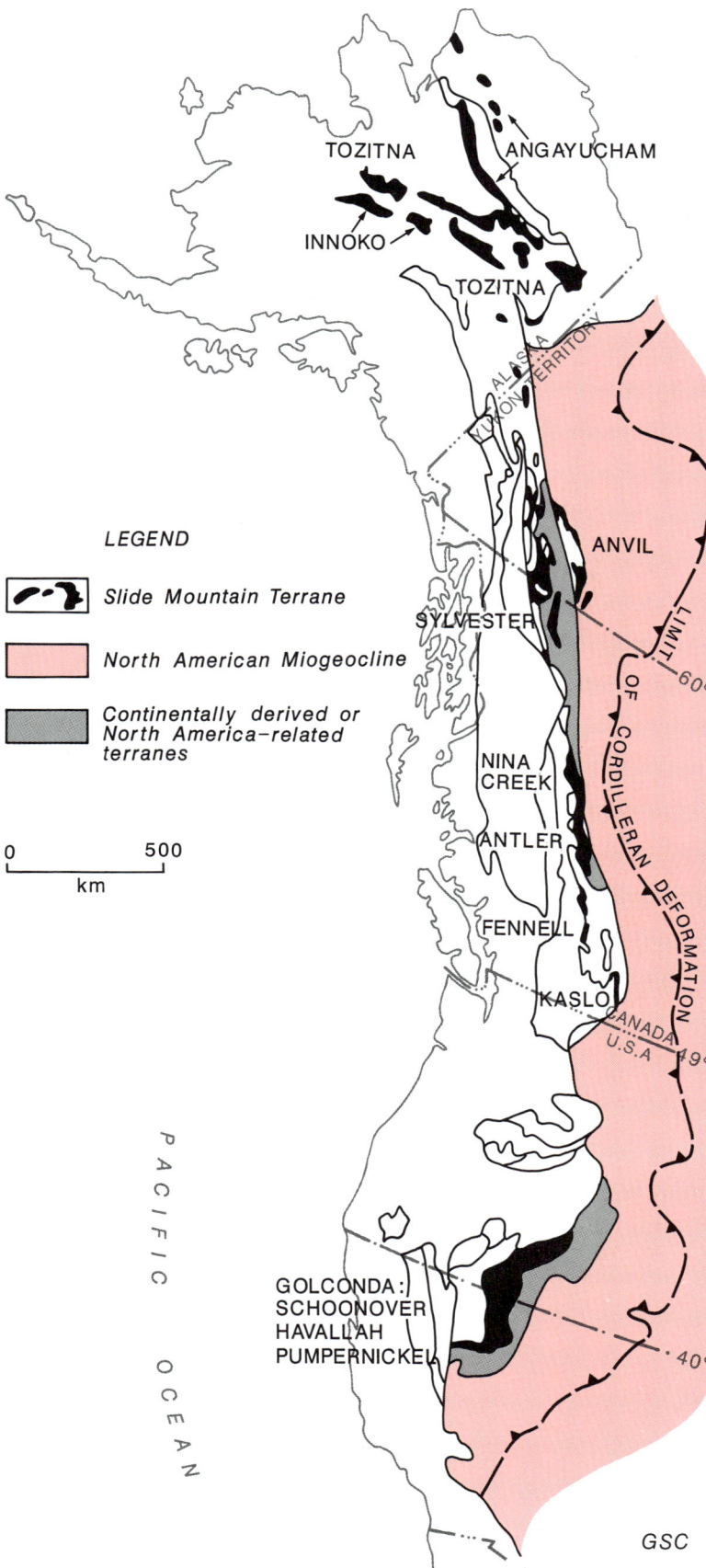

Figure 18.13. Distribution of Slide Mountain and similar oceanic terranes in the North American Cordillera.

of an early Mesozoic volcanic and plutonic island arc forming Quesnellia.

The emplacement of oceanic allochthons onto continental margins seems to have been common. Similarities can be drawn between Cordilleran examples and those in the Appalachians of Newfoundland (Williams and Hatcher, 1982), the Semail Nappe in Oman (Welland and Mitchell, 1977) and the Papuan ultramafic bodies in New Guinea (Davies and Smith, 1971). Although it seems likely that obduction resulted from convergence and collision processes, the evolution of obducted oceanic lithosphere prior to emplacement remains unclear. Local occurrences of blueschist minerals coupled with intense slicing are probably the products of subduction accretion rather than oceanic processes. Another view is that the character of terranes like the Slide Mountain, which have relatively undisrupted stratigraphy between faults, resulted from contraction in a back-arc basin.

Quesnellia, of well defined island arc affinity, separates the oceanic Cache Creek and Slide Mountain terranes. Except for the Harper Ranch Subterrane and possibly related Anarchist rocks in the Okanagan region, the stratigraphic record encompasses only Late Triassic through Middle Jurassic time. The terrane is most extensively exposed in Canada but rocks of similar age and lithology occur in the Huntington Terrane of eastern Oregon and in the Walker Lake Terrane of west-central Nevada and adjacent parts of California (Fig. 18.14).

Permian Tethyan faunas, characteristic of the Cache Creek Terrane, occur in exotic limestone blocks in the San Juan Terrane of northern Washington, in the Baker Terrane of east-central Oregon, in limestone blocks in the Hayfork Terrane of northwestern California, in the Foothills Terrane in the western Sierra Nevada and in the Peninsular Terrane of Alaska (Fig. 18.15). The Tethyan limestone occurs either in laterally extensive units that encompass a considerable age range, as in the Cache Creek, or in olistolithic blocks within younger radiolarian chert or fine grained clastic facies. The Cache Creek limestones locally demonstrate remarkably stable conditions from Mississippian through Permian time, a phenomenon perhaps possible only on a volcanic plateau within an ocean basin.

Cache Creek oceanic rocks include more mélange and are much more disrupted than those of the Slide Mountain Terrane. In the Cache Creek of central British Columbia, and correlative terranes in Oregon and northwestern California, Late Triassic radiometric ages have been obtained from blueschist minerals. The blueschist probably formed as a consequence of subduction that facilitated amalgamation of Stikinia, Quesnellia and the Cache Creek Terrane. In this model the complicated internal structure of the Cache Creek derives, in part, from its deformation in an accretionary prism setting.

Stikinia, bounded on the east by the Cache Creek Terrane, is the largest of the Cordilleran terranes. It extends for a short distance into Alaska and correlates in lithology and age with the Eastern Klamath Terrane in northern California (Fig. 18.14). The Eastern Klamath Terrane, however, lies east of equivalents of the Cache Creek Terrane. A volcanic-plutonic arc (Stuhini) formed on its northern and eastern margins during Late Triassic time. Lower to Middle Jurassic volcanic (Hazelton) and plutonic rocks define a complex of island arcs that almost surrounded the newly developing Bowser Basin. Their distribution suggests a tectonic setting akin to that of the Celebes arc of modern Indonesia (Hamilton, 1979).

In northern Stikinia a marked difference in structural style and metamorphism is observed between foliated and phyllitic Paleozoic rocks and weakly foliated upper Lower Triassic rocks; details of the inferred tectonism are lacking. A similar relationship has been reported in Quesnellia of the southern Cordillera (Read and Okulitch, 1977). Deformation and metamorphism that occurred in Early Triassic (?) time is of about the same age as that of the Sonoman Orogeny in Nevada which involved terranes possibly equivalent to the Slide Mountain Terrane and Quesnellia. One implication is that there may have been a loose association of Stikinia, Quesnellia and Slide Mountain Terrane in Early Triassic time. At least, they were all involved in a widespread, roughly coeval tectonic event.

The Alexander Terrane and Wrangellia have few if any obvious correlatives elsewhere in the Cordillera. Rocks typical of Wrangellia occur in eastern Oregon but underlie younger cover to the west. Recent observations in southeast Alaska of a late Paleozoic pluton cutting both the Alexander Terrane and Wrangellia suggest that they were together prior to Late Pennsylvanian time, or, that they were never separate terranes (R.B. Campbell, C.J. Dodds and G. Plafker, pers. comm., 1987).

EARLY JURASSIC TO PALEOGENE

The evolution of the Canadian Cordillera from the Early Jurassic to the Paleogene embraces the development of most of the region's main structural elements and marks a change from the dominance of disparate terranes to that of morphogeological belts resulting from amalgamation and accretion of terranes to the pre-mid-Jurassic continental margin. Within this broad concept, however, there are many problems which preclude unequivocal synthesis. These include the precise nature and timing of terrane amalgamation, the spatial relationship of tectonic phenomena in the suspect terranes with reference to those in ancestral North America, the temporal continuity of migrating tectonism versus tectonic pulses, the effects of oblique versus orthogonal collisions, the roles of Benioff as opposed to Ampferer subduction in tectonic processes and the clear identification of the tectonic processes that reorganized the northernmost Cordillera in the Late Jurassic and Cretaceous.

Early Jurassic - Volcanic island arcs, amalgamation of the Intermontane Superterrane and terminal phase of the miogeocline

In the Canadian Cordillera the Early Jurassic was a time of transition from terrane-specific volcanism, plutonism and sedimentation to the development of overlap assemblages in the Middle Jurassic. In Wrangellia the Late Triassic rifting of a Paleozoic marine volcanic arc was

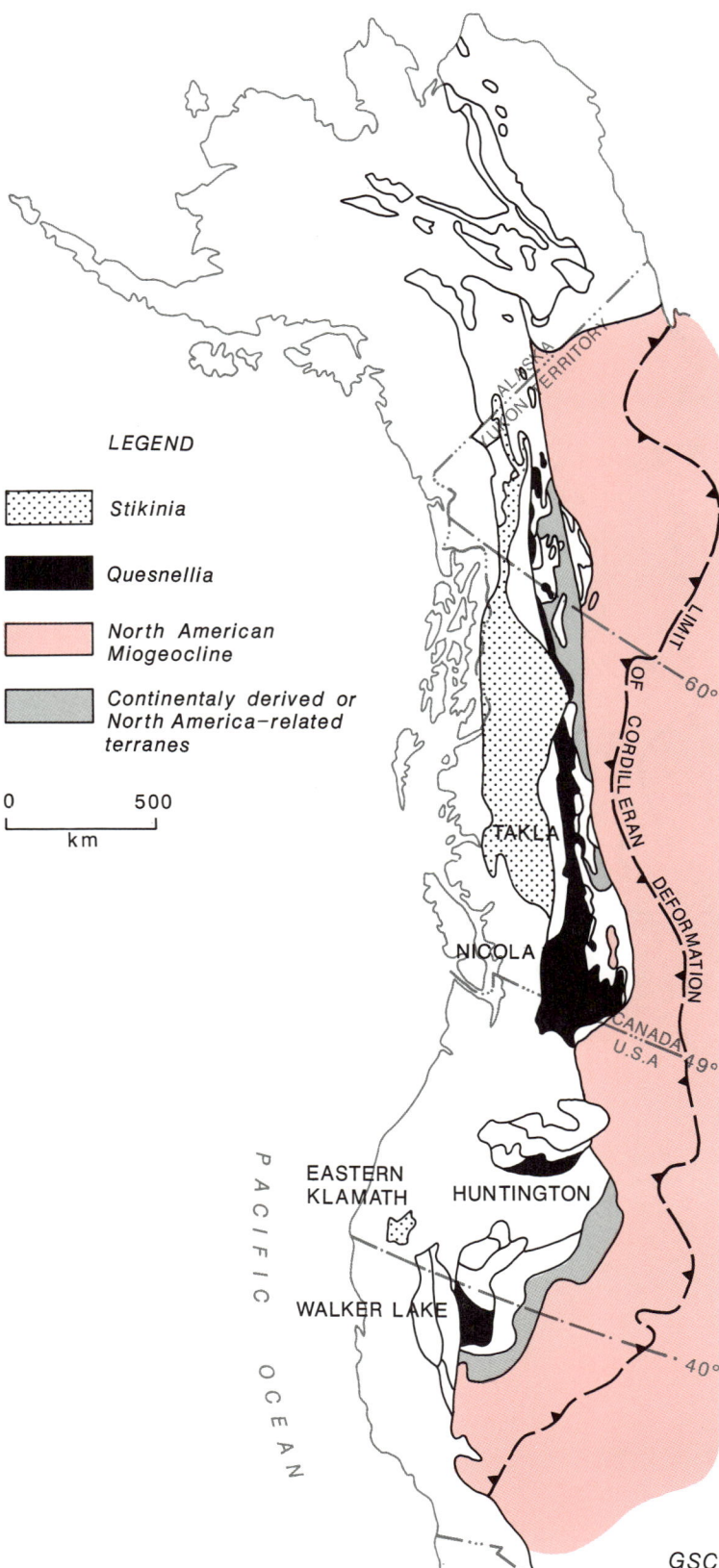

Figure 18.14. Distribution of Quesnellia and Stikinia and similar volcanic, island arc terranes in the North American Cordillera.

CHAPTER 18

succeeded, in Early Jurassic time, by the development of the largely subaerial Bonanza volcanic-plutonic arc on Vancouver Island and, in the Late Jurassic, on the Queen Charlotte Islands. Stikinia and Quesnellia were the sites of marine and nonmarine volcanic arcs (Hazelton and Nicola, respectively) with associated plutonic rocks until Middle Jurassic (Bajocian) time. In the Cache Creek Terrane of the Yukon Territory and northern British Columbia Lower Jurassic greywacke appears to have been derived from Quesnellia on the east and from Stikinia to the west. The youngest radiolarian chert recognized in the northern Cache Creek Terrane is of Late Triassic (Norian) age; shortly thereafter the oceanic setting of the Cache Creek Terrane was terminated. In the south, however, oceanic conditions persisted locally until Middle Jurassic time. On the central and southern miogeocline, condensed, incomplete successions of shale and some limestone accumulated in a stable cratonic setting.

Middle Jurassic to Early Cretaceous (Hauterivian) - accretion and orogeny

Although the Lower Triassic rocks provide hints of the amalgamation of Stikinia, Quesnellia, Cache Creek and Slide Mountain terranes into the Intermontane Superterrane, it was the Middle Jurassic Epoch that witnessed the onset of orogeny resulting in the formation of the Omineca Belt and the development of the Foreland Belt.

Island arc environments continued during the Middle Jurassic in northern Wrangellia before amalgamation with the Alexander Terrane, an event recorded by the Upper Jurassic to Lower Cretaceous Gambier overlap assemblage. The time of amalgamation or collision of the resulting Insular Superterrane with the Intermontane Superterrane is contentious but according to one view (Monger et al., 1982) probably occurred by mid-Cretaceous time. The similarity of the stratigraphy across the several small terranes of the southern Coast Belt, however, suggests that amalgamation may have occurred as early as Late Jurassic time.

Evidence for the collision of the Intermontane Superterrane with ancestral North America is contained in the record of sedimentation, deformation, plutonism and metamorphism. The best sedimentary and structural evidence for the timing of accretion comes from the region in and near the northern Bowser Basin where Bajocian marine and nonmarine conglomeratic strata are known to have been derived from uplifted Cache Creek Terrane to the northeast. Farther south the main influx of Cache Creek detritus into the Bowser Basin occurred in the Bathonian. The King Salmon Thrust Fault, carrying Cache Creek rocks in its hanging wall, and associated southwesterly directed structures, were cut by granitic plutons ranging in age from 147 to about 175 Ma. Continuing uplift of the Cache Creek Terrane and perhaps the northern fringe of Stikinia facilitated the progradation of a deltaic wedge over older, deep marine turbidites of the Bowser Basin during Late Jurassic time. The sedimentary history is similar along the eastern margin of the basin but there knowledge of the structural development is obscured by younger strike-slip faults.

The latitudinal position of the Bowser Basin relative to ancestral North America during the Middle and Late

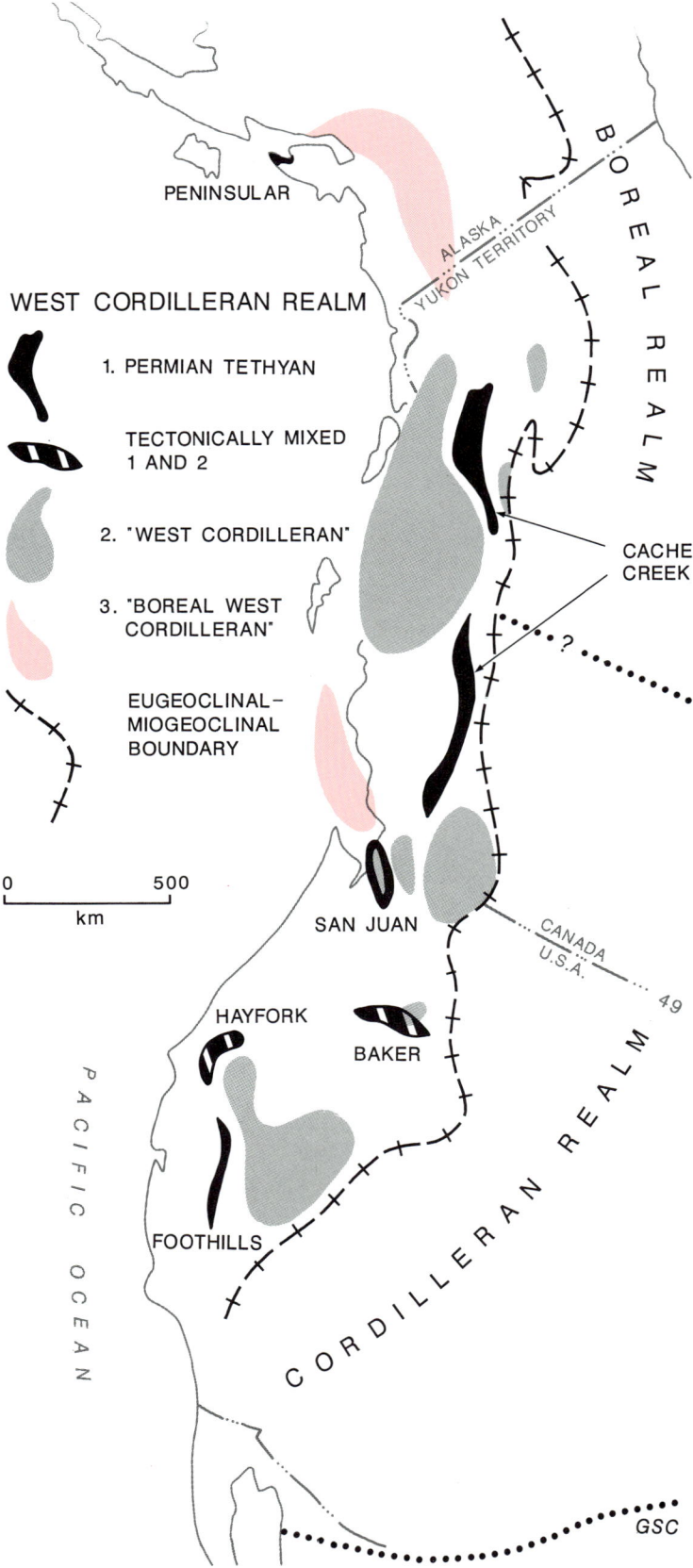

Figure 18.15. Distribution of Cache Creek and similar oceanic terranes in the North American Cordillera showing relationships of Permian fossil assemblages.

Jurassic is in doubt and the aforementioned events may reflect collision with the craton farther south. Indeed, current interpretations of paleomagnetic data suggest that the combined Intermontane and Insular superterranes (Baja, B.C.) were in more southerly paleolatitudes during mid-Cretaceous time than at present. Even though accretion with North America may have occurred at such latitudes it was not until early in the Tertiary Period that they reached their present position by coastwise translation. This interpretation suggests that either a northern extension of the Intermontane Superterrane, or an entirely different terrane, had collided with the craton during the Middle Jurassic and was subsequently displaced or replaced. Paleontological data support a northern hemisphere position for Stikinia as early as Pliensbachian time. Both interpretations, however, show that an important accretion event occurred during Middle Jurassic time; the resolution of this problem awaits further research.

Significant uplifts in the western part of southern Quesnellia are recorded by thick, easterly derived clastic rocks in the Relay Mountain and Ashcroft formations and by the evidence of termination of marine sedimentation throughout Quesnellia in Bajocian time. Collision of the Intermontane Superterrane with ancestral North America also resulted in the eastward emplacement of Quesnellia and probably the Slide Mountain Terrane onto the outer part of the miogeocline.

Dating of metamorphism, deformation and plutonism in the Omineca Belt is well constrained in the Columbia Mountains where highly deformed and regionally metamorphosed rocks (Quesnel and Kootenay terranes and the miogeocline) are cut by Middle Jurassic granitic plutons that have yielded radiometric ages ranging from about 160 to 180 Ma. These are the oldest plutons that clearly occur in all these terranes.

In the Foreland Belt the first direct evidence for clastic sediments derived from the west is the presence of clasts of radiolarian chert, from either or both of the Slide Mountain and Cache Creek terranes, in the Upper Jurassic to Lower Cretaceous Monteith Formation (M.E. McMechan, pers. comm., 1985). It is probable, however, that some Oxfordian clastics also were westerly derived. Uplift of the western miogeocline is clearly shown by the thick Jurassic-Cretaceous foredeep successions.

In the northern Yukon long established northern source areas for clastic sediments were replaced in Albian time by sources within the Cordillera to the south. Debate continues on the nature and timing of tectonic reorganization, however, recent paleomagnetic results obtained by Halgedahl and Jarrard (in press) support the model of counter-clockwise rotation of the Brooks Range away from the Arctic Archipelago and the consequent opening of Canada Basin sometime during Early to mid-Cretaceous time (Lawver and Scotese, 1990). The Albian age of thrusting in the Brooks Range possibly constrains the time of cessation of opening. It is suggested that the centre of rotation is expressed by the "Porcupine Virgation" a region of structural trend bifurcation and change in structural style west of Mackenzie Delta (see Chapter 17).

The eastern limit of Middle Jurassic tectonism is unknown, but no data show that it extended farther east than the Omineca Belt. The orogeny may have temporally overlapped with the classical Late Jurassic Nevadan Orogeny in eastern California (Schweickert, 1981) and with Middle Jurassic metamorphism, plutonism and deformation in the northern Klamath Mountains (Hill, 1985). The only other regions in the Cordillera that had significant Middle Jurassic plutonism are the Peninsular Terrane of the Aleutian Range and Talkeetna Mountains in southern Alaska (Hudson, 1983) and Wrangellia in the Queen Charlotte Islands. Except for the Omineca and eastern California regions none of the other examples involve miogeoclinal rocks.

Data obtained from equilibrium assemblages of metamorphic minerals show that a considerable increase in crustal thickness occurred in the Omineca Belt during Middle Jurassic deformation and metamorphism. Burial of Upper Proterozoic rocks to depths of greater than 25 km and heating at temperatures between about 550° and 650°C indicate the important role of compressional deformation in more than doubling the probable stratigraphic thickness (Archibald et al., 1983; Brown et al., 1986; Pigage, 1978, Simony et al., 1980; Evenchick, 1986; Mansy, 1986). A less significant part of the thickening resulted from the emplacement of Slide Mountain Terrane and, to a lesser extent, Quesnellia onto the miogeocline.

Two main hypotheses have been proposed to explain the early phase of Mesozoic structural development in the western part of the miogeoclinal wedge and, in particular, the westward verging structures that dominate the western part of the Omineca Belt. One proposes eastward obduction of the Slide Mountain Terrane and Quesnellia followed by the development of west-verging backfolds and back thrusts in supracrustal rocks balanced by eastward decoupling and underthrusting of basement rocks (Brown et al., 1986; Fig. 18.16A). Vergence changed from west to east following and in response to a change in configuration of the continental margin profile from west-sloping and thinning to east-sloping and thinning after obduction and structural thickening. The other hypothesis proposes obduction of the accreted terranes followed by tectonic wedging. In this case a wedge of accreted rocks was driven eastward between westward-verging, parautochthonous, supracrustal rocks delaminated from an autochthonous basement complex (Price, 1986; Fig. 18.16B). Vergence was determined by structural position above or below the wedge. In both models high-grade regional metamorphism was induced by structural thickening. In the Cassiar Mountains it has been demonstrated that eastward obduction of the Slide Mountain Terrane was accompanied by westward-verging duplex style thrust faulting in the underlying parautochthonous rocks (Harms, 1986). Westward-verging faults and folds persist to the deepest structural levels observed in the region.

Two axes of structural divergence formed, perhaps sequentially, during the early phase of Middle Jurassic to Early Cretaceous deformation. One occurs in the Intermontane Belt and separates the westward-verging structures of the Cache Creek oceanic rocks related to collision with Stikinia from the eastward-verging structures related to the emplacement of Quesnellia and the Slide Mountain Terrane (Fig. 18.17). The other occurs farther east in the Omineca Belt within the thickened prism of parautochthonous rocks (Fig. 18.17). In the southeastern Cordillera the axis of this structural divergence has been attributed variously to rotation of earlier formed easterly verging structures on the west side of the

CHAPTER 18

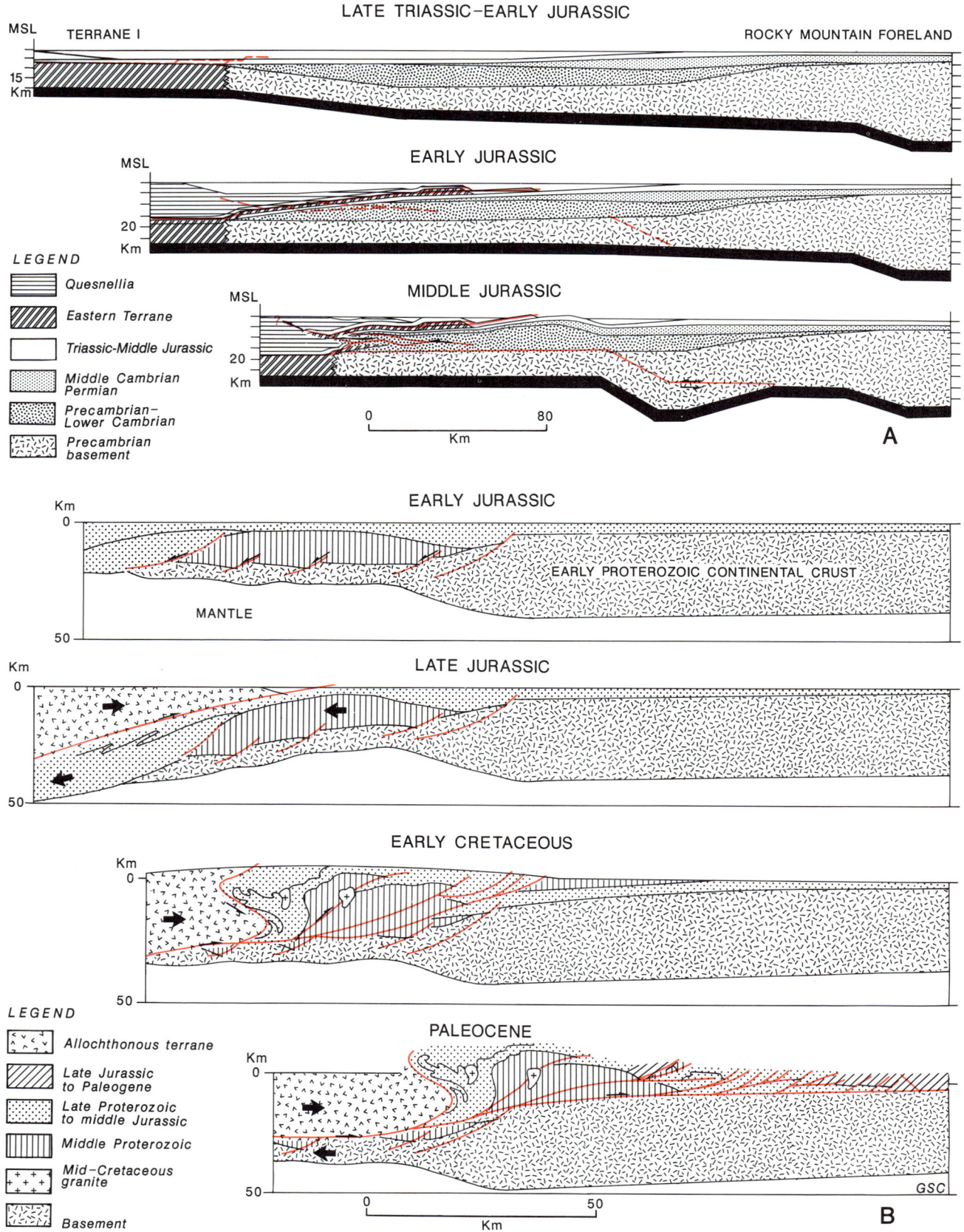

Figure 18.16. A. Model for Jurassic terrane collision involving obduction and eastward-underthrusting of sialic crust. After Brown et al. (1986). B. Model for Jurassic to Paleocene deformation involving obduction, wedging and delamination. After Price (1986).

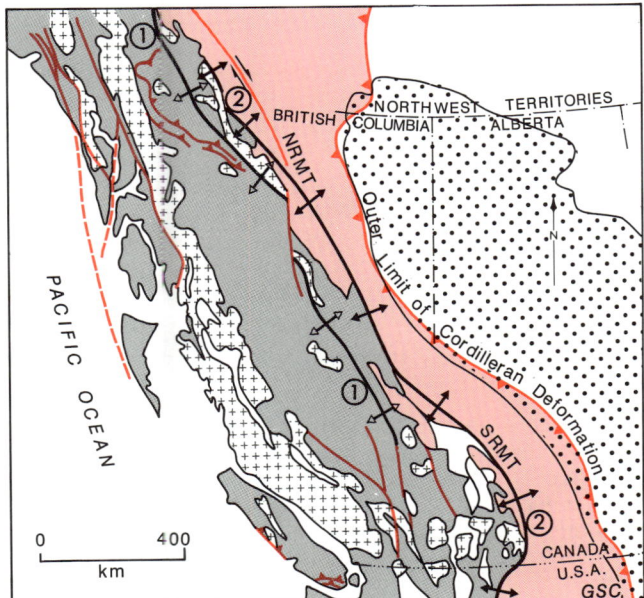

Figure 18.17. Axes of structural divergence in the central and eastern parts of the Canadian Cordillera; 1) Axis mainly between westward-verging rocks of the Cache Creek Terrane and eastward-verging rocks of Quesnellia. 2) Axis within miogeoclinal and structurally related rocks. NRMT-Northern Rocky Mountain Trench; SRMT-Southern Rocky Mountain Trench. Grey, rocks of terrane affinity; pink, rocks of miogeoclinal affinity; patterned area indicates foredeep deposits.

axis by wedging (Price, 1986) or to multiphase deformation (Brown and Tippett, 1978).

From a broader perspective the collision of terranes arising from closure of the Cache Creek and Slide Mountain oceans, and subsequent intraplate deformation necessitated the delamination and obduction of oceanic surface layers and ultramatic rocks from a crust the remainder of which was subducted (Fig. 18.18). Studies by Harms (1986) and Struik and Orchard (1985) indicate that the width of the Slide Mountain ocean was substantial and that subsequent contraction of obducted material began in late Paleozoic time. The polarity of inferred subduction zones and, indeed, their positions during this interval are not obvious. In southern British Columbia the Cache Creek mélange probably developed from an accretionary wedge related to eastward-directed subduction beneath Quesnellia. Middle Jurassic plutons that occur in Quesnellia and the pericratonic Kootenay Terrane (miogeocline) could be explained by this model. Similarly, in northern British Columbia the Cache Creek and overlying Lower Jurassic rocks have characteristics of an accretionary prism. There, evidence for Middle Jurassic plutonism in Quesnellia is sparse but Early Jurassic subvolcanic plutonism is widely documented.

Deformation and uplift of Cache Creek rocks in Middle Jurassic time north and east of the Bowser Basin was probably accomplished by intraplate deformation facilitated by subduction which generated a plutonic and volcanic arc along the northern and northeastern margin of Stikinia.

Middle to Late Jurassic tectonism of the Canadian Cordillera, particularly in the accreted terranes, has some similarity in age and style to that in the northern Klamath Mountains and Northern Sierra Terrane of California (Hill, 1985; Mortimer, 1985; Schweickert et al., 1984). In the western United States there is little evidence to suggest that deformation extended eastward into unequivocal North American strata until Early Cretaceous time (Armstrong, 1968) although uplift to the west is indicated by westerly derived sediments of the Oxfordian and younger Morrison Formation in Utah, Wyoming, Colorado and New Mexico.

Pre-Hauterivian to Barremian pedimentation

Clastic sedimentation (Kootenay, Bowser Lake and Relay Mountain assemblages) containing the record of Middle Jurassic to Early Cretaceous uplifts and including the first clastic-wedge in the Foreland Belt was succeeded by apparent Cordillera-wide erosion and pedimentation east of the Coast Belt. This event is demonstrated by an unconformity at the base of the middle clastic wedge in the southern Foreland Belt, beneath Hauterivian strata in the northern Yukon, under mid-Albian strata in northern British Columbia, directly below Hauterivian strata in central British Columbia and beneath Barremian strata in the Tyaughton-Methow Trough. Widespread uplift coincided with a lull in granitic plutonism and volcanism throughout the western Cordillera between 135 and 125 Ma (Armstrong, 1988) and a period of relatively slow motion of the North American Plate (Engebretson, 1982).

Early Cretaceous (Barremian) to Paleocene

A second clastic wedge (Blairmore Assemblage) was shed eastward into the foredeep of the emerging orogen between Barremian and Cenomanian time in response to another pulse of uplift in the Cordillera to the west. In the northernmost Cordillera, orogenic activity arising from the opening of the Canada Basin spread eastwards into the British and Barn mountains. The resulting Brooks Orocline thus confined clastic sedimentation to the Blow Trough west of the Mackenzie Delta wherein thick accumulations of flysch and molasse were deposited in response to continuing uplift associated with dextral strike-slip and normal faulting. Whereas most large-scale bends in structural grain in the North American Cordillera were controlled by the initial shape and thickness of the supracrustal rocks, the Brooks Orocline may be one of the few that formed as a result of rotation about a vertical or semi-vertical axis. During the same time thick clastic marine to nonmarine strata were shed westward from the Omineca Belt into Sustut and Skeena basins, and Tyaughton-Methow Trough. Equally thick successions of clastic rocks, locally with volcanics, were deposited in the regions of the Coast (Gambier Assemblage) and Insular (Honna Assemblage) belts. The first sediments in the latter belt derived from uplift of the Coast Belt to the east are of Albian to Coniacian age. Clastic-wedge sedimentation was coupled with uplift of the Omineca Belt which was the locus of deformation, widespread felsic S-type granitic plutonism characterized by rocks having $^{87}Sr/^{86}Sr$ ratios of 0.710 to 0.740, minor

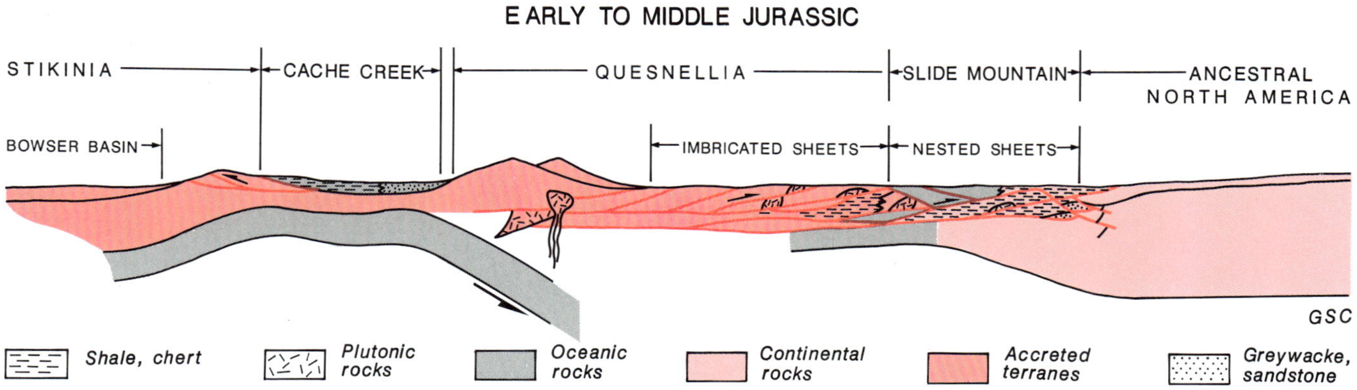

EARLY TO MIDDLE JURASSIC

Figure 18.18. Tectonic elements involved in Middle Jurassic collision, northern British Columbia. The Slide Mountain ocean may have closed by western subduction which generated Quesnellia during Late Triassic time. This subduction ceased before eastward subduction under Quesnellia which closed the Cache Creek ocean. Blueschist mineral assemblages at a number of localities in the Slide Mountain and Cache Creek terranes indicate that both were involved in subduction at some stage.

volcanism and regional metamorphism. The eastern limit of mid-Cretaceous plutonism is parallel with but much to the west of the eastern limit of Mackenzie Mountain and Rocky Mountain foreland deformation. Mid-Cretaceous plutonism, characterized by I-type granitic rocks with $^{87}Sr/^{86}Sr$ ratios of less than 0.706 (generally less than 0.7045) and associated with volcanism was extensive in the southern Coast Belt, where Cenomanian uplift provided a source of coarse clastic sediment for the Tyaughton-Methow Trough to the east and Insular Belt to the west. In the western part of the central Coast Belt west verging structures were synchronous with plutonism and regional metamorphism.

In the Foreland Belt the eastern limit of mid-Cretaceous deformation and the western limit of deposition are difficult to determine. Directly east of the Northern Rocky Mountain Trench deformation was accompanied by regional metamorphism dated at a minimum of 120 to 125 Ma. There, and in the Selwyn Mountains pre-mid-Cretaceous contraction of at least 50% is expressed by intense imbrication of lower Paleozoic strata. This imbrication, and overlap on thrust faults in northwestern Selwyn Mountains, indicate crustal shortening of possibly 200 km or more (S.P. Gordey and R.I. Thompson, pers. comm., 1987). In the Omineca Belt all regional contractional structures are cut by mid-Cretaceous plutons. No mid-Cretaceous structures are evident in the middle Albian part of the Sustut Group which was derived from the Omineca Belt.

In the Cassiar and Omineca mountains mid-Cretaceous deformation formed broad anticlinoria and synclinoria in previously deformed and regionally metamorphosed rocks. The structurally lowest rocks in the cores of the anticlinoria yield presumably reset mid-Cretaceous K-Ar ages whereas the flanks, with structurally highest rocks, yield earliest Cretaceous K-Ar ages. Rb-Sr dating of the structurally low rocks give ages as old as Middle Jurassic (Parrish, 1979). In the southeastern part of the Omineca Belt rocks that had been buried to depths of more than 20 km during Middle Jurassic tectonism were uplifted more than 13-14 km and intruded by mid-Cretaceous granite (Archibald et al., 1983). By about middle Albian time dextral strike-slip faulting began in the northern Cordillera (Gabrielse, 1985).

West-verging structures were generated in mid-Cretaceous time along the west side of the Coast Belt in southeastern Alaska and east of the Queen Charlotte Islands and on the west side of the Cascade segment of the Coast Belt including San Juan Islands whereas east verging thrusts developed on the east side of the Cascade Mountains.

By the beginning of Late Cretaceous time the Coast and Omineca belts were well established as uplifted metamorphic and plutonic welts. The Omineca Belt, for the most part, had undergone renewed uplift whereas the Coast Belt had just evolved from an island arc into a fully developed plutonic-metamorphic welt. The evolution of the Coast Belt has been related partly to collision between the Insular Superterrane and North America, which, by mid-Cretaceous time included the Intermontane Superterrane (Monger et al., 1982). Coeval, extensive plutonism, although of markedly different lithology and chemistry in the Coast and Omineca belts, raises questions concerning the collision mechanism. The suturing of the Insular and Intermontane superterranes along the axis of the Coast Belt could have been accommodated by eastward-dipping subduction between the two superterranes although the temporally closest arc volcanism (Gambier Assemblage) is Early Cretaceous in age. The magmatic arc with its ductile structures contrasts with much less penetrative westerly verging contraction structures that disrupt Wrangellia on eastern Vancouver Island and which may have formed during collision of the Insular and Intermontane superterranes. Another possibility is that much of the crustal thickening in the Coast Belt resulted from intraplate contraction and plutonism related to a subduction zone farther west. The parallelism of the eastern limit of granitic rocks in the Omineca Belt with the eastern margin of Cordilleran deformation suggests a coupled relationship in which the generation of plutons resulted from westward subduction of continental crust. The S-type nature of Omineca plutons points to a source within sialic material.

Early to mid-Cretaceous deformation and plutonism also occurred in the western United States with the emplacement of large granitic plutons in the Sierra Nevada and in Baja California and eastwardly verging folding and thrusting attributed to the Sevier Orogeny in Nevada and Utah (Armstrong, 1968). The interval spanning these events was marked by a relatively high velocity of the North American Plate in comparison with the preceding Valanginian to Barremian interval (Fig. 18.19B). The role of collision in the United States Cordillera is difficult to assess; speculations depend upon the paleogeographic position of the Canadian Intermontane Superterrane at that time.

Late Cretaceous to Paleocene time witnessed the deposition of the third and youngest of the clastic wedges (Brazeau Assemblage) in the foredeep of the Foreland Belt. The thickest and coarsest sediments occur in the upper part (Campanian to Paleocene). During Campanian time, uplift and deformation of the Coast Belt resulted in a reversal of sediment transport from westward to eastward in the Sustut Basin and from eastward to westward in the Insular Belt where a thick clastic wedge accumulated in Georgia Basin. This was a time of great crustal shortening in the southern Omineca and Foreland belts (up to 200 km) but considerably less than that in the northern Cordillera. Upward and eastward movement in the Coast Belt resulted in marked contraction of the supracrustal rocks in the Bowser Basin and Tyaughton-Methow Trough and was possibly related to westward thrusting in the Insular Belt. The mid-Cretaceous magmatic front lay mainly west of the Coast Range Megalineament in mid-Cretaceous time but moved to generally east of the megalineament in Late Cretaceous time. Throughout the Intermontane Belt block faulting was associated with local plutonism and volcanism. Explosive, felsic volcanism was particularly important during Campanian and Maastrichtian times, as shown by widespread tuff in the Sustut Basin and in the Foreland Belt (A.R. Sweet, pers. comm., 1984; Baadsgaard and Lerbekmo, 1982; Yorath and Cook, 1981). Abundant volcanics, in part alkaline in eastern exposures (Carmacks Volcanics) attest to widespread volcanism in southwestern Yukon Territory. Dextral transcurrent faulting probably occurred on many of the northwesterly trending faults although precise dating of the movements is not established. Accretionary wedges formed along the boundary between the Pacific and North American plates in the Gulf of Alaska. Structures in the central Coast Belt indicate a significant component of dextral shear between the Kula and North American plates.

Eocene - magmatism, uplift, extension, dextral strike-slip faulting

A short-lived (mainly 55-45 Ma) intense and widespread phase of granitic magmatism and uplift in the Coast and Omineca belts postdated the last main episode of contractional deformation and sedimentation east of the Insular Belt. Nonmarine sedimentation was restricted to linear basins along dextral transcurrent fault zones or occurred in extensional basins related to a dextral strain regime. Thick nonmarine to marine successions were deposited on the west flank of the Insular Belt and continued to form an accretionary wedge along a subduction zone in the Gulf of Alaska.

In the Omineca Belt of south-central British Columbia Eocene uplift was synchronous with volcanism and extension faulting. Locally, the amount of extension may have approached 100 km (Tempelman-Kluit and Parkinson, 1986). High heat flow is indicated by the widespread resetting of K-Ar ages in biotite and hornblende in structural culminations. Dextral transcurrent faulting continued along the Northern Rocky Mountain and Tintina trench systems and was important along the Fraser River system. In the Cassiar Mountains next to the Northern Rocky Mountain Trench, thrust faults of Late Cretaceous to Eocene age formed in response to compression along a left-stepping transcurrent fault (Evenchick, 1986).

The largest areas of Eocene granitic rocks are along the eastern part of the Coast Belt. They were emplaced between 62 and 48 Ma (Hollister, 1982) near the close of a period of rapid uplift of the Coast Belt (near 2.0 km/Ma).

One of the most remarkable elements in the Coast Belt is a foliated, steeply dipping tonalitic body about 10 to 15 km wide that lies immediately east of the Coast Range Megalineament (Work Channel lineament) and extends from near the British Columbia-Yukon border to south of Prince Rupert, a distance of more than 800 km (Brew and Ford, 1978; Woodsworth et al., 1983; Tectonic Assemblage Map 1712A, in pocket). Ages of the body range from Maastrichtian to Late Paleocene. Ductile structures along the western margin have been related to emplacement during deformation, perhaps involving strike-slip faulting (Woodsworth et al., 1983). Another possibility is that the tonalite represents an upturned slab originating at a deep crustal level in the Coast Plutonic Complex. An analog might be the tabular bodies of gneissic tonalite that occur at the base of an uptilted block of Archean crust along the Kapuskasing uplift in the Canadian Shield (Percival and Card, 1983).

The change from a contractional regime, presumably caused by orthogonal to oblique subduction of the Farallon Plate in the late Mesozoic and Paleocene, to one of uplift and extension east of the Insular Belt in the Eocene, has been attributed to changes in the relative motions among the Farallon, Kula, Pacific and North American plates (Engebretson, 1982; Fig. 18.19A,B). Prior to mid-Cretaceous time, relative movements of the Farallon and North American plates indicate oblique subduction with a component of sinistral shear. This argument has been used by Harper et al. (1985) to explain Jurassic sinistral offset along the Mojave-Sonora megashear in the southwestern United States and northern Mexico. During the mid-Cretaceous, relative movement of the two plates became essentially orthogonal. At about 85 Ma initiation of the Kula Plate with its dominantly northward movement relative to North America provided potential for dextral shear north of the ridge separating it from the Farallon Plate. With the demise of the Kula Plate at about 43 Ma the same result was obtained north of the triple junction between the Pacific, Farallon (Juan de Fuca) and North America plates.

There is considerable evidence in the Canadian Cordillera for dextral shear since mid-Cretaceous time, but, it may have been a minor component of oblique subduction until the effects of Kula Plate movement became important. The Paleocene and Eocene epochs, with fairly rapid west-to-southwest relative motion of the North American Plate, may have been an interval when the

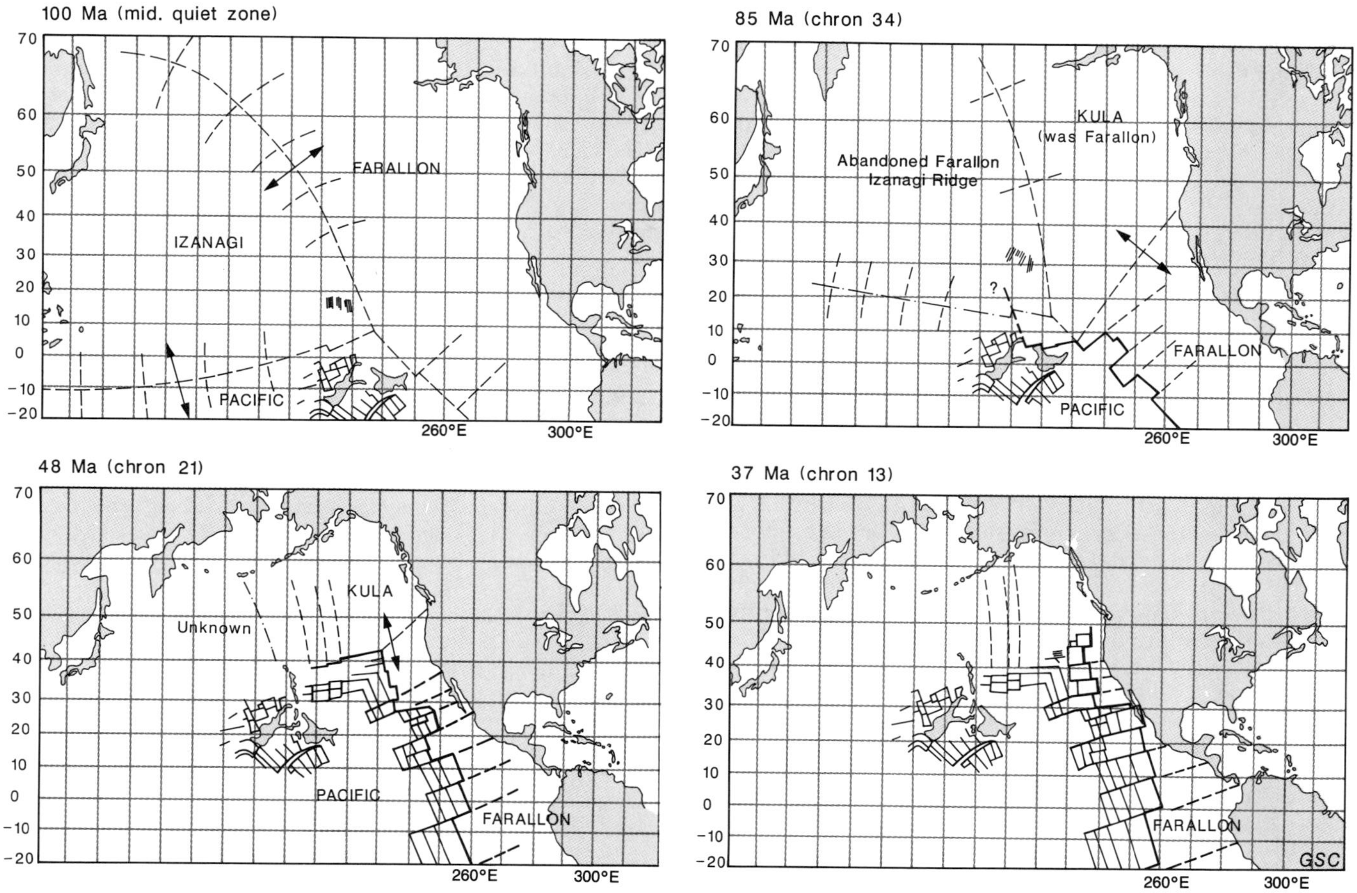

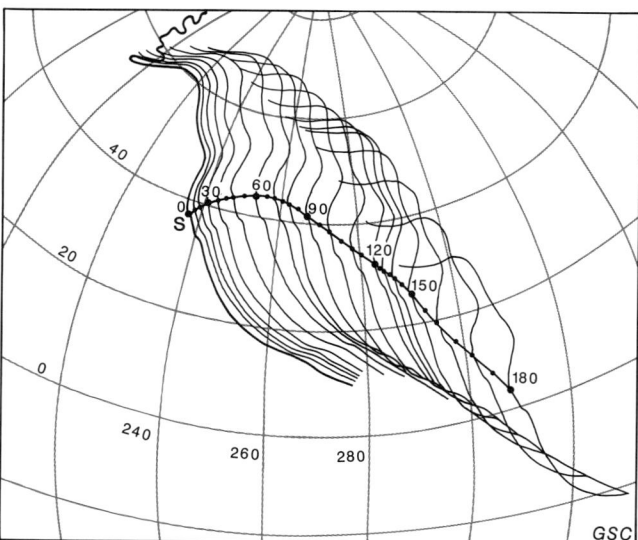

Figure 18.19. A. Plate reconstructions in fixed hotspot reference frame. B. Motion of North America with respect to hotspot reference for past 180 Ma at 10 Ma intervals. S= San Francisco. After Engebretson (1982).

component of dextral shear became important. Structural data and magmatic history suggest a dextral shear regime coupled with eastward subduction. Throughout the Jurassic to Eocene history of the Canadian Cordillera there is good correlation between episodes of deformation, plutonism and volcanism and changes in velocity or direction of North American Plate motion.

NEOGENE
Oligocene to Recent

Following cessation of the widespread Eocene events the remaining tectonic history of the Canadian Cordillera was determined largely by the interaction of the Pacific, Juan de Fuca and North American plates. As a consequence of northward motion of the Pacific Plate, sediment derived from the uplifted Coast and Insular belts has contributed to accretionary prisms adjacent to the Queen Charlotte Islands and in the Gulf of Alaska. Calc-alkaline volcanism occurred above the subducting slab in the Saint Elias Mountains. In the Insular Belt, Queen Charlotte Basin formed as a consequence of rifting associated with dextral motion along the Louscoone Inlet and Sandspit fault systems and through flexural subsidence due to the initiation of oblique subduction of the Pacific Plate beneath the Queen Charlotte Islands (Yorath and Chase, 1981; Yorath and Hyndman, 1983). North of the triple junction, which moved southerly along the continental margin from north of the Queen Charlotte Islands in the early Neogene to its present position between Queen Charlotte and Vancouver islands, northerly trending rifts related to the same regional stress system localized alkaline volcanism in the Stikine region. South of the triple junction two distinct volcanic-plutonic arcs formed above the subducting Juan de Fuca Plate; the Early to Late Miocene northwest-trending Pemberton Belt, and the Late Miocene to Recent more northerly trending Garibaldi Volcanic Belt (Cascade). In the interior of southern British Columbia flood basalts of about the same age as the Pemberton Belt volcanics reflect a back-arc setting. The conspicuous eastward-trending Anahim Volcanic Belt may define the trace of a mantle hotspot across central British Columbia where, from west to east, peralkaline volcanic centres become progressively younger. Rapid Pliocene-Recent uplift of the southern Coast Belt above the subducting Explorer and Juan de Fuca plates is attributed to thermal expansion and magma emplacement (Parrish, 1982).

Seismic reflection profiles across southern Vancouver Island reveal imbricate thrust faults related to Benioff-type subduction similar to those of Ampferer type in the Foreland Belt. In the Insular Belt Neogene contraction and uplift of Wrangellia were due to the underplating of oceanic materials. These emplacements require that considerable thicknesses of Wrangellian lithosphere must have been displaced so as to make room for the underplated crust. Moreover, the uplift of western Vancouver Island led to the local removal of approximately 9 km of Wrangellia above pervasive Early and Middle Jurassic granodiorite. Beneath the continental shelf and slope further similarities with Foreland Belt structures are expressed by seaward- and landward-verging thrust faults.

For the most part the modern geophysical signature of the Cordillera reflects its post-accretionary character and clearly defines the tectonophysical nature of the active convergent and transform components of the continental margin. There are two zones in the Cordillera across which regional geophysical properties change. The eastern zone is nearly coincident with the Southern Rocky Mountain Trench and, from east to west, is characterized by a 10 to 15 km shallowing of the Moho, an apparent 40 m/Wm^2 increase in surface heat flow, a 50 to 100 mGal increase in Bouguer anomaly values, a smoothing of long-wavelength magnetic anomalies and a pronounced increase in electrical conductivity. The western zone occurs in the Coast Belt, close to the Garibaldi (Cascade) volcanic arc. From east to west the crust thickens by about 20 km, surface heat flow declines by as much as 40 m/Wm^2, low level seismicity rises sharply within a horizontal distance of 20 km and the Bouguer anomaly increases by up to 150 mGal over a distance of about 80 km. A linear magnetic anomaly high of about 1000 nT is associated with the western zone. The region between these zones, the Omineca and Intermontane belts and the eastern part of the Coast Belt, has a relatively thin, less dense crust, high heat flow, attenuated long-wavelength magnetic anomalies, and high conductivity.

CONCLUSIONS AND COMPARISONS WITH OTHER OROGENS

The present architecture of the Canadian Cordillera is the product of an evolution that spans an interval of more than 1.6 Ga. The dominant elements of structure, metamorphism, plutonism and volcanism, however, are mainly the result of Mesozoic to Recent tectonism. Much of the orogen's complexity can be attributed to the widely disparate origins of the distinct component terranes and the timing and dynamics of their accretion to the continent. Although the general characteristics of the volcanic-arc and oceanic terranes are known the internal tectonic history of many are poorly understood. For example, the age and spatial relationships of volcanic island arcs that dominated the evolution of Stikinia throughout the late Paleozoic and early Mesozoic eras require much study. Almost nothing is known about the basement of Stikinia although initial strontium ratios suggest a crust not as evolved as that of ancestral North America. The extent of depositional basins within the small terranes in the southern Coast Belt and their interrelations, if any, have only recently become topics for detailed study. Similarly, the problems of movements between the displaced terranes and ancestral North America are now being addressed with the realization that paleogeographic restorations of the Cordilleran region must account not only for contractional deformation but for important and very large transcurrent displacements.

The evolution of the crustal structure of the Canadian Cordillera can be compared and contrasted with that of several other major orogenic belts (Fig. 18.20). Most, if not all models are based upon plate-tectonic principles as applied to a particular mountain system. The simplest setting is exemplified by intraoceanic subduction beneath the Mariana island arc in the western Pacific Ocean. Probably all of the better known ancient arcs in the Cordillera, however, had evolved considerably further than the Mariana arc before amalgamation and accretion.

The more evolved island arc system of the Japanese Archipelago provides a possible model for arc settings in Quesnellia, Stikinia and Wrangellia. No evidence has been found, however, to relate their initial origin to back-arc spreading away from a cratonic block. The relationship of the Slide Mountain Terrane to ancestral North America and to Quesnellia in late Paleozoic time may have approximated that of the Japan Basin to Asia and to the Japan island arc. It is probable, however, that Quesnellia in the late Paleozoic was not as evolved as the Japan arc and, indeed, may not hence become an important element until Late Triassic time.

An Andean model of oceanic subduction under a continental margin has been proposed for the western part of the Canadian Cordillera, particularly for the development of the Coast Plutonic Complex. In both cases it has been suggested that collision of terranes also played a role in formation of the volcanic-plutonic belts. At least in its present configuration the Andean orogen seems less complicated than the Canadian Cordilleran orogen, perhaps indicating that terrane collisions were considerably less important in the Andes.

The Canadian Cordillera can be compared and contrasted to the Alps and Himalayas, both attributed to the collision of cratonal blocks. Of note are the much greater volumes of granitic and plutonic rocks and the more widespread obduction of oceanic crust in the Cordillera. This partly reflects the contribution of island arcs in the Cordilleran accreted terranes and possibly the extent of uplift following obduction. The enormous area affected by Himalayan deformation perhaps results from the collision

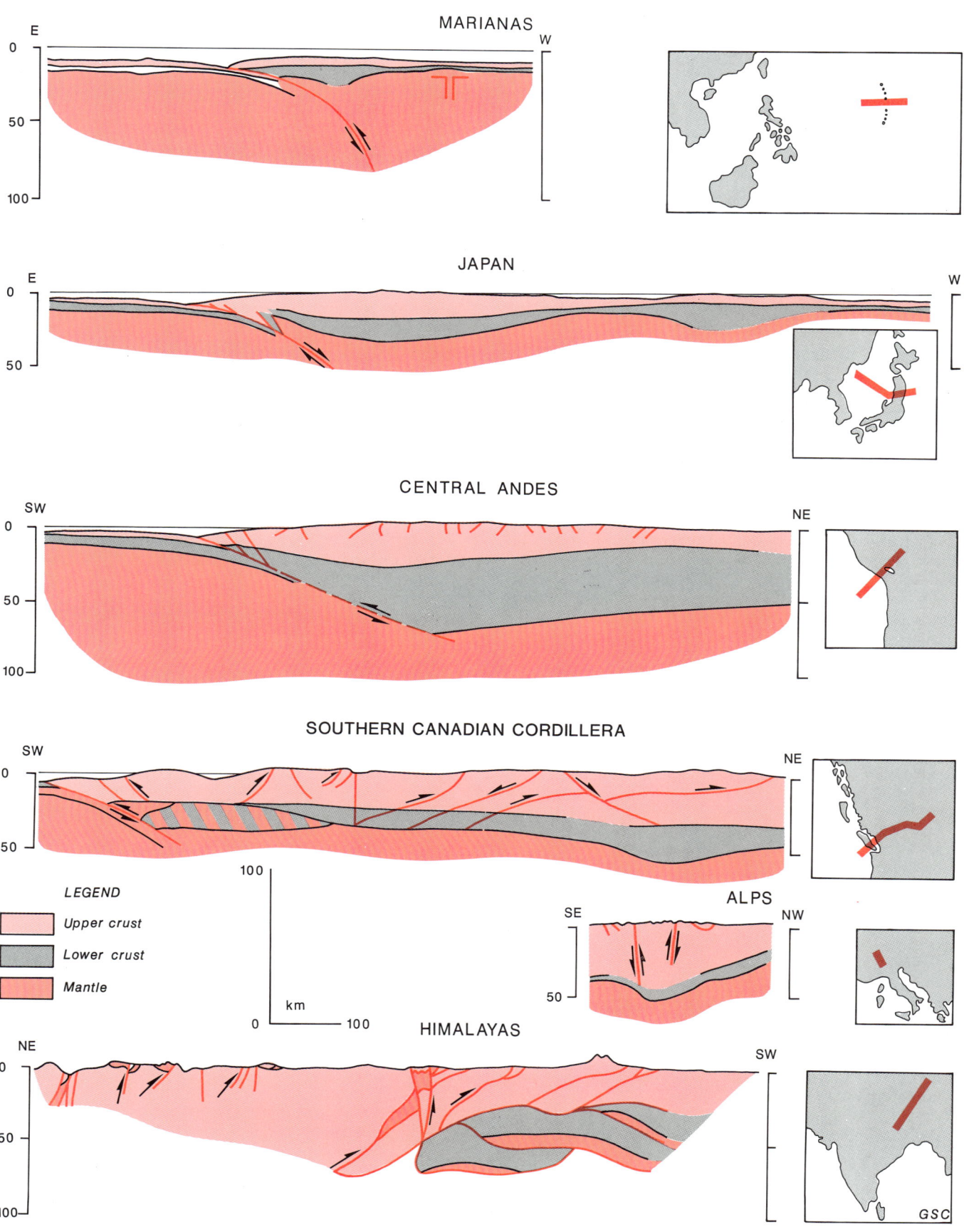

Figure 18.20. Crustal cross-sections of various orogenic belts comparing the styles and complexity of interaction between ocean-ocean (Marianas after Hussong and Uyeda, 1981; Katsumata and Sykes, 1969), ocean-arc (Japan after Murauchi and Yasui, 1968; Shiki and Misawa, 1982), ocean-continent (Andes after James, 1971; Ocola and Meyer, 1973; Laubacher, 1978; Isacks and Barazangi, 1977), arc(s)-continent (Canadian Cordillera after Monger et al., 1985), and continent-continent (Alps after Mueller et al., 1980 and Rybach et al., 1980; Himalayas after Burg and Chen, 1984; Hirn et al., 1984; Allegre et al., 1984). Compiled by S. Friday.

of two equally thick sialic crustal masses. In the Cordillera collisions involved the thick, sialic North American crust with much smaller and less buoyant island arc and oceanic terranes.

In conclusion, despite its complexity, the Canadian Cordilleran Orogen contains remarkably well exposed and preserved examples of many tectonic elements that contribute to the development of the earth's crust. These include assemblages representative of passive margins, volcanic, plutonic and sedimentary island arcs, ocean basins and foredeeps all involved variously in contractional, extensional and strike-slip deformation. The oceanic spreading centres and a host of related phenomena including the Juan de Fuca and Explorer subduction zones are among the world's most accessible and have great potential for studies of plate-tectonic processes and metallogenesis. The scope for detailed investigations of these subjects is exciting and unlimited.

REFERENCES

Aleinikoff, J.N., Foster, H.L., Nokleberg, W.J., and Dusel-Bacon, C.
1984: Isotopic evidence from detrital zircons for Early Proterozoic crustal material, east-central Alaska; in The United States Geological Survey in Alaska: Accomplishments during 1981, W.L. Coonrad and R.L. Elliott (ed.), United States Geological Survey Circular 868, p. 43-45.

Allegre, C.J., Courtillot, V., and Tapponnier, P.
1984: Structure and evolution of the Himalaya - Tibet orogenic belt; Nature, v. 307, p. 17-22.

Anderson, M.M. and King, A.F.
1981: Precambrian tillites of the Conception Group on the Avalon Peninsula, southeastern Newfoundland; in Earth's Pre-Pleistocene Glacial Record, M.J. Hambrey and W.B. Harland (ed.), Cambridge University Press, Cambridge, p. 760-767.

Archibald, D.A., Glover, J.K., Price, R.A., Farrar, E., and Carmichael, D.M.
1983: Geochronology and tectonic implications of magmatism and metamorphism, southern Kootenay Arc and neighbouring regions, southeastern British Columbia. Part I: Jurassic to mid-Cretaceous; Canadian Journal of Earth Sciences, v. 20, p. 1891-1913.

Armstrong, R.L.
1968: Sevier orogenic belt in Nevada and Utah; Geological Society of America Bulletin, v. 79, p. 429-458.
1988: Mesozoic and Early Cenozoic magmatic evolution of the Canadian Cordillera; in Processes in continental lithospheric deformation, S.P. Clark, Jr., B.C. Burchfiel, and J. Suppe (ed.), Geological Society of America, Special Paper 218, p. 55-91.

Baadsgaard, H. and Lerbekmo, J.F.
1982: Abstracts for Items NDS 126 and NDS 127, in Numerical Dating in Stratigraphy, G.S. Odin (ed.), Wiley Interscience, Chichester, v. 2, p. 795-797.

Bjorlykke, K.
1978: The eastern marginal zone of the Caledonide orogen in Norway; Geological Survey of Canada, Paper 78-13, p. 49-55.

Bond, C.B., Nickeson, P.A., and Kominz, M.A.
1984: Breakup of a supercontinent between 625 Ma and 555 Ma: new evidence and implications for continental histories; Earth and Planetary Science Letters, v. 70, p. 325-345.

Bond, G.C., Christie-Blick, N., Kominz, M.A., and Devlin, W.J.
1985: An early Cambrian rift to post-rift transition in the Cordillera of western North America; Nature, v. 315, p. 742-745.

Brew, D.A. and Ford, A.B.
1978: Megalineament in southeastern Alaska marks southwest edge of Coast Range batholithic complex; Canadian Journal of Earth Sciences, v. 15, p. 1763-1772.

Brown, D.A., Campbell, K.S.W., and Crook, K.A.W.
1968: The geological evolution of Australia and New Zealand; Pergamon Press, 409 p.

Brown, R.L. and Tippett, C.R.
1978: The Selkirk fan structure of the southeast Canadian Cordillera; Geological Society of America, Bulletin, v. 89, p. 548-558.

Brown, R.L., Journeay, J.M., Lane, L.S., Murphy, D.C., and Rees, C.J.
1986: Obduction, backfolding and piggy back thrusting in the metamorphic hinterland of the southeastern Canadian Cordillera; Journal of Structural Geology, v. 8, p. 255-268.

Brueckner, H.K. and Snyder, W.S.
1985: Structure of the Havallah sequence, Golconda allochthon, Nevada: Evidence for prolonged evolution of an accretionary prism; Geological Society of American Bulletin, v. 96, p. 1113-1130.

Burg, J.P. and Chen, G.M.
1984: Tectonics and structural zonation of southern Tibet, China; Nature, v. 311, p. 219-223.

Crittenden, M.D. Jr. and Wallace, C.A.
1973: Possible equivalents of the Belt Supergroup in Utah; in Belt Symposium, v. 1, Department of Geology, University of Idaho and Idaho Bureau of Mines and Geology, Moscow, Idaho, p. 116-138.

Crittenden, M.D. Jr., Christie-Blick, N., and Link, P.K.
1983: Evidence for two pulses of glaciation during the late Proterozoic in northern Utah and southeastern Idaho; Geological Society of America Bulletin, v. 94, p. 437-450.

Davies, H.L. and Smith, I.E.
1971: Geology of eastern Papua; Geological Society of America Bulletin, v. 82, p. 3299-3312.

Eisbacher, G.H.
1985: Late Proterozoic rifting, glacial sedimentation, and sedimentary cycles in the light of Windermere deposition, western Canada; in Paleogeography, Paleoclimatology and Paleoecology, v. 51, p. 231-254.

Engebretson, D.C.
1982: Relative motions between oceanic and continental plates in the Pacific Basin; Ph.D. thesis, Stanford University, 211 p.

Evenchick, C.A.
1986: Stratigraphy, metamorphism, structure and their tectonic implications in the Sifton and Deserters ranges, Cassiar and northern Rocky Mountains, northern British Columbia; Ph.D. thesis, Queen's University, Kingston, Ontario, 197 p.

Gabrielse, H.
1972: Younger Precambrian of the Canadian Cordillera; American Journal of Science, v. 272, p. 521-536.
1985: Major dextral transcurrent displacements along the Northern Rocky Mountain Trench and related lineaments in north-central British Columbia; Geological Society of America Bulletin, v. 96, p. 1-14.

Gabrielse, H., Loveridge, W.D., Sullivan, R.W., and Stevens, R.D.
1982: U-Pb measurements on zircon indicate middle Paleozoic plutonism in the Omineca Crystalline Belt, north-central British Columbia; in Current Research, Part C, Geological Survey of Canada, Paper 82-1C, p. 139-146.

Gabrielse, H., Snyder, S.W., and Stewart, J.H.
1983: Sonoma orogeny and Permian to Triassic tectonism in western North America; Geology, v. 11, p. 484-486.

Gehrels, G.E. and Saleeby, J.B.
1984: Paleozoic geologic history of the Alexander Terrane in S.E. Alaska, and comparisons with other orogenic belts; Geological Society of America, Abstracts with Programs, v. 16, p. 516.

Gehrels, G.E., Saleeby, J.B., and Berg, H.C.
1983: Preliminary description of the Klakas orogeny in the southern Alexander terrane, southeastern Alaska; in Pre-Jurassic Rocks in the Western North American Suspect Terranes, C.H. Stevens (ed.), Pacific Section, Society of Economic Paleontologists and Mineralogists, Los Angeles, California, p. 131-141.

Grow, J.A.
1981: Structure of the Atlantic Margin of the United States; in Geology of Passive Continental Margins, American Association of Petroleum Geologists, Education Course Note Series no. 19, p. 3-1 to 3-41.

Halgedahl, S. and Jarrard, R.
in press: Paleomagnetism of the Kuparuk River Formation from oriented drill core: Evidence for rotation of the North Slope block; in Alaskan North Slope Geology, I.L. Tailleur and P. Weimer (ed.), Los Angeles, Society of Economic Paleontologists and Mineralogists, Pacific Section.

Hambrey, M.J. and Harland, W.B. (ed.)
1981: Earth's pre-Pleistocene glacial record; Cambridge University Press, Cambridge.

Hamilton, W.
1979: Tectonics of the Indonesian Region; United States Geological Survey, Professional Paper 1078, 345 p.

Harms, T.A.
1986: Structural and tectonic analysis of the Sylvester Allochthon, northern British Columbia: Implications for paleogeography and accretion; Ph.D. thesis, University of Arizona, 80 p.

Harper, G.D., Saleeby, J.B., and Norman, E.A.S.
1985: Geometry and tectonic setting of sea-floor spreading for the Josephine Ophiolite and implications for Jurassic accretionary events along the California margin; in Tectonostratigraphic Terranes of the Circum-Pacific Regions, D.G. Howell (ed.), Circum-Pacific Council for Energy and Mineral Resources, Houston, Texas, p. 239-257.

Harris, H.L., Johnson, M.R.W., and Powell, D.
1978: The orthotectonic Caledonides (Moixes and Dalradians) of Scotland; in Caledonian-Appalachian Orogen of the North Atlantic Region, Geological Survey of Canada, Paper 78-13, p. 79-86.

Henriksen, N. and Higgins, A.K.
1976: East Greenland Caledonian fold belt; in Geology of Greenland, A. Escher and W.S. Watt (ed.), Geological Survey of Greenland, Copenhagen, 601 p.

Hill, B.L.
1985: Metamorphic, deformational, and temporal constraints on terrane assembly, northern Klamath Mountains, California; in Tectonostratigraphic Terranes of the Circum-Pacific Regions, D.G. Howell (ed.), Circum-Pacific Council for Energy and Mineral Resources, Houston, Texas, p. 173-186.

Hirn, A., Lepine, J-C., Jobert, G., Sapin, M., Wittlinger, G., Xin, X.Z., Yuan, G.E., Jing, W.X., Wen, T.J., Bai, X.S., Pandey, M.R., and Tater, J.M.
1984: Crustal structure and variability of the Himalayan border of Tibet; Nature, v. 307, p. 23-25.

Hollister, L.S.
1982: Metamorphic evidence for rapid (11 mm/yr.) uplift of a portion of the Central Gneiss Complex, Coast Mountains, B.C.; Canadian Mineralogist, v. 20, p. 319-332.

Hudson, T.
1983: Calc-alkaline plutonism along the Pacific rim of southern Alaska; in Circum-Pacific Plutonic Terranes, J.A. Roddick (ed.), Geological Society of America, Memoir 159, p. 159-170.

Hussong, D.M. and Uyeda, S.
1981: Tectonics in the Marianas Arc: results of recent studies including DSDP Leg 60; in Oceonologica ACTA, Proceedings 26th I.G. Congress, Geology of Continental Margins Symposium, Paris, p. 203-212.

Isacks, B.L. and Barazangi, M.
1977: Geometry of Benioff zones lateral segmentation and downwards banding of the subducted lithosphere; in Island Arcs, Deep Sea Trenches and Back-Arc Basins, Maurice Ewing Series, v. 1, M. Talwani and W.C. Pitman III (ed.), A.G.U. Washington, D.C., p. 99-114.

James, D.E.
1971: Plate Tectonic model for the evolution of the Central Andes; Geological Society of America Bulletin, v. 82, p. 3325-3346.

Kanasewich, E.R., Clowes, R.M., and McCloughan, C.H.
1968: A buried Precambrian rift in western Canada; Tectonophysics, v. 8, p. 513-527.

Katsumata, M. and Sykes, L.R.
1969: Seismicity and tectonics of the western Pacific: Izu-Mariana-Caroline and Ryuku Taiwan regions; Journal of Geophysical Research, v. 74, p. 5923-5948.

Ketner, K.B. and Smith, J.
1982: Mid-Paleozoic age of the Roberts thrust unsettled by new data from northern Nevada; Geology, v. 10, p. 298-303.

Klepacki, D.W. and Wheeler, J.O.
1985: Stratigraphic and structural relations of the Milford, Kaslo and Slocan groups, Goat Range, Lardeau and Nelson map areas, British Columbia; in Current Research, Part A, Geological Survey of Canada, Paper 85-1A, p. 277-286.

Laubacher, G.
1978: Géologie des Andes Pèruviennes; Travaux et Documents de L'Office de la Recherche Scientifique et Technique (Paris), no. 95, 217 p.

Lawver, L.A. and Scotese, C.R.
1990: A review of tectonic models for the evolution of the Canada Basin; in The Arctic Region, A. Grantz, G.L. Johnson and J.F. Sweeney (ed.), Geological Society of America, The Geology of North America, v. E.

Mansy, J.L.
1986: Géologie de la Chaine d' Omineca des rocheuses aux Plateaux Interieurs (Cordillere Canadienne), son evolution depuis le Precambrien; Doctorat d'Etat des Sciences Naturelle, University of Lille, France, 718 p.

McKenzie, D.F.
1978: Some remarks on the development of sedimentary basins; Earth and Planetary Science Letters, v. 40, p. 25-32.

McMannis, W.L.
1963: La Hood Formation - a coarse facies of the Belt Series in southwestern Montana; Geological Society of America Bulletin, v. 74, p. 407-436.

McMechan, M.E. and Price, R.A.
1982: Superimposed low-grade metamorphism in the Mount Fisher area, southeastern British Columbia - implications for the East Kootenay orogeny; Canadian Journal of Earth Sciences, v. 19, p. 476-489.

Miller, E.L., Bateson, J., Dixter, D., Dyer, J.R., Horbaugh, D., and Jones, D.L.
1981: Thrust emplacement of the Schoonover sequence, northern Independence Mountains, Nevada; Geological Society of America Bulletin, v. 92, p. 730-737.

Miller, E.L., Kanter, L.R., Larue, D.K., Turner, R.J., Murchey, B., and Jones, D.L.
1982: Structural fabric of the Golconda Allochthon, Antler Peak Quadrangle, Nevada: Progressive deformation of an oceanic sedimentary assemblage; Journal of Geophysical Research, v. 87, p. 3795-3804.

Miller, E.L., Holdsworth, B.K., Whiteford, W.B., and Rodgers, D.
1984: Stratigraphy and structure of the Schoonover sequence, northeastern Nevada: Implications for Paleozoic plate-margin tectonics; Geological Society of America Bulletin, v. 95, p. 1043-1076.

Monger, J.W.H., Price, R.A., and Tempelman-Kluit, D.J.
1982: Tectonic accretion and the origin of the two major metamorphic and plutonic welts in the Canadian Cordillera; Geology, v. 10, p. 70-75.

Monger, J.W.H., Clowes, R.M., Price, R.A., Simony, P.S., Riddihough, R.P., and Woodsworth, G.J.
1985: B-2 Juan de Fuca Plate to Alberta Plains; Geological Society of America, The Decade of North American Geology, Centennial Continent/Ocean Transect #7, 21 p., 2 sheets.

Mortimer, N.
1985: Structural and metamorphic aspects of Middle Jurassic terrane juxtaposition, northeastern Klamath Mountains, California; in Tectonostratigraphic Terranes of the Circum-Pacific Region, D.G. Howell (ed.), Circum Pacific Council for Energy and Mineral Resources, Houston, Texas, p. 201-214.

Mueller, S., Ansorge, J., Egloff, R. and Kissling, E.
1980: A crustal cross section along the Swiss geotraverse from the Rhinegraben to the Po Plain; Eclogae Geologicae Helvetiae, v. 73, p. 463-483.

Murauchi, S. and Yasui, M.
1968: Geophysical investigations in the seas around Japan; Kagaku, v. 38, no. 4, p. 192-200 (in Japanese).

Ocola, L.C. and Meyer, R.P.
1973: Crustal structure from the Pacific basin to the Brazilian Shield between 12° and 30° south latitude; Geological Society of America Bulletin, v. 84, p. 3387-3404.

Parrish, R.R.
1979: Geochronology and tectonics of the northern Wolverine Complex, British Columbia; Canadian Journal of Earth Sciences, v. 16, p. 1428-1438.

1982: Cenozoic thermal and tectonic history of the Coast Mountains of British Columbia as revealed by fission track and geological data and quantitative thermal models; Ph.D. thesis, University of British Columbia, Vancouver, 166 p.

Percival, J.A. and Card, K.D.
1983: Archean crust as revealed in the Kapuskasing uplift, Superior province, Canada; Geology, v. 11, p. 323-326.

Pigage, L.C.
1978: Metamorphism and deformation on the northeast margin of the Shuswap Metamorphic Complex Azure Lake, British Columbia; Ph.D. thesis, University of British Columbia, Vancouver, 289 p.

Poole, F.G.
1974: Flysch deposits of the Antler Foreland Basin, western United States, in Tectonics and Sedimentation, W.R. Dickinson (ed.), Society of Economic Paleontologists and Mineralogists, Special Publication no. 22, p. 58-82.

Price, R.A.
1986: The southeastern Canadian Cordillera: thrust faulting, tectonic wedging, and delamination of the lithosphere; Journal of Structural Geology, v. 8, p. 239-254.

Read, P.B. and Okulitch, A.V.
1977: The Triassic unconformity of south-central British Columbia; Canadian Journal of Earth Sciences, v. 14, p. 606-638.

Rybach, L., Mueller, S., Milnes, A.G., Ansorge, J., Bernoulli, D. and Frey, M.
1980: The Swiss geotraverse Basel-Chiasso - a review; Eclogae Geologicae Helvetiae, v. 73, p. 437-462.

Schwab, F.L.
1981: Late Precambrian tillites of the Appalachians; in Earth's pre-Pleistocene Glacial Record, M.J. Hambrey and W.B. Harland (ed.), Cambridge University Press, Cambridge, p. 751-755.

Schweickert, R.A.
1981: Tectonic evolution of the Sierra Nevada range; in The Geotectonic Development of California, W.G. Ernst (ed.), Prentice Hall, Englewood Cliffs, New Jersey, p. 87-131.

Schweickert, R.A., Harwood, D.S., Girty, G.H., and Hanson, R.E.
1984: Tectonic development of the Northern Sierra Terrane: an accreted late Paleozoic island arc and its basement; in Western Geological Excursions, Volume 4, J. Lintz, Jr. (ed.), Geological Society of America and the Department of Geological Sciences, Mackay School of Mines, University of Nevada-Reno, p. 1-65.

Sears, J.W. and Price, R.A.
1978: The Siberian connection: A case for Precambrian separation of the North American and Siberian cratons; Geology, v. 6, p. 267-270.

Shih-Fan, L.
1981: Sinian glacial deposits of Guizhou Province, China; in Earth's pre-Pleistocene Glacial Record, M.J. Hambrey and W.B. Harland (ed.), Cambridge University, Cambridge, p. 414-423.

Shiki, T. and Misawa, Y.
1982: Forearc geological structure of the Japanese Islands; in Trench-Forearc Geology, J.K. Leyett (ed.), Geological Society of London Special Publication No. 10, p. 63-73.

Silberling, N.J. and Jones, D.L.
1984: Lithotectonic terrane maps of the North American Cordillera; United States Geological Survey, Open-File Report 84-523.

Silberling, N.J. and Roberts, R.J.
1962: Pre-Tertiary stratigraphy and structure of northwestern Nevada; Geological Society of America, Special Paper 72, 58 p.

Simony, P.S., Ghent, E.D., Craw, D., Mitchell, W., Robbins, D.B.
1980: Structural and metamorphic evolution of northeast flank of Shuswap complex, southern Canoe River area, British Columbia; in Cordilleran Metamorphic Core Complexes, M.D. Crittenden, Jr., P.J. Coney and G.H. Davis (ed.), Geological Society of America, Memoir 153, p. 445-461.

Snyder, W.S. and Brueckner, H.K.
1983: Tectonic evolution of the Golconda Allochthon, Nevada: Problems and perspectives; in Pre-Jurassic Rocks in Western North American Suspect Terranes, C.H. Stevens (ed.), Pacific Section Society of Economic Paleontologists and Mineralogists, Los Angeles, p. 103-123.

Speed, R.C.
1984: Paleozoic and Mesozoic continental margin collision zone features, Mina to Candelaria, NV, Traverse; in Western Geological Excursions, Volume 4, J. Lintz (ed.), The Geological Society of America and the Department of Geological Sciences, Mackay School of Mines, University of Nevada, Reno, p. 66-80.

Stewart, J.H.
1972: Initial deposits of the Cordilleran geosyncline: Evidence of a late Precambrian (850 m.y.) continental separation; Geological Society of America Bulletin, v. 83, p. 1345-1360.

Stewart, J.H., MacMillan, J.R., Nichols, K.M., and Stevens, C.H.
1977: Deep water upper Paleozoic rocks in north-central Nevada - a study of the type area of the Havallah Formation; in Paleozoic Paleogeography of the western United States, J.H. Stewart, C.H. Stevens and A.E. Fritscho (ed.), Pacific Coast Paleogeography Symposium I, Los Angeles, Pacific Section Society of Economic Paleontologists and Mineralogists, p. 337-347.

Struik, L.C. and Orchard, M.J.
1985: Later Paleozoic conodonts from ribbon chert delineate imbricate thrusts within the Antler Formation of the Slide Mountain Terrane, central British Columbia; Geology, v. 13, p. 794-798.

Tempelman-Kluit, D.J. and Parkinson, D.
1986: Extension across the Eocene Okanagan crustal shear in southern British Columbia; Geology, v. 14, p. 318-321.

Von der Borch, C.C.
1980: Evolution of late Proterozoic to early Paleozoic Adelaide foldbelt, Australia: Comparisons with post-Permian rifts and passive margins; Tectonophysics, v. 70, p. 115-134.

Welland, M.J.P. and Mitchell, A.H.G.
1977: Emplacement of the Oman ophiolite: A mechanism related to subduction and collision; Geological Society of America Bulletin, v. 88, p. 1081-1088.

Williams, H. and Hatcher, R.D. Jr.
1982: Suspect terranes and accretionary history of the Appalachian orogen; Geology, v. 10, p. 530-536.

Woodsworth, G.J., Crawford, M.L., and Hollister, L.S.
1983: Metamorphism and structure of the Coast Plutonic Complex and adjacent belts, Prince Rupert and Terrace areas, British Columbia; Field Trip No. 14, Annual Meeting, Geological Association of Canada, Mineralogical Association of Canada and Canadian Geophysical Union, 41 p.

Yongji, M.
1981: Luoguan Tillite of the Sinian System of China; in Earth's pre-Pleistocene Glacial Record, M.J. Hambrey and S.B. Harland (ed.), Cambridge University Press, Cambridge, p. 402-413.

Yorath, C.J. and Chase, R.L.
1981: Tectonic history of the Queen Charlotte Islands and adjacent areas - a model; Canadian Journal of Earth Sciences, v. 18, p. 1717-1739.

Yorath, C.J. and Cook, D.G.
1981: Cretaceous and Tertiary stratigraphy and paleogeography, northern interior plains, District of Mackenzie; Geological Survey of Canada, Memoir 398.

Yorath, C.J. and Hyndman, R.D.
1983: Subsidence and thermal history of Queen Charlotte Basin; Canadian Journal of Earth Sciences, v. 20, p. 135-139.

Authors' addresses

H. Gabrielse
Cordilleran Division
Geological Survey of Canada
100 West Pender Street
Vancouver, British Columbia
V6B 1R8

C.J. Yorath
Pacific Geoscience Centre
9860 West Saanich Road
P.O. Box 6000
Sidney, British Columbia
V8L 4B2

Printed in Canada

Chapter 19

REGIONAL METALLOGENY

Summary
Part A. METALLOGENY OF THE CRATON AND DISPLACED, PERICRATONIC AND ACCRETED TERRANES
  Introduction
  Ancestral North American Miogeocline
    North American(?) Basement: Monashee Complex - pre-accretionary deposits
    Purcell-Wernecke Assemblage - pre-accretionary deposits
      Purcell Supergroup
      Wernecke Supergroup
      Muskwa Ranges Succession
      Mackenzie Mountains Assemblage - pre-accretionary deposits
      Windermere Supergroup - pre-accretionary deposits
      Cambrian to Middle Devonian passive margin - pre-accretionary deposits
        Carbonate-hosted Zn,Pb deposits
        Sedimentary-exhalative Zn,Pb deposits
        Skarn and replacement post-accretionary deposits
    Middle Devonian to Mississippian clastic wedge
  Displaced continental margin
    Cassiar Terrane - pre- and post-accretionary deposits
    Arctic Alaska Terrane
    Nisling Terrane
  Pericratonic terranes
    Kootenay Terrane
      Southeastern British Columbia
    Kootenay Arc
    Shuswap
      Barkerville Subterrane
      Nisutlin Subterrane
  Accreted terranes
    Intermontane Superterrane
      Slide Mountain Terrane
      Dorsey Terrane
      Quesnellia
      Cache Creek Terrane
      Stikinia
      Bridge River, Cadwallader, and Methow terranes

    Coast Plutonic Complex
      Pre-accretionary(?) deposits
        Massive sulphides in overlap assemblages
        Deposits related to accretionary plutonism
        Post-accretionary deposits
    Insular Superterrane
      Alexander Terrane
      Southern Wrangellia
        Pre-accretionary deposits
      Northern Wrangellia
      Pacific Rim and Crescent terranes

Part B. MINERAL DEPOSIT MODELS
  Precious metals (Au, Ag, platinum-group elements)
    Deposit types and their distribution
    Tectonic evolution and age of gold mineralization
  Volcanogenic massive sulphide deposits
  Porphyry deposits
  Skarn deposits
  Sedimentary exhalative deposits
  Conclusions
  References

Chapter 19

REGIONAL METALLOGENY

K.M. Dawson, A. Panteleyev, A. Sutherland Brown and G.J. Woodsworth

SUMMARY

The Canadian Cordillera is a region of great geological and metallogenic diversity. Just as each Cordilleran terrane preserves a stratigraphic record different from those of neighbouring terranes, characteristic suites of mineral deposits, as integral parts of their host terranes, reflect fundamental differences in their depositional environments. The miogeocline and displaced equivalents in the eastern Cordillera, as well as each of the terranes comprising the accreted collage of the western Cordillera, possess unique lithotectonic characteristics that are reflected in the types of mineral deposits they contain.

Predominantly stratiform deposits of Zn, Pb, Cu, Ba, and Fe and skarn deposits of W, Zn, Pb, Mo, and Sn are hosted by layered sedimentary strata of the ancestral North American miogeocline. The similar types of mineral deposits of displaced (Cassiar) and/or deformed (Kootenay, Nisling) continental margin terranes support their cratonal linkage.

Stikinia and Quesnellia, which together constitute the bulk of the Intermontane Superterrane, host a suite of mineral deposits typical of their predominantly calc-alkalic volcanic-arc composition: abundant porphyry Cu,Mo deposits, Cu, Zn volcanogenic massive sulphides, Cu and Au skarns, and Au,Ag veins. On the other hand, the ophiolitic Cache Creek and Slide Mountain terranes of the Intermontane Superterrane display distinctive kinds of mineral deposits typical of their oceanic origin: magmatic Cu,Ni, volcanogenic Cu,Zn and mesothermal Au veins, in addition to ultramafic pluton-related asbestos, jade, Cr and platinum group element (PGE) deposits.

The dominantly arc volcanic character of the diverse terranes of the Coast Belt is reflected in their metallogeny: volcanogenic Cu,Zn, porphyry Cu,Mo, mesothermal Au, and epithermal Au,Ag,Sb,Hg veins. The lithologies of both Wrangellia and the Alexander Terrane of the Insular Superterrane represent several depositional settings and host commensurate suites of mineral deposits including: platformal sediments with stratiform gypsum and barite; oceanic arc volcanics with volcanogenic Zn,Cu,Ag,Pb,Ba; porphyry Cu,Mo and Au veins; oceanic rift volcanics with volcanogenic Cu,Co,Ag, Mn, and magmatic Ni,Cu,PGE; and carbonate strata hosting Fe, Cu, and Au skarns.

In addition to the fundamental relationships between terrane crustal types and differences in origin, type, and distribution of contained mineral deposits, these mineral deposits can be classified further in relation to the time of terrane accretion as pre-accretionary, accretionary, and post-accretionary. Deposits formed prior to the amalgamation of the host terrane with others, and prior to subsequent accretion to the North American craton may have undergone significant modification and redistribution due to accretionary metamorphic, plutonic, and hydrothermal processes. Comparison with examples in the western United States, Mexico, and Australia demonstrates that not only are certain syngenetic pre-accretionary deposits uniquely associated with specific crustal types of terranes, e.g. manganiferous chert in oceanic terranes, and Kuroko-type volcanogenic massive sulphides in calc-alkaline island arcs, but also accretionary and post-accretionary deposits, regardless of age, correlate strongly with host and/or basement lithology.

Dawson, K. M., Panteleyev, A., Sutherland Brown, A., and Woodsworth, G. J.
1991: Regional metallogeny, Chapter 19 in Geology of the Cordilleran Orogen in Canada, H. Gabrielse and C. J. Yorath (ed.); Geological Survey of Canada, Geology of Canada, no. 4, p. 707-768 (also Geological Society of America, The Geology of North America, v. G-2)

CHAPTER 19

PART A. METALLOGENY OF THE CRATON AND DISPLACED, PERICRATONIC AND ACCRETED TERRANES

Introduction

The Canadian Cordillera is extensively mineralized with a variety of metallic, nonmetallic, and fossil fuel deposits. Exploitation of mineral resources began with the discovery of coal in 1835, followed by lode gold in 1851, and placer gold in 1857. Current annual production of mineral and fossil fuel commodities is close to 3500 million dollars.

The first comprehensive description of Cordilleran metallogeny was by Sutherland Brown et al. (1971). At that time, in anticipation of breakthroughs in the newly emerging plate tectonics concepts, Cordilleran deposits were described without elaborate tectonic synthesis in a simple, geosynclinal-miogeoclinal context. With the exception of the work of Ney (1966) no attempts were made to apply older schemes of geosynclinal evolution such as those devised in the Soviet Union and made popular by Bilibin (1955). Elsewhere, such as in the Canadian Appalachians, metallogenic interpretations based on Soviet-style vertical tectonic schemes were being applied with considerable success (McCartney and Potter, 1962; McCartney, 1965).

The widespread espousal of plate tectonic theory resulted in a proliferation of papers in the early 1970s relating mineral deposits and plate tectonism (Guild, 1971; Pereira and Dixon, 1971; Sawkins, 1972; Strong, 1976; Mitchell and Garson, 1976). Inevitably, metallogeny of the Canadian Cordillera also was considered within a plate tectonics context (Wolfhard and Ney, 1976; Sinclair et al., 1978; Tempelman-Kluit, 1981). Plate tectonics and metallogeny are inextricably interwoven, although plate tectonic theory has not been the panacea for all the shortcomings of earlier metallogenic analysis (Sangster, 1979). Nonetheless, terrane analysis, within a plate tectonic context, provides a powerful framework in which to consider mineral deposits (Sawkins, 1972, 1984). In this chapter, metallogenesis is considered within an implicit plate tectonic framework in which tectonic processes are expressed as pre-accretionary, accretionary, and post-accretionary events.

Currently over 10,400 mineral occurrences are documented in British Columbia in a mineral inventory file (MINFILE); 1367 additional deposits are listed in the 1987 Canmindex file for the Yukon Territory. At least 1400 of these mineral deposits, the locations of which are shown on GSC Map 1513A (Dawson, 1984), are economically significant. Their distribution clearly is not random; it is readily evident that concentrations of deposits and metals occur in certain geomorphological belts, terranes, and ages of host rocks (Sutherland Brown et al., 1971; Fig. 19.1, in pocket). A contour map of the abundance of deposits (Fig. 19.2) reveals that most of the western Cordillera is extensively mineralized. Clearly the south has been more intensively explored than the north and therefore has a greater number of known deposits. The areas with less than average abundance are: the Foreland Belt, the high-grade metamorphic core of the Coast Belt, the youngest continental volcanics (mainly plateau basalts of the Intermontane Belt), and the Mesozoic to early Cenozoic clastic basins. The latter areas are most important for their coal deposits (see Chapter 20).

Analyses of mineral deposit frequency distribution patterns (Fig. 19.3) reveal the following: 1) relative concentrations of given deposit types differ widely among the tectonic belts; 2) within the same belt metal deposits are not uniformly distributed, and most occur near the margins of the belts; 3) the greatest diversity of metal occurrences is in the Omineca Belt. Characteristic metals (Sinclair et al., 1978) in the belts, from west to east, are Cu and Au in the Insular Belt, Au in the Coast Belt, Mo and Cu in the Intermontane Belt, Pb, Ag, Zn, Au in the Omineca Belt, and Pb, Zn, Ag, W, Ba in the Foreland Belt.

Genetic types of deposits also display zoning, as shown by major occurrences in Figure 19.3 and Sutherland Brown (1976b). Each tectonic belt has one or more dominant type of deposit. Porphyry deposits are most abundant in the

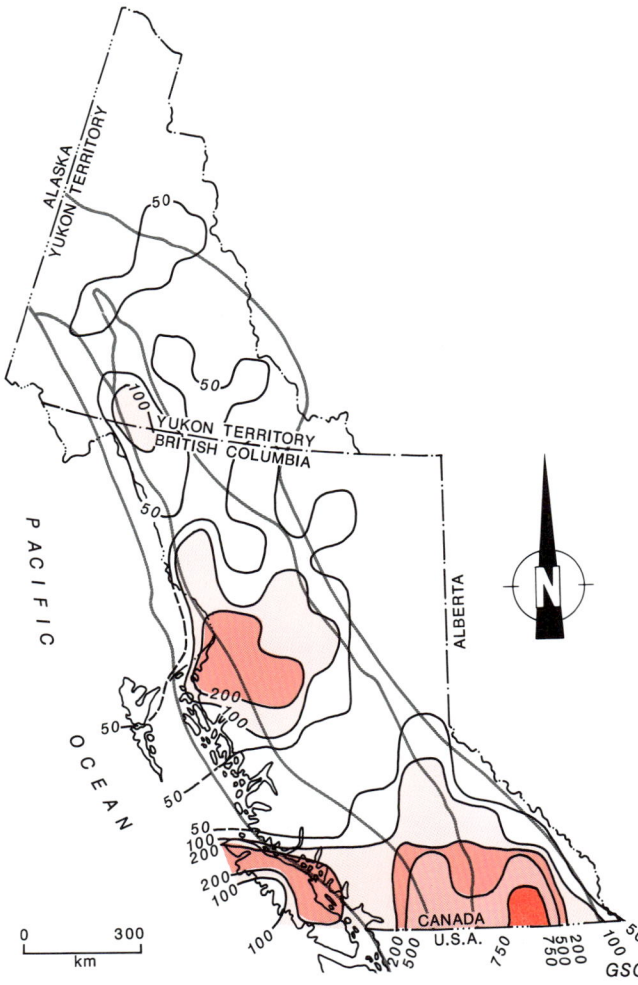

Figure 19.2. Distribution of mineral occurrences. Contours of the number of mineral inventory occurrences in each two-by-one degree 1:250 000 map area.

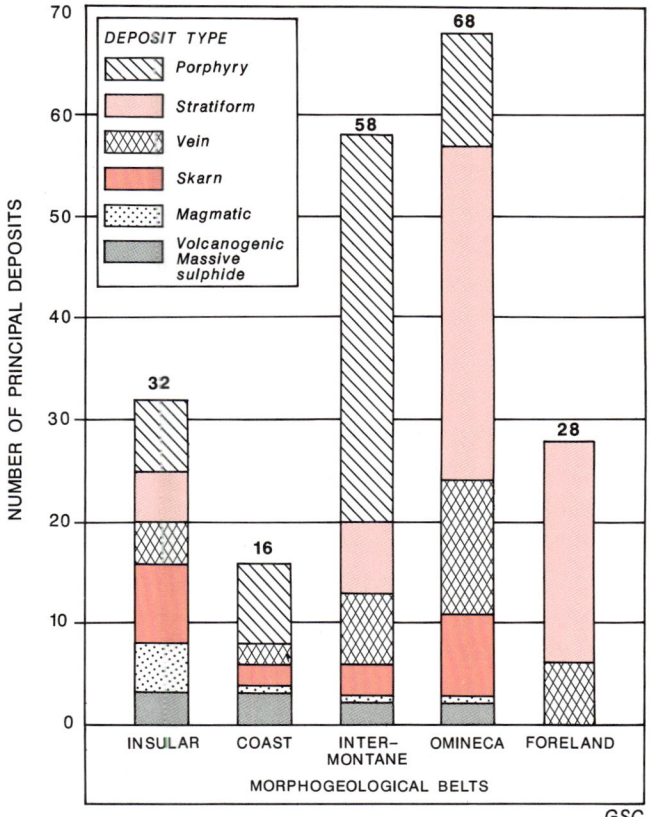

Figure 19.3. Distribution of classes of deposits by morphogeological belts. Medium and large sized deposits from Dawson (1984), Geological Survey of Canada, Map 1513A.

Intermontane Belt, skarns are important in the Insular and Omineca belts, and stratiform deposits in the Omineca and Foreland belts. In addition, skarn deposits display belt-specific metal zoning within the class itself with Fe; Fe, Cu; and Cu skarn deposits in the Insular Belt, Cu; Cu, Fe; and Au in the Intermontane Belt, and W, Mo, Cu, and minor Sn in the Omineca Belt (Tempelman-Kluit, 1981).

Consideration of the ages of mineralized host rocks from which ore has been produced reveals that there are preferred lithologies and periods of ore deposition. Distribution diagrams of ore production according to age of host rocks (Fig. 19.4) show intrusive rocks to be the most productive. This is because they host the large, low-grade bulk-mineable Cu,Mo porphyry deposits. When other deposits with higher, roughly equivalent unit values per-mined-tonne are considered, the most favourable host lithologies for mineralization are Triassic and Jurassic volcanic rocks. Proterozoic rocks appear to be productive, but this is greatly influenced by single-source production from the huge Sullivan Pb,Zn deposit. Paleozoic, Cretaceous and Cenozoic rocks, together with intrusive rocks hosting deposits other than porphyry-type ores have been moderately productive. Jurassic and younger sedimentary rocks have not been productive.

In this chapter the discussion of metallogeny is treated in two parts. Part A emphasizes the relationship of mineral deposits to terranes, and, where possible, refers to them in terms of pre-, syn- and post-accretionary, whereas Part B deals with models of genetically related types of mineral deposits.

Canadian Cordilleran mineral deposits herein are classified first by host terrane, and second by accretionary history. The former places the deposit in the appropriate lithotectonic context, the latter classification tends to juxtapose diverse deposits whose interrelationships may not have been otherwise examined.

Ancestral North American Miogeocline
North American (?) basement: Monashee Complex - pre-accretionary deposits

The Monashee Complex (Fig. 19.1, in pocket), which forms part of the Shuswap Metamorphic Complex in southeastern British Columbia, consists of core paragneiss dated at 2.8 and 2.2 Ga, that has been intruded by granitoid plutons dated at 2.1 and 1.96 Ga and which is overlain by heterogeneous paragneisses intruded by 750 or 770 Ma syenite (Okulitch, 1984; Parrish, Chapter 4). The upper paragneiss succession hosts several stratabound Zn,Pb prospects, which range from less than 1 million to 6.5 million tonnes of ore. Included are the Cottonbelt, River Jordan, Ruddock Creek, Colby, and Big Ledge prospects, which contain an average of 10% combined Zn + Pb as thin extensive sulphide layers folded and metamorphosed along with their predominantly calcareous and schistose host rocks (Høy, 1982b; Fyles, 1970).

Most of the Monashee Complex Zn,Pb prospects have been classified as 'sediment-hosted exhalative' by Sangster (1986), as have several significant producing mines and former producers hosted by similar shallow-water successions of carbonates and calcareous quartzites in Kootenay Arc and Purcell Anticlinorium that flank the Monashee Complex on the east and southeast. Although correlation of the Monashee paragneisses with rocks of either the Paleozoic Kootenay Arc or the Proterozoic Windermere and Purcell supergroups is possible (Okulitch, 1984),

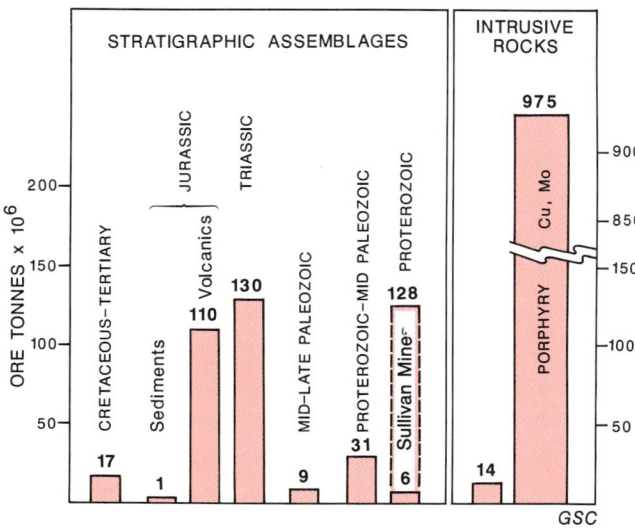

Figure 19.4. Ore production (tonnes) from various time-stratigraphic units and intrusive rocks, modified from McKechnie (1966).

definitive evidence is lacking. Correlation of Monashee paragneisses and contained stratabound Zn,Pb deposits with potentially equivalent rocks and sedimentary exhalative deposits to the east also remains speculative.

Nepheline syenite gneiss and locally associated carbonatite form concordant layers within the mantling paragneiss of the Monashee Complex at Frenchman Cap Dome. A formerly producing molybdenum mine with 1970-1974 production of about 190 000 tonnes of ore grading 1.82% Mo is associated with aplitic and pegmatitic segregations of the nepheline syenite gneiss at Mount Copeland (McMillan, 1973). Small amounts of pyrochlore and columbite-tantalite occur in carbonatites associated with the syenite gneiss (Høy and Pell, 1986; McMillan, 1973).

The close association of the Mo; Nb,Ta; and possibly Zn,Pb deposits with alkalic intrusions dated at 773 Ma by Okulitch et al. (1981) supports a genetic relationship of the deposits to a widespread rifting event contemporaneous with the onset of Windermere Supergroup deposition in the northern Cordillera (Armstrong et al., 1982; Jefferson and Parrish, 1989), and throughout the length of the evolving margin of western North America (Eisbacher, 1981). Mineral deposition in the Monashee Complex is interpreted to be rift-related, therefore pre-accretionary and probably synchronous with the separation of this segment of Precambrian crystalline basement and mantling Proterozoic sediments from the North American cratonal margin in Windermere time.

Purcell-Wernecke Assemblage - pre-accretionary deposits

Purcell Supergroup

Sedimentary rocks of the Middle Proterozoic Purcell Supergroup in the type area in southeastern British Columbia comprise a predominantly passive-margin(?) depositional sequence of fine grained basinal clastic rocks at least 11 km thick (Reesor, 1958; McMechan, 1980, 1981) in the west. They thin eastward to platformal sediments and are overlain by shallow marine and nonmarine rocks (Aitken and McMechan, Chapter 5). Formation of the continental margin and subsequent deposition of a prograding wedge of Purcell (Belt) sediments was attributed to a major rifting event in the Middle Proterozoic by Monger et al. (1972).

In the Middle Proterozoic lower and middle Aldridge Formation, facies changes, turbidite deposition, local sedimentary thickening and the accumulation of intraformational conglomerates were attributed by Høy (1982a) to rift-related synsedimentary faulting. Spatially related sulphide deposition and tourmaline alteration indicate that the principal northeast-trending faults and related structures served as conduits for important sedimentary exhalative mineralization of the Sullivan, North Star, and Kootenay King Pb,Zn,Ag deposits (Høy, 1982b; Hamilton et al., 1982; Fig. 19.1A, in pocket).

The Sullivan Mine, one of the world's largest base-metal deposits, has produced more than 112 million tonnes of ore and contains about 50 million tonnes of reserves of 6.1% Zn, 4.9% Pb and 37 g/t Ag. Hosted conformably by siltstones in a tectonically controlled sub-basin near the top of the lower Aldridge Formation, the 2000 x 1600 m orebody is primarily a massive synsedimentary lens of pyrrhotite-rich sulphides, up to 100 m thick in its western half. It grades upward and laterally eastward to laminated pyrrhotite-sphalerite-galena, and distally to delicately interlayered sulphide-siliciclastic beds. The orebody footwall, primarily intraformational conglomerate with lesser siltstone, wacke, and bedded sulphides, is cut by discordant breccia, veined by sulphides, and altered extensively in the west to a funnel-shaped zone of tourmaline. The latter is interpreted as a pre-ore alteration assemblage centred over the hydrothermal conduits. Local, discordant zones of albite, chlorite, pyrite, and carbonate in the footwall and a similar extensive zone in the hanging wall are post-ore, reflecting a progressive compositional change in exhalative hydrothermal fluids from early B, Fe, Mg-rich through Fe, Zn, Pb-rich ore stage to late Na-rich (Hamilton et al., 1982, 1983).

In the Clarke Range of southeastern British Columbia and southwestern Alberta predominantly clastic rocks of the Appekuny, Grinnell, and lower Siyeh formations of the lower Purcell Supergroup contain numerous minor occurrences of stratabound Cu(Ag) and less common Zn,Pb(Cu) (Fig. 19.1A, in pocket). Copper sulphide occurrences are most abundant in the Grinnell Formation, the stratigraphic equivalent of the Revett Formation in Montana (Price, 1964), which hosts the important Spar Lake Cu(Ag) deposit (Harrison, 1972). Erratically disseminated chalcocite, bornite, and less common copper sulphides, which typically occur in relatively permeable white quartz-arenite members of predominantly red argillite beds, were interpreted by Kirkham (1974) to represent late diagenetic mineralization of eolian beds in a sabkha sequence. Collins and Smith (1977), however, considered it to be the product of cyclically controlled redox conditions during short-lived lacustrine to fluvial episodes. Morton et al. (1974) related the source of the metal-bearing fluids to exhalative activity controlled by faults. Some local enrichment of stratabound Cu(Ag) mineralization has been ascribed to diorite sills contemporaneous with the Purcell lavas, and to hydrothermal activity adjacent to normal faults (Morton et al., 1974).

Widespread low-grade copper sulphide and malachite occurrences in a predominantly shallow-water clastic Middle Proterozoic sequence at Cap Mountain in the southern Franklin Mountains, Northwest Territories, were noted by Aitken et al. (1973) to resemble stratabound Cu occurrences in the Belt Supergroup of Montana. The Cap Mountain succession bears no lithological relationship to nearby strata of the Mackenzie Mountain Supergroup (see Chapter 5) but can be lithostratigraphically correlated with the Hornby Bay-Dismal Lakes succession of the Coppermine Homocline (Meijer-Drees, 1975) and is correlated temporally with older clastic Purcell Supergroup rocks in the southern Cordillera (Aitken and McMechan, Chapter 5). The aulacogen depositional environment proposed for the Hornby basin (Kerans et al., 1981; Dewey and Burke, 1973), in which an elongate rift opens westward to the continental margin, is similar to the setting of a structurally controlled Purcell embayment in the continental margin supported by Aitken and McMechan (Chapter 5). The markedly similar copper mineral occurrences of these two widely separated yet tectonically similar Proterozoic sequences supports a common, rift-related exhalative origin of the copper deposits.

Wernecke Supergroup

The Middle Proterozoic Wernecke Supergroup of the Wernecke and Ogilvie mountains of northern Yukon Territory is at least 14 km thick and consists dominantly of fine grained terrigenous clastics which grade upwards to carbonate strata (Delaney, 1978, 1981). It has been broadly correlated with the Purcell Supergroup (Young et al., 1979) and is considered older than 1.2 Ga on the basis of crosscutting breccia pipes, isotopically dated by Archer and Schmidt (1978) and Archer et al. (1986). Bell's (1982) proposal that the Wernecke Supergroup may be a displaced terrane incorporating continental margin sediments dextrally displaced in Late Proterozoic (Windermere) time is further expanded by Aitken and McMechan (Chapter 5) as the distal part of the Hornby Bay-Dismal Lakes miogeoclinal wedge.

The most significant mineral deposits hosted by rocks of the Wernecke Supergroup (Fig. 19.1A) are the more than 40 U occurrences, mainly brannerite, $(U,Ca,Fe)_3Ti_5O_{16}$, accompanied by several assemblages of Cu, Fe, U, Ba, Co, and Au minerals. They occur in narrow discontinuous vein-zones peripheral to metre- to kilometre-scale monolithic and heterolithic breccia bodies that commonly cut the lower parts of the sedimentary sequence (Archer and Schmidt, 1978; Bell, 1978, 1982). In addition, minor deposits consist of Pb and Zn sulphides localized in the carbonate matrices of breccias that cut dolomite of the upper Wernecke Supergroup in the western Ogilvie Mountains (Abbott, 1987). Possible origins of the breccias include fault crushing, hydrothermal stoping, gas streaming, and evaporite solution-collapse (Delaney, 1981; Bell, 1982). The latter mechanism may involve intrusion and collapse of evaporite diapirs (Bell, 1989) similar to that ascribed to breccias which host mainly Cu and Fe deposits in the Late Proterozoic Adelaidean geosyncline of South Australia (Dalgarno and Johnson, 1968). In addition, there may be an association with the large Olympic Dam Fe,Cu,U,Au breccia deposit adjacent to the Adelaidean geosyncline. The spatial association of breccias to major faults, lineaments, and mafic dykes, as well as the mesothermal character of the metallic and alteration mineral assemblages (e.g. Na and K-feldspar, hematite-magnetite, silica, chlorite, and carbonate), however, support a mineral origin by hydrothermal fluids of deep, possibly basement circulation. Accordingly, a pre-displacement age of U mineralization is proposed, related to a synsedimentary rifting event which occurred at about 1.2 Ga, prior to dextral displacement of the Wernecke Supergroup in Windermere time. Local remobilization of uranium within the occurrences took place during later deformation.

Also located near the top of the Wernecke Supergroup but possibly unrelated to breccias, is the Hart River massive sulphide deposit hosted by black argillite 50 m below pillow basalt flows (Abbott, 1987). It contains proven reserves of 525 000 tonnes grading 3.6% Zn, 1.45% Cu, 0.9% Pb, 46 g/t Ag, and 1.3 g/t Au (Morin, 1978). Both Hart River (1.24-1.28 Ga; Morin, 1978) and Sullivan (1.43 Ga; LeCouteur, 1979) are Middle Proterozoic products of exhalative activity within distal sedimentary facies possibly related to faulting (Aitken and McMechan, Chapter 5; Høy, 1982a).

Muskwa Ranges Succession

The Muskwa Ranges succession of northwestern British Columbia comprises a lower quartzite-carbonate succession of platformal character, 3.5 km thick, and an upper shaly flysch succession 2.5 km thick, which together are tentatively correlated with the Purcell (Belt) Supergroup (Bell, 1968; Aitken and McMechan, Chapter 5). Significant copper vein deposits are hosted by basinal clastic and impure carbonate rocks of the Aida and Gataga formations of the upper Muskwa Assemblage in the Racing River-Gataga River region (Taylor and Stott, 1973).

Twelve chalcopyrite-pyrite-quartz-ankerite vein deposits are known in the Racing River-Gataga River region, of which only the Magnum mine of Churchill Copper Corp. Ltd. (Fig. 19.1A) has produced (490 000 tonnes of about 3.3% Cu in 1971-74). The steeply dipping Magnum vein system is a zone 100 m wide which trends northeastward perpendicular to axes of overturned folds of the Aida Formation. The vein was intruded by a swarm of post-mineral diabase dykes and sills. Overlying Cambrian basal conglomerate contains clasts of mineralized vein material and diabase (Preto and Tidsbury, 1971; V.A. Preto, pers. comm., 1984; Carr, 1971).

The northeastward trending diabase dyke swarm crosscuts folds in the Muskwa strata, yet no diabase dykes are known to occur in Windermere strata to the west of the region. A pre-Windermere, late or post-Purcell age of dyke emplacement and mineralization is therefore indicated.

Mackenzie Mountains Assemblage - pre-accretionary deposits

An extensive pericratonic platformal succession composed of at least 4000 m of predominantly shallow water quartzite, shale, and partly stromatolitic carbonates (Aitken et al., 1978a; Gabrielse et al., 1973) was deposited in the Mackenzie Mountains region between about 1.2 and 0.8 Ga. The Mackenzie Mountains Supergroup, as originally proposed by Jefferson (1978) and Young et al. (1979), included both the Redstone River and Coppercap formations of the Rapitan Assemblage, i.e. the "copper cycle" of Aitken (1981) or Coates Lake Group of Jefferson and Ruelle (1986). As presently defined by Aitken (1981) this supergroup includes only the four unconformably underlying units, in ascending order Unit H_1, the Tsetzotene Formation, and Katherine, and Little Dal groups, and is constrained by two zircon dates of 1100-1175 Ma for basement and 778 Ma for diorite intruding the Little Dal Group (Jefferson and Parrish, 1989). Proterozoic rocks of this age are not known in the central and southern Canadian Cordillera, but correlation of the lower formations of the Mackenzie Mountains Supergroup with the Rae Group of the Coppermine Homocline, Northwest Territories, has been established by Young (1977) and Aitken et al. (1978b).

The most significant mineral deposits known in the Mackenzie Mountains Supergroup are the carbonate-hosted (Mississippi Valley-type) Zn,Pb deposits of the Gayna River region (Fig. 19.1A). The district comprises some 18 deposits and more than 100 occurrences. Some deposits exceed 1 million tonnes of up to 10% combined Zn + Pb. The entire camp has estimated potential resources of more than 50 million tonnes of greater than 5% combined Zn + Pb (Hewton, 1982).

Most Gayna River deposits are hosted by a dolostone member of the Grainstone Formation of the lower Little Dal Group (Aitken, 1981). They occur either as disseminations of pale sphalerite and lesser amounts of galena in the matrices and clasts of primary slump breccias developed over the flanks of large stromatolitic reefs or as generally richer concentrations of coloured sphalerite, sparry dolomite and calcite with lesser amounts of galena, pyrite, and barite as matrix fillings of secondary solution-collapse and fault-related crackle breccias (Hewton, 1982; Hardy, 1979). A major rifting event, which is suggested by the presence of widespread synsedimentary faults (Eisbacher, 1981; Jefferson and Parrish, 1989), may have begun during upper Little Dal Group sedimentation. Thus ore-controlling reefs and sedimentary breccias in the lower Little Dal Group at Gayna River may have formed in response to synsedimentary faulting at an early, incipient rifting stage of this event (ibid). Facies trends shown by Aitken (1981, Fig. 3.14) are parallel to dykes and normal faults.

Disseminated copper sulphides occur in minor abundance in several other units of the Mackenzie Mountains Supergroup: at Unit H_1 contacts with the Tsetzotene Formation, in carbonate units in the Katherine Group, and at the lower and upper contacts of the Gypsum Formation of the Little Dal Group (Jefferson, 1978). Several deposits and occurrences of carbonate-hosted Zn,Pb and Ag,Pb,Zn are known from dolostone units of the possibly correlative Pinguicula Group in the Wernecke Mountains.

Windermere Supergroup - pre-accretionary deposits

The dominantly clastic rocks of the Upper Proterozoic Windermere Supergroup, exposed almost continuously from Alaska, through Yukon Territory and British Columbia southward into California (Fig. 19.1A) contain the depositional record of a major rifting event along the western continental margin of North America (Stewart, 1972; Eisbacher, 1981; Young, 1982; Gabrielse and Campbell, Chapter 6). Lowest and easternmost units, i.e. the Rapitan Assemblage, best reflect a rift environment with rapid facies and thickness changes and a suite of rift-related igneous intrusions and extrusions whose isotopic ages centre on 770 Ma. Diamictite, in part glaciogenic, occurs at several localities and stratigraphic levels, notable at two well defined horizons in eastern Mackenzie Mountains, one of which also contains jaspilite-hematite iron-formation.

The lower Windermere Supergroup is characteristically a thick monotonous sequence of gritty feldspathic sandstone which gives way in upper and western formations to more lithologically diverse, carbonate-rich strata, representing a prograding wedge of sediments along the rifted cratonal margin (Gabrielse and Campbell, Chapter 6). Two cycles of mainly clastic strata were recognized by Pell and Simony (1987) in south-central British Columbia, whereas Eisbacher (1985) recognized three major shoaling-upwards cycles in the Mackenzie Mountains. Throughout the eastern Cordillera the Windermere Supergroup is unconformably overlain by Lower Cambrian sandstone, indicating a minimum age of about 570 Ma.

Significant stratabound copper deposits are hosted by rocks of the Coates Lake Group, an unconformity-bounded rift assemblage which occurs mainly in six synsedimentary fault-controlled depositional embayments over a 300 km long arcuate belt marking the eastern limit of Upper Proterozoic strata in the Mackenzie Mountains (Jefferson and Ruelle, 1986).

Basalt at the top of the Little Dal Formation reflects the onset of a major extensional tectonic regime. It is overlain unconformably by basalt, conglomerate and arenite which grade upward through clastic and carbonate strata to evaporites, carbonate conglomerate, and red mudstones of the Redstone River Formation (Gabrielse et al., 1973). These restricted units mark local uplift prior to subsidence along a fault-controlled Windermere basin margin (Eisbacher, 1985). An interval of evaporitic mudstones and algal laminites, transitional from underlying redbeds to basinal turbiditic limestone sequences of the Coppercap Formation (Gabrielse et al., 1973), includes up to eight repetitive sabkha sequences at Coates Lake (Fig. 19.1A) containing Cu, Fe, and less abundant Pb and Zn sulphides disseminated in algal limestone beds (Jefferson and Ruelle, 1986). The lowest copper-bearing bed in this transition zone, the only one that approaches economic proportions, averages 1 m thick and contains 37 million tonnes of 3.92% Cu and 11.3 g/t Ag within an area of about 12 km^2 (Ruelle, 1982). Sulphide assemblages in successively higher and more northeasterly mineralized beds in the transition zone are systematically depleted in Cu and enriched in Fe, Pb, and Zn (Jefferson and Ruelle, 1986).

The origin of stratabound copper deposits at Coates Lake has been addressed by several authors, as summarized by Jefferson and Ruelle (1986). The relationship of the copper-bearing sequence to extensional faulting was first proposed by Coates (1964), then expanded upon by Eisbacher (1977, 1981, 1985) who recognized the relationship of lithofacies to north- and northeast-trending faults. The depositional environment was related finally to transtensional faulting by Jefferson and Ruelle (1986). A relationship of the copper deposits to extensional tectonism through a source of copper in the rift-related Little Dal Group basalts has been proposed by Coates (1964), Kirkham (1973), Jefferson (1978, 1983), Ruelle (1982), and Brown and Chartrand (1983). The channelling of copper-bearing hydrothermal fluids by extensional faults was proposed by Helmstaedt et al. (1979). Most authors support copper transport by unfocused, upward-moving oxidized, chloride-rich solutions which deposited Cu-rich minerals prior to Fe- or Pb, Zn-rich, minerals by local reduction in algal carbonate beds. Subsequent diagenetic mineralization was supported by Brown (1978), Chartrand (1981) and Jefferson (1983). The timing of copper mineralization at Coates Lake was, as a maximum, the age of the onset of the major rift event dated at 778 Ma by Jefferson and Parrish (1989) and, as a minimum, the time of deposition of unconformably overlying Sayunei Formation conglomerate in which copper-mineralized clasts of the Redstone River Formation have been reported (Helmstaedt et al., 1979).

One of the largest hematite-jaspilite iron deposits in North America, plus numerous regional occurrences of related iron-formation, are hosted by glaciogenic clastic rocks of the Rapitan Group in the Mackenzie Mountains. The predominant 'proglacial' siltstone facies of the Sayunei Formation, 700 m thick, is overlain by an 'ice marginal' diamictite complex of the 500 m thick Shezal Formation (Eisbacher, 1985). The top of the Sayunei Formation in the Thundercloud Range and near Mountain River, and the base of the Shezal Formation near Snake River in the southern and northern Mackenzie Mountains, are characterized by extensive hematite-jaspilite iron-formation (Yeo, 1981, 1986) correlated by Young (1982) and Payne and Allison (1981) with laminated hematite-jasper iron-formation in siltstone and diamictite of the Upper Tindir Group near Tatonduk River in eastern Alaska.

The Crest iron deposit at Snake River (Fig. 19.1A), in its richest central part, contains 5.6 billion tonnes of iron-formation averaging 47.2% Fe (Stuart, 1963), whereas the regional resource is estimated to exceed 18.6 billion tonnes (Yeo, 1986). The main zone, about 150 m thick and 150 m above the base of the Shezal Formation, comprises more than 10 iron-rich subzones up to 24 m thick of either banded jasper hematite, nodular jasper hematite, or laminated maroon cherty hematite, which interfinger with siltstone and mixtite (Yeo, 1986). Gross (1965) proposed that the hematite and silica were deposited as a chemical sediment from fumarolic waters discharged into a marine basin along synsedimentary faults. A system of high-angle synsedimentary extensional faults has been advocated as the principal control both of lithofacies in the Sayunei and Shezal formations (Eisbacher, 1981; Yeo, 1981) and of synchronous mafic volcanic and exhalative activity within the Rapitan Group in western Yukon Territory (Yeo, 1986; Upitis, 1966) and eastern Alaska (Young, 1982). Yeo (1986) modified the exhalation and seawater dilution model of Gross (1965) and proposed transport of the iron- and silica-rich brines by currents generated by the thermal gradients between cold glacial and warm hydrothermal waters.

Magnetite-rich iron-formation is located in the Misinchinka Group of the Windermere Supergroup in the Pine Pass area of central British Columbia. Two beds of magnetite, with some hematite, interbedded with argillite, greywacke and jasperoid chert, have been regionally metamorphosed and isoclinally folded. Reported reserves at the Falcon prospect are 9 million tonnes of 34% Fe (Hancock, 1988).

Cambrian to Middle Devonian Passive Margin - pre-accretionary deposits

The Cambrian to Middle Devonian miogeoclinal sedimentary prism within the Foreland Belt of the eastern Cordillera consists of three main tectonic assemblages: a Middle Cambrian clastic rift sequence, a Paleozoic carbonate-clastic passive continental margin, and thinner platformal sequences (Gabrielse et al., Chapter 2). Mineral deposits, principally stratiform and skarn deposits of Zn, Pb, Ag, and W, are concentrated within carbonate shelf and, to a lesser extent, clastic basinal facies of the Paleozoic continental margin assemblage.

Sedimentary lithofacies exerted a primary control upon the localization of pre-accretionary sediment-hosted mineral deposits in the Foreland Belt. Carbonate-hosted Mississippi Valley-type (MV) Zn,Pb deposits occur commonly at the tectonically unstable western margin of lower Paleozoic platformal carbonate successions. Post-accretionary skarn and replacement deposits are located in essentially the same tectonic setting, where relatively thick and pure limestone beds were intruded by Upper Mesozoic to lower Cenozoic felsic granitoid rocks. Sedimentary exhalative (sedex) Zn,Pb deposits are hosted commonly by fine cherty, calcareous and/or carbonaceous clastic successions within linear sub-basins whose development was controlled by synsedimentary extensional block faulting near the shelf-slope transition.

Carbonate-hosted Zn,Pb deposits

Minor occurrences of sphalerite and galena are a common feature of carbonate rocks of all ages in the North American miogeocline. Other than the stratiform deposits of the Kootenay Arc, and skarn, replacement, and vein deposits, which are considered separately, almost all Cordilleran carbonate-hosted Zn,Pb deposits are of the Mississippi-Valley (MV) type, i.e. epigenetic ore and gangue minerals fill open spaces in pre-existing host dolostone (Sangster, 1976). Sphalerite, galena, and pyrite with dolomite, quartz, calcite, barite, and lesser amounts of gypsum, fluorite, chalcopyrite, and pyrobitumen fill vugs, pores, burrows, various sedimentary and tectonic breccias, and fractures in the dolostone host. Deposits are associated commonly with breccia zones and secondary dolomite, unconformities, reefs, carbonate-shale facies changes, basement highs and karst terrain (Sangster, 1984).

The only carbonate-hosted Zn,Pb deposits in the Canadian miogeocline with significant production, the Monarch, Kicking Horse, and Silver Giant mines, were discovered in the Middle Cambrian Cathedral Formation in the southern Rocky Mountains in the late nineteenth century. Discovery of important economic deposits in the Pine Point district of the Northwest Territories in the mid-1960s preceded the significant discovery, in 1969, of several deposits in the northern Rocky Mountains centred on Robb Lake. Subsequent intense exploration resulted in hundreds of discoveries in the Mackenzie, Selwyn, Ogilvie, and Richardson mountains, notably Godlin Lakes in 1972, Goz Creek in 1973, and Gayna River in 1974. Overviews of the deposits, their geological settings, and exploration histories are given by Dawson (1975), Sangster and Lancaster (1976), Macqueen (1976), and Brock (1976).

The principal Robb Lake deposits, the largest of six camps in the northern Rockies Zn,Pb district (Macqueen, 1976), contain an aggregate of 5.5 million tonnes of 7.3% Zn+Pb (Northern Miner, 30/01/1975). The deposits are confined to secondary breccias of either solution or tectonic origin within folded dolostones of the Middle Devonian Stone Formation (Taylor et al., 1975). The Bear-Twit deposit, the largest of the Godlin Lakes camp in the southern Mackenzie Mountains, contains 5 to 7 million tonnes of 7% Zn+Pb (Brock, 1975). Mineralized fault-breccias occur in dolostones of the Whittaker, Delorme, and Camsell formations of Late Silurian to Early Devonian age (Dawson, 1975). Elsewhere in the Godlin Lakes district and throughout the Mackenzie Mountains, several significant deposits are hosted by orange-weathering ferroan dolostones of the Lower Cambrian Sekwi Formation (Fritz, Chapter 7; Dawson, 1975). The Goz Creek deposits in the northern Mackenzie Mountains include a core zone of 1.4 million tonnes of 10% Zn+Pb (Northern Miner, 30/01/1975) within a potential ore zone of 9 to 11 million tonnes of 7% Zn+Pb (Northern Miner, 25/12/1975). The two main deposits and numerous occurrences extend for 8 km along the eastward strike of the Risky Formation of the Upper Proterozoic Backbone Ranges Group (Fritz, Chapter 7) as both stratigraphically and tectonically controlled mineral zones within pervasively silicified sandy dolostone (Dawson,

1975). An alternate, sedimentary exhalative origin of some of the Goz Creek deposits is supported, in part, by stratiform deposit morphology (J.W. Lydon, pers. comm., 1988).

In summary, Mississippi-Valley Zn,Pb deposits in the southern Rocky Mountains are concentrated mainly near the shelf margins of Middle Cambrian platformal dolostones (Høy, 1982b), whereas deposits in the northern Rocky Mountains are most commonly localized near carbonate-shale facies changes in Lower and Middle Devonian dolostones (Macqueen, 1976). The numerous carbonate-hosted Zn,Pb deposits and occurrences in the Mackenzie, Selwyn, Ogilvie, and Richardson mountains of the northern Cordillera are most commonly hosted by dolostones of Late Proterozoic and Early Cambrian age, but also occur in moderate numbers in several mainly shelf dolostone units ranging from Late Cambrian to Middle Devonian age (Brock, 1975, 1976), in which tectonic controls to mineralization predominated.

The time of Mississippi Valley-type mineral deposition is not known with certainty (Sangster, 1986). Apparent spatial relationship of some MV-type mineralization to extensional structures and processes, however, implies a timing of Zn,Pb deposition similar to that of major rifting events. Structures and processes include rift-induced synsedimentary and block faulting, uplift, basinal subsidence and resultant facies changes, reefal development, karstification, brecciation, and basinal brine migration. Major Windermere and younger extensional episodes correlative with MV-type mineralization in coeval or older carbonate rocks include: (a) synsedimentary rifting recorded by the deposition of the Coates Lake and Rapitan groups (Jefferson and Parrish, 1989); (b) the opening of Misty Creek Embayment in Early to Middle Cambrian time (Cecile and Norford, Chapter 7) preceding the opening of contiguous Selwyn Basin in Cambro-Ordovician time (Tempelman-Kluit, 1981); (c) Middle Cambrian rifting in the northern Rocky and Selwyn mountains; (d) the development of the Meilleur River Embayment and Root Basin in Early to Middle Ordovician time (Morrow, 1984); (e) Cambro-Ordovician and Devonian rifting in the Rocky Mountains and Selwyn Basin; and (f) the southward progradation of a clastic wedge following block uplift, granitoid plutonism, and alkalic volcanism accompanying a major Cordilleran rifting event in Late Devonian to Early Mississippian time (Gordey, Chapter 8).

The common association of hydrocarbons with Zn,Pb deposits in carbonate rocks has suggested a genetic relationship between MV mineralization and oil maturation, migration, and entrapment (Jackson and Beales, 1967). Oil presently trapped in Devonian reservoirs in the southern Interior Platform did not mature until requisite burial depths and maturation temperatures were attained in Cretaceous time, whereas Paleozoic and Proterozoic sedimentary rocks of the northern Rocky Mountains underwent burial metamorphism sufficient to produce oil maturation before the Cretaceous Period (Macqueen, 1976). In either case, Zn,Pb mineralization, if related genetically to oil maturation-migration processes, would be a relatively late-stage event.

Cambrian carbonate strata of the southern Rocky Mountains host large stratabound deposits of magnesite at Mount Brussilof, with total geological reserves of 30.7 million tonnes of 92-95% MgO (MacLean, 1989), and smaller deposits at Marysville, near Cranbrook. Bedded deposits of gypsum in Cambrian carbonate strata occur at Windermere Creek, Kootenay River and Lussier River (Leech, 1966).

Sedimentary-exhalative Zn-Pb deposits

The most significant stratiform shale-hosted sedimentary-exhalative Zn,Pb deposits of the Cambrian to Middle Devonian passive margin, termed "sedex" by Carne and Cathro (1982), are contained within the Anvil and Howards Pass districts, which occupy linear belts on opposite sides of Selwyn Basin in the northern Cordillera. Selwyn Basin was established as a major negative tectonic element by latest Proterozoic time, flanked to the north and east by the Ogilvie and Redstone Arches, and on the southeast by MacDonald Platform (Gabrielse, 1967). The western part of the basin was overridden by allochthonous upper Paleozoic and Mesozoic rocks (Tempelman-Kluit, 1979) and truncated by the Tintina Fault. Accordingly, age and stratigraphic correlation of the metamorphosed pelitic rocks hosting the Anvil district deposits, which lie directly east of allochthonous units and Tintina Fault, are not known with certainty. The tectonic setting, stratigraphy, and regional correlation of Cambrian to Silurian assemblages of Selwyn Basin are described by Fritz et al. (Chapter 7).

The first major sedimentary exhalative event is represented by the sedex deposits of the Anvil district. They occur in a 150 m succession that is located at the transition between the informally named Mount Mye formation and conformably overlying Vangorda formation which comprise at least 3 km of non-calcareous and calcareous phyllites. The five sub-units of the Vangorda Formation are correlated lithologically by Jennings and Jilson (1986) with Lower Cambrian limestone conglomerate and pelite mapped in eastern Selwyn Basin (Gordey, 1979, in press) and with the Cambro-Ordovician Rabbitkettle Formation in the same general area (Gabrielse et al., 1973). A 1 km-thick metabasaltic sequence, the informally named Menzies Creek formation (Jennings and Jilson, 1986), is interleaved with the upper shaly part of the Vangorda formation which contains Early Ordovician to Early Silurian graptolites (Gordey, 1983; Tempelman-Kluit, 1972), indicating a correlation with the Road River Group of Selwyn Basin.

Five stratiform pyritic Zn,Pb,Ag(Au,Cu,Ba) deposits and two stratiform pyritic Cu,Zn occurrences which extend over a strike length of 45 km constitute an aggregate premining reserve of 120 million tonnes of 5.6% Zn, 3.7% Pb, and 45-50 g/t Ag (Jennings and Jilson, 1986). Faro, the largest deposit, was the only one mined prior to 1990, when Vangorda and Grum started production. The seven deposits are hosted by a distinctive graphitic phyllite unit which serves as a district-wide metallotect. Local orebody lithofacies include basal graphitic and pyritic quartzite, central pyritic massive sulphide, and upper baritic, pyritic massive sulphide. Primary and deformational textures support a synsedimentary, pre-metamorphic age of mineralization. Coincidence of marked southwestward thickening in graphitic phyllite with a linear array of centres of alkaline basaltic volcanism indicates that rift-related synsedimentary faults may have served as conduits for sedimentary exhalative fluids, however demonstrable feeder zones have not been observed. The accumulation of stratiform sulphide deposits in lower Paleozoic basinal facies of the outer Cordilleran miogeocline and spatially associated basaltic volcanism apparently are related to episodic Middle Cambrian to Early Ordovician extension and synsedimentary block faulting. Elongate tectonic sub-basins were postulated to have confined exhalative metalliferous brines in a reducing environment where base metal sulphides were precipitated (Jennings and Jilson, 1986). Mineralization in the Anvil District was pre-accretionary with respect to the Mesozoic attachment of the allochthonous Slide Mountain and Kootenay terranes immediately to the southwest.

A second major sedimentary exhalative event was localized in eastern Selwyn Basin at Howards Pass where large stratiform bodies of Zn,Pb,Fe sulphides were deposited in Lower Silurian (Norford and Orchard, 1983) carbonaceous and limy mudstone and chert. Deep basinal clastic sedimentation adjacent to the Mackenzie carbonate platform was contemporaneous with two Early Paleozoic extensional episodes: the opening of Misty Creek Embayment in Early to Middle Cambrian time and the opening of Meilleur River Embayment in Early to Middle Ordovician time (Fritz et al., Chapter 7). Alkalic basaltic volcanism was widespread during the latter episode, notably in the Menzies Creek and upper Rabbitkettle formations.

Laminated sulphidic sediments of the XY, Anniv, and OP deposits of the Howards Pass district, which extend discontinuously southeastward for over 20 km, were localized within elongate sub-basinal depressions at the base of the slope 10 to 20 km southwest of the carbonate platform margin (Morganti, 1979, 1981). The XY mineralized zone, a lens up to 50 m thick decreasing gradually in thickness over its 3 to 4 km length, is composed of fine grained, well bedded sphalerite, galena, and pyrite, with traces of chalcopyrite, molybdenite, and other sulphides thinly interlaminated with carbonaceous and limy mudstone and chert (Morganti, 1979). The XY and Anniv deposits combined contain indicated reserves of 113 million tonnes of 5% Zn and 2% Pb, plus a further 363 million tonnes of inferred reserves at a similar grade (Placer Development Ltd. Annual Report, June, 1982).

The Howards Pass deposits were formed from Pb- and Zn-rich fluids which discharged episodically into a stable, starved marine basin during a period of restricted seawater circulation and resultant sulphidic, anoxic bottom waters. The anoxic conditions corresponded to an episode of rifting, volcanism, basinal subsidence, local marine transgression, and related hydrothermal activity (Goodfellow and Jonasson, 1986).

Skarn and replacement post-accretionary deposits

Skarn and replacement deposits in the Canadian Cordillera are abundant, diverse, and economically significant. Skarn deposits of the miogeocline are localized commonly where Middle Cretaceous, generally S-type granitoid plutons of the Selwyn, Cassiar, Tombstone, and Bayonne suites (Woodsworth et al., Chapter 15) discordantly intrude the lowest and(or) thickest limestone beds of an upper Proterozoic to lower Paleozoic shelf carbonate-pelite sequence. The broad thermal aureole at the contact between a Cretaceous quartz monzonite stock and a Lower Cambrian limestone is a typical setting.

A belt of W,Cu(Zn,Mo) skarns that follows an arcuate trend of generally small, mid-Cretaceous granitoid plutons from southeastern Yukon Territory and southwestern District of Mackenzie, Northwest Territories, northwestward to the Dublin Gulch district contains one of the world's largest reserves and resources of skarn tungsten, mainly in the Canada Tungsten mine at Tungsten, Northwest Territories, and in the Mactung deposit at Macmillan Pass, Yukon Territory, 175 km northwest of Tungsten. The large, high-grade E-zone orebody at Canada Tungsten (Mathieson and Clark, 1984) contains geological (i.e. total of proven, probable, and possible) reserves of 7 million tonnes of 1.5% WO_3 and 0.2% Cu (Canada Tungsten Mining Corp. Ltd., pers. comm., 1986). Scheelite orebodies are hosted by basinal facies of two outer-shelf limestone members of the Lower Cambrian Sekwi Formation (Fritz, Chapter 7; Blusson, 1968) which form of an overturned anticline whose lower limb is truncated by a quartz monzonite pluton spatially related to skarn ore. Similar shelf rocks describe an arcuate belt that flanks Selwyn Basin on the east and northeast and hosts the world-class Mactung deposit at Macmillan Pass. Geological reserves of the developed prospect are 32 million tonnes of 0.92% WO_3 (Atkinson and Baker, 1986). Two scheelite-chalcopyrite ore zones have replaced limestone breccia and interbedded limestone-pelite in a gently dipping, locally folded Cambro-Ordovician outer shelf sequence which is both flanked and inferred to be underlain by mid-Cretaceous granitoid stocks (Atkinson and Baker, 1986). Other potentially economic tungsten skarn deposits in this belt include, from southeast to northwest: Bailey, Baker, Lened, Clea, and Ray Gulch, whose reserves range between 1 and 5 million tonnes and grades between about 0.7 and 1.5% WO_3 (Dawson and Dick, 1978; Godwin et al., 1980; Glover and Burson, 1986; see Table 19.4 19.4,2 and Fig. 19-16).

Northern Cordilleran W,Cu(Zn,Mo) skarns are typically stratabound bodies adjacent to unaltered coarse grained porphyritic quartz monzonite or granodiorite plutons. Relatively deep-seated passive plutonic emplacement is evidenced by wide hornfelsed aureoles, the presence of migmatite and pegmatite, the chemically reduced states of the mineral assemblages (e.g. graphite, pyrite and high Fe^2/Fe^3 ratios) and the absence of brecciation, breccia pipes, intense stockwork fracturing and(or) dyke swarms. Prograde almandine garnet-hedenbergitic pyroxene-scheelite skarn has overprinted contact calc-silicate hornfels. Subsequent cooling and influx of meteoric water has caused hydrous retrograde alteration of the skarn to an assemblage of amphibole-biotite-chlorite with redistribution of scheelite, including both depletion and upgrading, and deposition of sulphides (Dawson, in Eckstrand, 1984, p. 55).

Significant tungsten was produced from skarn deposits in the Salmo area of southeastern British Columbia between 1943 and 1973. Four mines, the Emerald, Invincible, Feeney, and Dodger, produced a total of 1.4 million tonnes of ore at an average grade of 1.3% WO_3. Skarn orebodies were developed in limestone of the Lower Cambrian Laib Formation adjacent to granitoid stocks and dykes associated with the mid-Cretaceous Bayonne Batholith (Ball, 1954; Mulligan, 1984).

In addition to tungsten skarns, base-metal skarn and replacement deposits are the only other skarn subtype of economic significance in the North American miogeocline.

Numerous subeconomic Zn,Pb,Ag(Cu,W) skarns are adjacent to the Mount Billings Batholith in southeastern Yukon Territory and the Coal River stock in southwestern Northwest Territories in a tectonic setting essentially identical to that of the nearby, in part overlapping, belt of tungsten skarns described previously (Dawson and Dick, 1978; Fig. 19-16). Zn,Pb skarns in some cases developed along lithological and structural pathways more distant from an intrusion than W,Cu skarns. Some pass distally into replacement manto and chimney deposits that are significantly higher in Ag content and total ore tonnage than the skarn (Dawson, 1987). Sphalerite and the prograde skarn minerals hedenbergite and andradite are Fe-rich, whereas the retrograde skarn minerals epidote, actinolite, and chlorite are Mn-rich (Table 19.4.3).

Two relatively large and potentially economic Zn,Pb,Ag replacement deposits are located in southeastern Yukon Territory at Mount Hundere and Quartz Lake, whereas the main belt of such deposits (e.g. Midway, Silver Hart, Tintina Silver) lies 100 km westward within the Cassiar Terrane (Fig. 19.1B, in pocket). Both Mount Hundere and Quartz Lake stratabound Zn,Pb,Ag deposits replace carbonate members of a folded upper Proterozoic to Cambrian clastic-carbonate shelf sequence (Gabrielse, 1966; Gabrielse and Blusson, 1969; Vaillancourt, 1982a,b). The deposits exhibit marked stratigraphic and structural control of mineralization (Hamilton, 1982; Morin, 1981; Abbott, 1981). Both deposits are relatively distant from granitoid plutons but are closely associated with faults interpreted to have served as conduits to underlying plutons (Dawson and Dick, 1978; Hamilton, 1982; Morin, 1981).

Middle Devonian to Mississippian clastic wedge

Devono-Mississippian tectonism in the miogeocline involved local uplift and granitic intrusion in northern Yukon, volcanism in central Yukon and south-central British Columbia, and a dramatic change in sedimentation patterns throughout the northern Cordillera as shelf carbonate-clastic platforms were drowned and starved of clastic sediments before being inundated by fine- to coarse-grained, mainly turbiditic chert-rich clastics derived from the west and north (Gordey, Chapter 8). The abrupt change from passive-margin to variably coarsening-upward clastic sedimentation represented by the Earn Assemblage has been attributed to local block uplift as a consequence of regional extension or strike-slip faulting (Gordey, Chapter 8); ensialic arc magmatism, uplift and foreland clastic wedge deposition (Gabrielse et al., 1982); rifting, Mississippian volcanism, and graben sedimentation (Tempelman-Kluit, 1979); syndepositional extensional and (or) wrench faulting (Abbott, 1986); and Paleozoic sinistral transcurrent faulting along the cratonal margin (Eisbacher, 1983).

In the Macmillan Pass area, Yukon Territory, sedex barite occurrences (e.g. Cathy, Pete) and Zn,Pb,Ag,Ba deposits (e.g. Tom, Jason) occupy three or more stratigraphic levels, ranging from Middle to Late Devonian age, in the lower Earn Group and one Early Mississippian level (e.g. Tea) in the upper Earn Group. The lower Upper Devonian (Frasnian) barite interval is a metallotect of regional extent that corresponds to the principal stratiform Zn,Pb,Ag,Ba horizon at Macmillan Pass (Dawson and Orchard, 1982). Most sedex deposits are spatially related to syndepositional faults which bound a westerly trending rift-related trough, filled successively with dark siliceous shale and chert, turbiditic siltstone, shale, sandstone, and chert conglomerate, and silver-blue-weathering siliceous shale of the lower Earn Group (Abbott, 1986). Pyritic siliceous argillite of the uppermost (Frasnian) unit hosts the Tom West and Jason Main Zn,Pb,Ag,Ba deposits (Dawson and Orchard, 1982). The two structurally separated zones of the deformed Tom deposit, with an aggregate geological reserve of 15.7 million tonnes of 7.0% Zn, 4.6% Pb, and 49.1 g/t Ag, originally constituted one stratigraphically continuous body (McClay and Bidwell, 1986). The distribution of principal ore facies at Tom, including Cu,Ag-rich footwall stockwork overlain by Pb,Zn-rich massive sulphide facies grading upward and laterally to Zn,Fe-rich laminated sulphide facies, and distally to Ba-rich ore represents zonal deposition from low-temperature brines exhaled into an anoxic sub-basin (McClay and Bidwell, 1986; Carne, 1979; Large, 1983). The Jason Pb,Zn,Ag,Ba deposits, 5 km southwest of Tom, possess similar ore facies but are closer to the graben margin, being characterized by extensive slump and debris flows. Three ore zones contain an aggregate geological reserve of 14.1 million tonnes grading 7.09% Pb, 6.57% Zn, and 79.9 g/t Ag (Bailes et al., 1986; Fig. 19-1A).

In the Gataga district of northeastern British Columbia cherty clastic strata of the Earn Assemblage are underlain by shelf and basinal sedimentary rocks of Kechika Basin, a southeastern extension of Selwyn Basin (Fritz et al., Chapter 7). Eight sedex Zn,Pb,Ag,Ba and Ba deposits, which extend in a belt 180 km southeastward from Driftpile Creek to Akie River, are localized within inferred euxinic sub-basins in a structurally controlled trough, flanked in part by carbonate reefs (MacIntyre, 1982). The largest deposit in the district, the Cirque, contains reserves of 32.2 million tonnes of 7.9% Zn, 2.1% Pb, and 47.7 g/t Ag within Frasnian shales of the Earn Assemblage (Pigage, 1986). Massive baritic and pyritic ore facies, concentrated in the thicker western part of the orebody, thin and grade eastward to laminated pyritic facies with siliceous shale interbeds (Jefferson et al., 1983; Gorzynski, 1986).

Stratiform barite-rich deposits hosted by North American Devono-Mississippian clastic strata form part of the world's largest barite metallogenic province, which follows the eastern Cordilleran miogeocline from the Arctic to Mexico (Dawson, 1983). Marked similarities between these distinctive miogeoclinal deposits and barite-rich sedex deposits in Cassiar (e.g. Clear Lake, MM) and Kootenay (e.g. Rea, Homestake) terranes, reinforces lithological correlations between displaced and pericratonic terranes and the craton.

Displaced continental margin
Cassiar Terrane — pre- and post-accretionary deposits

Cassiar Terrane, together with Cariboo Subterrane, extends along the western margin of the North American miogeocline from the Cariboo district of east-central British Columbia 1350 km northwestward to central Yukon Territory (Fig. 19.1B). It comprises an Upper Proterozoic to Upper Triassic miogeoclinal succession of sedimentary strata similar to platformal- and shelf-facies rocks farther east. Cassiar Terrane was displaced as a coherent block from its original position by 500 to 1000 km of dextral displacement along the Tintina and Northern Rocky Mountain Trench fault systems in Late Mesozoic to Early Tertiary time (Gabrielse et al., Chapter 2).

A suite of early-formed stratiform and stratabound base-metal and barite deposits closely resembles equivalent deposits in the North American miogeocline and reinforces lithological correlations between the displaced terrane and the craton. Mississippi Valley-type Zn,Pb prospects at Wasi Lake and Atan Lake are hosted by Lower Cambrian carbonate strata, like the majority of their cratonal equivalents (e.g. Godlin Lakes). Sedimentary exhalative Zn,Pb,Ag,Ba and Ba deposits in the Pelly Mountains (e.g. Clear Lake, MM, GK) occur within Mississippian turbidites equivalent to the upper Earn Group which are the host rocks to sedex deposits near Macmillan Pass (Dawson and Orchard, 1982).

Cassiar Terrane is flanked on the west and intruded by calc-alkaline granite to granodiorite plutons of the Cassiar Suite. The extensive mid-Cretaceous suites of plutons in the Omineca Belt are characteristically S-type, felsic, and high in initial Sr ratios (Anderson, 1988; Armstrong, 1988). Accretion of the Intermontane Superterrane generated plutonism to the west of Cassiar Terrane, represented by the Middle Jurassic Three Sisters Suite (Woodsworth, Chapter 15). Subsequent overthrusting, thickening, westward subduction, metamorphism, and anatexis of cratonal sediments generated an extensive post-accretionary suite of mid-Cretaceous S-type granitoid plutons in the eastern Cordillera. Conversely, the mid-Cretaceous plutons of the Coast and Insular belts are I-type, hosted mainly by volcanic rocks and related to eastward subduction of oceanic rocks (Woodsworth, Chapter 15).

Two groups of Mo(W) and W(Mo) stockwork-skarn deposits typically display porphyry-type Mo(W) stockwork vein development in the associated plutons and W(Mo) skarn-type minerals in the adjacent limestone. The Red Mountain, Risby, and Stormy deposits are associated with the mid-Cretaceous Cassiar Suite plutons. The W and Mo minerals in these plutons which intrude displaced miogeoclinal strata of Cassiar Terrane are similar to those in the Selwyn Suite plutons which intrude North American strata directly to the east (Anderson, 1983) and are characteristic of S-type granitoid plutons throughout the Omineca Belt (Sinclair, 1986). A younger, Late Cretaceous group of Mo(W) deposits in the Cassiar district, including the Cassiar Moly, and Storie porphyry deposits and Lamb Mountain skarn deposit (Panteleyev, 1980), is hosted by felsic and porphyritic Needlepoint plutons of the post-tectonic Surprise Lake Suite (Woodsworth, Chapter 15).

A belt of economically significant replacement deposits hosted by lower Paleozoic carbonate/clastic successions of the Cassiar Terrane includes the Ketza River Au mine, the Midway, Silver Hart, and Tintina Silver developed Ag,Pb,Zn prospects, and the Ingenika prospect. Most of these deposits, in addition to the Mount Haskin and Mount Reed Mo,W stockwork-skarns and the Fiddler W,Cu,Sn greisen, are associated with a suite of small felsic stocks and dykes of mainly Eocene age. Both intrusion and related replacement mineralization were controlled by discordant north, northwest, and east-west faults in a transtensional environment ascribed to late movement on the Tintina, Kechika, and Cassiar faults (Abbott, 1983; Gabrielse, 1985). Significant similarities between these replacement deposits and the manto and chimney deposits of northeastern Mexico, in addition to morphology and mineralogy, include cratonal basement, post-orogenic F-rich granitoids and extensional setting (Dawson, 1987).

Arctic Alaska Terrane

Upper Proterozoic to lower Paleozoic sedimentary, volcanic and granitic rocks of cratonal character in northwestern Yukon Territory bordering Alaska (Fig. 19.1B) are bounded by the Kaltag-Porcupine and related faults and have undergone translation of an unknown amount relative to the North American craton (Gabrielse et al., Chapter 2). Clastic, pelitic and carbonate rocks are deformed, uplifted, and intruded by Early Devonian to Early Mississippian granitoid plutons, including the Fitton, Sedgewick, and Old Crow intrusions (Norris and Yorath, 1981), which are similar in composition to early Paleozoic granitoids of the Nisutlin Subterrane of Kootenay Terrane (Woodsworth et al., Chapter 15). A belt of small subeconomic W,Mo,U,Cu skarns extends northwestward from Mount Fitton to Mount Sedgwick, and small Au placers occur in creeks draining the intrusions. Small W,Cu,U,Sn skarns and W placers are peripheral to the Old Crow Batholith (Geological Survey of Canada, 1981).

Nisling Terrane

Nisling Terrane, in southwestern Yukon (Fig. 19.1C), is composed of two assemblages whose stratigraphy is similar to that of ancestral North America: the metamorphosed Nisling Assemblage of mainly Early Paleozoic and perhaps Late Proterozoic age, and the partly metamorphosed Nasina Assemblage of Cambro-Devonian age. The miogeoclinal aspect of both Nisling continental marginal siltstone and Nasina offshelf, partly carbonaceous shale, sandstone, and carbonate provides tentative correlation with Cassiar Terrane to the east.

The stratabound pre-metamorphic Lucky Joe Creek disseminated Cu prospect in Nasina metasedimentary rocks, although of uncertain origin, is similar to and probably correlative with deposits at Minto and Williams Creek in granitoid gneiss of the Stikine Terrane 150 to 200 km to the southeast (McClintock and Sinclair, 1986). The Lone Star Au,Ag vein deposit and several Au-quartz vein occurrences that may have served as a source of placer gold in the Klondike district are of uncertain age and origin (Debicki, 1984, 1985; Mortensen, 1984). Two small skarns, Hopkins Cu(Mo,Ag,Au) and Sekulmun Zn(Cu,Ag), hosted by carbonate units of the Nasina Assemblage adjacent to the mid-Cretaceous Nisling granodiorite pluton probably represent accretionary mineralization. The Pluto Mo,W porphyry prospect in the Klondike district is Eocene (Tempelman-Kluit, 1981a,b) and post-accretionary.

Pericratonic terranes
Kootenay Terrane
Southeastern British Columbia

The Kootenay Terrane is a discontinuous belt of metamorphosed and intensely deformed siliceous clastic, carbonate, volcanic, and plutonic rocks generally in thrust fault contact with ancestral North America on the east and oceanic volcanic terranes on the west (Fig. 19.1C). Some of the Proterozoic and lower Paleozoic rocks, notably the Eagle Bay Assemblage in Kootenay Arc and at Adams Lake in southeastern British Columbia, appear to be related stratigraphically to ancestral North America whereas younger rocks do not. Rocks of the Kootenay Arc and Shuswap regions of southeastern British Columbia, linked by common stratigraphy, are parts of Kootenay Terrane whereas related units in east-central British Columbia (Barkerville) and southwestern Yukon (Nisutlin) are assigned subterrane status (Wheeler, Chapter 8).

Kootenay Arc

Platformal Lower Cambrian carbonate strata in Kootenay Arc host a suite of pre-accretionary deformed stratiform Zn,Pb,Ag deposits of sedimentary exhalative origin (Sangster, 1986). Principal deposits include the important past producers Reeves Macdonald, Jersey, and HB to the south of Kootenay Lake, and the prospects Duncan Lake, Jackpot, Wigwam, and Mastodon, mainly in the northern Kootenay Arc (Høy, 1982a; Fyles, 1970). These stratiform mineral deposits were isoclinally folded and metamorphosed along with their host rocks prior to the deposition of unconformably overlying strata of the Carboniferous Milford Assemblage (Wheeler, Chapter 8).

An episode of Mesozoic regional metamorphism and deformation that affected rocks of Quesnellia, Slide Mountain and Kootenay terranes and ancestral North America culminated in the emplacement of the Nelson Suite of calc-alkalic granitoid plutons in Middle Jurassic time (Armstrong, 1988; Woodsworth et al., Chapter 15). Previously producing Ag,Pb,Zn vein and replacement deposits of the Riondel (Bluebell) (Høy, 1980; Ransom, 1977) and Ainsworth camps are interpreted by Andrew (1982), on the basis of galena-Pb isotopes, to be remobilized from Lower Cambrian stratiform Kootenay Arc deposits by the Nelson Batholith. A similar origin for Ag,Pb,Zn veins of the Ainsworth camp was proposed by Goldsmith and Sinclair (1983), based mainly upon geochemical data. The time of emplacement of the Nelson Suite is younger than the early Paleozoic deformation that affected Kootenay Arc rocks, yet is not considered here to be post-accretionary, because tectonic events related to this early Paleozoic period of deformation are not well understood. Nelson Suite plutonism coincides with the linkage of southern Quesnellia to ancestral North America (Armstrong, 1988) and therefore is considered to be accretionary. In addition to the Bluebell and Ainsworth deposits, important accretionary mineral deposits associated with the Nelson Batholith include Ag veins of the Slocan camp and Au veins of the Sheep Creek and Rossland camps, all hosted by rocks of Quesnellia.

Shuswap

In the Adams Lake district of the Shuswap region about 200 km northwest of Kootenay Lake metamorphosed phyllite, siltstone, mafic to felsic volcanics, and limestone in at least five imbricated west-verging thrust sheets are correlative in their lower part with the Eagle Bay Assemblage (Preto and Schiarizza, 1985) and in the upper, Mississippian, section, with the Milford Assemblage (Wheeler, Chapter 8) of Kootenay Arc. The four principal types of stratiform and stratabound mineral deposits hosted by rocks of the Eagle Bay Assemblage recognized by Preto and Schiarizza (1985) are: (1) stratiform Ag,Pb,Zn deposits in clastic metasedimentary sequences (e.g. Lucky Coon, Mosquito King); (2) stratabound volcanogenic Cu,Pb,Ag deposits in mafic metavolcanic rocks (e.g. Harper Creek); (3) volcanogenic massive baritic Au,Ag,Zn,Pb,Cu sulphide deposits in felsic to intermediate metavolcanics (e.g. Rea Gold, Samatosum, Homestake, Twin Mountain); and (4) volcanic-hosted stratabound pyritic U,F deposits (e.g. Rexspar).

The pre-deformational Rexspar deposit which is hosted by felsic metavolcanic and metaplutonic rocks (Preto, 1978), contains reserves of 1.1 million tonnes of 1.55% U_3O_8, 1.36 million tonnes of 30% CaF_2 (Northern Miner, 9/6/1977), as well as subeconomic amounts of rare earth elements, Th,Pb,Zn,Cu,Mo, and W.

Barkerville Subterrane

Barkerville Subterrane, in the Cariboo region of eastern British Columbia, includes possible Upper Proterozoic and lower Paleozoic continental shelf sediments and volcanics of unknown paleogeographic origin, which may be facies equivalents of the Cariboo Subterrane of Cassiar Terrane to the northeast. Probably correlative lower Paleozoic sedimentary-volcanic sequences that include the Jowett and Broadview formations of the Kootenay Arc, the Eagle Bay Formation at Adams Lake, and the Downey succession in the Cariboo region host similar Au- and Ag-rich stratabound mineral deposits (Struik, 1986).

Production from the important Au vein and replacement deposits of the Cariboo-Barkerville camp, which include the Cariboo Gold Quartz, Island Mountain, and Mosquito Creek mines, totals 38.1 tonnes of Au (Table 19.1). Principal pyritic Au replacement lodes at Mosquito Creek are conformable with folded contacts between marble and talc dolomite schist in the Lower Cambrian Downey Creek succession, and are interpreted to be synmetamorphic (F. Robert and B.E. Taylor, pers. comm., 1987). Replacement lodes are cut by two sets of post-tectonic veins. Pb-Pb and K-Ar isotopic data cited by Andrew et al. (1983) support a Middle to Late Jurassic/Early Cretaceous, syn- to post-metamorphic age of Au mineralization in the district.

Chlorite- to sillimanite-grade regional metamorphism affecting Barkerville Subterrane, which may range in age from Paleozoic to Cretaceous, is interpreted by Struik (1986) to have occurred mainly in early Mesozoic time and to have been related to the successive overthrusting of Cariboo Subterrane of Cassiar Terrane (Fig. 19.1B), Slide Mountain Terrane, and Quesnellia during the accretion of Intermontane Superterrane. The age of replacement- and vein-gold mineralization in the Cariboo-Barkerville district, although not well established, is interpreted to be accretionary with respect to the protracted metamorphic events that affected Barkerville Subterrane.

Another example of probable accretionary mineralization in Barkerville Subterrane is the concordant Eaglet fluorite deposit hosted by a Late Devonian (Mortensen et al., 1987) orthogneiss pluton, which intrudes metasedimentary rocks at Quesnel Lakes. Recent K/Ar dating of alteration muscovite yielded an Early Cretaceous age of mineralization (A. Panteleyev, pers. comm, 1991). In addition to reserves of 21.7 million tonnes of 11.5% CaF_2, the deposit contains subeconomic amounts of Mo,W,Pb,Zn, and Ag (Ball and Boggaram, 1985).

Nisutlin Subterrane

The allochthonous Nisutlin Subterrane (Fig. 19.1C), composed mainly of metamorphosed and intensely sheared sedimentary, volcanic, and intrusive rocks of Late Proterozoic, Paleozoic, and possibly Mesozoic age, forms extensive thrust sheets in southern and western Yukon Territory (Tempelman-Kluit, 1979). Mid- and upper Paleozoic rocks are correlated with those of the Kootenay Terrane in the southern Canadian Cordillera on the basis of lithological similarity and isotopic studies (Mortensen and Jilson, 1985; Wheeler, Chapter 8). Minor pre-accretionary mineralization is represented by a suite of Cu,Zn massive sulphide occurrences (e.g. Fyre Lake) of pre-metamorphic age and probable volcanic affinity which are hosted by quartzite, phyllite, and chlorite-muscovite schist in the southern Pelly Mountains. The Nisutlin Allochthon has been intruded by granitoid plutons and gneissic sills of the Devono-Mississippian Mink Creek and Simpson Ranges suites (Woodsworth et al., Chapter 15). The related Late Devonian Fiftymile Batholith in western Yukon (Tempelman-Kluit and Wanless, 1980) hosts several Ag,Pb,Zn,Au replacement and vein prospects (e.g. Connaught) and U,Th vein occurrences (e.g. Jove) of probable Paleozoic, pre-accretionary age. Similar Devono-Mississippian igneous activity took place in the Yukon-Tanana terrane of east-central Alaska (Aleinikoff et al., 1986).

The similar lithophile metallogeny and Late Devonian age of Rexspar (Bell, 1985) and Fiftymile pluton-related deposits support a common origin for their host granitoid rocks. An extensional tectonic environment of middle Paleozoic alkaline intrusions in ancestral North America and calc-alkaline intrusions within the Selkirk (Kootenay) Terrane, as proposed by Struik (1987), may apply to the longer belt of mid-Paleozoic plutonic rocks that extends from southeastern British Columbia to eastern Alaska (Gabrielse et al., 1982).

Post-accretionary mineralization in Kootenay Terrane is represented in the south by a W-skarn occurrence which is located at Silence Lake, near Clearwater, British Columbia and hosted by limestone of the Eagle Bay Assemblage adjacent to mid-Cretaceous and early Tertiary granitoid stocks (Dawson et al., 1983). The Taurus porphyry deposit in eastern Alaska contains estimated reserves of 450 million tonnes of 0.5% Cu and 0.07% Mo in Early Tertiary, post-accretionary granitoid plutons that intrude schist and gneiss of the Yukon-Tanana terrane (Nokleberg et al., 1987).

Accreted terranes
Intermontane Superterrane

Slide Mountain (Fig. 19.1D), Dorsey (Fig. 19.1D), Quesnellia (Fig. 19.1E), Cache Creek (Fig. 19.1D), and Stikinia (Fig. 19.1F) are accreted terranes with distinctive stratigraphic, paleontological and/or paleomagnetic signatures which indicate varying amounts of distant origin with respect to the ancient continental margin. All apparently originated as either ensimatic volcanic arc or oceanic/marginal basin deposits prior to their amalgamation, by the end of Triassic time, to form the composite Intermontane Superterrane (Gabrielse et al., Chapter 2). The Intermontane Superterrane was accreted to North America in the Jurassic Period and now mainly underlies the Intermontane Belt (for an alternative view see Chapter 3).

Slide Mountain Terrane

This easternmost accreted terrane forms a narrow, discontinuous belt extending 2000 km from southeastern British Columbia to northwestern Yukon Territory (Fig. 19.1D). The terrane is characterized by an ocean-basin assemblage of chert, fine grained clastics, tholeiitic volcanics, mafic to ultramafic intrusions, and carbonates which ranges in age from Devonian to Late Triassic. Less abundant coarse grained quartzose clastics, calc-alkaline volcanics, and granitoid plutons that support localized island arc and marginal basin environments were first juxtaposed with ophiolitic facies along thrust faults in Permian time (Gabrielse, Chapter 8), but most thrust faulting was post-Triassic in age. The stacked and imbricated package of thrust sheets was emplaced onto the North American continental margin and pericratonic terranes in Mesozoic time (Wheeler, Chapter 8).

The most economically significant mineral deposits in Slide Mountain Terrane are chrysotile asbestos stockworks in serpentinized ultramafic plutons at Cassiar, British Columbia, and Clinton Creek, Yukon Territory. The Cassiar deposit has produced 2.05 million tonnes of high-quality fibre from a total of 23.2 million tonnes of ore mined between 1953 and 1984 (Burgoyne, 1986). The adjacent McDame zone contains proven and probable reserves of 32 million tonnes of ore containing fibre of similar quality (Northern Miner, 12/12/1987). At Clinton Creek, northwest of Dawson, 0.94 million tonnes of fibre were produced from 15.9 million tonnes of ore mined between 1967 and 1978. Subeconomic asbestos prospects occur at Cassiar Creek near Dawson and east of Frances Lake. Gem-quality nephrite jade accompanies chrysotile at Cassiar mine and Frances Lake (Leaming, 1978). Chrysotile, along with serpentine, is generally considered to form early in the emplacement history of the host ultramafic pluton. Textural evidence cited by O'Hanley and Wicks (1987) favours formation of earliest fibre contemporaneously with emplacement of the Sylvester Allochthon at Cassiar. Limited $^{40}Ar/^{39}Ar$ dating of Cassiar nephrite indicates an Early Cretaceous, post-accretionary age of mineralization (T. Harms, pers. comm., 1987).

At Lang Creek near Cassiar a subeconomic Cu,Zn massive sulphide deposit occurs in tholeiitic basalt and cherty argillite near the base of the Sylvester Allochthon. The similar Chu Chua volcanogenic massive sulphide, located 80 km north of Kamloops, is hosted by pillowed tholeiitic basalt and cherty tuffite of the Fennell Formation. Geological reserves are 2.7 million tonnes of 2.0% Cu, 0.4% Zn, 8 g/t Ag, and 0.4 g/t Au (McMillan, 1980).

At Erickson Creek near Cassiar, Au,Ag-rich mesothermal quartz-carbonate-graphite veins are hosted by an assemblage of mafic volcanics, argillite, chert, and ultramafic rocks of the Sylvester Allochthon (Diakow and Panteleyev, 1981). A total of 5.74 tonnes of Au and 3.99 tonnes of Ag have been produced from 425 476 tonnes of ore mined up to October 31, 1987 (Annual Report, Total

Erickson Resources, 1987). Early Cretaceous K-Ar ages from several Au-quartz-sericite veins indicate that mineralization predated emplacement of the adjacent mid-Cretaceous Cassiar Batholith (Sketchley et al., 1986). Essentially identical galena-Pb isotope ratios from Erickson Creek Au veins and the nearby Lang Creek Cu,Zn massive sulphide (Hooper, 1984) support a host-rock source of Au. Mineralization is interpreted to be related to thermal and(or) structural events initiated by contraction which formed the synclinorium containing the Sylvester Allochthon and culminated in the emplacement of the Cassiar Batholith.

Post-accretionary mineralization in Slide Mountain Terrane is represented by the Windpass occurrence, 6 km north of Chu Chua Mountain. The magnetite-rich Au,Cu,Ag shear zone assemblage in Cretaceous quartz diorite, similar to the Chu Chua massive sulphide, probably reflects the same source in the host Fennell volcanic rocks. Gold placers at Manson Creek in central British Columbia have a potential source in mesothermal Au vein occurrences (e.g. Germansen Creek, Motherlode) in underlying Slide Mountain rocks.

Dorsey Terrane

Upper Paleozoic rocks of Dorsey Terrane, which straddle the Yukon-British Columbia border (Fig. 19.1D), overlie those of Slide Mountain Terrane in an unknown relationship. Thick ribbon chert sequences resemble rocks of Slide Mountain Terrane, but volcanics are scarce and ultramafic rocks are lacking. Quartzose clastic and thick, non-oceanic carbonate units suggest proximity to a cratonal area (Gabrielse, Chapter 8).

Assuming an accretionary history similar to that of Slide Mountain Terrane, the only pre-accretionary deposit recognized is the Bar or Smeg sedimentary exhalative Zn,Pb,Ag,Ba prospect 50 km northeast of Teslin, Yukon Territory. Exhalative chert, pyrite, and barite, with minor amounts of galena and sphalerite, are interbedded with a chert-argillite-siltstone-chert pebble conglomerate sequence reminiscent of upper Earn Group lithologies on the North American craton. Late Mississippian conodonts, Devono-Mississippian galena-Pb isotopes, and Mississippian barite-S isotopes confirm a Mississippian age of mineralization (K.M. Dawson, unpubl. data).

The most significant deposit in Dorsey Terrane is the large, low-grade Logtung porphyry W,Mo deposit with reserves of 162 million tonnes of 0.13% WO_3 and 0.03% Mo (Noble et al., 1984). This skarn W(Mo) and stockwork Mo,W mineralization, in and adjacent to mid-Cretaceous (109 ± 2 Ma; Sinclair, 1986) felsic granitoid plutons of the Cassiar Suite, is post-accretionary. Similar post-accretionary stockwork-skarn Mo,W mineralization at the Red Mountain, Stormy, and Risby deposits in Cassiar Terrane to the north suggests an affinity between the two terranes. Another post-accretionary pluton, the mid-Cretaceous Seagull Batholith, is unique in its high B,Be,F,Cl content (Dick and Hodgson, 1983) and associated Sn skarn and stockwork deposits (Mato et al., 1983). The presence of the Cassiar Suite of plutons in the Dorsey Terrane indicates that the terrane is structurally underlain by cratonal crust.

Quesnellia

Upper Paleozoic and lower Mesozoic volcanic, sedimentary, and plutonic rocks of Quesnellia are best developed in the southern Canadian Cordillera (Fig. 19.1E). Paleozoic strata, divisible into two subterranes of oceanic and arc affinity, are overlain unconformably by characteristic Quesnellia island-arc volcanic and sedimentary strata of the Upper Triassic and lowermost Jurassic Nicola Assemblage. Parallel facies belts in the latter define a west-facing arc which progresses compositionally from calc-alkaline in the west to alkaline in the east. Comagmatic granitoid plutons generally show the same chemical trend (Mortimer, 1987). Lower Mesozoic volcanic and sedimentary strata are overlain by Lower and Middle Jurassic clastic rocks.

Distinct facies belts are not seen in the central and northern Cordillera where Upper Triassic augite porphyries, characteristic of the eastern facies belt in the south, predominate. Volcanics are flanked on the east by black phyllite and intruded by calc-alkaline to alkaline comagmatic plutons of the Guichon and Copper Mountain suites. Quesnellia may have undergone Eocene extension in the southern Cordillera (Parrish, Chapter 17) but narrows to a discontinuous sliver north of 54°N latitude. The terrane is bounded by steeply dipping fault contacts with Cache Creek Terrane and Stikinia on the west, and on the east it stratigraphically and structurally overlies Slide Mountain, Cassiar, and Kootenay terranes.

Distinctive calc-alkaline to alkaline plutons of the Guichon and Copper Mountain suites in southern and central Quesnellia, and of the Klotassin and Long Lake suites in Yukon Territory are latest Triassic and Early Jurassic in age. They were emplaced, therefore, either before or during the amalgamation of the Intermontane Superterrane, but prior to its accretion to North America. The Guichon and Copper Mountain suites in southern Quesnellia are regarded as the cogenetic roots of the early Mesozoic volcanic arc of the Nicola Assemblage (Monger, Chapter 8). The deposits related to Early Jurassic and older members of these plutonic suites are pre-accretionary.

The large porphyry Cu,Mo deposits of the Highland Valley district are hosted by the calc-alkaline Guichon Creek Batholith, in association with the youngest, innermost, and most leucocratic phase and with late dyke swarms (Woodsworth et al., Chapter 15). The aggregate production, proven and probable reserves for the Valley Copper, Lornex, JA, Highmont, Bethlehem, and Krain deposits, almost 2 billion tonnes of 0.45% Cu equivalent (McMillan, 1985), constitute the largest porphyry Cu district in the Canadian Cordillera (Table 19-3). The Craigmont Cu,Fe skarn, 30 km south of Highland Valley, was developed in calcareous volcaniclastic and reefoid carbonate rocks of the western facies belt of the Nicola Assemblage, at their embayed contacts with the border phase of the Guichon Creek Batholith. Between 1961 and 1982, 33.4 million tonnes of chalcopyrite-magnetite ore were mined, at an average grade of 1.7% Cu and 19% Fe. (Table 19.4.1) Limited production was obtained from several small Cu,Fe skarns associated with the Iron Mask Batholith.

The alkalic Copper Mountain Stock and Iron Mask Batholith are cogenetic with, and probably subvolcanic to Nicola Assemblage eastern facies augite porphyry flows. The important alkalic porphyry-related Cu,Au,Ag mines of Copper Mountain-Ingerbelle and Afton are associated with the plutons near Princeton and Kamloops, respectively. The Copper Mountain-Ingerbelle deposits are hosted almost entirely by a narrow septum of fragmental andesitic volcanics between the Copper Mountain and Lost Horse stocks. Silica-deficient stockworks of chalcopyrite-pyrite, lesser amounts of bornite and hematite, and an alteration assemblage of biotite-albite-epidote-chlorite-K-feldspar-scapolite displays a skarn-like, zonal relationship to the similarly altered Lost Horse Stock (Fahrni et al., 1976). Conversely, the steeply dipping Afton chalcopyrite-bornite-pyrite-magnetite orebody is developed entirely within fractured dioritic and latitic porphyry phases of the composite alkalic Iron Mask Batholith. The deposit is distinguished by high Au (0.58 g/t) and Ag (4.19 g/t) contents, deep weathering, and economically recoverable supergene native Cu (Carr and Reed, 1976; Kwong, 1987; see Table 19.3).

The Alaskan-type zoned gabbro-ultramafic Tulameen complex intrudes, and may be subvolcanic to augite basalt of the western facies of the Nicola Assemblage (Findlay, 1969; Nixon and Rublee, 1988). Significant Pt,Pd occurrences near Tulameen, 25 km northwest of Princeton, accompany chromite in the layered dunitic core of the complex (St. Louis et al., 1986). Primary magmatic disseminations of titaniferous magnetite occur in pyroxenite of the complex at Lodestone Mountain (Eastwood, 1960).

At Hedley, significant Au-bearing skarns are developed at contacts between calcareous sediments of the Upper Triassic eastern sedimentary facies of the Nicola Assemblage and Early to Middle Jurassic Hedley Intrusions. The dioritic to gabbroic stocks, sills, and dykes are older than the adjacent Middle Jurassic Pennask and Similkameen plutons, and may reflect back-arc rifting relative to the Nicola arc to the west. Ray et al. (1987) related the coincidence of sedimentary facies changes and plutonism to an underlying basement flexure. The arsenical skarn deposits, which produced about 55 million grams of gold between 1902 and 1955, have been redeveloped recently as an open pit mine with surface mineable reserves of 9.0 million tonnes of ore grading 4.56 g/t Au (B.C. Mineral Exploration Review, Circular, 1988-1, see Table 19.4.5).

Skarn Au and Ag deposits, similar to those at Hedley but of less certain stratigraphic correlation and age, occur at Tillicum Mountain between Lower Arrow Lake and Slocan Lake in southeastern British Columbia. Deposits are hosted by calcareous beds in a metamorphosed volcanic-sedimentary pendant intruded by post-mineral plutons of mid-Cretaceous (Hyndman, 1968) to Eocene (Parrish, 1981) age. Hyndman (1968) correlated the siltstone-mafic volcanic-argillite-limestone sequence with the Pennsylvanian to Permian Milford and Kaslo groups. In the Nemo Lakes belt, south of Tillicum Mountain, Parrish (1981) correlated these rocks, in part, with upper Paleozoic strata included in Slide Mountain Terrane. The shoshonitic (high-K) basalts at Tillicum Mountain are correlated by Ray and Spence (1986), on a chemical basis, with volcanics of the Rossland Group and the eastern facies belt of the Nicola Assemblage of Quesnellia. Siliceous calc-silicate skarns with gold, pyrrhotite, pyrite, sphalerite, and galena are spatially related to undated monzodiorite porphyry sills and dykes interpreted, by Roberts and McClintock (1984), to be cogenetic with flows higher in the sequence. Production started in 1988 at the Heino-Money zone which, in conjunction with the East Ridge zones contains proven and drill indicated reserves of 730 500 tonnes grading 9.3 g/t Au (Esperanza Explorations Ltd., internal report, 1989).

The unusual, pre-accretionary Willa breccia pipe deposit on Aylwin Creek, 8 km south of Silverton, is hosted by an inlier of Rossland Group augite andesite flows, pyroclastics, and volcaniclastics within the northern part of the Nelson Batholith (Wong, 1985). A ring dyke of quartz latite porphyry which, with a younger feldspar porphyry stock, contains stockwork Au,Cu,Ag mineralization, yielded an Early Jurassic zircon U-Pb age of 194 Ma (Brown and Logan, 1988), substantially older than plutons of the enclosing Middle Jurassic Nelson Suite but coeval with Rossland Group volcanics. Felsic intrusions and sulphides are cut by an intrusive breccia pipe. Veins of precious metal-bearing pyrite-chalcopyrite ± pyrrhotite and magnetite are associated with a calc-silicate gangue assemblage (Wong, 1985). The West Zone contains proven reserves of 550 000 tonnes of 7 g/t Au, 1.04% Cu, and 8.6 g/t Ag (George Cross Newsletter, 5/1/1988).

In the Quesnel River region of central British Columbia, pre-accretionary Cu,Au and Au mineralization is associated with alkaline granitoid intrusions comagmatic with and subvolcanic to a succession of calc-alkalic to alkalic (shoshonitic) island arc basaltic rocks, that comprise the 5000 m thick upper portion of the Nicola-equivalent 'Quesnel River Group' (Morton, 1976; Bailey, 1978). Mount Polley (Cariboo Bell), a developed alkalic porphyry Cu,Au prospect 56 km northeast of Williams Lake, contains mining reserves of 48 million tonnes grading 0.38% Cu and 0.55 g/t Au (Imperial Metals Corp., Company Report, 1990). A subvolcanic syenite laccolith of earliest Middle Jurassic age was emplaced within a trachybasalt eruptive centre. Two breccia zones within the syenite contain disseminated magnetite-chalcopyrite-pyrite and pervasive potassic alteration (Hodgson et al., 1976).

The Quesnel River (QR) deposit, located 15 km northwest of Cariboo Bell, is an auriferous pyrite-epidote replacement of calcareous beds between alkalic augite porphyry flows and overlying argillite. Two zones of gold-pyrite-chalcopyrite within a halo of propylitic alteration surround a small composite diorite-monzodiorite stock of Early Jurassic (201 ± 7 Ma) age (Fox et al., 1987, see Table 19.4.5). A northwest-trending belt of Au-bearing zones east of Quesnel River also may be related to small alkalic stocks that intrude the underlying volcano-sedimentary sequence. Panteleyev (1987) recorded broad zones of propylitic alteration associated with a suite of Lower Jurassic (192-196 Ma) alkalic stocks south of Quesnel Lake. A black porphyroblastic Middle to Upper Triassic phyllite unit, basal to the Norian alkalic volcanics and flanking them on the east, hosts stratabound gold apparently unrelated to intrusions.

The Lorraine porphyry Cu,Au prospect, 56 km west of Germansen Landing (latitude 56°N), consists of two zones of chalcopyrite-bornite-magnetite disseminated in the Duckling Creek syenite pluton (Garnett, 1978). The Middle Jurassic deposit is younger than other alkalic porphyries of the Copper Mountain Suite, and the pluton is enclosed within the Early Jurassic calc-alkaline Hogem Batholith rather than cogenetic volcanic rocks. However, the pluton is assigned to the petrologically similar Copper Mountain Suite rather than to the accretionary Three Sisters Suite 200 km to the northwest (Woodsworth et al., Chapter 15).

The principal accretionary suite of plutons that intrudes Quesnellia is the calc-alkaline Middle Jurassic Nelson Suite, referred to previously under "Kootenay Terrane". In the Rossland district of southeastern British Columbia (Fyles, 1984), the Rossland Monzonite pluton, interpreted by Gilbert (1948), Thorpe (1967), and Little (1982) to be related to the Trail pluton of the Nelson Batholith, is closely associated with Au,Cu-bearing iron sulphide mineralization. Thorpe and Little (1973) related vein-mineral zonation and formation of skarn-mineral assemblages to pyrometasomatic mineralization genetically associated with the Rossland Monzonite. Production from the Rossland camp, which started in 1894, was second only to the Bridge River camp in British Columbia, with total production of 85.4 tonnes of Au from 5.6 million tonnes of ore at an average grade of 13 g/t Au, 17 g/t Ag, and 1% Cu (Table 19.1).

The Frasergold prospect near Eureka Peak, 25 km south of Quesnel Lake, is a zone of deformed quartz-carbonate-pyrite-Au veins interpreted by Bloodgood (1987) to have formed early in the structural history of the area as metamorphic segregations generated during the accretion of Quesnellia.

In the Slocan-Silverton district of southeastern British Columbia isoclinally folded beds of the Upper Triassic (Orchard, 1985) Slocan Group were intruded syn- to late-kinematically by the Nelson Batholith and related dykes and sills in Middle Jurassic time (Woodsworth et al., Chapter 15). Ag,Pb,Zn veins, related to the minor intrusions by Cairnes (1934) and Hedley (1952), follow faults and dykes which crosscut the folded argillite-quartzite-limestone-tuffite units.

Replacement orebodies in the Slocan district like Cork Province (Hedley, 1947) occur only in limestone. A district-wide deposit zonation with increasing depth, from

Table 19.1. Major Cordilleran gold producers 1894-1984

DEPOSIT OR CAMP	YEARS OF PRODUCTION	TONNES MILLED (10^6)	GOLD RECOVERED (TONNES)	HOST TERRANE/ SUBTERRANE	DEPOSIT TYPE
1. PLACER GOLD Yukon British Columbia	1885-1984 1857-1984	--- ---	353.934 164.861	Various	Placer Placer
2. BRIDGE RIVER CAMP Bralorne, Pioneer, Minto, Wayside	1899-1978	7.319	129.963	Bridge River, Cadwallader	Quartz veins - mesothermal, California Motherlode-type. 'Bralorne type': Au (As,W); 'Minto type': Au,Ag (As,Sb,Zn,Hg)
3. ROSSLAND CAMP Le Roi, Centre Star, War Eagle, Josie, etc.	1894-1974	5.625	85.387	Quesnellia	Fracture and fault-controlled sulphide veins; copper-gold mineralization with average 13 g/t Au, 17 g/t Ag and 1% Cu per tonne; Tertiary intrusive-related (Fyles, 1984)
4. PORTLAND CANAL-STEWART CAMP Silback-Premier Mine	1918-1976 1989-	4.237	56.117	Stikinia	Quartz-sulphide veins and stockworks; 'telescoped' epithermal deposit. 1270 tonnes silver produced. 1987 open pit reserves 5.8 Mt of 2.37 g/t gold, 92.2 g/t silver.
5. HEDLEY CAMP Nickel Plate, Hedley, Mascot, French, Good Hope, etc.	1904-1955 1987-	3.612	53.829	Quesnellia	Skarn. Structurally and lithologically controlled pyroxene calc-silicate zones with arsenopyrite, tellurides and late base metal sulphides related to Jurassic dykes and sills. Open pit reserves 10.8 Mt of 4.56 g/t Au.
6. CARIBOO-BARKERVILLE CAMP Cariboo Gold Quartz, Island Mountain (Aurum), Mosquito Creek	1933-1983	2.730	38.059	Kootenay/ Barkerville	Quartz veins and lithologically controlled massive, pyritic replacements.
7. BOUNDARY-GREENWOOD CAMP Phoenix, Motherlode, Greyhound, BC, Emma, Oro Denoro, etc.	1900-1978	31.795	32.680	Quesnellia	Mainly skarn. Intrusive-related or stratigraphically controlled calc-silicate copper sulphide-magnetite deposits; some genetically related porphyry copper and peripheral vein mineralization.
8. SHEEP CREEK & YMIR CAMPS Queen, Reno, Kootenay, Goodenough, Yankee Girl, etc.	1899-1943	2.585	28.926	Kootenay & Quesnellia	Quartz veins, mainly in faults cutting quartzite and hornfels. Possibly related to Middle Jurassic to mid-Cretaceous intrusions.
9. ISLAND COPPER	1971-1984*	168.761	19.920	Wrangellia	Porphyry Cu,Mo; calc-alkaline, dyke-related volcanic type. Mid- to late Jurassic. Island Intrusions of Vancouver plutonic suite.
10. COPPER MOUNTAIN Similkameen/Ingerbelle	1914-1984*	107.725	16.915	Quesnellia	Porphyry Cu; alkaline, Early Jurassic stock-related volcanic type. Copper Mountain plutonic suite.

* Mine in production 1987.

Ag,Pb through Pb,Zn,Ag to Zn was attributed by Hedley (1947) to the protracted thermal influence of the Nelson Batholith. Andrew et al. (1984) proposed that galena-Pb isotope data from Slocan deposits define a 160 Ma isochron representing mixing of upper and lower crustal Pb by Nelson Batholith magmatic activity. Mines in the Slocan district, between 1894 and 1950, produced 1780 tonnes of Ag, 192 130 tonnes of Pb, and 139 700 tonnes of Zn (Hedley, 1952; Table 19.4.3A).

In the Ymir-Sheep Creek district at the southeastern margin of the Nelson Batholith, folded Quesnellia strata of the Triassic Ymir Group and Lower Jurassic Rossland Formation (Little, 1960) host Au- and Ag-rich base-metal veins. As in the Slocan district, veins discordantly cut thermally metamorphosed strata adjacent to granitoid dykes related to the Nelson Batholith (Cockfield, 1936; McAllister, 1951; Høy and Andrew, 1988). Several small mines in the Ymir-Sheep Creek area produced a total of 29 tonnes of Au mainly in the periods 1896-1908 and 1934-1940 (McAllister, 1951; Table 19.1).

Post-accretionary Au, Ag, Mo, and U are associated with rocks of the Okanagan Subterrane of Quesnellia in the Beaverdell area of south-central British Columbia. Like Au deposits elsewhere in Quesnellia, Au-bearing base-metal veins of the Carmi deposit are associated with relatively old volcanic rocks of oceanic aspect, in this case the mafic volcanics of the Wallace Formation, which are correlated with the Permian part of the Anarchist Group (Little, 1961; Peatfield, 1978). Ag-rich Zn,Pb fissure veins of the Beaverdell (Highland Bell) mine occur mainly within quartz diorite and granodiorite of the probably Middle Jurassic West Kettle Batholith, similar to the Nelson Suite (Woodsworth et al., Chapter 15). The timing of Ag mineralization, however, is bracketed at 50 Ma by the K-Ar ages of pre- and post-mineral felsic dykes related to the quartz monzonitic Beaverdell stock (Christopher, 1975; Watson et al., 1982). Porphyritic alkalic to calc-alkalic stocks and dykes, which are associated with a disseminated Mo prospect and a larger Mo,U,F prospect 6 km northwest of Carmi, are also Eocene and probably related to the early Tertiary alkalic Coryell Suite.

Coryell stocks and calc-alkalic plutons of the Okanagan Composite Batholith underlie the significant Blizzard and Tyee basal U deposits 25 km northeast of Beaverdell. The Blizzard, with reserves of 4000 tonnes of U (Sawyer et al., 1981), is the largest U deposit in the Canadian Cordillera. These deposits, similar deposits near Hydraulic Lake, and deposits of the probably contiguous Spokane, Washington U district are related to U-rich lower Tertiary granitoid and related continental volcanic rocks. Secondary uranium minerals derived from uraniferous basement rocks are concentrated in paleochannel sediments where oxidation has been retarded by a capping of plateau basalt of the Chilcotin Assemblage (Bell, Chapter 20).

Paleozoic oceanic rocks of the Okanagan Subterrane lying both east and west of the Okanagan crustal shear zone in southern British Columbia (Tempelman-Kluit and Parkinson, 1986) host several small Au deposits of probable post-accretionary age. At Camp McKinney, 25 km northwest of Rock Creek, Au was recovered between 1894 and 1902 from base-metal veins in greenstone of the Permian Anarchist Group (Hedley, 1940). These were among the first producing lode mines in the province. At the Oliver, Fairview, Twin Lakes, and Dividend camps to the west of the gently westward-dipping Okanagan crustal shear, deformed precious- and base-metal-bearing quartz veins are hosted by the Kobau Group and the Old Tom and Shoemaker formations adjacent to Middle Jurassic and Early Cretaceous granitoid plutons. Total production from all camps prior to 1945 was in the order of 4000 kg of Au (Hedley and Watson, 1945). Age of vein mineralization is dated isotopically as dominantly Early Cretaceous and therefore probably related to post accretionary plutonism. Some epithermal Au deposits (e.g. Dusty Mac) are Eocene and possibly related to post-accretionary hydrothermal activity accompanying detachment faulting in an extensional environment (Parrish, Chapter 17).

In the Greenwood district of south-central British Columbia, significant Cu,Au,Ag mineralization accompanied the post-accretionary emplacement of granitoids of probably Middle Jurassic age into upper Paleozoic oceanic rocks of the Okanagan Subterrane and overlying Triassic sedimentary strata of Quesnellia. In summarizing production from 26 mines in the district from 1893 to 1985, Church (1986) noted that most of the Ag,Au,Pb,Zn vein deposits are hosted by granodiorite stocks of the Early Cretaceous Wallace Creek Batholith. Veins hosted by ultramafic rocks and schist, greenstone, and argillite of the Paleozoic Knob Hill and Attwood groups are usually localized close to granodiorite contacts. Most mineral production, however, has come from five Cu,Au,Ag skarn deposits hosted by limestone of the Middle to Upper Triassic (Little, 1983) Brooklyn Formation. Nearly all of the Cu (270 000 tonnes), 94% of the Au (36 tonnes), and 64% of the Ag (117 tonnes) produced in the Greenwood district were derived from the Phoenix, Mother Lode, Oro Denoro, Greyhound, and Marshall skarn deposits (Church, 1986; Tables 19.1 and 19.4.1). In a typical skarn deposit, pyrite, chalcopyrite and magnetite are associated with a garnet-rich assemblage of andradite, clinozoisite, diopside, and quartz (Carswell, 1957).

Significant amounts of Pt and Pd accompany Cu, Ag, and Au minerals in the Eocene Coryell intrusions in southeastern British Columbia (Drysdale, 1915). A prospect of current interest, the Sappho, 10 km south of Greenwood (Church, 1986) contains pyrite and chalcopyrite with Ag, Au, Pt and Pd associated with concentrations of shonkinite-pyroxenite marginal to the host augite syenite intrusions (Hulbert et al., 1988). The similar Maple Leaf, in the Franklin camp 80 km west-northwest of Nelson (O'Neill and Gunning, 1934), is associated with pre-Coryell, Late Jurassic intrusions (Keep, 1989).

Porphyry Cu,Mo stockwork veins at the Brenda deposit are notably younger and less intensely altered than nearby pre-accretionary porphyry deposits hosted by the Nicola Assemblage. Cu,Mo mineralization hosted by a quartz dioritic unit of the Early Jurassic Pennask Batholith has been interpreted by Soregaroli and Whitford (1976) to be of latest Jurassic age. Reserves at commencement of production in 1969 were 144 million tonnes of 0.183% Cu and 0.049% MoS_2. Production statistics to 1984 are given in Table 19.3.

Middle Cretaceous porphyry-type Mo mineralization at Boss Mountain in the Cariboo region of central British Columbia is superimposed upon the Early Jurassic Takomkane Batholith which has intruded volcanic rocks of the Nicola Assemblage. A sequence of molybdenite veins is hosted by felsic dykes and breccias related to the adjacent quartz monzonitic Boss Mountain stock (Soregaroli and Nelson, 1976). Between 1965 and 1982, 2.2 million tonnes of ore grading 0.7% Mo were produced (19.3).

Cache Creek Terrane

The Cache Creek Terrane is a disrupted assemblage of upper Paleozoic to lower Mesozoic predominantly oceanic rocks and subduction complexes that extends, within the Intermontane Belt, from central Yukon Territory to southern British Columbia (Fig. 19.1D). In northern British Columbia, large plutons of variably serpentinized ultramafic and mafic rock of the French Creek Subterrane (Monger, Chapter 8) contain mineral deposits typical of alpine-type ultramafic rocks in oceanic terranes and presumably are of pre-accretionary age. In the King Mountain (Cry Lake) area, nephrite jade has been mined from several small deposits hosted by fault-bounded serpentinized ultramafic rocks (Leaming, 1978). Also in the Cry Lake area, sub-economic chrysotile asbestos stockworks (e.g. Letain) and Cu,Ni sulphide disseminations (e.g. Pyrrhotite) occur in ultramafic rocks. Nephrite deposits and podiform chromite occurrences are hosted by ultramafic plutons of the Trembleur Intrusions at Mount Sydney Williams and Mount Ogden in central British Columbia.

The Kutcho Creek polymetallic, felsic-volcanic-hosted deposits in the Cry Lake area (Bridge et al., 1986) are the only significant volcanogenic massive sulphide deposits within the extensive Upper Triassic to Lower Jurassic arc volcanic rocks of the central and eastern Cordillera. They mark the unusual transition, in Late Triassic time, from Cache Creek tholeiites to immediately overlying calc-alkaline basalts and rhyolites (Thorstad and Gabrielse, 1986). Calc-alkaline arc volcanism may signal the onset of subduction of Stikinia beneath the Cache Creek Terrane at the time of their amalgamation (H. Gabrielse, pers. comm., 1988). Contemporaneous volcanogenic mineralization is, therefore, considered to have been early accretionary. The complex alkaline-subalkaline chemistry of Late Triassic island arc volcanism in the Cordillera has been attributed to oblique subduction by Monger (1977) and Paterson (1977), and to rapid subduction accompanied by rifting by Souther (1977). The Kutcho Creek Cu,Zn,Ag deposits are similar to Japanese Kuroko-type deposits whose calc-alkaline magmatic parentage was interpreted to be an integral part of a sequence of subduction-related arc volcanics (Bridge et al., 1986; Sato and Sasaki, 1973). Both the paucity of calc-alkaline volcanics and lack of volcanogenic massive sulphide deposits in Upper Triassic Cordilleran island arcs may be related to the subduction style. Size and grade of the Kutcho Creek deposits are given in Table 19.2.

The Granite Mountain diorite-quartz diorite pluton of the Guichon Suite between Williams Lake and Quesnel, intruded volcanic and sedimentary strata of the Cache Creek Terrane in earliest Jurassic time. At the Gibraltar porphyry Cu,Mo mine, multiple intrusion, deformation, greenschist facies metamorphism, and mineralization were interpreted to have taken place in a short time, even overlapping in age at about 204 Ma (Drummond et al., 1976). Late stages of Cu,Mo veining have undergone only minor deformation. Gibraltar mine Cu,Mo mineralization coincided with a period of intense northwestward trending shearing that probably was generated during the amalgamation of Cache Creek Terrane with Quesnellia and Stikinia to the east and west, respectively. Although other Guichon and Copper Mountain suite plutons of similar, even slightly younger age have been considered pre-accretionary, the well-documented continuum of events at Gibraltar mine supports an early accretionary age. Ore reserves at commencement of production in 1972 were 326.5 million tonnes of 0.37% Cu and 0.016% MoS_2 (Drummond et al., 1976; 19.3).

Significant post-accretionary Hg mineralization occurred within the Cache Creek Terrane along its faulted eastern boundary with Quesnellia in central British Columbia. A belt of 12 or more Hg occurrences extends from Pinchi Lake 100 km northwestward to Mount Ogden. It includes the Pinchi Lake mine, which produced 6 million kg of Hg between 1942 and 1975 (Canadian Minerals Yearbook, 1975), and the smaller Bralorne Takla mine. The Pinchi and subsidiary faults within the Pinchi Fault Zone host replacement lodes and breccia fillings of red cinnabar where massive limestone units are intersected. Serpentinized ultramafic rocks, chert, argillite, and greenstone along the same mineralized fault zones are extensively altered to an assemblage of Fe,Mg carbonates, quartz, mariposite, chlorite, and talc (Armstrong, 1949). Hydrothermal Hg mineralization was younger than both Upper Triassic amalgamation-related blueschists and Upper Cretaceous-Lower Tertiary conglomerate described by Paterson (1977), and may have coincided with the termination of a period of uplift, magmatism, and transcurrent faulting in Eocene-Oligocene time described by Gabrielse (1985).

Mesothermal Au,Ag-sulphide veins occur in geological settings essentially identical to those of Hg veins. In and adjacent to the Pinchi Fault Zone, the Lustdust (Kay), Indata Lake, and Snowbird prospects are associated with carbonatized fault zones in limestone. Coarse, angular placer Au from creeks draining the Pinchi fault zone indicates a local source (Armstrong, 1949). Placer Au deposits at Dease Lake and Wheaton Creek are underlain by rocks of the Cache Creek Terrane, but few lode Au occurrences are known. In the Atlin district in northwestern British Columbia there has been major placer Au production from creeks draining several mesothermal Au(Pb,Zn,Cu) vein prospects including Spruce Creek, Lakeview, Surprise, Yellowjacket, and Pictou. Au-bearing veins hosted by sheared and carbonatized greenstone and serpentinite yield galena-Pb isotopes concordant with the age of Au,Pb minerals adjacent to a Middle Jurassic granite stock, supporting a similar age for vein mineralization.

A significant Mo deposit and several W, Sn, and U occurrences are associated with felsic, post-tectonic plutons of the Surprise Lake Suite which intrude Cache Creek Assemblage rocks at Atlin. Late Cretaceous quartz monzonitic phases of the Surprise Lake pluton host the Adanac porphyry Mo deposit (Table 19.3), whereas Surprise Lake W,Cu,Sn greisen veins and W,Sn(Cu,Pb,Zn) skarns are localized at adjacent plutonic contacts with Cache Creek limestones (Christopher and Pinsent, 1982). Several U, Be, W, and F prospects characterize the unusual lithophile metallogeny of the plutonic suite (Ballantyne and Littlejohn, 1982). The adjoining Fourth of July Batholith, a diorite to granodiorite pluton of Middle

Table 19.2. Volcanogenic massive sulphide deposits

Name	Host Rock Lithology	Age	Terrane or Assemblage	Size Million Tonnes	Grade% Cu	Grade% Zn	Grade% Pb	Grade g/t Ag	Grade g/t Au	Fe Sulphide	Sulphate, other metals	Reference
Britannia	Gambier Gp.- andesite to dacite	Jurassic- Cretaceous	Coast Plutonic Complex; Gambier Gp.	47.8	2.8	0.26		3.8	0.3	py	Ba,Ah,Gyp	Payne et al., 1980
Westmin Lynx-Myra H-W	Sicker Gp.- basalt, rhyolite	Devonian	Wrangellia	6.1 13.8	1.4 2.2	7.6 5.3	1.2 0.3	106 37.7	2.1 2.4	py	Ba	Walker, 1983; company reports
Kutcho Creek Kutcho Zone Sumac West Zone Esso West Zone	'Kutcho Fm'- schist, rhyolite	Triassic	Cache Creek	17.0 10.0 1-1.5	1.62 1.0 ~3	2.32 1.2 ~4	0.06	29.2	0.3	py		Bridge et al., 1986
Tulsequah Chief	'Stikine volcanics' felsic volcs.	Paleozoic (Penn./ Permian?)	Stikinia	0.83 5.26	1.59 1.60	7.0 7.03	1.54 1.31	127 100	3.8 2.7	py	Ba	Production Redfern Res. Ann. Rept., 1989
Eskay Creek excluding 21A zone	mudstone, felsic volcanic dykes andesite	Jurassic	Stikinia	3.97				998	26	py	Pb,Zn	Company reports, 1990
Twin J	Sicker Gp.- basalt, rhyolite	Devonian	Wrangellia	0.6	3	0.9	0.08	~100	~2	py	Ba,Ah	Stevenson, 1945
Lara	Sicker Gp.- basalt, rhyolite	Devonian	Wrangellia	0.84	0.6	3.6	0.8	80.5	3.3	py		company reports, 1987
Seneca	Harrison Lake Fm.-andesite, rhyolite	Jurassic	Harrison	~0.9	0.84	5.2	0.2	55.5	1.1	py	Ba,Gyp	company reports, 1984
Homestake	Eagle Bay Fm.- schist, felsic volcanics	Devono- Missis- sippian	Kootenay	0.9	0.55	4	2.5	224	0.8	py	Ba	Høy and Goutier, 1986;
Rea and Samatosum	Eagle Bay Fm.- schist, felsic volc., phyllite chert, basalt	Devono- Missis- sippian	Kootenay	0.243 0.6	0.53 1.2	2.25 3.5	2.15 1.7	73.4 1100	6.5 1.8	py,arseno py	Ba	Høy and Goutier, 1986 company reports
Ecstall	schist-felsic? volcanics	Paleozoic?	Coast Plutonic Complex; meta- morphic pendant	4.5	0.8	2.3				py		Bacon, 1953
Windy-Craggy	'Tats Group'- basalt, siltstone, limestone	Late Triassic	Alexander	165.4	1.9	pres- ent		3.9	0.2 up to 12.8	po,py	.08%Co	Company reports, 1990
Greens Creek	argillite, chert, rhyolitic tuff	Late Triassic	Alexander	3.1	0.5	9.7	3.9	823	6.17	py,po	Ba	Berg, 1981 Buneltzen, et al, 1988
Anyox	basalt, siltstone	Triassic?	Coast Plutonic Complex; Stikinia?	25.1	1.4			9.5	0.17	po,py		Sharp, 1980; Alldrick, 1986
Granduc	'Unuk R. Fm.'- schist, andesite calcareous, siliceous seds.	Jurassic	Stikinia	32.5	1.93			7	0.13	po,py		Grove, 1986
Goldstream	Lardeau Group- phyllite, calc- areous and graphitic schist	Early to Middle Paleozoic	North America	3.2	4.5	3.1		20		po	mangani- ferous chert	Høy et al., 1984
Chu Chua	Fennell Formation basalt	Permian	Slide Mountain	~3	2	0.5		9 up to 3	0.5 0.1% Co	py	.05 to	Aggarwal & Nesbitt, 1984
Sunro	Metchosin Fm.- basalt, gabbro	Tertiary	Crescent	2.8	1.3			2	0.7	po,py		Stevenson, 1950

Main sources of information: Unpublished compilation by T. Høy (1984); McMillan et al. (1986).
Abbreviations used: po-pyrrhotite, py-pyrite, gyp-gypsum, arseno-arsenopyrite, Ba-barite, Ah-anhydrite, Co-cobalt

Table 19.3. Major porphyry copper and molybdenum deposits

DEPOSIT	MAP NO. FIG. 19.12	PRODUCTION YEARS	SIZE (MILLION TONNES) PRODUCTION TO 1984	RESERVES	ORE GRADE RECOVERED PLUS RESERVE GRADE Cu%	Mo%	Au g/t	Ag g/t
Bell (Newman)	1	1972-1982*	42	17	0.37	-	0.17 (.35)	1.0
Granisle		1966-1982	52	-	0.41	-	0.09 (.12)	1.12
Gibraltar	2	1972-1984*	152	163	0.35	0.004	- (.007)	1.03
Highland Valley	3	1962-1984*						
Lornex			228	349	0.39	0.013	0.0005(.006)	1.2
Bethlehem			106	38	0.50	-	0.012	0.1
Highmont			35	88	0.25	0.024	- (.004)	0.9
Valley Copper			17	559	0.48	-	0.014	
JA			-	286	0.43	0.017	-	
Brenda	4	1970-1984*	131	33	0.16	0.026	0.014 (0.031)	0.63
Island Copper	5	1971-1984*	114	143	0.45	0.01	0.11	0.63
Fish Lake	7			218	0.24	-	0.51	1.5
Poison Mountain	8	-	-	159	0.33	0.007	0.12 (0.3)	3.1
Catface	9	-	-	151	0.45	-	-	
Berg	10	-	-	238	0.39	0.03	- (.05)	5.0
Schaft Creek	14	-	-	910	0.3	0.025	(0.11 - .32)	1.5
Casino	16	-	-	162	0.37	0.023	- (.32)	1.75
Copper Mountain	18	1917-1962	31	-	0.9	-	0.19	3.9
Similkameen		1972-1984*	77	64	0.47	-	0.13	.63
Afton, Ajax, etc.	19	1977-1984*	15	56	0.75	-	0.44 (.60)	4.0
*Mount Polleg	20	-	-	48.3	0.38	-	.55	4.5
Stikine - (Galore Ck.)	22	-	-	125	1.06	-	- (.40)	7.7
Mt Milligan	22a	-	-	417	0.221	-	0.49	1.0
Endako	23	1965-1982*	139	141	-	0.08	-	
BC Moly - (Kitsault)	24	1967-1972	45	36	-	0.2	0.01	4.6
Boss Mountain	25	1965-1982	2.2	-	-	0.7	-	
				4.2	-	0.14	-	
Logtung	26	-	-	145	-	0.03	0.13% WO$_3$	
Glacier Gulch	27	-	-	30	-	0.16	0.06% WO$_3$	
Ajax	28	-	-	197	-	0.07	-	
Adanac	29	-	-	201	-	0.06	- (.01)	0.20
Red Mountain	30	-	-	187	-	0.10	-	
Quartz Hill (Alaska)	31	-	-	1207	-	0.13	-	

* denotes active mine
( ) denotes contained precious metal in geological reserves
sources: mine production statistics; Sinclair et al. (1982); unpublished company data.

Jurassic age (Mihalynuk et al., 1991), hosts the formerly producing Atlin Ruffner Ag,Pb,Zn vein deposit.

Near Cache Creek in southern British Columbia, the undeveloped Maggie porphyry Cu,Mo deposit occurs in a Paleocene granitoid stock concordant with the northwest trend of host rocks of the Cache Creek Assemblage (Miller, 1976). The Maggie stock may be related to a belt of similar isolated Cu,Mo porphyries that flanks the eastern margin of the Coast Plutonic Complex, but apparently is controlled primarily by steep faults bounding the southern extremity of the Cache Creek Terrane. The nearby Big Slide Au,Ag,Cu vein is associated with a diorite stock of unknown age and affinity.

Stikinia

Stikinia, the largest terrane in the Canadian Cordillera, underlies most of the Intermontane Belt and extends more than 1700 km from eastern Alaska to south-central British Columbia (Fig. 19.1F). Stikinia is a coherent stacked sequence of Devonian to Jurassic volcanic and sedimentary strata, and plutonic suites largely comagmatic with the arc-related volcanics. It includes the Devono-Permian Asitka, Triassic Stuhini, and Lower and Middle Jurassic Hazelton and Takwahoni assemblages, and is overlain by several post-accretionary volcanic and sedimentary sequences including the Skeena, Carmacks and Kamloops assemblages (Fig. 19.1F).

Volcanic and sedimentary strata of the Asitka Assemblage in northwestern Stikinia may represent an upper Paleozoic calc-alkaline volcanic arc with fringing carbonate banks and basinal carbonate, clastic, and chert sequences. Volcanogenic Au,Ag,Zn,Cu,Pb massive sulphide deposits of Kuroko type occur in the Tulsequah district (e.g. Tulsequah Chief, Polaris Taku; Table 19.2). Deformation preceded the resumption of volcanism in Late Triassic (Carnian to middle Norian) time (Monger, 1980). Similar subalkaline-to-alkaline volcanic-arc sequences, in which breccias dominate over flows, submarine tuff, limestone, and argillite, were deposited as the western Takla Group in eastern and northeastern Stikinia, the Stuhini Group in northern and northwestern parts of the terrane, and the Lewes River Group in the Whitehorse Trough.

The Golden Bear Au mine, 137 km west of Dease Lake in northwestern Stikinia, started production in 1989. Three mineralized zones include 1.3 million tonnes of proven and probable reserves, which grade 11 g/t Au (North American Metals Corp., 1987 Annual Report). Mesothermal Au-quartz veins within a north-trending fault zone 20 km long are hosted by silicified limestone, dolostone, and tuffite of the Permian Asitka Assemblage. The main period of mineralization has been interpreted, from isotopic studies, to have been Early Jurassic (Schroeter, 1987).

The Sustut Copper prospect in northeastern Stikinia is a large stratabound assemblage of copper sulphides and native copper in intermediate fragmental volcanic rocks of the Takla Group (Kirkham, 1970; Harper, 1977; Wilton and Sinclair, 1988). Similar Cu(Ag) deposits of either burial-metamorphic or diagenetic origin occur at White River, Yukon Territory, Coppermine River, Northwest Territories, and Keweenaw Peninsula, Michigan (Kirkham, 1984).

Upper Triassic augite porphyries of Stikinia petrologically resemble volcanics of the eastern Nicola Group of southern Quesnellia (Souther, 1977). Continuity of the two terranes is disrupted by the intervening Cache Creek Terrane, which is, according to Mortimer (1986), the remnant of at least 900 km of oceanic crust subducted eastward beneath Quesnellia. Polarity of the Stikinia arc is not known, but chemical and petrological affinities with Quesnellia imply a geometry similar to the westward-facing Nicola arc as proposed by Mortimer (1987).

Amalgamation of Stikinia with the Cache Creek Terrane on the east in Late Triassic time and consequent steepening of the eastward-dipping subduction zone and(or) initiation of subduction of the Cache Creek Terrane southwestward beneath Stikinia may have generated the shoshonitic (high-K) Polaris, Stikine, and Copper Mountain plutonic suites and comagmatic volcanics (Spence, 1985; de Rosen-Spence and Sinclair, 1988). Alaskan-type zoned mafic-ultramafic plutons of the Polaris Suite intrude coeval volcanics in Quesnellia. As in Quesnellia, comagmatic plutonic suites both coeval with and slightly younger than the upper Triassic volcanics host important porphyry-type deposits of the alkalic suite (Galore Creek Cu,Au,Ag) and calc-alkalic suite (Schaft Creek Cu,Mo; Table 19.3).

Uplift and plutonism in the Skeena and Stikine arches started in Late Triassic time and influenced Jurassic deposition in the intervening Bowser Basin, the Toodoggone Volcanic Belt, the southern flank of the Stikine Arch, the Hazelton belt on the Skeena Arch and along its flanks, and in the Whitehorse Trough lying northwest of the Stikine Arch. Lower Jurassic and lower Middle Jurassic rocks are marine to nonmarine, typically calc-alkaline pyroclastics, flows, and interbedded sediments, all of which were deposited as a series of basins, arches, and volcanic belts that possibly were interrelated as elements of one complex island arc environment (Tipper, Chapter 8).

Several calc-alkaline plutonic suites were comagmatic with arc volcanism of the Hazelton Assemblage in Early to early Middle Jurassic time, but the genetic relationship of volcanism and plutonism to either amalgamation or accretionary processes is not known. In southwestern Yukon Territory, calc-alkaline stocks and batholiths of the Klotassin and Long Lakes suites may have been comagmatic with Early Jurassic calc-alkaline volcanism in the Whitehorse Trough (Tempelman-Kluit, in press). Cu,Mo porphyry (e.g. Cash, Klazan, Revenue) and Au,Ag vein (e.g. Laforma) prospects are hosted by the Big Creek, Minto, and Granite plutons. Au(Ag,Pb,Zn) veins occur within the Klotassin Batholith in the Moosehorn Range.

The mainly volcanic Hazelton Assemblage was deposited on the Skeena Arch and along its flanks. The thickest and most felsic sections of the Lower Jurassic Telkwa Formation coincide with plutons of the Topley Suite, interpreted to be comagmatic roots of Telkwa volcanism by Tipper and Richards (1976). Significant mineralization is not known to have occurred with the Topley Suite. Deformed and metamorphosed Early and Middle Jurassic plutons along the eastern part of the Coast Plutonic Complex and the western margin of Stikinia at Stewart, Terrace, and Whitesail Lake are lithologically similar to the Topley Suite, and interpreted by Woodsworth et al. (Chapter 15) to have been comagmatic with the Telkwa Formation of the Hazelton Group. In the Stewart-Iskut River district the coeval and probably comagmatic Early Jurassic Texas Creek Granodiorite and related, mainly alkaline stocks and dykes are associated with significant Au,Ag(Cu,Mo) vein deposits, which include the operating mines Silbak-Premier, Big Missouri, and the developed prospects, Mount Johnny, Scottie, Sulphurets, and Snip (Alldrick et al., 1987; Britton and Alldrick, 1988). On the opposite side of Stikinia, along the southwestern flank of the Stikine Arch, calc-alkaline plutons of the Black Lake Suite intruded coeval and comagmatic Lower Jurassic Toodoggone volcanics and underlying Takla Group volcanics and sediments. The Toodoggone district is metallogenically similar to the Stewart-Iskut River district, with several significant epithermal Au,Ag vein deposits, including the Cheni mine and the developed Al, Mets, and Chapelle prospects, in addition to porphyry Cu,Mo and skarn Au,Cu prospects, all of which are of Early Jurassic age (Schroeter, 1983; Vulimiri et al., 1986).

Ag-rich Zn,Pb, barite, jasper deposits in the Alice Arm camp, including the formerly producing Dolly Varden and Torbrit mines, have been described as epithermal veins (Campbell, 1959). The deposits were recently reinterpreted (Devlin and Godwin, 1986) as structurally displaced portions of a once continuous volcanogenic massive sulphide layer in volcanic strata of the Lower Jurassic Hazelton Assemblage.

Calc-alkaline plutons of the predominantly Middle Jurassic Three Sisters Suite in the Skeena Arch intruded Cache Creek strata that were thrust southwestward over the amalgamated Intermontane Superterrane during its accretion to North America (Anderson, 1983). The Red-Chris porphyry Cu, Au prospect might be alkalic member of the Middle Jurassic Three Sisters Suite rather than the Copper Mountain Suite.

Felsic plutons of the Late Jurassic to Early Cretaceous Francois Lake Suite (Carter, 1982) intruded rocks of Stikinia, Quesnellia, and Cache Creek terranes along the eastern part of the Skeena Arch in central British Columbia. The post-accretionary plutons which range from 135 to 145 Ma (White et al., 1970) have few coeval plutons in the Canadian Cordillera and no comagmatic volcanics. The Endako quartz monzonitic pluton of the Francois Lake Suite hosts the Endako porphyry Mo mine, the largest Mo deposit in Canada, with 1971 reserves of 190 million tonnes of 0.15% MoS_2 (Dawson and Kimura, 1972; Table 19.3).

The Whitehorse Suite of post-tectonic, mid-Cretaceous, predominantly granodioritic plutons, which intruded the Lewes River and Laberge groups in Whitehorse Trough, was interpreted by Tempelman-Kluit (pers. comm., 1987) to have been comagmatic with subaerial volcanics of the Mount Nansen Group. Cu,Fe,Mo,Au,Ag skarns of the 30 km Whitehorse Copper Belt, are developed at re-entrants along the irregular, dioritic western contact of the Whitehorse pluton. Pyroxene and garnet-rich skarns formed after both limestone and dolostone of the Lewes River Group (Meinert, 1986). Total production from 11 deposits mined intermittently between 1898 and 1982 was 10 million tonnes of ore grading about 1.5% Cu and 0.7 g/t Au (Watson, 1984; Table 19.4.1).

The Late Cretaceous Bulkley Plutonic Suite forms a belt of small stocks and batholiths that trends northward along the Skeena Arch, localized by north-trending faults (Carter, 1982). These plutons may have been coeval with the Surprise Lake Suite and volcanics of the Carmacks Assemblage in northern Stikinia. Upper Cretaceous Brian Boru volcanics are interpreted to have been comagmatic with the Rocher Deboule stock (Sutherland Brown, 1960). The Bulkley Suite and comagmatic volcanics are interpreted by Woodsworth et al. (Chapter 15) to be high-level products of deep-seated plutonism and metamorphism in the Coast Belt to the west, representing a transpressive orogen controlled by an extensional stress field.

Two periods of Mo,W mineralization at the Glacier Gulch (Hudson Bay Mountain) porphyry prospect at Smithers are related to the intrusions of a sheet-like granodiorite body and a later quartz-porphyry plug (Kirkham, 1967; Bright and Jonson, 1976). The nearby Huckleberry and Ox Lake porphyry prospects, 80 km southwest of Houston, contain Cu and Mo minerals disseminated in granodiorite porphyry plugs and hornfelsed Hazelton tuffs (Carter, 1970; Sutherland Brown, 1969). Au,Ag(Cu,Pb,Zn) veins at the developed Dome Mountain prospect 20 km east of Smithers are interpreted (MacIntyre et al., 1987) to be contemporaneous with buried intrusions of the Bulkley Suite emplaced during early stages of folding of the lower Hazelton Group fragmental volcanic host rocks.

Plutons of the Eocene Nanika Suite, the most widespread in Stikinia, include the Nanika Intrusions south of Bowser Basin (Carter, 1982), and the Kastberg and Alice Arm intrusions east and west of the basin, respectively (Woodsworth et al., Chapter 15). The characteristically small, calc-alkaline granitic to granodioritic plutons, dykes, and sills were emplaced along steeply dipping, northward-trending faults related, in part, to the emplacement of the earlier Bulkley Suite. Nanika plutons south of Bowser Basin were interpreted by MacIntyre (1985) to be roots of deeply eroded coeval volcanics. Significant porphyry deposits associated with the Nanika Intrusions include the Berg Cu,Mo (Panteleyev, 1981), the Lucky Ship and Red Bird Mo prospects southwest of Houston, and the Mount Thomlinson Mo and Big Onion Cu,Mo prospects in the Hazelton and Smithers areas, respectively (Carter, 1982). Reserves for the Berg deposit are given in Table 19.3. Small, multi-stage felsic intrusions and related breccias of the Nanika Intrusions forcefully intruded and thermally metamorphosed Hazelton Assemblage host rocks. Sulphide minerals were preferentially deposited at their fractured and altered contacts. The Equity Silver (Goosly) mine is a vein stockwork that is, in part, massive semi-concordant orebodies of Ag,Cu,Au,Sb sulphide and sulphosalt minerals in pyroclastic and clastic rocks of the Cretaceous Skeena Assemblage. Adjacent gabbro, syenomonzonite, and somewhat younger quartz monzonite of the Goosly Lake Intrusions are subvolcanic to Eocene volcanic centres according to Church (1971), and genetically related to the mineralization. The previously producing Silver Queen (Nadina) Ag,Zn,Au,Cu,Pb vein deposit, 18 km south of Houston, is hosted by Late Cretaceous andesitic volcanics and bracketed by Eocene feldspar prophyry dykes similar to those at Goosly Lake (Leitch et al., 1990).

The Quanchus Suite in west-central British Columbia is similar to the Nanika Suite, which adjoins it on the west, but contains only minor porphyry Cu,Mo prospects (Woodsworth, Chapter 15). In the Fawnie Range, uplift and erosion of the Stuhini and Hazelton assemblages exposed felsic volcanic centres of the lower Tertiary Ootsa Lake Group and comagmatic plutons of the Quanchus Suite. The Wolf epithermal Au,Ag vein prospect was related by Andrew et al. (1986) to resurgent doming within a caldera, and the adjacent Capoose Lake epithermal Ag(Pb,Zn,Cu,Au) vein prospect was related by Church and Diakow (1982) to a subvolcanic porphyritic pluton.

The Eocene Alice Arm Intrusions extend northwestward along the eastern margin of the Coast Plutonic Complex from Terrace to the Stewart area as a group of widely separated, typically small granite to quartz monzonite porphyry stocks with significant associated porphyry molybdenum deposits. They have intruded clastic rocks of the Upper Jurassic Bowser Assemblage at or near the intersections of north-northwest and east-northeast faults (Carter, 1982). The previously producing British Columbia Molybdenum (Kitsault) mine contains evidence of multiple stages of quartz-molybdenite veining associated with a zoned quartz monzonite-quartz diorite stock (Steininger, 1985). Molybdenite mineralization is characteristically displayed around a core of potassic altered quartz monzonite porphyry and alaskite, whereas Ag,Pb,Zn veins are found peripheral to the stocks (Carter, 1982). Other significant porphyry Mo prospects in the Alice Arm camp include Bell Moly, Ajax, and Roundy Creek (Table 19-3).

Table 19.4. Major Cordilleran skarn and replacement deposits

1. Copper skarns Cu (Fe,Au,Ag,Mo)

Locality, status M-Mine, D-District/ PR-Prospect, PP-Past Producer, AC-Active	NTS	Years Prod'n	Metal produced tonnes, kilograms	Reserves M: mining; G: geological E: estimate	Terrane	Tect. ass; age; formation; lithology	Intrusive, Age	References
1. Craigmont M/PP	92I/02	1961-1982	402 714 t. Cu, minor Fe,Au,Ag	1.5 Mt 1.13% Cu(M)	Quesnellia	Nicola; lTr W.Nicola, lst, vclst.	granodior. Jur	Morrison, 1980
2. Greenwood D/PP	82E/02	1893-1976	269 604 t. Cu 35 984 kg Au 117 242 kg Ag	ca. 2 Mt 0.5% Cu (M)	Quesnellia	Nicola; m-lTr Brooklyn; lst.	granodior.	Church, 1986
3. Ingerbelle M/AC (see also porphyry copper deposits, Table 19.3)	92H/07	1972-pres.	228 132 t. Cu 9 932 kg Au 44 554 kg Ag (to 1982)	127 Mt 0.38% Cu .16 g/t Au .63 g/t Ag (1982, M)	Quesnellia	Nicola; lTr Nicola; andes.	monzonite, syenodior. eJur	Fahrni et al., 1976
4. Whitehorse Copper D/PP	105D/11	1898-1982	142 000 t. Cu 7 090 kg Au 90 000 kg Ag	2.6 Mt 1.11% Cu (M)	Stikinia	Stuhini; lTr Lewes River; lst, ark	diorite, granodior. mK	Watson, 1984 Meinert, 1986
5. Coast Copper D/PP	92L/06	1962-1971	42 300 t. Cu 3 840 kg Au 10 976 kg Ag		Wrangellia	Karmutsen; lTr Quatsino/Karm; lst, volc.	diorite gabbro eJur	BC Minfile, 1984 Meinert, 1986
6. Stikine Copper (Galore Creek) D/PR (see also Table 19.3)	104G/03	-	-	125 Mt 1.06% Cu 0.4 g/tAu 7.7 g/tAg (G)	Stikinia	Stuhini; lTr -; volc, sltst.	syenite eJur	Allen et al., 1976

2. Tungsten skarns W,Cu (Zn,Mo)

Locality, status M-Mine, D-District/ PR-Prospect, PP-Past Producer, AC-Active	NTS	Years Prod'n	Metal produced tonnes, kilograms	Reserves M: mining; G: geological E: estimate	Terrane	Tect. ass; age; formation; lithology	Intrusive, Age	References
1. Canada Tungsten M/PP	105H/16	1962-1986	40 087 t. WO_3 to 1986	1.53 Mt 1.24% WO_3 0.2% Cu(M)	North America	Gog; e.Camb.; Sekwi, lst.	qtz monz mK 90-92 Ma	Mathieson and Clark, 1984
2. Salmo D/PP	82F/03	1943-1973	8 056 t. WO_3	-	North America	Gog; e.Camb.; Laib, lst.	qtz monz mK	Mulligan, 1984
3. Macmillan Pass PR	105O/08			32 Mt 0.92% WO_3 (G)	North America	Gog; e.Camb.; Sekwi, lst	qtz monz mK (88 Ma)	Atkinson and Baker 1986
4. Risby PR	105F/14			2.7 Mt 0.81% WO_3	Cassiar	Gog; eCamb.; -; lst.	qtz monz mK8	The Northern Miner, July, 1982
5. Ray Gulch PR	106D/04			5.44 Mt 0.82 WO_3 (G)	North America	Hyland; Prot.-Camb, Grit; lst	granodior. uK	Lennan, 1986
6. Lened PR	105I/07			ca.150 000 t 1.2-1.3% WO_3 (E)	North America	Rocky Mtn; Camb.-Ord.; Rabbitkettle; lst	qtz monz mK 85-92 Ma	Glover and Burson, 1986
7. Baily PR	105A/15			405 455 t 1.0% WO_3(E)	North America	Rocky Mtn;Dev.; Unit 7; lst	granodior. mK	Dawson and Dick, 1978 Gabrielse, 1967 DIAND, 1981, p. 140
8. Baker PR	105H/16			120 000 t. 1.3-1.5%WO_3 (E)	North America	Gog; e.Camb.; Sekwi, lst	qtz monz mK	Canada Tungsten staff, 1986
9. Clea PR	105I/13			ca.100 000 t 1.5% WO_3 (E)	North America	Rocky Mtn; Camb.-Ord.; Rabbitkettle; lst	qtz monz mK (94 Ma)	Godwin et al., 1980

Table 19.4. Continued.

3. Zinc-lead-silver skarns and replacements Zn,Pb,Ag (Cu,Au,Cd)

3.A Southeastern Cordillera

Locality, status M-Mine, D-District/ PR-Prospect, PP-Past Producer, AC-Active	NTS	Years Prod'n	Metal produced tonnes,kilograms	Reserves M: mining; G: geological E: estimate	Terrane	Tect.ass, age; formation; lithology	Intrusive, Age	References
Riondell (Bluebell) M/PP	82F/10	1895-1971	248 871 t. Zn 233 736 t. Pb 228 t. Ag 2 855 t. Cu	-	Kootenay	Gog; e.Camb.; Badshot,lst	granodior. m.Jur.	Høy, 1982b
Mineral King M/PP	82K/08	1928-1968	90 190 t. Zn 37 292 t. Pb 27 000 t. Ag 662 t. Cu 314 t. Cd 56 516 t. $BaSO_4$	-	North America	Purcell; Late Prot.; Mt. Nelson,dolomite	not exposed	Fyles, 1959
Slocan-Sandon D/PP	82F/14	1894-1950	192 130 t. Pb 139 700 t. Zn 1 780 t. Ag (both replacement and vein deposits)	-	Quesnellia	Nicola; Late Trias.; Slocan; lst,argill.qtze	granodior M.Jur.	Hedley, 1952

3. Zinc-lead-silver skarns and replacements Zn,Pb,Ag (Cu,Sn,Au)

3.B Northeastern Cordillera

Locality, status M-Mine, D-District/ PR-Prospect, PP-Past Producer, AC-Active	NTS	Years Prod'n	Metal produced tonnes,kilograms	Reserves M: mining; G: geological E: estimate	Terrane	Tect.ass, age; formation; lithology	Intrusive, Age	References
1. Mt. Hundere AC	105A/10			5.2 Mt 13.3% Zn 5.3% Pb 63.8 g/t Ag(M)	North America	Hyland;e.Camb.; lst, phyll	not exposed mK(?)	Dawson and Dick, 1978 Abbott, 1981 Mining Review, 1990
2. Quartz Lake D/PR	95D/12			1.50 Mt 6.6% Zn 5.5% Pb 102 g/t Ag(M)	North America	Hyland;e.Camb.; limy qtze	not exposed	Morin, 1981 Vaillancourt, 1982a,b
3. Midway PR	104O/16			1.18 Mt 9.6% Zn 7.0% Pb 410 g/t Ag(M)	Cassiar	Cassiar; mDev.; McDame; lst	not exposed 1K(?)	Bradford and Godwin, 1988
4. Silver Hart PR	105B/06			97 070 t. 3.8% Zn 1.4% Pb 958 g/t Ag(M)	Cassiar	Gog; e.Camb.; Atan; lst, schist	qtz monz mK	Abbott, 1983 Buhlmann, pers. comm., 1988
5. Cassiar D/PR	104P/05			488 500 t. 4.4% Zn 5.3% Pb 168g/t Ag(G)	Cassiar	Gog; e.Camb.; Atan; lst. dol.	qtz monz uK (72 Ma)	Bloomer, internal report, Shell Canada Resources, 1981
6. Roy PR	95E/12			145 000 t. 3.11% Zn 5.24% Pb 147 g/t Ag 1.3 g/t Au(G)	North America	RockyMtn;l.Camb.; Rabbitkettle lst	granodior mK	Kukor, internal reports, Logan Mines, 1982
7. Tintina Silver D/PR	105G/03			91 000 t. 10% Zn 6% Pb 640 g/t Ag(E)	Cassiar	Gog; e.Camb.; Atan; lst	qtz dior eT (51 Ma)	Yukon MIR, 1976; Dawson (unpub. data)

Table 19.4. Continued.

4. Iron skarns Fe (Cu,Au,Ag)

Locality, status M-Mine, D-District/ PR-Prospect, PP-Past Producer, AC-Active	NTS	Years Prod'n	Metal produced tonnes,kilograms	Reserves M: mining; G: geological E: estimate	Terrane	Tect.ass, age; formation; lithology	Intrusive, Age	References
1. Texada D/PP	92F/15	1885-1976	10 Mt Fe (est.) 35 900 t. Cu 40 012 kg Ag 3 313 kg Au	-	Wrangellia	Karmutsen, lTr; Marble Bay/ Karm; lst, bslt	granodior, gabbro, eJur	Sangster, 1969 Ettlinger & Ray, 1988
2. Tasu M/PP	105C/16	1914-1982	5.5 Mt Fe 56 084 t. Cu 47 030 kg Ag 1 302 kg Au	5.46 Mt 45% Fe 0.3% Cu (1976)(M)	Wrangellia	Karmutsen, lTr; Kunga/Karm; lst, basalt	dior porph	Sutherland Brown, 1968
3. Jedway M/PP	103B/06	1962-1968	3.94 Mt Fe	-	Wrangellia	Karmutsen, lTr; Kunga/Karm; lst, basalt	diorite	Sutherland Brown, 1968
4. Brynnor (Kennedy Lake) M/PP	92F/03	1962-1970	3.0 Mt Fe	-	Wrangellia	Karmutsen; lTr; Quatsino; lst	granodior eJur	Sangster, 1969
5. Argonaut-Iron Hill M/PP	92F/13	1951-1957	2.03 Mt Fe	-	Wrangellia	Karmutsen; lTr; Quatsino/Karm; lst, basalt	granodior eJur	Sangster, 1969 Meinert, 1984
6. Merry Widow-Kingfisher D/PP	92L/06	1957-1967	1.68 Mt Fe	-	Wrangellia	Karmutsen; lTr; Quatsino,/Bonanza;lst/volc.	dior-gabbro eJur	Sangster, 1969
7. Nimpkish M/PP	92L/07	1959-1963	890 704 t. Fe	-	Wrangellia	Karmutsen, lTr; Quatsino/Karm; lst, volcs.	diorite eJur	Sangster, 1969
8. Coast Copper-Old Sport M/PP	92L/06	1962-1970	508 023 t. Fe 27 488 t. Cu 11 731 kg Ag 3 869 kg Au	816 480 t. 0.33% fe (M)	Wrangellia	Karmutsen; lTr; Quatsino/Karm; lst/andes.	diorite eJur	Sangster, 1969
9. Zeballos M/PP	92L/03	1962-1963	227 066 t. Fe	-	Wrangellia	Karmutsen; lTr; Quatsino/Bonanza;lst, volc.	diorite eJur	Sangster, 1969

5. Gold skarns Au (Ag,Cu)

Locality, status M-Mine, D-District/ PR-Prospect, PP-Past Producer, AC-Active	NTS	Years Prod'n	Metal produced tonnes,kilograms	Reserves M: mining; G: geological E: estimate	Terrane	Tect.ass, age; formation; lithology	Intrusive, Age	References
Hedley D/AC	92H/08	1902-1955 1987-	55 380 g Au 6 083 g Ag	10.8 Mt (M) 4.56 g/t Au	Quesnellia	Nicola; lTr; Sunnyside;lst	diorite, gabbro e-m Jur	Ray et al., 1987
Tillicum Mountain (Esperanza) D/AC	82K/04	1987-	-	730 500 t(M) 9.3 g/t Au	Quesnellia	Nicola;lTr-eJ; Rossland; siltst,volc.	syenodior eJur	Esperanza Exploration Interal Report, 1989
Texada (Marble Bay, Little Billie, etc.) D/PP	92F/15	1896-1952	2 425 kg Au 16 368 kg Ag 9 157 t. Cu	152 000 t 1.5% Cu 7.2 g/t Au	Wrangellia	Karmutsen, lTr; MarbleBay/Karm; lst, basalt	granodior, gabbro	Ettlinger & Ray, 1988 Mining Review, 1990
Banks Island (Tel, etc) PR	103G/08	-	-	90 700 t(M) 17.4 g/t Au	Alexander	Pz?; lst, sltst.	qtz diorite eK (120 Ma)	Ettlinger & Ray, 1988
QR (Quesnel River) PR	83A/12	-	-	1 m.t. (G) 6.5 g/t Au	Quesnellia	Nicola, lTr-eJ; Quesnel R; basalt, argill.	diorite-monzodior eJur(201Ma)	Fox, 1986

The Eocene Hyder pluton in the Stewart area north of Alice Arm is predominantly a biotite granodiorite (Smith, 1977) and related to an abundant Early Tertiary plutonic suite in the northeastern Coast Plutonic Complex (Woodsworth et al., Chapter 15). The pluton is spatially related to numerous Au,Ag,Pb,Zn vein deposits, including the significant past producing Prosperity-Porter Idaho, Silverado, and Indian mines (Alldrick, 1985; Alldrick et al., 1987).

The Eocene Babine Intrusions are small, distinctive biotite-feldspar granodiorite-to-diorite porphyry stocks and dykes in a fault-controlled northwest-trending belt in the Babine Lake area. The copper-bearing plutons, which include multiple intrusions of porphyry dykes and breccia (Kirkham, 1971), were regarded, by Carter (1982), as volcanic centres for extrusive equivalents in the area. Four significant porphyry Cu,Au,Ag deposits, including the producing Newman (Bell Copper) and previously producing Granisle mines and the developed Morrison and Old Fort prospects, demonstrate the annular zoning of sulphide minerals and potassic- through phyllic- to argillic-alteration assemblages (Carson and Jambor, 1974) and the high precious-metal contents characteristic of classic (non-plutonic) calc-alkaline porphyry deposits (Sinclair et al., 1982). Reserves are given in Table 19.3.

The recently closed Blackdome mine is situated near the south end of Stikinia. Epithermal, bonanza-type Au and Ag mineralization occurs in Upper Cretaceous to lower Tertiary volcanics of the Kamloops Assemblage (Faulkner, 1986).

Bridge River, Cadwallader, and Methow terranes

Several small terranes occur in the western part of the southern Intermontane Belt. The Bridge River Terrane consists of variably metamorphosed chert, argillite, basalt, alpine-type ultramafics, and minor carbonate and plutonic rocks that represent an accretionary prism and oceanic crust of Permian to Middle Jurassic age. The Cadwallader Terrane contains Upper Triassic island arc volcanics and clastics, suggested by Rusmore (1987) to have formed in the same basin as the Bridge River Group. These are overlain by Jurassic sedimentary and volcanic rocks and Jura-Cretaceous clastic sediments. The Methow Terrane contains Upper Triassic basalt overlain by Lower Jurassic arc clastics and volcanics and Jura-Cretaceous clastic wedges shed from Quesnellia. All three terranes contain economically important Au deposits related to initiation or reactivation of transcurrent faults of the Fraser and subsidiary fault systems.

Excluding placer deposits, the numerous vein deposits of the Bridge River camp (Fig. 19.5), particularly the Bralorne and Pioneer mines, have been the largest producers of Au in the Canadian Cordillera. From 1899 to 1978, 130 tonnes of Au and 31 tonnes of Ag were produced from 7.3 million tonnes of ore (Table 19.1).

The Bralorne and Pioneer veins are hosted by Permian diorite and Upper Triassic greenstone within a complex, northwest-trending fault system (the Bralorne fault zone). The stratigraphic assignment of these rocks is controversial (e.g. Rusmore, 1985; Leitch and Godwin, 1988), but there is some agreement that the rocks belong to the Cadwallader and/or Bridge River terranes. The veins, which are of mesothermal type, include minor pyrite, arsenopyrite, gold, scheelite, and stibnite and occupy en echelon tension fractures in the wall rocks (Cairnes, 1937; Joubin, 1948). The age of the mineralization has also been controversial, but recent U-Pb and K-Ar dating by Leitch and Godwin (1988) strongly indicates a Late Cretaceous age.

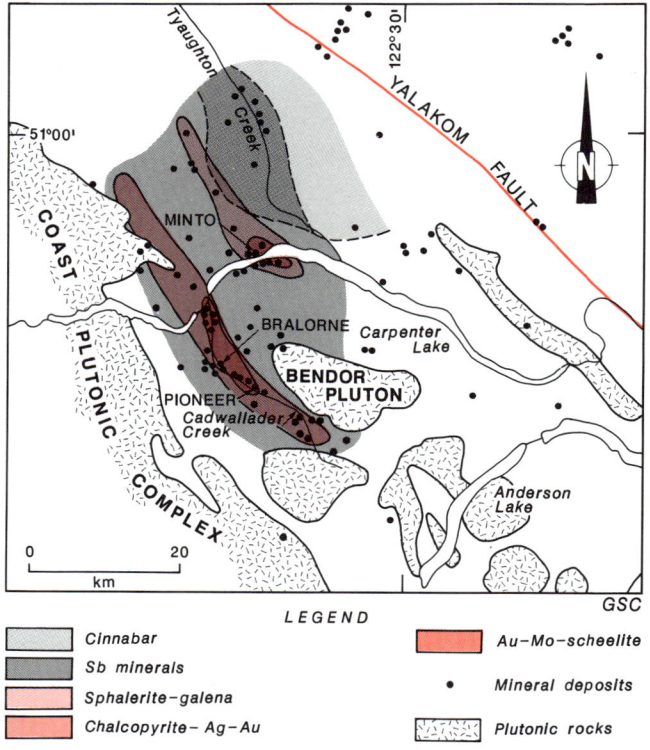

Figure 19.5. Zoning of mineral deposits in the Bridge River camp, British Columbia. After Woodsworth et al. (1977).

Woodsworth et al. (1977) described a distinctive zoning pattern in the Bridge River camp (Fig. 19.5). Two northwest-trending mineralized centres of Au mineralization are apparent. The western, centred around Bralorne, is characterized by high Au/Ag ratios. The eastern, centred around the formerly producing Minto mine, has low Au/Ag ratios. These two zones lie within a larger area of Sb mineralization. Northeast of the two centres, the Sb zone is partly overlapped by a Hg zone that extends across the Yalakom Fault and, in general, lacks Au deposits. Woodsworth et al. (1977) suggested that this zoning reflects proximity to Late Cretaceous to Early Tertiary plutons of the Coast Plutonic Complex, from which heat and water were derived. In western Stikinia near Terrace, the vein deposits show a similar zoning, with high-temperature deposits near plutons and low-temperature deposits farther away.

East of the Fraser River, stratified rocks of the Methow Terrane host a north-northwest-trending belt of Au deposits, including the former producer Carolin mine. Au mineralization occurs in and proximal to greenstones and clastic sediments of the Ladner Group and Spider Peak Formation (Ray, 1983; Ray et al., 1986). All known deposits are near the faulted contact of the Ladner Group with serpentinite, chert, argillite, and greenstone of the Bridge River Terrane. Mineralization is mesothermal, replacement-type characterized by pyrrhotite, pyrite, arsenopyrite, albitic alteration, and quartz veins. The age of mineralization is not known but appears to be contemporaneous with mid-Cretaceous to pre-Late Eocene folding (Ray et al., 1986) and hence may be roughly the same age as that in the Bridge River camp.

Coast Plutonic Complex

The Coast Plutonic Complex (Fig. 19.1G) is coincident throughout much of its length with the Coast Belt (Woodsworth et al., Chapter 15). The complex consists of discrete and coalescing plutons that intrude subgreenschist to amphibolite-facies volcanics, sediments and orthogneiss. The east and west boundaries are somewhat arbitrary, as there are no sharp changes in age, composition, or plutonic style vis a vis flanking regions.

Generally, the Coast Plutonic Complex may be viewed as a metamorphic and plutonic welt reflecting Middle Jurassic to Tertiary arc plutonism and volcanism and accretion of the Insular Superterrane to the Intermontane Superterrane. Superposed metamorphism and deformation and the paucity of fossils make determination of the protolith and age of metamorphic rocks difficult. However, the complex includes some strata that definitely belong to the Harrison, Chilliwack, and Wrangellia terranes in the south and to the Alexander, Taku, and Stikine terranes in the north. Most plutons and many metavolcanic rocks have arc affinities which reflect a long-lived accretionary history and, to a large extent, most mineral deposits in the Coast Plutonic Complex are in a sense accretionary in origin. This extended period of subduction, accretion and mineralization which contrasts with shorter accretionary events recognized in other Cordilleran terranes is an unique attribute of the Coast Plutonic Complex.

The main commodities in mineral deposits of the Coast Plutonic Complex are Au and Cu; Ni, Mo, and Pb,Zn are locally important (Woodsworth and Roddick, 1977). Compared with other areas of the Canadian Cordillera, the Coast Plutonic Complex has less-than-average mineral wealth, both in terms of number of known deposits and production to date (Fig. 19.1G; Sinclair et al., 1978). Nonetheless, the belt hosts one of the largest massive sulphide deposits in the Canadian Cordillera (e.g. Britannia). Furthermore, the important Au deposits of the Bridge River area and porphyry deposits in the Bulkley and Nanika suites east of the Coast Plutonic Complex may reflect Late Cretaceous to Early Tertiary metamorphic and plutonic events within the Coast Plutonic Complex (Woodsworth et al., 1977).

The distribution of mineral deposits in the Coast Plutonic Complex varies both along and across the regional grain. In a longitudinal direction, two moderately well mineralized segments (49-51°N and 52-55°N) are separated by a relatively barren area that is flanked on the east by a similarly barren part of the Intermontane Belt. In the northern mineralized segment, the area southwest of Work Channel Lineament (Chapter 17, Fig. 17.8) contains numerous sulphide deposits. The region northeast of the lineament hosts few deposits; the high grade metamorphic rocks of the Central Gneiss Complex are particularly barren.

Pre-accretionary(?) deposits

Several massive sulphide deposits of uncertain origin may predate the main Jura-Cretaceous accretionary events. In the Stewart-Iskut River area, polydeformed sediments and volcanics host the large Granduc massive sulphide deposit (Fig. 19.1G). Gypsiferous, graphitic and calcareous sediments and andesitic volcanics host concordant sulphide lenses dominated by pyrite, chalcopyrite, pyrrhotite, sphalerite, and galena (Grove, 1986). About 32.5 million tonnes averaging 1.93% Cu were mined (Table 19.2). The host rocks are of uncertain age, but they possibly correlate with Paleozoic or lower Mesozoic strata of Stikinia.

Several massive sulphide deposits in the Ecstall River area lie within highly deformed metamorphic rocks of the Work Channel Lineament (Gareau, 1988). The orebodies are steeply-plunging, pipe-like structures enclosed in quartz-sericite schist layers in amphibolite-facies metasediments and metavolcanics. Mineralization consists largely of pyrite with greatly subordinate sphalerite, chalcopyrite, and galena. Precious metal values are low. Mineralization clearly predated regional metamorphism and deformation. A volcanogenic origin is probable for these deposits, but both their age and mode of origin remain speculative. The Scotia deposit in the same belt to the north is similar in many respects to the Ecstall deposits, but is characterized by sphalerite-rich ore with relatively minor pyrite.

In the Anyox area, volcanogenic massive sulphide deposits include the former Hidden Creek mine that produced about 25 million tonnes grading 1.4% Cu (Table 19.2). The principal sulphide lenses in the area are pipe-like to sheet-like lenses in pillowed tholeiitic ridge basalt (Sharp, 1980). Grove (1986) correlated the pillow lavas with Middle Jurassic strata of the Hazelton Group, but a Triassic or Paleozoic age seem equally plausible.

The Seneca deposit in the Harrison Terrane of the southern Coast Belt is hosted by Middle Jurassic andesite, rhyolite, and argillaceous breccia of the Harrison Lake Formation. The fragmental, polymetallic sulphide ore is associated with gypsum and barite and is similar to the Kuroko-type deposits of Japan (Church et al., 1977; Table 19.2).

Massive sulphides in overlap assemblages

The Jura-Cretaceous Gambier Assemblage (Fig. 19.1F) is presumed to be an arc-related overlap assemblage which is the volcanic equivalent of many of the plutons within the Coast Plutonic Complex. The assemblage hosts the Britannia deposit, after Windy Craggy the largest known volcanogenic deposit in the Canadian Cordillera. Production from the various orebodies totalled about 48 million tonnes grading 2.8% Cu and 0.25% Zn (Table 19.2). The Kuroko-type orebodies are situated near the contact of dacitic tuff and breccia with overlying Albian(?) shales and were formed from hydrothermal fluids genetically related to the dacitic volcanism (Payne et al., 1980). Several periods of inhomogeneous strain produced the northwest-trending Britannia shear zone; some sulphides were remobilized into crosscutting quartz veins during and after deformation. The Nifty prospect northeast of Bella Coola is a massive sulphide deposit that also may be genetically related to Gambier volcanism.

The origin of the Northair deposits southwest of Pemberton is controversial. The deposits are in intermediate metavolcanic rocks which have been correlated with the Gambier Assemblage (Woodsworth et al, 1977; Miller and Sinclair, 1985). Sulphide minerals are mainly pyrite, sphalerite, galena, chalcopyrite, and tetrahedrite. Au is present in economically important amounts. The deposits have been interpreted by many geologists as veins formed during emplacement of plutons bounding the pendants (e.g. Little, 1974). Another explanation is that the deposits formed as distal volcanogenic exhalites during Early Cretaceous volcanism (Miller and Sinclair, 1985) with remobilization of sulphides during emplacement of nearby plutons.

Deposits related to accretionary plutonism

The majority of mineral deposits in the Coast Plutonic Complex are related to Jurassic to Tertiary "accretionary" plutonism. These deposits include auriferous veins, Au and Cu,Fe skarns, Ni,Cu magmatic deposits, and Cu,Mo porphyry bodies.

Veins show a great variety in geological setting, mineralogy, and economic importance. Gold, the main commodity in most veins, occurs both as the native metal and as tellurides. The only significant production has come from the Surf Inlet and Surf Point deposits (Fig. 19.1G). The Surf Inlet veins produced about 920 000 tonnes of ore averaging about 13 g/t Au. The veins occupy late tensional fractures subsidiary to a complex northwest-trending fault zone (Gill and Byers, 1948). At Surf Point they are localized along an arch of foliation planes in a small tonalite stock (Smith, 1947, 1948). At both deposits, veins are quartz with auriferous pyrite and minor chalcopyrite. K-Ar dating of vein sericite suggests that mineralization at Surf Point may be Late Cretaceous in age (P. van der Heyden, pers. comm., 1988). The age of the Surf Point stock is unknown but, on regional radiometric studies (P. van der Heyden, pers. comm., 1988), an Early Cretaceous age is possible.

Of the several porphyry deposits known in the southern Coast Plutonic Complex, the OK deposit near Powell River is the best-known. Chalcopyrite and minor molybdenite are associated with a quartz stockwork related to a dyke-like body of porphyritic leucogranodiorite (Meyer et al., 1976). Stratified rocks in the region are included in Wrangellia, suggesting that the OK deposit may be part of the suite of Upper Jurassic porphyry deposits on nearby Vancouver Island.

Skarn and related deposits are found mainly in the western Coast Plutonic Complex and include the economically significant Au skarns on Banks Island as well as small Fe and Cu,Pb,Zn,Ag deposits. The Banks Island deposits (Tel, Bob, Hepler) occur in marble and metasediments. Although the ages of these strata are unknown, the rocks likely are part of either the Alexander Terrane or Wrangellia. Enclosing plutonic rocks are Late Jurassic in age (P. van der Heyden, pers. comm., 1988). Mineralization is in skarn, and quartz-sulphide veins that are, in part, controlled by west to northwest-trending fracture zones (Ettlinger and Ray, 1988; Table 19.4.5).

The Giant Mascot mine near Hope in the southern Coast Plutonic Complex has been the only Ni producer in the Canadian Cordillera; 4.3 million tonnes of ore averaging about 0.63% Ni and 0.3% Cu were mined. The orebodies are concentrated in a crudely elliptical ultramafic complex about 2 km across enclosed in granodiorite to diorite of the Spuzzum Pluton (Vining, 1977). The ultramafic rocks are dominantly pyroxenite with lesser dunite and hornblendite. Unlike zoned Alaskan ultramafic complexes of the Polaris and Union Bay suites (Chapter 15), the Giant Mascot bodies contain abundant orthopyroxene. Sulphides are dominantly pyrrhotite with subordinate pyrite, pentlandite, chalcopyrite and other sulphides and form interstitial fillings among the silicates and massive sulphide bodies (Aho, 1956; Clark, 1969). There is little doubt that the ultramafic rocks and sulphide deposits are genetically related to the Early Cretaceous Spuzzum pluton. They may reflect reaction of tonalitic Spuzzum magma with pre-existing, mantle-derived ultramafic rocks which have been found in the region or, less likely, they may represent early cumulate phases of the Spuzzum Pluton.

Post-accretionary deposits

Few Eocene mineral deposits exist within the Coast Plutonic Complex, although porphyry deposits on its east flank (British Columbia Molybdenum) and vein deposits related to the Hyder Pluton near Stewart are clearly a consequence of Eocene plutonism in the complex.

The Oligocene to Miocene Chilliwack Suite of plutons extends from Washington northwest along the axis of the southern Coast Plutonic Complex (Chapter 15). Metallogenically important mineralization is associated with several of these high-level, calc-alkaline plutons. Au ± Ag mineralization in quartz veins in tension fractures is found along the west side of the Chilliwack Batholith and near small Miocene intrusions along Harrison Lake (e.g. Doctors Point) (Ray, 1986). Porphyry mineralization is found in several plutons. The best-known of these is the Salal Creek deposit (Stephens, 1972), well dated at 8 Ma. Molybdenite mineralization is concentrated in an arcuate zone centred on the contact between a fine grained granite core and a medium grained granite to granodiorite margin. Cu and Au values are very low. In contrast, the Franklin Glacier porphyry, dated at 7 Ma, contains pyrite and chalcopyrite with subordinate molybdenite.

The largest known post-accretionary porphyry deposit in the Coast Plutonic Complex is the world-class Quartz Hill deposit in southeastern Alaska. Stockwork and disseminated molybdenite mineralization is distributed throughout an Oligocene granite stock (Hudson et al., 1979). Reserves are estimated at about 1.3 billion tonnes averaging about 0.136% Mo (Bundtzen et al., 1984).

Insular Superterrane
Alexander Terrane

Two belts of Devonian to Permian, mainly marine volcanic and sedimentary rocks of the Alexander Terrane, one along its northeastern contact with Wrangellia and the other in the central part of the terrane in the Saint Elias Mountains, are deformed and regionally metamorphosed to lower greenschist facies (Campbell et al., Chapter 8; Fig. 19.1H). Granitoid plutons of the Late Pennsylvanian to Early Permian Icefield Ranges Suite are of unknown tectonic significance (Dodds and Campbell, 1988). Important mineral deposits are not known to occur in Paleozoic rocks in the Canadian part of Alexander Terrane.

Upper Triassic rocks in Alexander Terrane are best documented in the Tatshenshini River area near the developed Windy Craggy Cu,Co(Au,Ag,Zn) volcanogenic massive sulphide prospect. A basal limestone and turbiditic siltstone unit of unknown age and thickness is in fault contact with deformed Middle Devonian carbonates. It is overlain by and partly interfingered with a 1000 m thick lower division of tholeiitic, massive amygdaloidal to pillowed basalt flows, diorite sills, minor chert, and argillite. A middle division, 2000 m thick, which hosts the Windy Craggy orebody in its lower part, includes limy argillite, chert, calc-alkaline dioritic sills and dykes, mafic to intermediate, massive amygdaloidal to pillowed flows, tuff, agglomerate, and limestone. Conodonts from the middle division, including orebody host rocks, have determined its age to be Early Norian or mid-Late Triassic (M.J. Orchard, pers. comm., 1983). An upper volcanic division consists of at least 1500 m of calc-alkaline pillow basalts of Early to Middle Norian age (Gammon and Chandler, 1986; MacIntyre, 1984). The large cupriferous pyrite deposit, which contains reserves of 300 million tonnes of 1.5% Cu, 0.08% Co, and 1-3 g/t Au (Table 19.2), shows some similarities to Cyprus-type deposits, yet the dominantly alkaline to subalkaline compositions of the host volcanics (MacIntyre, 1986) and the abundant craton-derived clastic sediments are more characteristic of Besshi-type mineralization (Fox, 1984). Pb isotopes from Windy Craggy sphalerites plot within the field of the Besshi-type ores of the Sambagawa belt, Japan (Sato and Sasaki, 1976, 1980). The tectonic setting of the Upper Triassic rocks in the Tatshenshini River area, as deduced from adjacent miogeoclinal sediments, the progression from tholeiitic to calc-alkaline volcanism, turbiditic clastics, and Besshi-type mineralization, is interpreted to have been one of epicontinental rifting prior to amalgamation with Wrangellia in Middle Jurassic time.

The age of several stratiform polymetallic sulphide-barite occurrences in the Mount Henry Clay area, including the Low Herbert Cu,Zn(Ba,Ag,Au) and Glacier Creek, Alaska Ba,Zn,Cu,Ag,Pb prospects, is the same as that of the Upper Triassic volcano-sedimentary assemblage in the Tatshenshini River area, based on galena-Pb isotope and limited conodont data (Dawson, 1985). Unlike those at Windy Craggy, the host rocks are intermediate to felsic calc-alkaline volcanics, volcaniclastics, and equivalent quartzose phyllite and schist (MacIntyre and Schroeter, 1985). In southeastern Alaska, a related Upper Triassic metallogenic province of polymetallic massive sulphides in metamorphosed calc-alkaline volcano-sedimentary rocks, recognized by Berg (1981), may also include the Greens Creek Ag,Zn,Au,Pb,Cu mine on Admiralty Island and the Tel Au-rich base metal skarn on Banks Island in southernmost Alexander Terrane (Dawson, 1985).

In the Datlasaka Range of the eastern Tatshenshini map area, calcareous argillite, mudstone, and limestone of Norian age are interbedded with mafic volcanics and have been intruded by gabbro-diabase sills (Campbell et al., Chapter 8). Lenses of gypsum at the O'Connor River gypsum-anhydrite prospect are stratabound within deformed limestone and argillite beds (White, 1986a). In the Rainy Hollow area, limestone, argillite, and quartzite host several post-accretionary Ag,Cu(Au,Bi,Zn) skarns, including the previously producing Maid of Erin mine (Watson, 1948). The Upper Triassic sediments form a pendant within quartz diorite and granodiorite of the Three Guardsmen Batholith of predominantly Oligocene, therefore post-accretionary, age (Dodds and Campbell, 1988). A pluton in the southeastern tip of the batholith is of mid-Cretaceous age (MacKevett et al., 1974) indicating that the batholith has a composite age and was partly emplaced at the time of accretion.

Several skarn and vein deposits in southern Alexander Terrane are associated with Cretaceous granitoid plutons and are considered to be syn-accretionary in age. Included are the Cu,Au skarns of the Jumbo district, Prince of Wales Island (Kennedy, 1953), and possibly the Au skarns and veins on Banks Island (Ettlinger and Ray, 1988), mentioned previously under 'Coast Plutonic Complex'.

Few significant post-accretionary deposits are known in the Alexander Terrane (Fig. 19.1H). In addition to the Rainy Hollow Ag,Cu skarns are the Gold Cord Au vein occurrence in the same area, hosted by diorite of the Oligocene Tkope pluton, and the Souther porphyry Cu,Mo occurrence in a pluton subvolcanic to Wrangell volcanics.

Southern Wrangellia

Southern Wrangellia is, to a large degree, coextensive with the southern Insular Belt. Its eastern boundary with the Intermontane Superterrane, however, has been largely obliterated by plutons of the Coast Belt. On southern Vancouver Island the comparatively small Pacific Rim and Crescent terranes occur adjacent to and beneath Wrangellia (Fig. 19.1I).

Wrangellia is dominated by three thick, discrete volcanic piles, each separated by thinner platformal sedimentary sequences and penetrated by plutons that are comagmatic and coeval with the youngest volcanic sequence. The tectonic settings of the three superposed volcanic successions were, in time sequence: a primitive marine arc, a marine rift or backarc rift, and a mature emergent arc. The oldest volcanic succession is in the Sicker Group and is of Devonian age. It is dominated by fragmental basaltic andesites, but dacites and rhyolites occur amidst intermediate pillow lavas and epiclastic volcanic rocks in its upper accumulations. The volcanics are overlain by Carboniferous sedimentary strata characterized by bioclastic crinoidal limestone of the Buttle Lake Formation. These Paleozoic rocks have not been observed in the Queen Charlotte Islands. Minor folding, uplift, erosion, and shale deposition occurred prior to Late Triassic eruption of thick, uniform ferro-tholeiite pillow lavas of the Karmutsen Formation. These lavas in turn are overlain by shallow-water carbonate of the Quatsino Formation on Vancouver Island and the Kunga Formation in the Queen Charlotte Islands. The third volcanic sequence on Vancouver Island is the Bonanza Group of Early Jurassic age, comprising pyroclastic andesites to rhyolites, which in their upper part were subaerial in origin. In the Queen Charlotte Islands eruption of the marine and andesitic Yakoun Formation did not begin until the Middle Jurassic Epoch. Neither the bases nor the tops of these superposed piles have been recognized, yet their measured aggregate thickness is more than 12 km. The Island Intrusions on Vancouver Island and the Jurassic plutons on the Queen Charlotte Islands were comagmatic with the Jurassic volcanism.

Accretionary sedimentary sequences are not important volumetrically but are represented by the Lower Cretaceous Longarm Formation in the Queen Charlotte Islands and the Kyuquot Group on Vancouver Island. Post-accretionary clastic sequences are represented by the Queen Charlotte and Coal Harbour groups of mid-Cretaceous age, and the Nanaimo Group of Late Cretaceous age. Volcanism and plutonism re-occurred in mid- to late Paleogene and early Neogene times, resulting in the Catface Intrusions on Vancouver Island, the post-tectonic Queen Charlotte Suite, and bimodal Masset volcanics. An overlap sequence of mainly nonmarine clastics of Mio-Pliocene age, the Skonun Formation, overlies the Masset and the inferred boundary between Wrangellia and the Alexander Terrane on eastern Graham Island.

On southern Vancouver Island the Pacific Rim Terrane consists principally of Upper Triassic to Lower Cretaceous volcanic, metavolcanic, and metasedimentary rocks of the Pacific Rim Complex and Leech River Formation in the footwalls of the Westcoast and San Juan-Survey Mountain faults. The Crescent Terrane consists of marine pillow lavas of the Eocene Metchosin Complex and occurs adjacent to and beneath the Pacific Rim Terrane in the footwall of the Leech River Fault.

The diverse, voluminous, and long-lasting volcanism, related plutonism, intercalated reactive carbonates, and repeated tectonic activity provided a uniquely fertile environment for metallic mineralization. This is borne out in the southern Insular Belt by the high density of mineral occurrences. From an exposed area of about 50 000 km², 765 deposits are included in the MINDEP file (Sinclair et al., 1978). This amounts to 15.4 occurrences per 1000 km², the highest density in the Cordillera, and twice the average for the remainder of British Columbia. Most of the deposits are small but consistent with the metallogenic pattern evident in the distribution of major deposits types. Most of the deposits and the major examples of pre-accretionary types are dominated by concentrations of iron, copper, zinc, and molybdenum sulphides, precious metals, and magnetite in volcanic massive sulphide, skarn, and porphyry deposits. The most common setting of significant volcanic massive sulphides is in the upper part of the Sicker Group in or adjacent to rhyolites. Skarn deposits occur at or near the common contact between the Karmutsen Formation, Quatsino Formation, and Island Intrusions, and less commonly, between plutons that intruded either the Quatsino Formation and Bonanza Group or the Buttle Lake Formation. Small plutons of the Island Intrusions are the most common sites for porphyry deposits. Many vein and shear-zone deposits within Wrangellia, commonly sulphide-rich, also are clearly related to development of volcanic piles. Still others, which occupy young structures, are locally rich in quartz and ferroan dolomite and are considered to be post-accretionary deposits, which may in part involve the redistribution of earlier metals. Mineral deposits of accretionary age are virtually unknown in the southern Insular Belt, but post-accretionary deposits are common. Many are epigenetic precious-metal accumulations in young fault zones that cut Wrangellian rocks, and others notably in the Queen Charlotte Islands occur within sequences or plutons of post-accretionary origin. Also important are porphyry Cu,Mo, Cu,Au,Ag(Mo) or Au,Ag deposits in the Catface Suite. A few unique or ambiguous deposits are included in this category, such as a regolithically enriched skarn deposit, a germanium and kaolinite deposit in lignite and claystone, and a diffused auriferous quartz-ferroan dolomite stockwork in schistose Sicker rocks.

The Pacific Rim and Crescent terranes contain few occurrences; among them meta-turbidite-hosted gold veins in the Leech River Formation of the Pacific Rim Terrane and massive sulphides in shear zones in the Metchosin Complex of the Crescent Terrane.

Pre-accretionary deposits

Volcanogenic massive sulphide deposits are clearly an integral part of the host sequence. Significant examples occur in felsic volcanics of the upper part of the Sicker Group. Near-surface accumulations are contained in domal culminations of the Buttle Lake and Cowichan anticlinoria. Such deposits include the Westmin Lynx-Myra-Price and HW orebodies of the Buttle Lake Anticlinorium (Walker, 1983) and the Mount Sicker and Lara deposits of the Cowichan Anticlinorium. The Lynx-Myra-Price deposit is a single, thin, extensive but folded and disrupted orebody, which is segmented by erosion. It consists principally of pyrite, sphalerite, and chalcopyrite with minor galena. Ore with elevated silver and gold content commonly contains barite and is banded. The orebody occurs above a thin rhyolite unit in the upper part of the mine sequence and some 500 m below the base of the Buttle Lake Formation. The deposit has been mined since 1966 and in 1983 had accumulated production and reserves of approximately 6 million tonnes grading Cu: 1.5%, Zn: 7.6%, Pb: 1.1%, with about 109 g/t silver and 2 g/t gold (Fleming et al. 1983; Table 19.2).

The HW is a blind orebody (i.e. not exposed at the surface) which occurs about 500 m below the Lynx zone. It is a more continuous, thicker, and more pyritic but less disrupted orebody than the Lynx, and exhibits prominent lateral metal zonation. It occurs beneath a thick rhyolite unit and above a zone of intense sericitic alteration and stringer pyrite. Total HW reserves were quoted in 1983 as 15 million tonnes, with an average grade of Cu: 2.2%, Zn: 5.3%, Pb: 0.3% with about 37.7 g/t silver and 2.4 g/t gold (Fleming et al., 1983; Table 19.2). Mining of the HW zone began in 1985.

Two deposits of similar character occur in the Cowichan Anticlinorium; the formerly producing Mount Sicker (Twin J) mine and the newly discovered Lara prospect. These deposits, which are much more disrupted than the Buttle Lake orebodies, occur in schistose upper Sicker Group rocks adjacent to thrust faults. The Mount Sicker property was mined intermittently between 1898 and 1964, with total production of less than 300 000 tonnes of ore of a grade similar to that of the Buttle Lake deposits. The Lara prospect is currently being explored; reserves quoted in 1987 are 837 000 tonnes grading 0.61% Cu, 3.59% Zn, 0.81% Pb, with 89.5 g/t silver and 3.3 g/t gold (Table 19.2).

Skarn deposits rich in iron, some containing significant copper and precious metals, are common in southern Wrangellia. Most are near or partly in Jurassic plutons at the top of the Karmutsen Formation, replacing it and/or limestone of the Quatsino or Kunga formations. The orebodies are highly irregular and controlled by the distribution of favourable strata and igneous contacts as well as by pre-ore faults and breccia pipes which they crudely mimic in shape (Sutherland Brown, 1968, 1972). The ore minerals, magnetite and chalcopyrite, are contained normally in an envelope of silicate skarn minerals, principally epidote, chlorite, and andradite garnet with lesser actinolite, diopside, pyrite, and pyrrhotite (Sangster, 1969). Deposits of this type that have been mined include: Tasu and Jedway on the Queen Charlotte Islands, Argonaut, Coast Copper, Kennedy Lake, Merry Widow, Nimpkish, and Zeballos on Vancouver Island, and Texada on Texada Island (Fig. 19.1I, see Fig. 19.16). Some deposits with different characteristics from those described above include the smaller, generally pipe or vein-like copper-rich skarn ore zones such as Ikeda and Vananda, on Moresby and Texada islands, respectively. A few of the iron skarns, such as the Merry Widow, are sited at the contact of the Bonanza Group with the Quatsino Formation. Still other copper skarns replace limestone of the Buttle Lake Formation (i.e. the mined component of the Thistle deposit). Gold may be erratically distributed in copper- and sulphide-rich parts of some iron skarns (Meinert, 1984; Ettlinger and Ray, 1988).

Most of the iron skarn deposits supported short-lived mines of only a few million tonnes. The two largest deposits were Tasu, which between 1963 and 1982 produced 20.9 million tonnes containing 5.5 million tonnes of magnetite concentrate with 56,084 tonnes of copper, plus significant precious metals, and Texada, which produced somewhat less magnetite and Cu ore, mainly between 1957 and 1976 (Table 19.4.4).

Porphyry deposits of Jurassic age are not common in southern Wrangellia, although some pyritic stockworks in plutons of this age are known. The single known economic deposit is the important Island Copper mine on northern Vancouver Island (Northcote and Robinson, 1973). The orebody is cored by a 100 m wide quartz-feldspar dyke which dips 70 degrees to the northeast, and has intruded, brecciated, altered, and mineralized Bonanza pyroclastic andesites in a fairly symmetrical pattern (Cargill et al., 1976; Fleming, 1983). The dyke is similar to and projects toward a nearby pluton dated at 154 ± 4 Ma, which is thought to have been comagmatic with dacites of the upper part of the Bonanza Group (Northcote and Muller, 1972). Intensely pyrophyllitized breccia caps the deposit, and breccias with rotated fragments flanking the dyke grade outward into shatter breccias. The ore minerals consist chiefly of chalcopyrite and molybdenite with lesser magnetite, and form an orebody that is about twice the width of the dyke, which generally is less mineralized. Weathering and supergene enrichment are not significantly developed. Alteration extends outwards from a core in the dyke of weakly developed sericite through intense chlorite-sericite to biotite and ultimately epidote. The mine began production in 1971 with stated reserves of 257 million tonnes grading 0.52% Cu, 0.17% Mo with significant precious metals and high trace rhenium (Cargill et al., 1976; Table 19.3).

Several minor deposits of unique or ambiguous character, but which are metallogenically important, occur in southern Wrangellian rocks. These include three different deposits near Port Alberni, including: skarn in limestone and massive sulphides in mafic dykes at the Thistle mine, auriferous hematite at the Villalta prospect, and auriferous altered stockworks at the Debbie prospect. A group of small stratabound Cu(Ag,V) deposits near Menzies Bay, Vancouver Island and on Quadra Island in the upper Karmutsen Formation occur in minor interlava fossiliferous calcareous shale. They contain chalcocite, bornite, native copper, malachite, and a yellow vanadium mineral, volborthite mainly near the base of the shale and partly replacing fossils. The individual interlava shales are 1 to 2 m thick and only tens of metres in length. The deposits appear to be syngenetic to diagenetic in origin, with metals leached from the enclosing flows, which have high background copper (Northcote and Muller, 1972).

The Thistle mine consists of two deposits of apparently different type and possibly different age. One is a copper skarn and the other, massive to disseminated copper-iron sulphide bodies in mafic dykes. The locale is geologically complex, being on the intensely faulted southwestern flank of the Cowichan Anticlinorium, where the latter is cut by diabase dykes of Karmutsen affiliation. The mined ore was a Au-rich chalcopyrite skarn replacing Buttle Lake limestone. The sulphide deposits are contained in a series of steeply dipping sheeted dykes, which may have acted as eruptive conduits for the overlying Karmutsen lavas.

The two other unusual deposits seem to combine preaccretionary Wrangellian mineralization with post-accretionary enrichment or reconcentration. The Villalta is a small lens of auriferous hematite adjacent to thinly bedded carbonates and cherts of the Buttle Lake Formation, near both Island and Catface intrusions, and at the unconformity beneath the Nanaimo Group. The intensely sheared Paleozoic rocks show indications of early skarn mineralization. The origin of the Villalta deposit is unknown but two stages of mineralization followed by some supergene enrichment seem likely.

The Debbie prospect is a newly discovered deposit yielding high gold assays, but its geological setting is poorly known. The new 900 zone seems to consist of a dense stockwork of auriferous quartz-ferroan dolomite stringers in schistose upper Sicker Group tuffs and cherts near faults. Some of the pyritic cherts may have been exhalative in origin and have served as a protolith for post-accretionary Au concentration.

Post-accretionary deposits

Post-accretionary deposits consist mainly of epigenetic veins, stockworks or porphyries, but some unusual bedded types are found also. Veins and stockworks occur both in post-accretionary sequences and in Wrangellian rocks where they may be difficult to recognize. Unequivocal post-accretionary deposits are numerous in the Queen Charlotte Islands where they are found in the Masset and Skonun formations as well as in plutons of the Queen Charlotte Suite. They are somewhat rarer on Vancouver Island except in association with the Catface Suite. In the Queen Charlotte Islands these deposits are represented by the Inconspicuous prospect on northern Graham Island, Cinola deposit on southeastern Graham Island, and the April prospect on Lyell Island.

The Inconspicuous prospect is hosted by felsic lapilli tuff of the Masset Formation in a broad zone of argillic alteration enclosing small veinlets of chalcedony with minor pyrite and gold. The April prospect also occurs in the Masset Formation in marine mixed pyroclastic breccias near the Beresford Fault on northwestern Graham Island. It consists of a silicified and pyritized area with small gold-bearing veins.

Cinola, the largest and most fully explored example of a post-accretionary deposit on the Queen Charlotte Islands, occurs in the upper Tertiary Skonun Formation along the Sandspit Fault Zone. A northeast-dipping splay of this system juxtaposes Skonun conglomerate and sandstone in the hanging wall against the Queen Charlotte Group and unconformably overlying Masset Formation in the footwall. In its hanging wall the fault was the locus of intrusion of a quartz porphyry dyke up to 40 m wide. The rhyolite has been dated at 14 Ma (Champigny and Sinclair, 1982). Where thickest, the intrusion has an intensely silicified envelope surrounded by argillic alteration. The dyke and Skonun host have been subjected to multiple hydrothermal brecciation and silicification, the intensity of which correlates with gold grade (Tolbert et al., 1987). Drusy and banded veins of quartz and chalcedony with minor pyrite, marcasite, and rare free gold originated as gash fractures related to late movement on the fault. These veins also have been brecciated. The deposit has been interpreted by Tolbert et al. (1987) as an epithermal hotspring-type precious-metal system. The mineable reserves have been quoted as 24.8 million tonnes of 2.45 g/t Au (City Resources (Canada) Ltd. unpublisheb report, 1988).

On Vancouver Island vein deposits in post-accretionary rocks are best represented by the Domineer-Lakeview zone at Mount Washington near Comox (Muller and Carson, 1969) and the Privateer at Zeballos. The Domineer is a flat, wide, and seemingly continuous silicified vein-shear that cuts Nanaimo Group, Catface sills, Karmutsen Formation, and mixed breccias. The vein system is spatially associated with the Mount Washington porphyry Cu(Mo) deposit and with numerous mineralized breccia pipes and screens. The main metallic vein minerals are pyrite and arsenopyrite, but minor amounts of an extensive suite of sulphides, tellurides, realgar and orpiment exist. Reserves of 214 000 tonnes of 7.4 g/t Au have been outlined (Better Resources News Release, 1/1987).

Privateer was the largest and most productive of a group of small mines in and adjacent to the Zeballos stock of Catface Suite age. The Privateer veins, hosted by thermally metamorphosed Bonanza Group rocks as well as the stock, are steep, thin, ribbon quartz veins with fairly abundant pyrite and arsenopyrite, lesser amounts of sphalerite, chalcopyrite, galena and pyrrhotite, and rare free gold (Stevenson, 1950). The mine produced 282 000 tonnes containing 18.8 g/t of gold with some silver, lead, and zinc.

Vein deposits in young structures that disrupt Wrangellian host rocks in the Queen Charlotte Islands are typified by the Court prospect and by a host of small deposits associated with Tertiary fault zones on Vancouver Island. The Court, on southwestern Graham Island, consists of a zone of westerly trending small veins in an area of argillic alteration in the Yakoun Formation close to the Tertiary Masset volcanics. Sulphide minerals consist of pyrite and gold-bearing stibnite. On Vancouver Island many veins occur within and adjacent to the northeasterly dipping high-angle thrust faults which outline the Cowichan Lake Fault Zone. These faults served as conduits for the Catface Intrusions which were commonly emplaced within Nanaimo Group strata. The thrusts are cut by still younger, steep northerly trending faults. Both fault systems are characterized by areas of intense ferroan dolomite alteration and fissure veins of auriferous quartz and ferroan dolomite. Many of the vein deposits of the China Creek area near Port Alberni are of this type. Specifically the Victoria mine, which produced a small tonnage between 1934 and 1936, is now part of the Debbie property.

Porphyry copper deposits in plutons of the Eocene Catface Suite are moderately common, and those at Mount Washington and Catface Mountain have been extensively explored. The Mount Washington mine occurs northeast of the Domineer prospect, where sediments of the Nanaimo Group are intercalated with sills of quartz porphyry which were injected from the nearby Mackay Lake quartz diorite stock dated by K-Ar at 35 Ma. About 400 000 tonnes of ore were mined from two small pits in 1965 and 1966; an average of about 1.16% copper with minor precious metals was recovered.

The Catface deposit near Tofino is in and adjacent to a small pluton of quartz monzonite but centred on a slightly younger phase of porphyritic quartz diorite that forms an irregular dyke-like apophysis (McDougall, 1976). Much of the ore is in wall rocks of meta-basalt, believed to be part of the Karmutsen Formation. The orebody is sub-cylindrical and more regular in shape than the plutons. The deposit is deeply oxidized and somewhat enriched along fractures as a result of its location in a high rainfall area with steep relief. The random stockwork has been mineralized with quartz-sulphide veinlets and sulphide-filled fractures which contain an extensive suite of minerals low in iron; chalcopyrite exceeds pyrite, bornite and molybdenite are common, and secondary minerals are common locally, even deep within the body. At the completion of exploration the stated reserves were 181 million tonnes of 0.45% copper (McDougall, 1976).

A unique deposit of germanium and kaolinite occurs in a small basin of Upper Cretaceous, possibly Nanaimo Group rocks, on the mainland coast at Lang Bay near Powell River. The undisturbed basin contains strata of mudstone, sandstone, and conglomerate, with common lignite lenses. The lignite contains up to 139 g/t GeO_2, but recovery is difficult. The intercalated kaolinitic rocks are regarded as economically important (White, 1986b).

Northern Wrangellia

In southwestern Yukon Territory the Duke River Fault separates Wrangellia from Alexander Terrane on the southwest, and the Shakwak and Dalton faults separate it from the Gravina-Nutzotin overlap assemblage on the northeast. A smaller part of Wrangellia is faulted against the southwestern side of Alexander Terrane in the Saint Elias Mountains (Fig. 19.1I). Volcano-sedimentary oceanic-arc rocks of the upper Paleozoic Skolai Assemblage, equivalent to the upper Sicker Assemblage in southern Wrangellia (Monger, Chapter 8), were intruded by the Upper Triassic Quill Creek Complex (Hulbert et al., 1988). The overlying tholeiitic and partly subaerial Nikolai

Greenstone is overlain, in turn, by shallow-water sediments as in the correlative Upper Triassic Karmutsen Assemblage in southern Wrangellia.

In the Kluane Ranges a belt of Ni,Cu,Co,platinum-group-element occurrences, 130 km long including the formerly producing Wellgreen mine and the Canalask deposit, is hosted by volcaniclastic and clastic sedimentary rocks of the Pennsylvanian Station Creek Formation of the Skolai Assemblage (Campbell, 1960; Read and Monger, 1976). Massive pyrrhotite-pentlandite-chalcopyrite-magnetite lenses of magmatic segregation-type are preferentially developed at altered and faulted contacts with gabbroic marginal units of the ultramafic-mafic Quill Creek Complex. Lesser amounts of sulphides are disseminated in gabbro, peridotite, and Paleozoic host rocks (Campbell, 1976; Hulbert et al., 1988). Prior to production at Wellgreen mine in 1972-73, reserves were given as 669 150 tonnes of ore grading 2.04% Ni, 1.42% Cu, 0.07% Co, 1.2 g/t Pt, and 0.86 g/t Pd (Muller, 1967). The deposit is distinguished by its high content of platinum group elements relative to other Ni,Cu deposits, particularly the high content of the rare elements Os, Ir, Ru, Rh, and also by its unique Late Triassic, rather than Precambrian, age (Hulbert et al., 1988). The deposits clearly precede the accretion of the Insular Superterrane in Late Jurassic or mid-Cretaceous time.

Generally subeconomic Cu sulphide-native Cu deposits, which are abundant in predominantly subaerial tholeiitic basalt of the Karmutsen Assemblage of Wrangellia in southwestern Yukon Territory (Read and Monger, 1976), are equivalent to similar Cu deposits hosted by the Nikolai Greenstone in the Wrangell Mountains of eastern Alaska (MacKevett, 1976). At the Silver City Cu,Ag deposit in the White River area, Yukon, 352 km northwest of Whitehorse, stringers and disseminations of mainly native Cu and chalcocite with lesser amounts of bornite, chalcopyrite, and pyrite are stratabound in a glomeroporphyritic basalt unit (Sinclair et al., 1979). Similar fault-bounded occurrences of stringer Cu sulphides are found in the Quill Creek area and near Dezadeash Lake, where lenses of sulphides at the Johobo deposit yield locally high Cu and Ag assays (Carrière et al., 1981). A genetic relationship to the rich Cu deposits at Kennecott, Alaska in the immediately overlying Chitistone Limestone was proposed by Armstrong and MacKevett (1977). Carrière et al. (1981) noted that equivalent limestone of the sabkha type does not occur in Canadian Wrangellia. Stratiform gypsum in the equivalent Nizina Limestone at Bullion Creek is interpreted by Campbell et al. (Chapter 8) to be a structurally displaced segment of Wrangellia within Alexander Terrane.

Campbell (1981) has cited spatial and temporal relationships between the Quill Creek Complex and the Nikolai Greenstone in support of a genetic relationship between Cu,Ni and Cu sulphide-native Cu deposits. An origin of the Nikolai Greenstone, and the Karmutsen Assemblage in general, by rifting of a previously established Paleozoic arc as proposed by Jones et al. (1977), apparently supports the emplacement of comagmatic mafic-ultramafic plutons along steeply dipping extensional structures. As previously noted in the case of the Sustut Cu sulphide-native Cu deposit in Stikinia, the origin of this class of deposit is not well understood, but a relatively early, pre-metamorphic, i.e. pre-accretionary age of mineralization, is favoured.

The Baultoff porphyry Cu(Mo) prospect in eastern Alaska contains an estimated 240 million tonnes of 0.2% Cu and 0.01% Mo in Cretaceous quartz diorite, quartz diorite porphyry, and granite porphyry (Nokleberg et al., 1987) and is considered to be accretionary. The similar Cork Cu(Mo) porphyry prospect in the Quill Creek area is Oligocene (Christopher et al., 1972) and post-accretionary.

Pacific Rim and Crescent terranes

The Pacific Rim Terrane is sparsely mineralized, although the Leech River Formation contains a number of prospects, mostly lode gold in meta-turbidites. These are typified by the Valentine Mountain prospect, where two sets of veins cut Leech River slate and schist, an early ramifying set of small veins and a later, planar crosscutting set of richer tenor. Both sets contain sparse sulphides, mainly pyrite, with erratic but spectacular free gold. Erosion of such deposits probably produced the placer gold deposits in the Sombrio and Leech rivers.

The predominantly volcanic Crescent Terrane is moderately well mineralized and hosts one recently producing mine, several small mines, from which ore was shipped near the turn of the century, and many small prospects. The prospects show similar characteristics, most consisting of massive to disseminated sulphides in penecontemporaneous shears with associated hornblendic alteration of basalt or gabbro. The Sooke Gabbro also contains similar structurally controlled deposits of pyrrhotite and magnetite. The Sunro mine at Jordan River consisted of a series of lenses of massive pyrite and chalcopyrite in a steeply dipping shear which cuts basalts of the Metchosin Complex and intercalated Sooke Gabbro sills. Both rock types in the shear zone were amphibolitized prior to sulphide deposition. The mine produced 1.3 million tonnes of about 1% copper with minor precious metals between 1961 and 1968.

PART B. MINERAL DEPOSIT MODELS

Several types of mineral deposits of economic importance in the Canadian Cordillera are considered in terms of global genetic models. Precious metal veins of epithermal and mesothermal type are emphasized, in accord with current commodity values. Volcanogenic massive sulphide deposits of both polymetallic (Zn,Pb,Cu) and cupriferous pyrite (Cu,Zn) types are discussed. The economically significant granitoid-related suite of porphyries, skarns and replacements is treated in detail. Sedimentary exhalative deposits from three miogeoclinal sedimentary assemblages are compared. Several deposit types of greater metallogenic than current economic importance, including Mississippi Valley-type Zn,Pb, redbed-type Cu, mafic Ni,Cu,PGE and iron-formations are treated elsewhere, under their host terrane or tectonic assemblage.

Precious metals (Au, Ag, platinum-group elements)

Gold, historically the most important metal in the Cordillera, was first mined from bedrock sources in 1852 in the Queen Charlotte Islands. This was five years before the production of significant placer gold from a tributary of the Thompson River near Lytton. To the end of 1989, production of 1230 tonnes gold has been recorded - 690 tonnes from lode sources, mainly in British Columbia, the rest from placer workings. Reported placer output of 375 tonnes of gold from Yukon and 165 tonnes from British Columbia is dominated by production at the end of the 1800s from the Klondike gold fields of the Yukon and during the mid- to late-1800s from the Cariboo gold fields of central British Columbia (Table 19.1). The Yukon deposits were among the richest placer gold fields in the world, mainly because they escaped dispersal by glaciation. Placer and lode gold deposits are spatially related (Fig. 19.6 and 19.7) and many lode deposits were discovered as a consequence of exploration in old placer areas. Notably, only recently have potential bedrock sources or 'motherlode' for the Klondike deposits been identified (see 'Nisling Terrane', previously discussed in this chapter).

Lode production of gold and silver from mainly quartz veins or replacement deposits flourished from the turn of the century into the 1960s. Thereafter byproduct gold and silver from porphyry copper and other large base-metal mines accounted for over 80% of annual Cordilleran production. In 1987 about 55% of Cordilleran lode gold production of about 17 tonnes gold and 520 tonnes silver were byproducts. Porphyry copper deposits yield an average 0.07 grams gold and 0.7 grams silver per tonne, skarn Cu; Cu,Fe; and Fe ores recover 0.3 grams gold and 19 grams silver per tonne. Volcanogenic massive sulphide deposits in basaltic rocks yield 0.07 grams gold and 30 grams silver per tonne but those in rhyolitic host rocks produce 1.75 grams gold and 82 grams silver per tonne (see Table 19.2) Following the price increase of gold in 1979, there has been renewed interest in primary gold mining. Current exploration is focused on small, high-grade, bonanza-type veins and large tonnage, bulk mineable precious metal deposits.

More than 100 lode mines have operated since 1894 but nearly 80% of gold production has come from six major camps, each producing in excess of 1 million ounces gold (31.1 tonnes). The main camps shown in Table 19.1 are: Bridge River, Rossland, Portland Canal-Stewart (Premier mine), Hedley, Cariboo-Barkerville and Boundary-Greenwood. The major primary producers, Bridge River, Rossland, Cariboo-Barkerville, as well as Sheep Creek-Ymir, are structurally, and in part stratigraphically controlled, quartz vein and pyritic replacement deposits or cupriferous gold veins. Hedley is a skarn gold deposit. The Portland Canal-Stewart deposits, including the major gold-silver producer, the Silback-Premier Mine, are subvolcanic or epizonal pluton-related vein and stockwork deposits. The sources of gold in the Greenwood-Boundary camp are skarn copper and copper-magnetite deposits, their related porphyry copper deposits and peripheral veins. Major byproduct gold has come from the Island Copper porphyry copper-molybdenum mine, Copper Mountain-Similkameen and Bell copper deposits and the Westmin and Britannia volcanogenic massive sulphide deposits. These and other gold producers were tabulated and reviewed by Schroeter and Panteleyev (1986a,b) and described by Barr (1980).

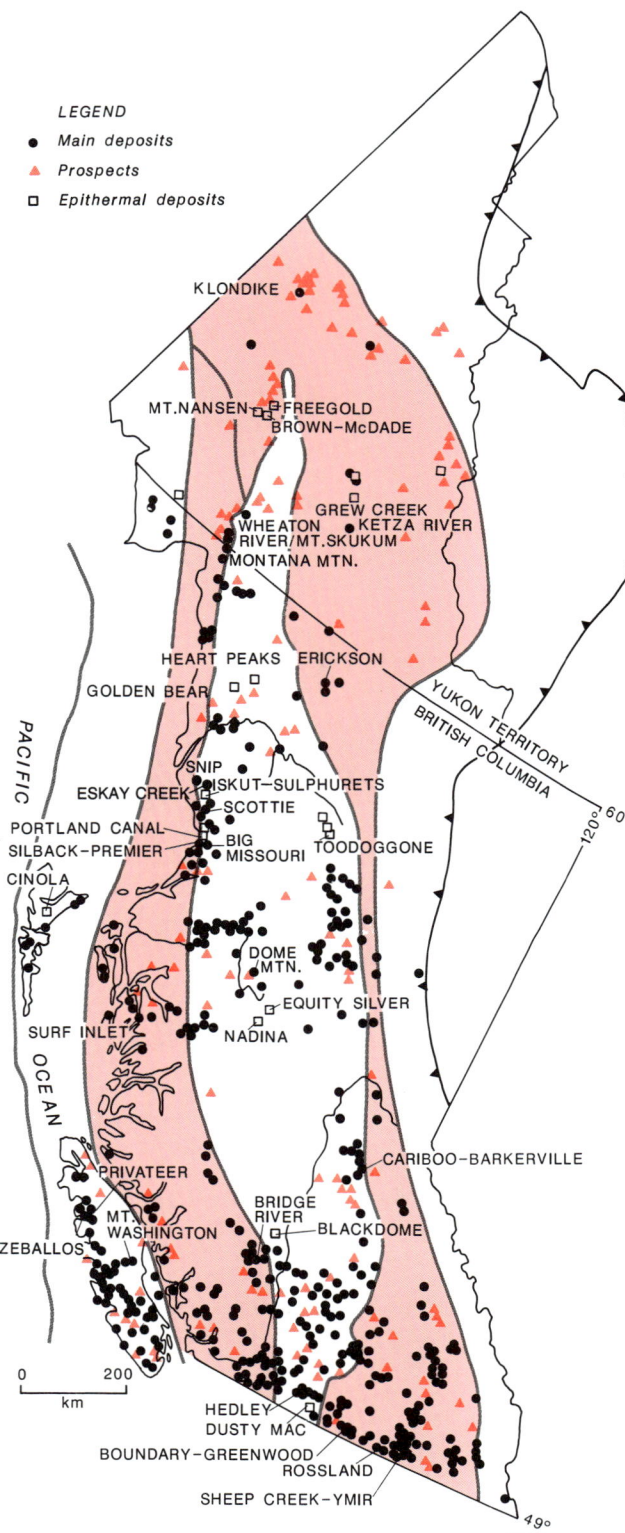

Figure 19.6. Lode gold deposits in the Cordillera.

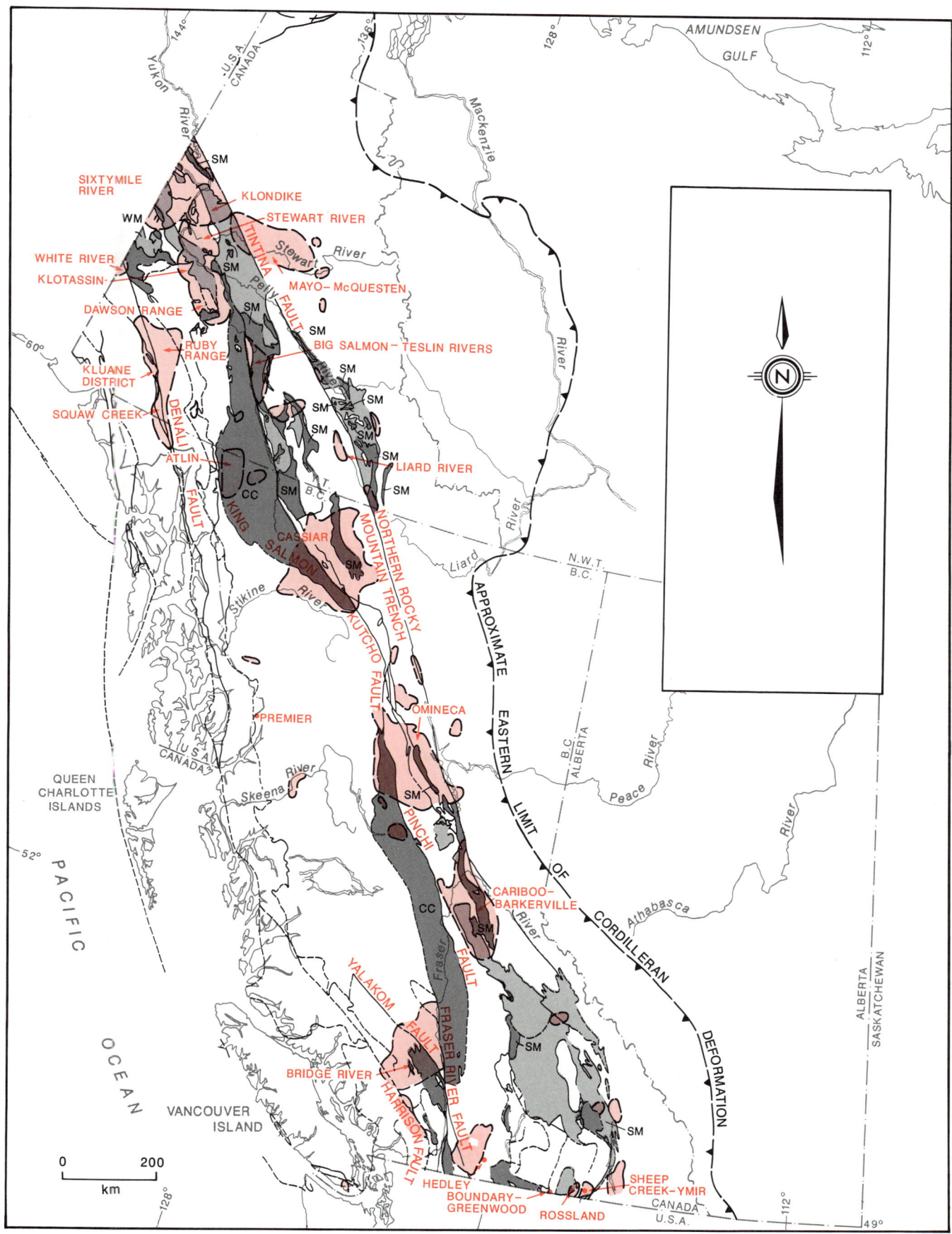

Figure 19.7. Placer gold areas (pink), major faults (red), major lode gold camps (dark grey) and oceanic terranes (light grey). CC, Cache Creek; SM, Slide Mountain; WM, Windy-McKinley; BR, Bridge River; MT, Methow.

Silver production from the Canadian Cordillera accounts for about one-third of annual Canadian production and is dominated by the Sullivan Pb,Zn,Ag deposit which has combined production and reserves totalling 11 000 tonnes of silver. Similar deposits in the Anvil district, Yukon, collectively contain close to 5000 tonnes Ag. Other major quantities of silver are recovered as byproducts from volcanogenic massive sulphide and porphyry copper deposits. The Westmin massive sulphide deposits and Highland Valley porphyry copper district contain silver reserves of about 1100 and 1700 tonnes, respectively. The major primary Ag resources (production and reserves) are vein deposits at the United Keno Hill mines, Yukon (5700 tonnes Ag) and Highland Bell (1000 tonnes Ag). Major, in part bulk-mineable, volcanic-hosted vein, stockwork and replacement deposits are Equity Silver (2750 tonnes Ag) and Silbak-Premier (1400 tonnes Ag).

Deposit types and their distribution

Precious metal deposits of diverse types formed throughout the evolution of the Cordillera. Two contrasting, fundamental genetic types are recognized — mesothermal and epithermal. Although the 'thermal' connotation in this classification proposed by Lindgren (1928), Nolan (1933) and others, is no longer necessarily valid, the two deposit types are now considered to reflect mineralization at different crustal depths involving different source fluids (Nesbitt et al., 1986). Mesothermal deposits are those formed from tectonically driven, large scale, deeply circulating fluid systems. The fluids tend to be alkaline to neutral pH, reduced, rich in CO_2, and highly evolved through isotopic exchange and interaction with wallrocks. Isotopic data show low ^{207}Pb values for ore minerals indicative of uniform, primitive lower crustal sources and relatively heavy ^{18}O fluid values (^{18}Osmow‰ from +3 to +16; Nesbitt et al., 1986).

Epithermal deposits tend to be related to late structural events during high level and subaerial magmatic activity. The hydrothermal solutions are shallowly circulating, meteoric-source, neutral to strongly acidic, oxygenated, dilute, low to moderate CO_2 content, commonly boiling fluids. Isotopic data generally show mixed-source, more radiogenic leads (Andrew, 1982) and light ^{18}O fluid values (^{18}Osmow‰ -14 to +3; Field and Fifarek, 1985).

Some precious metal deposits defy placement in this two-fold classification. Many deposits in mining camps and districts related to plutons, commonly display depth and thermal zoning and are transitional between mesothermal and epithermal types. In addition, the epithermal-mesothermal classification is best applied to vein, stockwork and breccia deposits; deposits of other morphologies such as skarns, mantos, replacements and detachment-type deposits require their own classifications and genetic models.

A pronounced clustering of various types of gold deposits is evident along the western margin of the Omineca Belt and along both the western and eastern margins of the Coast Belt (Fig. 19.6). A number of lode gold deposits and many of the major placer districts are associated with major structural breaks (Hodgson et al., 1982), along terrane boundaries of oceanic volcanic terranes such as Bridge River, Cache Creek and Slide Mountain (Fig. 19.7). These oceanic terranes consist of disrupted ophiolitic or composite volcanic arc-marine clastic rocks intruded by diorite or younger granitic intrusions. The gold deposits in these settings, such as in the Bridge River camp, Cassiar and Atlin areas, are mesothermal deposits, similar to the California Motherlode-type occurrences. They consist of structurally controlled, multiple, massive to ribboned, quartz-vein systems with considerable lateral and vertical extent. Gold-to-silver ratios are typically greater than one. Quartz vein mineralization with sparse sulphide minerals and minor arsenopyrite, scheelite and/or tellurides persist to depth with little vertical change in mineralogy or metal zoning. Typically veins have alteration envelopes with pyrite, Ca,Fe,Mg carbonate, quartz, sericite, fuchsite and, less commonly, albite and chlorite alteration. In a number of the districts Hg and Sb mineralization is found peripheral to gold deposits or occurs along strike in the major faults (McMillan et al., 1986).

The albitic, pyritic replacement zones in Jurassic clastic rocks at Carolin mine (Ray et al., 1986) are similar in many respects to mesothermal deposits near major faults and in or adjacent to oceanic terranes. At Carolin mine, initial host rock permeability appears to have controlled gold deposition rather than later quartz vein development. Similar lithologically controlled, structurally induced hydrothermal activity in Paleozoic metasedimentary rocks is responsible for the pyritic replacement orebodies at the Cariboo Gold Quartz deposits (Benedict, 1945). These deposits appear to be multiple-stage ores related first to Middle Jurassic sulphide replacements in marble and talc-dolomite schist, followed by superimposed Early Cretaceous quartz veins (Andrew et al., 1983; F. Robert and B. Taylor, pers. comm., 1986). Elsewhere, such as in the Rossland district and at Scottie mine near Stewart, mesothermal veins in volcanic rocks are characterized by high temperature Fe-Mg silicate alteration assemblages with pyrrhotite-rich, base-metal mineralization.

Other deposits associated with lode gold deposits in oceanic terranes are gold placers containing minor platinum group elements, polymetallic massive sulphides, asbestos deposits, most notably Cassiar mine, and small nephrite jade lenses. Of the more than 50 reported placer occurrences containing platinum group elements (Rublee, 1986), the only significant production has been 622 kg derived from the Alaskan-type Tulameen ultramafic complex between 1889 and 1936.

Deposits with epithermal or transitional characteristics are generally related to local structures in Mesozoic and younger volcanic fields or epizonal plutons in broad structural zones that are transverse to the predominant northwest Cordilleran trend. Two of the larger topographic and structurally disrupted zones or "arches" are the transverse Skeena and Stikine arches of northern British Columbia. They are broad belts of Late Mesozoic to Cenozoic, periodically reactivated deep fractures with mesozonal to epizonal plutonism. A similar deep-seated fracture zone with young intrusions has been noted in southeastern British Columbia by Høy (1982b). He described a long history of structurally constrained, syndepositional Proterozoic and Paleozoic base-metal mineralization followed by Mesozoic epigenetic, structurally controlled base and precious metal deposition.

Pluton-related auriferous polymetallic vein and skarn gold or copper-gold deposits are found commonly in

REGIONAL METALLOGENY

Mesozoic volcanic island arc rocks in the Insular and Intermontane belts, in the transverse structural zones, and in Tertiary extensional settings with epizonal plutonic and subvolcanic intrusions. Porphyry copper deposits and related breccia bodies in these geological settings commonly contain economically significant gold. A number of alkalic porphyry deposits associated with shoshonitic Triassic to Lower Jurassic arc rocks, and a few younger porphyry deposits associated with quartz-bearing stocks, are noted for their elevated gold and silver contents (see Table 19.3).

Cordilleran epithermal deposits are similar to those found elsewhere around the Pacific margin (Sillitoe, 1983; Bonham, 1985). Virtually all the deposits are volcanic-hosted. Some, such as Blackdome mine (250 000 tonnes ore, 25 grams per tonne gold) resemble the 'Tertiary type' or bonanza deposits of the southwestern United States (Hayba et al., 1985; Heald et al., 1987). Another newly discovered deposit, Cinola (28 million tonnes, 2.46 grams per tonne gold) is a Miocene or younger hotspring-related deposit (Tolbert et al., 1987) which formed, in part, contemporaneously with sedimentation of the host Skonun Formation (Christie, 1989). It is similar to the hotspring and geothermal deposits described by Berger (1985), Henley and Ellis (1983) and Henley (1985).

A notable difference to epithermal deposits in other regions is that a number of Cordilleran deposits (e.g. in Toodoggone and Stewart areas) are of Mesozoic age. They are hosted in Lower to Middle Jurassic calc-alkaline to alkaline (shoshonitic) andesites which reflect accreted island arc or back-arc depositional settings. Subaerial rocks in these environments are minor and mineralization in them is rare (e.g. Toodoggone). An epithermal model that outlines the origin of the deposits and their distribution in the strongly dissected, tectonically active Cordilleran accretionary setting has been described by Panteleyev (1986; Fig. 19.8). The model reviews the continuum of mineralization from hotspring deposits at surface to pluton-related deposits at depth and compares Cordilleran deposits to those described elsewhere (Berger, 1982, 1985).

Deposits not found to date are the sedimentary-hosted, large tonnage, disseminated 'Carlin-type' deposits in calcareous sedimentary rocks (Tooker, 1985), nor flat-fault-related 'detachment-type' deposits in metamorphic terranes (Wilkins, 1984). Recently the Dusty Mac and Camp McKinney deposits of southern Quesnellia have been proposed by Tempelman-Kluit and Parkinson (1986) to be related to Early Tertiary detachment faults (see Okanagan Subterrane, discussed previously in this chapter).

Tectonic evolution and age of gold mineralization

Figure 19.9 outlines development of gold deposits in the various morphogeological belts and terranes through time. Clearly, much of the gold in numerous vein and some breccia deposits (e.g. Willa), skarn (e.g. Hedley) and porphyry-copper-associated propylitic replacement bodies (e.g. QR) is related to latest Triassic to earliest Jurassic island arc and early accretionary magmatic activity. The intrusions in Paleozoic to Mesozoic oceanic rocks and the early Mesozoic volcanic island arcs that comprise the bulk of the Intermontane and Insular superterranes, are mainly subduction-related, I-type, commonly cupriferous, diorite

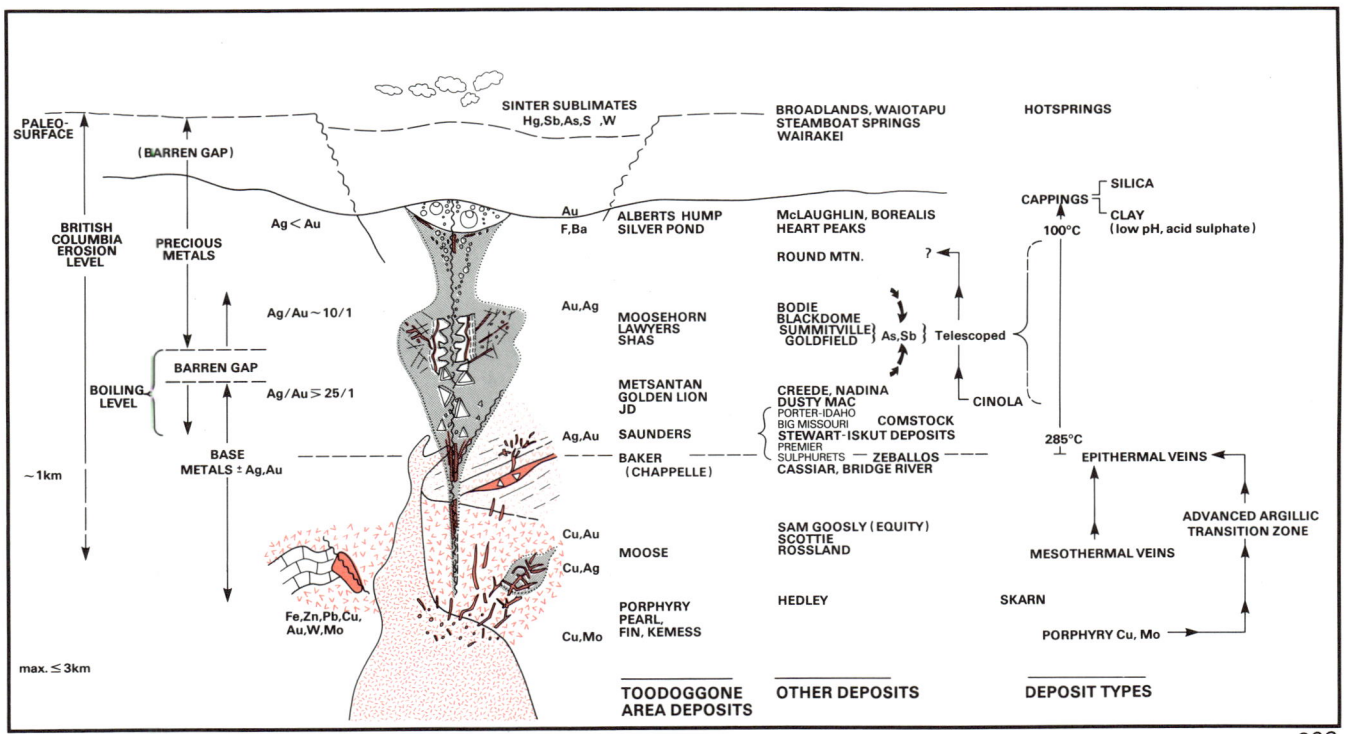

Figure 19.8. British Columbia epithermal deposits model depicting depth-zoning relationships between porphyry copper, skarn, transitional vein, epithermal and hotspring precious metal deposits; after Panteleyev (1986).

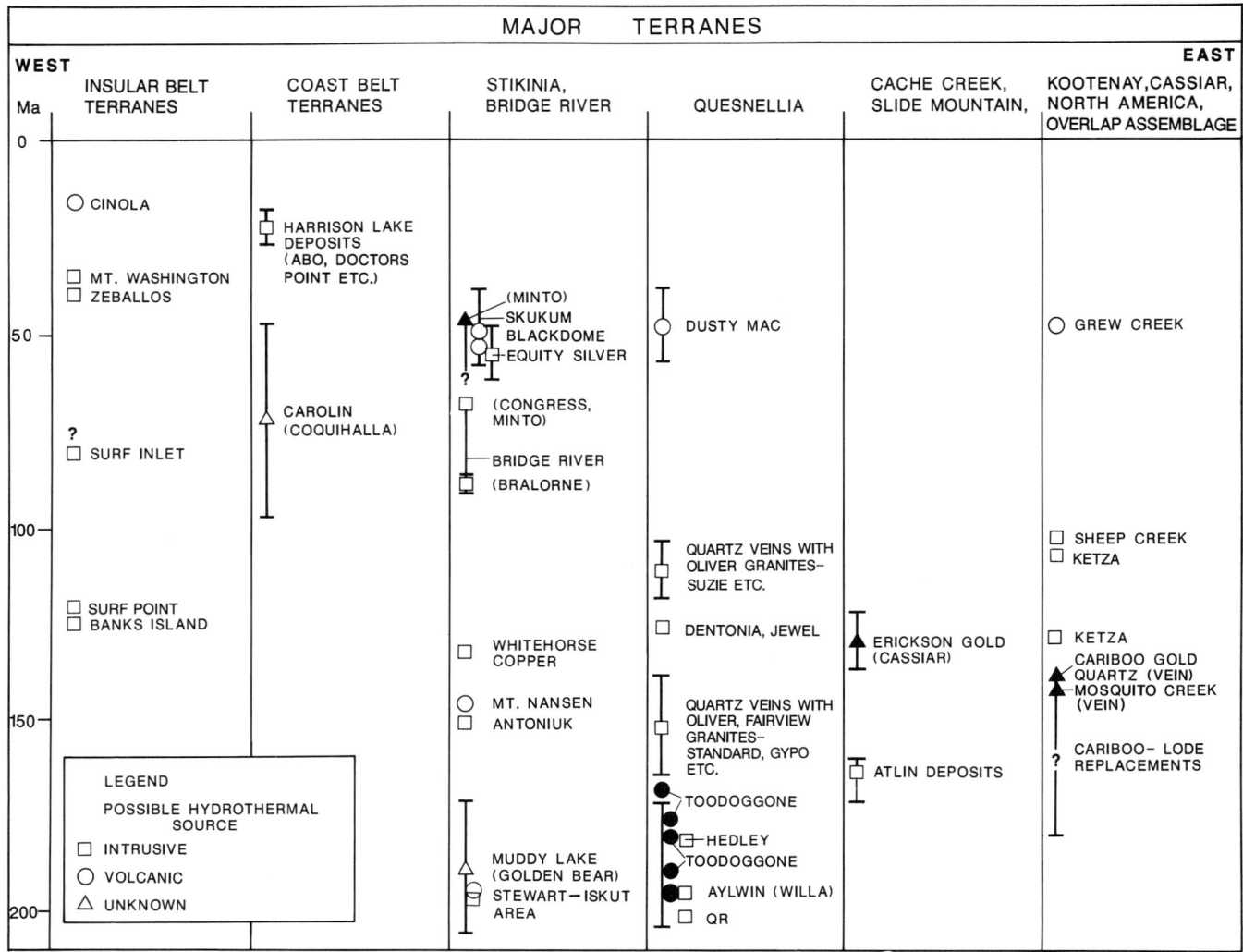

Figure 19.9. Radiometric age of precious metal mineralization in major terranes or terrane-overlapping rocks. Closed symbols - age of mineralization; open symbols - age of host rocks; bar shows possible range in age. Main sources of information: Ministry of Energy, Mines and Petroleum Resources - The University of British Columbia file of radiometric dates, R.L. Armstrong et al.; also Yukon Geology, v. 1, 1984.

to granodiorite plutons. Precious metal deposits in these settings are associated with both volcanic and mesozonal to epizonal-plutonic hydrothermal activity. The most notable older deposits in volcanic arc rocks are those of the Portland Canal-Stewart area (e.g. Silback-Premier Mine). Closer to the continental margin, coeval back-arc or marginal basin volcanism in the Toodoggone area deposited alkalic (shoshonitic) to calc-alkalic subaerial rocks with associated typical volcanic-hydrothermal epithermal precious metal deposits.

Numerous, generally small, gold- and silver-bearing polymetallic vein and replacement deposits formed in association with 150 to 100 million-year-old plutons. Examples are the many gold deposits of southern Quesnellia and elsewhere, such as those in the Oliver, Fairview, Sheep Creek and other small mining camps. Curiously, the core zones of the Omineca and Coast belts are not well mineralized with precious metals nor base metal deposits. This is possibly because the core areas are strongly uplifted and too deeply eroded. Instead, mineralization is best developed in or near the smaller, epizonal, Tertiary satellitic plutons along the margins of the tectonically disrupted crystalline belts.

The large Cordilleran terrane boundary faults, particularly those that juxtapose oceanic rocks with other terranes, are zones of subduction-related Jurassic and younger tectonic accretionary overlap. Some have large, post-mid-Cretaceous right lateral displacements. Many of the fault systems localized zones of large-scale hydrothermal fluid migration and contain important mesothermal gold deposits, related Hg deposits and As, Hg and Sb occurrences. The mineralizing fluids may have been derived from composite sources, interpreted to be mainly deeply circulating meteoric fluids by Nesbitt et al. (1986), but probably coupled with fluids from metamorphic dewatering and possibly some magmatic sources, as proposed for the Motherlode, California deposits by Bohlke and Kistler (1986). The oldest deposits are the Early to Middle Jurassic Muddy Lake (Golden Bear) and Atlin area deposits of northern Stikinia. Similar but somewhat younger deposits are the Middle Jurassic to Early Cretaceous Cariboo-Barkerville district deposits and Early Cretaceous Erickson Creek deposits formed during collision of the Intermontane Superterrane with North America. The major deposits of the eastern Cordillera are similar in tectonic setting, origin, and age to the Grass Valley, Motherlode and other deposits of the Sierra Nevada foothills. The California deposits range in age from 140 to 110 Ma, with mean and median ages between 120 and 115 Ma (Böhlke and Kistler, 1986). This suggests that major tectonic activity with a related gold-forming metallogenic epoch occurred along a large portion of the western margin of North America at the end of the Jurassic Period and during Early Cretaceous time. These metallogenic indications, along with some evidence of coeval metamorphic events, might signify major perturbations in Pacific (Farallon and Kula) and North American plate interactions following the Late Jurassic.

During Late Cretaceous time in southwestern British Columbia, initiation or reactivation of wrench faults in the Fraser and subsidiary fault systems along the eastern margin of the Coast Belt were associated with development of the largest Cordilleran gold deposits, the mesothermal deposits of Bridge River camp (see Bridge River, Cadwallader and Methow terranes, described previously in this chapter). Recent radiometric dating and isotopic studies suggest a protracted, episodic mineralizing event that coincides with emplacement of plutonic rocks in the Coast Belt during Late Cretaceous to Early Tertiary time (90-45 Ma; Leitch et al., 1988). A similar geological setting has been described 175 km to the southeast in the Coquihalla belt (Ray et al., 1986). There geological evidence suggests post-mid-Cretaceous and pre-Late Eocene mineralization.

Tertiary post-accretionary precious metal mineralization is related throughout the Cordillera to continental volcanism and epizonal to subvolcanic plutonism, generally in zones of extensional tectonism. Eocene volcanic-hosted examples are the Mount Skukum and Grew Creek deposits, Yukon; Blackdome bonanza-type epithermal deposits and the silver-rich Equity Silver porphyry-related transitional-type deposit (Cyr et al., 1984). One of the youngest major deposit is Cinola, an epithermal hotspring deposit hosted mainly by Tertiary clastic rocks and associated with a Miocene rhyolite dyke (Champigny and Sinclair, 1982; Christie, 1989). Other young intrusive-related deposits are quartz veins and breccia-stockwork deposits in or adjoining Oligocene intrusions near Zeballos and Mount Washington on Vancouver Island and the Oligocene-Miocene stocks near Harrison Lake (Ray, 1986). Other types of young, structurally controlled precious metal deposits are the Cretaceous silver veins in the rich Keno Hill district, Yukon and the Late Cretaceous or Tertiary silver-rich manto and chimney deposits of the northern Cordillera at Midway, and probably the Ketza manto deposit, Yukon Territory.

Volcanogenic massive sulphide deposits

Massive sulphide deposits (Fig. 19.10) in Paleozoic and Mesozoic marine volcanic terranes are an economically important class of deposit in the Cordillera (Thompson and Panteleyev, 1976). The deposits were among the earliest major mines developed in British Columbia, for example Mount Sicker (1898-1908), Anyox (1914-1935) and the long-lived Britannia mine (1905-1974). Massive sulphide deposits were the main source of Cordilleran copper production prior to development of porphyry copper deposits in the 1960s. They continue to be attractive exploration targets; recently discovered deposits include: Windy-Craggy, H-W, Kutcho Creek, Goldstream and Samatosum (see Table 19.2). Many of the Cordilleran deposits are large. Windy-Craggy, a 300 million tonne deposit, is a giant in a global context; others such as Britannia, Granduc, Anyox and the Westmin deposits, are equal in size to major Archean greenstone-hosted deposits of the Superior Province and elsewhere. Cordilleran deposits found prior to 1960 were generally regarded to be epigenetic shear-zone replacements. During the 1960s, interaction with Japanese, European and Australian geologists influenced genetic concepts (Oftedahl, 1958; Stanton, 1960; Tatsumi,

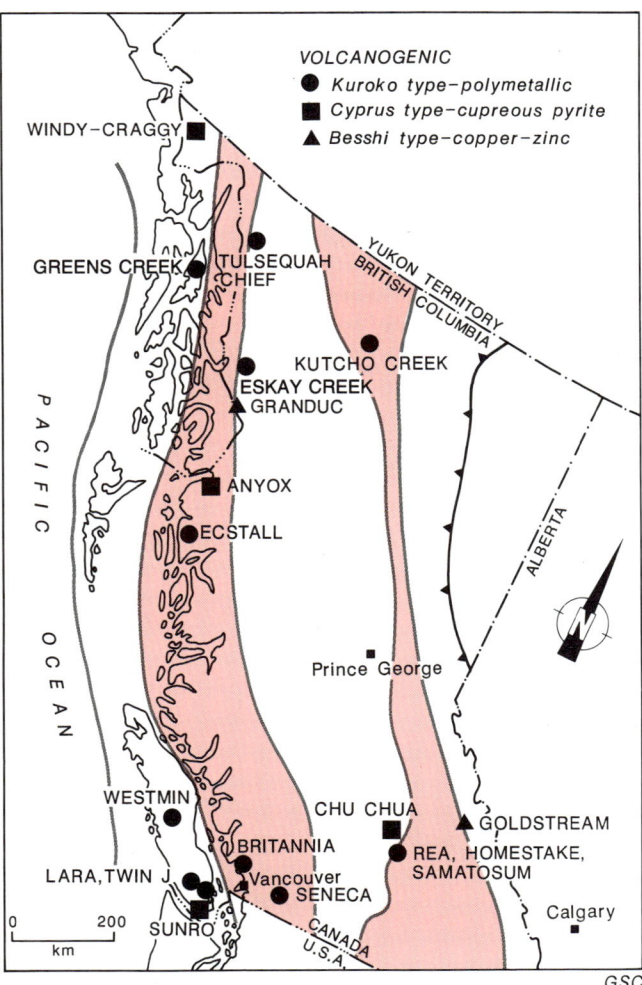

Figure 19.10. Volcanogenic massive sulphide deposits in the Cordillera.

1970). By the early 1970s the genetic connection between volcanism and ore deposition was firmly established (Sangster, 1972; Hutchinson, 1973, 1980; Sillitoe, 1972; Sato, 1974; Sawkins, 1976; Franklin et al., 1981). During the 1970s a number of Cordilleran deposits were discovered, guided in part by volcanogenic genetic concepts (Fig. 19.11 and 19.12). These new deposits include H-W (Westmin), DY, Kutcho Creek, Goldstream and Chu Chua (see Table 19.2).

Volcanic-related massive sulphide deposits differ from each other in detail but two fundamental types are recognized according to their host volcanic lithologies, metal contents (tenor) and tectonic setting. These correspond to the polymetallic, felsic-volcanic-hosted deposits and the cupriferous pyrite, mafic volcanic and sediment-hosted deposits (Table 19.2, Fig. 19.11 and 19.12). Alternatively, they are the Zn,Pb,Cu or Cu,Zn types of Franklin et al. (1981) and Lydon (1984, 1988).

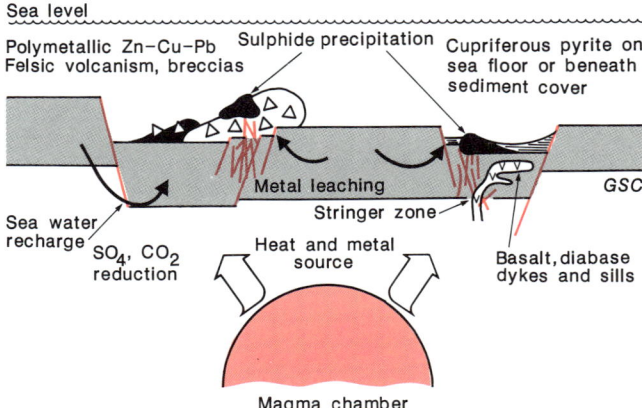

Figure 19.11. Volcanogenic massive sulphide genetic model, modified from Sawkins (1984), Hutchinson (1980), and others.

Polymetallic deposits are similar to the Tertiary Kuroko deposits of Japan (Ishihara, 1974; Lambert and Sato, 1974; Ohmoto and Skinner, 1983) and the Archean deposits in greenstone belts of the Canadian Shield (Sangster, 1972 and Franklin et al., 1981). The deposits are associated with felsic lavas, subaqueous pyroclastic rocks and autobrecciated domal rocks in differentiated, calc-alkaline andesitic or bimodal basalt-rhyolite volcanic suites formed in convergent, island arc settings. Deposits occur as combinations of massive sulphide lenses, breccias with sulphide clasts and vein stockworks. They are commonly associated with eruptive centres in structurally controlled submarine calderas (Scott, 1980). Ore sulphides are associated with silica and sericite alteration. Peripheral alteration zones are propylitic or zeolite-bearing and commonly contain pyrite, chlorite and clay minerals (montmorillonite and mixed-layer clays). Zoning of sulphide bodies is common. Copper-rich core zones and their underlying vein stockworks either replace or are overlain and flanked by zinc-rich zones. The zinc-rich ores are deposited from cooler, more oxidized fluids and are accompanied by abundant sulphate minerals (barite, anhydrite and gypsum), and rarely massive clay zones. Cordilleran examples of polymetallic deposits are Westmin, with layered massive sulphide, breccia and stockwork ores, Seneca with fragmental sulphide ores, and Britannia with extensive stockwork ores. Some deposits such as the zinc-rich Lara, the lead-zinc-bearing Eskay Creek and the tetrahedrite-bearing Samatosum deposits have notably elevated precious metal contents with up to 19 ppm gold and 1100 ppm silver, (see Table 19.2).

Cupriferous pyrite deposits are associated with marine basalts and their overlying fine grained clastic rocks. The setting, in predominantly volcanic terranes, resembles Cyprus-type deposits (Constantinou and Govett, 1973) and the modern day ocean-ridge environments (Scott, 1985). Deposits in volcanic-sedimentary and predominantly sedimentary terranes are similar to the Japanese Besshi deposits (Kanehira and Tatsumi, 1970; Fox, 1984) and the Kieslager deposits of Europe (Klau and Large, 1980;

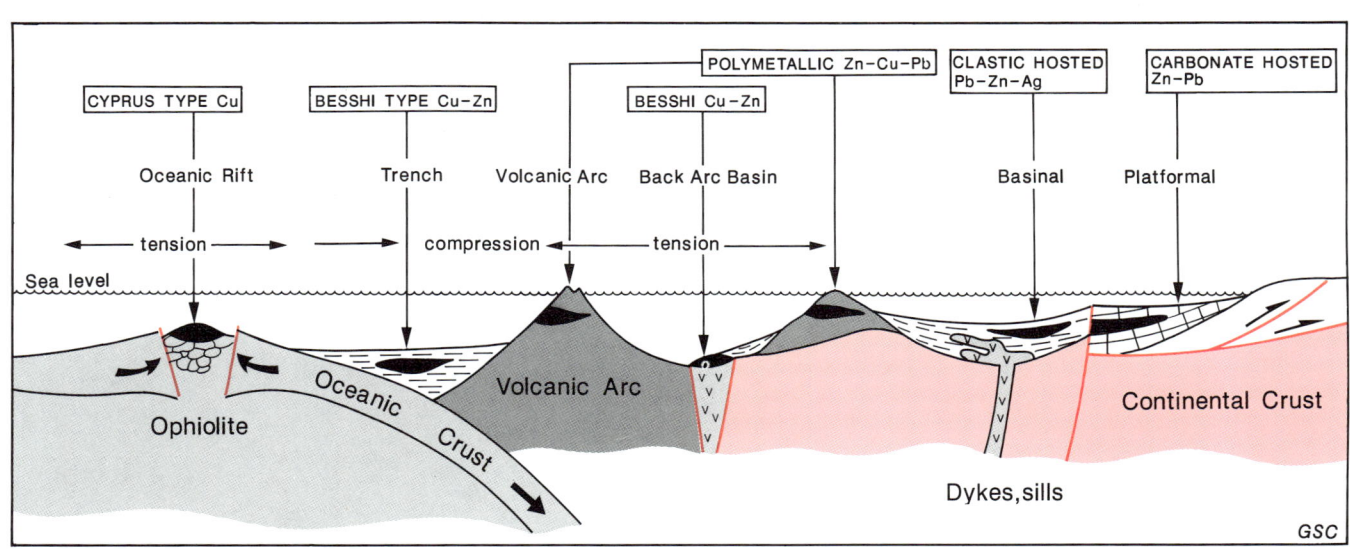

Figure 19.12. Tectonic environments of massive sulphide deposits; modified by T. Høy and others, after Hutchinson (1980).

Franklin et al., 1981). The deposits occur as pyritic and pyrrhotitic cupriferous and zinc-bearing massive sulphide lenses and stratabound sheets. Some deposits have underlying sulphide vein stockworks. Some metamorphosed and deformed deposits are markedly attenuated and rod-like in form. Compared to the polymetallic felsic-volcanic-hosted deposits, cupriferous pyrite deposits contain less lead but more cobalt (commonly up to 1000 ppm). They are generally associated with extensive magnesian chlorite alteration zones and have more abundant bedded silica and carbonate gangue rather than sulphate and clay minerals. Many deposits are spatially related to large linear structures; others are localized in fault blocks. In this interpreted extensional tectonic setting, rifting initiates submarine hydrothermal venting late in the volcanic cycle or following cessation of active volcanism. In mineralized sedimentary basins, sulphides were deposited in structural zones marked by abrupt facies changes from shallow to deep water deposits, major stratigraphic and lateral transitions from basaltic to fine grained black clastic sequences, presence of abundant basaltic or diabasic dykes and sills, and chemical changes in magmas from tholeiitic to calc-alkaline compositions. Differences between individual deposits are interpreted as due to variations in the amount of sedimentation that accompanied hydrothermal discharge and sulphide deposition either on the seafloor or under a thin cover of unconsolidated sediment (Fig. 19.11).

Sediment-starved oceanic spreading ridges with pillow basalts, sheet flow lavas, sheeted gabbroic subvolcanic intrusions and minor siliceous and oxidized ferruginous or manganiferous chemical sediments are interpreted as equivalent to the ophiolitic (i.e. oceanic crust) settings of the Cyprus, Oman, Newfoundland and other similar deposits (Spooner, 1980; Alabaster et al., 1982; Upadhyay and Strong, 1973). Alternatively, these basalt or basalt-sediment hosted deposits may have formed in off-ridge ocean seamounts (Alabaster et al., 1982) or in back-arc or inter-arc sites (Miyashiro, 1973).

Cordilleran deposits hosted by basaltic host rocks include the huge Windy-Craggy prospect in northwestern British Columbia. There, sulphide deposits in Triassic basalts and at basalt-argillite contacts contain approximately 300 million tonnes of massive to disseminated cupriferous pyrite. A footwall stringer zone has been reported to contain up to 12.8 ppm gold and approximately 1 per cent copper in a 38.7 m drill intercept (Gammon and Chandler, 1986). MacIntyre (1986) interpreted the tectonic setting of the Windy-Craggy deposit, based on host rock petrochemical data, to be an immature island arc or back-arc spreading regime, complete with abundant diorite (diabase?) dykes and sills. The similar Anyox deposits were interpreted by Sharp (1980) to be hosted by tholeiitic ridge basalts, and the Chu Chua deposit was thought by Aggarwal and Nesbitt (1984) to have formed in a Carboniferous ocean floor or oceanic island (seamount) setting. The small Sunro prospect is a rare example of a Tertiary deposit. It is associated with gabbroic intrusions in the recently accreted Eocene ophiolite suite of the Crescent Terrane.

Sedimentary and metasedimentary terranes with subordinate volcanic rocks contain deposits that have been compared to Besshi-type deposits. Goldstream (Høy et al., 1984) is a single massive sulphide layer 1 to 3 m thick in metamorphosed terrigenous clastic rocks deposited along the rifted Eocambrian or younger continental margin (Høy, 1979). Granduc mine comprises a series of concordant massive sulphide lenses in folded, laminated metavolcanic rocks containing calcareous and gypsiferous sediments that were deposited in a shallow water, island arc setting (Grove, 1986). A Recent geological environment similar to the Goldstream setting is found in the transform-fault-related Guaymas Basin in the Gulf of California (Lonsdale et al., 1980; Koski et al., 1985; Scott, 1985).

Active and recently extinct massive sulphide depositional sites are being investigated on modern spreading ridges in the northeast Pacific (Rona, 1984; Koski et al., 1988). The ridges mark the zone of interaction between the Pacific Plate and the actively subducting Juan de Fuca Plate. Following the discovery in 1979 of active 'black smokers' on the East Pacific Rise at 21°N (Francheteau et al., 1979), at least 8 sites of hydrothermal deposition have been noted on the Explorer and Juan de Fuca spreading ridge segments and seamounts off the British Columbia coast.

Most of the deposits are small massive sulphide spires or collapsed spires formed along submarine rift faults or caldera margins. The larger deposits are mounds 5 to 10 m thick covered by a blanket of unconsolidated sediment. Growth of sulphide mound-chimney deposits is largely due to sulphide accumulation by open space filling and replacement within the lens due to the impoundment and defocussing of hydrothermal fluids by previously deposited sulphides and anhydrite (Lydon, 1988). Along the southern Explorer Ridge, 60 deposits are known. Most exposed massive sulphide deposits are small; the largest one at 'Magic Mountain' is a sediment-covered sulphide mound measuring 300 m in diameter. In 1986 a very large sulphide mound was discovered in the thickly sedimented failed rift of Middle Valley on the northern Juan de Fuca Ridge (Davis et al., 1987). This mound has the potential to contain about 100 million tons of massive sulphide material (S. Scott, pers. comm. to A.P., 1987).

Porphyry deposits

Large tonnage, bulk-mineable, low-grade, disseminated Cu, Cu,Mo and Mo porphyries are the dominant deposit type hosted by granitic rocks. Cordilleran porphyry deposits (Fig. 19.13 and Table 19.3) contain about 60% of the copper and virtually all of the Mo resources of Canada. The deposits were first considered to be economic in the late 1950s. Production began in the Highland Valley in 1962. During the 1960s and 1970s a series of coincident economic conditions made exploitation of these very low grade deposits possible. Copper from porphyry copper deposits was the economically most important commodity produced until it was replaced by coal in the early 1980s. In recent years increased operating costs combined with decreased base-metal prices have allowed established mines to continue operation only if they have highly efficient mining practices, fully amortized capital assets and/or production of significant byproduct precious metals.

Cordilleran porphyry copper deposits, typically associated with intermediate to felsic, hypabyssal, porphyritic intrusive rocks are similar to those found elsewhere in the world (Titley and Hicks, 1966; Lowell and Guilbert, 1970; Gustafson and Hunt, 1975; Beane and Titley, 1981; Titley and Beane, 1981; and Titley, 1982). Molybdenum deposits

CHAPTER 19

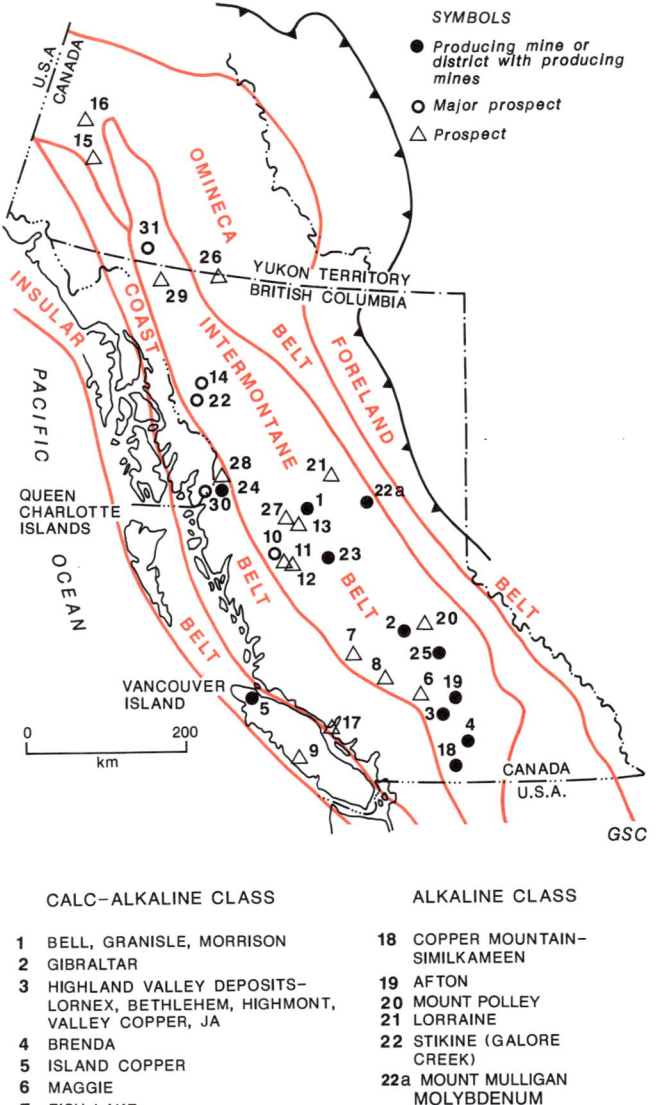

Figure 19.13. Major Cordilleran porphyry copper and molybdenum deposits.

CALC-ALKALINE CLASS

1. BELL, GRANISLE, MORRISON
2. GIBRALTAR
3. HIGHLAND VALLEY DEPOSITS—LORNEX, BETHLEHEM, HIGHMONT, VALLEY COPPER, JA
4. BRENDA
5. ISLAND COPPER
6. MAGGIE
7. FISH LAKE
8. POISON MOUNTAIN
9. CATFACE
10. BERG
11. HUCKLEBERRY
12. OX LAKE
13. BIG ONION
14. SCHAFT CREEK
15. MOUNT NANSEN
16. CASINO
17. OK

ALKALINE CLASS

18. COPPER MOUNTAIN-SIMILKAMEEN
19. AFTON
20. MOUNT POLLEY
21. LORRAINE
22. STIKINE (GALORE CREEK)
22a. MOUNT MULLIGAN MOLYBDENUM DEPOSITS
23. ENDAKO
24. BC MOLY (KITSAULT)
25. BOSS MOUNTAIN
26. LOGTUNG
27. GLACIER GULCH
28. AJAX
29. ADANAC
30. QUARTZ HILL (ALASKA)
31. RED MOUNTAIN

are similar to those described by Clark (1972); Mutschler et al. (1981); Westra and Keith (1981); and White et al. (1981). British Columbia and Yukon porphyry deposits have been extensively discussed by Sutherland Brown (1976a,c). Porphyry molybdenum deposits have been reviewed by Soregaroli and Sutherland Brown (1976) and porphyry copper deposits by McMillan and Panteleyev (1980).

Canadian Cordilleran perceptions about porphyry deposits were outlined in extensive reviews by Sutherland Brown (1976a,c). He proposed a three-fold classification of porphyry copper deposits according to their morphology. This classification describes 'phallic, volcanic and plutonic' porphyry deposits. The 'phallic' deposits resemble the conventional deposits of the southwestern United States described by Lowell and Guilbert (1970) and later called 'classic' type by McMillan and Panteleyev (1980). Sutherland Brown's morphological classification is based mainly on depth-zoning considerations but also considers a wide variety of ore-deposit characteristics (Fig. 19.14).

Distinct characteristics of Canadian Cordilleran deposits are as follows: supergene mineralization is absent, except in rare cases (Ney et al., 1976), volcanic rocks host most deposits and ore zones most commonly occur in the wall rocks and less commonly in the porphyries. The resulting alteration patterns are more akin to the wallrock porphyries described by Guilbert and Lowell (1974) rather than granitic-hosted deposits described by Lowell and Guilbert (1970). In many deposits biotite hornfels derived from both volcanic and sedimentary rocks is the most important ore host. The early developed hornfels, formed by thermal (contact) metamorphism, appears to be important in providing a brittle host rock that became highly fractured during later magma emplacement and mineralizing hydrothermal activity.

A small but important subgroup of porphyry copper deposits, the 'alkalic porphyry deposits', is associated with unusual silica-deficient magmas that give rise to alkalic gabbro to syenite porphyry intrusions (Barr et al., 1976). These deposits are characterized by intensely K-metasomatized volcanic host rocks composed of coarsely crystalline assemblages of orthoclase, biotite, garnet, anhydrite, and, less characteristically, pseudoleucite. The ore reserves in two alkalic porphyries, Ingerbelle and Stikine Copper, occur almost entirely within the volcano-sedimentary hostrocks as skarn mineral assemblages which show zonal relationships to adjacent alkalic intrusions (Fahrni et al., 1976; Allen et al., 1976). Abundant ore minerals are the copper-rich, iron-deficient sulphides chalcocite and bornite, as well as chalcopyrite; the deposits contain almost no molybdenum. Some of the porphyry deposits of the alkalic subclass have significantly higher-than-average copper and precious metal contents (see Table 19.3). For example, Stikine Copper (Galore Creek) contains 125 million tonnes of primary mineralization with 1.06% Cu, 0.4 grams gold and 7.7 grams silver per tonne. The elevated precious metal values for gold of roughly 0.2 to 0.7 grams per tonne and for silver of invariably greater than 2.35 grams per tonne, characterize the alkalic-type porphyry deposits and distinguish them from the other types of porphyry deposits (Sinclair et al., 1982).

Cordilleran Mo deposits described by Soregaroli and Sutherland Brown (1976) can be also considered to have two distinct modes of occurrence. One group of deposits, such as Glacier Gulch, are similar to Climax-type deposits (see White et al., 1981). In these deposits 'ore shells', some with appreciable W and Sn minerals, resemble inverted tea cups that overlie specific mineralizing intrusive phases in composite plutons. Variants of this type are deposits associated with small composite plutons in which vertically extensive annular ore shells have been developed in the wallrocks along the margin of one or more granodiorite to

REGIONAL METALLOGENY

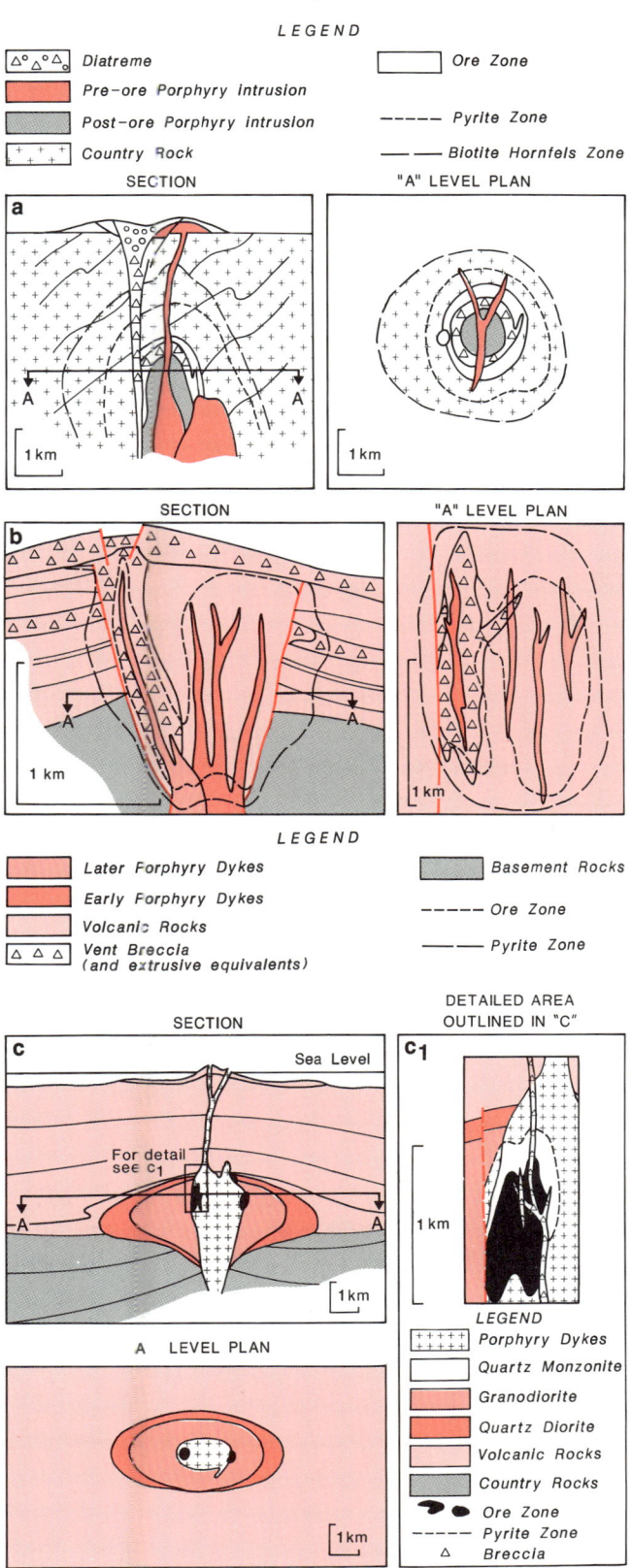

Figure 19.14. Models of porphyry copper deposits; modified from Sutherland Brown (1976c) and McMillan and Panteleyev (1980). The three-fold morphological and depth-zoning classification types are: a) classic, b) volcanic, and c) plutonic.

quartz monzonite intrusive phases. A number of these deposits is known in northwestern British Columbia; one example is BC Moly/Kitsault (Steininger, 1985). The other group of deposits is extensive, low-grade quartz-vein stockworks developed in relatively homogeneous, large, quartz monzonite to granodiorite plutons (e.g. Endako). Alternatively, some deposits are small, high-grade vein systems combined with stockworks and breccia zones within small plutons (e.g. Boss Mountain).

One of the more significant impacts of porphyry deposit metallogenic studies has been their influence on tectonic interpretations. This is an outgrowth of the extensive studies of the radiometric ages (Fig. 19.15) and investigations of isotopic characteristics and petrochemistry of plutons and their related ore deposits (Armstrong, 1988). However, no simple distribution of granites according to S-, I- nor A-type (Chappell and White, 1974; Ishihara, 1981; Pitcher, 1982; White and Chappell, 1983), nor any rigorous correlation among them and different ore types has been demonstrated. The long established generalized relationship between quartz monzonite and granodiorite-related Mo, W and Sn deposits in miogeoclinal rocks along the ancestral North American continental margin and Cu deposits in the volcanic-plutonic terranes to the west was reviewed by Sutherland Brown et al. (1971). In only a few cases are convincing genetic associations demonstrated between the more specific granite types and ore deposits. For example, in the northern Cordillera specialization of Mo, W and Sn with S-type granitoids has been described by Sinclair (1986). In Part A of this chapter a consistent association throughout the eastern Cordillera of skarn W with apparently S-type granites is described. Similarly, Anderson (1988) described a progression in the northern Cordillera from older island arc-related I-type granitoids, to mid-Cretaceous continent-margin S-types, followed by Late Cretaceous to Early Tertiary A-type granites. He related the petrogenetic changes to tectonically controlled, evolving stress regimes that tapped different magma sources during the protracted history of continental accretion and later disruption along the continental margin. Elsewhere, Griffiths and Godwin (1983) used Cu,Mo ratios and other petrogenic constraints to outline the relationship between Cu deposits and I-type anatectic volcanic arc or oceanic-source crustal melts, in contrast to Cu,Mo and Mo deposits associated with S-type and indeterminable mixed-source continental-margin magmas.

Ages of porphyry copper deposits and their host rocks were used by Godwin (1975) to indicate sources of mineralizing magmas within a plate tectonics scenario. The resulting hypothesis of imbricate, progressively younger outward-stepping subduction zones was provocative and influenced some early Cordilleran tectonic syntheses (Monger and Irving, 1980; Monger et al., 1982). More recently, analysis of spatial distribution of magmatic events summarized by Armstrong (1988) and others (Fig. 19.15), reveals a much more complicated pattern of successive, Andean-type magmatic arcs developed over large parts or throughout the length of the North American Cordillera.

The ages of porphyry Cu,Mo and Mo-bearing intrusions compared to all Cordilleran magmatic events (Fig. 19.4) clearly show a number of well defined mineralizing epochs (Christopher and Carter, 1976). In addition, many of the plutons with Cu or Cu,Mo deposits, and virtually all the alkaline intrusions with alkalic, gold-enriched copper

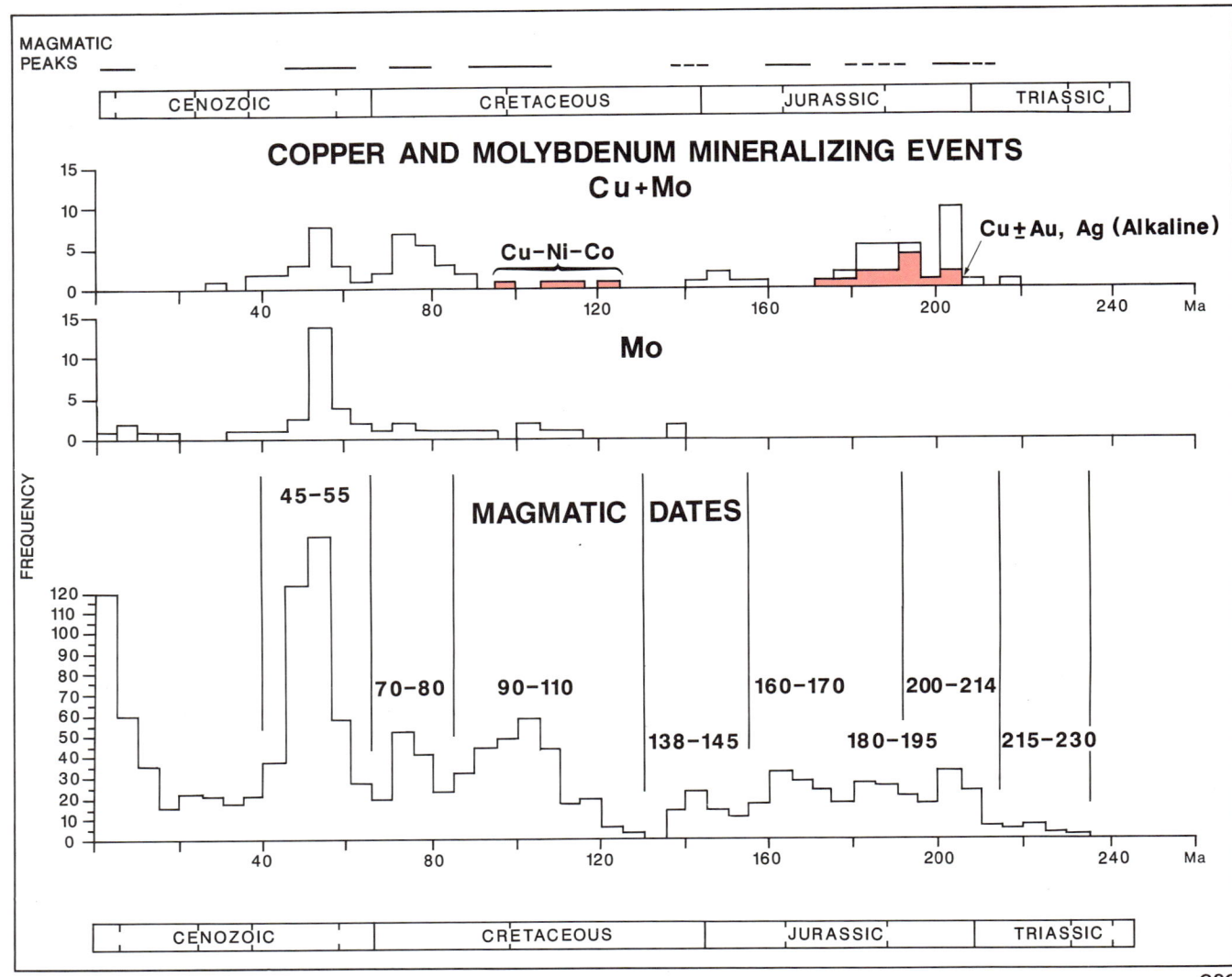

Figure 19.15. Radiometric ages of magmatic (mainly plutonic) events. Ages mainly from Armstrong (1988); hydrothermal events from Christopher and Carter (1976), Sinclair (1986), and other sources.

deposits are older than any purely Mo-bearing deposits. The early 218 to 179 Ma old copper-bearing deposits (Highland Valley, Stikine Copper, Schaft Creek, Cariboo-Bell) and others are related to volcanic arc magmatism in pre-accretionary settings which predate terrane amalgamation with North America. A few Lower to Upper Jurassic Cu,Mo deposits (Island Copper, Brenda) are known and can be considered to be accretionary, but there is an absence of copper-bearing deposits from 140 to 95 Ma. This gap coincides with the appearance of a few Mo deposits. Notably, one of the largest, Endako, with a radiometric age of about 140 Ma, is the oldest known Mo deposit in the southern Cordillera. A few older Mo prospects are known in Yukon Territory (Sinclair, 1986). The sole British Columbia nickel-cobalt mine, Giant Mascot, is also related to a Cretaceous accretionary plutonic event. The majority of the approximately 120 known Cu; Cu,Mo and Mo porphyry deposits are post-accretionary and were formed in the Late Cretaceous to Late Eocene (95 to 40 Ma); most of the deposits are Paleocene to Mid-Eocene (65 to 45 Ma). A few Cu,Mo and Mo deposits, mainly in the Insular and Coast belts, are Late Eocene to Oligocene in age (40-26 Ma). The youngest porphyry deposits are Mo deposits in the southern Coast Belt near Mount Waddington. These include the Late Miocene Salal (8 Ma) and Franklin Glacier (7 Ma) prospects. The presence of Mo-bearing plutons (with or without associated Cu) possibly signifies the tectonic evolution and maturation of the accreted continental margin. This was marked by the partial melting of evolved crust in underplated terranes containing subducted sediments and recycled, Mo-bearing continental detritus.

Skarn deposits

The Canadian Cordillera hosts several types of economically important skarn deposits distributed in several distinct metallogenic belts. Tungsten-rich skarns lie in North American miogeoclinal shelf limestone-pelite assemblages and displaced equivalents in the Omineca Belt of the eastern Cordillera. Tungsten-rich skarns are formed in a relatively deep, reducing environment, characteristically in a Lower Cambrian limestone adjacent to a mid-Cretaceous S-type quartz monzonitic pluton. Several developed prospects with large reserves define an arcuate belt which includes Canada Tungsten mine and flanks Selwyn Basin on the east and northeast (Fig. 19.16). Prior to its recent closure, Canada Tungsten mine was the largest W producer in the western world. Significant production during and shortly after World War II was obtained from four other W skarn deposits in the Salmo district of southeastern British Columbia (Table 19.4).

In the northern Cordillera, Zn,Pb,Ag (Cu,W,Sn,Au) skarn and replacement deposits are hosted by the same miogeoclinal sediments as W-rich skarns. Deposits occur in belts adjacent to the Mount Billings Batholith in southeastern Yukon, the Coal River stock in southwestern Northwest Territories and the Cassiar and Seagull batholiths in Cassiar Terrane (Fig. 19.16). The typical northern Cordilleran Zn-rich skarn and replacement deposit consists of a structurally controlled Fe- and Mn-rich mineral assemblage, formed at a relatively high level in a Cambrian to Devonian carbonate host, distal to a small felsic to dioritic stock which may not be exposed. Production has been attained from one Zn-rich skarn in the northern Cordillera, i.e. Mount Hundere, and, significant reserves have been developed in several deposits in both the craton and Cassiar Terrane (Table 19.4.3).

In the southeastern Canadian Cordillera, several replacement deposits have produced significant amounts of Ag, Pb and Zn, notably the mines of the Riondel and Slocan camps and the Mineral King (Table 19.4.3A). The Riondel (Bluebell, Ainsworth) and Slocan (Sandon) camps are adjacent to the Nelson Batholith, but are hosted by carbonate strata of both Kootenay and Quesnel terranes. The Mineral King is hosted by dolostone of the Mount Nelson Formation in the upper part of the Purcell Assemblage, with no associated intrusion exposed. In comparison with late to post-orogenic deposits of the northern Cordillera, these are more deformed and metamorphosed.

Skarn Cu deposits, important present and past producers of Cu, Au, Ag and Fe in the Cordillera, include Whitehorse Copper, Craigmont and parts of the Ingerbelle and Galore Creek porphyry copper deposits in the Intermontane Belt, Greenwood in the Omineca Belt and Coast Copper in the Insular Belt (Fig. 19.16 and Table 19.4.1). Unlike W- and Zn-rich skarns in miogeoclinal North American strata, Cu skarns are hosted by Upper Triassic limestones, dolostones and intermediate volcanic rocks of island arc affinity within the accreted Intermontane and Insular superterranes. Associated mafic to felsic alkaline to calc-alkaline intrusions are members of Jura-Cretaceous, I-type subduction-related plutonic suites which also commonly host major porphyry Cu and Cu,Mo deposits. Cu skarns tend to be large and relatively rich in Au and Ag, particularly those like Ingerbelle-Copper Mountain associated with porphyry Cu systems. Cu, Mo and Fe sulphides are concentrated closer to the intrusive contact than are Zn,Pb,Ag skarns, yet at a higher stratigraphic level than W-rich skarns and within a structurally controlled framework of fractures, stockworks and breccias.

Fe skarns in the Canadian Cordillera are part of a discontinuous belt that extends from northern California through western British Columbia into southeastern Alaska (Fig. 19.16). Important past producers include Tasu and Jedway on Queen Charlotte Islands, Kennedy Lake, Argonaut, Coast Copper, Merry Widow, Nimpkish and Zeballos on Vancouver Island, and several deposits on Texada Island. Deposit size ranges up to 30 million tonnes of 40-50% Fe as magnetite, hematite and martite (Table 19.4.4). All Canadian Cordilleran Fe skarns are hosted by Upper Triassic limestone and volcanic units in the upper part of the tholeiitic Karmutsen Assemblage of Wrangellia. Associated intrusions are Jurassic I-type granitoids of broad compositional range, from syenite to gabbro, and high Fe and Mg content relative to plutons associated with W, Zn and Cu skarns.

The economic significance of Au skarn deposits in the Canadian Cordillera (Ettlinger and Ray, 1988) was emphasized by the redevelopment, in 1987, of the Hedley district as an open-pit mine, the opening of the Tillicum Mountain deposit, and exploration activities centred on several Au skarn prospects, including ones on Texada and Banks islands. Important Au skarns, such as Hedley district, contain only Au as a recoverable metal, whereas other genetically related skarns in the district contain zones with W, Cu, Fe, Zn and Pb and could be considered subtypes of the gold skarns. Several significant Au skarns, including the Hedley district, Tillicum Mountain and possibly the QR deposit, are hosted by lower Mesozoic volcano-sedimentary assemblages of Quesnellia in association with comagmatic mafic to intermediate alkaline granitoid plutons. The previously producing Au,Ag,Cu,Fe skarns of Texada Island (Table 19.4.5) are related to adjacent Fe and Cu,Fe skarns. The Tel, Bob and other Au(Zn,Cu,Pb) prospects on Banks Island occur as skarns, replacements and veins within screens and roof pendants of metasedimentary rocks, possibly belonging to the Alexander Terrane, enclosed by Late Jurassic granitoid plutons of the Coast Plutonic Complex.

Sedimentary exhalative deposits

Significant stratiform sedimentary exhalative (sedex) Zn,Pb deposits occur in clastic facies of three assemblages in the North American miogeocline, i.e. the Sullivan mine in the Purcell Supergroup, deposits of the Howards Pass and Anvil districts in the Cambrian to Middle Devonian off-shelf, passive margin strata of Selwyn Basin and deposits of the Macmillan Pass and Gataga districts in the Middle Devonian to Mississippian clastic wedge. Geological characteristics of sedex deposits in general have been summarized by Large (1983), Gustafson and Williams (1981) and Morganti (1981). The orebodies are composed, at least in part, of bedded sulphides indicative of pre-lithification deposition and neither volcanic nor plutonic rocks are demonstrably related to ore formation. In comparison, stratabound deposits in carbonate-clastic shelf strata, e.g. Mississippi Valley-type Zn,Pb and Redstone-type Cu, may be superficially similar to sedex deposits but commonly exhibit non-bedded, discordant morphology on a local scale. Sedex deposits are localized within specific lithofacies of a sedimentary basin or sub-basin, commonly adjacent to or overlying major synsedimentary faults.

The Sullivan and Macmillan Pass-Gataga deposits, although of different ages, have several features in common. The depositional environment is a flysch basin

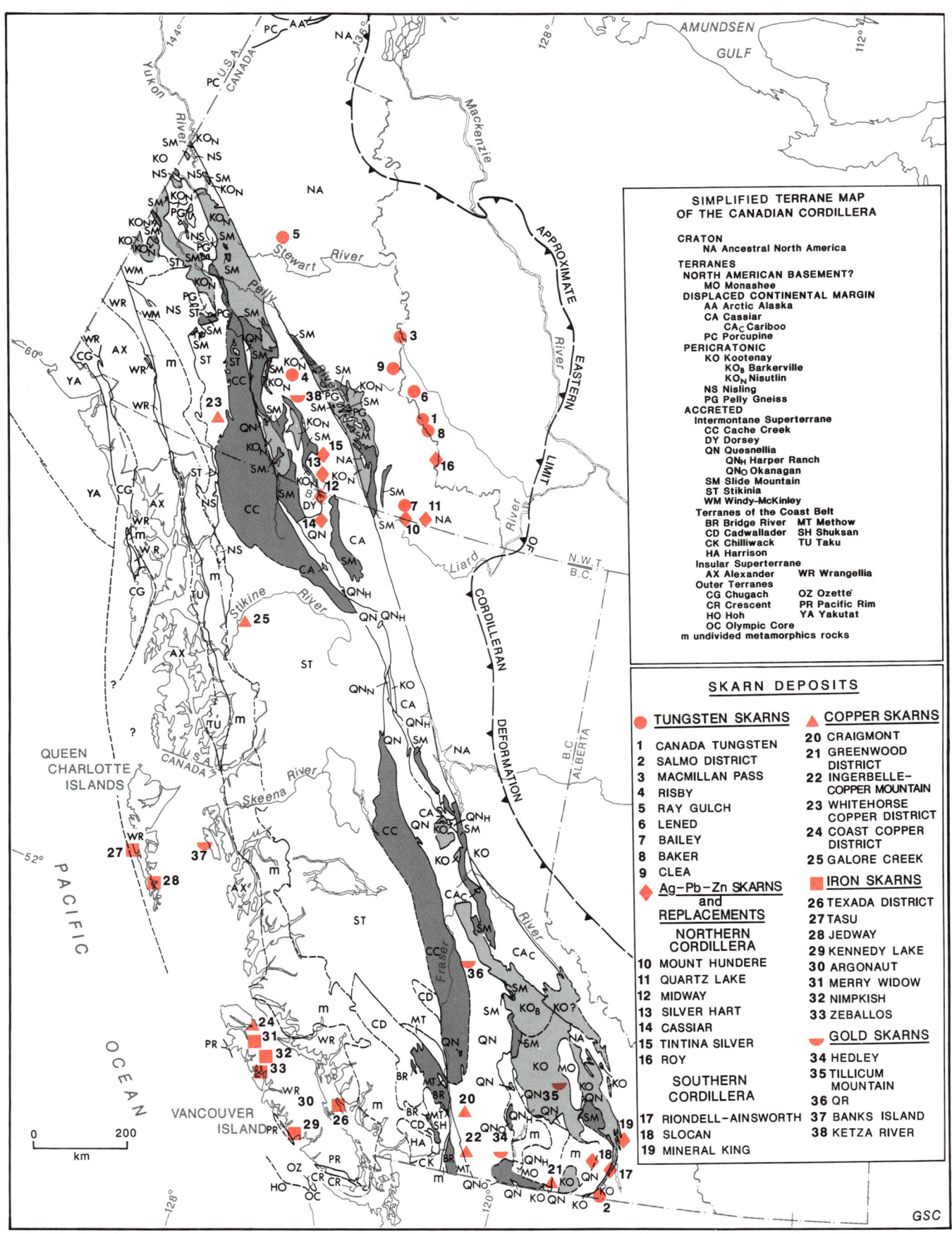

Figure 19.16. Skarn deposits in the Canadian Cordillera.

containing a thick succession of turbiditic siltstone, conglomerate, greywacke and mudstone (Morganti, 1981). Sulphides are contained within sub-basins related to synsedimentary grabens. Host turbidites are characterized by rapid changes in facies and thickness (Smith, pers. comm., 1978; Høy, 1982b). An altered and mineralized feeder zone lies adjacent to or underlies the centre of the deposit (Hamilton et al., 1983; McClay and Bidwell, 1986). Devono-Mississippian ore deposits, which are characterized by abundant barite, grade laterally to barite-rich, sulphide-deficient ore facies and distally to laminated baritite sequences. Collectively, these barite-sulphide facies constitute a metallotect of regional extent (Dawson, 1983). The Proterozoic Sullivan deposit, in contrast, contains only minor amounts of barite.

Turbidite-hosted sedex deposits show a clear genetic relationship to their tectono-stratigraphic setting. Zonation in mineralogy and stable isotopic composition of ore and gangue in relation to a feeder zone supports deposition by progressive oxidation of an exhaled, low temperature, reduced brine (Anger et al., 1966; Lydon et al., 1979; Nesbitt et al., 1986). Sulphur may be derived either from the hydrothermal fluid or the reduced, sulphur-rich seawater in sub-basins of restricted circulation (Goodfellow and Jonasson, 1984), or some combination of both. The underlying sediments are the most probable source of metals. This is supported by the conformability of galena-Pb isotopic data from all sedex deposits of Proterozoic to Devonian age to a growth curve representing isotopic evolution of upper crustal Pb in the North American miogeocline (Godwin et al., 1982). The spatial relationship of turbidite-hosted sedex deposits to faults has been emphasized by Abbott (1983) in the Macmillan Pass district, MacIntyre (1982) in the Gataga district and Høy (1982a) in the Sullivan district.

Sedex deposits of the Selwyn Basin type (Carne and Cathro, 1982), exemplified by deposits of the Howards Pass district, are localized within tectonically controlled sub-basins, seaward of major platforms and cratonal shelves, and dominated by fine grained carbonaceous to calcareous cherty clastic strata. Finely laminated sulphide deposits, unlike turbidite-hosted deposits, are low in Ba, Ag, Cu and Fe, and have simple sulphide mineralogy (Morganti, 1981). Sedex deposits of the Howards Pass district were deposited synchronously with Lower Silurian carbonaceous and calcareous mudstone and chert. Sedimentation was slightly younger than two early Paleozoic extensional events in eastern Selwyn Basin and was accompanied by widespread alkalic basaltic volcanism. Pb- and Zn-rich brines probably were discharged along basin marginal faults into sub-basins with restricted seawater circulation and resultant sulphidic, anoxic bottom waters (Goodfellow and Jonasson, 1986). Very large, finely laminated, carbonaceous Zn,Pb orebodies were subsequently deposited.

Zn,Pb,Ag(Cu,Ba) deposits of the Anvil district may be either distal volcanogenic or more proximal to an exhalative vent. Volcanic rocks are present at several horizons in the mineralized section, and graphitic ore lithofacies thicken markedly across a linear, basin-marginal array of volcanic centres (Jennings and Jilson, 1986).

CONCLUSIONS

Metallogenic analysis of cratonic, pericratonic and accreted terranes of the Canadian Cordillera demonstrates a consistent relationship between the lithotectonic character of host terranes and the type and composition of their typical suites of mineral deposits. Cratonic and pericratonic terranes of the Foreland and Omineca belts, for example, host stratiform deposits of Zn,Pb; Cu; Fe; and Ba and skarn deposits of W, Zn,Pb,Ag and Sn. Volcanic-arc lithologies of the accreted Intermontane Superterrane, by comparison, host porphyry-type Cu,Mo; Mo and Cu,Au,Ag and related skarn Cu,Fe and Cu,Au deposits. The probable derivation of exhalative brines and of anatectic, S-type granitoids associated with cratonic mineral deposits, from craton-derived clastic sequences contrasts with the oceanic and arc basement lithologies and subduction-related I-type granitoids related to porphyry and skarn deposits of the Intermontane Superterrane. The distinctive lithophile character of typical cratonic ore element assemblages, which may include the diagnostic lithophile elements W, Ba, U, Th, Nb, Ta, B and Be, probably reflects their origin in an upper crustal silicate protolith. In comparison, the mainly chalcophile ore element assemblages of typical Intermontane Belt deposits, that may include any of the chalcophile elements Cu, Hg, Ag, Au, Mo, Fe, As, and Sb, form sulphide ores genetically related to deep seated igneous rocks of lower crustal or oceanic parentage. Metallogeny, therefore, readily distinguishes Cordilleran cratonic terranes and their displaced equivalents from accreted arc and oceanic terranes.

Present knowledge of the time of tectonic, plutonic and metallogenic events, although imperfect, allows classification of mineral deposition according to the accretionary history of the host or immediately adjacent terrane. In terranes of cratonic affinity several periods of pre-accretionary mineralization may have been generated, in part, by rifting events recognized in Late Proterozoic and early Paleozoic time. Terrane displacements subsequent to rifting are not well understood. On the other hand lateral displacements on transcurrent faults are, at least locally, fairly well documented. Pre-accretionary mineral deposits are predominantly stratiform types, hosted by sedimentary and volcanic strata. A suite of significant porphyry-type deposits formed as a result of early synvolcanic plutonism in volcanic-arc terranes of the Intermontane Superterrane shortly before or during their amalgamation and before subsequent accretion.

Deposits of clearly accretionary timing are minor, partly because of the inherent rarity of metamorphogenic mineralization, but mainly due to the difficulty in unequivocally ascribing the generation of a plutonic suite and attendant mineralization to a specific accretionary event. Granitoid plutonism and related mineralization accompanied amalgamation of Quesnellia, Stikinia and Cache Creek terranes before accretion of the resultant Intermontane Superterrane to ancestral North America.

The majority of granitoid-related porphyry, skarn and vein deposits are, however, post-accretionary. Post-orogenic magmatic events, widespread in Middle Jurassic, mid- and Late Cretaceous and Early Tertiary times correspond to important post-accretionary metallogenic epochs that include, for example, W and Zn,Pb skarns of the Omineca

Belt, most Cu,Mo porphyries of the Intermontane Belt and precious metal veins throughout the Canadian Cordillera.

The Canadian Cordillera is well endowed with a variety of porphyry, stratiform, skarn and vein-type deposits, in addition to volcanogenic massive sulphide, magmatic and several other less common deposit types. Comparison of Canadian Cordilleran mineral deposits with well established global models has contributed to the understanding of deposit genesis, established relationships between regional metallogeny and the new plate tectonic-terrane accretion framework developed throughout this volume, and provided a powerful predictive tool for the explorationist.

REFERENCES

Abbott, G.
1981: A new geological map of Mt. Hundere and the area north; in Yukon Geology and Exploration 1979-80, Indian and Northern Affairs Canada, p. 45-50.
1983: Geology, Macmillan Fold Belt, 105O/SW and parts of 105P/SW; Indian and Northern Affairs Canada, Open File 1983.
1985: Silver-bearing veins and replacement deposits of the Rancheria district; in Yukon Exploration and Geology 1983, Indian and Northern Affairs Canada, p. 34-44.
1986: Devonian extension and wrench tectonics near Macmillan Pass, Yukon Territory, Canada; in The Genesis of Stratiform Hosted Lead and Zinc Deposits, R.J.W. Turner and M.T. Einaudi (ed.), Conference Proceedings, Stanford University Press, p. 85-89.
1987: Field Activities, 1986; in 1986 Yukon Mining and Exploration Overview, Mineral Resources Directorate, Northern Affairs Program, Yukon Department of Indian Affairs and Northern Development, p. 32-33.

Aggarwal, P.K. and Nesbitt, B.E.
1984: Geology and geochemistry of the Chu Chua massive sulphide deposit, British Columbia; Economic Geology, v. 79, p. 815-825.

Aho, A.E.
1956: Geology and genesis of ultrabasic nickel-copper-pyrrhotite deposits at the Pacific Nickel Property, southwestern British Columbia; Economic Geology, v. 51, p. 444-481.

Aitken, J.D.
1981: Stratigraphy and sedimentology of the Upper Proterozoic Little Dal Group, Mackenzie Mountains, Northwest Territories; in Proterozoic Basins of Canada, F.H.A. Campbell (ed.), Geological Survey of Canada, Paper 81-10, p. 47-71.

Aitken, J.D., Long, D.G.F., and Semikhatov, M.A.
1978a: Progress in Helikian stratigraphy, Mackenzie Mountains; in Current Research, Part A, Geological Survey of Canada, Paper 78-1A, p. 481-484.
1978b: Correlation of Helikian strata, Mackenzie Mountains-Brock Inlier-Victoria Island; in Current Research, Part A, Geological Survey of Canada, Paper 78-1A, p. 485-486.

Aitken, J.D., Macqueen, R.W., and Foscolos, A.E.
1973: A Proterozoic sedimentary succession with traces of copper mineralization, Cap Mountain, southern Franklin Mountains, District of Mackenzie (95 O); in Report of Activities, Part A, Geological Survey of Canada, Paper 73-1A, p. 243-246.

Alabaster, T., Pearce, J.A., and Malpas, J.
1982: The volcanic stratigraphy and petrogenesis of the Oman ophiolite complex; Contributions to Mineralogy and Petrology, v. 81, p. 168-183.

Aleinikoff, J.N., Dusel-Bacon, C., and Foster, H.L.
1986: Geochronology of augen gneiss and related rocks, Yukon-Tanana Terrane, east-central Alaska; Geological Society of America Bulletin, v. 97, p. 626-637.

Alldrick, D.J.
1985: Stratigraphy and petrology of the Stewart mining camp (104B/1); in Geological Fieldwork 1984, British Columbia Ministry of Energy, Mines and Petroleum Resources, Paper 1985-1, p. 316-341.
1986: Stratigraphy and structure in the Anyox area (103P/5); in Geological Fieldwork, British Columbia Ministry of Energy, Mines and Petroleum Resources, Paper 1986-1, p. 211-216.

Alldrick, D.J., Brown, D.A., Harakal, J.E., Mortensen, J.K., and Armstrong, R.L.
1987: Geochronology of the Stewart Mining Camp (104B/1); in Geological Fieldwork 1986, British Columbia Ministry of Energy, Mines and Petroleum Resources, Paper 1987-1, p. 81-92.

Allen, D.G., Panteleyev, A., and Armstrong, A.T.
1976: Galore Creek; in Porphyry Deposits of the Canadian Cordillera, A. Sutherland Brown, (ed.), Canadian Institute of Mining and Metallurgy, Special Volume 15, p. 402-414.

Anderson, R.G.
1983: Geology of the Hotailuh batholith and surrounding volcanic and sedimentary rocks, north-central British Columbia; Ph.D. thesis, Carleton University, Ottawa, 669 p.
1988: An overview of some Mesozoic and Tertiary plutonic suites and their associated mineralization in the northern Canadian Cordillera; in Recent Advances in the Geology of Granite-Related Mineral Deposits, R.P. Taylor and D.F. Strong (ed.), Canadian Institute of Mining and Metallurgy, Special Volume 39, p. 96-113.

Andrew, A.
1982: A lead isotope study of selected precious metal deposits in British Columbia; M.Sc. thesis, University of British Columbia, Vancouver, 80 p.

Andrew, A., Godwin, C.I., and Sinclair, A.J.
1983: Age and genesis of Cariboo gold mineralization determined by isotope methods (93H); in Geological Fieldwork 1982, British Columbia Ministry of Energy, Mines and Petroleum Resources, Paper 1983-1, p. 305-313.
1984: Mixing line isochrons: a new interpretation of galena lead isotope data from southeastern British Columbia; Economic Geology, v. 79, p. 919-932.

Andrew, K.P.E., Godwin, C.I., and Cann, R.M.
1986: Wolf epithermal precious metal vein prospect, central British Columbia; in Geological Fieldwork 1985, British Columbia Ministry of Energy, Mines and Petroleum Resources, Paper 1986-1, p. 317-320.

Anger, G., Nielsen H., Puchelt, H., and Ricke, W.
1966: Sulfur isotopes in the Rammelsberg ore deposit (Germany); Economic Geology, v. 61, p. 511-536.

Archer, A.R. and Schmidt, U.
1978: Mineralized breccias of Early Proterozoic age, Bonnet Plume River district, Yukon Territory; Canadian Institute of Mining and Metallurgy, Bulletin, v. 71, no. 796, p. 53-58.

Archer, A., Bell, R.T., and Thorpe, R.I.
1986: Age relationships from U-Th-Pb isotope studies of uranium mineralization in Wernecke breccias, Yukon Territory; in Current Research, Part A, Geological Survey of Canada, Paper 86-1A, p. 385-391.

Armstrong, A.K. and MacKevett, E.M.
1977: The Triassic Chitistone Limestone, Wrangell Mountains, Alaska; United States Geological Survey, Open File Report 77-217, 72 p.

Armstrong, J.E.
1949: Fort St. James map-area, Cassiar and Coast districts, British Columbia; Geological Survey of Canada, Memoir 252, 210 p.

Armstrong, R.L.
1988: Mesozoic and early Cenozoic magmatic evolution of the Canadian Cordillera; in Processes in Continental Lithospheric Deformation, S.P. Clark, Jr. (ed.), Geological Association of America, Special Paper 218, p. 55-91.

Armstrong, R.L., Eisbacher, G.H., and Evans, P.D.
1982: Age and stratigraphic-tectonic significance of Proterozoic diabase sheets, Mackenzie Mountains, northwestern Canada; Canadian Journal of Earth Sciences, v. 19, p. 316-323.

Atkinson, D. and Baker, D.J.
1986: Recent developments in the geologic understanding of Mactung; in Mineral Deposits of Northern Cordillera, J.A. Morin (ed.), Canadian Institute of Mining and Metallurgy, Special Volume 37, p. 234-244.

Bacon, W.R.
1953: Ecstall River; in British Columbia Department of Mines, Annual Report 1952, p. 79-81.

Bailes, R.J., Smee, B.W., Blackadar, D.W., and Gardner, W.D.
1986: Geology of the Jason lead-zinc-silver deposits, Macmillan Pass, eastern Yukon; in Mineral Deposits of Northern Cordillera, J.A. Morin (ed.), Canadian Institute of Mining and Metallurgy, Special Volume 37, p. 87-99.

Bailey, D.G.
1978: The geology of the Morehead Lake area, south-central British Columbia; Ph.D. thesis, Queen's University, Kingston, 198 p.

Ball, C.W.
1954: The Emerald, Feeney and Dodger tungsten ore-bodies, Salmo, British Columbia, Canada; Economic Geology, v. 49, p. 625-638.

Ball, C.W. and Boggaram, G.
1985: Geological investigation of the Eaglet fluorspar deposits; Mining Magazine, v. 152, p. 506-509.

Ballantyne, S.B. and Littlejohn, A.L.
1982: Uranium mineralization and lithogeochemistry of the Surprise Lake batholith, Atlin, British Columbia; in Uranium in Granites, Y.T. Maurice (ed.), Geological Survey of Canada, Paper 81-23, p. 145-155.

Barr, D.A.
1980: Gold in the Canadian Cordillera; Canadian Institute of Mining and Metallurgy, Bulletin, v. 73, no. 818, p. 59-76.

Barr, D.A., Fox, P.E., Northcote, K.E., and Preto, V.A.
1976: The alkaline suite porphyry deposits: a summary; in Porphyry Deposits of the Canadian Cordillera, A. Sutherland Brown (ed.), Canadian Institute of Mining and Metallurgy, Special Volume 15, p. 359-367.

Beane, R.E. and Titley, S.R.
1981: Porphyry copper deposits, Part II. Hydrothermal alteration and mineralization; in Economic Geology Seventy-Fifth Anniversary Volume, B.J. Skinner (ed.), Economic Geology Publishing Company, p. 235-269.

Bell, R.T.
1968: Proterozoic stratigraphy of northeastern British Columbia; Geological Survey of Canada, Paper 67-68, 75 p.
1978: Breccias and uranium mineralization in the Wernecke Mountains, Yukon Territory - a progress report; in Current Research, Part A, Geological Survey of Canada, Paper 78-1A, p. 317-322.
1982: Comments on the geology and uraniferous mineral occurrences of the Wernecke Mountains, Yukon and District of Mackenzie; in Current Research, Part B, Geological Survey of Canada, Paper 82-1B, p. 279-284.
1985: Overview of uranium in volcanic rocks of the Canadian Cordillera; in Uranium Deposits in Volcanic Rocks, Proceedings of International Atomic Energy Agency Technical Committee Meeting, El Paso, Texas, April, 1984, p. 319-335.
1989: A conceptual model for development of megabreccias and associated mineral deposits in Wernecke Mountains, Canada, Copperbelt, Zaire, and Flinders Range, Australia, in Uranium Resources and Geology of North America, Proceedings of IAEA workshop, Saskatoon, Sask., Canada, Aug, 1987 International Atomic Energy Agency, Vienna, IAEA TEC-DOC-500, p. 149-169.

Benedict, P.C.
1945: Structure at Island Mountain Mine, Wells, B.C.; Canadian Institute of Mining and Metallurgy, Transactions, v. 48, p. 755-770.

Berg, H.C.
1981: Upper Triassic volcanogenic massive-sulfide metallogenic province identified in southeastern Alaska; in The United States Geological Survey in Alaska: Accomplishments During 1979, N.R.D. Albert and T. Hudson (ed.), United States Geological Survey, Circular 823-B, p. B104-B108.

Berger, B.R.
1982: The geological attributes of Au-Ag-base metal epithermal deposits; in Characteristics of Mineral Deposit Occurrences, R.L. Erickson (compiler), United States Geological Survey, Open File Report 82-0795, p. 119-126.
1985: Geologic-geochemical features of hot-spring precious-metal deposits; in Geologic Characteristics of Sediment- and Volcanic-Hosted Disseminated Gold Deposits; Search For An Occurrence Model, E.W. Tooker (ed.), United States Geological Survey, Bulletin 1646, p. 47-53.

Bilibin, Y.A.
1955: Metallogenic provinces and metallogenic epochs; 1967 translation by E.A. Alexandrov, Geological Bulletin, Department of Geology, Queen's College Press, Flushing, New York, 35 p.

Bloodgood, M.A.
1987: Geology of the Triassic black phyllite in the Eureka Peak area, central British Columbia (93A/7); in Geological Fieldwork 1986, British Columbia Ministry of Energy, Mines and Petroleum Resources, Paper 1987-1, p. 135-142.

Blusson, S.L.
1968: Geology and tungsten deposits near the headwaters of Flat River, Yukon Territory and southwestern District of Mackenzie; Geological Survey of Canada, Paper 67-22, 77 p.

Böhlke, J.K. and Kistler, R.W.
1986: Rb-Sr, K-Ar, and stable isotope evidence for the ages and sources of fluid components of gold-bearing quartz veins in the Northern Sierra Nevada Foothills Metamorphic Belt; Economic Geology, v. 81, p. 296-322.

Bonham, H.F., Jr.
1985: Characteristics of bulk-minable gold-silver deposits in Cordilleran and island-arc settings; in Geologic Characteristics of Sediment- and Volcanic-Hosted Disseminated Gold Deposits; Search For An Occurrence Model, E.W. Tooker (ed.), United States Geological Survey, Bulletin 1646, p. 71-77.

Bradford, J.A. and Godwin, C.I.
1988: Midway silver-lead-zinc manto deposit, northern British Columbia; in Geological Fieldwork, 1987, British Columbia Ministry of Energy, Mines and Petroleum Resources, Paper 1988-1, p. 353-360.

Bridge, D.A., Marr, J.M., Hashimoto, K., Obara, M., Suzuki, R.
1986: Geology of the Kutcho Creek volcanogenic massive sulphide deposits, northern British Columbia; in Mineral Deposits of Northern Cordillera, J.A. Morin (ed.), Canadian Institute of Mining and Metallurgy, Special Volume 37, p. 115-128.

Bright, M.J. and Jonson, D.C.
1976: Glacier Gulch (Yorke-Hardy); in Porphyry Deposits of the Canadian Cordillera, A. Sutherland Brown (ed.), Canadian Institute of Mining and Metallurgy, Special Volume 15, p. 455-461.

Britton, J.M. and Alldrick, D.J.
1988: Sulphurets map area (104A/05W,12W; 104B/08E,09E); in Geological Fieldwork 1987, British Columbia Ministry of Energy, Mines and Petroleum Resources, Paper 1988-1, p. 199-209.

Britton, J.M. Webster, I.C.L. and Alldrick, D.J.
1989: Unuk map area (104 B/7E, 8W, 9W, 10E) in Geological Fieldwork 1988, British Columbia Ministry of Energy, Mines and Petroleum Resources, Paper 1989-1, p. 241-250.

Brock, J.S.
1975: Mining: Yukon's first industry; Western Miner, v. 48, no. 2, p. 63-66.
1976: Recent developments in the Selwyn-Mackenzie zinc-lead province, Yukon and Northwest Territories; Western Miner, v. 49, no. 3, p. 9-16.

Brown, A.C.
1978: Stratiform copper deposits-evidence for their post-sedimentary origin; Minerals Science and Engineering, v. 10, p. 172-181.

Brown, A.C. and Chartrand, F.M.
1983: Stratiform copper deposits and interactions with co-existing atmospheres, hydrospheres, biospheres and lithospheres; Precambrian Research, v. 20, p. 533-542.

Brown, D.A. and Logan, J.M.
1988: Geology and mineral evaluation of Kokanee Glacier Provincial Park, southeastern British Columbia (82F/06); in Geological Fieldwork 1987, Mineral Resources Division, British Columbia Ministry of Energy, Mines and Petroleum Resources, Paper 1988-1, p. 31-48.

Bundtzen, T.K., Eakins, G.R., Clough, J.G., Lueck, L.L., Green, C.B., Robinson, M.S., and Coleman, D.A.
1984: Alaska's mineral industry 1983; Alaska Division of Geological and Geophysical Surveys, Special Report 33, 56 p.

Bundtzen, T.K., Green, C.B., Peterson, R.J. and Seward, A.F.
1988: Alaska's mineral industry, 1987, Alaska Division of Geological and Geophysical Surveys, Special Report 41, 69 p.

Burgoyne, A.A.
1986: Geology and exploration, McDame asbestos deposit, Cassiar, B.C.; Canadian Institute of Mining and Metallurgy, Bulletin, v. 79, no. 889, p. 31-37.

Cairnes, C.E.
1934: Slocan Mining Camp, British Columbia; Geological Survey of Canada, Memoir 173, 137 p.
1937: Geology and mineral deposits of Bridge River mining camp, British Columbia; Geological Survey of Canada, Memoir 213, 140 p.

Campbell, F.A.
1959: The geology of Torbrit silver mine; Economic Geology, v. 54, p. 1461-1495.
1960: Nickel deposits in the Quill Creek and White River areas, Yukon; Canadian Institute of Mining and Metallurgy, Bulletin, v. 53, p. 953-959.

Campbell, S.W.
1976: Nickel-copper sulphide deposits in the Kluane Ranges, Yukon Territory; Department of Indian and Northern Affairs, Open File Report EGS 1976-10, 17 p.
1981: Geology and genesis of copper deposits and associated host rocks in and near the Quill Creek area, southwestern Yukon; Ph.D. thesis, University of British Columbia, Vancouver, 215 p.

Cargill, D.G., Lamb, J., Young, M.J., and Rugg, E.S.
1976: Island Copper; in Porphyry Deposits of the Canadian Cordillera, A. Sutherland Brown (ed.), Canadian Institute of Mining and Metallurgy, Special Volume 15, p. 206-218.

Carne, R.C.
1979: Upper Devonian stratiform barite-lead-zinc-silver mineralization at Tom claims, Macmillan Pass, Yukon Territory; M.Sc. thesis, University of British Columbia, Vancouver, 149 p.

Carne, R.C. and Cathro, R.J.
1982: Sedimentary exhalative (sedex) zinc-lead-silver deposits, northern Canadian Cordillera; Canadian Institute of Mining and Metallurgy, Bulletin, v. 75, no. 840, p. 66-78.

Carr, J.M.
1971: Geology of Churchill Copper Deposit; Canadian Institute of Mining and Metallurgy, Transactions, v. 74, p. 152-156.

Carr, J.M. and Reed, A.J.
1976: Afton: a supergene copper deposit; in Porphyry Deposits of the Canadian Cordillera, A. Sutherland Brown (ed.), Canadian Institute of Mining and Metallurgy, Special Volume 15, p. 376-387.

Carrière, J.J., Sinclair, W.D., and Kirkham, R.V.
1981: Copper deposits and occurrences in Yukon Territory; Geological Survey of Canada, Paper 81-12, 62 p.

Carson, D.J.T. and Jambor, J.L.
1974: Mineralogy, zonal relationships and economic significance of hydrothermal alteration at porphyry copper deposits, Babine Lake area, British Columbia; Canadian Institute of Mining and Metallurgy, Bulletin, v. 67, p. 110-133.

Carswell, H.T.
1957: The geology and ore deposits of the Summit camp, Boundary district, British Columbia; M.Sc. thesis, University of British Columbia, Vancouver, 80 p.

Carter, N.C.
1970: Len; in Geology, Exploration and Mining in British Columbia 1970, British Columbia Department of Mines and Petroleum Resources, p. 104-107.
1982: Porphyry copper and molybdenum deposits in west-central British Columbia; British Columbia Ministry of Energy, Mines and Petroleum Resources, Bulletin 64, 150 p.

Champigny, N. and Sinclair, A.J.
1982: The Cinola gold deposit, Queen Charlotte Islands, British Columbia; in Geology of Canadian Gold Deposits, R.W. Hodder and W. Petruk (ed.), Canadian Institute of Mining and Metallurgy, Special Volume 24, p. 243-254.

Chappell, B.W. and White, A.J.R.
1974: Two contrasting granite types; Pacific Geology, v. 8, p. 173-174.

Chartrand, F.
1981: Evolution diagenetique des depots stratiformes de cuivre du Lac Coates, ceinture cuprifere de Redstone, Territoires du N.O., Canada; M.Sc. thesis, Université de Montreal, Quebec, 137 p.

Christie, A.B.
1989: Cinola gold deposit, Queen Charlotte Islands (103F/9E); in Geological Fieldwork 1988, British Columbia Ministry of Energy, Mines and Petroleum Resources, Paper 1989-1, p. 423-428.

Christopher, P.A.
1975: Carmi-Beaverdell area (82E/6,11); in Geological Fieldwork 1975, British Columbia Department of Mines and Petroleum Resources, p. 27-31.

Christopher, P.A. and Carter, N.C.
1976: Metallogeny and metallogenic epochs for porphyry mineral deposits in the Canadian Cordillera; in Porphyry Deposits of the Canadian Cordillera, A. Sutherland Brown (ed.), Canadian Institute of Mining and Metallurgy, Special Volume 15, p. 64-71.

Christopher, P.A. and Pinsent, R.H.
1982: Geology of the Ruby Creek and Boulder Creek area near Atlin (104N/11W), notes to accompany Preliminary Map 52; British Columbia Ministry of Energy, Mines and Petroleum Resources, 10 p.

Christopher, P.A., White, W.H., and Harakal, J.E.
1972: Age of molybdenum and tungsten mineralization in northern British Columbia; Canadian Journal of Earth Sciences, v. 9, p. 1727-1734.

Church, B.N.
1971: Geology of the Owen Lake, Parrott Lakes, and Goosly Lake area; in Geology, Exploration and Mining in British Columbia 1970, British Columbia Department of Mines and Petroleum Resources, p. 119-125.
1986: Geological setting and mineralization in the Mount Attwood-Phoenix area of the Greenwood mining camp; British Columbia Ministry of Energy, Mines and Petroleum Resources, Paper 1986-2, 65 p.

Church, B.N. and Diakow, L.
1982: Geology and lithogeochemistry of the Capoose silver prospect (93F/3,6); in Geological Fieldwork 1981, British Columbia Ministry of Energy, Mines and Petroleum Resources, Paper 1982-1, p. 109-112.

Church, B.N., Pearson, D.E., and Preto, V.A.
1977: Volcanic suites of southwestern British Columbia; Geological Association of Canada/Society of Economic Geologists, Joint Annual Meeting, 1977, Field Trip No. 6: Guidebook, 24 p.

Clark, K.F.
1972: Stockwork molybdenum deposits in the Western Cordillera of North America; Economic Geology, v. 67, p. 731-758.

Clarke, W.E.
1969: Giant Mascot Mines Limited: geology and ore controls; Western Miner, v. 42, no. 6, p. 40-46.

Coates, J.A.
1964: The Redstone bedded copper deposit and a discussion on the origin of red bed copper deposits; M.Sc. thesis, University of British Columbia, Vancouver, 77 p.

Cockfield, W.E.
1936: Lode gold deposits of Ymir-Nelson area, British Columbia; Geological Survey of Canada, Memoir 191, 78 p.

Collins, J.A. and Smith, L.
1977: Genesis of cupriferous quartz arenite cycles in the Grinnell Formation (Spokane equivalent), Middle Proterozoic (Helikian) Belt-Purcell Supergroup, eastern Rocky Mountains, Canada; Bulletin of Canadian Petroleum Geology, v. 25, p. 713-735.

Constantinou, G. and Govett, G.J.S.
1973: Geology, geochemistry and genesis of Cyprus sulfide deposits; Economic Geology, v. 68, p. 843-858.

Cyr, J.B., Pease, R.B., and Schroeter, T.G.
1984: Geology and mineralization at Equity silver mine; Economic Geology, v. 79, p. 947-968.

Dalgarno, C.R. and Johnson, J.E.
1968: Diapiric structures and Late Precambrian-Early Cambrian sedimentation in Flinders Ranges, South Australia; in Diapirism and Diapirs, a Symposium, J. Braunstein and G.D. O'Brian (ed.), American Association of Petroleum Geologists, Memoir 8, p. 301-314.

Davis, E.E., Goodfellow, W.D., Bornhold, B.D., Adshead, J., Blaise, B., Villinger, H., and LeCheminant, G.M.
1987: Massive sulfides in a sedimented rift valley, northern Juan de Fuca Ridge; Earth and Planetary Science Letters, v. 82, p. 49-61.

Dawson, K.M.
1975: Carbonate-hosted zinc-lead deposits of the northern Canadian Cordillera; in Report of Activities, Part A, Geological Survey of Canada, Paper 75-1A, p. 239-241.
1983: A review of barite in the northern Canadian Cordillera; Canadian Institute of Mining and Metallurgy Symposium, Mineral Deposits of Northern Cordillera, Whitehorse, Program and Abstracts, p. 18.
1984: (Compiler) Mineral deposits and principal mineral occurrences of the Canadian Cordillera and adjacent parts of the United States of America; Geological Survey of Canada, Map 1513A.
1985: Review of metallogenic studies in Cassiar, Alexander and oceanic terranes; Cordilleran Geology and Exploration Roundup, Vancouver, Program and Abstracts.
1987: Silver-rich base metal skarn and replacement deposits of the northern Canadian Cordillera and Mexico; Northwest Mining Association 93rd Annual Convention, Spokane, Washington, Program and Abstracts, p. 28.

Dawson, K.M. and Dick, L.A.
1978: Regional metallogeny in the northern Cordillera: tungsten and base metal-bearing skarns in southeastern Yukon and southwestern Mackenzie; in Current Research, Part A, Geological Survey of Canada, Paper 78-1A, p. 289-292.

Dawson, K.M. and Kimura, E.T.
1972: Endako; in Copper and Molybdenum Deposits of the Western Cordillera, International Geological Congress, Montreal, 24th Session Guidebook for Field Excursion A09-C09, p. 36-47.

Dawson, K.M. and Orchard, M.J.
1982: Regional metallogeny of the northern Cordillera: biostratigraphy, correlation and metallogenic significance of bedded barite occurrences in eastern Yukon and western District of Mackenzie; in Current Research, Part C, Geological Survey of Canada, Paper 82-1C, p. 31-38.

Dawson, K.M., Friday, S.J., and McLaren, M.
1983: Mineralogical and chemical zonation in the Silence Lake tungsten skarn, Clearwater, British Columbia; Geological Association of Canada/Mineralogical Association of Canada/Canadian Geophysical Union, Joint Annual Meeting, Victoria, Program and Abstracts, p. A16.

Debicki, R.L.
1984: Bedrock geology and mineralization of the Klondike area (west), 115O/14,15 and 116B/2,3; Exploration and Geological Services Division, Yukon, Indian and Northern Affairs Canada, Open File.
1985: Bedrock geology and mineralization of the Klondike area (east), 115O/9,10,11,14,15,16 and 116B/2; Exploration and Geological Services Division, Yukon, Indian and Northern Affairs Canada, Open File.

Delaney, G.D.
1978: A progress report on stratigraphic investigations of the lowermost succession of Proterozoic rocks, northern Wernecke Mountains, Yukon Territory; Department of Indian and Northern Affairs, Open File EGS 1978-10.
1981: The mid-Proterozoic Wernecke Supergroup, Wernecke Mountains, Yukon Territory; in Proterozoic Basins of Canada, F.H.A. Campbell (ed.), Geological Survey of Canada, Paper 81-10, p. 1-23.

de Rosen-Spence, A. and Sinclair, A.J.
1988: Lower Jurassic volcanism of the Stikine Super-terrane; in Geological Fieldwork 1987, British Columbia Ministry of Energy, Mines and Petroleum Resources, Paper 1988-1, p. 211-216.

Devlin, B.D. and Godwin, C.I.
1986: Geology of the Dolly Varden camp, Alice Arm area (103P/11,12); in Geological Fieldwork 1985, British Columbia Ministry of Energy, Mines and Petroleum Resources, Paper 1986-1, p. 327-330.

Dewey, J.F. and Burke, K.
1973: Plume generated triple junctions; Eos (Transactions, American Geophysical Union), v. 54, p. 239.

Diakow, L.J. and Panteleyev, A.
1981: Cassiar gold deposits, McDame map-area (104P/4,5); in Geological Fieldwork 1980, British Columbia Ministry of Energy, Mines and Petroleum Resources, Paper 1981-1, p. 55-62.

Dick, L.A. and Hodgson, C.J.
1983: Contrasting environments of formation of W- and Sn-bearing skarns in the NE Canadian Cordillera; Geological Association of Canada/Mineralogical Association of Canada/Canadian Geophysical Union, Joint Annual Meeting, Victoria, B.C., Program and Abstracts, p. A17.

Dodds, C.J. and Campbell, R.B.
1988: Potassium-argon ages of mainly intrusive rocks in the Saint Elias Mountains, Yukon and British Columbia; Geological Survey of Canada, Paper 87-16, 43 p.

Drummond, A.D., Sutherland Brown, A., Young, R.J., and Tennant, S.J.
1976: Gibraltar-regional metamorphism, mineralization, hydrothermal alteration and structural development; in Porphyry Deposits of the Canadian Cordillera, A. Sutherland Brown (ed.), Canadian Institute of Mining and Metallurgy, Special Volume 15, p. 195-205.

Drysdale, C.W.
1915: Geology of the Franklin Mining Camp, British Columbia; Geological Survey of Canada, Memoir 56, 246 p.

Eastwood, G.E.P.
1960: Magnetite in Lodestone Mountain Stock; in British Columbia Minister of Mines, Annual Report 1959, p. 39-53.

Eckstrand, O.R. (ed.)
1984: Canadian mineral deposit types: a geological synopsis; Geological Survey of Canada, Economic Geology Report 36, 86 p.

Eisbacher, G.H.
1977: Tectono-stratigraphic framework of the Redstone Copper Belt, District of Mackenzie; in Report of Activities, Part A, Geological Survey of Canada, Paper 77-1A, p. 229-234.
1981: Sedimentary tectonics and glacial record in the Windermere Supergroup, Mackenzie Mountains, northwestern Canada; Geological Survey of Canada, Paper 80-27, 40 p.
1983: Devonian-Mississippian sinistral transcurrent faulting along the cratonic margin of western North America: A hypothesis; Geology, v. 11, p. 7-10.
1985: Late Proterozoic rifting, glacial sedimentation, and sedimentary cycles in the light of Windermere deposition, Western Canada; Palaeogeography, Palaeoclimatology, Palaeoecology, v. 51, p. 231-254.

Ettlinger, A.D. and Ray, G.E.
1988: Gold-enriched skarn deposits of British Columbia; in Geological Fieldwork 1987, British Columbia Ministry of Energy, Mines and Petroleum Resources, Paper 1988-1, p. 263-279.

Fahrni, K.C., Macauley, T.N., and Preto, V.A.G.
1976: Copper Mountain and Ingerbelle; in Porphyry Deposits of the Canadian Cordillera, A. Sutherland Brown (ed.), Canadian Institute of Mining and Metallurgy, Special Volume 15, p. 368-375.

Faulkner, E.L.
1986: Blackdome deposit (92O/7E, 8W); in Geological Fieldwork 1985, British Columbia Ministry of Energy, Mines and Petroleum Resources, Paper 1986-1, p. 106-109.

Field, C.W. and Fifarek, R.H.
1985: Light stable-isotope systematics in the epithermal environment; in Geology and Geochemistry of Epithermal Systems, B.R. Berger and P.M. Bethke (ed.), Society of Economic Geologists, Reviews in Economic Geology, v. 2, p. 99-128.

Findlay, D.C.
1969: Origin of the Tulameen ultramafic-gabbro complex, southern British Columbia; Canadian Journal of Earth Sciences, v. 6, p. 399-425.

Fleming, J.A.
1983: Island Copper; in Field Trip Guidebook, Vol. II, Fieldtrip 9, Geological Association of Canada, Victoria, 1983, p. 21-35.

Fleming, J.A., Walker, R., and Wilton, P.
1983: Mineral deposits of Vancouver Island: Westmin Resources (Au-Ag-Cu-Pb-Zn), Island Copper (Cu-Au-Mo), Argonaut (Fe); Geological Association of Canada, Victoria, 1983, Field Trip Guidebook, Vol. II, Fieldtrip 9, 41 p.

Fox, J.S.
1984: Besshi-type volcanogenic sulphide deposits - a review; Canadian Institute of Mining and Metallurgy, Bulletin, v. 77, no. 864, p. 57-68.

Fox, P.E., Cameron, R.S., and Hoffman, S.J.
1987: Geology and soil geochemistry of the Quesnel River gold deposit, British Columbia; Geoexpo/86, Exploration in the North American Cordillera, Association of Exploration Geochemists, Symposium Proceedings, Vancouver, May 12-14, 1986, p. 61-71.

Francheteau, J., Needham, H.D., Choukroune, P., Juteau, T., Séguret, M., Ballard, R.D., Fox, P.J., Normark, W., Carranza, A., Cordoba, D., Guerrero, J., Rangin, C., Bourgault, H., Cambon, P., and Hekinian, R.
1979: Massive deep-sea sulphide ore deposits discovered on the East Pacific Rise; Nature, v. 277, p. 523-528.

Franklin, J.M., Lydon, J.W., and Sangster, D.F.
1981: Volcanic-associated massive sulfide deposits; in Economic Geology, Seventy-Fifth Anniversary Volume, B.J. Skinner (ed.), Economic Geology Publishing Company, p. 485-627.

Fyles, J.T.
1960: Mineral King (Sheep Creek Mine Limited); British Columbia Ministry of Mines, Annual Report 1959, p. 74-89.
1970: The Jordan River area near Revelstoke, British Columbia; British Columbia Department of Mines and Petroleum Resources, Bulletin 57, 64 p.
1984: Geological setting of the Rossland Mining Camp; British Columbia Ministry of Energy, Mines and Petroleum Resources, Bulletin 74, 61 p.

Gabrielse, H.
1966: Watson Lake map-area, Yukon Territory; Geological Survey of Canada, Map 19-1966.
1967: Tectonic evolution of the northern Canadian Cordillera; Canadian Journal of Earth Sciences, v. 4, p. 271-298.
1985: Major dextral transcurrent displacements along the Northern Rocky Mountain Trench and related lineaments in north-central British Columbia; Geological Society of America Bulletin, v. 96, p. 1-14.

Gabrielse, H. and Blusson, S.L.
1969: Geology of Coal River map-area, Yukon Territory and District of Mackenzie (95D); Geological Survey of Canada, Paper 68-38, 22 p.

Gabrielse, H., Blusson, S.L., and Roddick, J.A.
1973: Geology of Flat River, Glacier Lake and Wrigley Lake map-areas, District of Mackenzie and Yukon Territory; Geological Survey of Canada, Memoir 366, Part 1, 153 p., Part 2, 268 p.

Gabrielse, H., Loveridge, W.D., Sullivan, R.W., and Stevens, R.D.
1982: U-Pb measurements of zircon indicate middle Paleozoic plutonism in the Omineca Crystalline Belt, north-central British Columbia; in Current Research, Part C, Geological Survey of Canada, Paper 82-1C, p. 139-146.

Gammon, J.B. and Chandler, T.E.
1986: Exploration of the Windy Craggy massive sulphide deposit, British Columbia, Canada; in Geology in the Real World - The Kingsley Dunham Volume, E.W. Nesbitt and I. Nichol (ed.), Institute of Mining and Metallurgy, p. 131-141.

Gareau, S.A.
1988: Preliminary study of the Work Channel lineament in the Ecstall River area, Coast Plutonic Complex, British Columbia; in Current Research, Part E, Geological Survey of Canada, Paper 88-1E, p. 49-55.

Garnett, J.A.
1978: Geology and mineral occurrences of the southern Hogem batholith; British Columbia Ministry of Mines and Petroleum Resources, Bulletin 70, 75 p.

Geological Survey of Canada
1981: Assessment of mineral and fuel resource potential of the proposed northern Yukon national park and adjacent areas (Phase 1); Geological Survey of Canada, Open File 760, 33 p.

Gilbert, G.
1948: Rossland Camp; in Structural Geology of Canadian Ore Deposits, Canadian Institute of Mining and Metallurgy, Jubilee Volume, p. 189-196.

Gill, J.E. and Byers, A.R.
1948: Surf Inlet and Pugsley mines; in Structural Geology of Canadian Ore Deposits, Canadian Institute of Mining and Metallurgy, Jubilee Volume, p. 99-104.

Glover, J.K. and Burson, M.J.
1986: Geology of the Lened tungsten skarn deposit, Logan Mountains, Northwest Territories; in Mineral Deposits of Northern Cordillera, J.A. Morin (ed.), Canadian Institute of Mining and Metallurgy, Special Volume 37, p. 255-265.

Godwin, C.I.
1975: Imbricate subduction zones and their relationship with Upper Cretaceous to Tertiary porphyry deposits in the Canadian Cordillera; Canadian Journal of Earth Sciences, v. 12, p. 1362-1378.

Godwin, C.I., Armstrong, R.L., and Tompson, K.M.
1980: K-Ar and Rb-Sr dating and the genesis of tungsten at the Clea tungsten skarn property, Selwyn Mountains, Yukon Territory; Canadian Institute of Mining and Metallurgy, Bulletin, v. 73, no. 821, p. 90-93.

Godwin, C.I., Sinclair, A.J., and Ryan, B.D.
1982: Lead isotope models for the genesis of carbonate-hosted Zn-Pb, shale-hosted Ba-Zn-Pb, and silver-rich deposits in the northern Canadian Cordillera; Economic Geology, v. 77, p. 82-94.

Goldsmith, L.B. and Sinclair, A.J.
1983: Spatial density of silver-lead-zinc-gold vein deposits in four mining camps in southeastern British Columbia (82F); in Geological Fieldwork 1982, British Columbia Ministry of Energy, Mines and Petroleum Resources, Paper 1983-1, p. 251-265.

Goodfellow, W.D. and Jonasson, I.R.
1984: Ocean stagnation and ventilation defined by $\delta^{34}S$ secular trends in pyrite and barite, Selwyn Basin, Yukon; Geology, v. 12, p. 583-586.
1986: Environment of formation of the Howards Pass (XY) Zn-Pb deposit, Selwyn Basin, Yukon; in Mineral Deposits of Northern Cordillera, J.A. Morin (ed.), Canadian Institute of Mining and Metallurgy, Special Volume 37, p. 19-50.

Gordey, S.P.
1979: Stratigraphy of southeastern Selwyn Basin in the Summit Lake area, Yukon Territory and Northwest Territories; in Current Research, Part A, Geological Survey of Canada, Paper 79-1A, p. 13-16.
1983: Thrust faults in the Anvil Range and a new look at the Anvil Range Group, south-central Yukon Territory; in Current Research, Part A, Geological Survey of Canada, Paper 83-1A, p. 225-227.
in press: Evolution of the northern Cordilleran miogeocline, Nahanni map-area (105I), Yukon Territory and District of Mackenzie; Geological Survey of Canada, Memoir 428.

Gorzynski, G.A.
1986: Geology and lithogeochemistry of the Cirque stratiform sediment-hosted Ba-Zn-Pb-Ag deposit, northeastern British Columbia; M.Sc. thesis, University of British Columbia, Vancouver, 129 p.

Griffiths, J.R. and Godwin, C.I.
1983: Metallogeny and tectonics of porphyry copper-molybdenum deposits in British Columbia; Canadian Journal of Earth Sciences, v. 20, p. 1000-1018.

Gross, G.A.
1965: Iron formation, Snake River area, Yukon and Northwest Territories; in Report of Activities: field 1964; Geological Survey of Canada, Paper 65-1, p. 143.

Grove, E.W.
1986: Geology and mineral deposits of the Unuk River-Salmon River-Anyox area; British Columbia Ministry of Energy, Mines and Petroleum Resources, Bulletin 63, 152 p.

Guilbert, J.M. and Lowell, J.D.
1974: Variations in zoning patterns in porphyry ore deposits; Canadian Institute of Mining and Metallurgy, Bulletin, v. 67, p. 99-109.

Guild, P.W.
1971: Metallogeny: a key to exploration; Mining Engineering, v. 23, p. 69-72.

Gustafson, L.B. and Hunt, J.P.
1975: The porphyry copper deposit at El Salvador, Chile; Economic Geology, v. 70, p. 857-912.

Gustafson, L.B. and Williams, N.
1981: Sediment-hosted stratiform deposits of copper, lead, and zinc; Economic Geology, Seventy-Fifth Anniversary Volume, B.J. Skinner (ed.), Economic Geology Publishing Company, p. 139-178.

Hamilton, J.M., Bishop, D.T., Morris, H.C., and Owens, O.E.
1982: Geology of Sullivan orebody, Kimberley, B.C. Canada; in Precambrian Sulphide Deposits, The H.S. Robinson Memorial Volume, R.W. Hutchinson, C.D. Spence and J.M. Franklin (ed.), Geological Association of Canada, Special Paper 25, p. 597-665.

Hamilton, J.M., Delaney, G.D., Hauser, R.L., and Ransom, P.W.
1983: Geology of the Sullivan Deposit, Kimberley, B.C., Canada; in Short Course in Sediment-hosted Stratiform Lead-Zinc Deposits, D.F. Sangster (ed.), Mineralogical Association of Canada, Short Course Handbook, v. 8, p. 31-84.

Hamilton, J.V.
1982: Geology of the north showing at the Mt. Hundere zinc-lead-silver skarn deposit, southeastern Yukon Territory; B.A.Sc. thesis, University of British Columbia, 73 p.

Hancock, K.D.
1988: Magnetite occurrences in British Columbia; in British Columbia Ministry of Energy, Mines and Petroleum Resources, Open File 1988-22, p. 131-132.

Hardy, J.L.
1979: Stratigraphy, brecciation and mineralization, Gayna River, N.W.T.; M.Sc. thesis, University of Toronto, Ontario.

Harper, G.
1977: Geology of the Sustut Copper deposit in B.C.; Canadian Institute of Mining and Metallurgy Bulletin, v. 70, no. 777, p. 97-104.

Harris, M.W.
1991: The Mt. Milligan porphyry system: a structural and genetic approach; CIM 93rd Annual General Meeting, Vancouver, Program and Abstracts, Canadian Institute of mining and metallurgy, Bulletin, v. 84, n. 947, p. 88.

Harrison, J.E.
1972: Precambrian Belt basin of northwestern United States: its geometry, sedimentation and copper occurrences; Geological Society of America Bulletin, v. 83, p. 1215-1240.

Hayba, D.O., Bethke, P.M., Heald, P., and Foley, N.K.
1985: Geologic, mineralogic, and geochemical characteristics of volcanic-hosted epithermal precious-metal deposits; in Geology and Geochemistry of Epithermal Systems, B.R. Berger and P.M. Bethke (ed.), Society of Economic Geologists, Reviews in Economic Geology, v. 2, p. 129-167.

Heald, P., Foley, N.K., and Hayba, D.O.
1987: Comparative anatomy of volcanic-hosted epithermal deposits: acid-sulfate and adularia-sericite types; Economic Geology, v. 82, p. 1-26.

Hedley, M.S.
1940: Geology of Camp McKinney and of the Cariboo-Amelia mine, Similkameen district; British Columbia Department of Mines, Bulletin 6, 39 p.
1947: Geology of Whitewater and Lucky Jim Mine areas, Slocan District; British Columbia Department of Mines, Bulletin 22, 54 p.
1952: Geology and ore deposits of the Sandon area, Slocan mining camp, British Columbia; British Columbia Department of Mines, Bulletin 29, 130 p.

Hedley, M.S. and Watson, K.De P.
1945: Lode-gold deposits, central southern British Columbia; British Columbia Department of Mines, Bulletin 20, Part III, 27 p.

Helmstaedt, H., Eisbacher, G.H., and McGregor, J.A.
1979: Copper mineralization near an intra-Rapitan unconformity, Nite copper prospect, Mackenzie Mountains, Northwest Territories, Canada; Canadian Journal of Earth Sciences, v. 16, p. 50-59.

Henley, R.W.
1985: The geothermal framework for epithermal deposits; in Geology and Geochemistry of Epithermal Systems, B.R. Berger and P.M. Bethke (ed.), Society of Economic Geologists, Reviews in Economic Geology, v. 2, p. 1-24.

Henley, R.W. and Ellis, A.J.
1983: Geothermal systems ancient and modern: a geochemical review; Earth-Science Reviews, v. 19, p. 1-50.

Hewton, R.S.
1982: Gayna River: a Proterozoic Mississippi Valley-type zinc-lead deposit; in Precambrian Sulphide Deposits, H.S. Robinson Memorial Volume, R.W. Hutchinson, C.D. Spence and J.M. Franklin (ed.), Geological Association of Canada, Special Paper 25, p. 667-700.

Hodgson, C.J., Bailes, R.J., and Verzosa, R.S.
1976: Cariboo-Bell; in Porphyry Deposits of the Canadian Cordillera, A. Sutherland Brown (ed.), Canadian Institute of Mining and Metallurgy, Special Volume 15, p. 388-396.

Hodgson, C.J., Chapman, R.S.G., and MacGeehan, P.J.
1982: Application of exploration criteria for gold deposits in the Superior Province of the Canadian Shield to gold exploration in the Cordillera; in Precious Metals in the Northern Cordillera, A.A. Levinson (ed.), Association of Exploration Geochemists, p. 173-206.

Hooper, D.
1984: A study of the gold-quartz veins at Erickson Gold Camp, Cassiar, north-central British Columbia; B.Sc. thesis, University of British Columbia, Vancouver, 96 p.

Høy, T.
1979: Geology of the Goldstream area; British Columbia Ministry of Energy, Mines and Petroleum Resources, Bulletin 71, 49 p.
1980: Geology of the Riondel area, Central Kootenay Arc, southeastern British Columbia; British Columbia Ministry of Energy, Mines and Petroleum Resources, Bulletin 73, 89 p.
1982a: The Purcell Supergroup in southeastern British Columbia: sedimentation, tectonics and stratiform lead-zinc deposits; in Precambrian Sulphide Deposits, The H.S. Robinson Memorial Volume, R.W. Hutchinson, C.D. Spence and J.M. Franklin (ed.), Geological Association of Canada, Special Paper 25, p. 127-147.
1982b: Stratigraphic and structural setting of stratabound lead-zinc deposits in southeastern B.C.; Canadian Institute of Mining and Metallurgy, Bulletin, v. 75, no. 840, p. 114-134.

Høy, T. and Andrew, K.
1988: Preliminary geology and geochemistry of the Elise Formation, Rossland Group, between Nelson and Ymir, southeastern British Columbia (82F/06); in Geological Fieldwork 1987, Mineral Resources Division, British Columbia Ministry of Energy, Mines and Petroleum Resources, Paper 1988-1, p. 19-30.

Høy, T. and Goutier, F.
1986: Rea Gold (Hilton) and Homestake volcanogenic sulphide-barite deposits southeastern British Columbia (82 M/4 W); in Geological Fieldwork 1985, Mineral Resources Division, British Columbia Ministry of Energy, Mines and Petroleum Resources, Paper 1986-1, p. 59-68.

Høy, T. and Pell, J.
1986: Carbonatites and associated alkalic rocks, Perry River and Mount Grace areas, Shuswap Complex, southeastern British Columbia; in Geological Fieldwork 1985, British Columbia Ministry of Energy, Mines and Petroleum Resources, Paper 1986-1, p. 69-87.

Høy, T., Gibson, G., and Berg, N.W.
1984: Copper-zinc deposits associated with basic volcanism, Goldstream area, southeastern British Columbia; Economic Geology, v. 79, p. 789-814.

Hudson, T., Smith, J.G., and Elliott, R.L.
1979: Petrology, composition, and age of intrusive rocks associated with the Quartz Hill molybdenite deposit, southeastern Alaska; Canadian Journal of Earth Sciences, v. 16, p. 1805-1822.

Hulbert, L.J., Duke, J.M., Eckstrand, O.R., Lydon, J.W., Scoates, R.F.J., Cabri, L.J., and Irvine, T.N.
1988: Geological environments of the platinum group elements; Cordilleran Section, Geological Association of Canada, Short Course Notes, 151 p.

Hutchinson, R.W.
1973: Volcanogenic sulfide deposits and their metallogenic significance; Economic Geology, v. 68, p. 1223-1246.

1980: Massive base metal sulphide deposits as guides to tectonic evolution; in The Continental Crust and its Mineral Deposits, D.W. Strangway (ed.), Geological Association of Canada, Special Paper 20, p. 659-684.

Hyndman, D.W.
1968: Petrology and structure of Nakusp map-area, British Columbia; Geological Survey of Canada, Bulletin 161, 95 p.

Ishihara, S. (Editor)
1974: Geology of Kuroko deposits; Society of Mining Geologists of Japan, Mining Geology, Special Issue 6, 435 p.
1981: The granitoid series and mineralization; Economic Geology, Seventy-Fifth Anniversary Volume, B.J. Skinner (ed.), p. 458-484.

Jackson, S.A. and Beales, F.W.
1967: An aspect of sedimentary basin evolution: the concentration of Mississippi Valley-type ores during late stages of diagenesis; Bulletin of Canadian Petroleum Geology, v. 15, p. 383-433.

Jefferson, C.W.
1978: Stratigraphy and sedimentology, Upper Proterozoic Redstone Copper Belt, Mackenzie Mountains, Northwest Territories - a preliminary report; in Mineral Industry Report for 1975, Northwest Territories, Indian and Northern Affairs, Economic Geology Series 1978-5, p. 157-169.
1983: The Upper Proterozoic Redstone Copper Belt, Mackenzie Mountains, N.W.T.; Ph.D. thesis, University of Western Ontario, London, 445 p.

Jefferson, C.W. and Parrish, R.R.
1989: Late Proterozoic stratigraphy, U-Pb zircon ages and rift tectonics, Mackenzie Mountains northwestern Canada; Canadian Journal of Earth Sciences, v. 26, n. 9, p. 1784-1801.

Jefferson C.W. and Ruelle, J.C.L.
1986: The Late Proterozoic Redstone Copper Belt, Mackenzie Mountains, Northwest Territories; in Mineral Deposits of Northern Cordillera, J.A. Morin (ed.), Canadian Institute of Mining and Metallurgy, Special Volume 37, p. 154-168.

Jefferson, C.W., Kilby, D.B., Pigage, L.C., and Roberts, W.J.
1983: The Cirque barite-zinc-lead deposits, northeastern British Columbia; in Short Course in Sediment-hosted Stratiform Lead-Zinc Deposits, D.F. Sangster (ed.), Mineralogical Association of Canada, Short Course Handbook, v. 8 (sic v.9), p. 121-140.

Jennings, D.S. and Jilson, G.A.
1986: Geology and sulphide deposits of Anvil Range, Yukon; in Mineral Deposits of Northern Cordillera, J.A. Morin (ed.), Canadian Institute of Mining and Metallurgy, Special Volume 37, p. 319-361.

Jones, D.L., Silberling, N.J., and Hillhouse, J.
1977: Wrangellia - A displaced terrane in northwestern North America; Canadian Journal of Earth Sciences, v. 14, p. 2565-2577.

Joubin, F.R.
1948: Bralorne and Pioneer mines; in Structural Geology of Canadian Ore Deposits, Jubilee Volume, Canadian Institute of Mining and Metallurgy, p. 168-177.

Kanasewich, E.R., Clowes, R.M., and McCloughan, C.H.
1969: A buried Precambrian rift in western Canada; Tectonophysics, v. 8, p. 513-527.

Kanehira, K. and Tatsumi, T.
1970: Bedded cupriferous iron sulphide deposits in Japan, a review; in Volcanism and Ore Genesis, T. Tatsumi (ed.), University of Tokyo Press, Tokyo, p. 51-76.

Keep, M.
1989: The geology and petrology of the Averill alkaline plutonic complex near Grand Farks, British Colombia; MSc. thesis, University of British Columbia, Vancouver.

Kennedy, G.C.
1953: Geology and mineral deposits of Jumbo basin, southeastern Alaska; United States Geological Survey, Professional Paper 251, 46 p.

Kerans, C., Ross, G.M., Donaldson, J.A., and Geldsetzer, H.J.
1981: Tectonism and depositional history of the Helikian Hornby Bay and Dismal Lakes Groups, District of Mackenzie; in Proterozoic Basins of Canada, F.H.A. Campbell (ed.), Geological Survey of Canada, Paper 81-10, p. 157-182.

Kirkham, R.V.
1967: Glacier Gulch; British Columbia Ministry of Mines, Annual Report 1966, p. 86-90.
1970: Certain copper deposits in Jurassic volcanic rocks of central British Columbia (93L,M, 94D, 103L); in Report of Activities, Part A, Geological Survey of Canada, Paper 70-1A, p. 96-97.

1971: Intermineral intrusions and their bearing on the origin of porphyry copper and molybdenum deposits; Economic Geology, v. 66, p. 1244-1249.
1973: Environments of formation of concordant and peneconcordant copper deposits in sedimentary sequences (abstract); Canadian Mineralogist, v. 12, p. 145-146.
1974: A synopsis of Canadian stratiform copper deposits in sedimentary sequences; Centenaire de la Societé Géologique de Belgique, Gisements Stratiformes et Provinces Cupriferes, Liege, p. 367-382.
1984: Volcanic redbed copper; in Canadian Mineral Deposit Types: A Geological Synopsis, O.R. Eckstrand (ed.), Geological Survey of Canada, Economic Geology Report 36, p. 37.

Klau, W. and Large, D.E.
1980: Submarine exhalative Cu-Pb-Zn deposits - a discussion of their classification and metallogenesis; Geologisches Jahrbuch, v. 40, p. 13-58.

Koski, R.A., Lonsdale, P.F., Shanks, W.C., Berndt, M.E., and Howe, S.S.
1985: Mineralogy and geochemistry of a sediment-hosted hydrothermal sulfide deposit from the Southern Trough of Guaymas Basin, Gulf of California; Journal of Geophysical Research, v. 90, p. 6695-6707.

Koski, R.A., Scott, S.D., Hannington, M.D., Delaney, J.R., and Tivey, M.K.
1988: Hydrothermal processes and massive sulphide deposits on the Juan de Fuca Ridge and other northeast Pacific spreading axes; in Geology and Resource Potential of the Continental Margin of Western North America and Adjacent Ocean Basins Beaufort Sea to Baja California, Circum-Pacific Council for Energy and Mineral Resources Earth Science Series, v. 6, p. 621-638.

Kwong, Y.T.J.
1987: Evolution of the Iron Mask Batholith and its associated copper mineralization; British Columbia Ministry of Energy, Mines and Petroleum Resources, Bulletin 77, 55 p.

Lambert, I.B. and Sato, T.
1974: The Kuroko and associated ore deposits of Japan: a review of their features and metallogenesis; Economic Geology, v. 69, p. 1215-1236.

Large, D.E.
1983: Sediment-hosted massive sulphide lead-zinc deposits: an empirical model; in Short Course in Sediment-Hosted Stratiform Lead-Zinc Deposits, D.F. Sangster (ed.), Mineralogical Association of Canada Handbook, v. 8 (sic v.9), p. 1-29.

Leaming, S.F.
1978: Jade in Canada; Geological Survey of Canada, Paper 78-19, 59 p.

LeCouteur, P.C.
1979: Age of the Sullivan lead-zinc deposit; in Evolution of the Cratonic Margin and Related Mineral Deposits, Geological Association of Canada, Cordilleran Section Symposium, Program and Abstracts, p. 19.

Leech, G.B.
1966: Kananaskis Lakes, West Half, NTS 82J; Geological Survey of Canada, Open File 634.

Lefebure, D.V. and Malott, M.L.
1990: Northwestern District; in Exploration in British Columbia 1989, British Columbia Geological Survey, Branch p. 29-42.

Leitch, C.H.B. and Godwin, C.I.
1988: Isotopic ages, wallrock chemistry and fluid inclusion data from the Bralorne gold vein deposit (92J/15W); in Geological Fieldwork 1987, British Columbia Ministry of Energy, Mines and Petroleum Resources, Paper 1988-1, p. 301-324.

Leitch, C.H.B., Dawson, K.M., and Godwin, C.I.
1988: Late Cretaceous-Early Tertiary gold mineralization: a galena lead isotope study of the Bridge River mining camp, southwestern British Columbia, Canada; Bicentennial Gold '88, Melbourne, Australia.

Leitch, C.H.B., Hood, C.T. Cheng, Xiao-lin and Sinclair, A.J.
1990: Geology of the Silver Queen mine area, Owen Lake, central British Columbia; in Geological Fieldwork 1989, British Columbia Geological Survey Branch, Paper 1990-1 p. 287-295.

Lennan, W.B.
1986: Ray Gulch tungsten skarn deposit, Dublin Gulch area, central Yukon; in Mineral Deposits of Northern Cordillera, J.A. Morin (ed.), Canadian Institute of Mining and Metallurgy, Special Volume 37, p. 245-254.

Lindgren, W.
1928: Mineral deposits; Third Edition, McGraw Hill, New York, 1049 p.

Little, H.W.
1960: Nelson map-area, west half, British Columbia (82F W1/2); Geological Survey of Canada, Memoir 308, 205 p.
1961: Geology of Kettle River, west half (82E), B.C.; Geological Survey of Canada, Map 15-1961.
1982: Geology of the Rossland-Trail map-area, British Columbia; Geological Survey of Canada, Paper 79-26, 38 p.
1983: Geology of the Greenwood map-area, British Columbia; Geological Survey of Canada, Paper 79-29, 37 p.

Little, L.M.
1974: The geology and mineralogy of the Brandywine property lead-zinc-gold-silver deposit, Brandywine map-area, south-western British Columbia; B.Sc. thesis, University of British Columbia, Vancouver, 96 p.

Lonsdale, P.F., Bischoff, J.L., Burns, V.M., Kastner, M., and Sweeney, R.E.
1980: A high-temperature hydrothermal deposit on the seabed at a Gulf of California spreading center; Earth and Planetary Science Letters, v. 49, p. 8-20.

Lowell, J.D. and Guilbert, J.M.
1970: Lateral and vertical alteration-mineralization zoning in porphyry ore deposits; Economic Geology, v. 65, p. 373-408.

Lydon, J.W.
1984: Ore deposit models - 8. Volcanogenic massive sulphide deposits, Part I: a descriptive model; Geoscience Canada, v. 11, p. 195-202.
1988: Ore deposit models #14. Volcanogenic massive sulphide deposits, Part 2: genetic models; Geoscience Canada, v. 15, p. 43-65.

Lydon, J.W., Lancaster, R.D., and Karkkainen, P.
1979: Genetic controls of Selwyn Basin stratiform barite/sphalerite/galena deposits: an investigation of the dominant barium mineralogy of the Tea Deposit, Yukon; in Current Research, Part B, Geological Survey of Canada, Paper 79-1B, p. 223-229.

MacIntyre, D.G.
1982: Geologic setting of recently discovered stratiform barite-sulphide deposits in northeast British Columbia; Canadian Institute of Mining and Metallurgy, v. 75, no. 840, p. 99-113.
1984: Geology of the Alsek-Tatshenshini Rivers area (114P); in Geological Fieldwork 1983, British Columbia Ministry of Energy, Mines and Petroleum Resources, Paper 1984-1, p. 173-184.
1985: Geology and mineral deposits of the Tahtsa Lake district, west-central British Columbia; British Columbia Ministry of Energy, Mines and Petroleum Resources, Bulletin 75, 82 p.
1986: The geochemistry of basalts hosting massive sulphide deposits, Alexander Terrane, northwest British Columbia (114P); in Geological Fieldwork 1985, British Columbia Ministry of Energy, Mines and Petroleum Resources, Paper 1986-1, p. 197-210.

MacIntyre, D.G. and Schroeter, T.G.
1985: Mineral occurrences in the Mount Henry Clay area (114P/7,8); in Geological Fieldwork 1984, British Columbia Ministry of Energy, Mines and Petroleum Resources, Paper 1985-1, p. 365-379.

MacIntyre, D.G., Brown, D., Desjardins, P., and Mallett, P.
1987: Babine Project (93L/10,15); in Geological Fieldwork 1986, British Columbia Ministry of Energy, Mines and Petroleum Resources, Paper 1987-1, p. 201-222.

MacKevett, E.M., Jr.
1976: Mineral deposits and occurrences in the McCarthy quadrangle, Alaska; United States Geological Survey, Miscellaneous Field Study Map, MF 773-B.

MacKevett, E.M., Jr., Robertson, E.C., and Winkler, G.R.
1974: Geology of the Skagway B-3 and B-4 quadrangles, southeastern Alaska; United States Geological Survey, Professional Paper 832, 33 p.

MacLean, M.E.
1989: Mount Brussilof magnesite project, southeast British Columbia (82J/13E); in Geological Fieldwork 1988, British Columbia Ministry of Energy, Mines and Petroleum Resourses, Paper 1989-1, p. 507-510.

Macqueen, R.W.
1976: Sediments, zinc and lead, Rocky Mountain Belt, Canadian Cordillera; Geoscience Canada, v. 3, , p. 71-81.

Mathieson G.A. and Clark, A.H.
1984: The Cantung E Zone scheelite skarn orebody, Tungsten, Northwest Territories: a revised genetic model; Economic Geology, v. 79, p. 883-901.

Mato, G., Ditson, G., and Godwin, C.I.
1983: Geology and geochronometry of tin mineralization associated with the Seagull batholith, south-central Yukon Territory; Canadian Institute of Mining and Metallurgy, Bulletin, v. 76, no. 854, p. 43-49.

McAllister, A.L.
1951: Ymir map-area, British Columbia; Geological Survey of Canada, Paper 51-4, 58 p.

McCartney, W.D.
1965: Metallogeny of post-Precambrian geosynclines; in Some Guides to Mineral Exploration, E.R.W. Neale (ed.), Geological Survey of Canada, Paper 65-6, p. 33-42.

McCartney, W.D. and Potter, R.R.
1962: Mineralization as related to structural deformation, igneous activity and sedimentation in folded geosynclines; Canadian Mining Journal, v. 83, no. 4, p. 83-87.

McClay, K.R. and Bidwell, G.E.
1986: Geology of the Tom deposit, Macmillan Pass, Yukon; in Mineral Deposits of Northern Cordillera, J.A. Morin (ed.), Canadian Institute of Mining and Metallurgy, Special Volume 37, p. 100-114.

McClintock, J.A. and Sinclair, W.D.
1986: Disseminated chalcopyrite in Nasina facies metamorphic rocks near Lucky Joe Creek west central Yukon; in Mineral Deposits of Northern Cordillera, J.A. Morin (ed.), Canadian Institute of Mining and Metallurgy, Special Volume 37, p. 169-177.

McDougall, J.J.
1976: Catface; in Porphyry Deposits of the Canadian Cordillera, A. Sutherland Brown (ed.), Canadian Institute of Mining Special Volume 15, p. 299-310.

McKechnie, N.D.
1966: Distribution of productive mineral deposits related to time-stratigraphic sequences in British Columbia; in Tectonic history and mineral deposits of the western Cordillera, Canadian Institute of Mining and Metallurgy, Special volume no. 8, p. 193-207.

McMechan, M.E.
1980: Stratigraphy, structure and tectonic implications of the Middle Proterozoic Purcell Supergroup in the Mount Fisher area, southeastern British Columbia; Ph.D. thesis, Queen's University, Kingston, Ontario, 279 p.
1981: The middle Proterozoic Purcell Supergroup in the southwestern Rocky and southeastern Purcell Mountains, British Columbia and the initiation of the Cordilleran miogeocline, southern Canada and adjacent United States; Bulletin of Canadian Petroleum Geology, v. 29, p. 583-621.

McMillan, W.J.
1973: Mount Copeland Mine; in Geology, Exploration and Mining in British Columbia 1973, British Columbia Ministry of Energy, Mines and Petroleum Resources, p. 104-113.
1980: CC prospect, Chu Chua Mountain (92P/8W); in Geological Fieldwork 1979, British Columbia Ministry of Energy, Mines and Petroleum Resources, Paper 1980-1, p. 37-48.
1985: Geology and ore deposits of the Highland Valley camp; Geological Association of Canada, Mineral Deposits Division, Field Guide and Reference Manual Series, No. 1, 121 p.

McMillan, W.J. and Panteleyev, A.
1980: Ore deposit models - 1. Porphyry copper deposits; Geoscience Canada, v. 7, p. 52-63.

McMillan, W.J., Panteleyev, A., and Høy, T.
1986: Mineral deposits in British Columbia: a review of their tectonic settings; in Geoexpo 86, Exploration in the North American Cordillera, Symposium Proceedings, I.L. Elliott and B.W. Smee (ed.), Association of Exploration Geochemists, p. 1-18.

Meijer-Drees, N.C.
1975: Geology of the lower Paleozoic formations in the subsurface of the Fort Simpson area, District of Mackenzie, N.W.T.; Geological Survey of Canada, Paper 74-40, 65 p.

Meinert, L.D.
1984: Mineralogy and petrology of iron skarns in western British Columbia, Canada; Economic Geology, v. 79, p. 869-882.
1986: Gold in skarns of the Whitehorse Copper Belt, southern Yukon; in Yukon Geology, v. 1, Exploration and Geological Services Division, Yukon, Indian and Northern Affairs Canada, p. 19-43.

Meyer, W., Gale, R.E., and Randall, A.W.
1976: O.K.; in Porphyry Deposits of the Canadian Cordillera, A. Sutherland Brown (ed.), Canadian Institute of Mining and Metallurgy, Special Volume 15, p. 311-316.

Mihalynuk, M.G., Mountjoy K.J., McMillan, W.J., Ash, C.H. and Hammack, J.L.
1991: Highlights of 1990 Fieldwork in the Atlin area; in Geologicial Fieldwork 1990, British Columbia Geological Survey Branch, Paper 1991-1 p. 145-152

Miller, D.C.
1976: Maggie; in Porphyry Deposits of the Canadian Cordillera, A. Sutherland Brown (ed.), Canadian Institute of Mining and Metallurgy, Special Volume 15, p. 329-335.

Miller, J.H.L. and Sinclair, A.J.
1985: Geology of the Callaghan Creek roof pendant (92J/3); in Geology in British Columbia 1977-1981, British Columbia Ministry of Energy, Mines and Petroleum Resources, p. 98-101.

Mitchell, A.H.G. and Garson, M.S.
1976: Mineralization at plate boundaries; Mineral Science Engineering, v. 8, no. 2, p. 129-169.

Miyashiro, A.
1973: The Troodos ophiolitic complex was probably formed in an island arc; Earth and Planetary Science Letters, v. 19, p. 218-224.

Monger, J.W.H.
1977: The Triassic Takla Group in McConnell Creek map-area, north-central British Columbia; Geological Survey of Canada, Paper 76-29, 45 p.
1980: Upper Triassic stratigraphy, Dease Lake and Tulsequah map areas, northwestern British Columbia; in Current Research, Part B, Geological Survey of Canada, Paper 80-1B, p. 1-9.

Monger, J.W.H. and Irving, E.
1980: Northward displacement of north-central British Columbia; Nature, v. 285, p. 289-294.

Monger, J.W.H., Souther, J.G., and Gabrielse, H.
1972: Evolution of the Canadian Cordillera: a plate-tectonic model; American Journal of Science, v. 272, p. 577-602.

Monger, J.W.H., Price, R.A., and Tempelman-Kluit, D.J.
1982: Tectonic accretion and the origin of the two major metamorphic and plutonic welts in the Canadian Cordillera; Geology, v. 10, p. 70-75.

Morganti, J.M.
1979: The geology and ore deposits of the Howards Pass area, Yukon and Northwest Territories: the origin of basinal sedimentary stratiform sulphide deposits; Ph.D. thesis, University of British Columbia, Vancouver, 317 p.
1981: Ore deposit models - 4. Sedimentary-type stratiform ore deposits: some models and a new classification; Geoscience Canada, v. 8, p. 65-75.

Morin, J.A.
1978: A preliminary report on Hart River (116A/10) - a Proterozoic massive sulphide deposit; in Mineral Industry Report 1977, Yukon Territory, EGS 1978-9, Indian and Northern Affairs Canada, p. 22-25.
1981: The McMillan deposit - a stratabound lead-zinc-silver deposit in sedimentary rocks of Upper Proterozoic age; in Yukon Geology and Exploration 1979-80, Department of Indian and Northern Affairs, p. 105-109.

Morrison, G.W.
1980: Stratigraphic control of Cu-Fe skarn ore distribution and genesis at Craigmont, British Columbia; Canadian Mining and Metallurgical Bulletin, v. 73, p. 109-123.

Morrow, D.W.
1984: Sedimentation in Root Basin and Prairie Creek Embayment-Siluro-Devonian, Northwest Territories; Bulletin of Canadian Petroleum Geology, v. 32, p. 162-189.

Mortensen, J.K.
1984: Bedrock geology of the Klondike placer gold district and evidence for structural controls on lode gold mineralization; in Symposium, Cordilleran Geology and Mineral Exploration: Status and Future Trends, Geological Association of Canada, Cordilleran Section Annual Meeting, Program and Abstracts, p. 27-29.

Mortensen, J.K. and Jilson, G.A.
1985: Evolution of the Yukon-Tanana Terrane: Evidence from southeastern Yukon Territory; Geology, v. 13, p. 806-810.

Mortensen, J.K., Montgomery, J.R., and Fillipone, J.
1987: U-Pb zircon, monazite and sphene ages for granitic orthogneiss of the Barkerville Terrane, east-central British Columbia; Canadian Journal of Earth Sciences, v. 24, p. 1261-1266.

Mortimer, N.
1986: Late Triassic, arc-related, potassic igneous rocks in the North American Cordillera; Geology, v. 14, p. 1035-1038.
1987: The Nicola Group: Late Triassic and Early Jurassic subduction-related volcanism in British Columbia; Canadian Journal of Earth Sciences, v. 24, p. 2521-2536.

Morton, R.D., Goble, R.J., and Fritz, P.
1974: The mineralogy, sulfur-isotope composition and origin of some copper deposits in the Belt Supergroup, southwest Alberta, Canada; Mineralium Deposita, v. 9, p. 223-241.

Morton, R.L.
1976: Alkalic volcanism and copper deposits of the Horsefly area, central British Columbia; Ph.D. thesis, Carleton University, Ottawa, 196 p.

Muller, J.E.
1967: Kluane Lake map-area, Yukon Territory (115G, 115F E1/2); Geological Survey of Canada, Memoir 340, 137 p.

Muller, J.E. and Carson, D.J.T.
1969: Geology and mineral deposits of Alberni map-area, British Columbia (92F); Geological Survey of Canada, Paper 68-50, 52 p.

Mulligan, R.
1984: Geology of Canadian tungsten occurrences; Geological Survey of Canada, Economic Geology Report 32, 121 p.

Mutschler, F.E., Wright, E.G., Ludington, S., and Abbott, J.T.
1981: Granite molybdenite systems; Economic Geology, v. 76, p. 874-897.

Nelson, J., Bellefontaine, K., Green, K. and MacLean, M.
1991: Regional geological mapping near the Mount Milligan copper-gold deposit (93K/16, 93N/1); in Geological Fieldwork 1990, British Columbia Geological Survey Branch, Paper 1991-1, p. 89-110.

Nesbitt, B.E., Murowchick, J.B., and Muehlenbachs, K.
1986: Dual origins of lode gold deposits in the Canadian Cordillera; Geology, v. 14, p. 506-509.

Ney, C.S.
1966: Distribution and genesis of copper deposits in British Columbia; in Tectonic History and Mineral Deposits of the Western Cordillera, Canadian Institute of Mining and Metallurgy, Special Volume 8, p. 295-303.

Ney, C.S., Cathro, R.J., Panteleyev, A., and Rotherham, D.C.
1976: Supergene copper mineralization; in Porphyry Copper Deposits of the Canadian Cordillera, A. Sutherland Brown (ed.), Canadian Institute of Mining and Metallurgy, Special Volume 15, p. 72-78.

Nixon, G.T. and Rublee, V.J.
1988: Alaskan-type ultramafic rocks in British Columbia: new concepts of the structure of the Tulameen Complex; in Geological Fieldwork 1987, Geological Survey Branch, British Columbia Ministry of Energy, Mines and Petroleum Resources, Paper 1988-1, p. 281-294.

Noble, S.R., Spooner, E.T.C., and Harris, F.R.
1984: The Logtung large tonnage, low-grade W (scheelite)-Mo porphyry deposit, south-central Yukon Territory; Economic Geology, v. 79, p. 848-868.

Nokleberg, W.J., Bundtzen, T.K., Berg, H.C., Brew, D.A., Grybeck, D., Robinson, M.S., Smith, T.E., and Yeend, W.
1987: Significant metalliferous lode deposits and placer districts of Alaska; United States Geological Survey, Bulletin 1786, 104 p.

Nolan, T.B.
1933: Epithermal precious-metal deposits; in Ore Deposits of the Western States, American Institute of Mining and Metallurgical Engineers, New York, Part VI, p. 623-640.

Norford, B.S. and Orchard, M.J.
1983: Early Silurian age of rocks hosting lead-zinc mineralization at Howards Pass, Yukon Territory and District of Mackenzie: local biostratigraphy of Road River Formation and Earn Group; Geological Survey of Canada, Paper 83-18, 35 p.

Norris, D.K. and Yorath, C.J.
1981: The North American Plate from the Arctic archipelago to the Romanzof Mountains; in The Ocean Basins and Margins, Volume 5, The Arctic Ocean, A.E.M. Nairn, M. Churkin, Jr. and F.G. Stehli (ed.), Plenum Press, New York, p. 37-103.

Northcote, K.E. and Muller, J.E.
1972: Volcanism, plutonism and mineralization: Vancouver Island; in Canadian Institute of Mining and Metallurgy, Bulletin, v. 65, no. 726, p. 49-57.

Northcote, K.E. and Robinson, W.C.
1973: Island Copper mine; in Geology, Exploration and Mining in British Columbia, 1972, British Columbia Department of Mines and Petroleum Resources, p. 293-303.

Oftedahl, C.
1958: A theory of exhalative-sedimentary ores; Geologiska Foereningen i Stockholm Forhandlingar, v. 80, part 1, p. 1-19.

O'Hanley, D.S. and Wicks, F.J.
1987: Structural control of serpentine textures in the Cassiar Mining Corporation's open-pit mine at Cassiar, British Columbia; Geological Association of Canada-Mineralogical Association of Canada, Annual Meeting, Saskatoon, Saskatchewan, Program and Abstracts, p. 77.

Ohmoto, H. and Skinner, B.J. (ed.)
1983: The Kuroko and related volcanogenic massive sulfide deposits; Economic Geology, Monograph 5, 604 p.

Okulitch, A.V.
1984: The role of the Shuswap Metamorphic Complex in Cordilleran tectonism: a review; Canadian Journal of Earth Sciences, v. 21, p. 1171-1193.

Okulitch, A.V., Loveridge, W.D., and Sullivan, R.W.
1981: Preliminary radiometric analyses of zircons from the Mount Copeland syenite gneiss, Shuswap Metamorphic Complex, British Columbia; in Current Research, Part A, Geological Survey of Canada, Paper 81-1A, p. 33-36.

O'Neill, J.J. and Gunning, H.C.
1934: Platinum and allied metal deposits of Canada; Canada Department of Mines, Economic Geology Series, no. 13, 165 p.

Orchard, M.J.
1985: Carboniferous, Permian and Triassic conodonts from the central Kootenay Arc, British Columbia: constraints on the age of the Milford, Kaslo and Slocan groups; in Current Research, Part A, Geological Survey of Canada, Paper 85-1A, p. 287-300.

Panteleyev, A.
1980: Cassiar map-area; in Geological Fieldwork 1979, British Columbia Ministry of Energy, Mines and Petroleum Resources, Paper 1980-1, p. 80-88.
1981: Berg porphyry copper-molybdenum deposit; British Columbia Ministry of Energy, Mines and Petroleum Resources, Bulletin 66, 158 p.
1986: Ore deposits #10. A Canadian Cordilleran model for epithermal gold-silver deposits; Geoscience Canada, v. 13, p. 101-111.
1987: Quesnel gold belt - alkalic volcanic terrane between Horsefly and Quesnel Lakes; in Geological Fieldwork 1986, British Columbia Ministry of Energy, Mines and Petroleum Resources, Paper 1987-1, p. 125-133.

Parrish, R.R.
1981: Geology of the Nemo Lakes belt, northern Valhalla Range, southeast British Columbia; Canadian Journal of Earth Sciences, v. 18, p. 944-958.

Paterson, I.A.
1977: The geology and evolution of the Pinchi Fault Zone at Pinchi Lake, central British Columbia; Canadian Journal of Earth Sciences, v. 14, p. 1324-1342.

Payne, J.G., Bratt, J.A., and Stone, B.G.
1980: Deformed Mesozoic volcanogenic Cu-Zn sulfide deposits in the Britannia district, British Columbia; Economic Geology, v. 75, p. 700-721.

Payne, M.W. and Allison, C.W.
1981: Paleozoic continental-margin sedimentation in east-central Alaska; Geology, v. 9, p. 274-279.

Peatfield, G.R.
1978: Geologic history and metallogeny of the 'Boundary District', southern British Columbia and northern Washington; Ph.D. thesis, Queen's University, Kingston, Ontario, 249 p.

Pell, J. and Simony, P.S.
1987: New correlations of Hadrynian strata, south-central British Columbia; Canadian Journal of Earth Sciences, v. 24, p. 302-313.

Pereira, J. and Dixon, C.J.
1971: Mineralisation and plate tectonics; Mineralium Deposita, v. 6, p. 404-405.

Pigage, L.C.
1986: Geology of the Cirque barite-zinc-lead-silver deposits, northeastern British Columbia; in Mineral Deposits of Northern Cordillera, J.A. Morin (ed.), Canadian Institute of Mining and Metallurgy, Special Volume 37, p. 71-86.

Pitcher, W.W.
1982: Granite type and tectonic environment; in Mountain Building Processes, K.J. Hsu (ed.), Academic Press, London, p. 19-40.

Preto, V.A.
1978: Setting and genesis of uranium mineralization at Rexspar; Canadian Institute of Mining and Metallurgy Bulletin, v. 71, no. 800, p. 82-88.

Preto, V.A. and Schiarizza, P.
1985: Geology and mineral deposits of the Adams Plateau-Clearwater region; in Field Guides to Geology and Mineral Deposits in the Southern Canadian Cordillera, D.J. Tempelman-Kluit (ed.), Geological Society of America, Cordilleran Section Meeting, Vancouver, Field Trip 16, p. 1-11.

Preto, V.A. and Tidsbury, A.D.
1971: Magnum Mine; in Geology, Exploration and Mining in British Columbia 1971, British Columbia Department of Mines and Petroleum Resources, p. 81-89.

Price, R.A.
1964: The Precambrian Purcell system in the Rocky Mountains of southern Alberta and British Columbia; Bulletin of Canadian Petroleum Geology, Field Conference Guidebook Issue, v. 12, p. 399-426.

Ransom, P.W.
1977: An outline of the geology of the Bluebell Mine, Riondel, B.C.; in Lead-Zinc Deposits of Southeastern British Columbia, T. Høy (ed.), Geological Association of Canada/Society of Economic Geologists, Joint Annual Meeting, Vancouver, Fieldtrip Guidebook No. 1, p. 44-51.

Ray, G.E.
1983: Carolin Mine - Coquihalla gold belt project (92H/6, 11); in Geological Fieldwork 1982, British Columbia Ministry of Energy, Mines and Petroleum Resources, Paper 1983-1, p. 62-84.
1986: Gold associated with a regionally developed mid-Tertiary plutonic event in the Harrison Lake area, southwestern British Columbia (92G/9; 92H/3,4,5,6,12); in Geological Fieldwork 1985, British Columbia Ministry of Energy, Mines and Petroleum Resources, Paper 1986-1, p. 95-97.

Ray, G.E. and Spence, A.
1986: The potassium-rich volcanic rocks at Tillicum Mountain - their geochemistry origin and regional significance; in Geological Fieldwork 1985, British Columbia Ministry of Energy, Mines and Petroleum Resources, Paper 1986-1, p. 45-49.

Ray, G.E., Dawson, G.L., and Simpson, R.
1987: The geology and controls of skarn mineralization in the Hedley gold camp, southern British Columbia (92H/8, 82E/5); in Geological Fieldwork 1986, British Columbia Ministry of Energy, Mines and Petroleum Resources, Paper 1987-1, p. 65-79.

Ray, G.E., Shearer, J.T., and Niels, R.J.E.
1986: The geology and geochemistry of the Carolin gold deposit, southwestern British Columbia, Canada; in Gold '86, An International Symposium on the Geology of Gold Deposits, Toronto, Proceedings Volume, A.J. Macdonald (ed.), p. 470-487.

Read, P.B. and Monger, J.W.H.
1976: Pre-Cenozoic volcanic assemblages of the Kluane and Alsek ranges, southwestern Yukon Territory; Geological Survey of Canada, Open File 381, 96 p.

Reesor, J.E.
1958: Dewar Creek map-area with special emphasis on the White Creek Batholith, British Columbia; Geological Survey of Canada, Memoir 292, 78 p.

Roberts, W.J. and McClintock, J.
1984: The Tillicum gold property; Western Miner, v. 57, no. 4, p. 29-31.

Rona, P.A.
1984: Hydrothermal mineralization at seafloor spreading centres; Earth-Science Reviews, v. 5, p. 191-212.

Rublee, V.J.
1986: Occurrence and distribution of platinum-group elements in British Columbia; British Columbia Ministry of Energy, Mines and Petroleum Resources, Open File 1986-7, 94 p.

Ruelle, J.C.L.
1982: Depositional environments and genesis of stratiform copper deposits of the Redstone Copper Belt, Mackenzie Mountains, N.W.T.; in Precambrian Sulphide Deposits, H.S. Robinson Memorial Volume, R.W. Hutchinson, C.D. Spence and J.M. Franklin (ed.), Geological Association of Canada, Special Paper 25, p. 701-737.

Rusmore, M.E.
1985: Geology and tectonic significance of the Upper Triassic Cadwallader Group and its bounding faults, southwestern British Columbia; Ph.D. thesis, University of Washington, Seattle, 174 p.
1987: Geology of the Cadwallader Group and the Intermontane-Insular superterrane boundary, southwestern British Columbia; Canadian Journal of Earth Sciences, v. 24, p. 2279-2291.

St. Louis, R.M., Nesbitt, B.E., and Morton, R.D.
1986: Geochemistry of platinum-group elements in the Tulameen Ultramafic Complex, southern British Columbia; Economic Geology, v. 81, p. 961-973.

Sangster, D.F.
1969: The contact metasomatic magnetite deposits of southwestern British Columbia; Geological Survey of Canada, Bulletin 172, 85 p.
1972: Precambrian volcanogenic massive sulphide deposits in Canada: a review; Geological Survey of Canada, Paper 72-22, 44 p.
1976: Carbonate-hosted lead-zinc deposits; in Handbook of Stratabound and Stratiform Ore Deposits, K.H. Wolf (ed.), v. 6, Elsevier Scientific Publishing Company, Amsterdam, p. 447-456.
1979: Plate tectonics and mineral deposits: a view from two perspectives; Geoscience Canada, v. 6, p. 185-188.
1984: Mississippi Valley lead-zinc; in Canadian Mineral Deposit Types: a Geological Synopsis, O.R. Eckstrand (ed.), Geological Survey of Canada, Economic Geology Report 36, p. 25.
1986: Classifications, distribution and grade-tonnage summaries of Canadian lead-zinc deposits; Geological Survey of Canada, Economic Geology Report 37, 68 p.

Sangster, D.F. and Lancaster, R.D.
1976: Geology of Canadian lead and zinc deposits, 2: The Mackenzie Valley lead-zinc district, Canada; in Report of Activities, Part A, Geological Survey of Canada, Paper 76-1A, p. 303-307.

Sato, T.
1974: Distribution and geological setting of the Kuroko deposits; Society of Mining Geologists Japan, Special Issue 6, p. 1-9.

Sato, K. and Sasaki, A.
1973: Lead isotopes of the black ore ("Kuroko") deposits from Japan; Economic Geology, v. 68, p. 547-552.
1976: Lead isotopic evidence on the genesis of pre-Cenozoic stratiform sulphide deposits in Japan; Geochemical Journal, v. 10, p. 197-203.
1980: Lead isotope features of the Besshi-type deposits and its bearing on the ore lead evolution; Geochemical Journal, v. 14, p. 303-315.

Sawkins, F.J.
1972: Sulfide ore deposits in relation to plate tectonics; Journal of Geology, v. 80, p. 377-397.
1976: Massive sulphide deposits in relation to geotectonics; in Metallogeny and Plate Tectonics, D.F. Strong (ed.), Geological Association of Canada, Special Paper 14, p. 221-240.
1984: Metal deposits in relation to plate tectonics; in Mineral and Rocks 17, P.J. Wyllie (ed.), Springer-Verlag, v. 17, 326 p.

Sawyer, D.A., Turner, A.T., Christopher, P.A., and Boyle, D.R.
1981: Basal type uranium deposits, Okanagan region, south central British Columbia; in Field Guides to Geology and Mineral Deposits, R.I. Thompson and D.G. Cook (ed.), Geological Association of Canada, Annual Meeting, Calgary, Alberta, p. 69-77.

Schroeter, T.G.
1983: Toodoggone River area (94E); in Geological Fieldwork 1982, British Columbia Ministry of Energy, Mines and Petroleum Resources, Paper 1983-1, p. 125-133.
1987: Golden Bear project; in Geological Fieldwork 1986, British Columbia Ministry of Energy, Mines and Petroleum Resources, Paper 1987-1, p. 103-109.

Schroeter, T.G. and Panteleyev, A.
1986a: Gold in British Columbia; British Columbia Ministry of Energy, Mines and Petroleum Resources, Preliminary Map no. 64, Map and Tabulation.
1986b: Lode gold-silver deposits in northwestern British Columbia; in Mineral Deposits of Northern Cordillera, J.A. Morin (ed.), Canadian Institute of Mining and Metallurgy, Special Volume 37, p. 178-190.

Scott, S.D.
1980: Geology and structural control of Kuroko-type massive sulphide deposits; in The Continental Crust and its Mineral Deposits, D.W. Strangway (ed.), Geological Association of Canada, Special Paper 20, p. 705-721.
1985: Seafloor polymetallic sulfide deposits: modern and ancient; Marine Mining, v. 5, p. 191-212.

Sharp, R.J.
1980: The geology, geochemistry and sulfur isotopes of the Anyox Massive Sulfide Deposits; M.Sc. thesis, University of Alberta, Edmonton, 211 p.

Sillitoe, R.H.
1972: Formation of certain massive sulfide deposits at sites of seafloor spreading; Institute of Mining and Metallurgy Transactions, v. 81, p. B141-148.

1983: Styles of low-grade gold potential of volcano-plutonic arcs; in Papers given at the precious-metals symposium, N.E. Kral (ed.), Sparks, Nevada, 1980; Nevada Bureau of Mines and Geology Report 36, p. 52-68.

Sinclair, A.J., Bentzen, A., and McLeod, J.A.
1979: Geology of the White River native copper deposit, Yukon Territory; Indian and Northern Affairs Canada, Publication No. QS-Y0001-000-EE-A1, 27 p.

Sinclair, A.J., Drummond, A.D., Carter, N.C., and Dawson, K.M.
1982: A preliminary analysis of gold and silver grades of porphyry-type deposits in western Canada; in Precious Metals in the Northern Cordillera, A.A. Levinson (ed.), Association of Exploration Geochemists, Symposium Proceedings, Vancouver, B.C., p. 157-172.

Sinclair, A.J., Wynne-Edwards, H.R., and Sutherland Brown, A.
1978: An analysis of distribution of mineral occurrences in British Columbia; British Columbia Ministry of Energy, Mines and Petroleum Resources, Bulletin 68, 125 p.

Sinclair, W.D.
1986: Molybdenum, tungsten and tin deposits and associated granitoid intrusions in the northern Canadian Cordillera and adjacent parts of Alaska; in Mineral Deposits of the Northern Cordillera, J.A. Morin (ed.), Canadian Institute of Mining and Metallurgy, Special Volume 37, p. 216-233.

Sketchley, D.A., Sinclair, A.J., and Godwin, C.I.
1986: Early Cretaceous gold-silver mineralization in the Sylvester allochthon, near Cassiar, north central British Columbia; Canadian Journal of Earth Sciences, v. 23, p. 1455-1458.

Smith, A.
1947: Control of ore by primary igneous structures, Porcher Island, British Columbia; Geological Society of America Bulletin, v. 58, p. 245-262.
1948: Surf Point and Edye Pass mines; in Structural Geology of Canadian Ore Deposits, Canadian Institute of Mining and Metallurgy, Jubilee Volume, p. 94-99.

Smith, J.G.
1977: Geology of the Ketchikan D-1 and Bradfield Canol A-1 Quadrangles, southeastern Alaska; United States Geological Survey, Bulletin 1425, 49 p.

Soregaroli, A.E. and Nelson, W.I.
1976: Boss Mountain; in Porphyry Deposits of the Canadian Cordillera, A. Sutherland Brown (ed.), Canadian Institute of Mining and Metallurgy, Special Volume 15, p. 432-443.

Soregaroli, A.E. and Sutherland Brown, A.
1976: Characteristics of Canadian Cordilleran molybdenum deposits; in Porphyry Deposits of the Canadian Cordillera, A. Sutherland Brown (ed.), Canadian Institute of Mining and Metallurgy, Special Volume 15, p. 417-431.

Soregaroli, A.E. and Whitford, D.F.
1976: Brenda; in Porphyry Deposits of the Canadian Cordillera, A. Sutherland Brown (ed.), Canadian Institute of Mining and Metallurgy, Special Volume 15, p. 186-194.

Souther, J.G.
1977: Volcanism and tectonic environments in the Canadian Cordillera - a second look; in Volcanic Regimes in Canada, W.R.A. Barager, L.C. Coleman and J.M. Hall (ed.), Geological Association of Canada, Special Paper 16, p. 3-24.

Spence, A.
1985: Shoshonites and associated rocks of central British Columbia; in Geological Fieldwork 1984, British Columbia Ministry of Energy, Mines and Petroleum Resources, Paper 1985-1, p. 426-442.

Spooner, E.T.C.
1980: Cu-pyrite mineralization and seawater convection in oceanic crust - the ophiolitic ore deposits of Cyprus; in The Continental Crust and its Mineral Deposits, D.W. Strangway (ed.), Geological Association of Canada, Special Paper 20, p. 685-704.

Stanton, R.L.
1960: General features of the conformable "pyritic" orebodies; Canadian Institute of Mining and Metallurgy Transactions, v. 63, p. 22-27.

Steininger, R.C.
1985: Geology of the Kitsault molybdenum deposit, British Columbia; Economic Geology, v. 80, p. 57-71.

Stephens, G.C.
1972: The geology of the Salal Creek pluton, southwestern British Columbia; Ph.D. thesis, Lehigh University, Bethlehem, Pennsylvania, 177 p.

Stevenson, J.S.
1945: Geology of the Twin "J" Mine in Transactions of the Canadian Institute of Mining and Metallurgy, v. 48, p. 294-308.

1950: Geology and mineral deposits of Zeballos area; British Columbia Department of Mines, Bulletin 27, 145 p.

Stewart, J.H.
1972: Initial deposits in the Cordilleran geosyncline: evidence of a Late Precambrian (<850 m.y.) continental separation; Geological Society of America Bulletin, v. 83, p. 1345-1360.

Strong, D.F., editor
1976: Metallogeny and plate tectonics; Geological Association of Canada, Special Paper No. 14, 660 p.

Struik, L.C.
1986: Imbricated terranes of the Cariboo gold belt with correlations and implications for tectonics in southeastern British Columbia; Canadian Journal of Earth Sciences, v. 23, p. 1047-1061.
1987: The ancient western North American margin: an Alpine rift model for the east-central Canadian Cordillera; Geological Survey of Canada, Paper 87-15, 19 p.

Stuart, R.A.
1963: Geology of the Snake River iron deposit; Department of Indian Affairs and Northern Development Assessment Files, Yellowknife, N.W.T., 18 p.

Sutherland Brown, A.
1960: Geology of the Rocher Deboule Range; British Columbia Department of Mines and Petroleum Resources, Bulletin 43, 78 p.
1968: Geology of the Queen Charlotte Islands, British Columbia; British Columbia Department of Mines and Petroleum Resources, Bulletin 54, 226 p.
1969: Ox (Ox Lake Property); in Geology, Exploration and Mining in British Columbia 1969, British Columbia Department of Mines and Petroleum Resources, p. 93-97.
1972: Texada; in Copper and Molybdenum Deposits of the Western Cordillera, International Geological Congress, 24th Session, Montreal, Guidebook A09-C09, p. 15-20.
1976a: Editor, Porphyry deposits of the Canadian Cordillera; Canadian Institute of Mining and Metallurgy, Special Volume 15, 510 p.
1976b: Metallic mineral resources of Canadian Cordillera; in Circum-Pacific Energy and Mineral Resources, M.T. Halbouty, J.C. Maher and H.M. Lian (ed.), American Association of Petroleum Geologists, Memoir 25, p. 503-508.
1976c: Morphology and classification; in Porphyry Deposits of the Canadian Cordillera, A. Sutherland Brown (ed.), Canadian Institute of Mining and Metallurgy, Special Volume 15, p. 44-51.

Sutherland Brown, A., Cathro, R.J., Panteleyev, A., and Ney, C.S.
1971: Metallogeny of the Canadian Cordillera; in Canadian Institute of Mining and Metallurgy, Bulletin v. 64, no. 709, p. 37-61, and Transactions v. 74, p. 121-145.

Tatsumi, T., (Editor)
1970: Volcanism and Ore Genesis; University of Tokyo Press, Tokyo, 488 p.

Taylor, G.C. and Stott, D.F.
1973: Tuchodi Lakes map-area, British Columbia; Geological Survey of Canada, Memoir 373, 37 p.

Taylor, G.C., Macqueen, R.W., and Thompson, R.I.
1975: Facies changes, breccias and mineralization in Devonian rocks of Rocky Mountains, northeastern British Columbia (94B,G,K,N); in Report of Activities, Part A, Geological Survey of Canada, Paper 75-1A, p. 577-585.

Tempelman-Kluit, D.J.
1972: Geology and origin of the Faro, Vangorda and Swim concordant zinc-lead deposits, central Yukon Territory; Geological Survey of Canada, Bulletin 208, 73 p.
1979: Transported cataclasite, ophiolite and granodiorite in Yukon: evidence of arc-continent collision; Geological Survey of Canada, Paper 79-14, 27 p.
1981a: Craig property description; in Yukon Geology and Exploration 1979-80; Department of Indian and Northern Affairs Canada, Geology Section, Whitehorse, Yukon, p. 225-230.
1981b: Geology and mineral deposits of southern Yukon; in Yukon Geology and Exploration 1979-80, Department of Indian and Northern Affairs, Whitehorse, Yukon, p. 7-31.
1981c: Pluto property description; in Yukon Geology and Exploration 1979-80; Department of Indian and Northern Affairs Canada, Geology Section, Whitehorse, Yukon, p. 288-289.
in press: Quiet Lake and Finlayson Lake map-areas, Yukon Territory; Geological Survey of Canada, Memoir.

Tempelman-Kluit, D. and Parkinson, D.
1986: Extension across the Eocene Okanagan crustal shear in southern British Columbia; Geology, v. 14, p. 318-321.

Tempelman-Kluit, D.J. and Wanless, R.K.
1980: Zircon ages for the Pelly Gneiss and Klotassin granodiorite in western Yukon; Canadian Journal of Earth Sciences, v. 17, p. 297-306.

Thompson, R.I. and Panteleyev, A.
1976: Stratabound mineral deposits of the Canadian Cordillera; in Handbook of Strata-bound and Stratiform Ore Deposits, Volume 5, K.H. Wolf (ed.), Elsevier Scientific Publishing Company, Amsterdam, p. 37-108.

Thorpe, R.I.
1967: Controls of hypogene sulphide zoning, Rossland, British Columbia; Ph.D. thesis, University of Wisconsin, Madison, 131 p.

Thorpe, R.I. and Little, H.W.
1973: The age of sulfide mineralization at Rossland, British Columbia: discussion; Economic Geology, v. 68, p. 1337-1340.

Thorstad, L.E. and Gabrielse, H.
1986: The Upper Triassic Kutcho Formation, Cassiar Mountains, north-central British Columbia; Geological Survey of Canada, Paper 86-16, 53 p.

Tipper, H.W. and Richards, T.A.
1976: Jurassic stratigraphy and history of north-central British Columbia; Geological Survey of Canada, Bulletin 270, 73 p.

Titley, S.R. (editor)
1982: Advances in geology of the porphyry copper deposits, southwestern North America; University of Arizona Press, Tucson, 560 p.

Titley, S.R. and Beane, R.E.
1981: Porphyry copper deposits. Part 1. Geologic settings, Petrology, and Tectogenesis; in Economic Geology, Seventy-Fifth Anniversary Volume, B.J. Skinner (ed.), Economic Geology Publishing Company, p. 214-235.

Titley, S.R. and Hicks, C.L. (editors)
1966: Geology of the porphyry copper deposits, southwestern North America; University of Arizona Press, Tucson, 287 p.

Tolbert, R.S., Baldys, C., Froc, N.V., and Watkins, T.A.
1987: The Graham Island (Cinola) deposit revisited - evidence for an epithermal hot-spring-type gold deposit; District 6 Annual Meeting, Vancouver, B.C., Program and Abstracts, Canadian Institute of Mining and Metallurgy, Bulletin, v. 80, no. 904, p. 30.

Tooker, E.W. (editor)
1985: Geologic characteristics of sediment- and volcanic-hosted disseminated gold deposits - search for an occurrence model; United States Geological Survey, Bulletin 1646.

Upadhyay, H.D. and Strong, D.F.
1973: Geological setting of the Betts Cove copper deposits, Newfoundland: an example of ophiolite sulfide mineralization; Economic Geology, v. 68, p. 161-167.

Upitis, U.
1966: The Rapitan Group, southwestern Mackenzie Mountains, Northwest Territories; M.Sc. thesis, McGill University, Montreal, Quebec, 70 p.

Vaillancourt, P.de G.
1982a: Geology of pyrite-sphalerite-galena concentrations in Proterozoic quartzite at Quartz Lake, southeastern Yukon; in Yukon Exploration and Geology 1982, Indian and Northern Affairs Canada, Exploration and Geological Services, Whitehorse, Yukon, p. 73-77.
1982b: Geology and genesis of pyrite-sphalerite-galena concentrations in Proterozoic quartzite at Quartz Lake, Yukon Territory; M.Sc. thesis, University of Western Ontario, London, 177 p.

Vining, M.R.
1977: The Spuzzum pluton northwest of Hope, B.C.; M.Sc. thesis, University of British Columbia, Vancouver, 147 p.

Vulimiri, M.R., Tegart, P., and Stammers, M.A.
1986: Lawyers gold-silver deposits, British Columbia; in Mineral Deposits of the Northern Cordillera, J.A. Morin (ed.), Canadian Institute of Mining and Metallurgy, Special Volume 37, p. 191-201.

Walker, R.R.
1983: Westmin Resources' massive sulfide deposits; in Field Trip Guidebook, vol. II, Field Trip 9, Geological Association of Canada, Victoria 1983, p. 5-19.

Watson, K.DeP.
1948: The Squaw Creek-Rainy Hollow area, northern British Columbia; British Columbia Department of Mines, Bulletin 25, 71 p.

Watson, P.H.
1984: The Whitehorse Copper Belt: 105D/11 - a compilation; Exploration and Geological Services Division, Yukon, Indian and Northern Affairs Canada, Open File Map.

Watson, P.H., Godwin, C.I., and Christopher, P.A.
1982: General geology and genesis of silver and gold veins in the Beaverdell area, south-central British Columbia; Canadian Journal of Earth Sciences, v. 19, p. 1264-1274.

Westra, G. and Keith, S.B.
1981: Classification and genesis of stockwork molybdenum deposits; Economic Geology, v. 76, p. 844-873.

White, A.J.R. and Chappell, B.W.
1983: Granitoid types and their distribution in the Lachlan Fold Belt, southeastern Australia; in Circum-Pacific Plutonic Terranes, J.A. Roddick (ed.), Geological Society of America, Memoir 159, p. 21-34.

White, G.V.
1986a: Preliminary report, O'Connor River gypsum deposit (114P/10E); in Geological Fieldwork 1985, British Columbia Ministry of Energy, Mines and Petroleum Resources, Paper 1986-1, p. 279-282.
1986b: Preliminary report, Lang Bay germanium prospect (92F/16W); in Geological Fieldwork 1985, British Columbia Ministry of Energy, Mines and Petroleum Resources, Paper 1986-1, p. 261-264.

White, W.H., Bookstrom, A.A., Kamilli, R.J., Ganster, M.W., Smith, R.P., Ranta, D.E., and Steininger, R.C.
1981: Character and origin of Climax-type molybdenum deposits; in Economic Geology, B.J. Skinner (ed.), Seventy-Fifth Anniversary Volume, Economic Geology Publishing Company, p. 270-316.

White, W.H., Sinclair, A.J., Harakal, J.E., and Dawson, K.M.
1970: Potassium-argon ages of Topley Intrusions near Endako, British Columbia; Canadian Journal of Earth Sciences, v. 7, p. 1172-1178.

Wilkins, J. Jr.
1984: The distribution of gold- and silver-bearing deposits in the Basin and Range Province, Western United States; in Gold and Silver Deposits of the Basin and Range Province, Western U.S.A., Joe Wilkins, Jr. (ed.), Arizona Geological Society Digest, v. 15, p. 1-27.

Wilton, D.H.C. and Sinclair, A.J.
1988: Ore petrology and genesis of a strata-bound disseminated copper deposit at Sustut, British Columbia; Economic Geology, v. 83, p. 30-45.

Wolfhard, M.R. and Ney, C.S.
1976: Metallogeny and plate tectonics in the Canadian Cordillera; in Metallogeny and Plate Tectonics, D.F. Strong (ed.), Geological Association of Canada, Special Paper 14, p. 359-392.

Wong, R.H.
1985: Assessment report of the 1984 diamond drilling program on the Rockland Group 8006 and Willa Group 8101 claims, Slocan Mining Division, B.C., NTS 82F/14W; British Columbia Ministry of Energy, Mines and Petroleum Resources, Assessment Report 13382.

Woodsworth, G.J. and Roddick, J.A.
1977: Mineralization in the Coast Plutonic Complex of British Columbia, south of latitude 55°N; in The Relations Between Granitoids and Associated Ore Deposits of the Circum-Pacific Region, J.A. Roddick and T.T. Khoo (ed.), Geological Society of Malaysia, Bulletin 9, p. 1-16.

Woodsworth, G.J., Pearson, D.E., and Sinclair, A.J.
1977: Metal distribution patterns across the eastern flank of the Coast Plutonic Complex, south-central British Columbia; Economic Geology, v. 72, p. 170-183.

Yeo, G.M.
1981: The Late Proterozoic Rapitan glaciation in the northern Cordillera; in Proterozoic Basins in Canada, F.H.A. Campbell (ed.), Geological Survey of Canada, Paper 81-10, p. 25-46.
1986: Iron-formation in the late Proterozoic Rapitan Group, Yukon and Northwest Territories; in Mineral Deposits of Northern Cordillera, J.A. Morin (ed.), Canadian Institute of Mining and Metallurgy, Special Volume 37, p. 142-153.

Young, G.M.
1977: Stratigraphic correlation of upper Proterozoic rocks of northwestern Canada; Canadian Journal of Earth Sciences, v. 14, p. 1771-1787.
1982: The late Proterozoic Tindir Group, east-central Alaska; evolution of a continental margin; Geological Society of America, Bulletin, v. 93, p. 759-783.

Young, G.M., Jefferson, C.W., Delaney, G.D., and Yeo, G.M.
1979: Middle and late Proterozoic evolution of the northern Canadian Cordillera and Shield; Geology, v. 7, p. 125-128.

CHAPTER 19

Authors' addresses

A. Sutherland Brown
546 Newport Ave.,
Victoria, B.C.
V8S 5C7

K.M. Dawson, G.J. Woodsworth
Geological Survey of Canada,
100 West Pender St.,
Vancouver, B.C.
V6B 1R8

A. Panteleyev
Geological Survey Branch
British Columbia Ministry of Energy, Mines and
Petroleum Resources,
106-553 Superior St.,
Victoria, B.C.
V8V 1X4

ADDENDA

Several significant mineral deposits were discovered since time of writing, the two largest of which are described briefly below. Some recently obtained isotopic ages of mineralization which bear upon metallogenic conclusions are also discussed.

At Eskay Creek, 70 km northwest of Stewart in Stikinia, an Au, Ag-rich pyritic VMS deposit occurs mainly in argillite, but also in footwall, rhyolitic volcaniclastics and hangingwall andesitic flows of the Mount Dilworth Formation of the Hazelton Group (Britton et al., 1989; see Table 19.2). Reserves in the VMS zone are 3.97 million tonnes grading 26 g/t Au, 998 g/t Ag plus values in Pb and Zn (Company reports, 1990). The VMS horizon also hosts discordant, high-grade Au, Ag veins (e.g. 21 zone) with probable geological reserves of 3.35 million tonnes of ore grading 19.2 g/t Au and 521.3 g/t Ag (Lefebure and Malott, 1990).

The Mount Milligan alkaline porphyry Cu, Au prospect, about 80 km south-southeast of Germansen Landing in Quesnellia, is associated with several small brecciated monzonite porphyry stocks and a laccolithic sill, probably comagmatic with the host augite porphyritic andesites of the Witch Lake Formation of the Upper Triassic to Lower Jurassic Nicola Assemblage (Nelson et al., 1991). The developed prospect contains geological reserves of 417 million tonnes of ore grading 0.22 % Cu and 0.4 g/t Au (Harris, 1991).

Recent British Columbia Geological Survey Branch K/Ar isotopic dating of alteration muscovite at the Eaglet concordant fluorite deposit in Barkerville Subterrane of Kootenay Terrane has resulted in its re-classification from pre-to late accretionary, similar to the nearby Cariboo Barkerville gold veins.

K/Ar dating of mesothermal Au-quartz veins in the Okanagan Subterrane of Quesnellia has demonstrated a predominantly Early Cretaceous, post-accretionary age of mineralization, although granitoid intrusions adjacent to the veins are interpreted to be dominantly Middle Jurassic and therefore accretionary. Nearby epithermal precious metal veins of Eocene age are possibly related to detachment faulting in an extensional setting.

Printed in Canada

Chapter 20

ENERGY AND GROUNDWATER RESOURCES OF THE CANADIAN CORDILLERA

Summary

Part A. PETROLEUM

C.J. Yorath, P.L. Gordy, and G.K. Williams
  Petroleum resources of the Foreland Belt
  Petroleum resources of the Intermontane Belt
  Petroleum resources of the Insular Belt

Part B. COAL

R.M. Bustin
  Foreland Belt
  Intermontane Belt
  Insular Belt
  Tertiary coal deposits
  Reserves and resources

Part C. URANIUM AND THORIUM

R.T. Bell
  Uranium
  Beaverdell area
  Clearwater area
  Okanagan 'young' or surficial
  uranium deposits
  Wernecke Mountains area
  Thorium

Part D. GEOTHERMAL ENERGY

J.G. Souther
  Target identification
  Target evaluation
  Present status of exploration
  Garibaldi Volcanic Belt
  Anahim Volcanic Belt
  Clearwater-Quesnel volcanic province
  Stikine Volcanic Belt
  Resource assessment
  Resource base
  Resource recovery
  Resource estimate

Part E. GROUNDWATER
E.C. Halstead
    Hydrogeological environment
    Groundwater in bedrock
    Igneous plutonic rocks
    Folded and faulted sedimentary rocks
    Flat-lying and gently dipping lava flows
    Groundwater in surficial deposits
    Coastal Lowland
    Interior valleys
    Permafrost
    References

Chapter 20

ENERGY AND GROUNDWATER RESOURCES OF THE CANADIAN CORDILLERA

C.J. Yorath, P.L. Gordy, G.K. Williams, R.M. Bustin,
R.T. Bell, J.G. Souther, and E.C. Halstead

SUMMARY

Petroleum resources of the Canadian Cordillera are mainly confined to the Foreland Belt where structural traps dominated by thrust faults and folds in Mesozoic and Paleozoic rocks form the main reservoirs. In the Insular Belt hydrocarbons have yet to be found, however, potential reservoirs are associated with rocks of mainly Tertiary age within structures which have arisen from interplate processes along an active continental margin. In the Foreland Belt established initial recoverable reserves of gas amount to 375×10^9 m^3; two oil fields together contain an estimated recoverable reserve of 65×10^6 m^3. In the Insular Belt estimates of potential recoverable reserves at an "average expectation" (34% level of probability) are: 265×10^9 m^3 for gas and 38.5×10^6 m^3 for oil.

Vast resources of bituminous and sub-bituminous coal and lignite occur in the Cordillera within strata of Jurassic, Cretaceous and Tertiary age. Upper Jurassic and Cretaceous coal deposits are part of regionally extensive clastic wedges that accumulated in the Foreland, Intermontane and Insular belts in response to tectonism, partly associated with accretion of suspect terranes. The most significant deposits and the only ones presently being mined are bituminous coals in the Foreland Belt. In the Intermontane Belt Jurassic and Lower Cretaceous(?) coal measures occur in the Whitehorse Trough, Bowser Basin and on the north flank of the Skeena Arch. The Whitehorse Trough and Groundhog coal measures of the Bowser Basin contain the only significant deposits of anthracite in the Cordillera. In the Insular Belt bituminous coal and anthracite lie within Jurassic rocks of the Queen Charlotte Islands and high volatile bituminous coal is found in Upper Cretaceous rocks of Vancouver Island. Tertiary deposits are widely distributed throughout the central and northern Cordillera and include large resources of bituminous and sub-bituminous coal and lignite. For the most part Tertiary coal measures are local deposits in fault-bounded basins; the Hat Creek deposit is 370 m thick and is one of the thickest coal deposits in the world. The total coal resources of the Cordillera are estimated at about 71 000 million tonnes.

The number of documented uranium occurrences in the Cordillera is small. Two uranium provinces are evident. The first is in the Okanagan region of the southern Omineca and Intermontane belts where deposits are contained in clastic rocks within paleodrainage valleys incised mainly into Cretaceous granite and associated metamorphic rocks. The second province is in the eastern Wernecke Mountains of the northern Foreland Belt where numerous high grade deposits of uranium occur in association with copper, gold and silver in megabreccias of Middle Proterozoic age. At present thorium is not exploited in the Cordillera although anomalously high concentrations are contained in some plutons of Late Cretaceous and Tertiary age in the northern Intermontane and Omineca belts and in the Bugaboo Mountains of the southern Omineca Belt.

High temperature geothermal systems, capable of generating electrical power, are associated with Neogene volcanism. Two hydrothermal systems have been identified in the Garibaldi (Cascade) Volcanic Belt and others may lie within the Anahim and Stikine volcanic belts. If these resources were ultimately proven and developed it is estimated that a power production rate of 100 megawatts would deplete the high-temperature hydrothermal resource in the three belts in about 50 years; the accompanying low-temperature systems would take nearly 150 years to be exhausted.

The variability of topography in the Cordillera has a profound effect on the initiation and distribution of precipitation. Together with the complexity of lithology these two factors influence the manner and degree to which groundwater flow systems are recharged. The mountains of the Insular Belt and the west flank of the Coast Belt are wet zones separated by a somewhat drier coastal trough. The Intermontane Belt of comparatively low topography is a dry zone bounded on the east by the wet west flank of the Rockies and mountain systems of the Omineca Belt. Within each of these regions the hydrodynamic characteristics of these groundwater systems depend on a wide range of criteria including surface slopes, lithology and structural style and the distribution of unconsolidated surficial deposits which are the major sources of groundwater supply.

Yorath, C.J., Gordy, P.L., Williams, G.K., Bustin, R.M., Bell, R.T., Souther, J.G., and Halstead, E.C.
1991: Energy and ground water resources of the Canadian Cordillera, Chapter 20 in Geology of the Cordilleran Orogen in Canada, H. Gabrielse and C.J. Yorath (ed.); Geological Survey of Canada, Geology of Canada, no. 4, p. 769-801 (also Geological Society of America, The Geology of North America, v. G-2)

CHAPTER 20

PART A. PETROLEUM

C.J. Yorath, P.L. Gordy and G.K. Williams

Petroleum exploration in the Cordillera, with rare exceptions, has been confined to the Foreland and Insular belts and only in the former region have discoveries been made (Fig. 20.1). The following account of Cordilleran petroleum resources is brief. For a more complete treatment of the subject the reader is referred to the volume on the Sedimentary Cover of the Craton in Canada (Geology of Canada, no. 5, Geological Society of America, DNAG vol. D-1).

Petroleum Resources of the Foreland Belt

Oil and gas resources in the Foreland Belt (Fig. 20.2) occur primarily in Paleozoic and Triassic carbonate and Cretaceous sandstone reservoirs and in structures mainly associated with thrust faults. The major accumulations, with the exception of the Turner Valley (Fig. 20.3) and Norman Wells oil fields, are mostly sour natural gas deposits estimated to amount to 375×10^9 m^3 of established initial recoverable reserves. Recoverable oil reserves for Turner Valley and Norman Wells are estimated to be 22 and 43×10^6 m^3 respectively (Alberta Energy Resources Conservation Board ST 85-18; British Columbia Ministry of Energy, Mines and Petroleum Resources, 1983; Kempthorne and Irish, 1981).

The Foothills structural belt of Alberta is coarsely divisible into an eastern and western zone according to the spacing, degree of stratigraphic separation and amount of displacement of listric thrust faults. Nearly all of the structural traps in the eastern zone are single (e.g. Pincher Creek and Jumping Pound fields) or stacked (e.g. Jumping Pound West and Stolberg fields) thrust sheets which, at the level of the Mississippian carbonate reservoirs, have small displacement (Fig. 20.2). Trap capacities depend upon the extent of reservoir development in carbonate rocks of the Rundle Group, strike length of thrust sheets, horizontal displacements and degree of vertical uplift above the regional Mississippian system gradient. Marine shales of the Jurassic Fernie Formation provide an effective top seal, and, because nearly all traps are fault closed, it is assumed that fault planes form seals for the eastern flanks of the structures (Fig. 20.3). Apparently most structures in the eastern Foothills with sufficient reservoir development are filled to capacity with gas.

Thrust faults of the western Foothills Belt generally have large displacements, in the order of many tens of kilometres, and extend deep into the section to dislocate both Mississippian and Devonian reservoir strata. Several thrust sheets involving Paleozoic rocks are commonly stacked in a broad anticlinal form and thus provide multiple trap and reservoir objectives (e.g. Waterton, Panther River and Moose Mountain fields; Fig. 20.4, 20.5; Plate 32). However, not all closed structures are filled to capacity, probably due to complex maturation and migration histories and/or lack of adequate seals. The thin but widely developed organic shale of the Exshaw Formation at the base of the Mississippian section is considered to be a principal source rock for Foothills accumulations.

In northeastern British Columbia the Foothills structural belt is less clearly defined; thrust faults with large displacements pass northwestward into large amplitude folds, blind thrusts, and subordinate thrusts with small displacement. Gas reservoir development is in Mississippian (Sikanni and Cypress fields) and Triassic (e.g. Sukunka field) carbonates as well as Lower Cretaceous (e.g. Buick Creek and Rigel fields) and Jurassic (Grizzly field) clastic rocks (Fig. 20.5). Lower Cretaceous clastics also yield significant oil production (e.g. Aitken Creek field). As in the Alberta Foothills, traps occur along the leading edges of thrust sheets and are associated with low relief anticlines. Stratigraphic traps may be important in the easternmost fields. Thirty-one gas fields and one oil field have been developed in this area.

Petroleum exploration in the outer Mackenzie arc has been much less successful, perhaps due to the fact that all potential reservoirs are breached at the surface in several Paleozoic outliers and along the many ranges of the Franklin Mountains. The structural style is expressed by large, en échelon arcuate upright folds and subordinate

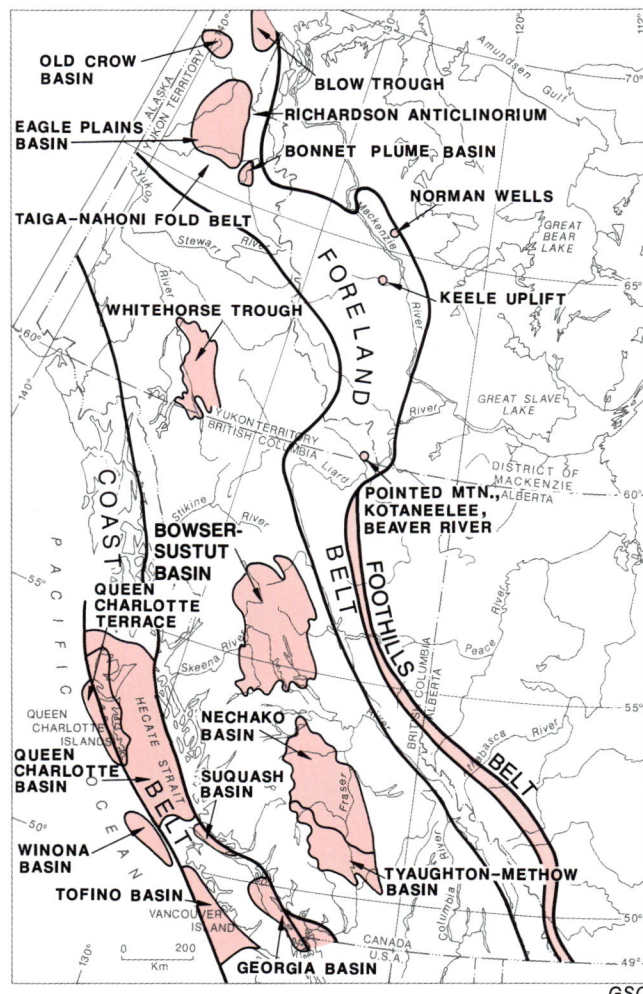

Figure 20.1. Distribution of prospective petroleum areas in the Canadian Cordillera. ('Bowser Basin' includes Sustut Basin)

thrust faults within a broad synclinorium. At the extreme southern end of the Mackenzie arc the Beaver River, Pointed Mountain and Kotaneelee gas accumulations are trapped in Middle Devonian carbonates disposed in large anticlines that may be associated with thrust faults (Fig. 20.1). Together these three fields contain estimated initial marketable gas reserves of 14.5 x 10^9 m³. Farther north, the Norman Wells oil field comprises a reef trap in the Middle Devonian Ramparts Formation carbonates on the west flank of the Norman Range in the Franklin Mountains. Source rocks for this field are considered to be the overlying and adjacent bituminous shales of the Devonian Canol and Hare Indian formations.

Although some 32 wildcat wells have been unsuccessful in finding additional Ramparts reef reservoirs, this type of play is by far the most prospective for additional oil reserves in the northern Cordillera. Nearby, other prospective targets are associated with the Keele Uplift where a non-commercial oil discovery was made in porous Cambrian to Ordovician dolomite immediately below the

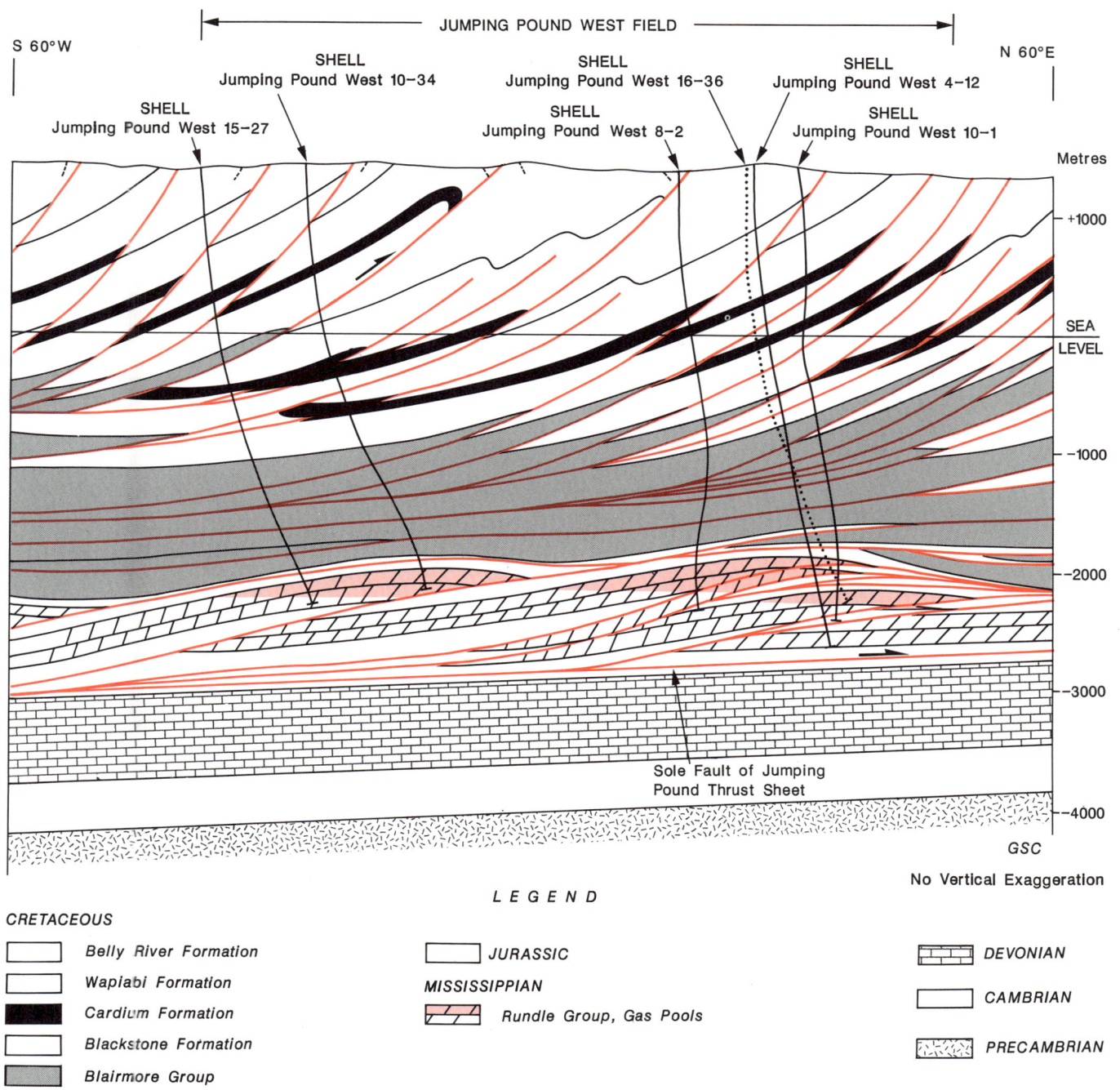

Figure 20.2. Structure-section through the Jumping Pound West field, Alberta (modified from F.R. Frey, Shell Canada Resources Ltd.). See Fig. 20.5B for location.

CHAPTER 20

sub-Cretaceous unconformity. In this case the source rocks are considered to be overlying Cretaceous shales which could provide petroleum to porous rocks ranging in age from Proterozoic to Devonian.

In several areas of the northern Cordillera the Upper Cambrian to Middle Devonian shelf carbonate to basinal shale transition is potentially an important exploration play. Carbonates at the bank edge could be reefoid and

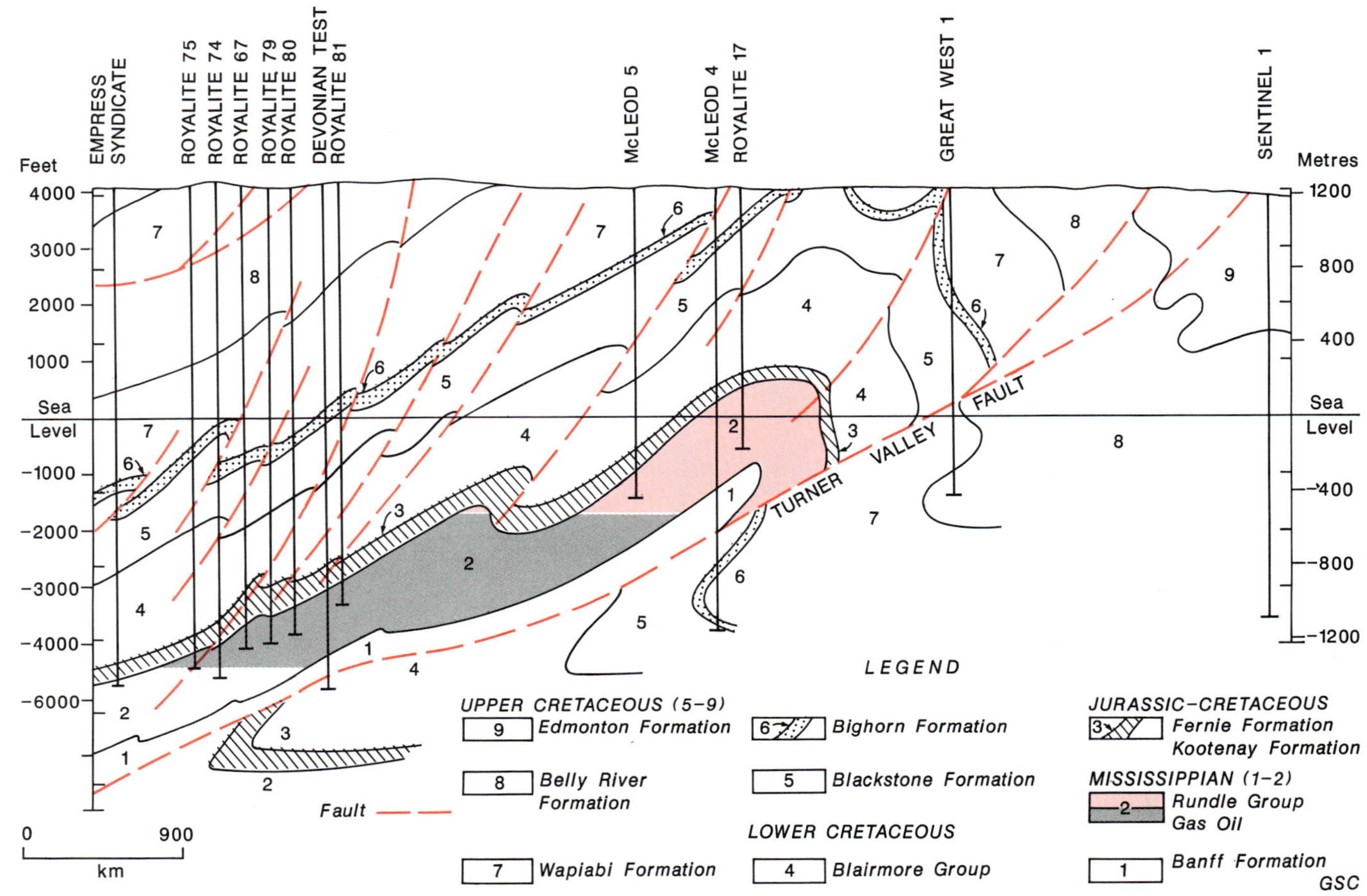

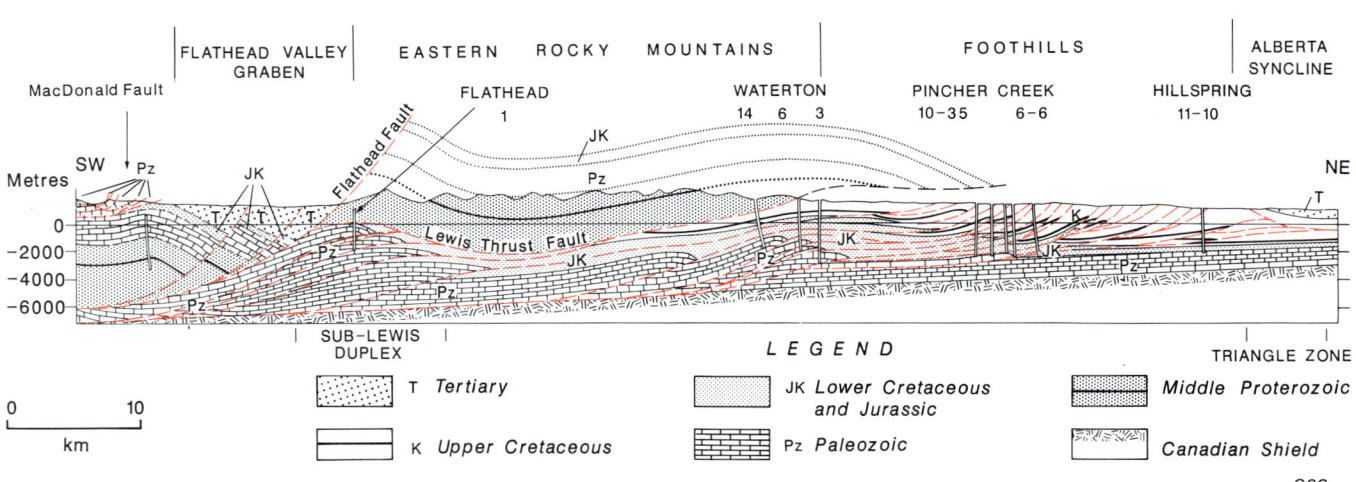

Figure 20.3. Structure-section through the central part of the Turner Valley oil and gas field, Alberta (modified from Gallup, 1951). See Fig. 20.5A for location.

Figure 20.4. Structure-section through Waterton and Pincher Creek gas fields, Alberta (after Balley et al., 1966). See Fig. 20.5B for location.

thus good reservoir rocks given adequate seals. Several exploratory wells, however, have failed to encounter reservoir facies. Moreover, due to high levels of thermal maturation as a consequence of deep burial, dry gas is probably the best that can be expected in much of the area.

The small Bonnet Plume Basin at the south end of the Richardson anticlinorium consists of about 1500 m of nonmarine Upper Cretaceous strata overlying the deeply eroded core of the anticlinorium. Some potential for gas exists in the Cretaceous strata, and if adequate Cretaceous or older source beds are found, reservoirs as old as Cambrian may contain hydrocarbons in much the same way as do Paleozoic strata of the Keele Uplift.

In Eagle Plains Basin, strata of every period except the Triassic are present; carbonates are dominant in the Upper Cambrian to Middle Devonian succession, whereas clastic strata predominate in the younger rocks. Oil and gas in significant quantities have been found in five of the 19 wells drilled in rocks of Carboniferous and Permian age. Gas recovery has been obtained from Lower Cretaceous strata, and promising reservoir rocks are present in Ordovician and Lower to Middle Devonian porous dolomite (Martin, 1973).

Old Crow Basin in the northwestern part of Yukon Territory, virtually unexplored as yet, is believed to contain about 1500 m of Mesozoic clastic rocks unconformably overlying up to 3000 m of Devonian to Permo-Carboniferous clastic and carbonate rocks. Both source and reservoir rocks may be present in stratigraphic and combination traps, but nothing is known of the basin's thermal and subsidence history. In Blow Trough in northern Yukon Territory structural deformation and widespread breaching of potential source and reservoir rocks have reduced the prospects of finding significant hydrocarbon reserves. To date three wells have been drilled in this area.

Petroleum Resources of the Intermontane Belt

The principal sedimentary basins of the Intermontane Belt are the Whitehorse Trough, Bowser, Nechako and Tyaughton-Methow basins. The Whitehorse Trough developed in an interarc position between Stikinia on the west and Quesnellia on the east. Its intensely deformed basin fill consists of Triassic to Lower Cretaceous limestones and marine and nonmarine clastic rocks of which the limestones of the Upper Triassic Lewes River Group are possibly the best reservoir rocks (Koch, 1973). Samples collected from the basin yielded small amounts of dry gas. The Bowser Basin, containing marine and nonmarine sediments of Middle Jurassic to Early Cretaceous age, is a basin developed upon Stikinia and has many of the attributes of a foreland basin. Upper Paleozoic limestone and reworked marine volcanics of the Hazelton Group, the latter tested by the only well drilled in the basin, are considered the primary exploration targets (Koch, 1973). In the Nechako and Tyaughton-Methow basins the primary exploration target for the three wells drilled was the clastic and limestone sequence of the Upper Triassic to Lower Jurassic Takla Group; oil staining was reported in Lower Cretaceous marine clastic beds (Koch, 1973).

In all of the Intermontane Belt basins anticlines are the most likely type of trap to be expected. Source rocks range in age from Early Permian to Late Jurassic and include black, richly organic shales in the Cache Creek, Laberge and Bowser Lake groups. Rocks of Permian, Jurassic and Cretaceous age could have good reservoir potential assuming favourable trapping conditions. Several adverse factors complicate exploration in these areas, not the least of which is the substantial degree of thermal metamorphism and intense deformation to which the rocks have been subjected, which probably has destroyed much of their petroleum.

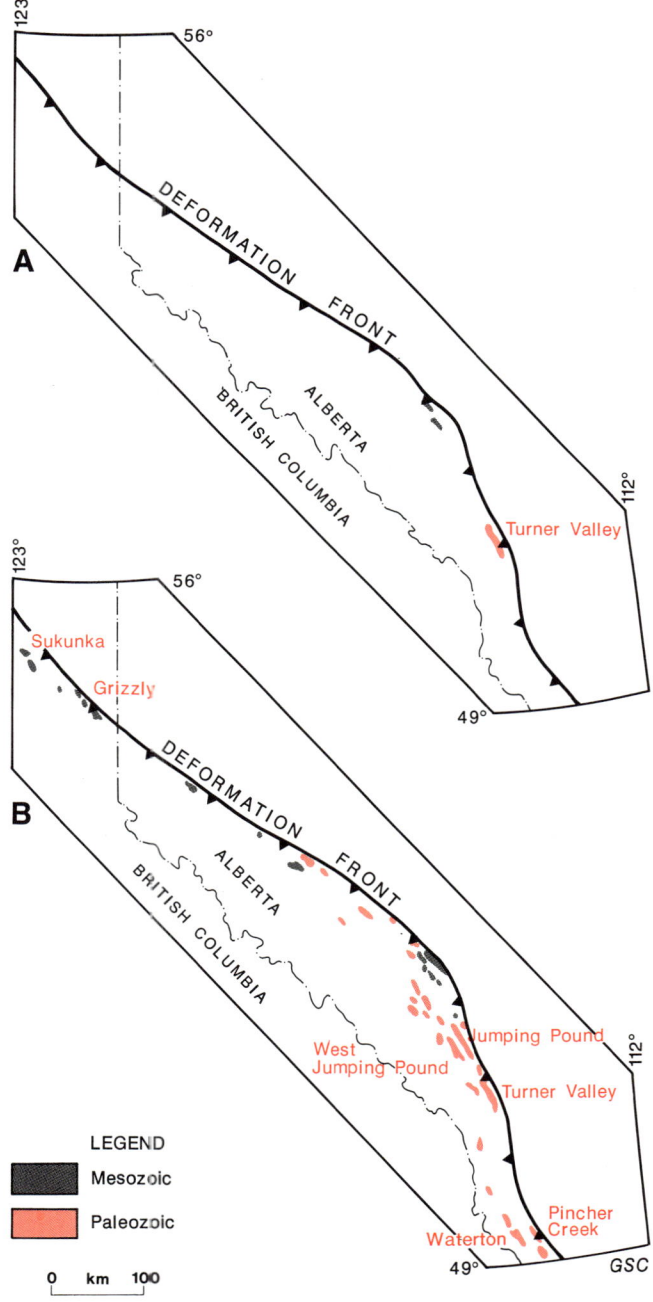

Figure 20.5. Distribution of principal oil (A) and gas (B) fields in the southern Canadian Cordillera (modified from Wallace-Dudley, 1982a,b).

Petroleum Resources of the Insular Belt

The principal petroleum prospective sedimentary basins of the Insular Belt are the Georgia Basin beneath the Strait of Georgia and adjacent Fraser Lowlands, Tofino Basin underlying the continental shelf off the west coast of Vancouver Island and the Queen Charlotte Basin lying beneath Queen Charlotte Sound, Hecate Strait and northwestern Queen Charlotte Islands. The Winona Basin, on the Explorer Plate adjacent to the northern Vancouver Island continental margin, and the Suquash Basin, on and near northeastern Vancouver Island, might conceivably contain hydrocarbons but are not considered prospective because of difficulty of access to the former and lack of sufficient sediment volume in the latter.

Eight wells have penetrated the thick nonmarine deltaic Tertiary succession in the Fraser Lowlands near the Georgia Basin, and one has been drilled into the Upper Cretaceous Nanaimo Group in the Gulf Islands. Currently, exploration of the Nanaimo Group, including slim-hole drilling is proceeding in the Nanaimo and Comox sub-basins on the east coast of Vancouver Island. In the Tertiary strata abundant reservoirs are available, but neither adequate source rocks nor traps have been found. The nonmarine character of the succession suggests that gas would be the principal hydrocarbon. The largely marine Nanaimo Group offers several opportunities for hydrocarbons in both structural and stratigraphic traps; however, porosities of surface rocks are generally 5% or less. Moreover, vitrinite reflectance data and stratigraphic analyses suggest that much of the basin sustained burial within low to moderate thermal regime.

Six wells have tested the Tertiary succession in the Tofino Basin (Shouldice, 1973). Gas shows were encountered in locally highly pressured Miocene and Pliocene marine strata. Potential reservoir rocks are commonly tight due to the clay-mineral plugging of initial porosity in the largely immature clastic sediments. Although abundant opportunities exist for large structural traps, adequate source rocks have yet to be encountered. Moreover, due to the basin's location above the descending Juan de Fuca Plate, the depressed thermal regime may have been insufficient for maturation beyond the level of immature gas.

Queen Charlotte Basin probably has the best potential for substantial petroleum resources in the Insular Belt. Fourteen wells have penetrated the Tertiary section both on northeastern Graham Island and offshore in Hecate Strait and Queen Charlotte Sound where one oil show was encountered in Pliocene nonmarine strata (Shouldice, 1973). Exploration ceased in the late 1960s, but considerable new geological and geophysical information has been acquired since then which, in part, has resulted in renewed interest on the part of petroleum companies.

Queen Charlotte Basin is believed to have formed through the combined processes of rifting and lithospheric flexure, and developed above and across the suture between Wrangellia and the Alexander Terrane (Yorath and Chase, 1981). The rifting model provides a favourable thermal regime and structural style for the maturation and entrapment of hydrocarbons of Tertiary age (Yorath and Hyndman, 1983). Source rocks for the basin may be found in the Queen Charlotte Terrace off the west coast of the Queen Charlotte Islands. The terrace is a folded sedimentary prism overlying the Pacific Plate which, adjacent to the Queen Charlotte Transform Fault, is moving northward at about 55 km/Ma. During the time of basin formation it may have been off the mouth of Queen Charlotte Sound. Additional sources of petroleum occur within Triassic and Jurassic rocks of the Queen Charlotte Islands (Macauley, 1983; Cameron and Tipper, 1981) which are believed to underlie the Tertiary sedimentary and volcanic succession of the basin. On the Queen Charlotte Islands petroleum occurs in vesicular basalt of Tertiary age and, thus, is likely to be found also in Tertiary sedimentary rocks offshore (Sutherland Brown et al., 1983).

Some hydrocarbon potential exists in the Haida and Honna formations of the Queen Charlotte Group assuming that adequate stratigraphic seals and source rocks can be found. Recent attempts to explore beneath the thick Tertiary lavas of the Masset Formation on the Queen Charlotte Islands have been unsuccessful.

PART B. COAL

R.M. Bustin

In the Canadian Cordillera coal deposits are widely distributed in strata ranging in age from Jurassic to Tertiary. Many have been extensively explored and some developed, but others, particularly in the northern and central Cordillera, have received only cursory examination. Most of the coal that is mined comes from Upper Jurassic and Lower Cretaceous coal measures in the Rocky Mountains and Foothills of the Foreland Belt where medium and low volatile bituminous coals are produced. Present coal production from the Cordillera is in the order of 20 million tonnes per year; additional mines are anticipated depending on market demand.

The distribution, age and rank of significant coal occurrences in the Canadian Cordillera are summarized in Figure 20.6. Cordilleran deposits can be divided into two broad stratigraphic categories: Upper Jurassic through Upper Cretaceous, and Tertiary deposits. Upper Jurassic to Upper Cretaceous measures occur in the Foreland, Intermontane and Insular belts. These coals form part of a regionally extensive molasse that accumulated in response to tectonism associated with the accretion of suspect terranes. In contrast Tertiary coals are for the most part local accumulations in fault-bounded basins which formed in response to extension related to right-lateral strike-slip faults. They occur throughout the central and northern parts, and locally in the western parts of the Cordillera.

Foreland Belt

Coal deposits of the Foreland Belt occur in the Front Ranges and Foothills of the Rocky Mountains and extend from the Canada-U.S.A. International border to north of Peace River. There are four principal coal-bearing sequences (Fig. 20.7): the Upper Jurassic to Lower Cretaceous(?) Mist Mountain Formation; the Aptian to Albian Gething Formation; the Albian Gates and Malcolm Creek formations and the uppermost Cretaceous to Paleocene Saunders Group. Collectively the coal measures form part of the foreland basin clastic wedge that was shed eastward from the emerging Omineca and Foreland belts in Late Jurassic through Tertiary time. The detailed stratigraphy of the measures reflect major and minor transgressions and regressions and intervals of widespread erosion (see Chapter 9).

In the Late Jurassic an extensive seaway occupied a marginal position adjacent to the emerging Cordilleran Orogen and was the site of deposition of marine and transitional marine sediments of the upper part of the Fernie Formation. Later, during the latest Jurassic to Early Cretaceous, easterly and northerly flowing rivers established deltas along the western margin of the seaway leading to the accumulation of thick, laterally extensive coal deposits of the Mist Mountain Formation of the Kootenay Group. North of the North Saskatchewan River equivalent strata of the Nikanassin Formation, and, farther north, the Minnes Group were deposited (Fig. 20.7). Coal occurs in each of these sequences but is thin and uneconomic.

The coal measures of the Mist Mountain Formation are up to 1000 m thick near Elkford British Columbia and thin to the east, both depositionally and as a result of pre-Cadomin (pre-Albian) erosion. Near Blairmore, Alberta, the formation is about 74 m thick; farther east it is absent. The number of mineable seams differs between individual deposits. The Crowsnest, Elk and Flathead coalfields contain in excess of 10 mineable seams with an aggregate thickness of 80 m. The thickest seams are about 15 m thick. East of the Lewis Thrust Fault the number of mineable seams is fewer, and at many localities, only one is present. The Cascade coalfield, in the Bow Valley near Canmore, Alberta, has between five and eight seams, ranging in thickness from 1.5 to 4 m.

The quality of Mist Mountain Formation coal is variable; locally it has a high ash content and near-surface measures are typically pervasively oxidized. Everywhere the coals are characterized by low sulphur content (less than 1.0%). In the Crowsnest, Elk and Flathead coalfields and nearby areas in the southeastern Cordillera the coals are of medium to low volatile bituminous rank. To the north the rank progressively increases and at Canmore, Alberta (Cascade Coalfield) it is semi-anthracite.

Coal has been mined at numerous localities throughout the Mist Mountain coalfields during the last 90 years. Present production is from the Crowsnest and Elk Valley coalfields where approximately 9 million tonnes of metallurgical and 3 million tonnes of thermal coal are mined annually.

In post-Valanginian time a major uplift resulted in erosion of the foredeep resulting in the progressive bevelling of the Kootenay, Nikanassin, and Minnes sediments and older strata to the east. Along the eastern margin of the Cordillera, conglomerate and sandstone of the Cadomin Formation were deposited on the erosion surface. A major southerly transgression of a boreal sea took place in Aptian to Albian time. South of this, in the Foothills between the Peace and Smoky rivers, coal measures of the Gething Formation (Aptian-earliest Albian) were deposited apparently as part of a delta complex (Stott, 1982). Farther to the south the Gething passes laterally into fine-grained alluvial plain, lacustrine and estuarine deposits of the Gladstone Formation (Stott, 1984) containing only thin uneconomic seams.

The coal measures of the Gething Formation have economic potential in a number of areas although they are presently not being exploited. Up to 50 seams occur in the Carbon Creek area where 12 seams are greater than 1.5 m thick and some are up to 4 m thick. Significant coal resources have been documented in the Dowling, Willow, Hasler and Norman creeks, Peace River, Pine Pass and Sukunka areas. In these areas the most important resources are restricted to one or two seams.

The rank of coal in the Gething Formation decreases progressively from the core of the Alberta syncline (semi-anthracite) to the western edge of the Foothills (high to low volatile bituminous) in response to shallower pre-orogenic depth of burial. Gething coal in the Peace River region was mined intermittently between 1908 and the late 1960s. Currently there is no significant production.

Transgression of the boreal sea continued southward as far as the town of Cadomin leading to deposition of the Moosebar Formation. The transgression was followed by a northward retreat of the shoreline and progradation of deltaic and alluvial coal measures of the Gates Formation. Economic coals accumulated from near Pine River in northeastern British Columbia to Waiparous Creek, 700 km to the south. Farther to the south, this stratigraphic interval is represented by part of the nonmarine Beaver Mines Formation which includes no commercial coal. In the plains to the east equivalent strata assigned to the Mannville Group contain thick coals beneath central Alberta.

The Gates Formation in the Sukunka-Monkman region is up to 263 m thick and locally includes up to 11 seams of coal. Several prospective mining areas which contain an average aggregate coal thickness of 22 m have been identified (Northeast Coal, 1977). Coal is presently being produced in the Mount Spieker-Bullmoose and Quintette areas (Fig. 20.6).

The Grande Cache Member of the Gates Formation is 130 m thick at Torrens Ridge near the British Columbia-Alberta border and thins southeastwards to 15 m at Waiparous Creek (McLean, 1982). Up to nine coal seams occur in the Grande Cache Member with the thickest being 6 m. Coals of the Grande Cache Member (included in the Luscar Group; see Holten and Mellon, 1972) have been mined since 1911, with the main production from the Luscar, Cadomin, Mountain Park, Nordegg and Grande Cache areas. Presently production is from the Luscar and Grande Cache areas.

The rank of coal in Gates Formation ranges from high volatile to low volatile bituminous. It is highest in the Alberta syncline-outer Foothills and decreases to the west in the inner Foothills.

The basal part of the Upper Cretaceous clastic wedge in the eastern Foreland Belt is composed of marine and transitional marine strata assigned to the Alberta and

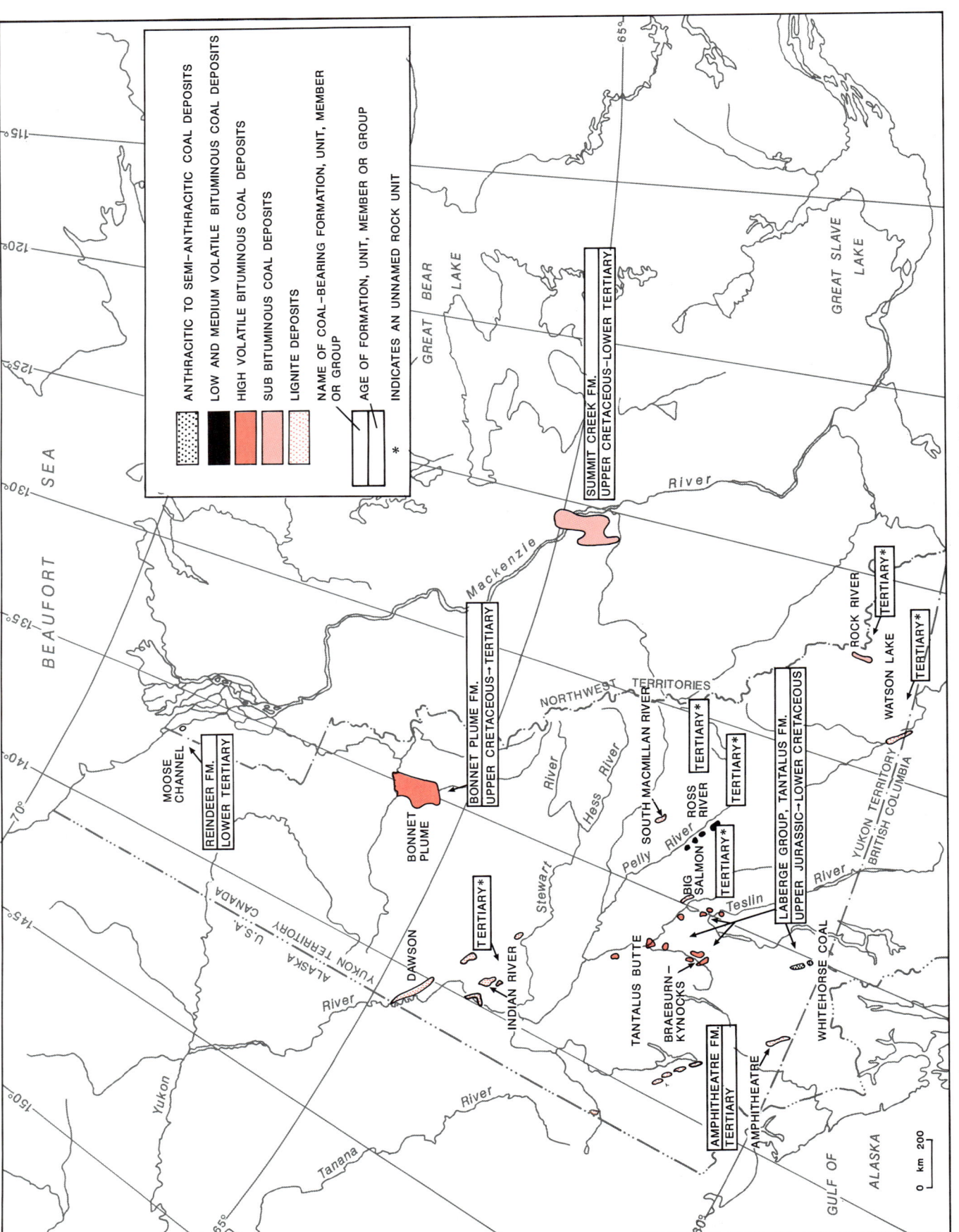

Figure 20.6. Distribution of coal deposits in the Canadian Cordillera.

ENERGY AND GROUNDWATER RESSOURCES OF THE CANADIAN CORDILLERA

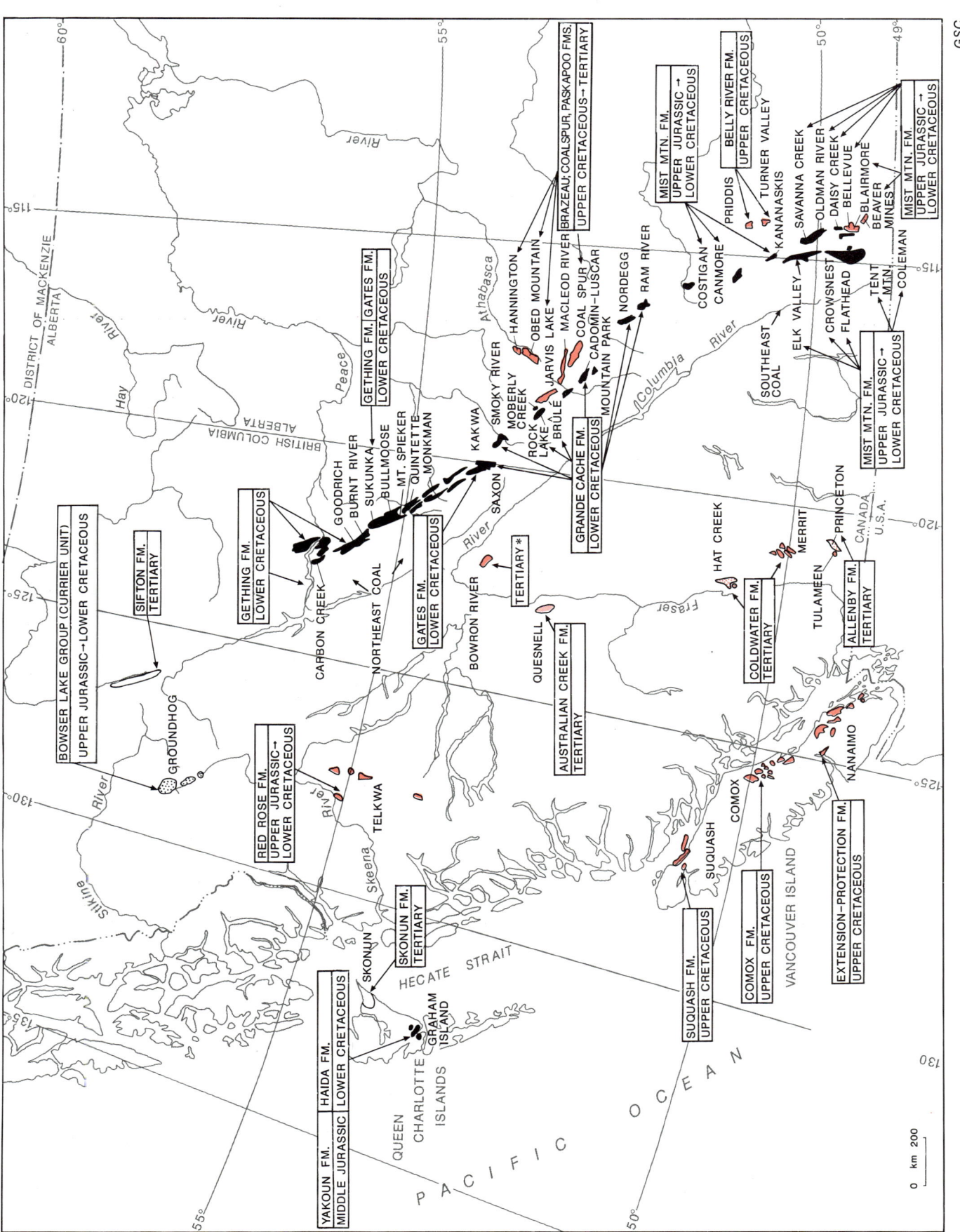

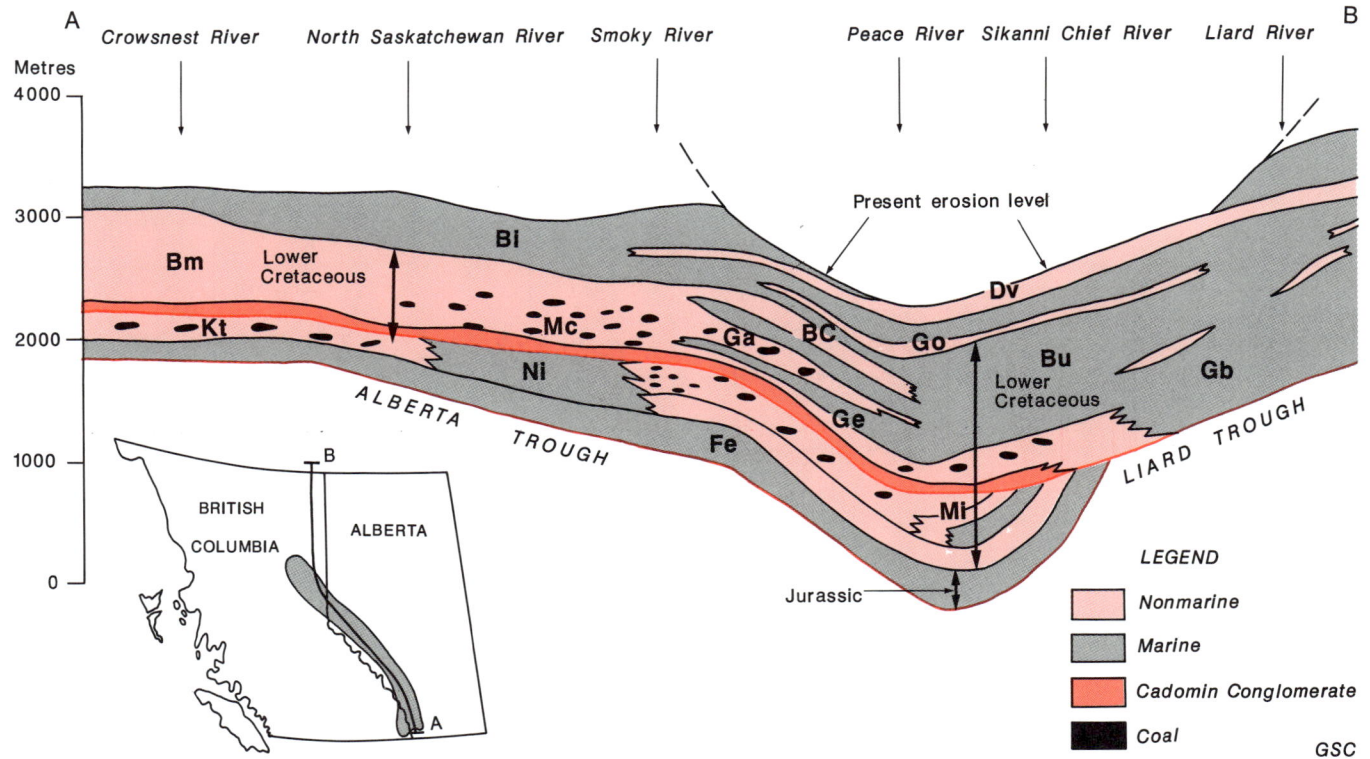

Figure 20.7. Schematic restored southeast-northwest structural cross-section along the Rocky Mountains and Foothills of Alberta and British Columbia (modified from Stott, 1974 and 1984).

Bl - Blackstone Formation
Dv - Dunvegan Formation
Bm - Blairmore Group
Mc - Malcolm Creek Formation
Go - Goodrich Formation
BC - Boulder Creek Formation
Ga - Gates Formation
Ge - Gething Formation
Bu - Buckinghorse Formation
Gb - Garbutt Formation
Kt - Kootenay Group
Ni - Nikanassin Formation
Mi - Minnes Group
Fe - Fernie Formation

Smoky groups. Overlying strata are a major regressive interval represented in the southern foothills by the Belly River, Bearpaw, St. Mary River and Willow Creek formations and, farther to the north by the Saunders Group and Wapiti Formation. Beneath the central Alberta plains equivalent strata are assigned to the Judith River and Bearpaw formations, and Edmonton Group. Deltaic and alluvial deposits of the Saunders Group, Wapiti Formation and Edmonton Group contain substantial resources of subbituminous and high volatile bituminous coal and locally high volatile bituminous coal occurs in the Belly River Formation.

During Late Cretaceous and Paleocene time the coal measures of the Foreland Belt were folded and faulted. Throughout the coalfields of the eastern Foreland Belt the structural style of the seams considerably influences the mineability of the coal. It is variably sheared and locally thickened or thinned. At Coal Mountain in the Crowsnest coalfield, for example, the basal seam of the Mist Mountain Formation, normally about 20 m thick, has been structurally thickened to more than 100 m in the hinge zones of synclines.

The Belly River Formation includes numerous thin seams, and locally seams up to 2.5 m thick of high volatile bituminous coal. Although some localities have been mined there has been no substantial production nor are major resources considered to exist. Beneath the plains area to the east equivalent strata of the Belly River Formation contain significant coal resources.

In the central Foothills the Saunders Group contains substantial coal resources in Maastrichtian to Paleocene strata of the Coalspur Formation in the Coal Valley, Brazeau River, Athabasca River and adjacent areas. The Coalspur Formation is approximately 250 m thick and includes at least eight zones of high volatile bituminous coal that in aggregate are about 48 m thick. From the Coal Valley area to the southwest there is marked decrease in thickness and number of seams. Coal deposits of the Saunders Group have been mined since early in this century and currently about 2 million tonnes per annum are produced in the Coal Valley area. In the northern Alberta plains major coal resources occur in equivalent strata of the Wapiti Formation, and to the east in the Edmonton Group.

Intermontane Belt

In the Intermontane Belt Upper Jurassic and Lower Cretaceous coal measures occur in three main areas: in the Whitehorse Trough in Yukon Territory, in the Groundhog coal area in Bowser Basin, and in the Telkwa coalfield on the north flank of the Skeena Arch. In these areas the coal is found in paralic deposits that characterized sedimentation throughout much of the Intermontane Belt during the Jurassic and Cretaceous periods and which were folded, faulted, and locally intruded during Late Cretaceous and Tertiary time (Tipper and Richards, 1976; Bustin and Moffat, 1983).

In the Whitehorse Trough coal is contained in the upper part of the Laberge Group (Jurassic) and Tantalus Formation (Upper Jurassic-Lower Cretaceous(?)) between Carmacks and Whitehorse (Wheeler, 1961). The upper Laberge Group is over 1000 m thick and up to eight coal seams have a cumulative thickness of 18 m (Milner and Craig, 1973). The overlying Tantalus Formation is 1500 m thick and locally includes at least three seams which, in aggregate, are up to 8 m thick. Coal measures of both the Laberge Group and Tantalus Formation are faulted, folded and locally cut by igneous intrusions. The coal is mainly of bituminous rank but is anthracite adjacent to intrusions. Coal has been periodically mined from the Tantalus Formation since 1905 and reached a maximum production of about 1700 tonnes per annum in 1967, mostly from the Carmacks area.

In the Groundhog coal area near the headwaters of the Skeena River coal forms part of the Currier Formation (Upper Jurassic-Lower Cretaceous) of the Bowser Lake Group. The Currier unit is between 400 and 600 m thick and includes 16 seams of anthracite and meta-anthracite each ranging from 0.5 to 4 m thick (Bustin and Moffat, 1983). The coal measures are tightly folded, locally cleaved and cut by quartz and carbonate veins. They have been extensively explored but as yet are not developed.

In the Telkwa, Bulkley River valley, Seeley Lake, and Kispiox areas on the north flank of the Skeena Arch coal occurs in the Red Rose Formation (Upper Jurassic-Lower Cretaceous) of the Skeena Group. The deposits are in the basal part of the formation and comprise up to 17 seams ranging from 1 to 7 m thick. They are partitioned into several segments by north-trending normal and reverse faults and are intruded by Tertiary volcanic rocks. The coal is mainly of high to medium volatile bituminous rank but is anthracite near intrusive rocks. Coal has been mined on a small scale since about 1900 mainly in the region of Goat Creek, and extensive exploration is currently under way.

Insular Belt

In the Insular Belt coal measures are found in the Queen Charlotte Islands (Sutherland Brown, 1968) and on Vancouver Island (Muller and Jeletzky, 1970). In the Queen Charlotte Islands bituminous coal and anthracite occur on Graham Island and Moresby Island in the Middle Jurassic Graham Island Formation of the Yakoun Group (Cameron and Tipper, 1985) and possibly in the Lower Cretaceous strata of the Haida Formation. As many as three coal seams up to 2.5 m thick exist near Kagan Bay on southern Graham Island but complex structure, uncertain stratigraphy and limited outcrop have inhibited evaluation of the deposits and thwarted early attempts at mining.

On Vancouver Island high volatile bituminous coal was mined for many years from the Nanaimo Group (Upper Santonian to Maastrichtian). The Nanaimo Group, up to 2450 m thick, is a succession of variously interbedded sandstone, conglomerate, shale and mudstone which accumulated in a post-accretionary basin on Wrangellia (Ward and Stanley, 1982; Muller and Jeletzky, 1970). The coal measures are preserved beneath the eastern coastal plain of Vancouver Island and on the adjacent Gulf Islands in three major coalfields; the Nanaimo, Cumberland, and Suquash fields. In the Nanaimo Coalfield three seams averaging 2 m thick occur in the Extension-Protection Formation (Campanian). In the Cumberland Coalfield three seams, also averaging 2 m thick, lie in the Comox Formation (Santonian). In the Suquash Coalfield up to nine coal zones have been recognized in the Suquash Formation (Campanian) but the seams are thin and of poor quality. Mining commenced in the Suquash Basin in 1836 and in the Comox and Nanaimo basins in 1852. In 1967 the last mine closed on Vancouver Island after a total production from all fields of about 72 million tonnes. Recent exploration indicates open pit reserves of 25.3 million tonnes at Quinsam Lake and Wolfe Mountain, both in the Comox Basin.

Tertiary coal deposits

Tertiary coal deposits of the Cordillera include large reserves of high volatile and sub-bituminous coal and lignite. The deposits are distributed throughout the central and northern Cordillera and locally on the Queen Charlotte Islands in the Insular Belt. With the exception of the Merritt, Tulameen, and Princeton coalfields in the southern Intermontane Belt none have been mined except on a local scale. In addition to the recognized coalfields (Fig. 20.6) thin or locally thick and discontinuous seams are found in almost all Tertiary sedimentary sequences in the Cordillera.

Tertiary coal deposits of the central and northern Cordillera are mainly in small isolated fault-bounded basins, many of which formed as a result of extension accompanying large scale, dextral, strike-slip fault displacements (Long, 1981). The Bonnet Plume Basin, at the south end of the Richardson Anticlinorium, formed during the Late Cretaceous and Paleocene epochs and is considered to have significant and potentially economic coal resources. The Bonnet Plume Formation, up to 1500 m thick, includes at least five seams of high volatile bituminous coal, each up to 15 m thick (McKinney, 1983). Within the Tintina Trench poorly exposed and steeply dipping lower Tertiary coal deposits are present at several localities, the most significant of which are in the Watson Lake, Ross River and Dawson areas (Hughes and Long, 1980). In the Dawson area several seams of Eocene lignite, each up to 13 m thick, were mined for local use at the turn of the century. In the Ross River area Paleocene(?) and Eocene strata include at least five seams of sub-bituminous to semi-anthracite coal cumulatively about 5 m thick. In the Watson Lake area five seams of lignite with an aggregate thickness of 15 m occur in Paleocene and Eocene strata. East of Watson Lake, in the Rock River area, significant thicknesses of Tertiary lignite (greater than 9 m) recently have been reported (Wright and Miller, 1983). Along the Denali Fault the Paleocene Amphitheatre Formation contains at least four seams of lignite that range up to 1.5 m thick (Muller, 1967).

Between the Fraser River-Straight Creek fault system and the Rocky Mountain Trench, Tertiary coal deposits lie within the Hat Creek, Tulameen, Princeton, Merritt, Quesnel and Bowron River coal basins. In the Bowron River basin Paleocene measures include at least three seams of high volatile bituminous rank, cumulatively about 10 m thick. In the Quesnel field coal is present in the Oligocene-Miocene Australian Creek Formation along the Fraser River Valley (Graham, 1978). It locally contains three seams of sub-bituminous coal each with reported thicknesses of up to 22 m. The Hat Creek, Merritt, Tulameen and Princeton coalfields in the south-central Cordillera are all of Eocene age although each deposit probably developed in isolated fault basins. The coal measures at the Hat Creek and Merritt coalfields are part of the Kamloops Group whereas those of the Princeton and Tulameen coalfields are assigned to the correlative Allenby Formation of the Princeton Group. In the Hat Creek field exceptionally thick high ash lignite and sub-bituminous coal measures of Late Eocene age are preserved in an elongate, narrow fault-bounded basin. Each of the two major deposits contains four coal zones with an aggregate thickness of 370 m, one of the thickest coal deposits in the world. The Merritt coal field was mined periodically between 1906 and 1962. At least seven seams have an aggregate thickness of about 9 m. In the Princeton Basin seven seams of sub-bituminous rank were mined between 1909 and 1961 during which time a total of 1.9 million tonnes was produced. In the Tulameen Basin coal was mined between 1919 and 1940, and in 1954. To date a total of about 2.2 million tonnes has been extracted. Two thick seams of high volatile bituminous coal occur in the Tulameen Basin; the upper is 15 to 20 m thick and the lower averages about 7.5 m thick.

On northern Graham Island in the Insular Belt lignite in Tertiary strata of the Skonun Formation outcrops in several isolated areas. At least 10 seams ranging from 30 cm to 4 m thick, are exposed at Skonun Point which, together with other occurrences, suggests the presence of significant resources which have yet to be evaluated.

In the northern Cordillera the Moose Channel coal deposits occur in the Aklak Member (Uppermost Cretaceous to Paleocene) of the Reindeer Formation. The Reindeer Formation is part of a thick succession of Upper Cretaceous and Tertiary deltaic sediments of the Beaufort-Mackenzie Basin. The Aklak Member is up to 580 m thick and is considered to include significant resources of sub-bituminous to high volatile bituminous coal at relatively shallow depths (Young, 1975). In the region of Coal Mine Lake two seams, 7 m and 3.7 m thick, were mined for domestic use at Aklavik between 1939 and 1958.

Along the Mackenzie Mountains front in the Fort Norman area, the uppermost Maastrichtian to Paleocene Summit Creek Formation hosts a coal-bearing clastic wedge (Yorath and Cook, 1981). The coal is of lignite to sub-bituminous rank with a speculative resource potential of 1.5 to 2.0 million tonnes. In the nearby Fort Norman area even greater resources may be present (B.D. Ricketts, pers. comm., 1985).

Reserves and resources

The total inferred coal resources of the Cordillera are estimated to be about 71 000 million tonnes of which 57 000 million tonnes are medium and low volatile bituminous coal and 14 000 million tonnes are high volatile bituminous and lower rank coal (Energy, Mines and Resources Canada, 1980). Table 20.1 shows reserves and resources for coal deposits in the Cordillera. The coal resources of many deposits are unknown and in others distinctions are not made between reserves and resources.

PART C. URANIUM AND THORIUM

R.T. Bell

Uranium and thorium occur in most kinds of rocks (Jones, 1990), but, the number of deposits and significant occurrences in the Canadian Cordillera is small (Fig. 20.8). Most of these occurrences developed following terrane accretion (Fig. 20.9). In addition to the known deposits in the Okanagan region and Wernecke Mountains, significant resources might be found in regions underlain by felsic volcanics and clastic deposits associated with late stage granitic intrusions.

Uranium
Beaverdell area

The largest uranium occurrence in the Canadian Cordillera is the Blizzard deposit (Fig. 20.10, 20.11) which contains about 4000 tonnes uranium at an average grade of 0.18% uranium (Sawyer et al., 1981; Boyle, 1982; Bates et al., 1980). Nearby are a few smaller deposits (Fig. 20.10), the largest of which is the Tyee deposit containing 650 tonnes uranium at an average grade of 0.03% uranium. The other deposits together contain approximately 1300 tonnes uranium. The discovery of this type of mineralization in British Columbia was based on a model of similar basal channel deposits in central and western Japan (Katayama et al., 1974).

The following description is mainly of the Blizzard and associated deposits, except where noted and follows those by Sawyer et al. (1981), and Boyle (1982). The deposits are contained in fluvial silt, sand, and gravels a few tens of metres thick within paleovalleys incised into mainly Cretaceous granite gneiss and schist on the west flank of the Omineca Belt. Generally the deposits are overlain and protected by Pliocene lavas of the Chilcotin Group. Underlying rocks are Paleogene continental volcanic and sedimentary rocks and the Eocene to Early Oligocene Coryell syenitic intrusions. Overall the sequences fine upwards

ENERGY AND GROUNDWATER RESSOURCES OF THE CANADIAN CORDILLERA

Table 20.1. Coal areas in the Canadian Cordillera

Age	Coal Field	Formation	Coal Rank	Reserves/Resources* (mega tonnes)	
British Columbia					
Tertiary	Skonun	Skonun Fm.	Lignite	0-1000	(c)
Albian-Turonian	Skonun	Haida Fm.			
Bajocian-Callovian	Graham Island	Yakoun Fm.	L.V.B. & M.V.B.	100-800	(c)
Campanian to Maastrichtian	Suquash	Suquash Fm.	H.V.B.	27	(v)
Santonian	Comox (Cumberland)	Comox Fm.	H.V.B.	30.48	
Campanian	Nanaimo	Extension-Protection Fm.	H.V.B.	9.1-64	(c)
U. Jur. to L. Cret.	Groundhog	Bowser Lake Gp.	Anth.	-	-
U. Jur. to L. Cret.	Telkwa	Red Rose Fm.	H.V.B. & M.V.B.	-	-
Paleocene	Bowron River	Unnamed	H.V.B. & M.V.B.	73.6	(v)
Oligocene	Quesnel	Australian Creek Fm.	Sub-bituminous		
Tertiary	Hat Creek	Coldwater Fm.	Lignite	809	(v)
Tertiary	Merritt	Coldwater Fm.	H.V.B.	18.2	(v)
Tertiary	Tulameen	Allenby Fm.	H.V.B.	37.3	(v)
Tertiary	Princeton	Allenby Fm.	Lignite	10.0	(v)
				Total: 1195.0 m. tonnes (v)	
				25 300.0 m. tonnes (c)	
B.C. (Southeast)					
U. Jur. to L. Cret	Elk Valley	Mist Mountain Fm.			
U. Jur. to L. Cret	Crowsnest	Mist Mountain Fm.			
U. Jur. to L. Cret	Flathead	Mist Mountain Fm.	L.V.B. & M.V.B.		
U. Jur. to L. Cret	Tent Mountain	Mist Mountain Fm.			
U. Jur. to L. Cret	Coleman*	Mist Mountain Fm.			
				Total: 575.0 m. tonnes (v)	
				8100.0 m. tonnes (c)	
B.C. (Northeast)					
L. Cret.	Carbon Creek	Gething Fm.	L.V.B. & M.V.B.	-	-
L. Cret.	Goodrich	Gething Fm.	L.V.B. & M.V.B.	-	-
L. Cret.	Burnt River	Gething Fm.	L.V.B. & M.V.B.	-	-
L. Cret.	Sukunka	Gething & Gates Fm.	L.V.B. & M.V.B.	-	-
L. Cret.	Mount Spieker	Gething & Gates Fm.	L.V.B. & M.V.B.	520	(c)
L. Cret.	Bullmoose	Gething & Gates Fm.	L.V.B. & M.V.B.	-	-
L. Cret.	Quintette	Gething & Gates Fm.	L.V.B. & M.V.B.	-	-
L. Cret.	Monkman	Gething & Gates Fm.	L.V.B. & M.V.B.	-	-
L. Cret.	Saxon	Gates Fm.	L.V.B. & M.V.B.		
Yukon and Northwest Territories					
L. Cret. to Paleocene	Bonnet Plume	Bonnet Plume Fm.	H.V.C. - "C"	650	(v)
Tertiary	Rock River	Unnamed	Subbituminous	-	-
U. Jur. to L. Cret.	Tantalus Butte	Laberge Group & Tantalus Fm.	H.V.B. - "A" & "B"	-	-
U. Jur. to L. Cret.	Whitehorse	Tantalus Fm.	Anthracite	-	-
Paleocene	Amphitheatre	Amphitheatre Fm.	Lignite	-	-
L. Cret. to Paleocene	Moose Channel	Reindeer Fm.	H.V.B.	-	-
Eocene	Dawson	Unnamed	Lignite	-	-
Eocene	Indian River	Unnamed	Lignite	-	-
Paleocene	Big Salmon	Unnamed	Lignite	-	-
Tertiary	S. Macmillan River	Unnamed	Lignite	-	-
Eocene	Ross River	Unnamed	L.V.B. to M.V.B.		
Maastrichtian-Paleocene	?	Summit Creek Fm.	Lignite-Subbituminous	1500-2000	
Alberta (Mountain Region)					
U. Jur. to L. Cret.	Beaver Mines	Mist Mtn. Fm.	H.V.B.	4	(v)
U. Jur. to L. Cret.	Bellevue	Mist Mtn. Fm.	H.V.B.	43	(v)
U. Jur. to L. Cret.	Blairmore	Mist Mtn. Fm.	H.V.B.	212	(v)
L. Cret.	Brule	Grande Cache Fm.	L.V.B.	13	(v)
L. Cret.	Cadomin-Luscar	Grande Cache Fm.	M.V.B.	198	(v)
U. Jur. to L. Cret.	Canmore	Mist Mtn. Fm.	L.V.B.	160	(v)
U. Jur. to L. Cret.	Coleman	Mist Mtn. Fm.	M.V.B.	76	(v)
U. Jur. to L. Cret.	Costigan	Mist Mtn. Fm.	M.V.B.	0	(v)
U. Jur. to L. Cret.	Daisy Creek	Mist Mtn. Fm.	M.V.B.	26	(v)
L. Cret.	Kakwa River	Grande Cache and Gates Fm.	L.V.B.	16	(v)
U. Jur. to E. Cret.	Kananaskis	Mist Mtn. Fm.	L.V.B.	25	(v)
L. Cret.	Moberly Creek	Grande Cache Fm.	M.V.B.	0	(v)
L. Cret.	Mountain Park	Grande Cache Fm.	H.B.B.	52	(v)
L. Cret.	Nordegg	Grande Cache Fm.	L.V.B.	77	(v)
U. Jur. to L. Cret.	Oldman River	Mist Mtn. Fm.	L.V.B.	107	(v)
L. Cret.	Ram River	Grande Cache Fm.	L.V.B.	140	(v)
U. Jur. to L. Cret.	Savanna Creek	Mist Mtn. Fm.	L.V.B.	40	(v)
L. Cret.	Smoky River	Grande Cache Fm.	L.V.B.	366	(v)
U. Jur. to L. Cret.	Tent Mtn.	Mist Mtn. Fm.	M.V.B.	32	(v)
Alberta (Foothills)					
L. Cret. to Tert.	Coalspur	Saunders Gp.	H.V.B.	275	(v)
L. Cret. to Tert.	Hannington	Saunders Gp.	H.V.B.	168	(v)
L. Cret. to Tert.	McLeod River	Saunders Gp.	H.V.B.	251	(v)
L. Cret. to Tert.	Obed Mountain	Saunders Gp.	H.V.B.	110	(v)
L. Cret. to Tert.	Priddis	Belly R. Gp.	H.V.B.	1	(v)
L. Cret.	Turner Valley	Belly R. Gp.	H.V.B.	7	(v)

Abbreviations:
H.V.B. - High Volatile Bituminous Coals; M.V.B. - Medium Volatile Bituminous Coals; L.V.B. - Low Volatile Bituminous Coals; Anth - Anthracitic Coals; (v) - Reserve Estimate; (c) - Resource Estimate; N.M. - Never Mined.
*Alberta included in reserve/resource calculations

CHAPTER 20

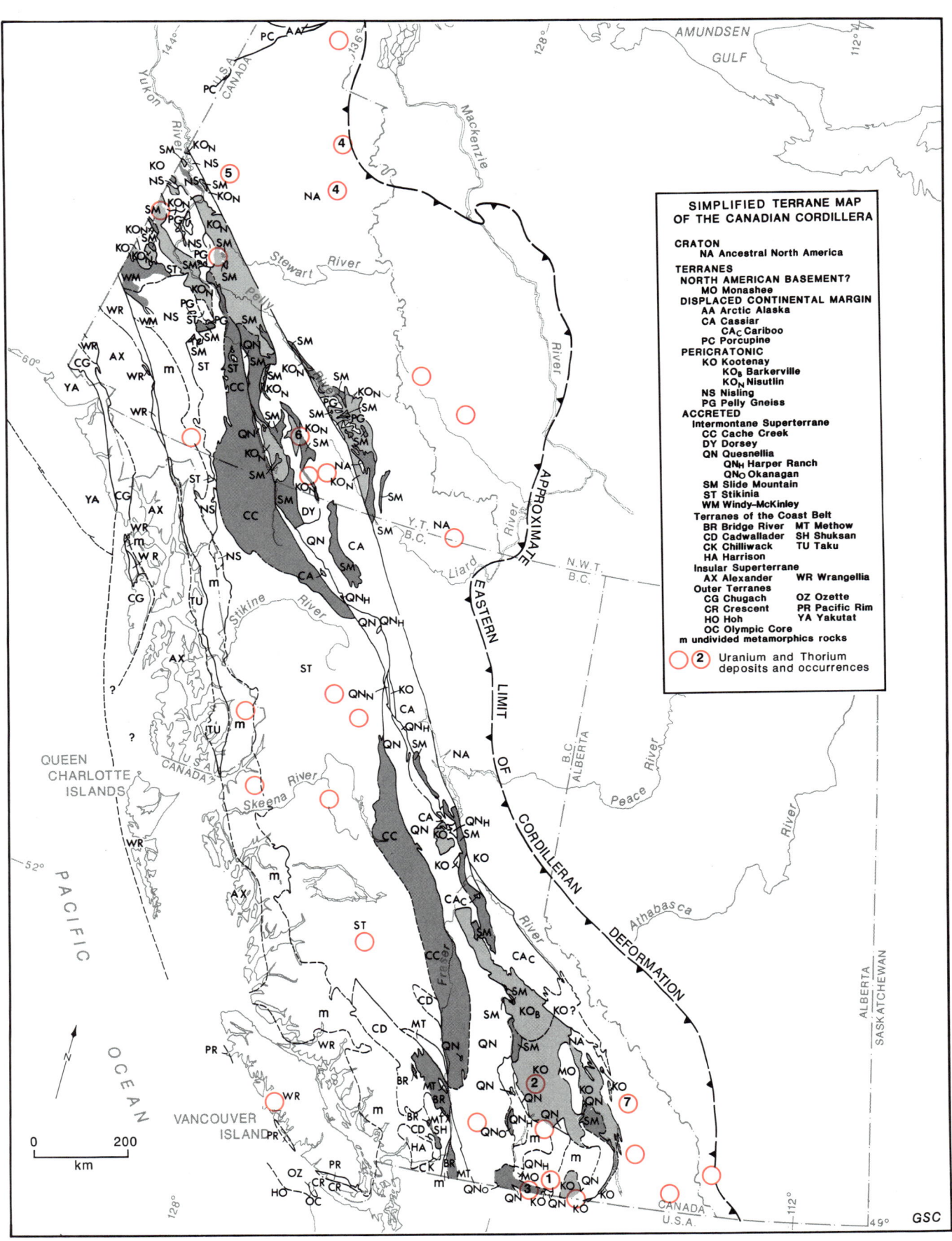

784

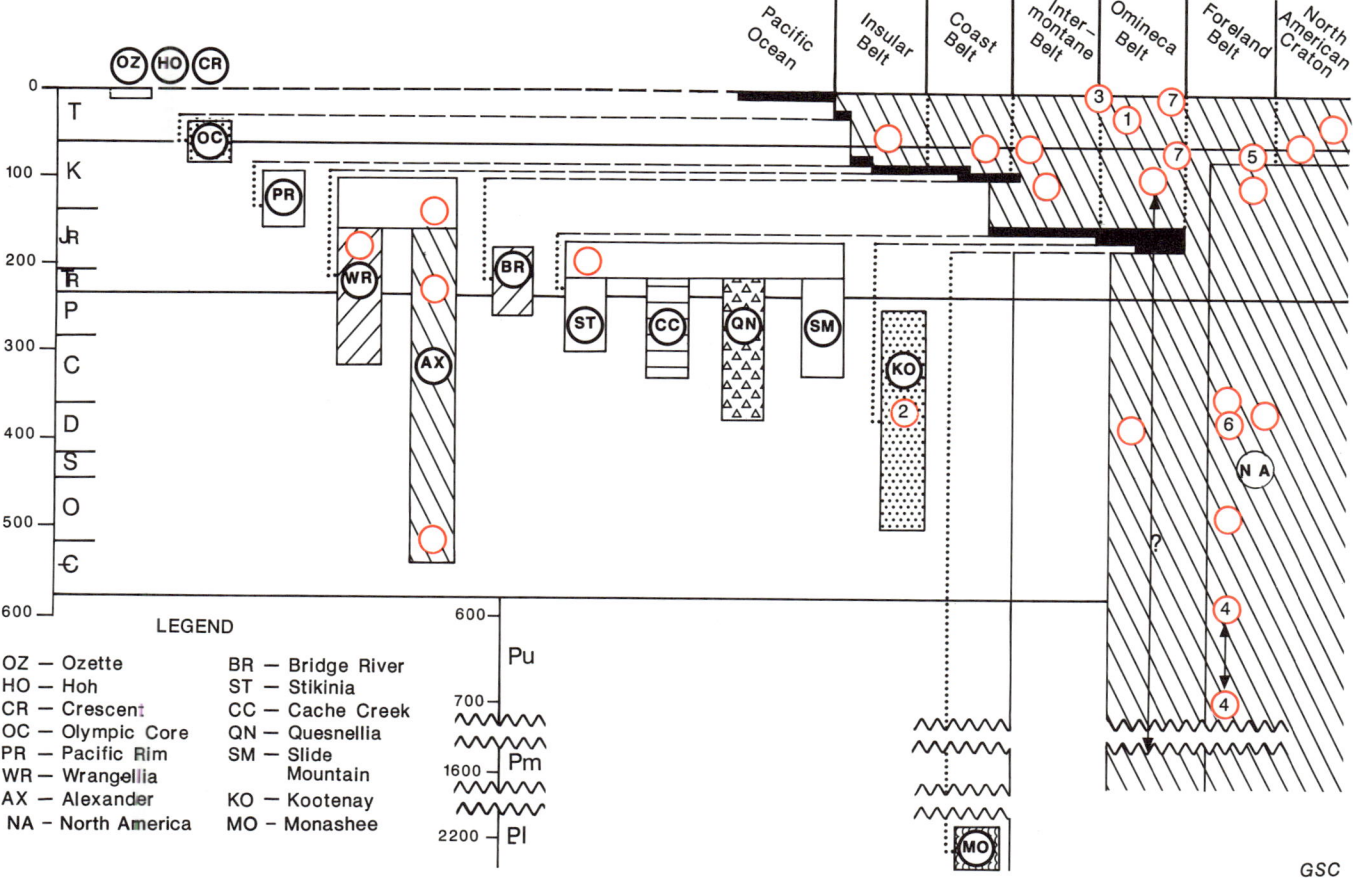

Figure 20.9. Spatial and temporal relationships of suspect terranes in Canadian Cordillera with selected groups of uranium occurrences and deposits. See Figure 20.8 for legend.

from coarse gravel at the base to fine sand and silt at the top. In some nearby paleovalleys similar sediments are interbedded with basaltic flows, implying that the sediments maybe as young as Early Pliocene.

Uranium is found as the yellow minerals saleeite and autunite (respectively magnesium and calcium uranyl/phosphate hydrates) in oxidized facies of the lower limonitized coarse sediments and in disseminated organic material in the reduced facies of the upper, fine sand and silt. Pitchblende and ningyoite (uranium calcium rare-earth phosphate hydrate) (Boyle, 1982) are the only uranium minerals identified in the reduced facies. Ningyoite (Boyle et al., 1981) also is reported in the Tyee deposit where marcasite acts as cement in the higher grade zones. Minor mineralization in the form of saleeite or autunite has been deposited in fractures in underlying regolithic granitic rocks and rarely in fractures in the overlying Chilcotin basalts. Minor mineralization also occurs in the silty matrix and in stoped sedimentary fragments in a basaltic breccia pipe at the north end of the Blizzard deposit. In British Columbia and Japan (Fig. 20.10, 20.11) the deposits are best developed in the uppermost levels of the paleodrainage system and, moreover, the richer deposits overlie granitic rocks. Additional prospects for this type of deposit lie in central British Columbia and in the southwestern part of Yukon Territory where similar conditions prevail: felsic igneous basement rocks, incised by late Tertiary fluvial channels and overlain by Neogene basalt.

Clearwater area

South of Clearwater, British Columbia, the small Rexspar deposit of fluorite, celestite and uranium-thorium minerals occurs in a sequence of metamorphosed feldspar-porphyritic trachyte and tuff on the western margin of the Kootenay Terrane (Preto, 1978; Morton et al., 1978; Bell, 1985). Stratigraphic tops and ages are uncertain in the structurally disrupted sequence. The trachyte may be older than Early Carboniferous and perhaps is related to the late mid-Devonian Mount Fowler granitic intrusion to the south. A single K-Ar date on the associated fluorphlogopite (Morton et al., 1978) suggests that mineralization can be not younger than Permian.

Figure 20.8. Terrane map of the Canadian Cordillera with selected groups of uranium occurrences and deposits: 1. Beaverdell area (Tyee, Blizzard, PNC deposits); 2. Clearwater area (Rexspar); 3. Okanagan area; 4. Wernecke Mountains; 5. Tombstone Mountains; 6. Pelly Mountains; 7. Bugaboo Creek.

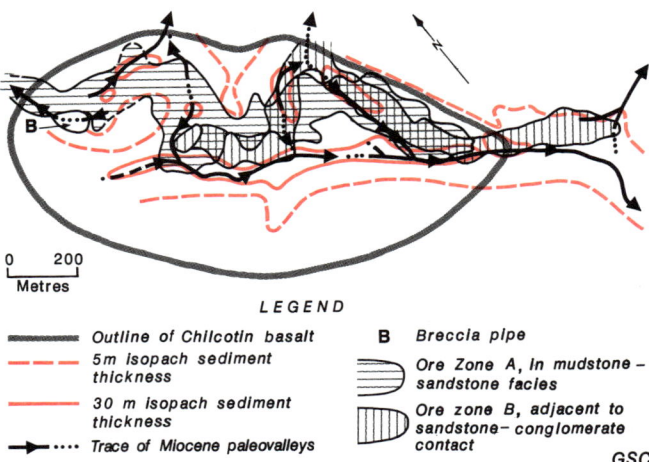

Figure 20.11. Plan view of Blizzard uranium deposit showing sediment thickness, main ore zones, outline of Chilcotin Basalt capping (after Norcen, 1979) and interpreted paleodrainage.

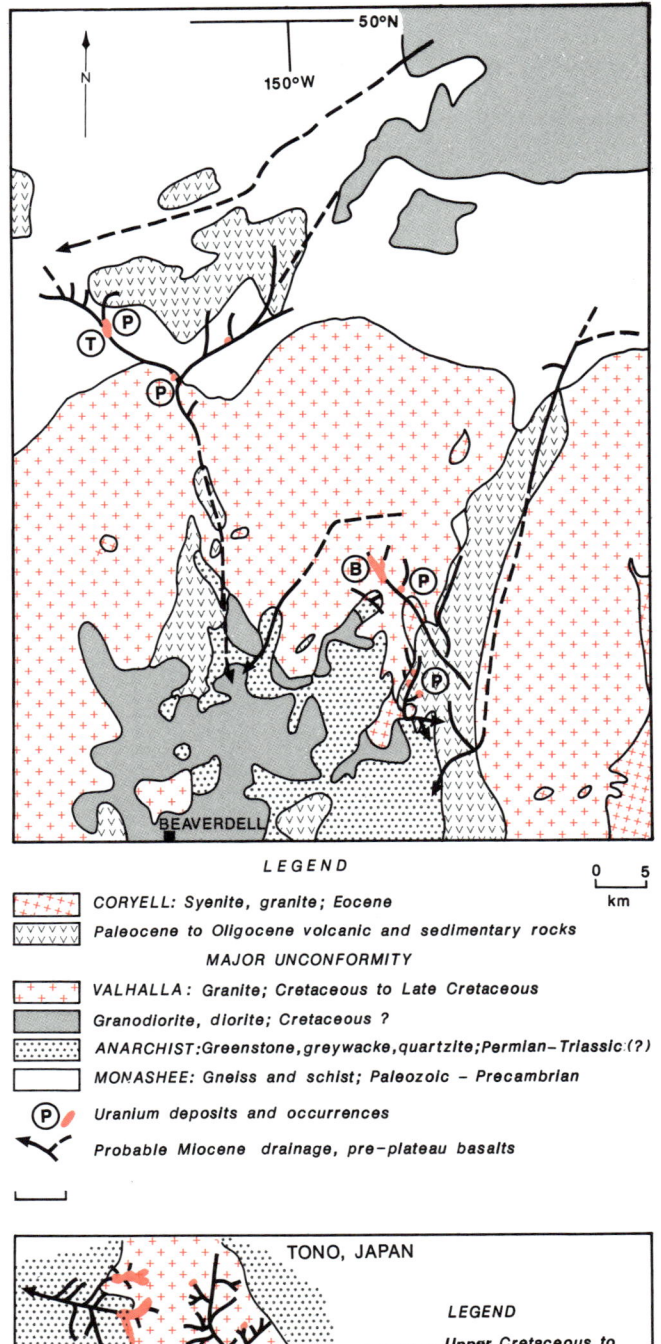

Figure 20.10. Basal channel uranium deposits on pre-Pliocene basement map, Beaverdell area (after Little, 1961 and 1957; and Boyle, 1982) and interpreted pre-Pliocene drainage with inset on Tono deposit in Japan (after Katayama et al., 1974). B-Blizzard, T-Tyee, and P-PNC deposits.

The Rexspar deposit includes three mineralized zones in sheared dark grey tuffaceous schist. They comprise disseminated uraninite, uranoan thorite and uranothorianite in a matrix of sericite, albite and quartz and accompanied by abundant coarse pyrite, fluorphlogopite, purple fluorite, celestite and calcite and minor bastnaesite, zircon, monazite and rutile. A fourth zone contains massive and disseminated pale green fluorite and celestite with pyrite, molybdenite and minor galena but only traces of uranium and thorium. Fluorite mineralization both preceded and followed uranium mineralization.

Assuming the section at Rexspar is upright, Preto (1978) postulated a syngenetic model wherein the deposition of sulphides, fluorite and uranium was due to late stage deuteric, volatile-rich fluids generated during volcanism. He discounted late hydrothermal events because there is no hydrothermal alteration in the schists beneath the host trachyte. On the basis of fluid inclusion studies Morton et al. (1978) suggested that uranium, thorium and rare earth elements were transported in an early hydrothermal system strongly charged with CO_2. On venting to the surface the sudden drop in P_{CO_2} would result in precipitation of the uranium minerals. The lack of hydrothermal alteration in the rocks beneath the trachytes is to be expected if the sequence is overturned (Bell, 1985). No analogue for the Rexspar deposit is known. Until structural and stratigraphic problems are resolved it will remain enigmatic.

The deposit contains recoverable reserves of about 700 tonnes uranium at grades of about 0.066% uranium (Preto, 1978; Bates et al., 1980). Additional resources are present in small deposits immediately to the east. Thorium recovery could equal that of uranium. Fluorite and celestite and rare earth elements may also prove to be economic.

Okanagan 'young' or surficial uranium deposits

In the Okanagan area, mainly between Kelowna and Osoyoos, there are about 40 uranium occurrences (Culbert et al., 1984) termed 'young' or 'surficial' in that they are forming at present in postglacial sediments. Because they are so young they have little or no gamma-active daughter products. They occur mainly in closed or cyclically closed alkaline lakes or playas, and to a lesser degree in association with organic-rich swamps in fluviatile systems. Factors important for this type of uranium deposit are a semi-arid climate in areas underlain by felsic igneous rocks. The underlying and adjacent rocks are Cretaceous and Tertiary granite and gneisses, and Eocene Marron volcanic members of the Penticton Group. The area is directly west of the Beaverdell uranium deposits. Uranium occurs in concentrations of up to a few thousand parts per million, and, being very loosely bonded, is easily remobilized.

The deposits are small, containing a few tens to a few hundreds of tonnes of uranium with grades in the order of 0.04% uranium; those known in the Cordillera aggregate about 800 tonnes uranium (Bates et al., 1980). Despite being small and of low grade, they could eventually prove to be economic insofar as they are numerous, are near the surface (less than 10 m depth) in unconsolidated sediments, and contain negligible daughter elements (Culbert et al., 1984).

Wernecke Mountains area

More than 40 groups of radioactive minerals occur in the northeastern Wernecke Mountains. A single uraniferous locality is known in the southern Richardson Mountains. The minerals are all associated with breccias with the following characteristics (Bell, 1982; 1986; 1989):

1) The breccia bodies are irregular to elongate and as much as 20 km long and 8 km wide.

2) They contain kilometre-size blocks of sedimentary rocks which have been transported upwards by as much as 10 km.

3) The blocks and matrix are of dolomitic sandstone and siltstone, argillite, phyllite, quartzite and dolomite and their altered protoliths with minor jaspilite, greenstone or gabbro fragments.

4) The breccias cut mainly the lower, entirely sedimentary, Wernecke Supergroup; they possibly cut younger Precambrian rocks west of the Wernecke Mountains but if so, it is likely due to later remobilization.

5) The breccia bodies contain irregular zones of intense alteration including carbonatization, hematization, albitization, silicification and chloritization, these also may be rebrecciated.

6) Mineralization includes: brannerite ± pitchblende ± gold; pitchblende + chalcopyrite + cobaltite ± iriginite; brannerite + monazite + pitchblende; chalcopyrite + pyrite + pyrrhotite ± chalcocite ± bornite ± cobaltite; cobaltite + brannerite; bournonite + chalcopyrite + tetrahedrite + galena + sphalerite + gold; and barite + magnetite; all are accompanied by hematite, hematite + magnetite, or siderite. (brannerite = uranium titanium oxide; iriginite = uranyl molybdenate hydrate.

7) Breccias are widespread west of the Wernecke Mountains but except for one locality in the southern Richardson Mountains and one in the Ogilvie Mountains they do not contain uraniferous minerals; most contain copper (Laznicka and Edwards, 1979) and minor cobalt.

The breccias appear to be mostly intrusive and diapiric. Intensive local metasomatism and the presence of a few rare-earth elements suggest that deep-seated carbonatites may be involved. At two localities euhedral magnetite-bearing breccia dykes with milled fragments are suggestive of diatremes.

Similar breccias (Bell, 1986; 1989; Bell and Jefferson, 1987) containing copper, cobalt, uranium and gold occur in the Copperbelt of Zaire (Lefebvre, 1980) and in South Australia (Dalgarno and Johnson, 1968; Roberts and Hudson, 1983). The origins of these breccias are not yet clearly understood but may relate to cryptic salk domes. The Wernecke area has good potential for U + Cu + Co ± Au although economic deposits have yet to be defined.

Thorium

At present the market for thorium is very small. There is some research on thorium as an energy metal; it is currently used in a small reactor in the Federal Republic of Germany and could be used in a modified CANDU reactor system.

Thorium occurs in the Rexspar deposit and is significant in the Tombstone Mountains in Yukon Territory (Olade and Goodfellow, 1978). Anomalously high concentrations occur in Late Cretaceous and Early Tertiary granitic plutons in the northern Intermontane Belt, in Carboniferous volcanic complexes in the Pelly Mountains (Bell, 1985), and along with U, Nb, Ta and REE in Recent placers in the Bugaboo Mountains.

PART D. GEOTHERMAL ENERGY

J.G. Souther

The natural heat within the earth's crust may be extracted in the form of hot water or steam and used, either directly for space or industrial process heating, or for the generation of electrical power. The type of application of geothermal energy depends primarily on the temperature of the resource and its proximity to market. Direct use can employ relatively low temperature fluids but the heat cannot be transported more than a few kilometres. The generation of electricity, which can be transmitted over great distances, requires relatively high temperatures. For a conventional low-pressure steam turbine a minimum temperature of 150°C is necessary to supply steam at 345 kPa. Binary-cycle generating systems, employing a working fluid with a lower boiling point than water, can operate at somewhat lower temperatures but power generation is generally not practical at temperatures much below 150°C.

In Canada geothermal energy is an undeveloped resource (Souther, 1981). Most of the potential is believed to lie in sedimentary basins where large volumes of warm brine, confirmed by petroleum exploration, could be tapped for direct heating use. High temperature geothermal systems, capable of power production, are confined to the Cordillera where hydrothermal activity is associated with Neogene volcanism. Two such systems, Meager Mountain and Mount Cayley, have been confirmed by drilling and several others are inferred from geological data. The following discussion is limited to those geothermal sites in western Canada that may be hot enough to produce electrical energy.

Target identification

The depth to which geothermal resources may be exploited is determined by the economic and technological limits of drilling (3 to 10 km). Geothermal gradients in the Canadian Cordillera are commonly about 20°C/km, thus only those areas where the geothermal gradient is anomalously high can be considered to have resource potential. Moreover, the technology to exploit hot dry-rock reservoirs is still in the experimental stage. Using existing technology geothermal energy can be recovered only from those reservoirs that are permeable to fluids and have sufficient intergranular water to supply the production wells. The geological environment most likely to meet these criteria is a convective hydrothermal system driven by heat from a shallow magma reservoir. Because heat is lost to the surface through time such systems have a finite life, in the order of a few million years after intrusion of the heat source.

In the Canadian Cordillera geologically young magmatic activity has been localized along the Neogene volcanic belts (Garibaldi, Anahim and Stikine belts; see Chapter 10). Large subvolcanic plutons within these belts are most likely to be associated with volcanoes that have a significant salic component because of the greater magma viscosity; mafic magma is less likely to form subvolcanic plutons. Evidence of associated hydrothermal activity is usually manifest in the presence of hotsprings. However, in regions of high rainfall and mountainous topography such as in the northern Cordillera, the ascent of hot water to the surface may be suppressed by deeply circulating cold groundwater.

Target evaluation

The initial evaluation of potential geothermal targets in Canada includes geological mapping to determine the composition, age, and volume of the associated volcano, and analysis of thermal spring waters to assess subsurface temperatures (Table 20.2) (Souther, 1975). Estimating reservoir temperatures using chemical geothermometers is based on the fact that groundwater equilibrates with minerals in the reservoir rock. The concentrations of various ions are temperature-dependent and the reactions are generally sluggish and endothermic. Thus water-rock equilibrium is established during prolonged residence time in the reservoir whereas the transit time back to a surface spring is commonly too short for re-equilibration to occur, allowing the water to retain some chemical "memory" of its deep, high-temperature environment.

The geometry of geothermal anomalies is estimated by various geophysical methods based on the relationship between temperature and the electrical properties of rock; conductivity increases with rising temperature and the presence of intergranular brine. Deep thermal structure can be investigated using active or passive telluric methods that measure earth currents induced by an external field, whereas the shallow thermal regime can be measured directly by d.c. resistivity.

Geochemical and geophysical models are tested by direct temperature measurement in shallow (100 to 500 m) diamond drill holes, providing temperature gradients that may be extrapolated to greater depth.

Table 20.2. Estimated subsurface temperatures for 58 thermal springs in the western Canadian Cordillera (for locations see Fig. 20.12)

Spring No.	Na-K-Ca T°C	SiO$_2$ T°C	Spring No.	Na-K-Ca T°	SiO$_2$ T°C
1	-15.4	50.7	30	28.5	109.3
2	16.8	92.9	31	46.3	119.4
3	-	-	32	49.0	67.4
4	94.0	70.6	33	48.7	77.9
5	0.1	26.5	34	153.7	92.6
6	162.3	73.8	35	86.7	141.7
7	89.2	91.8	36	-16.2	113.7
8	31.2	103.8	37	88.0	158.8
9	28.4	100.1	38	29.0	66.2
10	-20.6	41.9	39	172.5	168.1
11	-31.2	13.0	40	227.3	177.0
12	17.3	38.1	41	97.5	119.4
13	19.6	48.1	42	82.8	111.1
14	5.1	76.7	43	146.5	128.3
15	-3.4	53.1	44	88.6	112.8
16	-4.2	53.1	45	98.3	141.2
17	11.6	86.0	46	99.1	106.5
18	20.1	46.0	47	48.7	97.2
19	15.9	80.9	48	48.7	138.5
20	79.1	87.2	49	72.3	147.8
21	-2.9	65.4	50	187.0	129.0
22	67.4	55.4	51	93.0	136.8
23	5.2	67.4	52	228.3	186.7
24	24.7	65.1	53	59.1	123.2
25	26.3	98.3	54	69.5	117.0
26	78.7	117.0	55	38.2	129.6
27	54.3	104.5	56	89.7	141.2
28	37.7	102.5	57	169.2	145.3
29	73.9	128.3	58	167.4	143.3

Calculations based on: Fournier and Truesdell (1973); Fournier and Rowe (1966).

Present status of exploration

Water from most of the known thermal springs in the western Cordillera (Fig. 20.12) has been analyzed and estimates of subsurface temperature calculated by various geochemical thermometers (Table 20.2). Springs located outside the belts of Neogene volcanic activity appear to be discharging from hydraulically driven, isothermal systems related to deep circulation of meteoric water through porous fault zones and aquifers. Such systems are not believed to have potential for power production. The chemical geothermometry of water from springs associated with Neogene and younger volcanic centres predicts higher subsurface temperatures. This, plus relatively high concentrations of sulphur, boron and other elements of magmatic origin, suggests an association with convective hydrothermal systems driven by heat from subvolcanic intrusions.

Garibaldi Volcanic Belt

Most of the exploration for geothermal power sources in British Columbia has focused on the Garibaldi Volcanic Belt because of its proximity to large population centres in southwestern British Columbia. Meager Mountain near the northern end of the belt, is a composite andesite-dacite-rhyodacite dome with a volume of about 15 km^3 (Lewis and Souther, 1978; Read, 1979). The oldest rocks were erupted 1.9 Ma ago on the south side of the complex. Subsequent activity migrated progressively northward, culminating with the eruption of a dacite pumice plume (Bridge River ash) from a vent on the north flank 2490 ^{14}C years ago. Pebble Creek hot springs on the north and Meager Creek hot springs on the south issue at about 60°C. Geochemical thermometers predict subsurface temperatures of about 140°C but the chemistry of the two springs is distinctively different, suggesting that they are related to two separate flow systems (Clark et al., 1982). This is further indicated by the presence of distinct resistivity anomalies (25 to 30 ohm-m in a background of 200 to 1200 ohm-m) that are believed to define separate north and south reservoirs. A thermal gradient of over 200°C/km was obtained from a shallow borehole in the north reservoir and diamond drill holes in the south reservoir record gradients as high as 750°C/km and bottom-hole temperatures up to 202°C. Extensive exploration of the south reservoir by B.C. Hydro and Power Authority included the drilling of three large-diameter rotary wells 3000 to 3500 m deep, which encountered temperatures ranging from 233° to 264°C in granitic basement rocks beneath the volcanic complex (Moore et al., 1983). Low porosity and permeability in two of the holes precluded any sustained flow of steam but the third, (MC-1) produced a mixed flow of steam and water from a porous fault zone dipping beneath the northern margin of the volcanic complex.

South of Meager Mountain, in the central Garibaldi Belt, Mount Cayley, with an edifice volume of about 2 km^3, is a multiple plug dome of mainly dacite and rhyodacite (Souther, 1980; Souther and Dellechaie, 1984). An early central dome-building phase (3.8 Ma) was succeeded by the emplacement of two satellitic domes and effusion of andesite flows between 1.1 and 0.11 Ma. Each of these satellitic centres is associated with thermal springs and distinct resistivity anomalies. Chemical geothermometers from spring and borehole waters are inconsistent, predicting

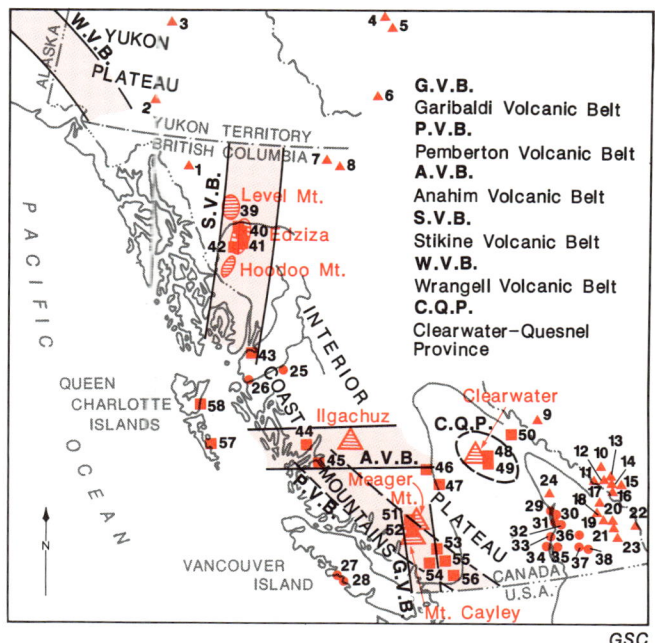

Figure 20.12. Locations of the principal thermal springs of western Canada with respect to Late Neogene and Quaternary volcanic belts. Class I springs are associated with deep flow systems in layered carbonate rock; Class II springs issue from fractures in granitic or metamorphic rocks of nonvolcanic regions; Class III springs are located in or near belts of Neogene and younger volcanic rocks.

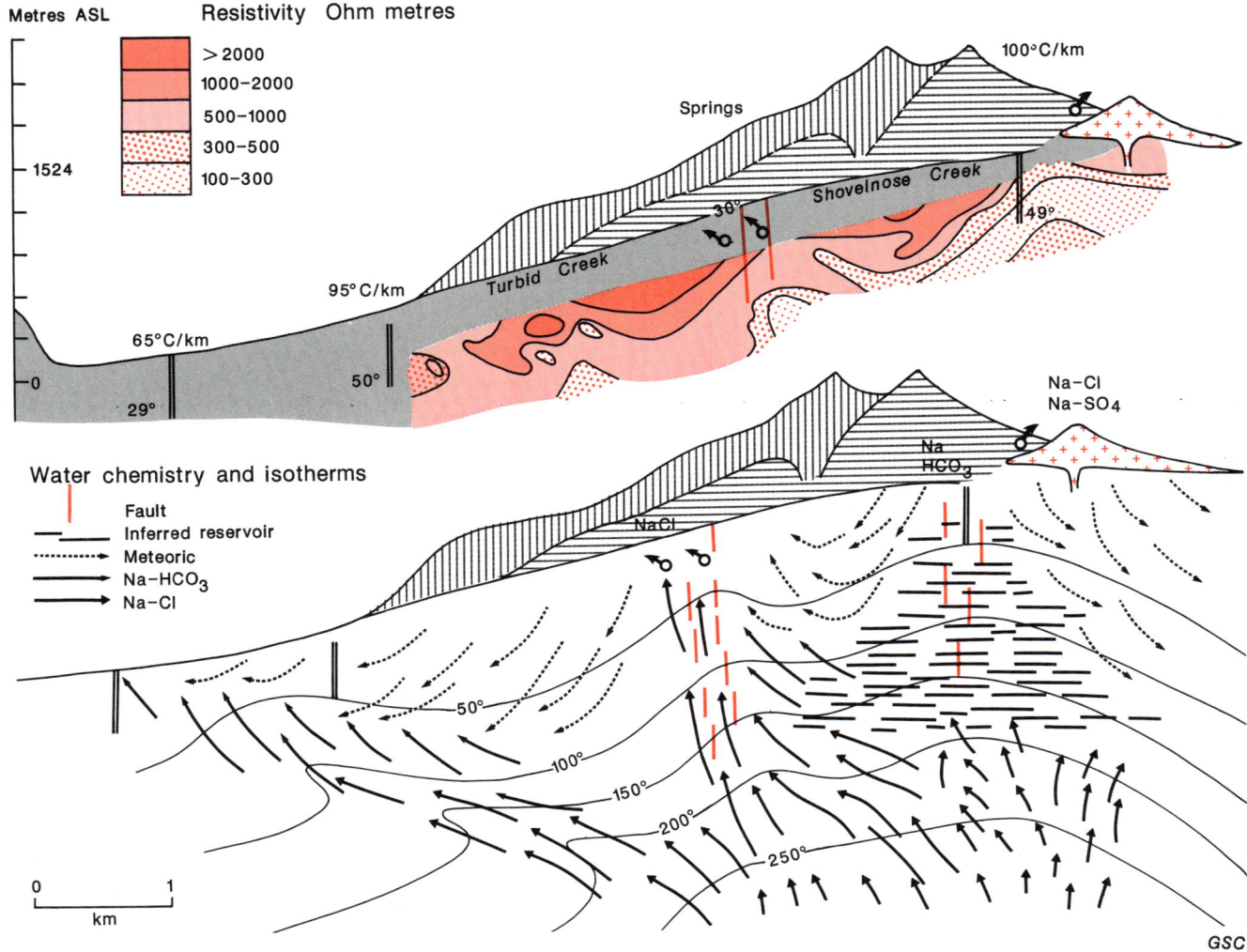

Figure 20.13. Schematic cross-section showing an interpretation of the thermal structure and flow regime beneath Mount Cayley, based on resistivity, drilling and geochemical surveys.

reservoir temperatures from 67° to 166°C. Temperature gradients from six diamond drill holes range from 48° to 105°C/km. An interpretation of exploration data from Mount Cayley (Fig. 20.13) suggests that a reservoir with temperatures above 150°C may be present at moderate depth.

Geophysical work and exploration drilling have not been undertaken elsewhere in the Garibaldi Belt but the juxtaposition of thermal springs with silicic volcanic piles suggests that geothermal reservoirs may be associated with Cauldron Dome, near Mount Cayley, and with the Mount Silverthrone complex in the central Coast Mountains (see Fig. 10.7). The latter is a dacitic cauldron subsidence complex, at least 16 km in diameter, that has been active within the last one million years. Its large size and the presence of thermal springs and extensive, potentially porous breccia make it an attractive target.

Anahim Volcanic Belt

Direct evidence of a convective hydrothermal system in the Anahim Belt has not been established but a broad thermal anomaly is indicated by the presence of hot springs in the western part of the belt (Souther, 1986). These appear to be associated with alkaline, subvolcanic plutons that have been exhumed in the deeply dissected Coast Mountains. Geochemical thermometry on the spring waters yields scattered results but the sulphate-water geothermometer predicts subsurface temperatures from 97° to 137°C for the Nascall, Eucott Bay and Talheo flow systems (Clark, pers. comm., 1985). A 400 m borehole in one of the subvolcanic plutons near Bella Bella yielded a linear, relatively high conductive gradient of 32.2°C/km (Nevin Sadlier-Brown Goodbrand Ltd., pers. comm., 1984).

No thermal springs are known in the central Anahim Belt but outflows of thermal water beneath the shield volcanoes would probably be dispersed by the outwardly dipping sheets of impervious lava interlayered with porous scoria (Souther, 1984). Magnetotelluric soundings in the Ilgachuz Range indicate the presence of a conductive subvolcanic zone. This, plus the presence of a caldera structure suggest that the central Ilgachuz may be underlain by a thermal reservoir.

Clearwater-Quesnel volcanic province

Volcanism in the Clearwater-Quesnel area culminated with the eruption of several basaltic Quaternary centres (Hickson and Souther, 1984). The area contains a few seeps of warm water but these show no direct relationship to the young cinder cones. However, cones in the Spanish Creek area are associated with phreatic explosion-pits, indicating that the eruption was accompanied by extensive groundwater heating. Moreover, the area has an anomalously high level of seismic activity and s-waves passing beneath the volcanic area are preferentially attenuated (G.C. Rogers, pers. comm., 1986). Thus, both the geological and geophysical data support the concept of a thermal anomaly, possibly a magma chamber, beneath part of the Clearwater-Quesnel volcanic province. On this basis it is considered to have some geothermal energy potential.

Stikine Volcanic Belt

The larger centres of the Stikine Volcanic Belt, Level Mountain, Mount Edziza and Hoodoo Mountain comprise bimodal, basalt-rhyolite assemblages that appear to be the product of crystal fractionation in crustal magma reservoirs (Souther, 1975; Souther and Hickson, 1984). They have been episodically active throughout Neogene time and into Quaternary time and the presence of associated thermal springs suggests that the central part of the belt is underlain by a broad thermal anomaly that may locally contain hot subvolcanic intrusions. An analysis of thermal data in the Stikine region (Jessop et al., 1984) suggests that a modest heat flow of 72mW/m^2 at Buckley Lake, west of Mount Edziza, may be explained by an accumulation of volcanically transported heat. The gradient at Buckley Lake is 33°C/km and 12 km southwest of Mount Edziza a gradient of 28°C/km was measured in a mineral exploration hole at Schaft Creek. The thermal gradient beneath the volcanic complexes is probably much higher. Hotsprings on the Edziza Complex discharge near the base of the volcanic pile and yield geochemical temperature estimates of about 73°C. An interpretation of the spring water chemistry by Piteau and Associates (1988) suggests that the flow systems have penetrated no more than 1000 m below the surface. Accordingly, based on the geother-mometer estimates, the geothermal gradient under the volcanic complex would be about 70°C/km. The geothermal energy potential of the area is predicated on the reasonable assumption that the shallow flow-systems, feeding existing springs, are underlain by deeper convective systems that are responsible for the high thermal gradients. Communication between the deep and shallow flow-systems is probably inhibited by the presence of an impervious zone, sealed by the precipitation of secondary minerals. Impervious cap-rocks are a common component of producing geothermal fields elsewhere (Rybach and Muffler, 1981) but confirmation of similar deep thermal reservoirs in the Stikine Volcanic Belt must await future geophysical surveys and ultimately exploration drilling.

Resource assessment

Estimating the geothermal energy potential of the Canadian Cordillera from the scant data available is necessarily speculative. However, the exercise can provide some reasonable upper limits on future energy budgets that may be expected from this resource. The scheme followed here is adapted from the system the United States Geological Survey uses which estimates the temperature and volume of each target area, calculates the total heat content or "resource base", and estimates the amount of recoverable energy (White and Williams, 1975). The parameters used in calculating the "resource base" in Canada are derived from experience at Meager Mountain, which is probably representative of other volcanic targets in the Canadian Cordillera (Fig. 20.14).

Resource base

The surface area of each anomaly is assumed to be equal to the area of the related volcanic edifice and its effective depth equal to the economic limit of drilling (3 km). Heat within the anomalous volume is distributed among four different regimes: (1) 30% hot dry rock with 0 permeability; (2) 20% high-temperature hydrothermal (250°C) with porosity 0.1; (3) 30% low-temperature hydrothermal (150°C) with porosity 0.2; (4) subeconomic heat (<100°C). Using the above criteria and a temperature drop (delta T) of 100°C the resource base for each 1 km^2 of surface anomaly is 25.7 megawatt-years distributed between the four thermal regimes as shown in Figure 20.14.

Resource recovery

The assumptions are made that none of the hot-dry-rock or very-low-grade (<100°C) heat is recoverable, that the high-temperature hydrothermal resource is recovered by a conventional low-pressure steam turbine, and that the low-temperature hydrothermal resource is recovered using a binary-cycle system capable of operating in the 150°C range.

The proportion of heat in the resource base that can be removed depends largely on the permeability of the reservoir rock. In an ideally permeable medium the recovery factor for thermal energy is about 50%. In the case of fracture permeability only that heat close to the fractures

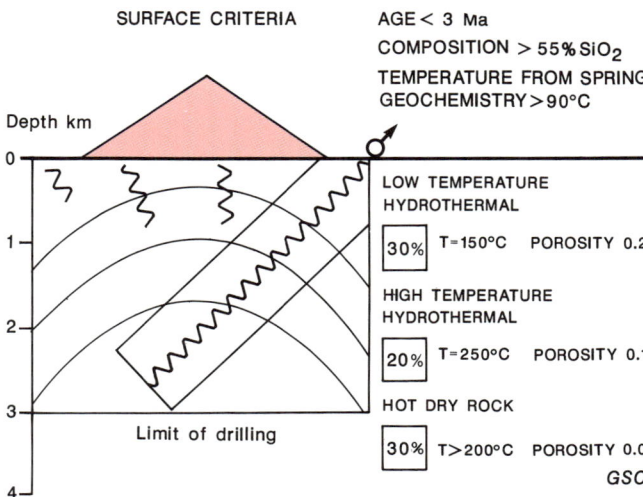

Figure 20.14. Model showing assumed distribution of heat among four thermal regimes underlying a geothermal anomaly related to a volcanic centre in western Canada.

can be conducted to the transporting fluid during the life of the reservoir and the recovery factor decreases with decreasing fracture density. Extrapolating from Meager Mountain, and the known surface geology of other centres, the type of permeability to be expected in the volcanogenic reservoirs of western Canada is probably due to fracturing of otherwise impervious rock. A recovery factor of 0.2 is herein assumed for the low temperature resource, which is commonly at lesser depths in more highly fractured rock, and 0.1 for the high temperature resource in less fractured rock. A further 50% energy loss occurs during the conversion of thermal energy to electricity (Fig. 20.15).

Resource estimate

The total recoverable geothermal energy in each of the three Neogene volcanic belts of western Canada has been calculated using the above criteria (Table 20.3). If the resource was produced at a rate of 1000 megawatts the high-temperature-hydrothermal systems would be depleted in about 50 years and the low-temperature-hydrothermal systems would last about three times as long. To put this in perspective the above hypothetical 1000 MW extraction rate assumed for western Canada compares to a total 1981 world generating capacity of 3300 MW from geothermal resources. Total energy consumption in British Columbia in 1981 was nearly 40 000 MW of which about 8000 MW was in the form of electricity. Thus, while the geothermal potential is dwarfed by total energy consumption it could make a significant contribution to the electrical component.

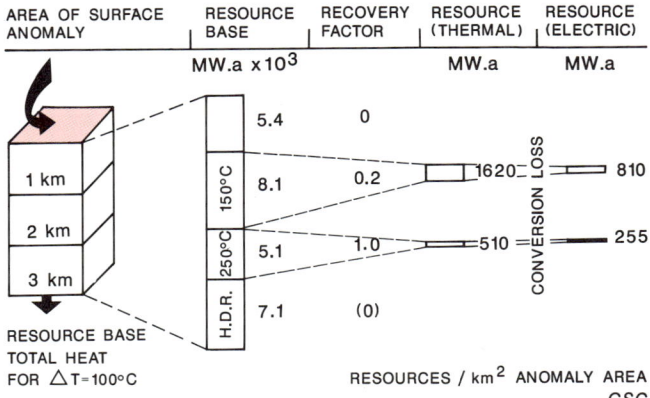

Figure 20.15. Schematic diagram showing the amount of recoverable energy that may be extracted from a 3 km³ resource base (1 km² anomaly) having the thermal structure depicted in Figure 20.14.

Table 20.3. Estimated recoverable energy from geothermal anomalies in the Neogene and younger volcanic belts of western Canada. For locations of volcanoes associated with anomalies see Chapter 10, Neogene Assemblages

Volcanic Belt	Anomaly Name (area in km²)		Recoverable energy (megawatt-years)	
			High Temperature	Low Temperature
Garibaldi	Meager Mtn. North	(18)	4 590	14 580
	Meager Mtn. South	(20)	5 100	16 200
	Mount Cayley	(16)	4 080	12 960
	Cauldron	(6)	1 530	4 860
	Mount Silverthrone	(20)	5 100	16 200
	SUBTOTAL		20 400	64 800
Anahim and Clearwater	Ilgachuz Range	(12)	3 060	9 720
	Clearwater	(12)	3 060	9 720
	SUBTOTAL		6 120	19 440
Stikine	Mount Edziza			
	Elwyn	(12)	3 060	9 720
	Taweh	(20)	5 100	16 200
	Mess	(25)	6 375	20 250
	Level Mountain	(10)	2 550	8 100
	Hoodoo Mountain	(16)	4 080	12 960
	SUBTOTAL		21 165	67 230
TOTAL ESTIMATED RECOVERABLE ENERGY			47 685	151 470

PART E. GROUNDWATER

E.C. Halstead

Hydrogeological environment

The geology and related topography of the Cordillera are important in the initiation and subsequent distribution of precipitation, the gathering of the resultant runoff into river systems, and the recharging of groundwater flow systems. For example, on some windward mountain slopes in British Columbia and the southwestern Yukon Territory, annual precipitation exceeds 3200 mm over large areas; in contrast, leeward slopes and interior valleys in the same region receive annual totals of less than 400 mm. Because the climate is affected by topography, great variations in the amount of precipitation are experienced due to differences in elevation, slope aspect, latitude and inland or coastal locations. These lead to climatic extremes expressed by continuous ice fields in the higher ranges, near-desert conditions in interior valleys, and discontinuous or continuous permafrost at higher latitudes and altitudes.

Precipitation is the primary source of all surface water and groundwater, and in a broad sense it is the result of the dominant types of air mass (Fig. 20.16). Within the coastal zone, the flow of air from the Pacific Ocean maintains relatively mild annual temperatures and wet winters due to frequent mid-latitude cyclonic storms. Polar continental and Arctic air masses commonly dominate in the northern regions and extend into interior valleys bringing colder temperatures and modest precipitation. In general, the mountains of the Insular Belt and the western flank of the Coast Mountains are wet zones, separated by a somewhat drier coastal trough. East of the Coast Mountains, the Interior Plateau and associated valleys are relatively dry and are bounded on the east by the interior wet belt, within which lies the less humid zone of the Rocky Mountain Trench. These contrasts have many implications in regard to ground and surface-water distribution as well as runoff patterns. Considering the Cordillera as a whole, about 89% of precipitation gathers as runoff (Canada Water Year Book, 1978-1979) but significant local departures are apparent. Within the Lillooet Basin of the Coast Mountains, 61% of the total annual precipitation is released from the basin as surface water runoff and 17% enters the groundwater system (Jamieson and Freeze, 1983). Lawson (1968) estimated recharge to local flow systems in the Trapping Creek basin of the Okanagan Highlands to be in the order of 54% of the total annual precipitation of 559 mm.

The potential for recharge via fracture zones in the Cordillera is high, especially in the Coast Belt where rainfall ranges from 800 mm to more than 2400 mm annually. For the past decade, water-levels in a number of unlined observation wells up to 200 m in depth, in plutonic and sedimentary rocks, have been monitored for seasonal hydraulic responses (Kohut et al., 1983). The results indicate that groundwater levels respond cyclically on a seasonal basis to climatic variations. In general, recharge occurs with a rise in water levels responding to fall and winter precipitation. A seasonal maximum is reached by mid-April and water levels decline during the drier summer and early fall months to reach a seasonal minimum between October and December. In groundwater recharge areas (topographic highs) recorded water levels fluctuate 2 to 6 m, showing asymmetric recovery and recession curves. In discharge areas (topographic lows) and deeper portions of the groundwater flow systems (greater than 75 m in

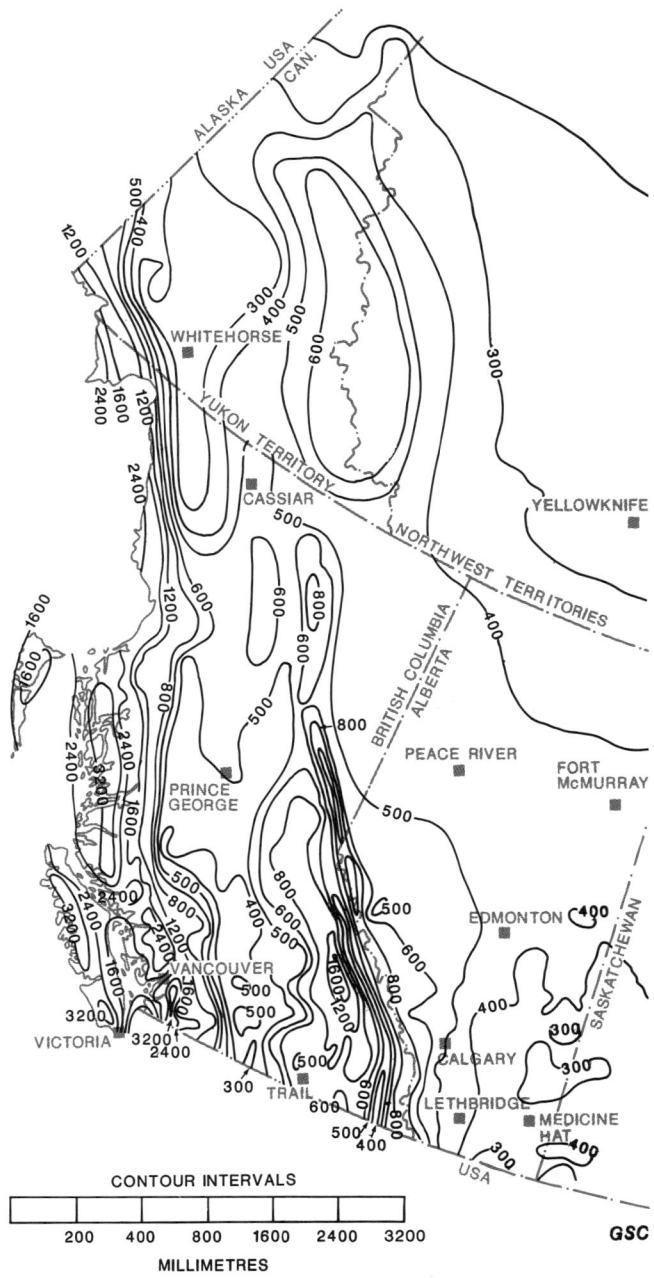

Figure 20.16. Annual precipitation rates in the Cordilleran region. Values are in millimetres per annum, based on the 30-year period 1941-1970.

depth), water levels fluctuate from 1 to 2 m showing nearly symmetrical recovery and recession curves. Seasonal maximum and minimum water levels in discharge areas lag several weeks behind the corresponding inflection points observed in recharge areas. Groundwater recharge may be significantly reduced during periods of below-normal precipitation but it is generally also limited by aquifer porosity and fracture conductivity. Similarly, during periods of excessive precipitation, recharge may not be enhanced due to physical limitations of the aquifers.

Water-level fluctuations monitored in observation wells penetrating water-table aquifers within surficial deposits of the Coastal Lowland show a significant rise often within hours of a heavy (10 mm) cumulative precipitation event. This type of water-level rise in part may be the result of air entrapment in the unsaturated zone where air pressures build up to values much greater than atmospheric (Freeze and Cherry, 1979). Confined aquifers are full of water, overlain by aquitards or less permeable materials, and water-level fluctuations respond only to changes in atmospheric pressure.

Groundwater in bedrock

The hydrogeological characteristics of several kinds of bedrock terranes are herein presented without reference to their age or origin.

Igneous plutonic rocks

The igneous plutonic rocks of the Cordillera are massive and generally homogeneous; hence intercrystalline porosities are rarely greater than 2%. A feature common to most plutons in the Coast Belt is the ubiquitous presence of well-developed joint systems. Roddick (1965) found, that as a rule, hornblende-rich rocks show closer spaced joints than biotite-rich plutonic rocks. Joints and fractures commonly extend only a few tens of metres but locally as much as a few hundred metres, below the ground surface. Major valleys are developed along large faults and shear zones. In prospecting for water supplies it is therefore likely that drilling at or near the valley bottoms will be more productive than drilling at or near the crests of hills or mounds.

The behaviour of groundwater in fractured media in the Insular Belt was investigated by Kohut et al. (1983), by monitoring the effects of pumping from two irrigation wells in plutonic rocks on southern Vancouver Island. The two wells were 200 mm in diameter, 67 and 179 m in depth, and were pumped at rates of 10.3 and 12.3 L/s for periods of 70 and 65 days, respectively; water levels were monitored in neighbouring wells in bedrock.

Pumping from these large capacity wells produced trough-like cones of depression of marked linearity (line sinks), which extended more than 1500 m from the pumped wells. The longitudinal axes of the drawdown troughs did not parallel the strike of regional fractures, but developed along trends coinciding with the locus of intersections of the two prominent bedrock joint sets, and bisected the angles between them. The drawdown troughs extended away from the pumped wells toward local recharge areas, in the direction of increasing regional potentiometric head. This observation in part may be misleading because the number of observation wells downgradient from the pumping wells was too small to enable delineation of the full extent and shape of the drawdown troughs. It is likely that the development of drawdown troughs in fractured bedrock is controlled by orientation, spacing, dimensional characteristics, degree of interconnection, and permeability of the fractures, in addition to the regional potentiometric gradient. The occurrence of the line-sink phenomena indicates that fracture anisotropy has a pronounced effect on the magnitude and direction of drawdown development around a pumped well in granitic terrane.

Folded and faulted sedimentary rocks

The main populated areas underlain by folded and faulted sedimentary rocks include all of the Foreland Belt, and a belt of similar rocks along the east side of Vancouver Island, extending southward from Campbell River to the Gulf Islands.

In hydrological terms the Foreland Belt can be regarded as consisting of a number of separate cells in which surface runoff plays a major role, with a smaller contribution by groundwater movement (Cherry et al., 1972). Transfer of subsurface water from one cell to the next either does not occur or is very limited in scale and dependent upon favourable structure and permeability. In several instances, circulation is both deep and fast enough to result in the discharge of geothermally heated water with high SO_4^{2-}.

Cherry et al. (1972) observed that the Rocky Mountains and Foothills could be divided into four hydrogeological zones characterized by the average slopes of their valleys' flanks. The alpine zones, with slopes greater than 16°, are the sites of numerous springs and seeps with highly variable yields and temperatures, quickly responding to precipitation. Groundwater quality is characterized by extremely low (100 mg/L) content of dissolved solids and by a predominance of Ca^{2+}, Mg^{2+} and HCO_3^-. Recharge areas are commonly devoid of vegetation or support only stunted growths of trees, shrubs and grasses, whereas discharge areas support lush phreatophytic vegetation.

Slopes between 4° and 16° display phenomena similar to those observed in alpine zones; spring flows and temperatures are less variable, however, and total solids concentrations are slightly higher. Flow systems in these zones appear to be less dynamic than in the alpine zone. The flow systems are deeper, more extensive and less subject to variations.

In the hilly zone, with slopes between 1° and 4°, springs are numerous, and discharge rates can be high (many hundreds of litres per minute) and steady. The major ions in the water are Ca^{2+}, Mg^{2+} and HCO_3^-, with minor amounts of Na^+ and SO_4^{2-}. Dissolved solid concentrations are in the range of 100 to 600 mg/L. The typical flow system is active, moderately deep (30 to 120 m), and moderately long and steady.

In the rolling type of hydrogeological zone, slopes range from 0.25° to 1°, spring discharge is generally 45 L/min or less, but is steady all year round. Dissolved solids concentrations are up to 800 mg/L. Na^+, SO_4^{2-} and Cl^- are important components of the groundwater. The flow systems are moderately deep, long and slow and are only moderately responsive to short-term changes in climate.

In the Nanaimo Group of eastern Vancouver Island and the adjacent Gulf Islands groundwater moves largely through fracture systems; the hydrogeological significance of the northwesterly trending folds and faults is not known. The widespread occurrence of groundwater in bedrock indicates that most groundwater motion is associated with joint systems rather than with discrete faults. Wells yield up to 4 L/s. Water quality is variable, sodium being the dominant cation in all water samples; highly mineralized groundwater is commonly found locally near faults. Dakin et al. (1983), investigating the groundwater regime of Mayne Island, found from isotope data that all groundwater on the island is of recent meteoric origin, and that local flow systems prevail, with no evidence of discharge from deep regional systems. The composition of the groundwater ranges from less than 500 mg/L dissolved solids content to more than 13 000 mg/L, with sodium and chloride the predominant ions. The salt content of the groundwater can be accounted for by the slow release, through molecular diffusion, of Na^+ and Cl^- from the low-permeability shales which are intersected by the fracture systems through which the groundwater flows. Under the present hydrological regime on the island, intrusion of ocean water and upward flow of brine from deep zones within the sedimentary basin are believed to be relatively unimportant contributors of salt to the groundwater flow systems. However, on Saltspring Island at least six saltwater seeps occur of probable deep but otherwise unknown origin.

Flat-lying and gently dipping lava flows

The Chilcotin Group lavas of the Intermontane Belt potentially constitute one of the more important bedrock aquifers in the Cordillera. Their internal structure allows relatively free circulation of groundwater, and springs are commonly found at the base of the lava flows. Because the lavas outcrop in a relatively dry zone where annual precipitation ranges between 30 and 50 cm (Fig. 20.17), available recharge is limited. Well yields of up to 1.6 L/s can be obtained from these rocks in some areas; a yield of 12 L/s has been recorded from one well.

Groundwater in surficial deposits

Unconsolidated surficial deposits overlying bedrock are the major sources of groundwater supply in the Cordillera. The geology of the surficial deposits is complex and varied, and the most important occurrences are in modern river valleys where they may reach thicknesses of several hundred metres. Large-capacity wells have been developed near rivers (Table 20.4).

Coastal Lowland

The Coastal Lowland (Fig. 20.18) includes the Fraser Lowland and low relief areas adjacent to the Strait of Georgia. Within this region glaciogenic deposits are divided into hydrostratigraphic units, some of which act as confining layers retarding groundwater movement.

In the Fraser Lowland, large areas are covered by fluvial sand and gravel deposited as outwash, deltaic or fan deposits. These sediments constitute water-table aquifers that are recharged annually by winter precipitation; during the drier summer months their discharge maintains the reduced flow of superimposed streams. Through many municipal, industrial and irrigation wells these aquifers commonly produce yields of 100 L/s or more. Due to short residence times within the aquifers, water qualities are good with dissolved-solid contents of less than 100 mg/L. Confined aquifers consisting of fluvial sand and gravel underlying stony clays, or in association with diamictites, also are present at greater depths. In these cases sodium and bicarbonate are the chief constituents of the water in which the dissolved solids can amount to 250 mg/L. Interconnected lenses of coarser material contained within silts, clays and fine sands that fill valleys to depths of more than 300 m, commonly yield flowing artesian groundwater. These deeper confined aquifers yield groundwater with dissolved solids as high as 1000 mg/L, of which sodium and chloride are the principal constituents. Because of the considerable demand for water imposed by agriculture, industry and municipalities in the Fraser Lowland, groundwater is an important resource; consumption during 1981 amounted to more than $29.0 \times 10^6 m^3$ (Halstead, 1986).

On the east coast of Vancouver Island, one of the more important surficial aquifers is a well-sorted, glaciofluvial, fine to coarse sand with minor gravel and silt, commonly exposed in wave-cut cliffs. Recharge occurs during midwinter and water qualities indicate dissolved solids in the range of 150 mg/L. Pumping tests of several municipal wells revealed transmissivity values (the product of hydraulic conductivity and aquifer thickness) slightly over 100 m^2/day, and hydraulic conductivities (a measure of the

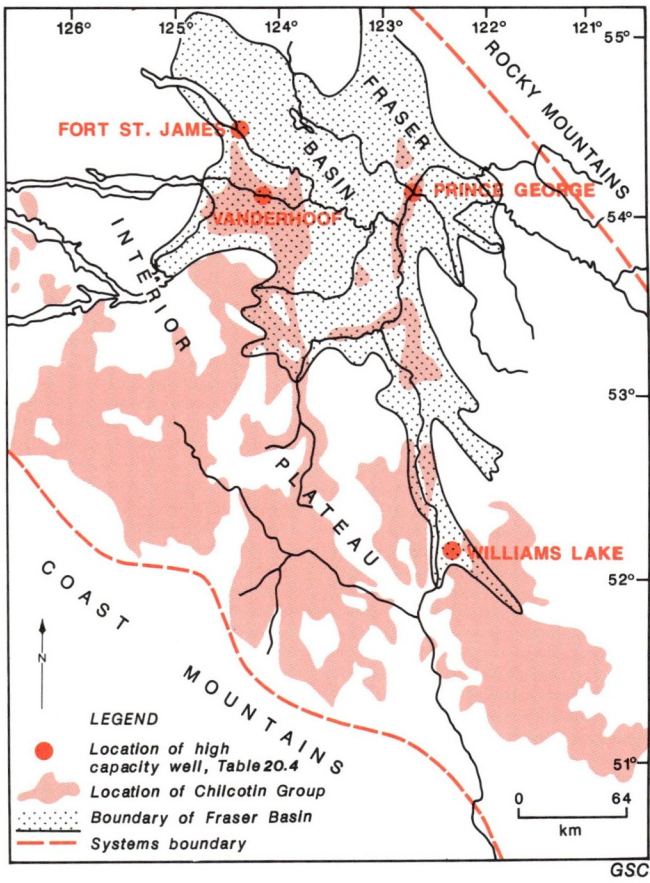

Figure 20.17. Capacity wells and distribution of the Chilcotin Group in the Interior Plateau, central British Columbia.

Table 20.4. High capacity wells in surficial deposits

Topographic Feature	Location	Aquifer (Lithology)	Average Well Depth(m)	Yield L/s	Transmissivity m²/s	Facility or Use
Coastal Plain	Duncan	unconfined sand and gravel	21	126	1.4×10^{-1}	Municipal Well
	Cassidy	unconfined sand and gravel	20	1264	7.0×10^{-2}	Industrial
	Parksville	sand in part confined	20		1.1×10^{-3}	Collector Wells, Municipal Wells and Springs
	Abbotsford	confined sand and gravel	80	158	4.9×10^{-2}	Trout Hatchery
	Inches Creek	unconfined alluvial sand and gravel	21	95	2.7×10^{-1}	Salmon Hatchery
Similkameen Valley	Keremeos	unconfined glaciofluvial sand and gravel	42	109	1.4×10^{-1}	Irrigation
Okanagan Valley	Armstrong	confined sand and gravel	350	31.5	2.2×10^{-3}	
	Okanagan Falls	glaciofluvial sand and gravel	128	75	2.5×10^{-3}	Municipal
Highland Valley	Copper Mines	confined glaciofluvial sand and gravel	90	31.5		Industrial
Columbia River Valley	Castlegar	confined glaciofluvial sand and gravel	44	139	1.7×10^{-1}	Municipal
Purcell Trench	Creston	unconfined alluvial sand and gravel	25	58		Municipal
	Cranbrook	unconfined alluvial sand and gravel	15	57	2.8×10^{-2}	Municipal
Southern Rocky Mountain Trench	Bull River	unconfined alluvial sand and gravel	22	126		Fish Hatchery
Interior Plateau	Williams Lake	confined glaciofluvial sand and gravel	60	126	8.9×10^{-2}	Municipal
	Prince George	unconfined fluvial sand and gravel	30	802	1.3×10^{-1}	Municipal Radial Collector Wells
and	Vanderhoof	confined glaciofluvial sand and gravel	185	70	1.4×10^{-2}	Municipal Well
Fraser Basin	Fort St. James	confined glaciofluvial sand and gravel	100	265	1.4×10^{-2}	Municipal Well
Liard Plain	Watson Lake	unconfined alluvial and glaciofluvial sand and gravel	21	64	1.4×10^{-3}	Municipal
	Whitehorse	alluvial sand and gravel	19	80	5.1×10^{-2}	Riverdale Water Supply
Yukon	Whitehorse	alluvial sand and gravel	10	4	8.3×10^{-3}	Porter Creek Water Supply
River	Whitehorse	confined alluvial sand and gravel	67	30	5.1×10^{-2}	Nursery Well
Valley	Dawson	alluvial sand and gravel	7	126		Municipal

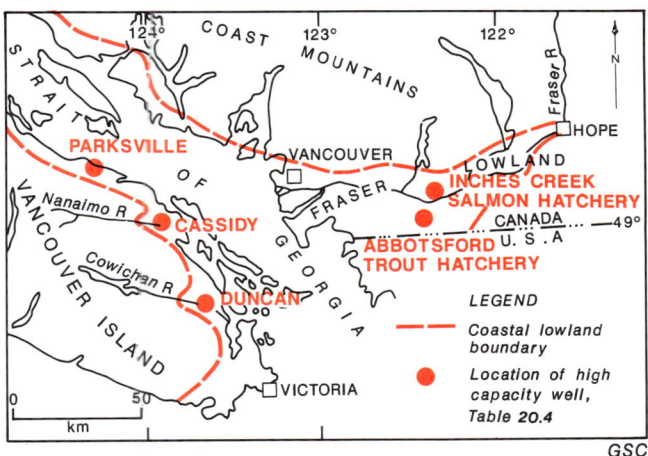

Figure 20.18. High capacity wells in the Coastal Lowland-Fraser Lowland area.

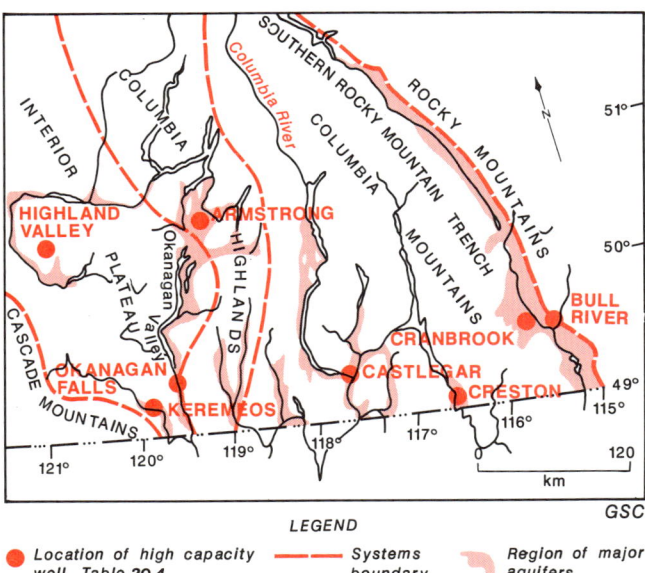

Figure 20.19. Map of the Interior Valleys showing the regions of major aquifers and the location of high capacity wells.

ease with which an aquifer will permit the movement of groundwater) of about 10^{-5} m/s.

Terraced fluvial and glaciofluvial deposits of coarse sand and gravel overlying till and occupying river valleys also constitute major water-table aquifers. Many examples are found near the Cowichan and Nanaimo rivers. The Cassidy aquifer underlies an area of about 8.6 km^2 within the Haslam Creek valley, a tributary of Nanaimo River. It supplies 1264 L/s for boiler feed at a nearby pulp mill. Along a stretch of the Cowichan River, 0.6 km long and 60 m wide, seven high-capacity wells produce yields ranging from 90 to 120 L/s; aquifer transmissivities range as high as 1.4×10^{-2} m/s.

Interior valleys

The Okanagan valley, in Canada, has a drainage area of 7700 km^2 (Fig. 20.19). The average annual precipitation over the basin is approximately 558 mm, with considerable losses to evaporation and evapotranspiration; 25 mm/year is estimated to be available for groundwater recharge (LeBreton, 1972), amounting to 5.4×10^6 m^3/a. Investigation of the groundwater potential of the north end of the valley in and near Armstrong revealed a valley fill sequence of silt and fine sand more than 550 m thick, with more permeable coarser materials constituting confined aquifers. The aquifers occur at depths ranging from 212 to 365 m, and yield 2.5 to 20 L/s. Water quality is characterized by up to 500 mg/L dissolved solids and the dominant ions include calcium, magnesium and bicarbonate.

Water-table aquifers in the upper part of the surficial deposits are of smaller areal extent and consist of alluvial fan deposits of sand and gravel near the valley sides. The unconfined O'Keefe aquifer occupies a tributary valley and consists of sand and gravel up to 150 m thick which supplies more than 75 L/s to individual wells.

Valley-fill deposits extending south from Skaha Lake to the Canada-U.S.A. International Boundary form an unconfined aquifer of coarse sand and gravel with exposed water tables forming lakes in the pitted outwash. These materials are recharged by the Okanagan River following snowmelt in the adjacent uplands. Irrigation and municipal wells can be developed to provide upwards of 75 L/s.

In the Columbia River valley, Purcell Trench and Southern Rocky Mountain Trench, two stratified glaciofluvial sequences, each separated by till, constitute the major aquifers. Recharge occurs annually following spring snowmelt and precipitation in the adjacent uplands. In some places, the confined aquifers accommodate long flow systems that yield artesian water with temperatures somewhat higher than the average annual air temperatures within the region.

The Fraser River Basin (Fig. 20.17) within the Intermontane Belt has considerable groundwater potential. The principal aquifers consist of fluvial and glaciofluvial sand and gravel occupying buried river channels cut into older tills or bedrock. These aquifers are overlain by lacustrine sediments deposited in ponded lakes that occupied the basin during deglaciation.

Many valleys of the Yukon Territory (Fig. 20.20) commonly contain two potential aquifers. The first aquifer is shallow modern alluvial sand and gravel associated with floodplains of the existing rivers, and the second is a deeper aquifer of glaciofluvial sand and gravel, and lenses of sand and gravel within glacial till, that fill the valleys. In some valleys, such as the Shakwak Valley, drilling has encountered coarse aquifer material underlying till at depths of more than 150 m.

Permafrost

Groundwater in permafrost areas differs in its distribution and hydrology from that in warmer climates because of the presence of frozen ground with very low permeability. In the discontinuous permafrost zone groundwater may be obtained from unfrozen zones. In the area of continuous permafrost, development of groundwater supplies

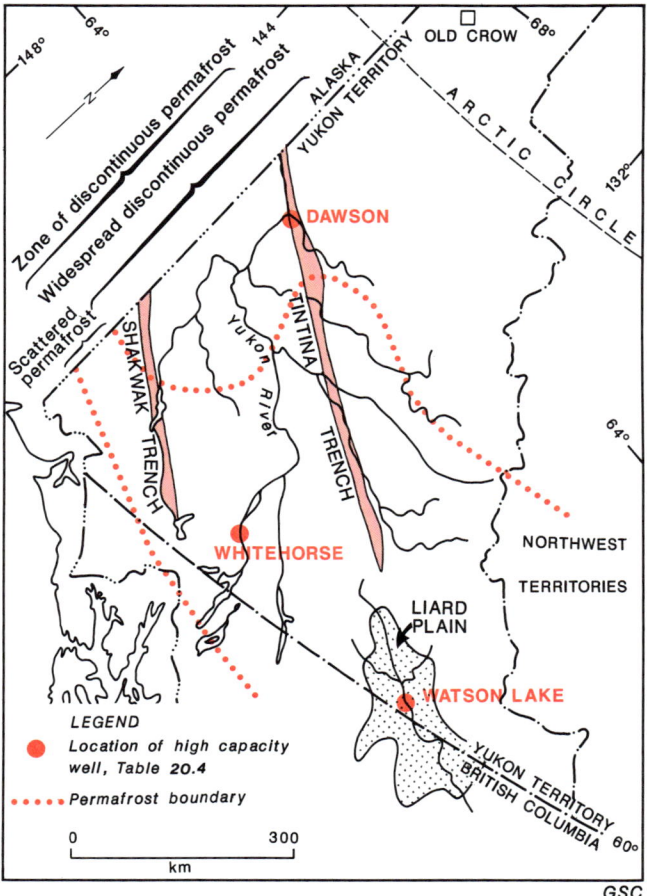

Figure 20.20. Map of the Yukon Territory showing the location of high capacity wells and permafrost boundaries.

is frequently impractical and may be impossible where the permafrost extends into bedrock.

The main effect of permafrost on the hydrology of an area is that it restricts the movement of groundwater. In regions of low relief this results in numerous lakes and swamps. Conversely, in areas of high relief runoff is often rapid due to lack of infiltration.

Groundwater in permafrost areas may occur as suprapermafrost water, intrapermafrost water or subpermafrost water. Suprapermafrost water is present nearly everywhere in permafrost regions during the summer thawing season. If thawing extends deep enough it can create an appreciable reservoir of groundwater perched upon the underlying frozen material. Included with the suprapermafrost water is groundwater that occurs beneath lakes and rivers. Some temporary water supplies are obtained during the summer from shallow wells dug into this active zone.

Subpermafrost water occurs beneath large areas of permafrost. Examples are flowing wells in several places along the Alaska Highway between Haines Junction and the Alaska Boundary. The highway follows the Shakwak Valley, the floor of which is underlain by extensive deposits of glaciofluvial sand and gravel. The water discharged by the flowing wells reportedly comes from beneath the permafrost. Another example was encountered during mining operations in the valley of Eldorado Creek near Dawson, when a 67 m shaft was sunk to bedrock through frozen muck, sand and gravel. Water encountered in the bedrock flowed at a rate of approximately 75 L/s. The mineral content of subpermafrost water is generally high.

Numerous springs, some thermal, are known in the northern Cordillera but little is known regarding their relation to permafrost. The most common host rocks of the larger springs are limestone and dolomite, consequently the waters are high in calcium and magnesium salts. Waters from many of the larger springs have a high sulphate content, suggesting solution of anhydrite or gypsum, or possibly oxidation of pyritiferous rocks. Chloride waters are also present; in most instances the source of the chloride is from solution of subsurface halite.

Old Crow, Yukon Territory, is the only Canadian community in an area of continuous permafrost that obtains a permanent water supply from subpermafrost groundwater; several others do so in the area of discontinuous permafrost. In the latter case these are situated near large streams or lakes, and obtain their water from wells in unfrozen material close to bodies of surface water. Dawson and Hay River both have water supplies of this type. Whitehorse obtains water from wells in addition to supplies from the Yukon River and uses the groundwater to warm the river water during cold periods, in order to prevent freezing in the distribution system.

REFERENCES

Alberta Energy Resources Conservation Board - Alberta
1981: Reserves of Coal Province of Alberta, ERCB 81-31; Energy Resources Conservation Board, Alberta Canada.

Armstrong, J.E.
1983: Environmental and engineering applications of the surficial geology of the Fraser Lowland, British Columbia; Geological Survey of Canada, Paper 83-23.

Bally, A.W., Gordy, P.L., and Stewart, G.A.
1966: Structure, seismic data, and orogenic evolution of southern Canadian Rocky Mountains; Bulletin of Canadian Petroleum Geology, v. 14, p. 337-381.

Bates, D.V., Murray, J.W., and Raudsepp, V.
1980: British Columbia - Royal commission of inquiry, health and environmental protection, uranium mining - Volume I Commissioners Report; Queen's Printer, Victoria, B.C., Canada, 328 pages and one map.

Bell, R.T.
1982: Comments on the geology and uraniferous mineral occurrences of the Wernecke Mountains, Yukon and District of Mackenzie; in Current Research, Part B, Geological Survey of Canada, Paper 82-1B, p. 279-284.

1985: Overview of uranium in volcanic rocks of the Canadian Cordillera; in Uranium in Volcanic Rocks, IAEA; TC-490, International Atomic Energy Agency, Vienna, Austria, p. 319-335.

Bell, R.T.
1986: Megabreccias in northeastern Wernecke Mountains, Yukon Territory; in Current Research, Part A, Geological Survey of Canada, Paper 86-1A, p. 375-384.

1989: A conceptual model for development of megabreccias and associated mineral deposits in Wernecke Mountains, Canada, Copperbelt, Zaire, and Flinders Range, Australia; in Uranium Resources and Geology of North America, International Atomic Energy Agency, Vienna, IAEA TEC-DOC-500, p. 149-169.

Bell, R.T. and Jefferson, C.W.
1987: An hypothesis for an Australian - Canadian connection in the Late Proterozoic and the birth of the Pacific Ocean; in International Congress of the Geology, Structure, Mineralisation and Economics of the Pacific Rim, proceedings of Pacific Rim Congress 87. The Australasian Institute of Mining and Metallurgy, Parkville, Victoria; p. 39-50.

Bevier, M.L.
1983: Regional stratigraphy and age of Chilcotin Group basalts, south-central British Columbia; Canadian Journal of Earth Sciences, v. 20, p. 515-524.

Boyle, D.R.
1982: The formation of basal-type uranium deposits in south-central British Columbia; Economic Geology, v. 77, p. 1176-1209.

Boyle, D.R., Littlejohn, A.L., Roberts, A.C., and Watson, D.M.
1981: Ningyoite in uranium deposits of the south-central British Columbia: first North American occurrence; Canadian Mineralogist, v. 19, p. 325-331.

British Columbia Ministry of Energy, Mines and Petroleum Resources
1983: Hydrocarbon and by-product reserves in British Columbia.

Bustin, R.M. and Moffat, I.
1983: Reconnaissance stratigraphy and structure, Groundhog coalfield, north-central British Columbia; Canadian Society of Petroleum Geology Bulletin, v. 31, p. 231-245.

Cameron, B.E.B. and Tipper, H.W.
1981: Jurassic biostratigraphy, stratigraphy and related hydrocarbon occurrences of Queen Charlotte Islands, British Columbia; in Current Research, Part A, Geological Survey of Canada, Paper 81-1A, p. 209-212.
1985: Jurassic stratigraphy of the Queen Charlotte Islands, British Columbia; Geological Survey of Canada, Bulletin 365.

Canada Water Year Book, Environment Canada
1978-1979: Hydrology of Permafrost Regions, p. 29-34.
1979-1980: Ground Water, p. 69-79.

Cherry, J.A., van Everdingen, R.O., Meneley, W.W., and Toth, J.
1972: Hydrogeology of Rocky Mountains and Interior Plains; International Geological Congress, Field Guide A 26.

Clague, J.J.
1975: Late Quaternary sediments and geomorphic history of the southern Rocky Mountain Trench, British Columbia; Canadian Journal of Earth Sciences, v. 12, p. 595-605.
1981: Late Quaternary geology and geochronology of British Columbia, Part 2, Summary and discussion of radiocarbon dated Quaternary history; Geological Survey of Canada, Paper 80-35.

Clark, I.D., Fritz, P., Michel, F.A., and Souther, J.G.
1982: Isotope hydrogeology and geothermometry of the Mount Meager geothermal area; Canadian Journal of Earth Sciences, v. 19, p. 1454-1473.

Culbert, R.R., Boyle, D.R., and Levinson, A.A.
1984: Surficial uranium deposits in Canada; in Surficial Uranium Deposits, P.D. Toens (ed.), International Atomic Energy Agency, Vienna, Austria, p. 179-191.

Dakin, R.A., Farvolden, R.N., Cherry, J.A., and Fritz, P.
1983: Origin of dissolved solids in groundwaters of Mayne Island, British Columbia; Journal of Hydrology, v. 63, p. 233-270.

Dalgarno, C.R. and Johnson, J.E.
1968: Diapiric structures and Late Precambrian - Early Cambrian sedimentation in Flinders Ranges, South Australia; in Diapirism and Diapirs, a Symposium, J. Braunstein and G.D. O'Brian (ed.), American Association of Petroleum Geologists, Memoir 8, p. 301-314.

Energy, Mines and Resources Canada
1980: Coal Resources and Reserves of Canada; Energy, Mines and Resources Canada Report ER-79-9, 37 p.

Foreraker, J.C., Livingston, E., and Brown, W.L.
in press: The contribution of groundwater to the development of the agricultural and industrial bases of British Columbia; Proceedings International Association Scientific Hydrology, Great Britain, 1985 (in press).

Fournier, R.O. and Rowe, J.J.
1966: Estimation of underground temperatures from silica content of water from hot springs and wet steam wells; American Journal of Science, v. 264, p. 685-697.

Fournier, R.O. and Truesdell, A.H.
1973: An empirical Na-K-Ca geothermometer for natural waters; Geochimica et Cosmochimica Acta, v. 37, p. 1255-1275.

Freeze, R.A. and Cherry, R.N.
1979: Groundwater; Prentice-Hall Inc., 604 p.

Gallup, W.B.
1951: Geology of Turner Valley oil and gas field, Alberta; American Association of Petroleum Geologists Bulletin, v. 35, p. 797-871.

Graham, P.S.W.
1978: Geology and coal resources of the Tertiary sediments, Quesnel-Prince George Area, British Columbia; Current Research, Part B, Geological Survey of Canada, Paper 78-1B, p. 59-63.

Halstead, E.C.
1986: Groundwater supply — Fraser Lowland, British Columbia; Environment Canada, National Hydrology Research Institute, Paper No. 26, Inland Waters Directorate Scientific Series No. 145, 80 p.

Hess, P.
1985: Groundwater use in Canada; Environment Canada, National Hydrology Research Institute.

Hickson, C.J. and Souther, J.G.
1984: Late Cenozoic volcanic rocks of the Clearwater-Wells Gray area, British Columbia; Canadian Journal of Earth Sciences, v. 21, p. 267-277.

Holten, M.E. and Mellon, G.B.
1972: Geology of the Luscar (Blairmore) coalbeds, central Alberta Foothills; in Proceedings First Geological Conference on Western Canadian Coal, G.B. Mellon, J.W. Kramer and E.J. Seagel (ed.), Research Council of Alberta Information Series No. 60, p. 124-136.

Hughes, J.D. and Long, D.G.F.
1980: Geology and coal resource potential of Early Tertiary strata along Tintina Trench, Yukon Territory; Geological Survey of Canada, Paper 79-32, 21 p.

Hydrological Atlas of Canada
1978: Fisheries and Environment Canada, 34 maps.

Indian and Northern Affairs Canada
1983: Yukon - Exploration and Geology 1982; Exploration and Geological Services, Indian and Northern Affairs Canada, Canadian Government Publishing Centre, Supply and Services Canada, Ottawa, Ontario, Canada, p. 259.

Jamieson, G.R. and Freeze, R.A.
1983: Determining hydraulic conductivity distributions in a mountainous area using mathematical modelling; Ground Water, v. 21, p. 168-177.

Jessop, A.M., Souther, J.G., Lewis, T.J., and Judge, A.S.
1984: Geothermal measurements in northern British Columbia and southern Yukon Territory; Canadian Journal of Earth Sciences, v. 21, p. 599-608.

Jones, L.D.
1990: Uranium and thorium occurrences in British Columbia, British Columbia, Geological Survey Branch, Open File 1990-32, 78 pages, 1 map and 1 MINFILE dataset disc.

Katayama, M., Kubo, K., and Hirono, S.
1974: Genesis of uranium deposits of the Tono mine, Japan; in Formation of Uranium Ore Deposits, IAEA-SM-183, International Atomic Energy Agency, Vienna, Austria, p. 437-452.

Kempthorne, R.H. and Irish, J.P.R.
1981: Norman Wells - A new look at one of Canada's largest oilfields; Journal of Petroleum Technology, 33 p.

Koch, N.G.
1973: Central Cordilleran region; in Future Petroleum Provinces of Canada, R.G. McCrossan (ed.), Canadian Society of Petroleum Geologists, Memoir 1, p. 37-72.

Kohut, A.P., Foweraker, J.C., Johanson, D.A., Tradewell, E.H., and Hodge, W.S.
1983: Pumping effects of wells in fractured granitic terrain; Ground Water, v. 21, p. 564-572.

Kohut, A.P., Hodge, W.S., Johanson, D.A., and Kalyna, D.
in press: Natural seasonal response of groundwater levels in fractured bedrock aquifers of the southern coastal region, British Columbia; International Association Scientific Hydrology Symposium, Montreal (in press).

Lawson, D.W.
1968: Groundwater flow systems in the crystalline rocks of the Okanagan Highland, British Columbia; Canadian Journal of Earth Sciences, v. 5, p. 813-824.

Laznicka, P. and Edwards, R.J.
1979: Dolores Creek, Yukon - a disseminated copper mineralization in sodic metasomatites; Economic Geology, v. 74, p. 1352-1370.

LeBreton, E.G.
1972: A hydrogeological study of the Okanagan River Basin, Canada - British Columbia Basin Agreement, Technical Supplement 11, 87 p.

Lefebvre, J.J.
1980: A propos de l'existence d'un "Wilflysch Katangien"; Annales de la Societe geologique de Belgique, t. 103, p. 1-13.

Lewis, T.J. and Souther, J.G.
1978: Meager Mountain, B.C. - a possible geothermal energy source; Earth Physics Branch, Energy, Mines and Resources Canada, Geothermal Series No. 9.

Little, H.W.
1957: Kettle River (east half), B.C.; Geological Survey of Canada, Map 6-1957 with marginal notes.
1961: Kettle River (west half), B.C.; Geological Survey of Canada, Map 15-1961 with marginal notes.

Long, D.G.F.
1981: Dextral strike slip faults in the Canadian Cordillera and depositional environments of related freshwater intermontane coal basins; in Sedimentation and Tectonics in Alluvial Basins, A.D. Miall (ed.), Geological Association of Canada, Special Paper 23, p. 153-186.

Macauley, G.
1983: Source rock - oil shale potential of the Jurassic Kunga Formation, Queen Charlotte Islands; Geological Survey of Canada, Open File 921.

Martin, H.L.
1973: Eagle Plain Basin; in Future Petroleum Provinces of Canada; R.G. McCrossan (ed.), Canadian Society of Petroleum Geologists, Memoir 1, p. 275-306.

McKinney, J.S.
1983: Bonnet Plume coalfield; Mineral Deposits of Northern Cordillera symposium, Canadian Institute of Mining and Metallurgy Abstracts, 23 p.

McLean, J.R.
1982: Lithostratigraphy of the Lower Cretaceous coal bearing sequence, Foothills of Alberta; Geological Survey of Canada, Paper 80-29, 46 p.

Milner, M.W. and Craig, D.B.
1973: Coal in the Yukon; Indian and Northern Affairs Ottawa, Document 061751.

Monger, J.W.H.
1978: Evolution of the Cordillera; Energy, Mines and Resources Canada, Geos, Fall 1978, p. 5-8.

Monger, J.W.H. and Berg, H.C.
1984: Lithotectonic terrane map of western Canada and southeastern Alaska; in Lithotectonic Terrane Map of the North America Cordillera, Part B, N.S. Silberling (ed.), Department of the Interior, United States Geological Survey, Miscellaneous Field Studies, Map MF-B, p. 1-11 and 41-44.

Moore, J.N., Admas, M.C., and Stauder, J.J.
1983: Geologic and geochemical investigations of the Meager Creek Geothermal System, British Columbia Canada; in Geothermal Resources: energy on tap!, Geothermal Resources Council, Transactions Volume 7, p. 1-6.

Morton, R.D., Aubut, A., and Gandhi, S.S.
1978: Fluid inclusion studies and genesis of the Rexspar uranium-fluorite deposit, Birch Island, British Columbia; in Current Research, Part B, Geological Survey of Canada, Paper 78-1B, p. 137-140.

Muller, J.E.
1967: Kluane Lake map area Yukon Territory; Geological Survey of Canada, Memoir 312, 137 p.

Muller, J.E. and Jeletzky, J.
1970: Geology of the Upper Cretaceous Nanaimo Group, Vancouver and Gulf Islands, British Columbia; Geological Survey of Canada, Paper 69-25, 77 p.

Norcen Energy Resources Ltd.
1979: Statement of evidence relating to summary of the geology of the Blizzard uranium deposit, Phase I, overview; presented by D.A. Sawyer, Norcen Energy Resources Ltd., September 1979, to British Columbia Royal Commission of Inquiry into Uranium Mining, Document 20255, p. 11 plus figures.

Northeast Coal
1977: Northeast Coal Study, Resources Subcommittee on Northeast Development; Ministry of Mines and Petroleum Resources, British Columbia, 50 p.

Olade, M.A. and Goodfellow, W.D.
1978: Lithogeochemistry and hydrogeochemistry of uranium and associated elements in the Tombstone batholith, Yukon, Canada; in Geochemical Exploration 1978, J.R. Watterson and P.K. Theobald (ed.), Proceedings of the Seventh International Geochemical Exploration Symposium, Association of Exploration Geochemists, p. 407-428.

Piteau and Associates
1988: Geochemistry and isotope hydrology of the Mount Edziza and Mess Creek geothermal waters; Geological Survey of Canada, Open File 1732.

Preto, V.A.
1978: Setting and genesis of uranium mineralization at Rexspar; Canadian Mining and Metallurgical Bulletin, v. 71, no. 800, p. 82-88.

Read, P.B.
1979: Geology, Meager Creek geothermal area, British Columbia; Geological Survey of Canada, Open File 603, map and descriptive notes.

Roberts, D.E. and Hudson, G.R.T.
1983: The Olympic Dam copper-uranium-gold deposit, Roxby Downs, south Australia; Economic Geology, v. 78, p. 799-822.

Roddick, J.A.
1965: Vancouver North, Coquitlam and Pitt Lake map areas, British Columbia; Geological Survey of Canada, Memoir 335.

Rybach, L. and Muffler, L.J.P. (editors)
1981: Geothermal Systems: Principles and case histories; John Wiley and Sons, New York - Toronto.

Sawyer, D.A., Turner, A.T., Christopher, P.A., and Boyle, D.R.
1981: Basal type uranium deposits, Okanagan region, south-central British Columbia; in Field Guides to Geology and Mineral Deposits, R.I. Thompson and D.G. Cook (ed.), Calgary '81, Annual Meeting, Geological Association of Canada, p. 69-77.

Shouldice, D.H.
1973: Western Canadian continental shelf; in Future Petroleum Provinces of Canada, R.G. McCrossan (ed.), Canadian Society of Petroleum Geologists, Memoir 1, p. 7-36.

Souther, J.G.
1975: Geothermal potential of western Canada; Proceedings, 2nd United Nations Symposium of the Development and Use of Geothermal Resources, v. 1, p. 259-267.
1980: Geothermal reconnaissance in the central Garibaldi Belt; in Current Research, Part A, Geological Survey of Canada, Paper 80-1A, p. 1-11.
1981: Canadian Geothermal Research Program; in Energy Resources of the Pacific Region, American Association of Petroleum Geologists, Studies in Geology no. 12, p. 391-400.
1984: The Ilgachuz Range, a peralkaline shield volcano in central British Columbia; in Current Research, Part A, Geological Survey of Canada, Paper 84-1A, p. 1-10.
1986: The western Anahim Belt: root zone of a peralkaline magma system; Canadian Journal of Earth Sciences, v. 23, p. 895-908.

Souther, J.G. and Dellechaie, F.
1984: Geothermal exploration at Mt. Cayley - a Quaternary volcano in southwestern British Columbia; in Transactions, Geothermal Resources Council, 1984 Annual Meeting, p. 463-468.

Souther, J.G. and Hickson, C.J.
1984: Crystal fractionation of the basalt comendite series of the Mount Edziza Volcanic Complex, British Columbia: major and trace elements; Journal of Volcanology of Geothermal Research, v. 21, p. 79-106.

Stott, D.F.
1974: Lower Cretaceous coal measures of the Foothills of west-central Alberta and northeastern British Columbia; Canadian Institute of Mining and Metallurgy Bulletin, v. 67, p. 87-100.
1982: Lower Cretaceous Fort St. John Group and Upper Cretaceous Dunvegan Formation of the Foothills and Plains of Alberta, British Columbia, District of Mackenzie and Yukon Territory; Geological Survey of Canada, Bulletin 328.
1984: Cretaceous sequences of the Foothills of the Canadian Rocky Mountains; in The Mesozoic of Middle North America, D.F. Stott and D.J. Glass (ed.), Canadian Society of Petroleum Geologists, Memoir 9, p. 85-108.

Sutherland Brown, A.
1968: Geology of the Queen Charlotte Islands, British Columbia; British Columbia Department of Mines and Petroleum Resources, Bulletin 54, 226 p.

Sutherland Brown, A., Yorath, C.J., and Tipper, H.W.
1983: Geology and tectonic history of the Queen Charlotte Islands; Geological Association of Canada, Mineralogical Association of Canada, Canadian Geophysical Union, Joint Annual Meeting, Victoria, B.C. Field Guide Book 8.

Tipper, H.W. and Richards, T.A.
1976: Jurassic stratigraphy and history of north-central British Columbia; Geological Survey of Canada, Bulletin 270, 73 p.

Wallace-Dudley, K.E.
1982a: Gas pools of western Canada; Geological Survey of Canada, Map 1558A.
1982b: Oil pools of western Canada; Geological Survey of Canada, Map 1559A.

Ward, P. and Stanley, K.O.
1982: The Haslam Formation: a Late Santonian - Early Campanian forearc basin deposit in the Insular Belt of southwestern British Columbia and adjacent Washington; Journal of Sedimentary Petrology, v. 52, p. 975-996.

Wheeler, J.O.
1961: Whitehorse map area Yukon Territory; Geological Survey of Canada, Memoir 312, p. 156.

White, D.F. and Williams, D.L. (editors)
1975: Assessment of geothermal resources of the United States; United States Geological Survey, Circular 726.

Wright, J. and Miller, D.
1983: Rock River Coal Basin; Mineral Deposits of Northern Cordillera - a symposium; Canadian Institute of Mining and Metallurgy, Abstracts, 23 p.

Yorath, C.J. and Chase, R.L.
1981: Tectonic history of the Queen Charlotte Islands and adjacent areas - a model; Canadian Journal of Earth Sciences, v. 18, p. 1717-1739.

Yorath, C.J. and Cook, D.G.
1981: Cretaceous and Tertiary stratigraphy and paleogeography, northern Interior Plains, District of Mackenzie; Geological Survey of Canada, Memoir 398.

Yorath, C.J. and Hyndman, R.D.
1983: Subsidence and thermal history of Queen Charlotte Basin; Canadian Journal of Earth Sciences, v. 20, p. 135-159.

Young, F.G.
1975: Upper Cretaceous stratigraphy, Yukon Coastal Plain and northwestern Mackenzie Delta; Geological Survey of Canada, Bulletin 249, 83 p.

Authors' addresses

C.J. Yorath
Pacific Geoscience Centre
9860 West Saanich Road
P.O. Box 6000
Sidney, British Columbia
V8L 4B2

J.G. Souther
Cordilleran Division
Geological Survey of Canada
100 West Pender Street
Vancouver, British Columbia
V6B 1R8

R.M. Bustin
Department of Geological Services
University of British Columbia
Vancouver, British Columbia
V6T 2B4

G.K. Williams
Institute of Sedimentary and
 Petroleum Geology
Geological Survey of Canada
3303-33rd Street N.W.
Calgary, Alberta
T2L 2A7

R.T. Bell
Geological Survey of Canada
601 Booth Street
Ottawa, Ontario
K1A 0E8

E.C. Halstead
4432 Lions Avenue
North Vancouver, British Columbia
V7R 3S6

P.L. Gordy
Box 118
Bowser, British Columbia
V0R 1G0

Printed in Canada

Chapter 21

NATURAL HAZARDS

Summary
Introduction
Earthquakes
Volcanism
Landslides
   Mitigation of landslide hazards
Snow avalanches
Floods
Tsunamis
Erosion
References

Chapter 21

NATURAL HAZARDS

John J. Clague

SUMMARY

A variety of geological and geomorphic processes are potentially hazardous to life and property in the Canadian Cordillera. Natural hazards in this region are mainly products of high relief, high precipitation, and earthquakes.

Earthquakes are common in the western Cordillera, especially on the ocean floor off the Queen Charlotte Islands and Vancouver Island, in and adjacent to the Strait of Georgia, and in the Saint Elias and Mackenzie mountains. Much of the seismic activity is closely associated with the interactions of lithospheric plates in the northeast Pacific Ocean. Although most large earthquakes are centred on the ocean floor far from major population centres, a magnitude 7-8+ event could occur close enough to Vancouver, Victoria, or other west coast communities to inflict major damage and some loss of life.

During the Quaternary Period, there have been scores of volcanic eruptions in the western Canadian Cordillera. Most Quaternary volcanic centres in this region are small and have formed through the eruption of low-viscosity, silica-poor magma. Future eruptions of this type probably would have only limited localized effects. In contrast, a few large composite volcanoes have erupted explosively during the Quaternary, producing lahars, pyroclastic flows and surges, as well as plumes of tephra that have blanketed large areas. The most recent eruption of this type occurred near the Alaska-Yukon boundary about 1200 years ago.

Destructive landslides are common in the Canadian Cordillera and are related to instabilities in steep rock slopes and to mountain torrent systems. Most landslide damage and loss of life has resulted from frequent rockfalls along transportation corridors, rare large rockslides and rock avalanches, debris flows that have crossed inhabited fans, and fine grained sediment flows and related lateral spreads. Submarine mass movements, especially those in fiords, also are potentially hazardous.

Avalanches of dry and wet snow occur throughout the mountains of western Canada during winter and spring, and have been responsible for the deaths of many people engaged in recreational and work-related activities. Avalanches occur both on open slopes and in ravines and gullies, and are favoured by certain meteorological conditions. In many, although not all, cases it is possible to identify areas prone to avalanching and thus to apply preventative measures to reduce hazard.

Most streams in the Canadian Cordillera periodically flood, either as a result of heavy rainfall, rapid melting of snowpacks, or the formation and breakup of ice jams. In addition, some short-lived large floods result from the sudden draining of moraine- and glacier-dammed lakes. Some low-lying areas along the British Columbia coast are flooded by the sea during severe storms, unusually high tides, or as a result of tsunamis. The tsunami generated by the 1964 Alaska earthquake heavily damaged parts of Port Alberni on Vancouver Island. Other distant or local earthquakes or large submarine landslides could trigger similar destructive tsunamis in the future.

Coastal and river-bank erosion and catastrophic gullying have damaged property in various parts of the Cordillera. Areas of troublesome coastal erosion include some shores along the Strait of Georgia and along Juan de Fuca and Hecate straits. The channels of many streams migrate or suddenly change position during periods of high water. Braided streams, which are common in and around glaciated mountain ranges, are especially susceptible to sudden shifts of channels, and consequently their floodplains are particularly hazardous sites for development.

INTRODUCTION

Earthquakes, volcanic eruptions, landslides, and floods have played a major role in shaping the landscape of the Canadian Cordillera. These natural processes also pose hazards to people and property in this region. Some of these hazards can be mitigated, but only after sufficient geological, geophysical, and historical data have been collected to appraise the probable location, magnitude, and frequency of future destructive events. Earthquakes, floods, and snow avalanches occur repeatedly, thus risk can be assessed by making observations over a relatively short period of time. Areas susceptible to volcanic eruptions and landslides also can be identified, although predicting exactly when and where such an event will occur is not yet possible. While in some cases the only option is to avoid a hazardous area, development may be possible after appropriate steps have been taken to reduce risk.

This chapter summarizes naturally occurring processes that are potentially dangerous to life and property in the Canadian Cordillera, namely earthquakes, volcanism, landslides, snow avalanches, floods, tsunamis, and erosion.

EARTHQUAKES

The incidence of earthquakes in parts of the Canadian Cordillera is high, reflecting the proximity of the region to lithospheric plate boundaries in the northeast Pacific

---

Clague, J. J.
1991: Natural hazards, Chapter 21 in Geology of the Cordilleran Orogen in Canada, H. Gabrielse and C. J. Yorath (ed.); Geological Survey of Canada, Geology of Canada, no. 4, p.803-815 (also Geological Society of America, The Geology of North America, v. G-2)

Ocean. Although most earthquakes are small and cause no damage, an average of two quakes greater than Richter magnitude 6.5 occur in western Canada every decade (Whitham and Hasegawa, 1975). Bearing in mind the San Fernando (California) earthquake of 1971 (Richter local magnitude = 6.4, 58 deaths, >$500 000 000 damage), the potential for earthquake damage in the region is considerable.

Most earthquakes in the Canadian Cordillera occur along the Queen Charlotte-Fairweather fault system (Pacific-America plate boundary) and along the system of offshore spreading ridges and associated fracture zones west of Vancouver Island (Pacific-Juan de Fuca-Explorer boundaries) (Fig. 21.1; Milne et al., 1978). Some of the largest historic earthquakes in Canada (to Richter magnitude 8) and greatest release of seismic energy have been on the Queen Charlotte fault.

Earthquakes also are common in a few areas east of these plate boundaries, notably coastal southwestern British Columbia, southwestern Yukon, and parts of eastern Yukon-western District of Mackenzie. The relationship of this seismicity to the present-day tectonic regime of the Canadian Cordillera is poorly known, although earthquakes in the southern Strait of Georgia-Puget Sound region may be related to northeasterly subduction of the Juan de Fuca plate beneath North America (Riddihough and Hyndman, 1976; Keen and Hyndman, 1979), and those in southwestern Yukon to subduction of the Pacific plate under Alaska and the adjacent Yukon (Plafker et al., 1978; Horner, 1983).

Isolated small and intermediate earthquakes have been recorded in almost all other parts of the Canadian Cordillera. Most of these appear to be random events that are not clearly associated with major geological or tectonic features, and for which a satisfactory explanation is not available. The total amount of energy released in these scattered earthquakes is miniscule in comparison to that generated by earthquakes at plate boundaries in the northeast Pacific Ocean.

The foregoing indicates that destructive earthquakes are most likely to occur in the westernmost Canadian Cordillera. The greatest hazard is in south coastal British Columbia, because it is here that most of the population of the region is located.

Earthquake damage is produced by direct ground motion and by secondary causes such as landslides and fires. Although probabilistic estimates of ground motion have been made for the Cordilleran region (Basham et al., 1985), the potential for secondary effects has not been systematically assessed. The extent of damage due to seismic shaking is closely related to magnitude, epicentral proximity, and focal depth. In addition, earth materials respond differently to shaking, thus geological factors are important determinants of earthquake damage. Structures located on Quaternary sediments generally are more prone to damage from shaking than structures on bedrock, although some dense compact sediments perform as well as rock in this respect. Saturated fine grained sediments may liquefy during a major earthquake, giving rise to destructive flows and spreads. Delta-front sediments off the mouth of Fraser River presumably could fail in this way (Luternauer and Finn, 1983). Other materials prone to failure during earthquakes are artificial fills and fine glaciolacustrine and glaciomarine deposits.

VOLCANISM

More than 100 volcanoes in the Canadian Cordillera have erupted one or more times during the Quaternary Period (i.e., the last two million years). Although there have been no eruptions during this century, one lava flow in a remote area near the Alaska-British Columbia boundary probably is less than 200 years old (Souther, 1977).

Late Tertiary and Quaternary volcanoes are concentrated in several linear belts (Fig. 21.2; Souther, 1977). This pattern is a product of lithospheric plate interactions in the northeast Pacific Ocean, perhaps coupled with the migration of continental crust over one or more hot spots in the mantle (Bevier et al., 1979; Rogers and Souther, 1983; Souther, 1986). It is likely that future eruptions also will occur within these belts, although their precise locations cannot be predicted.

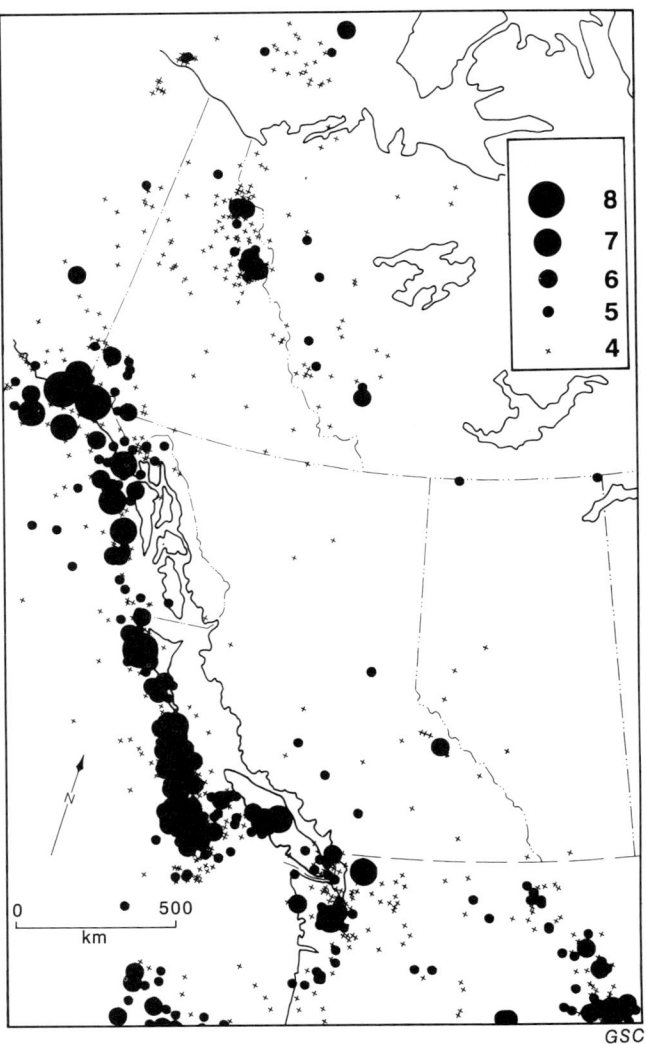

Figure 21.1. Epicentres of historical earthquakes (M≥4) in the Canadian Cordillera and adjacent areas (to 1986).

Most Quaternary eruptive centres have been the locus of a single pulse of activity during which one or more small pyroclastic cones were built and a small volume of basaltic lava erupted to form thin blocky flows. The past record indicates that an eruption of this type is likely to happen somewhere in the Cordillera in the next several centuries. Such an eruption probably would affect only a small area and would not be hazardous, unless it occurred in a populated region.

In contrast, several volcanic centres are large composite volcanoes that have erupted repeatedly during late Tertiary and Quaternary time (Fig. 21.3). Some, such as Mount Edziza, are formed mainly of basalt, whereas others, such as Mount Garibaldi, Mount Cayley, and Mount Meager, consist largely of dacite and andesite. The more siliceous volcanoes probably erupted explosively on several occasions during the Quaternary Period, with far-reaching effects. The youngest such activity occurred about 1200 years ago when a layer of tephra (volcanic ash) was deposited over a large area of the southern Yukon and western District of Mackenzie during an eruption from a vent near the Alaska-Yukon boundary (Bostock, 1952; Hughes et al., 1972; Lerbekmo et al., 1975). Other large explosive eruptions from the same volcano 1500-1900 BP, from Mount Meager about 2400 BP, Mount St. Helens, Washington, about 3400 BP, and Mount Mazama (Crater Lake, Oregon) about 6800 BP also produced extensive ash falls in western Canada (Bostock, 1952; Nasmith et al., 1967; Westgate and Dreimanis, 1967; Westgate et al., 1970; Fulton, 1971; Hughes et al., 1972; Lerbekmo et al., 1975; Mullineaux et al., 1975; Westgate, 1977; Mathewes and Westgate, 1980; Bacon, 1983). Although it had little impact on Canada, the May 1980 eruption of Mount St. Helens (Lipman and Mullineaux, 1981) was similar to these large explosive events and produced many of the same effects. Future eruptions of this type in British Columbia or the Yukon would likely damage property and crops and disrupt transportation in the Canadian Cordillera. This would result mainly from fallout of ash and dust over large areas and from flooding and aggradation in stream valleys surrounding the volcano. Lahars, pyroclastic flows, and pyroclastic surges would produce additional damage near the volcano.

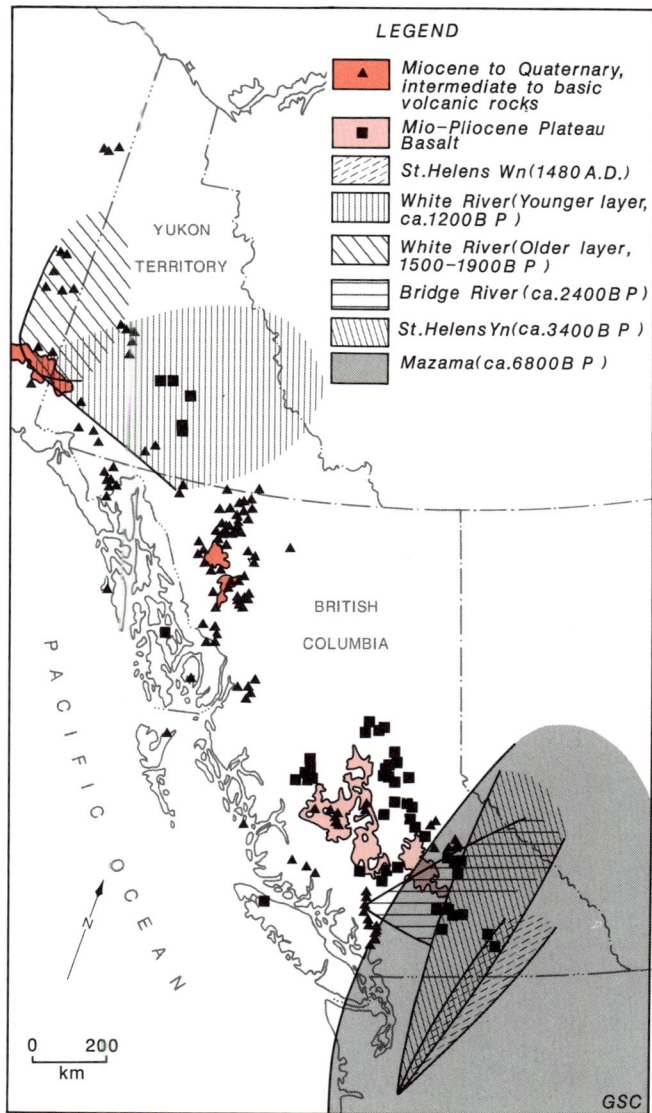

Figure 21.2. Late Tertiary and Quaternary volcanoes and Holocene tephras in the Canadian Cordillera. See text for references.

Figure 21.3. Late Pleistocene lava flow near Mount Garibaldi (view east towards Mount Price). This flow, which is probably 10 000-12 000 years old, formed the dam behind which Garibaldi Lake (background) has been ponded. Photo by Austin Post, United States Geological Survey; provided by W.H. Mathews, University of British Columbia. (GSC 205381)

LANDSLIDES

A wide variety of potentially destructive mass movements occur in the Canadian Cordillera (Eisbacher, 1979; Evans, 1982; Cruden, 1985). Most are complex and are related to instabilities in steep bedrock slopes and to high-gradient mountain streams. The occurrence and characteristics of mass movements are controlled by unique combinations of topography, geology, climate, seismicity, and human activity (e.g., mining, logging, urban and recreational development).

Landslides involving bedrock range from small falls and topples to large ($>10^6$ m^3) slumps and slides (for terminology see Varnes, 1978). *Rockfalls* are common on steep bedrock slopes in areas of intensely fractured or jointed rocks and are triggered by freeze-thaw activity, intense precipitation, and earthquakes (Peckover and Kerr, 1977; Piteau, 1977; Mathews, 1979). They regularly disrupt road and rail traffic and occasionally take human lives. The small size of these landslides (typically 10-1000 m^3) belies the fact that they are probably the most costly slope failures in the region. Aside from economic losses due to traffic delays, there is a considerable cost involved in scaling, blasting, and grouting threatening rock faces and in removing debris from roads and railways.

Rapid bedrock failures involving a sliding or flowing type of movement (*rockslides* and *rock avalanches*) are most common in areas of high relief where geological discontinuities (e.g., fractures, faults, bedding planes, cleavage, intrusive contacts) dip in the direction of the slope. In the Mackenzie and Rocky mountains, for example, these landslides are typically associated with dip slopes of limestone, dolostone, or quartzite. In the Coast and Insular mountains, on the other hand, they occur along structural and stratigraphic discontinuities in Quaternary volcanic rocks and older sheared granitic, metasedimentary, and metavolcanic rocks (Fig. 21.4). Large bedrock failures are triggered by excessive pore-water pressure, earthquakes, and human activity, among other things. Fundamental natural causes include erosion of slopes by streams and glaciers and the gradual destruction of cohesion along discontinuities by physical weathering and solution.

At least 17 large rockslides and rock avalanches have occurred in historical time in the Canadian Cordillera: at Rubble Creek (1855 or 1856), Britannia (1915), Hoodoo Mountain (1919), Mount Meager (1931, 1947, 1975, 1986), Pandemonium Creek (1959 or 1960), and Mount Cayley (1963, 1984), all in the Coast Mountains; at Frank (1903) and Brazeau Lake (1933) in the Rocky Mountains; at Kaouk River (1925 or 1926) and Mount Colonel Foster (1946) on Vancouver Island; near Hope in the Cascade Mountains (1965, Fig. 21.4); and southwest of Carlson Lake near the eastern front of the Mackenzie Mountains (1985) (for further information on these landslides, see: Kerr, 1948; Mokievsky-Zubok, 1977; Cruden and Krahn, 1978; Mathews and McTaggart, 1978; Moore and Mathews, 1978; Mathews, 1979; Clague and Souther, 1982; Cruden, 1982; Eisbacher, 1983; Evans, 1987; Evans et al., 1987). Although uncommon in comparison to small rockfalls and debris flows, rockslides and rock avalanches can be extremely destructive; they are responsible for about 40% (140) of the recorded landslide deaths in the Canadian Cordillera.

Figure 21.4. Hope slide, the largest historical rockslide in the Canadian Cordillera (view east). This photograph, taken shortly after the event in January 1965, shows Highway 3 southeast of Hope deeply buried by rock debris. Photo courtesy of *Vancouver Sun*.

Rotational slides (*slumps*) of a range of sizes are common in most parts of the Canadian Cordillera, occurring in both bedrock and Quaternary sediments. Rotational sliding often takes place in association with other types of mass movements. For example, many landslides involving Quaternary sediments are retrogressive slumps at their heads but translational slides or flows at their distal ends. Also, some debris flows are initiated when rain-swollen torrents are blocked by slumps.

Large deep-seated slope failures characterized by the slow downslope movement of internally broken rock masses along poorly defined rupture surfaces ("sagging slopes") are found in some areas of foliated metamorphic rocks, for example, in parts of the Columbia Mountains. An especially well documented example of such a failure is the Downie slide in Columbia River valley (Piteau et al., 1978; Brown and Psutka, 1980). Although initiated thousands of years ago, this and many other large slope sags in the Cordillera are still active, and any permanent structures located on them ultimately will be damaged or destroyed. In addition, parts of some sags may detach from the main mass of creeping debris and move rapidly downslope; such a possibility should be considered when siting communities and major developments such as hydroelectric reservoirs.

Flows of unconsolidated sediments or weathered bedrock involve the displacement of a mass of material as a viscous fluid. Most of those in the Canadian Cordillera are *debris flows* caused by the failure of water-saturated, heterogeneous Quaternary sediments (e.g., till, glaciofluvial deposits, pyroclastic deposits, colluvium). Debris flows are common in areas of high relief and abundant precipitation, for example the Coast and Insular mountains of western British Columbia. Most are triggered by intense fall, winter, and spring rainstorms (Eisbacher and Clague, 1984; VanDine, 1985; Church and Miles, 1987); some, however, are initiated by the sudden draining of moraine- or glacier-dammed lakes (Jackson, 1979; Blown and Church, 1985; Clague et al., 1985). Most debris flows are funnelled along steep valleys and ravines, although some occur on open slopes and are not confined to well defined channels. The former generally debouch onto fans or cones, which are the sites of much of the past destruction and loss of life due to mass movements in the Cordillera (Fig. 21.5); examples include Howe Sound in the southern Coast Mountains (Eisbacher, 1983; Jackson et al., 1985; VanDine, 1985) and Port Alice on Vancouver Island (Nasmith and Mercer, 1979). Although much smaller than large rockslides, debris flows are more frequent and are responsible for more than one-third (ca. 115) of all recorded landslide deaths in the Canadian Cordillera.

Flows and complex slides consisting mainly of silt and clay occur in areas of Pleistocene glaciolacustrine and glaciomarine sediments (Evans, 1982). Water is the critical determinant of stability of these fine sediments; failure commonly results from high pore pressures attributable to both natural causes such as heavy rains and rapid snowmelt and to human activity such as irrigation. In drier parts of the Cordillera, special care must be taken with surface water in areas where glaciolacustrine silt and clay form bluffs. A large landslide at Ashcroft in 1880 (Evans, 1984) and destructive landslides at Spences Bridge in 1905 (Drysdale, 1914) and Summerland in 1970 (Evans, 1982) resulted from infiltration of irrigation water into glaciolacustrine sediments; many other failures have had a similar origin. The addition of large amounts of water into these naturally dry sediments also causes piping (i.e., subsurface erosion by groundwater) and surface collapse (Cockfield and Buckham, 1956; Miller and Nyland, 1981).

Several cities and towns in the Canadian Cordillera (e.g., Vancouver, Kamloops, Prince George, Kitimat, Whitehorse) are located partly on glaciolacustrine or glaciomarine deposits, thus flows and related slides and lateral spreads pose a significant hazard to people and property. At least 25 people have died in landslides in these materials in the twentieth century, and property damage has been extensive.

Sediment flows also are common on slopes in permafrost areas. Failure occurs in zones of high pore pressure at the base of the active layer during summer thaw (Hughes et al., 1973; Mackay and Mathews, 1973; Rutter et al., 1973; McRoberts and Morgenstern, 1974). These flows are found on slopes as gentle as about 7° and are especially common along river and coastal bluffs underlain by ice-rich Quaternary sediments.

Large slow-moving flows in weathered volcanic and sedimentary rocks of Cretaceous and early Tertiary age occur in the Fraser River basin in south-central British Columbia (Fig. 21.6; Eisbacher, 1979; VanDine, 1980; Bovis, 1985). Perhaps the best known of these is Drynoch landslide near Spences Bridge, the toe of which is crossed by both the Trans-Canada Highway and the mainline of the Canadian Pacific Railway. Abundant montmorillonite (a clay mineral that swells when wetted) and elevated pore pressures are linked to the development of many of these landslides. Although many of the flows are presently active, they move very slowly (generally a few centimetres to perhaps a few metres per year) and consequently pose a hazard only to certain structures located on them (e.g., roads and rail lines).

Subaqueous mass movements similar to some terrestrial sediment flows are relatively common in British Columbia fiords, at several other sites along the coast, and in some lakes. Most occur at the fronts of active deltas and on steep slopes underlain by unconsolidated sediments (Prior and Bornhold, 1984, 1986; Prior et al., 1982, 1984, 1986). They involve artificial fills and deltaic, marine, and possibly glaciomarine sediments of very low strength and high water content. There have been several historical submarine slope failures in Howe Sound (Terzaghi, 1956; Prior et al., 1981) and Kitimat Arm (Luternauer and Swan, 1978; Swan and Luternauer, 1978; Prior et al., 1982, 1984); one in Kitimat Arm generated a train of sea waves that damaged port facilities at Kitimat (see "Tsunamis"). There also is evidence of large prehistoric failures at the front of the

Figure 21.5. Aftermath of a debris flow, Lions Bay, British Columbia. The debris flow swept across the cone on which the community is situated on February 11, 1983, causing considerable damage and killing two people. Lions Bay is located along Howe Sound approximately 20 km north-northwest of Vancouver. Photo courtesy of *Vancouver Sun*.

Figure 21.6. Large slow-moving flow near Pavilion, British Columbia ("Pavilion earthflow"; view southeast). This landslide occurred in Cretaceous sedimentary rocks and overlying Quaternary sediments. Photo by J.J. Clague. (GSC 205368)

Fraser River delta (Tiffin et al., 1971; Hamilton and Luternauer, 1983; Luternauer and Finn, 1983).

Mitigation of landslide hazards

The mitigation of landslide hazards follows three steps: hazard appraisal, application of "passive" countermeasures, and application of "active" countermeasures (Eisbacher and Clague, 1984). Hazard may be appraised by determining the age and frequency of past mass movements from geological and geomorphic evidence and from historical records, and by assessing ground conditions in an area where failure might occur. Once the hazard has been assessed, passive and active countermeasures may be taken to minimize danger to people and property. Passive measures are those which mitigate hazard without modifying the land surface; they include land-use zoning and monitoring. In contrast, active measures modify the landscape to lessen the likelihood of destructive landslides or to reduce their impact. Active measures against small rockfalls are mainly preventative engineering structures placed on and at the base of steep rock faces. Active measures against debris flows are directed towards the control and stabilization of debris in source areas; on fans they have primarily protective functions (Hungr et al., 1987).

Approaches that may be taken to reduce risk generally depend on the volume of the potential landslide and the probability of its occurrence (Eisbacher and Clague, 1984). In this context, preventative structures and protective forests are most appropriate for landslides of small size ($<10^3$ m^3) and high probability (<1-10 years average recurrence). More expensive protective and control structures are required for larger (10^3-10^6 m^3) landslides of moderate to high probability. These can be justified economically only if the value of the property to be protected exceeds the cost of the structures, or if human lives might be in jeopardy. Land-use restrictions generally are necessary to provide protection against large ($>10^6$ m^3) landslides that are likely to occur once every few hundred years or less; for rarer events, the risk generally is accepted, although it can be reduced through monitoring.

SNOW AVALANCHES

Avalanches of dry and wet snow, in some instances with dispersed sediment or freshly broken rock, are common in the mountains of western Canada and are a hazard to those who venture into these areas in winter and spring. They occur both on open slopes and in ravines and gullies. Areas of frequent avalanche activity on forested slopes are easily identified because they lack trees (Fig. 21.7); in contrast, avalanche areas above treeline may be more difficult to recognize.

Although they can occur almost anytime on slopes that are moderately steep and snow-covered, avalanches are favoured by certain weather conditions (Fitzharris and Schaerer, 1980). They are especially common during thaws or periods of rain after heavy snowfalls and when thick dry snow accumulates on old icy snow surfaces. In the latter instance, the buried icy surface is a plane of weakness along which overlying snow may easily slide.

In recent decades, avalanches have taken the lives of many people engaged in recreation activities in mountain

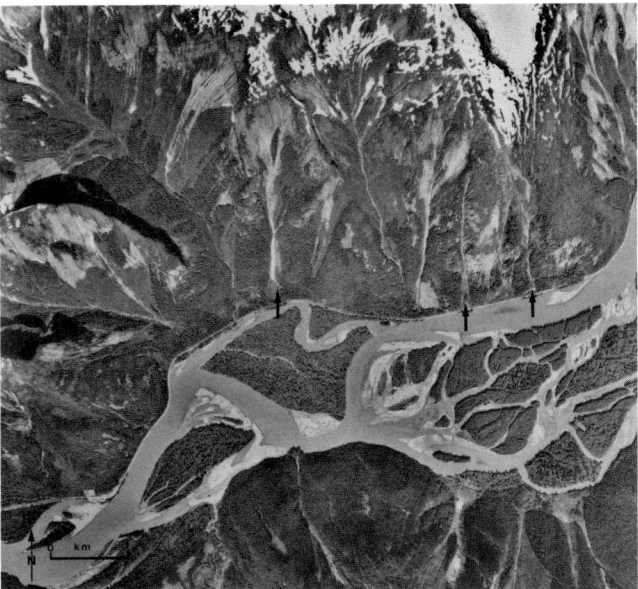

Figure 21.7. Typical mountain slopes subject to winter and spring avalanche activity (Skeena River valley west of Terrace, British Columbia). Areas lacking forest cover are periodically swept by snow avalanches. Some avalanche tracks extend to the valley floor and cross Highway 16 (examples are indicated by arrows). Province of British Columbia photo BC5111-220. (GSC 205387)

areas (Stethem and Schaerer, 1979, 1980). Avalanches also were responsible for disastrous mining-related accidents at Chilkoot Pass in 1898 and Granduc Mine in 1965 (Barker, 1977). They regularly disrupt road and rail traffic in some mountain passes and valleys (e.g., Rogers Pass in the Selkirk Mountains, Skeena River valley west of Terrace, and Fraser Canyon between Hope and Lytton) (Schaerer, 1962; Avalanche Task Force, 1974). Many employees of the Canadian Pacific Railway were killed by avalanches in Rogers Pass before the line was protected with snow sheds. Today, detailed records are kept of the frequency, size, and location of avalanches in Rogers Pass and its approaches, and winter weather conditions are monitored closely. When snowpack conditions on slopes overlooking the railway and the Trans-Canada Highway are dangerous, traffic is stopped, and potential avalanches are brought down by artillery fire (Schaerer, 1962). Mounds and barriers also have been constructed to slow down avalanches that otherwise might overwhelm the road and railway.

FLOODS

Although there has been very little loss of life due to floods in the Canadian Cordillera, property damage has been extensive and the cost of flood-control measures high. The most destructive historical flood in this region occurred in southwestern British Columbia in May and June of 1948 (Fig. 21.8). At that time, Fraser River broke its dykes in Fraser Lowland and flooded large tracts of agricultural land. As a result, 16 000 people were evacuated and 2000 homes damaged; total losses amounted to more than $20 000 000 (Barker, 1977). More recently, in October 1984, heavy warm rains melted large amounts of snow and ice in the southern Coast Mountains, causing severe flooding and extensive damage in Pemberton Valley.

Flooding in the southern Canadian Cordillera occurs during extended periods of warm weather in the spring and during periods of heavy rainfall. Large rivers generally crest in late spring as the winter snowpack melts. In contrast, many medium-size rivers have maximum flows during rainy periods in the fall. Streams with small catchments may flood at any time of the year.

Although rainstorm and snowmelt floods also occur in the northern Cordillera, another type of flood is perhaps more common there. In winter, rivers in the north freeze over, often to a depth of more than 1 m. During spring thaw, these rivers swell and their ice crusts are broken into sheets that float downstream. Channel constrictions and river bars may obstruct the free passage of the ice blocks and thus create large jams that force the river over its banks. When the flow becomes powerful enough to break the ice jams, the backed-up waters flood downstream. Considerable damage may be caused in this manner both by the water and by the debris and ice carried along with it. Similar floods occasionally occur when icings impede flow sufficiently to force a stream over its banks during spring thaw.

Some low-lying areas along the coast of British Columbia are flooded by the sea during severe storms, unusually high tides, or as a result of tsunamis. For example, exceptionally high spring tides raised the level of Fraser River during the 1948 flood, causing additional damage to parts of Fraser Lowland.

Finally, lakes impounded by present-day glaciers and by bulky Neoglacial end moraines may drain catastrophically to produce severe downstream floods (Fig. 21.9). There have been numerous historical jökulhlaups (glacier-outburst floods) in the Coast, Rocky, and Saint Elias mountains (Marcus, 1960; Mathews, 1965, 1973; Gilbert, 1972; Jackson, 1979; Clague, 1982b; Clague and Rampton, 1982; Clarke, 1982; Clarke and Waldron, 1984; Blown and Church, 1985; Jones et al., 1985; Ryder, 1985; Gilbert and Desloges, 1987). Floods from moraine-dammed lakes have occurred historically in the Selkirk and southern Coast mountains (S.G. Evans, Geological Survey of Canada, pers. comm., 1984; Blown and Church, 1985).

Figure 21.9. Nostetuko Lake and the breached late Neoglacial end moraine of Cumberland Glacier in the southern Coast Mountains of British Columbia (view southeast). About 6×10^6 m^3 of water were suddenly released from the lake in July 1983 when a series of waves overtopped the moraine. The escaping waters trenched the moraine, deposited the adjacent fan of coarse debris, and produced a flood that devastated the valley below. Photo by S.G. Evans. (GSC 204047-G)

Figure 21.8. Flooded terrain in eastern Fraser Lowland at the peak of the 1948 Fraser River flood (view north). Province of British Columbia photo BC581-54 (June 5, 1948). (GSC 205382)

TSUNAMIS

Tsunamis are sea waves produced by large-scale, short-duration disturbance of the ocean floor, most commonly by shallow submarine earthquakes, but occasionally by submarine mass movements or volcanic eruptions. Tsunamis have low amplitudes in the open sea, but may pile up to great heights (30 m or more) in shallow water, and consequently can cause considerable damage to coastal communities.

The most destructive, historical tsunami in British Columbia resulted from the Alaska earthquake of March 27, 1964 (Fig. 21.10; Wigen and White, 1964; Thomson, 1981). This earthquake generated a series of sea waves that moved radially outward from the epicentre near the head of Prince William Sound. Within a few hours, the waves reached the outer coasts of the Queen Charlotte Islands and Vancouver Island, damaging the communities of Port Alberni, Hot Springs Cove, and Zeballos. Port Alberni was hardest hit: 260 houses in this community were damaged, and the total loss was estimated to be $10 000 000.

Seismic sea waves comparable to those generated by the 1964 earthquake are rare. Of the 176 tsunamis recorded in the Pacific Ocean between 1900 and 1970, 35 caused damage near their sources, but only 9 resulted in widespread destruction (Thomson, 1981). Even fewer produced significant wave runups on British Columbia shores.

Tsunamis generated by distant earthquakes do not affect all parts of the British Columbia coast equally. For example, the 1964 tsunami was highly destructive at Port Alberni, but caused little damage at Tofino only 65 km away; it also had very little effect in protected waterways such as the Strait of Georgia. These differences are attributable to the morphological complexity of the British Columbia shoreline and the adjacent sea floor. Damage at Port Alberni in 1964 resulted, in large part, from the funnelling effects of the long narrow inlet up which the waves travelled.

Not all tsunamis in British Columbia are the result of distant earthquakes; some are produced by local quakes and submarine landslides. These locally generated tsunamis may damage areas that are unlikely to be affected by far-travelled seismic sea waves. For example, the Vancouver Island earthquake of June 23, 1946 (surface-wave magnitude = 7.2) produced sea waves in the Strait of Georgia and nearby inlets (Murty, 1977). At Sisters Rock near Texada Island, the main wave produced by this earthquake had an amplitude of over 2 m; a slightly smaller secondary wave was responsible for the death of one person near Mapleguard Point on eastern Vancouver Island. Another local tsunami caused about $600 000 damage to shore installations at Kitimat in April 1975 (Campbell and Skermer, 1975; Murty, 1979). This tsunami, unlike those in 1946 and 1964, was triggered by a submarine landslide.

EROSION

Coastal and river-bank erosion have caused some property damage in the Canadian Cordillera. Some reaches of shoreline in British Columbia have receded rapidly (up to a few metres per year) in historical time, necessitating the abandonment or relocation of homes and roads. Such erosion has been limited to relatively small areas along those parts of the coast bordered by thick Quaternary sediments, for example eastern Graham Island, eastern and southern Vancouver Island, and near Vancouver (Clague and Bornhold, 1980; Clague, 1982a).

During periods of high discharge, streams may erode their banks and, in some instances, occupy new channels. Erosion may be particularly severe at meander bends and other inflection points, and opposite the mouths of aggrading tributaries. Rapid lateral erosion may create steep banks which are prone to landsliding. Braided streams, which are common in and around glaciated mountain ranges, are especially susceptible to rapid shifts of channels; this, in conjunction with normal flooding, renders the floodplains of these streams hazardous sites for development.

Finally, anomalous surface runoff may rapidly incise loose Quaternary sediments and thus carve deep ravines or canyons. Such events, although rare, may occur in developed or developing areas and thus destroy valuable land or damage human works. Perhaps the most spectacular example was the formation in January 1935 of "Campus Canyon" at the University of British Columbia in Vancouver. Within a period of two days following a week of unprecedented snow and rain, a raging torrent in a formerly minor ravine removed about 100 000 m³ of sand and thus created a badland canyon of impressive proportions (Eisbacher and Clague, 1981). This catastrophic gullying was initiated when a large body of standing water was artificially diverted from the poorly drained University campus into the ravine.

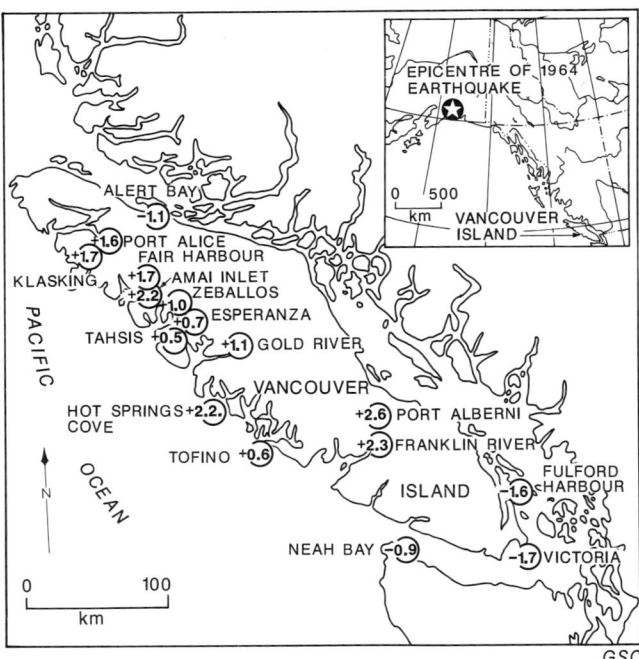

Figure 21.10. Sites in southwestern British Columbia that reported significant tsunami activity following the 1964 Alaska earthquake. Numbers are maximum wave heights in metres above or below higher high water level (positive values >HHW; negative values <HHW). The tsunami caused damage in Hot Springs Cove, Zeballos, and Port Alberni. Adapted from Wigen and White (1964).

REFERENCES

Avalanche Task Force
1974: Report on findings and recommendations to the Honourable Graham R. Lea, Minister of Highways, September 30, 1974; British Columbia Department of Highways, Victoria, 33 p.

Bacon, C.R.
1983: Eruptive history of Mount Mazama and Crater Lake caldera, Cascade Range, U.S.A.; Journal of Volcanology and Geothermal Research, v. 18, p. 57-115.

Barker, M.L.
1977: Natural resources of British Columbia and the Yukon; Douglas, David & Charles, North Vancouver, 155 p.

Basham, P.W., Weichert, D.H., Anglin, F.M., and Berry, M.J.
1985: New probabilistic strong seismic ground motion maps of Canada; Seismological Society of America Bulletin, v. 75, p. 563-595.

Bevier, M.L., Armstrong, R.L., and Souther, J.G.
1979: Miocene peralkaline volcanism in west-central British Columbia—Its temporal and plate tectonics setting; Geology, v. 7, p. 389-392.

Blown, I. and Church, M.
1985: Catastrophic lake drainage within the Homathko River basin, British Columbia; Canadian Geotechnical Journal, v. 22, p. 551-563.

Bostock, H.S.
1952: Geology of northwest Shakwak Valley, Yukon Territory; Geological Survey of Canada, Memoir 267, 54 p.

Bovis, M.J.
1985: Earthflows in the Interior Plateau, southwest British Columbia; Canadian Geotechnical Journal, v. 22, p. 313-334.

Brown, R.L. and Psutka, J.F.
1980: Structural and stratigraphic setting of the Downie slide, Columbia River valley, British Columbia; Canadian Journal of Earth Sciences, v. 17, p. 698-709.

Campbell, D.B. and Skermer, N.A.
1975: Report to British Columbia Water Resources Service on investigation of seawave at Kitimat, B.C.; Golder Associates, Vancouver, 9 p.

Church, M. and Miles, M.J.
1987: Meteorological antecedents to debris flow in southwestern British Columbia; Some case studies; in Debris Flows/Avalanches: Process, Recognition, and Mitigation, J.E. Costa and G.F. Wieczorek (ed.), Geological Society of America, Reviews in Engineering Geology, v. 7, p. 63-79.

Clague, J.J.
1982a: Erosion at Point Grey, British Columbia; Geoscience Canada, v. 9, p. 129-131.
1982b: The role of geomorphology in the identification and evaluation of natural hazards; in Applied Geomorphology, R.G. Craig and J.L. Craft (ed.), George Allen and Unwin, London, p. 17-43.

Clague, J.J. and Bornhold, B.D.
1980: Morphology and littoral processes of the Pacific coast of Canada; in The Coastline of Canada, S.B. McCann (ed.), Geological Survey of Canada, Paper 80-10, p. 339-380.

Clague, J.J. and Rampton, V.N.
1982: Neoglacial Lake Alsek; Canadian Journal of Earth Sciences, v. 19, p. 94-117.

Clague, J.J. and Souther, J.G.
1982: The Dusty Creek landslide on Mount Cayley, British Columbia; Canadian Journal of Earth Sciences, v. 19, p. 524-539.

Clague, J.J., Evans, S.G., and Blown, I.G.
1985: A debris flow triggered by the breaching of a moraine-dammed lake, Klattasine Creek, British Columbia; Canadian Journal of Earth Sciences, v. 22, p. 1492-1502.

Clarke, G.K.C.
1982: Glacier outburst floods from "Hazard Lake", Yukon Territory, and the problem of flood magnitude prediction; Journal of Glaciology, v. 28, p. 3-21.

Clarke, G.K.C. and Waldron, D.A.
1984: Simulation of the August 1979 sudden discharge of glacier-dammed Flood Lake, British Columbia; Canadian Journal of Earth Sciences, v. 21, p. 502-504.

Cockfield, W.E. and Buckham, A.F.
1956: Sinkhole erosion in the White Silts of Kamloops; Royal Society of Canada, Transactions, ser. 3, v. 40, sect. 4, p. 1-10.

Cruden, D.M.
1982: The Brazeau Lake slide, Jasper National Park, Alberta; Canadian Journal of Earth Sciences, v. 19, p. 975-981.
1985: Rock slope movements in the Canadian Cordillera; Canadian Geotechnical Journal, v. 22, p. 528-540.

Cruden, D.M. and Krahn, J.
1978: Frank rockslide, Alberta, Canada; in Rockslides and Avalanches, 1, Natural Phenomena, B. Voight (ed.), Elsevier Scientific Publishing Company, New York, p. 97-112.

Drysdale, C.W.
1914: Geology of the Thompson River valley below Kamloops Lake, B.C.; Geological Survey of Canada, Summary Report 1912, p. 115-150.

Eisbacher, G.H.
1979: First-order regionalization of landslide characteristics in the Canadian Cordillera; Geoscience Canada, v. 6, p. 69-79.
1983: Slope stability and mountain torrents, Fraser Lowlands and southern Coast Mountains, British Columbia; Geological Association of Canada, Mineralogical Association of Canada, Canadian Geophysical Union, Joint Annual Meeting (Victoria), Field Trip Guidebook, No. 15, 46 p.

Eisbacher, G.H. and Clague, J.J.
1981: Urban landslides in the vicinity of Vancouver, British Columbia, with special reference to the December 1979 rainstorm; Canadian Geotechnical Journal, v. 18, p. 205-216.
1984: Destructive mass movements in high mountains: hazard and management; Geological Survey of Canada, Paper 84-16, 230 p.

Evans, S.G.
1982: Landslides and surficial deposits in urban areas of British Columbia: a review; Canadian Geotechnical Journal, v. 19, p. 269-288.
1984: The 1880 landslide dam on Thompson River, near Ashcroft, British Columbia; in Current Research, Part A, Geological Survey of Canada, Paper 84-1A, p. 655-658.
1987: A rock avalanche from the peak of Mount Meager, British Columbia; in Current Research, Part A, Geological Survey of Canada, Paper 87-1A, p. 929-934.

Evans, S.G., Aitken, J.D., Wetmiller, R.J., and Horner, R.B.
1987: A rock avalanche triggered by the October 1985 North Nahanni earthquake, District of Mackenzie, N.W.T.; Canadian Journal of Earth Sciences, v. 24, p. 176-184.

Fitzharris, B.B. and Schaerer, P.A.
1980: Frequency of major avalanche winters; Journal of Glaciology, v. 26, p. 43-52.

Fulton, R.J.
1971: Radiocarbon geochronology of southern British Columbia; Geological Survey of Canada, Paper 71-37, 28 p.

Gilbert, R.
1972: Drainings of ice-dammed Summit Lake, British Columbia; Canada Department of Environment, Inland Waters Directorate, Water Resources Branch, Scientific Series, No. 20, 17 p.

Gilbert, R. and Desloges, J.R.
1987: Sediments of ice-dammed, self-draining Ape Lake, British Columbia; Canadian Journal of Earth Sciences, v. 24, p. 1735-1747.

Hamilton, T.S. and Luternauer, J.L.
1983: Evidence of seafloor instability in the south-central Strait of Georgia, British Columbia: a preliminary compilation; in Current Research, Part A, Geological Survey of Canada, Paper 83-1A, p. 417-421.

Horner, R.B.
1983: Seismicity in the St. Elias region of northwestern Canada and southeastern Alaska; Seismological Society of America Bulletin, v. 73, p. 1117-1137.

Hughes, O.L., Rampton, V.N., and Rutter, N.W.
1972: Quaternary geology and geomorphology, southern and central Yukon (northern Canada); 24th International Geological Congress (Montreal), Guidebook, Field Excursion A11, 59 p.

Hughes, O.L., Veillette, J.J., Pilon, J., Hanley, P.T., and van Everdingen, R.O.
1973: Terrain evaluation with respect to pipeline construction, Mackenzie Transportation Corridor, central part, lat. 64° to 68°N.; Environmental-Social Committee, Northern Pipelines, Task Force on Northern Oil Development, Report No. 73-37, 74 p.

Hungr, O., Morgan, G.C., VanDine, D.F., and Lister, D.R.
1987: Debris flow defenses in British Columbia; in Debris Flows/Avalanches: Process, Recognition, and Mitigation, J.E. Costa and G.F. Wieczorek (ed.), Geological Society of America, Reviews in Engineering Geology, v. 7, p. 201-222.

Jackson, L.E., Jr.
1979: A catastrophic glacial outburst flood (jökulhlaup) mechanism for debris flow generation at the Spiral Tunnels, Kicking Horse River basin, British Columbia; Canadian Geotechnical Journal, v. 16, p. 806-813.

Jackson, L.E., Church, M., Clague, J.J., and Eisbacher, G.H.
1985: Slope hazards in the southern Coast Mountains of British Columbia, Field Trip 4; in Field Guides to Geology and Mineral Deposits in the Southern Canadian Cordillera, D.J. Tempelman-Kluit (ed.), Geological Society of America, Cordilleran Section, 1985 Annual Meeting (Vancouver), Field Trip Guidebook, p. 4-1 - 4-34.

Jones, D.P., Ricker, K.E., Desloges, J.R., and Maxwell, M.
1985: Glacier outburst flood on the Noeick River: the draining of Ape Lake, British Columbia, October 20, 1984; Geological Survey of Canada, Open File 1139, 81 p.

Keen, C.E. and Hyndman, R.D.
1979: Geophysical review of the continental margins of eastern and western Canada; Canadian Journal of Earth Sciences, v. 16, p. 712-747.

Kerr, F.A.
1948: Lower Stikine and western Iskut River areas, British Columbia; Geological Survey of Canada, Memoir 246, 94 p.

Lerbekmo, J.F., Westgate, J.A., Smith, D.G.W., and Denton, G.H.
1975: New data on the character and history of the White River volcanic eruption, Alaska; in Quaternary Studies, R.P. Suggate and M.M. Cresswell (ed.), Royal Society of New Zealand, Bulletin 13, p. 203-209.

Lipman, P.W. and Mullineaux, D.R. (editors)
1981: The 1980 eruptions of Mount St. Helens, Washington; United States Geological Survey, Professional Paper 1250, 844 p.

Luternauer, J.L. and Finn, W.D.L.
1983: Stability of the Fraser Delta front; Canadian Geotechnical Journal, v. 20, p. 603-616.

Luternauer, J.L. and Swan, D.
1978: Kitimat submarine slump deposit(s): a preliminary report; in Current Research, Part A, Geological Survey of Canada, Paper 78-1A, p. 327-332.

Mackay, J.R. and Mathews, W.H.
1973: Geomorphology and Quaternary history of the Mackenzie River valley near Fort Good Hope, N.W.T., Canada; Canadian Journal of Earth Sciences, v. 10, p. 26-41.

Marcus, M.G.
1960: Periodic drainage of glacier-dammed Tulsequah Lake, British Columbia; Geographical Review, v. 50, p. 89-106.

Mathewes, R.W. and Westgate, J.A.
1980: Bridge River tephra: revised distribution and significance for detecting old carbon errors in radiocarbon dates of limnic sediments in southern British Columbia; Canadian Journal of Earth Sciences, v. 17, p. 1454-1461.

Mathews, W.H.
1965: Two self-dumping ice-dammed lakes in British Columbia; Geographical Review, v. 55, p. 46-52.
1973: Record of two jökullhlaups [sic]; International Association of Scientific Hydrology, Publication No. 95, p. 99-110.
1979: Landslides of central Vancouver Island and the 1946 earthquake; Seismological Society of America Bulletin, v. 69, p. 445-450.

Mathews, W.H. and McTaggart, K.C.
1978: Hope rockslides, British Columbia, Canada; in Rockslides and Avalanches, 1, Natural Phenomena, B. Voight (ed.), Elsevier Scientific Publishing Company, New York, p. 259-275.

McRoberts, E.C. and Morgenstern, N.R.
1974: The stability of thawing slopes; Canadian Geotechnical Journal, v. 11, p. 447-469.

Miller, G.E. and Nyland, D.
1981: Geological hazards and urban development of glacio-lacustrine silt deposits in the Penticton area, British Columbia; 34th Canadian Geotechnical Conference (Fredericton), Proceedings, 20 p.

Milne, W.G., Rogers, G.C., Riddihough, R.P., McMechan, G.A., and Hyndman, R.D.
1978: Seismicity of western Canada; Canadian Journal of Earth Sciences, v. 15, p. 1170-1193.

Mokievsky-Zubok, O.
1977: Glacier-caused slide near Pylon Peak, British Columbia; Canadian Journal of Earth Sciences, v. 14, p. 2657-2662.

Moore, D.P. and Mathews, W.H.
1978: The Rubble Creek landslide, southwestern British Columbia; Canadian Journal of Earth Sciences, v. 15, p. 1039-1052.

Mullineaux, D.R., Hyde, J.H., and Rubin, M.
1975: Widespread late glacial and postglacial tephra deposits from Mount St. Helens volcano, Washington; United States Geological Survey, Journal of Research, v. 3, p. 329-335.

Murty, T.S.
1977: Seismic sea waves—tsunamis; Fisheries Research Board of Canada, Bulletin 198, 337 p.

1979: Submarine slide-generated water waves in Kitimat Inlet, British Columbia; Journal of Geophysical Research, v. 84, p. 7777-7779.

Nasmith, H.W. and Mercer, A.G.
1979: Design of dykes to protect against debris flows at Port Alice, British Columbia; Canadian Geotechnical Journal, v. 16, p. 748-757.

Nasmith, H., Mathews, W.H., and Rouse, G.E.
1967: Bridge River ash and some other Recent ash beds in British Columbia; Canadian Journal of Earth Sciences, v. 4, p. 163-170.

Peckover, F.L. and Kerr, J.W.G.
1977: Treatment and maintenance of rock slopes on transportation routes; Canadian Geotechnical Journal, v. 14, p. 487-507.

Piteau, D.R.
1977: Regional slope-stability controls and engineering geology of the Fraser Canyon, British Columbia; in Landslides, D.R. Coates (ed.), Geological Society of America, Reviews in Engineering Geology, v. 3, p. 85-111.

Piteau, D.R., Mylrea, F.H., and Blown, I.G.
1978: Downie slide, Columbia River, British Columbia; in Rockslides and Avalanches, 1, Natural Phenomena, B. Voight (ed.), Elsevier Scientific Publishing Company, New York, p. 365-392.

Plafker, G., Hudson, T., Burns, T., and Rubin, M.
1978: Late Quaternary offsets along the Fairweather fault and crustal plate interactions in southern Alaska; Canadian Journal of Earth Sciences, v. 15, p. 805-816.

Prior, D.B. and Bornhold, B.D.
1984: Geomorphology of slope instability features of Squamish Harbour, Howe Sound, British Columbia; Geological Survey of Canada, Open File 1095.
1986: Sediment transport on subaqueous fan delta slopes, Britannia Beach, British Columbia; Geo-Marine Letters, v. 5, p. 217-224.

Prior, D.B., Bornhold, B.D., Coleman, J.M., and Bryant, W.R.
1982: Morphology of a submarine slide, Kitimat Arm, British Columbia; Geology, v. 10, p. 588-592.

Prior, D.B., Bornhold, B.D., and Johns, M.W.
1984: Depositional characteristics of a submarine debris flow; Journal of Geology, v. 92, p. 707-727.
1986: Active sand transport along a fjord-bottom channel, Bute Inlet, British Columbia; Geology, v. 14, p. 581-584.

Prior, D.B., Wiseman, W.J., and Gilbert, R.
1981: Submarine slope processes on a fan delta, Howe Sound, British Columbia; Geo-Marine Letters, v. 1, p. 85-90.

Riddihough, R.P. and Hyndman, R.D.
1976: Canada's active western margin—the case for subduction; Geoscience Canada, v. 3, p. 269-278.

Rogers, G.C. and Souther, J.G.
1983: Hotspots trace plate movements; GEOS, v. 12, no. 2, p. 10-13.

Rutter, N.W., Boydell, A.N., Savigny, K.W., and van Everdingen, R.O.
1973: Terrain evaluation with respect to pipeline construction, Mackenzie Transportation Corridor, southern part, lat. 60° to 64°N; Environmental-Social Committee, Northern Pipelines, Task Force on Northern Oil Development, Report No. 73-36, 135 p.

Ryder, J.M.
1985: Terrain inventory for the Stikine-Iskut area; British Columbia Ministry of Environment, MOE Technical Report 11, 85 p.

Schaerer, P.
1962: Avalanche defenses for the Trans-Canada Highway at Rogers Pass; National Research Council of Canada, Division of Building Research, NRC 7020 (DBR Technical Report No. 141), 52 p.

Souther, J.G.
1977: Volcanism and tectonic environments in the Canadian Cordillera— a second look; in Volcanic Regimes in Canada, W.R.A. Baragar, L.C. Coleman and J.M. Hall (ed.), Geological Association of Canada, Special Paper No. 16, p. 3-24.
1986: The western Anahim Belt: root zone of a peralkaline magma system; Canadian Journal of Earth Sciences, v. 23, p. 895-908.

Stethem, C.J. and Schaerer, P.A.
1979: Avalanche accidents in Canada. I. A selection of case histories of accidents, 1955 to 1976; National Research Council of Canada, Division of Building Research, NRCC 17292 (DBR Paper No. 834), 114 p.
1980: Avalanche accidents in Canada. II. A selection of case histories of accidents, 1943 to 1978; National Research Council of Canada, Division of Building Research, NRCC 18525 (DBR Paper No. 926), 75 p.

Swan, D. and Luternauer, J.L.
1978: Mosaic of side scan sonar records, northern Kitimat Arm, B.C.; Geological Survey of Canada, Open File 579.

Terzaghi, K.
1956: Varieties of submarine slope failures; 8th Texas Conference on Soil Mechanics and Foundation Engineering, Proceedings, 41 p.

Thomson, R.E.
1981: Oceanography of the British Columbia coast; Canada Department of Fisheries and Oceans, Canadian Special Publication of Fisheries and Aquatic Sciences 56, 291 p.

Tiffin, D.L., Murray, J.W., Mayers, I.R., and Garrison, R.E.
1971: Structure and origin of foreslope hills, Fraser Delta, British Columbia; Bulletin of Canadian Petroleum Geology, v. 19, p. 589-600.

VanDine, D.F.
1980: Engineering geology and geotechnical study of Drynoch landslide, British Columbia; Geological Survey of Canada, Paper 79-31, 34 p.
1985: Debris flows and debris torrents in the southern Canadian Cordillera; Canadian Geotechnical Journal, v. 22, p. 44-68.

Varnes, D.J.
1978: Slope movement types and processes; in Landslides, Analysis and Control, R.L. Schuster and R.J. Krizek (ed.), National Research Council, Transportation Research Board, Special Report 176, p. 12-33.

Westgate, J.A.
1977: Identification and significance of late Holocene tephra from Otter Creek, southern British Columbia, and localities in west-central Alberta; Canadian Journal of Earth Sciences, v. 14, p. 2593-2600.

Westgate, J.A. and Dreimanis, A.
1967: Volcanic ash layers of Recent age at Banff National Park, Alberta, Canada; Canadian Journal of Earth Sciences, v. 4, p. 155-162.

Westgate, J.A., Smith, D.G.W., and Tomlinson, M.
1970: Late Quaternary tephra layers in southwestern Canada; in Early Man and Environments in Northwest North America, R.A. Smith and J.W. Smith (ed.), University of Calgary, Archaeological Association, Calgary, p. 13-33.

Whitham, K. and Hasegawa, H.S.
1975: The estimation of seismic risk in Canada—a review; Canada Department of Energy, Mines and Resources, Earth Physics Branch Publications, v. 45, p. 137-162.

Wigen, S.O. and White, W.R.
1964: Tsunami of March 27-29, 1964, west coast of Canada; Canada Department of Mines and Technical Surveys, Ottawa, 6 p.

Authors' address

J.J. Clague
Terrain Sciences Division
Geological Survey of Canada
100 West Pender Street,
Vancouver, British Columbia
V6B 1R8

Printed in Canada

Chapter 22

OUTSTANDING PROBLEMS

Introduction
Miogeocline
    Character, age and correlation of the basement
    Nature of early rifting
    Late Proterozoic-Early Cambrian rifting
    Causes for rifting, volcanism and formation of deep basins
    Late Devonian-Mississippian tectonism
    Correlations with the Innuitian and Brooks Range orogens
Displaced terranes
Pericratonic terrane
Accreted terranes
    Basement
    Correlations and relationships
    Paleomagnetism
Tectonism
    Proterozoic tectonism
    Paleozoic tectonism
    Mesozoic and Cenozoic tectonism
Basin evolution
    Foreland Belt
    Intermontane, Coast and Insular belts
Biostratigraphy
    Magmatism
    Timing
    Settings
    Physical volcanology
    Ultimate origins
Regional metamorphism
Physiographic evolution and Quaternary geology
Natural hazards
Metallogeny

Chapter 22

OUTSTANDING PROBLEMS

H. Gabrielse and C.J. Yorath

Introduction

The synthesis of Cordilleran geology presented in this volume provides the reader with a summary of the current geoscience database and leading hypotheses derived therefrom. Some of the problems and questions that remain unanswered in this description of the evolution of the Canadian Cordillera are reviewed in this chapter and generally are the result of a lack of data. Many of the problems summarized below may be solved simply through detailed investigations in specific regions. Others demand greatly increased input from geophysical and geochemical studies. Finally there is a continuing need for integration of the complete spectrum of geoscience research to produce theoretical models of Cordilleran tectonic history for the benefit of resource exploration and evaluation and to provide guidance for future research.

Miogeocline

1. Character, age and correlation of the basement

The paucity of outcrops in most areas precludes adequate characterization of basement of the Cordilleran miogeocline. Those that have been studied in detail can be related only in part to crystalline rocks exposed in the Canadian Shield. Correlations of basement gneisses across the Southern Rocky Mountain Trench have been disputed. The source for voluminous feldspathic, quartzose grits in the Windermere Supergroup is unknown but preliminary dates in the northern Cordillera suggest a basement terrane somewhat older than the Thelon-Wopmay accreted terranes in the northwestern Canadian Shield. The nature and distribution of crystalline basement under the Selwyn and Kechika basins during Paleozoic time are controversial. If present, how much was it attenuated as a result of Late Proterozoic and early Paleozoic rifting?

2. Nature of early rifting

The configuration of the western margin of ancestral North America at the beginning of Purcell (Belt) deposition remains problematical. Firm correlation of the Muskwa Assemblage in the Northern Rocky Mountains with the Purcell-Wernecke Assemblage in the Purcell and Wernecke mountains is essential in this analysis. Were aulacogenes important elements in Middle Proterozoic deposition? Was the 'Mackenzie Mountains Supergroup' deposited south of its present distribution in the Mackenzie Mountains? One of the enigmas in Cordilleran geology is the great stratigraphic hiatus between the Purcell and Windermere supergroups in the southern Cordillera. The nature of the event or events that terminated Purcell-Wernecke deposition is essentially unknown except that it involved low-grade regional metamorphism and local deformation. Conceivably the western margin of ancestral North America could have evolved through at least two miogeoclinal regimes (Middle Proterozoic and Late Proterozoic through Paleozoic) each initiated by rifting episodes separated by about 800 Ma.

3. Late Proterozoic-Early Cambrian rifting

One of the main questions concerning the Windermere rifting event is its relationship to postulated Cambrian drifting almost 200 Ma later. Except for subsidence curves approximating those for theoretical thermal subsidence there is the general lack of data to support the notion of drifting in earliest Cambrian time. Of global importance is the correlation of the two Late Proterozoic glacial episodes recorded in the northern Cordillera with glaciations of Late Proterozoic age in many other parts of the world. The ultimate causes of Late Precambrian glaciations have not been demonstrated. These problems, and concepts of eustatic changes in sea level resulting from glaciations, offer exciting challenges.

4. Causes for rifting, volcanism and formation of deep basins

Throughout its history, at least until Mississippian time, the miogeocline was the site of minor but widespread volcanism near the boundary between the carbonate platform to the east and shale basins to the west. Rifting episodes during the Middle Cambrian and Middle Devonian led to the formation of fault blocks, at least in part parallel with the miogeoclinal trend. Volcanism took place locally at various times and was most widespread during the Middle Ordovician. Carbonatites were emplaced mainly in Devono-Mississippian time. Successions of Siluro-Devonian carbonate strata, locally exceeding 3 km in thickness, were deposited in Root Basin. All these processes occurred in a 'passive margin' environment. Can these various phenomena be related to one or more of sediment loading, crustal flexure or episodic rifting?

Gabrielse, H. and Yorath, C. J.
1991: Outstanding problems, Chapter 22 in Geology of the Cordilleran Orogen in Canada, H. Gabrielse and C. J. Yorath (ed.); Geological Survey of Canada, Geology of Canada, no. 4, p. 817-823 (also Geological Society of America, The Geology of North America, v. G-2)

5. Late Devonian-Mississippian tectonism

The relationships among Late Devonian and Mississippian alkalic volcanism, block faulting and clastic wedge deposition in the northern Cordillera are far from clear. The tectonic setting of these events, similar in large part to those associated with the Antler Orogeny in the southwestern United States, requires much further study.

6. Correlations with the Innuitian and Brooks Range orogens

Correlations of stratigraphy, structure and tectonic episodes among the Cordilleran Orogen and the Brooks Range and Innuitian orogens are speculative. Correlations of the Pinguicula, Lower Tindir, Windermere and Neruokpuk strata are in doubt which seriously inhibits understanding of Late Proterozoic paleogeography. Possible correlations between Northern Yukon geology and that in the Innuitian orogen are only general and speculative.

Displaced terranes

The Cassiar Terrane is clearly a fragment of ancestral North America, however, the amount of displacement from its original position remains controversial. Further study is needed on the bounding Tintina and Northern Rocky Mountain Trench faults and related structures farther west. The extent and time of displacement of the Arctic Alaska Terrane in relation to ancestral North America is disputed with widely conflicting estimates.

Pericratonic terrane

The main concern about the Kootenay Terrane and its subterranes is their relationship to ancestral North America. The Kootenay Terrane could be a relatively little displaced distal part of ancestral North America with which it has some close stratigraphic similarities but also some significant differences. Correlations of stratigraphy and tectonic events in the Kootenay Terrane require much clarification before meaningful comparisons can be made with ancestral North America. A possible relationship between Devono-Mississippian plutonism in the Kootenay Terrane and clastic wedge deposition in the miogeocline should be examined.

Accreted terranes

Concepts and analyses of terranes are only in their infancy and further studies undoubtedly will modify present assignments.

1. Basement

Except for the Slide Mountain, Cache Creek, Bridge River and Hozameen terranes, which appear to consist almost entirely of oceanic crust, little is known about basement rocks in most of the accreted terranes. Stratigraphic, geochronological and geochemical studies are needed in critical areas, particularly in the northeastern and western parts of Stikinia, parts of the Alexander Terrane and the southern regions of Quesnellia. The character and thickness of crust are important in assessing the role of buoyancy in collision processes.

2. Correlations and relationships

Problems of correlation require a great deal of biostratigraphic, petrographic and geochronological work. For example, the relationships among Stikinia and the Taku and Cadwallader terranes are unknown. The age span of the Nisling Terrane and its relationship to Stikinia is in considerable doubt. Structural, metamorphic and plutonic complexities make the unravelling of terrane relationships in the Coast Belt, particularly in the southern part, a major challenge. There is much controversy concerning the suture between the Intermontane and Insular superterranes. Were they loosely amalgamated even before Middle Jurassic tectonism occurred in the Omineca Belt?

A host of questions can be raised concerning the relationship between the Cache Creek and Slide Mountain terranes prior to the thrusting of the former onto Stikinia and the latter onto ancestral North America. Their paleogeographic positions, stratigraphy and tectonic histories are different, yet in the northern Cordillera, structures in the two terranes seem to verge outward from the Teslin Suture Zone.

3. Paleomagnetism

Although discrimination between terranes is based on stratigraphy it is paleomagnetism, supported by paleontology, that offers the greatest potential for determining the pre-amalgamation positions of terranes relative to ancestral North America. The favoured current view is that the accreted terranes, perhaps including parts of the pericratonic Kootenay Terrane and the displaced Cassiar Terrane (Baja British Columbia) moved south from near their present positions to more southerly latitudes in the northern hemisphere in post-Permian time. Then, between mid-Cretaceous and Eocene time, Baja British Columbia moved northward, perhaps as much as 2300 km to its present position. Much paleomagnetic work is required to help define the eastern limit of these displacements and more study needs to be devoted to the geological constraints on the paleomagnetic model.

Additional paleomagnetic data from all of the terranes would assist in evaluating their pre-amalgamation history. Throughout much of the Cordillera, particularly the Slide Mountain and Cache Creek terranes, no data are available for rocks magnetized before the mid-Cretaceous.

One of the most perplexing problems for paleomagnetic studies is the origin of the enigmatic, widespread mid-Cretaceous magnetic overprint. Was it entirely of mid-Cretaceous age?

Tectonism

The geological history of the Cordillera has been punctuated by a number of tectonic episodes indicated by magmatic, structural, metamorphic and sedimentological criteria.

1. Proterozoic tectonism

Little is known about the nature and extent of tectonism that ended deposition of the Purcell-Wernecke Assemblage. Do the scattered bits of information on deformation, metamorphism and plutonism represent only the fringe effects of a significant orogeny, the major evidence of which was removed by rifting during Late Proterozoic time?

2. Paleozoic tectonism

Most aspects related to the development of the Upper Devonian and Mississippian clastic wedge in the northern Cordillera remain speculative. The relationship of plutonism in the Kootenay Terrane with rifting and clastic wedge deposition in the distal parts of the miogeocline, the possible role of strike-slip faulting along the margin of ancestral North America, the problem of alkalic volcanism and the correlation with phenomena associated with the Antler Orogeny of the southwestern United States all demand further study.

The evolution of the Aklavik Arch, which reached a climax in Permian time and resulted in a widespread unconformity and the generation of coarse conglomerate in a northeast trending belt extending across the northern Yukon, presents an interesting problem. What caused the uplift and was there a relationship with the widespread sub-Permian unconformity that occurs throughout the length of the miogeocline in Canada?

Evidence for Permian tectonism is present also in the Slide Mountain Terrane as expressed by thrust faults, the presence of blueschist and eclogitic minerals, and the presence of granitic plutons. The relationship of these criteria to events in other terranes is unknown.

3. Mesozoic and Cenozoic tectonism

Much study has been devoted to the geometry and evolution of structures in the Omineca and Foreland belts. Far less is understood about the broader aspects of tectonism, for example, the role of terrane collisions, the nature and duration of intraplate deformation, the relationship of magmatism to deformation, the cause and effect of deformation and the development of clastic wedges, the relationship of plate convergence and episodic(?) tectonism, and the relationship between strike-slip faulting and relative plate movements. The interplay between subduction, volcanism, plutonism extension, and strike-slip faulting in Eocene time is controversial.

Although tectonism in the Omineca Belt during Middle Jurassic time and thrusting in the eastern Foreland Belt in Paleocene time seem to be well documented, the migration of deformation in time and space across the eastern Cordillera is little understood. Where were the eastern limits of deformations at various times? What were the relationships in time and space between westward and eastward verging structures? Can the age of individual thrust faults be determined with modern geochronological techniques?

Estimates of the magnitude of contractional, extensional and strike-slip displacements are, for the most part, poorly constrained. Estimates of shortening in the Mackenzie Mountains and central and eastern Northern Rocky Mountains are much less than those for the Southern Rocky Mountains. On the other hand, a great amount of shortening is suggested by deformation in the Selwyn Mountains and westernmost Northern Rocky Mountains. Most of these structures are older than those that effected the major contraction in the eastern Foreland Belt. The northern Cordillera, particularly, is lacking in detailed studies concerned with these problems.

Basin evolution

1. Foreland Belt

The exploration for petroleum and coal resources in the Foreland Belt has led to extensive research on foredeep clastic wedge deposits. Theoretical models relating basin evolution to crustal loading by thrust sheets and lithospheric flexuring are generally accepted. To what extend the stratigraphic record preserved in the foredeeps provides a history of eastern Cordilleran tectonism is, however, debatable. For example, do the Lower(?) Cretaceous clastics preserved only locally in the western Mackenzie Mountains represent an early clastic wedge that was once much more extensive and directly linked to Early Cretaceous deformation in the inner part of the northern Foreland Belt?

The evolution of miogeoclinal basins remains an important topic for research. Few detailed or comprehensive studies have been carried out in most of the miogeoclinal assemblages. Causes of basin configurations, effects of crustal attenuation, thermal and mechanical subsidence have been little explored. What led to the thick and widespread Triassic deposits in the Peace and Liard river regions? Was rifting responsible for the widespread occurrence of volcanic rocks in lower Paleozoic miogeoclinal strata and the locally thick deposits of Ordovician, Silurian and Devonian rocks? How important was Middle Cambrian rifting in the formation of the Richardson Trough, the Misty Creek Embayment and modification of the carbonate to shale boundary throughout the length of the Cordillera?

Analysis of Proterozoic and Lower Cambrian basins presents a host of questions. How was the sedimentation of the Gog and Hamill groups reflected in the presence of Montania and the Purcell Arch? Why did deposition of coarse, feldspathic grit continue much longer in Selwyn Basin and presumably Kechika Basin than in the Cassiar Terrane. How does the well displayed Upper Proterozoic stratigraphy of the Cariboo and Kaska mountains correlate with the Misinchinka and Miette groups of the Rocky Mountains? Correlation of the Coates Lake, Pinguicula, Tindir (at various places along the boundary between Alaska and Yukon Territory) and the Neruokpuk Group is problematical but essential to reconstruction of basin configuration in Late Proterozoic time. The original extent and controls for deposition of the 'Mackenzie Mountains Supergroup' are not well known. Finally, the depositional setting and configuration of the Purcell-Wernecke basin or basins have been studied only locally.

2. Intermontane, Coast and Insular belts

A wealth of information on terrane amalgamation, accretion and tectonism is contained in the stratigraphy and character of Mesozoic and Cenozoic basins west of the Omineca Belt. Most have been studied in only a reconnaissance manner, however, and many details of facies changes, sedimentary transport, correlations and environments of deposition are lacking. The source of Upper Tiassic(?) feldspathic grit interbedded with radiolarian chert in the upper part of the Cache Creek Group in northern British Columbia is a question. If the source was Quesnellia a link was established between the Cache Creek Terrane and Quesnellia by Late Triassic time. Separation of the mainly eastward-derived Inklin strata from the southwestward-derived Takwahoni strata, both of Early Jurassic age, is an important problem relative to the time of terrane amalgamation in the Whitehorse Trough. The initial configuration of the Bowser Basin, and its relationship with the Relay Mountain part of the Tyaughton Trough and with the volcanic-bearing Gambier Assemblage are poorly known. The histories and degree of stratigraphic correlations among the Dezadeash, Bowser, Relay Mountain, Gambier, Tyaughton and Methow troughs are fundamental to concepts of terrane amalgamation, accretion, and western Cordilleran paleogeography. The controls of sedimentation in the Sustut and Tantalus basins should be explored relative to transcurrent faulting. Similarly, the Queen Charlotte and Nanaimo basins in the Insular Belt contain a record of tectonism in the Coast Belt that needs further study. Stratigraphic, biostratigraphic and sedimentological studies of the Tofino and Winona basins are critical to understanding Cenozoic plate motions, margin uplift histories and accretionary wedge development.

The mechanisms of basin formation in strike-slip environments have been little studied. How did Eocene basins, some of apparently great thickness, evolve along strike slip faults?

Biostratigraphy

The past decade has witnessed an explosion of biostratigraphic data, in large part through the introduction of conodont and radiolaria dating techniques. Further studies should be directed at greatly increasing the database, particularly in the island-arc and oceanic terranes where macrofossil data are scarce, and at refining biostratigraphic dating by means of correlation between various kinds of fossils. The exquisite preservation of faunas including radiolaria, foraminifera, conodonts and ammonites in Mesozoic strata of the Queen Charlotte Islands offers great promise for refined biostratigraphic zonation. Future studies on endemism of various fossils should further help to constrain paleogeographic models of terrane distributions.

Magmatism

Problems concerning the nature and timing of volcanism and plutonism are fundamental to an understanding of Cordilleran evolution. Tertiary volcanics, because of their exposure and general lack of alteration, have received considerable study focused variously on physical form, stratigraphy, chemistry and evolution of volcanic centres in various parts of the Cordillera. Older volcanic assemblages have not received the same degree of investigation and in most areas have been described in only a cursory way. Few plutons or plutonic suites have received adequate study with modern geochronological and geochemical techniques and, indeed, many plutons have not been mapped except around their margins.

1. Timing

Many volcanic units and plutons are poorly constrained by radiometric or biostratigraphic data. The assignment of plutons to various suites and correlations of volcanic sequences is, therefore, commonly tentative. Durations of island-arc volcanism, the extent of individual arcs and the evolution of arc terranes is little known for Mesozoic and older rocks. In particular, the relationships of Lower Jurassic volcanic rocks around the periphery of the Bowser Basin, generally included in the Hazelton Group, are understood only locally. Few reliable ages are available for Proterozoic and Paleozoic volcanic rocks. Much more dating is needed for Paleozoic plutons in the Kootenay and Arctic Alaska terranes. Perhaps the greatest task will be dating plutons in the Coast Belt which range in age from Late Triassic to Cenozoic, and which, for Mesozoic plutons at least, have the same geochemical signature. Finally, dating of volcanic episodes should be related to the volcanic ashes in the several Cordilleran sedimentary basins.

2. Settings

Speculations on the tectonic settings of volcanic and plutonic rocks have been based on their chemistry with particular emphasis on the concentration of relatively stable, rare-earth elements. Very little chemical work has been done on most volcanic assemblages and plutonic suites, however, and thus there has been little debate regarding alternative environments suggested by geochemical models. For example, are geochemical variations within assemblages such as the Nicola, Takla, Stuhini and Hazelton as great as those between assemblages? What is the significance of the alkalic Copper Mountain Suite of plutons within the alkaline part of the Nicola arc volcanics? Are they coeval and a product of the western calc-alkaline to eastern alkaline polarity of the Nicola arc? Chemical polarities of most other volcanic island arc rocks in the Cordillera are unknown.

3. Physical volcanology

Few detailed studies with emphasis on facies and thickness changes have been carried out on volcanic assemblages in pre-Cenozoic rocks. Such studies, providing data on proximal to distal facies and their distributions relative to terrane boundaries may provide information on terrane linkages and strike-slip displacements. Mapping of a widespread Bajocian tuff unit which appears to be thickest in the Iskut River area on the west side of the Bowser Basin, but has been observed in the northern and eastern parts of the basin, should provide an excellent marker horizon in northwestern British Columbia and may constrain models of Cretaceous and Cenozoic deformation.

4. Ultimate origins

Many problems concerning the genesis of volcanic and plutonic rocks, not peculiar to the Canadian Cordillera, need to be addressed. Perhaps the most fundamental problem is the role of Benioff subduction, generally held to provide the mechanism for generation of island arc magmatic rocks. Were the mid-Cretaceous S-type granitic rocks in the Omineca Belt generated directly by Benioff subduction? If so, how can this process be reconciled with the generation of mid-Cretaceous granitic rocks in the Coast Belt? What was the role of crustal melting in each belt? Could crustal thickening by deformation in the Omineca Belt (Ampferer subduction) have resulted in melting of sialic crust and generation of S-type granites?

The widespread occurrence of Eocene volcanic and plutonic rocks requires explanation. Many occurrences in the Omineca and Intermontane belts are associated with tensional basins generated in a dextral strike-slip regime. Do their distributions and compositions, ranging from tholeiitic to calc-alkaline basalts to high potassic rhyolites require a direct link to Benioff subduction? Possibly, the volcanics in the rift basins have an origin similar to those in the rift valleys of Africa remote from Benioff subduction zones.

Regional metamorphism

Much of the information on metamorphic mineral assemblages in the Canadian Cordillera comes from studies concerned mainly with other subjects. As a result, data on metamorphic minerals are rarely comprehensive and in many cases are completely lacking. There is a great need for routine determinations of these minerals in mapping projects with accurate co-ordinates and identification of the rock units in which they were found. It is essential that these data be digitized to facilitate updating of the Cordilleran metamorphic map.

More carbon maturation and vitrinite reflectance data are required from areas having well calibrated mineralogical data on pressures and temperatures, especially in lowest chlorite zone rocks and zeolite-pumpellyite-prehnite facies. Data are sparse in the Bowser Basin, Rocky Mountains, Mackenzie Mountains and Vancouver Island.

Careful studies should be made of relationships between metamorphic isograds and terrane boundaries to determine the relative ages of the boundaries and metamorphism, the relationships of structural styles and metamorphism in adjacent terranes and the differences in P-T conditions. Investigations of this type are needed particularly in the southwestern Cordillera where the nature of terrane boundaries is the subject of considerable controversy.

Absolute and relative ages of metamorphism and deformation present some of the most difficult problems in Cordilleran geology. Careful integration of metamorphic, structural and radiometric dating techniques are required in the Coast and Omineca belts. Research combining P-T determinations from metamorphic mineral assemblages with radiometric ages to estimate depths of burial and rates of uplift are few and should be attempted in many more areas of complex tectonic history. There seems to be good potential for these kinds of studies in the northern Omineca and Coast belts where several ages of plutons cut earlier metamorphosed rocks and have metamorphic aureoles around them.

Physiographic evolution and Quaternary geology

No comprehensive studies of geomorphological evolution in the Canadian Cordillera have been made. Of particular interest is the history of landscape development from Late Miocene to Late Pleistocene time. During this interval the interplay between tectonism, glaciation, erosion and sedimentation should be resolvable. The significance and extent of various Cenozoic erosion surfaces and relationships to paleodrainage and sedimentation are little understood.

The ultimate origin of many spectacular lineaments in the western part of the Coast Belt and on Vancouver Island requires study. The origin of the Southern Rocky Mountain Trench is debatable. What controlled the linear distribution of the late normal faults that contributed to the present relief of the trench?

One of the most vexing problems of Quaternary research is the difficulty in correlation of rocks older than 50 000 years, the limit of reliable radiocarbon age determination. As a result, the chronology of older Pleistocene events is disputable.

Integration of Pleistocene events suffers from the lack of systematic terrain mapping. Much of northern and central British Columbia has undergone no regional terrain study.

Natural hazards

Studies of geological hazards in the Cordillera have received attention only in the past decade or so and only in relatively few regions. The most critical studies, concerning potential damage and loss of life, are those dealing with the nature and timing of large magnitude earthquakes. This research depends upon integration of a vast amount and variety of data including the results of historical seismic activity, enhanced monitoring of current seismic activity interpretations of the current plate-tectonic regime and related tectonic processes and theoretical models that may constrain hypotheses on the predictability of earthquakes.

Slope stability investigations have been restricted to localities of immediate concern and no regional assessments of hazards related to topography have been undertaken.

Metallogeny

Studies on the importance of tectonic setting for mineral deposits, the role of granitoid rocks in metal concentration, and mechanisms of the origin of strata-bound mineral deposits have received much attention in the last thirty years. There is a great need for ongoing stable lead isotope studies for dating mineral deposits and placing constraints on possible source rocks. Dating of wall-rock alterations is critical in research concerning correlations and genesis of deposits. For many mineral occurrences structural controls have been studied only locally and relations with regional structures are unknown or speculative. Models relating epithermal precious metal deposits to hydrothermal and hotspring systems hold great potential. The concentrations of zeolitic alterations for Mesozoic volcanic rocks such as those in the Hazelton Group need study with respect to mineral deposits.

Authors' addresses

H. Gabrielse
Cordilleran Division
Geological Survey of Canada
100 West Pender Street
Vancouver, British Columbia
V6B 1R8

C.J. Yorath
Pacific Geoscience Centre
9860 West Saanich Road
P.O. Box 6000
Sidney, British Columbia
V8L 4B2

Printed in Canada

INDEX

A

(A) subduction 683
A-E turbidites 103
A-type plutonism 526
absolute motion 441, 442
Acadian Faunal Province 156
Acado-Baltic Province 157
accreted terranes 26
    Devonian 210
    Ordovician – Silurian 195, 196
Active Formation 183
Adamant Pluton 509, 553
Adams Lake 147
Adams Plateau 601
Adelaide Basin 687
Adelaidean geosyncline 717
Aida Formation 113
Airy isostasy 47
Aitken Creek field 776
Aiyansh 391
Aklak Member 786
Aklavik Arch 255, 275, 825
Aklavik Arch Complex 639, 651, 652, 654, 664, 694
Aklavik Formation 275
Alaska Highway 9, 802
Alaskan-type ultramafic plutons 657
Alberta Basin 545
Alberta Plateau 645
Alberta Platform 224
Alberta Syncline 545, 641, 643
Albertella Zone 172, 173
Aldridge Formation 103, 104, 716
Alert Bay Volcanic Belt 373, 377, 382, 387, 485
Aleutian Arc 373
Aleutian Trench 44, 406, 439, 444
Alexander Terrane 31, 148, 183, 210, 277, 280, 310, 311, 312,
    467, 470, 472, 498, 559, 566, 578, 582, 692, 693, 697, 740, 824
Aley carbonatite 499, 547
Alice Arm Intrusions 518, 734
alkalic porphyry deposits 754
alkalic porphyry-related Cu, Au, Ag mines 727
Allenby Formation 786
Alligator Lake 394
Allison fault systems 600
Almstrom Creek Formation 275
Alps 684, 705
Alsek Province 395
Alsek River 148, 353, 395, 578, 580
Altyn Formation 104
America Plate 406, 439, 441, 443, 446
Ammerman 230, 233
Ammonites 37
Ampferer subduction 692, 826
amphibolite-facies Barrovian metamorphism 540
Amphitheatre Formation 353, 354, 377, 581, 785
Amundsen Embayment 463
Anabarites-Circotheca-Protohertzina Zone 157
Anahim Volcanic Belt 387, 388, 389, 439, 453, 462, 485,
    487, 705, 794
Anarchist Assemblage 290
Anarchist Group 292
ancestral Aklavik Arch 256, 261, 262
Ancestral North America 6, 26, 619, 624, 629, 688
ancestral North American miogeocline 185
anchizone 554

Ancient Wall Reef Complex 227
Andalusite zone 542
Andean orogen 705
Andean-type magmatic arcs 755
Anderson plains 116
Andinotype environment 526
Andinotype plutonism 526
Angayucham 695
Ankwell Member 475
Ankwell volcanics 557
Anne Creek Member 275
Anstey Pluton 610
anthracite 775, 785
Antler Flysch 242
Antler Formation 287, 288, 468
Antler Orogeny 695, 824
Antrim Formation 241
Anvil Allochthon 282, 283, 284, 285, 467, 548, 634, 636
Anvil Batholith 512, 549
Anvil district 720, 757, 759
Anvil Mining District 179
Anvil plutonic suite 343
Anvil Range 631
Aphebian 6
Aphelaspis Zone 179, 181
Apollo structure 585
Appekunny Formation 104, 105
Archibald Formation 292
Arcs Member 226
Arctic Alaska Terrane 26, 128, 262, 281, 723
Arctic Archipelago 639, 695
Arctic Innuitian region (Ellesmerian orogeny) 241
Arctic Red Formation 337
Arctomys Formation 181
Argonaut Mountain Fault 549, 551
Arnica formation 205
artesian groundwater 799
Ashcroft 539
Ashcroft Formation 290, 293, 601
Ashman Formation 346
Asitka Assemblage 556, 733
Asitka Group 75, 298, 299, 469
Askin Group 209, 239
Astoria fans 449
Athabasca River 640, 641, 643
Atlantic passive margin 690
Atlin 391, 601
Atlin Lake 596
Atna Peak 560
Attenborough Creek Formation 345
Attwood Formation 292
Atwell Stage 382
Au skarn deposits 727, 757
aulacogen 115, 116, 652, 685, 689, 823
Australian Creek Formation 785
autunite 786
Avalanche Formation 172
Avalanches 809
Axelgold Gabbro 70, 513

B

Baakan Formation 241
Babine Intrusions 518, 738
Back Range Fault 142
Backbone Ranges Formation 134, 137, 157, 160, 164

INDEX

Badshot Formation 161, 162, 168
Bahama Banks 296
Baja Alaska 65, 80
Baja British Columbia 63, 65, 71, 76, 79, 824
Baja California 73
Baker Terrane 696
Baldonnel Formation 271, 273
Baldy batholith 611
Balourdet Pluton 519, 554
Banff Formation 241, 249, 250, 251, 252, 253
barite 237, 695, 722, 723
Barkerville Subterrane 282, 618, 637, 724
Barn Mountains 157, 639, 653
Barnard Glacier pluton 500
Barr Complex 66
Barrovian assemblages 552, 559, 561, 562
Basement rocks 89, 91, 639, 645, 652, 683, 686, 699, 823, 824
Basin and Range Province 453
bathozone 551
Bathyuriscus-Elrathina Zone 172, 173
Baultoff porphyry 745
Bayonne Suite 512
Bear Province 93, 100
Bear Rock Formation 206
Bear Rock-Stone sequence 204
Bearpaw Formation 342, 781
Bearpaw Mountains igneous complexes 519
Bearpaw Ridge sodalite syenite 499
Beaufort Range 584
Beaufort Sea 5
Beaufort Shelf 235, 242, 275, 653, 664
Beaufort-Mackenzie Basin 786
Beaverfoot Formation 192
Beaver Mines Formation 781
Beaverdell area 786
Beggerlay Creek Pluton 501, 502
Belcourt Formation 251, 252, 259, 404
Bella Bella Complex 485
Bella Bella Formation 389
Belly River Formation 342, 781, 784
Belt-Purcell Basin 101, 685
Bendor Suite 523
Benioff (B) subduction 449, 450, 683, 692, 705, 826
Bennett Lake 483
Bennett Suite 519
bentonite 338, 480
Beresford Bay Fault 582
Besa River Assemblage 221, 230, 241, 252, 253
Besa River Formation 252, 254
Big Creek Fault 608, 635
Big Salmon Complex 140
Big Salmon Fault 636
Billhook Creek Formation 309, 350, 562
Biotite zone 541
Bison Creek Formation 181, 182
bituminous coal 775, 785
Black Dome 522
Black Lake Suite 507, 567
Black Stuart Formation 209
Black Stuart Group 239
Black Tusk 382
Blackie Formation 255
Blairmore Assemblage 701
Blairmore Group 337, 339, 340, 405, 546, 781
blastomylonites 588
Blow Trough 329, 337, 701, 779
Blueflower Formation 134, 137
Blueschist-facies 282, 539, 540, 556, 595, 695, 696
Blumberg Formation 357
Bock's Brook stock 525
Bocock Formation 271, 274

Bokan Mountain Granite 511
Bolaspidella Zone 172
Bonanza Group 73, 314, 315, 352, 462, 473, 476, 564
Bonaparte Fault 307
Bonaparte Subterrane 297
Bonnet Plume Formation 338, 341, 779, 785
Bonnia-Olenellus Zone 167
Border Ranges Fault 353, 373, 578, 580
boreal embayment 339, 340, 341
Boreal Realm 38
Boss Mountain 552, 729, 755
Boswell formation 284
Botanie Creek Fault 607
Bouguer anomaly 45, 47, 705
Bouleau rhyolite 345
Boundary Creek Formation 341
Bow platform 192
Bowen Island Group 561, 591
Bowron Basin 362
Bowron River coal basins 785
Bowser Basin 49, 277, 296, 298, 303, 304, 305, 306,
            330, 343, 345, 405, 557, 596, 603, 697, 698, 701, 703,
            775, 779, 826
Bowser Lake Assemblage 330, 345, 557
Bowser Lake Group 37, 305, 345, 557, 560, 785
Boya Formation 160
brannerite 791
Brazeau Assemblage 703
Brazeau Formation 342
Brewster Limestone Member 266, 268, 273
Brian Boru Formation 344, 478, 481
Bridge River Assemblage 308
Bridge River Complex 593, 594
Bridge River tephra 386
Bridge River Terrane 49, 277, 307, 308, 561, 563, 592,
            746, 751
Bridget Cove Volcanics 515
British Mountains 128, 157, 653, 654, 701
British-Barn Basin 157, 162, 169, 176
Broadview Formation 194, 195, 276, 281
Broken Skull Formation 179
Brokenback Hill Formation 309, 350
Brooks Orocline 653, 701
Brooks Peninsula 485
Brooks Range 83, 233, 639, 651, 653, 695, 699, 824
Brooks Range Geanticline 337
Brothers Peak Formation 349, 481, 557
Buchan-type metamorphism 552, 561, 562, 563, 564, 566
Buchias 353
Buck Creek volcanics 483
Buckinghorse Formation 547
Bug Creek Group 275, 333
Buick Creek gas field 776
Bulkley Plutonic Suite 344, 480, 516, 734
Bulkley River valley 785
Bulldog-Yellowjacket gneiss 92
Bullhead Group 339
Bullion Creek 745
Burgess Shale 8, 155
buried river channels 801
Burnaby Island plutons 510
Burrard Formation 358
Burgess Shale 8, 155
Bushy Point Pluton 522
Butedale plutons 522
Butler Range 608
Buttle Lake Formation 313, 564, 584
Byng Formation 143

C

Caborca Terrane 695
Cache Creek Assemblage 277, 294, 469, 556

INDEX

Cache Creek Group 471, 595, 599, 601, 825
Cache Creek mélange 461
Cache Creek Terrane 26, 31, 36, 277, 290, 292, 294, 295, 296,
    297, 298, 302, 306, 307, 406, 461, 469, 470,
    471, 472, 487, 539, 595, 596, 694, 696, 698, 730
Cache Creek Uplift 334, 337, 654
Cadomin Formation 339, 404, 781
Cadwallader Assemblage 277, 308
Cadwallader Group 307, 308, 593
Cadwallader Terrane 277, 308, 563
Cairn Formation 225, 226
Cake Hill Pluton 299, 501
Caledonides 687, 693
Calico Bluff Formation 233
California Motherlode 748
Calmar Formation 228
Cambrian zonal succession 156
Camp Cove Formation 309, 562
Camsell Formation 204
Canada Basin 48, 130, 664, 688, 699
Canadian Cordilleran Regional conductor 50, 51
Canadian Shield 49, 823
Canol Formation 230, 232
CANOL road 9
Canyon Creek Formation 183
Canyon Mountain Province 394
Cap Mountain succession 114, 684, 716
Cape Kellet Fault Zone 652
Captain Cove Pluton 67, 521
Carbon Creek area 781
carbon maturation 827
carbonatite 498, 499, 823
Cardium Formation 342
Cariboo Group 144, 145
Cariboo Mountains 140, 146, 161, 551, 555, 618, 620, 636, 655
Cariboo Subterrane 618, 722
Carmacks Assemblage 342, 344
Carmacks Batholith 508
Carmacks Group 343, 481
Carmanah Group 358, 563, 566
Carruthers Member 475
Cascade coal basin 547, 781
Cascade core zone 561, 562, 593
Cascade Mountains 42, 540
Cascade Segment of the Coast Belt 592
Cascade Suite 512, 515
Cascade Volcanic Belt 382, 449, 453, 485, 486
Cascadia Basin 449
Cassiar Batholith 513, 554
Cassiar mine 748
Cassiar Mountains 139, 615
Cassiar Platform 160, 167, 173, 181, 690
Cassiar Suite 512, 526
Cassiar Terrane 26, 33, 194, 209, 281, 285, 555, 608,
    615, 618, 688, 722, 824
Cassidy aquifer 801
Castle Mountain Fault 661
Catface Suite 353, 482, 517, 744
Cathedral Formation 719
Cauldron Dome 385, 793
cauldron subsidence complex 483
Cedar Cove Assemblage 312
Cedar District 358
Cedaria Zone 178, 181
celestite 789
Cenozoic plate tectonic regime 683
Central Coast Belt 559, 586
Central Gneiss Complex 523, 560, 588, 589
Central Tethyan Province 36
Challis-Absaroka arc 462, 483
Challis-Kamloops Volcanic Belt 481

Chamosite Member 114
Chancellor Formation 548
Chandindu area 629
Chapin Peak Formation 312
Chapperon Group 292, 468
Charles Camsell 8
Charles Formation 243
Charlie Lake Formation 271, 273
Chase Mountain 553
Chatanooga Formation 241
Chatham Sound 559
Chatham Strait 662
Chatham Strait Fault 446, 660, 661, 662
Cheakamus Formation 350
Cheekye Stage 382
chemical geothermometers 792
Cherry Creek Fault 607
Chilcotin Group 377, 378, 387, 388, 462, 486, 487, 558
Chilcotin River 409
Chilkat River fault 660
Chilliwack Assemblage 309
Chilliwack Batholith 377, 485, 524, 593
Chilliwack Group 309
Chilliwack Suite 524
Chilliwack Terrane 277, 309, 694
China Creek area 744
Chinle Formation of New Mexico 74
Chisana Formation 314
Chischa Formation 113
Chism Creek schist 563
Chitistone Limestone 314
Chlorite zone 541, 547
Chu Chua Formation 345
Chuaria sp. 119
Chuckanut Formation 358, 407, 564, 594
Chugach Terrane 353, 476, 578, 580
Chugach-Saint Elias Fault system 446
Chulitna Terrane 38
Church Mountain Thrust Fault 594
Churchill Province 93
Chutine area 298
Chuwanten Fault 593
Clachnacudainn Gneiss 500
Clark Range 102, 104, 105, 107, 108
clastic wedge 241, 683, 693, 701, 703, 775, 781
Clausen Formation 253, 254
Clearwater area 789
Clearwater basalt 397
Clearwater Valley 397
Clearwater-Quesnel Volcanic Province 395, 487, 794
Cleoniceras 348
climatic
    deterioration 424
    extremes 797
    perturbations 421
clockwise rotation 63, 67, 78
coal 329, 335, 337, 339, 340, 341, 342, 346, 347,
    348, 355, 357, 360, 362, 363, 539, 545, 775, 780, 781, 784
Coal Creek Dome 110, 547, 650, 651
Coal Glacier faults 662
Coal Harbour Group 357
Coal Mine Lake 786
Coal Mountain 784
Coal River stock 722
Coal Valley 784
Coalspur Formation 342, 784
Coast Belt 5, 23, 39, 42, 45, 49, 50, 51, 350, 539,
    558, 559, 567, 585, 586
Coast Mountains 42
Coast Plutonic Complex 488, 512, 519, 520, 559, 561, 739
Coast Range Megalineament 703

INDEX

coastal erosion 809
Coastal Lowland 798
Coates Lake diatreme 91
Coates Lake Group 81, 82, 94, 116, 133, 134,
135, 464, 717, 718
Coates Lake volcanics 462
Cold Fish volcanics 305
Coldwater Formation 345
Colorado Plateau 73
Columbia Basin 161, 168, 175, 183
Columbia Mountains 144, 147, 148, 607, 619, 699
Columbia River Fault Zone 550, 553, 608, 625, 665
Columbia River valley 801
Columbian Orogen 18
Columbus Fault 353, 580
Colville Hills 638, 646, 648
Combustion metamorphism 542
Comfortlessbreen Group 687
Comox Basin 357, 358
Comox Formation 785
Conception Group 687
conodont alteration index (CAI) 540, 545
Conodonts 31
Conophyton 107, 118
Constitution Formation 357, 566
Copp Creek stock 523
Copper Creek Formation 115
Copper Creek basalts 120
Copper Mountain 70, 73, 504, 526, 726, 727, 746, 757
Copper Mountain Stock 292, 727
Copper Mountain Suite 293, 504, 505
Coppercap Formation 717
Coppermine Group 463
Coppermine Homocline 114, 120, 684
Coppermine lavas 100, 115, 684
Coppermine River Group 685
Coppery Creek group 107, 109
Coquihalla Complex 378, 379, 485, 524
Cordilleran collage 462
Cordilleran Ice Sheet 421, 422, 431, 433
Cordilleran Orogen 5, 651, 824
core complex 667
core zones 539, 558, 586, 592, 607, 608, 618
Cornwallis limestone 312
Coronation Island 662
Coryell Suite 518
Coryell syenites 345
Costigan Member 229
Covada Group 281
Cow Head Group 157
Cowichan Anticlinorium 584
Cowichan Lake Fault 584
Cranbrook Formation 161, 162, 168, 169
Crater Lake, Oregon 811
Crepicephalus Zone 176, 178
Crescent Formation 358, 477
Crescent Terrane 53, 330, 358, 477, 564, 565, 566,
584, 585, 745
Crescent Thrust 585
Creston Formation 104, 105
Cretaceous magnetizations 78
Cretaceous Normal Superchron 83
Crillon layered gabbro 513
Crooked Amphibolite 619, 637
Cross kimberlite 499
Crossport sill 499
Crownite Formation 377
Crowsnest coalfield 781
Crowsnest Formation 83, 340, 479, 512
Crowsnest Pass 640
Crowsnest Volcanics 479

Crucible Dome 385
Crustal deformation related to glaciation 431
Cryptic metamorphism 540, 566
Cultus Assemblage 309
Cultus Formation 309
Cumberland coal field 785
Cunningham Formation 145
Curie isotherms 49
Currier Formation 348, 785
Custer Gneiss 307
Cyprus-type deposits 752

D

Dala River Pluton 521
Dalton Stage 382
Dana Facies 381
Darrington Phyllite 593
Dave Lord Intrusion 233
Dave Lord Pluton 499
Dave Lord Range 651
Dave Lord Uplift 654
Dawson Thrust Sheet 160, 629, 631, 650
Deadend Fault 648
Deadman plutons 514
Deadman River Formation 377
Debolt Formation 254
debris flows 812, 813
Decatur Terrane 357
Deep refraction and reflection seismology 39
Delorme sequence 199
Dempster Highway 133
Denali Fault System 42, 45, 353, 354, 559, 581,
660, 661, 662, 668
Descon Formation 31, 467
Deserters Gneiss 91
Deserters Range 91, 539, 548, 639
detachment faults 407, 639, 643, 645, 647, 649
Devastator 386
Devil's Claw unit 348
Devonian assemblages 196
Devonian-Mississippian clastics 230
Dewdney Creek Formation 309
Dewdney Creek Group 351
Dewdrop Flats Formation 345, 482
dextral strike-slip faulting 703
Dezadeash Formation 353, 477, 661, 826
Dezadeash Group 315
Dezadeash-Alsek valley 408
Dezadeash-Tyaughton-Methow Trough 462, 477
diamictite 131, 132, 133, 137, 139, 141, 147, 686
diapiric anticlines 654
diatreme 463, 466, 467, 498, 499
discharge areas 797, 798
Dismal Lakes Group 114
displaced terranes 694, 824
Dixon Entrance 439, 440, 445, 446
Dog Creek 422
Dogtooth Duplex 622
Dogtooth Range 146
Dome Creek Formation 168, 183
Donald Formation 169
Donjek River 608
Dorsey Assemblage 294
Dorsey Terrane 277, 294, 548, 636, 726
Douglas 156
Douglas Channel 559
Dowling 781
drainage diversions 412
Dryden Anticline 629
Dublin Gulch district 721
Dubose stock 523

828

Ducette Member	271
Duckling Creek Syenite	504, 505, 507, 727
ductile shear zones	588
Duke River	578
Duke River Fault	311, 353, 578, 579, 581, 661
Dumbell Mountain Pluton	501
Dunderbergia Zone	173, 179, 181, 182
Dunira Formation	312
Dunira Island	586
Dunvegan Formation	339, 340
Duo Lake Formation	178, 191
duplex	608, 611, 636, 699
Dusty River	395
Duvernay Formation	226

E

Eager Formation	168, 169
Eagle Bay Assemblage	281
Eagle Bay Formation	147, 183, 281, 288
Eagle Fold Belt	651, 652, 654
Eagle Granodiorite	601
Eagle Plain Group	341
Eagle Plains	329, 652
Eagle Plains Basin	779
Eagle Plutonic Complex	510, 556
Early to Middle Miocene shoreline	408
Earn Assemblage	230, 232, 235, 238, 239, 242, 254, 722
Earn Group	235, 254, 466, 722
earthquakes	584, 585, 662, 809, 810
earthquake fault-plane solutions	440
East Africa Rift Zone	239
East Fork Formation	341
East Pacific Rise	51
Eastern Klamath Terrane	696
Echooka Member	262, 273
eclogite	282, 542, 548
Ecstall Pluton	522, 559, 560
Ediacaran fauna	134, 138, 142, 144, 161
Edmonton Group	781, 784
Edziza-Spectrum Complex	389, 391
Eldorado Creek	802
electrical conductivity	49, 705
electrical power	791
Electromagnetic depth sounding	49
Elise Formation	290, 293
Elk Point Basin	207
Elk Valley coalfields	781
Ellesmerian Orogeny	235, 654, 695
Elvinia Zone	169, 178, 179, 181
Empetrum Formation	350
Endako Group	344, 482, 558
Endeavour Seamount	443
Endicott Group	128, 256
Entrance conglomerate	342
Eocene extension	290, 664
Eocene fault blocks	407, 408
Eocene rhyolite and basalt	343
epicontinental boreal sea	404
epidote-zoisite	541
epithermal	745, 748, 749, 750
epithermal veins	733
Erosion	816
Esbataottine Formation	191
Escalante Formation	358
Eskimo Lakes Arch	334
Eskimo Lakes Fault Zone	664
Espee Formation	139, 140, 145
Esplanade Thrust Fault	622
Etherington Formation	251, 252
Ettrain Formation	255
eugeosyncline	15, 18

Euler pole	64, 72, 76
eustatic sea-level lowering	431
evaporites	134
exhalative mineralization	716
exploration history,	9
Explorer Plate	373, 442, 443, 447, 449, 705
Explorer Ridge	443, 447, 753
explosion-pits	794
Exshaw Formation	230, 241, 243, 249, 251, 252, 776
extension, southern Foreland Belt	646
Extension-Protection Formation	358, 785
extensional basins	703

F

facies and zones, régional metamorphism	540
Fairchild Lake Group	110
Fairholme Group	224
Fairholme sequence	222, 224
Fairweather Fault	446, 580, 662, 683
Fairweather-Chatham Strait-Denali Fault systems	439
Fairweather-Queen Charlotte faults	662
Fairweather-Saint Elias faults	580
Fallotaspis zone	167
Fantasque Formation	261, 270
Farallon Plate	65, 73, 74, 79, 373, 377, 485, 486, 488, 703
fault-bounded basins	780, 785
fault-controlled basins	342, 344, 343
faults	578, 584, 651, 655, 660, 668
Fe skarns	757
Fennell Formation	287, 468
Fernie Basin	335, 640
Fernie Formation	276, 335
"Field Creek" volcanics	467
Fifty-Mile Batholith	282
Finlay Fault	599, 659
Finlay Ranges	554
Fire Lake Group	309, 350
Fish Lake stock	516
Fish River Group	341
Fish-Scale marker	340
fish-scale serpentinite	657
Fitton Intrusion	230, 233
Flathead coalfield	781
Flathead Fault	342, 640, 646
Flathead graben	407
Flathead Valley	640
Flett Formation	253, 254
flexural subsidence	704
floods	809
Flores volcanics	353, 358, 482, 517
Flume Formation	224, 225
fluorite	789
focal depth	810
Foothills Subprovince	638, 639, 641, 643, 645, 683, 780
Ford Lake Formation	254, 255
foredeep clastic wedge deposits	343, 346, 683, 699, 701, 825
Foreland Belt	5, 18, 31, 39, 41, 45, 48, 49, 50, 53, 333, 404, 406, 407, 539, 546, 548, 566, 698, 780
Fort Norman	786
Fort Simpson Formation	230
Fort St. John Basin	263
Fort St. John Group	339, 340
Fort Steele Formation	103
Fourth of July Batholith	519
Franciscan Complex	36, 476, 599
Francois Lake Suite	73, 507, 511
Franklin Glacier Complex	378, 380, 485
Franklin intrusions	80
Franklin Mountain Formation	178
Franklin Mountains	116, 638, 646, 648, 650, 684
Fraser Bend Formation	377

Fraser Fault System 607
Fraser Lowland 780, 799, 815
Fraser River 362, 407, 814
Fraser River Basin 801
Fraser River Fault Zone 593, 594, 659, 703
Fraser River Valley 785
Fraser River-Straight Creek Fault System 659
French Range Formation 294, 296
French Range Subterrane 296, 307
Frenchman Cap gneiss dome 92, 500, 608, 610
Freshwater Bay Formation 312
Funeral Formation 206
Funnel Creek Formation 164
fusulinids 28

G

Gabriel Peak Pluton 515
Gale Passage dyke swarm 389
Galena Bay stock 509, 526
Galton Range 107, 108
Gambier arc 462
Gambier Assemblage 309, 310, 312, 315, 330, 350, 353, 559, 701, 739, 826
Gambier Group 315, 350, 477, 561, 591
Gambier Island 591
Gambier overlap assemblage 698
Gametrail Formation 134, 137
Gamsby Metamorphic Complex 521, 560
Garibaldi Volcanic Belt 42, 373, 377, 378, 382, 383, 386, 485, 486, 705, 793
Garnet zone 541
gas 775, 776, 780
Gataga district 757
Gataga Formation 114
Gates Formation 781
Gateway Formation 108
geobarometry 540, 543, 551, 558, 608
geochemical temperature estimates 795
geochemical thermometers 792, 793, 794
geomagnetic field 63
geomagnetism 48, 49
geomorphic environments of New Guinea 405
George Formation 113
Georgia Basin 330, 357, 358, 376, 584, 780
geothermal anomalies 792
geothermal energy potential 795
geothermal gradients 539, 545, 792
geothermal reservoirs 793
geothermal systems 775
geothermally heated water 798
geothermometry 539, 540, 543, 551, 554, 558, 559, 608
Germansen batholith 556
Gething Formation 339, 781
Gil Island diorite 521
Gillespie Lake Group 110
glacial cycle 422, 431
glacial lakes 429
glaciation 137, 421, 424, 429
Glacier National Park 105
glacio-isostatic depression 431
glaciomarine sediments 429
Gladstone Formation 339, 781
glaucophane 309, 541
Glenogle Formation 193, 547
Glossopleura Zone 169
Gnat Lakes Ultramafite 501
Goat Creek coal 785
Gog Group 161, 167
Golata Formation 253, 254
Golconda Allochthon 695
Golden Embayment 223

Golden Horn Batholith 518
Goosly Lake Intrusions 483, 518
Gorman Creek Formation 335
Graham Island 42, 355, 565, 786
Graham Island Formation 785
Grainstone Formation, Little Dal Group 120, 718
Graminia Formation 229
Granby fault system 665
Grand Canyon of the Stikine River 556, 603
Grand Canyon Supergroup 685
Grande Cache Member 781
Granite Batholith 507
granulite 539, 542, 560
Gravina-Nutzotin Assemblage 309, 310, 312, 315, 351, 462, 477, 478, 479
Gravina-Nutzotin belt 559, 660
gravity and isostasy 45
Grayling Formation 264, 270, 273
Great Basin 156, 157, 690
Great Bear magmatic zone 91
Great Slave Lake Shear Zone 685
Great Valley sequence 38
Greenberry Formation 240
greenschist facies 539, 540, 541, 545, 546, 567
Greenwood 665, 729
Grenville Channel Fault 588
Grenville Loop 82, 120
Grey beds 276
Grinnell Formation 105, 716
"Grit Unit" 138
Grizzly field 776
Grizzy Bear Formation 205
Grotto Member 226
Groundhog coal area 346, 557, 775, 784, 785
groundwater flow systems 775, 797
groundwater in fractured media 798
groundwater in permafrost areas 801
groundwater potential 801
groundwater quality 798
groundwater recharge areas 797
groundwater supply 775
growth and decay of continental ice sheets 422
Guaymas Basin 753
Guichon Creek Batholith 70, 73, 292, 503, 595, 600, 726
Guichon Suite 293, 503
Gulf Islands 357, 583, 785, 798
Gulf of Alaska 662
Gulfian seaways 340, 341
Gull Lake Formation 164, 166
Gundahoo Fault 643
Guyet Formation 240
gypsum 271, 273, 314
Gypsum formation 118, 120

H

Hadrynian 6
Haida Formation 355, 565, 780, 785
Haig Brook Formation 104
Haines, Alaska 660, 662
Haines ultramafic complex 515
Halfway River 645
Hall Assemblage 277, 293
Hall Formation 293
Hamill Group 146, 161, 162, 169, 466, 689
Hamilton Island limestone 312
Hammond Terrane 693
Harbledown Formation 314, 315
Hare Indian Formation 230, 232, 777
Harper Ranch Assemblage 287, 290, 292, 468
Harper Ranch Formation 292
Harper Ranch Group 31, 290, 292, 309, 596, 601

Harper Ranch Subterrane	277, 290, 291, 294, 468, 696
Harper rift assemblage	132, 547
Harrison Fault	593, 594
Harrison Lake	308, 540, 561
Harrison Lake Assemblage	309, 315, 350
Harrison Lake Formation	309, 561, 739
Harrison Lake Terrane	277, 307, 308, 309
Harrison Lake-Lillooet River fault zone	561
Harrogate Formation	224
Hart River culmination	651
Hart River Formation	255
Hasen Creek Formation	314
Haslam Formation	358
Hasler Formation	781
Hat Creek	363, 407, 775, 785, 786
Hat Creek Graben	607
Havallah Formation	695
Hawkesbury warp	66
Hay River	225, 802
Hay River Platform	224
Hayfork Terrane	696
Hayhook event	650
Hazelton Assemblage	556, 557, 561, 733
Hazelton Group	73, 277, 298, 299, 304, 306, 461, 473, 555, 557, 605, 606
Hazen Trough	689
Heart Peaks shield volcano	391, 392, 394
heat flow	41, 453, 539, 703, 705, 795
Hecate Strait	373, 445, 582, 780
Hector Formation	144
Hedmark Group	687
Helikian	6
Hellroaring Creek stock	103, 499
Helm Formation	350
hematite-jaspilite iron deposits	718
Henry Creek Formation	113
Herchmer Fault	646
Hercynotype environment	526
Hesquiat Formation	358
Hess River Formation	166, 171, 178
heulandite-clinoptilolite	540, 545, 547
Hickman Batholith	502
high temperature geothermal systems	791
high volatile bituminous coal	775, 781, 784, 785
high-pressure and low-temperature metamorphism	539, 540
Highland Valley faults	600
Hillard Formation	169, 178
Himalayas	684, 705
historical earthquakes	810
Hobson Lake pluton	553
Hogem Batholith	501, 504, 513
Hollebeke Formation	224, 226
Holmes Creek Member	161
Holy Cross Mountain	143
Honna Assemblage	330
Honna Formation	356, 565, 780
Hoodoo Mountain	389, 391, 486, 795
Hope slide	812
Horetzky Dyke	523
Hornby Bay Group	114
Horsefeed Formation	294
Horseranch Fault	554
Horseranch Range	93, 140, 554, 668
Horsethief Creek Group	144, 146, 147, 162
hot spot	440, 447, 453, 487
hot springs	749, 792, 794
Hota Formation	168
Hotailuh Batholith	501, 510
Hound Island Volcanics	312
Howards Pass district	721, 757, 759
Howell Creek Suite	512, 526
Howson facies	475, 556
Hozameen Fault	593
Hozameen Group	308, 351, 593
Hubbard Fault	578, 580
Huckleberry Formation	465
Hudson Bay Mountain	734
Hugh Allan Creek-Mt. Blackman gneiss	92
Hughes Range	103, 104, 107, 161
Hume Formation	232
Hume-Dunedin sequence	223
Hungry Valley Fault	308, 592
Hunt Fork Formation, shale	232, 233
Huntingdon Formation	594
Huntington Terrane	696
Husky Creek Formation	115
Husky Formation	333
Hutshi Formation	479
Hutshi Group	343
Hyd Assemblage	312
Hyder pluton	738
hydrocarbon potential	780
hydrocarbons	315
hydrodynamic characteristics	775
hydrogeological zones	798
hydrostratigraphic units	799
hydrothermal resource	775
Hyland Assemblage	133
Hyland Group	138, 139
Hyland Highland	360

I

I-, S- and A-type plutonism	525, 526
I-type plutons	502, 519, 525, 526, 749, 757
Ice River Complex	499, 548
Icefield Ranges	353, 355
Icefield Ranges pelitic assemblage	312
Icefield Ranges Suite	312, 315, 500, 566
Ilgachuz Range	388, 389, 390, 485, 487, 794
Illinoian age	424
Illtyd Formation	164, 171
ilmenite-series plutons	525, 526
Imperial Assemblage	230, 232, 236, 255
Imperial Formation	233, 234, 235, 237, 547
Index Formation	183, 194, 195
Ingenika Fault	556
Ingenika Group	139
Ingta Member	159
initial 87Sr/86Sr ratios	497, 499, 507, 511, 512, 514, 524, 525
Inklin Formation	277, 297, 303, 307, 825
Inklinian Orogeny	6
Innoko Terrane, Alaska	695
Innuitian Orogen	130, 651, 653, 683, 824
Insular Belt	5, 23, 49, 51, 352, 406, 540, 563, 578, 714
Insular Superterrane	308, 310, 329, 462, 683, 698, 702
Interior Plains	646, 652
Interior Platform	41, 638
Intermontane Belt	5, 18, 21, 39, 41, 45, 49, 50, 51, 306, 377, 539, 567, 683, 784
Intermontane Superterrane	308, 310, 329, 462, 698, 699, 702, 703, 725
Irene volcanics	147, 462, 465
Ireton Formation	241
iriginite	791
Iron Mask Batholith	292, 504, 507, 727
iron skarns	743
iron-formation	137, 686, 745
Isaac Formation	145
Isaac Lake Synclinorium	618
Ishbel Assemblage	222, 249, 256, 259
Ishbel Trough	256
Iskut River	298, 556

I

Iskut Trench	409
Island Intrusions	315, 510
Island Mountain amphibolite	619
Island Suite	564
isostatic adjustments	431
isostatically depressed lowlands	429
Itcha Range	388, 389, 485, 487
Ivishak Member	273
Iyoukeen Assemblage	312

J

Jack Mountain Thrust Fault	308, 593
Jackass Mountain Group	351, 352, 405
Jackfish Gap Member	253
Jackson Unit	346
Jacutophyton	118
jadeite	297, 539, 541, 556
Japan Basin	705
Japanese Archipelago	683, 705
Jasper	144
Jervis Inlet	561
Johanson Lake	501
Johnson Canyon Formation	261
jökulhlaups	815
Jones Lake Formation	273
Jones Ridge Formation	133, 164, 169, 176
Jowett Formation	194, 467
Juan de Fuca	373, 704
Juan de Fuca Plate	42, 49, 51, 53, 388, 439, 441, 442, 443, 447, 449, 452, 488, 585, 705, 810
Juan de Fuca Plate subduction zone	449
Juan de Fuca Ridge	11, 44, 49, 373, 439, 441, 442, 443, 444, 447, 582, 753
Juan de Fuca-America Plate convergence	442
Juan de Fuca-Pacific spreading ridge	485, 486
Jubilee Formation	168, 183
Judith River, Formation	781
Jumping Pound field	776
Jumping Pound West field	776
Jungle Creek Assemblage	222, 256, 261, 262
Jungle Creek Formation	262

K

K-feldspar zone	542
Kaketsa Stock	501, 502
Kakisa Formation	228
Kakwa Platform	192
Kaltag-Porcupine Fault Zone	654, 664, 695
Kamik Formation	333
Kamloops Assemblage	342, 344, 345
Kamloops Group	345, 482, 786
Kamloops Lake	406
Kananaskis Formation	252
Kanayut Conglomerate	232, 233
Kandik Basin	337
Kaniksu Batholith	512
Kano Suite	524
Kapoose Formation	356
Karheen Assemblage	312
Karmutsen Formation	75, 314, 462, 470, 472, 563, 583
Karmutsen volcanics	564
Kasalka Group	344, 406, 478, 479, 480, 482
Kasiks pluton	560
Kaska Mountains	555, 608
Kaskaskia Sequence	223
Kaslo Formation	281
Kaslo Group	287, 288, 289, 293, 467, 468
Kastberg intrusions	397, 487, 518
Katherine Group	81, 116, 118
Kathul Greywacke	337
Kayak Formation	232, 255, 256
Kaza Group	144, 145, 551, 552, 618
Kechika Basin	160, 167, 173, 179, 689
Kechika Fault	362, 608, 655, 659
Kechika Group	167, 181, 193, 194, 466
Keele Formation	132, 137
Keele Range	654
Keele Uplift	777, 779
Keele-Kandik Basin	329, 337
Keen Creek assemblage	467
Keg River Barrier	208
Kekiktuk Formation	233, 255, 256
Keku Volcanics	312
Kettle domes	548
Kettle River Fault	665
Kettle River Formation	345
Kettle-Grand Forks Complex	93
Keweenawan lavas	463
Keweenawan rocks	81
kimberlite	467
Kindle Formation	261
King Island Pluton	388, 524
King Salmon Thrust Fault	304, 596, 599, 601, 603, 698
Kingak Formation	273, 275, 276, 333
Kingsvale Group	344, 481
Kintla Formation	80
Kishenehn Formation	342, 408
Kiskatinaw Formation	254
Kitchener Formation	105
Kitimat Trench	414
Kitsilano Formation	358
Kitsuns Creek	348
Klakas Orogeny	693
Klamath Mountains	75, 309, 699, 701
Klawak Formation	312
Klaza advance	422
Klondike goldfields	5, 411
Klondike Plateau	608, 668
Klondike schist	284
Klotassin batholith	507, 635
Klotassin Suite	507
Kluane Lake	581, 660, 661
Kluane Ranges	353, 660
Kluane Ranges Plutonic Suite	353, 512, 515
Kluane Schist	353, 661
Knob Hill Group	292, 468
Kobau Group	292, 468
Kohse Creek Volcanics	463, 464
Kootenay Arc	50, 162, 276, 281, 549, 550, 607, 611, 624, 625, 626, 629, 634, 637, 695, 724
Kootenay Assemblage	222
Kootenay Facies	381
Kootenay Group	335
Kootenay Lake	50, 147, 553, 625, 626, 628
Kootenay Lake High	625
Kootenay Terrane	26, 183, 194, 195, 210, 276, 281, 283, 288, 289, 467, 498, 611, 618, 619, 624, 688, 693, 694, 695, 824
Kotaneelee gas accumulations	776
Kotcho Formations	230
Kotsine facies	475
Kruger pluton	507
Kula Plate	63, 65, 73, 76, 373, 407, 488, 703
Kunga Formation	314, 315, 476, 583
Kuskanax Batholith	508, 526, 551, 553, 627
Kuskanax Suite	508, 526
Kuskanax-Clachnacudainn High	625
Kutchin Formation	202
Kutcho Assemblage	294, 296, 599, 657
Kutcho Fault	599, 657
Kutcho Formation	296, 299, 304, 471
Kyanite zone	542
Kyuquot Group	356

L

La France Creek group	109
La Perouse layered gabbro	513, 524, 525
La Perouse Suite	525
Laberge Group	298, 299, 303, 785
Lachlan Fold Belt I-type plutons	526
Lachlan Orogen	693
Ladner Assemblage	308
Ladner Group	308, 351
Ladybird Suite	517
Laib Formation	183, 721
Landry Formation	206
landslides	809, 812
Lardeau Group	146, 183, 194, 195, 288, 550, 552, 626
Lardeau Valley	625
Latham Pluton	501, 502
Lau Basin	684
laumontite	540, 547
Laurentian Shield	81
Laurentide Ice Sheet	422, 424
lawsonite	541, 556, 562, 566
lawsonite blueschist assemblages	541, 561, 562
layered gabbro	513
Ledbetter Formation	183
Leduc reefs	227
Leech River Fault	357, 358, 566
Leech River Formation	357
Leech River Schist	566
Leith Line	114
Level Mountain shield volcano	65, 389, 391, 392, 486, 795
Lewes River Group	296, 297, 298, 299, 307, 470, 471, 779
Lewis Thrust Sheet	104, 640
lherzolite nodules	391, 487
Liard Basin	695
Liard Depression	192, 689
Liard Formation	270, 271
Liard Plain region	360
Lightning Creek Anticlinorium	619
lignite	349, 542, 545, 775, 785
Lillooet Basin	797
Lillooet River	561
Lillooet River lineament	561
Lisburne Assemblage	221, 222, 233
Lisburne Group	255, 256, 262
LITHOPROBE	39, 450
lithospheric flexure	780
Little Bear Formation	341
Little Bell Member	275
Little Dal Group	81, 116, 118, 134, 135
Little River Fault	551, 552, 555
Livingstone Formation	250, 251, 252
Lizard Range	103, 107
Llama Member, Sulphur Mountain Formation	264, 266, 273
Lluvia Peak Pluton	523
Lokken batholith	507, 508
Lone Land Formation	114
Long Lake Suite	503, 507, 508, 733
long-wavelength magnetic anomalies	705
Longarm Assemblage	354, 356
Longarm Formation	354, 356, 583
Loomis Member, Mount Head Formation	254
Lopez Complex	357
Louis Creek Fault	601
Louscoone Inlet Fault Zone	582, 704
low temperature fluids	791
low volatile bituminous coals	780
Lowell Glacier	148
Ludington Formation	270, 271, 273
Lunar Creek	501
Luscar Group	339, 340, 781
Lyell Formation	181, 182
Lynx Group	181, 182

M

MacDonald Suite	499
Mackenzie Delta	51, 654, 664
Mackenzie Dyke Swarm	463
Mackenzie fold belt	629
Mackenzie Loop	82
Mackenzie Lowland	406
Mackenzie Mountains	49, 83, 91, 131, 133, 134, 638, 646, 650
Mackenzie Mountains Assemblage	116, 717
Mackenzie Mountains Supergroup	81, 94, 116, 133, 134, 461, 684, 685
Mackenzie Platform	157, 159, 160
Mackenzie River	406
MacLaren Metamorphic belt	661
Macmillan Pass	235, 632, 757
magnesite	162
magnetic anomaly	440, 442, 705
magnetite	141, 525
magnetite-rich iron-formation	719
magnetite-series	526
Magnetotelluric Depth Sounding	50
Mahto Formation	168
Maligne Formation	226
Malton Gneiss	91, 92, 620, 639
Malton Range	620
Manicouagan Impact Crater	74
Mannville Group	781
mantle hotspot	488, 705
Manuel Creek Formation	275
map unit H_1	116
Marama Formation	345, 482
Marble Canyon Formation	33
Marble Range Subterrane	297
Marblemount Quartz Diorite	501
Mariana island arc	705
Marmot Formation	191, 465, 467, 499
Marron Formation	345, 482, 518
Marsh Lake	596
Marten Conglomerate	288, 289
Martin Creek	333
Martin House Formation	337
Masset Formation	66, 373, 377, 380, 381, 485, 564, 565
Matanuska Assemblage	565
Matthews Fault	551
Mattson Assemblage	222, 249, 252, 253, 254, 255
Mattson Formation	254, 547
Maude Formation	315
Mayne Island ground water	799
Mazama ash	384
McBride map area	142
McCann Hill chert	241
McCarthy Formation	314
McCloud Limestone	309
Thust Fault	640
McConnell advance	394
McConnell Creek	487
McConnell Creek Volcanic Province	397
McConnell Thust Fault	640
McDougall-Segur conglomerate	405
McEvoy unit	348
McGuire Formation	333
McHardy Assemblage	289
McKay Group	182, 183, 193, 547
McKinley Terrane	284
McNaughton Formation	146, 161
Meager Creek hot springs	793
Meager Mountain	51, 380, 386, 485, 791, 793
Meager Mountain volcanic complexes	524

INDEX

Meilleur Member, Flett Formation 254
Meilleur River Embayment 191, 192, 689
mélange 297, 356, 476, 595, 599, 696, 701
Menard Creek ultramafic body 501
Merritt coal field 786
Mesilinka Pluton 504, 513
Mess Creek 414
Mestognathus 31
meta-anthracite 785
Metaline Formation 183
metallogeny 827
metamorphic culminations 539
Metchosin Assemblage 330
Metchosin Igneous Complex 66, 358
Methow Terrane 49, 277, 307, 308, 309, 561, 592, 826
Methow Trough 73, 592
Mica Dam Antiform 622
micro-seismicity 661
mid-ocean ridge 308
mid-ocean ridge basalts (MORB) 381, 461, 485, 488
Midas Formation 145, 161
Miette Group 142, 161
Miles Canyon 391
Milford Assemblage 281
Milford Group 276, 281, 288, 289, 468, 552, 553, 626, 627
Mine stock 553
Mineral deposit models 745
  Adanac porphyry Mo deposit 730
  Afton deposits 504, 727
  Ainsworth camp 724, 757
  Ajax prospect 734
  Al prospect 733
  Alice Arm camp 733, 734
  Anniv deposit 721
  Anyox deposit 739, 751, 753
  April prospect 744
  Argonaut deposit 743, 757
  Atlin Ruffner Ag,Pb,Zn vein 732
  Bailey deposit 721
  Baker deposit 721
  Banks Island deposits 652, 653, 740
  Bar prospect 726
  BC Moly deposit 755
  Bear-Twit deposit 719
  Beaverdell (Highland Bell) mine 729
  Bell Copper deposit 518, 746
  Bell Moly prospect 734
  Berg Cu,Mo prospect 734
  Besshi deposits 741, 752, 753
  Bethlehem deposits 726
  Big Ledge prospect 715
  Big Missouri 733
  Big Onion Cu,Mo prospect 734
  Big Slide Au,Ag,Cu vein 732
  Blackdome mine 738, 749, 751
  Blizzard basal U deposits 729, 786
  Bluebell 757
  Bob deposit 740
  Boundary-Greenwood camp 746
  Bralorne Takla mine 730, 738
  Brenda deposit 729, 756
  Bridge River camp 738, 748
  Brisco deposit 192
  Britannia deposit 739, 746, 751, 752
  British Columbia Molybdenum mine 734
  Camp McKinney deposit 749
  Canada Tungsten mine 166, 721, 757
  Canalask deposit 745
  Capoose Lake epithermal Ag prospect 734
  carbonate-hosted Zn,Pb deposits 719
  Cariboo Bell prospect 727, 756
  Cariboo Gold Quartz deposit 748
  Carmi deposit 729
  Carolin mine 738, 748
  Cash prospect 733
  Cassiar deposit 607, 608, 725
  Cassiar Moly deposit 239, 723
  Catface deposit 744
  Cathy 722
  Chappelle prospect 733
  Cheni mine 733
  chrysotile asbestos 296, 725, 730
  Chu Chua deposit 725, 752, 753
  Cinola deposit 744, 749, 751
  Cirque deposit 722
  Clea deposit 721
  Clear 722
  Clear Lake deposit 723
  Climax-type deposits 754
  Clinton Creek 725
  Coast Copper deposit 743, 757
  Colby prospect 715
  Connaught deposit 725
  Cork Province deposit 728
  Cottonbelt prospect 715
  Court deposit 744
  Craigmont Cu,Fe skarn 726, 757
  Crest iron deposit 719
  Dease Lake placer gold 730
  Debbie prospect 743
  Dividend camp 729
  Dodger mine 721
  Dolly Varden mine 733
  Dome Mountain prospect 734
  Domineer deposit 744
  Driftpile Creek deposits 722
  Duncan Lake deposit 724
  Dusty Mac deposit 749
  DY deposit 752
  Eaglet fluorite deposit 724
  Emerald mine 721
  Endako porphyry Mo mine 511, 734, 755, 756
  Equity Silver mine 734, 748, 751
  Erickson Creek deposits 725, 751
  Fairview camp 729, 750
  Falcon prospect 721
  Faro deposit 721
  Feeney mine 721
  Fiddler W,Cu,Sn greisen 723
  Franklin Glacier deposit 485, 524, 756
  Franklin Glacier porphyry 740
  Frasergold prospect 728
  Fyre Lake occurrence 725
  Galore Creek Cu,Au,Ag deposit 505, 733, 754, 757
  Gayna River deposits 717, 719
  Giant Mascot mine 516, 740, 756
  Gibraltar porphyry Cu,Mo mine 730
  GK deposit 721
  Glacier Creek prospect 741
  Glacier Gulch prospect 734, 754
  Godlin Lakes deposits 719
  Golden Bear Au prospect 733, 751
  Goldstream deposit 751, 752, 753
  Goz Creek deposits 159, 719
  Granduc massive sulphide deposit 739, 751, 753
  Granisle au, Mo deposit 518, 738
  Grass Valley deposit 751
  Greens Creek mine 741
  Grew Creek deposit 751
  Greyhound storn deposit 729
  H-W deposit 742, 751, 752
  Harper Creek 724

INDEX

Hart River deposit 233, 650, 717
HB deposit 724
Hedley camp 727, 746, 749, 757
Hepler deposit 740
Hidden Creek mine 739
Highland Valley district deposits 726, 748, 753, 756
Highmont deposit 726
Homestake deposit 722, 724
Hopkins skarn 723
Huckleberry prospect 734
HW deposit 742
Ikeda skarn deposit 743
Inconspicuous prospect 744
Indata Lake prospect 730
Indian mine 738
Ingenika prospect 723
Ingerbelle deposits 504, 727, 757
Invincible deposit 721
Island Copper mine 510, 743, 746, 756
Island Mountain 724
JA deposit 726
Jackpot deposit 724
Jason deposit 235, 722
Jedway deposit 511, 743, 757
Jersey deposit 724
Johobo deposit 745
Jove occurrence 725
Kennedy Lake deposit 743, 757
Keno Hill district 549, 751
Ketza mante deposit 751
Ketza River Au mine 723
Kicking Horse mine 719
Klazan prospect 733
Kootenay King deposit 716
Krain deposit 726
Kuroko deposits 297, 730, 739, 752
Kutcho Creek deposits 730, 751, 752
Laforma prospect 733
Lakeview prospect 730
Lamb Mountain skarn deposit 723
Lang Creek deposit 725
Lara deposit 742, 752
Lened deposit 721
Letain prospect 730
Logtung porphyry W, Mo deposit 726
Lone Star Au, Ag vein deposit 723
Lornex deposit 600, 726
Lorraine porphyry Cu, Au prospect 727
Low Herbert prospect 741
Lucky Coon deposit 724
Lucky Joe Creek Cu prospect 723
Lucky Ship prospect 734
Lustdust prospect 730
Mactung deposit 721
Maggie porphyry Cu, Mo deposit 732
Magnum mine 717
Maid of Erin mine 741
Maple Leaf prospect 729
Marshall skarn deposit 729
Mastodon deposit 724
mercury occurrences 730
Merry Widow deposit 743, 757
mets prospect 733
Midway deposit 722, 723, 751
Mineral King deposit 757
Minto deposit 723
MM 722
Monarch mine 719
Morrison prospect 738
Mosquito King deposit 724
Mother Lode deposit 729

Mount Brussilof deposits 720
Mount Copeland 93, 716
Mount Haskin skarn 723
Mount Hundere deposit 722
Mount Johnny 733
Mount Reed skarn 723
Mount Sicker deposit 742, 751
Mount Skukum deposit 751
Mount Thomlinson Mo prospect 734
Mount Washington porphyry Cu(Mo) deposit 744
nephrite jade 286, 730
Newman (Bell Copper) deposit 738
Nifty prospect 739
Nimpkish deposit 743, 757
North Star deposit 716
Northair deposits 740
O'Connor River gypsum-anhydrite prospect 741
OK deposit 740
Old Fort prospect 738
Oliver camp 729, 750
Olympic Dam deposit 717
OP deposit 721
Oro Denoro deposit 729
Ox Lake porphyry prospect 734
Pete occurrence 722
Phoenix deposit 729
Pictou prospect 730
Pioneer mine 738
Pluto Mo,W porphyry prospect 723
Polaris deposit 733
Portland Canal-Stewart camp 746
Premier mine 746
Privateer deposit 744
Prosperity-Porter Idaho mine 738
Pt,Pd occurrences 727, 729
QR deposit 749
Quartz Hill deposit 740
Quartz Lake deposit 722
Quesnel River (QR) deposit 727
Ray Gulch deposit 721
Rea deposit 722
Rea Gold deposit 724
Red Bird Mo prospect 734
Red Mountain deposit 723
Red-Chris prospect 734
Reeves Macdonald deposit 724
Revenue Cu, Mo prospect 733
Rexspar deposit 724, 781, 791
Riondel camp 724, 757
Risby deposit 723
River Jordan prospect 715
Robb Lake deposits 205, 719
Rossland camp 724, 728
Roundy Creek prospect 734
Ruddock Creek prospect 715
Salal Creek deposit 740, 756
Samatosum deposit 724, 751, 752
Sappho prospect 729
Schaft Creek Cu, Mo deposit 733, 756
Scottie mine 733, 748
Sekulmun skarn 723
Seneca deposit 739, 752
Sheep Creek camp 509, 724, 750
Sheep Creek-Ymir 746
Silback-Premier mine 733, 746, 748, 750
Silence Lake occurrence 725
Silver City Cu, Ag deposit 719, 745
Silver Hart deposit 722, 723
Silver Queen deposit 734
Silverado mine 738
Similkameen deposit 509, 746

835

Slocan-Silverton district 728
Snip deposit 733
Snowbird prospect 730
Spar Lake deposit 716
Spruce Creek prospect 730
Stikine Copper deposit 754
Storie porphyry deposit 723
Stormy deposit 723
Sullivan mine 716, 748, 757
Sulphurets 73
Sunro mine 745, 753
Surf Inlet deposit 740
Surf Point deposit 740
Surprise prospect 730
Surprise Lake W, Cu, Sn greisen veins 730
Sustut Copper prospect 733
Taku deposit 733
Tasu deposit 511, 743, 757
Taurus porphyry deposit 725
Tel Au-rich deposit 740, 741
Texada deposit 743
Texada Island deposits 757
Thistle deposit 743
Tillicum Mountain skarn 727, 757
Tintina Silver deposit 722, 723
Tom deposit 235, 722
Torbrit mine 733
Tulsequah Chief deposit 733
Twin Lakes camp 729
Twin Mountain deposit 724
Tyee deposit 786
United Keno Hill mines 748
uranium occurrences 775, 786, 789
Valentine Mountain prospect 745
Valley Copper deposit 726
Vananda skarn deposit 743
Vangorda deposit 720
Victoria mine 744
Villalta prospect 743
Wellgreen mine 745
Westmin massive sulphide deposits 742, 746, 748
   751, 752
Wheaton Creek placer Au deposits 730
Whitehorse Copper deposit 734, 757
Wigwam deposit 724
Willa breccia pipe deposit 727, 749
Williams Creek deposit 723
Windy Craggy (Cu, Co, Au, Ag, Zn) prospect 312, 741,
   751, 753
Wolf epithermal Au,Ag vein 734
XY deposit 721
Yellowjacket prospect 730
Ymir-Sheep Creek district 729
Zeballos deposit 743, 751, 757
MINFILE 714
Mink Creek Suite 499
Minnes Group 37, 335, 781
Minto Pluton 507, 508
Miocene and Pliocene glaciations 422
Miocene drainage 408, 409, 410
Miocene paleogeography 411
miogeocline 6, 15, 18, 49, 461, 646, 688, 690, 693,
   694, 697, 698, 699, 823
Misinchinka Group 141, 548, 719
Mission Canyon 243
Missisquoia Zone 181
Mississippi Valley-type Zn,Pb 719, 723, 745, 757
Mississippian carbonate reservoirs 776
Mist Mountain coalfields 781
Mist Mountain Formation 547, 781, 784
Mistaya Formation 181

Misty Creek Embayment 465, 654, 690
Moffat rhyolite 312
Mohican Formation 162
Mohorovicic discontinuity (Moho) 39, 47, 705
Molar-tooth structure 110, 113
molybdenum deposits 754
Monashee Complex 92, 109, 144, 162, 547, 549, 550,
   551, 553, 608, 610, 624, 665, 715
Monashee Décollement 550, 553, 608, 610, 622, 625, 665
Monashee Mountains 146, 620
Monashee Terrane 49, 498, 608
Monk Formation 147
Monster Formation 341
Montania 825
Monteith Formation 335, 699
Monteregian Intrusion 67
montmorillonite 545
Moose Channel coal deposits 786
Moose Channel Formation 545
Moose Dome 251
Moose Mountain fields 776
Moose Valley Fault 556
Moosebar Formation 781
Moosevale Formation 472
Morania 119
mordenite 557
Moresby Group 354
Moresby Island 355, 565
morphogeological belts 15, 17, 697
Morro Member 229
Moth Bay Pluton 522
Motherlode, California 729, 751
Mount Attwood Formation 281
Mount Baird Formation 206
Mount Barr Batholith 377
Mount Billings Batholith 722
Mount Brenner Pluton 514
Mount Burnham orthogneiss 284
Mount Caplice Stock 502
Mount Cayley 51, 385, 485, 791, 793, 811
Mount Constantine complex 500
Mount Copeland syenite gneiss 500
Mount Edziza 65, 391, 392, 422, 486, 795, 811
Mount Fee 385
Mount Fitton Pluton 499, 723
Mount Fowler Batholith 500
Mount Garibaldi 382, 485, 811
Mount Gilbert Pluton 522, 523
Mount Goodenough Formation 334
Mount Grace carbonatite 500
Mount Harper complex 499
Mount Harper rift event 650
Mount Harper volcanic complex 133, 462, 464
Mount Hawk Formation 226, 228, 230
Mount Head Formation 250, 251, 252
Mount Kindle Formation 187, 188
Mount Lloyd George 111, 114
Mount Logan 353, 578, 580, 581
Mount Logan Batholith 511
Mount Lytton Plutonic Complex 293, 352, 504, 592,
   595, 601, 607
Mount Mazama 811
Mount Meager 811
Mount Nansen Group 343, 406, 479, 512
Mount Nelson Formation 109
Mount Price 382, 384
Mount Raleigh 477, 561, 562, 563, 589
Mount Roberts Formation 281, 467
Mount Robson 168
Mount Rogers Formation 687
Mount Roosevelt 173

INDEX

Mount Saint Elias 353, 580
Mount Sedgwick 499
Mount Selwyn area 142
Mount Silverthrone Complex 380, 485, 794
Mount Spieker-Bullmoose 781
Mount St. Helens 811
Mount Stuart batholith 67, 71, 516
Mount Toby stock 509
Mount Washington 564, 744, 751
Mount Wilson Formation 192, 193
Mountain diatreme 196
Mowitch Formation 252, 261
Moyie Intrusions 103, 463
Moyie sills 499
Moyie Suite 499
Mt. Allan 405
Mt. Cairnes Gabbro-Greenstone Complex 311
Mudcracked formation 118
Muir-Chichagof Suite 515
Muncho-McConnell Formation 202, 204
Mural Formation 145, 161, 168
Murray Ridge Formation 275
Muskox Intrusion 463
Muskwa Anticlinorium 643
Muskwa Assemblage 100, 111, 823
Muskwa Ranges 646, 689
Muskwa Ranges succession 111, 684, 686, 717
Mysterious Creek Formation 309

N
Nahanni area 632
Nahanni earthquakes 45
Nahlin Basin 350
Nahlin Fault 597, 599, 601
Nahlin ultramafic body 294
Nahoni Range 651, 654
Nakina Formation 294
Nakina Subterrane 294, 296
Nanaimo Assemblage 330, 357
Nanaimo Basin 357
Nanaimo Group 38, 357, 564, 780, 785, 798, 826
Nanika intrusions 344, 518
Nanika Suite 518, 519, 734
Nanoose Uplift 470, 584
Nansen glacial advance 422
Narakay volcanics 115, 116
Narchilla Formation 138, 139, 159, 160
Nascall spring water system 794
Nasina Formation 284
Nasina Quartzite 608
Naswhito Creek Formation 345
Nation River Formation 230, 232, 241
native Cu 727
Natla Formation 206
Natural hazards 809, 827
Nazcha Formation 290, 293
Nazko Cone 485
Nechako Basin 346, 348, 606
Needle Peak Pluton 518
Needlepoint Pluton 516
Nehenta Formation 312
Nelson Batholith 509, 550
Nelson Suite 509, 526, 724
Nelway Formation 183
Nemo Lakes belt 287
Neruokpuk Formation 128, 157, 233, 256, 547, 654
Netalzul volcanics 347, 476, 557
Nevadan Orogeny 699
Nevadella Zone 161, 162, 166, 167, 168
New Guinea 404, 405, 406
Nicoamen Syncline 601
Nicol Creek Formation 108, 462

Nicola Assemblage 290, 556
Nicola batholith 293, 607
Nicola Group 277, 292, 299, 307, 461, 470, 471, 473, 595, 601
Nicola horst 556
Nicola-Princeton basins 363
Nig Creek Platform 263
Nikanassin Formation 335, 781
Nikolai Greenstone 314, 470, 472, 565
Nilkitkwa Depression 475
Nilkitkwa Formation 304, 305, 475, 557
Nimpkish deposit 743, 757
ningyoite 786
Nisku Formation 226
Nisling Range 462, 482
Nisling Range alaskite 479, 483, 519
Nisling Terrane 26, 148, 284, 298, 307, 548, 555, 608, 723, 824
Nisutlin Allochthon 282, 284, 467
Nisutlin Assemblage 282
Nisutlin batholith 608
Nisutlin Plateau 284
Nisutlin Subterrane 91, 148, 276, 634, 635, 693, 725
Nitinat fan 449
Nitinat Formation 210
Nizi Formation 285
Nizina Formation 314
Noatak Formation 232
Nome Lake Batholith 507
Nootka Fault 44, 442, 447, 449
Nordenskiold Dacite 508
Norman Wells 648
Norman Wells oil field 776
North American and Pacific crustal plates 5
North American craton 18
North American Faunal Province 156, 157
North Saskatchewan River 640, 641
North Slope Terrane 695
North Thompson 549
North Thompson Fault 551, 555
North Thompson River 362, 551
northern miogeocline 185
Northern Rocky Mountain Trench 139, 141, 362, 415, 549, 594, 608, 643, 655, 703
Northern Sierra Terrane 693, 701
Northern Tethyan Province 36
Nostetuko Lake flood 815
Nutzotin Mountains Sequence 353, 661

O
obduction 699, 701
oblique subduction 703, 704
ocean ridge 439
ocean-bottom seismographs 442
oceanic allochthons 696
oceanic crust 487
off-ridge seamounts 461
Ogilvie Formation 131, 638, 684, 689
Ogilvie Deflection 651
Ogilvie Mountains 132, 638, 639, 651
Oil and gas 342, 775, 779
Okanagan 291, 665, 790
Okanagan Complex 548, 555
Okanagan Composite Batholith 504, 509
Okanagan Highlands 408, 409, 797
Okanagan Subterrane 277, 290, 292, 294, 729
Okanagan Valley 607, 801
Okanagan Valley Fault 665
O'Keefe aquifer 801
Olalla Pluton 507
Old Crow Basin 779, 802
Old Crow Batholith 499, 664

INDEX

Old Crow intrusions 230, 233
Old Dominion Limestone 168
Old Tom Formation 292
Oldhamia 133, 157, 159
Oliver Pluton 509
Olympic Mountains 585
Olympus Sandstone Lentil 266
Omineca Belt 5, 18, 49, 51, 144, 406, 539, 548
555, 567, 607, 698, 714, 715
Omineca Mountains 139, 140, 608
One Tree Formation 356
Ootsa Lake Group 344, 482
Ootsa-Endako volcanic complexes 462
Operation Stikine 9
Ordovician assemblages 184
organic maturity 542, 544, 545
Osilinka Intrusions 504, 513
Ospika diatreme 499
Osprey Lake Pluton 510
Ottertail Formation 182
overlap assemblages 26, 580, 683, 697
Oweegee Peak 299, 469

P

Pacific Coast 51
Pacific Ocean floor 444
Pacific Orogen 18
Pacific Plate 42, 49, 53, 71, 72, 73, 439,
441, 446, 447, 488, 810
Pacific Plate subduction complexes 486
Pacific Rim 53
Pacific Rim Assemblage 330, 356
Pacific Rim Complex 356
Pacific Rim Terrane 36, 315, 330, 356, 476, 540, 564,
565, 566, 585, 745
Pacific-America motion 440
Pacific-Farallon spreading ridge 373
Paleogeomorphology 403, 404
paleomagnetic results 699
Paleomagnetic studies 63
Paleomagnetism 63, 67, 73, 824
Pali Dome 385
Palliser Formation 229, 250, 251, 252
Palliser sequence 223, 229
Pandora Peak Unit 357
Pangea 73
Panther River 776
Papuan ultramafic bodies 696
Parafusulina 31, 286, 695
Pardonet Formation 271, 273, 547
Park Ranges 548
Parson Bay Formation 314, 564
Parsons Assemblage 333
Pasayten Fault 592, 601
Pasayten Group 351, 352, 405
Paskapoo Formation 341, 342
passive margin 683, 685, 686, 688, 690
Pattison Pluton 519
Paul Revere Ridge 447
Pavilion earthfow 813
Pavilion Subterrane 297, 298
Peace River 161, 411, 642, 781
Peace River Arch 223, 252, 340, 690
Peace River Embayment 252
Pebble Creek hot springs 793
pedimentation 404, 701
pediplanation 403
Peechee Member 225
Peel Plain 406
Peel Plateau 406
Peel Platform 224

Peel Trough 337, 341
Pekisko Formation 250, 252, 253, 254
Pelly Gneiss 93, 282, 284, 499
Pelly Mountains 140, 237, 608, 615, 791
Pemberton 373, 377, 378, 485
Pemberton Diorite Complex 522
Pemberton Volcanic Belt 377, 381, 388, 485, 486, 705
Pend d'Oreille sequence 281
Peninsula Formation 309, 350
Peninsular Terrane 696, 699
Pennask Batholith 509
Penticton Group 345, 482, 558
Perdrix Formation 226
pericratonic terranes 26, 461, 694, 723, 824
permafrost 797, 801, 813
Permian tectonism 825
petrogenetic grid 543
petroleum 641, 776
petroleum exploration 776
petroleum resources 775
Peyto Formation 168
Peyto Member 161
Phillips Formation 108
phosphorite 338
Phroso Siltstone Member 264, 273
piedmont complexes 422
Piltz Peak Pluton 516
Pinchi Fault 290, 599, 657, 659
Pinchi Lake 555, 556, 595, 599
Pinchi Lake Hg occurrence 730
Pine Pass 642, 781
Pine River 781
Pinguicula Group 110, 121, 133, 547, 684
Pinnacles Dome 92, 553
pitchblende 786, 791
Pitman Batholith 504
Placentian Series 156, 157, 160
placer gold 730, 747
Plagiura-Poliella Zone 173
plate boundaries 810
Plateau Fault 646, 647, 649
Platyvillosus 33
Pleistocene glaciers 422
Plio-Pleistocene terrace gravels 411
plume and rift complexes 388
Pointed Mountain 776
polar wander loop 82
polar wander paths 64
Polaris Ultramafic Suite 500, 501
Polygnathus - Palmatolepis 31
Polymetamorphism 558
Ponder Pluton 523, 560
Porcupine Creek Anticlinorium 640
Porcupine Fault 664
Porcupine Hills Formation 341, 342
Porcupine River 664
Porcupine River basin 428
PORCUPINE TERRANE 131
Porcupine virgation 653, 699
porphyry copper deposits 754, 755
porphyry Cu, Mo deposits 726, 742
Port Refugio Formation 312
Porteau Pluton 70, 80, 522
Portrait Lake Formation 235
Precambrian basement 41, 49, 51
Precambrian plutonic rocks 498, 684
Precious metals 746
prehnite-pumpellyite 539, 540, 541, 542, 545, 547
prehnite-pumpellyite burial metamorphism 540
Premier Anticlinorium 618
Premier Porphyry 521

Presqu'ile Complex 206
Pressure stabilities of equilibria 544
Pressure-temperature diagrams 541
Prevost Formation 235, 241
Prichard Formation 105
Primary magnetizations 64
Prince Rupert shear zone 559
Princeton 407, 785
Princeton Basin 607, 786
Princeton coalfields 785
Princeton Group 786
Prometheus magnetic anomaly 585
Prophet Formation 253, 254, 255
Prophet Trough 242, 254
Prospect thrust sheet 78
Proterozoic igneous rocks 463
Prudential unit 348
Ptychaspis-Prosaukia Zone 180, 181, 183
pull-apart basins 607
pull-apart structures 461
pumpellyite 541, 542
Pumpernickel 695
Puppets Formation 312
Purcell (Belt) paleopoles 81
Purcell (Belt) Supergroup 101, 684
Purcell Anticlinorium 100, 147, 549, 625, 632, 646
Purcell Arch 825
Purcell Fault 548, 551
Purcell intrusives 499
Purcell Lava 80, 108, 462
Purcell Mountains 103, 107, 144, 146, 620, 622
Purcell Supergroup 147, 716
Purcell Thrust Fault 620, 622
Purcell Trench 801
Purcell-Wernecke Assemblage 100, 110, 716, 823
Pybus Assemblage 312
Pygodus - Periodon 31
Pyramid Mountain 395
Pyroxenite Creek ultramafite 515

Q

Quanchus Suite 518, 734
Quartet Group 110
Quartz Hill stock 525
Quaternary eruptive centres 811
Quaternary sedimentation and erosion 429
Quatsino Formation 314
Quatsino Sound 356
Queen Charlotte Basin 373, 564, 583, 704, 780
Queen Charlotte Fault 42, 45, 373, 439, 440, 444, 446
447, 582, 583, 662
Queen Charlotte Group 355, 582, 583
Queen Charlotte Islands 5, 36, 39, 354, 406, 431, 433, 439,
440, 444, 564, 578, 582, 785
Queen Charlotte Sound 42, 582, 780
Queen Charlotte Terrace 42, 445, 447, 583, 780
Queen Charlotte-Fairweather Fault 44, 49, 440, 810
Quesnel 407, 785
Quesnel Highlands 618, 636
Quesnel Lake 551, 552, 556, 619
Quesnel Lake Orthogneiss 500
Quesnel Lake volcaniclastics 70
Quesnellia 31, 33, 38, 49, 277, 288, 290, 291, 299, 302,
306, 307, 461, 467, 468, 469, 470, 472, 473, 552, 555,
595, 599, 637, 694, 696, 697, 698, 699, 726, 825
Quiet Lake 608
Quill Creek Complex 745
Quinsam Lake 785
Quintette 781
Quottoon Pluton 406, 523, 560, 588

R

Rabbitkettle Formation 178, 179, 191
Rackla dykes 498
Racklan event 6, 650
radiolarian chert 694, 698, 699
Rae Group 120, 684
Raft Batholith 611
Rainbow Range 387, 388, 389, 485, 487
Rainy Hollow area 741
Ramparts Formation 224, 232, 776
Randall Formation 312
Ranger Canyon Formation 259, 261
Rapid Creek Formation 338
Rapid Depression 341
Rapid Fault Array 51, 664
Rapitan Assemblage 718
Rapitan Group 81, 82, 110, 133, 134, 547, 718
Rat uplift 654
Rat River Formation 334
Recharge areas 797, 798, 799, 801
Red Rose Formation 348, 785
redbed-type Cu 745
redbeds 404
Redcap Mountain 588, 606
Redstone Plateau 647
Redstone River Formation 134, 717, 718
Redstone-type Cu 757
reference paleopoles 64
regional metamorphism 539, 827
Reid (Nisling) glacial advance 394
Reindeer Formation 341, 786
Relay Mountain Group 308, 351, 477, 826
remanent magnetization 63, 64
Rennell Sound Fault Zone 355, 356, 582, 583
replacement deposits 723
Revere-Dellwood fracture zone 442
reversal of the geomagnetic field 63, 73
Revett Formation 105, 716
Rexspar deposit 724, 789, 791
Reynolds Point Formation 82
Richardson Anticlinorium 338, 639, 651, 652, 654
Richardson Fault Array 639, 652, 654
Richardson Mountains 45, 48, 406, 639, 652, 694
Richardson Trough 157, 169, 652, 689
ridge propagation 444
rift basins 689
rifted continental margin 685, 686
rifted continental terrace wedge 487
rifting 461, 683, 705, 780, 823, 825
rifting and glaciation 686
Rigel fields 776
ring-dyke complex 511
Risky Formation 134, 137, 157
Road River Formation 171, 189, 190, 191, 203, 237
Road River Group 166
Robert Service Thrust Sheet 631
Roberts Mountain Allochthon 242
Roberts Mountain Thrust 695
Robson Basin 161, 168, 173, 181
Rocher Deboule stock 516
rock avalanches 812
Rock Creek Member 276
Rock River 785
Rockslide Formation 171, 172
rockslides 812
Rocky Mountain Trench 50, 51, 622, 634
Rocky Mountain Trough 338, 341, 342
Rocky Mountains 406, 638, 780
Rocky Mountains Assemblage 221
Rocky Mountains Subprovince 639, 641, 643

'Rocky Ridge' volcanic rocks 348, 557
Romanzof Mountains 233
Romanzof Uplift 653
Ronde Member 225, 228
Ronde-Kakisa sequence 222, 223, 228
Roosevelt facies 173
Roosevelt Graben 179
Roosville Formation 108, 109
Root Basin 689
Rosella Formation 167
Ross Creek Formation 256
Ross Lake Fault 351, 593, 659
Ross River 360, 655, 785
Rossland 473, 509, 746, 748
Rossland Group 290, 292, 293, 470
Rossland volcanics 473
rotation poles 443
rotational slides 812
rotations, magnetic 77, 78
Ruby Range granodiorite 518
Rundle Assemblage 221, 222, 250, 252, 253, 255
Rundle Group 241, 243, 250, 251, 252, 776
runoff 797
Rusty Shale formation 120

S

S-type granites 497, 526, 757, 690, 701, 755, 826
sabkha 314, 718
Sadlerochit Assemblage 262, 263
Sadlerochit Formation 262, 272, 273
Sagavanirktok River 653
Saginaw Bay Formation 312
Saint Elias Fault 580
Saint Elias Mountains ... 5, 148, 353, 408, 409, 566, 578, 660, 662
Saint Elias Plutonic Suite 353, 511, 566
Salal Creek Pluton 377, 378, 386, 485, 524
Salal Glacier Volcanic Complex 386
saleeite 786
Salmo 625, 626
Salmo district 757
Saltspring Island 799
Saltspring Intrusions 500
saltwater seeps 799
San Andreas Fault system 440
San Christoval plutons 510
San Juan Fault 357, 566
San Juan Islands 357, 564, 566
San Juan Terrane 696
Sanak-Baranof belt 662
Sandpile Group 194
Sandspit Fault System 582, 704
Sangamonian Stage 422, 425
Sans Sault Formation 337, 338
Saskatchewan 411
Sassenach Formation 229
Saukia Zone 179, 180, 181, 183
Saunders Group 781, 784
Savage Mountain Formation 472
Sayunei Formation 133, 137, 718
Schaffer Intrusion 233
schist 309
Scho Creek Member 275, 654
Schooler Creek Group 264
Schoonover Terrane 695
Schroeder Fault 629
Scrip Nappe 620
sea-level changes 431
secondary magnetizations 64
Sedgwick Intrusion 230, 233
sediment flows 813
sediment transport 703

Sedimentary exhalative deposits (sedex) 719, 720, 723, 745, 757
Seeley Lake 785
seeps, warm water 794
seismic reflection profiles 584, 585, 705
seismic refraction 39
seismicity 42, 705
Sekwi Formation 164, 166, 719
Selkirk Allochthon 144, 550, 610, 624, 665
Selkirk lavas 394
Selkirk Mountains 146, 162, 620, 624
Selwyn Basin 31, 157, 159, 160, 164, 465, 629, 654, 689, 690
Selwyn Gneiss 500
Selwyn Mountains 5, 138, 555, 638, 646, 650
Selwyn plutonic suite 512, 513, 526
Semail Nappe 696
Semenof Block 636
Semenof Fault 599, 636, 657
Semenof formation 284
semi-anthracite coal 781, 785
Sentinel Subterrane 294, 296
Settler Schist 307
Sevier Orogeny 703
Seward Glacier 662
Seward Suite 517
Shakwak Trench 660
Shakwak Valley 801, 802
Shames River fault zone 587
Sharon Creek Formation 194
Sharp Mountain Formation 337
Sheepbed Formation 134, 137
sheeted dykes 294
shelf to off-shelf facies boundary 694
Sheppard Formation 80, 107, 108
Sheppard Intrusions 517
Sheppard-Gateway Formation 109
Shezal Formation 133, 137, 719
shield volcanoes 409
Shingle Creek porphyry 518
Shoemaker Assemblage 292
Shoemaker Formation 31
Shonektaw Formation 290, 293
shortening, crustal 615, 631, 638, 646, 650, 702, 703
Shorts Creek Formation 345
Shublik Formation 272, 273, 274
Shuksan metabasalt 309
Shuksan Terrane 277, 307, 309, 562
Shuksan Thrust Fault 594
Shulaps ultramafic body 308, 563
Shunda Formation 250, 251, 252, 254
Shuswap Complex 550, 555
Shuswap Metamorphic Complex 109, 147
Sicker Assemblage 462, 470, 693
Sicker Group 210, 313, 470, 563
Sifton Antiform 554
Sifton Assemblage 330, 350
Sifton Basin 362
Sifton Fault 554, 555, 608
Sifton Formation 362
Sifton Pass 141
Sifton Ranges 91, 362, 554, 608
Sikanni 776
Siletz volcanics 64, 66
sillimanite 405
sillimanite-kyanite zone 542
sillimanite-muscovite zone 542
Silurian assemblages 184
Silver Creek Stock 377
Silverthrone Complex 485
Similkameen batholith 504, 509
Simpson Allochthon 282, 283, 467

INDEX

Simpson Peak batholith 507
Simpson Range Suite 282, 500
Sinwa Formation 297
Siyeh Formation 105, 107
Skagit Gneiss 307
Skaha Formation 345
Skaha Lake 801
skarn and replacement deposits 721, 727, 742, 756, 757, 758
Skeena Arch 304, 305, 306, 330, 344, 345, 596,
603, 605, 606, 775
Skeena Assemblage 308, 351, 357
Skeena Basin 404, 406, 701
Skeena Fold Belt 558
Skeena Group 346, 348, 557, 605
Skeena Mountains 5
Skeena River 560
Skidegate Formation 355
Skoki Formation 547
Skolai Group 314, 462, 470, 565
Skonun Formation 373, 375, 380, 388, 564, 565, 582, 786
Skukum Group 343, 483
Skukum Volcanic Complex 484
Skukum-Sloko igneous complexes 462, 482
Sky Pilot pendant 561, 589
Slater River Formation 338
Slats Creek Formation 169, 176
Slave structural Province 93
Slave Point Formation 224
Slide Mountain Allochthon 467
Slide Mountain Assemblage 284, 292, 294
Slide Mountain Group 284, 287
Slide Mountain Terrane 26, 31, 277, 281, 282, 284, 287,
288, 289, 290, 294, 461, 467, 469, 487, 498, 548,
550, 552, 618, 634, 636, 637, 697, 699, 725
Sliding Mountain 287, 288
Slocan 509, 757
Slocan Group 277, 290, 292, 293, 628
Slocan Lake Fault 550, 665
Slocan syncline 550, 553
Sloko Group 343, 483
slope failures 813
Smithers Formation 305, 475
Smoky Group 341, 781
Snake Indian Thrust Faults 640
Snake River 134, 137
Snow avalanches 814
Snowshoe Formation 144, 146
Snowshoe Group 282, 551
Sombre Formation 205
Sonoman Orogeny 695, 697
Sonomia 38, 308
Sooke Gabbro 358, 476, 517
Sophie Mountain Formation 343
South Fork Assemblage 342, 344
South Fork Volcanics 406, 419, 462, 514, 549
southern miogeocline 192
Southern Ogilvie Mountains 639, 650
Southern Rocky Mountain Trench 49, 144, 415, 548, 639,
640, 655, 690, 801
Southern Rocky Mountains 144, 639
Southesk Formation 225, 228
Sovanco fracture zone 442, 447
Spanish Creek area 794
Spatsizi Group 305
Spectrum Range 391, 486
speleothem deposition 425
Spences Bridge Group 307, 344, 352, 406, 478, 480, 557
Spences Bridge volcanism 462, 479
Spider Peak Group 308, 351
Spinel Fault 655
Spirit Fault 647

Spray Lakes Group 249, 252, 261
Spray River Assemblage 222, 263, 273, 274
Spray River Group 264
spreading rates 441, 442
spreading ridge 485, 753
spring water chemistry 795
Springbrook Formation 345
springs 798, 799, 802
Spuzzum Pluton 70, 80, 516, 562, 593
Squamish Granodiorite 522, 523, 561
St. Clare Province 394, 395, 396
St. Mary River Formation 342
St. Mary's fault 109
Standfast Creek Slide Fault 608, 626
Starbird Formation 224
Starlight Evaporite Member 266, 273
Station Creek Formation 314
staurolite 405, 542
Steele Creek complex 500
Steele Glacier 394
Stelkuz Formation 139, 140, 145, 160, 554
Step Formation 262
Stephens pluton 70
Stikine Arch 277, 303, 306, 345, 596, 693
Stikine Batholith 510
Stikine River 414
Stikine Suite 303, 501, 510
Stikine Volcanic Belt 373, 388, 389, 390, 391, 409,
453, 462, 486, 775, 795
Stikinia 26, 31, 38, 277, 298, 299, 302, 306, 307, 406, 461,
469, 471, 472, 473, 601, 693, 694,
696, 697, 698, 701, 732, 824
stilbite 540
stilpnomelane 541
Stoddart Group 243, 254
Stolberg field 776
Stone Formation 205, 719
Storelk Formation 252
Straight Creek Fault 351, 594, 659
Strait of Georgia 406, 428, 799
stratabound copper deposits 718
stratabound volcanogenic Cu, Pb, Ag deposits 724
stratabound Zn, Pb prospects 715
stratiform lead-zinc 235
stratiform Pb-Zn-barite 690
striated dropstones 132
$^{87}Sr/^{86}Sr$ initial ratios 504, 519, 525, 529
structural style 784
structural thickening 699
structural traps, petroleum 775, 776, 780
Stubbs Thrust Fault 288, 289, 629
Stuhini Assemblage 556
Stuhini Group 75, 298, 299, 305, 461, 470, 471, 472, 601
sub-bituminous coal 775, 785, 786
subaqueous mass movements 813
subduction complex 476, 548, 585
subduction zone 373, 439, 443, 488, 701
subgreenschist facies 541, 546, 566
subpermafrost water 802
subterranes 26, 296
Sukunka field 776, 781
Sukunka River 643
Sullivan Formation 181, 182
sulphide mound-chimney deposits 753
Sulphur Creek orthogneiss 500
Sulphur Mountain Formation 264, 266, 271
Summers Creek-Quilchena 600
Summit Creek Formation 341, 343, 786
Sunblood Formation 191
superterranes 26
suprapermafrost water 802

INDEX

Suquash Basin 358, 780, 785
Suquash Formation 785
surface runoff 798
surficial uranium deposits 790
Surprise Lake Suite 516, 526
Survey Mountain Fault 357
Survey Peak Formation 181, 182
Sustut Formation 701, 826
Sustut Basin 330, 349, 405, 406, 603, 703
Sustut Group 343, 346, 349, 556, 557, 702
Sustut-Skeena Basin 343
Swan Hills reefs 224, 225
Swannell Formation 139, 145, 554, 608
Swannell Ranges 553
Sweetgrass Hills 519
Sylvester Allochthon 285, 286, 467, 468, 550, 615, 636
Sylvester Group 277, 285
Symphysurina Zone 183

T

Tachek batholith 507
Taenicephalus Zone 179, 180, 181, 183
Taghanic Onlap 223, 226
Tahkandit Formation 262
Tahltanian Orogeny 6
Tahtsa Complex 521, 522
Tahtsa Lake 558, 560, 606
Tahtsa Lake caldera 480
Taiga Formation 176
Taiga-Nahoni fold belt 651
Takla belt 472
Takla Fault 606
Takla Group 290, 293, 294, 297, 299, 300, 470, 472, 779
Taku Assemblage 309, 310
Taku Terrane 277, 309, 310, 315, 580
Takwahoni Formation 277, 299, 303, 304
Talheo flow systems 794
Talkeetna Mountains 699
Talston 684
Tango Creek Formation 349, 557
Tantalus Formation 363, 406, 599, 785
Tarr Inlet suture zone 580
Tartu Facies 380
Tatchun Batholith 507, 508
Tatchun Belt 599
Tatla Lake 555
Tatla Lake Metamorphic Complex 516, 518, 539, 555, 558, 560, 606, 667
Tatlayoko Lake 591
Tatlmain Pluton 507, 508, 567
Tatonduk River 131
Tats Volcanic Complex 312, 472
Tatsieta Formation 202, 203
Tawu anticline 118
Tawuia sp. 119
Taylor Creek Group 308, 352
Taylor Flat Formation 254
Tchaikazan Fault 352, 591
tectonic assemblages 15
tectonic attenuation 80
tectonic wedging 699
Tegart Formation 193
Telford Formation 256, 261
Telkwa coalfield 784, 785
Telkwa Formation 304, 475, 556, 557, 605
Ten Mile Creek body 505
Ten Peak Pluton 515
terrace, B.C. 557, 606
terrane amalgamation 6, 825
terranes 6, 17, 561
Tertiary coal deposits 780, 785

Teslin Crossing stock 510
Teslin Fault 599, 657
Teslin Formation 296
Teslin Suture Zone 636, 657, 824
Teslin Trench 657
Tetcho Formation 230
Tethyan faunas 28, 277, 308, 693, 696
Tetsa Formation 113
Texas Creek Granodiorite 521, 733
Thelon Tectonic Zone 93, 684, 823
thermal maturation 779
thermal metamorphism 779
thermal springs 780, 792, 795
Thibert Fault 599, 607, 657
Thor-Odin Dome 553, 608
thorium 775, 791
Three Sisters Pluton 510
Three Sisters Suite 510, 525, 733
Thudaka pluton 513
Thundercloud dykes 498
Thundercloud Formation 134
tillite 137, 756, 776
Tillite Group 687
Tindir Group 131, 464
"Ting Creek" intrusion 519
Tintina Fault 42, 160, 629, 655, 668
Tintina Trench 138, 140, 360, 664, 703, 785
Tip Top Hill Formation 344, 478, 481
Tkope River Batholith 524
Tkope Suite 524
Toad Formation 270, 273
Tobermory Formation 252
Toby Formation 109, 147
Tochieka Gneiss 91
Todhunter Member 252
Tofino Basin 330, 358, 376, 407, 582, 585, 780, 826
Tombstone Mountain Formation 104
Tombstone Mountains 791
Tombstone Pluton 514
Tombstone Suite 512, 514, 526
Tombstone Thrust Sheet 631
tonalitic body, Coast Belt 703
Toodoggone 749, 750
Toodoggone volcanics 305, 473
Topley Intrusions 475, 507
Topley Suite 557
Toricelli Mountains 406
Totschunda Fault 581, 661, 662
Tozitna Terrane, slarka 695
trace fossils 139
Tracy Arm Terrane 307
Trail Pluton 509
Tranquille Formation 345, 482
transcurrent faults 407, 461, 606, 655, 703
transform boundary 662, 664
transform fault 439, 440
trap, petroleum 776, 780, 779
Trapping Creek basin 797
trellis drainage pattern 403, 406
Trevor Formation 338
triangle zone 641, 643
triple junction 442, 446
triple point 543
Triune Formation 194
Trout Creek assemblage 347
Trout River formation 230
Tsalkom Formation 287
Tsaydiz Formation 139, 145, 554
Tsaytis Pluton 523
Tsaytis River area 605, 606
Tsetso Formation 189

INDEX

Tsezotene Formation 81, 116, 118
Tsezotene Range 118
Tsezotene sill 81
Tshinakin limestone unit 183
Tsichu Formation 254
tsunamis 809, 816
Tuchodi Anticline 100, 645, 646
Tuchodi Formation 113
Tuchodi Lakes 111
Tuktoyaktuk Peninsula 652, 653
Tulameen Basin 786
Tulameen gabbro and ultramafic complex 501, 727
Tungsten-rich skarns 756
Turnagain complex 501
Turner Valley Formation 250, 251
Turtleback Complex 500
Tuttle Formation 230, 233
Tuya Basin 350
tuyas 350, 385, 395, 486
Tuzo Wilson Knolls 442, 444, 447
Tweedsmuir Glacier 148
Twitya Formation 133, 137
two-sided orogen 683
Tyaughton Group 308
Tyaughton Trough 477, 478, 826
Tyaughton-Methow Trough 308, 329, 343, 350, 351, 404, 405, 659, 701, 702, 703, 779
Tyrwhitt Formation 252

U

Ucluth Volcanics 356
Uinta Trough 685
ultramafic complexes 501
underplating 407
Union Bay Suite 515
Union Bay Ultramafic Suite 515
uranothorianite 789
Uslika Formation 293

V

Valdez Group 353, 580
Valhalla 548
Valhalla Complex 92, 517, 548, 550, 665
Valkyr shear zone 665
valley glaciers 422
valley incision 429
Valley-fill deposits 801
Vampire Formation 159, 160, 164
Van Creek Formation 107
Vancouver Island 5, 51, 356, 564, 578, 582, 583, 785, 798, 799
Vaseaux Formation 93
Vaseaux Lake 613
Vedder Complex 309, 594
Vedder discontinuity 594
Vega Siltstone Member 264, 273
Vera Formation 203
Verbeekinidae 31
Verde Creek Pluton 516
Virginian Ridge Assemblage 349, 352
Vital Fault 599
Vital Range 556
vitrinite reflectance 540, 545, 827
volcanic arcs 373, 461, 485, 488
volcanic eruptions 809
volcanic massive sulphides 742
Volcano Mountain 394
volcanogenic massive sulphide deposits 730, 745, 751

W

Waiparous Creek 781
wairakite 540
Wales Group 148, 467
Walker Lake Terrane 696
Wapiti Formation 342, 781, 784
Wapiti Group 545
Wapiti Platform 263
Wark and Colquitz complexes 508
Warneford River area 237
water-table aquifers 799, 801
Waterfowl Formation 182
Waters River Member 275
Waterton field 776
Waterton Formation 104
Waterton Park 104
Watson Lake 785
Watt Mountain Formation 223
Watt Mountain hiatus 223
Waucoban Series 156, 162
Waverly Formation 239, 240
Wells Creek volcanics 309
Wernecke Mountains 110, 133, 639, 650, 651, 775
Wernecke Mountains diatremes 498
Wernecke Supergroup 91, 110, 116, 133, 547, 684, 717, 791
West Alberta Ridge 223, 225
West Coast Complex 315
Westcoast Crystalline Complex 510, 564
Westcoast Suite 508
Weston Fault 141
Whatcom Basin 358
Whistler Member 264, 266, 273
White Channel gravels 411
White Creek Batholith 512
White Lake Basin 482
White Lake Formation 345, 558
White River 733
White River Ash 395, 486
White River Assemblage 284
Whitehorse 264, 802
Whitehorse Formation 264, 266, 273
Whitehorse Suite 512, 514, 734
Whitehorse Trough 296, 302, 303, 304, 307, 596, 603, 607, 657, 775, 779, 784, 785
Whitesail Formation 305
Whitestone River formations 338
Whitewater diorite 500
Whitewater Thrust Fault 288, 629
Whittaker Formation 188
Whudzi plutons 513
Wildhorse Batholith 504
Williston Lake 141, 608
Willow Creek Fault 685
Willow Creek Formation 342, 781
Wilson cycle 687
Wind River culmination 650, 651
Windermere rifting 823
Windermere Supergroup 109, 127, 132, 133, 461, 550, 553, 620, 685, 686, 718
Windy Range 551
Windy Terrane 284
Windy-McKinley Terrane 276, 284, 634
Winnifred Member of the Whitehorse Formation 266, 268
Winona Basin 447, 449, 780, 826
Winona Block 445, 447, 449
Winthrop Formation 352
Wisconsinan 422
Wokkpash Formation 204

Wolfe Mountain 785
Wolverine Metamorphic Complex 140, 287, 553, 554, 667
Wopmay Orogen 91, 93, 684
Work Channel Lineament 559, 560
Wrangell Lava 566, 581
Wrangell Mountains 73, 422
Wrangell Suite 524
Wrangell Volcanic Belt 377, 394, 395, 462, 486
Wrangell volcanoes 446
Wrangellia 26, 31, 35, 38, 53, 280, 281, 308, 310, 311, 312, 313, 315, 470, 472, 476, 498, 563, 564, 565, 566, 582, 693, 697, 699, 702, 741
Wrangellia (W1, W2, W3) 578
Wrangellian lithosphere 705

Y

Yahatinda Formation 223
Yakoun Formation 353, 354
Yakoun Group 308, 315, 473, 476, 785
Yakutat Assemblage 354
Yakutat Block 44, 446, 664
Yakutat Group 354
Yakutat Terrane 578, 580, 581, 683
Yalakom Fault 308, 351, 352, 593, 659, 667
Yankee Belle Formation 145
Yanks Peak Formation 145, 161
Yellow Aster Complex 309, 500, 594
Yellowstone-Snake River hot spot 453
Yohin Formation 253
Yukon Fault 131, 664
Yukon Group 284
Yukon Platform 157, 160, 164, 169, 176
Yukon River 664
Yukon Territory 51, 634, 733, 801
Yukon-Koyukuk Basin 664
Yukon-Tanana Terrane 148, 468, 693, 695
Yukon-Tatonduk area 131
Yusezyu Formation 138, 157

Z

zeolite facies 539, 540, 545, 546, 547, 556, 563
zeolitic alteration 302, 304
Zn, Pb deposits 719